Nuclear Data for Science and Technology

Proceedings of the International Conference

Antwerp 6 - 10 September 1982

Editor

K. H. Böckhoff

Central Bureau for Nuclear Measurements

Geel, Belgium

D. REIDEL PUBLISHING COMPANY

Dordrecht : Holland / Boston : U.S.A. / London : England

Library of Congress Cataloging in Publication Data
Main entry under title:

Nuclear data for science and technology.

English and French.
Proceedings of the Antwerp Conference 1982 on Nuclear Data for Science and Technology, organized by the Central Bureau for Nuclear Measurements, Commission of the European Communities, with the cooperation of the OECD Nuclear Energy Agency, Nuclear Data Committee, and the International Atomic Energy Agency, and sponsored by Fonds national de la recherche scientifique (Belgium) and others.
Includes bibliographical references and indexes.
1. Nuclear physics–Congresses. 2. Nuclear reactors–Congresses. 3. Nuclear engineering–Congresses. I. Böckhoff, K. H. II. Antwerp Conference on Nuclear Data for Science and Technology (1982). III. Commission of the European Communities. Central Bureau for Nuclear Measurements. IV. Fonds national de la recherche scientifique (Belgium)
QC770.N74 1983 539.7 83–2877
ISBN-13: 978-94-009-7101-1 e-ISBN-13: 978-94-009-7099-1
DOI: 10.1007/978-94-009-7099-1

Publication arrangements by
Commission of the European Communities
Directorate-General Information Market and Innovation, Luxembourg

EUR 8355

Softcover reprint of the hardcover 1st edition 1983

Published by D. Reidel Publishing Company
P.O. Box 17, 3300 AA Dordrecht, Holland

Sold and distributed in the U.S.A. and Canada
by Kluwer Boston Inc.,
190 Old Derby Street, Hingham, MA 02043, U.S.A.

In all other countries, sold and distributed
by Kluwer Academic Publishers Group,
P.O. Box 322, 3300 AH Dordrecht, Holland

D. Reidel Publishing Company is a member of the Kluwer Group

This conference was organized by

CENTRAL BUREAU FOR NUCLEAR MEASUREMENTS

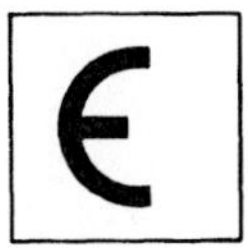

JOINT RESEARCH CENTRE

COMMISSION OF THE EUROPEAN COMMUNITIES

with the cooperation of

OECD NUCLEAR ENERGY AGENCY
NUCLEAR DATA COMMITTEE

INTERNATIONAL
ATOMIC ENERGY AGENCY

and sponsored by

NFWO NATIONAAL FONDS VOOR WETENSCHAPPELIJK ONDERZOEK (Belgium)

EUROPEAN PHYSICAL SOCIETY

UNIVERSITAIRE INSTELLING
ANTWERPEN

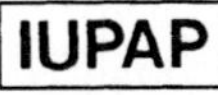

INTERNATIONAL UNION
OF PURE AND APPLIED PHYSICS

CITY OF ANTWERP

CONFERENCE ORGANISATION

Chairman K. H. Böckhoff C.E.C.

Local Organisation and Programme Committee

F. Corvi	C.E.C.	F. Poortmans	Belgium	E. Wattecamps	C.E.C.
P. De Bièvre	C.E.C.	G. Rohr	C.E.C.	H. Weigmann	C.E.C.
H. Knitter	C.E.C.	G. Vanpraet	Belgium		

International Programme Committee

B. Allen	Australia	H. Küsters	F.R. Germany	H. Motz	U.S.A.
J. Bouchard	France	H. Liskien	C.E.C.	J. Rosén	N.E.A.
C. Coceva	Italy	J. Lynn	U.K.	J. Rowlands	U.K.
J. Darvas	C.E.C.	C. Mahaux	Belgium	J. Schmidt	I.A.E.A.
A. Deruytter	C.E.C.	A. Michaudon	France	G. Stöcklin	F.R. Germany

International Advisors

V. Benzi	Italy	A. Ferguson	U.K.	R. Richmond	Switzerland
M. Bustraan	The Netherlands	E. Fort	France	A. Smith	U.S.A.
C. Campbell	U.K.	T. Fuketa	Japan	M. Sowerby	U.K.
R. Chrien	U.S.A.	S. Igarasi	Japan	H. Vonach	Austria
S. Cierjacks	F.R. Germany	J. Lachkar	France	S. Whetstone	U.S.A.
H. Condé	Sweden	F. Perey	U.S.A.	S. Yiftah	Israel
W. Cross	Canada	S. Qaim	F.R. Germany		

ACKNOWLEDGEMENT

This is, first of all to give recognition to the very valuable recommendations from the members of the afore mentioned advisory bodies and to Dr. B. Rose, concerning the organisation and the programme of the Conference.

The Conference Secretariat and the execution of the major part of the preparatory work for these Proceedings was assumed by Mrs. A. Faes, who mastered these very demanding tasks with great skill and personal engagement. She was efficiently assisted and complemented by Mrs. R. van Berkel. Their contributions during all phases of the Conference were essential and are sincerely appreciated.

Particular thanks are due to Prof. G. Vanpraet from the Rijksuniversitair Centrum Antwerpen (RUCA) for his outstanding engagement in the local organisation and the management in situ as well as for the support from his University with material and manpower.

Special acknowledgement is also owed to a group of collaborators who helped with great enthusiasm to prepare and run the Conference: Ms. H. Kerslake for her continuous voluntary and competent assistance, Dr. E. Wattecamps (poster and social arrangements), Mr. K. Bürkholz (finances), Mr. L. Sanspoux (general aid, RUCA) and the volunteers Mrs. E. Spanopoulou, Mrs. H. Allaert, Ms. P. Feyen and Mr. W. Nagel.

Further important contributions were given by Mr. H. Prins (graphical design work), Mr. U. Meloni (computer programmes), Mr. W. Schreiber and his workshop team (poster panel construction).

Mrs. L. Eisen from Directorate General XIII (Commission of the European Communities, Luxembourg) deserves thanks for carefully checking the manuscripts of the contributions during the Conference, for preparing the Index of Authors and for further editorial assistance.

The CINDA Index was provided by Drs. G. Coddens and A. Thompson from the Nuclear Energy Agency (OECD).

K. H. Böckhoff

PROLOGUE

The Antwerp Conference 1982 on Nuclear Data for Science and Technology which is subject of these proceedings, continues a long sequence of application oriented nuclear data conferences. Their origin may be dated back to the sixties when the Washington/ Knoxville series was launched in the USA with subsequent meetings in 1966, 1968, 1971, 1975 and 1979. IAEA contributed to the sequence with international conferences in 1966 (Paris) and 1970 (Helsinki). In the USSR meetings of similar scope were held within the Kiev series in 1971, 1973, 1975, 1977 and 1980. (Besides these data conferences there have been several nuclear physics oriented conferences, one of them also at Antwerp in 1965).

Based on the initiative of B. Rose and starting with the Harwell Conference in 1978, the previously largely uncorrelated series of nuclear data conferences was given a rigid scheme with respect to timing and location. It was agreed to hold annual meetings with the location cycled between Europe, the USA and the USSR. In the spirit of this agreement the Harwell Conference was followed in 1979 by the Knoxville Conference and in 1980 by the Kiev Conference. Once through with this cycle the turn was then again on Europe and upon request of the Nuclear Energy Agency Nuclear Data Committee, the Central Bureau for Nuclear Measurements accepted to organise the following Conference. There was at that time a broad consensus within the Western Nuclear Data and Reactor Physics Communities to enlarge the period between subsequent conferences of this type from one year to two years. The decision was then made to hold the Conference in 1982 in Antwerp and an agreement was achieved for a cooperation with NEANDC and IAEA and for sponsorships by the Belgian Nationaal Fonds voor Wetenschappelijk Onderzoek, the University of Antwerp, the European Physical Society, the International Union for Pure and Applied Physics and the City of Antwerp.

A Local and an International Programme Committee, the latter one extended by corresponding members, formed the advising body for the organisation and the programme of the Conference.

The scope of the Conference was decided to be similar to that of its predecessors, emphazising Nuclear Data and Neutron Physics which pertain to fission and fusion energy programmes, but considering also nuclear data for bio-medical purposes, astrophysics and solid state research. Underlying Physics, in particular the present picture of our understanding of neutron induced nuclear reactions completed this scope.

The Conference was attended by 270 scientists from 35 countries, among them for the first time also a delegation from the People's Republic of China.

A total of 229 contributions were presented, including 10 invited papers. The number of papers in this category has been drastically cut down, as compared to previous conferences.

For the first time in the history of such conferences, poster presentations were introduced. In fact, about 2/3 of all contributions were given as posters.
Taking this option, rather than that one of parallel sessions, was not an easy decision. The organizers of the Conference were well aware of an intrinsic concern about poster contributions, dominant within the scientific community, which still considers posters as third class contributions.

In the opinion of the organizers, however, poster presentations offer - if sufficient care is taken - an excellent chance for scientific communication, which may be better than that of the usually numerous, short oral presentations with their very limited discussion times. They address to a smaller audience of interested colleagues and discussions can be more profound and extended.

A number of measures have been taken to avoid discrimination of contributions with poster presentation against those with oral presentation.

- Both were considered as fully equivalent. Consequently texts of all these contributions entered into the Proceedings and were there reproduced without reference to their form of presentation at the Conference.
- All posters were on display during the whole period of the Conference.
- There have been five poster sessions of about one hour each, during which the stands of the indicated poster group had to be manned.
- The authors of contributed papers with poster presentation were invited to indicate themselves on a place above their poster two periods of minimum one hour each during which their stand would be manned in addition to the period prescribed by the organizers of the Conference.
- The spatial sequence of poster panels was interrupted by numerous white boards to ease discussions.
- Pigeon holes were provided for each participant to facilitate appointments on a personal base.

This specific form of the organization of a conference was an experiment, which - following the reactions after the conference - found a positive reception.
The future conferences will show whether the psychological barrier against poster presentations has become smaller.

SESSION CHAIRMEN

- OPENING SESSION AND CLOSING SESSION

  K. H. Böckhoff

- NEUTRON DATA OF TRANSACTINIUM ISOTOPES

  G. Campbell
  E. Fort

- NEUTRON DATA OF STRUCTURAL MATERIALS, COOLANTS AND SHIELDING OF FISSION REACTORS

  C. Coceva
  Y. Kikuchi

- NUCLEAR DATA PERTAINING TO FISSION REACTOR FUEL CYCLES AND FISSION PRODUCTS

  H. Küsters
  G. Vanpraet

- NUCLEAR DATA PERTAINING TO FUSION REACTORS AND INTENSE NEUTRON SOURCES

  J. Darvas
  H. Vonach

- NEUTRON STANDARD DATA

  H. Liskien

- THE UNDERSTANDING OF NUCLEAR REACTIONS AND NUCLEAR MODEL CALCULATIONS

  D. Seeliger
  E. Sharapov
  Sheng Qingbiao

- NEUTRON DATA ANALYSIS AND EVALUATIONS

  H. Mehta
  F. Perey

- UNDERLYING PHYSICS AND SUPPORTING DATA

  A. J. Deruytter
  J. Lachkar

- NUCLEAR DATA FOR BIOMEDICAL APPLICATIONS

  G. Stöcklin

- SOLID STATE, MOLECULAR AND ATOMIC EFFECTS ON NUCLEAR DATA

  M. Nève de Mévergnies

- GENERAL INTEREST

  J. A. Harvey
  J. J. Schmidt

TABLE OF CONTENTS

NEUTRON DATA OF STRUCTURAL MATERIALS, COOLANTS AND SHIELDING OF FISSION REACTORS

Invited Paper

Contributed Papers

NUCLEAR DATA PERTAINING TO FISSION REACTOR FUEL CYCLES AND FISSION PRODUCTS

Invited Paper

Contributed Papers

NUCLEAR DATA PERTAINING TO FUSION REACTORS AND INTENSE NEUTRON SOURCES

Invited Paper

Contributed Papers

NUCLEAR DATA FOR DOSIMETRY

Contributed Papers

NEUTRON STANDARD DATA

Contributed Papers

THE UNDERSTANDING OF NUCLEAR REACTIONS AND NUCLEAR MODEL CALCULATIONS

Invited Papers

Contributed Papers

NEUTRON DATA ANALYSIS AND EVALUATIONS

Contributed Papers

UNDERLYING PHYSICS AND SUPPORTING DATA

Invited Papers

Contributed Papers

NUCLEAR DATA FOR BIOMEDICAL APPLICATIONS

Contributed Papers

NUCLEAR DATA FOR THE UNDERSTANDING OF STELLAR NUCLEO-SYNTHESIS

Contributed Papers

SOLID STATE, MOLECULAR AND ATOMIC EFFECTS ON NUCLEAR DATA

Contributed Papers

GENERAL INTEREST

Invited Paper

Contributed Papers

Opening Session

OPENING ADDRESS

J. Dinkespiler,

Director General Joint Research Centre
Commission of the European Communities
Brussels

Mesdames, Messieurs,

Ce congrès international sur les données nucléaires pour la science et la technologie fait partie d'un cycle de congrès similaires sur les données nucléaires qui se tiennent successivement aux Etats-Unis d'Amérique, en URSS et en Europe. C'est à nouveau au tour de l'Europe de recevoir cette assemblée et vous découvrirez, je crois, qu'Anvers n'est pas un mauvais choix pour l'accueillir. Malheureusement, l'important programme scientifique et les activités sociales annexes ne vous laisseront que peu de temps pour profiter des ressources touristiques locales.

Le Comité des données nucléaires de l'Agence de l'Energie Nucléaire de l'OCDE a confié l'organisation de ce congrès au Bureau Central de Mesures nucléaires du Centre Commun de Recherche de la Commission des Communautés Européennes. En ma qualité de Directeur Général de ce Centre, j'ai l'agréable devoir de remercier les membres du Centre Belge d'Etudes Nucléaires de Mol et de l'Université d'Anvers pour leur participation active dans l'organisation de cette réunion. Je dois aussi spécialement remercier le Recteur de l'Université d'Anvers qui a mis à notre disposition ses remarquables installations pour conférences.

La coopération avec l'Agence de l'Energie Nucléaire à Paris et l'Agence Internationale de l'Energie Atomique à Vienne a été la clef permettant d'ouvrir ce congrès à tous les Etats Membres de ces organisations, lui donnant ainsi un véritable caractère international.

C'est un plaisir pour moi de souhaiter la bienvenue aux délégués de 33 pays parmi lesquels, pour la première fois dans l'histoire de ce congrès sur les données nucléaires, des délégués de la République Populaire de Chine.

Le grand nombre de participants à ce congrès prouve qu'il vient au bon moment et qu'il y a encore un intérêt mondial dans la recherche de données nucléaires, en dépit des irritations que le mot "nucléaire" suscite ici ou là. Je suis certes très impressionné par votre nombre. Je le suis encore plus par votre qualité et par votre importance.

Le développement de matériels nouveaux, de procédés nouveaux, de techniques nouvelles nécessite toujours, en plus de l'imagination et des finances, un cocktail où entrent pour une certaine fraction les lois de la physique et pour une autre fraction des données numériques. C'est ce cocktail qui permet de prévoir ce qui va se passer. A une époque où tout est cher et urgent, ce cocktail permet d'éviter les tâtonnements inutiles et coûteux.

Bien sûr, à part quelques progrès réalisés dans la compréhension des réacteurs nucléaires, ce n'est pas vous qui versez dans le shaker les lois de la physique. Mais vous êtes ceux qui y versent l'autre ingrédient fondamental : les données numériques. Et c'est pour cela que vous êtes, selon moi, des gens très importants.

Les données nucléaires ont joué un rôle absolument essentiel depuis 40 ans dans le développement des réacteurs à fission. A neutrons thermiques d'abord et maintenant à neutrons rapides, alors que le travail sur les réacteurs à fusion commence dans le monde entier. Elles continuent à être indispensables à l'établissement et à l'amélioration permanente de la sécurité de ces réacteurs. Elles seront aussi nécessaires aux études à conduire et aux décisions à prendre sur les différents cycles du combustible. La demande pour l'amélioration des données disponibles, et même pour de nouvelles données paraît en être le témoin.

Qu'en est-il des données nucléaires et que deviennent-elles? En arriverait-on à la fin, ayant atteint les objectifs? Des budgets doivent-ils y être consacrés? Je ne suis pas prophète et je n'oserais pas fournir des réponses à ces questions. Vous devez les proposer vous-mêmes.

Votre congrès est une étape importante sur la route, les données nucléaires progressent. Cette étape devrait vous montrer où vous êtes arrivés et ce que vous avez réalisé. Je doute qu'elle vous dise exactement jusqu'où vous devez aller, mais espérons qu'elle vous stimule et vous donne une nouvelle orientation pour l'avenir.

Je ne vais pas continuer à abuser du temps précieux de vos conférences.

Au nom de la Commission des Communautés Européennes, je déclare ouvert ce Congrès international sur les données nucléaires pour la Science et la Technologie.

Je vous souhaite un travail fructueux et aussi, bien sûr, un bon séjour à Anvers.

K. H. Böckhoff (ed.), Nuclear Data for Science and Technology, 3.

KEYNOTE SPEECH

B. Rose
United Kingdom

It is an honour and a responsibility to be invited to give the "key note talk" at this meeting on Nuclear Data for Application - though when I began to think about it, I became progressively less certain what the job entailed. It was clear that one thing to do was to tell you what fine fellows you all are - but you know that anyway. It was also clear that it would not be sensible to attempt a preview of the meeting itself - for that is already evident in the detailed, and perhaps rather daunting, programme which you already have before you.

After some thought it finally seemed appropriate to remind you first of all of the reasons why the meeting is being held at all - to put it into the context of other international meetings on related subject matter - and secondly to attempt a brief review of the present state of health of the nuclear data community in this part of the world, by which I mean Western Europe. No doubt other matters will be touched on before the end of my talk.

The particular characteristics of this meeting are that it is modest in size, the major group of participants is from Western Europe and that it is principally concerned with matters of intense practical importance - the nuclear physics of the fission energy programme. There are nevertheless major contributions from outside Western Europe and there are other important topics for discussion.

The great international jamborees like the "Atoms for Peace" conferences, the last of which was held in Geneva in 1971, are big prestige occasions, which last two weeks, cover almost everything under the sun and are attended by many thousands of people who come and go during the course of the meeting to attend the relatively small fraction of sessions that they can hope to understand.

The international neutron physics conferences began in Columbia University in 1957, followed by Antwerp '65, Budapest '72 and Lowell '76. These meetings, generally somewhat larger than the present one, are devoted principally to the development of the basic physics of neutron interactions and aim in principle at equiparticipation from all regions of the world.

There are also the so-called "Specialist Meetings" typically of around 50 people, where attention is given to a very narrow range of 'hot topics', i.e. those which at that specific time are ripe for detailed examination in depth. From such meetings important conclusions are drawn and recommendations for future actions are made.

Another group of meetings is that of the "Neutron Inter-laboratory Seminars", restricted to ~ 50-100 participants all from Western Europe, held roughly every 18 months, the purpose of which is to deal primarily with experimental neutron physics and to discuss in detail in smallish informal groups the dirty aspects of experimental physics which rarely get into the published papers.

In 1966 and 1970 meetings were held in Paris and Helsinki respectively whose primary concern was accurate neutron data for reactors. They were sponsored by the IAEA and, with attendances of ~ 600 and 200, attracted world-wide participation. The subject matter was very much the same as that with which we shall be dealing this week, though there was little attention given to other sorts of data or other kinds of application. There have been no succeeding meetings in this series.

The reason for this is partly due to the fact that, particularly in the U.S.A. but also in the U.S.S.R., there were a series of Nuclear Data Conferences devoted to the same topics and held at 2-4 year intervals. These, known as the Washington series and the Kiev series respectively, are primarily devoted to the needs of the regions in which they were held, though outside contributors and participants are welcomed and make important contributions.

It was therefore decided to set up a similar series in Western Europe with the idea of serving this community, particularly but far from exclusively; the attempt was also made to interleave them with the U.S.A. and U.S.S.R. meetings on a 3 year cycle to avoid clashes of timing and also to ensure that each region would have its own meeting on a regular basis. A deliberate attempt was also made to widen the scope of the meetings to include other types of application, as had been done for example in Washington in 1975.

The first such meeting was held at Harwell in 1978 and it was extremely successful in giving a general overview of the subject and its underlying theory and also of surveying related areas of potential practical importance and giving a good general perspective on work going on in this part of the world.

The advantages of having such a meeting in reasonable proximity to many groups of working scientists are fairly obvious. Participation is made easier by being less of a strain on travel budgets which are under continual review, usually downwards. If participation is reasonably high then members of various groups at all levels get to know each other and each other's programmes better, which greatly helps small groups, particularly when working on related subjects.

The widespread interest shown in the meeting is illustrated by the attendance - which at ~ 280 is a little more than at Harwell and comes from over 30 countries. However in my view one would not want anything much bigger than this, which seems to me a good size for such a meeting - it is sufficient to be representative of all work being done in the area and avoids the difficulties of many parallel sessions which would automatically follow with a larger attendance. I treasure the remark made many years ago by Alec Merrison that, if in attending an APS meeting - typically of 10 parallel sessions - you didn't go to any of the sessions, you only missed 10 % of the meeting. Another advantage of this size of meeting is that there is a good chance of being able to find someone you particularly wish to talk to - which is not the case with the bigger meetings.

The emphasis is somewhat different from that at Harwell where a conscious attempt was made to cover the whole field relating to applications and to the corresponding basic physics, and parallel sessions were held. This time rather fewer and rather shorter invited papers are being given and far more contributed papers have been accepted, though many of these will be presented in poster sessions rather than orally. All will appear in the proceedings however and in this way it has been possible to bring to the attention of the participants far more work in progress than was possible last time.

As most of you will be aware this year is the 50th anniversary of the discovery of the neutron and, to celebrate the fact, a conference entitled "The neutron and its applications" is being held in Cambridge next week. At that meeting, which is

K. H. Böckhoff (ed.), Nuclear Data for Science and Technology, 4–6.

primarily a celebration and a demonstration of the versatility of the neutron, only a relatively small amount of time is being devoted to the biggest single demonstration of the commercial importance of this peculiar particle. However it reminds one of the fallibility of judgement of great men, even in their own subject, for Rutherford clearly thought that there was no possibility of releasing nuclear energy on a significant scale, and discouraged speculation on the subject. I am also reminded of the British Astronomer Royal not so many years ago who declared space travel to be bunk, a few weeks before Sputnik I was launched.

Of course, the sort of physics at which Rutherford was so outstanding is not the sort of physics we are concerned with today. High precision was rarely of any consequence, for he was mainly concerned with new ideas where semi-quantitative results were quite sufficient to demonstrate the correctness of a fairly revolutionary concept. The classic case is the Rutherford scattering law which showed the presence of nuclei in atoms. The essential feature was that an appreciable number of alpha particles were scattered through wide angles, not that the scattering was proportional to cosec $^4\theta/2$. However such semi-quantitative experiments are rarely of any value in a technological context.

Modern experiments are usually extremely complex, and depend upon a most elaborate technological infrastructure, which the experimenter has to understand thoroughly if he is to achieve accurate results. The experiments themselves are extremely difficult - and their nature is such that one begins to wonder at times whether any claimed accuracies at the one percent level can be sustained in cross section measurements. These experiments frequently take years to carry out and, sometimes alas get stopped for financial reasons before the experimenter is really satisfied with them.

Modern theory is similarly placed, being extremely complex and with an elaborate infrastructure, though its infrastructure consists of hardware (in the form of computers), elaborate theoretical ideas and also, very often, elaborate computer codes coming from elsewhere.

There is also now a new breed of men, the evaluators, who can only become good at their trade after many years spent as experimenter or in an experimental environment but who also need a sound theoretical understanding. Evaluation plays such an important role that one is compelled nowadays to ask "Who will evaluate the evaluators?" - "quis custodiet ipsos custodios?".

So now I have fulfilled one of my tasks - by showing how difficult your work is I have told you what splendid chaps you all are.

The next question I should like to consider is the state of health of the nuclear data community. People have been concerned over this matter ever since the mid-sixties when it was widely felt that too much effort was going into the data side of the nuclear power business compared with effort on other subjects both inside and outside nuclear power. The analyses of the problem at that time and the subsequent fashion in spending cuts has lead to a steady reduction of the scale of effort when viewed world-wide. The important and worrying questions are of course whether this continuing reduction will continue or should continue.

There are two extremes of individuals' approach to worry. There was the case of the paranoic couple who described to a visiting friend various things that were bothering them, and capped them all by saying " - and in addition, the thrush in our garden is not nesting in the same place as she has done previously". "Don't worry", said the friend, "it is probably not the same bird this year". "We know it isn't", was the reply, "but, you see, we are not ready for a new thrush".

The other extreme is illustrated by the man who after visiting his doctor for some diagnosis or other, received a very full account of his condition. The doctor asked him if he was worried by the diagnosis and he said he wasn't. "In that case", said the doctor, "it is clear that you haven't understood the problem".

It is of course very important to try and steer an optimized "worrying course" rather well away from these two extremes. What I shall try to do is to induce what one could term "creative worry" among those who have the power of decision in funding and staffing levels, bearing in mind the old saying that a professor (or Director of Research) has as little time for research as an archbishop has for prayer.

First it is quite clear that the total requirements for data, as set by quite stringent criteria, do not get any less. The data needed for fast reactors as seen at present cover a much wider energy range and a greater variety of materials than the original data needed for thermal reactors, even though the accuracy requirements may not be so great in many cases. Despite the fact that the fast reactor projects as a whole in most countries are moving slowly, because of financial restraints and misplaced pressure from environmental groups, the next few years is the right time to press ahead with fulfillment of current data needs since in many cases they affect the detailed designs of reactors or plants which can still be modified. The extra time that may be available can be used to do better experimental work - it should not be wasted by attempting to make economies by cutting down scientific teams.

Mature projects - such as various thermal reactors - continue to throw up important and very difficult data matters. It is therefore impossible to believe that the still unfulfilled fast reactor projects will not disclose the need for new information 20-30 years from now.

If one looks at the fusion reactor programme, either the "conventional" magnetic confinement systems, or the more speculative inertial confinement ideas where one would need to work with states of matter at densities hitherto not experienced in the laboratory, let alone the factory it must be obvious that new and difficult data needs will arise. It is clear that, for very understandable reasons, the fusion community, particularly in Europe, has not fully addressed itself to the nuclear physics aspects of fusion. Nevertheless it is time now to construct a serious nuclear data programme to run in parallel with the other programmes on confinement and various engineering aspects if the whole fusion - power concept is to become credible in the early 21st century.

In all such work, it is most important not to take too narrow a view of when work should be stopped, because apparently sufficient is known to satisfy present needs. Many years ago in Harwell work on the glassification of nuclear wastes was stopped because the process would not be needed for many years and the remaining problems could be solved when the need arose - while work on disposal or storage sites was barely started. When the problems were taken up again, they proved, of course, to be still very difficult technically but in addition a strong environmental movement had grown up which regarded all things nuclear as its bête-noire. Just think what the political advantage to the nuclear power industry would have been if they had been able to say confidently that it had solved these problems already, and had not been forced back to the weak statement (which incidentally I still believe to be true) that it would be able to solve them in the future.

One matter that comes to mind in this connection is in the fusion reactor field, and the question of whether data should be actively collected to assist the evaluation of radiation damage experiments carried out at neutron energies of up to 40 MeV.

We thus face the necessity for maintaining data measuring teams and their experimental facilities in being for the foreseeable future. It has to be clearly understood that it will be quite impractical to consider dispersing existing teams on to other tasks and closing down experimental facilities for 10 years or so (even if fast reactors were to be so long delayed) and then to start them up afresh after a long delay. Expertise will quite litarelly be lost for decades. As an example of the problems to be faced in building new equipment one has only to consider the experience at Harwell over the past 10 years in rebuilding the linac there, even with the existing team kept in being; or in building up totally new facilities such as those at PTB, Braunschweig.

It is my belief that the existing experimental teams in Europe are on the average close to their lowest possible size for reasonably efficient working and often too small for the tasks demanded of them. Some are subcritical. One method of assessing this is to apply the bus test - to perform a Gedanken-Experiment in which a key member of the team falls under a bus and is killed - and to consider the effect on the work of the group. In far too many cases for comfort the effects would be very serious indeed as, even though people of the right intellectual capacity can in general be found, the years, sometimes decades, of experience lost will take nearly as long to recover. In other words, it is obvious that there is very little if any fat in the present system and to be in a healthy state, and able to stand up to the effects of real as opposed to Gedanken buses, illness, transfer to other jobs, premature retirement and so on teams need to be larger than they are at present, not smaller.

However even the maintenance of teams in being at their present level calls for not a passive but a very active policy, because it necessitates a positive attitude to recruitment. The age structure of the present nuclear data community is such that during the next ten years, many will retire and after that the loss rate will become very high indeed. Young people are needed now for two quite separate reasons. In the short term (i.e. the next 10 years) they will be a great stimulus to their older colleagues, who by the nature of things will be working more slowly (and, dare one say it, more conservatively) than when they were younger. For the longer term, they will be accumulating the experience necessary to prepare some of them to be good evaluators or for greater management responsibilities when the present generation of team leaders and research managers is turned out to grass.

It is therefore essential that means be found to recruit new workers to this field, despite the serious financial problems that face all funding agencies. I have in mind particularly though not exclusively the need to ensure the future of the teams at the dedicated reactor research institutes, because it is there that the greater part of the work is done in direct support of nuclear power programmes. This is not in any way to undervalue the contribution made by the University research departments which have been and surely will continue to be of great importance in all aspects of this work. However they are inevitably affected by different priorities arising from their teaching responsibilities, particularly in relation to research students on whom so much of the research work depends. Without the work done in University departments, it is possible to conceive of a system in which data needs would be met over a very long period of time and a comparatively healthy community exist - although it would not be nearly as vigorous as one would like. Without the dedicated Government-financed research teams, it is in my view not possible to conceive of such tasks being carried out.

I should now turn to some encouraging developments in Western Europe over the past few years. The first one relates to the high level of co-operation that has developed over the Joint European Evaluation of data for the production of a common data file. For many years there has been fairly good informal collaboration between the experimental groups but this has now been overtaken on the evaluation side by the formal co-operation on the Joint European Data file. I believe that this augurs well for data work in Western Europe at a time when there are difficulties in collaborating elsewhere - provided the national funding agencies do not think that they can cut their national effort still further.

The second is related to the NEA Data Bank at Saclay. The amalgamation of the computer library from Ispra and the Data Centre at Saclay produced the expected 10 % saving of running costs. Much more important than the financial savings have been the improvements in compilation of data, so that there is essentially now no backlog, and new data coming in is compiled within a matter of days.

These improvements have had a most powerful positive effect on the European evaluation already referred to - by providing strong technical back up in the holding of the files, production of new files in ENDFB format and their rapid distribution.

Data has been coming into the Data Bank at a fairly constant rate over the past few years and the question must be asked how long will this continue. Perhaps it is better to rephrase the question as to the continuing need for data and we shall be hearing plenty of answers to that during the next few days. The answers will I am sure be partly that totally new data will be called for, partly improved accuracy on older requests and partly by data produced from improved theory, the latter being based on data which itself is not directly relevant to any reactor system.

So far as improved accuracy on data is concerned, there is a paper towards the end of the meeting with a very provocative title which I hope will throw some light on what has to be done. However I am reminded of the story of an English tenor giving a recital in Italy and, after one long and particularly difficult aria, was rewarded with rapturous applause and a demand to sing it again - so he did. Once more the aria, the applause and a demand for a repeat. After the third time and a third request to do it again he pleaded that he was very tired and did not want to oblige. A voice called out from the audience "You will sing it again and again - until you get it right". Some of those working particularly in the important field of standards must feel rather like that man, - though the analogy is far from exact as the aria is different each time.

However there is much more to this field of work than mere repetition - even in the field of standards - as we shall be hearing over the next few days. It is clear that the work is tremendously challenging, to the theorist, to the experimentalist and to the technologist. We shall be hearing about many exciting new ideas and developments. I wish the conference well and stand down to let you get on with it.

Neutron Data of Transactinium Isotopes

^{238}U ISSUES RESOLVED AND UNRESOLVED*

G. de Saussure
Oak Ridge National Laboratory
Oak Ridge, Tennessee

A. B. Smith
Argonne National Laboratory
Argonne, Illinois

The interaction of 1 eV to 20 MeV neutrons with ^{238}U is discussed with emphasis on recently resolved and remaining issues relevant to both application need and physical understanding.

The apparent inability of older ^{238}U evaluations to predict the measured ^{238}U capture rate in thermal critical lattices has stimulated several recent precise measurements of the ^{238}U cross sections, reanalysis of older data, and improved evaluations. The uncertainty in the evaluated value of the parameters of the first few large resonances is now estimated to be smaller than 2% for the neutron widths and smaller than 3% for the capture widths. The recent evaluations predict satisfactorily the ^{238}U capture rate in thermal critical lattices.

In the region from 1.5 to 4 keV there are differences of the order of 15%, sometimes larger, between the values of the neutron widths of the main resonances reported by several experimenters or obtained by different evaluators. These systematic discrepancies have been examined by several authors but are not yet fully understood and deserve further research.

Above 4 keV there are only sparse results of resonance analysis and most evaluations adopt a statistical treatment of the resonance structure. However, an inspection of good resolution data suggests that the energy and neutron width of the most prominent resonances could be determined with reasonable accuracy at least up to 10 keV. Since these large levels control the resonance self-protection effect it appears that it would be desirable to include them explicitly in evaluations. Recent work indicates that the statistical treatment adopted in many evaluations may not account properly for resonance self-shielding. Some factors affecting the determination of the average properties of the resonance parameters will be discussed.

Above the inelastic-scattering threshold, energy-averaged neutron total, scattering, capture and fission cross sections are reviewed in a unified manner integrating measurement, calculation and evaluation. (n;n') and (n;2n') energy-transfer mechanisms are addressed. Particular attention is given to neutron capture, stressing precisions consistent with applied need. Fission properties are discussed including: prompt and delayed fission-neutron spectra and nubar, and fission-product yields. Physical understanding is assayed, with attention to compound-nucleus and direct-reaction mechanisms, and applications impact is illustrated in the context of fast-breeder-reactor performance.

[^{238}U, review of resolved- and unresolved-resonance and energy-averaged total, capture, scattering and fission cross sections and associated processes, E_n= 1 eV to 20 MeV.]

I. INTRODUCTION

Radiative capture, fission and scattering in ^{238}U have been recognized as central problems of nuclear technology since the first attempts to develop a fission-chain reaction in 1939. The reduction in capture achieved by self-shielding in lumped uranium provided the first solution to the problem of achieving criticality.[1] Over the subsequent 40 years our understanding of the complex nature of ^{238}U cross sections and associated properties has dramatically improved thanks to better measurement and analysis techniques and to more detailed nuclear models. Yet problems remain: inconsistencies among differential data, and between measured and computed integral parameters. These problems involve not only microscopic measurement and analysis but also the representation of the data in evaluated files and the processing and integral-calculation techniques.

This paper discusses the interaction of the neutron with ^{238}U over the region of primary applied interest (1 eV-20 MeV) with emphasis on recently resolved and remaining issues in the context of both application need and physical understanding. It is convenient to organize the discussion by incident-neutron energy; starting with the resolved and unresolved resonance regions and proceeding through the inelastic-scattering and fission thresholds to the energy-averaged MeV region. At the lowest energies the applied needs can be very specific, even to individual resonance properties. At higher energies the needs are more general and can be stated in broad energy averages.

*Research sponsored by the Division of Reactor Research and Technology, U.S. Department of Energy, under contract W-7405-ENG-26 with the Union Carbide Corporation and contract W-31-109-ENG-38 with the University of Chicago.

II. LOW-ENERGY AND RESONANCE REGION

II-A. PARAMETERS OF THE FIRST FEW LARGE RESONANCES

The most important neutron data for thermal-reactor applications are the cross sections in the thermal region (not discussed in this paper) and the parameters of the first few large ^{238}U resonances.

In a typical commercial PWR, about 20% of the neutrons are absorbed in ^{238}U. Table 1, from a recent paper of Tellier[2], gives the absorption in the first few resonances relative to the total absorption in ^{238}U; about 63% of the absorption is in the first five large levels below 100 eV. The apparent inability of ^{238}U evaluations before 1976 to predict correctly the capture in thermal lattices was the main subject of a "Seminar on ^{238}U Resonance Capture" held at Brookhaven in 1975.[3] The analyses and discussions presented at that seminar have stimulated several new measurements[4-9] and new evaluations of older data.[10,11] This effort has resulted in a great improvement in our knowledge of the parameters of the first few large resonances.

K. H. Böckhoff (ed.), Nuclear Data for Science and Technology, 9–20.

Table 1. Absorption rate in the first resonances of uranium 238 relatively to the total resonance absorption*

Resonance eV	6.67	20.9	36.7	66	80.7	102.6 116.4 189.6
Absorption %	27.7	15.2	11.9	5.2	2.8	6.7

*Reproduced from H. Tellier, Ref. 2.

Some measured[4-11] and evaluated[12-18] values of the parameters of the five large resonances below 100 eV are given in Table 2. Significant differences are observed between the pre-1976 evaluations and the more recent measurements and evaluations. The capture widths of the first three resonances are now estimated to be 10 to 15% lower than their pre-1976 values, and the neutron width of the 20.9 eV resonance has been increased by 10%. These changes resulted from new measurements and improved methods of analysis: most recent data are based on accurate shape analysis using a multilevel formalism. The importance of multilevel effects and truncation corrections had generally been underestimated in pre-1976 evaluations.[4,5,19,20]

The accuracies requested by reactor designers are 1 meV in the capture widths and 2,3, and 5%, respectively, in the neutron widths of the first three large resonances.[21] The spread among the values obtained in recent measurements suggests that these accuracies have now been met. Furthermore, calculations of resonance absorption in thermal lattices are now in reasonable agreement with the results of direct measurements.[2,20,22]

the estimated uncertainty in the evaluated parameters of the large levels below 100 eV and their correlations.[23]

II-B. NEUTRON WIDTHS OF THE RESONANCES BETWEEN 100 eV and 4 eV

It has been conventional in recent evaluations to represent the ^{238}U cross sections with resolved resonance parameters up to 4 keV and with unresolved (statistical) parameters above that energy. The boundary between the resolved and unresolved resonance ranges is to a large extent arbitrary and recent studies indicate that it would be both desirable and feasible to extend the resolved range to higher energies.[21,23] The resonances above 4 keV will be discussed in later sections.

An accurate knowledge of the parameters of the resonances in the resolved and lower unresolved ranges is particularly important for the correct calculation of the Doppler coefficient of reactivity and other safety parameters in fast reactors. Rowlands[24,25] and Salvatores et al.[26] have recently presented detailed studies of the sensitivities of the safety parameters to the resonance structures at different energies: in a typical fast reactor more than 95% of the Doppler effect in ^{238}U is contributed by the resonances below 25 keV.

In the past ten years or so several laboratories have undertaken major programs of measurement of the ^{238}U resonance parameters in the keV energy range.[5,7,8,11,27] Up to 1.5 keV the values of the neutron widths obtained from the different experiments are consistent to within a few percent. Above 1.5

Table 2. Comparison of ^{238}U resonance parameters

	6.67 eV		20.9 eV		36.7 eV		66.0 eV		80.7 eV	
	Γ_n(meV)	Γ_γ(meV)	Γ_n(meV)	Γ_γ(meV)	Γ_n(meV)	Γ_γ(meV)	Γ_n(meV)	Γ_γ(meV)	Γ_n(meV)	Γ_γ(meV)
Measurements										
Jackson and Lynn (1962)	1.52 ± 0.01	27.2 ± 0.4								
Rahn et al. (1972)	1.52 ± 0.05		8.5 ± 0.8	22.0 ± 3.0	38.0 ± 2.0	23.0 ± 2.0	26.0 ± 2.0	21.0 ± 2.0	1.71 ± 0.18	
Liou and Chrien (1977)	1.50 ± 0.03	21.8 ± 1.0	9.86 ± 0.5	23.5 ± 1.5	33.3 ± 1.2	23.6 ± 2.0	25.6 ± 1.8	22.2 ± 2.0	2.16 ± 0.18	23.7 ± 2.5
Olsen et al. (1977)	1.48 ± 0.03	23.0 ± 0.8	10.16 ± 0.21	22.8 ± 0.8	33.8 ± 0.7	22.9 ± 0.8	24.4 ± 0.5	23.2 ± 0.8	1.82 ± 0.05	24.3 ± 1.3
Haste and Moxon (1978)	1.507 ± 0.008	23.54 ± 0.53	10.17 ± 0.10	22.44 ± 0.43	34.41 ± 0.41	24.03 ± 0.87	25.64 ± 0.66	23.46 ± 0.50	1.794 ± 0.058	
Nakajima (1980)			10.09 ± 0.68		33.85 ± 0.73		25.42 ± 0.81		2.24 ± 0.18	
Poortmans et al. (1981)	1.50 ± 0.01	24.2 ± 0.6	10.2 ± 0.1	23.2 ± 0.6	34.1 ± 0.5	22.9 ± 0.3	23.9 ± 0.8	24.0 ± 0.4	1.81 ± 0.08	
Block et al. (1982)	1.51 ± 0.05	22.36 ± 0.66	9.39 ± 0.16	23.76 ± 0.31	31.47 ± 0.13	23.55 ± 0.13	27.54 ± 1.84	23.34 ± 0.56	1.92 ± 0.05	25.60 ± 0.94
Evaluations										
BNL-325 (1973)	1.52 ± 0.02	26 ± 2	8.5 ± 0.5	25 ± 3	31.0 ± 0.9	25 ± 2	25.0 ± 1.2	22 ± 2	2.0 ± 0.2	21 ± 15
Moxon (1974)	1.510 ± 0.009	26.9 ± 0.37	8.9 ± 0.175	25.7 ± 1.0	31.6 ± 0.5	26.55 ± 1.20	24.0 ± 0.4	23.56 ± 0.76	1.96 ± 0.07	21.17 ± 8.9
ENDF/B-IV (1975)	1.50	25.6	8.8	26.8	31.1	26.0	25.3	23.5	2.0	23.5
ENDF/B-V (1977)	1.510 ± 0.015	22.5 ± 0.6	10.12 ± 0.10	23.1 ± 0.5	33.9 ± 0.4	22.9 ± 0.3	24.6 ± 0.4	23.7 ± 0.3	1.91 ± 0.04	24.2 ± 1.2
KEDAK (1981)	1.495	23.7	9.94	23.67	33.64	23.64	24.61	23.69	1.907	24.17
BNL-325 (1981)	1.50 ± 0.02	22.8 ± 0.6	10.04 ± 0.20	23.5 ± 0.8	34.0 ± 0.4	23.5 ± 0.3	24.0 ± 1.0	23.6 ± 0.1	1.9 ± 0.1	25 ± 1
JENDL-2 (1981)	1.50	23.7	10.1	23.0	33.3	23.5	24.9	22.9	1.87	24.4

Though the spread among results from recent measurements is small compared to the accuracies needed for reactor design, these results nevertheless contain large inconsistencies with respect to the very small estimated uncertainties of the measurements. In particular the neutron widths of the 20.9 eV and 36.8 eV resonances obtained from the RPI self-indication data[9] are 4 to 5 standard deviations smaller than the widths obtained from the other, mostly transmission, data. These discrepancies imply unrecognized systematic errors which should perhaps be further investigated.

Finally, for several applications it would be desirable to have a covariance matrix which represents

keV the results show systematic differences which increase with energy and are of the order of 10 to 20% above 3 keV. This situation is illustrated in Fig. 1 where values of the local s-wave strength function obtained from five sets of data are compared.[15]

A great deal of effort was expended in trying to understand the sources of systematic discrepancies between various data sets, particularly by Derrien, Ribon, and L'Heriteau[28] who have reanalyzed several transmission measurements using the Saclay shape analysis code. An illustrative example is shown in Table 3, where the neutron widths for the large resonances between 1.45 and 1.76 keV are compared, as obtained by Rahn et al., [11] by Carraro and Kolar,[27] and as obtained from these two sets of transmission measurements with the Saclay shape analysis code.

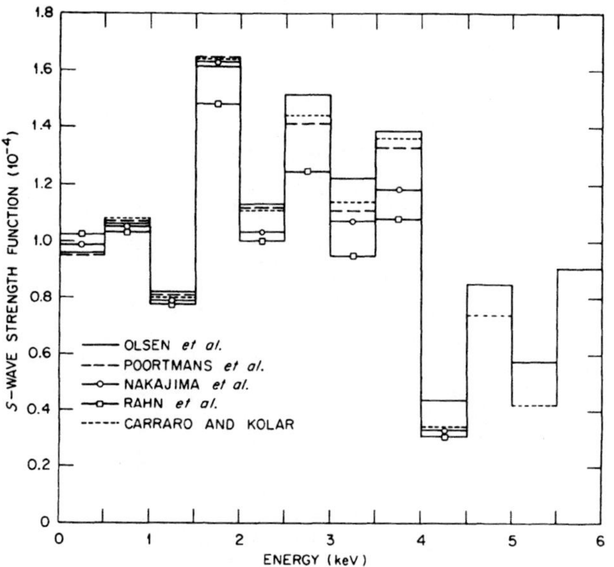

Fig. 1. Comparison of local s-wave strength function for 0.5 keV intervals for five sets of data.

Table 3. ^{238}U **neutron widths for large resonances between 1450 eV and 1760 eV***

Energy eV	Shape analysis of Geel data (2 thicknesses) Γ_n, meV	Shape analysis of Columbia data (3 thicknesses) Γ_n, meV	Geel published values Γ_n, meV	Columbia published values Γ_n, meV
1473.4	114 ± 2	108 ± 2	125 ± 8	125 ± 10
1522.3	215 ± 4	236 ± 3	260 ± 15	240 ± 15
1597.5	309 ± 6	352 ± 4	351 ± 40	355 ± 25
1622.3	97 ± 2	88 ± 2	116 ± 15	68 ± 14
1637.4	50 ± 1	46 ± 2	60 ± 5	50 ± 8
1662.0	201 ± 4	214 ± 4	241 ± 20	171 ± 20
1687.3	98 ± 2	97 ± 2	104 ± 9	92 ± 10
1709.0	81 ± 2	77 ± 2	94 ± 7	86 ± 8
1755.2	121 ± 3	116 ± 3	135 ± 10	105 ± 10
$\Sigma\Gamma_n$	1286	1334	1486	1292

*Reproduced from the paper of H. Derrien, Ref. 28.

The table illustrates that significant differences are obtained whether the same analysis is used on two different transmission measurements or whether the same transmission is analyzed by two different methods. Obviously the precise analysis of transmission or self-indication measurements in the keV energy range is still an unresolved issue. More investigation is required on multilevel effects, truncation corrections, Doppler and resolution broadening, and the effect on the interpretation of the data of level overlap, particularly the overlap between small unresolved p-wave levels and large s-wave levels. The importance of all these experimental problems increases with increasing incident-neutron energy.

II-C. CAPTURE WIDTHS AND AVERAGE PROPERTIES OF THE RESOLVED RESONANCE PARAMETERS

Capture widths are best obtained by an area analysis of the larger resonances below 1 keV: the results of shape analysis are less reliable because they are very sensitive to poorly understood multilevel effects and Doppler and resolution broadening kernels; the area of small resonances ($\Gamma_n \lesssim \Gamma_\gamma$) is insensitive to the value of the capture width; and above 1 keV the increasing resonance overlap perturbs area analysis.

Several recent studies suggest an average capture width of approximately 23.6 meV with a very narrow distribution.[15,29] The extensive and very thorough investigation of the Geel group[8] gives

$$\langle\Gamma_\gamma\rangle = 23.60 \text{ meV} \pm 0.11 \text{ meV (stat.)} \pm 0.50 \text{ meV (syst.)}$$

with a dispersion 0.67 ± 0.11 meV. It is important to remember, however, that the sample of resonances used in the determination of $\langle\Gamma_\gamma\rangle$ is selected (large Γ_n) and probably includes no p-wave level. Since the available transition levels are different, the p-wave average capture width could differ from that for s-wave levels.[23]

Poortmans et al.[8] have shown that the integral distribution of reduced neutron widths above 25 meV is consistent with a Porter-Thomas distribution. Below 25 meV corrections are needed for the admixture of p-wave levels and for missed small neutron widths. For the s-wave average level spacing up to 4.26 keV these authors obtain $\langle D\rangle$ = 21.7 ± 0.9 eV in good agreement with several previous determinations.[11,30]

As shown in Fig. 1, the local s-wave strength function undergoes large fluctuations, consistent with a Porter-Thomas reduced width distribution.[31] For the interval 0-4.26 keV Poortmans et al. obtain an average value of $S_o = (1.15 \pm 0.12) \cdot 10^{-4}$. Other values reported for this energy range[5,7,11,27] are consistent with this result, within the 10% estimated uncertainty.

The p-wave strength function is much more difficult to estimate from a study of the resolved resonances since an important fraction of the levels is not observed, even below 1 keV. Estimates range from the value $1.4 \cdot 10^{-4}$ by Rahn et al.[11] to the value $(2.3 \pm .5/.4) \cdot 10^{-4}$ by Corvi et al.[32]

The analysis of recent data suggests a value for the effective scattering radius of $(0.944 \pm 0.025) \cdot 10^{-12}$ cm.[15] This value is somewhat higher but consistent with the previously accepted value of 0.918.[14]

II-D. THE ^{238}U CAPTURE CROSS SECTION ABOVE 4 keV

A precise knowledge of the ^{238}U infinitely-dilute capture cross section in the range 100 eV to 100 keV is required for the calculation of k_{eff}, the breeding ratio and several safety parameters of fast reactors. Based on detailed sensitivity analyses by Salvatores et al.[26] and by Rowlands,[25] reactor designers have recently requested an accuracy of 3% over the entire range.[21]

From 20 to several-hundred keV the ENDF/B-V evaluation is representative of the capture cross section to within ± 5%, as discussed in Sec. III-B below.[33,34] Below 20 keV the direct capture cross section measurements have proven very difficult: since monoenergetic sources of neutrons are not available, measurements must be done by the time-of-flight technique and hence must rely on the detection of the prompt-capture-gamma radiation. The chief problems are the relatively low ^{238}U binding energy, the possibility of an incident neutron energy dependence of the capture-gamma-ray cascade and the unavoidable background due to the specific gamma activity of the ^{238}U decay chain. Furthermore, below 20 keV the ^{238}U cross sections exhibit a pronounced resonance structure; hence the measurements require appreciable self-protection and multiple scattering corrections. Moore et al.[35] have recently stressed the danger of using a statistical approach to make these corrections. The dispersion between the relatively few measurements in the 4-20 keV range is of the order of 10%.[36] To improve the accuracy new measurements should attempt to reduce the corrections for background, efficiency variation and multiple scattering by stressing new approaches and better techniques.

A promising approach is the combination of high resolution transmission with direct capture data. In Fig. 2 the result of a direct capture measurement[37] (heavy line) is compared to a calculation based on the

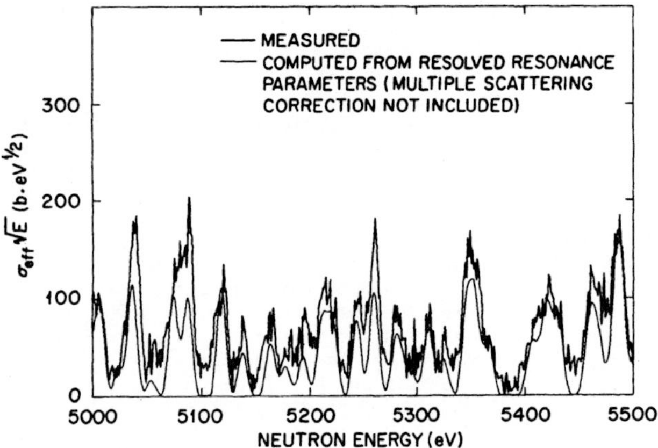

Fig. 2 Comparison of measured and computed effective capture cross section. The measurement was obtained by Yang et al.[37]; the computation is based on the resonance parameters obtained by Meszaros and Olsen[38] from the transmission measurements of Olsen et al.[5] A multiple scattering correction to the computation has not yet been included.

resonance parameters[38] obtained from transmission data.[5] This comparison is still preliminary because the multiple scattering correction has not been included in the calculation. The detailed comparison afforded by such studies may help identify sources of error.

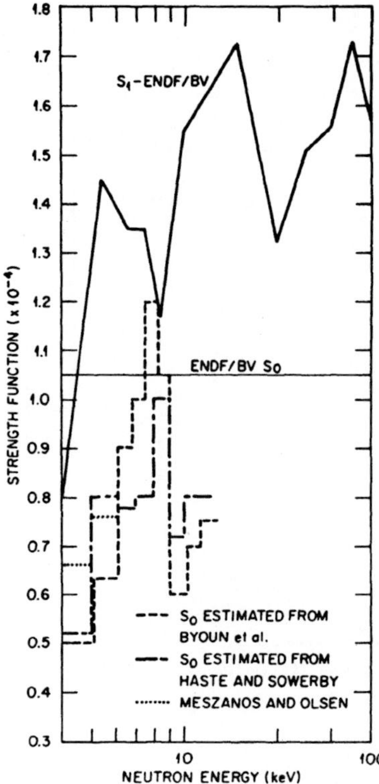

Fig. 3. Variation of strength functions with neutron energy. The ENDF/B-V s-wave strength function is assumed to be constant above 4 keV. The ENDF/B-V p-wave strength function fluctuates as shown. The histogram shows experimental estimates of the s-wave strength function.

II-E. UNRESOLVED RESONANCE PARAMETERS

Perhaps even more important than the infinitely dilute capture cross section, particularly below 10 keV, is the determination and representation of the resonance structure which allows the self-shielding factors to be calculated as a function of fuel composition and reactor temperature. Again, on the basis of the detailed sensitivity studies of Rowlands[25] and Salvatores et al.[26], the fast reactor designers need a 1% accuracy in the self-shielding factor as a function of dilution and 5% in the temperature derivative of the self-shielding factors.[21]

Several widely used evaluations (ENDF/B-V,[39] JENDL-2[18]) represent the ^{238}U cross section above 4 keV and up to 100 keV by specifying the probability distribution of unresolved (statistical) resonance parameters. In order to reproduce the observed fluctuations of the locally averaged capture cross sections, these evaluations were obtained by adjusting the energy dependence of a subset of "average" resonance parameters. For instance, in ENDF/B-V the s-wave strength function is taken to be constant and the p-wave strength function is fitted so as to reproduce the fluctuations of the the average capture data. In Fig. 3 the values assumed by ENDF/B-V for the s-wave and p-wave local strength functions up to 40 keV are compared with recent experimental determinations of the local s-wave strength function.[38,40] As the figure indicates, there is no experimental basis to take the s-wave strength function as constant above 4 keV. The ENDF/B-V procedure will reproduce the evaluated infinitely dilute capture cross section by construction, but there is no reason to expect that it will yield correct values for the self-shielding factors.[41-43] In fact, recent studies[44] indicate that when the ENDF/B-V unresolved resonance methodology is applied in the resolved range it yields self-shielding factors which may differ, locally, by more than 20% from the "correct" values, i.e. the values computed with the resolved resonance parameters. An example of this study is shown in Table 4.

Even when the self-shielding is averaged over 1 to 3 keV wide intervals, the self-shielding computed with the unresolved methodology may differ by 5% from that computed with resolved parameters. This is significant compared to the 1% accuracy requested by fast reactor designers.

To improve unresolved resonance representation, Sowerby[45] had already recommended in 1974 that evaluations should be made consistent not only with average capture data but with all available transmission and self-indication measurements as a function of sample thickness and temperature.

In the region below 10 (or perhaps 20) keV, where self-shielding effects are important, it is probably unreasonable to expect a statistical treatment to lead to the 1% requested accuracies so that an extension of the resolved range to these energies is desirable. With increasing incident neutron energy there is an increasing overlap of small levels, nevertheless an inspection of high reolution transmission and capture data suggests that the large resonances which are responsible for most of the self-shielding effects can be resolved and analyzed well above 10 keV. Efforts are now underway at ORNL to extend the resolved representation to 10 keV. Figure 2 shows an example from this study.

Table 4. ^{238}U self-shielded capture cross sections from 1 to 4 keV

EL-EH eV	$\sigma_n\gamma f$	T = 300° K, σ_o = 50 b $\sigma_n\gamma f^*$	Ratio
900 - 1100	1.535	1.257	1.22
1100 - 1400	1.001	1.000	1.00
1400 - 1600	1.026	1.059	0.97
1600 - 1800	1.303	1.068	1.22
1800 - 2100	0.818	0.941	0.87
2100 - 2400	1.081	1.042	1.04
2400 - 2600	1.169	0.919	1.27
2600 - 2900	0.865	0.955	0.91
2900 - 3100	0.646	0.827	0.78
3100 - 3400	1.123	1.133	0.99
3400 - 3600	0.933	0.823	1.13
3600 - 3800	0.843	0.795	1.06
3800 - 4000	0.856	0.836	1.02
900 - 2100	5.683	5.325	1.07
2100 - 3100	3.761	3.743	1.00
3100 - 4000	3.755	3.587	1.05
900 - 4000	13.199	12.655	1.04

The self-shielding factors were computed by the Bondarenko formula using either a statistical group average as is done in the unresolved region

$$f_\gamma = \frac{1}{<\sigma_n\gamma>} \frac{\left< \frac{\sigma_{n\gamma}}{\sigma_t + \sigma_o} \right>}{\left< \frac{1}{\sigma_t + \sigma_o} \right>}$$

or an energy average as is usually done in the resolved region.

$$f_\gamma{}^* = \frac{1}{<\sigma_{n\gamma}>} \frac{\int \frac{\sigma_{n\gamma}(E)dE}{\sigma_t(E) + \sigma_o}}{\int \frac{dE}{\sigma_t(E) + \sigma_o}}$$

II-F. THE C/E DISCREPANCY OF ^{28}c/^{49}f IN FAST ASSEMBLIES

A review of the ^{238}U unresolved issues is incomplete without a short discussion of the most elusive discrepancy affecting the U.S. fast reactor physics program: the persistent overprediction, by 5 to 10%, of the ^{238}U capture rate, ^{28}c, relative to the ^{239}Pu fission rate, ^{49}f, in typical fast reactor spectra. To the extent that the discrepancy points to inadequacy in the evaluation or treatment of the ^{238}U cross sections, it impacts on the calculation of the most important fast reactor performance parameters, such as the breeding ratio, k_{eff}, and the Doppler coefficient of reactivity.

A complete discussion of the issue is beyond the scope of this paper, but was the central theme of a recent specialist's workshop on "^{238}U Capture in Fast Reactors,"[46] where all the aspects of the problem were extensively reviewed but where no clear solution emerged. The discrepancy is often "blamed" on the ^{238}U differential capture data because these data exhibit a spread which suggests relatively large uncertainties. As previously explained, the direct measurements of capture have proven to be very difficult. Yet the recent evaluations of the ^{238}U capture cross sections are already lower than most basic data and a further lowering to resolve the discrepancy is not technically justified.[34]

It is relevant to note that roughly half the captures in fast reactor spectra are in the resonance region where self-shielding is important. Recent fast reactor testing indicates that the ^{28}c/^{49}f discrepancy decreases in harder spectra,[46,47] suggesting that the problem is indeed in the resonance region where, as previously stressed, the treatment of self-shielding needs further validation.

A possible conclusion of the workshop on "^{238}U Capture in Fast Reactors" is that it is difficult to ascribe the ^{28}c/^{49}f discrepancy to a single 5 to 10% problem but that it results more probably from several problems at the 2 to 3% levels, with possibly also the ^{239}Pu data and the direct measurements and their representation. In any case the important technological impact of the discrepancy justifies further work towards a better understanding of the ^{238}U resonance region.

III. ENERGY-AVERAGED REGION

Above the inelastic-neutron-scattering threshold (45 keV) the neutron cross sections and associated processes of ^{238}U follow a relatively energy-smooth behavior. There are some basic issues but the primary motivation is applied: for the design, optimization and operation of fast-breeder reactors (FBR). Recently that applied need has been reiterated by Lineberry et al.[48], as per Table 5.

Table 5. Energy-averaged ^{238}U FBR Data Needs

Reaction	Energy	Desired Accuracy (%)
(n;f)	> Threshold	5
(n;γ)	> 25 keV	2
(n;n')	> Threshold	8
Nu_{total}	> Threshold	1
$Nu_{delayed}$	> Threshold	3

These specifications have been essentially constant for more than two decades. There are additional specialized needs related to the fuel-cycle, diagnostic procedures and alternate-energy concepts.

III-A. TOTAL CROSS SECTIONS

These quantities are a governing factor in macroscopic transport and basic microscopic parameters. Their uncertainties are reflected throughout a comprehensive evaluation and they are a benchmark for model development. One would expect the ^{238}U energy-averaged total cross sections to be well established. An assessment of the situation is largely an exercise in geriatrics as only two sets of data have been reasonably reported since 1974 and only one of these extends above 100 keV.

The low-energy region (45-200 keV) is a concern in applications and microscopic measurements as self shielding effects can distort the cross sections by 5-8% depending upon the samples and energies involved.[49,50] The measured values can be experimentally corrected to infinitely-dilute cross sections using various sample sizes, or theoretically corrected using resonance models.[49] However, attention to self-shielding is clearly evident in only two of the reported data sets. At low energies, even in a broad average (e.g. 5 keV), the cross section displays fluctuations of up to 5% in magnitude.[50] The details are not well known but measurements made at single energies are questionable and the fluctuations will certainly be enhanced in isolated exit channels (e.g. capture and scattering channels).

From several-hundred keV to 10 MeV the data base is strong, with 8-10 comprehensive sets agreeing to within ≈ 1%. Fluctuations persist into this region but they are of very small magnitude. Above 10 MeV the experimental uncertainties increase, and above 15 MeV the data become sparse and even conflicting. The latter shortcomings are of minor note as it is a region where extrapolation from well established charged-particle reactions can be effective.

At the recent meeting on scattering from actinide nuclei, it was recommended that a rigorous evaluation of the ^{238}U total cross section be undertaken in an effort to establish the contemporary accuracy.[51] That recommendation has been pursued[52] with the results illustrated in Fig. 4. The evaluation indicates that the total cross section is known to fractional-percent accuracies with uncertainties of $\gtrsim$ 1% in only very limited regions and these are supported by comparisons with the independent ENDF/B-V evaluation. At the meeting[51] it was also suggested that a simple spherical optical model could effectively parameterize actinide total cross section.[53] Such a model is descriptive of the evaluated results (see Fig. 4) and is useful for interpolation, though it may have shortcomings in other contexts.

From the above it is concluded that: i) The total cross section is sufficiently known above several-hundred keV, and is even of a standard-reference quality. ii) At low energies uncertainties locally increase to several percent due to limited experimental information, fluctuations and self-shielding effects. The situation is a concern for applications, evaluation and model development. iii) Cross-section uncertainties increase above 15 MeV but there is no strong motivation for improvement.

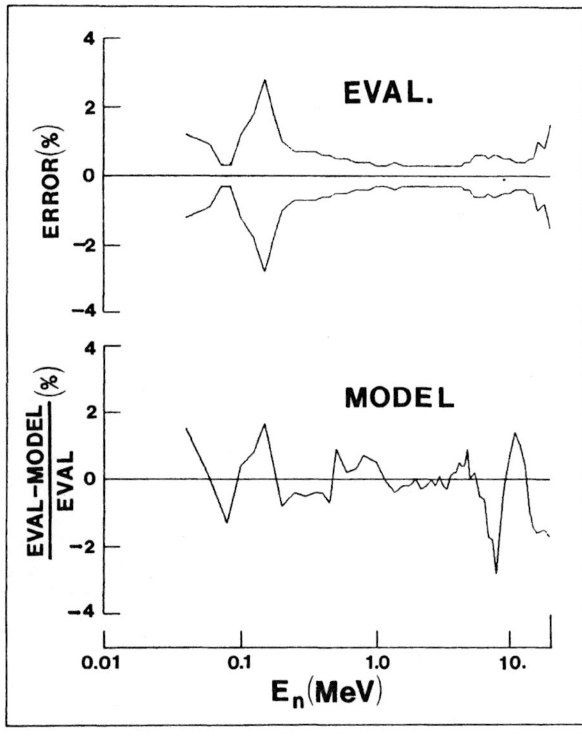

Fig. 4. ^{238}U total-cross-section-evaluation uncertainties (upper) and deviation of the evaluation from the model (lower).

III-B. RADIATIVE-CAPTURE CROSS SECTIONS

The $^{238}U(n;\gamma)$ process is one of the most sensitive and yet poorly known of FBR cross sections. For example, dk $\approx$ 0.2 $d\sigma^{48}$ and even the one-group cross sections, cited for a reference benchmark, vary by ± 3% (RMS).[54] The situation has recently been reviewed by Poenitz,[55] assessing measurement capabilities, data status and future options. The present remarks draw upon that review.

Direct-detection measurements have employed large-liquid-scintillation, energy-proportional and energy-weighting detectors. At low energies, there is little alternative to these direct-detection methods and they have also been used well into the MeV region. The scintillation tank is troubled by low binding energy and high backgrounds. Energy-weighting and energy-proportional devices avoid some problems. However, comparisons of results obtained with prompt-detection techniques continue to show discrepancies in both shape and magnitude of at least 5% and frequently above 8%. Moreover, these discrepancies are not decreasing with time. We may well be "bumping" into a technological limit.

Fortuitously, the ^{238}U (n;γ) process leads to ^{239}Np decay that can be precisely measured, particularly using the complementary decay of ^{243}Am.[56] When one seeks high accuracies, there are detailed problems in activity measurements but they are manageable. It has been demonstrated that activities induced in differential and integral irradiations can be determined to a consistency of $\approx$ 1%.

Flux determination is inherent to nearly all methods but flux measurements have greatly improved with the advent of such devices as the "black" detector[57] and now appear capable of 1-2% accuracies, as illustrated by the contemporary knowledge of the $^{235}U(n;f)$ cross section. Furthermore, such important applied parameters as $^{28}c/^{49}f$ are ratios requiring no explicit flux determination.

The contemporary data base, outlined in Fig. 5, remains essentially unchanged from the 1979 Knoxville Conference. There it was concluded that the uncertainties were ± 5% up to 0.5 MeV and ± 10% to 1.5 MeV.[34] The data base represents a number of measurements made with a variety of techniques. It can be hoped that systematic uncertainties will, at least in part, compensate one another. The above uncertainty estimates are supported by a recent evaluation,[58] correlating a number of standard cross sections (see Fig. 6). The new evaluation agrees with the prior ENDF/B-V to within 5-8%. If anything, the new evaluation tends toward locally larger cross sections, in contradiction to integral suggestions.

Clearly, the applied goal has not been reached. However, with diligence, 2% accuracies appear achievable at selected energies in the important region below 500 keV. The most promising technique appears to be activation.

The objective is essentially a "standard" accuracy that transcends the capabilities of model calculations but the latter can give guidance as to the relative energy dependence of future evaluations.

III-C. SCATTERING CROSS SECTIONS

Knowledge of ^{238}U elastic scattering in the few-MeV region has considerably improved with high resolution results such as shown in Fig. 7. Broader resolution measurements yield the angle-integrated elastic-scattering cross sections to ± 3%. These elastic-scattering results certainly meet the applied need in this energy region. Free adjustment of elastic scattering in the evaluation process is no longer a viable option. Together with well-known total and

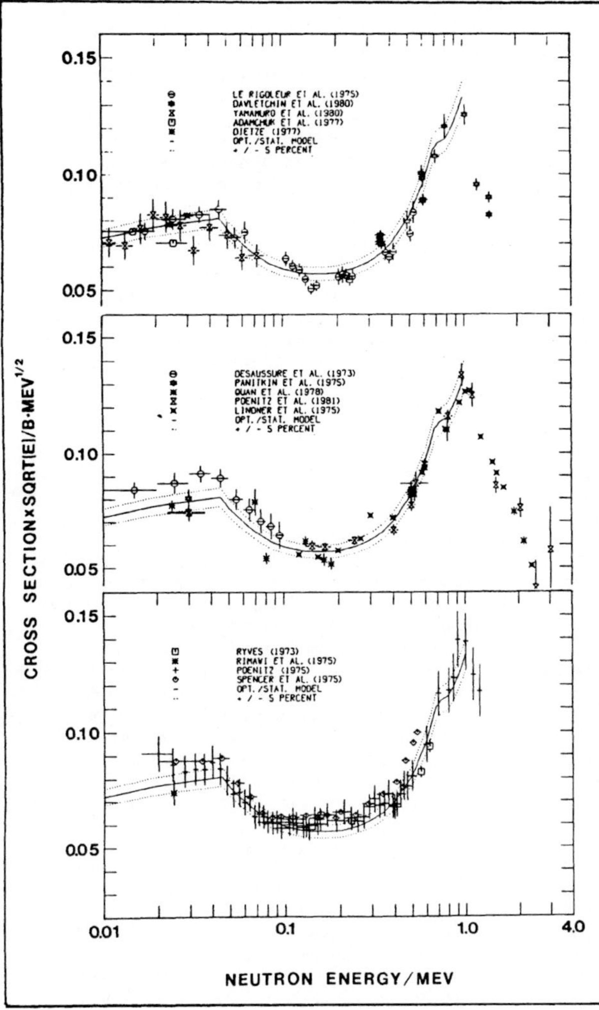

Fig. 5. $^{238}U(n;\gamma)$ cross sections. Symbols note measured values and curves model results, ± 5% (from ref. 55).

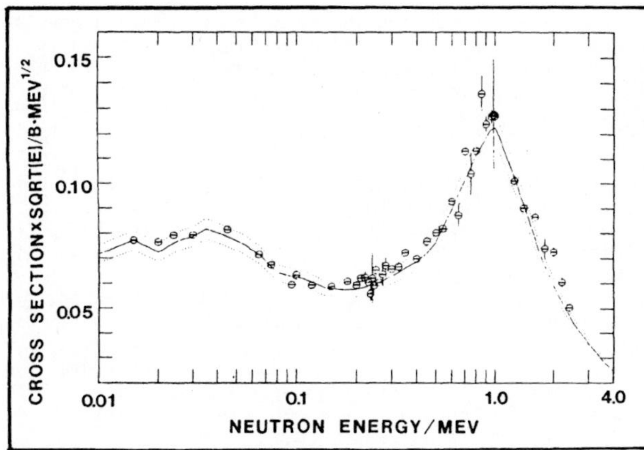

Fig. 6. Evaluated $^{238}U(n;\gamma)$ cross sections. Symbols note the evaluation of ref. 58 and curves that of ENDF/B-V (± 5%).

fission cross sections, these elastic-scattering results imply total inelastic-scattering cross sections to 8-10% accuracies.[60] Thus, above 1 MeV, inelastic-scattering uncertainties are associated with the distribution of the elements of the transfer matrix within the envelope of a reasonably well known total inelastic-scattering cross section.

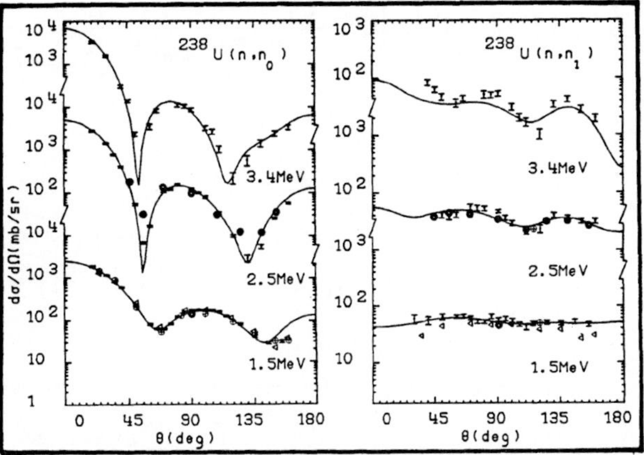

Fig. 7. Direct-scattering from the ^{238}U GSRB. Elastic = left. Inelastic (E_x = 45 keV) = right. Measurements (symbols) and calculations (curves) are shown. (from ref. 59)

The differential direct excitation of the ground-state-rotational band (GSRB) in the few-MeV region is well defined by measurements (see Fig. 7). While these results represent only modest energy transfer in FBR systems, and integral calculations conventionally assume isotropic inelastic neutron emission, they are an excellent foundation for the development of coupled-channels models[61] that provide quantitative extrapolation in energy and reaction type and guidance for evaluation. The models also give insight into the basic nature of the scattering process, particularly as to coupling schemes and the collective deformation.

Below 1 MeV scattering is dominated by contributions from the GSRB and the inelastic process is essentially due to the compound-nucleus (CN) interaction. It is a region of particular FBR importance, yet the experimental results are widely spread (see Fig. 8). At the lowest energies, inelastic measurements show an anisotropy not characteristic of the energy-averaged CN process[62] and there is evidence of considerable energy-dependent fluctuation. The latter could contribute to the evident differences between measured values. In this region theory is marginal and calculations characteristically under-predict the magnitudes of the inelastic cross sections. The effect is illustrated in Fig. 8 where calculated results (using the fluctuation correction of Engelbrecht and Wiedenmüller and the optical parameters of Lagrange[61,63]) are at the lower bound of both measurement and evaluation. The calculations are sensitive to the fluctuation correction. The shortcomings of the CN calculations are less evident in differential-cross-section comparisons (see Fig. 9). This example is at an important and experimentally favorable energy and the measured values should be better known.

A wealth of levels commences at $E_x \approx 0.7$ MeV with the onset of collective-vibrational-band structure. Inelastic scattering is reasonably known near the first few thresholds (see Fig. 10) but becomes increasingly uncertain with energy. The applied need for detailed level information is limited but it is important for the interpretation of (n;n',γ) measurements.[64,65] Level uncertainties tend to limit usefulness of that type of measurement to within

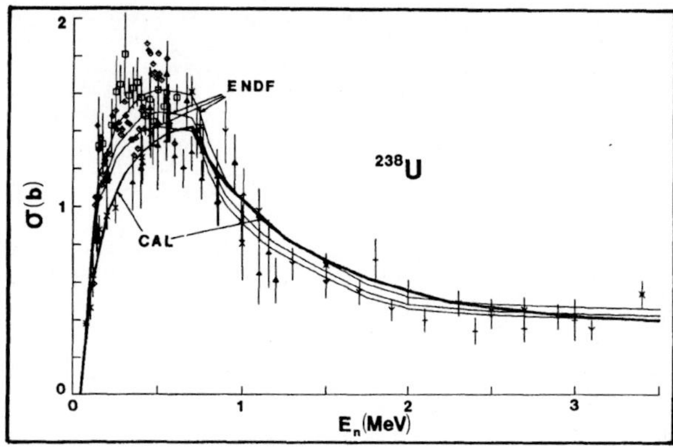

Fig. 8. Inelastic excitation of the 2+ (45 keV) rotational level. Symbols note measured values and curves the results of calculation and evaluation (± evaluation uncertainty).

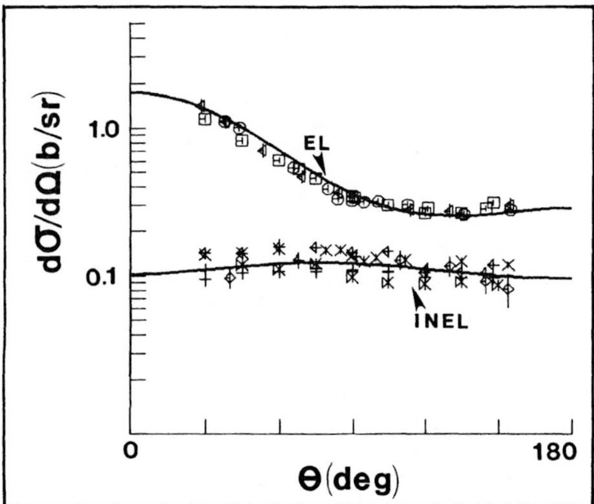

Fig. 9. Scattering of 0.55 MeV neutrons from ^{238}U. Symbols note experimental results and curves calculated values.

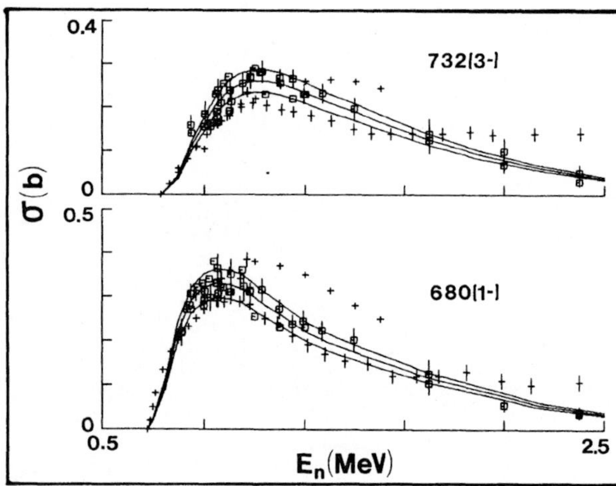

Fig. 10. Inelastic excitation of the first two members of the octupole band. Symbols note experimental values and curves ENDB/B-V (± 10%).

several-hundred keV of threshold. Recourse is often made to models, using statistical-level prescriptions, for the calculation of channel excitation and competition. The results are sensitive to the poorly known statistical-level properties.[66] In addition, there is the theoretical question of the direct excitation of vibrational bands and it has not been resolved in simpler contexts.[67] Like measurement, calculation is most productive near threshold.

Inelastic scattering due to levels at > 1 MeV appreciably affects the transfer matrix. Measurements are fragmentary and calculation faces the same problems encountered at lower energies (see above). Evaluators usually employ simple statistical models or pseudo levels, with magnitudes adjusted for consistency with integral results.

It has been pointed out[68] that the (n;γ,n') process, while small, can significantly affect the FBR transfer matrix. Evaluations generally ignore the process.

Some alternate-energy systems (e.g. fusion-fission hybrid) need high-energy scattering data. Such information is fragmentary and recourse should be made to model extrapolation from complementary and better known charged-particle processes.[69] High-energy inelastic-continuum spectra have been measured.[70] The results, complicated by fission and (n;2n') contributions, consist of CN and pre-CN components with appreciable angle-energy correlations that are not consistent with most evaluation formats. Thus evaluated high-energy inelastic spectra are generally model constructions adjusted for consistency with integral observation.[71]

It is concluded that direct scattering from the GSRB and the total inelastic cross section are sufficiently known above ≈ 1 MeV. Low-energy inelastic scattering ($\lesssim$ 0.7 MeV) remains experimentally and calculationally inadequately known but experimental resolution of the shortcomings appears feasible. Understanding of the excitation of the collective-vibrational bands is improving but above ≈ 1 MeV the situation remains uncertain and no experimental or calculational solution is clearly in sight. There are fundamental questions involving: level and decay properties, fluctuations, reaction mechanisms, coupling schemes and statistical-level properties.

III-D. FISSION CROSS SECTIONS

Immediately above the inelastic-scattering threshold, the ^{238}U(n;f) cross section displays a profusion of subthreshold resonances[72] that have been attributed to a partially damped vibrational level. This type of process has been considered and found to have little effect on FBR systems.[73]

From the subthreshold region, through the higher energies of applied interest, there have been a number of absolute and relative (to ^{235}U(n;f)) fission-cross-section measurements. The results, through 1977, have been reviewed and evaluated by Poenitz et al.[33] It was concluded that there was a good consistency between absolute and relative evaluated results (see Fig. 11). The ENDF/B-V evaluation is a logical combination of these two evaluations. The evaluation uncertainties are outlined in Table 6. They are generally smaller than needed for applications.

Subsequent to the above review, there have been a few new results. Most of these are reference values at isolated energies, which tend to support the prior conclusions. One new measurement was a comprehensive study of the ^{238}U/^{235}U(n;f) ratio from the subthreshold region to above 20 MeV.[72] These latter results are generally consistent with the evaluation of ref. 33 (see Fig. 12) excepting the

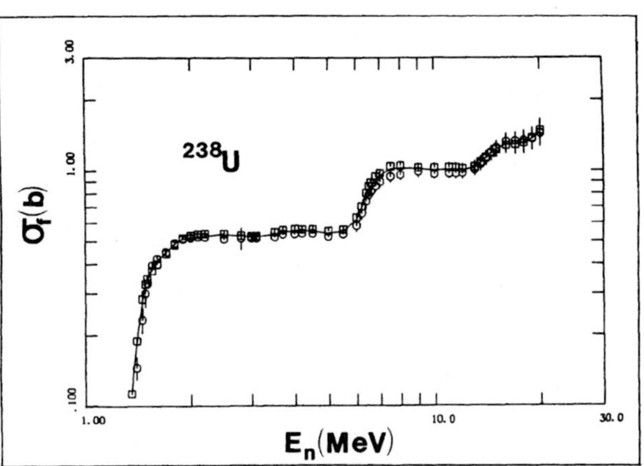

Fig. 11. Evaluated $^{238}U(n;f)$ cross sections from ref. 28. ○ = deduced from absolute data, □ = deduced from ratios to $^{235}U(n;f)$, ——— = composite evaluation.

Table 6. $^{238}U(n;f)$ evaluation uncertainties[33]

E_n(MeV)	$d\sigma/\sigma$ (%)
0.5	10.0
1.0	8.0
2.0	1.3
3.0	2.4
5.0	2.6
10.0	2.3
20.0	8.4

region just above 2 MeV where the cross sections are 4-5% larger than indicated by evaluation and other measurements. This is an energy-local discrepancy but it does imply ≈ 1.5% change in the fission-spectrum-integrated cross section and direct measurements of that quantity are now reported to the 1.5-2.0% uncertainty level.[74] It is also noted that the $^{28}f/^{49}f$ ratio, as measured in FBR cores, has characteristically been larger than calculated.[75]

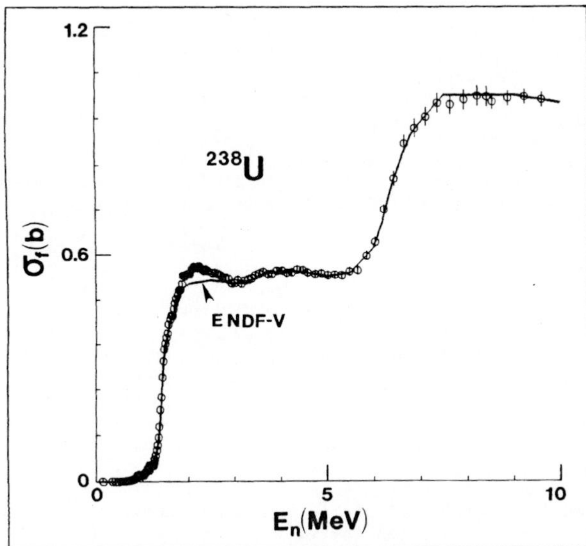

Fig. 12. Comparison of the $^{238}U(n;f)$ evaluation of ref. 33 (curve) with the measured values implied by ref. 72 (symbols, assuming the ENDF/B-V $^{235}U(n;f)$ cross section).

There remain considerable discrepancies above 15 MeV. These may be associated with fission-fragment anisotropies which are now becoming better known.[76] The above comments assume a relatively smooth energy dependence of the cross section above the first threshold. High-resolution measurements show that there are fluctuations of up to 5% in magnitude over the 2-4 MeV plateau region.[77] Such structure is not of direct applied interest nor is it represented in common evaluations; however, it is of concern in the use of the $^{238}U(n;f)$ cross section as a flux monitor.

Generally, the contemporary knowledge of the $^{238}U(n;f)$ cross section meets the applied need. It is disturbing that a recent quality measurement suggests a local discrepancy near 2 MeV and the situation in that limited region should be re-examined.

III-E. (n;2n) AND (n;3n') CROSS SECTIONS

The $^{238}U(n;2n')$ cross section has been determined using both direct-detection and activation techniques. A comprehensive data base has been accumulated over more than three decades. That knowledge has been reviewed and correlated with modern reference parameters by Kornilov et al.,[78] with the results summarized in Fig. 13. The experimental results give consistent coverage from threshold to 20 MeV. Kornilov et al. evaluated these data, guided by a statistical model (see curve of Fig. 13). The evaluation uncertainties are 5-10% in regions of appreciable cross section. A small artifact ("bump") near 12 MeV is within the evaluation uncertainty.

The $^{238}U(n;3n')$ cross sections can only be measured by direct detection. However, they are defined to 10-15%, over regions of appreciable magnitude, to 20 MeV and beyond.[79] The process is of little relevance to fission-energy systems.

The (n;2n') and (n;3n') emitted-neutron spectra have not been explicitly measured. However, the statistical model has been widely employed to obtain reasonable spectral estimates.[80]

Generally, FBR needs for (n;2n') and (n;3n') data appear reasonably met. There may be some unsatisfied requirements for higher-energy (n;xn') data (e.g. for electro-nuclear breeding, dosimetry, etc.)

III-F. PROMPT-FISSION-NEUTRON EMISSION

There have been a number of measurements of nubar of ^{238}U, using liquid-scintillation and thermalization detectors. The results span the range from threshold to 15 MeV with detail and accuracy as shown in Fig. 14. Many of the experimental uncertainties are in the 2-5% range and the normalization is conventionally relative to nubar of ^{252}Cf-spontaneous or ^{235}U-thermal fission. The overall energy dependence is well represented by the linear parameterization of Howerton[81] (Fig. 14, curve A, the small offset is due to differing ^{252}Cf reference values). Madland and Nix[82] have used a detailed evaporation model to describe the energy dependence (Fig. 14, curve C). The model is general and well verified with better-known actinide fission processes.

The experiments suggest a slight non-linear energy dependence of nubar with a transition point at ≈ 3 MeV. It is a small effect that appears to emerge from the assemblage of data. The evaluation of Vorob'eva et al.[83] portrays this non-linearity (Fig. 14, curve B). It has been suggested that the non-linearity reflects an interplay between fragment excitation and kinetic energies in the few-MeV region.[83]

Measurements of the ^{238}U prompt-fission-neutron spectrum are exceedingly difficult and the results

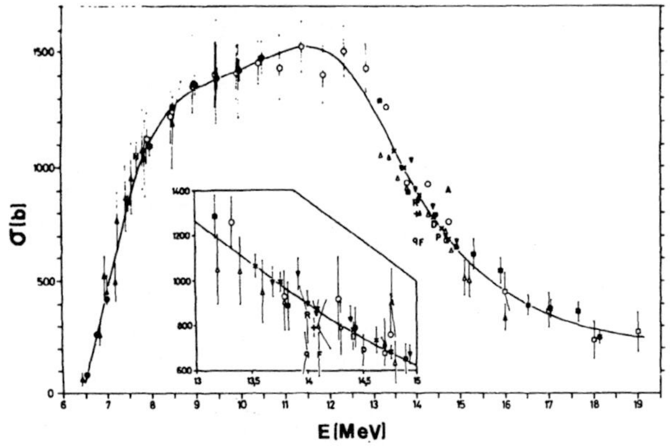

Fig. 13. Measured (symbols) and evaluated (curve) $^{238}U(n;2n')$ cross sections (taken from ref. 78).

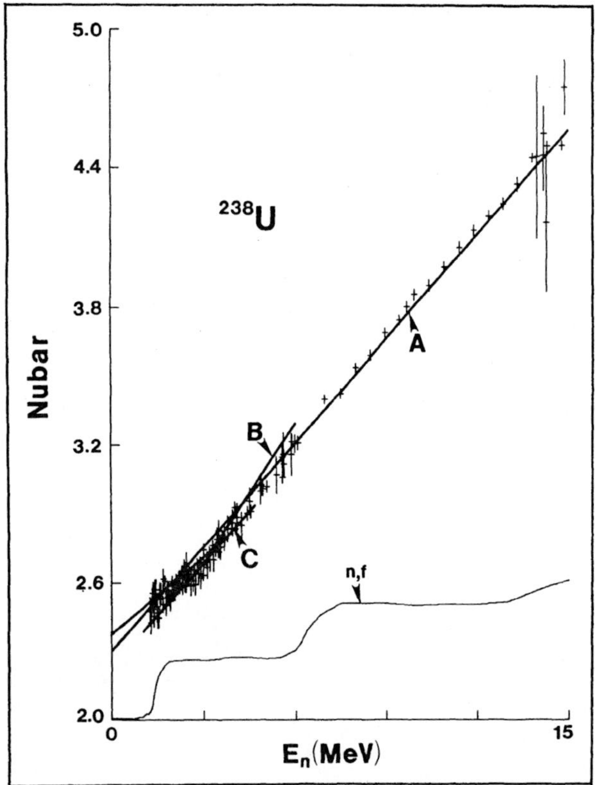

Fig. 14. Nubar of ^{238}U. Experimental results are indicated by +. Curves denote; A = ref. 81, B = ref. 83 and C = ref. 82.

not particularly definitive.[84] The experiments indicate an approximately Maxwellian shape with a temperature of ≈ 1.3 MeV near threshold, increasing linearly at the rate of 0.02 MeV/MeV.[84] It is perhaps more reliable to interpolate between far better known spectra using the model of Madland and Nix[82] which predicts an average energy of 2.041 MeV (at 2 MeV), a linear increase in average energy at the rate of 0.029 MeV/MeV over the several MeV region, and a spectrum somewhat differing from a Maxwellian.

The energy dependence of nubar appears adequately known for applied purposes. The normalization is uncertain by 2-3%, approximately half of which is due to the ^{252}Cf reference value. Detailed questions as to energy-dependent shape may be of basic interest but are of minor applied note.

III-G. DELAYED-FISSION-NEUTRON EMISSION

Reactivity is directly related to β_{eff}. Most important is the total delayed-neutron yield, with group yields and decay constants less critical factors. β_{eff} is not particularly sensitive to the spectrum; thus the spectrum requirements are a modest 20% in mean energy.[85]

Contemporary knowledge of the total delayed-neutron yield of ^{238}U is summarized in Fig. 15.

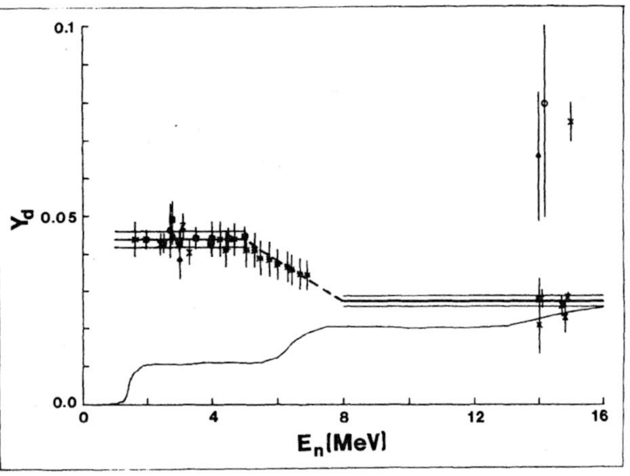

Fig. 15. ^{238}U total delayed-neutron yields. Symbols note measured values, curves the evaluation of Tuttle,[86] ± 5%.

It is not an auspicious display. Tuttle[86] suggests that "fast" and 14 MeV yields are uncertain by 2.3% and 2.9%, respectively, but only a single measurement (near 14 MeV) approaches those accuracies. It is theoretically suggested that the yield slowly rises from the first to the second threshold, then abruptly drops to the value observed near 14 MeV.[86,87] Only a single data set spans the transition region[88] and there is a void from 7-14 MeV. The single set indicates the onset of the transition well below the second threshold. This may reflect the effect of shell structure.[89] The contemporary knowledge of total delayed-neutron yields only marginally, if at all, meets the need and recent studies of noise in FBR systems suggest that microscopic data underpredict β_{eff} by 3-7%.[90] Measurements to the 2% level in the "fast" region are desirable and now more promising due to better known delayed-neutron spectra.

The six-group decay structure of Keepin et al.[91] is widely accepted for applications, though it is clearly an approximation of a much more complex phenomenon. Within this approximation the group abundances and decay constants warrant improvement. Recent results indicate that more of the delayed-neutron spectrum is below 100 keV than was previously thought.[92] These new results probably establish the average-gross-spectrum energy to 20% and, if so, the need is met. Major advances have been made in the understanding of nuclei far from stability using on-line mass separators (e.g. TRISTAN[93]). The results are of outstanding fundamental interest and, as they accumulate, will be combined to improve the gross delayed-neutron parameters of applied interest.

III-H. FISSION-PRODUCT YIELDS

^{238}U "fast" (e.g. FBR spectrum) fission-product yields are reasonably known.[94] Recently the explicit energy dependence has been better defined.[95] The character is isotope dependent but, generally, at the maxima of the yield peaks the yields decrease with energy at the rate of several percent per MeV and in the valley increase by several-tens percent. Detailed energy dependences are illustrated in Fig. 16. The log of the yield is essentially a linear function of energy with the slope of the valley yields changing as the fission thresholds are passed. A number of these yields are used as experimental indexes for burnup, fission-rate determinations, etc. and for constructing delayed-neutron parameters. Care must be taken to account for energy dependence and the requisite information is frequently unavailable in the general evaluated files.

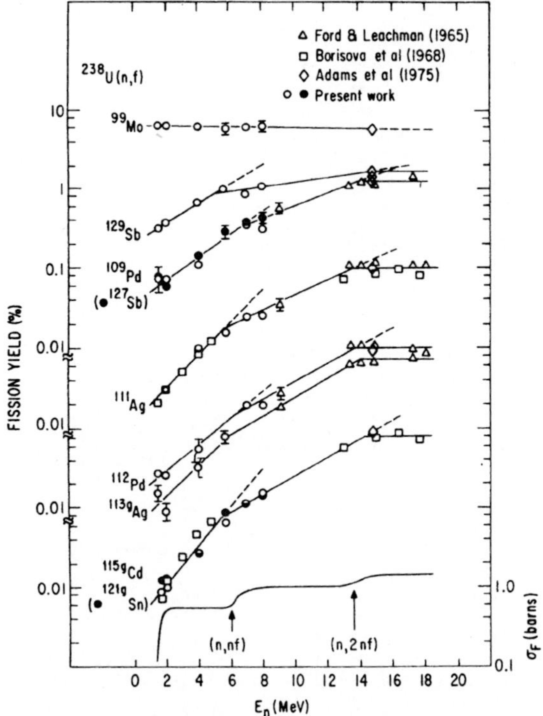

Fig. 16. Illustrative energy dependences of fission-product yields (from ref. 95).

IV. CONCLUDING COMMENTS

Looking back, recent measurements and their analyses have resulted in a great improvement in the values of the parameters of the first few prominent resonances. Remaining inconsistencies, although large with respect to the very small estimated experimental uncertainties, are small when compared to the applied accuracy goals, and absorption in thermal lattices is reasonably predicted. Persisting systematic discrepancies in the values reported for the neutron widths above 1.5 keV are disturbing and suggest the need for further work on multilevel resonance overlap effects as well as Doppler and resolution broadening. Dilute capture in the 4-20 keV region is uncertain by ± 10%. Above 4 keV evaluations conventionally employ unresolved resonance descriptions based upon energy dependent s_0 and s_1 estimates. This statistical approach may result in errors in the calculation of self-shielding factors which are large compared to the applied goal of 1% self-shielding accuracy. Improvement probably hinges upon the detailed study of, at least, the prominent resonances to 10 keV. Differences between measured and calculated integral values for the $^{28}c/^{49}f$ ratio persist and this is a key integral parameter. Energy-averaged total and (n;xn') cross sections and fission-product yields are reasonably in hand. There remain some detailed questions as to the fission cross section and prompt-fission-neutron emission; however, the broad aspects of both processes seem clear and future improvements considerably depend upon those of the relevant reference standards. Energy-averaged radiative-capture cross sections remain a serious problem and delayed-neutron emission is not sufficiently understood. In both of these cases experimental solutions appear feasible. There are large uncertainties in some aspects of the scattering process and their future resolution is not clear. Many of the outstanding energy-averaged problems involve accuracies that are inconsistent with a reasonable projection of present theoretical knowledge. An exception is the scattering process where improved basic understanding has the potential for sharp impact. Certain aspects of, particularly, the fission process raise basic questions and answers could significantly influence the applied data.

ACKNOWLEDGEMENTS

The authors ar indebted to Drs. P. Moldauer, R. Peelle, R. Perez and W. Poenitz for assistance in the preparation of this manuscript.

REFERENCES

1. L. Dresner, Resonance Absorption in Nuclear Reactors, Pergamon Press (1960).
2. H. Tellier, Proc. of the IAEA Consultants Mtg. on Uranium and Plutonium Isotope Resonance Parameters, INDC(NDS)-129/GJ, p. 7 (1982).
3. Seminar on ^{238}U Resonance Capture, Brookhaven Natl. Lab. Report, BNL-NCS-50451, (1975).
4. H. Liou and R. E. Chrien, Nucl. Sci. and Eng., 62 463 (1977).
5 D. K. Olsen et al. Nucl. Sci. and Eng., 62 479 (1977); also Nucl. Sci. and Eng., 69 202 (1979).
6. T. J. Haste and M. C. Moxon, Proc. Inter. Conf. Neutron Physics and Nuclear Data for Reactors and Other Applied Purposes, Harwell, CONF-780925, p. 337, IAEA (1979).
7. Y. Nakajima, J. Nucl. Energy, 7 25 (1980).
8 F. Poortmans et al., Proc. IAEA Consultants Mtg. on Uranium and Plutonium Isotope Resonance Parameters, INDC(NDS)-129/GJ, p. 112 (1982)
9 R. C. Block et al., Nucl. Sci. and Eng., 80 263 (1982).
10. H. E. Jackson and J. E. Lynn, Phys. Rev., 127 461 (1962).
11. F. Rahn et al., Phys. Rev., C6 1854 (1972).
12. S. F. Mughabhab and D. I. Garber, Neutron Cross Sections, Vol. I, Resonance Parameters, Brookhaven Natl. Lab. Report-325, Third Edition (1973).
13. M. C. Moxon, Evaluation of the ^{238}U Resonance Parameters, Specialists Mtg. on Resonance Parameters of Fertile Nuclei and ^{239}Pu, NEANDC(E) 163U, p. 73 (1975).
14. F. J. McCrosson, Brookhaven Natl. Lab. Report, BNL-NCS-50451, p. 22 (1975).
15. G. de Saussure et al., Progress in Nuclear Energy, Vol. 3, p. 87 (1979).
16. F. Froehner, reported by B. Goel and F. Weller, KFK Report, 2386/111 (1977).
17. S. F. Mughabhab and M. Divadeenam, Proc. IAEA Consultant Mtg. on Uranium and Plutonium Isotope Resonance Parameters, INDC(NDS)-129/GJ, p. 426 (1982).
18. T. Nakagawa et al., Proc. IAEA Consultant Mtg. on Uranium and Plutonium Isotope Resonance Parameters, INDC(NDS)-129/GJ, p. 282 (1982).
19. G. de Saussure et al., Nucl. Sci. and Eng., 61 496 (1976).

20. H. Tellier, Nucl. Sci. and Eng., 79 393 (1981).
21. J. L. Rowlands, Proc. IAEA Consultant Mtg. on Uranium and Plutonium Isotope Resonance Parameters, INDC(NDS)-129/GJ, p. 8 (1982).
22. J. Hardy, Benchmark Testing of ENDF/B-V Data for Thermal Reactors, to be published (1982).
23. F. Froehner, Proc. IAEA Consultant Mtg. on Uranium and Plutonium Isotope Resonance Parameters, INDC(NDS)-129/GJ, p. 15 (1982).
24. J. Rowlands, Brookhaven Natl. Lab. Report, BNL-NCS-51363, p. 23, Vol. 1 (1982).
25. J. Rowlands Proc. IAEA Consultant Mtg. on Uranium and Plutonium Isotope Resonance Parameters, INDC (NDS)-129/GJ p. 25 (1982).
26. M. Salvatores et al., Proc. IAEA Consultant Mtg. on Uranium and Plutonium Isotope Resonance Parameters, INDC(NDS)-129/GJ, p. 31 (1982).
27. G. Carraro and W. Kolar, Proc. Inter. Conf. Nuclear Data for Reactors, IAEA-CN-26/226, p. 403, Helsinki (1970).
28. J. P. L'Heriteau and P. Ribon, Conf. on Neut. Cross Sections and Technology, p. 438, CONF-710301, (1971); H. Derrien and P. Ribon, p. 63 in NEANDC(E) 163U (1975), see Ref. 13; H. Derrien p. 156, Brookhaven Natl. Lab. Report-NCS-50451 (1975), see Ref. 3.
29. F. Rahn and W. W. Havens, Jr., A Review of Total Radiation Widths of Neutron Resonances of ^{238}U, NEANDC(US)-179/U (1973).
30. G. A. Keyworth and M. S. Moore, Proc. Inter. Conf. Neutron Physics and Nuclear Data for Reactors and Other Applied Purposes, Harwell, CONF-780925, p. 241, IAEA (1979).
31. J. E. Lynn, Neutron Resonance Reactions p. 228, Clarendon Press, Oxford, (1968).
32. F. Corvi, G. Rohr, and H. Weigmann, Vol. 2 p. 733 Proc. Conf. on Nuclear Cross Sections and Technology, NBS Special Publication 425, CONF-75619216 (1975).
33. W. P. Poenitz et al., Argonne Natl. Lab. Report ANL/NDM-32, Argonne, (1977).
34. W. P. Poenitz, Proc. Conf. on Nuclear Cross Sections for Technology, NBS Spec. Pub. 594 p. 368 (1979).
35. M. S. Moore et al., Proc. IAEA Consultants Mtg. on Uranium and Plutonium Isotope Resonance Parameters, INDC(NDS)-129/GJ, p. 151 (1982).
36. N. Yamamuro et al., Nucl. Sci. and Tech. 17 582 (1980).
37. J. T. Yang et al., to be published (1982).
38. P. S. Meszaros and D. K. Olsen, Trans. Am. Nucl. Soc., 41 560 (1982).
39. ENDF/B-V file for ^{238}U (MAT-1398); E. Pennington, A. B. Smith, W. P. Poenitz, and R. Howerton principal evaluators.
40. N. J. Bee, Proc. IAEA Consultants Mtg. on Uranium and Plutonium Isotope Resonance Parameters, INDC(NDS)-129/GJ, p. 182 (1982).
41. G. de Saussure and R. G. Perez, Ann. Nucl. Energy, 9 79 (1982).
42. J. L. Munoz-Cobos and V. J. Serradell, Energia Nuclear, 134, 463 (1982).
43. S. Ganesan, Proc. IAEA Consultants Mtg. on Uranium and Plutonium Isotope Resonance Parameters, INDC(NDS)-129/GJ. p. 415 (1982).
44. R. B. Perez et al. Trans. Am. Nucl. Soc., 39 883 (1981).
45. M. G. Sowerby, Proc. Specialist Mtg. on Resonance Parameters of Fertile Nuclei and ^{239}Pu, NEANDC(E) 163U (1975).
46. Specialist's Workshop: ^{238}U Capture in Fast Reactors, (1982) Teton Village, Wyoming. Proceedings unpublished.
47. C. R. Weisbin et al., Meeting Cross Section Requirements for Nuclear Energy Design, Oak Ridge Natl. Lab. Report, ORNL/TM-822 (1982) to be published in Ann. Nucl. Energy.
48. M. Lineberry et al., Chapter prepared for NEACRP monograph on the Current Status of Fast-reactor Physics (1981).
49. W. Poenitz et al., Nucl. Sci. and Eng., 78 333 (1981).
50. D. Olsen et al., Oak Ridge Natl. Lab. Report, ORNL/TM-5915 (1977); see also ref. 5.
51. Proc. Specialists Mtg. on Fast-Neutron Scattering on Actinide Nuclei, Paris (1981), NEANDC-158U.
52. W. Poenitz and A. B. Smith, to be published.
53. P. Hodgson, Mtg. of ref. 51.
54. L. LeSage et al., Argonne Natl. Lab. Report, ANL-80-78 (1980).
55. W. Poenitz, Proc. Specialists Mtg. on Fast-Neutron Capture, Argonne (1982).
56. W. Poenitz et al., in press.
57. W. Poenitz, Nucl. Instr. and Methods, 109 413 (1973).
58. W. Poenitz, Brookhaven Natl. Lab. Report, NEANDC(US)-209 (1981).
59. G. Haouat et al., Nucl. Sci. and Eng., 81 491 (1982).
60. A. B. Smith et al., elsewhere in these proc.
61. C. Lagrange, Mtg. of ref. 51.
62. R. Winters et al., Nucl. Sci. and Eng., 78 147 (1981).
63. C. Englebrecht and H. Wiedenmüller, Phys. Rev., C8 859 (1973).
64. D. Olsen, Mtg. of ref. 51.
65. G. Kegel, ibid.
66. E. Arthur, ibid.
67. P. Guenther et al., to be pub. in Phys. Rev.
68. H. Hummell and W. Stacey, Nucl. Sci. and Eng., 50 397 (1973).
69. L. Hansen, Mtg. of ref. 51.
70. N. Kornilov et al., Proc. Kiev Conf. on Neut. Phys., 2 44 (1980).
71. C. Wong et al., Lawrence Livermore Natl. Lab. Report, UCRL-51144 (1972).
72. F. Difilippo et al., Nucl. Sci. and Eng., 68 43 (1978), also Phys. Rev., C21 1400 (1980).
73. D. Meneghetti, private communication.
74. V. Adamov et al., INDC(CCP)-180(L) (1982).
75. M. Lineberry et al., private communication.
76. J. Meadows, to be published.
77. S. Cierjacks, NBS Special Pub. 493 (1977).
78. N. Kornilov et al., INDC(CCP)-181/G (1982).
79. L. Veeser and E. Arthur, Proc. Inter. Conf. on Neut. Phys. and Nuclear Data for Reactors and Other Applied Purposes, Harwell, CONF-780925, IAEA (1978).
80. M. Craner et al., Nucl. Sci. and Eng., 59 395 (1976).
81. R. Howerton, Nucl. Sci. and Eng., 62 438 (1977).
82. D. Madland and J. Nix, Nucl. Sci. and Eng., 81 213 (1982).
83. V. Vorob'eva et al., UDC 539.173.84 (1981).
84. B. Zhuravlev et al., INDC(CCP)-154/L (1980)
85. Ph. Hammer, INDC(NDS)-107/G (1979).
86. R. Tuttle, ibid.
87. R. Moller and J. Nix, Phys. and Chem. of Fission, IAEA Press (1973).
88. M. Krick and A. Evans, Nucl. Sci. and Eng., 47 131 (1972).
89. B. Wilkins et al., Phys. Rev., C14, 1832 (1976).
90. E. Bennett, Argonne Natl. Lab. Report, ANL-81-72 (1981).
91. G. Keepin et al., Phys. Rev., 107 1049 (1957).
92. J. Williams, INDC(NDS)-107/G (1979).
93. R. Gill et al., Nucl. Instr. and Methods, 186 243 (1977).
94. E. Couche, At. Data Tables, 19 419 (1977).
95. J. Gindler et al., Nucl. Sci. and Eng., 70 101 (1979).

BESOINS EN DONNEES NUCLEAIRES POUR LES REACTEURS A NEUTRONS THERMIQUES

J. BOUCHARD*, C. GOLINELLI*, H. TELLIER**
Commissariat à l'Energie Atomique
Département des Réacteurs à eau

*CEN CADARACHE
B.P. n°1
13115 St Paul lez Durance

** CEN SACLAY
B.P. n° 2
91190 Gif-sur-Yvette

Les besoins en données nucléaires pour les réacteurs à neutrons thermiques sont conditionnés par la persistance de désaccords entre les calculs et les mesures de certains paramètres, par le bénéfice que l'on peut retirer d'une réduction des marges et par les modifications envisagées. Trois domaines particuliers sont abordés :
- la réduction de l'écart sur les coefficients de température par modification de la forme des sections efficaces de capture de l'U^{238} et de l'U^{235} et de fission de l'U^{235} dans le domaine thermique.
- l'augmentation de la précision des mesures cinétiques par une meilleure connaissance des données liées aux neutrons retardés.
- l'amelioration de la prévision de l'activité neutronique des combustibles usés par des mesures sur les sections efficaces de capture du ^{242}Pu et du ^{243}Am.

Introduction

L'exploitation industrielle des réacteurs à neutrons thermiques, essentiellement des réacteurs à eau ordinaire apporte depuis plusieurs années la confirmation que les calculs prévisionnels des performances réalisés à l'aide des codes de physique des réacteurs et des bibliothèques de données nucléaires se révèlent très corrects. Ceci ne doit pas complètement masquer le fait que certains ajustements sont appliqués dans ces calculs, soit au niveau des données, en particulier des sections efficaces exprimées par groupes d'énergies, soit directement sur le paramètre recherché. Il est probable également que des compensations de petites erreurs interviennent, ne serait-ce que pour atténuer l'effet d'incertitudes, associées à certaines données primordiales, qui sont trop importantes vis à vis de la précision requise sur les paramètres calculés.

L'état actuel des données, compte tenu des remarques précédentes,permet de satisfaire sans difficultés majeure les études de conception et de fonctionnement des réacteurs en service ou en projet. On peut néanmoins se poser trois questions :

a) Doit-on laisser sans réponse certains problèmes physiques soulevés par des constatations expérimentales plus ou moins en contradiction avec les prévisions théoriques appuyés sur la connaissance actuelle des données nucléaires ?

b) Indépendamment de ces cas particuliers, quel bénéfice pourrait-on retirer d'une amélioration des précisions sur les données nucléaires ?

c) Dans quelle mesure la situation actuelle pourra-t-elle satisfaire les demandes futures liées à des développements nouveaux, en particulier dans le cycle en pile des combustibles de réacteurs à eau ?

Les physiciens ne peuvent pas être indifférents à la première question. Les travaux correspondants ne sont pas nécessairement de première priorité, sauf lorsqu' ils concernent des données d'importance primordiale pour ce type de réacteurs comme les sections efficaces des ^{235}U et ^{238}U dans le domaine thermique, mises en cause pour tenter d'expliquer le désaccord permanent et historique sur les coefficients de température.

La seconde question a fait l'objet de nombreuses controverses. Dans un système extrêmement imbriqué où sont mélangés des approximations de méthodes de calcul, des incertitudes sur les données nucléaires, les résultats d'expériences intégrales et ceux de fonctionnement des réacteurs en service, il est pratiquement impossible de relier clairement l'amélioration de la connaissance des données, qui interviennent le plus en amont, à une économie des coûts d'investissement ou des coûts de cycles. Il y a toutefois quelques exceptions qui, pour les réacteurs actuellement en service ou en construction, sont liées aux marges prises en compte dans l'évaluation de certains paramètres de fonctionnement. Ces marges concernent des grandeurs qui ne sont pas directement accessibles par des mesures dans le réacteur, par exemple l'effet d'une éjection de barre ou la puissance résiduelle à des temps très courts, ou dont la mesure "in situ" est très imprécise. Elles ne sont pas toutes liées à des précisions insuffisantes sur les données nucléaires mais ces dernières peuvent intervenir de façon non négligeable comme, par exemple, à travers la détermination de la fraction effective de neutrons retardés.

Pour répondre à la dernière question il faut situer les modifications qui peuvent intervenir dans le futur de ces réacteurs. Indépendamment de la conception de nouveaux types de réacteurs à neutrons thermiques qui n'est pas prise en compte ici, on peut citer quelques évolutions prévisibles et concernant le coeur des réacteurs à eau :
- augmentation des taux de combustion
- allongement des cycles avec utilisation plus importante de poisons consommables
- modification du rapport de modération
- recyclage des noyaux lourds (uranium de retraitement, plutonium, autres actinides).

Certains de ces domaines ont déjà été couverts par des expériences intégrales, d'autres le seront prochainement. On peut déjà dire, que l'augmentation des taux de combustion, aussi qu'a fortiori le recyclage du plutonium nécessiteront un effort sur les sections efficaces des isotopes secondaires du plutonium et de l'américium sans préjuger des conclusions qui pourront conduire à l'expression des besoins nouveaux.
Dans ce domaine les désaccords persistants pour la prédiction des quantités de ^{244}Cm formées dans le combustible doivent être expliqués.

A travers ces questions ce sont donc trois problèmes

K. H. Böckhoff (ed.), Nuclear Data for Science and Technology, 21–28.

précis et d'actualité que nous allons évoquer plus en détail dans cet exposé. Ce n'est pas une remise en cause des besoins qui ont été exprimés de façon plus exhaustive lors de conférences précédentes /1, 2/ et qui demeurent pour beaucoup valables même si des progrès importants ont été réalisés, comme en témoigne le document de synthèse sur l'uranium 238 présenté aujourd'hui /3/. Pour bien marquer le fait que les trois problèmes exposés en détail ne représentent pas les seules préoccupations actuelles pour les réacteurs à eau, nous terminerons en mentionnant brièvement, à travers une revue des principaux paramètres de réacteurs, d'autres sujets de réflexion sur les besoins en données nucléaires.

1- Forme des sections efficaces des ^{235}U et ^{238}U dans le domaine thermique

Lorsqu'un désaccord est constaté entre des résultats de mesures intégrales précises et les valeurs annoncées par le calcul il est d'usage de pratiquer un "ajustement" qui peut être explicite, modification a posteriori du paramètre calculé, ou implicite, modification d'une ou plusieurs données en amont pour que le calcul donne directement la bonne valeur. Il est admis le plus souvent que cet ajustement doit permettre de pallier une précision insuffisante des données microscopiques ce qui revient à dire que l'on opère un meilleur choix des paramètres fondamentaux dans les marges d'incertitude expérimentale /4/. Ces ajustements portent normalement sur le niveau des données intégrées par groupes ou par domaines d'énergie.

Appliquée aux calculs de réactivité des coeurs de réacteurs à neutrons thermiques, cette démarche permet de fixer un choix de données cohérentes pour obtenir des valeurs précises des principaux paramètres, à l'exception du coefficient de température. Or il s'agit d'un paramètre important, qui intervient dans les études de conception, de fonctionnement et de sûreté. Depuis longtemps l'objectif fixé est de calculer ce paramètre avec une incertitude de $\pm$1 pcm/°C ($10^{-5}\frac{\delta k}{k}$ /°C). Les désaccords constatés entre prévisions théoriques et mesures intégrales se situent nettement au delà de cette marge d'incertitude, en moyenne de 3 à 5 pcm/°C, et portent sur un nombre considérable d'expériences dans des réacteurs de très faible puissance et dans les réacteurs des centrales électronucléaires.

Après avoir résumé la situation de ces comparaisons expérimentales et examiné la sensibilité de ce paramètre aux diverses sections efficaces nous évoquerons les possibilités d'explication suggérées par différents auteurs et étudiées en détail par ERRADI et SANTAMARINA /5/ en concluant sur le besoin d'amélioration de la connaissance des formes de sections efficaces.

1-1 Résumé des constatations faites à partir des expériences intégrales

Les expériences critiques permettent de réaliser des mesures très précises du coefficient de réactivité résultant d'une variation de température du modérateur. Toutefois ces expériences sont, pour les réacteurs à eau, généralement limitées au domaine de température inférieures à 90°C. Deux exceptions, l'expérience KRITZ en Suède /6/ qui a permis d'atteindre des températures un peu supérieures à 200°C pour un réseau complet et l'expérience CREOLE en France /7/ dans laquelle un assemblage était porté à près de 300°C.

Afin de disposer d'un jeu de valeurs cohérentes ces expériences ont été interprétées avec le code APOLLO /8/ et sa bibliothèque à 99 groupes dans laquelle ont été utilisées les sections efficaces ^{235}U et ^{238}U de ENDF/B IV modifiées légèrement en fonction des résultats d'ensemble des mesures de keff.

L'écart moyen relevé sur les expériences critiques à basses températures est de :

$$E - C = 4,2 \pm 2,0 \text{ pcm/°C}$$

Pour l'expérience KRITZ, avec un réseau UO_2, l'écart varie un peu avec le domaine de température :

Domaine de température	(E-C) pcm/°C
90 - 20	+ 5,0
160 - 90	+ 5,0
210 - 160	+ 3,3

Enfin dans l'expérience CREOLE l'interprétation conduit pour la transposition à des effets sur un réacteur de puissance aux écarts suivants :

Réseau	20°C	90°C	250°C	300°C
UO2	+ 6	+ 2	+ 2	- 1
UO2 + 1200 ppm Bore	+ 6	+ 2	0	+ 1
UO2 - PUO_2	+ 6	0	0	- 3

L'incertitude affectant ces valeurs d'écart est estimé à 1pcm/°C pour les températures basses et à 2 pcm/°C pour les températures les plus élevées.

On peut constater que la tendance à une variation de l'écart avec le domaine de température est nettement confirmée quel que soit le type de réseau considéré.

Les corrections dues aux effets des liaisons cristallines mises en évidence par une série de mesures par oscillation dans MINERVE réduisent les valeurs des écarts surtout à 20°C et à 90°C dans une moindre mesure.

Les mesures effectuées sur les réacteurs de puissance sont moins précises qu'en expériences critiques. L'interprétation des résultats est souvent complexe. Néanmoins l'accumulation d'un grand nombre d'essais de ce type permet de dégager des tendances. En comparaison avec des calculs effectués avec les mêmes données APOLLO qu'indiquées précedemment, ces tendances confirment une surestimation de 1 à 3 pcm/°C du coefficient de température dans le domaine de température du fonctionnement normal des réacteurs à eau pressurisée, c'est-à-dire autour de 300°C.

1-2 Sensibilité du coefficient de température aux sections efficaces

L'interprétation des expériences CREOLE fondée sur la bibliothèque des sections efficaces d'APOLLO permet de découpler les différents effets, tout particulièrement, ceux dûs à la densité d'eau et au spectre.

L'analyse qui a été faite conduit à suggérer que l'erreur sur le coefficient de température dans le domaine où la densité de l'eau varie peu est principalement liée aux effets de spectre, très sensibles aux formes des sections efficaces des ^{235}U et ^{238}U dans le domaine thermique.

Un programme complémentaire dans lequel seule la densité apparente de l'eau est changée par l'utilisation de surgaines autour du combustible confirme par ailleurs qu'à 300°C l'effet prépondérant est dû à la variation de la densité d'eau.

Les études de sensibilité montrent que :

- pour ^{235}U les modifications de niveau des sections efficaces dans le domaine thermique n'entraînent pas de variation importante du coefficient de température, contrairement à l'effet obtenu si on change leur forme

- pour ^{238}U alors que la plupart des bibliothèques de sections efficaces thermiques donne une capture en 1/v, il apparaît qu'également seule une modification de cette loi peut entraîner une variation sensible du coefficient de température.

- pour l'eau, les marges d'incertitude sont relativement étroites et leur effet sur le coefficient de température est faible.

- l'intégrale effective de résonance du ^{238}U joue un rôle non négligeable.

- les variations du η dans le domaine épithermique et de la section de fission par neutrons rapides du ^{238}U jouent un rôle négligeable.

- l'incertitude sur les sections de diffusion de l'eau intervient d'une manière importante et elle pourrait expliquer l'effet de densité d'eau à température élevée.

1-3 Explications physiques et besoins de données nucléaires associées

Depuis de nombreuses années le désaccord sur les coefficients de température a été attribué aux données nucléaires, la variété de réacteurs et de méthodes de calcul pour lesquels il était constaté ne permettant pas de suspecter d'autre cause.

EDENIUS, dans une étude fondée sur les résultats expérimentaux de KRITZ /9/, concluait que seule une modification de la section efficace de capture du ^{238}U dans le domaine thermique pouvait expliquer une part importante du désaccord.

ASKEW, à partir des résultats obtenus sur les AGR, avait suggéré que la forme du η de ^{235}U dans le domaine thermique pouvait être responsable de l'écart constaté /10/.

ERRADI, ayant regroupé l'interprétation des résultats de CREOLE avec celle des expériences antérieures /5/ a proposé une modification simultanée des formes des sections efficaces des ^{235}U et ^{238}U dans le domaine thermique, compatible avec les marges d'incertitude des données expérimentales et les modèles théoriques, mais à la limite de ces marges. Brièvement cette modification comprend :

* Pour le ^{238}U, l'introduction d'une résonance négative à - 0,005 ev, dont la largeur conserve la section de capture à 2200 m/s, mais qui fait décroître la section plus vite qu'une loi en 1/V. (Cf. Fig I).

* Pour le ^{235}U, la première grande résonance négative est déplacée à -0,85 ev et une petite résonance essentiellement de capture est introduite à -0,01 ev. La modification de η5 qui en résulte est montrée sur la figure 2. Les paramètres des autres résonances négatives sont légèrement modifiés pour conserver les sections efficaces moyennes.

Ces deux modifications permettent de récupérer l'essentiel du désaccord sur les coefficients de température puisque l'écart moyen pour les expériences critiques passe de 4,2 à 0,3 pcm/°C cependant que pour les résultats de CREOLE l'évolution est la suivante :

Réseau / Températures	UO_2		UO_2 + 1200 ppm B		$UO_2 - PuO_2$	
	A	B	A	B	A	B
20 - 90°C	+4,2	+0,4	+5,0	+1,4	+2,3	+1,5
90 - 245°C	+1,9	-0,5	+1,1	-1,3	+0,4	0
245 - 290°C	-1,0	-2,6	-0,3	-2,1	-2,2	-2,3

A - Jeu non modifié
B - Jeu modifié proposé par ERRADI

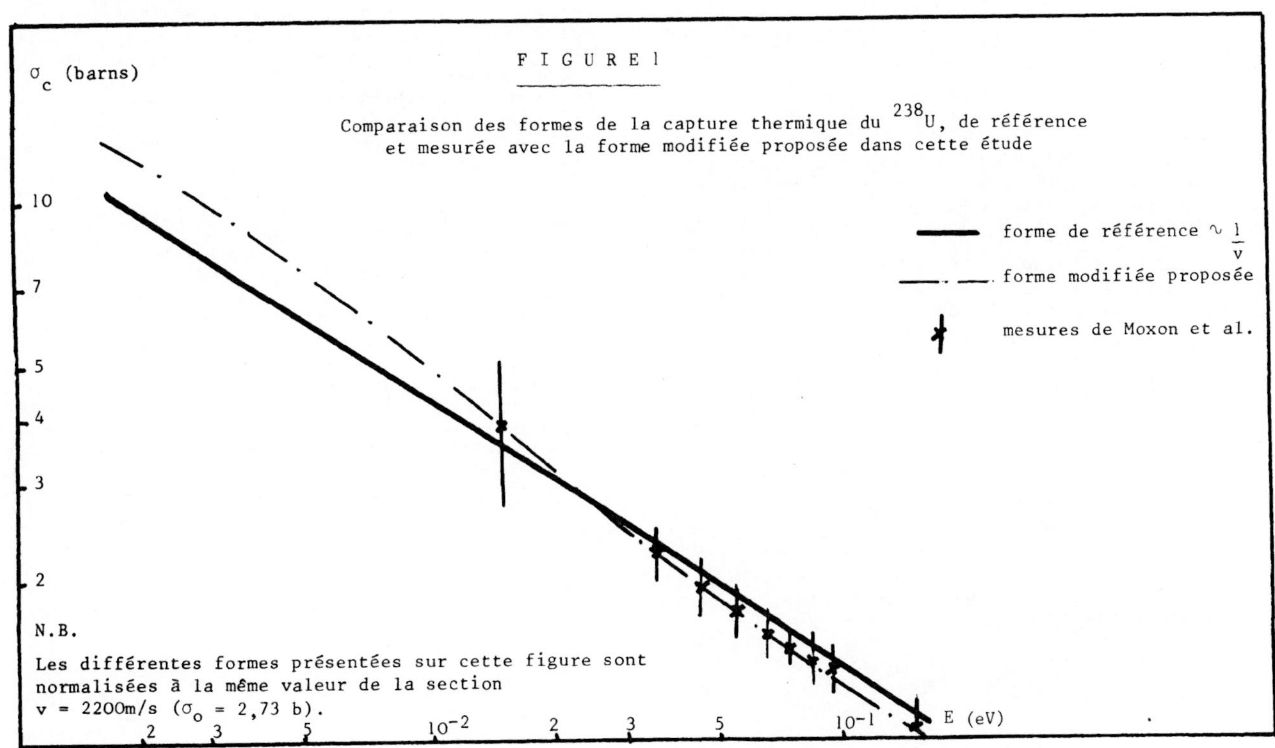

FIGURE I

Comparaison des formes de la capture thermique du ^{238}U, de référence et mesurée avec la forme modifiée proposée dans cette étude

N.B.
Les différentes formes présentées sur cette figure sont normalisées à la même valeur de la section v = 2200m/s (σ_o = 2,73 b).

FIGURE 2

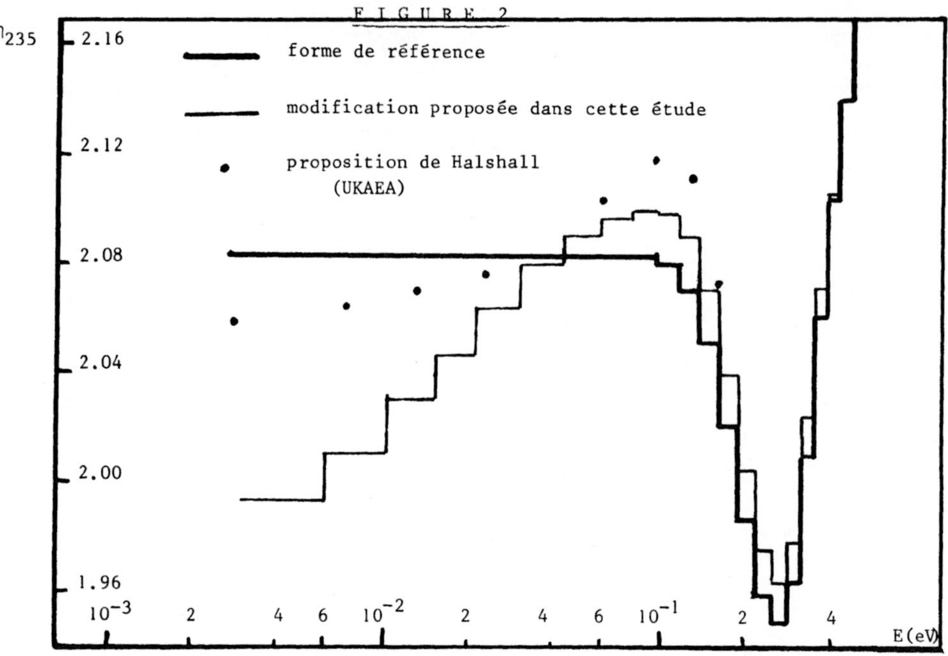

η Thermique de ^{235}U : Comparaison à la forme de référence des formes modifiées.

Les valeurs négatives dans le domaine des températures plus élevées restent attribuées à l'effet de densité d'eau.

Les modifications proposées par ERRADI, en principe compatibles avec la connaissance actuelle des données nucléaires, doivent être vérifiées ou améliorées à partir de mesures différentielles.

Les sections efficaces impliquées sont la capture de l'uranium 238, la capture et la fission (ou la fission et η) de l'uranium 235. La gamme d'énergie va de 1 eV à 0,002 eV. La partie à très basse énergie, qui est aussi la plus difficile à mesurer, est la plus intéressante car c'est elle qui est la plus sensible aux modifications proposées. Les précisions requises ne semblent pas inaccessibles. Il suffit que la précision sur la section efficace (ou sur η) soit meilleure que l'écart entre la valeur actuellement admise et la valeur modifiée. On peut se rendre compte de cette précision à l'aide des figures 1 et 2. Sur ces courbes sont également représentés une modification de η proposée par HALSHALL de l'UKAEA /11/ qui est du même type que celle d'ERRADI bien que d'amplitude moindre et le résultat d'une mesure non publiée de la capture du ^{238}U réalisée par MOXON de HARWELL /12/. Cette mesure ne permet pas de conclure car elle n'est pas assez précise et surtout ne couvre pas suffisamment les basses énergies. En résumé, il faudrait, à basse énergie, mesurer η avec une précision meilleure que 4 % et la capture de ^{238}U avec une précision meilleure que 10 %. La difficulté semble provenir de la gamme d'énergie à couvrir plutôt que de la précision à obtenir. Notons enfin qu'une mesure relative de la forme peut être suffisante, la normalisation pouvant être réalisée par les mesures intégrales.

2- Fraction effective de neutrons retardés

Dans le domaine des effets cinétiques, la précision requise pour les réacteurs est généralement exprimée sur la fraction effective de neutrons retardés qui est le paramètre intégral le plus important. Un objectif de 3 à 4 % est considéré comme raisonnable /13/. La connaissance détaillée des temps de vie et des rendements de chaque précurseur ainsi que celle du spectre des neutrons retardés est également nécessaire sans toutefois qu'il soit simple de fixer des objectifs de précision pour chacun de ces paramètres.

On considère souvent que l'état actuel des connaissances sur les neutrons retardés est suffisant pour les études de projet, de dimensionnement des systèmes de contrôle et de sécurité et pour l'application des critères de sûreté aux phases accidentelles. A notre avis, ce jugement doit être pondéré en tenant compte de trois constatations :

a) les erreurs sur les données nucléaires ne sont pas les seules sources d'incertitudes sur le βeff, le calcul de ce paramètre comportant des approximations, et il y a lieu de rechercher une précision meilleure que la fourchette de 3 à 4 % indiquée précédemment sur le rendement absolu de neutrons retardés (α ou $\beta = \frac{\alpha}{\nu}$) des principaux noyaux fissiles, ^{235}U et ^{239}Pu.

b) les recommandations des évaluations récentes conduisent pour les réacteurs à neutrons thermiques à des valeurs de βeff nettement plus élevées (5 à 6 %) que les anciennes valeurs de KEEPIN. Or beaucoup d'études sont encore fondées sur ces anciennes valeurs et la prise en compte des nouvelles dans l'interprétation des coefficients de réactivité remet partiellement en cause l'équilibre entre les résultats de mesures dynamiques et de mesures statiques.

c) une amélioration des données, en précision et en fiabilité, permettrait d'une part de réduire les marges sur des paramètres dont la vérification en réacteurs de puissance est effectuée par des mesures cinétiques, et d'autre part d'utiliser dans de meilleures conditions des mesures absolues de coefficients de réactivité pour l'analyse de certains problèmes physiques.

La valeur absolue du rendement des neutrons retardés des noyaux fissiles a fait l'objet depuis une quarantaine d'années de très nombreuses mesures. Les derniers résultats montrent encore dans le cas du ^{235}U une assez grande dispersion.

Auteur	Energie des neutrons incidents (MeV)	α
KEEPIN 57	Thermique	0,0158 ± 4,4 %
KEEPIN 57	1,45	0,0167 ± 4,4 %
CONANT 70	Thermique	0,0158 ± 6,3 %
CLIFFORD 72	0,63	0,0174 ± 4,6 %
KRICK 70	0,1 - 1,8	0,0172 ± 8 %
COX 79	0,96	0,0165 ± 6,3 %
"	1,97	0,01657 "
"	2,98	0,01696 "
"	3,98	0,01666 "
BESANT 77	Rapide	0,0164 ± 3,7 %

RUDSTAM (21) en sommant différents précurseurs séparés par voie chimique trouve une valeur inférieure :

0,0149 ± 6 % /14/

Les valeurs recommandées ont évolué depuis KEEPIN. COX (22) qui est à l'origine de la valeur ENDF/B IV donne 0,01668 ± 4,2 % pour un spectre de réacteur, TUTTLE (23) a analysé l'ensemble des valeurs publiées; il a corrigé les résultats en tenant compte de certaines erreurs systématiques dues à la méthode expérimentale ou au choix d'un étalon.

Il trouve pour un réacteur à neutrons thermiques 0,01654 ± 2,5 % et pour un réacteur à neutrons rapides 0,01714 ± 1,3 %.

KEEPIN donnait deux valeurs différentes pour les réacteurs à neutrons thermiques ou rapides.Les lois de la physique nucléaire tendent à montrer que α devait diminuer avec la dureté du spectre. La tendance actuelle serait de donner une valeur unique pour les différents types de réacteurs. C'est le cas de la recommandation de COX,et TUTTLE pondère les deux valeurs indiquées précedemment et propose 0,01697 ± 1,2 %.

L'écart entre les deux recommandations est faible mais les incertitudes indiquées sont très différentes. Il semble que TUTTLE qui applique une pondération statistique de tous les résultats soit optimiste, du fait d'erreurs systématiques provenant des méthodes de mesures.

A partir des précisions annoncés dans ENDF/B IV pour les différents noyaux fissiles on peut estimer l'incertitude du α moyen pour un réacteur à eau à 5 ou 6 %. Les 6 groupes de neutrons retardés ont été introduits par KEEPIN à partir d'un ajustement de 6 exponentielles sur la courbe de décroissance des neutrons retardés. Chacune des 6 familles est caractérisée par son rendement α_i. Lorsque le milieu multiplicateur est composé de plusieurs éléments fissiles, il faut pondérer les valeurs de α_i et de λi tout en conservant leur cohérence. Un effort peut être entrepris pour améliorer la connaissance des couples (αi, λi) en particulier sur les réacteurs de puissance qui évoluent tout au long du cycle.

La connaissance du spectre d'émission des neutrons retardés intervient dans les calculs de réacteur en raison d'une importance plus grande de ces neutrons qui sont en moyenne à plus basse énergie que les neutrons prompts. Compte tenu des mesures récentes il n'apparait pas de problème majeur sur cette connaissance.

En définitive, avec les erreurs introduites par les approximations des méthodes de calcul et la situation des données sur les neutrons retardés qui vient d'être bièvement résumée, la précision sur la connaissance du βeff pour un réacteur à neutrons thermiques se situe actuellement à environ 7 ou 8%, donc assez loin de l'objectif cité au début.

Par ailleurs, toujours dans l'état actuel, peu de mesures intégrales permettent d'assurer une vérification ou un recalage de ces valeurs. A notre point de vue il y a donc lieu de rechercher une amélioration par les deux voies, mesures des données fondamentales et mesures intégrales.

3- Capture des ^{242}Pu et ^{243}Am

L'augmentation des taux de combustion des combustibles de réacteurs à eau et l'éventualité d'un recyclage du plutonium font croître les teneurs en transplutoniens. L'effet sur le bilan en réactivité des isotopes de l'américium et du curium qui sont formés n'est pas trés important. Par contre la connaissance des concentrations de ces éléments l'est pour d'autres facteurs et en particulier pour l'émission neutronique des combustibles, essentiellement liée aux teneurs en curium. Dans ce domaine le curium 242 est responsable de l'émission à court terme et le curium 244 de l'émission à moyen terme. La chaîne de production des isotopes du curium est rappelée sur la figure 3.

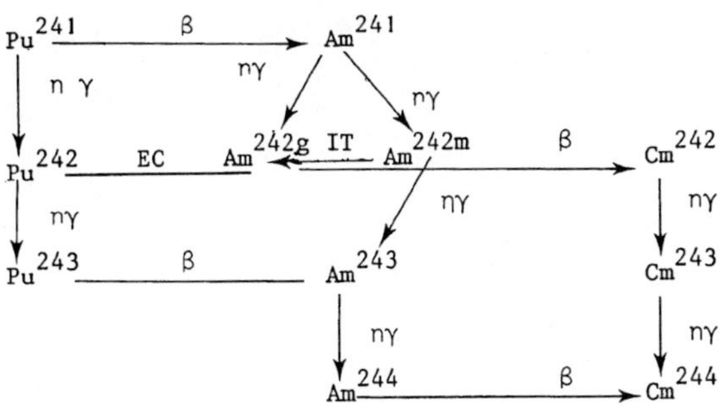

Figure 3 : Formation du ^{244}Cm : chaîne complète

Certaines périodes de décroissance sont courtes devant le temps d'irradiation du combustible et cette chaîne peut être simplifiée. En effet, le flux neutronique n'est pas suffisament élevé pour qu'il puisse y avoir des captures en quantité appréciable dans les isotopes à vie courte. Les calculs prévisionnels peuvent donc être réalisés à l'aide de la chaîne réduite représentée sur la figure 4. Dans ce schéma simplifié le curium 244 peut être produit par trois voies différentes, capture du curium 243 ou capture de l'américium 243 produit soit par le plutonium 242 soit par l'américium 242m. Dans un réacteur à eau, le taux de capture du curium 243 est négligeable et la production d'américium 243 par capture dans l'américium 242m relativement faible. Par conséquent c'est la voie ^{242}Pu - ^{243}Am qui est prépondérante. La teneur en curium 244 est donc essentiellement régie par les sections efficaces de capture du plutonium 242 et de l'américium 243.

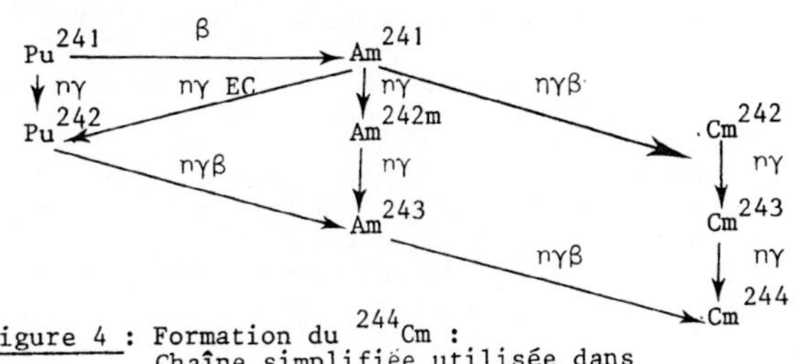

Figure 4 : Formation du ^{244}Cm : Chaîne simplifiée utilisée dans les calculs d'évolution

Les mesures de la composition chimique et isotopique de combustibles irradiés dans des réacteurs à eau ont montré que la teneur en curium 244 n'est pas prévue avec la précision souhaitée /15/. En effet les écarts entre le calcul et l'expérience sont surpérieurs à 30 %, le calcul sous-estimant la teneur en Curium 244. Il y a donc une erreur dans la chaîne de production du curium 244.

Afin de pouvoir chiffrer la précision avec laquelle il serait souhaitable de connaître les sections efficaces un calcul de sensibilité de la variation de teneur en curium 244 en fonction des variations des sections efficaces de capture du ^{242}Pu, de ^{243}Am et de ^{244}Cm a été réalisé. Les variations de sections efficaces sont supposées avoir lieu pour des domaines d'énergie correspondant aux neutrons thermiques, aux neutrons épithermiques et aux neutrons rapides. Le tableau suivant rassemble des variations relatives (en pour cent) de la teneur en curium 244 pour une augmentation de 10 % des sections efficaces de capture dans un spectre des réacteurs à eau ordinaire.

Energie / Isotope	Thermique	Résonances	Rapide
^{242}Pu	+ 0,4	+ 8	+ 0,2
^{243}Am	+ 1,3	+ 7	+ 0,1
^{244}Cm	- 0,1	- 0,7	- 0,1

On voit que le domaine des résonances à une part prépondérante aussi bien pour la capture du ^{242}Pu que pour celle du ^{243}Am. Pour ces deux noyaux les sections efficaces de capture à 2200m/s et les intégrales de résonances recommandées dans les évaluations de ces dernières années sont regroupées dans le tableau suivant :

		ENDF/B3	ENDF/B4	UKNDL 80	BNL 325
^{242}Pu	σ_γ	18,5	19,2	18,8	18,5 ± 0,4
	I_γ	892	1269	--	1130 ± 60
^{243}Am	σ_γ	181,9	74,8	180,4	79,3 ± 2
	I_γ	1367	1820	1355	1820 ± 70

Pour le ^{242}Pu la situation semble claire dans le domaine thermique ; par contre l'intégrale de résonance a notablement évolué mais la valeur haute qui est retenue dans ENDF/B IV va bien dans le sens souhaité pour la production du curium.

Pour le ^{243}Am le gros désaccord sur la section à 2200m/s est tranché actuellement en faveur de la valeur basse /16/, ce qui a aggravé le problème précité mais toutefois avec une incidence limitée. Dans le domaine des résonances la situation ne parait pas claire.

D'une façon générale il y a donc, avec les valeurs des évaluations récentes, une sous estimation de la production de ^{244}Cm dont la cause ne paraît pas pouvoir être ailleurs que dans les sections de capture de ^{242}Pu et ^{243}Am, essentiellement de ce dernier. Il est donc souhaitable qu'un nouvel effort soit fait, au niveau des mesures différentielles, pour tenter de résoudre cette incohérence.

4- Revue de quelques autres problemes de données nucléaires rencontrés dans la conception ou le fonctionnement des réacteurs à eau

Ces problèmes apparaissent le plus souvent comme la conséquence d'écarts constatés entre des prévisions théoriques fondées sur les données microscopiques et des résultats de mesures intégrales de paramètres de réacteurs.

4-1- Calculs des réacteurs au démarrage

Ces calculs font l'objet de nombreuses vérifications en expériences critiques ou lors de l'analyse des essais physiques au démarrage des réacteurs de puissance. En plus des écarts sur les coefficients de température exposés précédemment, nous pouvons évoquer quelques points qui nécessiteraient des éclaircissements ou des approfondissements :

- les résonances de capture de zirconium :
  La mauvaise connaissance des données fondamentales de ces résonances introduit une incertitude de 0,3 à 0,4 % $\delta k/k$ sur la réactivité du coeur. Une étude récente /17/ montre que ceci est dû essentiellement aux valeurs imprécises des largeurs moyennes de capture radiative $\Gamma\gamma$ pour les isotopes du zirconium.

Le désaccord apparent sur l'intégrale de résonance peut être chiffré en comparant les valeurs de deux évaluations : 1,66 barns avec la bibliothèque UKNDL 68 et 0,99 barn avec ENDF/B IV.

La répartition de l'effet de l'absorption du zirconium sur le facteur de multiplication effectif d'un réacteur à eau pressurisée est la suivante :

Bibliothèque	Domaines d'énergie			
	Thermique	Résonances	Rapide	Total
UKNDL 68	209	891	115	1215
ENDF/B IV	210	488	188	886

(effets exprimés en 10^{-5} $\delta k/k$)

Nous ne disposons pas actuellement de mesures intégrales permettant de séparer avec précision cet effet et un effort peut être fait dans ce domaine. Il parait également nécessaire de clarifier la situation des mesures différentielles pour les paramètres de ces résonances.

- Les liaisons cristallines de l'uranium dans l'oxyde.
  Ces liaisons influent sur le calcul de l'effet Doppler du combustible. De récentes expériences par oscillation d'échantillons d'UO_2 chauffés ont conduit à une température de DEBYE de 620°K. /7/.Si cette valeur est proche de ce qui était supposé pour l'UO_2, elle est en contradiction avec les interprétations précédentes qui conduisaient à penser que dans l'UO_2 les températures équivalentes étaient trés différentes pour l'uranium et l'oxygène, et que la correction sur l'effet Doppler des résonances du ^{238}U était négligeable /18/. L'effet de la différence éventuelle entre la température de DEBYE et la température thermodynamique intervient uniquement sur la différence de réactivité entre froid et chaud puisqu'il décroit rapidement avec la température. Si les mesures récentes étaient confirmées la correction correspondante atteindrait environ 0,2 % $\delta k/k$ pour un réacteur à eau. Il parait nécessaire de confirmer les mesures intégrales dans ce domaine des basses températures du combustible et d'examiner la possibilité d'expériences

fondamentales sur les liaisons cristallines dans l'UO_2.

- Les données de la thermalisation :
les paramètres de la thermalisation des neutrons dans l'eau sont une cause possible d'explication des écarts sur les effets de spectre (variations de température ou de rapport de modération). Les études de sensibilités portent généralement sur la comparaison de différents noyaux de thermalisation /9/. Il serait intéressant de confirmer la précision dans la connaissance des fréquences et des amplitudes des vibrations.

4-2 Calculs d'évolution

Ces calculs jouent un rôle important pour l'évolution de la réactivité, le repositionnement des combustibles, la composition finale des combustibles déchargés et la connaissance de leurs activités.

Par ailleurs c'est dans le domaine du cycle du combustible des réacteurs à eau que sont prévus actuellement plusieurs développements :

- allongement des cycles avec utilisation renforcée des poisons consommables
- augmentation des taux de combustion de rejet
- recyclage des matières fissiles (uranium et plutonium)

Le problème des données intervenant dans le calcul du curium a été discuté précédemment ; on peut également citer :

- les données intervenant dans la formation de plutonium. C'est essentiellement la capture résonnante du ^{238}U qui était mise en cause pour expliquer une légère dérive sur les prévisions de formation du plutonium. La situation s'est améliorée avec de récents développements /19/ et le bilan présenté à cette conférence /3/ permettra peut être de clore ce dossier.

- les isotopes pairs du gadolinium : l'importance des sections efficaces de capture des isotopes impairs (^{155}Gd et ^{157}Gd) conditionne l'efficacité de ce poison et les caractéristiques de son évolution dans les réacteurs à eau. La précision des données actuelles parait suffisante sous réserve que la forme de la section efficace dans le domaine où elle décroit rapidement soit bien prise en compte.

Par contre, après usure des isotopes impairs, la composition isotopique et l'effet résiduel en réactivité dépendent principalement des isotopes pairs pour lesquels les données ne sont pas connues avec une précision suffisante.

- les données d'évolution des isotopes mineurs de l'uranium et du plutonium : il s'agit principalement de la capture du ^{236}U qui joue un rôle important en cas de recyclage de l'uranium de retraitement /20/, et des sections efficaces intervenant dans la formation des ^{232}U, ^{236}Pu et ^{238}Pu dont les teneurs augmentent rapidement avec l'accroissement des taux de combustion des combustibles UO_2.

La connaissance des données des résonances de capture du ^{236}U doit être améliorée pour obtenir une bonne précision dans l'estimation de son effet en réactivité, de son évolution et donc du surenrichissement nécessaire pour les combustibles fabriqués avec de l'uranium de retraitement.

Conclusion

Si elle n'est pas impérative pour les calculs de conception et de fonctionnement des réacteurs à neutrons thermiques actuels, une amélioration de la connaissance de certaines données nucléaires peut à coup sûr être très profitable pour expliquer des désaccords physiques apparents, réduire progressivement des marges et faciliter l'introduction ultérieure de nouveaux développements. Il n'est pas évident que les trois points que nous avons choisi de mettre en lumière constituent aux yeux de tous des priorités incontestables, de même que la liste des autres problèmes que nous avons évoqués n'est certainement pas exhaustive. Nous avons choisi cette démarche pour éviter de noyer des problèmes précis et d'actualité dans le brouillard de longues listes de demandes avec des chiffres peu explicites et plus ou moins justifiables. Un progrès sensible sur l'un ou l'autre de ces points constituerait incontestablement un apport non négligeable dans le développement des réacteurs à neutrons thermiques.

REFERENCES

1 - ROWLANDS J.L : "Nuclear Data for Reactor Design, Operation and Safety" - Conference on Neutron Physics and Nuclear Data, HARWELL september 1978

2 - WESTON : Review of Cross Section Important to the Uranium-Plutonium Fuel Cycle in Thermal Reactors Brookaven (1979)

3 - DE SAUSSURE : ^{238}U, Issues Resolved and Unresolved (this meeting)

4 - REUSS P. : The Search of Tendancies. Thermal Reactor Benchmark Calculations, Techniques and Applications (Brookaven 1982)

5 - ERRADI L. : "Etude des effets de temperature dans les réseaux caractéristiques des réacteurs nucléaires de la filière à eau ordinaire" - Thèse Docteur cs Sciences, PARIS Orsay Février 1982

6 - PERSONN R. et al. : High Temperature Critical Experiments with H_2O - Moderated Fuel Assemblies In KRITZ Technical Meeting N°2/11 NUCLEX 72

7 - GOLINELLI C. et al : Température Coefficient and Doppler Effect Measurements.Topical Meeting on Advances in Reactor Physics and Shielding (Sun Valley 1980)

8 - HOFFMANN S. et al : Apollo Code multigroupe de résolution de l'équation de transport pour les neutrons thermiques et rapides. Note CEA N°1610 (Mars 1973)

9 - EDENIUS M. : "Studies of the Reactivity Temperature Coefficient in Light Water Reactors", Report AE.RF.76.3160, AB Atomenergie (1976)

10 - ASKEW : Thermal Reactor Temperature Coefficient Studies in the U.K. UKAEA Report AEEW-R-886 (1973)

11 - HALSHALL : Communication personnelle (UKAEA)

12 - MOXON M.C., JOLLY J.E. : "The Neutron Capture Cross - Section of U 238 from 0.01 ev" - IAEA Consultants Meeting on U/Pu Resonance Parameters, Vienne October (1981)

13 - Proceeding of a Panel on Fission Product Nuclear Data - Bologna,(November 1973)

14 - Proceedings of the Second Advisory Group Meeting on Fission Product Nuclear Data - Petten, (September 1977)

15 - DARROUZET M., GIACOMETTI A., ROBIN M. : Formation et disparition des actinides secondaires dans les réacteurs à eau et les réacteurs à neutrons rapides - Conférence on Neutron Physics and Nuclear Data, Harwell, September (1978)

16 - BENJAMIN : Status of Measured Neutron Cross Sections of Transactinium Isotopes for Thermal Reactors (Karlsruhe 1975)

17 - TEMPERVILLE : Communication personelle (CEA)

18 - BUTLAND : A note on crystalline binding in uranium dioxide and its effect on the Doppler broadening of Uranium Resonances. Annals of Nuclear Science and Engineering 1,575 (1974)

19 - TELLIER H. : Multilevel Effect in Uranium 238 and Thorium 232 Effective Neutron Capture Resonance Integrals.

20 - DARROUZET et al. : Recycling of the Reprocessed Uranium in PWRS (ENC Bruxelles 1982)

21 - RUDSTAM : Status of Delayed Neutron Data Petten 1977

22 - COX : Delayed Neutron Data - Review and Evaluation ANL/NIM 5 - 1974

23 - TUTTLE : Delayed Neutron Data for Reactor Physics Analysis - Nuclear Science and Engineering 56,37-71 (1975)

MODIFIED SCATTERING MATRICES TO IMPROVE TRANSPORT CALCULATIONS WITH APPROXIMATE FISSION-SOURCE MATRICES

U. Salmi, J.J.Wagschal, Atara Ya'ari and Y. Yeivin

Racah Institute of Physics, the Hebrew University
91904 Jerusalem, Israel

Several commonly used transport codes (ANISN and DOT for example) can only accept fission matrices given as outer products of $\nu\sigma^f$ and a single fission spectrum. This approximation of the transport equation results in unnecessary inaccuracies in the calculation of integral parameters of neutronic assemblies. To compensate for the difference between such an approximate fission matrix and the corresponding true fission matrix, we introduce a modified scattering matrix,defined as the difference between the true total-source matrix and the approximate fission matrix. Generally, these modified scattering matrices can be readily generated.

We have calculated integral parameters in benchmark critical assemblies and in fusion-reactor blankets. "Approximate" reactivities differ from their corresponding "true" values by several standard deviations of the experimental values, whereas the "modified" calculations practically coincide with the true ones. Relative deviations of approximate reaction rates from corresponding true values are 10% for critical assemblies, and 5% for the driven blankets.

[fission matrices, modified scattering matrices, neutronic-benchmark calculations, reactivity calculations, reaction-rate calculations]

Introduction

Calculating the reactivity, or other integral parameters such as reaction rates, in a given neutronic assembly, may seem a trivial problem. Nevertheless, the accuracy of the calculated values must be carefully considered. This is particularly true for benchmarks used for data testing and data adjustment. A seemingly negligible inaccuracy with respect to the calculated value itself, might turn out to be more significant with respect to the discrepancy between this value and the corresponding experimental value.

Fast reactor benchmark no.1 of CSEWG (the ENDF cross-section evaluation working group) is a bare Plutonium sphere, better known as JEZEBEL[1]. Even for such as simple an assembly there is a difference of 0.15% between the calculated values of k-effective obtained by LANL and ORNL in their recent ENDF/B-V data testing calculations.[2] This difference is about equal to the experimental uncertainty in k of JEZEBEL.[3] The corresponding calculated values of f28/f25 differ by more than twice the experimental uncertainty of this parameter.

The LANL and ORNL calculations are based on ENDF/B-V, and on similar cross-section processing methods.[2] Although the LANL multigroup library has a group structure different from that of the ORNL VITAMIN-E library,[4] the main difference between the libraries lies in the treatment of the fission spectrum.

In this paper we discuss the sensitivity of integral parameters to the fission-source matrices for different formulations of these matrices. In particular, it is demonstrated that even with the use of a single fission spectrum, quite good results can be obtained, provided the scattering matrix is appropriately modified.

The Modified Scattering Matrix

Some universally known and extensively used transport codes, such as ANISN,[5] XSDRNPM[6] and DOT,[7] can only use fission-source matrices which are given as an outer product of ($\nu\sigma^f$) and a fission spectrum (f). It is for this reason that the multigroup VITAMIN-E library, which is based on ENDF/B-V, has only one fission spectrum for each fissionable isotope, corresponding to 1 MeV incident neutrons, and identical to its thermal fission spectrum according to ENDF/B-V.

To overcome this limitation, we suggest to modify the scattering matrix in order to compensate for the difference between the correct fission matrix and any simplified fission matrix accepted by the above mentioned transport codes. We define a modified scattering matrix by

$$\hat{\tau}^s_{gg'} = \tau_{gg'} - f_g(\nu\sigma^f)_{g'}$$

where $\tau_{gg'}$ is the correct total-source matrix. This latter matrix should obviously include the correct fission matrix. Generally, such matrices can be easily generated. The AMPX-II system,[8] for instance, internally calculates a correct fission matrix, which however is not transmitted to the output.

The modfied scattering matrices may contain "upscattering" elements, and even negative elements, both of which are obviously devoid of any physical meaning. Nevertheless, these matrices are very effective as a practical device to eliminate the errors inherent in employing the outer-product fission matrices.

Critical Assembly Calculations

The reactivity, average neutron energy and the reaction-rate ratio f28/f25 have been calculated for a pure Pu-239 sphere of radius 6.3849 cm and density 15.61 gr/cm^3. This, of course, is a JEZEBEL-like assembly. Different combinations of fission and scattering matrices have been employed. All calculations were performed in an S16-P4 approximation, using 9-group cross sections based on ENDF/B-IV, and with 30 equally spaced radial intervals. The results of these calculations are listed in Table I.

K. H. Böckhoff (ed.), Nuclear Data for Science and Technology, 29-31.

Table I - Integral Parameter Calculations in a JEZEBEL-Like Assembly

integral parameter	reference value	scattering matrix	single fission spectrum		
			14 MeV	Average	thermal
(k-1)x1000	20.558	true modified	32.436 20.312	21.444 20.540	18.593 20.600
$\bar{E}$(MeV)	1.8165	true modified	2.0547 1.8116	1.8314 1.8162	1.7764 1.8174
f28/f25	0.2053	true modified	0.2324 0.2047	0.2070 0.2052	0.2005 0.2054

The second column ("reference value") lists the values of the three parameters calculated with the correct, energy-dependent fission spectrum, as specified in ENDF/B-IV, i.e. with the proper (irreducible) fission matrix. On the other hand, the entries in the last three columns of the table were calculated with single, energy-*independent* fission spectra. In other words, the fission matrices employed in these calculations were outer products of a single fission spectrum and the group-averaged values of $(\nu\sigma^f)$. The fourth column lists values calculated with the fission spectrum of 14 MeV neutrons, in the fifth column the listed values were obtained with the average fission spectrum

$$\bar{f}_g = \sum_{g'} \tau^f_{gg'} \phi_{g'} \Big/ \sum_{g'} (\nu\sigma^f)_{g'} \phi_{g'},$$

and the sixth-column values were calculated with the thermal fission spectrum.

We first consider the upper entries in the last three columns. These calculations were performed with the proper scattering matrix, and they thus appear in the lines designated "true" under the column heading "scattering matrix". For all three integral parameters, and just as was to be expected, the softer the fission spectrum, the lower the values obtained in these single-fission-spectrum calculations.

The lower entries in columns four, five and six, designated"modified", are the integral-parameter values calculated with the same outer-product fission matrices used to calculate the upper entries, but with the appropriately modified scattering matrices. The practical disappearance of the discrepancies between the single-fission-spectrum values and the corresponding reference values is rather striking. Such an improvement, however, is not surprising, since the sum of any of the modified scattering matrices and its corresponding outer-product fission matrix is obviously the correct total-source matrix.Had our Plutonium sphere been exactly critical, the discrepancies would have completely vanished. On the other hand, the tiny discrepancies which we still have, due to the slight super-criticality of our calculated assembly, would increase the farther would the assembly be from criticality.

It is worth our while to note that in formulations of the stationary transport equation other than the k-formulation, of which the asymptotic-inverse-period (α) formulation is the most common example, the modified scattering-matrix algorithm would always completely eliminate calculational errors due to outer-product fission matrices, irrespective of criticality. This, of course, is so because in these other formulations only the total-source matrix is needed, whereas in the k-formulation the scattering and fission matrices are needed separately.

Fusion-Reactor Blanket Calculations

Having realized the possible significance of the errors resulting from the use of approximate fission matrices in fast-reactor core calculations, we also wanted to examine the magnitude of these errors in other neutronic systems. To this end we have calculated the neutron field in a 20 cm thick U-238 slab with a 14 MeV neutron source on one face, which is representative of a breeding blanket around a fusion reactor. This assembly was calculated in the same approximation of the transport equation used to calculate the Plutonium sphere. The integral parameters studied were the average neutron energy and the U-238 fission rate.

Figure 1 show the behaviour of the two parameters across the slab.
These two functions were properly calculated, using the correct, energy-dependent fission matrix. We note that close to the irradiated face of the slab, the neutron spectrum is strongly influenced by the source neutrons, but as we go deeper into the slab, the spectrum softens appreciably, and reaches its asymptotic form about half way across the slab. And since U-238 fission is a threshold reaction, the behaviour of its spectrum-averaged cross section is very similar to that of $\langle E \rangle$

We have then calculated the fission rate using the approximate fission matrices (and the *unmodified* scattering matrix). The results of these calculations are shown in Figure 2.

The three curves represent the fission rates calculated with single-fission-spectrum (sfs), i.e.outer-product, fission matrices, for the fission spectrum of 14 MeV neutron, the neutron-flux averaged fission spectrum, and for the thermal fission spectrum. The figure depicts these fission rates divided by the correct (crt) fission rate which is given in Fig.1. We note that the approximate fission

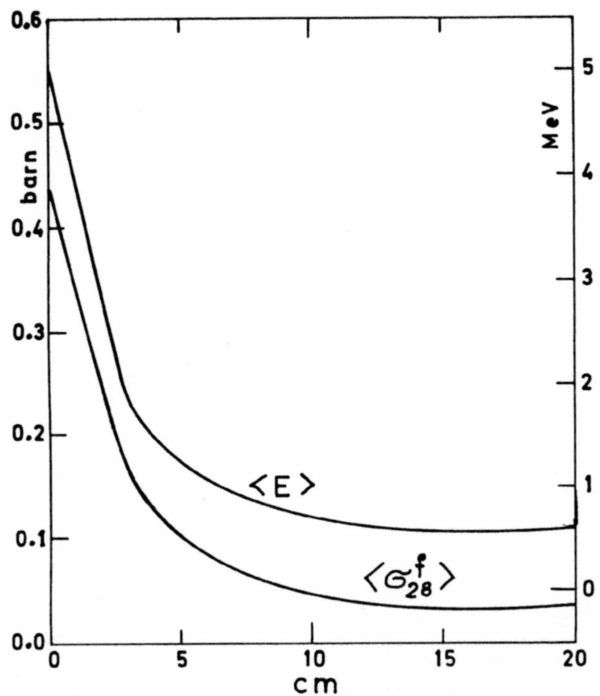

Fig. 1 The average neutron energy and fission rate accross a fusion-reactor U-238 blanket

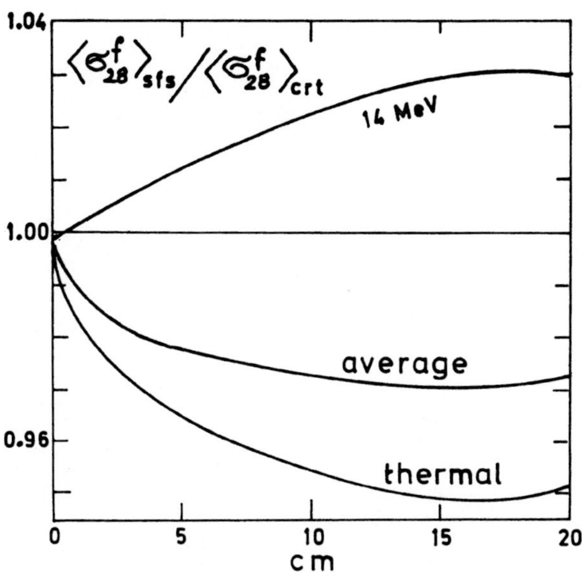

Fig. 2 The approximate fission rates relative to the correct rate, across a fusion-reactor U-238 blanket

rate calculated with the "standard"VITAMIN-E fission matrix (employing the thermal fission spectrum), deviates by up to 5% from the correct rate. This should certainly be of some consequence in calculating the power generated in the blanket.

Concluding Remarks

The foregoing results have an obvious and well-defined practical aspect. Beyond this, however, they also have a somewhat broader aspect, touching upon the general approach to data testing and data adjustment. It should be clear that we were not discussing cross-section errors, but rather what, in adjustment terminology, is referred to as "methods uncertainties". The subject of this paper could therefore be also defined as a possible source of uncertainty in cross-section processing. On the other hand, however, even though the calculational errors that we have indicated might not be of major importance, we have shown how they could be easily eliminated altogether. This should therefore always be taken care of in any case.

A final remark concerns the rather rudimentary assemblies which we have calculated to illustrate our argument. The same effects obviously exist in actual assemblies as well. As a matter of fact, this study was originally motivated by actual results from recent ENDF/B-V data testing. Unfortunately, however, these microscopic data files are not available outside the USA,which was a further reason not to illustrate our discussion by calculations of more realistic assemblies.

References

1. G.E. Hansen, H.C. Paxton, LA-UR-79-2923, Los Alamos National Laboratory (1979)

2. R.Q. Wright, "Fast Reactor Data Testing of ENDF/B-V",UCND-IC of November 17 1981, Oak Ridge National Laboratory

3. J.J. Wagschal, Atara Ya'ari, Nuclear Data in Science and Technology 2, 83, Paris (1973)

4. C.R. Weisbin, R.W. Roussin, J.J.Wagschal, J.E. White, R.Q. Wright, ORNL-5505, Oak Ridge National Laboratory (1979)

5. W.W. Engle, Jr., K-1693, Oak Ridge National Laboratory (1967)

6. L.M. Petrie, N.M. Greene, RSIC-PSR-63, ch.8, Oak Ridge National Laboratory(1978)

7. F.R. Mynatt, K-1694, Oak Ridge National Laboratory (1967)

8. N.M. Greene, J.L. Lucius, J.E. White, RSIC-PSR-63, ch.3, Oak Ridge National Laboratory (1978)

COMPARISONS OF CALCULATED SELF SHIELDING FACTORS WITH MEASURED VALUES FOR ^{239}Pu, ^{235}U, Fe and Na

S. Ganesan, M.M. Ramanadhan and V. Gopalakrishnan

Reactor Research Centre, Kalpakkam - 603 102
Tamil Nadu, INDIA

Using two different approaches, the self shielding factors were generated for ^{239}Pu, ^{235}Pu, Fe and Na and compared with available experimental values. The first method uses J* formalism of Hwang which employs ψ and χ approximations and additional approximations in the formulation and numerical evaluation of the J* integrals. The other method which uses Cullen's codes avoids these approximations. The present comparison studies support the recent investigations performed at Kalpakkam towards the objective of evolving a correct method of calculation of self shielding factors in the resonance region. The results show that at lower background dilutions there is a wide dispersion among the values of self shielding factors calculated by two different methods and those obtained experimentally. At higher dilutions the values of self shielding factors calculated by both the methods converge towards experimental values satisfactorily.

[Self shielding factors, theory, experiment, comparisons, ^{235}U, ^{239}Pu, Na, Fe]

Introduction

The central question about the data representations and their processing in the resonance region is whether or not the present approach leads to a correct estimate of resonance self-shielding [1-4]. Obviously, a direct comparison of calculated self shielding factors (SSFs) with measured values as a function of temperature, energy and background dilution will answer this question. The present work compares the SSFs calculated by J* method of Hwang [5] with those calculated by Cullen's codes [6-9] for ^{235}U and ^{239}Pu using ENDF/B-IV data. The reported experimental values [10-11] of SSFs have been used for comparison with calculated values. For ^{23}Na and Fe, the experimental values [10] of SSFs are compared with values calculated using Cullen's codes [6-9].

The Reported Experimental Values of SSFs

Arnaud et al. [10] have reported self shielding factor measurements for natural iron and ^{23}Na between 24 keV and 160 keV at 300°K. The experimental set up used the fast time of flight method and consisted of apparatus usually employed for measuring microscopic cross sections. The accuracy of the infinite dilution cross section is reported to be 6 %. Similarly Bakalov et al. [11] have carried out transmission and self indication measurements for ^{235}U and ^{239}Pu. For ^{235}U and ^{239}Pu isotopes the errors in the deduced values of SSFs for total reaction process are in the range 8 - 10 % in 1 - 21.5 keV region and 5 % below 1 keV for zero dilution cross sections. For fission process the errors in SSFs are about 5 % in 1 - 21.5 keV energy range and about 3 % at energies below 1 keV. The uncertainties decrease with increasing dilution. The principle behind the semi-empirical determination of self shielding factors is well documented in the literature [10-12].

The Two Calculated Methods

The first method for calculation of SSFs uses RAMBHA code [13] developed at Kalpakkam for the generation of group constants for fast reactor applications and employs mainly the J* formalism of Hwang [5] in the resonance region. The familiar line shape broadening functions ψ and χ are employed for temperature broadening of the resonance line shapes in this method. The J* method of course invokes the narrow resonance approximation. The approximations in the ψ and χ formalism have been well described by Cullen and Weisbin [14].

The second method which employs Cullen's codes LINEAR [6] - RECENT [7] - SIGMA1 [8] and GROUPIE [9] (LRSG codes) uses direct numerical broadening techniques to evaluate energy dependent broadened reaction rates at a higher temperature starting from the energy versus cross section file at a lower temperature. The self shielding factor is then calculated using direct numerical integration techniques for various background dilutions at a given temperature using Bondarenko's definition.

The energy dependent cross sections are obtained in a linear interpolable form by running the codes LINEAR and RECENT using ENDF/B-IV data.

The applicability of Cullen's codes is limited to resolved resonance region and that of RAMBHA to Single Level Breit Wegner formalism at present.

Approximations in Doppler Broadening

The SIGMA1 method uses exact free atom Doppler broadening equations

$$\sqrt{E}\,\sigma(E,T) = \frac{1}{2}\left(\frac{a}{\pi E}\right)^{1/2} \int_0^\infty [\sqrt{E_\nu}\,\sigma(E_\nu,o)] \times \left\{\exp[-a(\sqrt{E}-\sqrt{E_\nu})^2] - \exp[-a(\sqrt{E}+\sqrt{E_\nu})^2]\right\} dE_\nu \quad \ldots (1)$$

where E is the projectile energy, E_ν the relative energy of the neutron as 'seen' by the target, $a = A/kT$ where A is the mass of target atom in units of mass of the neutron, k the Boltzmann constant and T the temperature.

The SIGMA1 method assumes that the cross section is a function given by a table of cross section versus energy with linear-linear interpolation in energy and cross section between tabulated values. The Doppler broadening equation is numerically solved without any further assumptions.

On the other hand the approximations in ψ - χ formalisms are

1. the second exponential can be ignored compared to the first. This is valid for large $a\sqrt{E_\nu E}$ ($\sim aE$). Ignoring the second term results in larger reaction rates than could be obtained using the exact kernel.
2. The introduction of Doppler width's definition which amounts to the approximation:

$$\exp[-a(\sqrt{E}-\sqrt{E_\nu})^2] = \exp[-\frac{a}{4E}(E-E_\nu)^2] \quad \ldots (2)$$

3. The lower limit is extended from 0 to $-\infty$

$$\sqrt{E}\,\sigma(E,T_2) = \frac{1}{\Delta\sqrt{\pi}} \int_{-\infty}^{+\infty} \sqrt{E_\nu}\,\sigma(E_\nu,T_1) \times \exp\left\{-\left[\frac{(E_\nu - E)}{\Delta}\right]^2\right\} dE \quad \ldots (3)$$

K. H. Böckhoff (ed.), Nuclear Data for Science and Technology, 32–35.

4. The partial and total widths and wave numbers which are usually slowly varying functions of energy are taken to be energy independent, across a resonance.

Approximations in Calculation of SSFs

There are also differences in the way the Bondarenko self-shielded cross sections are generated in the two methods.

The Bondarenko self shielded cross sections are expressed as

$$\tilde{\Sigma}_x(\sigma_b,T) = \int_{E_G}^{E_{G+1}} \frac{\Sigma_x(E,T)\,S(E)dE}{(\Sigma_t(E,T)+\Sigma_b)^N} \quad ..... \quad (4)$$

where Σ_b is the macroscopic dilution cross section, the suffices x and t stand respectively for partial and total reaction process, S(E) is the weighting spectrum and N = 1 or 2 gives flux weighted or current weighted self shielded cross sections. Note that $\Sigma_x(E,T)$ is obtained by running SIGMA1 at temperature T.

Each of the functions $\Sigma_x(E)$, $\Sigma_t(E)$ and S(E) is assumed to be given as a table value $(E_k, f(E_k))$ with linear interpolation between tabulated values. Since no approximation is introduced in performing the integrals used to get Σ_x the accuracy of the results is guaranteed to be at least as good as the accuracies with which the cross sections $\Sigma_x(E)$ and $\Sigma_t(E)$ (Generated with SIGMA1 code) and the weighting spectrum are given by the user.

In the J★ method $\tilde{\Sigma}_t$ is obtained using the following expression

$$\tilde{\Sigma}_t = \sum_i S_i \frac{\Gamma_{t_i}\Sigma_b J^\star_{t_i}}{\Delta U\, E_{o_i}\, f} \quad ..... \quad (5)$$

where f is the familiar flux correction factor

$$f = \sum_i S_i - \frac{1}{\Delta U}\sum_j S_j \Gamma_{t_j} J^\star_{t_j} \quad ..... \quad (6)$$

and

$$\tilde{\Sigma}_x = \sum_i \frac{S_i \Sigma_b \Gamma_{x_i}}{\Delta U\, E_{o_i}\, f} J_{x_i} \quad ..... \quad (7)$$

The J★ integrals are defined as:

$$J^\star_t = \frac{1}{2}\int_{-\infty}^{+\infty} \frac{(\psi_i + a_i\chi_i)dx_i}{\beta_i + \psi_i + a_i\chi_i + \sum_{i'\neq i}(A_{i'}\psi_{i'} + B_{i'}\chi_{i'})} \quad ..... \quad (8)$$

and

$$J^\star_x = \frac{1}{2}\int_{-\infty}^{+\infty} \frac{\psi_i\, dx_i}{\beta_i + \psi_i + a_i\chi_i + \sum_{i'\neq i}(A_{i'}\psi_{i'} + B_{i'}\chi_{i'})} \quad ..... \quad (9)$$

Here i is the resonance index, E_{o_i} the resonance energy, Γ_{x_i} the reaction widths for process x and ΔU the lethargy width.

To save space the reader is referred to Section III of Ref.[15] for a derivation, and for a physical meaning of the symbols. The quantity J★ is calculated by efficient numerical methods, devised originally by Hwang, involving fixed point Gauss-Jacobi quadrature or asymptotic expressions depending upon Σ_b and the characteristics of the resonance. The introduction of J★ function involves the following assumptions:

The cross section is assumed to be represented as sums of single levels:

in $dU = \frac{dE}{E}$ the E is put = E_{o_i}; the weighting spectrum S(E) is assumed to be constant over the resonance = S_i and integration is extended to cover all lethargies $(-\infty, +\infty)$.

Results and Discussion

From the results of comparison studies typical of which are shown in Figs. 1 to 5 and in Table I we obtain the following conclusion:

1. For ^{235}U and ^{239}Pu, there is a wide dispersion among the values of SSFs calculated by RAMBHA code system and by LRSG code system at low background dilutions specially below 10-15 barns. The 'low' background dilution at which the discrepancy between LRSG and RAMBHA results became significant depended on the details of energy structure of the resonances and thus varied from group to group and from nuclide to nuclide. At higher dilutions the values of SSFs calculated by both the methods converge towards experimental values satisfactorily. The unacceptably large discrepancies in SSFs at low dilutions are of practical interest[16] only in situations involving pure fissionable isotopes below 100 eV.

2. For ^{23}Na and natural iron, the SSFs calculated by LRSG code system are larger than experimental values of Arnaud et al. The calculated values of SSFs in this case for capture process in ^{23}Na and Fe as well as the calculated values of SSFs for total reaction process for ^{239}Pu and ^{235}U were always larger than experimental values. For SSFs for fission process the trend in the disagreement was not uniform as shown in Table I.

Table I Comparison of SSFs for Fission Process: ^{239}Pu and ^{235}U

Energy region (eV) and Nuclide	σ_b (barns)	RAMBHA	LRSG	Exptl. (Bakalov et al. Ref.[11])
4.65-10.0	10	0.184	0.282	0.225
^{239}Pu	100	0.289	0.381	0.367
	1000	0.624	0.685	0.736
21.5 -46.5	10	0.174	0.225	0.260
^{239}Pu	100	0.332	0.387	0.399
	1000	0.705	0.729	0.778
4.65-10.0	10	0.458	0.328	0.379
^{235}U	100	0.533	0.491	0.559
	1000	0.807	0.791	0.838
21.5 -46.5	10	0.481	0.502	0.476
^{235}U	100	0.675	0.678	0.701
	1000	0.909	0.902	0.928

3. In the case of ^{239}Pu and ^{235}U, even at 'higher' dilutions the values of SSFs calculated by RAMBHA i.e. J★ method were found to be larger than the values obtained with Cullen's codes. This trend is in agreement with the calculations performed by independent set of codes VITAMIN-E and MC^2-2 method as communicated by Cullen[17] for ^{238}U for total reaction process.

4. The ignoring of flux correction factor in J★ method RAMBHA-A in Figs. 3 to 6 surprisingly brought good agreement between measured values of SSFs and calculated values of SSFs in many cases. However inclusion of flux correction factor in J★ method gives a more accurate

interpretation of evaluated data. The LRSG code system which even more accurately reproduces the contents of evaluations than J* method with inclusion of flux correction term gives values of SSFs which disagree with experimental values. There is a need to re-assess the experimental values of SSFs in all these cases. For instance the experimental value of SSF reported by Bakalov et al. for ^{239}Pu in 4.65 to 10 eV seems to show an unphysical result. The J* method without inclusion of flux correction term and the experimental value while agreeing well with each other yield a self shielded cross section which is below the minimum of cross section in this energy region. See Fig. 6. Physically the self shielded cross section for any σ_b must lie [17] between maximum and minimum of $\sigma(E,T)$ in a given energy group for a given temperature T.

Acknowledgement

The authors sincerely thank Dr. D.E. CULLEN, Nuclear Data Section, IAEA for providing us with the latest version of his LINEAR-RECENT-SIGMA1-GROUPIE code system which enabled the authors to perform the present study. A fruitful correspondence with him is gratefully acknowledged.

References

1. R.B. Perez et al., Trans. Am. Nucl. Soc. 39, 883 (1981).
2. S. Ganesan, 'On the Need for Changing ENDF/B Convention in the Representation of Cross Section in the Unresolved Resonance Region for Fertile and Fissile Nuclei, Proceedings of IAEA Consultants Meeting on Uranium and Plutonium Isotope Resonance Parameter's, 28 Sep.-2 Oct.1981, IAEA Report. D.E. Cullen (Ed) INDC(NDS)-129/GJ pp. 415-425, Vienna.
3. J.L. Munoz-Cobos, G. de Saussure, R.B. Perez, Trans. Am. Nucl. Soc. 38, 666 (1981).
4. G. de Saussure, R.B. Perez, Ann. Nucl. Energy 9, 79 (1982).
5. R.N. Hwang, Nucl. Sci. Eng., 52, 157 (1973).
6. D.E. Cullen, 'Program LINEAR' UCRL-50400, Vol.17, Part A, Lawrence Livermore Laboratory (1979).
7. D.E. Cullen, 'Program RECENT' UCRL-50400, Vol.17, Part C, Lawrence Livermore Laboratory, (1979).
8. D.E. Cullen, 'Program SIGMA1' UCRL-50400, Vol.17, Part B, Lawrence Livermore Laboratory (1979).
9. D.E. Cullen, 'Program GROUPIE' UCRL-50400,Vol.17, Part D, Lawrence Livermore Laboratory, (1979).
10. A. Arnaud et al., Nucl. Cross Sections Technology, Washington (1975) p. 961.
11. Bakalov et al., Nucl. Cross Section Technology, Knoxville (1979) p. 692.
12. S. Ganesan, 'Evaluation of Statistical Resonance Parameters for ^{232}Th in 4 to 41 keV Energy Region' RRC-42, INDC(IND)-26/GJ (1980) IAEA, Vienna.
13. S. Canesan et al., 'Development of a new fast reactor processing code RAMBHA at RRC' in Proceedings of the Workshop on Nuclear Data Evaluation, Processing and Testing, INDC(IND)-30 (1981), IAEA, Vienna.
14. D.E. Cullen, C.R. Weisbin, Nucl. Sci. Eng. 60, 199 (1976).
15. H. Henryson, B.J. Toppel, C.G. Stenberg, MC^2-2: A Code to Calculate Fast Neutron Spectra and Multigroup Cross Sections" ANL-8144(ENDF 239) Argonne National Laboratory (1976).
16. R.N. Hwang, Ann. Nucl. Energy, 9, 31 (1982).
17. D.E. Cullen, Private Communication (15 Feb. 1982).

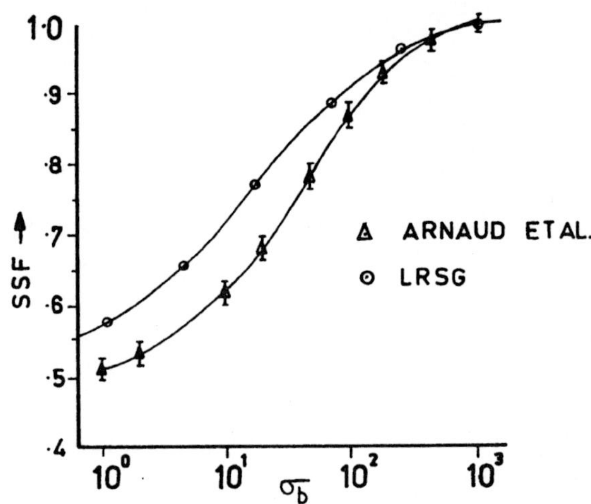

Fig. 1. SSFs for capture process for natural iron; T = 300K; Energy region: 24.75 - 40.75 keV

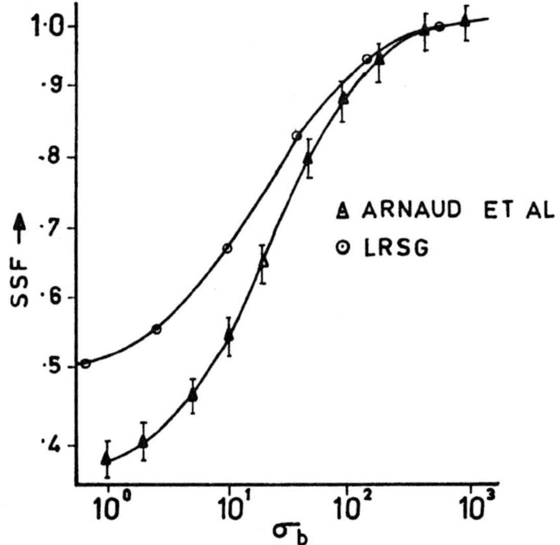

Fig. 2. SSFs for capture process for sodium; T = 300K; Energy region: 49.5 - 57.5 keV

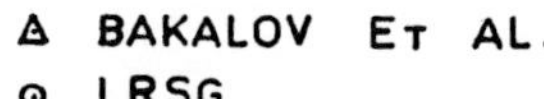

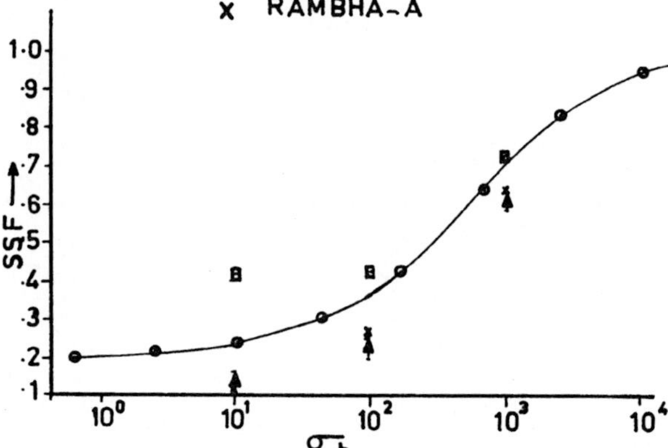

Fig. 3. SSFs for total reaction process for ^{239}Pu; T = 300K; Energy region: 10.0 - 21.5 eV. In RAMBHA-A, the factor f is unity in Eq.(5)

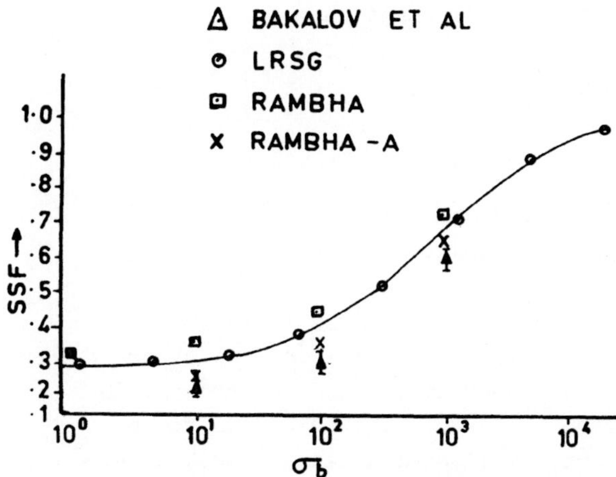

Fig. 4. SSFs for total reaction process for ^{239}Pu; T = 300K; Energy region: 4.65 - 10.0 eV. In RAMBHA-A, the factor f is unity.

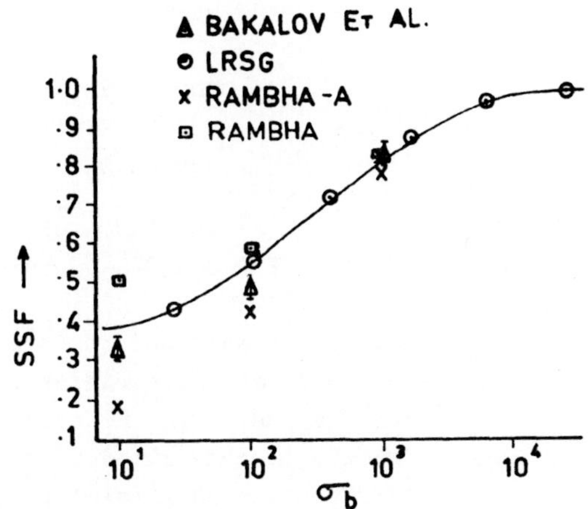

Fig. 5. SSFs for total reaction process for ^{235}U. T = 300K, Energy region 4.65 - 10.0 eV. In RABMHA-A, the factor f is unity.

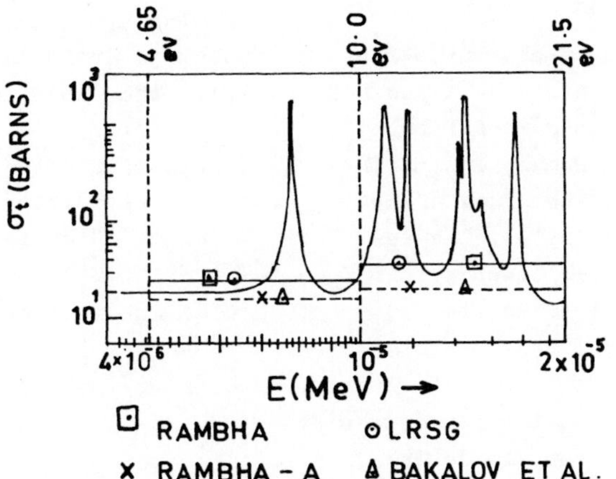

Fig. 6. Plot of σ_t at 300K for ^{239}Pu. The horizontal lines correspond to self-shielding group cross sections in the two groups for σ_b = 10 barns.

FAST NEUTRON INDUCED FISSION CROSS SECTION FOR Pu-239

Zhou Xianjian, Yan Wuguang, Zhou Huiming,
Deng Xinlu, Rong Chaofan, Wu Jingxia,
Sun Zhongfa, Ye Zhongyuan and Zhou Shuhua

Institute of Atomic Energy, Academia Sinica
P.O.Box 275-15 Beijing, People's Republic of China

The fission cross sections of Pu-239 in the neutron energy ranges 1.0--1.6 MeV and 3.4--5.6 MeV are measured absolutely. The monoenergetic neutrons are produced from $T(p,n)^3He$ and $D(d,n)^3He$ reactions on a 2.5 MV Van de Graaff accelerator. The neutrn fluence is determined with recoil proton telescope system and the number of fission events with fission ionization chamber filled with methan gas. The deposit of Pu-239 and the polyethylene film are put in a back-to-back geometry and the neutron fluence and fission fragments are measured simultaneously. The absolute values of fission cross section of Pu-239 are obtained and compared with other experimental and compiled results.

[^{239}Pu(n,f), fission cross sections, E_n=1.0--1.6, 3.4--5.6 MeV, nuclear data.]

Introduction

The fission cross section of plutonium-239 is one of the important nuclear data and has been measured by many authors in different laboratories. In the last decade due to the progress of techniques the accuracy of measurement was improved. But still, the results do not agree with each other within their apparent uncertainties. In view of this situation more careful measurements are needed.

We have measured the σ_f of Pu-239 in neutron energy range accesible by our Van de Graaff accelerator, i.e. 1.0--1.6 MeV using T(p,n) reaction and 3.4--5.6 MeV using D(d,n) reaction. The neutron fluence was determined by means of a recoil proton telescope, the fission fragments were recorded by an ionization chamber. We adopted the back-to-back concept, i.e., the telescope and the fission chamber were constructed together, the fission sample and the hydrogen radiator (thin plastic film) were placed close to each other, and the recoil protons and fission fragments were recorded simultaneously.

Experimental Method

The experimental set up including the electronics used was shown in Fig.1. A shallow fission chamber (that is, the effective thickness of the ionization gas layer was not thick enough to stop the fragments and the α particles as well) was used in order to improve the anti-α pile up characteristics. The chamber was filled with 1 atm pure CH_4 gas and covered with 0.5 mm cadmium sheet to avoid the thermal neutrons scattered in the experimental hall from striking the sample.

Sample of fissile material was prepared by electrodeposition of 1 mg Pu on a platinum backing (40 mm in diameter and 20 μm thick). The size of deposit Pu plot was a circle of 26 mm in diameter. The isotope composition of the sample was listed in Table I.

The recoil proton telescope contains a thin

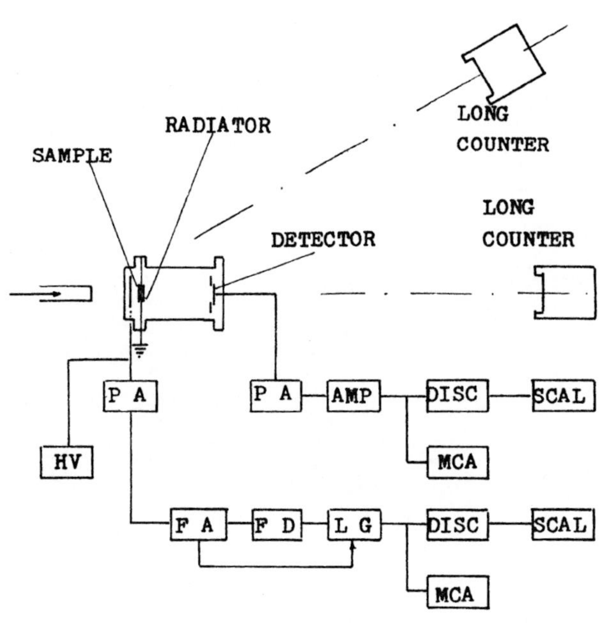

Fig.1 Experimental arrangement

K. H. Böckhoff (ed.), Nuclear Data for Science and Technology, 36–38.

Table I. Isotope composition of the sample

Isotope	Pu-239	Pu-240	Pu-241
Fgaction	96.9±0.8	3.0±0.5	0.1±0.1

plastic film as hydrogen radiator. The number of neutrons was determined by using a silicon semiconductor detector of surface-barrier type to detect the recoil protons. Two polyethylene films with thicknesses 5.3269 and 3.9358 mg/cm^2 were used in the 3.4--5.6 MeV energe range, and two polypropylene films with thicknesses 0.74345 and 0.47453 mg/cm^2 were used in the 1.0--1.6 MeV energe range. The ratio of atomic numbers of carbon to hydrogen for these two kinds of films was 1:2. The telescope was evacuated to 10^{-2} mm Hg in order to improve the resolution of recoil proton spectrum. A molybdenum plate was inserted between the radiator film and the semiconductor detector to stop the protons produced from (n,p) scattering of direct neutrons, and thus the background of recoil proton spectrum was measured. The plate was inserted or removed alternatively during measurement by a small motor controlled remotely. At the 0° and 30° directions with respect to deuteron beam two long neutron counters were settled to monitor the neutron yields. A shadow cone was inserted between the neutron source and fission chamber to mea- the fission background counts.

Obtaining the counts of fission events, recoil protons and their respective backgrounds, we could calculate the fission cross section directly.

Corrections and Uncertainties

The following Corrections and sources of errors have been considered:

(1) The neutron induced fission of the isotopes Pu-240 and Pu-241 contained in the plutonium sample were subtracted. The spontaneous fission of the Pu-240 was also considered. The quantity of this correction was about 4--5 %, the error brought in by this effect was 0.6 %.

(2) Uncertainty of the number of hydrogen nuclei.

For the measurement in 3.4--5.6 MeV region, the fraction of impurity contained in the film was smaller than 0.2 %,for the 1.0--1.6 MeV region it was smaller than 0.3 % The relative error of proton number due to weighing was 0.15 % for the 3.4--5.6 MeV region, and it was 1 % for the 1.0--1.6 MeV region. The deviation of the H/C ratio from 2 was less than 0.3 %. The uncertainty in the measurement of film diametre was 0.4 %. The error caused by the small variation of thickness along the film was 1 %. All the above gave the error in determination of proton number for measurements in low energy region 1.5 %, and in high energy region 1.15 %.

(3) Uncertainty of the number of Pu-239 atoms. The quantity of Pu-239 in the sample was determined with two methods. The -activity was measured absolutely in a low solid angle equipment, then the value of 2.5213 10^{18} Pu-239 atoms was obtained from the decay life times of varios Pu isotopes and the isotope composition of the sample. In the other hand, the absolute number of Pu atoms was determined by constant current Coulomb method and a value of 2.4876 10^{18} Pu-239 atoms was derived. This agrees well with the result from low solid angle measurement. We adopted the value given by constant current Coulomb method, the relative error was 1 %.

(4) The detection efficiency of fission ionization chamber.

The detection efficiency of fission chamber didn't reach 100 % because we set a threshold in discriminator to cut off the α pile up. The semiempirical formula N_f $B/(X_o-X)^2$ was used to fit the fission fragment pulse height spectrum in the near threshold region and then extrapolated to the zero threshold. The efficiency ε_f thus determined in our experiment was 94--96 %. Considering the deviation of fitting curve from the exact spectrum, the instability of the threshold and other sources of error,we gave the correction factor $(1-\varepsilon_f)$ an relative uncertainty 20 %, this

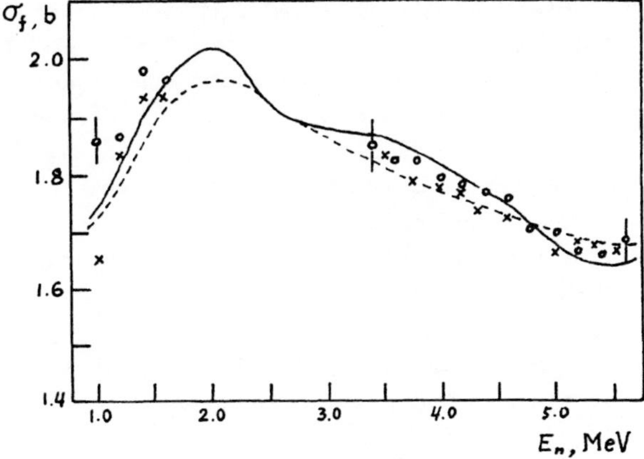

Fig.2. Fission cross section of Pu-239
----- ENDF/B-IV;
o Present data;
× Value from Kari;
—— Evaluated curve of Liu.

would bring a 1% error in fission cross section values.

(5) Correction for absorption in sample and fission fragment anisotropy.

Owing to the thickness of the fissile layer some of the fission fragments can not go out the sample and be recored. The fraction of the lost fragments Σ can be calculated according to the respective parameters. The Σ value obtained was 1.9 % at $E_n = 1$ MeV and 2.4% at $E_n = 4$ MeV.

(6) Correction of neutron scattering.

The neutrons scattered by material in or around the neutron source and detector assemblies can cause background fission events and recoil protons as well. This kind of background cannot be subtracted by inserting shadow cone and molybdenum plate. The amount of correction was 2 3 % in 1.0--1.6 MeV energy region and 0.1--0.2 % in the other region. The error of calculation was 5 %.

(7) Correction of neutron attenuation.

The transmission attenuation of neutron beam when it was passing through the backings could be calculated. The resulting correction was 0.5 %.

(8) Error of the efficiency factor of the telescope.

This error was 0.9 % which contains the error of the (n,p) scattering cross section and the error of geometric dimensions.

The overall uncertainties of σ_f was 2.3 % in 3.4--5.6 MeV region and 2.7 % in 1.0--1.6 MeV region.

Result and Comparison with other data

The values of σ_f obtained are shown in Fig.2 together with some other results for comparison. It can be seen that in 3.4--5.6 MeV neutron energy range our results are in good agreement with the compilation of Liu[1] and the experimental values of Kari[2]. In 1.2--1.6 MeV range,all of these results agree with each other within their uncertainties. At 1.0 MeV, however, our value is obviously higher than theirs, but agrees with early results given by Henkel[3] and Netter[4]. It is shown that so far the disagreement of Pu-239 fission cross section at 1.0 MeV neutron energy still exists.

References

1. Liu Jicai, Evaluation of Pu-239 fission cross section in 1 keV--20 MeV region, hsj-75005, 1975.
2. K.Kari, KFK-2673 (1978) p.59.
3. R.L.Henkel,LA-2114 (1957).
4. F.Netter, CEA-1913 (1961).

ON THE NEUTRON INELASTIC-SCATTERING CROSS SECTIONS OF ^{232}Th, ^{233}U, ^{235}U, ^{238}U, ^{239}Pu AND ^{240}Pu*

by

Alan B. Smith, P. T. Guenther and R. D. McKnight
Argonne National Laboratory
Argonne, Illinois

Differential-neutron-emission cross sections of ^{232}Th, ^{233}U, ^{235}U, ^{238}U, ^{239}Pu and ^{240}Pu are measured from $\approx$ 1.0 to 3.5 MeV with sufficient angle and magnitude detail to provide angle-integrated emission cross sections to $\lesssim$ 3% accuracies. Emitted-neutron resolutions are quantitatively defined and vary from $\approx$ 0.1 to 0.35 MeV. The experimental results are corrected for fission-neutron contributions to obtain pseudo-elastic-scattering cross sections which, together with independently measured neutron total cross sections, define the non-elastic cross sections to within specified energy resolutions. The latter results imply inelastic-neutron-scattering cross sections which are compared with quantities deduced from prominent evaluations. Good general agreement is noted for ^{232}Th, ^{233}U, ^{235}U, and ^{238}U inelastic-scattering cross sections. The agreement is poor in the case of ^{240}Pu and very poor for ^{239}Pu. The implications of the present results on the integral behavior of ^{239}Pu critical assemblies and FBR energy systems are discussed.

[^{232}Th, ^{233}U, ^{235}U, ^{238}U, ^{239}Pu, ^{240}Pu; neutron emission, nonelastic and inelastic cross sections, E_n=1.0-3.5 MeV]

I. INTRODUCTION

Inelastic-neutron scattering from fissile and fertile nuclei has long been of applied concern.[1] Direct measurements have led to partial understanding of inelastic-scattering from ^{232}Th and ^{238}U[2,3], fragmentary knowledge for ^{240}Pu[4] and very little information for ^{233}U, ^{235}U and ^{239}Pu.[5] In the few-MeV region inelastic-scattering cross sections of fertile nuclides are not known to better than 10% and to much lesser accuracies in fissile cases. These uncertainties considerably exceed fast-breeder-reactor (FBR) accuracy goals.[1] Appreciable improvement by direct measurement is exceedingly difficult, therefore alternate approaches are sought.

Few-MeV neutron total cross sections of ^{232}Th, ^{233}U, ^{235}U, ^{238}U, ^{239}Pu and ^{240}Pu are now known to $\lesssim$ 1.5%.[6] Radiative-capture cross sections are not as well known but very small. Fission cross sections are known to 2-3% and nubar to 1-2%. Elastic-scattering cross sections can be measured to accuracies of $\lesssim$ 3%. These considerations suggest the determination of the nonelastic cross sections and thus the total inelastic-scattering cross sections. The concept is not new[7] but contemporary knowledge and technologies offer new potential for improved understanding of inelastic-scattering cross sections for the excitation of all levels above $\approx$ 100 keV. These are the effective inelastic-scattering cross sections of primary applied interest. This paper reports measurements of non-elastic (and thus inelastic-scattering) cross sections of ^{232}Th, ^{233}U, ^{235}U, ^{238}U, ^{239}Pu and ^{240}Pu and explores the impact of the results on FBR performance characteristics.

II. EXPERIMENTAL METHODS

An essential factor in the measurements was the availability of metallic samples of > 95% isotopic purity. The sample properties are defined in ref. 8.

All of the measurements were made with the Argonne National Laboratory ten-detector time-of-flight apparatus.[9] The relative energy-dependent responses of the ten detectors were determined by observing neutrons emitted at the spontaneous fission of ^{252}Cf[10] and cross-correlated by observing neutrons scattered from carbon at a fixed angle. The angular distribution of neutrons scattered from carbon and the total cross section of carbon were measured at each energy. At the energies of the present work these two cross sections are essentially identical. The angular distribution was integrated to obtain the relative angle-integrated scattering cross section of carbon and the absolute sensitivity of the detector system adjusted to bring that value into agreement with the observed total cross section. This method is independent of a reference standard and is relatively insensitive to experimental perturbations. It implies actinide scattering cross sections relative to the observed total cross sections of carbon.

The measurement objective was the determination of pseudo-elastic-scattering cross sections to the best possible accuracies, not the optimization of energy resolution. The observed actinide-scattering cross sections were inclusive of inelastic-scattering contributions due to the excitation of very low-lying levels (not generally of applied interest). The scattered-neutron-resolution functions were carefully determined from the measured velocity spectra and varied from $\approx$ 100-350 keV depending upon the incident energies. Some of the samples were very radioactive (e.g. ^{233}U) and attention was given to the consequent time-uncorrelated backgrounds.

All of the actinide measurements included fission-neutron contributions which were large at some energies and angles (particularly for ^{233}U and ^{239}Pu). Corrections for this fission component were made in two ways. In the first, a Maxwellian (T=1.33 MeV) fission-neutron spectrum was fitted to the high-energy portion of the velocity spectra (to regions corresponding to energies greater than the incident energy) and subtracted from the spectral region of interest. Pseudo-elastic-scattering cross sections were deduced from the corrected velocity spectra and further corrected for multiple-event, beam attentuation and angular-resolution effects.

*This work supported by the U.S. Department of Energy.

K. H. Böckhoff (ed.), Nuclear Data for Science and Technology, 39–44.

In the second fission-neutron-correction method, differential emission cross sections, including the fission-neutron contribution, were deduced from the velocity spectra. The resulting emission cross sections were then corrected for fission-neutron, beam-attentuation, multiple-event and angular-resolution effects using a detailed Monte-Carlo simulation of the experiment. The simulation was based upon: the observed emission cross sections, well known total and fission cross sections[8,11] and on nubar (neutrons per fission) and its energy dependence.[12] The fission-neutron spectra were assumed to be of the Maxwellian form (T=1.33 MeV). Inelastic-scattering processes corresponding to the excitation of levels beyond the resolution function were considered. Contributions due to the neutron-capture cross sections were very small and were ignored. The above two methods led to similar results (frequently differing by < 1%). Generally, fission-neutron-correction uncertainties were not a major component of the overall experimental uncertainties.

More details of the experimental method are to be found in ref. 8.

III. EXPERIMENTAL RESULTS

The measurements were made at approximately six incident energies ($\approx$ 1.0 - 3.5 MeV) and at $\approx$ 20 scattering angles distributed between 20-160 degrees and known to $\approx$ 0.5 degree. Incident-neutron energy resolutions were 50-70 keV. The scattered neutron resolutions were larger, well defined and varied with target and incident energy.

Normalization of the actinide results was dependent upon the observation of differential-elastic-scattering from carbon. A representative carbon differential-scattering distribution is shown in Fig. 1. Eighty differential values were obtained at each incident energy providing a four-fold redundancy at each angle. The statistical uncertainties of the measured cross sections were generally $\lesssim$ 1% and the results were consistent to within that value. The angle-integrated elastic-scattering cross sections were obtained by fitting the observed distributions with 6th-order Legendre polynominal series. The resulting angle-integrated values were related to the measured neutron total cross sections to obtain the absolute normalization of the measurement system, as outlined above.

The actinide differential-scattering distributions were measured concurrently with the above carbon determinations. Forty differential values were obtained at each incident energy. The resulting differential-emission cross sections were corrected for fission-neutron and other perturbations, as outlined above, to obtain the pseudo-elastic-scattering cross sections inclusive of well defined low-energy inelastic excitations. Illustrative actinide pseudo-elastic-scattering cross sections are shown by the symbols of Fig. 1. The statistical uncertainties of the differential values were generally $\lesssim$ 1% and the systematic uncertainties due to calibration and correction procedures were $\approx$ 2-3%. The objective was the derivation of implied inelastic-scattering cross sections thus the measurement emphasis was on angle-integrated scattering cross sections. The latter were obtained by fitting the measured actinide differential distributions with 8th-order Legendre polynominal series. The quality of the fit to the measured data is illustrated by the curves of Fig. 1. The resulting angle-integrated actinide cross sections were generally known to $\approx$ 3%. The dominant contributions to this uncertainty were due to detector calibrations and scattering-angle determinations which were independent at each measurement energy. Thus, the observed consistency of the results with energy was gratifying.

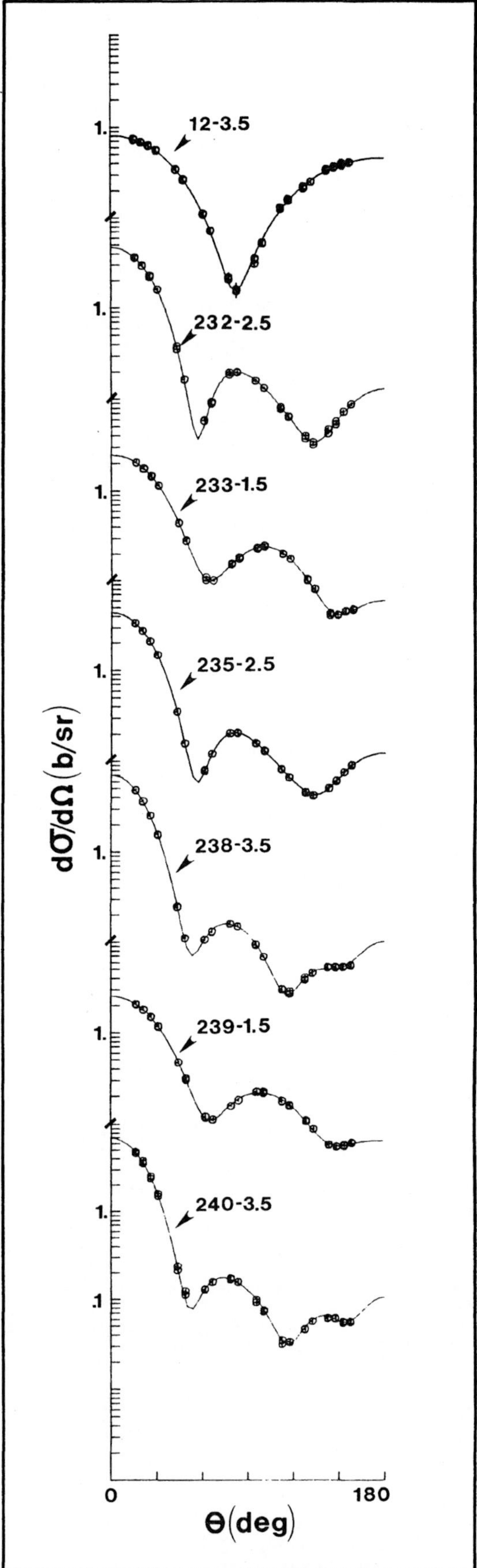

Fig. 1. Illustrative differential scattering cross sections. Measured values are indicated by symbols. Target mass and incident energy in (MeV) are noted for each distribution.

IV. DISCUSSION

The total neutron-inelastic-scattering cross sections, above well defined and low excitation energies, were deduced by subtracting the above measured values and the other partial cross sections from the neutron total cross section. The necessary neutron total cross sections were taken from results recently obtained at this laboratory.[6] At the relevant energies these total cross sections are known to ≈ 1 1/2%. The capture cross sections were taken from ENDF/B-V assuming a 10% uncertainty. The capture process is a minor consideration in the present work. The ^{232}Th and ^{238}U fission cross sections and associated uncertainties were taken from refs. 2 and 3. Those for ^{233}U and ^{235}U were taken from ref. 11 using the stated ^{235}U uncertainties and assuming that the ^{233}U/^{235}U fission ratio was known to 1 1/2%. ^{239}Pu and ^{240}Pu fission cross sections were taken from the ratio values relative to ^{235}U as given in ref. 13. Generally, these fission cross sections are consistent with those of ENDF/B-V, identically so in the case of ^{235}U. The results are total inelastic-scattering cross sections of primary FBR interest (i.e. those involving significant energy transfers). The respective numerical inelastic-scattering values implied by the present measurements are given in Table 1. They are compared with the quantities deduced from ENDF/B-V in Fig. 2.

The experimentally deduced (ED) ^{232}Th inelastic-scattering cross-section uncertainties are ≈ 6-8%, excepting the lowest-energy (0.93 MeV) value which involves the difference between two large and similar numbers. Generally, the ^{232}Th ED results are similar to those deduced from ENDF/B-V. At the lowest energies the present results are somewhat smaller than the evaluated quantities (and similar to the ^{238}U results, below). This suggests that the ENDF/B-V ^{232}Th evaluation underestimates the excitation of low-lying levels at incident energies of ≈ 1 MeV. The ^{238}U ED inelastic-scattering results are generally in agreement with the values derived from ENDF/B-V to within the experimental uncertainties of 7-9%. There is a similar agreement in the case of ^{235}U although the experimental uncertainties are somewhat larger (11-12%). The lowest-energy ^{235}U result may not be too meaningful as the fission cross section changes rapidly in that energy region. The generally good agreement between measured and evaluated ^{235}U results is encouraging as it supports the above treatment of the fission-neutron corrections and the ^{235}U fission cross sections are large thoughout the energy range of the present experiments. The uncertainties associated with the ^{233}U ED inelastic-scattering cross sections are relatively large (17-24%), reflecting the large fission cross section (and consequently small inelastic-scattering cross section). An additional experimental factor was the high radioactivity of the measurement sample. Despite these problems, the present results are consistent with those deduced from ENDF/B-V.

The uncertainties associated with the ^{240}Pu ED inelastic cross sections are relatively large (14-19%) but the results are 20-30% smaller than those deduced from ENDF/B-V, increasingly so with energy. This suggests that the evaluation may not have properly accounted for the direct excitation of the ground-state-rotational-band of this even-even nucleus. The ^{239}Pu ED inelastic-scattering cross sections are defined to within 15-20%. These are relatively large uncertainties but the experimental results lie 30-50% below those deduced from ENDF/B-V.

Table 1. Actinide Scattering Cross Sections (in barns)

E_n(MeV)			Th-232	U-233	U-235	U-238	Pu-239	Pu-240
0.93	E_x(keV)[a]	=	> 90	-----	> 90	> 90	> 90	-----
	Scat.[b]	=	6.96±0.108	-----	5.73±0.110	7.07±0.109	5.42±0.121	-----
	Exp.[c]	=	5.79±0.231	-----	5.02±0.200	6.12±0.240	4.93±0.246	-----
	Inel.[d]	=	1.17±0.255	-----	0.71±0.228	0.95±0.263	0.49±0.274	-----
1.27	E_x(keV)	=	> 120	> 120	> 120	> 120	> 120	> 120
	Scat.	=	6.71±0.103	4.66±0.110	5.48±0.107	6.89±0.105	5.09±0.115	5.26±0.111
	Exp.	=	4.54±0.159	4.05±0.162	4.13±0.124	4.79±0.167	4.01±0.160	4.23±0.150
	Inel.	=	2.17±0.189	0.61±0.195	1.35±0.164	2.10±0.197	1.08±0.197	1.03±0.193
1.49	E_x(keV)	=	> 130	> 130	> 130	> 130	> 130	> 130
	Scat.	=	6.75±0.104	4.82±0.113	5.51±0.107	6.65±0.106	5.07±0.120	5.28±0.100
	Exp.	=	4.28±0.128	4.01±0.110	4.11±0.123	4.46±0.134	4.10±0.120	4.19±0.120
	Inel.	=	2.47±0.164	0.81±0.157	1.40±0.163	2.19±0.171	0.97±0.169	1.09±0.156
1.85	E_x(keV)	=	> 130	> 130	> 130	> 130	> 130	> 130
	Scat.	=	6.98±0.107	5.10±0.118	5.75±0.112	6.68±0.108	5.18±0.121	5.45±0.119
	Exp.	=	4.29±0.129	4.05±0.118	4.23±0.127	4.28±0.128	4.10±0.120	4.11±0.130
	Inel.	=	2.69±0.167	1.05±0.166	1.52±0.169	2.40±0.167	1.08±0.169	1.34±0.176
2.55	E_x(keV)	=	> 200	> 200	> 300	> 200	> 300	> 200
	Scat.	=	7.44±0.114	5.64±0.128	6.30±0.120	7.05±0.115	5.67±0.124	5.86±0.116
	Exp.	=	4.51±0.190	4.57±0.155	4.55±0.176	4.46±0.172	4.47±0.180	4.40±0.130
	Inel.	=	2.93±0.221	1.07±0.204	1.75±0.213	2.59±0.206	1.20±0.218	1.45±0.176
3.55	E_x(keV)	=	> 300	> 350	> 300	> 300	> 300	> 300
	Scat.	=	7.88±0.121	6.26±0.132	6.76±0.124	7.44±0.121	6.14±0.128	6.31±0.147
	Exp.	=	4.95±0.136	5.13±0.170	5.14±0.153	5.03±0.164	5.11±0.220	4.97±0.120
	Inel.	=	2.93±0.182	1.13±0.215	1.62±0.196	2.41±0.203	1.03±0.258	1.34±0.189

a. Inelastic cross sections correspond to all excitations above this lower limit.

b. Scattering cross section ≡ $\sigma_t - \sigma_f - \sigma_{cap}$.

c. Measured values corresponding to excitations of less than the value of footnote a.

d. Implied inelastic cross section ≡ Scat. - Exp.

These are large discrepancies in an important cross section of an essential FBR material. It is difficult to attribute these discrepancies to experimental error as the six actinides were studied concurrently and the results for four of the actinides are in reasonable agreement with ENDF/B-V. ^{239}Pu and ^{233}U are odd nuclides with similar neutron total and fission cross sections and level structure. Thus, one might expect similar inelastic-scattering cross sections. That expectation is consistent with the present observations but very different from the conclusions drawn from the evaluation. In addition, the present ^{239}Pu inelastic-scattering results are reasonably consistent with the values calculated by Jary[14] and with the evaluation of Antsipov et al.[15] (see Fig. 2).

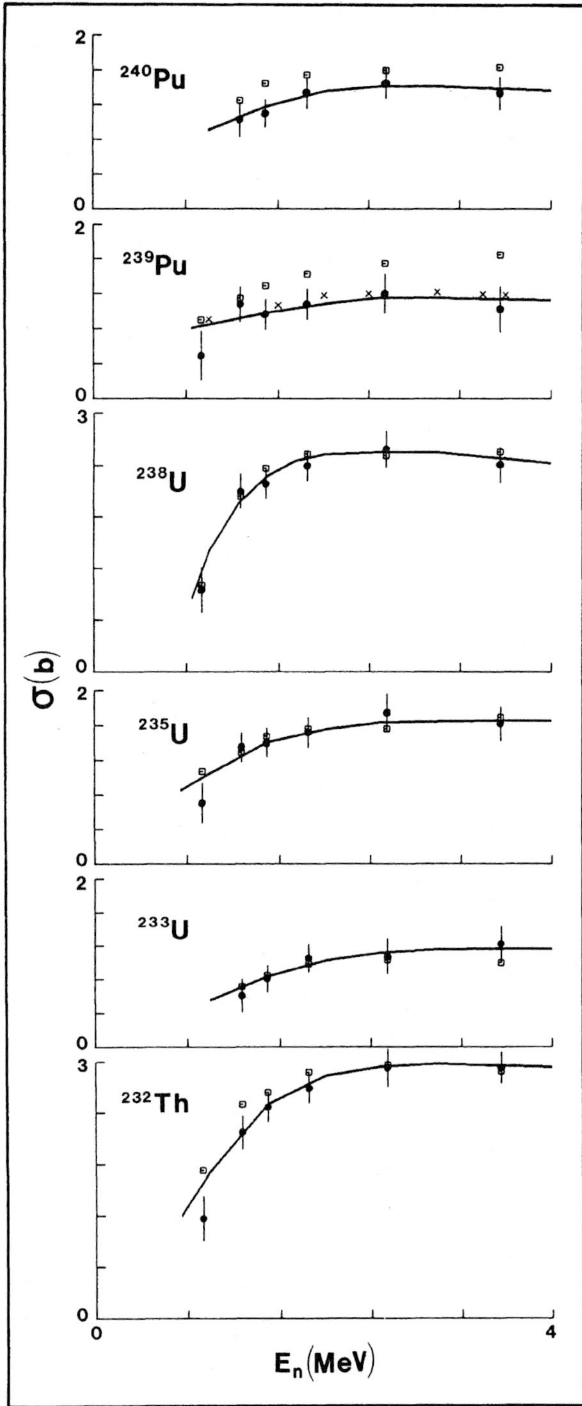

Fig. 2 Total-Inelastic-Scattering Cross Sections of ^{232}Th, ^{233}U, ^{235}U, ^{238}U, ^{239}Pu and ^{240}Pu. ● and ——— = present results. □ = ENDF/B-V.[11] × = evaluation of ref. 15.

V. INTEGRAL SENSITIVITY

The above ^{239}Pu inelastic-scattering descrepancy is disturbing as this is the primary FBR fissile nuclide. In order to determine the implications of the discrepancy a revised ^{239}Pu evaluated file was constructed. The revision used the inelastic-scattering cross sections of Antsipov et al.[15], somewhat normalized to explicitly agree with the total ^{239}Pu inelastic-scattering cross sections deduced from the present measurements. The remaining portions of the file were directly taken from ENDF/B-V[11] with only the elastic-scattering cross section adjusted to assure internal consistency. This revised file and ENDF/B-V were used to calculate key integral parameters in two fast critical assemblies: the bare plutonium (95% ^{239}Pu) sphere JEZEBEL[16] and the LMFBR benchmark assembly ZPR-6 Assembly 7.[17] The ^{239}Pu enrichment (^{239}Pu/total heavy metal) in ZPR-6/7 was 13%. These two integral test cases span an extreme range of ^{239}Pu enrichments. Calculations were performed using one-dimensional spherical benchmark models of these assemblies. For each assembly two sets of homogeneous cross sections with 29 broad energy groups (E_{top} = 14.19 MeV; $\Delta u \simeq 0.5$) were generated with the MC^2-2 code. One set used all ENDF/B-V data; the other set used the revised ^{239}Pu file (which was the ENDF/B-V file with compensating modifications to the inelastic- and elastic-scattering cross sections as described above) with ENDF/B-V data for all other nuclides.

The effects of these changes in the ^{239}Pu scattering data, obtained by direct calculations (with S_{16} transport theory for JEZEBEL and diffusion theory for ZPR-6 Assembly 7), are summarized in Table 2. The impact on integral parameters is more dramatic in JEZEBEL--increases eigenvalue by 0.5%; increases central reactivity worth of ^{238}U by 5%; increases the threshold fission rate of ^{238}U (relative to ^{239}Pu fission) by $\approx$ 4%; and decreases capture rates in ^{235}U, ^{238}U, and ^{239} Pu (relative to ^{239}Pu fission) by 3%, 3%, and 4%, respectively. The impact on integral parameters in the ZPR-6 Assembly 7 is much smaller--increases eigenvalue by 0.1%; changes the central reactivity worths of heavy metals by ± 0.1 to 0.2% and of scattering materials by + 0.5 to 1.0%; and increases ^{238}U fission (relative to ^{239}Pu) by 0.6%.

These changes observed in calculated integral parameters reflect the hardening of the calculated real flux spectra in these assemblies due to the decrease in the ^{239}Pu inelastic scattering cross section in the few-MeV range. As shown in Fig. 3 the impact on the calculated flux spectrum in JEZEBEL is significant. Large changes (2-6%) in group fluxes occur over the entire energy range from 100 keV to 10 MeV.

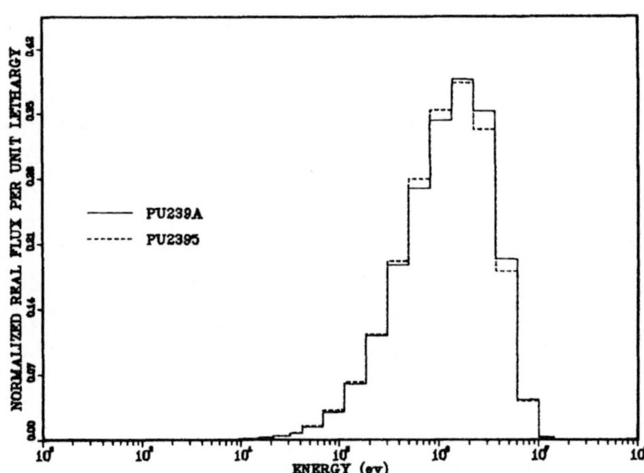

Fig. 3. Central real flux spectrum in JEZEBEL. ——— = present results. ------ = results obtained with ENDF/B-V. The group fluxes are normalized such that the sum of the group fluxes equals one.

Table 2. Impact of ^{239}Pu Inelastic-Scattering Data from Present Measurements on Fast Reactor Systems.

	JEZEBEL			ZPR-6 Assembly 7		
	Present σ^{49}_{inel}	ENDF/B-5 σ^{49}_{inel}	% Diff. Rel. to V5	Present σ^{49}_{inel}	ENDF/B-5 σ^{49}_{inel}	% Diff. Rel. to V5
Eigenvalue, k_{eff}	1.01169	1.00673	+0.49	0.98086	0.98002	+0.09
Beta-effective, β_{eff}	1.8192-3	1.8300-3	-0.59	3.4210-3	3.4165-3	+0.13
Prompt Neutron Lifetime, ℓ_p	3.5302-9	3.5997-9	-1.93	4.7534-7	4.7667-7	-0.28
Reactivity Factor, Ih/%δk/k	1.3718+3	1.3640+3	+0.57	9.5555+2	9.5602+2	-0.05
Central Material Reactivity Worths (10^{-5} δk/k/mole)						
ρ^{49}	3120.06	3113.67	+0.21	46.33	46.29	+0.09
ρ^{40}	2029.65	2011.27	+0.91			
ρ^{25}	1567.42	1576.66	-0.59	36.79	36.84	-0.14
ρ^{28}	213.57	203.95	+4.72	-2.575	-2.581	-0.22
ρ^{B10}				-30.90	-30.99	-0.29
ρ^{Na}				-0.204	-0.202	+0.84
ρ^{Fe}				-0.309	-0.307	+0.59
ρ^{Ni}				-0.517	-0.515	+0.52
Central Reaction Rate Ratios (relative to f^{49})						
f^{25}	0.7069	0.7091	-0.31	1.0645	1.0653	-0.07
f^{28}	0.1450	0.1398	+3.74	0.0243	0.0242	+0.59
f^{40}	0.6885	0.6809	+1.11	0.2103	0.2098	+0.22
f^{41}	0.9187	0.9225	-0.42	1.3997	1.4006	-0.07
c^{25}	0.0683	0.0704	-3.00	0.3231	0.3236	-0.17
c^{28}	0.0459	0.0473	-2.93	0.1537	0.1539	-0.10
c^{49}	0.0328	0.0342	-3.92	0.2896	0.2903	-0.26
$n,2n^{28}$	0.00769	0.00752	+2.22	0.00080	0.00080	0.00

The calculated parameters in Table 2 should not be compared directly with experimental values. As described above, these calculated values have been obtained from homogeneous spherical benchmark models. As such they require corrections for benchmark modeling and methods. However, the changes in the calculated values due to changes in the ^{239}Pu scattering data do reflect the changes in calculated-to-experimental (C/E) values for these assemblies. For JEZEBEL the C/E values for fission rate ratios are improved with the present ^{239}Pu scattering data. The $\approx$ 4% increase in f^{28}/f^{49} eliminates one-half of the C/E discrepancy obtained with ENDF/B-V for this parameter. However, the $\approx$ 0.5% increase in k_{eff} for JEZEBEL doubles the eigenvalue discrepancy obtained with ENDF/B-V (a bias most likely due primarily to the evaluation of the ^{239}Pu fission cross section). The observed changes in C/E values for ZPR-6 Assembly 7 are rather consistently both small and in the right direction.

It should be noted that current FBR designs have ^{239}Pu enrichments intermediate to the two assemblies discussed above. Current homogeneous designs (e.g., PHENIX[18]) have enrichments in the range of 15 - 20%, and current hetrogeneous designs (e.g., CRBR[19]) have enrichments in the range of $\approx$ 30%. These higher (than ZPR-6 Assembly 7) concentrations of ^{239}Pu will increase the effects of the present ^{239}Pu scattering data. These consequences could be $\approx$ 0.25% in eigenvalue, $\approx$ 1% on f^{28}/f^{49}, and $\approx$ 0.5% on α^{49} (c^{49}/f^{49}). These changes would impact several design specifications, including enrichment, reactivity swing, and burnup.

VI. SUMMARY

The present measurements reasonably define the pseudo-elastic-scattering cross sections of ^{232}Th, ^{233}U, ^{235}U, ^{238}U, ^{239}Pu and ^{240}Pu from $\approx$ 1.0-3.5 MeV. The results imply neutron total-inelastic-scattering cross sections of ^{232}Th, ^{233}U, ^{235}U and ^{238}U that are consistent with ENDF/B-V. This suggests that improved understanding of inelastic scattering from these four nuclides should emphasize the determination of energy transferred within a relatively well established total-inelastic-scattering cross section. The present measurements imply ^{239}Pu and ^{240}Pu effective total-inelastic-scattering cross sections 30-50% and 20-30% smaller than given by ENDF/B-V, respectively. These are large discrepancies in important FBR cross sections. The impact of the indicated reduction in the ^{239}Pu inelastic cross sections alone on key integral parameters can be large in contemporary high-enrichment FBR systems (e.g. 0.2-0.4% in k_{eff}). The integral effects will be enhanced by the complimentary decreases in ^{240}Pu inelastic scattering suggested by the present measurements, particularly in FBR systems with high ^{240}Pu content.

References

1. G. Palmiotti and M. Salvatores, Proc. Specialists Meeting on Fast-Neutron Scattering on Actinide Nuclei, Paris (1981) to be published.

2. W. Poenitz et al., Argonne National Laboratory Report, ANL/NDM-32 (1977).

3. J. Meadows et al., Argonne National Laboratory Report, ANL/NDM-35 (1978).

4. A. B. Smith et al., Nucl. Sci. and Eng. 47, 19 (1972).

5. See Proc. Specialist's Meeting on Fast-Neutron Scattering on Actinide Nuclei, Paris (1981) to be published.

6. W. Poenitz et al., Nucl. Sci. and Eng. 78, 333 (1981).

7. M. Walt, Proceeding Conference on Peaceful Uses of Atomic Energy 2, 18 (1955).

8. A. B. Smith and P. Guenther, Argonne National Laboratory Report, ANL/NDM-63 (1982).

9. A. B. Smith et al., Nucl. Instr. and Methods 50, 277 (1967).

10. A. B. Smith et al., Nucl. Instr. and Methods 140, 397 (1977).

11. Evaluated Nuclear Data File-B, Version-V, Brookhaven National Laboratory Report, ENDF-201 (1979), compiled by R. Kinsey.

12. R. Howerton, private communication (1982).

13. J. W. Meadows, Nucl. Sci. and Eng. 68, 360 (1978) and 79, 233 (1981).

14. J. Jary, as discussed at the meeting of ref. 5, above.

15. G. Antsipov et al., International Atomic Energy Agency Report, INDC(CCP)-116/CHJ (1981).

16. G. E. Hansen et al., Los Alamos Scientific Laboratory Report, LA-4208 (1969).

17. C. E. Till et al., Argonne National Laboratory Report, ANL-7910 (1972).

18. M. Robin et al., IAEA-SM-244(19), Aix-en-Provence (1980).

19. CRBRP Preliminary Safety Analysis Report, Project Management Corporation, Docket No. 50-537 (1980).

RECENT RESULTS OF NEUTRON INELASTIC SCATTERING TO HIGHER-EXCITED LEVELS IN ^{232}Th and ^{238}U

G.P. Couchell, C. Ciarcia, J.J. Egan, G.H.R. Kegel, A. Mittler
D.J. Pullen, W.A. Schier and J. Shao

Department of Physics and Applied Physics
University of Lowell
Lowell, MA 01854
United States of America

Neutron inelastic scattering cross sections for levels of ^{232}Th and ^{238}U between 680 and 1200 keV excitation energy have been measured in the neutron energy range 0.9-1.5 MeV. Nearly monoenergetic neutrons were generated using the ^{7}Li(p,n) reaction initiated by a pulsed proton beam produced by a Van de Graaff accelerator using terminal pulsing in conjunction with a Mobley bunching system. Scattered neutrons were observed with a time-of-flight spectrometer with experimental parameters chosen to yield an overall energy resolution of 12-15 keV for inelastically scattered neutrons. Level cross sections were deduced from measured 125°-differential scattering cross sections. Excitation functions were compared with our own previous (n,n′γ) measurements, recent theoretical calculations and the ENDF/B-V evaluations.

[^{232}Th(n,n′), ^{238}U(n,n′), $\sigma(E_n, \theta=125°, E_n=0.9\text{-}1.5$ MeV, neutron time-of-flight; deduced level cross sections, estimated E0 branching ratios]

Introduction

Inelastic scattering cross sections for fast neutrons have been measured by the neutron time-of-flight (TOF) method for levels near 1 MeV in excitation of the actinide nuclei ^{232}Th and ^{238}U. These cross sections are of considerable importance for calculating neutron slowing-down cross sections in fast reactor systems. We have previously studied[1,2] the same level cross sections by the (n,n′γ) pulsed-beam technique and can now make level-by-level comparisons between the two sets of measurements. Experimental parameters for the present (n,n′) TOF measurements were chosen to yield optimum energy resolution (12-15 keV overall) for scattered neutrons with outgoing energies between 200 keV and 400 keV; thus, these measurements determine only the low-energy portion of each level excitation function. An experiment designed to study the same level cross sections for higher outgoing neutron energies is now underway and results will be presented in a future communication.

For both ^{232}Th and ^{238}U we find that level cross sections determined by the (n,n′) TOF method are in good agreement with those deduced from our (n,n′γ) studies for levels whose decay modes and branching ratios are generally assumed to be well known. However, a number of investigators[3,4] have remarked on the difficulty in deducing neutron level cross sections in these two nuclei from (n,n′γ) production cross sections due to uncertainties in: 1) the decay schemes of a number of levels; 2) the multipolarity, and hence internal conversion correction, for many transitions; 3) electric monopole E0 transitions. D.K. Olsen has discussed many such ambiguous cases in a recent comprehensive review[5] of ^{232}Th and ^{238}U(n,n′γ) measurements.

States for which our own (n,n′) and (n,n′γ) level cross sections display significant disagreement are generally those whose decay properties are acknowledged to be uncertain. Thus, considering the good agreement between the two experiments except in cases where the (n,n′γ) method is assumed to have large uncertainties, it is reasonable to assume that the more direct (n,n′) measurements yield more reliable results. Based on this assumption we have used our two sets of measurements to estimate or to set limits on E0 branching ratios for those levels that can decay by electric monopole transitions according to most level classification schemes[6-8].

Our cross sections are also compared with those of the ENDF/B-V evaluations[9,10]. In many instances cross sections for several levels were combined for comparing with those for pseudolevels given in the evaluation. For ^{232}Th our results are generally in good agreement with ENDF/B-V, except near threshold where the evaluated cross sections are consistently higher than our measurements. The excitation function near threshold was estimated using our (n,n′γ) level cross sections, normalized in absolute magnitude to the (n,n′) TOF data in the region 200-400-keV above threshold. For ^{238}U no similar discrepancy in the threshold region was observed between our measurements and the evaluation. In fact most of our cross sections are in good agreement with those for the corresponding ENDF/B-V levels or pseudolevels. The most notable discrepancy is for the 827.1-keV(5$^-$) state, where our results are approximately a factor of 2 smaller than the evaluated cross sections.

Experimental Method

The experimental arrangement for the (n,n′) TOF measurements is shown in Figure 1. Neutrons were generated via the ^{7}Li(p,n)^{7}Be reaction. A fresh lithium target was prepared for each measurement in order to minimize effects due to target deterioration. The thickness was controlled during the evaporation by observing the neutron yield with a separate neutron TOF monitor positioned along the beam line about 4.8 m from the target. The thorium and uranium samples were disk-shaped, each mounted with its plane oriented at an angle of 52° relative to the 0° line. This orientation minimized flight-time dispersion for 1.0-1.5-MeV incident neutrons and 200-400-keV outgoing neutrons. Some of the relevant experimental parameters are listed in Table I.

Scattered neutrons were detected using a 4.5-inch diameter by 0.5-inch thick Pilot B plastic scintillator mounted on a 5-inch RCA 8854 low-noise phototube. This detector, denoted as the main detector in Fig. 1, was placed inside a lead shield which was imbedded in a shield containing a paraffin and lithium carbonate

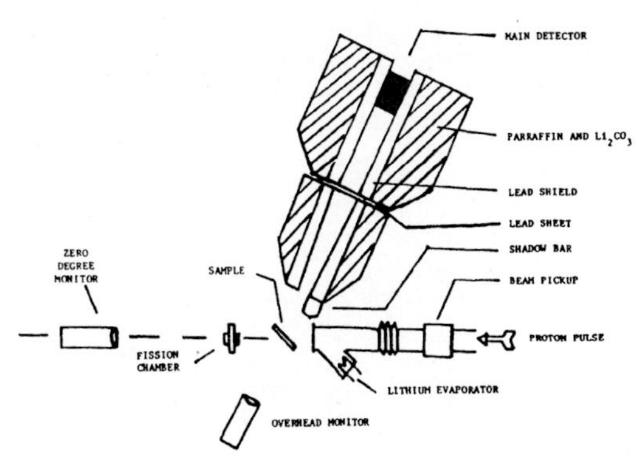

Fig. 1. Experimental arrangement.

Table I Experimental Parameters

Proton pulse width	<0.5 ns
Repetition frequency	5 MHz
Lithium target thickness	8 keV
Sample diameter	3.8 cm
Sample thickness: ^{232}Th	1.53 cm
^{238}U	1.03 cm
Sample mass: ^{232}Th	188 gm
^{238}U	218 gm
Target-to-sample distance	10 cm
Sample-to-detector distance	175 cm
Overall energy resolution	12-15 keV
Run duration	8-12 hr

mixture. A second lead-lined shield was placed in front of the one housing the detector. Further gamma-ray attenuation was achieved by inserting a 0.1-cm thick lead sheet between the two shields. The main detector was positioned at an angle θ=125° during the scattering measurements. The same detector was used to measure the neutron fluence incident on the sample. After removing the thorium or uranium scatterer, the incident neutron yield was measured with the main detector rotated to θ=0°. The 0°-direct measurement was normalized to the (n,n´) 125°-scattering measurement through an overhead monitor neutron TOF system consisting of a Pilot U scintillator mounted on a 2-inch phototube, located approximately at θ=30°. The efficiency of the main detector as a function of neutron energy was determined by comparing it with a uranium (93.1% ^{235}U enrichment by weight) fission chamber of known efficiency.

Data Reduction and Analysis

Figure 2 shows a typical (n,n´) scattered neutron spectrum. Because both the ^{232}Th and ^{238}U average level spacings in the excitation region of interest for these measurements is 20-25 keV, considerable attention has been devoted to the problem of spectrum unfolding. We have developed a two-step stripping procedure: 1) the generation of a simulated TOF response function for each neutron group of interest, 2) the approximation of the measured spectrum, in a least-squares fashion, by a superposition of response functions using the program TINA. A detailed description of the method for generating the response function is presented at this Conference in a separate contribution[11]. Generally, the detector time response functions were not symmetric about the mean, but displayed tails on the low-energy side that became more pronounced for low outgoing neutron energies. The excellent quality of fit achieved for an isolated level such as the 680 keV (1-) state in ^{238}U is shown in Figure 3. The lithium target profile was the only free parameter adjusted to achieve this single-level shape. Figure 4 shows two measured doublets in ^{232}Th and their decomposition using TINA. A precise knowledge of each nuclear level scheme was extremely helpful in analyzing more complex spectra as illustrated by Figure 5. The shape of the response function for each level as well as the two gamma-ray peaks was generated and kept fixed during the fitting procedure. The approximate peak locations were determined from the known level energies, and only slight adjustments in these positions were allowed in the search for the best fit to the measured spectrum. For levels separated by more than 5 keV, consistently better overall fits resulted when they were treated as individual levels; however, unless the energy separation was greater than 10 keV only the composite fit was considered reliable. The background subtraction entailed probably the most subjective part of spectrum stripping. Fortunately, aside from the prominent target- and sample-gamma-ray peaks, the background varied smoothly with energy; we have found that the combination of a constant plus an exponentially decaying term provide an excellent representation of the overall background.

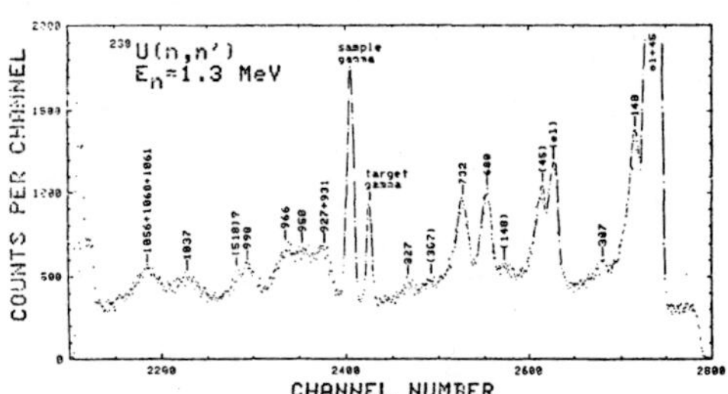

Fig. 2. Typical neutron TOF spectrum taken with the ^{238}U sample.

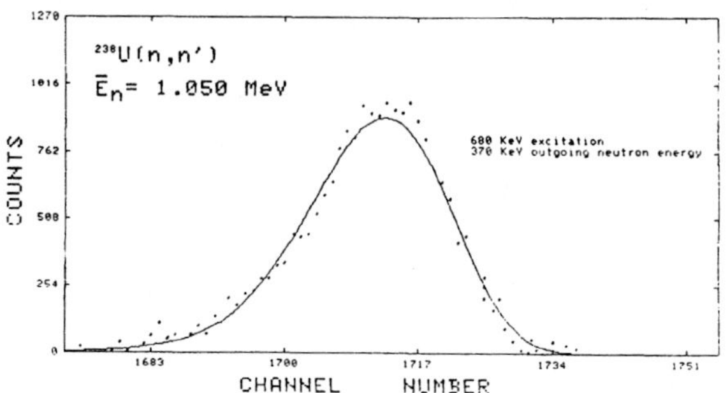

Fig. 3. Computer-generated time response function fitted to the isolated 680-keV state in ^{238}U.

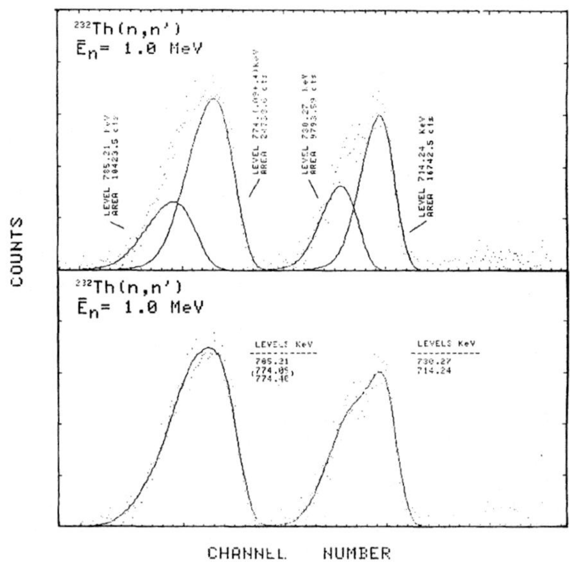

Fig. 4. Response functions fitted to two ^{232}Th doublets (bottom) and decomposition into individual levels (top).

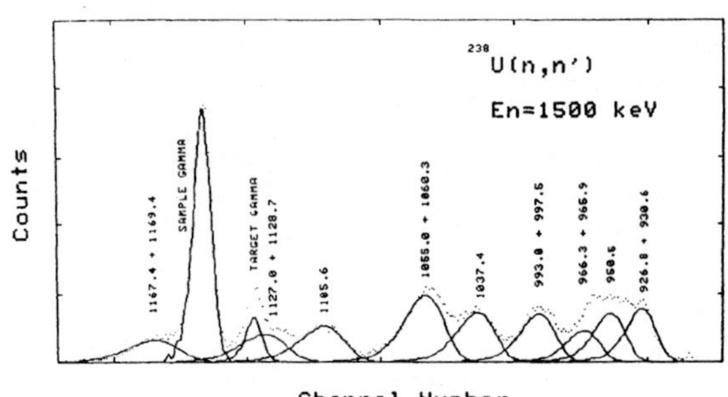

Fig. 5. Decomposition of a more complex ^{238}U(n,n´) TOF spectrum using weighted-least-squares fit program TINA.

The areas obtained from the stripping analysis were corrected for neutron attenuation and multiple scattering in the sample. The code IMBUI[12] was used to calculate single-scattering probabilities and time profiles; a companion program GAVEA, as yet unpublished, was used to calculate double scattering probabilities and time profiles. The double-to-single scattering probabilities were approximately 20% for both the ^{232}Th and ^{238}U measurements. These ratios were used to estimate the small correction due to higher-order scattering events.

The direct results of our (n,n´)TOF measurements are excitation functions of the 125°-differential scattering cross section of individual levels. We have approximated the angle-integrated level cross section by 4π times the differential cross section, on the assumption that the inelastic angular distribution is primarily a linear combination of $P_o(\cos\theta)$ and $P_2(\cos\theta)$ terms. This assumption is in excellent accord with angular distributions calculated for ^{232}Th and ^{238}U with the program CINDY[13], which show that corrections due to higher-order Legendre-polynomial terms are less than 1-2% for all low-spin ($J\leq6$) states above the ground-state rotational band in the energy region of interest for this study. Although a large direct-reaction component is not expected at these energies for the levels of higher rotational-vibrational bands in ^{232}Th and ^{238}U, there is little experimental evidence supporting the validity of our assumption of predominance of the compound-nucleus mechanism; thus, in the near future we plan to perform (n,n´) angular distribution measurements to test its validity.

Results and Discussions

Level schemes for ^{232}Th and ^{238}U are shown in Figures 6 a) and b), respectively. The levels are decomposed according to proposed[6-8] classifications of their collective band structure. The present study deduced level cross sections for a number of levels comprising the higher-excited rotational-vibrational bands depicted in the figure. For ^{232}Th we have obtained results for the 714-,730-,774-,785-,829-,873-,883-plus-889-, 960-,1054-,1073-plus-1078(0^+)-plus-1078(1^-)-,1094-, 1105-,1122-,1143-plus-1148- and the 1183-keV levels. Our results for ^{238}U include cross sections for the 680-,732-,827-,927-plus-931-,950-,966(2^+)-plus-966(7^-)-, 993-plus-998-,1037-,1055-plus-1060-,1106-,1127-plus-1129- and 1168-plus-1169-keV states. A complete tabulation of these measurements will be available through the U.S. National Nuclear Data Center. Some representative results for each nucleus are shown in Fig. 7. The solid curves represent theoretical calculations performed by E. Sheldon, entailing the unified statistical S-matrix formalism, and reported as a separate contribution[14] to this Conference.

We first consider the ^{232}Th results. The (n,n´) TOF cross sections for the 714-keV (1^-) level are somewhat lower than those measured in the (n,n´γ) study, suggesting this level may be fed from higher levels via low-energy transitions such as from the 774-keV(3^-) second member of the K=0^- octupole band. The cross sections for the combined 774-keV (2^+ and 3^-) levels are correspondingly higher in the (n,n´) experiment. However, the discrepancy in this doublet may be attributed in part to an E0 transition from the 774-keV(2^+) level to the 49-keV(2^+) state. The E0 branching ratio could be as small as 15% if low-energy feeding to the 714-keV level is invoked, and as high as 40% if the complete discrepancy is attributed to the electric monopole transition. The 730-keV (0^+) level can decay by an E0 transition to the ground state. The larger (n,n´) cross sections observed for this level indicate its E0 branching ratio may be as high as 50%. Similarly the discrepancy between (n,n´) and (n,n´γ) cross sections for the 873-keV (4^+) level may be attributed to an E0 transition to the 162-keV (4^+) member of the ground-state band. Its branching ratio could be as large as 60%. The fallacy of attributing all such discrepancies to E0 transitions is illustrated by the comparisons for the 1054-keV (2^-) lowest member of the K=2^- octupole band. Here the much larger cross sections measured in the present (n,n´) experiment have been attributed to unobserved decay branches in the (n,n´γ) study, since this level is assumed unable to decay by an electric monopole transition. The results for the 1074-keV (2^+) and the 1078-keV (1^- and 0^+) triplet are illustrative of the excellent agreement often observed between the two experiments.

Our ^{232}Th(n,n´) individual level cross sections have been suitably combined for comparison with the levels and pseudolevels used in the ENDF/B-V evaluation[9]. For the most part the present data are in good agreement with the evaluation for energies greater than 0.4-0.5 MeV above threshold. However, our results show large deviations from the evaluated cross sections at lower neutron energies. The discrepancies are even more marked in the (n,n´γ) results, which extend even closer to the threshold energies due to the greater sensitivity of this method. Comparison of our data with the theoretical calculations using the unified statistical S-matrix formalism indicates a very similar behavior. Agreement is generally rather good well above threshold, but the theory overestimates cross sections near threshold.

Results from the ^{238}U(n,n´) measurements are shown in the right-hand column of Fig. 7. The (n,n´) TOF cross sections for the 680-keV (1^-) and 732-keV (3^-) levels are in rather good agreement with those obtained in our earlier (n,n´γ) study. In both cases, however, there is evidence that the (n,n´γ) results increase relative to the (n,n´) data with increasing neutron energy, suggesting that each level is fed from higher excited states by transitions that were not included when level cross sections were inferred from our (n,n´γ) production cross section. It was surprising to observe that the (n,n´γ) cross sections for the 927-keV (0^+)-plus-931-keV (1^-) doublet were somewhat larger than the (n,n´) values. We had expected just the opposite result, since the 0^+ level was assumed to have an E0 decay branch to the ground state. A possible explanation for the observed behavior may be drawn from the comparison of the two experiments for the 950-keV (2^-) level. In this case the (n,n´) cross section is approximately a factor of 2 larger, although it is believed this level cannot decay by an E0 transition. It is likely that the discrepancy is due in part to an unobserved low-energy transition from the 950-keV level to the 931-keV ground-state member of the K=1^- octupole band. If one invokes this argument to decrease the measured (n,n´γ) cross section for the 927-plus-931-keV doublet by a corresponding amount, the result is almost in perfect agreement with the present (n,n´) data, suggesting there is no important E0 decay branch from the 927-keV state. On the other hand the 1037-keV (2^+) level could have an E0 branching ratio as large as 50%, if one attributes the discrepancy between our two experiments entirely to the electric monopole transition. The cross sections obtained for the 1055-plus-1060-keV (4^+ and 2^+) doublet offer another example of the very good agreement often observed between the two experiments.

Our (n,n´) TOF results are in excellent agreement with the ENDF/B-V evaluation for the 680- and 732-keV levels. The largest observed discrepancy with the evaluation occurred for the 827-keV state, where our cross sections are on the average more than a factor of 2 smaller. We observed somewhat better agreement with the 965-keV pseudolevel on combining our cross sections for the 927-,931-,950-,966-, and 998-keV states; in this case our measurements were approximately 25% higher than the evaluation. After combining our cross sections for the 1037-,1055- and 1060-keV levels, we observed excellent agreement with the values given for the 1048-keV pseudolevel. Our combined cross sections for the 1106-,1129-,1168- and 1169-keV states yielded results that were approximately 30% higher than those for the 1170-keV pseudolovel. In contrast to the case of ^{232}Th, we observed no systematic deviation in the threshold region of our ^{238}U(n,n´) cross sections relative to those of

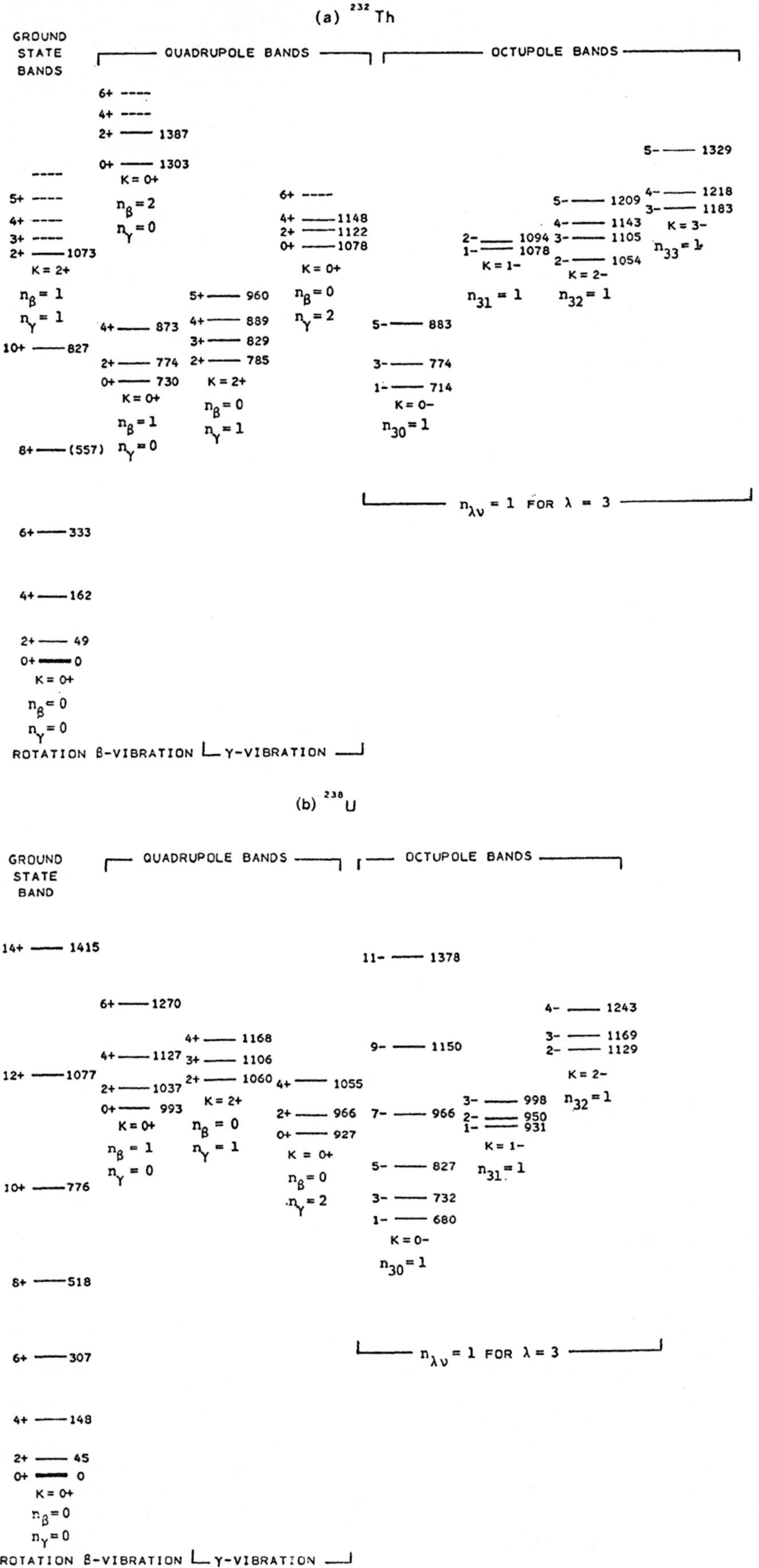

Fig. 6 Level schemes for ^{232}Th and ^{238}U with levels arranged according to their assumed[6-8] rotational-vibrational band classification

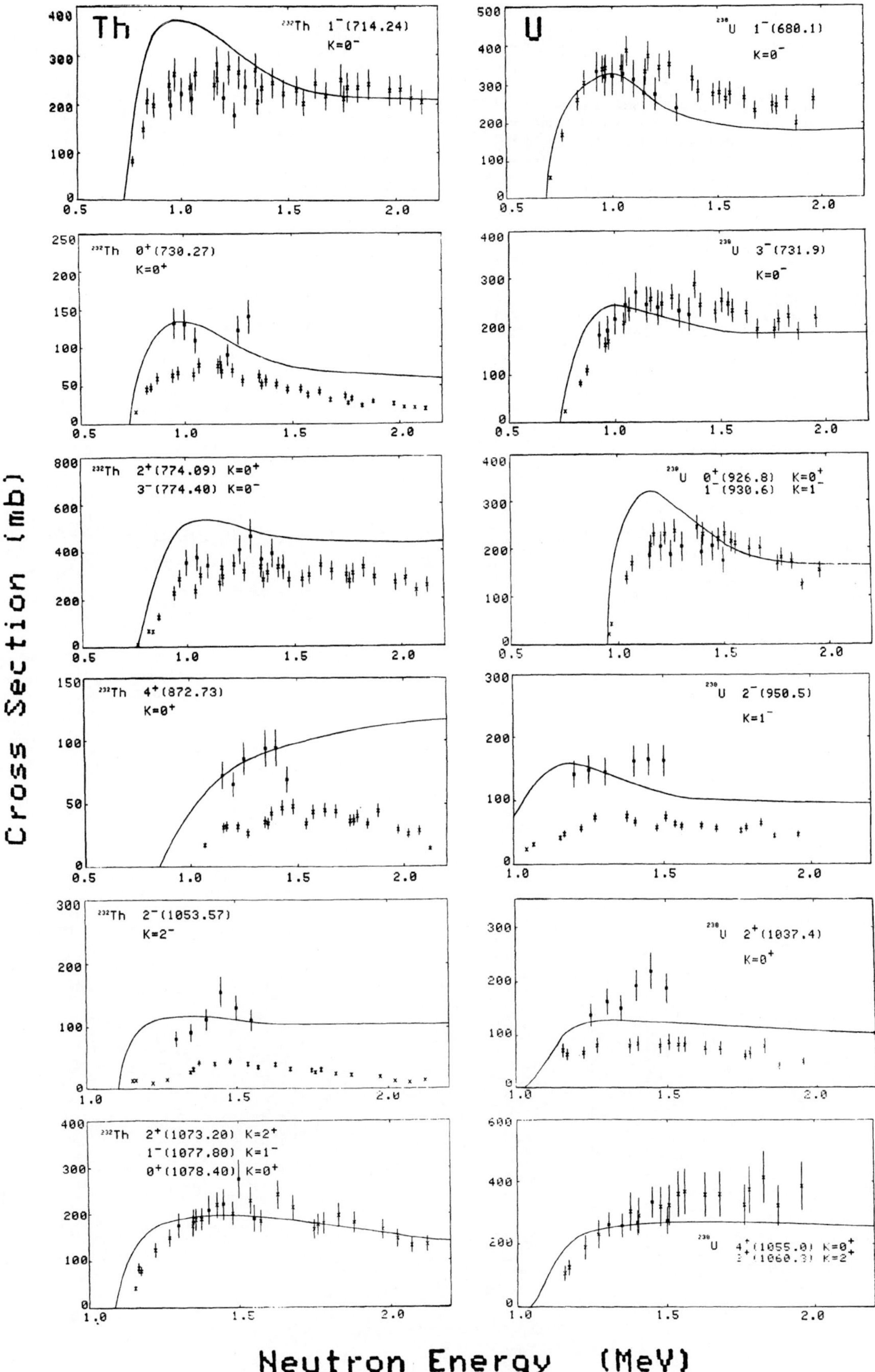

Fig. 7. Representative samples of present (n,n′) TOF level cross sections shown as solid squares along with previous (n,n′γ) measurements[1,2] represented by x's. Unified statistical model calculations[14] are indicated by solid curves.

ENDF/B-V. In most instances the theoretical predictions obtained using the unified S-matrix formalism are in good agreement with our $^{238}U(n,n')$ measurements.

Acknowledgements

The authors would like to express their appreciation to Dr. Eric Sheldon for valuable discussions and for his assistance with theoretical computations useful for both our (n,n') and (n,n'γ) experiments. We also express our gratitude to the U.S. Department of Energy for its continued support of this project.

References

1. A. Mittler, G.P. Couchell, W.A. Schier, S. Ashar J.H. Chang and A.T.Y. Wang, in Nuclear Cross Sections for Technology, ed. by J.L. Fowler, C.H. Johnson and C.D. Bowman, NBS Special Publication 594, U.S. Government Printing Office (1980) p. 680.

2. J.J. Egan, J.D. Menachery, G.H.R. Kegel and D.J. Pullen, in Nuclear Cross Sections for Technology, ed. by J.L. Fowler, C.H. Johnson and C.D. Bowman, NBS Special Publication 594, U.S. Government Printing Office (1980) p. 685.

3. W.R. McMurray, I.J. van Heerden, E. Barnard and D.T.L. Jones, Southern Universities Nuclear Institute Annual Research Report SUNI-41, 4 (1975).

   W.R. McMurray, I.J. van Heerden, E. Barnard and D.T.L. Jones, Southern Universities Nuclear Institute Annual Research Report SUNI-45, 5 (1976).

4. D.K. Olsen, G.L. Morgan and J.W. McConnell, "Measurement of $^{238}U(n,n'\gamma)$ and $^{7}Li(n,n'\gamma)$ Gamma-Ray Production Cross Sections", ORNL/TM-6832 (1979).

5. D.K. Olsen, in Proceedings of the OECD(NEANDC) Specialists' Meeting on Fast Neutron Scattering on Actinide Nuclei, Paris, (1981) (to be published).

6. M.R. Schmorak, Nucl. Data Sheets 20, 165 (1977).

7. Y.A. Ellis, Nucl. Data Sheets 21, 549 (1977).

8. D.W.S. Chan, J.J. Egan, A. Mittler and E. Sheldon, Phys. Rev. C, 26 (3), (1982).

9. J.W. Meadows, W.P. Poenitz, A.B. Smith, D.L. Smith, J.F. Whalen, R.J. Howerton, B.R. Leonard, G. DeSaussure, R.L. Macklin, G. Gwin, M.R. Bhat, Evaluated Nuclear Data File-B, Version V (ENDF/B-V), MAT 1390, Nat. Neutron Cross Section Center, (1979).

10. E. Pennington, A. Smith, W. Poenitz, M.R. Bhat, Evaluated Nuclear Data File-B, Version V (ENDF/B-V) MAT 1395, NNCSC, BNL (Upton, New York, (1977 & 1980)).

11. G.H.R. Kegel, C. Ciarcia, G.P. Couchell and J. Shao, this conference.

12. G.H.R. Kegel, Computer Phys. Commun. 24, 205 (1981).

13. E. Sheldon, private communication (1978).

14. E. Sheldon, this conference.

MESURES ABSOLUES DE ^{240}Pu(n,f), ^{242}Pu(n,f) ET ^{237}Np(n,f)
A L'ENERGIE INCIDENTE DE 2,5 MeV

M. Cancé, G. Grenier

Service de Physique Neutronique et Nucléaire
Centre d'Etudes de Bruyères-le-Châtel
B. P. N° 561
92542 MONTROUGE-CEDEX, France

Measurements of the absolute neutron fission cross section of ^{240}Pu, ^{242}Pu and ^{237}Np have been made at 2.5 MeV using a hybrid detector. The fission events were detected in an ionization chamber (2π) and the neutron flux was determined by a proton recoil telescope and a directional long counter. Our values are compared to previous data.

[^{240}Pu(n,f), ^{242}Pu(n,f), ^{237}Np(n,f), neutron fission cross sections, E_n = 2.5 MeV].

Introduction

Ces dernières années de nombreuses mesures de sections efficaces de fission ont été faites, mais très peu de mesures absolues. Ces dernières sont cependant indispensables aux évaluateurs. Des mesures absolues très précises peuvent être faites à 14 MeV en utilisant la méthode de la particule associée et la technique du temps de vol [1].

Comme il est important d'avoir des valeurs absolues à plus faible énergie,des mesures absolues ont été entreprises à 2,5 MeV. Cette énergie permet, d'une part de normaliser les mesures relatives dans de bonnes conditions, d'autre part de bénéficier avec la réaction T(p,n)^{3}He d'une source de neutrons monoénergétiques de bonne qualité.

Les mesures absolues à 2,5 MeV sont faites en utilisant un détecteur hybride [2] composé d'une chambre d'ionisation pour la détection de fissions et d'un télescope à protons de recul pour mesurer le flux de neutrons. Cette technique permet d'utiliser une deuxième référence pour la détermination du flux de neutrons : un long compteur directionnel (BF_3).

Les mesures absolues à 2,5 MeV présentées ici, sont celles relatives aux sections efficaces de fission de ^{240}Pu, ^{242}Pu et ^{237}Np.

Conditions expérimentales

Implantation

Le détecteur hybride et le long compteur directionnel sont placés à 0° respectivement à 10 cm et 250 cm. La transmission des neutrons à travers le détecteur hybride mesurée, est de l'ordre de 97%.

Source de neutrons

Les neutrons de 2,5 MeV sont produits par la réaction T(p,n)^{3}He avec un accélérateur Van de Graaff 4 MV. La cible est constituée d'un dépôt de titane tritié sur un support d'or. Le faisceau de protons utilisé est pulsé à une fréquence de 2,5 MHz et une largeur de bouffée d'environ 1 ns.

Détection

Le détecteur hybride schématisé sur la figure 1, est composé d'une chambre d'ionisation 2π, pour la détection des fissions, dans sa partie avant et d'un télescope à protons de recul, pour la mesure du flux de neutrons, dans sa partie arrière. Le dépôt fissile et le radiateur hydrogéné sont placés de part et d'autre d'un disque mince assurant la séparation entre la chambre d'ionisation et le télescope. Le gaz de remplissage de la chambre d'ionisation est un mélange d'argon (90%) et de méthane (10%). Le télescope à protons de recul est constitué d'un radiateur en polyéthylène de 10 μ d'épaisseur et de deux détecteurs ΔE et E, complètement désertés, respectivement de 50 μ et 100 μ d'épaisseur. L'analyse du nombre d'atomes d'hydrogène présents dans le radiateur a été faite par les laboratoires Huffman (Colorado). Le long compteur directionnel (BF_3) est utilisé comme deuxième référence pour la mesure du flux de neutrons.

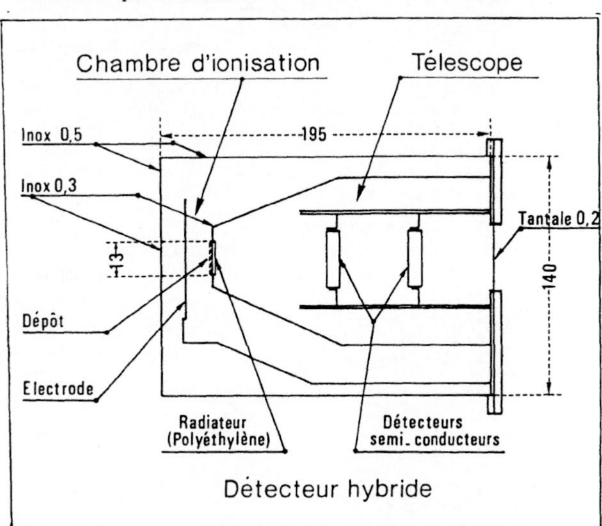

Fig. 1. Vue schématique du détecteur hybride.

K. H. Böckhoff (ed.), Nuclear Data for Science and Technology, 51–54.

Tableau 1. Compositions isotopiques et épaisseurs des dépôts.

Dépôt	Composition isotopique (at %)							Epaisseur µg/cm²
	^{237}Np	^{238}Pu	^{239}Pu	^{240}Pu	^{241}Pu	^{242}Pu	^{244}Pu	
^{237}Np	100							74,49
^{240}Pu		0,0109	0,6724	98,478	0,4701	0,3677	0,0006	44,33
^{242}Pu		0,0035	0,0286	0,0339	0,0648	99,8192		50,00

Tableau 2. Corrections.

EFFET	Corrections (%)		
	^{237}Np	^{240}Pu	^{242}Pu
Détecteur de fissions			
Extrapolation vers zéro du spectre d'amplitude de fission	1,084	1,27	0,94
Perte de fissions	0,6	0,35	0,35
Fissions dues aux autres isotopes		1,78	0,2
Fissions dues aux neutrons diffusés	4,7	4,7	4,7
Fissions obtenues sans tritium dans la cible	0,1	0,1	0,1
Télescope			
Evènements sans radiateur	0,9	8,2	0,93
Evènements sans tritium dans la cible	0,6	0,6	0,6
Atténuation des neutrons dans le support de Ta	1,5	1,5	1,5
Long compteur directionnel (BF_3)			
Fond sans tritium dans la cible	0,4	0,4	0,4
Fond avec barre d'ombre	3,05	3,05	3,05
Atténuation des neutrons :			
. dans l'air	2,7	2,7	2,7
. dans le détecteur hybride	2,7	2,7	2,7

Tableau 3. Erreurs.

EFFET	Erreur (%)		
	^{237}Np	^{240}Pu	^{242}Pu
Détecteur de fissions			
Extrapolation vers zéro du spectre d'amplitude de fission	0,1	0,1	0,1
Perte de fissions	0,1	0,03	0,03
Fissions dues aux autres isotopes		0,18	0,02
Fissions dues aux neutrons diffusés	0,5	0,5	0,5
Nombre d'atomes	0,6	0,3	0,5
Statistique	0,88	0,5	0,9
Facteurs géométriques	0,5	0,5	0,5
Télescope			
Evènements sans radiateur	0,9	0,82	0,86
Evènements sans tritium dans la cible	0,1	0,1	0,1
Atténuation des neutrons dans le support de Ta	0,1	0,1	0,1
$\sigma(n,p)$	0,75	0,75	0,75
Nombre d'atomes d'hydrogène	0,2	0,2	0,2
Efficacité	0,3	0,3	0,3
Statistique	0,4	0,2	0,3
Long compteur directionnel (BF_3)			
Fond avec barre d'ombre	0,3	0,3	0,3
Atténuation des neutrons :			
. dans l'air	0,3	0,3	0,3
. dans le détecteur hybride	0,3	0,3	0,3
Efficacité	2,5	2,5	2,5
Erreur finale			
BF_3	2,88	2,72	2,84
Télescope	1,86	1,52	1,76

Caractéristiques des dépôts fissiles

Les éléments fissiles ont été déposés sur un support de tantale de 0,3 mm d'épaisseur. Les dépôts de $^{240}PuF_3$ et $^{242}PuF_3$ ont été réalisés à Geel par évaporation sous vide. Le dépôt de $^{237}NpO_2$ a été fait à Bruyères-le-Châtel par "sputtering".

Les compositions isotopiques et les épaisseurs des dépôts sont données dans le Tableau I. Le nombre d'atomes a été obtenu par comptage alpha en géométrie définie en utilisant les périodes recommandées par Lorenz [3]. Celles-ci sont : 6550 ± 20 ans pour ^{240}Pu (3,76 ± 0,02) 10^6 ans pour ^{242}Pu et (2,14 ± 0,01) 10^6 ans pour ^{237}Np.

Acquisition

La méthode du temps de vol est utilisée de façon à ne prendre en compte que les évènements de fission corrélés avec la bouffée de neutrons. L'identification des protons de recul dans le télescope est faite à partir de l'acquisition biparamétrique des énergies déposées dans les deux détecteurs par le proton. Le bruit de fond est mesuré, d'une part en ôtant le radiateur, d'autre part avec une cible de titane non chargée en tritium.

Résultats et discussion

La nature et la grandeur des corrections effectuées sont résumées dans le Tableau II.

Le nombre de fissions, dues aux neutrons diffusés sur les matériaux de structure voisins du dépôt, donc non éliminées par la méthode du temps de vol, est calculé à l'aide d'un code Monte Carlo. La perte des fragments de fission dans le dépôt est aussi obtenue avec un programme Monte Carlo.

Toutes les erreurs contribuant à l'incertitude finale sur la section efficace sont données dans le Tableau III. Les erreurs sont ajoutées quadratiquement.

Les sections efficaces obtenues avec les deux références : télescope et long compteur directionnel sont données dans le Tableau IV.

Nos résultats sont comparés aux données antérieures sur les figures 2,3 et 4.

Comme pour chaque élément les valeurs obtenues proviennent de deux méthodes différentes, la valeur moyenne est la valeur moyenne pondérée. Celle-ci est définie par l'expression suivante :

$$\bar{a} = \frac{1}{\Sigma(\Delta a_i)^{-2}} \; \Sigma\, a_i(\Delta a_i)^{-2}$$

$a_i \pm \Delta a_i$ étant les valeurs individuelles.

Tableau 4 . Sections efficaces à E_n = 2,47 ± 0,03 MeV.

Isotope	σ(n,f) barn	
	BF_3	Télescope
^{237}Np	1,630 ± 0,047	1,646 ± 0,031
^{240}Pu	1,596 ± 0,043	1,667 ± 0,025
^{242}Pu	1,388 ± 0,040	1,391 ± 0,024

^{237}Np

Nos deux valeurs sont en très bon accord, puisque l'écart entre celle-ci est inférieur à 1%. La valeur moyenne pondérée de 1,64 b remplace et annule le résultat préliminaire de 1,66 b mentionné par Derrien dans son évaluation [4]. Cette valeur moyenne est très proche de la section efficace évaluée par Derrien de 1,653 b, la différence étant de l'ordre de 0,7%.

^{240}Pu

L'accord entre nos deux résultats est moins bon que pour le ^{237}Np. L'écart est en effet de l'ordre de 4%.

La valeur moyenne pondérée de 1,650 est cependant en très bon accord avec la mesure absolue de Kari et Cierjacks [8]. De plus, si cette valeur est comparée à une autre mesure récente de Behrens et al. [9], relative à ^{235}U, l'écart est seulement d'environ 2% .

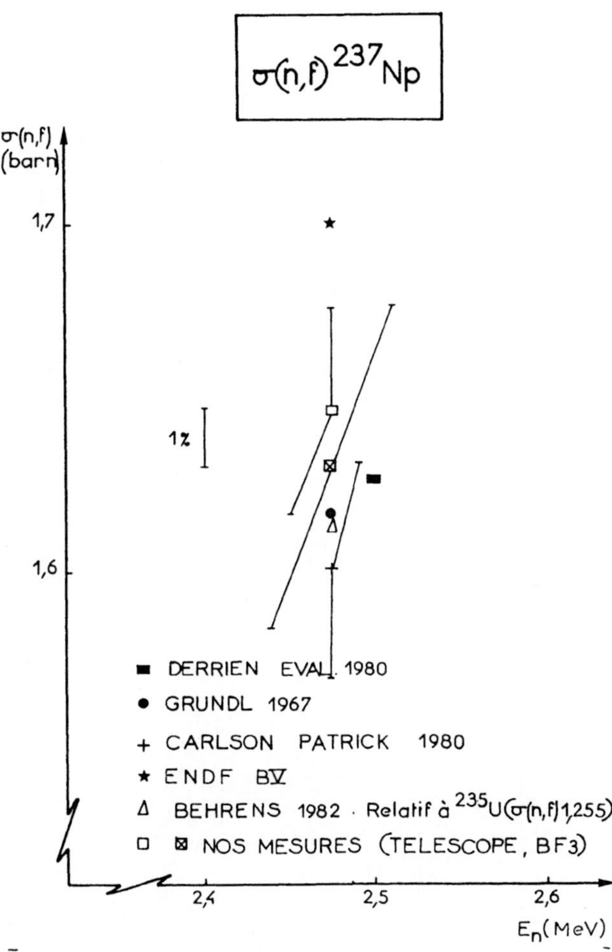

Fig. 2. Section efficace de fission de ^{237}Np entre 2,4 et 2,6 MeV.

^{242}Pu

Il y a un excellent accord entre nos deux résultats.

La valeur moyenne pondérée de 1,390 b est aussi en bon accord avec les mesures relatives récentes de Behrens et al. [9] et de Meadows [11]. Notre valeur confirme donc l'évaluation de Antsipov [13] qui repose pour une bonne part sur les résultats de Behrens et Meadows.

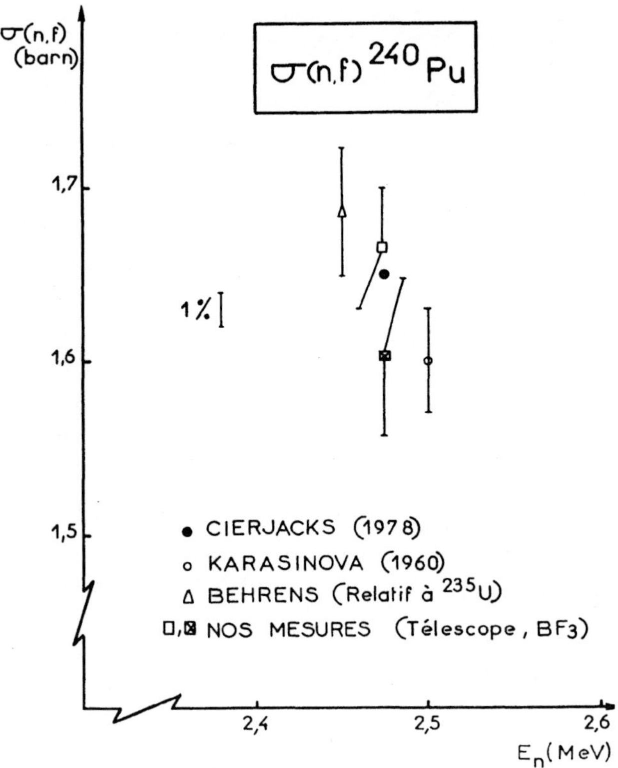

Fig. 3. Section efficace de fission de ^{240}Pu entre 2,4 et 2,6 MeV.

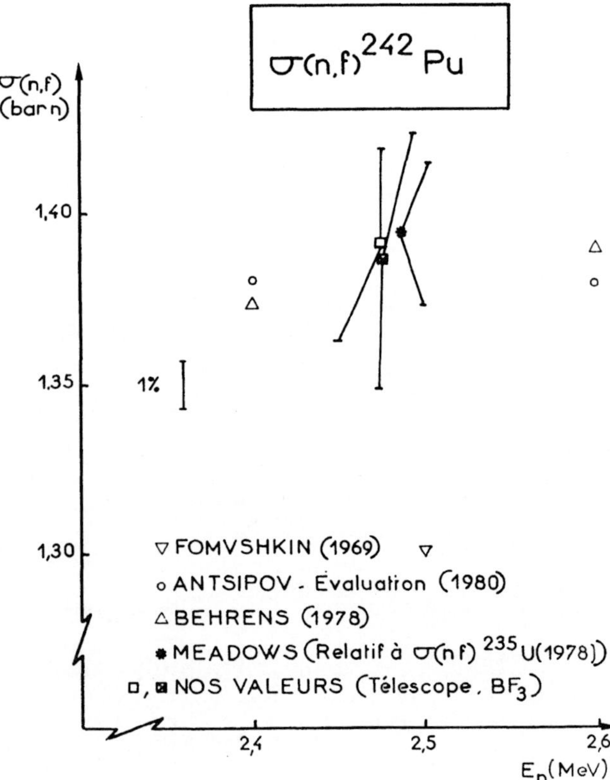

Fig. 4. Section efficace de fission de ^{242}Pu entre 2,4 et 2,6 MeV.

Références

1. M. Cancé, G. Grenier, Nucl. Sci. Eng. 68, 197 (1978).
2. M. Cancé, G. Grenier, Note CEA-N-2194, NEANDC (E) 214 "L" (1981).
3. A. Lorenz, INDC (NDS) 108/N (1979).
4. H. Derrien et al., INDC (FR) 42/L (1980).
5. J.W. Behrens et al., Nucl. Sci. Eng. 80, 393 (1981).
6. A.D. Carlson, B.H. Patrick, Proceedings of Knoxville Conference (1980).
7. J.A. Grundl, Nucl. Sci. Eng. 30, 39 (1967).
8. K. Kari, S. Cierjacks, Proceedings of Harwell Conference, 905 (1978).
9. J.W. Behrens et al., Nucl. Sci. Eng. 66, 433 (1978).
10. M.I. Kazarinova et al., Sov. At. Energy V, 125 (1960).
11. J.W. Meadows, Nucl. Sci. Eng. 68, 360 (1978).
12. Eh. F. Fomushkin et al., Yad. Fiz. 10, 917 (1969).
13. G.V. Antsipov et al., INDC (CCP)-150/LJH 1 (1980).

ABSOLUTE MEASUREMENTS OF ^{235}U AND ^{239}Pu FISSION CROSS SECTION INDUCED BY 14.7 MeV NEUTRONS

Li Jingwen, Li anli, Rong Chaofan, Ye Zhongyuan
Wu Jingxia and Hao Xiuhong

Institute of Atomic Energy, Academia Sinica
P.O.Box 275-15 Beijing, People's Republic of China

The cross section of uranium-235 and plutonium-239 fission induced by 14 7 MeV neutrons were measured using associated particle method with time correlation technique. The results obtained are 2.098±0.040 b and 2.532±0.050 b for U-235 and Pu-239 respectively. Comparison with data of other authors is also given.

[^{235}U(n,f), ^{239}Pu(n,f), fission cross sections, E_n=14.7 MeV, nuclear data.]

Introduction

The values of fission cross sections for neutron energies in the vicinity of 14 MeV are of significant importance in nuclear energy application. They are usually used as normalization values in the relative measurement of cross sections, so the accuracy needed is higher than that of other energy region. In fact, though, the results from different laboratories do not agree with each other. For instance, White[1] gave the value 2.17±0.04 b for U-235 fission at 14.1 MeV, but at 14.6 MeV, the data of 2.075±0.040 b and 2.063 ± 0.039 b were provided by Czirr[2] and Cance[3] respectively. The latter two values are about 5 % lower than the former one. The situation for σ_f of Pu-239 is similar. In order to eliminate the discrepancy mentioned above more careful experimental determinations are needed. We have measured the σ_f of U-235 and Pu-239 using associated particle method.

Experimental Arrangement

Fig.1 shows the device and the block diagram of electronic circuits used in our experiment.

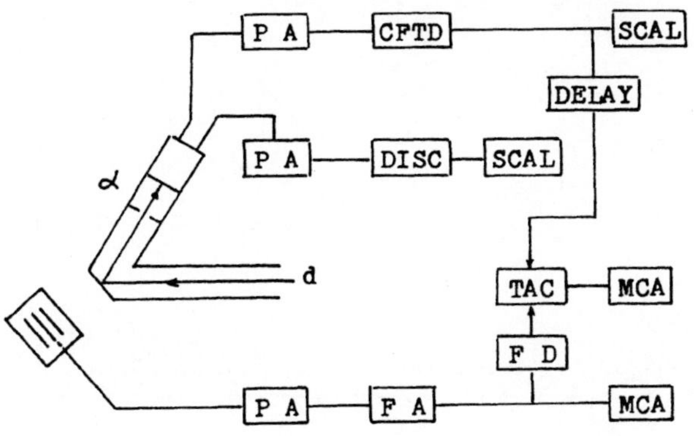

Fig. 1. Experimental arrangement

Neutron Source. Neutrons with energy 14.7 MeV were produced by T(d,n)^{4}He reaction on the 600 kV Kockcroft-Walton accelerator in IAE. The incident energy of deuteron beam was 220 keV. The recoil angle of ^{4}He particle was selected as 135°. In the direction of 40° with respect to the deuteron beam the corelated neutrons of 14.7 MeV were obtained.

Fission Chamber. The fission ionization chamber was a cylindrical one 100 mm in diameter 50 mm in length, which was composed of two identical halves. The incident direction of correlated neutron beam was along the axis of the chamber. The chamber was filled with methan gas of one atmosphere. The fast current collection was adopted.

Samples. The sample of uranium and plutonium were electrodeposited on platinum backings. The diameter of the deposited fissile material layer was 26 mm. Three methods were used to determine the quantity of uranium contained in the sample: by direct weighing, by alpha counting in a low solid angle equipment and by titration. For plutonium the alpha counting technique and the constant current Coulomb method were used. The details of preparation and quantitative analysis determination of samples had been described elsewhere[4,5]. The nonuniformity of the deposited layer was tested in following way: A small diaphragm was placed in front of the sample and the alpha particles were counted using a silicon detector of surface barrier type.

Measurement

The alpha particles produced in reaction T(d,n)^{4}He were detected by a thin film scin-

K. H. Böckhoff (ed.), Nuclear Data for Science and Technology, 55–57.

tillation counter consisted of a 100 μm thick plastic scintillation film and a fast photomultiplier of 56/DVP/03A type. The resolution time of the output alpha signal was smaller than 2 ns. Alpha detector was placed in the direction 135° with respect to the incident deuteron beam. The fission chamber was located 5 cm away from the T-Ti target at the correlated direction. This ensured that all the associated neutrons were strike on the U or Pu sample. The counting rate of alpha particles was restricted within 10^4 to 10^5/s through whole process of measurement. In order to determine the correct position of the neutron beam correlated with the 135° alpha particles we have used a small plastic scintillator as a neutron detector scanning along the horizontal and verticle directions to measure the coincidence rate with alpha particles.

The fission fragment pulses after discremination was fed to the time-to-amplitude converter (TAC) as its start signal. The duration of the TAC gate was 200 ns. The alpha particle pulses were fed into a constant fraction timing discriminator (CFTD) which gave a narrow output pulse used as stop signal for TAC after a delay of 100 ns. In the same time the CFTD gave a broad pulse fed to a scaler for counting the associated alpha particles. The output signals of the TAC were recorded in a multichannel analyzer directly. In Fig.2 a typical time distribution of fission pulses was shown. It can be seen that by use of TAC the correlated events were seperated clearly from the background. As for the background, one may notice that the average counting number per channel in the left side of the coincident peak is higher slightly than that in the right side region from the peak. This effect was caused by two facts, one was that the collecting time of electrons produced by fission fragments in ionization gas reveals a distribution which has rather long tail, the other was that in some cases the TAC, after being started by the fragment pulse, might be erroneously stopped due to a uncorrelated alpha pulse which preceded the right correlated one. So we take the average number of counts per channel in the righthand region with respect to the peak as the background counts per channel.

The main advantage of the associated particle method is that the neutron fluence could be determined directly and precisely. The inter-

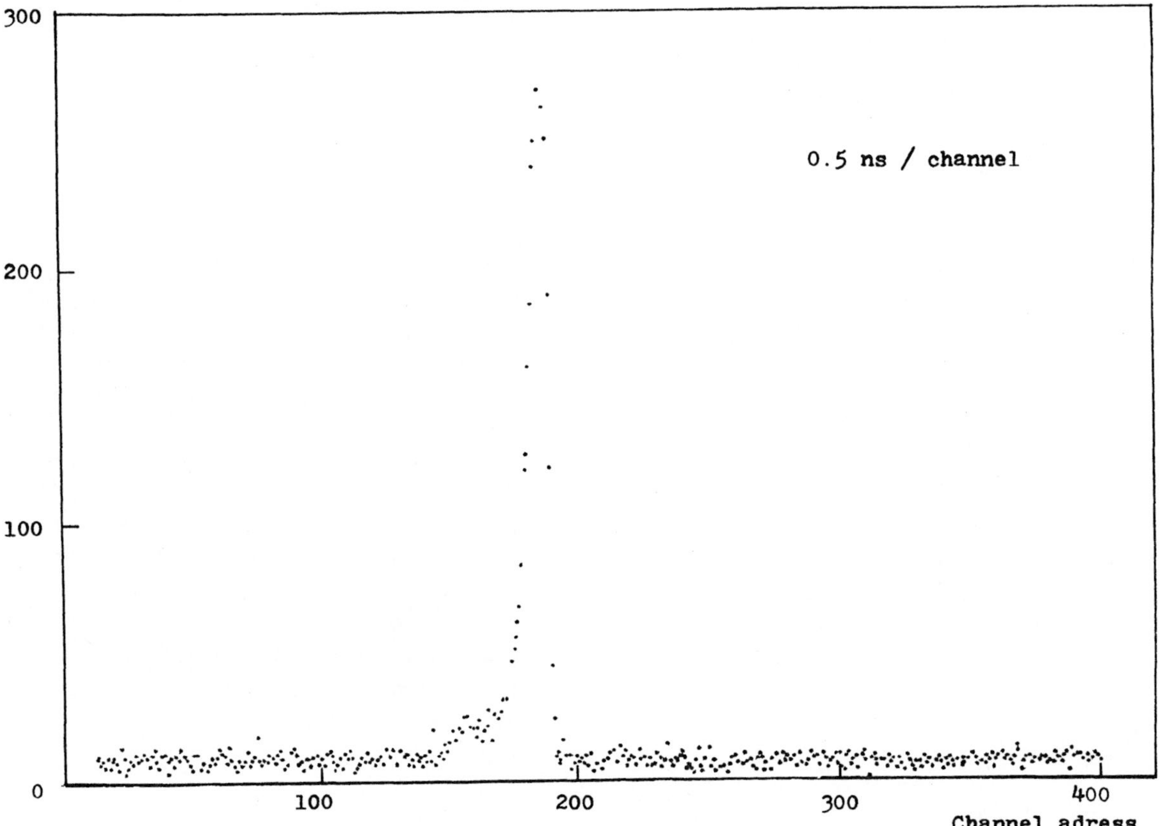

Fig. 2. Typical time distribution of fission pulses.

ference of all background fission events induced by scattered neutrons, by slowing down neutrons and by the neutrons from $D(d,n)^3He$ reaction as well as the false fission events due to alpha pile up were greatly supressed. Backgrounds of different origins could be subtracted directly in the time distribution. In calculating σ_f the detail knowledge of the solid angle and the other geometric factors were no longer needed.
The protons from the parasitic reaction D(d,p)T may produce pulses higher than the alpha's in the plastic scintillator. We used a seperate discriminator with the threshold higher than the endpoint of alpha pulse distribution to record these protons simultaneously and then subtracted from the total alpha counts. In order to reduce the yield of these protons the target used in the experiment was changed with new one every time if the proton yield exceed 1 % of the alpha's.

Corrections and Uncertainties

1. There are two factors which would reduce the efficiency of fission detector. The first effect is that the pulse height of some fragments outgoing from the sample layer would be lower than the setting threshold of the discriminator. The fraction of these events may be obtained by extrapolating the amplitude distribution of fragment pulses to zero threshold. The second effect is the self absorption of fragments in the sample layer. In the mean time it is necessary to take into account the absorption variation due to the momentum brought in by the incident neutrons and due to the anisotropic angular distribution of fission fragments.
2. The attenuation of neutrons
The target backing, water layer for cooling the target, the cadmium cover in front of the fission chamber, the front window of the chamber and the backings of the samples--all these caused some attenuation of neutron beam striking on the fissile layer. This correction was calculated according to transmission consideration . The amount resulted was 2.68 %.
3. Fission events due to the other isotopes contained in the sample such as ^{234}U, ^{238}U, ^{240}Pu, and ^{241}Pu were calculated directly according to the isotope composition shown in Table I. The contribution was 5.92 % and 3.00 % for U-235 and Pu-239 respectively.

Table I. Isotope composition of samples

Uranium	234	235	238
	1.0 0.1	90.5 0.9	8.5 0.8
Plutonium	239	240	241
	96.9 0.8	3.0 0.5	0.1 0.1

Results and Discussion

From our experiment it was obtained that the σ_f of U-235 induced by 14.7 MeV neutrons was equal to 2.098±0.040 b. Our result was in agreement with those of Czirr[2], Cance[3] and Alhazov[6] within standard error.As for the σ_f of Pu-239, the present result was equal to 2.532±0.050 b at 14.7 MeV.
The ratio of the σ_f(Pu-239) to σ_f(U-235) in present work was equal to 1.207 which agrees well with the previous result 1.196 obtained in a relative cross section measurement of our laboratory[5].
As for the compiled data, the ENDF/B-IV gave the values 2.218 b and 2.552 b for U-235 and Pu-239 respectively, but the ENDF/B-V gave the U-235 σ_f value 2.101 b which is in good agreement with present result.

References

1. P.H.White, J.Nucl.Energy A/B, 19,325(1965).
2. J.B.Czirr et al , Nucl. Sci. Eng.,57,18 (1975).
3. M.Cance et al., Nucl. Sci. Eng., 68,197 (1978).
4. Yan Wuguang et al., Keiji,No.2,133 (1975).
5. Den Xinlu et al., Keiji,No.1,12 (1981).
6. I.D Alhazov et al., Atomnaja energia, 47, 416 (1979).

MEASUREMENTS OF THE 14 MeV FISSION CROSS-SECTIONS FOR ^{235}U AND ^{239}Pu

M. Mahdavi and G. F. Knoll

The University of Michigan
Ann Arbor, Michigan 48109
USA

J. C. Robertson
The University of New Mexico
Albuquerque, New Mexico 87131
USA

The fission cross-sections of ^{235}U and ^{239}Pu have been measured at a neutron energy of 14.63 MeV. A 150 kV accelerator installed on a "low mass" floor at the center of a large laboratory provided a source of 14 MeV neutrons. Fission fragments passing through a limited solid angle aperture were registered on polyester track-etch films. The angular distribution of fission fragments for ^{235}U and ^{239}Pu were measured in a separate experiment. The masses of the foil deposits were determined by microbalance weighings and confirmed by thermal fission and alpha counting. The neutron flux was measured relative to $^{56}Fe(n,p)^{56}Mn$ and $^{27}Al(n,\alpha)^{24}Na$ reaction cross-sections. ^{56}Mn and ^{24}Na activities were measured in a $4\pi\beta$ gas flow proportional counter. Absolute efficiencies of the beta detector for counting these activities were separately measured using a $4\pi\beta$-γ coincidence counting technique. The precise neutron energy was directly measured using a silicon surface barrier detector. Values of the fission cross-sections obtained for ^{235}U and ^{239}Pu were 2.070$\pm$.046 and 2.44$\pm$.092 barns respectively with a corresponding cross-section ratio of 1.179$\pm$.028.

[Fission cross-sections, ^{235}U, ^{239}Pu, 14.63 MeV neutrons, $^{235}U(n,f)$, $^{239}Pu(n,f)$]

Introduction

Fission cross-sections of ^{235}U and ^{239}Pu are of particular interest since they are widely used as standards for relative cross-section measurements. In addition, reactor physicists often normalize the fast fission cross-section data in the 14 MeV region.

Although there have been many measurements covering the 14 MeV energy region for ^{235}U fission cross-section, there is disagreement between different experiments employing associated particle technique.[1-4] In addition, the absolute measurements using proton recoil monitor appear to be ~5% larger than the measurements employing associated particle method.[5] There are relatively few direct measurements of the ^{239}Pu fission cross-section in the 14 MeV energy region, characterized by a wide scatter in the data base.[6,7]

The present work utilizes an independent and unique experimental technique to measure the ^{235}U and ^{239}Pu fission cross-sections. The results of this work along with the most recent absolute measurements in the 14 MeV energy region help to establish ^{235}U and ^{239}Pu as primary standards.

Experimental Technique

A 150 kV accelerator installed on a "low mass" floor at the center of a large laboratory provided a source of 14.6 MeV (d,t) neutrons under conditions of low scattering. Irradiations were carried out at the University of Michigan Neutron Experimental Bay. The general source-detector geometry adopted for the fission cross-section measurements is shown in Fig. 1. Fission reaction rate measurements were carried out for several source-detector spacings which facilitated the determination of the room return contribution. Fission fragments passing through a limited solid angle aperture were recorded on polyester track-etch films. After exposure, these films were etched in 6.0 N KOH at 68°C for ~4.5 hours resulting in track sizes of approximately 14 μm. The films were then gridded and the accumulated tracks were displayed on the screen of a projection microscope and counted manually. Each film was counted by different individuals until the results agreed to within .25%. Limited solid angle counting avoids large angle fragment emission and scattering problems which affect the more common 2π geometry, but requires independent knowledge of the anisotropic fragment emission.

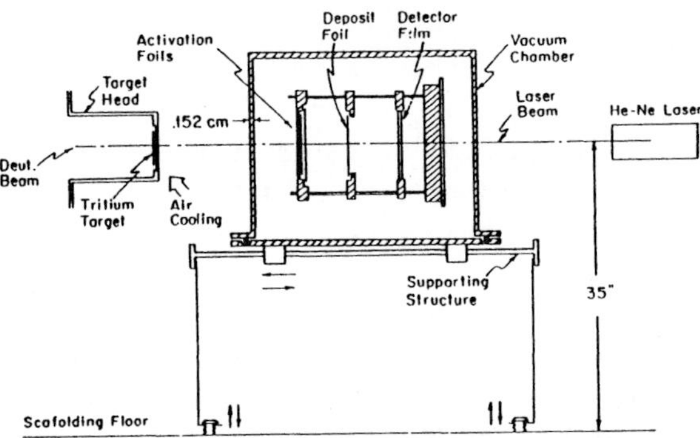

Fig. 1. Source-Detector Geometry for Fission Rate and Flux Measurements

The angular distribution of fission fragments for ^{235}U and ^{239}Pu were measured in a separate experiment. In this case, the 14 MeV neutron flux impinged on a target deposit which was defined by a circular mask of 6 mm in diameter. Fission fragments were detected by a polyester track-etch film which was shaped to cover a 110° arc on a sphere of 5.08 cm radius, centered on the deposit. The experimental assembly was then positioned inside a vacuum chamber. The etched and gridded films were later counted to give the measured fragment yield at angles ranging from 0° to 90° with respect to the incident neutron beam. A normalization run using thermal neutrons was used to account for positioning errors and variations in the solid angle with direction.

The neutron flux was measured relative to $^{56}Fe(n,p)^{56}Mn$ and $^{27}Al(n,\alpha)^{24}Na$ reactions. The activities were measured in a $4\pi\beta$ gas flow proportional counter. The use of iron foil as a primary flux standard is an important feature of this measurement, since the $^{56}Fe(n,p)$ cross-section has been determined absolutely with an accuracy of $\pm$1% as a part of an international 14 MeV flux comparison.[8]

In order to determine the absolute activity of the standard foil, the absolute efficiencies of the beta

K. H. Böckhoff (ed.), Nuclear Data for Science and Technology, 58–61.

detector for counting ^{56}Mn and ^{24}Na activities were measured in a separate experiment. The experimental technique employed a $4\pi\beta$-γ coincidence counting system to measure an uncorrected detector efficiency and a k-correction to account for the complex decay scheme of ^{56}Mn and ^{24}Na and non-ideal beta detector response. The decay spectrum corrections were derived from the results of efficiency extrapolation experiments employing the foil absorption method. These results were later confirmed by measuring the k-corrections using the self absorption method for efficiency extrapolation. The precise neutron energy was measured at the time of the final fission cross-section measurements. In this experiment, a silicon surface barrier detector was used as a neutron spectrometer. Neutron-induced reactions in silicon produce an energy spectrum consisting of a series of well defined peaks. The energies of these peaks are a means for direct measurement of neutron energy, provided the PHA is energy calibrated. To accomplish this calibration, we exploit the fact (from reaction kinematics) that for deuteron energies from 10 to at least 160 keV, the neutron energy is ~14 MeV at angles of around 96° to the incident deuteron beam.[9] Therefore, a measurement at 96° will produce a spectrum corresponding to ~14 MeV neutrons which can be reliably used to calibrate the analyzer. The mean energy at 0° to the incident beam can then be found from the shift of any prominent peak with respect to its original position in the 96° measurement. In this work, the shift of the prominent peak due to $^{28}Si(n,\alpha_0)^{25}Mg$(g.s.) was investigated.

Data Analysis and Results

Neutron Energy

The overlapped spectra resulting from measurements with the detector positioned at 0° and 96° to the incident deuteron beam is shown in Fig. 2. The energy calibration was obtained by utilizing the fact that the Q values of the two peaks α_0 and α_4, acquired at 96° to the incident beam, are separated by a known difference of ΔQ=1.9649±.0006 MeV. The shift of the α_0 peak centroid from channel c_0 to c_0' gives the mean neutron energy at 0°:

$$E_n(0°) = E_n(96°)+(\frac{\Delta Q}{c_0-c_4})(c_0'-c_4).$$

The centroid positions were determined by employing Non Linear Least Squares fitting (NLLS) to the peaks of interest.

Major corrections and uncertainties in the measured neutron energy were due to: 1) pulse height defect and resolution anomalies in the silicon detector; 2) the rapid fluctuation in the $^{28}Si(n,\alpha_0)$ cross-section as a function of energy, which alters the shape of the peak α_0' compared to α_0; 3) uncertainties in determining the centroid positions of the (n,α) peaks and the Q values of the reactions; 4) angular positioning of the SSB detector with respect to the deuteron beam and 5) short term electronic drifts. The mean neutron energy at 0° was determined to be 14.63±.02 MeV.

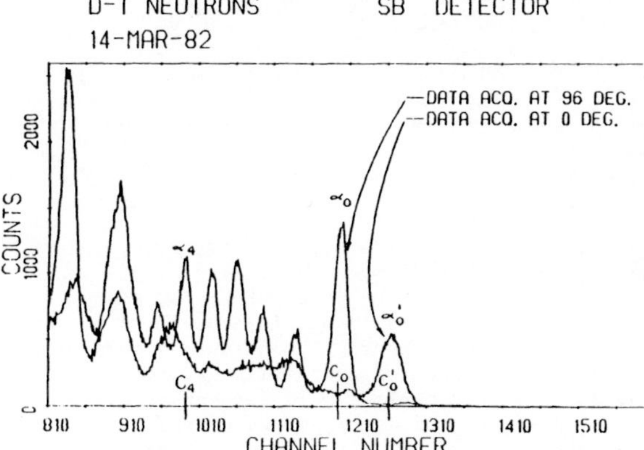

Fig. 2. Overlap of the Spectra Resulting from Measurements at 0° and 96°

Angular Distribution of Fission Fragments

The fission fragment angular distributions for ^{235}U and ^{239}Pu were measured absolutely. The fission fragments were measured at angles of 0°, 15°, 30°, 60°, 75° and 90° with respect to the incident neutron beam. At each angle, angular spreads of ±5° were counted. The data points (after thermal normalization) were least squares fitted to an expression of the form,

$$\omega(\theta) = 1 + A\cos^2\theta \ ,$$

where, θ = polar angle of fragment emission with respect to the incident neutron and A = measured anisotropy.

The errors in the individual data points were due to systematic counting error (~.5%), counting statistics and thermal normalization.

Several corrections were applied to the measured anisotropics: 1) A solid angle correction, due to the finite size of the detector (~1%); 2) the scattering correction, due to varying incident neutron angles (~2.8%); 3) a thermal fission correction, due to the contamination of anisotropy with thermal fissions (~.8%); and 4) a correction due to finite angular spreads (~<1%).

The final values of anisotropy for ^{235}U and ^{239}Pu were .426±.024 and .295±.024 respectively.

Absolute β-detector Efficiency

Absolute efficiencies of the $4\pi\beta$ proportional counter were measured for each one of 9 iron and 6 aluminum standard foils used in flux measurements. A block diagram of the $4\pi\beta$-γ coincidence counting configuration employed to measure the uncorrected efficiencies and k-corrections is shown in Fig. 3. The uncorrected efficiency is: $\varepsilon_{\beta,unc} = \frac{N_c}{N_\gamma}$, where N_c and N_γ are

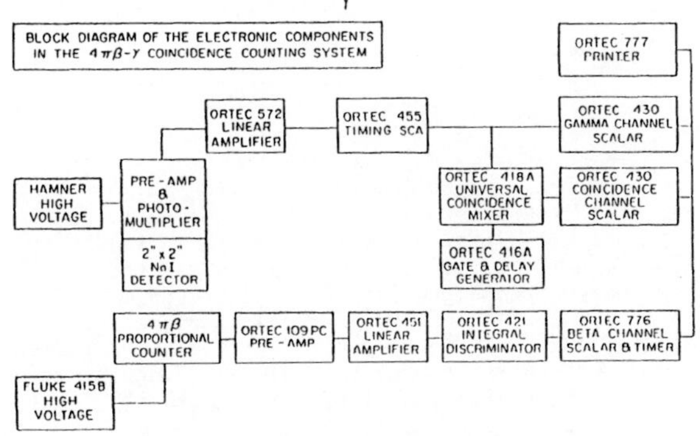

Fig. 3. Block Diagram of the $4\pi\beta$-γ Counting System

coincidence and gamma channel count rates corrected for background, dead time and chance coincidences (in the case of N_c). K-corrections were derived from the results of efficiency extrapolation experiments based on the expression,

$$\frac{N_c}{N_\gamma} = \frac{1}{N_0 k} N_\beta - (\frac{1-K}{K}) \ ,$$

where, N_c, N_γ and N_β are coincidence, gamma and beta channel count rates corrected for background, dead times and chance coincidences (in the case of N_c), N_0 is the absolute source disintegration rate and K is the decay spectrum correction factor. Efficiency variations were accomplished by employing foil absorption technique.[10] The resulting k-corrections for ^{56}Mn and ^{24}Na were .938±.016 and .981±.013 respectively. Measurements of k-corrections using the self absorption method confirmed the above results. The absolute β-detector efficiencies for irradiated iron foils (primary flux standards) are listed in Table I.

Table I Absolute Beta Detector Efficiency Results for ^{56}Mn

Foil Identification	Mass (gm)	$\varepsilon_\beta \pm \sigma_{\varepsilon\beta}$ (% error)
Fe-E-.091	.4366	.728±.008(1.0)
Fe-C-.091	.4327	.731±.006(.8)
Fe-D-.091	.4306	.730±.008(1.0)
Fe-B-.091	.4304	.729±.008(1.0)
Fe-A-.091	.4263	.731±.005(.7)
Fe-E-.051	.2414	.807±.008(1.0)
Fe-C-.051	.2411	.803±.006(.7)
Fe-D-.051	.2410	.809±.008(.9)

*Foil type-name-thickness (mm)

The quoted errors are mainly due to: 1) counting statistics; 2) efficiency extrapolation procedure and 3) small systematic uncertainties in the measurement of the k-factor.

Neutron Flux

The scalar neutron flux was derived from the activation rate equation, taking into consideration the time dependency of neutron flux over the exposure time. Variation in the neutron yield over time was due to tritium depletion or redistribution in the tritium-titanium target. Flux time dependence was monitored by employing a long counter set at a fixed position from the neutron source. The average neutron flux (over the exposure time) at the activation foil is then given by the expression,

$$\bar{\phi}_{foil} = \frac{\omega N_c(\Delta t)}{t_0 \cdot \varepsilon_\beta \cdot P \cdot M \cdot AN \cdot \bar{\sigma}_{act} \cdot e^{-\lambda t_0}(e^{-\lambda t_1} - e^{-\lambda t_2})} \; * \; \frac{\int_0^{t_0} f(t)dt}{\int_0^{t_0} e^{\lambda t} f(t)dt}$$

where, $N_c(\Delta t)$ = counts obtained during Δt, ω = atomic weight of the element, M = mass of the foil, ε_β = absolute beta detector efficiency, AN = Avgadro's number, t_0, t_1, and t_2 = irradiation duration, beginning of counting interval and end of counting interval, σ_{act} = activation cross-sections for $^{56}Fe(n,p)$ and $^{27}Al(n,\alpha)$ reactions over the D-T spectrum, and P = fractional isotopic abundance of target atoms. Time variation of flux is represented by the function f(t), resulting from a polynomial fit of the long counter versus time data.

The activation cross-sections employed in the above equation were obtained from cirtical evaluation of the most accurate existing data. $^{56}Fe(n,p)$ cross-section is measured absolutely to a high accuracy (±1%) as a result of an international effort to establish iron foil as a primary 14 MeV flux standard. The rest of the parameters in the above equation were independently measured in this work to a high accuracy.

To determine the scalar flux at the fission deposit, the ratio of the average flux per unit source strength at the deposit to that of the foil (see Fig. 1) had to be calculated. The results of average flux per unit source strength calculations are shown in Fig. 4 which shows a close approximation to point source-disk detector configuration for geometries employed in this work.

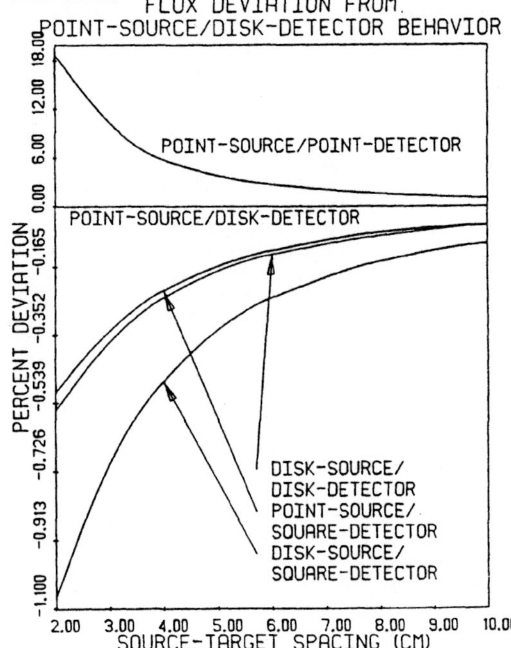

Fig. 4. Average Flux Departure from Point Source-Disk Detector Geometry

Total number of fission events

The total number of fission events was obtained from the track counts and the calculated detector efficiency.

The efficiency for detection of fission events at equal-area rings on the deposit is an anisotropy weighted numerical integration over the aperture solid angle (see Fig. 1). The corresponding values of efficiency is then numerically averaged over the target foil and weighted by the deposit density distribution.

In addition to imprecision always associated with Poisson distributed counts, an uncertainty of .25% due to scanner bias was assumed based on the average difference in repeated counts of the films by different scanners. Uncertainties due to track overlap were negligible. No uncertainty was associated with the discrimination against alpha and background pits. A registration efficiency of 100±0% has been assumed for the track-etch method of fission detection using limited solid angle angle counting based on the tests reported by Gilliam and Knoll.[11]

The uncertainties in the flux-efficiency values were mainly due to geometrical errors (~.24%) and uncertainties in the fission fragment anisotropy (~.5-.6%).

Target Foils

The masses of the foils were determined by the supplier, the Isotope Target Laboratory at Oak Ridge National Laboratory, by means of microbalance weighings.The relative isotopic contents were determined by mass spectrographic analyses also performed by ORNL. Confirmation of the mass assays by alpha and thermal fission counting were conducted at the National Bureau of Standards. The deposit mass uncertainties are an estimated ±.5% for ^{235}U and ±1.4% for ^{239}Pu.

Fission Cross-Section, Corrections and Uncertainties

The reaction rates measured led directly to the spectrum average cross section values. The total reaction rate measured at a given source-detector spacing is the sum of the direct streaming source neutrons and the room scattered neutrons.

The room scattered flux is proportional to the source strength as is the direct neutron flux. In a large room with the source at the center, the room scattered flux is very nearly constant over small displacements in the neighborhood of the source, while the direct streaming drops off approximately as $1/R_d^2$ where R_d is the source-target spacing. The integral fission cross section ($\bar{\sigma}_f$) can then be expressed as,

$$\bar{\sigma}_f = \bar{\sigma}_A - B \frac{1}{\varepsilon\phi}$$

where, $\bar{\sigma}_A$=apparent cross-section, B=the room scattered background rate, and $\varepsilon\phi$=average direct fluence efficiency at the target surface. A weighted linear least squares fitting to $\bar{\sigma}_A$ versus $1/\varepsilon\phi$ data results in $\bar{\sigma}_f$ as the y intercept and the background rate (B) as the slope.

Several corrections were applied to the $\bar{\sigma}_A$ data before attempting the fitting procedure. The major factors are compiled in Table II and are mainly self-explanatory. The sign attahced to each correction indicates

Table II Adjustment and Corrections with Residual Errors for ^{235}U and ^{239}Pu

Type of Perturbation	Correction(%) ^{235}U	^{239}Pu
Fragment Emission Anisotropy	-11.2±.48	- 8.3±.6
Angular Distribution Normalization to Lab	-1.92±.48	-1.90±.48
Scattering in Platinum Bakcings	-1.76±.4	-1.76±.4
Scattering in Other Structure	-1.03±.16	-1.03±.16
Foil Isotopic Composition	- .45±.1	-.024±.0067
$\frac{d\sigma}{dE_n}$ Adjustment for Fe(n,p) Reaction Cross-Section	+1.23±.06	+1.23±.06

*Note: correction is already included in the solid angle efficiency calculations.

the direction of the associated change in the cross-section value due to the particular adjustment applied. Table III shows a summary of major uncertainties in the measured fission cross-sections for ^{235}U and ^{239}Pu.

Table III Major Uncertainties in ^{235}U and ^{239}Pu Measurements

Source of Error	Uncertainty(%) ^{235}U	^{239}Pu
Fission Track Counting	.84	1.13
Fission Fragment Anisotropy	.55	.63
Angular Distribution Normalization to Lab	.48	.48
Total Scattering Perturbation	.43	.43
Total Geometric Error	.24	.24
Total Flux Uncertainty	1.5	1.4
Deposit Masses	.5	1.4

Finally, the corrected values of $\bar{\sigma}_A$ were fitted vs the $1/\varepsilon\phi$, accounting for the uncertainties associated with $\bar{\sigma}_A$ (Table III). Graphical representation of the above fitting procedure for ^{235}U is shown in Fig. 5.
Table IV lists the final cross section values and their ratio with the estimated residual errors.

Table IV ^{235}U and ^{239}Pu Fission Cross-Sections (barns) and Ratio

Neutron Energy (MeV)	$^{25}\sigma_f$	$^{49}\sigma_f$	$^{49}\sigma_f(n,f)/^{25}\sigma_f(n,f)$
14.63±.02	2.070±.046 (2.2%)	2.44±.092 (3.7%)	1.179±.028 (2.3%)

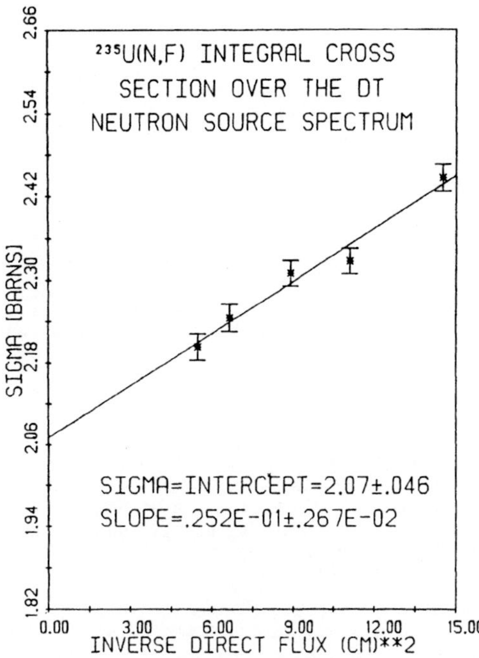

Fig. 5. Graphical Representation of ^{235}U Fission Cross-Section Extrapolation

Comparisons

The ^{235}U and ^{239}Pu fission cross-sections measured in this work are presented in Table V along with some of the more recent absolute measurements.

Table V ^{235}U and ^{239}Pu Fission Cross-Sections (barns)

E_n(MeV)	$^{25}\sigma_f$	$^{49}\sigma_f$	Author
14.63±.02	2.070±.046	2.44±.092	Present work, 82
14.60±.13	2.063±.039	2.29±.052	Cance, 78
14.70±.2	2.085±.023	2.505±.045	Adamov, 80

Acknowledgements

We wish to thank the Department of Energy for their support of this work. We would also like to thank Ken Zasadny, Jere Hassberger and Douglas Poland for their general assistance in various aspects of these measurements.

References

1. O. A. Wasson, "Neutron Cross-Section Standards", Proc. of the Intern'l Conf. on Nucl. Cross-Sections for Tech., Univ. of Tennessee, Knoxville, TN, 1979.
2. P. H. White, J. of Nucl. Energy A/B, 19,325(1965).
3. V. M. Adamov et al., see ref. 1, p 995 (1980).
4. M. Cance and G. Grenier, Nucl. Sci. and Eng., 68, 197 (1978).
5. K. Kari, Karlsuhe Nucl. Res. Center, Report KFK 2673 (1978).
6. V. M. Adamov et al., see ref. 3, p 990 (1978).
7. R. Arlet et al., see ref. 3, p 993 (1978).
8. T. B. Ryves and E. J. Axton, see ref. 1 (1980).
9. T. B. Ryves et al., Nucl. Inst. & Methods, 167, (1979) 449-453.
10. A. P. Baerg, Matrologia, 2, No. 1, 23 (1965).
11. D. M. Gilliam and G. F. Knoll, Trans. Am. Nucl. Soc., 13, 526 (1970).

STUDY OF RESONANCE NEUTRON CROSS SECTION STRUCTURE OF U-238 AND Pu-239

T. Bakalov, G. Ilchev, S. Toskov, Tran Khanh Mai, N. Janeva
Laboratory of Neutron Physics, Joint Institute for Nuclear Research, Dubna, USSR

A.A. Van'kov, Yu.V. Grigoriev, V.F. Ukraintsev
Physics Energy Institute, Obninsk, USSR

The results of investigation are reported of neutron cross section structure of U-238 and Pu-239 in the unresolved resonance energy region on the basis of transmission and self-indication analysis.

The results of self-shielding factors in total cross section and absorption cross section for U-238 and fission cross section for Pu-239 obtained are compared with the group constant system ABBN-78 data.

The resonance structure of neutron cross sections of U-238 and Pu-239 was investigated in the Laboratory of Neutron Physics (JINR). The transmission and self-indication measurements were carried out at the IBR-30 fast pulsed reactor using the time-of-flight spectrometer. The transmission

$$T(n) = \int_{\Delta u} f(u)\exp(-\sigma_t(u)n)\,du \Big/ \int_{\Delta u} f(u)\,du \qquad (1)$$

was measured on the 1000 m flight path with the ^{3}He proportional counter and the self-indication

$$T_f(n) = \int_{\Delta u} f(u)\sigma_f(u)\exp(-\sigma_t(u)n)\,du \Big/ \int_{\Delta u} f(u)\sigma_f(u)\,du \qquad (2)$$

on the 75 m and 250 m flight paths using the fission chamber. Here, f(u) are the neutron spectra in the lethargy interval ΔU, σ_x is the cross section of registered reaction of type X. In the former case the so called "reactor mode" was employed (pulse repetition rate - 5 Hz, pulse width - 70mcsec), the latter used the linear accelerator mode (100 Hz, 4 mcsec).

The time-depending reactor background was measured in the open beam and for each sample thickness by the resonance filter technique. Most of obtained experimental data were used in the development of the system of group constants ABBN-78 (see Ref.[1]).

Data on U-238

These experiments are described in Ref.[2], the later additional measurements on transmission function were carried out under better conditions.

Table 1 presents the comparison of obtained results -- average total cross sections and their resonance self-shielding factors (at room temperature) -- with the theoretical estimates from Ref.[1]. It is seen that in the energy region E > 2 keV (group 11-13) our experimental estimates for $\langle\sigma_t\rangle$ are by 1-2 barns higher than those reported in Ref.[1]. It is the same with the estimates obtained for self-shielding. For the energy range 1-2 keV (group 14) an agreement is achieved in estimates of $\langle\sigma_t\rangle$ and $f_t(\sigma_0)$ at $\sigma_0 \geqslant 10$ barns.

Table 1 Average group total cross sections for U-238 and its resonance self-shielding factors at T=300^0 as compared with calculated estimates from Ref.[1].

No. Gr.	E, keV	Experiment $\langle\sigma_t\rangle$	Experiment $f_t(0)$	Experiment $f_t(10)$	Estimates $\langle\sigma_t\rangle$	Estimates $f_t(0)$	Estimates $f_t(10)$
11	10-21.5	16.0±0.6	0.70±0.05	0.77	14.5	0.76	0.83
12	4.65-10	17.3±0.8	0.60±0.07	0.69	15.9	0.67	0.76
13	2.15-4.65	20.0±1.0	0.44±0.10	0.61	19.0	0.45	0.58
14	1-2.15	22.2±1.0	0.39±0.06	0.46	22.2	0.18	0.48

Table 2 gives a comparison between self-shielding factors of absorption cross sections obtained in our experiment(for the delution cross section equal to zero, "room" temperature) with estimates from Ref.[1] in both versions: average absorption cross sections are evaluated using nuclear data from microscopical experiments with and without correction on the basis of integral data. It was assumed that in the evaluation of $\langle\sigma_c\rangle$ for U-238 with correction using integral data and variation of selfshielding factors, one may achieve a better agreement between the corrected results and the initial microscopic data.

Table 2 Resonance self-shielding factors of absorption cross section for U-238

No. Gr.	Experiment	Estimate (Ref.[1]) without correction	Estimate(Ref.[1]) with correction
11	0.44±0.03	0.54	0.50
12	0.55±0.03	0.64	0.59
13	0.68±0.04	0.66	0.62

Pu-239 data

Fig. 1 presents a comparison of experimental and calculated data on transmission and self-indication in the fission of Pu-239. In the energy range E > 100 eV the calculation was performed using the method of statistical simulation (the Monte Carlo Method) of the neutron cross section resonance structure in the frame of multilevel R-matrix formalism (Ref.[5]).with variation of parameters dependent on the energy group number. The S-matrix was calculated taking into account the s- and p- resonances statistics. The capture channel contribution is taken into account in the Riech-Moore approximation. The initial values of average resonance parameters were written using the V.A. Konshin evaluation[6] for the fission channels in the states 0+ and 1+.

In the energy range E < 100 eV the multilevel resonance parameters are derived in the Adler's scheeme of s-matrix theory (Ref.[9]).

It can be seen that the calculated values, derived within the statistical model, are in good agreement with experimental ones on transmission and self-indication except for the self-indication point at a largest sample thickness.

The results of our experimental (Ref.[8]) and calculated data on self-shielding factors are presented in Table 3 as compared with the data evaluated in Ref.[1]. Our values of $f_f(\sigma_0)$ for Pu-239 are systematically higher than the same values from Ref.[1]. Good agreement for f_t values is seen.

It must be noted that our experiment gives the value of the total cross section between resonances about 1.5-2 barns higher than Derrien's data (Ref.[7]) for the energy range below 50 eV.

K. H. Böckhoff (ed.), Nuclear Data for Science and Technology, 62–64.

Table 3 Resonance self-shielding factors for Pu-239 in the energy region from 10 eV to 21.5 keV
A- experiments (Ref.[8]);
B- calculated values;
C- evaluated data ABBN (Ref.[1])

No. group	E		$f_f(\sigma_0)$		$f_t(\sigma_0)$	
			0	100	0	100
II	2I.5 - - I0 (KeV)	A B C	0.974 0.890 0.920	0.998 0.980 0.990	0.836 0.86I 0.945	0.986 0.978 0.995
I2	I0. - - 4.65	A B C	0.872 0.8I0 0.840	0.990 0.970 0.980	0.785 0.8I3 0.868	0.978 0.962 0.985
I3	4.65 - - 2.I5	A B C	0.765 0.700 0.640	0.950 0.930 0.930	0.728 0.689 0.739	0.924 0.990 0.94I
I4	2.I5 - - I.	A B C	0.688 0.6I7 0.508	0.920 0.895 0.88I	0.598 0.634 0.635	0.858 0.849 0.88I
I5	I000. - - 465. (eV)	A B C	0.598 0.46I 0.373	0.83I 0.764 0.786	0.386 0.455 0.467	0.696 0.679 0.72I
I6	465. - - 2I5.	A B C	0.54I 0.509 0.296	0.757 0.822 0.668	0.36I 0.408 0.36I	0.7I6 0.702 0.553
I7	2I5. - - I00.	A B C	0.435 0.378 0.205	0.664 0.709 0.527	0.269 0.292 0.279	0.52I 0.546 0.423
I8	I00. - - 46.5	A B C	0.300 0.308 0.I3I	0.658 0.583 0.383	0.I65 0.I79 0.I55	0.357 0.383 0.3I8
I9	46.5 - - 2I.5	A B C	0.235 0.258 0.I32	0.399 0.427 0.363	0.I62 0.2I4 0.235	0.304 0.306 0.359
20	2I.5 - - I0.	A B C	0.I67 0.I84 0.I49	0.366 0.375 0.372	0.I24 0.II5 0.II0	0.229 0.220 0.2I7

Conclusions

The results of self-indication measurements confirm the necessity of further experimental investigation of unresolved resonance structure of neutron cross-sections in order to obtain more precise values of self-shielding factors and other characteristics of neutron cross section structure in the heavy nuclei region.

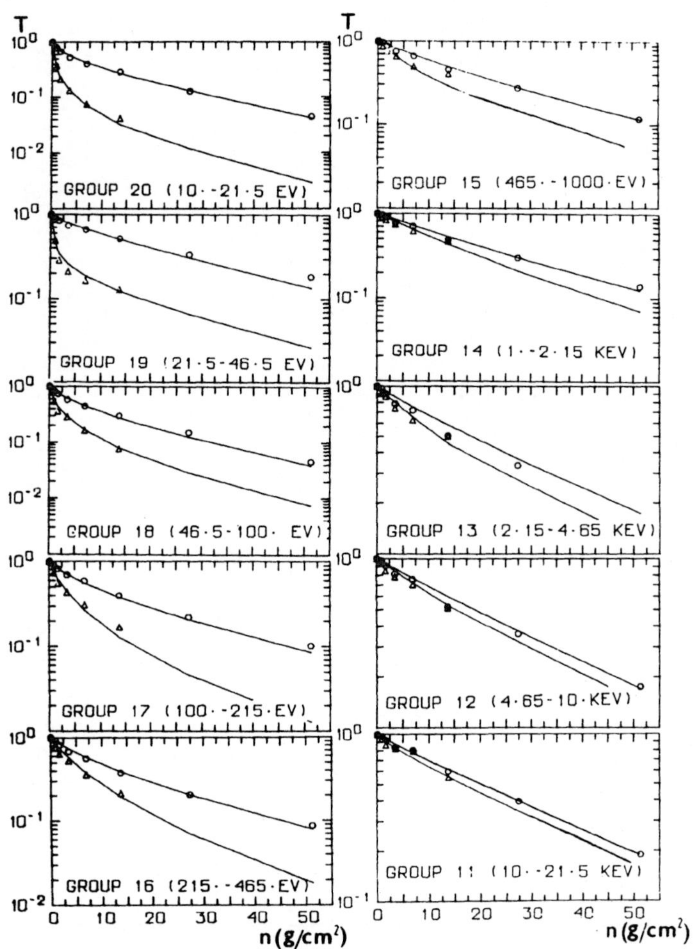

Fig. 1 Experimental and calculated values of transmission and self-indication for Pu-239 vs thickness; O- experimental values T_t Δ - experimental values T_f — calculated values

References

1. Gruppovye konstanty dlja rascheta reaktorov i zaschity, M., Eneroizdat, 1981. Auth.: Abagian L.P., Bazazjants N.O., Nikolaev M.N., Tsibulja A.M.
2. A.A.Van’kov, Yu.V.Grigoriev, M.N.Nikolaev, et al. Proc.of Int.Conf.Nuclear Data for Reactors, Helsinki, IAEA, 1970, 1, p.559.
3. A.A.Van’kov, Yu.V.Grigoriev, A.M.Tsibulja, B.Bemer, K.Dtse. Vsesojuznaja konferentsija po nejtronnoj fizike, Kiev, 1975, part 3, p.200.
4. A.A.Van’kov, A.M.Tsibulja, Tran Khanh Mai. Otsenka sechenija zakhvata i ego resonansnoj samoekranirovki dlja U-238 iz analiza eksperimentov po propuskaniju nejtronov v oblasti nerazreshennykh rezonansov. Preprint FEI-1005, Obninsk, 1980.
5. N.Kuindzhieva, N. Janeva. Atomnaja nauka i tekhnika, serija Jadernye konstanty, 1981, part 42(3), p.88.
6. G.V.Antsinov, V.A.Zenevich, V.A.Konshin. Voprosy atomnoj nauki i tekhniki, serija Jadernye konstanty, 1981, part 2(41) p.42.
7. Derrien H. Proc.of Int.Conf.Nuclear Data for Reactors, IAEA, Paris, 1967, v.2, p.195.
8. T.Bakalov et al. Proc.of Int.Conf. Nuclear Cross Sections for Technology, 22-26 October, Knoxville, 1979, p.692.
9. F.T.Adler, D.B. Adler In: Proc.Conf.Neutron Cross Sections and Tech., Washington, March 4-7, 1968, p.967.

NEUTRON TOTAL CROSS SECTION MEASUREMENTS OF ^{238}U AT THE Fe-FILTERED NEUTRON ENERGY BANDS IN keV REGION

I. Tsubone and Y. Kanda

Interdisciplinary Graduate School of Engineering Sciences
Kyushu University, Kasuga, Fukuoka, Japan

and

Y. Nakajima and Y. Furuta

Japan Atomic Energy Research Institute
Tokai-mura, Ibaraki-ken, Japan

The neutron total cross section of ^{238}U has been measured in the energy range from 24 keV to 1 MeV at a 100 m flight path station of the neutron time-of-flight facility of the Japan Atomic Energy Research Institute linac. Neutrons were filtered by a 30 cm-thick iron and detected by a NE-110 plastic scintillation counter. Attention was paid to taking into account the self-shielding effect below 200 keV. Fitting the measured total cross section with the average R-matrix, the neutron strength function for p- and d- wave were found to be $S_1 x10^4$=1.60±0.08 and $S_2 x10^4$=0.95±0.50, and the average distant level parameters for s- p- and d-wave to be R_0^∞=-0.081±0.005, R_1^∞=0.13±0.04, and R_2^∞=-0.01±0.68, respectively.

[^{238}U, total cross section, E_n=24 keV - 1 MeV, Fe-filtered neutron beam technique, time-of-flight, self-shielding effect, strength functions, average distance level parameters.]

Introduction

The accurate knowledge of the total cross section of ^{238}U is one of the most important nuclear data in the field of the fission reactor science and technology. Especially improved data in the keV region are desirable for fast reactor design and safety. There are, however, some experimental difficulties to obtain the accurate cross section in this energy region with white neutron sources. Firstly, fine energy-dependent structure complicates the measurements of the cross section by the self-shielding effect. Second, the time-dependent background is large and it is difficult to be determined accurately.

Recently, filtered neutron beam techniques have been used at reactor[1-4] and accelerator[7-10] facilities to obtain more accurate cross sections. This method is characterized with a large signal-to-background ratio, and has the advantage of both reducing the time dependent background and determining the background easily. Thus, the utilization of this method in time-of-flight measurements leads accurate cross sections, although the available energy is restricted and the counting rate decreases.

Some cross section measurements using this method have been reported. In these measurements, mainly 24 keV[5-9] and/or 82 keV bands[11] for an iron filter and a 144 keV band[6-7] for a silicon filter were utilized. In the present experiment, the iron-filter was used and 25 filtered energy bands were made to be available in the keV region as well the 24 keV and 82 keV bands. In the present work, transmission measurements were carried out on ^{238}U with this filtered neutron beam to obtain the accurate average total cross section.

Average resonance parameters are utilized for the calculation of partial cross sections. They are also important data for the interpretation of the neutron induced reactions. The reliable strength functions for higher partial waves can be derived from the total cross sections. In this paper, we present results of the total cross section measurements and of the average resonance parameters deduced from these data.

Experimental Procedure

Measurements were carried out with the neutron time-of-flight method using the Japan Atomic Energy Research Institute linac. Neutrons were produced by a pulsed electron beam of 120-MeV in a water-cooled tantalum target. The linac was operated at 300 pps of repetition rate with 30 ns electron burst width. The neutrons produced in the tantalum target were moderated by a boron-loaded polyethylene plate of 20 cm diameter x 5 cm thickness. The iron filter of 20 cm diameter x 30 cm thickness was located 7 m from the neutron source. The neutron beam filtered by the thick iron was collimated to 6 cm diameter at a sample position.

Samples were metallic disks of 8 cm diameter depleted uranium (99.95%). The transmission of four samples of different thicknesses, having 0.00628, 0.0259, 0.0495, and 0.0807 atoms/barn, were measured so as to correct the self-shielding effect. Samples were mounted on a sample changer placed at 7.5 m from the neutron target.

The neutron beam was directed to a 100 m station through an evacuated aluminium flight tube. Neutrons were detected by a NE-110 plastic-scintillation counter of 5.08 cm diameter x 1.27 cm thickness. The stop signals from the neutron detector were fed into a time analyzer via a dead time generating circuit which was used for generating the constant and definite dead time for the whole data acquisition system. Channel width of the analyzer was 25 ns. Cyclic measurements of neutron flux spectra were automatically repeated for two configurations of sample-in and sample-out to reduce the error by the neutron beam fluctuation. Data acquisition was made with an on-line computer system.

Transmission measurements were also carried out on polyethylene to examine the systematic error in the detection system. The measured total cross section of polyethylene was very close to the cross section composed of the cross sections of hydrogen and carbon in JENDL-1.[11] The neutron time-of-flight spectra measured with the sample of 0.0459 atoms/barn are shown in Fig. 1.

Analysis

Data Reduction

The dead time correction was made for measured spectra. A magnitude of this correction reached about 9 % change of cross section at the peak with the highest counting rate.

Backgrounds were determined by the least squares method. Off peak regions of the time spectra were fitted using a smooth curve. Uncertainties caused by background determination were negligible.

After the dead time correction and the subtraction of the background, the effective average total cross section $<\sigma eff(E,n)>$ was calculated by the equation

$$<\sigma eff(E,n)> = -\frac{1}{n} \ln(< T >) \qquad (3)$$

K. H. Böckhoff (ed.), Nuclear Data for Science and Technology, 65–68.

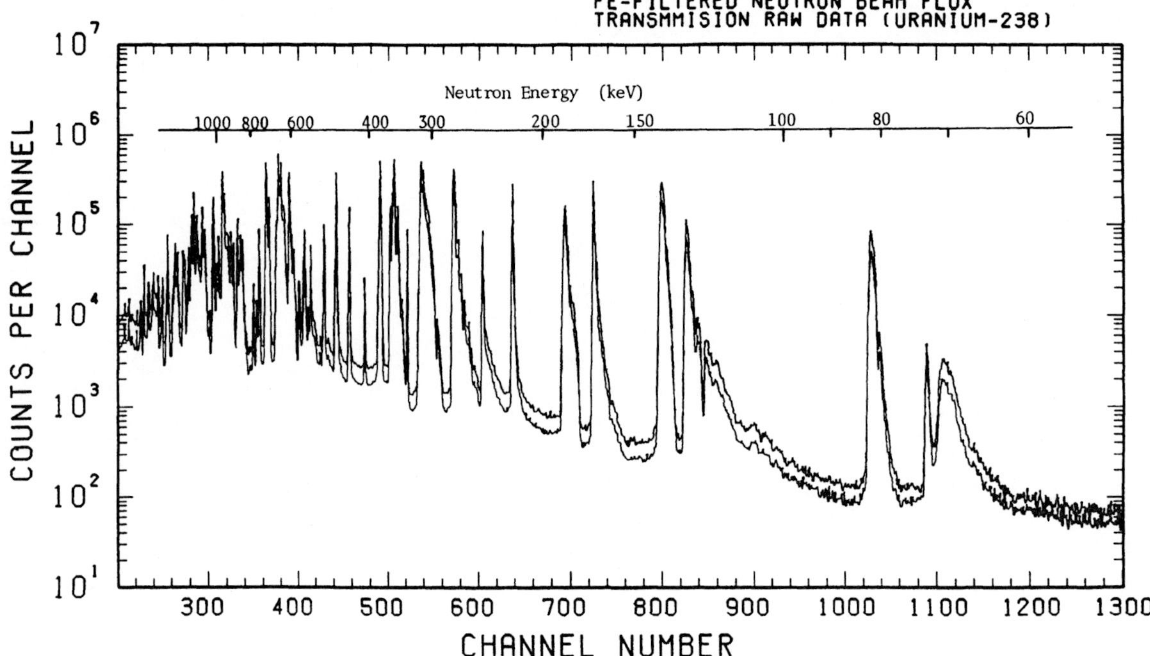

Fig. 1. Time spectra of iron filtered neutron beam. Upper and lower lines show the configurations of sample-out and sample-in, respectively.

at the filtered energy E for each sample, where n is sample thickness and $< T >$ is the measured average transmission.

In low keV region, the measured transmission yields $<\sigma_{eff}(E,n)>$ less than true average total cross section $<\sigma_t>$ by self-shielding effect, since fine energy-dependent resonance structure still exists. The average total cross section $<\sigma_t>$ is, however, given by

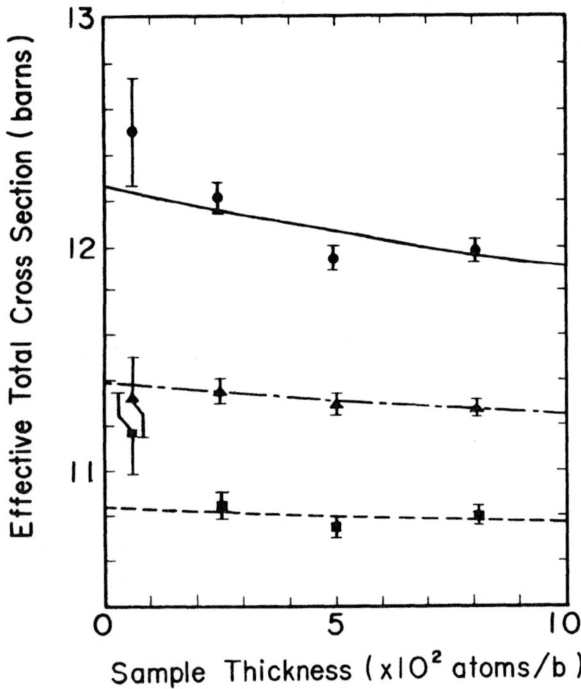

Fig. 2. Effective total cross section measured with four different sample thicknesses. Attenuation of the cross section is caused by the self-shielding effect. ●, ▲, and ■ are measured at 82 keV, 138 keV, and 183 keV, respectively.

$$<\sigma_t> = \lim_{n \to 0} <\sigma_{eff}(E,n)> \tag{4}$$

This implies use of sufficently thin samples, such as $(n \cdot \sigma_t << 1)$, and it takes extremely long counting time to measure the transmission of this sample with sufficient statistics. Consequently, by measuring $<\sigma_{eff}(E,n)>$ with different sample thickness and the $<\sigma_t>$ is obtained from extrapolating the values to zero sample thickness.

For the extrapolation, we simulated the self-shielding effect using the Monte-Carlo techniques. Similar techniques were also used by Tsang and Brugger,[5] Kobayashi et al.,[9] and Poenitz et al.[12] Calculations were carried out mainly based on the TIMS-1[13] code which was originated by Takano et al. for production of group constants of heavy elements. The result instructed that the calculated curve of the effective total cross section as a function of sample thickness was dependent on the average resonance parameter sets of input. Thus we found out that the function

$$A + B \cdot \exp(-C \cdot n) \tag{5}$$

well reproduced the form of the self-shielded cross section curve even if the input average resonance parameters and calculation conditions were changed, where A, B, and C were parameters.

The function, therefore, was fitted to the experimental data by the least squares method, and the value at n=0 was taken as the true average total cross section, $<\sigma_t>$. Examples of their self-shielding effect are shown in Fig. 2. Above 200 keV where the self-shielding is not significant, $<\sigma_t>$ were obtained by the weighted mean of measuerd values.

Data Interpretation

The average total cross section $<\sigma_t(\ell)>$ for incident neutrons of the orbital angular momentum ℓ is expressed by the R-matrix theory:

$$<\sigma_t(\ell)> = 2\pi\lambda^2(2\ell+1) \; [1 - \mathrm{Re}U(\ell)] \tag{6}$$

and

$$U(\ell) = \exp(-2i\phi_\ell) \frac{1 - P_\ell \cdot L_\ell^*}{1 - R_\ell \cdot L_\ell} . \tag{7}$$

In eq.(7),

$$\phi_\ell = \arctan(-j_\ell/n_\ell)$$
$$R_\ell = R_\ell^\infty + i\pi s_\ell$$
$$L_\ell = B_\ell + \ell + iP_\ell \ .$$

Here ϕ_ℓ is the phase shift for ℓ-wave neutrons scattered from a hard sphere, and j_ℓ and n_ℓ are the spherical Bessel and Neumann functions, respectively; R_ℓ^∞ represents the overall effect of far away levels and s_ℓ the pole strength function; B_ℓ is the modified level shift factor and P_ℓ the penetration factor. The pole strength function s_ℓ is related to experimental strength function S_ℓ:

$$S_\ell = \frac{2kR\cdot s_\ell}{\sqrt{E(eV)/1eV}} \qquad (8)$$

The cross section for each partial wave is characterized by two unknown parameters, R_ℓ and S_ℓ . Therefore, from the best fit of this formula to the measured cross sections by the least squares method,* S_ℓ and R_ℓ for s-,p-,d- and f-wave were extracted.

This method of the analysis is similar to those used by Uttly et al.[14] Camarda,[15] and Jain et al.[16]

The s-wave neutron strength function (S_0= 1.0 x10^{-4}) was fed to the program and treated as a fixed parameter. It is valid since the s-wave compound nucleus component is not so large in the present energy range and S_0 is reliably obtained from the analysis of individual resonances in the resolved resonance region, and there are only four points up to 100 keV in the present experiment.

Results and Discussions

The measured total cross section and the best fit are shown in Fig. 3 with statistical errors. Dashed lines marked S, P, D and F indicate the s-, p-, d- and f-wave components of the total cross section, respectively.

At the 24 keV energy band, our result is higher than that of Tsang and Brugger,[5] and is in very good agreement with ENDF/B-IV evaluated value.[17] In the whole energy range, our data have a tendency of being lower than those of Uttly et al.[14] and Poenitz et al. but are in good agreement with the JENDL-1[11] and ENDF/B-IV[17] evaluated data below 400 keV.

The average resonance parameters extracted are listed in Table.I. with the previous results. It was assumed, in the parameter search for higher partial wave, by Uttley et al.[14] that S_1= $S_3$and R_2^∞= R_3^∞= 0 and by Camarda[16] that S_0= $S_2$and R_0^∞ = R_2^∞ . It is caused by the reason that one expect the similar behavior between s- and d-wave parameters and/or p- and f-wave parameters as a function of mass number, and small shape elastic cross sections for d- and f-wave below 1 MeV. On the other hand, the parameters except S_0 were treated as free in the present analysis. The parameters resulted without any assumption show well the behavior between the partial wave parameters, which is mentioned above, but are not in agreement with Uttley et al.

The p-wave strength function S_1 is rather near to the value by Rahn et al.[18] which was derived by distinguishing the individual levels into p- and s-waves based on the known statistical behavior, and is also in very good agreement with the values by Moxon[19] and Tsang and Brugger[5]

In the case of ℓ = 0 , the potential scattering radius R' is related to the nuclear radius R :

$$R' = R\,(1 - R_\ell^\infty) \qquad (9)$$

where R was taken to 1.4$\cdot$ $A^{1/3}$ fm. Thus , R' was also determined to be 9.37±0.04 fm , which was in good agreement with the data by Tsang and Brugger,[5] and Olsen et al.[20]

In the present work, the transmission measurements, using the iron-filtered neutron beam which was available up to 1 MeV, were carried out on ^{238}U, and the accurate total cross sections were obtained at each filtered energy band. From the analysis by the average R-matrix, the consistent average partial wave parameters could be obtained, which will be some useful quantities to the nuclear data evaluation and the calculations in the technological important energy region for reactor physics.

* SALS, by T.Nakagawa and Y.Koyanagi, was used in this analysis.

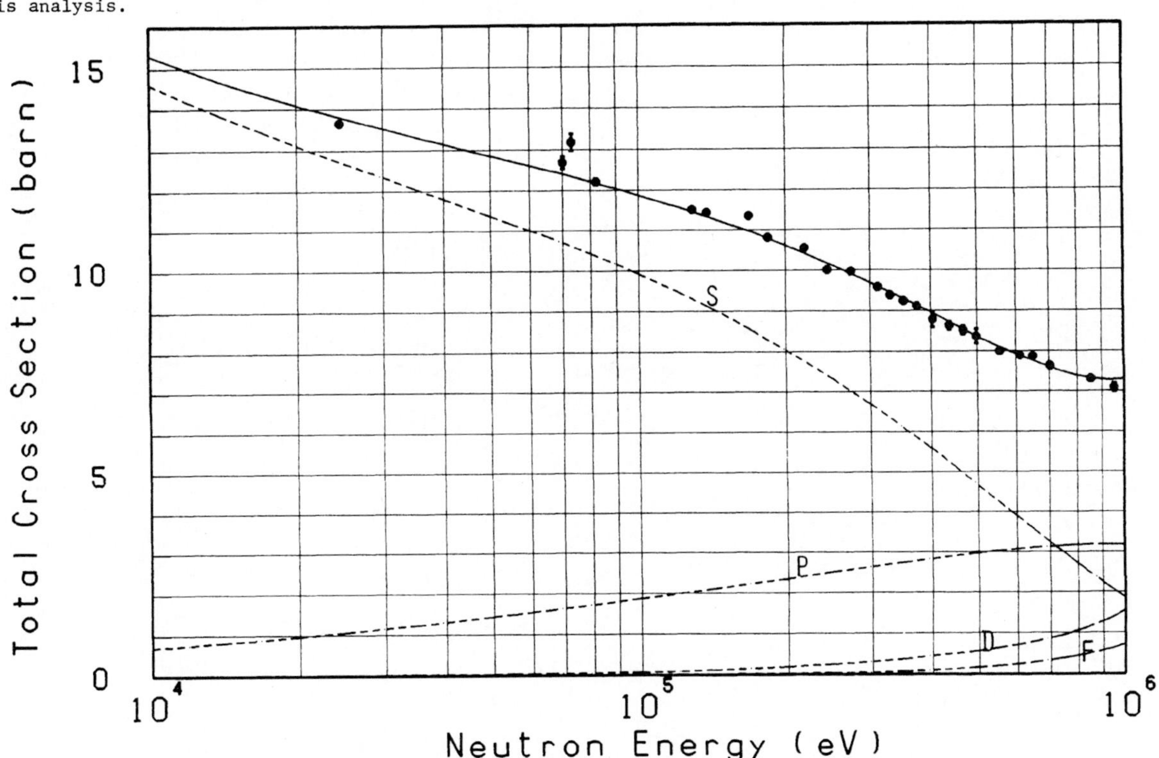

Fig. 3. Total cross section of ^{238}U and the best fit. Dashed lines marked S, P, D, and F indicate s-, p-, d- and f-wave components, respectively.

Table. I. Average resonance parameters for partial waves

	Moxon	Rahn	Tsang & brugger	Olsen	Uttley	Present
$S_0 x10^4$	1.3 ±0.3	1.08 ±0.10	0.943	0.968 ±0.035	1.0 *	1.0 *
R_0^∞					-0.098*±0.016	-0.087±0.005
R' (fm)		9.6	9.4	9.44 ±0.05	9.185**	9.37 ±0.04
$S_1 x10^4$	1.59±0.45	1.4	1.64		$2.47\ ^{+0.16}_{-0.28}$	1.60 ±0.08
R_1^∞					$0.06\ ^{+0.00}_{-0.04}$	0.13 ±0.04
$S_2 x10^4$					$0.05\ ^{+0.30}_{-0.40}$	0.90 ±0.50
R_2^∞					0.0 *	-0.01 ±0.68
$S_3 x10^4$					S_1 *	2.0 ±0.5
R_3^∞					0.0 *	0.1 ±1.9

* These values were assumed.
** This was calculated from $R = 1.35 \cdot A^{1/3}$

Acknowledgements

The authors wish to thank Dr. H. Takano for instructive discussions in the use of the TIMS-1 code and also to thank the operation crew of the JAERI linac for stable beam supplying.

References

1. Simpson O.D., Miller L.G.: Nucl. Instrum. Methods, 61, 245(1968)
2. Tsang F.Y., Brugger R.M.: Nucl. Instrum. Methods, 134, 441(1976)
3. Nakazawa M., et al.: UTNL-R-0062(1977)31
4. Robert A.K., Brugger, R.M.: Nucl. Instrum. Methods, 155, 307(1978)
5. Tsang F.Y., Brugger R.M.: Nucl. Sci. Eng., 72, 52(1979)
6. Tsang F.Y., Brugger R.M.: Nucl. Sci. Eng., 74, 34(1980)
7. Kobayashi K., et al.: Jour. Nucl. Sci. Technology, 12, 1(1975)
8. Harvey J.A.: keV neutron total cross section measurements at ORELA, Proc. ANS Topical Meeting on New Developments in Reactor Physics and Shielding Calculation, CONF-720901,(1972)
9. Kobayashi K., et al.: Nucl. Sci. Eng., 65, 347(1978)
10. Winters R.R., Morgan G.L.: Nucl. Sci. Eng., 78, 147(1981)
11. Igarasi S.,et al.: JENDL-1, JAERI 1261 (1979)
12. Poenitz W.P., et al.: Nucl. Sci. Eng. 78,333(1981)
13. Takano H., et al.: JAERI 1267(1981)
14. Uttly C.A., et al.: Proc. of the Conference on Nuclear Data for Reactors, Paris,1966,Vol. I, p.165.
15. Camarda H. S.: Phys. Rev. C9, 28(1974)
16. Jain A.P., et al.: Phys. Rev. 137, B83(1965)
17. BNL-17541, Brookhaven National Laboratory (1975)
18. Rahn F., et al.: Phys. Rev. C6, 54(1972)
19 Moxon M.C.: M.Sc Thesis(1968)
20. Olsen K.D., et al.: Nucl. Sci. Eng.,62, 479(1977)

THE ^{241}Pu(n,f) CROSS-SECTION AND ITS NORMALIZATION

C. Wagemans *

SCK/CEN Mol, Belgium and
Nuclear Physics Lab.,
Gent, Belgium

A.J. Deruytter

Commission of the European Communities
Joint Research Centre
CBNM, Geel, Belgium

A series of measurements of the ^{241}Pu(n,f) reaction cross-section has been performed at GELINA. The neutron energy range from 10^{-2} eV up to 10^5 eV was covered, allowing a normalization to the 2200 m/s reference cross-section. The neutron flux was determined via the ^{10}B(n,α)^{7}Li reaction. Back-to-back ^{241}Pu-^{10}B foils were viewed by surface barrier detectors, so the fission fragments counting rate and the neutron flux were measured simultaneously and from the same position in the neutron beam. The present σ_f-values are compared with previous results, giving special attention to the influence of the normalization. In particular, the situation of the secondary normalization integrals $\int_{12eV}^{20eV}\sigma_f dE$ and $\int_{0.1keV}^{1kev}\sigma_f dE$ is reviewed.

Introduction

A few years ago, we investigated the low-energy part of the ^{241}Pu fission cross-section, giving special attention to normalization problems [1]. At that time we observed important discrepancies between the scarce results reported. Moreover, the neutron energy intervals covered in these measurements were generally small. Especially, not a single experiment covering the neutron energy region from thermal up to 30 keV was available. Afterwards, two new measurements have been reported by Carlson et al. [2] and by Weston and Todd [3], which, unfortunately, were different by about 10 %. In view of the importance of the ^{241}Pu fission cross-section for the fast reactor development, we performed a new series of measurements at GELINA in an effort to improve this situation.

Experimental procedure

The measurements were performed at a 9.3 m flight path of the linac at CBNM, Geel (Belgium). Pulsed neutrons with a broad energy spectrum were produced by bombarding a natural uranium target with 100 MeV electrons, the spectrum being softened in a polyethylene moderator. Since we wanted to cover thermal neutron energies, no overlap filters could be used, hence the repetition frequency of GELINA had to be limited to 100 Hz to avoid overlap of neutrons from successive bursts. A neutron burst-width of 20 ns was selected in order to still obtain a sufficient energy-resolution in the keV-region. Additional measurements were performed at 9.3 m using a Cd-overlap filter and at 30.45 m using a ^{10}B-overlap filter. In these cases, GELINA was operated at a repetition frequency of 800 Hz and a burst width of 4 ns.

A ^{10}B-layer was mounted back-to-back with a ^{241}Pu-layer in the center of a large evacuated detection chamber (Ø = 50 cm). The ^{10}B(n,α)-particles and the ^{241}Pu(n,f)-fragments were detected by two collinear gold-silicon surface barrier detectors placed outside the neutron beam, with which the detector axis made an angle of 90°. After amplification, a fast timing signal was derived from each detector signal and sent into one of the two half-sections of a 16384 channels time-of-flight analyser with channel widths varying between 4 and 4096 ns. So the fission fragments and the ^{10}B(n,α)-particles were detected simultaneously and from the same position in the neutron beam. In the same conditions, a background determination was performed, with Cd-, Rh-, Au-, Co- and Mn-neutron filters, together with a permanent Al-neutron filter.

The ^{241}Pu layer was prepared by electrospraying of plutoniumacetate on an aluminium disk. The thickness of the deposit was 550 μg/cm^2 with an isotopic enrichment of 91.02 atom % of ^{241}Pu. The remaining Pu-isotopes were ^{238}Pu(0.03 %), ^{239}Pu(1.17 %), ^{240}Pu (4.48 %) and ^{242}Pu(3.30 %). The ^{10}B-layer was prepared by evaporating elemental boron in vacuum. In this way a homogeneous layer with a thickness of 184 μg/cm^2 and an isotopic enrichment of 93 % was obtained. Both targets were manufactored by the CBNM Sample Preparation Group.

Treatment of data

The data reduction was done using an updated version of the computer code ANGELA [4]. An analytical expression for the background was determined by fitting polynomes

$$y = \sum_{i=0}^{n} a_i T^i$$

to the background values obtained in the black resonances. Here T is the time-of-flight, y is the counting rate and a_i are constants. The n-value yielding the best χ^2-value was chosen.

With the assumption of a 1/v-shape for the ^{10}B(n,α) cross-section, the ratios of the corresponding fission and ^{10}B(n,α) time-of-flight spectra yield the (unnormalized) $\sigma_f(E)\sqrt{E}$ values through application of the formula:

$$\sigma_f(E)\sqrt{E} = k\,\frac{N_f(E) - B_f(E)}{N_n(E) - B_n(E)}$$

where k is a normalization constant; $N_f(E)$ and $B_f(E)$ the number of fission counts and the corresponding background at energy E; $N_n(E)$ and $B_n(E)$ the number of ^{10}B(n,α) counts and the corresponding background at energy E. These $\sigma_f\sqrt{E}$ values were corrected for the slight deviation from 1/v of the ^{10}B(n,α) cross-section making use of the analytical expression proposed by Sowerby et al. [5]. The data were also corrected for the ^{239}Pu(n,f) contribution, due to the ^{239}Pu-impurity in the target, for which purpose the σ_f-data of Wagemans et al. [7] were used.

The present fission cross-section data were normalized to the fission integral

$$\int_{0.02eV}^{0.10eV}\sigma_f dE = 60.11 \text{ b.eV}$$

determined previously [1] and corresponding to a 2200 m/s σ_f-value [6] of 1015 b. A normalization to this thermal σ_f-value via a linear least-squares fit of $\sigma_f\sqrt{E}$ from 0.02481 to 0.02586 eV, yielded a consistent normalization factor, but was only done for control purposes.

* Nationaal Fonds voor Wetenschappelijk Onderzoek

K. H. Böckhoff (ed.), Nuclear Data for Science and Technology, 69–73.

Results and discussion

The main goal of these experiments was to determine accurate average fission cross-section values directly normalized to the thermal reference value and covering a large energy interval. At present, data were obtained from 10 meV up to 30 keV, so for the first time for ^{241}Pu almost seven energy decades were covered in a single experiment. These results are graphically shown in Figs. 1-3. Numerical values are given in Table 1. From this table it is clear that the present results are in excellent agreement with our previous data [1]. This is also obvious from Fig. 1.

The uncertainty of the present data is mainly caused by five contributions: (i) The error on the normalization, which is determined by the accuracy of the 2200 m/s fission cross-section being 1015 ± 7 b according to Lemmel [6]. (ii) The error on the background determination for the fission as well as for the neutron flux spectra. Especially in the fission case, very stable measuring chains are required to avoid background variations due to the extremely high natural α-emission rate of the target resulting in large amplitude pile-up signals. In the present experiments, typical amplitudes of the background correction were 0.8 % at thermal and 5 % at 25 keV for the (n,f)-reaction, and 0.4 % and 1.5 % for the ^{10}B(n,α)-reaction. (iii) The error on the correction for impurities in the sample. Since the history of the sample is very well known, the error on this correction is very small and determined by the uncertainty of the σ_f-data [7] for ^{239}Pu. (iv) The uncertainty in the shape of the ^{10}B(n,α) cross-section. This effect mainly plays a role above 1 keV, since below that energy the shape has an almost perfect 1/v-behaviour. (v) The statistical accuracy of the data. By combining these contributions we obtain an overall accuracy of 2 % below 30 eV, 3 % below 1 keV and 4 % up to 30 keV.

In Table 1 the present results are compared with other experimental [1-3/8-13] and evaluated $\bar{\sigma}_f$-data in the same energy region. These data were mainly obtained from the OECD/NEA Data Bank in Saclay (France). This table reveals differences of up to 40 % between extreme experimental values. In order to investigate possible causes for these discrepancies, we summarized the main characteristics of these experiments and the normalization methods in Table 2.

The normalization methods are very different and sometimes rather peculiar. Nevertheless, they can only partially explain the differences observed in Table 1. This is illustrated with Table 3, in which the normalization integrals or cross-sections used in the various experiments are compared with the corresponding values obtained from the present measurements. The ratio of both varies between 0.974 and 1.080. In the same table the fission integrals from 12 to 20 eV are given and compared with the value of 1379 b.eV calculated from our data. For consistent and suitably normalized data, the difference Δ between both ratios should be very small. This is only true for ref. [1/3/10], which all three happen to be normalized in the thermal region. The present comparison illustrates the rather common underestimation of the errors introduced by the normalization procedure.

Another sometimes neglected feature is the influence of the neutron overlap filters used on the shape of the neutron flux. This is illustrated in Fig. 4, which shows the ^{10}B(n,α)-counting rate in function of the neutron energy in three cases: (a) without overlap filter, (b) with a Cd-overlap filter and (c) with

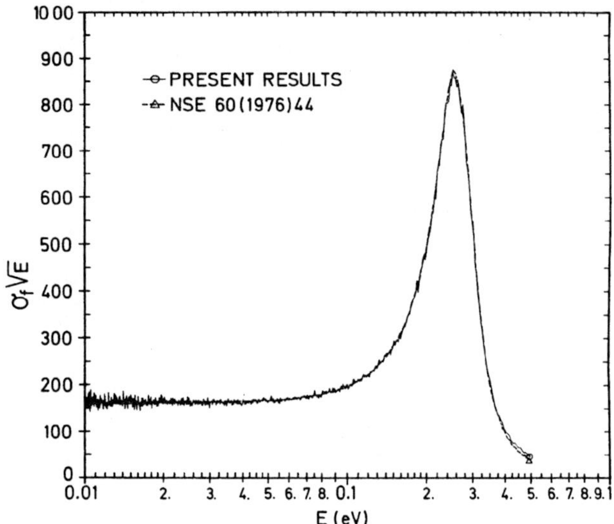

Fig. 1. Present $\sigma_f\sqrt{E}$-curve for ^{241}Pu from 0.01 to 0.5 eV (full line) compared with our previous data [1] (intermittent line)

Table 1 Comparison of the present $\bar{\sigma}_f$-values (barn) with other results

	EVALUATIONS		EXPERIMENTS									
E_n	ENDF-B4	JENDL2	Watanabe[8]	Moore[9]	James[10]	Simpson[11]	Migneco[12]	Blons[13]	Wagemans[1]	Carlson[2]	Weston[3]	This work
0.02-0.1 eV	751.2	753.4			721.2				751.3		745.5	751.3
0.1 -0.5	673.2	678.6	654.8		609.0				681.5		689.3	681.6
0.5 -1	40.3	40.3	34.8		35.9						36.7	(42.4)
1 -10	183.4	189.1	187.6		189.3				187.2		192.4	189.7
10 -20	143.3	140.9	156.9	152.3	153.3		146.8	146.3	150.3		151.5	151.7
20 -30	88.3	83.8	86.5	85.1	85.7		82.9	82.9	83.9	78.9	86.2	85.5
30 -40	48.0	48.8	47.1	48.7	46.6	38.7	46.3	46.5			49.2	49.6
40 -50	48.4	40.9	51.0	49.4	42.2	35.3	36.8	38.8			43.7	45.5
50 -100	40.1	38.0	51.4		51.6	39.6	39.7	36.4			40.6	39.6
0.1 -0.5 keV	24.3	23.2			24.7	27.8	21.0	21.0		22.1	24.4	23.6
0.5 -1	13.2	11.8			(11.6)	11.5	10.4	10.8		11.7	13.0	12.1
1 -5	7.14	6.96				6.58		6.70			7.37	7.21
5 -10	4.18	4.32				3.89		4.16			4.43	4.2
10 -20	3.02	3.37				3.36		3.31		3.02	3.39	3.09
20 -30	2.65	2.87				3.09		2.81		2.62	2.98	2.76

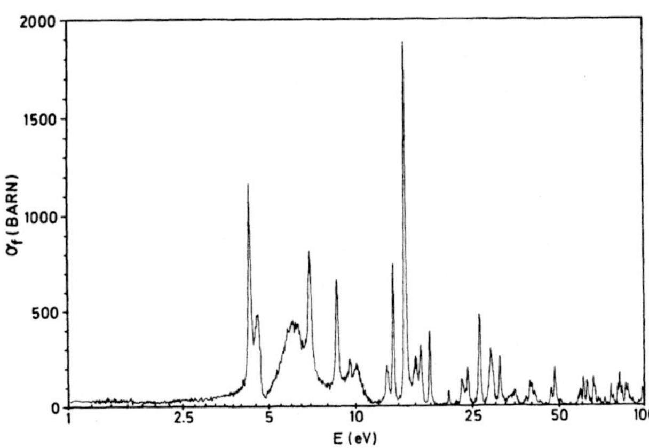

Fig. 2. [241]Pu fission cross-section in the neutron energy region from 1 to 100 eV

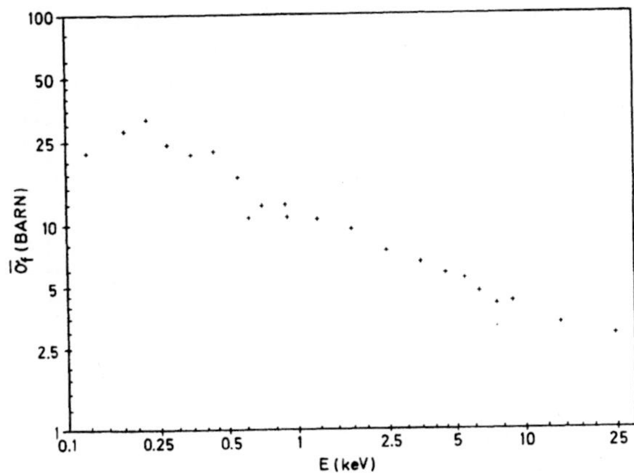

Fig. 3. $\bar{\sigma}_f$-values for [241]Pu in function of the neutron energy from 100 eV up to 30 keV

Table 2 Comparison of the Main Characteristics and the Normalization Methods Applied in the Considered [241]Pu Fission Cross-Section Measurements

References	Main Characteristics and Normalization Method
Watanabe [8]	σ_f normalized at a peak value of 406 b for the 6-eV resonance; chopper measurement, 8.5-m flight path; gas-scintillation chamber; neutron flux determined with BF_3 counters.
Moore [9]	σ_f normalized at a peak value of 406 b for the 6-eV resonance; Linac measurement; 10.56-m flight path; gas-scintillation chamber; neutron flux determined with BF_3 counters.
James [10]	σ_f normalized at 0.0253 eV to σ_f° = 1010 b via a linear least-squares fit of $\sigma_f\sqrt{E}$ from 0.02 to 0.05 eV; Linac booster measurement; 5- and 15-m flight path; silicon surface-barrier detectors; neutron flux determined with BF_3 counters.
Simpson [11]	nuclear explosion; solid state detectors; neutron flux determined with ^{6}Li foil viewed by solid state detectors.
Migneco [12]	σ_f normalized to $\int_{4.65eV}^{10eV} \sigma_f(E)\frac{dE}{E}$ = 193.6 b (Hennies [14]); Linac measurement; 30.617-m flight path; liquid scintillation detector; uncorrected for 0.87 % [239]Pu impurity; neutron flux determined with BF_3 counters.
Blons [12]	σ_f normalized to $\int_{20eV}^{70eV} \sigma_f(E)dE$ = 2367.5 b.eV (James [15]); Linac measurement; 10.89- and 50.07-m flight path; gaseous scintillator; uncorrected for 0.87 % [239]Pu contribution; neutron flux determined with BF_3 counters.
Wagemans [1]	σ_f normalized to σ_f° = (1015 ± 7) b by a linear least-squares fit of $\sigma_f\sqrt{E}$ from 0.02481 to 0.02586 eV; Linac measurement; 8.1-m flight path; silicon surface-barrier detectors; neutron flux determined with thin ^{10}B-foil viewed by a surface-barrier detector.
Carlson [2]	σ_f normalized to an average cross-section of 1015 b in the energy interval from 0.0203 to 0.0303 eV; Linac measurement; fission chamber; neutron flux determined with ^{6}Li glass scintillators. Data in keV-region obtained via two-step normalization.
Weston [3]	σ_f normalized to $\int_{0.02eV}^{0.03eV}\sigma_f\sqrt{E}dE$ = 1.603 b.$\sqrt{eV}$ which corresponds to σ_f° = 1007.8 b; Linac measurement; 20- and 85-m flight path; fission neutron scintillation detector; neutron flux determined with $^{10}BF_3$ ionization chamber. Data in keV-region obtained via two-step normalization.
This work	σ_f normalized to $\int_{0.02eV}^{0.10eV}\sigma_f dE$ = 60.11 b.eV (ref.[1]) which corresponds to σ_f° = 1015 b; Linac measurement; 9.3- and 30.45-m flight path; silicon surface-barrier detectors; neutron flux determined with thin ^{10}B-foil viewed by a surface-barrier detector.

Table 3 Comparison of the Original Normalization Integrals or Cross-Sections Used in the Experiments Considered with the Corresponding Values Deduced from the Present Work

Reference	Original normalization (O)	Present value(P)	Ratio P/O	$I_1 = \int_{12eV}^{20eV} \sigma_f dE$ (b.eV)	1379 b.eV/I_1	Δ(%) *
Watanabe[8]	σ_f(6 eV) = 406 b	430 b	1.059	1395	0.989	7
Moore[9]	σ_f(6 eV) = 406 b	430 b	1.059	1388	0.994	6.5
James[10]	σ_f^o = 1010 b	1015	1.005	1342	1.028	2.3
Migneco[12]	$\int_{4.65eV}^{10eV} \sigma_f dE/E$ = 193.6 b	188.5	0.974	1329	1.037	6.3
Blons[13]	$\int_{20eV}^{70eV} \sigma_f dE$ = 2367.5 b.eV	2557	1.080	1335	1.033	4.7
Wagemans[1]	σ_f^o = 1015	1015	1	1363	1.012	1.2
Carlson[2]	σ_f^o = 1015	1015	1	1294	1.066	6.6
Weston[3]	$\int_{0.02eV}^{0.03eV} \sigma_f \sqrt{E} dE$ = 1.603 b.$\sqrt{eV}$	1.638	1.022	1377	1.001	2.1
Weston[3]	σ_f^o = 1007.8	1015	1.007			
This work	σ_f^o = 1015	1015	1	1379	1	0

* $\Delta = | P/O - 1379/I_1 | \times 100$

a ^{10}B-overlap filter. In the three cases an Al-window was present in the beam, which is responsible for the absorption dips at 5.95 and 34.7 keV. A few conclusions can be drawn when intercomparing these three graphs. The first one is that Cd is not a suitable overlap filter when one wants to collect data between 40 eV and about 3 keV, because of the large number of secondary resonances appearing in this region. The second one is the drastic flux reduction below 100 eV when using a boron overlap filter, which deteriorates the signal-to-noise ratio in this energy region. Obviously, this energy region, and especially the region below 10 eV, is not the best place to normalize cross-sections if a boron-overlap filter has been used, which was e.g. the case for refs. [12/13].

As can be seen in Table 1, the present $\bar{\sigma}_f$-data are in fair agreement with the ENDF-B4 and the JENDL evaluations. Since the ENDF-B5 $\bar{\sigma}_f$-data were not available we performed a graphical comparison between our results and the ENDF-B5 curves reported by Weston and Wright [16]. Within the limited accuracy of this method, the agreement seems to be good.

In Table 4 the secondary normalization integrals $I_1 = \int_{12eV}^{20eV} \sigma_f dE$ and $I_2 = \int_{0.1keV}^{1keV} \sigma_f dE$ calculated from the various experimental and evaluated data files are compared. For I_1 the extreme values differ by almost 8 %; for I_2 this difference increases up to 20 %. Another important number is the ratio I_2/I_1, which in the case of consistent data sets, should remain constant within the combined errors. This is clearly the case for our data, those of James [10], Carlson er al. [2] and Weston and Todd [3], which were all obtained from measurements without overlap filter and normalized in the thermal region. In the case of Migneco et al. [12] and Blons [13], I_2/I_1 ratios are obtained which are about 10 % smaller. This might be a consequence of the use of a boron overlap filter in their experiments. Also in the case of the ENDF-B4 evaluation, a rather divergent I_2/I_1 ratio has been obtained.

In Table 5 the fission cross-section measurements normalized in the thermal region are all renormalized to a thermal fission cross-section of 1015 b, and the renormalized value I_1' and I_2' of the integrals mentioned above are given. Average values of

Table 4 Comparison of the values of the secondary normalization integrals $I_1 = \int_{12eV}^{20eV} \sigma_f dE$ and $I_2 = \int_{0.1keV}^{1keV} \sigma_f dE$

Reference	I_1(b.eV)	I_2(b.keV)	I_2/I_1
ENDF-B4	1308	16.35	12.50
JENDL2	1349	15.15	11.23
Watanabe[8]	1395		
Moore[9]	1388		
James[10]	1342	15.66	11.67
Simpson[11]		16.87	
Migneco[12]	1329	13.58	10.22
Blons[13]	1335	13.86	10.38
Wagemans[1]	1363		
Carlson[2]	1294	14.70	11.36
Weston[3]	1377	16.23	11.79
This work	1379	15.50	11.24
Exp. average	1356	15.20	11.11

$\int_{12eV}^{20eV} \sigma_f dE$ = 1354 b.eV and $\int_{0.1keV}^{1keV} \sigma_f dE$ = 15.57 b.keV with estimated errors of 2 % resp. 4 % are obtained.

Conclusion

In the present work, $\bar{\sigma}_f$-values for ^{241}Pu(n,f) in the neutron energy region from 10 meV up to 30 keV are reported which have been obtained in one single experiment. These data are in excellent agreement with our previous results [1] and in agreement with the results recently reported by Weston and Todd [3]. Possible explanations for the discrepancies with other literature values are discussed.

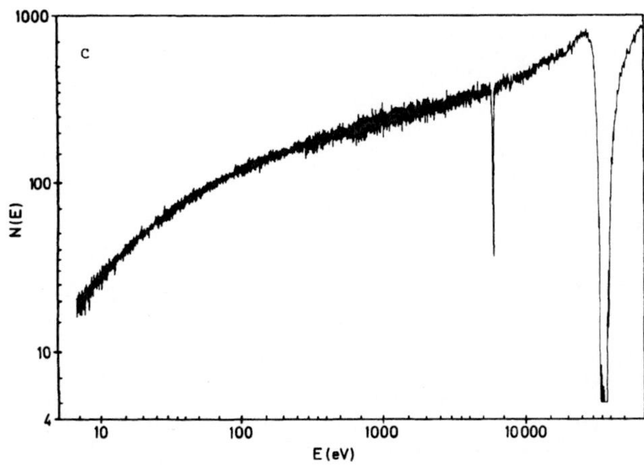

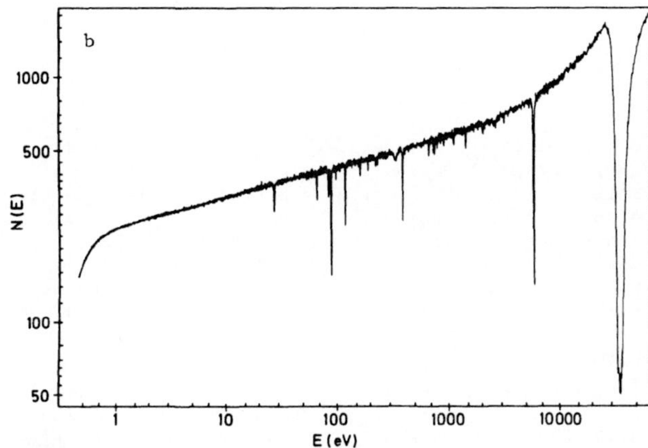

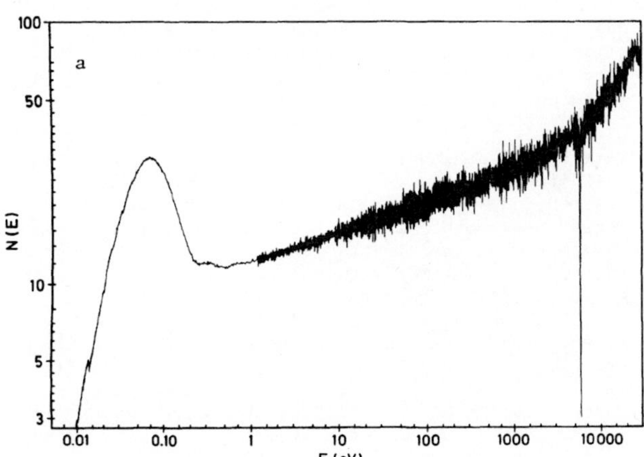

Fig. 4. $^{10}B(n,\alpha)$ counting rates as a function of the neutron energy (a) without overlap filter (b) with a cadmium overlap filter (c) with a boron overlap filter

Acknowledgements

Mr. R. Barthélémy and J. Van Gils are acknowledged for their help during the experiments and the analysis of the data.

Thanks are also due to Dr. G. Coddens from the NEA Data Bank in Saclay for providing us with integrated data from the ^{241}Pu files.

Table 5 Renormalized values of the fission integrals $I_1 = \int_{12eV}^{20eV} \sigma_f dE$ and $I_2 = \int_{0.1keV}^{1keV} \sigma_f dE$

Reference	Renormalization factor	I_1'(b.eV)	I_2'(b.keV)
James [10]	1.005	1349	15.74
Wagemans[1]	1	1363	
Carlson[2]	1	1294	14.70
Weston[3]	1.007	1387	16.34
This work	1	1379	15.50
Average		1354	15.57

References

1. C. Wagemans, A. Deruytter, Nucl. Sc. Eng. 60, 44 (1976)
2. G. Carlson, J. Behrens, J. Czirr, Nucl. Sc. Eng. 63, 149 (1977)
3. L. Weston, J. Todd, Nucl. Sc. Eng. 65, 454 (1978)
4. C. Bastian, Internal Report CBNM
5. M. Sowerby, B. Patrick, C. Uttley, K. Diment, Int. Conf. on Nucl. Data for Reactors, Helsinki, I, 161 (1969)
6. H. Lemmel, Proc. Conf. Nucl. Cross-Sections and Techn., Washington, NBS Special Publication 425, I, 286 (1975)
7. C. Wagemans, G. Coddens, H. Weigmann, R. Barthélémy, Ann. Nucl. En. 7, 495 (1980)
8. T. Watanabe, O. Simpson, Phys. Rev. B133, 390 (1964)
9. M. Moore, O. Simpson, T. Watanabe, J. Russell, R. Hockenbury, Phys. Rev. B135, 945 (1964)
10. G. James, Nucl. Phys. 65, 353 (1965)
11. O. Simpson, R. Fluharty, M. Moore, N. Marshall, B. Diven, A. Hemmendinger, Proc. Conf. Neutron Cross-Sections and Technology, Washington, II, 910 (1966)
12. E. Migneco, J. Theobald, J. Wartena, Proc. Conf. Nucl. Data for Reactors, Helsinki, I, 437 (1969)
13. J. Blons, Nucl. Sc. Eng. 51, 95 (1973)
14. H. Hennies, Proc. Conf. Nucl. Data for Reactors, Paris, II, 333 (1966)
15. G. James, Proc. Conf. Nucl. Data for Reactors, Helsinki, I, 267 (1969)
16. L. Weston, R. Wright, Proc. Conf. on Nucl. Cross-Sections for Techn., NBS Special Publication 594, 464 (1979)

NEUTRON INDUCED FISSION CROSS SECTION OF ^{244}Pu

M. S. Moore*, C. Budtz-Jørgensen, H.-H. Knitter, C. E. Olsen*, J.A. Wartena, and H. Weigmann

Commission of the European Communities
Joint Research Centre
Central Bureau for Nuclear Measurements
Geel, Belgium

The neutron induced fission cross section of ^{244}Pu has been measured in the energy range from 1 keV to 8 MeV relative to ^{235}U(n,f). The CBNM linear accelerator has been used as a pulsed neutron source. Fission events were detected with a parallel plate ionization chamber with ns-timing, thus providing an effective time-of-flights resolution of 0.4 ns/m.
In the energy range from 1 keV to 100 keV a number of narrow structures in the fission cross section are observed which are interpreted as being caused by class II states in the second minimum of the deformation potential. The conditions of coupling of the class II states with the class I compound nuclear resonances and the probable class II parameters are discussed.

Introduction

Among the actinides which show intermediate structure in the sub-barrier neutron induced fission cross section, ^{244}Pu is one of the worst known isotopes. In the sub-barrier region only one measurement exists performed in 1971 using an underground nuclear explosion as a pulsed neutron source [1]. For MeV neutron energies 2 more measurements exist [2-3], but the data are rather discrepant, the average cross section of ref.[2] being 20 to 30% higher than the other two data sets in most of the energy range. In order to improve the knowledge of systematic trends in fission properties of the actinides, we have performed a series of measurements of the ^{244}Pu fission cross section in the neutron energy range from 50 eV to 8 MeV.

Experimental Method

The experimental set up is described in detail elsewhere [4] and will only be summarized here.

Most of the measurements were performed at the neutron time-of-flight spectrometer of the 150 MeV electron linac of CBNM, Geel. The accelerator was operated with a burst width of 4.5 ns (F.W.H.M.). With a neutron flight path of 25m, typical figures for the overall energy resolution (F.W.H.M.) were 19 eV, 260 eV and 6.1 keV at 10 keV, 100 keV and 1 MeV neutron energy, respectively. Below 100 keV the largest contribution to this resolution is the uncertainty of the neutron moderation time in the polyethylene moderators.

Fission fragments were detected in a parallel plate ionization chamber similar to the one described by Knitter and Budtz-Jørgensen [5]. The Pu-samples had been prepared at Los Alamos National Laboratory by painting layers of 180 $\mu g/cm^2$ of PuO_2 (isotopic purity 98.58%) onto two 12.5 μm thick stainless steel backings. The diameter of the deposits was 10 cm. For the measurement of the neutron flux, a 390 $\mu g/cm^2$ $^{235}U_3O_8$ layer was separately placed into the fission chamber.

Data were accumulated for a total measuring time of about 750 h, including runs for background determination with "black resonance" filters of Co, Na and S.

In order to check the normalization of the cross section measured at the linear accelerator, two additional measurements of the fission cross section of ^{244}Pu at two neutron energies (2.0 and 2.5. MeV) were performed at the 7 MeV van de Graaff accelerator of CBNM. For these measurements 2.8 cm diameter samples of 150 $\mu g/cm^2$ PuO_2 and 213 $\mu g/cm^2$ $^{235}U_3O_8$ on 0.5 mm stainless steel backings were placed back to back in a small fission chamber. The chamber was placed in the zero degree direction such that the distance from the neutron producing target to the fission samples was 8 cm. Time-of-flight techniques were used to separate neutron induced fission from background events.

Data Analysis

Conversion of the measured fission yield to cross sections requires the knowledge of the "effective mass" of the Pu-samples, i.e. the product of the mass of ^{244}Pu times the efficiency of the ionization chamber for fission fragments at a given bias setting. It is obtained from the rate of spontaneous fission, provided the spontaneous fission half life of ^{244}Pu is known. Fields et.al. [6] give a value of

$$T_{1/2}\ (\text{S.F. } ^{244}\text{Pu}) = (6.55 \pm 0.32)\ .\ 10^{10}\ \text{y}$$

In the present investigation, the spontaneous fission half-life has been re-determined by spontaneous fission and α -counting as well as α-spectrometry measurements performed on the small diameter Pu-sample together with mass-spectrometric measurements performed at Los Alamos National Laboratory. Details are given in ref.[4]. The result obtained here is

$$T_{1/2}\ (\text{S.F. } ^{244}\text{Pu}) = (6.56 \pm 0.30)\ .\ 10^{10}\ \text{y}$$

in excellent agreement with the above value of Fields et. al.

The masses of the ^{235}U samples have been determined by α-counting and mass spectrometric analysis to an accuracy of 1%.

From the known effective masses of the ^{244}Pu and ^{235}U samples and the measured fission yield ratios, the ^{244} Pu fission cross section is obtained, using the ENDF/B-V data for the ^{235}U fission cross section above 25 keV neutron energy, and ^{235}U fission cross section integrals as measured by Wagemans and Deruytter[7] in the lower resonance region.

Results

The fission cross section of ^{244}Pu between 0.4 and 8 MeV neutron energy as obtained in the present work, is shown in fig.1. The error bars shown in fig.1 represent statistical uncertainties only. The additional normalization uncertainty of $\pm$ 5% is mainly due to the uncertainty in the "effective mass" of the Pu-sample and hence to the spontaneous fission half-life redetermined in the present experiment. The numerical data represented by the data points of fig.1 are available from the NEA data bank. The broken

* Los Alamos National Laboratory, Los Alamos, N.M., USA

K. H. Böckhoff (ed.), Nuclear Data for Science and Technology, 74–77.

line also shown in the figure represents a statistical model calculation which will be referred to in section 5.

Table 1 lists the measured ^{244}Pu fission cross sections averaged over a number of broad energy intervals.

Table 1: Average ^{244}Pu Fission Cross Section

Energy Interval [MeV]	Cross Section [b]	Statistical Uncertainty [b]	Total Uncertainty [b]
0.1 - 0.2	0.0088	0.0005	0.0007
0.2 - 0.3	0.0202	0.0005	0.0011
0.3 - 0.4	0.0442	0.0007	0.0023
0.4 - 0.5	0.0974	0.0010	0.0050
0.5 - 0.6	0.1955	0.0015	0.0099
0.6 - 0.7	0.4085	0.0025	0.021
0.7 - 0.8	0.7127	0.0038	0.036
0.8 - 1.0	1.188	0.0044	0.060
1.0 - 2.0	1.507	0.0035	0.075
2.0 - 3.0	1.389	0.0064	0.070
3.0 - 4.0	1.297	0.0090	0.065
4.0 - 5.0	1.162	0.012	0.059
5.0 - 6.0	1.220	0.020	0.064
6.0 - 7.0	1.711	0.035	0.092
7.0 - 8.0	1.899	0.057	0.111

In fig.2 the present data (upper curve) are compared to the cross section calculated from the σ (^{244}Pu)/ σ(^{235}U)- ratio data of Behrens et.al.[2] using the ENDF/B -V data for ^{235}U (lower curve). For the sake of clarity of the figure, error bars have been omitted from fig.2. As is seen from the figure, the shapes of the two cross section curves are very similar; especially below 0.8 MeV neutron energy structures tend to coincide. However, the absolute normalization of the present cross section is higher than the one of Behrens et. al. by about 12% on the average, which in most of the energy range of fig.2 clearly exceeds the combined uncertainties.

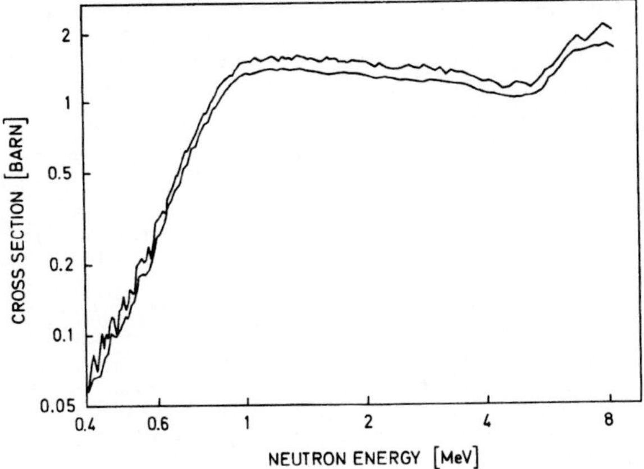

Fig. 2. Comparison of the present fission cross section (upper curve) to the one obtained from the σ(^{244}Pu)/σ(^{235}U) ratio data of Behrens et al.[2] with the aid of the ENDF/B-V data for ^{235}U(lower curve).

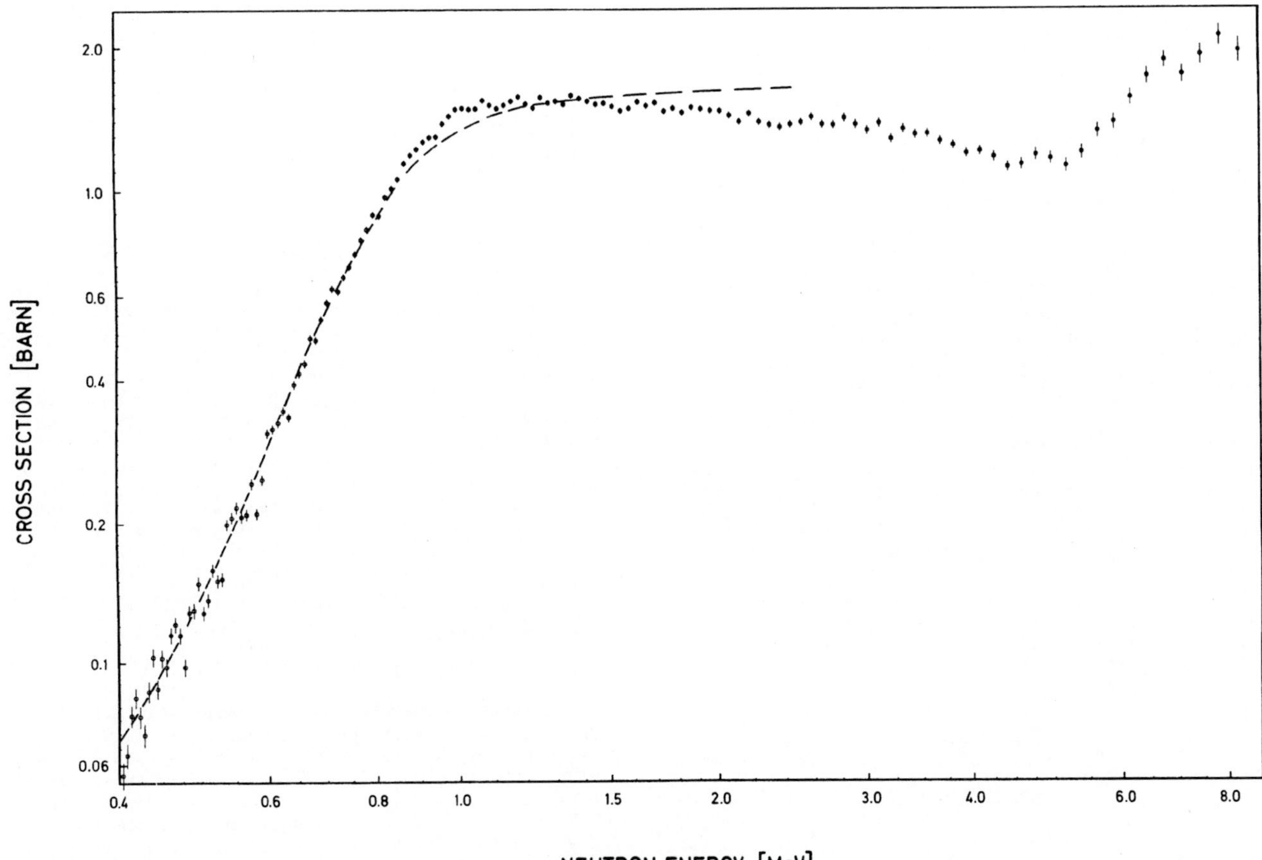

Fig. 1. The measured fission cross section of ^{244}Pu between 0.4 and 8 MeV neutron energy. The error bars represent statistical uncertainties only; a common normalization uncertainty of ±5% has to be added. The broken curve represents the statistical model calculation referred to in section 5.

It was this discrepancy which prompted us to perform the supplementary measurements at the van de Graaff described above. They confirmed the absolute normalization of the linear accelerator measurements to within 1.5%. In fact the data representing the present measurements in figs. 1 and 2 and in table 1 are based on a normalization which is the weighted average of the normalization of the linear accelerator - and van de Graaff measurements.

It should be noted that the measurements performed at the linear accelerator and the van de Graaff were essentially independent: neutron sources, experimental arrangements, detectors, electronics, and the Pu- and U-samples were all different. The only common feature was the method by which the masses of the Pu-samples have been determined : in both cases, it is based on the α-activity and mass-spectrometric measurements used to re-determine the ^{244}Pu spontaneous fission half-life. The resulting value, however, is in perfect agreement with the one given by Fields et. al. [6].

In the neutron energy range from 1 to 100 keV a number of narrow structures are observed in the ^{244}Pu fission cross section. Fig. 3 shows the fission cross section in several small energy intervals where the most prominent structures occur. In table 2 the fission area

$$A_f = \int \sigma_f \, dE$$

integrated over individual peaks in the fission cross section, is given for most of the observed structures. Apart from the peaks listed in the table, additional weak ones have been observed at 10.02, 11.04, 21.9, 28.2, 28.9, 35.0, 38.7, and 52.2 keV neutron energy. Their fission area was too small to be determined with sufficient confidence, i.e. the statistical error on the fission area was larger than a tolerated 25%. This required statistical accuracy defines a bias for "accepted" peaks which as a function of neutron energy E_n above 5 keV may be expressed as

$$\ln B = 0.45 \cdot \ln E_n + 0.58 \qquad (1)$$

where E_n is in keV and B in b.eV.

Table 2: Fission Areas of Resonances in the ^{244}Pu Fission Cross Section.

E_n[keV]	A_f[b.eV]*	Remarks
1.6504	35.8 ± 2.0	$g\,\Gamma n \Gamma_f/\Gamma$ = 14.4 meV
3.691	9.1 ± 0.9	$g\,\Gamma n \Gamma_f/\Gamma$ = 8.2 meV;doubl.?
5.422	4.9 ± 0.8	multiplet?
7.114	9.3 ± 1.1	doublet? degeneracy?
7.554	11.6 ± 1.2	multiplet?
11.30	7.7 ± 1.3	
12.00	12.7 ± 1.6	
14.42	37 ± 2.9	
15.58	11.9 ± 2.0	
17.85	32 ± 2.8	
18.26	11.9 ± 1.7	
32.7	19.4 ± 2.7	
48.9	13.3 ± 3.0	
50.8	9.7 ± 2.4	
55.8	15.3 ± 3.0	
58.5	21.9 ± 4.0	
59.7	20.4 ± 3.6	
76.4	26.8 ± 3.8	
77.2	17.2 ± 3.2	
82.2	22.5 ± 4.4	

* Statistical error only; additional normalization error is ± 5%.

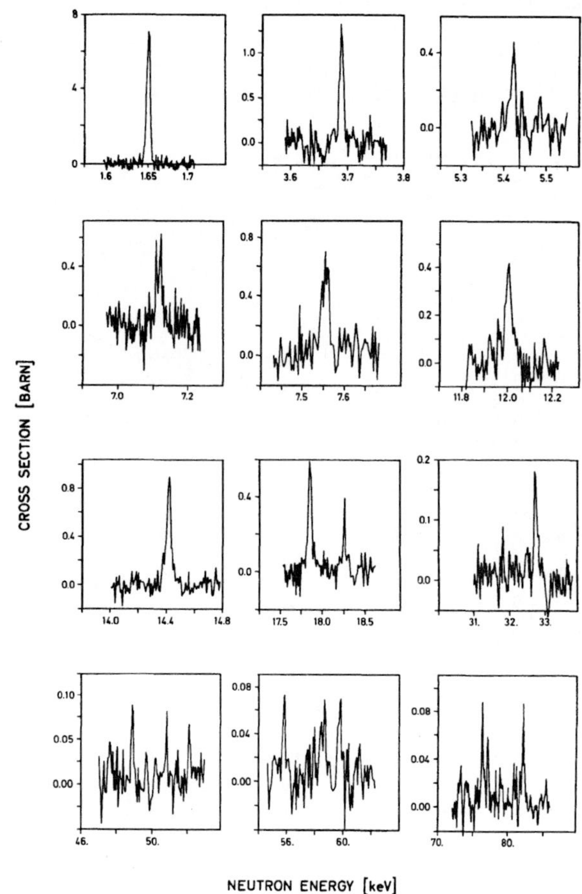

Fig. 3. The fission cross section of ^{244}Pu in selected energy intervals between 1 keV and 100 keV.

Discussion

A statistical model calculation has been performed to reproduce the measured fission cross section in the neutron energy range between 0.4 and 2.4 MeV. For this calculation, the compound nucleus formation cross section has been taken from a recent evaluation of Antsipov et al.[8] for ^{242}Pu. The fission probability has been calculated with a code developed by Jensen [9], the main feature of which is a deformation - dependent shell correction and appropriate collective enhancement factors of the various level densities.The best reproduction of the experimental fission cross section, indicated as the broken line in fig.1, has been obtained with the following barrier parameters :

height of inner barrier:	E_A = 5.67 MeV
curvature of inner barrier:	$\hbar\omega_A$= 0.64 MeV
height of outer barrier:	E_B = 5.33 MeV
curvature of outer barrier:	$\hbar\omega_B$= 0.52 MeV

The height and curvature of the higher (inner) barrier are quite well determined (to about 0.1 and 0.05 MeV respectively) by the position of the threshold and the slope of the cross section at threshold. The parameters of the lower (outer) barrier, however, are quite uncertain: they depend strongly on the assumed compound nuclear formation cross section and on the assumed density of transition states above the second saddle. For example, a change of the transition state density by a factor of two would call for a change in E_B by about 0.2 MeV.

Each of the structures in the ^{244}Pu fission cross section seen in fig.3 and listed in table 2 is assumed to be caused by an individual class II state in the second minimum of the deformation potential, very weakly coupled to neighbouring compound nuclear (class I) resonances. With the exception of the 1650.4 eV resonance which is discussed in detail in ref.[4], a possible class I fine structure is not sufficiently resolved to allow a detailed analysis. The only experimental observables are the total fission areas integrated over complete class II structures and the only way to compare them to a barrier model is on a statistical basis.

For this purpose a Monte-Carlo routine is used which generates sequences of class I and class II levels and their parameters (class II fission and coupling widths, class I neutron widths and fine structure fission widths) from the relevant distribution laws. S- and p-wave resonances have been taken into account with neutron strength functions of $1 \cdot 10^{-4}$ and $1.8 \cdot 10^{-4}$, respectively. From the sequence of resonances constructed this way the total fission area for each of the generated class II structures is calculated. The whole procedure is repeated a sufficient number of times to obtain meaningful averages and distributions.

Several such resonance simulations have been performed with different barrier parameters. In table 3 the results of three simulations are compared to the experimental data: the numbers of class II structures with fission areas larger than the above bias function equ.(1), and the average area of these structures are given for two neutron energy intervals. The figures given in parenthesis indicate the widths of the distributions of the respective quantities as obtained from the sampling procedure, expressed in standard deviations.

As is seen from the table, the average area of structures above bias is not very sensitive to the barrier parameters, but the number of these structures is : Simulation 1 which uses the barrier parameters as obtained from the statistical model calculation for the threshold region, gives much too few levels. For simulation 2, E_B has been reduced to E_B = 4.9 MeV; this is in line with the above statement that the parameters of barrier B are not well determined by the statistical model calculation. However, the number of predicted levels is still too small.A further reduction of E_B is not justified: It would lead to class II fission widths of more than 15 eV, in contradiction to the observation that most structures at lower energies (E_n < 30 keV, where the experimental resolution is sufficient to tell) consist of a small number (1 - 3) of fine structure resonances. Thus we have instead increased $\hbar\omega_A$ to $\hbar\omega_A$ = 0.7 MeV in simulation 3 which reproduces the experimentally observed number of structures above bias very well. We recall that E_B = 4.9 MeV is the smallest acceptable value, and that a larger E_B would call for a still larger $\hbar\omega_A$. Indeed, the value of $\hbar\omega_A$ = 0.7 MeV is closer to commonly accepted values [10] than the rather small $\hbar\omega_A$ which was necessary to reproduce the fission cross section in the threshold region.

The inconsistencies found here are similar to those observed for other nuclei [11]. They call for a detailed investigation and comparison of the statistical model codes used to describe fission cross section in the threshold region. On the other hand, the necessity to use different values for $\hbar\omega_A$ near threshold and in the sub-barrier region, may indicate that the "shape" of the first barrier is not well represented by the usually adopted inverted parabola.

References

1. G.F. Auchampangh, J.A. Farrell and D.W. Bergen, Nucl. Phys. A 171 (1971) 31.
2. J.W. Behrens, R.S. Newbury and J.W. Magana, Nucl. Sci. Eng. 64 (1978) 433.
3. B.M. Gokhberg, S.M. Dubrovina and V.A. Shigin,Y.F. 25 (1977)21;transl.Sov.J. Nucl.Phys.25 (1977) 11.
4. M.S. Moore, J.A. Wartena, H. Weigmann, H.-H. Knitter and C. Budtz-Jørgensen, to be published.
5. H.-H. Knitter and C. Budtz-Jørgensen, Proc. of Int. Conf. on Nuclear Cross Sections for Technology, Knoxville 1979, NBS Special Publication 594, Washington, D.C. 1980.
6. P.R. Fields, A.M. Friedman, J. Milsted, J. Lerner, C.M. Stevens, D. Metta, and W.K. Sabine, Nature 212 (1966) 131.
7. C. Wagemans and A.J. Deruytter, Ann. Nucl. Energy 3 (1976) 437.
8. G.V. Antsipov, L.A. Bakhanovich, A.R. Benderskij, V.A. Zenovick, A.B. Klepatskij, V.A. Konshin, and E. Sh. Sukhovitskij, report INDC (CCP)-150/LJH (1979) 42.
9. A.S. Jensen, Proc. of Int. Conf. on Neutron Physics and Nuclear Data, Harwell 1978,p.378.
10. S. Bjørnholm and J.E. Lynn, Rev. Mod. Phys. 52 (1980) 725.
11. H. Weigmann, in "Nuclear Theory for Applications-1980", report IAEA-SMR-68/I (1981) p.91.

The authors are indepted for the use of the target materials to the Office of Basic Energy Sciences, U.S. Department of Energy, through the transplutonium element production facilities of the Oak Ridge National Laboratory

Table 3: Monte-Carlo Simulations of Sub-barrier Fission Structures.

	E_A [MeV]	$\hbar\omega_A$ [MeV]	E_B [MeV]	$\hbar\omega_B$ [MeV]	average area [b · eV]		number of levels	
E_n [keV]					1 - 20	20 - 80	1 - 20	20 - 80
simul. 1	5.67	0.64	5.33	0.52	16 (15)	14 (3)	0.94 (1.0)	0.66 (0.7)
simul. 2	5.67	0.64	4.9	0.52	19 (20)	26 (11)	6.2 (2.4)	5.3 (2.1)
simul. 3	5.67	0.7	4.9	0.52	21 (17)	25 (7)	10.7 (2.8)	9.9 (2.8)
experim.					17	19	11	9

MESURE DE $\bar{\nu}_p$ et $\bar{E}_\gamma$ POUR LA FISSION DE ^{232}Th, ^{235}U ET ^{237}Np INDUITE PAR DES NEUTRONS D'ENERGIE COMPRISE ENTRE 1 ET 15 MeV

J. Fréhaut, A. Bertin, R. Bois

Service de Physique Neutronique et Nucléaire
Centre d'Etudes de Bruyères-le-Châtel
B.P. N° 561
92542 MONTROUGE-CEDEX, France

The average number of prompts neutrons, $\bar{\nu}_p$, and the average prompt gamma-ray energy, $\bar{E}_\gamma$, have been measured for the neutron induced fission of ^{232}Th, ^{235}U, and ^{237}Np in the energy range from 1 up to 15 MeV. Data were obtained relative to the corresponding quantities $\bar{\nu}_{Cf}$ and $\bar{E}_{\gamma Cf}$ for the ^{252}Cf spontaneous fission. For ^{235}U and ^{237}Np, $\bar{\nu}_p$ is approximately a linear function of energy with small changes in slope around 4 MeV and around 6 MeV and 12 MeV (second and third chance fission). For ^{232}Th the second chance fission results in a strong increase of $\bar{\nu}_p$, due to the increase of the fragment total kinetic energy with excitation energy for that nucleus. A linear correlation observed between $\bar{E}_\gamma$ and $\bar{\nu}_p$ is interpreted in terms of a competition between neutron and gamma-ray emission, connected with an increase of the average angular momentum of the fission fragments with excitation energy. The variance of the fission neutron multiplicity distributions is observed to be proportional to the excitation energy of the fissioning nucleus.

Introduction

Les mesures de $\bar{\nu}_p$ en fonction de l'énergie d'excitation ont jusqu'à présent surtout été réalisées sur les isotopes de l'uranium et du plutonium. Cependant, il s'avère nécessaire de disposer de mesures précises pour une plus large gamme de noyaux, afin de mieux étayer les évaluations semi-empiriques qui ont été proposées [1] ou de tester les modèles théoriques qui prédisent la variation de $\bar{\nu}_p$ en fonction de l'énergie d'excitation [2]. C'est dans cet esprit que nous avons réalisé des mesures pour la fission de ^{232}Th, ^{235}U et ^{237}Np induite par des neutrons d'énergie comprise entre 1,5 et 15 MeV.

D'autre part, nous avons mesuré simultanément, pour ces mêmes noyaux, les variations de l'énergie moyenne $\bar{E}_\gamma$ des rayons gamma prompts de fission en fonction de l' énergie d'excitation. A notre connaissance, aucune mesure n'avait été réalisée auparavant dans ce domaine. Des résultats partiels pour la mesure sur ^{237}Np ont déjà été publiés [3].

Méthode expérimentale

Les mesures ont été réalisées par la méthode de la chambre à fission associée à un gros scintillateur liquide chargé au gadolinium [4,5]. Deux expériences ont été réalisées. Dans la première, la chambre à fission contenait 100 mg de ^{237}Np et 100 mg de ^{235}U enrichi à 99,83% répartis sous forme d'oxydes en dépôts d'épaisseur 0,7 mg/cm^2 sur des disques de platine d'épaisseur 0,2 mm espacés de 1,5 mm. La chambre était remplie d'un mélange d'argon à 20% de méthane sous une pression de 3 bars. Dans la seconde expérience la chambre à fission contenait 1 g de ^{232}Th réparti en dépôts d'oxyde d'épaisseur 1 mg/cm^2 sur des disques de titane d'épaisseur 55 µ espacés de 0,5 mm. La chambre était remplie d'un mélange d'argon à 2% d'azote sous une pression de 7,5 bars. Les deux chambres contenaient en outre un dépôt de ^{252}Cf dont la fission spontanée a été utilisée comme référence pour les mesures de $\bar{\nu}_p$ et de $\bar{E}_\gamma$.

Les neutrons incidents ont été produits pour les deux expériences par les réactions T(p,n)^{3}He et D(d,n)^{3}He en utilisant des cibles gazeuses de 7,5 cm de longueur avec une pression comprise entre 1 et 2 bars et le faisceau de l'accélérateur tandem Van de Graaff de 7 MV. Le faisceau était pulsé, afin d'éliminer par la méthode du temps de vol les fissions induites par les neutrons issus de réactions secondaires dans les cibles.

Les valeurs relatives de $\bar{E}_\gamma$ ont été obtenues à partir de la mesure de l'aire du signal prompt délivré par le scintillateur liquide. Ce signal correspond à la détection du rayonnement gamma prompt de fission et des protons de recul créés par le ralentissement des neutrons de fission. Cette dernière composante, proportionnelle à $\bar{\nu}_p$, a été évaluée à partir des données de la réf. [6] et introduit au maximum une correction de 5% pour le rapport $\bar{E}_\gamma/\bar{E}_{\gamma Cf}$ de l'énergie gamma pour la fission provoquée des noyaux étudiés et la fission spontanée du ^{252}Cf.

Les valeurs expérimentales de $\bar{\nu}_p$ ont subi les corrections classiques [4,5] de bruit de fond, de temps mort (120 ns) et d'efficacité, déterminée en adoptant la valeur $\bar{\nu}_p$ = 3,732 ± 0,000 pour la fission spontanée du ^{252}Cf et en tenant compte de la différence des spectres en énergie des neutrons de fission. La contribution des rayons gamma retardés a été réduite en commençant le comptage des neutrons 600 ns après la fission [7]. Faute de données suffisantes, aucune correction n'a été effectuée pour la contribution résiduelle, que l'on peut estimer de l'ordre de - 0,2%. Cette contribution est d'ailleurs pratiquement compensée par l'effet lié à l'épaisseur des dépôts [8] dont on n'a pas tenu compte.

Résultats et Discussion

Les résultats sont donnés avec leur incertitude statistique dans le Tableau I et sont présentés sur les figures 1,2 et 3 respectivement pour ^{232}Th, ^{235}U et ^{237}Np. Il n'existait auparavant aucun résultat de précision comparable pour ^{232}Th. Les valeurs de $\bar{\nu}_p$ pour ^{235}U sont en bon accord avec nos mesures antérieures [4,7] et les autres résultats publiés [9]. Par contre les valeurs de $\bar{\nu}_p$ pour ^{237}Np sont inférieures de 3% aux deux séries de mesures existantes [10,11]. On retrouve, pour ces 2 noyaux la variation quasi linéaire de $\bar{\nu}_p$ en fonction de l'énergie E_n des neutrons incidents, caractéristique de tous les noyaux mesurés jusqu'à présent [9]. Par contre on observe pour ^{232}Th une remontée brutale de $\bar{\nu}_p$ lorsque la réaction (n,n'f) apparaît, avec une pente passant par un maximum de 0,35 neutron/MeV.

Pour les 3 noyaux, on remarque que $\bar{E}_\gamma$ augmente avec E_n en dessous du seuil de réaction (n,n'f) , puis

K. H. Böckhoff (ed.), Nuclear Data for Science and Technology, 78–81.

présente un plateau ou une diminution au voisinage de ce seuil. A plus haute énergie, on observe de nouveau une augmentation approximativement linéaire de $\bar{E}_\gamma$ avec E_n, suivie d'un autre plateau au voisinage du seuil de réaction (n,2n'f). Ces résultats pour $\bar{E}_\gamma$ peuvent être interprétés à partir de la compétition entre l'émission de neutrons et de rayons γ au cours de la désexcitation des fragments de fission. En dessous du seuil de réaction (n,n'f), on observe une relation linéaire entre $\bar{E}_\gamma$ et $\bar{\nu}_p$ et l'on obtient par la méthode des moindres carrés, en prenant une valeur de 7,06 MeV pour $\bar{E}_{\gamma Cf}$ [12] et en exprimant $\bar{E}_\gamma$ en MeV :

pour ^{232}Th $\bar{E}_\gamma = (0,64 \pm 0,07)\ \bar{\nu}_p + (4,48 \pm 0,20)$
pour ^{235}U $\bar{E}_\gamma = (0,98 \pm 0,08)\ \bar{\nu}_p + (4,37 \pm 0,21)$
pour ^{237}Np $\bar{E}_\gamma = (0,90 \pm 0,07)\ \bar{\nu}_p + (4,40 \pm 0,20)$

Le terme constant dans ces expressions correspond aux prévisions du modèle statistique [13]. Cependant, les neutrons de fission sont évaporés avec une énergie relativement faible (∿ 2 MeV) et ne réduisent pas de façon importante le spin des fragments. En fin de désexcitation, ceux-ci se retrouvent avec un moment angulaire relativement élevé et la probabilité d'émission de rayons γ devient importante, même si l'émission d'un neutron est encore énergétiquement possible. La relation linéaire qui existe entre $\bar{E}_\gamma$ et $\bar{\nu}_p$ traduit une augmentation de la valeur moyenne du spin des fragments avec l'énergie d'excitation. Nifenecker et Coll. avaient abouti à des conclusions similaires lors de l'étude de la fission spontanée du ^{252}Cf [6].

Au cours d'une réaction (n,n'f) une partie de l'énergie d'excitation du noyau composé est emmenée par le neutron évaporé avant la fission. Les fragments de fission émettent donc moins de neutrons et de rayons γ, d'où le plateau observé pour $\bar{E}_\gamma$ au voisinage du seuil de réaction (n,n'f). Par contre, on ne peut expérimentalement distinguer le neutron évaporé avant fission des neutrons de fission. Dans le cas de ^{235}U et ^{237}Np, l'augmentation de $\bar{E}_\gamma$ avec l'énergie d'excitation est approximativement compensée par une diminution de l'énergie cinétique $\bar{E}_K$ des fragments de fission : toute l'énergie fournie par le neutron incident sera utilisée pour l'émission de neutrons ; comme l'énergie de liaison d'un neutron dans les fragments et le noyau composé n'est pas très différente, on mesurera approximativement un même nombre moyen de neutrons pour la fission directe et la réaction (n,n'f). Pour ^{232}Th, il est bien connu [14] que $\bar{E}_K$ augmente avec l'énergie d'excitation avec une pente de 0,34 MeV/MeV. Compte tenu de l'augmentation de $\bar{E}_\gamma$, il n'y a que 55% de l'énergie fournie pour le neutron incident qui contribue à l'émission de neutrons de fission. Contrairement au cas précédent, on observera donc une différence de $\bar{\nu}_p$ d'environ 0,45 neutron entre la fission directe et la réaction (n,n'f), ce qui explique quantativement les résultats expérimentaux. Un raisonnement similaire dans le cas où $\bar{E}_K$ diminue quand l'énergie d'excitation augmente montre que $\bar{\nu}_p$ sera plus faible pour la réaction (n,n'f) : c'est vraisemblablement cet effet qui a été mis en évidence récemment pour la fission de ^{242m}Am [15].

Nous avons porté sur la figure 4 la variation de la variance σ^2 des distributions P(ν) des neutrons de

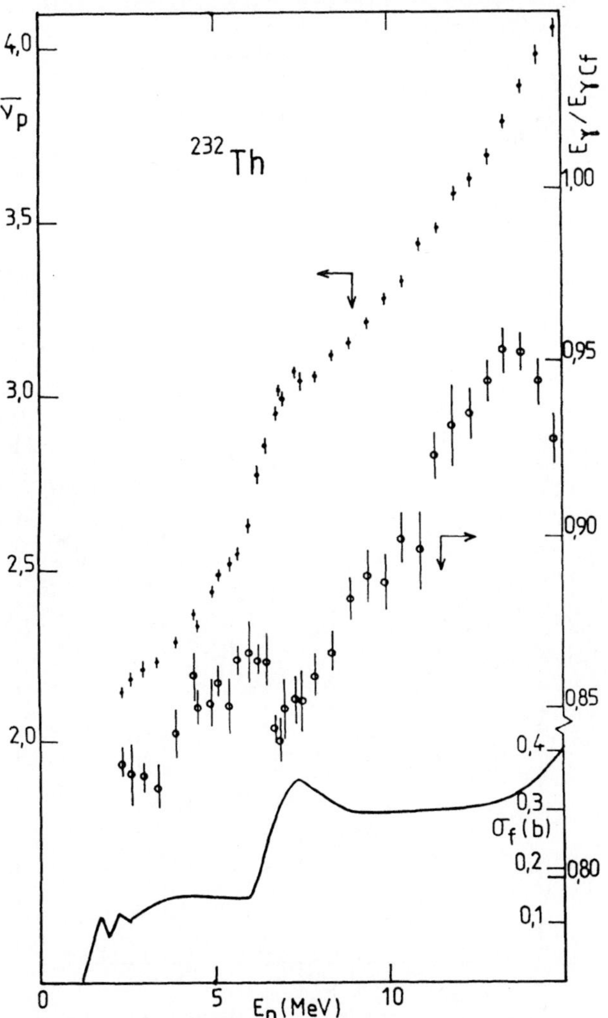

Fig. 1. Variation de $\bar{\nu}_p$, $\bar{E}_\gamma/\bar{E}_{\gamma Cf}$ et σ_F en fonction de l'énergie E_n des neutrons incidents pour la fission de ^{232}Th.

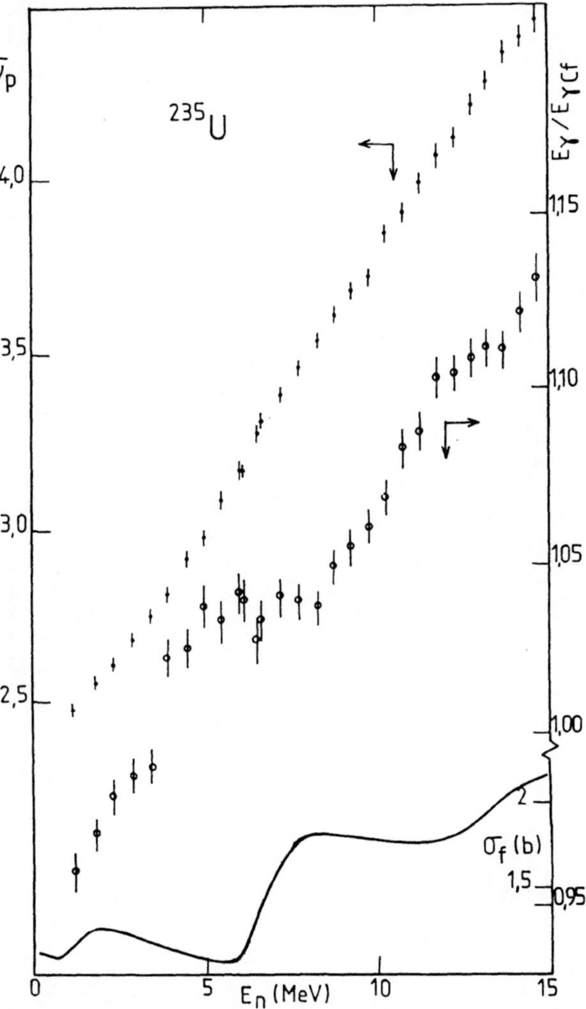

Fig. 2. Variation de $\bar{\nu}_p$, $\bar{E}_\gamma/\bar{E}_{\gamma Cf}$ et σ_F en fonction de l'énergie E_n des neutrons incidents pour la fission de ^{235}U.

fission en fonction de l'énergie des neutrons incidents pour les 3 noyaux étudiés. L'augmentation de σ^2 avec l'énergie d'excitation n'est pas surprenante, elle est liée à la nature statistique du phénomène de désexcitation des fragments de fission. Dans la réaction (n,n'f), le neutron évaporé avant fission diminue l'énergie d'excitation du noyau composé, qui émettra donc moins de neutrons dans la fission qui s'ensuit. La variance σ^2 sera donc plus faible pour cette réaction que pour la fission directe, d'où les plateaux observés entre 6 et 8 MeV sur la figure 4.

Nous avons porté sur la figure 5 la variation de σ^2 en fonction de $\bar{\nu}_p$ pour des énergies d'excitation inférieures au seuil de réaction (n,n'f) et pour l'ensemble des noyaux que nous avons étudiés jusqu'à présent. Pour tous ces noyaux sauf ^{232}Th, une seule droite d'équation $\sigma^2 = 0{,}21\ \bar{\nu}_p + 0{,}75$ ajuste les résultats expérimentaux. Pour ^{232}Th, on mesure des variances plus faibles que cette loi moyenne. Cela est probablement lié à l'observation d'une compétition neutron-gamma plus faible pour ce noyau.

Il faut également noter que l'ouverture de la voie (n,2n) peut avoir une influence sur $\bar{\nu}_p$. En effet, après émission d'un premier neutron, la probabilité de décroissance du noyau résiduel par fission - donc $\bar{\nu}_p$ - pourra dépendre fortement de son énergie d'excitation selon que l'émission d'un second neutron est énergétiquement possible ou non.

Au-dessus du seuil de réaction (n,2n'f) les structures

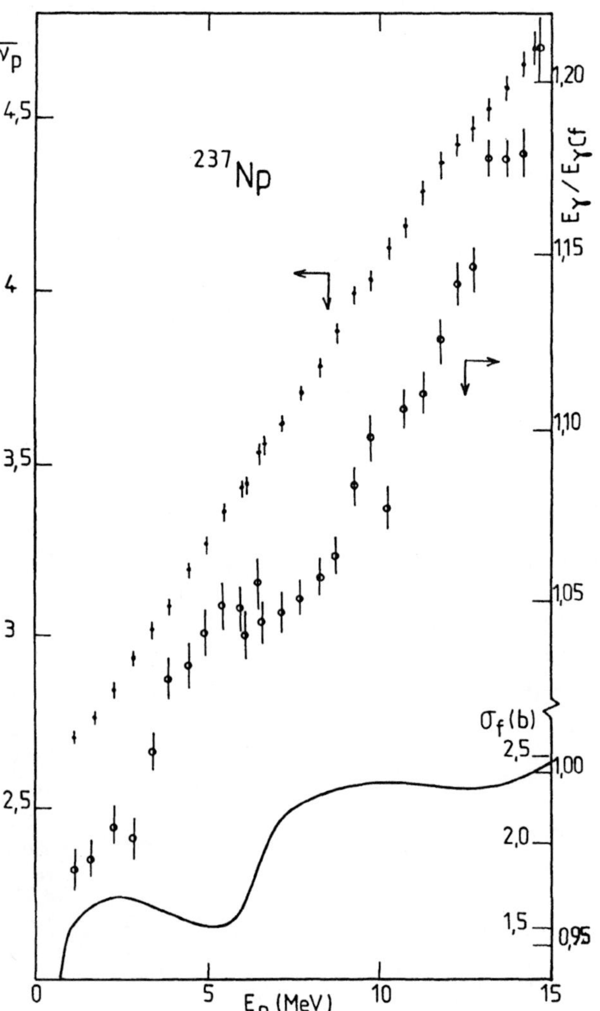

Fig. 3. Variation de $\bar{\nu}_p$, $\bar{E}_\gamma/\bar{E}_{\gamma Cf}$ et σ_F en fonction de l'énergie E_n des neutrons incidents pour la fission de ^{237}Np.

que l'on peut observer sur $\bar{\nu}_p$, $\bar{E}_\gamma$ et σ^2 peuvent s'interpréter de façon similaire aux structures observées au niveau de la réaction (n,n'f).

Enfin, pour tous les noyaux que nous avons étudiés jusqu'à présent (^{232}Th, ^{235}U, ^{238}U, ^{237}Np, ^{239}Pu, ^{240}Pu, ^{241}Pu) on observe une augmentation de la pente $d\bar{\nu}_p/dE_n$ au-dessus d'une énergie de neutrons incidents comprise entre $\sim$ 3,5 MeV et 4 MeV qui, pour l'instant, n'a pas reçu d'interprétation satisfaisante.

Références

1. R. Bois, J. Fréhaut, Rapport CEA-R-4791 (1976).
2. D.G. Madland, J.R. Nix, Nucl. Sci. Eng. 81 (1982) 213.
3. J. Fréhaut, R. Bois, A. Bertin, Rapport NEANDC (E) 219 "L" (1981).
4. M. Soleilhac, J. Fréhaut, J. Gauriau, J. Nucl. Energy 33 (1969) 257.
5. J. Fréhaut, G. Mosinski, R. Bois, M. Soleilhac, Rapport CEA-R-4626 (1974).
6. H. Nifenecker, C. Signarbieux, M. Ribrag, J.Poitou, J. Matuszek, Nucl. Phys. A189 (1972) 285.
7. J.W. Boldeman, J. Fréhaut, R.L. Walsh, Nucl. Sci. Eng. 63 (1977) 430.
8. J.W. Boldeman, J. Fréhaut, Nucl. Sci. Eng. 76 (1980) 49.
9. F. Manero, V.A. Konshin, Atomic Energy Review 10 (1972) 637.
10. L. Veeser, Phys. Rev. C17 (1978) 385.
11. V.G. Vorobeva, B.D. Kuzminov, V.V. Malinovsky, N.N. Semenova, INDC (CCP) 177/L (1982) 39 et INDC (CCP) 156/G (1980).

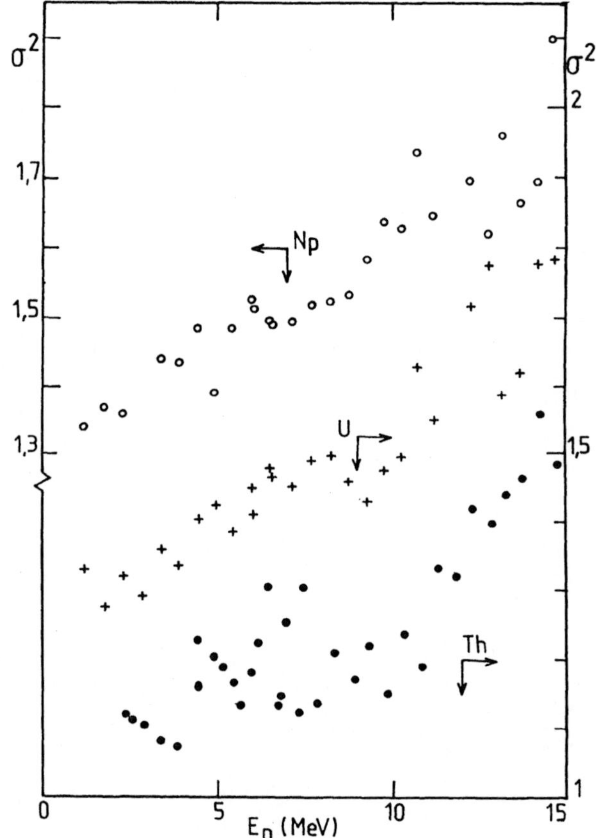

Fig. 4. Variation en fonction de l'énergie E_n des neutrons incidents de la variance des distributions P (ν) des neutrons de fission pour ^{232}Th, ^{235}U et ^{237}Np.

12. F. Pleasonton, R.L. Ferguson, H.W. Schmitt, Nucl. Phys. 231 (1973) 413.

13. R.B. Leachman, C.S. Kazek, Phys. Rev. 105 (1957) 1511.

14. J. Trochon, H. Abou-Yehia, F. Brisard, Y. Pranal, Nucl. Phys. A318 (1979) 63.

15. R.E. Howe, J.C. Browne, R.J. Dougan, R.J. Dupzyk, J.H. Landrum, Nucl. Sci. Eng. 77 (1981) 454.

Tableau I. Valeurs expérimentales de $\bar{\nu}_p$ et du rapport $R = \bar{E}_\gamma/\bar{E}_{\gamma Cf}$ pour la fission de ^{232}Th, ^{235}U et ^{237}Np induite par des neutrons d'énergie E_n comprise entre 1 et 15 MeV.

	^{232}Th			^{235}U		^{237}Np	
$E_n \pm \Delta E_n$ (MeV)	$\bar{\nu}_p \pm \Delta\bar{\nu}_p$	$R \pm \Delta R$	$E_n \pm \Delta E_n$ (MeV)	$\bar{\nu}_p \pm \Delta\bar{\nu}_p$	$R \pm \Delta R$	$\bar{\nu}_p \pm \Delta\bar{\nu}_p$	$R \pm \Delta R$
2,37 ± 0,02	2,146 ± 0,012	0,834 ± 0,004	1,14 ± 0,24	2,475 ± 0,018	0,960 ± 0,005	2,706 ± 0,021	0,972 ± 0,006
2,59 ± 0,08	2,184 ± 0,021	0,831 ± 0,009	1,73 ± 0,19	2,557 ± 0,017	0,971 ± 0,004	2,759 ± 0,020	0,975 ± 0,005
2,93 ± 0,02	2,215 ± 0,015	0,830 ± 0,004	2,30 ± 0,16	2,610 ± 0,019	0,982 ± 0,005	2,842 ± 0,022	0,985 ± 0,006
3,39 ± 0,06	2,236 ± 0,014	0,827 ± 0,006	2,85 ± 0,14	2,685 ± 0,019	0,988 ± 0,005	2,932 ± 0,022	0,981 ± 0,006
3,91 ± 0,06	2,289 ± 0,015	0,843 ± 0,007	3,38 ± 0,13	2,751 ± 0,021	0,991 ± 0,005	3,015 ± 0,025	1,007 ± 0,006
4,43 ± 0,05	2,369 ± 0,015	0,859 ± 0,007	3,91 ± 0,12	2,816 ± 0,022	1,023 ± 0,006	3,084 ± 0,024	1,027 ± 0,006
+,49 ± 0,12	2,338 ± 0,020	0,851 ± 0,006	4,43 ± 0,11	2,919 ± 0,022	1,026 ± 0,006	3,193 ± 0,025	1,031 ± 0,007
4,95 ± 0,05	2,440 ± 0,015	0,852 ± 0,007	4,95 ± 0,10	2,981 ± 0,023	1,038 ± 0,006	3,272 ± 0,025	1,041 ± 0,007
5,13 ± 0,09	2,490 ± 0,017	0,857 ± 0,005	5,47 ± 0,09	3,084 ± 0,022	1,036 ± 0,006	3,368 ± 0,025	1,049 ± 0,007
5,47 ± 0,05	2,519 ± 0,018	0,851 ± 0,008	5,99 ± 0,09	3,170 ± 0,023	1,042 ± 0,006	3,437 ± 0,025	1,048 ± 0,007
5,72 ± 0,07	2,547 ± 0,023	0,865 ± 0,004	6,50 ± 0,08	3,278 ± 0,025	1,028 ± 0,007	3,536 ± 0,028	1,056 ± 0,008
5,98 ± 0,04	2,623 ± 0,020	0,866 ± 0,009	6,03 ± 0,34	3,173 ± 0,021	1,040 ± 0,006	3,451 ± 0,023	1,040 ± 0,007
6,27 ± 0,06	2,776 ± 0,014	0,844 ± 0,004	6,61 ± 0,29	3,311 ± 0,021	1,034 ± 0,006	3,560 ± 0,022	1,044 ± 0,006
6,49 ± 0,04	2,849 ± 0,018	0,864 ± 0,008	7,17 ± 0,25	3,387 ± 0,019	1,040 ± 0,005	3,621 ± 0,021	1,047 ± 0,006
6,82 ± 0,05	2,776 ± 0,024	0,864 ± 0,004	7,71 ± 0,23	3,460 ± 0,020	1,039 ± 0,005	3,708 ± 0,022	1,052 ± 0,005
7,00 ± 0,04	2,984 ± 0,021	0,852 ± 0,009	8,23 ± 0,21	3,537 ± 0,021	1,038 ± 0,005	3,785 ± 0,023	1,058 ± 0,006
7,51 ± 0,04	3,035 ± 0,023	0,852 ± 0,009	8,75 ± 0,19	3,609 ± 0,023	1,049 ± 0,005	3,882 ± 0,025	1,064 ± 0,005
6,90 ± 0,20	3,015 ± 0,011	0,841 ± 0,006	9,26 ± 0,17	3,681 ± 0,022	1,055 ± 0,005	3,988 ± 0,025	1,084 ± 0,005
7,35 ± 0,25	3,066 ± 0,014	0,853 ± 0,007	9,77 ± 0,16	3,768 ± 0,025	1,061 ± 0,005	4,029 ± 0,032	1,098 ± 0,007
7,88 ± 0,22	3,055 ± 0,013	0,860 ± 0,007	10,27 ± 0,15	3,843 ± 0,026	1,069 ± 0,005	4,121 ± 0,029	1,078 ± 0,006
8,39 ± 0,20	3,115 ± 0,012	0,867 ± 0,006	10,76 ± 0,14	3,903 ± 0,029	1,083 ± 0,006	4,179 ± 0,028	1,107 ± 0,005
8,90 ± 0,18	3,150 ± 0,014	0,882 ± 0,006	11,26 ± 0,14	3,993 ± 0,029	1,087 ± 0,005	4,287 ± 0,032	1,111 ± 0,006
9,40 ± 0,17	3,211 ± 0,016	0,889 ± 0,007	11,75 ± 0,13	4,068 ± 0,035	1,103 ± 0,006	4,364 ± 0,039	1,125 ± 0,006
9,90 ± 0,16	3,278 ± 0,015	0,887 ± 0,008	12,24 ± 0,12	4,118 ± 0,030	1,104 ± 0,005	4,418 ± 0,032	1,142 ± 0,006
10,39 ± 0,15	3,329 ± 0,017	0,900 ± 0,007	12,72 ± 0,12	4,215 ± 0,031	1,108 ± 0,006	4,469 ± 0,034	1,146 ± 0,006
10,88 ± 0,14	3,441 ± 0,020	0,896 ± 0,011	13,21 ± 0,11	4,279 ± 0,027	1,111 ± 0,005	4,524 ± 0,031	1,179 ± 0,005
11,37 ± 0,13	3,487 ± 0,016	0,923 ± 0,007	13,69 ± 0,11	4,365 ± 0,036	1,110 ± 0,005	4,586 ± 0,033	1,179 ± 0,006
11,86 ± 0,13	3,586 ± 0,021	0,932 ± 0,012	14,18 ± 0,10	4,408 ± 0,032	1,121 ± 0,006	4,655 ± 0,037	1,180 ± 0,007
12,34 ± 0,12	3,623 ± 0,019	0,935 ± 0,007	14,66 ± 0,10	4,459 ± 0,040	1,132 ± 0,007	4,702 ± 0,047	1,210 ± 0,009
12,85 ± 0,11	3,692 ± 0,018	0,944 ± 0,006					
13,31 ± 0,11	3,792 ± 0,021	0,953 ± 0,007					
13,80 ± 0,10	3,891 ± 0,020	0,952 ± 0,006					
14,29 ± 0,10	3,978 ± 0,026	0,944 ± 0,006					
14,74 ± 0,10	4,061 ± 0,023	0,928 ± 0,007					

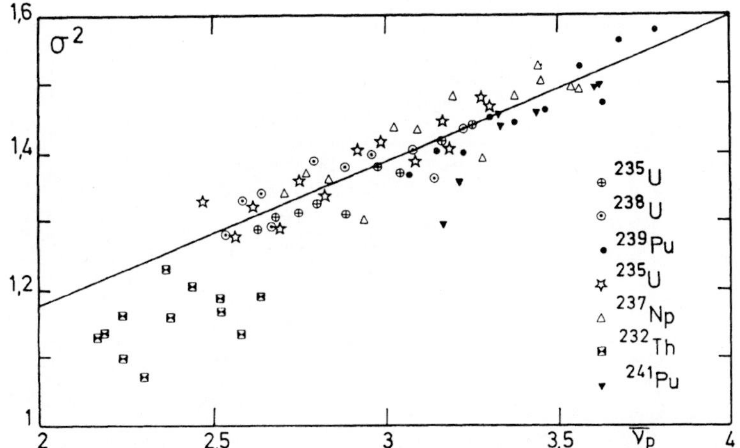

Fig. 5. Variation de la variance des distributions P(ν) des neutrons de fission pour différents noyaux en fonction du nombre moyen $\bar{\nu}_p$ de neutrons : ⊕ ⊙ • : réf. 4 ; ☆ △ ⊞ : ce travail ; ▼ : réf. 5. La droite a pour équation $\sigma^2 = 0,21\ \bar{\nu}_p + 0,75$.

Neutron Data of Structural Materials, Coolants and Shielding of Fission Reactors

CONVERGENCE OF INTEGRAL AND DIFFERENTIAL CROSS-SECTION DATA FOR STRUCTURAL MATERIALS

J. L. Rowlands, R. W. Smith, J. M. Stevenson and W. H. Taylor

United Kingdom Atomic Energy Authority
Atomic Energy Establishment, Winfrith
Dorchester, Dorset, UK

Nuclear data for structural materials are used in calculations of reactor fuel enrichment, burn-up reactivity variations, kinetics and safety characteristics, irradiation damage and activation of the materials, and the effectiveness of reflectors and shields. The structural materials of interest are those present in reactor cores, reflectors and shields, components in the coolant circuits and in fuel storage, transport and reprocessing facilities (in connection with criticality calculations). Correlations with integral measurements have been used extensively in the design of reactors, and in refining calculations of neutronics properties. Aspects of the present status of the integral and differential data are discussed.

Introduction.

A detailed representation of the nuclear cross-sections of structural materials is required for calculations of the neutronics characteristics of reactors (and related facilities) and for other characteristics, such as heating and irradiation damage effects. Particular characteristics for which nuclear data are required are the following:

(a) Neutron economy. Structural materials affect the neutron multiplication and conversion (or breeding) characteristics of reactor cores because they both scatter and absorb neutrons, thus influencing the neutron spectrum and leakage probability and the neutron balance.

(b) Temperature coefficients of reactivity and safety related characteristics. Changes in core temperatures result in changes in the spectrum and hence changes in the neutron balance. These changes can be influenced by structural material cross sections. Structural materials can contribute more directly to fast reactor temperature coefficients as a consequence of the Doppler effects in narrow (p-wave) resonances. In hypothetical severe accidents, structural material can be ejected from the core and the reactor power transient can depend on the reactivity worth of structural materials.

(c) Irradiation damage of structural materials. The maximum irradiation permitted for fuel assemblies can depend on the irradiation endurance of structural materials. Atomic displacements, associated with nuclear recoil in neutron-nuclear interactions, and helium formation from (n,α) reactions, cause swelling and creep of materials and changes in the metallurgical characteristics.

(d) Activation of the coolant circuit and of components. Trace quantities of structural materials can be transferred to the coolant circuit, irradiated in the core and deposited elsewhere in the circuit. This can result in radiation levels which make maintenance of components more complicated.

(e) Core reflectors and shielding. The effectiveness of core reflectors and shields, and optimisation of shield design, have implications for core neutron economy and irradiation doses on such components as pressure vessels. Heat deposition in reflectors and shields depends on γ-emission and interaction data, (with neutron scattering, and other interactions, making a smaller contribution to the heating). Reactor sizes and costs depend on shield optimisation.

(f) Criticality of storage, transport and re-processing plant units. There is an economic incentive to use fuel stores, transport flasks and reprocessing plant more efficiently. Prediction of the criticality of reactors is based on extensive integral correlations. The required reference integral measurements for these other applications are not so complete.

(g) Dosimetry reactions. To monitor the irradiation exposure of components, foils of materials which have suitable activation cross-sections and decay properties are used. Many of these dosimetry reactions are in structural materials.

Structural materials used in some current designs of reactor are:

(i) Water cooled thermal reactors. Zircalloy is used for fuel cladding and pressure tubes. In a typical PWR about 1% of fission neutrons are absorbed in the zircalloy cladding and a further 0.5% in the inconel grids which hold the fuel pins in a fuel assembly.

Zircalloy consists of about 98% zirconium with about 1.5% zinc and 0.3% (Fe + Cr + Ni) In Zircalloy-4 the Ni is absent. An alternative alloy which is used is zirconium with about 2% niobium. The zirconium must be refined to reduce the fraction of rare-earth elements such as hafnium, which are strong neutron absorbers. Calandria tubes have been made using several different alloys, including aluminium alloys.

K. H. Böckhoff (ed.), Nuclear Data for Science and Technology, 85–97.

(ii) Gas cooled thermal reactors. In the UK Advanced Gas Cooled Reactors the fuel is clad in stainless steel and the cluster of fuel pins which forms an assembly uses stainless steel grids, braces and tie bars. The fraction of fission neutrons absorbed in steel is about 10%. This compares with an absorption in graphite of about 4%.

Checks on the stainless steel composition and the level of trace impurities are made as part of the quality assurance procedures. These include reactivity worth measurements for samples of materials in a critical facility.

(iii) Sodium cooled fast reactors. Stainless steels and nimonic alloys are used for the fuel pin cladding and sub-assembly wrapper tubes. The fraction of fission neutrons absorbed in these is typically about 5%. A typical stainless steel composition is

Fe 65%, Cr 19% Ni 13%, Mn 1.6%, Mo 1.3%

but many different alloys are being used, and investigated, because of the requirement to find materials which can withstand long irradiations.

As will be shown later, the percentage reactivity effects of structural materials can be much larger than the fraction of fission neutrons absorbed.

The designs of fuel assemblies for current types of reactor have made extensive use of integral measurement correlations. These measurements (combined with a knowledge of the energy structure of the cross-sections) has proved sufficient to produce satisfactory designs. Why is nuclear data for structural materials still of concern?

For fast reactors the requirements are associated particularly with the fact that fuel irradiation endurance is limited by the endurance of structural materials. The evaluation of alternative structural materials and the refinement of designs to prolong fuel endurance and reduce costs is a major part of fast reactor development.

The irradiation characteristics of structural materials can be correlated with the irradiation dose (that is with the number of atomic displacements and with temperature). Helium formation via (n,α) reactions also contributes. However, precise relationships between changes in properties and the nuclear interactions have not yet been established and this has not yet been identified as requiring improved nuclear data.

Circuit activation from structural materials is of particular concern in water cooled reactors. By modelling the processes by which the coolant and components in the circuit become activated, ways of reducing the activity can more easily be developed. Designing the facilities for handling components, (and for maintenance), requires an ability to predict the activity levels. Reactor measurements are being successfully analysed using models of the processes involved and associated nuclear data.

The geometrical complexity of reflectors and shields, and the approximations made in representing this complexity in calculations, has meant that a high accuracy in the nuclear data was not warranted. With developments in the power of computers it becomes practical to make more accurate Monte Carlo calculations and to optimise designs further. More accurate nuclear data then becomes of value. Experiments using benchmark assemblies of shield materials provide valuable tests of nuclear data relevant to both shield and core design.

Absorption and Reactivity Effects of Structural Materials

In Table I the thermal Maxwellian values and resonance integrals of structural material absorption cross-sections are summarised, together with an indication of resonances below 1 KeV. For steels and aluminium alloys the resonance contribution to the absorption is less than 10% in a thermal reactor spectrum (Mg, Al, Cr, Fe, Ni). For zircalloy and Zr-Nb alloys the resonance contribution predominates. However, the fractional absorption in zircalloy is very small (∿1%) and the estimated uncertainties in the resonance integrals are acceptable. Variations in composition possibly result in comparable uncertainties.

Table II summarises components of the reactivity effect of steel in a fast reactor contributed by the constituent elements and separate reactions. The largest effects in Fe, and Cr arise from inelastic scattering and the capture effect of Mo exceeds that of Cr (although the relative atomic percentages are 1.3% and 19% respectively). There are many different steels being used in fast reactors and the relative contributions vary significantly. For example, the nimonic alloy PE16 is used in the UK PFR for subassembly wrapper tubes (Ni 43.5%, Fe 33.2%, Cr 16.5%, Mo 3.3%) and stellite pads (Co 67%, Cr 27%, W 5%) are used on the sides of the tubes.

The relative importance of (n,p), (n,α) and (n,γ) reactions in a fast reactor spectrum are illustrated in Table III.

TABLE III

Cross Sections Averaged in a Fast Reactor Spectrum

(millibarns)

Element	(n,γ)	(n,p)	(n,α)
Al	2.1	0.38	0.057
Cr	7.2	0.05	0.018
Mn	91.1	0.21	0.003
Fe	10.2	0.63	0.024
Co	56.1	0.14	0.016
Ni	16.6	7.57	0.465
Zr	20.8	0.26	0.024
Nb	271	0.04	0.006
Mo	156	0.06	0.014
(FGL5 data)			
Possible trace elements			
B	0.034	0.26	450
N	0.055	13.0	11.6

Some features of the absorption in structural materials in an LMFBR are:

(a) FE About 25% of the absorption is in the 1.15 KeV resonance.

(b) Ni About 30% of the absorption is in the (n,p) reaction.

(c) Mn About 60% of the absorption is in the two resonances at 337 eV and 1.1 KeV.

(d) Co Absorption in the 132 eV resonance predominates. Resonance shielding effects can be very large in some materials (eg stellite which is used for subassembly rubbing pads).

The high priority measurement requirements for structural materials in the NEANDC-NEACRP High Priority Measurement Request List are given in Table IV.

TABLE I

STRUCTURAL MATERIAL CROSS-SECTIONS AT LOW ENERGIES

Element	σ_o (2200 m/sec) barns	(s.d) (percent)	RI (0.5 eV) barns	(s.d) (percent)	Resonance Contribution to R.I. (approximate*)	Resonances below 1KeV
Mg	0.063	(5)	0.038	(10)	0.009	-
Al	0.234	(1)	0.16	(6)	0.05	-
Cr	3.1	(7)	1.7	(10)	0.3	-
Mn	13.3	(2)	14.0	(2)	8	337eV
Fe	2.55	(1)	1.4	(7)	0.3	(Fe58, 0.31% 230eV)
Co	37.2	(0.2)	74	(3)	53	132eV
Ni	4.6	(6)	2.2	(10)	0.1	-
Zr	0.185	(2)	1.10	(14)	1.0	Several in Zr91: Zr96
Nb	1.15	(4)	8.5	(6)	8.0	Several
Mo	2.65	(3)	22	(10)	21	Several
Sn	0.63	(2)	8.5	(3)	8.2	Several
(Hf)	102	(2)	2000	(5)	1950	Several

* Calculated as RI - 0.45 σ_o

A PWR Spectrum averaged value is approximately equal to 0.07 σ_o + 0.03 (RI)

Data taken from (a) BNL 325 Third Edition, Volume 1 (1973)
(b) Pisanko and Felorova, 1980 Kiev Conf. Vol. 3 p.270
(c) BNL-NCS-51388 (January 1981)

TABLE II

COMPONENTS OF THE REACTIVITY EFFECT OF STEEL IN A FAST REACTOR

(IN PERCENT)

Element	(Percent (by Atoms)	Absorption	Inelastic Moderation	Elastic Moderation	Leakage	Total
Cr	(19%)	-0.3	-0.7	-0.3	0.6	-0.7
Mn	(1.6%)	-0.3	-0.1		0.1	-0.3
Fe	(65%)	-1.7	-2.7	-0.6	1.8	-3.2
Ni	(13%)	-0.8	-0.4	-0.4	0.7	-0.9
Mo	(1.3%)	-0.5	-0.1		0.1	-0.5
TOTAL		-3.6	-3.9	-1.4	3.3	-5.6

(Calculated for the NEACRP LMFBR Benchmark Model using FGL5 data)

TABLE IV

HIGH PRIORITY MEASUREMENT REQUIREMENTS FOR STRUCTURAL MATERIALS

Reaction	Energy Ranges	Typical Requirement (percent)	Requester	Status (Geel 1977) (percent)
Capture				
Cr	0.1 - 100K	20	E	20
Fe	0.1K - 1M	5 - 10	E	10 - 20
Fe56	0.01 - 1M	10 - 15	US	
Ni	0.1K - 1M	10 - 20	E	10 - 20

Corresponding total cross-section measurements and resonance parameters analyses are needed. There are some large differences in the resonance parameters derived in different analyses and this could have implications for the calculation of resonance shielding effects. Resolution of some normalisation problems could reduce the uncertainties in the average infinite dilute cross-sections to about ±10% in the energy range up to 300 KeV and new evaluations are required.

Zr	Thermal RI	5 5	France) France)	Evaluations of existing data required

ORNL and BCMN resonance region measurements for Zr are being analysed by CNEN.

Scattering				
Fe(n,n')	Thr - 4M 4 - 15M	5 - 10 5 - 30	E E	20% to 5 MeV
Fe57(n,n')	Thr - 10M	20	US	
Ni(n,n')	Thr - 4M 4 - 15M	5 5 - 30	E E	20% to 5 MeV

High resolution scattering measurements can be used to determine the level spins for p- and d-wave resonances and these are needed to determine self-shielding factors. Data on scattering anisotropy are required for multiple scattering calculations.

The measurement requirements are all for fast reactor applications.
E denotes a combined West European request.

Status of Some Current Evaluated Nuclear Data Files

The evaluated nuclear data files considered here are those for the elements Fe, Cr and Ni in the libraries ENDF/B-V (1), JENDL-2B (2) and KEDAK-3 (3).

ENDF/B-V (1977)

Fe: A new evaluation by Fu and Perey. In the resonance range a new evaluation of the parameters has been made by Perey and Perey (Ref. 4). The high energy data have also been revised.

Cr and Ni. In the resonance region (to 640 KeV for Cr and 690 KeV for Ni) the data are from ENDF/B-2. The high energy data, inelastic scattering, (n,p), (n,α), etc have been re-evaluated.

JENDL-2B (1979)

The cross-sections for Fe, Cr and Ni at all energies have been re-evaluated (Ref. 5). In the resonance region the data are based on analyses published in 1977, and earlier. For Fe below 250 KeV the data for Fe56 and Fe57 are from the Perey and Perey evaluation and for Fe54 from the analysis by Pandey (Ref. 6). The data are therefore very similar to ENDF/B-V in this energy range. The resonance region evaluation for Cr extends to 300 KeV and for Ni to 400 KeV.

KEDAK-3 (Resonance region (1977), higher energies (1975)).

The evaluations in the resonance region were made by Froehner (Ref.7), the energy range being : Fe to 300 KeV, Cr to 200 KeV and Ni to 100 KeV. These evaluations result in different sets of resonance parameters to those in the ENDF/B-V and JENDL-2B evaluations.

An indication of the status of resonance data for Fe 56 is given in Table V.

TABLE V.

Two recent measurements for the 1.15 KeV resonance

	Γ_n	Γ_γ
AERE(1979)(Normalisation Au);	52.6 ± 0.16	614.8 ± 5.8
ORNL(1982)(Normalisation Au);	83.6 ± 1.7	(577)

The second measurement gives a capture area 50% larger than the first. How is this discrepancy to be resoved?

Capture Widths for s-wave Resonances (eV)

Energy (keV)	Perey and Perey	Froehner (1977)	Brusegan et al. (1979)
27.8	1.40 ±.1	1.25 ± .2	0.80 ± .20
73.95	0.73 ±.07	0.65 ± .15	0.57 ± .07
83.6	1.28 ±.13	0.58 ± .22	0.94 ± .13
129.8	1.0 ±.2	1.30 ± .40	
140.3	2.2 ±.2	1.48 ± .31	

Rohr (9) has recently described how some uncertainties in the capture can be reduced from about ± 20% to about ± 6%. New evaluated files incorporating the resonance parameter analyses which are now being carried out must be produced in order to provide data which approach the requested accuracies. Discrepancies between measured and calculated values for integral parameters will reflect the uncertainties in the current files and might help in the selection of parameter values.

In addition to the differences between the resonance parameters there are large differences between the mean resonance spacings found in different analyses. Smith and Story (10) conclude that a complete and consistent set of s-wave resonances can be derived from the measurements for iron isotopes. In deriving the fit it is necessary to take into account the energy dependence of the mean spacing. Their level spacings tend to be higher than those derived by other evaluators.

TABLE VI

Mean Resonance Spacings for Iron Isotopes

	Froehner et al (KEDAK-3)	Smith and Story
Fe54	20.4 ± 2.7	23.9 ± 0.8
Fe56	21.4 ± 1.9	27.0 ± 1.2
Fe57	6.5 ± 0.8	9.15± 0.5
Fe58	21.6 ± 5.6	29.1 ± 2.0

At the 1979 NEANDC Topical Discussion on Progress in the Field of Neutron Data of Structural Materials A B Smith et al described measurements of the total and scattering cross sections of Cr, Fe and Ni 60 in the energy range 1.0 to 4.5 MeV (11). The total cross sections were measured to accuracies of ∿ 1% and the scattering cross-sections to ∿ 5%. These accuracies meet the target requirements in the High Priority Request List. The authors of the paper report that these measurements suggest significant changes in the ENDF/B-IV File, resulting in changes of ∿ 0.15% in the Keff value of a typical LMFBR due to changes in the inelastic scattering of Fe alone. These measurements are probably too recent to have been fully taken into account in any current evaluated data file. The required new evaluations should include such recent high accuracy measurements.

Integral Measurements for Structural Materials

Integral measurements are considered in the following categories:

1) Activation (for those isotopes and reactions which result in radioactive products).

2) Compositional analysis of irradiated samples (eg to determine (n, α) reactions from the helium formed).

3) Reactivity perturbation effects of small samples, sample size effects and the temperature dependence of the reactivity effect.

4) Neutron balance measurements for simple critical test zones (ie by measuring or calculating all the other compoments of the neutron balance and hence deriving the fraction absorbed in the structural material).

5) Neutron spectrum measurements on critical test zones with and without the structural material present. Usually a substance such as carbon which has a well known scattering cross section is used in the test zone without the structural material to produce a similar overall spectrum.

6) Spectrum and attenuation measurements through thick samples and blocks of materials.

One type of data the measurements provide is the value of a spectrum averaged cross-section, (or cross-section ratio to some standard). The spectra include:

thermal Maxwellian
resonance integral in a (1/E)dE wieghting spectrum
fission spectrum
standard benchmark field
characteristic fast reactor spectra.

The spectra are either measured or calculated, with dosimetry reactions sometimes providing a check, or data for adjusting calculated spectra.

The spectrum measurements, and the more general reactivity and criticality measurements, do not provide simple integral data for a single reaction but a more complex dependence. These measurements can only be used either to provide an overall test of the differential nuclear data or to adjust the data. Some localised information can sometimes be obtained from a comparison of spectrum measurements with calculation, (for example, checking that the cross-sections through resonances are correctly represented in the nuclear data library used in the calculations).

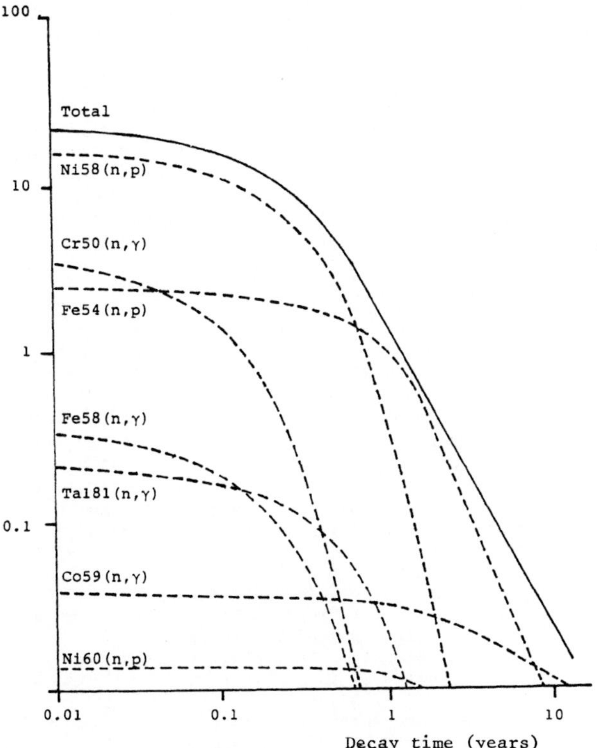

FIG. 1 - COMPONENTS OF THE ACTIVITY IN M316 STAINLESS STEEL

Measurements of the Spectrum Averaged Values of Activation Cross-Sections

Structural material reactions which produce radioactive products which emit strong gamma rays when they decay are important because the activity can present handling and maintenance problems. The activity can arise either from the direct irradiation of the components or from trace quantities of materials which are transported in the coolant, irradiated in the core and then deposited on the component. Components of the activity induced in stainless steel in a fast reactor spectrum are illustrated in Fig. 1.

Because of the strong activity induced, measurements of the spectrum averaged values of the cross-sections can be made by irradiating samples in zero power critical facilities, power reactors and other spectra followed by high resolution gamma spectrometry. Many of the reactions of concern are also suitable for flux dosimetry and have been extensively studied. In addition to measurements in reactor spectra measurements have been made in U235 and Cf252 fission spectra and standard benchmark fields, such as CFRMF, and the Mol ΣΣ facility (12). D L Smith (13) has made a comparison of measured and calculated values of fission spectrum averages of non-fission threshold reactions in the ENDF/B-V Dosimetry File. For the two threshold reactions of most interest, Fe54 (n,p) Mn54 and Ni58 (n,p) Co58 the agreement with the U235 and Cf252 fission spectrum averaged values is within $\pm$ 3%.

Structural Material Activation Measurements in Fast Reactor Spectra

Measurements of the radioactivity induced in structural materials in fast reactor spectra have been made in the ZEBRA critical facility. The measurements have been made in both core and breeder regions using different types of sample; sections of reactor components (such as fuel cans, subassembly wrapper tubes and stellite rubbing pads), thin foils of individual elements (such as Mn, Fe, Co, Ni etc) and thin plate samples of alloys of the types used in fast reactors (such as PE16, M316 and FV548) with the compositions checked by an independent chemical analysis. The advantages of using sections of reactor compoments are that resonance self-shielding and flux depression effects are reproduced, but the chemical composition is not known so accurately (particularly of impurities like Co) and the irregular geometries cannot be represented precisely in calculations. The compositions of the foils were known accurately and the geometry can be represented explicitly in the standard calculation models.

All irradiations were made at maximum power in Zebra, corresponding to peak fluxes of $\sim 2 \times 10^{10}$ n/cm^2/sec and over as long a period as practical in order that the residence time in a fast power reactor could be approached as closely as possible.

The reaction rates were determined by measuring the absolute levels of activities induced in the samples by γ-ray spectroscopy and these rates were expressed relative to the Pu239 fission rate measured nearby.

In order to present each of this wide variety of samples to a gamma ray spectrometer in the same solid angle geometry each sample was taken into solution and liquid sources prepared for presentation to the spectrometer in a fixed location (Fig. 2). This is a high resolution spectrometer which is calibrated by preparing sources containing standard source solutions from Amersham and presenting them in the same geometry. The measured spectrum from an M 316 steel sample is shown in Figure 3 where it will be seen that the predominant activities are from the reactions Fe54 (n,p) Mn54, Ni58 (n,p) Co 58, Co59 (n,γ) Co60, Ni60 (n,p) Co60. The spectrum from a sample irradiated in the PFR shows that the same activities are dominant.

The measurements were made in three different assemblies in ZEBRA, Core 14, BZB and BZD/3. Core 14 was a simulation of PFR which is a two zone core with Pu/Pu+U enrichments of 21% and 24% in the inner and outer core regions respectively. The core contained a number of control rod positions and experimental subassemblies. Irradiations of structural components were carried out near the core centre and in the outer region of the upper axial blanket. Assembly BZB was a larger two-zone conventional core, the enrichments of the inner and outer fuel zones being 16% and 21% respectively. There were several control-rod positions but in those near the core centre, the sample position for the irradiation, the absorber was fully withdrawn. Assembly BZD/3 was a simple annular core, representing a possible heterogeneous power-reactor design. There were no control rod simulations and the fuel enrichment of the single fissile zone was 24%. Irradiations were carried out in the fissile annulus near the peak flux and near the centre of the central breeder zone, in both cases close to the core mid-plane.

Table VII lists the spectrum averaged values of the 8 most important long lived steel activation reaction cross-sections measured using foils in the fissile regions of Core 14 and Assembly BZD/3. The activation measurements were made relative to Pu239 fission. Corrections have been applied for the flux fine structure through the plate cells in which the measurements are made to relate the flux at the position of the foil to the position of the Pu239 fission chamber. To obtain the reaction cross-section the ratio measurements have been scaled by multiplying by the FGL5 Pu239 fission cross-section (calculated with resonance shielding appropriate to the chamber position). The spectrum averaged value of the Pu239 fission cross-section in the FGL5 set agrees with the values in ENDF/B-V and JENDL-2 to within about $\pm$ 1% (for the spectra calculated for the NEACRP LMFBR Benchmark Calculational Model). The activation reaction rate measurements following all corrections have been made to an accuracy which is better than the target requirement of $\pm$ 10%. The calculated neutron spectra are given in Table VIII. Four of the reactions have been calculated using cross-section data from the ENDF/B-V Dosimetry Library. For these the agreement with the measured values is within about $\pm$ 10%. We can note, in particular, the good agreement obtained for Fe58 (n,γ). The remaining 4 have been calculated using FD5/FGL5 data. The agreement is not so close, but it is within about $\pm$ 20%, excepting for Mo98 (n,γ) for which the C/E value is 1.64. These integral measurements provide correction factors to be applied to FD5/FGL5 calculated values. For some reactions, Fe58 (n,γ), Co59 (n,γ), Ni60 (n,p) and Mo98 (n,γ), the C/E values are significantly different in fissile and fertile regions. The spectra in the fertile regions have a higher proportion of low energy neutrons and so these differences could imply errors in the energy shapes of the cross-sections. However, the calculated spectra for fertile regions are subject to larger uncertainties and the adjustment factors in these cases partly compensate for these uncertainties.

Measurements have also been made for a number of other reactions which result in the same active products, Mn55 (n,2n) Mn54, Fe54 (n,α) Cr51 and Co59 (n,p) Fe59. The results are given in Table IX. The contribution to the activity from these routes can generally be neglected. The reactions Mn55 (n,γ) Mn56 (2.576 hrs) and Ni58 (n,np) Co57 (271.4 days) have also been measured. In addition to structural material activation reactions, the following have been measured: Na23 (n,γ) Na24, K41 (n,p) A41, Zn64 (n,γ) Zn65, Cd114 (n,γ) Cd115 and Cd114 (n,γ) Cd115m. Full details are given in reference (14). In the measurements using steel samples in the Core 14 series there was a spread of up to 35% in activation rate values for some reactions. This could be due to differences in the compositions of samples from the nominal compositions for the alloys. Approximations in calculating flux fine structure effects could also contribute to the variations. The high

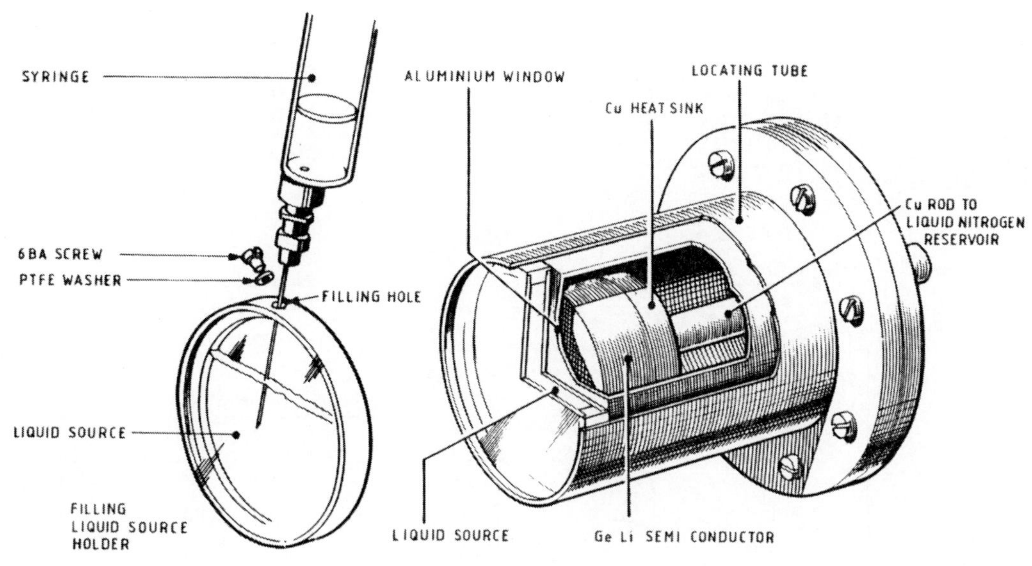

FIG. 2 - Ge Li GAMMA RAY SPECTOMETER

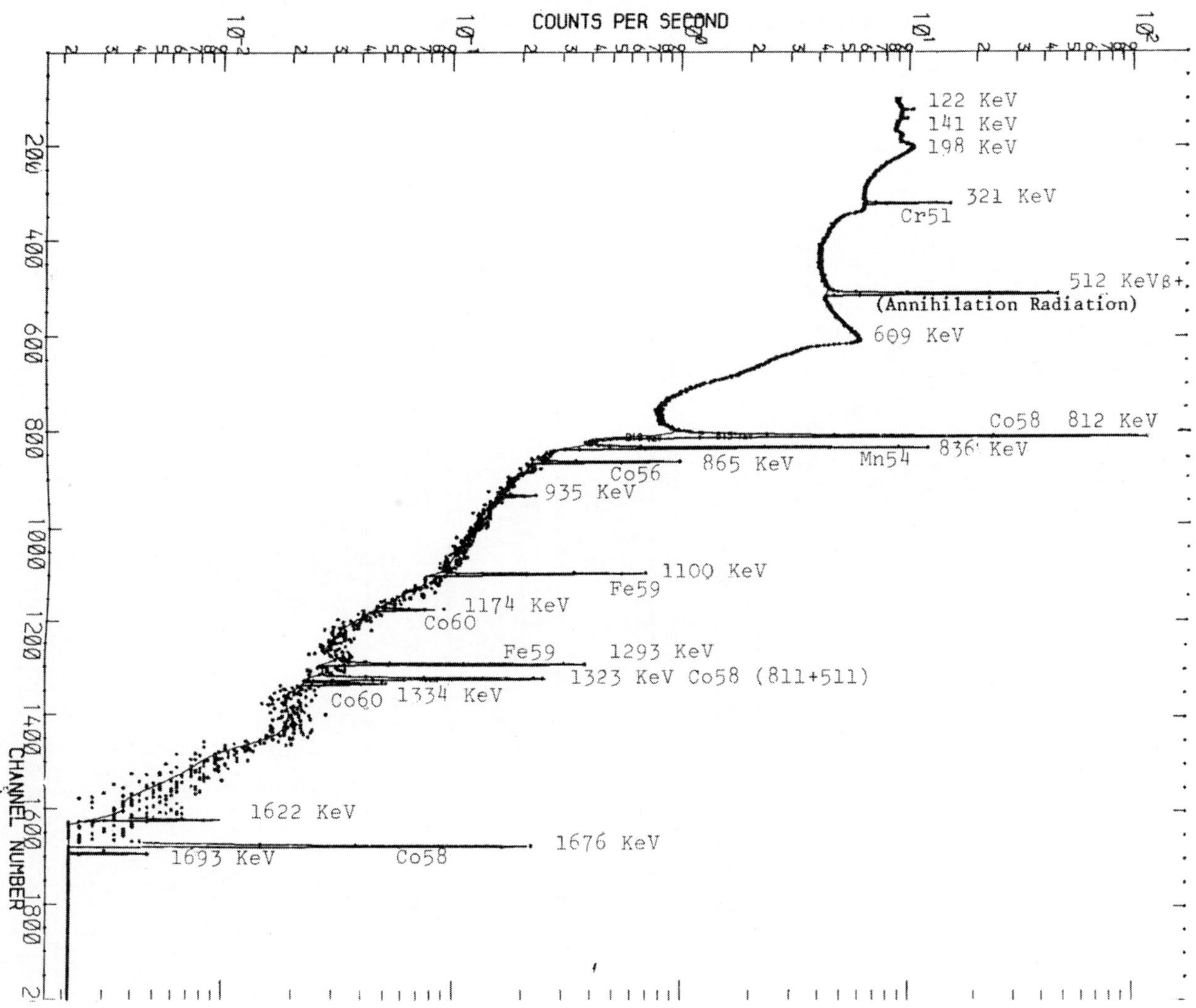

FIG. 3 - GAMMA-RAY SPECTRUM FROM AN IRRADIATED M.316 STEEL-SAMPLE

threshold reactions vary by ∿ 20% through a cell. For the BZD/3 experiments chemical analyses of the samples were made. These agreed with the stated compositions for the alloys. The samples were also located more precisely within the cell in a region midway between the fissile plates (rather than adjacent to them). The measurements for different samples were generally consistent both between samples and with the foil measurements.

TABLE VII

Measured and Calculated Values of Core Spectrum Averaged Cross-Sections for the Most Important Steel Activation Reactions

(Measurements made using foils in fissile regions of ZEBRA Assemblies)

Reaction	Measured cross-sections (in millibarns)				Ratio of calculation to measurement		
	Core 14		BZD/3		Data Source	Core 14	BZD/3
		s.d.		s.d.			
Cr50(n,γ)Cr51	30.8	±6%	28.5	±5%	FD5	0.80	0.79
Fe54(n,p)Mn54	10.5	±5%	12.35	±4%	ENDF/BV	0.90	0.95
Fe58(n,γ)Fe59	9.83	±6%	8.38	±5%	ENDF/BV	0.99	1.01
Co59(n,γ)Co60*	53.8	±5%	52.4	±5%	FD5	0.80	0.84
Ni58(n,p)Co58	14.4	±6%	16.86	±4%	ENDF/BV	0.88	0.92
Ni60(n,p)Co60	0.256	±11%	0.345	±5%	ENDF/BV	0.96	0.90
Mo98(n,γ)Mo99			76.3	±4%	FD5		1.64
Ta181(n,γ)Ta182*	664	±5%	680	±5%	FD5	1.13	0.95

(* Corrections for resonance shielding in the foil must be included in calculations)

TABLE VIII

Fissile Region Spectra in Cores 14 and BZD/3

Group	Lower Energy	Spectra	
		Core 14	BZD/3
1	10 MeV	12	15
2	6.065	232	293
3	3.679	1130	1428
4	2.231	3200	3902
5	1.353	4804	5432
6	0.821	5954	6461
7	0.498	10397	11140
8	0.302	10709	10913
9	0.183	12394	12684
10	0.111	12549	12323
11	67.4 KeV	10231	9819
12	40.9	7892	7383
13	25.0	5576	5166
14	15.0	4899	4621
15	9.12	3441	3060
16	5.53	1899	1674
17	3.35	1274	1032
18	2.03	429	372
19	1.23	1325	1087
20	748 eV	855	655
21	454	467	329
22	275	180	122
23	167	96	57
24	101	37	23
25	61.4	11	7
26	37.3	2	2
27	22.6	1	1

TABLE IX

Additional Activation Reactions Measured in the ZEBRA Programme

Reaction	Core	Cross-Section (millibarns)
Mn55(n,γ)Mn56	BZD/3	32.0
Mn55(n,2n)Mn54	Core 14	0.034
Fe54(n,α)Cr51	Core 14	0.080
	BZD/3	0.126
Ni58(n,np)Co57	BZD/3	0.036
Na23(n,γ)Na24	Core 14	1.19
K41(n,p)A41	BZD/3	0.278
Zn64(n,γ)Zn65	BZD/3	39.1
Cd114(n,γ)Cd115	BZD/3	88.0
115m	BZD/3	8.7

TABLE X

Ratio of ZEBRA Measurements to Fission Spectrum Averaged Values for Threshold Reactions (millibarns)

Reaction	Core 14	BZD/3	Fission Spectrum	Ratios	
				Core 14	BZD/3
Mn55(n,2n)Mn54	0.034	-	0.258 ± 5%	0.132	-
Fe54(n,p)Mn54	10.5	12.35	82.5 ± 6%	0.127	0.150
Fe54(n,α)Cr51	0.080	0.126			
Ni58(n,p)Co58	14.4	16.86	113 ± 6%	0.127	0.149
Ni58(n,np)Co57	-	0.036	0.24 ± 10%	-	0.150
Ni60(n,p)Co58	0.256	0.345	2.3 ± 15%	0.111	0.150

Fission spectrum averaged values from Wölfle and Qaim (submitted to Radiochem. Acta. Sept. 1979).

There are more recent evaluations for some of the reactions. Evaluations have been made at Petten for some corrosion products of stainless steel (15) and for molybdenum isotopes (16). The Petten evaluation for Cr50 (n,γ) is in better agreement with experiment than the FD5 value, C/E for the fissile region of Core 14 being 0.84.

Table X contains ratios of the values measured in ZEBRA to fission spectrum averaged values reported by Wölfle and Qaim (17).There is good consistency between the two sets of measurements. From these ratios a value for the fission spectrum averaged cross-section Fe54 (n,α) Cr51 can be obtained. The consistency between the measurements in the two cores is not good, the derived fission spectrum averaged values being 0.64 mb and 0.84 mb for Core 14 and BZD/3 respectively.

The integral activation measurements made in the fissile and fertile regions of these ZEBRA cores provide a test of the differential cross-section data to an accuracy which is better than the target requirement for activation predictions of ± 10%. When the spectrum averages of the differential cross-sections differ from the integral measurements adjustment factors can be applied. However, when the factors are different in fissile and fertile regions improved differential cross-sections (or spectrum calculation methods) are required to enable accurate predictions to be made for all regions of a fast reactor.

Integral Measurements of Total Absorption by the Null-Reactivity Technique

In these measurements a test zone is built within a critical assembly. The test zone must be sufficiently large for the spectrum at the centre to be the asymptotic spectrum for the test zone composition. The composition is chosen to have an infinite medium effective multiplication, $K_\infty = 1$. The value of K_∞ is measured by removing a small region (or lattice cell) at the centre of the test zone and observing the change in reactivity of the assembly. This is zero when $K_\infty = 1$ for the composition of the test zone, (provided that this is sufficiently large for asymptotic conditions to be established at the centre of the zone).

The test zone usually comprises:

(a) A fissile material, such as U235 or Pu239.

(b) A moderator having well known cross-sections, such as H_2O or graphite.

(c) An absorber with well known, or measurable absorption cross-sections such as boron or U238.

(d) The test substance for which the absorption cross-section is to be measured (such as iron).

The spectrum can be varied by changing the proportions of moderator and absorber to fissile materials.

The spectrum of the test zone is either measured or calculated, with the calculations being supplemented by dosimetry reaction rate ratio measurements. The spectrum measurement techniques include time of flight, proportional counter and solid state detectors. Reaction rate ratio measurements are made for the fission reactions and the measurable absorption reactions.

Two approaches are followed to derive the absorption cross-section of the test substance:

(i) From measurements and calculations of all the reactions (relative to the primary fission reaction) other than the absorption cross-section of the test material. The accuracy of the test material absorption cross-section depends on the accuracy of the K∞ measurement, of $\bar{\nu}$ calculations for the fissile materials, the reaction rate ratio measurements and calculations for reaction rate ratios (such as alpha for the fissile materials). The fraction of the fission neutrons absorbed in the structural material is, typically, 0.15 and the accuracy to which this can be derived using this method is about ± 10%.

(ii) From measurements made on two test zones, one containing the test substance and the other with the test substance replaced by moderator and absorber materials with well known and measurable cross-sections chosen to produce a similar neutron spectrum and absorption. Preferably a range of such test zones would be studied having different spectra. In this way the accuracy of the test material absorption can be derived to an accuracy of about 5%, this accuracy depending on the accuracy of K∞ measurements for the two test zones and of the relative absorption in the reference absorber.

Examples of Null-Reactivity Test Zones:

(a) RB-2 Reactor Central Test Zone (AGIP NUCLEARE)(18)
Mixtures of Microspheres of diameter ∼1mm.
UO_2 + C; Graphite, Borated Graphite, Structural Materials, Fe, Cr, Ni, Mo, Nb, Ta.

(b) PROTEUS Test Zones (19)
Pin geometry lattices.
$U\text{-}PuO_2$;UO_2 ; Fe and stainless steel
Cores 6, 7 and 10

(c) ZEBRA 8 Series of Test Zones (20)
Plate geometry cells.
Pu, U, Graphite, Stainless-steel, $Pu\text{-}UO_2$, UO_2
Cores 8A, 8B, 8C, 8F.

(d) ERMINE OAIO(Steel) and ONIO(Nickel)
C/E for CARNAVAL -IV data
Steel : -1 ± 10%; Nickel -4 ± 6%

A problem associated with this type of measurement is that, in order to get an accurate measurement of structural material absorption, a high proportion of the material must be included in the test zone and this results in stronger resonance shielding effects than are present in the normal LMFBR composition. The iron capture cross-section resonance shielding factors for two null-reactivity test zones containing iron (PROTEUS 10 and ZEBRA 8C) are compared with the values for a typical LMFBR in Table XI.

An example of the agreement obtained for the prediction of null-reactivity test zone K∞ values is illustrated by calculations made for the PROTEUS Test Zones 6, 7 and 10 using the adjusted data library FGL5 (with ENDF/BIV oxygen data) and using ENDF/BIV.

PROTEUS	Composition (pins)				K∞	
Test Zone	U/PuO_2	UO_2	SS	Fe	(C-E)/E FGL5	(Percent) ENDF/B-IV
6	1	1	-	-	0.7	-3.5
7	3	-	3	-	-0.2	-8.1
10	5	-	-	11	1.2	-8.5

The FGL5 results are broadly consistent with the uncertainties in the K∞ measurements.

TABLE XI

Iron Shielding Factors in Different Compositions

Group	Lower Energy	Shielding Factors		
		LMFBR	PROTEUS 10	ZEBRA 8C
	500 KeV			
9	302	0.99	0.96	0.96
10	183	0.94	0.87	0.91
11	111	0.88	0.80	0.84
12	67.4	0.77	0.67	0.72
13	40.9	0.77	0.54	0.63
14	24.8	1.10	0.95	1.19
15	15.0	1.00	0.93	1.02
16	9.12	0.95	0.84	0.91
17	5.53	0.94	0.92	0.97
18	2.04	1.08	0.96	0.98
19	1.23	1.01	0.99	1.01
20	0.748	0.86	0.41	0.48

Small Sample Reactivity Worth Measurements

Central small sample reactivity worth measurements have been made for structural materials in many fast reactor critical assemblies. The calculated worths are dependent on both the absorption and moderation cross-sections of the material. Inelastic scattering effects are particularly important. These are sensitive to uncertainties in the high energy cross-sections of other substances, such as U238 fission and the shape of the high energy neutron spectrum, as well as to the moderation cross-sections of the structural material. The accuracy which can be achieved in the measurements depends on the accuracy of determination of the reactivity scales. The reactivity scale can be determined by the kinetic response of the reactor, and hence depends on the accuracy of delayed neutron data. This reactivity scale uncertainty is typically ± 5%. The scale can also be determined by relating the measured reactivity worth of a small sample of the principal fissile isotope in the core to the calculated value. Again the potential accuracy is about ± 5%.

However, there are problems in the calculation of such reactivity worths. These problems include approximations in the treatment of sample size effects (resonance shielding and flux perturbation), the heterogeneity of the core in the region of the sample (eg plate heterogeneity) and uncertainties in the calculation of central values of the normal and adjoint flux (relative to the core averaged values). Discrepancies of ± 20% have been found in calculations of the worths of small fissile samples whereas agreement to within ± 5% is expected. However, good agreement has been found for some assemblies.

Calculations made using the JENDL-2B cross-section library are compared with the ZPPR-9 measured values in Table XII.

TABLE XII

Comparisons of Experimental and Calculated Small Sample Reactivity Worths

JENDL-2B Calculations for ZPPR-9

	C/E
Pu239	1.06
Fe	1.18
Ni	1.25
Cr	1.16
Mo	1.17
SS	1.15

(Reported in Proc. 1981 Seminar on Nuclear Data, Japanese Nuclear Data Committee INDC(JAP)68/G.p161)

These results are typical of the agreement currently obtained for such calculations. The discrepancies could be due to overestimates of the capture cross-sections of Fe,Cr and Ni by ~30% but this is not the only possible explanation.

Structural Material Doppler Effects

Measurements of small sample Doppler effects have been made for iron and stainless steel samples in the Japanese assemblies FCA-V and VI. These have been analysed by Takano and Inoue (21) using JENDL-2 and ENDF/BIV data. In the hard spectrum cores FCA-V-I and V-2 JENDL-2 calculations overestimate the measured effects in Fe and s.s. by almost a factor of 2 whereas the ENDF/B-IV calculations are in good agreement. In the softer spectrum core FCA-VI-2 the two data sets are in much closer agreement. The JENDL-2 calculations are in good agreement with the measurements and the ENDF/B-IV calculations are not inconsistent with the uncertainty range.

Integral Measurements of (n,p) and (n,α) Reactions

Spectrum averaged activation measurements of (n,p) cross-sections can potentially be made for all the naturally occurring isotopes of the elements Fe and Ni, for the mono-isotopic elements Co and Mn and for all isotopes of Cr excepting for Cr50 (4.35% abundant). However, for some of these radioactive products the decay modes might be too weak or they might be too short-lived to be detectable. The products of (n,p) reactions in Cr52, 53 and 54, Fe56 and 58 and Ni61, 62 and 64 are very short lived, and the activities from Mn55 and Fe57 would be difficult to detect.

Few (n,α) reactions in isotopes of Cr, Fe and Ni result in radioactive products, although Mn55 and Co59 (n,α) reactions result in active products and spectrum averaged measurements have been made for these. For Cr, Fe, Ni and other structural material elements, integral helium production cross-section measurements have been made in reactor spectra. The helium produced in samples following a long reactor irradiation is measured. Measurements were made by Freeman et al (22) in the Dounreay Experimental Power Reactor and by Weitman et al (23) under a boral shield in a thermal reactor. The accuracy of the measurements is typically ± 20%. Because the effective thresholds for the reactions are above about 3 MeV it was considered acceptable to derive fission spectrum averaged values from the reactor spectrum averaged values using threshold dosimetry monitors and calculated relationships. These are the values usually quoted. Measurements have also been made by Lippincott et al (24). Values are summarised in Table XIII which is taken from the compilation by Gryntakis (25). In such measurements care must be taken to ensure that the samples are sufficiently pure and free from small quantities of elements with very high (n,α) cross sections such as boron and nitrogen. As well as the measurements for pure elements, the structural materials used in power reactors have also been irradiated and analysed. This is necessary because of the possible contributions of trace elements and two stage reactions such as Ni58 (n,γ) Ni59 (n,α). Such reactions were observed by Weitman et al (23) in measurements made in a thermal reactor spectrum. It will be seen from Table XIII that the measurements have a spread of about ± 15% for Cr, ± 30% for Fe and ± 5% for Ni. Recent advances in mass spectrometric methods for measuring small qunatities of helium could now make more accurate measurements possible.

TABLE XIII

Fission Spectrum Averaged Values of Helium Production Cross-Sections

From Gryntakis (1979)

	(m.barns)	
Cr	0.187 ± 0.066	Freeman (1969)
	0.192	Weitman (1970)
	0.14	Lippincott (1975)
Fe	0.333 ± 0.10	Byrne (1965)
	0.198 ± 0.044	Freeman
	0.363	Weitman
	0.30	Lippincott
Ni	5.17 ± 0.66	Freeman
	4.89	Weitman
	4.78	Lippincott

Structural Material Transmission Measurements The Iron Block Experiment

Benchmark experiments for reactor shielding have been reviewed at a NEA Specialists Meeting (26). The experiments include measurements of transmission of neutrons through large blocks of iron. A fission plate provides the source of neutrons on one face of the iron block and flux spectra are measured at points through the block. In the ASPIS experiment (27) threshold detectors were used to measure the high energy flux, proton recoil spectrometers measure the spectra between about 7 KeV and about 4 MeV (with some data up to 10 MeV) and resonance sandwich foils measure the low energy flux. The threshold activation reactions used were Al27 (n,α) Na24, S32 (n,p) P32, In115 (n,n') In115m and Rh103 (n,n') Rh103m . The resonance sandwich foil reactions were Cd (Au(n,γ)),Cu63(n,γ)W186(n,γ)and Au(n,γ). A range of proton recoil detectors were used. For high energies (> 0.8 MeV) an NE213 organic scintillation counter was employed and for the range 7 KeV to 1.75 MeV seven different gas proportional counters were used. Spectrum measurements were made at 4 distances from the neutron source plate, 20.3 cm, 50.8 cm, 76.2 cm and 101.6 cm.

In the analysis of the ASPIS measurements by M C G Hall (28) a comparison was made between the adjusted iron cross-sections obtained by fitting to the ASPIS measurements and the adjustments to the iron data incorporated in the FGL5 cross-section set. These latter adjustments were based on measurements of the neutron balance, reactivity worths and spectra in critical compositions containing iron and so they are more sensitive to iron absorption and moderation cross-sections and not so sensitive to the transport cross-sections. The transmission through the iron block is more sensitive to minima in the total cross-section and provides data relating more specifically to iron.

Differences between the two sets of adjusted cross-sections might arise for one of the following reasons:

(a) The adjustments might partly compensate for approximations in the calculation methods, such as group averaging and the use of the transport approximation in the core physics calculations.

(b) The adjustments to individual cross-sections and energy ranges are not separately significant and it is only the combined set of adjustments which is significant. The different sensitivities of the two types of experiment might mean that adjustments made to fit one type do not significantly improve the accuracy of prediction of the other type.

(c) Some of the assumed uncertainties in the cross-sections might be unrealistic and this might constrain the adjustments to fit the two types of experiments differently.

The analyses of the ASPIS measurements were performed using discrete-ordinates multigroup transport codes. The total inelastic and absorption cross-sections of iron were adjusted in 100 groups.

The adjusted cross section set, FGL5, took into account integral measurements of k_{eff}, k_∞, central buckling, central reaction-rate-ratios, central spectra and central reactivity worths. The calculations were performed using collision-probability cell-averaging with anisotropic scattering approximated using the transport approximation. The scattering and absorption cross-sections of iron were adjusted in 10 energy groups followed by a smoothing of the adjustment factors. Both the ASPIS adjustments and the FGL5 adjustments were applied to the UK Nuclear Data Library File DFN-908.

Calculations of the neutron spectra in the iron block were made by M Hall using the three iron cross-section sets, DFN908, the ASPIS adjusted set and the FGL5 set. The McBEND Monte Carlo code was used, and the spectra were calculated in eleven energy groups (from 10 KeV to 15 MeV). Features of the results are:

1 In the top three energy groups the flux calculated using FGL5 iron cross-sections is the same as for DFN908. This is because these groups are outside the range of the ten group FGL5 adjustment scheme; this illustrates the different energy range of sensitivities of the integral data used to produce FGL5 and the ASPIS measurements.

2 There is quite good agreement between the calculations using FGL5 and the ASPIS adjusted cross-sections in groups four to eleven. In both cases the flux is reduced at energies above the inelastic threshold and increased at energies below. This is consistent with the increase in the inelastic cross-section found in both adjusted sets.

This agreement suggests that it should be possible to produce an adjusted cross-section set which takes into account both core physics and shielding integral data. It also gives confidence that adjusted cross-sections will have wide ranges of validity and need not be restricted to applications within the range of the types of integral data used to produce them.

In a recent NEACRP LMFBR Shielding Benchmark Intercomparison Study it was concluded that differences between calculations made by different participants are often related to the calculation methods, rather than the nuclear data. In particular, the methods used to produce multigroup data contribute to the differences.

Conclusions

Evaluated nuclear data files for structural materials are used in calculations of thermal and fast reactor neutronics characteristics (including kinetics and safety related properties), irradiation damage effects, material activation, effectiveness of reflectors and shields, energy deposition in structural materials and criticality of fuel stores, transport flasks and reprocessing plant. The high priority measurement requirements are for Cr, Fe and Ni capture and inelastic scattering cross-sections for fast reactor applications. New evaluated data files are required which take into account recent measurements of scattering cross sections at MeV energies (1 - 5 MeV) and improved resonance parameter analyses which resolve the large differences between current files.

Correlations with integral measurements have been used extensively in thermal and fast reactor design. Integral measurements of structural materials are also used as part of the quality assurance procedures.

Activation reaction measurements in reactor spectra provide the simplest type of integral data test and measurements have been made for all the important structural material activation reactions to accuracies of about ± 5%. The agreement obtained using evaluations which are in the ENDF/B-V Dosimetry File meets the requirements for the prediction of activation rates of ± 10%. Tests have not yet been made using the most up-to-date evaluations for all reactions of interest but it is probable that improved data are required for some reactions. In the interim, adjustment factors must be used, but these are different between core and blanket region spectra for some reactions.

The null-reactivity test zone measurements could help in making a choice between different resonance parameter analyses, if the differences cannot be resolved in other ways. The null-reactivity test zones can measure total absorption to an accuracy of about ± 5% although resonance shielding effects can be large.

Spectrum and attenuation measurements in large blocks of materials, such as the iron block experiment, provide an effective test of inelastic scattering and total cross-sections. However, sophisticated methods are required to treat resonance shielding effects (such as Monte-Carlo methods and detailed energy point representations of cross-sections).

(n,p) cross-sections. Many of the constituent isotope contributions could be measured by activation analysis although most of them produce short lived activities.

(n,α) cross-sections. Reactor measurements have been made of the helium production which is the property of interest. The quoted uncertainties and the spread in values is typically ± 20% to 30%. Techniques for measuring small fractions of helium have improved and more accurate measurements could now be made.

Encouraging consistency is found between iron block spectrum calculations made using the Fe data in the fast reactor core neutronics adjusted cross-section library FGL5, and the data adjusted to fit the ASPIS iron block measurements. This suggests that an adjusted library could be developed to fit both types of measurement, core measurements and shield transmission spectrum measurements.

References

1. R Kinsey (Compiler)
ENDF/B Summary Documentation,
BNL-NCS-17541 (ENDF-201), 3rd Edition (ENDF/B-V) (July 1979).

2. Y Kikuchi et al
J. Nucl. Sci. Technol., 17 (7) 567 (1980).

3. B Goel and B Krieg
Status of Nuclear Data Library KEDAK-3, October 1975. KFK 2234 (December 1975) (In addition reports on the later updates, see for example Reference 7).

4. C M Perey and F G Perey
Evaluation of Resonance Parameters for Neutron Interaction with Iron Isotopes for Energies up to 400 KeV. ORNL/TM-6405 (ENDF-298) (Sept 1980).

5. S Iijima, H Yamakoshi, N Sekine, Y Kikuchi and T Asami
Evaluation of Neutron Cross-Sections of Cr, Fe and Ni for JENDL-2, NEANDC Topical Discussion on Neutron Data of Structural Materials, CBNM-GEEL (September 1979)

6. M S Pandey and J B Gavg
High Resolution Total Neutron Cross-Section in Fe54 and Fe56, Proceedings of a Conference on Nuclear Cross-Sections and Technology, Washington, 1975, p 748, NBS Special Publication 425 (Volume 2).

7. F Fröhner
Evaluation of Resonance Data for Cr, Fe, Ni Below 300 KeV with Special Regard to Doppler Effect Calculation. Neutron Data of Structural Materials for Fast Reactors, Proceedings of a Specialists' Meeting held at CBNM, GEEL (5-8 December 1977).

8. A Brusegan, F Corvi, G Rohr, R Shelley and T Van der Veen
Neutron Capture Cross-Section Measurements of Fe56, Proceedings of an International Conference on Nuclear Cross-Sections for Technology, Knoxville, 1979, NBS Special Publication 594, p 163.

9. G Rohr
The Status of Structural Material Data in the Resolved Resonance Region, NEANDC/NEACRP Specialists' Meeting on Fast Neutron Capture Cross-Sections, Argonne, April 1982.

10. J S Story and R W Smith
Evaluating the Resonance Cross-Sections of the Structural Materials, Extract from Winter College Lectures, International Centre for Theoretical Physics, Trieste, 1982 (to be published).

11. A B Smith, P T Guenther and J F Whalen
Fast Neutron Total and Scattering Cross-Sections of Chromium, Iron and Ni60, NEANDC Topical Discussion on Neutron Data of Structural Materials, CBNM-GEEL (September 1979).

12. H J Holthenius, W L Zijp and W J Hoogendam
Consistency Between Data from the ENDF/BV Dosimetry File and Corresponding Experimental Data for Some Fast Neutron Reference Spectra, ECN Petten (August 1980).

13. D L Smith
Status of the Data Base and Some Integral-Differential Comparisons for Non-Fission Threshold Dosimetry Reactions from ENDF/B-V, NEANDC Topical Conference, AIX-EN-PROVENCE, April 1981 (Compiled by E Fort), NEANDC 150-U.

14. M F Murphy, J M Stevenson and W H Taylor
Comparison of Measured Neutron Reaction Rates in Fast Reactor Steels with Values Calculated using FD5 Data (to be published).

15. H Gruppelaar and H A J Van der Kamp
Evaluation of Neutron Cross-Sections for Some Corrosion Products of Stainless Steel, ECN, Petten.

16. H Gruppelaar and J W M Dekker
Evaluation and Adjustment of Radiative Capture Cross-Sections of Natural Molybdenum and the Stable Molybdenum Isotopes. Proc. of Specialists' Meeting, CBNM, GEEL (1977).

17. S M Qaim and R Wölfle
Measurement of (n, Charged Particle) Reaction Cross-Sections of Structural Materials for Fast Reactors, NEANDC Topical Discussion on Neutron Data of Structural Materials, CBNM-GEEL September 1979.

18. P Azzoni et al
Measurements of Structural Material Capture to U235 Fission Rate Ratios in Intermediate and Fast Spectra. Proc. Symposium on Fast Reactor Physics, Aix-en-Provence (September 1979).

19. S Seth et al
Measurements and Calculations of Integral Capture Cross-Sections of Structural Materials in Fast Reactor Spectra, p 53, Neutron Data for Structural Materials, Proc. Specialist Meeting, EEC Geel, (December 1977).

20. A M Broomfield et al
Measurements of K∞, Reaction Rates and Spectra in ZEBRA Plutonium Lattices, p 15, Proc. Int. Conf. Physics of Fast Reactor Operation and Design, London (June 1969).

21. H Takano and H Inoue
Doppler Effect of Structural Materials in Fast Reactors, Journal of Nuclear Science and Technology, 18(4), pp315-318 (April 1981).

22. N J Freeman et al
The (n,α) Cross-Sections for Natural Nickel, Iron, Chromium and Molybdenum in a Fission Neutron Spectrum, JNE 23, 713 (December 1969).

23. W Weitman et al
Trans. Am. Nucl. Soc 13, (2 papers) (1970).

24. E P Lippincott et al
Proc. 4th Conference on Nuclear Cross-Sections and Technology, Vol 1, Washington 1975.

25. E M Gryntakis
A Compilation of Fission Spectrum Average Neutron Cross-Sections for (n,α) Reactions, Jour. Radioanalytical Chem. 52, 219, (1979).

26. Nuclear Data and Benchmarks for Reactor Shielding Proc. Specialists' Meeting, NEA Paris (Oct 1980).

27. M D Carter et al
The ASPIS Iron Benchmark Experiment - Results and Calculation Model, Nuclear Data and Benchmarks for Reactor Shielding, Proc. Specialists' Meeting NEA Paris (October 1980).

28. M C G Hall
A Simple Comparison of Adjusted Data Sets Based on Measurements in ASPIS and in the Cores of Fast Reactor Criticals, Nuclear Data and Benchmarks for Reactor Shielding, Proc. Specialists' Meeting, NEA Paris (October 1980).

THE INTEGRAL CHECK OF NEUTRON CROSS SECTION DATA FOR REACTOR STRUCTURAL MATERIALS BY MEASUREMENT AND ANALYSIS OF NEUTRON SPECTRA

I. Kimura, S. A. Hayashi, K. Kobayashi, S. Yamamoto

Research Reactor Institute, Kyoto University
Kumatori-cho, Sennan-gun, Osaka-fu
590-04, Japan

Hiroshi Nishihara
Department of Nuclear Engineering, Kyoto University
Yoshidahonmachi, Sakyo-Ku, Kyoto-city
606, Japan

T. Mori and M. Nakagawa
Reactor System Laboratory, Japan Atomic Energy Reseach Institute
Tokai-mura, Naka-gun, Ibaraki-ken
319-11, Japan

In order to check neutron cross section data and group constants for some important structural materials for fast breeder reactors and controlled thermonuclear reactors, the energy spectrum of the neutrons from about 1 keV to a few MeV in a sample pile of assorted materials has been measured by the linac time-of-flight method and the results were compared with that theoretically predicted by one-dimensional transport calculation. To avoid geometrical complexity, we used spherical vessels in which elemental sample powders were packed. Each pile had a lead photoneutron target at the center. To clarify the spherical symmetry of the neutron transport in the pile, the spatial and angular distributions of the $^{58}Ni(n,p)^{58}Co$ and $^{197}Au(n,\gamma)^{198}Au$ reactions were measured for each pile. Group constants were produced by SUPERTOG-JR3 with the JENDL and ENDF/B-IV nuclear data files, and then the neutron transport calculations were carried out by DTF-IV or ANISN.

[Neutron spectrum, time-of-flight method, neutron cross sections, group constants, integral check, structural materials, fast breeder reactor, controlled thermonuclear reactor, transport calculation, photoneutron source, spatial and angular distributions, molybdenum, iron, nickel, chromium, stainless steel, niobium, titanium, electron linear accelerator]

Introduction

For the purpose of safe and economical design of fast breeder reactors (FBR) and controlled thermonuclear reactors (CTR) the neutron cross section data for structural materials are required with the accuracy of a few % to a few ten % in the relevant energy region[1]. Among the data, higher accuracy is requested for their total cross sections, and the value for iron is 1 - 5 %. Those for chromium and nickel are given to be 3 %. On the other hand, the requested accuracies to elastic scattering, inelastic scattering and radiative capture for iron are 5 - 10 %, 5 - 10 % and 5 - 21 %, respectively. However the accuracy to (n, p), (n,α) and (n,2n) reactions are requested slightly lower. In addition to these three elements, the reliable cross section data for molybdenum and manganese are requested for both FBR and CTR. From the standpoint of the CTR design, cross section data for more structural materials such as niobium, copper, titanium, aluminum, tungsten would be necessary[2].

At several nuclear data centers in the world, the neutron cross section data for important reactor materials are collected and are evaluated systematically, and thereby large scale evaluated nuclear data files or libraries are prepared for the reactor design. However, there frequently exist considerable large discrepancies among different evaluated nuclear data files. Even though the values in a few data files seem to agree satisfactorily each other, the accuracy of these values may be beyond the requested value. Therefore integral check or test of evaluated nuclear data files is often requested before they are practically utilized for the reactor design.

There are several methods for the integral check of evaluated nuclear data files. Critical experiments are suited for the integral check of nuclear data for fissile or fertile nuclides under the similar nuclear characteristics to real power reactors. However for the purpose of checking nuclear data for a single element, measurement of the energy spectrum of neutrons in a sample pile of a reactor material, especially a structural material, is superior to other methods, because this method has less ambiguity by its simple geometry and by its homogeneous pure constituents. The experimental success of this method is mainly due to the development of electron linear accelerators as a powerful tool of an intense pulsed neutron source for the neutron time-of-flight spectroscopy. At first, this method was started in the thermal neutron region and was applied to checking the models of the neutron scattering law. Then, several groups applied this method to the intermediate and fast neutron regions and tried to assess nuclear data of reactor materials in higher energy region[3~8]. Present authors also started the measurement and analysis of neutron spectrum in reactor materials about 15 years ago and have investigated more than 20 samples. The outline of our work has been published elsewhere[9~12].

In the present paper, newer results of integral check of nuclear data or neutron cross section data for some important structural materials for FBR and CTR are presented. Although the result for a molybdenum pile was already published very recently[13], it will be briefly reintroduced in this paper.

We made a sensitivity analysis code DTF-IV-KANDO for evaluating this method quantitatively, and its procedure and some results have been described elsewhere[12,14]. From this analysis, we can find that the sensitivities of the total cross section and the scattering cross section to neutron spectra are quite high and medium, respectively, but that of the capture cross section is much lower for reactor structural materials.

As same as the earlier works, we subsidiarily measured the angular and spatial distribution of neutron reaction rates around a photoneutron target in the sample pile by the activation method.

Experimental Arrangement and Procedure

Sample Piles

The purity of sample powders, the shape and the size of sample piles and the position and the direction mainly measured are tabulated in Table I. Each sample powder was packed into a spherical vessel of steel, except the titanium pile which was a 14-hedron.

K. H. Böckhoff (ed.), Nuclear Data for Science and Technology, 98–105.

Table I Sample piles for experimental study

Sample	Purity and density	Shape and size	Position and direction mainly measured
Borated graphite (Standard pile)	2.57% boron in C 1.65	Rectangular parallelepiped 70 cm x 70 cm x 80 cm	r = 22.5 cm, μ = 0.0
Molybdenum pile	>99.9% 2.15	Sphere of 60 cm in diameter	r = 15 cm, μ = 0.0
Iron pile	99.87% 3.63	Same above	Same above
Nickel pile	99.7% 2.94	Same above	Same above
Chromium pile	99.8% 3.95	Same above	Same above
Niobium pile	>99.8% 4.39	Sphere of 28 cm in diameter with/without lead reflector	r = 7 cm, μ = 0.0
Titanium pile	99.4% 1.20	14-hedron, in which a sphere 104 cm in diameter inscribes	r = 25 cm, μ = 0.0 r = 35 cm, μ = ±0.7

The size of the spherical vessel is the same for all cases except for the niobium pile which is much smaller than the others, We have measured the neutron spectra in a large rectangular parallelepiped iron pile[9] and a smaller iron pile surrounded by a lead reflector[11] before. But recently, we made a spherical pile in which iron powder was packed, and then the neutron spectrum in it was measured. Thereby, we have finished the fulfilment of the measurement and the analysis of the neutron spectra in all of main constituents of stainless steel (Fe, Ni, Cr and Mo) in almost the same experimental conditions.

Each pile has a cylindrical lead photoneutron target at the center and a reentrant hole from which the neutron beam with certain position and direction is extracted to the neutron flight path. When we measure background counts, a plug in the bottom of the reentrant hole is removed and then the hole becomes through.

In order to measure angular and spatial distribution of neutron reaction rates around the photoneutron target, each pile has several small radial holes around the target. We can set activation wires or foils, such as nickel and gold, in these holes.

Pulsed Neutron Source

Fast neutron pulses are generated with the electron linear accelerator of KURRI. The typical operation condition of the accelerator is as follows: electron energy: about 30 MeV; repetition rate: 250 pps; pulse width: 22, 33, or 47 ns; target current: about 700 mA at peak.

The spectrum of the neutrons directly emitted from the photoneutron target was measured by the time-of-flight method and the result is shown in Ref.15. The angular distribution of the photoneutrons was seen to be isotropic as shown before[12].

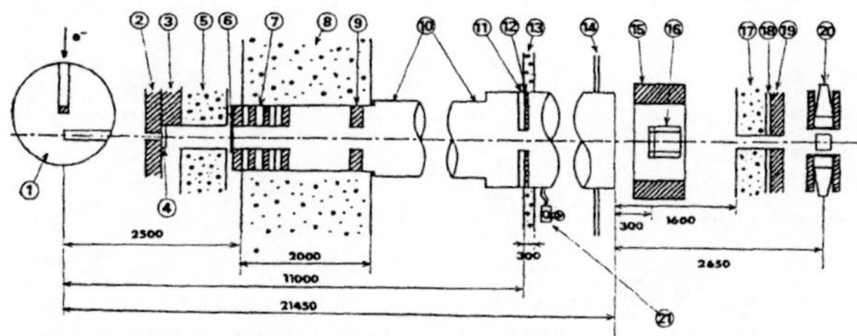

Fig. 1 Experimental arrangement of TOF experiment. (1) Sample pile, (2) Pb pre-collimator(22mm in diameter, 200mm thick), (3) Pb shield(200mm thick), (4) U filter(20mm thick), (5) heavy concrete shield (400mm thick), (6) Cd filter(0.5mm thick), (7) Pb and B_4C collimators((47mm Pb and 47mm B_4C)x4, 50mm in diameter), (8) concrete wall, (9) Pb collimator(100 mm in diameter, 200mm thick), (10) flight tube, (11) B_4C collimator(160mm in diameter,80mm thick), (12) Pb collimator(160mm in diameter, 60mm thick), (13) concrete wall, (14) wall of hut, (15) Pb shield for ^{6}Li glass scintillators, (16) ^{6}Li glass scintillation counter bank being removed when a ^{10}B-vaseline-NaI(Tl) counter is used, (17) heavy concrete shield(400mm thick), (18) B_4C shield(50 mm thick), (19) Pb shield (150mm thick), (20) ^{10}B-vaseline-NaI(Tl) counter, (21) rotary pump.

Collimators and Flight Tube

The general configuration of the collimators and the flight tube is almost the same as earlier works and is depicted in Fig.1. The total distance between the electron beam center and the front surface of the ^{6}Li glass scintillator is about 22 m. A natural uranium plate of 2 cm thick and a cadmium plate of 0.5 mm thick are used at the entrance of the flight tube against gamma flash and slow neutron backgrounds.

Neutron Detectors and Electronics

As the neutron detector, we used a bank of three ^{6}Li glass scintillation counters (NE 912, in 12.7 cm in diameter and 1.27 cm thick) and a ^{10}B-vaseline-NaI (Tl) counter. The structure and the characteristics of these detectors are given elsewhere[16]. The significance of using these two detectors is to avoid a systematic error which may be caused by the neutron detector system.

The relative detection efficiencies of the both detectors along with some correction factors such as the transmission function through the flight tube, were experimentally determined by making use of a standard neutron spectrum field with a borated graphite pile[17]. Very recently we encased this pile in a steel frame and recalibrated the efficiencies of both detectors. A ^{235}U fission chamber and a BF_3 counter are used for neutron monitoring.

The electronic circuits and the computer system for the measurement and the data analysis have been recently improved. The neutron signals are stored in a multi-channel analyzer, CANBERRA Series 88/MP, through a time analyzer unit, Oken S-1218. The obtained data is processed with the PDP-11/34 computer.

Measurements of Spatial Distributions of Neutron Reaction Rates

As given elsewhere[12], we always measured the spatial distributions of the ^{58}Ni(n,p)^{58}Co and ^{197}Au(n,γ)^{198}Au reactions in each pile.

Theoretical Calculation of Neutron Spectrum

Configuration and Assumption

In the theoretical calculation each pile was assumed to be spherically symmetric. This assumption was confirmed by the measurement of the spatial distributions of neutron reaction rate described above.

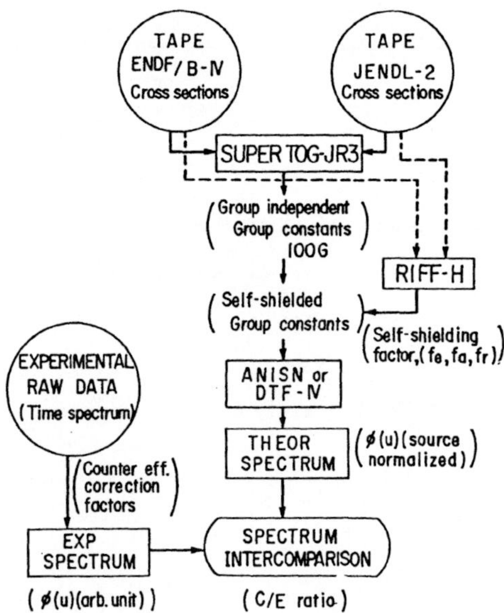

Fig.2 Flow chart of neutron spectrum calculation : production of self-shielded group constants from evaluated nuclear data files and neutron transport calculation.

We assumed the uniform and isotropic photoneutron source in the region of the lead target which was assumed to be spherical. The neutron spectrum in the lead target was taken from Ref.15. Usually the number of quadrature point, n, was taken to be 16, and the scattering anisotropy of the elastic scattering kernel was obtained by P_8 approximation.

Cross Section Libraries or Data Files and Group Constants

As shown in Fig.2, 100 group constants were produced by the SUPERTOG-JR3 code[18] from two evaluated nuclear data files ENDF/B-IV[19] and JENDL-2[20](or -1). In this calculation we assumed the weighting spectrum having the 1/E type below 100 keV and the fission neutron spectrum above it. The MAT numbers which we used in this work are tabulated in Table II.

On the other hand self-shielding factors for elastic scattering, absorption and removal cross sections were calculated by the RIFF-H code[21], which had been prepared by one of the authors. In this calculation, ultra fine mesh (0.0021 lethargy width) was taken. Multiplying the above group constants by these self-shielding factor, we obtained the self-shielded group constants for computer input. As the group constants for the reaction rate analysis we used the data in the ENDF/B-IV dosimetry file.

Table II List of MAT numbers used

Element	ENDF/B-IV	JENDL-1	JENDL-2
Mo	1287	1420 (1421-1427 for isotopes)	- - -
Fe	1192		2600
Ni	1190		2800
Cr	1191		2400
Ti	1286		- - -
Nb	1189		4193
Pb	1288		- - -

Computer Codes for Transport Calculation

The neutron transport calculation in the sample piles were carried out by the ANISN code[22] or the DTF-IV code[23] with the FACOM M200 of the Kyoto University Data Processing Center and with the same computer of the Data Computing Center of JAERI.

Results and Discussion

Molybdenum

The spatial distribution of the ^{58}Ni(n,p)^{58}Co reaction rate in the pile are shown in Fig.3. It can be seen that the distribution is almost symmetric around the lead target except for the backward direction ($\theta=135^{\circ}$). This result ensures the validity of the calculational model of spherical symmetry described in the previous section, at least in the forward hemisphere of the pile where there is the reentrant hole. A comparison between the measured and calculated distributions in Fig.3 shows fairly good agreement, although there is a little disagreement near the

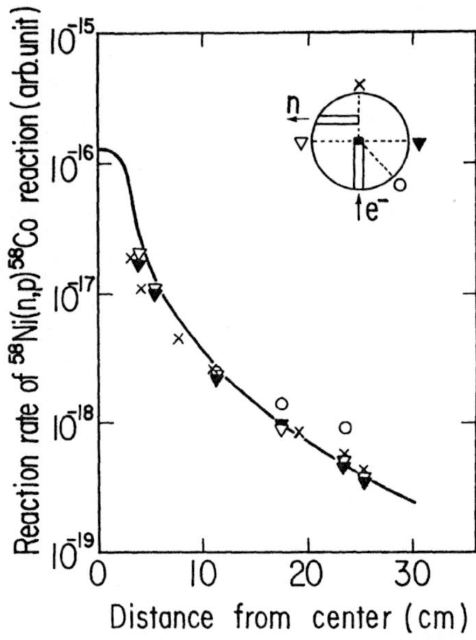

Fig.3 Spatial distribution of the reaction rate of ^{58}Ni(n, p)^{58}Co, comparing with the theoretical prediction shown by a curve.

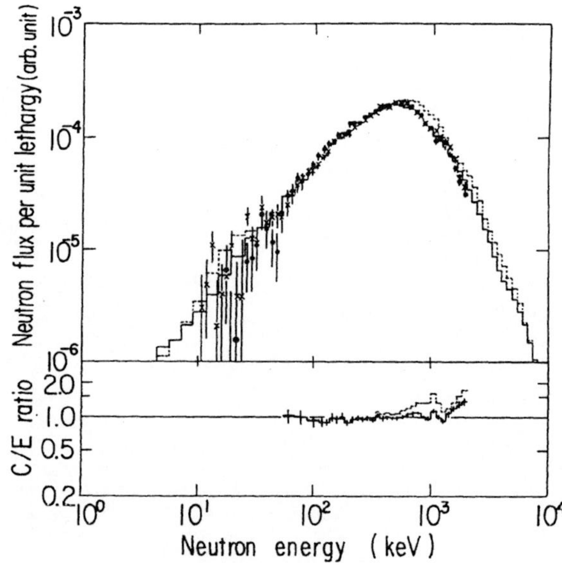

Fig.4 Neutron spectrum at r=15cm and μ=0 in the molybdenum pile : top.
• measured (^{6}Li glass), x measured (^{10}B-V-NaI),
—— calculated (JENDL-1), ---- calculated(ENDF/B-IV).
Rarios of the theoretical prediction to the experimental result (^{10}B-V-NaI) : bottom.
—— calculated (JENDL-1)/measured,
---- calculated (ENDF/B-IV)/measured.

center of the pile. Similar results were obtained for other piles.

Angular neutron spectrum at r=15 cm and μ=0 (θ= 0°) in the pile obtained in the TOF measurement are compared with theoretically predicted ones in Fig.4. Only the statistical uncertainty of the measured spectrum is indicated by an error bar in the figure. The experimental result is normalized to that calculated from the JENDL-1 data by the average flux from 740 keV keV to 1 MeV. Both of the two theoretical ones correspond to an unit neutron source. In this figure, the following can be seen.

(1) General agreement is obtained between the experimental results with the two kinds of neutron detectors.

(2) The measured spectrum is in good agreement with the predicted one from the JENDL-1 data.

(3) The prediction from the ENDF/B-IV data, on the other hand, gives higher flux by 30 % than the measured and also gives higher flux than that from the JENDL-1 data above several 100 keV, especially from 500 keV to 1.2 MeV.

From the result (1) mentioned above and the fact that the two detectors have quite different energy dependence of detection efficiency for neutrons, it can be concluded that the systematic error is not serious in the experimental results. The ratios of calculated spectra to the measured one with the ^{10}B-vaseline-NaI(Tl) counter are depicted on the lower side in Fig.4, which clarify the results (2) and (3) mentioned above.

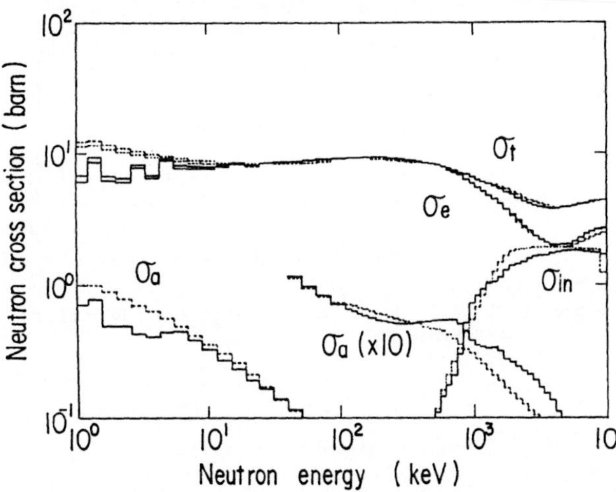

Fig.5 Comparison of the group constants for molybdenum produced by SUPERTOG-JR3.
JENDL-1, ENDF/B-IV.
σ_t : total cross section, σ_e : elastic scattering cross section, σ_i : inelastic scattering cross section, σ_c : capture cross section.

The group constants of total, capture, elastic scattering and inelastic scattering cross sections for molybdenum produced by the SUPERTOG-JR3 code are shown in Fig.5. There seems to be a little difference between the JENDL-1 data and the ENDF/B-IV data from this figure. The main difference between them exists in the treatment of inelastic scattering, in which no discretc level is considered in the ENDF/B-IV data, while 40 levels in natural molybdenum are in the JENDL-1 data. Furthermore, 99 levels are taken into account by use of the data for individual molybdenum isotopes in JENDL-1. This difference causes a large discrepancy in the energy distribution of inelastically scattered neutrons, as shown in Fig.6.

In order to identify the effect of partical cross section on predicted spectrum systematically the following group constants (A)-(E) are produced by substitution of partial cross sections from JENDL-1 into ENDF/B-IV :

(A) ENDF/B-IV (unchanged).

(B) The capture and inelastic scattering cross sections are taken from ENDF/B-IV, but the elastic scattering from JENDL-1.

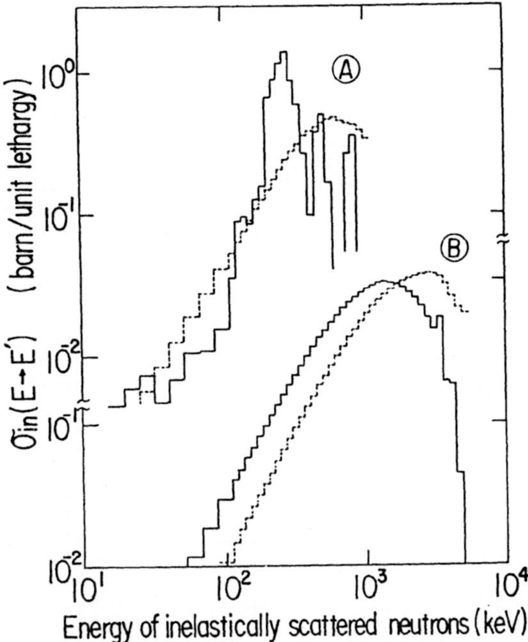

Fig.6 Comparison of energy distribution of inelastically scattered neutrons for molybdenum.
(A) : incident neutron energy ≃ 1 MeV,
(B) : incident neutron energy ≃ 5 MeV,
: JENDL-1,
: ENDF/B-IV.

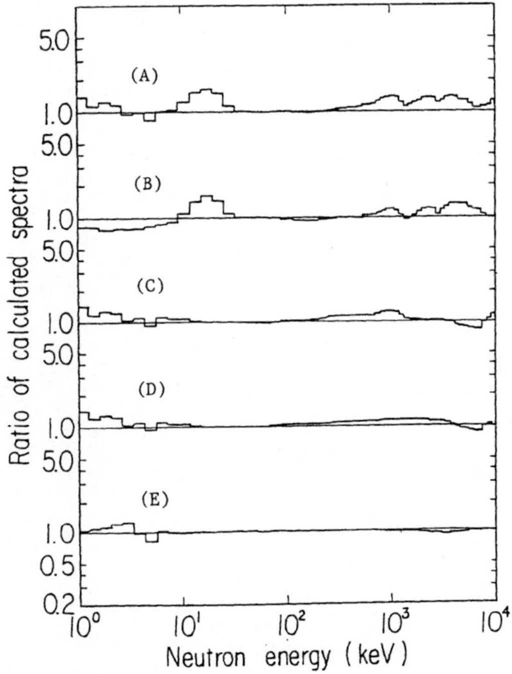

Fig.7 Ratios of neutron spectra calculated from five group constants to that from the JENDL-1 data. The explication of each capital is shown in the text.

(C) The capture, elastic and inelastic scattering cross sections are taken from ENDF/B-IV, but the energy distribution of inelastically scattered neutrons from JENDL-1.

(D) The capture and elastic scattering cross sections are taken from ENDF/B-IV, but the inelastic scattering cross section and the energy distribution of inelastically scattered neutrons are taken from JENDL-1.

(E) 99 discrete levels of inelastic scattering are taken from the data for individual molybdenum isotopes in JENDL-1 instead of 40 levels for natural molybdenum.

In the substitution, total cross sections are changed according to the change of substituted partial cross section. Ratios of the neutron spectra calculated from each of group constants (A)-(E) mentioned above to that from the JENDL-1 data were taken and each ratio corresponding to the group constants is shown in Fig.7 as (A)-(E), which we called the ratios -A to -E in the following discussion. This figure, as a whole, reveals the significance of the inelastic scattering on the spectrum above a few 100 keV, which is in particular shown by the fact that the structure of the ratio-A almost disappears in the ratios-C and -D. On the other hand, the ratio-B shows a similar variation with energy to that of the ratio-A, which indicates that the exchange of the elastic scattering cross section makes little change in the spectrum. Flatness in the ratio-D shows that the substitution of the inelastic scattering cross sections into ENDF/B-IV gives a very similar spectrum to that from the JENDL-1 data, and that the difference between the predictions from the JENDL-1 data and the ENDF/B-IV data is mainly ascribed to the difference in the inelastic scattering cross sections. The almost flat ratio-C shows that the difference in the energy distribution of inelastically scattered neutrons causes more than half of the change of the spectrum shape in the energy region around 1 MeV. Its effect appear even in the lower energy region around 20 keV, where statistical uncertainties in the experimental results are too large to make an accurate judgement on the cross section. It can be seen from the ratio-E that the calculation with the 99 levels of the inelastic scattering gives almost the same spectrum with the predicted one considering the 40 levels for natural molybdenum.

From these analyses, it can be seen that the inelastic scattering, especially the energy distribution of the inelastically scattered neutrons, is very important in the analysis of a neutron spectrum with a neutron source for which the energy distribution is peaked at a few MeV. The discrepancy between the measured spectrum and the theoretically predicted one from the ENDF/B-IV data is mainly due to the fact that the data for molybdenum has no discrete level for inelastic scattering of neutrons. On the contrary, the JENDL-1 data has the 40 discrete levels of inelastic scattering, which are sufficient for the present analysis, and the prediction from them is in good agreement with the experimental result.

Iron, Nickel and Chromium

The neutron spectra in the three major constituent elements of stainless steel were measured with almost the same experimental conditions as seen in Table I.

The experimentally obtained angular neutron spectra together with the theoretically predicted ones at r=15 cm and $\mu=0$ ($\theta=90^{\circ}$) for iron, nickel and chromium are shown in Figs.8, 10 and 12, respectively. The normalization between the measured spectrum and the predicted one was performed by the average flux in 300 keV - 1 MeV. In each figure, the ratio of the theoretically predicted spectrum to the experimentally ob-

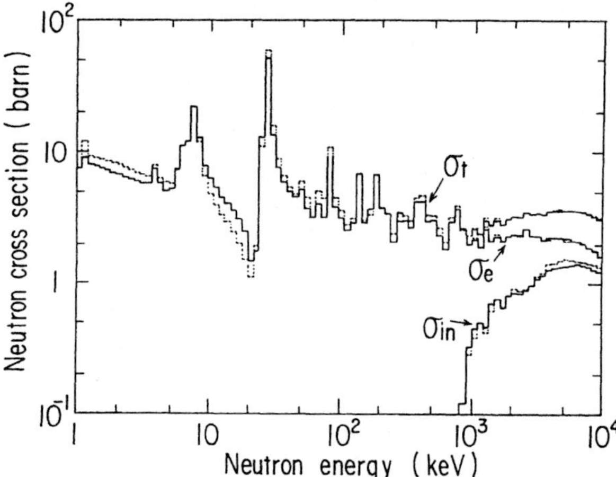

Fig.9 Comparison of the group constants for iron produced by SUPERTOG-JR3. All symbols are the same as in Fig.5, except
—┌— JENDL-2.

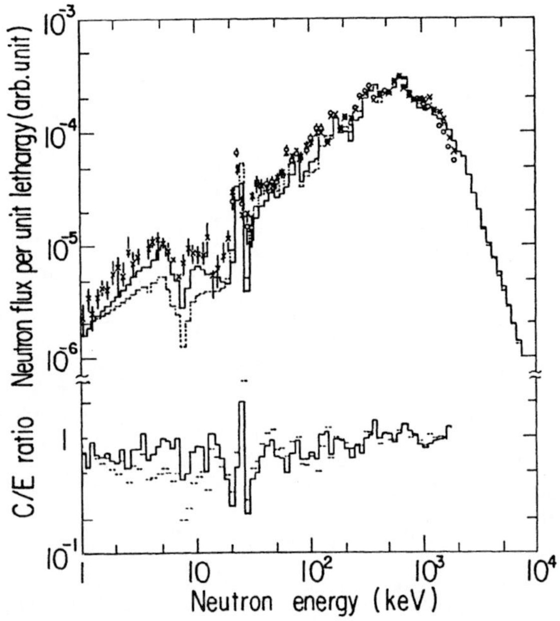

Fig.8 Neutron spectrum at r=15cm and μ=0 in the iron pile : top.
Ratio of the theoretical prediction to the experimental result (^{6}Li glass) : bottom.
All symbols are the same as in Fig.4, except
—┌— calculated (JENDL-2).

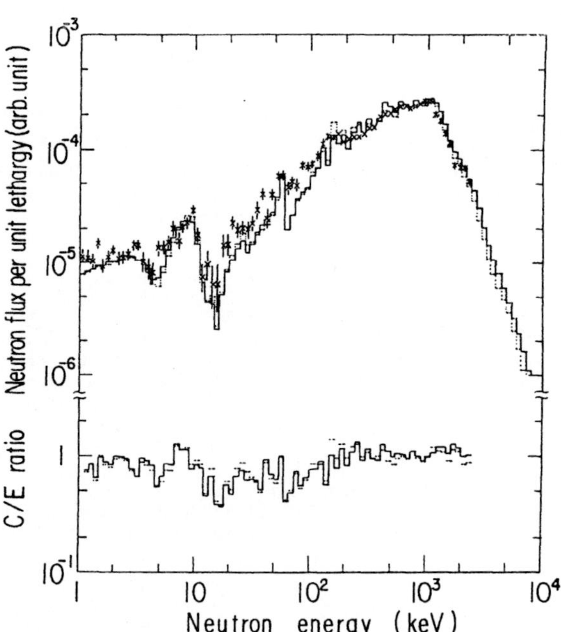

Fig.10 Neutron spectrum at r=15cm and μ=0 in the nickel pile : top.
Ratio of the theoretical prediction to the experimental result (^{6}Li glass) : bottom.
^ll symbols are the same as in Fig.8.

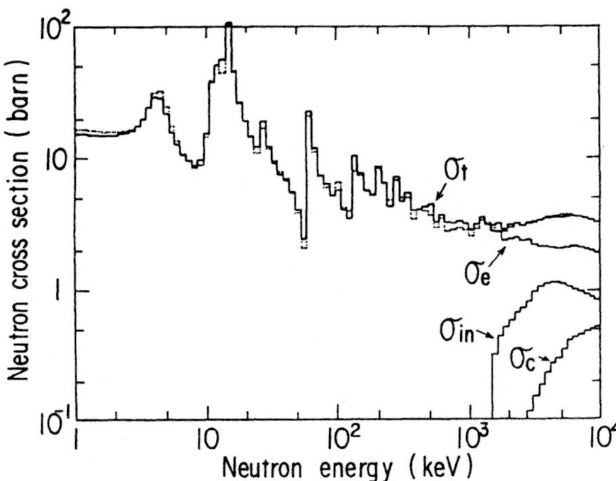

Fig.11 Comparison of the group constants for nickel produced by SUPERTOG-JR3. All symbols are the same as in Fig.9.

tained is also depicted below. In these cases, we took the data with the ^{6}Li glass scintillation counter bank to calculate the above ratio, because these data agree with those with the ^{10}B-vaseline-NaI(Tl) counter and have less statistical uncertainty.

The group constants of the total, elastic scattering and inelastic scattering cross sections for iron, nickel and chromium are shown in Figs.9, 11 and 13, respectively.

Throughout the neutron spectra in these three sample piles, it is noticeable that the measured spectra for both iron and nickel reasonably agree with the theoretically predicted in general from about 1 keV to about 2 MeV, while the measured spectrum for chromium starts exceeding the predicted one from about 100 keV and the difference between them gradually increases with decreasing energy.

For the spectrum in the iron pile, it can be seen that the C/E ratio for JENDL-2 is flatter than that for ENDF/B-IV in the keV region, especially below 20 keV. This is probably due to the difference of the total cross section between them. However, in 20 - 80 keV

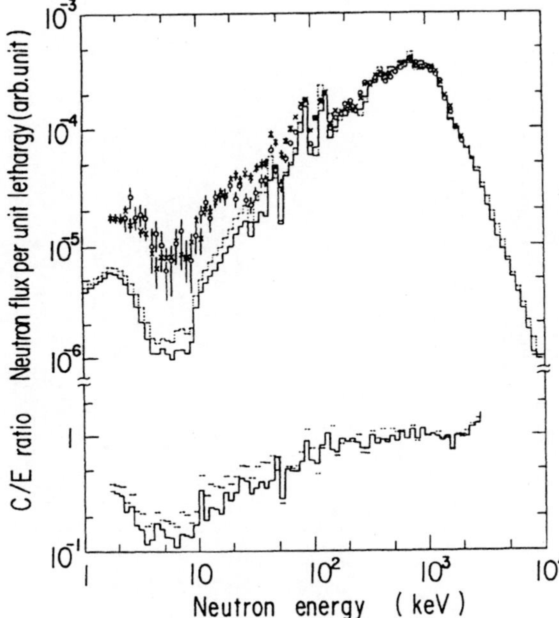

Fig.12 Neutron spectrum at r=15cm and μ=0 in the chromium pile : top.
Ratio of the theoretical prediction to the experimental result (^{6}Li glass) : bottom.
All symbols are the same as in Fig.8.

and 150 - 400 keV, the ENDF/B-IV spectrum is closer to the measured one than the JENDL-2 spectrum. Between the above two regions, the measured spectrum markedly exceeds the predicted ones. More investigation of the spectral shape in this region (20-400 keV), where the cross section for iron have a number of resonant structures, should be carried out in future.

Neutron spectra in large bulk iron assemblies had been measured by other groups[5,6] and the present authors[9] before. The present tendency of the relation between the measured spectrum and the predicted one is similar to that being found in our earlier work.

For the spectrum in the nickel pile, the agreement between the measured and calculated ones is best among the three piles, although the former slightly exceeds the latter at a few regions. The two spectra predicted with the ENDF/B-IV and JENDL-2 agree with each other better than the cases for iron and chromium. This is probably due to the fact that the group constants of these two nuclear data files for nickel agree with each other in the relevant energy region, as seen in Fig.11.

For the neutron spectrum in the chromium pile, a satisfactory agreement can be seen above 100 keV. But the measured spectrum markedly exceeds the predicted ones, as described above. In several keV, this difference reaches a factor of 5 - 6. The ENDF/B-IV data give 30 - 40 % higher spectrum which is closer to the measured spectrum than that from the JENDL-2 data. The reason which causes the above discrepancy has not been found yet. The precise cross section data around 100 keV for chromium are requred to solve this problem in future. As shown in the next section for titanium, not only more precise but also higher resolution cross section data must be obtained.

For the neutron spectrum in stainless steel, however, most of the above discrepancies which appear in elemental pile would be considerably smeared out as seen in the earlier work by the present authors[9].

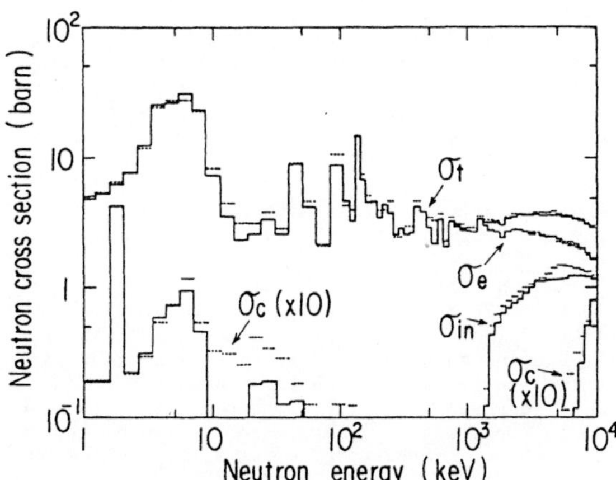

Fig.13 Comparison of the group constants for chromium produced by SUPERTOG-JR3. All symbols are the same as in Fig.9.

Titanium

Angular neutron spectra at r=25 cm and μ=0 in the pile obtained by the TOF measurement is shown in Fig. 14, in which the measured and calculated spectra are normalized in the energy range from 1 MeV to 1.5 MeV. The prediction from ENDF/B-IV with the self-shielding factors, which is shown by the solid line in the figure, agrees in general with the measured spectrum. However, this prediction underestimates the flux from 60 keV to about 200 keV. The disagreement is also seen around 20 keV.

Experimental reevaluation of total cross sections was also carried out through the transmission measurement by the linac TOF method. A measurement against a silicon-filtered neutron beam of energy near 147 keV was made as well as that against a quasi-white neutron beam. The details of the transmission measurement will

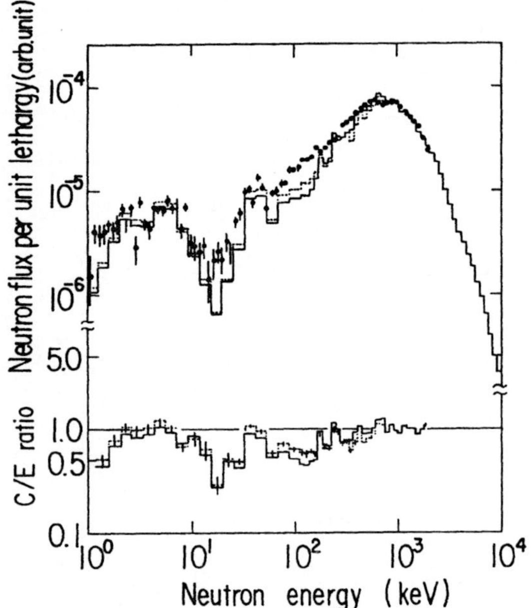

Fig.14 Neutron spectrum at r=25cm and μ=0 in the titanium pile : top.
measured (^{10}B-vaseline-NaI(Tl) counter)
calculated with ENDF/B-IV (MAT=1286),
calculated with the adjusted group constants.
Ratio of the theoretical prediction to the experimental result : bottom.
calculated (ENDF/B-IV)/measured,
calculated (modified)/measured.

be published soon[23]. The obtained cross sections are averaged with a 1/E weighting spectrum and shown in Fig.15, where they are compared with the group constants from ENDF/B-IV. The present measurement gives higher values of the cross section by about 30 % for neutrons below 200 keV, and smaller values by 10 % above 200 keV. Below 80 keV, present results contain uncertainty of about 10 % due to that in the background subtraction, which is not included in the errors in the figure. The weighted-average total cross section obtained with the silicon-filtered neutron beam, is 2.10 ± 0.05 barn at 147 keV and supports the results of quasi-white source measurement, of which the weighted-average one is 2.15 ± 0.06 barn. On the other hand, the ENDF/B-IV data gives 1.62 barn, which is smaller by 25 % than the present measurements.

We tried to adjust the group constants by exchanging the total cross section in ENDF/B-IV for the present measured ones. The other partial cross sections (σ_a, σ_e, and σ_{in}) were also changed by the same rate of change as the total cross section. The spec-

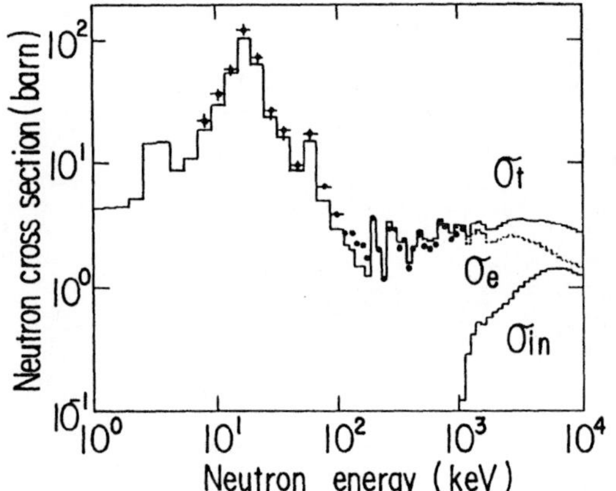

Fig.15 Comparison of the group constants for titanium produced by SUPERTOG-JR3 with ENDF/B-IV and the newly measured total cross sections(•).

trum predicted from the adjusted group constants is depicted in Fig.14, as a dotted line. It can be seen that the adjusted group constants improve the disagreement between the results of measurement and calculation around 200 keV. However, a notable discrepancy still remains there..

As was mentioned above for molybdenum, the inelastic scattering, especially the energy distribution of neutrons inelastically scattered, often has an important role in the prediction of spectrum. The ENDF/B-IV data for titanium contains only one discrete level of inelastic scattering, while recent experiment and evaluation have given more information such as 12 or more levels[25,26]. We tried to modify the ENDF/B-IV data by substituting the 12 levels with keeping the total cross section nearly constant. However, little change was observed in the new spectrum with these modified constants from that with the original ENDF/B-IV. This is due to the fact that, in the ENDF/B-IV, the only one level(Q=0.98 MeV) is dominant in the inelasic scattering. In addition to the energy distribution the values of inelastic scattering cross section might be invalid near the threshold, about 1 MeV. The sensitivity analysis by the DTF-IV-KANDO code, however, shows very low sensitivity for the inelastic scattering cross section; the sensitivity coefficient of the inelastic scattering cross section from 1 to 2 MeV to the neutron flux around 200 keV is about 0.016, which means only 1.6 % increase of the inelastic scattering cross section there.

It can be found that the disagreement between the measured spectrum and the calculated one is mainly due to abrupt decreases observed at 224 keV and 166 keV in in the latter one. These decreases can be ascribed to large leakage through the cross section minima in the energy group just above. The discussion above suggests the problem in the total cross sections around 200 keV, i.e. the elastic scattering cross sections there. This statement seems to contradict the fact that we tried to measure the total cross section itself in the present work. However, our measurement had the energy resolution of about 1 %, which meant about 1 % ambiguity in the determination of neutron energy. The effective group constants of elastic removal, i.e. the group constants with the self-shielding factors, which are produced from ENDF/B-IV for example, are changed by several % or more, when the neu-

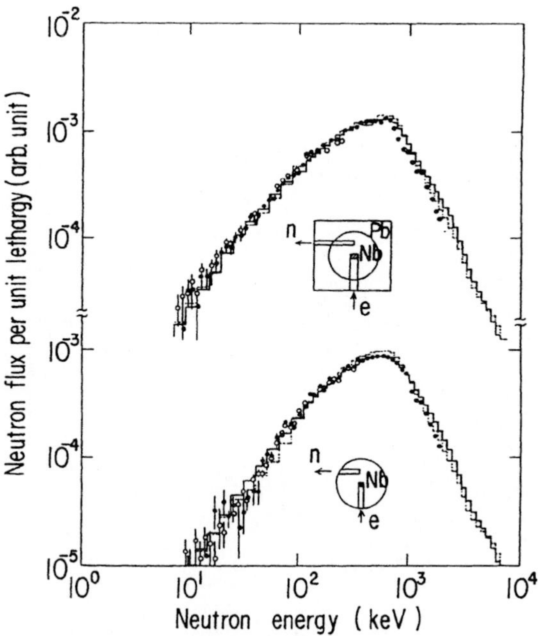

Fig.16 Neutron spectra at r=7cm and μ=0 in the niobium pile surrounded by the lead reflector (top) and that in the same pile bare-reflected (bottom).
measured (^{10}B-vaseling-NaI(Tl) counter),
calculated with JENDL-2,
calculated with ENDF/B-IV.

tron energy is changed only by 1 %. The sensitivity analysis shows that a 10 % increase of the elastic removal cross section of the 45th group (just above the decrease of the neutron flux at 166 keV), for example, results in a 4 % increase and a 3.6 % decrease of the neutron fluxes of the 46th and 45th energy groups, respectively, without any increase of the elastic scattering of the 45th group constants. These facts show the necessity of another measurement of the total cross section for titanium in the 100 keV region with higher resolution, and furthermore, might show the necessity of improvement in treatment of elastic cross sections near the cross section minima, especially just below them.

Niobium Pile with and without Lead Reflector

The spectra at r=7 cm and μ=0 in the niobium pile with and without a lead reflector are shown in Fig.16, where the calculated and measured spectra are normalized in the energy range from 100 to 300 keV. In this energy range, the two predictions from JENDL-2 and ENDF/B-IV reasonably agree. And both of these predictions are generally in good agreement with the measured spectra in both cases.

Concluding Remarks

The intercomparison of the experimentally obtained and theoretically calculated spectra of neutrons in the sample piles of the reactor structural materials has been demonstrated to check their cross section data integrally. This method seems to be very powerful to assess the scattering cross sections, both inelastic and elastic scattering cross sections at least for reactor structural materials, however the sensitivity to the capture cross section is rather small.

We have carried out a systematic study for main structural materials for FBR and CTR, such as molybdenum, iron, nickel, chromium, titanium and niobium, and then we have obtained new results for each element, though many of them are still to be discussed in future. Among them, the discrepancies between the measured and predicted spectra in the chromium and titanium piles are more noticeable than the others.

We adopted new Japanese nuclear data library, JENDL-2 for many cases, and ENDF/B-IV for comparison. More integral checks of other evaluated nuclear data with our neutron spectrum data are expected.

Furthermore, experimental work for other reactor materials will be continuously carried out at KURRI. For example, the neutron spectrum study in the copper pile is now being achieved.

Acknowledgement

This work has been supported by the Grant in Aid for Fundamental Research of the Ministry of Education, Science and Culture of Japan, by the Visiting Researchers Program of Research Reactor Institute, Kyoto University and by the Cooperative Research Program with JAERI.

The authors wish to express their sincere thanks to the staff of the Nuclear Data Center of JAERI and to the members of the Japanese Nuclear Data Committee, especially Drs. Y. Kikuchi, T. Asami and S. Iijima for the valuable discussion on the cross section data. They would like to acknowledge the cooperation and fruitful discussion of Mr. Y. Fujita of KURRI.

References

1. D. W. Muir(ed.), World Request List for Nuclear Data, INDC(SEC)-73/URSF, IAEA(1979).
2. O. N. Jarvis, European Appl. Res. Rept., Nucl. Sci. Technol., 3, Nos.1 and 2, pp.127-352 (1981) also EUR 7145 EN (1981).
3. M. S. Coates, et al., AERE-R 5364 (1968).
4. A. E. Profio, et al., Nucl. Sci. Eng., 35, 91 (1969).
5. R. J. Cerbone, GA-9149, Vol.1 (1969).
6. B. K. Malaviya, et al., Nucl. Sci. Eng., 47, 329 (1972).
7. N. N. Kaushal, et al., ibid., 49, 330 (1972).
8. A. J. H. Goddard, et al., Annals of Nucl. Sci. Eng., 1, 139 (1974).
9. I. Kimura, et al., NBS Sp. Publ. 425, Vol.1, p.184 (1975).
10. Hiroshi Nishihara, et al., J. Nucl. Sci. Technol., 14, 426 (1977).
11. I. Kimura, et al., ibid., 15, 183 (1978).
12. I. Kimura, et al., NBS Sp. Publ. 598, p.265 (1980).
13. T. Mori, et al., Annu. Rep. Res. Reactor Inst. Kyoto Univ., 13, 42 (1980).
14. T. Mori, et al., J. Nucl. Sci. Technol., 19, 427 (1982).
15. S. A. Hayashi, et al., Annu. Rep. Res. Reactor Inst. Kyoto Univ., 13, 23 (1980).
16. I. Kimura and S. Yamamoto, KURRI-TR-89, p.108 (1971).
17. I. Kimura, et al., Nucl. Instr. Meth., 137, 85 (1976).
18. K. Koyama, et al., JAERI M 7155 (1977).
19. National Neutron Cross Section Center, BNL "Evaluated Nuclear Data File IV" (data tape).
20. Nuclear Data Center, JAERI "Japanese Evaluated Nuclear Data Library" (data tape).
21. M. Nakagawa, unpublished.
22. W. W. Engle, Jr., K-1693 (1967).
23. K. D. Lathrop, LA-3373 (1965).
24. T. Mori, et al., to be published.
25. S. Philips, et al., ANL/NDM-28 (1977).
26. P. Guenther, et al., ANL/NDM-31 (1977).

INTEGRAL CAPTURE CROSS SECTION MEASUREMENTS OF SOME STRUCTURAL MATERIALS IN A FAST SPECTRUM

P. Azzoni, A. Salomoni
C.R.E. "E. Clementel"-ENEA
Via Mazzini, 2 - Bologna
Italy

C. Giuliani, R. Marvasi
AGIP NUCLEARE - Laboratori di
Medicina - Bologna
Italy

The integral capture cross section of iron, chromium, nickel and stainless steel, normalized to the integral fission cross section of ^{235}U, have been measured in fast neutron spectrum systems, using the fast-thermal reactor RB-2. The null reactivity method was used to obtain the K_∞ of the systems and the reaction rates were measured by ^{10}B capture and ^{235}U fission chambers. The results of the experiments provide a sensitive integral test of multigroup cross section sets.

[Iron, Chromium, Nickel, Stainless Steel, capture cross section].

Introduction

The structural materials used in the core of commercial fast reactors of current design are stainless steel and nimonic alloys. The principal elements in these steels are iron, chromium and nickel.

Nuclear data files for these materials are required from which multigroup cross section sets are derived in order to perform neutronic design calculations.

The reliability of these nuclear data can be tested against results of integral experiments in critical and sub-critical assemblies. To this end, an attempt was made at ENEA-Bologna - in the frame of ENEA-CEA cooperation in the Fast Reactor field - with the aim of improving our knowledge of neutron capture by structural materials.

The integral experiments were designed to provide a sensitive test of the (material capture/^{235}U fission) cross section ratio, R^m, for iron, chromium, nickel and stainless steel:

$$R^m=\sigma_c^m/\sigma_f^5=\int dE\ \sigma_c^m(E).\phi_\infty(E)\Big/\int dE\ \sigma_f^5(E).\phi_\infty(E) \qquad (1)$$

in "clean" neutron spectrum.

A first series of measurements was carried out in 1976-78 in epithermal spectra tailored to produce the bulk of the structural material capture events in the 100 eV - 100 KeV energy range. The results of those measurements were presented at the 1977 Geel NEANDC/ /NEACRP meeting[1] and at the 1979 Aix-en-Provence Symposium on Fast Reactor Physics[2].

This paper reports on the results obtained in 1980 in a fast neutron spectrum similar to that of a large fast breeder reactor.

General Features of the Experiments

The design of the experiments was based on the simple consideration that a comparison between theory and experiment can be established if the measurements are carried out in situations easily reproduced by calculation, so that discrepancies outside the overall experimental error should be attributed to deficiencies in the nuclear data.

Because of the difficulties in effecting direct measurements, the (σ_c^m/σ_f^5) cross section ratio has been derived from the neutron balance equation in a medium into which the structural material was inserted.

More precisely, the experiment was carried out in a medium simulating an infinite homogeneous critical reactor. In this way the neutron flux is only a function of neutron energy and can be reliably calculated.

The homogeneity of the medium considerably reduces the difficulties related to neutron flux computation.

The neutron absorbers in the medium, other than the structural material under study, were chosen from among those having "known" cross sections.

Derivation of the (σ_c^m/σ_f^5) Cross Section Ratio

In an infinite homogeneous medium, in which the structural material, (m), is inserted, the neutron balance equation is:

$$K_\infty=\frac{S_i\int dE\ \nu^i(E).\Sigma_f^i(E).\phi_\infty(E)}{S_i\int dE\ \Sigma_a^i(E).\phi_\infty(E)}=\frac{S_i(\nu^i.N^i.\sigma_f^i)}{S_i(N^i.\ \sigma_a^i)} \qquad (2)$$

where ν^i is the average number of neutrons of the element (i), $\phi_\infty(E)$ the "asymptotic" neutron flux and the σ are the flux-averaged microscopic cross sections; S is the summation symbol.

If it is supposed that a suitable amount of ^{10}B is added to the medium, fuelled with ^{235}U, until unit $K_\infty = 1$ is reached, then eq. (2) may be written as:

$$K_\infty = \frac{\nu}{1+\alpha^5+(X.R)^{10B}+(X.R)^m+E} = 1 \qquad (3)$$

$$\text{where}\quad \nu=\frac{S_i(\nu^i\ N^i.\sigma_f^i)}{(N.\sigma_f)^5}\ ;\ \alpha^5=(\frac{\sigma_c}{\sigma_f})^5;$$

$$X^i=\frac{N^i}{N^5}\ ;\ R^i=\frac{\sigma_c^i}{\sigma_f^5}$$

The term E takes the absorptions (other than Boron-10, structural material and Uranium-235) into account.

A similar equation holds in a critical medium made up of the same elements as the previous medium, other than structural material ("reference medium", r):

$$K_\infty^r = \frac{\nu_r}{1+\sigma_r^5+(X.R)_r^{10B}+E_r} = 1 \qquad (4)$$

By combining Eq.s (3) and (4), Eq. (1) becomes:

$$R^m=(\frac{\sigma_c^m}{\sigma_f^5}) = \frac{1}{X^m}\Big\{(\frac{\nu}{K_\infty}) - d\ (\frac{\nu}{K_\infty})_r + (X.R)^{10B} - $$

$$- d(X.R)_r^{10B} + d-1 + E - d.E_r\Big\} \qquad (5)$$

$$\text{where}\quad d=\frac{\alpha^5}{\alpha_r^5}$$

Measurements, performed with ^{10}B capture and ^{235}U fission miniaturized chambers, provide the values of the R^{10B} cross section ratios; the atom concentrations of the infinite homogeneous critical systems are inferred from K_∞ measurements, made by means of the "null reactivity technique"[3].

Theory is involved only to provide small contributions

K. H. Böckhoff (ed.), Nuclear Data for Science and Technology, 106–109.

of less important absorption reactions as well as calculations of the (α^5/α^5_r) ratios and the ν values.

The reliability of Eq. (5) to determine the R^m value rests on the fulfillment of strict requirements such as, (1) an infinite medium simulation as exact as possible and (2) good agreement between the asymptotic spectra of the "reference medium" and the medium with structural material.

The first requirement was met by means of optimization of the design parameters of the experimental facility while the second was satisfied through atom densities of the elements bound to give α^5_r and R_r^{10B} calculated values in the reference medium very near (±1%) to those in the medium containing the structural material.

Design of the Experiment

- Infinite homogeneous boundary conditions

As mentioned above, the measured and calculated parameters involved in Eq. (5) should be obtained for an infinite critical homogeneous medium.

The "infinite medium" conditions were simulated in the central cavity of the RB-2 Reactor[4].

This cavity was surrounded by homogeneous fast and heterogeneous "buffer" regions. These zones effectively coupled the fast to the thermalized spectrum of the outer "driver" region of the reactor, which maintained its criticality and provided the source of thermal neutrons to the inner fast region.

The materials with "known" cross sections utilized in the homogeneous media were:enriched (∿93% atom percent of ^{235}U) Uranium dioxide, graphite and Boron-10.

To minimize the influence of heterogeneities on the results, the media were made up of a loose mixture of UO_2 and nuclear grade graphite particles coated with pyrocarbon to an overall average diameter of 1000μm. A fraction of the graphite particles was loaded with Boron (as B_4C) to an enrichment of ∿12 w/o (weight percent of Boron). The structural material was added to the mixture as 750 μm dia particles (iron, nickel and stainless steel) or as roughly uniform size grains (chromium). The mixture components could be recovered after use by the "heavy liquids" separation technique. Satisfactory homogeneity of the media was achieved by means of an air-jet mixer specially conceived for this purpose[5].

- Composition of the infinite homogeneous critical medium

As mentioned above, the atom densities of the elements in an infinite homogeneous critical medium were obtained by means of the "null reactivity technique" which provides a method of determining the $(^{10}B/^{235}U)$ atom density ratio that makes $K_\infty = 1$.

The basis of the "null reactivity technique" is that zero change in reactivity, i.e. zero change in K_∞, is produced by removing a sample from an infinite homogeneous critical system. Conversely, if the $(^{10}B/^{235}U)$ atom ratio of a sample with suitable boundary conditions is adjusted until zero reactivity change occurs on removing it from the critical reactor, then K_∞ of the sample is unity.

In accordance with this, determination of the "null reactivity" composition involves essentially measurements of the "reactivity worth" of samples of varying boron contents, from which the "null reactivity" $(^{10}B/^{235}U)$ atom ratio can be deduced by linear interpolation.

In the present experiments each set of samples was characterized by the same quantity of fissile and structural material, when present. The increase of borated graphite microspheres was counterbalanced by a decrease in graphite microspheres to keep the total amount of graphite quite constant.

The "reactivity worth" of the sample was inferred by the Fourier analysis of the neutron density modulation,induced by the oscillation of a 60 mm - dia and 4200 mm - length column containing both the test and the empty sample. The first harmonic amplitude was normalized to the mean power level. The selected measurement parameters were: quasi square oscillation with 60 sec period, 1.8 sec sample transit time, 1500 mm stroke, 2 W reactor power, 100 mm sample height. The sample positioning was accurate to ±0.4 mm and maximum acceleration was below 0.2 g. Further information has been given in previous papers[6,7].

Flux modulation was measured by means of three independent counting rate chains. The three values ranged within the experimental error of (A/P):$\pm 2\cdot 10^{-4}$, corresponding to $dK_{eff}/K_{eff} = 0.1\cdot 10^{-5}$ of the reactor.

The natural Boron content in the borated graphite microspheres was chemically measured in four laboratories, following a technique described by P.L. Buldini and D. Ferri[8]; the resulting average value was (12.75±0.20) w/o weight percent.

Particular care was devoted to sample preparation in order to reduce the influence of the Boron content uncertainties. The guideline of the process was the utilization of most of the borated microspheres of the "reference samples",used in the "null reactivity measurements", in making up the mictures chich include the structural material under study. In this way, propagation of the Boron content error was greatly reduced[9].

The reactivity worth (A/P) of the measured samples was plotted against the (borated graphite/Uranium dioxide) microsphere weight ratios and least square linear fits were performed. Subtraction of the (A/P) value, measured while oscillating the column with the two empty samples, and utilization of the Boron content of the borated microspheres yielded the "null reactivity" $(^{10}B/^{235}U)$ atom ratio, X^{10B}.

The X^{10B} values given in Table I, together with the $(\text{structural}/^{235}U)$ atom ratio,are effected by an overall experimental error (1σ) ranging between 0.2%-0.3%.

The $(\text{graphite}/^{235}U)$ atom ratio of the reference sample was equal to 50, while the ratio varied between 34 and 37 for the other samples.

Table I - $(^{235}U/\text{structural material})$ and $(^{10}B/^{235}U)$ atom density ratios in the "null reactivity media"

Structural material	N^{235U}/N^m	N^{10B}/N^{235U}
Fe 1(°)	0.04435±0.8%	0.6156±0.2%
Fe 1.3(°)	0.03665±0.5%	0.6074±0.3%
Cr	0.04343±0.8%	0.5855±0.2%
Ni	0.04668±1.3%	0.4760±0.2%
SS(*)	0.04418±0.6%	0.5452±0.2%
reference medium	--	0.6206±0.2%

(°) Media with different concentrations of iron

(*) SS means Stainless Steel

- Reaction rate measurements

The ratio of the spectrum - averaged ^{10}B capture to ^{235}U fission cross section, R^{10B}, in the "null reactivity" samples was measured by means of ^{10}B capture (∿50 μgr ^{10}B coating) and ^{235}U fission (∿50 μgr ^{235}U coating) chambers (4 mm - dia, 10 mm active length), the spectrum in the thermal column of the reactor being used for normalization.

All measurements were intercalibrated with high statistical accuracy by continuous monitoring of the reactor power,using two ^{235}U fission chambers. To reduce background noise as well as signals from spurious reactions, the ^{10}B - coated chamber discriminator was set to reject the pulses corresponding to ^7Li of the reaction $^{10}B(n,\alpha)^7Li$. Dummy chambers - i.e. without active coating - were used to evaluate the magnitude of residual signals from reactions with the chamber materials. The measured effects were found to be negligible both in the fast and thermal spectrum.

The emission spectrum of the ^{10}B chamber above the selected bias was verified as being the same in the test zone and the thermal column, allowing the use of a single constant to convert measured pulse rates into Boron capture events. This constant and similarly the constant correlating ^{235}U fission to ^{235}U - coated chamber pulse rate, cancelled each other out in the $(^{10}B/^{235}U)$ cross section ratio evaluation.

The dead times of the ^{10}B and ^{235}U chambers and associated equipment were verified as being equal (3.7 ± 0.5 μs). By adjusting the reactor power to the specific chamber type and location, it was possible to obtain a close compensation of counting rate losses, yielding an overall dead-time correction factor of 0.997.

The chamber's self-shielding and the local flux perturbation factors were verified as being significant (0.980) only for the ^{10}B chamber in thermal flux. Particular care was devoted in positioning the chamber in the thermal column, because the gradient flux was such as to give a counting rate variation of $\pm 0.4\%/1$ mm.

In order to verify the flatness of the neutron spectrum in the central sample, ^{235}U fission reaction rate measurements were made at different radial and axial positions of the fast region.

Experimental tests confirmed the calculated predictions that "null reactivity" samples did not change the shape of the neutron energy flux in the central cavity of the reactor.

The experimental value of the $(^{10}B/^{235}U)$ cross section ratio in the central cavity (local value) was $1.638 \pm 0.7\%$ (1σ). The quoted uncertainty includes all known sources of experimental errors together with those on the 2200 m/s neutron reference cross sections.

Results

Because of the perturbation of the direct and adjoint fluxes in the central "null reactivity" samples, due to the influence of the surrounding regions, the statement that zero change in reactivity is produced by a removal of the sample of unit K_∞ should be taken as approximately true.

As a consequence, to reduce experimental measurements to the ideal infinite medium situation in order to allow easier comparison of the final results with theory, the necessary corrections were calculated by the two-dimensional (r,z) multigroup transport theory, using the DOT 3.5[10] and the MICRO[11] codes.
A 25-group cross section set was obtained from ENDF-B/IV, handled by the MC^2-II code[12], and from the CARNAVAL-IV set, handled by the HETAIRE code[13]. Cross sections for the driver region were obtained by means of the GGC-IV code[14].

Detailed calculations of the flux mismatch factors have shown that the corrections needed are small and, moreover, are not very sensitive to the nuclear data used.

The experimental values of the $R^m = \sigma_c^m/\sigma_f^5$ ratios , as inferred from Eq. (5), are presented in Table II with the overall uncertainty, (1σ), resulting from a quadratic combination of the errors of the measured quantities. The same table lists the calculated cross section ratios.

Additional calculations were made for iron. Substituting the capture cross section data of the ENDF-B/IV by the data supplied by F. Corvi[15] in the 1 KeV-100 KeV energy range, a new multigroup cross section set was calculated. The values of the $(\sigma_c^{Fe}/\sigma_f^5)$ cross section ratios obtained by using this set are shown in Table II.

Table II - (σ_c^m/σ_f^5) cross section ratios in the "null reactivity media"

Structural material	Experimental values ($x10^3$)		Calculated values ($x10^3$)		
	BIV(*)	CIV(*)	BIV	CIV	BIV_c(°)
Fe 1	3.50 ± 0.14	3.00 ± 0.12	4.67	2.96	3.90
Fe 1.3	3.59 ± 0.16	3.21 ± 0.14	4.53	2.84	3.75
Cr	4.86 ± 0.15	4.81 ± 0.14	8.13	4.50	-
Ni	10.71 ± 0.32	10.92 ± 0.33	11.56	9.67	-
SS	8.21 ± 0.25	7.79 ± 0.23	9.77	7.73	-

(*) Absorption reactions, E, and mismatch correction factors calculated by using ENDF-B/IV (BIV) and CARNAVAL-IV (CIV) data, respectively.

(°) Substitution of the iron capture cross section data of the ENDF-B/IV with the data supplied by F. Corvi[15], in the 1 KeV - 100 KeV energy range.

Conclusions

From Equation (5) it may be deduced that the influence of the calculated parameters keeping account of the less important absorption reactions, as well as of the ratio (α^5/α_r^5) and values of ν, become important when neutron captures of the structural material are low. This is due to the fact that the X^{10B} parameters have approximately the same value.

This situation occurs when iron is under study. The difference between parameters calculated by using ENDF-B/IV and CARNAVAL-IV data, respectively, is responsible for the difference in the $(\sigma_c^{Fe}/\sigma_f^5)$ experimental values, as is shown in the first two columns of Table II.

Comparison of the (σ_c^m/σ_f^5) ratios with predictions made by using ENDF-B/IV shows up significant discrepancies for all materials except nickel. The iron capture cross section data, supplied by F. Corvi, considerably improved the $(\sigma_c^{Fe}/\sigma_f^5)$ calculated values.

The situation in the case of the predictions made by using CARNAVAL-IV data is quite satisfactory, even though anomalous behaviour can be seen in the iron results.

Acknowledgements

Mr. G. Brighenti, Mr. M.L. Chiodi, Mr. B. Camiscioni, Mr. S. Gualdalini and Mr. L. Strazzari of AGIP Nucleare, Laboratori di Medicina, Mr. M. Piani of C.R.E. "E. Clementel", ENEA, Mr. R. Bruschi, V. Casasanta and T. Ferrari of C.S.N. Casaccia, ENEA, contributed greatly to the experimental work. The nuclear calculations were carried out by Mr. C. Petrella of C.R.E. "E. Clementel", ENEA. Mr. A. Garagnani,with the cooperation of Mr. A. Messora of C.R.E. "E. Clementel", ENEA and Mr. G. Beghelli of AGIP Nucleare, Laboratori di Medicina, was responsible for the fabrication of the homogeneous media. Many thanks are due to Mr. R. Martinelli for his helpful suggestions.

References

1. Neutron data of Structural Materials for Fast Reactor, Proceedings of a Specialist's Meeting, Geel - Belgium, 5-8 Dec. 1977, Pergamon Press (1979).

2. Fast Reactor Physics 1979, Proceedings of an International Symposium on Fast Reactor Physics, Aix-en-Provence, France, 24-28 Sept. 1979, IAEA (1980).

3. F. Benedetti et al.: Energia Nucleare (Italy) 24,4,199 (1978).

4. F. Benedetti et al.: Energia Nucleare (Italy) 24,2,73 (1977).

5. F.Lanza et al.: EUR-5143i, IRC; Ispra Establishment (1974).

6. P. Azzoni et al.: Energia Nucleare (Italy) 24,8/9,411 (1977).

7. P. Azzoni et al.: Nuclear Science and Engineering 76,1,70 (1980).

8. P.L. Buldini and D. Ferri: Microchemical Journal 25,143 (1980).

9. A. D'Angelo: Private Communication, ENEA (Italy).

10. RSIC Computer Code Collection, ORNL/TM-4280 (1977).

11. P. Azzoni and M. Galli: "RIT/FIS-LFSR(78)2, CNEN- - Bologna (Italy).

12. H. Henryson II et al.: ANL-8144, June 1976.

13. Private Communication: CEA (France).

14. J. Adir and D. Lathrop: GA-9021 (October 1968).

15. F. Corvi: B.C.M.N. Euratom, Geel (Belgium)-Private Communication.

ETUDE EXPERIMENTALE DE LA CAPTURE DES MATERIAUX DE STRUCTURE DANS DES RESEAUX A NEUTRONS RAPIDES PAR LA METHODE K∞ = 1

M. Darrouzet*, E. Fort**, G. Rimpault**, L. Martin Deidier*

Commissariat à l'Energie Atomique
* Département des Réacteurs à Eau
** Département des Réacteurs à Neutrons Rapides

Centre d'Etudes Nucléaires de Cadarache
B.P. N° 1 - 13115 - SAINT PAUL LEZ DURANCE
FRANCE

RESUME

Les matériaux de structure influent de façon importante sur le calcul du bilan en réactivité et le gain de surgénération des réacteurs de puissance de la filière à neutrons rapides.
A l'heure actuelle, la connaissance de la capture de l'acier dans la gamme de spectre intéressant ces réacteurs est très imparfaite et les incertitudes sont encore plus importantes pour les différents composants de l'acier pris individuellement (Fe, Cr, Ni, Mo...).
Afin de ramener ces incertitudes à un niveau raisonnable pour les projets des futurs grands réacteurs de la filière, un programme expérimental basé sur la méthode K∞ = 1 a été lancé au C.E.A.
Deux coeurs, fortement chargés en matériaux de structure, acier pour l'un, nickel pour l'autre, ont été réalisés dans l'assemblage couplé thermique-rapide ERMINE de la pile MINERVE. La mesure du bilan neutronique de leur cellule a permis d'obtenir directement des informations sur la capture des deux matériaux précédents avec les précisions requises (meilleures que 10 %).
Dans ces mêmes réseaux, la capture des autres éléments entrant dans la composition des aciers ou présents sous forme d'impuretés a été obtenue par la méthode du remplacement de cellules. Une comparaison des résultats obtenus avec les valeurs calculées à l'aide du formulaire CARNAVAL IV est présentée.

Introduction

A l'heure actuelle, un des facteurs d'incertitude les plus importants sur le calcul du bilan en réactivité des coeurs des réacteurs à neutrons rapides réside dans la mauvaise connaissance de la capture des matériaux de structure dans cette gamme de spectre [1]. On estime ainsi à environ 15 à 20 % la précision sur le calcul de l'effet de l'acier avec la version en cours du formulaire CARNAVAL (version IV) utilisé au CEA pour les projets de réacteurs à neutrons rapides [2]. Il en résulte une incertitude de plus de 700 pcm sur la détermination de la masse critique.

Afin de diminuer cette incertitude un programme de mesures intégrales a été décidé sur les réacteurs MINERVE (expériences ERMINE) et RB2 (dans le cadre de la collaboration CNEN/CEA [3]). Il est basé essentiellement sur des mesures de bilan neutronique dans des réseaux fortement chargés en matériaux de structure, les autres techniques de mesures (irradiation, oscillation) présentant d'importantes difficultés de mise en oeuvre ou d'interprétation [4].

Deux réseaux OA 10 et ON 10, contenant respectivement de l'acier et du nickel, ont été réalisés dans l'assemblage couplé thermique-rapide ERMINE. Les résultats obtenus ont permis, par interprétation du bilan neutronique des cellules, de déterminer la capture de ces deux matériaux avec une précision répondant aux besoins des projets de réacteurs (meilleure que 10 %).

Réseaux expérimentaux

1. Cellules

Les réseaux OA 10 et ON 10 réalisés dans l'assemblage ERMINE sont constitués de réglettes de simulation MASURCA, cylindriques ou carrées, de 12,7 mm de diamètre ou côté, rassemblées dans des tubes de structure en acier inoxydable.

Le pas des deux réseaux est de 26,5 mm.

Le combustible est constitué par de l'oxyde d'uranium enrichi à 27 % en ^{235}U, associé à une réglette d'oxyde d'uranium appauvri ou naturel afin d'obtenir un K∞ proche de l'unité.

Les matériaux de structure sont constitués par 2 réglettes d'acier pour OA 10 et deux réglettes de nickel pour ON 10.

Une coupe horizontale des cellules de ces 2 réseaux est présentée sur la figure 1. Leurs principales caractéristiques neutroniques calculées en mode fondamental avec le formulaire CARNAVAL sont données dans le tableau I.

Dans les deux cas, le profil de la sensibilité du K* à la capture des matériaux de structure est très voisin de celui caractéristique de la zone interne de SUPER PHENIX, ce qui élimine tout problème au niveau de la transposition des résultats expérimentaux [4].

K. H. Böckhoff (ed.), Nuclear Data for Science and Technology, 110–114.

COUPE HORIZONTALE DES CELLULES DES RESEAUX
OA 10 ET ON 10

FIGURE 1

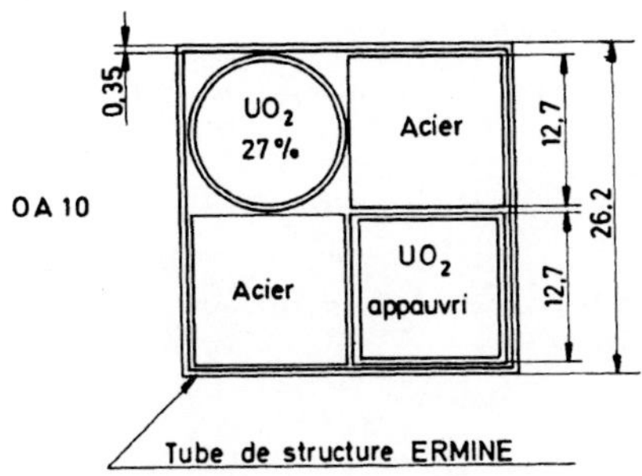

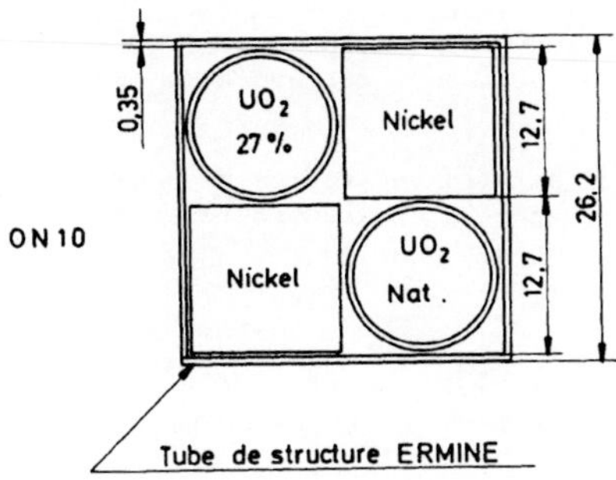

TABLEAU I : PARAMETRES NEUTRONIQUES EN MODE FONDAMENTAL DES RESEAUX OA10 et ON 10

	OA 10	ON 10
$K\infty$	1.01731	0.99368
K^*	1.01931	0.99333
$r = \frac{\nu \Sigma f}{\xi \Sigma s}$	0.259	0.243
$\sigma c5/\sigma f5$	0.314	0.323
$\sigma c8/\sigma f5$	0.145	0.141
$\sigma f8/\sigma f5$	0.0138	0.0142
$\sigma cFe/\sigma f5$	0.00288	0.00430
$\sigma cCr/\sigma f5$	0.00446	0.00457
$\sigma cNi/\sigma f5$	0.0107	0.00833

2. Assemblages

L'étude des deux réseaux OA 10 et ON 10 a été effectuée dans l'expérience ERMINE, version couplée thermique-rapide du réacteur MINERVE [5]. Dans cette configuration, le réseau à neutrons rapides central est entouré d'une zone de transition permettant d'obtenir au centre de la zone de mesure un spectre neutronique se rapprochant au mieux du mode fondamental, tout en limitant le volume de la zone centrale.

Dans le cas des réseaux OA 10 et ON 10, la zone d'adaptation était constituée par de l'oxyde d'uranium enrichi à 3,5 % et de l'uranium métallique appauvri. Une coupe horizontale de l'assemblage ERMINE est présentée sur la figure 2.

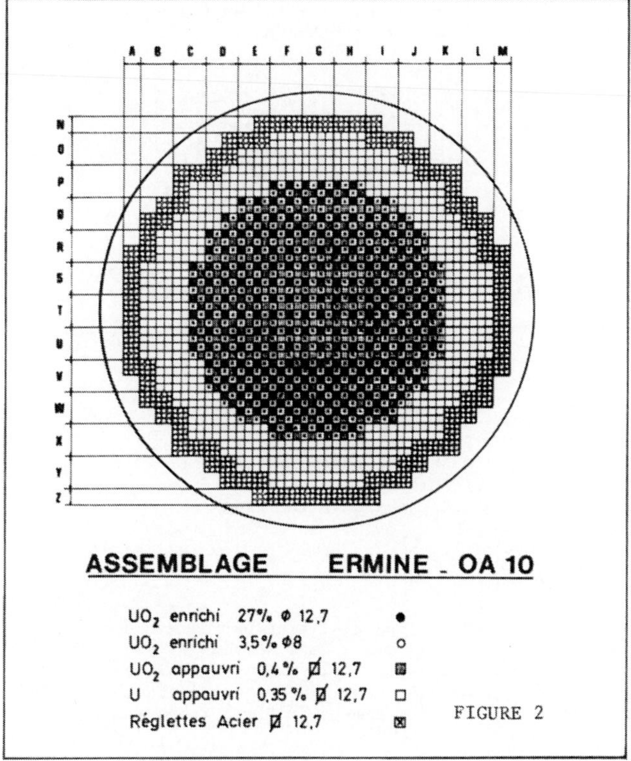

FIGURE 2

Mesures de taux de réaction

1. Mesures d'indices par chambres à fission

Les taux de fission qui ont été mesurés dans les réseaux OA 10 et ON 10 sont ceux de ^{233}U, ^{235}U, ^{238}U, ^{239}Pu.

Les mesures ont été effectuées à l'aide de chambres à fission miniatures, cylindriques de 4 mm de diamètre. Les chambres étaient introduites à l'intérieur de réglettes d'acier ou de nickel percées d'un trou Ø 6,5 mm.

Pour ramener les indices mesurés en mode fondamental, on applique un certain nombre de facteurs calculés afin de tenir compte :

- de la perturbation créée dans la cellule par l'enlèvement d'une partie de l'acier ou du nickel pour le passage de la chambre.

- du passage de l'indice mesuré dans le diluant à l'indice dans le combustible. Cette correction dite "d'hétérogénéité" est obtenue expérimentalement à l'aide de détecteurs feuilles (1/10 ou 2/10 mm d'épaisseur) positionnés dans les différentes régions de la cellule.

- de l'écart existant au centre de la zone de mesure avec le spectre en mode fondamental (correction spatiale).

Les résultats obtenus dans les 2 réseaux OA 10 et ON 10 sont présentés dans le tableau II sous forme d'une comparaison expérience-calcul avec les valeurs prédites par le formulaire CARNAVAL IV.

TABLEAU II : RESULTATS DES MESURES D'INDICES DE SPECTRE

E-C/C	OA 10	ON 10
$\sigma f3/\sigma f5$	- 1.8 ± 1.4	- 1.6 ± 1.4
$\sigma f8/\sigma f5$	+ 10.0 ± 1.4	+ 8.1 ± 1.1
$\sigma f9/\sigma f5$	- 1.5 ± 1.0	- 2.2 ± 1.0
$\sigma c8/\sigma f5$	- 0.5 ± 1.5	- 0.5 ± 1.6

On constate un désaccord assez important (+ 10 %) pour l'indice $\sigma f8/\sigma f5$ dans ce type de réseau fortement chargé en matériaux de structure.

2. Mesures de la capture de l'uranium 238 par détecteurs

Les mesures ont été réalisées par la méthode classique du comptage des γ émis lors de la décroissance du ^{239}U, avec étalonnage préalable en colonne thermique. Les détecteurs utilisés sont en uranium appauvri à 400 ppm en ^{235}U, de 2/10 mm d'épaisseur.

Les corrections appliquées aux valeurs mesurées sont les suivantes :

- correction spatiale.

- correction de dilution tenant compte de la dilution différente de l'isotope entre le milieu et le détecteur.

- correction de position tenant compte de la position exacte du détecteur dans la cellule.

Nous donnons dans le tableau II l'écart expérience-calcul sur l'indice $\sigma c8/\sigma f5$ qui est bien prédit par le formulaire CARNAVAL dans la limite des barres d'erreurs expérimentales.

Mesures du bilan en réactivité

Dans les réseaux à $K\infty$ voisin de 1, le bilan en réactivité K+ est aisément obtenu à partir d'une mesure de l'effet en réactivité de la cellule [6].

$$K+ = \frac{< \phi^*, P\phi >}{< \phi^*, K\phi >}$$ avec P = Production, K = Absorption + Transferts

Deux types de normalisation peuvent être utilisées :

- soit à l'effet du combustible de la cellule ; on écrit alors :

$$K+ - 1 = m \frac{\delta K \text{ cellule}}{\delta Kc}$$

avec δKc, effet en réactivité du combustible présent dans la cellule,
m, coefficient calculé, peu sensible aux sections efficaces et à la méthode de calcul utilisées [7].

- soit à l'effet d'une source de neutrons, ce qui revient à écrire, toutes simplifications effectuées [8] :

$$\frac{K+ - 1}{K+} = \frac{\delta K \text{ cellule}}{\delta K \text{ source}} \times \frac{Ne}{P}$$

avec δK_{source}, effet de la source de neutrons

Ne, nombre de neutrons émis par la source

P, production de la cellule, obtenue expérimentalement à partir d'une mesure absolue du taux de fission du ^{235}U dans la cellule et de l'indice $\sigma f8/\sigma f5$ également mesurée.

1. Techniques expérimentales

- L'effet en réactivité de la cellule du réseau est obtenu en comparant, par la méthode des oscillations, une cellule de référence de composition bien connue à un "vide de cellule". Celui-ci est créé en écartant artificiellement les 2 parties du chargement de la canne d'oscillations. Pour les 2 réseaux, la cellule de référence était constituée de 4 cellules élémentaires, soit 53 mm de côté sur 100 mm de hauteur environ.

- L'effet du combustible de normalisation, c'est-à-dire du ^{235}U, a été mesuré dans chaque réseau par oscillations, en utilisant la méthode du "remplacement de combustible" qui permet d'obtenir une bonne précision sur l'effet cherché tout en ne perturbant pas trop neutroniquement le réseau de référence. On a procédé par variations de densité ou d'enrichissement en utilisant des réglettes de diamètre moitié (ϕ 6.35 mm).

- L'effet de la source de neutrons (^{252}Cf) a été mesuré par oscillations en opérant à différentes puissances du réacteur de façon à obtenir par extrapolation à puissance nulle l'effet direct des neutrons émis.

2. Interprétation des mesures de vide de cellule

Diverses corrections doivent être appliquées à l'effet mesuré de la cellule avant de pouvoir comparer cet effet à la valeur prédite par le calcul :

- correction de composition pour ramener la cellule expérimentale à une composition identique à celle de la cellule calculée. On utilise pour cette correction les valeurs mesurées par remplacement de combustible des effets du ^{235}U et du ^{238}U ainsi que l'effet de l'acier inoxydable mesuré par oscillations.

- correction de gradient afin de corriger la mesure de l'effet de "vide de cellule" des termes de fuite.

- correction spatiale.

3. Résultats

Nous donnons dans le tableau III, les valeurs expérimentales du K+ des réseaux OA 10 et ON 10 obtenues par les deux méthodes : normalisation à l'effet du combustible et normalisation à l'effet de la source. La cohérence des résultats est très bonne.

TABLEAU III : RESULTATS DES MESURES DE BILAN EN REACTIVITE K+

Résultats en pcm	OA 10	ON 10
K+ - 1 par normalisat. à l'effet du combust.	+ 2740±180	+ 510±130
K+ - 1 par normalisat. à l'effet d'une source	+ 2495±160	+ 615±135
Valeur moyenne	+ 2600±145	+ 560±125
E - C	+ 860±145	+1190±125

Nous présentons dans le même tableau la comparaison expérience-calcul avec les valeurs obtenues à l'aide du formulaire CARNAVAL IV. On constate que les calculs sous-estiment d'environ 1000 pcm le bilan en réactivité pour les 2 réseaux.

Interprétation en bilan direct

Pour procéder à une interprétation du bilan en taux de réaction, il est nécessaire de passer de la valeur mesurée du K+ à la valeur du K* par l'intermédiaire d'un facteur α calculé tel que

$$\alpha = \frac{K^* - 1}{K^+ - 1}$$

Le coefficient α étant peu sensible aux sections efficaces utilisées dans la gamme des coeurs à $K\infty = 1$, on n'introduit pas d'erreurs supplémentaires sur le K* par cette méthode.

Les valeurs de K* ainsi obtenues sont ensuite corrigées des écarts observés entre valeur expérimentale et valeur calculée sur les taux de réaction mesurés par chambres à fission ou détecteurs (en pondérant ces écarts par la part que prennent les différentes réactions dans le bilan de la cellule).

L'écart résiduel subsistant entre la valeur mesurée et corrigée du K* et sa valeur calculée peut alors être imputé aux taux de réaction non mesurés dans la cellule : capture de l'acier pour OA 10 et capture du nickel pour ON 10.

Pour réaliser cette opération, nous procédons en différenciant l'expression du K* :

$$K+ = \frac{\sum_i N_i \gamma_i \sigma f_i}{\sum_i N_i (\sigma c_i + \sigma f_i)} =$$

$$\frac{\gamma_5 N_5 \sigma f_5}{N_5 \sigma f_5} \times \frac{1 + \sum_i \frac{Ni}{N5}\frac{\gamma i}{\gamma 5}\frac{\sigma fi}{\sigma f5}}{1 + \sum_i \frac{Ni}{N5}\frac{\sigma ci}{\sigma f5} + \sum_i \frac{Ni}{N5}\frac{\sigma fi}{\sigma f5}}$$

$$\frac{\Delta K^*}{K^*} = \frac{\Delta\gamma 5}{\gamma 5} + \sum_i \frac{\Delta(\gamma i/\gamma 5)}{\gamma i/\gamma 5}\frac{pi}{P} - \sum_i \frac{\Delta(\sigma ci/\sigma f5)}{\sigma ci/\sigma f5}\frac{ci}{A}$$

$$+ \sum_i \frac{\Delta(\sigma fi/\sigma f5)}{\sigma fi/\sigma f5}\left(\frac{pi}{P} - \frac{fi}{A}\right)$$

P et A représentent la production et l'absorption totales dans la cellule ; pi, fi, ci représentent la part dans le bilan de la production, de la fission et de la capture de l'isotope i.

$$\frac{\Delta K^*}{K} \quad \frac{\Delta\ (\sigma fi/\sigma f5)}{\sigma fi/\sigma f5} \quad \text{et} \quad \frac{\Delta\ (\sigma ci/\sigma f5)}{\sigma ci/\sigma f5}$$

sont assimilées aux écarts expérience-calcul respectivement sur le K* et sur les indices mesurés préalablement.

La décomposition du bilan des cellules OA 10 et ON 10 est indiquée dans le tableau IV.

TABLEAU IV : DECOMPOSITION DU BILAN DES RESEAUX OA 10 et ON 10 (normalisé au total des productions = 1)

Bilan en pcm	OA 10	ON 10
Fission ^{235}U	36 600	36 950
Capture ^{235}U	11 500	11 900
Fission ^{238}U	3 700	3 400
Capture ^{238}U	38 400	33 950
Capture Acier	7 800	1 750
Capture nickel	-	12 800
Capture oxygène	150	150
Absorption totale	98 200	100 900

L'interprétation en bilan direct et les tendances obtenues sur la capture de l'acier et du nickel sont présentées dans le tableau V.

TABLEAU V : INTERPRETATION DU BILAN EN TAUX DE REACTION

	OA 10	ON 10
Ecart exp-calc sur K*	+ 940±160 pcm	+ 1270±130 pcm
Correction des écarts sur les taux de réaction	- 870±760 pcm	- 750±790 pcm
Part des matériaux de structure dans le bilan	Acier : 7780 pcm	Nickel : 12800 pcm
Tendance sur les matériaux de structure (E-C/C)	Acier : -1 ± 10 %	Nickel : - 4 ± 6 %

On constate que dans la limite des barres d'erreur expérimentale, la capture de ces deux matériaux est correctement prédite par le formulaire CARNAVAL IV.

Conclusions

La réalisation des réseaux à K∞ voisin de 1, OA 10 et ON 10, dans l'assemblage couplé thermique-rapide a permis d'obtenir des informations très intéressantes sur la capture des matériaux de structure dans la gamme de spectre des réacteurs de la filière à neutrons rapides.

A partir des mesures de bilan en réactivité K+ par la méthode du "vide de cellule", on a pu mettre en évidence l'absence de désaccord avec le formulaire CARNAVAL pour la capture de l'acier inoxydable et du nickel. Les précisions obtenues sur ces deux paramètres, respectivement de ± 10 % et ± 6 %, s'inscrivent parfaitement dans les objectifs visés pour les projets des futurs grands réacteurs.

Par ailleurs, dans ces mêmes réseaux, on a procédé également à des mesures de remplacement de cellules afin d'étudier l'ensemble des éléments entrant dans la composition des aciers ou présents sous forme d'impuretés (Fe, Cr, Ni, Mo, Mn, Ti, Co).

Différents types d'acier prévus pour le développement de la filière à neutrons rapides ont également été étudiés de façon à pouvoir conserver à l'avenir une précision correcte sur la capture des matériaux de structure quelle que soit l'évolution de leur composition.

Références

1. P. HAMMER, "Nuclear data needs for Plutonium breeders", Knoxville, Int. Conf. on Nuclear Cross Section for Technology (1979).

2. J.P. CHAUDAT and al., "Data adjustments for fast reactor design", San Fransisco, ANS Winter Meeting (1977).

3. J.P. AZZONI and al., Nuclear Energy, 24, 411, (1977).

4. M. DARROUZET and al., "Experimental study of structural materials capture on fast breeder spectra", NEACRP (1980).

5. J. BOUCHARD and al., CEA-R-4246, Commissariat à l'Energie Atomique (1971).

6. D. DE LAPPARENT, J. BOUCHARD, BIST 170, 31, Commissariat à l'Energie Atomique (1972).

7. J.P. CHAUDAT and al., CEA-R-4552, Commissariat à l'Energie Atomique (1971).

8. M. DARROUZET, L. MARTIN DEIDIER, CEA-N-2063, Commissariat à l'Energie Atomique (1978).

AJUSTEMENT DES DONNEES NUCLEAIRES DE MATERIAUX DE STRUCTURE A PARTIR D'EXPERIENCES INTEGRALES SPECIFIQUES

J.P. TRAPP

DRNR/SEDC/SPNR/CPR
CENTRE D'ETUDES NUCLEAIRES DE CADARACHE
B.P. N°1
F.13115 - ST PAUL LEZ DURANCE
FRANCE

An adjustement concerning Ni, Cr, Fe and Na cross-section has been performed using propagation experiments on pure sodium and stainless-steel sodium mixtures media.

A concise analysis on these experiments and on the adjustement process is presented. A clear tendency appears concerning a slight reduction on the whole cross-sections. This adjustement is used in the new version D1 of the neutron propagation formulary Propane.

Introduction

Après une revue sommaire des différentes expériences de propagation effectuées à Cadarache et à Casaccia sur des massifs sodium et acier-sodium, on montrera la représentativité de ces expériences vis à vis des paramètres d'intérêt dans les réacteurs rapides de puissance (flux thermique, activation sodium secondaire).

L'interprétation de ces expériences permet un ajustement des sections efficaces du Nickel, du Chrome, du Fer et du sodium qui conduit à une réduction des écarts expérience-calcul constatés.

Expériences intégrales de propagation

Des expériences de propagation ont été effectuées tant à Cadarache sur un massif de sodium pur qu'à Casaccia sur un massif de sodium pur et des mélanges acier-sodium en proportions variables.

Le tableau I donne un résumé des caracteristiques essentielles de ces expériences. Une description détaillée de ces expériences peut être trouvée en /1/.

Dans ces expériences, deux types de source neutronique, conduisant à des spectres différents, ont été utilisés:

- spectre sortant du réflecteur,
- spectre sortant d'un milieu couverture UO_2-Na.

Des mesures par proton de recul permettent de déterminer le spectre source à l'entrée du massif de propagation.

Représentativité des expériences vis à vis des configurations de projet

Le problème de la représentativité des réponses expérimentales par rapport aux paramètres projet a été étudié /2/ et /3/.

On a comparé l'atténuation du flux thermique équivalent avec la réponse Au (n,γ) dans deux cas :

- configuration projet avec protection neutronique en acier-sodium (180 cm) suivie de 250 cm de sodium.

- configuration expérimentale Harmonie : 250 cm de sodium pur.

De la même façon a comparé l'atténuation du flux total avec la réponse expérimentale Mn (n,γ) dans le cas de la PNL projet et de l'expérience Tapiro acier-sodium (50% acier - 50% sodium).

D'une manière générale, on observera une très bonne représentativité des paramètres projet et des réponses expérimentales ; les détecteurs utilisant la réaction (n,γ) sont de bons intégrateurs de flux.

Les profils de sensibilité permettent aussi de juger de la représentativité des paramètres d'intérêt projet et des réponses expérimentales aux mêmes types de sections efficaces et en particulier aux mêmes zones d'énergie.

On notera que :

- Le profil de sensibilité de la réponse Au (n,γ) aux sections efficaces du sodium au voisinage de l'échangeur d'un réacteur rapide, et,

- le profil correspondant dans le cas de l'expérience Harmonie,

sont extrêmement voisins en particulier au-dessous de 1 Kev, domaine d'énergie qui représente la part prépondérante de la sensibilité.

De l'ensemble de ces études, il ressort que les conclusions extraites des expériences-écarts E-C/C, ajustement des sections efficaces résultats de la réduction de ces écarts - sont tranposables aux calculs de projet sans introduire d'erreurs systématiques.

Analyse des expériences : écarts E-C/C

De nombreux tests ont permis de discriminer dans les écarts E-C/C la part provenant des sections efficaces de celle provenant des méthodes (modélisation, effets de source, pondération de sections efficaces, option de calcul...).
Sommairement, la méthode utilisée consiste à calculer avec le code de transport, deux dimensions DOT 3.5, avec le formulaire Propane D_o :

- une source de volume dans le coeur et les couvertures des réacteurs Harmonie et Tapiro avec un calcul de k effectif.

K. H. Böckhoff (ed.), Nuclear Data for Science and Technology, 115–119.

- un calcul de propagation dans un milieu homogène utilisant cette source de volume permet de calculer les flux neutroniques aux points de mesure et l'activation des détecteurs intégraux utilisés.

Dans les massifs acier-sodium, milieu hétérogène, des calculs 2D et de nombreux tests ont permis de déterminer les facteurs d'hétérogénéité /4/.

Le tableau II donne les valeurs des écarts calcul-expérience ; il s'agit des écarts sur l'atténuation entre deux points de mesure (1/3 - 3/6 signifie qu'il s'agit du rapport entre les positions de mesure 1 et 3 ou 3 et 6). Les distances de propagation sont 100 et 150 cm pour Harmonie, 20-40-60 cm pour les massifs Tapiro.

Les incertitudes sur les valeurs des écarts expérience-calcul tiennent compte de manière globale :

- des incertitudes expérimentales (tableau I),

- des incertitudes sur la méthode d'analyse et le schéma de calcul.

On remarquera que les valeurs des écarts E-C/C sont :

- relativement faibles pour le massif de sodium d'Harmonie,

- plus importants en ce qui concerne le massif de sodium Tapiro.

- d'autant plus grand que le pourcentage d'acier est fort dans le cas des mélanges acier-sodium.

- les valeurs de E-C/C sont, en général, supérieures aux incertitudes et justifient donc d'un ajustement des sections efficaces utilisées dans le formulaire Propane Do.

Processus d'ajustement

a) Principe et Méthodes

L'ajustement des sections efficaces des matériaux de structure : Ni, Cr, Fe, Na doit réduire les écarts calculs-expériences constatés sur l'atténuation entre deux points successifs de la propagation jusqu'à une valeur proche des incertitudes associées à ces écarts.

On utilise la méthode statistique qui est basée sur l'hypothèse suivante :

- les incertitudes tant sur les valeurs expérimentales que sur les sections efficaces sont d'origine aléatoire.

Il en résulte que :

- le nombre d'expériences disponibles peut être inférieur, souvent de beaucoup, au nombre de paramètres à ajuster.

Cette méthode utilise le système de codes SAMPO/5/ qui permet :

- le calcul des coefficients de sensibilité aux sections efficaces partielles de fonctionnelles telles que dose ou taux de réactions ou rapport de doses ou rapport de taux entre deux points de l'espace (code ROSCOFF).

- les variations à appliquer aux sections partielles (σ_a, σ_{in}, σ_{tot}) pour réduire les écarts E-C/C (CORDAJ). Ce code permet de tenir compte des éventuelles corrélations entre données intégrales et microscopiques ou des corrélations en énergie des sections efficaces. Il calcule la déviation standard d'un paramètre intégral R, due à une matrice de dispersion B qui contient les informations relatives aux différentes corrélations.

Il permet en outre de calculer la nouvelle matrice de dispersion après ajustement des sections efficaces et effectue un test de X2 sur les ajustements effectués.

Les incertitudes sur les données de base et leurs corrélations proviennent pour le Fe et le Na, des valeurs données par ORNL /6/ mises en forme à Cadarache. Pour le Ni et le Cr, elles proviennent de /7/ pour les incertitudes ; aucune information sur les corrélations n'est disponible.

b) Tests effectués

De nombreux tests ont été effectués :

- sur la cohérence des expériences entre elles au moyen du test de X^2.

La figure montre les résultats de ces tests concernant soit, seules les expériences de propagation en sodium, soit seules les expériences de propagation dans l'acier-sodium, soit, l'ensemble des expériences.

On remarque, que pour tous ces essais, la probabilité de l'ajustement est supérieure à 50%.

- Influence des incertitudes initiales sur les données de base.

Une augmentation des incertitudes sur les sections du Ni et du Cr, toutes choses égales par ailleurs, conduit à :

. peu de modifications des résultats du test de X^2
. une augmentation non négligeable des $\Delta\sigma/\sigma$ avec l'accroissement des incertitudes initiales,
. des valeurs de E-C/C après ajustement peu différent.

- Influence des incertitudes sur les valeurs initiales de E-C/C.

De faibles effets ont été constatés pour ces tests.

- Influence de la prise en compte de corrélations en énergie pour le Nickel et le Chrome.

La réalité physique se trouvant quelque part entre les deux situations extrêmes suivantes :

. Pas de corrélation ,
. Corrélations totales.

La prise en compte de corrélations en énergie de type arbitraire et de la forme :

$$t = 1 - \frac{/I-J/ \times 3}{45}$$

avec $J \leqslant I + 10$ (I et J étant les indices correspondants aux groupes d'énergie), permet de répartir l'ajustement sur l'ensemble des 4 éléments à ajuster alors que sans corrélation (pour le Ni et le Cr) l'ajustement porte essentiellement sur le Fer et le Na.

Ce type de corrélations apparaît de toute façon plus réaliste que les situations extrêmes ci-dessus ($t = 0$ ou $t = 1$).

Vérifications de l'ajustement

Une expérience supplémentaire sur un massif 75% acier 25% Na (avec un spectre neutronique type coeur au lieu de type couverture) a été effectuée. Ces résultats n'ont pas été utilisés dans le processus d'ajustement.

Le recalcul de cette expérience avec les sections efficaces ajustées fait apparaître une amélioration importante des résultats comme le montre le tableau VI.

On constate, dans le passage de ENDFB IV à ENDFB V, pour le Fer et le Na, une tendance opposée à celle indiquée par les expériences intégrales à savoir une augmentation des flux. Voir aussi /9/.

On peut affirmer que l'utilisation des sections efficaces du Fer et du Na - ENDFB V - dans l'interprétation des expériences Harmonie (250 cm de propagation) et TSF (400 cm de propagation) auraient eu pour effet d'accroître les écarts expérience-calcul initiaux constatés.

Conclusions

L'ajustement des sections efficaces des matériaux de structure utilisées dans le formulaire Propane Do servant au calcul des protections neutroniques de la filière rapide française a été effectué à l'aide d'expériences de propagation dans des milieux sodium et acier-sodium en proporitons variables.

L'ajustement permet de réduire les écarts E-C/C observés lors de l'interprétation et conduit à une diminution générale des sections efficaces du Ni, Cr, Fe, Na et a été vérifié sur des expériences indépendantes du processus d'ajustement. Celui-ci présente une tendance opposée à celle observée quand on passe de ENDFB IV à ENDFB V (pour le Fer et le Na), ce qui toutefois ne paraît pas être de nature à remettre en cause l'ajustement effectué.

Références

1. J.C. ESTIOT, A. DE CARLI et J.P. TRAPP, Results of Neutron propagation in steel-sodium mixtures with various spectre on Harmonie et Tapiro.

2. M. SALVATORES, Neutron transport in structural materials and shielding design Knoxville. Octobre 1979.

3. C. PALMIOTTI - M. SALVATORES, Utilisation d'expériences intégrales pour la réduction des incertitudes affectant les paramètres projet dans les calculs de protection, Paris Octobre 1980.

4. M. CARTA, A. DE CARLI, V. RADO (Italie), M. SALVATORES, J.P. TRAPP (France), Expériences de propagation dans les milieux acier-sodium à Tapiro, Paris Octobre 1980.

5. J.C. ESTIOT, G. PALMIOTTO, M. SALVATORES, SAMPO. Un système de code pour les analyses de sensibilité et de perturbations à différents ordres de peu d'approximation, Paris Octobre 1980.

6. J.D. DRISCHLER, C.R. WEISEIN, Compilation of multigroup cross-sections covariance matrices for several important reactor materials, ORNL 5318 (1979).

7. J.S. SCHMIDT KFK 1966.

8. J.C. ESTIOT, Qualification de la méthode d'ajustement des données nucléaires utilisées pour l'optimisation du formulaire Propane, Paris, Octobre 1980.

9. M.CARTA, F. D'AVERSA, A. DE CARLI, P. PALMIOTTI, M. SALVATORES, J.P. TRAPP, Résultats d'un calcul étalon pour la comparaison de données de protection dans les laboratoires des pays de l'OCDE, Paris , juillet 1982.

TABLEAU I

		Milieux de propagation	Dimensions	Source Neutronique	Dispositif de Mesures	Mesures	Précision des Mesures
Mélange Acier-Socium	Tapiro	Acier-sodium . 25 % Acier . 50 % Acier . 75 % Acier Barreaux d'acier entourés de Na	Cuves parallélépipédiques Côté = 100 cm H = 150 cm Longueur de propagation utilisable 130 cm	Réacteur Tapiro . couverture UO_2-NA ou . Milieu coeur (R1)	13 canaux sur l'axe	Détecteur: Na/Cd Mn/Cd Au/Cd S Rh	± 15 % S Rh ± 15 % pour les autres
Sodium	Tapiro	Na 100 %	idem ci-dessus	Réacteur Tapiro . Réflecteur cuivre . Couverture UO_2-Na+25 cm	6 canaux sur l'axe	Protons de recul + Détecteurs inégraux: Na/Cd - Mn/Cd Au/Cd - S - Rh	idem
	Harmonie	Na 100 %	Cylindre ϕ = 150 cm H = 300 cm long. utile de propagation: 250 cm	Réacteur Harmonie Réflecteur Acier 25 cm Couverture UO_2-Na=25 cm	6 canaux dans le sodium + 1 canal sous la cuve	Protons de recul + Détecteurs intégraux (idem)	± 15 % S Rh ± 5 % pour les autres

TABLEAU II : Valeurs E-C/C (%) des expériences utilisées dans l'ajustement

Expérience			$\frac{E-C}{C}$ (%)	Incertitude (%)
Harmonie Na	Avec couverture	S 1/3	14.4	± 25
		Na/Cd 1/3	- 13.5	± 15
		Na/Cd 3/6	+ 6.5	idem
		Mn/Cd 1/3	- 17.0	idem
		Mn/Cd 3/6	- 11.8	idem
	Sans couverture	Rh 1/3	+ 8	± 25
		S 1/3	- 40	idem
		Na/Cd 1/3	- 19.9	± 15
		Na/Cd 3/6	+ 0.5	idem
		Mn/Cd 1/3	- 22.7	idem
		Mn/Cd 3/6	- 15.5	idem
		Au/Cd 1/3	- 10.4	idem
		Au/Cd 3/6	+ 0.5	idem
Tapiro Na	Avec couverture	S 1/2	- 11.6	± 25
		Rn 1/4	- 35.4	idem
		Na/Cd 1/6	- 25.1	± 30
		Mn/Cd 1/6	- 31.2	± 30
	Sans couverture	S 1/2	+ 9.5	± 25
		Rn 1/4	- 28.0	idem
		Na/Cd 1/6	- 30.9	± 30
		Mn/Cd 1/6	- 37.0	idem
		Au/Cd 1/6	- 21	idem

Expérience		$\frac{E-C}{C}$ (%)	Incertitude (%)
Tapiro 25 % Acier 75 % Na	S 1/5	- 38.1	± 25
	Rh 1/6	- 26.1	idem
	Na/Cd 1/6	- 27.7	± 15
	Na/Cd 7/10	- 9.4	idem
	Au/Cd 1/6	- 11.6	idem
	Au/Cd 7/10	- 2.4	idem
Tapiro 50 % Acier 50 % Na	S 1/7	- 16.0	± 25
	Rn 1/8	- 50.7	idem
	Na/Cd 1/8	- 28.5	± 15
	Au/Cd 1/8	- 14.6	idem
	Mn/Cd 1/8	- 18.5	idem
Tapiro. R1 50 % Acier 50 % Na	S 1/6	- 30.2	± 25
	Na/Cd 1/9	- 19.8	± 15
	Na/Cd 10/13	- 11.6	idem
	Mn/Cd 1/9	- 11.0	idem
	Mn/Cd 10/13	- 9.4	idem
	Au/Cd 1/9	- 6	idem
	Au/Cd 10/13	+ 0.5	idem
Tapiro 75 % Acier 25 % Na	S 1/5	- 40.6	± 25
	Rh 1/6	- 41.9	idem
	Na/Cd 1/6	- 31.9	± 20
	Mn/Cd 1/6	- 32.6	idem
	Au/Cd 1/6	- 22.2	idem

TABLEAU III : Variations des sections efficaces et déviations standards initiales

Domaine Energétique	Chrome						Sodium					
	Inélastique		Absorption		Totale		Inélastique		Absorption		Totale	
	$\Delta\sigma/\sigma$ (%)	Dev. Std. (%)	$\Delta\sigma/\sigma$ (%)	Dev. Std. (%)	$\Delta\sigma/\sigma$ (%)	Dev. Std. (%)	$\Delta\sigma/\sigma$ (%)	Dev. Std. (%)	$\Delta\sigma/\sigma$ (%)	Dev. Std. (%)	$\Delta\sigma/\sigma$ (%)	Dev. Std. (%)
En > 1.65 Mev	0.5 à 1.0	7.5 à 12.5	< 0.1	7.5 à 12.5	1 à 9.3	4 à 10	7.1 à 7.5	22 à 43	< 1.0	36 à 50.	1 à 1.40	2.7 à 3.2
0.82 < En ≤ 1.65 Mev	1 à 2.0	12.5 à 22	< 0.1	12.5 à 15.	3.5 à 10	4 à 10	5.5	17 à 20.	< 0.1	36.	1.3	2.6
183 < En ≤ 820 Kev	1.6	20.	< 2.0	20.	3.5 à 9.7	4 à 10	< 5.0	16 à 45	< 0.6	38 à 50	1.8 à 2.3	3.1 à 4.8
67. < En ≤ 183 Kev			2 à 10	20 à 10	3 à 4.0	5.0			0.5 à 0.6	16 à 27.	2.4 à 5.2	3.4 à 5.7
9.1 < En ≤ 67 Kev			10 à 20	30.0	2 à 2.5	5.0			< 1.0	13 à 15.	7 à 9.	6.5 à 8.3
2.74 Kev < En ≤ 9.1 Kev			13.	20.0	1.5 à 2	5.0			1.3	10 à 13	6 à 7.5	5.7 à 6.3
0.454 < En ≤ 2.74			8 à 14.0	20.	1.2 à 1.5	5.0			< 1.2	7 à 10	6.	4 à 6.5
En ≤ 454 ev			0.8 à 6.6	7.0	< 1	5.0			< 0.5	< 5.4	< 6.4	4.5 à 7.0

TABLEAU IV : Variations des sections efficaces et déviations standards résiduelles

Domaine Energétique	Fer						Nickel					
	Inélastique		Absorption		Totale		Inélastique		Absorption		Totale	
	$\Delta\sigma/\sigma$ (%)	Dev. Std. (%)	$\Delta\sigma/\sigma$ (%)	Dev. Std. (%)	$\Delta\sigma/\sigma$ (%)	Dev. Std. (%)	$\Delta\sigma/\sigma$ (%)	Dev. Std. (%)	$\Delta\sigma/\sigma$ (%)	Dev. Std. (%)	$\Delta\sigma/\sigma$ (%)	Dev. Std. (%)
En < 1.65 Kev	0.1 à 2.5	8 à 14	0.2	20	< 0.5	20	1.2	20	< 0.5	20	< 2.3	4 à 7.5
0.82 ≤ En ≤ 1.65 Kev	< 1.0	6 à 7	0.2	15 à 18	< 0.3	1.5	1.1	17.5	< 0.6	17 à 25	1.2 à 2.6	4 à 7.5
183 Kev < En ≤ 820 Kev			< 1.4	16 à 20	< 0.5	1.95 à 2.5			0.5 à 2.0	17.5	1.2 à 10.0	4 à 10
67 Kev < En ≤ 183 Kev			< 1	11 à 15	< 0.4	1.2 à 2			< 0.4	5.0	10 à 14	10.
9.1 Kev < En ≤ 67 Kev			< 0.8	10.	1 à 2	2.9 à 5			< 0.3	5.0	3 à 5.0	4.0
2.74 Kev < En ≤ 9.1 Kev			1 à 4	12 à 14	1.4 à 1.8	5 à 6			< 0.3	5.0	2 à 3.0	4.0
0.454 Kev < En ≤ 2.74 Kev			4.8 à 12	15 à 19	< 1.4	2.8 à 4.4			< 0.4	5.0	1.4 à 1.8	4.0
En ≤ 454 ev			< 3.6	< 13	< 0.3	< 2.0			< 0.4	5.0	0.3 à 1.2	4.0

TABLEAU V

Résultat de l'Ajustement

	Det	Distance dans le sodium (cm)	E-C/C (%) avant	Incertitude (%)	E-C/C(%) après recalcul de l'expérience	E-C/C(%) après recalcul de l'expérience	Incertitude (%)	E-C/C (%) avant	Distance dans le mélange (cm)	Détecteur	Exp
avec couverture	S	120	-23	± 25	-19	8.6	± 25	19.0	50	5	Tapiro 50% Acier-50%Na
		120	15.6	± 15	11.5	77.0	"	100	57	Rh	
	Na/Cd	170	16.2	"	8.2						
		270	8.6	"	-7.0	26.0	± 15	37.9	57	Na/Cd	
		120	20.6	"	12.0						
	Mn/Cd	170	22.1	"	10.0						
		270	36.6	"	17.0	13.4	"	25.7	57	Mn/Cd	
sans couverture	S	120	67	± 25	55.0						
	Rh	70	-7.4	± 25	-13.0	10.2	"	19.2	57.0	An/Cd	
		120	24.8	± 15	19.9						
	Na/Cd	270	24.2	"	6.0						
		120	29.4	"	19.1	30.0	25	40.9	42	S	Tapiro 50%acier 50%Na (R_1)
	Mn/Cd	270	53.7	"	28.3						
	An/Cd	120	11.7	"	7.7	8.1	15	22.0	65.0	Na/Cd	
	An/Cd	270	12.3	"	4.7	9.9	"	38.0	130.0		
sans couverture	S	35cm	-5.5	± 25	-6.8	-0.1	"	13.0	65.0	Mn/Cd	
	Rh	70cm	43.8	"	38.0	-2.4	"	24.0	150.0		
	Na/Cd	120	45.0	± 35	40.7	2.1	"	6.3	65.0	An/Cd	
	Mn/Cd	120	89.4	"	47.8	-19.0	"	0.8	130.0		
	An/Cd	120	25.8	"	22.9						
	S	48.0	61.6	± 25	47.9	53.0	± 25	68.3	39.0	S	Tapiro 75%acier 25% Na
	Rh	59.0	35.3	± 25	24.1	56.0	"	71.8	51.5	Rh	
	Na/Cd	89.0	44.2	± 15	41.0	30.2	± 15	45.2	51.5	Na/Cd	
		110.0	59.1	"	46.0	31.5	"	47.8	51.5	Mn/Cd	
	An/Cd	89.0	19.5	"	18.0	25.0	"	35.0	51.5	An/Cd	
		110.0	26.0	"	14.0						

TABLEAU VI : Massif 75 % Acier - 25 % Na Coeur R1

Détecteur	Distance dans le milieu (cm)	E-C/C (%) avant	Incertitude (%)	E-C/C (%) après ajustement
	61.6	44.1	15	12.5
Na/Cd	122.3	144.3	"	53.2
	61/6	38.7	"	14.4
Mn/Cd	134.6	173.9	"	75.8
	61.6	22.1	"	-4.5
An/Cd	134.6	116.1	"	32.8

FIGURE 1

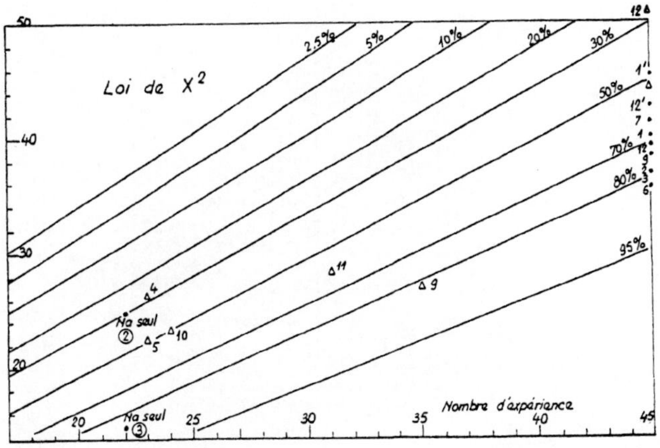

ADJUSTMENT OF NEUTRON MULTIGROUP CROSS SECTIONS WITH ERROR COVARIANCE MATRICES TO DEEP PENETRATION INTEGRAL EXPERIMENTS

G. Hehn, R.-D. Bächle, G. Pfister, M. Mattes

Universität Stuttgart
Institut für Kernenergetik und Energiesysteme (IKE)
Postfach 80 11 40, D-7000 Stuttgart 80
Federal Republic of Germany

W. Matthes

Joint Research Center Euratom
I-21020 Ispra (Varese)
Italy

For radiation protection and reactor safety improved accuracies are required in analysis of radiation penetration. For checking the consistency of neutron cross-sections with integral data of deep penetration experiments we have designed the computer program ADJUST-EUR, which operates within the modular system RSYST. To adjust partial group cross-sections an iterative procedure is applied based on the least-square method and using the results of one- and twodimensional transport calculations with linear perturbation theory. The iron experiments ASPIS/AEE-Winfrith and EURACOS/Euratom-Ispra are analysed. The adjustment of the inelastic and elastic iron cross-sections is made in the 100 group EURLIB structure using the newest error covariance information.

Radiation shielding, iron, group cross-sections, adjustment, computer program, integral experiments.

Need of group data adjustment

For determination of target quantities in radiation shielding like biological dose, gamma heating, neutron activation, or radiation damage special group data libraries are required which allow reliable calculations of neutron and gamma fluxes in deep penetration of material. The special features of group data for radiation shielding had always much profit from measurements of integral experiments. Modern adjustment of group cross-section libraries can make use of following important advantages:

- the availability of comprehensive sensitivity studies, showing the details of data requirements needed.
- the availability of error covariance information with correlation across energy range and reaction type, and last not least
- the available results of clean, deep penetration, integral experiments, designed for single materials in nearly onedimensional geometry.

The need for detailed consistency checks with integral measurements have been worked out quite clearly in the course of two shielding benchmark exercises proposed by NEACRP for power reactors of fast breeder and of pressurized water type /1/, /2/. The need for adjustment of group cross-sections is confined to very few elements with iron and sodium as the most important ones. Extensive sensitivity studies with complete error analysis show up the reaction types and energy regions, which produce the largest portion of nuclear data uncertainties.

For integral checks of iron group data several benchmark experiments with medium and deep penetration depth are now available. The Karlsruhe measurements on iron spheres surrounding a strong Cf-252 source realize exactly onedimensional neutron fields, which can be perfectly analyzed /3/. But with a largest thickness of 40 cm of iron the penetration depth is relatively small, so that only the inelastic cross-section can be tested. Higher sensitivities to the inelastic and elastic cross-sections are given by deep penetration experiments like ASPIS of AEE-Winfrith and recently EURACOS of Euratom JRC-Ispra

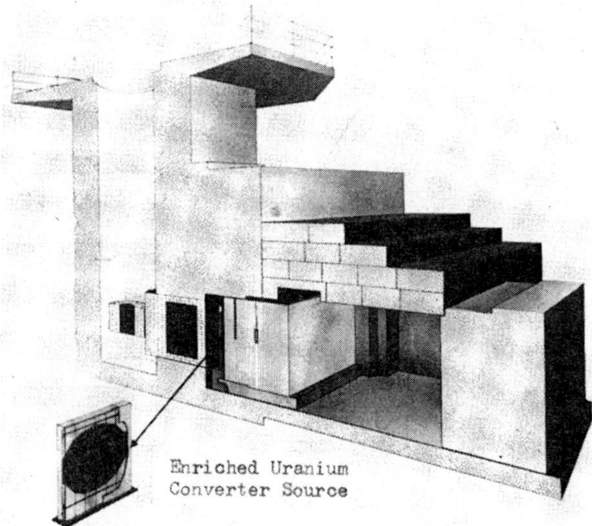

Fig. 1a. The iron deep penetration experiment EURACOS
EURACOS = Enriched URAnium COnverter Source

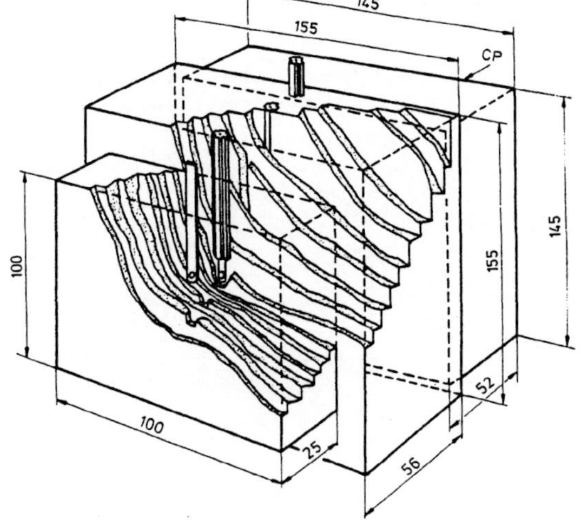

Fig. 1b. Arrangement of the iron blocks of EURACOS. The converter plate and thermal column of the reactor are on the back side. The detectors along the central axis are inserted from above /5/.

K. H. Böckhoff (ed.), Nuclear Data for Science and Technology, 120–126.

/4/, /5/. They consist of a large iron block irradiated by a converter plate placed at the end of a thermal column of a nuclear reactor as shown in figure 1 schematically for EURACOS. A large variety of detectors are applied to measure the neutron field for penetration depth up to 1 m of iron and more. The main problem is, that the measured counting rates at different positions along the central axis of the iron block do not agree sufficiently with their calculations based on available cross section sets.

In the past consistency tests between integral measurements and group cross sections were performed primarily in few group structures used for reactor design. For the most part adjustments of few group data were done without or with a relatively poor information of cross-section covariances. But recently the error covariance data have been improved considerably /6/. New target accuracies of reactor safety and radiation protection require shielding calculations with multigroup cross-sections.

Refined calculation methods of neutron transport promise a better separation between uncertainties from group data and from the transport method. A straight forward "least-square" fit with a "global detector" option reduces the effort for multigroup adjustment appreciably /7/. Further developments in this field can give more support for evaluating basic cross-sections and appropriate error covariance informations with the help of integral experiments.

Adjustment procedure and program

The least-square method, as applied to the adjustment of group cross-sections, has been published in detail /7/, /8/, /9/. The main result can be summarized: If $M_n (n=1,2,\ldots N)$ are measured counting rates and $C_n(\sigma^{(i)})$ the corresponding calculated values, based on a given cross-section set $\sigma^{(i)}$, an updated (improved) cross-section set $\sigma^{(i+1)}$ is obtained by the iteration

$$\sigma^{(i+1)} = \sigma^{(i)} + \delta\sigma \tag{1}$$

where the correction term $\delta\sigma$ depends on the difference between measurement and calculation $M_n - C_n\,(\sigma^{(i)})$ and on the sensitivity matrix $\partial C_n/\partial\sigma^{(i)}$. The evaluation of the matrix however needs a high calculation effort, especially for many measurements. For this reason a further simplifying assumption is made, which allows us to rewrite (1) in the form:

$$\sigma^{(i+1)} = \sigma^{(o)} + C_{ov} \cdot (\partial C_w / \partial\sigma^{(i)}), \tag{2}$$

where one sensitivity coefficient only has to be calculated /7/. Here C_{ov} is the error covariance matrix of the initial $\sigma^{(o)}$ and C_w is a reaction rate of one "global detector", whose space and energy dependent reaction cross-section is a linearly weighted average of the difference between measurement and calculation $M_n - C_n\,(\sigma^{(i)})$.

The outlined procedure for multigroup adjustment has been programed in modular form in the code ADJUST-EUR. It runs within the code system RSYST /10/ and uses the onedimensional S_N-transport code ANISN /11/ as RSYST-module called SN1D and the module SWAN derived from the linear perturbation code SWANLAKE /12/. The handling of different codes with large data input and output is simplified herewith appreciably. In RSYST each operation such as forward or adjoint S_N-calculations or determination of cross-section sensitivities is performed by a certain module, which obtains all data required from a data pool and writes its results back into the pool. The data are stored in blocks containing e.g. cross-sections, neutron fluxes, sensitivities or

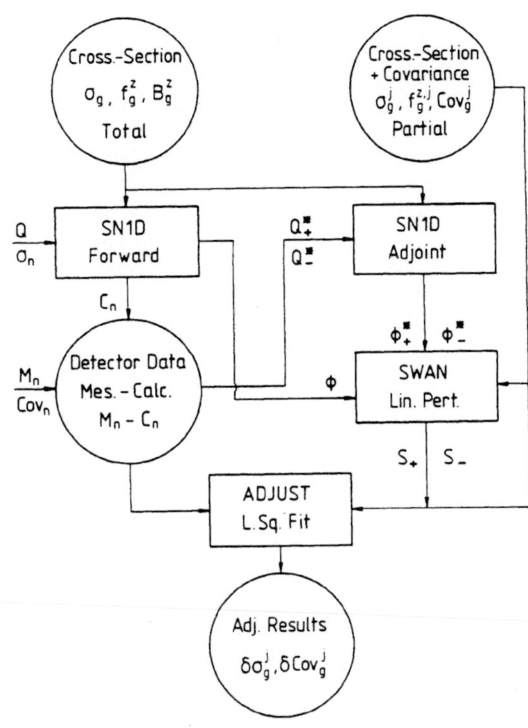

Fig. 2. Adjustment procedure applied in ADJUST-EUR for one iteration step. Circles indicate data blocks and rectangles are modules of RSYST. Only the main modules are shown.

even input sequences. The scheme of the adjustment procedure of the module sequence ADJUST-EUR is shown in figure 2. The circles represent the data blocks and the modules are indicated by rectangular fields. The consistency test with integral measurements is performed in the 100 group structure used in the EURLIB library /13/.

For calculating the neutron field with high precision along the central axis of the iron block illustrated in figure 1, we need in addition to the reference group cross-section set σ_g a modifying self-shielding factor f_g^z, which cares for a shift in the resonance flux weighting in function of penetration depth z. Likewise any transvers leakage of neutrons has to be regarded carefully by bucklings depending on energy group g and penetration depth z. With this nuclear data effort and a good representation of the fission source in the converter plate the detector counting rates are calculated for comparison with their measured values. Separated into positive and negative differences of the measured and calculated counting rates (M_n-C_n), two adjoined transport calculations are performed followed by the determination of sensitivity profiles S. For the latter all partial cross-sections σ_g^j are needed with the appropriate self-shielding factors $f_g^{z,j}$. Finally the adjustment of the reference group cross-section is accomplished using the error covariance information of partial cross-sections Cov_g^j and of integral measurements Cov_n. The total procedure is applied in several iteration steps in the code, so that any shortcomings of linear perturbation theory are eliminated completely. In the global detector option of the code the adjustment is restricted to partial cross-sections. A second version is in progress, using the Newton iteration method with the aim to adjust the cross-section covariance data, too.

Refined group data for ASPIS and EURACOS

For precise calculation of the detector counting rates in large iron blocks the available multigroup cross-sections like EURLIB (100 groups) or VITAMIN-C (171 groups) /14/ turned out to be totally insufficient. The main problem is caused by the group structure between 70 keV and 3 MeV, whereas the representation of the large cross-section minimum at 24 keV is perfect in both libraries. As a first step we designed a special fine group structure, which takes into account all important iron minima in this energy range, to open the streaming path for neutrons in deep penetration. Figure 3 shows the energy details of the iron total cross-section in a point data plot /15/ compared with the respective VITAMIN-C and special fine group data. Both group data are $1/\sigma_T$-weighted.

Besides the minima problem we tried also to reduce the sensitivity of the fine group data to the flux weighting applied. We were quite successful for important groups with high neutron fluxes and diminished all abnormal high sensitivities inherent in some VITAMIN-C groups appreciably, as can be seen from figure 4. Here the relative differences are plotted between the total group cross-section of iron weighted with 1/E and $1/\sigma_T$. For some groups, which are placed on one side of large resonances, the weighting effect is still high, but the general importance of such groups is low in neutron transport.

Our processing of refined group cross-sections is shown in figure 5. Starting with the best available evaluated point cross-section information, fine group data are produced with the code NJOY /16/ for all nuclides in the ASPIS and EURACOS experiments. The flux weighting

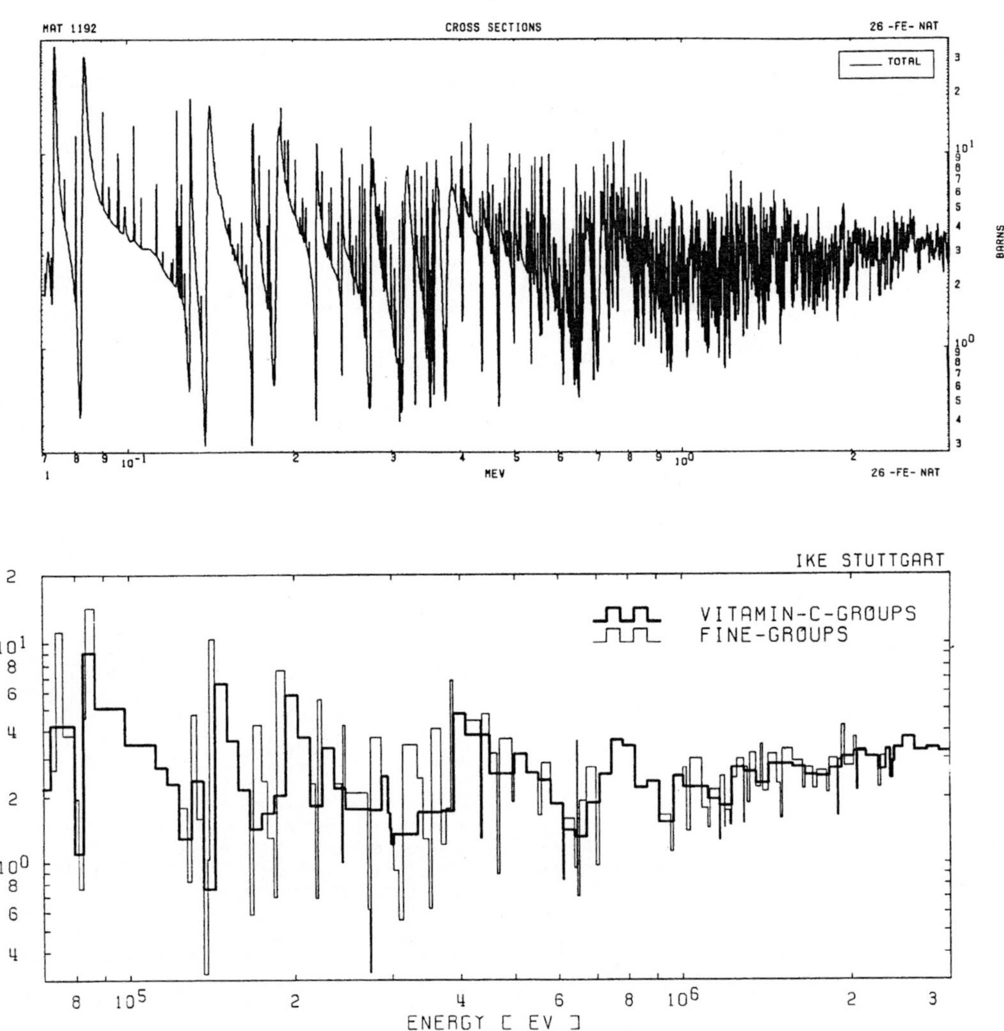

Fig. 3. Total neutron cross section of iron /15/. Comparison of point data (above) and their representation in two energy group structures: VITAMIN - C and our fine group structure (below). The latter shows the special treatment of the important cross-section minima. The flux weighting is $1/\sigma_T$ in both cases.

w(E) of the iron point data has been $1/\sigma_T$ and flat weighting of the other nuclides with small importance. Special care has been taken of a sufficiently accurate integration in the resonance region. The numerical results of the scattering matrix elements were checked with the respective total scattering to correct any numerical error. With these fine group data onedimensional transport calculations were performed for both experiments resulting in weighting functions $\emptyset_k^z$ for collapsing all partial cross-sections to the EURLIB group structure. These weighting fluxes depend on penetration depth in iron as can be seen from figure 6. The coarse shape of the flux spectrum brings a large reduction of the MeV-region beyond half a meter of iron. The fine structure of the flux shows a typical filter effect of the iron resonances, being slowly increased with depth. The transport along the iron minima in the upper keV-range becomes effective for penetration depth beyond half a meter of iron. The effective iron cross-sections are reduced with penetration depth. This effect can be separated into a reference cross-section with locally dependent self-shielding factors. Such factors for both scattering cross-sections are shown in figure 7 averaged over the second, third, and fourth iron zone of ASPIS. Reference cross-sections are the $1/\sigma_T$-weighted values of the first iron zone.

Since in the adjustment procedure used several forward and adjoint transport calculations must be performed in multigroup version, we have to apply one-dimensional codes like ANISN. The small transvers outstreaming must be accounted by the buckling concept. Our processing of refined buckling data is shown in figure 5. For each experiment like ASPIS and EURACOS a twodimensional S_N-calculation is run with the code DOT4.2 /17/. Hereby the refined group cross-sections are used in EURLIB structure, regarding all effects of the resonances with penetration depth. The radial dependence of the group fluxes $\emptyset_g(r,z)$ are least-square fitted to get buckling factors depending on group and penetration depth. The latter dependence has been averaged in four iron zones for both experiments. The same zone averaging has been performed with the self-shielding factors, which modify the reference cross-sections of iron with penetration depth. The adjustment procedure pertains to partial cross-sections of the reference group data only, e.g. the inelastic and elastic cross-sections of iron. The decisive question is, which precision can be achieved with the buckling and self-shielding corrections in calculating the detector counting rates? The precision of the buckling correction for ASPIS is given in figure 8. At the positions of the three detectors S-32, In-115, and Rh-103 the relative differences of the counting rates calculated by DOT4.2 and SN1D with bucklings vary between -1.1 % and 0.7 %. These values are small, if compared with the relative differences between measured and calculated counting rates of figures 9, 10, 11.

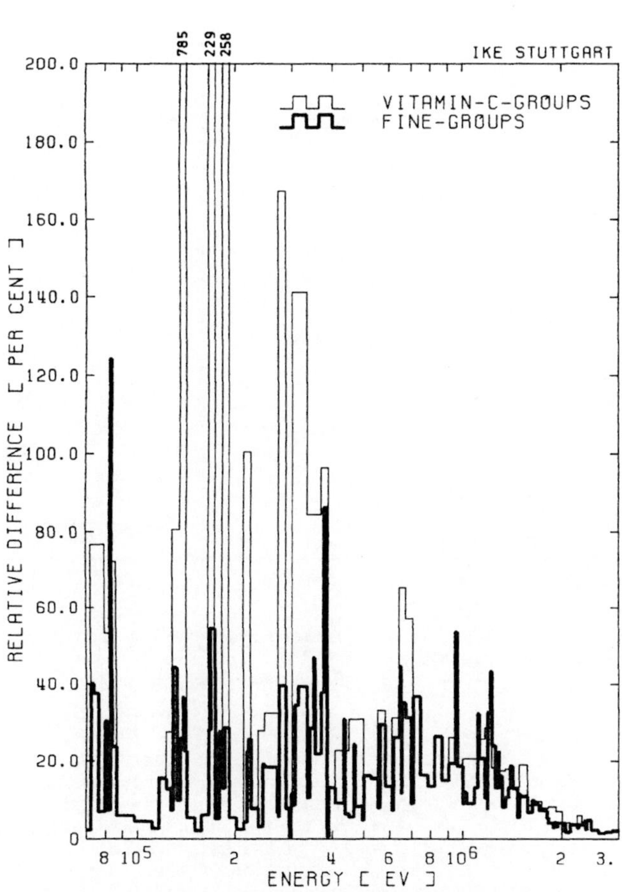

Fig. 4. Relative difference of the iron total group cross-section with extremely different flux weighting e.g. flat 1/E-weighting and $1/\sigma_T$-weighting.

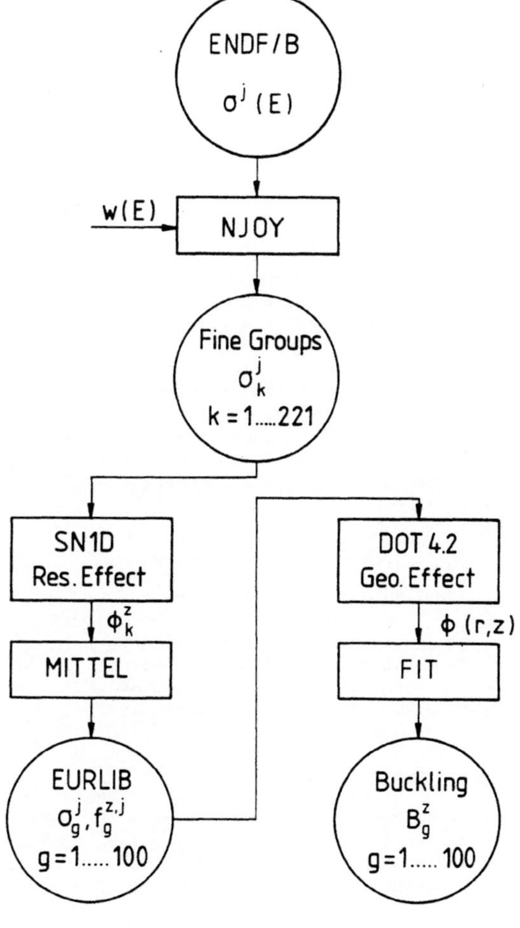

Fig. 5. Processing of refined group data in EURLIB structure for evaluation of benchmark experiments with deep penetration: partial group cross sections σ_g^j, self - shielding factors $f_g^{z,j}$ and bucklings B_g^z.

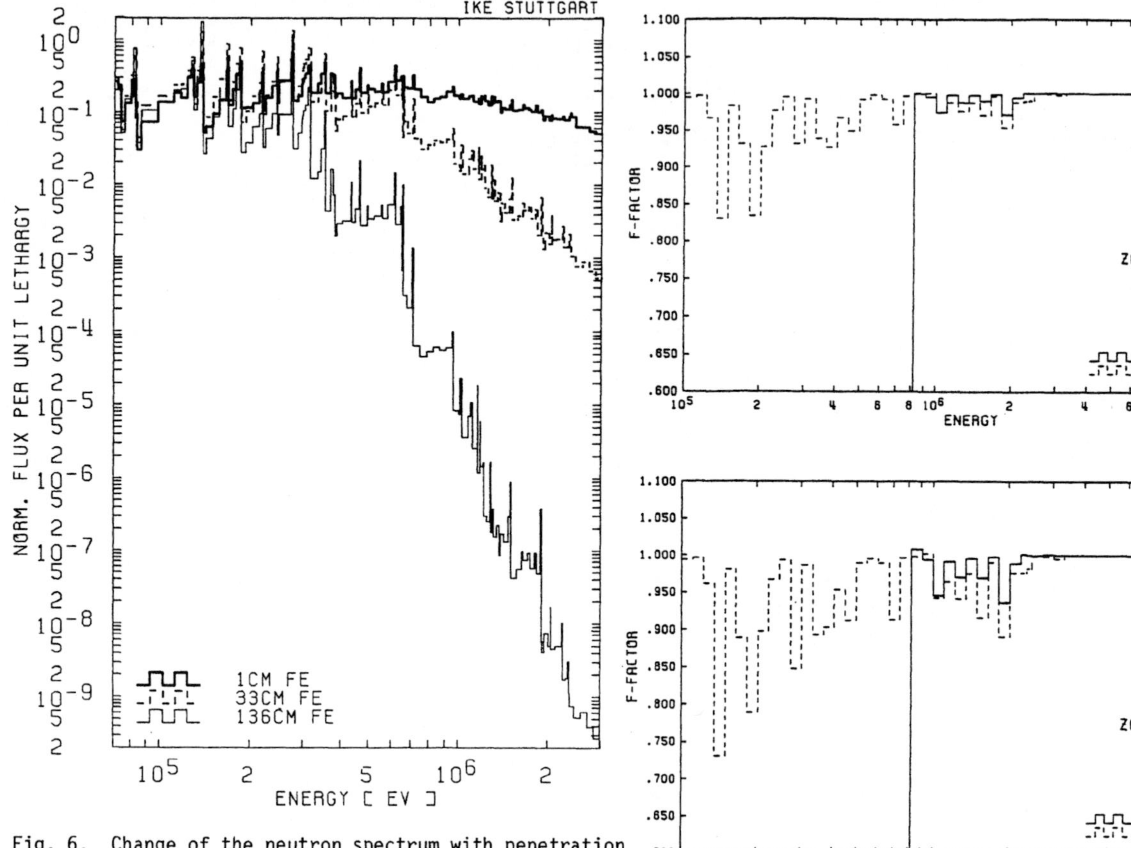

Fig. 6. Change of the neutron spectrum with penetration depth in iron, shown in the resonance energy region for ASPIS.

It should be remarked, that our perfect zone and group dependent buckling values were achieved by additional fitting by hand. We observed two effects, which contribut to the buckling. The values are very sensitive to the local transvers streaming and to the radial distribution of the fission source in the converter plate. For EURACOS this distribution has been recently least-square fitted to a cosine-distribution based on a set of separate measurements, when the iron block was removed.

Results and further work

A detailed analysis has been performed to calculate the penetration of fission neutrons through the large iron block of ASPIS /4/. Comparing to the integral measurements we get the following results: The calculated flux of fast neutrons above 1 MeV is correct up to a penetration depth of 4o cm of iron, but for deeper penetration we get an increasing underestimation as can be seen from figure 9 for the $^{32}S(n,p)$ - reaction rate. A similar result follows for the ^{115}In (n, n') - reaction shown in figure 1o. In this reaction also neutrons from the upper keV-region are included. But quite different are the results for the $^{103}Rh(n,n')$ - reaction with the lowest threshold given in figure 11. Here we have a strong overestimation up to 6o cm of iron and then for extreme deep penetration over 1 m again a significant underestimation. The calculations are based on primary group cross-sections in VITAMIN-C structure and twodimensional transport calculations with collapsed energy groups in EURLIB structure.

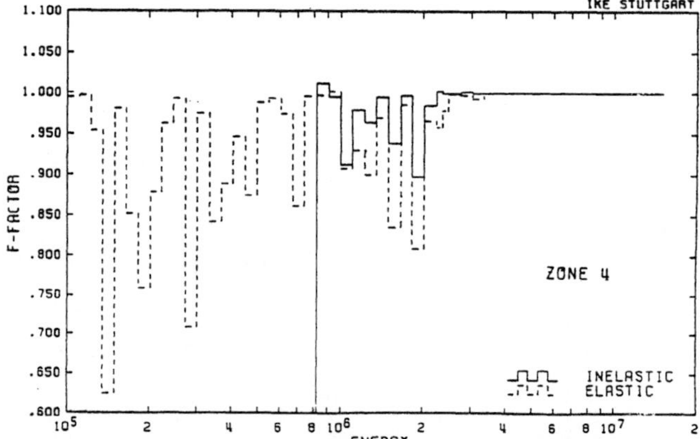

Fig. 7. Self-shielding factors averaged for iron zone 2 (7.1 - 32.9 cm), zone 3 (32.9 - 72.9 cm) and zone 4 (72.9 - 137.16 cm) of ASPIS.

To describe the spectral shift with penetration depth the group collapsing to the 1oo group structure was performed in four iron zones separated at iron depth of 7.1cm, 32.9cm and 72.9cm. We started to improve consistency with the sulphur reaction alone and got mainly a reduction of the inelastic scattering cross-section of iron. By adding the In- and Rh-measurements with their additional information pertaining to the keV-energy range, we achieved the results plotted in figure 12. The reduction of the inelastic and elastic scattering cross-sections, shown as relative difference between original and adjusted values, are proportional primarily to the

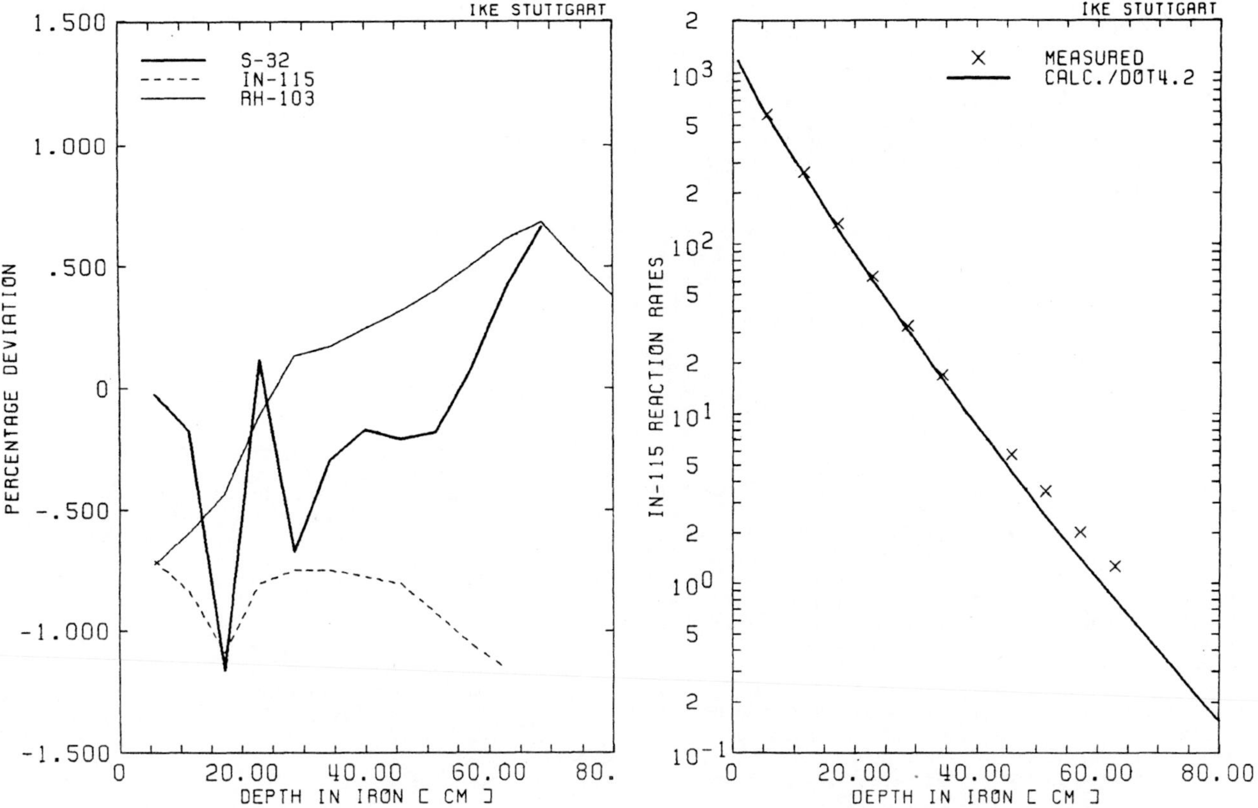

Fig. 8. Precision of the buckling correction in onedimensional transport calculations. Relative difference of the counting rates in two- and onedimensional calculation at detector positions for ASPIS.

Fig. 10. ^{115}In(n, n') reaction rates measured and calculated in function of penetration depth in iron for ASPIS

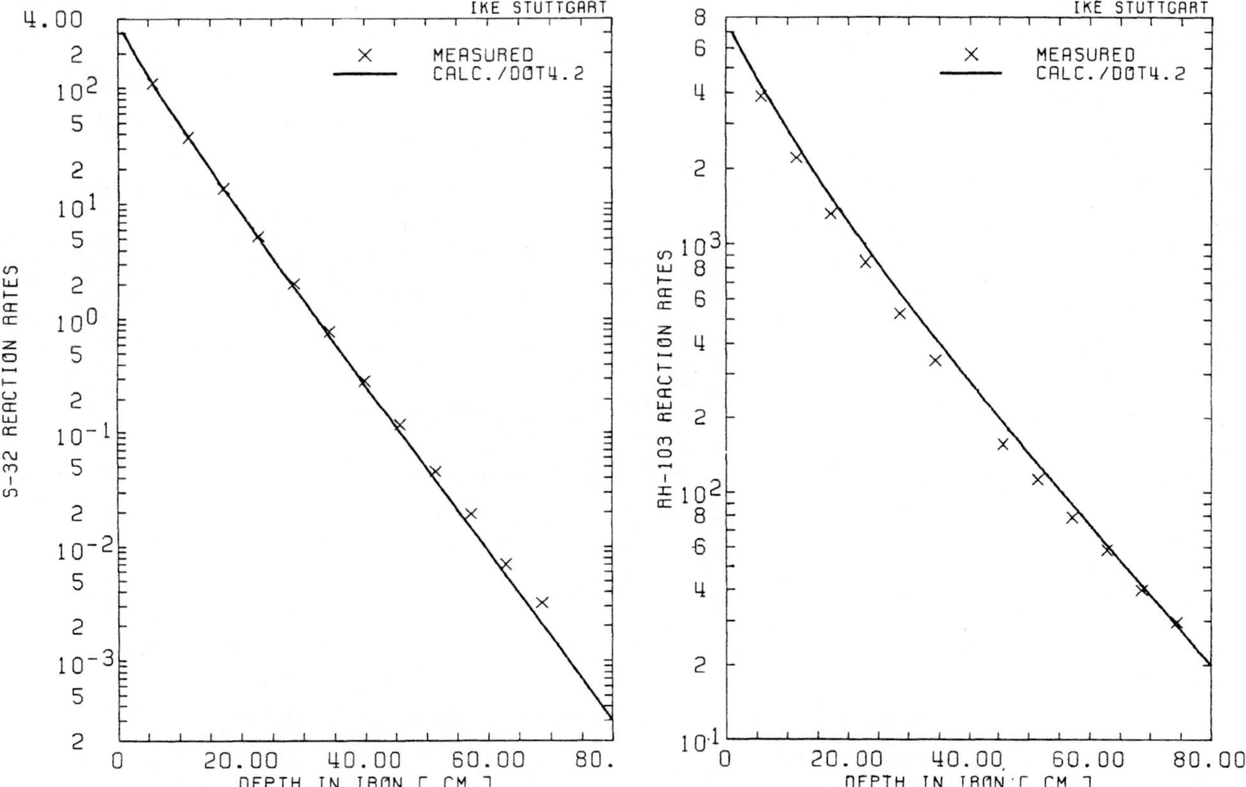

Fig. 9. ^{32}S(n,p) reaction rates measured and calculated in function of penetration depth in iron for ASPIS.

Fig. 11. ^{103}Rh(n,n') reaction rates measured and calculated in function of penetration depth in iron for ASPIS

sensitivity profiles of the "global detector" with maxima in the lower MeV- and upper keV- energy range and secondly to the relative standard deviations of the cross-sections as given in figure 13. So the increasing adjustment values at the high energy end is a consequence of the high uncertainty of both cross-sections in this energy region. The same holds for the inelastic scattering cross-section at threshold energy.

Work is in progress to reduce the remaining discrepancies between measured and calculated integral quantities at deep penetration with the help of our specially designed fine group structure reported. Important measurements at intermediate and low energy regions including the results of proton recoil measurements will be added to the "global detector", which will increase the sensitivity of the elastic cross-section in the keV-energy region of importance to deep penetration in iron.

In parallel with the reevaluation of the ASPIS experiment with the help of improved iron cross-sections including detailed covariance information the main work is directed now to the evaluation of EURACOS. Since the radius of the converter plate is smaller, we get higher buckling corrections. The self-shielding factors are similar for equivalent penetration depth in iron. The cross-section adjustment of EURACOS will be included in the "global detector" results of ASPIS.

In a further step the correct Newton iteration procedure of equation (1) will be used with the advantage of simultaneous adjustment of the error covariance matrices, which aren't possible in the "global detector" approximation of equation (2). Finally a fine group adjustment is planned applying energy block consistency, the results of which can be of special interest for the evaluation of basic inelastic and elastic iron cross-sections.

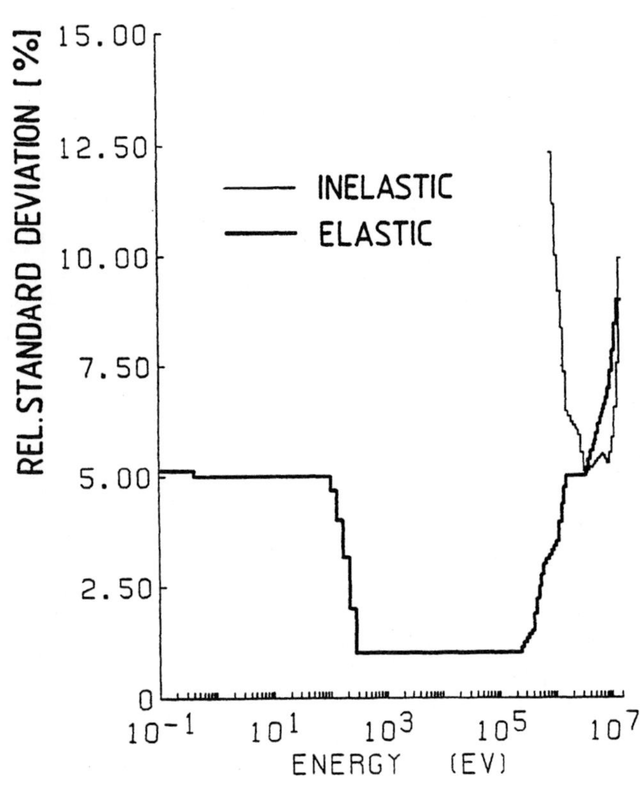

Fig. 13. Relative standard deviation of the iron inelastic and elastic scattering cross-section in EURLIB group structure from ENDF/B-5 /6/.

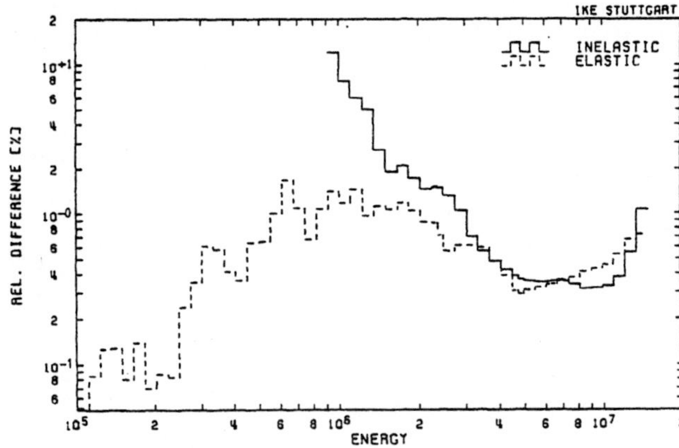

Fig. 12. Relative difference between original group cross-sections and adjusted group data for the iron inelastic and elastic scattering cross-sections. Improved consistency with the S-32, In-115, and Rh-103 detector measurements of ASPIS. Global detector version of ADJUST-EUR.

References

1. G. Palmiotti, M. Salvatores NEACRP-A-503 (Draft), (1982)

2. G. Hehn NEACRP-A-504 (Draft), (1982)

3. H. Werle, H. Bluhm, G. Fieg, F. Kappler, D. Kuhn, M. Lalovic, KFK 2219, (1975)

4. M. M. Chestnutt, A.K.McCracken, NEA Spec. Meeting on Nuclear Data, Paris, (1980)

5. G. Perlini, G. Gonano, NEA Spec. Meeting on Nuclear Data, Paris, (1980)

6. D. W. Muir, R.J. LaBauve, LA-8733-MS (1981)

7. W. Matthes, NEA Spec. Meeting on Nuclear Data, Paris, (1980)

8. A. Pazy, G. Rakavy, Y. Reiss, J.J. Wagschal, A. Ya'ari, Y. Yeivin, NSE, 55, (1974)

9. Y. Yeivin, W. Matthes, EUR-6041-en, (1978)

10. R. Rühle, ATKE26, (1975)

11. W.W. Engle, USAEC, K-1693, (1967)

12. D.E. Bartine, F.R. Mynatt, E.M. Oblow, ORNL-TM-3809, (1973)

13. G. Hehn, M. Mattes, K. Al Malah, G. Kicherer NEA Spec. Meeting on Nuclear Data, Paris, (1980)

14. R.W. Roussin, C.R. Weisbin, J.E. White, N.M. Green, R. O. Wright, J.B. Wright, ORNL/RSIC-37, (1978)

15. D. Garber, BNL-NCS-17541 (ENDF-201) (1975)

16. R.E. Mac Farlane, R.J. Barret, D.W. Muir, R.M. Boicourt, LA-7584-M, (1978)

17. W.A. Rhoades, D.B. Simpson, R.L. Child, W.W. Engle, ORNL/TM-6529 (1979)

^{54}Fe NEUTRON CAPTURE CROSS SECTION

A. Brusegan, F. Corvi, G. Rohr, R. Shelley, T. van der Veen, C. Van der Vorst

Commission of the European Communities
Joint Research Centre
Central Bureau for Nuclear Measurements
Geel, Belgium

B.J. Allen

AAEC - Australian Atomic Energy Commission Research Establishment
Lucas Heights Research Laboratories
Sutherland, NSW 2232, Australia

At the Geel electron Linac, neutron capture cross section measurements have been performed on an enriched ^{54}Fe oxide sample in the range from 0.3 to 500 keV. The nominal resolution was 0.086 ns/m. The analysis of the data has been completed up to 200 keV, both with an area code and a R-matrix shape fitting code. Resonance parameters and capture areas are given and the experimental evidence of valence neutron effects is discussed.

[^{54}Fe(n,γ), 1-200 keV, E_o, A_γ, Γ_n, Γ_γ]

Introduction

At the fourth International Symposium on Neutron Capture Gamma-Ray Spectroscopy [1] preliminary results of ^{54}Fe(n,γ) cross section measurements performed at CBNM Geel, were presented and resonance parameter correlations and asymmetries in the resonance shape were discussed for s-wave resonances in the range from 7 to 300 keV.

The important reduction of the capture widths observed in comparison with the previous measurements of Allen et al. [2], was attributed to the experimental set up in use at Geel, i.e. to the absence of fluorine in the C_6D_6 liquid scintillator detectors and to the minimized amount of material in the vicinity of the scintillator. This results in a much smaller prompt background correction (PBG) for sample scattered neutrons. As a consequence, the strong initial state correlation between the s-wave reduced neutron widths and the radiation widths $\rho_I(\Gamma_n^o,\Gamma_\gamma)$ shown by Allen [2] was seen to be reduced and heavily dependent on the very broad 188 keV resonance and could not confirm the expected valence neutron capture effect.

The valence correlation $\rho_\nu(\Gamma_\gamma^\nu,\Gamma_\gamma)$ between the predicted valence and the observed radiative widths lead to similar conclusions. A new high resolution measurement has been performed with the intention to confirm the previous observations and to extend the resonance parameter investigation to the non s-wave resonances. A general discussion about the discrepancies in the structural material data, related to systematic errors resulting from different detector systems and normalization procedures, has recently been presented by G. Rohr at the Specialist Meeting on Fast Neutron Capture Cross Sections held in Argonne [3] where the CBNM data for ^{54}Fe are taken from the analysis of the first measurement.

Experimental details and data normalization procedure

Neutron capture γ-rays were detected by a pair of 7.6 cm thick and 10.2 cm diameter C_6D_6 liquid scintillator detectors placed at the 58 m measuring station of the 150 MeV Geel Linac, which was running at 5 ns burst width and 800 Hz repetition rate. The monitors of the experiment were a 0.05 cm thick ^{6}Li glass detector placed permanently in the neutron beam, 63 cm in front of the sample, and BF_3 detectors. The sample, consisting of 66.2 g iron oxide enriched to 97.69 % ^{54}Fe ($9.92\cdot10^{-3}$ nuclei/b), was canned in a 0.03 cm thick Al container with 8 cm diameter. For each detected capture event the time information was analysed by a 4 ns digital time coder built at CBNM [4] and the amplitude information by an Analog to Digital Converter (ADC) over 256 channels. The linearity of the detecting system was measured with several γ-ray sources and the stability of the gain regularly controlled with the Compton edge of the ^{238}Pu-^{13}C(α,nγ)^{16}O source (6.13 MeV γ-ray energy). No capture events were accepted below 150 keV electron energy. A standalone multiparameter system CORSYS [5] controlled by a GA 16/460 minicomputer (320 kB of main memory) provided the acquisition for the capture event informations: the data were weighted on-line according to the associated pulse height (total energy weighting technique) and stored in 16 k time of flight channels. Unweighted, weighted and weighted squared spectra, the latter containing the statistical error information, were recorded in the memory of the minicomputer.

For the neutron flux determination the experimental conditions were taken as similar as possible to the capture yield measurements by replacing the ^{54}Fe sample in front of the detectors with a 0.3 cm thick $^{10}B_4C$ slab. A fast single channel analyzer was set on the amplified anode signal of the EMI photomultipliers of the detectors in order to accept only those events associated with the 478 keV photons of the ^{10}B(n,$\alpha\gamma$)^{7}Li reaction.

The background for the flux was determined with the same detectors and a carbon scatterer 0.35 cm thick. Both for the flux and its background, the time of flight spectra were recorded in 16 k channels as in the capture yield run.

The ^{54}Fe capture data have been normalized to the transmission result for the 1.15 keV resonance of ^{56}Fe ($\Gamma_n = (58.9 \pm 2.0)\cdot10^{-3}$ eV, $\Gamma_\gamma = 0.610 \pm 0.060$ eV)[3-6]. This latter isotope is present as a 2.25 % impurity in the sample. The problems connected to the application of the weighting technique for resonances with different γ-ray spectral hardness have been discussed elsewhere [3] and are still relevant. However, since the average hardness of the γ-ray spectra, obtained by the mean weights listed in Table 1 for the ^{54}Fe, corresponds very well to that of the 1.15 keV ^{56}Fe resonance (see also Fig. 2 in Ref. [3]), normalization to the capture area of this resonance overcomes these problems.

Due to the large scattering to capture ratio ($> 10^4$) in s-wave resonances of ^{54}Fe, the problem associated with the capture, in the detector, of neutrons scattered by the sample (PBG) is crucial. A preliminary value for the neutron sensitivity $\epsilon_n/\epsilon_\gamma = (1.50 \pm 0.75)\cdot10^{-4}$, not including any energy dependence, has been determined by the comparison of the count rate from a 0.8 cm thick graphite disk to that from a 0.01 cm thick Au sample. The background was measured both with a 0.1 cm carbon scatterer and with no sample in the beam.

Data analysis

For the flux spectra, the background measured with a graphite scatterer contributed up to 3 % of the observed count rate. The background of the weighted

K. H. Böckhoff (ed.), Nuclear Data for Science and Technology, 127–130.

capture yields has been determined at the minima between well separated resonances and described with a monotonic function from 0.3 to 350 keV.

The "capture cross section" spectrum, shown in Fig. 1 for the neutron energy range from 90 to 180 keV, was obtained from the ratio of the background corrected capture and flux yields performed channel by channel and includes a proper energy dependent "efficiency" factor for the boron slab and the normalization factor. No corrections have been applied for doppler and resolution broadening, multiple scattering and prompt background contributions.

The presence of large s-wave resonances in ^{54}Fe compels the use of the R-Matrix formalism to correctly reproduce the very complicated self-screening and multiple scattering effect. Hence s-wave resonances up to 100 keV have been analyzed with a modified version of the R-Matrix Reich-Moore shape analysis code Fanac [7].

At higher energy the Reich-Moore formalism has been replaced by a 2-channel R-Matrix formalism which considers the neutron and one capture channel. With this one can reproduce more accurately the asymmetric shape of the s-wave resonances at 130, 147 and 174 keV, seen in Fig. 1.

An example of the importance of the multiple scattering correction and the prompt background contribution to the primary yield is shown in Fig. 2 for the 7.8 keV s-wave resonance.

Non s-wave resonances have been analyzed with the single level Breit-Wigner shape analysis procedure included in Fanac and, for well isolated resonances, the area code TACASI [8]. The comparison shows that below 30 keV a discrepancy between the capture areas of about 3 % exists, increasing in some cases to 10 % at higher energy. This was considered to be caused by the use of a gaussian resolution function, in the shape analysis code, which is inadequate for the

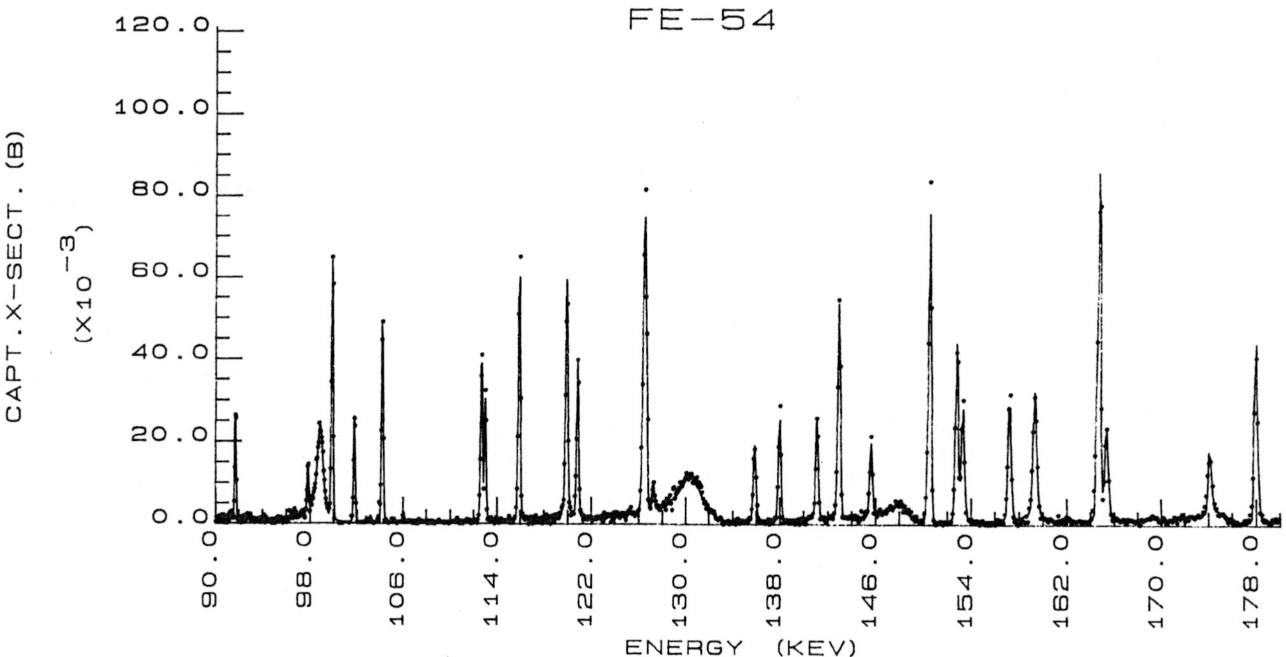

Fig. 1 The ^{54}Fe "capture cross section" vs neutron energy from 90 to 180 keV

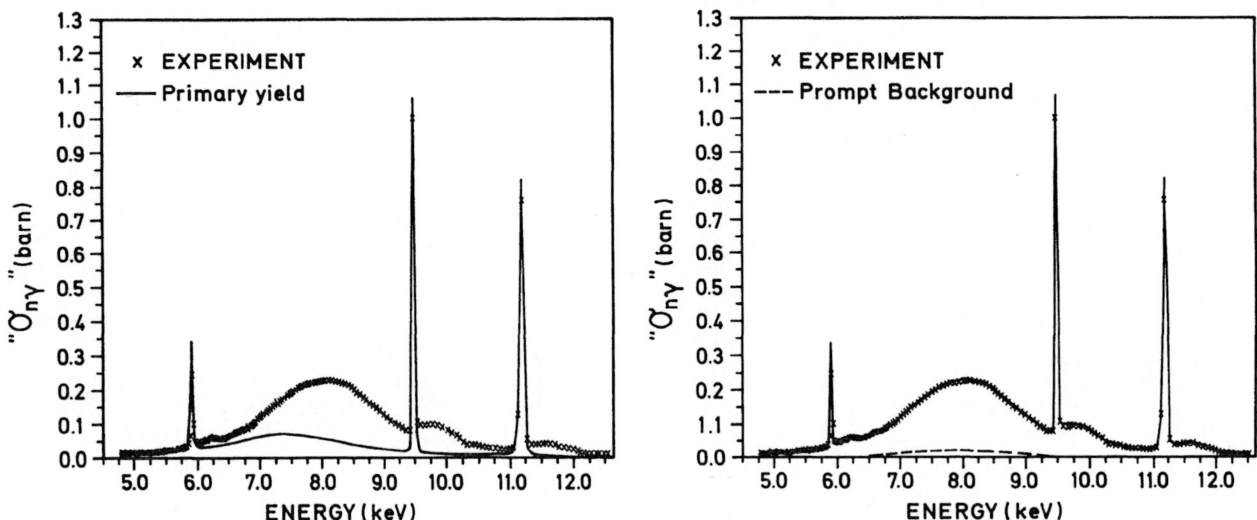

Fig. 2 The multiple scattering correction and the prompt background contribution to the primary yield for the 7.8 keV s-wave resonance in ^{54}Fe

Table I Capture areas and resonance parameters of ^{54}Fe in the range 1 to 200 keV from the present work compared to the results of ORNL-AAEC [2]

PRESENT RESULTS						ORNL-AAEC RESULTS b)
E_o (keV)	$\bar{w}$	$g\Gamma_n\Gamma_\gamma/\Gamma$ (eV)	$g\Gamma_n$ (eV)	$g\Gamma_\gamma$ (eV)	J^π	$g\Gamma_n\Gamma_\gamma/\Gamma$ (eV)
3.099	23.	0.0028 ± 0.0003				0.0030
7.80	44.	1.74 ± 0.11	1040. ± 100.	1.74 ± 0.11	$1/2^+$	1.8 ± 0.4
9.486	30.	0.55 ± 0.03	1.2 ± 0.3	1.03 ± 0.10		0.55
11.181	40.	0.77 ± 0.04	7.7 ± 0.8	0.86 ± 0.05		0.69
13.581	31.	0.035 ± 0.003				0.034
14.465	30.	0.62 ± 0.02	1.4 ± 0.5	1.12 ± 0.07		0.62
19.278	31.	0.050 ± 0.004				0.047
23.03	32.	0.37 ± 0.02				0.39
28.22	32.	0.16 ± 0.01				0.17
30.66	29.	0.83 ± 0.03	7.7 ± 1.0	0.93 ± 0.04		0.96
35.26	27.	0.23 ± 0.02				0.26
38.44	28.	0.83 ± 0.03				0.92
39.14	29.	0.77 ± 0.04	17. ± 2.	0.81 ± 0.04		0.82
41.20	31.	0.024 ± 0.006				0.028
50.14	29.	0.056 ± 0.07				0.075
51.58	35.	0.36 ± 0.03	6. ± 2.	0.38 ± 0.04		0.36
52.84	41.	1.50 ± 0.30	1978. ± 1.	1.50 ± 0.30	$1/2^+$	2.4 ± 0.4
53.59	24.	0.52 ± 0.04	17. ± 3.	0.54 ± 0.04		0.60
55.07	29.	0.62 ± 0.05				0.68
55.43	29.	0.55 ± 0.05	32. ± 3.	0.56 ± 0.05		0.68
59.20	29.	0.37 ± 0.03				
68.76	25.	0.24 ± 0.03				0.31
71.81	26.	0.28 ± 0.15	1676. ± 1.	0.28 ± 0.15	$1/2^+$	1.32 ± 0.26
75.80	29.	0.65 ± 0.03				0.76
77.21	27.	1.29 ± 0.06	4. ± 2.	1.91 ± 0.14		1.62
81.28	38.	0.29 ± 0.03				0.30 ± 0.03
83.19	26.	0.96 ± 0.07				1.27
83.48	30.	0.36 ± 0.05				0.45 ± 0.03
87.33	32.	0.39 ± 0.04				0.50
91.67	28.	0.25 ± 0.03				
97.77	35.	0.11 ± 0.02				0.24 ± 0.03
98.80	48.	1.20 ± 0.25	511. ± 5.	1.20 ± 0.25	$1/2^+$	1.65 ± 0.25
99.84	24.	0.73 ± 0.06				0.79
101.71	21.	0.29 ± 0.03				0.35 ± 0.02
104.18	24.	0.60 ± 0.06				0.79
112.63	28.	0.64 ± 0.07				0.72
112.96	30.	0.42 ± 0.06				0.56
115.85	31.	1.06 ± 0.07	26. ± 5.	1.10 ± 0.07		1.21
119.78	49.	1.01 ± 0.09	26. ± 8.	1.05 ± 0.10		1.11
120.81	25.	0.72 ± 0.11	40. ± 10.	0.73 ± 0.11		0.89
126.4 a)	29.	2.5 ± 0.4	60. ± 10.			2.59
127.15	36.	0.12 ± 0.04				
130.00	40.	2.9 ± 0.9	3470. ± 10.	2.9 ± 0.9	$1/2^+$	3.22 ± 0.64
135.81	31.	0.50 ± 0.05	80. ± 10.	0.50 ± 0.05		0.69
137.93	37.	0.61 ± 0.05				0.96
141.01	25.	0.59 ± 0.07				0.48
142.82	25.	1.30 ± 0.12				1.72
145.51	30.	0.48 ± 0.07				0.56
147.49	37.	0.54 ± 0.49	3780. ± 15.	0.54 ± 0.49	$1/2^+$	2.31 ± 0.46
150.47	43.	2.10 ± 0.17				2.88
152.71	27.	1.28 ± 0.15				1.78
153.22	25.	0.92 ± 0.11	60. ± 10.	0.93 ± 0.12		1.14
157.18	33.	1.06 ± 0.10				1.45
159.29	40.	1.39 ± 0.17	100. ± 10.	1.41 ± 0.17		1.89
164.64	35.	3.56 ± 0.25	105. ± 10.	3.69 ± 0.27		4.12
165.33	26.	0.87 ± 0.14	75. ± 10.	0.88 ± 0.14		0.81
174.07	32.	0.79 ± 0.16				0.82
174.09		0.58 ± 0.72	4590. ± 30.	0.58 ± 0.72	$1/2^+$	3.5 ± 1.1
177.94	37.	1.70 ± 0.20				1.88
182.38	28.	1.02 ± 0.16	150. ± 15.	1.03 ± 0.16		1.20
188.00		< 0.9	38170. ± 100.	< 0.9	$1/2^+$	10.0 ± 4.0
189.09	28.	0.27 ± 0.09				0.49 ± 0.03
192.12	31.	1.07 ± 0.14				1.33
194.05	45.	1.03 ± 0.20				1.14
194.56	39.	1.67 ± 0.21	100. ± 15.	1.70 ± 0.22		1.69
197.70	26.	0.85 ± 0.13				1.23

a) Doublet
b) Statistical and background errors are less than 5 % unless the error is quoted

reproduction of the asymmetric profiles generated by our Linac target-moderator configuration.

The implementation of the asymmetric resolution function in the Fanac code is underway.

Results

In Table I, our results for ^{54}Fe are listed together with the capture areas ($g\Gamma_n\Gamma_\gamma/\Gamma$) of Allen et al. [2].

The quoted neutron widths for the known s-wave resonances are taken from the work of Cornelis et al. [9], except for the 7.8 keV. In this case and for all the non s-wave resonances, the given $g\Gamma_n$ values are those of Pandey et al. [10]. Non s-wave resonances have been treated as p-wave resonances taking the g factor equal 1 and resonances where $g\Gamma_n < g\Gamma_\gamma$ were analyzed assuming $\Gamma_\gamma = 1$ eV. The results of the s-wave resonances are corrected for the prompt background by subtracting its contribution to the radiative widths. A more precise correction procedure is in preparation.

The overal accuracy on the resonance energies is taken to be 0.1 % for p-wave and up to 0.3 % for the large s-wave resonances. The calculated errors for the capture areas include the statistical plus the following systematic contributions: 5 % assumed on the background; up to 5 % related to the discrepancy of the results between the shape and the area analysis. Finally, the results of the s-wave resonances include the 50 % uncertainty on the prompt background correction. The normalization error, estimated to be 5 %, is not included for individual resonances. Also no uncertainty has been introduced for the differences in spectrum shape discussed earlier.

The present results for p-wave resonances are systematically lower than those of ORNL-AAEC [2]. For a selected set of resonances, which have a good statistical accuracy and which do not overlap with s-wave resonances, the discrepancies on the capture areas are on average 4 % below 30 keV and 17 % for resonances at higher energy.

For the same selected set of resonances, the ratio of the present data to those of our first ^{54}Fe run (partly published in Ref. [3]) is 1.04 independent of the neutron energy range.

A similar comparison of the PBG uncorrected s-wave capture widths to those published in Grenoble [1] shows a good agreement below 130 keV and a large deviation at higher energy. However a 40 % reduction for the uncorrected Γ_γ of the 188 keV resonance can be explained by a change of only 4 % in the background. As mentioned in the introduction the initial state correlation is strongly dependent on the parameters of this resonance. Therefore in view of the increase in the neutron sensitivity compared to our previous measurement, which will decrease the total radiative width, and the large uncertainty due to the background this correlation becomes very doubtful.

Acknowledgements

The authors would like to thank Mr. A. De Keyser from CBNM for his indispensable help with the data acquisition system and ORNL for the loan of the ^{54}Fe sample.

References

1. A. Brusegan, F. Corvi, G. Rohr, R. Shelley, T. van der Veen, B.J. Allen, Proc. Int. Conf. on Neutron-Capture Gamma-Ray Spectroscopy and Related Topics, Grenoble, 7-11 September 1981, p. 408
2. B.J. Allen, A.R. de L. Musgrove, J.W. Boldeman, R.L. Macklin, AAEC/E403, (1977) and Proc. Spec. Meeting, CBNM, Geel, 1977, p. 447
3. G. Rohr, Proc. Spec. Meeting on Fast Neutron Capture Cross Sections, Argonne, 1982
4. S. de Jonge, Fast Time Coder Type 7602D, Internal Report GE/IN/72/82, CBNM Geel
5. A. De Keyser, The Correlative Systems Data Acquisition System Documentation GE/TN/DE/85/82, CBNM Geel and Correlative Systems Reference Manual
6. A. Brusegan, F. Corvi, G. Rohr, R. Shelley, T. van der Veen, Proc. Int. Conf. on Nuclear Cross Section for Technology, Knoxville, 1979, p. 163
7. F.H. Fröhner, 1976, Report KFK 2129
8. F.H. Fröhner, 1966, Report GA-6906
9. E.M. Cornelis, C.R. Jungmann, L. Mewissen, F. Poortmans, Proc. Int. Conf. on Nuclear Cross Section for Technology, Knoxville, 1979, p. 159
10. M.S. Pandey, J.B. Gary, Proc. Int. Conf. on Nuclear Cross Section and Technology, Washington, D.C., 1975, p. 748

HIGH RESOLUTION NEUTRON CAPTURE CROSS SECTION MEASUREMENTS OF ^{56}Fe

F. Corvi, A. Brusegan, R. Buyl, G. Rohr, R. Shelley, T. van der Veen

Commission of the European Communities
Joint Research Centre
Central Bureau for Nuclear Measurements
Geel, Belgium

Capture cross section measurements with a nominal resolution of 0.075 ns/m have been carried out on an enriched ^{56}Fe sample at the Geel electron linac. The data have been analysed up to 256 keV both with an area code and an R-matrix shape fitting code. Capture areas and resonance parameters have been deduced and compared to previous results.

[$^{56}Fe(n,\gamma)$, 1-256 keV, E_o, A_γ, Γ_n, Γ_γ]

Introduction

Preliminary data based on ^{56}Fe neutron capture measurements performed at CBNM, Geel, covering the energy range from 1 to 100 keV, were reported at the Knoxville Conference [1]. Comparison with previous results showed that the Geel capture areas, represented by the quantity ($g\Gamma_n\Gamma_\gamma/\Gamma$), were in good agreement with Harwell's data [2] but were systematically about 15% lower than the Oak Ridge values [3,4]. This important discrepancy was ascribed to a different procedure of normalizing the data. A more extensive inter-comparison of ^{56}Fe capture data, including also the results of RPI [5] and Karlsruhe [6] has been recently carried out by G. Rohr [7] in an effort to investigate the possible systematic errors related to given detector systems and normalization methods.

The measurements presented at Knoxville, which will be referred to as Run 1 in the following, have also been analysed by M.C. Moxon [8] with his R-matrix fitting code REFIT, yielding resonance parameters in agreement with those of ref.[1]. More recently, the capture measurements have been completely repeated with the following improvements or extensions :

a) the relative neutron flux was measured from 1 to 400 keV neutron energy. In fact, it was the failure to correctly determine the flux at high energy that prevented the determination of resonance data above 100 keV in Run 1;
b) the data were weighted according to the corrected weighting function, referred to as "NEW W(E)" in ref.[1];
c) the threshold of acceptance of pulses from the scintillators was lowered to 150 keV electron energy as compared to 300 keV used in Run 1. Though a lower threshold results in a lower signal-to-background ratio, it has the advantage of reducing a possible dependence of the detector sensitivity on the shape of the capture γ-ray spectrum;
d) the scattered neutron sensitivity was measured relative to the capture detection efficiency and the s-wave capture areas and resonance parameters were corrected for that contribution;
e) the data were analysed not only with the area code TACASI, as in ref.[1], but also with the R-matrix multilevel code FANAC, which was adapted to our high resolution spectra. In fact, s-wave resonances can be correctly analysed only with such type of codes.

All results given in this paper are based on this new measurement campaign which will be referred to as Run 2 in the following. For the reasons listed above we believe that this set of data should replace the preliminary results of ref.[1].

Experimental Method

The experimental conditions of Run 2 were to a large extent similar to that of Run 1. Basically, the measurements were performed at the 150 MeV Geel electron linac, at a flight distance of 58 m. The linac was operated to provide 4.4 ns wide bursts of electrons of 100 MeV energy with a repetition frequency of 800 Hertz. Since the peak current was 10 A, the average beam power was about 3.6 kWatt. Neutrons impinging on the sample were only those coming from the 4 cm thick polyethilene moderator, while neutrons and gammas coming directly from the uranium target were blocked by a copper and lead shadow bar. The addition of copper to the lead bar used in Run 1 was found necessary in order to eliminate some structure on the high energy side of the neutron flux due to the shape of the lead cross section. The sample, on loan from ORNL, consisted of iron oxide enriched to 99.93% ^{56}Fe, packed in a thin aluminium container of 8 cm diameter. The sample thickness was 0.015 atoms/barn. The detectors were two C_6D_6 liquid scintillators encapsulated in thin aluminium containers of 10.2 cm diameter and 7.6 cm height. Events were weighted according to their amplitude information in order to achieve a detector efficiency proportional only to the total γ-ray energy emitted in the capture process. The weighted counting rate was sorted out as an 8K time-of-flight spectrum with 4 ns minimum channel width, covering the energy range from 1 keV to about 600 keV. The relative neutron flux was measured with a ^6Li-glass 0.5 mm thick below 100 keV and with a multi-plate ^{235}U fission chamber above 100 keV. The relative flux from the fission chamber was normalized to the Li-glass flux in the overlapping region from 60 to 100 keV. A description of the fission chamber can be found in ref.[9].

After background subtraction, correction for the neutron flux and proper normalization, the weighted counting rate can be reduced to capture cross section. This is shown in fig. 1, where $\sigma(n,\gamma)$ for ^{56}Fe is plotted vs neutron energy in the range from 175 to 265 keV. It should be noted that the plotted $\sigma(n,\gamma)$ is still Doppler and resolution broadened and it is not corrected for multiple scattering and prompt background contributions which are particularly important around the large s-wave resonances.

A measurement was carried out to determine the ratio ϵ_n/ϵ_c, i.e. the scattered neutron sensitivity of the detector system relative to the capture efficiency. The count rate from a 8 mm thick graphite disk was compared to that from a 0.1 mm thick gold sample after subtraction of the open beam background. A preliminary value $\epsilon_n/\epsilon_c = (1.5 \pm .75)\times10^{-4}$ was found in the region below 100 keV. This value was used to correct for the scattered neutron contribution in all s-wave resonances analysed, independent of their energy.

Data Normalization and Weighting Method

Similarly to ref.[1] the data were normalized to the 1.15 keV ^{56}Fe resonance assumed to have a neutron width $\Gamma_n = 58 \pm 3$ meV or equivalently a value $g\Gamma_n\Gamma_\gamma/\Gamma = 53 \pm 3$ meV (taking $\Gamma_\gamma \simeq 0.60$ eV). Such a value was deduced from a transmission measurement carried out at Geel on a 2 mm thick metallic sample of natural iron and was found in agreement with a thermal energy calibration which yielded $\Gamma_n = 56 \pm 6$ meV. More recently the figure of 58 meV has been confirmed by a new transmission measurement on a 1mm thick sample.[14]

K. H. Böckhoff (ed.), Nuclear Data for Science and Technology, 131–134.

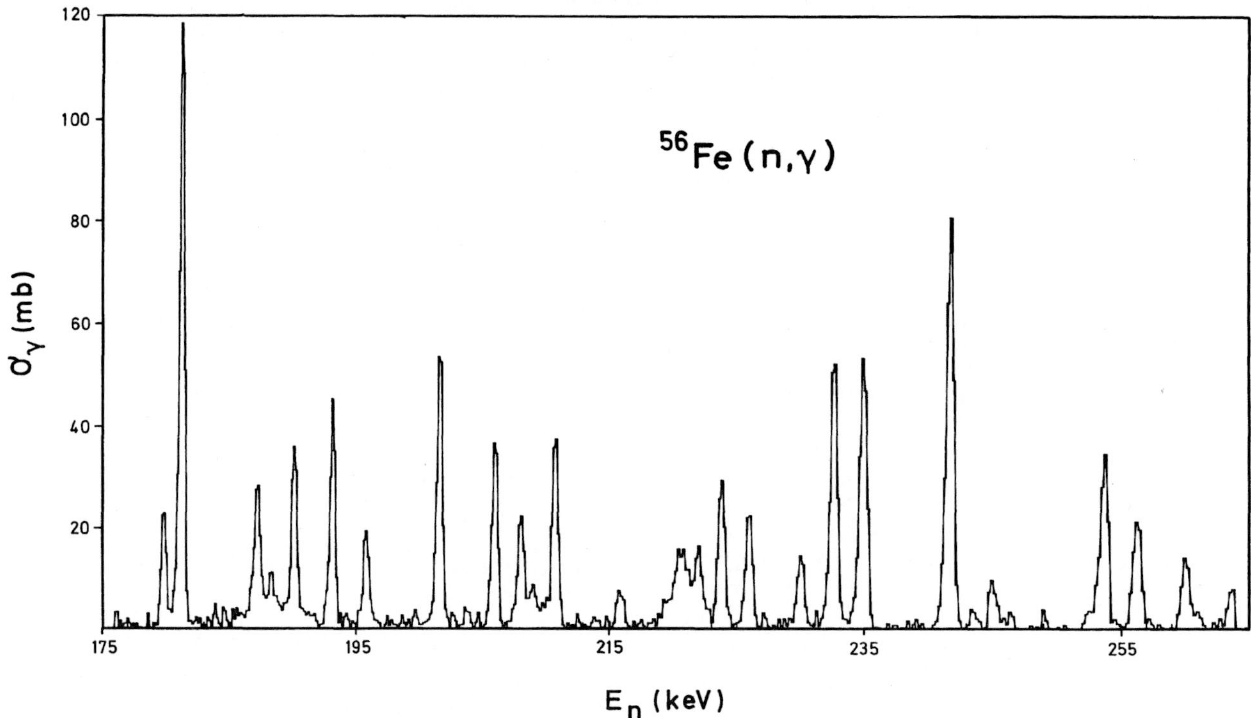

Fig. 1. The ^{56}Fe capture cross section vs neutron energy from 175 to 265 keV

A problem directly connected with the normalization is the validity of the weighting method in the case of ^{56}Fe and of structural materials in general. A discussion on this point can be found in ref. [1]. To summarize, the limited number of primary transitions of the ^{56}Fe capture spectrum (about 20 according to ref. [10]) and the large average strength of the highest energy ones, gives rise to considerable variation of the spectrum shape of the resonances. If, as a first approximation, we represent this spectrum shape, or better its hardness, with the average weight $\bar{w}$, we find that this quantity can change up to a factor 2 for different resonances. In this situation it is clear that normalizing to a given resonance is only correct if it can be proved that the detector efficiency is independent of spectrum shape. To check that we have carried out some further calibrations besides those already described in ref. [1]: in particular we have concentrated on measuring Fe capture relative to Ag and Au capture rates at both thermal and resonance energies. The results are summarized in Table I. At thermal energy both natural and enriched Fe samples were compared to the Au standard. The known thermal Fe capture cross sections, listed under R, were taken from ref. [11]. The saturated resonance technique was applied in the epithermal region, making use of the Ag and Au resonances at 5.2 and 4.9 eV, respectively: the two normalization constants obtained differed by 2.8 % and their average was taken. After correction for the relative neutron flux, measured always with a ^6Li glass, a value $A_\gamma = g\Gamma_n\Gamma_\gamma/\Gamma = 66.2 \pm 2.8$ meV was derived for the 1.15 keV resonance, corresponding to $\Gamma_n = 74.3 \pm 3$ meV. This result agrees well with a previous value of $A_\gamma = 65$ meV obtained by normalizing to saturated Ag resonances between 16 and 71 eV when use was made of the "New" weighting function [7]. The measured area is 25 % larger than the value obtained from transmission and listed under R in Table I. A similar trend is apparent at thermal energy though here the effect is limited to 8-10 %. This difference of the relative efficiency ϵ between thermal and resonance energies is really not to be understood since the γ-ray spectrum of the 1.15 keV resonance is very similar [10] to the thermal one. Recently Macklin [12] has also measured the $E_n = 1.15$ keV capture area using C_6F_6 detectors and calibrating to the 4.9 eV gold resonance. He has found $g\Gamma_n\Gamma_\gamma/\Gamma = 73.0 \pm 0.5$ meV, a value which is 38 % higher than our transmission data !

To conclude, the application of the weighting method to structural materials gives problems which are far from being solved: the increase of the efficiency with the hardness of the spectrum seems well established but its exact amount varies widely according to the particular measurement. The consequence of this effect on ^{56}Fe capture will be in practice an under-estimation of the capture areas for the resonances with softer spectra since the normalizing resonance at 1.15 keV has a rather hard spectrum (average weight $\bar{w} = 26$). If we make the (arbitrary) assumption that the efficiency increase is proportional to the increase in average weight $\bar{w}$, we can estimate that the effect on the softer resonances will be about 60 % of that found in the case of Ag-Au calibration. Therefore it will be of the order of 6 % if one follows the thermal results and 15 % when relying on the resonance calibration.

Analysis and Results

The data were analysed both with the area programme TACASI and with the R-matrix shape fitting code FANAC. This last code was used mainly for the analysis of s-waves and of groups of unresolved resonances. The results obtained with the two programmes were compared by analysing some well defined and isolated resonances: it was found that the capture areas given by FANAC are systematically slightly lower than those of TACASI by an amount which increases with neutron energy up to values of about 10 % in some cases. We ascribed this effect to the inadequacy of the gaussian resolution function used in FANAC to fit asymmetric profiles as those produced by our present target-moderator configuration [13]. A modification of the programme is underway to cope with this effect. For this reason all the isolated resonances have still been analysed with TACASI.

The results obtained are listed in Table II aside those of the ORNL evaluation of F. Perey [4]. The values of J, ℓ and Γ_n or Γ_γ (whichever is the larger) from this evaluation have been taken as fixed input parameters in the TACASI and FANAC programmes. The values of Γ_γ or Γ_n (whichever is the lower) derived from the codes and listed in the 3rd and 4th column of Table II depend on the assumed parameters and in particular on the resonance spin J. They should therefore be taken with some caution, particularly

Table I Results of calibration tests of capture in natural and enriched Fe samples relative to capture in Ag and Au at thermal and resonance energies

Calibration Energy	Calibration Element	Sample Thickness	Fe Sample Nucl.	Thickness	R (refer.value)	C (value from calibrations)	ϵ = C/R
E_{th}=0.025 eV	Au	0.1 mm	Fe	0.5, 1 mm	σ_{th}= 2.56 ± .03 b	σ_{th}= 2.82 ± 0.8 b	1.10 ± .04
E_{th}=0.025 eV	Au	0.1 mm	^{56}Fe	0.5 mm	σ_{th}= 2.59 ± .14 b	σ_{th}= 2.79 ± .08 b	1.08 ± .07
E_o = 4.9 eV	Au	0.1 mm	^{56}Fe	0.5 mm	A_γ = 53 ± 3 meV	A_γ = 66.2 ± 2.8 meV	1.25 ± .09
E_o = 5.2 eV	Ag	0.2 mm					

when the J value proposed in ref. [4] is not based on experimental data.

The given uncertainties of the capture areas are composed of the statistical plus the following systematic errors: 5 % error on the determination of the background line; an error on the relative neutron flux increasing from 0 at 1 keV to 5 % at 100 keV and kept at a constant value of 5 % above. For resonances analysed with FANAC an additional 5 % uncertainty is taken because of the discrepancy with TACASI. Moreover, large s-waves have 50 % uncertainty on the scattered neutron sensitivity. All these errors have been quadratically combined. The error on normalization, which is estimated at about 5 %, is not included in those of the single resonances. Finally, no uncertainty has been introduced for the difference in spectrum shape because of the unsettled situation of this problem. A comparison of the present results to those from previous works leads to the following remarks:

a) the present capture areas in the region up to 100 keV are on average about 6 % larger than those[1] of Run 1;
b) the present values of the radiative widths Γ_γ for the nine s-wave resonances analysed are considerably smaller than those of the Perey evaluation [4];
c) over the whole range analysed the present capture areas for p- and d-waves are on average 11 % lower than the corresponding ones of refs. [3-4];
d) the present capture areas are on average only 4 % lower than those recently measured by the Karlsruhe group [15] in the range from 20 to 113 keV.

It should be noted that the data sets of the Oak Ridge and Karlsruhe groups were both normalized to Au capture. Therefore the sign of the discrepancies confirms the results of the calibration of Table I though their absolute amount indicates that this effect is not too large.

To investigate further the dependence of the efficiency of the capture detectors on the spectral shape, we have plotted in Fig. 2 the total radiation width Γ_γ versus the average weight $\bar{w}$ for p- and d-waves separately. Though the spread of the data is too large to draw any definite conclusion, there seems to be no systematic increase of Γ_γ with $\bar{w}$ in the case of p-waves. On the other hand, there is a certain evidence of such an increase for d-waves. Large values of $\bar{w}$ are usually an indication of strong high energy primary transitions which can contribute considerably to Γ_γ. Therefore the increase noticed is explainable though it is not clear why it should be limited to d-waves alone. Finally, from the data of Table II one can calculate the averages and standard deviations of Γ_γ and the related effective number of degrees of freedom ν_{eff} for different ℓ values:

$$<\Gamma_\gamma> = 0.85 \pm 0.41 \ (\nu_{eff} = 9.6) \quad \text{for s-waves}$$
$$<\Gamma_\gamma> = 0.50 \mp 0.18 \ (\nu_{eff} = 17.1) \quad \text{for p-waves}$$
$$<\Gamma_\gamma> = 0.73 \mp 0.25 \ (\nu_{eff} = 18.5) \quad \text{for d-waves}$$

These values compare well with the results of the statistical model calculation reported in ref. [16].

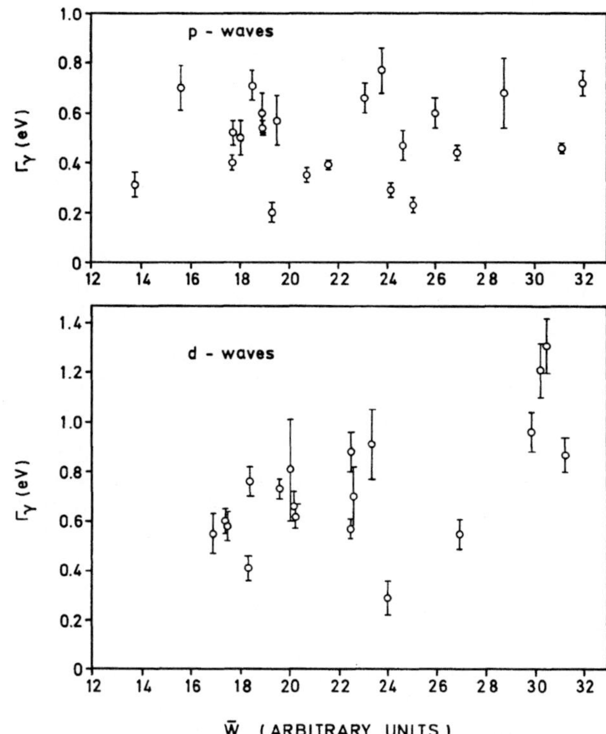

Fig. 2. Values of the total radiation width Γ_γ vs the average weight $\bar{w}$ for p- and d-waves separately

References

1. A. Brusegan, F. Corvi, G. Rohr, R. Shelley, T. van der Veen, Proc. Int. Conf. on Neutron Cross Section for Technology, Knoxville, 1979, NBS-SP.594, p. 163
2. D.B. Gayther, B.W. Thomas, B. Thom, M.C. Moxon, Neutron Data of Structural Materials for Fast Reactors, Geel, 1977, Pergamon Press, p. 547
3. B.J. Allen, A.R. de L. Musgrove, J.W. Boldeman, M.J. Kenny, R.L. Macklin, Nucl. Phys. A269 (1976) 408
4. F.G. Perey, G.T. Chapman, W.E. Kinney, C.M.Perey, Neutron Data of Structural Materials for Fast Reactors, Geel 1977, Pergamon Press, p.530
5. R.N. Hockenbury, Z.M. Bartolome, J.J. Tatarozuk, W.R. Moyer, R.C. Block, Phys.Rev. 178(1969)1746
6. A. Ernst, D. Kompe, F.H. Fröhner, Proc. Int. Conf. on Nuclear Data for Reactors, Helsinki, 1970, Vol.I, p. 633
7. G. Rohr, Proc. Spec. Meeting on Fast Neutron Capture Cross Sections, Argonne, 1982
8. M.C. Moxon, U.K. Nuclear Data Progr. Rep. (Jan.-Dec.1980), NEANDC(E)222, Vol.8, p.22
9. F. Corvi, L. Calabretta, M. Merla, T. van der Veen, M.S. Moore, Proc. Spec. Meeting on Fast Neutron Capture Cross Sections, Argonne,1982
10. R.E. Chrien, M.R. Bhat, O.A. Wasson, Phys. Rev. C1 (1970) 973
11. S.F. Mughabghab, M. Divadeenam, N.E. Holden, Neutron Cross Sections, Vol.1, Part A, Academic Press, 1981
12. R.L. Macklin, private communication
13. A. Bignami, C. Coceva, R. Simonini, Rep. EUR 5157 e, (1974)
14. A. Brusegan, private communication
15. F. Käppeler, K. Wisshak, L.D. Hong, to be published
16. G. Rohr, Neutron Capture Gamma-Ray Spectroscopy, Plenum Press, 1979, p. 734

Table II Capture areas and resonance parameters of ^{56}Fe in the range 1 to 256 keV from the present work compared to the data of the Oak Ridge evaluation

	Present results			ORNL evaluation				
E_o(keV)	$g\Gamma_n\Gamma_\gamma/\Gamma$ (eV)	Γ_n (eV)	Γ_γ (eV)	$g\Gamma_n\Gamma_\gamma/\Gamma$ (eV)	Γ_n (eV)	Γ_γ(eV)	J	l
1.152	0.053 ± 0.003	0.058 ± 0.003		0.055 ± 0.004	0.060 ± 0.004	0.60 ± .06	1/2	1
2.353	0.00040 ± 0.00008	0.00020 ± 0.00005		0.0004 ± 0.0001	0.00020 ± 0.00006	0.84 ±	3/2	2
12.463	0.0027 ± 0.0007	0.0028 ± 0.0007		0.0023 ± 0.0002	0.0023 ± 0.0003	0.54	1/2	1
17.771	0.014 ± 0.002	0.014 ± 0.002		0.019 ± 0.002	0.019 ± 0.002	0.54	1/2	1
20.194	0.0080 ± 0.0017	0.0040 ± 0.0008		0.0094 ± 0.0009	0.0047 ± 0.0005	0.84 ±	3/2	2
22.815	0.17 ± 0.01	0.24 ± 0.02		0.18 ± 0.02	0.27 ± 0.06	0.54	1/2	1
27.74	0.82 ± 0.12		0.82 ± 0.12		1520. ± 30.	1.4 ± .1	1/2	0
34.26	0.58 ± 0.03		0.46 ± 0.01	0.64 ± 0.05	0.79 ± 0.30	0.54	3/2	1
36.75	0.26 ± 0.01	0.096 ± 0.006		0.28 ± 0.02	0.11 ± 0.01	0.84	5/2	2
38.45	0.35 ± 0.02	0.25 ± 0.02		0.40 ± 0.04	0.32 ± 0.08	0.54	3/2	1
46.09	0.52 ± 0.03		0.54 ± 0.03	0.50 ± 0.05	10. ± 3.	0.53 ± .05	1/2	1
52.18	0.75 ± 0.04		0.39 ± 0.02	0.81 ± 0.08	12. ± 1.	0.42 ± .04	3/2	1
53.60	0.37 ± 0.02		0.57 ± 0.05	0.40 ± 0.04	1.0 ± 0.4	0.67 ± .17	1/2	1
53.73	0.03 ± 0.01			0.11 ± 0.04	1.9 ± 0.3	0.12 ± .05	1/2	1
59.28	0.79 ± 0.04		0.44 ± 0.03	0.87 ± 0.09	4.0 ± 0.5	0.49 ± .05	3/2	1
63.52	0.63 ± 0.04		0.52 ± 0.05	0.65 ± 0.07	0.80 ± 0.16	0.55 ± .13	3/2	1
73.04	0.69 ± 0.04		0.72 ± 0.05	0.70 ± 0.07	20. ± 4.	0.72 ± .07	1/2	1
74.06	0.52 ± 0.07		0.52 ± 0.07	0.73 ± 0.07	535. ± 10.	0.73 ± .07	1/2	0
77.14	0.27 ± 0.02		0.29 ± 0.03	0.30 ± 0.03	3.6 ± 0.5	0.33 ± .03	1/2	1
80.91	1.98 ± 0.11		0.73 ± 0.04	2.04 ± 0.20	7.0 ± 0.7	0.74 ± .08	5/2	2
83.64	0.45 ± 0.09		0.45 ± 0.09	1.28 ± 0.13	1250. ± 50.	1.28 ± .13	1/2	0
90.42	0.78 ± 0.05		0.40 ± 0.03	0.89 ± 0.09	14. ± 1.	0.46 ± .05	3/2	1
92.77	0.98 ± 0.06		0.71 ± 0.06	0.93 ± 0.09	1.6 ± 0.3	0.65 ± .11	3/2	2
92.99	0.52 ± 0.04	0.51 ± 0.07		0.53 ± 0.05	0.52 ± 0.15	0.54	3/2	1
96.27	0.49 ± 0.08	0.21 ± 0.04		1.26 ± 0.13	0.67 ± 0.4	1.1 ± .9	5/2	2
96.44	0.52 ± 0.06		0.86 ± 0.17	0.3 ± 0.2	1.3 ± 1.1	0.4 ± .3	1/2	1
96.69	1.18 ± 0.11		0.77 ± 0.09	0.7 ± 0.4	2.5 ± 0.3	0.4 ± .2	3/2	1
102.78	0.69 ± 0.06		0.35 ± 0.03	0.71 ± 0.07	21. ± 1.	0.36 ± .04	3/2	1
103.16	0.73 ± 0.06		0.70 ± 0.12	0.8 ± 0.1	0.76 ± 0.24	0.84	3/2	2
106.02	1.33 ± 0.08		0.87 ± 0.07	1.55 ± 0.16	2.8 ± 0.3	1.07 ± .17	3/2	2
112.81	1.04 ± 0.07		0.57 ± 0.04	1.17 ± 0.12	5.5 ± 0.5	0.65 ± .07	3/2	2
121.13	0.04 ± 0.03	0.04 ± 0.03		0.03 ± 0.01	0.026 ± 0.005	0.54	1/2	1
122.69	0.15 ± 0.05	0.09 ± 0.03		0.15 ± 0.02	0.090 ± 0.011	0.54	3/2	1
122.92	0.54 ± 0.04		0.27 ± 0.03	0.54 ± 0.05	46. ± 2.	0.27 ± .03	3/2	1
124.29	0.60 ± 0.05		0.66 ± 0.06	0.63 ± 0.06	7.5 ± 1.	0.68 ± .07	1/2	1
125.28	1.16 ± 0.08		0.61 ± 0.04	1.27 ± 0.13	10. ± 1.	0.65 ± .07	3/2	2
130.05	0.57 ± 0.08		0.57 ± 0.08	1.1 ± 0.1	500. ± 50.	1.10 ± .10	1/2	0
130.37	0.70 ± 0.06	0.60 ± 0.09		0.79 ± 0.08	0.75 ± 0.22	0.84	3/2	2
140.56	1.40 ± 0.26		1.40 ± 0.26	2.19 ± 0.22	2700. ± 100.	2.2 ± .2	1/2	0
141.38	0.67 ± 0.07		0.81 ± 0.21	0.68 ± 0.07	0.57 ± 0.17	0.84	3/2	2
142.55	0.57 ± 0.05		0.57 ± 0.10	0.55 ± 0.06	0.56 ± 0.17	0.54	3/2	1
150.03	0.25 ± 0.04	0.16 ± 0.03		0.29 ± 0.03	0.20 ± 0.03	0.54	3/2	1
154.08	0.59 ± 0.06	0.45 ± 0.07		0.56 ± 0.06	0.42 ± 0.09	0.84	3/2	2
161.91	1.14 ± 0.09		0.62 ± 0.05	1.14 ± 0.11	6.5 ± 1.	0.62 ± .06	3/2	2
169.24	1.10 ± 0.16		1.10 ± 0.16	1.1 ± 0.1	1000. ± 100.	1.1 ± .1	1/2	0
169.30	1.58 ± 0.20		0.91 ± 0.14	2.0 ± 0.2	6. ± 3.	1.2 ± .15	3/2	2
173.30	0.26 ± 0.05	0.17 ± 0.04		0.56 ± 0.06	0.58 ± 0.17	0.54	3/2	1
173.82	0.45 ± 0.06		0.23 ± 0.03	0.36 ± 0.04	42. ± 1.	0.18 ± .02	3/2	1
175.94	0.06 ± 0.03							
179.92	0.48 ± 0.07		0.50 ± 0.07	0.53 ± 0.05	13. ± 3.	0.55 ± .05	1/2	1
181.31	2.73 ± 0.23		0.96 ± 0.08	3.21 ± 0.32	17. ± 2.	1.1 ± .1	5/2	2
187.28	0.50 ± 0.10		0.29 ± 0.07	0.60 ± 0.06	1.7 ± 1.	0.33 ± .04	3/2	2
188.02	0.72 ± 0.31		0.72 ± 0.31	2.98 ± 0.30	3600. ± 100.	3.0 ± .3	1/2	0
188.36	0.08 ± 0.04							
190.14	0.80 ± 0.09		0.41 ± 0.05	1.04 ± 0.10	15. ± 7.	0.53 ± .05	3/2	2
193.17	1.16 ± 0.10		0.60 ± 0.05	1.13 ± 0.11	20. ± 10.	0.57 ± .06	3/2	2
195.95	0.59 ± 0.08		0.60 ± 0.08	0.69 ± 0.07	66. ± 8.	0.70 ± .07	1/2	1
201.78	1.69 ± 0.14		0.88 ± 0.08	1.95 ± 0.20	24. ± 2.	1.0 ± .1	3/2	2
203.91	0.08 ± 0.03							
206.14	1.11 ± 0.10		0.55 ± 0.08	1.45 ± 0.15	1.14 ± 0.35	0.84	5/2	2
208.20	0.68 ± 0.08		0.70 ± 0.09	1.19 ± 0.12	22. ± 2.	1.26 ± .13	1/2	1
209.08	0.20 ± 0.05	0.12 ± 0.04		0.27 ± 0.03	0.18 ± 0.03	0.54	3/2	1
210.14	0.12 ± 0.04	0.06 ± 0.03		0.39 ± 0.04	0.25 ± 0.04	0.84	3/2	2
210.91	1.08 ± 0.11		0.58 ± 0.06	1.33 ± 0.13	9. ± 1.	0.72 ± .07	3/2	2
216.04	0.16 ± 0.04	0.22 ± 0.08		0.40 ± 0.06	1.5 ± 0.5	0.54	1/2	1
220.90	1.55 ± 0.20		1.55 ± 0.20	2.16 ± 0.22	1150. ± 50.	2.16 ± .2	1/2	0
222.11	0.39 ± 0.08		0.20 ± 0.04	0.64 ± 0.06	8. ± 3.	0.34 ± .04	3/2	1
223.89	1.10 ± 0.11			1.32				
226.07	0.92 ± 0.11		0.47 ± 0.06	1.00 ± 0.10	28. ± 2.	0.51 ± .05	3/2	1
230.13	0.59 ± 0.08		0.31 ± 0.05	0.62 ± 0.06	7.5 ± 0.5	0.33 ± .03	3/2	1
232.76	2.29 ± 0.19		1.21 ± 0.11	2.67 ± 0.27	20. ± 1.	1.4 ± .2	3/2	2
235.09	2.19 ± 0.17		0.76 ± 0.06	2.51 ± 0.25	20. ± 1.	0.87 ± .09	5/2	2
241.86	3.47 ± 0.26		1.31 ± 0.11	3.84 ± 0.38	10. ± 1.	1.47 ± .15	5/2	2
243.61	0.16 ± 0.50	0.10 ± 0.04		0.14 ± 0.01	0.08 ± 0.01	0.54	3/2	1
245.25	0.52 ± 0.09		0.52 ± 0.09	0.80 ± 0.08	600. ± 20.	0.80 ± .08	1/2	0
246.60	0.09 ± 0.05							
252.59	0.17 ± 0.09	0.10 ± 0.05		0.24 ± 0.02	0.15 ± 0.02	0.54	3/2	1
253.83	1.62 ± 0.17		0.55 ± 0.06	1.53 ± 0.15	28. ± 2.	0.52 ± .05	5/2	2
256.43	1.22 ± 0.23		0.68 ± 0.14	1.08 ± 0.11	6. ± 1.	0.60 ± .06	3/2	1

NEUTRON RESONANCE STRUCTURE OF ^{54}Fe AND ^{56}Fe FROM HIGH RESOLUTION TOTAL CROSS SECTION EXPERIMENTS

E.M. CORNELIS

University of Antwerp, RUCA, B-2000 Antwerp, Belgium

L. MEWISSEN and F. POORTMANS

S.C.K.-C.E.N., B-2400 Mol, Belgium

High-resolution total neutron cross section measurements have been performed on samples of enriched isotopes (on loan from ORNL) of ^{54}Fe and ^{56}Fe in the energy range from 240 keV up to 19 MeV. The experiments were performed on the 400 meter flight path at the Gelina facility with a burst width of 4.5 nsec. The resonance parameter analysis is completed for ^{56}Fe up to 850 keV.

Introduction

The knowledge of the neutron-resonance structure of the structural materials of a fast reactor is requested for the calculation of resonance shielding effects. At the previous Nuclear Data Conference[1] we have presented the results from an analysis of the total cross section measurements of ^{54}Fe and ^{56}Fe up to 300 keV. The present paper describes a new series of measurements with an improved time-of-flight resolution, covering the energy range from 240 keV up to 19 MeV. The resonance parameter analysis is completed for ^{56}Fe up to the inelastic threshold around 850 keV. The analysis of the ^{54}Fe data is in progress.

Experimental Details

The experiments were performed at the neutron time-of-flight spectrometer of the 150 MeV linear accelerator of CBNM, Geel. The burst width was 4.5 nsec (F.W.H.M.) and the repetition rate 800 Hz. We have used the direct, unmoderated beam from the neutron producing target. Two filters were placed in the neutron beam: a 37 g/cm^2 thick uranium filter to reduce the effect of the gamma-flash in the detector and a 0.4 g/cm^2 thick $^{10}B_4C$ filter to absorb low-energy neutrons.

The total flight path length between neutron target and detector was 387.713 ± 0.012 meter. The uncertainty on this flight path length takes into account the uncertainties in the effective centers for neutron production and detection.

The time-of-flight analyzer consisted of a time digitizer with minimum channel width of 4 nsec. interfaced via a double data buffer (2 x 128 words, each 32 bits) to a computer HP 2113-E with 96K words of memory available for data storage. The data taking system could handle a count rate of more than 10^4 events per second and the dead-time after each event was only 540 ns. The maximum value for the dead-time correction was 10% without sample in the beam.

The neutrons were detected by proton recoil in a NE 110 scintillator. The dimensions of this detector were 15.24 x 15.24 x 2.54 cm. Four small photomultiplier tubes, type RCA 4516, each having a diameter of 1.9 cm were mounted side-on to the scintillator, out off the neutron beam. The construction of the detector was made such as to keep capture of the incident neutrons by the construction materials very low.

In order to optimise the signal to background ratio, we divided the pulse-height range into three windows and stored only those time-of-flight events for which the pulse-height was within the required window. In this way, the only background component of importance was a nearly constant one due to various activities in the detector room and to long-lived radiations from the neutron source. This background was determined at long flight times, shortly before the arrival of the next neutron burst. The ratio background to signal was of the order of 0.1% between 2 MeV and 700 keV and increased with decreasing energy : 0.3% at 500 keV, 1.5% at 300 keV and finally 5% at 240 keV.

The samples were prepared from oxide powder (Fe_2O_3) canned in aluminium tubes with windows of 0.3 mm thick. The sample diameters were 35 mm and the thicknesses were :

^{54}Fe	(97.68%)	6.983 10^{-2}	Fe atomes per barn
^{56}Fe	(99.93%)	7.810 10^{-2}	Fe atomes per barn

The data were corrected for the total cross section of oxygen which we have measured with the same experimental conditions as for the iron-oxyde measurements.

The neutron energy resolution width ΔE(F.W.H.M.) is given by the following expression :

$$\left(\frac{\Delta E}{E}\right)^2 = \left(2\,\frac{\Delta L}{L}\right)^2 + \left(2\,\frac{\Delta t}{t}\right)^2 = (0.020 + 0.18\ E) \times 10^{-6}$$

In this expression L is the flight path length, t is the total flight time of neutrons with energy E and ΔL and Δt are the corresponding uncertainties. E is expressed in MeV. Typical values for ΔE (FWHM) were : 64 eV at 250 eV, 166 eV at 500 keV and 450 eV at 1 MeV.

The resolution function was carefully checked by shape-fitting very narrow resonances.

Analysis

The neutron widths were determined from a shape analysis of the total cross-section data using the Reich-Moore multilevel code MULTI[2].

The spin and parity of sufficiently broad resonances (resonance width not much smaller than the resolution width) could be assigned in the following way :

1. s-wave resonances were easely detected because of the strong resonance-potential interference effect.

K. H. Böckhoff (ed.), Nuclear Data for Science and Technology, 135–138.

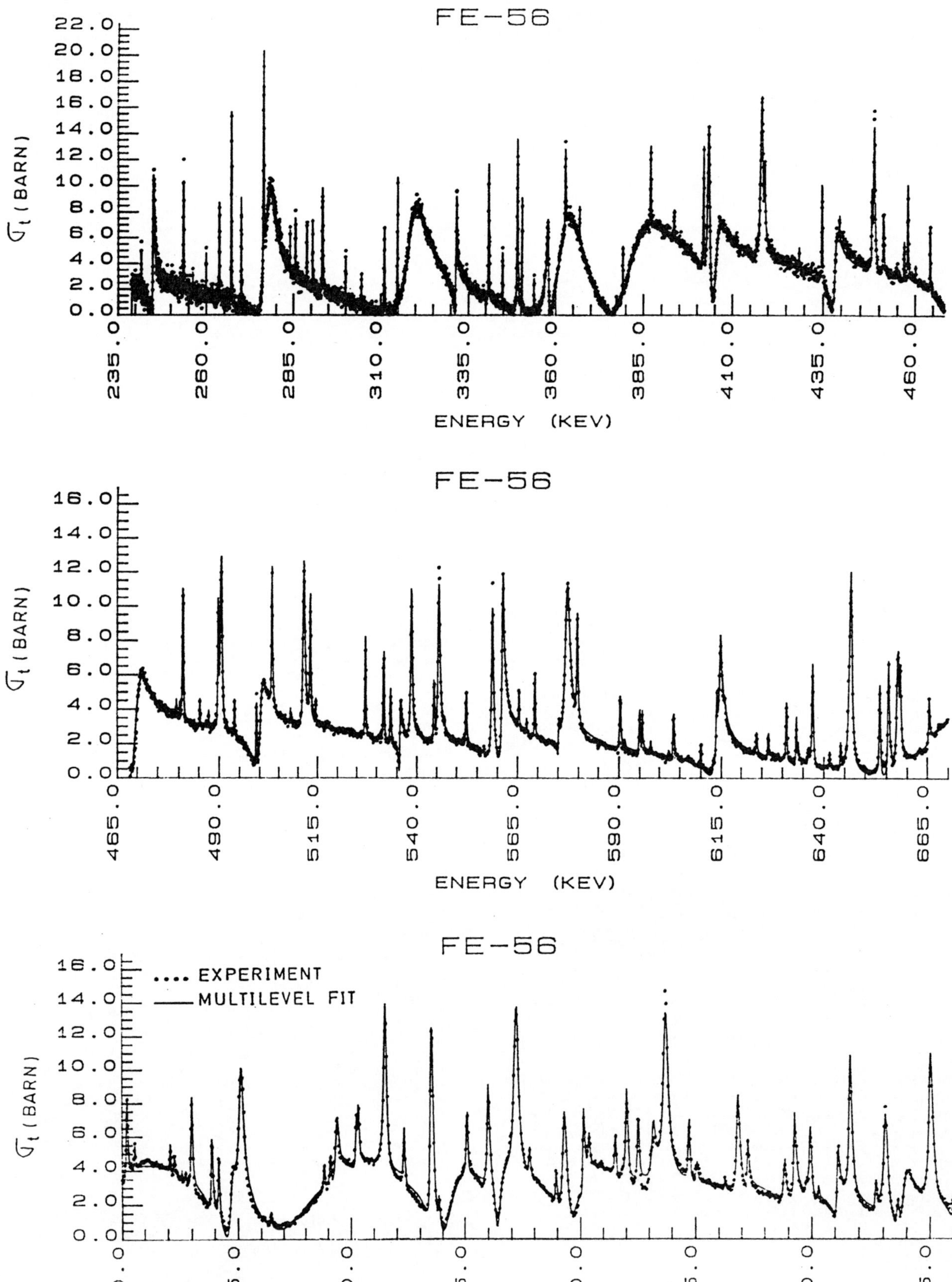

Fig. 1 : ^{56}Fe total cross section, experiment and R-matrix fit.

2. assignment of l=1 or l=2 could be done on basis of the resonance shape: l=1 resonances show assymetric shape due to interference with p-wave potential scattering, whereas l=2 resonances do not.
3. spin assignment was based on the peak total cross section.
4. in certain cases l- and J-assignment was possible on the basis of observed resonance-resonance interference.

Results for ^{56}Fe

At present we can only give the results from the analysis of the ^{56}Fe data. The analysis of the ^{54}Fe data is not completed yet.

Fig. 1 shows the experimental total cross section data together with the R-matrix fit. The best overall fit was obtained for a channel radius R = 5 fm. In total 188 resonances were analyzed; for 80 among them the spin and parity was determined. Fig. 2 shows the cumulative sums of reduced neutron widths versus neutron energy for the various spin groups.

For the energy range between 240 keV and 850 keV we obtain the following value for the s-wave strength function :

$$S_o = 2.6 \begin{smallmatrix} + 0.9 \\ - 0.6 \end{smallmatrix} \times 10^{-4} \qquad \text{(24 resonances)}$$

For the p-wave and d-wave resonances, the spin and parity was assigned for 2/3 of the total strength. So if we increase the assigned strength for each spin group by 50%, we obtain the following values for the p-wave and d-wave strength functions :

$$S_1 = 5.0 \begin{smallmatrix} + 1.5 \\ - 1.1 \end{smallmatrix} \times 10^{-5} \qquad \text{(30 assigned p-wave resonances)}$$

$$S_2 = 2.8 \begin{smallmatrix} + 0.9 \\ - 0.7 \end{smallmatrix} \times 10^{-4} \qquad \text{(26 assigned d-wave resonances)}$$

Comparison with other results

Below 400 keV we have compared our results with the ENDF-B/V evaluation[3] and above 450 keV with the results published by S. Cierjacks and I. Schouky[4]. In both references, resonance spins and parities were determined from the analysis of the angular distribution of elastically scattered neutrons.

From this comparison, the following conclusion can be drawn :

1. What the neutron widths of the s-wave resonances concern, our results agree reasonably well with the values of reference 3 below 400 keV but there are important descrepancies between our results and those of reference 4 above 450 keV.
2. What spin and parity assignments of the p-wave and d-wave resonances concern, we are not in agreement neither with the results of reference 3 below 400 keV nor with those of reference 4 above 450 keV.

Fig. 3 shows, as an example, our total cross section data in a small energy range around 350 keV together with the values calculated with the present resonance parameter values and with those of reference 3.

References

1. E.M. Cornelis, C.R. Jungmann, L. Mewissen, F. Poortmans, Proceedings of the Int. Conf. on Nuclear Cross Sections for Technology, Knoxville, Oct.22-26, 1979, NBS Special Publication 594, 159 (1980)
2. G.F. Auchampaugh, Report LA-5473-MS (1974)
3. C.M. Perey and F.G. Perey, Report ORNL/TM-6405 ENDF-298 (1980)
4. S. Cierjacks and I. Schouky, Proc. Int. Conf. on Neutron Physics and Nuclear Data for Reactors, Harwell, Sept. 1978, p. 187.

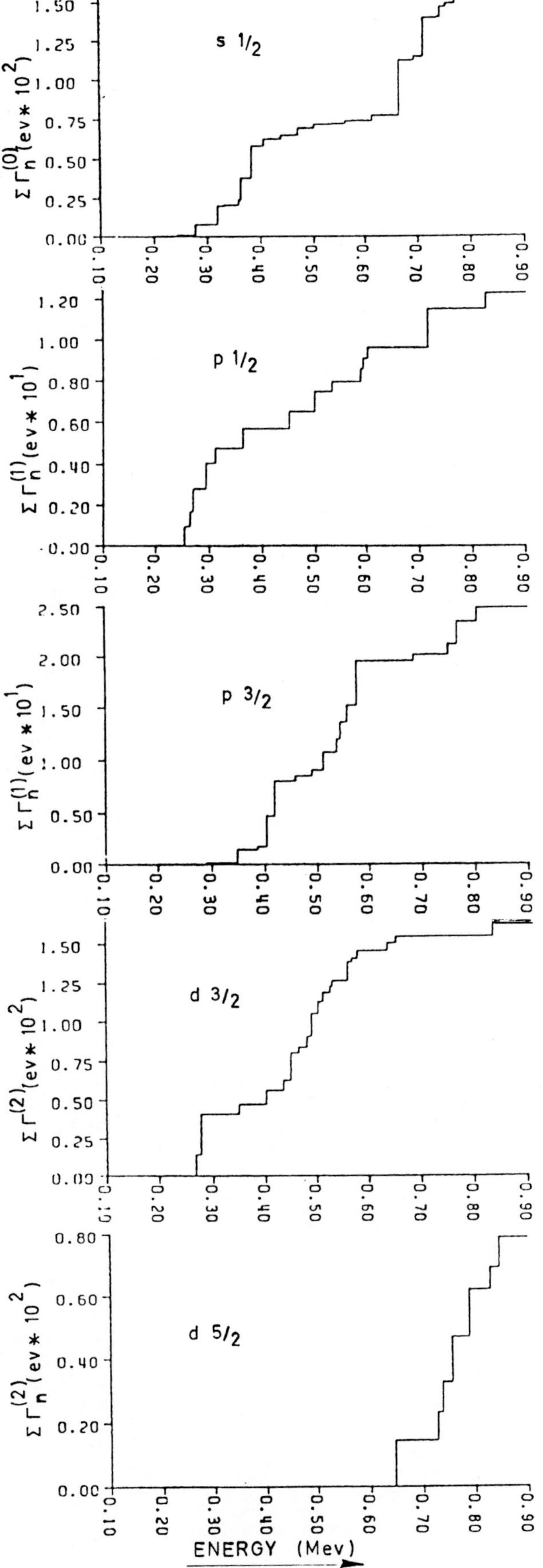

Fig. 2: cumulative sums of reduced neutron widths.

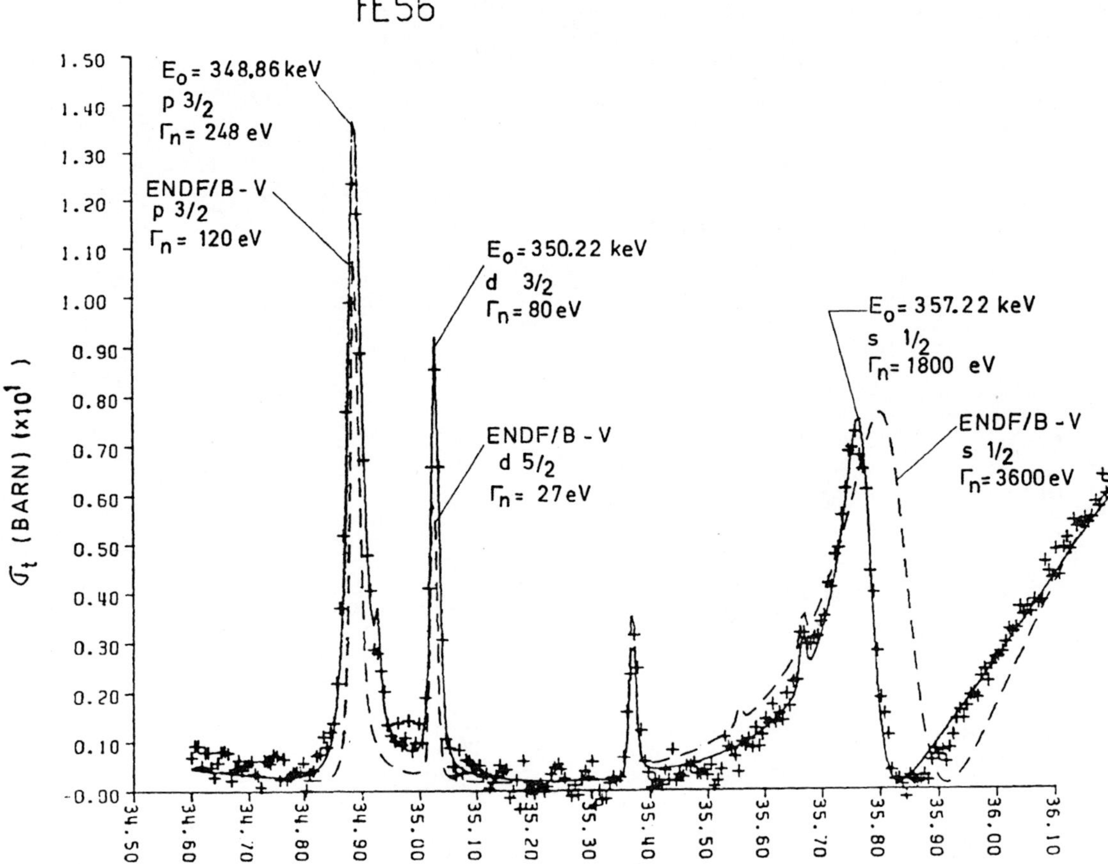

Fig. 3 : experimental total cross section of ^{56}Fe together with calculated values from present resonance parameters and from ENDF-B/V parameters.

RESONANCE PARAMETERS OF ^{57}Fe

G. Rohr, A. Brusegan, F. Corvi, R. Shelley, T. van der Veen, C. Van de Vorst

Commission of the European Communities
Joint Research Centre
Central Bureau for Nuclear Measurements
Geel, Belgium

High resolution neutron transmission and capture measurements have been performed at the 150 MeV Geel linac with a 4 ns pulse width. An enriched ^{57}Fe-oxide sample has been employed at 50 m and 60 m, covering energy ranges up to 100 keV and 600 keV for transmission and capture measurements respectively. The resonance capture data have been normalized to the resonance parameters of the 1.6 keV transmission results. The data sets have been analysed by means of modified versions of FANAL, FANAC and TACASI computer programs and compared with existing data.

[$^{57}Fe(n,\gamma)$, σ_{tot}, 1-200 keV, E_o, A_γ, $g\Gamma_\gamma$, $g\Gamma_n$]

Introduction

Within the framework of the structural materials programme at CBNM Geel, high resolution neutron transmission and capture measurements have been performed at the 150 MeV linac (GELINA) with enriched ^{57}Fe-oxide samples. The neutron energy range covered was up to 100 keV for transmission and 600 keV for capture, although analyses presented here are restricted to 61 keV and 200 keV respectively and are performed with a multi-level Reich-Moore formalism including the elastic and one inelastic channel. This, together with the low threshold of inelastic scattering (14.4 keV) for spin J=1 resonances, enables the determination of the inelastic scattering cross section by means of transmission measurements [1]. The resonance parameters obtained from the transmission measurement are compared with KFK [1] and the capture results are compared with data published by ORNL [2]. A more extensive inter-comparison for weak resonances has recently been performed by G. Rohr [3] to investigate the possible systematic errors related to particular detector systems and normalization methods.

Experimental Procedure

Utilizing the 4 ns bursts of electrons, with a repetition frequency of 800 Hz, from GELINA the measurements were performed at neutron flight paths of 50 m and 60 m for transmission and capture respectively. Two enriched ^{57}Fe-oxide samples, on loan from ONRL, were canned in 0.3 mm aluminium with a diameter of 80 mm and placed back to back in the neutron beam. The respective purities of the samples were 93.4 % and 90.4 % giving a total ^{57}Fe thickness of $8.976\cdot 10^{-3}$ atoms/barn.

For the transmission measurement two 4" x 3" NaI(Tl) detectors were used together with a 5 mm $^{10}B_4C$ slab. The background was determined with the black resonance technique (and corrected in order to allow for the neutron flux reduction due to the filters). For capture two cylindrical aluminium canned C_6D_6 liquid scintillators of 10.2 cm diameter and 7.5 cm thickness were independently connected to EMI photomultipliers. By reducing the amount of material in the vicinity of the scintillator to a minimum these detectors are characterized by their low prompt sensitivity to sample scattered neutrons. The so-called weighting technique was used whereby capture events were weighted according to their amplitude information in order to achieve a detector efficiency proportional to the total gamma ray energy released in the capture process. For each event the amplitude and time of flight information were analyzed over 256 and 16 K channels respectively. By storing this information in list-mode on a disk, with 2 files, connected to an IBM 370/138 main computer the data have been sorted and weighted according to the associated pulse height using the two files in ping-pong [4]. Amplitude calibration and the gain of the amplifiers have been regularly controlled by the Compton edge of the ^{238}Pu $^{13}C(\alpha,n\gamma)^{16}O$ 6.13MeV γ-rays and by suppressing the first 3 amplitude channels a pulse height amplitude bias of 150 keV was set. Monitoring the shape of the neutron flux was done by keeping a ^6Li-loaded glass scintillator, 0.5 mm thickness and 10 cm diameter, permanently in the neutron beam 63 cm in front of the sample. The relative neutron flux was measured with a $^{10}B_4C$ slab detector of 3 mm thickness using the C_6D_6 detectors to view the 478 keV photons associated with the neutron absorption $^{10}B(n,\alpha\gamma)^7Li$ reaction. Comparison was also made with the flux measured by a ^6Li glass detector below 100keV and with that measured by a multi-plate ^{235}U fission chamber above 100 keV and agreement within 3 % was obtained in the energy range up to 400 keV.

The neutron sensitivity of the capture detector has been measured by comparing directly the count rate from an 8 mm thick graphite disc with that from a 0.1 mm thick Au sample. The background has been measured both under an open-beam condition (no sample in place) and with a 1 mm carbon disc. A preliminary energy independent value of the neutron sensitivity $\epsilon_n/\epsilon_\gamma = (1.5 \pm 0.75) \times 10^{-4}$ has been used to correct for the scattered neutron contribution in all s-wave resonances within our range of analysis.

Normalization and Weighting

With our sample containing some 8 % ^{56}Fe as an impurity we initially attempted to normalize the ^{57}Fe capture data to the capture area of the 1.15 keV resonance of ^{56}Fe, as was done in ref. [5]. Furthermore, the 1.63 keV resonance of ^{57}Fe is also an almost pure capture resonance ($\Gamma_n \ll \Gamma_\gamma$) and can equally well be used for normalization as the capture area can be determined very precisely in a transmission measurement. As shown in Table II our transmission measurement results in $g\Gamma_n$ = 53.3 meV and $g\Gamma_\gamma$ = 1.25 eV for this resonance. In addition, normalization of the capture data to resonances of Ag below 70 eV (black resonance technique) was performed and found to be consistant with our transmission result for the 1.6 keV resonance but in considerable disagreement with known transmission results for the 1.15 keV resonance of ^{56}Fe. It can be seen in Table I that the average weight $\bar{w}$, i.e. hardness of the gamma spectrum, for these two resonances is very different. By the use of different weighting functions we have attempted to obtain a consistent set of results between capture and transmission.

Four weighting functions (WF) are plotted in Fig. 1. The 'OLD' WF is calculated with a Monte Carlo based computer code developed at Cadarache and Karlsruhe [6] and was in excellent agreement with WFs obtained at Harwell using the routine GAMOC [7] and at Oak Ridge, where a code based on an analytical method has been developed [8]. All three codes calculate the most probable energy loss of electrons in the scintillator. The 'NEW' WF has been calculated using an increase in this energy loss obtained by employing newly evaluated data from Atomic Data [9]. The 'GAB' WF includes γ-absorption in the sample and scintillator containers by assuming an absorption of 40 % for a 330 keV γ-ray, which corresponds to the absorption in the Fe oxide samples used at CBNM or in 1 mm of Ag. Finally, for the sake of comparison, the extreme limit of a linear (LIN) WF has been included. In Table I capture results, normalized to Ag, obtained from the different WFs are listed for

K. H. Böckhoff (ed.), Nuclear Data for Science and Technology, 139–142.

Table I Weighting function tests compared with transmission measurement results

Weighting Function	^{57}Fe 1.6 keV $g\Gamma_n\Gamma_\gamma/\Gamma$ (meV)	^{56}Fe 1.15 keV $g\Gamma_n\Gamma_\gamma/\Gamma$ (meV)
OLD	54.	70.
NEW	51.5 ($\bar{w}$ = 20.5)	65. ($\bar{w}$ = 28.9)
GAB	49.	61.
LIN	41.	49.
TRA	51.1	53.

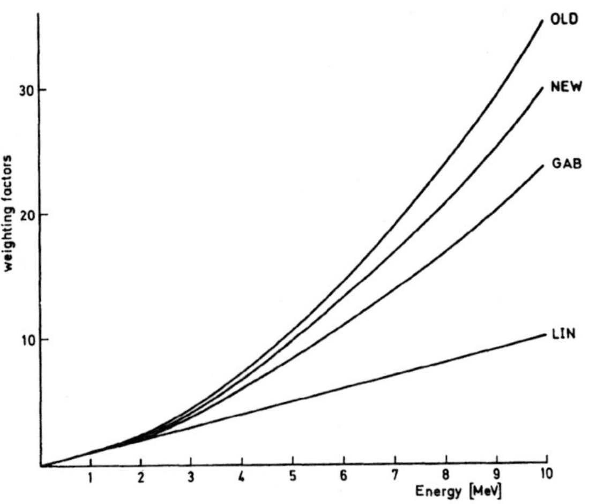

Fig.1. C_6D_6 capture detector weighting functions.

^{56}Fe(1.15 keV) and ^{57}Fe (1.63 keV). The transmission (TRA) values for these two resonances are also given.

The excellent agreement in the case of ^{57}Fe (1.63 keV) obtained between the transmission value and the capture result using the 'NEW' weighting function can readily be seen. However, to get a similar agreement for ^{56}Fe (1.15 keV) it becomes necessary to utilise a WF that is nearer to the linear one. There is no single WF for which the transmission results of both these resonances can be reproduced. Therefore, one may at least conclude that the weighting method can be used confidently for resonances with an average weight up to $\bar{w}$ = 20.5. The method fails for the ^{56}Fe (1.15 keV) resonance where $\bar{w}$ = 28.9, leading to a 22.6 % overestimation of the capture area calculated with the 'NEW' weighting function. In order to determine the upper limit of the validity of the pulse height weighting technique and to study the behaviour of the detector beyond this limit, more pure capture resonances with different $\bar{w}$ have to be measured. But there is a possibility of getting this information by comparing results from other capture detectors which do not depend so strongly on the spectral shape [3].

In view of the above discussion and the fact that the average weight for nearly 100 resonances of ^{57}Fe is 24.1 ± 5.3, we have normalized the capture data of the present work to the resonance parameters obtained from our tranmission measurement of the 1.63 keV resonance. The average weight $\bar{w}$ for the resonances is included in Table II in order to eventually correct the results for those with a hard spectrum.

Analysis

After background subtraction, correction for the neutron flux and proper normalization the weighted counting rate can be reduced to the capture cross section. This is shown in Fig. 2, where $\sigma(n,\gamma)$ for ^{57}Fe is plotted vs neutron energy in the range from 23 to 44 keV. It should be noted that the plotted $\sigma(n,\gamma)$ is still Doppler and resolution broadened and it is not corrected for multiple scattering and prompt background contributions, which are particularly important around the large s-wave resonances. Analysis of the capture data has been performed with the capture area fitting program TACASI and with the R-matrix shape fitting program FANAC, both of which have been modified at CBNM. The former code has not been utilized for s-wave resonances nor for regions in the spectrum where resonances are poorly resolved. The results obtained for resonances below about 50 keV show that these two programs are consistent to within 3 % uncertainty at low energies but as one increases in energy they tend to diverge. In the higher energy range of our analysis the capture area given by TACASI is as much as 10 % above the equivalent results from FANAC whose gaussian resolution function is unable to fit the asymmetric profiles as produced by our target-moderator configuration. A modification of FANAC is underway to cope with this effect. The broad s-wave resonances above 90 keV have been a particular source of problems during the analysis of the capture data. Their low signal to background ratio means that the background determination in this region is very critical for an accurate determination of the resonance parameters. In fact, several such resonances known from transmission can hardly be seen in capture and even the relatively large structure seen at 110 keV has the added complication of being two overlapping s-waves resonances, with the same spin, and hence cannot be satisfactorily described by the Reich-Moore formalism. Above 200 keV the statistics of the capture experiment become rather poor and no worthwhile analysis could be contemplated. The preliminary transmission data analysis has been done up to 61 keV with a modified FANAL program and is still in progress.

Results

Table II gives our neutron capture measurement results, obtained from analyses with the NEW weighting function, for 120 resonances alongside those published by KFK [1] (transmission) and ORNL [2] (capture). The values of J^π and large neutron widths from KFK have been taken as fixed input data for our analysis programs, except for where values are shown in brackets - these are preliminary results from our transmission measurement. For all p-wave resonances g is assumed to equal 1 and in cases where $\Gamma_n < \Gamma_\gamma$ a fixed value of 0.5 eV has been taken for the radiation width. A comparison of the two capture area data sets, for well resolved p-wave resonances where our statistics are good, is best done in three energy ranges. Firstly below 20 keV, where two large s-wave structures are predominant and ten resonances are compared, our results are on average 15 % higher than those of ORNL. Between 20 keV and 50 keV agreement within 2 % is achieved, while over the rest of the energy range analysed, where as mentioned earlier broad s-wave resonances, poorer statistics and also a higher level density have made the analysis rather difficult, our results average 7 % higher. The agreement between ORNL and Geel averaged over the whole energy range is within 8 % although this figure would be somewhat improved were it not for the large low energy discrepancy which may be attributable to an incorrectly determined neutron flux [3]. It should also be noted that the ORNL data have since been increased by 3.6 % [10], thus bringing both data sets to within 5 % of each other. Our uncertainties include statistical and systematic errors due to the background, neutron sensitivity, relative flux and inconsistancy of our two analysis programs as discussed earlier, but not for the difference in spectrum shape because of the unsettled situation of this problem.

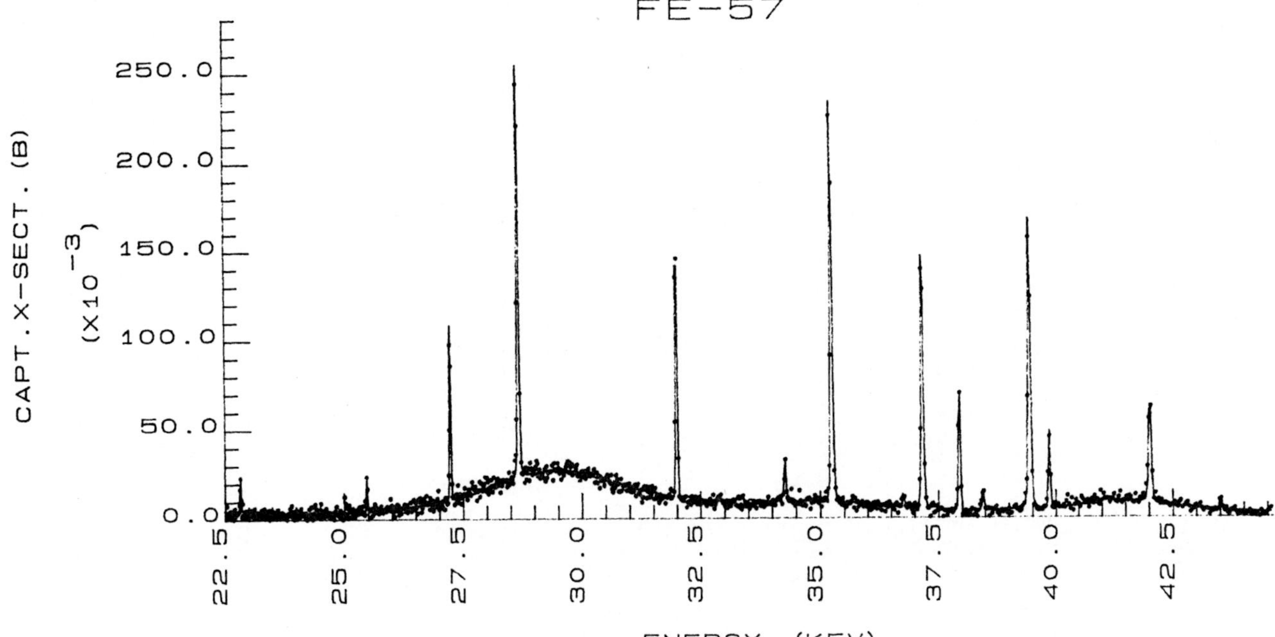

Fig. 2. Neutron capture cross section fitted by the R-matrix shape code FANAC over the energy range 23 to 44 keV

References

1. G. Rohr, K.N. Müller, Z. Physik 227, (1969) 1
2. B.J. Allen et al. Proc. NEANDC/NEACRP Spec. Meet.on Neutron Data of Struct.Mat.for Fast Reactors, Geel 1977
3. G. Rohr, Proc. Spec. Meeting on Fast Neutron Capture Cross Sections, Argonne, 1982 to be published
4. T. Babeliowsky et al. Acquisition system for two parameter measurements with high data rate (to be published)
5. A. Brusegan et al. Proc. Int. Conf. on Neutron Cross Sections for Technology, Knoxville 1979 NBS-SP 594 p.163
6. C. Le Rigoleur, A. Arnaud, Proc. NEANDC/NEACRP Spec. Meeting, Karlsruhe 1973, report KFK 2046(1975)
7. R.B. Thom, U.K. Nuclear Data Progress Report (1979), NEANDC(E)212, Vol.8, p.22
8. R.L. Macklin, J.H. Gibbons, Phys. Rev. 159 (1967) 1009
9. L. Pages et al. Atomic Data 4, (1972) 1
10. B.J. Allen, Private communication

Table II ^{57}Fe resonance parameters up to 200 keV

PRESENT CAPTURE RESULTS					TRANSMISSION - KFK			OAK RIDGE
E_o(keV)	$\bar{w}$	$g\Gamma_n\Gamma_\gamma/\Gamma$(eV)	$g\Gamma_n$(eV)	$g\Gamma_\gamma$(eV)	$g\Gamma_n$(eV)	$g\Gamma_{n'}$(eV)	J^π	$g\Gamma_n\Gamma_\gamma/\Gamma$(eV)
1.630	20	(0.051 ± 0.0026)	(0.0533 ± 0.0022)	(1.25 ± 0.3)				
3.964		0.195 ± 0.02		0.78 ± 0.08	(255.3)		0^+	0.17 ± 0.02
4.753	19	0.043 ± 0.002	0.047 ± 0.002					0.035
6.215	24	0.86 ± 0.08		0.86 ± 0.08	(286.)		1^+	0.69 ± 0.23
7.227	32	0.42 ± 0.04		0.78 ± 0.08	(0.92)			0.29
7.944	23	0.16 ± 0.01	0.230 ± 0.015					0.13
8.533	21	0.008 ± 0.002	0.008 ± 0.002					0.006
9.213	20	0.028 ± 0.002	0.030 ± 0.002					0.023
12.10	18	0.013 ± 0.001	0.013 ± 0.001					0.013 ± 0.001
12.87	24	0.39 ± 0.02		0.57 ± 0.02	(1.2)			0.34
13.31	20	0.083 ± 0.01	0.10 ± 0.01					0.087
13.99	20	0.70 ± 0.04		0.75 ± 0.04	(9.6)		(1^+)	0.56
14.14		0.004 ± 0.003	0.004 ± 0.003					
18.08	20	0.21 ± 0.015	0.36 ± 0.25					0.19
18.25	23	0.51 ± 0.03		0.60 ± 0.03	(3.5)			0.40
18.73	19	0.037 ± 0.004	0.040 ± 0.004					0.034
21.08	22	0.28 ± 0.02		0.31 ± 0.02	(2.5)			0.25
21.32	24	0.54 ± 0.04		0.64 ± 0.04	(3.4)			0.48
21.37	21	0.10 ± 0.015	0.12 ± 0.18					0.11
25.46	19	0.014 ± 0.003	0.014 ± 0.003					
27.22	21	0.10 ± 0.01	0.12 ± 0.01					0.12
28.67	18	0.42 ± 0.03		0.45 ± 0.03	(7.6)			0.42
28.94	23	2.65 ± 0.50		2.65 ± 0.50	(2600.)	(385.)	1^+	3.34 ± 0.67
31.99	21	0.23 ± 0.02	0.41 ± 0.04					0.22
35.24	20	0.54 ± 0.03		0.57 ± 0.03				0.44
37.16	21	0.34 ± 0.02		0.37 ± 0.02	(4.2)			0.34
37.94	19	0.13 ± 0.02	0.17 ± 0.2					0.15
39.44	19	0.45 ± 0.03		0.47 ± 0.03	(10.7)			0.45
39.85	22	0.09 ± 0.01	0.11 ± 0.01					0.11
41.44	23	3.55 ± 0.75		3.55 ± 0.75	(558.)	(1375.)	1^+	2.8 ± 0.6
42.02	19	0.18 ± 0.02		0.18 ± 0.02	(7.1)			0.19
43.51	21	0.016 ± 0.006	0.017 ± 0.006					0.035 ± 0.004
47.14	21	0.85 ± 0.11		0.85 ± 0.11	(310.)	(36.)	1^+	0.67 ± 0.7
50.02	23	0.12 ± 0.01		0.13 ± 0.01	(2.9)			0.14
51.01	20	0.41 ± 0.03		0.44 ± 0.03				0.33

(Table II - cont.)

52.80	26	0.10 ± 0.03	0.12 ± 0.04					0.097 ± 0.015
52.94	22	0.43 ± 0.05		0.45 ± 0.05	(11.2)			0.37
56.00		0.60 ± 0.20		0.60 ± 0.20	(1690.)		0^+	1.0 ± 0.30
56.27	23	0.38 ± 0.04		0.39 ± 0.04				0.24 ± 0.12
58.67	22	0.037 ± 0.01	0.04 ± 0.01					
58.93	24	0.16 ± 0.03						0.18
61.00		1.70 ± 0.26		1.70 ± 0.26	(2235.)	(225.)	1^+	1.18 ± 0.28
62.00	26	0.05 ± 0.03	0.05 ± 0.03					
62.37	24	0.25 ± 0.03		0.26 ± 0.03				0.24
63.12	20	0.13 ± 0.02		0.14 ± 0.02				0.13 ± 0.01
64.05	29	0.40 ± 0.03		0.41 ± 0.03				0.31
64.29	23	0.08 ± 0.02						0.14 ± 0.01
66.90	20	0.63 ± 0.08		0.80 ± 0.10				0.54
68.19	21	0.29 ± 0.03		0.29 ± 0.03				0.23 ± 0.03
71.55	28	0.36 ± 0.03		0.36 ± 0.03				0.25 ± 0.02
72.83	26	0.33 ± 0.03		0.35 ± 0.03				0.32 ± 0.03
74.40		0.10 ± 0.03		0.10 ± 0.03				0.19 ± 0.04
77.16		0.95 ± 0.20		0.95 ± 0.20	1463.	563.	1^+	0.76 ± 0.22
80.48	19	0.27 ± 0.04		0.31 ± 0.04				0.27
83.32	31	0.06 ± 0.03	0.07 ± 0.03					0.12 ± 0.02
84.66	22	0.22 ± 0.04						0.22 ± 0.02
87.24	30	0.05 ± 0.02	0.06 ± 0.02					
88.28	23	0.30 ± 0.02		0.30 ± 0.02				0.25 ± 0.02
89.98	21	0.21 ± 0.03		0.21 ± 0.03				
91.34	26	0.05 ± 0.03	0.06 ± 0.03					
93.57	25	0.72 ± 0.15		0.72 ± 0.15				0.64 ± 0.15
93.70		0.35 ± 0.15		0.35 ± 0.15	150.	150.	1^+	≤ 0.45
93.97		1.00 ± 0.20		1.00 ± 0.20				0.41 ± 0.10
94.90	30	0.24 ± 0.02						0.27 ± 0.02
96.49	19	0.76 ± 0.06		0.77 ± 0.06				0.62 ± 0.06
98.10	26	0.06 ± 0.02	0.07 ± 0.02					
99.02	23	0.20 ± 0.03						0.20 ± 0.05
99.87	24	0.18 ± 0.03						
101.22	20	0.42 ± 0.04		0.44 ± 0.04				0.33
101.92	21	0.48 ± 0.05		0.50 ± 0.05				0.51
103.97		0.11 ± 0.03						
105.21	28	0.14 ± 0.04						
105.87		0.15 ± 0.04						0.25 ± 0.03
106.83	23	0.88 ± 0.07		0.89 ± 0.07				0.84
107.60		0.05 ± 0.03		0.05 ± 0.03				
108.28	24	0.45 ± 0.07		0.45 ± 0.07				0.47
109.60		2.25 ± 1.13		2.25 ± 1.13	1725.	< 150.	1^+	1.27 ± 0.31
110.15		0.75 ± 0.38		0.75 ± 0.38	900.	1163.	1^+	0.33 ± 0.17
111.30	34	0.12 ± 0.04		0.12 ± 0.04				0.21 ± 0.03
113.00	27	0.90 ± 0.10		0.90 ± 0.10				0.72 ± 0.05
113.58		0.04 ± 0.02	0.04 ± 0.02					
114.68	25	0.54 ± 0.05		0.56 ± 0.05				0.45 ± 0.04
119.47	30	0.38 ± 0.04		0.38 ± 0.04				0.48
121.70	30	0.35 ± 0.04		0.35 ± 0.04				0.35 ± 0.05
123.83	38	0.18 ± 0.03		0.18 ± 0.03				0.35
125.00		1.50 ± 0.75		1.50 ± 0.75	1125.	750.	1^+	0.45 ± 0.23
125.08	25	1.00 ± 0.20		1.00 ± 0.20				0.53
125.30		0.10 ± 0.05		0.10 ± 0.05				0.6
126.00		0.30 ± 0.15		0.30 ± 0.15	625.		0^+	0.33 ± 0.17
126.99	30	0.31 ± 0.04		0.33 ± 0.04				0.34 ± 0.03
129.50		2.25 ± 1.13		2.25 ± 1.13	3150.	6000.	1^+	0.26 ± 0.16
130.50	23	0.39 ± 0.05		0.39 ± 0.05				0.45 ± 0.05
132.10	20	0.77 ± 0.07		0.78 ± 0.07				0.72
134.35	35	0.42 ± 0.06		0.43 ± 0.06				0.28 ± 0.03
134.50		0.95 ± 0.25		0.95 ± 0.25	825.		0^+	1.00 ± 0.30
135.28	28	0.15 ± 0.04						
141.10		0.57 ± 0.20		0.57 ± 0.20	375.		0^+	0.85 ± 0.03
149.25	21	0.48 ± 0.07		0.48 ± 0.07				0.32 ± 0.05
150.25	21	0.15 ± 0.04						
151.06	27	0.40 ± 0.06		0.4 ± 0.06				0.33 ± 0.08
154.97	32	0.15 ± 0.04						
159.22	28	0.15 ± 0.04						
160.28	19	0.40 ± 0.06		0.40 ± 0.06				0.22 ± 0.06
166.53		0.15 ± 0.08						
167.00		1.12 ± 0.56		1.12 ± 0.56	825.	675.	1^+	0.87 ± 0.22
167.80	32	0.49 ± 0.10		0.49 ± 0.10				0.72
168.82	24	0.20 ± 0.05		0.20 ± 0.05				0.70
168.90		1.50 ± 0.75		1.50 ± 0.75	1275.	900.	1^+	0.22 ± 0.13
169.31	31	0.35 ± 0.08		0.35 ± 0.08				
176.00					175.		0^+	0.3 ± 0.20
180.35	33	0.60 ± 0.09		0.60 ± 0.09				0.61 ± 0.04
185.00		2.25 ± 1.13		2.25 ± 1.13	2625.	< 300.	1^+	0.87 ± 0.44
186.18	28	0.4 ± 0.07		0.40 ± 0.07				0.46 ± 0.06
187.20	25	0.33 ± 0.06		0.33 ± 0.06				0.41 ± 0.05
189.00		0.38 ± 0.19		0.38 ± 0.19	800.		0^+	0.25 ± 0.13
194.25	25	0.90 ± 0.15		0.90 ± 0.15				0.69 ± 0.06
197.30	50	0.65 ± 0.20		0.65 ± 0.20				
198.90	18	0.30 ± 0.10		0.30 ± 0.10				
200.10	28	0.25 ± 0.10		0.25 ± 0.10				

TOTAL CROSS SECTION MEASUREMENTS OF THERMAL AND 24 keV NEUTRONS FOR CRYSTALLINE MATERIALS

O. Aizawa, T. Matsumoto

Atomic Energy Research Laboratory
Musashi Institute of Technology
Ozenji 971, Aso-ku, Kawasaki-shi,
Kanagawa-ken, 215 Japan

and

H. Kadotani

Century Research Center Corp.
Nihonbashi-Honmachi 3-2,
Chuo-ku, Tokyo, 103 Japan

Total cross sections of some reactor materials were measured in the energy range from 0.001 eV to 0.3 eV by means of the time-of-flight transmission method, and also measured at the monochromatic energy of 24 keV from an iron filter. The experiments in the thermal energy range were performed for zirconium, niobium, magnesium and silicon with a chopper TOF facility of the Musashi Institute of Technology Reactor. Inelastic and elastic scattering cross sections for these samples were calculated with the THRUSH and UNCLE-TOM code, respectively. The experiments were compared with those calculations. In the whole energy range, good agreement was obtained except for a little difference for the individual data in some energy range. The experiments of 24 keV neutrons were performed for beryllium, carbon, zirconium, bismuth, niobium and molybdenum. The total cross sections were determined to an overall accuracy of ±0.3~0.5%.

[time-of-flight, total cross section, thermal neutron, crystalline materials, zirconium, niobium, magnesium, silicon, filtered beam, 24 keV neutrons]

1. Introduction

The total cross sections of low energy neutrons still play an important role in designing experimental facilities as well as power reactors. Especially, the total cross sections of metals in the polycrystalline state have not been measured systematically. Therefore, further measurements are needed for the following reasons: (1) The total cross sections of some reactor materials have not yet been measured in some energy ranges. (2) Some of the total cross sections in BNL-325 have been measured in phase of solid metal with a preferential orientation of the grains. (3) The comparison of the experimental cross sections with the calculations is not throughly performed. In the present measurement, the crystalline materials of Zr, Nb, Mg and Si were selected for the above reasons.

The precise measurements of total cross section around 24 keV are useful to provide a precise bench-mark to which other conventional cross sections can be compared and normalized. A bench-mark for total cross sections at 24 keV is given for Be, C, Zr, Bi, Nb and Mo with an overall accuracy of ±0.3~0.5%.

2. Total Cross Section Measurements of Thermal Neutrons by TOF Method

2.1 Time-of-Flight Experimental Facility

The experiment was performed with the chopper and TOF facility of Musashi Institute of Technology Reactor (TRIGA-II, 100 kW). The outline of the experimental arrangement is shown in Fig. 1. The two kinds of filtered beams were used for the TOF experiment: The first one is a beam filtered by a large nearly-perfect single-crystal of about 52 cm thick silicon-rod, which is used as a very effective thermal neutron bandpass filter.[1] The second one is a cold neutron beam below 5 meV, which is transmitted by 15 cm thick polycrystalline beryllium block. They are called "Si-beam" and "Be-beam" in this paper, respectively.

The BN (boron-nitride) rotor of the chopper is 12 cm in diameter and 12 cm in height and has nine slits of 0.4 x 7.0 cm and is connected with a driving motor. In the present experiment two rotation frequencies are chosen: 2000 rpm and 3800 rpm, where neutron burst widths (FWHM) are 280 μsec and 170 μsec, respectively. The flight path can be set up to 7.2 m. In the present measurements a distance of 6.25 m was chosen. The typical energy resolutions are 5.5% for 0.002 eV neutrons at 2000 rpm and 7.5% for 0.01 eV neutrons at 3800 rpm.

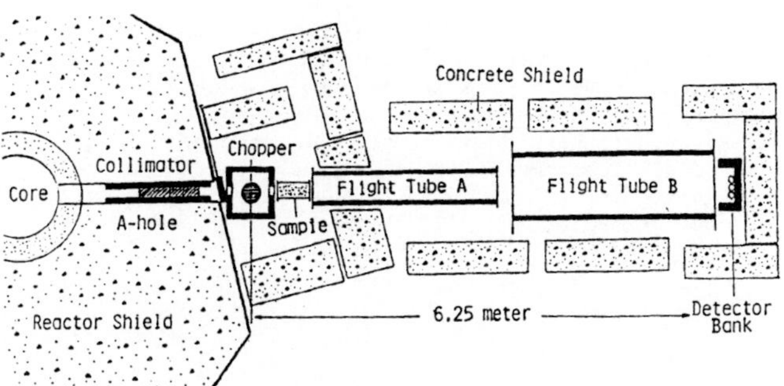

Fig. 1 Arrangement of Time-of-Flight Facility

2.2 Samples for the Measurements

The measurements were performed for four kinds of elements, i.e. zirconium, niobium, magnesium and silicon. The cylindrical samples were prepared in forms of powder, slug and clush. Table I shows the samples measured in the present experiment. Almost all of these samples were packed with the canning machine.

Table I. Samples for Total Cross Section Measurements of Thermal Neutrons

Element	Form or Phase	Purity (%)	Density (g/cm^3)	Dimension Dia. (cm)	Length (cm)
Zr	slug	99.8	3.00	9.8	18.0
	solid	99.9	6.67	5.5	8.0
Nb	clush	99.9	4.98	9.8	18.0
Mg	powder	99.0	1.00	9.8	36.0
Si	rod	99.95	2.34	6.1	22.0 30.2
	powder	98.0	1.19	9.8	30.0 36.0

2.3 Experiment and Data Analysis

The total cross sections were measured with the "Si-beam" and the "Be-beam" except for Si-rod sample. The measurement for Si-rod sample was performed as follows: At first no filter was inserted in the experimental hole. This is called the "complete open beam" in this paper. A time-of-flight spectrum was measured in the "complete open beam" (Spectrum:A).

K. H. Böckhoff (ed.), Nuclear Data for Science and Technology, 143–146.

After a Si-filter of 30.2 cm in thickness was inserted at the position of collimator, a TOF spectrum was measured (Spectrum:B). With an additional Si-filter of 22.0 cm in thickness, a TOF spectrum was again measured (Spectrum:C). The total cross section of Si-rod was reduced from the ratios of Spectrum:A/Spectrum:B, Spectrum:B/Spectrum:C and Spectrum:A/Spectrum:C. In the case of the "Be-beam", the 52.2 cm Si-rod sample was set at the sample position in Fig. 1. Examples of the TOF spectra are shown in Fig. 2 for the "Si-beam" and the "Be-beam". The background spectra of the "Si-beam" and the "Be-beam" were measured with a Cd-sheet instead of the samples. The background neutrons are mainly caused from fast neutrons passing through the BN rotor, and it can be seen from Fig.3(b) and (c) that the background neutrons decrease to about 1/100 in the case of the "Si-beam" in comparison with the "complete open beam". Another background is caused by the overlap as shown in Fig. 3(d) for higher rotation rate of the chopper. The dotted line indicates the overlapping situation. It can be seen in Fig. 3(e) that the background is underestimated in the case of the Cd-sheet transmission. As this background could not be experimentally subtracted, the summation of this and the fast neutron background was fitted to the function of $f(t)=-a[(t-b)^2-c]^2+d$. An example of the background correction is shown in Fig. 4. The present background correction was verified using the well-known cross section of beryllium.[2)]

The cross section for each channel is derived by the following equation:

$$\sigma_i = \frac{1}{N\cdot d}\ln\frac{C_o^i - B_o^i}{C_s^i - B_s^i},$$

where N is the number density of atoms in the sample, d the thickness of sample, C_o^i i-channel counts for no sample, C_s^i i-channel counts for sample-in, B_o^i i-channel background counts for no sample, B_s^i i-channel background counts for sample-in. As the counts of the lower energy region were very small, they were bunched in several channels.

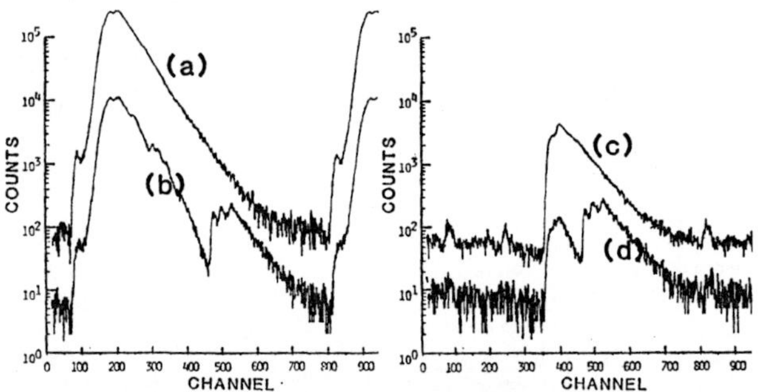

Fig. 2 Raw data of time-of-flight spectrum
(a) "Si-beam" at 2000 rpm (1ch.=20μsec)
(b) Magnesium transmission spectrum for "Si-beam"
(c) "Be-beam" at 2000 rpm (1ch.=20μsec)
(d) Magnesium transmission spectrum for "Be-beam"

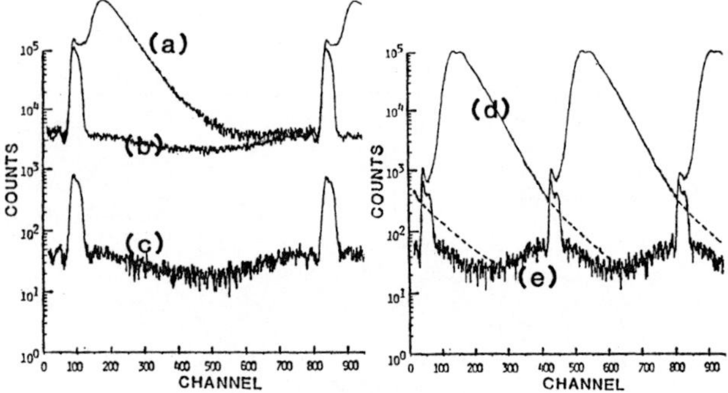

Fig. 3 Raw data of background spectrum
(a) "complete open beam" at 2000 rpm (1ch.=20μsec)
(b) Background spectrum of "complete open beam"
(c) Background spectrum of "Si-beam" at 2000 rpm
(d) "Si-beam" at 3800 rpm (1ch.=20μsec)
(e) Background spectrum of "Si-beam" at 3800 rpm

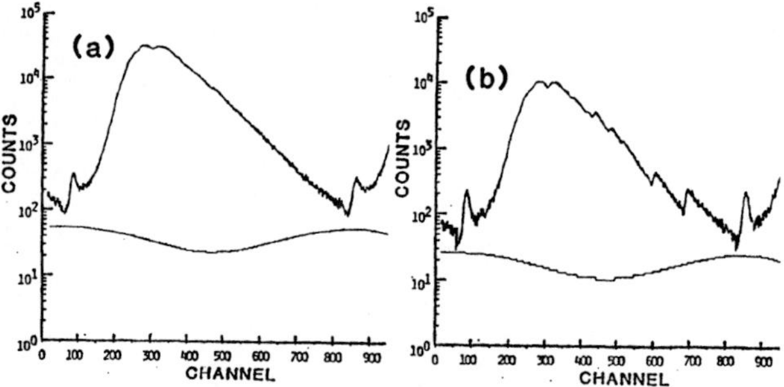

Fig. 4 An example of background-fitting correction
(a) "Si-beam" at 3800 rpm (1ch.=10μsec)
(b) Transmission spectrum for powdered silicon

2.4 Calculations

In the present calculation, the UNCLE-TOM[3)] code and the THRUSH[4)] code were used. The former code calculates zero-phonon elastic scattering cross sections for crystals of face-centered cubic, body-centered cubic and hexagonal structures. The latter is developed by one of the authors for computation of scattering kernels by means of the phonon expansion method using an incoherent approximation. In the present calculation, a Debye frequency spectrum was assumed, and the 1/v absorption cross section was added to the scattering cross section. The nuclear and physical constants used for the calculations are shown in Table II.

Table II. Physical Constants for Measured Samples

	Crystal Structure	Lattice Const. (Å)	Debye Temp. (°K)	σ_s (b)	b_{coh} (fm)	σ_{incoh} (b)	$\sigma_{\gamma o}$ (b)
Zr	hcp	a=3.23 c=5.14	310	6.40	7.0	0.15	0.185
Nb	bcc	a=3.29	252	6.37	7.11	0.006	1.15
Mg	hcp	a=3.21 c=5.21	342	3.41	5.38	0.046	0.063
Si	dia-mond	a=5.43	640	2.04	4.15	0.009	0.171

2.5 Comparison and Discussion

(1) Zirconium

The measurement was performed for two different phases, i.e. zirconium slug and zirconium metal in solid phase. The measured cross section for the Zr-slug is in good agreement with the calculation in the energy range above 2.5 meV where the elastic scattering is dominant. The inelastic scattering cross section below 2.5 meV is about 50% higher than the calculation as shown in Fig. 5(a). However, the measured cross section for Zr-metal in solid phase is in good agreement with the calculation in the whole energy range as shown in Fig. 5(b). The disagreement for Zr-slug data may be due to the amorphous component in Zr-slug.

Furthermore, it is clear that solid-phase Zr-metal has no preferential orientation of grains, because the measured elastic scattering cross sections show good agreement with the calculation.

(2) Niobium

In Fig. 6 the total cross section for clushed-form niobium is compared with the calculation. These curves are similar but not identical. Comparing them with the preliminary results obtained by the authors for solid-phase niobium metal, we note that the shape of the elastic cross section for clushed-form is normal, while that for solid-phase shows a unique peak at higher-energy side of each elastic peak.

The fact that the first elastic peak diminished in the clushed-form sample can be explained in terms of non-random orientation of the grains. Therefore, it is necessary to get powdered niobium in order to precisely measure the first elastic peak.

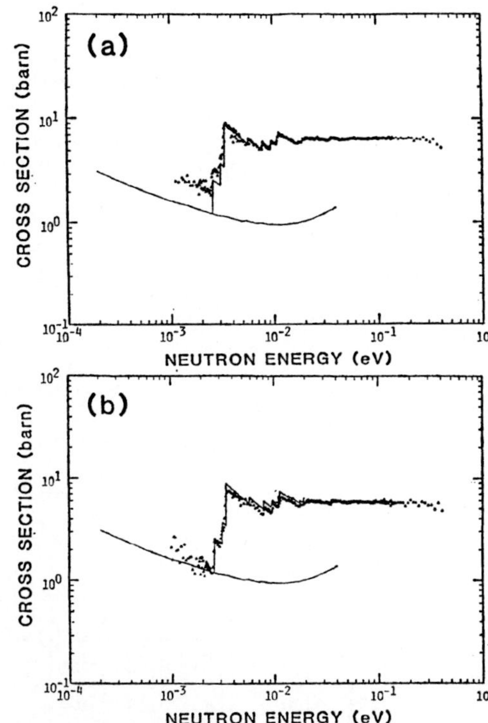

Fig. 5 Total cross section of zirconium
(a) Zr-slug (sponge)
(b) Zr-metal in solid phase

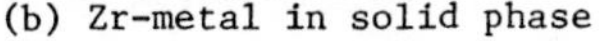

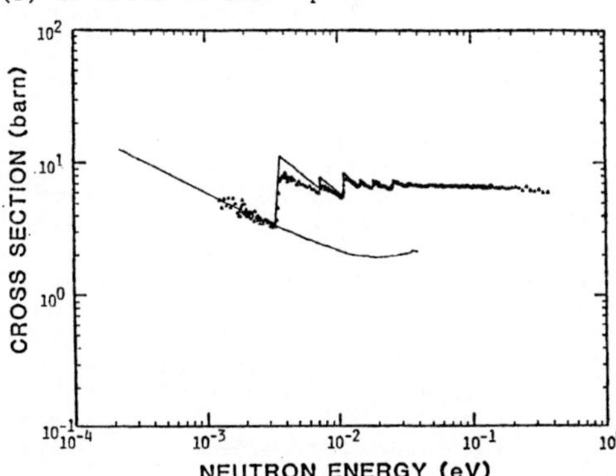

Fig. 6 Total cross section of niobium

(3) Magnesium

The powdered magnesium sample is used in the present experiment. The experimental result is compared with the calculation in Fig. 7. It shows that the measured elastic scattering cross section is in good agreement with the calculation, but the inelastic scattering cross sections are about 20% higher than the calculation below 2.5 meV. It is supposed that the disagreement is caused by the amorphous component in powdered magnesium.

(4) Silicon

The inelastic scattering cross section measured by using a single crystal rod is compared with the calculation in Fig. 8(a). It can be seen that the results from the "Be-beam" experiment below 5 meV are in good agreement with the calculation. There is some discrepancy around 0.01 eV, but it is clear that better agreement is obtained at higher energies in comparison with the results previously obtained by Brugger and Yelon.[1)]

The experimental result for Si-powder is compared with calculation in Fig. 8(b). From this figure the experimental result of elastic cross section is in good agreement with the calculations, but there still exists difference in the inelastic scattering cross sections at energies below 2 meV. The difference is supposed to be originated from the same reason as above.

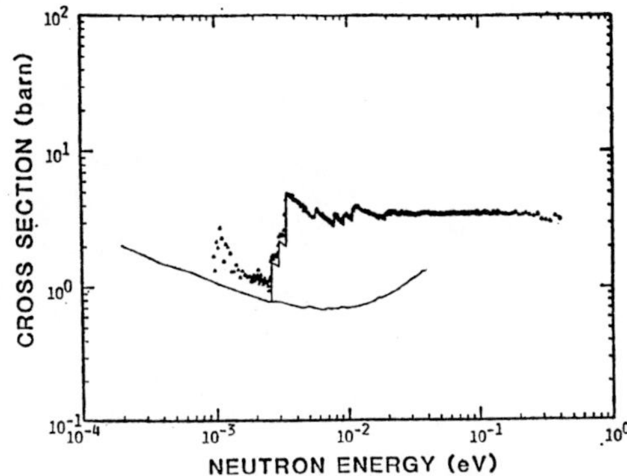

Fig. 7 Total cross section of magnesium

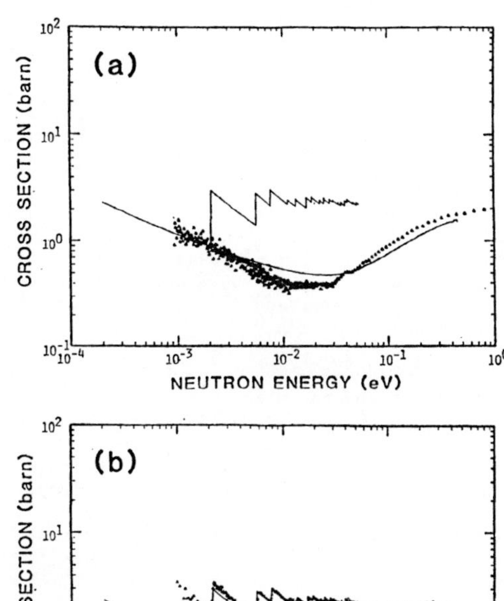

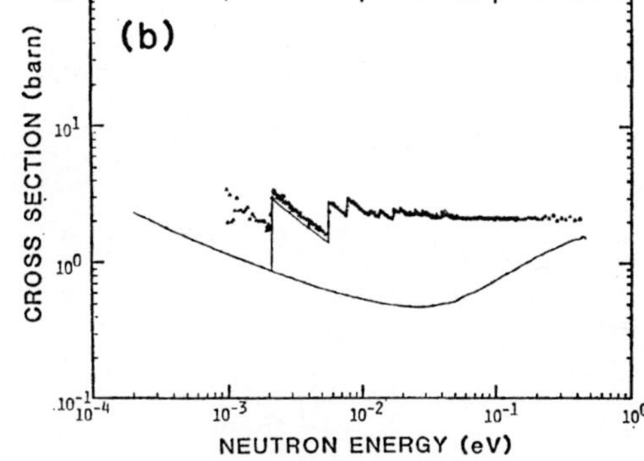

Fig. 8 Total cross section of silicon
(a) Single crystal silicon-rod
(b) Powdered silicon

3. Total Cross Section Measurements of 24 keV Neutrons Using Iron-Filtered Beam

3.1 Experimental Arrangement

The iron filter constructed for the Musashi Institute of Technology Reactor is illustrated in Fig. 9. The filter was developed for the 'B-hole', which penetrates the biological shielding.

The collimator and the filter are located at a distance of 1.2 m from the graphite reflector to reduce neutron activation. The section of lead and polyethylene collimator, 60 cm long and 6 cm in inner-diameter, forms the first section (A). The next three filter sections consist of 30 cm thick aluminium, 40 cm thick iron and 10 cm thick sulfur zones. The last section (C) is a lithium-tile collimator of 10 cm long and 5 cm in inner-diameter. The transmission samples and the Ti-filter for background-run were inserted to the position (D), and the other collimator (E) of 5 cm in inner-diameter was set outside the 'B-hole'. A proton recoil counter (LND 27024, hydrogen 1 atm.) was used as a neutron detector, which was set at the position (F), 1 m apart from the sample position.

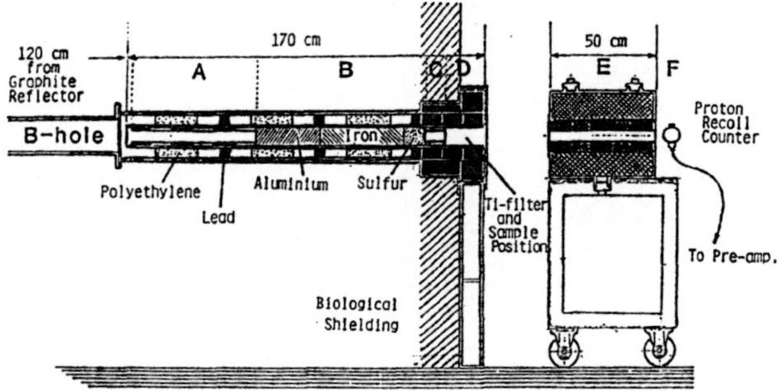

Fig. 9 Geometrical arrangement of filtered beam facility

3.2 Samples for the Measurements

The beryllium sample is a plate of high-quality metal (5.08 cm x 5.08 cm square and 1 cm thickness). Each plate was weighed, and its density was determined to be 1.866 ± 0.002 g/cm^3. The carbon sample is of reactor-grade graphite (total impurity: 30 ppm). The density was determined to be 1.753 ± 0.005 g/cm^3 by the same method. The zirconium, niobium and molybdenum samples are polished disks (5.5 cm dia. and 2 cm thickness). The densities were determined to be 6.67 ± 0.02, 8.79 ± 0.02 and 10.35 ± 0.03 g/cm^3, respectively. The bismuth sample is in form of fused and polished block (4.95 x 4.95 x 19.95 cm), and its density was determined to be 9.88 ± 0.01 g/cm^3.

3.3 Energy Calibration and Data Reduction

The proton recoil counter which contained a small amount of ^{3}He gas (0.1 mmHg) was irradiated in the thermal column. The peak at 764 keV due to the ^{3}He (n,p) reaction was observed. The amplifier gain was increased to bring the 24 keV shoulder to a proper position. From the ratio between these two amplifier gains we have determined the absolute energy scale. The position of pulse-height which corresponds to 24 keV neutrons is shown with an arrow in Fig. 10(a).

Four spectra were measured for each sample: i.e. without sample in the beam (Fig. 10(a)), with only Ti-filter (Fig. 10(b)), with sample in the beam (Fig. 10(c)) and with the sample and Ti-filter both in the beam (Fig. 10(d)).

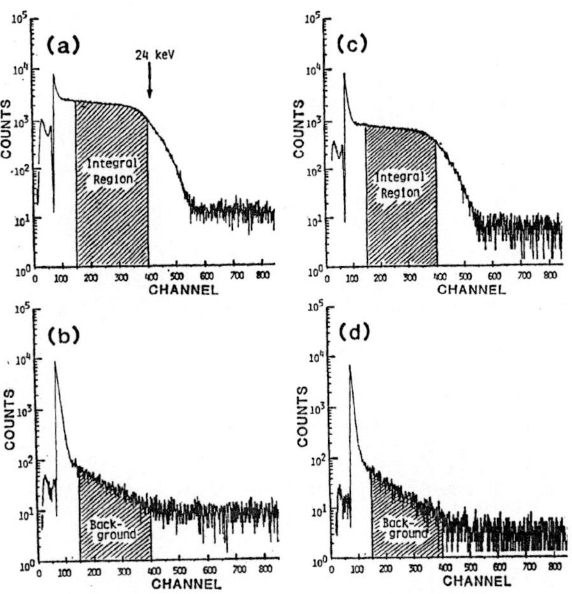

Fig. 10 Raw data of filtered beam experiments
(a) without sample in the beam
(b) with only Ti-filter
(c) with sample (Be 2 cm) in the beam
(d) with the sample and Ti-filter in the beam

3.4 Evaluation of Errors

An integral between 150 ch. and 400 ch. of the pulse-height spectrum was finally adopted as the measured cross section in the present experiment. The ambiguity of the integrating region introduced an uncertainty in cross section. It was estimated to be 0.05~0.1%. The counting statistics of 0.1~0.4% were the largest error in the cross section.

3.5 Results and Comparison

The results of the experiment are presented in Table III. All errors listed are standard deviation. Comparing the data from other sources, we note the evaluated cross section based on ENDF/B[5] is 0.8% larger than the experimental result for Be, and is in good agreement with the result for C. The results of this experiment for Be and C are 0.4~0.7% smaller than the LINAC data of Block et al.[6]

Table III. Results of Experiments

Element	Total cross section (barn)	Total Error (%)	* Partial Error (a) (%)	(b) (%)	ENDF/B 5) Block 6) (barn)
Be	5.88 ± 0.02	0.34	0.12	0.10	5.928 5) 5.903 6)
C	4.65 ± 0.02	0.35	0.21	0.28	4.651 5) 4.684 6)
Zr	7.69 ± 0.03	0.40	0.26	0.30	———
Bi	10.54 ± 0.03	0.30	0.28	0.10	———
Nb	7.54 ± 0.03	0.46	0.40	0.22	———
Mo	7.63 ± 0.04	0.55	0.47	0.29	———

*(a) Statistical error (b) Sample thickness

4. Summary and Conclusion

In the thermal energy region, the total cross sections were re-measured for crystalline materials of Zr, Nb, Mg and Si. The present experiment supplies new data for the energy regions which have not been covered. They were compared with the calculations, and it was found that the structures due to coherent elastic scattering could be successfully measured by means of powdered samples, however, the inelastic scattering cross sections could not be exactly measured due to the amorphous component in powdered samples. On the other hand, the metal samples in solid phase are suitable for the measurements of inelastic scattering cross sections.

For 24 keV neutrons, a bench-mark is proposed with an accuracy of 0.3~0.5% for the total cross sections of Be, C, Zr, Bi, Nb and Mo, using the iron-filtered beam method.

Acknowledgement

This work was supported by a grant from the Ministry of Education, Science and Culture: Grant-in-Aid for Scientific Research No. 53358086 and No. 56580162. The authors are indebted to Prof. E. Arai of Tokyo Institute of Technology and to Prof. K. Kanda of Kyoto University Research Reactor Institute for critical reading of the manuscript.

References

1. R. M. Brugger and W. Yelon, Proceedings of the Conference of Neutron Scattering, Vol.II, 1117 (1976)
2. K. Kanda, H. Kadotani and O. Aizawa, Nucl. Sci. Tech., 12, 601 (1975)
3. S. Iijima, private communication
4. H. Kadotani, JAERI-M-8927 (1980)
5. ENDF/B Summary Documentation, BNL-NCS-17541(ENDF-201), Third Edi. (ENDF/B-V), Edited by R. Kinsey
6. R. C. Block, Y. Fujita, K. Kobayashi and T. Osaki, Nucl. Sci. Tech., 12, 1 (1975)

SEARCH FOR GAS PRODUCING REACTIONS IN THERMAL REACTORS

P.D'hondt, C.Wagemans, E.Allaert, A.De Clercq

Nuclear Physics Laboratory, B-9000 Gent and SCK-CEN, B-2400 Mol, Belgium

A.Emsallem

Institut de Physique Nucléaire, F-69621 Villeurbanne (Lyon), France

R.Brissot

Institut Laue-Langevin, 156X Centre de Tri, F-38042 Grenoble, France

Four types of gas producing reactions induced by thermal neutrons have been studied at an intense thermal neutron beam of the Grenoble High Flux Reactor.

(i) hydrogen, tritium and helium production during the ternary fission process for several actinides.

(ii) hydrogen and helium production via (n_{th},p) and (n_{th},α) reactions on structural materials like V, Zn and Mo.

(iii) helium production via (n_{th},α) reactions on fission products like Ru, Pd, Cd and Sn.

(iv) helium production via (n_{th},α) reactions on actinides from ^{232}Th to ^{239}Pu.

Gas producing cross-sections ranging from a few barn to smaller than one microbarn have been observed

[NUCLEAR REACTIONS : ^{233}U(n,f), ^{235}U(n,f), ^{237}Np(n,f), ^{239}Pu(n,f), ^{50}V(n,p), ^{65}Zn(n,α), ^{95}Mo(n,α), ^{97}Mo(n,α) ^{101}Ru(n,α), ^{105}Pd(n,α), ^{113}Cd(n,α), ^{115}Sn(n,α), ^{232}Th(n,α), ^{233}U(n,α), ^{235}U(n,α), ^{238}U(n,α), ^{239}Pu(n,α), E=thermal; yield of ternary particles, σ(n,α).]

Intoduction

Up to now, most of the information available on gas producing reactions in nuclear reactors has been obtained via fast neutron induced reactions. However, also thermal neutrons are inducing a large variety of gas producing reactions in thermal fission reactors. Since very little information is available for thermal neutron induced (n,p) and (n,α) reactions on actinides, on fission products and on structural materials, giving rise to hydrogen and helium production, we performed a series of (n_{th},p) and (n_{th},α) measurements on such nuclei taking profit of the excellent thermal neutron beam characteristics available at the Grenoble High Flux Reactor. For the actinides we also studied the hydrogen, tritium and helium production during the ternary fission process.

Experimental conditions.

The detection chamber was installed at the 87m curved neutron guide of the High Flux Reactor at the ILL (Grenoble). The very intense neutron beam ($\simeq 10^9$ neutrons/cm^2.sec) leaving this tube has a ratio of about 10^6 for the number of slow neutrons to that of epithermal and fast neutrons. In addition to this the direct γ-ray flux from the reactor is also reduced by a factor of about 10^6.

The targets studied, inclined at an angle of 30° with respect to the incident neutron beam, are viewed either by a gold-silicon surface barrier detector, or by a telescope assembly consisting of two gold-silicon surface barrier detectors, placed outside the neutron beam. The use of a telescope assembly has several advantages. When operating the two detectors in an anti-coïncidence mode, it considerably reduces the background in the low energetic part of the spectrum which permits a accurate study of (n,α) reactions with low Q_α -values. When using the two detectors in a coïncidence mode, it permits to identify the particles studied based on the difference in energy loss in the thin transmission detector for different kind of particles with the same incident energy, removing the ambiguity on the nature of the observed particles. Most of the targets were prepared at the Central Bureau for Nuclear Measurements at Geel. When available, highly isotopic enriched target materials were used.

The energy calibration of the detector was done based on the well-known energies of the α-particles emitted during the radioactive decay of various uranium isotopes. The ^{6}Li(n_{th},α)t (E_α=2.055 MeV, E_t=2.727 MeV) and the ^{143}Nd(n_{th},α) (E_α=9.45 MeV) reactions were used as well. The neutron flux calibration was done via the thermal neutron induced fission of ^{235}U (σ_f°=587.6b[1])

The experimental cross-section values are deduced using the relation

$$\sigma_x = \frac{N_u}{N_x}\frac{C_x}{C_u}\ 587.6\ \text{barn}$$

were N stands for the number of atoms per cm^2 and C_x and C_u are respectively the counting rates corrected for background for the reaction studied and for ^{235}U(n_{th},f).

Measurements and Results

We will report first on our results obtained for the ternary fission process for several actinides. For the thermal neutron induced ternary fission of ^{235}U we have studied the Z=1 and Z=2 particles. The energy distributions for protons, tritons and α-particles are given in fig.1. The measured relative probabilities and B/LRA (i.e. the average number of fission reactions needed to emit one ternary α-particle) are given in table I. Taking into account the σ_f° value for the ^{235}U(n_{th},f) reaction[1], these probabilities can be converted into cross-section values. In table I we also report the B/LRA values which we obtained, relative to this of the ^{235}U(n_{th},f) reaction, for the thermal neutron induced fission of ^{233}U, ^{237}Np and ^{239}Pu. Also these numbers have been converted into cross-section values taking into account the known σ_f° values for these reactions.

Table I: Relative probabilities for Z=1 and Z=2 particles emitted during the ^{235}U(n_{th},f) and B/LRA values

	RELATIVE PROBABILITIES	B/LRA	CROSS-SECTION (barn)
^{235}U(n_{th},f)	p:0.96±0.01		(9.6±0.3) 10^{-3}
	t:6.27±0.03		(62.7±1.5) 10^{-3}
	α: 100	589±9	(1.00±0.02)
^{233}U$(n_{th}$,LRA)		461±15	(1.15±0.04)
^{237}Np$(n_{th}$,LRA)		493±20	(40±9) 10^{-6}
^{239}Pu$(n_{th}$,LRA)		450±15	(1.65±0.05)

A second group of results deals with (n_{th},α) reactions on actinides. For the fissile nuclei the (n_{th},α) particles are emitted in the presence of an intense background due to the thermal neutron induced ternary fission process for which we already have given some results. Taking into account the Q_α-values for the reactions studied, the statistical accuracy of the present data and the energy resolution of the telescope, we determined an upper limit of 300μb for the ^{233}U(n_{th},α), 100μb for the ^{235}U(n_{th},α) and 370μb for the ^{239}Pu(n_{th},α)

K. H. Böckhoff (ed.), Nuclear Data for Science and Technology, 147–149.

reaction cross-section . In addition to these nuclei we also studied the (n_{th},α) reaction on two non-fissile targets (^{232}Th and ^{238}U) with very small σ_f^o values respectively 3μb and 2.7μb and, consequently, a very small ternary fission background. Fig.2. shows

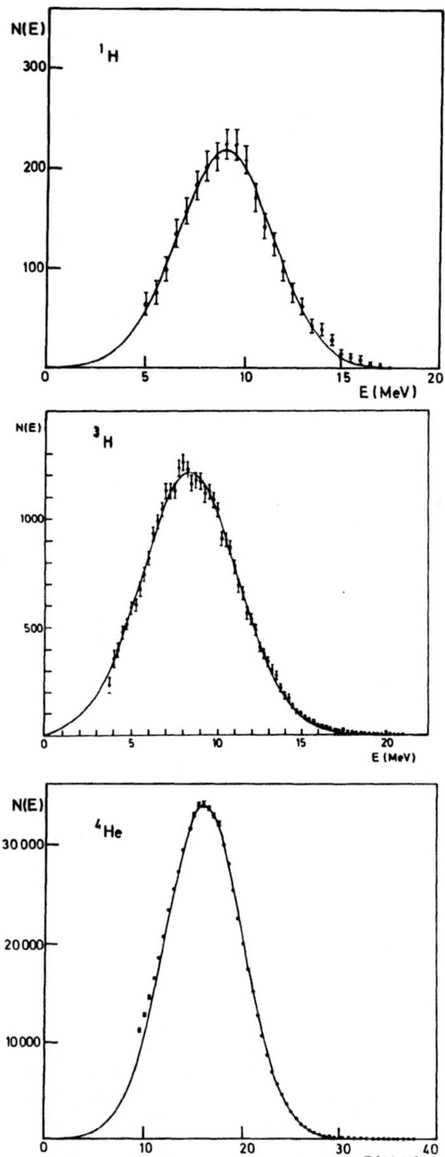

Fig.1. Energy distributions for the ^{1}H, ^{3}H and ^{4}He particles emitted during the $^{235}U(n_{th},f)$ reaction.

the identified energy distribution obtained for the $^{238}U(n_{th},\alpha)$ reaction; also the corresponding background run is shown. For the ^{238}U case the ^{212}Po and ^{216}Po α-lines (daughter products of ^{232}Th impurities in the vacuum chamber) are observed in the (n_{th},α) spectrum and in the background run. The intensity of the α-line at 8.55 MeV corresponds to a $\sigma(n_{th},\alpha)$ value of 1.5±0.5 μb. For the ^{232}Th case the observation of a (n_{th},α) line (which is expected at E_α=8.266 MeV) is strongly hindered by the detection of the radioactive decay α-particles emitted by the ^{232}Th target. Nevertheless, the statistical accuracy of the present data allowed us to calculate an upper limit of 1μb for the $^{232}Th(n_{th},\alpha)$ reaction. Our experimental values and the previously reported values for the (n_{th},α) cross-sections on actinides are summerized in table II. It is clear that the present results are lower (in some cases several orders of magnitude) than all previous data.

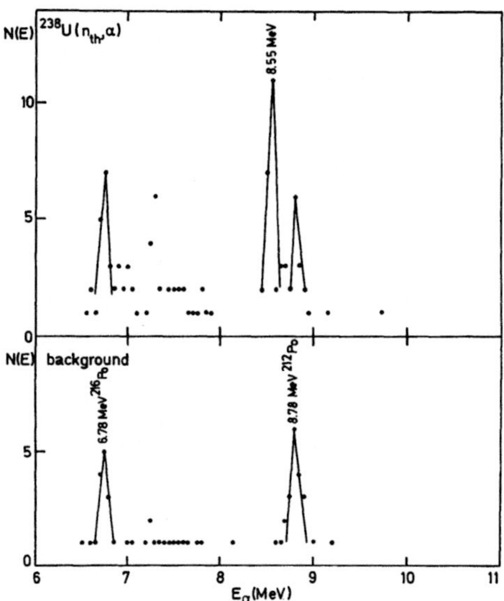

Fig.2. Energy distribution of the identified $^{238}U(n_{th},\alpha)$ particles and the corresponding background run.

Table II: A survey of experimental (n_{th},α) cross-section values (μb) reported in the actinide region.

Ref.	^{232}Th	^{233}U	^{235}U	^{238}U	^{239}Pu
4			3x10^4		
5			3x10^3		
6			⩽(5±2)x10^4		
7					2x10^4
8				⩽100	
9				1.3±0.6	
10			⩽330		
This work	⩽1	⩽300	⩽100	1.5±0.5	⩽370

The (n_{th},p) and (n_{th},α) reactions on structural materials are very important data from technological point of view. This is best illustrated by the surprisingly high cross-section value (13.1b)[11] for the $^{59}Ni(n_{th},\alpha)$ reaction. We will report here on the results which we we obtained up to now in this field. Fig.3. shows the

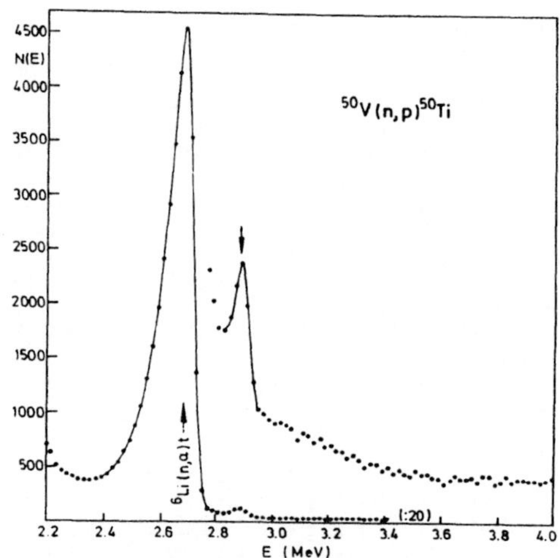

Fig.3. Energy spectrum for the $^{50}V(n_{th},p)$ reaction.

energy spectrum of the charged particles obtained after neutron irradiation of a $^{50}V_2O_5$ target.Besides a prominent peak due to the $^{6}Li(n_{th},\alpha)t$ reaction on Li impurities in the target,a well resolved proton peak at (2.920±0.015) MeV is observed.The intensity of this proton line corresponds to a $\sigma(n_{th},p)$ value of 400μb. For the $^{97}Mo(n_{th},\alpha)$ reaction we observed a small peak at the expected energy of E_α=5.15 MeV with a corresponding $\sigma(n_{th},\alpha)$ value of 0.44±0.18 μb.For the $^{95}Mo(n_{th},\alpha)$ reaction a cross-section of 30 μb has been observed [12]. Preliminar results have been obtained for the ^{65}Zn (n_{th},α) reaction,where we observed a peak at the expected energy of 6.03 MeV with a cross-section value of 0.15 b.

As a last point of the present work we have studied the (n_{th},α) reaction on some selected fission products.Up to now mainly light fission products and a few "valley" isotopes have been investigated.For a given Z,the isotope with the highest calculated Q_α-value was chosen (when available) since there the highest σ_α-value might be expected.This correlation is illustrated in table III by comparing the ^{95}Mo and ^{97}Mo data.Our results obtained for the $^{105}Pd(n_{th},\alpha)$ reaction are shown in fig.4.We observe a small peak at the expected energy with a corresponding $\sigma(n_{th},\alpha)$ value of 0.5 μb.We have chosen this example to demonstrate that even that small cross-sections can be measured with the present technique.For the $^{115}Sn(n_{th},\alpha)$ reaction we also observed a (n_{th},α) peak at the expected energy with a corresponding $\sigma(n_{th},\alpha)$ value of 58±15 μb.For the $^{101}Ru(n_{th},\alpha)$ and the $^{113}Cd(n_{th},\alpha)$ reaction no indication of an α-peak has been observed.Taking into account the statistical accuracy of the corresponding data,an upper limit of respectively 0.15 μb and 1 μb could be obtained.In table III the (n_{th},p) and (n_{th},α) cross-section values for structural materials and the fission products mentioned above are summerised.

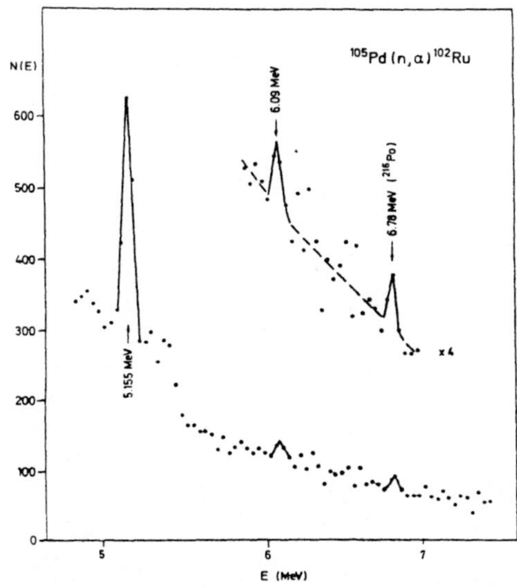

Fig.4.Energy spectrum for the $^{105}Pd(n_{th},\alpha)$ reaction.

Conclusion

Our systematic study of (n_{th},α) and (n_{th},p) reactions has shown that these reactions only give rise to μb-cross-section values for the actinides and fission products studied.For structural materials much higher values have been found which might create mechanical problems when using some of these materials in reactor walls or for canning purposes.Also in the thermal neutron induced ternary fission of the fissile isotopes, large gas producing cross-sections have been found.So we can conclude that hydrogen,tritium and helium production during the ternary fission process on fissile nuclei and via thermal neutron induced (n,α) and (n,p) reactions on some structural materials are not negligible sources of gas production in thermal reactors.

Table III:Results for (n_{th},p) and (n_{th},α) cross-sections (μb) for selected structural materials and fission products.

Nucleus	Reaction	E_α or E_p (MeV)	σ_{exp} (μb)
^{50}V	(n,p)	2.92	400±20
^{65}Zn	(n,α)	6.03	$0.15x10^6$ °
^{95}Mo	(n,α)	6.14	30±4
^{97}Mo	(n,α)	5.15	0.44±0.18
^{101}Ru	(n,α)	5.57	≤0.15
^{105}Pd	(n,α)	6.09	0.54±0.18
^{113}Cd	(n,α)	4.77	≤1
^{115}Sn	(n,α)	5.99	58±15

° preliminar result

Acknownledgements

Prof.Dr.A.J.Deruytter and Dr.M.Nève de Mévergnies are acknowledged for their continuous interest in this work. This research was supported by the Interuniversitair Instituut voor Kernwetenschappen-Nationaal Fonds voor Wetenschappelyk Onderzoek.

References

1. A.Deruytter,J.Spaepen and P.Pelfer,J.Nucl.Energ.27, 645 (1973).
2. P.D'hondt,C.Wagemans,G.Barreau and A.J.Deruytter, Ann.Nucl.En.7,367 (1980)
3. C.Lederer and V.Shirley,Table of Isotopes(7thEd), 1978.
4. M.Sowinsky,M.Dakowski and H.Piekarz,Phys.Lett.6, 321, (1963)
5. J.Chwaszczewska,M.Dakowski,T.Krogulski,E.Piasecki, M.Sowinsky,A.Stegner and J.Tys,Acta Phys.Pol.35, 187, (1969)
6. C.Wagemans and A.J.Deruytter,Z.PhysA275,149, (1975)
7. V.N.Andreev,Bull.Ac.Sc.USSR,Vol25,n°1,116, (1961)
8. I.Almodovar,J.Cantarell and H.Bielen,Z.Phys.177, 451, (1964)
9. M.Asghar,A.Emsallem,R.Chéry,C.Wagemans,P.D'hondt and A.J.Deruytter,Nucl.Phys.A259,429, (1976).
10. P.D'hondt,A.De Clercq,A.J.Deruytter,C.Wagemans,M. Asghar and A.Emsallem,Nucl.Phys.A303,275, (1978).
11. M.Ashar,A.Emsallem and N.Sjöstrand,Z.Phys.A282, 375, (1977).
12. A.Emsallem and M.Asghar,Z.Phys.A275,157, (1975).

STUDY OF THE $^{22}Na(n,p)^{22}Ne$ REACTION IN THE NEUTRON ENERGY RANGE UP TO 1 keV

Yu.M.Gledenov, J.Kvitek, S.Marinova, Yu.P.Popov, J.Rigol
V.I.Salatsky

Laboratory of Neutron Physics, Joint Institute for Nuclear Research, Dubna, USSR

The $^{22}Na(n,p)^{22}Ne$ reaction has been investigated at a neutron energy up to 1000 eV by the time-of-flight method at IBR-30 reactor of JINR. A neutron resonance was found in the reaction cross section. By fit, the parameters assigned to the resonance were E_0=145±11 eV; Γ_n^0 =1.6±0.2 eV; Γ_{p1} =114±20 eV. The presence of this resonance explains large values for the thermal cross section of this reaction.

NUCLEAR REACTIONS $^{22}Na(n,p)^{22}Ne$; $E_n \approx$ 0.01-1000 eV, measured $\sigma(E)$, ^{22}Na neutron resonance, deduced Γ_p, Γ_n .

In the capture of slow s-neutron by the radioactive nucleus ^{22}Na ($T_{1/2}$=2.6 Y) the compound nucleus ^{23}Na with spin 5/2+ or 7/2+ at an excitation energy above neutron binding energy (B_n=12416.7± 2 keV, Ref.[1]) is formed. This excited state may decay by a proton (Q_p= =3624.4 keV) or α -particle (Q_α =1951 keV) emission.

The $^{22}Na(n,p)^{22}Ne$ reaction was earlier studied with thermal neutrons. As shown in Refs.[2,3] the main contribution into cross section is made by the proton transition to the first excited state of ^{22}Ne: $\sigma_{n,p1}$= =(30.6±2.6).10^3 b. A weak proton transition to the ground state of ^{22}Ne with an intensity ratio of I_{P0}/I_{P1}= =(7.4±0.2).10^{-3} was found in Ref.[3]. This work gives also the upper limit for the cross section $\sigma_{np_1} \leq$11 b for the isotope ^{22}Na at the neutron energy E_n=2 keV.

The investigation of the excited states of the ^{23}Na nucleus is possible by means of the reaction ($^{19}F+\alpha$). Several resonances were found in the energy region that corresponds to the neutron binding energy at the nucleus ^{22}Na. The positions of the excited states of ^{23}Na are determined with an uncertainty ± 4 keV and thus one may assume that the resonances at E_α =2354 keV or 2360 keV in the $^{19}F(\alpha,p)$ reaction might be responcible for the thermal cross section of the $^{22}Na(n,p)$ reaction (see the Table).

This work was carried out in order to obtain information about the energy dependence of the $^{22}Na(n,p)^{22}Ne$ reaction cross section as well as to look for the resonance responeible for the thermal cross section.

The measurements were carried out with the time-of-flight neutron spectrometer at the pulsed reactor IBR-30 of the Laboratory of Neutron Physics, JINR (Ref.[6]) both in the reactor and booster modes. Three runs were made at resolutions of 4.5; 0.2 and 3.0 μs/m. The target with an activity of about 400 μCi was prepared by the deposition of NaCl dissolved in HCl on the polycarbonate film 4 μm thick. A silicon surface-threshold detector with an area ob 8 cm^2 and a sensitive layer 200 μm thick was used for the registration of protons. The detector was placed inside an evacuated chamber at a distance of 40 mm from the target. The measurements with Li^6 and B^{10} targets were made for the calibration of amplitude spectra and normalization of cross sections of the investigated reaction. The two-dimensional information about the energy of registered particles and about the neutron time-of-flight was written on the magnetic tape of the PDP-11/20 measuring module. In the data processing we used energy spectra of charged particles in different time windows corresponding to definite neutron energies. One of the spectra at E_n=0.02-0.08 eV is shown in Fig. 1. The strong peak corresponds to the p_1-transition with the energy 2.25 MeV. The p_0-transition at 3.47 MeV is much weaker. The peak at 2.73 MeV is due to tritons from the $^6Li(n,T)^4He$ reaction on lithium admixture in the Na target.

The time-of-flight spectrum of the proton yield was obtained by selecting in the amplitude spectrum the window corresponding to p_1 peak. This spectrum was used to build up the dependence of the $^{22}Na(n,p_1)$ reaction cross section on neutron energy (Fig.2). The experimental points are marked differently to show the results obtained in different runs. The absolute normalization was made over the reaction thermal cross section from Ref.[3], which coincided with the absorption total cross section from Ref.[7]. One may see in Fig.2 the earlier unknown resonance at $E_n \simeq$ 150 eV. Solid curve is the cross section calculated with Breit-Wigner formula using the following parameters obtained by fitting: E_0=145±11 eV; Γ_n^0 =1.6±0.2 eV; g =4/7 and Γ_{P1} = =114±20 eV. Here E_0, Γ_n^0 , Γ_p are the position, reduced neutron width, proton neutron width, respectively. On the choice of the statfactor g see below. Dashed line is the dependence $\sigma \sim 1/v$ normalized over the thermal point.

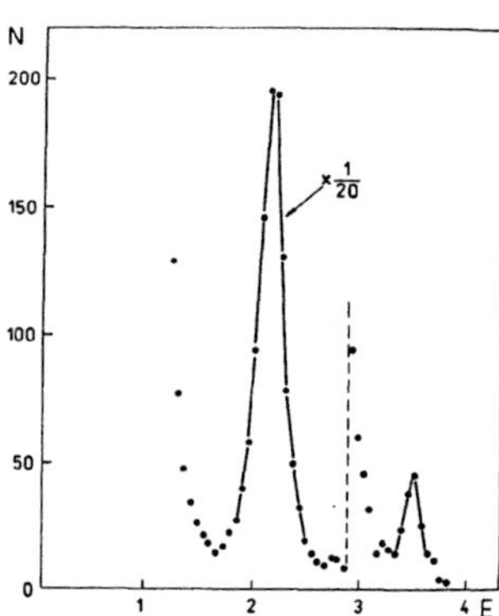

Fig. 1. Spectrum of charged particles. N - number of particles, E - energy of particles in MeV.

For 17 experimental points the value of χ^2 appeared to be equal to 17.5. It was not possible to describe the cross section curve with the help of the p-resonance. In particular, the cross section curve described with a sum of the p-resonance at 145 eV and of contribution $1/v$ from the s-resonance gives $\sigma_{np} \simeq$100 b at E_n= 2 keV which is in contradiction with the upper estimate of $\sigma_{np} \leq$11 b (Ref.[3].

Since the resonance observed determines the thermal cross section, then the ratio of widths of decay through different channels may be found near the thermal neutron energy, where the statistical accuracy of the measurement is much higher. Our results in comparison with other data are summarized in the Table. The upper three lines correspond to the $^{22}Na(n,p)$ reaction, the other three lines -- to the $^{19}F(\alpha,p)$ reaction. The comparison of these data allows one to assume that the resonance at E_n=145 eV most probably corresponds to the 2360 keV resonance. The comparison of ratios Γ_{P0}/Γ_{P1} for these resonances serves also

K. H. Böckhoff (ed.), Nuclear Data for Science and Technology, 150–151

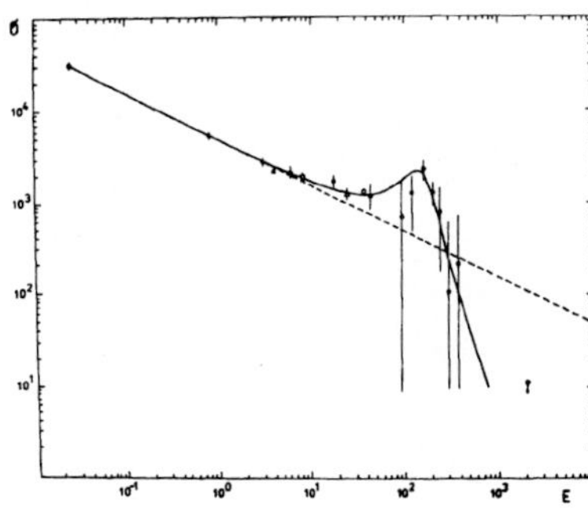

Fig. 2. Cross section of the $^{22}Na(n,p)^{22}Ne$ reaction. σ - in barns, E - neutron energy in MeV.

to the advantage of such conclusion. So, under this assumption and with more accurate values of B_n, one may make more precise the excitation energy scale for the compound nucleus ^{23}Na in the $^{19}F(\alpha,p)$ reaction. However, in comparison of our results with data from Ref.[4] a question arises whether such comparison of excited states is unambiguous. According to Ref.[4] the ra-

Table

E_{exp} keV	E_n keV	E_α keV	Γ_{p0}/Γ_{p1}	$\Gamma_{\alpha 0}/\Gamma_{p1}$	Ref. R
12416.8	0.15±0.01	-	$(6\pm2).10^{-3}$	$\ll 1$	p.p.
- " -	4±4	-	-	$(1\pm1.5).10^{-3}$	/2/
-" -	1	-	$(7.4\pm0.2).10^{-3}$	-	/3/
12407	$(-5)^x$	2354±4	0.31	-	/5/
12412	$(0.15)^x$	2360±4	0.03	$\simeq 1$	-"-
12447	$(35)^x$	2402±2	0.38		-"-

^{x}in brackets the possible energies of neutron resonances are indicated, if our resonance at E_n=147 eV corresponds to E_α=2360 keV resonance in the $^{19}F(\alpha,p)$ reaction.

tio of parameters for the E_α=2360 keV resonance is $(2J+1)\cdot\Gamma_{p1}\cdot\Gamma_{\alpha 0}/\Gamma$ =230 eV, from where it follows that $\Gamma_{\alpha 0} \simeq \Gamma_{p1}$. At the same time as it follows from our data $\Gamma_{\alpha 0} \ll \Gamma_{p1}$ which is in agreement with the estimate $\Gamma_{\alpha 0}/\Gamma_{p1} \lesssim 10^{-3}$ derived in Ref.[2]. It may occur that the E_n=145 eV resonance is poorly excited in the $^{19}F(\alpha,p)$ reaction due to its small α-width, and the resonances from Ref.[5] are outside the investigated excitation energy region.

No definite assignment is made for the spin of the resonance at E_n=147 eV. But from a comparison of experimental ratio $\Gamma_{p0}/\Gamma_{p1} = 6.10^{-3}$ with average proton widths calculated in the frame of the optical model-- $\bar{\Gamma}_{p0}/\bar{\Gamma}_{p1}$ $(J^\pi=5/2^+)$=0.3 and $\bar{\Gamma}_{p0}/\bar{\Gamma}_{p1}$ $(J^\pi=7/2^+)$=0.01-- it is seen that the most probable is the value $J^\pi=7/2^+$ which we used in the fitting of neutron resonance parameters. The extrapolation of the obtained curve for the cross section of the $^{22}Na(n,p)$ reaction to 2 keV gives σ_{np} (E_n=2 keV)=0.7 b which is not in contradiction with the experimental upper estimate $\lesssim 11$ (Ref.[3]).

The measurements of the cross section in the energy range up to 1 keV allowed us to obtain a real estimate of the capture resonance integral $I_p = \int_{0.5}\sigma_{np}\frac{dE}{E} \simeq \int_{0.5}^{1000}\sigma_{np}\,dE/E$ = 2.5×10^4 b. It appeared possible because the next resonance seems to be higher enough (see the Table) and the resonance contribution into the resonance integral decreases rapidly with increasing E_n. The value of I_p must coincide with the total integral of absorption, since for thermal neutrons the total cross section of absorption practically coincides with σ_{np}. But the value of $I=16.10^4$ b given in the Table of Cross Sections in Ref.[7] much exceeds our results. However, the depletion value obtained in Ref.[8] confirmed that the measured effective cross section for ^{22}Na was not significantly influenced by epicadmium neutrons.

References

1. V.A.Kravtsov. Massy atomov i energii sviazi jader. Atomizdat, M., 1974.
2. R.Ehehalt, H.Morinaga, Y.Shida. Z.Naturforsch., 1971, Bd.26a, p.590.
3. J.Kvitek e.a. Z.Physik, 1981, v.A299, p.187.
4. J.Kuperus. Physica, 1965, v.31, p.1603.
5. L.Van der Zwan, K.W.Geiger. Nucl.Phys., 1977, v.A284, p.109.
6. I.M.Frank. Particles and Nucleus, 1972, v.2, part 4, p.807.
7. Neutron Cross Sections. BNL-325, Third.Ed., v.1, 1973.
8. R.D.Werner, D.C.Santry. J.Nucl.Energy, 1972, v.26, p.403.

GAMMA-RAYS FROM CAPTURE OF 400-keV NEUTRONS

N. Yamamuro, M. Igashira, T. Maruyama, K. Hashimoto and H. Kitazawa

Research Laboratory for Nuclear Reactors,
Tokyo Institute of Technology
O-okayama, Meguro-ku,
Tokyo, Japan.

Gamma-ray spectra following neutron capture in Nb, Mo and Sn have been measured at the neutron energy of 400 keV. A pulsed 3-MV Pelletron accelerator provided proton bursts of 1.5-ns width at 2-MHz repetition rate, and neutrons were produced by the ^{7}Li(p,n) reaction. A 7.5 cmϕ × 15 cm NaI(Tl) detector centered in an annular NaI(Tl)-crystal, which was surrounded by a heavy shield consisting of lead, boric acid, and paraffin, was used as a gamma-ray detector. A sample was located at 15 cm from the neutron source, and the distance between the sample and the detector was 110 cm. The axis of the detector made an angle of 125° with respect to the proton-beam direction. The capture gamma-ray spectra were obtained after background subtraction, spectrum unfolding, and correction for the gamma-ray self-absorption in the sample. A comparison between the present and the other experimental data and the calculated spectra based on a statistical model was carried out.

[^{93}Nb(n,γ), Mo(n,γ), Sn(n,γ), capture gamma-ray spectra, En=400 keV, high energy gamma-ray transition, statistical model]

Introduction

Gamma-ray spectra from the neutron capture reaction in the keV-region are one of tools to comprehend the process of gamma-ray transition from excited states to low-lying levels of the compound nucleus. The spectrum frequently shows the enhancement of high energy component or of the discrete lines corresponding to the transitions leading to near ground state, which cannot be described with the statistical processes of the nuclear reaction. Several non-statistical models for the nuclear reaction were proposed to understand the phenomena, and the more accurate spectrum measurement is requested to give the accurate information on the neutron capture. On the other hand, gamma-ray spectra following keV-neutron capture are important nuclear data for shielding design, for heating calculation, and for material damage estimation in the nuclear reactors. Although the keV-neutron capture gamma-ray spectra were compiled[1], more experimental studies on the gamma-ray spectrum are necessary to precisely evaluate the problems on the design of a fast nuclear reactor or of a fusion reactor.

In the present study, gamma-rays from capture of 400-keV neutrons were measured with an anti-Compton NaI spectrometer under the good signal-background ratio and the unfolded experimental spectra were compared with the other ones and with a theoretical calculation.

Experimental Procedures

The pulsed neutrons were produced by the ^{7}Li(p,n) reaction induced by proton burst of 1.5-ns width at 2-MHz repetition rate from the 3-MV Pelletron accelerator. A sample was placed at 15 cm from the neutron source and a 7.5 cmϕ × 15 cm NaI detector was located at about 110 cm from the sample. The sample was a natural metal cylinder with 4.5 cmϕ × 4.5 cm for Nb and Mo and 4.5 cmϕ × 5.0 cm for Sn. The detector was centered in an annular NaI crystal, 25.4 cm in outer-diameter and 28 cm in length. The detector system operated as the anti-Compton gamma-ray spectrometer and was shielded with 10-cm thick lead, 35-cm thick boricacid and paraffin. The detector and the shield were mounted on a goniometer and the detector system was placed at an angle of 125° with respect to the proton beam in the present experiment. The experimental arrangement in the present study is shown in Fig. 1.

The response functions of the anti-Comption NaI detector were determined with the several calibrated gamma-ray sources, ^{24}Na and Am-Be sources, and the monoenergetic gamma-rays from the ^{19}F(p,αγ), ^{9}Be(p,γ) and ^{27}Al(p,γ) reactions. Since the 80% of Compton scattering events, 90% of single escape peak and almost all double escape peak can be removed by the anti-coincidence detection with the annular NaI scintillator, the response functions shown in Fig. 2 were obtained. A response matrix, which is necessary for unfolding the observed pulse height specturm, was composed by interpolating the resopnse functions determined experimentally.

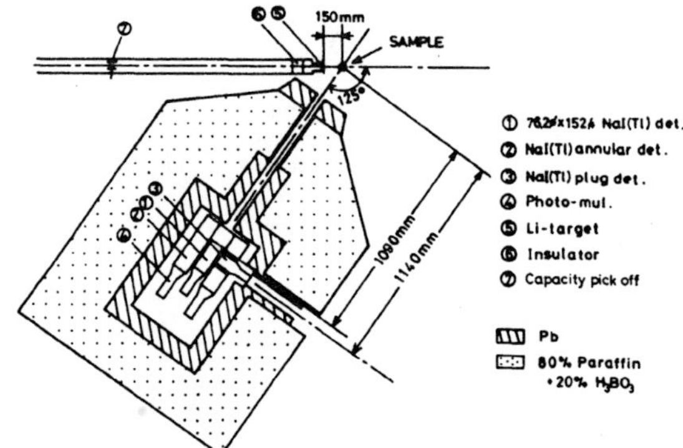

Fig. 1. Experimental arrangement.

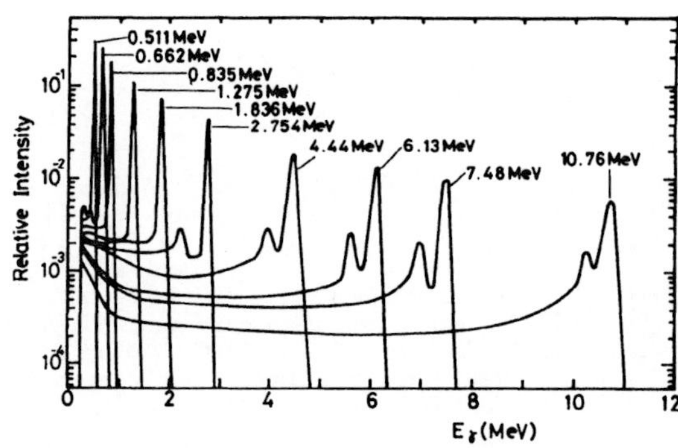

Fig. 2. Experimental response functions of the NaI detector.

The signals of the central NaI detector were fed into a linear amplifier and were recorded in a CAMAC controled data processing system[2] through a linear gate and stretcher and an ADC. The time-of-flight (TOF) spectrum of the signals from the detector was measured using the reference signal from a beam pickoff cylinder. Fig. 3 shows the TOF spectrum for the Nb(n,γ) reaction. The time windows were placed at A and C in Fig. 3, and the pulse height spectra corresponding to the time independent background and to the gamma-rays

K. H. Böckhoff (ed.), Nuclear Data for Science and Technology, 152–155.

from the Nb(n,γ) reaction were obtaind for both windows respectively, as shown in Fig. 4. In the spectrum for the constant background, we can see the contributions from the $^{127}I(n,\gamma)$ reaction, but their amounts were small except the 6.8 MeV peak. Similar TOF spectra were obtained for Mo and Sn. The neutron energy used in the experiment was measured by the TOF method using a thin ^{6}Li glass scintillation detector, resulting in 420 ± 20 keV for Nb and Mo and 400 ± 20 keV for Sn.

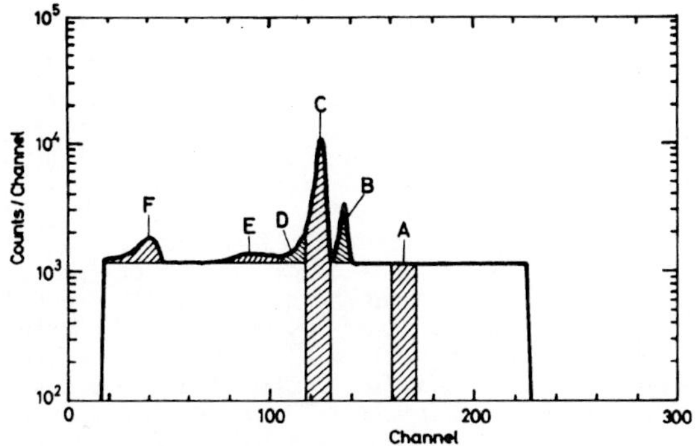

Fig. 3. TOF spectrum. Notations in the figure indicate the following; A: constant background, B: gamma-ray scattered with sample, C: capture gamma-ray, D and E: capture gamma-ray due to scattered neutron, F: neutron scattered with sample.

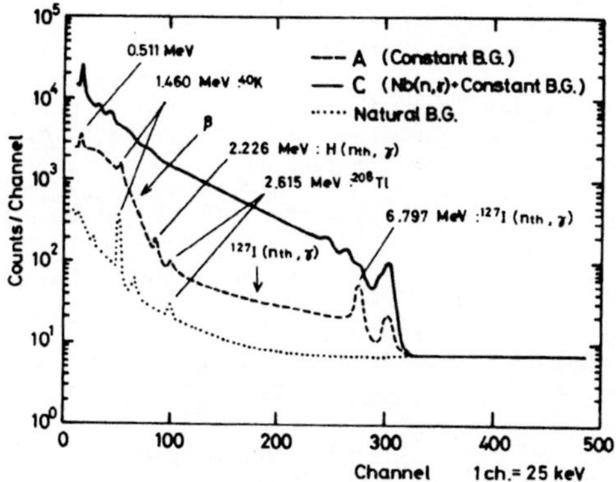

Fig. 4. Pulse height spectrum.

Experimental Results

The pulse height spectrum from the (n,γ) reaction was obtained by subtracting the time independent background shown by the dashed line in Fig. 4 from the spectrum shown by the solid line in the same figure. To determine the gamma-ray energy distribution from the (n,γ) reaction, two data processes were taken. Firstly, the spectrum was calculated with the FERDOR code[3], which unfolded the observed pulse height spectrum by using the response matrix of the detector determined previously. Secondly, the spectrum was corrected for the self-absorption of gamma-rays in the sample with the same procedure as the first process using the response spectrum that shows the emitted gamma-ray spectrum from the sample when the monoenergetic gamma-rays were homogeneously produced over the whole volume of sample.

The capture gamma-ray spectra for Nb, Mo and Sn at the neutron energy of 400 or 420 keV are shown in Figs.5 through 8. In the present study, the thick samples were used and the neutron flux incident upon the sample was not measured, so that the spectra are shown with the relative intensity. The errors of the intensity was estimated to be 7 ~ 8 % below 4 MeV and ~ 10 % between 4 and 6 MeV. The comparisons with the data of Barrett et al.[4] for 430 keV neutrons are illustrated in Figs. 5 through 7. While the experimental arrangement, the detector resolution and the method of data reduction are different from each other, the intensity distributions show the general consistency between both data. Above 5 MeV, we can find structures or discrete lines due to the transitions from the capturing states to the low-lying levels, although we notice the some energy shift between both spectra for Mo(n,γ) reaction. There are a few discrete lines below 2 MeV which are probably gamma-rays from the

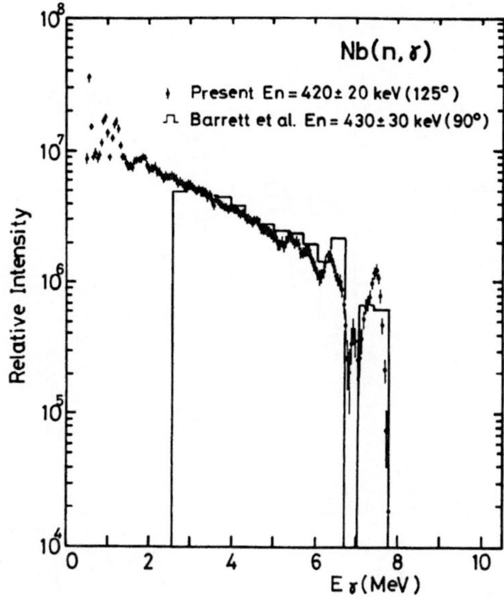

Fig. 5. Gamma-ray spectra from the Nb(n,γ) reaction.

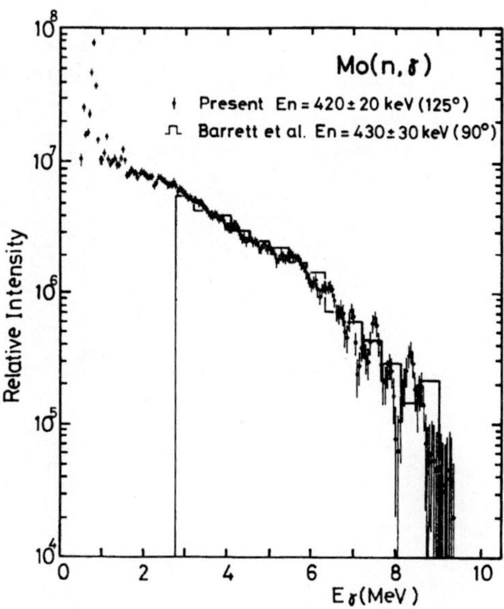

Fig. 6. Gamma-ray spectra from the Mo(n,γ) reaction.

transitions between the lower excited levels and they are not shown in Barrett et al's data. Morgan and Newman[5] have measured the gamma-ray spectra from Mo(n,xγ) reaction in the study of gamma-ray production cross sections at ORELA. The comparison can be made between our data and ORELA data for neutron energy of 0.2 ~ 0.6 MeV and shows the excellent agreement over the whole gamma-ray energy region, as shown in Fig. 8, and the present data reveal the small statistical uncertainty. From the facts described above, it would indicate that the confidence of the present spectra is placed in the relative intensity distribution, except above 8 MeV in the Sn(n,γ) spectrum where the statistics of experimental data was poor.

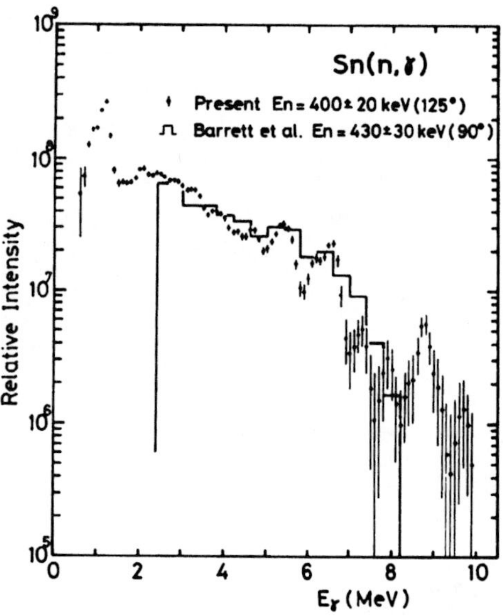

Fig. 7. Gamma-ray spectra from the Sn(n,γ) reaction.

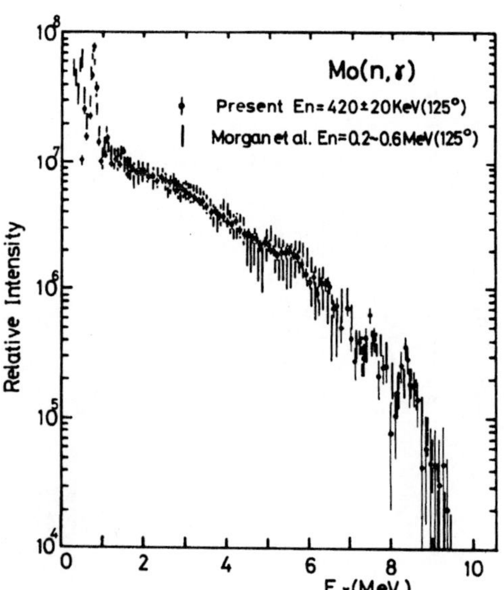

Fig. 8. Gamma-ray spectra from the Mo(n,γ) reaction. To compare obviously, data are plotted at the suitable interval of points.

Theoretical Calculations

A statistical model calculation was carried out to explore the reaction processes for keV-neutron capture in nuclei of the mass region A=90 ~ 120. The theoretical models adopted are the spherical optical model and the modified Hauser-Feshbach method, where the sum of the partial cross sections calculated with the statistical model or with the other nuclear model keeps to be equal to the total cross section obtained from the optical model calculation[6]. The energy spectrum of the emitted gamma-rays can be calculated by the population probability of states and the branching ratio of each state to the low-lying levels. Above the neutron separation energy of the compound nucleus, the branch to the states of the target nucleus opened by the neutron scattering is taken into account. The gamma-ray spectra and cross sections for the ^{93}Nb(n,γ) reaction were calculated using the CASTHY computer code devised with the modified method of neutron cross section calculations described above[6]. The optical potential parameters of Igarashi et al.[7] and the level density parameters of Reffo et al.[8] were used. The Brink-Axel type profile function with the giant dipole resonance parameters refered from the table of Berman[9] is adopted to calculate the gamma-ray transmission coefficient. In the calculation of gamma-ray energy distribution the scheme of low-lying levels and the density of continuum levels of the compound nucleus are important. However, the number of low-lying levels that can be read in the present code is limited to 30. In this calculation, levels above 1.35 MeV are assumed to be overlapping and 30 levels from ground to 1.33 MeV are selected as the discrete levels of ^{94}Nb. This assumption will introduce some uncertainty into the calculated spectrum of high energy region.

The calculated result for ^{93}Nb(n,γ) reaction gives, as shown in Fig. 9, the excellent agreement with the experimental one in the continuum region from 2 to 5 MeV and the contribution from the direct transition to the discrete levels at 6.4 MeV and 7.5 MeV, where the experimental spectrum also shows the peaks. The disagreement is found in the region from 5 to 6 MeV, where the gamma-rays due to the transition to the

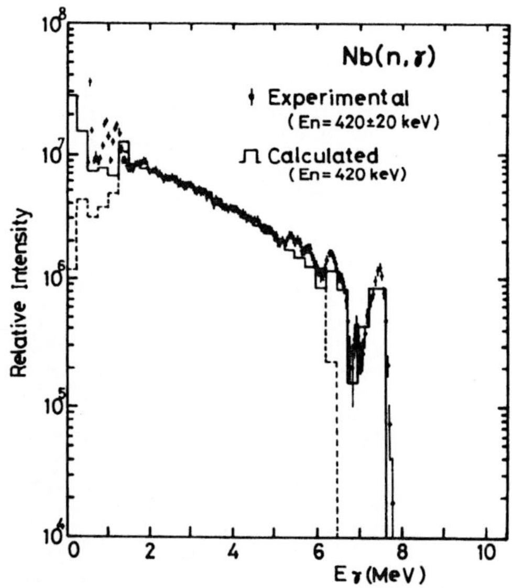

Fig. 9. Comparison between experimental and theoretical spectra for the Nb(n,γ) reaction. Dashed lines show the spectrum due to transitions in the continuum region.

discrete levels do not exist in the calculated spectrum because of the no precise data of discrete levels above 1.35 MeV and from the limit in the present code as mentioned above. Below 1.5 MeV, the calculation cannot represent the peaks, probably due to the gamma-ray transitions between the discrete levels.

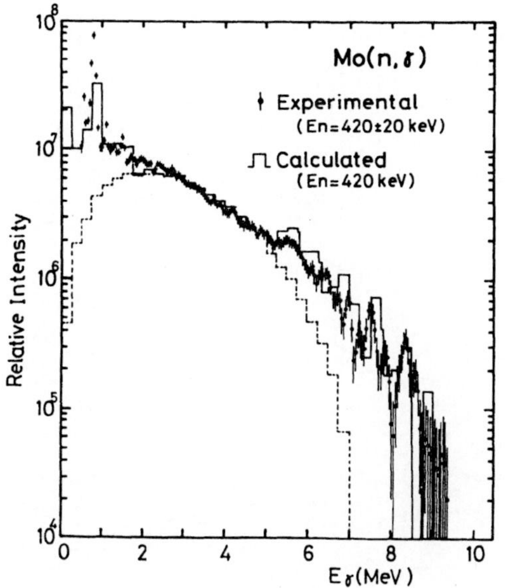

Fig. 10. Comparison between experimental and theoretical spectra for the Mo(n,γ) reaction. Dashed lines show the spectrum due to transitions in the continuum region.

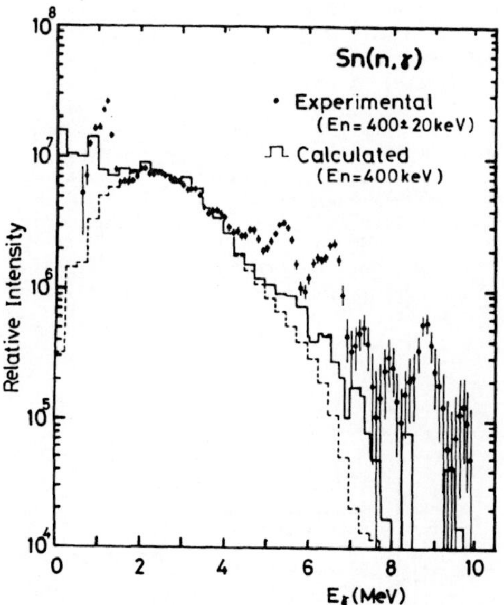

Fig. 11. Comparison between experimental and theoretical spectra for the Sn(n,γ) reaction. Dashed lines show the spectrum due to transitions in the continuum region.

The calculations of gamma-ray spectra from Mo(n,γ) and Sn(n,γ) reactions with the CASTHY code were performed and the results are shown in Figs. 10 and 11 respectively, for the comparison with the experimental spectra. In these calculations, the optical potential parameters of Igarashi et al.[7], the Brink-Axel type profile function and the level density parameters of Gilbert-Cameron[10] for the stable isotopes of natural element were used. The gamma-ray spectra for each isotope were summed up with the weight, which was proportional to the product of the number of photon per neutron absorption, the compound formation cross section and the abundance ratio of isotope. The fair agreement between experimental and calculated spectrum is found over the whole gamma-ray energy region for Mo(n,γ) spectrum. While for Sn(n,γ) spetrum, the agreement with the experimental one was not good except from 1.5 to 4 MeV. There are the enhanced transitions above 4 MeV in the experimental spectrum.

Conclusion

Gamma-ray spectra following neutron capture in Nb, Mo and Sn have been measured at the neutron energy of 400 KeV by the time of flight method and the anti-Compton spectroscopy. The observed pulse height spectra were unfolded by the responses of the detector and the correction of the gamma-ray self-absorption in the sample was made to obtain the energy distribution of capture gamma-ray. The reliability of present result is proved from the fact that the good agreement exists with the other experimental spectrum, especially with Morgan and Newman's result of Mo(n,γ) spectrum. The theoretical calculation based on the statistical model was carried out for the comparion with the experimental spectrum. The good agreement is obtained for ^{93}Nb(n,γ) and Mo(n,γ) except the lower energy region and some resonances at higher energy region. However, the disagreement exists above 4 MeV in the Sn(n,γ) spectrum. To make clear the cause of the disagreement, the more detailed examinations of the statistical model calculation and trials of non-statistical model calculation seem to be necessary.

References

1. J.R. Bird, B.J. Allen, I. Bergqvist and J.A. Biggerstaff, Nucl. Data Tables, 11, 433 (1973).
2. T. Emoto and N. Yamamuro, Bull. Research Lab. Nucl. Reactors, 4, 1 (1979).
3. H. Kendrick and S.M. Sperling, An Introduction to the Principles and Use of the FERDOR Unfolding Code, GA-9882 (1970).
4. R.F. Barrett, K.H. Bray, B.J. Allen and M.J. Kenny, Nucl. Phys. A278, 204 (1977).
5. G.L. Morgan and E. Newman, ORNL-TM-5097 (1975).
6. S. Igarashi, J. Nucl. Sci. Technol. 12, 67 (1975).
7. S. Igarashi et al., Evaluation of Fission Product Nuclear Data for Fast Reactor, JAERI-M 5752 (1974).
8. G. Reffo, F. Fabbri, K. Wisshak and F. Käppeler, Nucl. Sci. Eng. 80, 630 (1982).
9. B.L. Berman, Atlas of Photoneutron Cross Sections obtained with Monoenergetic Photons, UCRL-78482 (1976).
10. A. Gilbert and A.G.W. Cameron, Can. J. Phys. 43, 1446 (1965).

MEASUREMENT OF (n,α) CROSS-SECTIONS FOR Cr, Fe AND Ni AT 14 MeV NEUTRON ENERGY

E. Wattecamps, H. Liskien and F. Arnotte

Commission of the European Communities, Joint Research Centre
Central Bureau for Nuclear Measurements, B-2440, Geel, Belgium

Helium production cross-sections for the main constituents of stainless steel (Cr, Fe, Ni) have recently been published for neutron energies between 5 and 10 MeV. The α-particles were detected with a multi-angle telescope and cross-section data relative to the well known n-p scattering cross-section were deduced. Those measurements have been performed now also at 14 MeV. At this energy the background condition had to be improved by changing the neutron collimation, by reducing the sensitive volume of the ΔE-proportional counters, and by replacing remaining low-Z material inside the chamber by tantalum. Listing mode data acquisition is used and α-particle identification is performed by transforming the observed (ΔE,E) signal into a (MZ^2,E) signal. The measurements yield the angle-differential cross-sections for five fixed angles. Angle-integrated cross-sections are compared with the few results available in literature.

[measurement, (n,α), cross-sections, Cr, Fe, Ni, 14 MeV, neutrons]

Introduction

Cross-section data for (n,α) reactions on structural materials of fission and fusion reactors are requested. WRENDA [1] contains 25 requests for the main constituents of stainless steel Cr, Fe and Ni. The majority of the requested data (priority one or two) are for the helium production cross-sections. These are used in calculations of the helium accumulation to estimate radiation damage effects. Accuracies of 10 to 20% are demanded and the energy range of interest is from threshold to 15 MeV.

Most known (n,α) cross-section data are measured by activation techniques, but of the 13 isotopes involved in Cr, Fe and Ni only six lead to measurable radioactive isotopes. Therefore, (n,α)-cross-sections for elemental Cr, Fe and Ni are rather scarce. Farrar et al.[2] and Kneff et al.[3] reported results obtained by post-irradiation high sensitivity gas mass spectrometry to determine the total amount of helium in a sample. However, due to the necessary high-intense neutron source, this method is at present restricted to 14 MeV. Determination of 14 MeV cross-sections by prompt α-particle detection was performed by Dolya et al.[4-5] and by Grimes et al.[6]. The latter experiment uses a magnetic quadrupole spectrometer. Recently, in this laboratory Paulsen et al. measured cross-sections from 5 to 10 MeV with a multitelescope device [7-8]. To extend these measurements and to compare (n,α)measurements made by different groups with very different techniques it was decided to use the device of Paulsen et al. for measurements also at 14 MeV. In principle the technique delivers the double differential helium generation cross-section relative to the well-known differential n-p scattering cross-section [9].

Experimental set-up and measurements

The multi-telescope system, already used for experiments with neutrons of 5 to 10 MeV energy, was described earlier [7-8]. The modified set-up around the target is illustrated in fig. 1, the multi-telescope-chamber is drawn in fig. 2 and an outline of the electronic system with the ND 6660 data acquisition and analysis system is given in fig. 3.

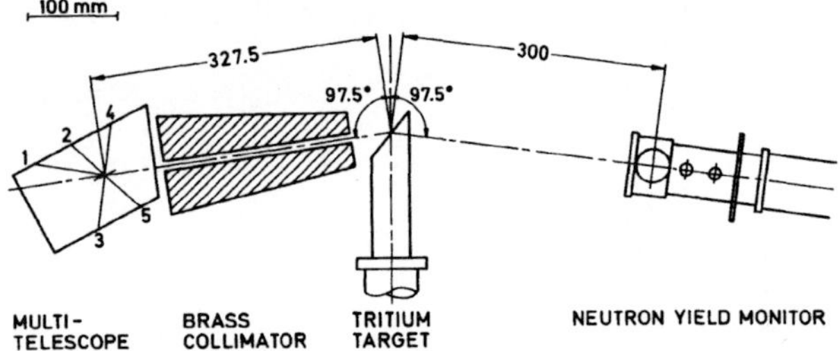

Fig. 1. The experimental set-up around the target.

The D_2 gas target was replaced by a solid TiT target of 3 mg/cm^2 on a Cu-backing, tilted by 45° relative to the deuteron beam. The deuterons had an energy of 1 MeV and were completely stopped in this thick TiT layer. The multi-telescope was set under 97°. This choice reduces the neutron energy spread to $\pm$ 0.1 MeV. Running the CN-type Van de Graaff accelerator at a d.c. current of 20 μA yields $3\cdot10^8$ n/s/sr of 14.1 MeV. This rather low yield is to be compared with $7.5\cdot10^8$ n/s/sr of 8 MeV at 0°from a 5 μA deuteron beam on a deuterium gas target as used in previous experiments[7-8]. A brass collimator of 21.5 cm length is placed between the target and the chamber to shield the surface barrier detectors against the direct neutron beam. The neutron yield monitor is a proton-recoil telescope at approximately 97° on the opposite side of the target.

The multi-telescope (see fig. 2) comprises five telescopes. Each telescope has two multi-wire proportional counters in front of a total energy solid state detector. All three are connected to a triple coincidence system with 250 ns time width. The mean angle for trajectories from the central sample, a disc of 25 mm diameter, to the disc shaped solid state detector of 24 mm diameter is 14, 51, 79, 109 and 141°. Four samples on a sample-changer can be slided sequentially to the center of the chamber :

- the sample under investigation,Cr, Fe or Ni of approximately 3 mg/cm^2 ;
- a polyethylene foil for n-p scattering rate measurements under identical conditions;
- a Ta sample for background determination;
- a ^{241}Am source for pulse height and solid-angle calibration.

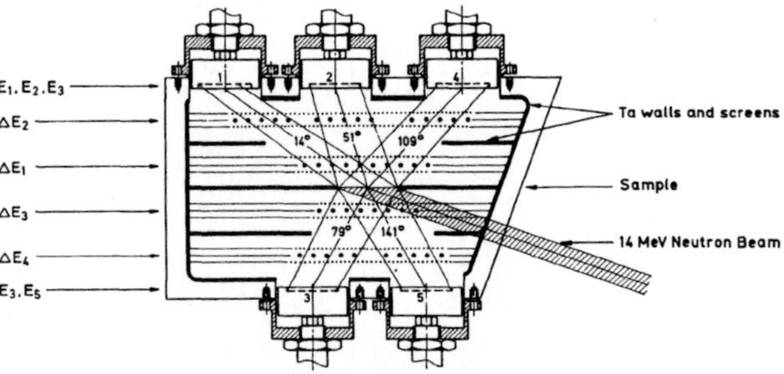

Fig. 2. The multi-telescope chamber.

At first the multi-telescope used for the 5 - 10 MeV range [8] was also tried at 14 MeV, but the larger background level had to be investigated and reduced by a sequence of actions. Tantalum apertures were inserted between the proportional counters, and the frames carrying the multi-wire electrodes were also made from tantalum. In addition the outlining of all inner surfaces of the chamber consists now entirely out of 0.5 mm tantalum sheets. The sensitive volume of the proportional counters was drastically reduced.

K. H. Böckhoff (ed.), Nuclear Data for Science and Technology, 156–158.

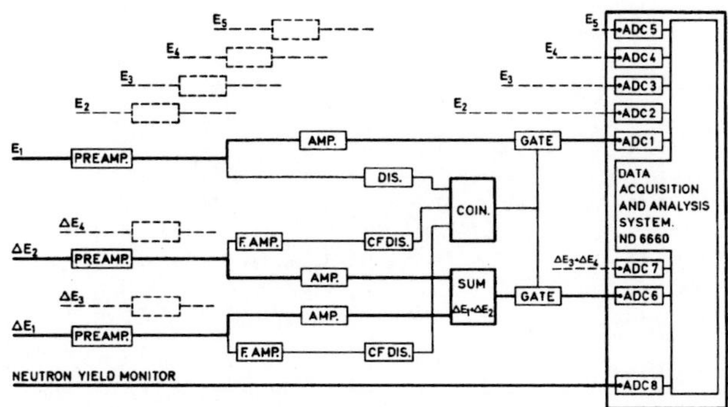

Fig. 3. Outline of the electronic- and data acquisition system.

Their area is now confined to the geometrical minimum, while their thickness was brought down to 6 mm. This and the use of 10 μm diameter counting wires (gold-plated tungsten) resulted in faster rise times. The time spread between coincident proportional counter events from α-particles of ^{241}Am is 30 ns FWHM. However, a reduction of the time resolution of the triple coincidence from 250 ns to 125 ns did not reduce the background accordingly and the remaining background is therefore attributed to α-particles from the surface barrier detectors crossing the telescope in the opposite direction. As can be seen in fig. 4a column one and two, the background below 4 MeV α-particle energy is many times the size of the net signal and this limits the applications of the system.

As may be seen from fig. 2 and 3, the pair of proportional counters ΔE_1, ΔE_2 is common to E_1, E_2 and E_4; whereas the pair ΔE_3, ΔE_4 is common to E_3 and E_5. The pulse heights from proportional counters belonging to the same pair are added. This sum-signal and a signal from a relevant solid state detector, if coincident within 250 ns, form a valid event to be converted and stored. Data acquisition is performed with a ND 6660 system in listing mode on tape. In addition the events are stored by increment mode in active memory (5 x 64 x 64 channels) with means for display of any two-dimensional spectrum. ADC 8 is dedicated to simultaneous fluence monitoring. This procedure combines the advantage of critical data appraisal already during acquisition, but leaves flexibility for various analyses after completion of acquisition.

The presented cross-section data are the results of foreground runs for Cr, Fe and Ni, which lasted 14, 19, and 14h, respectively. The poor signal-to-background ratio made it necessary that about the same times had to be devoted to background determination. The runs with the polyethylene reference sample were typically not longer than 0.5 h. The counts of the neutron yield monitor was used to normalize fore- and background events of X(n,α) and H(n,p) measurements to the same fluence of 14 MeV neutrons.

Data Analysis and Results

The α-particles have to be discriminated against other charged particles. To reduce the arbitrariness of this separation the measured spectra in ΔE,E variables are transformed [10] off-line in spectra versus the variables $M \cdot Z^2$,E. These spectra are reduced to one-dimensional energy spectra (64 channels) by integration between energy independent $M \cdot Z^2$ limits. The energy channels are furthermore collapsed to 8 groups of 8 channels each to reduce statistical fluctuations. As an example spectra obtained for Cr, Fe and Ni with detector 3

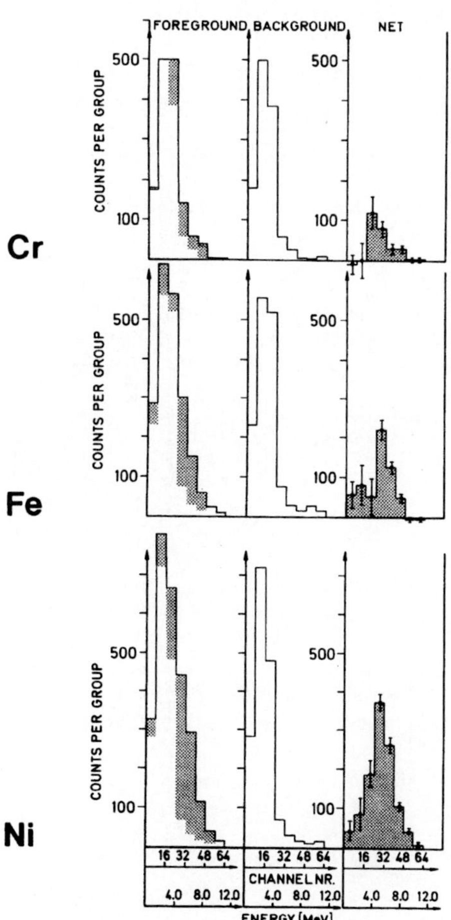

Fig. 4 a.
Foreground-, background- and net α-particle spectra of detector 3.

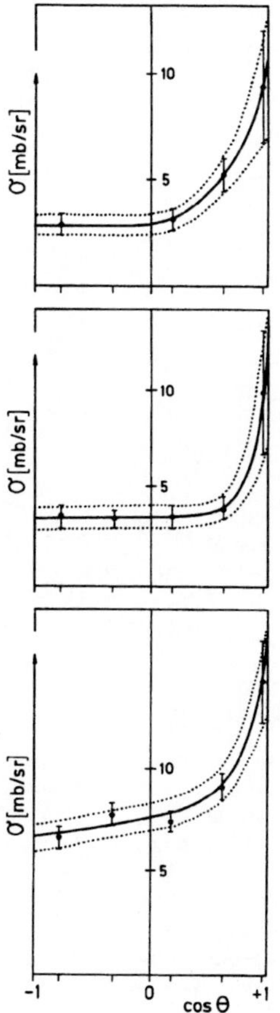

Fig. 4 b
Angular distribution of α-particle production.

REFERENCE	E [MeV]	σ [mb]
DOLYA et al. [4]	14.7	44 ± 4
GRIMES et al. [6]	15	38 ± 6
KNEFF et al. [3]	14.8	34 ± 4
ENDF/B-V [11]	14.0	65.9
KEDAK [12]	14.0	52
THIS WORK	14.1	46 ± 9

REFERENCE	E [MeV]	σ [mb]
DOLYA et al. [5]	14.7	54 ± 5
GRIMES et al. [6]	15	43 ± 7
KNEFF et al. [3]	14.8	48 ± 3
ENDF/B-V [11]	14.0	41.7
KEDAK [12]	14.0	82
THIS WORK	14.1	45 ± 9

REFERENCE	E [MeV]	σ [mb]
DOLYA et al. [5]	14.7	129 ± 12
GRIMES et al. [6]	15	97 ± 16
KNEFF et al. [3]	14.8	100 ± 7
ENDF/B-V [11]	14.0	146
KEDAK [12]	14.0	95
THIS WORK	14.1	95 ± 9

Fig. 4 c
Angle-integrated (n,α) cross-section data.

Table 1. Table of constants and uncertainties

$\left[\frac{d\sigma(\Theta_1)}{d\Omega}\right]_H$ = 207.1 mb/sr ± 1.0 %				
N_H = 111.3 ± 0.4 %	N_{Cr} = 3.069 ± 0.5 %	N_{Fe} = 3.058 ± 0.7 %	N_{Ni} = 3.212 ± 1 % ★
	$C_{1,H,Cr}$ ± 1.0 %	$C_{1,H,Fe}$ ± 1.0 %	$C_{1,H,Ni}$ ± 1.0 %
	$M_{H,Cr}$ ± 0.6 %	$M_{H,Fe}$ ± 0.6 %	$M_{H,N}$ ± 0.6 %
	M_{Cr} ± 0.1 %	M_{Fe} ± 0.1	M_{Ni} ± 0.1 %
	$C_{1,Cr}$ ± 30 %	$C_{1,Fe}$ ± 33 %	$C_{1,Ni}$ ± 15 %
Ω_1/Ω_2 = 0.125 ± 1.3 %	$C_{2,Cr}$ ± 15 %	$C_{2,Fe}$ ± 11 %	$C_{2,Ni}$ ± 7 %
Ω_1/Ω_3 = 0.215 ± 1.3 %	$C_{3,Cr}$ ± 20 %	$C_{3,Fe}$ ± 10 %	$C_{3,Ni}$ ± 6 %
Ω_1/Ω_4 = 0.309 ± 1.3 %	$C_{4,Cr}$ —	$C_{4,Fe}$ ± 15 %	$C_{4,Ni}$ ± 7 %
Ω_1/Ω_5 = 0.236 ± 1.3 %	$C_{5,Cr}$ ± 21 %	$C_{5,Fe}$ ± 17 %	$C_{5,Ni}$ ± 9 %

★number of atoms per cm^2 to multiply by 10^{19}

under 79° are given in fig. 4a. The pulse height scale is known from the previous calibration with the 5.5 MeV α-particles of ^{241}Am taking into account the energy loss in the counting gas. The dotted area is the difference between foreground and background. Similar spectra are available for all angles.

Net counts are obtained by simple summing over the energy groups. The uncertainties are mainly determined by the poor statistics and the unfavourable foreground-to-background ratio, especially for the low energy groups. For the conversion from the measured net counts $C_{i,X}$ to angular-differential cross-sections the following equation is used :

$$\left[\frac{d\sigma(\Theta_i)}{d\Omega}\right]_X = \left[\frac{d\sigma(\Theta_1)}{d\Omega}\right]_H \cdot \frac{N_H}{N_X} \cdot \frac{C_{i,X}}{C_{1,H,X}} \cdot \frac{\Omega_1}{\Omega_i} \cdot \frac{M_{H,X}}{M_X}$$

with :

$\left[\frac{d\sigma(\Theta_i)}{d\Omega}\right]_X$	angular-differential cross-section of element X, at angle Θ_i, in mb/sr
$\left[\frac{d\sigma(\Theta_1)}{d\Omega}\right]_H$	angular-differential cross-section for elastic scattering on hydrogen, at angle Θ_1, in mb/sr
N_H / N_X	ratio of the number of hydrogen atoms to the number of atoms of element X on the irradiated samples
$C_{i,X}$ / $C_{1,H,X}$	ratio of the counts of X(n,α) and H(n,p) reactions observed at angle Θ_i and Θ_1 respectively
Ω_1 / Ω_i	ratio of solid angles
$M_{H,X}$ / M_X	ratio of fluence monitor counts of foreground runs of H(n,p) and X(n,α) runs

The obtained angular-differential cross-sections are given graphically in fig. 4b. The uncertainties shown do not include components which are common to all differential cross-sections of a given reaction (see Table 1).

The smooth eye guide curves drawn through the data of fig. 4b were used to integrate the differential cross-section over cos Θ. The resulting (n,α)cross-sections are listed in fig. 4c. The uncertainty of these values is the quadratic sum of those components which are common to a given reaction, and a component which results from the pair of dotted curves as shown in fig. 4b. Our results agree well within the uncertainty margins with measured data of G.P. Dolya et al. [4-5], S.M. Grimes [6], and D.W. Kneff [3]. The large recommended values of ENDF/B-V for Cr and Ni are surprising, whereas the agreement in the case of Fe is satisfactory. The recommended values of KEDAK [12] agree with our data in the cases of Cr and Ni, but the iron value seems very large.

Measurements of helium generation cross-sections are feasible at 14 MeV with the multi-telescope, though low neutron yield and relatively large background imposed some adaptations. Results are consistent with data obtained by other techniques.

Many stimulating and clarifying discussions with A. Paulsen are gratefully aknowledged.

References

1. WRENDA 81/82, edited by N. Dayday, INDC(SEC)-78/ URSF, July 1981.
2. H. Farrar IV and D.W. Kneff, Trans. Am. Nucl. Soc. 28, 197 (1978).
3. D.W. Kneff, B.M. Oliver, M.M. Nakata, H. Farrar IV, BNL-NCS-27800 or INDC(USA)-83 U, May 1980,p. 181.
4. G.P. Dolya, V.P. Bozhko, V.Ya. Golovnya, A.P. Klyucharev, A.T. Tutubalin, Proc. 2nd All-Union Conf. on Neutron Physics, Kiev, 28th May-1st June 1973, part 3, p. 131.
5. G.P. Dolya, A.P. Klyucharev, V.P. Bozhko, V.Ya. Golovnya, A.J. Kachan, A.I. Tutubalin, Neitronnaya Fizika, Part 4, p. 173-179. Proc. 3rd All-Union Conf. on Neutron Physics, Kiev, 9th-13th June 1975.
6. S.M. Grimes, R.C. Haight, K.R. Alvar, H.H. Barschall, R.R. Borchers, Phys.Rev. C 19, Nr. 6,p. 2127,June 1979.
7. A. Paulsen, H. Liskien, F. Arnotte, R.Widera. Proc. Int. Conf. on Nuclear Cross Sections for Technology p. 844, Knoxville, Oct. 22-26, 1979.
8. A. Paulsen, H. Liskien, F. Arnotte and R. Widera, Nuclear Science and Engineering, 78, 377-385 (1981).
9. J.C. Hopkins and G. Breit, Nucl. Data Tables, A9, 137 (1971).
10. H. Liskien and A. Paulsen, Ann. of Nucl. Energy Vol. 8, p. 423-429 (1981).
11. ENDF/V Library - Gas Production File,
Cr. Evaluation by A.Prince and T. Burrow, BNL, Dec. 1977.
Fe. Evaluation by C.Y. Fu, ORNL, Nov. 1979.
Ni. Evaluation by M. Divadeenam, BNL, Mar. 1977.
12. B. Goel, KEDAK 3, KFK 2386/II, Mar. 1977.

FAST NEUTRON INTERACTION WITH CHROMIUM-52 AND MOLYBDENUM-92

I.A.Korzh, V.A.Mishchenko, M.V.Pasechnik, N.M.Pravdivy

Institute for Nuclear Research of the Ukrainian Academy of Sciences, prosp. Nauki 119, 252028 Kiev-28, USSR

Experimental differential cross sections of elastic and inelastic neutron scattering by chromium-52 and molybdenum-92 nuclei in the energy range 1.5-7.0 MeV are given. Energy dependences of the total cross sections and integrated cross sections of the elastic scattering and inelastic scattering accompanied by the excitation of the lowest two levels are presented. The experimental data are analysed using the spherical optical, statistical, and coupled-channels models. An essential change of the scattering mechanism is shown to occur in the investigated energy range.

[^{52}Cr(n,n), (n,n'), ^{92}Mo(n,n), (n,n'), E_n=1.5-7.0 MeV, TOF, scattering cross sections, spherical optical, statistical, and coupled-channels model analysis]

Introduction

Extensive applications of chromium and molybdenum in modern and, very likely, future fission and fussion nuclear installations determine the technological value of the systematic and sufficiently accurate data on the neutron interaction cross sections for these elements and their isotopes. Moreover, molybdenum isotopes are among the most abundant fission fragments. From the view point of physics it is of interest to study the dynamics of changes in the neutron scattering mechanism in the MeV range for the even-even medium weight nuclei.

For the neutron energies up to 4 MeV are available relatively numerous data on the elastic scattering cross sections [1-13] and inelastic scattering cross sections [11-19] with excitation of chromium-52 discrete levels. However, the discrepancy between these data are so significant, that they exceed the experimental errors (especially below 3 MeV). Under such conditions, new experimental results can considerably modify the evaluated data. An example is the appreciable change of the evaluated neutron inelastic scattering cross sections for chromium isotopes in the fifth version [20] of the ENDF/B system as compared to the forth version [21].

The information about neutron scattering cross sections in the energy range above 4 MeV is much more scarce due to experimental difficulties in obtaining such data. For this energy range there are only the data on neutron elastic scattering cross sections in the papers [10,22,23] and the data on inelastic scattering cross sections for chromium-52 in the papers [18,22,23], but these were obtained only at several points of the energy range.

Even fewer are the data on the neutron elastic [24-27] and inelastic [26] scattering cross sections for molybdenum-92.

Therefore new measurements of the neutron elastic and inelastic scattering cross sections for these nuclei would be useful.

We have measured differential cross sections of the neutron elastic scattering and inelastic scattering with excitation of the lowest two levels for chromium-52 and molybdenum-92 at the energies 1.5, 2.0, 2.5, 3.0, 5.0, 6.0, and 7.0 MeV. The data on the chromium-52 cross sections have been published occasionally [11-14,23]; we give a systematic presentation here. The presented data on the molybdenum-92 cross sections are published for the first time. The experimental data are analysed using the spherical optical, statistical, and coupled-channels models. To make the analysis complete, we have used also the energy dependences of the total cross sections and the integrated elastic and inelastic scattering cross sections for the nuclei under consideration in the energy range 0.5-9.0 MeV.

Experimental technique

Neutron elastic and inelastic scattering cross sections were measured by means of a high resolution time-of-flight fast neutron spectrometer with the use of a pulsed accelerator EG-5. Neutrons from the reactions T(p,n)^{3}He and D(d,n)^{3}He on the solid Ti-T and Ti-D targets with the energy spreads ±(90-140) keV were scattered from the cylindrical scatterers, which have been located at the distance 10 cm from the target and consisted of highly isotope enriched powders: 99.5 % of chromium-52 (mass 41 g) and 92.2 % of molybdenum-92 (mass 55 g), respectively.

The scattered neutrons were registered at the flight paths up to 3 m and for 9-15 angles in the range 20-150° with a scintillation detector, equipped with n- -discrimination and mounted in a massive shield collimator supplied with an additive wedge shield.

To normalize the inelastic scattering cross sections we measured the neutron scattering by hydrogen contained in a small diameter polythene sample of the mass 1.3 g.

The neutron flux monitoring was realised with a long counter, time-of-flight detector, oriented at the angle 30° relative to the primary particle beam direction, and current integrator.

A detailed description of the neutron spectrometer and the measurement technique is given in the papers [28,29].

Experimental results

Neutron elastic and inelastic scattering differential cross sections measured are presented in Figs. 1 and 2. These are corrected analytically for the neutron flux attenuation in the sample, anisotropy of the neutron yield from the targets, experimental angular resolution, and multiple neutron scattering in the sample. The errors given in the Figs. are total and include the experimental, normalization, and correction errors.

K. H. Böckhoff (ed.), Nuclear Data for Science and Technology, 159-162.

For the energies of the experiment, the most differential cross sections were obtained for the first time.

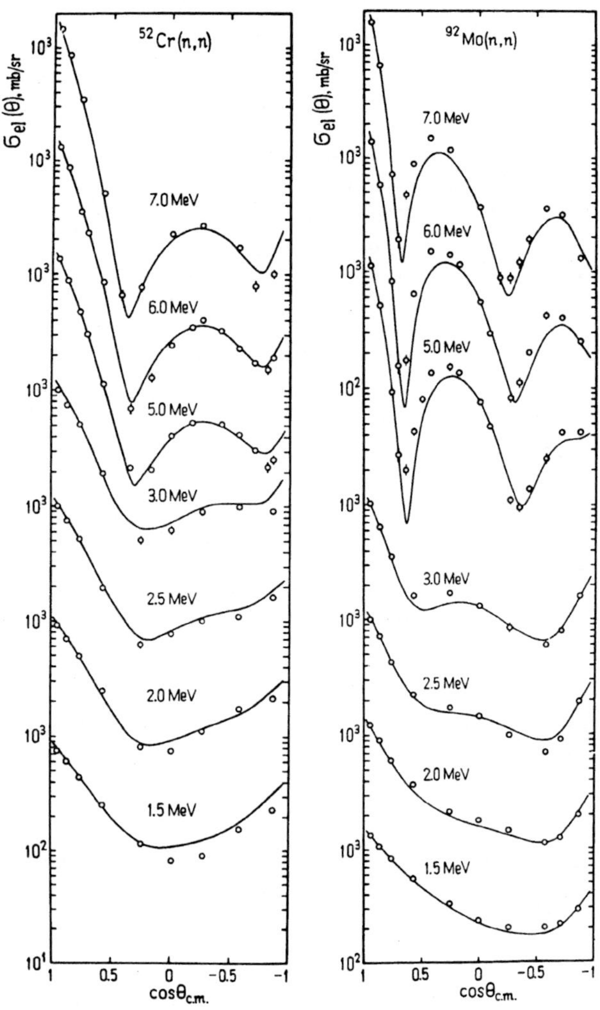

Fig.1. Differential cross sections of the neutron elastic scattering by chromium-52 and molybdenum-92 for the indicated energies: experiment (points) and the calculations using the spherical optical model and statistical model taking into account the level-widths fluctuations (curves).

For the comparison and analysis, Figs. 3 and 4 show energy dependences of the total cross sections and the integrated elastic and inelastic scattering cross sections, including our data and the data published by other authors. Our data for both isotopes under investigation are in good agreement with the most data of the other authors.

Theoretical analysis

The experimental data were analysed using the spherical optical, statistical, and coupled-channels models[34].

The shape elastic scattering differential cross sections and the transmission coefficients to be applied for the calculation of compound cross sections were obtained using the optical model with the spherical potential of the following form [35]:

$$V(r)=-V_c f(r)-iW_c g(r)+V_{so}\left(\frac{\hbar}{m_\pi c}\right)^2 \frac{1}{r}\frac{df(r)}{dr}\vec{\sigma}\cdot\vec{\ell},\quad (1)$$

where

$$f(r)=\left[1+\exp\left(\frac{r-R}{a}\right)\right]^{-1},\quad g(r)=\exp\left[-\left(\frac{r-R}{b}\right)^2\right],$$

$$R=r_o A^{1/3},$$

and the averaged potential parameter set [36]

$$V_c=(48.7-0.33E)\text{ MeV},\ W_c=(7.2+0.66E)\text{ MeV},$$
$$V_{so}=7.5\text{ MeV},\ a=0.65\text{ Fm},\ b=0.98\text{ Fm},$$
$$r_o=1.25\text{ Fm}.\quad (2)$$

When carrying out the calculations, we made use also of the optical potential, in which the form factor of the imaginary potential

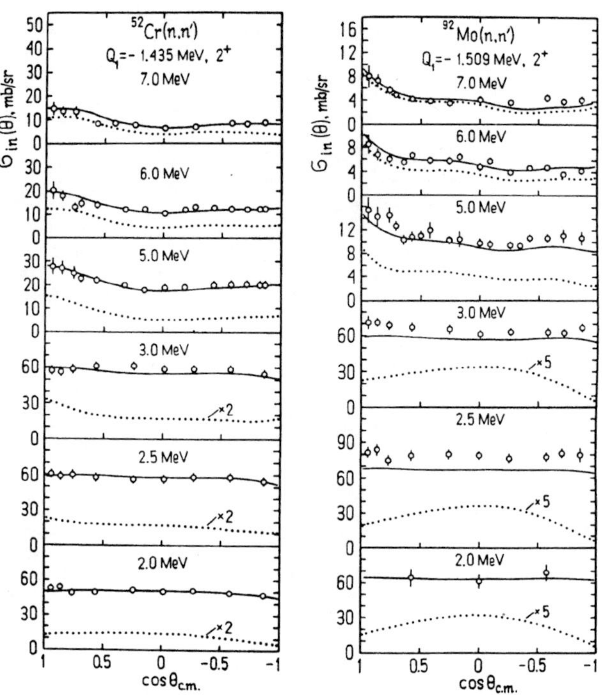

Fig.2. Differential cross sections of the neutron inelastic scattering by chromium-52 and molybdenum-92 with excitation of their first levels for the indicated energies: experiment (points) and the calculations using i) the coupled-channels model and statistical model taking into account the level-widths fluctuations (solid curves) and ii) only the coupled-channels model (dot curves).

was taken to be a derivative of the real potential form factor; in this case the diffuseness parameter was reduced by a factor of 2.3.

The direct inelastic scattering cross sections were calculated using the coupled-channels model [37] under assumption, that the nature of the lowest nuclear levels is vibrational and only the first excited level is strongly coupled to the ground state. The computations were carried out using the programm [38] with the averaged potential parameters (2) (except W_c, reduced by 20 % in order to obtain the same σ_t, as in the spherical optical model). The quadrupole deformation coeffitients β_2 were taken equal to 0.23 for chromium-52 and 0.116 for molybdenum-92 [39].

The compound scattering cross sections were

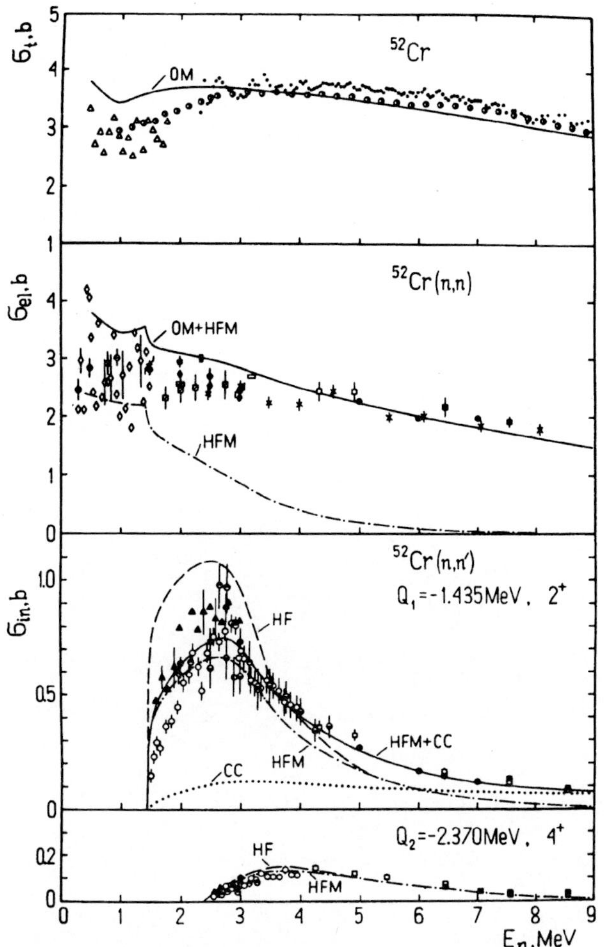

Fig.3. Energy dependences of the total cross sections and integrated neutron cross sections of the elastic scattering and inelastic scattering with excitation of the lowest two levels of chromium-52 in the energy range 0.5-9.0 MeV. The points show the experimental cross sections: ▮ - [1], ⊖ - [2], ⊗ - [3], ⊞ - [4], ◊ - [5], ▭ - [6], + - [7], ◆ - [8], ⊠ - [9], × - [10], ● - [11-14; 23], ◒ - [15], ▲ - [16], ◐ - [17], ◓ - [18], ○ - [19], □ , ■ - [22], △ - [30], • - [31], ◑ - [32]. All data on the total cross sections and elastic scattering cross sections (except the results of the papers [11-14,23] and [24] - ■) were obtained for the natural chromium. The curves show results of the theoretical calculations: using the spherical optical model (OM), coupled-channels model (CC), and statistical model both ignoring (HF) and allowing for (HFM) the level-widths fluctuations.

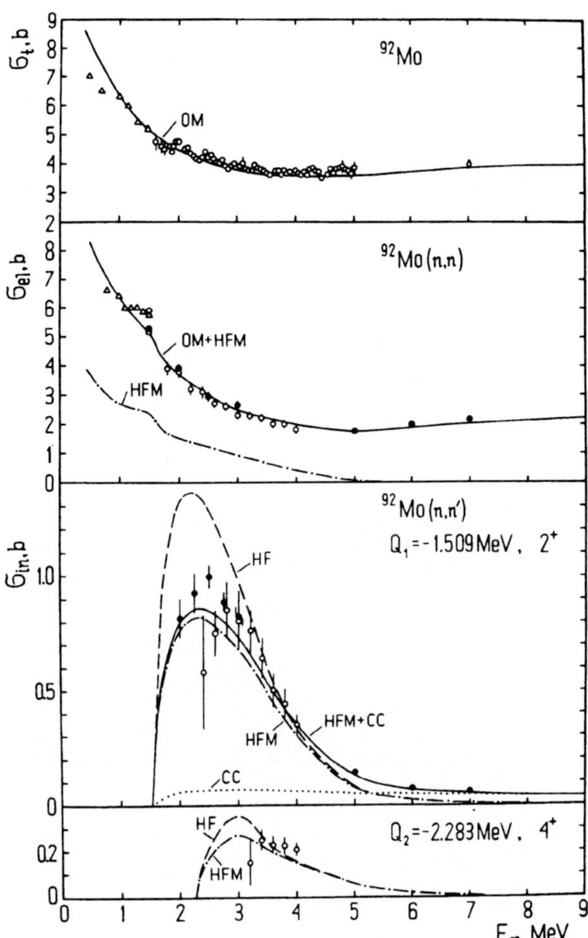

Fig.4. The same, as in Fig.3, for molybdenum-92. The points show the experimental cross sections: ● - this paper, △ - [24], □ - [25], ○ - [26], ◊ - [33].

calculated by means of the method, described in the papers [14,42] and based on the use of the statistical model in neglect of [40] and taking into account [41] the level-widths fluctuations.

In the statistical model calculations we have taken into account the known discrete levels of the nuclei under investigation, taken from the compilation [43]. The higher levels with the unknown characteristics were allowed for statistically using the Fermi-gas model with a "back-shift". The level density in the continuum level spectra were calculated using the formula [44]

$$\rho(U,i'')=\frac{2i''+1}{24\sqrt{2}\,a^{1/4}U^{5/4}\,\sigma^3}\exp\left[2\sqrt{aU}-\frac{(i''+1/2)}{2\sigma^2}\right],\quad(3)$$

where $U=E-\Delta$, a and Δ are the level density parameters, σ is the spin cut-off parameter, which is related to a, U, and nucleus mass A as $\sigma^2=0.146\sqrt{aU}\,A^{2/3}$ [45]. The parameters a and Δ were taken from the paper [46], but renormalized, since we made use of an alternate expression for the level density.

The results calculated in the above manner are given in Figs.1-4 in order to be compared with the experimental data. The agreement between the calculated and measured cross sections is fairly good both for the differential and integrated cross sections in the case, when the level-widths fluctuations were allowed for in the statistical model.

This agreement enables one to make conclusions about the role of direct and compound processes in fast neutron scattering by the nuclei under investigation. The relative contributions of the direct and compound processes into the summary scattering cross sections are essentially changed as the neutron energy changes. For example, the compound components for chromium-52, which are equal to a half of the summary elastic scattering cross sections and 80 % of the summary inelastic scattering cross sections at the initial part of the energy range of interest, are practically negligible at the energy 8 MeV. For molybdenum-92 these become negligible for the energies as low as 4-5 MeV.

The analysis carried out not only allows to make conclusions about the role of the diverse mechanisms in the fast neutron scattering, but also indicated the possibility of using the applied theoretical models for predicting the cross sections of fast neutron scattering by the nuclei under investigation.

References

1. О.А.Сальников, Атом. эн. 3, I06 (I957).
2. W.B.Gilboy, J.H.Towle, Nucl. Phys. 42, 86 (1963).
3. И.А.Корж, Н.Т.Скляр, И.А.Тоцкий, Укр. физ. ж. 9, 577 (I964); M.V.Pasechnik, I.A.Korzh, I.E.Kashuba, I.A.Totsky, In: Nucl. Struct. Study with Neutrons, Antwerpen, 1965, p.573.
4. L.Ya.Kazakova, V.E.Kolesov, V.I.Popov, O.A.Salnikov, V.M.Sluchevskaja, V.I.Trikova, In: Nucl. Struct. Study with Neutrons, Antwerpen, 1965, p.576.
5. A.B.Smith, P.T.Guenther, In: Neutron Cross Sections, BNL-325, Second Ed., vol.IIA, Suppl. nr.2, Brookhaven, 1966.
6. R.L.Becker, W.G.Guindon, G.J.Smith, Nucl. Phys., 89, 154 (1966).
7. И.А.Корж, В.А.Мищенко, М.В.Пасечник, Н.М.Правдивый, И.Е.Санжур, И.А.Тоцкий, Укр. физ. ж. I2, I57I (I967).
8. М.В.Пасечник, И.А.Корж, И.Е.Кашуба, В.А.Мищенко, Н.М.Правдивый, И.Е.Санжур, Яд. физ. II, 958 (I970).
9. B.Holmqvist, S.G.Johansson, M.Salama, T.Wiedling, AE-385, Studsvik (Sweden), 1970.
10. B.Holmqvist, T.Wiedling, AE-430, Studsvik (Sweden), 1971.
11. И.А.Корж, В.А.Мищенко, Э.Н.Можжухин, М.В.Пасечник, Н.М.Правдивый, И.Е.Санжур, В кн.: Нейтронная физика, М., ЦНИИатоминформ, I976, ч.IУ, с.220.
12. И.А.Корж, В.А.Мищенко, Э.Н.Можжухин, М.В.Пасечник, Н.М.Правдивый, Яд. физ. 26, II5I (I977).
13. M.V.Pasechnik, I.A.Korzh, E.N.Mozhzhukhin, In: Nucl. Cross Sections for Technology, U.S. Dep. of Comm., Nat. Bureau of Stand. Spec. Publ. nr.594,1980, p.893.
14. И.А.Корж, И.Е.Кашуба, А.А.Голубова, В кн.: Нейтронная физика, М., ЦНИИатоминформ, I976, ч.IУ, с.203.
15. D.M.Van Patter, N.Nath, S.M.Shafroth, S.S.Malik, M.A.Rothman, Phys. Rev. 128, 1246 (1962).
16. Д.Л.Бродер, В.Е.Колесов, А.И.Лашук, И.П.Садохин, А.Г.Довбенко, Атом. эн. I6, I03 (I964).
17. М.Б.Федоров, Т.И.Яковенко, В кн.: Нейтронная физика, Обнинск, изд.ФЭИ, I974, ч.III, с.56.
18. E.Almen-Ramström, AE-503, Studsvik (Sweden), 1975.
19. P.T.Karatzas, G.P.Couchel, B.K.Barnes, L.E.Beghian, P.Harihar, A.Mittler, D.J.Pullen, E.Sheldon, N.B.Sullivan, Nucl. Sci. Eng. 67, 34 (1978).
20. A.Prince, T.W.Burrows, Evaluation of Natural Chromium Neutron Cross Sections for ENDF/B-V, BNL-NCS-51152 (ENDF-286), BNL, Upton, N.Y., 1979.
21. A.Prince, Evaluation of Chromium Neutron and Gamma Production Cross Sections for ENDF/B-IV, BNL-NCS-50593 (ENDF-246), BNL, Upton, N.Y., 1976.
22. W.E.Kinney, F.G.Perey, ORNL-4806, Oak Ridge, 1974.
23. И.А.Корж, В.А.Мищенко, Э.Н.Можжухин, Н.М.Правдивый, Яд. физ. 35, I097 (I982).
24. P.Lambropoulos, P.Guenther, A.B.Smith, J.Whalen, Nucl. Phys. A201, 1 (1973).
25. F.D.McDaniel, J.D.Brandenberger, G.P.Glasgow, H.G.Leichton, Phys. Rev. C10, 1087 (1974).
26. A.B.Smith, P.Guenther, J.Whalen, Nucl. Phys. A244, 213 (1975).
27. J.Rapaport, T.S.Cheema, D.E.Bainum, R.W.Finlay, J.D.Carlson, Nucl. Phys. A313, 1 (1979).
28. В.В.Жук, А.А.Козарь, И.А.Корж, В.А.Мищенко, Э.Н.Можжухин, Н.С.Назаров, М.В.Пасечник, В.С.Подобайло, Н.М.Правдивый, И.Е.Санжур, И.А.Тоцкий, В кн.: Нейтронная физика, Обнинск, изд.ФЭИ, I974, ч.IУ, с.203.
29. И.А.Корж, В.А.Мищенко, И.Е.Санжур, Укр. физ. ж. 25, I09 (I980).
30. C.T.Hibdon, Phys. Rev. 108, 414 (1957).
31. D.G.Foster, Jr., D.W.Glasgow, Phys. Rev. C3, 576 (1971); C3, 604 (1971).
32. C.M.Neustead, S.Cierjacks, KFK-2060, Karlsruhe, 1974.
33. D.I.Garber, R.R.Kinsey, BNL-325, Third Ed., TID-4500, Brookhaven, BNL, 1976.
34. И.А.Корж, В.А.Мищенко, Э.Н.Можжухин, М.В.Пасечник, Н.М.Правдивый, И.Е.Санжур, Яд. физ. 3I, I3 (I980).
35. F.Bjorklund, S.Fernbach, Phys. Rev. 109, 1295 (1958).
36. М.В.Пасечник, И.А.Корж, И.Е.Кашуба, В кн.: Нейтронная физика, Киев, Наукова думка, I972, ч.I, с.253.
37. T.Tamura, Rev. Mod. Phys. 37, 679 (1965).
38. А.В.Игнатюк, В.П.Лунев, В.С.Шорин,В кн.: Вопр.атом. науки и техн., сер. Яд.конст. М., ЦНИИатоминформ, I974, вып.I3, с.59.
39. P.H.Stelson, L.Grodzins, Nucl. Data A1, 21 (1965).
40. W.Hauser, H.Feshbach, Phys. Rev. 87, 366 (1952).
41. P.A.Moldauer, Phys. Rev. B135, 642 (1964); B136, 947 (1964); Rev. Mod. Phys. 36, 1079 (1964).
42. Г.В.Анципов, В.А.Коньшин, В.П.Коренной, Е.Ш.Суховицкий, В кн.:Вопр. атом. науки и техн., сер. Яд.конст., М., Атомиздат, I975, вып.20, с.I64.
43. C.M.Lederer, V.S.Shirley (Eds.), Table of Isotopes, XVII Ed., N.Y.-London, John Wiley and Sohns, Inc., 1978.
44. A.Gilbert, A.G.W.Cameron, Canad. J. Phys. 43, 1446 (1965).
45. U.Faccini, E.Saetta, En. Nucleare, 15, 54 (1968).
46. W.Dilg, W.Schantl, H.Vonach, M.Uhl, Nucl. Phys. A217, 269 (1973).

Nuclear Data Pertaining to Fission Reactor Fuel Cycles and Fission Products

NUCLEAR DATA NEEDS FOR URANIUM-PLUTONIUM FUEL CYCLE DEVELOPMENT

H. Umezawa

Japan Atomic Energy Research Institute
Tokai Research Establishment
Tokai-mura, Naka-gun, Ibaraki-ken 319-11
Japan

Potential needs of nuclear data have been surveyed through the uranium-plutonium fuel cycle. A wide variety of data were used in operation of fuel cycle facilities and its relating activities. Those were decay data, cross section data and fission-product yield data. Although the nuclear data were used mostly as physical constants in analyzing results of measurements and calculations made, the users did not give a first priority to their requirements on the nuclear data. It is necessary to disseminate uniquely evaluated data for securing uniformity of data-treatment basis of the users. A covariance file of the evaluated cross section data was requested by workers involved in reactor dosimetry studies and in actinides incineration studies for waste management. Also, in the various parts of the fuel cycle equally required were some evaluated computer codes including reasonably organized nuclear data set which enable one to predict the amounts of actinides and fission-product nuclides in irradiated nuclear fuels.

[survey, requirements, cross section, half-life, fission-yield, nuclear decay, nuclear fuel cycle, facilities, operation, waste, safeguards]

Introduction

A large number of Light Water Reactors (LWR) have already been in operation, supplying a considerable fraction of electricity, and Fast Breeder Reactors (FBR), and perhaps thermal reactors utilizing recycled plutonium too, will come forward for significant energy sources in near future. In this context, nuclear fuel cycle of uranium and plutonium is of great importance and only realistic fuel cycle at present.

The entire fuel cycle consists of various processes: Mining and milling of uranium ore, enrichment of ^{235}U, fabrication of uranium fuel, reactor operation, spent fuel storage and reprocessing, and fabrication of plutonium-uranium mixed-oxide fuel. Waste management and environmental protection are connected to all of those processes. Transportation of nuclear material, particularly of spent fuel from reactors, is also an important part to be considered. Besides, safeguards of nuclear material for diversion from any process in the fuel cycle to non-peaceful uses must be effectively implemented. A simplified flow diagram of the uranium-plutonium fuel cycle is shown in Fig. 1.

The main purposes of nuclear data activity organized so far are to provide a sufficient basis for designing and developing nuclear reactors and exclusively concerned with cross section data, mostly for neutrons. However, effective operation of nuclear facilities out of the reactors has become a key issue, as all of the fuel cycle activity has been developed and continuous supply of fresh fuels to the nuclear power plants and adequate management of spent fuels from those are inevitably required. Nuclear data activity has, therefore, to be developed by taking such a view into consideration.

Taking cognizance of the present stage of fuel cycle development in Japan, the Japanese Nuclear Data Committee (JNDC) had organized a working group to make a survey of up-to-date needs of nuclear data in operation of nuclear fuel facilities. The group was composed of some nuclear data evaluators and a good many of data users who were dealing with daily operation of such facilities. The group examined extensively how the nuclear data are being used in various phases of the fuel cycle activity and sought to recommend what activity on nuclear data is needed to support the fuel cycle development.

Needs of a wide variety of data were identified. Those were decay data, cross section data and fission-product yield data. Although those nuclear data were used mostly as physical constants that should be accurate enough, in analyzing results of measurements and calculations made in the operation of fuel cycle facilities, the users did not, in general, give a first priority to their requirements on the nuclear data. It implies that dissemination activity of evaluated nuclear data is particularly important.

A covariance file of the evaluated cross section data was requested by workers involved in reactor dosimetry studies and in actinides incineration studies for waste management. Also in various parts of the fuel cycle equally required were some evaluated computer codes including reasonably organized nuclear data set which enable one to predict correctly the amounts of actinides and fission products in irradiated nuclear fuels.

This paper describes, mostly based on the deliberations of the JNDC working group on the fuel cycle nuclear data[1], the uses of the nuclear data in operation of the nuclear fuel facilities and its relating activities, and it summarizes requirements of nuclear data.

Use of Nuclear Data in the Operation of Fuel Cycle Facilities

Mining and milling of uranium ore

Facilities for the mining and milling of uranium ore have been operated essentially the same as those for ores of other kinds of elements, but that the facilities subject to uranium in this case. From the crushed ore, uranium is leached most commonly with sulphuric acid. Nitric acid, hydrochloric acid or sodium carbonate is also used instead, if it is more adequate to the uranium compound in the ore. The learched uranium is purified by means of chemical precipitation, ion exchange and solvent extraction, and it is converted to uranium oxide of reactor grade.

Through the whole process, isotopic measurements are not necessarily required and process control analysis of nuclear material is concerned with only its chemical form and quantity. No nuclear data have been necessarily referred to the treatment of the measurement results.

Release of radioactive nuclides from the waste debris is a problem from the point of view of environmental protection. Decay data of the daughter nuclides of uranium are, however, well known to evaluate the environmental impact, and the environmental problems as well as any operational aspects of the mining and milling of uranium ore would not be influenced by improving nuclear data from present day forth.

Enrichment of ^{235}U

Gaseous diffusion, gas centrifuge, gas nozzle, chemical diffusion and laser excitation have been developed

K. H. Böckhoff (ed.), Nuclear Data for Science and Technology, 165–169.

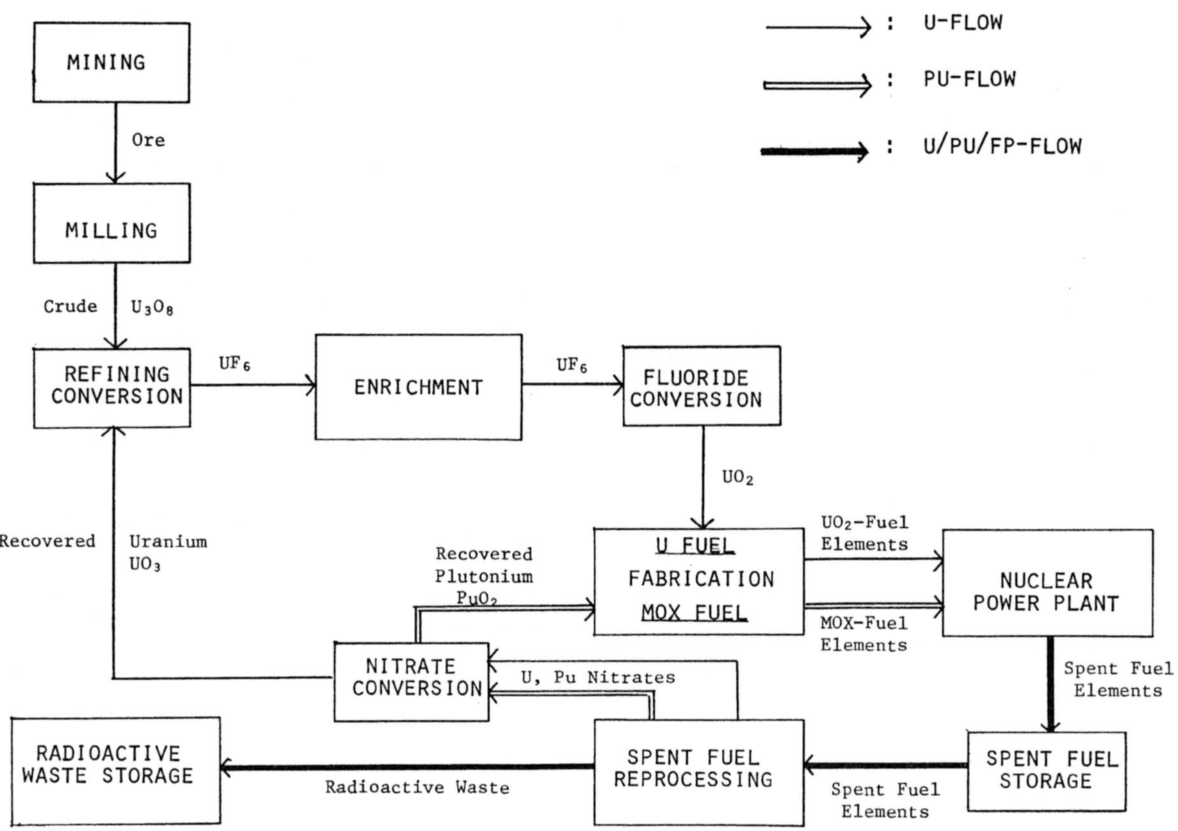

Fig. 1. A simplified flow diagram of the uranium-plutonium fuel cycle.

to make uranium enriched in ^{235}U. In Japan, a pilot plant of gas centrifuge has been constructed and operated.

Uranium is prepared in the form of uranium hexafluoride gas from uranium dioxide by two-step fluorination with hydrofluoric acid first to obtain uranium tetrafluoride then with fluorine to convert it to the hexafluoride. The uranium hexafluoride gas is fed to a cascade of a large number of centrifuge in which ^{235}U is separated from ^{238}U according to the mass difference.

In operation of the enrichment facilities, isotopic measurements of uranium are essential and must be carried out at many points in the process to control the plant operation conditions and to establish material accountancy in the plant, but nuclear data have not been used in connection with the operation.

Once uranium recovered from spent fuel reprocessing is recycled to the enrichment process, however, measurements of ^{232}U that is formed by successive (n,2n) reactions in the fuel, and evaluation of the radiation dose of its daughter nuclides become serious problems. Nuclear data of the decay-product series of ^{232}U are again known enough to make such evaluation.

On the other hand, ^{232}U can be hardly measured by mass but radioactivity because of its extremely low concentration and high specific activity, and correlation between contents of ^{232}U and ^{235}U or ^{236}U would be useful to estimate the amounts of ^{232}U in such recycled uranium. It would be necessary to give adequate cross section data and decay data of nuclides involved for predicting formation of ^{232}U in spent fuels.

Fabrication of uranium fuels

The uranium hexafluoride gas enriched in ^{235}U is reacted with pure water and then treated with ammonium hydroxide to obtain ammonium diuranate. Then, it is filtrated, dried and calcined to convert it to uranium dioxide powder. Enrichment, impurities, uranium content and specific surface area are to be controlled in that process. From the powder, uranium dioxide pellets are manufactured by slugging,pressing, sintering and grinding. Of the pellets, density, diameters, impurities, enrichment, uranium content, moisture and cracks are to be examined. Those pellets are packed in zircaloy sheaths and those are bundled up to fabricate fuel assemblies.

Uranium fuel fabrication plants have been in very routine operation and material accountancy system in those facilities have been designed to meet the requirements of quality control, process control, radiation safety control, property management and safeguarding of nuclear material. The basis of the system is to measure flow and inventory of nuclear material at measurement points and to establish material balance for a given period within a material balance area set up in the facilities. Most of the measurements are weighing and chemical analysis by taking samples, though non-destructive assay by gamma-ray spectrum measurements for enrichment and active neutron interrogation for ^{235}U content made are of nuclear techniques. Calibration of the non-destructive assay equipment is used to be carried out with a reference sample whose value is determined by precise chemical analysis. Accuracy and precision of the present system of accountancy for and control of nuclear material applied to the facilities of uranium fuel fabrication are seemed to achieve the present objectives. Immediate improvement of nuclear data is not required.

If the system would be developed to make real-time inventory control in future, non-destructive assay equipment might be more installed and nuclear data would be used to develop and operate such advanced equipment. If recovered uranium from reprocessing of spent fuels would be refabricated, decay data of ^{232}U

and its daughter nuclides might be needed in the measurement and control of those contaminations.

Reprocessing of spent fuels

For reprocessing spent fuels from LWR, the PUREX (Plutonium and Uranium Recovery by EXtraction) process has been only operated as experienced plants. Spent fuel of LWR of burnup 30,000MWD/t, for example, contains fission products of 3%, plutonium of 1%, neptunium of 0.05% and transplutonium elements of 0.01% in addition to about 94% of uranium remained in the fuel. After mechanical chopping the fuel rods, the oxide fuel is leached with hot nitric acid. Uranium and plutonium are extracted into organic solvent including tributyl phosphate (TBP) as extracting reagent from the nitric acid solution, remaining other actinides and fission products in the aqueous phase that comes to high radioactive aqueous waste (HAW). The uranium and plutonium are, then, mutually separated in the second extraction cycle by controlling the oxidation state of plutonium. The uranium is further purified, concentrated and calcined to produce uranium trioxide that is stored in the product storage area and shipped to reenrichment or refabrication. The plutonium is purified and concentrated as plutonium product nitrate solution, or it is further converted to plutonium dioxide by precipitation and calcination.

When spent fuels are reprocessed, it is desirable to know the inventory of the fission products and actinides in the fuels in connection with safety control. The nuclides of interest are as follows:
(1) With respect to the problems of environmental release, ^{3}H, ^{14}C, ^{85}Kr, ^{129}I and aerosols of plutonium and transplutonium elements are taken into consideration.
(2) In connection with process control, ^{106}Ru, ^{95}Zr, ^{95}Nb, ^{90}Sr, ^{134,137}Cs are important for radioactivity inventory. Chemical concentration of zirconium, molybdenum, ruthenium and rhodium is interested in relation to formation of insoluble particles and phase instability in extraction process.
(3) For the purposes of criticality monitoring in process, ^{238}Pu, ^{240}Pu, ^{242}Cm and ^{244}Cm are of importance as neutron sources and yields of (α,n) reaction are also to be considered in the same respect. For evaluating criticality in the process of the bulk nuclear facilities, fission cross sections of the fissile materials involved, namely ^{235}U, ^{239}Pu and ^{241}Pu, and capture cross sections of the neutron absorbing material used, such as boron, cadmium, gadolinium or hafnium, are to be calculated under all possible conditions in the process.

The reprocessing plant would receive a variety of spent fuels in its exposures and cooling, and the inventory of the nuclides mentioned above has to be estimated by calculation. Simple and reliable computer code is quite desirable for the plant operators. Nuclear data aspects concerning control and measurements of plutonium are discussed in following sections.

Fabrication of MOX fuels

The fuel of mixed-oxide (MOX) of uranium and plutonium is manufactured to use for FBR and thermal reactors which are LWR and heavy water reactors for plutonium utilization. The plutonium content of the FBR fuel is 15-30% and of thermal reactors 0.5-6%. The mixed-oxide is usually prepared by mixing uranium dioxide powder and plutonium dioxide powder. On the other hand, mixing in the form of nitrate solutions of uranium and plutonium is being developed and it may provide more homogeneous composition of the products. In this case, the same conversion process to that at the end of reprocessing is made to obtain the mixed-oxide powder. Manufacture of pellets from the MOX powder and fabrication of fuel assemblies are also similarly carried out as those of the uranium fuels, except that all materials in this process should be contained within glove boxes.

Nuclides included in the process are mainly ^{234}U, ^{235}U and ^{238}U of uranium and ^{238}Pu, ^{239}Pu, ^{240}Pu, ^{241}Pu and ^{242}Pu of plutonium, but ^{241}Am is continuously formed by the beta decay of ^{241}Pu. If recycled uranium and plutonium would be used in future, ^{232}U, ^{236}Pu and the daughter nuclides of those should be measured and controlled.

In this process, nuclear data are used in connection with both quality assurance of products and process safety control. For the former, impurities of fuel pellets are always expressed as boron equivalent concentrations and neutron capture cross section of each of the impurities is referred to.

Reactivity of the MOX fuel is controlled by the content of fissile plutonium nuclides, namely ^{239}Pu and ^{241}Pu. When isotopic composition of plutonium is measured by the use of a mass spectrometer, such an equipment is, in general, calibrated by a reference standard material for which decay correction is used to be done for a period of several years or, often, more than ten years according to the half-life data. Among those data, the half-life of ^{241}Pu is rather short and has not been well determined yet, so that errors due to the decay correction exceed sometimes precision attained in relative measurements of the isotopic ratios with the mass spectrometer.

When alpha spectrum measurement or gamma-ray spectrometry was applied to the isotope analysis, data obtained from the measurements have to be analyzed based on the half-lives and the branching ratios of measured alpha and gamma emmisions of the relevant nuclides of plutonium and some daughters like ^{241}Am.

Problems with respect to safety control are evaluation of radiation dose at all working areas and criticality in the process. As to the radiation dose, gamma activities of ^{241}Pu and ^{241}Am and neutron fluence from (α,n) reactions are of main concern. Neutron cross sections of nuclides concerned under the possible critical conditions in the process are needed to evaluate criticality safety. Evaluation of neutron source intensity of materials in the process is also useful in relation to criticality monitoring as well as in the reprocessing plant.

Reactor operation

Apart from the nuclear data needs for the design and development of reactor core, reactor dosimetry is one of the topics much relating to nuclear data. Recent trends in reactor dosimetry studies are critical examination of evaluated neutron fluence and neutron spectrum. Measured data of irradiation effects of reactor component material have been much accumulated, and systematic discrepancy among those of different reactors and of different irradiation positions has been recognized. These facts lead to requirements of reevaluation of the dosimetry approaches and reevaluation of the values of neutron dose obtained from the approaches.

At the same time, early constructed LWRs have been operated long close to the reactor lifetime and timely improvement is necessary of high-precision evaluation of the data of irradiation damage and integral neutron dose in the pressure vessels of such reactors.

So far data treatment procedures have been mostly established and uncertainties associated with the input data of reaction rates which are errors of the dosimetry cross sections and estimated neutron spectra, are being examined. Covariance file must be made available for the evaluated dosimetry cross sections.

Non-destructive measurements of burnup and accumulation of plutonium in irradiated fuels are interested from the point of view of making control and management of the power reactor fuels. Measured are gamma activities of fission products: ^{140}Ba-^{140}La, ^{103}Ru, ^{106}Ru-^{106}Rh, ^{134}Cs, ^{137}Cs, ^{144}Ce-^{144}Pr, and ^{154}Eu. Correlation between the activity or activity ratio of those nuclides and parameters of interest is determined in reference to chemical analysis results obtained from post-irradiation examination or reprocessing of the measured fuel or equivalent ones to it. Based on the correlation thus obtained, non-destructive assay for burnup and plutonium content of spent fuels can be made. For the analysis and correction of the measured data in connection with irradiation history and cooling period, yield data of those nuclides in fissions of ^{235}U, ^{238}U, ^{239}Pu, ^{240}Pu and ^{241}Pu, half-life data of those, and gamma-ray energy and intensity data are necessary to be given, although most of those data have been already measured with accuracy and the evaluation of the data may be done without much difficulty.

Another problem concerning to reactor would be of decommissioning. Using the Japan Power Demonstration Reactor (JPDR) as an object, studies on the decommissioning of LWR are going to be made in Japan. After taking all spent fuels out of reactor, long-lived neutron-induced radioactivities in structural materials have to be evaluated. Main sources of the activities are ^{55}Fe, ^{60}Co and ^{63}Ni in steel and ^{39}Ar, ^{41}Ca, ^{55}Fe, ^{60}Co, ^{151}Sm, ^{152}Eu and ^{154}Eu in concrete. In addition, ^{14}C, ^{59}Ni, ^{93}Mo, ^{94}Nb and ^{108m}Ag might be somewhat formed in such structural materials. Even though accuracy required in the radioactivity inventory in the structural materials is not high and the presently available nuclear data of cross section and decay perhaps meet the requirements, methods to calculate the production rates of those nuclides in such complex materials have to be more developed.

Spent fuel: transportation and shielding

Spent fuels from nuclear power plants must be stored at reactor site at least for some months and transported to other storage area away from reactor or a reprocessing plant. Cask to contain the spent fuels should have enough shielding ability against the intense radiation of the fission products contained.

For the purposes of shielding analysis, the intensity of radiation source in the spent fuel, that is of neutrons and gamma-rays, has to be evaluated with confidence. Elements to be considered as to neutron sources are spontaneous fissions of some actinide nuclides and (α,n) reactions on light nuclides, such as ^{16}O, ^{18}O and ^{13}C, with alpha particles emitted by heavy actinide nuclides. Cross section data of those reactions are not satisfactory at present and desired is determination of the cross section from 0.1MeV to 8MeV at energy resolution of about 0.1MeV.

Methods to evaluate the energy spectra of neutrons generated from the (α,n) reactions are also to be further developed. Moreover, important is multiplication of neutrons thus generated in such a system. That problem could be reduced to the requirements for the fission and capture cross sections and estimating density of the fissile material and neutron absorbing material.

It is obvious that main sources of gamma radiations are the fission products. Fission yield data of the product nuclides and nuclear decay data of those are used to calculate the fission-product inventory in the spent fuel. The requirements of the nuclear data in this case are the similar to that for non-destructive assay of spent fuel discussed previously in reactor operation.

Waste management

A variety of radioactive waste is generated from different stages of the fuel cycle. In addition to the waste from operations of the fuel cycle facilities, the maintenance and deconstruction of those would produce waste, and the waste treatment itself might also generate secondary waste.

Very low level waste could be released to environment after evaluating its environmental influence. A lot of low level waste arises from reactor operation, that contains neutron-induced activities of iron, nickel and cobalt nuclides and of the fission-product strontium and cesium nuclides. Because of little heat generation the waste is solidified with cement, asphalt or plastics, and those would be given to sea or land disposal after establishing confident conditions.

High level waste that contains uranium, neptunium, plutonium, americium, curium and the fission products, arise from reprocessing of spent fuel. Since there are long-lived alpha radioactive nuclides as well as beta-gamma-active fission products, it is necessary to control against the decay heat and radiations and to isolate from human environment for very long time, although the amounts of the high level waste might not be so much.

An alternate approach studied is nuclear incineration of the long-lived actinides waste. For reactor physics studies to make assesment on technical feasibility of the nuclear incineration process, fission and capture cross sections for neutrons and decay constants of the actinides nuclides involved in the successive neutron capture and decay chain from ^{235}U to ^{245}Cm must be provided. Particularly the data of ^{237}Np, ^{238}Pu, ^{241}Am and ^{243}Am have not been determined well enough and required to be determined as precisely as those of ^{235}U and ^{239}Pu for FBR fuel. The nuclear data required are fission and capture cross sections and delayed neutron characteristics.

Considering difficulties associated with the measurements of differential cross sections on such alpha radioactive nuclides of actinides, group constants of the cross section could be evaluated and adjusted by using integral measurement data. Procedures to make the adjustment have been developed, and information on errors of the evaluated cross section data is necessary. Especially covariance is indispensable.

Waste of the plutonium-use-facilities is contaminated with alpha active material, but it does not bear much of beta and gamma activities. It is treated as the similar to the high level waste described above.

Reactor-structural materials activated by neurton exposure are the waste of decommissioning of reactor that might be classified either into high level or low level waste contaminated with beta and gamma activities corresponding to its radiation level. In a case where the plutonium-use-facilities are decommissioned, most of the waste must be contaminated with alpha activities and treated as alpha contaminated waste.

From reactor and reprocessing facilities, gaseous radioactive materials, such as ^{3}H, ^{14}C, ^{85}Kr, ^{127}I and ^{129}I, and aerosol of some fission products and alpha radioactive nuclides of the actinides are discharged. Evaluation of the rate of discharge of those activities is important for environmental safety assesment.

In order to develop and establish adequate management system of the radioactive waste, it is necessary to evaluate various factors of the waste: Generation and inventory, heat generation, shield calculation, criticality either in the treatment process or in the disposal, and long-term safety analysis. The requirements involve complete preparation of basic decay data of the actinide nuclides as well as the daughter nuclides in the radioactive decay chains and those of the fission products.

Safeguards

In 1977, requests of nuclear data for safeguards purposes were investigated through questionaires sent to 183 world-wide experts concerned with the field. Answers of 18 among 52 correspondences have contributed to the entries of the questionaire. Fuketa has examined the correspondences and reported in the last International Conference on Nuclear Physics and Nuclear Data for Reactor and other Applied Purposes, Harwell 1978[2]. The situation of utilization of the nuclear data was as follows:

(1) Implementation of safeguards in the most important at present, and there may not wait for improving nuclear data if those are involved in the safeguards

elements.
(2) In the presently adopted measurements methods and equipment which use necessarily nuclear data, many factors out of the nuclear data are dominant and concern might not focussed on the nuclear data.
(3) If uncertainties associated with half-livies of ^{238}Pu and ^{241}Pu could be reduced, for instance, calorimetry of plutonium might more accurately carried out. Those techniques, however, have not been demonstrated yet as practical use.
(4) Improvement of nuclear decay data of plutonium nuclides, such as gamma-ray energy and branching ratio, may contribute to reduce error level of measurements of isotopic composition of plutonium by gamma-ray spectrometry. But it could be alternated by calibration using better physical standards, even though such standard material might be hardly available. Besides, the other factors than the nuclear data still have affected dominantly to the measurement results.

It is true yet at the present time that the attitude of the workers in the field of safeguards of nuclear material is as such that their higher priority of improvement of factors involved should be given to other factors than the nuclear data. They are, however, expecting always to adopt the best nuclear data to their uses for the measurements, evaluation and control of nuclear materials for safeguards. Those data are of a wide variety: Half-lives and decay energy and intensity of the nuclides of uranium, plutonium and some fission products; the data of spontaneous fission; yield data of the fission products; Fission and capture cross section data of the nuclides of uranium, plutonium and some fission products. A detailed list of the nuclear data uses for the safeguards technology has been prepared by Lammer[3].

Summary of Requirements and Conclusion

Although the nuclear data are not connected with routine operation of the facilities where only fresh uranium deals with, measurements of materials and evaluation of parameters in the operation of the plutonium-use-facilities have close relation to the nuclear data. In the fields of waste management and other aspects of fuel cycle deployment the nuclear data are concerned with at any rate.

The nuclides involved are of actinides, particularly of uranium and plutonium, and the fission products. The data used are cross sections, half-lives, energy and intensity of nuclear radiations, neutron emission, and fission-product yields[1]. Those data are commonly used in various stages of the uranium and plutonium fuel cycle that is presently developed, even if there are a number of factors which affect the results of measurements and calculations to be made, other than the nuclear data. In such a case, improvement of the nuclear data might not directly result in reduction of errors associated with the measurements.

Therefore, the users of nuclear data who deal with the fuel cycle facilities operation and not familiar with measurements and evaluation of nuclear data, are rather indifferent to adoption of the best evaluated values of nuclear data and, often, puzzled which ones to take among those appeared in different compilations, in spite of using in fact the nuclear data as physical constants which must be always accurate enough, affecting at least one order less than the other factors. Adopting different values of nuclear data causes, in fact, inconsistency between results obtained from measurements made at different points in the fuel cycle facilities. Lammer has also pointed out this situation with illustrative examples on the data of half-lives and neutron emissions of some actinide nuclides[3].

Based upon the facts just pointed out, it may be quite important to unify the results of data evaluation and recommend the evaluated and unified data to the users spreaded over the fuel cycle activities. Adequate dissemination activity carrying out these objectives should be encouraged.

Requests of cross sections for studying reactor dosimetry and reactor physics characteristics of actinide fuel would be coped with extension or upgrade of the evaluated cross section data file that is existing.

There were needs to estimate the amounts of fission products and heavy nuclides in the spent fuels with accuracy by simple calculation code. For those purposes, ORIGEN or some other computer codes have been available, but caluculated results, often, differed considerably from measured ones.This might not result from choise of the code. Computation procedures involved in those codes have been well developed and problems are in the input-data preparations involved. Necessary data for the calculation are cross sections, fission-product yields, half-lives, and nuclear decay data including energy, intensity and branching ratio of radiations emitted.

In addition to elaborate the base data file for the calculation codes, it might be very much helpful to provide such a code system involving necessary data library and indicating confidence limit to the results obtained by the code in connection with its applicability range. Along this line JNDC has organized a new working group on evaluation for generation and depletion of nuclides whose activity is to examine input data file for such a code and to evaluate the whole performance of the code on LWR fuels. Other activities to reach this objective should also be encouraged.

References

1. H. Umezawa, K. Hisatake (Editors), JAERI-M 9993, NEANDC(J)79/U, INDC(JAP)66/L (1982).
2. T. Fuketa, Proceedings of on International Conference on Neutron Physics and Nuclear Data for Reactor and other Applied Purposes, pp. 922-940, Harwell, September 1978.
3. M. Lammer, INDC/P(81)-24.

DONNEES DE BASE DANS LE CYCLE DU COMBUSTIBLE DES REACTEURS A NEUTRONS RAPIDES
BILAN ET PERSPECTIVE

L. COSTA - G. GRANGET - F. JOSSO

Centre d'études nucléaires de Cadarache
B.P. N° 1, 13115 Saint Paul lez Durance France

Résumé

On fait le bilan des données nucléaires nécessaires aux études relatives au cycle des combustibles des réacteurs à neutrons rapides.

A partir des précisions obtenues actuellement sur ces données, des calculs de sensibilité ont permis de définir les priorités pour les mesures expérimentales et les évaluations futures.

I - Introduction

On peut considérer le cycle du combustible des réacteurs à neutrons rapides comme un ensemble de transformations physiques, nucléaires, chimiques. Il s'agit pour nous de caractériser le combustible et les structures en ce qui concerne les transformations nucléaires et donc d'établir un bilan de masse et de radioactivité aux différents postes du cycle. La connaissance des données nucléaires de base de l'ensemble des nuclides mis en jeu est l'élément essentiel pour ce type de calcul.

Le nombre de nuclides mis en jeu, étant à priori très important, il est nécessaire de définir des priorités d'études pour la connaissance de ces données.

On essaie de montrer ici, à l'aide d'exemples, la démarche permettant d'établir leur importance relative, compte tenu des contraintes aux divers postes du cycle.

II - La cycle du combustible : principales grandeurs physiques et données de base

II.1 - Le cycle du combustible

On peut définir le cycle du combustible par ses principaux postes et leurs caractéristiques fonctionnelles.

* Usine de fabrication : mise sous forme d'assemblages d'oxyde d'uranium et plutonium et d'acier inactif.

* Réacteur : production d'énergie par fission, transmutations nucléaires par captures neutroniques.

* Usine de retraitement : transformations chimiques pour la séparation des différents composants et l'extraction du plutonium.

* Stockage des déchets : conditionnement des éléments (produits de fission, actinides, structures acier) autres que uranium et plutonium.

* Transport et manutention : assurent la liaison entre les divers postes pour les combustibles neufs ou irradiés et les déchets

II.2 - Principales grandeurs physiques

La définition d'une stratégie générale pour le cycle nécessite les études suivantes :

* Dimensionnement des usines.

* Mise au point des procédés.

* Optimisation de l'ensemble.

Soit dans le cadre d'études de prospective, soit dans une optique de projet on doit établir, à chaque poste du cycle, un bilan de masse et un bilan de radioactivité.

- Bilan de masse

On doit prendre en compte aussi bien d'un point de vue chimique qu'isotopique les nuclides que l'on peut regrouper dans les principales familles suivantes :

* Uranium plutonium.

* Actinides (autres que U et PU).

* Produits de fission.

* Matériaux de structure.

Ces bilans de masse isotopique et chimique sont nécessaires pour résoudre les problèmes de :

* Criticité.

* Balance plutonium.

* Mise au point des procédés de retraitement et de stockage.

* Dimensionnement en général de l'ensemble des usines.

- Bilan en radioactivité

Il s'agit de calculer les intensités et les spectres des différents rayonnements X, γ, α, β, n. Ces rayonnements, émis par les combustibles neufs ou irradiés, sont déterminants pour obtenir :

K. H. Böckhoff (ed.), Nuclear Data for Science and Technology, 170–174.

* La puissance thermique hors irradiation.

* Les débits de dose et les épaisseurs de protection biologique nécessaires.

* Le niveau de contamination.

II.3 - Les données nucléaires de base

En amont des calculs d'estimation de l'ensemble des grandeurs physiques que l'on vient d'examiner se situent les données nucléaires caractéristiques de chaque nuclide :

* Les valeurs des sections efficaces en fonction de l'énergie des neutrons :

$\sigma_{capture}$ (E), $\sigma_{fission}$ (E), σ_{n2n} (E)...

* Les schémas de formation par radioactivité ou absorption neutronique.

* Les périodes de décroissance radioactive.

* Les rendements de fission.

* La nature, l'intensité et le spectre énergétique des rayonnements émis.

En considérant les quatre grandes familles de nuclides, on arrive à un nombre très important (de l'ordre de 700 nuclides).

U, PU : de l'ordre de 16
autres actinides : de l'ordre de 30
Produits de fission : de l'ordre de 650
Produits d'activation : de l'ordre de 15

Il ne serait pas réaliste de faire ici l'inventaire complet des données nucléaires relatives à chacun de ces nuclides, mais, après un bilan rapide de l'état actuel, on donnera les grandes lignes de notre étude :

* Définition systématique de l'importance de chaque nuclide.
* Définition des priorités pour l'évaluation des données nucléaires.

III - Incertitudes - Bilan actuel

On résume dans les tableaux suivants (1 à 4) les nuclides dont il est nécessaire de prévoir la concentration ou l'intensité des rayonnements émis.

Pour chacun d'eux, on précise l'incertitude que l'on affecte aux calculs de leur concentration qui est estimée soit :

* A partir des incertitudes affectées aux données nucléaires de base spécifiques de chacun d'eux et à l'aide d'un calcul de sensibilité.

* A partir de comparaisons calcul-expérience effectuées sur le cycle des réacteurs RAPSODIE et PHENIX.

On peut constater que certains nuclides à période de décroissance très grande ne sont pas affectés d'incertitude ; leur importance apparaît dans la perspective de stockage des déchets nucléaires et ces nuclides sont l'objet des études actuelles.

IV - Systématique d'étude - Priorités d'évaluation

IV.1 - Systématique d'étude

En fonction des contraintes et des grandeurs physiques que l'on doit estimer à chaque poste, on doit définir :

* Pour les différents nuclides une importance relative.
* Pour chaque nuclide la ou les données nucléaires prépondérantes.

Le schéma ci-dessous résume la systématique d'étude qui permet de définir les importances relatives des différents nuclides et les priorités pour les évaluations des données nucléaires.

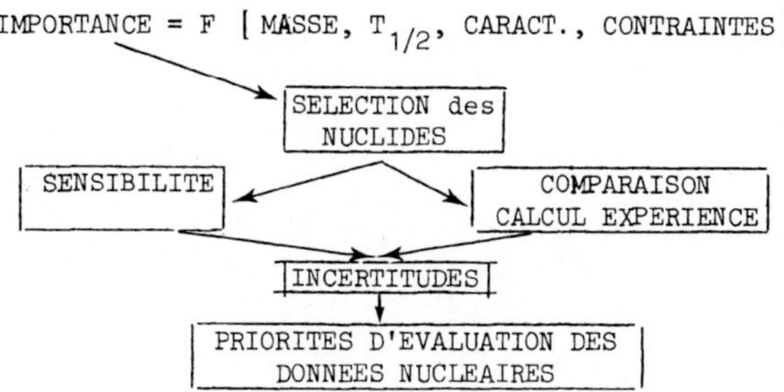

IV.2 - Exemples d'application

On illustre par quelques exemples, choisis aux différents postes du cycle, l'application de la méthode d'étude et les conséquences pour les évaluations des données nucléaires.

- Exemple 1 : usine de fabrication

* Contraintes : protection biologique et débit de dose.

* Nuclides prépondérants :
Tl 208 (γ dur 2,6 Mev) Am 241 (γ mou 0,06 Mev)

* Sensibilités :
Concentrations en PU 236 et Pu 241 initiales

* Priorités d'évaluation :
Tl 208 : σn2n de U 238
σn2n de U 237
et rapport de branchement { U 236 M / U237 F

Am 241 : période radioactive du PU 241
spectre γ de Am 241 et coefficients d'interaction avec la matière.

* Incertitudes sur les concentrations :
Pu 236 ± 30 % Am 241 ± 5 %

- Exemple 2 : réacteur

* Contrainte : puissance thermique résiduelle pour des temps courts après arrêt.

* Nuclides prépondérants : - P.F à vie courte
- U 239 et Np 239

* Données : - rendement de fission : Y
- énergie moyenne des β : E_β
- énergie γ : E_γ

- Exemple 3 : transport réacteur-retraitement

* Contraintes : puissance thermique résiduelle
protections biologiques

* Sensibilité : P.F ($\bar{\beta}$, γ) 270 P.F
Actinides (α et neutrons) Cm 242 et Cm 244
Acier (γ) Co 60

100 % sur intensité γ ⇒ environ 2 cm de plomb .
Variation de 200 kev autour de 1 Mev ⇒ environ 4 cm de Pb (discrétisation plus ou moins fine du spectre γ).

- Exemple 4 : usine de retraitement

En plus des problèmes de criticité ou de puissance thermique et de protection biologique déjà examinés, on peut noter un problème spécifique dû au procédé.

* Concentration en platinoides : Pd, Ru, Rh
* Données : rendement (Y)
* Incertitudes : ∿ 15 %

- Exemple 5 : stockage de déchets

Les nuclides importants ont été repertoriés. On peut citer quelques exemples importants :

* Produits d'activation des matériaux de structures Ni 63, Ni 59, Mo 93.
* Actinides
* Un produit de fission : Ho 166 M
* Données : σ_c pour Ni, Actinides et Ho 166 M;

V - Conclusion

Nous avons considéré le cycle du combustible des réacteurs à neutrons rapides dans son ensemble et défini la systématique d'étude pour établir les priorités d'évaluation des données nucléaires relatives à l'ensemble des nuclides. L'analyse du bilan actuel des incertitudes et les quelques exemples d'application de la méthode représentatifs d'une étude plus générale permettent de préciser les grandes lignes des études futures dans le domaine des données nucléaires :

* Pour les nuclides "majeurs" U235, U238, Pu239 à Pu242 qui sont importants pour la criticité et la puissance des réacteurs, une précision de 1 % est nécessaire sur les sections efficaces de capture et de fission. Ces nuclides font l'objet d'études particulières par les physiciens des coeurs de réacteur.

* Pour l'ensemble des autres nuclides (Actinides, produits de fission, produits d'activation, on peut se satisfaire d'une précision de l'ordre de 10 % sur les grandeurs physiques qui les caractérisent aux divers postes du cycle hors réacteur ; concentration, puissance thermique, intensité et spectre des rayonnements émis. Les tableaux du paragraphe III et les quelques schémas d'étude de sensibilité permettant de définir les domaines sur lesquels devra porter l'effort.

* Produits de fission à vie très courte ou très longue : rayonnements, énergie émise, sections de capture.

* Actinides importants par leur concentration ou par leur rôle intermédiaire ; en plus de ceux déjà cités ici, on devra faire un effort particulier sur les données nucléaires de ceux à vie très longue ou de numéro atomique supérieur au curium 244.

* Produits d'activation : on peut retenir ici des évaluations de sections efficaces de capture pour la formation des Ni59, Ni63, et du Mo93.

Tableau 1

PRODUITS DE FISSION
OPTIQUE RETRAITEMENT
INCERTITUDE ± 10 % A ± 15 %

ZN 72 +	KR 80	ZR 90 +	RH 102M	IN 115M	TE 122 +	CS 132	FR 141 +	EU 151 +
GA 72 +	KR 82 +	ZR 91 +	RH 102F	IN 115F +	TE 124 +	CS 133 +	PR 143 +	EU 153 +
GA 73	KR 83M	ZR 92 +	RH 103M	IN 117M	TE 125M +	CS 134 +	PR 144	EU 154 +
GE 72 +	KR 83F +	ZR 93 +	RH 103F +	IN 117F	TE 125F +	CS 135 +	PR 145 +	EU 155 +
GE 73 +	KR 84 +	ZR 94 +	RH 105 +	IN 118M	TE 126 +	CS 136 +	PR 146	EU 156 +
GE 74 +	KR 85M	ZR 95 +	RH 106F	SN 115 +	TE 127M +	CS 137 +	ND 143 +	EU 157 +
GE 75	KR 85F +	ZR 96 +	RH 107	SN 117 +	TE 127F +	CS 138	ND 144 +	EU 158
GE 76 +	KR 86 +	ZR 97 +	RH 109	SN 118 +	TE 128 +	BA 134 +	ND 145 +	GD 154
GE 77 +	KR 87	NB 93M	PD 102	SN 119M	TE 129M +	BA 135	ND 146 +	GD 155 +
GE 78	KR 88	NB 93F	PD 105 +	SN 119F +	TE 129F +	BA 136 +	ND 147 +	GD 156 +
AS 75 +	RB 85 +	NB 95M	PD 106 +	SN 120 +	TE 130 +	BA 137 +	ND 148 +	GD 157 +
AS 76 +	RB 86 +	NB 95F +	PD 107 +	SN 121M	TE 131M +	BA 138 +	ND 149	GD 158 +
AS 77 +	RB 87 +	NB 96	PD 108 +	SN 121F +	TE 131F +	BA 139	ND 150 +	GD 159 +
AS 78	RB 88	NB 97	PD 109 +	SN 122 +	TE 132 +	BA 140 +	FM 147 +	GD 160 +
SE 76 +	RB 89	NB 98	PD 110 +	SN 123M	TE 133M	LA 139 +	PM 148M +	TB 159 +
SE 77 +	SR 86 +	MO 95 +	PD 111M	SN 123F +	TE 134	LA 140 +	PM 148F +	TB 160 +
SE 78 +	SR 88 +	MO 96 +	PD 112 +	SN 124 +	I 127 +	LA 141	PM 149 +	TB 161 +
SE 79 +	SR 89 +	MO 97 +	AG 107 +	SN 125 +	I 128	LA 142	PM 150	TB 162M
SE 80 +	SR 90 +	MO 98 +	AG 109 +	SN 126 +	I 129 +	CE 140 +	PM 151 +	DY 160 +
SE 81M	SR 91 +	MO 99 +	AG 111 +	SN 127	I 130 +	CE 141 +	SM 147 +	DY 161 +
SE 81F	SR 92	MO 100 +	AG 112	SN 128	I 131 +	CE 142 +	SM 148 +	DY 162 +
SE 82	Y 89 +	MO 101	AG 113	SB 121 +	I 132	CE 143 +	SM 149 +	DY 163 +
SE 83	Y 90M	TC 98	AG 115	SB 122 +	I 133 +	CE 144 +	SM 150 +	DY 164
BR 79	Y 90F +	TC 99M +	CD 111 +	SB 123 +	I 134		SM 151 +	HO 165 +
BR 80M	Y 91M	TC 99F +	CD 112 +	SB 124 +	I 135 +		SM 152 +	
BR 80F	Y 91F +	TC 101	CD 113 +	SB 125 +	XE 128 +		SM 153 +	
BR 81 +	Y 92	RU 98	CD 114 +	SB 126 +	XE 129 +		SM 154 +	
BR 82 +	Y 93 +	RU 99	CD 115M +	SB 127 +	XE 130 +		SM 155	
BR 83	Y 94	RU 100 +	CD 115F +	SB 128 +	XE 131M		SM 156 +	
BR 84		RU 101 +	CD 116 +	SB 129	XE 131F +			
		RU 102 +	CD 117M	SB 130	XE 132 +			
		RU 103 +	CD 117F	SB 131	XE 133M			
		RU 104 +	CD 118		XE 133F +			
		RU 105 +			XE 134 +			
		RU 106 +			XE 135M			
					XE 135F +			
					XE 136 +			
					XE 138			

(+) Rendement (Y) + Capture

Tableau II
Actinides

Nuclide	Période (ans)	Incertitude en %	Nuclide	Période (ans)	Incertitude
PO 209	102		Pu 239	24110	± 1
Ra 226	1600		Pu 240	6553	± 1
Ra 228	5.75		Pu 241	14.70	± 1
Ac 227	21.77		Pu 242	376000	± 1
Th 228	1.91		Pu 244	8.26×10^7	± 10
Th 229	7340		Am 241	432.6	± 5
Th 230	77000		Am 242M	152.0	± 20
Th 232	1.41×10^{10}		Am 243	7380	± 40
Pa 231	32760		Cm 243	28.50	± 40
U 232	71.70	± 40	Cm 244	18.11	± 40
U 233	159200	± 30	Cm 245	8500	± 40
U 234	244500	± 20	Cm 246	4730	
U 235	7.04×10^8	± 1	Cm 247	1.56×10^7	
U 236	2.341×10^7	± 20	Cm 248	339700	
U 238	4.468×10^9	± 1	Cm 250	6900	
NP 236F	115000	± 30	BK 247	1380	
NP 237	2.14×10^6	± 30	CF 249	350.6	
Pu 236	2.851	± 40	CF 250	13.08	
Pu 238	87.74	± 10	CF 251	895.0	

Tableau III
Produits de fission - Optique déchets de classement

Nuclide	Période (ans)	Incertitude en %	Nuclide	Période (ans)	Incertitude
SE 79	65000		Rh 102	2.9	
KR 85	10.7		Pd 107	$6.15\ 10^6$	
Rb 87	$4.8\ 10^{10}$		Ag 108M	127	
Sr 90	29.1		SN 121M	55	
Zr 93	$1.5\ 10^6$		SN 126	100000	
Nb 91	10000		I 129	$1.6\ 10^7$	
Nb 92	$3.2\ 10^7$	± 20 %	Cs 134	2	± 20 %
Nb 93M	16.4		Cs 135	$2.3\ 10^6$	
Nb 94F	20300		Cs 137	30	
Tc 97	$2.6\ 10^6$		Sm 146	$7.4\ 10^7$	
Tc 98	$4.2\ 10^6$		Sm 147	$1.1\ 10^{11}$	
Tc 99	213000		Sm 148	$8.\ 10^{15}$	
Rh 101	3.3		Sm 149	$4.\ 10^{14}$	
Sm 151	90		Gd 150	$1.2\ 10^5$	
Eu 150M	34		Tb 157	150	
Eu 152	13.3		Tb 158	150	
Eu 154	8.8		Dy 154	$1.\ 10^7$	
Eu 155	4.9		Ho 163	10	
Gd 148	93		Ho 166M	1200	

Tableau IV
Produits d'activation

Nuclide	Période (ans)	Incertitude en %	Nuclide	Période (ans)	Incertitude
H 3	12.4	± 501	Ar 39	269	
Be 10	$1.6\ 10^6$		Ar 42	33	
C 14	5730	± 300	Fe 55	2.7	± 100
Na 22	2.6	± 30	Co 60	5.3	± 100
Ni 59	75000	± 200	Mo 93	3500	± 200
Ni 63	100	± 200			

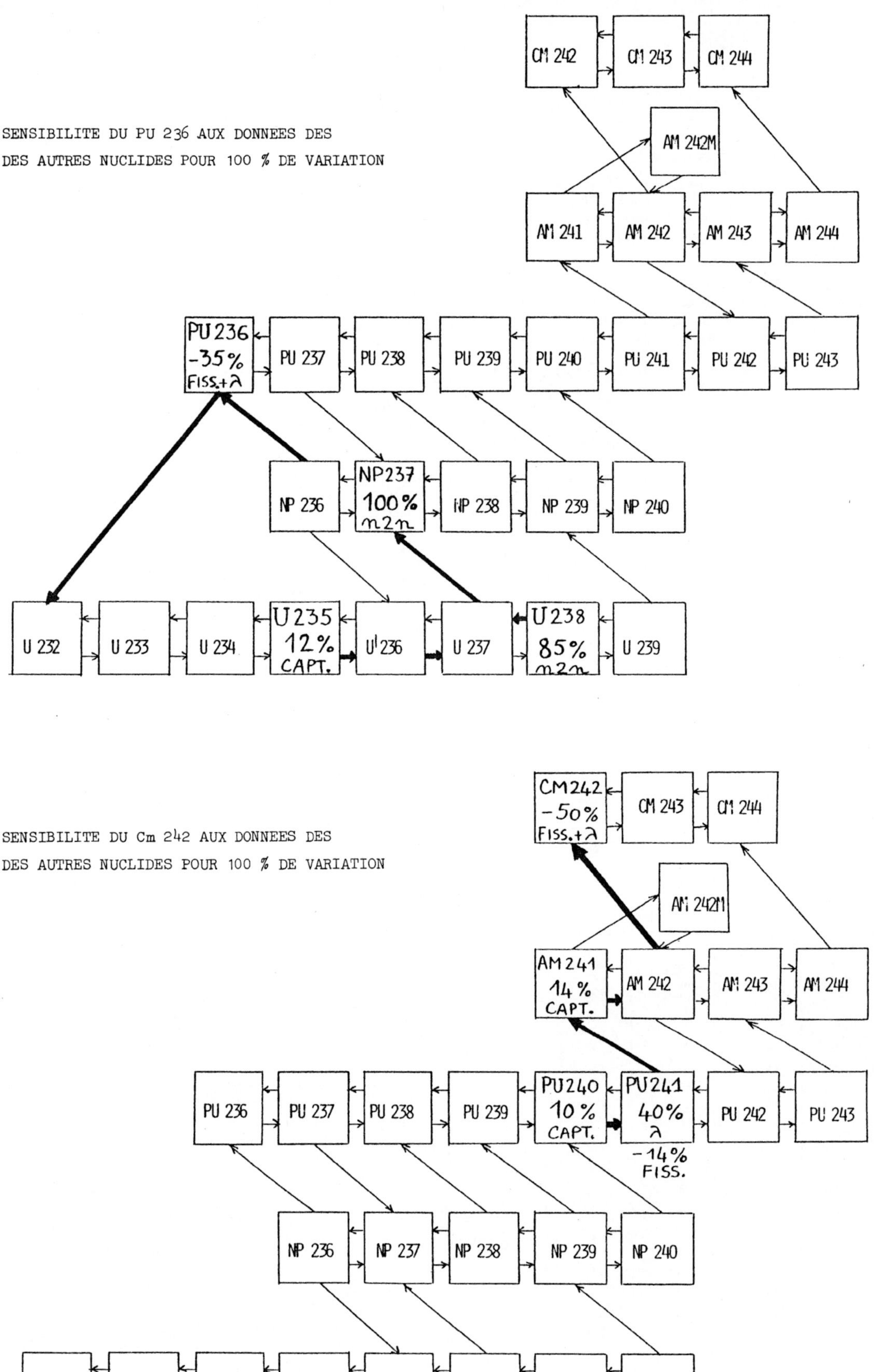

SENSIBILITE DU PU 236 AUX DONNEES DES
DES AUTRES NUCLIDES POUR 100 % DE VARIATION

SENSIBILITE DU Cm 242 AUX DONNEES DES
DES AUTRES NUCLIDES POUR 100 % DE VARIATION

THE TACO EXPERIMENT FOR THE DETERMINATION OF INTEGRAL NEUTRON CROSS-SECTIONS IN A FAST REACTOR

A. Cricchio, R. Ernstberger, L. Koch, R. Wellum
Commission of the European Communities
Joint Research Centre
Karlsruhe Establishment
European Institute for Transuranium Elements
Postfach 2266, D-7500 Karlsruhe
Federal Republic of Germany

For the determination of integral neutron cross-sections, actinides and fission product nuclides have been irradiated in the Rapsodie Fortissima reactor. A normalisation procedure was sucessfully applied which eliminated uncertainties from losses of material before and during the dissolution stages. After irradiation, the nuclides were separated and isotope ratios were measured by mass-spectrometry. Integral neutron absorption, fission and capture cross-sections were determined and are presented for 4 actinide nuclides, and capture cross-sections are given for 15 fission products. Values calculated from microscopic neutron cross-section data are presented and compared.

{ Integral neutron cross-sections, fast reactor, ^{233}U, ^{237}Np, ^{241}Am, ^{243}Am, fission product cross-sections, ^{95}Mo, ^{97}Mo, ^{98}Mo, ^{100}Mo, ^{106}Pd, ^{108}Pd, ^{110}Pd, ^{133}Cs, ^{139}La, ^{141}Pr, ^{143}Nd, ^{144}Nd, ^{146}Nd, ^{148}Nd, ^{149}Sm, mass-spectrometer, KEDAK}

Introduction

The aims of the TACO(TAUX de COMBUSTION) experiment were the measurement of integral neutron cross-sections and fission yields of actinides in a fast reactor (RAPSODIE). The results of the fission yield measurements have been reported previously[1] and the present investigation concentrated on the determination of the integral neutron absorption, capture and fission cross-sections of the nuclides ^{233}U, ^{237}Np, ^{241}Am, ^{243}Am.

For the fission yield determinations corrections for the burn-up of the fission products were needed. A number of selected fission products were therefore irradiated at the same time as the actinides so that their neutron capture cross-sections could be measured.

Experimental

Actinide and fission-product nuclides irradiated are given in table I. Each nuclide to be irradiated was dissolved in nitric acid. Aliquots of the solutions were placed in thin-walled aluminium capsules and dried under vacuum. The nitrates were then decomposed by heating. The capsules were closed under vacuum and collapsed when atmospheric pressure was restored. The capsules were folded and placed in stainless steel cans of 4.3 mm o.d. and 30 mm length which were in turn sealed by electron beam welding. Aliquots of each sample were kept and stored for subsequent analysis.

The irradiation took place in RAPSODIE FORTISSIMA at Cadarache. The capsules were loaded in 8 standard fuel pins. The irradiation started in February 1971 and was completed in June 1972 giving a total irradiation equivalent to 263.5 days at full power.

Principle of Analysis

Complete recovery of the irradiated nuclides and the product material from the irradiation cans is not necessary as the concentration of nuclides analysed is expressed relative to the total number of atoms in the capsule before and after irradiation. Thus if the absorption cross-section of a nuclide is σ_a, the integrated flux ϕt, the number of atoms before irradiation No and the number after irradiation N, then

$$\frac{N}{\Sigma(No)_i} = \frac{No}{\Sigma(No)_i} \cdot \exp(-\sigma_a \phi t) \qquad 1)$$

The total number of atoms before irradiation is expressed as $\Sigma(No)_i$ where i indicates that all isotopes of the irradiated element are to be included in the summation. After irradiation, the total number of atoms is expressed as ΣNi and includes daughter nuclides produced by neutron reactions and subsequent decays. For irradiated actinides, the contribution from fission must be included and the total number of atoms before and after irradiation can be expressed as

$$\Sigma(No)_i = \Sigma Ni + N(Nd)/y \qquad 2)$$

where N(Nd) is the number of neodymium-148 atoms and y its fission yield. (Nd-148 is a convenient isotope for the determination of the number of fissions because its fission yield remains relatively constant from one actinide to another.) Equation 1) can now be expressed with the terms normalised to the total number of atoms before and after irradiation:

$$\frac{N}{\Sigma Ni + N(Nd)/y} = \frac{No}{\Sigma(No)_i} \cdot \exp(-\sigma_a \phi t) \qquad 3)$$

i.e.
$$\sigma_a \phi t = -\log \frac{N\, \Sigma(No)_i}{No\,(\Sigma Ni + N(Nd)/y)} \qquad 4)$$

The simplest application of this formula is for a nuclide which by neutron capture produces an isotope of the parent element:

$$^{A}_{Z}X(n,\gamma)\,^{A+1}_{Z}X$$

In this case a measurement of isotope ratios before and after irradiation determines the neutron capture cross-section. If however, the product of neutron capture, ^{A+1}X, decays by beta or alpha emission, then the ratio of the parent nuclide and decay products must be determined. The method of choice was mass-spectrometric isotope dilution because of its precision, wide applicability and relative lack of interferences.

By similar arguments to those above, the capture cross-section can be expressed as:

$$\sigma_c \phi t = \frac{\frac{^{A+1}N}{^{A}N} \cdot \exp(-^{A}\sigma_a \phi t) - \frac{^{A+1}No}{^{A}No} \cdot \exp(-^{A+1}\sigma_a \phi t)}{\exp(-^{A}\sigma_a \phi t) - \exp(-^{A+1}\sigma_a \phi t)} \cdot (^{A+1}\sigma_a - {}^{A}\sigma_a)\phi t \qquad 5)$$

and for actinides the fission cross-section as

$$\sigma_f \phi t = \frac{\sigma_a \phi t\, \Sigma(No)_i \;/\; (1 + y \cdot \Sigma Ni / N(Nd))}{No\;(1 - \exp(-\sigma_a \phi t))} \qquad 6)$$

The cross-sections of reactions such as (n,2n), (n,3n), (n,p) and (n,α) are low compared with σ_c and σ_f and therefore an independent check of the neutron capture cross-section is obtained:

$$\sigma_c \phi t = \sigma_a \phi t - \sigma_f \phi t \qquad 7)$$

K. H. Böckhoff (ed.), Nuclear Data for Science and Technology, 175–177.

Table I Actinide and fission-product nuclides irradiated in RAPSODIE

^{233}U	^{237}Np	^{241}Am	^{243}Am			
^{95}Mo	^{106}Pd	^{133}Cs	^{139}La	^{141}Pr	^{143}Nd	^{149}Sm
^{97}Mo	^{108}Pd				^{144}Nd	
^{98}Mo	^{110}Pd				^{146}Nd	
^{100}Mo					^{148}Nd	

Table II Neutron energy spectrum of RAPSODIE

Energy range		Neutrons (%)
0.748-1.23	kev	0.03
1.23 -2.04	kev	0.115
2.04 -3.36	kev	0.165
3.36 -5.53	kev	0.207
5.53 -9.12	kev	0.418
9.12 -15.0	kev	0.809
15.0 -24.8	kev	2.00
24.8 -40.9	kev	3.03
40.9 -67.4	kev	4.44
67.4 -111	kev	7.14
111 -183	kev	9.89
183 -302	kev	14.2
302 -498	kev	13.5
498 -821	kev	17.4
.821 -1.35	Mev	10.6
1.35 -2.23	Mev	8.69
2.23 -3.68	Mev	5.18
3.68 -14.5	Mev	2.17

Separation and Isotope Ratio Measurements

The capsules were dissolved individually in nitric acid in the presence of mercuric nitrate to aid the dissolution. Spikes were added where necessary to aliquots of the solution and after preliminary pre-conditioning steps to ensure isotopic exchange between spikes and sample nuclides, separations were carried out to remove aluminium and possible isobaric interferences for the mass-spectrometric measurements.

The separation procedures chosen employed ion-exchange. The resultant solutions containing the purified nuclides of interest were evaporated to dryness, taken up in dilute nitric acid and dried on the filament of the mass-spectrometer.

Mass-spectrometry was carried out using a Varian MAT CH5 spectrometer with rhenium filaments.

Duplicate aliquots were analysed from each capsule and, with the exception of ^{243}Am for which insufficient material was available, two capsules of each actinide were irradiated. This allowed estimations to be made of the measurement errors. For certain of the fission product nuclides in particular, interferences due to natural background from the filaments were observed. Corrections were applied based on selected masses, naturally abundant but depleted in the irradiated material.

Results and Discussion

The cross-sections obtained from the experiment are integral neutron cross-sections specific to the irradiation conditions in the RAPSODIE reactor.

It was of interest to compare these integral cross-sections with those calculated from microscopic neutron cross-sections even though the neutron energy spectrum at the irradiation position was not known very accurately. The results (tables III,IV) are given here together with calculated cross-sections[2] for the following actinides: ^{233}U, ^{237}Np, ^{241}Am and ^{243}Am and in table V for certain fission products. The calculated cross-sections were obtained using the microscopic cross-section library KEDAK[3,4,5] and the pertinent neutron spectrum[6] (table II).

Standard deviations, expressed in percent of the measured values, were calculated either from duplicate or triplicate capsule results and are also inserted in tables III and IV for the actinides. (Errors are not given for the fission-product results as only one capsule was irradiated for each fission-product nuclide). Neutron capture cross-sections were calculated directly (eq. 5) and also from the absorption and fission cross-sections (eq. 7). Comparison of the values from the two methods demonstrates the internal consistency of the measurements. Three capture cross-sections were determined for ^{241}Am: for the reactions (a) $^{241}Am(n,\gamma)^{242m}Am$, (b) $^{241}Am(n,\gamma)^{242}Am \xrightarrow{\varepsilon c} {}^{242}Pu$ and (c) $^{241}Am(n,\gamma)^{242}Am \xrightarrow{\beta} {}^{242}Cm$. The daughter products measured were ^{242m}Am, ^{242}Pu and ^{238}Pu (produced by decay of ^{242}Cm) respectively.

Table III Integral absorption and fission cross-sections of actinides in RAPSODIE (barns). Percentage standard deviations given in brackets.

Isotope	Absorption		Fission	
	σ(exp)	σ(calc)	σ(exp)	σ(calc)
^{233}U	2.47 (5)	2.29	2.31 (6)	2.12
^{237}Np	1.28 (7)	1.26	0.65 (7)	0.60
^{241}Am	1.62 (9)	1.42	0.70 (7)	0.52
^{243}Am	1.48 (-)	1.08	0.45 (-)	0.47

Table IV Integral capture cross-sections of actinides in RAPSODIE (barns). Percentage standard deviations given in brackets.

Isotope	σ_c(exp)	σ_c(calc)	σ_c(eq. 7)
^{233}U	0.155(.2)	0.155	0.156
^{237}Np	0.64 (7)	0.66	0.63
^{241}Am	1.05 (7)	0.89	0.92
	0.17 (12)	(^{242m}Am)	
	0.15 (16)	(^{242}Pu)	
	0.73 (15)	(^{242}Cm)	
^{243}Am	1.0 (-)	0.61	1.03

Table V Integral cross-sections of fission products irradiated in RAPSODIE (barns)

Isotope	σ_c(exp)	σ_c(calc)
^{95}Mo	0.070	0.133
^{97}Mo	0.158	0.128
^{98}Mo	0.015	0.038
^{100}Mo	0.092	0.044
^{106}Pd	0.094	0.097
^{133}Cs	0.208	0.180
^{139}La	0.030	0.012
^{141}Pr	0.056	0.039
^{143}Nd	0.140	0.129
^{144}Nd	0.045	0.039
^{146}Nd	0.058	0.052
^{148}Nd	0.072	0.067
^{150}Nd	0.084	0.077
^{149}Sm	0.759	0.787

Conclusions

A method was developed and sucessfully employed whereby only nuclide ratios have to be measured thus avoiding errors due to material losses during and after dissolution. The measurements were performed using mass-spectrometric techniques almost exclusively.

One of the original aims of the experiment was the determination of fission yields - the results have been reported previously[1] - and only a limited amount of effort was given to characterising the neutron energy spectrum. The comparison of the experimental results with those calculated from microscopic cross-sections is therefore only of limited value. The experience acquired in this experiment will be used in a follow-up experiment planned in KNK II under circumstances where the neutron flux can be well characterised.

References

1. L. Koch et.al., European Applied Reports, vol. 3, no. 1 and 2 (1981)

2. I. Broeders, KfK, Karlsruhe, F.D.R., private communication

3. B. Krieg, KfK-1725 (1973)

4. F.H. Fröhner, B. Goel, U. Fischer, F. Jahn, paper A-16, this conference

5. H. Gruppelaar, NEANDC/NEACRP Specialists' Meeting, Argonne 20-23 April 1982, ECN-82-045

6. M. Robin, CEA-Cadarache, France, private communication

INTEGRAL EXPERIMENTS TO MEASURE THE PRODUCTION RATES OF ^{242}Cm AND ^{244}Cm IN FAST REACTOR SPECTRA

J.M. Stevenson, A.D. Knipe, D.W. Sweet

Reactor Physics Division, AEE, Winfrith, Dorchester, Dorset, England

R.A.P. Wiltshire, K.M. Glover, B. Whittaker

Chemistry Division, AERE, Harwell, Oxfordshire, England

Integral production cross-sections for ^{242}Cm and ^{244}Cm have been experimentally determined in the zero power fast reactor ZEBRA. Sensitive radiochemical methods have been used to determine the curium produced in milligram samples of irradiated ^{241}Am and ^{243}Am. The experimental results are compared with values calculated using neutron spectra from ZEBRA standard analysis methods and americium capture cross-sections from recent Harwell differential evaluations. The comparisons provide valuable information on the accuracies of these capture data, which were largely based on theoretical models.

Introduction

The accurate prediction of the neutron source strength in irradiated subassemblies in a fast power reactor is important both for the interpretation of subcritical monitoring during refuelling operations and for the design of the fuel handling route. This neutron source is dominated by spontaneous neutrons from ^{242}Cm and ^{244}Cm which are formed by neutron capture in ^{241}Am and ^{243}Am, respectively. Lack of suitable target material makes the measurement of differential cross-sections of these higher actinides difficult, and considerable reliance has to be placed on theoretical models, as in recent Harwell evaluations of the data for these isotopes (e.g. Reference 1).

One-group curium production cross-sections have now been determined by measuring the curium produced following irradiation of small samples of ^{241}Am and ^{243}Am in a representative range of spectra in four assemblies of the zero power fast reactor ZEBRA. The curium produced was chemically separated and measured by alpha counting and alpha spectrometry techniques and the integral production rate for each sample was obtained. The measurement of ^{239}Pu fission rates at the sample positions provided the data for the fluence received during irradiation.

The measured production cross-sections have been compared with values calculated using 37-group fluxes obtained using standard ZEBRA analysis methods, and 37-group americium capture cross-sections deduced from the Harwell evaluations.

Measurements

The core and breeder regions of ZEBRA assemblies consist of plate cells of the type shown in Figure 1. The cells are contained inside vertical steel sheaths approximately 51mm square. The individual plates are 3.17 or 6.35mm thick and the plutonium, UO_2 and sodium plates are clad in steel. Americium samples of 6-12mgs $^{241}AmO_2$ and 10-15mgs $^{243}AmO_2$, contained in small aluminium ampoules, were introduced into the cells in 12.7mm thick aluminium sample holders (see Figure 2), which replaced two sodium plates. In the first three assemblies, the samples were introduced at the centres of the cylindrical cores. The last assembly contained an annular core and samples were introduced at the mid-core radius and at the middle and edge of the central breeder island. The samples were irradiated at maximum power in ZEBRA (peak flux ~2 x 10^{16} n/cm^2/sec) in 8 hour units over periods of up to 40 days.

Because of very low yields of curium isotopes produced in the irradiation (0.1-1.0ppm by activity) it was necessary to chemically separate the curium product from americium, before any estimation of curium could be made. To avoid the possibility of ^{242}Cm growing into ^{241}Am because of the presence of any ^{242m}Am contamination, isotopically pure ^{241}Am, milked from ^{241}Pu, was used for the irradiations. The ^{243}Am irradiated was exhaustively purified from curium isotopes and the low level of ^{242}Cm present from the ^{242m}Am decay measured; any effects on the estimation of chemical yield because of ^{242}Cm in-growth were minimised by over-spiking.

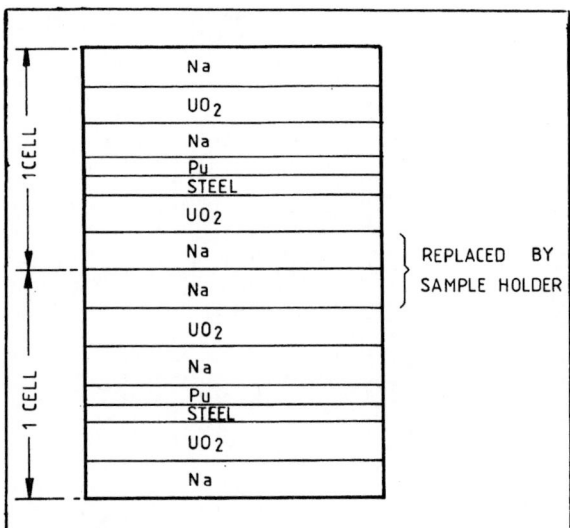

Fig. 1. Layout of two cells from ZEBRA 21.

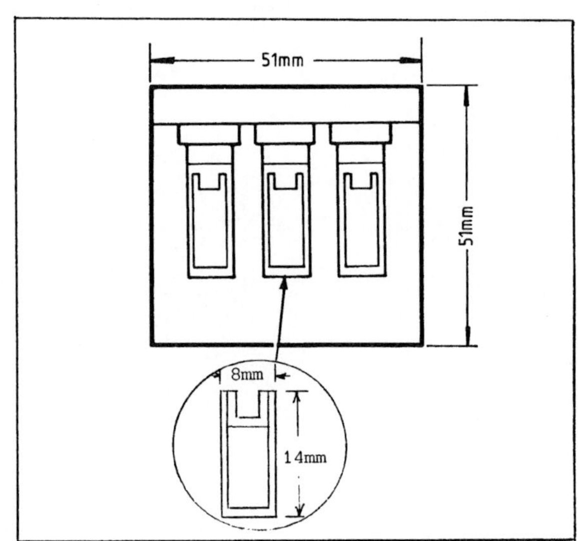

Fig. 2. Aluminium plate and ampoule used to insert 6 —15mgs AmO_2 into cell.

K. H. Böckhoff (ed.), Nuclear Data for Science and Technology, 178–180.

Following irradiation the samples of americium oxide were removed from the sample holder and the ampoules and contents were dissolved in hydrochloric acid. Americium and curium were co-precipitated by addition of excess NaOH; the aluminium remained in solution as the soluble aluminate. The Am/Cm hydroxide was recovered by centrifugation and redissolved in hydrochloric acid before being made up to volume in a volumetric flask previously spiked with a known activity of curium. Curium-244 was used as a spike to measure chemical yields in the ^{241}Am irradiations and ^{242}Cm was used similarly for the ^{243}Am samples. The quantity of americium irradiated was measured to an accuracy of ±0.5% by taking known aliquots from the volumetric flasks and preparing them as thin alpha sources for gross alpha counting in an absolute low geometry counter. Chemical separation of the curium from americium was achieved using cation exchange columns heated at 50°C, eluting with 0.4M alpha-hydroxyisobutyric acid adjusted to pH4.0. Full details are given in Reference 2.

The final curium product was prepared as a thin source on a tantalum tray using tetraethylene glycol as a spreading agent. Quantitative and qualitative analyses of curium isotopes were made using gross alpha counting and alpha spectrometry techniques.

The curium yields were converted to production cross-section using integrated fluxes at the sample positions. These flux levels were deduced from measured ^{239}Pu fission rates and one-group ^{239}Pu fission cross-sections calculated as described in Section 3. The fission rates were determined either after the irradiation via ^{239}Pu foils introduced into the sample position and subsequently gamma counted for fission products using calibrated detector, or during the irradiation by using calibrated ^{239}Pu fission chambers introduced into the plate cells near the sample.

The measured one-group production cross-sections are shown in Table 1.

Estimates of uncertainty in these measurements have been identified as follows:

(1) radiochemical data (random) ±1-2% for ^{241}Am, ±3-4% for ^{243}Am;

(2) radiochemical data (systematic) ±2% for ^{241}Am, ±5% for ^{243}Am;

(3) fluence ±5% for core samples, ±7% for breeder samples.

Combining these errors gives an uncertainty in the measured capture cross-sections of 6% to 8% for ^{242}Am and 8% to 9% for ^{243}Am.

Calculated Cross-Sections and Comparison With Experiment

Calculated one-group cross-sections for capture in ^{241}Am and ^{243}Am and for fission in ^{239}Pu were obtained by the standard methods for analysing ZEBRA assemblies. Fluxes from 37-group diffusion-theory models, weighted to allow for the flux fine structure in the plate cell, were combined with microscopic cross-sections.

Cell cross-sections for the core and breeder zones were prepared by the collision probability code MURAL (Reference 3), and the 2240 group FGL5 data set (Reference 4). The plate cells were represented in slab geometry with the plutonium and uranium regions represented separately and the materials between smeared together. The single buckling search option, appropriate to a large region of the cell, was used. Cell-average macroscopic cross-sections, microscopic cross-sections for each region and flux fine structures through the cell, were output in the 37-groups of the FD5 scheme. More details of the cell calculations are given in reference 4.

The cell-average macroscopic cross-sections were used with the diffusion-theory code TIGAR - generally in 3D because of the complexity of the ZEBRA assemblies. The resulting 37-group fluxes at the sample positions in the reactor were corrected for the fine structure variation of the flux in the cell and combined with the appropriate microscopic cross-sections. The broad-group data for capture in ^{241}Am and ^{243}Am were produced from the Harwell evaluations which were almost entirely based on theoretical models. The derivation of the data for ^{241}Am is described in reference 1 where the capture cross-sections are in reasonable agreement with the limited number of differential measurements. The ^{239}Pu fission cross-sections for the sample or chamber location were from the MURAL calculation.

Three corrections have been applied to the calculated one-group cross-sections:-

(i) The errors associated with the use of diffusion theory for the flux calculations have been estimated by comparing transport theory and diffusion theory in BZ geometry. The only significant

Sample Location	^{242}Cm Production			^{244}Cm Production		
	Experimental (E)	Calculated (C)	C/E	Experimental (E)	Calculated (C)	C/E
ZEBRA 12 Core Centre	1.48	1.28	0.86	-	-	-
ZEBRA 14 Core Centre	1.28	1.16	0.91	1.61	1.60	0.99
ZEBRA 16 Core Centre	-	-	-	1.99	1.72	0.86
ZEBRA 21 Mid-Core Radius	1.18	1.03	0.87	1.34	1.43	1.06
Edge of Breeder Island	2.20	1.88	0.86	2.79	2.58	0.92
Near Centre of Breeder Island	3.92	3.13	0.80	-	-	-
Centre of Breeder Island	4.07	3.32	0.82	4.85	4.52	0.93
Mean and Standard Deviation			0.85 ±0.04			0.95 ±0.08

Table 1 Experimental and Calculated One-Group Cross-Sections for Curium Production

difference was at the edge of the breeder in ZEBRA 21 where the americium capture cross-sections from diffusion theory have been increased by 1.5%.

(ii) Self-shielding effects in the americium samples have been estimated using ^{235}U capture resonance data. The calculated one-group capture cross-sections at the centre and edge of the breeder island of ZEBRA 21 have been reduced by 2% and 1%, respectively. For the core positions the effects are negligible (<0.5%).

(iii) The perturbation effects caused by replacing sodium plates in the standard cells by the aluminium sample holders and the special fission chambers have been studied using MURAL and small corrections (~1.5%) have been applied to the calculated capture cross-sections. The corrections to the ^{239}Pu fission cross-sections were negligible.

The calculated one-group cross-sections for curium production are also given in Table 1. For ^{242}Cm, the calculated capture cross-sections for ^{241}Am have been multiplied by 0.70 to allow for capture events which lead to ^{242m}Am, and for ^{242}Am decaying by electron capture to yield ^{242}Pu. There is an uncertainty of a few percent in this factor.

The C/E ratios vary between 0.80 and 0.91 for ^{242}Cm production and 0.86 and 1.06 for ^{244}Cm production, showing much better agreement than with earlier americium group cross-sections. The mean ratios are 0.85 and 0.95 with standard deviations of 0.04 and 0.08. These mean C/E ratios can be used as bias factors for curium production rates calculated using methods and data similar to those for the ZEBRA analysis.

Conclusions

Sensitive radiochemical measurements have provided direct determinations of very small amounts of ^{242}Cm and ^{244}Cm produced by neutron capture in ^{241}Am and ^{243}Am in a range of fast reactor spectra in ZEBRA. The results have been converted to one-group production cross-sections, which have been compared with calculated values obtained using neutron spectra from standard ZEBRA analysis methods, together with capture cross-sections from recent differential evaluations of americium data. The calculated to experimental ratios show means of 0.85 for ^{242}Cm production and 0.95 for ^{244}Cm production. The uncertainties on these ratios are 5% and 8%, respectively. These data give valuable support to the evaluations which, because of shortage of suitable target material for differential cross-section measurements, were almost entirely based on theoretical models.

The mean ratios can be applied as bias factors to calculated production rates of the curium isotopes which dominate the neutron source in irradiated sub-assemblies of fast power reactors.

References

1. J.E. Lynn, B.R. Patrick, M.G. Sowerby, E.M. Bowey, Progress in Nuclear Energy, 5, 255 (1980).
2. R.A.P. Wiltshire, K.M. Glover, Journal of Radioanalytical Chemistry, 64, 47 (1981).
3. J.D. McDougall, International Symposium on Physics of Fast Reactors, Tokyo, Paper A32 (October 1973).
4. J.L. Rowlands, C.J. Dean, J.D. McDougall, R.W. Smith, International Symposium on Physics of Fast Reactors, Tokyo, Paper A30 (October 1973).
5. M.J. Grimstone, J.M. Stevenson, R. Bohme, E.A. Fischer, Fast Reactor Physics 2, 401 (197).

DETERMINATION EXPERIMENTALE DES SECTIONS EFFICACES DES ISOTOPES DE Pu, Am, Cm
DANS UN SPECTRE DE NEUTRONS DE REACTEUR A EAU

M. Darrouzet, A. Giacometti, R. Girieud, M. Robin

Commissariat à l'Energie Atomique
Département des Réacteurs à Eau

Centre d'Etudes Nucléaires de Cadarache
B.P. N° 1 - 13115 - SAINT PAUL LEZ DURANCE
FRANCE

ABSTRACT

The experiment described here determines the effective capture cross-sections of the principal isotopes of Pu, Am and Cm in a LWR neutron spectrum from the formation by capture under irradiation of the sucessor isotope of the studied isotope.
The system, representative of assemblies used in PWR, was irradiated in the swimming-pool reactor MELUSINE (CEN de GRENOBLE).
It consisted of a 5 x 5 square lattice with PWR 17 x 17 fuel type (enriched UO_2, zircaloy cladding).
To minimize neutron flux perturbations, some pellets of the central pin were slightly doped (∿ 0.1 mass per cent) with a particular isotope to be studied.
In this paper, we present : experimental apparatus - study of the method to determine effective capture cross-sections - transposition of the results to a PWR reference lattice - comparison with values calculated with French library.
The results concerned the following isotopes : ^{240}Pu, ^{241}Pu, ^{242}Pu, ^{241}Am, ^{243}Am, ^{244}Cm.

1. Introduction

Le but de l'expérience décrite ici est d'améliorer la connaissance des sections efficaces de capture des principaux isotopes du plutonium, de l'américium et du curium. Ce travail a été réalisé dans le cadre d'un contrat avec la commission des Communautés Européennes.

Les sections de capture sont déterminées à partir de la formation du nucléide "fils", au cours d'une irradiation dans un spectre neutronique semblable à celui qui règne dans le combustible des réacteurs PWR.

On décrit sommairement le dispositif expérimental, irradié dans le réacteur piscine de Grenoble, et le programme post-irradiatoire qui permet de déterminer la variation du rapport des nombres d'atomes des nucléides "fils" et "père". On en déduit facilement les taux de capture dans les conditions expérimentales.

L'étape suivante de l'interprétation consiste à vérifier que le spectre neutronique, réalisé dans le réacteur expérimental, est voisin de celui existant dans un réacteur PWR, et à transposer les taux de réaction dans ce dernier spectre, pour obtenir ainsi des sections efficaces caractéristiques d'un réseau PWR bien défini.

Les résultats expérimentaux sont ensuite comparés aux valeurs correspondantes obtenues avec la bibliothèque française de sections efficaces.

2. Description de l'expérience SHERWOOD

2.1 Description du dispositif expérimental

Le coeur du réacteur est constitué d'assemblages du type MTR. Le dispositif d'irradiation a été introduit dans un créneau entré en périphérie du coeur, en remplacement d'un assemblage standard.

Il est composé d'un assemblage de 5 x 5 crayons de type PWR 17 x 17 (UO_2 enrichi et gainage zircaloy) dans un boîtier en aluminium. Sur la figure 1 est représenté, le coeur du réacteur et la position du dispositif d'irradiation.

La température de l'eau du coeur étant de 32°C, le pas des crayons a été réduit afin de conserver le rapport de modération des réacteurs de puissance. L'enrichissement de l'UO_2 retenu (2,8 %) a permis d'éviter l'emploi d'un système de refroidissement supplémentaire.

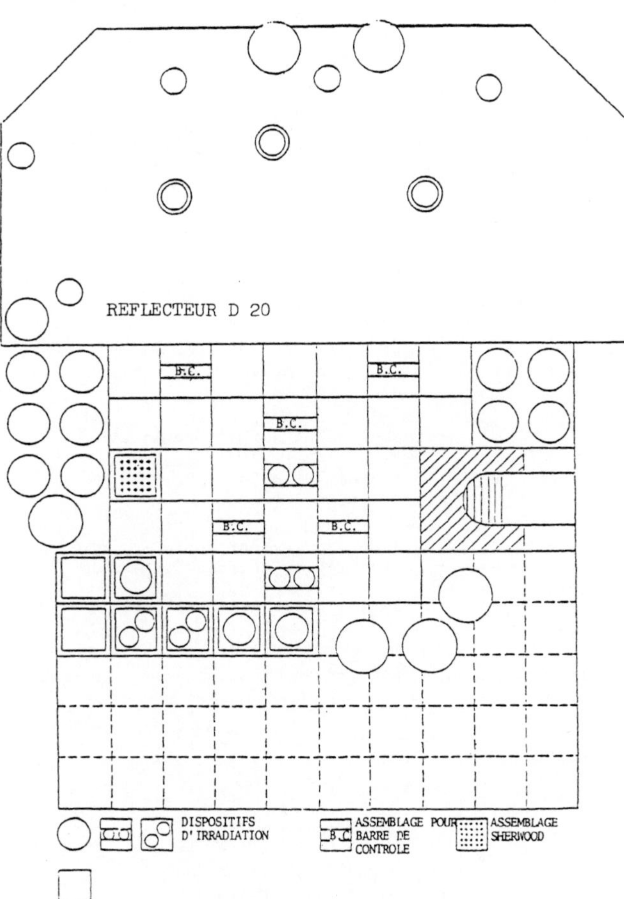

Fig. 1 Réacteur Melusine
Configuration pour l'irradiation Sherwood

K. H. Böckhoff (ed.), Nuclear Data for Science and Technology, 181–184.

Les nucléides, dont on veut mesurer les sections efficaces de capture, sont introduits au centre du réseau 5 x 5, sous forme de traces (environ 0.1 % en masse) dans certaines pastilles d'UO_2 du crayon central.

2.2 Conditions d'irradiation

L'irradiation a eu lieu en Juin et Juillet 1979 représentant 1021.5 heures de fonctionnement du réacteur à pleine puissance. Le taux de combustion maximum atteint par le crayon central a été de 1000 MWj/t environ.

2.3 Mesures post-irradiatoires

Une spectrométrie γ du crayon central a permis d'obtenir la répartition de la fluence le long du crayon.

Les taux de réactions sont déduits de la variation de la teneur en isotope "fils", ce qui implique des analyses sur des pastilles irradiées et des pastilles témoins. Les différentes concentrations sont obtenues par spectrométries de masse, α ou γ.

Les mesures ont portées sur des pastilles d'UO_2 pur et dopées en ^{239}Pu, ^{242}Pu, ^{241}Am, ^{243}Am, ^{244}Cm. 3 pastilles irradiées et 2 témoins ont été analysés sur le lot de pastilles d'UO_2 pur, et 2 irradiées et 1 témoin pour chaque lot de pastilles dopées.

3. Détermination des sections efficaces

3.1 Détermination des taux de réaction dans les échantillons.

Le taux de capture (σ_{ci} τ_i) de l'isotope A_i peut être relié à la formation de l'isotope A_{i+1} par une relation simple de la forme :

$$\frac{N_{i+1}(\tau_i)}{N_i(\tau_i)} = \sigma_{ci}\,\tau_i \times f_o(\sigma_i\,\tau_i)$$

avec les notations suivantes :

- σ_{ci} — section efficace de capture de l'isotope A_i au début de l'irradiation
- τ_i — fluence reçue par l'échantillon
- $N_i(\tau_i)$ — nombre de noyaux de l'isotope A_i en fin d'irradiation

Le terme $f_o(\sigma_i\,\tau_i)$ est un facteur correctif qui prend en compte toutes les autres réactions susceptibles de faire varier N_i et N_{i+1} et l'évolution des sections efficaces en cours d'irradiation.

Si l'isotope A_{i+1} est présent initialement, on peut écrire :

$$\frac{N_{i+1}(\tau_i)}{N_i(\tau_i)} - \frac{N_{i+1}(0)}{N_i(0)} = \sigma_{ci}\,\tau_i \times f_o(\sigma_i\,\tau_i)$$

et par un raisonnement analogue, on a :

$$\frac{N_{i+2}(\tau_i)}{N_{i+1}(\tau_i)} = \frac{1}{2}\,\sigma_{ci+1}\,\tau_i \times f_1(\sigma_i\,\tau_i)$$

L'obtention des taux de réactions de capture dans chaque échantillon suppose donc simplement la connaissance du rapport de concentrations adéquat et d'un facteur correctif.

Si l'on veut pouvoir comparer les résultats entre échantillons différents il est, de plus, nécessaire de connaître la répartition de la fluence, afin de ramener les taux de réactions à une fluence de référence.

Les rapports isotopiques sont déterminés par spectrométrie de masse après dissolution des échantillons.

Les facteurs correctifs, proche de l'unité, sont obtenus en simulant dans un calcul d'évolution tous les phénomènes intervenant sur la concentration des isotopes A_i et A_{i+1}.

3.2 Calcul des facteurs correctifs $f(\sigma_i\,\tau)$

Le calcul est effectué à l'aide du module EVOGENE du système KAFKA mis au point au CEA. EVOGENE est un code d'évolution ponctuelle permettant de calculer les concentrations de tous les isotopes lourds et d'un certain nombre de produits de fission pendant l'irradiation du combustible.

Il utilise des sections efficaces condensées à 1 groupe d'énergie, paramétrées en fonction de la fluence, à partir des résultats du code de calcul neutronique APOLLO [2].

Le module Apollo du système NEPTUNE [3] utilisé, résoud l'équation intégrale du transport avec un découpage en 99 groupes d'énergie. En option MULTICELLULE, il permet la prise en compte des interactions neutroniques entre "cellules" de nature différentes.

Le maillage adopté permet de représenter les 25 cellules du réseau 5 x 5, l'environnement du réseau étant constitué de milieux homogènes équivalents.

On en déduit
$$f(\sigma_i\,\tau) = \frac{1}{\sigma_{ci}\,\tau}\;\frac{N_{i+1}(\tau)}{N_i(\tau)}$$

3.3 Détermination de la fluence

La répartition de la fluence entre les différentes pastilles est déduite des mesures de gammamétrie faites sur le crayon central, en effectuant un lissage sur les valeurs obtenues sur les pastilles d'UO_2 pur.

La fluence absolue, nécessaire au calcul des facteurs correctifs, est obtenue à partir de la valeur moyenne des taux de fission du ^{235}U des 3 pastilles d'UO_2 pur.

4. Transposition des résultats à un réseau PWR 17 x 17

Le but du programme expérimental est la détermination d'un jeu de sections efficaces caractéristiques d'un réseau PWR.

Le spectre neutronique obtenu dans le crayon central est aussi proche que possible du spectre souhaité. Néanmoins, il faut prendre en compte les perturbations induites au niveau des échantillons (densité, diamètre, etc...), l'effet de l'environnement sur le réseau 5 x 5, les différences spécifiques aux deux types de réseau (5 x 5 à pas réduit, PWR 17 x 17).

Les taux de réactions obtenus ayant été ramenés à une même fluence, déduite du taux de fission moyen de ^{235}U, on détermine un jeu de rapports de sections efficaces σ_{ci}/σ_{f5}.

D'autre part, le même code de calcul neutronique (APOLLO) a été utilisé pour calculer les sections efficaces à 1 groupe d'énergie dans le crayon central du réseau 5 x 5 et dans le réseau PWR 17 x 17.

La transposition des résultats sera donc assurée par une formulation du type :

$$\frac{\sigma_{ci}}{\sigma f_5} = \left[\frac{\sigma_{ci}}{\sigma f_5}\right] \text{exp.} \times \left[\frac{\dfrac{\sigma_{ci}}{\sigma_{f5}}\ \text{PWR 17 x 17}}{\dfrac{\sigma_{ci}}{\sigma f_5}\ \text{échant.}}\right]_{\text{calcul}}$$

Les effets de spectre sur le crayon central dus à l'environnement ont été étudiés en comparant, les flux en 4 groupes d'énergie du calcul du dispositif et du crayon central en réseau infini.

Les bornes sont les suivantes

ϕ_1 entre 0.9 MeV et 10 MeV

ϕ_2 entre 5 keV et 0.9 MeV

ϕ_3 entre 0.625 eV et 5 keV

ϕ_4 entre 0 et 0.625 eV

L'effet de l'environnement apparaît dans le tableau ci-dessous :

	ϕ_1/ϕ_4	ϕ_2/ϕ_4	ϕ_3/ϕ_4
Dispositif réel	1.7	2.6	1.8
Réseau infini	1.7	2.4	1.5

Le même code de calcul est utilisé pour calculer le réseau de type PWR 17 x 17 en milieu infini, d'enrichissement 2.8 %.

La bonne similitude entre les deux spectres permet la transposition des résultats.

On dispose ainsi des sections efficaces de capture de ^{235}U et ^{238}U par analyses sur les pastilles d'UO_2 pur. La précision des analyses a permis, en outre, d'en déduire les sections efficaces des ^{239}Pu et ^{240}Pu.

Des mesures sur les pastilles dopées en ^{239}Pu on obtient les sections de ^{240}Pu et ^{241}Pu.

Les autres sections sont déduites des mesures sur les pastilles dopées en isotope de même nom : ^{242}Pu, ^{241}Am, ^{243}Am, et ^{244}Cm.

5. Ecarts expérience - bibliothèque française

La bonne cohérence des taux de réaction mesurés et calculés pour les isotopes les mieux connus (^{235}U, ^{238}U, ^{239}Pu) montrent que le calcul du spectre est correct.

Dans le tableau I sont donnés les écarts expérience-calcul en utilisant la bibliothèque française ainsi qu'un jeu d'ajustement sur les isotopes mineurs déduits d'études précédentes.[4] [5]

TABLEAU I

COMPARAISONS DES SECTIONS EFFICACES DE CAPTURE (NORMALISEES A σ_{f5}) DANS LE CRAYON CENTRAL, MESUREES ET CALCULEES

Section de capture	Valeur expérimentale	Ecart avec la bibliothèque franc. (E-C/C en %)
240 Pu	3.35 ± 0.07	0.1
241 Pu	0.853 ± 0.029	- 1.0
242 Pu	0.416 ± 0.019	- 18.
241 Am	2.28 ± 0.06	2.0
243 Am	0.982 ± 0.025	14.
244 Cm	0.227 ± 0.015	- 5.0

6. Conclusion

Cette expérience a permis de déterminer les sections efficaces de capture des transuraniens dans un spectre de réacteur PWR. Les grandes options retenues se sont révélées bonnes : irradiation dans un réacteur de recherche très souple d'utilisation, réalisation d'un réseau de 5 x 5 crayons à pas réduit pour retrouver de bonnes conditions neutroniques, étude individuelle des sections efficaces de capture par addition à de l'UO_2 pur d'une faible quantité de chacun des isotopes étudiés.

Les résultats obtenus sont utiles à tous ceux qui sont concernés par le cycle du combustible. Les sections efficaces intégrales obtenues ici sont directement utilisables dans certains cas caractérisés par des conditions neutroniques voisines du réseau de référence étudié. Elles sont plus particulièrement destinées à la qualification des bibliothèques de sections efficaces utilisées dans les codes de calcul des réacteurs à eau.

Remerciements

Les auteurs de ce document tiennent à remercier tout spécialement les Communautés Européennes qui ont contribuées à la réalisation de cette expérience. Ils remercient également de leurs contributions, le personnel du réacteur MELUSINE, le laboratoire COMIR du Département des Réacteurs à Eau ainsi que la Section d'Etudes Analytiques du Cycle du Combustible, la Section d'Etudes et d'Analyse Isotopique et Nucléaire du Département de Chimie Appliquée et d'Etudes Analytiques.

Références

[1] J. PINEL
KAFKA - Communication personnelle.

[2] A. HOFFMANN et Al.
APOLLO - Code multigroupe de résolution de l'équation du transport pour les neutrons thermiques et rapides
Note CEA N-1610 (1973).

[3] J. BOUCHARD, A. KAVENOKY, P. REUSS
Le développement et la qualification du système NEPTUNE de calcul des réacteurs à eau.
European Nucléar Conférence -
HAMBOURG (Mai 1979).

[4] A. GIACOMETTI
Etude neutronique des noyaux lourds formés dans le cycle du combustible des réacteurs nucléaires
Thèse - Orsay (Mai 1978).

[5] M. DARROUZET, A. GIACOMETTI, M. ROBIN
Formation et disparition des actinides secondaires dans les réacteurs à eau et à neutrons rapides.
International Conference on Neutron Physics and Nuclear Data for Reactors (Harwell, Septembre 1978).

REACTOR IRRADIATIONS OF ^{242}Pu, COMPARISON OF MEASURED AND CALCULATED YIELDS OF ^{244}Pu, ^{243}Am AND ^{244}Cm, AND STUDY OF THE FISSION PRODUCT YIELDS

Ch. DE RAEDT, P. DE REGGE

S.C.K./C.E.N., B-2400 Mol, Belgium

T. BABELIOWSKY and E. WATTECAMPS

C.E.C.-J.R.C. Central Bureau for Nuclear Measurements
B-2440 Geel, Belgium

To find the optimum conditions for the production of ^{244}Pu and to provide a test and/or correction factors for theoretical predictions, samples of approximately 1 mg ^{242}Pu-oxide were irradiated and quantitatively analysed at the S.C.K./C.E.N. in Mol on behalf of C.B.N.M. Some important parameters were investigated by calculation and by experiment, namely the irradiation time, the neutron flux level and the neutron spectrum. The yield of ^{244}Pu, ^{243}Am and ^{244}Cm, the depletion in ^{242}Pu, the yield of fission products and the material balance are linked, and are compared with calculation. These comparisons show that predictions made with unadapted cross-section sets overestimate the ^{244}Pu yield. From the analysis of some fission products it was possible to deduce some mass chain yields for the fission of ^{245}Cm.

Introduction

^{244}Pu can be used as a spike for the absolute determination, by mass spectrometric isotope dilution, of the various plutonium isotopes present in e.g. irradiated nuclear fuel. In the frame of the C.B.N.M. feasibility study of a mass separator, the production of such ^{244}Pu spikes was envisaged.

^{244}Pu is produced by neutron irradiation of other plutonium isotopes according to the scheme illustrated in fig. 1.

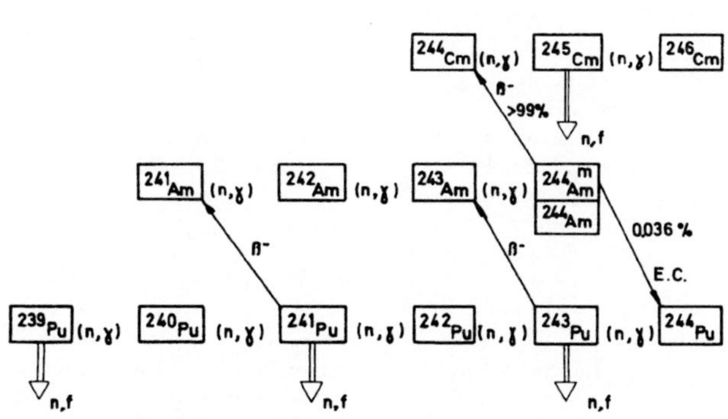

Fig. 1. Formation schemes of ^{244}Pu and of the other heavy nuclides considered in this paper

Very high neutron fluxes are needed in view of the competing transmutations neutron capture versus natural β^- decay (with a half-life of only 4.96 h) in ^{243}Pu. Also the subsidiary formation of ^{244}Pu through americium remains small because of the branching ratio of ^{244m}Am with only 0.036 % E.C. The main target nuclide for the practical production of ^{244}Pu is ^{242}Pu.

With a view to examining the possibilities offered by BR2 (S.C.K./C.E.N., Mol) of producing ^{244}Pu in mg quantities per year, sixteen (= eight duplicate, for the sake of redundancy) small samples, each of approximately 1 mg PuO_2, highly enriched in ^{242}Pu, were irradiated in capsules in BR2.

The composition of the samples and the main irradiation characteristics are indicated in table I. Relatively short and long irradiation times were chosen in reactor channels with mainly thermal (H1 channel) and mixed thermal-epithermal (C41 channel) neutron flux spectra.

Table I Characteristics of the samples and the irradiations

Sample N°	^{242}Pu/Pu (at %)	Irrad. position in BR2: Channel	Th. flux (n/cm^2 s)	Epith. flux (n/cm^2 s)	Irrad. time (days)
74, 75	99.744	H1	$7.60\ 10^{14}$ ± 5 %	$0.243\ 10^{14}$ ± 10 %	88.4
82, 90	87.778				
71, 76	99.744	C41	$2.64\ 10^{14}$ ± 5 %	$0.422\ 10^{14}$ ± 10 %	
88, 81	87.778				
72, 80	99.744	H1	$6.74\ 10^{14}$ ± 5 %	$0.216\ 10^{14}$ ± 10 %	176.7
85, 87	87.778				
77, 78	99.744	C41	$2.57\ 10^{14}$ ± 5 %	$0.411\ 10^{14}$ ± 10 %	
84, 86	87.778				

The samples were of two types, as to their isotopic composition. In the present paper only the samples with 99.744 at % ^{242}Pu/Pu will be further considered (samples N°s 71-80). The PuO_2 powder was poured into thin aluminium tubes of 2 mm inner diameter which were introduced into quartz ampoules of 40 mm length and 5 mm outer diameter. Four such ampoules, together with fluence monitors, fit in a BR2 aluminium capsule, of which four were thus irradiated.

After the irradiations, which took place in 1980-1981, the samples were discharged and analysed.

K. H. Böckhoff (ed.), Nuclear Data for Science and Technology, 185–189.

The radiochemical techniques utilized and the comparison of experimental and calculated yields will be discussed in the present paper.

Analysis of the irradiated samples. Radiochemical techniques and experimental results

After the irradiations the quartz ampoules containing the samples were opened and the contents were dissolved in a boiling mixture of nitric and hydrofluoric acid under reflux. The solution was transferred into a 50 ml flask and weighed in a shielded cell. All subsequent aliquotations and dilutions are done by weight.

A number of chemical separation methods have been combined to achieve the separations required for this study by the use of a single Dowex 1-x4 anion exchange column. Actinides and rare earths are retained in 1 M HNO_3 - 90 % ethanol while other fission products are washed out. The rare earths are eluted to a large extent with a 0.5 M NH_4 SCN-80 % methanol mixture; Cm and Am are separately eluted in 5 M HNO_3 - 50 % methanol. Plutonium is finally obtained by washing the column with 0.35 M HNO_3. A decontamination of 10^3 for most elements was obtained for the Pu fraction while decontamination of Am from Cm exceeds 10^2. Decontamination of Cm from certain fission products, particularly Eu and Ru was only 5 to 10 but sufficient for the measurements intended.

A general scheme of the radiochemical analysis applied is shown in figure 2. For the determination of the heavy isotopes and a selected number of fission products, essentially three techniques have been used. The selection of a particular technique for a particular isotope was motivated by the required accuracy and the specific radiochemical characteristics of the nuclide.

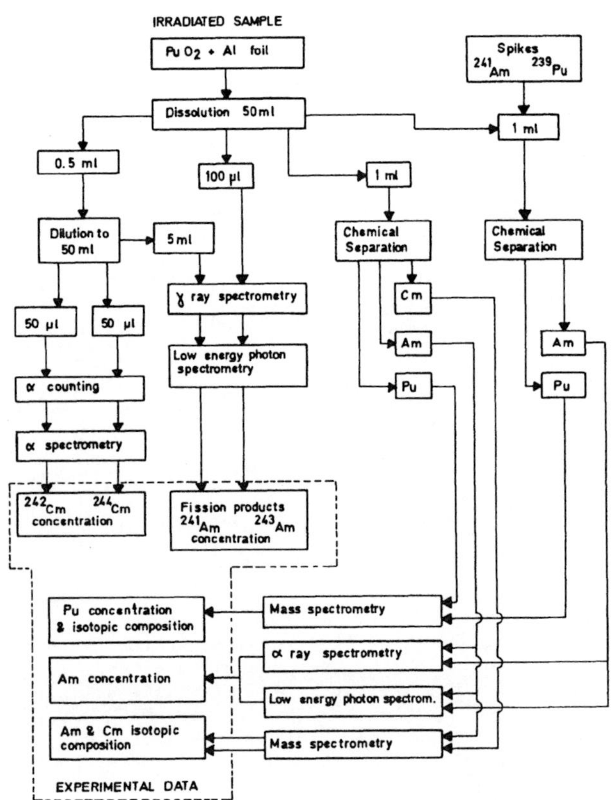

Fig. 2. General scheme for the radiochemical analysis of an irradiated PuO_2 sample.

The fission products have been determined without any chemical separation by gamma-ray spectrometry. The spectrometer has been calibrated with reference samples of similar geometry.

Low energy photon spectrometry of the samples permitted also the determination of ^{241}Am and ^{243}Am, using a reference source of ^{241}Am for calibration. A reference solution of ^{243}Am has been prepared by isotope dilution techniques, using mass spectrometry and alpha spectrometry. Measurements of the reference ^{243}Am sources with a calibrated low energy photon spectrometer allow the calculation of a branching factor of about 0.70 for the 74.7 KeV gamma-ray of ^{243}Am, which value has been used in this work. To obtain a higher precision on the ^{243}Am concentration, particularly in the "long irradiation" samples, the ^{243}Am concentration was also determined by isotope dilution with ^{241}Am as a spike. After the chemical separation of the americium, both gamma-ray and alpha-ray spectrometry were used for the measurement of the ^{243}Am/^{241}Am ratio. ^{242m}Am has been determined in a few samples by mass spectrometric isotopic analysis, but turned out to have a very low content and therefore to be of little relevance to the study.

Alpha spectrometry combined with absolute alpha counting permitted the determination of the ^{242}Cm and ^{244}Cm concentrations in the samples. The other curium isotopes have been determined in selected samples by mass spectrometric isotopic analysis relative to ^{244}Cm.

The plutonium isotopic composition was measured by mass spectrometry (alpha spectrometry for the ^{238}Pu). The plutonium concentration was determined by mass spectrometric isotope dilution after chemical separation, using a ^{239}Pu spike solution obtained by dissolution of a plutonium metal reference material.

The concentrations of the main transuranium nuclides thus determined are indicated in table II per atom initial ^{242}Pu in the samples. The concentrations of the other americium and curium isotopes measured were significantly smaller than those of ^{243}Am and ^{244}Cm.

Table II Experimental results (atoms per atom initial ^{242}Pu)

Nuclide	Sample 74 Sample 75 Average H1, short	Sample 76* C41, short	Sample 72 Sample 80 Average H1, long	Sample 77 Sample 78 Average C41, long
^{242}Pu	0.770 0.657 0.713	0.768	0.610 0.610 0.610	0.649 0.662 0.655
^{244}Pu	0.000286 0.000323 0.000305	0.000194	0.000535 0.000527 0.000531	0.000315 0.000325 0.000320
^{243}Am	0.134 0.150 0.142	0.144	0.146 0.156 0.151	0.143 0.150 0.146
^{244}Cm	0.0672 0.0786 0.0729	0.0543	0.163 0.168 0.165	0.122 0.122 0.122

*The results for sample 71 were discarded as during the dissolution operations contamination and loss of part of the precipitate occurred.

As can be seen from table II, ^{244}Pu quantities of 0.2 ... 0.5 μg per mg target ^{242}Pu have been formed; the amount of ^{244}Pu increases with increasing irradiation time (see further, fig. 4); more ^{244}Pu is formed in the H1 (thermal flux) than in the C41 (thermal-epithermal flux) channel. The following error margins are to be ascribed to the measured values (including calibration uncertainties on the reference materials) : ^{242}Pu : 1 %; ^{244}Pu : 3 %; ^{243}Am : 3 % ; ^{244}Cm : 2 %. For all samples, the material balance of all heavy nuclides plus fission products measured amounted to 94 ... 98 % of the initial amount of plutonium (except for sample 75 : 90 %). The apparent loss of material might be explained by stoichiometric effects, moisture present during the initial plutonium weighing and/or impurities in the target samples.

Calculated quantities of transuranium nuclides

By solving the Bateman equations (ref. 1) corresponding to the scheme given in fig. 1, the quantities of ^{242}Pu, ^{244}Pu, ^{243}Am and ^{244}Cm remaining or formed can be calculated, as a function of the irradiation time. The reaction rates per atom intervening in these equations were calculated in the two-group scheme :

$$\int_0^\infty \sigma(E)\ \emptyset(E)dE \approx \sigma_{2200} \cdot \emptyset_{th} + RI \cdot \emptyset_{epi} ,$$

where : σ_{2200} : 2200 m/s cross-section

RI : resonance integral

$\emptyset_{th}$: conventional thermal flux $(v_o \int_o^{0.5} n(E)dE;\ v_o = 2200$ m/s)

$\emptyset_{epi}$: epithermal flux, per unit lethargy.

The values of $\emptyset_{th}$ and $\emptyset_{epi}$ were taken from table I. As starting point for the calculations, the 2200 m/s cross-sections and the resonance integrals (infinite dilution) tabulated in table III were adopted.

Table III Nuclear constants adopted for the initial calculations

Nuclide	$\sigma_{2200\ n,\gamma}$ (barns)	$\sigma_{2200\ n,f}$ (barns)	$RI_{n,\gamma}$ (barns)	$RI_{n,f}$ (barns)
^{242}Pu	18.5 (± 0.4)	0	1170 (± 100)	5.0 (± 5.0)
^{243}Pu*	60 (± 30)	180 (± 16)	264 (± 132)	542 (± 50)
^{244}Pu	1.7 (± 0.2)	0	43 (± 4)	0
^{243}Am	75.2 + 4.1** (± 6.0)	0	1709 + 111 (± 500)	0
^{244}Cm	13.9 (± 4.0)	1.2 (± 0.5)	650 (± 50)	12.5 (± 2.5)

*$\lambda_{^{243}Pu} = 3.886\ 10^{-5} s^{-1}$

**1st term : $^{243}Am(n,\gamma)^{244m}Am$;
2nd term : $^{243}Am(n,\gamma)^{244g}Am$

The $\sigma_{2200\ n,\gamma}$ values of all five nuclides, the $RI_{n,\gamma}$ values of ^{244}Pu, ^{243}Am and ^{244}Cm as well as the $\sigma_{2200\ n,f}$ and $RI_{n,f}$ values of ^{242}Pu and ^{244}Cm correspond to those given in BNL 325, 3rd ed. (ref. 2). The $\sigma_{2200\ n,f}$, $RI_{n,\gamma}$ and $RI_{n,f}$ values of ^{243}Pu were taken from refs 3 to 6, while $RI_{n,\gamma}$ of ^{242}Pu is the average of the values given in ref. 7. The error margins indicated between brackets correspond approximately to the widest spread of values found in the literature.

The results are illustrated in fig. 3 where the ratios nuclide quantity calculated/nuclide quantity measured are indicated (values "O"). As can be seen the agreement is in general rather bad, with a discrepancy of up to a factor of nearly two for ^{244}Cm in the C41 channel.

Adaptation of the nuclear constants and irradiation data to improve the agreement with the measured values

A first observation that can be made concerning the "O" values on fig. 3 is that the calculations tend to overestimate the nuclear transmutation reactions : the calculated ^{244}Pu, ^{243}Am and ^{244}Cm quantities are higher than the measured ones while the calculated ^{242}Pu left is generally lower. A much better agreement would be obtained by decreasing in a substantial manner the flux (thermal, epithermal or both) values cited in table I and used in the calculations. The error margins attributed to these flux values are however too narrow to enable an important amelioration of the agreement calculations/measurements.

The capture cross-section of ^{242}Pu was therefore investigated first. Although the targets (PuO_2) only weighed approximately 1 mg each, a very strong self-shielding occurs in the 2.67 eV resonance of ^{242}Pu, which is responsible for the major part of the captures in ^{242}Pu (about 80 % , for infinite dilution, in a C channel of BR2). Adopting realistic estimates of the density and of the geometry of the PuO_2 in the target ampoules, self-shielding factors of 0.2 ... 0.7 were calculated for the resonance integral (e.g. an infinite slab of 0.1 mm thickness and density 3 g/cm^3 yields a self-shielding factor of the resonance integral equal to about 0.4; the same self-shielding factor is valid for isolated microspheres of 0.15 mm diameter and 6 g/cm^3 density).
Starting from the cross-section values indicated in table III (leading to the "O" ratios in fig. 3, as mentioned above), the yields of the four heavy nuclides concerned were re-calculated, applying self-shielding factors to the capture resonance integral of ^{242}Pu varying with steps of 0.05 between 0.75 and 0.40. The best agreement between calculations and measurements was obtained for 0.55. The results are indicated in fig. 3 in the columns marked "a".

Keeping these $RI_{effective}$ values, the thermal flux values given in table I were increased next by 5 % : the ratios "b" on fig. 3 were thus obtained.

Next the thermal capture cross-section for the reaction $^{243}Am(n,\gamma)^{244m}Am$ was increased by 5 % (while maintaining the modifications introduced above), yielding the ratios "c" on fig. 3.

Finally, in order to obtain calculated ^{244}Pu quantities more symmetrically spread around the measured values, the thermal capture cross-section of ^{243}Pu was increased by 5 % (keeping all modifications introduced hitherto), resulting in the ratios "d" on fig. 3.

The last two modifications were within the cross - section error margins generally quoted.

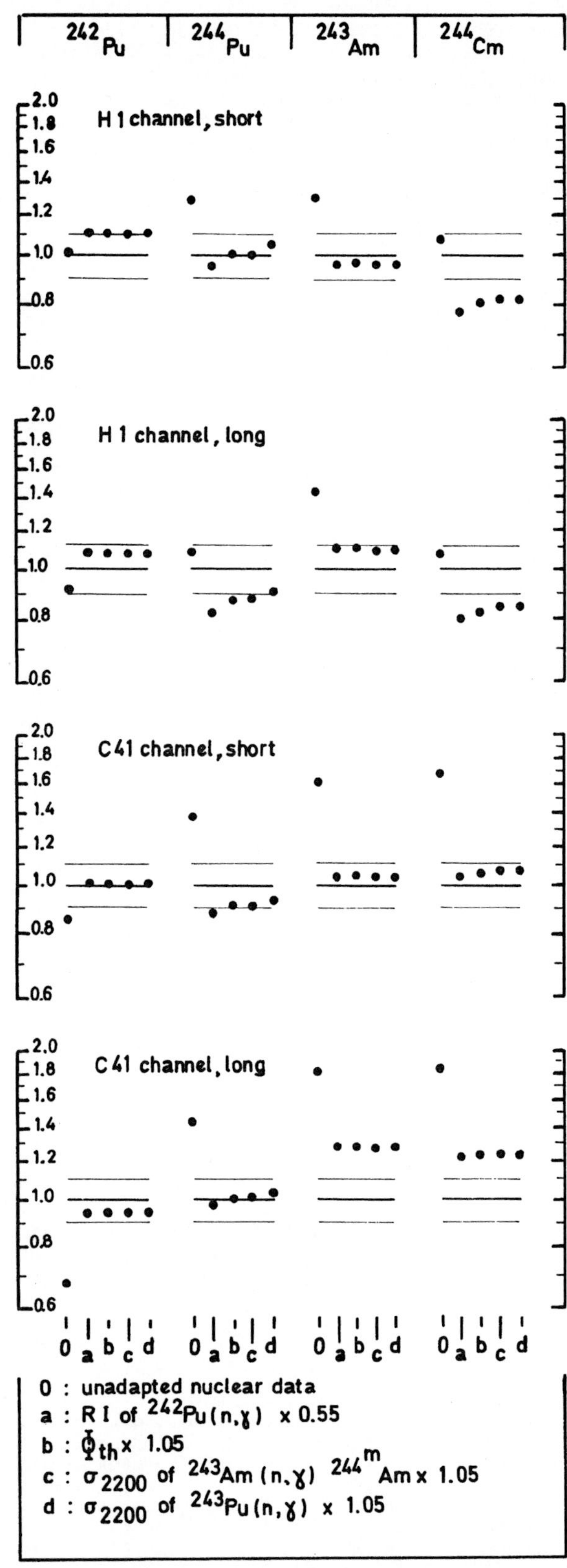

Fig. 3. Ratios of computed to measured quantities of ^{242}Pu, ^{244}Pu, ^{243}Am and ^{244}Cm. Influence of the modifications applied to the nuclear constants and irradiation data used in subsequent calculations.

A least square error can be defined :

$$E^2 = \sum_{i=1}^{4} \sum_{j=1}^{4} \left[\left(\frac{\text{calc.}}{\text{meas.}}\right)_{i,j} - 1\right]^2$$

where i (1...4) : irradiation type : H1 and C41, both short and long

j (1...4) : nuclide type : ^{242}Pu, ^{244}Pu, ^{243}Am, ^{244}Cm.

E^2 decreases as follows : 3.054 (0) → 0.311 (a) → 0.267 (b) → 0.252 (c) → 0.246 (d).

As can be seen - also from fig. 3 - by far the most important improvement is due to the taking into account of the resonance self-shielding of the ^{242}Pu capture. The ratios calculation/measurement "d" may be considered as rather satisfactory, except for ^{244}Cm.

Several other modifications of cross-sections were tried but all led either to negligible improvements of the agreement calculation/measurement or to improvements for some nuclides with corresponding degradation for other nuclides. In any case, it should be kept in mind that some of the correction factors applied, viz those to the RI of ^{242}Pu and to the thermal flux, are not necessarily identical for all four irradiations (H1 and C41, both short and long).

With the adapted cross-section and flux values (corresponding to "d" on fig. 3) the disappearance or formation curves of the four heavy nuclides considered were drawn (fig. 4) for the two typical irradiation positions in BR2, viz H1 and C41. The measured values were added to the figure (only those for the long irradiation time).

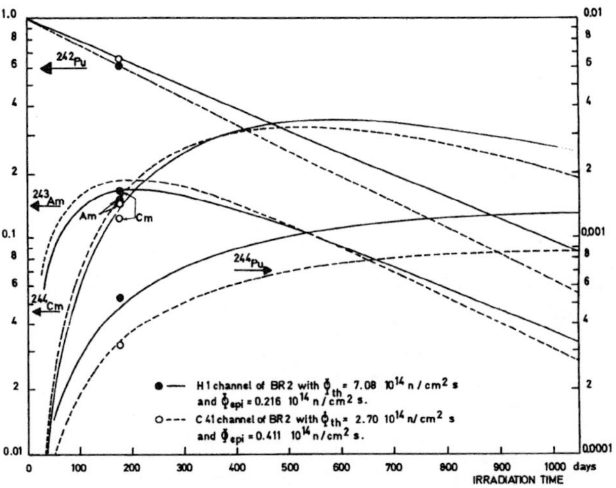

Fig. 4. Calculated and measured yields of ^{242}Pu, ^{244}Pu, ^{243}Am and ^{244}Cm per atom ^{242}Pu initially present, as a function of irradiation time

Origin of the fission products and estimation of some mass chain yields for ^{245}Cm

The fission products in the samples are mainly formed by the fission of the three fissionable heavy nuclides ^{241}Pu, ^{243}Pu and ^{245}Cm (see fig. 1). The small quantity of ^{240}Pu initially present in the target material is transformed into ^{241}Pu and ^{241}Pu disappears by fission and neutron capture to ^{242}Pu with relative reaction rates of 1 and 0.36.

This total burnout of the minor isotopes occurs in the first few weeks of the irradiation. The number of fissions can be calculated from the amount of minor isotopes in the base material.

The ratio of the (n,γ) to (n,f) cross-sections of ^{243}Pu is about 0.333, which means that the number of fissions due to ^{243}Pu is about three times the number of ^{244}Pu atoms formed. The fission rate of ^{243}Pu is proportional to the neutron flux and is uniform over the entire irradiation period.

From the Bateman equations it can be deduced that the ratio ^{245}Cm/^{244}Cm reaches a constant value after a short initiation period. The isotope ^{245}Cm disappears by fission and by neutron capture to ^{246}Cm. The transformation of ^{246}Cm into ^{247}Cm is very slow due to its low capture cross-section. The amount of ^{246}Cm in the irradiated samples can thus be used as a measure for the disappearance of ^{245}Cm by fission using a value of 5.88 for the ratio of the (n,f)/(n,γ) reaction rates as derived from the neutron flux and cross-section values given in table I and ref. 2, respectively. The fission rate as a function of the irradiation time has been calculated from the Bateman equations while the number of fissions in the samples has been derived from the experimental data.
For the short irradiation in the H1 channel (sample 75) the total number of fissions per 10^6 initial Pu atoms is 7510, with 23.7 % due to ^{241}Pu, 12.8 % due to ^{243}Pu and 63.5 % due to ^{245}Cm. For the long irradiation in the H1 channel (sample 72) the total number of fissions per 10^6 initial Pu atoms is 24351, with 7.3 % due to ^{241}Pu, 6.5 % due to ^{243}Pu and 86.2 % due to ^{245}Cm.

A number of radioactive fission products have been measured in the samples. Applying the appropriate in-pile and out-of-pile decay corrections - based on the fission rate in the sample as a function of irradiation time - the total number of nuclides for different mass chains has been calculated. Their ratio to the total number of fissions in the sample provides an average fission yield, which, in view of the distribution of the fissions, can be used as an estimate of the mass chain yields for the fission of ^{245}Cm. Preliminary results are shown in table IV. Refined calculations and further measurements of the isotopic ratios in other samples will be carried out to improve the data.

Table IV Estimated mass chain yields in percent for the thermal fission of ^{245}Cm. Comparison with other fissile isotopes.

Mass	Sample 75	Sample 72	^{235}U	^{239}Pu	^{241}Pu
95	3.0	2.6	6.54	4.99	4.03
103	6.3	5.7	3.04	6.38	6.57
106	6.6	7.5	0.46	4.40	6.12
141	5.8	5.2	5.84	5.27	4.89
144	4.7	4.6	5.54	3.78	4.17

Conclusions

The irradiation of mg samples of $^{242}PuO_2$ in BR2 has demonstrated that ^{244}Pu quantities of the order of 1 μg per mg target ^{242}Pu can be produced in the high flux channels. Even for such small samples account has to be taken, when calculating the production of transuranium nuclides, of the strong self-shielding of the 2.67 eV resonance of ^{242}Pu. In this way, a satisfactory agreement between calculated and measured yields of the heavy nuclides is obtained, except for ^{244}Cm.

By analysing the fission products an estimate of some mass chain yields for the thermal fission of ^{245}Cm has been made.

The theoretical analysis of the PuO_2 samples with 87.778 at % ^{242}Pu/Pu will be carried out in the near future.

References

1. H. Bateman. The solution of a system of differential equations occurring in the theory of radio-active transformations. Proceedings of the Cambridge Philosophical Society 15, p. 423-27 (1910).
2. S.F. Mughabghab, D.I. Garber. Neutron cross sections, Volume I, Resonance parameters. BNL 325, third edition, volume I, June 1973.
3. R.W. Benjamin. Status of measured cross sections of transactinium isotopes for thermal reactors. Review paper N° B1. Transactinium isotope nuclear data (TND). Vol. II. Proceedings of an advisory group meeting on transactinium isotope nuclear data, held at Karlsruhe 3-7 November 1975. IAEA-186.
4. R.W. Benjamin, F.J. McCrosson, T.C. Gorrell, V.D. Vandervelde. Consistent set of heavy actinide multigroup cross-sections. DP-1394, December 1975.
5. R.W. Benjamin. Nuclear data for actinide production and depletion calculations. DP-MS-78-39, 1978.
6. E. Menapace, G. Oliva, L. Tondinelli. Preliminary sensitivities for transcurium isotope build-up in thermal reactors. 1st technical meeting on the nuclear transmutation of actinides. Ispra 16-18 April 1977. EUR 5897.
7. E.M. Gryntakis, J.I. Kim. A compilation of resonance integrals. Part II, Z=53-100. Journal of radioanalytical chemistry, vol.42 (1978).

Acknowledgements

The authors thank B. Rose for suggesting this study, F. Hendrickx for the sample preparation, J.M. Gandolfo, W. Boeykens and P. Vanmechelen who were responsible for the BR2 irradiations and neutron flux dosimetry, as well as R. Boden and D. Huys for the radiochemical analyses. They are also indebted to A. Paulsen for his continuous interest in the work.

EXPERIMENTAL VALIDATION OF IRRADIATED FUEL INVENTORIES CALCULATED BY THE FISPIN CODE FUELS

S. Pickles and H. J. Powell

Dounreay Nuclear Power Development Establishment
Dounreay
Thurso, Caithness
SCOTLAND

Post-irradiation examination work on fuels irradiated in the Prototype Fast Reactor (PFR) includes measurements of activities or concentrations of several isotopes. Control of the reprocessing plant requires numerous analyses of samples. The measurement techniques used include alpha spectrometry and mass spectrometry for the actinides, gamma spectrometry for radioactive fission products and mass spectrometry for stable fission products, including the gases krypton and xenon.

Comparison of these measurements with predicted values affords the possibility of validating parts of the calculational procedure, particularly the FISPIN code(1), its data libraries and data derived from other sources which form input to the code. This paper will show that despite the present limited scope of such comparisons, the results are distinctly encouraging, especially for the actinides.

Introduction

The calculation of the composition of irradiated fuel is subject to a number of uncertainties and approximations, in respect of prime and derived nuclear data, calculational methods and models and possibly most importantly, from the practical realisation of the calculated design and operational specification.

In the UKAEA, fast reactor calculations are performed with the aid of a data retrieval system which includes the adjusted nuclear data set FD/4 and a number of codes for producing detailed neutronics calculations for complex and detailed reactor models. The output from such calculations can then be used with the FISPIN code to predict changes in fuel composition with irradiation.

Cross sections for nuclear reactions are spectrum dependent. For the FISPIN calculations reported herein a set of one group values derived for a notional fast reactor spectrum has been used. If this were to prove a significant source of error, recourse would have to be made to available multigroup cross-sectional data.

Although FISPIN can be applied to as detailed a model as desired, from a fraction of a fuel pellet to a whole reactor core, very detailed calculations require a considerable computing resource and for this paper a relatively coarse (smeared) representation has been used. It is recognised that this will introduce a further source of error in the comparison between the results presented here and the measured data. In order to minimise this error, the one group cross sections for the important actinides were replaced by values appropriate to the region under consideration.

Overall, the sources of error or uncertainty in calculating the compositions and activities of irradiated fuel may be summarised as:

1. Prime nuclear data.
2. Reactor modelling - division into zones of smeared 'uniform' composition
3. Neutronics calculations

and for the FISPIN calculations reported here:

4. Smearing over much larger regions.
5. Use of one-group, irradiation independent cross sections.
6. Decay data uncertainties.
7. Fission yield uncertainties

Numerous evaluations have been made of the accuracy of individual items and groups of data relevant to irradiated fuel calculations. The most comprehensive survey, by Tobias (2), concentrates on the relevance to decay heat. Though many of the quantities have been determined with reasonable precision others are subject to large possible errors and some are no more than estimates based on theoretical predictions. Fortunately, not all the errors will operate in the same direction, and with 90 actinides and 825 fission product isotopes in the respective libraries a substantial measure of error cancellation can be expected.

It is therefore of greatest importance that experimental evidence should be obtained to compare with the calculations. Data now emerging from the post irradiation examination and the reprocessing of PFR fuel is beginning to make such comparisons possible, and some of the early findings are presented in this document.

Fuel Samples and Irradiation Assessment

These early results are derived almost entirely from a group of experimental fuel pins incorporating non-standard but well characterised fuel of high fissile content. The fuel cluster had been irradiated from the start of the commissioning runs of the reactor, and so had been subjected to an abnormal degree of power cycling.

Evaluation of the fuel irradiation was based on the experimental measurements of reactor power and corresponding times, and the physics calculations for each fuel loading pattern of the reactor, related to a 3-dimensional spatial mesh and a total reactor power of 600 MW.

FISPIN calculations were used to obtain details of the irradiated fuel composition. To simplify the calculation, the complex power cycling pattern was equated to five constant power runs giving the correct integrated power-time product. The 600 MW fluxes were adjusted to the equivalent values for each of the mean powers of the smoothed irradiation programme. Good agreement was obtained between the reactor operations calculation and FISPIN calculations using the standard PFR centre plane or region-averaged cross section libraries and the appropriate fluxes.

Analytical Techniques

The FISPIN output for the cooling time at which the analyses were performed lists 54 elements, of which almost all are present in several isotopic forms. Obviously it would be straining the limits of

K. H. Böckhoff (ed.), Nuclear Data for Science and Technology, 190–195.

analytical possibility to attempt a full analysis of every component.

In practice, the analyses were restricted to:

i. High resolution gamma spectrometry on the intact fuel pin.
ii. Mass spectrometry on the plenum gases.
iii. Mass spectrometry on solutions of fuel samples dissolved in aqua regia - evaluations of U and Pu isotopes and of Nd isotopes for burn-up measurement.
iv. Alpha spectrometry on the above solutions - supplementary information on actinide compositions.
v. Gamma spectrometry on the solutions - giving a measure of some fission product concentrations.

(i) The gamma scan of the intact fuel pin is useful in showing the movement of fission products where this occurs. Absolute measurements of fission product masses are not really practicable, and concentration comparisons between different isotopes are spoiled by differential self-shielding, resonance and interference effects.

The results for the non-mobile isotopes however provide a useful check of the burn-up profile.

(ii) The plenum gases contain the fission products krypton and xenon, though only about 30% to 85% of the total, depending on the burn-up, density and temperature of the fuel, is released to the plenum during irradiation.

Mass spectrometry gives the isotopic compositions, which can be compared with calculated values, but the ratio of krypton to xenon is normally a little higher than in the total fission products because of differential release from the fuel.

(iii) Mass spectrometry of the dissolved fuel samples gives the absolute concentrations of the main isotopes of uranium and plutonium. The presence of a large amount of U238 prevents measurement of Pu238. The measurements of the four stable neodymium isotopes (Nd 143, 145, 146 and 148) in the same solutions are related to the U + Pu concentrations to determine burn-up.

In the standard burn-up calculations at Dounreay the neodymium fission yields used are those determined in DFR, related to relative fission ie rates at mid-burn-up composition. Use of the Crouch-2 fission yields[(3)] with the integrated relative fission rates given by FISPIN in conjunction with the analytical neodymium measurements leads to an approximately 2% incrase in the burn-up evaluation. The Crouch-2 yields have been used for the analytical burn-up values in the comparisons which follow.

(iv) Alpha spectrometry is substantially less precise than mass spectrometry, but is useful in distinguishing between different nuclides with the same isotope mass.

Measurements were made on the fuel solutions. The peak at 5.5 MeV gave the sum of the Pu238 and the Am241 while that at 5.15 MeV gave the Pu239 + Pu240. A peak at 6.0 MeV was taken to represent Cm242, but probably swamped a small peak at 5.8 MeV resulting from Cm243 and Cm244.

Alpha spectrometry on the plutonium after separation from other actinides by extraction gave a value for Pu238 alone and hence, by difference, for Am241.

(v) The fission products that could be measured by gamma spectrometry on the fuel solutions were limited by the cooling time of the fuel (433 days) and by the method of preparing the solutions.

These were made by dissolving 1 cm lengths of fuel pin, including cladding, in aqua regia by prolonged boiling, in the course of which much of the ruthenium present would be lost as the very volatile ruthenium tetroxide.

The fission products of which measurements were made included Zr95, Nb95, Cs134, Cs137, Ce144, Pr144, Sn125 and Eu154.

It was not possible to weigh the irradiated fuel before dissolution, but the mass spectrometer gave the absolute concentrations in atoms/ml of uranium, plutonium and neodymium. The equivalent initial fuel concentration was obtained by adding the fissions/ml calculated from the neodymium yields to the U+Pu atoms/ml, and the other activities were related to this figure.

In addition to the measurements on the experimental pin cluster, results are becoming available from the routine operation of the PFR reprocessing plant. The most useful information will come from statistical evaluation of the results from a long period of operation, but it was thought worth while including here a comparison of the isotopic composition of the plutonium product from a standard outer core sub-assembly with the corresponding FISPIN calculation.

Results

Burn-Up The neodymium yields used in the burn-up calculations were taken from the FISPIN calculation. This code derives burn-up from the total number of fissions and works out the fission product composition from the Crouch-2 yield sets and the integrated fission rates for a group of actinides specified in the actinide library as contributing significantly to fission.

Because of the unusual initial composition of the fuel and the extra actinides in this category, the standard library was augmented by the allocation of yield sets to U234 and U236 and to the americium and curium isotopes.

For each fuel sample duplicate measurements were made of the yields of the four stable neodymium isotopes (143, 145, 146 and 148) and burn-ups were calculated using the Crouch-2 and the DFR yield values.

These are compared with the calculated burn-ups in Table I.

Table I Comparison of Burn-ups Determined by Neodymium Analysis and Physics Calculations

Sample	Calculated Burnup	Nd + crouch 2		Nd + DFR Yields	
		BU%	SD*(%)	BU%	SD*(%)
1	4.55	4.63	1.3%	4.54	1.3%
2	6.13	6.18	1.5%	6.06	2.1%
4	8.32	8.06	1.6%	7.91	2.5%
5	8.76	8.61	1.2%	8.45	2.6%
7	8.01	7.65	2.6%	7.50	3.4%
9	6.28	6.31	2.2%	6.19	2.9%

* Duplicate samples were analysed by mass spectrometry for four neodymium isotopes. Each burn-up figure quoted is therefore the mean of eight measurements.

Gamma scan although the gamma scan of the intact fuel pin does not give absolute or even relative values of the concentrations of different isotopes, when applied to fission products that are known not to migrate to different parts of the fuel it can give a warning of bulk fuel movement should this occur, and it can provide a useful check of the burn-up profile in a stable fuel column.

In Table II the gamma scan measurements have been divided by the burn-ups interpolated from the graph of neodymium analyses and Crouch-2 yields. If no fuel or isotope movement takes place the ratio should be constant along the length of the pin. This is illustrated in the Table by normalising the ratios for each isotope to the mean values and noting the standard deviations. Except in the case of niobium-95 for which the measurements were marred by severe peak interference, the results show encouraging endorsement of the calculated profile.

Table II High Resolution Gamma Scan of Intact Fuel Pin

Burn-up	Gamma Activity per Unit Burn-up							
	Ce 144		Pr 144		Zr 95		Nb 95	
	A	B	A	B	A	B	A	B
4.84	2.19E5	1.08	2.33E6	1.02	4.19E4	1.07	1.16E5	1.26
6.43	2.05E5	1.01	2.24E6	0.98	3.81E4	0.97	1.13E5	1.23
7.36	1.96E5	0.97	2.27E6	0.99	3.79E4	0.96	1.08E5	1.17
8.18	2.07E5	1.02	2.30E6	1.01	3.91E4	0.99	9.81E4	1.06
8.63	2.03E5	1.00	2.22E6	0.97	3.77E4	0.96	1.10E5	1.19
8.30	2.06E5	1.02	2.31E6	1.01	4.20E4	1.07	7.34E4	0.79
7.54	1.91E5	0.94	2.43E6	1.06	4.16E4	1.06	7.84E4	0.85
6.99	1.96E5	0.97	2.29E6	1.00	4.01E4	1.02	5.8E4	0.63
6.16	2.01E5	0.99	2.18E6	0.95	3.94E4	0.91	7.71E4	0.83
Mean	2.03E5	1.00	2.28E6	1.00	3.94E4	1.00	9.26E4	1.00
St Dev	0.08E5	4.02%	0.07E6	3.15%	2.19E3	5.57%	0.93E5	22.8%

A - Measured B - Relative to mean

<u>Actinides</u> The isotopic composition of fission product elements should be essentially the same for all samples with the same irradiation and cooling history but different fluxes and hence different burn-ups. Because of the complication introduced by breeding, this is not true for the actinides and so separate calculations were undertaken for each sample. Spatial variations in neutron energy spectrum were ignored. The measured and calculated isotopic compositions of the experimental fuel sections are compared in Table III. The measured and calculated isotopic compositions of the experimental fuel sections are compared in Table III.

Reprocessing separates not only the uranium and fission products but also the higher actinides from the plutonium. This simplifies the analysis of the plutonium product and should improve the accuracy, but it is difficult to believe that the closeness of the agreement between the analysed and calculated isotopic compositions of the plutonium product from a full sub-assembly (including integral breeder) shown in Table IV is not to some extent fortuitous. The results shown in Table IV do cast doubts on those of Table III, where poor prediction of the higher plutonium isotopes was found.

Table III Heavy Element Compositions - Comparison of Experimental and Calculated Values

Isotope	Ratio of CALCULATED/EXPERIMENTAL values					
	Burn-up % (Nd analysis, Crouch-2 Yields)					
	4.63	6.18	8.06	8.61	7.65	6.31
	MASS COMPARISON					
A	<u>Comparison with Mass Spectrometry Measurements</u>					
U235	0.996	0.997	1.001	1.002	1.026	1.004
U238	1.039	1.037	1.035	1.036	1.028	1.033
Pu239	0.997	0.996	0.999	0.999	1.004	1.001
Pu240	0.890	0.926	0.901	0.914	0.879	0.883
Pu241*	0.8	0.9	0.8	0.9	0.9	0.8
Pu242*	0.8	-	-	-	1.1	0.9
	ACTIVITY COMPARISON					
B	<u>Comparison with Alpha Spectrometry Measurements</u>					
Pu239 +) Pu240)	1.004	-	1.033	0.946	0..994	1.003
Pu238	0.69	-	1.26	0.85	0.85	0.76
Am241	1.04	-	0.73	1.28	1.6	1.75
Pu238 +) Am241)	0.86	-	0.97	0.98	1.05	1.06
Cm242, 3,4	1.10	-	0.94	0.97	0.92	0.95

* Analytical measurements quoted to only one significant figure.

Table IV Reprocessed Plutionium Composition

Isotope	Calculated/Experimental Masses*
Pu 238	No analysis
Pu 239	1.0001
Pu 240	0.998
Pu 241	0.961
Pu 242	0.967

* Results from a standard outer core sub-asssembly. Calculations include 3% of breeder plutonium from integral breeders.

Experimental measurements by mass spectrometry did not include Pu238, for which the calculated concentration was 0.14%.

<u>Gamma Spectrometry on Solutions of Fuel Sections</u> The gamma emissions from the six fuel sections were analysed by high resolution gamma spectrometry, and concentrations were reported for nine fission product isotopes. The technique does not yield a very high precision - in particular the niobium peak is subject to interference and there is a chronic imbalance between Ce144 and Pr144 which should have virtually identical disintegration rates in view of the very short half life of the praeseodymium daughter (17 minutes).

Nevertheless, the ratios of Calculated/Experimental activities for the positionally stable isotopes listed in Table V are commendably close to unity, and the means for the sets of twelve measurements (duplicate determinations on each of the six samples) have standard deviations no greater than would be claimed for the experimental technique.

The measurements of the ruthenium-106 were considerably lower than calculated, as expected from the method of preparing the solutions, and are of no significance; the caesium isotopes show C/E ratios which vary widely with the location of the sample, and confirm the migration of this element to the cooler parts of the fuel.

As the samples were not uniformly distributed along the fuel length, an attempt was made to assess the overall C/E ratio for the migratory isotopes by graphical integration. This worked fairly well for Cs134 where the duplicate determinations were in reasonably good agreement, but there were such wide differences in some of the Cs137 duplicates that the graphical integration became somewhat arbitrary.

Table V Variation of Fission Product Concentrations with Burn-up

Sample No	Burn-up	Calculated/Experimental Activities						
		Zr95	Nb95	Ru106	Sb125	Ce144	Pr144	
1	4.63%	1.20	1.55	2.22	-	1.17	1.40	1.2
		1.20	1.50	2.03	5.20	1.07	1.40	1.31
2	6.18%	0.96	1.34	2.93	2.76	0.95	1.30	1.36
		1.04	1.31	2.73	3.29	0.93	1.08	1.15
4	8.06%	0.89	2.07	3.91	5.11	0.77	0.93	1.20
		0.98	2.35	4.58	1.11	0.88	0.97	1.10
5	8.61%	1.08	1.61	3.18	2.75	1.09	1.23	1.14
		1.12	1.44	3.77	2.13	1.10	1.25	1.14
7	7.65%	1.02	1.40	4.18	3.62	1.02	1.19	1.16
		0.97	1.30	4.00	3.37	1.02	1.15	1.12
9	6.31%	1.05	1.32	1.97	3.36	0.82	1.02	1.25
		1.05	1.52	2.09	0.96	0.96	1.17	1.21
Mean		1.05	1.56	3.13	3.06	0.98	1.17	1.20
Std Dev		8.9%	21%	30%	45%	12%	13%	6.6%

NOTE: Analyses were obtained by gamma spectrometry on solutions of fuel pin sections. Average accuracy claimed for the method is ± 20%.

Included with the mobile isotopes in Table VI is the non-mobile europium-154. This isotope is produced mainly by neutron capture of Eu153; the capture rate is presumably strongly spectrum-dependent and since only a single one-group cross section is available in the FISPIN libraries close agreement with experimental measurement cannot be expected. As seen from Table VI the production varies from 0.2 times the calculated value at the ends of the fuel column to 4 times at the centre, the integrated measured values showing over twice the calculated integral.

Table VI Variation of Caesium and Europium activities with Burn-up and with Distance from Centre Plane

Sample	Axial Position 1% of 1/2 fuel length	Burn-up	Activities - Ci/kg fuel		
			Cs134	Cs137	Eu154
1	89.4%	4.63%	17.9	107	0.414
			18.2	122	-
2	62.8%	6.18%	40.8	250	1.36
			43.0	263	-
4	22.8%	8.06%	54.4	322	9.07
			50.2	775	9.58
5	2.8%	8.61%	28.2	174	11.6
			27.0	270	12.3
7	-41.2%	7.65%	36.0	211	9.39
			39.1	227	7.83
9	-77.2%	6.31%	52.6	311	1.86
			52.6	647	-
Mean by graphical integration			41.0	343	5.58
Calculated mean			41.5	273	2.04
CALCULATED/EXPERIMENTAL MEAN			1.01	0.80	0.37

<u>Fission Product Gases</u> The isotopic compositions of krypton and xenon as determined by mass spectrometric analysis of the plenum gases are compared with the calculated values in Table VII. In addition to the experimental fuel mainly considered in this work, four pins from a reference driver sub-assembly were examined, and the results included in the comparisons.

Table VII Krypton and Xenon in Plenums

Quantity	Ratios - CALCULATED/EXPERIMENTAL Values				
	Experimental Fuel	Standard Sub-assembly			
		Pin 1	Pin 2	Pin 3	Pin 4
Total Kr	1.18	3.40	3.32	3.47	3.17
Total Xe	1.24	3.33	3.32	3.27	3.06
Ratio Xe/Kr	1.05	0.98	1.00	0.98	0.97
ISOTOPIC COMPOSITION (atom %)					
Kr 83	0.984	0.993	0.980	0.992	0.985
Kr 84	1.017	1.019	1.019	1.022	1.021
Kr 85	0.982	1.013	0.993	0.995	1.011
Kr 86	0.993	0.987	0.995	0.988	0.989
Xe 128	0.66	0.65	0.45	0.65	0.76
Xe 130	0.36	0.44	0.38	0.38	0.45
Xe 131	1.050	1.043	1.040	1.047	1.042
Xe 132	0.999	0.999	0.997	1.00	0.997
Xe 134	0.965	0.952	0.952	0.950	0.952
Xe 136	1.018	1.031	1.034	1.030	1.033

Discussion

Burn-up Agreement between calculated burn ups and those derived from neodymium measurements is very satisfactory. The Crouch-2 yield set gives slightly smaller errors and standard deviations than those measured in DFR. However for either yield set the difference between calculated and experimental burn up is small compared with experimental and calculational uncertainties of about 5%.

The burn-up profiles obtained from the physics calculation and the neodymium analyses are confirmed by the gamma scans for Ce 144, Pr 144 and Zr 95, assuming that these isotopes do not migrate along the fuel column. The standard deviations of 4.0%, 3.2% and 5.6% respectively are very reasonable, having regard to the experimental accuracy of the measurements. Although it is appreciated that peak interference lessens the accuracy of the niobium-95 measurements it is not clear why the relative activity should fall off so strongly in the lower half of the fuel column.

Actinides The mass comparisons in Table IIIA show good agreement between calculated and experimental values for U235 (C/E mean = 1.004 ± .011) and for Pu 239 (C/E mean = 0.999 ± .003) but worse agreement in the case of Pu240 (C/E mean = 0.900 ± .018) and of U238 (C/E mean = 1.035 ± .018) and of U238 (C/E mean = 1.035 ± .004).

The experimental results suggest that the capture rate in both U238 and Pu239 have been underestimated in the calculations. This is consistent with a neutron spectrum of lower average energy than assumed but there is no obvious reason for a softer spectrum in an enriched cluster than in a standard sub- assembly.

The alpha spectrometry measurements (Table IIB) are much less precise than the mass spectrometry, but give a measure of some high activity, low number density actinides not evaluated by the latter technique. In general, the ratios of Calculated/Experimental values are very much closer to unity and show very much less scatter than would be expected from the analysts' modest claim of ± 20% accuracy. The average of the 5.15 MeV activities (Pu 239 + Pu 240) had a C/E = 0.996 ± .03, for 5.5 MeV (Pu 238 + Am 241) C/E was 0.984 ± .08, and for 6.0 MeV (curium isotopes) the C/E was 0.976 ± .07. The separation of Am 241 and Pu 238 gave more variable answers, but as the results were quoted to only two significant figures, and the americium was derived by difference, the agreement with calculation remains surprisingly good.

The comparison of the isotopic composition (by mass spectrometry) of the plutonium product from bulk reprocessing of standard driver fuel with the calculated composition shows remarkably good agreement (Table IV), well within the tolerance of the measurements. Indeed, if the results are rounded to 0.1% total plutonium the agreement is exact.

Fission Products It has been possible to attempt evaluation of only a few of the many fission product isotopes, these being the main gamma emitters at the time of measurement. Fortunately most of the beta emitters are associated with gamma emission from precursors or daughters, so that accurate prediction of the gamma activities would give grounds for confidence in beta activity and total fission product power predictions.

Unfortunately, fission product migration along the fuel pins and complications of chemistry of fuel dissolution reduce the value of some of the measurements. Most importantly, caesium migrates from the hottest regions of the fuel, and its evaluation involves integration over the length of the fuel pin. In the present work no data were available for the ends of the fuel column or the adjacent breeder and plenum regions. Also, ruthenium in a highly oxidising environment forms the volatile oxide RuO_4 (B.Pt. 40 C) and much of it would be expected to be lost when the fuel was dissolved by boiling in aqua regia. Other fission product elements including Sn, Sb, Pd, Mo, Nb, form oxides with low solubility or give unstable solutions in nitric acid, resulting in partial precipitation, but most of them make only a minor contribution to the total fission product power. An exception is niobium, which is almost completely insoluble, though its high specific activity (Nb95) and very low mass concentration lead to a substantial activity in solution. Fortunately its precursor Zr 95 with which it is effectively in equilibrium can be measured fairly accurately.

The results in Table V show good agreement with calculations in the cases of Zr95 and Ce144, though the analyses for Pr144 were generally less than predicted and on average some 20% below the corresponding Ce144 values. Since on theoretical grounds these isotopes should be in equilibrium and have the same disintegration rates, a weakness in the analytical measurements is indicated, and this is being investigated.

The measured activities of Nb95, Ru106 and Sb125 were much lower than calculated, as implied by the foregoing discussion of the dissolution process.

In an attempt to assess the caesium isotopes graphical integrations were made over the length of the fuel pin, and these were compared with integration combining FISPIN calculations with the (analytical, Crouch-2) burn-up profile in Table VI.

The results were very satisfactory for Cs134 (C/E = 1.01) but not so good for Cs137. This appears to be the result of measurement difficulties as for two of the samples the duplicates differed by over 2:1.

The graphical integration method was used to study the production of Eu154. This isotope, which because of its relatively long half life and high energy gamma emission is frequently observed in longer-cooled fuels, is formed in the reactor by n,gamma reaction from Eu153. The reaction is energy-dependent, and as only a single one-group cross section, non-specific to reactor or region, is available in the FISPIN libraries, and as the formation is (apart from a very small primary fission yield) related to the concentration of Eu153, it is not to be expected that the concentration of Eu154 would be directly proportional to burn-up.

The measurements of europium-154 are compared with FISPIN calculations in Table VI. Although the FISPIN values are an order of magnitude higher than the calculated Eu154 primary fission yields (Crouch-2) they are considerably lower than the observed values in the central region of the fuel and averaged over the length of the fuel column.

Fortunately the europium is a minor and relatively unimportant constituent of the fission products, but the foregoing observations indicate scope for improving the FISPIN libraries with respect to the neutron capture cross sections of fission products.

Fission Gases A fraction of the total production of the noble gases krypton and xenon is released to the plenum during irradiation, and can be collected and analysed during post-irradiation examination. The remainder of the gases comes off during fuel dissolution and is not easily measured.

Results are available for one pin from the experimental cluster and four pins from a standard, reference driver fuel sub-assembly, and these are reviewed in Table VII. The calculated values used in the comparisons were based on mean burn-ups obtained from the physics calculation in the case of the driver sub-assembly and on the neodymium analysis (Crouch yields) plus physics profile in the case of the experimental fuel.

The first section of the Table shows that whereas over 80% of the gases were released to the plenum from the experimental fuel, the driver fuel, at much lower rating and burn-up, released only about a third of the calculated production.

The xenon:krypton ratio is marginally higher than unity for the experimental fuel. If any significance could be read into this it could be supposed that the calculated ratio is 5% too high, and that the lower ratio for the driver fuel indicates slower diffusion of xenon from the 60+% retained in the fuel. It would be unwise to give much weight to this postulate on the basis of a single observation, but similar observations were made on fuel irradiated in the experimental Dounreay Fast Reactor (4).

The isotopic compositions shown in the second part of the Table indicate agreement to within 2% - well within the tolerance of the measurements - in the case of krypton, but rather poor agreement in the case of the minor zenon isotopes Xe128 and Xe130. The consistent 4-5% overestimate of Xe131 is possibly significant also, and would agree with the similarly higher Xe/Kr ratio for 80% vis a vis 30% gas release.

Overall Assessment

In general, the comparisons of the experimental measurements with the calculated values are extremely encouraging. Because the analyses so far available relate mainly to a somewhat unusual experimental fuel that was not uniquely specified in the reactor physics model, a lower calculational accuracy was achieved (particularly with respect to burn-up) than may be expected for standard fuel sub-assemblies, but even so the agreement was well within the tolerance of the analyses for all major components.

The actinide results are particularly pleasing in view of the extended chains of reactions by which some are produced. The small discrepancies in the U238 and Pu240 results may be explainable by an overestimate of the hardness of the neutron spectrum, and the excellence of the agreements for Pu239 and Cm242 is very reassuring in view of the doubts that had been expressed about the accuracy of the relevant nuclear data. These calculations encourage considerable confidence in calculations of actinide decay power.

It would be rash to claim that the agreement obtained for a few fission product isotopes provides a blanket endorsement of all fission product and decay power calculations - in particular two nuclides which make important contributions to decay power at decay times of a few months to a few years, ie strontium and ruthenium, have not been measured - but it would appear that where comparisons are available the calculated values using the existing FISPIN libraries are at least as good as any experimental measurements. (An exception must be made for isotopes formed from other fission product isotopes by neutron capture, but these account for only a very small part of the total fission products).

Two useful validations apply to praeseodymium, which tends to dominate the shielding requirements for medium-cooled fuel, and to the noble gases, which have implications for pin design and for stack discharges. A possible 5% overestimate of Xe 131 will have no design implications.

A full validation of decay power calculations can probably be achieved only by means of calorimetric measurements on irradiated fuel. The most than can be said at the present time is that the actinide power calculations seem to be highly satisfactory, and there is nothing in the relatively few comparisons that are available to suggest that the fission product calculations may not be equally good overall.

Conclusions

Calculations of the composition and activity of irradiated fuel (fast reactor) have been vindicated to well within the limits of experimental accuracy for almost all the quantities which have been measured. Results for the actinides were particularly good, and there seems no justification for applying an additional safety margin to the FISPIN values of decay power for plant design purposes.

Though good agreement with calculations was also obtained for important fission products, there is not yet sufficient information to give such an unequivocal endorsement of the overall fission product decay power calculation.

There are indications of weaknesses in the FISPIN data relating to the production of some minor fission product isotopes, notably Xe128, Xe130 and Eu154. The work has also revealed scope for improvement in analytical techniques, particularly in the field of high resolution gamma spectrometry, in the hope of resolving the discrepancies in the cerium and praeseodymium measurements, and of reducing peak interference and improving accuracy for Nb95 and Cs137.

References

1. R F Burstall FISPIN - A computer code for nuclide inventory calculations. ND-R-328(R) October 1979

2. A Tobias Decay heat Progress in Nuclear Energy 5, 1 - 93, Pergamon Press Ltd, 1980

3. EAC Crouch Fission Product yields from neutron-induced fission. Atomic data and nuclear data tables 19, 417 (1977).

4. H J Powell Behaviour and Chemical State of Irradiated Ceramic Fuels. IAEA, Vienna 1974.

A CRITICAL REVIEW OF RESONANCE INTEGRALS AND POSTIRRADIATION FUEL ANALYSIS FOR IMPORTANT ISOTOPES OF AM AND CM

B. Goel and U. Fischer

Institut für Neutronenphysik und Reaktortechnik
Kernforschungszentrum Karlsruhe GmbH
P.O. Box 3640, D-7500 Karlsruhe
Federal Republic of Germany

During the course of KEDAK-4 evaluation for ^{241}Am, ^{242m}Am, ^{243}Am and ^{244}Cm a wide range in the published values for the resonance integrals was observed. For actinide isotopes, due to the presence of strong subcadmium resonances the measured values of resonance integrals are very sensitive to the effective cutoff energy. The resonance integrals for ^{241}Am for example, change by a factor of about 2.5 by changing the cutoff energy from 0.3 eV to 0.6 eV. For ^{241}Am measured resonance integrals are consistent with each other and those calculated from differential data if proper account of the Cd-cutoff energy is made. For ^{243}Am there is a discrepancy of about 20% between the capture resonance integral value calculated from differential data and those measured in a reactor. The validity of the KEDAK-4 data as used in the burnup code KORIGEN to predict the isotopic formation of higher actinides in the spent fuel of thermal reactors has been demonstrated.

[^{241}Am, ^{242m}Am, ^{243}Am, ^{244}Cm fission and capture resonance integral, postirradiation fuel analysis]

Introduction

An efficient radioactive waste management stipulates an accurate knowledge of the inpile and out of pile characteristics of the burnt fuel. The isotopic composition of the fuel is to be calculated as a function of reactor history. The accuracy of these calculations depends upon the calculational models of the reactor codes and the quality of the nuclear data used. For a variety of applications in the field of thermal reactors it is sufficient to know the effective cross sections, which in turn can be derived from the knowledge of thermal cross sections and the resonance integrals. In this paper we examine the present status of the resonance integrals of the isotopes of Am and ^{244}Cm. A comparison of the measured resonance integrals with those calculated from the evaluated neutron data files, such as KEDAK-4, UKNDL and ENDF/BV, is also done wherever the data are available. The compilation of Gryntakis and Kim/1/ with some additions from CINDA form the basis of this survey. The results of the post irradiation fuel analysis which can be an important tool to check the quality of the data used in the burnup calculations are also critically examined in this report.

Resonance Integrals

During the course of the evaluation/2/ of the neutron data for the actinide isotopes a wide range in the published values for the resonance integrals was observed. The published resonance integrals can be classified in 4 different categories:

a) Calculated from the differential data
b) Semiempirical nuclear systematics
c) Direct measurement using more or less pure isotopes
d) Resonance Integrals derived from the isotope production measurements in the reactor fuel.

In this paper the data belonging to the categories a) and b) will not be discussed as it is assumed that the current evaluations include all the scrutinised information about the differential data and nuclear systematics. The measurement of resonance integrals is done using the Cd-difference technique. The probe is protected from the thermal neutron flux with a Cd-cover. The thickness of the Cd-cover determines the lowest energy of neutrons reaching the probe. If strong resonances are present in the region of Cd-cutoff energy the measured resonance integral will strongly depend on the thickness of the cadmium sheet. The measurement and the interpretation of the resonance integrals in the actinide region is hampered due to the presence of strong resonances (Fig.1) near the Cd-cutoff energy. Thus a precise knowledge of the Cd-cutoff energy is a prerequisit for a proper evaluation of the

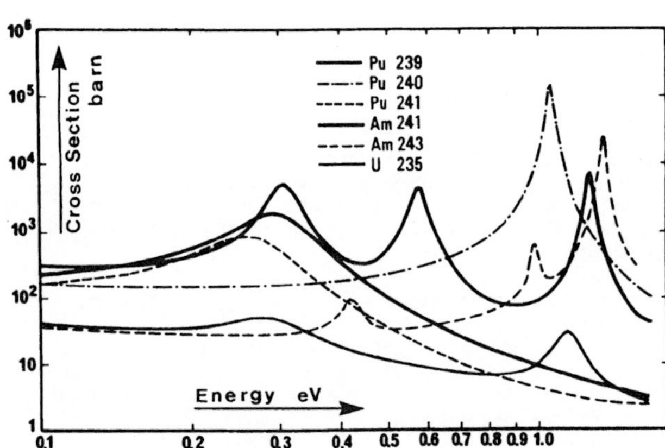

Fig. 1 Capture cross sections for different actinides.

resonance integral. This information, however, is often missing in the publications. The measurements of the type d) are further complicated due to the mutual resonance shielding i.e. the presence of isotopes with strong resonance in or near the sample causes the shape of neutron flux to deviate from the assumed 1/E behaviour. The production of Am and Cm isotopes in the reactor fuel is a result of multiple neutron absorption which takes place in competetion with the different decay modes of the intermediate isotopes(Fig. 2). Though some of the data uncertainties may compensate each other, others may accumulate to give a larger uncertainty of the measured value.

K. H. Böckhoff (ed.), Nuclear Data for Science and Technology, 196–201.

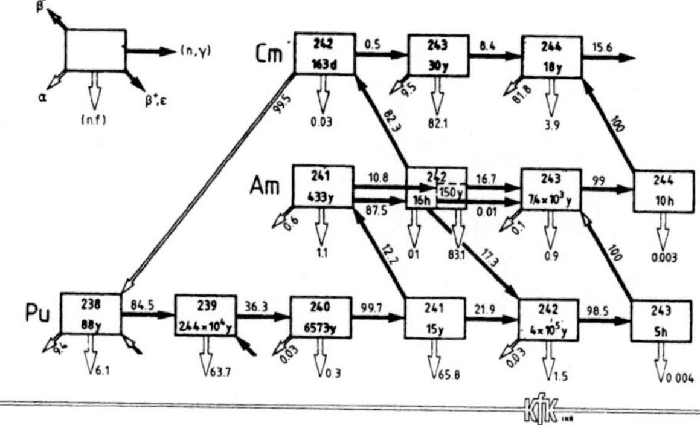

Fig. 2 Isotope Production in a Thermal Reactor ($\phi = 10^{13} n/cm^2.sec$)

In the following we examine the experimental data for different isotopes.

^{241}Am:

For this isotope there are a number of measurements. Due to the two strong subcadmium resonances at 0.308 eV and 0.577 eV the measured value depends strongly on the cadmium cutoff energy and consequently the range in the published values is large. The resonance integrals change by more than a factor of 2 by varying the Cd-cutoff energy from 0.2 eV to 0.7 eV. A similar effect was observed and discussed previously/3/ for the thermal cross section of ^{241}Am. Table I reproduces the results of different measurements for resonance integrals for ^{241}Am.

TABLE I Measured resonance integrals for ^{241}Am

Author	year	E_{Cd}	capture	fission
Bak +/4/	1967	(0.3)	2400 ±200	21 ± 2
Schuman/5/	1969		1100 ±100	
Eberle+/6/	1972		1390	21
Harbour+/7/	1973	0.369	1538 ±135	
Zhuravlev+/8/	1975			27.7 ±1.6
Gavrilov+/9/	1977	0.68	1800 ±100	22.5 ±0.25

Now let us examine the different data critically. Bak et al./4/ have measured the capture resonance integrals for ^{241}Am and ^{243}Am and fission integral for ^{241}Am. For the measurement of ^{241}Am resonance integrals a "pure" sample was used but in the publication no information about the purity of the probe is given. Information about the cutoff energy is also missing. In Ref. /3/ it was shown that the effective cutoff energy in this experiment is likely to be arround 0.3 eV. Thus both resonances at 0.308 eV and 0.577 eV would contribute to the resonance integral. The value of both capture and fission resonance integrals are consistent with this cutoff energy (Fig. 3).

The aim of Schuman's work/5/ was to produce samples rich in ^{242}mAm. Pure samples of ^{241}Am were irradiated in high flux core positions of ETR. The values given for resonance integrals are not corrected for variable Cd-shields. The value obtained for isomeric ratio in this experiment is 0.773. This is much lower than the value obtained in other experiments and presently recommended value of the isomeric ratio. For these reasons the results of this experiment are to be discarded.

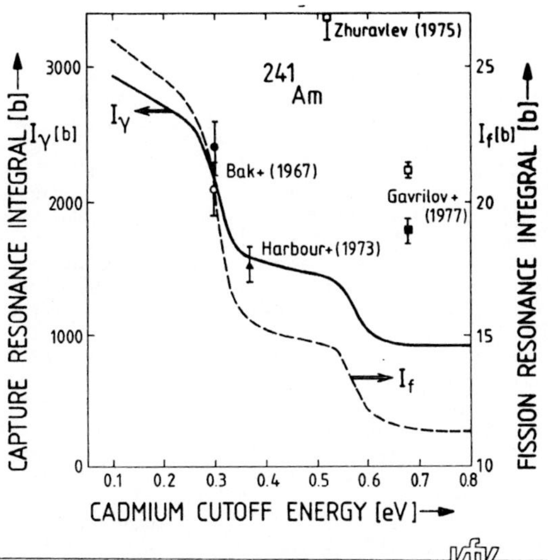

Fig. 3 Resonance Integrals for ^{241}Am as a Function of Cd-Cutoff Energy

Full Symbols are for Capture, Empty Symbols are for Fission.

Eberle et al./6/ is an example of the category d) measurements. Pu-Al-aloy containing 1.5% Pu was irradiated in a reactor until all the Pluonium was burnt and the production of different actinides was measured. The initial composition of the sample was 90.91% ^{239}Pu, 8.22% ^{240}Pu, 0.83% ^{241}Pu and 0.04% ^{242}Pu. The nuclear data were taken from the literature and adjusted to reproduce experimentally determined isotopic compositions of the irradiated probe. Moreover, the ratio of thermal to epithermal flux was also varied to obtain an optimal agreement between experiment and calculations. In most cases the isotope production was a result of multiple neutron absorption in competion with different decay modes of the intermediate nuclei. Thus an accurate knowledge of constants for all intermediate processes is required to interprete the experiment propperly. This type of experiment is of only a limited use to extract cross sections. Though it can be an important tool to check the consistency and the accuracy of the nuclear data for a definite purpose.

The best available measurement for the capture resonance integral of ^{241}Am is that of Harbour et al./7/. They have measured the equivalent thermal capture cross section and capture resonance integral on a highly pure (> 99.9% ^{241}Am) probe and have carefully examined the effect of epithermal to thermal flux ratio and the cadmium cutoff energy on the measured data. They give a value of 1538 ± 135 b for a Cd-cutoff energy of 0.369 eV. The value calculated form KEDAK-4 data for this cutoff energy is 1549 b which is in excellent agreement with the result of Harbour et al.

Zhuravlev et. al./8/ have measured the fission resonance integrals for a number of isotopes ranging from ^{239}Pu to ^{249}Cf. Though the details of the calculation are not published corrections for various effects have been applied to the data. Nevertheless their results for ^{241}Am with strong subcadmium resonances are up to a factor of 2 higher than the values obtained from differential data. The Cd-cutoff energy is given to be 0.52 eV. Even reducing this energy to an extremely low value will not match their results to the differential data. We believe that there is some unrecognised error in this experiment and discard their data from present evaluation.

Gavrilov et al./9/ have measured the capture and fission resonance integrals for ^{241}Am and ^{243}Am. They have also used almost pure isotopes and the thickness of the Cd-shield is given to be 1 mm which corresponds to a cadmium cutoff energy of 0.68 eV for pure 1/v absorbers. Since the investigated isotopes have strong subcadmium resonances the effective cutoff energy is likely to be lower than this value. The authors claim a good agreement between their results and those of Harbour et al/7/. This again is not consistent with the cutoff energy of 0.68 eV. As is seen in Fig. 3 capture integral reduces by about 40% by changing the cutoff energy from 0.369 eV to 0.68 eV. Gavrilov et al.'s value on the contrary is 17% larger than that of Harbour et al. The authors have observed subcadmium neutrons while checking the method on ^{239}Pu. These were then avoided by reducing the detector size. Since the subcadmium resonance in ^{239}Pu is at a lower energy than those of ^{241}Am and ^{243}Am neutrons which do not appreciably effect the fission integral for ^{239}Pu may still enhance the resonance integral for Am isotopes.

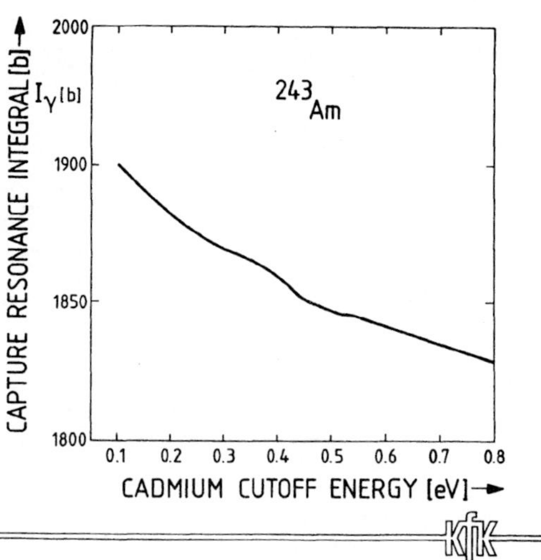

Fig. 4 Resonance Integrals for ^{243}Am as a Function of Cadmium Cutoff Energy

^{242m}Am

For this isotope there are 6 entries in Ref. /1/. Out of these only two are of the category c), namely Schuman/5/ for the absorption resonance integral and Zhuravlev et al./8/ for the fission resonance integrals. Schuman did this measurement concurrent to that of ^{241}Am. The value derived for the ^{242}mAm depends strongly on the cross sections adopted for ^{241}Am. Since these are in large error the same is also true for his value of 7000 ± 2000 b for the absorption resonance integral of ^{242}mAm. Thus the only reliable data for the capture resonance integral for this isotope are due to differential measurements of the capture cross section. The value derived from the KEDAK-4 evaluation is 280 b for the capture integral and 1909 b for the absorption integral.

For the fission resonance integral the only measurement is that of Zhuravlev et al/8/. They give a value of 2260 ± 100 b for ^{242}mAm. This value is about 40% higher than the value of 1629 b derived from KEDAK-4 which is based upon the differential data. The sample contained only about 1% of ^{242}mAm. Thus the contribution to the fission count rate of the probe due to fission in ^{241}Am and ^{243}Am, in spite of their comparitively low cross sections, is of the same order as that due to ^{242}mAm. Their value for fission integral of ^{241}Am as shown above is too large. If a smaller value for fission integral of ^{241}Am is taken to extract the fission integral of ^{242}mAm its value will be still larger. This discrepancy can not be presently resolved. It is recommended to use the value derived from the differential data.

^{243}Am

Table II gives experimental data for ^{243}Am. It is seen that the results of different authors for capture integral agree well within the experimental uncertainties. Fig. 4 shows that the resonance integral for this isotope, in spite of a resonance at 0.4 eV, is not very much dependant on the cadmium cutoff energy. The value of the capture resonance integral for ^{243}Am based on differential data (KEDAK-4) however is 1847 which is about 20% lower than the results of the integral mesurements. An increase in the resonance parameters to match this value can not be justified at present. An explanation for this discrepancy has yet to be found.

TABLE II Measured resonance integrals for ^{243}Am

Author	year	capture	fission
Butler +/10/	1957	2290 ± 50	
Bak +/4/	1967	2300 ± 200	
Folger +/11/	1968	2250 ± 50	
Zhuravlev+/8/	1975		9 ± 1
Gavrilov+/9/	1977	2210 ± 150	17.1 ± 1.3

For the fission resonance integral of ^{243}Am the differential information is lacking and the two integral measurements differ by a factor of 2. For the reasons given above we discard the data of Zhuravlev et al./8/ and recommend the results of Gavrilov et al./9/.

^{244}Cm

For this isotope there are 6 category c) measurements for resonance integrals. The eldest measurement is due to Folger et al./11/. Folger et al. and Smith et al./12/ are twin papers on this subject. The value measured in Ref. /11/ is 700 b. Ref. /12/ uses a value of 631 b for the analysis of their experiment. A reason of this inconsistency is not given in the publications. Both these values are, nevertheless, consistent with the result of Schuman/13/. Since the first resonance for this isotope is at 7.67 eV the results are not affected by the variable Cd-shields. Schuman reports a large uncertainty in the ^{245}Cm content of the sample. It is not clear whether this is included in the uncertainty of ± 50 b or not.

The experiment of Thompson et al./14/ was essentially performed to measure the fission resonance integrals of Cm isotopes. It was done in two parts, short irradiation measurement to determine the fission integral with respect to ^{235}U and long irradiation measurement to obtain a consistent set of cross sections for Cm isotopes. In short irradiation measurement samples with 93 - 95%

enriched ^{244}Cm were irradiated in Cd-capsules together with ^{235}U for about 5 hours. The buildup of ^{131}I was measured with γ-spectroscopy thus obtaining a direct measurement of fission integral with respect to ^{235}U. In the long irradiation experiment samples with different compositions were irradiated long enough to transmute a substantial amount of the probe. From the isotope production calculations the value of capture integral of 650±50 b for ^{244}Cm could be confirmed. The value of fission integral was found to be 12.5 ± 2.5 b. The paper states that this value of resonance integral has been confirmed by fission chamber measurements with a 99.02% ^{244}Cm sample. This measurement with 99.02% ^{244}Cm sample was performed by Benjamin et al./15/. The measurement was done relative to ^{235}U with fission chambers and with solid state track detectors. The value of fission resonance integral for ^{244}Cm is given to be 18 ± 1 b. We recommend this value as the latest result from SRL for the fission resonance integral of ^{244}Cm.

Gavrilov and Goncharov/16/ have published a short note indicating the measurement of capture resonance integrals for Cm isotopes. The work of Zhuravlev et al. has already been discussed. Table III shows results of different authors. It is seen that capture integral values of different authors agree well with each other.

TABLE III Measured resonance integrals for ^{244}Cm

Author	year	capture	fission
Folger +/11/	1968	700	
Schuman /13/	1969	650 ± 50	
Thompson+/14/	1971	650 ± 50	12.5 ± 2.5
Benjamin+/15/	1972		18.0 ± 1
Zhuravlev+/8/	1975		13.4 ± 1.5
Gavrilov+ /16/	1978	626 ± 53	

Comparison of Different Evaluations

In Table IV and V the resonance integrals calculated from different libraries of evaluated data are compared. The resonance integrals for UKNDL and JENDL are due to Mattes/17/. It is seen that the agreement among the different data is good. Except for ^{241}Am, where there is a sufficient experimental data base the agreement is rather decieving and reflects the limited data base. For example for the evaluation of ^{243}Am data in the resonance region there is practically only one publication, namely that of Simpson et al./18/, on which different evaluations are based. The measured capture resonance integral for ^{243}Am is about 20% larger than the values given in Table IV. The KEDAK value of the capture resonance integral for ^{244}Cm is higher than other evaluations. This is due to the inclusion of the data of Kalebin et al./19/. Including these data in the resonance analysis leads to a larger value for neutron width of the first few resonances in ^{244}Cm (see Ref. 2).

TABLE IV Capture resonance integrals from different evaluations

	^{241}Am	^{242m}Am	^{243}Am	^{244}Cm
KEDAK-4	1452	280	1847	637
UKNDL	1415		1846	
JENDL		207		593
ENDF/B V	1420	286	1818	595

TABLE V Fission resonance integrals from different evaluations

	^{241}Am	^{242m}Am	^{243}Am	^{244}Cm
KEDAK-4	14.8	1629	4.4	18.1
UKNDL	15.02		5.95	
JENDL		1575		17.8
ENDF/B V	13.4	1886	6.15	18.7

Post Irradiation Fuel analysis

As has been stated before the isotopic composition of the spent fuel can be used as a tool to check the data used in burnup calculations. Whether this is a good check or not depends upon:
a) the quality of calculational methods and
b) the reliablity of the experimental results.

At KfK the isotopic composition of the spent fuel is calculated using the computer code KORIGEN/20/. This is an improved version of the well known code ORIGEN/21/. The improvements concern physics of the code as well as the improved and extended data base. The main feature of these improvements is that, the effective cross sections are not derived from the thermal cross sections and resonance integrals rather they are calculated from a 83 group (30 thermal and 53 epithermal groups) cross section set. This cross section set is in turn generated form the differential data library KEDAK-4. The effective cross sections are calculated at each burnup stage using the current neutron spectrum obtained with time dependant cell calculations. This enables to take account of effects like resonance shielding, spectrum hardening etc. as a function of burnup in a consistent manner. To calculate the isotopic composition of the spent fuel a good knowledge of its irradiation history is essential. This is very often an object of commercial secret. Therefore, the analysis is limited to those cases where we have an access to the irradiation data. Post irradiation fuel analysis was performed for different reactors. In this report the results for a prototype 350 MWe PW-reactor KWO are presented. The experimental composition and the irradiation data are taken from Ref. /22/ and /23/.

The validity of the code has been checked by reproducing the measured isotopic composition of the major actinides i. e. for the isotopes of U and Pu for different reactors. Fig. 5 and 6 compare the experimental and calculated concentrations of isotopes ^{241}Pu and ^{242}Pu in the spent fuel of the reactor KWO.

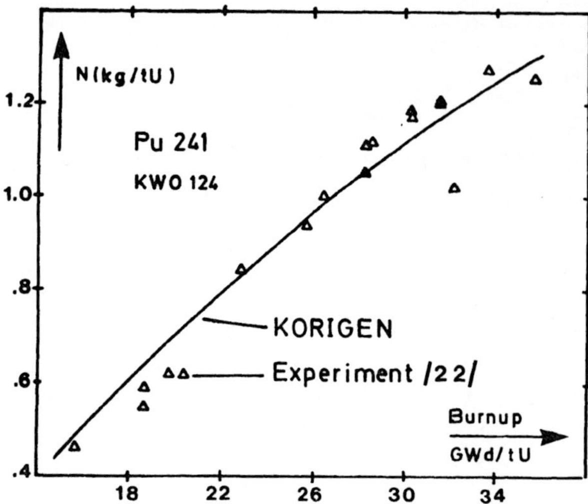

Fig. 5 ^{241}Pu concentration in the spent fuel of PW reactor KWO

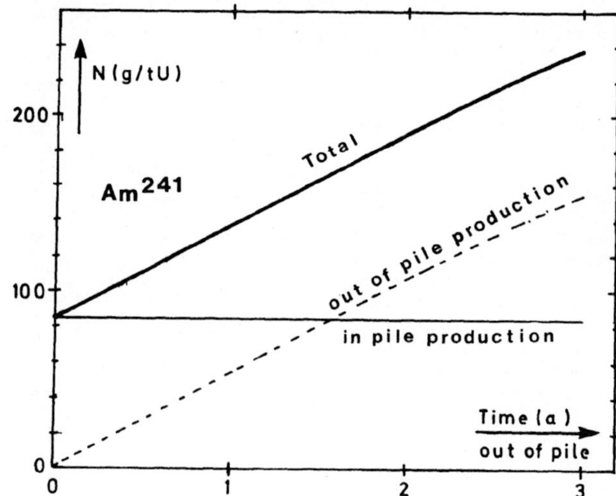

Fig. 7 Out of pile accumulation of ^{241} Am in spent fuel

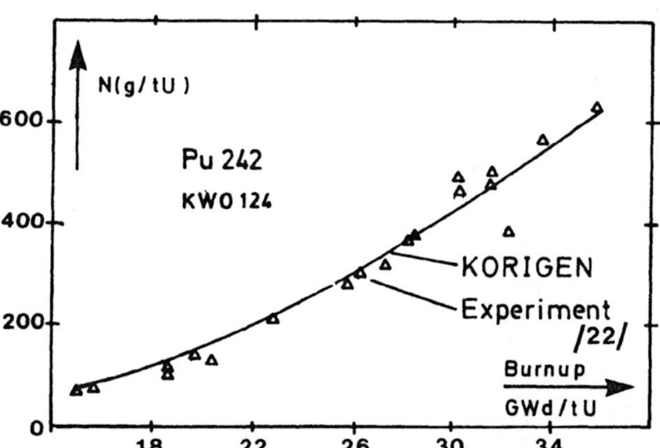

Fig. 6 ^{242}Pu concentration in the spent fuel of PW reactor KWO

The measurement of the isotopic compostion of the spent fuel is done after a cooling period whose length is determined by technical and managerial problems. Generally it is of the order of 2 to 3 years. During this period ^{242}Cm with its half life of 162 days has decayed to ^{238}Pu. A subtantial amount of ^{241}Am has been generated through the decay of ^{241}Pu with the half life of 14.35 ± 0.02 years/24/. The measurement of isotopic concentration of Am and Cm isotopes is performed with the methods of α-spectroscopy and mass analysis. The α-peaks of ^{241}Am are overshadowed by 238Pu, which is formed in the reactor mainly via the reaction ^{237}Np(n,γ)^{238}Np--β-->^{238}Pu and the decay of ^{242}Cm as mentioned above. This is the main source of error in the determination of ^{241}Am concentration as it is obtained by measuring the ^{238}Pu-^{241}Am peak before and after the extraction of Pu from the spent fuel. This is a difference of approximately equal numbers and accordingly the error in the ^{241}Am measurement is large. More over the account of he ^{241}Am formed due to the out of pile decay of ^{241}Pu has to be made. Fig. 7 shows the out of pile increase in the ^{241}Am concentration in the spent fuel due to the decay of ^{241}Pu. It is seen that at the time of experimental analysis the ^{241}Am produced in the reactor accounts for only 25% of the measured ^{241}Am concentration. Due to the uncertainty in the measured ^{241}Am concentration which is of the same order this isotope is not suitable for data check.

^{243}Am is measured with mass analysis relative to ^{241}Am and thus the error in the ^{241}Am concentration also dominates the error in the ^{243}Am concentration. Only recently a measurement/25/ using the isotope dilution technique for the determination of the concentrations of Am isotopes in the spent fuel has become available. This makes the ^{243}Am concentration free from the errors in the ^{241}Am concentration. Fig. 8 compares the results of isotope dilution measurement/25/, which is in good agreement with the earlier measurement based on α-spectroscopy/22/, with the results of the code KORIGEN. A good agreement is observed. A similar good agreement is observed for ^{244}Cm (Fig. 9). Since more than 99% of ^{244}Cm is built via ^{242}Pu and ^{243}Am the good agreement between calculated and measured concentration of Pu isotopes as well as ^{243}Am and ^{244}Cm indicate the validity of the data for the involved processes, namely the neutron cross sections for ^{242}Pu(n,γ), ^{243}Am(n,γ) and ^{244}Cm(n,abs.). To verify the same for ^{241}Am it is necessary that isotopic analysis of the spent fuel is made soon after irradiation i.e. before the ^{241}Am formed out of pile dominates the experimental value.

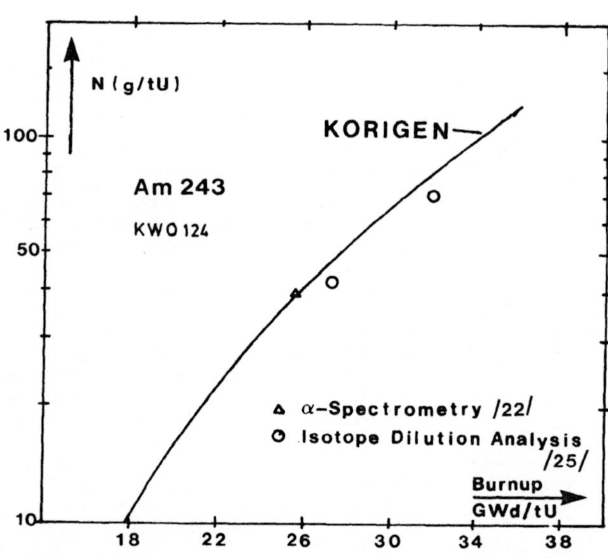

Fig. 8 ^{243}Am concentration in the spent fuel of KWO

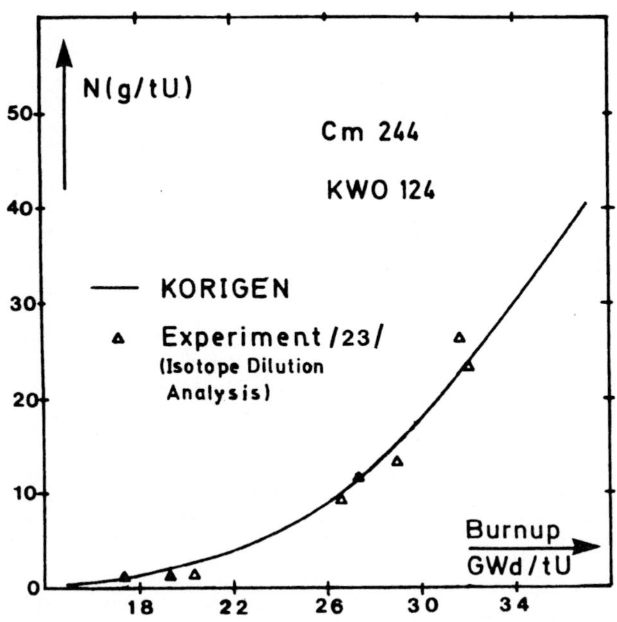

Fig. 9 ^{244}Cm concentration in the spent fuel of KWO

Conclusion and Recommendations

The validity of the KEDAK-4 data as used in the burnup code KORIGEN to predict the isotopic formation of higher actinides in the spent fuel of thermal reactors has been demonstrated in the preceeding chapter.

For ^{241}Am different evaluations are consistent with each other and with integral data except that of Zhuravlev et al./8/.

For ^{242m}Am the data basis is poor but the importance of this isotope in burnup calculation is low.

For ^{243}Am there is a discrepancy of about 20% between the capture resonance integral value calculated from differential data and those measured in a reactor. New measurement seems necessary to resolve this discrepancy.

For ^{244}Cm the data are in good shape. The recent evaluation of differential data (KEDAK-4) give about 5% higher value for the capture resonance integral of ^{244}Cm. This value is supported by the recent measurement of Gavrilov and Goncharov/16/ and the post irradiation fuel analysis calculations.

Acknowledgement

The authors thank Dr. Küsters for his support and many extensive and fruitful discussions.

References

1 E.M. Gryntakis, J.I. Kim, J. Radioanal. Chem. 42 (1978) 181
2 F.H. Fröhner, B. Goel, U. Fischer, H. Jahn, This meeting, Paper A-16
3 B. Goel, Proc. of the Specialist's Meeting on Nuclear Data of Plutonium and Americium Isotopes for Reactor Applications, Brookhaven, Nov. 20-21, 1978, (BNL-50991) p. 177
4 M.A. Bak, A.S. Krivokhatskii, K.A. Petrzhak, Yu.G. Petov, Yu.F. Romanov, E.A. Shlyamin, Soviet At. Energy, 23 (1967) 1059
5 R.P. Schuman, WASH-1136, 1969, Pages 51 and 53
6 S.H. Eberle, H.J. Bleyl, H. Braun, A.V. Baeckmann, L. Koch, Report KfK-1453 (EUR-4726d), 1972, p. 1
7 R.M. Harbour, K.W. Macmurdo, F.J. McCrosson Nucl. Sci. Eng. 50 (1973) 364
8 K.D. Zhuravlev, N.I. Kroskin, A.P. Ujetvjerikov, At. Engery (USSR), 39 (1975) 285
9 V.D. Gavrilov, V.A. Goncharov, V.V. Ivanenko, V.N. Kustov, V.P. Smirkov, Sov. At. Energy 41 (1977) 808
10 J.P. Butler, M. Lounsbury, J.S. Merrit, Can. J. Phys. 35 (1957) 147
11 R.L. Folger, J.A. Smith, L.C. Brown, R.F. Overman, H.P. Holcomb, 68 Wash, 2 (1968) 1279
12 J.A. Smith, C.J. Banick, R.L. Folger, H.P. Holcomb, I.B. Richter, 68 Wash, 2 (1968) 1285
13 R.P. Schuman, WASH-1396 (1969) p. 54
14 M.C. Thompson, M.L. Hyder, R.J. Reuland, J. Inorg. Nucl. Chem., 33 (1971) 1553
15 R.W. Benjamin, K.M. Macmurdo, J.D. Spencer, Nucl. Sci. Eng. 47 (1972) 203
16 V.D. Gavrilov, V.A. Goncharov, Sov. At. Energy 44 (1978) 274
17 M. Mattes, Private Communication (1982)
18 O.D. Simpson, F.B. Simpson, J.A. Harvey, G.C. Slaughter, R.W. Benjamin, C.E. Ahlfeld, Nucl. Sci. Eng. 55 (1974) 273
19 S.M. Kalebin, IAEA 186 (1976) Vol. III, p. 121
20 U. Fischer, H.W. Wiese, KfK-3014 (to be published)
21 M.J. Bell, ORIGEN - The ORNL Isotope Generation and Depletion Code, ORNL-4628, May 1973
22 P. Barbero et al., Report EUR 5605e (1979)
23 M. Wantschik, KfK 3316 (1982)
24 H. Ottmar, KfK 3149 (1981)
25 B. Ganser, KfK 3380 (1982)

ANALYSIS OF NEUTRON CROSS SECTIONS FOR THE FORMATION OF PU236 and CO58,60 IN BOTH THERMAL AND FAST REACTORS

H.W. Wiese, U. Fischer, B. Goel

Kernforschungszentrum Karlsruhe, GmBH
Institut für Neutronenphysik und Reaktortechnik

P.O.Box 3640, D-75 Karlsruhe
Federal Republic of Germany

For safe handling of irradiated nuclear fuel and of activated structural components the radiation of these materials has to be well predicted. Recently shielding problems have gained interest which arise from the intense high-energy γ-ray of Tl208 in the decay chain of Pu236 in Pu from LWRs in Fast Reactor fuel elements and from the Co-isotopes 58 and 60 as activation products in structural materials.
The formation of Pu236 in LWRs strongly depends on the (n,2n)-cross-section of Np237. The fission spectrum averaged cross-section $\langle\sigma_{n,2n}(Np237)\rangle_\chi$ = 3.35 mb as calculated from 1977 KEDAK differential data is about a factor two larger than that deduced from older evaluations and integral experiments. By consistent use of the KEDAK value in burnup calculations good agreement is achieved between measured and calculated Pu236 concentrations in burnt PWR-TRINO-fuel.
The formation of Co58,60 is due to(n,p)-reactions in Ni58,60 respectively and to (n,γ)-reactions in Co59. With effective cross-sections based on KEDAK and ENDF/B-IV data the formation of Co in FBRs and LWRs and the role of the initial Co in the structural material is analysed. In FBR stainless steel,the Co production from Ni58 is most dominant. In LWR zircaloy, the ratio of the Co formation from Ni and Co59 varies from about 7:3 to 9:1 depending on the initial Co59 concentration.

Background of Pu236 Formation Analysis

Tl208, a daughter product at the end of the decay chain of Pu236, emits hard γ-radiation of 2.6 MeV. Due to this radiation, there may arise additional shielding problems in handling Fast Reactor fuel elements, fabricated with reprocessed LWR-plutonium. To estimate the γ-dose rate of such fuel elements, it is important to predict reliably the amount of Pu236 in plutonium discharged from Thermal Reactors.

Formation of Pu236 in Thermal Reactors

In Thermal Reactors, Pu236 is produced mainly via the following nuclear processes:

i) U235(n,γ) U236(n,γ) U237 $\searrow\beta^-$ Np237(n,2n)Np236 $\xrightarrow{\beta^-}$ Pu236

ii) U238(n,2n)U237 $\nearrow\beta^-$

For a power reactor of the German Biblis-type (PWR, 1300 MW_e) fuelled with 3.2 w/o enriched uranium, irradiated to a burnup of 30 GWd/tU, it was shown by means of a sensitivity study using the KfK-code KORIGEN /1/, that route i) contributes ca. 80%, and route ii) ca. 20% to the production rate of Pu236. The production rate of Pu236 passes the bottleneck Np237(n,2n) Np236. Furthermore, the formation of Pu236 depends almost linearly on the cross-section of this process. Thus the attention focusses on the (n,2n) cross-section of Np237.

Differential Data

At present there exist no direct measurements of $\sigma_{n,2n}(E)$ of Np237. There are, however, a few measurements of the formation cross-section of Pu236 /2-5/ in the energy range 14-15 MeV, and, in one case at 9.6 MeV. The precursor of Pu236, Np236 occurs in two isomeric states, a long-lived one ($T_{1/2}=1.15\cdot10^5$ a/6/) and a short-lived one ($T_{1/2}$=22.5 h /6/). Only the short-lived state of Np236 is relevant for the formation of Pu236, it decays at a fraction of 48% /7/ by β-emission to Pu236 and at a fraction of 52% by EC to U236. The total (n,2n)cross-section of Np237 thus can be obtained from the measured formation cross-section of Pu236 by correcting the last one for the β^-/EC ratio and the isomer ratio of Np236.

The isomer ratio $\sigma_{n,2n}$(Np237→Np236(s))/$\sigma_{n,2n}$(Np237→Np236(s+l)), where Np236(s) designates the short-lived, Np236(l) the long-lived state of Np236, has been measured by Landrum et al. /2/ and Myers et al. /8/ in underground thermonuclear tests. They exposed samples of Np237 to intense fluxes of high-energy neutrons; the spectrum was believed to be peaked at 14 MeV. Myers et al. give 0.35 ± 0.05 for the ratio Np236(l)/Np236(s), that is 0.75 for the isomer ratio, Landrum et al. give 0.76 ± 0.03 for the ratio Np236 (s)/Pu236, that is 0.725 for the isomer ratio.
If the isomer ratio is not strongly energy dependent in the vicinity of 14 MeV, we have adequate experimental information on the total (n,2n)-cross-section of Np237 in this energy region. At lower energies, where the major contribution to the spectrum averaged formation cross-section of Pu236 comes from, one has to rely on model calculations. Since theoretical models of (n,2n)reactions have been tested on neighbouring nuclei,cf. e.g/24/, such calculations are believed to be accurate within 10%.

Integral Data

The threshold of the reaction Np237(n,2n) Np236 is at E_s=6.8 MeV. Above this energy the neutron spectrum in a Thermal Reactor can be well approximated by the fission spectrum $\chi(E)$.
The (n,2n)cross-section of Np237, averaged over the reactor neutron spectrum thus can be written as

$$\langle\sigma_{n,2n}\rangle_\phi = f_\phi \cdot \langle\sigma_{n,2n}\rangle_\chi \qquad (1)$$

with the fission spectrum averaged cross-section

$$\langle\sigma_{n,2n}\rangle_\chi = \int_0^\infty \sigma_{n,2n}(E)\chi(E)dE / \int_0^\infty \chi(E)\,dE$$

and the spectral factor

$$f_\phi = \frac{\int_0^\infty \chi(E)\,dE}{\int_{E_s}^\infty \chi(E)\,dE} \cdot \frac{\int_{E_s}^\infty \phi(E)dE}{\int_0^\infty \phi(E)dE}$$

$$= 75.19 \cdot \phi(E > E_s)/\phi_{tot}$$

K. H. Böckhoff (ed.), Nuclear Data for Science and Technology, 202–205.

using a Cranberg fission spectrum. The ratio of the flux above the threshold E_s to the total flux has to be calculated for the given reactor by spectral codes. An early evaluation of $<\sigma_{n,2n}>_\chi$ of Np237 by Pearlstein in 1965 yielded 1.3 mb using a Cranberg fission spectrum weighting /9/. In 1974 Paulson and Hennelly presented a value of 1.46 mb for $<\sigma_{n,2n}>_\chi$ (Np237)/10/. This value was deduced from an integral experiment, where the ratio of the concentrations of Pu236 and Pu238 of several samples irradiated in a heavy water moderated reactor at Savannah River was measured by α-spectrometry. The (n,2n)cross-section was adjusted until agreement was observed between the calculated and the measured ratio of Pu236/Pu238. Thus the value for $<\sigma_{n,2n}(Np237)>_\chi$ deduced by Paulson and Hennelly from integral experiment and that one evaluated by Pearlstein from differential data agreed satisfactorily.

There are recent evaluations of the differential (n,2n) cross-section of Np237, e.g. that of Jary as cited by Patrick /11/, Caner et al. /12/, Fort /13/, which are consistent with the experimental differential data cited above. These evaluations yield a fission spectrum averaged cross-section which is approximately twice as large as the value deduced by Paulson and Hennelly. The evaluation of Caner et al. /12/ for the German KEDAK-File (Fig.1), based on systematics and the measured values of Ref./2/ and /3/ yields a fission spectrum averaged cross-section of 3.35 mb. This value is by a factor 2.3 larger than that of Paulson and Hennelly.

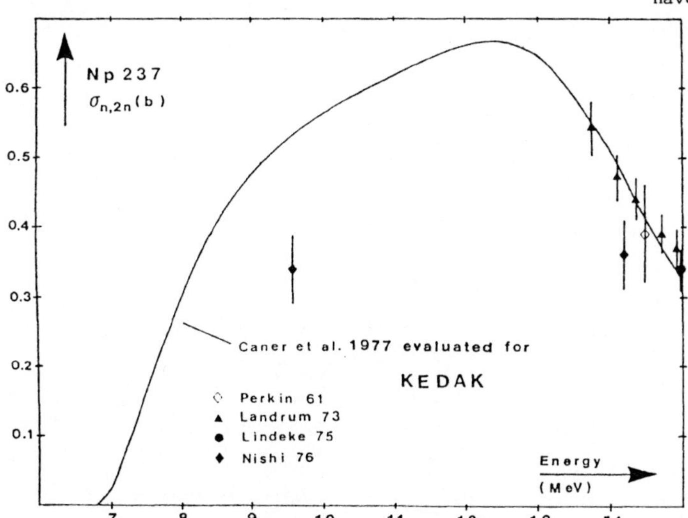

Fig.1: Evaluated and Measured (n,2n)-Cross-Sections for Np237

Integral Data Versus Differential Data

It has been argued by Patrick /11/, that this discrepancy may disappear, assuming that the isomer ratio of Np236 is energy-dependent. If the isomer ratio with decreasing energy tends to zero at the threshold, as proposed by Fort /13/, the contribution of $\sigma_{n,2n}(E)$ of Np237 to the formation cross-section of Pu236 in the energy range 6.8 to about 12 MeV would be reduced considerably. There is indeed no evidence, that the isomer ratio should be energy-independent as it is assumed so far in the calculations for the formation of Pu236. There is, however, some uncertainty about the dependence of the isomer ratio r(E) as a function of neutron energy. As shown by Fort by model calculations /13/ there are two possible curves for r(E), depending on the spin assignments of the ground resp. isomeric state of Np236. Clearly, more experimental information is needed in the energy range 6.8 to 12 MeV, where the major contribution to $<\sigma_{n,2n}>_\chi$ comes from. The experiments should be based on neutron counting methods to obtain directly the total (n,2n)cross-section, as proposed by Patrick /11/.

Post Irradiation Analysis: Calculation and Experiment

Further experimental integral information on Pu236 formation is obtained from experimental post irradiation analysis of pellets from the PWR Trino Vercellese (250 MWe) /14/. In the frame of this analysis, the concentration of Pu236 has been measured by means of α-spectrometry at the Joint Research Centre at Ispra. This experiment represents a good base to check the nuclear data relevant for the formation of Pu236 using the KfK-burnup code KORIGEN, since the concentrations of U-, Pu-, Am- and Cm-isotopes are well reproduced by KORIGEN-calculations, if actinide cross-sections from the KEDAK-library are used /1/.

To get the neutron spectrum averaged cross-section $<\sigma_{n,2n}>_\phi$ from the fission spectrum averaged cross-section $<\sigma_{n,2n}>_\chi$ (see.eq.(1) to (3)), the ratio $\phi(E > 6.8\ MeV)/\phi_{tot}$ has to be determined for the Trino reactor. This was done in a cell-calculation for the Trino PWR using a modified version of the code HAMMER /16/. The ratio $\Phi(E > 6.8\ MeV)/\Phi_{tot}$ was obtained from this calculation as $5.0\cdot10^{-3}$.

Using the Pearlstein value for $<\sigma_{n,2n}>_\chi$ of Np237 the calculated concentrations of Pu236 for the Trino pellets are to a factor 2.7 lower than the measured ones (Fig.2). Nearly the same holds if the Paulson/Hennelly value is used (Fig.2). Using however the KEDAK value of 3.35 mb results in a slight overprediction in comparison with measured concentrations of Pu236 (Fig.2). It should be mentioned, that the KEDAK value is not adjusted. In these calculations we have used the energy-independent isomer ratio of Np237 as given by Landrum /2/. If the energy-dependence of the isomer ratio as suggested by Fort /13/ is used, the spectrum averaged formation cross-section of Pu236 changes by roughly +25 resp. -30%.

Discussion of Uncertainties

There are three principle sources of uncertainties. The first of these is the fission spectrum averaged formation cross-section of Pu236. The uncertainty in the evaluated $\sigma_{n,2n}$(Np237) data is given to be ± 10%. The contribution to the uncertainty due to the energy dependence of the isomer ratio may amount to about 30%. The second source of possible errors lies in the neutron flux above the threshold of Np237 (n,2n)Np236 compared to the total flux, calculated by the HAMMER code. It is difficult to estimate the accuracy of the calculated high-energy part of the neutron spectrum. The last source of uncertainties certainly is the experimental uncertainty in

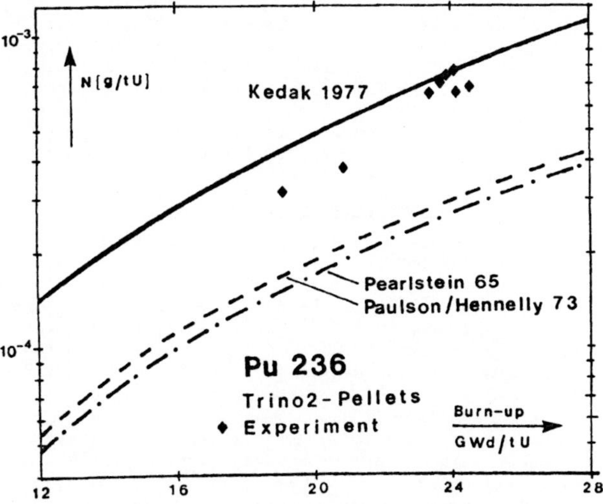

Fig.2: Comparison of Calculated and Measured Concentrations of Pu236 in Trino Irradiated Fuel

determining the concentration of Pu236. This experimental error is given to be 10%. As an upper limit the uncertainty in determining the formation of Pu236 is estimated to be ± 30%.

Conclusions

The fission spectrum averaged cross-section $\langle \sigma_{n,2n}(Np\ 237) \rangle_\phi$ = 3.35 mb, based on the KEDAK-evaluation of Caner et al. for $\sigma_{n,2n}(E)$, is suitable to calculate the formation of Pu236 in PWR-fuel. For fuel from the TRINO-PWR a slight overprediction in the Pu236 concentrations in comparison with experimental results is achieved. For the γ-radiation from FBR fuel elements thus conservative estimates will be obtained. Shielding of these fuel elements has to be reinvestigated.

Background of Co Formation Anlysis

In structural materials of fission reactors(stainless steel in Fast Breeder Reactors and zircaloy in Light Water Reactors), radioactive Co, namely Co58 and Co60, is generated during irradiation. Co58 decays by electron capture with a half-life of 71d and emits predominantly γ-rays of .81 MeV energy, Co60 decays by ß-transition with a half-life of 5.26a mainly emitting γ-rays of 1.17 MeV. These hard γ-rays may neccesitate expensive shielding measures, especially if handling of structural components after short cooling times, during which only a small amount of Co58 could decay, is intended.

For the Karlsruhe Fast Test Reactor KNK-II a stainless steel test rig for irradiation purposes is under discussion. It is foreseen to be handled within a period of up to ten days after unloading from the reactor. Since the only stable Co isotope Co59 which principally undergoes (n,γ) and (n,2n) reactions to form Co60 and Co58, respectively, is always present in stainless steel as an impurity, the question arose, whether it is necessary and reasonable to choose a special kind of stainless steel with a low initial content of Co in order to keep the resulting γ-radiations small.

Production of Co58,60

The principal routes of production of Co58,60 are

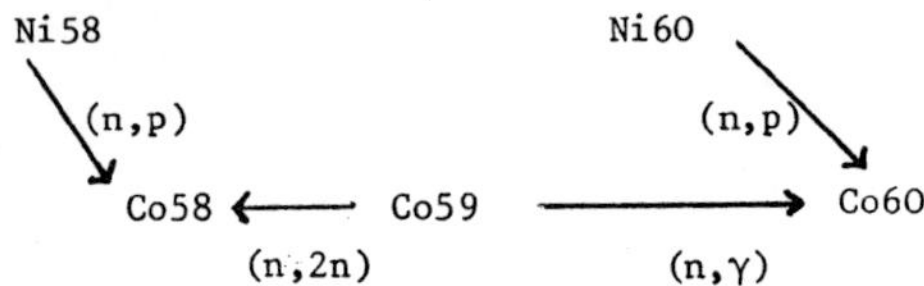

In addition to the already mentioned formation of Co58, 60, from Co59, there are (n,p)-processes on Ni58 and Ni60 leading to Co58 and Co60, respectively. Because of the thresholds of the latter reactions of about 1 MeV and 4 MeV only a relatively small number of direct or slightly degraded fission neutrons contribute to the production of Co from Ni. Since the (n,2n)-reaction on Co59 has a threshold at about 10 MeV, there will be no significant built-up of Co58 from Co59 in fission reactors. This situation will be different in fusion reactors where one has to do with 14.1 MeV (D,T)-neutrons. The probability of an (n,γ)-process on Co59 generally decreases with increasing energy; thus, qualitatively, the role of the latter reaction will be more important in Thermal Reactors.

Status of Data

For the quantitative analysis of Co58,60 formation KEDAK /17/ and ENDF data were used. The evaluation for Ni58(n,p)Co58 and Ni60 (n,p)Co60 for KEDAK was completed in 1975. In the meantime the data for Ni58(n,p)Co58 from ANL /18/ have become available. These data are lower than those from the KEDAK evaluation. On the other hand the fission spectrum averaged cross-section calculated from KEDAK (105. mb) is in excellent agreement with that from ENDF/B-V (105.1 mb) and the result of an integral measurement (108.5 mb ± 5%) /19/. The situation for Ni60(n,p)Co60 is less satisfactory. Except for some new measurements at 14 MeV no experimental data were published after the close of the KEDAK evaluation. The fission average cross-section from KEDAK is 2.1 mb. The same from ENDF/B-V is 2.61 mb and the only measurement /20/ is 1.69 ±.18 mb. The data for Co59(n,γ)Co60 are taken from ENBF/B-IV which seems to be the same as ENBF/B-V. A new evaluation has not yet been made.

Reactor Spectrum Averaged Cross-Sections

For a comparison of the routes of production of Co58 and Co60 both in Fast and Thermal Reactors the relevant neutron spectrum averaged cross-sections were calculated from KEDAK data using the Group Cross Section Generation Code MIGROS-3 /29/. The KNK-II midplane neutron spectrum in the center of the test rig was determined by a 2d-Diffusion calculation /22/. For comparison also a PWR spectrum /23/ and the SNR 300 outer-core spectrum was applied. Fig.3 shows these $E \cdot \phi(E)$ spectra with $\int \phi(E) \cdot dE = 1$ normalization together with the (n,p)-cross-section of Ni58.

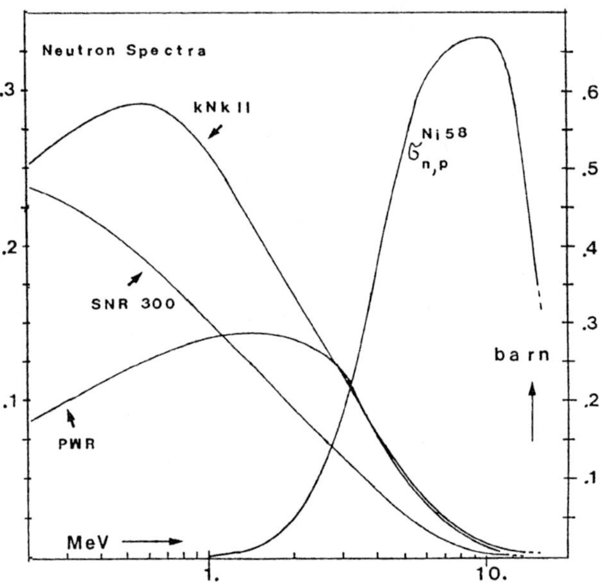

Fig.3: Normalized Fast and Thermal Reactor Neutron Spectra, and $\sigma_{n,p}(E)$ of Ni58

The spectrum of the compact KNK-II core with its comparatively low U238 content and therefore lower degradation by inelastic scattering is much harder than that of the larger and more U238 containing SNR 300 core. The difference in the spectra is directly reflected in the averaged (n,p)-cross-sections of Ni in Table I

Table I: Reactor Neutron Spectrum Averaged Cross-Sections /mb/

Reactor	Ni58 $\langle \sigma_{n,p} \rangle$	Ni60 $\langle \sigma_{n,p} \rangle$	Co59 $\langle \sigma_{n,\gamma} \rangle$
KNK-II	36.	.75	13.
SNR 300	20.	.41	42.
PWR	35.	.83	7000.

$\sigma_{n,\gamma}$ is larger in the SNR 300 than in KNK-II by a factor of ca. 3. In the PWR spectrum $\sigma_{n,\gamma}$ (Co59) reaches 7b. This value is taken from ORNL calculations /15/, since the KEDAK Co data were converted from ENDF/B-IV. The PWR spectrum averaged (n,p)-cross-sections of Ni are comparable to those calculated for KNK-II. This is due to the fact that in the considered PWR the neutron spectrum for E > 2 MeV is similar to that of KNK-II - see Fig.3.

Quantification of Sources of Co58,60

On the basis of the Ni58,60 and Co59 contents in structural materials as given in Table II for the stainless steel type 1.4948 (as an example) and for zircaloy the specific total Co activities and the contributions due to the initial Ni58, 60 and Co 59 contents were calculated - Table III.

For FBRs a 360d irradiation in a constant neutron flux density of $\phi = 2 \cdot 10^{15}$ n/cm² sec, for the PWR 1000d with $\phi = 3 \cdot 10^{14}$ n/cm² sec was assumed.

Table II: Ni and Co Contents in Structural Materials (Examples)

Material	Ni58	w/o Ni60	Co59
SS 1.4948	7.42	2.53	4.2
Zircaloy	3.12	1.19	.0032 to 0.21

Table III: Total Co Activities and Contributions from Ni58, 60 and Co59

Reactor	Ci/kg Str.Mat.	Ni58	% from Ni60	Co59
KNK-II	1458.	99.8	.1	.1
SNR 300	815.	99.2	.1	.7
PWR[a]	125.	71.3	.2	28.5
PWR[b]	95.	94.0	.3	5.7

a) .021 w/o Co b) .0032 w/o Co

The main result is: In FBR stainless steel 1.4948 the predominant source of radioactive Co is Ni58. The contribution of Co59 is < 1% already if a maximum initial Co59 content of .2 w/o is considered. A reduction of the initial Co will not affect the total activity significantly. In PWR zircaloy the contributions from Ni58 and Co59 roughly behave as 7:3 and 9:1 at maximum and minimum initial Co content, respectively. Here the total activity considerably depends on the initial Co, a limitation of the latter is reasonable. In both cases FBR and PWR structural material, the contribution of Ni60 is very small: <1%. The factor of about 10 between the total FBR and PWR Co activities is mainly caused by the same factor in the neutron flux densities. Attention is drawn to the about 2:1 ratio of the KNK-II and SNR 300 Co activities. This is mainly due to different neutron spectra and the resulting about 2:1 ratio in the (n,p)-cross-sections of Ni58.
In the context of differential nuclear data the following can be stated: Most important for the Co formation both in Thermal and Fast Reactors is $\sigma_{n,p}(E)$ of Ni58. This cross-section, being a dosimetry cross-section where more stringent accuracy requirements are made, is sufficiently well known. In Thermal Reactors $\sigma_{n,\gamma}(E)$ of Co59 plays a certain role. With limited initial Co-contents in PWR structural materials the present uncertainties seem to be tolerable. This will be further investigated.

Acknowledgement

The authors wish to thank Dr. H. Küsters for supporting the work and for helpful discussions.

References

1. U.Fischer,H.W.Wiese, KfK-3014 (in preparation)
2. J.H.Landrum. R.J. Nagle, M.Lindner, Phys. Rev.C, 8 (1973) 1938
3. K.Lindeke, S.Specht, H.J.Born; Phys. Rev. C 12 (1975) 1507
4. J.L.Perkin, R.F.Coleman, Journ. Nucl. Eng. 14 (1961)69
5. T.Nishi, I.Fujiwara, N.Imanishi, INDC (JAP)28L,20 (1975)
6. H.Seelmann-Eggebert, G.Pfennig, H.Münzel, H.Kleve-Nebenius; Karlsruher Nuklidkarte, 5.Auflage, 1981
7. M.R.Schmorak, Nucl. Dat. Sheets 20, (1977)165
8. W.A.Myers, M.Lindner, R.S.Newbury, Journ. Inorg. Chem. 37, 637-639 (1975)
9. S. Pearlstein, Nucl. Sci.Eng. 23,(1965)238
10. C.K.Paulson, E.J.Hennelly, Nucl. Sci.Eng.55, (1974)24
11. B.H.Patrick, NEANDC -124A, INDC(UK)-33/6,42
12. M.Caner, S.Wechsler, S.Yiftah, IA-1346(1977)
13. E.Fort; NEANDC 150-U. p.84
E.Fort,H.Derrien,J.P.Doat; this conference
14. P.Barbero et al.; EUR-5605e (1977)
15. M.J.Bell;ORNL-4628 (1973)
16. J.E.Suich, H.C.Honeck; DP-1064(1967)
17. B.Goel; KfK-2386/II (1977)
18. D.L.Smith, I.W.Meadows; Nucl. Sci.Eng 58 (1975) 314
19. W. McElroy, A.Fabry and B. Magurno in a private communication, reported at 3rd ASTM-EURATOM Dosimetry Symposium. Ispra 1979
20. F.Nasyrov; Sov. At. En. 25(1968)1251
21. I.Broeders, B.Krieg; KfK-2388 (1977)
22. W.Väth, I.Broeders; private communication
23. A.G.Croff, M.A.Bjerke, G.W.Morrison, L.M.Petrie; ORNL/TM-6051 (1978)
24. M.Segev, M.Caner; Ann.Nucl.Eng. 5(1978)239

NEUTRON INDUCED FISSION CROSS SECTION OF ^{238}PU IN THE ENERGY RANGE FROM 5 eV TO 10 MeV

C. Budtz-Jørgensen, H.-H. Knitter and D.L. Smith *

Commission of the European Communities
Joint Research Centre
Central Bureau for Nuclear Measurements
Geel, Belgium

The fission cross section of ^{238}Pu was measured in the neutron energy range from 5 eV to 10 MeV. Several methods of neutron production were employed using the Van de Graaff and the electron linear accelerator of the CBNM. The neutron induced fission events were detected with a specially designed ionization chamber that permitted discrimination against the high α-activity of the ^{238}Pu sample. The fission cross section was measured relative to the ^{235}U(n,f) cross section above 100 keV and at lower neutron energies relative to the ^{6}Li(n,t)^{4}He cross section shape, however normalized between 7.8 eV and 11.0 eV to the known resonance fission integral of ^{235}U.

The present data provide unique new information above 5 MeV covering the second chance fission threshold. In the resonance region the existence of intermediate structures as observed in a previous nuclear explosion measurements was confirmed.

[reactions ^{238}Pu(n,f), ^{235}U(n,f), ^{6}Li(n,t)^{4}He, cross sections $\sigma^{38}_{n,f}(E)$ 5 eV $< E_n <$ 10 MeV, fission areas, fission widths 5 eV $< E_n <$ 500 eV]

Introduction

The neutron induced fission cross section of ^{238}Pu is requested with an accuracy of 15-20% below 15 MeV [1]. The ^{238}Pu isotope is abundant to 2% in light water reactor plutonium, but a recently-proposed proliferation-resistant fuel cycle envisages concentrations of ^{238}Pu in the range of 5-10%. Concentrations at the 5% level would render the fuel useless for weapon applications because the heat released by the high α-particle activity of ^{238}Pu would raise the temperature at the surface of a minimum-critical-mass sphere to ~ 875°C which is above the melting point for plutonium [2].

In addition ^{238}Pu has applications as one of the most favourable isotopic heat sources. In processing large quantities of the material it is favourable to have a knowledge of its fission cross section so that problems of nuclear criticallity may be envisaged.

Measurements of the ^{238}Pu fission cross sections are complicated by the relatively short alpha and spontaneous fission half lives of 87.5 and $5\cdot10^{10}$ a, respectively. Hence, special detection methods have to be employed to cope with the intense sample activities. There exist two measurements [3-4] covering neutron energies from some eV to a few MeV. Both employed nuclear explosion sources which have the advantage of extremely high neutron intensities. Therefore the inherent radioactivity of ^{238}Pu was negligible compared to the neutron-induced reaction rates. However this type of measurements encountered severe difficulties in the past. This was the case for ^{241}Am where a nuclear explosion measurement [5] yielded values for the subthreshold fission cross sections nearly two orders of magnitude larger than those found in recent laboratory measurements.[6]

A number of measurements [7-10] of the ^{238}Pu(n,f) cross section have been made employing mono-energetic fast neutrons from electrostatic accelerators. Generally the reported data sets are in reasonable agreement within the requested 15-20% accuracy level. However above ~ 3 MeV, there are only a few data points available and no information has been obtained about the fission cross section in the region of the second chance fission threshold and at higher energies.

The present measurements determined the ^{238}Pu(n,f) cross section from ~ 5 eV to ~ 10 MeV. The fast region above ~ 0.15 MeV was investigated at the Van de Graaff accelerator, whereas a range from ~ 5 eV to ~ 2 MeV was covered at the electron linear accelerator (GELINA) of CBNM.

Experimental Details

The experimental techniques and procedures were very similar to those employed in a recent measurement [11] of the ^{240}Pu(n,f) cross section. Therefore only the essentials will be reported below.

Figure 1 shows schematically the experimental arrangement at the Van de Graaff. A fission fragment detector was positioned at zero degree with respect to the incident ion beam with the two fissile layers located at a distance d ~ 6 cm, from the neutron producing target. A neutron time-of-flight detector was installed with L=11.3 m from the target and was used to monitor neutron production, neutron energy and energy spread during the measurement. The Van de Graaff accelerator was operated in the pulsed mode giving an average current of 4 μA, pulse width of 1.5 ns and a repetition frequency of 2.5 MHz. Conventional time-of-flight technique was used to suppress the time uncorrelated background. The reactions ^{7}Li(p,n)^{7}Be, T(p,n)^{3}He and D(d,n)^{3}He were used to produce neutrons between 0.15 and 1.0 MeV, 1.0 and 4.0 MeV and between 4.0 and 10.0 MeV, respectively.

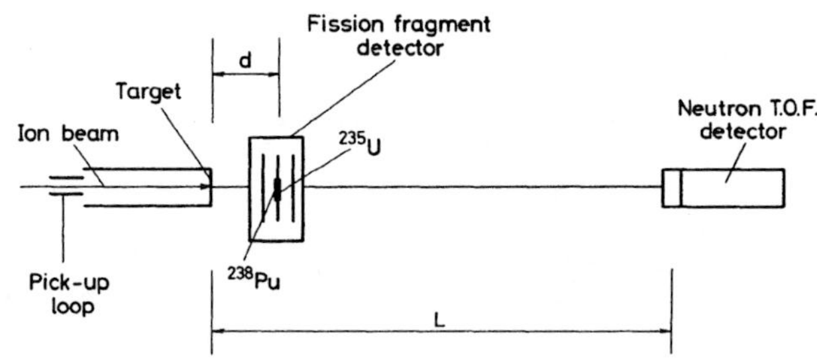

Fig. 1. Schematic representation of experimental set-up at the Van de Graaff.

The experimental set-up at GELINA is shown schematically in fig. 2. The fission chamber was exposed to the neutron beam at a distance of (8.392±0.010)m from the neutron-producing target. The neutron beam was collimated to a diameter of 3 cm. GELINA was operated with a pulse width of 4.4 ns, a repetition frequency of 800 Hz, and an average electron beam power of 4 kW at 100 MeV. A 0.1 g/cm^2 ^{10}B cut-off filter was used to remove neutrons with flight-times longer than the time between two successive bursts. The neutron energy spectrum extended from several eV to several MeV.

The fission layers of ^{235}U and ^{238}Pu were mounted back-to-back in an ionization chamber as shown in fig. 1. Both fissile layers had diameters of 28 mm, and one consisted of evaporated ^{235}UF$_4$ and the other of hard fired ^{238}Pu-oxide on 0.5 mm thick stainless steel backings. The metrological data of these layers are given in table I. The masses of the layers were determined by low-geometry alpha counting using the isotopic composition of the material as determined by mass-spectrometry.

* Visiting scientist from Argonne National Laboratory Argonne, Illinois 60439, U.S.A.

K. H. Böckhoff (ed.), Nuclear Data for Science and Technology, 206–210.

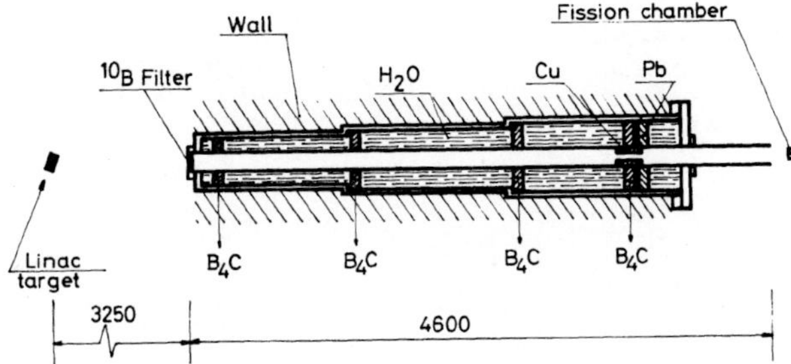

Fig. 2. Sketch of the experimental arrangement at GELINA. Dimensions are in millimeters.

Although the ^{238}Pu sample had a mass of only 50 μg it had an alpha activity of $3.10^{+7}s^{-1}$. Therefore the special detector-design as described in ref. [11] was used in order to reduce the effects of pulse pile-up due to alpha's. The detector signals could be used both for ns timing and for fission fragment energy determination. The ^{235}U fission fragments were detected with a conventional parallel plate ionization chamber. Fig. 3 displays typical pulse height spectra for both ^{238}Pu and ^{235}U obtained from the neutron induced fission at the Van de Graaff. Events uncorrelated with the neutron bursts have been subtracted. The detection efficiency for fission fragments was 96% for both detectors. In the case of ^{238}Pu the loss of fission fragments is entirely due to the electronic threshold since losses in the 7.75 $\mu g/cm^2$ thick and visually homogeneous plutonium deposit were negligible.

Table I : Metrological data of the fission layers

^{235}U weight:(591.7±3.0)μg	Pu weight:(53.24±0.53)μg
^{234}U : 0.1776 (Atom %)	^{238}Pu : 89.63 (Atom %)
^{235}U : 99.1896 (Atom %)	^{239}Pu : 9.71 (Atom %)
^{236}U : 0.0310 (Atom %)	^{240}Pu : 0.634 (Atom %)
^{238}U : 0.6018 (Atom %)	^{241}Pu : 0.0204 (Atom %)
	^{242}Pu : 0.0069 (Atom %)

Measurements, Analysis, and Results

At the Van de Graaff

The neutron energy range from 150 keV to 10 MeV was covered by 89 fission cross section ratio measurements. The measuring time per point varied between one and four hours as required in order to obtain statistical accuracy of 2 to 3 % for the ^{238}Pu fission fragment counts. The fission cross section σ_{Pu} (E) of ^{238}Pu was calculated according to the following expression :

$$\sigma_{Pu}(E)= [\frac{N_U}{N_{Pu}} \cdot C_g \cdot \frac{\epsilon_U}{\epsilon_{Pu}} \frac{F_{Pu}(E)}{F_U(E)} -P_9 \cdot R_9(E)-P_1 \cdot R_1(E)]\sigma_U(E)$$

where N_U and N_{Pu} are the number of ^{235}U and Pu nuclei in the respective samples. $F_{Pu}(E)$ and $F_U(E)$ are the number of detected fission events at neutron energy E in the plutonium and uranium detector respectively. $\sigma_U(E)$ is the ^{235}U standard fission cross section. The numerical values were obtained from the ENDF/B-V file[12]. ϵ_U and ϵ_{Pu} are the detection efficiencies for fission fragments in the respective detectors including the effect of the absorption of fragments in the fissile layers. ϵ_{Pu} also accounts for the loss of fission fragments due to the reaction kinematics. C_g corrects for the difference in geometry of the two fission layers with respect to the neutron target; it also includes the change of the neutron fluence through the fissile layers by the backing material. P_9 and P_1 are the relative amounts of ^{239}Pu and ^{241}Pu in the sample. R_9 and R_1 are the ratios of the fission cross sections of these isotopes to the fission cross section of ^{235}U. The numerical values for R_9 and R_1 were deduced from plots given in ref. [13,14,9,15] respectively. The relative amounts of ^{239}Pu and ^{241}Pu are below 10% and the correction leads to only a small uncertainty in the ^{238}Pu cross section above the fission threshold, even though an uncertainty of 10% is assumed for the cross section ratios $R_1(E)$ and $R_9(E)$. However, this correction is of considerable importance in the low energy region of the present measurements where the fission cross section for ^{238}Pu becomes smaller than for ^{239}Pu and ^{241}Pu.

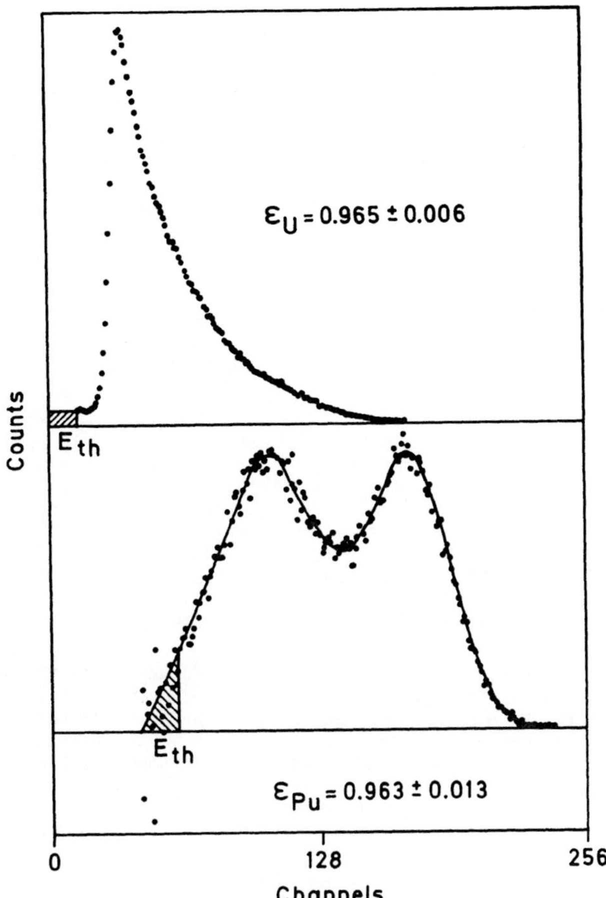

Fig. 3. Neutron induced fission-fragment energy spectra. Upper part ^{235}U, lower part ^{238}Pu.

When using the $^7Li(p,n)^7Be$ source reaction, corrections were made for the influence of a weak second neutron group which corresponds to the population of the first excited state of ^{7}Be. This correction increases from zero at the threshold for the production of this neutron group up to 3% at the point where the neutron energy for the ground state group is 1 MeV.

For the measurements made with the $D(d,n)^3He$ neutron source, additional background measurements were performed without deuterium in the gas target cell in order to determine the influence of secondary neutrons not produced by the interaction between deuterons. For deuteron energies higher than 5 MeV, secondary neutrons are produced in the gas cell from the D(d,np)D break-up reaction. The ratio of (d,np) to (d,n) neutron yield at zero degree as well as the energy distribution of the break-up neutrons have been investigated in recent works [16-17]. Using these results, the measured fission cross section ratios were corrected for fission events induced by the break-up neutrons.

The representative statistical errors for $F_{Pu}(E)$ and $F_U(E)$ and other relevant error sources, with their typical values in percent are listed in table II.

In fig. 4, comparison is made of the fission cross sections of ^{238}Pu obtained at the Van de Graaff with measurements of Ermagambetov et al.[8-9], Silbert et al.[3]

Barton et al.[7] and Fomushkin et al.[10]. These sets of data agree with each other within the experimental errors. No agreement was found with the data of Butler et al.[18].

At the Linac

The fission chamber was mounted ~ 8.4 m from the GELINA target. TOF spectra were recorded simultaneously for both ^{238}Pu and ^{235}U for a period of 100 beam hours. The neutron TOF information was obtained from a 4 ns time coder and was, together with the pulse height of the fission pulses, stored event by event on magnetic tape. The correction for the large amount of ^{239}Pu in the ^{238}Pu sample was made from a subsequent TOF measurement, where the ^{238}Pu sample was substituted by an isotopically pure > 99% ^{239}Pu sample. Normalization to the ^{238}Pu measurement was readily done using prominent ^{239}Pu resonances observed in both TOF spectra.

The fission cross section ratio of ^{238}Pu to ^{235}U for neutron energies above 100 keV was formed directly from the two respective TOF spectra. The ratio was normalized in the neutron range between 1.1 and 2.1 MeV to a value of 1.716±0.058 as found in the Van de Graaff measurements. The results obtained at the Van de Graaff and at GELINA were consistent within their uncertainties, as can be seen from fig. 4, where the ^{238}Pu fission cross sections were calculated from the ratios using the ENDF-B/V values for the ^{235}U cross sections.

The ^{235}U fission cross section below 100 keV is not useful as a standard due to its resonance structure. Therefore a measurement was made of the shape of the GELINA neutron spectrum relative to the ^{6}Li(n,t)^{4}He standard cross section. An ionization chamber similar to the one used for the fission measurement was loaded with two ^{6}LiF layers of 75 μg/cm^2 thickness each. The detector was placed at the position of the

Table II : Error sources and their typical values (in percent)

Error source	Typical Value
Random	
$F_U(E)$	1.5
$F_{Pu}(E)$	3.0
Systematic	
ϵ_U	0.7
ϵ_{Pu}	1.6
N_U	0.5
N_{Pu}	1.0
C_g	0.6
^{239}Pu contamination	1 - 4
Normalization at GELINA	3.4
Neutron flux at GELINA	1 - 5

fission chamber and TOF spectra for the forward and backward hemisphere with respect to the neutron beam were recorded simultaneously. Averaging the two spectra eliminated the effect of angular distributions and the neutron spectrum was found using the ENDF-B/V values for the angle integrated ^{6}Li(n,t)^{4}He cross section. Below 100 keV the shape could be well represented by

$$\Phi(E)dE = E^{-\alpha} \cdot \exp(-\frac{\beta}{\sqrt{E}})\, dE$$

$$\alpha = 0.874 - 2.249 \cdot 10^{-4} \cdot \sqrt{E}, \quad \beta = 2.856\ \sqrt{eV}$$

with E in eV. The exponential term describes the

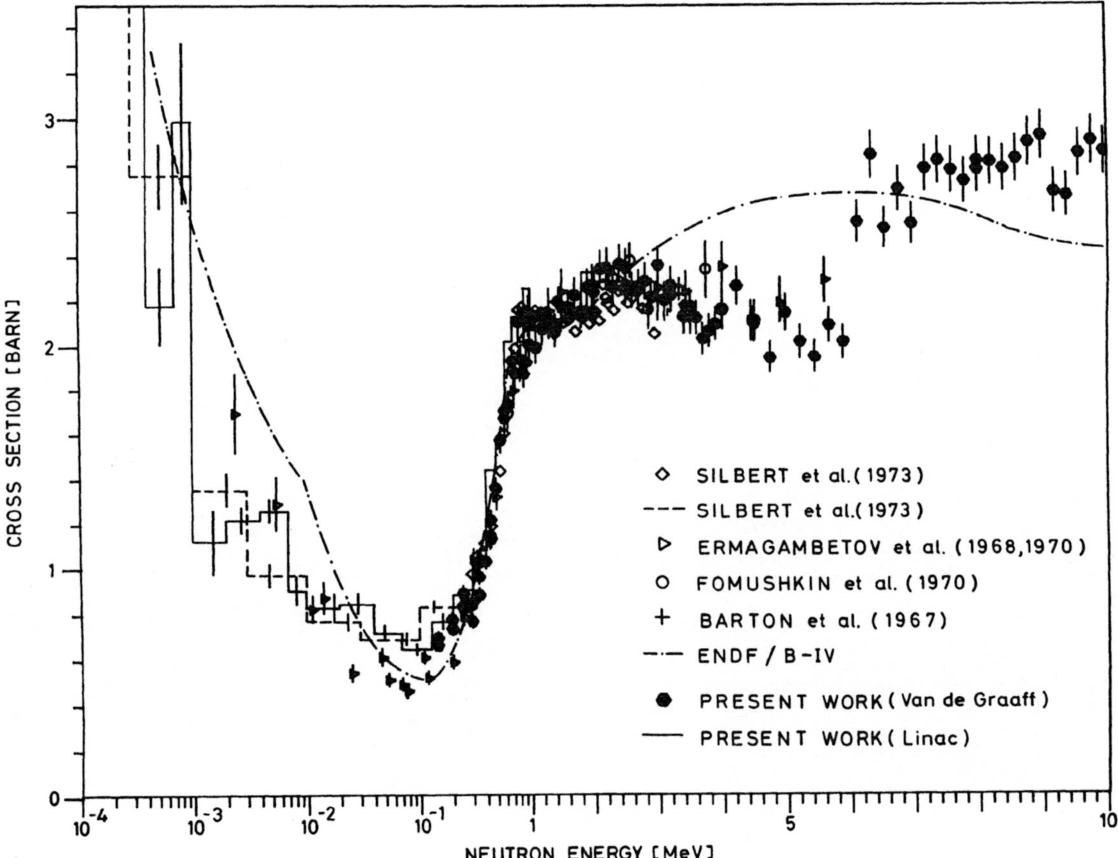

Fig. 4. Neutron-induced fission cross section of ^{238}Pu from 400 eV to 10 MeV. Note that the energy scale is logarithmic below 1 MeV and linear above.

effect of the [10]B cut-off filter. The background was determined to be < 2% from black resonance measurements using sodium, cobalt, and gold filters. The absolute neutron flux to which the fission samples were exposed was found from the observed number of fission events between 7.8 and 11.0 eV in the [235]U spectrum and by normalizing to :

$$\int_{7.8\,eV}^{11\,eV} \sigma_f^{35}(E)dE = 240.0 \text{ barn.eV}$$

in accordance with ref.[19]. From the neutron spectrum shape, and the above normalization the [238]Pu fission cross section below 100 keV was calculated in a straight forward manner. Fig. 4 shows the obtained [238]Pu fission cross sections averaged over a number of broad energy intervals down to 400 eV. Good agreement within the stated uncertainties of ~ 5% is found with the nuclear explosion data of Silbert et al.[3]. The mono-energetic data of Ermagambetov et al.[9] are in the energy range from 20 to 200 keV ~ 15% lower than the present results. Below 20 keV the average cross sections from the ENDF/B-IV file are about 20-50% higher than the present measurement.

The cross sections measured in the range from 5 eV to 1.0 keV are shown with the full energy resolution in fig. 5. The energy resolution below ~ 50 eV is mainly determined by Doppler broadening, whereas at higher energies it is determined by the neutron slowing-down time in the GELINA moderators. Due to the rather poor counting statistics which could be obtained with the 50 μg [238]Pu sample and the correction for the [239]Pu content, only resonances with a fission area larger than 5 b. eV could be safely identified. This sensitivity was an order of magnitude worse than in the measurement reported by Silbert et al.[3]. However the present energy resolution in the resonance region is about a factor two better than the one employed in the nuclear explosion measurement. Moreover, comparisons are valuable, considering the very different experimental conditions for the two measurements.

Table III lists the observed fission-resonance energies and areas below 500 eV: Generally the agreement with the data of Silbert et al.[3] is quite good, although the latter fission areas are on the average about 10% larger than the present values. The two

Table III : Resonance Parameters

E_0 (eV)	Fission Area (b . eV)	Γ_f (meV)	
9.97	13.8±1.3	6.3±0.6	a
18.54	21.1±1.7	0.9±0.19	
70.34	22.3±3.0	6.7±1.6	
84.32	46.9±4.2	2.7±0.3	
110.3	24.3±3.1	5.1±0.9	
113.8	47.3±3.8	5.5±0.7	
118.9	20.9±3.9	1.3±0.28	
122.7	113.4±5.8	8.2±1.1	
151.6	93.6±5.8	8.3±1.1	
176.9	30.0±4.7	39.3±14	
182.9	57.9±5.5	6.3±1.0	
192.5	216.7±8.9	117 ±90	
216.1	97.2±6.9	21.5±4.4	
244.8	36.9±7.4	21.6±8.1	
251.9	120.4±8.2	51 ±27	
285.7	309.9±12.7	3280 ±400	b
289.4	114.8±8.1	20.5±4.5	
300.5	355.2±13.5	90 ±65	
305.5	89.1±7.4	92 ±45	
312.4	73.4±8.2	- - -	c
320.7	100.5±8.1	9.5±1.3	
326.9	69.2±7.2	15.2±2.5	
420.9	103.4±8.8	23 ±8	
427.8	45.5±7.0	9.2±2.5	
450.1	38.3±6.5	72 ±48	
464.0	15.6±4.5	3.2±1.1	
473.6	10.9±5.9	3.3±1.4	
497.3	4.3±4.2	2.8±2.7	

a Based on the reduced neutron width Γ_n^0 reported in ref.[23]. All other resonances have been determined unless stated, using the capture areas from ref.[20]. The gamma width Γ_γ was assumed to be 34 meV for each resonance.

b From Breit-Wigner shape fit.

c Not reported in ref.[3].

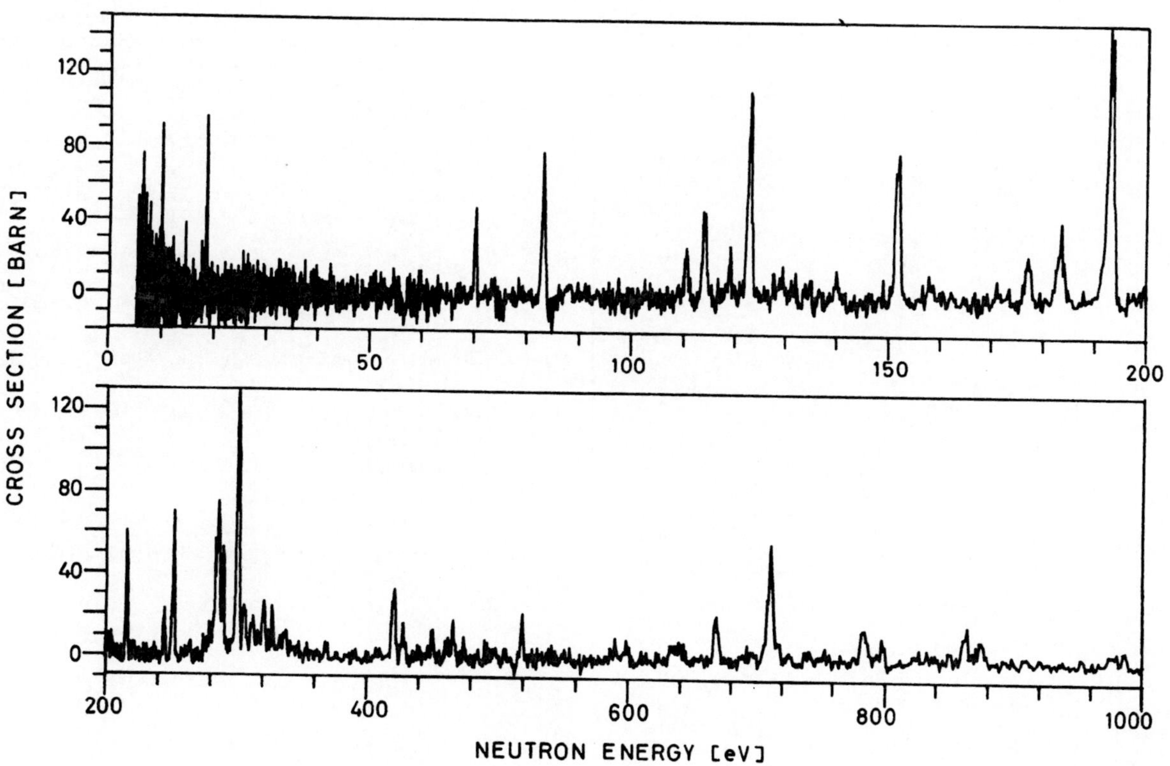

Fig. 5. Neutron-induced fission cross section of [238]Pu from 5 eV to 1 keV.

data sets differ most (> 30%) in resonances where the capture area , as measured by Silbert et al.[20], is large compared to the fission area. This could indicate that the fission detectors used in the nuclear explosion experiments were more sensitive to capture γ-rays than assumed. Good agreement is found with the fission areas of Stubbins et al.[21] who reported values for resonances up to 122 eV. For the four resonances below 100 eV observed by both Gerasimov [22] and us, we obtain areas about twice as large as those of Gerasimov.

Also listed in table III are the fission widths as calculated from the observed fission areas and the capture areas reported by Silbert et al.[20], except for the 9.97 eV resonance where the reduced neutron width Γ°_n = 0.066±0.004 meV reported by T.E. Young et al.[23] has been used. A gamma width Γ_γ = 34 meV was assumed throughout.

The present results confirm the marked enhancement of the fission widths between 200 and 400 eV. The most striking feature is the behaviour of the 285-eV resonance which has a width 2-3 times larger than the experimental resolution of 1.5 eV at this energy. Fig. 6 shows the 285-eV resonance together with the neighbouring 289-eV resonance. The latter has a width which conforms well with the experimental resolution. A single-level Breit-Wigner fit to the 285-eV resonance yielded : Γ_f = 3.28±0.40 eV and Γ°_n = 1.61±0.08 meV in surprisingly good agreement with Silbert et al.[20] who reported : Γ_f = 3.5±0.5 eV and Γ°_n = 1.54±0.57 meV. The nuclear explosion measurement had at ~ 300 eV an energy resolution of 2.9 eV, nearly the same as the width of the 285-eV resonance. Bjørnholm and Lynn [24] have recently interpreted the 285-eV resonance as essentially a class-I resonance close to a class-II state and having a large coupling matrix element. Their analysis was based on the data of Silbert et al.[20]. Due to the excellent agreement with the present data, the numerical results for the fission width $\Gamma_{\lambda II}(f)$ = 50 eV and for the coupling width $\Gamma_{\lambda II}(c)$ = 4.5 eV will remain unchanged.

Intermediate structures were also found at higher energies and could be identified up to neutron energoes of 20 keV. A level spacing $D_{II} \sim 1$ keV for the class-II levels in the second minimum of the fission barrier was deduced, confirming the observations of Silbert et al.[3].

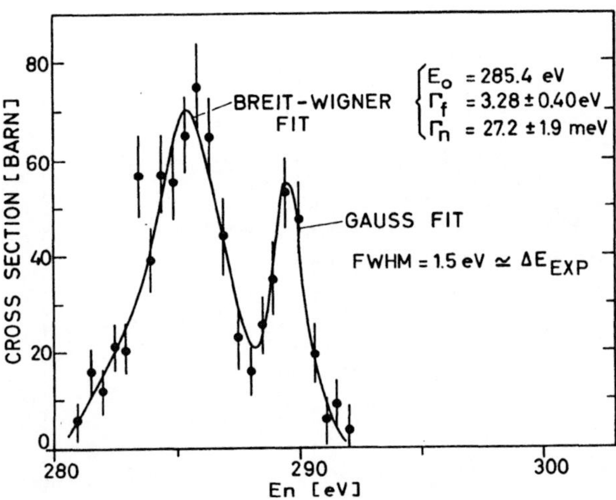

Fig. 6. Shape analysis of the 285-eV resonance.

Acknowledgement

The authors are indebted to Mrs. K.M. Glover, AERE - Harwell, U.K. and Mr. J. Van Audenhove, CBNM - Geel, Belgium for the preparation of the ^{238}Pu-sample and for the ^{235}U- and ^{6}LiF-samples respectively. The authors also like to thank Prof. Dr. A.J. Deruytter for the careful study of the manuscript.

References

1. D.W. Muir,"World Request List for Nuclear Data-WRENDA-80", FNDC(SEC)-73/URSF, IAEA Vienna (1979) page 102.
2. "AGNS Proposes Proliferation Resistant Fuel Cycle", Nuclear Industry, Sept. 1978, Page 24.
3. M.G. Silbert, A. Moat, T.E. Young, Nucl. Sci. Eng. 52, 176(1973).
4. D.M. Drake, C.D. Bowman, M.S. Coops and R.W. Hoff, "Fission Cross Sections from Pommard", P.A. Seeger, Ed., LA-4420, p.101,(1970).
5. P.A. Seeger, A. Hemmendinger and B.C. Diven, Nucl. Phys., A96, 605(1967).
6. H.-H. Knitter and C. Budtz-Jørgensen, Atom-Kernenergie, 33, 205 (1979).
7. D.M. Barton and P.G. Koontz, Phys. Rev., 162, 1070 (1967).
8. S.B. Ermagambetov and G.N. Smirenkin, Sov. At. Energy, 25, 1364 (1968).
9. S.B. Ermagambetov and G.N. Smirenkin, JETP Letters, 9, 309 (1969).
10. E.F. Fomushkin and E.K. Gutnikova, Sov. J. Nucl. Phys., 10, 529(1970).
11. C. Budtz-Jørgensen and H.-H. Knitter, Nucl. Sci. Eng. 79, 380 (1981).
12. "Evaluated Nuclear Data File ENDF/B-V", National Nuclear Data Centre, Brookhaven National Laboratory (1979).
13. W.P. Poenitz and P.T. Guenther, Proc. of NEANDC/NEACRP Specialists Meeting on Fast Neutron Fission Cross Sections of ^{233}U, ^{235}U, ^{238}U and ^{239}Pu, Argonne National Laboratory, Illinois, U.S.A. Suppl. NEANDC (US) 199/L (1976).
14. L.W. Weston, Proc. of the Specialists Meeting on Nuclear Data of Plutonium and Americium Isotopes for Reactor Applications, Brookhaven National Laboratory, Upton, N.Y., USA, NEANDC L-116, Nov. 1978, Page 1.
15. B.H. Patrick, Proc. of the Specialists Meeting on Nuclear Data of Plutonium and Americium Isotopes for Reactor Applications, Brookhaven National Laboratory, Upton, N.Y., USA, NEANDC L-116, Nov. 1978, Page 133.
16. D.L. Smith and J.W. Meadows, Argonne National Laboratory Report ANL/NDM-9 (1974).
17. J.W. Meadows and D.L. Smith, Argonne National Laboratory Report ANL/NDM-53 (1980).
18. D.K. Butler, R.K. Sjoblom, Bull. Amer. Phys. Soc. 8, 369 (1963).
19. C. Wagemans and A.J. Deruytter, Ann. N. Energy, 3, 437 (1976).
20. M.G. Silbert and J.R. Berreth, Nucl. Sci. Eng. 52, 187 (1973).
21. W.F. Stubbins, C.D. Bowman, G.F. Auchampaugh and M.S. Coops, Phys. Rev., 154, 1111 (1967).
22. V.F. Gerasimov, Sov. J. Nucl. Phys., 4, 706 (1967).
23. T.E. Young, F.B. Simpson, J.R. Berreth and M.S. Coops, Nucl. Sci. Eng., 30, 355 (1967).
24. S. Bjørnholm and J.E. Lynn, Rev. Mod. Phys., 52, 862 (1980).

NEUTRON CROSS SECTION EVALUATION FOR ^{241}Am, ^{242m}Am, ^{243}Am AND ^{244}Cm

F.H. Fröhner, B. Goel, U. Fischer and H. Jahn

Kernforschungszentrum Karlsruhe
Institut für Neutronenphysik und Reaktortechnik
Postfach 3640, D-7500 Karlsruhe
West Germany

Neutron cross sections for ^{241}Am, ^{242m}Am, ^{243}Am and ^{244}Cm were evaluated for the KEDAK file. Gaps and deficiencies in differential data necessitated heavy use of model calculations and integral data, especially in the thermal range and above the resolved resonance region. Special features of the evaluation such as treatment of bound levels, reevaluation of resonance parameters including Monte Carlo generation of missing resonance widths, determination of level statistics and of optical-model and fission barrier parameters are briefly discussed. Recommended thermal cross sections and level-statistical parameters are given. Results are compared with some recent measurements and evaluations.

Introduction

The nuclides ^{241,242m,243}Am and ^{244}Cm are produced in thermal and fast reactors in quantities sufficient to influence core neutronics and especially the radiation hazard of spent reactor fuel. Their fission and capture cross sections are needed with 10-30 % accuracy for core design and production studies of hazardous nuclides (e.g. the neutron emitters ^{244}Cm and ^{252}Cf) and with 20-50 % accuracy for more general studies of transplutonium nuclide production in the context of fuel recycling or nuclear incineration.

When the present KEDAK evaluation was begun, fission cross sections had been measured for all four nuclides considered here and at most energies needed for KEDAK (1 meV to 15 MeV), but thermal values for ^{241}Am were wildly discrepant, some of the subthreshold data for ^{242}Am and ^{243}Am looked quite dubious and gaps existed especially for ^{243}Am and ^{244}Cm. Capture cross sections were known for ^{241}Am up to 350 keV but for the other three nuclides only at thermal. Total cross sections, essential for a consistent evaluation, existed only in the resolved resonance range for ^{241}Am, ^{243}Am and ^{244}Cm. Elastic and inelastic scattering data were totally lacking. With such a deficient data base the evaluator must rely heavily on nuclear systematics and reaction theory.

Theoretical Tools

At thermal energies and in the resolved resonance region the multi-level Breit-Wigner resonance representation was used (in the ENDF sense, i. e. with single-level representation of fission and capture), unknown level spins and fission channel interference precluding a more rigorous resonance description. The unknown distant levels were described in level-statistical approximation and one bound level was determined per nuclide so as to ensure the correct thermal cross sections (see [1]). For each nuclide a complete set of point cross sections was generated coherently (i. e. for all reaction types from the same parameters) for zero temperature with the multi-level program STRUMA.

Above the resolved resonance region cross sections were computed coherently with the Hauser-Feshbach formalism including width fluctuation corrections. Neutron transmission coefficients were obtained basically from a spherical optical potential for actinides[2] that fitted total and scattering cross sections of ^{238}U below 15 MeV. The real and imaginary parts of this potential have Woods-Saxon and derivative Woods-Saxon radial dependence, respectively. The potential depths, mean radii and diffuseness parameters for the real and imaginary part are as follows,

$$V = 47.01 \text{ MeV} - 0.267\, E - 0.00118 \text{ MeV}^{-1} E^2$$

$$R_r = 1.21 \text{ fm } A^{1/3} , \quad a_r = 0.66 \text{ fm} ,$$

$$W = 9.0 \text{ MeV} - 0.53\, E$$

$$R_i = 1.30 \text{ fm } A^{1/3} , \quad a_i = 0.48 \text{ fm} .$$

Capture transmission coefficients were calculated with giant dipole resonance profiles[3]. Fission transmission coefficients for double-hump fission barriers were obtained by fits to fission cross section data. Width fluctuation corrections were calculated in the approximation of Tepel et al.[4].

Toward the resolved resonance region these more global calculations (with a modified version of the code HAUSER*4 [5]) were replaced by more specific ones with the newly developped program FITACS. This code adjusts level-statistical parameters for l=1,2,3,4 by a simultaneous fit to all angle-integrated average cross section data (total, capture, fission, scattering) available for a given nuclide up to 100-200 keV, ensuring compatibility with strength functions, level densities, average radiation and fission widths found from resolved-resonance parameters (see [6]). Inelastic scattering is fully taken into account. Width fluctuation corrections are calculated according to Moldauer[7]. The necessary s-wave strength functions and level spacings, corrected for missing levels, were obtained from resolved resonances as maximum-likelihood estimates with the code STARA[8].

Thermal Cross Sections

The biggest problem was encountered for ^{241}Am where effective capture cross sections measured with the Cd ratio method are severely affected by two resonances straddling the Cd cutoff which is at 0.5-0.6 eV depending on Cd thickness. Ignoring inadequately documented values we base our recommendation mainly on accurate total cross sections (Adamchuk et al.[9] and Kalebin et al.[10]), on the absorption data of Weston and Todd[11]) which in the resonance region yielded resonance parameters in excellent agreement with the transmission measurement of Derrien and Lucas[12], and on recent capture data at 14.75 meV by Wisshak et al.[13]. Time-of-flight data by Bowman et al.[14] and Browne et al.[15] were utilised for ^{242}Am and "barn book" values[16] for ^{243}Am. The total cross section recommended for ^{244}Cm is that of Berreth et al.[17], the fission cross section is a weighted average of irradiation results by Thompson et al.[18], Benjamin et al.[19] and Zhuravlev et al.[20], and the capture cross

K. H. Böckhoff (ed.), Nuclear Data for Science and Technology, 211–214.

section was deduced from data of Folger et al.[21], Thompson et al.[18] and Gavrilov and Goncharov[22]. The recommended thermal cross sections are listed in Table I (see [23] for more detail).

Table I - Recommended Thermal Cross Sections

Target Nucleus	σ_t (b)	σ_γ (b)	σ_f (b)	σ_n (b)
^{241}Am	625±16	610±19	3.15±0.16	12±3
^{242m}Am	8250±900	1400±860	6840±180	10±5
^{243}Am	85±5	79.3±2.0	0	5.7±4.5
^{244}Cm	23±3	14.4±1.1	1.03±0.18	7.57±3.20

Resolved Resonance Region

The resonance parameters of ^{241}Am were taken as weighted averages over all available values[12,24-27]. The absorption results of Weston and Todd[11] were renormalised to our recommended thermal cross sections which meant essentially an increase of neutron widths by about 4 %. First it was feared that this would destroy the good agreement with the transmission results of Derrien and Lucas [12]. Although for many of the weak levels this happened to some extent, the agreement for the strong levels was mostly improved. A mean value of Γ = 45 ± 2 meV was determined from the low-lieing resonances with the most accurate widths, and this average was used to recalculate the fission widths of Gayther and Thomas [27] who had used a radiation width of 40 meV. Again agreement with Ref. [12] was improved. Missing fission widths were generated by Monte Carlo sampling from the χ^2-distribution with Γ = 0.23 meV and ν = 4 reported in [12]. Fig. 1 shows the calculated fission cross section in the thermal region and across the first two resonances together with experimental data and other evaluations.

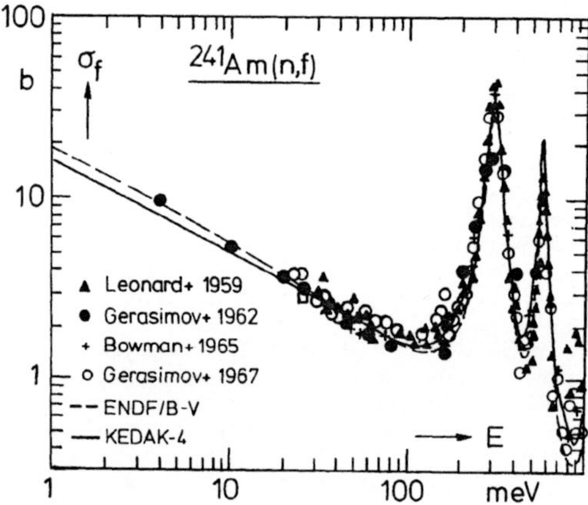

Fig. 1 - Measured and calculated fission cross sections for ^{241}Am+n below 1 eV

For ^{242} Am parameters were only available for the first six resonances[14]. Since no radiation widths had been measured we adopted Moore's level-statistical estimate, 49 meV [28].

For ^{243}Am the "barn book" values[16] (based essentially on the results of Simpson et al.[42]) were adopted except for a redetermination of the bound level.

Evaluation of ^{244}Cm resonance parameters[17,29-32] required reassessment of the radiation widths. From the information on the first two levels a weighted average of Γ = 36.2 ± 1.1 meV was deduced. All parameter sets were then reevaluated with this value in such a way that transmission dip areas remained unchanged while the change of capture and fission peak areas was minimal. Unphysical fluctuations of Γ were thus removed e. g. from the results of [32].

Unresolved Resonance Region

For ^{241}Am a FITACS fit to all available capture data[11,25,31] (renormalised where necessary) and to the level-statistical parameters from resolved resonances produced the parameters shown in Table II. Fission barrier parameters were adjusted so as to be compatible with the average fission width[12] in the resolved resonance region and to fit the best fission data[33-37] in the fission threshold region above 200 keV. Fig. 1 shows the fitted capture data.

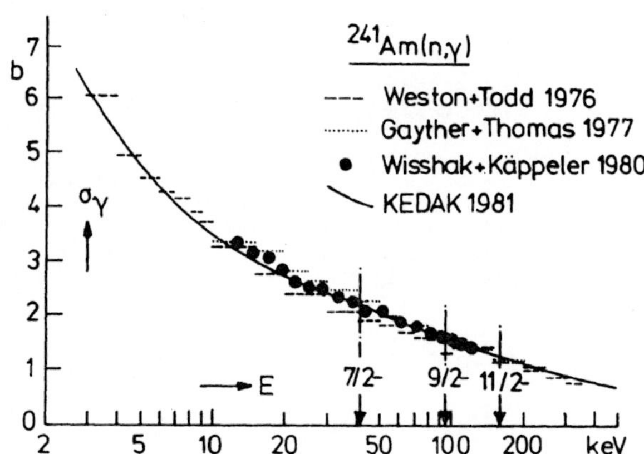

Fig. 2 - Measured and fitted capture cross sections for ^{241}Am in the unresolved resonance region. The theoretical curve is discontinuous at inelastic thresholds indicated by arrows and characteristics of residual excited levels.

With the final parameters a coherent set of point cross sections was generated up to 200 keV. Above this energy the results from HAUSER*4 calculations were adopted, with neutron transmission coefficients from the optical model[2] described earlier. Both codes gave similar results in the overlap region so that matching was unproblematic. Fig. 3 shows the fission cross section obtained together with the most recent data and an evaluation by Fort et al.[33].

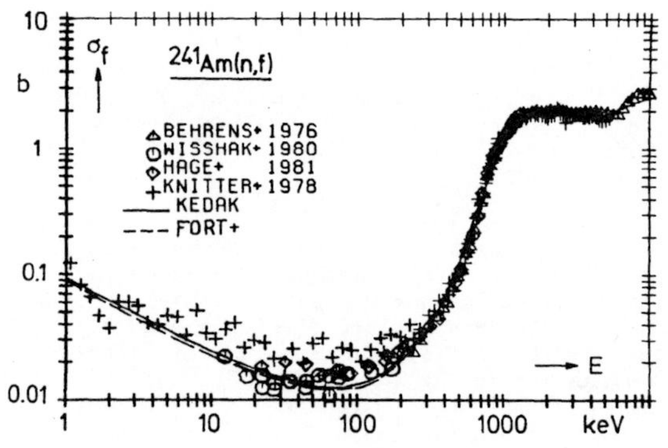

Fig. 3 - The most recent fission data for ^{241}Am+n and calculated curves

The very low subthreshold fission cross section is mainly determined by the average fission width in the resolved resonance region. It supports the results of Shpak[34] and more recent data[35-38] and agrees well with the evaluation of Fort et al.[33] but is totally incompatible with older bomb shot results[39]. When the transmission results of Philipps and Howe[40] became available they agreed well with the HAUSER*4 results, strengthening confidence in the optical potential used[2] (see Fig. 4).

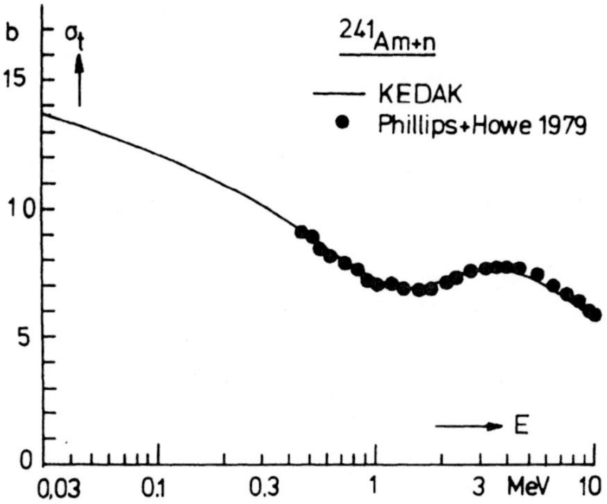

Fig. 4 - Calculated (predicted) and measured total cross section for ^{241}Am+n

A statistical (STARA) analysis of the six known ^{242m}Am resonances gives a mean level spacing which is almost certainly too high because the resonances overlap badly and probably contain unresolved doublets. The Gilbert-Cameron composite level density formula[41] yields an estimate of D = 0.3 eV. The s-wave strength function is obtainable from the average fission cross sections measured by Bowman et al.[14] and Browne et al.[15]. Below 100 keV FITACS point data were generated with the parameters in Table II, above this energy the HAUSER*4 calculations were adopted.

The s-wave strength function of ^{243}Am listed in Table II is a compromise between the STARA result and the value from our actinide optical potential[2]. The mean radiation width is that of the resolved resonances[40] and the fission barrier parameters were chosen so as to reproduce Behrens' fission data[41]. Point cross section were calculated up to 10 keV with FITACS, above this energy with HAUSER*4 (both calculations gave almost identical results below 30 keV). Fig. 5 shows the capture cross sections thus obtained together with the ENDF/B-V evaluation and new data of Wisshak and Käppeler[44].

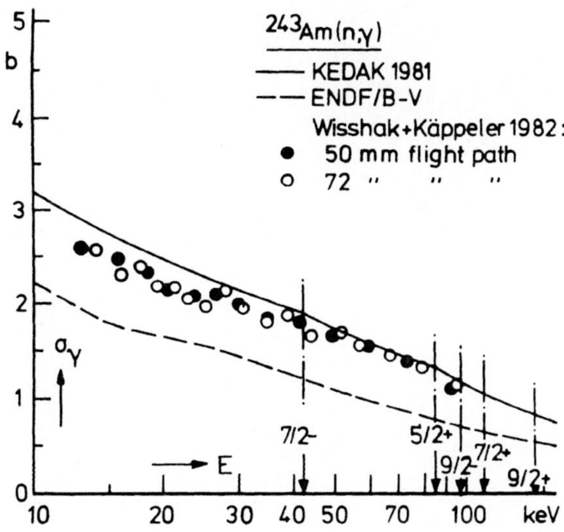

Fig. 5 - Calculated (predicted) and measured capture cross sections for ^{243}Am

Table II - Level-statistical Parameters (s-, p- and d-wave level spacings are related by Bethe's spin dependence[41])

Target Nucleus	ℓ	S_ℓ (10^{-4})	$\bar{\Gamma}_\gamma$ (meV)	D (eV)
^{241}Am	0	1.05	45	0.58
	1	2.14	45	
	2	1.21	45	
^{242m}Am	0	1.40	49	0.30
	1	2.15	49	
	2	1.15	49	
^{243}Am	0	1.05	39	0.71
	1	2.25	39	
	2	1.10	39	
^{244}Cm	0	1.14	36.2	11.9
	1	2.30	36.2	
	2	1.17	36.2	

These KEDAK evaluations were extensively checked against resonance integrals and by comparison of burn-up calculations with irradiation results, see [43].

Conclusions

The KEDAK evaluation for four actinides is based to a large extent on consequent utilisation of nuclear theory. All cross sections were coherently calculated from suitably parametrised reaction formalisms. The various tests and experience with newly published data seem to indicate that with our optical potential for actinides total cross sections can be predicted with about 2-5 % uncertainty and capture cross sections below a few hundred keV with perhaps 5-20 %, at least in cases where good resonance data define the radiation width and the level density reasonably well. In cases where the data base is poor, as for ^{242}Am, the latter margin may be somewhat larger, but still the giant dipole resonance model[3] seems to yield very reasonable radiation strengths.

Acknowledgement

We thank all colleagues at KfK who helped to create the evaluated files, especially Mrs. B. Krieg and E. Stein. Discussions with E. Fort, H. Derrien, E. Menapace, M. Stefanon, D. Gayther, B. Patrick and L. Weston are gratefully acknowledged.

References

1. F.H. Fröhner, Proc. Conf. on Nucl. Data Eval. Meth. and Proced. 1980, BNL-NCS-51363 (1981)
2. U. Fischer, KfK 2907, Karlsruhe (1980)
3. J.A. Holmes, S.E. Woosley, W.A. Fowler and B.A. Zimmermann, At. and Nucl. Data Tables 18(1976)306
4. J.W. Tepel, H.M. Hoffman and H.A. Weidenmüller, Phys. Letters 49B(1974)1
5. F.M. Mann, HEDL-TME 76-80, Hanford (1976)
6. F.H. Fröhner, B. Goel and U. Fischer, Proc. Meeting on Fast-Neutron Capture Cross Sections, ANL 1982 (in print)
7. P. Moldauer, Nucl. Phys. A344(1980)185
8. F.H. Fröhner, Proc. Meeting on Neutron Cross Sections of Fission Product Nuclei, Bologna 1979, RIT/FIS-LDN(80)1, p. 145
9. Y.B. Adamchuk et al., Conf. on Peaceful Uses of Atomic Energy, Geneva (1955), vol. 4, p. 216
10. S.M. Kalebin, V.S. Artamonov, R.N. Ivanov, G.V. Pukolaine, T.S. Belanova, A.G. Kolesov, V.A. Safonov, Sov. At. Energy 40(1976)373
11. L.W. Weston and J.H. Todd, Nucl. Sci. Eng. 61(1976)356

12. H. Derrien and B. Lucas, Nucl. Cross Sections and Technol., Washington D.C., NBS SP 425(1975)637

13. K. Wisshak, J. Wickenhauser, F. Käppeler, G. Reffo and F. Fabbri, Nucl. Sci. Eng. 81(1982)396

14. C.D. Bowman, G.F. Auchampaugh, S.C. Fultz and R.W. Hobbs, Phys. Rev. 166(1968)1219

15. J.C. Browne, R.E. Howe, R.J. Dougan, R.J. Dupsyk and J.H. Landrum, Conf. on Neutron Physics and Nucl. Data, Harwell (1978), p. 887

16. S. Mughabghab and D.I. Garber, BNL 325, 3rd ed., Brookhaven (1973)

17. J.R. Berreth, F.B. Simpson and B.C. Rusche, Nucl. Sci. Eng. 49(1972)145

18. M.C. Thompson, M.L. Hyder and R.J. Reuland, J. Inorg. Nucl. Chem. 33(1971)1553

19. R.W. Benjamin, F.J. McCrosson and W.E. Gettys, DP-1447, Savannah River (1977)

20. K.D. Zhuravlev, N.I. Kroshkin and A.P. Chetverikov, Sov. At. Energy 39(1976)907; K.D Zhuravlev and N.I. Kroshkin, INDC(CCP)-68G, Vienna (1976)

21. R.L. Folger, J.A. Smith, L.C. Brown, R. F. Overman and H.P. Holcomb, Nucl. Cross Sections and Technol., NBS SP 299, Washington D.C. (1968) II p. 1279; J.A. Smith, C.J. Banick, R.L. Folger, H.P. Holcomb and J.B. Richter, ibid. p. 1285

22. V.D. Gavrilov and V.A. Goncharov, Sov. At. Energy 44(1978)274

23. F. Fröhner, B. Goel, H. Jahn, U. Fischer, KfK 3415, Karlsruhe (in print)

24. C.D. Bowman. M.S. Coops, G.F. Auchampaugh and S.C. Fultz, Phys. Rev. 137(1965)B326

25. V.F. Gerasimov, Sov. J. Nucl. Phys. 4(1975)706

26. T.S. Belanova, A.G. Kolesov, V.M. Nikolaev, V.A. Safonov, V.Ya. Gabeskiriya, V.A. Poruchikov, S.M. Kalebin, R.J. Ivanov and O.M. Gudkov, Sov. At. Energy 38(1975)33

27. D.B. Gayther and B.W. Thomas, Proc. Kiev Conf. on Neutron Physics, vol. 3, p. 3, Moscow (1977)

28. M.S. Moore, Conf. on Neutron Phys. and Nucl. Data, Harwell (1978), p. 313

29. R.E.Cote, R.F. Barnes and H. Diamond, Phys. Rev. 134B(1964)1281

30. M.S. Moore and G.A. Keyworth, Phys. Rev. C3(1971)1656

31. S.M. Kalebin, Transactin. Isot. Nucl. Data, IAEA 186, Vienna(1976), vol. II, p. 121

32. O.D. Simpson, F.B. Simpson and T.E. Young, ANCR-1088, Idaho Falls (1972)

33. E. Fort, M. Darrouzet, H. Derrien, P. Hammer and L. Martin-Deidier, Nucl. Cross Sections for Technology 1979, NBS SP 594, Knoxville(1980)636

34. D.L. Shpak, Y.B. Ortopenko and G.N. Smirenkin, Sov. Phys. JETP 10(1969)175

35. K. Wisshak and F. Käppeler, Nucl. Sci. Eng. 6(1980)148

36. H.H. Knitter and C. Budtz-Jorgensen, Atomkernenergie-Kerntechnik 33(1979)205

37. W. Hage, K. Wisshak and F. Käppeler, Nucl. Sci. Eng. 78(1981)248

38. J.W. Behrens and J.C. Browne, UCID-17324, Lawrence Livermore Laboratory (1976)

39. P.A. Seeger, A. Hemmendinger and B.C. Diven, Nucl. Phys. A96(1967)605

40. T.W. Phillips and R.E. Howe, Nucl. Sci. Eng. 69(1979)375

41. A. Gilbert and A.G.W. Cameron, Can. J. Phy. 43(1965)1446

42. O.D. Simpson, F.B. Simpson, J.A. Harvey, G.G. Slaughter, R.W. Benjamin and C.E. Ahlfeld, Nucl. Sci. Eng. 55(1974)

43. J.W. Behrens, UCID-17504, Livermore (1977)

44. K. Wisshak and F. Käppeler, this conference

45. B. Goel and U. Fischer, this conference

THE NEUTRON CAPTURE CROSS SECTION OF ^{243}Am IN THE ENERGY RANGE FROM 5 TO 250 keV

K. Wisshak and F. Käppeler
Kernforschungszentrum Karlsruhe GmbH
Institut für Angewandte Kernphysik
P.O.B. 3640, D-7500 Karlsruhe
Federal Republic of Germany

The neutron capture cross section of ^{243}Am was measured in the energy range from 10 to 250 keV using ^{197}Au as a standard. Kinematically collimated neutrons were produced via the ^{7}Li(p,n) and the T(p,n) reaction with the Karlsruhe 3 MV pulsed Van de Graaff accelerator. Capture events were detected by two Moxon-Rae detectors with graphite and bismuth-graphite converters. Fission events were observed by a NE 213 liquid scintillator with pulse shape discriminator. A ^{243}Am sample of ∿1 g was positioned at flight paths as short as 5-7 cm from the neutron target in order to obtain a signal-to-background ratio of the order of one. The preliminary results for the capture cross section of ^{243}Am show good agreement with the KEDAK 4 evaluation, while discrepancies of 20-30 % are found to ENDF/B-V.

[^{243}Am(n,γ), capture cross section, actinides, E_n = 10-250 keV]

Introduction

Associated with the further development of fast reactors, the problems of fuel handling, reprocessing, and waste disposal have received an increasing concern in the past years. This caused a demand for improved neutron cross sections of minor actinide isotopes in the keV range.

Moxon-Rae detectors in connection with kinematically collimated neutrons from (p,n) reactions and the use of very short flight paths proved to be a suited method to measure capture cross sections of these highly radioactive actinides in the keV range. Measurements have been performed for ^{240,242}Pu and ^{241}Am using ^{197}Au and ^{238}U as standards[1,2,3]. We have completed this series by investigating a sample of ^{243}Am. The high energy of the gamma radiation emitted by the sample material causes severe problems in capture cross section measurements. Therefore no data have been published for this isotope in the keV range so far.

One of the largest systematic uncertainty in measurements with Moxon-Rae detectors is due to the fact that the efficiency deviates slightly from the ideal shape (which increases linearly with gamma-ray energy). To reduce this uncertainty we used in the present experiment two Moxon-Rae detectors with different converter materials so that we could check the efficiency correction.

Experimental Method

Similar to earlier work by Macklin et al.[4], the principle of the experimental method is to use (p,n) reactions on light nuclei for neutron production at proton energies just above the reaction threshold. In this case the neutrons are kinematically collimated by the center of mass motion of the compound nucleus. All neutrons are emitted within a cone in a forward direction, the opening angle of which is determined by the proton energy. Therefore, further collimation of neutrons is not necessary and flight paths as short as 5 to 10 cm can be used in the experiments. The capture and fission events can be observed by detectors placed at backward angles completely outside the neutron cone.

Details of the experimental technique have been published in Refs. 1, 2 and 3, and therefore only a brief description is given here.

The experimental set-up is shown schematically in Fig. 1. The pulsed proton beam of the Karlsruhe 3 MV Van de Graaff accelerator hits a water cooled ^{7}Li or ^{3}H target producing kinematically collimated neutrons in the energy range 5-100 keV and 50-250 keV, respectively. Two Moxon-Rae detectors with graphite and bismuth graphite converter served for the detection of capture events. As the Moxon-Rae detector cannot distinguish between gamma rays from capture and fission, we had to use a second detector (sensitive to fission events only) for correction. This was a NE-213 liquid scintillator, which was operated with a pulse shape discriminator, to separate fission neutrons from gamma-ray background. From the data recorded with this detector, it will be possible to determine the fission cross section of ^{243}Am, too.

All detectors are located at backward angles of 120 deg with respect to the beam axis, completely outside of the neutron cone. They are shielded with lead against the prompt gamma radiation from the neutron target. The intense gamma radiation from the ^{243}Am sample itself was attenuated by lead absorbers of variable thickness in front of the Moxon-Rae detectors.

Four samples are mounted in a low mass sample changer and cycled automatically into the measuring position:

1) ^{243}Am: A pellet of 0.970 g $^{243}AmO_2$ was prepared with a diameter of 13.7 mm. After a sintering procedure the pellet was welded into a 0.15 mm thick stainless steel canning.
2) ^{197}Au: A 1 mm thick sample with the same 13.7 mm diameter was used as a standard.
3) graphite: The thickness of this sample was adjusted to give about the same scattering yield as the gold sample.
4) ^{235}U: This sample was used for normalization of the fission correction.

K. H. Böckhoff (ed.), Nuclear Data for Science and Technology, 215–217.

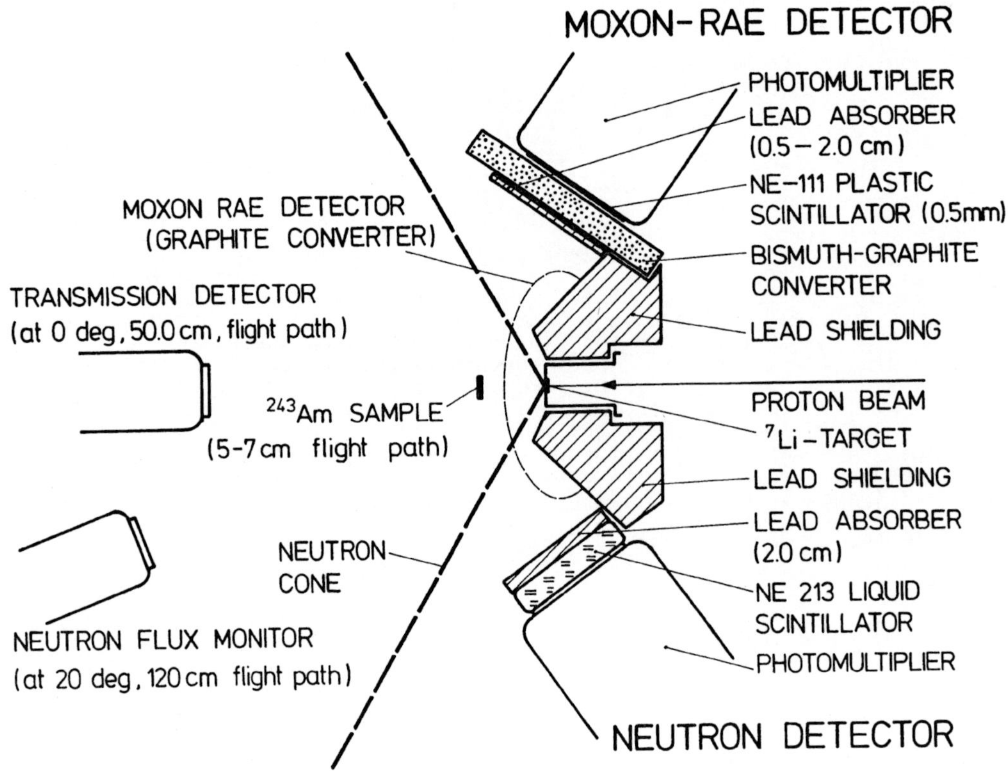

Fig. 1 Schematic view of the experimental set-up for the capture cross section measurement on ^{243}Am.

In order to obtain a similar time dependent background, all samples were canned in the same way.

Systematic uncertainties were studied in detail in several runs with modified experimental conditions. The essential parameters of these measurements are compiled in Table I. As an additional check, the cross section ratio $\sigma_\gamma(^{238}U)/\sigma_\gamma(^{197}Au)$ has been measured for the different lead absorbers and also without lead absorber.

To demonstrate the experimental signal to background ratio, Fig. 2 shows the time-of-flight spectrum measured with the bismuth-graphite converter in run 3, where a lead shielding of only 0.7 mm was used. The neutron flux at the sample position is large enough that with 1 g of sample material sufficient statistical accuracy is obtained in 1.5 days (~9 h per sample). The spectrum of

TABLE I Parameters for the Individual Measurements of the Neutron Capture Cross Section Ratio $\sigma_\gamma(^{243}Am)/\sigma_\gamma(^{197}Au)$

Run	Lead Shielding of Moxon-Rae Detectors [cm]	Flight Path [mm]	Neutron Energy Range [keV]
1	2.0	52	10-90
2	1.0	52	10-90
3	0.7	52	10-90
4	0.5	52	10-90
5	2.0	71	10-90
6	1.0	71	10-90
7	2.0	65	50-250
8	1.0	65	50-250

Fig. 2 can be evaluated with confidence in the energy range from 20-90 keV. At lower energies the shape of the cross section is

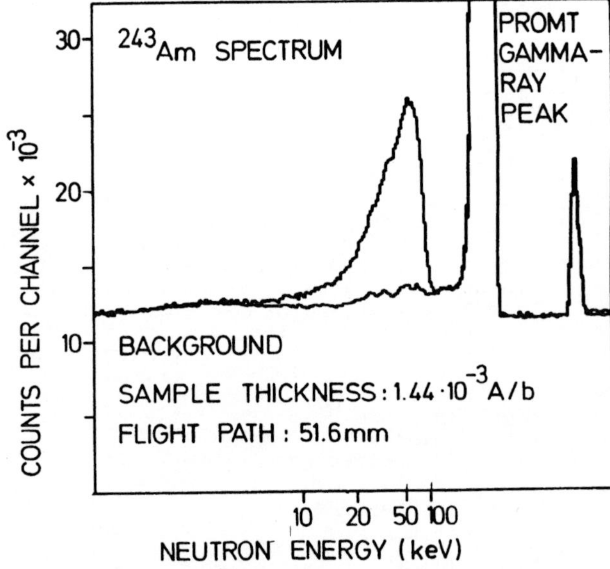

Fig. 2 Experimental TOF spectrum of the ^{243}Am sample measured by a Moxon-Rae detector with bismuth-graphite converter and with 0.7 cm thick lead shielding. A constant value as been added to the background spectrum obtained with the graphite sample to account for the time-independent background from the ^{243}Am decay (the peak right of the gamma peak is caused by a diaphragm in front of the Li-target).

determined from the measurements with thicker lead absorbers which therefore exhibit a much better signal-to-background ratio.

Data Analysis and Results

The data analysis has been described in detail in Refs.1,2,3. After subtraction of the time dependent and time independent background the TOF spectra taken with the ^{243}Am sample were corrected for fission events. As the fission cross section of ^{243}Am proved to be very small, only ~1 % of the observed net count rate was due to fission. A correction for capture in isotopic impurities was neglected as the sample was enriched to 99.82 % and 0.17 % from the remaining abundance is ^{241}Am, which has about the same capture cross section. The correction for multiple scattering and self shielding was determined using the SESH code[5]. Gamma-ray self absorption in the gold sample has been measured using samples with different thickness. The respective correction for the ^{243}Am sample was then estimated from the theoretical total energy absorption coefficients given in Ref. 6.

Finally the data were corrected for the deviation of the detector efficiency from a linear increase with gamma-ray energy. This requires knowledge of two quantities: (i) the shape of the capture gamma-ray spectrum of sample and reference sample and (ii)the shape of the efficiency curve for the particular converter material. Then the respective correction factors can be calculated according to the formula given in Ref. 7.

The capture gamma-ray spectra were calculated in the framework of the statistical model and the optical model. A detailed description of the method is given in Refs. 8,9. As neutron capture in ^{243}Am is very similar to capture in ^{241}Am (Ref. 10) we used the capture gamma-ray spectra for ^{241}Am as given in Ref. 9 to evaluate the correction. The shape of the detector efficiency of the different converter materials was evaluated in Ref. 11.

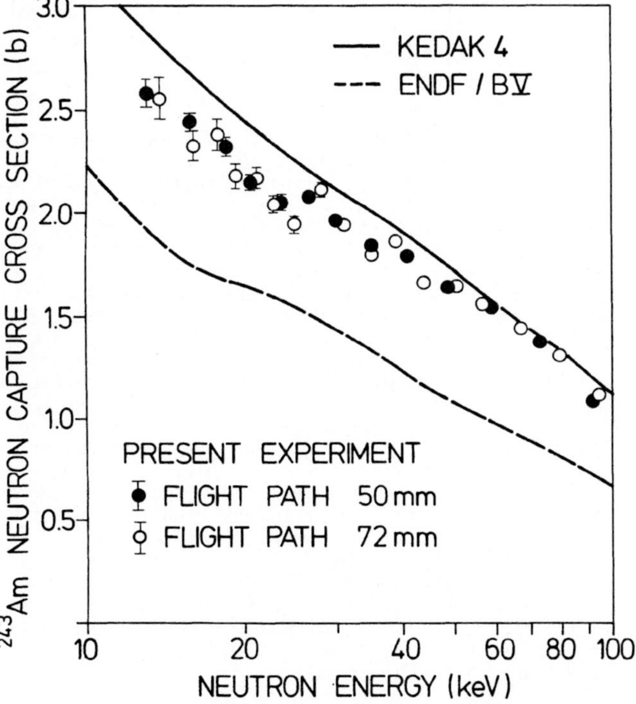

Fig. 3 Preliminary results for the neutron capture cross section of ^{243}Am. The error bars represent the statistical uncertainties. A comparison is made to evaluated files.

The experimental cross section ratio has to be normalized by a factor of 0.973 for measurements with the graphite converter, whereas the correction proved to be negligible for measurements with the bismuth-graphite converter. The influence of the lead absorber in front of the detectors was studied by measuring the ratio $\sigma_\gamma(^{238}U)/\sigma_\gamma(Au)$. We found that this ratio agrees within 0.3 % for all thicknesses of lead absorbers used. As the capture gamma-ray spectrum of ^{238}U and ^{241}Am is very similar[9,12] it was therefore evident that also the measured cross section ratio $\sigma_\gamma(^{243}Am)/\sigma_\gamma(Au)$ is independent of the thickness of the lead absorbers.

The results of the six independent runs performed in the energy range 10-100 keV(see Table I) differ by less than ± 2 % and within each run the cross section ratio obtained by the two detectors scattered always less than ±1.2%. These differences are well within the systematic uncertainty due to background subtraction.

Preliminary results for the capture cross section of ^{243}Am are given in Fig. 3. For easier survey the four runs performed with different lead absorbers at 50 mm flight path and the two runs measured a 72 mm flight path have been combined in one data set. In the energy range 5-10 keV and 100-250 keV data analysis is still in progress.The cross section of Fig. 3 was obtained by converting the experimental ratios to absolute values using the ENDF/B-V gold cross section. The plotted error bars represent only the statistical uncertainties.The systematic uncertainties are not yet analyzed in detail but will be of the order of ~5 %. A comparison to recent evaluations shows good agreement with KEDAK 4 while a discrepancy of 20-30 % is found to ENDF/B-V.

References

1 K. Wisshak and F. Käppeler, Nucl. Sci. Eng. 66, 363 (1978).

2 K. Wisshak and F. Käppeler, Nucl. Sci. Eng. 69, 39 (1979).

3 K. Wisshak and F. Käppeler, Nucl. Sci. Eng. 76, 148 (1980).

4 R.L. Macklin, J.H. Gibbons, and T. Inada, Nucl. Phys., 43, 353 (1963).

5 F.H. Fröhner, GA-8380 Gulf General Atomic (1968).

6 E. Storm and H.J. Israel, Nucl. Data Tables A 7, 565 (1970).

7 K. Wisshak, F. Käppeler, G. Reffo, and F. Fabbri, Proc. of the NEANDC/NEACRP Specialists Meeting on Fast Neutron Capture Cross Sections, Argonne, April 20-23, 1982, to be published.

8 G. Reffo, F. Fabbri, K. Wisshak and F. Käppeler, Nucl. Sci. Eng.80, 630 (1982).

9 K. Wisshak, J. Wickenhauser, F. Käppeler, G. Reffo and F. Fabbri, Nucl. Sci. Eng. 81, 396 (1982).

10 G. Reffo, private communications, 1982.

11 K. Wisshak and F. Käppeler, Nucl. Sci. Eng. 77, 58 (1981).

12 G. Reffo, F. Fabbri, K. Wisshak, and F. Käppeler, submitted for publication to Nucl. Sci. Eng.

NEUTRON-INDUCED FISSION CROSS SECTION MEASUREMENTS AND CALCULATIONS OF SELECTED TRANSPLUTONIC ISOTOPES†

R. M. White

Lawrence Livermore National Laboratory
Livermore, CA 94550

J. C. Browne

Los Alamos National Laboratory
Los Alamos, NM 87545

The neutron-induced fission cross sections of ^{242m}Am and ^{245}Cm have been measured over an energy range of 10^{-4} eV to ~20 MeV in a series of experiments at three facilities during the past several years. The combined results of these measurements, in which only sub-milligram quantities of enriched isotopes were used, yield cross sections with uncertainties of approximately 5% below 10 MeV relative to the ^{235}U standard cross section used to normalize the data. We summarize the resonance analysis of the ^{242m}Am(n,f) cross section in the eV region. Hauser-Feshbach statistical calculations of the detailed fission cross sections of ^{235}U and ^{245}Cm have been carried out over the energy region from 0.1 to 5 MeV and these results are compared with our experimental data.

[^{242m}Am(n,f), ^{245}Cm(n,f), fission cross section measurements, $E_n = 10^{-3}$ eV to 20 MeV; ^{235}U(n,f), ^{245}Cm(n,f), Hauser-Feshbach statistical calculations, E_n = 100 keV to 5 MeV]

Introduction

A series of measurements of the neutron-induced fission cross sections of ^{242m}Am and ^{245}Cm have been carried out at three facilities during the past several years (see Table I). These measurements span an energy range of 10^{-4} to $2x10^{+7}$ eV. In this report we summarize the results of the low energy ^{242m}Am resonance analysis and then outline the results and comparisons of our high energy (MeV) measurements with previously reported data. Hauser-Feshbach statistical calculations of the fission cross section over the energy region of the fission neutron spectrum are discussed and results of these calculations for the ^{235}U and ^{245}Cm fission cross sections are presented.

Table I. Summary of experimental measurements on the neutron-induced fission cross sections of ^{242m}Am and ^{245}Cm at various facilities since 1977.

YEAR	FACILITY	ISOTOPE	ENERGY RANGE (eV)
1977	LLNL-Linac†	^{242m}Am	10^{-2} – $2x10^{+7}$
1979	LLNL-Linac	^{242m}Am	10^{-4} – $2x10^{+7}$
		^{245}Cm††	10^{-4} – $2x10^{+7}$
1980	LANL-WNR‡	^{242m}Am	$\sim10^{+6}$ – $\sim10^{+7}$
		^{245}Cm	$\sim10^{+6}$ – $\sim10^{+7}$
1980	LLNL-ICT‡‡	^{242m}Am	14.1 MeV
		^{245}Cm	14.1 MeV

†Livermore 100-MeV Electron Linac, Ref. 1.
††Livermore 100-MeV Electron Linac, Ref. 2.
‡Los Alamos Weapons Neutron Research Facility
‡‡Livermore Insulated Core Transformer Accelerator

Experimental Techniques

The experimental techniques of the 1977 and 1979 Linac measurements, as well as preliminary statistical analysis of the low energy resonance parameters of both ^{242m}Am and ^{245}Cm, have been reported previously.[1,2] The ^{242m}Am and ^{245}Cm fission samples used in these measurements were prepared by the LLNL Nuclear Chemistry Division and were electroplated on 0.05mm thick hemispherically shaped ionization chambers. As the samples were extremely radioactive with α-decay, the hemispherical geometry helped to differentiate between the α-decay and fission signals by limiting the path-length of alphas and fission fragments through the gas in the fission chambers. Table II gives the pertinent data on the fission samples used.

Table II. Masses of ^{242m}Am and ^{245}Cm used in 1979 and 1980 measurements.†

ISOTOPE	SAMPLE	MASS (μg)	STATISTICAL	SYSTEMATIC
^{242m}Am	#1	207.3	± 1.0%‡	——
^{245}Cm	#1	195.3	± 0.4%	~ 2%
	#2	201.0	± 0.5%	~ 2%

Note: While these samples were isotopically enriched to >99%, there existed isotopic contamination which gave rise to spontaneous fission backgrounds which were random in time and could therefore be subtracted as constant backgrounds. The spontaneous fission rates were approximately as follows:

^{242m}Am #1 – 3.2 spontaneous fissions/sec
^{245}Cm #1 – 0.75 " " "
^{245}Cm #2 – 9.0 " " "

†The 1977 linac measurement (see Table I) used a mass of ~800μg of ^{242m}Am.
‡This number represents the total (statistical plus systematic) error on the mass assay of ^{242m}Am.

Until 1977, the only high energy data on ^{242m}Am were those of Seeger *et al.*[3] and Bowman *et al.*[4] In 1977 Browne *et al.*[1] measured the ^{242m}Am(n,f) cross section to 20 MeV. We repeated this measurement in 1979 (see Table I) with a new sample and in the same experiment measured the ^{245}Cm(n,f) cross section using two different samples (see Table II). Additional measurements of the ^{242m}Am(n,f) and ^{245}Cm(n,f) cross sections were subsequently carried out at the Los Alamos Weapons Neutron Research (WNR) facility (see Table I) to verify the Livermore Linac results in the ~1 to ~10 MeV region. Both facilities provided a 'white' source of neutrons and standard time-of-flight techniques were used to determine incident neutron energies. For these high energy measurements and for thermal energies, the cross sections were measured relative to the cross section of ^{235}U. In the resonance region, the neutron flux shape was measured with thin lithium glass detectors and the relative cross sections obtained were then normalized to the thermal data. An overlap check with the high energy data, measured independently with respect to ^{235}U, was then made in each case. Table III summarizes the errors for the ^{242m}Am(n,f) and ^{245}Cm(n,f) final data sets relative to the ENDF/B-V ^{235}U(n,f) cross section used to normalize these data.

Resonance Analysis of ^{242m}Am(n,f)

Results of a Breit-Wigner sum-of-single-level

†Work performed under the auspices of the U.S. Department of Energy by the Lawrence Livermore National Laboratory under contract number W-7405-ENG-48.

K. H. Böckhoff (ed.), Nuclear Data for Science and Technology, 218–221.

Table III. Summary of statistical and systematic† errors for $^{242m}Am(n,f)$ and $^{245}Cm(n,f)$ cross section measurements at the Livermore Linac.

STATISTICAL ERROR			SYSTEMATIC ERROR			
Energy	^{242m}Am	^{245}Cm		^{242m}Am	^{245}Cm	^{235}U
Thermal	0.5%	0.7%	Mass	± 1%	±2 %	± 2%
1 keV	3.5%	3.5%	Eff.‡	± 1%	$^{+1.5\%}_{-2.5\%}$	± 1%
1 MeV	0.5%	1.1%				
14 MeV	2.0%	4.5%				

†The known systematic errors evolve from the measured masses of ^{242m}Am, ^{245}Cm, and the ^{235}U 'standard' as well as from the efficiency of the corresponding fission chamber of each sample.
‡Estimated uncertainty in efficiency from fission pulse-height distribution and calculation of percent of fission fragments lost in sample deposit.

analysis of 48 fission resonances in the $^{242m}Am(n,f)$ cross section up to 20 eV are summarized in Ref. 1. These data show a complicated structure with many levels and a lack of any obvious interferences. This implies a large number of Bohr transition states (fission channels) open for this nucleus since the level-level interference is essentially nonexistent in the case of many channels. This is further supported by the fact that the distribution of fission widths for these levels follows a general chi-square distribution having at least 10 degrees of freedom. It is therefore more meaningful to fit these data with a sum of single levels than with a multilevel (R-matrix) approach which allows for interference between levels and which would require many channels per level to fit these data.

The complicated resonance structure in the $^{242m}Am(n,f)$ cross section in some regions between 1 and 20 eV requires a 'synthesis' of enough levels to fit the magnitude of the cross section while at the same time preserving its shape. Therefore, the average level spacing, <D>, which we determined to be 0.4 eV, is a maximum value since we have almost certainly missed narrow resonances in this region. When compared with even-even fissioning systems, the ^{243}Am nucleus should be expected to have a reduced average level spacing because the unpaired proton allows population of additional intrinsic excitations without the expenditure of energy necessary to first break a nucleon-nucleon pair. Because of the higher level density, the probability of the first resonance occurring lower in neutron energy is more likely. The first observed resonance in the $^{242m}Am(n,f)$ cross section[5] occurs at E_n=0.178 eV. The low energy of this first resonance together with the fact that it also has a very large reduced neutron width gives ^{242m}Am the largest thermal fission cross section known.

Data Comparison - High Energy Region

The high energy portions (10 keV to 10 MeV) of the 1977 and 1979 Linac measurements of the $^{242m}Am(n,f)$ cross section are plotted for comparison in Fig. 1. These were independent experiments using different samples of ^{242m}Am and the agreement in these two data sets is excellent. Also included in Fig. 1 are the recently published results of Fomushkin *et al.*[6]

The high energy portion (10 keV to 10 MeV) of the 1979 Linac measurement on the $^{245}Cm(n,f)$ cross section is given in Fig. 2 along with the previously measured $^{245}Cm(n,f)$ data of Moore and Keyworth[7] which were derived from a nuclear explosion used as a pulsed neutron source. From ~30 eV to 100 keV their data are in very good agreement with the present measurement and above 100 keV their data agree fairly well in shape with the present results.

Because of the poorer statistical quality of the 'white' neutron source data above 10 MeV, an

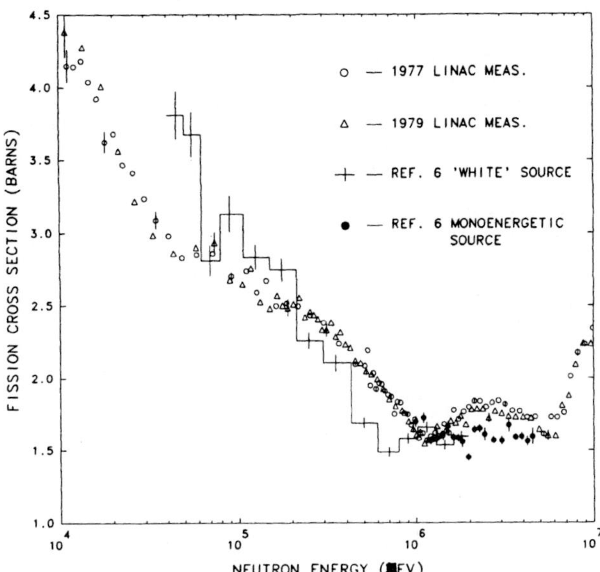

Figure 1. Comparison of the 1977 and 1979 Linac measurements of the $^{242m}Am(n,f)$ cross section. Also included are the recent data of Fomushkin *et al.*[6]

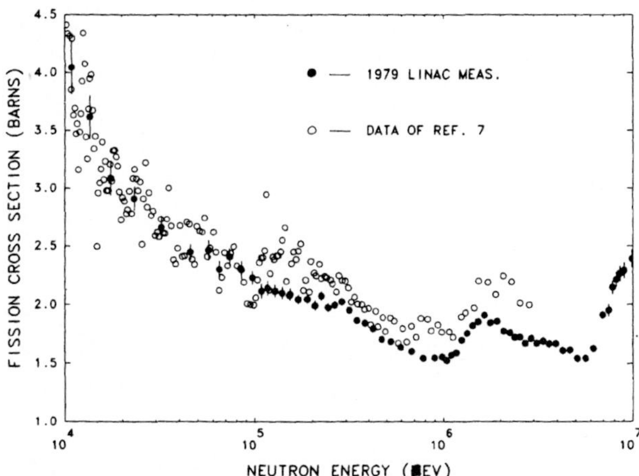

Figure 2. Comparison of our 1979 Linac measurement of the high energy $^{245}Cm(n,f)$ cross section with the data of Moore and Keyworth.[7]

independent monoenergetic measurement, using the same fission chambers, was conducted at the LLNL-ICT facility (see Table I). At the ICT, a pulsed 400-keV deuteron beam was used with a tritiated target to produce an intense flux of neutrons via the $^3H(d,n)^4He$ reaction. Collimating the neutron flux at 90° to the incident deutron beam produced a 14.1 MeV neutron beam of minimum energy spread. Time-of-flight techniques were employed to enhance the signal-to-background (spontaneous fission) ratio. The values obtained for the fission cross sections (relative to ^{235}U) at 14.1 MeV, along with the 14.8-MeV results of the $^{242m}Am(n,f)$ measurement of Ref. 6, are given in Table IV. As can be seen in the table, the 14-MeV cross section data for $^{242m}Am(n,f)$ are in good agreement. For both ^{242m}Am and ^{245}Cm these data also agree with our Linac measurements within the experimental uncertainty.

Table IV. Results and comparison of Livermore-ICT measurement of $^{242m}Am(n,f)$ and $^{245}Cm(n,f)$ cross sections at E_N=14.1 MeV and $^{242m}Am(n,f)$ cross section of Ref. 6 at E_N=14.8 MeV.

ISOTOPE	PRESENT WORK	ERROR		REF.6	ERROR
	$\sigma(n,f)$	Stat.	Sys.	$\sigma(n,f)$	Stat.
^{242m}Am	2.41 barns	5.1%	~5%	2.31 barns	2.2%
^{245}Cm	2.55 barns	2.8%	~5%	—	—

Calculation of the Fission Cross Section

Our present effort involves calculating these fission cross sections over the energy region of the fission neutron spectrum (~100 keV to ~5 Mev). In this energy region many of the fissile actinides have 'macroscopic' structure clearly seen in the $^{235}U(n,f)$ and $^{245}Cm(n,f)$ cross sections shown below. Bjørnholm and Lynn[8] have determined trends in fission barrier parameters over the actinide region. We have written a statistical code, FISCAL, utilizing their results, to investigate this structure as well as the general shape and magnitude of the fission cross section in this energy region.

For neutron-induced reactions, proceeding through channel x, we write the cross section as follows:

$$\sigma_{n,x}(E) = \sum_{J^\pi} \sigma_{cn}^{J^\pi}(E) \frac{T_x^{J^\pi}}{\sum_c T_c^{J^\pi}}$$

where $\sigma_{cn}^{J^\pi}(E)$ is the compound nucleus formation cross section (for a given J^π) written:

$$\sigma_{cn}^{J^\pi}(E) = \pi \lambda^2 g_J \sum_{L,S} T_{L,S}^{J^\pi}(E)$$

and J^π is the total compound nucleus angular momentum and parity, S is the channel spin (neutron+target) and L is the orbital angular momentum. For the neutron transmission coefficients we used the Moldauer form[9] suggested in Bjørnholm and Lynn. T_X is the total transmission coefficient for the decay of the compound nucleus through channel x and $\sum T_c$ is the sum of the total transmission coefficients for decay through all channels. The modes of decay important to this work are (n,n), (n,n′), (n,γ), and (n,f). Above a few hundred keV the competing decay modes are mainly the inelastic and fission channels. The total transmission coefficient for elastic/inelastic scattering is written:

$$T_n^{J^\pi}(E) = \sum_{i,L,S} T_{L,S}^{J^\pi}(E_{CM}-\varepsilon_i) + \sum_{I,L,S} \int_{E_d}^{E_{CM}} T_{L,S}^{J^\pi}(E_{CM}-\varepsilon)\rho(\varepsilon,I^\pi)\, d\varepsilon$$

where E_{CM} is center-of-mass energy, ε_i is the energy of the ith discrete level in the residual nucleus, and $\rho(\varepsilon,I^\pi)$ is the level density (used above the known discrete levels) of the residual nucleus. For the gamma-decay channel we used the total transmission coefficient given in Ref. 8. For the energy region considered here, the total transmission coefficient for fission is given by:

$$T_f^{J^\pi} = \frac{T_A^{J^\pi} \cdot T_B^{J^\pi}}{T_A^{J^\pi} + T_B^{J^\pi}}$$

where A and B are the inner and outer barriers of the double-humped fission barier and

$$T_A^{J^\pi} = \sum_i \frac{1}{1+e^{-2\pi(E_x-E_A-\varepsilon_i)/\hbar\omega_A}} + \int_{E_d}^{\infty} \frac{\rho(\varepsilon,J^\pi)}{1+e^{-2\pi(E_x-E_A-\varepsilon)/\hbar\omega_A}}\, d\varepsilon$$

where the sum is over discrete barrier levels (fission channels) and the integral is the contribution from higher levels. E_A is the barrier height, E_X is the excitation of the compound nucleus and a similar expression holds for barrier B.

Calculation of the $^{235}U(n,f)$ Cross Section

To test our code and to explore the sensitivity of the calculated cross section to the various input parameters, we chose to start with the best-known fission cross section, $^{235}U(n,f)$, and have used the recent evaluation of Poenitz[10] as our standard reference data. For the inelastic channel, 93 (discrete) levels[11,12] up to 1 MeV in excitation were included for the residual ^{235}U nucleus. Not all of these levels (particularly above 500 keV) have known spin and parity but a plot of the spin distribution of the known (and tentatively known) spins yielded a spin dispersion coefficient consistent with σ=4. Therefore, levels of unknown spin were assigned spins consistent with a spin dispersion coefficient σ=4. Above 1 MeV, a level density representation of the Gilbert-Cameron constant temperature form,[13]

$$\rho(E,I^\pi) = C(2I+1)e^{-(I+\frac{1}{2})^2/2\sigma^2} e^{E/\theta}$$

was employed. Values for C, θ, and σ were taken from Ref. 8 as were the parameters for the gamma channel which employed the same form.

For the fisson channel, the barrier reference parameters, E_A=5.63 MeV, $\hbar\omega_A$=1.04 MeV, E_B=5.53 MeV, $\hbar\omega_B$=0.6 MeV, of Ref. 8 were used. The fission barrier level densities were represented by the constant temperature form given above, with several temperature regions (see. Ref. 8) employed for each barrier. However, in order to calculate the detailed cross section, it was found necessary to replace the 'shape' of the barrier level densities represented by the several constant temperature regions with a level density function which maintained that general shape and which had a continuous first derivative.[14] The resulting calculation for the $^{235}U(n,f)$ cross section is plotted in Fig. 3. No width fluctuation correction, which will reduce the cross section below ~1 MeV in the fission channel,[15] is yet incorporated in this calculation.

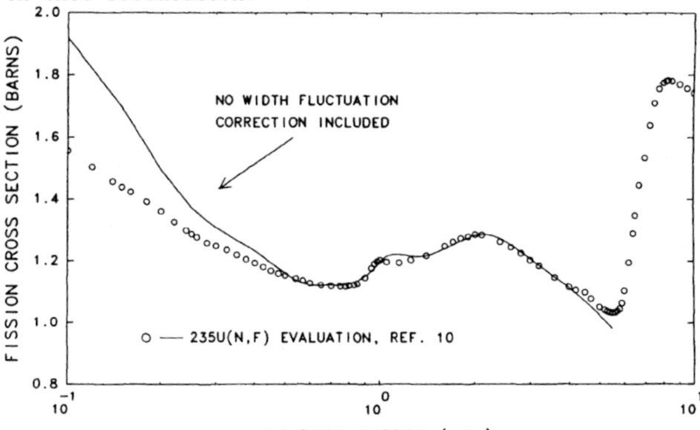

Figure 3. Hauser-Feshbach calculations of the $^{235}U(n,f)$ cross section compared to the evaluation of Poenitz.[10]

The step in the $^{235}U(n,f)$ cross section between 800 keV and 1 Mev is about 8% and, while the exact shape is not well known, the fact that this 'macroscopic' structure really exists is clearly seen in the various fission data sets on ^{235}U (see for example Ref. 10). This structure appears in the calculation because the decreasing fission cross section up to ~1 MeV is governed mainly by the increasing inelastic scattering cross section which is competing with fission. The shape of the increasing inelastic cross section with energy is governed by the increasing number of levels in the residual ^{235}U nucleus to which the neutron can inelastically scatter. Therefore, if the fission cross section is to be calculated correctly in this region, it is very important to represent the discrete levels in ^{235}U up to 1 MeV as complete as possible. From ~1.5 to ~5 MeV, the level density in the residual ^{235}U nucleus appears to be represented satisfactorily by a constant temperature form using the parameters of Bjørnholm and Lynn. However, if our representation of the discrete levels is reasonable, then there exists structure in the level density in the 1 MeV region of excitation in ^{235}U in order for our discrete representation to 'tie onto' the constant temperature form which fits the data at higher energies. This structure is seen, to a greater or lesser degree, in the level density representations of Bjørnholm and Lynn in the 1-3 MeV region of excitation above the fission barriers in ^{236}U. Without this structure in the level density our calculation would not reproduce the detailed ^{235}U(n,f) cross section.

Calculation of the ^{245}Cm(n,f) Cross Section

For the initial calculation of the ^{245}Cm(n,f) cross section, we maintained the same level density parameters as in the ^{235}U(n,f) calculation and used the barrier parameters, E_A=5.7 MeV, $\hbar\omega_A$=1.04, E_B=4.2 MeV, $\hbar\omega_B$=0.6 MeV, of Bjørnholm and Lynn for ^{246}Cm. With these parameters, we calculated the cross section to be in qualitative agreement with the magnitude of the measured cross section (~1.7 barns) in the MeV region. However, the detailed shape was completely wrong. We should not expect the level densities used for the ^{235}U calculation to fit the detail of the ^{245}Cm(n,f) cross section. Since less is known about the discrete levels up to ~1 MeV in ^{245}Cm than in ^{235}U, we allowed the shape of the level density in the residual ^{245}Cm nucleus to vary while maintaining the same fission barrier level densities as used in the ^{235}U calculation. The height of the outer barrier, E_B, in ^{246}Cm was also changed to 4.5 MeV to fit the measured cross section better. The results of the final calculation are shown in Fig. 4 with no width fluctuation correction yet included in the code. As in the case for the ^{235}U(n,f) calculations, we could only fit the ^{245}Cm(n,f) data by allowing some structure in the level density representation of the residual ^{245}Cm nucleus at an excitation energy of ~1 to 2 MeV.

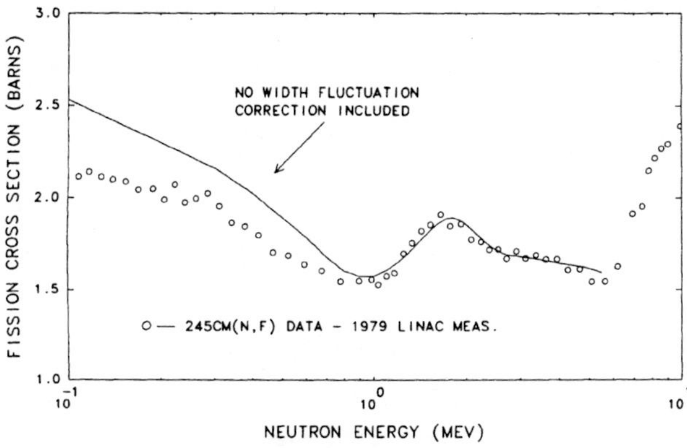

Figure 4. Hauser-Feshbach calculation of the ^{245}Cm(n,f) cross section compared to our 1979 Linac measurement.

Conclusions

We have completed measurements of the ^{242m}Am and ^{245}Cm fission cross sections from thermal energies to ~20 MeV in a series of experiments at several facilities. These measurements were accomplished to good accuracy with sub-milligram quantities of isotopically enriched ^{242m}Am and ^{245}Cm.

Hauser-Feshbach calculations have been carried out which reasonably represent the detailed ^{235}U and ^{245}Cm fission cross sections in the energy range from 1 to 5 MeV. Our calculations do not yet include a width fluctuation correction nor do they include a second chance fission calculation. We have been guided in this calculational effort mainly by parameterizations of Bjørnholm and Lynn. Our calculations indicate that, in order to reproduce the detailed cross section, we require level density functions which have continuous first derivatives and which contain some structure in both the residual and compound nucleus in the lower energy region of excitation.

While we conclude from these calculations that the 'macroscopic' structure seen in the 1 to 2 MeV region of these fission cross sections is most likely caused by structure in the level densities, we cannot positively rule out the possibility of this structure entering through the fission (barrier) transmission coefficients. If indeed this were the case, i.e., that the macroscopic structure were really intermediate structure, our present method of calculating the total fission transmission would not produce it. It seems unlikely that such intermediate structure should be present at an excitation of 2 to 3 MeV above the fission barriers.

There exist two important areas in which our calculations are being improved beyond the existing limitations discussed above. The first is in the calculation of the neutron transmission coefficients and the compound nucleus formation cross section. Neutron transmission coefficients calculated from a deformed optical potential which will fit the total cross sections over a range of actinides and over the energy region of interest here should improve the energy dependence of these coefficients and give a more physically realistic compound reaction cross section. The second area of improvement and probably the more important one is to calculate correctly the level density (intrinsic and collective) to at least a few MeV both at ground state deformation for the target nucleus and at deformations corresponding to the two barriers in the compound nucleus. It is essential that these density calculations be included if we attempt to use statistical calculations to accurately predict those cross sections for which measurements are not now feasible.

Acknowledgments

We appreciate the efforts of R. W. Hoff and the LLNL Nuclear Chemistry Division for their contributions to the preparation of the transplutonic fission samples. Our thanks are also due to G. F. Auchampaugh and P. W. Lisowski who collaborated in the WNR measurement. R. E. Howe contributed greatly to all phases of the measurements at both laboratories. Discussions with H. Marshall Blann, D. G. Madland and E. D. Arthur were very helpful toward our calculational effort. Finally, special appreciation is due to R. E. Strout, II of LLNL whose programming ability made difficult calculations a joy to perform.

References

1. J. C. Browne, *et al.*, "Fission Cross Section for ^{242m}Am," Neutron Physics and Nuclear Data for Reactors and other Applied Purposes, AERE, Harwell, 887 (1978).
2. R. M. White, *et al.*, "Fission Cross Section of ^{245}Cm from 10^{-3}eV to 10^{4}eV," Nuclear Cross Sections for Technology, NBS Special Publication 594, (Knoxville, TN, 1979), p. 496.
3. P. A. Seeger, *et al.*, Nucl. Phys. A96, 605 (1967).
4. C. D. Bowman, *et al.*, Phys. Rev. 166, 1219 (1968).
5. J. C. Browne and R. M. White, "The ^{242m}Am Fission Cross Section," to be published.
6. E. F. Fomushkin, *et al.*, Yadernaya Fizika 33, 620 (1981).
7. M. S. Moore and G. A. Keyworth, Phys. Rev. C 3, 1656 (1971).
8. S. Bjørnholm and J. E. Lynn, Rev. Mod. Phys. 52, 725 (1980).
9. P. A. Moldauer, Phys. Rev. 157, 907 (1967).
10. W. P. Poenitz, "Evaluation ^{235}U(n,f) Between 100 keV and 20 MeV," Argonne National Laboratory Report, ANL/NDM-45 (1979).
11. C. M. Lederer and V. S. Shirley, eds., Table of Isotopes, 7th Edition, John Wiley and Sons, New York, NY (1978).
12. M. R. Schmorak, Nuclear Data Sheets 21, 117 (1977).
13. A. Gilbert and A. G. W. Cameron, Can. J. Phys. 43, 1446 (1965).
14. F. N. Fritsch and R. E. Carlson, "Monotone Piecewise Cubic Interpolation," SIAM J. Numer. Anal. 17, 238 (1980).
15. E. D. Arthur, "Use of the Statistical Model for the Calculation of Compound Nucleus Contributions to the Inelastic Scattering of Actinide Nuclei," Los Alamos National Laboratory Report, LA-UR 81-3497 (1981).

AVERAGE CAPTURE CROSS SECTION OF THE FISSION PRODUCT NUCLEI ^{104,105,106,108,110}Pd

E. Cornelis and G.J. Vanpraet

R.U.C.A. University of Antwerp, Belgium

C. Bastian, G. Rohr, R. Shelley and T. van der Veen

CEC - JRC, Geel Establishment
Central Bureau for Nuclear Measurements, B-2440, Geel, Belgium

Neutron capture cross section measurements on enriched stable Pd isotopes have been performed at the 30 m station of Gelina in the energy range of 10 eV up to 600 keV. The neutron flux shape was determined with a 0.5 mm ^{6}Li-glass scintillator and a 0.6 mm thick $^{10}B_4C$-slab. The time dependent background was evaluated by using the yields from a 0.5 mm ^{208}Pb capture sample. The present data analysis covers the energy range between 10 keV and 300 keV.

[Average Capture Cross Section ^{104,105,106,108,110}Pd]

Introduction

Neutron cross sections of fission products are of great importance for predicting long term characteristics of fast reactors. Among these nuclei, seven Palladium isotopes are listed on the ENDF/B-V data files for fission product nuclides as important absorbers for fast and thermal reactors [1]. In this paper we present capture cross sections for $^{104-105-106-108-110}$Pd in regard to this poisoning effect. As in a large fast reactor, the energy range below several keV is important and should not be neglected, these measurements have been performed in the energy range of 4 eV up to 600 keV. The present data analysis however covers the energy range between 3 keV and 300 keV. Preliminary results for ^{105}Pd and ^{108}Pd have been discussed elsewhere [2].

Experimental Technique

The C.B.N.M. 150 MeV electron linear accelerator at Geel, GELINA [3], has been used with the characteristics listed in Table 1 for this experiment.

Table 1. Experimental Parameters

burst width		4 ns
repetition rate		800 Hz
electron energy		100 MeV
beam power		5 kW
moderator thickness		4 cm
gamma flash filter	lead	30 cm
	copper	20 cm
time overlap filter		5.68×10^{-3} ^{10}B at/b

Intense neutron bursts are produced in a mercury cooled natural U-target through a photonuclear reaction, and moderated by two halfmoon shaped polyethylene discs, placed one on top and one below the target. To suppress the detector paralyzing γ-flash effect of the bremsstrahlung, blocks of copper and lead are shielding the U-target in its horizontal plane through the beam line.

Enriched samples of ^{104,105,106,108,110}Pd were investigated at the 30 m time-of-flight station linked to the target bunker by an evacuated flight path tube, equipped with neutron beam collimators. The beam line was perpendicular to the target moderator that is entirely viewed by the capture sample. The detection of prompt γ-rays following neutron capture is carried out by two cylindrical C_6D_6 liquid scintillators, of 10.2 cm diameter and 7.5 cm thickness, each, coupled to an EMI photomultiplier. This capture detector is characterized by a low prompt sensitivity to sample scattered neutrons. This has been achieved by reducing the amount of materials around the capture detector to a minimum. The distance from the neutron source moderator assembly was determined to be 28.398 m. The electronic signals, carrying time and amplitude information are registered with the time-of-flight start signals derived from the electron bursts and sorted electronically in a Nuclear Data ND 6600 acquisition system linked to a C.B.N.M. built 4ns digital time coder [4]. This two-parameter experiment is required by the Mayer-Leibnitz weighting technique [5]. In this method, the detector events are assigned an importance proportional to the energy of detected photons and is independent of the gamma decay mode. This results in a detector response proportional to the total energy released in the capture process.

Amplitude calibration and amplifiers gain have been controlled regularly with the Compton edge from the Pu $^{13}C(\alpha,n)^{16}O^*$ 6.13 MeV γ-ray. A 200 keV bias was set for the pulse height amplitude by suppressing the first 4 channels. For the weighting, the events are sorted in 16 PH-groups of 8K TOF each.

Monitoring the shape of the neutron flux was done by putting permanently in the neutron beam, 75.0 cm in front of the capture sample, a ^{6}Li-loaded glass scintillator of 0.05 cm thickness and 10 cm diameter. It was contained in a thin Al-foil cylinder acting as a reflector and viewed by 2 EMI photomultipliers. The relative neutron flux was measured with a $^{10}B_4C$ slab, 0.06 cm thick and of 8 cm diameter, used as a capture sample by measuring the 478 keV photons associated with the neutron absorption $^{10}B(n,\alpha\gamma)^7Li$ reaction.

Analysis

Some effort has been devoted in obtaining a critical evaluation of the time dependent background with black resonance filters S, Na, Co, W and Ag and the yield obtained with a 0.05 cm thick ^{208}Pb sample of 8 cm diameter. For the latter case it was found that the actual background was higher when compared to the results obtained with the thick sulphur filter. This filter depleted the incident neutron flux by an average of about 30 % outside the 102 keV resonance. Fig. 1 shows typical results obtained for ^{105}Pd in the upper curve and for ^{208}Pb on the lower curve both normalized per monitor count. In Fig. 2, three different spectra are shown after subtracting the yield per monitor count obtained with ^{208}Pb; the upper curve is the ^{105}Pd capture yield without black resonance filters; the middle one with the Na filter and the bottom one was obtained with all filters. For this case the background evaluation by using the black resonance technique is off by 7 % compared to that obtained with the lead sample at the sulphur resonance. The channel numbers cover the energy range from 600 keV down to 10 keV. Similar figures for ^{108}Pd are shown in Figs. 3 and 4. For this case a discrepancy of about 35 % for this energy region was found. Palladium sample characteristics are given in Table 2.

Normalization

To normalize the capture data, the value of $2g\Gamma_n = (6.70 \pm 0.04)$ meV for the resonance at 55.2 eV in ^{105}Pd was taken from recent C.B.N.M. transmission data [6]. Hence, corrections for γ absorption of

K. H. Böckhoff (ed.), Nuclear Data for Science and Technology, 222–225.

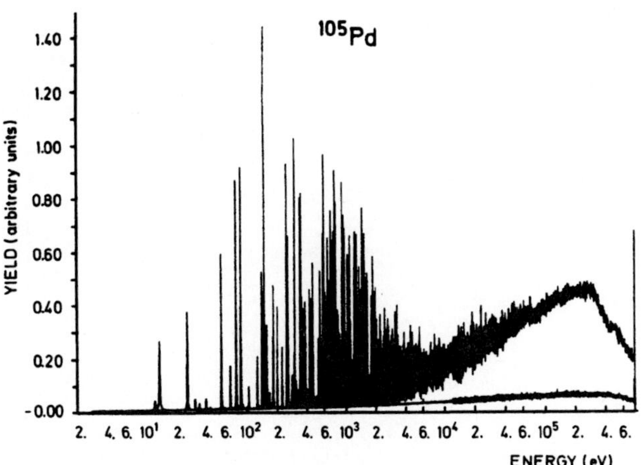

Fig. 1. Capture yield and background obtained for ^{105}Pd as a function of energy

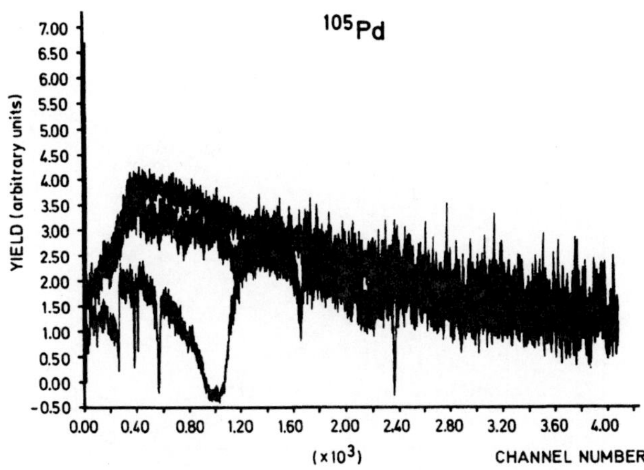

Fig. 2. Capture yield in ^{105}Pd obtained without filter, with Na filter and with all black resonance filters respectively top, middle and bottom curve after being corrected for ^{208}Pb background

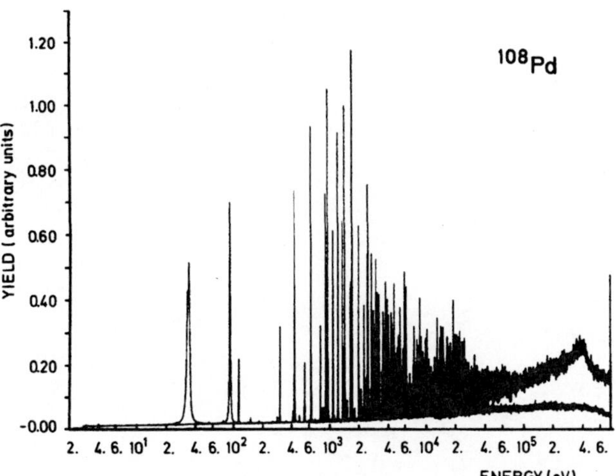

Fig. 3. idem as for Fig. 1, but now for ^{108}Pd

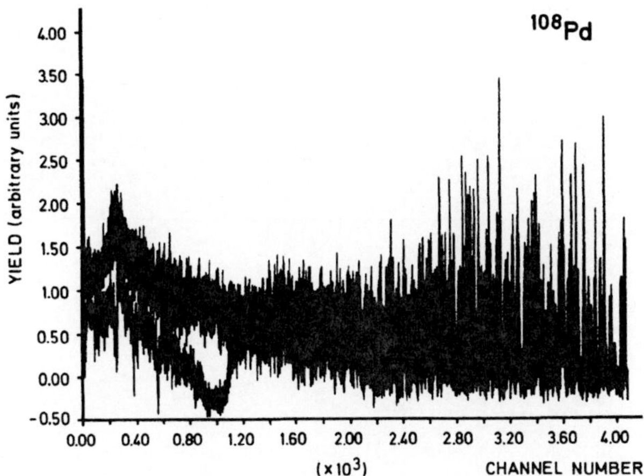

Fig. 4. idem as for Fig. 2, but now for ^{108}Pd

Table 2. Sample characteristics

Pd isotope	% abund.	^{105}Pd contamination	thickness (at/barn)x10^3
104	95.25	2.64	1.077
105	97.38	--	0.894
106	98.48	0.79	2.250
108	98.88	0.22	2.461
110	97.73	0.36	1.470

photons in the capture sample can be disregarded. Further confidence in the normalization procedure was obtained by cross calibration to the results obtained with a 0.05 cm thick Au sample, using the ENDF/B-V file at 200 keV. The agreement of both calibration procedures was found to be within 3 %. For the even Pd isotopes the normalization was only carried out relative to the standard cross section of Au. No corrections for self-protection and multiple scattering effects have been applied, but have been calculated. Since our samples are at least three times thinner than those used at O.R.N.L., these effects can only contribute for a small possible systematic error. Statistically good data have been obtained with these relative thin Pd-samples, especially for ^{105}Pd. Average capture cross sections are shown in the histograms of Figs. 5, 6, 7, 8 and 9. The statistical and systematic errors to which the background uncertainty is the main contributor, are estimated to be 5 % for ^{105}Pd and 10 % for the even isotopes. The numerical results for the other data corrections, gamma energy attenuation, and dead time correction are given in Table 3.

Results and Conclusions

Numerical results are given in Table 4. For ^{105}Pd, the O.R.N.L. results [7] show a systematic deviation relative to the C.B.N.M. data of approximately + 10 % below 10 keV neutron energy whereas at higher energies the data are in agreement within ± 5 % except for the last energy bin. The average capture cross sections show a systematic decrease with the atomic weight for the even isotopes. Compared to the O.R.N.L. results, we have found very similar systematic deviations as for ^{105}Pd:

^{104}Pd: approximately + 10 % below 15 keV and at higher energies agreement within ± 6 %
^{106}Pd: deviation of + 20 % below 15 keV and approximately + 7 % up to 150 keV
^{108}Pd: deviation of 10 % below 15 keV, reducing to 20 % for higher energies
^{110}Pd: approximately 30 % deviation over the whole energy range considered.

No explainable evidence for the observed discrepancy could be found on the base of our experimental investigations, except that with the use of the black resonance technique, the Geel data approach the values of O.R.N.L. The smooth curves in Figs. 5-9 describe our capture cross sections based on the statistical theory in terms of average values of neutron resonance paramaters. The results are discussed elsewhere [8].

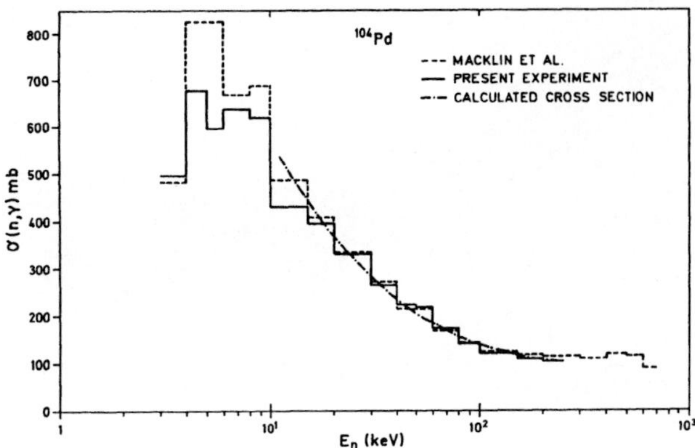

Fig. 5. Average Capture Cross Section as histogram for ^{104}Pd

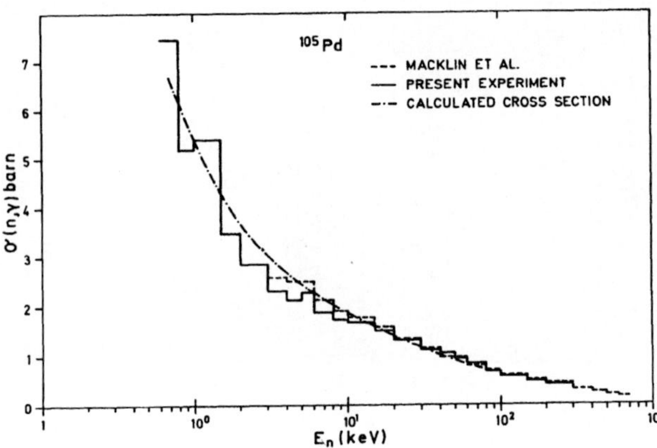

Fig. 6. Average Capture Cross Section as histogram for ^{105}Pd

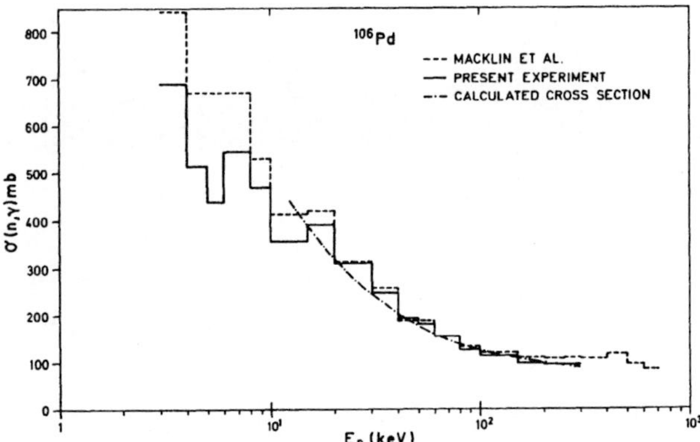

Fig. 7. Average Capture Cross Section as histogram for ^{106}Pd

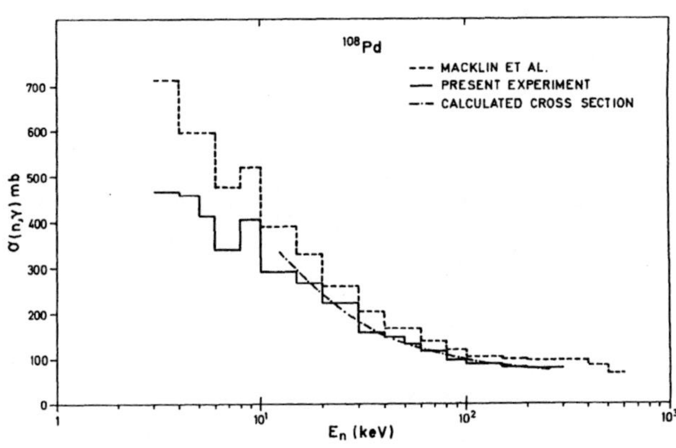

Fig. 8. Average Capture Cross Section as histogram for ^{108}Pd

Table 3. Corrections

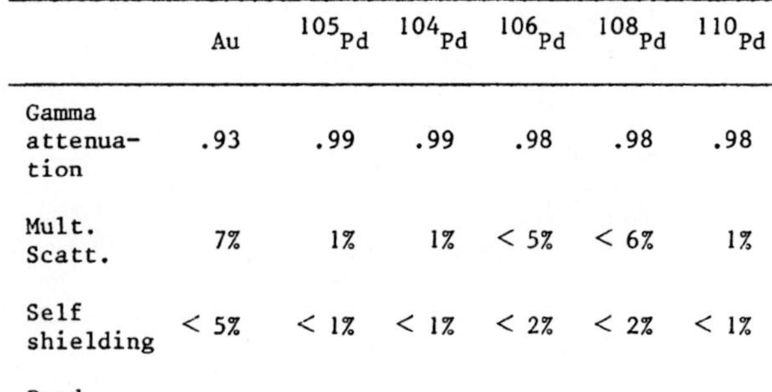

	Au	^{105}Pd	^{104}Pd	^{106}Pd	^{108}Pd	^{110}Pd
Gamma attenuation	.93	.99	.99	.98	.98	.98
Mult. Scatt.	7%	1%	1%	< 5%	< 6%	1%
Self shielding	< 5%	< 1%	< 1%	< 2%	< 2%	< 1%
Dead time	< 6%	< 4%	< 2%	< 2%	< 2%	< 2%

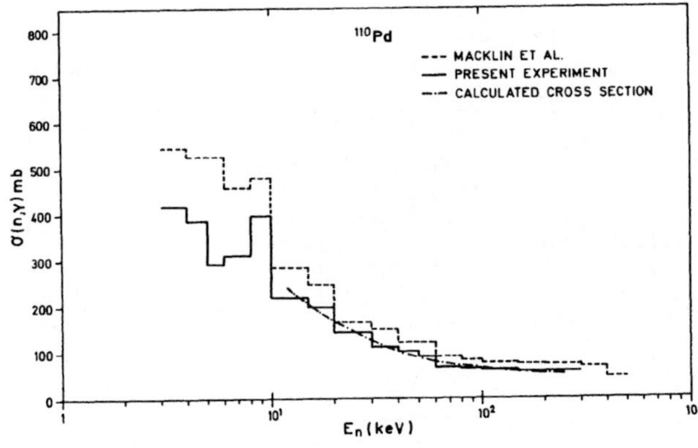

Fig. 9. Average Capture Cross Section as histogram for ^{110}Pd

Table 4. Average cross section (barn)

Energy (eV)	^{104}Pd	^{105}Pd	^{106}Pd	^{108}Pd	^{110}Pd
3000 - 4000	0.496	2.34	0.676	0.467	0.416
4000 - 5000	0.681	2.17	0.503	0.459	0.504
5000 - 6000	0.596	2.33	0.424	0.415	0.292
6000 - 8000	0.636	1.92	0.534	0.339	0.310
8000 - 10000	0.619	1.76	0.459	0.405	0.395
10000 - 15000	0.432	1.72	0.345	0.291	0.219
15000 - 20000	0.395	1.56	0.381	0.268	0.198
20000 - 30000	0.334	1.37	0.301	0.223	0.142
30000 - 40000	0.265	1.18	0.240	0.158	0.111
40000 - 50000	0.223	1.08	0.185	0.148	0.101
50000 - 60000	0.217	0.978	0.174	0.134	0.090
60000 - 80000	0.173	0.852	0.149	0.117	0.067
80000 - 100000	0.142	0.714	0.124	0.098	0.065
100000 - 150000	0.123	0.603	0.110	0.089	0.062
150000 - 200000	0.109	0.493	0.094	0.082	0.057
200000 - 300000	0.104	0.416	0.092	0.080	0.057

References

1. R.E. Schenter et al., Proc. Specialists' Meeting on Neutron Cross Sections of Fission Product Nuclei, Bologna 1979, p. 253.

2. G. Rohr, C. Bastian, E. Cornelis, R. Shelley, T. van der Veen, G. Vanpraet, Proc. Spec. Meeting on Fast Neutron Capture Cross Sections, Argonne, 1982, to be published

3. J. Salomé, K.H. Böckhoff, Proc. Knoxville Conf. on Nuclear Cross Sections for Technology, 1979, p. 534

4. S. de Jonge, Fast Time Coder Type 7602D, Internal Report GE/IN/DE/72/81, C.B.N.M. Geel

5. R.L. Macklin, J.H. Gibbons, Phys. Rev. 159, (1967) 1009

6. P. Staveloz et al., Proc. Specialists' Meeting on Neutron Cross Sections of Fission Product Nuclei, Bologna 1979, p. 53

7. R.L. Macklin, J. Halperin, R.R. Winters, Nucl. Sc. and Eng. 71 (1979) 182; and R.L. Macklin, Private Communication Nov. 7, 1980 concerning corrections to publication N.S.E. 71 (1979) 205

8. to be published

NEUTRON RADIATIVE CAPTURE AND TRANSMISSION MEASUREMENTS OF ^{107}Ag AND ^{109}Ag.

M. Mizumoto, M. Sugimoto, Y. Nakajima
M. Ohkubo, Y. Furuta and Y. Kawarasaki

Japan Atomic Energy Research Institute
Tokai-mura, Ibaraki-ken, Japan

Neutron capture and total cross sections of ^{107}Ag and ^{109}Ag have been measured using the time-of-flight facility at the Japan Atomic Energy Research Institute electron linear accelerator. Capture measurements were carried out with a 3500 l large liquid scintillation detector at 52 m flight path and transmission measurements were made with ^{6}Li-glass detectors at 56 m and 190 m flight paths. Resonance energies and neutron widths were determined up to 7 keV both for ^{107}Ag and ^{109}Ag. The s-wave strength functions and average level spacings were found to be S_0 = (0.42 ± 0.05) x 10^{-4}, D_0 = 22 ± 2 eV for ^{107}Ag and S_0 = (0.44 ± 0.05) x 10^{-4}, D_0 = 21 ± 2 eV for ^{109}Ag. The average capture cross sections were deduced in the energy range of 3.2 to 700 keV with an estimated accuracy of 4 to 8 %. The data to 100 keV were well fitted with the average resonance parameters. Recent evaluated data for ^{109}Ag disagree significantly with our results.

[^{107,109}Ag(n,n),(n,γ), E_n = 2-700 keV, measured $\sigma_t(E)$, $\sigma_{n\gamma}(E)$, deduced resonance parameters, E_n, $2g\Gamma_n$, s-wave strength functions, average level spacings]

Introduction

The accurate neutron cross sections of the silver isotopes are, as significant fission products, very important for the design of power reactors. In particular, the capture cross sections for ^{109}Ag have been officially requested with priority 1 with an accuracy of 10 % in the energy range of 10 keV to 1 MeV for calculations of reactivity and burn up in fast reactors.[1] The previous data for this isotope by Weston et al.[2] and Kononov et al.[3] are different from each other by a factor of two. The recent evaluations[4-6] are also influenced by either of these two results. On the other hand, average resonance parameters can be used to calculate the average cross sections in the low keV region and provide a valuable consistency check with the measured cross sections. Resonance parameters for Ag isotopes were previously obtained below 2.6 keV for separated isotopes and below 4.0 keV for natural element by several authors.[7-11] Below about 1 keV, previous data show the substantial difference between the s-wave strength functions for ^{107}Ag and ^{109}Ag.

In the present experiment, the neutron capture and transmissions were measured to obtain accurate capture cross sections and average resonance parameters. Our capture cross sections for ^{109}Ag, in good agreement with the recent ORELA data,[12] lie between the two discrepant data by Weston et al.[2] and Kononov et al.[3] Resonance energies and neutron widths for a large number of resonances were obtained up to 7 keV both for ^{107}Ag and ^{109}Ag, resulting in the very similar values for the s-wave strength functions and average level spacings. The capture cross sections for ^{109}Ag, calculated with our average resonance parameters, is found to be consistent with the measured cross section. In this measurement, the experimental uncertainties were expressed with the covariance matrix in order to meet the requirement for the evaluation.

Experiment

The experimental procedure will be briefed below since it has been described in previous publication.[13] Measurements were carried out with the neutron time-of-flight spectrometer at the 120 MeV JAERI electron linear accelerator. Pulsed neutrons were produced in a laminated water-cooled Ta target and moderated by 5 cm thick boron-loaded polyethylene. The linac was operated at 150 pps and 300 pps with electron burst widths of 80 ns and 20 ns, respectively.

Samples enriched in ^{107}Ag and ^{109}Ag were provided in the form of metallic powder from the Isotope Division of the Oak Ridge National Laboratory. Table I lists the isotope abundances and thicknesses of the samples. The impurities in both samples were negligible. The samples were contained in Al cases 7 cm in diameter.

Capture events were detected by observing prompt γ-rays with a large liquid scintillation tank. It is located at 52.95 m from neutron producing target. The neutron beam was collimated to 6 cm diameter at the sample position. The neutron flux shape was measured with a ^{10}B-NaI detector. The relative efficiency of the flux detector was calculated using evaluated ^{10}B(n,αγ) cross sections.[14] The corrections for neutron multiple-scattering and self-shielding in the flux detector were estimated with a Monte-Carlo method. Beam filters were inserted permanently in the neutron beam; 5 mm boron nitride to absorb low energy neutrons, 8 mm Pb to reduce γ-flash and 8 mm metallic Na encapsulated in Al container to normalize backgrounds at the 2.85 keV resonance. For background determination, notch filters were used, consisting of thick Al and Co.

The transmissions were measured either with an 11.1 cm diameter x 0.635 cm thick ^{6}Li-glass detector at 56.32 m or with five 11.1 cm diameter x 1.27 cm thick ^{6}Li- glass detectors at 191.49 m. The neutron beam was collimated to 3.5 cm at the sample position.

The energy calibration was made using the energies of sharp resonances present in the filters; 5.903 keV of Al, 53.191 keV of Na, and 3.36 keV and 71.191 keV of Pb.

Table I Isotopic abundances and thicknesses of samples

Isotope	Isotopic abundance (%)	Tickneses Thick (atoms/barn)	Thin
^{107}Ag	98.22	0.0170	0.00728
^{109}Ag	99.32	0.0177	0.00683

Analysis

The capture count rates were converted to relative capture cross sections by using conventional techniques; corrections for dead time losses, background subtractions by black resonance method and corrections for the resonance self-shielding and multiple-scattering. The normalization of the cross sections was made by the saturated resonance technique, using the resonance in ^{107}Ag at 16.30 eV and in ^{109}Ag at 30.40 eV. The scattered neutron sensitivity of the detector was negligible and was not corrected for.

The uncertainties in the measured cross sections were devided into statistical and systematical errors. The statistical errors range from 0.4 % at 4 keV to 4 % at 700 keV after the data were averaged over

K. H. Böckhoff (ed.), Nuclear Data for Science and Technology, 226–229.

Table II The sources of systematical uncertainty in the measured capture cross sections.

Source of uncertainty	Neutron energy region (keV) 10 - 20 (%)	80 - 100 (%)	300 - 400 (%)
Background of capture yield	1.4	2.5	1.0
Normalization	2.2	2.2	2.2
Background of neutron flux	<0.1	0.6	1.4
Relative efficiency of flux detector	1.0	1.4	2.6
Sample thickness	2.0	2.0	2.0
Multiple scattering and selfshielding	2.4	2.6	2.8

Table III A part of the correlation matrix (x100) for capture cross sections of Ag.

Group number	Energy (keV)	$\Delta\sigma$ (%)	10	11	12	13	14	15	16	17
10	120-100	6.08	100							
11	100- 77	6.14	68	100						
12	77- 60	6.06	67	69	100					
13	60- 47	6.11	67	69	68	100				
14	47- 36	6.19	72	75	73	77	100			
15	36- 22	6.18	65	69	62	69	71	100		
16	22- 17	6.15	64	67	66	67	68	68	100	
17	17- 13	6.33	58	61	60	59	68	64	63	100

suitable energy interval. Information on the sources of systematical errors is given in Table II. They are considered to be correlated from one energy to another. Because they change slowly with the neutron energy, these errors are given at a limited number of energies. For the sake of simplicity, the shapes of backgrounds and sample thickness corrections were expressed analytically in the calculation of covariance matrix. The correlation matrix for the $^{10}B(n,\alpha\gamma)$ cross section as a standard was taken from the error file of ENDF/B-V.[14] Table III gives, an example, a part of our estimated systematical errors and correlation matrix. The detailed description for these error estimations will be given elswhere.[15]

The transmissions were analyzed with the multi-level Breit-Wigner formula incorporated in a least squares fitting program,[16] which calculated effects due to Doppler and resolution broadening. An example of experimental and calculated transmission data is shown in Fig. 1. Due to insufficient statistics in the data taken at 190 m, many weak resonances were apparently missed. Radiation widths Γ_γ were taken constant to be 140 meV for ^{107}Ag and 130 meV for ^{109}Ag by averaging values given in BNL-325.[17] The statistical factor g=0.5 was assumed. Below 2.0 keV, reliable resonance parameters are available in the previous works.[17] In this experiment, only resonances, of which parameters are discrepant over quoted uceratintíes, were analyzed. For the missing resonance analysis, the Porter-Thomas distribution was assumed to fit the distribution of the measured neutron widths. Since the silver isotopes are located near a peak in the mass dependence of the p-wave strength function, many p-wave levels are expected to be observed. The conditional probabilities, that the resonances with given $g\Gamma_n$ values were s-wave or p-wave, were calculated by the Bayes' thorem and used as the weighting factors for the distribution.

The cumulative sum of observed levels between 2 to 7 keV is plotted for ^{107}Ag in Fig. 2. The cumulative sum of $g\Gamma_n^0$ is also plotted in Fig. 3. The similar plots for ^{109}Ag are omitted. The $g\Gamma_n^0$ plot is relatively insensitive to the missed weak resonances. The slopes in the figures give our estimated values of average level spacings and s-wave strength functions, after missing resonances were taken into account.

Results and discussions

The average capture cross sections are shown in Figs. 4 and 5 for ^{107}Ag and ^{109}Ag, respectively. Several previous experimental and evaluated data available for comparison are also shown in the figures. Our data on ^{107}Ag are in good agreement with the data of Weston et al.[2] over the whole energy range and with the data of Macklin et al.[18] and Chaubey et al.[19] at 24 keV. The data of Kononov et al.[3] have steeper energy dependence and are 25 % higher at 3C keV compared with our results. The evaluated data of

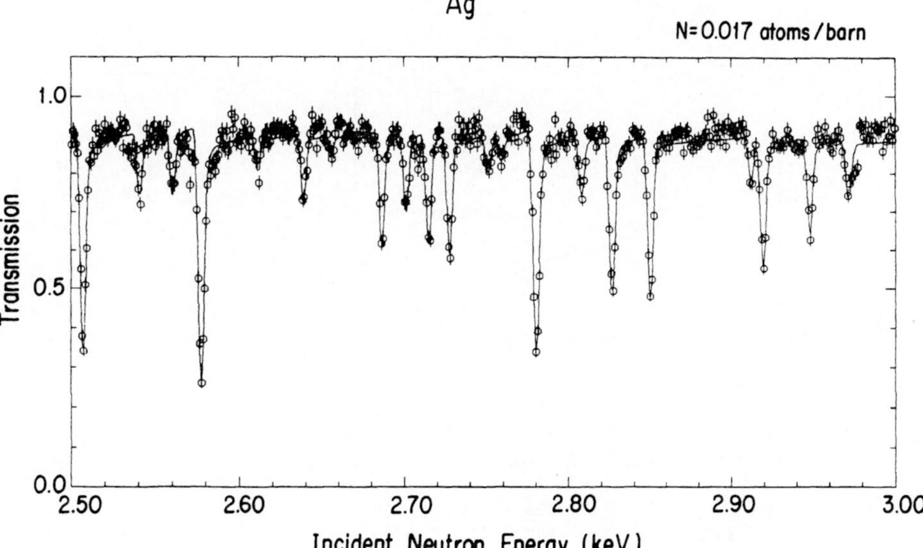

Fig. 1. Example of resonance analysis of the transmission data. The solid line is the calculated curve with a least squares fitting program.

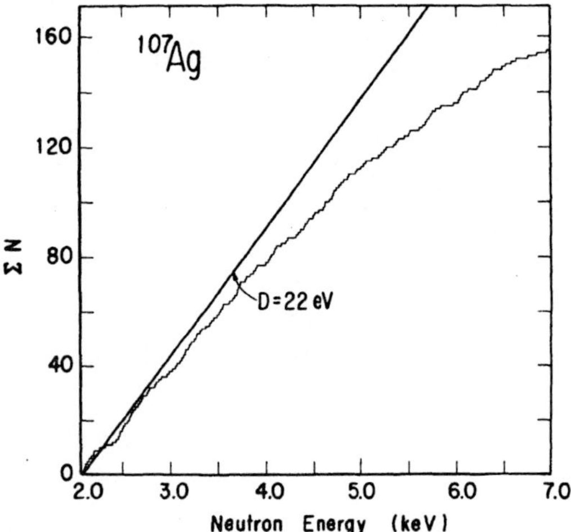

Fig. 2. The cumulative sum of observed levels vs energy for ^{107}Ag.

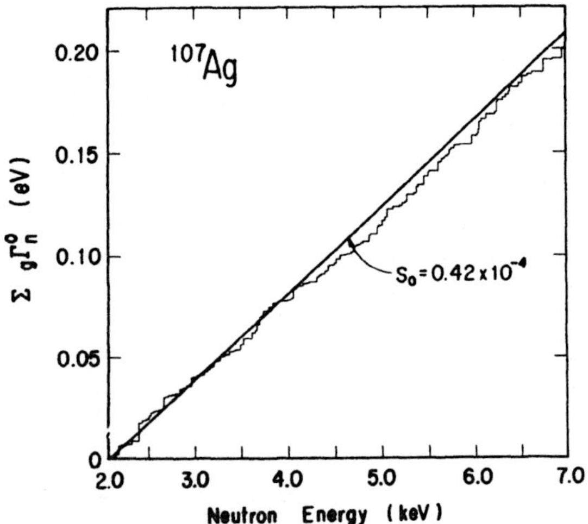

Fig. 3. The cumulative sum of $g\Gamma_n^0$ vs energy for ^{107}Ag.

RCN-2,[4] which are based on the calculation of strength function model, are more than 20 % higher in the energy region above 100 keV and the data of ENDF/B-V,[6] which are based on the adjustment of integral and differential data, are also systematically higher by about 5 to 15 %.

For ^{109}Ag, as already mentioned, the two existing experimental data disagree seriously with each other. The data of JENDL-1 evaluation[5] rely on the higher values of Weston et al.[2] It has been pointed out that the data of Kononov et al.[3] are almost certainly too low because the capture cross section of natural silver, calculated from their ^{107}Ag and ^{109}Ag data, are about 16 % lower than the data for natural silver by Kompe.[20] On the other hand, the integral data from the central reactivity worth measurements[4] suggested that the data of Weston et al.[2] were too high.

The previous values of the s-wave strength functions are given in Table IV. Our resonance energies are systematically higher than the high resolution data by Garg et al.[7] In the energy region below 2.0 keV, overall agreement between our resonance parameters and old data is obtained. Below about 1 keV, the marked difference between s-wave strength functions for ^{107}Ag and ^{109}Ag was observed previously, which was not expected from the statistical model. In our experiment covering wider energy range, the very similar values for both isotopes were obtained. The recent ORELA capture cross section data[12] were parametrized to extract capture areas between 2.65 and 7 kev, using our neutron widths for sample-thickness correction. Their data are more sensitive to weak resonances.

Table V shows the average resonance parameters for evaluations of capture cross sections for ^{107}Ag and

Table IV Resonance parameters of ^{nat}Ag, ^{107}Ag and ^{109}Ag.

Author	Isotope	Energy range (eV)	$S_0 \times 10^4$ ^{107}Ag	^{109}Ag
de Barros	^{nat}Ag	291- 753	$0.37^{+0.11}_{-0.08}$	$0.75^{+0.22}_{-0.16}$
Pattenden	^{107}Ag ^{109}Ag	41-2665	0.35±0.07	0.46±0.09
Chrien	^{109}Ag	30- 317		0.54±0.10
Muradyan	^{107}Ag ^{109}Ag	5.2- 915	$0.43^{+0.17}_{-0.12}$	$0.83^{+0.23}_{-0.19}$
Garg	^{nat}Ag	100-3996	0.48±0.04	
Present	^{107}Ag ^{109}Ag	2000-7000	0.42±0.05	0.44±0.05

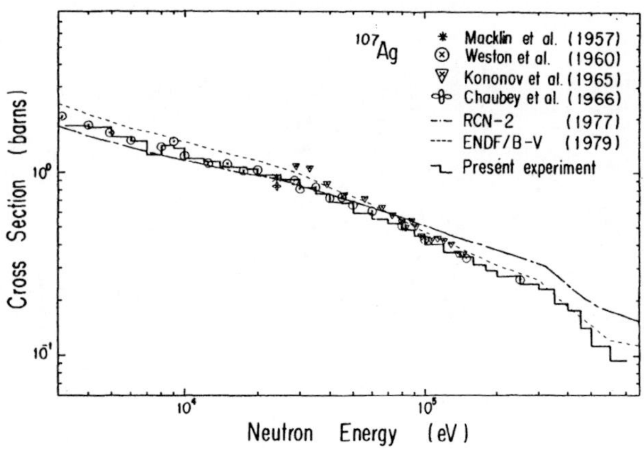

Fig. 4. The neutron capture cross sections of ^{107}Ag in the energy region from 3 to 800 keV.

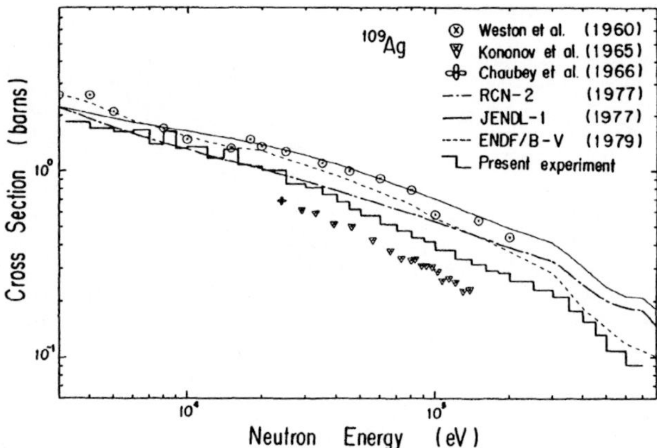

Fig. 5. The neutron capture cross sections of ^{109}Ag in the energy region from 3 to 800 keV.

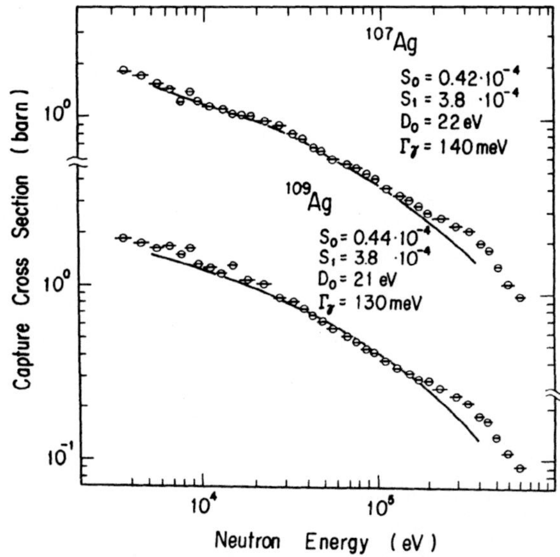

Fig. 6. The measured capture cross sections for ^{107}Ag and ^{109}Ag are compared with the calculated curves. The p-wave strength functions and radiation widths are taken from the BNL-325 fourth edition.[17]

^{109}Ag. The value of the p-wave strength functions 3.8 x 10^{-4} are from BNL-325 fourth edition based on the total cross section experiments by Camarda.[21] In the all evaluations which adopted the values by Weston et al.,[2] they took either large s-wave strength functions or small average level spacings in order to reproduce large cross section. As shown in Fig. 6, calculated curves for ^{107}Ag and ^{109} Ag, using our average resonance parameters, are in good agreement with the measured average capture cross sections below 100 keV. In the region above 100 keV, the d-wave contribution, which is not included in this calculation, may become significant.

Table V Resonance parameters for evaluations of ^{107}Ag and ^{109}Ag average capture cross sections.

Isotope	Data file	S_0x10^4	S_1x10^4	D_0 (eV)	Γ_γ (meV)
^{107}Ag	RCN-2	0.37	3.8	19	140
	BNL-325	0.38	3.8		140
	Present	0.42	(3.8)	22	(140)
^{109}Ag	RCN-2	0.6	3.8	17.5	129
	JENDL-1	0.354	5.26	12.7	130
	BNL-325	0.46	3.8		130
	Present	0.44	(3.8)	21	(130)

Summary

The average capture cross sections for ^{107}Ag and ^{109}Ag from 3.2 to 700 keV have been obtained with an estimated accuracy of 4 to 7 %, which are in good agreement with the recent ORELA results.[12] The present experiments have also provided neutron resonance parameters up to 7 keV both for ^{107}Ag and ^{109}Ag. Our extracted average resonance parameters in combination with the evaluated p-wave strength functions, can describe well the observed average capture cross sections in the energy region below 100 keV.

The authors wish to thank Mr. T. Shoji for his technical assistance and Drs. S. Tanaka and Y. Kikuchi for their critical reading of the mamuscript. They are indebted to Dr. R. L. Macklin for helpful discussions of his results for Ag isotopes. The contribution of the operation crew of linac is acknowledged.

References

1. N. DayDay (ed.), WRENDA 81/82, World Request List for Nuclear Data, IAEA- INDC(SEC)-78/URSF (1981)
2. L.W. Weston et al., Ann. Phys. 10, 477 (1960)
3. V.N. Kononov et al., First IAEA Conf. on Nuclear Data on Reactors, Paris, 1966, Vol. 1, p.496
4. H. Gruppelar, Tables of RCN-2 Fission Product Nuclides, ECN-33 (1977)
5. S. Iijima et al., J. Nucl. Sci. Tech. 14, 161 (1977), Y. Kikuchi et al., Neutron Cross Sections of 28 Fission Product Nuclides adopted in JENDL-1, (1981)
6. R.E. Schenter and T.R. England, ENDF/B-V Fission Product Cross Section Evaluations in Proc. Specialists' Meeting on Neutron Cross Sections of Fission Product Nuclei, Bologna, 1979, NEANDC(E) 209 "L" p. 253 (1980)
7. J.B. Garg, J. Rainwater and W.W. Havens,Jr., Phys. Rev. 137, B547 (1965)
8. G.V. Muradian and Yu.V. Adamchuk, First IAEA Conf. on Nuclear Data for Reactors, Paris, 1966, Vol. 1, p. 79
9. R.E. Chrien, Phys. Rev. 141, 1129 (1966)
10. N.J. Pattenden and J.E. Jolly, AERE-PR/NP16 (1969)
11. S. de Barros et al., Nucl. Phys. A131, 305 (1969)
12. R.L. Macklin, Private communication (1982)
13. M. Mizumoto et al., J. Nucl. Sci. Tech. 16, 711 (1979)
14. J.D. Smith, ORNL-TM 7221 (1980)
15. M. Sugimoto et al., to be published
16. G. de Saussure,D.K. Olsen and R.B. Berez ORNL-TM 6286 (1978)
17. S.F. Mughabahb, M. Divadeenam and N.E. Holden Neutron Cross Sections, Vol. 1A, Academic Press, New York (1981)
18. R.L. Macklin, N.H. Lazer and W.S. Lyon, Phys. Rev. 107, 504 (1957)
19. A.K. Chaubey and M.L. Sehgal, Phys. Rev. 152, 1055 (1966)
20. D. Kompe, Nucl. Phys. A133, 513 (1969)
21. H.S. Camarda, Phys. Rev. C9, 28 (1974)

CAPTURE CROSS SECTION, ENERGY DEPENDENCE OF TOTAL CROSS SECTION OF ^{152m}Eu ISOMER WITH $T_{1/2}$=9.3 h FOR THERMAL NEUTRONS

Pshenichniy V.A., Vertebnyi V.P., Gritzay E.A.

Institute for Nuclear Research
Academy of Sciences Ukr.S.S.R.,
252028 Kiev, U.S.S.R.

The capture cross section and energy dependence of total cross section of ^{152m}Eu isomer with $T_{1/2}$=9.3 h for thermal neutrons are given. For neutrons with E_n=0.0253 eV σ_{tot}=(68000±14000) barn. The measurements have been carried out at WWR-M-reactor of Institute for Nuclear Research in Kiev.

/ ^{152m}Eu, total cross section, capture cross section, decay constant, time-of-flight method, activation method, ^{151}Eu, isomeric ratio /

The study of interactions of neutrons with isomeric nuclei (isomeric pair) is interesting for enchacing our understanding of level density of compound nuclei with different quantum characteristics. At present there are five isomeric pair for which thermal cross sections have been measured both for metastable and ground states. These are ^{58}Co, ^{60}Co, ^{104}Rh, ^{148}Pm /1/, ^{165}Dy /2/. For all pairs with an exception of the last pair, the value of capture cross section proved to be larger for metastable state than for ground one. Therefore it is interesting to find out this trend for ^{152}Eu isomeric pair.

At WWR-M reactor of Kiev Institute for Nuclear Research a few measurements of thermal total cross section values have been carried out for ^{152m}Eu isomer / 3,4 /. For this aim the changes in transmission of the specimen containing ^{152m}Eu were observed during several days by the time-of-flight method. These changes are small as a result of low concentration of isomeric nuclei; there are also much ^{151}Eu that were irradiated by thermal neutrons in reactor to get europium-152 isomer. As a result of three runs of measurements the energy dependence of total cross section of ^{152m}Eu has been obtained in the energy range (0.013÷4) eV. The time resolution is about 4μs/m. This dependence is presented in Fig.1, where the straight line shows the 1/V-law for the total cross section versus neutron energy (upper scale): σ_{tot}=(68000±14000) barn at E_n=0.0253 eV.

Bearing in mind the small observed effect that was prescribed to ^{152m}Eu, we have measured the thermal capture cross section for the europium-152 isomer by activation method with the Ge(Li)-spectrometer. The yield of ^{152m}Eu nuclei relative to ^{152g}Eu was determined at different fluencies of thermal neutrons. It depends on decay constants of both kinds of ^{152}Eu nuclei as well as on the capture cross section of europium-152 isomer.

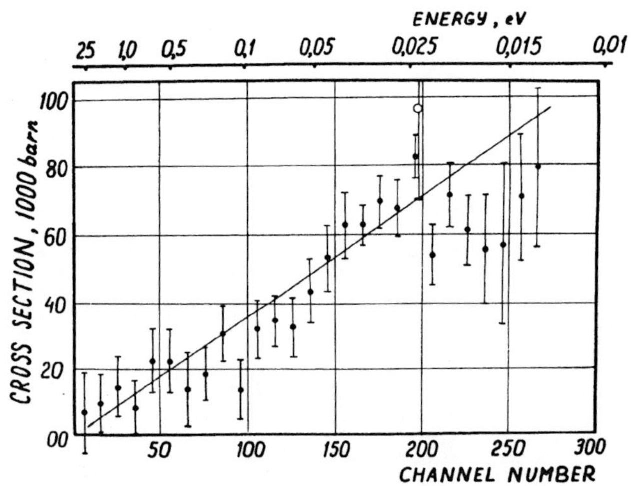

Fig.1. The energy dependence of total cross section of ^{152m}Eu ($T_{1/2}$=9.3 h) for thermal neutrons (filled circles-for total cross section by time-of-flight method; the empty circle-for capture cross section by activation method).

For neutron density flux less than $5*10^{13}cm^{-2}s^{-1}$ the isomeric yield ratio is

$$\frac{n_m}{n_g}=\gamma\cdot\frac{1}{g}\cdot\frac{1-e^{-x}}{x}\quad, \qquad (1)$$

K. H. Böckhoff (ed.), Nuclear Data for Science and Technology, 230–232.

where $x = \lambda_m t + (\sigma_m - \sigma_o)\Phi t$,

$g = 1 - \frac{1}{2}[\lambda_g t + (\sigma_g - \sigma_o)\Phi t]$,

n_m and n_g, λ_m and λ_g, σ_m and σ_g are the yields, decay constants, thermal capture cross section of ^{152m}Eu and ^{152g}Eu, respectively;

σ_o is the thermal total cross section of ^{151}Eu;

Φ is the neutron flux density;

t is the irradiation time.

The value γ has been obtained by measuring the isomeric yield ratio at the low neutron flux Φ, that is $\Phi t(\sigma_m - \sigma_o) \ll \lambda_m t$.

For thermal neutron flux monitoring the polyethylene samples have been taken (diameter 5mm, thickness up to 3mm) with a few milligram of cobalt-59. The cobalt weight uncertainty is 3%. Similar specimens with europium-151 (about $0.3*10^{-6}$gram) were used for the isomeric yield ratio measurements. After the irradiating of the europium samples in reactor the measurements of the gamma-ray spectra of the isomeric europium-152 pair were performed with the Ge(Li) detector. Then after one or two weeks the measurements of ^{152g}Eu were carried out with the same detector and at the same conditions. The gamma-ray transitions of ^{152m,g}Eu decay were observed at energies 121.78, 344.30, 841.68 and 867.37, 963.36 and 964.01 keV. Thus we have performed four runs for the isomeric yield ratio on each sample. Five samples were irradiated in thermal neutron flux and three ones done in the epicadmium neutron spectrum.

Formula 1 allows to obtain the value $x - \lambda_m t = (\sigma_m - \sigma_o)\Phi t$ that for the each sample is presented at Fig.2 (A - the thermal neutron flux, B - the epicadmium one). The slope of line 1 gives the ^{152m}Eu capture cross section to by (97000±28000) barn. The empty circle at Fig.1 marks this value. The both above mentioned cross sections are in agreement within the limits of estimated accuracy. This agreement proves that the obtained energy dependence of the total cross section is true and its value for E_n=0.0253 eV is equal to (68000±14000) barn. Besides the results of the measurements of the samples irradiated by the epicadmium neutrons allow to estimate the upper limit of the ^{152m}Eu resonance integral: $I <$ 100000 barn (Fig.2B). The slope of line 2 on Fig.2A indicates the upper limit (10000 barn) of the possible systimatic error that follows from the data obtained from three epicadmium samples (the empty circles at Fig.2A and the filled ones at Fig.2B).

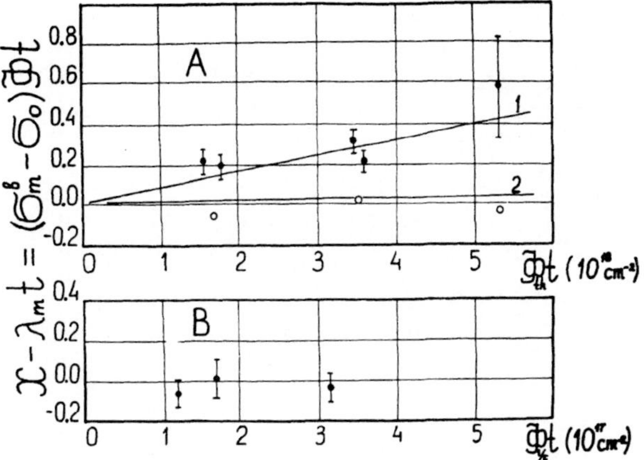

Fig.2. The experimental dependence of value $x - \lambda_m t$ related with the isomeric yield is presented versus the neutron fluency (A - thermal neutrons, B - epicadmium neutrons).

So we have obtained the sixth isomeric pair for which the thermal cross sections have been measured both for the metastable and ground states. For ^{152g}Eu the total cross section at E_n=0.0253 eV is (13120±600) barn /5/. The observed trend /2/ is confirmed for the europium-152 pair.

Other values obtained at these experiments are the isomer decay constant λ_m, an isomer ratio r_{th} for ^{151}Eu and the **relation** r_{ep}/r_{th} for the epicadmium neutrons and the thermal ones. λ_m has been obtained from the 62 measurements for 17 samples and is equal to (0.07486±0.00015) h^{-1}. It gives the half-life time of ^{152m}Eu $T_{1/2}$=(9.260± ±0.019) h that is in a good agreement with the value (9.302±0.049) h /6/. The isomeric ratio r_{th} has been obtained by Poormans et. al. /7/ and is equal to 0.35±0.015. Using the absolute intensity of the gamma-ray transitions /6/ we estimated r_{th}=0.40±0.02 and r_{ep}/r_{th}=1.01±0.01 compared with the value 0.97±0.01 /8/.

References

1. S.F. Mughabghab, D.J. Garber, BNL-325 3rd Ed., 1 (1973).
2. T. Sekine, H. Babe, J. Inorg. Nucl. Chem. 43, 1107 (1981).

3. В.А. Пшеничный, А.И. Блановский, П.Н. Ворона, Препринт КИЯИ-79-14 (1979).

4. В.А. Пшеничный, В.П. Вертебный, Е.А. Грицай, В.И. Кришталь, П.Н. Ворона, В.Г. Кривенко, Нейтронная физика. Материалы 5-ой Всесоюзной конференции, Киев-1980, ч.2, 131, Москва (1980).

5. В.П. Вертебный, П.Н. Ворона, А.И. Кальченко, В.Г. Кривенко, Препринт КИЯИ-75-14 (1975).

6. Table of Isotopes, eds. C.M. Lederer, V.S. Shirley. New York (1978).

7. F. Poortmans, J. Girlea, A. Fabry, Nucl. Phys. A172, 489 (1971).

8. B. Keisch, Phys. Rev. 129, 769 (1963).

MESURES PAR ACTIVATION D'ISOTOPES SEPARES DE PRODUITS DE FISSION DANS DES SPECTRES DE REACTEURS A NEUTRONS RAPIDES

L. Martin Deidier, M. Darrouzet

Commissariat à l'Energie Atomique
Département des Réacteurs à Eau

Centre d'Etudes Nucléaires de Cadarache
B.P. N° 1 - 13115 - SAINT PAUL LEZ DURANCE
FRANCE

RESUME

Dans le cadre du programme expérimental mis en oeuvre au C.E.A. pour améliorer la connaissance de la capture des produits de fission dans la gamme de spectre des réacteurs à neutrons rapides, les sections efficaces intégrales de capture d'un certain nombre d'isotopes ont été obtenues par la technique d'activation.
Les mesures ont été réalisées sur l'expérience critique couplé thermique-rapide ERMINE, dans deux réseaux à neutrons rapides de spectre neutronique représentatif des réacteurs PHENIX et SUPER PHENIX.
Douze nuclides, figurant parmi les produits de fission jouant un rôle important dans la capture globale ont été étudiés : ^{98}Mo, ^{100}Mo, ^{102}Ru, ^{104}Ru, ^{108}Pd, ^{139}La, ^{141}Pr, ^{142}Ce, ^{146}Nd, ^{148}Nd, ^{150}Nd et ^{152}Sm.
Les résultats obtenus sont comparés avec les valeurs calculées à l'aide du formulaire CARNAVAL IV, les données de base des produits de fission résultant des travaux d'évaluation effectués en commun par le CNEN et le CEA.

Introduction

Depuis 1972 un important programme expérimental a été mis en oeuvre au CEA afin d'améliorer la connaissance de la capture des produits de fission dans les réacteurs à neutrons rapides [1].

En complément des mesures directes de l'effet global des produits de fission par comparaison d'aiguilles de combustibles vierge et irradié [2], on a procédé également à des mesures sur les principaux isotopes séparés.

Dans ce cadre, la technique d'activation est intéressante pour obtenir des informations sur les sections de capture de nuclides donnant naissance à des isotopes de période courte (comprise entre quelques dizaines de minutes et quelques dizaines de jours) dont l'activité peut facilement être mesurée par spectrométrie γ.

Cette technique ne nécessitant que de faibles irradiations, elle peut être facilement mise en oeuvre sur expérience critique et, de ce fait, a été utilisée dans l'assemblage couplé thermique-rapide ERMINE au cours de deux campagnes expérimentales en 1975 et 1978. Les mesures ont été réalisées dans deux réseaux à neutrons rapides RONA 3, de spectre caractéristique de SUPER PHENIX et ZONA 1, de spectre caractéristique de PHENIX.

Au cours des deux campagnes, 12 produits de fission ont été étudiés, parmi les isotopes jouant un rôle important dans la capture globale. Les résultats obtenus ont été utilisés dans l'ajustement du formulaire CARNAVAL IV pour le calcul des réacteurs à neutrons rapides au CEA.

Réseaux expérimentaux

Les réseaux à neutrons rapides étudiés dans l'assemblage ERMINE (Version couplé thermique-rapide de la pile MINERVE) sont réalisées à l'aide de réglettes de simulation de type MASURCA. Ces réglettes sont de géométrie cylindrique ou carrée, de 12.7 mm de diamètre ou de côté.

RONA 3 est un réseau à combustible oxyde d'uranium et à diluant sodium + acier. L'enrichissement en U^{235} moyen sur la cellule est de 13,5 %.

Le pas de ce réseau est de 26,5 mm et son spectre est intermédiaire entre ceux des 2 zones du coeur de SUPERPHENIX.

ZONA 1 est un réseau à combustible oxyde mixte d'uranium-plutonium et à diluant Sodium + Acier. La teneur moyenne en plutonium est de 20 % et la composition isotopique de ce dernier est représentative d'un combustible de réacteur à eau légère retraité.

Le pas du réseau ZONA 1 est de 53 mm et, par son spectre, il est intermédiaire entre les deux zones du coeur de PHENIX.

Une coupe horizontale des cellules RONA 3 et ZONA 1 est présentée à la figure 1.

Dans la configuration ERMINE, le réseau à neutrons rapides central est entouré d'une zone de transition permettant d'obtenir au centre de la zone de mesure un spectre neutronique se rapprochant au mieux du mode fondamental, tout en limitant le volume de la zone centrale [4].

Principe de la méthode

Le principe recherché dans ce type de mesure par activation est de déterminer directement un rapport de sections efficaces de capture à partir d'un rapport de taux de comptage γ, en éliminant deux paramètres très difficiles à évaluer :

- le flux dans le réacteur pendant l'irradiation,
- l'efficacité absolue de l'appareillage de comptage.

K. H. Böckhoff (ed.), Nuclear Data for Science and Technology, 233–236.

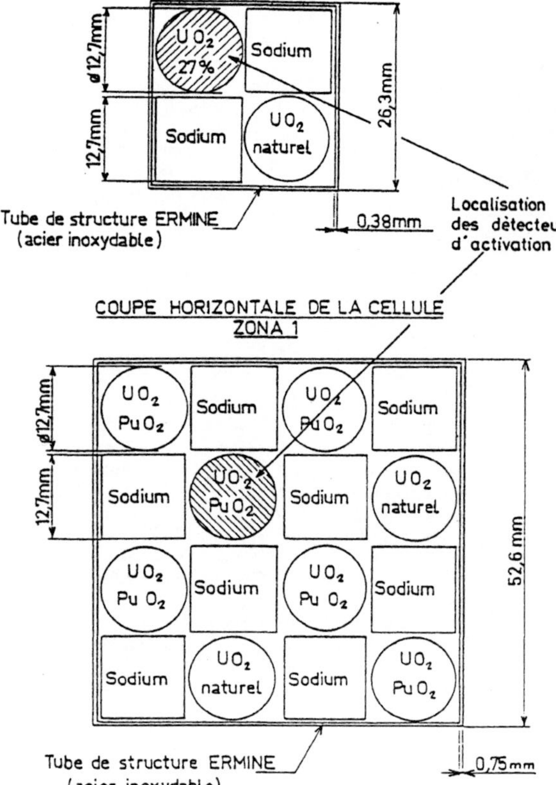

Figure 1 : Coupe horizontale des cellules RONA 3 et ZONA 1

Ainsi, pour un corps A irradié, on peut écrire l'activité du corps A + 1 produit par capture neutronique :

$$\lambda_{A+1} N_{A+1} = N_A \sigma_{cA} \cdot \phi \cdot (1 - e^{-\lambda_{A+1} T})$$

avec N_A et N_{A+1} : Nombre d'atomes du corps irradié et de son descendant par capture neutronique

λ_{A+1} : Constante de décroissance du corps A + 1

σ_{cA} : Section efficace de capture du corps A

ϕ et T : Flux et durée de l'irradiation.

(On néglige la disparition par capture du corps A+1 devant sa disparition par décroissance radioactive, ainsi que la disparition du corps A pendant l'irradiation).

L'indice de capture entre deux corps A et B s'écrit alors :

$$\frac{\sigma_{cA}}{\sigma_{cB}} = \rho_{AB} \frac{C_{A+1}}{C_{B+1}} \frac{N_B}{N_A} \frac{(1 - e^{-\lambda_{B+1} T})}{(1 - e^{-\lambda_{A+1} T})}$$

C représente le taux de comptage γ.

ρ_{AB} représente l'efficacité relative de l'appareillage de comptage pour les rayonnements γ d'énergie différente émis par les corps A+1 et B+1.

Dans notre cas, nous appliquons cette méthode pour déterminer la section de capture des différents produits de fission, normalisée à la section de capture du ^{238}U.

Technique expérimentale

1. Echantillons d'activation

Les isotopes de produits de fission qui ont été étudiés sont : ^{98}Mo et ^{100}Mo, ^{102}Ru et ^{104}Ru, ^{108}Pd, ^{139}La, ^{141}Pr, ^{142}Ce, ^{146}Nd, ^{148}Nd et ^{150}Nd, ^{152}Sm.

Les produits se présentaient sous forme de poudre métallique ou oxyde, d'une pureté supérieure à 90 % pour l'isotope étudié. Les poudres étaient placées dans des boîtiers en aluminium de 12 mm de diamètre, de 3 mm de hauteur, à raison d'environ 50 mg par boîtier (excepté pour le ^{141}Pr où la quantité était d'environ 400 mg).

Pour la mesure du ^{238}U on a utilisé un détecteur d'uranium métallique appauvri à 400 ppm en ^{235}U, de 2/10 mm d'épaisseur, placé dans un boîtier identique à celui utilisé pour les produits de fission.

2. Conditions d'irradiation

Pour l'irradiation, les détecteurs étaient placés dans la cellule centrale du réseau, au plan de flux maximum, et intercalés directement entre des pastilles combustibles d'oxyde d'uranium enrichi à 27 % (RONA 3) ou d'oxyde mixte d'uranium-plutonium (ZONA 1).

La durée d'irradiation était de 2 heures environ, pour un flux de l'ordre de 10^7 n/cm². sec.

3. Installation de comptage

Les mesures de spectrométrie γ ont été réalisées à l'aide d'un détecteur Ge-Li ayant une résolution d'environ 2 KeV.

La durée des comptages était déterminée de façon à obtenir une précision statistique meilleure que ± 2 % sur l'aire du pic γ principal analysé (correction de bruit de fond effectuée).

Pour chaque échantillon une dizaine de comptages successifs ont été effectués.

Nous présentons sur la figure 2 un exemple (^{105}Ru) d'un spectre γ enregistré.

L'efficacité relative de l'installation de comptage en fonction de l'énergie a été obtenue à l'aide d'une multisource ^{133}Ba + ^{152}Eu et qui permettait de couvrir la gamme 80 KeV - 1400 KeV.

Méthode d'interprétation

1. Coefficients correctifs appliqués aux taux de comptage mesurés

Différentes corrections doivent être appliquées aux taux de comptage mesurés avant de procéder à la détermination des indices de capture.

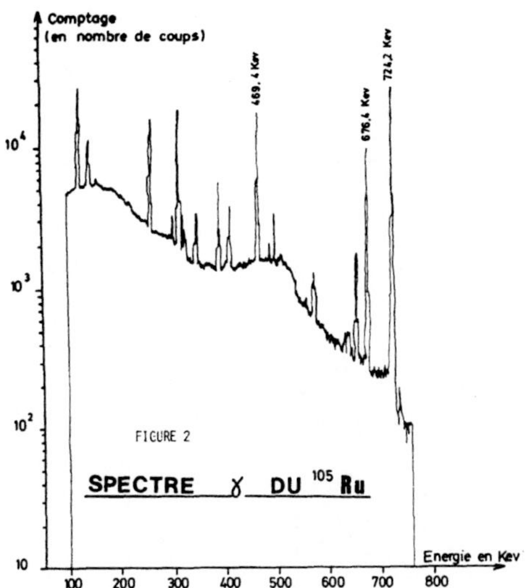

FIGURE 2

SPECTRE γ DU ^{105}Ru

a) Correction d'efficacité de l'installation de comptage

Cette correction, rendue nécessaire par la variation de réponse du détecteur Ge-Li suivant l'énergie du pic γ analysé, est déduite des étalonnages réalisés à l'aide de la multisource ^{133}Ba + ^{152}Eu.

La précision obtenue sur l'efficacité relative par rapport au rayonnement γ de 74 KeV du ^{239}U est comprise entre 2 et 3 % sur toute la gamme d'énergie examinée (80 KeV - 1400 KeV).

b) Correction d'absorption du boîtier

Elle est due à la variation en fonction de l'énergie de l'absorption du rayonnement γ par le couvercle des boîtiers renfermant les produits étudiés.

Cette correction est déterminée expérimentalement par l'intermédiaire de la multisource ^{133}Ba + ^{152}Eu placée successivement dans des boîtiers avec et sans couvercle. Sa valeur est faible (quelques %) et la précision sur sa détermination est estimée à ± 0,5 %.

c) Correction d'auto-absorption

Cette correction prend en compte l'absorption par le corps lui-même du rayonnement γ émis.

Pour les poudres des produits de fission, l'auto-absorption est faible et a été estimée par une formulation d'atténuation en ligne droite en utilisant les coefficients d'atténuation issus de la référence [5].

Dans le cas du détecteur d'uranium métallique l'auto-absorption est beaucoup plus importante et a été calculée à l'aide de code de Monte-Carlo MERCURE IV [6].

La précision sur cette correction est d'environ ± 0,5 % pour les poudres et de ± 3 % dans le cas du détecteur métallique.

Nous donnons à titre d'exemple dans le tableau I les valeurs des taux de comptage mesurés et des différentes corrections pour le cas du ^{104}Ru et du ^{238}U.

Le passage à la valeur finale de l'activité puis à l'indice de capture est réalisé en utilisant les intensités des raies γ analysées, les constantes de décroissance et les facteurs de branchement dans le cas de chaînes radioactives plus complexes. (^{98}Mo, ^{100}Mo, ^{108}Pd, ^{150}Nd) de la référence [7].

2. Interprétation des indices de capture mesurés

Un certain nombre de corrections sont effectuées sur les indices de capture mesurés afin de les ramener en mode fondamental et de prendre en compte les différentes perturbations occasionnées par la présence des détecteurs dans la cellule. Ces corrections sont classiques pour les mesures par détecteurs effectués à ERMINE et ont été détaillées à de nombreuses reprises en particulier dans [8].

a) Correction spatiale afin de tenir compte de l'écart existant au centre de la zone de mesure avec le spectre en mode fondamental.

b) Correction de dilution tenant compte de la dilution différente de l'isotope mesuré entre le détecteur et le milieu.

c) Correction de perturbation de spectre et de flux occasionné par la présence du détecteur.

d) Correction d'hétérogénéité pour passer au combustible moyen de la cellule.

Toutes ces corrections sont faibles comme on peut le constater dans l'exemple présenté dans le tableau II (Indice capture ^{104}Ru/capture ^{238}U).

Résultats et comparaison expérience-calcul

Les résultats obtenus au cours des deux campagnes expérimentales dans les réseaux ZONA 1 et RONA 3 sont donnés dans le tableau III. Les précisions obtenues sont comprises dans tous les cas entre ± 4 et ± 7 % ce qui met en valeur l'interêt de cette méthode expérimentale.

Les indicdes de capture mesurés sont comparés aux valeurs obtenues avec le formulaire CARNAVAL IV [3] utilisé au C.E.A. pour le calcul des réacteurs à neutrons rapides, les données sur les produits de fission résultant de l'évolution commune CNEN/CEA [9].

On constate une très bonne cohérence entre les résultats obtenus sur un même isotope lors des deux campagnes de 1975 et 1978 et également une bonne cohérence entre les écarts expérience-calcul obtenus dans les réseaux RONA 3 et ZONA 1 pour la plupart des isotopes (exceptés pour ^{108}Pd et ^{150}Nd).

Les principales tendances montrent :
- une forte sous-estimation par le calcul de la capture du ^{142}Ce du ^{152}Sm dans les 2 réseaux et du ^{108}Pd dans le réseau RONA 3.
- une surestimation importante par le calcul du ^{139}La, du ^{146}Nd dans les 2 réseaux et du ^{150}Nd dans ZONA 1.

Dans tous les autres cas, compte tenu des barres d'erreurs expérimentales, les désaccords expérience-calcul sont au maximum de quelques pour-cents.

TABLEAU I : FACTEURS CORRECTIFS APPLIQUES AU TAUX DE COMPTAGE DES DETECTEURS ^{238}U et ^{104} Ru IRRADIES DANS LE RESEAU RONA 3

	Energie du pic γ analysé	Correction d'efficac. du cristal	Correction d'absorpt. du boîtier	Correction d'auto.-abs. du détect.	Intensité de la raie γ analysée (%)	Comptages corrigés par sec et mg	Valeur moyenne	Rapport des taux de comptage
^{238}U nº1	74.67 KeV	1. (valeur de normalisat.)	1. (valeur de normalisat.)	2.092±0.063	74.4 ± 0.8	9.474±0.384	9.513 ± 0.380	0.335 ± 0.015
^{238}U nº2				2.114±0.063		9.568±0.388		
^{238}U nº3				2.097±0.063		9.522±0.386		
^{238}U nº4				2.017±0.061		9.489±0.384		
^{104}Ru nº1	469.4 KeV	1.957±0.039	0.962±0.005	1.002±0.002	17.54 ± 0.2	3.194±0.093	3.192 ± 0.067	
	676.4 KeV	2.927±0.058	0.957±0.005	1.001±0.002	15.65 ± 0.4	3.213±0.114		
	724.2 KeV	3.156±0.063	0.956±0.005	1.001±0.002	47.28 ± 0.4	3.181±0.072		
^{104}Ru nº2	469.4 KeV	1.957±0.039	0.962±0.005	1.002±0.002	17.54 ± 0.2	3.160±0.081		
	676.4 KeV	2.927±0.058	0.957±0.005	1.001±0.002	15.65 ± 0.4	3.224±0.108		
	724.2 KeV	3.156±0.063	0.956±0.005	1.001±0.002	47.28 ± 0.4	3.205±0.073		

TABLEAU II : EXEMPLE D'INTERPRETATION D'UN INDICE DE CAPTURE DANS LE RESEAU RONA 3 ($^{104}Ru/^{238}U$)

Correction spatiale	0.994
Correction de perturbation de spectre	0.988
Correction de dilution pour ^{128}Ru	1.036
Correction de dilution pour ^{104}Ra	< 1.003
Correction d'hétérogénéité	1.009

TABLEAU III : COMPARAISON EXPERIENCE-CALCUL (CARNAVAL IV) POUR LES INDICES DE CAPTURE (E-C/C EN %)

ISOTOPES	ANNEE DE MESURE	RESEAU RONA 3	RESEAU ZONA 1
^{98}Mo	1975	+ 4.3 ± 5.0	+ 2.9 ± 4.9
^{100}Mo	1975	-	- 5.9 ± 5.0
^{102}Ru	1975 1978	- 6.5 ± 4.4 - 6.2 ± 4.3	-
^{104}Ru	1975 1978	- 3.9 ± 4.2 - 5.0 ± 4.2	- 8.0 ± 4.3 - 6.8 ± 4.1
^{108}Pd	1978	+16.2 ± 7.3	+ 0.6 ± 5.2
^{139}La	1978	-10.6 ± 4.0	-21.7 ± 3.6
^{141}Pr	1975	+ 1.5 ± 5.2	-
^{142}Ce	1978	+44.3 ± 6.5	+46.9 ± 7.0
^{146}Nd	1978	-15.6 ± 4.1	-12.4 ± 4.2
^{148}Nd	1978	+ 7.8 ± 5.6	+ 7.7 ± 5.2
^{150}Nd	1978	+ 9.0 ± 5.5	-19.6 ± 4.3
^{152}Sm	1978	+44.4 ± 7.7	+ 96 ± 10

Conclusion

Dans le cadre du programme français sur la mesure de la capture des produits de fission, les résultats obtenus par la méthode d'activation se sont révélés très intéressants. Cette technique qui permet la mesure de la section de capture de nucleides donnant naissance à des isotopes de période courte est tout à fait complémentaire aux mesures par irradiation mises en oeuvre en particulier en France sur le réacteur PHENIX (expérience PROFIL {10}).

C'est ainsi qu'à l'aide de ces 2 méthodes et de l'a méthode d'oscillations, les sections de capture de la quasi totalité des 35 principaux produits intervenant dans la capture globale ont pu être mesurées {2}.

D'un point de vue expérimental l'intérêt de cette méthode est aussi évident car elle ne met en jeu que de faibles quantités de produits (quelques dizaines de mg) et ne nécessite que de faibles irradiations. Elle peut donc être mis en oeuvre sans difficulté sur expérience critique et permet d'obtenir des informations sur les sections de capture avec de bonnes précisions (± 5 % en moyenne).

Références

1. L. MARTIN DEIDIER and al., Harwell, Proc. Int. Conf. on Neutron Physics and Nuclear Data, 1016 (178).
2. L. MARTIN DEIDIER, CEA-R-5023, Commissariat à l'Energie Atomique (1979)
3. J.P. CHAUDAT ans al., "Data adjustments for fast reactor design", San Fransisco, ANS Winter Meeting (1977).
4. J. BOUCHARD and al., CEA-R-4246, Commissariat à l'Energie Atomique (1971).
5. T. ROCKWELL, "Reactor sheelding desing manual", D. Van Nostram Compagny.
6. C. DEVILLERS, CEA-R-3264, Commissariat à l'Energie Atomique (1967).
7. J. BLACHOT and al., CEA-N-1822, Commissariat à l'Energie Atomique (1975)
8. M. DARROUZET, L. MARTIN DEIDIER, CEA-N-2063, Commissariat à l'Energie Atomique (1978)
9. M. DARROUZET and al., Harwell, Proc. Int. Conf. on Neutron Physics and Nuclear Data, 1146 (1978).
10. A. GIACOMETTI and al., Aix-en-Provence, Proc. Int. Conf. on Fast Reactor Physics, I, 521, (1979).

MEASUREMENTS OF FISSION-PRODUCT DECAY HEAT FOR FAST REACTORS

M. Akiyama and S. An

Nuclear Engineering Research Laboratory
Faculty of Engineering, the University of Tokyo
Tokai-mura, Ibaraki-ken, Japan

Time dependent beta- and gamma-ray energy release rates from the decay of fission products (i.e., decay heat) have been measured following fast neutron-fissions of ^{233}U, ^{235}U, ^{239}Pu, ^{238}U, natural uranium, and ^{232}Th for cooling times between 10 and $\sim 10^5$ seconds using a radiation spectrometory method. Experimental results were compared with three summation calculations using alternate fission-product nuclear data libraries, which are ENDF/B-IV, TASAKA and JNDC libraries. As the results of the comparisons, it appeared that the values calculated using the JNDC fission-product nuclear data library were in well agreement with the present data. The present results were also compared with other experimental data.

[^{233}U, ^{235}U, ^{239}Pu, ^{238}U, natural uranium, $^{232}Th(n,f)$, decay heat, gamma energy release rate, beta energy release rate, NaI(Tℓ) detector, beta-ray spectrometer, summation calculation, fission-product nuclear data library]

Introduction

There were about four experimental data of the decay heat for fast reactors at that time when a new experimental program was made to measure the decay heat following the fast-neutron fissions of ^{235}U, ^{238}U, ^{239}Pu, etc. before six years at the University of Tokyo. Johnston measured the decay heat following ^{239}Pu fast fissions in the Dounreay Fast Reactor[1]. The PuO_2 samples were irradiated intermittently over periods of about 80 and 125 days and measurements made for cooling times of 40 - 150 days using a Calvet microcalorimeter. The beta and gamma decay heat were also measured separately by using a uranium gamma-ray absorber. Analytical expressions for the burst functions of beta, gamma and total decay heat were derived from those results. Costa, Rastoin and de Tourreil measured the decay heat following the fast fissions of ^{235}U and ^{239}Pu for decay times of 21 - 86 days using a Calvet microcalorimeter[2]. A composite fuel pin was irradiated in the Rapsodie Fast Reactor over a period of 3 years. The burst functions of the decay heat of ^{235}U and ^{239}Pu were obtained from those results. Fisher and Engle were measured the gamma decay heat following the fast fissions of ^{232}Th, ^{233}U, ^{235}U, ^{238}U and ^{239}Pu for cooling times of 0.2 - 45 seconds[3]. Each sample was irradiated in fission burst in the Godiva II critical assembly. The fission -product gamma-ray energy spectra were observed with a NaI total absorption spectrometer for gamma energies between 0.12 and 6.5 MeV. The spectral data were integrated over energy to obtain the burst functions of the gamma decay heat. Overall uncertainties of 12% were quoted for all the fissile isotopes, except for Pu, of which an uncertainty of 23% was given. Bunney and Sam measured the fission-product gamma-ray energy spectra for cooling times between 15 minutes and 72 hours following the fast fissions of ^{235}U and ^{238}U[4]. The samples were irradiated in a fast neutron flux produced by (γ,n) and (γ,pn) reactions in a water-cooled neutron converter in the beam of the 45 MeV electron linear accelerator. The neutrons had a spectrum that was similar to that of fission process. The gamma-ray spectral data were observed with a 5-in. dia. × 5-in. long NaI detector for gamma energies between 0.06 and 5.1 MeV. The gamma decay heat could be obtained by integrating the spectral data over energy. Uncertainties were about 15%.

In safety analysis of sodium cooled reactors, the knowlege of the residual power after reactor shutdown is important with cooling time ranging from 10 to 10^5 seconds[5]. Two calorimetric measurements by Johnston and Costa et al. observed the decay heat for longer cooling times than that of the requirements. Fisher and Engle and Bunney and Sam measured only the gamma decay heat, and those data of ^{235}U presented large values compared with the decay heat data for the thermal fissions of ^{235}U measured by Peelle et al.[6], and had large experimental uncertainties.

As mentioned above, our knowledge of the decay heat of the fission products for the fast fissions was insufficient at the beginning of the present experiments. After that, the beta decay heat had been measured following the fast fissions of ^{235}U and ^{239}Pu by Murphy et al.[7]. The samples were irradiate in the zero power fast reactor, Zebra, for time periods of 10^5 seconds. Measurements of the beta decay heat were made with a plastic scintillation spectrometer for cooling times of $20 \sim 3 \times 10^7$ seconds. The random errors of the measurements were assigned at 1% for the cooling times between 20 and 10^4 seconds, $1 \sim 2.5\%$ for $10^4 \sim 10^6$ seconds, and $2.5 \sim 5\%$ for $10^6 \sim 3 \times 10^7$ seconds. The systematic uncertainties were estimated at 2.5% for the cooling times $t \leq 10^6$ seconds, and 3.1% for longer times.

Experiments

There were two purposes of the present experiments[8] to observe the decay heat. One was to provide basic experimental data for the decay heat of fission products created during the fast fission process required for safety analysis of sodium cooled fast reactors. Other purpose was to verify the fission -product (FP) nuclear data libraries used in summation calculations of the decay heat by comparing calculated results with the present experimental results and other data. For safety analysis, the important cooling times after reactor shutdown are between 10 and 10^5 seconds. A calorimetric method for measurements of the decay heat has the advantage of giving the total decay heat in a single measurement, but it is not suitable for measurements of short cooling time of the decay heat because its response time is slow. On the other hand, a radiation spectrometry method has a rapid response, and gives more information, that is beta and gamma spectral data to be used for detailed checking summation calculations. Therefore, a radiation spectrometry method was adopted in the present experiments.

The samples of ^{233}U, ^{235}U and ^{239}Pu consisted of about 1.6 mg of the metallic fissile materials electrodeposited on 18 mm dia., 0.1 mm thick titanium foils. The diameter of the electrodeposited area was 10 mm. The ^{233}U sample was 99.44% enriched in mass number 233 isotopes, the ^{235}U sample was 97.78% enriched in mass number 235 isotopes, and the pultonium was 99.113% enriched in mass number 239 isotopes. Each sample was covered with thin myler film (10μm thick), fixed with a polyvinyl chloride ring (inner dia. 14 mm, outer dia. 20 mm and 2 mm thick) and sealed by adhesives. The samples were also packed in a thin polyethylene sack (30μm thick). The samples of ^{238}U, natural uranium (nat.U) and ^{232}Th were metallic foils of which diameter was 12.7 mm and thicknesses were 0.025, 0.15 and 0.10 mm, respectively. Each foil was covered with thin polyethylene film.

The samples were irradiated at the center of the

K. H. Böckhoff (ed.), Nuclear Data for Science and Technology, 237–244.

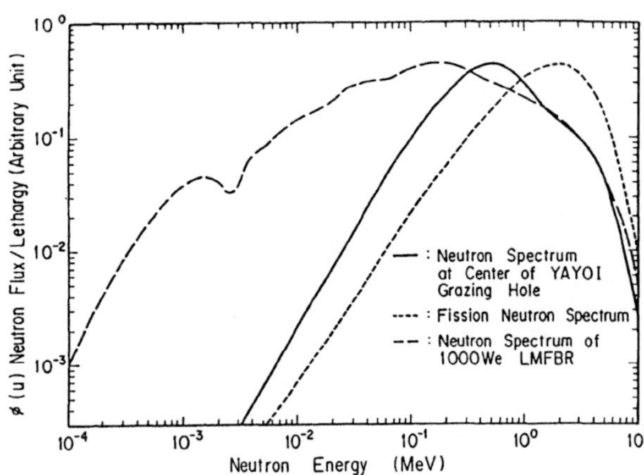

Fig. 1. Neutron energy spectrum at the center of the grazing hole of YAYOI Reactor where the samples were irradiated. Comparison of the spectrum with fission neutron spectrum and neutron energy spectrum of 100MWe sodium cooled fast breeder reactor calculated by Okrent (see Ref. 9)

grazing hole of the fast neutron source reactor YAYOI for time periods of 10 and 100 seconds for the measurements of the gamma decay heat, and 10, 60 and 300 seconds for the measurements of the beta decay heat. The samples were transported pneumatically to the irradiation position and rapidly recovered to the counting area following irradiations. The neutron energy spectrum at the irradiation position is presented in Fig. 1. The figure also shows the fission neutron spectrum and the neutron spectrum[9] of a sodium cooled fast reactor.

The time dependent beta- and gamma-ray pulse height distributions were measured separately at 34 time regions between 11 and 26000 seconds for ^{233}U, ^{235}U and ^{239}Pu samples. The gamma-ray pulse height data were measured for ^{238}U, nat.U and ^{232}Th. Because the fair amount of betas emitted from fission products was absorbed in the sample foils for ^{238}U, nat.U and ^{232}Th, and it was very difficult to correct the self -absorption effects in the measured beta-ray spectra, the beta-decay heat was not measured for ^{238}U, nat.U and ^{232}Th. A gamma-ray detector used was a 76.2 mm dia. × 76.2 mm long NaI(Tℓ) scintillation spectrometer which was contacted with a 30 mm thick polyethylene disk at it's front face to prevent beta -ray from entering into the scintillator. The measured energy region was between 0.1 and 5.0 MeV. The response function of the spectrometer was obtained experimentally for gamma-ray energy between 0.06 and 5.0 MeV using gamma-ray standard sources and sources made by neutron reactions. A beta-ray detector was a well type plastic scintillator which was 50.8 mm dia. × 50.8 mm long and had a cylindrical well of 16mm dia. and 20mm depth at a front face of the scintillator. It was contacted with a transmission type proportional counter which was 150 × 150 × 20 mm^3 and 60 mm dia. windows at the center of both sides. The windows were covered with poly-para-xylene films electrodeposited by gold (total thickness 0.15 mg/cm^2). The counter used a gas mixture of 90% argon and 10% methane. It served to effectively discriminate between betas and gammas since the probability of a gamma interaction producing a pulse was extremely low. The measured energy region was between 0.3 and 8.0 MeV. The responses of the beta-ray spectrometer were observed using monoenergetic electron beams with energies between 0.3 and 2.7 MeV taken from the iron -free beta-ray spectrometer at the Institute for Nuclear Study, the University of Tokyo, and beta-ray standard sources. The response function of the spectrometer was determined in the energy range between 0.2 and 8.0 MeV from the measured responses.

The number of fissions in each irradiated sample was determined from measurements of the gamma activity of ^{97}Zr using a calibrated Ge detector. The measured pulse height data were corrected for backgrounds, and unfolded by the FERDOR code[10] with the use of the response function of the beta- or gamma-ray spectrometer. The beta or gamma energy release data were derived to integrate the unfolded spectra over energy, as follows,

$$E_{tot} = \frac{1}{n_f}\int_{E_{min}}^{E_{max}} E\cdot\phi(E)dE \qquad (1)$$

where, n_f denotes the number of fissions per second. For the beta-ray data, the upper limit, E_{max} = 8 MeV, applied. The lower limit of the measured spectra was restricted to 0.3 MeV by the internal bias level of the detection system. The contribution from betas with energies lower than 0.3 MeV was estimated by extrapolating the unfolded spectra to zero energy. For the gamma-ray data, the upper limit, E_{max} = 5 MeV, and the lower limit, E_{min} = 0.1 MeV, applied. The contribution of low energy photons to the energy release was less than 0.5% and was neglected.

The energy release data in eq.(1) are used to obtain the pulse function f(t) of the decay heat, which is defined as the rate of energy release t seconds following a fission. If the waiting time, T_{wait} is longer than the irradiation period plus the counting time ($T_{irrad}+T_{count}$), and the cooling time, t is measured between the midpoints of irradiation and counting intervals, the pulse function can be estimated from the energy release data divided by the irradiation time, as follows,

$$f[T_{wait}+0.5(T_{irrad}+T_{count})] \simeq E_{tot}(T_{irrad}, T_{wait}, T_{count})/T_{irrad} \qquad (2)$$

If $T_{wait} \lesssim (T_{irrad}+T_{count})$, eq.(2) does not hold true. For this condition, the pulse function is determined from the right side of eq.(2) multiplied by a collection factor η which is obtained from the summation calculations[8].

$$f[T_{wait}+0.5(T_{irrad}+T_{count})] = \eta\cdot E_{tot}(T_{irrad}, T_{wait}, T_{count})/T_{irrad} \qquad (3)$$

$$\eta = \frac{f_{cal}.[T_{wait}+0.5(T_{irrad}+T_{count})}{F_{cal}.(T_{irrad}, T_{wait}+0.5T_{count})/T_{irrad}} \qquad (4)$$

where, f_{cal} is a result at cooling time $[T_{wait}+0.5(T_{irrad}+T_{count})]$ of the summation calculation based on a condition of instantaneous fission

Table I Comparison of the characteristics of three FP decay data libraries

	JNDC	TASAKA	ENDF/B-IV
Total number of nuclides	1172	1170	1114
Number of stable nuclides	140	159	113
Number of unstable nuclides	1031	1011	1001
Number of nuclides with known decay energies	749(87*)	392	207
Number of nuclides with estimated decay energies	283	619	794

* Number of nuclides which have Q_β-values larger than 5MeV, and of which decay energies are replaced by the estimated values

pulse, and F_{cal} is a result at cooling time ($T_{wait}+0.5T_{count}$) of the summation calculation based on a condition of T_{irrad} seconds irradiation at constant rate of one fission per second. The pulse functions were obtained for the beta decay heat of ^{233}U, ^{235}U, and ^{239}Pu, and for the gamma decay heat of ^{233}U, ^{235}U, ^{239}Pu, ^{238}U, nat.U and ^{232}Th by the use of eq.(2) or eq.(3), from the present experiments. The uncertainties (1σ) were about 5% for the beta decay heat. For the gamma decay heat, the uncertainties (1σ) were about 5% for ^{235}U and ^{239}Pu, about 6.5% for ^{233}U, ^{238}U and nat.U and 8% for ^{232}Th.

Comparisons of present data with calculations and other data

The present data were compared with summation calculations and other data. The pulse function, f(t), decrease with t roughly as t^{-1}. Hence for presentation purposes, it has become common practice to illustrate the pulse function as $t \times f(t)$. The pulse functions of the beta and gamma decay heat are presented in figures in this format for comparison with three alternate summation calculations and with other experimental data only for the gamma decay heat. The summation calculations were carried out using the DCHAIN code[11] with three alternate FP nuclear data libraries, JNDC FP nuclear data library[12], TASAKA FP nuclear data library[13] and ENDF/B-IV FP nuclear data library[14]. The characteristics of these libraries are summarized in Table I.

The beta-ray data for ^{233}U are presented in Fig. 2. The results of the summation calculations using JNDC FP nuclear data library agree very well with the present experimental data. The calculated results using TASAKA or ENDF/B-IV FP nuclear data libraries show slightly larger values than that of the experimental data except for the short cooling time. The gamma-ray data for ^{233}U are presented in Fig. 3. The present data agree very well with the summation calculations using JNDC FP nuclear data library for the cooling times between 20 and 300 seconds. For the cooling times between 300 and 2000 seconds, the present data are larger than the results of the summation calculations using any libraries. For the cooling times between 2000 and 5000 seconds, the present data agree with the calculated results using JNDC FP nuclear data library. The present data are between the calculated results using JNDC and TASAKA or ENDF/B-IV FP nuclear data libraries for $t > 5000$ seconds. The data of Fisher and Engle are about 27% larger than the present data.

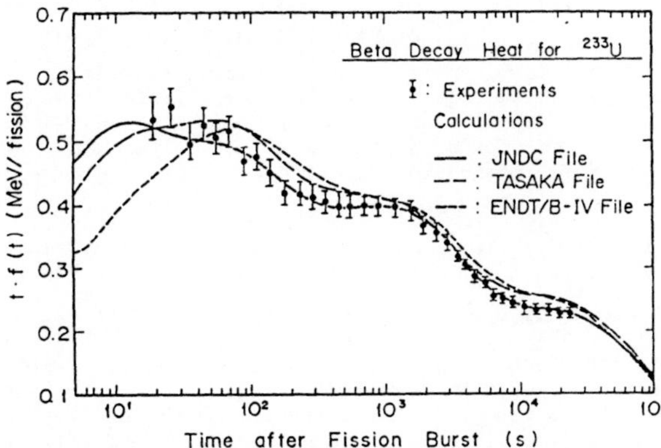

Fig. 2. Beta energy emission rate following an instantaneous pulse of fast-neutron fissions of ^{233}U. The abscissa, t, is the time after a pulse of fissions. The ordinate is a quantity derived by multiplying the pulse function, f(t), by t as described in the text. The solid circles represent the present data. The calculations were carried out using three FP nuclear data libraries by DCHAIN code. The solid line is the result using JNDC FP nuclear data library. The broken line is the result using TASAKA FP nuclear data library. The dotted line is the result using ENDF/B-IV FP nuclear data library.

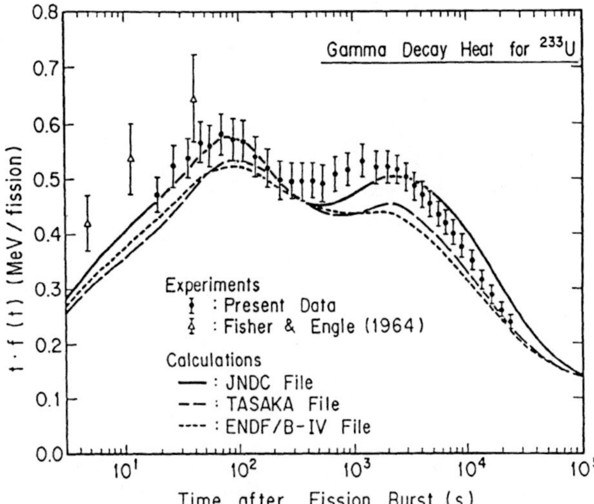

Fig. 3. Gamma energy emission rate following an instantaneous pulse of fast-neutron fissions of ^{233}U The axis are the same as Fig. 2. The solid circles represent the present data, and the open triangles represent the data of Fisher and Engle (see Ref. 3). The calculations are represented in the same manner as Fig. 2.

The beta-ray data for ^{235}U are presented in Fig. 4. The comparison of calculation using JNDC FP nuclear data library with the present data is good for cooling times between 20 and 5×10^3 seconds, but the calculated results are slightly smaller than the present data for $t > 5 \times 10^3$ seconds. Other two calculations are larger than the present data. The gamma-ray data for ^{235}U are shown in Fig. 5. The present data agree well with the calculated results using JNDC FP nuclear data library for the cooling times $t < 2000$ seconds and approach the values calculated using TASAKA FP nuclear data library for $t > 2000$ seconds. The data of Fisher and Engle are about 27% larger than the present data. The data of Bunney and Sam show about 17∿52% large values to be compared with the present data.

The beta-ray data for ^{239}Pu are presented in Fig. 6. The results of the summation calculation using JNDC FP nuclear data library agree very well with the present experimental data as same as the comparisons for ^{233}U. The calculated values using TASAKA or ENDF/B-IV FP nuclear data libraries are larger than the present data. The gamma-ray data are shown in Fig. 7. The results of the summation calculation using JNDC FP nuclear data library agree well with the present data for $t < 300$ seconds, and for $t > 2000$ seconds. For the cooling times between 300 and 2000 seconds, the present data are larger than the results of the summation calculations using any libraries. The data of Fisher and Engle are about 50% larger than the present data.

The gamma-ray data for ^{238}U are presented in Fig. 8. The present data agree with three summation calculations within assigned uncertainties for the cooling times $t < 300$ seconds. The calculation using JNDC FP nuclear data library agree well with the present data for $t > 3000$ seconds. For the cooling times between 300 and 3000 seconds, the present data are larger than the calculated values using any libraries. The data of Fisher and Engle which are not drawn are about 30% larger than the present data. The data of Bunney and Sam show 3∿40% large values to be compared with the present data. The gamma ray data for nat.U shown in Fig. 9 have the same behavior as the data for ^{238}U.

The gamma-ray data for ^{232}Th are shown in Fig. 10. The present data agree well with the calculation using JNDC FP nuclear data library for $t < 200$ seconds. But the experimental data show different behavior from the results of the summation calculations using

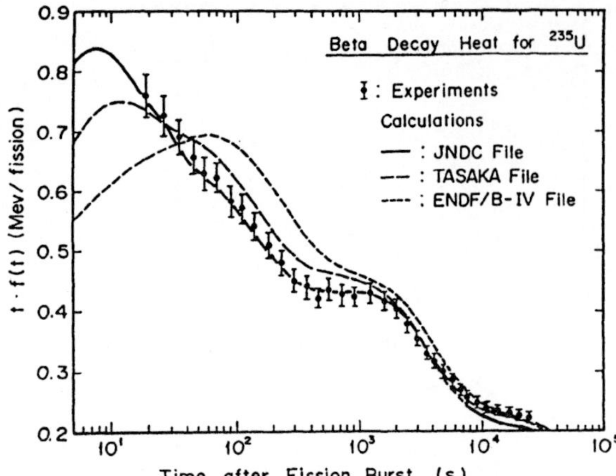

Fig. 4. Beta energy emission rate following an instantaneous pulse of fast-neutron fissions of ^{235}U. The solid circles represent the present data. The calculations are represented in the same manner as Fig. 2.

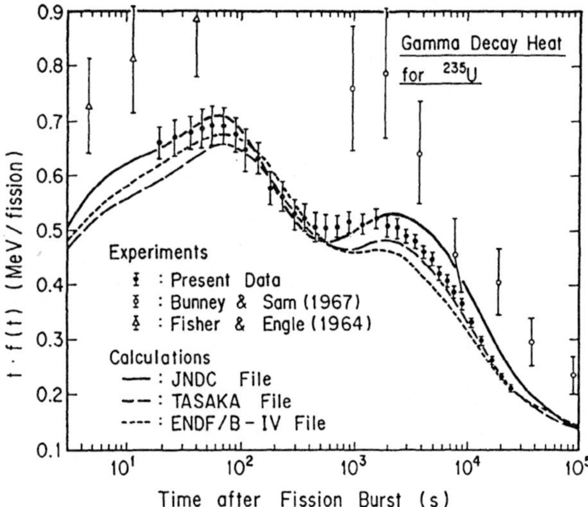

Fig. 5. Gamma energy emission rate following an instantaneous pulse of fast-neutron fissions of ^{235}U. The solid circles represent the present data. The open triangles are the data of Fisher and Engle (see Ref. 3). The open circles are the data of Bunney and Sam (see Ref. 4). The calculations are represented in the same manner as Fig. 2.

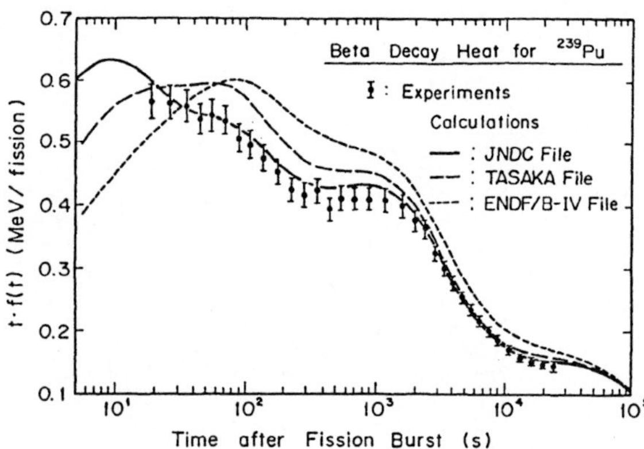

Fig. 6. Beta energy emission rate following an instantaneous pulse of fast-neutron fissions of ^{239}Pu. The solid circles represent the present data. The calculations are represented in the same manner as Fig. 2.

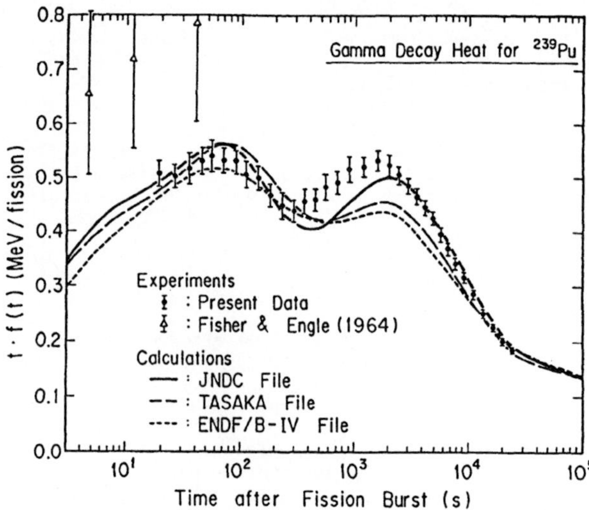

Fig. 7. Gamma energy emission rate following an instantaneous pulse of fast-neutron fissions of ^{239}Pu. The solid circles are the present data, and the open triangles are the data of Fisher and Engle (see Ref. 3). The calculations are represented in the same manner as Fig. 2.

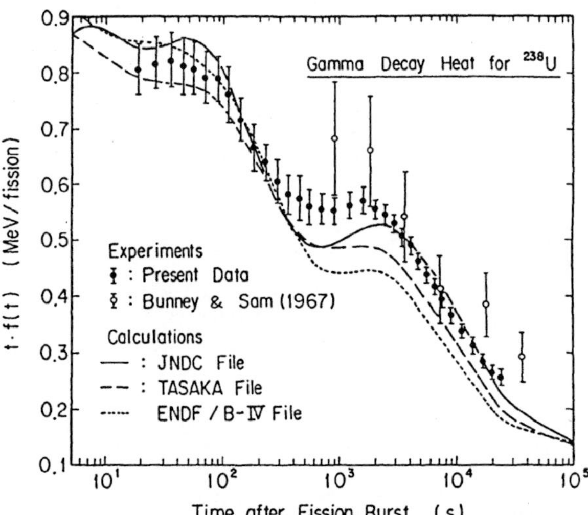

Fig. 8. Gamma energy emission rate following an instantaneous pulse of fast-neutron fissions of ^{238}U. The solid circles are the present data and the open circles are the data of Bunney and Sam (see Ref. 4). The calculations are represented in the same manner as Fig. 2.

any libraries for $t > 200$ seconds. The data of Fisher and Engle are about 30% larger than the present data.

As the results of these comparisons mentioned above, one may observe that the summation calculations for the beta decay heat using JNDC FP nuclear data library agree very well with the present data. For the gamma decay heat, the calculated results using JNDC FP nuclear data library are in better agreement with the present data than that of calculations using other libraries, although there are disagreements for the cooling times between 700 and 2000 seconds. These disagreements are in commmon with all fission types and with three summation calculations, and reasons of the disagreements are not clear. These suggest that further study is needed in this cooling time interval of the gamma decay heat.

The decay heat data following an irradiation period of 10^5 seconds were derived from the present data in order to compare with the beta-ray data of Murphy et al., as follows,

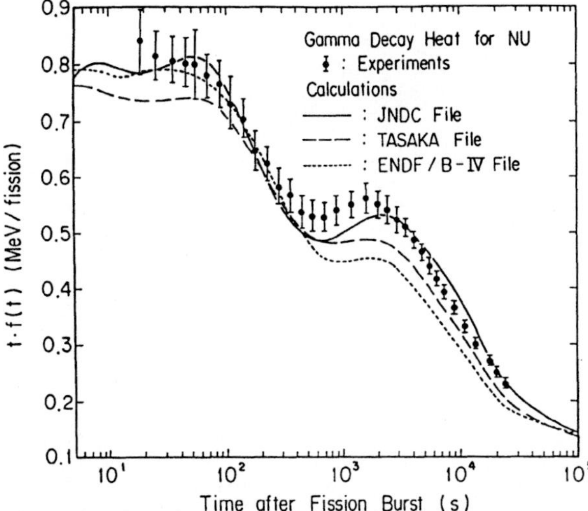

Fig. 9. Gamma energy emission rate following an instantaneous pulse of fast-neutron fissions of nat.U. The solid circles are the present data. The calculations are represented in the same manner as Fig. 2.

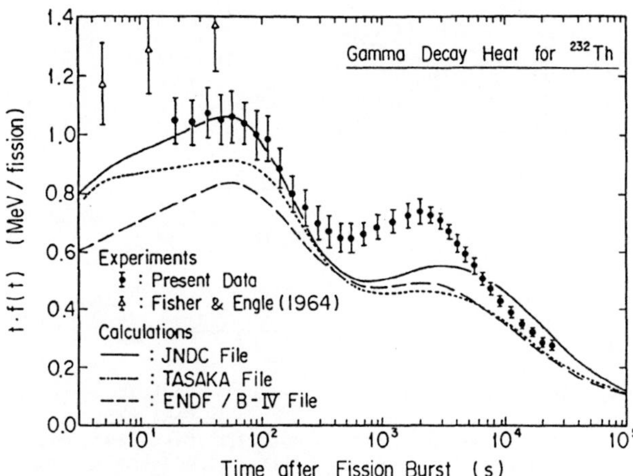

Fig. 10. Gamma energy emission rate following an instantaneous pulse of fast-neutron fissions of ^{232}Th. The solid circles are the present data, and the open triangles are the data of Fisher and Engle (see Ref. 3). The calculations are represented in the same manner as Fig. 2.

$$F(10^5,t) = \int_t^{10^5+t} f(\tau)d\tau \qquad (3)$$

The values for ^{235}U and ^{239}Pu obtained from the present data are shown in Fig. 11 and Fig. 12 compared with the data of Murphy et al., and the result of the summation calculation using JNDC FP nuclear data library. For both ^{235}U and ^{239}Pu, the data of Murphy et al. agree very well with the values obtained from the present data. The summation calculation is slightly smaller than these experimental values for ^{235}U, but agrees within the assigned experimental uncertainties. The uncertainties of the data of Murphy et al. were assigned about 5%. The agreement between the present data and the data of Murphy et al. and slight underestimation of the summation calculation suggest that a cause of the disagreement for $t > 5 \times 10^5$ seconds may be due to the summation calculation in the comparison of the beta-ray data for ^{235}U with summation calculation in Fig. 4.

In summation calculations for ^{235}U, there are small discrepancies between results for thermal-neutron fissions and for fast-neutron fissions for cooling time $t < 10^5$ seconds. But results for 14 MeV neutron fissions are about 30% smaller than results for

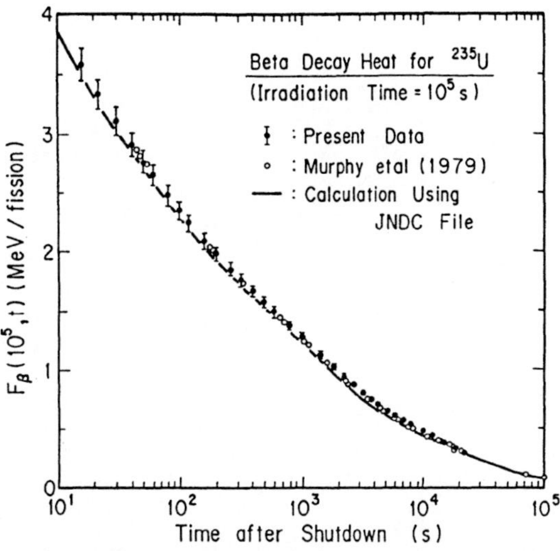

Fig. 11. Beta energy emission rate following a 10^5 sec irraciation of fast-neutron fissions of ^{235}U at constant fission rate without neutron capture in fission products. The solid circles are the results derived from the present data. The open circles represent the data of Murphy et al. (see Ref. 7). The calculation was carried out using JNDC FP nuclear data library.

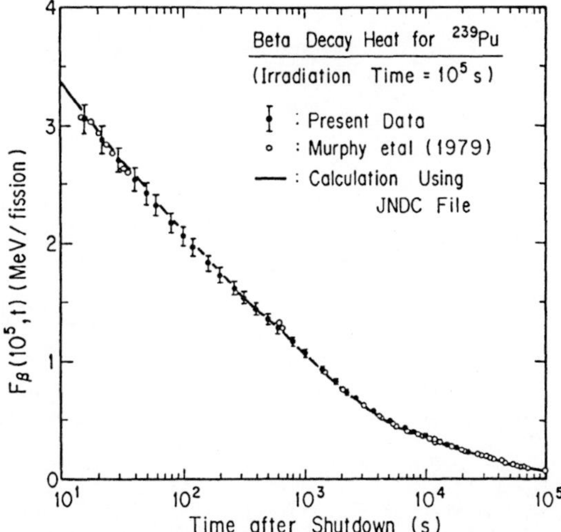

Fig. 12. Beta energy emission rate following a 10^5 sec irradiation of fast-neutron fissions of ^{239}Pu at constant fission rate without neutron capture in fission products. The open circles are the results derived from the present data. The open circles represent the data of Murphy et al. (see Ref. 7). The calculation was carried out using JNDC FP nuclear data library.

thermal- or fast-neutron fissions. The calculated results of the pulse functions are shown in Fig. 13 in the same format as Fig. 2. The comparisons were made between the present data and the data of ORNL[15] in order to confirm neutron energy dependence of beta and gamma energy emission rates. The beta -ray data for ^{235}U are presented in Fig. 14. The present data agree very well with the data of ORNL, and with the results of the summation calculation for fast-neutron fissions using JNDC FP nuclear data library within the assigned experimental uncertainties. The gamma-ray data for ^{235}U are shown in Fig. 15. The present data are in agreement with the data of ORNL for $t < 300$ seconds, but there is a disagreement for $t > 300$ seconds. The reason of the disagreement is not known. The beta- and gamma-ray data for ^{239}Pu are presented in Fig. 16 and Fig. 17, respectively. For both beta- and gamma-ray data, the present data

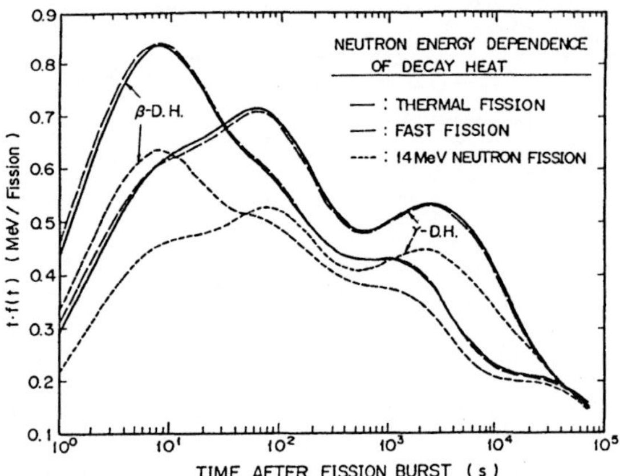

Fig. 13. Neutron energy dependence of beta and gamma energy emission rate following an instantaneous pulse of ^{235}U fissions. The solid lines are the results for thermal-fissions, the broken lines are the results for fast-fissions and the dotted lines are the results for 14MeV neutron fissions. These results were calculated using JNDC FP nuclear data library by DCHAIN code.

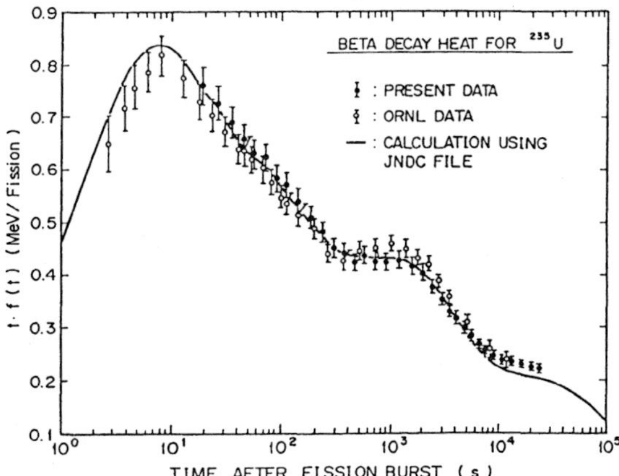

Fig. 14. Beta energy emission rate following an instantaneous pulse of ^{235}U fissions. The solid circles are the present data for fast-neutron fissions. The open circles are the data of Dickens, et al. (see Ref. 15) for thermal-neutron fissions. The calculation was carried out using JNDC FP nuclear data library for fast-neutron fissions of ^{235}U by DCHAIN code.

agree very well with the data of ORNL, and are in agreement with the results of the summation calculations using JNDC FP nuclear data library except for the gamma-ray data for the cooling times between 300 and 2000 seconds. In this cooling time interval, the calculation underestimates the experimental data about 12%.

As the comparisons of the present data with the data of ORNL mentioned above, it is confirmed experimentally that there are small effects for the neutron energy dependence of the decay heat of fission products to be created by fissions for neutrons with energy spectra softer than the fission neutron spectrum. But further study is needed because the behaviors of the decay heat of fission products created by fissions for neutrons with hard energy spectra are very interesting for assessments of FP nuclear data libraries used by summation calculations and relating a safety analysis of fusion-fission hybrid reactors as a future energy system.

The decay heat curves for an infinite period of irradiation were determined from the present pulse functions, as follows,

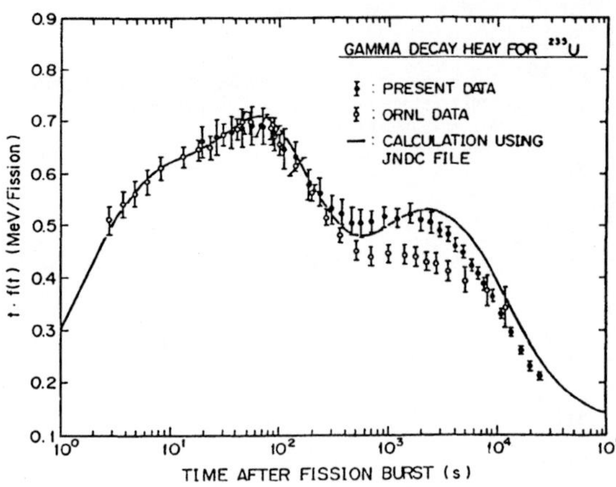

Fig. 15. Gamma energy emission rate following an instantaneous pulse of ^{235}U fissions. The solid circles are the present data for fast-neutron fissions. The open circles are the data of Dickens, et al. (see Ref. 15) for thermal-neutron fissions. The calculation was carried out using JNDC FP nuclear data library for fast-neutron of ^{235}U by DCHAIN code.

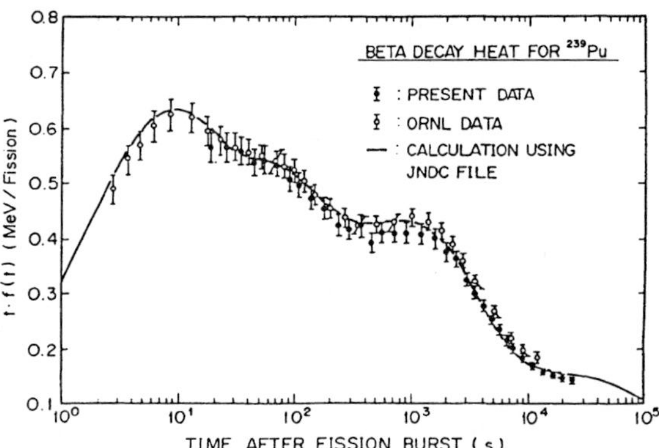

Fig. 16. Beta energy emission rate following an instantaneous pulse of ^{239}Pu fissions. The solid circles are the present data for fast-neutron fissions. The open circles are the data of ORNL (see Ref. 15) for thermal-neutron fissions. The calculation was carried out using JNDC FP nuclear data library for fast-neutron fissions of ^{239}Pu by DCHAIN code.

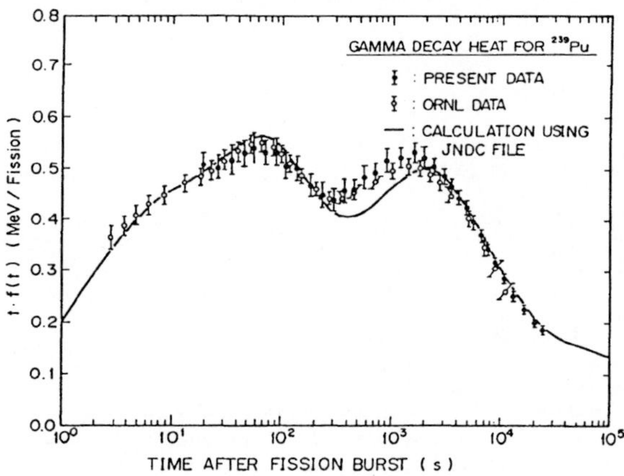

Fig. 17. Gamma energy emission rate following an instantaneous pulse of ^{239}Pu fissions. The solid circles are the present data for fast-neutron fissions. The open circles are the data of ORNL (see Ref. 15) for thermal-neutron fissions. The calculation was carried out using JNDC FP nuclear data library for fast-neutron fissions of ^{239}Pu by DCHAIN code.

$$F(\infty,t) = \int_t^\infty f(\tau)d\tau \qquad (4)$$

In actual practice, 10^{13} seconds is taken to be an infinite period of irradiation. The decay heat curves for ^{235}U and ^{239}Pu derived from the present data are shown in Fig. 18 and Fig. 19 compared with the data of ORNL, the calculated results using JNDC FP nuclear data, and the new ANS standard[16]. The data of ORNL for ^{235}U shown in Fig. 18 are somewhat smaller than the present data, because the gamma-ray data of ORNL in the pulse function has small values for $t > 300$ seconds shown in Fig. 15. But two data agree within the experimental uncertainties. The calculations are in well agreement with the present data, and the new ANS standard has about 2.5% larger values than the present data. The summation calculations agree very well with the present data and the data of ORNL shown in Fig. 19. But the new ANS standard are about 2.5% larger than the present data.

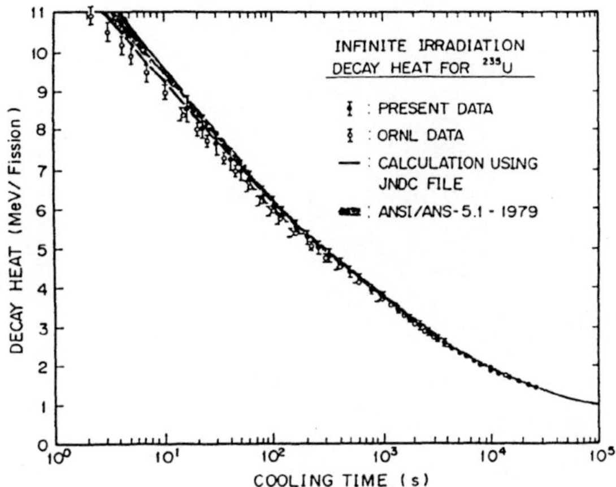

Fig. 18. Fission-product decay heat following an infinite period of irradiation for ^{235}U. The solid circles are the present data, and the open circles are the data of ORNL (see Ref. 15). The summation calculation was carried out using JNDC FP nuclear data library for fast-neutron fissions of ^{235}U by DCHAIN code drawn by the solid line. The dotted line represents the new ANS standard.

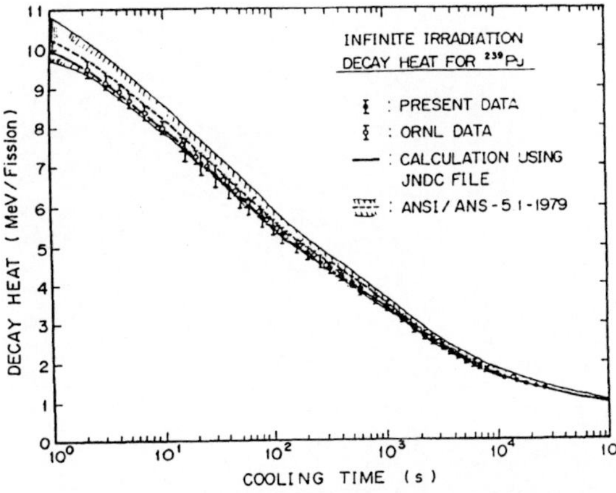

Fig. 19. Fission-product decay heat following an infinite period of irradiation for ^{239}Pu. The solid circles are the present data, and the open circles are the data of ORNL (see Ref. 15). The summation calculation was carried out using JNDC FP nuclear data library for fast-neutron fissions of ^{239}Pu by DCHAIN code drawn by the solid line. The dotted line represents the new ANS standard.

Conclusions

The decay heat data following fast-neutron fissions were measured for ^{233}U, ^{235}U and ^{239}Pu for the cooling times between 16 and 26000 seconds within the required accuracy[5]. The gamma-ray energy release data were also obtained for ^{238}U, nat.U and ^{232}Th. As the comparison of the present data and other data with the summation calculations, it is obvious that the present data agree very well with the data of Murphy, et al. and the data of ORNL for thermal neutron fissions, and the results calculated using JNDC FP nuclear data library are in well agreement with those three experimental data.

Acknowledgment

We wish to express their thanks to K. Sakata, T. Ida and K. Furuta for help with the experiments. We are also indebted to Y. Oka, M. Nakazawa, H. Hashikura and the member of reactor operating and maintenance section of the University of Tokyo for performing the experiments. We are also indebted to K. Tasaka, H. Ihara and the members of the Decay Heat Evaluation Working Group of the Japanese Nuclear Data Committee for their useful comments in performing the study. We also wish to acknowledge the expert assistance of J. Okugawa in typing the manuscript.

The experiment was financially supported by Power Reactor and Nuclear Fuel Development Corporation.

References

1. K. Johnston, J. Nucl. Energy 19, 527 (1965).
2. L. Costa, J. Rastoin, R. de Tourreil, J. Nucl. Energy 26, 431 (1972).
3. P. C. Fisher, L. B. Engle, Phys. Rev. 134, B796 (1964).
4. L. R. Bunney, D. Sam, Nucl. Sci. Eng. 29, 432 (1967).
5. C. Devillers, Proc. of the Second Advisory Group Meeting on Fission Product Nuclear Data, Petten, 1977 IAEA-213 (1978), Vol. I, P.61.
6. R. W. Peelle, R. C. Maienschein, W. Zobel, T. A. Love, "The Spectra of Gamma Rays Associated with Thermal Neutron Fission of ^{235}U" in Pile Neutron Research in Physics, IAEA (1962).
7. M. F. Murphy, W. H. Taylor, D. W. Sweet, M. R. March, "Experiments to Determine the Rate of Beta Energy Release Following Fission of ^{239}Pu and ^{235}U in a Fast Reactor", AEEW-R 1212 (1979).
8. M. Akiyama, K. Furuta, T. Ida, K. Sakata, S. An, J. At. Energy Soc. Japan 24 no.9 (to be published). ibid., 24 no.10 (to be published).
9. D. Okrent, Power Reactor Technol. 7, 107 (1964).
10. H. Kendrick, S. M. Sperling, "An Introduction to the Principles and use of the FERDOR Unfolding Code", GA-9882 (1970).
11. K. Tasaka, "DCHAIN: Code for Analysis of Build-up and Decay of Nuclides", JAERI 1250 (1977).
12. T. Yamamoto, M. Akiyama, Z. Matumoto, R. Nakasima, "JNDC FP Decay Data File", JAERI-M9357 (1981). H. Ihara, Z. Matumoto, K. Tasaka, M. Akiyama, T. Yoshida, R. Nakasima, "JNDC FP Decay and Yield Data", JAERI-M9715 (1981).
13. K. Tasaka, "Nuclear Data Library of Fission Products for Decay Power Calculation", NUREG/CR-0705 (1979).
14. T. R. England, R. E. Schenter, "ENDF/B-IV Fission-Product Files: Summary of Mojor Nuclide Data", LA-6116-MS (1975).

15. J. K. Dickens, J. F. Emery, T. A. Love, J. W. McConnell, K. J. Northcutt, R. W. Peelle, H. Weaver, "Fission-Product Energy Release for Times Following Thermal-Neutron Fission of ^{235}U Between 2 and 14000 Seconds", ORNL/NUREG-14 (1977).
J. K. Dickens, J. F. Emery, T. A. Love, J. W. McConnell, K. J. Northcutt, P. W. Peelle, H. Weaver, "Fission-Product Energy Release for Times Following Thermal-Neutron Fission of ^{239}Pu Between 2 and 14000 Seconds", ORNL/NUREG-34 (1978).
J. K. Dickens, T. A. Love, J. W. McConnell, R. W. Peelle, Nucl. Sci. Eng. 74, 106 (1980).
J. K. Dickens, T. A. Love, J. W. McConnell, R. W. Peelle, ibid., 78, 126 (1981).

16. "American National Standard for Decay Heat Power in Light Water Reactors", ANSI/ANS-5.1-1979, American Nuclear Society (1979).

A BRIEF SURVEY OF EXPERIMENTAL AND THEORETICAL DATA ON FISSION PRODUCT DECAY HEAT FROM U235 AND PU239

M. F. James

United Kingdom Atomic Energy Authority
Atomic Energy Establishment, Winfrith
Dorchester, Dorset, U.K.

A brief survey is made of calculations and measurements of decay heat for U235 and Pu239. Systematic differences exist between results from different laboratories, and between them and summation calculations. If the ratio of Pu239 to U235 decay heat is considered, most of these differences are removed, but there remains a discrepancy of about 8% between on the one hand results from two laboratories, and, on the other, results from one laboratory and summation calculations. Uncertainties large enough to allow for these discrepancies should be given to decay heat recommendations.

Introduction.

Decay heat from fission products has been of great importance since the early days of reactor design, but in the last few years work on this topic has burgeoned throughout the world: an excellent review has been published by Tobias (Ref. 1). There are three main reasons for this recent increase in effort. First, several very careful measurements of total decay heat or of its beta and gamma components have been made in a number of different laboratories. Secondly, computer power has increased sufficiently for rapid summation calculations to be feasible; and thirdly, better data for the libraries used by these summation calculations are available from measurements of half-lives, fission yields, and decay schemes of a large fraction of all fission products.

However, it is the aim of this paper to emphasise that there are still unresolved discrepancies between calculation and some measurements, and between different measurements, and to suggest that the recommended uncertainties in predicted decay heat should reflect these discrepancies.

Comparison of Measurements and Calculations.

Decay heat measurements have used a variety of irradiation times (I) and cooling times (t), making immediate intercomparisons difficult. Several techniques have been used to aid comparison, and we discuss each in turn.

Intercomparison using C/E Values

Each measurement can be compared with the results of a summation calculation for the appropriate I and t, and then values of the ratio C/E of calculation to experiment can be compared.

Figures 1 - 4 show C/E ratios for several experiments, in each case for both U235 and Pu239 total decay power. The calculations were made with the summation programmes FISPIN (Ref. 2) or FISP (Ref. 3), using the UK fission product decay data library UKFPDD2 (Ref. 4) and fission product yields Crouch 3I (Ref.5, 6).

Figs 1-4 (opposite) Values of Calculation/Experiment→ (C/E) for :(1) ORNL, (2) LASL, and (3 and 4) IRT Measurements

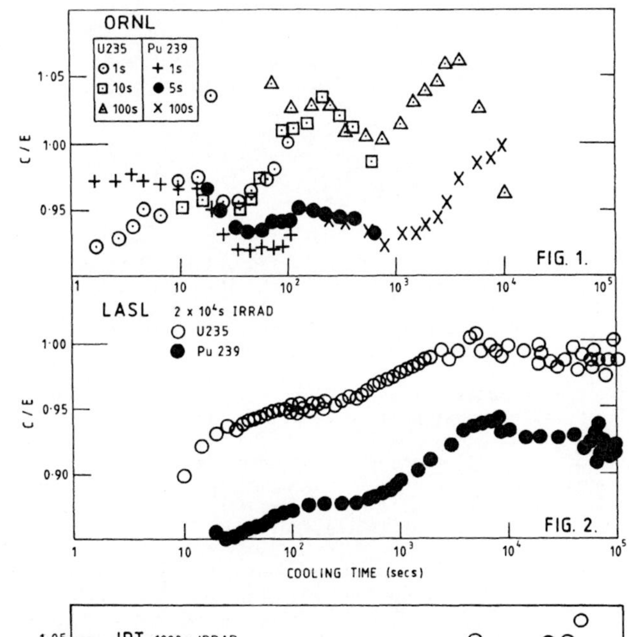

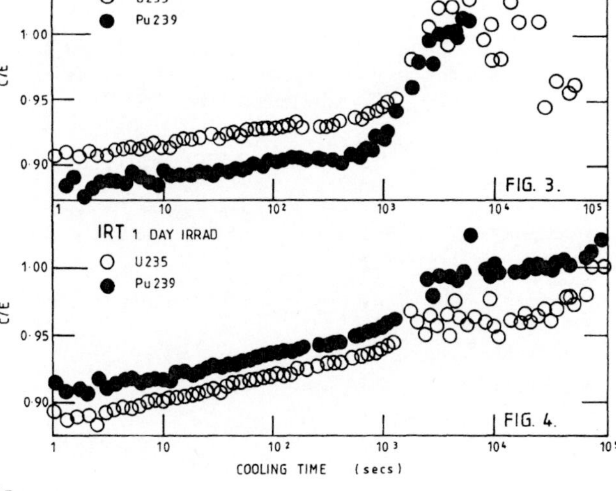

K. H. Böckhoff (ed.), Nuclear Data for Science and Technology, 245–248.

Fig. 1, for the ORNL measurements by Dickens et al (Refs. 7, 8), show good agreement for t > 20 secs for both U235 and Pu239, but with C/E values about 8% higher for the former than for the latter.

Fig. 2, for the LASL measurements by Yarnell and Bendt (Refs. 9, 10) show good agreement for U235, but C/E values no higher than 0.95 for Pu239: again there is a difference of about 8% between the two fissile nuclides.

Figs. 3 and 4, for two of the IRT measurements by Friesehahn et al (Ref. 11), show poorer agreement between calculation and experiment; but there is no such large discrepancy between Pu239 and U235.

Other measurements that have been compared with calculation, but which are not shown here because of lack of space include the beta heating measurements at Winfrith (Ref. 12) and the total decay power measurements at the French CEA laboratories (Ref. 13); it is clear from all these that, whatever deficiencies there may be in the data libraries used in summation calculations, there are systematic differences between different experiments. These differences are greatest for gamma heat measurements.

Comparison of Measurements by Synthesis of Different Irradiations

The decay power t seconds after an irradiation of 1 fission per second lasting for I seconds is:

$$M(I,t) = \int_{t}^{I+t} m(x)dx$$

where m(x) is the decay power x seconds after a single fission. (The small effect of neutron capture is ignored).

Therefore,

$$M(nI,t) = \sum_{K=0}^{n-1} M(I,t+KI)$$

Consequently values for one irradiation can be built up by summing data from a shorter irradiation, if the smaller value of I is a factor of the larger, and if the cooling times after the shorter irradiation extend to more than the longer irradiation.

To overcome experimental fluctuations, and to allow interpolation at different cooling times, a sufficiently smooth curve must be fitted to the results of the shorter irradiation; this was done first with a least squares cubic spline programme written at Harwell, and secondly by fitting a sum of exponentials. The latter fitting was made by Tobias (Ref. 14). Very similar results were obtained with the two types of fit.

Some results are shown in Figure 5, comparing LASL measurements (I = 20,000 s) with values computed from short irradiation ORNL measurements, and in Figure 6, comparing IRT measurements (I = 1000 sec) with ORNL values.

It is apparent that there are considerable differences between the different measurements: LASL results are up to 8% higher than ORNL, and IRT and ORNL results differ by 8 - 12%.

Reduction of Measurements to Burst Functions or to Infinite Irradiation

If $I \ll t$, $M(I,t) \simeq Im(t+ I/2)$, and so the 'burst function', m(t), giving the decay power after a single fission, may be estimated.

Conversely, the decay heat after an infinite irradiation;

$M(t) \equiv M(\infty,t)$

may be calculated from

$M(t) = M(I,t) + M(I + t)$

where, for I >> t, the second term on the right is a small correction that can be estimated from summation calculations.

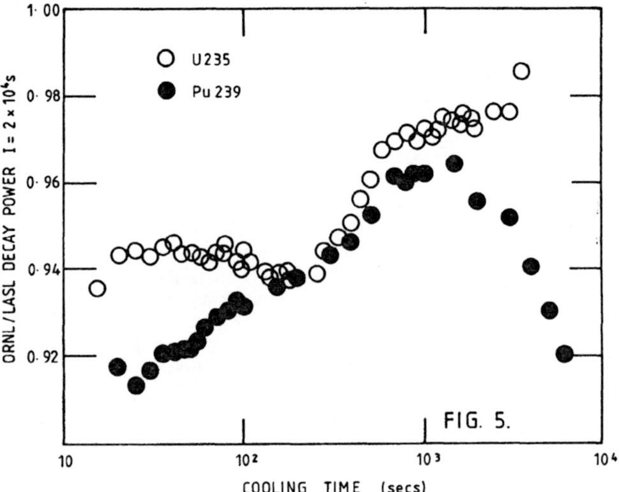

Fig 5 Comparison of LASL Measurements (I=20000 SEC) With Values Computed from ORNL Measurements

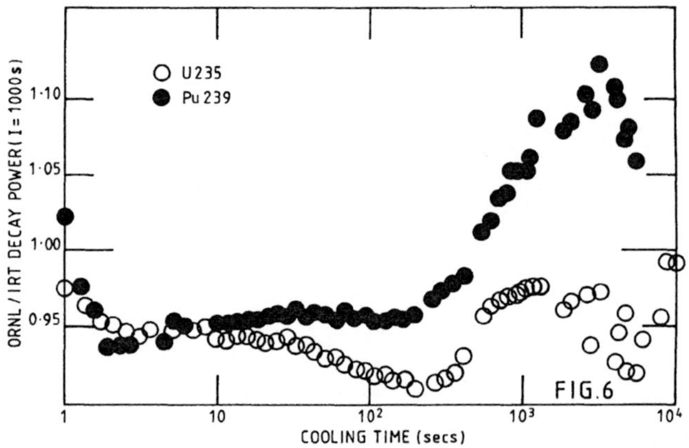

Fig 6 Comparison of IRT Measurements for an Irradiation of 1000 Sec with Values Computed from ORNL Measurements

Both m(t) and M(t) can be calculated very simply from the fits of M(I,t) to sums of exponentials, as described in the last sub-section, and this method has been used to derive burst functions from all the different total decay heat measurements.

Calculation of Definite Integrals of the Burst Function

Dickens et al (Ref 7, 8) have obtained partial integrals

$$\int_{t_i}^{t_{i+1}} m(t)dt,$$

for sets of times t_i, from the LASL, ORNL and IRT measurements for U235 and Pu239. These integrals were also computed using Tobias' exponential fits; agreement with the published results in Refs. 7 and 8 was very good.

Pu239/U235 Decay Heat Ratios.

Besides comparing measurements and calculations of Pu239 and U235 decay heat, it is also useful to compare calculated and experimental values of the ratio of Pu239 to U235 decay heat because there is considerable correlation between the data for the two nuclides (and of course for any other fissile nuclides). That this is so for calculations was pointed out by Trapp and Spinrad (15): although the yields are of course different for the two, the fission product decay data (half lives, energies and branching ratios) are common, although some chains are much more significant for one nuclide than for the other.

For measurements, one expects that some of the systematic errors of a particular method, (e.g. in determining the number of fissions; in detector efficiency) are common to each nuclide, and will thus tend to cancel out in the ratio.

For convenience, we write:

$$R(I,t) = \frac{M(I,t) \text{ for Pu239}}{M(I,t) \text{ for U235}}$$

and
$$r(t) = \frac{m(t) \text{ for Pu239}}{m(t) \text{ for U235}}$$

Fig. 7 shows R(20,000 sec, t) comparing measurements from LASL (Refs. 9, 10), with calculation and with values derived from measurements at other irradiation times at ORNL (Refs. 7, 8) and at IRT (Ref. 11), by the methods described in Section 2.2.

Figure 8 shows r(t), the ratio of the burst functions, from calculations and from the measurements at LASL (Refs. 9, 10), ORNL (Refs. 7, 8), and IRT (Ref. 11).

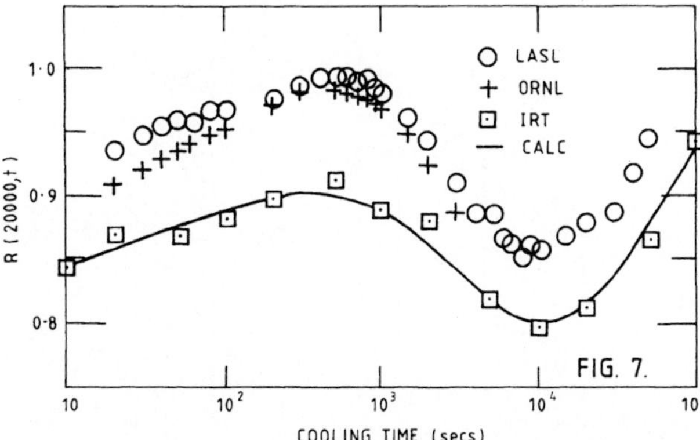

Fig 7 A Comparison of Calculated, Measured and Derived Values for R(20000,t)=(Pu 239 Decay Power)/(U235 Decay Power) After 20000 Sec Irradiations

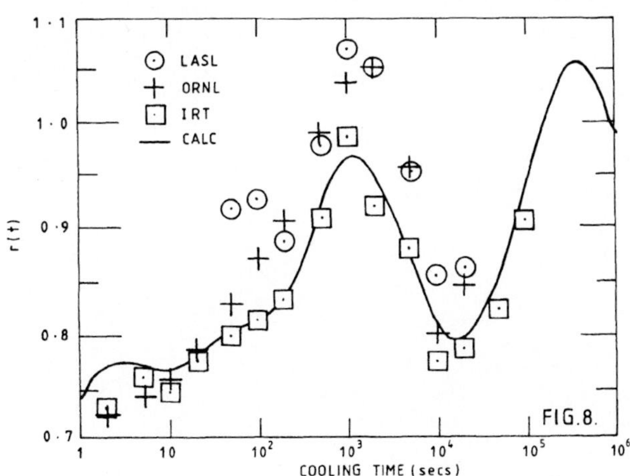

These two graphs show much the same picture. Much of the systematic differences between different experiments have indeed cancelled out as would be expected.

The ratios fall clearly into two groups. The first contains the calculated values and those from IRT, in remarkably good agreement; the second, about 8% higher, contains the LASL and ORNL measurements.

To explain the 8% discrepancy, we could make either of two extreme hypotheses, or a combination of them:

(i) The LASL and ORNL results are correct, so that r(t) reaches ∿ 1.05 at t ∿ 1000 secs. Then the agreement between calculation and the IRT results is probably coincidental, and there are large errors in the data libraries. But note that the largest sensitivity of the decay heat ratio in this time range to any single datum is $\sim 4 \times 10^{-2}$, so that to explain an error of 8% would need an error of about 200% in any one item; even if several items were erroneous, large errors would be required to explain the discrepancy. If reasonable uncertainties are assumed in the yield and decay data, the statistical uncertainty is only about 2% (1 standard deviation).

(ii) The ratios from summation calculations and from IRT measurements are correct (to perhaps ∿ 2%). Then there must be systematic errors between the Pu239 and the U235 measurements for each of the LASL and ORNL experiments, and of about the same magnitude (∿ 8%) for each laboratory. Since the discrepancy is nearly constant over a wide range of times (20 - 10^5 sec), perhaps the most likely source of error would be in the estimated number of fissions; an error in detector efficiency might be expected to have an effect that would be dependent on time, as the beta and gamma spectra change.

At present, however, it is not possible to decide between these extreme hypotheses, but only to emphasise the existence of the discrepancy.

It should be noted that the beta heating measurements and the preliminary results of gamma heating measurements for irradiations of 1 day and longer carried out at Winfrith show C/E values for the Pu239/U235 ratio that are close to unity.

Conclusions.

We have shown how different measurements of decay heat may be compared; and how these comparisons indicate systematic differences between them. Some may be due to errors in estimating the numbers of fissions, and some (a time dependent error) may be due to errors in estimating the efficiency of the detector or calorimeter.

Most of these differences disappear if the ratio of Pu239 to U235 decay power is considered. But an 8% discrepancy remains between two measurements and calculation and one measurement.

Until these discrepancies are resolved, care should be taken not to assume too small uncertainties in recommended decay heat values. It is also not necessarily the case that the Pu239 values are more uncertain than those for U235.

← Fig 8 A Comparison of Calculated and Experimental Values for r(t) =(Pu239 Decay Power)/(U235 Decay Power) for Single Fissions

Acknowledgement.

Much of this work would have been impossible without the exponential fits made to different sets of data by A Tobias, who readily made them available to the author.

I also wish to thank A Tobias, J S Story and J L Rowlands for invaluable discussions.

References.

1. A. Tobias, Prog. in Nucl. Energy 5 (1), (1980)

2. R. F. Burstall, UKAEA Report ND-R-328(R) (1979)

3. A. Tobias, CEGB Report RD/B/N4303 (1978)

4. A. Tobias and B. S. J. Davies, CEGB Report RD/B/N4942 (1980)

5. E. A. C. Crouch, At. Data and Nucl. Data Tables 19, 417 (1977) and INDC (NDS)-113/G+P (1980)

6. B. S. J Davies, A. Tobias, M. F. James, A. L. Nichols and E. A. C. Crouch, The UK Chemical Nuclear Data Libraries: Evaluated Nuclear Decay Data for Reactor Applications. This Conference.

7. J. K. Dickens, T. A. Love, J. W. McConnell and R. W. Peelle, Nucl.Sci. and Eng. 74, 106 (1980)

8. J. K Dickens, T. A. Love, J. W. McConnell and R. W. Peelle, Nucl. Sci. and Eng. 78, 126 (1981)

9. J. L. Yarnell and P. J. Bendt, LA-NUREG-6713 (1977)

10. J. L. Yarnell and P. J. Bendt, LA-7452-MS, NUREG/CR-0349 (1978)

11. S. J. Friesenhahn and N. A. Lurie, IRT-0034-004 (1977)

12. M. F. Murphy, W. H. Taylor, D. W. Sweet and M. R. March, UKAEA Report AEEW-R1212 (1979) and Private Communication

13. M. Lott, G. Lhiaubet, F. Dufreche and R. de Tourreil, J. Nucl. Energy 27, 597 (1973)

14. A. Tobias, CEGB Report RD/B/5079N81 (1981) and private communication.

15. T. J. Trapp and B. I. Spinrad, OSU-NE-7801 (1978).

CALCULS DE PUISSANCE RESIDUELLE A L'AIDE DE LA BIBLIOTHEQUE CEA DE DONNEES RADIOACTIVES ET DU CODE PEPIN

B. DUCHEMIN(1), J. BLACHOT(2), B. NIMAL(1), J.C. NIMAL(1), J.P. VEILLAUT(1)

COMMISSARIAT A L'ENERGIE ATOMIQUE FRANCE

(1) SERMA/LEPF/ - CEN-SACLAY - 91191 GIF SUR YVETTE

(2) DRF/CPN - CEN-GRENOBLE - 38041 GRENOBLE CEDEX

The CEA radioactivity data bank, has been updated mainly from ENSDF and from some recent experimental results. As already described in Harwell, this library contains the decay data for about 700 fission products (F.P.), 220 actinides and more than 1400 other nuclides. A comparison between our data and ENDF/B5 is shown for the fission products.

The fission products part of this library is currently used for shielding and decay heat calculations with the PEPIN code. Direct summation calculations and spectral comparisons of the available experiments. (Dickens, Lott, Yarnell...) and other recent calculations is made for thermal fission of ^{235}U and ^{239}Pu using our data bank as input.

[RADIOACTIVITY, Data bank, Actinides, Fissions products, Decay heat, Summation calculation, pulse irradiation, comparison with experiments.]

Introduction

La nécessité de disposer d'une bibliothèque de données spectroscopiques pouvant être utilisées dans de nombreuses applications, par exemple celles liées à la puissance résiduelle des réacteurs ou aux calculs des doses d'irradiation a été ressentie depuis longtemps. Dans ce travail, après avoir rappelé brièvement les développements récents de la bibliothèque C.E.A., nous définirons succinctement les autres données nécessaires à nos calculs, nous développerons la manière dont nous étudions la puissance résiduelle par la méthode de sommation et nous montrerons comment la comparaison entre nos résultats et d'autres soit théoriques soit expérimentaux peut fournir de nombreuses informations dans le but d'améliorer encore ces résultats.

1 - Les données d'entrée

A) Les données de décroissance :

La bibliothèque

Le CEA a développé depuis de nombreuses années une bibliothèque sur les données de décroissance radioactive. Lors du dernier Congrès de Harwell (1), la description du type de format des données, de leurs sources, des méthodes d'évaluation avait été présentée.

Cette bibliothèque est accessible par le réseau CISI, et une copie sur bande magnétique peut aussi être obtenue du NEA data Bank (2).

La partie produits de fission (P.F.) de cette bibliothèque comprend :

- 120 noyaux stables
- 417 nucléides avec des données expérimentales complètes permettant de déduire les énergies moyennes bêta et gamma.
- 162 nucléides avec des données expérimentales partiellement ou complètement manquantes

Origine des données

La mise à jour de la bibliothèque s'est poursuivie d'une façon continue depuis Harwell. Tous les six mois les données récentes du Fichier ENSDF (3) ont été traduites dans le format de notre fichier. Les masses A=139, 75, 116, 125, 113, 78 (4), résultant de toutes les dernières évaluations du réseau ENSDF faisaient partie du fichier ENSDF de février 1982. En raison du cycle de renouvellement encore trop long des données ENSDF, nous avons introduit des résultats récents sur les produits de fission à période courte. En particulier la 4e conférence sur les noyaux loin de la stabilité (5) nous a apporté beaucoup d'informations nouvelles.

Raies X et électrons

Elles sont calculées par un programme dérivé du code MEDLIST (6).

Le programme utilise d'une part des informations nucléaires stockées dans le fichier, et d'autre part des données atomiques qui sont inclues dans le programme. Les données venant de ENSDF possèdent les coefficients de conversion internes des transitions gamma de multipolarité connue, et les rapports de capture électronique. Les données atomiques nécessaires au calcul sont les énergies, les rapports d'intensité et les rendements de fluorescence (7).

K. H. Böckhoff (ed.), Nuclear Data for Science and Technology, 249–256.

Energies moyennes bêta et gamma

Pour les noyaux où le schéma de désintégration est connu, les énergies moyennes bêta et gamma sont dérivées d'une méthode déjà présentée (13). Il a été remarqué (12) que lorsque les énergies bêta étaient déduites du schéma de niveau elles étaient quelquefois trop élevées, car des transitions gamma de hautes énergies n'avaient pas été mesurées. Nous avons comparé dans le tableau I nos valeurs, souvent déduites de travaux récents, avec les résultats de Rudstam (12) et de ENDF/B V.

Table I - Comparaison des énergies moyennes bêta

Nucléide	Réf.12 (MeV)	ENDF/B V (MeV)	Fichier CEA (MeV)
^{86}Br	1.92 ± 0.11	1.78	1.94
^{87}Br	1.20 ± 0.06	2.50	1.72
^{88}Br	1.63 ± 0.07	2.54	2.44
^{89}Br	2.32 ± 0.12		2.93
^{89}Kr	1.38 ± 0.04	1.35	1.36
^{89}Rb	1.08 ± 0.03	1.03	1.02
^{91}Kr	1.53 ± 0.07	1.94	1.96
^{91}Rb	1.36 ± 0.07	1.50	1.58
^{92}Kr	2.67 ± 0.47	2.37	2.02
^{92}Rb	3.46 ± 0.23	3.48	3.52
^{93}Kr	2.70 ± 0.21	2.34	2.78
^{93}Rb	2.59 ± 0.14	2.61	2.73
^{93}Sr	0.90 ± 0.03	0.70	0.70
^{94}Sr	0.91 ± 0.03	0.90	0.90
^{94}Y	1.78 ± 0.07	1.80	1.69
^{95}Sr	1.92 ± 0.09		2.10
^{95}Y	1.52 ± 0.07	1.48	1.54
^{96}Y(6s)	2.83 ± 0.15	3.15	3.18
^{134}Sb(10s)	2.38 ± 0.12	2.80	2.95
^{136}I(48s)	2.11 ± 0.12	2.13	2.24
^{136}I(83s)	1.97 ± 0.08	1.97	2.0
^{137}I	2.25 ± 0.11		1.9
^{137}Xe	1.73 ± 0.03	1.77	1.79
^{138}I	2.27 ± 0.13		3.0
^{138}Xe	0.80 ± 0.10	0.64	0.68
^{138}Cs(32mn)	1.22 ± 0.12	1.20	1.21
^{139}I	2.01 ± 0.29		2.20
^{139}Xe	1.72 ± 0.06	1.70	1.74
^{139}Cs	1.73 ± 0.05	1.69	1.71
^{140}Xe	1.52 ± 0.05	1.18	1.27
^{140}Cs	1.89 ± 0.04	1.65	1.75
^{141}Xe	1.96 ± 0.11	2.35	2.28
^{141}Cs	1.68 ± 0.07	1.91	1.82
^{142}Cs	2.50 ± 1.01	2.50	2.12

Energie des noyaux à spectre inconnu

Les spectres de certains nucléides de la bibliothèque n'ont pas encore été mesurés expérimentalement. Les énergies moyennes bêta et gamma de ces nucléides sont estimées à partir des valeurs des noyaux connus en constituant une droite de régression par rapport au Q de la réaction.

B) Rendements de fission

Les rendements indépendants de fission Y_{if} (Z,A) sont calculés pour chaque nucléide en utilisant la méthode de Wahl (9)

$$Y_{if}(Z,A) = \frac{Y_f(A)*KP_f(Z,A)*NF_f(A)}{\sqrt{\pi C}} \int_{Z-0,5}^{Z+0,5} \exp\left[-\frac{(Z-ZP)^2}{C}\right] dZ$$

Z = numéro atomique

A = masse atomique

$Y_f(A)$ rendement cumulé de la chaîne A d'après RIDER (8)

ZP(A) valeur la plus probable du Z de la chaîne A d'après RIDER (8)

NF_f coefficient de normalisation

C paramètre d'ouverture de la gaussienne relié à σ par la relation :

$$\sigma = \left[\frac{C}{2} - \frac{1}{12}\right]^{1/2}$$

quelque soit le type de fission σ a une valeur constante de 0,56.

$KP_f(Z,A)$ coefficient traduisant les corrections dues aux effets de parité entre les nombres de protons et de neutrons. Ils sont donnés dans le tableau II d'après (9) :

Table II - Valeurs de KP_f (Z,A)

Protons	Neutrons	KP_f (Z,A)
pair	pair	EOZ*EON
pair	impair	EOZ/EON
impair	pair	EON/EOZ
impair	impair	1/(EOZ*EON)

Les valeurs de EOZ et EON sont déduites de (9) et données dans le tableau III.

Table III - Valeurs de EOZ et EON

Isotope fissile	Type de fission	EOZ	EON
^{235}U	thermique	1.26	1.07
^{235}U	rapide	1.15	1.05
^{238}U	rapide	1.10	1.05
^{239}Pu	thermique	1.07	1.
^{239}Pu	rapide	1.	1.
^{240}Pu	rapide	1.	1.
^{241}Pu	thermique	1.	1.
^{241}Pu	rapide	1.	1.
^{242}Pu	rapide	1.	1.
^{232}Th	rapide	1.25	1.08
^{233}U	thermique	1.30	1.07
^{233}U	rapide	1.10	1.05

Cas des isomères

Lorsqu'un nucléide (Z,A) possède un isomère, $Y_{if}(Z,A)$ a été distribué sur les isomères suivant un modèle statistique développé par Madland (10). Les valeurs des spins et parité nécessaires au calcul ont été prises dans la bibliothèque.

C) Données de capture

Notre méthode nécessite la connaissance des sections efficaces de capture (thermique, intégrale de résonance et rapide) et des rapports de branchements isomériques par capture. Les valeurs proviennent de diverses sources (11).

2 - Application aux calculs de puissance résiduelle

L'ensemble des données définies aux paragraphes précédents est utilisé pour calculer tous les résultats liés à la chaleur résiduelle due aux produits de fission après irradiation du combustible.

A) Méthode utilisée

Après avoir défini les données spécifiques à chaque étude particulière (flux thermique maxwellien, rapport épithermique sur thermique, quantités de fission distribuées sur les 12 espèces fissiles possibles...), notre code PEPIN résoud analytiquement les équations de Batemann :

$$\frac{dN_i\ (t)}{dt} = b_i + \sum_{j=1}^{i-1} C_{ij}\ N_j\ (t) - C_{ii}\ N_i\ (t)$$

avec :

b_i : nombre de noyaux i créés par fission
$= N_f\ \sigma_f\ \varphi Y_{if}$

C_{ij} : nombre de noyaux i créés par décroissance ou capture
$= \lambda_{ij} + \sigma_{ij}\ \varphi \qquad \forall\ j \neq i$

C_{ii} : nombre de noyaux i disparus par décroissance et capture
$= \lambda_i + \sigma_i\ \varphi$

En faisant l'hypothèse que le diagramme de marche du système irradiant peut être approché par un histogramme à palier constant (b_i = constante pour un palier), cette équation a des solutions analytiques bien connues.

Cette méthode, qui est aussi utilisée dans d'autres codes de ce type, par exemple CINDER (14), nous semble préférable à la méthode matricielle utilisée par exemple par ORIGEN (15), car elle donne une beaucoup plus grande souplesse dans les applications, permettant de traiter aussi bien les temps d'irradiation ou de refroidissement très courts que des temps d'irradiation ou de refroidissement très longs. Les instabilités numériques inhérentes à cette méthode sont maintenant bien connues et éliminées simplement.

B) Comparaison avec l'expérience et d'autres calculs

Pour étudier le comportement de l'ensemble code PEPIN et bibliothèque à 699 nucléides, nous avons comparé nos résultats avec les calculs effectués d'une part avec notre ancienne bibliothèque, d'autre part à l'aide du code CINDER et de la bibliothèque ENDF/B V (16) et les expériences effectuées par DICKENS (17), LOTT (18), et YARNELL et BENDT (19) pour les deux principaux corps fissiles ^{235}U et ^{239}Pu.

Fission élémentaire

La figure 1 montre que la puissance résiduelle totale est dans le cas ^{235}U très bien reproduite par les calculs avec une légère amélioration par rapport à B V entre 10 et 200 sec. La figure 2 relative à la puissance totale dans le cas ^{239}Pu montre qu'un déficit subsiste entre 200 et 3000 sec en particulier. Les figures 3 et 4 relatives à la puissance bêta respectivement sur ^{235}U et ^{239}Pu montre un comportement similaire, la puissance bêta étant surestimée entre 10 et 100 sec de manière plus importante dans le cas ^{235}U. Les figures 5 et 6 relatives à la puissance gamma montrent un comportement similaire jusqu'à 300 sec, la puissance gamma calculée étant plus faible que les expériences. Par contre entre 300 et 2000 sec le comportement est très différent entre l'^{235}U et le ^{239}Pu. Un tel comportement devrait permettre de déterminer, dans cette région, les noyaux sur lesquels un effort expérimental est encore à faire.

Tous ces résultats sont en légère amélioration par rapport à BV et en parfait accord avec les travaux récents tels que ceux de Rudstam (12) qui montrent que l'énergie bêta est en général surestimée dans les bibliothèques actuelles.

Irradiation infinie

Toutes les observations trouvées dans le cas de la fission élémentaire se retrouvent dans le cas de l'irradiation infinie. Dans les figures 7 et 8 concernant la puissance totale dans les cas U^{235} et Pu^{239} respectivement, nous avons également reporté le standard ANS 79 (20). Nos résultats sont toujours inférieurs au standard ANS. Dans les figures 9 et 10 la puissance bêta est supérieure aux temps courts dans les deux cas. Dans les figures 11 et 12 la puissance gamma est inférieure à l'expérience ce qui confirme bien que l'énergie bêta est en général surestimée.

Irradiation de durée définie

Dans les figures 13 et 14 nos résultats sont comparés à une irradiation de 2. 10^4 sec faite par Yarnell et Bendt (19). Les calculs de sommation montrent dans ces cas un déficit (faible pour U^{235}, plus grand pour Pu^{239}) par rapport à l'expérience. Ce résultat est également trouvé par les calculs utilisant ENDF/B V (19).

Spectres

Dans le but d'obtenir des informations complémentaires, nous avons comparé les calculs de spectres gamma effectués dans le cas de la fission élémentaire dans un découpage à 17 groupes en énergie avec les données DICKENS (17). Les figures 15 et 16 donnent un exemple de cette comparaison pour des temps de refroidissement de 12,7 et 1000 sec respectivement dans le cas ^{235}U. Il résulte de cette étude que les spectres calculés sont plus mous que les spectres expérimentaux pour les temps de refroidissement très courts, par contre pour $t_{réf} \gg 1000$ sec l'accord est satisfaisant sauf pour les gamma de haute énergie.

Conclusion

L'évolution constante des données des bibliothèques spectroscopiques nous paraît très importante à suivre par les utilisateurs même ceux qui font uniquement des calculs globaux. En effet, les études présentées ici montrent bien que, d'une part l'utilisation des données récentes fournit dans la plupart des cas, des résultats beaucoup plus proches de l'expérience, mais que d'autre part des études complémentaires sont nécessaires pour expliquer les divergences qui subsistent : surestimation des énergies bêta aux temps courts, déficit de la puissance totale dans le cas ^{239}Pu, etc......

Remerciements

Nous tenons à remercier M. C. FICHE qui a développé les programmes de gestion de la bibliothèque et du calcul des X.

Références

1 J. BLACHOT, C. FICHE - Harwell 1978 p. 216

2 C. FICHE - Bibliothèque de Radioactivité du CEA et les programmes EDIBIB, ISOTAB, TRIGAL (à paraître)

3 ENSDF evaluated nuclear structure data file Edité par le Neutron Nuclear Data Centre, Brookhaven National Laboratory pour le compte du Réseau International de Structure Nucléaire et Radioactivité

4 Nuclear Data Sheets - Vol. 32 n° 1,2,4 (1981)

5 4th International Conference on Nuclei far from stability Helsingor Danmark 7-13 juin 1981

6 W.B. EWBANK - Communication privée (1979)

7 J. BLACHOT, C. FICHE - Annales de physique Vol. 6 S (1981)

8 B.F. RIDER - Nedo 12154-3C (1981)

9 A.C. WAHL - Journal of Radioanalytical Chemistry Vol. 55 N° 1 (1980)

10 D.G. MADLAND - LA.6595.MS (ENDF-241)(1976)-

11 W.H. WALKER - AECL 3037 (1969) et Chart of the Nuclides (1977)

12 K. ALEKLETT, G. RUDSTAM - Nucl. Sci. Eng. 80, 74 (1982)

13 J. BLACHOT - Petten (1977) -

14 T.R. ENGLAND, W.B. WILSON, et M.G. STAMATELATOS LA 6746 MS (1976)

15 M.J. BELL ORNL 4628 (1973)

16 R.J. LABAUVE, T.R. ENGLAND, D.C. GEORGE LA 9090 MS (1981)

17 J.K. DICKENS et al. ORNL-NUREG 39 (1978) et ORNL-NUREG 66 (1980)

18 M. LOTT et al. - J.N.E. 27 p. 597 (1973) et C. FICHE NEACRP/L 212 (1976)

19 J.L. YARNELL, P.J. BENDT - LA-NUREG 6773 (1977) et LA 7452 MS (1978)

20 ANSI/ANS 5.1 (1979)

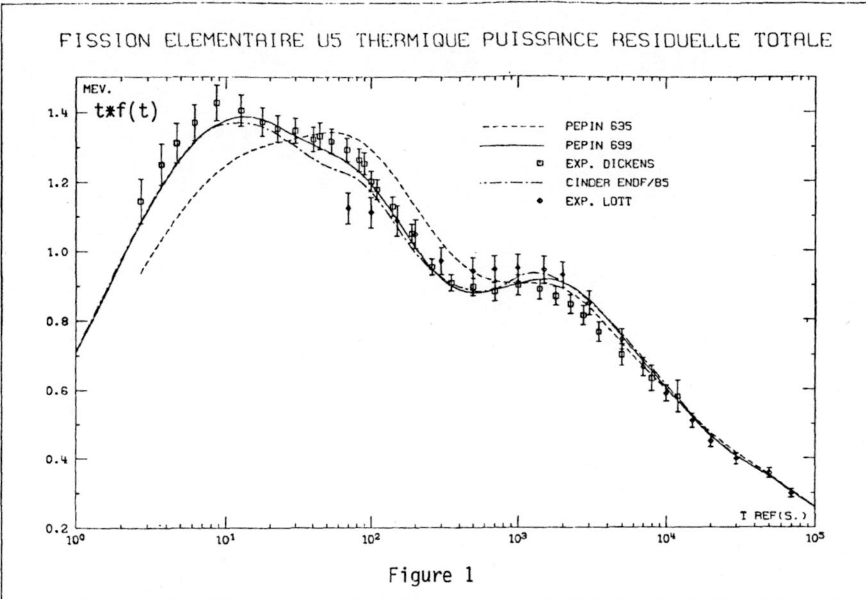

Figure 1

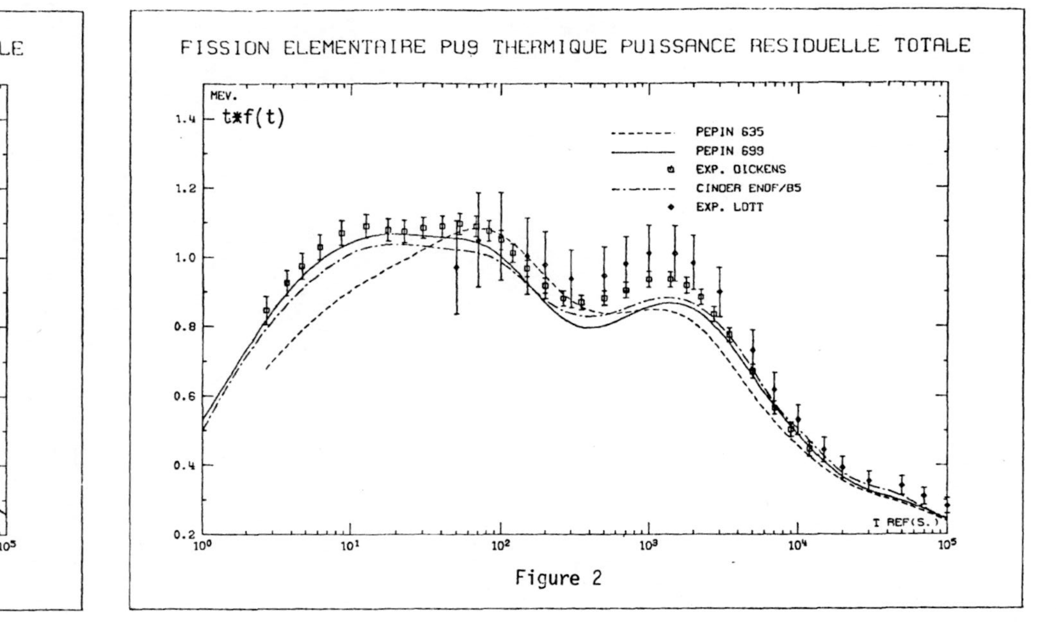

Figure 2

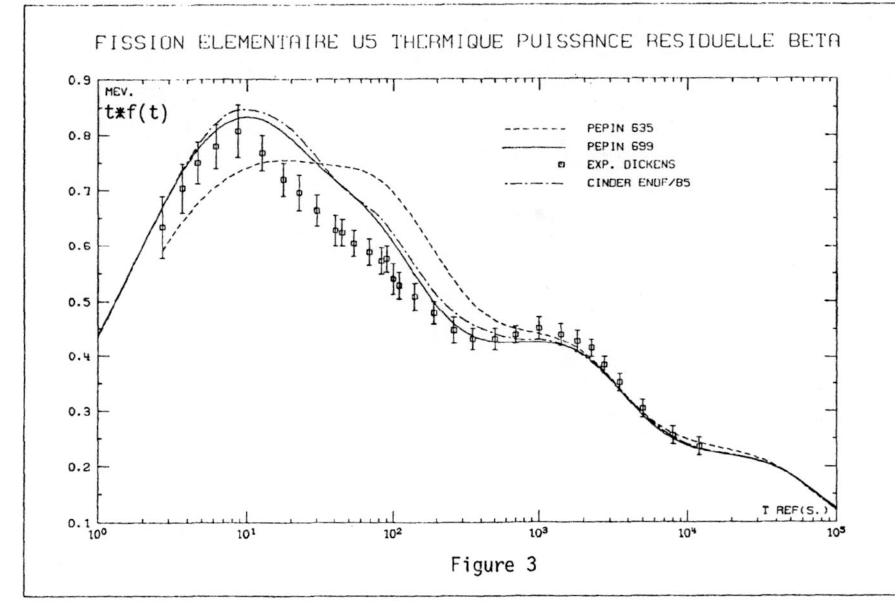

Figure 3

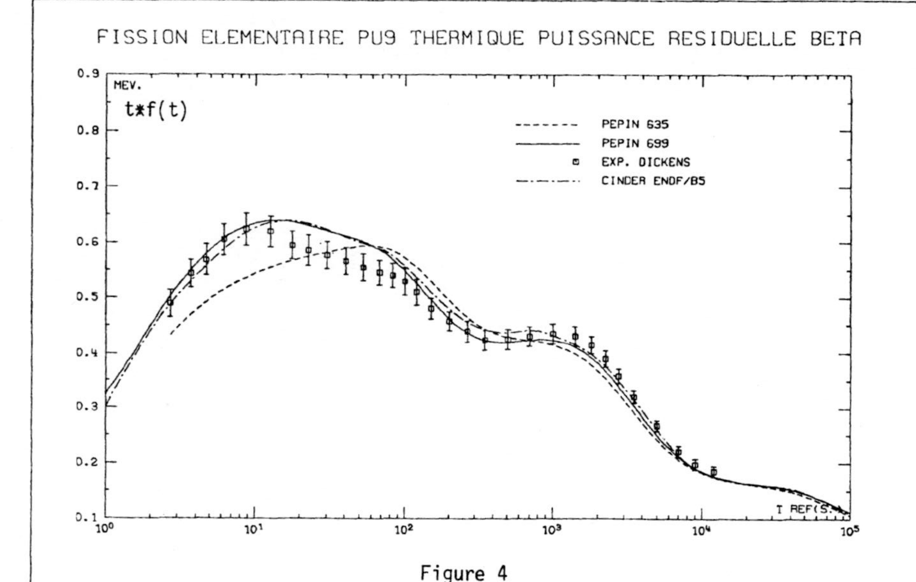

Figure 4

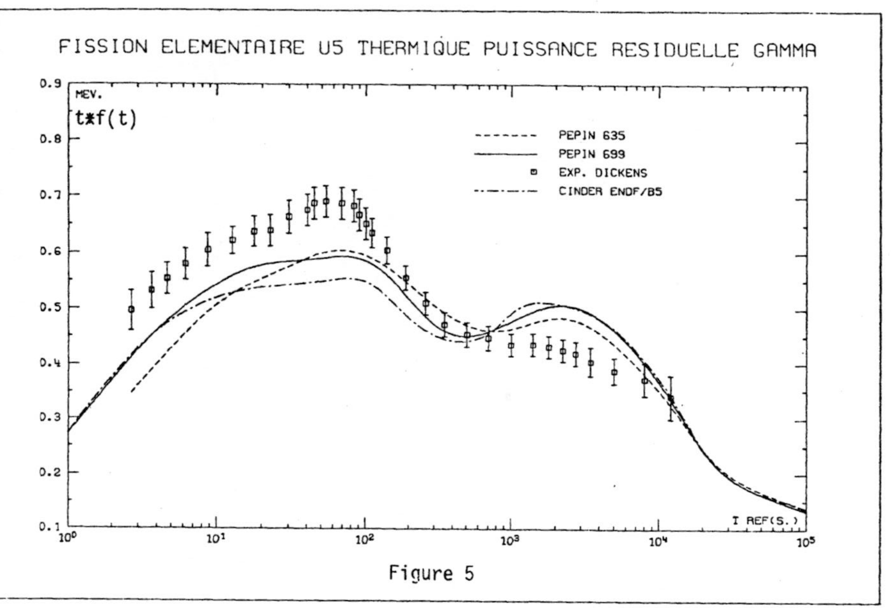

Figure 5

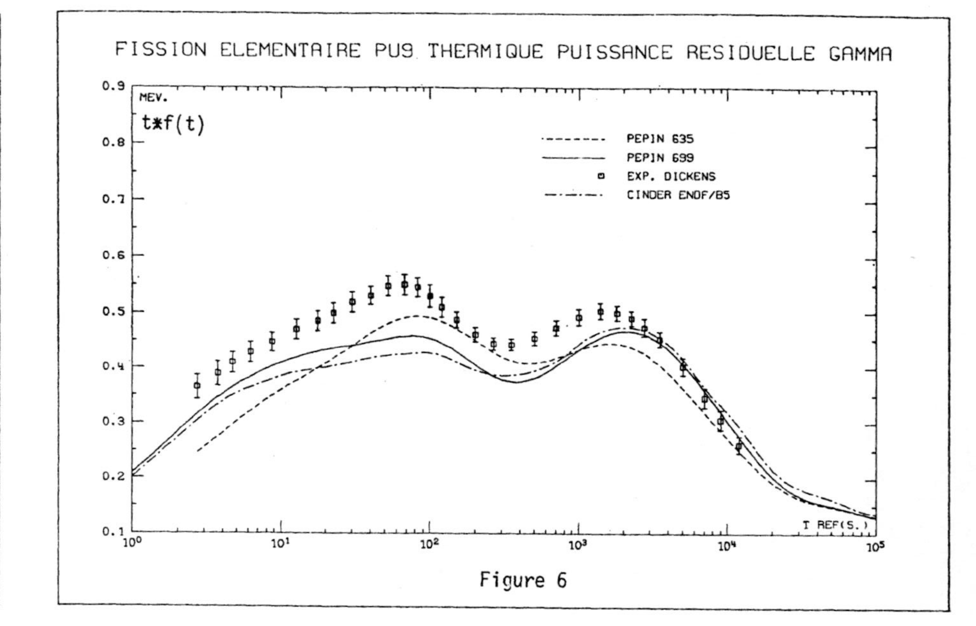

Figure 6

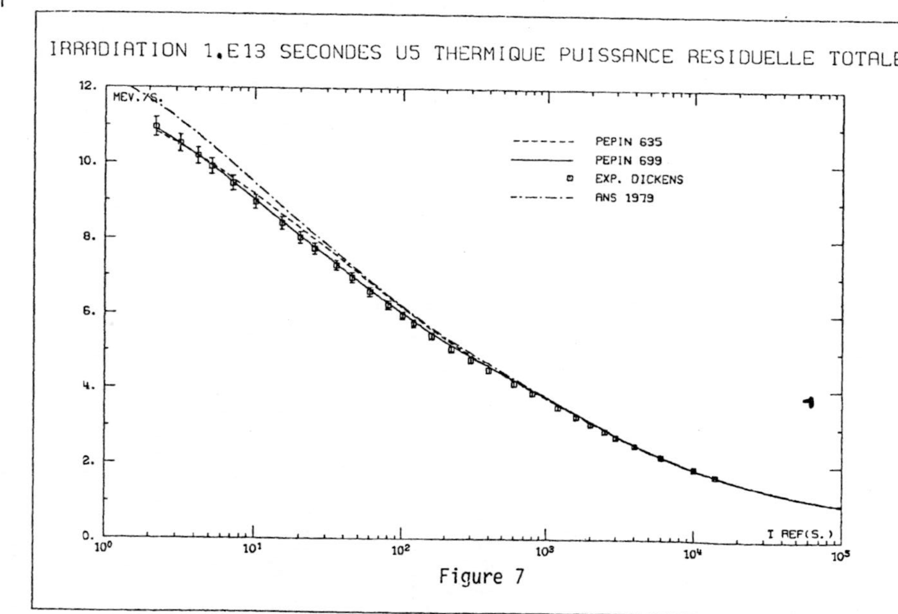

Figure 7

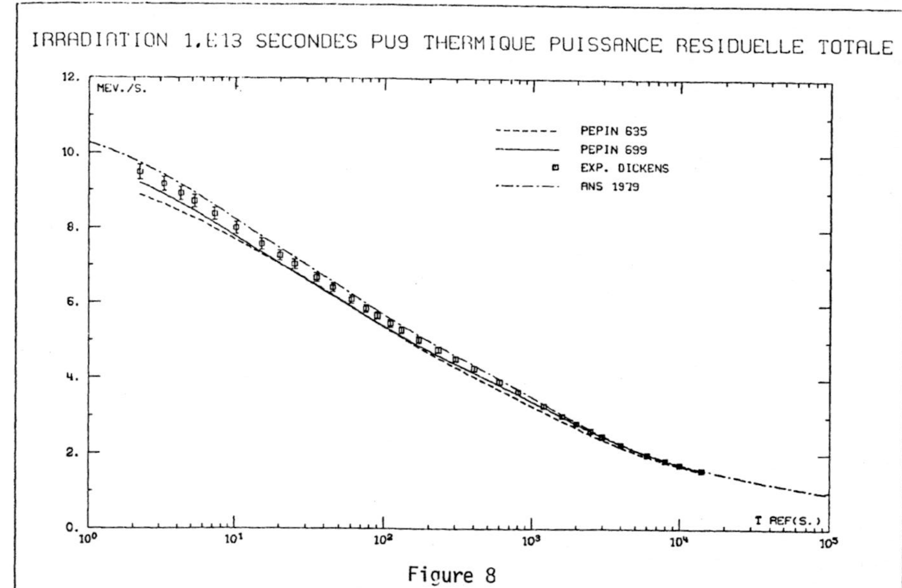

Figure 8

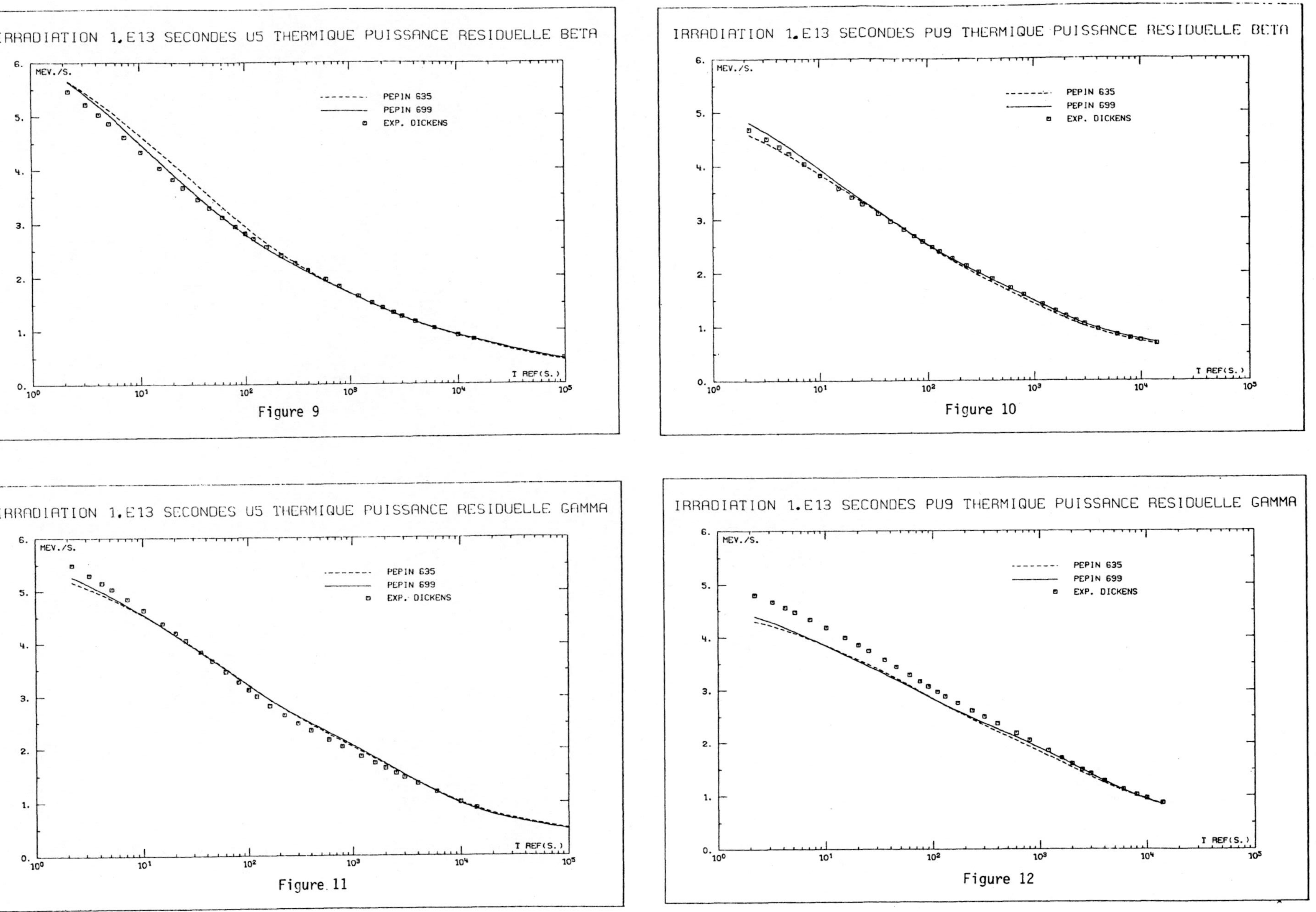

Figure 9

Figure 10

Figure 11

Figure 12

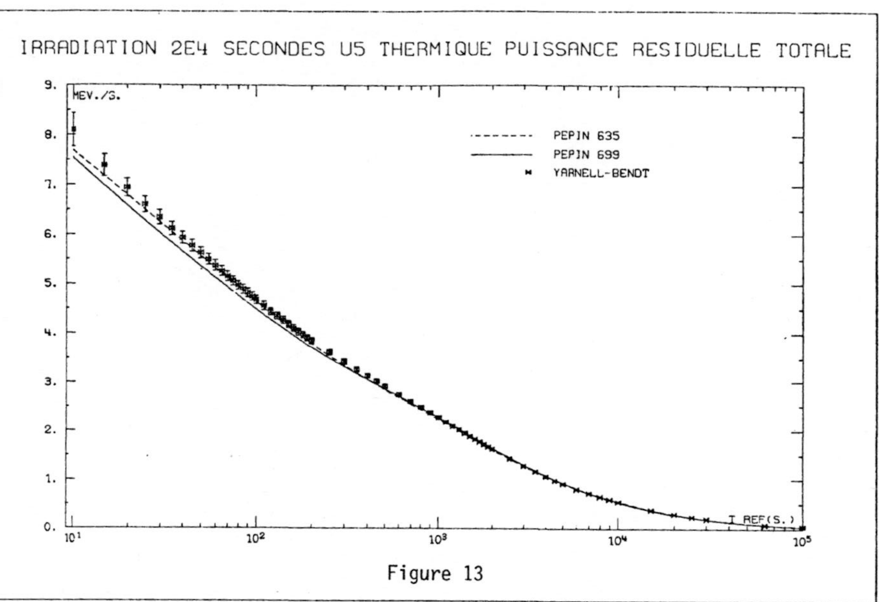
IRRADIATION 2E4 SECONDES U5 THERMIQUE PUISSANCE RESIDUELLE TOTALE
MEV./S.
PEPIN 635
PEPIN 699
YARNELL-BENDT
T REF(S.)

Figure 13

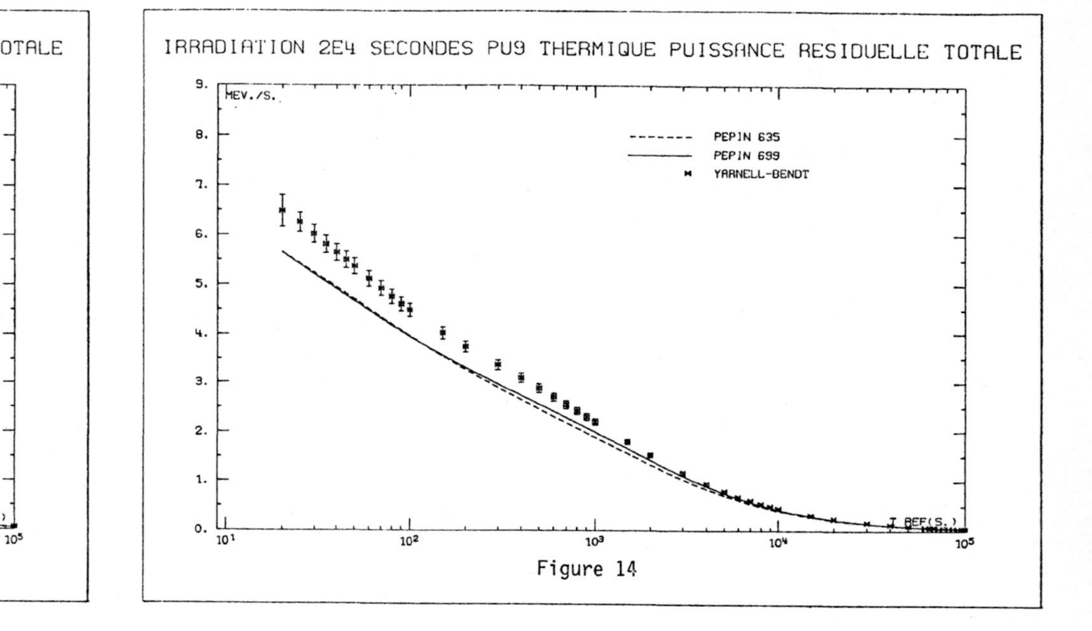
IRRADIATION 2E4 SECONDES PU9 THERMIQUE PUISSANCE RESIDUELLE TOTALE
MEV./S.
PEPIN 635
PEPIN 699
YARNELL-BENDT
T REF(S.)

Figure 14

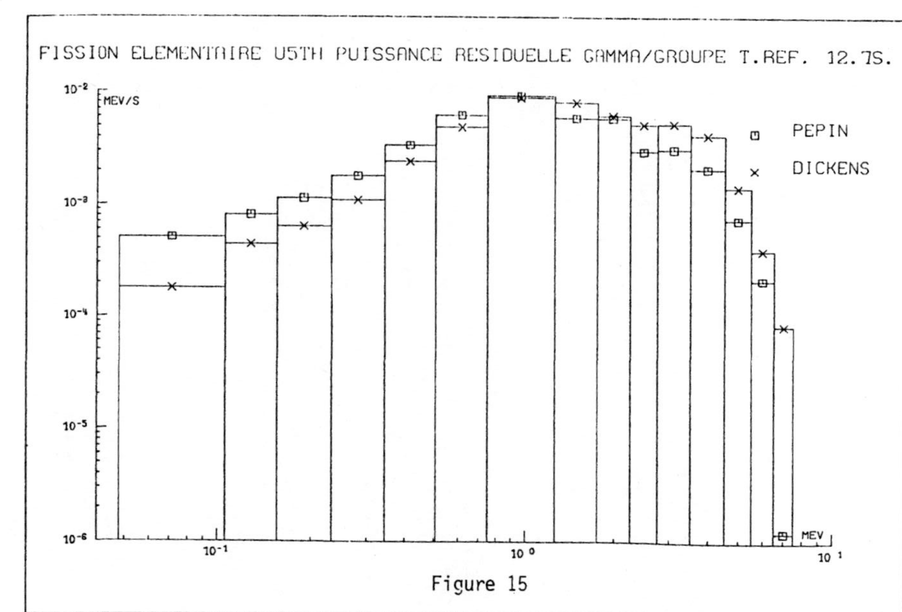
FISSION ELEMENTAIRE U5TH PUISSANCE RESIDUELLE GAMMA/GROUPE T.REF. 12.7S.
MEV/S
PEPIN
DICKENS
MEV

Figure 15

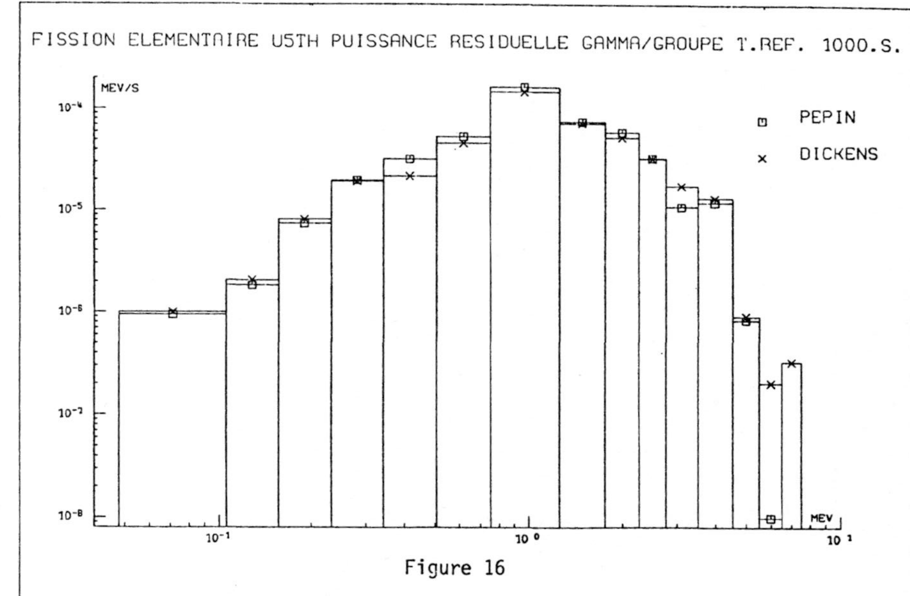
FISSION ELEMENTAIRE U5TH PUISSANCE RESIDUELLE GAMMA/GROUPE T.REF. 1000.S.
MEV/S
PEPIN
DICKENS
MEV

Figure 16

ENERGY DEPOSITION DATA FOR ^{233}U, ^{235}U, ^{239}Pu, ^{241}Pu and ^{252}Cf FISSION FRAGMENTS IN ANY SUBSTANCE

P. Dickstein, Y. Laichter, and N.H. Shafrir

Department of Nuclear Engineering
Technion - Israel Institute of Technology
IL-32000 Haifa, Israel

Analysis of extensive experimental range and energy loss data of ^{252}Cf-fission fragments in gases and solids, representing a wide range of atomic numbers throughout the periodic system reveals clearly the existence of Z_2 oscillations in the stopping process. A modified LSS expression which is based on these results enables, together with a suitable interpolation procedure, the prediction of range and energy loss data for ^{252}Cf fragments in any substance, taking into account the oscillatory behaviour of the stopping process. These data can be scaled to fission fragments of other fissile material of technological interest, using the few existing experimental results for these materials and a suitable scaling procedure.

[Nuclear fission, ^{252}Cf, ^{233}U, ^{235}U, ^{239}Pu, ^{241}Pu fission fragments - Z_2 oscillations, ranges, stopping cross sections]

Introduction

Range and energy loss predictions of fission fragments in gaseous and solid stopping media of any atomic number are of considerable interest for the study of the mechanism of the interaction of energetic heavy ions with matter. Another aspect is the study of the kinetic energy properties of the fragments, which leads to a better understanding of the fission process. Energy loss predictions for fission fragments are also of practical importance in a variety of fields such as direct energy conversion, radiation induced ionization and non-thermal ionization in magnetohydrodynamics. The inital energies associated with fission fragments and their mass range relate them to the medium velocity region of the order of $v_o Z_1^{2/3}$, where Z_1 is the nuclear charge of the fragment and v_o the Bohr velocity in the hydrogen atom. Existing theoretical treatments of the stopping process in the velocity region show very poor agreement with experiments[1].

In a recent experimental study accurate range data were determined for a large number of individual fission fragments[2,3] as well as accurate energy loss data for mean fission fragments[1,4] in gases and solids, representing a wide range of atomic numbers throughout the periodic system. The results revealed clearly the existence of Z_2-oscillations in the stopping process[4]. Based on these results a modified Lindhard, Sharff and Schiøtt (LSS) formulation has been suggested which enables, together with a suitable interpolation procedure, the prediction of range and energy loss data of ^{252}Cf fragments in any substance, taking into account the oscillatory behaviour of the stopping process. This data can be scaled to fission fragments of other fissile material of technological interest, using the few existing experimental results for these materials and a suitable scaling procedure.

Fission Fragment Ranges and Energy Loss - LSS Theory and Experimental Results.

In the fission fragment energy domain most of the energy will be lost by inelastic ion-electron collisions, resulting in ionization and excitation of the stopping medium, but with decreasing velocity, the contribution of quasi-elastic ion-atom collisions to the stopping process will be more significant. The LSS theory[5] is based on a statistical approach. In this theory are defined, respectively, reduced range and energy expressions as follows:

$$\rho = \pi \alpha^2 N \gamma R_{int} \tag{1}$$

$$\varepsilon = \frac{\alpha A_2}{Z_1 Z_2 e^2 (A_1 + A_2)} E \tag{2}$$

where R_{int} and E are the integral path length and energy in conventional units, A_1 and Z_1 are the mass and atomic number of the particle, A_2 and Z_2 are the mass and atomic number of the stopping medium, N the atomic density and e the electronic charge. The screening parameter α is given by:

$$\alpha = 0.8853\, \alpha_o\, (Z_1^{2/3} + Z_2^{2/3})^{-1/2} \tag{3}$$

and γ is given by:

$$\gamma = 4\, A_1 A_1\, (A_1 + A_2)^2 \tag{4}$$

For fission fragments, energy loss at high energies is due almost entirely to electronic collisions for which the specific energy loss per unit path length is given by:

$$(\partial\varepsilon/\partial\rho)_e = k\varepsilon^{1/2} \tag{5}$$

where

$$k = \xi_e \frac{0.0793\, Z_1^{1/2} Z_2^{1/2} (A_1+A_2)^{3/2}}{(Z_1^{2/3}+Z_2^{2/3})^{3/4}\, A_1^{3/2} A_2^{1/2}} \tag{6}$$

ξ_e is estimated by LSS to be roughly

$$\xi_e = Z_1^{1/6} \tag{7}$$

The total reduced stopping power will be the sum of the electronic and nuclear stopping contributions

$$(\partial\varepsilon/\partial\rho)_t = (\partial\varepsilon/\partial\rho)_e + (\partial\varepsilon/\partial\rho)_n \tag{8}$$

Integrating eq. (8), the range-energy relation is obtained:

$$\rho(\varepsilon) = \frac{2}{k}\left[\varepsilon^{1/2} - \frac{1}{2} k\, \Delta\, (k,\varepsilon)\right] \tag{9}$$

where the nuclear stopping term is given by LSS[5]. After inserting the numerical constants we obtain:

$$R_{int} = \frac{\left[\varepsilon^{1/2} - \frac{k}{2}\Delta(k,\varepsilon)\right]\, (Z_1^{2/3} + Z_2^{2/3})\, (A_1+A_2)^2}{83000\, A_1 k} \tag{10}$$

This equation has to be solved for k by an iterative process because $\Delta(k,\varepsilon)$ is not known analytically. The most probable nuclear change Z_p was taken as representative value for Z_1[6]. While from theoretical approaches, like above, the integrated range R_{int} is obtained, experimental determinations result in the projected range R_p. In order to make possible comparison between theoretical and experimental quantities the latter have to be corrected accordingly[5].

It is well known from the literature and has been shown by accurate range measurements of individual ^{252}Cf fission fragments in a large variety of stopping

K. H. Böckhoff (ed.), Nuclear Data for Science and Technology, 257–260.

media, that there are large discrepancies between the ranges calculated according to eq. (9) and the experimental data[3,7]. The same applies also to experimental energy loss data[1,4]. In order to overcome this discrepancy, Laichter and Shafrir have replaced eq. (7) by[3,4]:

$$\xi_e = Z_1^{\,x} \tag{11}$$

The x values were determined to satisfy eqs. (8) and (9) respectively, using experimental energy loss and range data.

In basic theories dealing with the slowing down of energetic heavy particles in matter, a smooth trend is assumed for the dependence of the stopping process on the atomic number Z_2 of the medium. However it is well known that there is considerable structure in the Z_2-dependence for light ions at about 1 MeV[8]. Recently the existence of Z_2-oscillations has also been demonstrated in the stopping of fission fragments[4] as well as for other heavy ions up to uranium[9].

Representative range results versus Z_2 for the ^{252}Cf-fission fragments ^{105}Ru and ^{149}Nd, shown in fig. 1, reveal clearly the existence of Z_2-oscillations in the stopping of fission fragments[4].

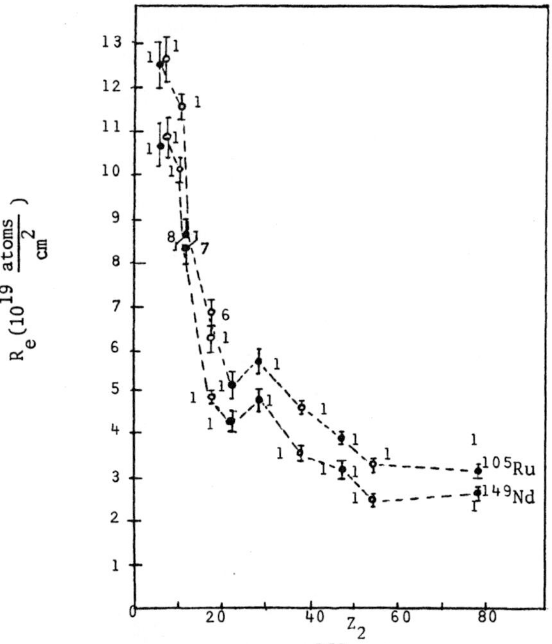

Fig. 1. Electronic ranges of ^{252}Cf fission fragments in solids and gases.

- ● - in solids
- o - in gases
- 1 - from Laichter and Shafrir[4]
- 6 - from Pickering and Alexander[7]
- 7 - from Birgul, et al.[22]
- 8 - from Viola, et al.[31]

In order to emphasise the oscillatory behaviour of the stopping process which may be attributed to the shell structure of the stopping atoms, electronic range data R_e in units of atoms/cm^2 are used. The R_e values are obtained by adding the calculated LSS nuclear range[5] to the total range value. Recently, significant improvements have been made to the calculation of nuclear stopping[10/11]. However, in the velocity domain in question, the nuclear stopping component is very small and thus the accuracy of this correction is of minor importance. In fig. 2 are shown representative x values versus Z_2 for the ranges of the fragment mass A_1 = 142 and A_1 = 105. In the same figure are also shown the x values for the mean heavy - M_H and mean light - M_L fission fragments of ^{252}Cf which are almost identical to A_1 = 142 and A_1 = 105, respectively. ΔE indicates the energy degradation of the initially unattenuated fragments.

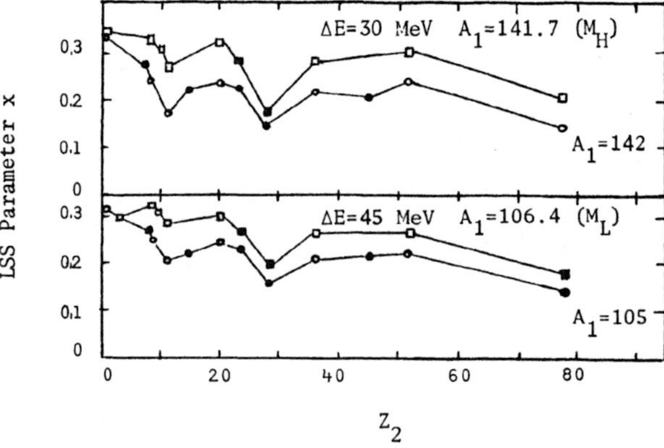

Fig. 2. Modified LSS-parameter-x versus atomic number of stopping medium - Z_2 for

- □ - thickness traversed in gases
- ■ - thickness traversed in solids
- o - ranges in gases
- ● - ranges in solids

We note that x fluctuates with Z_2 and that larger values for x are obtained from energy loss data than from range data. This indicates that x is also energy dependent[4].

Generation of Fission Fragment Range and Energy Loss Data

^{252}Cf FISSION FRAGMENT RANGES

From extensive range measurements of 22 individual ^{252}Cf fission fragments in 13 stopping media, as well as from additional experimental data taken from the literature[3], experimentally fitted x values versus Z_2 are obtained (see previous section). Plotting the x values versus fragment identity as defined by Z_p for any of the stopping media in which the range measurements have been performed shows a linear dependence. In fig. 3 representative results are given for Ti (Z_2 = 22) and Xe (Z_2 = 54).

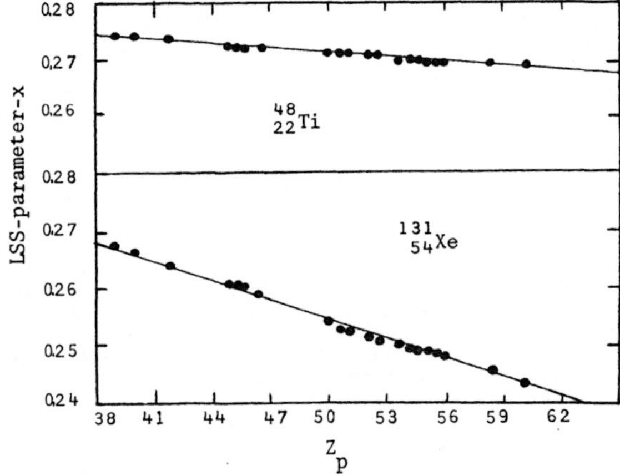

Fig. 3. Modified LSS-parameter-x versus most probable charge Z_p of fission fragment.

From interpolation between two nearby experimentally determined x values which represent the stopping process of a certain fission fragment (Z_p) in two different media, the x value for this fragement in any medium is found. Repeating this procedure for all measured fission fragments enables one to plot additional straight lines, like in fig. 3. In this way x values are found for the calculation of the range of any ^{252}Cf fragment in any stopping medium.

FISSION FRAGMENT RANGES OF OTHER FISSILE NUCLEI

Range data for individual ^{233}U, ^{235}U and ^{239}Pu fission fragments have been determined much less extensively than for ^{252}Cf fragments[12-21]. There do not seem to exist at all range data for ^{241}Pu fragments. We base the calculation of range data for the thermal fission of these four fissile nuclei on the existing extensive range data of ^{252}Cf fragments and the interpolation procedure described above as follows: The range of any individual fission fragment arising from the fissile nuclei in question is found by making use of the LSS-formulation as given in eq. (9), introducing the proper energy and most probable nuclear change Z_p[8] of the fragment. The k value to be used in the equation is found by inserting into eq. (11) the same x value as found for the respective ^{252}Cf fragment. In table I an illustrative sample for such a calculation is given for the range of ^{140}Ba in aluminum, and the results compared with existing experimental data. It can be seen that the ranges calculated by this method deviate only a few per cent from the experimental ranges, the deviation being within the experimental error.

Table I Calculated and Experimental Ranges of the Fission Fragment ^{140}Ba in Aluminum.

Fissioning nuclide	E (MeV)	Z_p 6)	R_{cal} (mg/cm²)	R_{exp}
^{233}U	67.7	55.02	2.71	2.92 10)
^{235}U	70.4	54.60	2.81	2.98 11)
^{239}Pu	74.5	55.34	2.89	3.04 10)
^{241}Pu	76.3	54.55	2.90	-
^{252}Cf	80.5	54.69	2.98	3.03 18)

FISSION FRAGMENT ENERGY LOSS

Existing experimental energy loss data for ^{252}Cf-fission fragments are almost exclusively for M_H and M_L[4,23-26] and the overall mean fragment M[1,27-29]. It has recently been shown that only minor changes in the mass and nuclear charge of the mean fragments occur while they are slowed down in matter[20]. The measurements are usually performed by determining the energy degradation ΔE of the fragment for a certain thickness of the stopping medium traversed. A second order polynom is then fitted to the experimental points as follows:

$$E_1 = E_1^0 + a_1 x_m + a_2 x_m^2 \quad (12)$$

where E_1^0 is the unattenuated fission fragment kinetic energy and E_1 the kinetic energy of the fragment after traversing thickness x_m. By derivation of eq. (12) the total atomic cross section is calculated by:

$$S = A_2/N^0 \; (dE/dx_m) \quad (13)$$

where N^0 is the Avogadro number. In order to isolate the pure electronic stopping effect, S is corrected for the nuclear stopping effect by substracting the calculated nuclear stopping component, as in the case for range data[5]. In figs. 4 and 5 are shown the resulting electronic cross section versus Z_2, for M_H-fragments at 0.5 MeV/u and 0.2 MeV/u and for M_L-fragments at 0.9 MeV/u and 0.5 MeV/u, respectively.

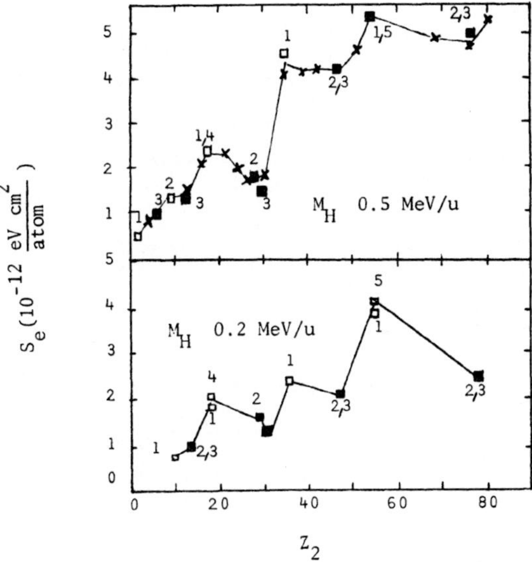

Fig. 4. Electronic stopping cross section Se versus Z2 for 0.5 MeV/u and 0.2 MeV/u M_H fragments from ^{252}Cf.

■ - in solids

□ - in gases

1 - from Laichter and Shafrir[2]

2 - from Forgue and Kahn[23]

3 - from Muller and Gonnenwein[24]

4 - from Bertin, et al.[25]

5 - from Kahn and Forgue[26]

x - values for Xe[23,27]

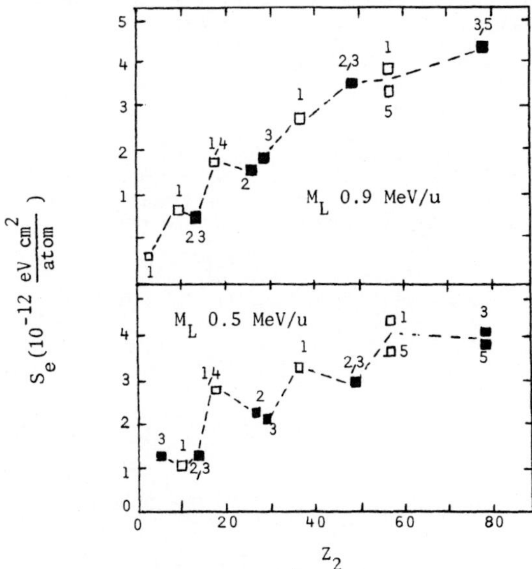

Fig. 5. Electronic stopping cross section Se versus atomic number of stopping medium - Z_2, for 0.2 MeV/u and 0.5 MeV/u M_L - fission fragment from ^{252}Cf.

For explanation of symbols and references, see text of fig. 4.

From these figures it is clearly seen that the Se values show a considerable structure in the Z_2 dependence. Plotting the Se values versus E_1/A_1 for all the stopping media in which energy loss experiments have been performed for M_L, M_H and M, enables interpolation for Se in any stopping medium Z_2.

In table II are given the mass and kinetic energy of mean fission fragments.

Table II Mass and Kinetic Energy of Mean Fission Fragments

Fissioning nuclide	E_H^0 (MeV)	M_H	Z_H	E_L^0 (MeV)	M_L	Z_L
^{233}U	67.7	138.2	54.26	99.9	93.3	37.68
^{235}U	68.4	138.6	53.63	99.8	94.9	38.04
^{239}Pu	73.2	138.1	54.53	98.8	100.5	40.45
^{241}Pu	76.3	138.8	54.07	103.2	100.4	40.12
^{252}Cf	79.3	141.7	55.39	104.1	106.4	41.58

From the knowledge of the Se values of the ^{252}Cf fragments and table II, Se values for fission fragments of other fissile nuclei can be found by using the following expression:

$$Se = \frac{[Se(E/A)^{1/2} Z_p^2]_{Cf}}{(E/A)^{1/2} Z_p^2} \qquad (14)$$

Adding to the calculated Se values the nuclear stopping component as described before[5], S values for mean fission fragments of various fissile materials in all stopping media can be found.

Conclusion

The Z_2 oscillations of the stopping cross sections and ranges of fission fragments have been clearly demonstrated. These oscillations set a limit to the accuracy one can expect from heavy ion range and stopping power calculations when using a smooth Z_2-dependence. Thus the use of any basic theory for the prediction of range and energy loss data, even after applying experimental fitting factors, is only justified when these oscillations are taken into account, as has been applied in this paper. Further investigation of the Z_2- oscillatory behaviour in the stopping of fission fragments and inclusion of their effect into an experimentally modified stopping expression are in progress.

References

1. M. Hakim and N.H. Shafrir, Can. J. Phys. 49, 3024 (1971).
2. Y. Laichter and N.H. Shafrir, Nuc. Instr. and Meth. 177, 459 (1980).
3. Y. Laichter and N.H. Shafrir, Nucl. Phys. A371, 45 (1981).
4. Y. Laichter and N.H. Shafrir, Nucl. Phys. A, to be published (1982).
5. J. Lindhard, M. Scharff and H.E. Schiott, Mat. Fys. Medd. Dan. Vid. Selsk 33, no. 14 (1963).
6. D.R. Nethaway, Tables of Values of Z_p, the Most Probable Charge in Fission, Lawrence Livermore Laboratory, Report UCRL-51640 (1974).
7. M. Pickering and J.M. Alexander, Phys. Rev. C6, 332 (1972).
8. J.F. Ziegler and W.K. Chi, At. Nucl. Data 13, 463 (1974).
9. H. Geissel, Y. Laichter, W.F.W. Schneider and P. Armbuster, G.S.I. - Nachrichten 2-82 (unpublished).
10. W.D. Wilson, L.G. Haggmark and J.P. Biersack, Phys . Rev., 15B, 2458 (1977).
11. J. Ziegler, Appl. Phys. Lett. 31, 544 (1977).
12. S.P. Dange, H.C. Jain, S.B. Manohar, K. Satyprakash, M.V. Ramaniah, A. Ramaswami and K. Regan, in: Physics and Chemistry of Fission (IAEA, Vienna, p. 741, 1969).
13. F. Demichelis, R. Liscia and A. Tartaglia, Il Nuovo Cimento, 10B, No. 2, 1972.
14. B. Chingalia, F. Demichelis and A. Tartaglia, Il Nuovo Cimento, Serie I, 4, 1185 (1970).
15. N.K. Aras, M.P. Menon and G.E. Gordon, Nucl. Phys. 69, 337 (1965).
16. O. Selig and R. Sizmann, Nukleonik 8, 303 (1966).
17. J.M. Slexander and M.F. Gazdik, Phys. Rev. 120, 874 (1960).
18. S. Hontzeas and H. Blok, Physica Scripta 4, 229 (1971).
19. M. Hollstein and H. Munzel, Radiochimica Acta 5, 195 (1966).
20. M. Hollstein and H. Munzel, Radiochimica Acta 6, 9 (1966).
21. G. Rohde and H. Munzel, Radiochimica Acta 8, 96 (1967).
22. O. Birgul, I. Olmez and N.K. Aras, Radiochimica Acta 18, 198 (1972).
23. V. Forgue and S. Kahn, Nucl. Inst. and Meth. 91, 357 (1971).
24. R. Muller and F. Gonnenwein, Nucl. Inst. and Meth. 91, 357 (1971).
25. A. Bertin, M. Bruno, G. Vannini, A. Vitale and M. Forte, Phys. Letters 43A, 231 (1973).
26. S. Kahn and V. Forgue, Phys. Rev. 163, 290 (1967).
27. P.M. Mulas and R.C. Axtmann, Phys. Rev. 145, 296 (1966).
28. M. Hakim and N.H. Shafrir, Nucl. Sci. Eng. 48, 72 (1972).
29. R.C. Axtmann and J.T. Sears, Nucl. Sci. Eng. 23, 299 (1965).
30. Y. Laichter, H. Geissel and N.H. Shafrir, Nucl. Inst. and Meth. 194, 45 (1982).
31. V.E. Viola, G.E. Gordon and W.B. Waters, Research in Nuclear Chemistry, Annual Progress report, Maryland University, MNC-4028-0012 (1973).

HIGH RESOLUTION MEASUREMENTS OF DELAYED NEUTRON EMISSION SPECTRA FROM FISSION PRODUCTS

T.-R. Yeh, D. D. Clark, and G. Scharff-Goldhaber

Ward Laboratory, Cornell University
Ithaca, N.Y. 14853 U.S.A.

R. E. Chrien, L.-J. Yuan, M. Shmid, and R. L. Gill

Brookhaven National Laboratory
Upton, N.Y. 11973 U.S.A.

A. E. Evans

Los Alamos National Laboratory
Los Alamos, N.M. 87545 U.S.A.

H. Dautet and J. Lee

McGill University
Montreal, Quebec H3A ZT8 Canada

Low-energy delayed neutron spectra of rubidium fission products have been measured in exploratory experiments with the time-of-flight technique using ^{3}He gas scintillator and ^{6}Li glass scintillator neutron detectors. The principal aim has been to investigate the feasibility of high resolution experiments capable of yielding level parameter data on neutron-rich far-unstable nuclides. Such data could be of use in constructing neutron cross section estimates for fission products. FWHM energy resolutions of ∿10% have been obtained for neutrons of 10 keV and further tests indicate resolutions approaching 1% could be achieved. Several peaks have been found in ^{95}Rb and ^{97}Rb. Survey measurements of the spectra of $^{93-97}$Rb and $^{143-146}$Cs have also been carried out with a Cuttler-Shalev detector. Both experiments used the TRISTAN facility at BNL.

[radioactivity, delayed neutrons, $^{93-97}$Rb, $^{143-146}$Cs, T_n spectra, time-of-flight, ^{3}He ion chamber]

Introduction

Quite apart from the practical value of data on fission-product delayed neutrons for reactor applications, the phenomenon affords unique empirical clues to neutron cross sections of far-unstable nuclides, including even cross sections of nuclei in excited states. The key point is to exploit the inverse relationship between neutron emission and absorption by compound-nuclear levels above the neutron binding energy. We report here exploratory high resolution measurements of the energy spectra of delayed neutrons (DN) of energies between 1 and 100 keV emitted in decay of mass-separated fission products. One purpose is to provide spectra to lower energies than previously observed, but the major motivation is to explore the feasibility of obtaining data of sufficient quality and quantity to allow construction of empirically based cross sections for very neutron-rich nuclides. Such cross sections are of interest for astrophysics and for nuclear theory.

Fig. 1 illustrates the DN process. The partial yield P_n^i, the fraction of neutrons emitted per beta decay that leave GC in state i, is governed by the combined probabilities of beta decay and neutron emission. The groundbreaking and fruitful work of Kratz and his collaborators (e.g. Ref. 1) has shown that for properties averaged over spans of NE levels it is useful to treat the former by the beta strength function and the latter by neutron transmission factors based on optical models. (See also Schenter et al.[2]) In the present work, however, the aim is to obtain level parameters of a number of individual isolated NE levels. Such parameters are known for stable nuclides from cross section studies. By extrapolation from those studies, expected values of NE level spacings D and widths Γ_n lead to the conclusion that DN energy spectra at keV energies will show clearly resolvable peaks associated with single NE levels. With high resolution, peak analyses could even yield estimates of widths of levels with large Γ_n. Even partial sets of data on level spacings and widths can cast useful light on neutron cross section estimates and can complement work such as that in Ref. 2.

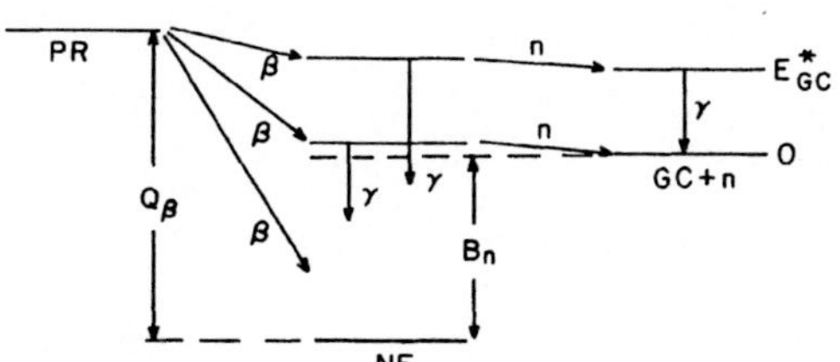

Fig. 1. The possible branchings in decay of a delayed neutron precursor. PR = precursor, NE = neutron emitter, GC = grandchild nuclide.

Experimental Method

The experiments were performed at Brookhaven National Laboratory. The source of precursors was the TRISTAN facility, which has been described elsewhere.[3] Its general features are shown in Fig. 2. A beam of about 10^{11} neutrons/cm^{2}sec is incident on a heated graphite target impregnated with several grams of enriched ^{235}U. For the experiments reported here, a positive surface ionization source was in use, and Rb and Cs isotopes were the principal available precursors. Magnets and lenses shown in Fig. 2 focused the ions onto a moving aluminized Mylar tape at the 0^0 port. Ion intensity varied with mass number and operating history but ranged as high as several × 10^5 per second.

The best method for high resolution spectroscopy of keV neutrons is time of flight (TOF). We have tried

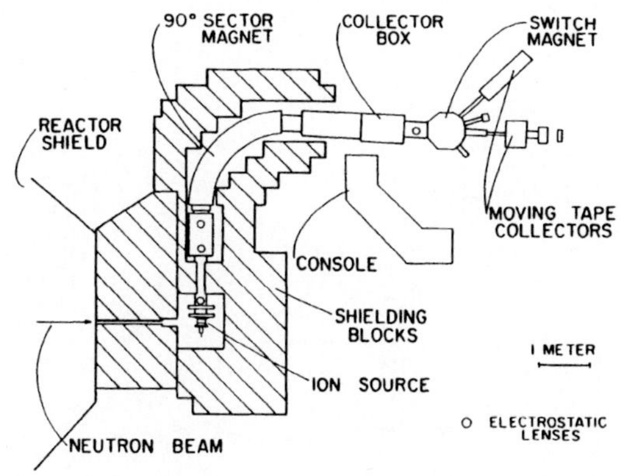

Fig 2. Layout of the on-line mass separator TRISTAN at the Brookhaven High Flux Beam Reactor (HFBR).

K. H. Böckhoff (ed.), Nuclear Data for Science and Technology, 261–264.

TOF with several detectors but have also employed a Cuttler-Shalev detector for surveys at lower resolution (to be described later). For the TOF system the populating beta ray and the emitted neutron provided the timing signals. For counting rate reasons betas were used for stops. They were detected by a telescope consisting of a thin and a thick fast plastic scintillator, each viewed by an RCA 8850 phototube. The telescope was for determination of beta spectra when appropriate. For the neutrons, three different scintillation detectors have been tested; each could be connected to provide a fast timing signal for TOF and a slower signal for pulse height analysis.

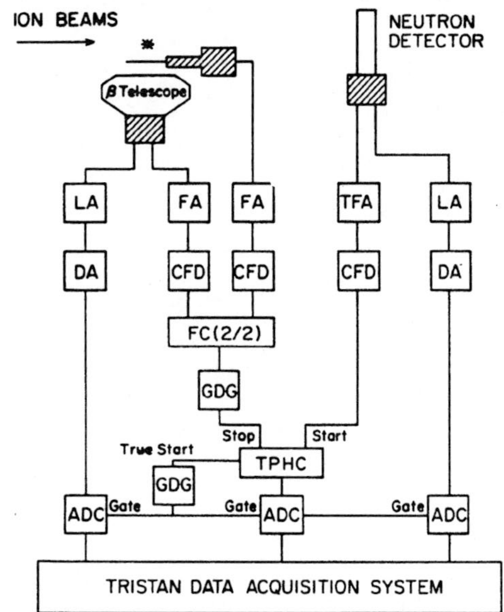

Fig. 3. Electronics block diagram for TOF system. FA = fast amplifier, CFD = constant fraction discriminator, FC = fast coincidence, TFA = timing filter amplifier, LA = linear amplifier, DA = delay amplifier, GDG = gate and delay generator, TPHC = time-to-pulse-height converter, and ADC = analog-to-digital converter.

Data acquisition was in event mode with three digitized parameters: time of flight, neutron detector pulse height, and thick beta detector pulse height. A block diagram is given in Fig. 3. The versatile computer-based acquisition and analysis systems that were used are described in Ref. 3.

In our application the neutron detector has to be sensitive to 1 to 100 keV neutrons, be capable of timing to 2 ns or better, and have dimensions along and transverse to the flight path axis such that geometrical contributions to the system energy resolution are suitably small. In addition, the detector should be insensitive to gamma rays (or permit good pulse shape discrimination), have high efficiency, and subtend a large solid angle. No detector can satisfy all these criteria. We have tested three, all scintillators, two of ^{6}Li glass and one with ^{3}He high pressure gas. Chronologically the first was a 2-mm thick 114-mm diameter disk of NE908 viewed edge-on by three RCA C31024 phototubes in a two out of three majority coincidence mode for timing and a sum of all three for pulse height analysis. The second was the high pressure gas detector, designed originally by A. E. Evans for another experiment, which had been loaded with ∿134 atm ^{3}He and ∿2.7 atm Xe (to enhance scintillations) and was viewed by an Amperex XP-2000 phototube. Because of the high pressure, this unit was in the form of a cylinder with a hemispherical cap on one end and a thick glass window on the other; the internal diameter was 44.5 mm and overall internal length 80 mm. The third detector, a recent loan from Argonne National Laboratory, was a 9.5-mm thick 127-mm diameter disk of NE912 viewed broadside by a selected RCA 4522 phototube. All three detectors are more sensitive to thermal neutrons than to those of the desired energies; hence boral or Cd shielding was required to reduce the contribution by room neutrons to random coincidences. The ^{3}He detector is the least gamma-sensitive of the three and the most neutron-efficient. Although its geometry is awkward for high resolution, it proved very useful for exploratory runs; see Fig. 4.

The beta telescope can in principle be used to determine which state in GC is populated by the neutrons in a given TOF peak (by finding the endpoint energy) as well as to measure the beam intensity of the

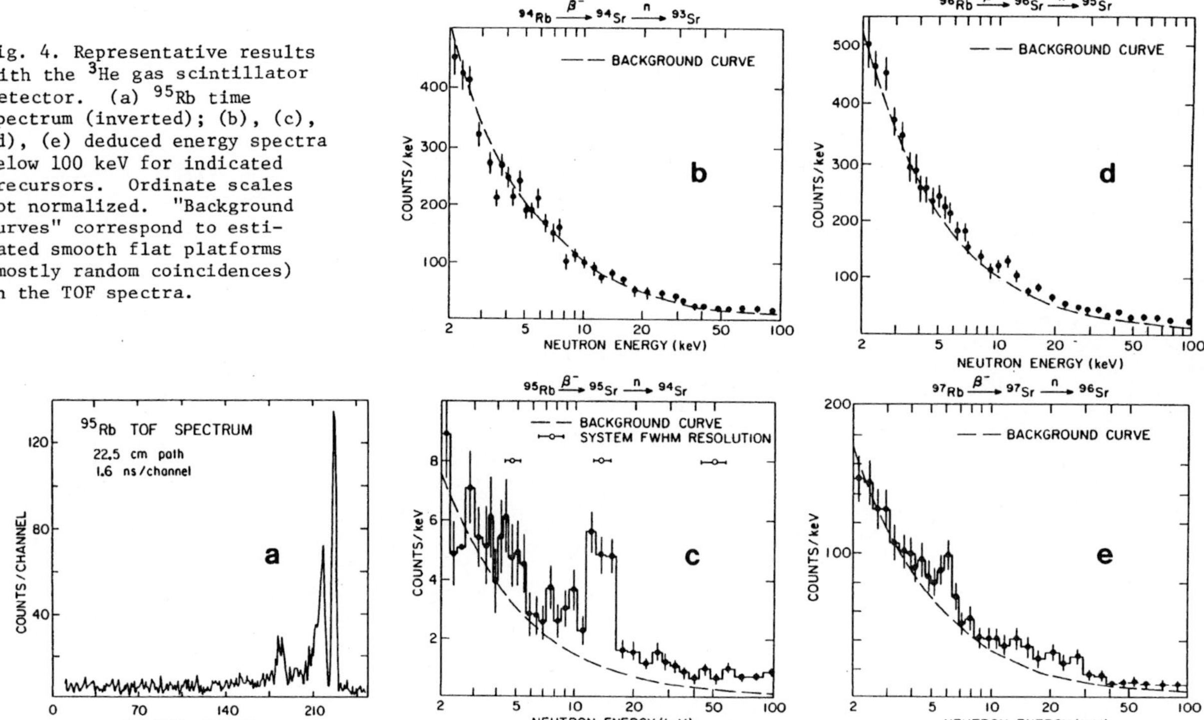

Fig. 4. Representative results with the ^{3}He gas scintillator detector. (a) ^{95}Rb time spectrum (inverted); (b), (c), (d), (e) deduced energy spectra below 100 keV for indicated precursors. Ordinate scales not normalized. "Background curves" correspond to estimated smooth flat platforms (mostly random coincidences) in the TOF spectra.

desired isobar in the mass chain at which TRISTAN is being operated. However, in practice it is usually much more satisfactory for both purposes to record the gamma spectra in coincidence with TOF events by using an auxiliary Ge(Li) detector. This approach requires knowledge of several decay schemes but is easier because analysis of beta spectra is avoided.

The flight path length is chosen to balance the conflicting criteria of counting rate and energy resolution. We have used values from 22.5 to 35 cm, measured from the moving tape to the center of the neutron detector.

Calibration and testing of this type of TOF system present several problems. It is not appropriate to use running time at the separator for the extended tests needed to optimize timing resolution, detector geometry, flight path, collimators, and so forth. We have adopted ^{252}Cf as a convenient source for off-line tests. Its (prompt) fission neutron spectrum is a standard and prompt gamma rays can substitute for beta rays at the beta detector (using the thick detector only). Selected regions of the TOF spectrum can be used to gate the neutron detector pulse height spectrum and thereby study the variation with energy of detector response. The fact that several gammas and neutrons are emitted together complicates the analysis but can be turned to diagnostic advantage; e.g., the gamma response of the neutron detector and the neutron response of the gamma (beta) detector can be isolated. On the other hand, the ^{252}Cf spectrum is not very well known at lower energies and it has no sharp peaks to allow tests of the TOF system energy resolution.

Data Analysis

The conversion of a TOF spectrum to an energy spectrum is straightforward. One question is whether to convert before or after subtracting the platform, due to random coincidences and other causes, on which the TOF peaks ride. As seen in Fig. 4, we choose to convert first to get a more instructive result. Energy bins were combined after the conversion to obtain bin widths appropriately sized with respect to the energy resolution function of the system.

Calibrations of the electronic system including timing were done by standard methods. The relative efficiency vs energy curves for the detectors were calculated from the known cross sections of Li and He. For the gas scintillator, the absolute efficiency was measured with the scandium and iron filtered beams[4] of the HFBR. Au foils were used to obtain the beam intensities. The detector, as expected, showed a 1/v variation between the 2 keV and 24 keV points.

The overall energy resolution of the TOF system is a key parameter. The better it is, the larger the sample of levels for which we will be able to obtain useful data. The system resolution is determined by Δt, the full width at half maximum (FWHM) of the electronics, including detector properties, and by $\Delta \ell$, the FWHM geometrical uncertainty in the flight path ℓ. The two contributions are independent, so they combine quadratically by the formula

$$\frac{\Delta T_n}{T_n} = \frac{2}{\ell}\sqrt{(\Delta \ell)^2 + (v\Delta t)^2}$$

where T_n and v are the kinetic energy and velocity of the neutron. Fig. 5 gives the calculated FWHM system resolution in percent for the three detectors we have used and for two different values of Δt. For these calculations we have taken $\Delta \ell$ to be one-half the detector thickness in the direction of the central flight path. The gas scintillator detector was used with its cylindrical axis perpendicular to the flight path, so in its case a weighted average thickness was employed. The differing variation with energy of the two contributions to the resolution is evident in the figure: the geometrical term dominates for the gas scintillator, and timing term for the 2-mm lithium glass. The figure is for ℓ = 25 cm; for a 50 cm path the ordinate scale would be divided by two. With the gas scintillator the typical Δt has been no better than 6 to 8 ns, but with the 9.5-mm lithium glass we have reached <3 ns in off-line tests. We anticipate being able to have better than 5% system energy resolution at 100 keV and down to about 1% below 10 keV in future experiments.

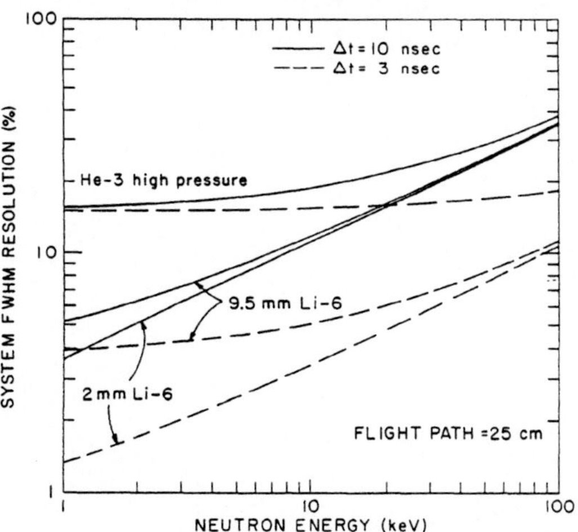

Fig. 5. Calculated TOF system FWHM energy resolution for three different detectors. See text for details.

Results

Fig. 4 displays some representative results from the current exploratory phase of our program. The data were taken with the ^3He gas scintillator detector. Fig. 4(a) is a raw TOF spectrum for ^{95}Rb; from right to left, the three peaks are due to gammas, to neutrons above about 100 keV, and, around channel 185, to neutrons of about 13 keV. Fig. 4(c) is the corresponding energy spectrum below 100 keV (note the logarithmic energy scale). The 13-keV peak, which has been reported previously by the Kratz group[5] in both Cuttler-Shalev and TOF experiments, is the strongest but clearly not the only low energy peak in the spectrum. Subsequent runs not only confirmed the existence of several peaks but also suggested that the strong peak might be as low as 11.5 or 12 keV. That conclusion should remain tentative, however, until confirmed by new runs in which some possible remaining systematic errors are eliminated.

Figs. 4(b), 4(d) and 4(e) are energy spectra taken in sequence with constant system parameters. The path here was 33 cm. ^{97}Rb clearly displays a peak at ~6.5 keV, but there is less evidence of structure in ^{94}Rb and ^{96}Rb since the GC nuclide is odd A and the levels near its ground state are more closely spaced. A higher resolution run would probably bring out some structure, however.

In a separate experiment on ^{95}Rb, the absolute P_n of the "13"-keV peak was found to be (0.50 ± 0.15)%. The absolute efficiency of the gas scintillator detector, determined as described earlier, was used, and the absolute intensity of ^{95}Rb ions was found from the counting rate in a calibrated Ge detector of 352-keV gamma rays whose absolute intensity in the decay scheme had been reported in Ref. 1. This P_n value agrees with an estimate using the shape of the energy spectrum and the accepted value of 8.9% for the total P_n of ^{95}Rb.

Cuttler-Shalev Detector Surveys

Surveys of $^{93-97}Rb$ and $^{143-146}Cs$ delayed neutron spectra were also carried out with a Cuttler-Shalev ionization chamber. This detector covers a larger energy range than our TOF system but has poorer energy resolution at the low end -- the FHWM of our detector was ~13 keV for thermal neutrons. Although the uncertainties introduced in unfolding low energy peaks from the thermal peak can be large, Cuttler-Shalev spectra are useful for our purposes because

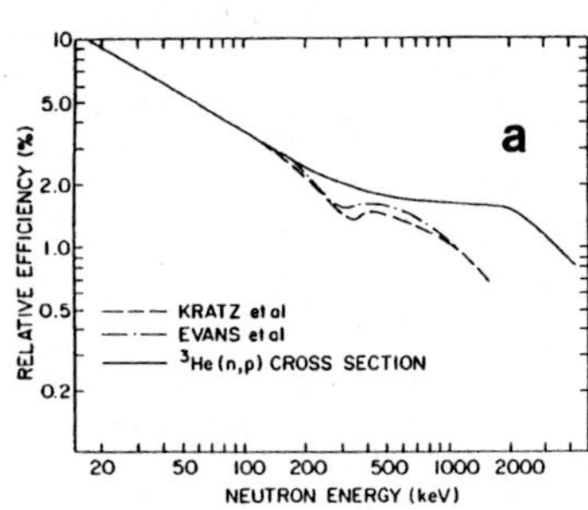

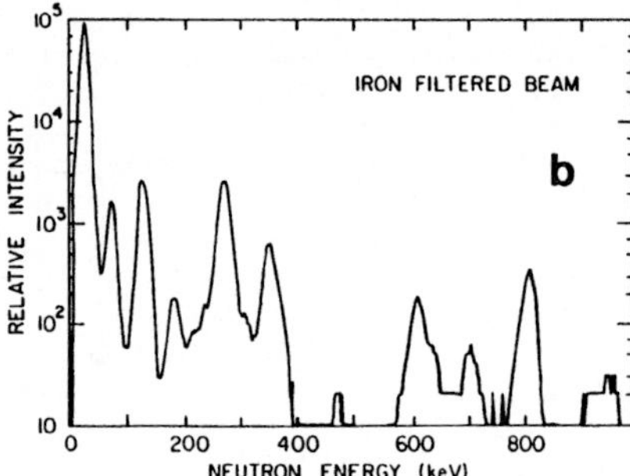

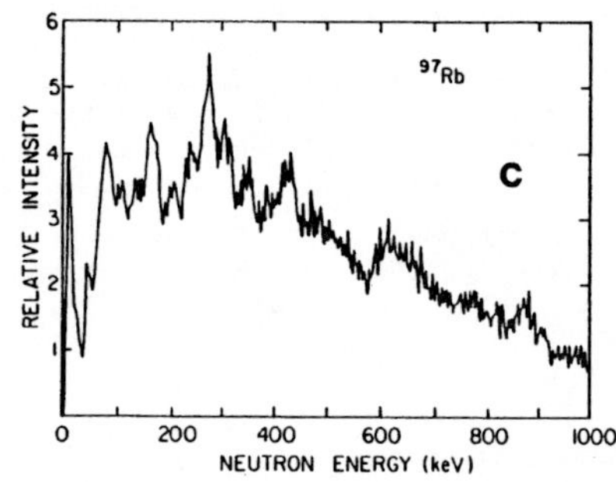

Fig 6. Cuttler-Shalev detector data. (a) Relative efficiency vs energy. (b) Energy spectrum of the 24-keV iron-filtered beam of the HFBR (note the logarithmic intensity scale). (c) DN energy spectrum of ^{97}Rb.

they can reveal whether or not there is much intensity in the low energy region.

Data were taken at the 0^0 port of TRISTAN, and were corrected for gamma pile-up, detector response function, and relative efficiency vs energy by standard methods such as those described in Ref. 6. The absolute efficiency at 24 keV was measured in the iron filtered beam of the HFBR. The energy calibration was done with a pulse generator based on the Q-value of the ^{3}He(n,p) reaction. Fig. 6(b) is the resulting energy spectrum of the filtered beam, and Fig. 6(c) is an example of the delayed neutron spectra we obtained. It shows a peak at very low energy for ^{97}Rb which probably corresponds to the more clearly resolved 6.5-keV peak in Fig. 4(e).

Discussion

Delayed neutron emission from fission products is a unique probe into level parameters of far-unstable neutron-rich nuclides. With sufficient data it should be possible to "measure" neutron cross sections for such nuclei by exploiting the inverse relation between emission and absorption. In a recent interesting study, Fogelberg et al.[7] have shown that levels in ^{87}Kr observed as resonances in the ^{86}Kr + n cross section correspond in one-to-one fashion with peaks in the DN spectrum of ^{87}Br that would be expected when spin and parity considerations are taken into account. (This is one of the only two cases of precursors for which the GC nuclide is stable and therefor for which the inverse reactions can be studied experimentally.) To make real use of this potential probe, however, the DN spectra must be investigated with the highest available resolution. The purpose of our experiments has been to explore the possibilities of the time-of-flight method. We have extended measurements to below 5 keV with ∿10% resolution. We expect to achieve improvements to the level of 1% FWHM in a second stage of experiments and hope to identify and characterize a number of unbound levels for use in cross section estimates.

Acknowledgment

This research was supported by the United States Department of Energy.

References

1. K.-L. Kratz, A. Schröder, H. Ohm, M. Zendel, H. Gabelmann, W. Ziegert, P. Peuser, G. Jung, B. Pfeiffer, K. D. Wünsch, H. Wollnik, C. Ristori, J. Crancon, Z. Phys. A306, 239 (1982).

2. R. E. Schenter, R. M. Mann, R. A. Warner, P. L. Reeder, this converence.

3. R. L. Gill, M. L. Stelts, R. E. Chrien, V. Manzella, H. Liou, Nucl. Instrum. Methods 186, 243 (1981).

4. R. C. Greenwood, R. E. Chrien, Nucl. Instrum. Methods 138, 125 (1976).

5. K.-L. Kratz, H. Ohm, A. Schröder, H. Gabelmann, W. Ziegert, H. V. Klapdor, J. Metzinger, T. Oda, B. Pfeiffer, G. Jung, L. Alquist, G. I. Crawford, Proc. 4th Int. Conf. on Nuclei Far from Stability, CERN 81-09, p. 317 (1981).

6. A. E. Evans, M. S. Krick, Nucl. Sci. Eng. 62, 652 (1977), H. Franz, W. Rudolph, H. Ohm, K.-L. Kratz, G. Hermann, F. M. Nuh, D. R. Slaughter, S. G. Prussin, Nucl. Instrum. Methods 144, 253 (1977).

7. B. Fogelberg, J. A. Harvey, R. L. Macklin, S. Raman, P. H. Stelson, in CERN 81-09, p. 339.

DELAYED NEUTRON SPECTRAL MEASUREMENTS AND COVARIANCE ERROR ANALYSIS FOR FAST FISSION IN ^{235}U AND ^{239}Pu

J. Walker, D.R. Weaver and J.G. Owen

Birmingham Radiation Centre
University of Birmingham
Birmingham, UK

A gridded ^{3}He ionisation chamber has been used to measure near-equilibrium delayed neutron spectra from fast fission in U-235 and Pu-239; experiments have been conducted with incident neutrons of various energies from 0.5 MeV to 7.5 MeV. An extensive analysis of errors has been made and a method devised for the calculation of the full covariance error matrix for the unfolded spectra, including contributions from both counting statistics and uncertainties in the response of the spectrometer.

Introduction

The importance of accurate delayed neutron data in predicting the kinetic response of nuclear reactors, particularly the fast reactor under fault conditions, has been recognised for several years[1]. Unfortunately the existing information is sparse, with only a few measurements with primary neutron energies in the fast region[2,3]. Disagreements also exist between spectra accumulated by various authors; particular emphasis has therefore been placed in this work on the calculation of good error information on the results and this has led to the generation of full covariance error information on all unfolded spectra. The error contributions from both counting statistics and uncertainties in the spectrometer's response function were included in the calculation of the covariance matrices.

Spectrum Measurements

A gridded ^{3}He spectrometer of the Shalev type[4] was used in these measurements. It contained 4.2 atmospheres of ^{3}He, 2.1 atmospheres of argon, and 0.3 atmospheres of methane. The resolution (FWHM) was 14.6 keV for thermal neutrons, and 27 keV at 1 MeV. In some of the later experiments a second ^{3}He spectrometer, filled to a total pressure of 10 atmospheres, was also employed. This second detector had a resolution of 11.6 keV for thermal neutrons, but the results taken with this detector have yet to be analysed because the response function measurements were not made until after the delayed neutron runs.

Pulse rise time discrimination was used in order to minimise the interference in the spectrometer's pulse-height distribution caused by recoil ^{3}He ions. The recoiling ^{3}He nucleus is more highly ionising than either the proton or the triton, and these recoil pulses therefore on the whole have short rise-times. By measuring and recording both the rise-time and pulse height parameters of the pulse we were able to discriminate against these recoil pulses during the analysis of the data.

Delayed neutron experiments have been performed on the Birmingham Radiation Centre's Dynamitron accelerator, and on the Tandem and IBIS accelerators at AERE Harwell. Experimental parameters are listed in Table I.

Obviously it is impossible to detect delayed neutrons in the presence of source neutrons and those emitted at the time of fission; the accelerator beams were therefore pulsed on and off. On the Dynamitron, pulsing was achieved by the application of an electrostatic potential to deflection plates in the high voltage terminal; on the Tandem a post-acceleration mechanical chopper was employed, whilst on the IBIS post-acceleration electrostatic deflection was applied before the beam entered a bending magnet. In all cases the reduction in primary neutron source strength from the beam-on to beam-off condition was better than 10^6:1.

Table I The Delayed Neutron Experiments

Location	Fissile sample	Mean primary neutron energy (MeV)	Neutron-producing target
Birmingham Dynamitron	U-235 (105g, 96.54% U-235)	0.49 0.94 1.44 1.76 6.0	Thick titanium-tritium or titanium-deuterium solid targets
Harwell Tandem	U-235 (214g, 96.54% U-235)	7.5 6.0	Deuterium gas target, 10 cm, 2 atmospheres
Harwell IBIS	Pu-239 (63g, 100%)	2.0	Tritium gas target, 5 cm, 1 atmosphere
		0.5	Titanium-tritium solid target

The timing cycle chosen was 0.8 seconds beam on and 1.0 seconds beam off. After the beam was turned off a delay of 0.1 seconds was imposed before counting began to allow recovery by the detector from the severe overload condition whilst the beam was on. The delay also allowed time for the primary neutrons scattered from the walls of the beam room to die away. Such a timing cycle produces a near-equilibrium distribution of the delayed neutron groups, as defined by the Keepin six-group model[5].

Spectrum Analysis

The spectrometer response to a mono-energetic neutron flux was represented by two Gaussians and a wall effect contribution:

$$Y(E) = P(1)\{\exp(-\tfrac{1}{2}(\frac{E-E_n}{P(2)})^2) + P(3)\exp(-\tfrac{1}{2}(\frac{E-(E_n-P(4))}{P(5)})^2) + P(6)\omega(E)\} \quad (1)$$

K. H. Böckhoff (ed.), Nuclear Data for Science and Technology, 265–267.

where Y(E) is the magnitude of the pulse height distribution at an energy E

E_n is the energy of the incident neutron

ω(E) is the wall effect prediction

P(1) is a normalising factor

and P(2) to P(6) are terms determined from a non-linear least squares fit to the data.

The response was measured with near monoenergetic neutron beams at several incident neutron energies, and at each was fitted to equation 1 by means of the non-linear least squares program. The variations in the terms P(2) to P(6) with incident neutron energy were plotted and fitted by polynomials of up to fourth order. Thus it was possible to predict the shape of the spectrometer's response to any incident neutron energy by first interpolating with the polynomials to derive the parameters P(2) to P(6), and then using these in equation 1 to give the pulse-height distribution.

The spectrometer efficiency was measured by comparison with a de Pangher precision long counter[6], cross-calibrated with the one at the National Physical Laboratory, Teddington, England.

In any neutron measurement, the relationship between the measured pulse-height distribution, $\underline{C}$, and the true neutron spectrum $\underline{\Phi}$ is:

$$\underline{C} = \underline{R} \cdot \underline{\Phi} \qquad (2)$$

where $\underline{R}$ is the response function of the detector. A computer code has been written which makes a guess at the true flux distribution (initially by dividing the measured pulse-height distribution by the detection efficiency), multiplies this by the response function (generated using equation 1) and compares the result, by means of a chi-squared test, with the measured pulse height distribution. If suitable agreement does not exist between the experimental and predicted values, a better approximation to $\underline{\Phi}$ is calculated and the procedure continued until the change in chi-squared between two successive iterations is less than some pre-determined value (usually 2%) when the flux distribution, $\underline{\Phi}$, is determined.

The determination of the errors on the final spectrum is clearly important, especially in view of the low counting rates which are inherent in these experiments. The method employed in determining the errors has been described in detail elsewhere (as has the calibration of the spectrometer and the spectrum unfolding procedure[7]), but the basis of the operation is to vary each parameter within the unfolding operation by a known amount, and then repeat the unfolding. Thus each channel of the pulse-height distribution is in turn changed by 1 standard deviation and the unfolding repeated in each case. Similarly, each parameter which is used to define the response function, both in shape and efficiency, is varied, and a revised unfolded distribution produced. In this manner a difference matrix $\underline{D}$ between each of the unfolded cases and the unperturbed result can be generated, and the error on the unfolded spectrum $\underline{V}_y$ then calculated from the relationship:

$$\underline{V}_y = \underline{D}^T \underline{V}_x \underline{D} \qquad (3)$$

where $\underline{V}_x$ is the covariance matrix of the input parameters. (It is important to note that the method of determining the covariance matrix is independent of the method employed to unfold the data; it merely requires that several unfolding runs are performed).

Delayed Neutron Spectra

The delayed neutron spectrum from ^{235}U following fast fission induced by neutrons of 0.94 MeV and 1.76 MeV are shown in Figures 1 and 2 respectively; the errors shown are the square root of the variance terms from the diagonal of the covariance error matrix. Typically these errors are ± 10% at low energies, ± 12% at 500 keV, and ± 25% at 1 MeV. Other spectra obtained from the experiments in Birmingham at different incident energies are similar although not identical and are known to approximately the same accuracy.

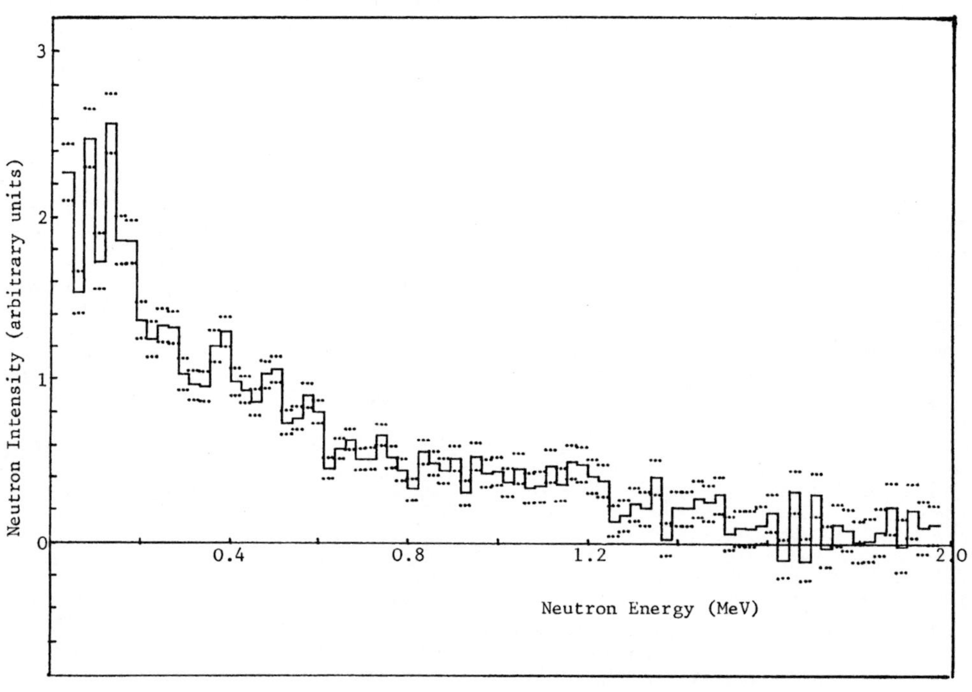

Figure 1: Delayed neutron spectrum for fast fission of ^{235}U produced by primary neutrons of mean energy 0.94 MeV. Each energy bin is 23.5 keV wide. Errors are indicated as $\pm$ 1 standard deviation.

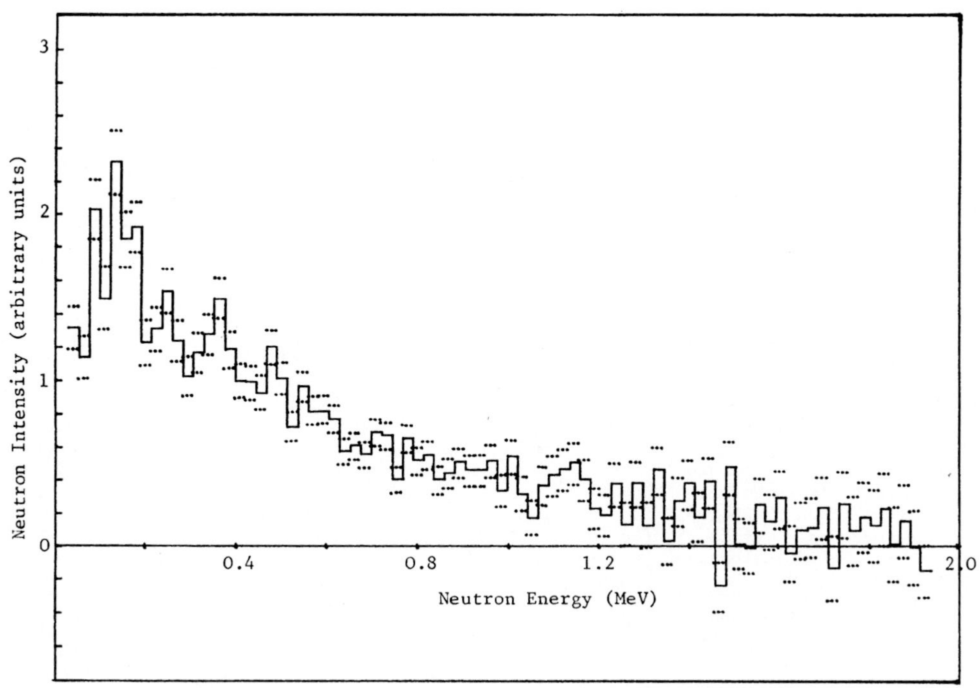

Figure 2: Delayed neutron spectrum for fast fission of ^{235}U produced by primary neutrons of mean energy 1.76 MeV. Each energy bin is 23.4 keV wide. Errors are indicated as $\pm$ 1 standard deviation.

Results obtained at UKAEA Harwell have greater statistical errors owing to a number of operational considerations. However, for the ^{239}Pu measurements the second spectrometer was employed and these data have yet to be analysed. We also hope to develop a method whereby data previously discarded as being heavily contaminated with recoil ^{3}He pulses can be analysed to give useful information.

Conclusions

Delayed neutron spectra from ^{235}U and ^{239}Pu have been measured with a high-resolution neutron spectrometer at a number of primary fission energies; all spectra show a distinct peak structure. Full covariance error information is also known for these spectra. It is only through a knowledge of this full error information that the true error on any derived parameters (such as the mean energy of the spectrum) can be determined. Covariances are also necessary if the errors on the distribution following some operation (such as normalisation) are to be known. Thus a knowledge of these error data will allow sensible comparisons to be made between spectra obtained by different workers, and it is hoped that some of the confusion with regard to delayed neutron spectra will be resolved. A more detailed account of how the covariance analysis applied here may be adapted to other situations is in preparation[8].

Acknowledgements

We wish to thank AERE Harwell for the loan of the uranium and plutonium samples and for the provision of accelerator time, and the Science and Engineering Research Council for financial support.

References

1. D. Saphier, D. Ilberg, S. Shalev and S. Yiftah, Nucl. Sci. Eng. 62, 660 (1977).
2. A.E. Evans and M.S. Krick, Nucl. Sci. Eng. 62 652 (1977).
3. G.W. Eccleston and G.L. Woodruff, Nucl. Sci. Eng. 62, 636 (1977).
4. S. Shalev and J.M. Cuttler, Nucl. Sci. Eng. 51 52 (1973).
5. G.R. Keepin, Physics of Nuclear Kinetics, Addison Wesley, Reading, Mass. (1965).
6. J. de Pangher and L.L. Nichols, Battelle-Northwest Laboratory Report, BNWL-260, Richland, Washington, USA (1966).
7. J.G. Owen, D.R. Weaver and J. Walker, Nucl. Inst. Methods 188, 579 (1981).
8. D.R. Weaver, "A Covariance Cookbook", Birmingham Radiation Centre Report, Birmingham University, Birmingham, England, BRC82/02. In preparation.

INTEGRAL MEASUREMENTS OF DELAYED NEUTRON AVERAGE ENERGIES FOR ^{235}U

S. Synetos[+]

77,3rd September Str
Athens 104
Greece

[+]Work carried out as part of the Ph D. thesis of the author at the :

University of London Reactor Centre
Imperial College, University of London

The average energy of delayed neutrons from ^{235}U thermal fission is determined using a count rate ratio technique. The experimental set-up consists of a fast pneumatic system to transfer the sample to and from the irradiation position, situated in a thermal column of the University of London Reactor. The delayed neutron decay curve is obtained in turn by three detectors, consisting of a BF_3 counter surrounded by polythene cylinders of diameters 2, 5 and 9 inch respectively.
The calibration curves of efficiency ratio versus energy for each pair of detectors are constructed using a Monte Carlo calculation, and corrected by counting neutron sources of known energy spectra (Am-Li and ^{252}Cf). The calibrated system is then used to obtain the average energy of the equilibrium delayed neutron emission (from the ratios of total counts), and the average energy of each delayed neutron group (from the ratios of counts due to each group). The average energy for the equilibrium delayed neutron emission obtained was 490±100 KeV, while for the first five groups of the Keepin group structure the values obtained were 250±12 KeV, 670±26 KeV, 300±28 KeV, 841±218 KeV and 385±193 KeV respectively. The above values show reasonable agreement with the literature values.

(average energy, delayed neutrons, ^{235}U thermal fission, count rate ratio, polythene cylinder, Monte Carlo, group structure)

1. Introduction

In recent years a lot of effort has been devoted to the measurement of the energy spectra of delayed neutrons[1,2,3]. This is done in response to the increased demand for a more complete and accurate set of data imposed by their users. Especially fast breeder reactors seem to be particularly sensitive to the details of the delayed neutron (d.n.) spectra used for their analysis[4].
Delayed neutron spectrum data come from two sources: spectroscopy of unseparated samples of fission products to provide integral data of use to reactor calculations and, spectral studies of separated d.n. precursors whose results apart from being of use to theoretical interpretations of d.n. emission, can be combined to support the integral measurements.
In the present paper we concentrate on integral d.n. spectroscopic data, and report measured average d.n. energies (E_{av}) for thermal fission of ^{235}U. A count rate ratio technique is employed to measure the average energy of the equilibrium d.n.emission, as well as E_{av} for each delayed group.
The question arises now whether the measurement of average d.n. energies is justified. We argue that a precise determination of E_{av} (to within 40 KeV) would give an indication of the energy spectrum under examination[3]. The determination of E_{av} per delayed group, already very important in reactor calculations[5], will give an independent check of the average energies and the energy spectra of different precursors, as well as reported spectra corresponding to the six half life groups.
The accuracy requirements for E_{av} vary according to their use. For reactor calculations the requirement at a recent IAEA meeting[6] was set to ±20%. For the group E_{av} the ratio technique can give better accuracy than the summation methods using experimental data or analytical representations, due to generally insufficient known d.n. yields and individual precursor abundancies.

2. Description of the method

The method employed in this work is based on the "ring ratio" technique proposed by Reeder et al[7,8].
A long BF_3 counter surrounded in turn by three polythene cylinders of various thicknesses is used to measure ratios of counts, normalized to the same number of d.n. emitted from the sample. If the shape of the efficiency curve for each moderator thickness is different, then the ratio of efficiency for a pair of cylinders would be a function of energy. Thus, the ratio of counts of any neutron emitting source obtained with such a pair of cylinders would be a function of the energy of the neutron emission. A preliminary calculation showed that polythene cylinders of diameters 9, 5 and 2 inches surrounding the BF_3 detector, would have quite different efficiency curves.
Calibration curves for two pairs of cylinders were obtained using a Monte Carlo calculation. The energy region of interest (20 KeV to 2 MeV) was divided into 10 intervals, and the efficiency of each cylinder $\varepsilon(E_i)$ was calculated as the number of neutrons absorbed in the detector per source neutron emitted with energy within the boundaries of group i.
The calibration curves based on the above calculation do not accurately represent the calibration curves needed for sources having a spectrum of neutron energies (e.g. delayed neutron sample). In order to take account of spectral sources we proceeded as follows:
The effective efficiency of each cylinder to a calibrated Am-Li source, whose spectrum is known to be similar to the equilibrium d.n. spectrum, was calculated from the relation:

$$\varepsilon_{eff} = \frac{\sum_i N(E_i)\varepsilon(E_i)}{\sum_i N(E_i)} \qquad (1)$$

where $\varepsilon(E_i)$ is the efficiency function of each cylinder and $N(E_i)$ the source spectrum. ε_{eff} was then related to the true average energy of the source, defined as:

$$E_{av} = \frac{\sum_i N(E_i)E_i}{\sum_i N(E_i)} \qquad (2)$$

The resulting effective efficiency ratios were then normalized to the measured efficiency ratios (for the Am-Li source) for each pair of cylinders. We thus end up with calibration curves normalized to the Am-Li source counting, and exhibiting one point (Am-Li counting) corrected for spectral sources. Counting of a calibrated Cf source provides a check of the normalization, and a second point in the calibration curves, corrected for spectral sources. The d.n. average energy can then be deduced from the experimentally obtained count ratios and the calibration curves obtained from the above analysis.
It should be noted that the technique described above gives correct E_{av} values under certain conditions. The following analysis is performed:
The effective efficiency ratio R_{eff} for a spectrum N(E) is:

$$R_{eff} = \frac{\int N(E)\varepsilon_1(E)dE}{\int N(E)\varepsilon_2(E)dE} \qquad (3)$$

where ε_2 refers to the reference, and ε_1 to some other detector. Expanding the efficiency ratio in a power series: $\varepsilon_1(E) / \varepsilon_2(E) = a + bE + cE^2 + \ldots$ (4)
and inserting (4) in (3) we get:

K. H. Böckhoff (ed.), Nuclear Data for Science and Technology, 268–271.

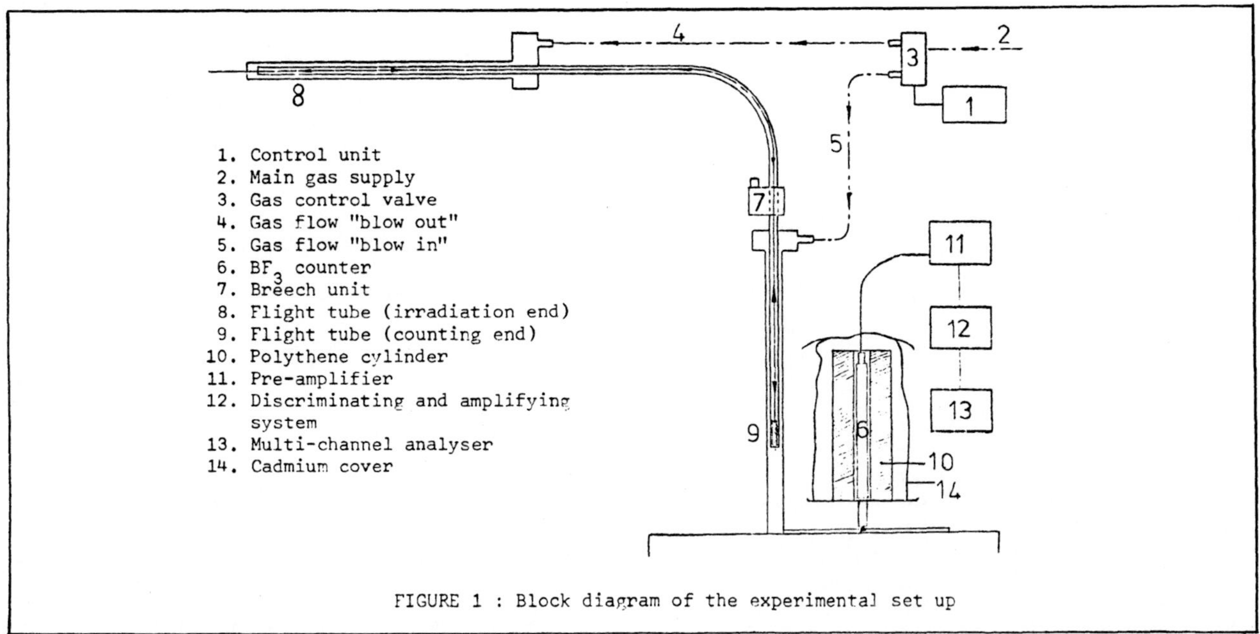

FIGURE 1 : Block diagram of the experimental set up

$$R_{eff} = a+b\frac{\int N(E)\varepsilon_1(E)EdE}{\int N(E)\varepsilon_2(E)dE} + c\frac{\int N(E)\varepsilon_1(E)E^2dE}{\int N(E)\varepsilon_2(E)dE}+.. \quad (5)$$

In order for R_{eff} to be a function of E_{av} only it is necessary that:

$$c = 0 \quad (6)$$

and

$$\varepsilon_2 = \text{const.} \quad (7)$$

In this case eq. (5) gives: $R_{eff} = a+bE_{av}$ (8)
and R_{eff} is a function of E_{av} only, and does not depend on any other attribute of the spectrum. Thus, for the method to be valid a reference detector exhibiting an efficiency function constant with energy should be employed.

In this work we used the calibration curves to obtain the average energy of each d.n. group. The time dependent decay curves obtained with each cylinder thickness were analysed using the least squares technique. The group half lives were fixed to those found in (9) and (10), or to Keepin's values. The following equation was fitted to the decay curve for each cylinder j:

$$C_j(t) = by_d \sum_{k=1}^{m} C_{mk} \sum_i P_{ij} e^{-\lambda_i t} + \delta(t) \quad (9)$$

Where:

y_d = absolute delayed neutron yield per fission
b = calibration factor of monitor counts to number of fissions, constant for all three detectors
C_{mk} = monitor counts per sec for the kth run
bC_{mk} = n_{fk} = fissions per sec for the kth run
P_{ij} = $\alpha_i \varepsilon_{ij}$ = relative yield x detector efficiency for detector j and delayed group i
λ_i = the group i decay constant (fixed)
$\delta(t)$= the statistical discrepancy of the count rate from its expectation value

The problem (9) is a linear one. From the least squares fit we get $b\varepsilon_{ij}\alpha_i$ for the three detectors. Taking into account that b and α_i are the same for each detector, the ratios of efficiencies per group can be obtained, and through the calibration curves we get the average energy per group.

3. Experimental details

3a. Description of the experimental set up

For the d.n. measurements the detector-moderator assembly has to be coupled to a sample transfer system. Such a system has been used in (9) for d.n. yield measurements. Figure 1 shows the experimental set up. The moderator consists of polythene cylinders hollowed around their axis to accommodate the long BF_3 counters. Scattering has been kept to a minimum by using as little an amount of structural material as possible.
Background was kept at reasonable levels by enclosing the counting end of the system in a Cd sheet (Fig. 1). Room return was checked by keeping the assembly away from walls or the floor.

The experimental procedure was very similar to that followed in the yield experiment (9), and the same set of electronics consisting of an amplifier, a discriminator and a multi-channel analyser was used. The sample consisting of a .2g metal foil (95.54% in ^{235}U), is shuttling between the irradiation position and the counting end. The irradiation position is at the centre of a spherical cavity in one of the thermal columns of the CONSORT reactor. Cadmium ratio measurements have shown the neutron temperature at the irradiation position to be in the order of $300^0 \pm 10^0$K. The cycle times used were 10min irradiation and 10min counting. At the end of the counting the activity had practically reached background levels.

3b. Calibration of the detectors

The absolute efficiency of each of the three detectors was determined using calibrated Am-Li and Cf neutron sources. Although the measurement of the neutron sources is straight forward, a number of corrections has to be applied, so that the result is directly comparable to ratios obtained using the ^{235}U sample.

(i). Volume correction

The effect of the large volume Am-Li source was investigated as described in (9). It was found that a correction factor of 1.014±.007 has to be applied to the Am-Li measured efficiency in order to make it equivalent to the efficiency of an equivalent point source (e.g. ^{235}U foil). Furthermore, it was found that the correction factor is independent of polythene thickness. Thus, no correction was applied to the efficiency ratios, but an uncertainty of ±1% is introduced.

(ii). Anisotropy correction

The Am-Li source used in the calibration (supplied by the National Physical Laboratory) exhibits an anisotropy in its neutron emission around its axis. The N.P.L. supplied us with the relative emission rates for 10^0 intervals from the cylindrical axis. These data were used in a calculation to find the difference between the efficiency of the detectors for an isotropic source from that of an anisotropic.

The anisotropy factors were calculated as:

1.031±1.5% for the 2 inch cylinder
1.018 ± .9% for the 5 inch cylinder
1.010 ± .5% for the 9 inch cylinder

Uncertainties were taken as 50% of the correction. Self-absorption in the sources was found to be negligible.

3c. The experiments

What we want to get from our experiments are the following data: d.n. decay curves following thermal fission of ^{235}U, obtained with each of the three detectors, and count ratios for the two neutron sources Am-Li and Cf, for the same source-detector distance.

Cylinder diameter / Counts	2 inch	5 inch	9 inch
Reference Am-Be count rate (cp 60 sec)	269	7,096	9,182
Am-Be net count rate when ^{235}U samples were counted (cp 60 sec)	281	7,253	9,268
Background (cp 60 sec)	47	83	66
Monitor 2 average (of all cycles)	1,333,745	1,327,534	1,327,981
Number of cycles	30	9	15
Net total counts (due to d.n.)	30,026	92,085	78,225
Normalize to 30 runs and 9 inch cylinder monitor average	29,896	307,053	156,450
Normalize to Am-Be net count rate at time of calibration	28,619	300,406	155,000
Efficiencies			
Am-Li source	.00033±.6%	.00287±.6%	.00145±.7%
Cf source	.00011±.6%	.00232±.5%	.00220±.5%

TABLE I: Results of the delayed neutron and source counting for the three cylinders

A LABEN multi-channel analyser was used to register the decay curves in 2048 time channels. The number of runs for each detector varied inversely proportional to its efficiency, so that the obtained counting statistics would be of the same order for all three detectors. A monitor counter (fission chamber) was operating during the irradiations, giving thus a means of normalizing the total d.n. counts obtained with each detector, to the same flux, and therefore to the same number of fissions. An Am-Be neutron source was counted before each experiment to ensure the reproducibility of the electronics settings. The results of the experiments are shown in table I, along with some experimental details.

4. Calculational details

4a. Calculation of the efficiency functions of the detectors

The efficiency functions of the three detectors were calculated using the Monte Carlo code MONK[12]. A point source located at the actual centre of the ^{235}U foil was assumed for the calculation. The efficiency for each cylinder was calculated as the number of neutrons absorbed in the detector per source neutron emitted with energy within the boundaries of group i. We assume for the moment that every neutron absorbed in the BF_3 is actually counted. Later on we correct for this assumption. The efficiency curves are shown in Fig. 2. Their shape is generally in agreement with that of polythene spheres of the same diameter[13]. $\varepsilon(E)$ for the 5 inch cylinder is almost constant as a function of energy, while that of the 9 inch cylinder is a linear function of energy. On the contrary the 2 inch cylinder exhibits an $\varepsilon(E)$ linear with respect to lnE (lethargy).

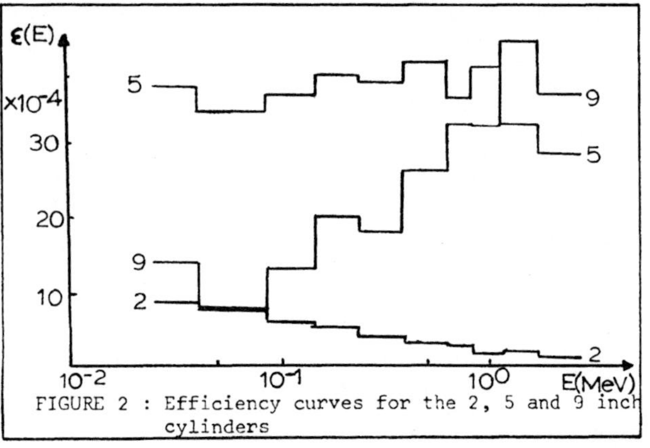

FIGURE 2 : Efficiency curves for the 2, 5 and 9 inch cylinders

Cylinder diameter / Efficiency	2 inch	5 inch	9 inch
Experimental	.00033±.6%	.00287±.6%	.00145 ±.7%
Anisotropy correction	x .97±1.5%	x .982±.9%	x .99±.5%
Volume correction	x1.014±.7%	x1.014±.7%	x1.014±.7%
Result	.000325±1.8%	.00286±1.3%	.00146±1.1%
Calculated	.00040	.00379	.00241
C/E value	1.231	1.325	1.651

TABLE II: Comparison of the calculated and experimental efficiencies for the Am-Li source

4b. Construction of the calibration curves

We are now equipped with enough data to construct the count ratio versus neutron energy curves. The data come from two sources: experimental determination of the count ratios for the Am-Li and Cf sources, and calculation of the efficiency ratios for monoenergetic sources. These data are combined as described in section 2. Table II shows the Am-Li efficiency for each polythene cylinder, first as measured and second as calculated using eq.(1). For the Am-Li and Cf sources the spectra given in references (14) and (15) were used. It is apparent that the calculation overpredicts the efficiency of all detectors. Discrepancies towards the same direction have been reported in (13) and (16). What is disturbing in our case is that the overprediction increases with increasing polythene thickness. One possible explanation is that the thermalization of neutrons in the polythene is underestimated in the MONK calculations. This would result in a decrease of the number of neutrons absorbed in the polythene and a hardening of the spectrum of neutrons entering the BF_3 detector region. Thus, the efficiency of the 9 inch cylinder as calculated by MONK would increase, while that of the 2 inch cylinder would decrease. For the purpose of our analysis we assumed that the shape of the efficiency functions as calculated by the Monte Carlo code was correct, but we multiplied them by a normalizing factor, so that the calculated efficiency for the Am-Li source would agree with the experimental one.

The calibration curves thus produced are depicted in Figures 3 and 4.

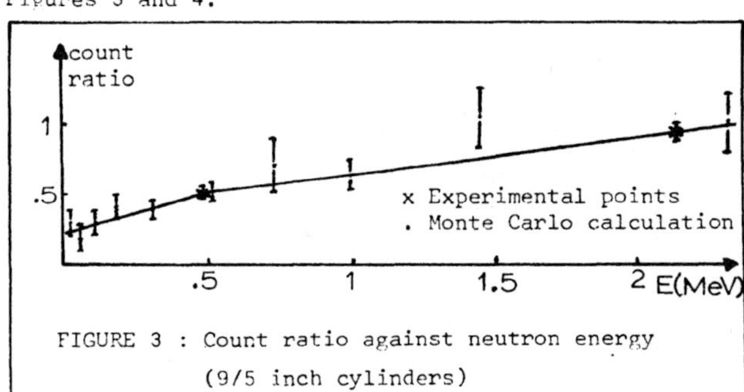

FIGURE 3 : Count ratio against neutron energy (9/5 inch cylinders)

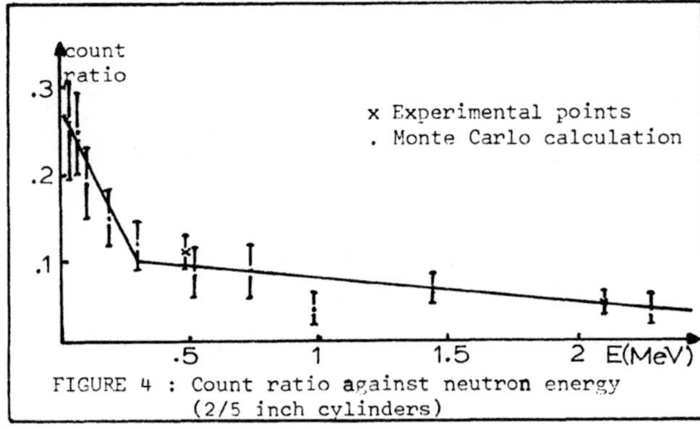

FIGURE 4 : Count ratio against neutron energy (2/5 inch cylinders)

5. Results

The average energy for the equilibrium d.n. emission is deduced from the measured count ratios and the curves of Fig. 3 to 4. From the 9/5 inch curve we get 480 KeV and from the 2/5 inch curve we get 500 KeV. The average for E_{av} is calculated to 490±100 KeV, where the uncertainty includes statistical and systematic errors.

We must though qualify the above result with the statement that it is normalized to the Am-Li source counting, and thus depends on the spectrum assumed for the source. The value of E_{av} obtained here can be compared with E_{av} = 540 KeV obtained by Reeder et al[7] using a similar technique. Reeders' et al reviewed measurement[8] gives an even higher value for E_{av}. On the other hand, calculations of E_{av} based on d.n. spectra measured by Saphier et al[1] and Eccleston and Woodruff[2] gave average energies 462 KeV and 405 KeV respectively.

The ring ratio against neutron energy curves of Fig. 3 and 4 were used for the derivation of the average energy for each d.n. group. Eq. (9) was fitted to the d.n. decay curves obtained with each cylinder using the least squares method, as described in (9). The group half lives were fixed to those reported in (9). For the sake of comparison with already published data of group average energies, we obtained a second set of data with the group half lives fixed to the values of Keepin[11]. The obtained ratios of efficiencies per group and the deduced E_{av} are given in Table III. The uncertainties in the ratios represent the uncetrainties in the least squares fit and in the normalization. The uncertainties in E_{av} are then deduced from the calibration curves of Fig. 3 and 4. A weighted average value for each group is calculated from the results obtained using each of the two ratios. The result is shown in Table III for both Synetos' and Williams'[9] and Keepin's[11] group structures. E_{av} for groups 1, 2 and 3 are identical for ours and Keepin's group structures. The differences in E_{av} for the other two groups reflect the differences between the two group structures. The higher E_{av} for our group 5 (which includes part of group 6) compared with Keepin's group 5 indicates that group 6 has a harder spectrum. This agrees with the result of Reeder et al[7]. Our results show a good agreement with those of Evans and Krick[17], where a trend towards higher average energies for the shorter lived d.n. groups is reported.

Group	9/5 inch	E_{av}	2/5 inch	E_{av}	weighted average E_{av} (KeV)
λ_i=Synetos[9]					
1	.407±5%	290±20	.140±5.5%	235±15	255±12
2	.548±3%	600±60	.081±5.2%	900±110	670±26
3	.425±15%	320±110	.112±17%	280±30	284±29
4	.587±11%	740±250	.083±17%	870±370	781±207
5	.572±28%	660±420	.084±45%	860±500	740±322
λ_i=Keepin[11]					
1	.403±5%	285±20	.141±5.6%	230±15	250±12
2	.547±3%	600±60	.081±5%	900±110	670±26
3	.455±10%	360±70	.108±12%	290±30	300±28
4	.608±11%	810±260	.080±19%	910±400	841±218
5	.547±39%	600±400	.099±53%	320±220	385±193

TABLE III: Group count ratios and average energies

6. Summary and conclusions

The average energy of d.n. from ^{235}U was determined using a count rate ratio technique. The experimental set up consisted of a BF_3 counter surrounded by polythene cylinders of diameters 2, 5 and 9 inch respectively. The calibration curves of count ratio versus neutron energy were constructed using sources of known spectra (Am-Li and Cf) and their detailed shape was established with the help of a Monte Carlo calculation of the efficiency functions of the detectors. The ratios of counts from the d.n. sample were measured with the three detectors,

Reference \ Group	1	2	3	4	5	6
Burgy et al[18]	300±60	670±10	650±90	910±90	400±70	-
Hughes et al[19]	250±60	560±60	430±60	620±60	420±60	-
Batchelor et al[20]	250±20	460±10	410±20	450±20	-	-
Fieg[21]	280±30	480±50	450±50	430±40	-	-
Reeder et al[7]	150	460	>690	540	520	>610
This work(Keepin's group structure)	250±12	670±26	300±28	841±218	385±193	>385

TABLE IV: Average energies (KeV) of delayed neutron groups in thermal neutron fission of ^{235}U

and E_{av} was deduced from the calibration curves. The value obtained was 490±100 KeV. This value agrees within the uncertainties with that obtained by Reeder et al[7] using a similar ratio technique. Comparison of our value with values of E_{av} obtained from the analytical representation of the d.n. spectra of Saphier et al[1] and Eccleston and Woodruff[2] indicates that both srectra might be too soft.

Average energies of d.n. groups were also obtained. The decay curves measured with the three detectors were analysed using the least squares method, and the ratios of counts due to each of five groups were obtained. E_{av} was then deduced from the calibration curves. Table IV presents a comparison of the group E_{av} obtained in this work with previously published data. Our result confirms that group 1 has the softest spectrum. Our result is in good agreement with that of Hughes et al[19]. It disagrees with most of the other measurements, especially for group 3 which appears quite soft in our measurement. The uncertainties associated with groups 4 and 5 appear high in our calculation, and this reflects the big uncertainties in the parameters α_i and λ_i fitted to groups 4 and 5. It is suggested that irradiation-counting cycles accentuating the contribution of groups 4 and 5 would improve the accuracy here.

Acknowledgements

The author wishes to express his appreciation for the help and advice received from the Operation Staff and especially Dr. J.G. Williams at the University of London Reactor Centre,and from Messrs D. Langford and K. Shuttler at A.W.R.E. in running the Monte Carlo code.

References

1. Saphier D. et al, Nucl.Sc. and Eng.62(1977) p.660
2. Eccleston G.W. and Woodruff G.L., Nucl.Sc. and Eng. 62(1977) p.636
3. Kratz K.L., in Proceedings of the consultants' meeting on delayed neutron properties, Vienna, March 1979, INDC(NDS)-107/G+ Special, p.103
4. Saphier D. and Yiftah S.,Nucl.Sc. and Eng.42(1970) p.272
5. Hammer Ph.,same as (3), p.1
6. Conclusions and Recommendations, same as (3), p.277
7. Reeder P.L. et al, Phys. Rev. C 15(1977) p.2098
8. Reeder P.L. and Warner R.A., same as (3), p.239
9. Synetos S. and Williams J.G., same as (3), p.183
10. Synetos S., Ph.D. thesis, Imperial College, University of London (1979)
11. Keepin G.R., Physics of Nuclear Kinetics, Addison-Wesley Publishing Co Inc., Reading, Massachusetts (1965)
12. Sherriffs V.S.W., "MONK-a general purpose Monte Carlo neutronics program",SRD R86, UKAEA (1978)
13. Bricka M. et al, IAEA-SM-167/19 (1971)
14. Werle H., Kernforschungszentrum Karlsruhe (1970) translated as ORNL-tr-2415
15. Lorch E.A., Int. J. Appl. Rad. Isot., 24(1973) p.590
16. Maerker R.E. et al, ORNL-TM-3451 and 3465 (1971)
17. Evans A.E. and Krick M.S., ANS trans. 23(1976) p.491
18. Burgy M. et al, Phys. Rev., 70(1946) p.104
19. Hughes D.J. et al, Phys. Rev., 73(1948) p.111
20. Batchelor R. and McHyder H.R., J. Nucl. En.,3(1956)p.7
21. Fieg G., J. Nucl. En., 26(1972)p.585

COMPILATION OF NEUTRON PRECURSOR DATA

F. M. Mann, M. Schreiber and R. E. Schenter

Westinghouse Hanford Company
Hanford Engineering Development Laboratory
Richland, Washington 99352, U.S.A.

T. R. England

Los Alamos National Laboratory
Los Alamos, New Mexico 87544, U.S.A.

Delayed neutron precursor data (yields and spectra) are being compiled in preparation of a new ENDF/B evaluation of delayed neutron data. Precursor P_n values have been updated using ENDF/B-VM yields, ENDF/B-V decay data, and consensus P_n values. Weighted averages of these P_n values will be presented along with their effect on fission yields, P_n systematics, and model code calculations.

[delayed neutron - P_n compilation]

Introduction

Delayed neutron data play an important role in reactor design and operation. These data are used to predict the effective delayed neutron fraction for the reactor β_{eff}, to predict the effective delayed neutron decay constant λ_{eff}, and to detect cladding failure. According to Hammer,[1] the most stringent requirements result from the need to know β_{eff} to interpret critical experiments for control and reactivities, sample reactivity worth, and other parameters.

The delayed neutron spectrum can be obtained by summing over each of the neutron precursors:

$$\phi_d(E,t) = \sum_i Y_n^i(t)\, X_d^i(E) = \sum_i P_n^i\, Y_f^i(t)\, X_d^i(E) \qquad (1)$$

where ϕ_d is the delayed neutron spectrum, Y_n^i is the neutron yield for precursor i, X_d^i is the normalized delayed neutron spectrum for precursor i, P_n^i is the probability of delayed neutron emission, and Y_f^i is the precursor yield from the fission process. As the half-lives of the parents of the precursor are short compared to that of the precursor, the time dependences of Y_n^i and Y_f^i are exp $(-\lambda_i t)$ where λ_i is the decay constant for precursor i.

Based on the need to know β_{eff} to 3% (1σ), Hammer states that the absolute delayed neutron yield

$$Y = \iint \phi_d(E,t)\, dE\, dt = \sum_i P_n^i \int Y_f^i(t)\, dt \qquad (2)$$

must be known to 1.5% for ^{235}U, ^{238}U and ^{239}Pu. Larger uncertainties (7% for ^{240}Pu and ^{241}Pu and 20% for ^{242}Pu and ^{241}Am) are acceptable for the higher actinides.

Although the total delayed neutron yield has been measured for many isotopes (see Tuttle),[2] the recent experimental work on precursors allows summation calculations such as Equation 2 to approach (and hopefully soon to be lower than) the accuracy of such direct measurements. This report discusses the status of P_n measurements and their implications. In future papers the time and energy dependence will be discussed. Some preliminary results have already been presented.[3]

P_n Compilation

The literature has been searched for all P_n measurements.[6-28] See the review by Rudstam[4] for a discussion of experimental methods and problems of P_n determinations and the review by Kratz[5] for a discussion of neutron spectra.

As some measurements provide relative values, these have been updated using modern reference values. For values given in neutrons per fission, ENDF/B-VM fission yields[29] were used (ENDF/B-VM fission yields based only on calculations were arbitrarily given 1000% uncertainties). For values normalized to gamma radiation, new normalizations were obtained using the decay data in the ENDF/B-V Fission Product File. Values normalized against other P_n values were renormalized by using the weighted average of the updated reference precursor.

Such corrections are particularly important for P_n values derived from measurements given in neutrons per fission. In many cases, estimated fission yields were used originally to normalize the data. For those values normalized to other P_n values, some reference P_n values have changed over the years generating significant differences.

This compilation which is available upon request differs from that given by Rudstam by the inclusion of the final (rather than preliminary) values from SOLIS[27] and OSIRIS,[21-22] by the use of renormalized ORSAY-74[20] data, and by the use of different reference values. The ORSAY-74 data were renormalized by the use of modern ^9Li data. Note that some new preliminary values from SOLAR measurements cast doubt on previous A=97 and 98 values.

Recommended Values

In past compilations, best values were obtained by calculating the weighted average for each precursor after disregarding outlying experiments. The uncertainty for the weighted average of each precursor was increased by the reduced chi square value for that precursor. Table I presents the weighted precursor average from this compilation as well as the 1979 Rudstam compilation.

This procedure ignores the fact that the measurements are not independent. Certainly for a given paper, the same detectors and efficiency determinations are used. Similar correlations might be expected for papers from the same laboratory.

To determine all the correlations is impossible because of the lack of information in the published papers. Therefore, simpler correlations were introduced through the use of normalization factors for each laboratory (or each paper, depending upon the analysis). The P_n values and normalization factors were determined using nontraditional constrained least squares techniques with the model equations being

$$E_{ik} \pm \Delta\varepsilon_{ik} = b_i * P_k$$

and

$$1 \pm \Delta \qquad = b_i$$

where E_{ik} and ε_{ik} are the experimental values and uncertainty for precursor k as measured by laboratory

K. H. Böckhoff (ed.), Nuclear Data for Science and Technology, 272–276.

Table I

P_n VALUES (%)
RELATIVE UNCERTAINTIES ARE GIVEN IN PERCENT

PRECURSOR	RUDSTAM-79	RECOMMENDED	COMMENTS
31-Ga-79	0.094 (17)	0.100 (11)	1 MEASUREMENT
31-Ga-80	0.80 (8)	0.86 (8)	1 MEASUREMENT
31-Ga-81	11.9 (8)	12.2 (9)	1 MEASUREMENT
31-Ga-82	21.9 (11)	21.6 (11)	1 MEASUREMENT
31-Ga-83	—	43.9 (17)	1 MEASUREMENT
33-As-84	0.09 (56)	0.078 (50)	1 MEASUREMENT
33-As-85	50.0 (100)	[9.1 (10)]	DISCREPANT DATA $n/10^4f$
33-As-86	12.0 (33)	[0.53 (19)]	$n/10^4f$
33-As-87	44.0 (32)	[2.2 (45)]	$n/10^4f$
34-Se-87	0.19 (16)	0.18 (15)	
34-Se-88	0.6 (150)	0.96 (16)	DISCREPANT DATA
34-Se-89	5.0 (30)	7.5 (30)	1 MEASUREMENT
35-Br-87	2.38 (3)	2.58 (8)	
35-Br-88	6.7 (3)	6.35 (8)	
35-Br-89	13.5 (19)	14.2 (7)	
35-Br-90	21.2 (11)	24.9 (10)	
35-Br-91	10.9 (17)	18.3 (12)	DISCREPANT DATA
35-Br-92	22.0 (29)	[1.13 (35)]	$n/10^4f$
36-Kr-92	0.033 (9)	0.031 (15)	
36-Kr-93	1.96 (9)	1.93 (10)	
36-Kr-94	5.7 (39)	6.1 (39)	1 MEASUREMENT
37-Rb-92	0.0116 (10)	0.0100 (6)	
37-Rb-93	1.39 (13)	1.36 (6)	DISCREPANT DATA
37-Rb-94	10.4 (11)	10.2 (6)	
37-Rb-95	8.8 (7)	8.60 (7)	
37-Rb-96	14.2 (10)	14.2 (6)	
37-Rb-97	28.0 (21)	26.9 (7)	DISCREPANT DATA
37-Rb-98	16.0 (6)	13.4 (8)	DISCREPANT DATA
37-Rb-99	15.0 (20)	13.4 (14)	1 MEASUREMENT
38-Sr-97	0.27 (33)	0.24 (33)+	1 MEASUREMENT
38-Sr-98	0.36 (31)	0.32 (31)+	1 MEASUREMENT
38-Sr-99	3.4 (71)	3.6 (65)	1 MEASUREMENT
39-Y-97	0.06 (33)	0.05 (17)+	1 MEASUREMENT
39-Y-98	3.4 (29)	3.0 (28)+	1 MEASUREMENT
39-Y-99	1.2 (67)	1.3 (67)	1 MEASUREMENT
49-In-127	0.65 (8)	0.69 (10)	1 MEASUREMENT
49-In-128	0.057 (14)	0.060 (15)	1 MEASUREMENT
49-In-129	3.5 (14)	3.0 (20)	
49-In-129*	—	0.25 (21)	1 MEASUREMENT
50-Sn-134	17.0 (41)	18.0 (41)	1 MEASUREMENT
51-Sb-134	0.108 (7)	0.117 (12)	DISCREPANT DATA
51-Sb-135	15.6 (13)	20.8 (12)	DISCREPANT DATA
51-Sb-136	23.0 (35)	[0.57 (11)]	$n/10^4f$
52-Te-136	0.9 (44)	1.14 (57)	
52-Te-137	2.5 (23)	2.7 (20)	1 MEASUREMENT
52-Te-138	6.3 (33)	6.7 (33)	1 MEASUREMENT
53-I-137	6.6 (9)	7.1 (9)	DISCREPANT DATA
53-I-138	5.3 (4)	5.5 (14)	DISCREPANT DATA
53-I-139	9.4 (5)	9.9 (8)	
53-I-140	23.0 (26)	9.4 (11)	
53-I-141	39.0 (33)	21.7 (22)	
54-Xe-141	0.044 (25)	0.041 (11)	
54-Xe-142	0.42 (10)	0.39 (12)	
55-Cs-141	0.036 (58)	0.034 (9)	DISCREPANT DATA
55-Cs-142	0.091 (3)	0.096 (8)	DISCREPANT DATA
55-Cs-143	1.68 (10)	1.63 (6)	
55-Cs-144	3.0 (20)	3.17 (7)	DISCREPANT DATA
55-Cs-145	13.3 (13)	13.8 (5)	DISCREPANT DATA
55-Cs-146	13.2 (5)	13.4 (9)	DISCREPANT DATA
55-Cs-147	25.4 (13)	26.2 (14)	1 MEASUREMENT
56-Ba-147	5.2 (10)	4.7 (11)	1 MEASUREMENT
56-Ba-148	23.9 (9)	21.5 (10)	1 MEASUREMENT
57-La-147	0.50 (34)	0.45 (34)	1 MEASUREMENT

+NEW PRELIMINARY PNL DATA NOT INCLUDED

(paper) i, b_i is the normalization (bias) factor for laboratory i, P_k are the recommended P_n values for precursor k, and Δ is the average standard deviation of the normalization factors. By transforming the data into logarithmns, linear techniques would be used and both recommended P_n values and normalization values were restructed to be positive definite. The results are insensitive to the starting value of Δ and the Δ iterations converge quickly.

Such a procedure is very sensitive to values far from the average. Thus experiments having values differing by greater than 5 standard deviations from the fitted value have had their uncertainties increased by 10.

In addition, for those isotopes (^{85}As, ^{86}As, ^{92}Br and ^{136}Te) where the P_n values are derived only from neutron per fission determinations and estimated ENDF/B-VM fission yields, the original neutrons per fission values were used, not the derived P_n values.

Except for the ORSAY papers, the normalization factors for each paper were not statistically different from the factors for that laboratory. Also, except for ORSAY-74[20] and TRISTRAN-69,[28] the normalization factors were near unity. Hence for the final fit each laboratory (except ORSAY) had one normalization factor and the a priori normalization uncertainties for ORSAY-74 and TRISTRAN-69 were increased by a factor of 5.

The results from this final fit are shown in Table I. Table II shows the normalization factors for this final fit and for the case in which each paper is treated separately. Although in general this procedure gives similar values to the precursor weighted average method, for a few precursors (^{85}As, ^{134}Sb, ^{135}Sb and ^{138}I) there are significant differences.

Table II

LABORATORY BIAS FACTORS

Laboratory	Number of Measurements	(Bias - 1.0)* 100 Based on Each Paper	Based on Each Laboratory
Ariel-75 (Ref. 6)	8	-3 ±8	1 ±8
Harwell-68 (Ref. 7)	3	8 ±15	3 ±12
-71 (Ref. 8)	2	2 ±15	3 ±12
Lohengrin-75 (Ref. 9)	8	-8 ±6	-6 ±6
-78 (Ref. 10)	7	-14 ±10	-6 ±6
Mainz-70 (Ref. 11)	2	2 ±16	-5 ±8
-72 (Ref. 12)	11	-2 ±8	-5 ±8
-73 (Ref. 13)	2	-3 ±13	-5 ±8
-74 (Ref. 14)	9	-7 ±11	-5 ±8
-75 (Ref. 15)	4	0 ±11	-5 ±8
Mol-71 (Ref. 17)	5	6 ±10	9 ±9
Orsay-69 (Ref. 19)	6	-5 ±8	-4 ±8
-74 (Ref. 20)	8	24 ±10	32 ±11*
ORSIS-80 (Ref. 21)	19	-4 ±5	-2 ±6
-80A (Ref. 22)	6	1 ±8	-2 ±6
OSTIS-79 (Ref. 23)	9	-6 ±5	-3 ±6
Russia-64 (Ref. 24)	5	-3 ±13	-3 ±12
SOLAR-77 (Ref. 25)	14	-2 ±6	0 ±6
-80 (Ref. 26)	12	-3 ±5	0 ±6
SOLIS-81 (Ref. 27)	12	8 ±6	11 ±8
TRISTRAN-69 (Ref. 28)	8	19 ±11	29 ±14*

*Orsay-74 and TRISTRAN had apriori bias uncertainties a factor of 5 greater than other laboratories.

Applications

For three precursors which have estimated ENDF/B-VM fission yields, there exists both neutrons per fission values as well as direct P_n determinations. Thus the cumulative fission yield can be determined and is shown in Table III. The ^{140}I cumulative fission yield as determined by delayed neutron data is significantly different from that estimated by ENDF/B-VM. Table IV presents the precursors which have values only from neutron per fission determinations.

Kratz and Herrmann[30] and Amiel and Feldstein[31] have suggested a correlation between P_n values and the available neutron energy. Amiel and Feldstein assume

$$P_n = a(Q_\beta - S_n)^b \tag{3}$$

Table III

FISSION YIELDS DETERMINED BY DELAYED NEUTRON DATA

Precursor	Delayed Neutron Derived	Estimated ENDF/B-VM
^{91}Br	(.172 ±.038)%	(.233 ±.149)%
^{140}I	(.367 ±.111)%	(.112 ±.071)%
^{141}I	(.063 ±.024)%	(.041 ±.026)%

Table IV

PN VALUES USING ESTIMATED ENDF/B-VM YIELDS

Precursor	$n/10^4f$	Estimated Yields (%) ENDF/B-VM+	Estimated Pn (%)
^{85}As	9.1 ±0.9	.137	66.0
^{86}As	0.53 ±0.10	.062	8.5
^{87}As	2.2 ±1.0	.053	42.0
^{92}Br	1.13 ±0.40	.0266	43.0
^{136}Sb	0.57 ±0.06	.0242	24.0

+Estimated uncertainty is >64%.

where Q_β (MeV) is the maximum energy for β decay, S_n is the neutron separation energy of the daughter nucleus, and a and b are free parameters. Figure 1 displays the recommended P_n values along with the correlation shown in Equation 1. Masses were taken from the 1981 Interim Mass Table of Wapstra[32] whenever possible (40 of the 58 cases). If not possible, the mass predictions of Liran and Zeldes[33] were used. Kratz and Herrmann assume

$$P_n = d\,[(Q_\beta - S_n)/(Q_\beta - C)]^e \qquad (4)$$

where C is a cut-off parameter depending upon the type of precursor nucleus:

$C = 0$ even-even precursor

$= 13/A^{1/2}$ odd precursor

$= 26/A^{1/2}$ odd-odd precursor

and d and e are free parameters.

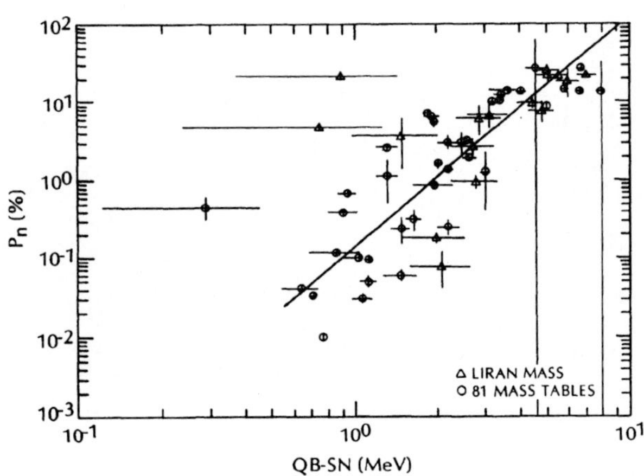

Fig. 1 - P_n Correlation of Amiel and Feldstein (Equation 3).

Figure 2 displays this correlation. Note for both correlation formulas, ^{147}Ba, ^{148}Ba and ^{147}La seem to be about a factor of 100 too high. This effect has already been noted in the original paper[27] (SOLIS-81) and in the Rudstam 1979 review.[4] Table V displays the parameters and resulting reduced chi square from least square fits to the recommended P_n values (except for ^{147}Ba, ^{148}Ba and ^{147}La). Uncertainties in the masses were propagated through to the P_n values during the fitting process. Of the formulas examined (including Equation 4 with C = 0 or using C determined by the type of daughter nucleus), Equation 4 provided the best fit, with 56% of the P_n's within a factor of 2 of their measured values and 84% within a factor of 4. Table VI presents the cases which are greater than a factor of 4. Although some of these cases are suspect because only one measurement exists or because there are discrepant data, some cases (e.g., ^{87}Br and ^{88}Br) are well established, suggesting large enhancements. The statistical model code BETA (without any parameter adjustments)[34] results in P_n predictions that are only slightly poorer than those of Equation 4.

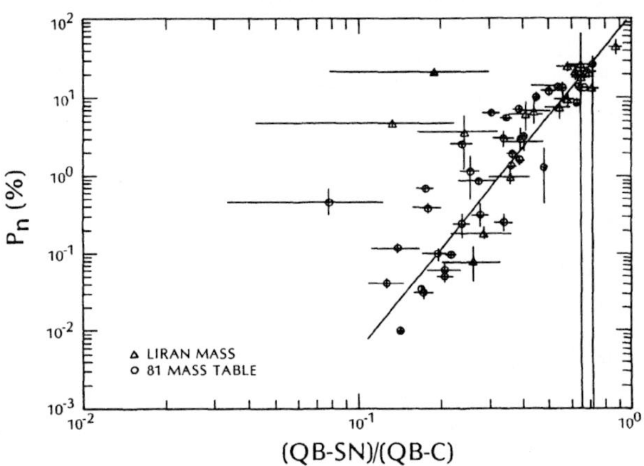

Fig. 2 - P_n Correlation of Kratz and Herrmann (Equation 4).

Table V

CORRELATION FIT TO PN'S

Parameter*	1982 Masses Only	Masses All
a	0.115	0.118
b	3.02	2.99
X^2/ν	39.0	27.0
d	123.0	122.0
e	4.37	4.36
X^2/ν	17.0	11.0

*See Equations 1 and 2.

Summary

An updated list of P_n values has been compiled. Using these data in a least-squares adjustment technique, new recommended P_n values have been derived. The new P_n values were used to derive fission yields for ^{91}Br, ^{140}I and ^{141}I. They were also used to determine a correlation equation for unmeasured P_n values.

Table VI

P_n VALUES WITH LARGE DISCREPANCIES WITH THE FIT

PRECURSOR	MEASURED P_n (%)	FITTED P_n (%)	RATIO	COMMENT
$^{84}As^+$	0.078 ± 0.039	0.368	0.21	1 MEASUREMENT
^{87}Br	2.58 ± 0.21	0.25	10.3	
^{88}Br	6.35 ± 0.51	0.735	8.6	
$^{99}Sr^+$	3.6 ± 2.3	0.27	13.3	
^{127}In	0.69 ± 0.07	0.064	10.8	1 MEASUREMENT
^{129}In	0.25 ± 0.10	1.14	0.22	1 MEASUREMENT
^{134}Sb	0.117 ± 0.014	0.023	5.1	DISCREPANT DATA
^{138}I	5.5 ± 0.8	1.3	4.2	DISCREPANT DATA
^{142}Xe	0.39 ± 0.05	0.070	5.6	
$^{147}Ba^+$	4.7 ± 0.5	0.019	247.0	1 MEASUREMENT
$^{148}Ba^+$	21.5 ± 2.2	0.088	244.0	1 MEASUREMENT
^{147}La	0.45 ± 0.15	0.002	236.0	1 MEASUREMENT

+ USED 1 OR MORE MASSES FROM LIRAN AND ZELDES.

References

1. Ph. Hammer, "Review of the Requirements of Delayed Neutron Data for the Design, Operation, Dynamics and Safety of Fast Breeder and Thermal Power Reactors," in Proceedings of the Consultants Meeting on Delayed Neutron Properties, Vienna, Austria, March 1979, (INDC NDS-107/G + Special), International Atomic Energy Agency (1979).

2. R. J. Tuttle, "Review of Delayed Neutron Yields in Nuclear Fission," in Proceedings of the Consultants Meeting on Delayed Neutron Properties, Vienna, Austria, March 1979, (INDC NDS-107/G + Special), International Atomic Energy Agency (1979).

3. T. R. England, W. B. Wilson, R. E. Schenter and F. M. Mann, "Delayed Neutron Spectral Calculation Using Augmented ENDF/B-V Data," Trans. Am. Nucl. Soc., 41 (1982) 567.

4. G. Rudstam, "Review of Delayed Neutron Branching Ratios," in Proceedings of the Consultants Meeting on Delayed Neutron Properties, Vienna, Austria, March 1979, (INDC NDS-107/G + Special), International Atomic Energy Agency (1979).

5. K. L. Kratz, "Review of Delayed Neutron Energy Spectra," in Proceedings of the Consultants Meeting on Delayed Neutron Properties, Vienna, Austria, March 1979, (INDC NDS-107/G + Special) International Atomic Energy Agency (1979).

6. M. Asghar, J. Crancon, J. P. Gautheron and C. Ristori, "Delayed Neutron Emission Probabilities of $^{92,93}Kr$, $^{92,93}Rb$, $^{141,142}Xe$ and $^{141,142}Cs$ Precursors," J. Inorg. Nucl. Chem., 37 (1975) 1563.

7. L. Tomlinson and M. H. Hurdis, "Delayed Neutron Precursors - II: Antimony and Arsenic Precursors Separated Chemically," J. Inorg. Nucl. Chem., 30 (1968) 1649.

8. L. Tomlinson and M. H. Hurdis, "Delayed Neutron Precursors - IV, ^{87}Se, ^{88}Se and ^{89}Se; Half Lives, Neutron Emission Probabilities and Fission Yields," J. Inorg. Nucl. Chem., 3 (1971) 3609.

9. M. Asghar, J. P. Gautheron, G. Bailleul, J. P. Bocquet, J. Grief, H. Schrader, G. Siegert, C. Ristori, J. Crancon and G. I. Crawford, "The P Values of the ^{235}U (n_{th},f) Produced Precursors in the Mass Chains 90, 91, 93-95, 99, 134 and 137-139," Nucl. Phys., A247 (1975) 376.

10. J. Crancon, C. Ristori, H. Ohm, W. Rudolph, K.-L. Kratz and M. Asghar, "Half-Lives on P_n Values of Delayed Neutron Precursors in the Mass Chains 85-87, 92, 135, 136 and 145," Nucl. Phys., A287 (1978) 45.

11. J. V. Kratz and G. Herrmann, "Half-Lives, Fission Yields, and Neutron Emission Probabilities of ^{87}Se and ^{88}Se, and Evidence for ^{87}As," J. Inorg. Nucl. Chem., 32 (1970) 3713.

12. Von H.-D. Schüssler and G. Herrmann, "Hauptkornptonen unter den Vorläugern verzögerter Neutronen bei der Spaltung von Uran - 235 durch thermische Neutronen," Radiochem. Acta., 18 (1972) 123.

13. J. V. Kratz, H. Franz and G. Herrmann, "Delayed Neutrons from Arsenic Isotopes, ^{84}As, ^{85}As and ^{86}As," J. Inorg. Nucl. Chem., 35 (1973) 1407.

14. K. L. Kratz and G. Herrmann, "Delayed Neutron Emission from Short-Lived Br and I Isotopes," Nucl. Phys., A229 (1974) 179.

15. W. Rudolph, K. L. Kratz and G. Herrmann, "Half-Lives, Fission Yields and Neutron Emission Probabilities of Neutron-Rich Antimony Isotopes," J. Inorg. Nucl. Chem., 39 (1977) 753.

16. P. Peuser, H. Otto, M. Weiss, G. Nyman, E. Roeckl, J. Bonn, L. von Reisky and C. Spath, "Half-Lives, Neutron Emission Probabilities, and Fission Yields of Neutron-Rich Rubidium Isotopes in the Mass Region A=96 to A=100," Z. Physik., A289 (1979) 219.

17. P. del Marmol and D. C. Perriocos, "Identification of ^{88}Se and Search for Delayed Neutron Emission from ^{87}Se and ^{88}Se," J. Inorg. Nucl. Chem., 32 (1970) 705.

18. P. del Marmol, P. Fettweiss and D. C. Perricos, "On the Delayed Neutron Yields of the Longer-Lived Halogen Precursors in Thermal Neutron Fission of ^{235}U," Radiochem. Acta., 16 (1974) 4.

19. I. Amarel, H. Gauvin and A. Johnson, "Delayed Neutron Emission Probabilities of Rb and Cs Precursors: The Half-Life of ^{97}Rb," J. Inorg. Nucl. Chem., 31 (1969) 577.

20. E. Roeckl, P. F. Dittner, R. Klapisch, C. Thibault, C. Rigaud and R. Prisels, "Delayed Neutron Emission from the Decay of Neutron Rich Rb and Cs Isotopes," Nucl. Phys., A222 (1974) 621.

21. E. Lund, P. Hoff, K. Alekett, O. Glomset and G. Rudstam, "Delayed Neutron Emission Probabilities of Gallium, Bromine, Rubidium, Indium, Antimony, Iodine and Cesium Precursors," Z. Physik., A294 (1980) 233.

22. K. Alekett, P. Hoff, E. Lund and G. Rudstam, "Delayed Neutron Emission Probabilities of the Precursors, $^{89,90,91}Br$ and $^{139,140,141}I$," Z. Physik., A295 (1980) 331.

23. C. Ristori, J. Crancon, K. D. Wünsch, G. Jung, R. Decker and K. L. Kratz, "Half-Lives and Delayed Neutron Emission Probabilities of Short-Lived Rb and Cs Precursors," Z. Physik., A290 (1979) 311.

24. P. M. Aron, O. I. Kostochkin, K. A. Petrzhak and V. I. Shpakov, "Probability of Delayed Neutron Emission from Halogens," Soviet Journal of Atomic Energy, 16 (1964) 447, original Russian article Atomnaya Energyia, 16 (1963) 368.

25. P. L. Reeder, J. F. Wright and L. J. Alquist, "Delayed Neutron Emission Probabilities of Separated Isotopes of Br, Rb, I and Cs," Phys. Rev., C15 (1977) 2108.

26. P. L. Reeder and R. A. Warner, "Delayed Neutron Emission Probabilities of Rb and Cs Precursors Measured by Both Ion and Beta Counting Techniques," PNL-SA-8766, Pacific Northwest Laboratory, Richland, WA 99352.

27. G. Engler and E. Ne´Eman, "Delayed Neutron Emission Probabilities and Half-Lives of Rb, Sr, Y, In, Cs, Ba and La Precursors with A=93-98, A=127-131 and A=142-148," Nucl. Phys., A367 (1981) 29.

28. W. L. Talbert, Jr., A. B. Tucker and G. M. Day, "Delayed Neutron Emission in the Decays of Short-Lived Separated Isotopes of Gaseous Fission Products," Phys. Rev., 177 (1969) 1805.

29. B. F. Rider, T. R. England, D. G. Madland, J. R. Liaw and R. E. Schenter, "Evaluation of Fission Product Yields for the U.S. National Nuclear Data Files," Proceedings of the Conference on Nuclear Data Evaluation Methods and Procedures, BNL-NCS-51363, Brookhaven National Laboratory, Upton, NY.

30. K. L. Kratz and G. Herrmann, "Systematics of Neutron Emission Probabilities from Delayed Neutron Precursors," Z. Physik., 263 (1973) 435.

31. S. Amiel and H. Feldstein, "A Semi-Empirical Treatment of Neutron Emission Probabilities from Delayed Neutron Precursors," Phys. Lett., 31B (1970) 59.

32. This preliminary mass adjustment is available from the National Nuclear Data Center, Brookhaven National Laboratory, Upton, NY.

33. S. Liran and N. Zeldes, "A Semiempirical Shell-Model Formula," Atomic and Nuclear Data Tables, 17 (1976) 431.

34. F. M. Mann, C. Dunn and R. E. Schenter, "Beta Decay Properties Using a Statistical Model," Phys. Rev., C25 (1982) 524.

THE WORK OF THE IAEA COORDINATED RESEARCH PROGRAMME ON THE MEASUREMENT AND EVALUATION OF TRANSACTINIUM ISOTOPE NUCLEAR DECAY DATA

C. W. Reich

Idaho National Engineering Laboratory
EG&G Idaho, Inc.
Idaho Falls, Idaho 83415 U.S.A.

In 1977, the IAEA organized a Coordinated Research Program to address the needs for precise actinide-isotope decay data identified at the first Advisory Group Meeting on Transactinium Isotope Nuclear Data, held in Karlsruhe in 1975. During the years of its existence, this CRP has made significant strides toward achieving the goals outlined at Karlsruhe. In this paper, we discuss the make-up of the CRP and its work in the areas of decay-data evaluation and measurement. The objectives of the evaluation effort and some of the results to date are summarized. The measurement activity being carried out within the various participating laboratories is presented. Finally, the significant accomplishments resulting from the work of the CRP participants are discussed, together with those tasks which remain to be done in order to be fully responsive to the goals of the Program as envisioned at Karlsruhe.

1. Introduction

Over the past decade or so, as their relevance to many different areas of science and technology has received increasingly wide recognition, data on the decay properties of radioactive nuclides have come to represent a significant subset of that body of information referred to as "Nuclear Data." Because of the mechanism through which they are most commonly produced, the fission-product nuclides have always occupied a special position in fission-reactor research and technology; and the importance of their decay data has long been recognized. Similarly, with the expanding emphasis on fast-reactor technology and all aspects of the associated fuel cycle, the need for reliable, accurate decay data on a number of isotopes of the actinide elements is being increasingly recognized.

However, in spite of the wide applicability of decay data in many areas of technology, the primary motivation for their measurement has historically been their importance for basic low-energy nuclear-structure physics. As a consequence of the extensive measurement activity in this area over the past 30 years or so, there presently exists a considerable volume of rather detailed data on the decay properties of unstable nuclei. However, due in part to the basic orientation and interests of most of the measurers of these data, the existence of this large quantity of measured information does not necessarily mean that the data needs within any given area are satisfied. To properly assess the adequacy of the existing decay-data base for a specific application requires a careful evaluation of this information in the light of a specific set of user needs, some of which can be rather highly specialized.

In 1975, the IAEA, in cooperation with the OECD Nuclear Energy Agency, convened an Advisory Group Meeting on Transactinium Isotope Nuclear Data[1] (TND) at Karlsruhe, FRG. The principal purpose of this meeting was to survey the requirements for and the status of transactinium ($Z \geq 90$) isotope nuclear data relevant to fission-reactor research and technology and to formulate recommendations and measures for the coordination of future work. One of the problem areas addressed at this meeting was the status of the decay data (half-lives, alpha and gamma intensities) for the transactinium nuclides. It was found that the accuracy of many of these data was not adequate to satisfy a number of needs in such areas of reactor technology as safeguards, fuel assay, sample-mass determination and standards preparation. A list of these important transactinium isotopes and the accuracy requirements for their decay data was drawn up. Further, it was recommended (General Recommendation 4 in Ref.[1]) that an internationally coordinated program of decay-data measurement and evaluation be set up to satisfy these identified data needs.

In response to this recommendation, the IAEA Nuclear Data Section subsequently organized a Coordinated Research Program (CRP) on the measurement and evaluation of transactinium-isotope nuclear decay data. Five established groups experienced in decay-data measurement agreed to participate in the work of this CRP. The first meeting of the participants in this program was held in April, 1978 in Vienna, where the objectives were discussed and the work initiated. Subsequent annual meetings have been held to monitor progress and plan future work. In this paper, we discuss the various activities of the CRP, its achievements as of May, 1982 and some of the work that remains to be done.

2. Composition and Activities of the Coordinated Research Program

The representatives of the various participating laboratories in the Coordinated Research Program are as follows:

Laboratory (Nation or Organization, Representative)

Central Bureau for Nuclear Measurements (CEC, R. Vaninbroukx)

Laboratoire de Métrologie des Rayonnements Ionisants (France, F. Lagoutine, G. Malet, J. Legrand)

Japan Atomic Energy Research Institute (Japan, H. Umezawa)

Atomic Energy Research Establishment Harwell (United Kingdom, A. J. Fudge)

Idaho National Engineering Laboratory (United States of America, C. W. Reich)

A. Lorenz, of the IAEA Nuclear Data Section, serves as project officer coordinating the program. In addition V. M. Kulakov of the Kurchatov Atomic Energy Institute (U.S.S.R.) and A. L. Nichols of AEE Winfrith (U.K.) have participated as observers at several meetings of the CRP.

Within the U.S., work at several laboratories is relevant to the objectives of the CRP. In addition to the γ-ray emission-probability (absolute-intensity) measurements being carried out at Idaho National Engineering Laboratory (INEL), absolute alpha-transition-intensity measurements are being done by I. Ahmad at Argonne National Laboratory (ANL). Valuable input to the CRP is also being provided by the U.S. Half-Life Evaluation Committee. This Committee was set up several years prior to the inception of the IAEA CRP to address the poor status of the half-life data on the important Pu isotopes (primarily ^{239,240,241}Pu). The laboratories (and representatives) making up this Committee are: Mound Laboratory (W. Strohm, Chairman); Argonne National Laboratory (A. Jaffey); Los Alamos National Laboratory (J. E. Rein); Lawrence Livermore National Laboratory (A. Prindle); National Bureau of Standards (L. Lucas); and Rocky Flats Laboratory (R. Carpenter).

The initial basis for planning the work of the CRP was provided by the decay-data requirements identified at the Advisory Group Meeting in Karlsruhe. These requirements are summarized in Table I. From this information, supplemented by additional data needs known to various of the participants, a measurement

K. H. Böckhoff (ed.), Nuclear Data for Science and Technology, 277–280.

program was agreed upon and initiated. At the Second IAEA Advisory Group Meeting on Transactinium Isotope Nuclear Data,[2] held in Cadarache in 1979, this request list was extensively reviewed and several modifications to it were made. In some instances, the required accuracies had been achieved and the related data requests could thus be regarded as having been satisfied. In others, however, additional data needs were identified or substantially increased accuracy requirements were requested. A tabulation of this revised listing of the requirements and status of transactinium isotope decay data can be found in Ref.[3].

3. Decay-Data Evaluation Activities Within the CRP

At the first CRP Participants meeting, the preparation of a list of "recommended" decay data was undertaken. While this list was intended initially to serve primarily as a working file for the use of the program participants, subsequently it was thought that it would also eventually be made available to interested users of transactinium-isotope decay data.

The first phase of this decay-data evaluation involved the preparation of a preliminary list of half-life values and branching-fraction data for approximately 40 nuclides, ranging from ^{228}Th through ^{253}Es. The data for this first-draft listing of half-lives were taken largely from the file of actinide-isotope decay data prepared at INEL for eventual incorporation into the ENDF/B-V Actinide File. In subsequent meetings, this list was refined and expanded and now contains information for essentially all the nuclides in the important actinide decay chains. The sources of data for this list have been expanded to include the Evaluated Nuclear Structure Data File (ENSDF), the U.K. Chemical Nuclear Data Committee Heavy Element Decay Data File (UKHEDD-1) and recent experimental results, as well as the INEL decay-data file.

Periodically, updated versions of this half-life data list have been issued, the most recent being that contained in Ref.[4]. Those nuclides that are presently included in this list are given in Table II.

After the basic content of the half-life file was established, the thrust of the evaluation effort was directed toward the preparation of lists of evaluated alpha-transition and gamma-ray energy and intensity values. Here, the desired nuclide coverage is much smaller than that of the half-life list and is generally restricted to data on the more prominent transitions from each nuclide. The nuclides included in these two lists are shown in Table III.

4. Decay-Data Measurements Within the CRP

The requested accuracies (cf. Table I and Ref.[3]) of many of the data, especially the intensity values, are quite high; and to achieve them represents a challenging experimental problem. In this regard, the existence of this Coordinated Research Program has

Table I. Summary of Nuclear Decay-Data Requirements for Actinides (Extracted from Proceedings of 1975 Karlsruhe TND Meeting, IAEA-186, Ref.[1]).

Nuclide	Data type	Required Accuracy	Needed for
U 234	I_α	1%	mass determination, fuel assay
	I_γ	5%	
U 235	$T_{1/2}$	1%	mass determination, fuel assay
	I_α	1%	
	I_γ	1%	
U 236	$T_{1/2}$	1%	mass determination
U 238	I_α	1%	mass determination
Np 237	I_α	1%	mass determination
Pu 238	$T_{1/2}$	0.5%	thermal and fast reactors, destructive fuel assay, calorimetry
		0.02%	high-precision mass standard
	I_α	0.1%	mass determination
	I_γ	1%	non-destructive fuel assay
Pu 239	$T_{1/2}$	0.5%	thermal reactor application, fast reactor fuel analysis
		0.2%	destructive fuel assay, mass determination
	I_α	1%	mass determination
	I_γ	1%	non-destructive fuel assay
Pu 240	$T_{1/2}$	1%	fast reactor fuel analysis
		0.2%	mass determination, calorimetry
	I_α	0.2%	mass determination
	I_γ	1%	non-destructive fuel assay
Pu 241	$T_{1/2}$	1%	thermal and fast reactor applications, destructive fuel assay
	I_γ	1%	non-destructive fuel assay
Pu 242	$T_{1/2}$	1%	mass determination
	I_α	4%	mass determination
Am 241	$T_{1/2}$	1%	fast reactor application
	I_γ	1%	intensity standard
Cm 242	$T_{1/2}$	0.1%	destructive fuel assay (decay correction)
	$T_{1/2}$ (S.F.)	3%	fast reactor fuel handling
Cm 244	$T_{1/2}$ (S.F.)	3%	fast reactor fuel handling
Cf 252	$T_{1/2}$	0.2%	destructive fuel assay (decay correction)

Table II. Nuclides included in the list of proposed recommended heavy-element radionuclide half-life and branching-fraction data.

Hg	206
Tl	206,206m,207,207m,208,209,210
Pb	205,209,210,211,212,214
Bi	210,210m,211,212,212m,$212m_2$,213,214,215
Po	209,210,211,211m,212,212m,213,214,215,216,218
At	215,217,218,219
Rn	217,218,219,220,222
Fr	221,223
Ra	223,224,225,226,228
Ac	225,227,228
Th	227,228,229,230,231,232,233,234,235
Pa	231,232,233,234,234m,235
U	232,233,234,235,235m,236,237,238,239,240
Np	236,236m,237,238,239,240,240m,241
Pu	236,237,238,239,240,241,242,243,244,245,246
Am	240,241,242,242m,243,244,244m,245,246,246m
Cm	241,242,243,244,245,246,247,248,249,250
Bk	249,250
Cf	249,250,251,252,253
Es	253

not only helped focus the existing capabilities of the participating laboratories on these measurements but also encouraged the development of such capabilities at other laboratories. This occurrence, together with the systematic production of additional accurately measured decay data, may represent one of the more significant accomplishments of this CRP.

It has been a little more than four years now since the first meeting of the participants in this CRP. During this period a great deal of progress has been made, although much still remains to be done to complete the measurement program initially envisioned. The present status of the measurements is summarized in Table IV.

5. Conclusions

During its existence, the IAEA Coordinated Research Program on the Measurement and Evaluation of Transactinium Isotope Nuclear Decay Data has accomplished a number of its original goals. Briefly summarized, these include the following:

1. the evaluation of the accuracy requirements for decay data established by the users at the Advisory Group Meetings on Transactinium Isotope Nuclear Data at Karlsruhe (1975) and Cadarache (1979). These requirements were sorted out into one of the following three categories: (i) those which are satisfied by the presently available data; (ii) those which are beyond the capabilities of present measurement techniques; and (iii) those which have not yet been satisfied but which are achievable with present experimental capabilities;
2. determination of the status of the existing data, including keeping abreast of the results of new measurements as they become available;
3. identification of those unsatisfied decay-data needs which can be met;
4. coordination of existing measurement capability to obtain the required data; and

Table III. Summary of nuclides included in the lists of recommended E_α/I_α and E_γ/I_γ data presently being reviewed by the CRP participants.

Nuclide	α-transition list	γ-transition list
^{228}Th	yes	yes
^{229}Th	yes	yes
^{230}Th	yes	no
^{233}Th	no	yes
^{231}Pa	yes	yes
^{233}Pa	no	yes
^{232}U	yes	yes
^{233}U	yes	yes
^{234}U	yes	yes
^{235}U	yes	yes
^{236}U	yes	yes
^{237}U	no	yes
^{238}U	yes	no
^{239}U	no	yes
^{237}Np	yes	yes
^{239}Np	no	yes
^{236}Pu	yes	no
^{238}Pu	yes	yes
^{239}Pu	yes	yes
^{240}Pu	yes	yes
^{241}Pu	yes	yes
^{242}Pu	yes	no
^{241}Am	yes	yes
^{243}Am	yes	no
^{243}Cm	yes	no
^{245}Cm	yes	no
^{246}Cm	yes	no
^{252}Cf	yes	no

5. completion of measurements of some of these data.

The CRP participants, through their broad experience in nuclear-data measurement, have critically evaluated the experimental feasibility of achieving the required data accuracies. Through their access to and familiarity with several national decay-data files, they have unified the information from a number of different sources and prepared a list of evaluated data which correctly reflects the status of the relevant data.

With the possible exception of completing the measurements presently in progress or planned, the most important task that remains to be done is the preparation of a final set of recommended values for the users of transactinium-isotope decay data. This will require critical evaluation not only of the data produced by the members of this CRP but also of data produced elsewhere. Throughout the above discussion, because of its emphasis on the activities of the CRP, not much mention has been made of measurement results in the field of transactinium-isotope decay data not done within the CRP. However, a significant amount of such information has been produced during recent years by groups that have no active involvement in the activities of this Program. The CRP participants have attempted to keep abreast of such measurement results and, where appropriate, have incorporated them into the data evaluations discussed in Section 3 above.

Of equal importance, careful consideration needs to be given to the data generated by the CRP participants themselves. One significant aspect of the program plan of the CRP was to have a number of the quantities measured independently by several of the participating laboratories. The evaluation of the potentially differing results in these cases, to produce for each quantity a single recommended value, will be an especially important task and one which the CRP participants should be uniquely qualified to perform. This reconciliation process, properly carried out using all the available information, should greatly enhance the credibility of the recommended data and

Table IV. Summary of the status, as of May, 1982, of the measurements being carried out by the various laboratories participating in the Coordinated Research Program.

Isotope and Quantity	Participating Laboratory						
	CBNM	HARWELL	JAERI	LMRI	ANL	INEL	U.S. Half-life Evaluation Comm.
^{228}Th I_γ [a]	completed					in progress	
^{231}Th I_γ	completed						
^{231}Pa I_γ		in progress					
^{233}Pa I_γ		completed[b]				completed	
^{232}U I_γ		in progress				in progress	
^{233}U I_γ		in progress				in progress	
I_α					completed		
^{234}U I_γ		in progress					
^{235}U I_γ	completed	in progress				in progress	
^{236}U I_γ		in progress					
^{237}Np I_γ		completed					
^{237}Pu $T_{1/2}$			completed				
^{238}Pu I_γ	completed[c]			completed		completed	
I_α	in progress				completed		
I_{LX}	completed						
^{239}Pu $T_{1/2}$	completed						completed
I_γ			in progress[d]	completed		completed	
I_α					completed		
^{240}Pu $T_{1/2}$							in progress
$T_{1/2}$(S.F.)	completed						
I_γ				in progress		completed	
I_α					completed		
^{241}Pu $T_{1/2}$	in progress	in progress					completed[e]
$T_{1/2}$ (α)	completed						
^{241}Am I_γ	in progress			completed			
^{242}Cm $T_{1/2}$			completed				
$T_{1/2}$(S.F.)			in progress				

[a] Including selected members of its decay chain.

[b] Includes relative, as well as absolute, values.

[c] Value for the 43.5-keV gamma ray only.

[d] H. Umezawa has brought to our attention work by Yoshizawa et al. of Hiroshima University.

[e] Results from two of the participating laboratories have been published separately; the committee has not yet adopted a recommended value.

hence enhance their acceptance by the international user community. Such a set of recommended decay data, based on careful evaluation of a reliable base of experimental information, will be the final product of this Coordinated Research Program.

Acknowledgements

Helpful comments from the other participants in this IAEA Coordinated Research Program in the preparation of this manuscript are gratefully acknowledged. The work at INEL has been supported by the U. S. Department of Energy under DOE Contract No. DE-AC07-76ID01570.

References

1. "Transactinium Isotope Nuclear Data," IAEA-186, Vol. I (IAEA, Vienna, 1976).
2. "Transactinium Isotope Nuclear Data -- 1979," IAEA-TECDOC-232 (IAEA, Vienna, 1980).
3. "Third Coordinated Research Meeting on the Measurement and Evaluation of Transactinium Isotope Nuclear Data, Vienna, 12-13 June 1980, Summary Report," INDC (NDS)-118/NE (October, 1980).
4. "Proposed Recommended List of Heavy Element Radionuclide Decay Data : Part I Half-Lives (December 1981 Edition); Part II Provisional List of Alpha Spectra (December 1981 Edition)," INDC (NDS)-127/NE (December, 1981).

THE UK CHEMICAL NUCLEAR DATA LIBRARIES: EVALUATED NUCLEAR DECAY DATA FOR REACTOR APPLICATIONS

B.S.J. Davies, A. Tobias
CEGB,
Berkeley Nuclear Laboratories
Berkeley, Glos, UK

M.F. James, A.L. Nichols
AEE Winfrith,
Dorchester,
Dorset, UK

E.A.C. Crouch
AERE Harwell,
Didcot,
Oxfordshire, UK

The United Kingdom Chemical Nuclear Data Committee has the responsibility for the development and maintenance of libraries of evaluated nuclear decay and fission yield data for use in the UK nuclear power programme. Currently the libraries contain decay data for 91 activation products (UKPADD-1), 125 heavy element nuclides (UKHEDD-1) and 855 fission product nuclides (UKFPDD-2) together with fission yield data (CROUCH 3I). The present contents and the methods of file construction are described together with the available means of producing suitable data tabulations.

The fission product data have been recently improved and tested in decay heat calculations. The calculated results have been compared with experimental values of beta, gamma and total heating.

Introduction

The United Kingdom Chemical Nuclear Data Committee (UKCNDC) was constituted to identify the chemical nuclear data needs of the nuclear energy programme in the United Kingdom and to remedy important deficiencies[1]. The role of this committee has included the co-ordination of specific laboratory measurements and data evaluations. The latter has included the establishment of a number of specific data libraries by the Data Library Subcommittee of the UKCNDC. Four data libraries have been assembled and are revised at regular intervals:

(i) fission product yields,
(ii) fission product decay data, excluding delayed neutron data,
(iii) activation product decay data,
(iv) heavy element decay data.

The first three of these libraries are in ENDF/B-IV format as recommended by the IAEA panel on fission product nuclear data[2,3]. For the heavy element decay data ENDF/B-V format has been used to permit the inclusion of spontaneous fission decay data including half-lives, $\bar{\nu}$ values and prompt gamma and neutron spectra.

The UKCNDC Data Libraries

Fission Yield Data

The fission product yield data were evaluated by Crouch. Firstly, cumulative yields were evaluated from experimental values[4]. These values were then adjusted to produce a set which is consistent with certain physical constraints:

(i) cumulative chain yields sum to 2.0,
(ii) conservation of nucleons,
(iii) conservation of charge,
(iv) if the fissioning nucleus has atomic number Z_f then the pair of elements with atomic numbers z and Z_f-z have equal yields,
(v) the sum of the fractional independent yields along a mass chain is 1.0.

Having produced an adjusted set by this method, an attempt was made to split the yields of individual nuclides between ground and metastable states. This was done by using the ratios in USENDF/B-IV[5].

By this means consistent yield sets have been produced[6] for the following fissioning nuclides and incident neutron energies:

^{232}Th, ^{238}U	(fast and 14.7 Mev)
^{233}U, ^{235}U	(thermal, fast and 14.7 MeV)
^{239}Pu, ^{241}Pu	(thermal, fast)
^{240}Pu	(fast)

The resulting file (CROUCH3I) is intended mainly for use in calculating decay heat or similar quantities involving many fission product species. When individual fission product yields are required, for example for burn-up determination, the data in reference 4 are recommended.

Fission Product Decay Data

The fission product decay data library (UKFPDD-1) has been described previously[1]. As improved measurements of these data have been reported, revisions have been made to the half-life values and radiation spectra in this library. These new data were taken either from the 1980 version of the Evaluated Nuclear Structure Data File[7] or directly from the published literature. The resultant file (UKFPDD-2) has been described by Tobias and Davies[8]. It contains data on 855 nuclides of which 736 are radioactive, and line spectra are given for 390.

Activation Product Decay Data

The currently recommended decay data for nuclides formed by neutron activation of reactor structural materials is that described by Nichols[9] (UKPADD-1). Comprehensive decay data for 91 nuclides have been evaluated, all of which have detailed radiation spectra. Major efforts are under way to expand this library to over 400 activation product nuclides that will cover decay data requirements for possible fusion as well as fission reactor systems.

Heavy Element Decay Data

This set of decay data covers the transactinides, produced by the neutron irradiation of nuclear fuel, and their subsequent heavy element decay products. The contents of this library (UKHEDD-1) have been described by Nichols and James[10]. 125 radionuclides are included, of which 122 have evaluated spectral data. This library has been completed recently with the addition of the spontaneous fission decay data, including $\bar{\nu}$ values and prompt radiation spectra. ENDF/B-V format has been used for these data.

Data Retrieval

While the files described above contain a great deal of information in a format which is suitable for input to a computer, it is desirable to have a system for extracting sub-sets of data, which are needed for particular purposes, and writing these sub-sets in a conveniently readable form. A suite of computer programs for this purpose has been written by Tobias[11]. The nuclides may be defined by stating ranges of atomic number, mass number and half-life; ranges of line energy and intensity may be stated for

K. H. Böckhoff (ed.), Nuclear Data for Science and Technology, 281–283.

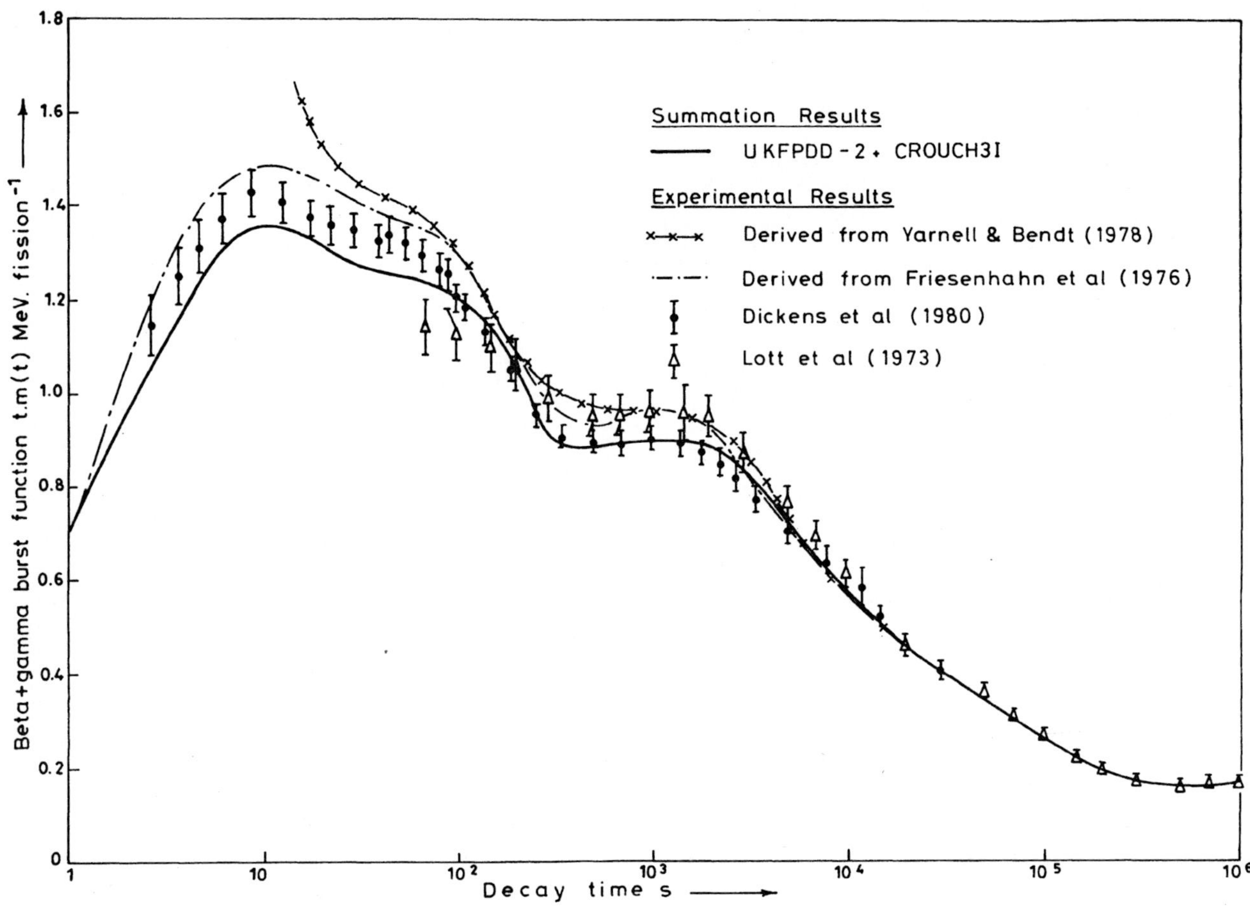

Fig. 1. Fission Product Beta + Gamma Burst Function - ^{235}U Thermal Fission

the radiation of interest. The data thus obtained may then be presented in a number of ways:

(i) data on any radiation type may be listed for the selected nuclides,

(ii) gamma radiation energies and intensities may be listed in order of energy,

(iii) beta radiation spectral shapes may be calculated and plotted.

Data Testing

Extensive checks have been made on the data, including tests of the consistency and completeness of decay data, intercomparisons with other equivalent libraries and comparisons of calculated values of decay heat with measured values. These exercises have pin pointed specific errors which have been subsequently eliminated.

Any serious problems encountered during the evaluation are described in the comments section associated with each nuclide: if the resultant data contain outstanding problems a statement may be made that improved measurements are required.

The fission product decay data and associated yield data were designed to be used in decay heat calculations which may be compared with experimental values. Examples of such a comparison are given in Figures 1 and 2. In these figures calculated decay heat values following an instantaneous fission pulse in ^{235}U are compared with values derived from experiment[12-17]. Remembering that comparisons based on very short irradiation times such as this are very stringent, it is evident from the figures that agreement between calculation and experiment is good. A more detailed discussion involving other irradiation times and other fissile nuclides is given by Tobias[18].

Conclusion

Extensive sets of evaluated data covering fission product yields and decay data for fission products, activation products and heavy elements have been assembled. A suite of computer codes exists for extracting the data and presenting them in a convenient and readable form. As part of the quality assurance checks extensive comparisons have been made between experimental and calculated decay heat values.

Acknowledgements

The authors gratefully acknowledge the comments of members of the Data Library Subcommittee and the UK Chemical Nuclear Data Committee.

This paper is published by permission of the Central Electricity Generating Board and the United Kingdom Atomic Energy Authority.

References

1. B.S.J. Davies, E.A.C. Crouch, M.F. James, A.L. Nichols, J.R. Parkinson, A. Tobias, D.G. Vallis, Int. Conf. Neutron Physics and Nuclear Data for Reactor Applications, Harwell (1978).

2. IAEA 'Fission Product Nuclear Data', Bologna, 1973, IAEA-169 (1974).

3. IAEA 'Fission Product Nuclear Data', Petten, 1977, IAEA-213 (1978).

4. E.A.C. Crouch, At. Data and Nuc. Data Tables, 19 419 (1977).

5. T.R. England, R.E. Schenter, LA-6116-MS, Los Alamos Scientific Lab (1975).

6. IAEA 'Progress in Fission Product Nuclear Data', INDC(NDS)-113/G+P (1980).

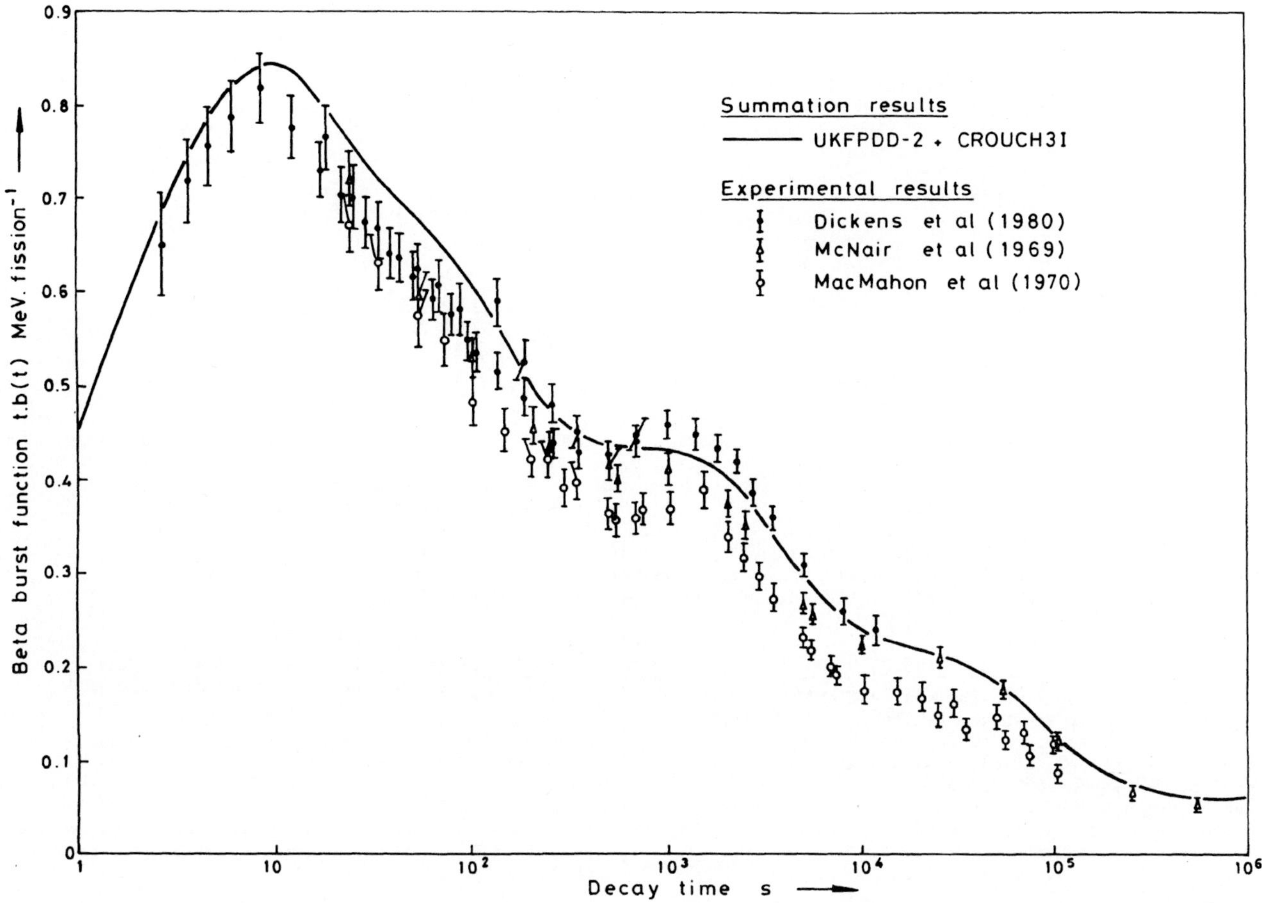

Fig. 2. Fission Product Beta Burst Function - ^{235}U Thermal Fission

7. W.B. Ewbank, M.R. Schmorak, F.E. Bertrand, M. Feliciano, D.J. Horen, ORNL 5054 Oak Ridge National Lab (1975).

8. A. Tobias, B.S.J. Davies, RD/B/N4942, CEGB (1980).

9. A.L. Nichols, AERE R-8903, UKAEA (1977).

10. A.L. Nichols, M.F. James, AEEW R-1407, UKAEA (1981).

11. A. Tobias, RD/B/5170N81, CEGB (1981).

12. J.K. Dickens, T.A. Love, J.W. McConnell, R.W. Peele, Nuc. Sci. Eng. 74 106 (1980).

13. A. McNair, R.L.G. Keith, J. Nuc. Energy 23 697 (1969).

14. T.D. MacMahon, R. Wellum, H.W. Wilson, J. Nucl. Energy 24 493 (1970).

15. J.L. Yarnell, P.J. Bendt, LA-7452-MS, Los Alamos Scientific Lab (1978).

16. S.J. Friesenhahn, N.A. Lurie, V.C. Rogers, N. Vagelatos, EPRI-NP-180, Electric Power Res. Inst. (1976).

17. M. Lott, G. Lhiaubet, F. du Freche, R. de Tourreil, J. Nuc. Energy 27 597 (1973).

18. A. Tobias, RD/B/N4949, CEGB (1980).

COMPUTER INTERPRETATION, ANALYSIS AND EVALUATION OF NUCLEAR DECAY SCHEMES

G. Evangelides

Chemical Engineering and Chemical Technology Department
Imperial College of Science and Technology
Prince Consort Road, London S.W.7
United Kingdom

Research Sponsored by the United Kingdom Atomic Energy Authority (U.K.A.E.A.)
Atomic Energy Research Establishment (A.E.R.E.)
Harwell, Near Didcot, OXON OX11 ORA
United Kingdom

Currently in the Telecommunications, Installations and Network Support Group
Science and Engineering Research Council (S.E.R.C.)
Daresbury Laboratory, Daresbury, Warrington WA4 4AD
United Kingdom

The CASCADE computer program, originally written for radiochemical purposes, was developed and greatly extended for use in the interpretation, analysis and evaluation of nuclear decay schemes. The results are recorded in ordered single or coincidence radiation catalogues. These can be re-ordered in terms of decay mode, half-life, ascending or descending energy, intensity, etc. The results can also be provided in standard ENDF/B-IV or V format for use by application programs, e.g. for use in the study of decay heat and shielding calculations of irradiated nuclear fuel. The source of the nuclear decay schemes for this study was the Evaluated Nuclear Structure Data Files, and a program called FELIX was written to perform the conversion.

Introduction

The CASCADE computer program was originally written by D.G. Vallis[1], for radiochemical purposes. It has been developed and greatly extended, by the author[2,3] for use in the interpretation, analysis and evaluation of nuclear decay schemes. It analyses any given nuclide and produces a list of cascades of effectively coincident radiations, hence the name of the program. From these, each component radiation with its energy and intensity is evaluated and the results are recorded in a catalogue of single radiations. This catalogue can be sorted, and/or merged with existing radiation catalogues, hence building up a radiation catalogue library. Such a library can be regularly maintained or updated, with data added or deleted by using the program CASCAT[4], written for this purpose.

The cascade radiations can also be analysed for the production of coincidence radiation catalogues. These consist mostly of two-way coincidences, but may include some three or more way coincidences, where more than one X-ray or Auger electron of identical energy are in coincidence with another radiation. A program called CONSORT[5] was written for sorting and merging these coincidence radiation catalogues with existing catalogues, or a coincidence radiation library.

Both the CASCAT and CONSORT programs depend on the use of the IBM SORT/MERGE[6] utilities for sorting and merging the radiation data which exist as random access data sets on disc.

The results of the decay scheme evaluations by CASCADE can also be produced in standard ENDF/B-IV and V formats[7,8]. The present version of CASCADE will produce 'partial' ENDF/B output, that is to say that one ENDF/B output file is produced per decay mode. The full ENDF/B implementation requires only one file for all decay modes of each nuclide. Hence a program, named AMALGAM[9], was written to amalgamate these 'partial' files into one.

The decay scheme data for input to CASCADE can be obtained from a variety of sources. The format that was devised for these data, is very similar in structure to the diagramatic representations of nuclear decay schemes as found, for example, in the Table of Isotopes[10]. However, in a project of this nature a generally available computerised input source is required. Hence the Evaluated Nuclear Structure Data Files (ENSDF)[11] were chosen as the basic input source. A computer program named FELIX[12] was written to convert the ENSDF compilations into a form suitable for CASCADE. In FELIX provisions had to be made to overcome some of the errors and inconsistencies, some systematic, which had been discovered in the ENSDF compilations.

Later versions of the ENSDF compilations have had many of these errors or inconsistencies removed. A notable improvement was the inclusion of the parent nuclide description in most of the decay data files. The latest version used in this project had been released in July 1979 and was very kindly provided to the author by the Nuclear Data Bank at Gif-sur-Yvette, France, while on a visit there.

Nuclear Decay Scheme Data

The nuclear decay scheme data used by CASCADE have been obtained from the Evaluated Nuclear Stucture Data Files (ENSDF). These consist of single record descriptors of 80 bytes length (i.e. card image format), starting with the nuclide identification and followed by such records as the parent nuclide description, Q-value and normalisation factor records, if any, and a record for each relevant energy level or emitted radiation. These may be interspersed with relevant comment records and other secondary information, such as theoretical and/or experimental internal conversion coefficients. These are given as secondary or continuation records.

There are a number of different possible record types. Each type is defined by a characteristic symbol placed in column 8. For example the letters A, B and E define alpha, negatron and positron or electron capture transition records respectively. These are the primary transitions, and they must follow the level record (L) to which they apply. The gamma transitions (G) should then follow. Comment records can be included with these and a letter C is placed in column 7, with the record type symbol in column 8, as before. It is also possible to have records in non-standard format. In which case, a record with the letter F in column 7 is used to define the format. This must be placed before the first occurrence of the record type to which it refers. A non blank character, in column 6, is used to define continuation records.

For efficient programming considerations it was found necessary for the type of records present in any particular nuclide to be known in advance of the evaluation by FELIX. Hence the ENSDF compilations had to be reorganised by FELIX to include 'header' records. These 'header' records consist of the parent and daughter nuclides, the number or records in the data scheme file, and a list of the type of records to follow, using the above mentioned symbols.

K. H. Böckhoff (ed.), Nuclear Data for Science and Technology, 284–286.

The FELIX program has two internal steps. The first is to reorganise the data and the second to produce the nuclear decay schemes in a form suitable for CASCADE. However, once the ENSDF compilations have been reorganised, as stated, then there is no need to repeat this step on any subsequent run.

It was unfortunate that there were a number of errors and inconsistencies in earlier versions of the ENSDF compilations, and provisions had to be made in FELIX to overcome these. For example, in a number of cases the parent record was either missing or incorrect. FELIX attempts to correct these as follows.

If the parent record was missing the decay mode was taken from the primary transition records, if any. Where the parent record was included, but was given the same as the daughter nuclide, then the daughter nuclide was checked for primary transition records, and if found, the decay mode was taken from these suitably modifying the parent nuclide description in the output data set. Otherwise, the nuclide was taken as an isomeric transition. Where the parent record differed from the daughter nuclide, and the decay mode thus obtained did not match with the primary transition, then the decay mode defined by the parent record was taken to be correct. It was found in practice, that these could account for most of the parent - daughter nuclide discrepancies.

In later versions of the ENSDF compilations the missing parent nuclide descriptor records had in most cases been included. A check is still made by FELIX for any discrepancies.

The 'CASCADE' Program

The CASCADE program was written in a variable dimension format, for efficient use of computer storage. There is no inherent limit to the number of energy levels in a given nuclide, nor to the number of decay schemes that can be analysed in a single execution of the program. The limiting factor is the maximum amount of storage and/or cpu time that is made available to the program.

Should the amount of storage made available to CASCADE not be sufficient, then some of the problem size parameters, which are defined on input by the user, such as the maximum number of energy levels, maximum number of cascades per nuclide, or the maximum number of coincidence radiations per cascade may be reduced. The choice for which of these parameters is to be reduced first is also defined on input, by the user.

CASCADE will calculate all cascades above a minimum intensity, which is specified on input. If the number of cascades being evaluated is about to exceed the maximum number permitted for the given storage space allocation, then CASCADE will increase the value of the minimum cascade to be evaluated and repeat the calculations. Up to eight reiterations of these calculations are permitted. The algorithm used to increase the minimum cascade intensity allows for quick convergence, and in the vast majority of cases, no more than three iterations are required.

The internal conversion coefficients, supplied on input could be supplemented, on request, by the theoretical internal conversion coefficients data from the Hager and Seltzer tables[12a] for atomic numbers 30 to 103 inclusive, or from the I. Band et al.[14] tables for atomic numbers 3 to 30 inclusive. Alternatively these theoretical tables could be used exclusively. The choice can also be made regarding which table is to be used in preference, when the atomic number is common to both.

Both theoretical and experimental internal conversion coefficients may be present in the ENSDF compilations. FELIX can select either, or a combination, of these.

Average negatron or positron energies can be calculated from the Widman et al.[15] tables of average to end-point energies, or from theoretical considerations, using A. Tobias'[16] program. K-electron capture to positron ratios could be used from the Zweifel[17] tabulations.

The 'CASCADE' Radiation Catalogues

The results of the CASCADE analysis can be recorded in a catalogue of emitted radiations consisting of the parent nuclide description, including half-life and metastable state number, its decay mode, emitted radiations given by type, energy and intensity, the computing parameters used and date of calculation. For beta particles both the end-point and average energies are recorded, and for alpha or heavy particle emission both the total transition energy and energy after recoil are given. These radiation catalogues can be ordered according to the above mentioned parameters, such as half-life, decay mode, in ascending or descending energies or intensities etc., or by user preference, using the IBM supplied SORT/MERGE utilities and the listing and editing code CASCAT.

Similar catalogues can be produced for the coincidences between the evaluated radiations, and these too can be sorted and merged using the IBM utilities and the listing code CONSORT.

ENDF/B-IV and V Formatted Output

The results of the CASCADE decay scheme analysis and evaluation can also be produced in standard ENDF/B-IV and V formats. The present version of CASCADE will produce 'partial ENDF/B' output, that is to say that one ENDF/B output file is produced per decay mode. The full ENDF/B implementation requires only one file for all decay modes of each nuclide. Hence a program, named AMALGAM, was written to amalgamate these 'partial' files into one.

The amalgamation of the 'partial' ENDF/B output has been prevented from being performed within the CASCADE program itself, due to computer storage restrictions, and due to the fact that the ENSDF compilations are ordered according to the mass of the daughter nuclide, instead of the parent. Thus a substantial amount of reorganisation would have had to be performed, by FELIX, in order to collate all the different decay modes into a single nuclear decay scheme for each required nuclide.

For the ENDF/B-V format the isomeric level number of isomeric nuclides is also required. This can only be correctly interpreted from the 'Adopted Decay' schemes in the ENSDF data, but since these do not, in general, contain primary or secondary transition propabilities, it was felt unnecessary to perform complete evaluation of these. However, the decay scheme data and the adopted decay schemes are summarised onto a separate stream, giving details of the nuclide and the level lifetimes, when known. This summary list is used by the AMALGAM program to correct the value of the isomeric transition level number.

'Non decay scheme' data are still interpreted by FELIX but placed on an alternative output stream. If any decay schemes are found to be internally inconsistent, then these are separated out from the rest of the data. It would be up to the evaluator to correct these manually.

The present version of CASCADE does not include specific uncertainty data for the ENDF/B output formats, except for half-life of the parent and daughter nuclides. An uncertainty value of 10% is given to those nuclides for which no half-life uncertainty is provided by the ENSDF data. An overall value is given to the rest of the ENDF/B data. This was arbitrarily chosen as 5% plus an additional contribution imposed by the computational interpretation performed by FELIX. This additional error is of the order of 0.0001% but can be as high as a few percent in inconsistent decay schemes.

The option of including more detailed uncertainty data, in the future, was left open for both FELIX and CASCADE programs.

Examples of 'partial' ENDF/B-IV and V output have been supplied to Dr A.L. Nichols for comparison against the United Kingdom Heavy Element and Fission Product Decay Data, UKHEDD-1 and UKPFDD-1A, respectively. Dr A.L. Nichols has reported in a recent private communication, that these are, in general, in good agreement with each other.

In a few cases where differences have been observed, these have been found to be attributed to evaluator preferences. Often related to difference in the choice of internal conversion coefficients, and especially in the absence of these in the ENSDF compilations.

Applications

The radiation catalogues or the ENDF/B output data files can be used with an inventory code such as FISPIN-5[18] to produce a table of neutron and gamma fluxes for decay heat calculations of irradiated nuclear fuel under specified irradiation conditions, The results can then be analysed by a neutron and photon transport code such as ANISN-W[19], and used, for example, to determine the shielding requirements and desired configuration of irradiated fuel transportation flasks. Sensitivity analysis programs could be included to determine which nuclides provide the most significant contribution to the radiation shielding problem.

References

1. CASCADE - D.G. Vallis, AWRE Report No. O 45/74, November 1974.
2. CASCADE - G. Evangelides, to be published, CASCADE Reference Manual, AERE-R9611.
3. CADCADE - G. Evangelides, to be published, CASCADE User's Manual, AERE-R9612.
4. CASCAT - G. Evangelides, to be published, AERE-R9613.
5. CONSORT - G. Evangelides. Intended for publication.
6. IBM OS SORT/MERGE Program, Program Number 360S-SM-023.
7. A. Tobias, C.E.G.B., ENDF/B-IV Format, RD/B/M3733, 1976.
8. A. Tobias, C.E.G.B., ENDF/B-V Format, RD/B/N4423, November 1978.
9. AMALGAM - G. Evangelides. Intended for publication.
10. Table of Isotopes, 7th Edition. Edited by C. Michael Lederer and Virginia S. Shirley, Lawrence Berkeley Laboratory, University of California, John Wiley & Sons Publication.
11. ENSDF - W.B. Ewbank and Marcel R. Schmorak, ORNL-5054/R1, February 1978.
12. FELIX - G. Evangelides, to be published. FELIX User's and Reference Manual, AERE-R9612.

12a R.S. Hager, E.C. Seltzer, Nucl. Data A9, 119 (1971).

14. I. Band, M.A. Listengarten and M.B. Trzhaskovskkaya. LIJaF-145 (translated by I.A.E.A.) April 1976, INDC(CCP)-89/N.
15. J.C. Widman, J. Mantel, N.H. Horwitz and E.R. Pownser, Int. Journal of Applied Rad. and Isotopes, 9, 1, (1968).
16. A. Tobias, C.E.G.B., RD/B/N3740, August 1976.
17. P.F. Zweifel, Phys. Rev. 107, 329 (1957).
18. FISPIN-5 - R.F. Burstall, G.N. Nash, J.T. WOrswick, UKAEA Risley, May 1979 ND-M778(R).
19. ANISN-W - R.G. Soltez, R.K. Disney, WANL-PR(LL)-034, Vol. 4, August 1970.

DECAY SCHEME DATA FOR ^{239}U, ^{154}Eu AND $^{140}Ba/^{140}La$

S.P. Holloway, J.B. Olomo*, T.D. Mac Mahon

Reactor Centre, Imperial College at Silwood Park,
Ascot, Berkshire, SL5 7PY, England.

B.W. Hooton**

Atomic Energy Research Establishment
Harwell, Oxon, OX11 ORA, England.

Measurements of the major β^- branching intensities in the decay of ^{239}U to ^{239}Np have been made by comparing γ-spectra obtained from ^{239}U β^- decay and ^{243}Am α decay which both populate levels in ^{239}Np. This comparison method eliminates the need for accurate internal conversion data, and the β^- branching intensities are much more precise than those currently quoted in the literature. The half-life of the fission product ^{154}Eu is being determined by monitoring the emission rate of the 1.27 MeV γ-ray relative to the emission rates of the 1.17 MeV and 1.33 MeV γ-rays of ^{60}Co in two mixed $^{154}Eu/^{60}Co$ sources using a Ge(Li) detector. The best estimate of the ^{154}Eu half-life is now 9.19 ± 0.24 y. Absolute emission probabilities of 21 γ-rays in the fission product pair ^{140}Ba and ^{140}La have been determined using $4\pi\beta$-γ coincidence techniques and Ge(Li) γ-ray spectroscopy.

(RADIOACTIVITY ^{239}U, ^{243}Am; measured I_γ, deduced ^{239}U I_β. ^{154}Eu; measured $T_{\frac{1}{2}}$. ^{140}Ba, ^{140}La; measured I_γ.)

^{239}U Introduction

^{239}U is the most important of the Actinide isotopes responsible for short-term decay heat production at the beginning of irradiated fuel handling and storage. A detailed knowledge of the nuclear decay scheme of this isotope is therefore required if computational methods are to successfully predict the ^{239}U contribution to the total decay heat output. However, the most recent Nuclear Data Sheets(1) quotes a 10% uncertainty in the intensity of the major ^{239}U β^- branch feeding the 74.67 keV level in ^{239}Np and a 5% uncertainty in the intensity of the accompanying major γ transition which decays to the ground state. These large uncertainties reflect the experimental difficulty of resolving a multiple branch β^- decay by β^- spectroscopy and the fact that inference of the β^- branching ratios from the associated γ decay data is only possible when this data itself is complemented by accurate internal conversion electron data.

An experiment has therefore been carried out to determine the major β^- branching ratios in the decay of ^{239}U by comparing γ-spectra from ^{239}U β^- decay and ^{243}Am α decay which both populate levels in ^{239}Np, as shown in fig. 1. This comparison method eliminates the need for accurate internal conversion data.

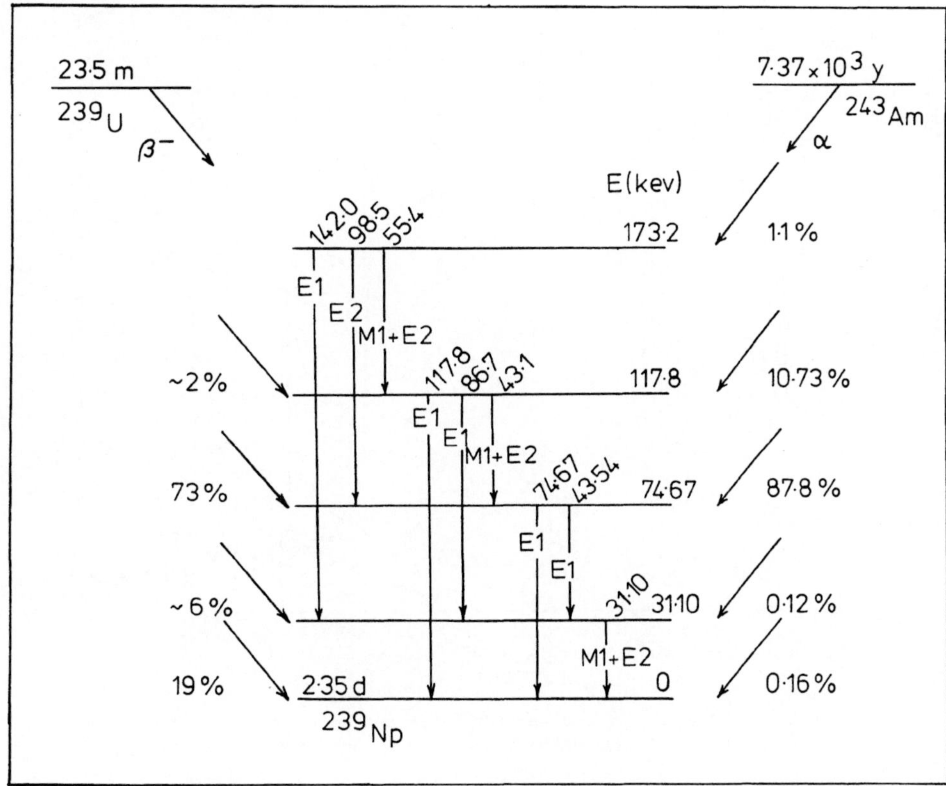

Fig. 1. ^{239}Np partial level scheme

* Present address: Department of Physics, University of Ife, Ile-Ife, Nigeria.

** Present address: UKAEA Headquarters London SU1Y 4QP, England

K. H. Böckhoff (ed.), Nuclear Data for Science and Technology, 287–290.

Experimental Method

The measured intensity of a particular γ-ray in the decay scheme from ^{239}U β^- decay is given by

$$I_\gamma(\beta^-) = No(\beta^-)\beta_\gamma F_\gamma E_\gamma \qquad (1)$$

where $No(\beta^-)$ is the absolute β^- decay source strength, β_γ is the unknown β^- branching intensity feeding the level from which this γ-transition originates, F_γ is the fraction of level decays which produce this particular γ-ray and E_γ is the absolute γ-ray detection efficiency. In general F_γ is unknown because of internal conversion and γ-branching but since the mode of decay is independent of the mode of formation a similar experiment using ^{243}Am α decay would give

$$I_\gamma(\alpha) = No(\alpha)\alpha_\gamma F_\gamma E_\gamma \qquad (2)$$

where $No(\alpha)$ is the absolute α activity and α_γ is the well-known α branching ratio. If these two measurements are combined then β_γ can be obtained and an independent measurement of $F_\gamma E_\gamma$ is not required. Correction terms, to allow for the population of levels by the decay of higher excited states, have been determined using the experimental internal conversion coefficient data of Engelkemeir(2).

^{239}U sources were produced by the reaction $^{238}U(n,\gamma)^{239}U$. After irradiating a few mg of uranium nitrate in solution (uranium content = 99.999% ^{238}U) at the University of London 100 kW research reactor a small volume of the active solution was placed on a thin VNYS disc and rapidly dried under an infra-red lamp before transferring to the counting position of a planar intrinsic Ge detector for γ-ray analysis. Fig. 2 shows a typical low-energy γ-spectrum obtained from a ^{239}U source. It was essential to count for only a short period of time to reduce interference from the 2.35d β^- emitting ^{239}Np daughter. All ^{239}U sources produced were free from fission product activity. A 4πβ-γ coincidence system(3) was then used to determine the absolute β^- activity of the ^{239}Np daughter from which it was possible to infer the original ^{239}U source β^- activity.

A low geometry Si surface barrier α detection system with accurately known fractional solid angle for α detection was used to determine the absolute activity of the ^{243}Am disc source.

Relative γ-ray intensities for transitions common to both β^- and α decay modes, and the absolute intensity of the 74.67 keV γ-transition fed by the major α and β^- branches, were also determined from the experimental data.

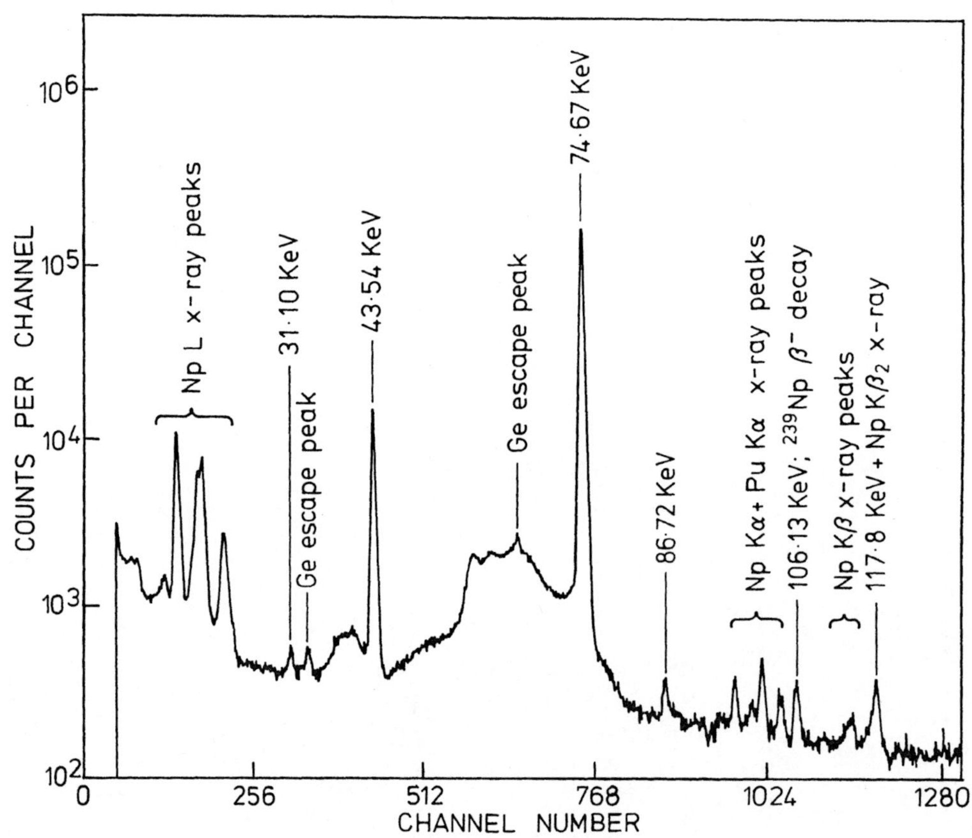

Fig. 2. ^{239}U low energy γ-spectrum obtained using a 1500 mm^2 x 10 mm deep planar intrinsic Ge detector

Results

Table I lists the calculated intensities of the ^{239}U β^- branches feeding the various energy levels, E, in the level scheme of fig. 1. The β^- branch to the ground state was determined by making all ^{239}U β^- transitions sum to 100%. The values of the ^{239}U β^- branching intensities obtained by this method are much more precise than the current published values with the uncertainty in the major β^- branch intensity reduced by one order of magnitude.

Table I Absolute β^- branching intensities from ^{239}U decay

E_β (keV)	E (keV)	Absolute β^- intensity (%)	
		Present work	Nuclear Data Sheets(1)
1168	117.8	1.8 ± 0.2	≃ 2
1211	74.67	67 ± 1	73 ± 10
1254	31.10	11 ± 3	≃ 6
1285	0	20 ± 3	19 ± 8

Table II γ-ray intensities from ^{239}U β^- decay

Energy (keV)	Relative γ-ray intensity	
	Present Work	Nuclear Data Sheets(1)
31.10	0.16 ± 0.02	-
43.54	9.1 ± 0.2	8.9 ± 0.3
74.67	100	100
86.7	0.11 ± 0.01	0.12
Absolute $I_{\gamma 74.67}$	= 0.47 ± 0.01 Present Work	
	= 0.48 ± 0.02 Nuclear Data Sheets	
	= 0.50 ± 0.05 Table of(4) Isotopes	

Table III γ-ray intensities from ^{243}Am α decay

Energy (keV)	Relative γ-ray intensity	
	Present Work	Nuclear Data Sheets(5)
31.10	0.075 ± 0.003	0.11 ± 0.01
43.54	9.1 ± 0.2	8.4 ± 0.6
74.67	100	100
86.7	0.49 ± 0.01	0.51 ± 0.05
117.8	0.78 ± 0.17	0.84 ± 0.12
142.0	0.18 ± 0.01	0.19 ± 0.02
Absolute $I_{\gamma 74.67}$	= 0.67 ± 0.01 Present Work	
	= 0.66 ± 0.03 Nuclear Data Sheets	

The relative intensities of the selected γ-ray transitions from ^{239}U β^- decay and ^{243}Am α decay used to determine the β^- branching intensities are listed in Tables II and III respectively. The absolute intensity of the major 74.67 keV γ-transition common to both ^{239}U decay and ^{243}Am decay is also given for each decay mode; in each case the absolute intensity of the 74.67 keV γ transition is in good agreement with the current published values.

^{154}Eu - Introduction

In the latest edition of the 'Table of Isotopes'[4] two values of the ^{154}Eu half-life are quoted: 8.5 ± 0.5 y and 16 y. To resolve this discrepancy a long-term experiment has been started in which the ^{154}Eu half-life is being measured relative to that of ^{60}Co. Two mixed sealed sources of ^{154}Eu and ^{60}Co have been prepared. The first source 'A' contains about 6 kBq of ^{154}Eu prepared by the ^{154}Sm (p,n)^{154}Eu reaction, together with approximately 29 kBq of ^{60}Co. Source 'B' contains about 43 kBq of ^{154}Eu prepared from the $^{153}Eu(n,\gamma)^{154}Eu$ reaction using 99.45% enriched $^{153}Eu_2O_3$, together with approximately 63 kBq of ^{60}Co. The sources are counted at intervals on an 86 cc, 12.5% Ge(Li) detector with 2.1 keV FWHM resolution, at a source-detector distance of 3 cm. The gamma ray spectrum in the region 1.1 to 1.4 MeV is recorded in each case, and the peak areas corresponding to the 1.17, 1.27 and 1.33 MeV gamma rays of ^{60}Co, ^{154}Eu and ^{60}Co respectively are determined.

Analysis of Data and Results

Let
$$R_1 = \frac{\text{area of 1.27 MeV } ^{154}\text{Eu peak}}{\text{area of 1.17 MeV } ^{60}\text{Co peak}} \quad (3)$$

and
$$R_2 = \frac{\text{area of 1.27 MeV } ^{154}\text{Eu peak}}{\text{area of 1.33 MeV } ^{60}\text{Co peak}} \quad (4).$$

Then
$$\ln R = t(\lambda_c - \lambda_E) + \ln R_o \quad (5)$$

where R stands for R_1 or R_2 and R_o is the initial value of R_1 or R_2; λ_c and λ_E are the decay constants of ^{60}Co and ^{154}Eu, respectively; and t is time. A plot of ln R versus t therefore gives a straight line of slope $(\lambda_c - \lambda_E)$. It should be noted here that the efficiency of the gamma ray detector does not enter into the calculation. It is only necessary that the relative efficiencies of the detector at the three energies concerned remain fixed.

It is planned to continue this experiment for several years, but an interim evaluation of the data so far accumulated leads to the results shown in Table IV. These results are the slopes of straight lines fitted by a least squares technique to R_1 and R_2 values obtained from the two sources over a period of 21 months.

Table IV ^{154}Eu half-life data

Source	Ratio	Slope of fitted line
A	R_1	0.0583 y^{-1}
A	R_2	0.0570 y^{-1}
B	R_1	0.0549 y^{-1}
B	R_2	0.0542 y^{-1}
Mean Slope	:	0.0561 ± 0.0019 y^{-1}
^{154}Eu $T_{\frac{1}{2}}$	:	9.19 ± 0.24 y

The ^{154}Eu half-life shown in Table IV is derived from the mean slope using a value of 0.1315 y^{-1} for the ^{60}Co decay constant[6]. It should be emphasized that the half-life quoted in Table IV should not be regarded as the final result of this experiment. In particular, it can be seen that the two sources give rise to fitted lines with slightly different slopes, and this requires further investigation.

^{140}Ba, ^{140}La Gamma Ray Emission Rates

Absolute gamma ray emission rates in the fission product pair ^{140}Ba, ^{140}La have been determined by a method which has been fully described in an earlier paper[7]. The method involves two separate techniques:

(a) measurement of the absolute disintegration

rate, N_o, of a radioisotope source using a high precision $4\pi\beta$-γ coincidence system;

(b) measurement of gamma ray emission rates, A_γ, of the same source by means of a well calibrated Ge(Li) spectrometer.

Absolute gamma ray emission rates are then given by

$$I_\gamma = A_\gamma/N_o \tag{6}$$

The results of the application of this method to sources of ^{140}Ba and ^{140}La in transient equilibrium and to sources of pure ^{140}La are shown in Table V and are compared to the data of Debertin et al[8].

Table V Absolute Gamma Ray Emission Rates in ^{140}Ba, ^{140}La. (in percent)

Nuclide	Gamma Energy keV	Debertin et al.[8]	This Work
Ba	132.84		0.47 ±0.01
	162.64	6.21 ±0.08	6.18 ±0.08
	304.84	4.30 ±0.05	4.35 ±0.05
	423.70	3.15 ±0.04	3.11 ±0.02
	437.55	1.93 ±0.04	1.92 ±0.02
	537.32	24.39 ±0.22	23.96 ±0.22
La	131.12		0.609±0.013
	241.97	0.473±0.028	0.492±0.012
	266.55	0.452±0.025	0.486±0.011
	328.77	20.74 ±0.18	20.48 ±0.21
	432.53	2.99 ±0.04	2.95 ±0.03
	487.03	45.94 ±0.38	45.52 ±0.37
	751.83	4.41 ±0.04	4.22 ±0.04
	815.80	23.64 ±0.17	23.02 ±0.18
	867.82	5.59 ±0.05	5.62 ±0.05
	919.63	2.68 ±0.03	2.55 ±0.03
	925.24	7.05 ±0.08	7.08 ±0.06
	951.1	0.539±0.019	0.559±0.021
	1596.49	95.40 ±0.08	95.42 ±0.06
	2348.8	0.846±0.017	0.839±0.012
	2521.7	3.43 ±0.08	3.40 ±0.05

The precision of the value of $I_\gamma(1596)$ was increased by requiring that all transitions to the ground state of ^{140}Ce should total 100%.

i.e. $I_\gamma(1596) + I_\gamma(2349) + I_\gamma(2522) + 0.34 = 100.00$

where the factor 0.34 represents the contributions of weak gamma rays and conversion electrons.

References

1. M.R. Schmorak, Nucl. Data Sheets 21, 168 (1977).
2. D. Engelkemeir, Phys. Rev. 181, 1675 (1968).
3. A.P. Baerg, Metrologia 2, 23 (1966).
4. Table of Isotopes, 7th ed., edited by C.M. Lederer and V. Shirley (Wiley, New York, 1978).
5. M.R. Schmorak, Nucl. Data Sheets 21, 177 (1977).
6. W.L. Zijp, J.H. Baard, Nuclear Data Guide for Reactor Neutron Metrology, Part 1, Netherlands Energy Research Foundation, ECN-70, (1979).
7. J.B. Olomo, T.D. Mac Mahon, Nucl. Energy, 20, 237-240, (1981).
8. K. Debertin, U. Schotzig, K.F. Walz, Nucl. Sci. Eng., 64, 784-6, (1977).

Nuclear Data Pertaining to Fusion Reactors and Intense Neutron Sources

TRITIUM BREEDING IN FUSION REACTORS

Mohamed A. Abdou

Fusion Power Program
Argonne National Laboratory
Argonne, Illinois 60439

Key technological problems that influence tritium breeding in fusion blankets are reviewed. The breeding potential of candidate materials is evaluated and compared to the tritium breeding requirements. The sensitivity of tritium breeding to design and nuclear data parameters is reviewed. A framework for an integrated approach to improve tritium breeding prediction is discussed with emphasis on nuclear data requirements.

1. Introduction

The first generation of commercial fusion reactors will operate on the DT cycle. In comparison with other fusion cycles, the DT cycle has technological requirements that are more attainable in the near term and power reactor economics that are more favorable. A self-sustaining DT fusion reactor must breed tritium. In all fusion reactor concepts, this is accomplished in a lithium-containing blanket that circumscribes the plasma.

The feasibility of producing tritium with an adequate breeding ratio TBR was a serious issue in the 1950's and early 1960's. The release of reasonable data for the ^{7}Li (n, n'α)t reaction led to estimates of TBR > 1 in the 1960's. The early generation[1,2] of conceptual designs predicted TBR of ~ 1.3 to 1.5 for natural-lithium blankets with full coverage. Early sensitivity studies[3] showed that the TBR in the natural lithium system was not overly sensitive to variations in cross sections. Consequently, concerns about attaining adequate TBR were alleviated and tritium breeding studies were not given high priority for most of the 1970's.

The safety problems of liquid lithium received a great deal of attention in the past several years. The STARFIRE[4] and DEMO[5] studies investigated breeder blanket concepts based on solid lithium compounds such as Li_2O, $LiAlO_2$, and Li_2SiO_3. Solid breeders appear now to be the leading candidates worldwide[6] for fusion blankets. However, the feasibility of solid breeders has not yet been established. Achieving adequate tritium breeding and acceptable tritium recovery from the blanket are the two most critical issues for solid breeders. Thus, the past three years have witnessed a serious interest in the tritium breeding issues.

This review paper is concerned with tritium breeding in fusion blankets. While the discussion of nuclear data is the primary motivation for this paper, the author avoids the conventional approach of providing a simple list of recommendations for a few cross sections to be measured. Rather, the paper is written to help nuclear-data specialists develop an appreciation of the many technological variables of the tritium breeding issue in fusion blankets. A selected number of the most promising design concepts for the breeding blanket are described. The breeding potential of candidate breeding materials and blanket designs is evaluated and compared to the tritium breeding requirements. The sensitivity of tritium breeding to design and nuclear data parameters is reviewed. Finally, a framework for an integrated approach to improve tritium breeding prediction is discussed.

There are many areas of fusion applications where nuclear data is important. Some of these are indicated in Table I. This paper is focused only on tritium breeding. The nuclear data requirements for other areas have been reviewed by a number of authors (see for example references 9, 42, 43, and 46).

The plans for the major fusion programs in the world call for a tritium-producing blanket in a fusion engineering device whose construction will start in the early 1990's. However, the requirements for improvements in the nuclear data base for tritium breeding must be satisfied much sooner—over the next several years. Accurate data is required now because it constitutes an important part of the input to evaluating the feasibility of the many design concepts and candidate breeding materials that are now under consideration.

2. Breeding Blanket Design and Materials

The primary functions of the blanket in a fusion reactor system is to convert the energy of the fusion neutrons into sensible heat and to breed tritium. The present focus of the blanket development is on defining a compatible combination of materials that can be integrated into a functional design. This design must serve the following requirements:

K. H. Böckhoff (ed.), Nuclear Data for Science and Technology, 293–312.

Adequate tritium production
Acceptable tritium recovery
Efficient heat conversion and recovery
Acceptable lifetime
Acceptable safety and environmental impact
Maintainable system

The blanket contains a number of materials, including a lithium-bearing material for tritium breeding and a structural material for containing the coolant. Some breeding materials require a neutron multiplier in order to achieve adequate breeding ratio. Lithium is not a very efficient moderator, and in some blanket concepts, another material is used for neutron slowing down. A reflector is normally incorpoated at the outer end of the blanket. Table II lists the candidate blanket materials that are now under consideration in fusion development.

A large number of blanket designs have been proposed and evaluated in design studies. The promising concepts can generally be classified into the following types:

1. Liquid Metal

   a. liquid metal as breeder and coolant
   b. liquid metal as breeder and a separate coolant

2. Solid Breeder (with separate coolant).

The candidate liquid metals are lithium, $Li_{17}Pb_{83}$ and $LiBi_5Pb_4$. They can be used as both the breeder and coolant or as the breeder alone with a separate coolant. $LiBi_5Pb_4$ and $Li_{17}Pb_{83}$ are very similar in many respects, particularly the neutronics characteristics. $LiBi_5Pb_4$ has the advantage of a lower melting point but the large polonium production in bismuth makes it less favorable. Water cannot be used as a coolant with liquid lithium because of their strong chemical reactivity and unacceptable safety consequences. The reactivity of water with $Li_{17}Pb_{83}$ is much lower than with liquid lithium, and water cooling in $Li_{17}Pb_{83}$ is believed to be acceptable. Sodium and helium are other candidate coolants for liquid metal breeders.

Figure 1 is a cross section of a representative liquid metal blanket with a separate coolant. The dimensions in the figure are approximately to-scale and are based on the $Li_{17}Pb_{83}$ blanket design developed in the STARFIRE/DEMO[5] study.

There are a number of critical issues for the liquid metal blankets, as shown in Table III. The very strong reactivity of lithium with air and water is an important safety issue. For both lithium and lithium-lead, the compatibility with the structural material and MHD effects are key issues. Resolving the compatibility issue may require the development of a structural alloy other than austenitic stainless steel, which is believed to be a costly and time-consuming development.

The above issues have led recently to serious consideration of solid breeders. The most promising solid breeders are Li_2O and the ternary ceramics: $LiAlO_2$, Li_2SiO_3 and Li_2ZrO_3. The most critical issues for solid breeders are achieving adequate tritium breeding and acceptable tritium recovery (see Table III). The tritium breeding issue will be discussed in the following sections. It remains to be demonstrated that the steady-state tritium inventory in solid breeders can be kept acceptably low (< 10 kg) under irradiation. Both helium and water have been proposed for use as coolants with solid breeders. It has been shown[11] that pressurized water is superior to helium under present materials and reactor design constraints. The loss of radiation-attenuating space associated with the use of helium in the inboard blanket results in a large economic penalty. Furthermore, the higher operating temperature required for helium requires a structural material other than austenitic stainless steel. The void space and the additional increase in the structural material volume required for helium also reduce the achievable tritium breeding ratio.

Lithium oxide is presently considered to be the most promising solid breeder, primarily because its breeding potential is higher than that for the ternary ceramics. One of the key problems with Li_2O is its reactivity with water to form the highly corrosive LiOH. The ternary ceramics (e.g., $LiAlO_2$) are more chemically stable but their tritium breeding is questionable, as discussed later.

Figure 2 shows a cross section of a typical design for a solid breeder blanket. The details are based on the Li_2O design for the STARFIRE/DEMO.[5] Figure 3 is a vertical cross section of a tokamak and shows the interrelation between the blanket and other components.

3. Tritium Breeding Requirements

The tritium breeding ratio is defined as $T = N^+/N^-$, where N^+ is the rate of tritium production in the system (normally the blanket) and N^- is the rate of burning tritium in the plasma. T must exceed unity by a margin (G) to cover 1) losses and radioactive decay during the period between production and use, and 2) supplying inventory for startup of other fusion reactors. Detailed expressions were derived[7,8] to correlate T, doubling time (t_d), and inventory. The expression has the general form

$$T = 1 + \frac{ID}{N^- t_d} \qquad [1]$$

where I is the total tritium inventory and D is a complex function that accounts for radioactive decay and other losses. The magnitude of the total tritium inventory is determined by the tritium inventories in the breeding blanket (I_B), fueling and exhaust systems (I_F), storage (I_S), and startup (I_O), respectively. At present, there are large uncertainties concerning the magnitude of the tritium inventory achievable in fusion reactor systems. The tritium inventory in the blanket will be < 1 kg for liquid lithium systems, but may be between 1 and 10 kg or greater in solid breeders because their tritium release characteristics cannot be quantified at present. The magnitude of I_F depends strongly on many plasma-performance and exhaust-system operating parameters such as the fractional tritium burnup in the plasma. I_S is determined by the amount of fuel required in reserve to guard against a temporary malfunction of the tritium recovery system. I_O is the tritium inventory that needs to be accumulated in order to start up a new reactor, and it is also a function of I_B, I_F and I_S.

Figure 4 shows the required tritium breeding ratio as a function of the fractional tritium burnup in the plasma and the doubling time assuming the blanket tritium inventory is 10 kg. The required tritium breeding ratio increases very rapidly as the fractional burnup decreases to ~ 1% and as t_d becomes very short. This is one of the reasons why achieving a fractional burnup of > 5% has been identified as one of the key goals for the impurity control and exhaust system development. The doubling-time goal is obviously related to the desired growth rate for fusion power. The historical growth of the power industry has been a doubling time of ~ 10 year. For the first generation of fusion power reactors, a short doubling time (< 5 y) is highly desirable.

The required tritium breeding ratio (T_0) in an actual operating reactor should thus be

$$T_0 = 1 + G \quad [2]$$

where G is the margin discussed above. For a fractional burnup of ~ 5% and doubling time of ~ 5 y the required T_0 is ~ 1.05. It should be clearly realized that if a fractional burnup of only ~ 1% is achievable, then the required T_0 is ~ 1.2 -- a very difficult breeding requirement to achieve as will become evident in later sections.

T_0 as defined in Eq. 2 is the actual net tritium breeding ratio that must be obtained in an operating fusion power reactor. The more difficult task is establishing the goal for the tritium breeding ratio (T_D) that must be achieved in a conceptual design. Clearly, T_D must exceed T_0 by an allowance (Δ) for the uncertainties in estimating T_0, i.e.,

$$T_D = 1 + G + \Delta \quad [3]$$

The sources of uncertainties in estimating Δ are numerous, but can be broadly classified into three areas: 1) reactor design definition, 2) neutronics calculations, and 3) nuclear data. A few comments on these are in order.

The tritium breeding ratio is sensitive to many of the design features of the fusion reactor. Some of the important features are: 1) the materials, geometry and size of the in-vessel components (e.g., limiter or divertor for impurity control) that intercept the fusion neutrons before entering the blanket; 2) materials, volume fraction and distributions of structural materials and coolants in the first wall and blanket; 3) size and locations of the many void regions that penetrate the blanket and serve important functions such as plasma heating, plasma-current drive, vacuum pumping and fueling; 4) the presence of non-breeding regions in the blanket, which are reserved for presently ill-defined requirements such as passive coils for plasma stabilization; and 5) overall plasma characteristics and reactor configuration, including fusion neutron source distribution, shape of first wall, modularity of components, etc. Fusion is still in a somewhat early stage of research and development with active conceptualization of reactor designs. Technology choices and design concept selections have not been made for many of the reactor systems. These selections will not be made for a number of years to come because a number of issues are yet to be resolved experimentally and analytically. The achievable tritium breeding ratio is certainly one of the key considerations in the selection among concepts.

The dependence of the tritium breeding ratio on technology and design concept choices can be quite large, 50% in some instances. Obviously, one cannot design a breeding blanket with a breeding potential that accommodates these large variations. Since the breeding potential for candidate breeding materials will be (and is already being) performed for various technology and design concept choices, we will require that T_D includes only 2% allowance for design definition. This is merely enough to account for those additional design details that cannot be developed at present for a given conceptual reactor design.

Neutronics calculations of the breeding ratio in a given system are subject to a number of uncertainties. Assuming that the designer uses the best methods and codes available, there are two sources of errors. First, geometrical modeling of the fusion reactor configuration entails some approximations that are necessary to make the problem practical from computer storage and computing-time viewpoints. Second, there are errors that are inherent in all calculational methods and codes for a variety of reasons such as those related to numerical techniques, averaging, and/or discrete treatment of continuous variables. Reducing the errors due to calculations to < 2% appears to be a very difficult goal.[9]

The third source of uncertainty in estimating the breeding ratio comes from errors in nuclear data (e.g., cross sections and energy and angular distribution of secondary neutrons). These include errors arising from the accuracy of measurements, representation of parameters in data files, and processing of the data into a form suitable for use in radiation transport codes. Judging from experience in the fusion program and the estimated sensitivity of T to variations in nuclear data (discussed later), it is not unreasonable to require accuracies in nuclear data that result in an error in T of no more than 1 to 2%.

From the above discussions, the allowance Δ in Eq. 3 required to compensate for a possible shortfall in tritium breeding due to the combined effects of uncertainties in design definition, calculations, and nuclear data is ~ 5 to 6%. Hence, the tritium breeding ratio required in a fusion reactor design must be ~ 1.1 in order for the design concept to have a high potential of achieving self-sustaining DT fusion power economy.

The possibility of operating fusion reactors with a tritium breeding ratio, T < 1, by using a tritium concentration in the plasma of <50%, has been suggested. However, it has been found[10] that the required magnetic field and reactor size have to be increased rather quickly as T decreases for the same reactor power. Thus, small decrements in T appear very costly in terms of the required plasma performance and the increase in the cost of electricity.

4. Breeding Potential of Candidate Materials

As discussed earlier, many blanket design concepts and a number of breeding materials are being considered for fusion reactors. One of the primary selection criteria is achieving an adequate breeding ratio. In this section, we examine the breeding potential of candidate breeding materials.

Based on the discussions in Section 2, the most promising blanket concepts are shown in Table IV. The breeding ratios obtainable with these concepts cover the range of breeding ratios potentially achievable in fusion blankets. There are many possible variations (e.g., helium instead of water coolant) on these designs. However, the four blanket concepts considered in Table IV are sufficient for our purpose here of examining the tritium breeding issues. These concpets were examined in detail in a recent study[5], and the neutronics calculations performed by Jung are documented in reference 12.

Table V describes the dimensions and materials used for calculating the breeding ratio for the four blanket options of Table IV. These four designs represent self-consistent choices for the four most promising breeding materials: Li, 17Li-83Pb, Li_2O and $LiAlO_2$. The neutron transport calculations were performed with the MORSE Monte Carlo code in a 3-D toroidal geometry (see Figure 5) that models the STARFIRE/DEMO design. A neutron source distribution that accounts for the spatial distribution of the DT reaction rate within the plasma was used. Nuclear data parameters are based on ENDF/BIV. To be consistent with recent experiments,[17-23,47] the $^7Li(n,n'\alpha)t$ reaction rate was reduced by 15% for the purpose of calculating the breeding ratio.

The tritium breeding ratio calculated for the four designs are shown in Table VI. These results are for a full blanket coverage, i.e., the first wall/blanket completely circumscribes the plasma with no allowance for penetrations. Thus, the breeding ratios indicated in Table VI represent an upper bound on the breeding potential for the four breeding materials without using a neutron multiplier. Below we discuss design and material variations that can affect T, the possibility of using a neutron multiplier, and the breeding characteristics of each breeding material.

4.1 Effects of Penetrations on T

A variety of penetrations must be accommodated in the blanket system. These can be classified into a) major penetrations, and b) normal penetrations. The major penetrations are large in size and their functional requirements do not permit substantial modifications in their shape. Examples of major penetrations are (see Figure 3): 1) auxiliary heating ducts, presently neutral beams and rf are the leading contenders; 2) vacuum pumping ducts; and 3) active impurity control, e.g., a divertor or limiter. Besides being large in number and size, these major penetrations have openings in the first wall in direct visibility to the plasma neutrons. On the other hand, normal penetrations are small or moderate in size, and they are amenable to substantial shaping of their path inside the blanket and bulk shield. Among the penetrations in this category are the gaps between blanket and shield modules, holes for diagnostics and refueling, and a variety of gaps for clearances around plugs and pipes. For gas-cooled blankets the coolant passages will represent additional radiation streaming paths.

The change of the tritium breeding ratio due to the penetrations, ΔT^P, consists of three components: ΔT^S, ΔT^V and ΔT^L due to variation in neutron spectrum, reduction in the volume of breeding zone, and enhanced loss of neutrons, respectively. Radiation streaming enables high energy neutrons to reach deeper regions of the blanket. This tends to increase the $^7Li(n,n'\alpha)t$ reaction rate. The reduction in the volume of the breeding zone depends strongly on the type of penetration. In recent designs, this volume reduction is ~ 5 to 10%. The relative reduction in the breeding ratio can be smaller or slightly larger than the relative reduction in the breeding volume, depending on the exact characteristics of the penetrations. Ide et al.[13] showed that the reduction in T due to the presence of typical neutral beam injectors is ~ 2%. Jung[12] estimates a reduction in T due to the presence of a limiter impurity control system of ~ 5%. A divertor system[14] results in a larger reduction in T than in the case of the limiter.

The reduction in the tritium breeding ratio due to penetrations in a practical power reactor system is > 5% but does not appear to exceed 10%. This reduction is not overwhelming considering the nature of the major penetrations. However, this reduction can be crucial for breeding materials with a marginal breeding potential.

4.2 Breeding in the Inboard Blanket

In magnetic confinement fusion reactors, there are strong incentives for reducing the total thickness of the blanket and bulk shield, Δ_{BS}, to maximize the reactor power for a given magnetic field. In tokamaks, the reactor performance and economics are particularly sensitive to Δ_{BS} in the inboard region (see figure 3). Lithium-containing materials are not very efficient in radiation attenuation. Therefore, it was proposed[16] that the portion of the breeding blanket in the inboard region (i.e., sector 1 in figure 5) be replaced by more efficient shielding materials. Obviously, this is acceptable only if attaining an adequate T is possible without the inboard blanket. However, the saving in Δ_{BS} is so important[15] that the feasibility of eliminating the breeding blanket from the inboard region must be considered as a factor in selection of breeding materials. Table VII shows the tritium breeding ratios for the case of when the inboard blanket is eliminated and the limiter impurity control penetration is accounted for. The reduction in the breeding ratio ranges from ~ 11% in the Li_2O case to ~ 15% in the Li-Pb system. Of this reduction, ~ 5% is due to the limiter and the balance is due to elimination of the inboard blanket.

4.3 Fast and Thermal Systems

Blanket concepts can be generally classified into "fast" and "thermal" systems. In the fast system, high energy neutrons in the 4 to 15 MeV range contribute significantly to breeding via the $^7Li(n,n'\alpha)t$ reaction. In the thermal system, neutrons are slowed down by elements other than ^7Li (before or after entering the breeding material region) so that essentially all the tritium breeding is achieved by low energy neutrons, inducing the reaction $^6Li(n,\alpha)t$.

Of the four systems considered in Tables V and VI, lithium and Li_2O represent a "fast" system, while 17Li-83Pb and $LiAlO_2$ are representative of a "thermal" system. Figures 6 and 7 show typical neutron spectra in both types of blanket. The neutron moderating power of lithium is not very strong. Therefore, the high energy neutrons penetrate into the deeper regions of the blanket, resulting in a significant $^7Li(n,n'\alpha)t$ reaction rate. In a liquid lithium blanket, T_7 is ~ 30% of T. Therefore, the relative error in T is approximately one-third of the relative error in the magnitude of the $^7Li(n,n'\alpha)t$ reaction cross section. The contribution of ^7Li to the tritium breeding ratio in Li_2O is ~ 24%. The reduction of T_7 in Li_2O relative to that in Li is due to scattering of high energy neutrons with oxygen in the Li_2O and the water coolant.

In the absence of a neutron multiplier, the contribution of ^7Li is an absolute necessity to achieve T > 1 in Li and Li_2O. Studies have shown that the natural isotopic composition of lithium (92.5% ^7Li and 7.5% ^6Li) yields nearly optimum breeding ratio with no benefit from enrichment in ^6Li. If a neutron multiplier is incorporated in Li and Li_2O blankets, the neutronics characteristics of these fast systems change to those of a thermal system. The increase in the (n,2n) reaction rate in the neutron multiplier is always associated with a reduction in the $^7Li(n,n'\alpha)t$ reaction rate. Thus the enhancement of T by using a neutron multiplier tends to be modest for the Li and Li_2O systems as will be discussed shortly.

Liquid 17Li-83Pb has one of the highest breeding potentials because of the large neutron multiplication in the breeding material itself through the Pb(n,2n) and (n,3n) reactions. The $^7Li(n,n'\alpha)t$ reaction rate is reduced to an insignificant contribution (see Table VI) due to the rapid slowing down of high energy neutrons by inelastic scattering in Pb. However, the neutron multiplication via the Pb(n,2n) is large enough to result in a T_6 that is much greater than the breeding ratio achievable in the Li and Li_2O systems. Since the number density of lithium in Li-Pb is very low (about an order of magnitude lower than in Li and Li_2O), isotopic enrichment of ^6Li is crucial to attaining a high tritium breeding ratio. The lithium in the Li-Pb system used in Tables V-VII is enriched to 90% ^6Li. Most of tritium breeding comes from neutrons in the low eV range. The soft spectrum that prevails in such a thermal system requires particular care in the neutronics treatment. This system also is more sensitive to the amount and type of structural material and coolant used.

The tritium breeding characteristics are similar for blankets using the ternary ceramics such as $LiAlO_2$, Li_2SiO_3, Li_2ZrO_3 and Li_2TiO_3. The strong neutron moderation by the non-lithium elements renders ^7Li useless for tritium production (see Table VI). In the absence of a neutron multiplier, T is always < 1 for these ceramic breeders. Hence, the viability of solid breeders (other than Li_2O) from tritium breeding viewpoint is unavoidably dependent on the ability to use a neutron multiplier.

4.4 Neutron Multipliers

Enhancing the tritium production capability of the blanket by incorporating a neutron multiplier has been examined in a number of design studies. The subject has been treated in some detail in references 4, 5, 12, 24, 25 and 27. The results of these studies are surprising. Finding an acceptable neutron multiplier appears to be extremely difficult.

Table VIII lists the candidate neutron multipliers and presents some of the key properties relevant to their utilization in fusion blankets. Fissionable materials are excluded since hybrid concepts are not presently included in the main-line fusion development.

A good neutron multiplier must have a large (n,2n) and/or (n,3n) cross section, with a threshold much lower than 10 MeV. It must also have a relatively small parasitic absorption over the entire energy range of 0 to 15 MeV. It is desirable that the inelastic cross section be relatively small. Lead, bismuth and zirconium appear to be the only high-Z materials that are potentially useful neutron multipliers. Compounds such as PbO, PbBi and Zr_5Pb_3 need to be considered for engineering reasons. Low-Z materials typically have low (n,2n) cross sections with the well-known exception of beryllium.

In order to select a material for any component of the blanket it must satisfy a number of engineering criteria in addition to possessing acceptable neutronics characteristics. Some of these engineering criteria are availability, cost, fabricability, and compatibility with coolant and structure.

Lead and bismuth have the most attractive neutronics characteristics except for beryllium. Zirconium has an (n,2n) cross section that is about a factor of 3 lower than that for Pb and Bi. The key engineering problem for lead and bismuth is their low melting point (327 and 271°C, respectively). To keep the material solid during reactor operation, its melting point must be higher than the coolant temperature. The exit temperature for a helium coolant is typically > 450°C. The lowest operating temperature in a blanket design is achievable with pressurized water, which yields an acceptable thermal conversion efficiency with a maximum temperature of ~ 320°C. Thus, the use of lead and bismuth in solid form in fusion power reactors is not practical. For near-term experimental reactors such as INTOR,[14,25] power conversion is not an objective and the maximum coolant temperature is ~ 100°C. In this case, lead or bismuth can be used in solid form. However, the poor thermal conductivity of lead limits the maximum thickness of the neutron multiplier zone to ~ 5 cm for a neutron wall load of ~ 1 MW/m^2. The allowable thickness is even smaller for bismuth. It should be noted that the significant production of the α-emitter ^{210}Po makes bismuth an undesirable material in a power reactor system.

The only practical possibility for using lead or bismuth in a fusion power reactor blanket is in liquid form. However, liquid lead and bismuth are not compatible with candidate structural materials. To limit corrosion of stainless steel, a maximum operating temperature of ~ 450°C has to be imposed. With such temperature limit and at a nuetron wall load of ~ 3 MW/m^2, the low thermal conductivity of lead limits the maximum spacing between coolant tubes to ~ 2 cm. This reduces considerably the benefits of lead as a neutron multiplier because of the parasitic neutron absorption in the coolant structure. The situation is even less attractive for bismuth and PbBi. Lead oxide has a much higher melting point (888°C). However, its lower thermal conductivity and the presence of oxygen make PbO inferior to lead as a neutron multiplier.

The intermetallic compound Zr_5Pb_3 was considered in the STARFIRE study because it has a higher melting point (~ 1400°C) than lead and a greater neutron multiplication than zirconium. However, as we will show shortly, the breeding enhancement achievable with Zr_5Pb_3 is not great.

Beryllium is very attractive neutronically because its (n,2n) cross section is large over a wide energy range (~ 2 to 15 MeV) and its parasitic neutron absorption is small. There are two serious issues in using beryllium in fusion reactors. The first is the limited known resources of beryllium. The estimated reserves for beryllium are 2.5 x 10^7 kg in the U.S. and ~ 3.8 x 10^8 kg in the world. A 5-cm-thick beryllium zone in a typical tokamak power reactor require ~ 7 x 10^4 kg of beryllium, i.e., ~ 0.3% of the U.S. reserves. Thus, beryllium reserves cannot surtain a full fusion power economy. However, the beryllium burnup over the reactor life (~ 40 y) is ~ 10%. Assuming recycling of beryllium is feasible, beryllium can be used in tens of fusion reactors without exhausting a significant fraction of beryllium reserves. The second serious issue is that beryllium is toxic. This toxicity makes beryllium difficult to handle and introduces additional factors into safety and environmental considerations for fusion reactors. The cost of beryllium is high, but this is more than offset by its excellent additional energy multiplication. Helium generation rate in beryllium is extremely large, typically ~ 8000 appm/yr in a power reactor. This requires using beryllium at only ~ 60 to 70% of its theoretical density in order to accomodate swelling. This requirement increases the effective fraction of structural mateiral and reduces slightly the net neutron multiplication benefit. The Be(n,t) reaction with a threshold of 11.6 MeV and a cross section of ~ 20 mb at 14 MeV resuls in a significant tritium production. The tritium inventory trapped in beryllium is a cause for concern. This inventory is estimated[12] to be in the range of 2 to 3 kg after five years of operation at 3 MW/m^2 neutron wall load in a typical power reactor.

Gohar[24,4] examined the effect of various neutron multipliers on the tritium breeding ratio in a blanket that uses $LiAlO_2$ as a breeding material. He performed one-dimensional calculations using the first wall/-blanket composition shown in Table IX. A neutron multiplier zone was used behind the first wall and preceeding the breeding material region. The coolant and structure were homogenized with the neutron multiplier. Figure 8 shows the tritium breeding ratio as a function of the neutron multiplier zone thickness for a number of neutron multipliers. A few important observations can be made on the results in this figure. The breeding ratio shows a maximum at a multiplier zone thickness of ~ 5 to 8 cm. The reason is that as the multiplier thickness increases the (n,2n) reaction rate increases but a larger fraction of the neutrons is thermalized and absorbed in the structure before reaching $LiAlO_2$. While Figure 8 indicates a comparable neutronics performance for lead and beryllium, it must be realized that the practicality of the lead system is questionable as discussed above. The very severe limitations on the spacing between coolant channels for lead will substantially degrade the obtainable breeding ratio to much lower values than those indicated in Figure 8. The maximum one-dimensional breeding ratio shown in Figure 8 for blankets with Be and Pb is 1.18, which is lower than that obtainable for Li and Li_2O without a neutron multiplier.

Beryllium has been shown[12] to be the only non-fissionable neutron multiplier capable of significantly increasing the breeding ratio in lithium and lithium-oxide blankets. (Using lead in a Li-Pb compound results in high T as discussed earlier but such concept is classified in another category.) The neutron multiplier can be incorporated in the blanket in a number of possible arrangements as illustrated in Figure 9. Optimum use of beryllium appears to be not immediately behind the first wall but behind several centimeters of Li or Li_2O, particularly for a thick (1-2 cm) first wall.

4.5 Conclusions on Breeding Potential

In a previous section, we indicated that the tritium breeding ratio required in an operating reactor is

$$T_0 \sim 1.05$$

We also indicated that present uncertainties in design definition, calculational methods, and nuclear data make it necessary to demand that prudent designs of fusion blankets performed at present should require that the design be capable of yielding a breeding ratio

$$T_D \sim 1.1.$$

Breeding materials and design concepts that are limited to $T < T_0$ should be rejected. In addition, if their maximum predicted breeding ratio (T_m) is such that $T_0 < T_m < T_D$, the development of these materials and concepts will entail a high risk of not achieving the tritium breeding requirement. It should not be inferred that the success of those materials and concepts that are now predicted to yield a breeding ratio slightly higher than T_D is guaranteed.

A large number of tritium breeding studies have been performed over the past 15 years. The most recent of these were summarized in this section. These studies are not exhaustive. However, they do permit a judgement on the T_m that can probably be achieved with each breeding material in a fusion power reactor. This judgement is based on all the neutronics, engineering, technology, and economics considerations known to us today. However, this judgement cannot now be conclusive and may have to be revised in the future if dramatic changes occur in fusion reactor designs.

Table X shows the estimated T_m achievable with present designs for liquid lithium, $Li_{17}Pb_{83}$, Li_2O and $LiAlO_2$. The highest breeding potential is offered by $Li_{17}Pb_{83}$ with a T_m of ~ 1.3. This margin is large enough to judge tritium breeding feasibility with $Li_{17}Pb_{83}$ to be certain. T_m for Li and Li_2O is 1.15 and 1.1 respectively. The risk of attaining the breeding feasibility is low to medium. Beryllium can be used to enhance their tritium breeding capability. Development of $LiAlO_2$ appears to involve a very high risk. All previous optimization studies for $LiAlO_2$, using the best neutron multipliers, yield a tritium breeding ratio that is < 1.1. The same appears to be true for the other ternary ceramics such as Li_2SiO_3 and Li_2TiO_3.

The feasibility of solid breeders with respect to tritium breeding cannot be assured at present. Since solid breeders offer many attractive features for fusion reactors, programs to resolve the tritium breeding issue should receive high priority.

5. Sensitivity to Data Uncertainties

Sensitivity analyses provide an important input to defining and prioritizing the nuclear data requirements. The objective is to estimate the sensitivity of a specific nuclear response to uncertainties in various cross sections and secondary neutron energy and angular distributions. The results of the sensitivity analysis can be used to define the accuracies in nuclear data parameters required to meet an accuracy goal for a nuclear response (e.g., tritium breeding ratio). In addition, if the uncertainties in the available nuclear data base are well defined, sensitivity analysis is a useful tool in defining the accuracy of the tritium breeding ratio estimated for a conceptual design.

The number of sensitivity studies on fusion blankets is rather limited.[28-39] However, reliable tools have been developed, and the recent availability of quantitative information on nuclear data uncertainties in ENDF/B files should encourage additional sensitivity studies in the future. From the available literature, we summarize below some of the results on the sensitivity of the tritium breeding ratio to uncertainties in nuclear data.

This section would have been more easily written if sensitivity studies were available for the most recent blanket designs discussed in earlier sections. Since this is not the case, we have to summarize portions of sensitivity studies performed for earlier versions of blanket designs based, in some cases, on cross-section sets that have undergone some changes in recent years. However, the results are useful in pointing out some important trends. We will select results from studies performed for fast and thermal systems and with and without neutron multipliers to show the differences in sensitivity to nuclear data.

Alsmiller et al.[35] compared the cross-section sensitivity for the two blanket systems described in Table XI. The first (ORNL) is based on natural lithium with vanadium or niobium as a structural material. Note that the amount of structure indicated is much less than that predicted for the more recent designs. This system can be classified as a "fast" system in which ^7Li provides large contribution to tritium breeding. With niobium, $T_6 = 0.9$, $T_7 = 0.6$; while with vanadium, $T_6 = 0.94$ and $T_7 = 0.61$. The second (LASL) uses beryllium as a neutron multiplier and lithium as a breeding material. The spectrum in the breeding zone is much softer than in the first design, resulting in T_6 = 0.91 and T_7 of only 0.13. Table XII lists the cross section uncertainties assumed by Alsmiller et al. and also specifies the corresponding cross section varied to maintain a constant total cross section. The estimated uncertainties in the breeding ratio are given in Table XIII. Note that an earlier study by the same authors showed that, of all the ^{7}Li partial cross sections only the ^{7}Li(n,n'α)t sensitivity was appreciable. The conclusions can be summarized as follows:

- The available ^{6}Li(n,α)t cross section introduces an uncertainty of less than a few tenths of a percent in the breeding ratio for both the fast and thermal designs considered.

- The sensitivity profiles for ^{7}Li(n,n'α)t are not very system dependent and have a shape approximately similar to that of the cross section itself. However, the resultant uncertainty in the magnitude of the tritium breeding ratio is system dependent (since T_7/T is system dependent). A 20% change in the ^{7}Li(n,n'α)t cross section results in ~ 6% change in the breeding ratio in a fast system and ~ 0 to 2% change in a thermal system.

- The uncertainties in the carbon elastic and inelastic cross sections indicated in Table XII result in an uncertainty in the breeding ratio of 0.6% in the LASL design, which uses graphite as a moderator.

Pelloni[39] has performed a recent sensitivity study for a variation of two blanket design options considered for INTOR.[26]. The first design is based on Li_2SiO_3 as a breeding material and lead as a neutron multiplier. The second uses $Li_{17}Pb_{83}$ as a breeding material. Both concepts have water coolant (D_2O and H_2O in different regions) and stainless steel 304. The dimensions and material compositions used by Pelloni in 1-D calculations are shown in Table XIV. Both systems are of the thermal type, with T_7 = 0.013. The ^{6}Li(n,α)t reaction rate per fusion neutron is T_6 = 1.16 for Li_2SiO_3 and T_6 = 1.32 for $Li_{17}Pb_{83}$. The sensitivity to the ^{7}Li data is not important in this system. Table XV gives the percentage change in the tritium breeding ratio due to 1% change in the partial cross sections of ^{6}Li, Pb, Fe, H, O and Al.

As expected, Table XV shows that the tritium breeding ratio increases with increasing the (n,2n) and/or decreasing the (n,γ) cross sections. The breeding ratio is most sensitive in both systems to the Pb(n,2n) cross section. The change in the breeding ratio is ~ 15 to 20% of the change in the Pb(n,2n) cross section. For the Li_2SiO_3 system, the next most important reactions are the (n,2n) and (n,γ) in the elements of stainless steel, particularly iron. The changes in the breeding ratio are + 0.06% and -0.05% due to 1% increase in the (n,2n) and (n,γ) cross sections, respectively. The results of Pelloni indicate that the sensitivity to ^{6}Li is very small.

Considering typical uncertainties in lead cross sections of ~ 7% for (n,2n) and ~ 10% for elastic and inelastic cross sections, the corresponding uncertainty in the breeding ratio is ~ 1.4%. Typical uncertainties in iron cross sections of ~ 15% in elastic, inelastic, and (n,2n) and ~ 10% in (n,γ) give an uncertainty in the breeding ratio of ~ 0.6% to ~ 1.5%. Since the uncertainties in the (n,2n) and (n,γ) cross sections are not correlated, they both can be lower or higher or one of them higher and the other lower than the presently available values. This, of course, affects the estimate of uncertainty in the breeding ratio.

In addition to the uncertainties in the cross sections discussed above, errors in the secondary neutron energy and angular distributions are of serious concern. Obviously, the neutron emitting reactions of all materials that contribute significantly to determining the neutron transport in the blanket can potentially affect the breeding ratio. Of particular importance are the secondary neutron energy distributions for ^{7}Li(n,n'α)t and for the (n,2n) reaction in neutron multipliers (e.g., Be and Pb). Unfortunately, very few studies have been performed in this area and deriving quantitative conclusions on the uncertainty in the breeding ratio due to errors in the secondary neutron energy and angular distributions is difficult at present. It should be noted, however, that these errors are generally much larger than those in the reaction cross sections and are thus likely to contribute more to the uncertainty in the breeding ratio.

Steiner and Tobias[28] examined the sensitivity of the breeding ratio to the energy distributions of the secondary neutrons produced by the ^{7}Li(n,n'α)t reaction. They varied the nuclear temperature, θ(E), in an evaporation model that describes this energy distribution by ± 50%. Increasing θ(E) by 50% results in a harder spectrum and an increase in T_7 by 2.8%. The harder spectrum increases the fraction of T_7 contributed by the secondary neutrons from the ^{7}Li(n,n'α)t reaction from 0.15 to 0.18. Decreasing θ(E) produces a softer spectrum and decreases T_7 by ~ 7%. The tritium production rate in ^{6}Li remained unchanged. Markovsii[40] obtained similar results. The importance of the accuracy in the energy distribution of the ^{7}Li(n,n'α)t is limited to non-thermal blankets.

The accuracy of T is affected by the uncertainty in the energy distributions of the secondary neutrons of the (n,2n) reaction when a significant fraction of these secondary neutrons have an energy well above the threshold for the (n,2n) or the ^{7}Li(n,n'α)t reactions. The effect can be expected to be the largest for beryllium, whose (n,2n) reaction has a threshold energy of only 2.8 MeV. Soran et al.[41] examined the sensistivity of T in a thermal-type blanket (similar in many respects to the LASL design discussed earlier) to uncertainties in the Be(n,2n) secondary energy distribution. They considered two data sets in which the average energy of the secondary neutrons are ~ 4 MeV and ~ 7 MeV. They found ~ 6% difference in the breeding ratio.

6. Recommended Effort

A serious R&D effort is necessary to demonstrate the feasibility of achieving adequate tritium breeding in fusion blankets. While some present blanket concepts such as $Li_{17}Pb_{83}$ provide adequate assurance of the feasibility of tritium breeding, a number of other concepts such as solid breeders have too small a breeding margin to assure success. In addition to tritium breeding, the selection of materials and design concepts depends on a number of critical requirements discussed in an earlier section (e.g., tritium release). Therefore, the fusion blanket R&D program has to investigate the critical issues such as tritium breeding and tritium release (see Table III) simultaneously for a number of materials and design concepts. Thus, a neutronics R&D program is a key part of the critical path leading to the selection of a viable blanket design.

Areas for the tritium breeding R&D neutronics effort are summarized in Table XVI. The effort should focus on:

A. Reducing the uncertainties in predicting T.
B. Improving the predictability of the uncertainty in T.

The two areas are, of course, interrelated. For example, the effort to improve the predictability of the uncertainty in T will lead to a better definition of areas where effort is needed to reduce uncertainties.

There are three sources of error that contribute to the uncertainty in T:

1. Design Definition
2. Calculations
3. Nuclear Data

The uncertainties of the design definition comes from the many yet unresolved technology choices for reactor components that can impact tritium breeding (e.g., neutral beams vs. rf and limiter vs. divertor) and from the lack of some engineering details in present conceptual designs. The present overall R&D strategy for fusion is such that design definition should continually improve over the next several years. Some technology choices are likely to remain open for sometime, however, and the neutronics analyst has to develop ways to cope with the situation.

The second source of uncertainty comes from the calculations. The critical phase of the calculations is neutron transport, i.e., predicting the neutron flux. There are some inherent errors in methods/codes due to numerical techniques, averaging, discrete treatment of continuous variables, etc. There are also errors introduced by the approximations made by the radiation transport analyst in describing the problem. The most common of these relates to the approximations in the geometrical modeling of the physical system. These approximations are often motivated by cost considerations, but sometimes are dictated by limitations on the input description for the transport code. These topics were discussed by several authors (see for example reference 9). The discrete ordinates method is normally used for 1-D and 2-D problems while Monte Carlo is the method of choice for 3-D problems. An important need that has been identified is to improve Monte Carlo codes to facilitate input specifications for geometric representation of specialized fusion geometries. For example, the input description for representing tori in both tokamaks and particular types of major penetrations in tokamaks and mirrors can be simplified.

The uncertainties from nuclear data have been discussed in earlier sections. The following section will discuss the improvement needs for nuclear data. It should be noted that in addition to errors in basic data, errors can also be introduced in data representation and processing.

One of the most serious shortcomings at present is the lack of adequate predictions of the uncertainties in the tritium breeding ratio. The only effort in this area has been limited to a number of sensitivity studies that were reviewed in an earlier section. It is presently very difficult to define quantitatively the accuracy needs in specific cross sections, energy spectra, etc. to meet the stated accuracy goal for the breeding ratio. This situation must be remedied by high priority R&D. The two key elements of this R&D should be integral breeder neutronics experiments and sensitivity analysis.

The types of integral experiments of immediate need are:

- "Clean" experiments that provide definitive information about the uncertainties in specific cross sections in the energy range 0-14 MeV.

- Simple engineering experiments that include key engineering features but the configuration is simple enough to be accurately calculated. These experiments should be performed with a 14 MeV neutron source and will provide an estimate of the difference between measured and calculated values for tritium production.

"Mock-up" experiments may ultimately be required to validate the neutronics design and performance prediction. However, the geometrical complexity involved in a realistic mock-up makes it unlikely that discrepancies between calculations and measurement can be resolved. Furthermore, mock-up experiments will encounter serious difficulties in attempting to realistically simulate a fusion (volumetric) neutron source. Therefore, the role of mock-up experiments in fusion blanket neutronics remains to be evaluated. It must be noted that all types of integral experiments will fall short of direct experimental demonstration of the achievable breeding ratio in an actual reactor, where penetrations and other geometrical effects will affect the tritium production. Extrapolation will, of course, have to be resorted to. However, such extrapolation will be adequate only if the tritium breeding margin exceeds the estimated uncertainties.

A number of very simple integral experiments have been performed and were reviewed by Jarvis.[42] The Japanese fusion program has made a significant step forward in the area of integral experiments. An intense 14 MeV neutron source was constructed in a versatile facility dedicated for fusion neutronics. The facility is now operational and initial experiments[45] have begun. The United States program is in the stage of planning for integral breeder neutronics experiments.

7. Nuclear Data Needs

The fusion program has been fortunate to have the extensive methods and data information base that has been developed over the past thirty years in the fission and weapons program. In the late 1960's and early 70's, the additional data needs for fusion in terms of materials, energy range, type of data, etc. were defined. Significant progress has been made over the past decade in satisfying many of these needs.

The nuclear data program should be aimed at two areas: 1) improving the broad data base for candidate fusion materials, and 2) satisfying high priority requests for high accuracy measurements in specific materials. A program to improve the overall data base is necessary for several reasons. There is presently a diversity of candidate materials and design concepts. The probability of selecting a number of materials has changed and will continue to change over the years. Furthermore, the sensitivity of estimated parameters such as the tritium breeding ratio are very system dependent. As the characteristics of the system change and become clearer in the future, it is likely that our priority list of very specific needs will be different.

Satisfying high priority requests for high accuracy measurements in specific materials should continue to be a part of the nuclear data effort. However, the list of priority requests must always be kept under review. Furthermore, methods for generating the priority requests should be improved. More specifically, integral neutronics experiments and sensitivity studies should be relied upon in developing priorities and accuracy requirements. Several reviews of the data needs were performed recently (see, for example references 42-44, 9, 46). We briefly review below the highlights of the data needs for tritium breeding.

Table XVII lists the elements of importance for tritium breeding. The author attempted to limit the number of materials, but the list remains long because of the large number of tritium breeding blanket concepts that are still under very active consideration in fusion development. For each material, the probability of eventually using it in a fusion reactor is given on three scales: high, medium, and low. This probability is derived based on all presently known considerations of engineering feasibility issues. A priority for the data needs for each material is given on a scale from 1 to 3 with 1 designating the highest priority. The priority for the data needs of each material combines considerations of a) the probability of eventually using the material in the blanket, b) the effect of the material on the tritium production rate, and c) the uncertainties in the presently available data. For example, hydrogen has a very high probability of use as an element of the water coolant, and it does affect significantly tritium production, but the accuracies of available data are excellent. Therefore, the priority for the data needs of hydrogen is very low. The materials included in Table XVII cover not only the blanket materials but also those in components that affect tritium production in the blanket, such as the first wall, limiter, rf.

The type of data needed are those that affect the neutron transport, i.e., 1) total, 2) elastic, 3) inelastic, 4) neutron emission [σ_{em}, $P(E')$, $d\sigma/d\Omega$], and 5) total parasitic neutron absorption; as well as the tritium-producing reaction cross sections, $^6Li(n,\alpha)t$ and $^7Li(n,n'\alpha)t$. Some of this data and other types of data are, of course, needed for nuclear heating, radiation damage, etc.

The energy range of interest extends from thermal energies to at least the average energy of the D-T neutrons, i.e., 14 MeV. It should also be noted that there is a considerable width in the spectrum of the source neutrons emitted from D-T plasmas significantly heated or driven by injecting energetic deuterons. Therefore, the high energy limit for nuclear data should extend to ~ 15 or 16 MeV. Sensitivity studies indicate that the 9 to 15 MeV range is the most important for predicting the neutron transport. Available data in this range are less adequate than those below 9 MeV. The low energy range, 0 to 100 keV, is very important for many of the recent breeding blanket concepts, particularly the "thermal" type discussed earlier. In some of these concepts, ~ 80% of the tritium production is contributed by neutrons of energies < 4 eV, inducing the $^6Li(n,\alpha)t$ reaction.

The accuracy goal for predicting the tritium breeding ratio is 2%. Thus, the accuracy of nuclear data should be such that the contribution of nuclear data uncertainties to the error in tritium breeding ratio estimates is ~ 1% or less. Specific accuracy requirements for material, reaction, energy range, etc. can be derived only from future integral breeding experiments and sensitivity studies. From the limited studies we have, it appears that an accuracy of ~ 5 to 10% in data for neutron transport may be sufficient. Exceptions can be noted for several key blanket materials such as beryllium and lead where accuracies of ~ 3% are required.[44]

^6Li and ^7Li are obviously critical to tritium breeding. The status of the $^6Li(n,\alpha)t$ cross section data appears adequate. As discussed earlier, available data appear to introduce an uncertainty of less than a few tenths of a percent in the tritium breeding ratio for present candidate designs. The importance of ^7Li depends critically on the blanket system considered. In the thermal system, ^7Li contribution is almost negligible. However, for systems such as those with natural lithium and Li_2O, the $^7Li(n,n'\alpha)t$ contribution can be more than one-third the breeding ratio. The available accuracies for the $^7Li(n,n'\alpha)t$ cross section and secondary neutron energy distribution are not adequate. The required accuracy in this cross section is ~ 3%.

Acknowledgements

The author thanks his colleagues, Drs. J. Jung and Y. Gohar, for many useful comments. Dr. Jung has graciously provided the author with some important results from his recent work prior to this publication. The efforts of Ms. L. Legerski in typing and organizing this manuscript are greatly appreciated.

References

1. B. Badger, et al., "UWMAK-I," UWFDM-68, University of Wisconsin (1973).
2. D. Steiner, "Analysis of a Benchmark Calculation of Tritium Breeding in a Fusion Reactor Blanket- The U.S. Contribution," ORNL-TM-4177, Oak Ridge National Laboratory (1973).
3. D. Steiner, M. Tobias, Nucl. Fusion 14, 2 (1974) 153.
4. C. Baker, et al., "STARFIRE-A Commercial Tokamak Fusion Power Plant Study," ANL/FPP-80-1, Argonne National Laboratory (1980).
5. M. Abdou, et al., "A Demonstration Tokamak Power Plant Study-Interim Report," ANL/FPP/TM-154, Argonne National Laboratory (1982).
6. "International Tokamak Reactor - Phase I," International Atomic Energy Agency, Vienna (1982).
7. W. F. Vogelsang, "Breeding Ratio, Inventory, and Doubling Time in a D-T Fusion Reactor," Nucl. Techn. 15 470 (1972).
8. R. Clemmer, "TCODE - A Computer Code for Analysis of Tritium and Vacuum Systems for Tokamak Fusion Reactors," ANL/FPP/TM-110, Argonne National Laboratory (1978).
9. M. A. Abdou, "Problems of Fusion Reactor Shielding," GTFR-10, Georgia Institute of Technology (1979).
10. J. Gilligan, K. Evans, Personal Communication (1982).
11. M. A. Abdou, D. Graumann, "The Choice of Coolant in Commercial Tokamak Power Plants," Proc. 4th Topical Meeting on the Technology of Controlled Nuclear Fusion, CONF-801011, Vol. III, p. 1740 (1980).
12. J. Jung, M. Abdou, "Breeding Potential of Candidate Breeding Materials," Nucl. Tech. (to be published).
13. T. Ide, Y. Seki, H. Aida, "Effects of Neutron Streaming through Injection Ports on Neutronics Characteristics of a Fusion Reactor," Proc. 2nd

Topical Meeting on the Technology of Controlled Nuclear Fusion, CONF-760935-P3, p. 395 (1976).

14. W. M. Stacey, Jr., et al., "U.S. Contribution to the International Tokamak Reactor Phase-I Study," US-INTOR/81-1, Georgia Institute of Technology (1981).

15. M. A. Abdou, "Radiation Considerations for Superconducting Fusion Magnets," J. Nucl. Mater. 72, 147-167 (1978).

16. M. Abdou, "Nuclear Design of the Blanket/Shield System for a Tokamak Experimental Power Reactor," Nucl. Techn. 29, 7 (1976).

17. D. W. Muir, M. E. Wyman, "Neutronic Analysis of a Tritium-Production Integral Experiment," Proc. Symp. Technology of Controlled Thermonuclear Fusion Experiments and the Engineering Aspects of Fusion Reactors, CONF-72111, p. 910 (1974).

18. H. Bachmann, U. Fritsher, F. W. Kappler, D. Rusch, H. Werle, H. W. Wiesse, "Neutron Spectra and Tritium Production Measurements in a Lithium Sphere to Check Fusion Reactor Blanket Calculations," Nucl. Sci. Eng. 67, 74 (1978).

19. W. A. Ruepke, "The Consistency of Differential and Integral Thermonuclear Neutronics Data," LA-7067-T, Los Alamos National Laboratory (1978).

20. A. Hemmendinger, C. E. Ragan, J. M. Wallace, Nucl. Sci. Eng. 70, 274 (1979).

21. D. L. Smith, M. M. Bretscher, J. W. Meadows, "Measurement of the Cross Section for the ^{7}Li(n,n't)^{4}He Reaction in the Energy Range 7-9 MeV" Nucl. Sci. Eng., 78, 359 (1981).

22. M. T. Swinhoe, C. A. Uttley, "Tritium Breeding in Fusion," Proc. Conf. Nuclear Cross Sections for Technology," NBS Special Publication 594, p. 246, National Bureau of Standards (1980).

23. P. G. Young, "Evaluation of n + ^{7}Li Reactions Using Variance-Covariance Techniques," Trans. Am. Nucl. Soc. 39, 272 (1981).

24. Y. Gohar, M. Abdou, "Neutronic Optimization of Solid Breeder Blankets for STARFIRE Design," Proc. 4th Topical Meeting on the Technology of Controlled Nuclear Fusion, CONF-801011, Vol. II, p. 628 (1980).

25. R. W. Conn, G. L. Kulcinski, "UWMAK-II," UWFDM-112, University of Wisconsin (1975).

26. "International Tokamak Reactor - Phase I," International Atomic Energy Agency, Vienna (1982).

27. M. Abdou, L. Wittenberg, C. W. Maynard, "A Fusion Design Study of Non-Mobile Blankets with Low Lithium and Tritium Inventories," Nucl. Techn. 26, 400 (1975).

28. D. Steiner, M. Tobias, "Cross-Section Sensitivity of Tritium breeding in a Fusion Reactor Blanket: Effects of Uncertainties in Cross-Sections of ^{6}Li, ^{7}Li, and ^{93}Nb," Nucl. Fusion 14, 153 (1974).

29. D. E. Bartine, E. M. Oblow, F. R. Mynatt, "Radiation-Transport Cross-Section Sensitivity Analysis - A General Approach Illustrated for a Thermonuclear Source in Air," Nucl. Sci. Eng. 55, 147 (1974).

30. S. A. W. Gerstl, D. J. Dudziak, D. W. Muir, "Cross-Section Sensitivity and Uncertainty Analysis with Application to a Fusion Reactor," Nucl. Sci. Eng. 62, 137 (1977).

31. R. Conn, W. M. Stacey, "Variational Method for Controlled Thermonuclear Reactor Blanket Studies," Nucl. Fusion 13, 185 (1973).

32. R. G. Alsmiller, J. Barish, T. A. Gabriel, R. T. Santoro, "Comparison of Cross-Section Sensitivities of Breeding Ratios in Fusion-Reactor Blankets," Trans. Am. Nucl. Soc. 19, 463 (1974).

33. D. E. Bartine, R. G. Alsmiller, C. M. Oblow, F. R. Mynatt, "Cross-Section Sensitivity of Breeding Ratio in a Fusion Reactor Blanket," Nucl. Sci. Eng. 53, 304 (1974).

34. S. A. W. Gerstl, "Blanket Design and Cross-Section Sensitivity Calculations Based on Perturbation Methods," Proc. 1st Topical Meeting on the Technology of Controlled Nuclear Fusion, CONF-740-402-P2, p. 136 (1974).

35. R. G. Alsmiller, R. T. Santoro, J. Barish, T. A. Gabriel, "Comparison of the Cross-Section Sensitivity of the Tritium Breeding Ratio in Various Fusion-Reactor Blankets," Nucl. Sci. Eng. 57, 122 (1975).

36. R. G. Alsmiller, J. Barish, C. R. Weisbein, "Uncertainties in Calculating Heating and Radiation Damage in the Toroidal Field Coil of a Tokamak Experimental Power Reactor Due to Neutron Cross-Section Errors," Nucl. Techn. 34, 376 (1977).

37. B. Arcipiani, G. Palmiotti, M. Salvatores, "Neutron Heating Sensitivity to Cross-Section Variations in a Controlled Thermonuclear Reactor Blanket," Nucl. Sci. Eng. 65, 540 (1978).

38. T. Wu, C. W. Maynard, "The Application of Uncertainty Analysis in Conceptual Fusion Reactor Design," Proc. of a Seminar-Workshop on Theory and Application of Sensitivity and Uncertainty Analysis, ORNL/RSIC-42, Oak Ridge National Laboratory (1979).

39. S. Pelloni, "Cross-Section Sensitivity and Uncertainty Analysis for European INTOR and U.S. FED Designs," GA-A16685 (1982).

40. D. V. Markovsii, et al., "Influence of Neutron Constants on the Behaviour of a Thermonuclear Reactor Blanket, IAE 2579 (1975).

41. P. D. Soran, et al., "Effect of Be (n,2n) Multigroup Treatment on Theta Pinch Blanket Neutronics, Proc. 1st Topical Meeting on the Technology of Controlled Nuclear Fusion, CONF-740402-P2, p. 172 (1974).

42. O. N. Jarvis, "Nuclear Data Needs for Fusion Reactors," European Appl. Res. Rep., Nucl. Sci. Techn. 3, 127 (1981).

43. "Nuclear Data for Fusion Reactor Technology," IAEA-TECDOC-223, International Atomic Energy Agency (1979).

44. E. T. Cheng, et al., "Magnetic Fusion Energy Program Nuclear Data Needs," GA-NDN-82-005 (1982).

45. T. Nakamura, et al., "Integral Experiments on Lithium Oxide Spherical Assembly with Graphite Reflector and on Duct Streaming," 3rd IAEA Technical Committee Meeting and Workshop on Fusion Reactor Design and Technology, Tokyo (1981).

46. G. Constantine, "Nuclear Data Requirements for Fusion Reactor Design - Neutronics Design, Blanket Neutronics, and Tritium Breeding," Nuclear Data for Fusion Reactor Technology, IAEA-TECDOC-223, International Atomic Energy Agency (1979).

47. H. Liskiel, et al., "Determination of $^{7}Li(n,n't)^{4}He$ Cross Sections," Proceedings of this conference.

Table I. Important Areas of Nuclear Data for Fusion

Applications	Comments
Fuel Cycle	Charged particle reactions for DT, DD and advanced fuels.
Tritium Breeding	$^{6}Li(n,\alpha)t$, $^{7}Li(n,n'\alpha)t$, neutron emission and parasitic absorption for blanket materials.
Nuclear Heating	Neutron and gamma transport neutron reaction kinetics, reaction Q-values.
Radiation Damage	DPA, gas production, transmutation.
Induced Activity	Transmutation reactions, decay schemes.
Radiation Shielding	Shielding of components and personnel; neutron and gamma transport, induced activity, heating.

Table II. Candidate First-Wall/Blanket Materials

Breeding Materials	Coolants	Structure	Neutron Multiplier
Liquid Metals	Water	Austenitic Stainless Steel	Be
Li	(H_2O, D_2O)	Ferritic Steels	BeO
Li-Pb	Liquid Metals	Nickel-base alloys	Pb
Li-Pb-Bi	Li	Refractory Alloys (e.g., V)	PbO
Intermetallic Compounds	Li-Pb		Bi
Li_7Pb_2	Li-Pb-Bi		Zr
Ceramic	Gases		Zr_5Pb_3
Li_2O	He		PbBi
$LiAlO_2$	Steam		
Li_2SiO_3			
Li_2ZrO_3			
Li_2TiO_3			

Table III. Critical Issues for the Breeder Blanket Options

1. LIQUID METALS

A. Lithium

- Safety
  Consequences of Lithium Fire
- MHD Effects
- Compatibility with Structural Materials

B. 17Li-83Pb

- Compatibility with Structural Materials
- For Li-Pb Cooling
  - MHD Effects
  - Tritium Processing and Containment
- For Water Cooling
  - Safety - large scale expulsion of Li-Pb from the blanket in off-normal Li-Pb/water contact
  - Tritium permeation to water as a result of tritium low solubility/high partial pressure
- For Sodium Cooling
  - Safety - chemical reactivity of sodium with water and air
  - MHD Effects

2. SOLID BREEDERS

- Blanket Tritium Inventory
  Particularly, effects of radiation on tritium inventory.
- Design Practicability
  A number of design problems related to maintaining the low-thermal-conductivity breeder material within the required narrow temperature range and controlling the temperature gradient at the breeder/structure/coolant interfaces.
- For $LiAlO_2$: Tritium Breeding
  Achieving a net tritium breeding ratio greater than one.
- For Li_2O: Reactivity with Water to Form LiOH
  - Difficulties in obtaining and maintaining high purity Li_2O.
  - Consequences of corrosive effects of LiOH under off-normal events involving breeder/coolant interaction.

Table IV. Promising Blanket Concepts Under Active Consideration

	Solid Breeder		Liquid Metal Breeder	
Breeder	Li_2O	$LiAlO_2$ (Li_2SiO_3)	Li	17Li-83Pb
Coolant	H_2O	H_2O	Li	(Na, Li-Pb, H_2O) ?
Structure	SS	SS	V (FS, SS?)	V, FS ?
Neutron Multiplier	?	Be	-	-

SS: Austenitic stainless steel, FS: Ferritic steel, V: Vanadium alloy.

Table V.
Description of Four Blanket Designs for Tritium Breeding Calculations.

Thickness	Armor 1 cm	First Wall 1 cm	Blanket 68 cm
Lithium	FS[a]	35% H_2O + 65% FS	90% Li* + 10% FS
17Li-83Pb	FS	35% H_2O + 65% FS	85% $Li_{17}Pb_{83}$** + 10% FS + 5% H_2O
Li_2O	SS[b]	35% H_2O + 65% SS	90% Li_2O* (70% dense) + 5% SS + 5% H_2O
$LiAlO_2$	SS	35% H_2O + 65% SS	90% $LiAlO_2$*** (70% desne) + 5% SS + 5% H_2O

*Natural lithium enrichment.
**90% ^6Li enrichment.
***60% ^6Li enrichment.

[a]FS = Ferritic Steel
[b]SS = Stainless Steel

Table VI.
Tritium Breeding Ratio Potential
(Full breeding blanket coverage with no penetrations.)

Breeding Ratio	Liquid Breeder Lithium[a]	17Li-83Pb[b]	Solid Breeder Li_2O[a]	$LiAlO_2$[c]
T_6	0.89	1.48	0.90	0.85
T_7	0.36	.002	0.29	0.03
$T = T_6 + T_7$	1.25	1.48	1.19	0.88

[a]Natural enrichment.
[b]90% ^6Li enrichment.
[c]60% ^6Li enrichment (no neutron multiplier).

Table VII.
Tritium Breeding Ratio with and without Inboard Blanket and Limiter

	Liquid Breeder		Solid Breeder	
	Lithium[a]	17Li-83Pb[b]	Li_2O[a]	$LiAlO_2$[c]
Full Blanket Coverage, No Penetrations	1.25	1.48	1.19	0.88
No Inboard Blanket, Limiter Penetration Included	1.09	1.26	1.06	0.79

[a]Natural enrichment.
[b]90% ^{6}Li enrichment.
[c]60% ^{6}Li enrichment (no neutron multiplier).

Table VIII. Properties of Candidate Neutron Multiplier Materials

Material	Be	BeO	Pb	PbO	Bi	Zr_5Pb_3	PbBi
Density, g/cm^3	1.85	2.96	11.34	9.53	9.8	8.93	10.46
Atoms or molecules/ cm^3, x 10^{-24}	0.1236	0.07127	0.03348	0.02571	0.02824	0.004680	0.03047
$\sigma(n,2n)$ at 14 MeV, barns	0.5	0.5	2.2	2.2	2.2	9.2	2.2
$\Sigma(n,2n)$ at 14 MeV, cm cm^{-1}	0.0618	0.256	0.0737	0.0565	0.0621	0.0431	0.0670
(n,2n)threshold, MeV	1.868	1.868	6.765	6.675	7.442	6.765	6.765
$\sigma(n,\gamma)$ at 0.0253 eV, barns	0.0095	0.0095	0.17	0.17	0.034	1.41	0.094
Radioactivity							
Isotopes	^{10}Be	^{10}Be	^{205}Pb	^{205}Pb	^{210}Po	^{93}Zr, ^{205}Pb	^{205}Pb, ^{210}Po
Melting point, °C	1278	2520	327.5	888	271.3	1400	125
Thermal Conductivity[a] at 25°C, W/m-°K	201	216[b]	35.3	2.8	7.92[c]	---	2.3[d]

[a]At 25°C.
[b]Pure beryllium oxide, hot pressed.
[c]Polycrystalline.
[d]At 200°C.

Table IX.
Blanket Parameters for Internally-Cooled Neutron Multiplier Concepts

Zone Description	Zone Thickness cm	Zone Composition, Vol. % Internally-Cooled
First Wall	1	50% PCA 50% H_2O
Neutron Multiplier	Variable	85% neutron multiplier 10% PCA 5% H_2O
Tritium Breeder	50	80% $LiAlO_2$ breeder[a] 10% PCA 5% H_2O 5% He purge
Reflector	15	50% carbon 25% PCA 25% H_2O

[a]90% ^{6}Li.

Table X.
Estimated Tritium Breeding Ratio Likely to be Achievable in Fusion Power Reactors

Breeding Material	Breeding Ratio	Comment
Liquid Lithium	1.15	Low Risk
$Li_{17}Pb_{83}$	1.3	Attractive
Solid Breeders		
Li_2O	1.1	Medium Risk
Li_2O (+Be)	1.3	Attractive
$LiAlO_2$	0.8	Impossible
$LiAlO_2$ (+Zr_5Pb_3)	1.04	Rejected
$LiAlO_2$ (+Be)	1.08	High Risk

Table XI.
Description of Two Blanket Concepts for Sensitivity.
(Analysis performed by Alsmiller, et al[35])

Zone	ORNL[a]		LASL[b]	
	Outer Radius[c] (cm)	Composition[d,e]	Outer Radius (cm)	Composition[d]
1	280	Plasma	15	Plasma
2	350	Vacuum	29.9	Vacuum
3	350.25	Structure	30	Al_2O_3
4	380.25	99% Li + 1% Structure	30.4	Nb
5	380.5	Structure	40	66% Li + 25% Be + 4% Al_2O_3 + 5% Nb
6	420.5	Graphite	62	86% C + 10% Li + 2% Al_2O_3 + 2% Nb
7	420.75	Structure	68	92% Li (95% ^6Li) + 4% Al_2O_3 + 4% Nb
8	450.75	99% Li + 1% Structure	70	25% Al_2O_3
9	451	Structure	110	67% Cu + 33% Al_2O_3
10			115	Ti
11			117	25% Al_2O_3
12			155	98% Li + 2% Nb

[a]Oak Ridge National Laboratory blanket design.[3]
[b]Los Alamos Scientific Laboratory theta-pinch reactor blanket design.[4]
[c]Calculations performed in 1-D cylindrical geometry.
[d]All composition percentages are by volume and all lithium is natural lithium except where noted.
[e]Structure is niobium or vanadium.

Table XII.
Cross-Section Uncertainties Assumed for Various Partial Cross Sections (Analysis performed by Alsmiller, et al[35])

Element	Cross-Section Type Varied	Energy Range (MeV)	Percent Increase in Varied Cross Section δC	Cross Section Type Varied to Compensate[a]
^6Li	$\Sigma_{(n,t)\alpha}$	$< 1 \times 10^{-7}$	0.5	$\Sigma_{Total\ Collision}$
		$10^{-7} - 10^{-2}$	1.0	
		$10^{-2} - 10^{-1}$	1.0 - 2.0[b]	$\Sigma_{Elastic}$
		$10^{-1} - 3 \times 10^{-1}$	5.0	
		$3 \times 10^{-1} - 5 \times 10^{-1}$	5.0 - 10[b]	
		$5 \times 10^{-1} - 7 \times 10^{-1}$	10 - 15	
		$7 \times 10^{-1} - 1 \times 10^{0}$	15	
		$1 \times 10^{0} - 1 \times 10^{0}$	15 - 10[b]	
		$1.7 \times 10^{0} - 1.4 \times 10^{1}$	10	
^7Li	$\Sigma_{(n,n')\alpha,t}$	All Energies	20	$\Sigma_{Elastic}$
C	$\Sigma_{Elastic}$	< 4.8	3.0	$\Sigma_{Total\ Collision}$
		4.8 - 9.0	5.0	$\Sigma_{Inelastic}$
		9.0 - 15	15	
Be	$\Sigma_{(n,2n')}$	1.85 - 6.4	10	$\Sigma_{Elastic}$
		6.4 - 14	10 - 15[b]	
F	$\Sigma_{Total\ Collision}$	All Energies	20	

[a]See discussion in text.
[b]Linear interpolation was used between the values shown.

Table XIII.
Breeding-Ratio Uncertainties Due to the Cross Section Variations Shown in Table XII.

Element	Breeding Material j	$\frac{R_j}{R}$ (in percent) ORNL Design With Nb	ORNL Design With V	LASL Design
^6Li	^6Li	0.14	0.07	0.19
	^7Li	- 0.01	- 0.01	0.00
	^6Li and ^7Li	0.13	0.06	0.19
^7Li	^6Li	0.26	0.35	0.02
	^7Li	5.60	5.56	2.42
	^6Li and ^7Li	5.86	5.91	2.44
C	^6Li			- 0.82
	^7Li			0.24
	^6Li and ^7Li			- 0.58
Be	^6Li			2.51
	^7Li			- 0.18
	^6Li and ^7Li			2.33

Table XIV.
INTOR Blanket Models used by Pelloni[39]

Li_2SiO_3 Blanket			$Li_{17}Pb_{83}$ Blanket		
Zone	Thickness cm	Composition[a]	Zone	Thickness cm	Composition[a]
1	220	Plasma	1	220	Plasma
2	30	Vacuum	2	30	Vacuum
3	13	100% SS	3	3	80% Al + 20% H_2O
4	9.1	60.7% Pb + 7.8% D_2O + 10% SS + 21.5% void	4	40	80% $Li_{17}Pb_{83}$ + 10% H_2O + 10% SS
5	6.2	28.6% Li_2SiO_3 + 22.1% H_2O + 11.2% SS + 16.6% He + 21.5% void	5	12	100% SS
6	6.2	24.5% Li_2SiO_3 + 27.9% H_2O + 11.1% SS + 15% He + 21.5% void			
7	6.2	12.3% Li_2SiO_3 + 29.7% H_2O + 11% SS + 25.5% He + 21.5% void			
8	12	100% SS			

[a]Lithium is enriched to 30% ^6Li.

Table XV.
Sensitivites[39] of Tritium Production Rate to Partial Cross Section Uncertainties for the Two INTOR Designs of Table XIV.
(Percent per 1% Cross Section Increase)

	Li_2SiO_3 Concept			$Li_{17}Pb_{83}$ Concept					
Cross Section Type	Pb	Fe	^6Li	Pb	Fe	^6Li	H	O	Al
(n,n)	0.000	0.011	--	0.029	0.016	--	0.057	0.003	-0.006
(n,n´)	-0.001	-0.000	--	-0.013	-0.012	--	--	-0.011	-0.057
(n,2n)	0.195	0.059	--	0.149	0.008	--	--	--	0.003
(n,γ)	--	-0.045	--	--	-0.078	--	--	--	--
(n,t)	--	--	0.001	--	--	0.006	--	--	--

Table XVI. Needed Neutronics Effort on Tritium Breeding

A. Reduce Uncertainties in Predicting T

1. Design Definition
   - Narrow Materials and Design Concepts
   - Greater Engineering Details
2. Calculations
   - Improve Methods (neutron transport)
   - More Detailed Modeling
3. Nuclear Data
   - Measurements
   - Evaluation
   - Data Representation and Processing

B. Improve Predictability of Uncertainty in T

1. Integral Experiments
   - Basic Experiments
   - Engineering Experiments
   - Mockup
2. Sensitivity Analysis
   - Improve Method
   - Perform Sensitivity Studies (geometry, material composition, cross sections, secondary neutron spectra, etc.)

Table XVII.
Probability of Use and Priority of Data Needs for Elements of Importance to Tritium Breeding

Element	Probability of Use	Priority for Data Needs	Comments
Lithium/Lithium Compounds			
Lithium	H	1	Definite
Lead	H	1	Li-Pb
Oxygen	H	3	Li_2O
Aluminum	M	1	$LiAlO_2$
Silicon	M	1	Li_2SiO_3
Zirconium	M	1	Li_2ZrO_3
Titanium	L	1	Li_2TiO_3
Neutron Multiplier			
Beryllium	H	1	
Lead	M	1	
Bismuth	L	2	Bi-Pb
Zirconium	L	3	Zr_5Pb_3
Coolant			
Helium	M	3	
Hydrogen	H	3	H_2O
Oxygen	H	3	H_2O
Structural Material			
Iron	H	1	SS
Chromium	H	2	SS
Nickel	H	2	SS
Manganese	H	2	SS
Vanadium	H	1	
Niobium	L	3	
Moderator/Reflector			SS, H_2O
Carbon	M	3	
In-Vessel Components			
Copper	H	2	
Vanadium	M	2	
Niobium	M	2	
Tungsten	M	3	
Tantalum	L	3	

[a] Judgement on the probability of using the material in fusion blankets: H=high, M=medium, L=low.
[b] Priority for data needs takes into account of (a) the probability of using the material, (b) effect on tritium breeding, and (c) uncertainties in presently available data.

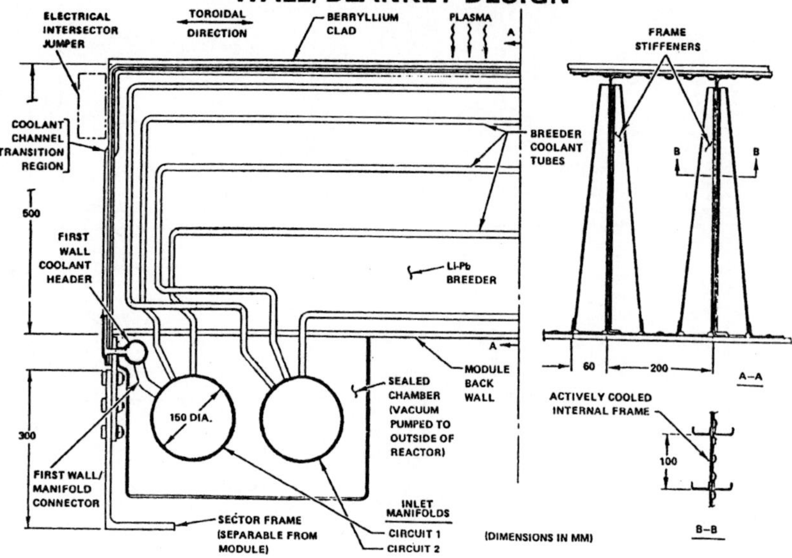

Fig. 1 17Li-83Pb liquid alloy breeder first wall/ blanket design

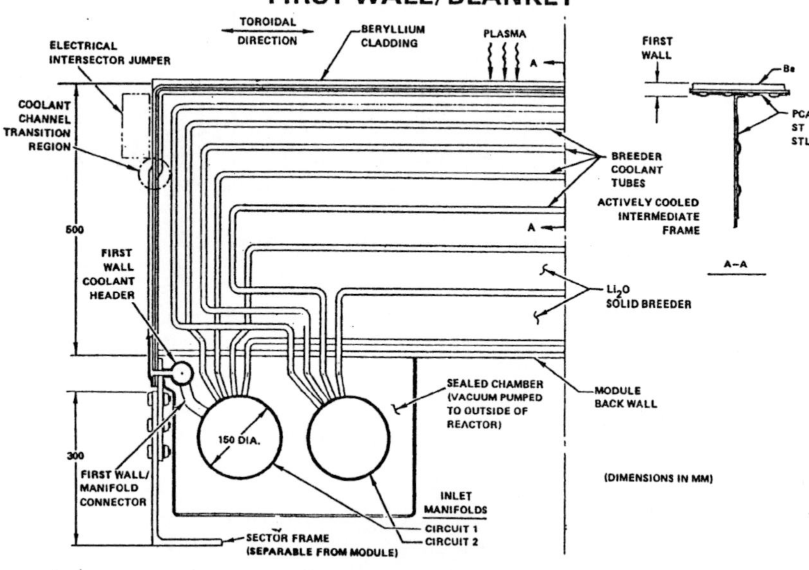

Fig. 2 Li_2O solid breeder reference design first wall/blanket

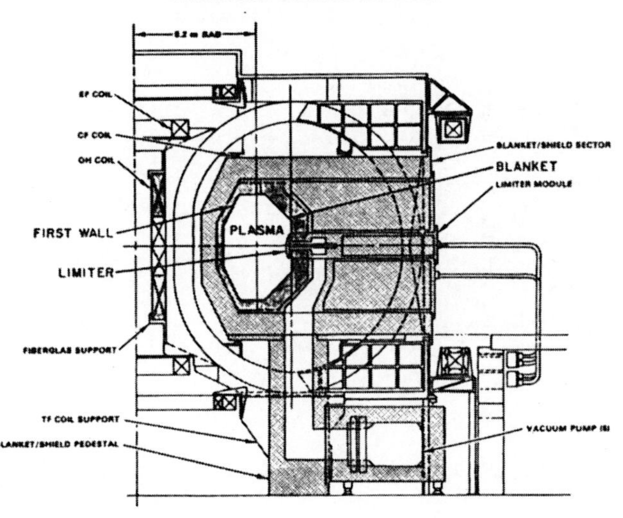

Fig. 3 DEMO reference design

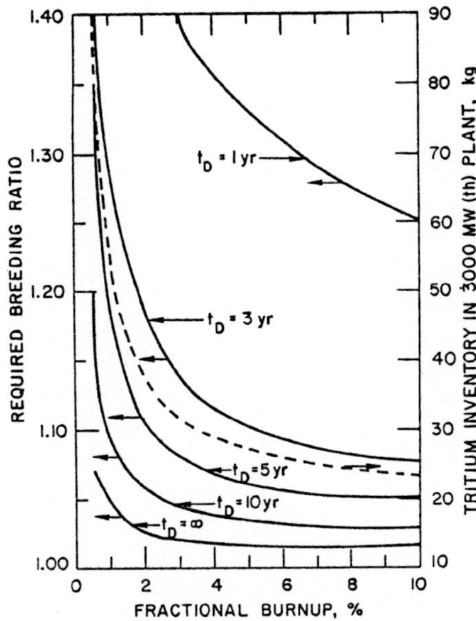

Fig. 4 Effect of fractional burnup and doubling time upon required breeding ratio

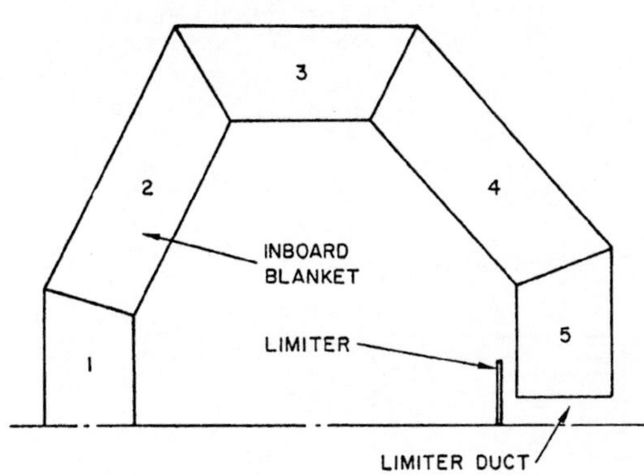

Fig. 5 Vertical cross section of the toroidal geometry used for Monte Carlo calculations

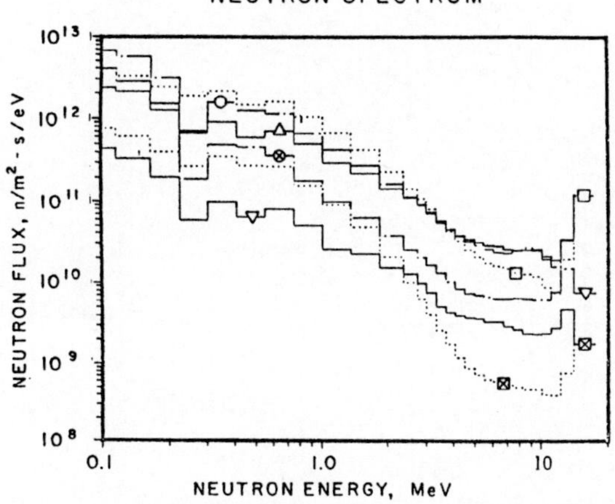

Fig. 6a Neutron spectrum in three blanket systems: Li_2O, Li and 17Li-83Pb. The systems are described in Table V. (Energy range 0.1 - 15 MeV)

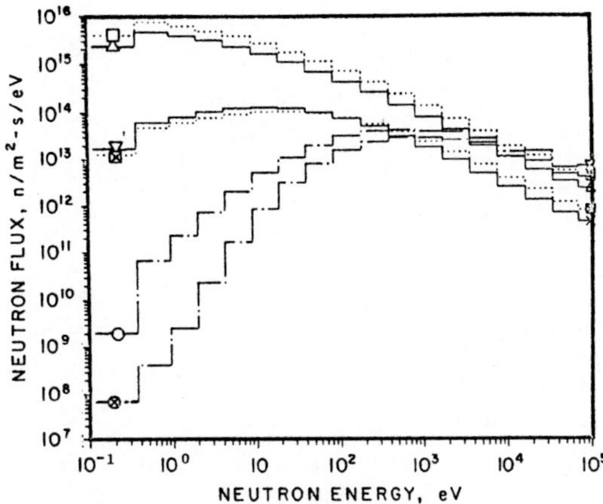

Fig. 6b Neutron spectrum in three blanket systems: Li_2O, Li and 17Li-83Pb. The systems are described in Table V. (Energy range 10^{-1} - 10^{-5} eV)

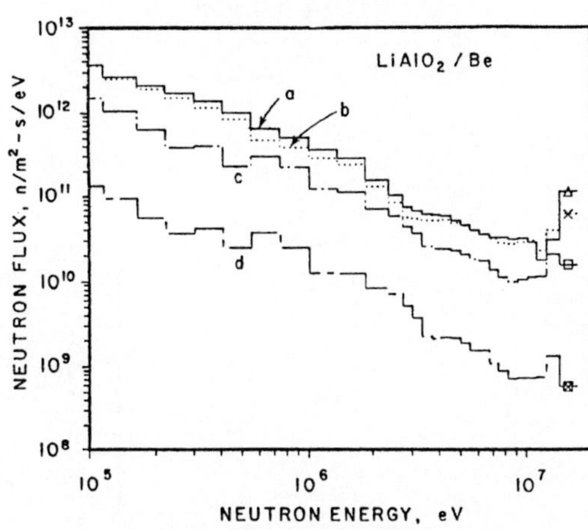

Fig. 7a Neutron spectrum in $LiA1O_2$ blanket with beryllium multiplier. The system consists of 1 cm first wall (Zone 1), 4 cm $LiA1O_2$ (Zone 2), 8 cm Be (Zone 3), 20 cm $LiA1O_2$ (Zone 4). Curves a and b are for the first and last points, respectively, in Zone 2. Curves c and d are for the first and mid-point of Zone 4. (Energy range 0.1 - 15 MeV)

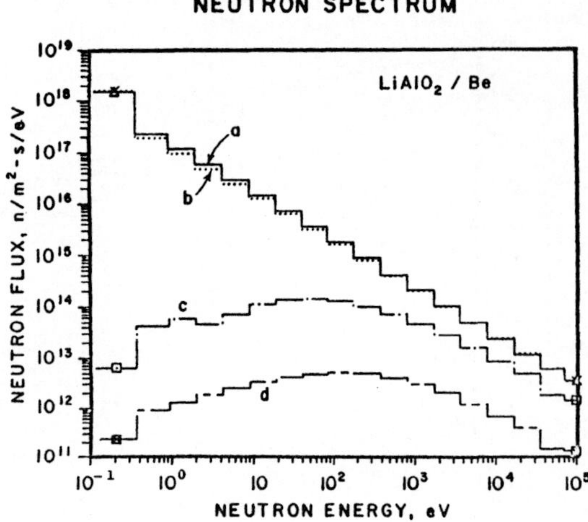

Fig. 7b Neutron spectrum in $LiAlO_2$ blanket with beryllium multiplier. The system consists of 1 cm first wall (Zone 1), 4 cm $LiAlO_2$ (Zone 2), 8 cm Be (Zone 3), 20 cm $LiAlO_2$ (Zone 4). Curves a and b are for the first and last points, respectively, in Zone 2. Curves c and d are for the first and mid-point of Zone 4. (Energy range 10^{-1} - 10^{-5} eV)

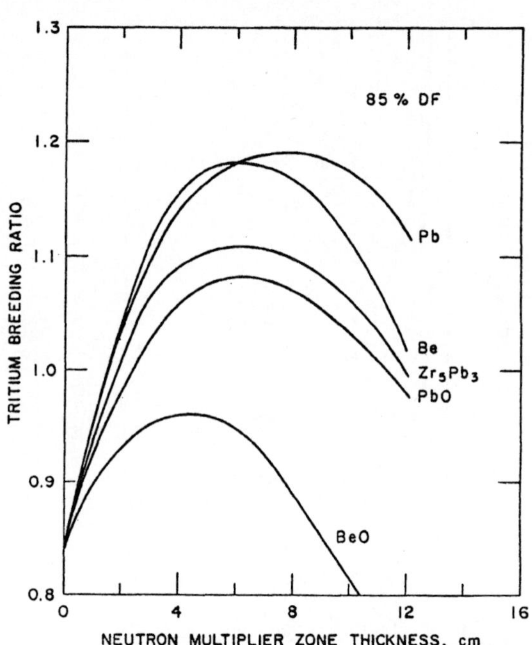

Fig. 8 Tritium breeding ratio from the separate zone blanket option with different neutron multipliers (internally cooled) with $LiAlO_2$ (90 % ^6Li) breeder, H_2O coolant and PCA structure

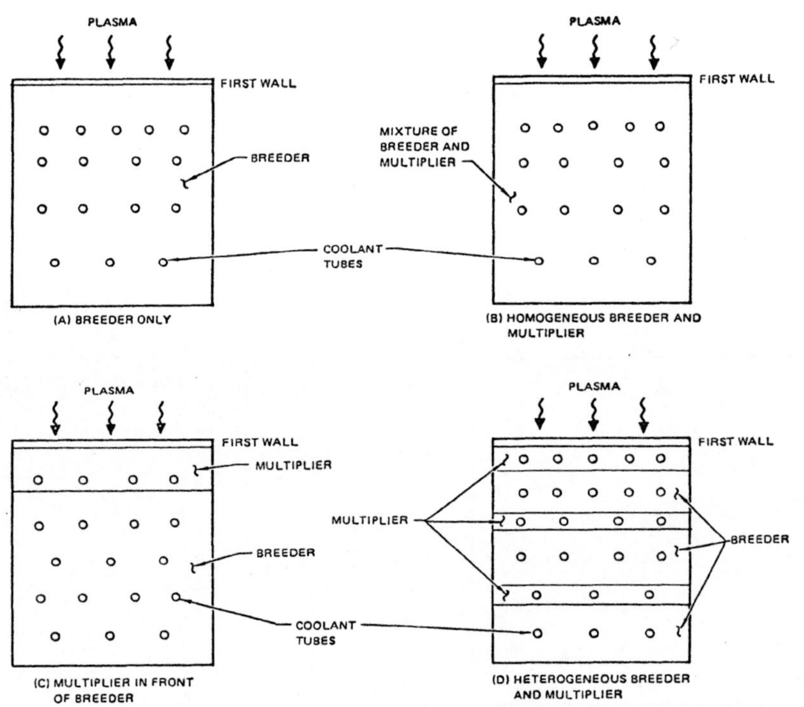

Fig. 9 Possible arrangements of solid breeder and solid neutron multiplier

FUSION REACTION MEASUREMENTS IN THE PRINCETON LARGE TOKAMAK

J.D. Strachan, R.E. Chrien(1) and W.W. Heidbrink

Plasma Physics Laboratory
Princeton University
Princeton, New Jersey 08544
U.S.A.

The d(d,n)^{3}He, d(d,p)t, d(t,n)α, and d(^{3}He,p)α fusion reactions have been measured by detection of the 2.5 MeV neutron, 3 MeV proton, 14 MeV neutron and 14.7 MeV proton emissions, respectively. These measurements provide information on the energy, density, and confinement properties of energetic ions in the tokamak plasma.

Introduction

During the last ten years, the fusion yield from tokamak experimental devices [1-7] has increased about one order of magnitude per year (Fig. 1). One reason for this growth is that heating schemes such as Neutral Beam Injection [8] and Ion Cyclotron Resonant Heating [9] have become available to create energetic ions within the plasma. Another reason is that as tokamaks have become larger, they are better able to confine energetic ions [10,11]. A consequence of the increasing reaction rates is a corresponding increase in plasma diagnostic information that is gained from the fusion reaction products.

One issue addressed by fusion reaction measurements is whether the reacting ions are part of a thermal distribution or of a non-thermal velocity "tail". Measurement of the energy spectrum of the fusion reaction products yields the relative energy of the reacting ions, while the magnitude of the emission is related also to the number density of the reactants.

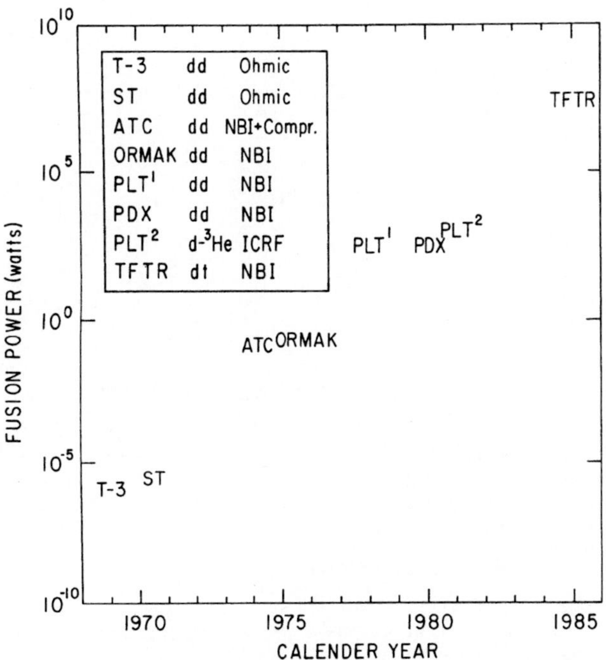

Fig. 1. The fusion yields (expressed as peak power produced)from various tokamaks: T-3 [1], ST [2], ATC [3], ORMAK [4], PLT^1 [5], PDX [6], and PLT^2 [7]. The TFTR power is the breakeven projection. Indicated in the insert is the fusion reaction and the form of the heating: OHMIC implies joule heating from the plasma current, NBI implies neutral beam heating, COMPR implies major radial plasma compression, and ICRF implies ion cyclotron resonant heating.

In this manner, d-d fusion reactions during injection of energetic hydrogen neutrals into a deuterium plasma ($H^\circ \rightarrow D^+$), during wave (ion cyclotron) heating of ^{3}He in a predominately deuterium plasma (^{3}He minority heating), and during ohmic heating (joule heating produced by the plasma current) have been identified as occuring in the bulk plasma. In contrast, measurements show that d-d reactions during D° neutral beam injection, during D minority ion cyclotron heating, and during lower hybrid wave heating are produced by energetic tail distributions.

Another issue addressed by fusion reaction measurements is the confinement properties of energetic ions as they slow down and burn up. The confinement can: a) be longer than the slowing-down time in which case the burnup is determined by the electron temperature of the plasma, b) be limited by the intersection of classical ion orbits with the walls in which case the burnup is determined by the plasma current, or c) be limited by anomalous processes (ion loss caused by a plasma instability or imperfect magnetic field topology) in which case the burnup is determined by the characteristics of the instability. In this paper, we illustrate the use of fusion reaction diagnostics by describing measurements of the energetic ion confinement in the Princeton Large Tokamak (PLT) and Poloidal Diverter Experiment (PDX), which currently are the largest producers of fusion reactions among tokamak devices. These studies rely on nuclear cross-section data primarily in the design and evaluation of detectors.

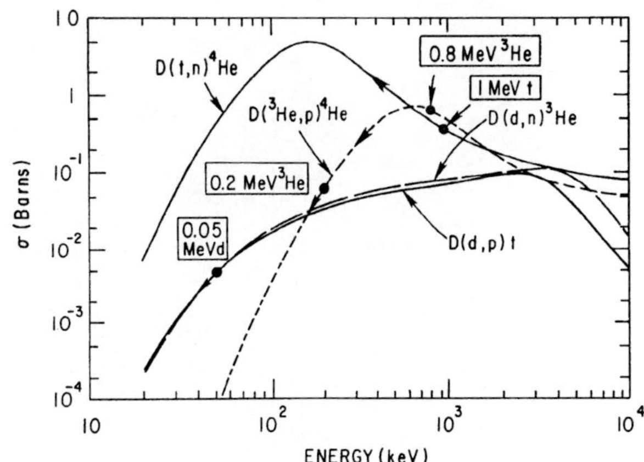

Fig. 2. Cross sections for the d(d,n)^{3}He, d(^{3}He,p)α, d(d,p)t, and d(t,n)α fusion reactions.
The 1 MeV triton and 0.8 MeV ^{3}He ions are produced by the d-d reactions and slow down through the peaks of the dt and d-^{3}He cross sections. The 0.05 MeV deuterons are produced by neutral beam injection. The 0.2 MeV ^{3}He ions are produced by ion cyclotron resonance heating.

K. H. Böckhoff (ed.), Nuclear Data for Science and Technology, 313–317.

Fusion Reactions in Tokamaks

The available energetic ions are able to cause several fusion reactions (Fig. 2). The $d(d,n)^3He$ and $d(d,p)t$ fusion reactions occur at their largest levels during high power (6 MW) neutral beam injection of 50 keV deuterons. After crossing the tokamak magnetic fields as neutral particles, the deuterons undergo charge exchange reactions with plasma ions and become 50 keV ions which are subject to the confinement characteristics of the tokamak. In a plasma with an electron temperature of 2 keV and an electron density of 3×10^{13} cm^{-3}, a typical deuteron slows down through Coulomb collisions and joins the thermal ion population in about 30 msec. Since the deuterons are on a steep part of the cross section (Fig. 2), most d-d reactions occur before the deuterons lose much energy and the cross-section weighted slowing-down time is about 10 msec.

The $d(^3He,p)\alpha$ fusion reactions occur at their largest levels during application of strong (MW) ICRF heating to a predominately deuterium plasma containing about 10% ^3He ions. When the frequency of the radio waves (25 MHz) equals the frequency of ^3He cyclotron motion at the plasma center, the ^3He ions are accelerated by the waves. Simultaneously, the ions lose energy through collisions with the background plasma. The result is a distribution function that is continuous up to 0.2 - 0.4 MeV energies. Again the cross-section weighted slowing-down time is about 10 msec.

D-^3He fusion reactions are also produced by the 0.8 MeV ^3He ion created by the $d(d,n)^3He$ fusion reaction. In this case, the ^3He ion is born near the plasma center and slows down through the maximum of the d-^3He cross section in a cross-section weighted slowing-down time of about 20 msec. Similarly, $d(t,n)\alpha$ fusion reactions are produced by the 1 MeV triton created in the $d(d,p)t$ fusion reaction. In this case, the 1 MeV triton is born at an energy well above the peak of the d-t cross section (Fig. 2) and the slowing down duration to the peak can be as long as 200 msec.

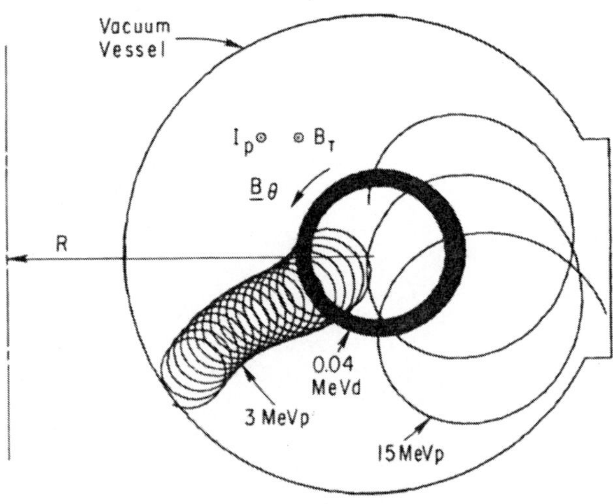

Fig. 3. Poloidal cross section of the tokamak indicating the torus major radius R, the orientation of the toroidal magnetic field, the toroidal plasma current and the poloidal magnetic field induced by the plasma current. Indicated are the poloidal projection of particle orbits for a 0.04 MeV deuteron, a 15 MeV proton, and a 1 MeV triton (or 3 MeV proton or 3 MeV alpha). Each particle was born with v-toroidal/v = 0.6 into a plasma with a 22 k Gauss toroidal magnetic field and a 400 kA plasma current.

Measurement of the emission levels, time evolution, spectra, and spatial origin of the fusion reaction products yield information on the confinement of 0.05 MeVd, 0.2 MeV ^3He, 0.8 MeV ^3He, and 1 MeV t plasma ions. These confinement properties are important in evaluating magnetic confinement schemes, auxiliary heating effectiveness, and processes which might influence the 3 MeV alphas in an ignited device.

Confinement Theory

Since the tokamak is a closed magnetic field system in the form of a torus, ions experience vertical curvature and grad B drifts as they travel along the curved magnetic field lines. Ions are prevented from drifting out of the machine by the poloidal magnetic field produced by the toroidal plasma current (Fig. 3). This field makes the net field (and particle orbits) helical. Thus, ions spend time both above and below the horizontal midplane so that the vertical drift causes displacement towards the plasma edge half the time and displacement towards the plasma center the other half; the net radial displacement in a full orbit is zero. Since the vertical drifts depend upon particle energy, more plasma current is required to confine more energetic ions. The energetic ions can exist at any pitch angle (direction of the particle velocity with respect to the magnetic field) and orbits for a particular pitch angle (ratio of v-toroidal/v) are shown in Fig. 3. For a specific device with a fixed maximum magnetic field and plasma current, some ions (such as 0.05 MeV d and 0.2 MeV ^3He ions in PLT) will be confined for all pitch angles, more energetic ions will be confined for some pitch angles and will intersect the plasma walls for other pitch angles (0.8 MeV ^3He and 1 MeV t ions are on such partially confined orbits in PLT), and still more energetic ions will leave the plasma for all pitch angles (15 MeV protons in PLT are all on unconfined orbits).

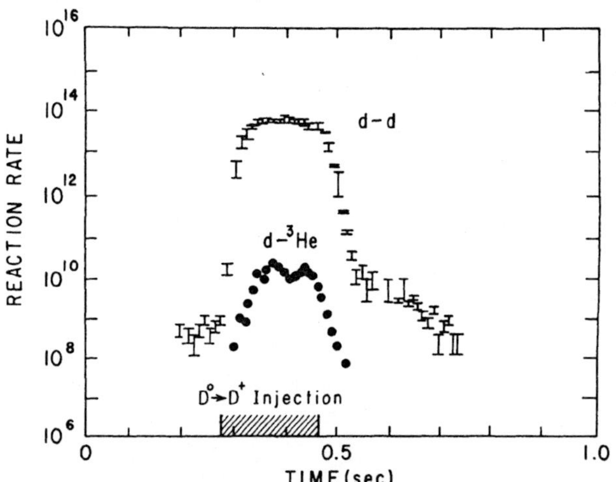

Fig. 4. Time evolution of the d-d fusion reaction rate measured by the 2.5 MeV neutron emission as detected by an array of moderated BF_3 proportional counters. The neutron emission increases five orders of magnitude when the 50 keV deuterium neutral beams (4 MW) are applied and decreases exponentially when the beams are turned off. The 15 MeV proton emission is caused by the burnup of the 0.8 MeV ^3He fusion product and is detected by a surface barrier detector located at the plasma edge.

Unconfined Orbits

All of the 15 MeV protons and many of the 3 MeV protons leave the plasma on classical orbits which intersect the walls (Fig. 3). These escaping particles are detected using surface barrier detectors located outside the plasma so that energy and time resolution of the $d(d,p)t$ and $d(^3He,p)\alpha$ reactions can be made. Spectral measurements of the 3 MeV protons [12] show negligible energy losses prior to detection, indicating that detected particles escape the plasma

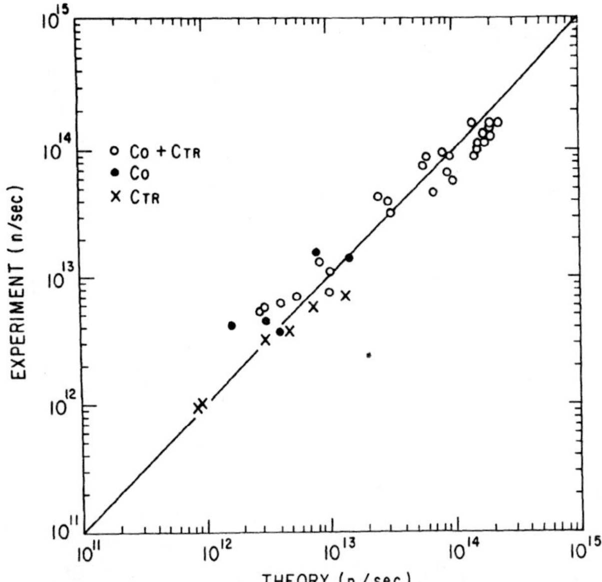

Fig. 5. Comparison of the experimental and theoretically predicted 2.5 MeV neutron emission from the d(d,n)^{3}He fusion reaction during neutral beam injection on PLT for a wide range of plasma and neutral beam conditions. The neutron calibration is good to about 30%. The ability to calculate the neutron yield depends upon experimental input in the form of temperature and density profiles which make the calculation good to about a factor of two.

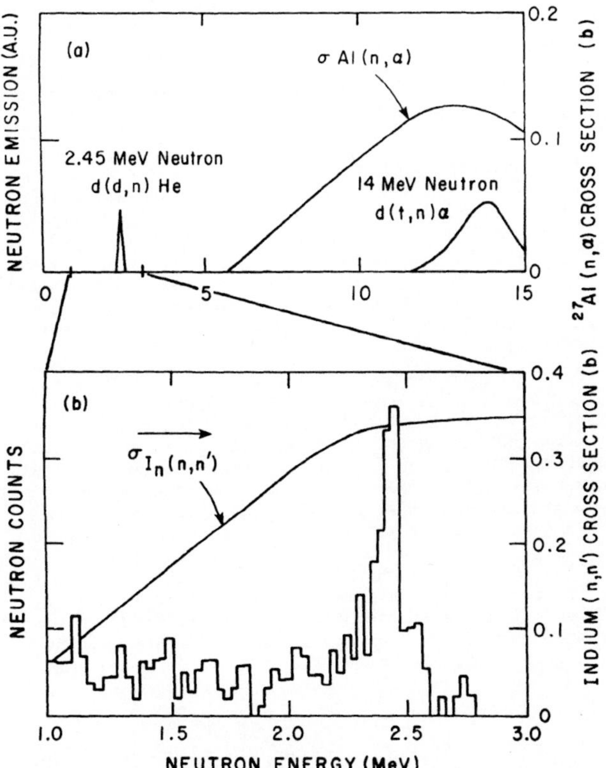

Fig. 6. Threshold reactions used for calibrating the PLT neutron emissions. The In(n,n') reaction eliminates many of the scattered neutrons and is shown in relation to the measured 2.5 MeV neutron spectrum from PLT [14,15]. The Al(n,α) reaction is sensitive to 14 MeV neutrons and insensitive to 2.5 MeV neutrons. In general (n,2n) reactions are not useful due to complications from (γ,n) reactions produced by runaway electrons [16].

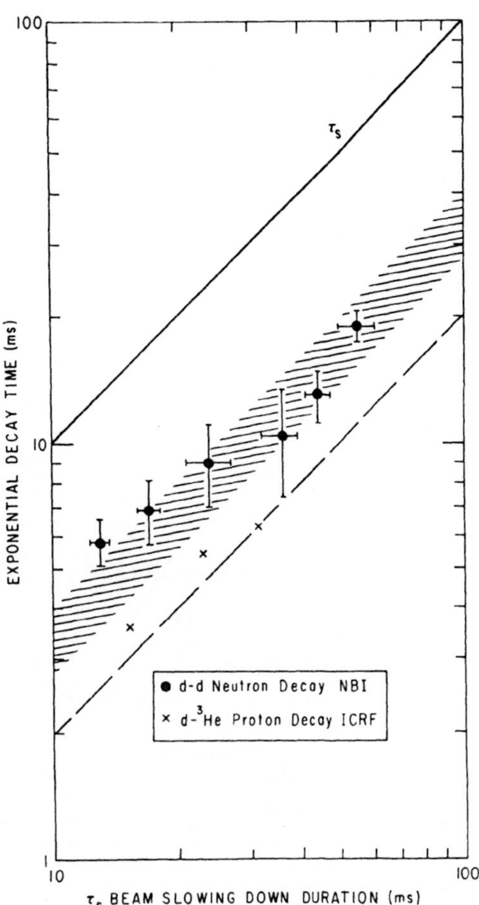

Fig. 7. The measured exponential reaction rate decay time (upon completion of the auxiliary heating) as a function of the calculated total slowing down duration. For d-d reactions from neutral beam injection (•), that time is expected to be about 1/3 of the slowing-down duration (cross-section weighted slowing-down time is shown as the shaded region). For d-^{3}He reactions from ICRF (x), that time is expected to be about 1/5 of the slowing-down duration (dashed line is the d-^{3}He cross-section weighted slowing-down time).

promptly on classical orbits. The poloidal distribution of the escaping 15 MeV protons was determined by the ^{56}Fe(p,n)^{56} Co vessel activation [13] and was concentrated below and outside the torus minor axis in agreement with classical calculations.

Confined Orbits

The confinement of the 0.05 MeV deuterons is examined by measurement of the time evolution of the 2.5 MeV neutron emission from the d(d,n)^{3}He fusion reaction [5] (Fig. 4). The magnitude agrees with that calculated for injected neutral beam ions which slow down by collision with the plasma electrons (Fig. 5). The principal experimental difficulty is the absolute calibration of the 2.5 MeV neutron emission. On PLT, the calibration has been accomplished using the ^{115}In(n,n')^{115m}In reaction [14]. Threshold reactions are especially useful since neutrons which scatter off the PLT machine structure and have degraded energies are of reduced importance (Fig. 6). After the neutral beams are turned off (Fig. 4), the neutron emission decays exponentially at a rate consistent with the cross-section weighted slowing-down time (Fig. 7).

The 0.2 MeV ^{3}He ion behavior is determined by measurement of the escaping 15 MeV proton (Fig. 3). The Doppler broadening of the 15 MeV protons (Fig. 8) indicates that the ^{3}He ions are in the range of 0.2 -

0.4 MeV. The ICRF accelerates the ^3He ions about 100 eV during each transit through the cyclotron resonance layer. (Because the tokamak toroidal field varies inversely with major radius, the ^3He cyclotron motion and the pump waves only resonate in a narrow

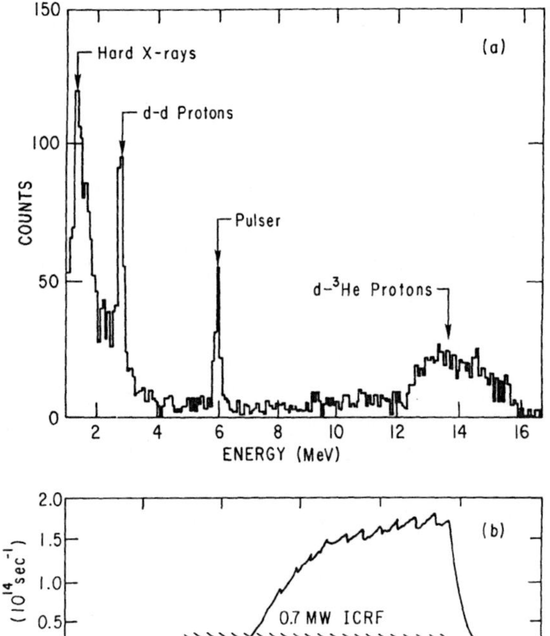

Fig. 8a). Proton spectrum during ICRF heating indicating a narrow 3 MeV proton signal, and a 2 MeV wide 15 MeV proton signal. Direct heating of the ^3He ions to 0.2 - 0.4 MeV by the ICRF accounts for the large Doppler width of the 15 MeV proton emission. In contrast, the plasma deuterons are heated by collisions with energetic ^3He ions to average energies of several keV and so produce a relatively narrow 3 MeV proton spectral line.

Fig. 8b). Time evolution of the d-^3He reaction indicating a slow rise to peak and an exponential decay after the rf is turned off. The 10% oscillations are caused by a plasma instability called an "internal disruption".

layer that passes vertically through the center of the machine). The delay between the application of the rf and the peak in the d-^3He rate is the time required for ^3He ions (gaining 100 eV per transit) to accelerate to 0.2 MeV and indicates that these ions must be confined about 50 msec. After the rf is turned off, the d-^3He rate decays exponentially with a time constant consistent with the cross-section weighted slowing-down time.

Generally, then, the lifetime of the energetic 0.05 MeV d and 0.2 MeV ^3He ions is determined primarily by collisional cooling with the background plasma. Occasionally, however, plasma instabilities can cause sudden losses of the energetic ions that induce modulations in the fusion emission rates. An example can be seen in Fig. 8b, with the 10% drops in the d-^3He rate occuring at "internal disruptions" (events which apparently redistribute but leave the ^3He ions contained inside the plasma). Another, more serious event is shown in Fig. 9, where with intense neutral beam injection at low toroidal magnetic fields, the d-d emission rate drops by 30% in 0.5 msec every 3 msec, indicating a large loss of energetic ions from the plasma [17]. Detailed studies of such events necessitate 2.5 MeV neutron detectors with high efficiency, spatial resolution, and MHz frequency response.

Partially Confined Orbits

The confinement of 0.8 MeV ^3He and 1 MeV t ions produced by the d-d reactions can be monitored by their subsequent burnup through the d-^3He and d-t reactions (Figs. 4 and 10) [18,19] since their burnup is about 10^4 times greater inside the plasma than in the vessel walls. Again, the principal experimental difficulty is the relative calibration of the d-^3He or d-t rate to the d-d rate.

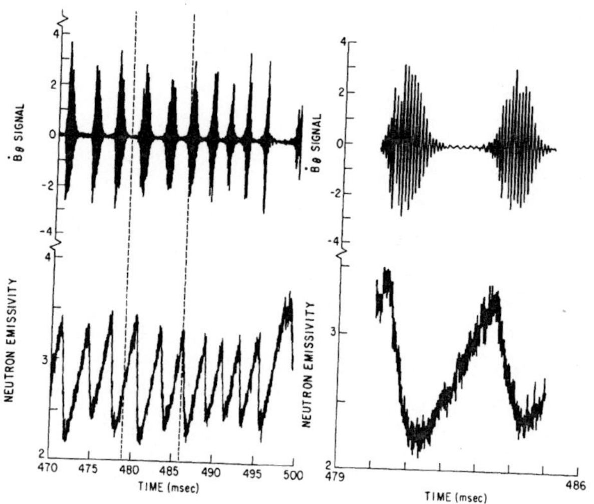

Fig. 9. Time evolution of the d-d neutron emission through a plasma instability which occurs at low toroidal magnetic fields during intense neutral beam injection. The 30% drop in neutron emission is an indication that many recently injected beam ions are lost from the plasma. Measurements of poloidal magnetic field perturbations at the edge of the plasma indicates a coincident 15 kHz oscillation that is sometimes also seen on the neutron emission.

The confinement of the 0.8 MeV ^3He ions can be nearly classical when the plasma has a high toroidal magnetic field. The burnup decreases when the plasma current decreases (Fig. 11), indicating that a major loss is by the classical intersection of orbits with the walls. Furthermore, there is a 10-20 msec delay between the burnup of the ^3He ion and its product (Fig. 12), which about equals the cross-section weighted ^3He slowing-down time.

On the other hand, the confinement of 1 MeV tritons and 0.8 MeV ^3He ions at lower toroidal magnetic fields (Fig. 13) is less than expected for classical confinement and burnup. This effect could be due to plasma instabilities (as in Fig. 9) or to incorrect calculations of classical processes which often assume small particle gyroradii. Development of 14 MeV neutron detectors [20] featuring > 10^4 cps time resolution and good rejection of hard x rays, epithermal neutrons, and 2.5 MeV neutrons (1 14 MeV neutron per 10^4 2.5 MeV neutrons) would help resolve the issue. Spectral resolution is also desirable. At present, the best measurements make use of the Al(n,α) threshold reaction (Fig. 6), which yield a good dt/dd relative calibration but no time resolution.

Acknowledgements

The authors thank the PLT and PDX experimental teams for their support. This work is supported by DoE Contract # DE-AC02-76-CHO-3073.

(1) Present Address:

Los Alamos National Laboratory
CTR 3
Los Alamos, New Mexico 87545

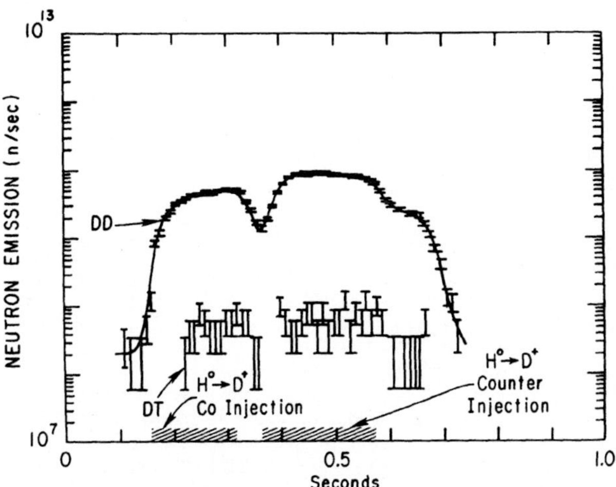

Fig. 10. 2.5 MeV neutron emission from the d(d,n)^{3}He fusion reaction during neutral beam injection on PLT as measured by moderated BF_3 counters. About one half of the 1 MeV tritons from the d(d,p)t reactions are confined in the plasma and < 1% burnup as they slow down through the maximum of the d-t cross-section. The 14 MeV neutron emission from the d-t reaction was measured with an NE213 scintillator.

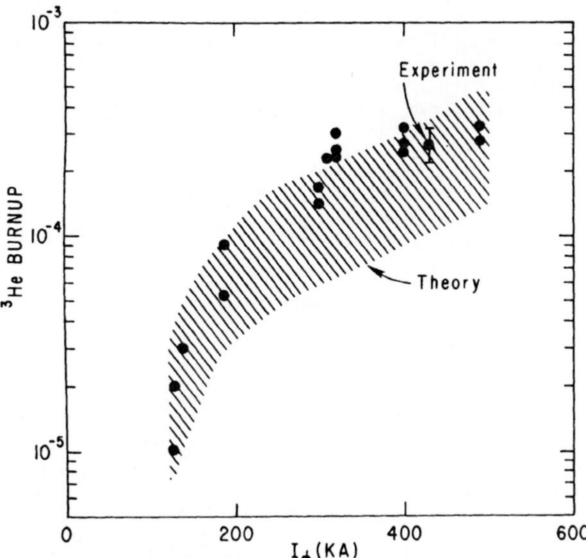

Fig. 11. Variation in the 0.8 MeV ^{3}He ion burnup with plasma current indicating agreement with a classical picture of energetic ion confinement. The confinement depends strongly upon plasma current since the energy dependent vertical particle drifts are compensated in a tokamak by the action of the poloidal magnetic field induced by the plasma current.

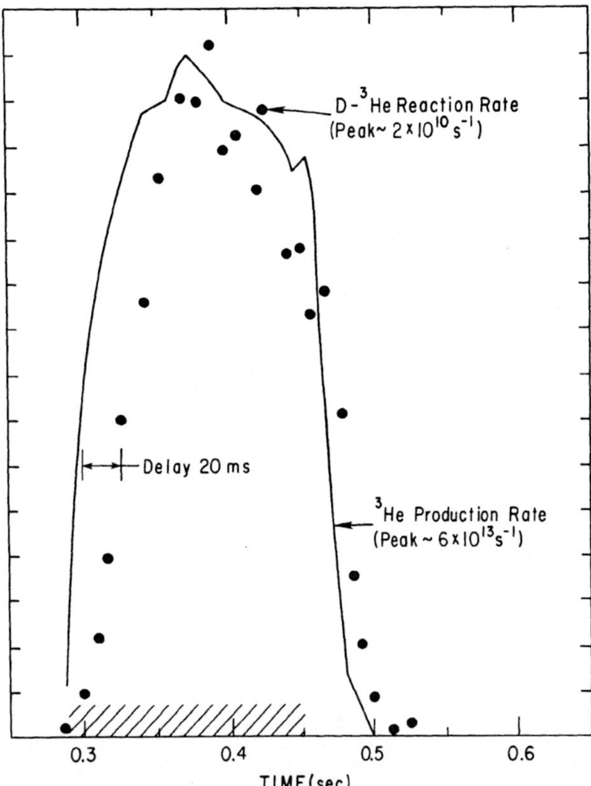

Fig. 12. Time evolution of the 0.8 MeV ^{3}He birth rate (solid line) as measured by detection of the 2.5 MeV neutrons produced in the d(d,n)^{3}He reaction during deuterium neutral beam injection (shaded region). The ^{3}He burnup rate (●) measured by detection of the 15 MeV proton emission is delayed with respect to the birth rate because ^{3}He ions take 10-20 msec to slow down to the maximum of the d-^{3}He cross section.

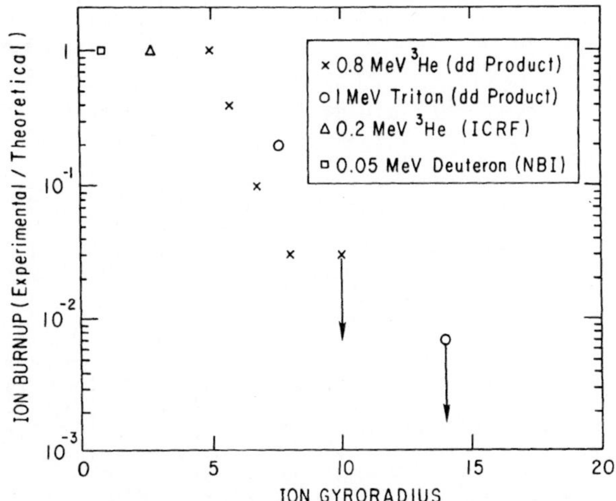

Fig. 13. Comparison of the experimental burnup to theoretical calculations for the (□) 0.05 MeV d, (Δ) 0.2 MeV ^{3}He (X) 0.8 MeV ^{3}He, and (o) 1 MeV t as a function of ion gyroradius. The scan in gyroradius is obtained by varying the toroidal magnetic field.

References

1) L.A. Artsimovich, et al., Sov. Phys. JETP 34, 306 (1972)
2) W. Stodiek, et al., "Plasma Phys. and Contr. Nucl. Fus. Res." Vol. 1 465 (1971) Madison
3) H. Eubank, "Course on Plasma Diagnostics and Data Acquisition Systems" (Varenna, 1975) 377
4) L. Berry, et al., ORNL-TM-5057 (1975)
5) J.D. Strachan, et al., Nucl. Fus. 21, 67 (1981)
6) K. Bol, et al., Proc. IAEA Tech. Comm. Meeting on Divertors and Imp. Control (Garching, 1981) 20
7) R.E. Chrien and J.D. Strachan - to be published
8) H. Eubank, et al., "Plasma Phys. and Contr. Nucl. Fusion Res." (Proc. 7th Int. Conf. Insbruk, 1978)
9) J. Hosea, et al., "Plasma Phys. and Contr. Nucl. Fus. Res." (Proc. 8th Int. Conf., Brussels, (1980)
10) L.A. Artsimovich, Nucl. Fus. 12, 215 (1972)
11) H.P. Furth, Nucl. Fus. 15, 487 (1975)
12) R.E. Chrien and J.D. Strachan, PPPL-1931 (1982)
13) R.E. Chrien and al., Phys. Rev. Lett. 46, 535 (1981)
14) G. Zankl, et al., Nucl. Instr. and Meth. 185, 321 (1981)
15) J.D. Strachan, et al., Nature 279, 626 (1979)
16) J.D. Strachan et al., Nucl. Fus. 17, 140 (1977)
17) M. Bell, et al., Proc. of the European Conf. on Plasma Phys. and Contr. Fus. (Moscow, 1981)
18) P. Colestock, et al., Phys. Rev. Lett. 43, 768 (1979)
19) W.W. Heidbrink, et al., PPPL-1937 (1982)
20) R.E. Chrien and J.D. Strachan, Rev. Sci. Instr. 51, 1638 (1980)

D(t,α)n REACTION CROSS SECTIONS AT LOW ENERGY

Nelson Jarmie, Ronald E. Brown, and R. A. Hardekopf
Los Alamos National Laboratory
Los Alamos, NM 87545 USA

Abstract

Because of discrepancies in previous measurements of the D(t,α)n reactions at low energies, we have developed a low-energy fusion cross-section facility (LEFCS) to remeasure the D(t,α) and other few-nucleon reactions. The experimental equipment features a windowless cryogenic target, a precision beam-intensity calorimenter, a 10 to 120 keV accelerator producing negative tritium ions, an accurate target gas flow and temperature system, and a tritium gas-handling system. The target density and geometry are calibrated with the D(p,p)D reaction using 10-MeV protons from the Los Alamos Van-de-Graaff. Presented are 16 D(t,α)n total cross sections from 8.3 to 78.1 keV equivalent deuteron energy. The relative error is 0.5% and the absolute error is 1.4% for most energies. For the lowest energy both uncertainties rise to about 4.7%. We will also attempt to measure the $D(t,\gamma)^5He$ cross section. We have made preliminary background studies with the LEFCS target and a bismuth germanate counter with encouraging results.

*Work performed under the auspices of the U.S. Department of Energy, Contract W7405-ENG-36.

I. INTRODUCTION

The D(t,α)n reaction has long been of interest to physicists and engineers. At low energies the reaction is dominated by a strong sharp isolated $3/2^+$ threshold resonance in the ^5He compound system at a deuteron equivalent bombarding energy of 107 keV. The peak cross section is 5 barns even though energy of the system is well below the Coulomb barrier. With a 17.6 MeV Q-value, this reaction has been widely used as a source of high energy neutrons, and will certainly be the reaction used to fuel the first magnetic and inertial fusion reactors that are expected to reach energetic break-even and eventually to provide significant amounts of energy for commercial use. The actual plasma operating range in these reactors is expected to be from 1 to 30 keV, which just corresponds[1] to a triton bombarding energy range of 10 to 120 keV that we study in this report.

Design studies involving the D + T reaction have used cross sections measured 30 years ago by Arnold et al.[2] and Conner et al.[3] who claim accuracies of a few percent. Recent awareness of Russian experiments by Katsaurov[4] and Kobsev,[5] and of similar experiments done with essentially the same equipment as Arnold and Conner led to the conclusion[1] that there may be systematic errors as large as 50% in the T(d,α)n cross sections. These discrepancies have motivated the present study.

Accurate measurement of the bombarding energy is difficult at low energies, and this is suspected to be the main cause of the cross-section discrepancies. Because the cross section is falling in steep exponential, slight energy shifts can produce a large error in the cross-section magnitude. Using the cross-section parameterization as a function of energy from Refs. 6 and 7, one can calculate, for example, that at 20 keV, a shift dE of only 0.5 keV in the bombarding energy will produce a 10% change in the cross section. At the lower energies, the fractional cross-section error varies as $dE/E^{3/2}$, so that the effect increases as the energy decreases. Accordingly, great care has been taken in this experiment to insure an accurate knowledge of the target-center beam energy.

We report in this paper the measurement of 16 values of the D(t,α)n total cross section from 8.3 to 78 keV equivalent-deuteron bombarding energy with relative errors less than 1% and absolute errors less than 1.5% except for the lowest energies. The bombarding energies have uncertainties in the range of 5 to 15 eV. The measurements were made at the Los Alamos Low Energy Fusion Cross Section (LEFCS) facility.

In Section II we will describe the details of the experimental equipment. In Section III, we discuss the D(t,α) experimental procedure, explain backgrounds and various corrections, and show examples of the raw data. Section IV describes various error contributions. In Section V, we present graphs of the final data, compare our results with previous work, and mention the possibility of measuring other reactions at low energy with our facility.

Descriptions about the progress of this experiment were given in Knoxville, Tennessee, in 1979[8] and in Denton, Texas, in 1980.[9]

II EXPERIMENTAL EQUIPMENT

A. Overview

The design of the experiment has been dominated by the need for accurate knowledge of the bombarding energy, detection of as low a counting rate as possible, and the elimination of as many sources of systematic error as possible. A schematic of the equipment is shown in Fig. 1. A cryogenic windowless deuterium-gas target was chosen to avoid energy-loss uncertainty in exit foils. A 10.5-K temperature of the target gas gives the maximum deuterium density compatible with an appropriate energy loss in the target. The target gas, after leaving the target, is pumped (frozen) by nearby 4-K surfaces. Accurate control of the target density is maintained by precise control of the continous flow of deuterium (at 5cc/min.) and temperature at the reaction volume. Alpha particles from the reaction come out of the cryogenic target region and are detected by standard solid-state surface-barrier detectors. The target-density exit-geometry product is calibrated using a 10.04-MeV proton beam from the Los Alamos Van de Graaff; the necessary D(p,p)D accurate cross sections were measured by us in a separate experiment using the Los Alamos 30" precision scattering chamber.

The 10 to 120 keV triton beam undergoes a large amount of charge exchange in passing through the target, so that it is necessary to use a calorimeter to measure the beam intensity with an accuracy of less than 0.5%. An extensive tritium handling system is also necessary to be able to use tritium in the ion source. We use a negative ion beam to eliminate unwanted molecular species and to reduce problems associated with slit-edge scattering. We typically used a very stable beam of 1-5 μA with 99% transmission through a 2.4-mm target aperture. The critical beam energy was measured with a high-voltage registor stack that had a calibration traceable to a primary voltage standard at the National Bureau of Standards.

K. H. Böckhoff (ed.), Nuclear Data for Science and Technology, 318–325.

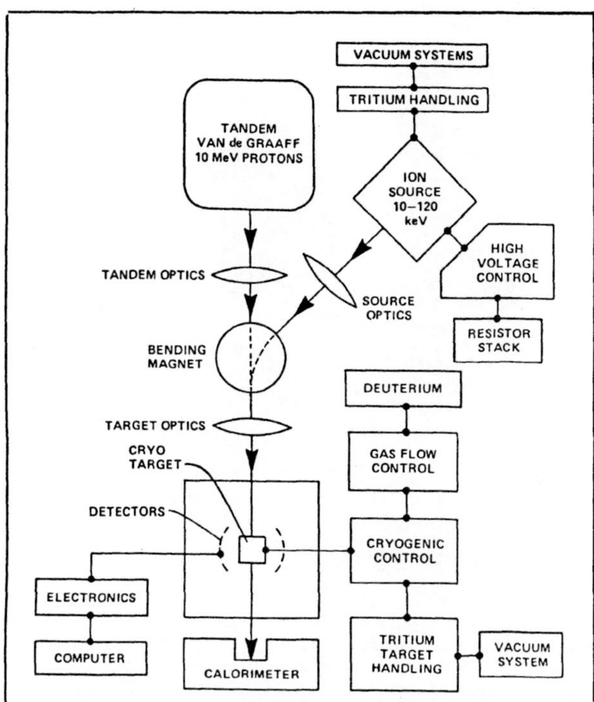

Fig. 1. Schematic of the Los Alamos Low Energy Fusion Cross Section facility (LEFCS).

An experiment measuring the time of flight of neutral beam pulses created by photodetachment of the negative beam by a pulsed Nd:YAG laser is in progress[9-10]; this will help to confirm the absolute energy and to study beam energy losses and widths.

Of the many souces of error considered, the relative error of the D(t,α)n cross section (averaging about 0.6%) was dominated by fluctuations in the target density (flow and temperature) counting statistics; and the repeatability of the calorimeter measurement; the latter two error sources were important especially at low energies. The absolute scale error was largely due to the uncertainty of the D(p,p)D calibration cross section (1.2%).

B. Low Energy Accelerator Beam Optics and Associated Equipment

The 10 to 120 keV dual-polarity ion source and injector system[9] was built to Los Alamos specifications by General Ionex Corporation. The beam originates in a standard duoplasmatron and 30-kV extraction lens, which can be operated to extract either positive or negative ions used as we did in this experiment. Although positive beams of greater intensity can be produced, the negative beam is preferred for tritium because neither the crossed-field analyzer nor the inflection magnet can separate T^+ ions from $^3H^+$, which might also be extracted from a positive-ion duoplasmatron.

The accelerating tube is a low-gradient multi-electrode structure providing low optical-strength coupling to ground potential. The final beam energy is determined by a highly stable 0 to 120 kV power supply between the duoplasmatron and ground. This system allows the beam energy to be altered without affecting the extraction mechanism, and thus the beam current available and the terminal focusing are essentially independent of beam energy over a broad range. The voltage of the source table was monitored by a precision 10^5-to-1 resistor stack, a Spellman Model HVD-200, 200-kV voltage divider. The beam, which is diverging as it exits the accelerator, is focused by an einzel lens to a minimum beam waist at a position beyond the exit of a 45° sector double-focusing analyzing magnet. The magnet bends the ion beam onto the target-chamber axis, which is also the axis of the beam line from the tandem Van de Graaff accelerator.

The beam size is delimited by a set of 4-way slits at the waist position and then focused into the target chamber by another einzel lens. Negative hydrogen beams of 50-μA intensity and 15-mrad divergence have been obtained through a 3-mm-diam aperture. The intensities are less for deuterium and tritium beams and for energies below 20 keV. Beam intensity is also lost constraining the beam emittance so that high transmission is obtained through the target, which is essentially a 2.4 mm-diam tube 7 cm long. The good transmission is necessary for a stable temperature of the 10.5-K target and for an absolute determination of the cross section. Transmissions of 100 or better were obtained for the low-energy beams and a transmission of 1000 was typical for the 10-MeV Van de Graaff beam. Typical beam currents were, for the low energy runs, from 1 to 5 μA and for the 10.04-MeV calibration runs, from 1 to 2 μA. Most of the time the beam was very stable with no significant degradation of either intensity or transmission over several hours. Contributions to the energy width of the beam come from ripple and instability of the source high voltage, plasma effects in the duoplasmatron, and energy straggling in the target gas. The total spread (FWHM) varies from 17 eV at E_t = 12.5 keV to 35 eV at 117 keV, both values with an uncertainty of 4 eV.

The tritium gas used in the ion source is recirculated with a turbopump and mercury booster diffusion pump. Details of the recirculating and handling are shown in Fig. 2 and given in Ref. 11. When running, the amount of tritium in circulation is about 10 to 20 standard cc.

C. Target, Detectors, and Target Chamber

Figure 3 shows a schematic drawing of the target and detector system. The chamber has an inside diameter of 38.7 cm. A gas target without windows in the beam entrance and exit is used to avoid uncertainties in the reaction energy, which is felt to have caused much of the disagreement among the earlier experiments. Indeed, to avoid a significant energy loss, the central target density was only 1.4 x 10^{16} atoms cm^{-3}, which corresponds to a pressure of 7.6 millitorr. Deuterium gas, precooled to 75 K, flows slowly at a rate of 5 standard $cm^3 min^{-1}$ into a copper cooling pot kept at 10.5 K. The cooled gas enters the reaction region and flows out of the target to be pumped by 4-K surfaces

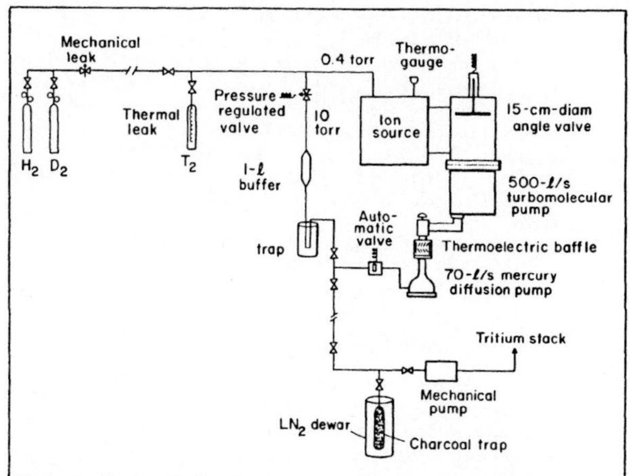

Fig. 2. Ion source gas handling-system. In operation, the tritium recirculates, the system is isolated from the charcoal trap and mechanical pump and operates at the accelerating voltage. Not shown is a LN_2 trap between the mercury pump and the turbopump.

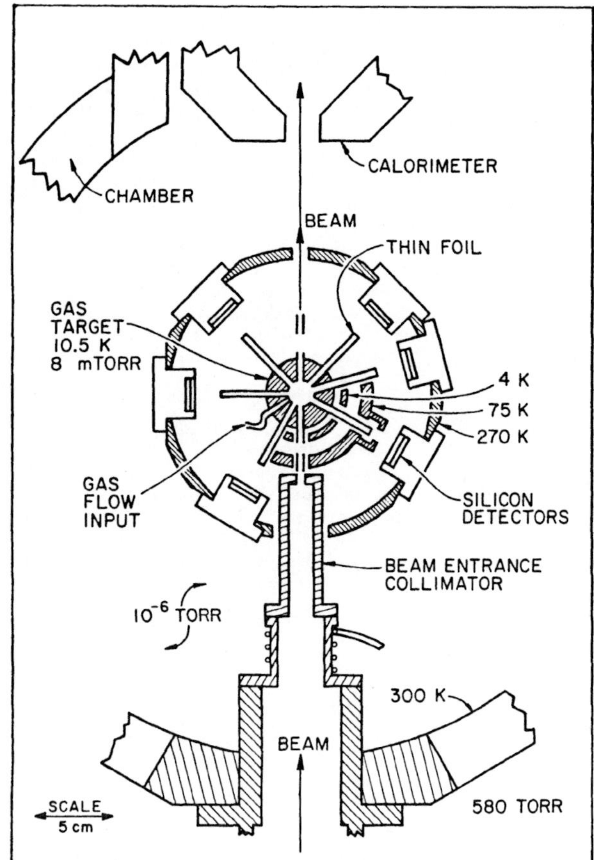

Fig. 3. Gas target and detector system. For clarity 4- and 75-K cylindrical shields are shown in one section only. The entrance collimator and the target are electrically isolated. Several secondary electron suppressors are not shown.

around the target. About 98% of the exit gas is frozen out, the rest is pumped by a 500 ℓ turbopump. The reaction products, in this case 4-MeV alpha particles, leave via six tubes capped by thin 30 $\mu g/cm^2$ stretched polypropylene foils[12] and are detected by 50-micron-thick silicon surface-barrier detectors. The exit geometry is determined by apertures at the foil end of the tube and where the tube meets the reaction area, and has a typical geometry factor[13] of 40 10^{-5} cm sr. The angular acceptance of one of the collimator tubes is 3.2° FWHM. Care must be taken to insure the target chamber is completely leak tight; a small leak can deposit enough solid oxygen and nitrogen on the thin foil surfaces to seriously degrade the energy of the alpha particles.

The 10.5-K target gas was chosen to be near the (9.5 K) freezing point of deuterium at the pressure used. For a given target configuration, cold target gas increases the density and decreases density fluctuations. Figure 4 shows the target pot and the cryogenic shield surfaces. The target system is supported by a cryostat containing a liquid nitrogen reservoir and a 5ℓ-liquid helium tank. The target pot is connected to the liquid helium tank supported by a stainless steel temperature gradient rod, to which is attached heating resistors that regulate the target temperature.

In addition to the 2.4-mm-diam beam entrance and exit beam collimators, detector collimators are placed at angles of 45°L, 45°R, 75°R, 90°L, 120°R, and 150°L, where L and R stand for left and right of the beam axis as indicated in Fig. 3. The range of collimator angles and the two symmetric 45° ports were useful in diagnosing target operation and

Fig. 4. View of the target pot and exit tubes. The beam enters the hole in the front of the pot, which normally is inside the 4 K shield in the middle, which in turn fits into the 75 K shield on the left.

increasing the data-taking rate because of the isotropy (in the center-of-mass system) of the D(t,α)n reaction . The fusion reaction under study has a large positive Q-value (17.6 MeV), which allows thin foils to be used over the detection ports to reduce gas leakage without appreciably slowing down the reaction products (∿4 MeV α). Silicon surface-barrier detectors are mounted in a support ring that is cooled by a thermoelectric device. Both ΔE and E detectors can be used in coincidence to reduce backgrounds and aid in particle identification when necessary (we used the coincidence and mass identification for the high-energy calibration experiment). Data-taking time periods were measured with a BNC Model 7010 digital delay generator. An on-line computer is used to store the spectra and record most of the experimental data and parameters.

D. Target Density Calibration

A vital element of our technique is the ability to measure the density of scattering centers as seen by the different detectors. To do this we used a 10.04-MeV proton beam from the Los Alamos Tandem Van de Graaff. The protons scatter off the deuterium target while the target is in the identical state used in the low-energy runs. The absolute D(p,p)D cross section was measured at 10.04 MeV by Kocher and Clegg[14] at Wisconsin with 0.5 to 2% uncertainties; so 10.04 MeV was chosen for the calibration.

As our work proceeded, it became clear that we needed to verify Kocher & Clegg's work, and if possible, improve the accuracy. In a separate experiment using the Los Alamos 30-inch precision scattering chamber,[15] we remeasured the D(p,p)D cross sections. This study will be described in a more detailed paper to be published later. The results are shown in Fig. 5, which plots the ratio of the Wisconson work to our measurements and also the unpublished accurate measurements made at the University of Minnesota.[16] Absolute error bars are shown and are dominated by the quoted errors of the Wisconsin data. The relative error of our D(p,p)D cross sections at 10.04 MeV was about 0.4% with the main contributions from the statistical counting error (0.3%), and the background subtraction (0.25%). The scale error was about 0.7% with the main contributions from gas purity (0.4%), absolute error of the p+p elastic scattering calibrating cross section[15] (0.49%), and an external error due to fluctuations of the measured values from several runs (0.5%). The total absolute error was 1.0%.

The above D(p,p)D cross sections were then used in a high energy (10.04 MeV) calibration of the LEFCS cryogenic target. The result is the value for each angle of the product Gn of the mean particle density n and the geometry factor G. A study of the results

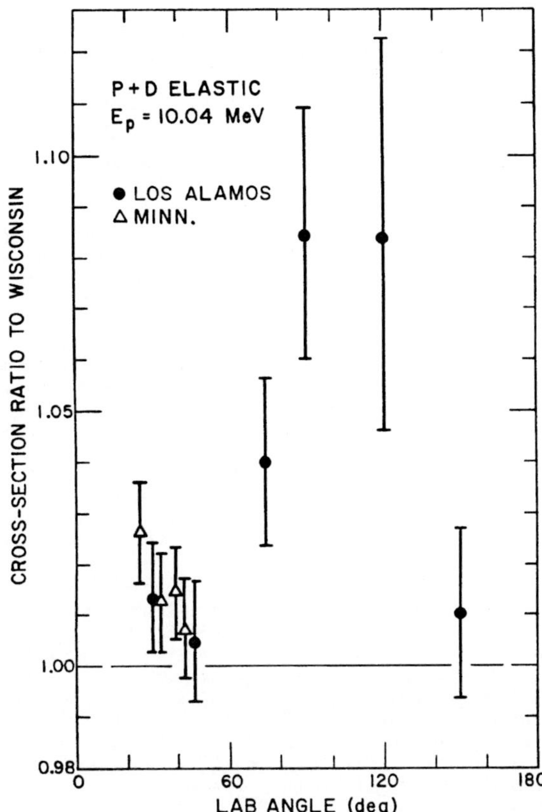

Fig. 5. Comparison of D(p,p)D cross sections at 10.04 MeV between the work of Kocher and Clegg (Ref. 14) at Wisconsin, and those of Los Alamos (present experiment) and Minnesota (Ref. 16).

at each angle showed, somewhat unexpectedly, that Gn remained constant over angle well within the errors of the measurement. We had suspected that the highest and lowest angles might have a lower Gn than the central angles, since we reasoned that the detectors of the extreme angles observe a target region closer to the beam tubes where the target gas flows out. An angle-averaged Gn (about 58 x 10^{11} cm^{-2} sr) was then used in the D(t,α)n measurement. The total error in Gn, 1.2%, was the quadratic sum of a relative error of 0.5% and a scale error of 0.8%.

The slopes of Gn versus flow and temperature were also measured in the 10.04 MeV experiment, so that corrections for small excursions in the operating region could be made.

E. Target Gas Flow and Temperature Measurement

The target temperature (T ≃ 10.5K) was measured with a germanium resistor attached to the target within a few mm of the reaction region. The resistance R of the germanium was measured by a standard four-wire system, with a 10μA constant current source and a digital voltmeter whose output was read by the on-line computer every 12 seconds. With a rate dR/dT of 35 Ω K^{-1}, we found the measurement very sensitive and also very reproducible. The temperature was stable enough that only infrequent manual adjustment of the heating-resistor current was necessary. The flow of deuterium gas (5 cm^3 min^{-1}) was determined by adjustment of a fine needle valve fed by a buffer volume whose constant pressure was regulated by an automatic feedback control system. The analog output of a Hastings model TNALL 50P flowmeter was read by a digital voltmeter whose output, in turn, was read by the computer every 12 seconds. The flow was also very stable and reproducible. Values of both the flow and temperature usually changed less than a percent of their initial values over several hours. These measurements were among the least troublesome parts of the experiment. No attempt was made to determine accurate absolute values of the parameters, as only the reproducibility matters from run to run and between the calibration experiment and the D(t,α)n runs.

F. Calorimeter

Construction and operation of the beam calorimeter that determines the beam intensity were based on a Swiss design.[17] This will be reported in more detail in a later paper. The beam collecting cup was kept close to room temperature by a constant flow of heat to a cooler. This heat loss was matched by the heat output of a transistor attached to the beam cup. The transistor current value was converted to a digital output that was scaled and read by the computer. The heat deposited by the tritium beam was rapidly and accurately compensated by a decrease in the transistor current. The measurement then, consisted of an integration of the output count for a given time before and after the beam is on (baseline runs) and during the irradiation. The calorimeter in addition automatically compensated for errors due to a change in room temperature.

The beam cup is also a faraday cup for calibration purposes. We calibrated the calorimeter by two methods: 1) by comparison with the integrated current of a beam of 3 MeV protons (which has no significant charge exchange), and 2) using the heat output of a precision resistor embedded in the beam cup. The results are shown in Fig. 6. In the beam power range from 20 to 800 mW the two methods agree very well and give a calibration constant $\mathcal{R}$, of 97.80 ± 0.08% μJ/output count. The limit at high power is just the cooling power limit of the Peltier cooler. At low powers the measurement becomes the difference of two large similar numbers; measurements can only be made during periods of a very steady beam-off baseline. A practical lower limit for our system is 5 to 10 mW. It must be remembered that the beam particle flux is determined from the ratio of the calorimeter output to the absolute beam energy so that the accuracy of the measured particle flux depends also on the beam-energy accuracy.

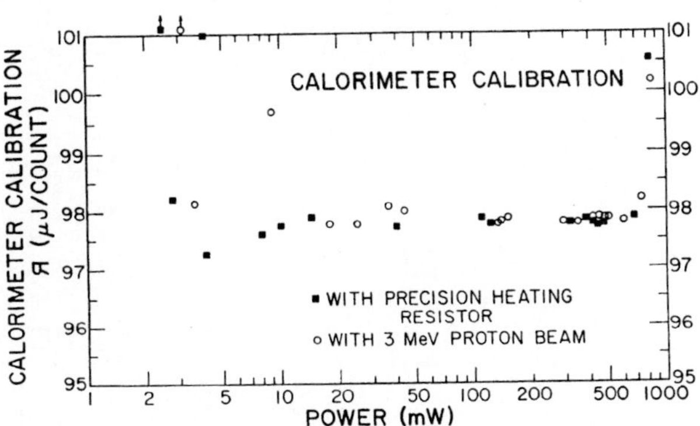

Fig. 6. Chart of the calorimeter calibration, $\mathcal{R}$, vs the power deposited in the calorimeter.

III. EXPERIMENTAL PROCEDURE, DATA SPECTRA

The procedure for the D(t,α)n experimental runs begins several days ahead, turning the power on critical electrical components: flow system, calorimeter, pulse electronics, CAMAC electronics, and target temperature. On run day after filling the liquid helium tank and letting the system come to equilibrium (about 1 hr), the tritium beam is tuned through the target (for example -3 μA). If the transmission is better than 100 (less than 30 nA hitting the target) and the beam is steady, the flow of deuterium is let into the target. The charge

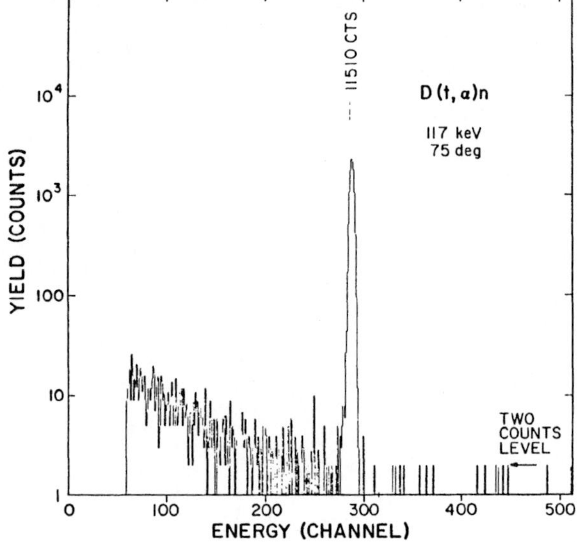

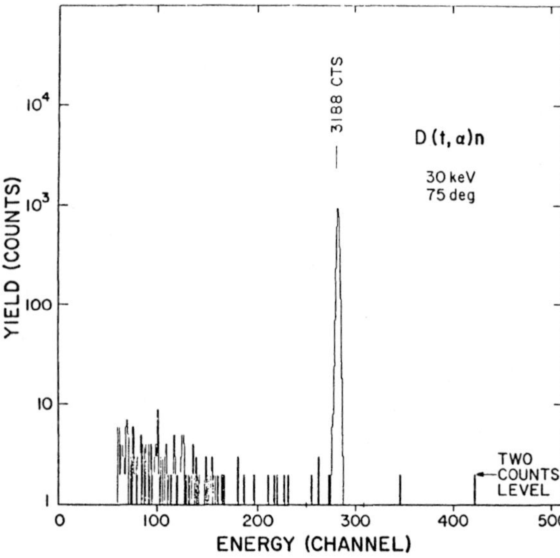

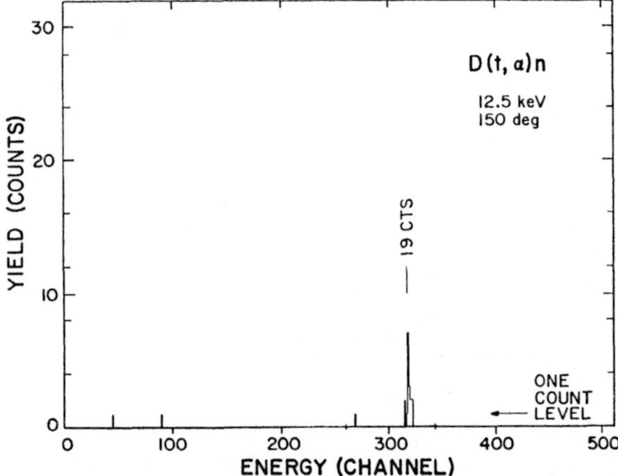

Fig. 7. Examples of raw data spectra at E_t = 12.5, 30, and 117 keV. Running times to produce these spectra were 64, 86, and 13 minutes respectively.

exchange of the beam in the target-gas is such that about -6 μA appears on the target body and +3 μA on the calorimeter faraday cup. The sum of the currents on the target body and faraday cup was close to the value determined by the calorimeter, but was always off by about 5% probably because of component of neutral beam entering the target. The "sum-of-currents" value was used to monitor the proper operation of the calorimeter. The beam is then stopped. When the beam energy, accelerating voltage calorimeter, flow, and target temperature are stable and steady and at their proper values, a beam-off calorimeter baseline run is taken for a few minutes. The beam is then put through, and after all values are again steady, data taking begins. The run is stopped after one hour or earlier if satisfactory counting statistics are achieved. Another baseline run is then taken. To go to another energy the flow would be stopped, the transmission would be checked for the current energy, and the whole process repeated. At least two runs were taken for each energy, and if any anomaly was observed in the energy, flow, temperature, or other parameter, the run was discarded. Every 24 hours the solid deuterium would be warmed and pumped away, and the liquid helium tank would be refilled.

Figure 7 shows representative examples of the raw data. Except for a few low channel counts at the highest energies, all the background was associated with the target gas struck by the beam, presumably D(t,n) neutrons. Even at the lowest energy, E_t=12.5 keV, the background was not a problem. We were limited at low energies by both the rapidly disappearing count rate, and by the increasing inaccuracy of the calorimeter. The limit at high energy was voltage breakdown of the source above 117 keV.

An overall view of the facility is shown in Fig. 8.

IV. ERRORS

Many sources of error were studied. In this discussion the "scale error" is defined such that (scale error)^2+ (relative error)^2= (absolute error)^2. Several errors were correlated: for example a shift in beam energy changes the energy loss in the target, the heat deposited in the calorimeter, and the count yield because of the steep slope of the cross section. Fortunately the correlated errors were small, usually less than 0.3%, and were arbitrarily folded into the scale error. There was no significant error contribution from neglecting the correlations. The following error sources and errors were studied as part of the relative error: 1) counting statistics, 2) beam intensity (calorimeter), 3) dead-time correction, 4) beam direction fluctuation, 5) fluctuations in the target density (flow and temperature), 6) stopping power (relative error), 7) beam energy fluctuations, and 8) beam lost hitting target. The following parameters were studied for contributions to the scale error: 9) absolute source voltage, 10) true beam angle and exit port angle, 11) calorimeter calibration, 12) resistor stack calibration, 13) stopping power (scale error), 14) source voltage shift (plasma), 15) target gas length, 16) calibration uncertainty (in Gn), and 17) the assumption of G constant with angle. The dominant errors are given in Table I, that shows a total scale error of 1.2% mostly from the Gn calibration, and a relative error, for most energies, of 0.5%. The relative error rises sharply at the lowest energies becoming 4.6% at an equivalent deuteron energy of 8.3 keV. Repeat measurements at one energy produced an external error in agreement with the internal total relative error. In the results below, please note that the error of the energy scale has been included in the error of the cross section.

V. RESULTS AND DISCUSSION

The D(t,α)n excitation function is given in Fig. 9, which shows the range of cross sections measured: from about 0.53 millibarns to 3.7 barns. The error bars are smaller than the plotting

Fig. 8. Overall view of the LEFCS facility. The source is on the right, with high voltage and beam optics controls next to it. R. Martinez is lowering the target and its cryostat into the chamber. Not easily visible are the tandem beam tube, tritium handling system and hood, and the flow control system.

TABLE I, SIGNIFICANT ERROR CONTRIBUTIONS, $T(D,N)^4He$

SCALE ERROR = 1.3%			FROM 9 SOURCES, LARGELY FROM THE D(P,P)D CALIBRATION; THE OTHER 8 SOURCES RARELY ABOVE 0.1%	
RELATIVE ERROR:			FROM 7 SOURCES; ONLY COUNTING STATISTICS (Y) AND BEAM INTENSITY (CALORIMETER, C_0) ABOVE 0.3%	
E_D (KEV)	Y (%)	C_0 (%)	TOTAL RELATIVE (%)	TOTAL ABSOLUTE (%)
8.3	4.5	0.5	4.6	4.8
10.0	2.3	0.5	2.3	2.7
13.3	1.7	0.5	1.7	2.1
16.7	0.7	0.4	0.9	1.5
20.0 – 12 PTS – 78.1	0.4	0.12	0.5	1.4%

symbols. For a more meaningful study and for comparison to other experiments Fig. 10 gives the data as the astrophysical S-factor (MeV-b):

$$\sigma(E) = S(E)\ (1/E)\ \exp\ (-1.0873/\ E^{-\frac{1}{2}}\ )$$

which extracts from the cross section expected dependencies on the wavelength of the beam and the Coulomb barrier penetration. The symbol E is the energy in MeV, and σ is the total cross section in barns. We also convert to equivalent laboratory deuteron energy for the abscissa. For clarity, comparison of our data with other results are shown in two graphs. The cross section now peaks at about 85 keV instead of 107 keV because of the unfolding of the steeply falling exponential penetration factor. Our data are the solid dots and have error bars that represent the absolute errors to compare with other work whose data also show absolute error bars. Our data show a provocative shape for the lowest four or

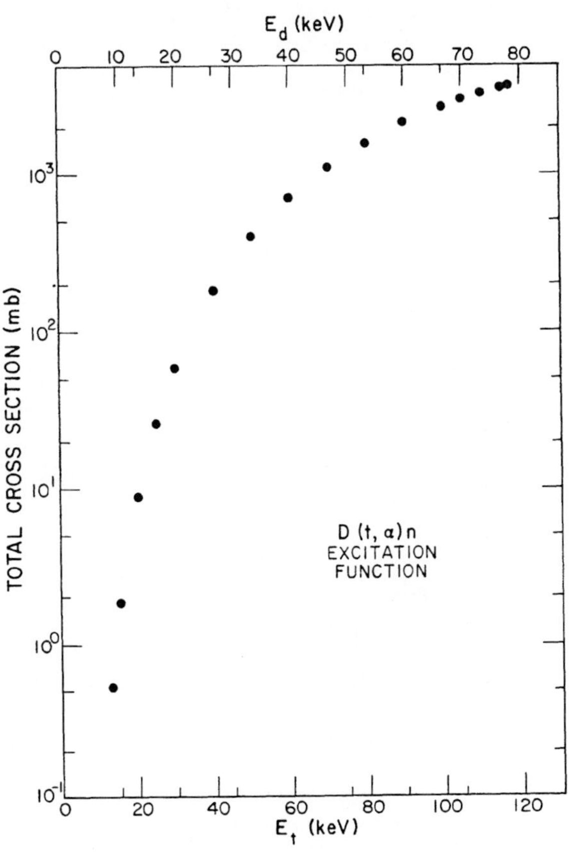

Fig. 9. Excitation function for the D(t,α)n total cross section. Absolute error bars are smaller than the plotting symbol. The unlabled marks on the cross section scale are "2" and "5". The equivalent deuteron bombarding energy is at the top of the figure.

five energies. A $\chi 2$ study of the fluctuations of these points about a reasonably smooth curve predicts one chance in 20 that the points would be distributed as they are. Since we know of no physical support for

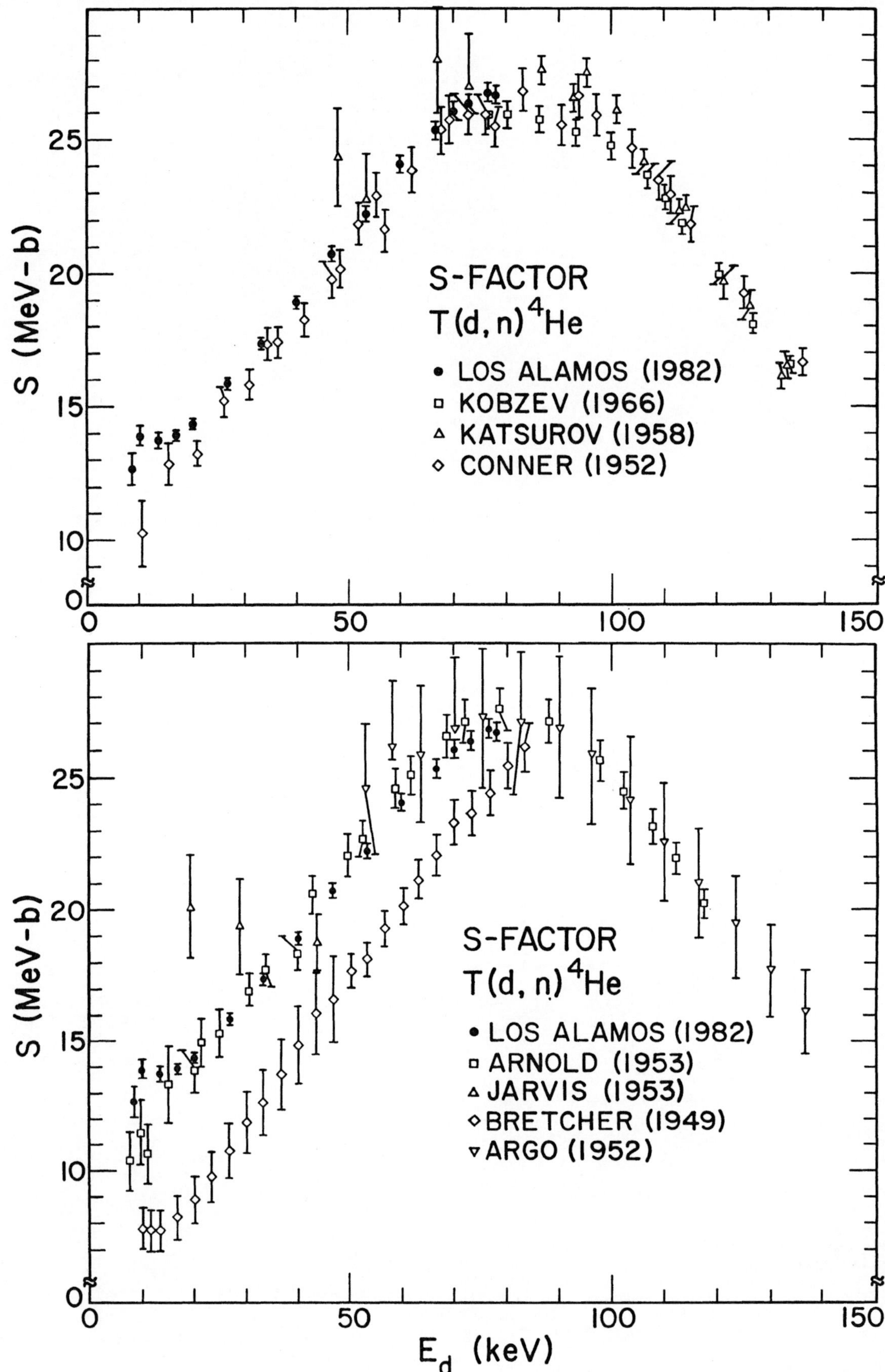

Fig. 10. The Los Alamos data plotted as the astrophysical S-Factor vs the equivalent deuteron bombarding energy, and compared with Arnold (Ref.2), Jarvis (Ref. 19), Bretcher (Ref. 18), and Argo (Ref. 20), Kobsev (Ref. 5), Katsaurov (Ref. 4) and Conner (Ref. 3). Only a selection of Arnold's points are plotted. The error bars represent the absolute error.

structure in that energy region, we conclude that the structure seen is a stochastic fluctuation.

Tables of cross sections and reactivity currently used are based upon the data of Arnold[2] and Conner.[3] These data fall lower than our values at the lowest energies but agree very well with most of the present data, indicating that the prediction of a possible large systematic uncertainty is not borne out. The cross sections of Katsurov[4] at lower energies appear to be too high. Bretcher[18] and Jarvis[19] are in marked disagreement with us. All of the data on the high energy side of the resonance appear to be in good agreement.

Analysis of our data is now under way with phenomenological models. A simple single-level resonance model fits very well. Comparison with an R-matrix analysis[21] of previous data of this and the other mass 5 channels indicates that our data lie 7 to 10% higher than the R-matrix prediction. A fit including our data is now underway.[22] One of the main conflicts in fitting is with the n + α total cross section of Shamu and Jenkin.[23] There is a preliminary indication that with a more careful interpretation of cross-section and energy uncertainties the conflict may fade. Also, the R-matrix analysis should shed light on low-energy extrapolation techniques used in astrophysical calculations.

In the future, we plan to study at these low energies: the D(d,p)T, D(d,^{3}He)n, and T(t,α)2n reactions. We will also attempt to make a reliable measurement of the D(t,γ)^{5}He cross section, poorly known[24-30] because of its branching ratio of 10^{-4} to 10^{-5} of the α + n channel. Recently, using the LEFCS target and a bismuth germanate gamma detector, we studied the backgrounds expected, with encouraging results.

Acknowledgements

Many people have been of great help to us in this work, especially Rudy Martinez who with diligence and care, has helped to assemble and construct much of the experiment. Ben Roybal, Gerald G. Ohlsen, F. David Correll, and Richard Hiebert have also made important contributions. We greatly appreciate discussions with Gerald M. Hale and Jim Toevs. We are grateful to P. J. Spellman and his collaborators at the Sandia Laboratory for the calibration of our resistor stack and digital voltmeters.

REFERENCES

1. N. Jarmie, Nucl. Sci. and Eng., 78, 404 (1981).

2. W. R. Arnold, J. A. Phillips, G. A. Sawyer, E. J. Stovall, Jr., and J. L. Tuck, Phys. Rev., 93, 483 (1954), see also by the same authors, "Absolute Cross Section for the Reactions T(d,n)He^4 from 10 to 120 keV," LA-1479, Los Alamos National Laboratory (1953).

3. J. P. Conner, T. W. Bonner, and J. R. Smith, Phys. Rev., 88, 468 (1952).

4. L. N. Katsaurov, "Investigation of the Reaction D(t,n)^{4}He by the Thin-Target Method in the Energy Region from 40 to 750 keV," Akad, Nauk, USSR, Trudy, Fizicheskii Inst., 14, 224 (1962); see also E. M. Balabanov, I. Ia. Barit, L. N. Katsaurov, I. M. Frank, and I. V. Shtranikh, "Measurement of the Effective Cross Section for the D(t,n)^{4}He Reaction in the Deuteron Energy Range 40-730 keV," Supp. 5., Sov. J. At. Energy, Atomnaya Energiya, P. 43 (1957).

5. A. P. Kobzev, V. I. Salatskii, and S. A. Telezhnikov, Sov. J. Nucl. Phys., 3, 774 (1966).

6. L. A. Artsimovich, "Controlled Thermonuclear Reactions," in Russian: Fiz Matgiz, p, 7, Moscow (1961); English translation: Gordon and Breach, p. 2, New York (1964).

7. L. Stewart and G. M. Hale, "The T(d,n)He and T(t,2n) Cross Sections at Los Energies," LA-58280MS, Los Alamos National Laboratory (Jan. 1975).

8. N. Jarmie, R. A. Hardekof, R. E. Brown, F. D. Correll, and G. G. Ohlsen, "Nuclear Cross Sections for Technology," J. L. Fowler and C. H. Johnson, NBS Special Publication #594, Sept. 1980 (Knoxville Conference) p. 733.

9. R. A. Hardekopf, R. E. Brown, F. D. Correll, N. Jarmie, and D. A. Clark, IEEE on Nucl. Sci. vol. NS-28, No. 2, p.1339, April 1981, (Denton, Conf.).

10. G. G. Ohlsen, R. E. Brown, N. Jarmie, R. A. Hardekopf, S. E. King, F. D. Correll, and D. A. Clark, Bull. Am. Phys. Soc. 24, 822 (1979).

11. R. A. Hardekopf, G. G. Ohlsen, R. V. Poore, and N. Jarmie, Phys. Rev. C13, 2127 (1976) and R. A. Hardekopf, Proc. of the Fourth Int. Symp. on Polarization Phenomena in Nuclear Reactions, Zürich, ed. by W. Gruebler and V. König (Birkhäuser, Basel, 1976) p. 865.

12. D. W. Barrus, and R. L. Blake, Rev. Sci. Instr. 48, 116 (1977).

13. E. A. Silverstein, Nucl. Instrum. Methods 4, 53 (1959).

14. D. C. Kocher and T. B. Clegg, Nuc. Phys. A132, 455 (1969).

15. N. Jarmie, J. H. Jett, J. L. Detch, Jr., and R. L. Hutson, Phys. Rev. C3, 10 (1971).

16. R. E. Brown, private communication.

17. C. H. Thomann and J. E. Benn, Nucl. Instrum. and Methods, 138, 293 (1976).

18. E. Bretcher and A. P. French, Phys. Rev. 75, 1154 (1949).

19. R. G. Jarvis and D. Roaf, Proc. Roy. Soc., A218, 432 (1953).

20. H. V. Argo, R. F. Taschek, H. M. Agnew, A. Hemmendinger, and W. T. Leland, Phys. Rev. 87, 612 (1952).

21. G. M. Hale and D. C. Dodder, "Nuclear Cross Sections for Technology" J. L. Fowler and C. H. Johnson, NBS Special Publication #594, Sept. 1980 (Knoxville Conference) p.650.

22. G. M. Hale, Private Communication.

23. R. E. Shamu and J. G. Jenkin, Phys. Rev. 135, B99 (1964).

24. A. Kosiara and H. B. Willard, Phys. Lett. 32B, 99 (1970).

25. W. Buss, W. Del Blanco, H. Wäffler, and B. Zeigler, Nucl. Phys. A112, 47 (1968).

26. V. M. Bezotosnyi, v. a. Zhmailo, L. M. Surov, M. S. Shvetsov, Sov. J. of Nucl. Phys. 10, 127 (1970) [Yad. Fiz. 10, 225 (1969)].

27. W. Buss, H. Wäffler, and B. Ziegler, Phys. Lett. 4, 198 (1963).

28. T. Hoang and R. Smith, Nucl. Phys. 83, 9 (1966).

29. J. H. Coon and R. W. Davis, Bull. Am. Phys. Soc. 4, 366 (1959).

30. P. Batay-Csorba, Thesis, Calf. Inst. of Tech. (1975).

CHARGED-PARTICLE ELASTIC CROSS SECTIONS

G. M. Hale and D. C. Dodder

Theoretical Division, Los Alamos National Laboratory
Los Alamos, New Mexico 87545 USA

J. C. DeVeaux

214 Nuclear Engineering Laboratory
University of Illinois
Urbana, Illinois 61801 USA

Modern treatments of energy loss in plasmas through elastic scattering of energetic ions require complete knowledge of charged-particle elastic cross sections. R-matrix theory provides an explicit separation of nuclear and Coulomb effects in these cross sections, and gives reasonable extrapolations to small angles and low energies, where data may be scarce. We outline the calculation of charged-particle elastic cross sections from R-matrix parameters, and give examples for d-T, d-α, and t-α scattering, obtained from comprehensive analyses of reactions in the ^{5}He, ^{6}Li, and ^{7}Li compound systems. Expansion coefficients for an exact polynomial representation for the difference of the scattering and Rutherford cross sections (σ_{NI}) are given for d-T scattering. Integral quantities involving σ_{NI} calculated from the present cross sections disagree substantially in some cases near resonances with a recent Livermore evaluation.

[Charged-particle elastic cross sections calculated from R-matrix theory. d-T, d-α, t-α scattering at energies below 5 MeV. Polynomial representation and integrals for σ_{NI}]

Introduction

Charged-particle elastic (CPE) scattering is an important energy-loss mechanism in ionized plasmas. Traditionally, the slowing-down treatment has taken into account only the effects of Rutherford (or "pure Coulomb") scattering, in which the energy loss comes from summing over large numbers of small-angle deflections. This approximation is valid in most cases at low energies, where the Coulomb amplitude dominates the nuclear amplitudes. However, at higher energies, where energy losses are more substantial, nuclear amplitudes affect the large-angle cross sections significantly, and enter even at small angles through interference with the Coulomb amplitude. It is important, therefore, to have a method for describing all the components of CPE cross sections, which provides reasonable extrapolations to regions not accessible to measurements (usually small angles and low energies).

R-matrix theory[1] is highly suitable for such a description, since, as has been pointed out at previous Cross Section meetings[2-4], it allows a parametric treatment of short-ranged (nuclear) effects, while accounting exactly for long-ranged effects, such as the Coulomb and angular momentum barriers. Thus, the relationship between the Coulomb and nuclear parts of the cross section as a function of energy and angle is constrained by a theory that embodies the fundamental properties of nuclear interactions and ensures the correct limiting behavior at small angles and low energies, where the long-ranged effects dominate.

In the following section, we outline the calculation of CPE cross sections from R-matrix parameters, then show for a few important cases how these calculated cross sections behave as functions of energy and angle and how they compare with experimental data. In Section III, we develop an exact polynomial representation for the difference of the CPE and Rutherford cross sections (σ_{NI}), and illustrate the behavior of this cross section and some of its expansion coefficients. Applications involving integrals of σ_{NI} are discussed in Section IV, where results based on the present R-matrix cross sections are compared with another recent evaluation. The final section summarizes the important points of the previous discussion and gives conclusions.

R-matrix Description of CPE Scattering

Elements of the R matrix are given by projections of essentially the Green's function operator for the internal Hamiltonian of a system of interacting particles on the two-body channel surface of the system[2]. They can be expressed as

$$R_{c'c} = \sum_{\lambda} \frac{\gamma_{c'\lambda}\gamma_{c\lambda}}{\varepsilon_\lambda - \varepsilon} \quad , \tag{1}$$

where the $\gamma_{c\lambda}$ and ε_λ are reduced-width amplitudes and eigenenergies, respectively, for states $|\lambda)$ having logarithmic derivative projections B_c on the channel surface defined by radii $r_c = a_c$. The channel label $|c) = |\alpha J s \ell)$ denotes a 2-body arrangement α of the system of particles having quantum numbers J,s,ℓ associated with total angular momentum, total spin, and orbital angular momentum, respectively.

In order to calculate observables, such as cross sections, of a scattering process, it is necessary to have matrix elements of the transition operator. These are related to the R matrix by [2,5]

$$T_{c'c} = e^{i\omega_{c'}} P_{c'}^{1/2} [R^{-1} - (L-B)]^{-1}_{c'c} P_c^{1/2} e^{i\omega_c} - e^{2i(\omega_c - \phi_c)} \delta_{c'c} \tag{2}$$

where B is the diagonal matrix of boundary-condition numbers B_c and L is a diagonal matrix containing logarithmic derivatives of channel outgoing spherical Coulomb waves, which can be expressed as

$$L_c = S_c + iP_c \quad , \tag{3}$$

in terms of the channel shift and penetrabiltiy functions, S_c and P_c. The quantities ω_c and $-\phi_c$ are phase shifts for Coulomb and hard-sphere scattering, respectively. Note that because of the matrix inversions in (2), any element of T depends, in general, on all the elements of R, and vice versa.

Observables for elastic scattering involve only the T-matrix elements for which $\alpha' = \alpha$. If these are denoted by

$$T_{c'c}(\alpha'=\alpha) = t^J_{s'\ell',s\ell} \ ,$$

then the elastic scattering cross section for charged particles can be written as the sum of three terms:

$$\sigma(\mu) = \sigma_R(\mu) + \sigma_N(\mu) + \sigma_I(\mu) \quad , \tag{4}$$

K. H. Böckhoff (ed.), Nuclear Data for Science and Technology, 326–330.

where for $\mu = \cos(\theta_{CM})$, the "pure Coulomb," or Rutherford scattering cross section is

$$\sigma_R(\mu) = \left[\frac{\eta}{k(1-\mu)}\right]^2 \quad , \tag{5}$$

the "pure nuclear" cross section is

$$\sigma_N(\mu) = [(2s_1+1)(2s_2+1)k^2]^{-1} \sum_L \sum_{s's} (-1)^{s'-s} \sum_{J_1 J_2 \ell_1' \ell_2' \ell_1 \ell_2} \bar{Z}(\ell_1' J_1 \ell_2' J_2, s'L) \; \bar{Z}(\ell_1 J_1 \ell_2 J_2 sL)$$

$$\mathrm{Re}\left(t^{J_1}_{s'\ell_1', s\ell_1} \, t^{J_2 *}_{s'\ell_2' s\ell_2}\right) P_L(\mu) \tag{6}$$

and the Coulomb-nuclear interference cross section is

$$\sigma_I(\mu) = - \frac{2\eta}{(2s_1+1)(2s_2+1)k^2(1-\mu)} \; \mathrm{Re}\Big[e^{i\eta \ln \frac{1}{2}(1-\mu)}$$

$$\sum_{Js\ell} (2J+1) t^J_{s\ell, s\ell} \, P_\ell(\mu)\Big] \quad . \tag{7}$$

In Eqs. (5)-(7), k is the center-of-mass wave number and η the Coulomb parameter for a distinguishable pair of interacting particles, s_1 and s_2 are their spins, $\bar{Z}$ is a modified Racah coefficient as defined in Ref. 5, and $P_n(\mu)$ is the nth order Legendre polynomial.

Elastic scattering cross sections calculated from relations (4)-(7) are shown in Figs. 1-3 for the case of d-T, d-α, and t-α scattering. These are the primary cross sections that are required when considering the slowing down of fast ions in a d-T plasma. The R-matrix parameters used in these calculations came from comprehensive studies of reactions in the ^{5}He, ^{6}Li, and ^{7}Li systems, some of which have been described in Refs. 2 and 3. The term "comprehensive" implies that many measurements for other reactions and observables were included in the analyses along with elastic scattering cross section data. Data for other reactions influence the elastic T-matrix elements through the relation between R and T noted above in connection with Eq. (2).

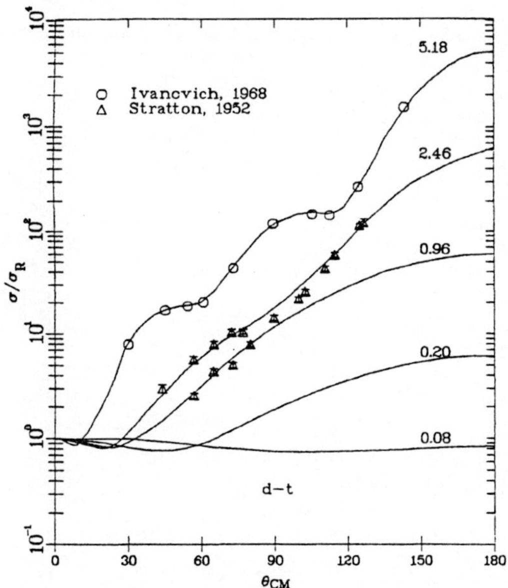

Fig. 1. Calculated and measured cross-section ratios to Rutherford scattering for d-T. The data (Φ) at .96 and 2.46 MeV are those of Stratton[6]; those (Φ) at 5.18 MeV are from Ivanovich[7].

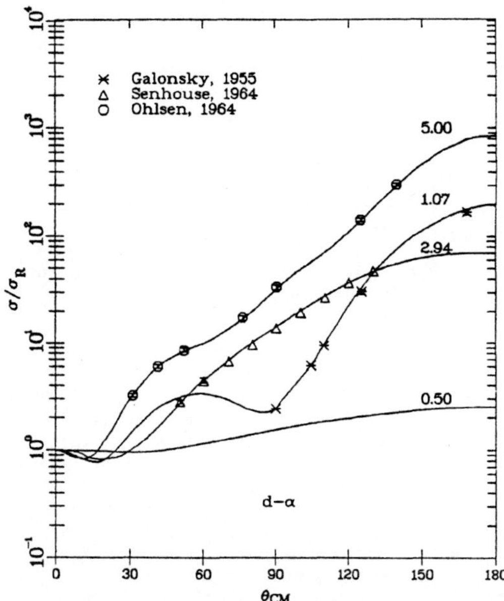

Fig. 2. Calculated and measured cross-section ratios to Rutherford scattering for d-α. The data at 1.07 MeV are those of Galonsky[8]; those at 2.94 MeV are from Senhouse[9]; those at 5.00 MeV are from Ohlsen[10].

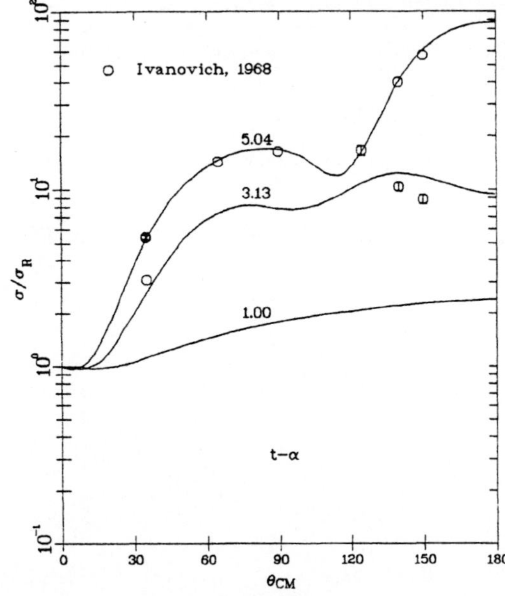

Fig. 3. Calculated and measured cross-section ratios to Rutherford scattering for t-α. The data at 3.13 and 5.04 MeV are those of Ivanovich[7].

The scattering cross sections in Figs. 1-3 are plotted as ratios to the Rutherford cross section, so that the effect of the nuclear amplitudes can easily be seen as deviations from unity. The onset of significant deviations from pure Coulomb scattering occurs at fairly low energies for all three reactions, but especially for d-T (Fig. 1). This is due mainly to the presence of prominant resonances in the reactions: a relatively broad $3/2^+$ resonance in d-T at E_d = 200 keV (109 keV in the T(d,n)α reaction), a narrow 3^+ resonance in d-α at E_d = 1.07 MeV, and a relatively narrow $7/2^-$ resonance in t-α at E_t = 3.85 MeV. In all three cases, the calculations indicate substantial deviations from pure Coulomb scattering at energies below the lowest energy where measurements have been made.

Fig. 1 shows predictions for d-T at 80 and 200 keV, then comparisons with the data of Stratton[6] at .96 MeV (the lowest energy measurements) and 2.46 MeV, and with those of Ivanovich[7] at 5.18 MeV. The agreement of the calculation with the data is fair at the lower two energies and good at the upper energy. The $3/2^+$ resonance gives strong contributions from the interference cross sections in the calculations for d-T, causing the interference minimum that diminishes and moves to smaller angles as the energy increases. The interference is so strong at 80 keV, however, that the scattering cross section falls below the Rutherford cross section at all angles greater than 30°.

Fig. 2 shows predictions for d-α at 500 keV and comparisons with the data of Galonsky[8] at 1.07 MeV, of Senhouse[9] at 2.94 MeV, and of Ohlsen[10] at 5.00 MeV. There is good agreement between the calculations and the measurements in this energy range. The D-wave resonance at 1.07 MeV causes more structure in the angular distribution and a deeper interference minimum at this energy than at nearby energies.

Fig. 3 shows predictions for t-α at 1 MeV and comparisons with measurements of Ivanovich[7] at 3.13 and 5.04 MeV. The agreement between the calculation and the data is poor at 3.13 MeV and good at 5.04 MeV. In this case, the energy of the resonance is relatively high, so that deviations from pure Coulomb scattering are less dramatic at low energies than for the other two reactions.

Exact Polynomial Representation for σ_{NI}

From Eqs. (5)-(7), it can be seen that the sum of the nuclear and interference cross sections,

$$\sigma_{NI}(\mu) = \sigma_N(\mu) + \sigma_I(\mu) \quad ,$$

can be expanded in Legendre polynomials, according to

$$\sigma_{NI}(\mu) = -\frac{2\eta}{1-\mu}\,\mathrm{Re}[e^{i\eta \ln \frac{1}{2}(1-\mu)} \sum_{\ell=0}^{\ell mx} \frac{2\ell+1}{2}\, a_\ell P_\ell(\mu)] + \sum_{\ell=0}^{2\ell mx} \frac{2\ell+1}{2}\, b_\ell P_\ell(\mu) \quad , \tag{8}$$

with

$$\frac{2\ell+1}{2}\, a_\ell = [(2s_1+1)(2s_2+1)k^2]^{-1} \sum_{Js} (2J+1)\, t^J_{s\ell,s\ell}, \tag{9}$$

and

$$\frac{2\ell+1}{2}\, b_\ell = [2s_1+1)(2s_2+1)k^2]^{-1} \sum_{s's} (-1)^{s'-s} \sum_{J_1J_2\ell_1'\ell_2'\ell_1\ell_2} \bar{Z}(\ell_1'J_1\ell_2'J_2,s'\ell)\, \bar{Z}(\ell_1J_1\ell_2J_2,s\ell)\, \mathrm{Re}(t^{J_1}_{s'\ell_1',s\ell_1}\, t^{J_2 *}_{s'\ell_2',s\ell_2}) \tag{10}$$

The sums in Eqs. (8)-(10) are limited by neglecting partial waves for nuclear scattering above ℓmx. As can be seen from Eqs. (9) and (10), the coefficients a_ℓ are complex (as are the t's), the b_ℓ are real, and the two sets are not independent. Therefore, it is not possible to fit experimental data directly in terms of a_ℓ and b_ℓ coefficients treated as independent parameters. An intermediate step, such as an R-matrix or phase-shift analysis, is necessary to establish the relation among the a's and b's through the t's, and in fact, to establish unitary relations among the t's themselves.

The exact polynomial expansion described above is one option allowed for representing elastic scattering cross sections in a new ENDF charged-particle format[11]. The format simply specifies a tabulation of the a_ℓ and b_ℓ coefficients as a function of energy. Files of the coefficients calculated from R-matrix parameters are available at Los Alamos National Laboratory for the reactions and energy ranges listed in Table I. From these coefficients, elastic scattering cross sections can be constructed over the entire angular range for most of the reactions among light charged particles from protons through alpha particles.

Table I Charged-Particle Elastic Cross Sections Available from Los Alamos R-matrix Analyses

Mat #	Reaction	Energy Range (MeV)
2002	p-p	0-20
2004	p-T	0-11
2005	p-^{3}He	0-10
2006	p-^{4}He	0-20
3003	d-d	0-10
3004	d-T	0-8
3005	d-^{3}He	0-8
3006	d-^{4}He	0-10
4004	t-t	0-3.5
4006	t-^{4}He	0-14
5006	^{3}He-^{4}He	0-12

The behavior of the expansion coefficients a_ℓ and b_ℓ as a function of energy is illustrated in Figs. 4 and 5 for d-T scattering. Fig. 4 shows the important b_ℓ coefficients for expanding the nuclear cross section at energies below 5 MeV. The low-energy d-T S-wave resonance is clearly evident as a peak in the b_0 coefficient at 200 keV. The real and imaginary parts of the a_ℓ coefficients are shown in Fig. 5, where again, the effect of the S-wave resonance is seen at low energies in a_0.

The separation of CPE cross sections into components given by R-matrix calculations allows one to see an interesting feature not normally visible in experimental data: the rapid oscillations at small angles

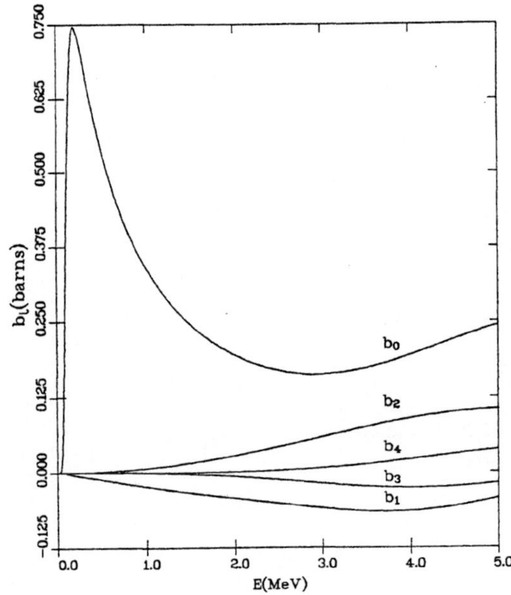

Fig. 4. Significant b_ℓ coefficients for d-T scattering at energies below 5 MeV.

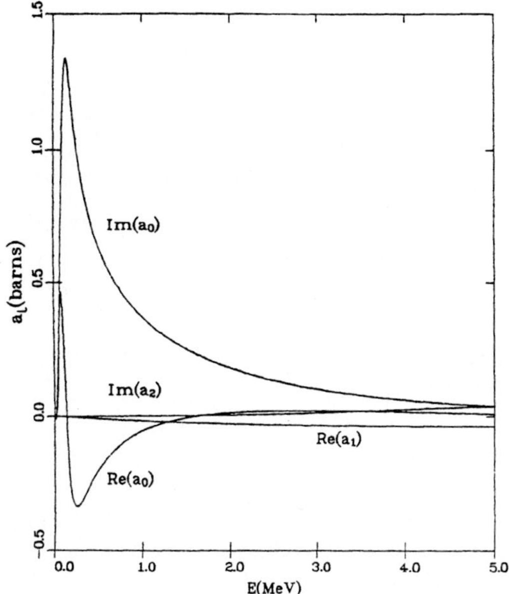

Fig. 5. Significant a_ℓ coefficients for d-T scattering at energies below 5 MeV.

in the interference cross section that are masked by the Rutherford cross section. These are illustrated in Fig. 6 by plotting $(1-\mu)\sigma_{NI}(\mu)$ for d-T scattering at 200 keV (Multiplying $\sigma_{NI}(\mu)$ by $(1-\mu)$ removes the pole at $\mu = 1$ in $\sigma_I(\mu)$; see Eq. 7.) Such oscillations are important in practice only at angles for which $\sigma(\mu) \neq \sigma_R(\mu)$ (in this case, $\theta_{CM} > 5°$), where they sometimes appear in the data as "Coulomb-nuclear interference" minima, but they introduce the mathematical difficulty that integrals of $\sigma_{NI}(\mu)$ over the entire angular range are undefined. We will return to this point in the next section.

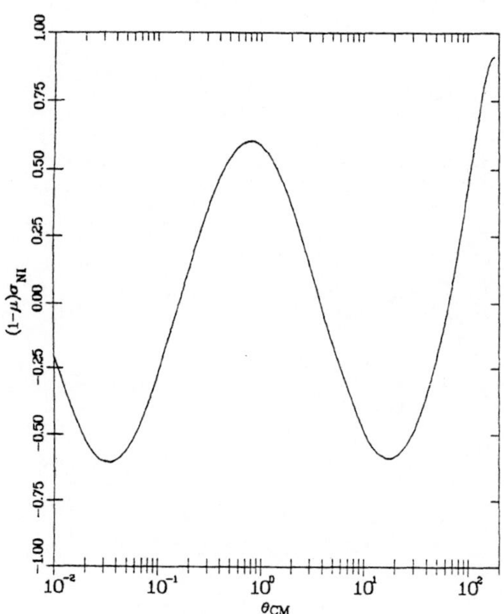

Fig. 6. Oscillatory behavior of $(1-\mu)\sigma_{NI}(\mu)$ for d-T scattering at 200 keV.

Integrals of σ_{NI}

In many CPE energy-loss calculations, the Rutherford cross section is treated with a continuous slowing-down model, while the remainder of the cross section, $\sigma_{NI}(\mu)$, is used in a discrete-event, Monte-Carlo transport framework. In such applications, integrals of $\sigma_{NI}(\mu)$ with other functions of μ are required, in principle, over the entire angular range. If σ_{NI} is to have the physical significance of a cross section, its integral over all angles must exist and be positive definite. The nuclear part of σ_{NI} satisfies these conditions, with $\sigma_N(\text{int.}) = 2\pi b_0$; but, as was mentioned at the end of the last section, the integral of σ_I over all angles is undefined due to divergent oscillations as $\mu \to 1$. Therefore, it is necessary to introduce a small-angle cutoff into the integrals of $\sigma_{NI}(\mu)$ in order to define a value for the integrated cross section, which in conventional useage should be positive.

For these reasons, integrals of σ_{NI} defined by Perkins and Cullen[12] are cut off at $\theta_{CM} = 20°$ (μ=.94), or at the largest angle (if > 20°) where σ_{NI} crosses through zero and remains positive. For the d-T cross section shown in Fig. 6, the cutoff angle according to their prescription would be 66°. Such an integral for d-T at 80 keV presumably would be zero, since σ_{NI} remains negative at angles out to 180°. In all cases, this prescription neglects the "interference" region of the cross section, where $\sigma/\sigma_R = 1 + \sigma_{NI}/\sigma_R < 1$.

However, in order to compare with the work of Ref. 12, we have calculated from our R-matrix cross sections the following integrals for $\sigma_{NI}(\mu)$:

$$\text{Integrated cross section} = \sigma_{NI}(\text{int.}) = 2\pi \int_{-1}^{\mu_c} \sigma_{NI}(\mu)d\mu$$

$$\text{Reaction rate} = \langle\sigma v\rangle = 2\pi \int_{-1}^{\mu_c} v_{rel}\sigma_{NI}(\mu)d\mu = v_{rel}\sigma_{NI}(\text{int.}),$$

$$\text{Frac. energy loss/collision} = \langle\frac{\Delta E}{E}\rangle = \frac{4\pi m_1 m_2}{(m_1+m_2)^2} \frac{1}{\sigma_{NI}(\text{int.})} \int_{-1}^{\mu_c} (1-\mu)\,\sigma_{NI}(\mu)\,d\mu,$$

$$\text{Frac. energy loss/unit path length} = \langle\frac{\Delta E}{E}\rangle\, \sigma_{NI}(\text{int.}),$$

Ave. Lab. Scatt. Cosine =

$$\langle\mu_{Lab}\rangle = \frac{2\pi}{\sigma_{NI}(\text{int.})} \int_{-1}^{\mu_c} \frac{m_1+m_2\mu}{\sqrt{m_1^2+2m_1m_2\mu+m_2^2}}\, \sigma_{NI}(\mu)d\mu,$$

with the cutoff $\mu_c = \min(.94, \mu_0)$, $\sigma_{NI}(\mu_0) = 0$. The last four quantities are plotted for d-T scattering as dashed curves in Fig. 7, compared to solid curves representing the evaluation of Perkins and Cullen[12]. The agreement is fairly good at energies above 2 MeV, but there are substantial differences at low energies in the region of the S-wave resonance, where our calculations give much larger reaction rates and energy losses. This trend is also evident in comparisons for other reactions where the differences are not quite as large, but largest near the resonances.

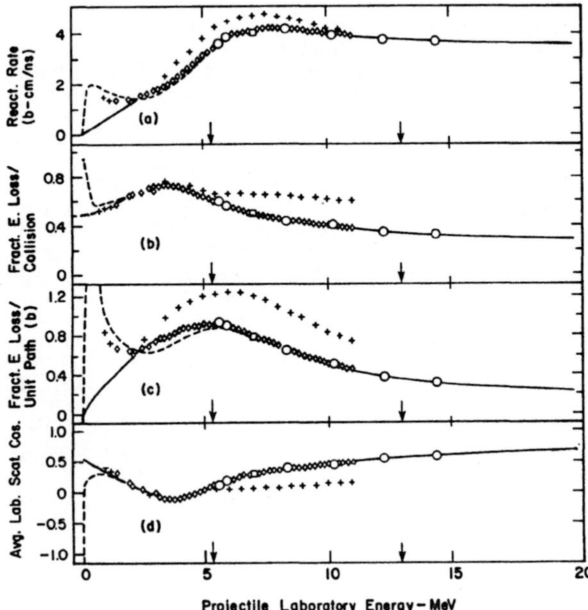

Fig. 7. Integrals of σ_{NI} for d-T scattering, as described in the text. The dashed curves are from the present calculations, the solid curves are evaluations of Perkins and Cullen[12], and the symbols represent measurements.

Summary and Conclusions

Calculating CPE cross sections from R-matrix parameters allows the separation of σ_{NI} into its constituent nuclear and Coulomb-nuclear-interference parts. The nuclear part has a well-behaved Legendre expansion with properties identical to those used to represent neutron cross sections. The interference part also has a Legendre expansion, but exhibits diverging, increasingly rapid oscillations as $\mu \to 1$. In practice, these oscillations are important only at angles for which σ_{NI}/σ_R differs noticeably from zero, where they cause measureable "interference minima" in the cross sections.

Coefficients for the Legendre expansion of σ_{NI}, tabulated vs energy in a new ENDF format, are available for most of the reactions involving protons through ^{4}He. These coefficients are calculated from R-matrix parameters resulting from comprehensive studies of few-nucleon systems done at Los Alamos over the past several years. The R-matrix calculations generally agree well with existing cross-section measurements, and provide reliable extrapolations to energies and angles where measurements are not available.

For the important case of d-T scattering, for instance, cross-section measurements do not extend to energies low enough to see the major effects of the strong $3/2^+$ S-wave resonance. These effects are clearly evident in the low-energy calculations, however, as peaks in the a_ℓ and b_ℓ expansion coefficients, strong interference terms in the cross section, and pronouced deviations from Rutherford scattering at backward angles. The effects of the resonance are also responsible for large differences at low energies between calculated integrals of σ_{NI} and a recent evaluation of Perkins and Cullen.

The oscillatory behavior of σ_I at small angles prevents defining an integrated cross section for σ_{NI} without introducing a small-angle cutoff. If the cutoff is chosen so that σ_{NI}(int.) is positive definite, as is required by its use in some slowing-down calculations, contributions to the integral from the interference region are neglected. These difficulties suggest that an improved method of dealing with CPE cross sections in slowing-down calculations is required. One possibility is to apply the continuous slowing-down treatment to the sum of the Rutherford and interference cross sections, leaving the well-behaved nuclear cross section to be used in Monte-Carlo transport. Another is to use the present (σ_R, σ_{NI}) separation, but place the cutoff angle where σ begins to deviate from σ_R. This prescription does not guarantee a positive-definite integrated cross section for σ_{NI}, however, and would require physical interpretation in the Monte-Carlo transport of a cross section that is sometimes negative with respect to Rutherford scattering.

The present calculations provide the separation and detail necessary to be useful, regardless of the methods finally chosen to apply charged-particle elastic cross sections to the important problems of fusion energy design.

References

1. E. P. Wigner and L. Eisenbud, Phys. Rev. 72, 29 (1974).
2. G. M. Hale and D. C. Dodder, Proc. Conf. Nuclear Cross Sections for Technology, Knoxville 1979 (NBS Spec. Pub. 594), p. 650 (1980).
3. G. M. Hale, Proc. Symp. Neutron Standards and Applications, Gaithersburg,1977 (NBS Special Pub. 493), p. 30 (1977).
4. G. M. Hale, Proc. Conf. Nuclear Data Evaluation Methods and Procedures, Brookhaven 1980 (BNL-NCS-51363, Vol. II), p. 509 (1981).
5. A. M. Lane and R. G. Thomas, Rev. Mod. Phys. 30, 257 (1958).
6. W. R. Stratton, G. D. Freier, G. R. Keepin, D. Rankin, and T. F. Stratton, Phys. Rev. 88, 257 (1952).
7. M. Ivanovich, P. G. Young, and G. G. Ohlsen, Nucl. Phys. A110, 441 (1968).
8. A. Galonsky, R. A. Douglas, W. Haeberli, M. T. McEllistrem, and H. T. Richards, Phys. Rev. 98, 586 (1955).
9. L. S. Senhouse and T. A. Tombrello, Nucl. Phys. 57, 624 (1964).
10. G. G. Ohlsen and P. G. Young, Nucl. Phys. 52, 134 (1964).
11. R. E. MacFarlane, G. M. Hale, and P. G. Young, "Charged Particle Format," ENDF draft report (1981).
12. S. T. Perkins and D. E. Cullen, Nucl. Sci. Eng. 77, 20 (1981).

CROSS SECTION SENSITIVITY STUDY FOR FUSION BLANKETS INCORPORATING LEAD NEUTRON MULTIPLIER

S. Pelloni

Swiss Federal Institute for Reactor Research (EIR)
CH-5303 Wuerenlingen
Switzerland

E.T. Cheng

General Atomic Company
P.O. Box 81608
San Diego, California
USA 92138

In the recent European INTOR design, lead has been considered for incorporation in the blanket as either an explicit or implicit neutron multiplier. The blanket employs either Li_2SiO_3 or $Li_{17}Pb_{83}$ as tritium breeding material. Nucleonic analysis was performed for this blanket using the DLC37 and DLC41 cross section libraries. The reaction rates were estimated using the reaction cross sections provided with both libraries. In addition to that, they were estimated using the MACKLIB-IV response library. The calculated tritium breeding ratio was found to be 5 % less and 15 % more in the calculations with DLC41 and DLC41 plus MACKLIB-IV libraries, respectively, than in the calculation with the DLC37 library. The Fe, Pb, and Li cross sections given by the ENDF/B-IV and V were reviewed. A sensitivity study of these cross section uncertainties shows that the tritium breeding ratio is relatively insensitive to the above mentioned partial cross sections. The calculated tritium breeding ratio can be known within $\pm$ 2 %.

Introduction

In the nuclear design of a fusion reactor, it is important that the nucleonic parameters be calculated within reasonable accuracy. These parameters include tritium breeding ratio, blanket energy multiplication, and radiation damage parameters in various reactor components. The most detrimental radiation damage parameters are probably those associated with damage to the toroidal field magnets, namely radiation dose at the insulators and atomic displacement in the copper stabilizer. Damage to the insulators cannot be recovered, although damage to the copper stabilizer can partially be recovered upon annealing. Accurate prediction of tritium breeding is necessary in fusion reactor design because of the requirement for self-sustaining tritium production and inventory control. However, nucleonic parameter calculations are often sensitive to the method employed and to uncertainties in the nuclear data. In this paper, we are mainly concerned with calculational uncertainties of some nucleonic parameters due to nuclear data uncertainties.

In this study we have chosen the International Tokamak Reactor (INTOR) being designed by the European Community in order to investigate the above-mentioned uncertainties.

The purpose of the INTOR is to demonstrate the feasibility of fusion reactor engineering performance, which includes the production of power and tritium in the blanket, in addition to creating and maintaining a burning plasma, and operation and protection of the superconducting magnet in a fusion neutron environment. The European Community INTOR design has been considering lead for incorporation in the blanket as either an explicit or implicit neutron multiplier. The blanket employs either Li_2SiO_3 or $Li_{17}Pb_{83}$ as tritium breeding material. In the Li_2SiO_3 concept, the first wall consists of a 1.3 cm thick plate of stainless steel SS-304. The neutron multiplier Pb is inserted in front of the blanket, which is subdivided into three regions, each containing the breeding material Li_2SiO_3 with different densities. Since the neutron multiplier, Pb, is incorporated in the breeder, the $Li_{17}Pb_{83}$ design uses a thicker first wall (3 cm) than does the Li_2SiO_3 design (1.3 cm). The first wall consists of aluminum and water. The breeding blanket itself is 40 cm thick. In both cases, water acts as coolant and stainless steel as structural material. The reflector is 12 cm thick and consists of stainless steel SS-304.

Nucleonic Aspect

The nucleonic aspects of the European Community INTOR (INTOR-EC) design have recently been studied and discussed in Ref.[1] Two blanket options have been selected for the INTOR-EC design. One of them employs solid lead as the neutron multiplier and Li_2SiO_3 as the breeding material. The other employs $Li_{17}Pb_{83}$ as both neutron multiplier and tritium breeding material. In both designs the neutronics model consists of the important components of a fusion reactor, namely the plasma and vacuum chamber, first wall, breeding blanket, and reflector. The model and material compositions are displayed respectively for the two blanket designs in Tables I and II. Neutronics calculations were performed using the one dimensional discrete-ordinates transport code ANISN (Ref. [2]) with P_3S_8 approximation, in cylindrical geometry. An albedo equal to 0.3 for all energy groups on the external boundary of the reflector was used to simulate the effect of the shield.

The Li_2SiO_3 concept was computed using first the coupled 100-group neutron and 21-group gamma ray nuclear data based on the DLC-37 library (Ref. [3]), and second, the coupled 171-group neutron and 36-group gamma ray nuclear data based on the DLC-41 library (Ref. [4]). Both libraries were collapsed into coupled 25 neutron and 21 gamma-ray groups, using an 1/E weighting spectrum. The group energy boundaries are the same as given in Ref. [5]. In each case, a specific INTOR working library which contains only the needed materials (namely Ni, Cr, Fe, Pb, D, O, Si, ^6Li, ^7Li) was created. The reaction rates were estimated using the reaction cross sections provided with both libraries and in addition using the MACKLIB-IV (Ref.[6]) response library with the flux calculation from the DLC-41 library. The alternate blanket concept, using $Li_{17}Pb_{83}$ as the tritium breeding material, was computed with the DLC-37 library. In both blankets the tritium breeding ratio was studied as a function of ^6Li enrichment in the breeder. Furthermore, the Fe, Pb, and Li cross sections given by the ENDF/B-IV and V were reviewed and the influence of their uncertainties on the tritium breeding ratio was studied.

Some Results from the Calculations

In Table III, the tritium breeding ratios for the Li_2SiO_3 blanket are displayed for the three cases using different nuclear data sets. Note that natural lithium is used in Li_2SiO_3.

K. H. Böckhoff (ed.), Nuclear Data for Science and Technology, 331–334.

Table I The Li_2SiO_3 INTOR-EC Neutronics Model Concept

Zone No.	Component	Zone Thickness (m)	Composition
1	Plasma	2.20	Vacuum
2	Vacuum	0.30	Vacuum
3	First wall	0.013	100% SS-304
4	Neutron multiplier	0.091	60.7% Pb 7.8% D_2O 10.0% SS-304 21.5% void
5	Tritium breeding blanket	0.062	28.6% Li_2SiO_3 22.1% H_2O 11.2% SS-304 16.6% He 21.5% void
6	Tritium breeding blanket	0.062	24.5% Li_2SiO_3 27.9% H_2O 11.1% SS-304 15% He 21.5% void
7	Tritium breeding blanket	0.062	12.3% Li_2SiO_3 29.7% H_2O 11.0% SS-304 25.5% He 21.5% void
8	Reflector	0.12	100% SS-304

Table II The $Li_{17}Pb_{83}$ INTOR-EC Neutronics Model Concept

Zone No.	Component	Zone Thickness (m)	Composition
1	Plasma	2.20	Vacuum
2	Vacuum	0.30	Vacuum
3	First wall	0.03	80% Al 20% H_2O
4	Breeding blanket	0.40	80% $Li_{17}Pb_{83}$ 10% H_2O 10% SS-316
5	Reflector	0.12	100% SS-304

Table III Tritium Breeding Ratio for the Li_2SiO_3 Blanket Calculated with the DLC-37, DLC-41, and DLC-41+MACKLIB-IV Libraries

Reaction	DLC-41 + MACKLIB-IV	DLC-37	DLC-41
$^6Li(n,\alpha)$	1.238	1.075	1.020
$^7Li(n,n')\alpha$	0.017	0.017	0.017
Total	1.255	1.092	1.037

The calculated tritium breeding ratio shown in Table III is mainly contributed by the $^6Li(n,\alpha)$ reaction and is found to be 5 % less and 15 % more in the calculations with DLC-41 and DLC-41 plus MACKLIB-IV libraries than in the calculation with the DLC-37 library. The contribution due to the $^7Li(n,n')\alpha$ reaction is in very good agreement in the three cases and is small because the multiplier Pb was incorporated in the design. The density of lead is larger than that of ^7Li and the Pb(n,2n) cross section in the 10 MeV energy range of interest is much higher than the $^7Li(n,n')\alpha$ cross section. The differences between the three calculations due to the $^6Li(n,\alpha)$ reaction, as Table III shows, can cause serious troubles in tritium breeding ratio evaluation. It is found the differences arise from the low energy neutrons in the range between 4 and 0.02 eV, which contribute about 80 % of the total breeding ratio. This is because the spectrum in the INTOR concept is soft. The contributions of each energy group to the tritium production due to the $^6Li(n,\alpha)$ reaction are listed in Table IV. The agreement between flux calculations using the DLC-37 and DLC-41 libraries is within 10 % except the last energy group (0.4 to 0.022 eV), where the discrepancy is larger. This alone could not explain satisfactorily the large differences in tritium breeding ratio between the DLC-37 and DLC-41 plus MACKLIB-IV libraries. These are due to the $^6Li(n,\alpha)$ reaction cross sections in the low energy range between 4 and 0.022 eV. The $^6Li(n,\alpha)$ cross section in the range 0.4 to 0.022 eV is about 50 % higher in the MACKLIB-IV library than in the DLC-37. The same nuclear data in the DLC-41 case is in between.

Furthermore, the neutron balance analysis for the system shows that when using the DLC-41 plus MACKLIB-IV library, the tritium breeding tends to be overpredicted and therefore the MACKLIB-IV activity cross section of the reaction $^6Li(n,\alpha)$ in the energy range between 4 and 0.02 eV is not reliable. The total gain of neutrons due to the source and the (n,2n) reactions is smaller than the total loss due to the (n,p),(n,α) and (n,γ) reactions. The difference between lost (1.70) and gain (1.45) is about 0.25 and, as expected, corresponds roughly to the tritium breeding discrepancy with the DLC-37 library.

The tritium breeding in the alternate design using $Li_{17}Pb_{83}$ as the tritium breeding material is studied in the following with the neutronics calculations performed only using the DLC-37 library. However, the implication from the above analysis applies to the $Li_{17}Pb_{83}$ blanket as well.

In Fig. 1 we plot the tritium breeding ratio as a function of ^6Li enrichment in lithium for the $Li_{17}Pb_{83}$ blanket and the Li_2SiO_3 blanket. The tritium breeding ratios for the Li_2SiO_3 blanket were also obtained using the DLC-37 library. These results seem to be in good agreement with those presented in Ref.[1]. In the $Li_{17}Pb_{83}$ blanket, the maximum tritium breeding ratio increases with the enrichment of ^6Li in lithium up to about 30 %. As already pointed out in Ref.[1], the tritium breeding ratio remains almost unchanged for enrichments above the 30 % value. This is primarily due to the moderating effect of water, which favors $^6Li(n,\alpha)$ reaction in such a way that all the available neutrons are already absorbed by ^6Li, even if its concentration in the breeder is small. In the Li_2SiO_3 design, the tritium production rate increases steadily with the ^6Li enrichment in lithium, since the breeding region is only 18.6 cm thick compared to that in the $Li_{17}Pb_{83}$ design which is about 40 cm thick.

Sensitivity and Uncertainty Study

The sensitivity and uncertainty analysis of the INTOR blankets were performed using 30 % ^6Li in the breeder as the reference cases. The materials of most interest are Pb, Fe, and ^6Li in the Li_2SiO_3 design and Pb, Fe, ^6Li, H, and O in the $Li_{17}Pb_{83}$ design, respectively. The ^7Li can be neglected since about 99 % of the tritium production is contributed from the $^6Li(n,\alpha)$ reaction. The contributions to the total uncertainty of the tritium breeding ratio due to Cr, Al, and Ni are similar to those due to Fe; they are less important and can also be neglected. Water (that is H and O) seems to be important in the lithium-lead blanket because of its neutron slowing down properties and will be discussed here. The method of the cross section sensitivity study employed here is based on the linear perturbation theory as described in Ref.[7]. The calculations of the sensitivities were performed using the ORNL developed sensitivity analysis code, SWANLAKE (Ref.[8]). The direct and adjoint flux distributions required to perform the sensitivity calculations were computed using ANISN (Ref.[2]). We recall at this point that the agreement of the $^6Li(n,\alpha)$ reaction rates in the direct and adjoint calculations are excellent, within 1 %.

Table IV Tritium Production by Energy Group for Calculations using DLC-37, DLC-41, and DLC-41 + MACKLIB-IV Libraries

Energy Group	DLC-37	DLC-41 + MACKLIB-IV	DLC-41
1	5.49174E-05 (a)	5.37736E-05	6.91075E-05
2	1.03696E-05	1.09533E-05	1.36334E-05
3	5.33995E-06	4.66788E-06	6.01342E-06
4	5.05888E-06	4.48707E-06	5.79944E-06
5	5.31828E-06	5.13779E-06	6.64589E-06
6	5.37325E-06	5.32616E-06	6.86301E-06
7	5.54604E-06	6.13108E-06	7.93648E-06
8	7.37450E-06	7.30909E-06	9.54709E-06
9	8.95342E-06	8.39406E-06	1.10246E-05
10	8.60496E-06	8.24168E-06	1.09580E-05
11	2.02109E-05	1.95476E-05	2.62026E-05
12	2.86364E-05	2.74894E-05	3.58671E-05
13	4.68540E-05	4.59363E-05	4.84098E-05
14	8.59495E-05	8.57795E-05	8.57795E-05
15	4.87816E-04	5.38839E-04	5.36613E-04
16	7.71251E-04	9.40307E-04	9.25557E-04
17	8.27853E-04	8.14378E-04	8.16567E-04
18	6.31603E-03	6.27025E-03	6.30342E-03
19	3.07366E-03	3.11898E-03	3.12300E-03
20	6.89372E-04	6.76872E-03	6.81210E-03
21	1.95010E-02	1.90800E-02	1.92005E-02
22	5.55231E-02	5.38045E-02	5.45416E-02
23	1.39825E-01	1.36296E-01	1.37173E-01
24	2.66909E-01	3.89991E-01	2.59996E-01
25	5.80606E-01	6.19914E-01	5.30040E-01
Total	1.07483E+00	1.23781E+00	1.01981E+00

(a) 5.49174E-05 means 5.49174×10^{-5}.

Cross Section Uncertainties

The cross sections of the mentioned materials given by the ENDF/B-IV and V data files were reviewed.

1. Lead. All the uncertainties in nuclear data are taken from Ref.[9]. The Pb(n,2n) reaction cross sections are known to ± 7 % in the energy range between 14 and 12 MeV and to $\pm$ 5 % below 12 MeV. Elastic and inelastic scattering cross sections are known within $\pm$ 10 %.
2. Iron. According to Ref.[10], the important cross sections like elastic and inelastic scattering are measured within $\pm$ 10 % to 15 % between 0 and 14 MeV; (Fe(n,γ) reaction cross sections are known within $\pm$ 10 % between 3 and 3000 keV and within $\pm$ 20 % above 3 MeV; Fe(n,2n) cross sections within $\pm$ 15 % above threshold energy.
3. Lithium-6. The ^{6}Li(n,α)T reaction is well known. The uncertainty in the tritium production cross section should be $\pm$ 5 % or less in the whole energy range according to Ref.[10]. As pointed out in the previous section, the discrepancy between the MACKLIB-IV and the DLC-37 libraries of ^{6}Li(n,α) cross sections in the low energy groups is larger than 50 %. This is due to the unreliability of the MACKLIB-IV rather than to the cross section uncertainty itself.
4. Hydrogen and Oxygen. According to Ref.[10], elastic scattering cross sections are assumed to be known very accurately to about $\pm$ 1 %.

Some Results

The overall uncertainty of the tritium breeding ratio in the INTOR blankets due to the cross section uncertainties are small and similar for the two designs. It is mainly contributed from the Pb(n,2n) reaction in the energy range between 14 and 12 MeV. The sensitivities of the tritium breeding ratio due to the Pb(n,2n) and Fe(n,2n) reactions are positive, since more neutrons become available in the system if these reaction cross sections become larger and therefore

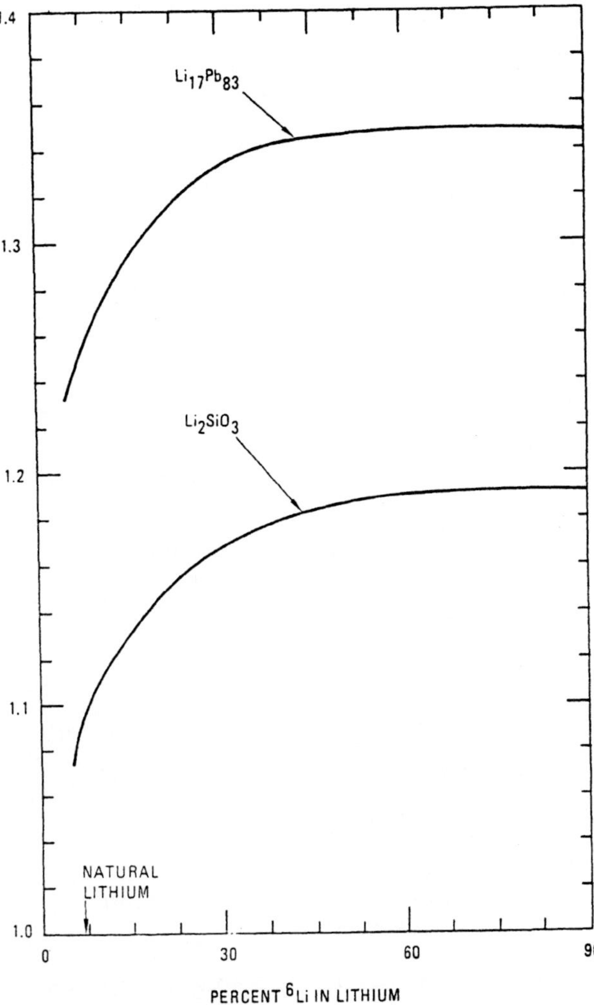

Fig. 1 Tritium breeding ratio as a function of ^{6}Li enrichment in lithium for INTOR-EC $Li_{17}Pb_{83}$ and Li_2SiO_3 blankets

the tritium production rate increases. The contributions due to reactions like Fe(n,γ) and Pb(n,γ) are on the contrary negative, because neutrons are destroyed if these cross sections increase. Increasing the ^{6}Li(n,α) cross section, the flux decreases and the negative flux change tends to compensate for the positive changes of the cross section. Therefore, the sensitivity due to ^{6}Li is very small and can be neglected. If the uncertainties of the cross sections described in Section "Cross Section Uncertainties" are coupled with the above calculated sensitivities, the uncertainty of the tritium breeding ratio can be assessed. It is within $\pm$ 2 % in both blanket designs. The most uncertainty ($\sim$1.7 %) is due to the Pb(n,2n) reaction.

Conclusions and Recommendations

Neutronics and cross section sensitivity studies were performed for the European Community International Tokamak Reactor design. From the results of the studies presented above, some conclusions can be reached and recommendations made about future neutronic calculations and cross section measurements.

The nucleonic analyses show clearly that the discrepancies of the ^{6}Li(n,α)T reaction cross sections between the DLC-37, DLC-41, and MACKLIB-IV libraries can cause serious trouble in the evaluation of tritium production in the INTOR. It was found that the MACKLIB-IV cross sections between 0.02 and 4 eV are not reliable. The $^{6}Li(n,\alpha)T$ reaction cross sections in the above energy range should be accurately collapsed using a representative weighting spectrum. In general, greater care must be given to preparation of broad group cross section sets for fusion, especially in the thermal energy region for blankets containing moderator materials. The tritium production in the INTOR blanket is found to be insensitive to the nuclear data uncertainties in most blanket materials except lead. The uncertainty of tritium production due to nuclear data uncertainties in lead is about $\pm$ 2 % and is primarily due to the Pb(n,2n) reaction cross section in the energy range between 12 and 15 MeV. In order to reduce the uncertainty of tritium breeding ratio below $\pm$ 1 %, the mentioned cross section should be known within $\pm$ 3 %. Since the uncertainty associated with the Pb(n,2n) cross section is presently $\pm$ 7 %, resulting in a breeding ratio uncertainty of $\pm$ 1.7 %, additional measurements appear to be needed. The uncertainty associated with the ^{6}Li and ^{7}Li cross sections are $\pm$ 5 % and $\pm$ 10 %, which result in $\pm$ 0.1 % and ~ 0.0 % uncertainty in the tritium breeding ratio. Thus these cross sections are adequately known to insure $\pm$ 1 % uncertainty on the INTOR-EC tritium breeding ratio.

References

1. C. Ponti and M. Rieger, "Chapter IX. Tritium Producing Blanket", European Community INTOR Design Study, March 1981.
2. W.W. Engle Jr., "A User's Manual for ANISN", K-1693, Oak Ridge Gaseous Diffusion Plant (1967).
3. D.M. Plaster et al., "Coupled 100 Group Neutron and Group Gamma-Ray Cross Sections for EPR Calculations, ORNL-TM-4873, Oak Ridge National Laboratory (1975).
4. "VITAMIN-C: 171 Neutron, 36 Gamma Ray Group Cross Sections for Fusion and LMFBR Calculations", RSIC, Oak Ridge National Laboratory, DLC-41 (1978).
5. E.T. Cheng et al., Nucl. Technol. 45, 77 (1979).
6. Y. Gohar and M.A. Abdou, "MACKLIB-IV, A Library of Nuclear Response Functions Generated with the MACK-IV Computer Program from ENDF/B-IV", ANL/FPP/TM-106, Argonne National Laboratory (1978).
7. S. Pelloni, Cross Section Sensitivity and Uncertainty Analysis for European Intor and U.S. TED Designs, GA-A16685, General Atomic Company, San Diego (1982)
8. D.E. Bartine, F.R. Mynatt, E.M. Oblow, "SWANLAKE: A Computer Code Utilizing ANISN Radiation Transport Calculations for Cross Section Sensitivity Analysis", ORNL-TM-3809 (May 1973).
9. J. Frehaut and G. Mosinski, 5th International Symposium on the Interaction of Fast Neutrons with Nuclei, Gaussig, DDR (November 17-21, 1975).
10. E.T. Cheng and D.R. Mathews, "The Influence of Nuclear Data Uncertainties on Thorium Fusion-Fission Hybrid Blanket Nucleonic Performance", Proceedings of the International Conference on Nuclear Cross Sections for Technology, Munich, October 22-26, 1979.

INTEGRAL EXPERIMENTS TO CHECK NUCLEAR DATA AND CALCULATIONAL METHODS FOR CTR BLANKETS

P. Cloth, V. Drüke, D. Filges, R. Hecker

Institut für Reaktorentwicklung
Kernforschungsanlage Jülich GmbH
Postfach 1913
D-5170 Jülich 1, Germany

Association EURATOM-KFA

ABSTRACT

The main objective of our work was comparison of experimental results with assessment of the theoretical models in order to obtain validation of nuclear data and calculational methods. The limitations and recent enhancements of the present blanket experiments are discussed. The measuring precision of several blanket parameters, e.g. the most prominent one - the tritium production rate - is principally limited by the precision of the neutron-source strength. The determination of this value has normally an uncertainty of at least 15 %. Consequently we try to reduce this error to about 5 % by development of a source-strength measurement technique utilizing coincidence counting of the neutrons and the associated alpha particles of the DT reaction used for neutron production.

INTRODUCTION

A series of neutronic model experiments on CTR blankets has been performed at KFA during the last years. Although these experiments were related to the study of the basic blanket parameters - mainly tritium production rates and to special development of appropriate measuring techniques - ,they were considered to be predominantly benchmark experiments for testing nuclear data and calculational methods. The benchmarks presented here are examples of comparing experimental results with theory and they illustrate the state of accuracy reached. It turns out that the overall uncertainty of the measurements is not better than 15-20 % and that the theoretical predictions are in agreement within this error margin. This is, however, unsatisfactory in a theoretical determination of the main blanket parameters e.g. the tritium breeding rate. Enhancement of the measuring techniques is therefore envisaged.

BENCHMARK EXPERIMENTS

Lithium Metal Blanket

The design of blanket models has always been chosen taking into account the properties of the current CTR concepts and the existing computational models. In our first experiment, the structure was for simplicity reasons restricted to lithium metal, as lithium in any form is the most important material for tritium breeding blankets. The toroidal geometry of CTR blankets has been approximated by a hollow cylinder. The compromise between a small size, desirable for technical reasons, and a representative model was found by a parametric study using the simple one-dimensional transport code ANISN /1/. The outer diameter of a homogeneous lithium hollow cylinder with a constant inner diameter of 20 cm was varied. The axial dimension was assumed to be always equal to the diameter. The single dimension of the ANISN code was taken to be the radial direction, whereas the axial leakage was considered by a buckling correction. As a result of this study, a hollow lithium cylinder of 120 cm diameter and height was chosen with an inner diameter of 20 cm /2/. The total mass of 690 kg of natural lithium was contained in a 2 mm thick stainless steel vessel. To allow for measurements at various points within this metal block, a system of radial channels has been provided (see Fig. 1). The DT-fusion plasma was simulated by the tritium target of a deuteron accelerator. The target was positioned within the axial channel on the center line of the blanket cylinder.

A comparison of the space-dependent tritium production rate evaluated by the Monte Carlo code MORSE /3/ was made for a typical section of the model blanket. The influence of the stainless-steel containment and the experimental channels was taken into account in the calculations using the 05R-geometry

K. H. Böckhoff (ed.), Nuclear Data for Science and Technology, 335–338.

package supplied with the MORSE code.

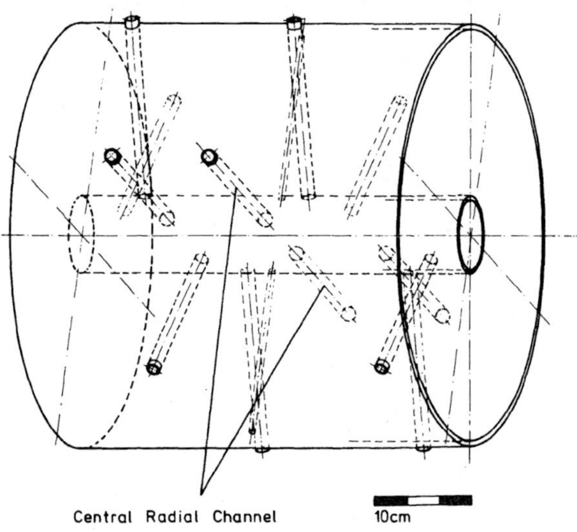

Figure 1

Li-Metal Blanket Model
(Steel Container with Measuring Channels)

The liquid scintillation counting of tritium extracted from lithium carbonate probes was shown to give the best results. The measured curve of tritium production in natural lithium is nearly identical with the calculated one (see Fig. 2). The absolute calibration of the liquid scintillation curve fits the calculated values better than expected from the estimated uncertainty of 15 % of this measurement. The strange increase of the experimental curve near the outer surface of the blanket vessel - not reproduced by the Monte Carlo calculation - is generated by low-energy neutrons backscattered from the concrete walls surrounding the experiment. These neutrons are absorbed by ^{6}Li within a few centimeters of penetration.

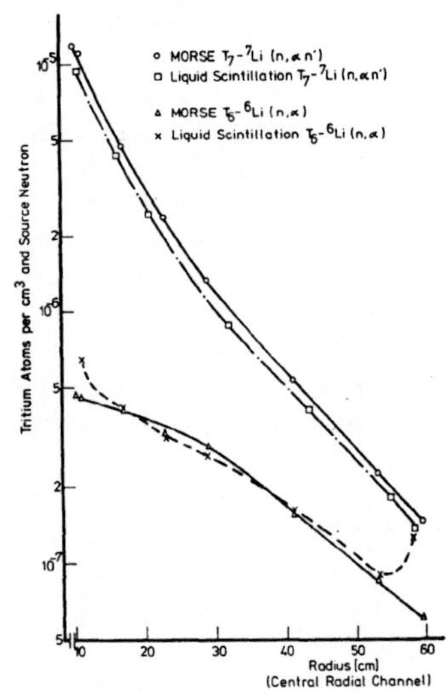

Figure 2

Comparison of Measured and Calculated Results of the Li-Metal Experiment

The measurement described above was obtained with probes containing the lithium isotopes in high abundance (95.62 % ^{6}Li and 99.99 % ^{7}Li, respectively).

Lithium-Aluminate Blanket

During systematic experimental studies on different concepts of the fusion reactor blanket a check of the theoretical predictions concerning the properties of a typical representative of solid state blankets was made. The usage of solid state materials for fusion reactor blankets has several advantages as compared to the liquid metal blanket, that enhance the feasibility of this blanket type /4/. One of the major advantages of the solid state blanket is the low lithium, and thus the low tritium inventory that can be achieved.

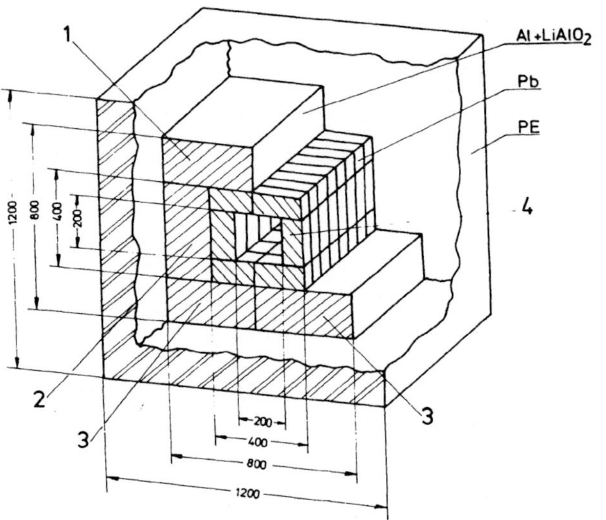

Figure 3

Li-Aluminate Blanket Model
(Schematic Drawing of the Layer Arrangement)

Among the big number of proposed materials for a solid state blanket /5,6/ only the lithium aluminates are considered in this study. The test facility (see Fig. 3) is composed of several material layers, the 20 by 20 cm neutron-source channel is made up of lead briques forming a layer of 10 cm thickness. The following 20 cm layer is piled up by aluminum boxes filled with lithium aluminate. The whole is surrounded by polyethylene to cubic dimensions of 120 cm. For measuring purposes two channels of 4 cm width penetrate the lithium aluminate and the polyethylene starting from the lead surface and extending to the outer surface. The space left between the inserted detectors is filled with suitable cylindrical bodies containing the appropriate materials.

The tritium measuring techniques was again the liquid-scintillation method which uses Li_2CO_3 from which the generated tritium is separated chemically as an aqueous solution /7/. The results of the measurements of the ^6Li-tritium breeding rate are depicted in Fig. 4 together with the calculated curve.

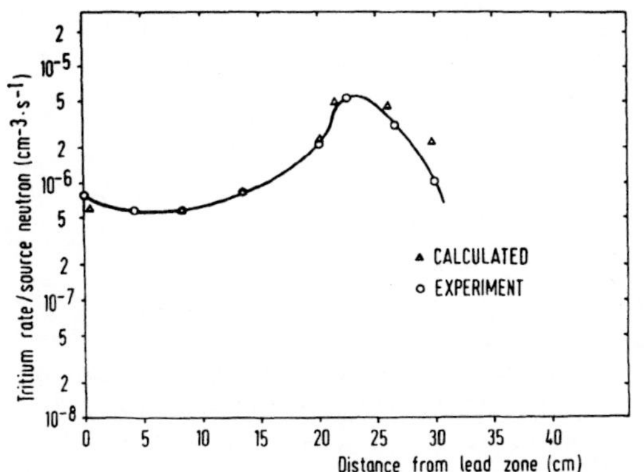

Figure 4

Comparison of Measured and Calculated Results of the Aluminate Blanket Experiment

Neutron Multiplication Experiment

Beryllium is used as an efficient neutron multiplier in several fusion reactor blanket designs. Therefore also measurements of the neutron multiplication in beryllium produced by 14-MeV neutrons were carried out to check basic nuclear data. The measurements were made in rectangular geometry as a function of beryllium thicknesses of 8 and 12 cm. The experimental values of the neutron multiplication were found to be 25 % lower than the calculated values for both thicknesses. The low value of the multiplication casts doubts to the feasibility of certain blanket designs using beryllium as a neutron multiplier to yield sufficient tritium breeding ratios.

A 14-MeV neutron source was surrounded by an 8 to 12 cm thick layer of beryllium. The total number of neutrons leaking out of the beryllium was detected in a large hydrogenous moderator system such as polyethylene (Fig. 5). Ideally, neutron multiplication can be determined as the ratio of total neutron absorptions in an infinite moderator with and without beryllium by means of 1/v detectors such as manganese wires or BF_3 counters. In practice, however, there is a small leakage from the moderator assembly of finite thickness, which can be corrected for. Since polyethylene (containing mainly hydrogen) is a 1/v absorber, the volume-integrated 1/v detector response is directly proportional to the total neutron absorption in the moderator. The results are given in Table I.

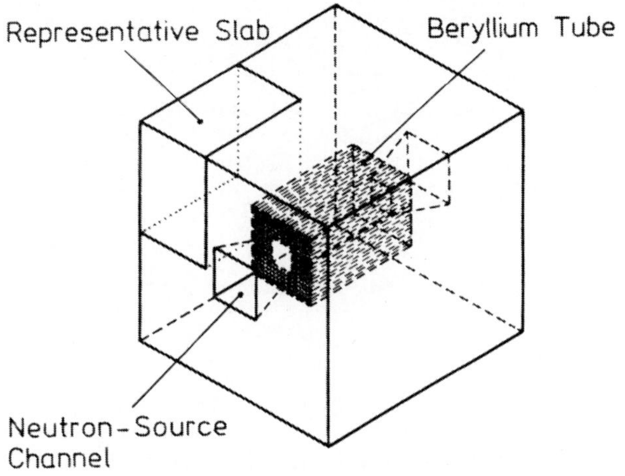

Figure 5

Experimental Setup for Neutron-Multiplikation Measurement of Beryllium

Table I Corrected Multiplication

Be thickness	experimental	calculated	experimental/calculated
8 cm Be	1.35	1.79	0.75
12 cm Be	1.58	2.03	0.78

ENHANCEMENT OF BENCHMARKS

Enhancement of experimental benchmarks as they are described above requires knowledge of the main sources of error in the measurements. Most of the values obtained in these measurements have to be related to the absolute strength of the neutron source used in the experiments. It was found in our series of benchmark measurements that the uncertainty of the source strength was usually dominant when foil-activation technique was used. Improvement of program and nuclear data tests by means of experimental benchmarks can therefore only be achieved by development of more precise source-strength measuring techniques.

Source Strength Determination

Determination of the neutron-generator output can be done by activation methods or by counting of the associated particles, i.e. by counting the α-particles which are produced simultaneously with the neutrons in the t(d,n)α reaction. By this method the absolute neutron output of the target can be determined. Fig. 6 shows a sketch of the experimental assembly. For α-detection a thin plastic scintillator NE 102A of 50μm thickness is placed at the end of a tube facing the target at an angle of 176° to the incident deuteron beam. A thin aluminum foil in front of the detector eliminates scattered deuterons, tritons, and photons as well. Most of the detected signals originate from the desired reaction:

$$d + t \rightarrow \alpha + n + 17.588 \text{ MeV} \quad (1)$$

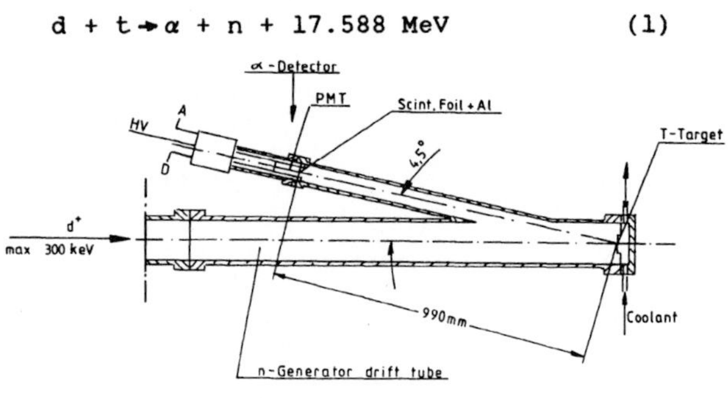

Figure 6

Arrangement for Associated Particle Measurement at a DT-Neuton Generator

Since the target contains tritium, a certain amount of of ^{3}He is obtained after some time by radioactive decay. Therefore the reaction

$$d + {}^3He \rightarrow \alpha + p + 18.34 \text{ MeV} \quad (2)$$

is also possible as well as the following reactions:

$$d + d \rightarrow {}^3He + n + 3.27 \text{ MeV and} \quad (3)$$

$$d + d \rightarrow t + p + 4.04 \text{ MeV.} \quad (4)$$

In the last reaction (4) the proton can be detected very well and, therefore, the fraction of dd neutrons can be determined. The second reaction (2) gives α-particles with an energy similar to that from the desired reaction number 1. This "α-background effect" can be reduced in fairly new targets to 1 % correction, if the energy of the incident deuterons is below 200 keV.

The source-strength determination by means of α counting is in our case possible up to a neutron yield of $5 \cdot 10^9$ n/sec. The extrapolation to higher neutron yields is done by simultaneously using α counting and activation foils or fission chambers.

The total error is estimated to be 5 %, as summed up from count losses of 2 %, contribution of parasitic α particles by 1 %, determination error of solid angle 2 % and counting statistics of 0,5 %. Absorption of neutrons in the target material and jacket are not considered.

The other possible method to determine the neutron output as was performed earlier is the foil activation method. We used the reaction $^{19}F(n,2n)\,^{18}F$. This reaction has a threshold energy of 12 MeV, therefore, low energy neutrons or dd neutrons are not measured. The sources of uncertainty in this method comprise counting statistics (0,5 %), determination of the activity standard (2 %) and the cross section of the (n,2n) reaction (10 %). The effect of scattered neutrons with an energy above 12 MeV has to be calculated as a correction factor. It might be that the difference observed between the two described methods of nearly 15 % can be explained by this effect.

REFERENCES

/1/ W. Engle
Report-K-1693, 1967

/2/ R. Herzing, L. Kuypers, P. Cloth, D. Filges, R. Hecker, N. Kirch
Nucl. Eng., Vol. 60 (2), 1976

/3/ E.A. Straker et al.
ORNL-4585, 1970

/4/ A. Mohsin
Jül-1705, Feb. 1981

/5/ J.R. Powell
CONF-750989, Vol. III, 1976

/6/ R.W. Conn, G. Kulcinski and C. Maynard
Nuclear Engineering and Design, 39, 1976, 5.44

/7/ R. Dierckx
Nucl. Instr. Meth. , 107, 307, 1973

COMPARISON OF LOS ALAMOS MATXS, VITAMIN-C AND DLC-37 MULTIGROUP LIBRARIES FOR A REFERENCE FUSION HYBRID BLANKET

D. J. Dudziak*, J. Stepanek, W. T. Urban** and G. Friedrich

Swiss Federal Institute for Reactor Research,
5303 Würenlingen, Switzerland

* On leave from Los Alamos National Laboratory,
Los Alamos, USA

** Los Alamos National Laboratory, Los Alamos, USA

The helium-cooled, thorium-metal based blanket of the tandem mirror hybrid reactor is used as a reference blanket for comparisons among the Los Alamos NJOY-produced fusion library collapsed into the 30 neutron- and 12 gamma-group MATXS library structure, the VITAMIN-C (DLC-41) library with 171 neutron- and 36 gamma-group structure, the VITAMIN-C library collapsed into the Los Alamos 30/12 group structure using a 1/E spectrum, and a 25 neutron- and 21 gamma-group cross-section set collapsed from the EPR (DLC-37) 100 neutron- and 21 gamma-group set using a 1/E spectrum. The NJOY-produced library is collapsed using a fusion-fission-1/E-Maxwellian spectrum weighting. The Los Alamos one-dimensional S_N transport code ONEDANT and the EIR one-dimensional "surface-flux" transport code SURCU were used for the calculations. Studies of modelling parameters such as order of approximation in quadratures and scattering were also performed. The reactions ^{6}Li (n,t), ^{7}Li (n,n't), Th (n,γ), Th (n,2n), Th (n,3n), Th (n,f), Fe (n,γ), Cr (n,γ) and Ni (n,γ) are compared for the beginning of life. Detailed specifications of the model and data are given, providing a preliminary proposed hybrid reactor blanket benchmark.

Introduction

Prior to embarking on a fusion/fission hybrid blanket design study at the Eidg. Institut für Reaktorforschung (EIR), a review was made of previous similar studies, especially the nucleonics aspects. After a ^{232}Th-^{233}U cycle was selected for the initial conceptual design at EIR, an evaluation of available nuclear data sets was undertaken. The gas-cooled ^{232}Th metal, Li_2O blanket composition used in a tandem mirror hybrid study by Lawrence Livermore Laboratory and General Atomic Company (GA) was selected as a base case for the nuclear data intercomparison. Using a reference blanket configuration that was well defined by Cheng[1], a series of transport calculations were performed with two ENDF/B-IV based data sets available at EIR (the 171 neutron- and 36 gamma-group VITAMIN-C, and the 30 neutron- and 12 gamma-group Los Alamos MATXS libraries [2,3]) and one data set at Los Alamos National Laboratory (LANL), the 30 neutron- and 12 gamma-group ENDF/B-V based MATXS5 library. These calculations were then compared with earlier ones for the same configuration by Cheng[1], who used the EPR (DLC-37) 100 neutron- and 21 gamma-group library [4] after collapsing it into a 25 neutron- and 21 gamma-group structure used at GA.

In addition to simply intercomparing data sets for a well defined computational model, several aspects of the model itself were analysed to determine their effect on the integral parameters of interest, which were principally tritium producing reaction rates in Li and various reaction rates in ^{232}Th. Variations in Fe, Ni and Cr reaction rates were computed, but the differences of 10 % or less found are not significant in the context of systems characteristics such as materials damage correlations, transmutation and activation, where errors of this magnitude are masked by other uncertainties. However, the indirect effects of the Fe, Ni and Cr reaction rates are, of course, incorporated implicitly in the fusile and fissile material breeding rates via their affecting neutron balance and spectrum.

Modelling parameters that were varied included order of angular quadrature and scattering approximations, spatial mesh refinement, and the numerical method for solution of the linear Boltzmann transport equation (i.e., discrete ordinates and surface flux methods). The analysis of computational modelling effects was performed with a view toward use of the model as a future benchmark. Thus, an attempt has been made to define the configuration, atom densities, and other properties of the computational model in sufficient detail to provide the basis for a future hybrid reactor benchmark specification.

Thorium Fast Fission Hybrid Blanket Specification

The thorium fast fission hybrid blanket consists of a 0.01-m Inconel-718 first wall, a 0.12-m thorium metal zone, a 0.4-m Li_2O breeding zone and a 0.1-m Inconel-718 reflector/hot shield. At the center is located the D-T plasma that generates 14-MeV fusion neutrons. Its radius is 0.9 m.

Between the plasma zone and the first wall a vacuum zone of thickness 0.1 m is located. The detailed configuration as well as the material compositions are given in Table I.

Table I Thorium Fusion-Fission Hybrid Blanket Configuration and Material Compositions

Zone	Thickness (m)	Composition	Remarks
1	0.90	Vacuum	Plasma
2	0.10	Vacuum	Vacuum
3	0.01	Inconel-718	First wall
4	0.12	6,3% Inconel-718 + 10% helium+ 83,7% thorium	^{233}U production
5	0.40	6,3% Inconel-718 + 10% helium+ 83,7% Li_2O	Tritium breeding
6	0.28	Inconel-718	Reflector/hot shield

Table II shows the atom number densities as used in all the nucleonic calculations. They are given to five significant figures for consistency, recognizing that actual engineering materials and dimensions are known generally to at best three significant figures.

The nucleonic calculations were performed using the one-dimensional discrete-ordinates transport code ONEDANT [5] and the surface-flux transport code SURCU[6]. The calculations were performed in cylindrical geometry considering an infinite height of the cylinder; i.e., buckling equal to zero, and a "GRAY" boundary condition at the external boundary with albedo equal to 0.3.

K. H. Böckhoff (ed.), Nuclear Data for Science and Technology, 339–344.

Sample calculations with the reference 171 neutron-group library showed the reaction rates to be insensitive to the right boundary condition, within four significant figures, when albedos of 0.0 and 0.3 were compared. The inhomogenious source was placed in the energy group encompassing 14.06 MeV, and was distributed uniformly over the spatial extent of the plasma region; i.e., 0.90-m radius. Results presented in the next section are given for various specified angular approximations and different spatial meshes. However, for the ONEDANT discrete-ordinates calculations the angular quadrature sets were always the generalized quadrature GQ_N sets from the program library (cf. Ref. 5, p. V-32 for discussion of GQ_N sets), as recommended in the code user's manual. Sample cases run with the Gaussian P_N quadrature set revealed insignificant differences for the integral quantities examined in this paper. Also, the ONEDANT spatial mesh was composed of coarse-mesh intervals identical to the "zones" defined in Table I, except for the ^{233}U-production zone 4, which was further subdivided into seven coarse-mesh intervals. Fine-meshes are then always equally spaced within a coarse-mesh interval by ONEDANT. Except for one direct comparison with Cheng's calculation, both the pointwise flux convergence and outer iteration convergence criteria were 10^{-4} (i.e., EPSI=EPSO=10^{-4} in ONEDANT).

The SURCU code is based on the DP_N (double P_N) surface flux integral transport method. The space is subdivided into homogeneous space intervals. The spatial distribution of the fluxes and sources in each space interval is represented by a legendre polynomial expansion, while double P_N (DP_N) spherical harmonics orthogonal in angular half-spaces are used to approximate the incoming and outgoing surface fluxes at the interfaces between space intervals. In the calculations the coarse space mesh distribution was chosen the same as the zones defined in Table I. Then the thorium and lithium zones and reflectors were subdivided into 2, 4 and 8 intervals. In some calculations also the plasma zones and first wall were subdivided into 2 intervals. Whereas P_0 to P_3 anisotropy of scattering was considered, a DP_2 to DP_4 approximation was used to approximate the angular distribution of the surface fluxes.

All computations at EIR and LANL employed fission neutron spectra from ENDF/B-V, specifically those for 14-MeV neutrons incident on ^{232}Th. Values of χ are given in Table III along with the group structures for the standard Los Alamos 30 neutron- and 12 gamma-group libraries. The χ values in 171 groups corresponding to the DLC-41 structure are given in Table IV. The calculation by Cheng used 25-group χ values for thermal fission of ^{235}U.

Table II Hybrid Blanket Atom Number Densities by Zone

Zone	Material	Nr. Density (atoms/b·cm)
3	Fe	2.2179E-2
	Cr	1.5628E-2
	Ni	4.4128E-2
4	Fe	1.3973E-3
	Cr	9.8456E-4
	Ni	2.7801E-3
	^{232}Th	2.5425E-2
5	Fe	1.3973E-3
	Cr	9.8456E-4
	Ni	2.7801E-3
	^{6}Li	5.0940E-3
	^{7}Li	6.3540E-2
	O	3.4317E-2
6	Fe	2.2179E-2
	Cr	1.5628E-2
	Ni	4.4128E-2

Table III Los Alamos 30/12 Energy Group Structure, and Fission Neutron Spectrum for 14-MeV Fission of ^{232}Th

(Group Boundaries in Electron Volts)

Neutrons

Group,g	Maximum Energy	χ_g
1	1.70000E+07	0.69572E-4
2	1.50000E+07	0.17738E-3
3	1.35000E+07	0.49943E-3
4	1.20000E+07	0.21855E-2
5	1.00000E+07	0.97425E-2
6	7.79000E+06	0.26611E-1
7	6.07000E+06	0.12780
8	3.68000E+06	0.10502
9	2.86500E+06	0.11613
10	2.23200E+06	0.11550
11	1.73800E+06	0.11693
12	1.35300E+06	0.16306
13	8.23000E+05	0.10136
14	5.00000E+05	0.56628E-1
15	3.03000E+05	0.29374E-1
16	1.84000E+05	0.22203E-1
17	6.76000E+04	0.51951E-2
18	2.48000E+04	0.11740E-2
19	9.12000E+03	0.26366E-3
20	3.35000E+03	0.58743E-4
21	1.23500E+03	0.13177E-4
22	4.54000E+02	0.29372E-5
23	1.67000E+02	0.65549E-6
24	6.14000E+01	0.14606E-6
25	2.26000E+01	0.32616E-7
26	8.32000E+00	0.72885E-8
27	3.06000E+00	0.16227E-8
28	1.13000E+00	0.36545E-9
29	4.14000E-01	0.80965E-10
30	1.52000E-01	0.0
EMIN	1.39000E-04	

Gamma Rays

Group,g	Maximum Energy
31	2.00000E+07
32	9.00000E+06
33	8.00000E+06
34	7.00000E+06
35	6.00000E+06
36	5.00000E+06
37	4.00000E+06
38	3.00000E+06
39	2.00000E+06
40	1.00000E+06
41	5.00000E+05
42	1.00000E+05
EMIN	1.00000E+04

Table IV Fission Spectrum for 171 Neutron Energy Groups of the DLC-41 Library

g	χ_g	g	χ_g	g	χ_g
1	4.55060E-05	16	2.03550E-03	31	3.21290E-02
2	2.63820E-05	17	2.55980E-03	32	3.63790E-02
3	4.35440E-05	18	3.16850E-03	33	4.00740E-02
4	3.11180E-05	19	3.86360E-03	34	2.12140E-02
5	3.91000E-05	20	4.64550E-03	35	2.18790E-02
6	4.89130E-05	21	5.51100E-03	36	2.24170E-02
7	6.03790E-05	22	2.04610E-03	37	2.28540E-02
8	1.64730E-04	23	4.41460E-03	38	2.31960E-02
9	2.43160E-04	24	7.46890E-03	39	2.34280E-02
10	3.51120E-04	25	8.54310E-03	40	1.56960E-02
11	4.94240E-04	26	9.66830E-03	41	3.95360E-03
12	6.82590E-04	27	1.08250E-02	42	3.92810E-03
13	9.23280E-04	28	1.20090E-02	43	7.87650E-03
14	1.22330E-03	29	1.31980E-02	44	1.57340E-02
15	1.59180E-03	30	1.43780E-02	45	2.35520E-02

Table IV (cont'd)

g	X_g	g	X_g	g	X_g
46	2.34130E-02	88	2.59330E-03	130	1.04760E-05
47	2.32030E-02	89	2.42230E-03	131	9.02040E-06
48	2.29090E-02	90	2.26040E-03	132	4.02560E-06
49	2.25380E-02	91	2.10990E-03	133	3.73590E-06
50	2.21000E-02	92	1.97020E-03	134	6.68380E-06
51	2.16310E-02	93	1.83470E-03	135	5.75380E-06
52	2.11160E-02	94	1.71380E-03	136	1.11260E-05
53	2.05630E-02	95	1.59450E-03	137	7.64800E-06
54	1.99250E-02	96	1.48910E-03	138	5.25760E-06
55	1.93540E-02	97	1.38500E-03	139	3.61410E-06
56	1.86560E-02	98	1.29200E-03	140	2.48420E-06
57	1.80430E-02	99	1.20400E-03	141	1.70760E-06
58	1.73450E-02	100	1.11880E-03	142	1.17360E-06
59	3.26520E-02	101	1.04380E-03	143	8.06770E-07
60	1.28210E-02	102	2.30310E-03	144	5.54430E-07
61	1.71360E-02	103	1.92390E-03	145	3.81020E-07
62	1.39750E-02	104	6.45080E-04	146	2.61960E-07
63	1.33250E-02	105	4.72670E-04	147	1.80000E-07
64	1.26830E-02	106	1.14640E-03	148	1.23720E-07
65	1.20600E-02	107	6.80020E-04	149	8.50350E-08
66	1.14540E-02	108	1.50820E-03	150	5.84450E-08
67	1.08600E-02	109	5.36930E-04	151	4.01660E-08
68	1.02930E-02	110	7.73460E-04	152	2.76100E-08
69	9.74180E-03	111	6.43450E-04	153	1.89750E-08
70	9.21140E-03	112	7.23160E-04	154	1.30410E-08
71	8.70040E-03	113	2.56990E-04	155	8.96100E-09
72	8.21020E-03	114	3.30100E-04	156	6.16110E-09
73	7.74470E-03	115	1.42920E-04	157	4.23370E-09
74	1.41680E-02	116	8.78200E-05	158	2.90980E-09
75	1.25410E-02	117	1.15980E-04	159	2.00000E-09
76	5.70910E-03	118	5.48780E-05	160	1.37450E-09
77	5.36050E-03	119	5.28880E-05	161	9.44760E-10
78	9.74990E-03	120	1.47380E-04	162	6.49230E-10
79	8.56730E-03	121	2.11850E-04	163	4.46260E-10
80	9.19690E-04	122	3.22160E-04	164	3.06690E-10
81	3.43690E-04	123	2.21850E-04	165	2.10770E-10
82	7.07020E-04	124	1.52790E-04	166	1.44910E-10
83	1.90760E-03	125	1.05150E-04	167	9.95680E-11
84	3.63200E-03	126	7.23480E-05	168	6.84330E-11
85	6.57610E-03	127	4.97660E-05	169	4.70360E-11
86	2.97130E-03	128	2.20420E-05	170	9.25990E-11
87	2.77630E-03	129	1.21690E-05	171	1.30070E-11

Two basic nuclear data multigroup libraries were used in this study. First, the 171 neutron- and 36 gamma-group VITAMIN-C (DLC-41) library [2] was used directly (uncollapsed) as a reference calculation, and then collapsed to the standard Los Alamos 30/12-group structure using a 1/E weighting spectrum. This latter spectrum was chosen to be consistent with Cheng's collapsing procedure for the DLC-37 library, where he used a 1/E weighting spectrum. Another study of collapsing spectra effects for the INTOR-EC blanket designs [7] showed, however, a difference of only ∿ 1 % in tritium breeding ratio between the DLC-41 library collapsed with a 1/E spectrum versus the standard Los Alamos spectrum, with the Los Alamos spectrum giving results closer to the reference 207-group calculations. The DLC-41 library was produced from ENDF/B-IV.

The second base multigroup libraries used were the Los Alamos 30/12-group MATXS libraries derived from ENDF/B-IV, a LANL evaluation of ^{6}Li, and ENDF/B-V, in all cases processed by the NJOY code into a fine-group structure and then collapsed using the standard Los Alamos fusion-fission-1/E-Maxwellian weighting spectrum[7]. Response function vectors for ^{6}Li(n,t), ^{7}Li(n,n't), ^{232}Th(n,γ), ^{232}Th(n,2n), ^{232}Th(n,3n), Fe(n,γ), Fe(n,2n), Ni(n,γ), Ni(n,2n), Cr(n,γ) and Cr(n,2n) were processed in the same way as the cross sections and transfer matrices for the transport calculations; i.e., with the same weighting functions for collapsing.

Table V Neutron 25 Energy Group Structure in eV

Group, g	Maximum Energy	Corresponding Fine Groups in 100 Group Structure [a]
1	1.4918 (+7)	1
2	1.3499 (+7)	2
3	1.2214 (+7)	3
4	1.1052 (+7)	4
5	1.0000 (+7)	5
6	9.0480 (+6)	6
7	8.1873 (+6)	7
8	7.4082 (+6)	8
9	6.7032 (+6)	9
10	6.0653 (+6)	10
11	5.4881 (+6)	11 - 12
12	4.4933 (+6)	13 - 14
13	3.6788 (+6)	15 - 16
14	3.0119 (+6)	17 - 18
15	2.4660 (+6)	19 - 24
16	1.3534 (+6)	25 - 30
17	7.4274 (+5)	31 - 36
18	4.0762 (+5)	37 - 45
19	1.6573 (+5)	46 - 54
20	3.1828 (+4)	55 - 63
21	3.3546 (+3)	64 - 72
22	3.5358 (+2)	73 - 81
23	3.7267 (+1)	82 - 90
24	3.9279 (+0)	91 - 99
25	4.1399 (-1)	100
EMIN	2.2000 (-2)	

[a] The original DLC-37 library group structure

Results obtained by Cheng were computed with the S_N code ANISN[9]. All his multigroup nuclear data except ^{232}Th came from the DLC-37 (transport)[4] and MACKLIB (response function)[10] libraries, which he collapsed into a 25/21 group structure (cf Table V for the neutron-group structure). Data for ^{232}Th from DLC-41 were collapsed into the same 25/21 group structure by Cheng[1]. Thus, all his multigroup data are ENDF/B-IV based, and collapsed using a 1/E weighting spectrum.

All cross sections used for the analyses were infinitely dilute (resonance unshielded) values. Note from Tables III and V that the lower bounds of the thermal groups varied considerably. However, for the hybrid-system analysed in this paper, in contrast to other fusion blankets[7], responses in the thermal group were a minor contribution to the total values.

Analysis of Results

Prior to attempting an intercomparison of responses among various multigroup data sets, it was necessary to establish the convergence of the transport calculations with respect to order of angular approximation and spatial mesh and order of expansion of the cross section scattering matrices. One such study was undertaken with the ONEDANT code, using mainly the Los Alamos 30/12-group MATXS library based upon ENDF/B-IV and a LANL evaluation of ^{6}Li that uses a pseudo-level formalism to better approximate secondary neutron energy spectra. Another was performed with the surface-flux transport code SURCU, using the 171 neutron-group DLC-41 library, with no gamma-ray transport considered and no kerma calculations.

Results of the ONEDANT calculations are presented in Table VI, where reaction rates in Li and Th are presented, along with neutron and gamma-ray kerma. Reaction rates for (n,γ) and (n,2n) reactions in Fe, Ni and Cr were also determined, and were found to agree to $<$ 10 % in all cases. The first set of calculations examined the order of scattering, holding the other parameters fixed; viz, S_4 and 40 mesh intervals.

Table VI Convergence Study with ONEDANT Using LANL (MATXS) Library. Pointwise convergence is 10^{-4}.

S_N	P_N	Number of Intervals per Zone and Total	Reaction Rates* $^6Li(n,t)$	$^7Li(n,n't)$	Breeding Ratio	$^{232}Th(n,\gamma)$	$^{232}Th(n,f)$	$^{232}Th(n,2n)$	$^{232}Th(n,3n)$	Energy Multiplicat. Factor: $E_T/14.06$
	0		0.8663	0.0624	0.9288	0.9088	0.1081	0.3505	0.1117	3.3303
4	1^+	1/1/1/8/20/9//40	0.8992	0.0819	0.9811	0.8391	0.1036	0.3376	0.1088	3.2274
	3		0.8975	0.0822	0.9797	0.8388	0.1035	0.3370	1.1086	3.2250
6	3	1/1/1/8/20/9//40	0.8955	0.0859	0.9814	0.8317	0.1025	0.3336	0.1076	3.2054
8	3	1/1/1/8/20/9//40	0.8952	0.0868	0.9820	0.8298	0.1022	0.3327	0.1074	-
4	3	10/1/2/12/30/14//69	0.8975	0.0836	0.9812	0.8380	0.1034	0.3368	0.1087	3.2231
6	3	10/1/2/12/30/14//69	0.8957	0.0874	0.9831	0.8311	0.1024	0.3335	0.1077	3.2041
8	3	10/1/2/12/30/19//69	0.8953	0.0883	0.9837	0.8291	0.1022	0.3326	0.1075	3.1990
8	3	20/1/2/12/30/14//79	0.8953	0.0883	0.9837	0.8291	0.1022	0.3326	0.1075	3.1990

S_N	P_N	Number of Intervals per Zone and Total	Neutron Energy Deposition* (MeV) FW	FISS	TBREED	REFL	Gamma-Ray Energy Deposition* (MeV) FW	FISS	TBREED	REFL
	0		0.8226	26.464	5.874	-0.0004	0.9077	11.723	0.8210	0.2124
4	1	1/1/1/8/20/9//40	0.7421	25.181	6.413	0.0056	0.8049	10.960	0.9577	0.3037
	3		0.7422	25.151	6.406	0.0074	0.8067	10.961	0.9566	0.3131
6	3	1/1/1/8/20/9//40	0.7349	24.913	6.456	0.0080	0.8106	10.856	0.9712	0.3188
8	3	1/1/1/8/20/9//40	0.7326	24.853	6.470	0.0082	0.8145	10.825	0.9755	0.3205
4	3	10/1/2/12/30/14//69	0.7304	25.132	6.430	0.0077	0.8085	10.933	0.9628	0.3132
6	3	10/1/2/12/30/14//69	0.7213	24.902	6.482	0.0084	0.8115	10.828	0.9780	0.3192
8	3	10/1/2/12/30/14//69	0.7192	24.840	6.495	0.0085	0.8152	10.797	0.9821	0.3210

* Per D-T neutron; + P_1 cross sections not transport corrected

As has been shown in numerous such studies, P_0 is clearly inadequate, whereas P_1 is sufficient for blanket (but not necessarily shielding) computations in the context of design studies, with error <<1 %. Also, the S_N-order study, not surprisingly, showed S_6 to be adequate for engineering design calculations (<<1%), and even S_4 is within ∿1 % of the S_8 results except for the $^7Li(n,n't)$ reactions at deeper penetration distances. Refinement of the spatial mesh, mostly in the plasma, thorium and Li_2O zones, showed small changes in going from 40 to 69 total intervals. Except for $^7Li(n,n't)$ reactions again, changes were generally <1 %. Further refinement of the mesh in the plasma-neutron source region from 10 to 20 intervals showed no effect on the integral quantities. Most sensitive to the order of scattering and quadrature is the energy deposition (actually kerma) in the reflector, where both neutron and gamma-ray penetration is greatest.

Negative neutron kerma for S_4P_0 occurs because of negative kerma factors, not transport calculational errors (i.e., fluxes are positive). These negative kerma factors arise because of inconsistencies with respect to energy conservation between neutron reactions and gamma-ray production data in the ENDF/B data files. In order to force integral energy conservation, the NJOY code first computes gamma-ray production cross sections, and then computes neutron kerma factors by the balance between total energy release and gamma-ray energy production. As a result, negative neutron kerma factors sometimes occur, [12] and the spatial distribution of the energy deposition has an inherent residual error. Thus, the MATXS library is the only one in this study that can provide correct total energy deposition and multiplication factors, because the other libraries contain neutron kerma factors and gamma-ray production data computed independently.

Results of the SURCU calculations are summarized in Table VII.The $^6Li(n,t)$, $^7Li(n,n't)$, breeding ratio, $^{232}Th(n,\gamma)$, $^{232}Th(n,f)$, $^{232}Th(n,2n)$ and $^{232}Th(n,3n)$ reaction rates are presented. P_0, P_1 and P_3 approximation of the neutron scattering combined with DP_2, DP_3 and DP_4 angular surface flux approximation and P_0 and P_1 spatial flux approximation were used for comparison. Although the P_0 scattering approximation is rather inadequate, it was used in order to show the convergence of the scattering approximation in connection with P_1 and P_3 approximations. As expected, the conclusion is the same as those for ONEDANT calculations; namely that P_1 approximation is adequate for this type of blanket calculation. Comparing the DP_2 and DP_3 calculations for 1/1/1/4/4/4 spatial intervals per zone, one sees that the DP_N angular approximation of surface fluxes at the mesh boundaries converges extremely fast. Already DP_2 approximation is adequate. Another SURCU advantage is that the spatial flux approximation using the Legendre polynomials leads to adequate results already for 1/1/1/4/4/4 spatial intervals per zone. This results, in comparison to S_N theory, in a smaller number of unknowns that have to be solved for a given accuracy of the expected results. Therefore, the code is very quick, needs generally less memory, and the resulting system of equations can be solved simpler. On the other hand, the disadvantage in comparison to the S_N methods is that the method is much more complicated than S_N theory, so that its generalization to additional geometries and programing the method needs more effort.

After establishing the convergence adequacy, an intercomparison was then undertaken among the libraries, as presented in Table VIII. In the second column are the values as computed by Cheng, using a pointwise convergence criteria of $2x10^{-2}$ in ANISN. This calculation was then repeated at EIR with ONEDANT, but using the MATXS library for transport and response data. Both the $2x10^{-2}$ and the default 10^{-4} convergence criteria were used, with results shown in columns four and five of Table VIII. Differences are seen in the third and fourth significant figures; hence, the default convergence was used in all other calculations to allow intercomparison to four figures. In either case the comparison with Cheng's values show differences of up to 6 %, apparently due to the different nuclear data. In order to extend the analysis to the DLC-41 library, those data collapsed to the Los Alamos group structure, but with the same 1/E weighting spectrum used by Cheng, were used in an otherwise identical ONEDANT calculation. No consistent pattern

Table VII Convergence Study with SURCU Using 171 Neutron + 36 Gamma Groups DLC-41 Library. Pointwise convergence is 10^{-4}.

DP_N	P_N	P_M	Number of Intervals per Zone and Total	$^6Li(n,t)$	$^7Li(n,n't)$	Breeding Ratio	$^{232}Th(n,\gamma)$	$^{232}Th(n,f)$	$^{232}Th(n,2n)$	$^{232}Th(n,3n)$
2	0	0	1/1/1/1/1/1//6	0.8175	0.0711	0.8284	0.4726	0.1001	0.3416	0.1033
			1/1/1/1/1/1//6	0.9330	0.0553	0.9883	0.8524	0.1077	0.3588	0.1071
2	0	1	1/1/1/2/2/2//9	0.8591	0.0522	0.9113	0.8916	0.1076	0.3586	0.1065
			1/1/1/4/4/4//15	0.8494	0.0511	0.9005	0.8980	0.1079	0.3594	0.1066
			1/1/1/1/1/1//6	0.9451	0.0924	1.0375	0.7608	0.0991	0.3329	0.0995
			1/1/1/2/2/2//9	0.8939	0.0858	0.9797	0.8072	0.1018	0.3421	0.1024
2	1	1	1/1/1/4/4/4//15	0.8877	0.0861	0.9738	0.8130	0.1022	0.3433	0.1029
			1/1/1/8/8/8//27	0.8864	0.0870	0.9734	0.8129	0.1021	0.3428	0.1028
			1/1/2/4/4/4//16	0.8877	0.0860	0.9737	0.8132	0.1022	0.3434	0.1029
			2/2/2/4/4/4//18	0.8884	0.0858	0.9742	0.8140	0.1024	0.3439	0.1031
3	1	1	1/1/1/4/4/4//15	0.8874	0.0862	0.9736	0.8126	0.1021	0.3430	0.1028
			1/1/1/8/8/8//27	0.8839	0.0881	0.9720	0.8123	0.1020	0.3420	0.1028
4	3	1	2/2/2/8/8/8//30	0.8844	0.0877	0.9721	0.8130	0.1021	0.3426	0.1029
			2/2/2/16/16/16//54	0.8841	0.0881	0.9721	0.8131	0.1021	0.3426	0.1029

Table VIII Comparison of the ONEDANT Results with Cheng's Results Using LANL (MATXS) and VITAMIN-C (DLC-41) Library. S_6P_3 calculations.

Reaction Rates	Cheng's Values (a) ($\varepsilon=2x10^{-2}$)	LANL ENDF/B-IV Library DLC-41 Th232 Response Functions (a)	LANL ENDF/B-IV Library, Li and Th Functions (a) ($\varepsilon=2x10^{-2}$)	($\varepsilon=10^{-4}$)	DLC-41 1/E Collapsed LANL Resp.F. (a) (c)	DLC-41 1/E Collapsed 1/E Resp.F. (a)	DLC-41 171 Gr. (a)	(b)	ENDF/B-V Response: ENDF/B-IV (b)	ENDF/B-V Response: DLC-41 (b)	ENDF/B-V Response: ENDF/B-V (b)
$^6Li(n,t)$	0.8708	-	0.8962	0.8955	0.8902	-	0.8771	0.8776	0.9334	-	0.9118
$^7Li(n,n't)$	0.0911	-	0.0859_3	0.0859_2	0.0877_2	-	0.0895_0	0.0910_7	0.0956_0	-	0.0956_4
Tritium Breeding Ratio	0.9619	-	0.9822	0.9814	0.9779	-	0.9666	0.9687	1.0290	-	1.0075
$^{232}Th(n,\gamma)$	0.8473	0.8154	0.8307	0.8317	0.8281	0.8117	0.8040	0.8034	0.9050	0.8859	0.8278
$^{232}Th(n,f)$	0.1039	0.1010	0.1024	0.1025	0.1019	0.1005	0.1002	0.1002	0.1087	0.1074	0.1074
$^{232}Th(n,2n)$	0.3304	0.3388	0.3336	0.3336	0.3311	0.3364	0.3363	0.3362	0.3566	0.3612	0.3337
$^{232}Th(n,3n)$	0.1119	0.1017	0.1076	0.1076	0.1064	0.1006	0.1002	0.1003	0.1100	0.1043	0.1291

(a) Using 40 mesh intervals as in Cheng's original calculation.
(b) Using 69 mesh intervals.
(c) S_6P_5 calculations yielded the identical values to 4 significant figures.

of change was observed, with the tritium breeding ratio closer to Cheng's DLC-37 based calculations and the ^{232}Th reaction rates more deviant except for the (n,2n). One other set of calculations using Cheng's model, except with the uncollapsed DLC-41 library, was performed. They showed good agreement in tritium breeding with Cheng (∿ 0,5 %), but large inconsistencies in ^{232}Th reaction rates; e.g. ∿10 % and 5% differences in (n,3n) and (n,γ) rates, respectively. Thus, all three libraries (MATXS, DLC-41, and DLC-37/MACKLIB), although derived mostly from the same ENDF/B-IV file, give largely divergent results.

A final set of calculations was undertaken to evaluate the variations in the above-mentioned reaction rates when the ENDF/B-V data were used. The standard MATXS5 library in the Los Alamos 30/12 group structure was used. These data are collapsed from a fine-group library with the same weighting function as the (ENDF/B-IV) MATXS library. In order to evaluate separately the effects of ENDF/B-V transport cross sections and response functions, calculations were performed with both MATXS and MATXS5, as well as collapsed DLC-41, response functions sets. As can be seen in Table VIII, there remain significant differences among the ENDF/B-V results for the three response functions. For the two sets of ENDF/B-IV response functions, LANL and collapsed DLC-41, the thorium reaction rates show parallel trends to those for transport calculations with the LANL ENDF/B-IV library, as should be expected. However, the consistent ENDF/B-V case (last column in table VIII) shows significant deviations from all the other results, with a higher tritium breeding ratio (4.5 % and 2.6 % higher respectively, than the 171 neutron-group DLC-41 and 30 neutron-group LANL results from ENDF/B-IV). Likewise, the ENDF/B-V thorium fission and (n,3n) reaction rates are considerably higher, and the (n,γ) rates are between those for the DLC-41 and LANL ENDF/B-IV multigroup libraries. If one adapts the consistent ENDF/B-V results as a reference, the best overall agreement in tritium breeding ratio and ^{232}Th reaction rates among the ENDF/B-IV data sets is shown by the consistent LANL MATXS multigroup library.

Conclusions

An extensive comparison of important blanket reaction rates has been made for a fusion-fission hybrid reactor based on the $^{232}Th/^{233}U$ fuel cycle. Generally minor variations were found between sets of transport calculations performed with two quite different numerical methods (viz, surface flux integral transport and discrete-ordinates integrodifferential), when identical nuclear data sets are employed. Convergence studies with the two methods showed that S_6P_3 in discrete ordinates, and DP_2 and P_3 in the surface flux method, were more than adequate for design studies; i.e., $<$ 1 % error.

Detailed comparisons of the MATXS, DLC-41 and DLC-37/ MACKLIB libraries showed large deviations, up to 10 % in Li and ^{232}Th reaction rates, even though all these multigroup libraries are based primarily on ENDF/B-IV pointwise data. Comparisons between the MATXS transport cross-section library and the DLC-41 transport cross-section library collapsed to the MATXS group structure with a 1/E weighting function, show only minor differences. Thus, where the same response functions are used, the weighting function for the transport cross-sections does not seem to be a significant source of error (<1 %). Larger discrepancies occur due to the use of MATXS versus DLC-41 response functions in these cases. Also, comparisons of ^{232}Th reaction rates, using the DLC-41 transport cross-sections and response functions, in 171 groups versus collapsed to 30 groups with a 1/E spectrum, show good agreement.

Consistent computations using the MATXS5 library based on ENDF/B-V data show much higher tritium breeding ratios than those with any of the libraries based on ENDF/B-IV. Also, a comparison of ^{232}Th reaction rates shows the effects of the ENDF/B-V evaluation are considerable; i.e., fission and (n,3n) values much higher and (n,γ) values slightly lower[12]. The increased tritium breeding may then be due more to a neutron richer system rather than increased tritium-producing cross-sections in ENDF/B-V.

Another source of variation in the reaction rates between the present calculations and those of Cheng is the choice of fission neutron yield spectra. However, test calculations show this to be a very minor effect, as could be expected from the low fission rate and consequently small reaction rate from secondary fission neutrons.

Acknowledgement

The authors gratefully acknowledge the invaluable programming efforts of Colin E. Higgs in adapting ONEDANT to the EIR computer, and the extensive help from K. Herda in processing nuclear data and computer code input.
The cooperative efforts of Dr. E.T. Cheng of General Atomic Company, in making his unpublished results available to us, were essential for this paper and are greatly appreciated.

References

1. E.T. Cheng, Private communication (16 Feb. 1982). See also E.T. Cheng and D.R. Mathews "The Influence of Nuclear Data Uncertainties on Thorium Fusion-Fission Hybrid Blanket Nucleonic Performance", Proc. Int. Conf.on Nuclear Cross-Sections for Technology, Knoxville, TN, 22-26 October 1979

2. VITAMIN-C, 171 Neutron, 36 Gamma-Ray Group Cross-Sections for Fusion and LMFBR Calculations", Radiation Shielding Information Center Data Library DLC-41 (1978)

3. R.J. Barrett and R.E. Mac Farlane, "The MATXS-TRANSX System and the CLAW-IV Nuclear Data Library", Proc. Int. Conf. on Nucl. Cross-Sections for Technology, Knoxville, TN, 22-26 October 1979, NBS Spec. Pub. 594

4. D.M. Plaster, et al. "Coupled 100 Group Neutron and 21 Group Gamma-Ray Cross-Sections for EPR Calculations", Oak Ridge National Laboratory report ORNL-TM-4872 (1975)

5. R.D. O'Dell, F.W. Brinkley, Jr., and D.R. Marr, "User's Manual for ONEDANT: A Code Package for One-Dimensional, Diffusion-Accelerated, Neutral-Particle Transport", Los Alamos National Laboratory report LA-9184-M (1982)

6. J. Stepanek, "The DP_N Surface Flux Integral Neutron Transport Method for Slab Geometry", Nucl. Sci. and Eng. 78, 53-65 (1981). (See also EIR-Report No. 365 for slab, spherical and cylindrical geometry)

7. S. Pelloni, J. Stepanek and D. Dudziak, "Comparison of Different Fusion Nuclear Data Libraries using the European INTOR Blanket Designs", Swiss Federal Institute for Reactor Research report (to be published)

8. R.E. Mac Farlane, D.W. Muir and R.M. Boicourt", The NJOY Nuclear Data Processing System; Vol.I: User's Manual", Los Alamos Scientific Laboratory report LA-9303-M (1982)

9. W.W. Engle, Jr., "A User's Manual for ANISN", Oak Ridge Gaseous Diffusion Plant report K-1693 (1967)

10. M.A. Abdou and R.W. Roussin, "MACKLIB-100 Group Neutron Fluence-to-Kerma Factors and Reaction Cross-Sections", Oak Ridge National Laboratory report ORNL-TM-3995 (1974)

11. E.T. Cheng and D.R. Mathews, "The Influence of Nuclear Data Uncertainties on Thorium Fusion-Fission Hybrid Blanket Nucleonic Performance", Proc. Int. Conf. on Nucl. Cross-Sections for Technology, Knoxville, TN, 22-26 Oct. 1979

12. R. E. MacFarlane, "Energy Balance of ENDF/B-V," Trans. Am. Nucl. Soc., Vol. 33, p. 681 (1979).

TWO-DIMENSIONAL CROSS-SECTION AND SED UNCERTAINTY ANALYSIS FOR THE FUSION ENGINEERING DEVICE (FED)

M. J. Embrechts and W. T. Urban

Theoretical Division
University of California
Los Alamos National Laboratory
P. O. Box 1663
Los Alamos, New Mexico 87545 USA

Donald J. Dudziak

EIR - Switzerland
Eidg. Institut fur Reaktorforschung
CH-5303 Würenlingen
Switzerland

The theory of two-dimensional cross-section and secondary-energy-distribution (SED) sensitivity was implemented by developing a two-dimensional sensitivity and uncertainty analysis code, SENSIT-2D. Analyses of the Fusion Engineering Design (FED) conceptual inboard shield indicate that, although the calculated uncertainties in the 2-D model are of the same order of magnitude as those resulting from the 1-D model, there might be severe differences. The more complex the geometry, the more compulsory a 2-D analysis becomes. Specific results show that the uncertainty for the integral heating of the toroidal field (TF) coil for the FED is 114.6%. The main contributors to the cross-section uncertainty are chromium and iron. Contributions to the total uncertainty were smaller for nickel, copper, hydrogen and carbon. All analyses were performed with the Los Alamos 42-group cross-section library generated from ENDF/B-V data, and the COVFILS covariance matrix library. The large uncertainties due to chromium result mainly from large covariances for the chromium total and elastic scattering cross sections.

Introduction

The calculation of nuclear design parameters is often limited in accuracy due to errors resulting from the calculational methods and the cross-section data utilized. Continuous developments in the calculational methods and computer codes accentuate the importance to improve the cross-section data. Cross-section sensitivity analysis[1,2] has become a widely applied design tool which allows the user to pinpoint the most crucial partial cross sections in a specific energy range for a particular design. A cross-section sensitivity analysis leads to a broader interpretation of the results from a transport calculation and, in conjunction with an uncertainty analysis, provides a way to estimate errors for a particular response (e.g. nuclear heating, radiation damage, etc.). Consecutively, it can be determined where improvements in the nuclear data or in the design are needed to reduce that uncertainty. Secondary-energy-distribution sensitivity and uncertainty analysis[3,4] becomes important for the study of threshold reactions, e.g., ^{7}Li(n;n',α)T, while design sensitivity analysis[5] can lead to a significant reduction in the number of transport calculations required during the optimization process for a particular design. The complicated geometry of a fusion reactor makes the capability to undertake two-dimensional sensitivity analysis in a toroidally symmetric system highly desirable. Therefore, a two-dimensional sensitivity and uncertainty analysis code, SENSIT-2D,[6,7] which can mock-up a toroidal geometry, was developed and applied to the FED.

Theory

The theory behind cross-section sensitivity and uncertainty analysis has been explained by several authors[1,2] and involves the use of covariance matrices and sensitivity profiles. The variance of the response corresponding to two correlated cross-section uncertainties can be expressed by

$$\left[\frac{\Delta I}{I}\right]^2 = \sum_{ij} P_{\Sigma_i} P_{\Sigma_j} \frac{\mathrm{Cov}(\Sigma_i,\Sigma_j)}{\Sigma_i \Sigma_j} , \qquad (1)$$

where I is the integral response, Σ_i and Σ_j are cross sections, P_{Σ_i} and P_{Σ_j} are sensitivity profiles of the integral response with respect to Σ_i and Σ_j, and $\mathrm{Cov}(\Sigma_i,\Sigma_j)$ is the matrix of the covariance data. The first part in the above expression ($P_{\Sigma_i}P_{\Sigma_j}$) is highly problem dependent, while the second part requires cross-section uncertainty information and is problem independent.

The sensitivity profiles are defined as the fractional change of the integral response per fractional change of cross-section, normalized per lethargy width. Sensitivity profiles can be calculated based on forward and adjoint angular fluxes. Cross-section covariance data describe the uncertainties in multigroup cross-section data and the correlation between those uncertainties. A non-zero non-diagonal covariance matrix element indicates that there was a common reason why an uncertainty in two different cross sections was introduced. Multigroup covariance data are ordered in covariance matrices. Such a covariance matrix contains GMAX rows and GMAX columns, where GMAX is the number of energy groups. A covariance matrix can contain covariance data of a particular partial cross section with itself over an energy range, with a different partial cross section for the same element, or with a partial cross section for a different element. A 30-group covariance library, COVFILS, based on ENDF/B-V which contains most of the elements commonly used in fusion reactor shielding has been constructed by Muir and LaBauve.[8] The covariance data in this library were processed by using the NJOY code.[9]

By introducing the "hot-cold" and "forward-backward" concepts, Gerstl[3,4] developed a framework which leads to expressions similar to Eq. (1) for calculating uncertainties due to uncertainties in secondary energy (SED) and secondary angular (SAD) distributions. For a SED uncertainty calculation a median energy group which distributes the secondary energy neutrons in a "hot" and a "cold" part is defined as the group into which the median energy of the secondary energy distribution falls. An integral SED profile is then defined which can be hot or cold depending whether the response is more sensitive to the "hot" secondary neutrons than to the "cold" ones. The equivalent of the covariance data in Eq. (1) are the fractional uncertainties, which are related with the uncertainty of

K. H. Böckhoff (ed.), Nuclear Data for Science and Technology, 345–348.

the spectral shape for the secondary energy distribution. The fractional uncertainties for some materials have been evaluated.[10] All secondary particle production processes for a particular element are presently treated as one single process. The simplicity of the "hot-cold" concept and the fact that all secondary particle production processes for a particular element are treated as one single process might be causes for an underestimation of the SED uncertainties.

In principle similar algorithms for calculating uncertainties due to uncertainties in the secondary angular distribution can be constructed using the "forward-backward" formalism. However, the necessary data haven't been evaluated yet.

SENSIT-2D

SENSIT-2D, which is an extension of the one-dimensional sensitivity code SENSIT,[11] has the capability for cross-section sensitivity and uncertainty analysis, SED sensitivity and uncertainty analysis and design sensitivity analysis. The algorithms utilized are based on first-order generalized perturbation theory. The code allows (x,y) or (r,z) geometry options and accepts group-dependent quadrature sets. It is intended for use with the two-dimensional, multigroup, discrete-ordinates transport code TRIDENT-CTR.[12] The triangular mesh used by TRIDENT-CTR allows unique modeling capabilities that are applicable to fusion reactor configurations, and thus SENSIT-2D can also analyze these configurations. The forward and adjoint angular fluxes generated by TRIDENT-CTR are required input to SENSIT-2D. Because the number of angular fluxes can be voluminous, a sophisticated data management scheme was necessary for the code to keep the execution time and memory requirements within reasonable limits.

The computational outline for a two-dimensional sensitivity analysis with SENSIT-2D is shown in Fig. 1. There are three major parts: a cross section preparation block which prepares cross sections and covariance data, a transport block which includes a forward and an adjoint transport calculation and a postprocessing of the angular fluxes, and a sensitivity block, which performs the sensitivity and uncertainty calculations.

Cross-Section and SED Uncertainty Analysis for the FED

Two-dimensional cross-section and SED sensitivity and uncertainty analyses of the integral heating in the inboard toroidal field coil of the FED conceptual design were performed. The TRIDENT-CTR forward and adjoint calculations used 30 neutrons and 12 gamma-ray groups, a P_3 scattering approximation and a group dependent quadrature set. The cross-section library used was based on ENDF/B-V data. The calculational model is shown in Fig. 2 and was based on the latest FED data available at the start of this effort.

The SENSIT-2D calculations were performed using the COVFILS[8] covariance matrix library which is based on ENDF/B-V data. Covariance data for copper are based on those for iron. The SED data used in this analysis were obtained from Reference 10. Because the elastic scattering cross-section data for hydrogen are well known and due to the lack of SED data for hydrogen, hydrogen was not considered for calculating uncertainties resulting from uncertainties in the SEDs. It was assumed that the uncertainties for a particular element in the various SS 316 zones are fully correlated, while all other uncertainties are assumed to be uncorrelated.

Results from this study are shown in Table I. The overall uncertainty for the heating of the TF coil (at one standard deviation) is 115% and is in reasonable agreement with the 132% overall uncertainty obtained from a one-dimensional analysis reported elsewhere.[13] It can be concluded that the cross-section uncertainties (predicted to be 113%) tend to be more important than the SED uncertainties (20%). The major contributors to the cross-section uncertainty are chromium with relative standard deviations of 96.7% and 47.3%, respectively. The corresponding values for SED uncertainties are 4.9% and 18.4%. Contributions to the total uncertainty were smaller for nickel, copper, hydrogen and carbon. A more detailed look at the computer listings generated by this analysis reveals that the largest uncertainties are produced by uncertainties in the total and the elastic scattering cross sections for chromium, and that the heating is less sensitive to chromium than to iron. This indicates that the calculated uncertainty is largely due to the fact that chromium has very large covariances. A re-evaluation of the covariance data for chromium is therefore recommended. If new covariance data would not reduce the predicted uncertainty, new experiments for measuring the chromium cross sections are suggested.

The SED uncertainties, although less relevant to the overall predicted uncertainty, tend to become more important in the outboard shield. A possible explanation for this behavior is related with the fact that the heating in the TF-coil will be very sensitive to

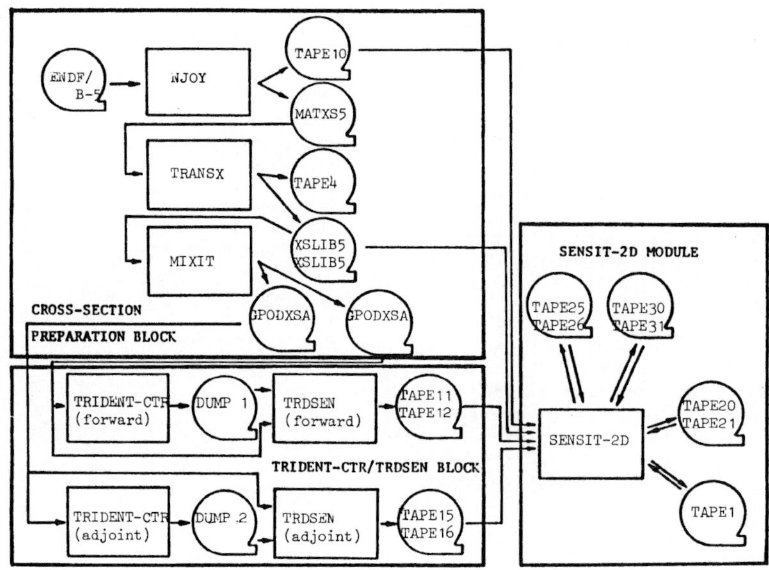

Fig. 1. Computational outline for a two-dimensional sensitivity analysis with SENSIT-2D.

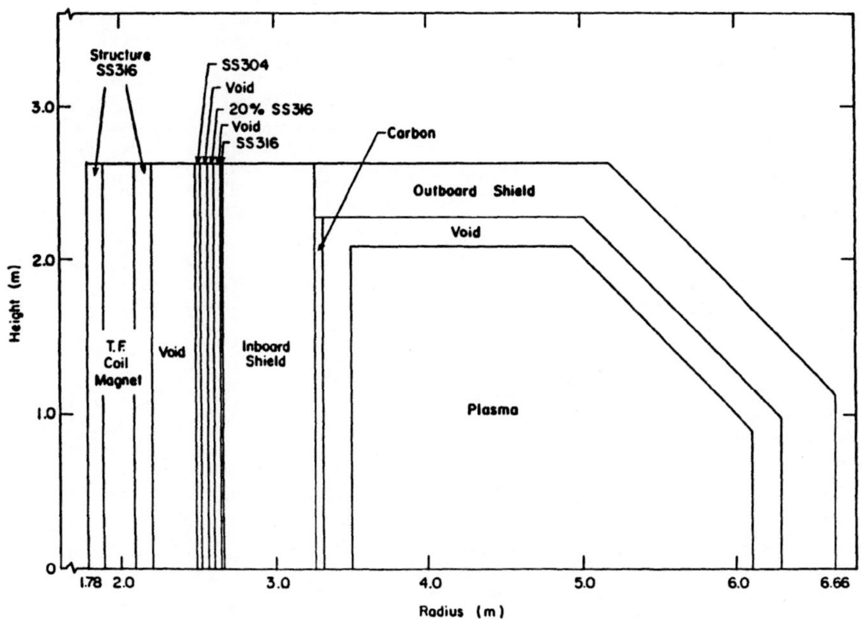

Fig. 2. FED calculational model.

TABLE I

PREDICTED UNCERTAINTIES (STANDARD DEVIATION) DUE TO ESTIMATED SED AND CROSS-SECTION UNCERTAINTIES FOR THE HEATING IN THE TF COIL FROM A TWO-DIMENSIONAL ANALYSIS

Cross Section Material	Zone	SED Uncertainties in % $\left[\frac{\Delta R}{R}\right]_{Mat,region}$	$\left[\frac{\Delta R}{R}\right]^a_{Mat}$	XS Uncertainties in % $\left[\frac{\Delta R}{R}\right]_{Mat,region}$	$\left[\frac{\Delta R}{R}\right]^a_{Mat}$
Cr	SS316	3.8		60.0	
	TFCOIL	0.2		34.5	
	SS304	0.1		4.5	
	SS312	0.0	4.9	1.1	96.7
	ISDLC	0.2		2.2	
	ISDLB	0.8		33.3	
	ISDLA	3.0		58.5	
Fe	SS316	14.8		18.9	
	TFCOIL	0.1		10.4	
	SS304	0.0		2.2	
	SS312	0.2	18.4	0.7	47.3
	ISDLC	0.5		4.4	
	ISDLB	2.7		23.6	
	ISDLA	10.8		34.5	
Ni	SS316	1.5		18.6	
	TFCOIL	0.7		11.8	
	SS304	0.0		0.9	
	SS312	0.0	4.3	0.4	31.4
	ISDLC	0.0		1.3	
	ISDLB	0.4		13.4	
	ISDLA	1.2		18.0	
H	TFCOIL	-		1.7	
	ISDLC	-	-	6.0	7.2
	ISDLB	-		3.7	
	ISDLA	-		0.5	
O	TFCOIL	0.1		-	
	ISDLC	0.2	0.3	-	-
	ISDLB	0.1		-	
	ISDLA	0.1		-	
C	TFCOIL	0.0		0.1	
	C-region	0.3	0.3	3.2	3.2
Cu	TFCOIL	2.9	2.9	10.1	10.1
Total[a]			19.7		112.9
Total uncertainty due to cross-section and SED uncertainties = 114.6%[a]					

[a]Quadratic Sums.

back-scattering in this region. The SED uncertainties in the outboard shield were different from those obtained with the one-dimensional model. More relevant than the uncertainty of the integral heating of the TF coil would be the uncertainty of the hot-spot heating of the TF coil. However, the results of the hot-spot analysis are very similar to those reported in Table I, (less than 5% difference).

Conclusions

The two-dimensional sensitivity and uncertainty analysis code, SENSIT-2D, was applied to the FED. The overall uncertainty for the heating of the TF-coil was found to be 115% and mainly due to cross-section uncertainties, while the SED uncertainties were less important. The largest uncertainties were caused by uncertainties in the chromium and iron cross section. A re-evaluation of the covariance data for chromium might lead to smaller predicted uncertainties.

References

1. S. A. W. Gerstl, Donald J. Dudziak, and D. W. Muir, "Cross-Section Sensitivity and Uncertainty Analysis with Application to a Fusion Reactor," Nuclear Science and Engineering, 62, 137 (1977).

2. D. J. Dudziak, "Cross-Section Sensitivity and Uncertainty Analysis for Fusion Reactors (a Review)," Proceedings of the Advisory Group Meeting on Nuclear Data for Fusion Reactor Technology, IAEA, Vienna, Austria, 11-15 December 1978.

3. S. A. W. Gerstl, "Sensitivity Profiles for Secondary Energy and Angular Distributions," Proc. Fifth Int. Conf. Reactor Shielding, 18-23 April 1977, Knoxville, TN, Proceedings edited by R. W. Roussin et al., Science Press, Princeton (1978).

4. S. A. W. Gerstl, "Uncertainty Analysis for Secondary Energy Distributions," Proc. Seminar-Workshop on Theory and Application of Sensitivity and Uncertainty Analysis, 22-24 August 1978, Oak Ridge, TN, Proceedings edited by C. R. Weisbin et al., ORNL/RSIC-42 (February 1979).

5. E. L. Simmons, Donald J. Dudziak, and S. A. W. Gerstl, "Nuclear Design Sensitivity Analysis for a Fusion Reactor," Nuclear Technology, 34, 317 (1977).

6. M. J. Embrechts, "Two-Dimensional Cross-Section Sensitivity and Uncertainty Analysis for Fusion Reactor Blankets," Los Alamos National Laboratory report LA-9232-T (1982).

7. M. J. Embrechts, "SENSIT-2D: A Two-Dimensional Cross-Section Sensitivity and Uncertainty Analysis Code," Los Alamos National Laboratory report LA-9515-MS (1982).

8. D. W. Muir and R. J. LaBauve, "COVFILS A 30-Group Covariance Library Based on ENDF/B-V," Los Alamos National Laboratory report LA-8733-MS (1981).

9. R. E. MacFarlane, D. W. Muir and R. M. Boicourt, "The NJOY Nuclear Data Processing System, Volume I: User's Manual," Los Alamos National Laboratory report LA-9303-M (1982).

10. S. A. W. Gerstl, R. J. LaBauve and P. G. Young, "A Comprehensive Neutron Cross-Section and Secondary Energy Distribution Uncertainty Analysis for a Fusion Reactor," Los Alamos Scientific Laboratory report LA-8333-MS (1980).

11. S. A. W. Gerstl, "SENSIT: A Cross-Section and Design Sensitivity and Uncertainty Analysis Code," Los Alamos Scientific Laboratory report LA-8498-MS (1980).

12. T. J. Seed, "TRIDENT-CTR User's Manual," Los Alamos National Laboratory report LA-7835-M (1979).

13. D. J. Dudziak, R. D. O'Dell, and R. E. Alcouffe, "Transport and Reactor Theory October 1 - December 31, 1981," Los Alamos National Laboratory report LA-9336-PR (1982).

DETERMINATION OF $^7Li(n,n't)^4He$ CROSS SECTIONS

H. Liskien

Commission of the European Communities, Joint Research Center
Central Bureau for Nuclear Measurements, B-2440 Geel, Belgium

R. Wölfle and S.M. Qaim

Institut für Chemie 1 (Nuklearchemie), Kernforschungsanlage Jülich GmbH
D-5170 Jülich, Federal Republic of Germany

Cross sections for the tritium-breeding reaction $^7Li(n,n't)^4He$ have been determined in view of their importance for the breeding factor of future fusion reactors and due to existing discrepancies. More than thirty samples have been irradiated with quasi-monoenergetic neutrons in the energy range from threshold to 8 MeV [D(d,n)] and from 13 to 16 MeV [T(d,n)]. Samples consisted of about 0.7 g metallic lithium enriched in ^7Li. The neutron fluences (10^{10} to 10^{11} cm^{-2}) have been determined relative to the 180° differential n-p scattering cross sections. The irradiated samples were transferred to a vacuum system and converted at 600-700°C by carrier hydrogen to lithium hydride (LiT). Thereafter, the temperature was raised to 800°C, the hydrogen carrier pumped off and aliquots collected in a 0.5 l proportional counting tube which was then filled up with methane and connected to a low background anticoincidence counting system. The measured cross sections have uncertainties of 6 to 7 %. They are higher than recent Harwell results but are consistent with the results of ANL at 7 and 8 MeV.

[fusion, tritium breeding $^7Li(n,n't)^4He$ reaction, cross sections, E_n = 5 to 16 MeV]

Introduction

Tritium needed in future D-T fusion reactors has to be bred from lithium via the reactions $^6Li(n,t)^4He$ and $^7Li(n,n't)^4He$ and it is the latter reaction which allows tritium breeding ratios ⩾ 1. Below the (n,2n)-threshold of 8.3 MeV the tritium production cross section of ^7Li is essentially equal to the inelastic scattering cross section. Only the first excited level of ^7Li is particle-stable, but neutrons populating this level are typically not separated from ground state neutrons in TOF-experiments [1]. However, a determination as difference of the total and the elastic scattering cross section leads to large uncertainties. A determination by observing the inelastic neutrons [1] or the emitted tritons [2] yields only results of modest accuracy. This is due to the fact that the observed energy spectra of the emitted particles have to be extrapolated to zero energy and an integration over the emission angle has to be performed. The most accurate method is certainly the activation method using quantitative tritium extraction from irradiated bulk samples and activity counting.

Integral experiments [3-5] led to the suspicion that the $^7Li(n,n't)^4He$ cross section as predicted by ENDF/B-IV (=ENDF/B-V) [6] is too high. Indeed, a recent activation experiment conducted at AERE Harwell [7] resulted in cross sections about 26 % lower than the ENDF/B. This prompted to start two other activation experiments. One performed at Argonne National Laboratory [8] was restricted to measurements at 7, 8, and 9 MeV and yielded cross sections about 17 % lower than ENDF/B. The other work is described here.

Irradiations

Quasi-monoenergetic neutrons were produced via the reactions $D(d,n)^3He$ and $T(d,n)^4He$ employing the 7 MV Van de Graaff accelerator of CBNM. For all irradiations titanium occluded targets on 0.3 mm silver backings were used. For irradiations in the 5 to 8 MeV neutron energy range targets of about 2 mg/cm^2 D-Ti were bombarded with deuteron currents between 8 and 18 μA. Samples were mounted at a distance of 77.5 mm (source to middle plane of the sample) under 0° with respect to the deuteron beam. The ion energy was adapted to obtain under 0° the wanted neutron energy. The irradiation time per sample varied between 28 and 53 h.

Irradiations in the 13 to 16 MeV neutron energy range were performed simultaneously by making use of the dependance of the neutron energy on the emission angle. A target of about 1 mg/cm^2 T-Ti was bombarded with a deuteron current of 25 μA for 160 h. The deuteron energy was adjusted to 1133 keV and samples were mounted at various angles between 46° and 134° at two distances, 94.2 and 102.2 mm.

Samples

All samples were made from the same badge of metallic lithium enriched in ^7Li to (99.99 ± 0.01) atom %. After machining to the nominal dimensions of 20 mm Ø and 4 mm thickness the samples were cleaned, weighed and transferred to an argon atmosphere. Typical weights were 660 mg with variations as small as ± 20 mg. Etching with methanol and encapsulation in 0.25 mm thick aluminium container occurred in the argon atmosphere. The loss in weight by the etching procedure has been determined to be 2 mg.

Tritium production by fast neutrons on ^6Li may be neglected due to the high enrichment of the lithium used. But the effect of the remaining ^6Li may be strongly enhanced by thermal (σ_{2200} = 936 b) and epithermal neutrons. Therefore we irradiated at three neutron energies (5.23, 6.49, and 7.73 MeV) additionally also samples enriched in ^6Li to (95.7 ± 0.5) atom %. However, the effect of ^6Li, when reduced to the high degree of ^6Li depletion as used in our ^7Li samples, was found to be negligible.

Neutron Fluence Determination

The neutron fluences seen by each sample were determined relative to the 180° n-p scattering cross section [9] employing a proton recoil telescope counter [10] positioned under 0° at 500 mm distance from the source as shown in Fig. 1. A typical proton recoil spectrum is given in Fig. 2. This telescope counter has been used successfully in the past at two international fluence intercomparisons [11-12], at 5 and 14 MeV, which adds additional confidence in the accuracy of the fluence determination. In case of DD neutrons the fluence at the position of the hydrogen radiator foil had to be corrected for the transmission of the sample. The transmission was calculated from the ^7Li total cross section but was also verified experimentally at three neutron energies using the same set-up and a spare sample. During some of the long irradiation runs the data acquisition for the proton recoil counter was interrupted by accident and the fluence during these interruptions had to be calculated from the ion beam charge assuming proportionality between charge and neutron production. An extra uncertainty has been added for such periods.

From the fluence at the position of the radiator foil the fluence averaged over the sample volume was de-

K. H. Böckhoff (ed.), Nuclear Data for Science and Technology, 349–352.

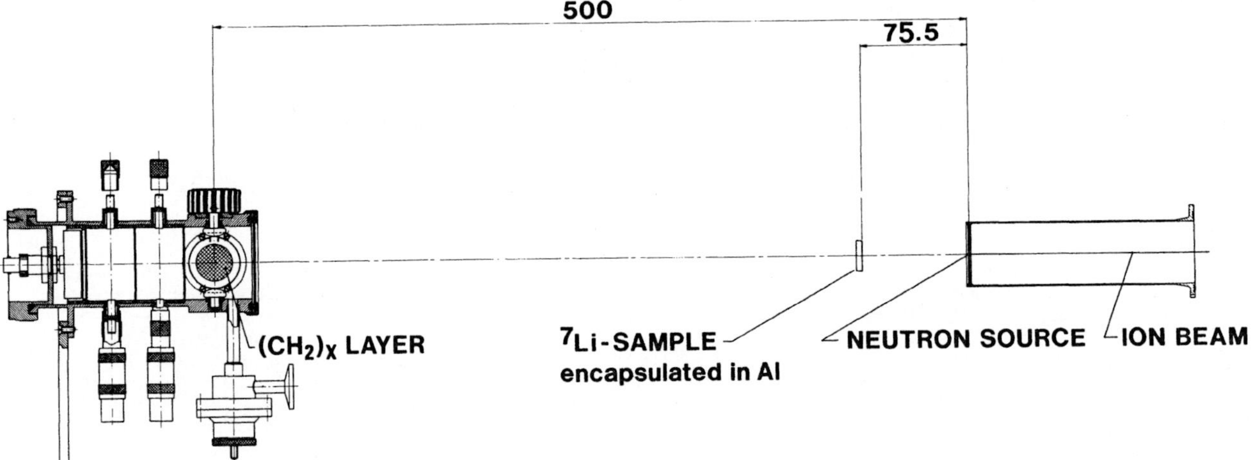

Fig. 1 The irradiation geometry as used for the energy range from threshold to 8 MeV together with the proton recoil telescope counter used for fluence determinations. In case of T(d,n) neutrons (13 to 16 MeV) samples were mounted at various angles between 46 and 134° at distances of 92 or 100 mm.

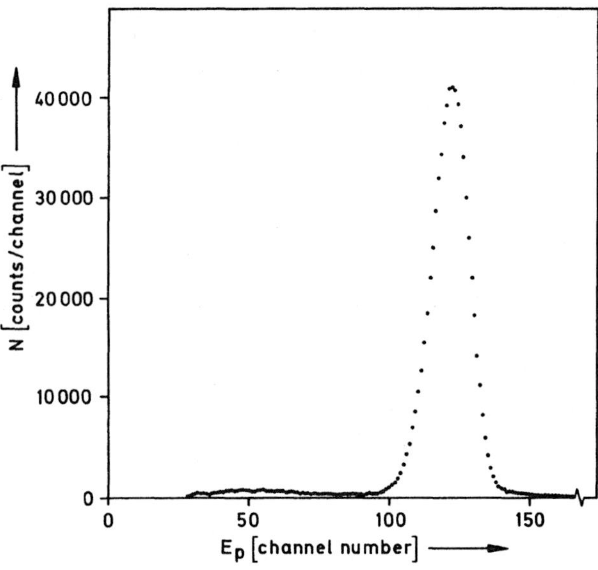

Fig. 2. The proton recoil spectrum as observed at 6 MeV neutron energy.

termined taking into account the geometry and the angular distribution of the source neutrons [13]. Corrections for neutron scattering and absorption within the sample were performed. The spectrum of secondary neutrons was determined with a calibrated NE-213 TOF spectrometer and was found to be consistent with results which were deduced from the low energy tail of the observed proton recoil spectra. The effect of these secondary neutrons with lower energy for the tritium breeding cross section is essentially zero below 6.5 MeV primary neutron energy and increases to about 3.5 % at 8 MeV.

Tritium Extraction and Counting

The tritium ($T_{1/2}$ = 12.35 y) produced in each lithium sample was separated in a vacuum apparatus and its activity determined in a proportional gas counter. A sketch of the vacuum apparatus used for the extraction of tritium is shown in Fig. 3.

Each irradiated sample was opened in an argon atmosphere and then transferred to the stainless steel tube. After evacuation a stoichiometric amount of H_2 carrier was introduced which brought the pressure to 340 Torr. The temperature was then raised to 600-700°C and maintained till the reaction to LiH (LiT) had ceased. Thereafter the temperature was raised to 800°C and the diffusion pump connected to the reaction vessel. The hydrogen carrier together with tritium (HT) was extracted into the upper part of the Töpler pump, which forwarded it into the combined volumes of bulb A and the counting tube. In order to ensure homogeneous mixing of tritium with carrier residues left in connecting lines and not fully absorbed by the lithium, the collected gas was expanded several times into bulb B and pumped back into bulb A and counting tube. The aliquot of the collected gas in the 500 ml proportional counting tube was then separated from bulb A, and the tube filled with methane to 760 Torr. Tritium (HT) was counted in an anticoincidence mode. The efficiency of the counting system was determined using tritium gas standards supplied by the CEA-Bureau National de Métrologie, Saclay, France. Tritium extraction and counting cycle was repeated with further hydrogen carrier. In most of the experiments the yield of the recovered hydrogen was $>$ 98 %. In general $>$ 97 % tritium activity was recovered already in the first extraction cycle. The system has been successfully used in extensive studies on (n,t) reactions[14-16].

The removal of the irradiated lithium from the Al capsule and its subsequent transfer to the vacuum system did not present any difficulty. The amount of lithium sticking to aluminium was determined gravimetrically and was generally $<$ 0.5 %. In a few cases, however, the samples were strongly corroded. The results from those samples were not reliable and were therefore discarded. Two further points needed consideration :

i) Loss of tritium from irradiated lithium on opening the capsule

ii) Recoil tritium which sticks to aluminium.

Thermodynamic considerations and our experience have shown (ref. 14, 15) that tritium produced via a nuclear reaction remains attached to metallic lithium as LiT. However, as a further check we transferred one of the irradiated capsules without opening to the vacuum system, filled up with H_2 carrier to 760 Torr, drilled a small hole into the Al capsule using a special manipulator and extracted the gases with heating into the counting tube. Upon counting no significant amounts of tritium could be detected.

The loss of tritium due to recoil in aluminium could be estimated theoretically from the shape of the tritium spectrum [2] and its range in lithium [17]. This amounted to $<$ 1 %. Experimentally in a few cases the aluminium capsule was degassed separately in the vacuum apparatus and the tritium content determined. It was invariably $<$ 1 %.

The various uncertainties associated with the tritium extraction and counting are discussed below.

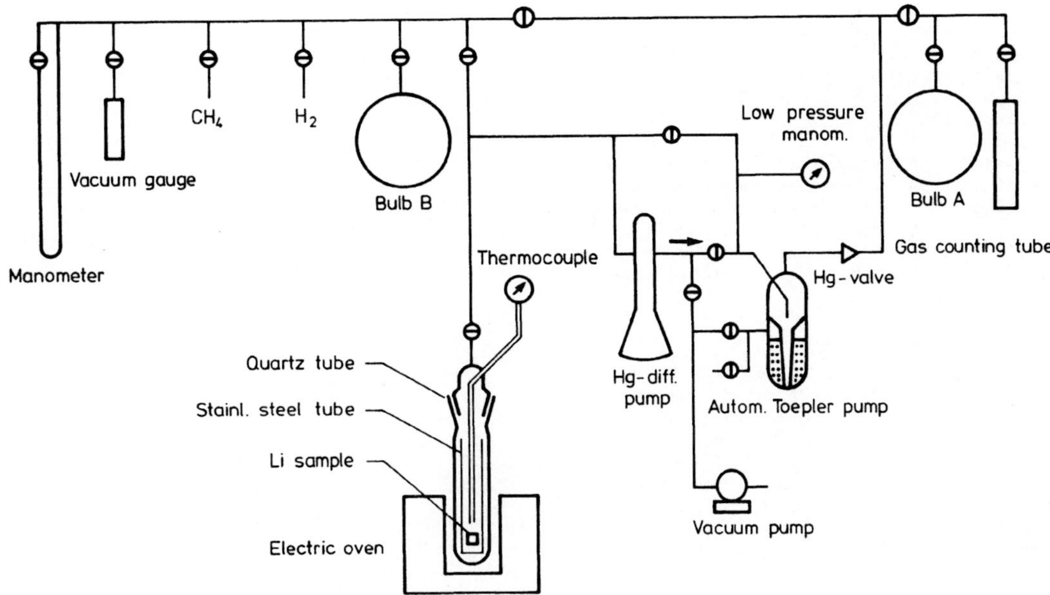

Fig. 3. Schematic diagram of vacuum apparatus for the extraction of tritium from irradiated lithium.

Results

The final results of the present work are given in numerical form in Table I. E denotes the average neutron energy. Its uncertainty never exceeds 25 keV and is due to uncertainties in the ion energy, the target thickness, the stopping power of the target material for deuterons [17], and the irradiation angle. ΔE, the quoted neutron energy spread, covers the full range of the spectrum seen by the sample. $\Delta\sigma$ are 1σ uncertainties obtained by quadratic summation of the various uncertainty contributions. Typical values (1σ) for these contributions are summarized in Table II together with indications whether a specific contribution is common to all or to a subset of the samples. This allows the construction of a correlation or covariance matrix.

Table I. Final results of this work.

E(MeV)	ΔE(MeV)	σ(mb)	Δσ(mb)
4.99	± 0.23	95	± 6
5.23	± 0.24	137	± 10
5.39	± 0.18	200	± 12
5.49	± 0.18	269	± 16
5.61	± 0.17	297	± 20
5.74	± 0.16	316	± 18
5.74	± 0.16	399	± 23
5.99	± 0.16	352	± 21
6.23	± 0.15	369	± 23
6.49	± 0.18	361	± 21
6.73	± 0.14	359	± 22
6.98	± 0.14	367	± 24
7.23	± 0.13	369	± 24
7.23	± 0.13	388	± 25
7.48	± 0.13	364	± 24
7.48	± 0.13	352	± 23
7.73	± 0.13	351	± 25
13.01	± 0.12	322	± 20
13.51	± 0.17	292	± 19
13.77	± 0.20	306	± 19
14.02	± 0.24	271	± 17
14.77	± 0.28	263	± 17
15.03	± 0.31	256	± 16
15.27	± 0.32	269	± 17
15.78	± 0.31	246	± 16
16.03	± 0.33	253	± 16

Table II. Uncertainty contributions (in %) and their origin. Remarks refer to existing correlations.

Origin	DD	DT
sample weight	0.5	0.5
isotopic sample composition	0.1[a)]	0.1[a)]
external geometry	2.9	2.0 - 2.1
effect of spurious neutrons	0.1 - 1.2	0.1
scattering within the sample	1.0	1.0
internal telescope geometry	1.3[a)]	1.3[a)]
hydrogen content of radiators	1.5[a)]	1.5[a)]
proton recoils	1.0 - 3.0	1.0[b)]
n-p scattering cross section	1.0 - 1.2[c)]	2.0[b)c)]
sample transmission	0.5[d)]	—
angular distribution of source	0.1	3.0
flux interpolation	0.0 - 2.6	—
tritium extraction	2.0	2.0
calibration of tritium counter	3.0[a)]	3.0[a)]
tritium counting	0.8 - 4.1	0.9
tritium half-life	0.2[a)]	0.2[a)]

a) common to all data points
b) common to all DT data points
c) see ENDF/B MAT-1301 MF-33 MT-2
d) see ENDF/B MAT-1272 MF-33 MT-1

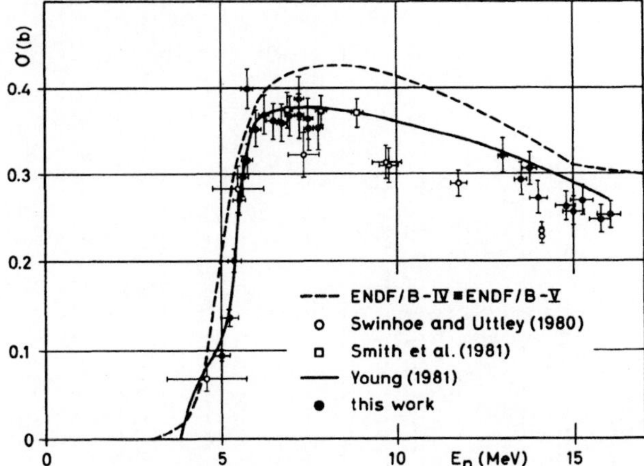

Fig.4. Results of this work and other recent activation data compared with ENDF/B and a recent LANL evaluation.

Conclusions

The results of this work are compared in Fig. 4 with the ENDF/B evaluation [6] and other recent activation data. Our results are higher than the Harwell results [7] but are in good agreement at 7 and 8 MeV with the Argonne values [8]. They are also consistent with a recent Los Alamos evaluation [18] in which variance-covariance analyses are performed for each of the major relevant cross section types under the constraint that all partial cross sections sum up to the total cross section. In fact agreement with this evaluation is excellent in the steep rise near threshold, while there is a trend to slightly larger evaluated values for higher neutron energies.

Acknowledgements

The authors wish to thank E. Freistedt for the careful production and assay of the lithium samples. The isotopic analysis has been performed by M. Gallet. Thanks are also due to R. Widera, F. Arnotte and the accelerator staff of CBNM's Van de Graaff group for the help given during the very time-consuming irradiations. We are grateful to Prof. G. Stöcklin for his encouragement and for stimulating discussions on tritium chemistry.

References

1. P.W. Lisowski, G.W. Auchampaugh, D.M. Drake, M. Drosg, G. Haouat, N.W. Hill, and L. Nilsson, Report LA-8342(1980).
2. H. Liskien and A. Paulsen, Annals of Nuclear Energy 8, 423 (1981).
3. R. Herzing, L. Kuijpers, P. Cloth, D. Filges, R. Hecker, and N. Kirch, Nucl. Sc. and Eng. 63, 341 (1977).
4. H. Bachmann, U. Fritscher, F.W. Kappler, D. Rusch, H. Werle, and H.W. Wiese, Nucl. Sc. and Eng. 67, 74 (1978).
5. A. Hemmendinger, C.E. Ragan, and J.M. Wallace, Nucl. Sc. and Eng. 70, 274 (1979).
6. "Evaluated Nuclear Data File ENDF/B-V", documented in Report ENDF-102, BNL-NCS-50496 "Data Formats and Procedures for ENDF" (1979).
7. M.T. Swinhoe and C.A. Uttley, Report AERE-R-9929 (1980).
8. D.L. Smith, M.M. Bretcher, and J.W. Meadows, Nucl. Sc. and Eng. 78, 359 (1981).
9. J.C. Hopkins and G. Breit, Nucl. Data Tables A9, 137 (1971).
10. H. Liskien, Proc. Symp. Neutron Standards and Flux Normalization,Argonne, page 343 (1970)
11. V.D. Huynh, Metrologia 16, 31 (1980).
12. H. Liskien and R. Widera, Report GE/R/VG/36 (1981).
13. H. Liskien and A. Paulsen, Nucl. Data Tables 11, 569 (1973).
14. S.M. Qaim, R.Wölfle and G. Stöcklin, J. Inorg. Nucl. Chem. 36, 3639 (1974).
15. S.M. Qaim, R.Wölfle and G. Stöcklin, J. Radioanalyt. Chem. 30, 35 (1976).
16. S.M. Qaim and R.Wölfle, Nucl. Phys. A295, 150 (1978).
17. K.H. Andersen and J.F. Ziegler, "Hydrogen Stopping Powers and Ranges in All Elements" Pergamon Press (1977).
18. P.C. Young, Report LA-8874-PR, page 2 (1981).

$^6Li(n\ n)$ CROSS SECTION IN THE ENERGY RANGE FROM 600 eV TO 80 keV

V.P.Alfimenkov, S.B.Borzakov, Vo Van Thuan, L.B.Pikelner, E.I.Sharapov

Laboratory of Neutron Physics, Joint Institute for Nuclear Research, Dubna, USSR

The $^6Li(n,n)$ cross section value and its dependence on neutron energy in the 0.6-80 keV region are measured at the IBR-30 pulsed reactor by the time-of-flight method. A constant value of 0.72±0.02 barn was found for the scattering cross section below 10 keV. The energy dependence of the $^6Li(n,n)$ cross section together with other data were analysed in the frame of R-matrix formalism to clarify the properties of levels of the ^7Li nucleus.

[$^6Li(n,n)$, total scattering cross section, 0.6-80 keV, R-matrix analysis]

The elastic scattering $^6Li(n,n)^6Li$ and the $^6Li(n,\alpha)T$ reaction are the dominant processes in the $(n,^6Li)$ interaction up to 2 MeV neutron energy. The (n,α) cross section as well as the total cross section were studied in many works (see e.g. Refs.[1-3]). However, the (n,n) cross section data are poor, the only known data on differential cross sections from Ref.[4] and Ref.[5] are in disagreement below 100 keV. This paper presents experimental data and results on total scattering cross section of neutrons by ^6Li.

The measurements were performed at the IBR-30 pulsed reactor with a time-of-flight resolution of 10 ns/m at the 500 m flight path. The 4π-neutron detector with the $ZnS(Ag)^{10}B$-plexiglass sandwiches shown in Fig.1 was used.

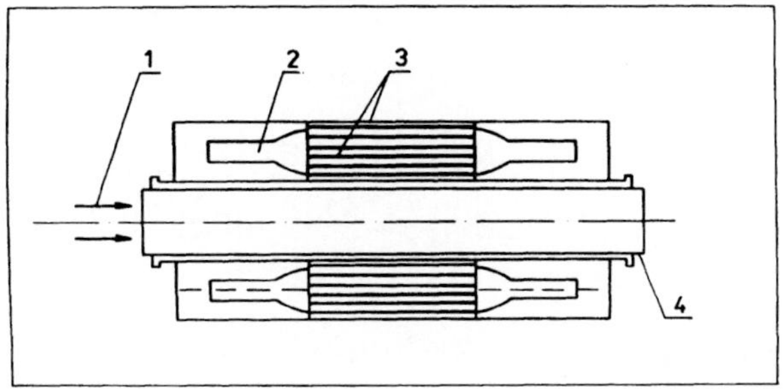

Fig. 1. Neutron detector. 1 - direction of neutron beam, 2 - photomultipliers, 3 - scintillator sandwiches, 4 - evacuated tube.

The $^6Li(n,n)$ cross section was measured relative to that known from Ref.[6] for ^7Li. Thin metallic samples 180 mm in diameter were used ($n=1.627 \times 10^{22}$ ^6Li nucl/cm^2 and $n=1.353 \times 10^{22}$ ^7Li nucl/cm^2). The walls of sample containers were made from lavsan sheets 15 μm thick. The ^6Li and ^7Li samples were placed in turn into the beam and the measurements with an empty container were used as background ones (the difference of real background measurements with black resonance filters for ^6Li and ^7Li was less than 3%). Fig. 2 presents the time-of-flight spectra, each of them being obtained during 10 hrs. In the determination of scattering cross section the following corrections were made : for the absorption of incident neutrons by ^6Li (from 3% at E_n=3.0 keV to 0.5% at E_n=60 keV) and for the absorption after multiple scattering (from 7% at E_n=3.0 keV to 1.3% at E_n=60 keV). Our scattering cross section results are shown in Fig. 3 (black points).

Below 10 keV our cross section is constant. Here the value σ_n =0.72±0.02 b is in agreement with σ_n = =0.76±0.04 b reported by Asami and Moxon[4]. Using the known relations $\sigma_n = (2|a_+|^2 + |a_-|^2)4\pi/3$, $a_{coh} = (2\,Re\,a_+ + Re\,a_-)/3$ and the corresponding data from Refs.[1,7] the (n^6Li) scattering lengths $a_\pm$ were refined. The indices ± correspond to channel spin values $J_\pm = I \pm 1/2 = 3/2$ or 1/2. The imaginary part of $a_\pm$ calculated from the optical theorem

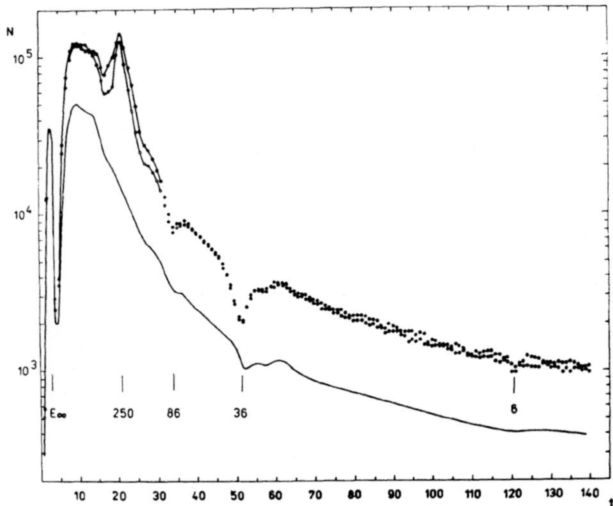

Fig. 2. Time-of-flight spectra (the detector number of counts N is in logarithmical scale). ^7Li - open points, ^6Li - black points, the curve - background, digitals - energy in keV.

and data of Glättli et al[1]. Thus we obtain

$$a_-(J=1/2)=(4.00\pm0.06)-i(0.53\pm0.02)\ \text{fm}$$

$$a_+(J=3/2)=(0.65\pm0.03)-i(0.07\pm0.01)\ \text{fm}$$

This means strong spin dependence of neutron scattering lengths of ^6Li, which is known already from Ref.[4]. However, the spin dependence of (n^6Li) scattering persists even at energies higher than 100 keV, e.g. we calculated the value $(a_+)_{eff}$=0.9 ±0.2 fm in the energy interval 0.1-0.5 MeV from the interference term between the p-wave 252 keV resonance and the s-wave potential scattering using data of Lane et al.[5].

The strong absorption of neutrons by ^6Li is usually interpreted as being due to the bound level in ^7Li with energy 6.60 MeV and $J = 1/2^+$. No such positive parity state has been directly identified as yet. Additional information on this problem may follow from the energy dependence of the s-wave neutron cross section for ^6Li. With this aim we have performed the analysis of the following data: $\sigma_n(E)$ of present work, $\sigma_\alpha(E)$ of Gayther[2] and $\sigma_t(E)$ of Meadows and Whalen[3]. In order to separate the s-wave contribution to cross sections it was necessary to describe very accurately the 252 keV p-wave neutron resonance which was the dominant feature in the (n^6Li) cross sections. We have tried the two level R-matrix formalism first, because both the known level 6.68 MeV and the p-resonance (the level 7.46 MeV) have the same spin and parity $J = 5/2^-$. However, it was impossible to achieve good fit to σ_n and σ_α cross sections simultaneously without changing the known from Ref.[8] parameters of the 6.68 MeV level. Therefore, the analysis was performed in one-level two-channel R-matrix approach to the p-wave resonance and in the phenomenological approach to the energy dependence of s-wave cross sections. The fol-

K. H. Böckhoff (ed.), Nuclear Data for Science and Technology, 353–355.

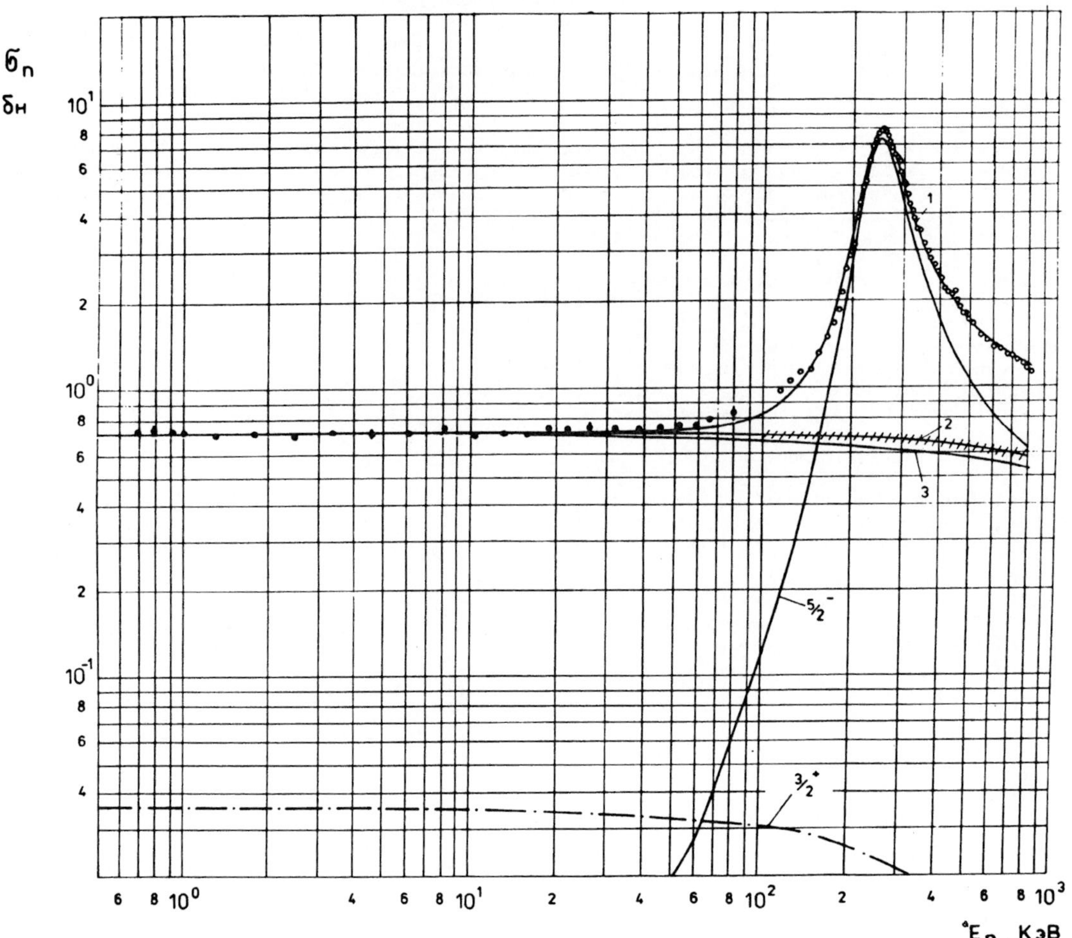

Fig. 3. Scattering cross section $\sigma_n(E)$ for ^{6}Li vs neutron energy. Black points - this work, open circles - the difference $(\sigma_t - \sigma_\alpha)$ (σ_t from Ref.[3], σ_α -- Ref.[2]), for the curves see description in the text.

lowing formulas were used in the least squares fit:

$$\sigma_n = \pi \lambda\!\!\!^-{}^2 \frac{\Gamma_n[\Gamma\cos\varphi - 2(E_\lambda+\Delta_\lambda-E)\sin 2\varphi]}{(E_\lambda+\Delta_\lambda-E)^2+\Gamma^2/4} + \tag{1}$$

$$+ 4\pi\lambda\!\!\!^-{}^2\sin 2\varphi + 0.72 + \delta\cdot E \quad ,$$

$$\sigma_\alpha = \pi\lambda\!\!\!^-{}^2 \frac{\Gamma_n\Gamma_\alpha}{(E_\lambda+\Delta_\lambda-E)^2+\Gamma^2/4} + \tag{2}$$

$$+ \frac{149.5}{\sqrt{10^3\cdot E}} - 0.025 - \beta\cdot\sqrt{E} \quad .$$

Here, neutron energy E is in keV, $\lambda\!\!\!^- = \lambda/2\pi$ in 10^{-12} cm and σ in barn. The formulas contain the R-matrix resonance term with traditional parameters $\Gamma_c = 2\gamma_c^2 P_c$, $E_\lambda = E_0 + \sum_c \gamma_c^2 S_c$, Δ_λ with the zero value for the boundary conditions parameter B and the p-wave potential phase $\varphi = kR_n - \mathrm{arctg}\, kR_n$. The phenomenological parameter δ describes the deviation of s-wave scattering cross section from the constant value 0.72 b. The analogous parameter β describes the deviation of the reaction cross section from $1/v$ behaviour. According to Shapiro[9] $\beta > 0$ in the case of a bound s-level present, and $\beta < 0$ in the case of a far away s-wave neutron resonance. The constant term (-0.025) b was calculated using the known Shapiro expression and thermal neutron data from Ref.[1]. The results of least squares fits with different values of channel radii R_n and R_α are given in the Table.

Table. Results of the least squares fit to (n^{6}Li) cross sections in the energy range from 0 to 800 keV

	R_n=3.88 fm R_α=2.5 fm	R_n=3.88 fm R_α=4.2 fm	R_n=4.1 fm R_α=4.4 fm
E_λ keV	-1837(14)	-801(6)	-625(5)
γ_n^2 keV	1161(8)	1106(8)	930(6)
γ_α^2 keV	410(4)	22.8(3)	22.3(3)
Γ_n keV	101	96	93
Γ_α keV	39	37	36
$\beta\cdot10^3$ b.keV$^{-1/2}$	-2.0(2)	-2.2(2)	-2.2(2)
$\delta\cdot10^4$ b.keV^{-1}	-1.6(2)	-1.6(2)	-1.8(2)

The data summarized in the Table show that the uncertainty in the choice of the channel radii affects little the Γ_n and Γ_α widths of the 252 keV p-resonance (their values are given at resonance energy) as

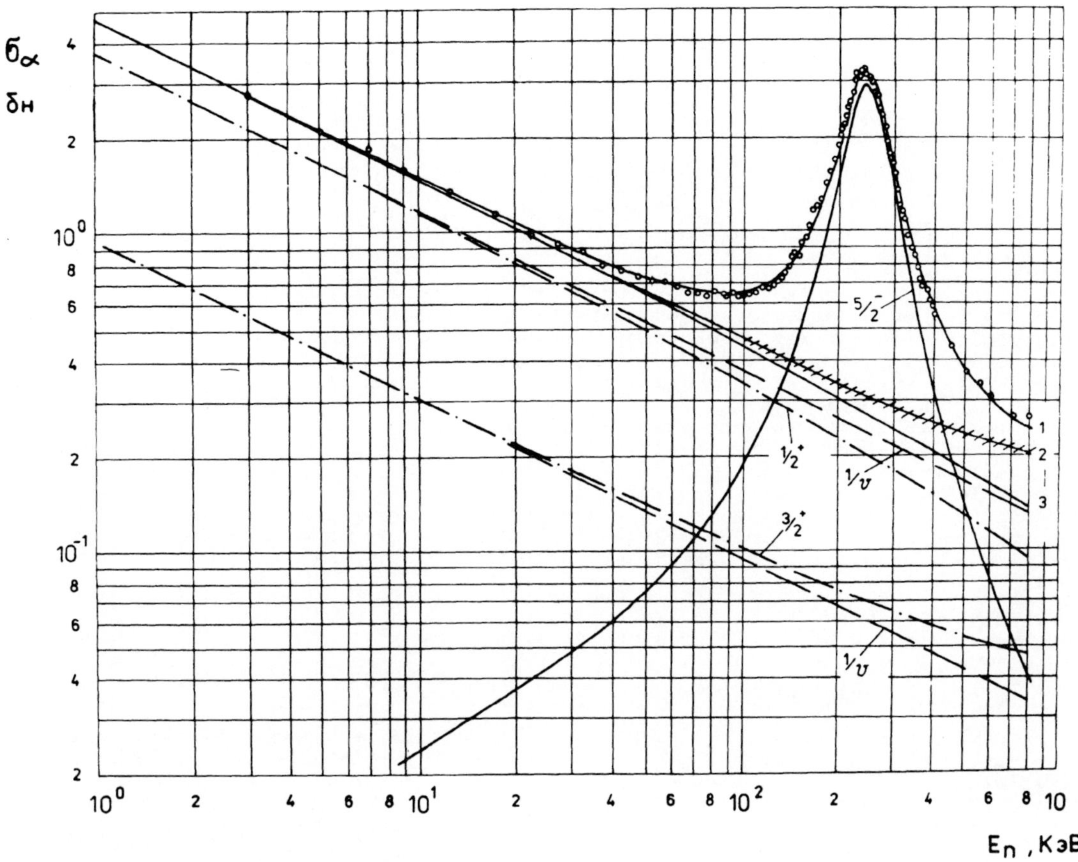

Fig. 4. Reaction cross section $\sigma_\alpha(E)$ for ^{6}Li. Points -- Ref.[2], for the curves see description in the text.

well as the parameters δ and β . The fits are near-equally good and we recommend the parameters of the last column having a slightly lesser value of χ^2. The new result of this analysis is the negative sign of the β -coefficient (the "experimental" s-wave reaction cross section is higher than the cross section due to the $1/v$ law).

In Figs. 3 and 4 our calculated curves are marked by 1. Small deviations between the calculated and experimental curves near the centre of E_0 are due to the $\sim$ 7 keV shift between the energy scales of data of Ref.[2] and Ref.[3]. The s-wave contributions obtained after the 252 keV resonance subtraction are marked by 2. They have the uncertainty intervals as obtained from the least squares fit errors. The curves 2 should be compared with curves 3, which were obtained for the s-wave cross sections under assumption of the existence of the bound level $E_0^{c.m.}(J=1/2^+)=-693$ keV, $\gamma_n^2=400$ keV, $\gamma_\alpha^2=1998$ keV and the level ($R_n=3.88$ fm, $R_\alpha=2.5$ fm) $E_0^{c.m.}(J=3/2^+)=2998$ keV, $\gamma_n^2=2733$ keV, $\gamma_\alpha^2=220$ keV above the neutron threshold. The two components of the curve 3 corresponding to spin channels J=1/2 and 3/2 are also shown. One can see that the "experimental" curves 2 do not coincide with curves 3 and the deviation cannot be explained by the influence of the known p-resonances at higher energy.

One can obtain a full agreement with curves 2, if the hypothesis of the bound level is left behind and instead the empirical $1/v$ cross section in the channel with $J=1/2^+$ and the level $E_0^{c.m.}=1420$ keV, $\gamma_n^2=903$ keV, $\gamma_\alpha^2=19$ keV in the channel with $J=3/2^+$ are used. In this case one should look for other physical explanations of the large σ_α cross section, e.g. that of Weigmann and Manakos[10] which consists in a direct reaction mechanism for ^{6}Li(nα)T due to the deuteron exchange.

The conclusion is that the energy dependence of the (n^{6}Li) cross sections points to the absence of a positive parity level in the (n+^{6}Li) system below neutron threshold.

References

1. H.Glättli et al. Phys.Rev.Lett.,40, 748(1978).
2. D.B.Gayther. Ann.of Nucl.En.,4,515 (1977).
3. J.W.Meadows, J.F.Whalen.Nucl.Sci.Eng.,48,221 (1972).
4. A.Asami, M.C.Moxon. Nucl.Data for Reactors, Vienna, IAEA,1,153 (1970).
5. R.O.Lane et al. Phys.Rev., B136, 1710 (1964).
6. V.P.Alfimenkov et al. Yad.Fiz.35, 542 (1982).
7. S.W.Peterson, H.G.Smith. J.Phys.Soc., 17 Suppl., 335 (1962).
8. R.J.Spiger, T.A.Tombrello. Phys.Rev., 163, 964 (1967).
9. F.L.Shapiro. Neutron Physics, Nauka, Moscow, 1, 211 (1976).
10. H.Weigmann, P.Manakos. Z.fur Phys., A289, 383 (1979).

CROSS SECTIONS FOR HYDROGEN AND HELIUM PRODUCING REACTIONS INDUCED BY FAST NEUTRONS ON POTENTIAL FIRST WALL MATERIALS OF FUSION REACTOR TECHNOLOGY

S.M. Qaim, R. Wölfle, and G. Stöcklin

Institut für Chemie 1 (Nuklearchemie), Kernforschungsanlage Jülich GmbH, D-5170 Jülich, Federal Republic of Germany

Cross sections were measured for the $^{61}Ni(n,p)^{61}Co$ and $^{58}Ni(n,\alpha)^{55}Fe$ reactions in the energy region of 5 to 10 MeV, using radiochemical separations and high resolution γ- and x-ray spectroscopy, respectively. An analysis of the systematic trends in the [(n,d)+(n,n'p)+(n,pn)] reaction cross sections at 14.7 ± 0.3 MeV was carried out. Cross sections were also measured for several (n,p), [(n,d)+(n,n'p) +(n,pn)], (n,α) and (n,n'α) reactions induced by a 30 MeV d(Be) break-up neutron spectrum, similar to the one expected in Fusion Materials Irradiation Test Facility (FMIT). Up to about 10 MeV incident neutron energy hydrogen and helium production is caused by (n,p) and (n,α) reactions, respectively. At 14-15 MeV the contributions of (n,n'x) processes are appreciable, the (n,n'p) process being exceptionally strong for target nuclei with neutron separation energies higher than the proton separation energies. With 30 MeV d(Be) break-up neutron spectrum, the contributions of (n,n'x) processes are significant for all the target nuclei.

[$^{61}Ni(n,p)$, $^{58}Ni(n,\alpha)$, measured σ radiochemically, E_n = 5-10 MeV; Systematics of (n,d)+(n,n'p)+(n,pn) reaction cross sections at 14.7 MeV; (n,x) and (n,n'x) processes with 30 MeV d(Be) break-up neutron spectrum, measured σ.]

Introduction

Studies of fast neutron induced hydrogen and helium producing nuclear reactions on potential first wall materials of fusion reactor technology (cf. 1) are of considerable significance for calculating nuclear heating, induced radioactivity and, above all, radiation damage effects originating from gas formation (cf. 2). Investigations with 14-15 MeV neutrons have shown (cf. 3-8) that both (n,x) and (n,n'x) type reactions, where x denotes a proton or an α-particle, contribute to hydrogen and helium formation. Due to their highly negative Q-values the (n,n'x) type processes, however, are expected to be less important at E_n < 14 MeV but more important at energies above 14 MeV. In this paper we report on our recent cross section measurements on some gas producing reactions induced by neutrons in three energy regions, namely 5 to 10 MeV, 14-15 MeV, and a 30 MeV d(Be) break-up neutron spectrum.

Experimental

Cross sections were measured by the activation technique using highly enriched isotopes (wherever necessary) as target materials and high-resolution γ- or x-ray spectroscopy. In general the activation product was separated radiochemically and a thin source suitable for counting was prepared.

Quasi monoenergetic neutrons in the energy range of 5 to 10 MeV were produced via the $D(d,n)^3He$ reaction. For this purpose a 3.7 cm x 2.0 cm ϕ cell filled with deuterium gas at a pressure of 1.8 bar was used. The front window of the cell consisted of a 7 μm Havar foil and the beam stop at the back a 100 μm Mo sheet which was cooled by a stream of air. The energy of the deuteron beam from our compact cyclotron CV28 could be varied between 2 and 7 MeV and beam currents of about 5 μA incident on the target were used. All the irradiations were done in the 0° direction; the energies of the incident neutrons could thus be calculated accurately[9]. The effect of break-up neutrons, which becomes appreciable at $E_d \geq 5$ MeV, was corrected for by performing at each deuteron energy a blank irradiation without any deuterium gas in the cell. The shape of the neutron spectrum was checked qualitatively using an NE 213 scintillator. The neutron flux densities were determined using the $^{58}Ni(n,p)^{58}Co$ (ref. 10) and $^{56}Fe(n,p)^{56}Mn$ (ref. 11) monitor reactions for $^{58}Ni(n,\alpha)^{55}Fe$ and $^{61}Ni(n,p)^{61}Co$ reactions, respectively.

The 14.7 ± 0.3 MeV neutrons were produced via the $T(d,n)^4He$ reaction in a Dynagen neutron generator and the irradiations were performed invariably in the 0° direction (cf. 12). The flux densities were determined using the $^{27}Al(n,\alpha)^{24}Na$ monitor reaction (σ = 115 ± 3 mb).

The deuteron break-up braod neutron spectrum was produced by bombarding a 1 cm thick Be target with 30 MeV deuterons at the Jülich isochronous cyclotron (JULIC) at beam currents of about 3 μA. The shape of the spectrum in the forward direction is known (cf. 13). We characterized it by multiple foil activation technique (cf. 14) using the spectrum unfolding code SAND II. Neutron fluences were calculated from the charge collected at the Be-converter as well as via the $^{27}Al(n,\alpha)^{24}Na$ monitor reaction ($\bar{\sigma}$ = 45 ± 8 mb).

Results and Discussion

E_n = 5 to 10 MeV

Two excitation functions have been measured for the first time and the results are shown in Figs. 1 and 2. Previous investigations on those reactions had been limited to 14-15 MeV region (cf. 4,6,8). We measured the $^{61}Ni(n,p)^{61}Co$ cross sections using a 92.92 % enriched sample as target material. In investigations on the $^{58}Ni(n,\alpha)^{55}Fe$ reaction, however, about 10 g natural nickel samples were irradiated and in each case the activation product ^{55}Fe ($T_{1/2}$: 2.7 y) was separated radiochemically. The final source for counting the 5.8 keV x-rays was about 1 mg/cm² thick.

K. H. Böckhoff (ed.), Nuclear Data for Science and Technology, 356–359.

The shape of the excitation function of the $^{61}Ni(n,p)^{61}Co$ reaction (Fig. 1) is similar to those of the (n,p) reactions on ^{58}Ni and ^{60}Ni (cf.10); the reaction thresholds and the magnitudes of the cross sections at a given incident neutron energy are, however, different due to different reaction energies involved. The activation product of the $^{61}Ni[(n,d)+(n,n'p)+(n,pn)]^{60}Co$ reaction could not be detected at $E_n \leq 10$ MeV.

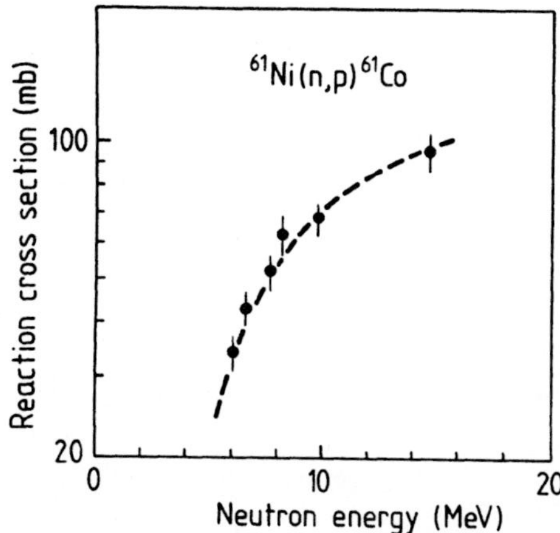

Fig. 1. Excitation function of the $^{61}Ni(n,p)^{61}Co$ reaction.

The excitation function of the $^{58}Ni(n,\alpha)^{55}Fe$ reaction (Fig. 2) is somewhat similar to that for ^{nat}Ni deduced by α-particle detection[15]. At 14-15 MeV the results for ^{58}Ni obtained by various techniques, viz. activation[4], charged particle detection[7] and mass spectrometry[8] are in good agreement. Our present activation measurements give the first helium production results in the energy region from 5 to 10 MeV for an individual isotope of nickel. These results should allow an estimation of the helium production rates from the other isotopes of nickel as well.

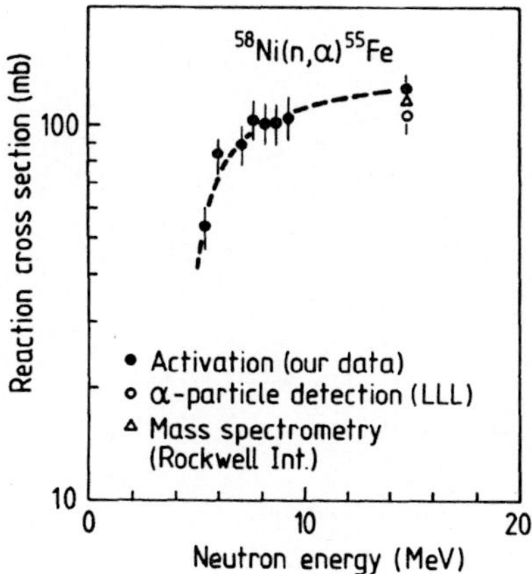

Fig. 2. Excitation function of the $^{58}Ni(n,\alpha)^{55}Fe$ reaction.

$E_n = 14.7 \pm 0.3$ MeV

In continuation of our earlier studies[3,4] on [(n,d)+(n,n'p)+(n,pn)] reactions at 14.7 ± 0.3 MeV we recently reported several new measurements[5] and completed an analysis of the systematic trends in those cross sections. The results are reproduced in Fig. 3. Evidently the cross sections are exceptionally high for target nuclei with neutron separation energies higher than proton separation energies ($S_n > S_p$). Since the (n,d) cross section is almost constant (≈ 10 mb) for all the nuclei in the mass region A = 27 to 100 (cf. 6,7) and the (n,pn) process, i.e. the emission of a neutron after the first chance emission of a proton, is expected to be relatively less probable, the (n,n'p) process appears to be exceptionally strong. For nuclei with $S_n < S_p$, on the other hand, the combined [(n,d)+(n,n'p)+(n,pn)] cross section is relatively small and amounts to ∿ 20 % of the (n,p) cross section.

By plotting the difference between the proton and neutron separation energies (S_p-S_n) against the [(n,d)+(n,n'p)+(n,pn)] reaction cross section it was observed[5] that the cross section decreases with the increasing S_p-S_n. The correlation emerges better if (S_p-S_n)/[(N-Z)/A] is plotted against the cross section, as depicted in Fig. 4. The dip relates to the cross section for nuclei with rather large values of the asymmetry parameter (N-Z)/A. For highly neutron deficient target nuclei (left hand side of Fig. 4) the cross section is rather high, the dominating contribution being from the (n,n'p) process. For the other target nuclei, however (right hand side of Fig. 4), the cross section is relatively small, and radiochemical methods of measurement are more advantageous.

As far as helium production is considered, the major source is the (n,α) reaction. Our studies have shown[3] that the contribution of the (n,n'α) process amounts to 10-15 %.

Break-up Neutron Spectrum

Cross sections of some (n,p) and [(n,d)+(n,n'p)+(n,pn)] reactions induced by the 30 MeV d(Be) break-up neutron spectrum[13,14] are given in Table I. In contrast to the 14.7 MeV data, the latter processes are fairly strong both for target nuclei with $S_p > S_n$ and $S_p < S_n$, and thus would contribute appreciably to hydrogen formation if an FMIT type neutron spectrum were to be used for radiation damage studies.

Some (n,α) and (n,n'α) reaction cross sections were also determined and are given in Table II. Apparently the contribution of (n,n'α) process is significant.

Conclusions

Up to about 10 MeV incident neutron energy hydrogen and helium production is caused by (n,p) and (n,α) reactions, respectively. At 14-15 MeV the contributions of (n,n'x) processes are appreciable, the (n,n'p) process being exceptionally strong for target nuclei with neutron separation energies higher than the proton separation energies. With 30 MeV d(Be) break-up neutron spectrum, the contributions of (n,n'x) processes are significant.

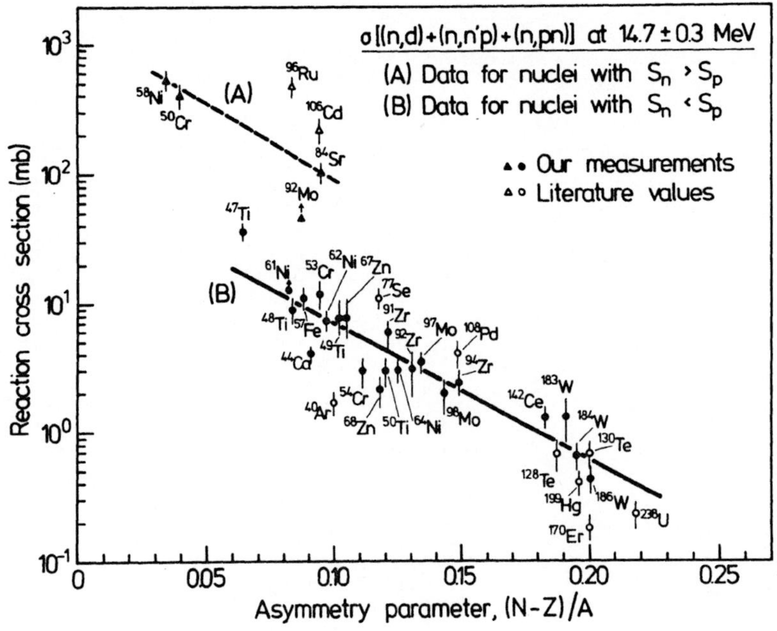

Fig. 3 Systematics of $[(n,d)+(n,n'p)+(n,pn)]$ cross sections induced by 14.7 ± 0.3 MeV neutrons.

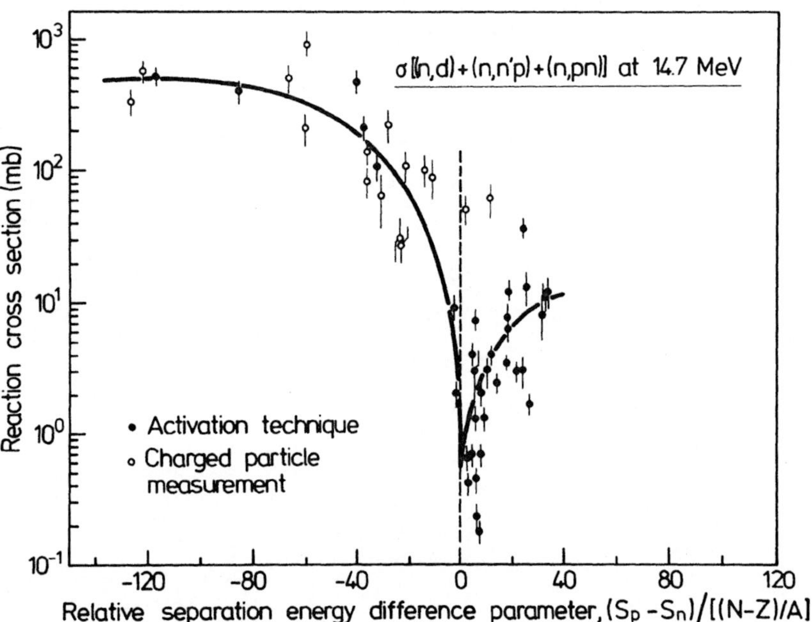

Fig. 4 $\sigma[(n,d)+(n,n'p)+(n,pn)]$ as a function of relative separation energy difference parameter $(S_p-S_n)/[(N-Z)/A]$. For negative values of the parameter (highly neutron deficient isotopes) the (n,n'p) process is the dominating mode of deexcitation.

Table I. Cross Sections for some (n,p) and [(n,d)+(n,n'p)+(n,pn)] Processes Induced by 30 MeV d(Be) Break-up Neutron Spectrum

Target nuclide	% enrichment	S_p-S_n (MeV)	Cross section (mb)	
			(n,p)	[(n,d)+(n,n'p)+(n,pn)]
^{47}Ti	79.5	1.58	61 ± 10	42 ± 8
^{48}Ti	99.13	-0.18	19 ± 5	25 ± 5
^{50}Ti	68.35	1.23	4.3 ± 0.7	7.3 ± 1.4
^{53}Cr	96.98	3.19	17 ± 4	25 ± 6
^{58}Ni	natural (67.76)	-4.05	340 ± 60	250 ± 50
^{96}Mo	96.8	0.15	6.4 ± 1.4	13 ± 3

Table II. Cross Sections for some (n,α) and (n,n'α) Processes Induced by 30 MeV d(Be) Break-up Neutron Spectrum

Target nuclide	% enrichment	Cross section (mb)	
		(n,α)	(n,n'α)
^{51}V	99.98	5.5±1.0	3.5±0.8
^{59}Co	natural (100)	9.5±2.0	
^{65}Cu	99.70		5.1±1.3

References

1. R.W. Conn, J. Nucl. Materials 76&77, 103 (1978)

2. S.M. Qaim, Proc. Advisory Group Meeting on Nuclear Data for Fusion Reactor Technology, Vienna, December 1978 (IAEA-TECDOC-223, 1979) p. 75

3. S.M. Qaim, G. Stöcklin, Proc. 8th Symp. on Fusion Technology, Noordwijkerhout (The Netherlands) June 1974 (EUR 5182e, 1974) p. 939

4. S.M. Qaim, N.I. Molla, Proc. 9th Symp. on Fusion Technology, Garmisch Partenkirchen, June 1976 (Pergamon Press, Oxford 1976) p. 589

5. S.M. Qaim, Nucl. Phys. A382, 255 (1982)

6. S.M. Grimes, R.C. Haight, J.D. Anderson, K.R. Alvar, R.R. Borchers, Proc. Symp. on Neutron Cross Sections from 10 to 40 MeV, Brookhaven, July 1977 (BNL-NCS-50681, 1977) p. 297

7. S.M. Grimes, R.C. Haight, K.R. Alvar, H.H. Barschall, R.R. Borchers, Phys. Rev. C19, 2127 (1979)

8. D.W. Kneff, B.M. Oliver, M.M. Nakata, H. Farrar IV, Proc. Symp. on Neutron Cross Sections from 10 to 50 MeV, Brookhaven, May 1980 (BNL-NCS-51245, 1980) p. 289

9. H. Liskien, A. Paulsen, Nuclear Data Tables 11, 569 (1973)

10. "ENDF/B-IV Dosimetry File", (Ed. B.A. Magurno), BNL-NCS-50446, Brookhaven National Laboratory (1975)

11. A. Paulsen, R. Widera, F. Arnotte, H. Liskien, Nucl. Sci. Engineering 72, 113 (1979)

12. S.M. Qaim, G. Stöcklin, J. Inorg. Nucl. Chem. 35, 19 (1973)

13. D.R. Nethaway, R.A. van Konynenburg, M.W. Guinan, L.R. Greenwood, Proc. Symp. on Neutron Cross Sections from 10 to 40 MeV, Brookhaven, July 1977 (BNL-NCS-50681, 1977) p. 135

14. S.M. Qaim, S. Khatun, R. Wölfle, Proc. Symp. on Neutron Cross Sections from 10 to 50 MeV, Brookhaven, May 1980 (BNL-NCS-51245, 1980) p. 539

15. A. Paulsen, H. Liskien, F. Arnotte, R. Widera, Nucl. Sci. Engineering 78, 377 (1981)

MEASUREMENT OF DOUBLE DIFFERENTIAL NEUTRON EMISSION CROSS SECTIONS WITH 14 MeV SOURCE FOR D, Li, Be, C, O, Al, Cr, Fe, Ni, Mo, Cu, Nb AND Pb.

A. Takahashi, J. Yamamoto, T. Murakami, K. Oshima,
H. Oda, K. Fujimoto and K. Sumita

Department of Nuclear Engineering,
Osaka University
Yamadaoka 2-1, Suita, Osaka, Japan

Secondary neutron spectra from ring samples of fusion reactor candidate elements were measured for 6 to 12 angle-points by means of time-of-flight technique. Data with good statistics and excellent energy-angle-resolutions were obtained by using relatively long flight path (9.5 m) and intense source of 14 MeV neutrons. Large discrepancies were found in comparison with ENDF/B-IV data, for many elements and in the range 5 to 13 MeV where measured secondary neutron specta showed characteristic structures by level excitations and a great deal of anisotropies. This shows predominant contributions of direct reaction processes for non-elastic neutrons, even upto high excitation levels.

(14 MeV neutron source, double differential emission cross section, fusion reactor materials, ring sample, 9.5 m TOF, comparison with ENDF/B-IV, discrepancies)

Introduction

The necessity to use double differential cross section (DDX) data (energy-angle-distribution data of secondary neutrons) in calculating anisotropic neutron transport in media surrounding D-T plasma has been pointed out at the IAEA Advisory Group Meeting on Nuclear Data for Fusion [1] in 1978, and in other reports[2,3,4]. To estimate precisely reaction rates such as tritium production in fusion reactors, the use of DDX in neutron transport calculation is recommended. Application for shielding calculation is also effective. Data of DDX-type are directly applicable to transport calculations [2,4], since the Boltzmann equation contains DDX-type scattering kernel.

Measurements of DDX for elements and isotopes of fusion reactor candidate materials have been carried out in several laboratories. Hermsdorf et al[5] measured DDXs for 34 elements with 14 MeV source, in the ranges 2 to 14.6 MeV and 40 to 150 degree scattering angle. Their enormous data, however, are not for forward scattering angles at which anisotropic scattering becomes important, and are not with sufficient energy-resolution enough to clarify reaction mechanism in the range of several MeV to 14 MeV for emitted neutrons. Drake et al [6] measured DDX for Be-9 at incident neutron energies of 5.9, 10.1 and 14.2 MeV, and reported that the ENDF/B cross sections strongly overemphasized the low-lying states in Be-9. The series experiments [7] at Triangle Universities Nuclear Laboratory have been to measure single differential cross sections for elastic and discrete-level-inelastic scattering. Data for high excitation levels and DDX could not be measured due to the continuous energy spectrum component of source, in the lower energy range, produced by their FN tandem Van de Graaf accelerator.

At the Intense 14 MeV Neutron Source Facility [8] (OKTAVIAN) of Osaka University, measurements of DDXs with 14 MeV incident energy have been carried out for many elements of fusion reactor candidate materials. Some results are presented in the following, in comparison with ENDF/B-IV[10]. Innovation in the present experiment is due to attaining DDX measurement with good energy-resolution for many scattering angles and without losing good statistics.

Experimental

Secondary neutron spectra from a ring sample were measured with varying scattering angle by means of time-of-flight technique. The experimental arrangement is shown in Fig.1. Deuteron ion beam with 1.5 ns pulse width in f.w.h.m. and 16 mA peak current by OKTAVIAN accelerator [8] bombards the 10 Ci TiT target (2 cm in diameter) to generate D-T neutrons. Angular distribution of D-T neutron yield from the target assembly was derived by the measurement of (n,alpha) reaction rates of aluminum foils. Relation between mean source neutron energy and angle was obtained by the TOF data of elastic scattering for lead and carbon ring-samples, as shown in Fig.2.

Scattering samples adopted were in ring shape, instead of short cylinders as used in the conventional goniometer-TOF experiment, so that the neutron flight path and the detector position were fixed to the direction of the accelerator beam line. Ring samples have 10 cm major radii and 1 cm minor radii for materials except

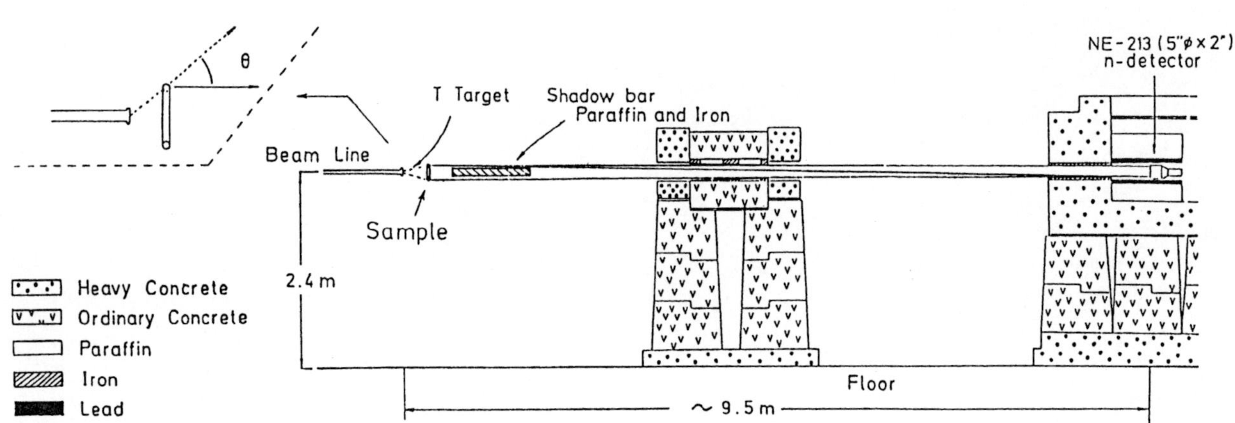

Fig.1 Schematic arrangement of ring-sample-TOF experiment

K. H. Böckhoff (ed.), Nuclear Data for Science and Technology, 360–367.

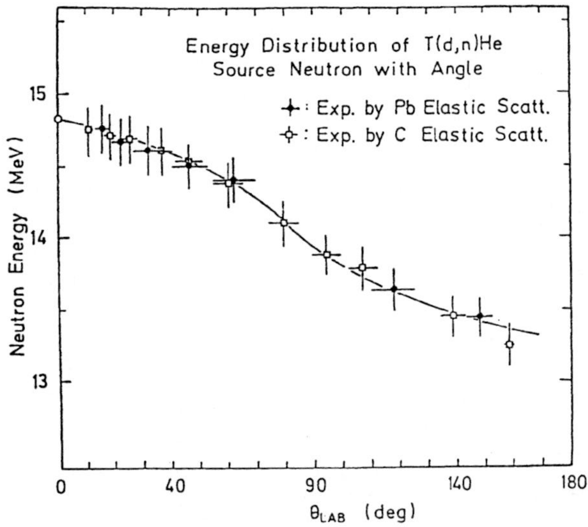

Fig.2 Relation of source energy vs. scattering angle

Be, Cr and Mo which have 2 cm x 2 cm cross sections.

Emitted neutrons from a ring sample were detected with a 5"dia.2" NE213 liquid scintillator placed at 9.5 m distance from the target. Source D-T neutrons viewing directly the detector were shadowed with a bar of iron-praffin mixture (9 cm in diameter and 110 cm in length). Room-returned neutrons were shielded with the pre-collimator and main-collimator shadowing system, as shown in Fig.1. Scattering angle was changed with distance between the target and a ring sample, so that source neutron energy varied from 14.8 MeV (0 degree) to 13.5 MeV (150 degree) corresponding to laboratory scattering angles (see Fig.2). This source condition was taken into account when comparing with ENDF/B data. TOF measurement was possible in the range 11 to 160 degree of laboratory scattering angle. Background run was done by removing a ring sample. Data normalization among a series of TOF runs was performed by counting source D-T neutrons with a 2"dia.2" NE213 detector positioned at 90 degree against the accelerator beam line and at 2.65 m from the target.

The used electronic circuit was a conventional one in TOF measurement. Elimination of gamma-ray counting was done with two parallel pulse shape discrimination circuits with "low" and "high" gain DDL amplifier, respectively, for the main detector. Neutron-gamma discrimination for the monitor detector was single. Energy-resolution of the present experiment was about 2% (0.17 MeV at 14.8 MeV).

Energy-dependent efficiency of the main detector was determined by the TOF experiment with Cf-252 fission neutron source for the range 0.5 to 8 MeV, and in addition by the TOF measurement of $^{1}H(n,n)$ scattering peaks for the range 6 to 14 MeV ; i.e., fission neutron spectrum and differential elastic scattering cross sections of hydrogen were used as standard. This experimental determination of detector efficiency was carried out after every series of runs for a ring sample. The $^{1}H(n,n)$ scattering experiment with a polyethylene ring of 0.7 cm minor radius and 10 cm major radius was also utilized to determine DDX values in barn/steradian/MeV for all ring samples, by normalizing with monitor counts and $^{1}H(n,n)$ standard cross sections.

One TOF run with ring sample took about an hour. DDX measurements were carried out for samples of D (D_2O in glass), Li (natural lithium in stainless steel can), Be, C, O (H_2O in glass), Al, Cr, Fe, Ni, Mo, Cu, Nb and Pb. All samples are in natural abundance. Special background runs were done with empty cans for H_2O, D_2O and Li. Present experiment for these three elements (D, O, Li) is preliminary since there remain problems in samples. Correction for multiple scattering in a ring is not yet done in the present results. However, the difference between the two experimental results with 1 and 1.5 cm minor radii for Cu was within statistical errors.

DDX from ENDF/B

By using the DDX processing module of the transport code system NITRAN [2,4] DDX data set with 135 energy-groups and 64 angle-points were produced from ENDF/B-IV data library.[10] The detail of calculational procedure is written in Ref.2. For both of the elastic scattering and the inelastic scattering with discrete level excitation, angular distribution data (single differential cross sections) were used with scattering kinematics to generate double differential cross sections for every reaction channel. For inelastic scattering with continuum level-excitation and (n,2n) reaction, isotropies in the laboratory system were assumed if angular distributions were not given. Then the energy-distribution data (renormalized to the unit of barns/MeV) were simply devided by 4π to generate DDX. Thus DDXs of total neutron emission were produced by gathering DDXs over all reaction channels for which any single differential data were available in ENDF/B-IV. For carbon, ENDF/B-V data were used.

Results and discussions

Experimental results of DDXs in comparison with ENDF/B-IV are shown in the following, from light element to heavy one. However, the results for D, Li and O will be shown at the end since they are preliminary. Experimental points are so many that only selected data of figures can be shown. Measured DDX data have sufficient energy-resolutions enough to deduce angle-dependent (single differential) level excitation cross sections for some of particular excitations, even for heavy elements. In the present paper, however, only elastic and few discrete-level inelastic cross sections of single differential are derived by integrating peaks in spectra. Comparisons with ENDF/B-IV are therefore mostly in DDX.

The lower energy source component produced by scattering of D-T neutrons in the target support will contaminate energy spectra of DDXs, somewhat significantly for extremely forward angles (11 to 15 degree), but only slightly for other angles. To perform exact

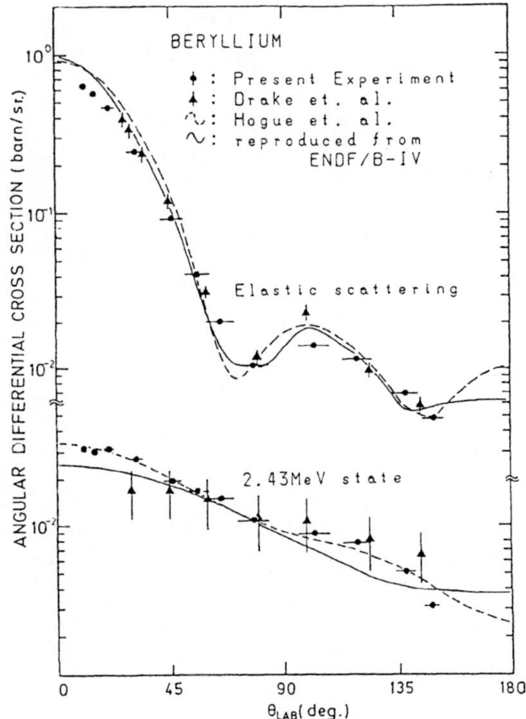

Fig.3 Angular distributions of elastic and 2.43 MeV state inelastic scattering, for Be

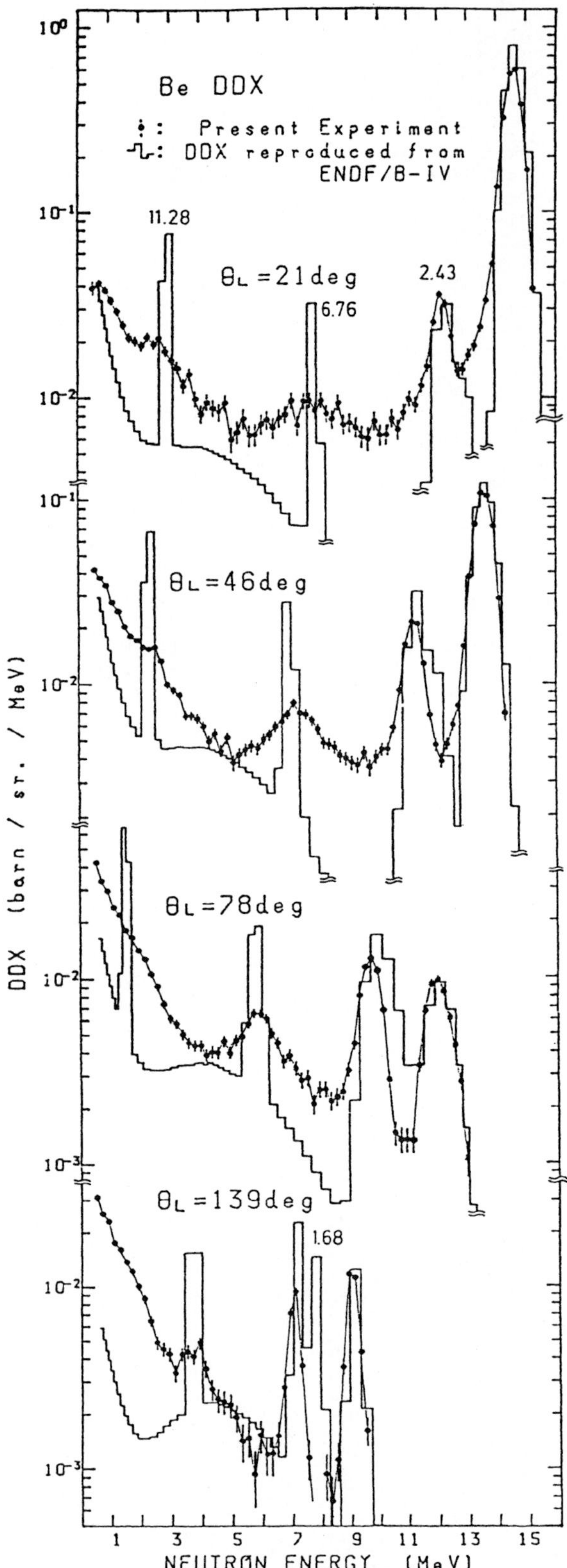

Fig.4 Selected DDXs for Be

comparisons the measured source spectra from the target assembly should be used in broadening calculated DDX data from ENDF/B. Within the discussions of the present paper, this correction can be neglected and the broadening of calculated DDX has been done with Gaussians of measured energy-angle resolution factors which are shown by error bars in Fig.2.

Beryllium : DDXs were measured at laboratory scattering angles of 11, 15, 21, 32, 46, 56, 65, 78, 103, 120,

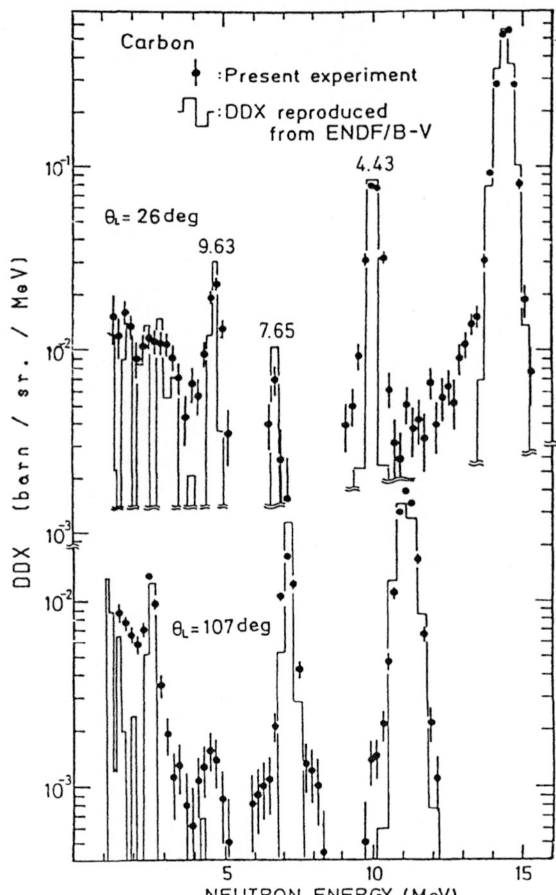

Fig.5 Selected DDXs for carbon

139 and 149 degree and for energy range 0.3 to 15 MeV. Angular distributions of elastic scattering and the second level (2.43 MeV state) inelastic scattering are shown in Fig.3, in comparison with ENDF/B-IV and other measurements. Selected DDXs are shown in Fig.4. For elastic scattering, present results are smaller by about 20% than others in forward angles (11 to 21 degree) where the curve by Hogue[7] is extrapolated with Legendre fit. As for discrete-level inelastic scattering, data for 1.68, 2.43, 6.76 and 11.28 MeV state are available in ENDF/B-IV and the 2.43 MeV state is treated as anisotropic, while others as isotropic. As clearly seen in measured DDXs (Fig.4), the 1.68 MeV state was not observed. The present results for the 2.43 MeV state are in very good agreement with those by Hogue et al[7] and considerably larger than ENDF/B-IV in forward angles. In DDXs, large discrepancies between measured and ENDF/B-IV data are seen in the energy range below the 2.43 MeV state peak. As pointed out by Drake et al [6], the "low-lying states" (6.76 and 11.28 MeV states) are overemphasized. The (n,2n) process of Be-9 in ENDF/B-IV is treated mostly via time-sequential process. The first emitted neutrons with 6.76 and 11.28 MeV excitations can be treated as inelastic scattering, then they must be in sharp peaks in energy spectra of DDXs. The present results, however, show that the (n,2n) process is not so frequently time-sequential but simultaneous emission of two neutrons may be more frequent. In any way, kinematics of the (n,2n) reaction should be taken into account in DDX caluculation, since the angular-dependence of emitted spectra is so different from ENDF/B-IV.

Carbon : DDXs were measured at 11, 19, 26, 37, 46, 60, 80, 94, 107, 139 and 159 degree and for the secondary neutron energy range 1 to 15 MeV. Elastic scattering and inelastic scattering for three discrete-levels (4.43, 7.65 and 9.64 MeV states) could be resolved as shown in Fig.5 for selected DDXs. In Fig.6 angular differential cross sections for these four

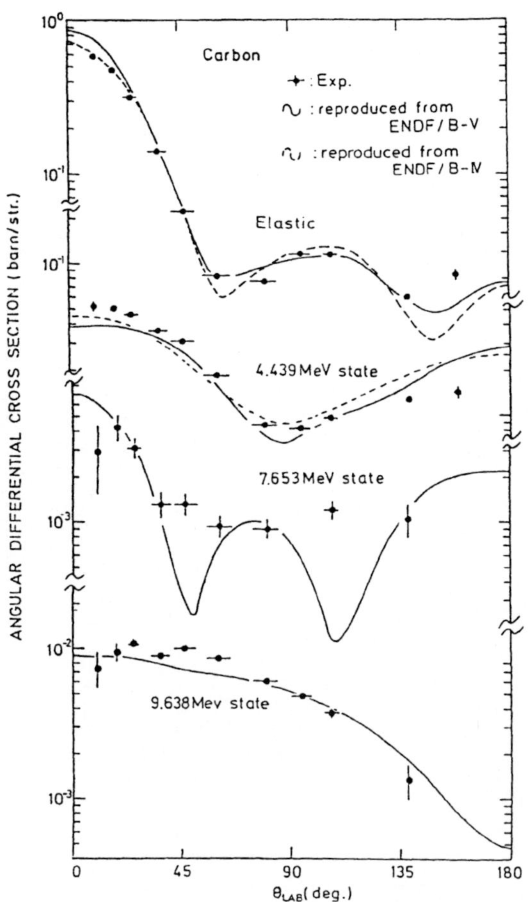

Fig.6 Angular distributions of elastic and 3 inelastic scattering cross sections, for carbon

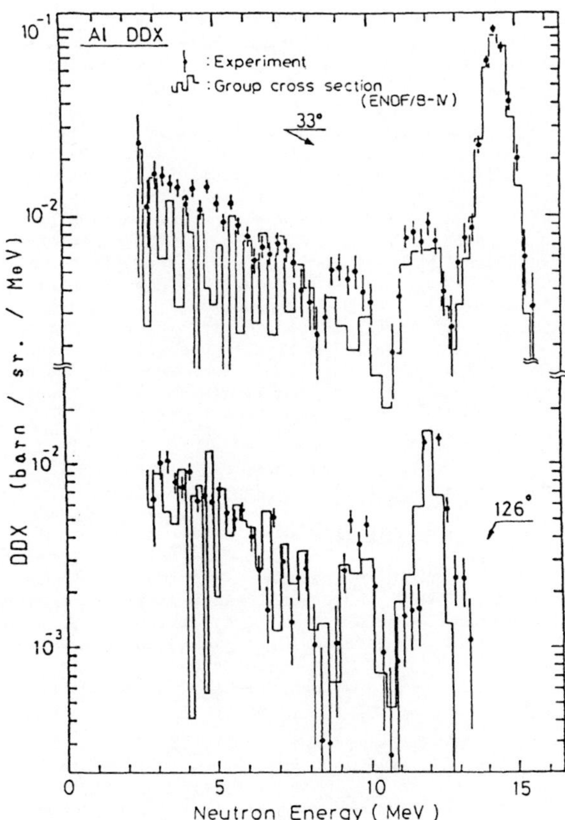

Fig.7 Selected DDXs for Al

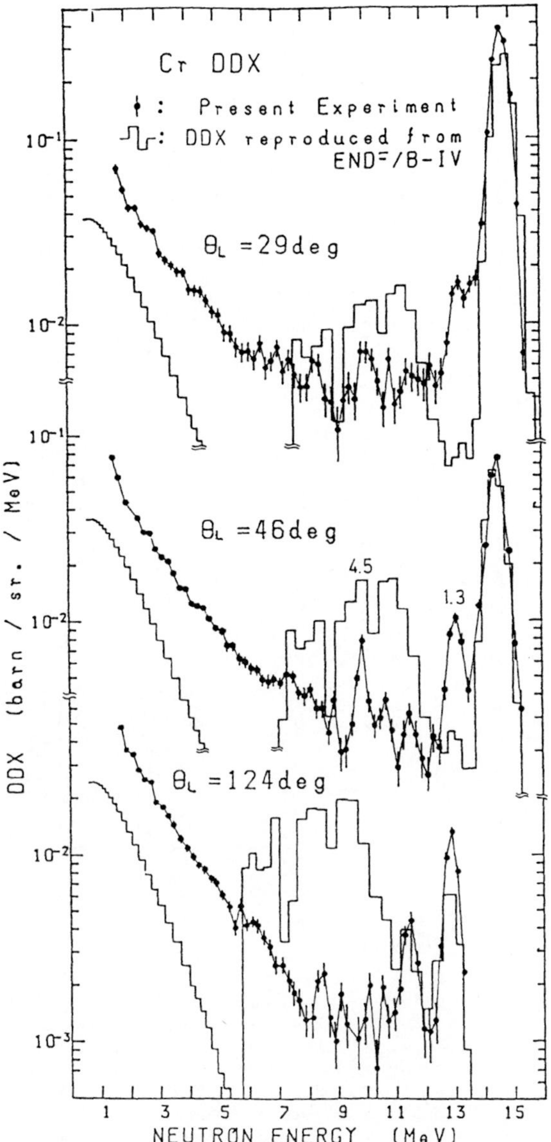

Fig.8 Selected DDXs for Cr

states are shown in comparison with ENDF/B-IV & -V data. In ENDF/B-IV 7.65 and 9.64 MeV states are not evaluated. As a whole, the present results are in good agreement with ENDF/B-V though the results for 4.43 MeV state in forward angles are larger by 20 to 30 % than ENDF/B and the two dips at 45 and 110 degree for 7.65 MeV state were not observed in experiment. The ENDF/B-IV data for carbon could not follow the experiment in the range below the 4.43 MeV state peak, since evaporation spectra are given. Experimental points in DDXs between the elastic and the 4.43 MeV state peak for 25 degree are due to the effect of the tail in source neutron spectrum. In addition to the three excitation states ENDF/B-V gives real levels of 10.30 and 10.84 MeV and pseudo-levels from 11.25 to 14.75 MeV with 0.5 MeV interval. Therefore DDXs from ENDF/B-V draw many peaks in the range below the 9.64 MeV state peak. In this range agreement with the present experiment is not good. ENDF/B-V data for carbon, however, are evaluated well as a whole.

Aluminum : DDXs were measured at 33, 43, 59, 107 and 126 degree and for the range 2 to 15 MeV. Results are shown in Fig.7. As a whole, agreement is good. Peaks by discrete-level inelastic scattering, measured in experiment in the range 7 to about 13 MeV, are well reproduced by ENDF/B-IV data. In the range below 7 MeV, where many pseudo-levels with 0.5 MeV interval are assumed in ENDF/B-IV, experimental results in forward angles seem to be larger in average.

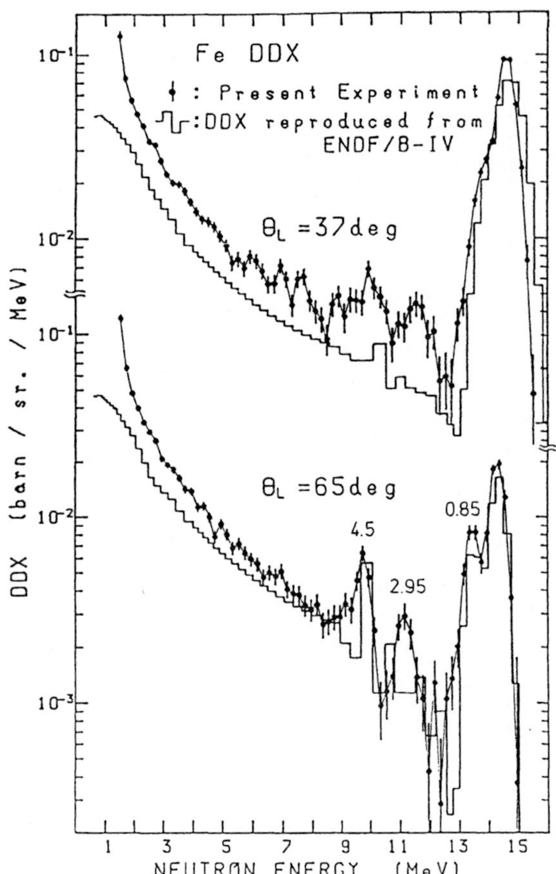

Fig.9 Selected DDXs for Fe

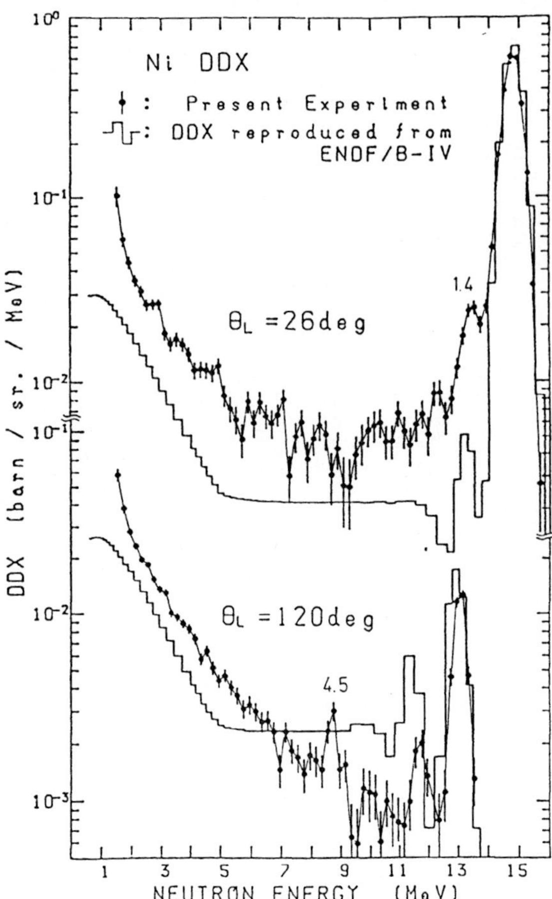

Fig.10 Selected DDXs for Ni

Chromium : DDXs were measured at 19, 29, 37, 46, 60, 73, 80, 103, 124, 143 and 159 degree and for the range 1 to 15 MeV. Obtained angular distribution of elastic scattering is in good agreement with ENDF/B-IV data, except 73 and 124 degree at which ENDF/B-IV data have dips. In Fig.8 selected DDXs are shown. In the range from 9 to 14 MeV for forward angles, 5 peaks by discrete-level-inelastic scattering are observed. Among them there exist two dominant peaks which correspond to "1.3" and "4.5 MeV state", respectively. Fine structures (small peaks) are seen down to about 5 MeV for secondary neutron energy. DDXs reproduced from ENDF/B-IV differ too far from experiments. For natural chromium, 40 discrete level-cross-sections are given isotropically in ENDF/B-IV and are forming the blocks of peaks as seen in the range 7 to 14 MeV. Very small cross section is given for continuum-inelastic scattering. In the lower energy range than 6 or 7 MeV, the spectra are occupied by (n,2n) neutrons. The present experiment shows that discrete-level-inelastic scattering is strongly anisotropic and continuous into the lower energy range. ENDF/B-IV data for Cr must be changed drastically. Comparison with ENDF/B-V should be done.

Iron : Selected DDXs are shown in Fig.9. DDXs were measured at 18, 29, 37, 51, 65, 80, 100, 120, 139 and 159 degree. As a whole, the present results are in fairly good agreement with ENDF/B-IV data. The experimental results however show stronger angular-dependence than ENDF/B-IV in which inelastic scattering is treated anisotropically for 26 discrete-levels, in the range 10 to 14 MeV. Three inelastic peaks than elastic are resolved in experiment. These correspond to 0.846, 2.95(2.939, 2.957) and 4.5 (4.45, 4.50) MeV state. Discrepancies are large at the latter two peaks

Nickel : DDXs were measured at 18, 26, 32, 41, 51, 64, 80, 101, 120, 143 and 161 degree. Selected DDXs are shown in Fig.10, in comparison with ENDF/B-IV. Elastic scattering cross sections in backward angles are significantly smaller than ENDF/B-IV. Upto 3.04 MeV state ENDF/B-IV gives 15 discrete-level-inelastic cross sections as isotropic distributions. Flat distribution from about 5 to 12 MeV in ENDF/B-IV curve (see Fig.10) shows continuum-inelastic scattering by precompound model which is treated as isotropic. Below about 5 MeV (n,2n) distribution continues. Experimental DDXs show structures by discrete level excitations in the range 4 to 14 MeV where angular dependencies are large and ENDF/B-IV data do not follow the experiment. In experiment dominant inelastic peaks are seen at 1.4 (1.33, 1.45) and "4.5" MeV state. The experimental results show that the precompound model is not successful and inelastic scattering via direct reactions with many discrete-level excitations should be taken into account with anisotropies. Moreover the present experiment shows anisotropy exists even in the lower energy range than about 4 MeV.

Molybdenum : DDXs were measured at 18, 29, 37, 46, 69, 79, 120 and 159 degree. Deduced differential elastic scattering cross sections were in good agreement with those of ENDF/B-IV, except 46 degree. Selected DDXs are shown in Fig.11. Non-elastic neutrons in ENDF/B-IV are from both of continuum-inelastic (precompound model) and (n,2n) reaction. In the (n,2n) region (below about 5 MeV) the present experiments are in very good agreement with ENDF/B-IV. On the contrary, in the continuum-inelastic scattering region 5 to 14 MeV discrepancies are large. Experiment shows many discrete-level excitations and strong anisotropy. Dominant excitations are around at 2.3 MeV state.

Copper : DDXs were measured at 11, 18, 29, 41, 51, 62, 79, 103, 120, 139 and 159 degree. Differential elastic (including inelastic scattering upto 1.865 MeV state) cross section and inelastic scattering cross section of "1.3 MeV state" (summation of 0.962, 1.165, 1.325 and 1.482 MeV) are shown in Fig.12. Agreements

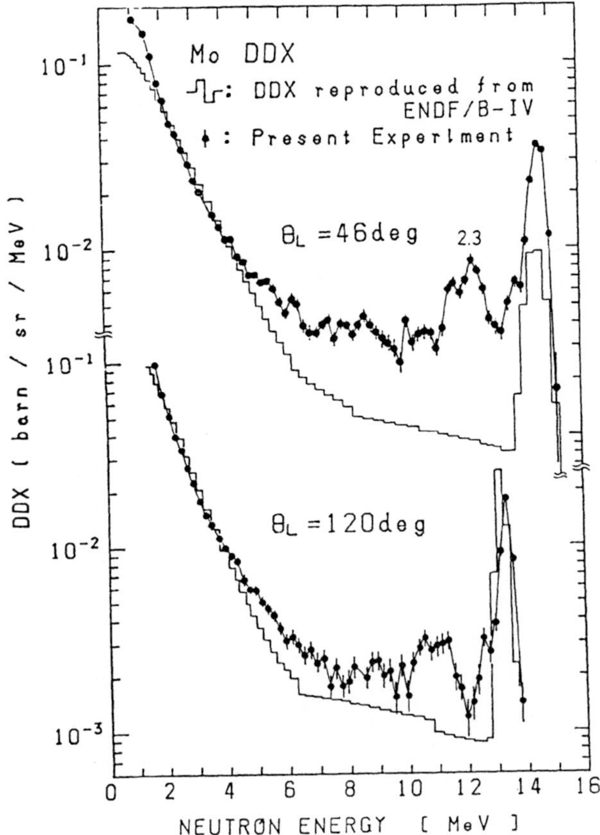

Fig.11 Selected DDXs for Mo

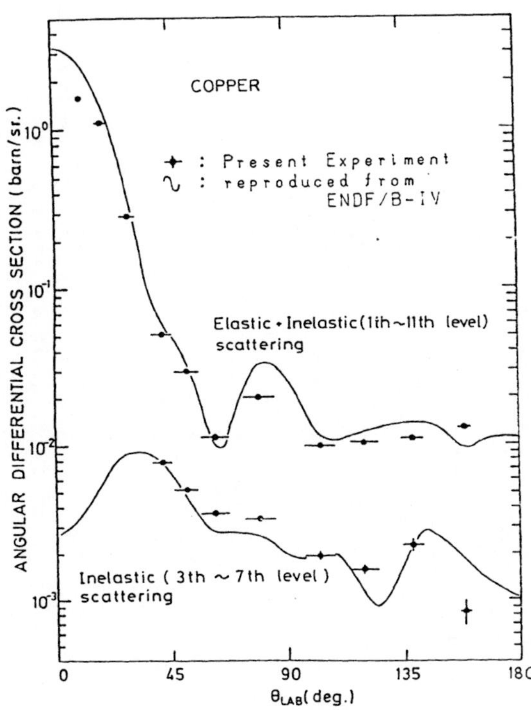

Fig.12 Angular distributions of elastic and "1.3 MeV" state inelastic scattering cross section, for Cu

with ENDF/B-IV are good, except 11 degree for elastic scattering. Selected DDXs are shown in Fig.13. Large discrepancies are found between experiments and ENDF/B-IV data. In ENDF/B-IV, discrete-level inelastic cross sections are given upto 1.865 MeV state and evaporation spectra are given for continuum inelastic scattering

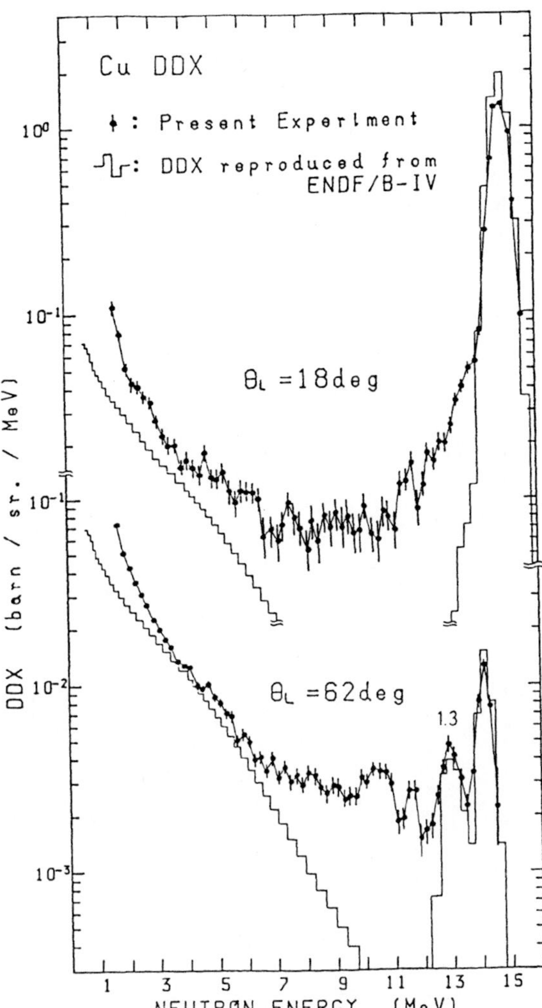

Fig.13 Selected DDXs for Cu

(Q-value = -1.9 MeV) and for (n,2n) reaction (Q-value = -9.9 MeV). The experimental results show that discrete-level-inelastic scattering with strong anisotropies should be treated upto about 7 MeV state (Q-value = -7 MeV) and evaporation model for this region is not successful at all. Many small peaks observed in experimental spectra of DDXs may be attributed to direct reaction processes with many excitation levels. Dominant excitations not evaluated in ENDF/B-IV are at "2.5" and "3.6" MeV state.

Niobium : DDXs were measured at 15, 23, 30, 37, 53, 70, 78, 102, 115, 130 and 143 degree and for the range 0.5 to 15 MeV. Selected DDXs are shown in Fig.14 in comparison with ENDF/B-IV. Discrepancies are large particularly at forward angles. In the range 5 to 14 MeV where the precompound model is used in ENDF/B-IV, measured spectra show very strong anisotropy and structures with discrete-level excitations. As already discussed for nickel and molybdenum, precompound models adopted in ENDF/B-IV can not reproduce experimental DDXs at all. Experimental spectra in the range below 5 MeV, where (n,2n) reaction contributes mainly, are harder than ENDF/B-IV data.

Lead : DDXs were measured at 15, 21, 27, 31, 39, 47, 50, 57, 69, 78, 103, 122, 137 and 154 degree and for the range 0.5 to 15 MeV. In Fig. 15 are shown selected DDXs. For discrete-level-inelastic scattering ENDF/B-IV gives 35 states upto 4.339 MeV and among them cross sections at 14 MeV incident energy are given for 11 states. As shown in Fig.15, excitations at 2.6 (sum of 2.615, 2.624 and 2.634 MeV state) and 4.3 MeV (sum of 4.076, 4.288 and 4.339 MeV) are dominant in ENDF/B-IV data. In experiment two peaks are seen corresponding to the two states. Differential cross

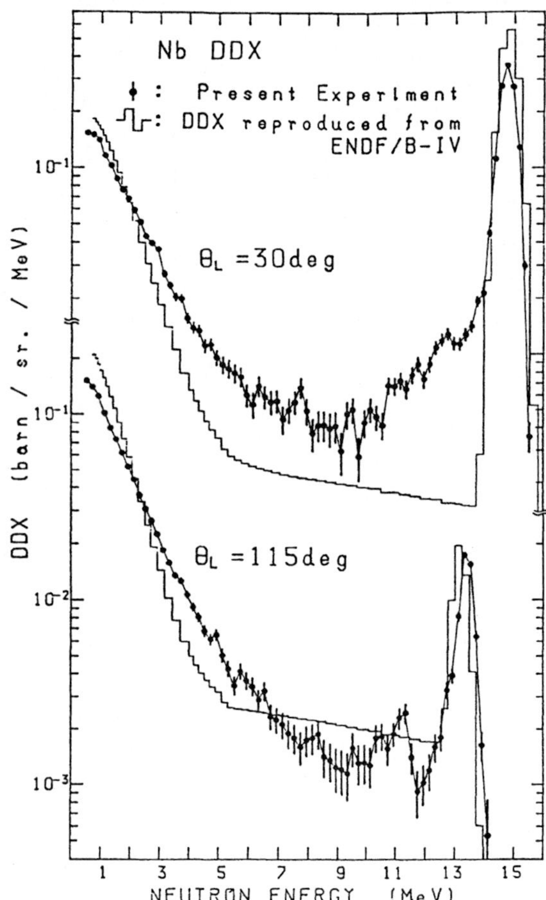

Fig.14 Selected DDXs for Nb

sections for these resolved level-inelastic scattering are shown in Fig.16 in comparison with ENDF/B-IV data. Agreements are well for elastic and the "2.6 MeV state", while the measured data are significantly larger for the "4.3 MeV" state. Above the 4.339 MeV state ENDF/B-IV treats inelastic scattering as "continuum", however the experimental spectra show the existence of discrete-level excitations and considerable anisotropy in the range 5 to 10 MeV. Below about 5 MeV, secondary neutron spectra are formed mainly by (n,2n) reaction. The present experimental results show that (n,2n) cross section at 14 MeV is significantly larger than that of ENDF/B-IV. This result is in consistent with the lead sphere experiment to have measured neutron multiplication. That experiment gave larger multiplication by about 20% than the calculation with ENDF/B-IV data. Therefore the tritium breeding with lead multiplier will become hopeful.

Deuterium : Measured DDX at 25 degree is shown in Fig.17 in comparison with that reproduced from ENDF/B-IV. At elastic scattering peaks agreements are very good, but for (n,2n) neutrons discrepancies are drastic. In the calculation isotropic distribution is assumed for (n,2n) secondary neutrons, in the laboratory system. The experimental result show that this assumption is quite incorrect; naturally (n,2n) neutrons can not appear in the energy range higher than elastic peaks, so that correct kinematics of the (n,2n) reaction must be taken into account.

Lithium : DDXs were measured at 15, 30, 40, 135 and 145 degree and for the range 0.4 to 15 MeV. Used sample is a brick 5x5 cm and 20 cm long so that angular resolution is poorer than other experiments. Examples are shown in Fig.18. Disagreements with ENDF/B-IV are seen in the energy ranges less than elastic peaks. Dominant excitation of the second level (4.63 MeV state, tritium producing reaction) of Li-7 is clearly seen. In ENDF/B-IV inelastic scattering except the first

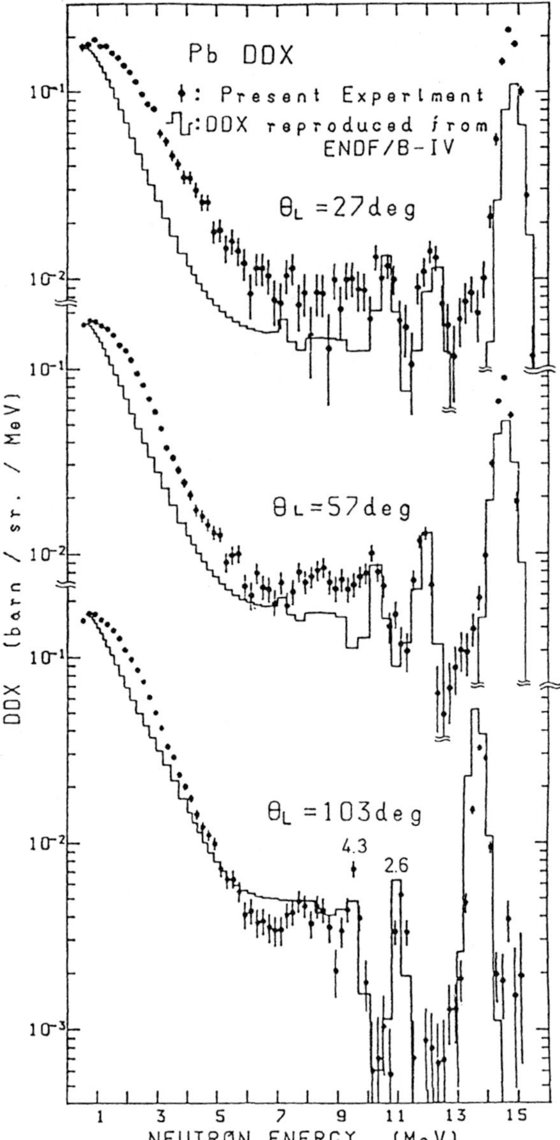

Fig.15 Selected DDXs for Pb

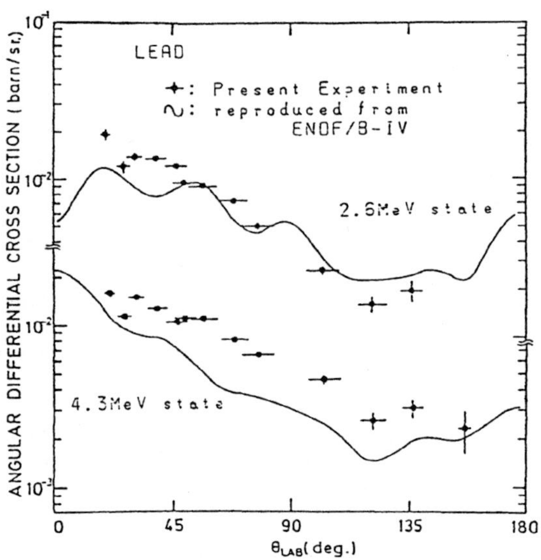

Fig.16 Angular distributions of "2.6" and "4.3 MeV" state inelastic scattering cross sections, for Pb

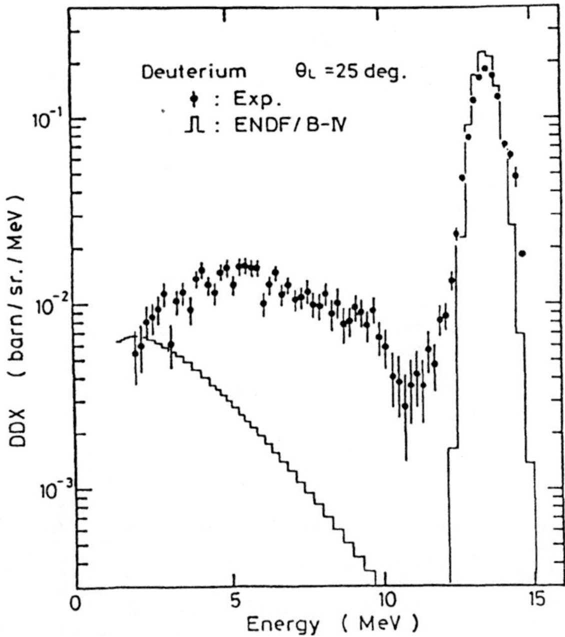

Fig.17 DDX of deuterium at 25 degree

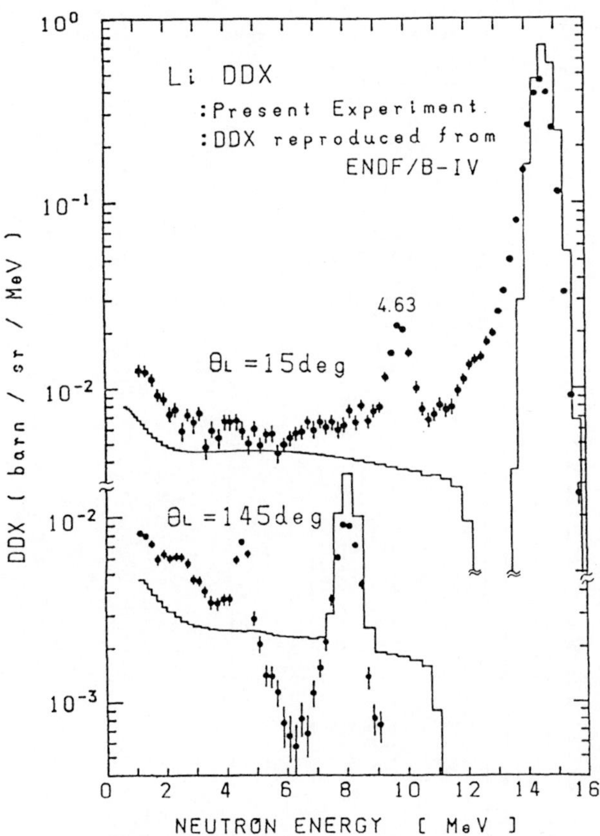

Fig.18 Selected DDXs for natural lithium

level (0.748 MeV state) is treated as isotropic in continuum. However the experiment shows that discrete-inelastic scattering with anisotropy should be included upto high excitation levels. Correct kinematics of (n,n't) of Li-7 must be treated, as in the case of (n,2n) of deuterium.

Oxygen : DDXs were measured at 25, 80, 115, 136 and 145 degree. Measurement between 25 and 80 degree was difficult due to dominant peak of hydrogen scattering - of sample water. Agreement with ENDF/B-IV data was fairly good. Example at 136 degree is shown in Fig.19.

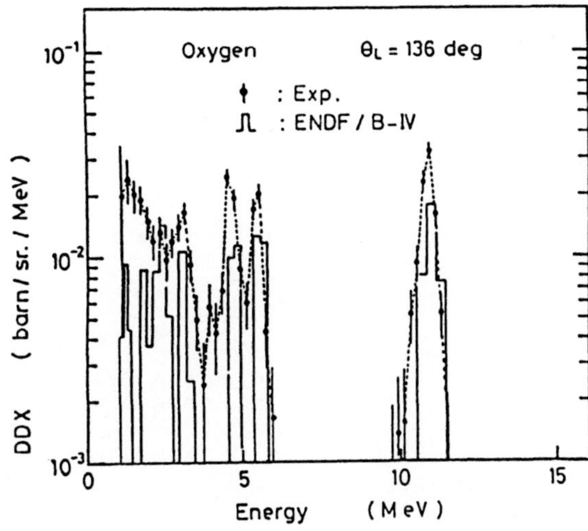

Fig.19 Example of DDX for oxygen

Disagreement in the lower energy range than 3 MeV may be due to multiple scattering effect of hydrogen which has large cross section.

Concluding remarks

The measured DDX data for many elements have not yet been fully analyzed. Further analysis will be done for emission cross sections, differential level-excitation cross sections and anisotropies. The present comparison with ENDF/B-IV has shown that there exist large discrepancies, particularly in the range 5 to 14 MeV, where many direct reactions with discrete level excitations should be taken into account. Even in the low energy range where important (n,2n) reactions exist, e.g. for Be and Pb, disagreements are large. Comparison with advanced nuclear data libraries, e.g. ENDF/B-V and others, should be done. Within the analyses of the present work, ENDF/B-IV data are well evaluated for oxygen and aluminum and fairly well for iron. Experiment for other elements than the present report is under way.

Acknowledgment : The present work has been supported by Grant-in-Aid for Scientific Research Promotion from the Ministry of Education, Science and Culture of Japan. The authors wish to thank post-graduate students (Mr. M. Ueda, Mr. M. Fukazawa & Mr. Y. Yanagi) for their assistances in the experiment and analysis.

References

1) INDC(NDS)-101/LF (1979)
2) Takahashi, A., Rusch, D.: KfK-2832/I,II (1979)
3) Jarvis, O. N.: J. Europian Sci. Technol., 3[1,2] (1981)
4) Yamamoto, J., et al.: J. Nucl. Sci. Technol., 19, 276 (1982)
5) Hermsdorf, D., et al.: ZfK-277(U) (1975)
6) Drake, D. M., et al: Nucl. Sci. Eng., 63,401 (1977)
7) Hogue, H. H., et al.: ibid., 69, 22 (1979)
8) Sumita, K., et al.: 12th SOFT at Julich, Sept. 1982
9) Takahashi, A., et al.: ibid.
10) ENDF/B summary documentation, BNL-NCS-17451 (1975)
D; MAT=1120, by Leonard, B. R., et al. (BNW)
Li; MAT=1271 & 1272, by Hale, G. M., et al (LASL)
Be; MAT=1289, MOD2, by Howerton, R. J., et al. (LLL)
C; MAT=1306, by Fu, C. Y., et al. (ORNL) ; ENDF/B-V
O; MAT=1276, MOD2, by Young, P. G., et al. (LASL)
Al; MAT=1193, by Young,P. G., et al. (LASL)
Cr; MAT=1191, by Prince, A. (BNL)
Fe; MAT=1192, by Perey, F. G., et al. (ORNL)
Ni; MAT=1190, by Bhat, M. R. (BNL)
Mo; MAT=1287, by Howerton, R. J. (LLL)
Cu; MAT=1295, MOD1, by Drake, M. K., et al. (SAI)
Nb; MAT=1189, by Howerton, R. J., et al.(LLL)
Pb; MAT=1288, MOD5, by Fu, F. G., et al.(ORNL)

MEASUREMENT AND EVALUATION OF (n,t) CROSS SECTIONS

Z. T. Bődy, F. Cserpák, J. Csikai, S. Sudár

Institute of Experimental Physics
Kossuth University, Debrecen, Hungary

K. Mihály

Teaching Reactor, Technical University, Budapest
Hungary

Cross sections for (n,t) reaction were measured on ^{9}Be and ^{40}Ca by tritium beta counting and activation methods around 14 MeV. Results were compared with the Hauser-Feshbach model. Compilation and evaluation of the available data at 14 MeV have been carried out. The possible use of the empirical formulae for the description of cross sections were studied.

[^{9}Be, $^{40}Ca(n,t)$, cross sections, E_n= 13.55-14.7 MeV, comparision with theory]

Introduction

In addition to the basic nuclear research, the knowledge of (n,t) cross sections are important for the estimation of tritium concentration produced in some structural materials of both fission and fusion reactors. Neutron energies of interest cover the range from thermal to about 14 MeV.

Recently the (n,t) cross sections have been investigated by the activation and tritium beta counting techniques mainly at Jülich, Zagreb and Debrecen. Qaim and Stöcklin [1] found that a maximum appears in the $\sigma_{n,t}$ values at Z=26, and an (N-Z)/A dependence exists for Z > 22. Similarly to Qaim's observation we also found the (N-Z)/A dependence for odd target nuclei, but the $\sigma_{n,t}$ data were higher than for even nuclei by one order of magnitude [2]. Further measurements are needed, however, to clear up the systematics in the $\sigma_{n,t}$ values. The aim of this work is the complete the (n,t) cross sections and to study the present states of data around 14 MeV.

Compilation and evaluation of $\sigma_{n,t}$ data

A compilation and evaluation programme has been iniciated to survey the available data for the (n,t) cross sections around 14 MeV. Some previous results are the following:

1) The vast majority of the experiment was devoted to the lightest elements (Li, Be, B). E.g. about the third of the measurements concerns the $^{6}Li(n,t)$ process if one take into account the data at all energies.
2) The experimental data for other elements are very scarce and contradictory. The maximum discrepancy can reach three order of magnitude (^{32}S, ^{40}Ca). Using the activation method the interfering reactions on the impurities can strongly influence the (n,t) cross sections. E.g. in the case of ^{32}S and ^{40}Ca the ^{31}P and ^{39}K contaminations, respectively must be less than ~0.01 %.
3) Neglecting the cross sections integrated for a broad energy range (e.g. the fission average cross sections of Heinrich or those measured by Qaim using deuteron

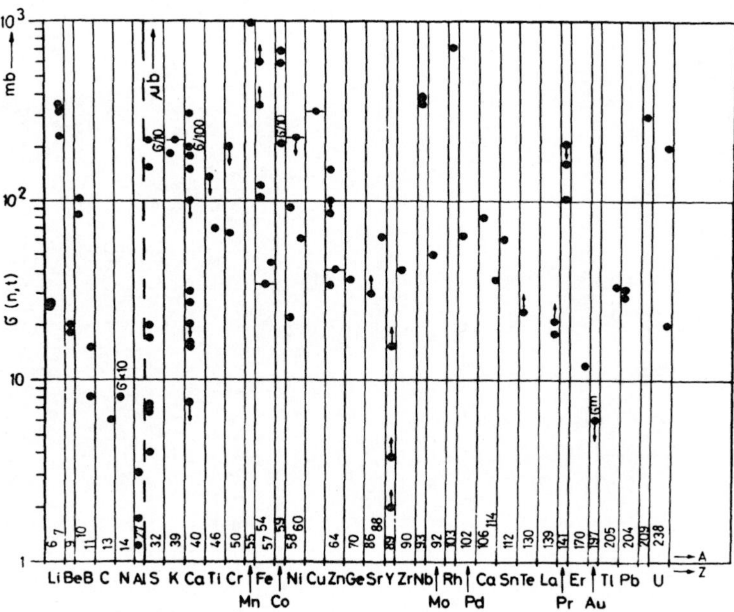

Fig. 1. Cross sections for (n,t) reactions at 14 MeV.

K. H. Böckhoff (ed.), Nuclear Data for Science and Technology, 368–370.

break-up neutrons) the following statements can be made for the atomic number range $10 \leqq Z \leqq 93$: Measurements were performed only for thirty elements, among these in four cases are data for two isotopes. There are wide atomic number ranges (Z = 33-37, 53-56, 62-67 and 69-78) where no experimental data at all. Excitation functions measured at least two energies were determined only in 8 cases. In the $10 \leqq Z \leqq 93$ region 54 cross section values were determined for the $14.2 \leqq E_n \leqq 14.9$ MeV energy range. In addition to this in 7 cases upper limits were given and in 8 cases partial cross sections for the ground or metastable states.

4) Data available in the energy range mentioned above are presented in Fig. 1. Fig. 2 shows that the $\sigma_{n,t}$ values as a function of (N-Z)/A are separated into two groups for even and odd target nuclei.

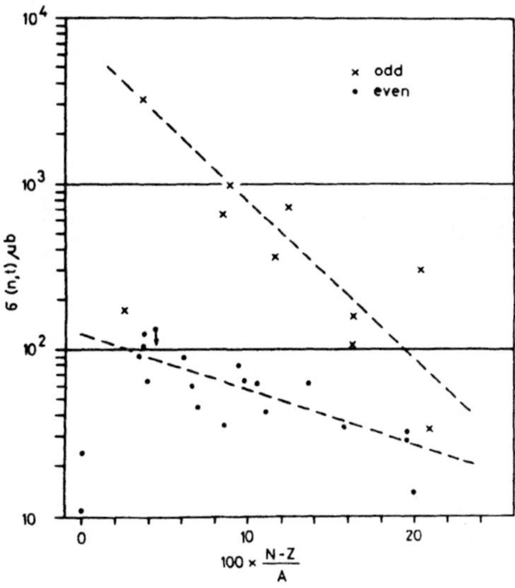

Fig. 2. Dependence of $\sigma_{n,t}$ on (N-Z)/A

For Z > 20 the experimental data are inside in the following intervals:

$$76\ e^{-9.2\frac{N-Z}{A}} \leqq \sigma_{n,t} \leqq 200\ e^{-8.3\frac{N-Z}{A}} \quad (\mu b)$$

for even target nuclei,

$$4500\ e^{-23.5\frac{N-Z}{A}} \leqq \sigma_{n,t} \leqq 5400\ e^{-13.9\frac{N-Z}{A}} \quad (\mu b)$$

for odd nuclei. These limits show that both the cross sections and the slope are higher for odd nuclei than for even ones. There are two similar empirical formulae given by Qaim and Stöcklin [1]

$$\sigma_{n,t} = 4.52(A^{1/3}+1)^2\ e^{-10\frac{N-Z}{A}} \quad (\mu b)$$

as well as Woo and Salaita [3]

$$\sigma_{n,t} = 7.684(A^{1/3}+1)^2\ e^{-13\frac{N-Z}{A}} \quad (\mu b)$$

for the calculation of the $\sigma_{n,t}$ values. Both formulae are good only for even nuclei where their qualities are about the same. However, for the range $20 < Z \leqq 44$ the Qaim's formula is better (in a 4:7 progression) while for the range $Z > 44$ the Woo's expression is preferable (in a 1:6 progression).

In addition to the compilation and evaluation of the (n,t) cross sections some new data were determined for ^{9}Be and ^{40}Ca around 14 MeV.

Experimental Procedure

For the determination of (n,t) cross sections both the tritium beta counting (Be) and the activation methods (Ca) were used. Berillium samples of 1.5-2.3g were irradiated simultaneously for 20h, placing them into different directions to the analysed deuterium beam of 170 keV to change the neutron energy around 14 MeV. The average neutron flux has been determined by the ^{93}Nb(n,2n)^{92m}Nb and ^{197}Au(n,2n)^{196}Au reactions covering the sample with Nb and Au foils on both sides. The details of the preparation of samples, and of the degassing procedure as well as the activity measurements have been published previously [4]. In the case of Ca samples of natural isotopic composition have been used. Discs of 19 mm diam. were pressed of spectral pure $CaCO_3$ powder. The thickness of the Ca samples were about 1.0 g/cm^2. For Ca the ^{14}N(n,2n) reactions have been used as flux monitors and its cross sections were accepted for standardization. The nitrogen monitors were prepared from cellulose nitrate and $C_6H_{12}N_4$ compound. The purity of the samples has been controlled by X-ray fluorescence analysis. During the measurement special attention was paid to the possible contribution of the high energy gamma's by annihilation to the 511 keV peak area.

Results and Discussion

Cross section of ^{9}B(n,t)^{7}Li reaction were measured at six energies between 13.5 and 14.7 MeV. Results are summarized in Table I

Table I ^{9}Be(n,t) cross sections

E_n (MeV)	σ (mb)
14.70	22.8 ± 3.4
14.63	18.9 ± 4
14.39	22.0 ± 2.8
14.10	19.1 ± 2.6
13.83	16 ± 2.5
13.55	15.5 ± 2.5

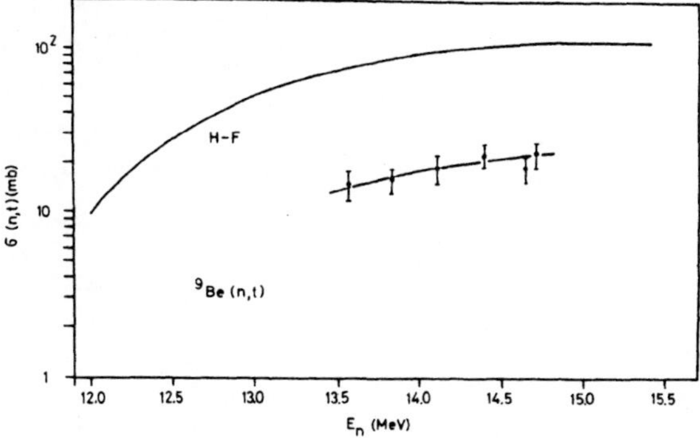

Fig. 3. Measured and calculated cross sections for ^{9}Be(n,t) reaction

and plotted in Fig. 3. The errors given in Table I are due the following sources: determination of neutron flux, counting statistics and efficiency, degassing efficiency. For comparision the excitation function between 12 and 15.5 MeV was calculated using the Hauser-Feshbach model. The details of the calculation has been described in a previous paper [2]. As it can be seen in Fig. 3 the shape of the curve can be well fitted to the experimental points but there is a factor of five in the absolute values.

In the case of Ca the $^{40}Ca(n,t)^{38m}K$ reaction was investigated at 14.7 MeV. A value of 3.5 ± 1.4 μb was found in this experiment which is higher with about two order of magnitude than that predicted the Hauser-Feshbach model.

The results obtained for the $^{9}Be(n,t)$ reaction is in good agreement with the literature data [4,5]. In the case of calcium a large spread exists in the measured values around 14 MeV.

Acknowledgements

The authors are grateful to Mrs. Dr. Juhász for her kind help during the measurements. Two of us (Z.T. Bődy and K. Mihály) wish to acknowledge the financial support of the International Atomic Energy Agency, Vienna under contract No. 3114/RB.

References

1. S.M. Qaim, G. Stöcklin, Nucl. Phys. A257 233 (1976)
2. S. Sudár, J. Csikai, Nucl. Phys. A319 157 (1979)
3. T.W. Woo, G.N. Salaita, Proc. Int. Conf. on Nucl. Cross Sections for Techn. Knoxwille TN, Oct. 22-26, GC12 853(1979)
4. M.E. Wyman, E.M. Fryer, M.M. Thorge, Phys. Rev. 112, 1264 (1958)
5. T. Biró, S. Sudár, Z. Miligy, Z. Dezső, J. Csikai, Inorg. nucl. Chem. 37, 1583 (1975)

THE $(n,{}^3He)$ REACTIONS AT $E_n \sim 14$ MeV

Đuro Miljanić and Mile Zadro
"Ruđer Bošković" Institute, POB 1016, 41001 Zagreb, Yugoslavia

The forward part of the ${}^{40}Ca(n,{}^3He_o){}^{38}Ar$ reaction angular distribution has been measured at E_n = 14.6 MeV using standard counter telescope technique. The ${}^{40}Ca(n,{}^3He){}^{38}Ar$ reaction cross section deduced from these measurements is much higher than previously measured $(n,{}^3He)$ reaction cross sections for some other nuclei. This difference could be explained mainly by the Coulomb effects.

[Nuclear reaction ${}^{40}Ca(n,{}^3He)$, E = 14.6 MeV, measured $\sigma(E_{{}^3He}, \theta)$, DWBA calculations]

Introduction

The data on $(n,{}^3He)$ reactions for neutron energies around 14 MeV are scarce. All $(n,{}^3He)$ reaction cross sections were up to now measured by activation technique (see e.g. Refs. [1,2]). Measured cross sections, having values of few μb, are the lowest cross sections of all 14 MeV neutron induced (n,x) reactions (x- nuclei with A≤4). A statistical model analysis[3] of existing data gave theoretical cross section values lower than the experimental ones. It probably indicates that these processes are mainly direct, which would be in accord with the results from different (p,t), $({}^3He,n)$ and other two nucleon transfer reaction studies. Present measurement of the ${}^{40}Ca(n,{}^3He){}^{38}Ar$ reaction was undertaken in order to shed some light on these processes. The ${}^{40}Ca$ nucleus was a natural choice not only because of its attractive structure, but mainly because of relatively low $(n,{}^3He)$ reaction Q value (Q = -6.99 MeV). It should be mentioned here, that the angular distribution of inverse ${}^{38}Ar({}^3He, n_o){}^{40}Ca$ reaction was measured at ${}^3He$ energies of 11.5 MeV[4] and 13.5 MeV[5]. The angular distributions display a pronounced stripping pattern. With the help of detailed balance principle one can deduce from the data that the total cross section for the ${}^{40}Ca(n,{}^3He_o){}^{38}Ar$ reaction at neutron energies of 18.1 and 20.0 MeV is of the order of 1 mb. This value is two orders of magnitude higher than any other $(n,{}^3He)$ reaction cross section measured with activation technique at $E_n \sim 14.6$ MeV.

Experimental Method

Present measurements were carried out at the "Ruđer Bošković" Institute neutron generator. 14.6 MeV neutron flux on the target was 2-5 · 10^7 n cm^{-1} s^{-1}. The target was a thin (2.4 mg cm^{-2}) layer of calcium fluoride evaporated on aluminum backing. The detection and identification of ${}^3He$ ions was done with the use of counter telescope consisted of a large photodiode (E-detector) and of three hydrogen filled proportional counters[6]. The cross sections were measured for eight setting angles in the 0° to 50° range.

Results

The only contributions observed in this experiment were those from the ground state transition. Measured ${}^{40}Ca(n,{}^3He_o){}^{38}Ar$ cross sections are given in Fig. 1. Error bars represent statistical errors only. Curves are the results of DWBA calculations normalized to the first experimental point. Full line represents calculation with only $(d^2_{3/2})$ configuration taken into account and dash one is obtained by including a $(f^2_{7/2})$ component. From the experimental results shown here one can estimate that the ${}^{40}Ca(n,{}^3He){}^{38}Ar$ reaction cross section is larger than 0.2 mb.

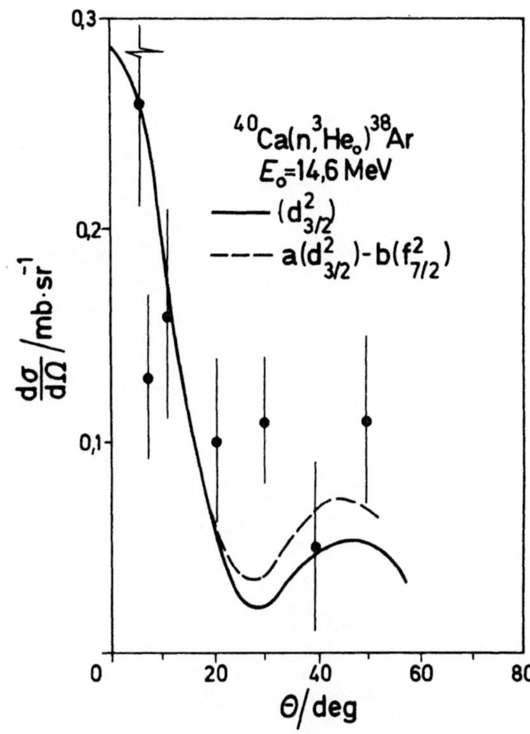

Fig. 1. Differential cross sections for the ${}^{40}Ca(n,{}^3He_o){}^{38}Ar$ reaction at E_n = 14.6 MeV

Discussion

There is a large difference between the cross section for ${}^{40}Ca(n,{}^3He){}^{38}Ar$ reaction and other measured $(n,{}^3He)$ cross sections[2]. This difference could be explained by Coulomb effects. Namely, due to large negative Q-values the energies of outgoing ${}^3He$ from all $(n,{}^3He)$ reactions at 14.6 MeV are below Coulomb barrier. This is shown in Fig. 2. Full line represents Coulomb barrier height for ${}^3He$ as a function of the charge of the nucleus, Z, and the points are centrum of mass energies for ${}^3He$ channel at E_n = 14.6 MeV. For a given chemical element only the isotope with the least negative $(n,{}^3He)$ Q value is shown. It is obvious that the $(n,{}^3He)$ reaction cross sections at E_n = 14.6 MeV are governed mainly by the Coulomb penetrabilities.
Because of that one can conclude that the cross sections for the lightest isotopes of even Z nuclei

K. H. Böckhoff (ed.), Nuclear Data for Science and Technology, 371–372.

will be larger than those for their odd Z neighbours. It could also be expected that the (n,^{3}He) reaction cross sections for ^{32}S and ^{36}Ar nuclei would be of the same order of magnitude as the one for ^{40}Ca, and all of them larger than the cross section for any other stable nucleus. And finally, (n,^{3}He) reactions have a minor role among the helium producing 14 MeV neutron induced reactions.

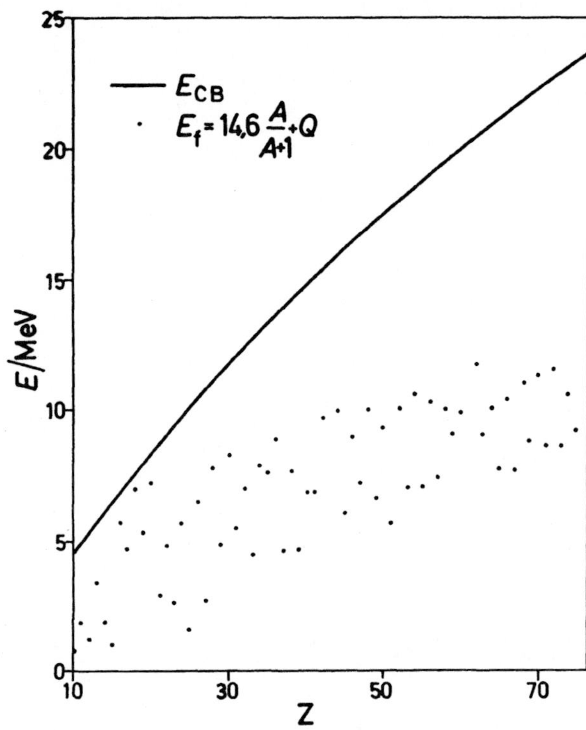

Fig. 2. Coulomb barrier heights for ^{3}He particles and c.m. energies available in ^{3}He channel for the (n,^{3}He) reactions at E_n = 14.6 MeV.

References

1. S.M. Qaim, J. Inorg. Nucl. Chem. 36, 239 (1974)
2. S.M. Qaim, Radiochim. Acta 25, 13 (1978)
3. S.M. Qaim, H.V. Klapdor and H. Reiss, Phys. Rev. C22, 1371 (1980)
4. W.P. Alford, R.A. Lindgren, D. Elmore and R.N. Boyd, Phys. Lett. B46, 356 (1973)
5. W. Bohne, K.D. Büchs, H. Fuchs, K. Grabisch, D. Hilscher, U. Janetzki, U. Jahnke, H. Kluge, T.G. Mastersn and H. Morgenstern, Nucl. Phys. A284, 14 (1977)
6. Đ. Miljanić, M. Furić and V. Valković, Nucl. Phys. A148, 312 (1970)

MEASUREMENT OF THE CROSS SECTIONS OF GAMMA RAYS PRODUCED IN THE INELASTIC INTERACTION OF 14.2 MeV NEUTRON WITH Fe, Ni, Cu, Pb AND Bi

Shi Xiamin, Wu Yongshun, Shen Ronglin, Xing Jinqiang, Ding Dazhao

Institute of Atomic Energy
P.O.Box 275, Beijing
People's Republic of China

The differential cross sections of the discrete gamma rays at 55° produced in the inelastic interaction of 14.2 MeV neutron with Fe, Ni, Cu, Pb and Bi have been measured using a heavy complete shielded Ge(Li) detector. TOF technique is used with the zero time picked by associated alpha particle in $T(d,n)^4He$ reaction. Tens of discrete gamma rays for each sample are identified and analysed. The energy level schemes and their transitions for various residual nuclei have been deduced. For determining the production cross sections, the correction factors of incident neutron attenuation and multiscattering in sample are calculated by Monte-Carlo method. The lowest limit of cross section measured in this experiment is about 0.2 mb/sr. The reliability of data have been checked in comparison with estimated total inelastic secttering cross sections.

[Gamma ray production cross section; Fe, Ni, Cu, Pb, Bi; En = 14.2 MeV; level schemes.]

Introduction

The inelastic scattering cross section and the gamma ray spectra produced in fast neutron interaction with nuclei are of importance to the understanding of neutron transportation in material and shielding study. At the meantime it provides the knowledge of energy level schemes of residual nuclei and the understanding to the nuclear reaction mechanism. In present measurement the cross sections of discrete gamma rays produced in inelastic interaction of 14.2 MeV neutron with Fe, Ni, Cu, Pb and Bi have been determined using a complex shielded Ge(Li) detector with TOF technique with the zero time picked by alpha particle in $T(d,n)^4He$ reaction. Due to the high energy resolution and low background of this detecting system, it is possible to measure the de-excitation gamma rays in different reaction channels of different isotopes simultaneously and the lowest limit of cross section which could be measured in this experiment is 0.2 mb/sr. Several weak transitions which haven't been reported in fast neutron interaction before have been identified in this experiment. Most of gamma transitions are from the (n,n') and (n,2n) reaction channels. Using the experimental data the level schemes for different residual nuclei have been deduced.

Experimental Method

The details of the experimental set up has been discribed in ref[1]. The neutron is produced by 200 KeV deuteron on a T-Ti target at a 600 KeV Cockroft-Walton high tension set. A Au-Si surface barrier detector at 90° with respect to deuteron beam is used to pick up the zero time signal for the TOF spectrometer and to determine the neutron number on the sample, while a smaller Au-Si detector at 135° is used as the monitor of the neutron flux. The samples are made of pure metallic disk with diameter of 100-130 mm and thickness ranging from 5-15 mm according to different material. The distance between neutron source and sample is 20 cm and all neutron correlated with 90° alpha particles detected by Au-Si detector were intercepted by the sample. The energy of incident neutron is 14.2 MeV with energy spread of 0.1 MeV which is determined by the T-Ti target thickness and the solid angle of 90° Au-Si detector.

A 40 cm^3 coaxial Ge(Li) detector enclosed in heavy complete shielding is used to detect the gamma ray yields at 55° with respect to neutron beam. The experimental set up and the block diagram of TOF spectrometer are shown in Fig.1(a) and 1(b) schematically. The time resolution of the TOF spectrometer is typically 4.1 ns at the energy selection threshold of 400 KeV. The energy resolution of Ge(Li) detector is 3.0 KeV for 1332 KeV gamma ray.

K. H. Böckhoff (ed.), Nuclear Data for Science and Technology, 373–376.

The relative efficiency of full energy peak and double escaping peak of Ge(Li) detector were determined by the "double line method" in gamma ray energy region from 0.2 to 10 MeV and then normalized with a set of standard gamma ray sources[2].

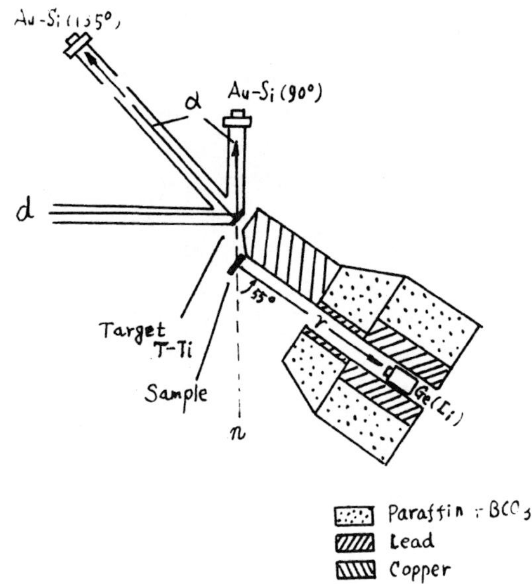

Fig.1(a)

The time spectrum with the sample out has been measured with good statistics and it appears a flat accident coincidence shape, in contrast with the time spectrum with the sample in, no gamma ray peak in the limit of statistics has been observed at the time region where the gamma rays peak should be presented. This ensures that the error in reaction induced gamma ray yields obtained by subtracting the spectra gated by background time window from the spectra gated by effect time window will be less than 3% generally.

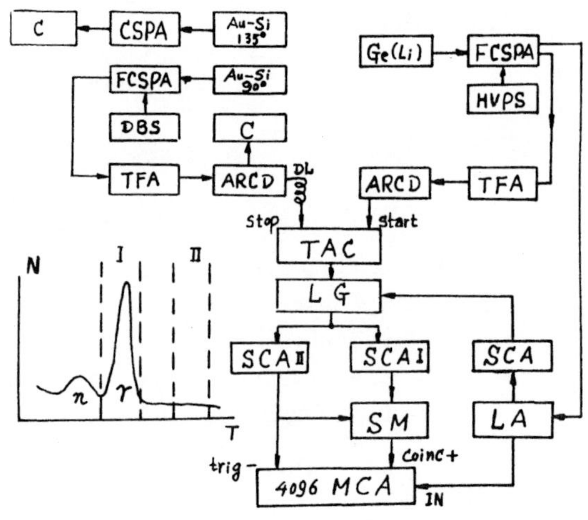

Fig.1(b)

Following formula is used to calculate the production cross section of gamma rays:

$$\frac{d\sigma}{d\Omega}(55°) = \frac{N}{4\pi\phi_n f T \varepsilon n}$$

where N--the counts in full energy peak (or in double escaping peak), ϕ_n -- the total incident neutron number on the sample, correlated with alpha particles, f--the correction factor of neutron attenuation and multi-scattering in sample, calculated by Monte-Carlo method, T--the average penetrating coefficient of gamma rays in sample, ε --the efficiency of Ge(Li) detector and n--the number of sample nuclei per cm^2 at the direction of incident neutron.

The estimated errors from different factors are shown in Table 1.

Table 1

statistics of the gamma ray peak counts	1-50%(dependent upon the peak counts)
subtraction of background under gamma ray peak	3%
background difference due to sample in and sample out	3%
correction of dead time	1%
difference of the effect and background window in time spectrum	1%
neutron flux	2%
factor f and T	3%
detector efficiency	4%
distance between sample and detector	1%
number of sample nuclei	5%

Results and discussion

The details of the experimental results have been published in ref(3)(4). 27 discrete gamma rays from $^{56}Fe(n,n'\gamma)$ and/or $^{56}Fe(n,2n)$ have been observed and their production cross sections have been determined, while for ^{54}Fe--3, for ^{57}Fe--2, for ^{58}Ni--8, for ^{60}Ni--5, for ^{62}Ni--2, for ^{63}Cu--18, for ^{65}Cu--11, for ^{206}Pb--12, for ^{207}Pb--7, for ^{208}Pb--13 and for ^{209}Bi--30 discrete gamma rays have been observed.

The level schemes for ^{56}Fe, ^{208}Bi deduced from this measurement are shown in Fig. 2(a) and 2(b).

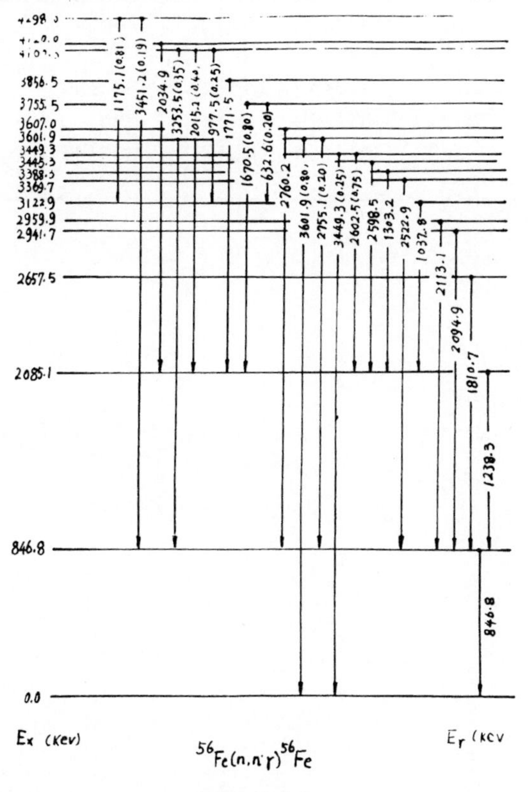

Fig. 2(a)

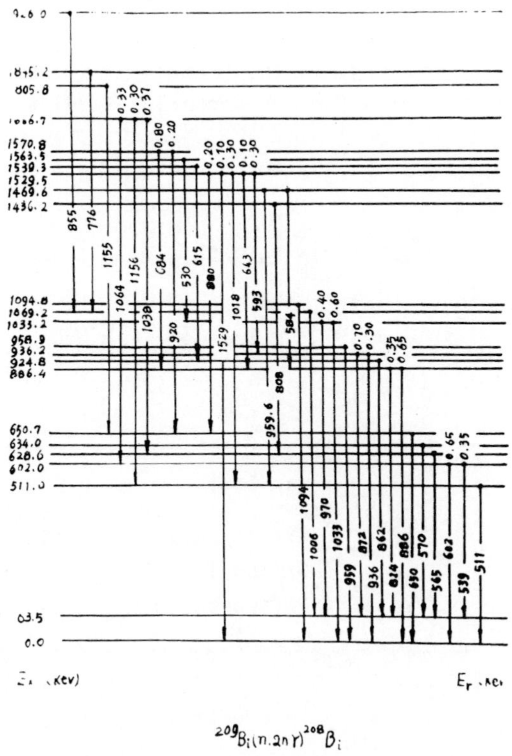

$^{209}Bi(n,2n\gamma)^{208}Bi$

Fig. 2(b)

There was only a little work on Ni, Cu, Bi while for Fe and Pb only several strong gamma ray transitions have been measured by some groups. The relative intensities of various gamma transitions for Fe and Pb are shown in Table 2. The overall agreement between present measurement with those of (5)(9) is quite well. This is probably due to the similarity of the experimental method adopted in these measurements.

Table 2

	E_γ (KeV) ref	Present work	(5)	(6)	(7)	(8)	(9)
Fe	847	1.000	1.000	1.000	1.000	1.000	
	1038	0.103	0.105	-----	0.110	0.095	
	1238	0.500	0.490	0.290	0.400	0.670	
	1811	0.075	0.076	0.072	0.095	0.092	
Pb	538	0.38		0.31	-----		-----
	570	1.22		1.98	-----		0.98
	804	1.00		1.00	1.00		1.00
	890	0.64		0.45	0.58		0.57
	1046/1060/1090	0.77		0.86	1.22		0.87
	2620	0.53		0.45	0.45		0.49

In order to check the reliability of present measurement we have calculated the low limit of inelastic scattering cross section using the measured gamma ray production cross sections with the corresponding level schemes and by assumption of $\sigma = 4\pi d\sigma/d\Omega(55^{o})$. On the other hand the inelastic scattering cross section could be estimated from

$$\sigma_{in}=\sigma_{non}-\sigma_{n,2n}-\sigma_{n,\alpha}-\sigma_{n,p}-\sigma_{n,d}-\sigma_{n,n\alpha}$$

for heavy nuclei, such as Pb and Bi, $\sigma_{n,\alpha}$, $\sigma_{n,p}$, $\sigma_{n,d}$ and $\sigma_{n,n\alpha}$ are negligible.

Using various evaluated or measured cross sections the σ_{in} could be deduced. The comparison of the estimated σ_{in} with the low limit of σ_{in} calculated by present measurement is shown in Table 3 for ^{56}Fe, ^{58}Ni, ^{60}Ni, ^{63}Cu, ^{65}Cu and ^{209}Bi, while for Pb, it is impossible to distinguish the (n,n'γ) of ^{207}Pb from (n,2n) of ^{208}Pb and (n,n'γ) of ^{206}Pb from (n,2n) of ^{207}Pb. The agreement between estimated σ_{in} with calculated low limit of σ_{in} by present measurement seems reasonable.

Table 3

nucleus	^{56}Fe	^{58}Ni	^{60}Ni	^{63}Cu	^{65}Cu	^{208}Bi
σ_{non} (mb) (estimated)	1360±30	1380±30	1380±30	1490±30	1490±30	2580±30
σ_{in} (mb) (low limit)	659±47	302±80	707±44	631±18	448±31	280±30
σ_{in} (mb) (present work)	613±48	320±30	766±72	596±64	463±63	295±43

Reference

(1) Shen Ronglin et al. Chinese Journal of Nuclear Physics 4 59(1982)(in Chinese)

(2) Xing Jinqiang et al. to be published in Nuclear Technique (in Chinese)

(3) Shi Xiamin et al. Chinese Journal of Nuclear Physics 4 121(1982)(in Chinese)

(4) Wu Yongshun et al. to be published in Chinese Journal of Nuclear Physics (in Chinese)

(5) J.Lachkar et al. Nucl. Sci. Eng. 55 168 (1974)

(6) B. Jonsson et al. Ark. Fys. 39 298(1968)

(7) F. C. Engesser et al. J. Nucl. Energy 21 487(1967)

(8) V. J. Orphan et al. Nucl. Sci. Eng. 57 309(1975)

(9) G. Clayeux et al. CBA-R-3807(1969)

THE GAMMA RAYS ASSOCIATED WITH THE INELASTIC SCATTERING OF 14 MeV NEUTRONS IN LARGE SAMPLES OF IRON AND CONCRETE

A.J. Cox and B. Al-Shalabi

Department of Physics, University of Aston in Birmingham
Gosta Green, Birmingham. B4 7ET

Iron and concrete will be used in fusion reactors and multiple scattering processes are important. The angular distribution of gamma rays following inelastic scattering of 14 MeV neutrons in iron and concrete have been measured. An associated particle time of flight system was used to gate the gamma ray signals. The 14 MeV neutrons were produced by the $T(d,n)^4He$ reaction. The neutron flux was monitored by counting the alpha particles associated with the neutrons. The gamma rays were detected by a NaI(Tl) scintillator mounted on a 56 AVP photo-multiplier tube. The samples varied in thickness from 2cms to 10.5cms. The differential cross-sections between 20° and 90° were measured. The increase in cross-sections σ due to multiple scattering effects was found to obey the law $\sigma = \sigma_0 e^{\alpha x}$ where x is the sample thickness in mean free paths and α has an average value of 0.35 (mean free paths)$^{-1}$. The concrete results show that the angular distributions are strongly anisotropic.

Introduction

The first fusion reactors will produce 14 MeV neutrons via the $T(d,n)^4He$ reaction. These neutrons will interact not only with the blanket containing lithium but also with the structural materials of the reactor. The most common structural materials are likely to be steel and concrete. These 14 MeV neutrons will inevitably produce large fluxes of gamma rays from neutron inelastic scattering with the iron and concrete. The gamma rays can cause problems associated with the biological shielding of fusion reactors and with the interactions of the gamma rays themselves with reactor materials. As the structures will be large, multiple scattering will play an important role in gamma ray production. The effect of multiple scattering on the gamma ray production cross-sections is to cause an increase in the gamma ray yield due to an effective increase in the neutron flux. Information on the dependance of the gamma ray angular distributions and the emergent gamma ray spectra is of considerable interest and can be obtained by the study of the inelastic scattering of neutrons in large slabs. In the present work the angular distributions and energy spectra of the emergent gamma rays from slabs of various thicknesses have been measured. An associated particle time of flight technique was used to gate the gamma ray signals in order to reduce the background. The results have been analysed using a semi-empirical model based on the Fermi-age theory.

The experimental arrangement is shown in Figure 1. The neutrons were produced using the $T(d,n)^4He$ reaction, with the deuterons being accelerated by a SAMES type J accelerator and by the 3 MeV Dynamitron at the Joint Aston and Birmingham Universities' Radiation Centre. The deuteron beam was incident on to a rotating water cooled tritiated titanium target and the resultant alpha particles were used to produce a zero time signal. The neutrons were inelastically scattered from the scattering sample and the resultant gamma rays detected by the NaI(Tl) detector located 1.39m from the scattering sample. The gamma ray detector was shielded by means of concentric cylinders of paraffin wax, boric oxide and lead. The 76.2mm NaI(Tl) crystal was mounted via a converging perspex light pipe (angle of convergence 9°) on to a 56 AVP photo-multiplier tube fitted with a μ-metal shield. The detector was also shielded from the neutron source by means of a paraffin wax

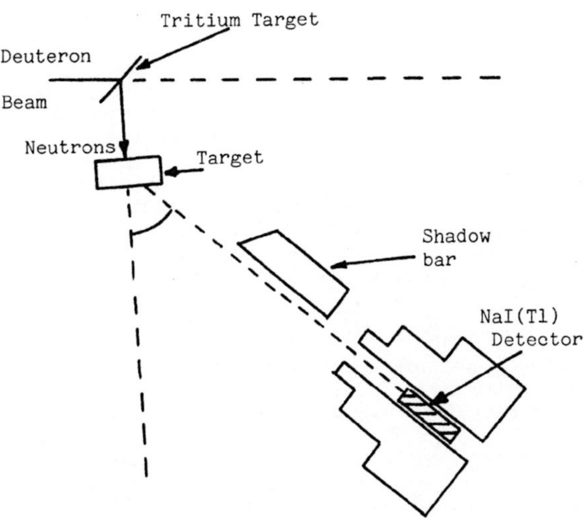

Fig 1 : Experimental Arrangement

shadow bar. The emitted alpha particles were detected by a NE 102a plastic scintillator sheet (0.5mm thick) mounted on a 56 AVP photo-multiplier tube at 90° to the incident deuteron beam. The detector was shielded from particles emitted from the tritium target and from elastically scattered deuterons by an Al foil 0.0044mm thick. An aperture placed in front of the detector gave an angular resolution of $\pm$ 4°. In order to reduce the background in the gamma ray detector an electronic gating system based on the time of production of the alpha particle was used. This is shown schematically in Figure 2. Two signals were taken from the gamma ray photo-multiplier tube, one from the anode which provided the timing signal and the other from the twelfth dynode which provided the linear signal. The timing signal was fed into the constant fraction discriminator, whose output was used as a start pulse for the time to pulse height coverter. The output from the alpha particle detector, after shaping and delay in the gate and delay generator, was passed into a 100 MHz discriminator. The negative

K. H. Böckhoff (ed.), Nuclear Data for Science and Technology, 377–379.

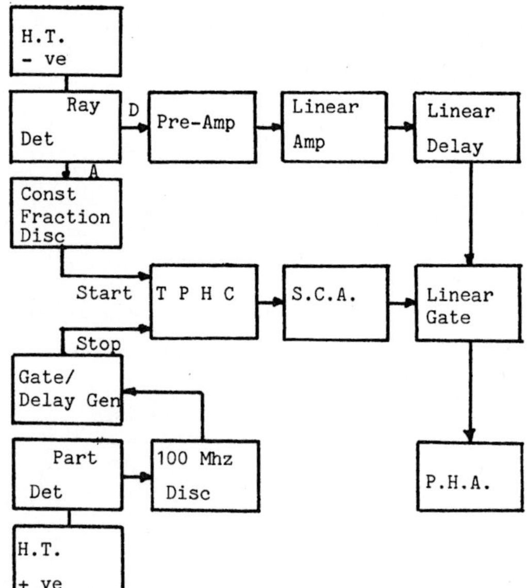

Fig 2 : Electronics System

output of this was used as the stop signal for the time to pulse height converter while the positive output was fed directly into a scaler, allowing the number of alpha particle pulses and hence the number of neutrons to be measured. The time to pulse height converter produced a signal which was proportional to the time difference between the two input signals. This output was fed to a single channel analyser, the window of which was set to allow through only those pulses which corresponded to the correct gamma ray arrival time. The output of the single channel analyser was then used to open the linear gate allowing the linear signal from the gamma ray detector dynode to be recorded on the pulse height analyser. The overall time resolution of the system was 3.5 n-seconds which enabled the gamma ray and neutron pulses to be totally resolved.

Experimental Procedure

The iron scattering samples were in the form of rectangular slabs 120mm x 220mm with various thicknesses ranging from 20mm to 76mm.

The concrete samples used were ordinary concrete (type 5) which had been ultrasonically treated to increase their density to about 2500kgm/m^3. This treatment has no effect on the chemical composition of concrete. Concrete has a very complicated chemical composition consisting of 15.1% hydrogen, 55.7% oxygen, 3.2% aluminium, 14.9% silicon and 3.6% calcium constituting about 93% of the concrete composition (1, 2). These percentages were used in the calculation of the differential cross-sections. The samples used were rectangular plates with dimensions 230 x 150mm. The thickness of each plate was 25mm. The different samples studied consisted of 2, 5, 8 and 10 plates giving a sample thickness of 50, 120, 200 and 250mm respectively.

The samples were positioned in the cone of neutrons associated with the detected alpha particles which lies between 78° and 92° to the incident deuteron beam. The samples were oriented at 45° to the incident neutrons and the absorption for different neutron paths through the sample was calculated numerically. The incident neutron flux was estimated from the integrated counts in the alpha monitor. This was corrected for the background radiation caused by the activation of the target assembly. The alpha monitor count rate was corrected for neutron absorption in the target assembly. In each case readings were taken at angular positions between 0° and 90°.

Results

A typical iron spectrum is shown in Figure 3. Two

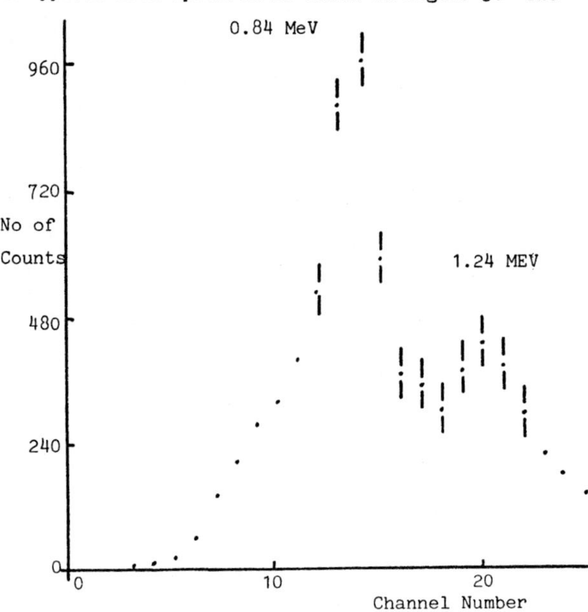

Fig 3 : Gamma Ray Spectrum from Iron Sample at 50°.

lines can be seen in the spectrum at 0.84 MeV and 1.24 MeV which are identified as the gamma rays from the first two excited states in ^{56}Fe. Figure 4 shows

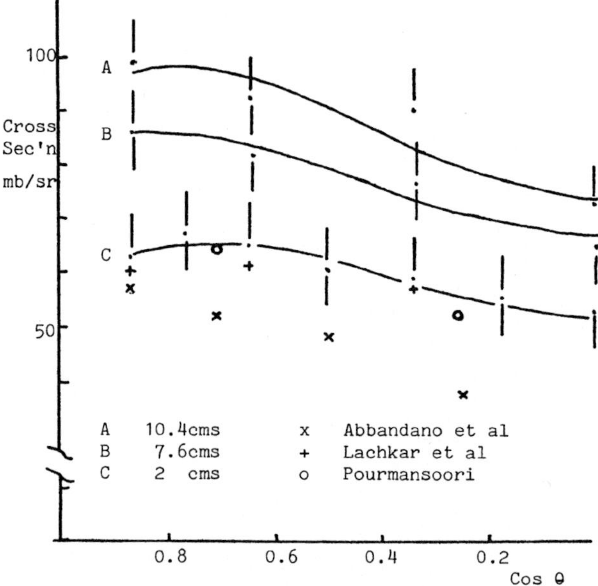

Fig 4 : Angular Distribution for 0.845 MeV Gamma Rays.

the angular distributions for the 0.84 MeV gamma ray for the various ample thicknesses. The differential cross-sections were first corrected for contributions from competing reactions and then fitted by the least squares method to an even order Legendre Polynomial series. The results obtained by other authors (3, 4, 5) are also shown in this Figure. It is seen that there is general agreement in the results. The general equation representing the Legendre Polynomial for the thin sample is :-

$$\sigma(\theta) = A_o + A_2P_2(\theta) + A_4P_4(\theta) + \dots$$

The solid lines in Figure 4 are the results of these fits.

The results for concrete are given in Table 1 in the form of differential cross-sections for the production gamma rays of a given energy between 1 and 6 MeV in integral steps of 1 MeV. Figure 5 shows the angular

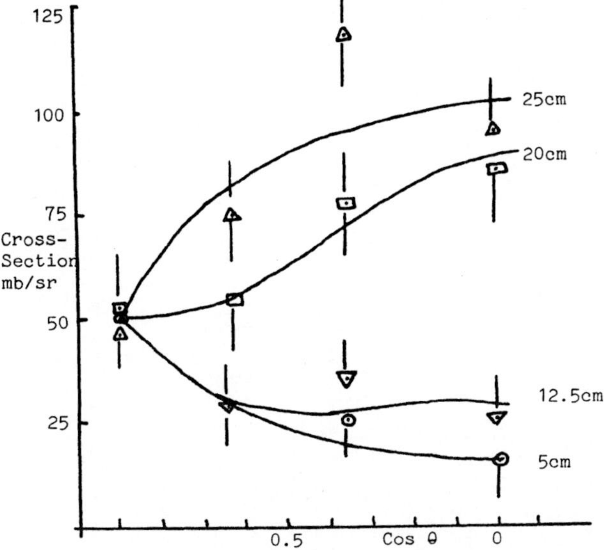

Fig 5 : Angular Distribution for 2-3 MeV Gamma Rays from Concrete.

distribution for the production of 2-3 MeV gamma rays from concrete samples of different thicknesses. In contradiction to the normal assumption the angular distribution is very anisotropic and the anisotropy increases with increasing sample thickness. The solid line in the Figure is a least squares fit of the experimental results using the modified Legendre Polynomials :-

$$\sigma\ (\theta) = B_0 + B_2 \cos^2\theta + B_4 \cos^4\theta + \dots$$

In each case the uncertainties were calculated by combining the component errors in quadrature. There is a general increase in the effective gamma ray production cross-section as the thickness of the sample increases due to multiple scattering effects and in the region covered by the present work this increase is described for both iron and concrete by the equation :-

$$\sigma = \sigma_0 \exp\ \{\alpha x\} \qquad \dots (1)$$

where σ is the effective cross-section for a sample x mean free paths thick and σ_0 is the cross-section for zero thickness. Fits of the experimental results yield a value for α of 0.35 ± 0.1 per mean free path.

Theoretical Interpretation of Results

Mostly, workers use Day's (6) suggestion that where the neutron transmission is greater than 70% then the attenuation of the neutron beam and the increase in effective cross-section due to multiple scattering compensate each other. In the present work the thickness and dimensions of the scattering sample do not allow the use of Day's criterion to be valid. Because of this a semi-phenomenological model was used to compare with the experimental results. This is based on the use of the continuous slowing down model and the Fermi-age equation to determine the spatial distribution of neutron energies inside the sample. This distribution was then used to predict the gamma yield from each position. The total yield of a particular gamma ray was then found by summing this over the whole sample. The yield was then subjected to the same correction as the experimental readings to convert it to the count rate it would produce in the gamma ray detector. This theory consistently underestimates the experimental values. The most likely explanation for this is that the theory takes account of the gamma ray yield after one neutron inelastic collision only, whereas gamma rays can be produced by more than one inelastic collision. This is emphasised because the discrepancy increases with increasing sample thickness as would be expected. The effect of neglecting all but the first neutron inelastic collision is partially compensated by neglecting the neutron attenuation and for the 20mm thick iron sample this compensation is totally effective when the transmission factor is 70% or more. This confirms Day's criterion.

The results of the work show the effect of multiple scattering on the gamma ray yield after neutron inelastic scattering in samples of differing thickness. This variation is described by the empirical relation shown at equation 1 for the range of thicknesses considered.

Table 1

The Differential Cross-Section for the Production of Gamma Rays at 90° in (mb/sr).

Energy Range MeV	Sample Thickness cms 5	12.5	20	25
1 - 2	11.3	31.2	75.3	77.9
2 - 3	14.1	25.6	78.6	96.8
3 - 4	13.6	32.2	64.6	70.7
4 - 5	11.8	15.6	21.1	49.8
5 - 6	15.4	16.2	24.9	30.7

All cross-sections are accurate to ± 11%.

References

1. Neville A.M., Properties of Concrete, Pitman - London, 1963.
2. Jaeger R.C. (Editor), Engineering Compendium on Radiation Materials Vol. II, Springer-Verlag, Berlin - Germany 1975.
3. Engeser F.C. and Thompson W.E., J. Nucl. Energy 21, 487, 1967.
4. Jonsson B., Nyberg K, and Bergquist I., Arkiv For Fysik 32, 295, 1969.
5. Lachkar J., Sigaud J., Patin Y., and Haugat G., Nuc. Sci. Eng. 55, 168, 1974.
6. Day R.B., Phys. Rev. 102, 767, 1956.

A BENCHMARK EXPERIMENT FOR FAST NEUTRON TRANSPORT IN GRAPHITE

H. Takahashi, and K. Sugiyama

Department of Nuclear Engineering
Tohoku University
Aoba, Sendai, 980
Japan

Benchmark experiments for high energy neutron transport up to about 33 MeV in graphite have been performed using fast neutrons from thick beryllium target bomberded by 35 MeV pulsed proton beam. Cubic graphite sample located at 13 m from the neutron source is viewed by a NE-213 scintillation counter. The experimental results are compared with Monte Carlo calculations to test a carbon data set of the DLC-58 cross section library.

[carbon, neutron scattering, integral cross sections, DLC-58, E_n= 6-33 MeV, time-of-flight, NE-213]

Introduction

In the fields concerning with fusion reactor or fast neutron therapy[1], neutron cross sections for energies up to about 40 MeV are necessary for shielding calculations and radiation damage studies. The cross sections based on model calculations such as optical model and the intranuclear-cascade-evaporation model have been widely used because of lack of measured data. It should be made to test these calculated cross sections of neutrons for reliable applications. There is an experiment which provides verification of the accuracy of the available neutron cross sections for use in transport calculations.
Present paper describes the benchmark experiments for high energy neutron transport up to about 33 Mev in graphite. With a NE-213 scintillation spectrometer, three sets of the measurements have been made;
1) transmission time spectra, 2) scattering time spectra, and 3) angle dependent pulse height spectra of secondary neutrons from the graphite assemblies.
Neutron detection efficiency and response matrix of the organic scintillator for unfolding are used on basis of previous results by present authors[2].
Comparisons between present experiments and calculations by Monte Carlo method with the carbon cross sections of the DLC-58 library are presented.

Experimental Arrangement

The neutrons were produced by the Be(p,n) reaction with thick target at Tohoku University Cyclotron. Time width of proton beam was 800 ps, and the energy resolution of the beam was analysed about 1/1000 by an analyser magnet[3]. Table 1 shows the contribution to T-O-F resolution for 30 MeV neutrons in the case of 13 m flight path. An over-all time(energy) resolution less than 1.1 ns (360 keV) has been obtained. No gamma flash was viewed with a neutron detector by a concrete shield of 1 m thickness. Schematic diagram of the integral experiments is illustrated in Fig. 1. Neutrons emitted from the assembly were detected with a cylindrical NE-213 scintillation counter of 5 inch diameter and 2 inch long which was followed by a neutron-gamma discrimination circuit. Flight time of neutrons with respect to RF signal of a beam chopper was sent to a time-to-amplitude convertor.

Table I T-O-F Resolution

	Time Resolution	Energy Resolution
Detector thickness	0.7 ns	230 keV
Target thickness	0.15	50
Beam energy spread	0.1	32
Beam time spread	0.8	260
TOTAL	1.1 ns	360 keV

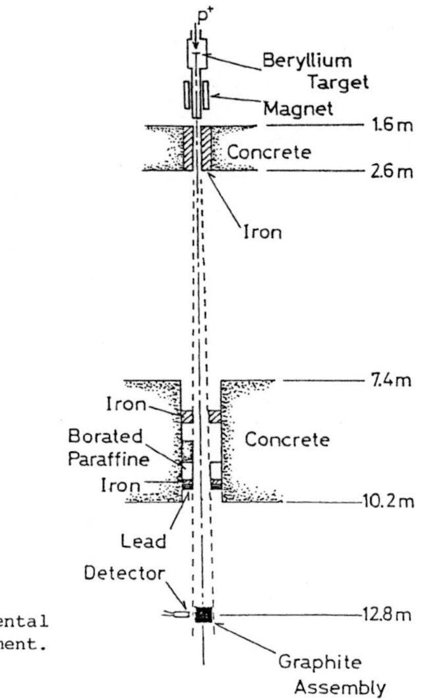

Fig. 1. Experimental arrangement.

Integral Experimental

A. Transmission Experiments
Measurements of transmission neutrons from graphite assemblies of thickness 10, 20 and 40 cm were performed. The neutron detector was placed at 13.05 m from neutron source. Bias of the detector was set to 1.32 ^{22}Na Light Unit. Energy spectra obtained from the time spectra are shown in Fig. 2 along with the values of ENDF/B-IV.

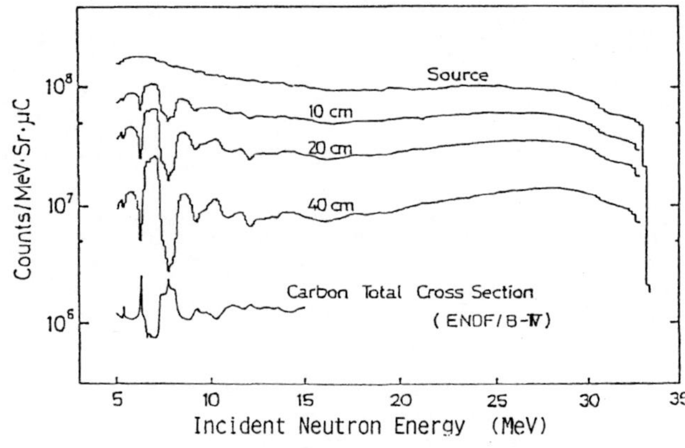

Fig. 2. Transmission neutron spectra from graphite.

K. H. Böckhoff (ed.), Nuclear Data for Science and Technology, 380–382.

B. Scattering Experiments

Neutron scattered from the graphite assembly of 30 cm cube were detected at the angles of 45°, 90° and 135° as shown in Fig. 3. Beam profil of incident neutrons was checked with a NE-213 scintillator of smaller size. Obtained neutron spectra after the time-to-energy conversion are shown in Fig. 4.

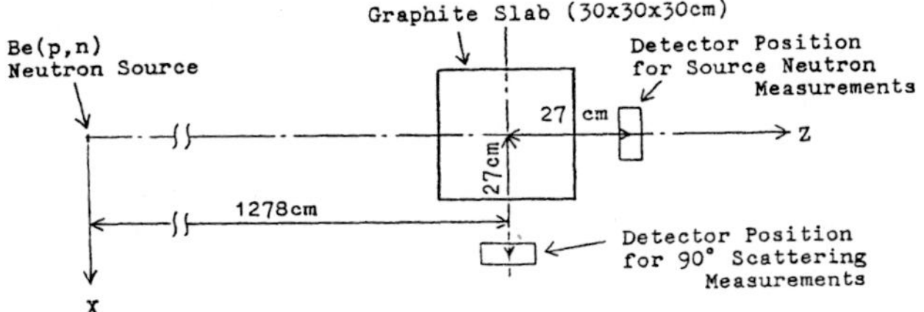

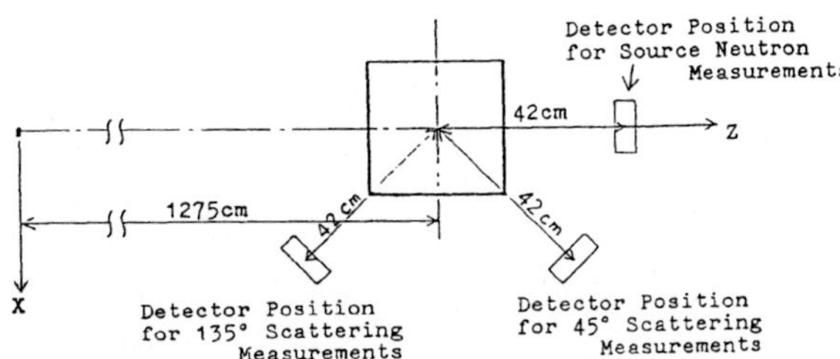

Fig. 3. Arrangements of detector and assembly for the scattering integral experiments.

Calculational Model

In the analysis of present experiment, calculations of neutron transport in the graphite were performed using the multigroup Monte Carlo code MORSE[4] with the cross section data library DLC-58 (47 neutron groups, 21 gamma-ray groups)[5] which extends from thermal neutron energy to 60 MeV. For neutron energy above 14.9 MeV, Legendre expansion of P5 is used, and differential cross sections are deduced from optical model calculation for the elastic scattering and from intra-nuclear-cascade-evapolation model for the nonelastic scattering. Below 14.9 MeV, Legendre expansion of P3 is used and the cross sections are from the Radiation Shielding Infomation Center's fusion energy cross section library.

Source neutrons incident to the face of the assembly in each energy group were given an initial age Tg determined by

$$Tg = \frac{L}{V_g^u} + \left(\frac{L}{V_g^l} - \frac{L}{V_g^u} \right) \times R$$

where L = distance between neutron source and assembly,

V_g^u = neutron velocity of the upper energy group g,

V_g^l = neutron velocity of the lower energy group g,

and R = uniformly distributed random number on the interval (0,1).

Neutron fluxes are calculated using next-flight estimation to a point detector. Both time of arrival and particle energy group are scored at the detector point. The time bins used to score the particles correspond to the arrival time of neutron which has the energy of group boundary. Neutron counts were calculated as $\Phi\varepsilon A$ where Φ, ε and A are the neutron fluence, the energy dependent detector efficiency and the detector geometrical cross-section, respectively. Amounts of source neutrons are given as $S\varepsilon A/4\pi r^2$, where S and r are the energy spectrum of source neutrons, and the distance from source to detector, respectively.

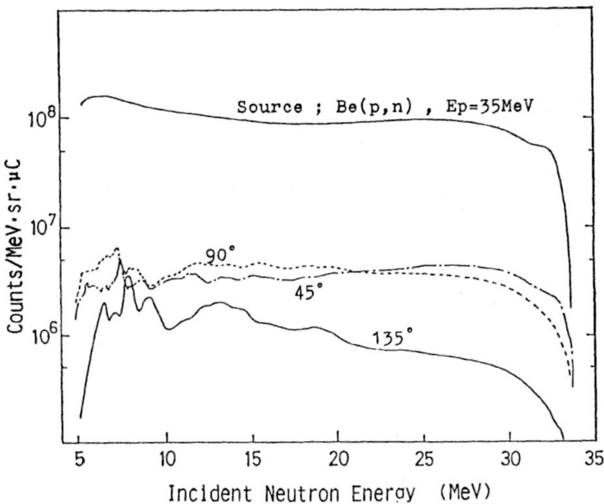

Fig. 4 Scattering neutron spectra from graphite assembly of 30 cm cube.

Results and Discussion

Present results on the carbon are summarized in the figures 5, 6, and 7. These results include comparisons of the measured integral data and the calculated, and also unfolded differential secondary neutron data.

Transmissions

In Fig. 5, comparisons of the final results for transmission integral experiments with the calculations are indicated. The values are given in terms of the ratio of real transmission rate to source neutrons with calculated values of the ratio of uncollided neutrons. Discrepancies of 3 to 35 % appear over the incident neutron energy range of 6 to 33 MeV. It reveals that more than 50% of the detector counts are due to uncollided neutrons for 10 cm thick assembly.

For 20 and 40 cm thick assemblies, opposit tedency beween the measured and the calculated are found out in the both sides of 14.9 MeV. Below 14.9 MeV, a similar tedency to the above has previously been reported[6].

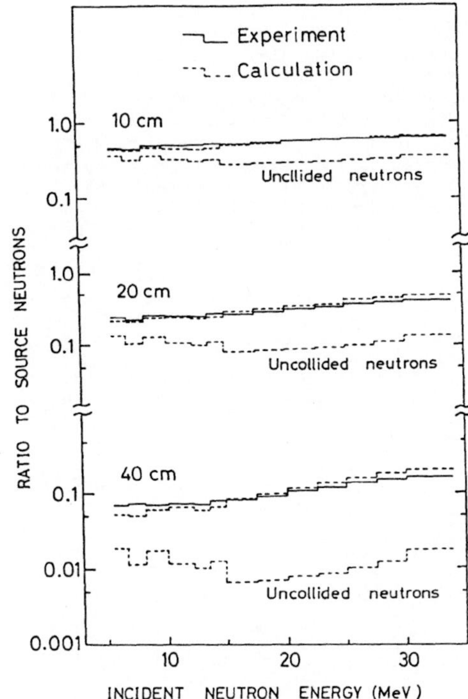

Fig. 5. Comparison of measured and calculated neutron transmission spectra through graphite.

Scattering Integral Results

In Fig. 6, discrepancies are 24 to 40% above 15 except for 135° where are remarkable discrepancie

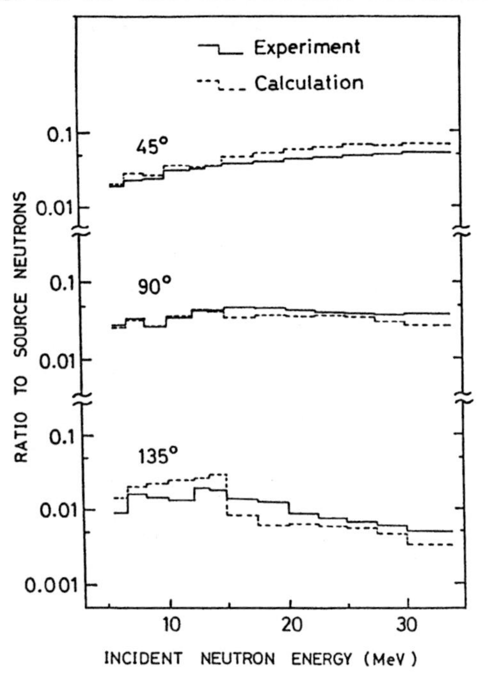

Fig. 6. Comparison of measured and calculated energy spectra of scattering neutrons from graphite assembly of 30 cm cube.

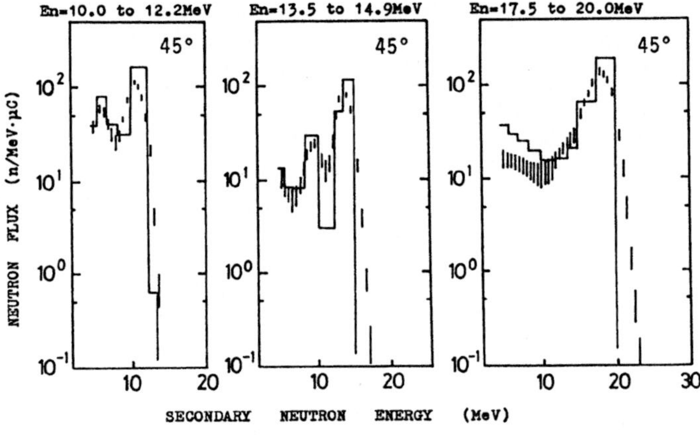

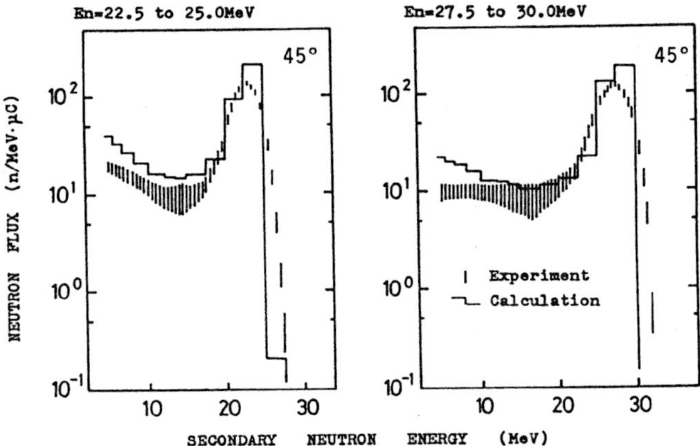

Results of secondary emitted neutrons

Unfolded results of differential energy spectra of the secondary neutrons are illustrated in Fig. 7 for the comparisons of the measured and the calculated. The data are given in terms of neutron fluences per unit energy which are incident to the neutron detector when the integrated proton charge on the target is 1 μC. The measued energy spectrum of the secondary neutrons is unfolded from the detected pulse-height distribution where time widths of incident neutrons are taken in corresponding to the energy bin of the DLC-58 library. In the secondary neutrons emitted to 45°, there are fairly agreements of the measured and the calculated below 14.9 MeV, except for highest energy peak and low energy tail in the region above 14.9 MeV. In the data for 90° and 135°, the results show significant discrepancies over the entire energies of incident and secondary emitted neutrons. The calculated results give negative fluxes and high energy peak which are physically unexpectable.

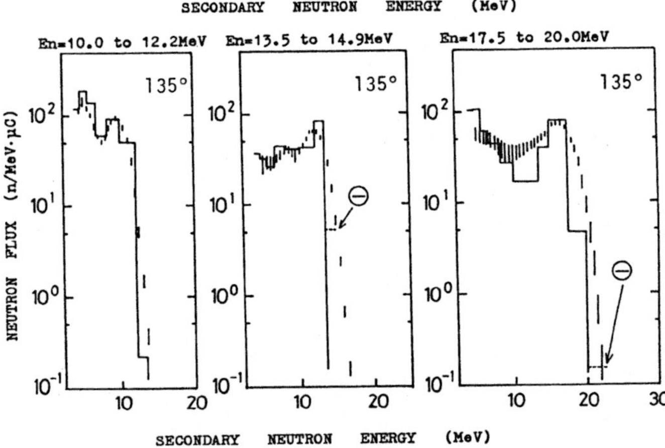

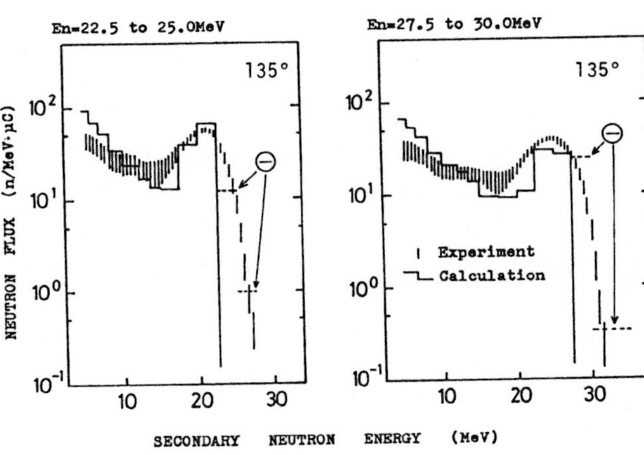

Fig. 7. Measured and calculated secondary neutron energy spectra from graphite. Incident energy is indicated by E_n.

References

1. BNL-NCS-50681(1977), BNL-NCS-51245(1980) Brookhaven National Laboratory, and for example, NIRS--M-19 (1977), Inst. Radiological Science, Japan.
2. H. Takahashi, H. Uchida, K. Sugiyama, Ann. Report 1980, Cyclotron and RI Center, Tohoku University.
3. H. Orihara and T. Murakami, Nucl. Instr. and Meth., 188, 15 (1981).
4. M.B. Emmett, ORNL-4972(1975), Oak Ridge National Laboratory.
5. R.G. Alsmiller, Jr., J. Barish, Nucl. Sci. Eng., 69, 378 (1979).
6. S.N. Cramer, E.M. Oblow, Nucl. Sci. Eng., 58, 33 (1975).

DIFFERENTIAL NEUTRON PRODUCTION CROSS SECTIONS FOR 590 MeV PROTONS

S. Cierjacks, Y. Hino, S.D. Howe†, F. Raupp and L. Buth††

Kernforschungszentrum Karlsruhe
Institut für Kernphysik
Postfach 3640, D-7500 Karlsruhe
Federal Republic of Germany

Double differential neutron production cross sections have been measured for C, Al, Fe, Nb, In, Ta, Pb and U. The measurements were carried out by the time-of-flight technique using a NE-213 liquid scintillation detector and the 590 MeV pulsed proton beam from the SIN cyclotron. The cross section data were taken at emission angles of 30°, 90° and 150°. Energy- and angle-dependent data are presented and compared with predictions based on the 3-dimensional "High Energy Nucleon-Meson Transport Code, HETC".

(C(p,xn), Al(p,xn), Fe(p,xn), Nb(p,xn), In(p,xn), Ta(p,xn), Pb(p,xn), U(p,xn), double differential neutron production cross sections, E_p=590 MeV, spallation neutron sources)

Introduction

The use of high-current, medium-energy particle accelerators as spallation neutron sources for solid and liquid state physics investigations and nuclear research has gained widespread interest in recent years. Design of such neutron sources depend strongly on the detailed knowledge of neutron and charged particle spectra emitted from the spallation target. Several groups in different laboratories have experimentally investigated the differential and integral yields and spectra obtained from proton bombardment of thick spallation targets. Comparisons between such measurements and predicted values calculated by the intranuclear cascade-evaporation model have revealed a considerable degree of disagreement. Concerning differential neutron production cross sections derived from thin target measurements comparisons have shown reasonable agreement with calculated values for the evaporation region of the spectrum. For energies greater than ∿20 MeV, however, considerable discrepancies were observed between calculations and measurements.

In order to investigate the existing discrepancies more systematically, accurate experimental data for simple target configurations are most desirable. Consequently, we have performed a series of measurements to determine absolute differential neutron and charged particle production cross sections from thin sample bombardment of 590 MeV protons over a wide range of masses. In this paper, preliminary results of neutron production cross section measurements at laboratory angles of 30°, 90° and 150° for targets of carbon, aluminum, iron, niobium, indium, tantalum, lead and uranium are reported. Some comparisons with model predictions are also given.

Experimental Details

The measurements were performed using the 590 MeV proton beam from the SIN cyclotron. The cyclotron provided proton beam pulses of 0.2 ns width at a frequency of 16.84 MHz. The beam was focused to about 2 cm diameter onto the thin plate metal targets. For the uranium target, a plate of depleted material (99.8% ^{238}U), 5 x 10 cm^2 by 5 mm thick, was used. The other targets were natural elements of high purity metal, 10 x 10 cm^2 by 5 mm thick.

A detailed description of the target-detector arrangement and the electronics used in the measurements has been given eleswhere[1]. In brief, neutrons emitted from the target at 30°, 90° and 150° were detected at the exit of a ∿1 m long iron collimator. The principal detector was a 3 cm thick, 4.5 cm diameter, NE-213 liquid scintillator employing n-γ pulse-shape discrimination. In order to eliminate pulses from charged particles, also produced in the target, a 5 mm thick plastic scintillator in front of the main detector was used as a veto counter. Background measurements were performed with the target opposite the collimator entrance removed.

Data accumulation was accomplished in a 4-parameter mode, employing a PDP 11/60 on-line computer. Each event was recorded with respect to its flight time, its pulse height from the liquid scintillator and its n-γ "pulse-shape discrimination time". The pulse height signal from the liquid scintillator was split into two amplifier channels, one with ten times the gain of the other. This was necessary to cover the large dynamic range and the expanded threshold region. The latter was important for the accurate determination of the detector threshold throughout the experiment which is an important prerequisite for a precise detection efficiency determination. The contents of the 4 ADC's were stored event by event on magnetic tape for subsequent off-line data processing.

The proton beam current was measured throughout the experiment by a proton beam monitor. This monitor consisted of a thin polyethylene scatterer placed in the incident proton beam far up-stream of the neutron producing target. Scattered protons were detected by a pair of thin plastic scintillators which operated in coincidence. The monitor was calibrated with respect to absolute proton flux by counting individual protons in the direct beam involving an auxiliary thin plactic detectors at extremely reduced proton beam current.

Data Analysis

The analysis of the neutron data began with the separation of events into neutron- and γ-events by a consideration of 2-dimensional arrays of pulse height versus "pulse-shape discrimination time". Excluding γ-events from further analysis, the neutron events from the corresponding background runs were then subtracted. These data were subsequently sorted into suitable time-of-flight bins and their corresponding energies calculated relativistically according to the time of occurence of the prompt γ-peak. In the time of flight measurements, a single overlap in part of the neutron time-of-flight spectrum was admitted, in order to optimize timing resolution with low detector threshold. Separation of the response due to high energy neutrons from that due to low energy neutrons was achieved by extrapolation of the high energy pulse height response down to the bias level. The error

† Present address; Los Alamos Scientific Laboratory, Los Alamos, New Mexico, U.S.A.

†† Institut für Neutronenphysik und Reaktortechnik, Kernforschungszentrum Karlsruhe, F. R. Germany

K. H. Böckhoff (ed.), Nuclear Data for Science and Technology, 383–386.

associated with this separation is small because of the largely different shapes of the corresponding distributions. The contents of each time bin were integrated and the results divided by the efficiency of the NE-213 neutron detector. The data were finally scaled by the solid angle subtended by the detector, the energy bin-width, the dead time correction factor and the incident proton flux to produce the double differential neutron production cross sections. The required detection efficiencies up to 400 MeV were calculated by the Monte Carlo code of Cecil et al.[2]. The results of this code in the range below 400 MeV have been carefully verified from experiments using the associated particle method[3]. For neutron energies in the range 400 to 600 MeV, experimentally determined detection efficiencies of Howe[4] were employed, which are significantly higher than those predicted by Cecil's code.

Results and Discussion

The results of the measurements are shown in Figs. 1 to 3 in terms of differential cross sections as a function of the energy of the emitted neutrons. In these figures the results from the various targets at a certain laboratory angle are combined in one graph. The cross section curves are seen to be all very much alike showing a more or less pronounced broad maximum around 1 MeV, then a rapid fall to about 15 MeV, and a different shape above that energy with a broad shoulder around 100 MeV. The two different components in the spectra are attributed to evaporation neutrons dominating in the spectrum shape below 15 MeV and cascade neutrons determining the shape above this energy. The absolute cross section values are seen to increase with increasing mass number of the target nucleus and the fraction of the high energy component tends to increase with decreasing mass number for all three angles.

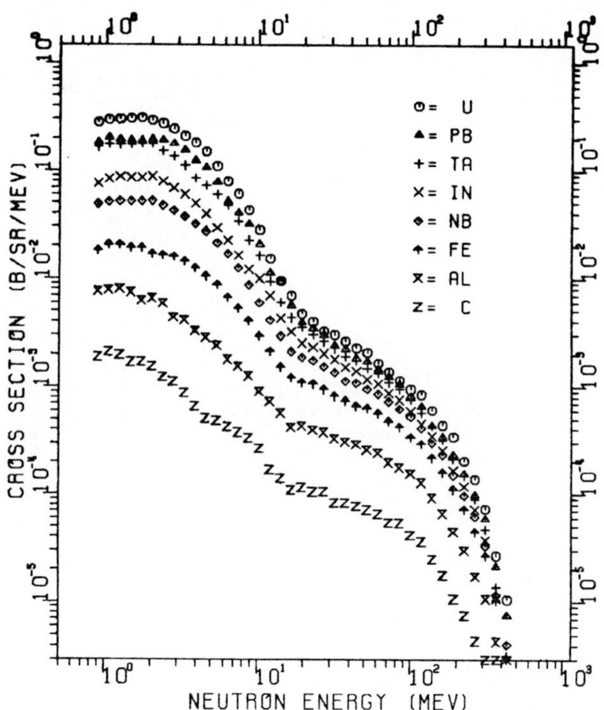

Fig. 2. Same as Fig. 1 except for a laboratory angle of 90°.

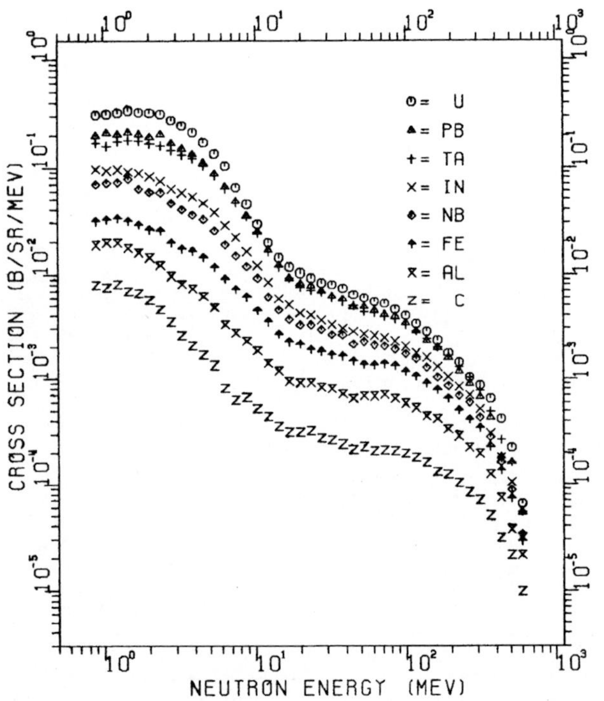

Fig. 1 Differential cross sections for neutrons emitted from various thin metal targets at a laboratory angle of 30°. The character size indicates the statistical uncertainty.

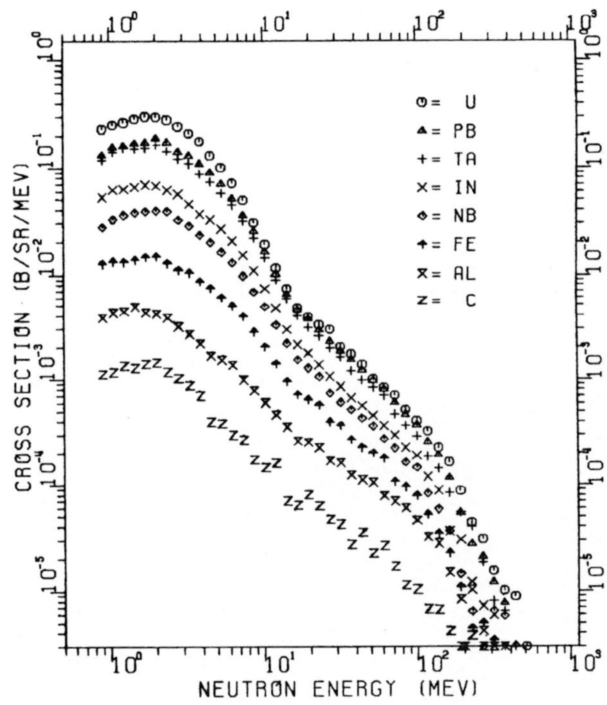

Fig. 3. Same as Fig. 1 except for a laboratory angle of 150°.

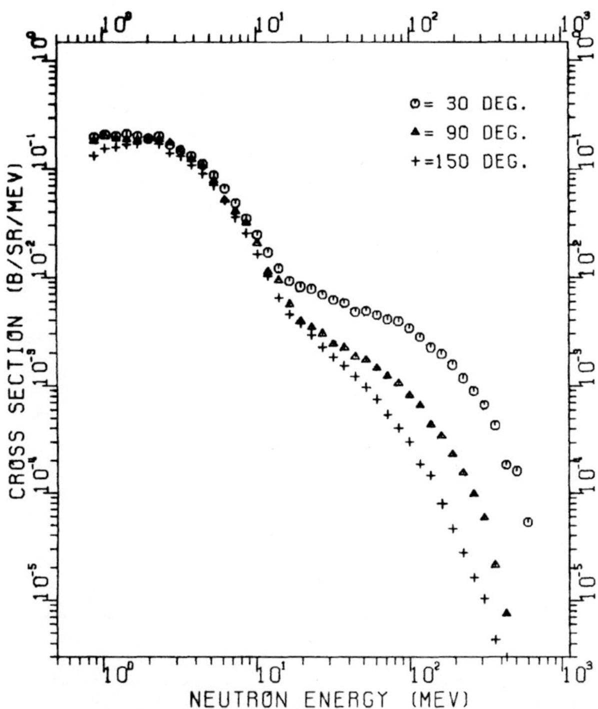

Fig. 4. Differential cross sections for neutrons emitted from a thin lead sample at laboratory angles of 30°, 90° and 150°.

In Fig. 4 the results for one target nucleus (lead) at the three different angles are combined. It is interesting to note that below 15 MeV the cross sections for all three laboratory angles coincide over the whole range, except for the 150°-data where the cross sections below 2 MeV are slightly smaller than for the other two angles. This observation is strongly indicative of an isotropic angular distribution for evaporation neutrons. In the high energy region, cross sections are found to decrease more rapidly with energy for increasing laboratory angles. This is in qualitative accordance with the intranuclear cascade process which is strongly forward peaked. The general features of the angular dependent spectra for the other heavy and medium weight target nuclei (down to niobium) are similar as for lead with an almost angle-independent evaporation cross section component and a strongly forward peaked cascade contribution.

Especially for the light element targets such as carbon and aluminum, the cross sections below 15 MeV vary also significantly with laboratory angle, showing the larger cross section values at forward angles. A possible explanation for this behavior might be due to the non-applicability of the compound mechanism for these light nuclei. In this context, however, the angular-dependent low energy cross sections for iron cannot be easily understood.

In Fig. 5 the total cross sections (integrated over all energies from 0.9 to 590 MeV) are shown as a function of mass number. The results are shown separtely for the three laboratory angles of 30°, 90° and 150°. The cross sections increase almost linearly with increasing mass number, but different slopes are observed for the different angles with smaller values for smaller emission angles. This is explained by the changing ratios of cascade neutrons to evaporation neutrons in the corresponding spectra.

Model Calculations

Calculations of neutron production from thin targets of lead and uranium have been performed by Armstrong and Filges using the HETC computer code[5,6]. Comparisons of the code's predictions for lead at 30°, 90° and 150° are shown in Figs. 6 to 8.

Calculations of the differential neutron production cross sections were made for three different choices of the level density parameter B_0, i. e. for values of $B_0=8.0$, $B_0=14.0$ and $B_0=$ variable. Changes of the B_0 parameter are only sensitive to the evaporation part of the cross sections and leave the cascade region (above 15 MeV) practically unchanged. But also below 15 MeV, only slight changes in the peak cross sections around 1 MeV and in the width of the evaporation spectrum occur for the adopted choices of B_0. From Figs. 6 to 8, it can be seen that the calculations tend to overpredict the measured results for lead in

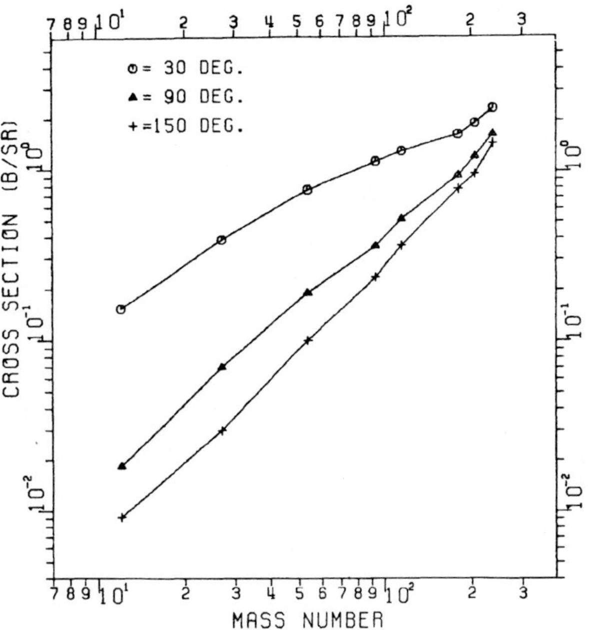

Fig. 5. Total cross sections (integrated over all energies from 0.9 to 590 MeV) for neutrons emitted from various thin metal targets at laboratory angles of 30°, 90° and 150° plotted as a function of the target mass number.

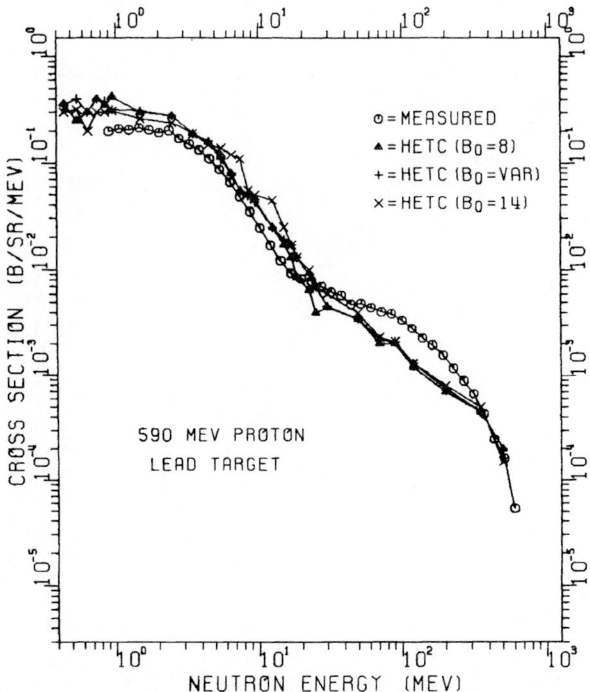

Fig. 6. Comparison of measured and calculated differential cross sections for neutrons emitted from lead at a laboratory angle of 30°. For explanation of the three choices for the B_0-parameter in the calculations see text.

the evaporation region up to a factor 1.5 throughout the energy range from 0.9 to 15 MeV with some inevitable changes between the three calculated curves. Above 15 MeV, however, the code underpredicts the cross section and for 90° and 150°, the difference between the measured and calculated results increases rapidly with increasing energy (for the 90° results,the calculated value at 200 MeV is almost a factor of 4 smaller than the corresponding measured cross section). For 30°, the calculated energy dependence disagrees with the measured curve in the range from 20 to 350 MeV, but joins favourably at the energies above ∿350 MeV. Comparisons of code predictions with experimental data for uranium have also been made[5,6], and showed a similar behavior as in the lead case. Large differences, especially for the high energy cross sections were also observed in recent comparisons of HETC code calculations with differential neutron production cross sections measured at 800 MeV in Los Alamos[7,8]. Consequently, we feel that model parameters used in the HETC code should be carefully reexamined.

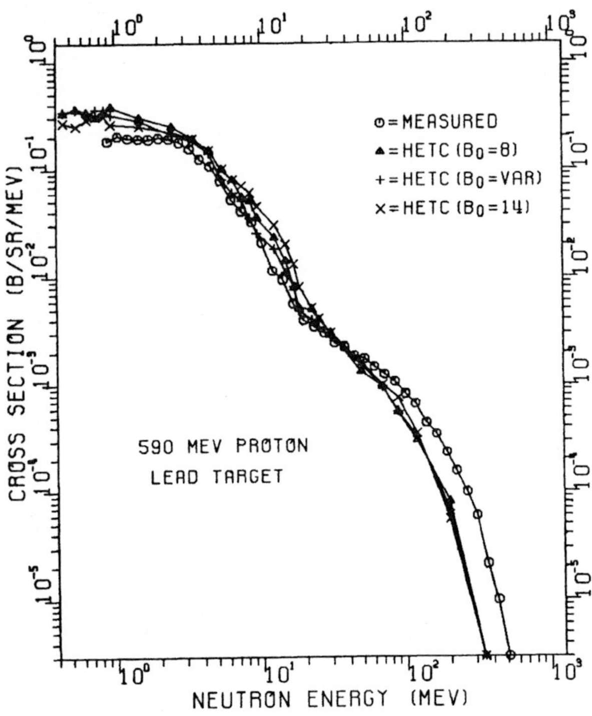

Fig. 7. Same as Fig. 6 except for a laboratory angle of 90°.

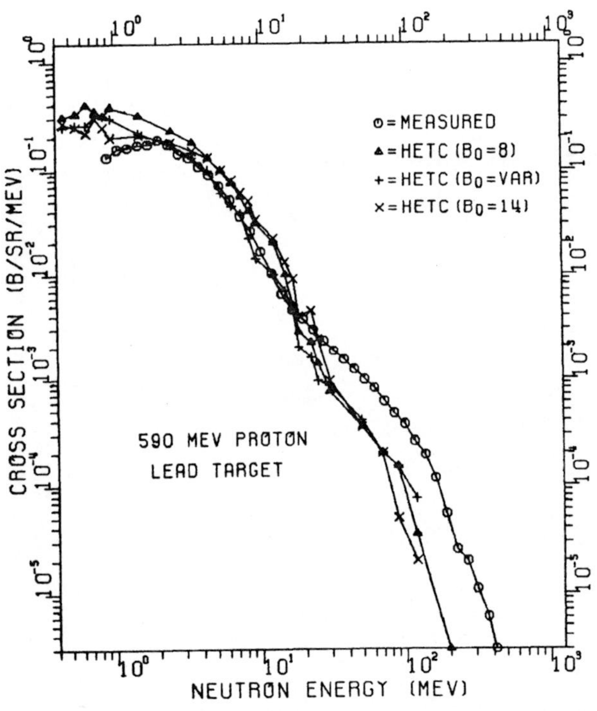

Fig. 8. Same as Fig. 6 except for a laboratory angle of 150°.

Conclusion

In conclusion, we have determined the double differential neutron production cross sections for 590 MeV proton bombardment of various thin metal targets at laboratory angles of 30°, 90° and 150°. All of these data clearly revealed a two-component structure with contributions from evaporaion processes and intranuclear cascade reactions. For heavy and medium weight nuclei evaporation data are strongly indicative of an isotropic angular distribution in the laboratory system. Data dominated by cascade processes are strongly forward peaked, and the fraction of cascade neutrons from 590 MeV protons tend to increase with decraesing target mass numbers.

Comparisons between experiments and calculations by the HETC computer code show a significant underprediction of the cross sections by the code at energies above 15 MeV, but also a systematic overprediction of the data in the evaporation energy range, at least for lead and uranium targets. The present comparisons indicate that model parameters used in evaporation and intranuclear cascade neutron cross section calculations need review. In the near future we intend to supplement the neutron data and the recently evaluated charged particle results[9] by additional charged particle data taken already for three additional laboratory angles between 23° and 90°. Such additional data can provide a more complete experimental data set for the suggested model parameter adjustments.

The authers would like to acknowledge the help of the SIN staff, especially Dr. W. Fischer and Dr. C. Tschalär. We are also indepted to Dr. T.W. Armstrong and Dr. D. Filges for the various discussions and for providing their HETC code calculations.

References

1. S. Cierjacks, F. Raupp, S.D. Howe, Y. Hino, M.T. Swinhoe, M.T. Rainbow and L. Buth., ICANS-V, Proc. 5th Meeting of the International Coll. on Advanced Neutron Source, G.S. Bauer and D. Filges (Eds), Jül-Conf-45, p.215-240, 1981

2. R.A. Cecil, B.D. Anderson and R. Mady, Nucl. Instr. and Meth., 161, 439 (1979)

3. S. Cierjacks, M.T. Swinhoe, L. Buth, S.D. Howe, F. Raupp, H. Schmitt and L. Lehmann, Nucl. Instr. and Meth. 192, 407 (1982)

4. S.D. Howe, Los Alamos National Laboratory, Private Communication 1982

5. T.W. Armstrong and B.L. Colborn, "Note on Calculated Results for Particle Spectra Produced in Thin Pb and U Targets and Comparison with Experimental Data", 1982 (unpublished)

6. D. Filges, "Calculated Neutron and Charged Particle Spectra in Thin Pb and U Targets and Comparison with Experimental Data", 1982 (unpublished)

7. S.D. Howe, PhD Thesis, Kansas State University, 1980

8. T.W. Armstrong, Private communication, 1981

9. S.D. Howe, S. Cierjacks, Y. Hino, F. Raupp, M.T. Rainbow, M.T. Swinhoe and L. Buth, ICANS-V, Proc. 5th Meeting on the International Coll. on Advanced Neutron Sources, G.S. Bauer and D. Filges (Eds.), Jül-Conf.-45, p. 313-332, 1981

MODEL PREDICTIONS OF DIFFERENTIAL NEUTRON PRODUCTION CROSS SECTIONS FOR LEAD AND URANIUM BOMBARDED BY 590-MEV PROTONS

T.W. Armstrong* and D. Filges

Institut für Reaktorentwicklung
Kernforschungsanlage Jülich GmbH
Postfach 1913
D-5170 Jülich 1, Germany

*KFA Consultant, P.O. Box 2807
La Jolla, California 92038, USA

ABSTRACT

The intranuclear-cascade-evaporation model has been used to predict the neutron production from nonelastic interactions of protons with lead and uranium. These model predictions for a 590 MeV proton beam were compared with measurements recently made at the SIN facility. In general, for both lead and uranium, the model predicts approximately the correct neutron production in the evaporation region, but over-predicts the neutrons around 15 MeV, whereas the high-energy (>100 MeV) neutrons are substantially underrepresented. The neutron production in the evaporation region proved to be sensitive to the assumed level-density parameter in the model. The charged pion-production cross section was also measured and found to be in agreement with SIN measurements.

INTRODUCTION

As part of a study to asses the accuracy of state-of-the art high-energy models for spallation neutron source applications, the intranuclear-cascade-evaporation model has been used to predict the double differential neutron-production cross sections for nonelastic collisions of protons with lead and uranium nuclei. The purpose here is to present some comparisons of model validation calculations with experimental data for neutron spectra produced in thin targets.

The calculations are made using the intranuclear-cascade-evaporation model contained in the HETC code. The cases considered are 590-MeV protons on Pb and U-238 target nuclei. The emphasis is on the angular dependent neutron spectra produced. The results on proton production are similar.

The measured neutron spectra which are compared with here were kindly provided by S. Cierjacks of KfK Karlsruhe, and are unpublished data from experiments performed at SIN. (The experimental method was summarized by Cierjacks, et.al. at ICANS-V /1/.). Cierjacks has indicated /2/ that the normalization of the measured data is to be checked in further experiments at SIN, so the comparisons here should be regarded as preliminary at present.

In the series of neutron measurements made by Cierjacks, et.al. data were taken for a range of target materials (C, Al, Fe, Nb, In, Ta, Pb, and U) and 5 angles of emission. At present, analyzed data are available to us for Pb and U targets at three angles (30, 90, and 150 degrees), and the model calculations are made only for these cases.

NEUTRON SPECTRA COMPARISONS

Figures 1-6 show comparisons of the present calculations and the KfK measurements for neutron spectra at 30°, 90° and 150° from thin lead and uranium targets bombard 590-MeV protons. The lead calculations were made using the evaporation model (EVAP-4) contained in the HETC code with the following modifications: (a) updated mass data, (b) changes for non-isotropic particle emission, and (c) a level density parameter of $B_o = 14$, rather than the standard value of $B_o = 8$ usually used in EVAP-4. For the uranium calculations the standard Rutherford and Appleton Laboratory fission model with $B_o = 14$ was used.

To show better the low-energy neutron comparisons in the evaporation region, Figures 7 and 8 give the low-energy (< 10 MeV) neutron part of spectra with a linear scale. The calculated spectra here are averaged over all emission angles.

K. H. Böckhoff (ed.), Nuclear Data for Science and Technology, 387–390.

The basic conclusions we draw from these comparisons are: (a) For uranium, there is rather good agreement in the evaporation region of the spectrum (few MeV and below). The magnitudes of the evaporation peaks agree within 25 %. The evaporation neutron maximum is lower in the calculations (1 MeV calculated vs. 2 MeV measured). In the "region of overlap" of the high-energy part of the evaporation spectrum and where the cascade production begins to dominate (i.e., in the energy range 10-25 MeV), the calculated results are higher, by as much as a factor of 3 at 10 MeV. The high-energy part of the spectrum (> 50 MeV) is underestimated by the calculations, by a factor of 3 for small (e.g., 30°) angles, with much worse agreement at the higher angles.
(b) For lead, the agreement is similar, except the agreement in the evaporation region is worse than for uranium.

To check the sensitivity of the predictions of the evaporation model to assumed input for specifying the level densities, additional calculations were made for the lead target for two cases: (a) a level density parameter of $B_o = 8$, which is the value assumed in the standard version of the code, and (b) a "variable" B_o, which is based on level density measurements and varies with the mass number of the residual nucleus at each step of the evaporation calculation.

In principle, the variable B_o method is better because it takes into account that lead is in the region of "magic" or "near-magic" nuclei, so on theoretical grounds the level densities should be lower in this narrow A region. (In actual code applications, this may not, of course, lead to improvements because the evaporation model used contains numerous inter-related approximations.)

The evaporation peak is about the same for $B_o = 8$ as when a variable B_o is used, with both in worse agreement with experiment than for $B_o = 14$. The variable B_o case gives a lower number of neutrons in the high-energy region of the evaporation spectrum (10 MeV), as indicated in Table I, which is in better agreement with the measurements than for B_o=8 or 14.

Table I Comparison of Calculated Neutron Production at Low Energies for Different Cases (a)

Case	Neutrons Produced ≤10 MeV, per Collision	Neutron Spectrum (b)	
		0.9 - 1.0 MeV	9 - 10 MeV
Pb, B_o = 8	11.8	2.69	0.27
Pb, B_o = variable	9.4	2.37	0.19
Pb, B_o = 14	10.0	2.02	0.32
U, B_o = 14	11.7	2.34	0.35

(a) Pb Cases calculated using EVAP, U case using RAL model.

(b) Neutrons per MeV per nonelastic collision.

PION PRODUCTION

Pion production is relatively small for the 590-MeV proton beam considered here, and the statistics on the pion spectra from the present calculations are not adequate to allow detailed comparisons with experimental data. However, we have compared the calculated pion production cross sections with the measured values of Crawford, et.al. made at SIN /3/ (Table II), and the agreement is reasonable good.

Table II Comparison of Pion Production Cross Section (in Millibarns) for Pb Target and 590 MeV Protons

	Calculated		Measured
	E_π >	$E_\pi \geq 25$ MeV	E_π > 25 MeV
π^+	140 ± 28	118 ± 35	86 ± 11
π^o	159 ± 32	105 ± 32	——
π^-	150 ± 30	43 ± 13	41.5 ± 5.1

CONCLUSIONS

The major deficiency of the present model is considered to be the underestimate of the high-energy neutrons and protons. The comparisons here are with preliminary experimental data, and with only a small part of the KfK data which have been taken, so the magnitude of the experimental/theoretical differences may change if further comparisons are made. However, there is enough evidence from these, and other comparisons which have been made, to believe that the difference, at least at large angles, is real, even though the magnitude may be considered still questionable, as indicated by comparisons of the KfK data with other measurements made below.

There are model changes which would, we believe, improve the high-energy particle production. In the meantime, this problem must be kept in mind in making predictions related to SNQ design.

The differences between calculation and measurements at evaporation energies, while much smaller, are also appreciable, and further investigations to improve the evaporation model predictions are needed. However, it would be better to have comparisons with the other target materials which have been measured (and, if possible, new measurements for other beam energies) before starting to change the model. Modifying the evaporation model is simpler than changing the intranuclear cascade model because it is much less complex in theory and programming, and candidate improvements are relatively easier to identify and test.

Barashenkov, et.al. quote results from measurements made using the 660 MeV proton beam at Dubna for the angular distribution of all neutrons and protons produced above 60 MeV from a thin U-238 target /4/. We have integrated the KfK measured neutron spectra for this energy range. It is interesting to note that the forward directions (< 90°) the KfK data for neutrons only agree with the Dubna data for neutrons + protons. (The calculations show that neutrons and protons are produced at roughly equal magnitudes at these energies.) Also, for the forward direction, the calculations here agree with the Dubna data but not with the KfK data.

REFERENCES

1. S. Cierjacks, et.al.,"High-Energy Particle Spectra from Spallation Targets", Proceedings of the 5th Meeting of the International Collaboration on Advanced Neutron Sources", G.S. Bauer and D. Filges (Eds.),22-26 June 1981, Jülich, Jül-Conf-45 (October 1981)

2. S. Cierjacks, private communication, 1981

3. J.W. Crawford, et.al.,"Measurement of Cross Sections and Asymmetry Parameters for the Production of Charged Pions from Various Nuclei by 585 MeV Protons", Phys. Rev. C 22 1184, (1980)

4. V.S. Barashenkov, et.al.,"Interaction of Particles and Nuclei of High and Ultrahigh Energy with Nuclei", Sov. Phys.-Usp. 16,31 (1973)

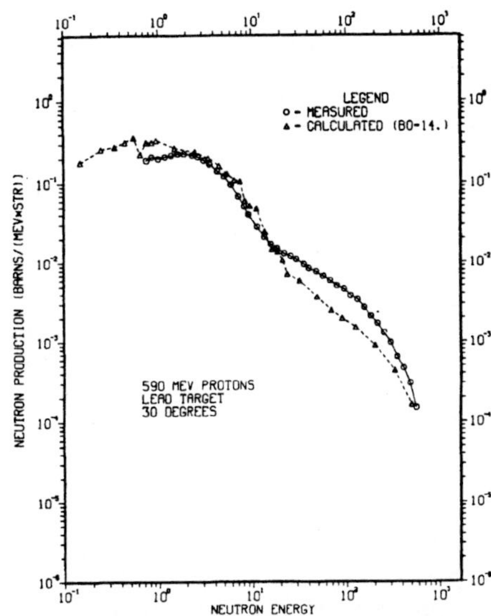

Figure 1. Comparison of calculated and KfK measured neutron spectra at 30° from a thin lead target bombarded by 590-Mev protons.

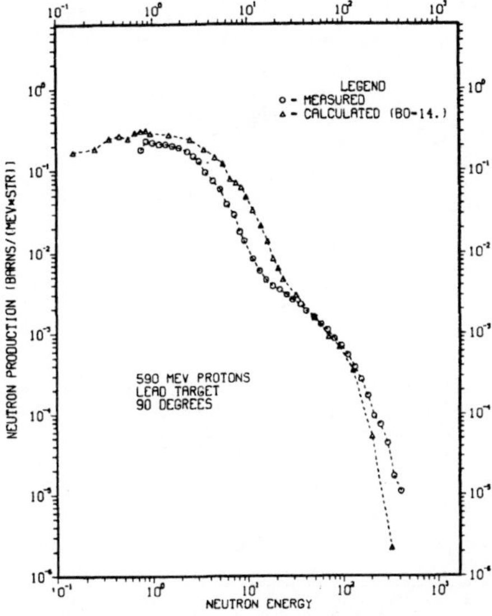

Figure 2. Comparison of calculated and KfK measured neutron spectra at 90° from a thin lead target bombarded by 590-MeV protons.

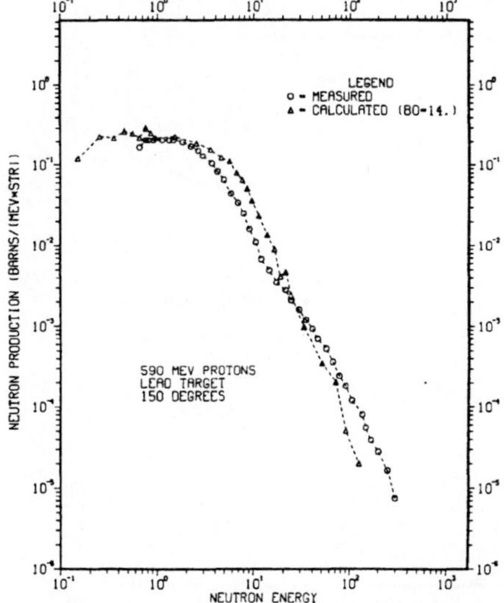

Figure 3. Comparison of calculated and KfK measured neutron spectra at 150° from a thin lead target bombarded by 590-MeV protons.

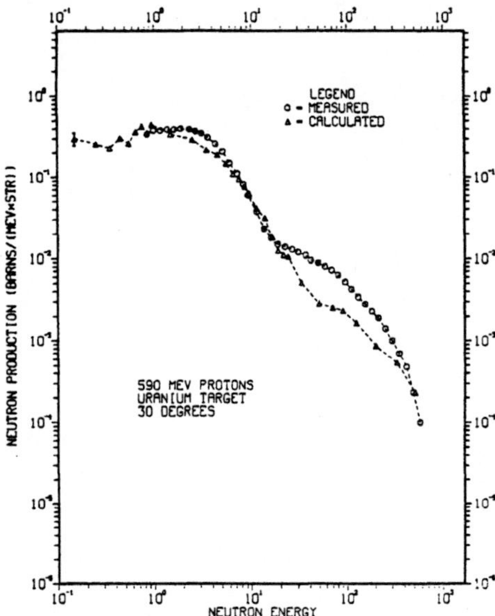

Figure 4. Comparison of calculated and KfK measured neutron spectra at 30° from uranium target bombarded by 590-MeV protons.

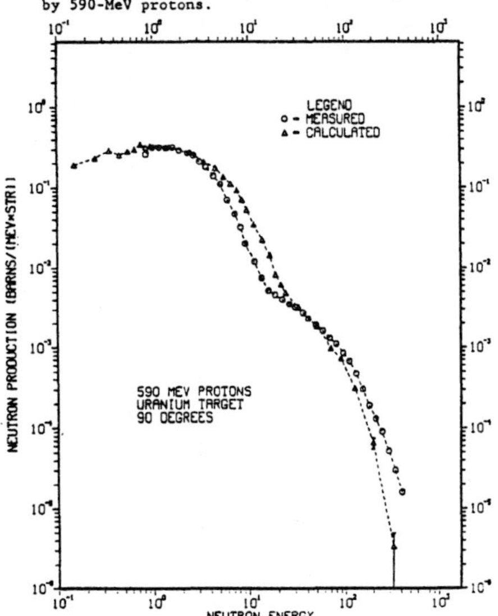

Figure 5. Comparison of calculated and KfK measured neutron spectra at 90° from uranium target bombarded by 590-MeV protons.

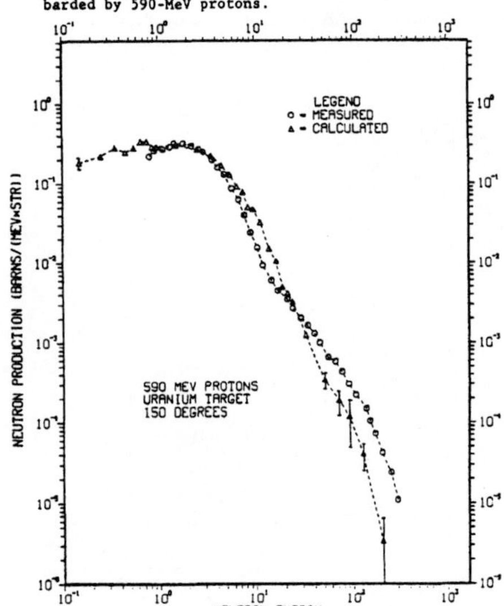

Figure 6. Comparison of calculated and KfK measured neutron spectra at 150° from uranium target bombarded by 590-MeV protons.

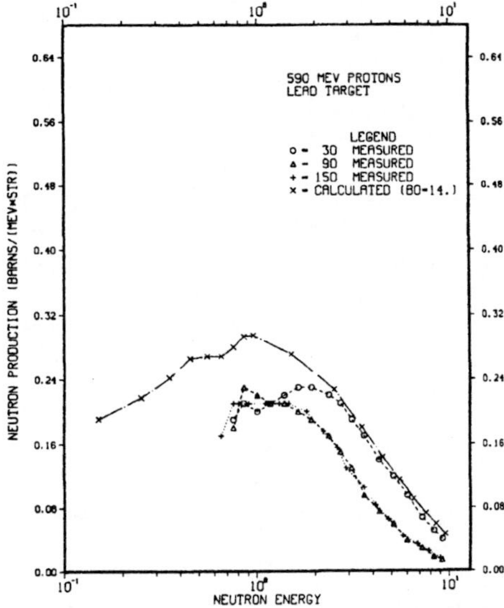

Figure 7. Comparison of calculated and KfK measured neutron spectra at low energies from a thin lead target bombarded by 590-MeV protons. The calculated spectrum is averaged over all emission angles.

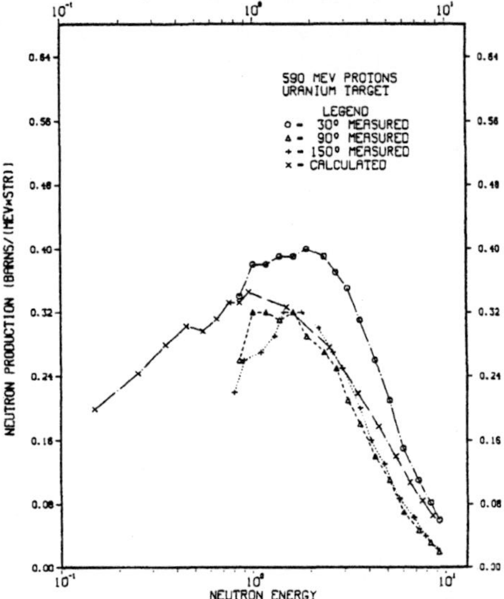

Figure 8. Comparison of calculated and KfK measured neutron spectra at low energies from a thin uranium target bombarded by 590-MeV protons. The calculated spectrum is averaged over all emission angles.

MODEL PREDICTIONS OF NEUTRON AND ISOTOPE PRODUCTION FROM PROTON INDUCED FISSION AT HIGH ENERGIES

T.W. Armstrong*, D. Filges, P. Cloth

Institut für Reaktorentwicklung
Kernforschungsanlage Jülich GmbH
Postfach 1913
D-5170 Jülich 1, Germany

*KFA Consultant, P.O. Box 2807
La Jolla, California 92038, USA

ABSTRACT

Calculations have been carried out to compare the results predicted by two models based on the intranuclear cascade plus evaporation with fission competition method for treating high-energy fission induced by hadrons. The models used have been developed recently at the Oak Ridge National Laboratory and the Rutherford and Appleton Laboratory. The motivation for the present calculations was to determine a suitable model for use in thick-target radiation transport predictions for spallation neutron source applications. The comparisons are for thin targets and some special cases are determined for the BNL-Cosmotron experiments for thick targets.

INTRODUCTION

For high-energy hadrons incident on high-mass target materials (e.g., uranium), the target nucleus may undergo fission during de-excitation. Recently, there have been several high-energy fission models developed /1-4/ for the hadronic transport code HETC /5/, which is used for spallation neutron source studies. The fundamental basis of all developed models is the statistical model of fission developed by Fong /6/. The objective here is to investigate the appropriate high-energy fission model for routine HETC calculations for spallation neutron source applications. All of the calculations were made using the Rutherford and Appleton Laboratory (RAL) high-energy fission model developed by Atchison /1/. Results for fission cross sections, neutron multiplicities and spectra fission fragment recoil energies, and residual mass distributions are compared with available results for the same cases computed by the Oak Ridge National Laboratory (ORNL) model developed by Alsmiller, et.al. /2/. The comparisons in this paper are for thin targets and some special cases are determined for the BNL-Cosmotron /7/ experiments for thick targets.

HIGH-ENERGY FISSION PROCESS

Without fission, the spallation collisions can be treated as a two-step process: (a) an intranuclear cascade, described by a series of independent particle-particle de-excitation by a series of particle emissions, which can be described by an evaporation model. For very heavy nuclei, there is competition between evaporation and fission at each step of the de-excitation sequence. The probability of fission at some step during de-excitation in high-energy (> 100 MeV) collisions is proportional to Z^2/A of the target nucleus. For example, $\sigma_f/\sigma_t = 0.05$ for lead and $\sigma_f/\sigma_t = 0.8$ for uranium.

If fission occurs, the following quantities are expected to be different compared to spallation without fission: (a) energy deposition, (b) residual mass distributions, (c) neutron multiplicity and (d) neutron spectrum.

The high-energy fission models developed differ in the approximations made in arriving at a practical implementation of the above general expression and in the physical data used. Therefore, the model comparisons here are in terms of "output" rather than assessment of the intermediate physics. Characterizations of the different models are given in Ref. /8/. It may be mentioned, that the parameter B_0 in the level density formula has an important influence on neutron production. This is in detail described in Ref. /9/.

K. H. Böckhoff (ed.), Nuclear Data for Science and Technology, 391–394.

THIN TARGET COMPARISONS

Figure 1 shows the fission cross sections calculated by the RAL Fission Model and compared with results from Alsmiller et.al. /2/ with the ORNL Model. The fission cross section predicted by the RAL model is about 15-20 % lower than for the ORNL model for beam energies below 1 GeV, and the energy dependence of the cross section predicted above 1 GeV appears to be different for the two models. Also shown in Figure 1 are the nonelastic cross sections from the two calculations, which are in agreement, as expected, since both calculations use the same intranuclear cascade model.

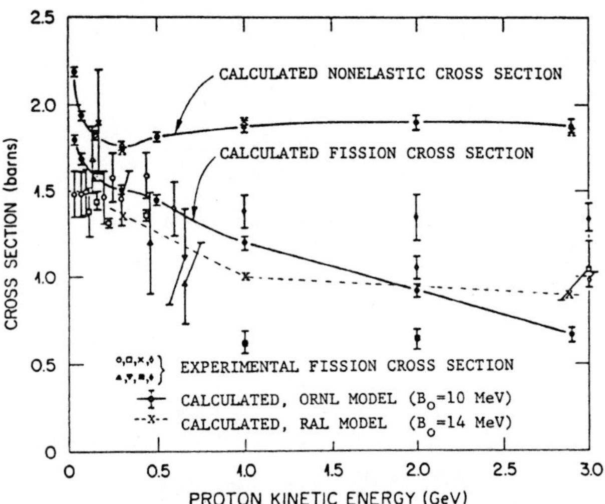

Fig. 1: Comparison of calculated cross sections using the RAL and ORNL fission models for the case of protons incident on thin U-238 target. References for experimental data given in /2/

To consider the influence of B_0, we have calculated the low-energy neutron production spectrum for 1-GeV protons using the RAL model (Figure 2). The integral neutron production below 12.5 MeV is 20 % higher for B_0 = 8 than for B_0 = 14. With fission included, the neutron production is about 8 % higher, and there is evidence of some "spectral hardening" for neutron energies above about 2 MeV. Neutron production above this energy is important because this corresponds approximately to the energy threshold for neutron induced fission for U-238 in thick U-238 targets providing a larger source of neutrons that can cause multiplication via low-energy fissioning.

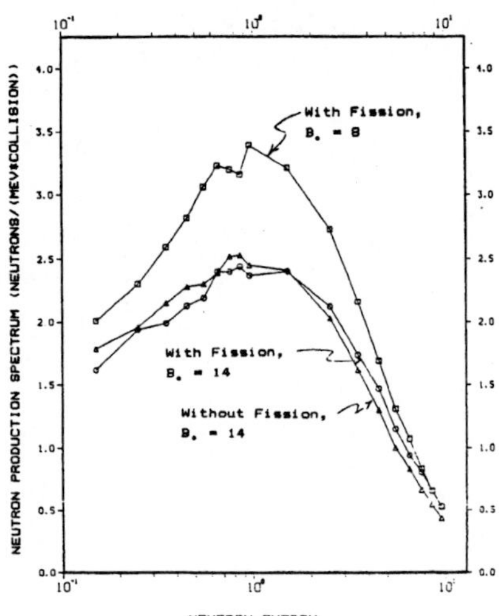

Fig. 2: Low-energy neutron production spectrum calculated using RAL fission model, 1-GeV proton beam on thin U-238 target

There are large differences in energy deposition for spallation collision with and without high-energy fissioning (20-30 MeV per collision). If fission takes place, the "local" energy deposition at the collision site is mainly from the kinetic energy of the fission fragments which, as in low-energy fission, is expected to be about 170 MeV per fission. The fission fragment kinetic energies predicted high-energy fission models is important predicting the target heating for spallation neutron sources. A comparison of the (total) fission fragment energies calculated using the RAL and ORNL models is given in Figure 3.

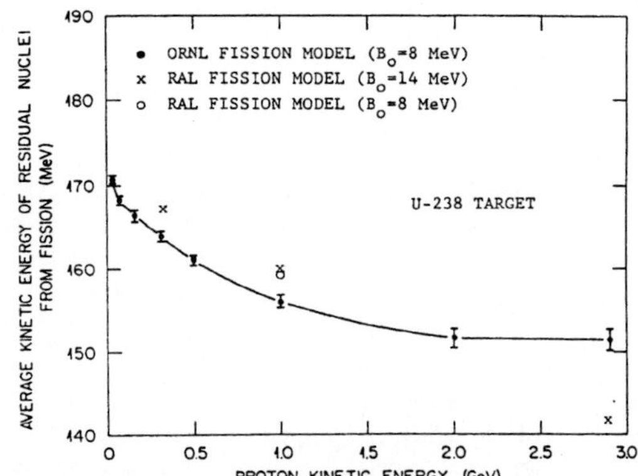

Fig. 3: Comparison of fission fragment energies predicted by the RAL and ORNL fission models for protons incident on thin U-238 target

Mass distributions for isotope production were calculated using the RAL model for proton beams having kinetic energies of 0.3, 1.0 and 2.9 GeV and are shown in Figure 4. The points shown are averages over ΔA = 5 intervals. The normalization is per nonelastic proton-uranium collision, which can be converted from yields to production cross sections by multiplying by the computed total nonelastic cross section.

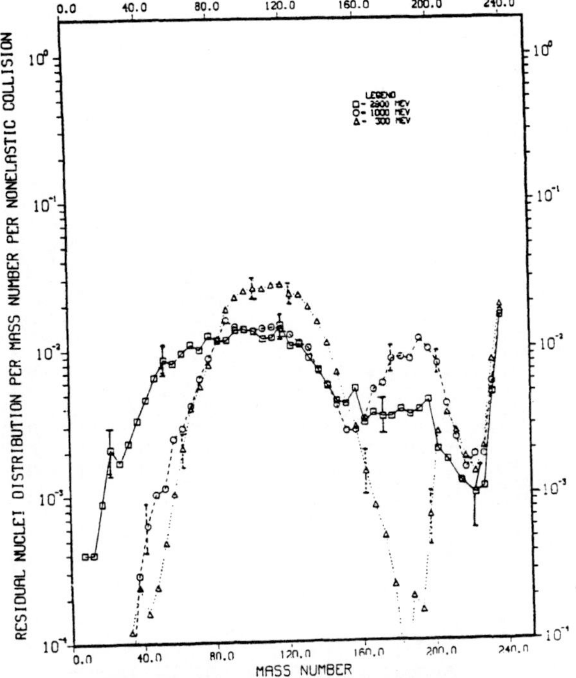

Fig. 4: Mass distribution predicted by RAL high-energy fission model for 300, 1000, and 2900 MeV protons on thin U-238 target

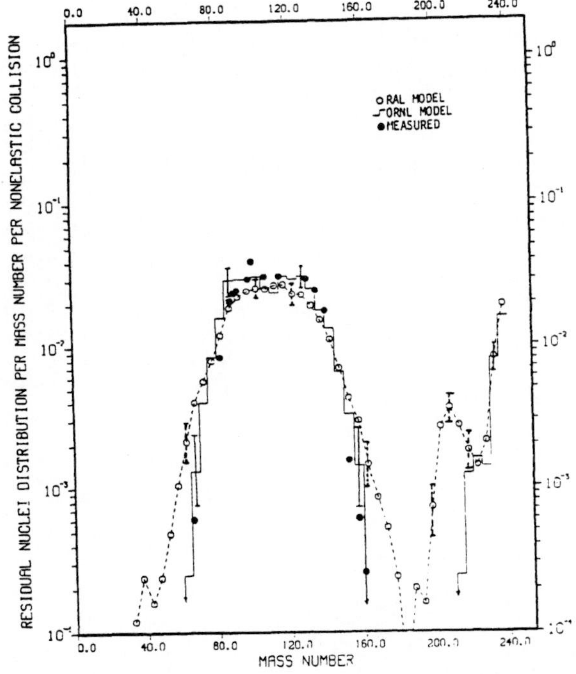

Fig. 5: Comparison of mass distributions computed using RAL model, from ORNL model calculations /11/, and from measurements of Stevenson, et al. /10/ for 300 MeV protons on thin U-238 target

In Figure 5 results from the RAL model are compared with ORNL model predictions and measured data. The calculations are also averaged over ΔA = 5 intervals. The normalization of the measured data of Stevenson, et.al. /10/ at 300 MeV is taken from the ORNL paper /2/, in which the area under the experimental points in the mass region from 60 to 160 was normalized to be the same as the area under the ORNL calculated histogram in this mass region. In the vicinity of the peaks of the fission fragment mass distributions the RAL model seems to predict a somewhat wider fission fragment distribution.

The RAL model predicts three peaks in the mass distribution Fig. 4 and 5: the fission fragment peak at A 110, the spallation peak at A = 238, and an intermediate peak at A = 200. This intermediate peak apparently results from spallation products which "survive" de-excitation through the mass region of high-fission probability into a lower mass region where further de-excitation by neutron emission is much more likely than fission.

THICK TARGET COMPARISONS

According to the BNL-Cosmotron experiments /7/ calculations were made at incident proton beam energies 590, 960 and 1470 MeV. The measured and calculated quantity is the neutron capture rate in the H_2O tank. A variety of B_0 values was applied to calculations with RAL- and ORNL-HEF models. The last column of Table I shows the ratios of experimental and calculational results.

Table I: Comparison with Cosmotron Data (Fraser et.al. /7/) Depleted Uranium Target, H_2O Captures per Incident Proton

Incident Proton Energy (MeV)	Experiment	HEF	B_0 MeV	Theory	Theory vs. Experiment
540	15.1+0.8	ORNL	10*	15.2±0.8	1.07
		RAL	14	13.5±1.1	0.81
960	32.3+1.6	ORNL	10*	33.7±1.0	1.04
		ORNL	10**	35.2±0.6	1.09
		RAL	10	30.9±0.3	0.96
		RAL	8	32.6±0.3	1.00
		RAL	14	29.0±0.3	0.89
1470	44.8+0.2	ORNL	10*	53.6±1.5	1.20
		RAL	14	46.0±0.4	1.03

* Calculations of Alsmiller et.al., Ref. /11/
**KFA-IRE calculations using uranium cross sections with self-shielding corrections

With B_0 = 10 MeV RAL- and ORNL-HEF models give very similar results, which are in good agreement with experimental results (the "standard" RAL model B_0 = 14 MeV /1/ underestimates the experiment) for all incident proton energies upto 1 GeV. At energies above 1 GeV the deviation from experiment is higher and significant in case of the ORNL-HEF model ("standard", B_0 = 10 MeV /11/), the RAL model however, meets the experiments even at B_0 = 14 MeV. Additional information of neutron production and neutron-reaction rates about the previous calculations given in Reference /9/.

Calculations were also done for BNL-Cosmotron setups for the spatial dependence and thermal neutron peak fluxes at three proton beam energies (590, 960, and 1470 MeV) using HETC-, MORSE-CG-, and SIMPEL-spallation computer code system at KFA-IRE as described in Ref. 12. In Fig. 6 the thermal peak fluxes for neutrons (n cm^{-2} s^{-1}) per proton are plotted as a function of proton beam energy for lead and uranium target with B_0 = 14.

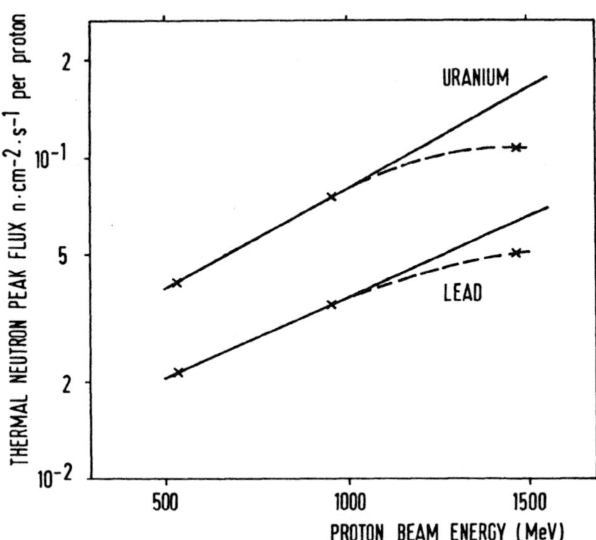

Fig. 6: Thermal neutron peak fluxes per proton as a function of proton beam energy (540, 960, 1470 MeV) for lead and uranium target with B_0 = 14 MeV

The uranium target system gives twice the thermal neutron peak flux of the lead system. The peak fluxes depend linearly on the incident proton beam energy upto 1 GeV. For higher energies there is only a weak increase of the neutron flux because of the spatial spreading out of the cascades.

REFERENCES

/1/ F. Atchison, Meeting on Targets for Neutron Beam Spallation Sources, G. Bauer (Editor), KFA-Jülich, FRG, 11-12 June 1979, Jül-Conf-34, January 1980

/2/ F.S. Alsmiller, et.al., ORNL-TM-7528 March 1981

/3/ H. Takahashi, to be published in Proc. of Symp. on Neutron Cross Sections 10-50 MeV, Brookhaven National Laboratory, 12-14 May 1980

/4/ Y. Nakahara, Proc. of 4th Meeting of International Collaboration on Advanced Neutron Sources (ICANS-IV), KEK National Laboratory for High-Energy Physics, Tsukuba, Japan, 20-24 October 1980

/5/ T.W. Armstrong and K.C. Chandler, Nucl. Sci. and Engr. 49, 110 (1972)

/6/ P. Fong, Gordon and Beach Science Publishers, New York, 1969

/7/ J.S. Fraser, et.al., Phys. in Canada 21, 17 (1965)

/8/ T.W. Armstrong, D. Filges, Proceedings of the 5th Meeting of the International Collaboration on Advanced Neutron Sources (Eds. G.Bauer and D.Filges), Jül-Conf-45 Jülich, June 22-26, 1981

/9/ T.W. Armstrong, P. Cloth, D. Filges, R.D. Neef, Proceedings of the 5th Meeting of the International Collaboration on Advanced Neutron Sources (Eds. G.Bauer and D.Filges), Jül-Conf-45, Jülich June 22-26, 1981

/10/ P.C. Stevenson, et.al., Phys. Rev. 111, 886 (1958)

/11/ R.G. Alsmiller Jr., T.A. Gabriel, J. Barish, F.S. Alsmiller, ORNL/TM-7527 (1981)

/12/ T.W. Armstrong, P. Cloth, D. Filges, R.D. Neef, Jül-Spez-120, Juli 1981

Nuclear Data for Dosimetry

COMPARISON OF MEASURED AND EVALUATED SPECTRUM-AVERAGED CROSS-SECTION DATA FOR REACTOR NEUTRON METROLOGY

W.L. Zijp, H.Ch. Rieffe and H.J. Nolthenius

Netherlands Energy Research Foundation ECN,
P.O. Box 1, 1755 ZG Petten, The Netherlands.

For several reference neutron spectra the measured spectrum-averaged cross-section values are compared with the spectrum-averaged cross-section values, derived from evaluated cross-section data files (ENDF/B-V reaction dosimetry file, and the DOSCROS81 library).

Apart from the discrepancy between these values themselves, we have also considered the consistency between the experimental and the evaluated cross-section values, taking into account the available information on the standard deviations in the averaged cross-section values. The standard deviations in the calculated cross-section values have been derived by taking into account the available information in files 32 and 33 of the ENDF/B-V dosimetry tile.
Remarkable discrepancies between measured and calculated spectrum-averaged cross-sections have been observed.

Introduction

The information presented here is an updated version of earlier studies made in 1979 and 1981 at ECN on the same topic.
This comparison of measured and calculated spectrum-averaged cross-section values for reactor neutron metrology reactions compares experimental values (determined by means of the activation technique) with calculated values (derived from the evaluated cross-section library DOSCROS[1], which is mainly based on the ENDF/B-V dosimetry file.

For the consideration of the so-called consistency between measured and evaluated spectrum-averaged cross-sections one needs to take into account the standard deviation in the experimental value, and the standard deviation in the calculated value (based in principle on the variance-covariance information for both the spectrum and the cross-section data). However, variance-covariance information of reference spectra is hardly available.

Approach

In our consideration the following reference neutron fluence rate spectra have been considered.

- The spontaneous fission neutron spectrum of ^{252}Cf.
- The thermal neutron fission spectrum of ^{235}U.
- The neutron spectrum in the Coupled Fast Reactivity Measurement Facility (CFRMF) at Idaho Falls, USA.
- The spectrum in the 10 per cent Enriched Uranium Cylindrical Critical Assembly (BIG-TEN).
- The spectrum of the Coupled Thermal/Fast Uranium and Boron Carbide Spherical Assembly (SIGMA-SIGMA) at Mol, Belgium.
- The spectrum of the NISUS facility at London (similar to SIGMA-SIGMA).

The numerical spectrum information and the experimental reaction rates, and also the spectrum covariance information were taken from the most recent published data, available to the authors early 1982.

For the reactions of interest we used the cross-section data as present in the DOSCROS81 library[1] in a 640 groups structure of the SAND-II type. These data were derived from the ENDF/B dosimetry file and supplemented with data from INDL/V. In both cases the conversion from the ENDF/B format to the format of the DOSCROS81 library was performed with the program ENTOSAN[2].

We were able to take into account the covariance information available in the ENDF/B-V dosimetry files, not only with respect to the major region of unresolved resonances (file 33), but also with respect to the region with resolved resonance peaks (file 32).

Quality of data

In this comparison we considered the following aspects (see Fig. 1):

The *imprecision*, defined as the statistical uncertainty, in terms of the standard deviation for both the measured and calculated average cross-section values. For reasons of easy comparison, values quoted are relative standard deviations, i.e. coefficients of variation, v. In formula:

$$v_m = \frac{\sqrt{\text{var}\ \alpha_m}}{\alpha_m} \quad \text{and} \quad v_c = \frac{\sqrt{\text{var}\ \alpha_c}}{\alpha_c}$$

where:

α_m = measured average cross-section;
α_c = calculated average cross-section;
var α_m = variance of measured values;
var α_c = variance of calculated values.

The *discrepancy*, defined as the relative difference Δ between measured and calculated average cross-section values. In formula:

$$\Delta = \frac{\alpha_m - \alpha_c}{\alpha_c}$$

The *consistency*, generally defined as the chi-square value found with the formula:

$$\chi^2 = \frac{(\alpha_m - \alpha_m)^2}{\text{var}\ \alpha_m + \text{var}\ \alpha_c}$$

In the attached tables a summary is presented of some main results obtained thus far in the study of uncertainties, discrepancies, and consistencies involved in reactor neutron metrology reactions. The following code is used to visualize the present situation.

quality	symbol	uncertainty (in per cent)	discrepancy (in per cent)	consistency
very good	++	$0 < v \leq 2$	$0 < \lvert\Delta\rvert \leq 2$	$0 < \chi^2 \leq 1,5$
good	+	$2 < v \leq 4$	$2 < \lvert\Delta\rvert \leq 4$	$1,5 < \chi^2 \leq 3$
moderate	0	$4 < v \leq 6$	$4 < \lvert\Delta\rvert \leq 6$	$3 < \chi^2 \leq 4,5$
bad	–	$6 < v \leq 8$	$6 < \lvert\Delta\rvert \leq 8$	$4,5 < \chi^2 \leq 6$
very bad	––	$8 < v$	$8 < \lvert\Delta\rvert$	$6 < \chi^2$

Some general conclusions

1. For most widely used neutron metrology reactions, the standard deviations of the measured spectrum-averaged cross-sections are in many cases larger than 4 per cent.
2. The standard deviations of the calculated spectrum-averaged cross-sections are much larger than the standard deviations of the experimental values. This is an indication that the uncertainties in the ENDF/B-V dosimetry file may have been estimated somewhat too large.

K. H. Böckhoff (ed.), Nuclear Data for Science and Technology, 397–399.

3. Large discrepancies between measured and calculated spectrum-averaged cross-sections are mainly observed for the category of (n,γ) and (n,2n) reactions.

4. For the ^{252}Cf spectrum and for the CFRMF spectrum many values of the χ^2-parameter for the consistency are classified as very good, due to large values for the imprecisions.

5. The situation with respect to imprecision, discrepancy and consistency is most favourable for CFRMF.

6. For various spectra large positive discrepancies are observed for the reaction ^{10}B(n,He).

7. For various spectra large negative discrepancies are observed for ^{47}Ti(n,p). The ENDF/B-V cross-section values seem to be about 20 per cent too high.

8. Sometimes remarkable discrepancies are also observed for ^{197}Au(n,2n) and ^{232}Th(n,f).

Final remark: More detailed information including numerical values is available from the authors as ECN-report with restricted distribution.

References

1. W.L. Zijp, H.J. Nolthenius, and H.Ch. Rieffe, "Cross-section library DOSCROS81 (in a 640 group structure of the SAND-II type)", Report ECN-111 (Netherlands Energy Research Foundation ECN, Petten, December 1981).

2. H.Ch. Rieffe, H.J. Nolthenius, and W.L. Zijp, "ENTOSAN. A program for the calculation of fine group cross section values from ENDF/B data", Report ECN-93 (Netherlands Energy Research Foundation ECN, Petten, April 1981).

Table I Qualification for ^{252}Cf spectrum.

reaction	imprecision		discrepancy	consistency
	measured v_m	calculated v_c	Δ	χ^2
^{19}F (n,2n)^{18}F	--		--	--
^{23}Na(n,γ)^{24}Na	0	--	--	+
^{24}Mg(n,p)^{24}Na	0		--	-
^{27}Al(n,p)^{27}Mg	+	0	0	++
^{27}Al(n,α)^{24}Na	+	0	0	++
^{32}S (n,p)^{32}P	0		0	++
^{46}Ti(n,p)^{46}Sc	+	--	+	++
^{47}Ti(n,p)^{47}Sc	+	--	--	0
^{48}Ti(n,p)^{48}Sc	+	--	+	++
^{54}Fe(n,p)^{54}Mn	+	+	+	++
^{55}Mn(n,2n)^{54}Mn	--	--	--	+
^{55}Mn(n,γ)^{56}Mn	-		0	++
^{56}Fe(n,p)^{56}Mn	+	0	+	++
^{58}Ni(n,p)^{58}Co	+	-	+	++
^{59}Co(n,γ)^{60}Co	0	--	--	++
^{59}Co(n,α)^{56}Mn	-	0	++	++
^{59}Co(n,2n)^{58}Co	--	--	--	-
^{59}Co(n,p)^{59}Fe	0		-	+
^{63}Cu(n,γ)^{64}Cu	-	--	--	--
^{63}Cu(n,α)^{60}Co	+	0	-	++
^{63}Cu(n,2n)^{62}Cu	--		--	--
^{64}Zn(n,p)^{64}Cu	+		++	++
^{75}As(n,γ)^{76}As	-		--	--
^{93}Nb(n,2n)^{92}Nb	0		--	+
^{98}Mo(n,γ)^{99}Mo	0		0	++
^{100}Mo(n,γ)^{101}Mo	-		--	+
^{103}Rh(n,n')^{103}Rh^m	0		-	+
^{115}In(n,n')^{115}In^m	+	--	--	++
^{115}In(n,γ)^{116}In^m	+	0	+	++
^{181}Ta(n,γ)^{182}Ta	0		--	-
^{197}Au(n,γ)^{198}Au	+	--	++	++
^{197}Au(n,2n)^{196}Au	+		+	++
^{232}Th(n,f)	--	0	--	++
^{235}U (n,f)	+	++	+	++
^{238}U (n,f)	+	++	++	++
^{237}Np(n,f)	+	--	++	++
^{239}Pu(n,f)	+		++	++

Table II Qualification for ^{235}U spectrum.

reaction	imprecision		discrepancy	consistency
	measured v_m	calculated v_c	Δ	χ^2
^{24}Mg(n,p)^{24}Na	0		0	++
^{27}Al(n,p)^{27}Mg	0	0	--	+
^{27}Al(n,α)^{24}Na	0	0	++	++
^{32}S (n,p)^{32}P	0	--	0	++
^{46}Ti(n,p)^{46}Sc	-	--	--	++
^{47}Ti(n,p)^{47}Sc	-	--	--	++
^{48}Ti(n,p)^{48}Sc	0	--	--	++
^{52}Cr(n,p)^{52}V	--		--	-
^{54}Fe(n,p)^{54}Mn	0	+	++	++
^{55}Mn(n,2n)^{54}Mn	-	--	--	+
^{56}Fe(n,p)^{56}Mn	0	0	++	++
^{58}Ni(n,2n)^{57}Ni	0	--	--	--
^{58}Ni(n,p)^{58}Co	0	-	-	++
^{59}Co(n,α)^{56}Mn	-	0	++	++
^{63}Cu(n,γ)^{64}Cu	--	--	-	++
^{63}Cu(n,α)^{60}Co	--	0	-	++
^{93}Nb(n,n')^{93}Nb^m	-		--	--
^{115}In(n,n')^{115}In^m	0	--	--	++
^{115}In(n,γ)^{116}In^m	0	0	-	++
^{127}I (n,2n)^{126}I	-	--	--	++
^{197}Au(n,γ)^{198}Au	0	--	+	++
^{197}Au(n,2n)^{196}Au	0		--	--
^{232}Th(n,f)	-	0	--	+
^{235}U (n,f)	+	++	+	++
^{238}U (n,f)	+	++	+	++
^{237}Np(n,f)	+	--	++	++
^{239}Pu(n,f)	+		++	++

Table III Qualification for CFRMF spectrum.

reaction	imprecision		discrepancy	consistency
	meas. v_m	calc. v_c	Δ	χ^2
^{6}Li(n,α)^{3}H	+	++	-	0
^{10}B (n,α)^{7}Be	+	++	--	--
^{27}Al(n,p)^{27}Mg	+	0	-	++
^{27}Al(n,α)^{24}Na	+	0	+	++
^{45}Sc(n,γ)^{46}Sc	+	--	+	++
^{46}Ti(n,p)^{46}Sc	+	--	-	++
^{47}Ti(n,p)^{47}Sc	0	--	--	0
^{48}Ti(n,p)^{48}Sc	+	--	--	++
^{54}Fe(n,p)^{54}Mn	+	+	0	+
^{58}Fe(n,γ)^{59}Fe	+	--	-	++
^{59}Co(n,γ)^{60}Co	+	--	0	++
^{58}Ni(n,p)^{58}Co	+	-	++	++
^{63}Cu(n,γ)^{64}Cu	-	--	++	++
^{115}In(n,n')^{115}In^m	+	--	++	++
^{115}In(n,γ)^{116}In^m	+	+	+	++
^{197}Au(n,γ)^{198}Au	+	+	--	0
^{232}Th(n,f)	0	0	+	++
^{232}Th(n,γ)^{233}Th	+	--	--	++
^{235}U (n,f)	+	+	++	++
^{238}U (n,f)	+	++	0	+
^{238}U (n,γ)^{239}U	+	++	++	++
^{237}Np(n,f)	+	--	-	++
^{239}Pu(n,f)	+		++	++

v_m = imprecision of measured data;
v_c = imprecision of calculated data;
Δ = discrepancy;
χ^2 = consistency.

Table IV Qualification for ΣΣ spectrum.

reaction	imprecision		discrepancy	consistency
	measured v_m	calculated v_c	Δ	χ^2
^{27}Al(n,p)^{27}Mg	--	0	0	++
^{27}Al(n,α)^{24}Na	+	0	--	0
^{55}Mn(n,γ)^{56}Mn	0		--	0
^{56}Fe(n,p)^{56}Mn	+	0	+	++
^{58}Ni(n,p)^{58}Co	+	-	++	++
^{63}Cu(n,γ)^{64}Cu	0	--	+	++
^{115}In(n,n')^{115}In^m	+	--	++	++
^{115}In(n,γ)^{116}In^m	+	+	--	+
^{197}Au(n,γ)^{198}Au	+	+	--	--
^{235}U (n,f)	+	++	++	++
^{238}U (n,f)	+	++	+	++
^{238}U (n,γ)^{239}U	0	++	--	--
^{237}Np(n,f)	+	--	-	++
^{239}Pu(n,f)	+	++	++	++

v_m = imprecision of measured value;
v_c = imprecision of calculated value;
Δ = discrepancy;
χ^2 = consistency.

Table V Qualification for BIG-TEN spectrum.

reaction	imprecision meas-ured v_m	imprecision calcu-lated v_c	discre-pancy Δ	consis-tency χ^2
$^{6}Li(n,\alpha)^{3}H$	++	++	0	--
$^{10}B(n,\alpha)^{7}Be$	++	++	--	--
$^{27}Al(n,p)^{27}Mg$	+	0	--	--
$^{27}Al(n,\alpha)^{24}Na$	+	0	--	--
$^{45}Sc(n,\gamma)^{46}Sc$	++	--	0	++
$^{46}Ti(n,p)^{46}Sc$	++	--	++	++
$^{47}Ti(n,p)^{47}Sc$	+	--	--	+
$^{48}Ti(n,p)^{48}Sc$	+	--	--	++
$^{54}Fe(n,p)^{54}Mn$	+	+	+	++
$^{55}Mn(n,\gamma)^{56}Mn$	+		-	--
$^{56}Fe(n,p)^{56}Mn$	+	0	--	--
$^{58}Fe(n,\gamma)^{59}Fe$	+	--	--	++
$^{58}Ni(n,p)^{58}Co$	++	-	++	++
$^{59}Co(n,\gamma)^{60}Co$	++	--	+	++
$^{63}Cu(n,\gamma)^{64}Cu$	0	--	++	++
$^{115}In(n,\gamma)^{116}In^{m}$	++	0	--	-
$^{115}In(n,n')^{115}In^{m}$	++	--	+	++
$^{197}Au(n,\gamma)^{198}Au$	++	0	-	+
$^{197}Au(n,2n)^{196}Au$	--		--	--
$^{233}U(n,f)$	+		+	+
$^{235}U(n,f)$	++	+	++	++
$^{237}Np(n,f)$	++	--	--	++
$^{238}U(n,\gamma)^{239}U$	+	++	++	++
$^{238}U(n,f)$	++	++	-	--
$^{239}Pu(n,f)$	++		++	++

v_m = imprecision of measured value;
v_c = imprecision of calculated value;
Δ = discrepancy;
χ^2 = consistency.

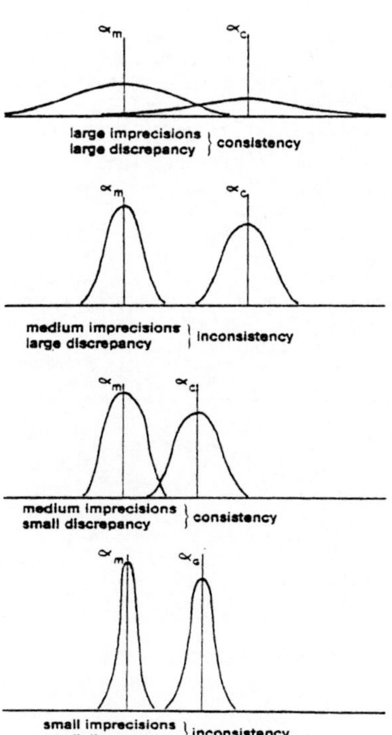

Fig. 1 RELATION BETWEEN IMPRECISION, DISCREPANCY AND CONSISTENCY

Table VI Qualification for NISUS spectrum.

reaction	imprecision meas-ured v_m	imprecision calcu-lated v_c	discre-pancy Δ	consis-tency χ^2
$^{10}B(n,\alpha)^{7}Be$	-	++	--	-
$^{24}Mg(n,p)^{24}Na$	+		++	++
$^{27}Al(n,p)^{27}Mg$	+	0	--	+
$^{27}Al(n,\alpha)^{24}Na$	+	0	0	++
$^{56}Fe(n,p)^{56}Mn$	+	0	++	++
$^{58}Ni(n,p)^{58}Co$	+	-	-	++
$^{64}Zn(n,p)^{64}Cu$	+		--	--
$^{103}Rh(n,n')^{103}Rh^{m}$	+		+	++
$^{115}In(n,\gamma)^{116}In^{m}$	+	+	--	0
$^{115}In(n,n')^{115}In^{m}$	+	--	-	++
$^{197}Au(n,\gamma)^{198}Au$	+	+	--	--
$^{235}U(n,f)$	++	++	++	++
$^{237}Np(n,f)$	++	--	--	++
$^{238}U(n,f)$	++	++	-	--
$^{239}Pu(n,f)$	++		++	+

Table VII Summary of discrepancies and consistencies.

reaction	^{252}Cf Δ	^{252}Cf χ^2	^{235}U Δ	^{235}U χ^2	CFRMF Δ	CFRMF χ^2	ΣΣ Δ	ΣΣ χ^2	NISUS Δ	NISUS χ^2	BIG-TEN Δ	BIG-TEN χ^2
$^{6}Li(n,\alpha)$					-	0					0	--
$^{10}B(n,\alpha)$					--	--			--	-	--	--
$^{19}F(n,2n)$	--	--										
$^{23}Na(n,\gamma)$	--	+										
$^{24}Mg(n,p)$	--	-	0	++					++	++		
$^{27}Al(n,p)$	0	++	--	+	-	++	0	++	--	+	--	--
$^{27}Al(n,\alpha)$	0	++	++	++	+	++	--	0	0	++	--	--
$^{32}S(n,p)$	0	++	0	++								
$^{45}Sc(n,\gamma)$					+	++					0	++
$^{46}Ti(n,p)$	+	++	--	++	-	++					++	++
$^{47}Ti(n,p)$	--	0	--	++	--	0					--	+
$^{48}Ti(n,p)$	+	++	--	++	--	++					--	++
$^{52}Cr(n,p)$			--	-								
$^{54}Fe(n,p)$	+	++	++	++	0	+					+	++
$^{55}Mn(n,2n)$	--	+	--	+								
$^{55}Mn(n,\gamma)$	0	++					--	0			-	--
$^{56}Fe(n,p)$	+	++	++	++			+	++	++	++	--	--
$^{58}Fe(n,\gamma)$					-	++					--	++
$^{58}Ni(n,2n)$			--	--								
$^{58}Ni(n,p)$	+	++	-	++	++	++	++	++	-	++	++	++
$^{59}Co(n,\gamma)$	--	++			0	++					+	++
$^{59}Co(n,\alpha)$	++	++	++	++								
$^{59}Co(n,2n)$	--	-										
$^{59}Co(n,p)$	-	+										
$^{63}Cu(n,\gamma)$	--	--	-	++	++	++	+	++			++	++
$^{63}Cu(n,\alpha)$	-	++	-	++								
$^{63}Cu(n,2n)$	--	--										
$^{64}Zn(n,p)$	++	++							--	--		
$^{75}As(n,\gamma)$	--	--										
$^{93}Nb(n,2n)$	--	+										
$^{93}Nb(n,n')$			--	--								
$^{98}Mo(n,\gamma)$	0	++										
$^{100}Mo(n,\gamma)$	--	+										
$^{103}Rh(n,n')$	-	+							+	++		
$^{115}In(n,n')$	--	++	--	++	++	++	++	++	-	++	+	++
$^{115}In(n,\gamma)$	+	++	-	++	+	++	--	+	--	0	--	-
$^{127}I(n,2n)$			--	++								
$^{181}Ta(n,\gamma)$	--	-										
$^{197}Au(n,\gamma)$	++	++	+	++	--	0	--	--	--	--	-	+
$^{197}Au(n,2n)$	+	++	--	--							--	--
$^{232}Th(n,f)$	--	++	--	+	+	++						
$^{232}Th(n,\gamma)$					--	++						
$^{233}U(n,f)$											+	+
$^{235}U(n,f)$	+	++	+	++	++	++	++	++	++	++	++	++
$^{238}U(n,f)$	++	++	+	++	0	+	+	++		--	-	--
$^{238}U(n,\gamma)$					++	++	--	--			++	++
$^{237}Np(n,f)$	++	++	++	++	-	++	-	++	--	++	--	++
$^{239}Pu(n,f)$	++	++	++	++	++	++	++	++	++	+	++	++

PRECISE MEASUREMENT OF CROSS SECTIONS FOR THE REACTIONS $^{90}Zr(n,2n)^{89}Zr$ AND $^{58}Ni(n,2n)^{57}Ni$ FROM THRESHOLD TO 20 MeV

G. Winkler, A. Pavlik, and H. Vonach

Institut für Radiumforschung und Kernphysik der Universität Wien und der Österr. Akademie der Wissenschaften, Boltzmanngasse 3, A-1090 Wien, Austria

A. Paulsen and H. Liskien

Commission of the European Communities, Joint Research Center, Central Bureau for Nuclear Measurements, Steenweg op Retie, B-2440 Geel, Belgium

The excitation functions for the reactions $^{90}Zr(n,2n)^{89}Zr$ and $^{58}Ni(n,2n)^{57}Ni$ were measured from 12.3 to 19.6 MeV in steps of 0.2 - 0.8 MeV using activation techniques. The reaction $T(d,n)^4He$ was used to produce the neutrons. Neutron fluences were determined by means of a proton-recoil telescope at 0° and the differential neutron production cross sections. Additionally, in the energy range 13.4 - 14.8 MeV measurements were performed in smaller energy steps (0.02 - 0.15 MeV) relative to well known cross sections for the reference reaction $^{27}Al(n,\alpha)^{24}Na$. Activity measurements were performed using a 12.7 cm x 12.7 cm NaI(Tl) well-type detector. The new cross section data were evaluated together with other experimental data from the literature to produce an improved set of evaluated group cross sections and their uncertainties for the reactions $^{90}Zr(n,2n)^{89}Zr$ and $^{58}Ni(n,2n)^{57}Ni$, including covariance information. Spectrum-averaged cross sections for the ^{252}Cf spontaneous-fission neutron field and the ^{235}U fission spectrum were calculated on this basis and compared with available experimental integral data.

[$^{90}Zr(n,2n)^{89}Zr$, $^{58}Ni(n,2n)^{57}Ni$, excitation functions, E_n = threshold to 20 MeV, ^{252}Cf- and ^{235}U-fission spectrum averaged cross sections, reactor dosimetry, fluence monitor, threshold reactions, spectrum unfolding]

Introduction

High-threshold reactions have become important for activation-detector dosimetry applications for fast reactors and proposed fusion energy devices. Priority 1 requests can be found both for the $^{90}Zr(n,2n)^{89}Zr$ and the $^{58}Ni(n,2n)^{57}Ni$ reaction in WRENDA 81/82,[1] asking for an accuracy of at least ±5% for the energy range from threshold to 30 MeV. So far the reaction $^{90}Zr(n,2n)^{89}Zr$, which is often used in spectrum unfolding techniques, and has been used several times as a reference for the determination of other cross sections, has been classified as a category-II reaction in reactor dosimetry,[2] in particular due to an inconsistency with measured spectrum-averaged data.[2,3] The reaction $^{58}Ni(n,2n)^{57}Ni$ has been classified as a category-I reaction; but above 15 MeV the knowledge of the excitation function is not satisfactory, there are differences >30% between the results of different authors, questioning the reliability of existing evaluations.[4,5] Therefore an effort was undertaken to precisely remeasure the excitation functions of the above reactions up to 20 MeV.

Experimental procedure

Two series of neutron irradiations were carried out using the $T(d,n)^4He$ reaction to produce the neutrons. In the energy range 13.4 - 14.8 MeV irradiations were performed at the Cockcroft-Walton-type neutron generator at the Institut für Radiumforschung und Kernphysik (IRK), Vienna; the energy of the incident deuterons (analysed d^+ beam) was 225 ± 5 keV. In the range 12.3 - 19.5 MeV irradiations were conducted using the 7 MV Van de Graaff accelerator at the Central Bureau for Nuclear Measurements (CBNM), Geel, employing a deuteron beam energy of 3.179 MeV which resulted in an average effective deuteron energy of 3.000 ± 0.015 MeV. Air-jet cooled solid state Ti-T targets (IRK) and Y-T targets (CBNM) were used in a low-mass target construction. For irradiation rectangular-shaped (16 x 8 x 1 mm^3) Zr and Ni samples (natural isotopic composition, purity > 99.8% and > 99.98%) were fastened to the inner side of a 1 mm thick aluminium ring (Ø 20 cm) with the target at the center (Fig. 1). They were arranged in steps of 6° in pairs positioned symmetrically at angles θ and - θ to cancel angular adjustment uncertainties. During the irradiations at IRK each sample was packed together with Al sandwich foils to measure the neutron fluence using the well known cross sections for the reaction $^{27}Al(n,\alpha)^{24}Na$ as evaluated by Tagesen and Vonach.[6] During the irradiation at CBNM the 0° neutron fluence was measured by means of a proton-recoil telescope with an uncertainty of ± 2.5%. Special attention was paid to a reliable energy scale. For the IRK irradiations in the 14 MeV region the procedure to establish the energy scale [7] was checked in the course of an international comparison [8] using the ratio of the specific activities in Zr and Nb samples. [9] As at an incident deuteron energy of ∿ 3 MeV (CBNM) an uncertainty in the source position in the direction of the ion beam seriously affects the average neutron energy, especially at 90°, activity ratios $^{89}Zr/^{24}Na$ induced in samples and attached monitor foils were used to transfer the energy scale from the IRK to the CBNM facility to establish the irradiation geometry precisely.

The ^{24}Na, ^{89g}Zr and ^{57}Ni γ-ray activities of the irradiated samples and monitor foils were measured absolutely at IRK with a 12.7 x 12.7 cm NaI(Tl) well-type detector. [10,11] The integral counting efficiency (discrimination level 22.1 keV) of this detector was known with an accuracy of 0.5% for monoenergetic γ-radiation and thin sources. The entire procedure and the code used for 4πγ-detector efficiency calculations for complex decay schemes is described by Winkler and

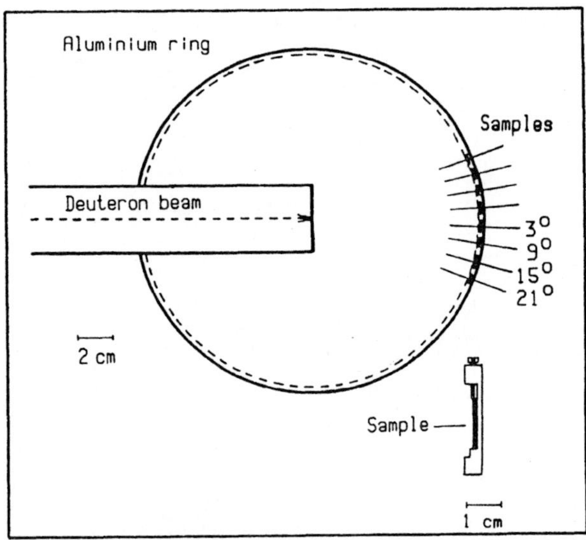

Fig. 1. Irradiation geometry and cross section of sample-holder ring with sample

Table I The principal sources of uncertainty in the $^{90}Zr(n,2n)^{89}Zr$ and $^{58}Ni(n,2n)^{57}Ni$ cross sections

	Source of error	Resulting uncertainty (%)	
		$^{90}Zr(n,2n)$	$^{58}Ni(n,2n)$
Proton-recoil telescope as neutron fluence monitor	Counting statistics	0.2 - 4.0	2.8 - 36.6
	Irradiation geometry for the samples	0.1 - 0.4	0.1 - 0.4
	Sample mass including purity	0.1	0.1
	Isotope content	0.2	0.2
	Detector efficiency	0.6	0.6
	Half-life of reaction product	< 0.01	0.3 - 0.4
	Half-life of interfering ^{57}Co	--	< 0.01
	Half-life of interfering ^{58g}Co	--	0.01 - 0.1
	Irradiation time, dead time	0.04	0.04
	Differential $T(d,n)^4He$ cross section	2.0	2.0
	Fluence determination with proton-recoil telescope	2.5	2.5
	Correction for absorption and scattering by sample-holder ring and samples between source and telescope	0.35	0.35
	Correction for neutron absorption in target backing (forward angles)	≤ 1.2	≤ 6.0
	Correction for neutron absorption in beam line (backward angles)	0.1 - 0.9	0.2
	Correction for neutron scattering effects due to target backing	<0.1 - 1.2	<0.1 - 0.2
	Correction for absorption, scattering and backscattering by samples and sample-holder ring	0.1	0.2
	Correction for relating the measured cross section to the average neutron energy	0.1 - 0.4	0.1 - 0.2
Relative to $^{27}Al(n,\alpha)^{24}Na$ cross sections	Counting statistics	0.3 - 0.6	0.8 - 2.7[a]
	Masses for sample and reference including purity	0.2	0.2
	Isotope content	0.2	0.2
	Detector efficiency for sample radiation	0.6	0.6
	Detector efficiency for ^{24}Na radiation	0.3	0.3
	Half-life of reaction product	0.15	0.2
	Half-life of interfering ^{57}Co	--	0.01
	Half-life of interfering ^{58g}Co	--	0.1
	Half-life of ^{24}Na	< 0.1	< 0.1
	Flux gradient in sample (experimentally determined)	~ 0.1	~ 0.1
	Correction for (elastic) neutron scattering by samples and sample-holder	< 0.1	< 0.1
	Correction for neutron scattering effects due to target backing	< 0.1	< 0.1
	Room return background	0.1	0.1
	Correction for relating the measured cross section to the average neutron energy	0.1	0.1
	Reference cross section for $^{27}Al(n,\alpha)^{24}Na$	0.5 - 1.3	0.5 - 1.3

a) 7.4% for irradiation at CBNM

Table II Measured $^{90}Zr(n,2n)^{89}Zr$ cross sections

	Average energy (MeV)	Energy resolution (MeV)	Total cross section (mb)
Proton-recoil telescope (Irradiation at CBNM)	12.335 ± 0.015	0.065	26.3 ± 1.4
	12.530 ± 0.020	0.080	75.0 ± 3.1
	13.015 ± 0.025	0.110	253.1 ± 8.8
	13.615 ± 0.030	0.130	472 ± 16
	14.310 ± 0.030	0.150	693 ± 23
	15.080 ± 0.035	0.160	874 ± 30
	15.895 ± 0.040	0.165	993 ± 35
	16.720 ± 0.040	0.170	1086 ± 36
	17.510 ± 0.035	0.170	1137 ± 38
	18.230 ± 0.025	0.170	1166 ± 39
	18.825 ± 0.030	0.175	1197 ± 40
	19.265 ± 0.025	0.215	1207 ± 40
	19.525 ± 0.020	0.225	1219 ± 41
Relative $^{27}Al(n,\alpha)^{24}Na$ cross sections (Irradiation at IRK)	13.435 ± 0.015	0.160	406.7 ± 6.2
	13.475 ± 0.015	0.150	416.2 ± 5.0
	13.515 ± 0.010	0.140	427.8 ± 4.8
	13.620 ± 0.010	0.120	468.1 ± 4.6
	13.745 ± 0.010	0.100	511.1 ± 5.2
	13.815 ± 0.010	0.090	536.2 ± 5.3
	13.885 ± 0.010	0.080	559.0 ± 5.4
	14.035 ± 0.010	0.080	600.9 ± 5.9
	14.190 ± 0.010	0.100	647.5 ± 6.5
	14.265 ± 0.010	0.110	674.7 ± 6.9
	14.345 ± 0.010	0.130	693.4 ± 6.9
	14.480 ± 0.015	0.170	734.3 ± 7.3
	14.610 ± 0.015	0.200	763.8 ± 7.6
	14.660 ± 0.015	0.210	775.0 ± 7.6
	14.710 ± 0.015	0.220	784.6 ± 7.5
	14.780 ± 0.015	0.240	805.1 ± 8.1
	14.830 ± 0.015	0.270	814.9 ± 9.1

Table III Measured $^{58}Ni(n,2n)^{57}Ni$ cross sections

	Average energy (MeV)	Energy resolution (MeV)	Total cross section (mb)
Proton-recoil telescope (Irradiation at CBNM)	13.305 ± 0.025	0.115	9.06 ± 3.33
	13.955 ± 0.030	0.130	21.43 ± 3.00
	14.690 ± 0.030	0.150	38.06 ± 3.08
	15.485 ± 0.035	0.160	51.92 ± 4.60
	16.305 ± 0.040	0.165	58.58 ± 3.09
	17.120 ± 0.040	0.170	65.41 ± 2.90
	17.880 ± 0.035	0.170	69.23 ± 2.97
	18.540 ± 0.030	0.175	72.16 ± 3.01
	19.065 ± 0.025	0.205	75.70 ± 3.16
	19.415 ± 0.025	0.220	78.70 ± 3.34
	19.570 ± 0.020	0.225	79.25 ± 3.42
Relative $^{27}Al(n,\alpha)^{24}Na$ cross sections (Irradiation at IRK)	13.470 ± 0.015	0.160	11.24 ± 0.33
	13.565 ± 0.015	0.130	12.98 ± 0.40
	13.620 ± 0.010	0.120	14.20 ± 0.36
	13.680 ± 0.010	0.110	15.08 ± 0.40
	13.815 ± 0.010	0.090	18.45 ± 0.36
	13.960 ± 0.010	0.080	21.37 ± 0.43
	14.035 ± 0.010	0.080	23.41 ± 0.46
	14.115 ± 0.010	0.090	24.77 ± 0.35
	14.265 ± 0.010	0.110	28.78 ± 0.59
	14.415 ± 0.010	0.150	31.47 ± 0.56
	14.480 ± 0.015	0.170	33.24 ± 0.49
	14.550 ± 0.015	0.180	34.84 ± 0.67
	14.660 ± 0.015	0.210	36.62 ± 0.47
	14.690 ± 0.030 a)	0.150	37.65 ± 2.82
	14.750 ± 0.015	0.230	39.05 ± 0.62
	14.780 ± 0.015	0.240	39.63 ± 0.52
	14.810 ± 0.015	0.250	40.01 ± 0.65
	14.830 ± 0.015	0.270	40.55 ± 0.54

a) Irradiation at CBNM

Pavlik,[11] considering all relevant error contributions and correlations amongst decay data and partial efficiencies for γ- and electron-radiation. A correction for efficiency increase due to Compton scattering in thicker samples was accounted for. Decay schemes and half-lives were taken from Ref. 12 or 13, if available therein; the half-life value 14.98 ± 0.02 h was used for ^{24}Na.[7] The gamma spectra of the irradiated samples were checked for purity by means of a Ge(Li) detector. Longer-lived interfering activities (^{90}Y, ^{95}Zr, ^{95}Nb and ^{57}Co, ^{58g}Co, ^{60}Co in the case of ^{89}Zr and ^{57}Ni, respectively) were corrected for. For more experimental details see Ref. 7.

Data reduction and results

Table I summarises the principal sources of uncertainties taken into account and provides a survey of the relevant procedures and corrections during the reduction of the experimental data (for details see Ref. 7). So far as available energy- and angle-dependent cross sections were used for absorption- and scattering corrections; in particular neutrons inscattered by the target backing and shifted in energy were considered. The final results are listed in Tables II and III. The neutron energy distribution profile was calculated for every experimental point to determine the energy spread and average energy. As energy resolution ± 1/2 (FWHM)-values are given in the tables. The uncertainty in the average energy was estimated on the basis of uncertainties in the effective deuteron energy, the Q-value of the T(d,n) reaction, stopping power data, target layer thicknesses, tritium loading ratio and angular position of the sample [7]. The specification of the total cross section uncertainty does not include the uncertainty of the average energy. As only the activity of ^{89g}Zr was measured the experimentally determined quantities $\sigma_g + 0.938\ \sigma_m$ (see Ref. 12) were converted to the total (n,2n) cross sections in the tables using isomeric ratios from the literature (see Ref. 7). The correction amounted to 0.4 - 1.2%. The uncertainty components for each point (standard or effective standard deviations) used to give the total uncertainties are available in the form of error matrices from the authors (for ^{90}Zr(n,2n) see Ref. 7) distinguishing between uncertainties uncorrelated or correlated over the energy range of the measurements.

Discussion

The results of this work for $^{90}Zr(n,2n)^{89}Zr$ are compared with the results of the evaluation by Tagesen et al. [14] in Fig. 2. This evaluation was updated [7] including the new results according to the procedure described in Ref. 14, chapter II, but restricting the input data to purely experimental values. The results of the new evaluation which demonstrate the significant improvement of the data base are shown in Table IV and Fig. 2 (For the correlation matrix of the group cross sections see Ref. 7).

The $^{58}Ni(n,2n)^{57}Ni$ cross sections were reevaluated in the same manner [14] including the new results and critically reviewing the available data from the literature, renormalizing if necessary, to account for adjustments in standard cross sections and decay schemes. The results are given in Table V and Fig. 3 together with the new experimental data, the ENDF/B-V evaluation [4], and results from a model calculation.[15] The values from ENDF/B-V are too low above 15 MeV. Details of the new $^{58}Ni(n,2n)^{57}Ni$ evaluation (including the correlation matrix for the group cross sections) will be published elsewhere.

Spectrum-averaged cross sections $\langle\sigma\rangle$ for the ^{252}Cf spontaneous-fission neutron field and the ^{235}U fission spectrum were calculated using the differential cross section data from the new evaluations (Tables IV and V) and compared with integral data (Table VI). Spectra parameters were taken from Ref.16. The uncertainties of the calculated $\langle\sigma\rangle$ data originate from the errors of the evaluated cross sections only, considering correlations. Obviously measurements of the integral response with a series of high-threshold reactions for which differential cross section data were carefully determined may serve to better define the high-energy part of the technologically important fission spectra.

Acknowledgment

The authors gratefully acknowledge the assistance of Dr. S. Tagesen in using his computer program for the cross section evaluations.

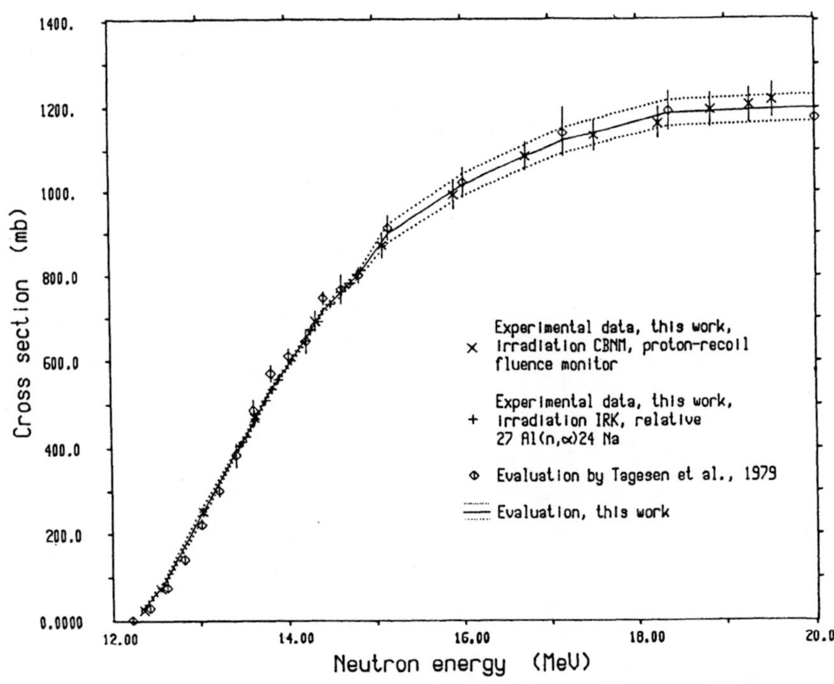

Fig. 2. Excitation function for the reaction $^{90}Zr(n,2n)^{89}Zr$

Table IV

Evaluated group cross sections for the reaction $^{90}Zr(n,2n)^{89}Zr$

Energy range (MeV)	Cross section (mb)
12.20 - 12.40	26.3 ± 4.0
12.40 - 12.70	82.3 ± 8.8
12.70 - 13.30	247.6 ± 12.5
13.30 - 13.50	392.0 ± 6.8
13.50 - 13.70	461.4 ± 8.8
13.70 - 13.90	535.5 ± 7.0
13.90 - 14.10	591.3 ± 6.4
14.10 - 14.30	652.2 ± 8.9
14.30 - 14.50	717.8 ± 6.8
14.50 - 14.70	763.8 ± 10.4
14.70 - 14.90	805.0 ± 8.0
14.90 - 15.40	900.6 ± 21.6
15.40 - 16.60	1016.6 ± 25.8
16.60 - 17.70	1124.9 ± 31.6
17.70 - 19.00	1189.3 ± 29.2
19.00 - 21.00	1200.1 ± 30.2

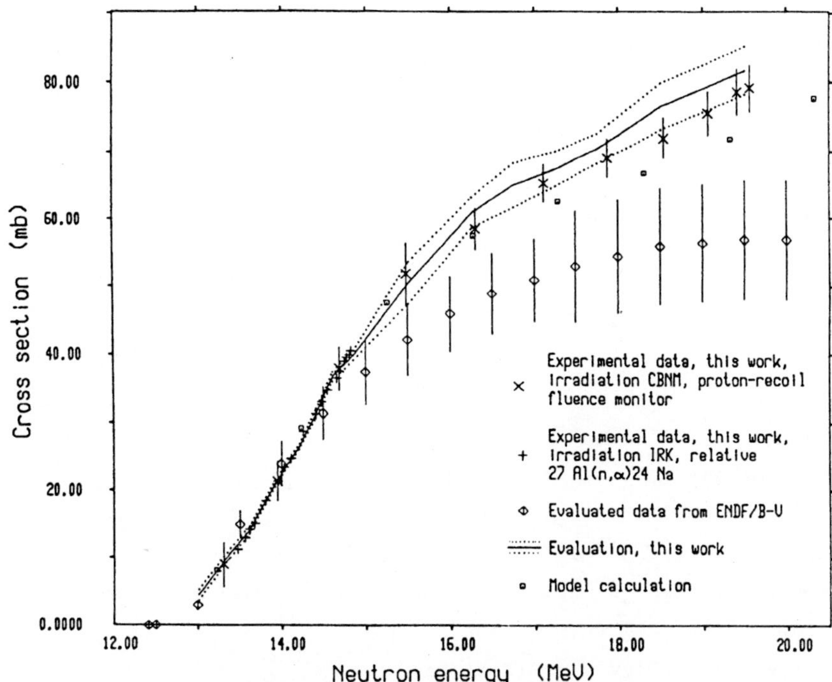

Fig. 3. Excitation function for the reaction $^{58}Ni(n,2n)^{57}Ni$

Table V

Evaluated group cross sections for the reaction $^{58}Ni(n,2n)^{57}Ni$

Energy range (MeV)	Cross section (mb)
12.75 - 13.25	4.54 ± 0.80
13.25 - 13.50	10.78 ± 0.69
13.50 - 13.70	14.12 ± 0.58
13.70 - 13.90	18.59 ± 0.54
13.90 - 14.10	22.88 ± 0.55
14.10 - 14.30	26.59 ± 0.45
14.30 - 14.50	31.49 ± 0.72
14.50 - 14.70	36.86 ± 0.66
14.70 - 15.00	39.94 ± 0.67
15.00 - 16.00	50.63 ± 3.10
16.00 - 16.50	60.99 ± 2.27
16.50 - 17.00	65.17 ± 3.33
17.00 - 17.50	67.60 ± 2.64
17.50 - 18.00	70.64 ± 2.19
18.00 - 19.00	76.68 ± 3.31
19.00 - 20.00	81.79 ± 3.42

Table VI Comparison of calculated and experimental spectrum-averaged cross sections <σ> for the ^{252}Cf spontaneous-fission neutron field and the ^{235}U fission spectrum

	Spectrum	<σ> for $^{90}Zr(n,2n)^{89}Zr$ (mb)	<σ> for $^{58}Ni(n,2n)^{57}Ni$ (mb)
^{252}Cf	Maxwellian (<E> = 2.13 MeV) and segment correction	0.207 ± 0.004	0.00856 ± 0.00040
	Maxwellian (<E> = 2.13 MeV) without segment correction	0.268 ± 0.005 [a]	0.0113 ± 0.0003
	Experimental	0.267 ± 0.015 (Ref. 17) 0.221 ± 0.006 (Ref. 18)	0.0101 ± 0.0012 (Ref. 18)
^{235}U	Maxwellian (<E> = 1.97 MeV) and segment correction	0.083 ± 0.002	0.00330 ± 0.00040
	Maxwellian (<E> = 1.97 MeV) without segment correction	0.131 ± 0.003 [b]	0.00532 ± 0.00025 [b]
	Experimental	0.247 ± 0.017 (Ref. 3)	0.00577 ± 0.00031 (Ref. 3) 0.0036 ± 0.0002 (Ref. 19)

[a] Variation of the average energy by ± 0.02 MeV changes the spectrum-averaged cross section by about ± 8.5%
[b] Result of an evaluation

References

1. N. Dayday, ed., WRENDA 81/82 World Request List for Nuclear Data, INDC(SEC)-78/URSF, IAEA Nuclear Data Section (Vienna 1981)
2. M.F. Vlasov, A. Fabry, W.N. McElroy, Status of Neutron Cross Sections for Reactor Dosimetry, INDC(NDS)-84/L+M, IAEA Nuclear Data Section (Vienna 1977)
3. A. Fabry et al., in "Neutron Cross Sections for Reactor Dosimetry", ed. M.F. Vlasov, IAEA-208, vol. 1, p. 233, IAEA Nuclear Data Section (Vienna 1978)
4. ENDF/B-V Dosimetry Files, Mod. 2, National Nuclear Data Center, Brookhaven National Laboratory (1981)
5. L. Adamski, M. Herman, A. Marcinkowski, INDC(POL)-8/L, IAEA Nuclear Data Section (Vienna 1977)
6. S. Tagesen, H. Vonach, Physics Data 13-3, Fachinformationszentrum Karlsruhe (1981)
7. A. Pavlik, G. Winkler, H. Vonach, A. Paulsen, H. Liskien, J. Phys.G: Nucl. Phys. (in print)
8. V.E. Lewis, National Physical Laboratory Teddington, private communication (1982)
9. V.E. Lewis, K.J. Zieba, Nucl. Instr. 174 (1980) 141
10. W. Mannhart, H. Vonach, Nucl. Instr. 136 (1976) 109
11. G. Winkler, A. Pavlik, Int. J. Appl. Radiat. Isotopes (to be published)
12. C.M. Lederer, V.S. Shirley, Table of Isotopes 7th edn. (Wiley-Interscience, New York 1978)
13. Nuclear Standards File INDC-36/LN, IAEA Nuclear Data Section, Vienna 1981
14. S. Tagesen, H. Vonach, B. Strohmaier, Physics Data 13-1, Fachinformationszentrum Karlsruhe (1979)
15. B. Strohmaier, M. Uhl, this conference
16. J. Grundl, C. Eisenhauer, ibid. Ref. 3, p.53
17. Z. Dezsö, J. Csikai, Proc. 4th All Union Conf. on Neutron Physics, Kiev, 22-26 April 1977, vol. 3, p. 32
18. W. Mannhart, Proc. 4th ASTM-EURATOM Symp. on Reactor Dosimetry, 22-26 March 1982, US Nat. Bureau of Standards
19. K. Kobayashi et al., NEANDC(J) 67U, 42, cited in D. Smith, NEANDC 150-U, p. 113 (1981)

MEASUREMENT OF THE CROSS SECTION RATIOS FOR THE REACTIONS $^{90}Zr(n,2n)^{89m+g}Zr$ by $^{93}Nb(n,2n)^{92m}Nb$, $^{63}Cu(n,2n)^{62}Cu$ by $^{27}Al(n,p)^{27}Mg$ and $^{27}Al(n,\alpha)^{24}Na$ by $^{27}Al(n,p)^{27}Mg$ FOR THE PURPOSE OF NEUTRON SPECTROMETRY AROUND 14 MeV

A. CHIADLI, A. AIT HADDOU AND M. VIENNOT
Laboratoire de Physique Nucléaire, Faculté des Sciences Rabat-Morocco.

G. PAIĆ
Rudjer Bošković Institute, POB 1016, 41001 Zagreb, Yugoslavia

The cross sections ratios for the reactions $^{90}Zr(n,2n)^{89m+g}Zr$ by $^{93}Nb(n,2n)^{92m}Nb$, $^{63}Cu(n,2n)^{62}Cu$ by $^{27}Al(n,p)^{27}Mg$ and $^{27}Al(n,\alpha)^{24}Na$ by $^{27}Al(n,p)^{27}Mg$ have been measured by activation method for the purpose of neutron spectrometry around 14 MeV. A good agreement with calculated neutrons energy is obtained in the case of $^{90}Zr(n,2n)^{89m+g}Zr$ by $^{93}Nb(n,2n)^{92m}Nb$ ratio while the $^{27}Al(n,\alpha)^{24}Na$ by $^{27}Al(n,p)^{27}Mg$ cross sections ratio has been found to be sensitive to the detector-sample distance.
|NUCLEAR REACTIONS, $^{90}Zr(n,2n)^{89m+g}Zr$, $^{93}Nb(n,2n)^{92m}Nb$, $^{63}Cu(n,2n)^{62}Cu$, $^{27}Al(n,p)^{27}Mg$, $^{27}Al(n,\alpha)$ measured σ ratios by activation;

There are several methods to determine the energy of neutrons. Usually the methods are rather complicated and involve sophisticated equipments.
Therefore one is still searching for a simple spectrometric method applicable in laboratories working on measurements of total cross section by activation method.
Lately, such methods were proposed[1,2].
They are based on the fact that if one measures the ratio of two cross sections in function of energy and if that ratio is a steep function of the neutron energy one may determine the neutron energy with considerable precision.
Two pairs of reactions have been proposed so far: the $^{93}Nb(n,2n)^{92m}Nb$, $^{90}Zr(n,2n)^{89m+g}Zr$, and the $^{63}Cu(n,2n)^{62}Cu$, $^{27}Al(n,p)^{27}Mg$ reactions.
We have tested the two pairs in our laboratory where we have made very careful calculations of the energy emitted by our neutron generator[3].
To determine the neutrons energy we have used the existing evaluation for the $^{90}Zr(n,2n)^{89m+g}Zr$ reaction[4], the data of Nethaway[5] for the $^{93}Nb(n,2n)^{92m}Nb$, the $^{27}Al(n,\alpha)$ data evaluated by Vonach[6] and the $^{27}Al(n,p)$ evaluation[7].

Experiment

The irradiation was done at angles 0 and 105 degrees with respect to the deuterons beam of the $T(d,n)\alpha$ neutron generator at an accelerating voltage of 300 KV. The samples were thin foils of 1 cm diameter and placed at 11 cm from the center of the Ti-T target.
The irradiated samples were measured with a Ge(Li) detector in conjunction with a 4098 channels analyzer.

Calculation of the neutron energy

The neutron energy was calculated allowing for the following considerations:

i) The angle of emission of the neutron.

ii) The existence of two groups of neutron energies due to the presence of atomic and molecular ions in the deuterons beam that is not analyzed.

iii) The energy spread due to the profile of the tritium in the titanium layer. We used a target of nominal titanium thickness of 200 $\mu g/cm^2$ and our measurements[3] show that the bulk of tritium is concentrated in a layer starting at 60 $\mu g/cm^2$ from the air-titanium interface and ending at 30 $\mu g/cm^2$ from the copper-titanium interface.

K. H. Böckhoff (ed.), Nuclear Data for Science and Technology, 404–405.

iv) The energy spread due to the final angular apertures spanned by the beam spot and samples dimensions.

Results

Ratio $^{63}Cu(n,2n)/^{27}Al(n,p)$

We have tried to determine the energy by the above ratio proposed by Jarjis[2)] but the results were neither consistent nor reproducible. The reason for that is apparently due to the large sensitivity of the annihilation process of β^+ used to determine $^{63}Cu(n,2n)$ cross section on the spatial arrangements of masses during the measurements. We therefore do not think this ratio to be a practical method.

Ratio $^{90}Zr(n,2n)^{89m+g}Zr/^{93}Nb(n,2n)^{92m}Nb$

The ratio of the $^{90}Zr(n,2n)/^{93}Nb(n,2n)$ cross sections yields a value of 1,88 $\pm$ 0,07 at 0 degree and 1,03 $\pm$ 0,08 at 105°. If one uses the cross section data of references 4) and 5) one obtains that these values correspond to neutron energies of 14,81 $\pm$ 0,09 MeV and 13,99 $\pm$ 0,01 MeV respectively.

Our calculation gives at the same angles, energies of 13,83 MeV and 14,93 MeV for an accelerating voltage of 300 kV.

We observe a very satisfactory agreement indicating that the existing cross section data or at least their ratio is sufficiently well determined to be used for spectrometry uses. It would be very useful if one quoted together with cross section data around 14 MeV the value of the ratio for these two reactions.

Ratio $^{27}Al(n,\alpha)/^{27}Al(n,p)$

The ratio $^{27}Al(n,\alpha)/^{27}Al(n,p)$ does not change very rapidly be we wanted to check our data with the existing evaluation[6,7)] giving $\sigma(n,\alpha)$ = 112,3 mb and $\sigma(n,p)$ = 71,4 mb at 14,83 MeV.

Our measurement done placing the sample at 15 cm from the window of the Ge(Li) detector, to avoid the influence of the coincidence due to the gamma cascade in ^{24}Na, yields a ratio of 1,75 $\pm$ 0,06. This ratio is in clear disagreement with the value 1,57 $\pm$ 0,03 extracted from the references 6) and 7). Presently we do not have an explanation for this disagreement with literature data but it is interesting to note that when placing the irradiated Al field against the window of the Ge(Li) counter the ratio obtained was 1,61 $\pm$ 0,06 in agreement with the ratio of the evaluation values.

The decrease in the ratio is due to the true coincidences in the detector for the ^{24}Na gammas.

It was not possible to find data on the sample-detector distance in older measurements but possibly the very important monitor reaction $^{27}Al(n,\alpha)$ should be reevaluated striking out measurements done close to the detector.

References

1) V.E. Lewis and K. Zieba, Nucl. Instr. and Methods 174 (1980) 141.

2) R.A. Jarjis, Nucl. Instr. and Methods 184 (1981) 439.

3) A. Chiadli, M.Sc. Thesis, Faculté des Sciences, Rabat Morocco 1982.

4) S. Tagesen and H. Vonach in Neutron Physics and Nuclear Data Proceeding of an International Conference, Harwell 1978 (O.E.C.D. Nuclear Energy Agency 1978).

5) D.R. Nethaway, J. Inorg. Chem. 40 (1978) 1285.

6) H. Vonach, INDC/NEANDC, Nucl. Standard File 1980 version. IAEA NUCLEAR DATA SECTION, May 1981.

7) P.G. Young, D.G. Forster, J.R. Dist ENDF/B Material no 6313. P.C to the normalisation and standards subcommittee in Feb 1979.

EXCITATION FUNCTIONS OF (N,P) REACTIONS IN THE REGION 13.75 TO 15 MeV FOR Ti, Fe AND Ni ISOTOPES

M. VIENNOT, A. AIT HADDOU, A. CHIADLI
Laboratoire de Physique Nucléaire, Faculté des Sciences
Rabat - Morocco

G. PAIĆ
Rudjer Bošković Institute, POB 1016, 41001 Zagreb, Yugoslavia

Cross sections for (n,p) reactions induced by neutrons from 13.75 to 15 MeV on $^{46,48,50}Ti$, $^{54,56,57}Fe$, $^{58,60,61,62}Ni$ have been measured by the activation technique, relative to the $^{27}Al(n,\alpha)$ and $^{27}Al(n,p)$ excitation functions.

The samples are made of natural pure powder for the elements studied, mixed Al_2O_3, and shaped as pellets, sealed in plastic foils. They are placed at several angles (from 0 to 120 degrees) at a distance of 11 cm from the center of the target.

The gamma-rays are detected using a Ge(Li) detector, the samples in contact with window of the detector.

The following corrections are applied: self absorption, coincidence losses (gamma-rays cascade coincidence) and (n,np) interfering reactions.

The measurement of energy trends of cross sections separated isotopes around 14 MeV is only seldom done because of the reduction in flux one has to put with to achieve a reasonable energy resolution of the inconing neutrons.

To achieve a reasonable activity one has to use important quantities for the measured sample what in turn makes the use of the separated isotopes prohibitively expensive.

We have therefore measured the excitation function for the (n,p) reactions on Ti, Fe, and Ni of natural composition in the range of energy available on a d-T neutron generator with a high voltage of 400 kV. In cases of adjacent isotopes the interferences of the (n,np) reaction was taken into account using literature values[1)] for the corrections.

The correction was attempted only when the contribution of the interfering reaction did not represent more than 25% of the measured yield. For this reason we were not able to extract the cross section for $^{47}Ti(n,p)$.

We have carefully established the energy of the neutrons at the angles measured[2)] taking into account the thickness of the Ti-T target the excitation function of the d+T reaction of the D^+/D_2^+ ratio in our unanalyzed deuterons beam.

The measurements have been carried at angles of 0, 30, 60, 75, 90, 105, 120 degrees.

The samples were placed at a distance of 11 cm from the center of the target. The energy quoted in the results correspond to a weighted average calculated taking into account the contribution of the D^+ and D_2^+ ions and assuming that the excitation function of the reaction studied may be approximated by a linear function inside the interval of the neutron energies considered at one angle.

The gamma rays were detected using a 60 cm^3 Ge(Li) detector shielded with lead.

The counting was performed with the irradiated samples in contact with the window of the Ge(Li) detector. In such a geometry one may have large distortion due to gamma-cascades since the probability of a true coincidence occuring is not negligible. Increasing the distance makes counting statistics poor in most cases.

We therefore chose to measure the importance of the corrections measuring for one sample the ratio in the cross section obtained relative to the monitor reaction (Al(n,p) or Al(n,α)) on the window of the detector and when measuring at 5 cm from the detector. This correction was then applied to the measurements done at other energies E_n.

We have measured the correction described above for each isotope and each reaction measured. This is necessary since several factors may influence the correction.

i) In case of β emitters there is possibility that γ rays also cause true coincidences if their energy

K. H. Böckhoff (ed.), Nuclear Data for Science and Technology, 406–408.

is sufficiently high to enter the detector.

ii) The number of γ rays in cascade varies from isotope to isotope.

iii) The angular correlation between γ rays is function of the spin and parity of levels involved.

Corrections for losses of radioactive nuclei by ejection of recoil particles were not necessary since the samples were sealed in plastic foils.

Corrections for self absorption in the sample were made. The targets were of the mixed powder type. Al_2O_3 powder was mixed with the element studied and pressed with a hydraulic press to obtain pellets of ⌀ = 16 mm and thickness 3 mm.

The composition of the targets and their masses are shown in table I.

The irradiation time was chosen in function of the decay time of the product measured but in no cases exceeded 2 hours. The decay parameters used in the calculation of the cross sections are shown in table II[5)].

To extract the (n,p) cross section in cases when the (n,np) reaction on the adjacent isotope could interfere, we subtracted the literature value[1)] using a single value for the whole energy interval. If limited to cases when the interference is small it should not introduce errors larger than 2% in the cross section. The result obtained for the cross sections at different energies are shown in table IV.

Table I. Characteristics of the targets

Target	Composition	Masses
$^{46,47,50}Ti$	pure	3 g
$^{56,57}Fe$	pure	2 g
^{54}Fe	Fe_2O_3	10 g
$^{58,61}Ni$	pure	3 g
$^{60,62}Ni$	pure	5 g
^{27}Al	Al_2O_3	

Table II : (n,p) nuclear reactions parameters.

A(n,p)B	$T_{1/2}$	a	E(Kev)	Iγ
$^{46}Ti/^{46}Sc$	83,80d	8%	889,25	100%
$^{47}Ti/^{47}Sc$	3,422d	7,4%	159,4	68,5%
$^{48}Ti/^{48}Sc$	43,67 h	73,7%	983,5	100%
$^{50}Ti/^{50}Sc$	1,7 m	5,2%	523,5	86%
$^{54}/^{54}Mn$	312,2d	5,8%	835	100%
$^{56}Fe/^{56}Mn$	2,58 h	91,8%	1811	27,19%
$^{57}Fe/^{57}Mn$	1,54m	2,15%	122	10,3%
$^{58}Ni/^{58m+g}Co$	70,8 d	67,76%	811	99,5%
$^{60}Ni/^{60m}Co$	10,47m	26,1%	56,8	2%
$^{61}Ni/^{61}Co$	1,6h	1,16%	67,5	86%
$^{62}Ni/^{62m}Co$	13,91m	3,59%	1163,5	69,0%
$^{62}Ni/^{62g}Co$	1,5m	3,59%	1129	13,1%

Table III: Cross section for monitor reactions

a) $^{27}Al(n,p)^{27}Mg$ b) $^{27}Al(n,\alpha)^{24}Na$

E_n(MeV)	σ_a (6)	σ_b (7)
13,77	79,8	122,725 ± 0,6
13,93	78	123,020 ± 0,5
14,11	76,8	121,907 ± 0,5
14,30	75,4	119,970 ± 0,5
14,47	73,8	116,046 ± 0,6
14,73	72	113,100 ± 0,4
14,83	71,2	112,307 ± 0,6

Table IV: Cross sections in millibarns at different neutron energies

Nuclear reactions	E_n :	13,77	13,93	14,11	14,30	14,47	14,73	14,83
$^{46}Ti(n,p)^{46}Sc$		310 ± 33	297 ± 38	310 ± 33	306 ± 37	242 ± 33	306 ± 34	275 ± 39
$^{48}Ti(n,p)^{48}Sc$		51 ± 3	53 ± 3	55 ± 3	53 ± 3	55 ± 3	61 ± 3	56 ± 3
$^{50}Ti(n,p)^{50}Sc$		8,4 ± 0,7	8,6 ± 0,8	10,3 ± 0,9	9,8 ± 0,9	9,8 ± 0,9	11,7 ± 1,0	12,6 ± 1,1
$^{54}Fe(n,p)^{54}Mn$		411 ± 26		433 ± 26		366 ± 24	346 ± 22	314 ± 20
$^{56}Fe(n,p)^{56}Mn$		109 ± 7	127 ± 8	109 ± 7	119 ± 7	109 ± 7	104 ± 6	105 ± 6
$^{57}Fe(n,p)^{57}Mn$		96 ± 8	96 ± 8	90 ± 8	93 ± 8	83 ± 7	89 ± 8	92 ± 8
$^{58}Ni(n,p)^{58m+g}Co$		389 ± 14		363 ± 13	332 ± 12	308 ± 11	276 ± 10	269 ± 9
$^{60}Ni(n,p)^{60m}Co$		40 ± 4		33 ± 3		39 ± 4	29 ± 3	28 ± 3
$^{61}Ni(n,p)^{61}Co$		27 ± 5	32 ± 6	44 ± 8	54 ± 10	58 ± 10	84 ± 15	93 ± 15
$^{62}Ni(n,p)^{62m}Co$		14 ± 2		14 ± 2		17 ± 2	18 ± 2	19 ± 2
$^{62}Ni(n,p)^{62g}Co$		20 ± 2		17 ± 2		20 ± 2	16 ± 2	22 ± 2

References

1) S.M. Qaim CRC Handbook Spectroscopy Vol. III (1981)

2) A. Chiadli Thèse de Diplôme d'Etudes Supérieures de 3ième cycle, Faculty of Sciences, Rabat, Morocco 1982.

3) W.W. Osterhage Journal of Radioanalytical Chemistry 56 (1980) 277.

4) S.M. Qaim and al Proc. 9th Symposium Fus. Technology, Garmisch FRG (1976).

5) C.M. Ledered, V.S. Shirley Table of Isotopes Seventh Edition (1978).

6) Evaluation I.A.E.A. 1979 Young, Foster, Dist.

7) S. Tagesen, H. Vonach to be published INDC/NEANDC 1980.

THE SPECTRUM-AVERAGE CROSS SECTION RATIO OF $^{63}Cu(n,\alpha)^{60}Co$ - TO - $^{27}Al(n,\alpha)^{24}Na$ IN A THICK-TARGET $^{9}Be(d,n)^{10}B$ NEUTRON SPECTRUM

H. Liskien, D.L. Smith*, and R. Widera

Central Bureau for Nuclear Measurements
Steenweg naar Retie, B - 2440 Geel, Belgium

A Cu-Al alloy sample has been irradiated in a potential 'benchmark' neutron field which was produced by bombarding a thick Be-target with 7 MeV deuterons. Due to the use of an alloy, and the close proximity of the involved γ-energies (^{24}Na : 1368 keV; ^{60}Co : 1333 keV), a very accurate value of 0.539 ± 0.009 was obtained for the spectrum average cross section ratio of $^{63}Cu(n,\alpha)^{60}Co$ - to - $^{27}Al(n,\alpha)^{24}Na$. This experimental value is compared with calculated values as derived from differential cross section evaluations and spectrum information, and it is found to support the recent $^{63}Cu(n,\alpha)^{60}Co$ evaluation of Winkler et al.[1].

[Reactor dosimetry, $^{63}Cu(n,\alpha)$, $^{27}Al(n,\alpha)$, cross section ratio, $^{9}Be(d,n)$ spectrum]

Introduction

$^{63}Cu(n,\alpha)$ leading to ^{60}Co with a half-life of 5.3 years is one of the rare threshold reactions which allow long-term fast-fluence integration. The cross section for this reaction appeared on discrepancy lists until recently when the work of Winkler et al. [1,2] apparently offered a solution to the discrepancy problem. On the other hand $^{27}Al(n,\alpha)^{24}Na$ is considered to be one of the best known reactions applied in dosimetry [3,4]. Ratio measurements of $^{63}Cu(n,\alpha)^{60}Co$-to-$^{27}Al(n,\alpha)^{24}Na$ can be performed with high accuracy if Cu-Al-alloy is used [5]. To see whether the evaluation of Winkler et al.[1] is superior to the ENDF/B-V evaluation[3] we have performed such a ratio measurement in a potential 'benchmark' neutron field [6] produced by bombarding a thick Be-target with 7 MeV deuterons.

Experimental Details

A Cu-Al (95 - 5 weight %) alloy sample of 2 cm diameter and 0.5 cm thickness was mounted in the zero-degree direction at 8 cm distance from the source and was irradiated for about T = 8 hours at a target current of 10 μA. This current was sufficiently stable to insure that no correction factor for the saturation term $(1 - e^{-\lambda_{Na} \cdot T})$ had to be applied. The use of an alloy and the similarity of the thresholds for the two involved reactions made it unnecessary to apply a correction for neutron absorption and scattering in the sample. The irradiation geometry and non-uniformity in the aluminium distribution within the alloy were responsible for some inhomogeneity in the activity distribution within the sample. Therefore, separate counting of the ^{24}Na and ^{60}Co activity with a calibrated Ge-Li detector was performed with first one side of the sample and then the other directed towards the Ge-Li crystal. The counting of ^{24}Na via the 1368 keV γ-line was started about 12 hours after the end of irradiation, and was continued for nearly five half lives. The decay was found to be consistent with a recently reported half life of (14.96 ± 0.01) h[7]. ^{60}Co activity, with a half-life of (1925 ± 1)d[7], was determined via the 1333 keV γ-line. Counting was started about one week later when the ^{24}Na activity had nearly disappeared. The close proximity in energy of the two lines essentially eliminates differences due to γ self-absorption and minimizes the error in the efficiency ratio. Corrections for losses in the full-energy peaks due to coincident γ-transitions were applied. Finally, we find for the cross section ratio :

$$\frac{\langle\sigma_{Cu}\rangle}{\langle\sigma_{Al}\rangle} = \frac{C_{Co}}{C_{Na}} \cdot \frac{\epsilon_{Na}}{\epsilon_{Co}} \cdot \frac{N_{Al}}{N_{Cu}} \cdot \frac{(1-e^{-\lambda_{Na} \cdot T})}{\lambda_{Cu} \cdot T} = 0.539 \pm 0.009$$

The quoted uncertainty contains the following contributions :
Count rates C at the end of irradiation (1.4%), detector efficiencies ϵ (0.5%), numbers of irradiated nuclei N (0.8%), time factors (0.2%).

Comparison with Calculated Values

This rather accurate ratio can be compared with values derived from evaluated cross sections and the Be(d,n) spectrum. We have calculated four such values as given in the Table I below.

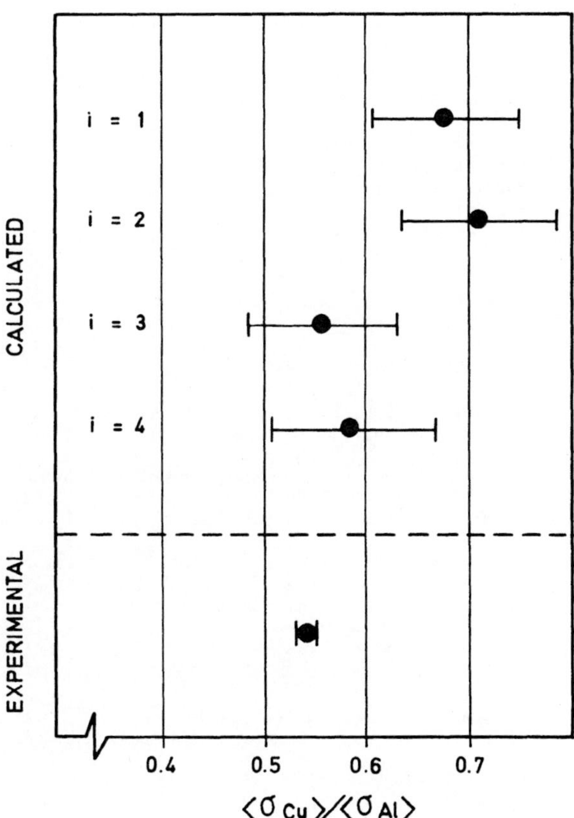

Fig. 1. Comparison of calculated and measured cross section ratios

*Permanent address: Applied Physics Division, Building 314,Argonne National Laboratory,Argonne,Ill.60439 USA

K. H. Böckhoff (ed.), Nuclear Data for Science and Technology, 409–410.

Table I   Calculated Results

i	$^{63}Cu(n,\alpha)^{60}Co$ evaluation	$^{27}Al(n,\alpha)^{24}Na$ evaluation	Ratio $<\sigma_{Cu}>/<\sigma_{Al}>$
1	ENDF/B-V a)	ENDF/B-V c)	0.677 ± 0.077
2	ENDF/B-V a)	Tagesen and Vonach d)	0.710 ± 0.079
3	Winkler et al. b)	ENDF/B-V c)	0.559 ± 0.076
4	Winkler et al. b)	Tagesen and Vonach d)	0.587 ± 0.079

a) Ref. 3 (MAT=6435, MF=3+33, MT=107)   b) Ref. 1
c) Ref. 3 (MAT=6313, MF=3+33, MT=107)   d) Ref. 4

The quoted uncertainties for the calculated ratios $<\sigma_{Cu}>/<\sigma_{Al}>$ and the quoted correlation matrix have been derived making full use of the covariance formalism. The details of these error calculations are documented in a separate report[8].

Table II   Correlation Matrix of Calculated Results

i = \ j =	1	2	3	4
1	1.00	0.89	0.83	0.72
2	0.89	1.00	0.75	0.83
3	0.83	0.75	1.00	0.92
4	0.72	0.83	0.92	1.00

Other evaluations for these reactions could not be considered due to the lack of complete uncertainty information. A comparison of the measured and calculated cross section ratios appears in Fig. 1. The present experiment clearly supports the recent $^{63}Cu(n,\alpha)^{60}Co$ evaluation of Winkler et al.[1], but it does not favor either $^{27}Al(n,\alpha)^{24}Na$ evaluation [3,4] over the other.

References

1. G. Winkler, D.L. Smith, and J.W. Meadows, Proc. International Conference on Nuclear Cross Sections for Technology, Knoxville, 1979, NBS Special Publication 594, page 199.

2. G. Winkler, D.L. Smith, and J.W. Meadows, Nucl. Sci. Eng. 76, 30 (1980).

3. "Evaluated Nuclear Data File - ENDF/B-V", documented in Report ENDF-102, BNL-NCS-50496 "Data Formats and Procedures for ENDF" (1979).

4. S. Tagesen and H. Vonach, "Evaluation of the Cross-Sections for the Reaction $^{27}Al(n,\alpha)^{24}Na$", to be published in Physics Data.

5. H. Liskien, A. Paulsen, and R. Widera, Journ. Nucl. Energy 27, 39 (1973).

6. A. Crametz, H.-H. Knitter, and D.L. Smith, "Thick Target $^{9}Be(d,n)^{10}B$ Neutron Spectrum", this conf.

7. H. Houtermans, O. Milosevic, and F. Reichel, Int. J. Appl. Radiat. Isotopes 31, 153 (1980).

8. D.L. Smith and H. Liskien, "Error Analysis for the Calculated Spectrum-Average Cross Section Ratio of $^{63}Cu(n,\alpha)^{60}Co$ - to - $^{27}Al(n,\alpha)^{24}Na$ in a Thick Target $^{9}Be(d,n)^{10}B$ Spectrum", to be published in European Applied Research Reports.

MEASUREMENTS OF CROSS SECTIONS FOR THE (n,2n) REACTION OF ^{58}Ni, ^{93}Nb, ^{181}Ta and ^{197}Au

Lu Hanlin, Huang Jianzhou, Fan Peiguo, Cui Yunfeng and Zhao Wenrong

Institute of Atomic Energy, Academia Sinica
P.O.Box 275-3 Beijing, China

Cross sections for the ^{58}Ni(n,2n)^{57}Ni, ^{93}Nb(n,2n)^{92m}Nb, ^{181}Ta(n,2n)^{180m}Ta and ^{197}Au (n,2n)^{196}Au reactions have been measured by the activation technique in the neutron energy range from 12 to 18 MeV. The measurements were relative to the known cross sections of the ^{27}Al(n,α)^{24}Na or ^{56}Fe(n,p)^{56}Mn reaction at 14.61 MeV. The results measured were compared with existing data. The recommended curves are also given.

[^{58}Ni(n,2n)^{57}Ni, ^{93}Nb(n,2n)^{92m}Nb, ^{181}Ta(n,2n)^{180m}Ta, ^{197}Au(n,2n)^{196}Au, cross sections, E_n=12--18 MeV, activation method, nuclear data.]

Introduction

Neutron threshold reaction cross sections are widely applied in the measurement of the neutron flux, fluence and spectrum. If the excitation functions for threshold reactions are well known, the data analysis would provide accurate information about the irradiating neutron spectrum above the threshold. In the present work the cross sections for the (n,2n) reaction on ^{58}Ni, ^{93}Nb, ^{181}Ta and ^{197}Au have been measured in the neutron energy range from 12.3 to 18.3 MeV, which are used for fast neutron flux determination. The ^{27}Al(n,α)^{24}Na and ^{56}Fe(n,p)^{56}Mn reactions were chosen to monitor the neutron fluence on the site of the sample in question at 14.61 MeV.
Existing data were collected and evaluated to obtain the recommended curves shown in the figures.

Experiment

Irradiation: The 12.3--18.3 MeV neutron irradiations were performed at 600 kV Cockcroft-Walton accelerator and 2.5 MV Van de Graaff accelerator of IAE.
The samples mounted on the surface of a ring (either 5, or 7.5 or 10 cm radius) centered at the D-T neutron source were irradiated at various positions. Therefore simultaneous irradiations were made in the neutron energy range of 12.3--18.3 MeV. The neutron flux density was closely monitored with an associated particle system or a Si detector. This made possible to make the correction for the small variations of neutron yield. The irradiations usually lasted for 5--20 hrs.
The neutron fluence at 14.61 MeV was determined relative to the ^{27}Al(n,α)^{24}Na or ^{56}Fe (n,p)^{56}Mn cross sections of 117.5(2.9) mb or 108.0(2.7) mb. Those at the rest neutron energies were obtained from the known neutron angular distributions taken from our group and H.Liskien[1].
The sample used were metal disks 1--2 cm in diameter and 0.02--1 mm thick. The sample employed for the absolute measurement were sandwiched between two Al or Fe foils and placed at 45° with respect to the deuteron beam.
Activity measurement: The radioactivity measurements after the irradiation were carried out with a calibrated 80x80 mm^3 NaI(Tl) and a 136 cm^3 Ge(Li) gamma-ray detector. The detectors were connected with a pulse height analyser or an on-line minicomputer.

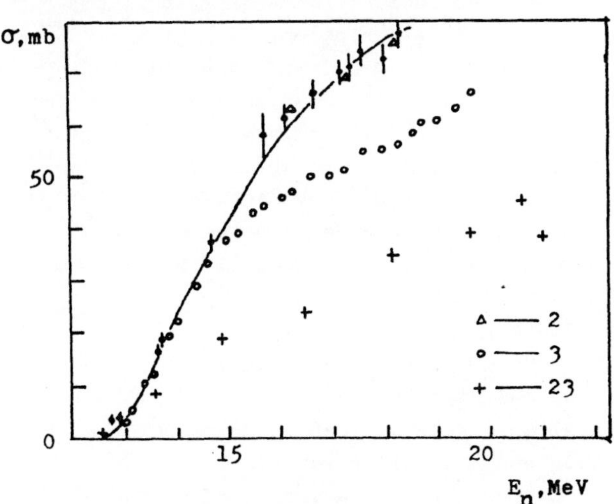

Fig.1. ^{58}Ni(n,2n)^{57}Ni reaction

K. H. Böckhoff (ed.), Nuclear Data for Science and Technology, 411–413.

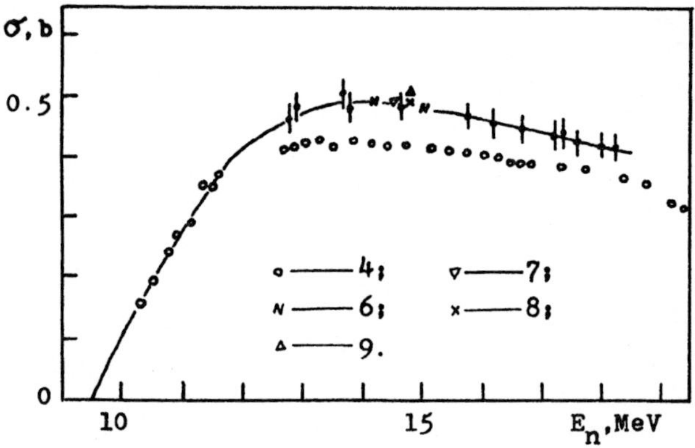

Fig.2. $^{93}Nb(n,2n)^{92m}Nb$ reaction

Corrections were made for self-absorption in the samples and coincident effects in the detectors.

Results and Discussion

The cross sections for the (n,2n) reactions obtained are shown in Figs.1--4. Values of present work were marked with black circles together with error bars. Some of the existing data from other laboratories as well as our recommended curves are also plotted in the Figs. for comparison. Because of the limited space the Figs. do not contain all data available in the literature.

The uncertainty of the measurement was estimated to be 3.5--5 %, except for the result of the $^{197}Au(n,2n)^{196}Au$ reaction. The important sources of error were from the standard cross section adopted; absolute activity determination; neutron angular distribution; statistics of the activity measurement; neutron absorption and scattering, etc.

$^{58}Ni(n,2n)^{57}Ni$ reaction:

Fig.1 shows the comparison with other data. Ours agree with B.P.Bayhurst et al(75)[2], but higher than those of A.Paulsen et al.(65)[3].

$^{93}Nb(n,2n)^{92m}Nb$ reaction:

Fig.2 shows the measured cross sections of this reaction. The cross section of the formation of a long lived isomeric state in ^{92}Nb was not included. Our results differ from A. Paulsen et al.(70)[4] and Wen-Deh Lu et al.(70)[5] but agree with D.R.Nethaway et al.(72)[6], E.T. Bramlitt et al.(62)[7], D.G.Vallis (66)[8] and S. M.Qaim et al.(71)[9].

$^{181}Ta(n,2n)^{180m}Ta$ reaction:

The cross sections measured for this reaction are presented in Fig.3 together with other available data. It can be seen that the absolute cross sections reported, however, differ from each other by a factor of 3. The discrepancy may be mainly due to the different decay schemes used. The cross sections determined for this reaction depends stronly on the decay scheme.

In the measurement of J.S.Brzosko et al.(69)[10] the ^{180m}Ta nucleus was identified by X-radiation accompanying its decay. It was assumed in calculating the cross sections that there are 0.56±0.06 KX-ray quanta from ^{180}W and ^{180}Hf for each decay of ^{180m}Ta. Thus, the absolute cross section obtained at 14.2 MeV was 1930(210) mb. Additonal measurements performed with 14.2 MeV neutrons using the β-particle counting method yielded σ_m=1350(100) mb using the decay scheme of H.N.Brown et al(51)[11] and σ_m=2180(150) mb on the basis of the G.J. Gallayber (62)[12] scheme.

In the vicinity of 14.MeV J.S.Brzosko (69), R.J.Prestwood (61)[13], M.Bormann (68)[14] and T. B.Ryves (80)[15] made use of the decay scheme of T.B.Ryves (80)[16] to obtain the absolute cross sections. They are 1558,1286,1335 and 1307 mb, respectively. All cross sections but that of J.B.Brzosko agree well with the present work. Also, there is the value of 2740 (30) mb given by A.Poularikas (60)[17]. This is the highest one among the all available data.

$^{197}Au(n,2n)^{196}Au$ reaction:

Fig.4 shows the measured cross sections which are the sum of the 6.18 day and 9.7 hr ^{196}Au isomers. Obvious discrepancies appear in the 14--15 MeV neutron energy range. It seems that there exist large differences between the prompt neutron detection measurements of the earliest data[18,19] and the activation measurements. Hence some authors consider that there are some hitherto undetected decay modes, i.e. isomeric states, which could be responsible for this discrepancy. The existing data which have been published since 1975 show that the cross sections of J.Frehaut(75)[20]

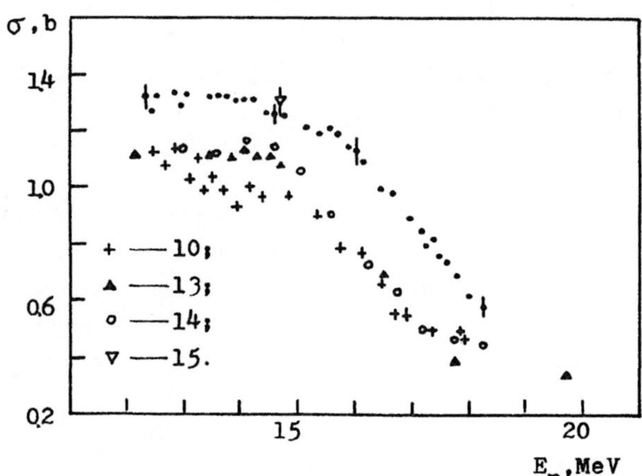

Fig.3. $^{181}Ta(n,2n)^{180m}Ta$ reaction

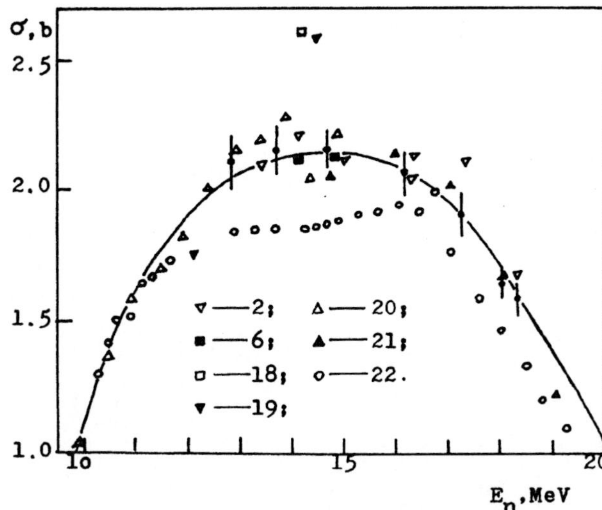

Fig.4. $^{197}Au(n,2n)^{196}Au$ reaction

and L.R.Veeser (77)[21] by the prompt neutron detection technique and of B.P.Bayhurst (75) by the activation technique are in good agreement with our results, except for that of A.Paulsen (75)[22].

The results for the $^{197}Au(n,3n)^{195}Au$ reaction at 18.20 MeV were obtained. The value,636(47) mb, was based on the cross section of the $^{197}Au(n,2n)^{196}Au$ reaction measured in this work.

Acknowledgement

The authors wish to thank Wang Dahai, Li Jizhou, Ma Hongchang and Chen Baolin for their participating also in this work.

References

1. H.Liskien et al., Nucl Data Tables, 11,569 (1973).
2. B.P.Bayhurst et al., Phys.Rev., C12,451(1975).
3. A.Paulsen et al., Nucleonik, 7, 117 (1965).
4. A.Paulsen et al., Z.Phys , 238,23 (1970).
5. Wen-Deh Lu et al., Phys.Rev., C1, 350 (1970).
6. D.R.Nethaway et al., Nucl.Phys., A190,635 (1972).
7. E.T.Bramlitt et al., J.Nucl.Energy, 24, 1321 (1962).
8. D.G.Vallis, AWRE-0-76/66 (1966).
9. S.M.Qaim et al., Chem.Nucl.Data Measurement and Application (1971) BNES, p.121.
10. J.S.Brzosko et al., Nucl Phys.,123, 603 (1969).
11. H.N.Brown et al., Phys.Rev., 84, 292 (1951).
12. C.J.Gallagher et al., Nucl.Phys., 33,285 (1962).
13. R.J.Prestwood et al., Phys.Rev., 121,1438 (1961).
14. M.Bormann et al., Nucl.Phys., A115, 309 (1968).
15. T.B.Ryves et al., J.Phys.G, 6, 771 (1980).
16. T.B.Ryves, J.Phys.G,6, 763 (1980).
17. A.Poularikas et al., J.Inorg.Nucl.Chem., 13, 196 (1960).
18. J.Ashby et al., Phys.Rev., 111, 616(1958).
19. D.S.Mather, AWRE-0-72/72 (1972).
20. J.Frehaut et al., Nucl.Cross Sections & Tech., NBS Spec.Pub., 425,vol.2, p.855 (1975).
21. L.R.Veeser et al., Phys.Rev., C16, 1792 (1977).
22. A.Paulsen et al., Atomkern Energie, 25, 34 (1975).
23. J.M.F.Jerenymo, Nucl.Phys., 47, 157(1963).

STUDY OF EXCITATION FUNCTIONS AROUND 14 MeV NEUTRON ENERGY

J. Csikai

Institute of Experimental Physics
Kossuth University, Debrecen, Hungary

A simple method has been developed for the determination of neutron energy in the interval of 12.2-15.5 MeV using the cross section ratio of the $^{90}Zr(n,2n)^{89}Zr$ and $^{93}Nb(n,2n)^{92m}Nb$ reactions. The error of this method does not exceed 50 keV for neutrons produced in D+T reaction using thick tritium target and 200 keV D^+ beam. This method needs to use only either a Ge(Li) or a NaI(Tl) gamma spectrometer. In the knowledge of the neutron energy emitted in different directions to the D^+ beam the excitation functions of the following reactions were measured: $^{27}Al(n,\alpha)$, $^{27}Al(n,p)$, $^{46}Ti(n,2n)$, $^{58}Ni(n,2n)$, $^{65}Cu(n,2n)$, $^{90}Zr(n,2n)$, $^{93}Nb(n,2n)$, $^{181}Ta(n,2n)$ and $^{197}Au(n,2n)$. Results were compared with the theoretical models.

[Activation cross sections, E_n= 13.5-14.78 MeV, neutron energy, reaction models.]

Introduction

The nuclear technology related to the design construction and operation of fission and fusion reactors needs a number of accurate reaction cross section data around 14 MeV (see Refs. [1,2]). The primary neutron spectrum to be expected from a D+T plasma of a few ten keV temperature is in the energy range which can be produced by the small neutron generators [3]. A survey [4] on the present status of 14 MeV neutron data indicates that a large spread exists in the cross sections measured in different laboratories. Most recently the International Atomic Energy Agency has initiated a co-ordinated research programme to improve the accuracy of the cross sections around 14 MeV.

The excitation functions for many threshold reactions vary significantly around 14 MeV, so the inconsistencies of the data may be caused by the unnormalized neutron energy. The neutron energy depends on the bombarding deuteron energy (E_d) and on the emission angle (ϑ). If E_d= 200 keV, the angular variation of energy is between 13.5 and 14.8 MeV and so, the cross section depends on the position and the dimension of the sample. This is the reason why the average energy of neutrons emitted in different directions should be determined before measuring the cross sections. The energy spread of neutrons at a definite angle arises mainly from the stopping and scattering of deuterons in the target [5]. In the angular interval from 93 to 103° the energy spread of neutrons does not exceed a few ten keV and the average energy of neutrons at 98° in a good approximation remains constant about 14.1 MeV independently of the deuteron energy below 500 keV. So, the data measured around 90° seems to be acceptable for normalization of the cross section curves.

The principle of the method for the determination of neutron energy described in this paper is based on the fact that the shapes of the excitation functions for the $^{90}Zr(n,2n)$ and $^{93}Nb(n,2n)^{92m}Nb$ reactions deviate significantly and so their ratio depends strongly on the neutron energy. After determining the neutron energy as a function of emission angle, the cross sections around 14 MeV have been determined for some reactions using the same arrangement as for the Nb and Zr.

Experimental considerations and procedure

Zirconium and niobium foils of 10 mm diam. and 0.1 mm thickness were irradiated simultaneously in a scattering free target arrangement shown in Fig. 1 with an angular

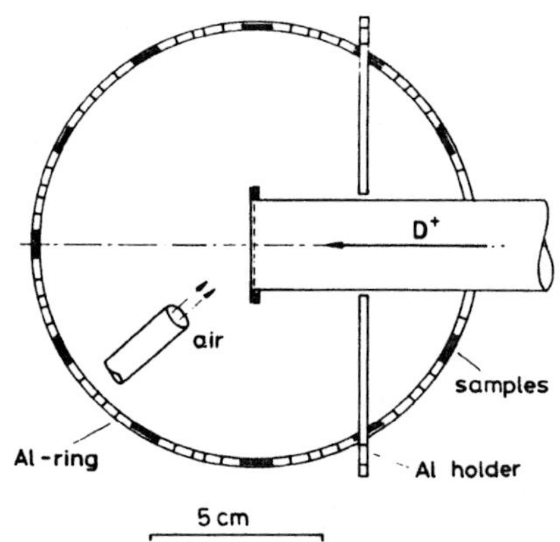

Fig. 1. Geometrical arrangement for the irradiation of the samples.

resolution of 1.6×10^{-2} sr. The details of the target holder and the cooling system have been described in an earlier paper [6]. Neutrons were produced in the D+T reaction using 0.3 mA analysed D^+ beam of 175 keV and 210 keV energies. During the relatively long about 3 hours irradiation time the constant yield of neutrons was assured by using a wobbling target system. After the irradiation of the Zr and Nb foils placed at 0, 30, 60, 90, 120 and 150° to the direction of the D+T-beam the intensities of the gamma lines in the decay of ^{89}Zr (511, 909 keV) and ^{92m}Nb (934 keV) were measured by a Ge(Li) and a NaI(Tl) detector. Typical gamma spectra are presented in Fig. 2.
If the half-lives are relatively long compared to the irradiation time the change in the position of the beam on the target surface can be neglected and so, we have the following simple relation between the cross section ratios and the measured intensities (I) as a function of neutron emission angle (ϑ):

K. H. Böckhoff (ed.), Nuclear Data for Science and Technology, 414–417.

$$R(\vartheta) = \frac{\sigma_{Zr}(\vartheta)}{\sigma_{Zr}(90^o)} \cdot \frac{\sigma_{Nb}(90^o)}{\sigma_{Nb}(\vartheta)} = \frac{I_{Zr}(\vartheta)}{I_{Zr}(90^o)} \cdot$$

$$\cdot \frac{m_{Zr}(90^o)}{m_{Zr}(\vartheta)} \cdot \frac{E_{Zr}(90^o)}{E_{Zr}(\vartheta)} \cdot \frac{I_{Nb}(90^o)}{I_{Nb}(\vartheta)} \cdot \frac{m_{Nb}(\vartheta)}{m_{Nb}(90^o)} \cdot$$

$$\cdot \frac{E_{Nb}(\vartheta)}{E_{Nb}(90^o)}$$

where

$E_x(\vartheta) = \left[e^{-\lambda_x t_1} - e^{-\lambda_x t_2}\right]$, $t_2 - t_1$ is the measuring time after the end of activation,

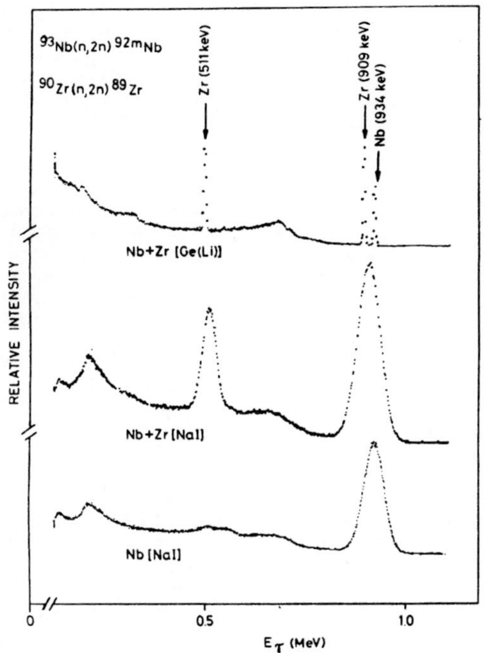

Fig. 2. Gamma spectra of ^{89}Zr and ^{92m}Nb

m_x is the mass of the foil placed at different angles to the D^+ beam.
The available data show that the excitation function of the $^{93}Nb(n,2n)^{92m}Nb$ reaction has a flat shape and so the $\sigma_{Nb}(E_n)$ is constant in the energy interval of 13.5-15 MeV. Using the following designation

$I_o(\vartheta) = I_x(\vartheta)m_x(\vartheta)E_x(\vartheta)$ we have

$$R(\vartheta) = \frac{Zr(\vartheta)}{Zr(90^o)} = \frac{I_{o,Zr}(\vartheta)}{I_{o,Zr}(90^o)} \cdot \frac{I_{o,Nb}(90^o)}{I_{o,Nb}(\vartheta)}$$

If the cross section curves for the standard reaction would change only the relative values to the 90° should be known. This means that for the determination of $R(\vartheta)$ function only the relative intensities of gamma lines should be measured.

Using a Ge(Li) spectrometer the 909 and 934 keV lines emitted in the decay of ^{89}Zr and ^{92m}Nb, respectively can be separated. In the decay of the ^{92m}Nb we have an additional line to the 934 keV one at 912 keV with a well known intensity which can be used for the determination of the relative efficiency of the Ge(Li) detector.

If a NaI(Tl) crystal is used the 909 and 934 keV lines cannot be separated, therefore, the ratio of the peak areas of the 511 and 909 keV lines should be measured (see Fig.2). This method gives a possibility for the separation of the intensities of these two lines in the decay of ^{89}Zr and ^{92m}Nb measured simultaneously.

Cross section curves relative to the $^{27}Al(n,\alpha)$ excitation function were measured for a few reactions indicated in Table I. The nuclear

Table I Nuclear parameters for reactions to be investigated

Reaction	$T_{1/2}$	E(MeV)	I (%)
$^{27}Al(n,\alpha)$	15 h	1.368 1.730	100 99.85
$^{27}Al(n,p)$	9.48 m	0.843 1.014	72 28
$^{46}Ti(n,2n)$	3.08 h	0.511	172
$^{58}Ni(n,2n)$	36 h	0.511	99
$^{65}Cu(n,2n)$	12.8 h	0.511	37
$^{90}Zr(n,2n)$	78.4 h	0.909	99
$^{93}Nb(n,2n)$	10.16 d	0.934	95.5
$^{181}Ta(n,2n)$	8.15 h	β^-(GM) 0.71 0.61	13
$^{197}Au(n,2n)$	6.18 d	0.333 0.356	24.4 93.6

data accepted in this experiment were taken from Refs. [7,8]. The relative efficiency of the Ge(Li) and the NaI(Tl) detectors have been determined by using standard sources and the relative intensities of the gamma lines from the irradiated samples accepting the literature data for the decay schemes.

Results and Discussion

Accepting the most recent literature data [9,10,11] the relative excitation function $R = \sigma(\vartheta)/\sigma(90^o) = \sigma(E_n)/\sigma(14.1)$ has been determined for the $^{90}Zr(n,2n)$ reaction. As it can be seen in Fig.3 the $R(E_n)$ function can

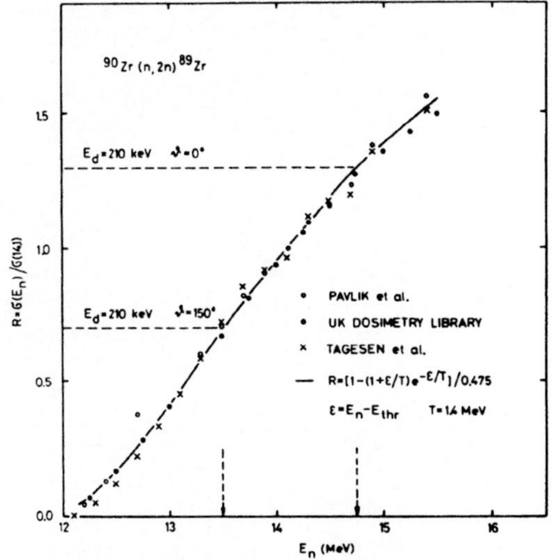

Fig. 3. Excitation function of $^{90}Zr(n,2n)^{89}Zr$ reaction normalized to the data obtained at 14.1 MeV

be well approximated by the Weisskopf formula

Table II Neutron energies vs. emission angle for D+T reaction

Angle	E_n (MeV)		
	Raics[5]	Present experiment	
	E_d= 175 keV	E_d = 170 keV	E_d = 210 keV
0	14.80 ± 0.17	14.70 ± 0.030	14.78 ± 0.055
30	14.70 ± 0.15	14.62 ± 0.025	14.66 ± 0.040
60	14.45 ± 0.12	14.39 ± 0.015	14.39 ± 0.025
90	14.12 ± 0.08	14.10 ± 0.010	14.10 ± 0.010
120	13.75 ± 0.10	13.82 ± 0.015	13.77 ± 0.015
150	13.52 ± 0.12	13.55 ± 0.015	13.50 ± 0.015
180	13.41 ± 0.13	-	-

$$R(E_n) = \left[1-(1+\varepsilon/T)e^{-\varepsilon/T}\right]/0.475$$

where T= 1.4 MeV and E_{thr} = 11.86 MeV.

On the basis of the measured R values for the $^{90}Zr(n,2n)$ reaction the average neutron energies as a function of emission angle were determined at E_d= 170 and 210 keV. Results are summarized in Table II. For comparision the values calculated by Raics [5] on the basis of "stopping and impinging spectra" at E_d= 175 keV are also given. The good agreement between the measured and calculated values shows that the average neutron energy at a given angle depends mostly on the stopping of deuterons in the thick tritium target and on the solid angle applied for the irradiation. The deviations in forward and backward directions can be explained by the scattering of deuterons in the thick target. The error of this method is determined by the shape of the $^{90}Zr(n,2n)$ cross section curve but does not exceed 50 keV, while the sensitivity dR/dE_n between 13 and 15 MeV is about 46 %/MeV. The R values can be determined by experiment with an accuracy of better than 1 % and so the error in the determination of the neutron energy at 14.1 MeV is about 20 keV.

Cross section values obtained in this experiment are summarized in Table III. Relative excitation functions were determined in the interval of 13.5-14.8 MeV accepting a value of 122.0 mb for the $^{27}Al(n,\alpha)^{24}Na$ reaction [9] at 14.1 MeV. The $\sigma(E_n)/\sigma(14.1)$ values are constant for $^{93}Nb(n,2n)^{92m}Nb$, $^{181}Ta(n,2n)^{180m}Ta$ and $^{197}Au(n,2n)^{196}Au$ reactions within the ± 0.9 %, ± 2.2 % and ±1.3 % errors of the measurements, respectively. These reaction can be used as flux monitors in the energy interval indicated above. The data for $^{27}Al(n,\alpha)$ reaction are in good agreement with those given by Pavlik et al.[9] except the point at 13.5 MeV. The absolute values of cross sections for $^{27}Al(n,p)$ reaction can be well described by the Hauser--Feshbach model [12]. The details of calculation were described in a previous paper [13]. The shapes of excitation functions of (n,2n) reactions are in good agreement with the Weisskopf estimate using T= 1.0 MeV for ^{46}Ti, ^{93}Nb, ^{181}Ta, ^{197}Au and T=1.4 MeV for ^{58}Ni, ^{65}Cu, ^{90}Zr, respectively, so that by fitting the measured and calculated values at 14 MeV, $\sigma_{n,2n}$ data for different energies can also be determined. This procedure was used in the case of $^{90}Zr(n,2n)$ reaction (see Table III). For comparision the measured and calculated values of the cross sections are plotted in Fig.4. As a conclusion of the results is that the small neutron generators can be used to study the shapes of the excitation functions around 14 MeV and through it to check the applicability of the reaction models as well as to determine standard data.

The author is grateful to Mrs. Dr. Juhász for her kind assistance during the measurements.

Table III Cross sections determined in the present experiment

Reaction \ Energy (MeV)	13.5	13.77	14.1	14.39	14.66	14.78
$^{27}Al(n,\alpha)$	128.3± 2.6	125.3± 2.5	122.0	117.6± 2.4	113.2± 2.3	113.3± 2.3
$^{27}Al(n,p)$	102.0± 5.6	97.0± 5.3	88.0± 4.8	84.0± 4.6	76.0± 4.2	73.0± 4.0
$^{46}Ti(n,2n)$	2.2± 0.2	7.5± 0.6	17.7± 1.3	30.9± 2.3	42.0± 3.2	50.2± 3.8
$^{58}Ni(n,2n)$	16.5± 1.0	19.2± 1.2	25.0± 1.5	29.4± 1.8	32.0± 1.9	34.8± 2.1
$^{65}Cu(n,2n)$	840 ± 50	862 ± 52	905 ± 54	955 ± 57	975 ± 59	980 ± 59
$^{93}Nb(n,2n)$	453 ± 29.5	456 ± 30	461 ± 30	455 ± 30	458 ± 30	462 ± 30
$^{181}Ta(n,2n)$	2210 ±192	2167 ±189	2148 ±187	2163 ±188	2154 ±187	2131 ±185
$^{197}Au(n,2n)$	2056 ±140	2087 ±142	2079 ±141	2092 ±142	2087 ±142	2094 ±142
$^{90}Zr(n,2n)$	451	546	655 ± 20	743	819	852

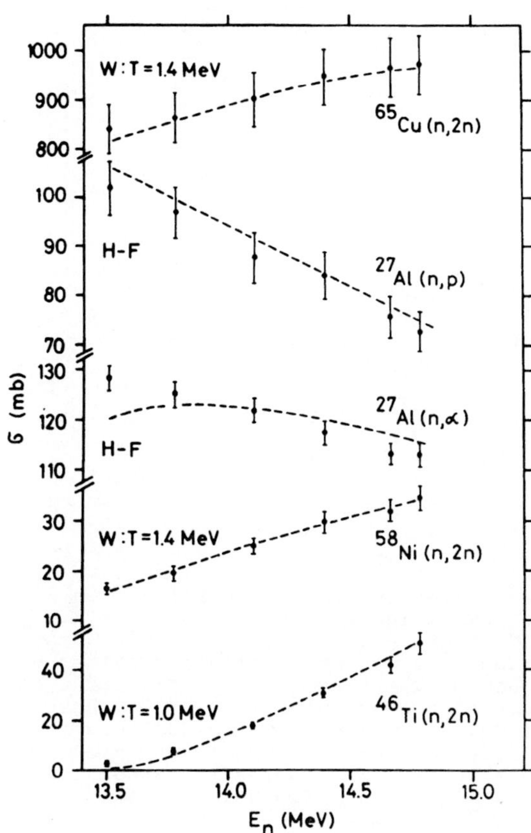

Fig.4. Measured and calculated (dashed curves) excitation functions. (W and H-F denote the Weisskopf and Hauser-Feshbach model, respectively.)

References

1. D.W. Muir, WRENDA 78/80, INDC SEC-73/URSF IAEA Vienna (1979)

2. N. DayDay, WRENDA 81/82, INDC(SEC)--78/URSF, IAEA, Vienna (1981)

3. H. Liskien, IAEA-TECDOC-223, Vienna (1978)

4. J. Csikai, IAEA-SMR-68/I, p.215, Trieste (1981)

5. P. Raics, Thesis, Kossuth University, Debrecen (1978)

6. G. Pető, J. Csikai, V. Long, S. Mukherjee, J. Bánhalmi, Z. Miligy, Acta Phys. Slovaca 25, 185 (1975)

7. C.M. Lederer, V.S. Shirley, Table of Isotopes, Seventh Edition. John Wiley and Sons, INC., New York (1978)

8. G. Erdtmann, W. Soyka, Die Gamma-Linien der Radionuklide. Jül-1003-Ac (1973)

9. A. Pavlik, G. Winkler, H. Vonach, Nucl. Phys. G. (to be published)

10. S. Tagesen, H. Vonach, B. Strohmaier, Physics Data 13-1 Fachinformationszentrum Karlsruhe, ISSN 0344-8401 (1979)

11. U.K. Nuclear Data Library (1982)

12. W. Hauser, M. Feshbach, Phys. Rev. 87, 366 (1952)

13. S. Sudár, J. Csikai, Nucl. Phys. A319, 157 (1978)

MEASUREMENT OF AVERAGE CROSS SECTIONS FOR ^{252}Cf NEUTRONS

Z. Dezső and J. Csikai
Institute of Experimental Physics
Kossuth University,
H-4001 Debrecen, P.O.Box 105. Hungary

Average cross sections for fourteen reactions have been measured relative to the ^{115}In(n,n')^{115m}In reaction in the spontaneous fission neutron field of ^{252}Cf. Irradiations have been done in a scattering free arrangement in the open air using a small Cf-neutron source and thin samples. Results are compared with previously measured $\langle\sigma\rangle$ data in this laboratory and those found in the literature.

[A(n,x)B reactions, cross sections, ^{252}Cf spectrum]

Introduction

The accurate knowledge of Californium-252 spectrum averaged cross sections is of great importance especially for reactor neutron dosimetry applications. Since the spontaneous fission neutron field of ^{252}Cf is one of the most well known standard neutron field, integral cross section measurements can be used for the validation and further improvement of energy-dependent cross sections used in reactor metrology.

The present work is a continuation of our earlier cross section measurements but with improved experimental conditions especially that of the Cf-source and the irradiation circumtances concerned. Since the accurate neutron source strength determination has not yet been available, measurements were made relative to the ^{115}In(n,n')^{115m}In reaction for which there are several measured data from different laboratories and they are in good agreement. The primary objective of this set of experiments was to investigate the applicability of a small source - both in size and in source strength - for accurate $\langle\sigma\rangle$ measurements. Since the establishment of the proper experimental conditions is still in progress detailed analysis of uncertainties are not discussed here only the statistical errors are quoted thus the data presented here can be considered as preliminary results.

Experimental Method

The californium source used in this experiment contains about 40 μg of ^{252}Cf encapsulated in a stainless steel cylinder 7.5 mm in diameter and 14 mm in height. The dimension of the ^{252}Cf itself is ∅ 3x3.5 mm^3 as reported in the certificate of manufacture. The exact location of the source in the container has been investigated throught the activity distribution measurement of 0.2 mm thick aluminium foils. The source was covered with the foil on its cylindrical side for activation and after irradiation it was cut into 1 mm wide strips and then counted with a GM-counter. The activity distribution showed a sharp maximum at 5 mm measured from the bottom of the source. The 86 percent of the total activity induced in the foil was found. within the lower 10 mm wide zones . Therefore samples of this size have been used for cross section measurements.

Irradiations have been performed in the open air about 12 m above the ground level and 4 m above the roof of the building. The source was suspended using steel wire of 0.8 mm in diameter. All metallic samples were 0.2 mm in thickness and were tightly wound on the surface of the ^{252}Cf-source. To provide a permanent irradiation geometry throughout the time of the activation, small pieces of adhesive tapes have also been applied. Si, S and Sr in powder form have been used and were encapsulated in 30 μ thick polyethylene foils having the same size as the metal samples. Special care was taken to avoid the disturbing effect of the air also encapsulated for the irradiation geometry.

The time of irradiations were set according to the half-life of the investigated radionuclide and to the estimated induced activity. The longest irradiation lasted 15 days. Induced γ-activities were measured by a Ge(Li) detector but in some cases a window-mounted NaI(Tl) detector was used for better counting statistics. The relative photopeak efficiency of the detectors have been obtained with standard γ-ray sources. In all cases the samples were placed nearest to the detector. In the case of ^{32}S(n,p)^{32}P cross section measurement the β^--activity of the sample was measured by an end window GM-counter. To correct for counting efficiency and self-absorption the activity of the sulfur sample was compared with the ^{27}Mg-activity of an aluminium sample irradiated with Cf-neutrons and having the same size and weight as the sulfur sample.

Since very thin and identical samples have been used in all cases only small corrections for self absorption differences and for dead time losses had to be applied in the present experiment.

Results and Discussion

The average cross sections measured in this experiment together with those obtained in other works are summarized in Table I. Nuclear data used for data evaluation are presented in Table II.

Since the measured cross section data for ^{115}In(n,n') reaction obtained in different laboratories are in very good agreement after renormalization an evaluated value of 200 mb has been accepted as monitor cross section in the present experiment.

K. H. Böckhoff (ed.), Nuclear Data for Science and Technology, 418–420.

Table I Measured average cross sections in mb

Reaction	This work	Other work	Reference
$^{27}Al(n,p)^{27}Mg$	4.84(2.9)	4.891(3.67) 4.90 (6.53) 5.11 (8.41) 4.7 (7.87)	1 2 3 4
$^{28}Si(n,p)^{28}Al$	7.12(3.3)	9.66 (5.69)	3
$^{32}S(n,p)^{32}P$	68.4(0.5)	72.52(4.08) 72.4 (6.63)	1 2
$^{46}Ti(n,p)^{46}Sc$	13.6(8.9)	13.8 (2.17) 13.4 (8.21) 12.4 (9.68)	5 3 6
$^{47}Ti(n,p)^{47}Sc$	19.4(0.5)	18.9 (2.12) 22.0 (4.03) 20.3 (5.42)	5 3 6
$^{54}Fe(n,p)^{54}Mn$	87.8(1.0)	87.63(4.97) 87.0 (6.21) 84.6 (2.36) 92.5 (5.41)	1 2 5 3
$^{56}Fe(n,p)^{56}Mn$	1.40(1.2)	1.44 (4.86) 1.43 (5.59) 1.450(2.41) 1.18 (6.78) 1.45 (4.14) 1.09 (6.24)	1 2 5 6 3 4
$^{58}Ni(n,p)^{58}Co$	118.8(0.3)	118.5(3.45) 119 (5.04) 118 (2.54) 113.4(4.23) 105 (4.76) 118.8(4.55) 95 (4.74)	1 2 5 3 6 7 4
$^{64}Zn(n,p)^{64}Cu$	41.3(7)	41.84(4.18) 41.7 (6.47) 39.4 (2.54) 46.4 (4.96) 36.2 (4.14)	1 2 5 3 4
$^{84}Sr(n,\gamma)^{85m}Sr$	35.4(6.5)		
$^{93}Nb(n,\alpha)^{90m}Y$	0.028(6)		
$^{113}In(n,n')^{113m}In$	164.3(1.1)	169.7 (4.77) 170 (5.88) 160 (2.50) 178 (4.27) 168 (5.36)	1 2 8 9 4
$^{181}Ta(n,\gamma)^{182}Ta$	89.25(1.2	119.9(5.42) 105.5(5.78)	9 10
$^{197}Au(n,\gamma)^{198}Au$	77.0 (0.1	76.2(2.36) 119.1(4.37) 79.9(3.63) 95.5(2.41) 78 (3.85)	8 9 10 11 4

A comparison of $<\sigma>$ data of this work with our earlier measurements indicates that for non-threshold reactions the earlie data appear somewhat higher than the presen which is probably due to the thermal neutron background in the laboratory. On the other hand the good agreement for threshold reactions implies that the effect of source encapsulation on the neutron spectrum is not considerable at least when the total uncertainty of the cross section measurement is not below 4-5 percent.

The results of several irradiations and activity measurements on the same reaction indicate that the irradiation geometry should be further improved to achive a better reproducibility than the presently available ± 2 percent. Since for many reactions concidered as Category I. and II. reactions in neutron dosimetry reasonable counting statistic can be obtained even with small ^{252}Cf-sources it is hoped that $<\sigma>$ data with uncertainty below 3 percent can be measured. This goal in accuracy is obviously needed for checking energy dependent cross sections as can be seen from Table 3. where a few reactions important in reactor metrology are listed for a comparision of measured and calculated $<\sigma>$ values using different $\sigma(E)$ data sets.

Table II Reactions and nuclear parameters

Reaction	Isotopic abund. %	Half-life	γ-Ray Detect. (keV)	γ-Ray Intens.
$^{27}Al(n,p)^{27}Mg$	100	9.462min	843.76	71.8
$^{28}Si(n,p)^{28}Al$	92.23	2.2405 "	1778.70	100
$^{32}S(n,p)^{32}P$	95.02	14.29d	-	-
$^{46}Ti(n,p)^{46}Sc$	8.1	83.83d	889.25	99.9840
$^{47}Ti(n,p)^{47}Sc$	7.4	3.351d	159.38	68.0
$^{54}Fe(n,p)^{54}Mn$	5.8	312.5d	834.83	99.9760
$^{56}Fe(n,p)^{56}Mn$	91.8	2.5785h	846.75	98.9
$^{58}Ni(n,p)^{58}Co$	68.27	9.15h	810.757	99.4
$^{64}Zn(n,p)^{64}Cu$	48.6	12.701h	1345.9	0.49
$^{84}Sr(n,\gamma)^{85m}Sr$	0.56	67.3min	231.68	84.5
$^{93}Nb(n,\alpha)^{98m}Y$	100	3.19h	202.53 479.51	91 97.5
$^{113}In(n,n')^{113m}In$	4.3	99.48min	391.7	64.90
$^{115}In(n,n')^{115m}In$	95.7	4.486h	336.24	46.7
$^{181}Ta(n,\gamma)^{182}Ta$	99.988	114.41d	1121.28 1221.42	35.30 27.17
$^{197}Au(n,\gamma)^{198}Au$	100	2.696d	411.80	95.404

Table III Measured and Calculated $<\sigma>$ in mb in the ^{252}Cf Fission Neutron Spectrum

Reaction	Measured	Calculated Origin of $\sigma(E)$ ENDF/B-IV	Calculated Origin of $\sigma(E)$ ENDF/B-V
$^{197}Au(n,\gamma)$	76.2-95.5	79.91	76.50
$^{46}Ti(n,p)$	12.4-13.8	12.52	13.46
$^{47}Ti(n,p)$	18.9-22.0	23.85	24.06
$^{115}In(n,n')$	186-201	175.5	181.9
$^{56}Fe(n,p)$	1.18-1.45	1.476	1.50
$^{58}Ni(n,p)$	95-118.8	115.0	116.5

References

1. K. Kobayashi, I. Kimura, W. Mannhart, J. Nucl. Sci. Technol. 19, 341(1982)
2. K. Kobayashi, I. Kimura, Proc. Third ASTM-EURATOM Symp. Reactor Dosimetry, Ispra, EUR-6813, Vol II., p.1004(1980)
3. Z. Dezső, J. Csikai, Proc. VII. Symp. on Interactions of Fast Neutrons with Nuclei. Gaussig, 1977, p. 44.
4. H. Benabdallah, G. Paic, J. Csikai, this conference
5. W.G. Alberts, E. Günther, M. Matzke, G. Rassl, Proc. First ASTM-EURATOM Symp. Reactor Dosimetry, Petten, EUR 5667 e/f, Part I., 131 (1977)
6. G.J. Kirouac, H.M. Eiland, C.J. Slavik, Proc. Topl. Mtg. Irr. Experimentation in Fast Reactors, Jackson Lake, CONF-730910, 412 (1973)
7. V. Spiegel, C.M. Eisenhauer, J.A. Grundl, G.C. Martin, Proc. Second ASTM-EURATOM Symp. Reactor Dosimetry, Palo Alto, NUREG/CP-0004, Vol2., 959(1978)
8. W. Mannhart, W.G. Alberts, Nucl. Sci. Eng. 69, 333 (1979)
9. Z. Dezső, J. Csikai, Proc. IV. All Union Conf. Neutron Physics, Kiev, Vol III., 32 (1977) Atomizdat
10. L. Green, Nucl. Sci. Eng. 58 361(1975)
11. H. Pauw, A.H.W. Aten, J. Nucl. Energy, 25, 457 (1971)

MEASUREMENT OF SOME AVERAGE CROSS SECTIONS FOR ^{252}Cf NEUTRONS

H. Benabdallah

Laboratoire de Physique Nucléaire, Faculté de Science,
Université Mohammed V, Rabat, Maroc

G. Paic

Institut Ruder Boskovic, Zagreb, Yugoslavia

J. Csikai

Institute of Experimental Physics,
Kossuth University, Debrecen, Hungary

The average cross sections for the reactions $^{115}In(n,n')^{115m}In$, $^{115}In(n,\gamma)^{116m}In$, $^{113}In(n,n')^{113m}In$, $^{197}Au(n,\gamma)^{198}Au$, $^{111}Cd(n,n')^{111m}Cd$ + $^{110}Cd(n,\gamma)^{111m}Cd$, $^{27}Al(n,p)^{27}Mg$, $^{64}Zn(n,p)^{64}Cu$, $^{68}Zn(n,\gamma)^{69m}Zn$, $^{138}Ba(n,\gamma)^{139}Ba$, $^{135}Ba(n,n')^{135m}Ba$ + $^{134}Ba(n,\gamma)^{135m}Ba$, $^{58}Ni(p,n)^{58}Co$ and $^{87}Sr(n,n')^{87m}Sr$ + $^{86}Sr(n,\gamma)^{87m}Sr$ have been measured in the spontaneous fission spectrum of neutrons from a small ^{252}Cf source. The results are in general agreement with other data in the literature. In the cases where energy differential cross sections were available the average cross sections have been calculated using T = 1.42 MeV Maxwellian temperature. The comparison of measured and calculated values indicates strong disagreements in the cases of the $^{56}Fe(n,p)$, $^{138}Ba(n,\gamma)$ and $^{135}Ba(n,n')$ + $^{134}Ba(n,\gamma)$ reactions.

[Reaction cross sections, ^{252}Cf neutron spectrum, comparision with the calculated values.]

Introduction

The energy spectrum of spontaneous fission neutrons emitted from ^{252}Cf has reached the status of a standard neutron field [1].

The shape of the neutron spectrum is agreed upon to be basically a Maxwellian with a temperature of 1.42 MeV, corresponding to an average energy of 2.13 MeV. Grundl and Eisenhauer [2,3] propose continuous segment corrections to the basic shape while the recent results of Blinov et al. [4] suggest that in the region from 1 keV to 1 MeV neutron energy no corrections to a Maxwellian is necessary. The spectrum of californium neutrons is known with an accuracy that it can be used to check the energy differential cross sections for neutron induced reactions in the same energy range. Therefore the use of a ^{252}Cf source in conjunction with a well calibrated detector set up makes possible integral tests of the energy differential cross section data. This comparison follows from the definition of the average cross section measured in a ^{252}Cf spectrum:

$$\langle \sigma \rangle = \frac{\int_0^\infty N(E)\ \sigma(E)dE}{N(E)dE} \qquad (1)$$

From the relation (1) it is obvious that if the spectrum shape N(E) is accurately known, the comparison between the measured $\langle\sigma\rangle$ and the one calculated via (1) shall bear on the knowledge of $\sigma(E)$. The determination of $\langle\sigma\rangle$ is a relatively simple experiment many small laboratories can afford but surprisingly the data are still rather scarce and scattered in results.

To be able to study the actual situation in the nuclear data for fission neutrons we have measured the average cross sections for the reactions shown in Table I.

Experimental Method

The source of ^{252}Cf of 20 μg ^{252}Cf located at the Laboratoire de Physique Nucléaire in Rabat has been used. The source strength was declared by the manufacturer (Amersham) to be $4.6 \times 10^7\ s^{-1} \pm 1.5$ % on 31 December 1975. To determine the strength of the source at the time of irradiation a half life of 2.638 years[5] was used.

The ^{252}Cf under the chemical form of oxide is doubly encapsuled in welded stainless steel capsules. The thicknesses of the walls of each capsule in 0.8 mm. The outside dimensions of the source cylinder is 7.8 mm in diameter and 10 mm in height.

For irradiations a special set up was constructed out of thin aluminium, allowing for irradiations at a distance of 4.3 ± 0.15 mm from the vertical axis running through the source in the median plane of the cylinder. The targets were discs of 10 mm in diameter ranging in thickness from 0.1 to 0.7 mm.

To determine the neutron fluence at the target position, the yield of the reaction $^{115}In(n,n')$ was measured in function of the source-target distance. It was found that the neutron fluence could be satisfactorily calculated assuming a point source and $1/r_{eff}^2$ dependence, where r_{eff} was found to be 5.2 ± 0.12 mm for sample thickness 0.1 mm. During irradiation the set up was hung in air 6 meters above ground and from any wall, minimizing thus the scattered neutron background.

The thermal neutron background was determined by the measurement of the $^{115}In(n,\gamma)$ reaction with and without a cadmium foil. The two yields, after corrections for distance and attenuation of the neutrons in the cadmium foil, were equal within the statistical error. So, no corrections were introduced for the thermal neutrons in the measured (n,γ) reactions.

K. H. Böckhoff (ed.), Nuclear Data for Science and Technology, 421–424.

Table I Parameters of the reactions studied

Target	Reaction	$T_{1/2}$	Atomic mass	Isotop. abund. (%)	γ-ray detect. (keV)	Intens. of the γ (%)
In	$^{115}In(n,n')^{115m}In$	4.486h	114.82	95.7	336.0	45.9
	$^{115}In(n,\gamma)^{116m}In$	54.3 mn	114.82	95.7	1293.4	84.6
	$^{113}In(n,n')^{113}In$	99.47mn	114.82	4.3	391.72	64
Au	$^{197}Au(n,\gamma)^{198}Au$	2.697d	196.967	100	411.794	95.5
Cd	$^{111}Cd(n,n')^{111m}Cd$	48.6 mn	112.4	12.8	245.35	94.2
	$^{110}Cd(n,\gamma)^{111m}Cd$	48.6 mn	112.4	12.4	245.35	94.2
Al	$^{27}Al(n,p)^{27}Mg$	9.462mn	26.98154	100	843.7	73
Zn	$^{64}Zn(n,p)^{64}Cu$	12.699h	65.38	48.9	511.006	38.6
	$^{68}Zn(n,\gamma)^{69m}Zn$	13.76h	65.38	18.6	438.7	94.8
BaO	$^{138}Ba(n,\gamma)^{139}Ba$	82.9mn	137.34	71.9	165.85	22
	$^{134}Ba(n,\gamma)^{135m}Ba$	28.7h	137.34	2.4	268.1	16
	$^{135}Ba(n,n')^{135m}Ba$	28.7h	137.34	6.5	268.1	16
Ni	$^{58}Ni(n,p)^{58}Co$	70.78d	58.7	67.76	810.74	99.44
Fe	$^{56}Fe(n,p)^{56}Mn$	2.58h	55.847	91.7	846.74	98.87
$[Sr(OH)_2, 8H_2O]$	$^{87}Sr(n,n')^{87m}Sr$	2.805h	87.62	7.0	388.4	82
	$^{86}Sr(n,\gamma)^{87m}Sr$	2.805h	87.62	9.9	388.4	82

The activity was measured with a 67 cm^3 GeLi detector. The samples have been placed at 6.5 mm from the detector window. Special corrections were introduced to take care of the counting geometry which was different from that used in the measurement of the efficiency. The detector efficiency has been determined with an uncertainty of ± 1.5 %. The gamma ray absorption in the samples was determined using the measured attenuation coefficients. The dead time corrections were negligibly small for the low intensity neutron source.

Results and Discussion

The list of the reactions studied and the nuclear parameters used are presented in Table I. The nuclear parameters were taken from the Table of Isotopes [6].

The average cross sections measured in the present experiment are shown in Table II together with the available literature data for the same reactions. The errors quoted in our results include only the counting statistics and the uncertainties in the r_{eff}. The uncertainty in the source strength ~ 1.5 % and in the detector efficiency ~ 1.5 % should be taken into account to obtain the total errors in Table II. Since the statistical error quoted is much larger than the latter two so, the overall uncertainty is almost the same as presented in Table II.

The cross sections for reactions on Cd and Sr refer to average values of the form:

$$<\sigma> = \frac{n_1 <\sigma_1> + n_2 <\sigma_2>}{n_1 + n_2}$$

where n_1 and n_2 are the respective number of nuclides of each isotopes and $<\sigma_1>$ and $<\sigma_2>$ are the average cross sections for each reactions.

As it can be seen in Table II there are very few data even for the useful dosimetric reactions like the one on indium.

The present results are in general agreement with the literature data except for some given in Ref. [10,11].

It is gratifying to see that in the case of the reactions on indium and gold the data measured at different laboratories are coherent enought to warrant the calculation of the mean values.

Rejecting the incoherent data [10,11,12] the mean values we obtained are given in Table III.

Comparing the mean values in Table III with our values from Table II, we find that only in the case of the $^{115}In(n,\gamma)$ reaction is our value outside the 68% error quoted.

The values of $<\sigma>$ have been calculated according to the relation (1) using the $\sigma(E)$ from the evaluated data files ENDF/BIV and ENDF/BV as well as the results of Smith and Meadows [14], accepting T = 1.42 MeV for the Maxwellian temperature.

The results for the calculated values $<\sigma>_{calc}$ compared with the measured ones $<\sigma>_{meas}$ are shown in Table IV for those reactions where $\sigma(E)$ data were available. When two sources for $\sigma(E)$ have been used both values of $<\sigma>_{calc}$ are given. In the last column the ratio $<\sigma>_{meas}/<\sigma>_{calc}$

Table II Average cross sections $\langle\sigma\rangle$ in mbarns

Reactions	Present work	Ref. 7	Ref. 9	Ref.8	Refs. 10,11	Ref. 12	Ref. 13
$^{115}In(n,n')^{115m}In$	196 ±8	198 ±5	202.2±12.0	-	199.4 ±10.5*	188 ±8	195 ±5
$^{115}In(n,\gamma)^{116m}In$	115.6 ±5	-	-	-	131.6 ± 5.7	125.0±4.3	123.8±3.6*
$^{113}In(n,n')^{113m}In$	168 ±9	-	-	-	178.3 ± 7.6	-	162.3±4.1*
$^{197}Au(n,\gamma)^{198}Au$	78 ±3	-	-	79.9±2.9	119.1 ± 5.2	95.5±2.3	76.2±1.8
$^{111}Cd(n,n')^{111m}Cd$ $^{100}Cd(n,\gamma)^{111m}Cd$	110.6 ±4	-	-	-	204 ± 7	-	-
$^{27}Al(n,p)^{27}Mg$	4.7 ±0.37	-	-	-	5.04±0.42*	-	-
$^{64}Zn(n,p)^{64}Cu$	36.2 ±1.5	40 ±1	-	-	46.4 ±2.3	-	-
$^{68}Zn(n,\gamma)^{69m}Zn$	1.85±0.12	-	-	-	-	-	-
$^{138}Ba(n,\gamma)^{139}Ba$	1.30±0.26	-	-	-	3.8 ±0.4	-	-
$^{135}Ba(n,n')^{135m}Ba$ $^{134}Ba(n,\gamma)^{135m}Ba$	180.9 ±11	-	-	-	255 ±28	-	-
$^{58}Ni(n,p)^{56}Co$	95.0 ±4.5	118 ±3	105 ±5	-	110.7 ±4.8*	-	-
$^{56}Fe(n,p)^{58}Mn$	1.09±0.068	1.450±0.035	1.18±0.08	-	-	-	-
$^{87}Sr(n,n')^{87m}Sr$ $^{86}Sr(n,\gamma)^{87m}Sr$	130 ±8	-	-	-	182 ±22	-	-

*renormalized in accordance with data from Table I.

has been shown. This ratio should give a measure of the accuracy of the agreement between the integral measurement and the data on $\sigma(E)$. It was pointed out by Smith [15] that the average cross sections

Table III Mean values of $\langle\sigma\rangle$

Reaction	$\langle\sigma\rangle$ mb
$^{115}In(n,n')^{115m}In$	196.4 ± 5.0
$^{115}In(n,\gamma)^{116m}In$	124.0 ± 3.5
$^{113}In(n,n')^{113m}In$	169.5 ± 5.5
$^{197}Au(n,\gamma)^{198}Au$	78.0 ± 2.0

may be insensitive to certain variations in the shape of the excitation functions and that therefore the requisite data base for nuclear energy applications cannot be established by means of integral measurements alone. However, if the agreement between $\langle\sigma\rangle_{calc}$ and $\langle\sigma\rangle_{meas}$ may not be always significant, the disagreement most certainly is indicative.

The errors to the ratios in Table IV is between 10 and 15 percent due to both uncertainties in $\langle\sigma\rangle_{meas}$ and $\sigma(E)$. Therefore on the basis of table IV, the following conclusions can be drawn:

i) When two sets of $\sigma(E)$ were used the ANL data gives a ratio closer to one;
ii) In the case of $^{113}In(n,n')$ reaction all measured values of $\langle\sigma\rangle$ are higher than the calculated ones. The disagreement may suggest that the shape of $\sigma(E)$ curve is not well known.

Table IV Comparison of measured average cross sections with calculated ones using T = 1.42 MeV

Reactions	$\langle\sigma\rangle$ meas (mb)	$\langle\sigma\rangle$ calc (mb)	Origin of $\sigma(E)$	$\langle\sigma\rangle_{meas}/\langle\sigma\rangle_{calc}$
$^{115}In(n,n')$	196	183	ENDF/B-V, ANL	1.07
$^{115}In(n,\gamma)$	116	130	ENDF/B-IV	0.89
$^{113}In(n,n')$	168	143.5	ANL	1.17
$^{197}Au(n,\gamma)$	78	76.5	ENDF/B-V	1.02
$^{58}Ni(n,p)$	95.1	116.5	ENDF/B-V	0.82
$^{58}Ni(n,p)$	95.1	102	ANL	0.93
$^{27}Al(n,p)$	4.7	5.18	ENDF/B-V	0.91
$^{27}Al(n,p)$	4.7	4.78	ANL	0.98
$^{64}Zn(n,p)$	36.2	37	ANL	0.98
$^{56}Fe(n,p)$	1.1	1.5	ENDF/B-V	0.74
$^{56}Fe(n,p)$	1.1	1.43	ANL	0.76
$^{138}Ba(n,\gamma)$	1.3	2.2	ENDF/B-V	0.59
$^{135}Ba(n,n')$+ $^{134}Ba(n,\gamma)$	181	230	ENDF/B-V	0.79

iii) The $\langle\sigma\rangle_{calc}$ values for $^{56}Fe(n,p)$ seem to be too high compared with the our measured data as well as with the one from Ref. [9] while the result of Alberts [7] is in agreement with $\langle\sigma\rangle_{calc}$.

iv) The data for reactions ^{138}Ba, ^{134}Ba and ^{135}Ba yield $\langle\sigma\rangle_{calc}$ values which are much higher than those measured in our experiment. The lack of more integral data on the same reaction precludes a definite statement about the $\sigma(E)$ in these cases.

Acknowledgments

The help of Drs. Z. Dezső and G. Pető in the early stage of the experiment is gratefully acknowledged. Our thanks go also to the staff of the Nuclear Data Section of the International Atomic Energy Agency for providing us with the necessary data on cross sections.

References

1. M.F. Vlasov, Ed. "IAEA Consultants Meeting on Integral Cross Section Measurements in Standard Neutron Fields" Vienna 15-19 Nov. 1976 IND (NDS) - 81/L+M IAEA Nuclear Data Section Vienna (1977)

2. J.A. Grundl and C.M. Eisenhauer "Fission Spectrum Neutrons for Cross Sections Validation and Neutron Flux Transfer" Proc. Conf. Nuclear Cross Sections and Technology, Washington D.C. March 3-7, 1975, NBS Special Publication 425, vol. 1, p. 250 National Bureau of Standards (1975). Also Proc. First ASTM-EURATOM Symposium Reactor Dosimetry, Petten, Sept. 22-26, 1975, EUR 5667 e/f Part. 1, p. 425, commision of the European Communities, Luxembourg (1977)

3. J.A. Grundl, V. Spiegel, C.M. Eisenhauer, II H.T. Heaton, D.M. Gilliam and J. Bigelow, Nuclear Technology 32, 315 (1977)

4. M.V. Blinov, V.A. Vitenko, V.I. Krevitsch, Proc. IXth International Symposium on the Interaction of Fast Neutrons with Nuclei, Nov. 26-30, 1979 Gaussig Edit. D. Seeliger and S. Unholzer INDCC (GDR) - 13/G+Sp (1979)

5. V. Spiegel, Nucl. Sc. Eng. 53, 326 (1974)

6. C. Lederer and V.S. Shirley, Table of Isotopes, Seventh Edition, Wiley Interscience 1978. (1978)

7. W.G. Alberts, E. Günther, M. Matzke and G. Rassi (1975) Symp. Reactor Dosimetry Petten September 22-26, 1975 EUR 5667 e/f part. I. p. 131 Commission of the European Communities, Luxembourg (1977)

8. L. Green, Nucl. Sci. Eng. 58, 361 (1975)

9. G.J. Kirouac, H.M. Eiland and C.J. Slavik (1973), Proc. Topl. Mtg. Irradiation Experimentation in Fast Reactors, Jackson Lake Lodge, Wyoming, September 10-12, 1973 CONF-730910, P. 412 U.S. Atomic Energy Commission 1973

10. Z. Dezső and J. Csikai (1977) Proc. IV. All Union Conf. Neutron Physics, Kiev. April 22-26, 1977 Vol. III p.32 Atomizdat Moscow (1977)

11. J. Csikai and Z. Dezső, Ann. Nucl. En. 3, 527 (1977)

12. H. Pauw and A.H.W. Aten, J. Nucl. Energy 25, 457. (1971)

13. W. Mannhart and W.G. Alberts, Nucl. Sci. Eng. 69, 333 (1979)

14. D.L. Smith and J.W. Meadows ANL/NDM-14 and ANL/NDM-10 Argonne National Laboratory (1975)

15. D.L. Smith ANL/NDM-30 Argonne National Laboratory (1977)

INTEGRAL REACTION RATE MEASUREMENTS IN ^{252}Cf AND ^{235}U FISSION SPECTRA

G. P. Lamaze, E. D. McGarry, and F. J. Schima

National Bureau of Standards
Washington, D. C. 20234

In support of the light water reactor-pressure vessel (LWR-PV) surveillance dosimetry program established by the U.S. Nuclear Regulatory Commission, the National Bureau of Standards is undertaking a series of measurements to provide a physical basis for neutron dosimetry standards. Reaction rate measurements have been made with both ^{252}Cf and ^{235}U fission neutron fields. The following reactions have been measured through an activation technique: ^{115}In(n,n'), ^{58}Ni(n,p) and ^{54}Fe(n,p).

The neutron emission rate of the ^{252}Cf source has been measured with the $MnSO_4$ bath technique, thus permitting a direct measure of the spectrum averaged integral cross section for Cf ($\bar{\sigma}_{Cf}$). The source strength of the ^{235}U fission neutron field has been measured relative to the ^{252}Cf neutron field using the ^{115}In(n,n') reaction. All measurementsof cross sections in the ^{235}U fission field are therefore relative measurements.

[^{115}In(n,n'), ^{54}Fe(n,p), ^{58}Ni(n,p), ^{235}U and ^{252}Cf fission neutrons, spectrum averaged cross sections, neutron dosimetry]

An important method of fast neutron dosimetry is the use of radiometric monitors for measuring neutron fluence and fluence rate. By selecting a set of reactions each of which is sensitive to a different neutron energy range, information on both the shape and magnitude of the fluence spectrum is obtained. For example, this technique is applied in a light water reactor pressure vessel (LWR-PV) dosimetry program established by the U.S. Nuclear Regulatory Commission. More complete information of the application of this technique can be found in a series of ASTM standards.[1-5]

In order to obtain a neutron fluence from a measured reaction rate, the spectrum averaged cross section, $\bar{\sigma}$, must be obtained from

$$\bar{\sigma} = \frac{\int \sigma(E)\phi(E)dE}{\int \phi(E)dE}$$

The differential cross section, $\sigma(E)$, is obtained from an evaluated nuclear data file, such as ENDF/B, which is based on both experimental data and calculations. In a complex system, such as a reactor, the differential fluence rate $\phi(E)$ is usually obtained by using a transport code to calculate the neutron energy and spatial distribution. However, if the neutron source and the radiometric monitor consititute a relatively well understood system, a measure of $\bar{\sigma}$ becomes a validation of the knowledge of $\sigma(E)$. An examination of recent literature,[6-8] shows that even in pure fission fields, discrepancies exist between calculated and measured values of $\bar{\sigma}$. This paper reports new values of $\bar{\sigma}$ for ^{115}In(n,n')^{115m}In in ^{252}Cf fission spectrum and preliminary values of $\bar{\sigma}$ for ^{54}Fe(n,p)^{54}Mn and ^{58}Ni(n,p)^{58}Co for both ^{235}U and ^{252}Cf fission spectra.

Experimental Procedure

A) ^{252}Cf Fission Fluences

The NBS ^{252}Cf irradiation facilities have been described in detail.[9] Briefly, the facility used in this experiment consists of a californium source made up of a CfO_2 bead in an aluminum pellet singly encapsulated in a thin-walled stainless steel cylinder. The source is raised into the irradiation position between nearly identical foils to be irradiated. In this compensated beam geometry, the first-order distance error is associated only with the separation of detectors; the uncertainty in source deposit to foil distance becomes second order. Measurements in compensated beam geometry involve the average of the response of the two foils. The samples are commercially available, high purity, natural metal foils having 12.7 mm diameters and range in thickness from .13 to .25 mm.

The source strength (neutron emission rate) is determined by a manganous sulfate bath technique in which the Cf source emission is compared to that of NBS-I, a standard Ra-Be neutron source. The neutron fluence incident on the foils is obtained through knowledge of the source strength, average foil distance, and length of time of the irradiation. Corrections are made for neutron scattering, the decay of the ^{252}Cf (for long irradiations), saturation of activity, and attenuation of gammas in the foils.

B) ^{235}U Fission Fluences

The ^{235}U fission irradiations are obtained in the NBS cavity fission source.[10,11] This source operates at the center of a 30-cm diameter spherical cavity located at the center of the NBS Research Reactor graphite thermal column. The source-detector capsule consists of two coaxial source disks of ^{235}U metal (16 mm dia. x 0.13 mm thick) placed outside of a cadmium box that encloses passive detectors for exposure. The cavity and source detector arrangement is shown in Fig. 1.

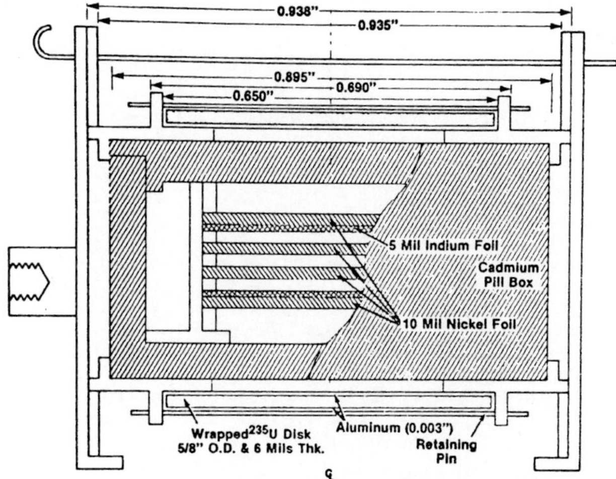

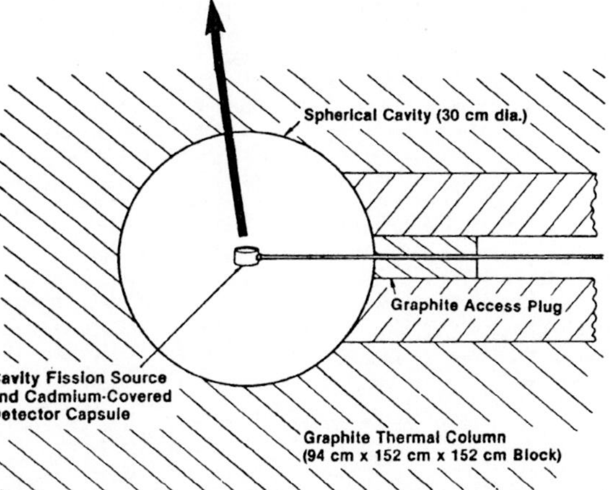

Fig. 1. U-235 Cavity Fission Source. Upper view: relative positions of neutron sensor foils, fission-disks and cadmium enclosure. Lower view: the upper assembly in the thermal column of the NBS reactor.

K. H. Böckhoff (ed.), Nuclear Data for Science and Technology, 425–428.

The fluence rate gradient between the two fission disks has been measured by simultaneously irradiating 13 nickel foils each .076 mm thick, the total mass thus approximates the total mass of a typical loading. The results of this measurement are given in Fig. 2. The fluence in the cavity is measured indirectly by using the $^{115}In(n,n')^{115m}In$ reaction. The indium is first irradiated in a ^{252}Cf field and the count rate at the end of irradiation (counts $s^{-1}g^{-1}$) measured. The gamma detector is then effectively calibrated in terms of fluence rate/count rate ($ncm^{-2}s^{-1}$/counts $s^{-1}g^{-1}$). Indium foils are then irradiated in the ^{235}U fission field to measure the fluence rate. The reactor power is also monitored with a small fission chamber in the cavity to record relative changes in the fluence rate during the irradiation. A correction is made to take into account the difference of the $^{115}In(n,n')$ cross section in the two fields using

$$\left[\bar{\sigma}_{Cf}\ (^{115}In(n,n'))/\bar{\sigma}_{25}\ (^{115}In(n,n'))\right] = 1.05$$

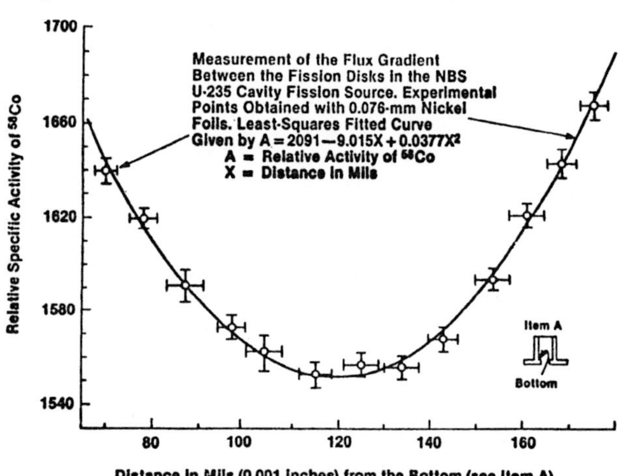

Fig. 2. Fluence rate gradient in the NBS Cavity Fission Source

C) Activity Measurements

In this work, the neutron induced radioactivities in nickel, iron and indium foils were assayed by measurement of the gamma-ray emission with an efficiency-calibrated Ge(Li) detector. The detector used is a 60-cm^3 co-axial diode with resolutions of 0.80 keV at 122 keV and 1.77 keV at 1332 keV. The efficiency calibration procedure is described in a preliminary report by A. T. Hershfeld et al.[12] and the subsequent report by B. M. Coursey et al.[13]

A computer-based data acquisition system[14] was used that was programmed to make periodic evaluation of the pre-selected gamma-ray emission rates. The peak fitting procedure is described in a report by F. J. Schima et al.[15] Data of each run were adjusted for decay to the time of the end of the neutron irradiation. The emission rate determination was obtained from the average of these periodic measurements. The decay parameters (the half-life and the gamma-ray probability per decay), were obtained from; D. D. Hoppes et al.[16] for ^{58}Co and ^{54}Mn, and the Nuclear Data Sheets[17] for ^{115m}In. Small corrections were made to the emission rates due to the finite size of the sources and the self-attenuation of the gamma-ray in the respective foil. Corrections were made for the decay within the measurement time for the ^{115m}In only. All measurements were performed at source distances for which calibrations exist.

Uncertainty Estimates

A) ^{252}Cf Measurements

The principal sources of systematic uncertainty in the ^{252}Cf cross section measurements are uncertainties in the source strength, distance measurements, neutron scattering, and gamma-ray counting efficiency. Table I summarizes the uncertainties of the three cross sections measured in the Cf field. The source strength has an uncertainty of 1.1% (1σ) for all measurements. For the ^{54}Fe and ^{58}Ni irradiations, the stem holding the ^{252}Cf source was slightly bent giving a source displacement uncertainty of about 1.6 mm. This source position uncertainty contributes a 0.7% (1σ) uncertainty to the fluence measurements. For the ^{115}In measurements, a different source was used and the source positioning contributes only a 0.4% uncertainty to the overall measurement.

Scattering effects are a function of the response range of the detector and must be considered individually. Eisenhauer et al.[18] describe the procedure used here to obtain the scattering corrections. Reaction rates measured in the perturbed spectrum should be divided by a correction factor, C_D, in order to infer the reaction rate in a free-field Cf spectrum. Since the response ranges of the $^{54}Fe(n,p)$ and $^{58}Ni(n,p)$ reactions are so similar, the scattering corrections have been assumed to be the same. Table II gives the reaction rate correction C_D due to scattering in various structures. The uncertainty in this correction is 1.0% (1σ), e.g., $C_{Ni} = 1.004 \pm .010$.

B) ^{235}U Fission Neutron Measurements

The uncertainty estimates for the $^{58}Ni(n,p)$ and $^{54}Fe(n,p)$ cross sections in ^{235}U cavity fission source are given in Table III. The principal source of error arises from the method of indirect measurement of the fluence rate. The principal contributors to this are the Cf source strength uncertainty and uncertainty in σ_{Cf}/σ_{U25} for In(n,n').

$$\left[\frac{\delta(\sigma_{Cf}/\sigma_U)}{\sigma_{Cf}/\sigma_U}\right]^2 = \sum_i \left[\mu_i^U\,\sigma_i^U/\sigma_U - \mu_i^{Cf}\,\sigma_i^{Cf}/\sigma_{Cf}\right]^2\left(\frac{\delta S_i}{S_i}\right)^2$$

$$+ \sum_i (1 - \sigma_i^U/\sigma_U)^2\ (\mu_i^U)^2\left(\frac{\delta\psi_i^U}{\psi_i^U}\right)^2$$

$$+ \sum_i (1 - \sigma_i^{Cf}/\sigma_{Cf})^2\ (\mu_i^{Cf})^2\left(\frac{\delta\psi_i^{Cf}}{\psi_i^{Cf}}\right)^2$$

where $\mu_i^U = \psi_i^U \Delta Ei$ and $\mu_i^{Cf} = \psi_i^{Cf}\Delta Ei$ are the ^{235}U and ^{252}Cf group spectra and $S_i = \sigma_i/\sigma_Q$ is the relative group averaged cross section for ^{115}In. This expression was evaluated numerically by selecting seven energy groups between 0.4 eV and 12 MeV and using available differential data. The sum of the spectrum uncertainty terms (second and third terms) is $2.3x10^{-4}$. The first term (cross section uncertainty term) was evaluated by arbitrarily taking 20 to 30% uncertainties in the differential cross section values. The value of this term is $0.2x10^{-4}$ indicating that the flux transfer between similar spectra is not sensitive to uncertainties in differential cross section data. The overall measurement errors in the ^{235}U fission spectrum measurement -- 2.7% for Ni and 3.1% for Fe -- can be reduced by establishing better scattering corrections based on Monte-Carlo calculations, and by irradiating to higher fluences and establishing longer counting intervals for Fe. New scattering calculations are already in progress.

Results

A) ^{252}Cf Fission Neutron Field

1) $^{115}In(n,n')^{115m}In$

The ^{115}In cross section was measured three times with two different pairs of .25 mm natural indium foils. To obtain the ^{115}In cross section, a natural abundance = .957, a $P\gamma$ = .459 and $T_{1/2}$ = 4.486h were used. The irradiation times varied from 48 to 64.5 hours to produce a nearly saturated activity. The counting statistics of each pair of foils were about 1% and the standard deviation of three measurements was also 1%. The final value for the $^{115}In(n,n')^{115m}In$ cross section was 196±4 mb.

Table I

Uncertainties for Cf Neutron Field Measurements

	$^{54}Fe(n,p)$	$^{58}Ni(n,p)$	$^{115}In(n,n')$
Source Strength of NBS-I	0.9%	0.9%	0.9%
Ratio of Cf to NBS-I	0.6	0.6	0.6
Source to Foil Distance	0.7	0.7	0.4
Statistics	0.2	0.1	0.9
Efficiency	0.8	0.8	0.8
Scattering	1.0	1.0	1.0
TOTAL	1.8%	1.8%	1.9%

Table II

Corrections for Scattering in NBS Foil Holder Arrangement

	C_{Ni}	C_{Fe}	C_{In}
Source Capsule	.992	.992	1.000
Source Support	1.002	1.002	1.002
Detector Support	1.004	1.004	1.004
Detector Holder	1.002	1.002	1.002
Foil Backing	1.002	1.005	1.001
Foil	1.002	1.001	1.001
TOTAL	1.004	1.006	1.01

Table III

Summary of Uncertainties for ^{235}U Neutron Field Measurements

Source		Contribution (%)	
Source Strength of NBS-I		0.9	
Ratio of Cf to NBS-I		0.6	
Cf Source to In Foil Distance		0.4	
Statistics of In Activity Counting		0.5	
σ_{Cf}/σ_{U25} of In(n,n')		1.6	
Scattering Correction for In at Cf		1.0	
Scattering Correction for In in Cavity		0.8	
Total Uncertainty in Fluence		2.4	
	^{54}Fe		^{58}Ni
Efficiency of γ Counting	0.75		0.75
Statistics of γ Counting	1.6		0.17
Flux Gradient in Cavity	0.4		0.4
Scattering Correction in Cavity	0.8		0.8
Total Uncertainty in σ	3.1%		2.7%

Table IV

Comparison of Present Results with Previous Results and Calculations

Cf Fission Neutron Field	This Work	Calculated[a]	Previous Work[b]
$^{115}In(n,n')^{115m}In$	196 ± 4 mb	182 ± 1.5 mb	195 ± 5 mb
$^{54}Fe(n,p)^{54}Mn$	89 ± 2	88.3 ± 1.2	87 ± 2.4
$^{58}Ni(n,p)^{58}Co$	121 ± 2	114 ± 1.5	118 ± 3
^{235}U Fission Neutron Field			
$^{58}Ni(n,p)^{58}Co$	109 ± 3	101 ± 2.6	109 ± 8
$^{54}Fe(n,p)^{54}Mn$	79 ± 3	77.8 ± 2.2	83 ± 4

[a]NBS evaluated fission spectra[20] and ENDF/B-V cross sections; errors are for spectrum uncertainties only.

[b]NBS evaluation of observed cross sections recorded in NBS compendium.[10]

2) $^{58}Ni(n,p)^{58}Co$

One pair of .25 mm natural nickel foils was irradiated for a total time of 22.798 days. This irradiation was interrupted twice for a total of 104 minutes and the decay of the ^{58}Co during these periods taken into account. The decay of the Cf source during this long irradiation is also taken into account. To obtain the cross section, values of the natural abundance = .6827, $P\gamma$ = 0.994, and $T_{1/2}$ = 70.78 days were used. Since this measurement has not yet been repeated, the value reported here must be considered preliminary. The value obtained is 121±2 mb.

3) $^{54}Fe(n,p)^{54}Mn$

The $^{54}Fe(n,p)^{54}Mn$ irradiation occured simultaneously with the $^{58}Ni(n,p)^{58}Co$ irradiation and the above irradiation times apply here. The natural iron foils of 99.64% purity were .13 mm thick and were backed by .75 mm thick titanium foils. The titanium results are not part of this report. Values of the natural abundance = .058, $P\gamma$ = .9997, and $T_{1/2}$ = 312.02 days were used to obtain the cross section. Since this measurement has not been repeated, the value reported here must be considered preliminary. The value obtained is 89±2mb.

B) ^{235}U Fission Neutron Field

1) $^{58}Ni(n,p)^{58}Co$

Two nickel irradiations have been made. In the first, two .25 mm natural metal foils were irradiated about 30 hours in a total fluence of 2.11×10^{15} n/cm^2. In the second irradiation, two .09 mm natural metal foils were irradiated about 55 hours in a total fluence of 3.00×10^{15} n/cm^2. The result from these four foils gives a value of 109 mb with a standard deviation of 0.9% indicating good reproducibility of the method. Combining all sources of uncertainty (see Table III) in quadrature gives as a final value of the cross section 109±3 mb.

2) $^{54}Fe(n,p)^{54}Mn$

The iron cross section value obtained in this experiment is the result of only one irradiation and as such must be considered preliminary. Two natural iron foils .13 mm thick were irradiated for about 29 hours in a total fluence of 2.17×10^{15} n/cm^2. Taking the average value of the two foils and combining all sources of uncertainty (see Table III) in quadrature gives a value of 79±3 mb for the $^{54}Fe(n,p)$ cross section. It should be noted that there is a 1.7% uncertainty assigned to the isotopic abundance of ^{54}Fe in natural iron. This uncertainty has not been included in the uncertainty of this measurement. If natural iron samples are used for neutron dosimetry this factor is eliminated in the application of the method.

Discussion

Table IV compares the present results with the NBS Compendium[10] which gives calculated values of the integral cross sections and an evaluation of observed values of these cross sections. The calculations use the NBS evaluation[20] of the ^{235}U and ^{252}Cf fission neutron spectra and ENDF/B-V cross sections. All of the present results are in good agreement with the observed values, as reported in the NBS compendium. Furthermore, the nickel and iron results are in agreement with the accurate, $^{58}Ni(n,p)/^{54}Fe(n,p)$, cross-section-ratio measurements of Fleming and Spiegel.[19] They report a ratio of 1.346±0.03 for the ^{235}U fission neutron spectrum and a value 1.326±0.03 for the ^{252}Cf spectrum. For comparison, the present measurements yield ratios of 1.38 and 1.36, respectively. The agreement is within the combined uncertainties of the present measurements.

While the $^{54}Fe(n,p)$ results agree with the calculated values for both of the mentioned fission neutron spectra, the calculated results for $^{58}Ni(n,p)$ are low by 6-8% in both fields and the $^{115}In(n,n')$ results are low by 8% for the ^{252}Cf field. These integral measurements cannot, of course, indicate whether the problems are associated with ENDF/B-V cross sections or the fission spectra evaluation, or both. However, since the cross section shape for both the $^{54}Fe(n,p)$ and $^{58}Ni(n,p)$ reactions are so similar, the calculated cross section ratios should be less sensitive to the shape of the fission neutron spectrum than to the evaluated cross section.

Acknowledgements

The authors thank C. M. Eisenhauer for the calculations of the scattering corrections and D. D. Hoppes for aiding in determinations of the gamma counting efficiency. The patience of Donna Shipe and Gloria Wiersma in the preparation of the manuscript is greatly appreciated.

References

1. Annual Book of ASTM Standards, Part 45, E706-81, Standard Master Matrix for Light-Water Reactor Pressure Vessel Surveillance Standards.
2. Ibid, E261-77, Standard Practice for Measuring neutron Flux, Fluence, and Spectra by Radioactivation Techniques.
3. Ibid, E844-81, Sensor Set Design and Irradiation for Reactor Surveillance.
4. Ibid, E263-77, Determining Fast-Neutron Flux by Radioactivation of Iron.
5. Ibid, E264-77, Determining Fast-Neutron Flux by Radioactivation of Nickel.
6. W. L. Zijp, H. Ch. Rieffe, H. J. Nolthenius, Proc. of the Fourth ASTM-EURATOM Symposium, Gaithersburg, MD (1982), p. 725.
7. W. Mannhart, Proceedings of the Fourth ASTM-EURATOM Symposium, Gaithersburg, MD (1982), p. 637.
8. R. J. LaBauve, D. G. Madland, submitted to American Nuclear Society, 1982 Annual Meeting, June 6-11, 1982, Los Angeles, CA.
9. J. A. Grundl, V. Spiegel, C. M. Eisenhauer, H. T. Heaton II, D. M. Gilliam, J. Bigelow, Nuc. Tech. 32 (1977).
10. J. A. Grundl, et al., Compendium of Benchmark Neutron Fields for Reactor Dosimetry, LWR Pressure Vessel Dosimetry Quarterly Progress Report, NUREG/CR-0551 (1978) (calculated cross-sections updated to ENDF/B-V).
11. E. D. McGarry, G. P. Lamaze, C. M. Eisenhauer, D. M. Gilliam, F. J. Schima, Proceedings of the Fourth ASTM-EURATOM Symposium, Gaithersburg, MD (1982), p. 597.
12. A. T. Hirshfeld, D. D. Hoppes and F. J. Schima, Proc. ERDA Symp. on X- and gamma-ray sources and applications, NTIS CONF-760539, 90 (1971).
13. B. M. Coursey, D. D. Hoppes and F. J. Schima, Nucl. Inst. and Meth. 193, 1 (1982).
14. F. J. Schima, Computers in activation analysis and gamma-ray spectroscopy, CONF-780421, 416 (1978).
15. F. J. Schima, D. D. Hoppes and A. T. Hirshfeld, Computers in activation analysis and gamma-ray spectroscopy, CONF-780421, 177 (1978).
16. D. D. Hoppes, J. M. R. Hutchinson, F. J. Schima and M. P. Unterweger, National Bureau of Standards, Special Publication 626, 85 (1982).
17. Nuclear Data Sheets 30, 455 (1980).
18. C. M. Eisenhauer, D. M. Gilliam, J. A. Grundl, V. Spiegel, Proceedings of the Second ASTM-EURATOM Symposium on Reactor Dosimetry, Palo Alto, CA (1977) p. 1177.
19. R. Fleming, V. Spiegel, Proc. of the Second ASTM-EURATOM Symposium on Reactor Dosimetry, Palo Alto, CA (1977) p. 953.
20. J. Grundl, C. Eisenhauer, Proc. of the First ASTM-EURATOM Symposium on Reactor Dosimetry, Petten (1975) p. 425.

MEASUREMENT AND EVALUATION OF INTEGRAL DATA IN THE CF-252 NEUTRON FIELD

W. Mannhart

Physikalisch-Technische Bundesanstalt
3300 Braunschweig, FR Germany

Average neutron cross sections were determined in the neutron field of spontaneous fission of ^{252}Cf. The reactions were: ^{19}F(n,2n), ^{24}Mg(n,p), ^{27}Al(n,p), ^{55}Mn(n,2n), ^{59}Co(n,p), ^{59}Co(n,α), ^{59}Co(n,2n), ^{58}Ni(n,2n), ^{63}Cu(n,γ), ^{63}Cu(n,α), ^{63}Cu(n,2n) and ^{90}Zr(n,2n). Conventional integral testing was applied to check the consistency of the experimental data with those calculated based on ENDF/B-V cross section data and various experimental and theoretical spectrum representations. Generalized least-squares techniques were used in the evaluation of a consistent set of integral data with regard to a complete covariance matrix.

[^{19}F(n,2n), ^{24}Mg(n,p), ^{27}Al(n,p), ^{55}Mn(n,2n), ^{59}Co(n,p), ^{59}Co(n,α), ^{59}Co(n,2n), ^{58}Ni(n,2n), ^{63}Cu(n,γ), ^{63}Cu(n,α), ^{63}Cu(n,2n), ^{90}Zr(n,2n), fission-spectrum averaged cross sections, ^{252}Cf, testing of ENDF/B-V, generalized least-squares techniques]

Introduction

Major nuclear data requirements deal with relating nuclear data to integral quantities. These quantities are, for example, reaction rates for radiation damage estimates or k_{eff} values and breeding ratios for fast reactor configurations and others. From this point of view, integral parameters play an important role in the evaluation of nuclear data as well as in the application of these data. Integral data can be roughly separated into two types: simple integral data and more complex ones. The terminology of "simple" integral data is chosen to distinguish those integral data depending alone on a specific neutron cross section, from more complex quantities which are influenced by a variety of cross section data. To the class of simple integral data belong integral reaction rates or spectrum-averaged cross sections (reaction rates normalized with the neutron flux density) in well-known neutron fields. This class mainly includes integral data in neutron benchmark fields such as fission spectra and a few others ($\Sigma\Sigma$, CFRMF). In this context, the neutron field of the spontaneous fission of ^{252}Cf is of particular importance. Compared with other "driven" benchmarks, the neutron spectrum is very clean, does not require extensive corrections for spectrum perturbation effects and, most important of all, the neutron spectrum is independent of any neutron cross section data. All this qualifies integral data measured in the neutron field of ^{252}Cf as candidates for inclusion (together with energy-dependent cross section measurements) in the evaluation of nuclear data files.

The present work continues work begun previously[1-3] on spectrum-averaged cross section data in the neutron field of spontaneous fission of ^{252}Cf. Special emphasis is given to the measurement of high-threshold (n,2n)-reactions with the aim of obtaining better information about the shape of the spectrum at high neutron energies. Conventional integral testing of evaluated energy-dependent cross section data is applied with various experimental and theoretical representations of the ^{252}Cf neutron spectrum. Particular attention is paid to the inclusion of covariance information available, to render the consistency checks of calculated and measured integral data as trustworthy as possible. For the high-threshold reactions, the integral testing is presently of more use in testing the spectrum representations than for giving information about the quality of energy-dependent cross section data. Finally the result of an evaluation of a best set of ^{252}Cf spectrum-averaged cross sections based on relative as well as on absolute experimental data is given. The data of this evaluation, performed with generalized least-square techniques, may serve as a basis for future inclusion of these data in evaluated nuclear data files.

Experiment

The neutron source and the experimental procedure were almost the same as in Ref. 3, and for this reason, only a brief description is given here. The ^{252}Cf neutron source was encapsulated in a double zircaloy cylinder with outer dimensions of 10 mm in diameter and 10 mm in height. The wall thickness of the cylinder was 1.5 mm. Spectrum perturbations due to the encapsulation were of the order of 1 %, somewhat dependent on the neutron energy. The neutron source strength was approximately 2.0×10^8 s^{-1} on July 1, 1981. A few experiments were performed with another neutron source of the same size but containing two milligrams of recently-produced californium. In these cases the neutron source strength was 4.7×10^9 s^{-1} on April 1, 1982. Irradiations were performed in a low-scattering arrangement. The samples were high-purity metallic foils with the exception of the ^{19}F(n,2n) reaction where teflon was used. Irradiations were performed using the sandwich method i.e. each sample was placed between two foils acting as neutron flux density monitors. Activities were determined with a large-volume Ge(Li) detector of a nominal efficiency of 28 %. The activities of the short-lived reaction products ^{18}F, ^{27}Mg and ^{62}Cu with half-lives of 109.8 m, 9.462 m and 9.74 m, respectively, were followed up to seven half-lives to identify the reaction products uniquely and to correct for other contributions. In the case of ^{27}Mg, the identification was relatively easy following the 843.8 keV gamma ray line. The strong positron emitters ^{18}F and ^{62}Cu were analyzed via the 511 keV line of positron annihilation. For ^{18}F, a small constant portion was found which could be identified as belonging to the detector background. For ^{62}Cu, in addition, a second half-life was found in the decay of the 511 keV gamma line. The component resulted from ^{64}Cu produced via the ^{63}Cu(n,γ) and ^{65}Cu(n,2n) reactions.

The experimental results are shown in Table I. Those data measured with the strong neutron source are especially marked. All data were normalized with the cross sections of the reactions ^{27}Al(n,α), ^{58}Ni(n,p) and ^{115}In(n,n'), respectively, used as neutron flux density monitor reactions. The corresponding values were taken from earlier experiments[2] and were 1.006 mb for ^{27}Al(n,α), 118 mb for ^{58}Ni(n,p) and 195 mb for ^{115}In(n,n'). Neither reaction, ^{63}Cu(n,γ)^{64}Cu or ^{65}Cu(n,2n)^{64}Cu can be experimentally separated. The spectrum-averaged cross section of ^{63}Cu(n,γ) was obtained by subtracting the portion weighted with the isotopic abundances which belongs to ^{65}Cu(n,2n) from the common response of both reactions. For these purposes, the spectrum-averaged cross section of ^{65}Cu(n,2n) was calculated based on ENDF/B-V data and was taken as 0.656 mb. In obtaining the cross section of the reaction ^{63}Cu(n,γ), the correction was 2.7 %. The resulting spectrum-averaged cross section of 10.39 mb is in good agreement with the earlier experiment of Green[8]. It is also fairly consistent with the calculated value of 9.649 mb based on ENDF/B-V cross section data and the NBS evaluation of the neutron spectrum (details below).

The present data are compared with those from other experiments. In many cases only a single previous experiment is available. Even when some of the present

K. H. Böckhoff (ed.), Nuclear Data for Science and Technology, 429–435.

Table I Experimental ^{252}Cf Spectrum-Averaged Cross Sections (in mb)

Reaction	Present Work	Other Experiments	
$^{19}F(n,2n)^{18}F$	$(1.63 \pm 0.05)\times10^{-2}$ a)	$(1.08 \pm 0.16)\times10^{-2}$	Ref.4
$^{24}Mg(n,p)^{24}Na$	2.01 ± 0.06	1.94 ± 0.09	Ref.5
$^{27}Al(n,p)^{27}Mg$	4.80 ± 0.09 a)	4.89 ± 0.18	Ref.5
		5.11 ± 0.43	Ref.4
$^{55}Mn(n,2n)^{54}Mn$	0.408 ± 0.009	0.58 ± 0.14	Ref.4
$^{59}Co(n,p)^{59}Fe$	1.68 ± 0.04	1.96 ± 0.10	Ref.4
$^{59}Co(n,\alpha)^{56}Mn$	0.222 ± 0.004 a)b)	0.218 ± 0.014	Ref.5
		0.20 ± 0.01	Ref.4
		0.20 ± 0.01	Ref.6
$^{59}Co(n,2n)^{58}Co$	0.406 ± 0.010	0.57 ± 0.06	Ref.7
$^{58}Ni(n,2n)^{57}Ni$	$(8.94 \pm 0.28)\times10^{-3}$ a)b)		
$^{63}Cu(n,\gamma)^{64}Cu$	10.39 ± 0.30 a)	10.95 ± 0.51	Ref.8
		17.6 ± 1.4 c)	Ref.9
$^{63}Cu(n,\alpha)^{60}Co$	0.671 ± 0.018	0.709 ± 0.017	Ref.10
$^{63}Cu(n,2n)^{62}Cu$	0.183 ± 0.007 a)	0.30 ± 0.03	Ref.7
$^{90}Zr(n,2n)^{89}Zr$	0.221 ± 0.006	0.267 ± 0.015	Ref.4

a) Measured with a recent neutron source of a strength of $4.7\times10^{9}\ s^{-1}$

b) Replaces a previous experiment (Ref.3)

c) $^{63}Cu(n,\gamma)^{64}Cu + ^{65}Cu(n,2n)^{64}Cu$

data are fairly consistent with other previous experiments, a general comparison indicates that there is an urgent need for additional confirmatory experiments, particularly to solve the large discrepancies seen between the present data and those from Ref. 4 and Ref. 7. In the case of the reaction $^{63}Cu(n,\alpha)$ the earlier experiment of Winkler et al.[10] was very carefully performed. Here, a re-measurement is being carried out to resolve the difference of 5.7 % between both experiments which is outside the error bars of both experiments.

Spectrum Representation of ^{252}Cf

The availability of recent experimental[11] and theoretical[14] data for the neutron spectrum of ^{252}Cf seems to be worthy of consideration in the integral testing procedures outlined below. A commonly used parametric representation of fission spectra is the Maxwellian distribution:

$$\chi(E)\equiv\chi^{M}(E,kT)=\frac{2}{\sqrt{\pi}}\frac{1}{(kT)^{3/2}}\sqrt{E}\ \exp(-E/kT) \qquad (1)$$

with $\chi(E)$ being a normalized spectral distribution. For ^{252}Cf, the temperature parameter is at present assumed to be kT = 1.42 MeV, corresponding to an average energy <E> of 2.13 MeV. The Maxwellian is a typical center-of-mass system representation and therefore somewhat unphysical in describing the spectrum in the laboratory system due to its neglecting the motion of fission fragments during the emission of neutrons. Nevertheless, in the analysis of experimental spectra, the Maxwellian is mostly used for its simplicity. A recent TOF experiment[11] of the ^{252}Cf neutron spectrum analyzed in terms of a Maxwellian is given by:

$$\chi^{EXP}(E)=1.095\ \chi^{M}(E,1.355\ \text{MeV}) \quad \text{with } 3\,\text{MeV}\leq E\leq 13\,\text{MeV} \qquad (2)$$

As the detector system was calibrated, this experiment is an absolute one. The normalization constant of 1.095 is based on a weighted average of experimental normalization factors at discrete energies between 3 MeV and 13 MeV. These factors (Fig. 5 of Ref. 11) indicate some structure in this energy range which, however, is probably due to problems in the efficiency calibration which still have to be solved. A constant factor has therefore been chosen instead of a function which is questionable in its physical meaning. Another representation of the ^{252}Cf neutron spectrum is given by an evaluation performed at the National Bureau of Standards[12]:

$$\chi^{NBS}(E)=\chi^{M}(E,1.42\ \text{MeV})\ \mu(E) \qquad (3)$$

Here the spectrum is described by a Maxwellian of <E> = 2.13 MeV corrected in five continuous segments with $\mu(E)$. Between 1.5 MeV and 6 MeV, Eq. (3) is essentially dominated by the pure Maxwellian. Above 6 MeV, the spectrum of Eq. (3) is decreased compared to the Maxwellian. Recent experimental investigations of the high-energetic part of the ^{252}Cf neutron spectrum[13] have confirmed the representation of Eq. (3) up to 20 MeV.

The theoretical work of Madland and Nix[14] describing the ^{252}Cf neutron spectrum is based on nuclear evaporation theory. It takes into account the center-of-mass motion of both fission fragments, the cooling of the fragments in the subsequent emission of neutrons and the energy dependence of the cross section for the inverse process of compound nucleus formation. A great advantage here is that the formalism simultaneously yields expressions for the neutron spectrum as well as for $\bar{\nu}$, the average number of neutrons emitted in spontaneous fission. The energy dependence of the compound nucleus cross section is based on optical model parameters from Becchetti and Greenless[15]. The high energy tail of this theoretical neutron spectrum depends mainly on two quantities: the average energy release and the nuclear level density parameter. The energy release for the division of the compound nucleus in a pair of fission fragments is the difference between the mass of the compound nucleus and the masses of both fission fragments. This quantity must be averaged over all fission-fragment mass and charge distributions. In the original work[14], Myers'mass formula[16] has been used in the calculation of the energy release. Replacing this formula by another mass formula of Moeller and Nix[17] results in a decrease of the average energy release and decreases the spectrum at high neutron energies. This can clearly be seen in Fig. 1. The figure neglects the part of the neutron spectrum below 2 MeV as this is of minor importance here. The decrease of the neutron spectrum of the NBS evaluation compared with the Maxwellian at high neutron energies is easily recognizable. The same is valid for the theoretical result with the Moeller-Nix mass formula instead of the mass formula of Myers originally quoted. A further decrease of the theoretical spectrum at high neutron energies could be obtained by increasing the nuclear level density parameter above the presently used value of

$$a = A/11 \quad \text{MeV}^{-1} \qquad (4)$$

(A being the mass number of the fissioning nucleus). However, before doing so, the physical evidence must be carefully examined. The experimental data of Ref. 11 show distinctly parallel behaviour to the NBS evaluation, even when extending the high energy trend of the evaluation to lower energies.

Integral Testing at Medium Neutron Energies

Conventional integral testing is one of the first steps in combining the information available on energy-

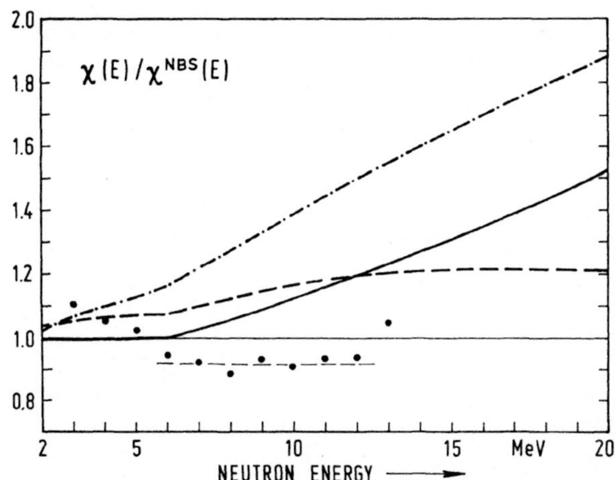

Fig. 1. Ratio of various neutron spectrum representations of ^{252}Cf relative to the NBS evaluation (Eq. 3). The solid curve is for a pure Maxwellian with kT = 1.42 MeV. The dashed curve is for the Madland-Nix formalism with Becchetti-Greenless potential and the mass formula of Moeller-Nix, the dot-dashed curve is for the same formalism with the mass formula of Myers. The points correspond to the experimental data of Ref. 11. The discontinuity of all curves at 6 MeV is due to the correction function of the NBS evaluation.

dependent cross sections with directly measured integral quantities. Provided that the integral measurements have been carefully carried out, the comparison between $\langle\sigma\rangle_{exp}$ and $\langle\sigma\rangle_{calc}$ allows conclusions to be drawn as regards the quality of energy-dependent cross section data as well as the adequacy of the spectrum representation of the benchmark neutron field used. Both types of conclusions are strongly interrelated, i.e., a specific conclusion always requires that its counterpart is free from any doubt. Nevertheless, even when this condition cannot be perfectly fulfilled, the calculated/experimental (C/E) ratio is of some use. In such cases, the C/E value of a specific reaction cannot be unambigiously interpreted, of course, but global trends can be recognized by regarding a bulk of C/E values representing various reactions.

All the above-mentioned spectrum representations were used in the comparison between measured and calculated integral ^{252}Cf data. The energy-dependent cross section data are based mainly on ENDF/B-V. In a few cases other evaluations[19-23] were used. The results are given in Table II. These reactions were selected with main energy responses in the range between 3 MeV and 13 MeV covered by the experiment of Eq. (2). In Column 2 and 3 of Table II the percental response of each reaction below 3 MeV and above 13 MeV is given. As these values never exceed 5 %, corrections in Column 5 arising from the normalization not fulfilled by Eq. (2) could be neglected. In the case of the spectrum-averaged cross sections calculated with the NBS evaluation, uncertainties are given. They were calculated with

$$\text{Rel.Var}(\langle\sigma\rangle) = P_\sigma^t\, C_\sigma\, P_\sigma \;+\; P_\phi^t\, C_\phi\, P_\phi \qquad (5)$$

C being the relative covariance matrix of the energy-dependent cross section data or of the spectral neutron flux density. The column vectors P contain sensitivity coefficients. The superscript t indicates a transpose. Details are explained elsewhere[24]. The covariance file of ENDF/B-V was used in generating[25] the final covariance matrices of most σ(E) data. In some cases, the information was directly taken from the literature. The evaluation of the second term of Eq. (5) presents basic difficulties. At present, no covariance matrix of the ^{252}Cf neutron spectrum is available, with the exception of an attempt to generate this matrix for a Maxwellian with kT = 1.42 MeV by attributing a 2 % uncertainty to the temperature parameter[26]. This result is far from satisfactory. The evaluation of the uncertainty component of

Table II Comparison between Experiment and Calculation (Neutron energy range: 3 to 13 MeV)

Reaction	Response (in %)		$\langle\sigma\rangle_{exp}$ [a] (in mb)	$\langle\sigma\rangle_{calc}$ (in mb)					
				Spectral Representation:					σ(E)-data
	below 3 MeV	above 13 MeV		Experiment[c] Böttger et al.	NBS[d] Evaluation	Maxwellian $\langle E\rangle$=2.13 MeV	Theory 1[f] $\langle E\rangle$=2.2167 MeV	Theory 2[g] $\langle E\rangle$=2.2791 MeV	from:
^{24}Mg(n,p)	0.0	2.6	2.01(6)[b]	2.049	2.159(88)[e]	2.335	2.449	2.825	Ref. 18
^{27}Al(n,p)	0.7	0.4	4.80(9)	5.015	5.14(29)	5.274	5.617	6.126	ENDF/B-V
^{27}Al(n,α)	0.0	3.2	1.006(22)	1.004	1.059(58)	1.158	1.208	1.406	ENDF/B-V
				0.959	1.012(29)	1.107	1.155	1.345	Ref. 19
^{46}Ti(n,p)	0.1	0.6	13.8(3)	13.13	13.5(17)	13.81	14.72	16.05	ENDF/B-V
^{48}Ti(n,p)	0.0	4.4	0.42(1)	0.3884	0.409(42)	0.4457	0.4653	0.5399	ENDF/B-V
^{56}Fe(n,p)	0.0	2.1	1.450(35)	1.348	1.414(63)	1.504	1.586	1.803	ENDF/B-V
^{59}Co(n,p)	2.9	0.8	1.68(4)	1.695	1.733	1.779	1.893	2.063	Ref.20,21
^{59}Co(n,α)	0.0	3.5	0.222(4)	0.2051	0.2162(91)	0.2348	0.2455	0.2842	ENDF/B-V
^{63}Cu(n,α)	0.0	1.7	0.671(18)	0.7242	0.758(40)	0.8066	0.8499	0.9652	ENDF/B-V
				0.6473	0.676(38)	0.7145	0.7547	0.8516	Ref. 22

a) This work and Ref. 2

b) Read as 2.01 ± 0.06

c) Ref. 11 and eq. (2)

d) Ref. 12

e) Uncertainty due to the σ(E)-data; without the uncertainty of the spectral representation

f) Madland-Nix formalism: Becchetti-Greenless potential, Möller/Nix mass formula

g) Madland-Nix formalism: Becchetti-Greenless potential, Myers mass formula

the neutron spectrum has therefore been consistently neglected, keeping in mind that such an artificial covariance matrix is of little use. The propagated relative uncertainties of Eq. (5) depend only weakly on details of the spectrum representation. Similar values can therefore be attributed to the data of the other spectrum representations given in Table II.

Table II shows that the recent cross section evaluations of $^{27}Al(n,\alpha)$[19] and of $^{63}Cu(n,\alpha)$[22] are superior to the older ENDF/B-V evaluations. The $\sigma(E)$ data of $^{59}Co(n,p)$ are based on the data of Ref. 20 up to 10 MeV. Above 10 MeV, data were taken from Ref. 21. Their influence on the results given here was a weak one, as a small portion of the total response (5.4 %) was above 10 MeV. The C/E values of the reactions of Table II for various spectrum representations are shown in Fig. 2. For $^{27}Al(n,\alpha)$ and $^{63}Cu(n,\alpha)$, the values of Fig. 2. are based on $\sigma(E)$ data taken from Ref. 19 and Ref. 22, respectively, instead from ENDF/B-V. The 90 % energy response ranges extend from 3.51 MeV to 9.84 MeV for $^{27}Al(n,p)$, on the left-hand side of Fig. 2., and from 6.57 MeV to 12.10 MeV for $^{24}Mg(n,p)$, on the right-hand side of the figure.

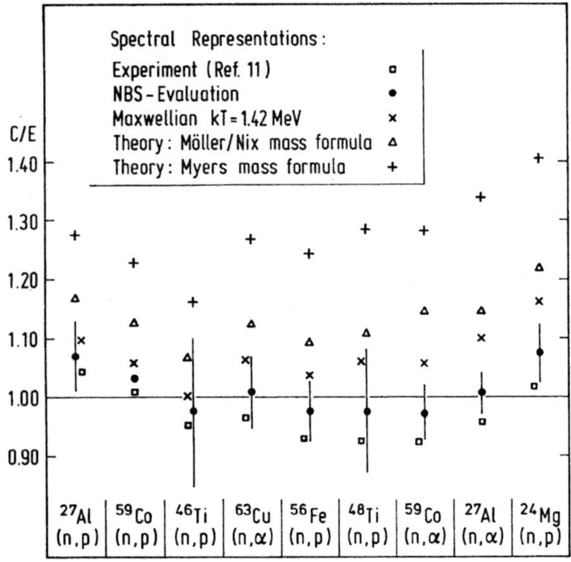

Fig. 2. Ratio of calculated ^{252}Cf spectrum-averaged cross sections relative to the experimental values for various spectrum representations. The error bars given for the data of the NBS evaluation are the quadratically combined uncertainties of the energy-dependent cross section data and of the experiments. The reactions are ordered according to their energy response ranges with energy responses at lower energies on the left-hand side and at higher energies on the right-hand side of the figure.

In the region of energy responses at lower energies (to the left in the figure), a fair consistency between experimental data and those calculated with the spectral representations of Eq. (2), Eq. (3) and that of a pure Maxwellian (kT = 1.42 MeV) is shown. The calculated data based on Eq. (2) show a tendency to be somewhat lower (between 3 % and 5 %) than those based on the NBS evaluation of the neutron spectrum. This effect is mainly due to the normalization factor of Eq. (2) whose relative uncertainty is presently estimated as being of the order of 5 %. At higher neutron energies (right-hand side of Fig. 2.), the consistency mentioned of the data calculated with Eqs. (2) and (3) continues, but the data of the pure Maxwellian show a clear increase, indicating an overestimation of the neutron spectrum by the pure Maxwellian. The data of the theory of Madland and Nix calculated with the Moeller-Nix mass formula show a similar trend. Compared with the NBS evaluation, these data are between 9 % and 14 % higher, which is in most cases outside the given error bars. The integral values calculated with the original theory of Madland and Nix[14] by using the mass formula of Myers, exceed the experimental data by 20 % to 40 %. This trend increases in the higher neutron energies, as shown in the next section.

Integral Testing at High Neutron Energies

The crucial test of which of the above-mentioned spectrum representations is most adequate in describing the prompt fission neutron spectrum of ^{252}Cf occurs at high neutron energies. There are two disadvantages in such a test. First, the energy-dependent cross section data of high-threshold reactions are not always as well-established as the data of the reactions given in Table II. Secondly, due to the small number of fission neutrons at high energies, the measured integral responses are weak, resulting in larger uncertainties of the measured data. On the other hand, the integral responses become very sensitive as regards the shape of the neutron spectrum at high energies. The results obtained with different spectrum representations differ by more than 20 % which in most cases is sufficient to recognize general trends.

In Table III the high-threshold reactions used in this test and their 90 % energy response ranges are given. The reactions cover the neutron energy range between 9 MeV and 18 MeV. The calculated integral data are based on the same spectrum representations as before, with the exception of data calculated with Eq. (2) not covering this energy range. Similar to before, the data are shown in Fig. 3. in a C/E representation.

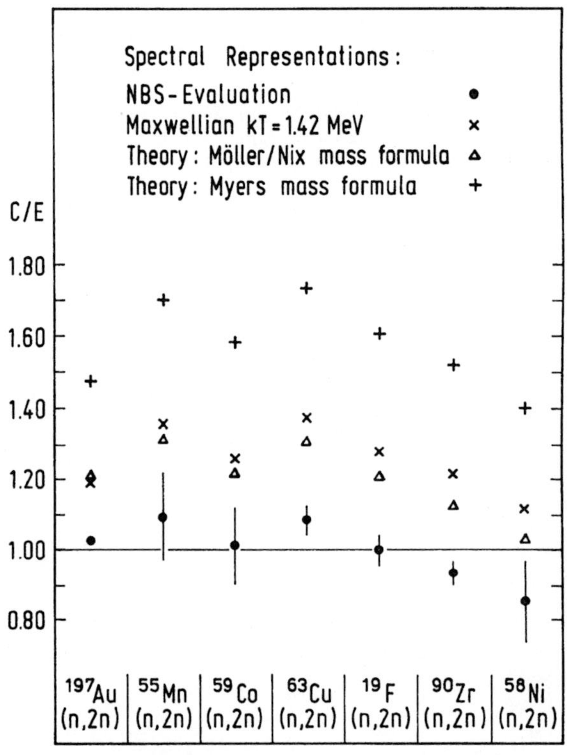

Fig. 3. Identical to Fig. 2 for high-threshold (n,2n) reactions

Taking into account the error bars we see that only one spectral representation, the NBS evaluation, achieves a full consistency between experimental and calculated integral data. As expected from the data shown in Fig. 1., at high neutron energies the integral value calculated with the Madland-Nix formalism based on the Moeller-Nix mass formula becomes lower compared with

Table III Comparison between Experiment and Calculation (High-threshold (n,2n)-reactions)

Reaction	90 % Energy Response Range (in MeV)		$<\sigma>_{exp}$ (in mb)	$<\sigma>_{calc.}$ (in mb)				
				Spectral Representation:				σ(E)-data
	from	to		NBS[c] Evaluation	Maxwellian $<E>$= 2.13 MeV	Theory 1[e] $<E>$= 2.2167 MeV	Theory 2[f] $<E>$= 2.2791 MeV	from
^{197}Au(n,2n)	8.91	14.1	5.50(14)[a][b]	5.646	6.551	6.642	8.120	ENDF/B-V
^{55}Mn(n,2n)	11.2	16.4	0.408(9)	0.446(56)[d]	0.5535	0.5355	0.6940	ENDF/B-V
^{59}Co(n,2n)	11.3	16.5	0.406(10)	0.410(43)	0.5118	0.4934	0.6420	ENDF/B-V
^{63}Cu(n,2n)	12.0	17.3	0.183(7)	0.1981(33)	0.2523	0.2390	0.3168	Ref. 18
^{19}F(n,2n)	12.0	17.6	$1.63(5)\cdot10^{-2}$	$1.626(43)\cdot10^{-2}$	$2.084\cdot10^{-2}$	$1.964\cdot10^{-2}$	$2.616\cdot10^{-2}$	Ref. 23
^{90}Zr(n,2n)	12.8	17.9	0.221(6)	0.2069(42)	0.2690	0.2488	0.3355	Ref. 18
^{58}Ni(n,2n)	13.1	18.0	$8.94(28)\cdot10^{-3}$	$7.60(83)\cdot10^{-3}$	$9.966\cdot10^{-3}$	$9.208\cdot10^{-3}$	$1.252\cdot10^{-2}$	ENDF/B-V

a) Ref. 2
b) Read as 5.50 ± 0.14
c) Ref. 12
d) Uncertainty due to the σ(E)-data; without the uncertainty of the spectral representation
e) Madland-Nix formalism: Becchetti-Greenless potential, Möller/Nix mass formula
f) Madland-Nix formalism: Becchetti-Greenless potential, Myers mass formula

that calculated with the pure Maxwellian. However, neither spectrum representation is consistent with the experimental data.

The influence of energy-dependent cross section data on the result of the test of Fig. 3 should not be neglected here. On the basis of the example of the reaction ^{59}Co(n,2n), Fig. 4 shows the shape of the excitation function and the response function of this reaction for ^{252}Cf. One recognizes that the integral of the response depends strongly on the exact slope of the cross section curve near threshold. Unfortunately, the neutron energy region below 14 MeV which here mainly determines the value of the integral response, is a source of problems as regards the measurement of energy-dependent cross section data. Most excitation functions are therefore not so well-established in this region as they are chiefly based on model calculations. For a single neutron reaction, all this can reduce the information of the consistency test of Table III and Fig. 3, of course. However, the fact that the information is obtained from a variety of different neutron reactions in Fig. 3 is believed to be a sufficient basis for the conclusions drawn.

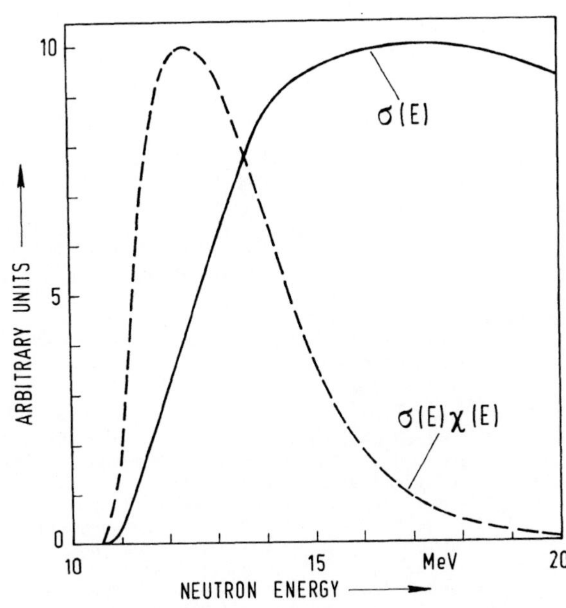

Fig. 4. Energy-dependent cross section and response in the ^{252}Cf neutron spectrum (NBS evaluation) for the reaction ^{59}Co(n,2n) in arbitrary units. Note that the response of such a high-threshold reaction is mainly dominated by the slope of the excitation function near threshold.

Evaluation of Integral Data

The aim of an evaluation is to combine the information from various experiments to obtain a result which can most likely be used as a basis for a further processing of these data. The present evaluation updates the result of a previous similar evaluation[27]. Generalized least-squares techniques were used which allow the successive addition of new data to the result of a previous evaluation. The methodology is described elsewhere[27,28]. The solution to the minimization of least squares for the present application is briefly summarized.

The (column) vector P stands for the result of an input evaluation of spectrum-averaged cross sections with a covariance matrix M. The vector D is a new set of data with a corresponding covariance matrix V. The elements of vector D are related to vector P in the form of absolute data or ratios. The combination of D and P results for the output of the evaluation P' and its covariance matrix M' in the matrix equations given by:

$$P' = P + A\,(N+V)^{-1}\,(D-D_P) \qquad (6)$$

$$M' = M - A\,(N+V)^{-1}\,A^t \qquad (7)$$

where $A = M\,G^t$ and $N = G\,A$. The superscript t indicates a transpose. The matrix G is the sensitivity matrix of partial derivatives of the elements of D_p with respect to the elements of P. The vector D_p contains elements corresponding to D calculated with the input data P.

Table IV Evaluation of ^{252}Cf Spectrum-Averaged Cross Sections

Reaction	<σ> (mb)	Rel. Std.Dev. %	Correlation Matrix (x 100)																					
Mg-24(n,p)	1.898	4.2	100																					
Al-27(n,p)	4.825	3.2	25	100																				
Al-27(n,α)	1.004	1.9	40	51	100																			
S -32(n,p)	71.20	3.7	28	30	44	100																		
Ti-46(n,p)	13.92	2.1	32	42	72	36	100																	
Ti-47(n,p)	19.04	2.0	34	44	75	37	74	100																
Ti-48(n,p)	0.4202	2.2	32	40	67	33	75	68	100															
V -51(n,p)	0.7092	8.0	6	16	19	8	16	17	16	100														
Fe-54(n,p)	85.58	2.0	36	44	74	38	65	68	62	17	100													
Fe-56(n,p)	1.446	2.1	35	42	72	37	63	66	59	16	69	100												
Ni-58(n,p)	115.0	1.7	37	49	82	42	76	80	70	18	81	71	100											
Co-59(n,α)	0.2165	6.2	12	20	26	15	21	22	20	10	22	21	24	100										
Cu-63(n,α)	0.7089	2.3	3	4	7	4	6	6	6	2	6	6	11	2	100									
Zn-64(n,p)	39.81	2.2	32	43	69	35	60	63	56	17	62	60	67	23	6	100								
In-113(n,n')	160.8	2.0	38	47	74	37	65	68	66	19	68	65	73	24	6	62	100							
In-115(n,γ)	124.6	2.6	29	36	58	29	51	54	52	15	53	52	57	19	5	49	71	100						
In-115(n,n')	196.3	2.0	40	48	76	38	67	70	70	19	70	68	75	24	7	63	87	69	100					
Au-197(n,γ)	76.17	2.0	36	46	74	36	65	67	64	19	68	65	73	24	6	62	86	69	86	100				
Au-197(n,2n)	5.461	2.2	33	45	69	33	59	63	59	20	63	60	67	24	6	58	78	61	78	84	100			
U -235(n,f)	1208	1.4	5	6	11	6	9	10	8	2	9	9	16	3	25	9	9	7	10	9	9	100		
Np-237(n,f)	1339	2.0	3	4	7	4	6	6	6	2	6	6	11	2	17	6	6	5	7	6	6	65	100	
U -238(n,f)	319.7	1.8	3	4	6	3	5	6	5	1	5	5	9	2	15	5	5	4	6	6	5	88	68	100
Pu-239(n,f)	1803	1.5	4	6	9	5	8	8	8	2	8	8	14	3	22	8	8	7	9	8	8	74	58	70 100

The input of the evaluation is based on a careful analysis of various experiments as regards their covariances[27]. In contrast to the previous evaluation, the recent data sets of Ref. 5 and Ref. 10 have been included. A few minor modifications were applied[3] to improve the overall consistency of the evaluation. The generation of the covariance matrices is shown with full details in Refs. 5 and 28. The output of the evaluation is given in Table IV. The evaluation resulted in a value of χ^2 of 25.7 with 25 degrees of freedom. Table IV shows that for most of the reactions considered the evaluation resulted in relative standard deviations smaller than 2.3 %. The few data with substantially higher standard deviations are mainly due to data based on a single experiment only. The data of Table I are not at present contained in the evaluation of Table IV. It is planned to include these data in the evaluation in a future step after the generation of a detailed covariance matrix not given here.

Looking at details of the correlation matrix of Table IV, one recognizes that the correlation between fission and non-fission reactions is very weak. This indicates that the experimental methods for determining the average cross sections of both types of reaction are different and relatively independent. There has been a recent indication[19] that the spectrum-averaged cross section ratios of ^{238}U(n,f)/^{235}U(n,f) and ^{239}Pu(n,f)/^{235}U(n,f) taken from Ref. 30 and included in the present evaluation may be inconsistent. This will be taken into consideration and will probably only partially influence the result of the evaluation of Table IV, as the last four rows of Table IV are chiefly concerned, and the impact on the other data of the evaluation should remain small due to the weak intercorrelations mentioned.

Data of the type shown in Table IV present the possibility of being used in neutron cross section adjustment procedures following the principles of Eqs. (6) and (7). The definition of adjustment given by Rowlands[31] is: "A way of predicting reactor properties by taking into account both energy-dependent cross section data and integral nuclear data measurements". It is mainly a question of at what stage integral data should be included in the evaluation of nuclear data. Considering a unique feedback of integral data to evaluated microscopic data as imperative, substantially reduces the amount of integral data which can be regarded as candidates for inclusion in the evaluation of nuclear data files such as ENDF/B. Only simple, well-defined integral data can be included in the first derivation of nuclear data files to keep the information contained in such basic data files largely free from data which depend on quantities not directly measured. Assuming that the ^{252}Cf neutron spectrum is sufficiently well-established, integral ^{252}Cf data can be regarded as broad resolution experiments compared with the narrow resolution experiments of direct energy-dependent cross section measurements. Under such circumstances, it is usually desirable to take such data into consideration in the evaluation of data files. As long as the condition of the best knowledge of the neutron spectrum is not valid for ^{252}Cf, the integral data may be of use together with other more complex integral data in the adjustment of secondary application-oriented cross section libraries.

A portion of the data in Table IV has already been used for such applications. The integral data were used for a variance reduction of fission spectra and nuclear cross section data to establish better parameters for the analysis of the neutron spectrum of the Oak Ridge Poolside Critical Assembly (PCA) with the aim of a more precise prediction of radiation damage in reactor pressure vessels[29]. Other integral data measured in the Coupled Fast Reactivity Measurements Facility (CFRMF)[32] were used to adjust the neutron spectrum of this facility as well as cross section data of important neutron dosimeter reactions[33].

One of the problems in adjusting is that, parallel to a cross section adjustment, there is always an implicit adjustment of the neutron spectrum. This may be of no disadvantage in cases concerning the CFRMF neutron field where the spectrum information available is mainly based on neutron transport calculations. In the case of ^{252}Cf, the neutron spectrum can be directly measured, which makes it undesirable that in an adjusting procedure, this spectrum is re-adjusted. It has to be proved that the procedure applied by Gruppelaar and Dragt[34] in correcting the adjusted spectrum back to the original one is mathematically sufficiently consistent to be applied to a future adjustment of integral ^{252}Cf data together with energy-dependent cross section data. At present, another essential condition for such an adjustment is not fulfilled in that of no reasonable covariance matrix of the ^{252}Cf neutron spectrum is available.

Summary and Conclusions

New integral data in the form of ^{252}Cf spectrum-averaged cross sections have been measured, special emphasis being placed on high-threshold (n,2n) reactions. The experimental data were fairly consistent with those calculated based on ENDF/B-V data. At present, only one spectrum representation, the NBS evaluation, has been found which fulfills the consistency statement mentioned at medium as well as at high neutron energies. This fact has partially been confirmed by a recent spectrum measurement. A promising attempt to describe the neutron spectrum by a recent evaporation model has been noted. Whether the parameters of this model can be further adjusted to yield a better consistency with the present data and with recent spectrum experiments still remains to be investigated. Generalized least-squares techniques were used in the evaluation of a consistent set of integral ^{252}Cf data. The impact of these data on future neutron cross section evaluations has been briefly reviewed.

Note added in proof:

The final analysis of the experiment of Ref. 11 resulted in a normalization factor of 1.08. This value differs only slightly from 1.095 used in the present work (Eq. 2).

References

1. W. Mannhart, Nucl. Sci. Eng. 77, 40 (1981)
2. W. Mannhart, W.G. Alberts, Nucl. Sci. Eng. 69, 333 (1979)
3. W. Mannhart, Proc. 4th ASTM-EURATOM Symposium on Rector Dosimetry, Washington, March 22-26, 1982 (to be published)
4. Z. Deszö, J. Csikai, Proc. IV All Union Conf. Neutron Physics, Vol. 3, p. 32, Atomizdat, Moscow (1977)
5. K. Kobayashi, I. Kimura, W. Mannhart, J. Nucl. Sci. Techn. 19, 341 (1982)
6. G.J. Kirouac, H.M. Eiland, C.J. Slavik, CONF-730910, p. 412 (1973)
7. Z. Deszö, J. Csikai, INDC(NDS)-103/M, p. 176 (1979)
8. L. Green, Nucl. Sci. Eng. 58, 361 (1975)
9. M. Buczkó, Z.T. Bödy, J. Csikai, Z. Deszö, S. Juhasz, H.M. Al-Mundheri, G. Petö, M. Vârnagy, CONF-760436, Vol. II, p. IV-19 (1979)
10. G. Winkler, V. Spiegel, C.M. Eisenhauer, D.L. Smith, Nucl. Sci. Eng. 78, 415 (1981)
11. R. Böttger, H. Klein, A. Chalupka, B. Strohmaier (this conference)
12. J. Grundl, C.M. Eisenhauer, Tech. Doc. IAEA-208, Vol. I, p. 53 (1978)
13. H. Maerten, D. Seeliger, B. Stobinski, INDC(GDR)-17/L (May 1982)
14. D.G. Madland, J.R. Nix, Nucl. Sci. Eng. 81, 213 (1982)
15. F.D. Becchetti, G.W. Greenless, Phys. Rev. 182, 1190 (1969)
16. W.D. Myers, "Droplet Model of Atomic Nuclei", IFI/Plenum, New York (1977)
17. P. Moeller, J.R. Nix, Nucl. Phys. A361, 117 (1981)
18. S. Tagesen, H. Vonach, B. Strohmaier, Physics Data 13-1 (1979)
19. S. Tagesen, H. Vonach, Physics Data 13-3 (1981)
20. D.L. Smith, J.W. Meadows, Nucl. Sci. Eng. 60, 187 (1976) and Report ANL/NDM-13 (June 1975)
21. G. Vasiliu, S. Mateescu, INDC(NDS)-103/M, p. 26 (1979)
22. G. Winkler, D.L. Smith, J.W. Meadows, Nucl. Sci. Eng. 76, 30 (1980)
23. B. Strohmaier, S. Tagesen, H. Vonach, Physics Data 13-2 (1980)
24. W. Mannhart, IAEA-TECDOC-263, 71 (April 1982)
25. W. Mannhart, IAEA-TECDOC-263, 47 (April 1982)
26. J.J. Wagschal, B.L. Broadhead, R.E. Maerker, NBS Spec. Publ. 594, 956 (1980)
27. W. Mannhart, F.G. Perey, EUR 6813, Vol. II, p. 1016 (1980)
28. W. Mannhart, Report PTB-FMRB-84 (June 1981)
29. J.J. Wagschal, R.E. Maerker, B.L. Broadhead, Proc. 4th ASTM-EURATOM Symposium on Reactor Dosimetry, Washington, March 22-26, 1982
30. D.M. Gilliam, C.M. Eisenhauer, H.T. Heaton, J.A. Grundl, NBS Spec. Publ. 425, 270 (1975)
31. J.L. Rowlands, BNL-NCS-51363, p. 23 (March 1981)
32. J.W. Rogers, D.A. Millsap, Y.D. Harker, Nucl. Techn. 25, 330 (1975)
33. R.A. Anderl, Y.D. Harker, D.A. Millsap, J.W. Rogers, J.M. Ryskamp, Proc. 4th ASTM-EURATOM Symposium on Reactor Dosimetry, Washington, March 22-26, 1982 (to be published)
    R.A. Anderl, BNL-NCS-51363, p. 91 (March 1981)
34. H. Gruppelaar, J.B. Dragt, BNL-NCS-51363, p. 133 (March 1981)

UNCERTAINTY ANALYSIS OF BENCHMARK DOSIMETRY MEASUREMENTS*

J. J. Wagschal

Racah Institute of Physics
The Hebrew University of Jerusalem
91904 Jerusalem, Israel

R. E. Maerker and B. L. Broadhead
Engineering Physics Division
Oak Ridge National Laboratory
Oak Ridge, TN 37830, U.S.A.

The joint consistency of a series of measured quantities and their calculated counterparts is defined. Ten measurements of average cross-sections or average cross-section ratios in benchmark ^{252}Cf spontaneous fission neutron fields are analyzed. The consistency of calculations based on a pure Maxwellian fission spectrum and ENDF/B-V dosimetry cross-sections, and of the corresponding measured values is being tested. Dosimetry cross-section covariance matrices based on ENDF/B-V, and covariances of the measured values are used in the analysis. Clear evidence of an inconsistency is presented and discussed. Removing one measurement results in a consistent data set.

[^{252}Cf neutron field, data consistency, LWR-PV surveillance dosimetry, chi square, ENDF/B-V, uncertainty analysis]

Introduction

The work presented here is part of the development of a consistent data base to be incorporated in the LEPRICON[1,2] methodology for estimating and reducing uncertainties in neutron fluence spectra within the pressure vessel of a PWR.

The ^{252}Cf spontaneous fission neutron field was recognized as a benchmark neutron field for pressure vessel surveillance dosimetry and is well documented[3]. Progress in integral average neutron cross-sections in ^{252}Cf benchmark fields and their accuracy is reported from time to time[4,5,6]. In these papers we find measured values of average cross-sections, their uncertainty, and recently even their covariances. One can also find "best" measured average cross-sections, that were derived by properly combining values from different measurements, and their correlations. Usually one also finds comparisons of measured and calculated average cross-sections based on various evaluated cross-section libraries. From such comparisons conclusions are drawn on the consistency of measured and calculated values for each reaction average cross-section separately.

Since different measured average cross-sections are frequently correlated and since the calculated average cross-sections are correlated through the fission spectrum and possibly also through correlations between the differential activation cross-sections, a joint consistency test has to be formulated and employed to such data. In the present work such a joint uncertainty analysis is presented in detail.

Ten different response measurements in ^{252}Cf benchmark fields were selected for this work. The experiments are carefully analyzed for possible correlations and a correlation matrix is derived. This is followed by a description of the calculated counterparts and their covariance. The discrepancies between the calculated and measured values and the covariance of these discrepancies are used next to define the joint consistency. An inconsistency is found in the data, its source is traced and a criterion for rejecting less consistent data in then described and used. The resulting consistent data base is finally discussed and some concluding remarks are made on the status of the data.

*This work was sponsored by the Electric Power Research Institute under research project 1399.

Selected measurements in ^{252}Cf fields

Ten "clean" measurements spanning the median response energy range in ^{252}Cf from 1.68 MeV to 8.7 MeV have been selected for this analysis. The same responses are used in the PCA[7] and PSF[8] experiments, and are thus of interest to the LWR-PV surveillance program.

Five measurements were performed at the Physikalisch-Technische Bundesanstalt in Germany and are documented in Refs[9,10]. Four measurements involving fission cross-sections were performed at the U.S. National Bureau of Standards. These measurements have been carefully reanalyzed and have a detailed list of uncertainty sources and contributions[11]. A more recent measurement of $^{63}Cu(n,\alpha)$ in the NBS ^{252}Cf field[12] is also included in our analysis.

In earlier work[2] an ad hoc assumption was made that the uncertainties of the NBS and PTB measurements are correlated via the source intensity uncertainty. This point has recently been checked with the experimentalists[13]. It is clear now that the NBS and PTB ^{252}Cf sources have not been calibrated against each other nor against a common third source[14], and therefore there is no correlation between the two source strengths uncertainty estimates.

The Winkler $^{63}Cu(n,\alpha)$ average cross-section measurement might have been a coupling link between the NBS fission cross-section measurements and the PTB activation measurements. The principal sources of uncertainty in the $^{63}Cu(n,\alpha)$ experiment were given by Winkler[12], and are reproduced here in Table I. Except for item No. 1, the counting statistics uncertainty, all other uncertainty sources could be correlated to the other NBS measurements or to the PTB measurements. In particular, the gamma-ray detector efficiency uncertainty could have been common to all activation measurements. However, since different special techniques were employed by Winkler[15] and by the PTB experimentalists for the activity measurements ($4\pi\gamma$ - counting and comparing dissolved sample solutions to standard solutions respectively), we conclude that the average $^{63}Cu(n,\alpha)$ corss-section measurement uncertainty is not correlated to the PTB measurements uncertainties.

On the other hand, the uncertainties of the $^{63}Cu(n,\alpha)$ and the $^{235}U(n,f)$ average cross-section measurements are indeed correlated through the common NBS source strength uncertainty and through uncertainties in corrections applied for neutron scattering due to the source capsule and support structure. The 1.1% for the source strength determination in both measurements

K. H. Böckhoff (ed.), Nuclear Data for Science and Technology, 436–440.

are fully correlated. Out of the 1.0% (see Table I Item #5) uncertainty due to scattering corrections, 0.71% are correlated to the corresponding uncertainties in the $^{235}U(n,f)$ measurement (the 1% quoted includes also scattering in the sample support structure in the $^{63}Cu(n,\alpha)$ measurement).

In the $^{63}Cu(n,\alpha)$ measurement Winkler has a 1.0% uncertainty due to corrections applied for neutron scattering, absorption and geometry due to the sample. The $^{235}U(n,f)$ measurement has a 0.8% uncertainty due to scattering corrections in the platinum backing of the sample. Since the samples and the analytical correction methods were different, no correlation is introduced. Even the uncertainty due to distance measurements (Table I, item 4) is not correlated to the "geometry" uncertainty in the earlier $^{235}U(n,f)$ measurement since a new computerized measuring device was used in the $^{63}Cu(n,\alpha)$ measurement[16].

Table I. The principal sources of uncertainty in the measured ^{252}Cf spontaneous fission neutron spectrum averaged cross section for the reaction $^{63}Cu(n,\alpha)^{60}Co$

No.	Source of Error	Resulting uncertainty (%)
1	Counting statistics and reproducibility of sample activity measurement	0.2
2	Gamma-ray detector efficiency	1.5
3	Source strength (neutron emission rate)	1.1
4	Distance measurements	0.3
5	Correction for neutron scattering due to the source capsule, source and sample support structures	1.0
6	Activation decay rate factor (G)	0.1
7	Correction for neutron scattering, absorption and geometry due to the sample	1.0

The description of the measurements analyzed, their median response energy, the measured values and the corresponding relative standard deviations are given in the first six columns of Table II. The full correlation matrix of these measurements is given in Table III.

Since for a detailed analysis of benchmark experiments a full description of the calculations and their uncertainties is also required, the next section is devoted to these calculations.

Calculation of the ^{252}Cf field experiments

Values of average cross-sections $\bar{\sigma}$, or values of the ratios $\bar{\sigma}_\alpha / \bar{\sigma}_\beta$, were calculated by folding the SAND-II group representation of the ENDF/B-V dosimetry file cross-sections[17] with a normalized ^{252}Cf spectrum, in the same group structure, calculated from a Maxwellian distribution[18] with a temperature of 1.42 MeV. The uncertainties in these calculations involve the uncertainties in the ENDF/B-V activation cross-sections, the uncertainty in the shape of the ^{252}Cf spontaneous fission spectrum and possibly uncertainties in the numerical procedure used to calculate $\bar{\sigma}$.

The uncertainties of the fission cross-sections and of the threshold activation cross-sections were processed into a 24 group variance-covariance representation. The covariances of the fission cross-sections $^{235}U(n,f)$, $^{238}U(n,f)$, $^{237}Np(n,f)$ and $^{239}Pu(n,f)$ were processed from the general purpose file of ENDF/B-V using the PUFF-2[19] code. Except for the cross-material covariances between the $^{239}Pu(n,f)$ and $^{235}U(n,f)$ reactions, there are no cross-material covariances in the ENDF/B-V general purpose file for the above mentioned fission cross-sections. Using comments of ENDF/B-V evaluators, cross-material covariances between $^{238}U(n,f)$ and $^{235}U(n,f)$ and between $^{238}U(n,f)$ and $^{239}Pu(n,f)$ were estimated[2] and used in our analysis. The covariances for the non-fission threshold reactions were read and collapsed by PUFF-2 directly from the 620 group dosimetry file of ENDF/B-V. There were no cross-material covariances for these cross-sections in the file.

The covariance matrix for the ^{252}Cf fission spectrum was also prepared in a 24 group structure. An uncertainty of 2% in the average energy and a Maxwellian shape were assumed[2].

The uncertainties in the numerical procedures used to calculate $\bar{\sigma}$ were found to be negligible by using a few different numerical schemes[20].

The uncertainties in the calculated responses due to propagated uncertainties in parameters (i.e., activation cross-sections and fission spectrum) are obtained by use of sensitivites, which in this simple case are of the form

$$S_{ig} = \frac{\sigma_{ig}\chi_g}{\bar{\sigma}_i} \tag{1}$$

Here S_{ig} denotes the sensitivity of $\bar{\sigma}_i$ (average cross-section of response number i) to the cross-section σ_{ig} or to the spectrum χ_g in the energy range $E_{g-1} \leq E \leq E_g$. The relative covariance matrix of the calculated responses is thus

$$\hat{C}_{\bar{r}\bar{r}} = S\hat{C}_{\sigma\sigma}S^+ + S\hat{C}_{\chi\chi}S^+ \tag{2}$$

where S^+ denotes the transpose of the sensitivity matrix, $\hat{C}_{\sigma\sigma}$ and $\hat{C}_{\chi\chi}$ denote the relative covariance matrices of the activation cross-sections and of the ^{252}Cf spontaneous fission spectrum respectively.

The vector of the calculated values of all responses is given in the seventh column of Table III. The respective relative standard deviations (i.e., the square roots of the diagonal elements of $C_{\bar{r}\bar{r}}$), are given in column eight of Table III.

The consistency of the measured and calculated data described earlier will be tested and discussed in the following section.

Initial consistency analysis

The consistency of a series of N measured values of responses, $r \equiv (r_i)$, $i = 1,2,...N$ and of their corresponding calculated values $\bar{r} \equiv (\bar{r}_i)$, $i=1,2,..,N$, is obtained by testing the discrepancy vector $d \equiv d_i = \bar{r}_i - r_i$, $i = 1,2,...N$ and its covariance matrix C_{dd}. The covariance matrix C_{dd} is, of course, given by

$$C_{dd} = C_{rr} + C_{\bar{r}\bar{r}} \quad , \tag{3}$$

since in our case there are no apriori correlations between the parameters and the responses. The conventional measure of the consistency of a set of measured values and their calculated counterparts is the so called "chi square of the data", namely,

$$\chi^2 = d^+ C_{dd}^{-1} d \tag{4}$$

Table II. Description of the Measurements and the corresponding Calculations

Experiment No.	Source	Average Cross-Section (or ratio)	Median Response Energy (MeV)	Measurement: Value (barns)	Measurement: Relative Standard Deviation (%)	Calculation: Value (barns)	Calculation: Relative Standard Deviation (%)	Discrepancy: Value divided by calculated value	Discrepancy: Standard Deviation divided by calculated value (%)	Individual χ^2
1	NBS	$^{235}U(n,f)$	1.68	1.205	2.10	1.239	1.96	-2.74	2.83	0.94
2	NBS	$^{238}U(n,f)/^{235}U(n,f)$	2.8*	0.2644+	1.06	0.2532	1.85	4.42	2.16	4.21
3	NBS	$^{237}Np(n,f)/^{235}U(n,f)$	2.1*	1.105+	2.08	1.093	9.38	1.10	9.62	0.01
4	NBS	$^{239}Pu(n,f)/^{235}U(n,f)$	1.77*	1.500+	1.33	1.447	1.99	3.66	2.42	2.29
5	PTB	$^{27}Al(n,\alpha)$	8.7	0.001006	2.14	0.001157	10.15	-13.05	10.32	1.60
6	PTB	$^{58}Ni(n,p)$	4.3	0.118	2.35	0.1146	7.34	2.97	7.73	0.15
7	PTB	$^{115}In(n,n')$	2.7	0.1964	2.28	0.1820	12.01	7.91	12.26	0.42
8	PTB	$^{54}Fe(n,p)$	4.5	0.0846	2.36	0.0889	4.91	-4.82	5.40	0.80
9	PTB	$^{46}Ti(n,p)$	6.0	0.0138	2.37	0.01381	13.80	-0.07	14.00	0.00
10	NBS	$^{63}Cu(n,\alpha)$	8.3	0.000709	2.37	0.000807	9.55	-12.1	9.77	1.54

*Median response energy of the numerator

+The ratios are pure numbers

Table III. Correlation Matrix of Measurements in Standard ^{252}Cf Spontaneous Fission Fields, (Blank Elements are Equal to Zero)

No.	Measurement	Source	Correlation Matrix									
1	$^{235}U(n,f)$	NBS	1.0	0.40	-.09	-.19						0.34
2	$^{238}U(n,f)/^{235}U(n,f)$	NBS	0.40	1.0	0.25	0.13						
3	$^{237}Np(n,f)/^{235}U(n,f)$	NBS	-.09	0.25	1.0	0.32						
4	$^{239}Pu(n,f)/^{235}U(n,f)$	NBS	-.19	0.13	0.32	1.0						
5	$^{27}Al(n,\alpha)$	PTB					1.0	0.75	0.77	0.74	0.74	
6	$^{58}Ni(n,p)$	PTB					0.75	1.0	0.70	0.68	0.68	
7	$^{115}In(n,n')$	PTB					0.77	0.70	1.0	0.70	0.70	
8	$^{54}Fe(n,p)$	PTB					0.74	0.68	0.70	1.0	0.67	
9	$^{46}Ti(n,p)$	PTB					0.74	0.68	0.70	0.67	1.0	
10	$^{63}Cu(n,\alpha)$	NBS	0.34									1.0

If $\chi^2 \lesssim N$ the data are considered to be consistent or, in other words, if χ^2/N ("chi square per degree of freedom") is less than or about equal to one, the data are considered consistent.

It is convenient to present the discrepancies (d) relative to their corresponding measured, r_i, or calculated values, $\bar{r}_i$. The values of $-d_i/\bar{r}_i$ are given in the ninth column of Table III. The corresponding standard deviations, also divided by the respective $\bar{r}_i$, are given in the tenth column of Table III.

The value of χ^2 when only one response is considered at a time (alongside with the relevant parameters used in its calculation) is given in the last column of Table III for each measurement under the heading "individual χ^2". As long as the discrepancy between the measured and calculated values of a response is less than or equal to the standard deviation of the discrepancy, the individual χ^2 is less than one. This means that the measured value of the particular response is consistent with the measured values of the relevant parameters.

The value of $\chi^2/10$ considering all responses in Table III is 2.45 indicating an inconsistency. This inconsistency will be the subject of the next section.

Systematic consistency analysis

The rather high value of χ^2/N for all measurements considered in this analysis indicates an inconsistency in the data (responses=integral and parameters= differential). The following procedure introduced by Yeivin et. al.[21] is applied to the current data in order to point out which response is "least" consistent with the rest of the data.

Out of the N measurements considered one measurement is temporarily suspended, and the value of χ^2 per degree of freedom is calculated for the remaining measurements. This is repeated for all measurements in turn, each time suspending a different measurement and reinstating the previous one. The particular measurement the suspension of which resulted in the lowest χ^2 per degree of freedom, is obviously least consistent with the rest of the data. After the definite identification of the least consistent measurement it is permanently rejected. This process can now be repeated for the N-1 measurements left, resulting in the permanent rejection of another measurement. Repeating this rejection technique for the N-2, N-3,... remaining measurements results in the establishment of a hierarchy of consistent measurements.

In Table IV values of χ^2 per degree of freedom obtained in such a rejection process for the ^{252}Cf fields measurements are presented. The least consistent measurement is No. 2, namely the ratio measurement of $^{238}U(n,f)/^{235}U(n,f)$. The value of $\chi^2/9$ for any nine (out of ten) measurements in the ^{252}Cf fields is always higher than 2.2 provided that measurement No. 2 is included. However, the exclusion of measurement No. 2 immediately reduces the $\chi^2/9$ value to 0.73. Since the remaining measurements are consistent now, the process could have been stopped at this point. It was continued nevertheless, and the resulting "optimal" consistent set of measurements is read from Table IV as follows: 9, 3, 8, 10, 5, 7, 6, 1, 4, 2. The value of χ^2/n is plotted as function of n for this optimal sequence in Fig. 1. The probability associated with the χ^2 values and n, $P(\chi^2,n)$, is also plotted in Fig. 1 as function of n. Note the change in slope when measurement No. 4 is included as ninth measurement, and the sharp change in the slope when the "inconsistent" measurement is included.

Table IV. Selection of Consistent Measurements by Successive Eliminations of Integral Measurements

Each entry indicates the value of χ^2/I when measurement No. J is eliminated from the array of the I+1 remaining measurements. At each step the "worst" response is permanently eliminated. Thus, for instance in the column I=9, the lowest $\chi^2/9$ value is obtained when response J=2 is eliminated. Out of the 9 responses left (after J=2 was permanently eliminated) one response is eliminated at a time and $\chi^2/8$ values are compared leading to the permanent elimination of J=4.

J \ I	9	8	7	6	5	4	3	2	1
1	2.7234	.7467	.5121						
2	.7336								
3	2.6873	.8253	.6281	.5918	.5584	.5541	.6155	.4385	$.0000_3$
4	2.5585	.5497							
5	2.2860	.7516	.5458	.5079	.4848	.4693			
6	2.6719	.7316	.5209	.4711					
7	2.6580	.7586	.5521	.5087	.4501				
8	2.6276	.8124	.6139	.5819	.5585	.5582	.5829	.0065	
9	2.7216	.7942	.5924	.5545	.5057	.4844	.5483	.4088	.0130
10	2.2256	.7848	.5855	.5335	.5012	.4792	.2993		

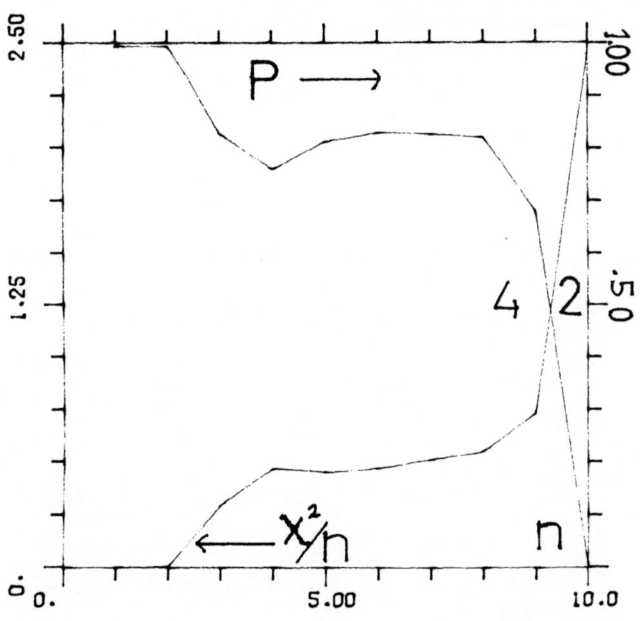

Figure 1. Consistency analysis
χ^2/n and associated probability $P(\chi^2,n)$ as function of n, the number of experiments analyzed. The sequence of measurements (from left to right is the optimal sequence, namely, those measurements which have a minimal value of χ^2/n for given n.

Although one might have guessed that measurement No. 2 is the least consistent measurement by examining the individual χ^2 values (Table II), the sequence of "least consistent measurements" is different from a similar sequence based only on the individual values. This is the result of the correlations which are not reflected in the analysis of one measurement at a time.

Cross sections, spectra or integral measurements, all contribute to χ^2 which is a measure of the joint consistency of the data considered. It is also interesting to see what the various data subsets contribute to χ^2. This will be the subject of the next section.

Contributions to χ^2

The measure of consistency, χ^2, introduced in Eq.(4) and used throughout this work, involved only the knowledge of the discrepancies between measured and calculated values of integral responses, $d = \bar{r} - r$, and the covariance of these discrepancies, C_{dd}. As is well known[2] this χ^2 value turns out to be the minimal value of the quadratic form

$$Q(\alpha',r')=(\alpha'-\alpha)^{\dagger}C_{\alpha\alpha}^{-1}(\alpha'-\alpha)+(\bar{r}'(\alpha')-r)^{\dagger}C_{rr}^{-1}(\bar{r}'(\alpha')-r), \quad (5)$$

which is the "loss function" introduced in generalized-least-square-cross-section-adjustment[22] procedures. Here the vector α denotes all differential parameters (cross sections, spectra, biases etc.) involved in the calculation of the integral responses, $\bar{r}(\alpha)$. The primed quantities are the respective adjusted quantities. The optimal adjustment is the one for which Q is minimal. Thus, χ^2 can be decomposed to contributions from the various data. Each contribution measures, in units of the respective uncertainties, the deviation between the original measured value and the value obtained when making all data, parameters and responses (calculated from these parameters) as consistent as possible. In other words, a high contribution to χ^2 indicates that a particular datum has to be substantially moved away from its original experimental value.

The measure of consistency of all ten benchmark experiments analyzed here gave a χ^2 value of 24.5 (2.45 per degree of freedom). The contributions to χ^2 from the integral response measurements, from the ^{252}Cf fission spectrum and from the various activation cross-sections as listed in column one of Table V, are given in the second column. The contributions of the ^{235}U, ^{238}U and ^{239}Pu fission cross-sections are lumped together since we included cross-material covariances for these cross-sections.

The contribution of the ^{252}Cf fission spectrum is rather small, but the contribution from the two high median-response-energy activation cross-sections,

$^{27}Al(n,\alpha)$ and $^{63}Cu(n,\alpha)$, is substantial. When only nine response measurements are tested for their consistency, but always including the "inconsistent" measurement No. 2, a similar pattern is observed as can be seen in columns 3,4,5 of Table V. However, when the "inconsistent" measurement, No. 2, is excluded the χ^2 contributions are quite different. In the last column of Table V we see that the ^{252}Cf fission spectrum has to be modified. We also note that the joint contribution of all (but No. 2) integral responses is small and so are the contributions of all activation cross-sections. This pattern is not changed when a smaller number of measurements is tested for consistency as long as measurement No. 2 is excluded. (Table VI).

The ^{252}Cf fission spectrum has to be softened by lowering its temperature (Θ=1.42 MeV) by slightly more than one standard deviation. This softening of the spectrum, reduces the two high ^{252}Cf fission spectrum averaged $^{27}Al(n,\alpha)$ and $^{63}Cu(n,\alpha)$ response values, and the respective differential cross-sections have to be only slightly modified. The softening is consistent with Mannhart's[5] results prefering the NBS segmented spectrum[3].

Table V. Contributions to χ^2

The elimination of an integral measurement does not drastically change in general the contributions to χ^2 . However, the elimination of measurement No. 2 susbstantially changes the Contributions to χ^2 .

I(J)	10(-)	9(1)	9(4)	9(10)	9(2)
χ^2/N	2.4526	2.7234	2.5585	2.2256	.7336
Integral	12.232	12.523	11.461	8.4012	1.4083
$^{252}Cf(\chi)$	.0466	.0459	.0417	.0637	1.3119
^{235}U, ^{238}U, $^{239}Pu(n,f)$	4.4633	4.1239	3.684	4.1361	1.8057
($^{235}U(n,f)$)	(45)	(45)	(35)	(35)	(3)
$^{237}Np(n,f)$	.3239	.3062	.2981	.2717	.0000
$^{27}Al(n,\alpha)$	2.9367	2.9435	2.9830	5.1304	.3996
$^{58}Ni(n,p)$	.4323	.4319	.4296	.3289	.6761
$^{115}In(n,n')$	.5848	.5849	.5857	.6212	.5166
$^{46}Ti(n,p)$	.0306	.0304	.0293	.0003	.2370
$^{54}Fe(n,p)$	.6020	.6034	.6119	1.0759	.0693
$^{63}Cu(n,\alpha)$	2.8739	2.9179	2.9031	---	.1780

Conclusions

In this work it was clearly demonstrated that the measured and calculated data of nine experiments in the ^{252}Cf benchmark neutron field are jointly consistent. This consistency reflects the measured responses, their covariance and the ENDF/B-V based information actually used for the differential activation cross-sections and their covariance. The use of a more recent evaluation of activation cross-sections[23], this time with many cross-material correlations, calls for a new consistency test reflecting the new information.

Measurement No. 2 is not consistent with the rest of the data. Using its information with the current data base for spectral unfolding is not advised. However, it should be stressed that the rejection of the $^{238}U(n,f)/^{235}U(n,f)$ experiment does not necessarily mean that there is anything wrong with the measured value. All ingredients involved, namely, the measured value, the experimental correlations with all other measurements, the cross sections and their auto and cross-material covariances have to be reexamined.

Table VI. Contributions to χ^2

After the elimination of Measurement No. 2, the contributions to χ^2 are small and they change gradually with the elimination of additional measurements.

I(J...)	9(2)	8(2,4)	7(2,4,1)	6(1,2,3,4)
χ^2/N	.7336	.5497	.5121	.5918
Integral	1.4083	.6023	.1045	.1028
$^{252}Cf(\chi)$	1.3119	1.3252	1.3694	1.3741
^{235}U, ^{238}U, $^{239}Pu(n,f)$	1.8057	.4133	.0014	--
$^{237}Np(n,f)$	.0000	.0014	.0307	--
$^{27}Al(n,\alpha)$	.3996	.3911	.3639	.3610
$^{58}Ni(n,p)$	.6761	.6777	.6833	.6839
$^{115}In(n,n')$	.5166	.5162	.5149	.5147
$^{46}Ti(n,p)$	.2370	.2387	.2452	.2458
$^{54}Fe(n,p)$	.0693	.0676	.0622	.0617
$^{63}Cu(n,\alpha)$	.1780	.1641	.2090	.2066

References

1. J.J. Wagschal, R.E. Maerker and B.L. Broadhead, "Surveillance dosimetry: achievements and disappointments", in Fourth ASTM-EURATOM Symposium on Reactor Dosimetry, Washington, D.C. (March 1982).

2. R.E. Maerker, J.J. Wagschal and B.L. Broadhead, "Developement and demonstration of an advanced methodology for LWR dosimetry applications", EPRI NP-2188 (December 1981).

3. J.A. Grundl, C. Eisenhauer and E.D. McGarry, "Benchmark neutron fields for pressure vessel surveillance dosimetry", in NUREG/CR-0551 (December 1978).

4. W. Mannhart, Nucl. Sci., Eng. 77, 40 (1981).

5. W. Mannhart, "Average neutron cross-sections in the CF-252 benchmark field", in Fourth ASTM-EURATOM Symposium on Reactor Dosimetry, Washington, D.C. (March 1982).

6. K. Kobayashi, I. Kimura and W. Mannhart, Journal of Nucl. Sci. and Tech., 19, 341 (1982).

7. F.W. Stallman, F.B.K. Kam, J.F. Eastham and C.A. Baldwin, "Reactor calculation 'Benchmark' PCA blind test results", ORNL/NUREG/TM-428 (1981).

8. R.E. Maerker and M.L. Williams, "Comparison of calculations with neutron dosimetry measurements performed at the Oak Ridge Poolside Facility", Trans. Am. Nucl. Soc. 39, 812 (1981).

9. W.G. Alberts, E. Gunther, M. Matzke and G. Rassl in "Proceedings of the First ASTM-EURATOM Symposium on Reactor Dosimetry", Petten, EUR 5667 elf, 1, 131, (1977).

10. W. Mannhart, W.G. Alberts, Nucl. Sci. Eng. 69, 333 (1979).

11. J.J. Wagschal, R.E. Maerker and D.M. Gilliam in "Proceedings of the Third ASTM-EURATOM Symposium on Reactor Dosimetry", Ispra, EUR 6813, Vol. II, 683 (1980).

12. G. Winkler, V. Spiegel, C.M. Eisenhauer, D.L. Smith, Nucl. Sci. Eng. 78, 415 (1981).

13. V. Spiegel, private communication (June 1981).

A NEW COVARIANCE FORMULATION INCLUDING THE ENERGY RESOLUTION CONCEPT FOR THE NEUTRON DOSIMETRY

M. Nakazawa

Nuclear Engineering Research Laboratry,
University of Tokyo
Tokai-mura, Ibaraki-ken, JAPAN

A new formulation of the covariance matrix has been proposed for the uncertainty representation of neutron cross sections and spectra especially required for the dosimetry studies. Main feature of this new formulation is the inclusion of the random error using the energy resolution concept in addition to the usual systematic error. Further discussions have been made on the interpolation error and the group averaging error that can be also written basing on the new formulation of the covariance matrix.

[Neutron Dosimetry, Uncertainty Analysis, Covariance Matrix, Energy Resolution, Random Error, Interpolation Error, Dosimetry File, Unfolding]

Introduction

From a point of view to complete the uncertainty analysis of the reactor neutron dosimetry method, the covariance matrix informations have been currently required for three kinds of basic quantities such as the measured reaction-rate values, corresponding neutron cross-sections and intial spectra for the unfolding calculations. On the definition of the covariance matrix and also on its evaluation procedures, there have been thought little problems for the discrete quantities like the reaction-rate data. For the non-discrete, function-like quantities such as the cross-section curves $\sigma(E)$ and the neutron spectrum $\phi(E)$, however, there are some problems how to represent their random (uncorrelative) error in the covariance matrix.

It should be commented that the following representation of the random error of $\sigma(E)$ by a degree of ε,

$$\overline{\Delta\sigma(E)\Delta\sigma(E')} = \varepsilon^2 \cdot \sigma(E)\sigma(E')\delta(E-E') \qquad (1)$$

where $\delta(E-E')$ is the delta function, is not correct even if it seems as a natural extension from the formula of the discrete quantities, because some wrong and confusing formula have been led using the eq.(1) such as, relative error of I,

$$\frac{\Delta I}{I} = \varepsilon \sqrt{\frac{<\sigma^2>}{<\sigma>^2\ (E_2-E_1)}} \qquad (2)$$

when the reaction rate value of $I = \int_{E_1}^{E_2} \sigma(E)dE$ is calculated. In the eq.(2), the right side has a strange dimension which can not be understandable, then the eq.(1) should be concluded as a mistake. This problem can be resolved through introducing the energy resolution concept which is shown in the next session.

Some additional uncertainties appeared in the cross-section production and evaluation procedures have been considered such as the interpolation error and the averaging error due to the variations of the weighting function, which should be represented also in the form of covariance matrix. Further discussions have been made on the transformability of the covariance matrix to the other energy group structures to represent it, that is not in a case of the present dosimetry file of ENDF/B-V.

Random Error

The random error is caused mainly by the stochastic properties of the radiation counting processes and it is found an important correlation of this random error in the cross-section curves $\sigma(E)$ and/or neutron spectra $\phi(E)$ to the energy resolution (or the uncertainty of the neutron energy) δ in the form of

$$\frac{\Delta\phi}{\phi} \times \sqrt{\delta} = \text{const.} \qquad (3)$$

and

$$\frac{\Delta\sigma}{\sigma} \times \sqrt{\delta} = \text{const.} \qquad (4)$$

where σ and ϕ (also their uncertainty $\Delta\sigma$ and $\Delta\phi$) are assumed to be represented with an energy resolution of the width δ , that can be simply demonstrated as is shown in the following subsection of (1) ad (2).

(1) The neutron flux $\phi(\delta)$ represented with the energy resolution δ is proportional to the detector counting C, then the uncertainty of $\phi(\delta)$ is given like

$$\frac{\Delta\phi(\delta)}{\phi(\delta)} = \frac{\Delta C}{C} = \frac{\sqrt{C}}{C} = \frac{1}{\sqrt{C}} \propto \frac{1}{\sqrt{\phi(\delta)}} \quad . \qquad (5)$$

Using the definition of $\phi(\delta)$ of

$$\phi(\delta) \equiv \int_E^{E+\delta} \phi(E)dE \doteq \phi(E) \cdot \delta \qquad (6)$$

we can obtain the relation of

$$\frac{\Delta\phi(\delta)}{\phi(\delta)} \times \sqrt{\delta} \propto \frac{1}{\sqrt{\phi(E)}} = \text{const.} \qquad (7)$$

(2) When the neutron cross-section σ is measured using the neutron flux $\phi(\delta)$ such as

$$\sigma = \frac{C}{N \cdot \phi(\delta)} \qquad (8)$$

where C is a detector counting and N is a number of target nucleus. The random error of σ can be given as

$$\frac{\Delta\sigma}{\sigma} = \frac{\Delta C}{C} + \frac{\Delta\phi(\delta)}{\phi(\delta)} \qquad (9)$$

neglecting the uncertainty of N and using the relations of

$$\frac{\Delta C}{C} = \frac{1}{\sqrt{C}} \propto \frac{1}{\sqrt{N\sigma\,\phi\cdot\delta}} \qquad (10)$$

$$\frac{\Delta\phi(\delta)}{\phi(\delta)} \propto \frac{1}{\sqrt{\delta}} \qquad (11)$$

we can obtain the final equation of

$$\frac{\Delta\sigma}{\sigma} \times \sqrt{\delta} = \text{const.} \qquad (\text{QED})$$

Using the above relations of eq.s (3) and (4), we can obtain an important conclusion that the random error of the neutron cross-section curves and the neutron spectra goes to infinitely large when the value of the energy resolution δ is made close to zero. This is the cause of the mistake of the eq.(1), because the value of ε can not be finite in the eq.(1) neglecting the energy resolution. Then instead of eq.(1) the correct covariance matrix can be given like

K. H. Böckhoff (ed.), Nuclear Data for Science and Technology, 441–443.

$$\overline{\Delta\sigma(E)\Delta\sigma(E')} = \varepsilon^2 \cdot \sigma(E)\,\sigma(E')\Delta(E-E') \qquad (1)'$$

where $\sigma(E)$ and $\Delta\sigma(E)$ are represented with the energy resolution δ and

$$\Delta(E-E') = \begin{cases} 1 , & |E-E'| \le \frac{\delta}{2} \\ 0 , & \text{otherwise.} \end{cases}$$

And the eq.(2) could be changed as

$$\frac{\Delta I}{I} \times \sqrt{E_2-E_1} = \varepsilon \cdot \sqrt{\delta \cdot \frac{\langle\sigma^2\rangle}{\langle\sigma\rangle^2}} \left(\doteqdot \varepsilon\sqrt{\delta} \right) \qquad (2)'$$

that is very interesting to see the conservation rule of the value of relative deviation multiplied by the root of the energy resolution.

On the properties of the random error, which is given in the eq.s (3) and (4) and could be called as the "Experimental Uncertainty Principle", further three comments are also interesting. At first, considering the random error of the cross section curves at the finite energy resolution, we can permit the existences of some small areal cross-sections such as the pygmy resonances in our present evaluation values, because the variations of the cross-section value at each energy can be severely fluctuated where the energy resolution of our measurements is improved. Secondly, this "Experimental Uncertainty Principle" can be considered equivalent to the uncertainty reduction rule by the smoothing procedures, which has the same formula as the eq.s (3) and (4) when the value of δ is thought as the smoothing width. Finally, the covariance term between the cross-section and the energy, i.e. $\overline{\Delta\sigma(E) \cdot \Delta E}$, can be obtained from these relations, which should be considered in the further calculation, especially in its uncertainty estimation.

Interpolation Error

The mathematical procedures of the interpolation and the fitting are often applied to obtain the continuous curves from several data at the discrete energy points in the cross-section evaluation and/or in the experimental efficiency curve estimation. For example, consider the interpolation of the two data, $(x_1, f_1 \pm \Delta f_1)$ and $(x_2, f_2 \pm \Delta f_2)$, by the linear function of $f(x) = ax+b$, we can obtain the values and uncertainties of two parameters of a and b using the input data. There are, however, some underestimation of the uncertainty when the interpolation error is missed, i.e. the uncertainty of the interpolated function at the midpoint of x_1 and x_2 is given like

$$\overline{\left(\Delta f(x)\right)^2} = \frac{1}{4}\left\{ \overline{(\Delta f_1)^2} + \overline{(\Delta f_2)^2} \right\} \doteqdot \frac{1}{2}\overline{\left(\Delta f_1\right)^2} \qquad (12)$$

where $\Delta f_1 \doteqdot \Delta f_2$ is assumed for simplicity, and it is very strange and also unreasonable that this uncertainty Δf is less than the original raw data uncertainty Δf_1 and Δf_2, which should be considered due to the neglection of the interpolation error. This interpolation error $\Delta f_i(x)$ can be given in the following form of

$$\Delta f_i(x) = \frac{(x-x_1)\cdot(x-x_2)}{2} \cdot f''(\xi) \qquad (13)$$

where ξ is a proper value between x_1 and x_2 depending on the value of x.

Practical example is shown on the interpolation error for the estimation of the γ-ray full-energy peak effciency values $\varepsilon(E)$ of the Germanium detector. In this case, simple fitting formula of $\log \varepsilon(E) = a \log E + b$ is sometimes applied and Fig. 1 shows the typical deviations between this formula and experiments, Second derivative values of $\log \varepsilon(E)$ have been obtained like Fig. 2 using the eye-guided line in Fig. 1. Using this data, two numerical examples of the interpolation error are shown here,

(1) if E_1 = 661 KeV (Cs-137) and
E_2 = 845 KeV (Mn-54),

then

$\Delta i(\ln \varepsilon(E)) \doteqdot 0.34$ %.

(2) If E_1 = 285 KeV (Hg-203) and
E_2 = 661 KeV (Cs-137),

then

$-5.8\ \% \le \Delta i\ (\ln \varepsilon(E)) \le 3.5\ \%$

depending on the choice of ξ values.

From these analysis, the ratio of E_1/E_2, which are the gamma-ray energies of standard sources, should be recommended less than 1.2 in order to decrease the interpolation error lower than 0.5 % in the case of 55-57 cc coaxial Germanium detectors.

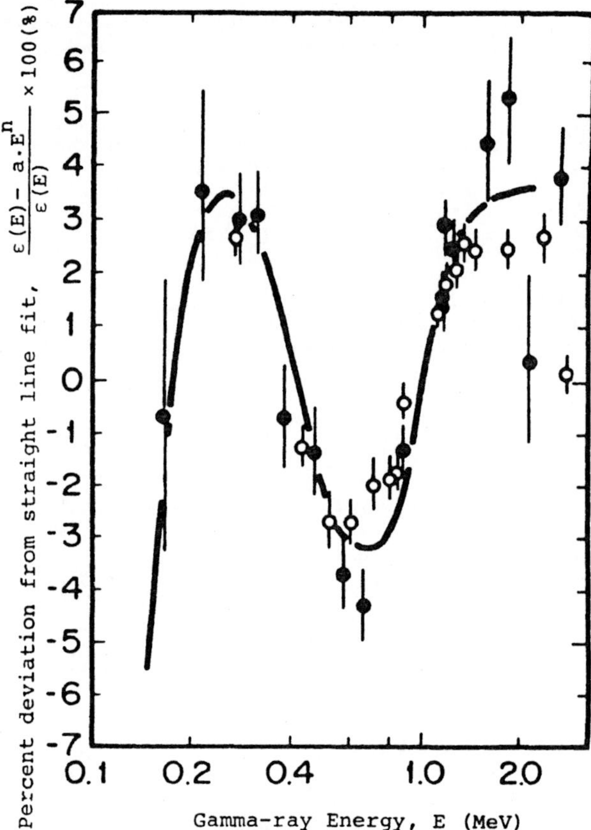

Fig. 1. Typical Gamma-ray Full-Energy Peak Efficiency Curves $\varepsilon(E)$ for the 55-57cc Ge (Li) detector
ϕ , by Gehrke et.al.[1] (55cc)
ϕ , by Yoshizawa et.al.[2] (57cc)
∿ , Semiempirical fitting data

As is shown in the above example, it is usually a main issue how to estimate the range of the value $f''(\xi)$ for the evaluation of this interpolation error. And the uncertainty of these interpolated curves should be considered depending on the energy resolution of the input data, then the range of the value $f''(\xi)$ has also the same dependency on the energy resolution as is shown in the eq.s (3) and (4). Further comment could be made on the correlation between the interpolation error at x and x', i.e. $\overline{\Delta fi(x)\Delta fi(x')}$, which will be produced depending on the energy resolution similar to the eq.(1)'.

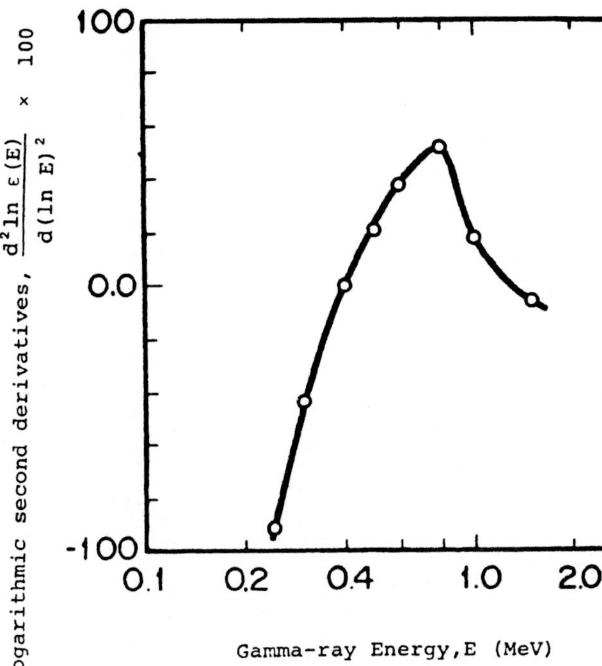

Fig. 2. Logarithmic Second Derivatives of the Gamma-ray Full-Energy Peak Efficiency Curves ε(E) of the 55-57cc coaxial Ge (Li) Detectors for the Estimation of the Interpolation Error.

Averaging Error

The averaging procedure to obtain the group averaged cross-section value is thought equivalent to the modification of the energy resolution to describe those values. There are two kinds of uncertainties in this group averaged cross-section, one is due to the original uncertainty of the micro cross-section, and another is new one due to the uncertainties of the weighting neutron spectra, which can be called the additional uncertainty due to the averaging procedure itself.[3]

Then it is also required to give the covariance matrix of the weighting spectrum, where the stochastic properties of the neutron spectra should be also considered such as the eq.(3). At present, there are, however, confessed we can not give detail and correct covariance matrix of the neutron spectra due to the difficulties of estimation procedures, even if some trials are reported very recently for the simple standard neutron field,[4] through combining the sensitivity coefficients with the covariance matrix of the group averaged cross-section, which is incomplete because of a neglect of the averaging error.

As a summary of these considerations, the covariance matrix of the group averaged cross-section should be given in a form of

$$\overline{\Delta\sigma g\, \Delta\sigma k} = \sigma g \cdot \sigma k \left\{ \varepsilon^2(g) \cdot \delta gk + \varepsilon(g,k) \right\} \qquad (14)$$

where $\varepsilon^2(g)$, δgk and $\varepsilon(g,k)$ are respectively the random error at the group g, Kronecker's delta function and the other uncertainty term. The random error of this eq.(14) should include the three components; the stochastic random error of the original micro-cross section, the interpolation error in the evaluation and the random error component of the averaging procedure due to the random error of the neutron spectra. On the transformation of the energy group structure of the covariance matrix, the random error component $\varepsilon^2(g)\delta gk$ should be transformed separately from the smooth component of the $\varepsilon(g,k)$, in order to conserve the value of $\varepsilon(g)\sqrt{\Delta ug}$ that is shown in the eq.(4).

Summary & Conclusion

The concept of the energy resolution has been found essentially important to describe the random error component of the uncertainties in the cross-section curve and neutron spectra for the reactor neutron dosimetry. A new formulation to represent the covariance matrix of the cross-section curves has been proposed with a special consideration to the error-propagation rule for the stochastic random uncertainty using the energy resolution concept, where the new concept of the "Experimental Uncertainty Principle" has been discussed. Two other components have been pointed out to be affecting the random error of the practical group-averaged cross-section values, they are the interpolation error and the averaging error, which are expected to be studied further using the energy resolution concept. As a conclusive remark, uncertainty evaluations of the neutron cross-section values should be made with a distinction between the random error and the other uncertainty, and should be filed seperately because their error-propagation rules are severely different from each other.

References

1. R.J. Gehrke, R.G. Helmer, R.C. Greenwood, "Precise Relative γ-ray Intensities for Calibration of Ge Semiconductor Detectors", Nucl. Instr. & Method, 147, 405 (1977)

2. Y. Yoshizawa, Y. Iwata et.al., "Precision Measurements of Gamma-ray Intensities", JAERI-M 8769 (1980)

3. M. Nakazawa, A. Sekiguchi, "Several Applications of J1-Unfolding Method of Multiple Foil Data to Reactor Neutron Dosimetry", EUR 6813 Vol.1, 751 (1980)

4. J.M. Ryskamp, R.A. Anderl, B.L. Broadhead, W.E. Ford III, J.L. Lucius, J.H. Marable, J.J. Wagschal, "Sensitivity and Uncertainty Analysis of the coupled Fast Reactivity Measurements Facility Central Flux Spectrum", Nucl. Technology 57, 20 (1982)

Neutron Standard Data

STUDIES OF THE $^6Li(n,t)^4He$ REACTION

H. Condé

National Defence Research Institute
Box 27322, S-102 54 Stockholm
Sweden

T. Andersson[x)], L. Nilsson and C. Nordborg[xx)]

Tandem Accelerator Laboratory
Box 533, S-751 21 Uppsala
Sweden

Measurements have been made on the $^6Li(n,t)^4He$ reaction and on its inverse the $T(\alpha,{}^6Li)n$ reaction in the neutron energy region between 200 keV and 3.5 MeV.

Angular distributions of the tritons and the ^{6}Li-ions have been measured at together six different energies for the $^6Li(n,t)^4He$ and the $T(\alpha,{}^6Li)n$ reactions, respectively. The results show a significant change in the angular distributions between 2.8 and 3.5 MeV of neutron energy.

The $^6Li(n,t)$ cross section has been calculated from measurement of the excitation function for the $T(\alpha,{}^6Li)n$ reaction relative to the elastic α-t scattering. The measurement was made at an angle of 3° in the laboratory system and at alpha energies from 11.5 to 17.0 MeV. In a separate measurement the α-t differential scattering cross sections were measured at alpha energies between 11.5 and 17 MeV.

1 Introduction

The $^6Li(n,t)^4He$ reaction mechanism is not very well understood (1, 2, 3, 4).

Theoretical calculations of the (n,t) cross section by the R-matrix formalism (2) are hampered by the not well known resonance structure of ^{7}Li. A large p-wave ($5/2^-$) resonance dominates the reaction in the few hundred keV region. At lower energies (<100 keV) the cross section follows essentially a 1/v-dependence, which imply a large s-wave resonance. This resonance has not been observed experimentally.

Hale (2) made a R-matrix analysis of the ^{7}Li system considering the t+^{4}He, n+^{6}Li and n+$^6Li^*$ reaction channels. A good fit to differential and integrated cross sections of the $^4He(t,t)^4He$, $^6Li(n,t)^4He$ and $^6Li(n,n)^6Li$ reactions were obtained assuming resonances with total angular momentum and parity (J^P) of $7/2^-$, $5/2^-$, $5/2^-$, $7/2^-$, $3/2^-$ and $3/2^+$. The two first levels are bound and the third level shows up prominently at E_n=240 keV. The $3/2^-$ level gives a shoulder to the $^6Li(n,t)$ cross section at about 2-3 MeV and $3/2^+$ is a broad s-wave resonance assumed at high energy. The 1/v cross section at low energies is due to distant background levels having $J^P=1/2^+$ and the broad $3/2^+$ resonance. Results below 400 keV obtained by Brown et al (5) from a measurement on the inverse reaction $^4He(t,{}^6Li)n$ are in good agreement with the calculations by Hale while cross section data obained from measurements on the $^4He(t,n)^6Li$ reaction by Drosg et al (4) deviates with about 30 % above 500 keV with the same calculations.

Polarization measurements by Glättli et al (6) at low energies point at a dominance of the 1/2 spin channel in favor of the 3/2 assumed in R-matrix calculations at these energies.

Present addresses:

x) Swedish Steel AB, Luleå, Sweden

xx) NEA Data Bank, Gif-sur-Yvette, France

Weigman and Manakos (3) suggested a deutron exchange process to explain the low energy (n,t) cross section with the majority of the reaction in the 1/2 spin channel. This process is consistent with the energy polarization results (6) and avoid postulating unobserved s-wave resonances.

A B Smith et al (1) analysed their total and elastic n+^{6}Li cross sections using a simplified R-matrix formalism but still allowing for off-diagonal elements in the background matrix between l=0 neutron and alpha channels as Hale (2) had done. This modification provided a reaction mechanism not associated with explicitly defined resonances, e.g. the deutron-exchange mechanism. To fit the observed elastic scattering data a broad $1/2^-$ level has to be introduced at about 4 MeV.

In general more experimental data on the ^{7}Li system are needed, in particular in the few MeV region, to get a base for better understanding of the reaction mechanism.

The present study gives information on the ^{7}Li system in the excitation energy region below 11 MeV of relevance for R-matrix calculations. The study includes measurements of triton angular distributions and excitation functions of the $^6Li(n,t)^4He$ reaction. The results have been obtained by studying the $^6Li(n,t)^4He$ reaction as well as the inverse reaction $T(\alpha,{}^6Li)n$. The measurements were performed at the 12 MeV tandem accelerator at TLU, Uppsala, Sweden.

2 The $^6Li(n,t)^4He$ reaction experiment

2.1 Experimental procedure and data processing

The experimental arrangement of the $^6Li(n,t)$ measurement has been described earlier (7).

The outgoing triton and alpha particles were recorded with solid state detectors at four angles simultaneously in a thin-walled scattering chamber.

Angular distributions for the tritons formed in the $^6Li(n,t)^4He$ reaction have been recorded at three incident neutron energies 2.04, 2.75 and 3.46 MeV. The neutrons produced by the $^7Li(p,n)$ reaction in evaporated Li-metal targets (2 mg/cm^2) had an energy spread of about ±70 keV. Neutrons of two different energies are simultaneously produced by the $^7Li(p,n)$ reaction. The relative intensity of the low energy neutrons (energies 1.58, 2.30 and 3.02 MeV) amount to about 10 %.

K. H. Böckhoff (ed.), Nuclear Data for Science and Technology, 447–450.

The yields of the alpha particles and tritons in each run were determined from the corresponding peaks in the pulse-height spectra of the solid state detectors. The yields divided by the solid angels gave a relative measure of the (n,t) differential angular cross section. Consideration was taken to the finite geometry, when calculating the solid angles and mean emission angles of the tritons and alphas.

Corrections of the order of 1-3 % were applied for the low energy neutrons from the $^{7}Li(p,n)$ neutron producing reaction utilizing angular distributions from Overlay et al (8) and of less than 0.5 % for multiple scattering according to Marion and Zimmermann (9).

2.2 Results

The results in the center-of-mass system of the relative (n,t) angular differential cross sections at 2.04, 2.75 and 3.46 MeV neutron energy from the $^{6}Li(n,t)^{4}He$ reaction measurement are given in table I and figure 1.

The errors in the cross sections are the sum of the statistical uncertainties (∿5 %) and the uncertainties in the calculation of the solid angles (0-10 %).

3 The $T(\alpha,{}^{6}Li)n$ reaction experiment

3.1 Experimental procedure and data processing

The experimental arrangement has been described earlier (9).

The target in the $T(\alpha,{}^{6}Li)n$ reaction experiment consisted of tritium adsorbed in a thin titanium layer evaporated onto a nickel backing (1.1 mg/cm^{2}). The alpha beam entered the target through the nickel backing and the outgoing ^{6}Li ions, emitted at angles near 0° in the laboratory system, were recorded by a combined magnetic spectrometer-surface barrier detection system.

Because the ^{6}Li ions are emitted in a narrow forward cone (<15°) the experiment was done with good angular resolution (<0.5^{o}) obtained by a narrow a carefully aligned slit arrangement.

Two different types of runs were made. One to obtain the angular distributions of the ^{6}Li ions at three incident alpha energies 14, 15 and 16 MeV and one to obtain the excitation function between 11.9 and 17 MeV.

The ^{6}Li yield was in both runs measured relative to the elastic α-t scattering cross section. The recoiling tritons were detected at one fixed angle relative to the incoming alpha beam by the use of a solid state detector. This angle was determined in a separate measurement using the H(p,p)H reaction. The target was CH_2 evaporated on a thin carbon foil. The angle (∿24°) was calculated from the energy shift of the scattered protons relative to that of the incident proton beam.

When turning the heavy magnet of the spectrometer in the angle interval 0-15° a small shift was observed in the relative counting rate of scattered protons from the H(p,p)H and C(p,p)C reactions. This was due to a small shift in angle (<1°) between in the incoming beam and the solid state detector monitor when the magnet was turned. A correction factor to the monitor counting rate of between 0 and 10 % at different angles was estimated.

The α-t differential scattering cross sections were measured separately at incident alpha energies between 11.5 and 17 MeV. The same tritium target as in the $T(\alpha,{}^{6}Li)n$ reaction experiment was placed in a scattering chamber. The recoiling tritons were counted at three angles (22, 24 and 26°) by solid state detectors relative to the integrated proton beam. The results show two resonances at alpha energies of about 11.5 and 16.5 MeV. No significant changes in the relative energy dependence of the differential cross sections was observed over the studied angular interval.

To obtain the ^{6}Li yield at one incident alpha energy and ^{6}Li emission angle the magnetic field of the spectrometer has to be changed over a certain range. This range was wider towards larger ^{6}Li angles. Thus, two settings of the magnet field was required at the lowest ^{6}Li emission angles while the number of settings increased to about ten at the highest angles.

The angular distributions were obtained by measuring the yields of the ^{6}Li ions relative to the α-t scattering in 2° steps starting at 1° and ending at a maximum angle (<15°) depending on the indicent alpha energy. The two different energy groups of ^{6}Li ions, which are produced in the $T(\alpha,{}^{6}Li)n$ reaction, corresponding to forward and backward emission in the center-of-mass system, were measured. After correcting for the shift in angle with the position of the magnet, the data were converted to the center-of-mass system and transformed to $^{6}Li(n,t)^{4}He$ cross sections.

The excitation function was determined from a measurement of the ^{6}Li yield at 3° in the laboratory system relative to the α-t scattering at 15 alpha bombarding energies between 11.9 and 17 MeV.

The ^{6}Li ions have a charge distribution dominated by Li^{3+}. The Li^{3+}/Li^{2+} ratio is about 20 and ions with unit charge are negligible (10). A correction was applied for the charge distribution.

After correcting the data for the energy dependence of the α-t scattering and multiple scattering (<0.1 %) they were converted to the center-of-mass system and transformed to $^{6}Li(n,t)^{4}He$ cross sections.

Table I. Relative triton angular distribution of the $^{6}Li(n,t)^{4}He$ reaction in the center-of-mass system

E_n = 1.30 MeV			E_n = 1.82 MeV		
θ cm	dσ/dΩ	Δ(dσ/dΩ)	θ cm	dσ/dΩ	Δ(dσ/dΩ)
7.5^{o}	1.00	±0.05	6.7^{o}	1.00	±0.05
23.0	0.93	0.04	20.4	0.96	0.05
39.7	0.79	0.04	34.9	0.88	0.05
59.7	0.85	0.04	51.2	0.81	0.05
134.3	0.69	0.04	72.4	1.17	0.10
150.0	0.70	0.04	125.6	0.79	0.05
150.4	0.60	0.04	142.8	0.74	0.05
163.0	0.59	0.04	155.1	0.67	0.05
163.1	0.65	0.04	165.6	0.74	0.05
174.5	0.68	0.04	170.5	0.84	0.05

E_n = 2.04 MeV			E_n = 2.32 MeV		
θ cm	dσ/dΩ	Δ(dσ/dΩ)	θ cm	dσ/dΩ	Δ(dσ/dΩ)
13.6^{o}	1.00	±0.03	18.8^{o}	1.00	±0.05
37.3	0.98	0.12	32.1	1.13	0.05
58.0	1.26	0.15	46.5	1.15	0.05
71.2	1.28	0.04	75.0	1.41	0.07
75.1	1.38	0.04	125.1	1.20	0.06
119.3	1.13	0.13	147.5	0.95	0.05
141.0	0.02	0.12	158.0	0.96	0.05
165.7	1.00	0.02	167.2	0.91	0.05

E_n = 2.75 MeV			E_n = 3.46 MeV		
θ cm	dσ/dΩ	Δ(dσ/dΩ)	θ cm	dσ/dΩ	Δ(dσ/dΩ)
13.6^{o}	1.03	±0.03	14.1^{o}	1.05	±0.04
13.8	0.97	0.02	14.1	0.95	0.03
37.8	1.31	0.03	38.4	0.89	0.04
58.8	1.69	0.04	59.7	0.78	0.03
90.0	1.42	0.03	90.7	0.66	0.04
118.1	1.12	0.03	117.1	0.64	0.03
140.1	1.05	0.03	139.5	0.71	0.03
165.2	1.06	0.02	165.0	0.84	0.03
			165.0	0.82	0.03

3.2 Results

The angular triton distributions of the $^6Li(n,t)^4He$ reactions in the center-of-mass system at 1.30, 1.82 and 2.32 MeV neutron energy obtained from the measurement on the $T(\alpha, {}^6Li)n$ reaction are given in table I and figure 1. The angular spread of the measurements increased in the center-of-mass system towards emission angles of 90°. Whereas significant the angular spread is shown in figure 1.

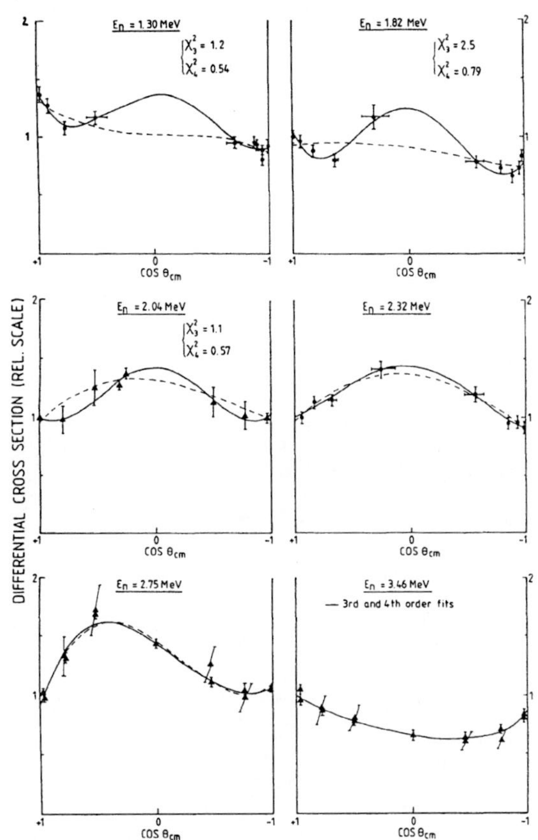

Figure 1. Triton angular distributions in the $^6Li(n,t)^4He$ reactions. Dashed and full-lines give the 3rd and 4th order Legendre polynomial fits, respectively.

The errors include statistical uncertainties (3-5 %) and an uncertainty in the correction for the angular shift with the magnet position (0-2 %).

Third and fourth order Legendre polynomials were fitted by a least square procedure to the experimental data and are shown as dotted and full-lines, respectively, in figure 1. In the figure are also given the χ^2-values for the fitted distributions.

The angular integrated cross sections ($4\pi B_o$) of the $^6Li(n,t)^4He$ reaction between 240 and 2 800 keV were calculated from the experimental data, obtained from the $T(\alpha, {}^6Li)n$ measurement at 3°, utilizing the energy dependence of Legendre coefficients observed in the present experiment and by Overlay et al (8). The data were normalised to ENDF/B V cross sections at 240, 290 and 340 keV. Results are given in table II and figure 2 which are based on both a third- and fourth-order Legendre fit to the angular distribution data.

Table II. The angular integrated cross section of the $^6Li(n,t)^4He$ reaction. The data are normalised to ENDF/B-V at 240, 290 and 340 keV

E_n (keV)	σ(barn) Third order Legendre fit (see text)	Fourth order Legendre fit (see text)
240	2.91 ± 0.17	
290	1.82 ± 0.09	
290	1.69 ± 0.09	
340	0.87 ± 0.05	
440	0.34 ± 0.03	
545	0.32 ± 0.03	
646	0.24 ± 0.02	
801	0.24 ± 0.02	
1 052	0.23 ± 0.02	
1 308	0.24 ± 0.03	0.26 ± 0.03
1 564	0.17 ± 0.02	0.19 ± 0.02
1 820	0.15 ± 0.03	0.16 ± 0.03
2 071	0.17 ± 0.04	0.17 ± 0.04
2 327	0.19 ± 0.04	0.18 ± 0.04
2 583	0.14 ± 0.02	0.15 ± 0.02
2 834	0.16 ± 0.02	0.17 ± 0.02

Due to straggling of the alpha particles passing the nickel foil and of the subsequent ^6Li ion passing the titanium layer a total energy spread of the ^6Li ions of the order of 100 keV was obtained. Thus, ENDF/B-V data weighted with a Gaussian neutron energy distribution (FWHM 25 keV) were used for normalisation and are shown as a dotted line in figure 2.

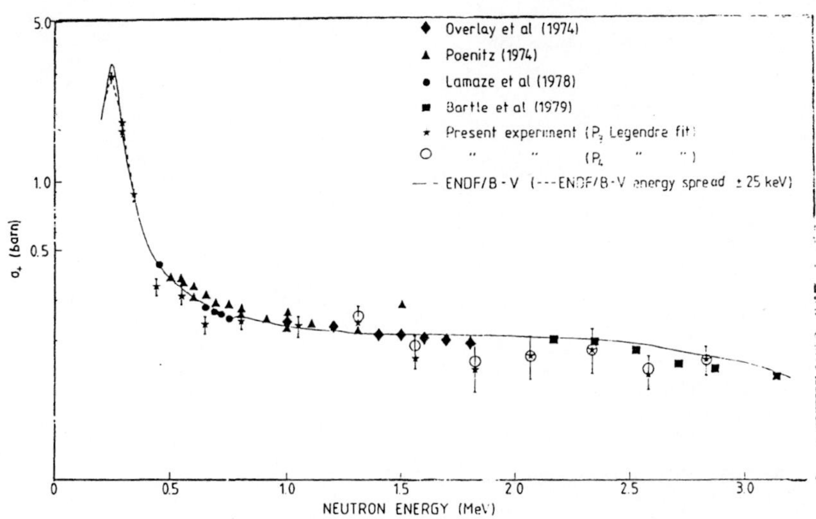

Figure 2. The $^6Li(n,t)$ angular integrated cross section. The present results are normalised to ENDF/B-V data at 240, 290 and 340 keV.

The errors of the integrated cross sections were calculated as the square root of the square sum of the statistical uncertainty (2-4 %), the relative uncertainty in the T(α,α)T cross section (0-10 %), the charge distribution correction uncertainty (<0.5 %), the uncertainties in the conversions to the center-of-mass system and $^{6}Li(n,t)$ cross sections (<2 %), and finally the uncertainty of the angular distributions (5-20 %).

4 Discussions

The experimental techniques used in the present measurement of the angular distributions of the ^{6}Li-ions from the $T(\alpha,{}^{6}Li)n$ reactions involve many experimental difficulties. One reason for this is the narrow cone (<15°) in which the ^{6}Li-ions are emitted in the α-energy interval of interest for this study. Extreme good angular resolution is required to study in particular the ^{6}Li intensity close to zero and maximum emission angles of the ^{6}Li-ions. The angular resolution is limited by the requirement to have a reasonable signal counting rate. Though, the experimental conditions of the present equipment was optimized a fairly narrow range of ^{6}Li emission angles could be studied with an increasing uncertainty of the values close to cosθ=0 (see Figure 1, E_n=1.30, 1.82 and 2.32 MeV).

The angular distributions are almost equally well fitted to 3rd and 4th order Legendre polynomials with the exception of the angular distribution at E_n=1.82 MeV. At this energy a 4th order Legendre polynomial is a better fit, but this is much depending on a single value at cosθ 0.3 which has a fairly large uncertainty.

The angular distributions change drastically between 2 and 3.5 MeV in which energy region $7/2^-$ and $3/2^-$ levels in ^{7}Li have been reported (2).

The observed $^{6}Li(n,t)$ integrated cross section (Figure 2) agrees fairly well with recent experiments. However, in the energy region from 1.5 to 2.5 MeV the present results are somewhat low. The large errors of the values in this energy region is mainly due to the uncertainties in the angular distributions.

5 Acknowledgement

The authors want to thank Dr L G Strömberg who participated in the early phase of the experimental work.

References

1 A B Smith, P T Guenther and J F Wahlen. Nucl. Phys. A273 (1982) 305

2 G M Hale, Proc. Symp. on Neutron Standards and Applications, Nat. Bureau of Standards Pub. NBS-493 (1977) 30

3 H Weigmann and P Manakos, Z. Phys. A289 (1979) 383

4 M Drosg, D M Drake, R A Hardekopf and G M Hale, Private communication

5 R E Brown, G G Ohlsen, R F Haglund Jr and N Jarmie, Phys. Rev., C16 (1977) 513

6 G Glättli, A Abragam, G Bacchella, M Fourmond, P Meriel, J Piesvaux and M Pinot, Phys. Rev. Lett. 40 (1978) 748

7 T Andersson, C Nordborg, L Nilsson, H Condé and L G Strömberg, Proc. Int. Conf. on Neutron Physics and Nuclear Data, Harwell 1978, p. 746

8 J C Overlay, R M Sealock and D H Ehlers, Nucl. Phys. A221 (1974) 573

9 J B Marion and Zimmermann, Nucl. Instr. and Methods, 51 (1967) 93

10 J B Marion and F C Young, Nucl. Reaction Analysis, North-Holland Publ. Company 1968

11 W P Poentiz, Z Phys 268 (1974) 359

12 G P Lamaze, R A Schrack and O A Wasson, Nucl. Sci. Eng. 68 (1978) 183

13 C M Bartle, Nucl. Phys. A330 (1979) 1

DETERMINATION OF THE $^6Li(n,t)^4He$ STANDARD CROSS SECTION

H.-H. Knitter, C. Budtz-Jørgensen and D.L. Smith*

Commission of the European Communities
Joint Research Centre
Central Bureau for Nuclear Measurements
Geel, Belgium

The neutron total and differential neutron elastic scattering cross sections of ^6Li were measured at the Van de Graaff with monoenergetic neutron beam and time-of-flight techniques in the neutron energy range from 80 keV to 3.0 MeV, yielding the angle integrated $^6Li(n,t)^4He$ cross section from 80 keV to 500 keV with an accuracy ranging from 2.5% to 6.5%. Angular distributions of the tritons with respect to the incident neutron beam direction were measured with the pulsed white spectrum neutron source GELINA in the neutron energy range from 10 eV to 330 keV. Both experiments were combined to give the absolute differential $^6Li(n,t)^4He$ cross section in the debated region of the resonance. A comparison of recent cross section shape measurements in the resonance region, which were normalized at low neutron energies to the $^6Li(n,t)^4He$ cross section itself, shows good agreement with the present cross section measurements for which the absolute scale does not depend from other experiments. A further comparison of the present differential (n,t) cross section for zero and 180 degrees shows excellent agreement with the same quantity determined by Brown et al. from the measurement of the zero degree laboratory ^6Li yield of the inverse reaction $^3H(^4He,^6Li)n$, which was normalized relative to the ^3H-^4He elastic scattering cross sections.

[reactions $^6Li(n,n)^6Li$, $^6Li(n,t)^4He$; cross sections $\sigma_{n,T}$, $\sigma_{n,n}(E,\vartheta)$ 80 keV $< E_n <$ 3 Mev; $\sigma_{n,t}(E,\vartheta)$ 10 eV $< E_n <$ 500 keV]

Introduction

The $^6Li(n,t)^4He$ reaction plays an important role in nuclear technology, e.g. as a widespread standard for neutron fluence determination and as a nonradiative capture reaction for neutrons in shielding applications. The present investigation was undertaken within the first context.

The $^6Li(n,t)^4He$ reaction possesses several favourable features for applications as a standard : Below 10 keV neutron energy the cross section exhibits a 1/v - behaviour with a large thermal cross section of 936mb [1]. The substantial positive Q-value ensures that the tritons and alpha particles can be readily detected. The reaction shows also a smooth behaviour up to neutron energies of 10 to 15 MeV [2] except for the region of the broad 250 keV resonance. Up to 1.72 MeV incident neutron energy the only open reaction channels are the (n,t) and (n,n) channels. The neutron capture process shows only a very small negligible cross section [3]. Therefore also the interpreting model can be kept comparatively simple.

In applications the $^6Li(n,t)^4He$ reaction is used e.g. for reactor neutron spectrometry up to several MeV neutron energy as a standard [4]. The main problem areas for the application of the $^6Li(n,t)^4He$ reaction as a standard were laying in the discrepant region for the angle integrated cross section around the 250 keV resonance and above, as well as in the angular distributions of the tritons with respect to the incident neutron beam direction from very low energy neutrons of $\approx$ 10 eV and upwards.

For these reasons we performed measurements of the neutron total and elastic neutron scattering cross sections between 80 keV and 3 MeV to determine parameters which describe the angle integrated $^6Li(n,t)^4He$ cross section. These measurements were only published in form of a laboratory report [5] and therefore some of the results are represented here together with new and comprehensive trition angular distribution measurements performed in the neutron energy range from 10 eV to 330 keV.

The Angle-integrated $^6Li(n,t)^4He$ Cross Section

The neutron total and differential elastic scattering cross sections were measured at the CBNM's Van de Graaff laboratory in the incident neutron energy range from 80 keV to 3.0 MeV using mono-energetic neutron beams and time-of-flight techniques. The details of the experiment can be found in ref.[5]. The typical over-all uncertainty for the total cross section is smaller than 1.7%. The representative error on the differential neutron elastic scattering cross sections is 4.7%, but for some cases at energies around the resonance the error is somewhat larger.

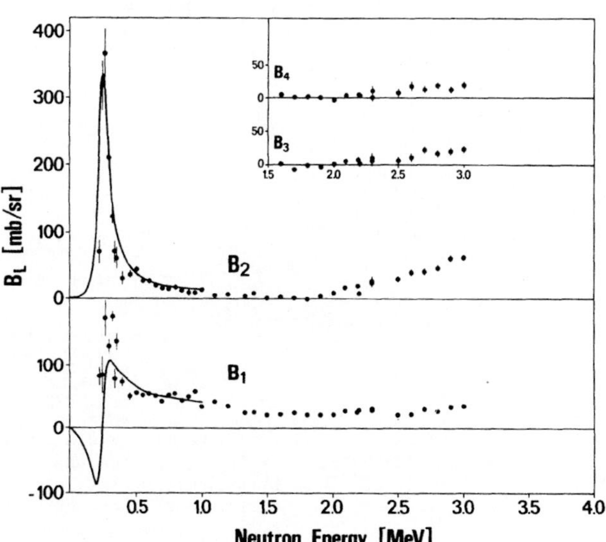

Fig. 1. Centre of mass Legendre polynomial expansion coefficients of the neutron elastic scattering angular distributions of ^6Li. Solid lines are calculated curves.

From the elastic neutron scattering angular distributions it can be seen that below 1.5 MeV incident neutron energy only s- and p- wave neutrons are interacting, since in this region the centre-of-mass Legendre coefficients with $L > 2$ are zero [6]. Interference of possible p-wave resonances at higher energies with the present resonance can only be very small since the

* Visiting Scientist from Argonne National Laboratory Argonne, Illinois 60439, U.S.A.

K. H. Böckhoff (ed.), Nuclear Data for Science and Technology, 451–455.

B_2-coefficient approaches zero in the neutron energy region from 1 to 1.5 MeV as one can see from fig. 1. The known $5/2^-$-resonance below the neutron binding energy, at an excitation energy of 6.64 MeV of the ^{7}Li-nucleus, has an extremely small reduced neutron width of γ_n^2 = (0.00 ± 0.01) MeV as determined by Spiger and Tombrello [7]. It follows that all the transition probabilities will be negligibly small for processes containing this neutron channel. In addition this implies that the $5/2^-$-resonance below the neutron binding energy can not give a contribution to the (n,t) and (n,n)-processes in the region above the neutron binding energy. Due to the above mentioned reasons, a single isolated resonance is considered and for the representation of the total and the integrated elastic scattering cross section a single level two channel Breit-Wigner formula plus linear background terms are used. These background terms account for the potential scattering and for an eventual contribution from distant levels. The $1/\sqrt{E}$ dependence of the (n,t)-cross section [8] is also included. These terms lead to the equations for the neutron total and integrated elastic scattering cross section of ^{6}Li

$$\sigma_T(E) = a_1 + a_2E + \frac{0.14956}{\sqrt{E}} + \frac{\pi\lambda^2(2J_\circ+1)}{(2i+1)(2I+1)} \cdot \frac{\Gamma_n^2 + \Gamma_n \cdot \Gamma_a}{(E_\circ+\Delta-E)^2+1/4(\Gamma_n+\Gamma_a)^2} \tag{1}$$

$$\sigma_{n,n}(E) = a_3 + a_4E + \frac{\pi\lambda^2(2J_\circ+1)}{(2i+1)(2I+1)} \cdot \frac{\Gamma_n^2}{(E_\circ+\Delta-E)^2+1/4(\Gamma_n+\Gamma_a)^2} \tag{2}$$

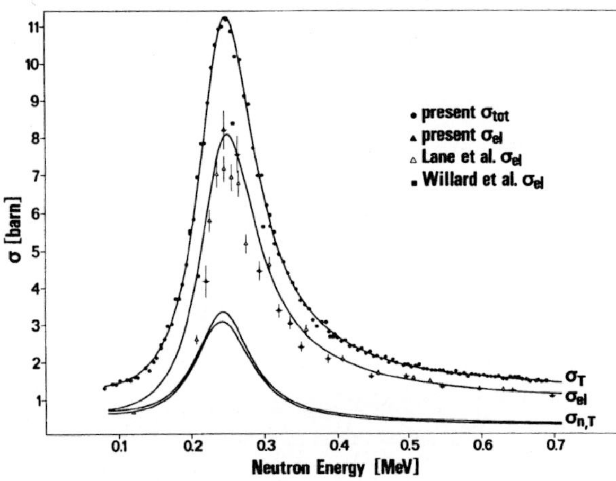

Fig. 2. The neutron total, integrated elastic and $(n,t)^4$He cross section of ^{6}Li are plotted versus the incident neutron energy. The full circles represent the measured values of the neutron total cross section of the present experiment. The full triangles are the measured values of the integrated neutron elastic scattering cross section from this work. The open triangles and the full squares are the measured values of the integrated neutron elastic scattering cross sections of ref.[9-10] and ref.[11] respectively. The upper two full lines are obtained by a simultaneous fit through the experimental total and integrated elastic scattering cross sections as described in the text. The two lowest full lines give the limits for the ^{6}Li$(n,t)^4$He cross section resulting from this work.

with $\Gamma_n = 2\gamma_n^2 P_n(E)$ and $\Delta = -\gamma^2(S_n(E)-S_n(E_\circ))$. γ_n^2 is the reduced neutron width, $P_n(E)$ and $S_n(E)$ are the neutron penetration and shift function respectively. In accordance with the findings of ref.[12] and with the fact that the Legendre coefficients with L > 2 are zero, only the neutron subchannel with a channel spin 3/2 and orbital angular momentum l was used. For the calculation of $P_n(E)$ and $S_n(E)$ an interaction radius of 3.89 fm was used. A simultaneous fit to the total and integrated cross section was made using all the cross section data up to 1.5 MeV minimizing the expression.

$$\chi^2 = \sum_i \left\{\frac{\sigma_T(E_i)_{exp.} - \sigma_T(E_i)_{cal.}}{\Delta\sigma_T(E_i)_{exp.}}\right\}^2 + \sum_j \left\{\frac{\sigma_{n,n}(E_j)_{exp.} - \sigma_{n,n}(E_j)_{cal.}}{\Delta\sigma_{n,n}(E_j)_{exp.}}\right\}^2 \tag{3}$$

This procedure allows a determination of the unknown parameters contained in equation (1) and (2). The following numerical results were obtained :

a_1= (0.616±0.020)b a_2= (0.209±0.015)b MeV^{-1}
a_3= (0.631±0.049)b a_4= (0.090±0.048)b MeV^{-1}
γ_n^2= (1.082±0.017)MeV Γ_a= (0.0357±0.0015)MeV
$E_\circ$= (0.2502±0.0007)MeV

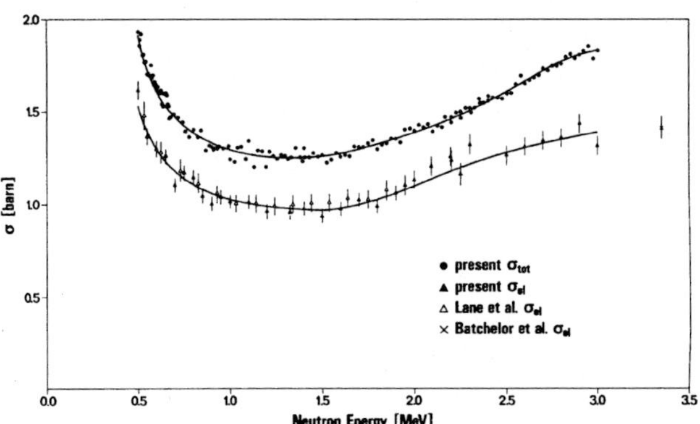

Fig. 3. The neutron total and integrated elastic scattering cross sections of ^{6}Li are plotted versus the incident neutron energy. Δ and χ are data of ref. 10 and 13 respectively, ▴and ● are the elastic scattering and neutron total cross sections of the present work.

The curves obtained by the fitting procedure for the total cross section and for the integrated elastic scattering cross section are plotted as full lines in fig.2 and fig.3 till 1.5 MeV incident neutron energy. Above 1.5 MeV the full lines represent eye-guide curves. The shape of the total cross section peak as given by the experimental points is very well represented by the calculated curve. The maximum of the total cross section has a value of (11.27 ± 0.12)b and appears at the incident neutron energy of (247 ± 3)keV.

The B_2 centre-of-mass Legendre polynomial expansion coefficient of the elastic neutron scattering angular distributions can be calculated from the parameters obtained by the fit without additional assumption. By specifying the general formula for the Legendre coefficients given in ref.[14] to this specific case one obtains :

$$B_2(E) = \frac{\lambda^2}{(2i+1)(2I+1)} \bar{Z}^2(1,J_\circ,1,J_\circ/s,2) \cdot \frac{\Gamma_n^2/4}{(E_\circ+\Delta-E)^2+1/4(\Gamma_n+\Gamma_a)^2} \tag{4}$$

The calculated values of the B_2-coefficient are plotted as the full line in fig. 1. As can be seen, the agreement between the calculated and observed values of the B_2-coefficient is good. The B_1-coefficient cannot be calculated from the above parameters without additional assumptions about the potential scattering phase shift for the channel spin 3/2 and the orbital angular momentum zero. However the information contained in B_1 could be used as well in a comprehensive R-matrix fit.

From the parameters obtained by the fit the $^6Li(n,t)^4He$ cross section was calculated. The errors of the (n,t) cross section were obtained from error propagation calculations using the parameters and their errors as obtained from the least quares fit. The 1σ-limits for the $^6Li(n,t)^4He$ cross section are given by the corresponding two full lines in fig.2 and 4. The $^6Li(n,t)^4He$

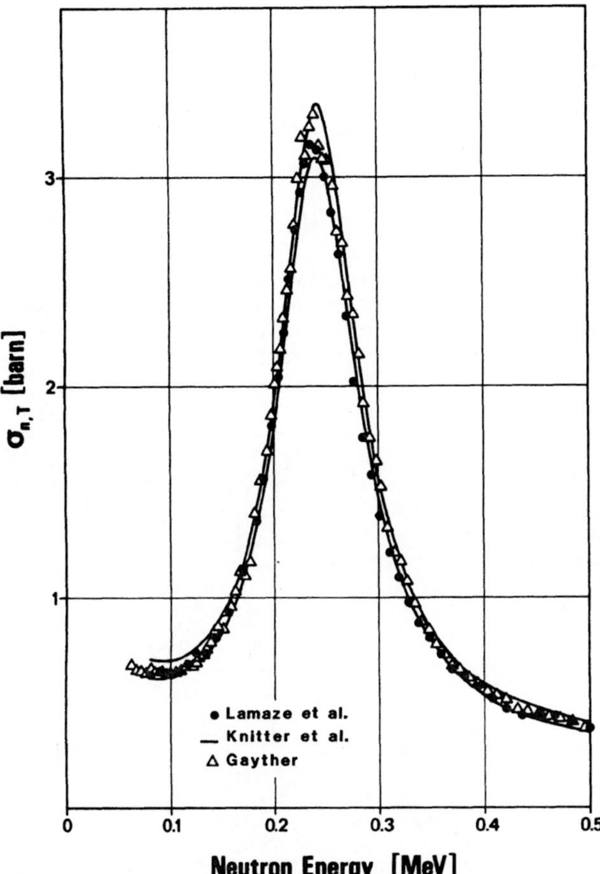

Fig. 4. $^6Li(n,t)^4He$ cross section is plotted versus the incident neutron energy. • and Δ are results of ref.[15] and [16] respectively. The area between the two full lines is the 1σ-area of the cross section as determined by the present experiment.

cross sections and their errors are given in tabular form in ref.[5] in convenient neutron energy steps between 85 keV and 500 keV incident neutron energy. The errors for the $^6Li(n,t)^4He$ cross section are about 4% in the top of the resonance and elsewhere between 2.5 and 6.5%. In fig. 4 the present (n,t) cross section results are compared with recent shape measurements of this quantity which were normalized in the 2 to 10 keV range of the $^6Li(n,t)^4He$ reaction itself. Good agreement is found with these measurements of Lamaze et al. [15] and Gayther [16]. The resonance integral between 100 and 500 keV was evaluated by Derrien et al.[17] who got values of 0.488 b•MeV and 0.486 b•MeV from the present work and from the comprehensive R-matrix evaluation of Hale [18] respectively. This is a difference of only 0.4%. The present cross section at the top of the resonance is 2.5% lower than the one of Hale [18] and 1.7% lower than the one obtained by Guenther et al.[19] from a similar anyalysis than the present one.

Angular Distribution Measurements

The angular distribution measurements were performed at the Geel electron linear accelerator (GELINA) using a neutron flight path of 31.66m length. The detector was an ionization chamber as shown in fig. 5. The detector is able to furnish signals, induced by charged particles emitted from the cathode plane and stopped in the space between cathode and grid, which contain in principle information on the energy, on the cos ϑ of the emission angle of the particle with respect to the normal on the cathode, and on ionization properties of a specific particle type with the detector

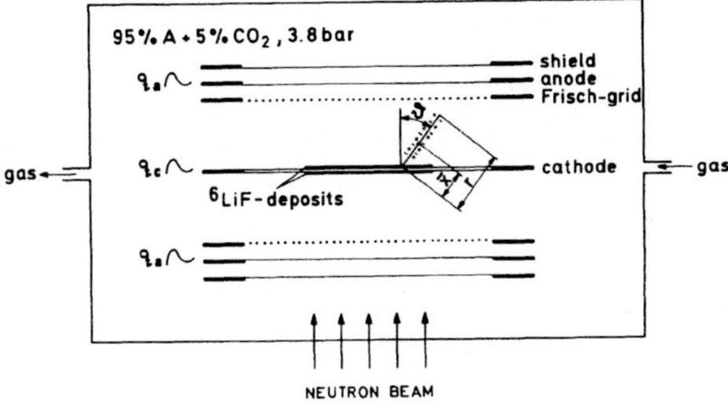

Fig. 5. Schematic representation of the ionization detector.

gas. Also the time of the particle emission can be obtained within a few ns or better [20], so that one can perform neutron time-of-flight experiments. The anode gives a signal q_a which is proportional to the particle energy E,

$$q_a = C_a \circ E \tag{5}$$

whereas the signal of the cathode q is given by

$$q_c = C_c \cdot E \circ \left(1 - \frac{\bar{X}(Z,A,E)}{d} \cdot \cos\vartheta\right) \tag{6}$$

where $\bar{X}(A,Z,E)$ is the distance from the beginning to the centre of gravity of the charge distribution of the ion trace. A and Z are the atomic and charge number of the particle respectively. The anode and cathode signals of one event together with $\bar{X}$ allow the determination of cos ϑ of the particle trace. Details of detector and experiment will be available in a further publication [21]. Fig. 6 shows a 3-dimensional spectrum of the anode signals q_a versus the ratio of the cathode and anode signals q_c/q_a for events which correspond to an incident neutron energy bin between 100 keV and 300 keV. It demonstrates that tritons and alpha particles can be readily distinguished by this detector. The cos ϑ-information is contained in q_c/q_a.

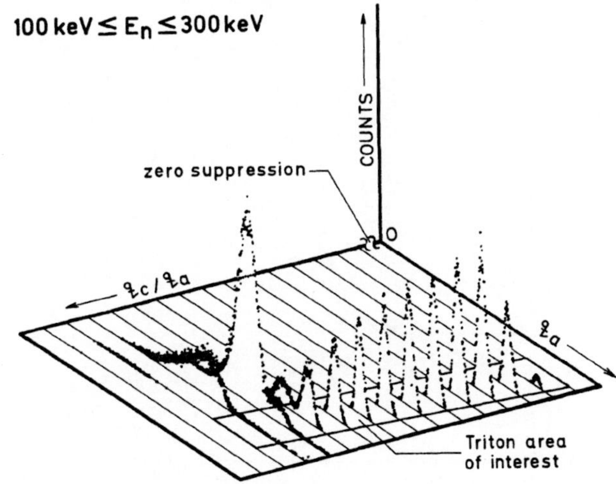

Fig. 6. The number of events induced by neutrons between e.g. 100 keV and 300 keV are plotted versus the particle energy q_a and versus the ratio of the cathode to anode signal q_c/q_a.

Triton angular distributions for 46 neutron energy bins are evaluated from the collected data in the incident neutron energy range from 10 eV to 330 keV. Fits with a Legendre polynomial expansion to the term with L = 2 are made. Higher terms are not needed in the present neutron energy range. The coefficients are normalized with the L = 0 coefficient and are shown in fig. 7, together with the corresponding

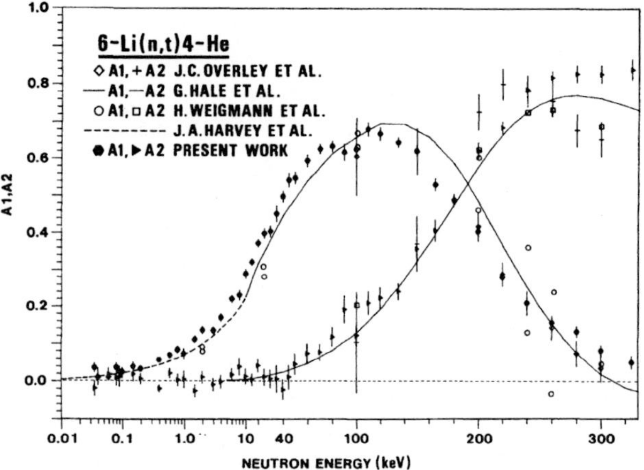

Fig. 7. The first and second order normalized centre-of-mass Legendre polynomial expansion coefficients are plotted versus the incident laboratory neutron energy.

values measured by Overley et al.[22] and values calculated with the deuteron exchange model by Weigmann et al.[23]. The full curves represent the results of the comprehensive R-matrix evaluation of Hale[18]. The details of the experiment, the evaluation procedure and the numerical values will be given in a forthcoming publication [21].

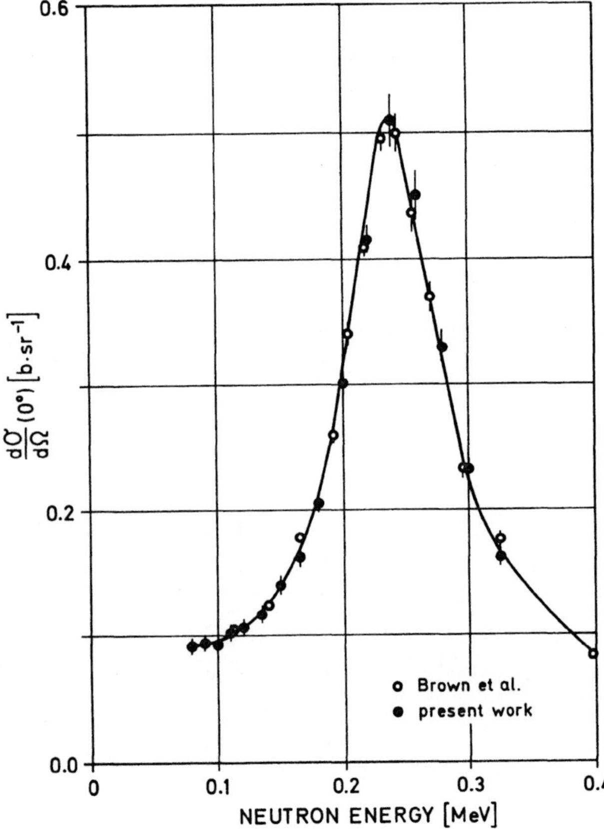

Fig. 8. Differential cross sections for zero degree of the ^{6}Li(n,t)^{4}He reaction are plotted versus the incident neutron energy.

Absolute Differential Cross Sections

The relative angular distributions obtained at GELINA can be brought on an absolute scale by the integrated ^{6}Li(n,t)^{4}He cross sections obtained in the measurements at the Van de Graaff. The necessary parameters to obtain the angle integrated ^{6}Li(n,t)^{4}He cross section as function of the neutron energy $\sigma_o(E)$ are given above.

The differential cross section as a function of energy E and angle ϑ is then given by

$$\frac{d\sigma}{d\Omega}(E,\vartheta) = \frac{\sigma_o(E)}{4\pi}\,[\,1+A_1(E)P_1(\vartheta)+A_2(E)P_2(\vartheta)] \qquad (5)$$

where $A_1(E)$ and $A_2(E)$ are the measured Legendre coefficients and $P_1(\vartheta)$ and $P_2(\vartheta)$ are the corresponding Legendre polynomials. The only systematic angular distribution measurements were done by Overley et al. [21] and a comparison in the overlapping region is made in fig. 7.

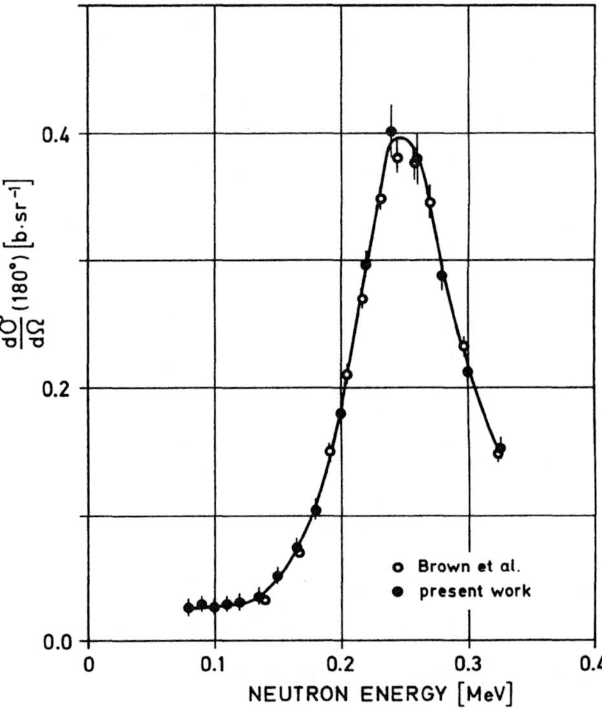

Fig. 9. Differential cross sections for 180 degrees of the ^{6}Li(n,t)^{4}He reaction are plotted versus the incident neutron energy.

Differential cross section measurements for the specific angles of 0° and 180° in the region of the 250 keV resonance were made by Brown et al.[24]. They measured the 0° and 180° cross section of the inverse reaction ^{3}H(α,^{6}Li)n relative to the cross section of the elastically recoiling tritons at 15°. A comparison of the present zero and 180 degrees differential cross sections obtained with equation (5) from the results of the Van de Graaff measurements [$\sigma_o(E)$] and from the GELINA measurements [$A_1(E)$ and $A_2(E)$] with the same cross sections of Brown et al.[24] is made in figs. 8 and 9 respectively. The agreement between the two sets of differential cross section data is excellent within their quoted errors.

Recently also Drosg et al.[25] reported some results of differential cross section measurements, by the inverse reaction ^{4}He(t,n)^{6}Li. However, those of the reported zero degree cross sections which fall in the region of the resonance do not give any new information, since they were used to calibrate the cross section scale and the energy scale with respect to the results of Brown et al.[24].

Conclusions

The neutron total and elastic scattering cross section measurements using mono-energetic neutron beam- and time-of-flight techniques delivered the angle integrated $^6Li(n,t)^4He$ standard cross section [5] in the debated neutron energy range from 80 keV to 500 keV with an accuracy between 2.5% and 6.5%. The normalization in the present experiment for the $^6Li(n,t)^4He$ cross sections depends on the metrological data of the ^6Li- and CH_2-samples used in the total and scattering cross section measurements and on the n-p differential cross section. The CBNM-measurements [5] are in agreement with those of Lamaze et al.[15]. Also the data of Gayther [16], which became available in the meantime, are in good agreement with the CBNM-data. Both groups normalized their cross section shape at low neutron energies to the value of the $^6Li(n,t)^4He$ cross section itself. The zero and 180 degrees differential cross sections of the present angular distribution measurements are also in excellent agreement with the corresponding values obtained by Brown et al.[24] by measuring the ^6Li-particles emitted in the inverse reaction and normalizing with respect to the differential triton-helium elastic scattering cross section. This is especially remarkable, since both experiments are of very complementary character.

All these experiments with different origins for the absolute cross section scale are in agreement within their quoted errors and therefore they give confidence in the results. The differential and angle integrated $^6Li(n,t)^4He$ cross section is now known with confidence to within a few percent in this difficult neutron energy region which covers the 250-keV resonance, and can be used in applications within these uncertainty limits. A proper evaluation could probably improve the picture presented here.

Acknowledgement

The authors wish to thank J. Van Audenhove, E. Freistedt and W. Lycke from the CBNM's sample preparation group for the careful preparation and definition of the ^6Li samples and Dr. P. De Bièvre from the mass spectroscopic group for the isotopic analysis of the Li-material. The authors are also very indepted to Messrs. H. Bax and R. Vogt for the technical support given during the periods of execution and evaluation of the experiments.

References

1. A.B. Smith and W.P. Poenitz, ENDC/NEANDC Nuclear Standard Files, Version 1980, INDC-36/LN (1980)
2. E.D. Pendlebury, AWRE 0-60/64 (1964)
3. F.C. Barker, Phil. Mag. 2, 780 (1957)
4. G. De Leeuw-Gierts and S. De Leeuw, S.C.K./C.E.N. Mol Report BLG-501 (1975)
5. H.-H. Knitter, C. Budtz-Jørgensen, M. Mailly and R. Vogt, EUR 5726 e (1977)
6. J.M. Blatt and L.C. Biedenharn, Rev.Mod.Phys.24, 258 (1952)
7. R.J. Spiger and T.A. Trombello, Phys.Rev.163, 964 (1967)
8. L. Stewart, Proc. of a Symposium on Neutron Standard Reference Data, IAEA Panels Proc.Series(1972)
9. R.O. Lane, A.S. Langsdorf Jr., J.E. Monahan, A.J. Elwyn, Report ANL-6172 (1960)
10. R.O. Lane, A.S. Langsdorf Jr., J.E. Monahan, A.J. Elwyn, Ann. Phys. 12, 135 (1961)
11. H.B. Willard, J.K. Bair, J.D. Kington, H.O. Cohn, Phys. Rev., 101, 765 (1956)
12. Y.S. Selim, Bull. Faculty of Science Alexandria University, Vol. II, 81 (1965)
13. R. Batchelor and J.H. Towle, Nucl. Phys. 47, 385 (1963)
14. H.B. Willard, L.C. Biedenharn, P. Huber and E. Baumgartner, Fast Neutron Physics (J.B. Marion and J.L. Fowler, Edts.) Interscience Publisher, Inc. New York
15. G.P. Lamaze, O.A. Wasson, R.A. Schrack and A.D. Carlson, Proc. of the Intern. Conf. on the Interactions of Neutrons with Nuclei, Lowell, Mass. U.S.A. 6-9 July 1976, Vol. 2, p. 1341
16. D.B. Gayther, Ann. Nucl. Energy, 4, 515 (1977)
17. H. Derrien and L. Edvardson, Proc. of Intern. Spec. Symposium on Nuclear Standards and Applications, held at the National Bureau of Standards, Gaithersburg, M.D., March 28-31, 1977, NBS Special Publication 493, p. 14
18. G.M. Hale, Proc. of Intern. Specialists' Symposium on Neutron Standards and Applications, held at the National Bureau of Standards, Gaithersburg, M.D., March 28-31, 1977, NBS Special Publication 493, p.3
19. P. Guenther, A. Smith and J. Whalen, ANL/NDM-52 (1980)
20. H.-H. Knitter and C. Budtz-Jørgensen,Proc. of the Intern. Conf. on Nuclear Cross Sections for Technology, Knoxville, Tenn.,U.S.A., Oct. 1979, NBS Special Publication 594, p.947
21. H.-H. Knitter, C. Budtz-Jørgensen, D.L. Smith and D. Marletta, accepted by Nucl. Sci. Eng.
22. J.C. Overley, R.M. Sealock and D.M. Ehlers, Nucl. Phys. A221, 573 (1974)
23. H. Weigmann and P. Manakos Z. Physik A289, 383 (1979)
24. R.E. Brown, G.G. Ohlsen, R.F. Haglund, Jr. and N. Jarmie, Phys. Rev. 16C , 513 (1977)
25. M. Drosg, D.M. Drake, R.A. Hardekopf, and G.M. Hale, LA-9129 MS

MEASUREMENT OF THE $^{235}U(n,f)$ CROSS SECTION FROM 0.3 TO 3.0 MeV USING THE NBS ELECTRON LINAC

A. D. Carlson and J. W. Behrens

National Bureau of Standards
Washington, D.C. 20234, U.S.A.

Progress is reported on a measurement of the $^{235}U(n,f)$ cross section from 0.3 to 3.0 MeV using the NBS electron linac as a pulsed neutron source. Fission events were detected using a parallel-plate, ionization fission chamber located 69.5 m from the neutron-producing tungsten target and containing 170.9 ± 2.1 mg of ^{235}U. The absolute neutron flux was measured with the NBS black neutron detector, located at 200.4 m and coaxial with the fission chamber. A Monte Carlo program was used to calculate the neutron detection efficiency of the black detector. The present paper gives preliminary results. The fission cross section shape is shown compared to that of the ENDF/B-V $^{235}U(n,f)$ cross section. Absolute values await a complete analysis of the experimental data and the measurement of the black detector efficiency at 2.6 MeV using the associated-particle technique at the NBS Van de Graaff. Accuracies of ± 2% are expected in the final analysis of this $^{235}U(n,f)$ cross section measurement.

[black neutron detector, electron linac, E_n = 0.3-3.0 MeV, fission cross section, ionization fission chamber, $^{235}U(n,f)$]

Introduction

The neutron-induced fission cross section of ^{235}U has long been a recognized standard for the neutron energy range above 100 keV. For example, most actinide fission cross sections are measured relative to the fission cross section of ^{235}U in the MeV range. Considerable effort has been spent during the past decade to accurately determine these fission cross section ratios.[1] So, efforts must continue to improve the accuracy to which the ^{235}U standard is known. The present measurement was designed to cover the neutron energy range from 0.3 to 3.0 MeV and it is part of a series of measurements at the U.S. National Bureau of Standards (NBS) to improve the accuracy of this important cross section. The present experiment used an ionization fission chamber and the NBS black neutron detector at the NBS linac neutron time-of-flight facility. A similar measurement,[2] conducted earlier using these detectors at the NBS 3-MV Van de Graaff, covered the neutron energy range from 0.2 to 1.2 MeV. There have been other measurements of the $^{235}U(n,f)$ cross section in this energy range. A summary of earlier work is given by Poenitz.[3]

When finalized the present measurement will be absolute since the total ^{235}U mass as well as the incident neutron fluence were measured. The present paper gives preliminary results. The cross section shape is shown compared to that of the ENDF/B-V $^{235}U(n,f)$ cross section.

Experimental Details

The present measurement was conducted with a parallel-plate ionization fission chamber located at the 69.5-m station of the 200-m neutron time-of-flight tube at the NBS linac. The neutron fluence which passed through the fission chamber was determined with a large plastic scintillation detector, *i.e.*, the NBS black neutron detector, at the 200-m station. The linac was operated at 720 Hz with an electron pulse width of ~ 30 ns to produce neutrons by the (γ,n) reaction in a thick, water-cooled tungsten target. The neutron beam was collimated so that the fission chamber could view the entire neutron source. However, for the black detector a small portion (< 1%) of the target could not be viewed due to a penumbra effect. For the measurements a 0.358 g/cm^2 boron-10 overlap filter and 1.59 cm of uranium for gamma flash reduction were used. The fission chamber contained 170.9 ± 2.1 mg of ^{235}U, as determined earlier by Wasson and Meier,[4] and is shown schematically in Fig. 1. The enrichment of the uranium was 99.912 at.% ^{235}U. The fission chamber which was fabricated at the Lawrence Livermore National Laboratory (LLNL) consists of ten deposits of ^{235}U, each having an areal density of approximately 100 μg/cm^2. The 10 cm x 18 cm rectangular deposits were painted[5] onto both sides of five aluminum backings of thickness 2.5 x 10^{-3} cm. The five deposit-carrying plates were alternated at 1 cm spacings with six blank aluminum plates. This arrangement was supported inside a 25 cm diameter aluminum cylinder with a 50 x 10^{-3} cm wall thickness. The cylinder was 25 cm in height.

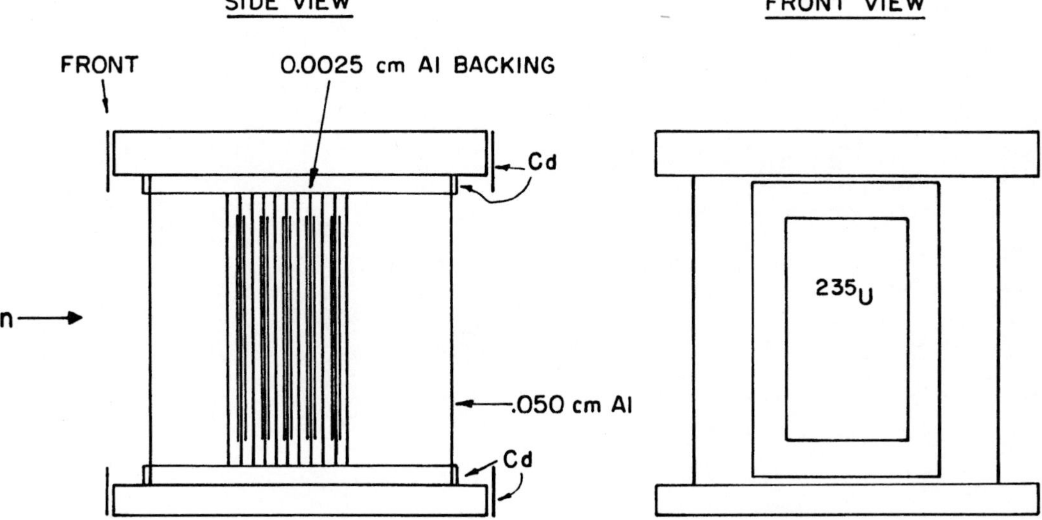

Fig. 1. Schematic drawing of the ^{235}U fission chamber.

K. H. Böckhoff (ed.), Nuclear Data for Science and Technology, 456–459.

The fission chamber gas filling consisted of a mixture of approximately 96% argon and 4% carbon dioxide at atmospheric pressure. The fission chamber was operated with a voltage of 400 volts. The preamplifier was a current preamplifier designed by Millard[6] and the main shaping amplifier had a time constant of 2 μs. A typical pulse-height distribution from the fission chamber is shown in Fig. 2 for $E_n \cong 1000$ keV. With the peak in channel 70 and a bias setting of channel 14 the fission chamber has an inefficiency of ~ 1%.[4] The fission yield produced by neutron scattering from the material in the fission chamber was calculated by Czirr[7] to be 0.2%/b of aluminum and is 0.6 - 1% in the 0.3 to 3.0-MeV neutron energy range.

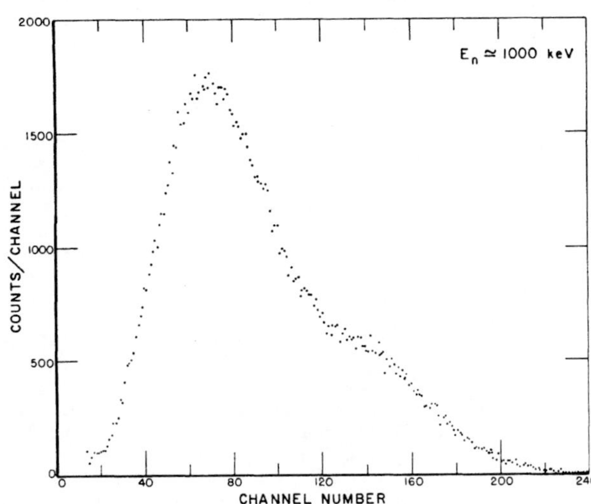

Fig. 2. Typical pulse-height distribution from the ^{235}U fission chamber at $E_n \cong 1000$ keV.

The neutron detector, shown schematically in Fig. 3, consisted of a large plastic scintillator mounted onto a single photomultiplier tube. The scintillator was 12.5-cm dia x 19-cm long. A reentrant hole 5 cm in diameter and 2.5 cm deep increases the neutron detection efficiency of the detector.[8,9] This detector has been referred to as a "black" neutron detector since it has a high neutron interaction probability which leads to a high neutron detection efficiency in the energy range of our measurement. The neutron beam was collimated to a diameter of 1.281 cm incident within the reentrant hole. This beam diameter was chosen to lower the counting rate in the black detector so that the dead-time correction factors would be no greater than 1.30. A computer control system was utilized which only allowed data accumulation when the electron beam current was within a narrow range. This condition reduced the error in the dead time correction. The black detector events were delayed in this experiment so they all occurred later in time than the fission detector events. This procedure eliminated dead time shadowing losses to the fission events from the black detector events so that less time was required to obtain the required statistical accuracy. A Monte Carlo program was used to calculate the neutron detection efficiency of the black detector. These efficiency calculations were experimentally verified by Meier[10] at the NBS Van de Graaff over the neutron energy range from 500 to 900 keV using the associated-particle method to measure the neutron flux. A similar measurement of the black detector efficiency at 2.6 MeV is planned at the NBS Van de Graaff. At the NBS work

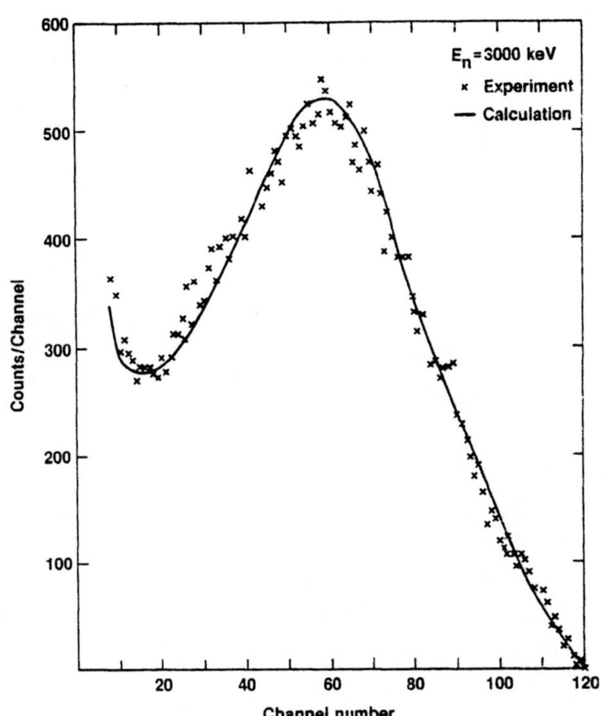

Fig. 4. Typical pulse-height distributions from the black detector and the Monte Carlo calculation at E_n = 3000 keV.

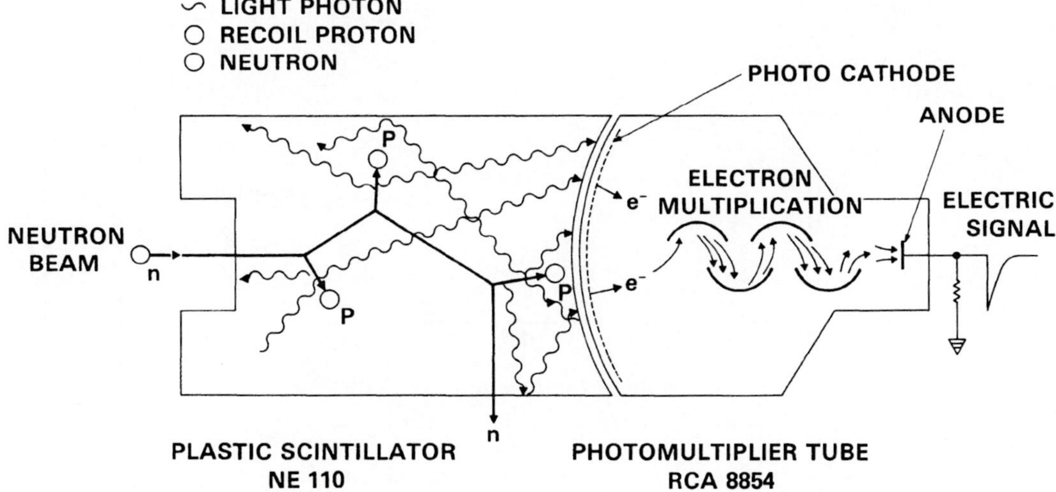

Fig. 3. Schematic drawing of the NBS black neutron detector.

is currently underway to remeasure the light tables used in the Monte Carlo calculations.[11] Further refinements include updating the cross section libraries used in the calculations. A typical pulse-height distribution from the black detector is shown in Fig. 4 for $E_n \cong 3000$ keV. The smooth curve represents the calculated pulse-height distribution.

The two parameter data, *i.e.*, pulse-height and time-of-flight, were simultaneously accumulated from the fission chamber and the black detector. Data were sorted into arrays of 512 pulse-height channels x 1024 time-of-flight channels for the fission chamber and 1024 pulse-height channels x 512 time-of-flight channels for the black detector. Time channels were 16 ns/channels for the fission chamber and 64 ns/channels for the black detector. A tagging system was used for the fission chamber and the black detector so that the same ADC and time digitizer could be used for both detector systems. The gamma flash from the tungsten target was the timing reference used for definition of the neutron energy scale for both detectors. The fission chamber was not off-gated during the gamma flash period and the time-of-flight spectra for this detector contain the profile of the gamma flash. The black detector, on the other hand, was pulsed off during the gamma flash. Tests with a radioactive gamma-ray source and a pulser indicated that the electronic system for the black detector had completely recovered from the pulsing off system before the data-taking time window. Periodic measurements of the black detector gamma flash channel were made by disconnecting the pulsing off system and including the gamma flash data in the data-taking time window.

For both detectors the background was measured at 250 keV with a black lithium filter. The fission chamber shielding reduced the background to a value within statistics of the ambient (late gate). For the black detector various configurations and amounts of lead and polyethylene were tried in an attempt to reduce the background to a negligible amount. The final configuration of the shield around the detector contained at least 20 cm of Pb and 20 cm of polyethylene. The minimum distance between the center of the detector and the shield was 20 cm. However, the time dependent background was not negligible. It was surmised that the remaining source of background was a result of neutrons which scattered from the detector, moderated and captured in the shield materials. The resulting gamma rays are detected with high efficiency in the black detector. This source of background was eliminated electronically by selecting a data taking energy interval which includes the maximum linac produced neutron energy down to below 300 keV and accumulating data in a one stop per start mode. These conditions eliminate background from detector scattered neutrons since a neutron which scatters from the detector is "detected" by the detector and stops the data taking system for that linac pulse. Thus if a background event results from that neutron, it will not be recorded since the data acquisition system will be dead.

Since the time dependent background at 250 keV is negligible for both detectors and it is expected that the percentage background will decrease with increasing neutron energy, only ambient backgrounds need to be subtracted from the data for the two detectors.

Results and Discussion

Over twenty-five days of data were collected in this measurement. The counting rate in the fission chamber was the limiting factor which caused this long data collection time. However, this detector was chosen primarily because it has been well characterized.[4] The raw data were taken in fifty runs which were completed June 1982. Figure 5 shows the shape data compared to the ENDF/B-V ^{235}U(n,f) cross section. The present work, normalized to ENDF/B-V in the 1.500 to 1.725 MeV interval, shows seven arbitrarily chosen energy intervals in the 0.3 to 3.0 MeV energy range with $\Delta E/E = 15\%$. Approximately half of the fifty runs were added together to get the values shown in Fig. 5. In the final analysis all fifty runs will be analyzed separately to better account for minute changes in energy scales deduced from gamma flash timing, dead-time corrections, and ambient background subtractions. The data, shown in Fig. 5, have been corrected for the transmission of neutrons through the materials which separate the ^{235}U deposits and the black detector, *i.e.*, air, windows, and aluminum. Error bars represent the statistical uncertainty of the data expressed as one standard deviation. At present the data are to be viewed as preliminary. Figure 5 shows that the present data agree reasonably well with ENDF/B-V in shape. Once finalized the data will be absolute. Accuracies of ± 2% are expected in the final analysis of this ^{235}U(n,f) cross section measurement. Final values await a complete analysis of the experimental data.

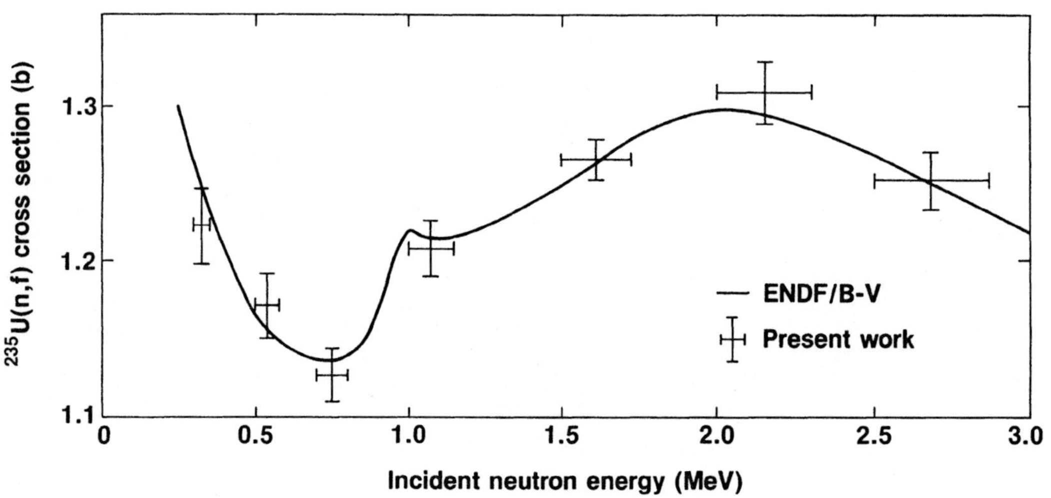

Fig. 5. The ^{235}U(n,f) cross section over the incident-neutron energy range from 0.3 to 3.0 MeV. Present work is normalized to ENDF/B-V in the 1.5 to 1.725 MeV energy interval.

References

1. J.W. Behrens *et al.*, Nucl. Sci. Eng. 63, 250 (1977); 66, 205 (1978); 66, 433 (1978); 68, 128 (1978); 77, 444 (1981); 80, 393 (1982); 81, 512 (1982).

2. O.A. Wasson, M.M. Meier, and K.C. Duvall, Nucl. Sci. Eng. 81, 196 (1982).

3. W.P. Poenitz, "Evaluation of ^{235}U(n,f) Between 100 keV and 20 MeV," ANL/NDM-45, Argonne National Laboratory (1979).

4. O.A. Wasson and M.M. Meier, Nucl. Instrum. Methods 190, 571 (1981).

5. J.W. Behrens, "Preparation of Fission Foils for Fission Ionization Chambers Using a Painting Technique," UCRL-51476, Lawrence Livermore National Laboratory (1973); see also, J.W. Behrens, Nucl. Instrum. Methods 200, to be published September 1, 1982.

6. J.K. Millard, "An Investigation of Broadband Current Preamplification for Obtaining Simultaneous High-Resolution Energy and Time Information from Nuclear Radiation Detectors, Thesis, ORNL-TM-3252, Oak Ridge National Laboratory (1971).

7. J.B. Czirr, Lawrence Livermore National Laboratory, private communication (1976).

8. W.P. Poenitz, "The Black Neutron Detector," ANL-7915, Argonne National Laboratory (1972).

9. G.P. Lamaze, M.M. Meier, and O.A. Wasson, "A Black Detector for 250 keV-1000 keV Neutrons, Proc. Conf. Nuclear Cross Sections and Technology, March 3-7, 1975, Washington, D.C., NBS Special Publication 425, 1, p. 73 (National Bureau of Standards, 1975).

10. M.M. Meier, "Associated Particle Methods," Proc. Symp. Neutron Standards and Applications, March 28-31, 1977, Gaithersburg, Maryland, NBS Special Publication 493, p. 221 (National Bureau Standards, 1977).

11. M. Dias, National Bureau of Standards, private communication (1982).

FISSION FRAGMENT ANGULAR DISTRIBUTION DATA FOR NEUTRON INDUCED FISSION OF ^{235}U

S.S. Kapoor, K.N. Iyengar, D.M. Nadkarni and V.S. Ramamurthy

Nuclear Physics Division
Bhabha Atomic Research Centre
Trombay, Bombay 400 085, INDIA

The available fragment anisotropy data in the neutron induced fission of ^{235}U for neutron energies up to 23 MeV have been compiled to evaluate this data for the preparation of a standard file for the fragment anisotropies. The compiled data are shown in Figs. 1 and 2. The data are being analysed within the framework of the statistical theory, taking into account multiple chance fission, to carry out data evaluation and to deduce the energy dependence of K_o^2. In particular, the expected shifting of the transition state shape from the location of the outer barrier to that corresponding to liquid drop model due to the washing out of shell effects will be tested by the analysis of the data.

[$^{235}U(n,f)$, fission fragment anisotropies, K_o^2 vs. excitation Energy]

Introduction

The fission fragment angular distributions provide important information on the quantum states available at the saddle point to the fissioning nucleus which in turn provides a basis for a theoretical understanding of the fission process. A knowledge of the fragment angular distributions is also important in evaluating those experimental fission cross section measurements in which fission fragments were detected at specified angles and not in a 2π geometry. In the case of neutron induced fission of an even-odd target nucleus like ^{235}U, at neutron energies exceeding a few MeV the number of K-states at the saddle point become sufficiently large for a statistical description [1] to be valid. In this region the fragment anisotropies depend on the value of a parameter K_o^2 (the variance of the assumed Gaussian distribution in K), which is related to the effective moment of inertia (and hence the nuclear shape) at the fission barrier. It has been pointed out earlier [2] that for nuclei exhibiting a double-humped barrier shape, the transition state is expected to gradually shift from the position of the outer barrier (barrier II) to the liquid drop model shape, with an increase in the excitation energy, as a result of the washing out of the nuclear shell effects with excitation energy. Some evidence of this feature in the experimental fragment anisotropy data for ^{242}Pu fissioning nucleus was also pointed out earlier [2]. In this work, we have looked into this feature of the expected excitation energy dependence of the fission transition state in the fragment anisotropy data of the fast neutron induced fission of ^{235}U. The present analysis provides evidence for the presence of the above feature expected in the case of all nuclei with double-humped barriers. This suggests that evaluation of the fission data for nuclei with double-humped fission barrier shape should incorporate the above important feature resulting from the gradual washing out of shell effects.

Description of the Data

The measured fragment anisotropies W(0°)/W(90°) in the neutron induced fission of ^{235}U in the neutron energy range of 0 to 23 MeV taken from Refs. [3-17] are given in Figs. 1 and 2. Results of recent measurements of J.W. Meadows and C. Budtz-Jørgensen obtained through private communication [17] have also been included in the figures. Figs. 1 and 2 provide a summary of the present status of the measured fragment

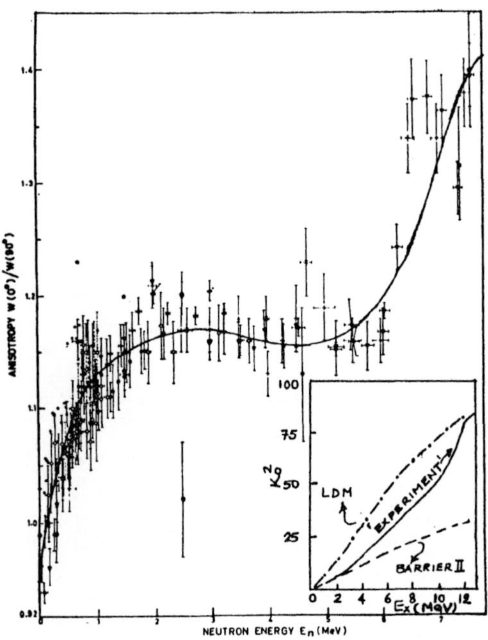

Fig. 1. The anisotropy W(0°)/W(90°) versus incident neutron energy (E_n) from Refs. [3-17] and the continuous line shows that resulting from the variation of K_o^2 vs. E_x shown in inset.

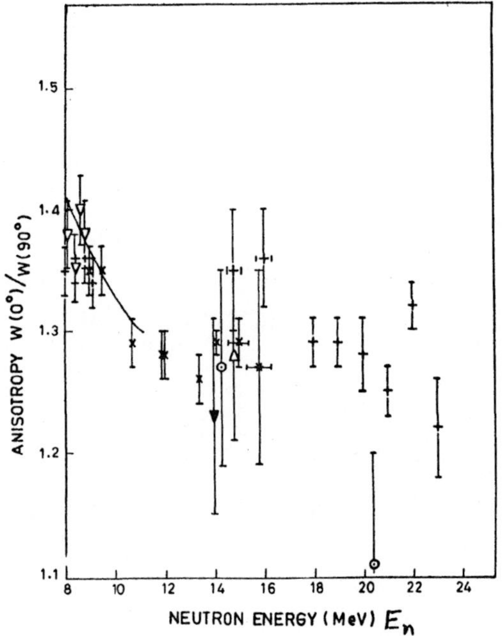

Fig. 2. The anisotropy W(0°)/W(90°) versus E_n from Refs. [3-17] and the continuous line shows that resulting from the variation of K_o^2 vs. E_x shown in inset of Fig. 1.

K. H. Böckhoff (ed.), Nuclear Data for Science and Technology, 460–461.

anisotropies for different neutron energies for the case of the target nucleus ^{235}U, resulting from thorough literature survey. Although the average trend of the variation of the anisotropy with the neutron energy is clearly brought out in the figures the different measurements of the anisotropy are generally in agreement with each other to only within a value of 0.05. Comparing the two most recent measurements [16-17] the anisotropies of Ref.[17] appear to be systematically lower by a few per cent as compared to those of Ref.[16]. For example, at 1040 keV and 300 keV where both have reported measurements, the values reported in Ref.[17] are 1.098 and 1.017 as compared to 1.170 and 1.040 of Ref.[16]. Further careful measurements at these two energies will be useful in resolving the above small discrepancy between the above two recent measurements.

Analysis of the Data and Discussion

We have carried out statistical model analysis of the compiled fragment anisotropy data (Figs. 1 and 2) on the basis of the statistical theory [1]. Below the second chance fission threshold, the theory predicts an angular distribution of the form

$$W(\theta) = (1 + a\cos^2\theta)/(1 + a/3)$$

where a is related to the quantity $p = \ell^2_{max}/4K^2_o$, ℓ_{max} being the maximum angular momentum brought in by the neutron in a sharp cut-off model and K^2_o is a parameter which is the product of the effective moment of inertia J_{eff} and temperature T of the saddle point nucleus. With the opening up of the multichance fission, one has to also consider the angular distributions of multiple chance fission fragments. In addition, it is known that for incident neutron energies of about 14 MeV, significant pre-equilibrium effects set in leading to uncertainties in the angular momentum distribution and excitation energy of the fissioning nucleus. We have therefore restricted the present analysis to the range $0 < E_n < 11$ MeV where third chance fission is also not open.

In the present analysis, the first chance fission anisotropy of the ^{236}U fission fragments for neutron energies up to 11 MeV have been deduced, from the experimental data on Γ_f/Γ_n and anisotropies for neutron induced fission of ^{235}U and ^{234}U. In this way, the observed anisotropies of ^{236}U have been corrected for the second chance fission in a nearly model independent way. The maximum angular momentum ℓ_{max} is obtained from the relationship given by Blann [18].

The continuous line in Figs. 1 and 2 represents the values of the anisotropy resulting from the variation of K^2_o versus E_x shown in the inset in Fig. 1 as solid line. Also shown in the inset, are the expected values of K^2_o versus E_x for the second barrier shape ((dashed line) and the shape predicted by the liquid drop model (LDM) (dash-dot line). For the LDM shape the shell and pairing corrections have been included as in Ref.[2]. Thus, the above analysis demonstrates that the fission of ^{236}U exhibits the same features as was earlier pointed out for the ^{242}Pu fission, namely, that due to washing out of shell effects the shape of the transition state shifts from the second barrier shape to the LDM shape. This feature is expected to be common to all actinide nuclei with double-humped fission barriers and should be considered in any fission data evaluation.

References

1. I. Halpern, V.M. Strutinsky, in Proceedings of the Second International Conference on the Peaceful Uses of Atomic Energy, Geneva, 1958 (United Nations, Geneva, Switzerland, 1958)
2. V.S. Ramamurthy, S.S. Kapoor, S.K. Kataria, Phys. Rev. Lett. 25, 386 (1970)
3. J.E. Brolley, Jr., W.C. Dickinson, R.L. Henkel, Phys. Rev. 99 (1), 159 (1955)
4. J.E. Brolley, Jr., W.C. Dickinson, Phys. Rev. 94 (3), 640 (1954)
5. A.N. Protopopov, V.P. Eismont, Soviet Phys. JETP 7, 173 (1958)
6. J.E. Simmons, R.L. Henkel, Phys. Rev. 120 (1), 198 (1960)
7. L.N. Blumberg Ph.D. Thesis, University of Columbia, 1962
8. R.W. Lamphere, Oak Ridge National Laboratory, Report ORNL-3306 (1962)
9. R.B. Leachman, L. Blumberg, Phys. Rev. 137 (4B), 814 (1965)
10. V.G. Nesterov, G.N. Smirenkin, D.L. Shpak, Soviet Journ. Nuclear Physics 4 (5), 713 (1967)
11. D.M. Nadkarni, S.S. Kapoor, P.N. Rama Rao, Proc. Nuclear Physics and Solid State Physics, Symp. Bombay, Nucl. Phys. II 133 (1968)
12. G.N. Smirenkin, D.L. Shpak, Yu B. Ostapenko, B.I. Furson, Sov. Phys. JETP Lett. 11, 333 (1970)
13. S.T. Hsue, G.F. Knoll, J. Meadows, Nuclear Science and Engineering 66, 24 (1978)
14. N.K. Chaudhuri, V. Natarajan, R. Sampathkumar, M.L. Sagu, R.H. Iyer, Nuclear Tracks 3, 69 (1979)
15. S. Ahmad, M.M. Islam, A.H. Khan, M. Khaliquzzaman, M. Hussin, M.A. Rahman, Nucl. Sci. and Engineering 71-72, 208 (1979)
16. A.R. de L. Musgrove, J.W. Boldeman, J.L. Cook, D.W. Lang, E.K. Rose, R.L. Walsh, J. Caruana, J.N. Mathur, Jour. Phys. G: Nuclear Physics 7, 549 (1981)
17. J.W. Meadows, C. Budtz-Jørgensen, Private Communication 1981
18. M. Blann, Phys. Rev. 21C, 1770 (1980)

MEASUREMENTS OF NEUTRON CAPTURE CROSS SECTIONS
FOR ^{197}Au AND ^{169}Tm BETWEEN 0.1-1.5 MeV

Chen Ying, Jiang Songshen, Luo Dexing, Zhu Shengyun and Zhou Zuying

Institute of Atomic Energy, Academia Sinica
P.O.Box 275-15 Beijing, China

The neutron capture cross sections for ^{197}Au and ^{169}Tm have been measured by activation technique in the energy range from 0.1 to 1.5 MeV. Absolute measurements were performed for ^{197}Au at the energies 0.462, 0.991, 1.190 and 1.490 MeV. The neutron flux was determined by the hydrogeneous gas proportional counters. The $^{169}Tm(n,\gamma)^{170}Tm$ cross sections were measured relative to $^{197}Au(N,\gamma)^{198}Au$ cross sections.

[$^{197}Au(N,\gamma)^{198}Au$, $^{169}Tm(N,\gamma)^{170}Tm$, capture cross sections, E_n=0.1--1.5 MeV, nuclear data.]

I. Measurements of ^{197}Au Capture cross sections

The discrepancies can be seen in Fig.1 of the existing ^{197}Au neutron capture cross sections, most of which were determined relatively. Our measurement was undertaken to determine by activation method ^{197}Au neutron capture cross sections from 0.1 to 1.5 MeV with an accuracy of 4 to 6 %.

Experimental Procedure

The neutrons were produced by the $T(p,n)^3He$ reaction using protons from the IAE 2.5 MV Van de Graaff accelerator. The target was cooled by a spray of water and air. The sample was a disk 0.3 mm thick by 19 mm diameter. Each sample to be measured was sandwiched between two other 0.1 mm gold foils and then placed in a 0.3 mm thick cadmium case.

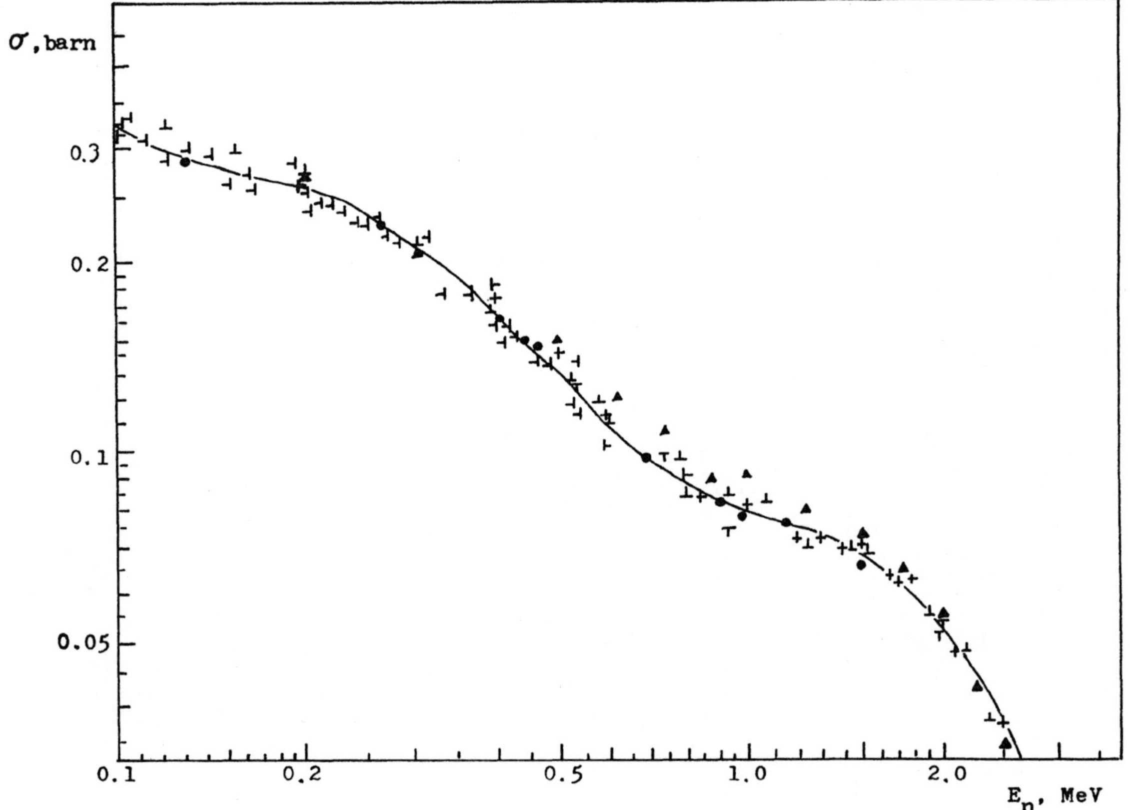

Fig.1. Neutron capture cross sections of ^{197}Au

⊥ —3; + —4; ⊢ —5; ▲ —6; ⊤ —7; ⊣ —8; —— ENDF/B-IV; ● —present work.

K. H. Böckhoff (ed.), Nuclear Data for Science and Technology, 462–464.

Table I. Angular distribution of $T(p,n)^3He$ reaction

Proton energy	1.3 MeV				1.8 MeV			
Angle	0°	25°	50°	75°	0°	25°	50°	75°
Present results	1	0.843	0.649	0.617	1	0.907	0.728	0.529
Liskien's value	1	0.816	0.609	0.580	1	0.900	0.706	0.464
Present work/ Liskien's value	1	1.03	1.07	1.06	1	1.01	1.03	1.14

Absolute measurements were performed at neutron energies of 0.462, 0.991, 1.190 and 1.49 MeV with the energy spreads 0.028, 0.022, 0.043 and 0.04 MeV respectively. Neutron flux was measured by hydrogeneous gas proportional counters. The distance between the sample and the target was 70 mm. Two identical sample packets 2 m from the source were used to determine the scattered neutron background. Two long counters were used for monitoring the flux variation during the run.
In the relative measurements the irradiation at 7 cm from the source was simultaneously carried out at 0°, 25°, 50° and 75° with respect to the proton beam. For E_p 1.8 MeV the neutron energies at the sites of the sample were 0.991, 0.903, 0.682 and 0.443 MeV with the energy spreads 0.022, 0.061, 0.079 and 0.064 MeV respectively. For E_p 1.3 MeV they were 0.462, 0.405, 0.266 and 0.131 MeV with the energy spreads 0.028, 0.045, 0.051 and 0.035 MeV respectively. The background measurement and flux monitoring were performed in the same way as in absolute measurements. The neutron capture cross sections at 0° were taken from the absolute measurements mentioned above. The capture cross sections at other angles were determined in terms of neutron source angular distributions and activity ratios of 0° to the angles.
Neutron source angular distributions were determined with hydrogeneous gas proportional counters. Listed in Table I are the results together with H.Liskien's recommended values[1]. Our results are considered more accurate than H.Liskien's data based on some long counter measurements in 1950s.
The radioactivity of ^{198}Au was determined by a 3"x3" well type NaI(Tl) scintillation counter. The absolute efficiency was calibrated by 4 - coincidence method. Corrections were made for the sample thickness and radial variation.

Corrections

A Monte Carlo calculation was performed to estimate the elastic and inelastic neutron scattering from the sample packet and sample holder, multiple scattering in the Cd-Au packet and neutron attenuation in the target and sample packet. Total correction ranged from 3 to 10 % ,depending on the neutron energies. The main sources of the correction were from neutron elastic scattering and neutron attenuation.

Results and Discussion

Our experimental results are shown in Fig.1. The accuracy of the absolute measurements was believed to be less than 4 %. While that of the relative measurements was 5--6 %. Inspection of the figure shows our results fall into good agreement with most of the published data.

II. Measurements of ^{169}Tm Capture cross sections

Only available data in this energy region were that of S.Joly[7] published during our experiment.

Experimental Procedure

^{169}Tm neutron capture cross sections were measured relative to recommended ^{197}Au capture cross sections[9]. The irradiation was performed at 0°. The distance between the sample and target was 3 cm. Neutrons were produced via the $T(p,n)^3He$ and $^7Li(p,n)^7Be$

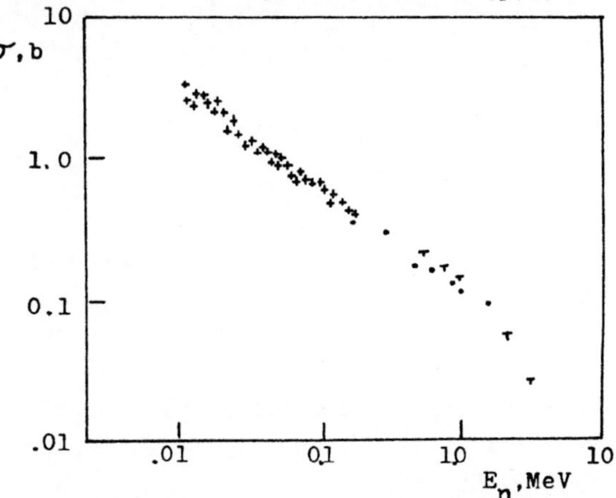

Fig.2. ^{169}Tm neutron capture cross sections T —— 7; + —— 10; • —— present work.

reactions. 0.158,0.283,0.454,0.560,0.84,0.97 and 1.47 MeV energies were chosen with the energy spreads 0.027,0.045,0.036,0.022,0.18, 0.16 and 0.12 MeV respectively.

The sample disk 20 mm in diameter and 0.3 mm thick (about 100 mg/cm^2) was prepared by pressing the Tm_2O_3 powder in a die. The sample disk was then placed into a small alluminium case and covered with a 10 μm thick alluminium foil. This kind of alluminium wrapper can prevent the Tm_2O_3 sample from crashing and permit a large fraction of β-particles passing through the alluminium foil.

β-activity was determined by a scintillation plastic anti-coincidence β counter. The efficiency of the counter was calibrated by the so called "simulated source technique". The simulated sources were a set of disks containing a known quantity of ^{170}Tm embeded uniformly in Tm_2O_3 with the form identical to the sample. The process of mixing ^{170}Tm with Tm_2O_3 was as follows. The standard solution of 71.4 mg ^{170}Tm was added to the solution of HNO_3 containing 10 g Tm_2O_3. The resultant solution was uniformly stirred, then heated slightly and dried. Thus the Tm_2O_3 powder having the specific activity was obtained. The detector efficiency calibrated with this technique is more accurate than the existing methods.

Correction and Results

The same Monte Carlo programme was used to estimate the correction due to neutron scattering and attenuation. The total amount of correction was about 2 %.

Our results together with other published data are shown in Fig.2. The accuracy of our measurement is 6--8 %, while J.H.Gibbons[10] 10 % and S.Joly[7] 6--15 %. At 0.16 MeV our value is fairly well connected with that of J.H.Gibbons. In energy range 0.5--1.5 MeV the results of S.Joly are about 20 % higher than ours.

References

1. Horst Liskien et al.,Nuclear Data Tables, 11,569 (1973).
2. Chen Ying et al., Chinese Jour.of Nuclear Physics 3,375 (1981).
3. M.Lindner et al., Nucl.Sci.Eng.,59,381 (1976).
4. W.P.Poenitz, Nucl.Sci.Eng.,57,300 (1975).
5. M.P.Fricke et al., Nucl.Data for Reactors vol.2,265 (1970).
6. A.Paulsen et al., Atomkern Energie,26,80 (1975).
7. S.Joly et al., Nucl.Sci.Eng., 70,53 (1979).
8. C.Le Rigolear et al., CEA-N-1662 (1973).
9. Jiang Songsheng et al., "Evaluation of 0.1--3.0 MeV ^{197}Au neutron capture cross sections", unpublished.
10. J.H.Gibbons et al., Phys.Rev.,122,182 (1961).

Investigation of the Prompt-Neutron Spectrum for Spontaneously-Fissioning ^{252}Cf*

by

W. P. Poenitz
Argonne National Laboratory
Argonne, IL, USA

T. Tamura
Research Participant from Tohoku University
Sendai, Japan

The prompt-fission-neutron spectrum of ^{252}Cf was investigated. The spectrum was measured with Black Neutron Detectors which have a well known efficiency. Considerations of various issues in such measurements lead to an experiment in which a time-calibration pulser, a random pulser, the neutron detector time-of-flight spectrum, the pulse-shape-discriminator gamma time-of-flight spectrum, and the detector-response spectra were simultaneously recorded for the prompt-fission neutrons, transmission through carbon, and shadowbars in a total-cross-section-type measurement. Corrections and associated uncertainties were applied for a large variety of effects which may have been overlooked in many of the previously reported measurements. Preliminary results indicate deviations from a Maxwellian shape toward a Watt-spectrum shape. Agreement is good with the shape differences relative to a Maxwellian from the recent theoretical calculation by Madland and Nix, however, a lower average energy was found.

I. Introduction

The prompt-fission-neutron spectrum of ^{252}Cf is of considerable interest for the interpretation of the fission process as well as for practical applications. It was proposed as a standard for emission spectra measurements (1,2), it is used for detector calibrations (3,4), and it has a direct impact upon the interpretation of $\bar{\nu}$ measurements (5). Substantial experimental effort has been devoted to its determination (see Ref. 6), but the available results are quite often contradictory. Prior to the IAEA-Consultants' Meeting in 1971 (1), two distinctly different groups of values for the average temperature, T, of an assumed Maxwellian spectrum existed (~ 1.41 MeV and 1.58 MeV, respectively). More recent measurements tend to confirm a value of 1.42 MeV, corresponding to an average energy of 2.13 MeV (7-9). However, the post 1971 values are spread over the range from 1.18 to 1.57 MeV (see the summary table in Ref. 6).

Agreement with the Maxwellian-spectrum shape was stressed in the more recent work (7,8), but both neutron excesses and deficiencies at higher or lower energies were noted in the post 1971 measurements (10-12). The apparent satisfaction over finding agreement with a Maxwellian-spectrum shape is in itself surprising: The clear improvement of the basic concept for the derivation of a fission-neutron spectrum shape, that is, taking into account the center-of-mass motion of the fission fragments (Watt-spectrum), suggests that agreement with a Maxwellian-spectrum shape should be viewed with suspicion.

Substantial criticism was directed toward past experimental effort during a recent workshop on fission-neutron spectra (13), more specific critical remarks were expressed in a recent review (6). The unsatisfactory situation is also reflected by the continuing presence of the ^{252}Cf spectrum as a high-priority item on nuclear-data-request lists (14,15). The present measurements were undertaken in order to provide improved data on the ^{252}Cf-fission-neutron spectrum for practical applications as well as for the testing of nuclear model predictions. Consideration of past experiments suggested that there was substantial room for improvememts in measurement technique and analysis. A major advantage of the present measurements might well be the use of neutron detectors with well-known efficiencies. However, additional physical, experimental, and instrumental effects which might have been neglected in previous work were investigated and corrected. The results presented here are considered preliminary pending further planned improvements in the measurements and analysis.

II. The Fission Source and Detector

The ^{252}Cf source used in the present measurements was one which was available, but not specifically prepared for this experiment. It consisted of a ^{252}Cf deposit on a 0.0254-cm-thick-platinum disc of 0.953-cm diameter. The source strength was determined by comparison with a ^{252}Cf-reference source. The latter was obtained by vacuum-self-deposition and its absolute fission rate was determined by low-geometry fission-fragment counting. The comparison between the ^{252}Cf source and the reference deposit was made by relative neutron-emission rate measurements using a Marion counter. The source strength was found to be $1.43 \cdot 10^5$ fissions per second at the end of the present measurements.

A gas-scintillation counter was used for the detection of fission events. The counter was described previously (16) but was further modified in order to reduce neutron transmission and scattering corrections. The detector has a cylindrical shape with 16 cm height and a diameter of 22 cm. Neutron entry and exit windows with a radius of 13 cm at the top and bottom of the counter consisted of ~ 0.0036-cm-thick aluminum foil. The ^{252}Cf source was mounted on the inside of one of these windows on the center axis of the detector. A mixture of 85% argon and 15% nitrogen was used as the scintillation gas. Four photo-multipliers were mounted on the outside of the cylindrical surface. The high voltage on each photomultiplier was adjusted to result in approximately the same pulse height spectrum. The lengths of the cables between the anode output of each photo-multiplier and a fast linear mixer were adjusted to result in the same delay for the prompt gamma peak in the time-of-flight spectrum.

The pulse-height spectrum obtained with this fission chamber for the ^{252}Cf-reference source is shown in Fig. 1. The efficiency was determined to be

K. H. Böckhoff (ed.), Nuclear Data for Science and Technology, 465–472.

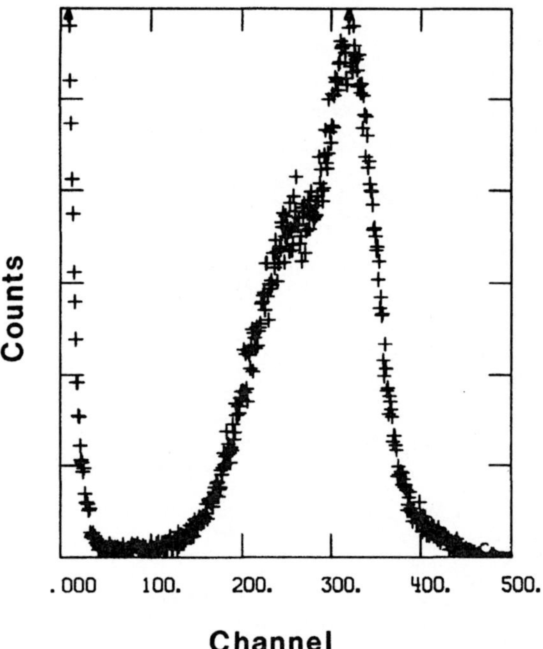

Fig. 1. The Pulse-height Spectrum Obtained with the Gas-scintillation Counter for the ^{252}Cf-Reference Deposit.

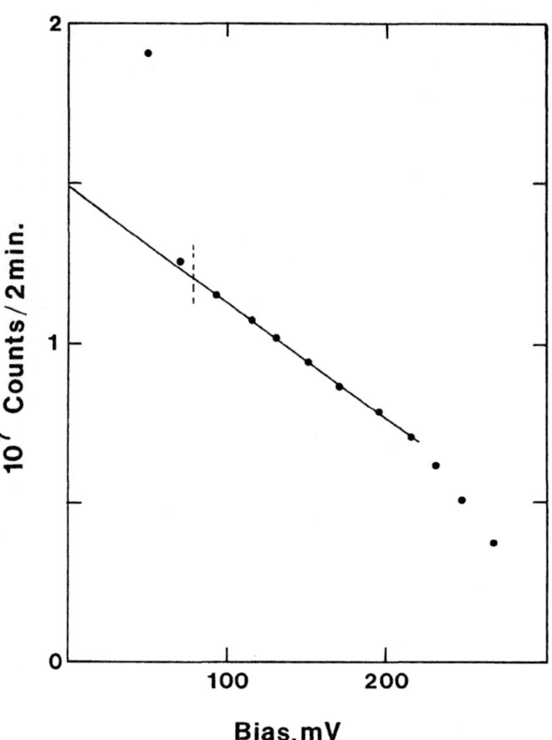

Fig. 2. The Count Rate Obtained with the Gas-scintillation Counter for the ^{252}Cf Source as a Function of the Fast-Trigger Threshold.

≈ 100%. The efficiency for the ^{252}Cf source used in the neutron-spectrum measurements is determined by the losses below the alpha-pile-up and total-fragment absorption. It was determined by measuring the count rate as a function of the fast-discriminator threshold (see Fig. 2), with subsequent extrapolation to zero pulse height. This measurement and the known source strength resulted in the determination of 20 ± 2% losses below the alpha-pile-up and 9 ± 3% total fragment absorption. This shows that substantial fragment energy was lost in the deposit. Fig. 3 shows the fragment-energy spectra obtained for the two deposits perpendicular to the source backings with a surface-barrier detector which supports this conclusion.

The possibility of ^{252}Cf migrated within the chamber was considered and determined at the end of the experiment after removal of the source. It was found as $\lesssim 4 \cdot 10^{-3}$%.

III. The Neutron Detectors and Their Efficiencies

Two Black Neutron Detectors (BND) were used in the present experiments (17,18). The smaller was a cylindrical liquid scintillator with a radius of 7.62 cm and a height of 17.78 cm, and with a cylindrical neutron-entrance channel of 1.9 cm radius and 4.76 cm length. An RCA 8854 multiplier was used for this detector. The larger BND was a cylindrical liquid scintillator with a radius of 10 cm and a height of 37 cm, and with a neutron entrance channel of 1.9 cm radius and 14 cm length. Four 58 DVP photomultipliers were used for this detector. A liquid scintillator medium was formulated for this experiment which had a higher hydrogen content, and appeared to have improved pulse-shape discrimination features relative to NE213.

The efficiency of a BND is ~ 100% based on its design principle, and the deviation from 100% efficiency can be accurately calculated by Monte Carlo techniques. Because this correction and its uncertainty depend mainly on well-known quantities

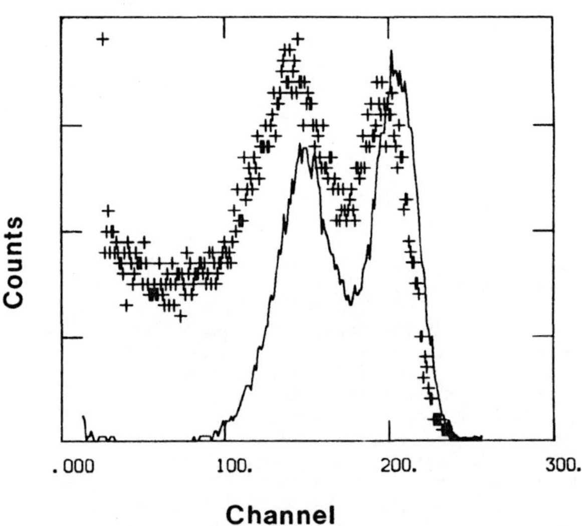

Fig. 3. The Pulse-height Spectra Obtained with a Surface-barrier Detector at 0° for the ^{252}Cf Source and the ^{252}Cf-Reference Deposit.

(transmission through the detector, hydrogen and carbon cross sections, and the detector geometry and composition), this type of detector has a well known efficiency and low uncertainties ($\lesssim$ 1 - 2%).

BND's are now used in many laboratories (16-22). Comparisons were made with the associated-partical technique (19,21) and agreement was found within 1.2% and 0.7% at 0.5 - 0.9 MeV and 14.1 MeV, respectively. Indirect support for the $\lesssim$ 1% uncertainty level can be found in recent ^{235}U (n,f) cross section measurements (23-25), by comparing results obtained with BND's and with proton-recoil counters or the associated-partical technique.

The smaller detector was used for measurements between ~ 200 keV and 4 MeV where it had an efficiency between 98 and 83%. The larger detector was used between 600 keV and 10 MeV with corresponding efficiencies between 96 and 77%.

IV. The Energy Scale and Resolution

The energy determination and verification is one of the most important factors in ^{252}Cf-spectrum measurements. Energy uncertainties can cause substantially larger spectrum uncertainties than the uncertainties of other quantities. The effects of energy resolution, (which unfortunately is often mistaken for energy uncertainty), can be corrected for, thus reducing the uncertainty of the measured spectrum.

The energy scale in a time-of-flight experiment is based upon three experimentally determined parameters: the time-scaling channel width, Δt, the channel of the center of the prompt-gamma peak, G, and the flight path, ℓ. The energy is determined from

$$E = E_0((1 - \ell^2/c^2T^2)^{-1/2} - 1) \qquad (1)$$

where E_0 is the neutron rest energy, c is the velocity of light and ℓ is the flight path between the source and the front of the detector. Physical processes within the detector will cause an increase of the time at which the neutron will be recorded. The average time given by the center of channel, N, is

$$T + \tau + \tau' = \sum_{N}^{G} \Delta t_i + \frac{\ell}{c} + \frac{H}{c'} \quad . \qquad (2)$$

This assumes that the photomultiplier is at the end of the detector which has a height, H, and a medium, with a light velocity c'. Thus, the additional right-hand terms of Eq. 2 correct for the time the gammas require to be detected. τ is the average time a neutron spends in the detector until the light produced by it exceeds the threshold of the discriminator, and τ' corrects for the time the scintillation light requires to travel the remaining distance to the photomultiplier. The individual relative channel width, Δt_i, can be determined with a random pulser; the average channel width can be determined with a time-calibration pulser. The flight path can be determined within ~ 0.5 - 1.0 mm, and thus contributes little to the energy uncertainties. However, differences in the detection mechanisms for neutrons and gammas might cause larger uncertainties. Because of the sensitivity of the neutron spectrum to energy uncertainties, verification of the energy scale appears mandatory.

The energy resolution is governed by a variety of physical and instrumental effects. These can be grouped into three major components:

a) the time resolution of the electronic components and of the fission chamber, which can be approximated by a Gaussian shape,

b) the time-scaling channel width, which is well defined, and

c) the time-response function of the neutron detector, which is determined by the physical processes within the scintillator.

The component, a, can be obtained from the width of the gamma peak, taking into account that the channel width is already involved. The component, c, can be obtained from Monte Carlo calculations which at the same time yield the average time, τ, the neutron spends in the detector, in order to be detected. Simultaneous verification of the energy scale and the energy resolution can be obtained by measurement of the total neutron cross section of carbon with its well-known resonances.

V. The Experimental Set-up

A schematic of the experimental set-up for the measurement with the smaller BND is shown in Fig. 4. The gas-scintillation counter (GSC) with the ^{252}Cf source (S) was in the cave of a 4π-neutron shield (S-S1 and S-S2) built with lithium-and baron-loaded plastic blocks. A rear-neutron-exit channel

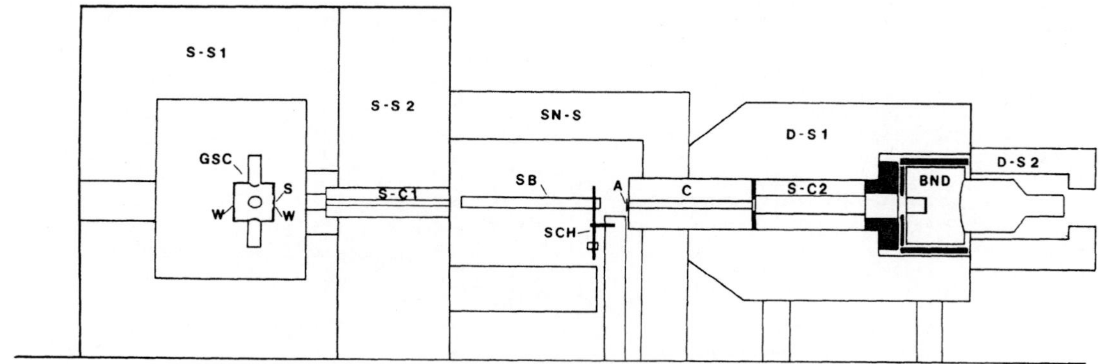

Fig. 4. Schematic of the Experimental Set-up. GSC is the Gas-Scintillation Counter with the ^{252}Cf Source (S) and the Neutron-Entry and Exit Windows. BND is the Black Neutron Detector. SCH is the Eight-Positions-Sample Changer with one of the Three Shadow-bars (SB) and one of the Two Carbon Samples Indicated in the Figure.C is the Main Collimator which Limits the Space Angle of the Source Neutrons for the BND. S-C1 and S-C2 are Shield Collimators. S-S1 (Lithium- and Boron-Loaded Plastic) and S-S2 (Lithium-Loaded Plastic) Form a 4π-Source-Neutron Shield. SN-S is a Neutron Shield for the Neutrons Scattered in the Shadow-bars.D-S1 (A Mixture of Lithiumcarbonate, Water and Lead-pellets) and D-S2 (Lead and Lithium-Loaded Plastic) is a Neutron Detector Shield. A Indicates the Au-Gamma Filter. Black Areas Indicate Lead Shielding.

assured that neutrons scattered within this shield could not reach the neutron detector. Background from neutrons and gammas leaking from the source shield, and scattered within the room, which might still penetrate the detector shield (D-S1 and D-S2), was determined with a shadow-bar (plexiglas). The latter had less than 0.1% neutron transmission. An additional 4π - neutron shield (SN-S) was built around the location of the shadow-bar in order to reduce any possible effect from the neutrons scattered in the shadow-bar or the carbon samples. An eight-position sample changer which moved in a rapid stepping motion was used in order to interchange 3 holes, 3 shadow-bars, and 2 carbon samples for the simultaneous measurement of the fission-neutron spectrum, the carbon transmission, and the shadow-bar background. The first shield collimator (S-C1) was inserted in order to reduce the number of neutrons which impinge on the sample changer, and the number of neutrons which would scatter closer to the neutron detector.

The main collimator, which limits the space angle of the source neutrons detected by the BND, is labeled C in Fig. 4. A second shield collimator (S-C2) reduces the number of neutrons which are scattered in the main collimator and arrive at the neutron detector or near to it. A 0.05-cm-thick gold filter was inserted in front of the main collimator in order to reduce the intensity of the rather soft delayed gammas emitted from the fission fragments. A 0.076-cm-thick-lead filter was placed at the end of the neutron-entrance channel of the BND for the same purpose.

The area between the ^{252}Cf source and the neutron detector was designed to reduce uncertainties in the corrections required for a variety of effects, and to permit an easy Monte Carlo simulation of the spectrum measurements. Neutron inscatter from the main collimator was estimated to be $< 0.1\%$. Neutron inscattering from the first shield collimator was estimated to be $\lesssim 0.2\%$. An additional measurement was made without this shield collimator and without the sample changer.

The detector was located behind the neutron-detector shield (D-S1) and the flight path was extended to ~ 3.50 m for the measurement with the larger BND. The anode signals from the four photomultipliers of the gas-scintillation counter were added in a fast mixer, clipped to ~ 20 nsec, and amplified with a fast-linear amplifier. The stop signal was obtained with a constant-fraction discriminator. The anode signals of the neutron-detector photomultiplier were split and one branch was amplified with a fast-linear amplifier. This signal was used to obtain the neutron-detector-start signal with a constant-fraction discriminator. The second branch was used to obtain a gamma identification signal from a pulse-shape discriminator. A dynode output from the photomultiplier was amplified and an energy signal was obtained from a subsequent linear-gate and stretcher.

The start and stop signals of a time-calibration pulser were added to the start and stop signals from the neutron detector and the fission counter, respectively, with an or-gate. In addition, a random-pulser signal was added to the start signal in the or-gate. The output signals of the or-gates were used to start and stop a time-to-amplitude converter (TAC) using a time range of 1000 nsec. The TAC signals and the pulse-height signals of the neutron detector were converted with amplitude-to-digital converters and recorded with an on-line computer. The origin of the various signals was identified with tag-bits which were obtained from secondary outputs of the respective triggers or the photodiodes on the sample changer. Thus, computer-input words were identified as neutron-detector event, neutron-detector-gamma event, random-pulser event, or time-calibration-pulser event for the fission-neutron spectrum, the carbon transmission, or the shadow-bar background.

The simultaneous measurement of these quantities provides a variety of benefits for the interpretation of the experiment. The gamma spectrum can be investigated for structure which might be present (to some extent) as residual background in the neutron spectrum. The shadow-bar-background spectrum indicates the success of the shielding against background from neutron and gamma inscattering from the room. The carbon cross section derived from the measured transmission and the ^{252}Cf spectrum provides a check of the energy scale and of the resolution. The random-pulser spectrum eliminates differential nonlinearity (but, unfortunately it too adds to the statistical uncertainty).

VI. Measurements and Corrections

The results presented here were obtained from three measurement sets. One was obtained with the smaller BND and a flight path of ~ 2.6 m, yielding data in the energy range of 0.2 - 4.0 MeV. A second set was obtained with the larger BND and a flight path of ~ 3.5 m, yielding data in the energy range of 0.7 - 10.0 MeV. The third data set was obtained with the smaller BND, but without the simultaneous measurement of the carbon cross section and without use of the random pulser. Also, the first shield collimator (S-C1 in Fig. 4) was not used in generating this last set. Furthermore, the gold filter was $\sim 70\%$ thicker, and a smaller fission chamber was used. Fig. 5 shows the neutron-time-of-flight spectrum obtained with the smaller detector. The

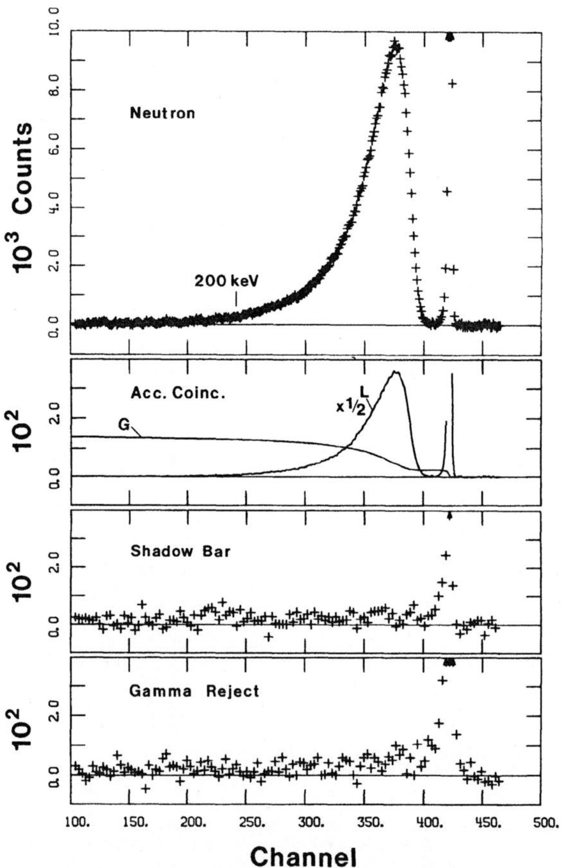

Fig. 5. The Time-of-Flight Spectra obtained with the Smaller Black Neutron Detector and a Flight Path of 2.58 m. Ambient Background has been Subtracted from all Spectra. The Shadowbar Background and the Accidental Coincidence Spectra were Subtracted (Added) from the Neutron Spectrum.

background spectra shown in the same figure were subtracted. Figs. 6 and 7 show the total cross section values obtained with the smaller and the larger BND's respectively. The statistical uncertainties of the cross sections are larger than for the measured spectra not only because of the lower number of counts in the transmission spectra, but also because only two carbon samples were used while there were three positions without samples. The smooth cross section curves in Figs. 6 and 7 were obtained by averaging the calculated transmission through the carbon samples using the ENDF/B-V cross section and a Maxwellian spectrum with the resolution of the present experiment. The carbon total cross section was then derived from this calculated transmission.

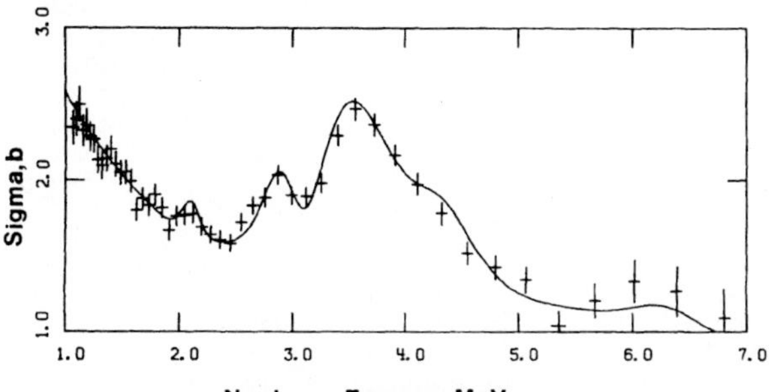

Fig. 6. The Total Cross Section of Carbon Obtained with the Smaller Black Neutron Detector and a Flight Path of 2.58 m. Data were Used only Below 4 MeV.

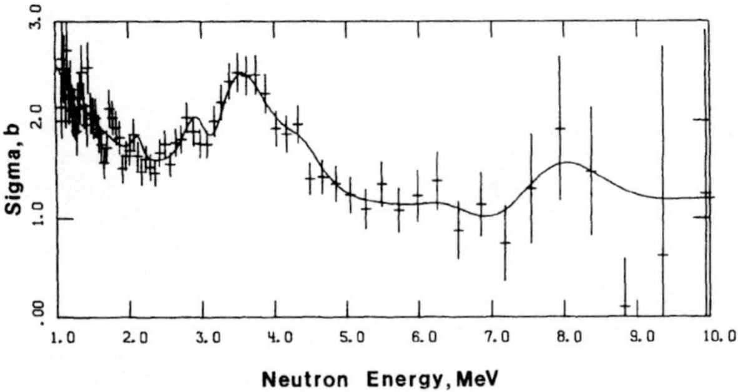

Fig. 7. The Total Cross Section of Carbon Obtained with the Larger Black Neutron Detector and a Flight Path of 3.47 m.

The random-fission-event stop pulses lead to accidental-coincidence losses and gains in the measured time-of-flight spectra. The required corrections were obtained from the associated probabilities:

The accidental-coincidence gains in channel i are given by

$$N_{acc}(i) = \sum_{\ell=i+1}^{\ell max} \ell^{-N_f t}\left(1 - \ell^{-N_f \Delta t}\right) N(\ell) \qquad (3)$$

where N_f is the stop rate, and t is the time associated with channel i. Thus, EXP $(-N_f t)$ is the probability that a random stop event does not occur in t, and 1 - EXP $(-N_f \Delta t)$ is the probability that at least one occurs within (the channel width) Δt. Such accidental-coincidence gains ("background") were corrected in some of the previous work (see for example Refs. 10, 26) but a wrong formula was used. The accidental-coincidence gains are, by necessity, the results of accidental-coincidence losses elsewhere in the spectrum. For, the channel, i, associated with the time, t, between the neutron-detector event and the stop pulse, these accidental-coincidence losses are given by

$$N_{acc}(i) = N(i)\left(1 - \ell^{-N_f t}\right) \quad . \qquad (4)$$

Both equations neglect a secondary effect which is caused by the dead time of the trigger in the stop branch. This can be shown to be negligible for the present experiment. The accidental-coincidence losses were not corrected in most previously reported ^{252}Cf-spectrum measurements. These corrections were derived independently, however, the problem was previously recognized by Chalupka (27). His formulas are different but they should yield the same result. The formulas for accidental coincidences require the knowledge of the true spectrum, thus, iterative steps were applied. The number of accidental gains is, of course, equal to the number of losses. Losses increase for higher-energy neutrons while the gains are predominant in the low-energy part of the spectrum. (See Fig. 5).

Total-fission-fragment absorption is predominant at angles near 90° relative to the axis perpendicular to the deposit backing. This causes a hardening of the measured spectrum. The required correction can be calculated based upon the kinetics involved. In order to reduce the uncertainty of this correction, measurements of the ratio of the spectra at 0° and near 90° were made. These measurements were made with a flight path of $\approx$ 1.7 m and the ^{252}Cf deposit mounted in the center of a fission chamber. The corresponding ratio was calculated with the known parameters of the light and heavy fission fragments (26). Best agreement was found for a total-fission-fragment absorption of ~ 7% (see Fig. 8). The required correction (spectrum without absorption versus spectrum with ~ 7% fission-fragment absorption) was ~ 5%.

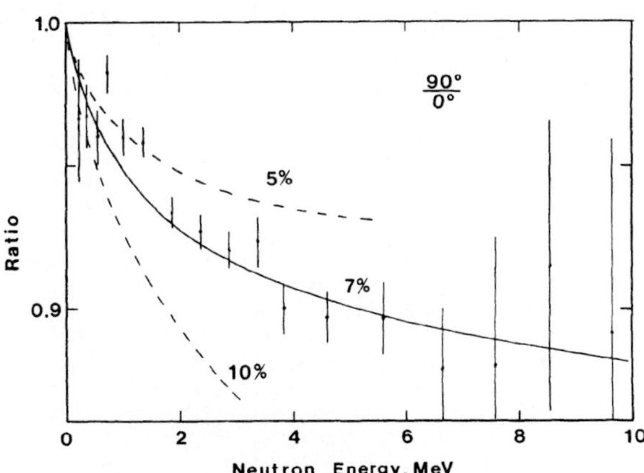

Fig. 8. The Ratio of the Fission-neutron Spectra Observed at ~ 90° and at 0° to the Deposit. The Calculated Curves Indicate Various Amounts of Fission-fragment Absorption at ~ 90°.

Measurements of the spectrum would also be affected, if neutron emision occurs after fragments were already slowed down in the deposit backing. In order to check this, effect, the ratio of the spectra measured at 0° and at 180° to the deposit axis was derived. The result is shown in Fig. 9 and it supports the previous conclusions that neutron emission occurs within 10^{-15} sec. of the fission process (6).

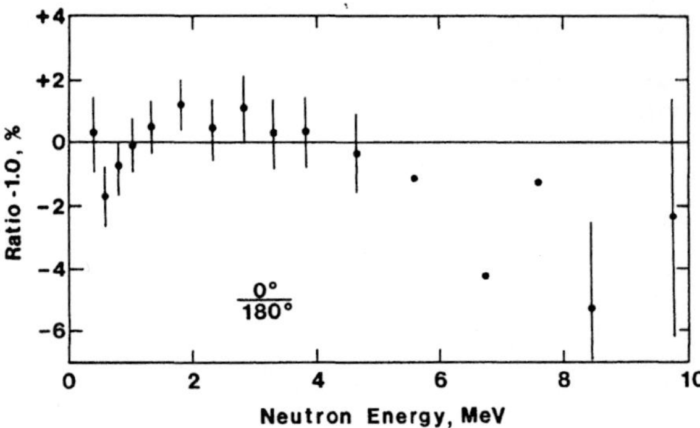

Fig. 9. The Difference from Unity in Percent of the Ratio of the Fission-neutron Spectra Observed at 0° and 180°.

Neutron transmission through the deposit backing, the air of the flight path, the gold filter, and the edges of the collimeter changed the measured spectrum. The corrections were suficiently small and/ or the cross sections well enough known to result in insignificant uncertainties for the corrected spectrum. Some of the corrections required for elastic and inelastic scattering events involve energy transfer and result in an experimental spectrum with more low-energy neutrons than the original spectrum. These effects were substantially reduced in the present measurements due to the design of the experiment. Neutrons which scattered in the ≈ 0.05 grams of aluminum forming windows of the GSC and in the ≈ 0.4 grams of the platinum backing could reach the detector. Only the scattering in the platinum backing was significant. The effect due to elastic scattering was corrected by using Legendre-coefficients for elemental platinum (28,29) and taking into account the increase of the interaction probability for neutrons emitted toward 90°. Data on the inelastic scattering in platinum are sparse. Only a few levels (< 1 MeV) were resolved with some cross section data available (29). For inelastic scattering to higher levels an evaporation spectrum was used. Scattering from the two collimators was estimated to be < 0.1% from the collimator in front of the detector and ≲ 0.2% from the collimator in front of the source. Because little energy transfer is involved, these corrections have been deferred but will be included in a Monte-Carlo simulation of the experiment at a later time.

Neutrons which penetrate the detector or which are scattered without being detected might return to the detector after scattering in the photomultiplier or other materials surrounding the neutron detector. The additional neutron-time-of-flight implies a softer measured spectrum. There is an additional advantage in using a BND because this correction is proportional to $1 - \varepsilon_n$, where ε_n is the neutron detector efficiency, which is high for the BND. This correction was < 1% for the present experiment.

Delayed gammas from the fission products are rather soft (30) and were mostly eliminated with the gamma discriminator, and the gamma absorption filters. Support for this was found in consideration of the residual background between the gamma peak and the neutrons in the time-of-flight spectra (see below).

The possibility of neutron production by the alphas emitted in ^{252}Cf decay was considered. The threshold for detecting the fission fragments was set above the alpha-pile-up level, however, some alpha-pile-up trigger pulses cannot be excluded. The (σ,n) threhsold in platinum is high enough so that neutron-production should not occur. The possibility of neutron production from other parts of the detector was checked with a strong ^{241}Am alpha source. No effect was found.

The (n, 2n) reaction in platinum contributes some neutrons to the spectrum. The required correction was calculated and found to be negligible (γ,n), (γ,2n) was estimated and also found to be negligible.

After the corrections for accidental coincidences, shadow-bar background and ambient background subtraction were applied, the residual background in the areas between the gamma peak and the neutrons, and below the neutron detection threshold were found to be very small in two of the spectra. The difference for the neutron spectra between not subtracting this residual background or subtracting it by linear interpolation between the two background ranges was ≲ 0.3% at the lowest and higher energies. For the third spectrum this background was somewhat higher. Ultimately, the residual background was subtracted by linear interpolation.

The agreement between all three measured spectra in respective overlap ranges was good.

VII. Results and Discussion

The results from the present measurements were reduced to a lesser number of energies then obtained with the TOF spectra which were identical for the three spectra. The procedure of reducing to this energy grid was described previously (31). A Maxwellian spectrum shape was used for extrapolation to the grid energies. This results in reduced statistical uncertainties and unchanged systematic uncertainties. The three spectra were than normalized to one another in the overlap range and averaged. The final result was normalized to the same number of neutrons in a Maxwellian spectrum with T=1.42 MeV within the range of the present measurements. The difference of the present results from a Maxwellian of T= 1.42 MeV is shown in Fig. 10. The difference of a Watt spectrum from a Maxwellian spectrum of the same average energy (2.13 MeV) is also shown in Fig. 10. The present result deviates from a Maxwellian-spectrum shape toward the Watt-spectrum shape. Madland and Nix (38) recently develped a theoretical approach for calculating fission properties. The ^{252}Cf-fission spectrum obtained from these calculations has an average energy of 2.2791 MeV (38). By using an improved mass formula, a spectrum with a lower average energy of 2.2167 MeV was calculated (39). This spectrum is also shown in Fig. 10 relative to a Maxwellian-Spectra of T = 1.42 MeV.

The average energy of the present result for the ^{252}Cf spectrum was obtained by approximating the low-and-high-energy regions not included in the present measurement with a Maxwellian spectrum of T = 1.42 MeV. The result for the average energy of the present measurement is 2.159 MeV. Fig. 11 shows the present results relative to a Maxwellian spectrum of the same average energy (2.159 MeV). The Watt spectrum is also shown for this average energy. The two spectra calculated by Madland and Nix (38,39) are shown relative to Maxwellian spectra of the respective average energies of 2.2167 MeV and 2.2791 MeV. This shows that the spectra differences relative to Maxwellian spectra of the same average energy are in first order independent of the average energy. Thus, we can conclude that our data support the shape difference between the spectrum calculated by Madland and Nix and a Maxwellian spectrum.

The present data were nevertheless fitted with a Maxwellian sprectrum using the variance-covariance matrix of the data. This led to a temperature of 1.439 MeV which corresponds to the identical average energy as determined from the experimental data.

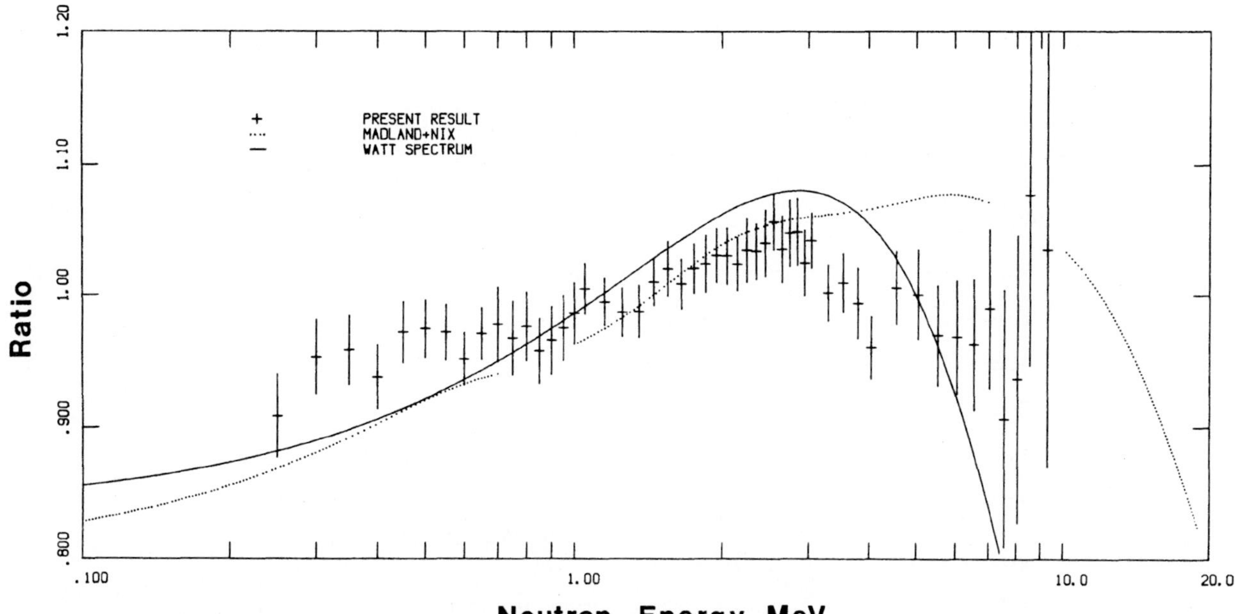

Fig. 10. The Ratio of the Present Results for the ^{252}Cf-Fission-Neutron Spectrum to a Maxwellian with a Temperature of 1.42 MeV ($\bar{E}$ = 2.13 MeV). The Ratios of a Watt Spectrum with the same Average Energy, and of the Theoretical Calculations by Madland and Nix to the Maxwellian are also shown. The Spectrum by Madland and Nix has an Average Energy of 2.2167 MeV.

The present result for the average energy is compared in Table 1 with results from other TOF measurements in the post-1971 area. The present value for the average energy is higher than those obtained in the last three measurements. Complete descriptions of these experiments are not yet available. It is more surprising that the other experiments listed in Table 1 did not result in more substantial differences. This may be due to partial compensation of effects overlooked in these experiments.

The present reuslts are preliminary, pending improvements for some of the corrections and a possible improvement in measurement technique involving a better ^{252}Cf source (and backing) and better statistical accuracy.

Acknowledgements

The present work was supported by the U.S. Department of Energy. Valuable discussions with Drs. A. B. Smith and J. M. Meadows were appreciated. The least-squares fit of the present data and their variance-covariance matrix to a Maxwellian spectrum was carried out by Dr. D. L. Smith.

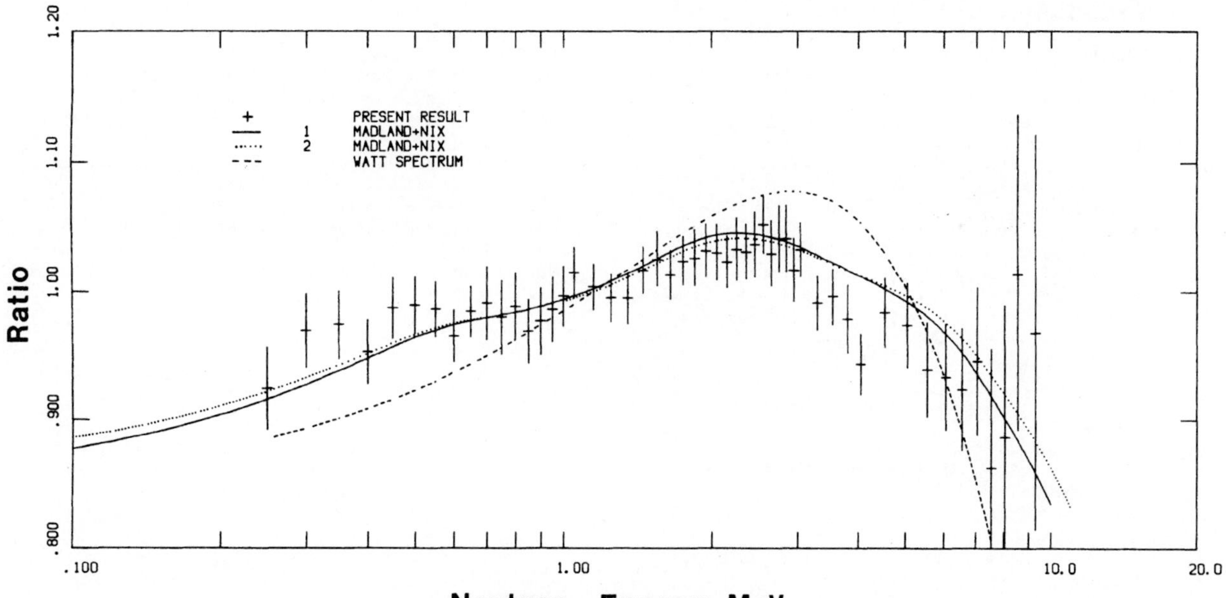

Fig. 11. The Ratio of the Present Experimental Results to a Maxwellian with the same Average Energy (2.159 MeV). The Ratio of the Watt Spectrum is also Shown for the Same Average Energy. The Calculated Spectra by Madland and Nix are Shown Relative to Maxwellian Spectra of the Same Average Energies of 2.2167 MeV and 2.2791 MeV.

Table 1. Comparison of the Present Result for the Average Energy with other Post 1971 Time-of-Flight Measurements

Reference		Year	$T_{MaxW.}$, MeV	$\bar{E}$, MeV
Green et al.	(10)	1973	1.406 ± 0.015	2.105 ± 0.01
Knitter et al.	(32)	1973	1.42 ± 0.05	2.13 ± 0.08
Kotelnikova et al.	(33)	1976	1.46 ± 0.02	(2.19 ± 0.03)
Batenkov et al.	(34)	1975	1.40	
Blinov et al.	(35)	1977	1.41 ± 0.03	2.12
Nefedov et al.	(36)	1977	1.28	(1.92)
Nefedov et al. (Starostov et at.)	(11)	1978 1981	1.43 ± 0.02	(2.15 ± 0.03)
Bertin		1978	(1.51)	2.27 ± 0.02
Blinov et al.	(7)	1979	1.42	
Bodeman et al.	(8)	1979	1.424 ± 0.013	2.136 ± 0.02
Mon Jiangshen et al.	(9)	1981	1.416 ± 0.023	
Present Result		1982	1.439 ± 0.010	2.159

References

1. Proc. IAEA Consultants' Meeting on Prompt Fission Neutron Spectra, Vienna, 1971.
2. Proc. IAEA Panel on Neutron Standard Reference Data, Vienna, 1974.
3. A. B. Smith et al., NIM 140, 397 (1977).
4. V. A. Vukolov et al., Proc. Conf. Neutr. Physics, Vol. 6, 265, Kiev 1976.
5. J. R. Smith, Summary of a Workshop on Fission Spectra, Los Alamos Report LA-7739-C (ENDF-278), 1978.
6. M. V. Blinov, Proc. IAEA Consultants' Meeting on Neutron Source Properties, Debrecen, p. 79 (1980).
7. M. V. Blinov, Proc. IX Intern., Symp. on the Interactions of Fast Neutrons with Nuclei, Rossendorf (1980).
8. J. W. Boldeman et al., ANS Transactions 32, 733 (1979).
9. Mon Jiangshen et al., Chin. I. of Nucl. Phys. 3, 163 (1981).
10. L. Green et al., NSE 50, 257 (1973).
11. B. I. Starostov et al., INDC (CCP)-1641L (1981).
12. See figures in Ref. 6.
13. Summary of a Workshop on Fission Spectra, Los Alamos Report LA-7739-C (ENDF-278), 1978.
14. Compilation of Requests for Nuclear Data, DOE/NDC-22/U (1981).
15. WRENDA 81/82, World Request List for Nuclear Data, INDC (SEC)-78/URSF (1981).
16. W. P. Poenitz, NSE 53, 370 (1974).
17. W. P. Poenitz, NIM 109, 413 (1973).
18. W. P. Poenitz, Argonne National Laboratory Report, ANL-7915 (1972).
19. G. P. Lamaze et al., NBS Special Publ. 425, Vol. I, 73 (1975), M. M. Meier, NBS Special Publ. 493, 221 (1977).
20. C. Renner, Oak Ridge National Laboratory, private communication (1978).
21. G. Grenier, 3rd. Conf. on Neut. Dos. (1977).
22. Tohoku Univ., Sendai, Japan (1981).
23. W. P. Poenitz, NSE 64, 894 (1977).
24. A. O. Wasson, to be published in NSE (1982).
25. R. Arlt, et al., Tech., Univ. Dresden Reports, 05-43 (1980).
26. H. R. Bowman et al., Phys. Rev. 126, 2120 (1962).
27. A. Chalupka, NIM 165, 103 (1979).
28. A. Langsdorf et al., Argonne National Laboratory, ANL-5567 (1956).
29. A. B. Smith et al., Argonne National Laboratory, ANL-7363 (1967).
30. S. A. E. Johansson, NP 64, 147 (1965).
31. W. P. Poenitz, Proc. Conf. on Nuclear Data Evaluation Methods and Procedures, NEANDC(US)-209, Vol. 1, p. 249 (1980).
32. H. H. Knitter et al., Atomkernenergie 22, 84 (1973).
33. G. V. Kotelnikova et al., INDC (CCP)-81/U (1976).
34. O. I. Batenkov et al., Conf. Neutron Physics, Kiev, Vol. 5, p. 114 (1975).
35. M. V. Blinov et al., NBS Spec. Publ. 493, 194 (1977).
36. V. N. Nefedov et al., Conf. Neutron Physics, Kiev, Vol. 3, p. 205 (1977).
37. A. Bertin et al., CEA-R-4895 (1975).
38. D. G. Madland and J. R. Nix, NSE 81, 213 (1982).
39. D. G. Madland, private communication (1982).

CALCULATION OF THE PROMPT NEUTRON SPECTRUM AND AVERAGE PROMPT NEUTRON MULTIPLICITY FOR THE SPONTANEOUS FISSION OF ^{252}Cf

D. G. Madland and J. R. Nix

Theoretical Division, Los Alamos National Laboratory
Los Alamos, New Mexico 87545, USA

We calculate the prompt fission neutron spectrum and average prompt neutron multiplicity for the spontaneous fission of ^{252}Cf. Our calculations are based upon conventional nuclear-evaporation theory and account for the effects of (1) the motion of the fission fragments, (2) the distribution of fission-fragment residual nuclear temperature, and (3) the energy dependence of the cross section for the inverse process of compound-nucleus formation. Because of increased accuracy requirements due to the use of the ^{252}Cf(sf) reaction as a standard, we perform our calculations with greater accuracy than previously. In particular, we calculate the integral for the average energy release without approximation and with the use of more recent mass sources. Following our theoretical calculations, we perform a least-squares adjustment to a specific experimental spectrum and present the comparison.

[^{252}Cf(sf); calculated prompt fission neutron spectrum N(E), average prompt neutron multiplicity $\bar{\nu}_p$, and $\bar{\nu}_p(A_H)$]

Introduction

On the basis of new developments[1] in the theory of the prompt fission neutron spectrum N(E) and average prompt neutron multiplicity $\bar{\nu}_p$, we calculate these quantities for the spontaneous fission of ^{252}Cf. We study this particular reaction because it is used as a standard in many measurements and applications of neutron physics. The new developments are based upon conventional nuclear-evaporation theory and account for the effects of (1) the motion of the fission fragments, (2) the distribution of fission-fragment residual nuclear temperature, and (3) the energy dependence of the cross section for the inverse process of compound-nucleus formation.

As an approximation to the result of Terrell[2], we take the residual nuclear-temperature distribution to be triangular in shape, extending linearly from zero to a maximum value T_m. For some of our purposes we calculate the compound-nucleus cross section from the optical model, whereas in other cases we use a constant cross section and readjust the value of the nuclear level-density parameter to simulate the energy dependence. The value of T_m is determined from the average energy release, the total average fission-fragment kinetic energy, and the level-density parameter of the Fermi-gas model.

Whereas for ^{252}Cf spontaneous fission the total average fission-fragment kinetic energy is a measured quantity and the Fermi-gas level-density parameter is inferred from measurements, the average energy release must be calculated. Previously we have calculated this quantity by use of a seven-point approximation[1] to the integral of the energy release over the fission-fragment mass and charge distributions, using measured or systematic masses of the 1977 Wapstra-Bos evaluation[3] when they exist and otherwise the droplet-model mass formula of Myers[4]. Here we replace these with the new 1981 Wapstra-Bos evaluation[5] and the new macroscopic-microscopic mass formula of Möller and Nix[6]. We then perform the integration for the average energy release without approximation. An identical set of changes is made in the integration for the average fission-fragment neutron separation energy, which is required in the calculation of the average prompt neutron multiplicity. With these improvements, we calculate the prompt fission spectrum N(E), the average prompt neutron multiplicity $\bar{\nu}_p$, and its decomposition into $\bar{\nu}_p(A_H)$, where A_H is the mass number of the heavy fragment.

We first summarize our theory of N(E) and present calculations for it. We then describe our formulation and calculations of $\bar{\nu}_p$ and $\bar{\nu}_p(A_H)$. Taking into account the constraints that operate between N(E) and $\bar{\nu}_p$, we finally perform a least-squares adjustment to a specific experimental spectrum. Some of the results presented here have already appeared in Refs. 1 and 7.

Prompt Fission Neutron Spectrum

To summarize our calculation[1] of the prompt fission neutron energy spectrum N(E), it is given by the average of the spectra calculated for neutron emission from the light L and heavy H average fission fragments, namely

$$N(E) = \tfrac{1}{2}[N(E,E_f^L,\sigma_c^L) + N(E,E_f^H,\sigma_c^H)] \ , \tag{1}$$

where E is the laboratory neutron energy, E_f is the average kinetic energy per nucleon of a moving fission fragment, and σ_c is the cross section for the inverse process of compound-nucleus formation. The spectrum due to a moving fission fragment is determined from nuclear-evaporation theory by integrating over a triangular distribution of fission-fragment residual nuclear temperature that extends linearly from zero to a maximum value T_m and transforming to the laboratory system, which gives

$$N(E,E_f,\sigma_c) = \frac{1}{2\sqrt{E_f}T_m^2}\int_{(\sqrt{E}-\sqrt{E_f})^2}^{(\sqrt{E}+\sqrt{E_f})^2} \sigma_c(\varepsilon)\sqrt{\varepsilon}\,d\varepsilon \times\int_0^{T_m} k(T)\,T\exp(-\varepsilon/T)\,dT \ . \tag{2}$$

In this equation ε is the center-of-mass neutron energy, T is the fission-fragment residual nuclear temperature, and k(T) is the temperature-dependent normalization constant for the corresponding center-of-mass spectrum.

If the cross section $\sigma_c(\varepsilon)$ for the inverse process of compound-nucleus formation is assumed constant, Eq. (2) reduces to the closed four-term expression

$$N(E,E_f) = \frac{1}{3(E_fT_m)^{1/2}}[u_2^{3/2}E_1(u_2) - u_1^{3/2}E_1(u_1) + \gamma(3/2,u_2) - \gamma(3/2,u_1)] \ , \tag{3}$$

where

$$u_1 = (\sqrt{E}-\sqrt{E_f})^2/T_m \ ,$$

$$u_2 = (\sqrt{E}+\sqrt{E_f})^2/T_m \ ,$$

$$E_1(x) = \int_x^\infty \frac{\exp(-u)}{u}\,du$$

K. H. Böckhoff (ed.), Nuclear Data for Science and Technology, 473–478.

is the exponential integral[8], and

$$\gamma(a,x) = \int_0^x u^{a-1} \exp(-u)\, du$$

is the incomplete gamma function[8]. The spectrum given by Eq. (2) is obtained by numerical integration, whereas the spectrum given by Eq. (3) is obtained directly, since the exponential integral and incomplete gamma function are standard computer program library functions.

We showed in Ref. 1 that the energy-dependent cross-section spectrum given by Eqs. (1) and (2) is more accurate than the constant cross-section spectrum given by Eqs. (1) and (3). However, the energy dependence of $\sigma_c(\varepsilon)$ can be simulated by readjusting the value of the nuclear level-density parameter so that a constant cross-section calculation reproduces reasonably well the energy-dependent cross-section calculation.

Spectra calculated from Eqs. (1) and (2) or Eqs. (1) and (3) depend primarily upon the values of three constants, namely the average kinetic energies per nucleon E_f^L and E_f^H of the average light and heavy fragments, respectively, and the maximum temperature T_m of the distribution of fission-fragment residual nuclear temperature.

The values of E_f^L and E_f^H are obtained by use of momentum conservation from the total average fission-fragment kinetic energy $\langle E_f^{tot}\rangle$, the mass number A of the compound nucleus undergoing fission, and the average mass numbers A_L and A_H of the light and heavy fragments, respectively. In this work, as in Ref. 1, we use the values $\langle E_f^{tot}\rangle = 185.9$ MeV, $A_L = 108$, and $A_H = 144$ that are obtained from the measurements of Unik et al.[9].

The value of T_m is obtained from the observation of Terrell[2] that in the triangular approximation T_m is related to the initial total average fission-fragment excitation energy $\langle E^*\rangle$ approximately by

$$T_m = (\langle E^*\rangle/a)^{1/2} , \tag{4}$$

where a is the nuclear level-density parameter. For spontaneous fission, $\langle E^*\rangle$ is given by

$$\langle E^*\rangle = \langle E_r\rangle - \langle E_f^{tot}\rangle , \tag{5}$$

where $\langle E_r\rangle$ is the average energy release in fission.

The energy release for division into a given pair of fission fragments is the difference between the ground-state mass of the fissioning compound nucleus and the ground-state masses of the two fission fragments. The average energy release $\langle E_r\rangle$ is then given by the integral of the product of this energy difference and the fission-fragment mass and charge distributions, taken over all possible mass and charge divisions.

In Ref. 1 we evaluated the integral for $\langle E_r\rangle$ by a seven-point approximation to this integral that is centered about the average values of the distributions, as illustrated in Fig. 1 of Ref. 1. In applying this approximation, the average or central fragments are obtained from the measurements of Unik et al.[9], which yield $^{108}_{42}$Mo and $^{144}_{56}$Ba for the ^{252}Cf(sf) reaction. The required energy differences are then obtained using experimental or derived systematic masses when they exist and otherwise a mass formula. The resulting value of $\langle E_r\rangle$ is used in Eq. (5) to obtain the initial total average fission-fragment excitation energy $\langle E^*\rangle$.

The remaining quantity required to evaluate Eq. (4) for T_m is the nuclear level-density parameter a. In Ref. 1 we used the value

$$a = A/(11 \text{ MeV}) \tag{6}$$

for energy-dependent cross-section calculations and

$$a_{eff} = A/(10 \text{ MeV}) \tag{7}$$

for constant cross-section calculations that simulate the energy dependence.

We now discuss four calculations of the prompt fission neutron spectrum that have been performed using the seven-point approximation. Due to space limitations we do not show comparisons of these calculations with experimental data, but instead present the essential results in the first four lines of Table I.

The first line of Table I gives the results for the energy-dependent cross-section calculation obtained using a value of $\langle E_r\rangle$ determined from the 1977 Wapstra-Bos[3] evaluation for eight of the required masses and the mass formula of Myers[4] for the remaining seven. The optical-model potential of Becchetti and Greenlees[10] is used to calculate $\sigma_c(\varepsilon)$ and the level-density parameter is given by Eq. (6). This spectrum is identical to that calculated and compared in Refs. 1 and 7 to experiments #1 and #7 of Boldeman et al.[11,12]. The first moment or average energy $\langle E\rangle$ of the spectrum is 2.279 MeV, which is 143 keV larger than the average value of 2.136 ± 0.020 MeV obtained by Boldeman et al.[11] in a Maxwellian fit to the data from several experiments. As shown in Figs. 27-30 of Ref. 1, the calculated spectrum is somewhat harder than the experimental spectra in the region above ~ 2 MeV and somewhat softer in the region below ~ 2 MeV. As explained in Ref. 1, the discrepancy may be due to a value of the level-density parameter a that is somewhat low.

The second line of Table I gives the results for the constant cross-section calculation. This spectrum, illustrated in Ref. 7, is very similar to the first spectrum, having an average energy $\langle E\rangle = 2.306$ MeV that is only 27 keV larger than that of the first spectrum.

The third and fourth lines of Table I correspond, respectively, to the first and second lines except that new sources of masses are used in the calculation of $\langle E_r\rangle$. These are the 1981 Wapstra-Bos evaluation[5], from which ten of the required masses are obtained, and the new mass formula of Möller and Nix[6], from which the remaining five masses are obtained. As Table I shows, the new value of $\langle E_r\rangle$ is reduced by 2.827 MeV, or 1.3%. This produces corresponding reductions of 8.4% in the excitation energy $\langle E^*\rangle$ and 4.2% in the maximum temperature T_m, which reduces the average energies of the third and fourth spectra by 66 keV relative to those of the first and second. The average energy of 2.213 MeV for the third spectrum is 77 keV larger than the fitted Maxwellian value of Boldeman et al.[11]. However, while it is true that the third spectrum agrees with experiment[11,12] better than do the other three calculations discussed, the value of the average prompt neutron multiplicity $\bar{\nu}_p$ that we simultaneously calculate is in this case significantly smaller than experimental values, as discussed in the next section. In an attempt to resolve this discrepancy, we next improve upon the seven-point approximation to the integral for the average energy release $\langle E_r\rangle$ by performing the full integration without approximation, repeating both the energy-dependent and constant cross-section calculations of the spectrum.

Table I Summary for $^{252}Cf(sf)$ of the Calculated Prompt Fission Neutron Spectrum and Average Prompt Neutron Multiplicity*

Integration	Mass Source	$\langle E_r \rangle$ (MeV)	$\sigma_c(\varepsilon)$	Level Density Parameter (1/MeV)	$\langle E \rangle$ (MeV)	$\langle S_n \rangle$ (MeV)	$\bar{\nu}_p$
seven-point	W-B 77[a] Myers[b]	219.408	B-G[e]	A/11	2.279	5.473	3.803
seven-point	W-B 77 Myers	219.408	const	A/10	2.306	5.473	3.788
seven-point	W-B 81[c] M-N[d]	216.581	B-G	A/11	2.213	5.233	3.554
seven-point	W-B 81 M-N	216.581	const	A/10	2.240	5.233	3.540
full	W-B 81 M-N	218.886	B-G	A/11	2.267	5.439	3.737
full	W-B 81 M-N	218.886	const	A/10	2.294	5.439	3.723 3.714[f]
full	W-B 81 M-N	218.886	B-G	A/9.6	2.168	5.439	3.791
full	W-B 81 M-N	218.886	const	A/8.4	2.167	5.439	3.792 3.783[f]

*In obtaining a mass value, we use the indicated experimental mass evaluation if possible, and the indicated mass formula otherwise; the level-density parameter is either a for energy-dependent cross-section calculations or a_{eff} for constant cross-section calculations; unless otherwise noted, $\bar{\nu}_p$ is calculated using Eq. (12).

[a] The 1977 Wapstra-Bos mass evaluation (Ref. 3).
[b] The droplet-model mass formula of Myers (Ref. 4).
[c] The 1981 Wapstra-Bos mass evaluation (Ref. 5).
[d] The macroscopic-microscopic mass formula of Möller and Nix (Ref. 6).
[e] Calculated using the optical-model potential of Becchetti and Greenlees (Ref. 10).
[f] Calculated using Eq. (14).

The average energy release in fission $\langle E_r \rangle$ is given exactly by

$$\langle E_r \rangle = \frac{\sum_{A_H} Y(A_H)\, E_r(A_H)}{\sum_{A_H} Y(A_H)} , \qquad (8)$$

where $Y(A_H)$ is the fission-fragment mass yield distribution, A_H is the heavy-fragment mass number, and $E_r(A_H)$ is the average energy release for a given mass division. It is, in turn, obtained by summing the contributions from all participating charge divisions, namely

$$E_r(A_H) = \frac{\sum_{Z_H} \rho(Z_H,A_H)\, E_r(Z_H,A_H)}{\sum_{Z_H} \rho(Z_H,A_H)} , \qquad (9)$$

where $\rho(Z_H,A_H)$ is the heavy fission-fragment charge distribution, Z_H is the heavy-fragment atomic number, and $E_r(Z_H,A_H)$ is the energy release for a given mass and charge division.

We use the fission-fragment mass-yield distribution $Y(A_H)$ measured by Weber et al.[13] and assume the fission-fragment charge distribution $\rho(Z_H,A_H)$ to be of Gaussian form,

$$\rho(Z_H,A_H) = \frac{1}{(2\pi\sigma_z^2)^{1/2}} \exp[-(Z_H - ZP_H)^2/(2\sigma_z^2)] , \qquad (10)$$

with the most probable heavy fragment charge ZP_H given by

$$\frac{ZP_H + c}{A_H} = \frac{Z}{A} = \frac{ZP_L - c}{A_L} \qquad (11)$$

In this equation, we use the value of 0.5 charge units determined by Unik et al.[9] for the charge division parameter c, except for symmetric fission where c = 0. We also use a value of 0.5 charge units for the width σ_z, which is approximately mid-range in the set of values determined by Wahl[14] in studies of fission-product charge distributions.

With these parameters, we perform the full integration and obtain a value for $\langle E_r \rangle$ of 218.886 MeV. This value is stable to within ± 0.025% (± 55 keV) for a change of ± 0.05 charge units in c, and is stable to within ± 0.1% (± 220 keV) for a change of ± 0.1 charge units in σ_z. These ranges are representative of the accuracy with which c and σ_z are known for ^{252}Cf spontaneous fission.

The new value of $\langle E_r \rangle$ lies between the two previous values obtained with the seven-point approximation. Our calculations of the spectrum corresponding to those obtained with the seven-point approximation, but using the full integration instead, are summarized in the fifth and sixth lines of Table I and are compared to experiment in Figs. 1 and 2. These figures show that, while both calculations agree fairly well with experiment #7 of Boldeman et al.[11,12], the energy-dependent cross-section calculation is preferred. However, this spectrum is itself somewhat hard in the tail region and somewhat soft in the peak region. The average energy $\langle E \rangle$ is 2.267 MeV, a value which is 131 keV larger than the average value obtained by Boldeman et al.[11] in a Maxwellian fit to several experiments. It thus appears that some further adjustment is necessary in our calculations of the spectrum. The clue to this adjustment is found by studying also the average prompt neutron multiplicity $\bar{\nu}_p$, which we calculate simultaneously with N(E) and to which we now turn our attention.

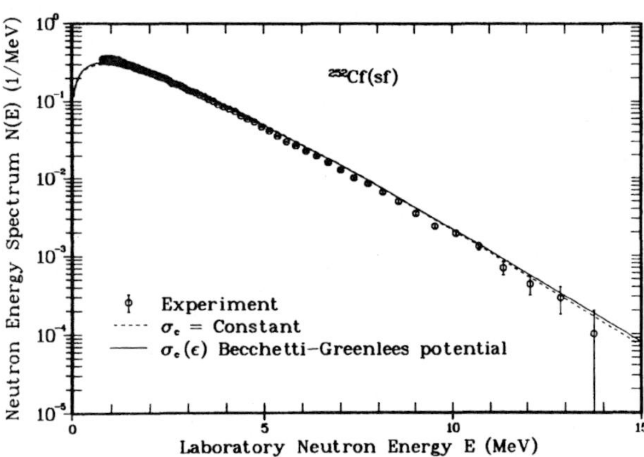

Fig. 1. Prompt fission neutron spectrum for the spontaneous fission of ^{252}Cf. The dashed curve gives the spectrum calculated with Eqs. (1) and (3) for a constant cross section, using T_m = 1.144 MeV resulting from Eq. (7), whereas the solid curve gives the spectrum calculated with Eqs. (1) and (2) for an energy-dependent cross section, using T_m = 1.200 MeV resulting from Eq. (6). For both calculated spectra the values of E_f^L and E_f^H are 0.984 MeV and 0.553 MeV, respectively. The experimental data are from experiment #7 of Boldeman et al. (Refs. 11, 12).

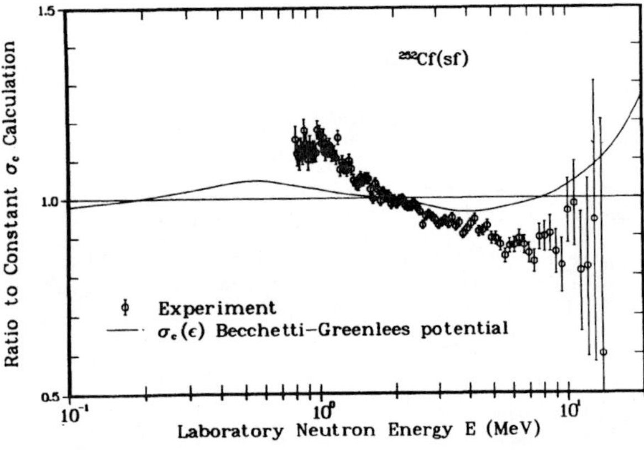

Fig. 2. Ratio of the spectrum calculated using energy-dependent cross sections and the experimental spectrum to the spectrum calculated using a constant cross section, corresponding to the curves shown in Fig. 1.

Average Prompt Neutron Multiplicity

The average prompt neutron multiplicity is given in Ref. 1 by

$$\bar{\nu}_p = \frac{\langle E^* \rangle - \langle E_\gamma^{tot} \rangle}{\langle S_n \rangle + \langle \varepsilon \rangle} , \qquad (12)$$

where $\langle E_\gamma^{tot} \rangle$ is the measured total average prompt gamma energy, $\langle S_n \rangle$ is the average fission-fragment neutron separation energy, and $\langle \varepsilon \rangle$ is the average center-of-mass energy of the emitted neutrons. For the energy-dependent cross-section calculation, $\langle \varepsilon \rangle$ is calculated numerically using the center-of-mass spectrum corresponding to Eqs. (1) and (2), whereas for the constant cross-section calculation $\langle \varepsilon \rangle$ is given by $(4/3)T_m$.

For spontaneous fission, the total average fission-fragment excitation energy $\langle E^* \rangle$ is given by Eq. (5). Thus, for a fixed value of $\langle E_f^{tot} \rangle$, $\bar{\nu}_p$ is very sensitive to the average energy release $\langle E_r \rangle$. It is also sensitive to the value of $\langle S_n \rangle$ since the average center-of-mass neutron energy $\langle \varepsilon \rangle$ is only about $0.2\langle S_n \rangle$. Moreover, because only $\langle \varepsilon \rangle$ in Eq. (12) depends on the level-density parameter, $\bar{\nu}_p$ is largely insensitive to the value of a. This is in contrast to Eq. (4) for T_m, which is sensitive to both $\langle E_r \rangle$ and a for a fixed value of $\langle E_f^{tot} \rangle$.

Thus, unsatisfactory agreement between calculated and measured N(E) means that $\langle E_r \rangle$ and/or a are in error, whereas unsatisfactory agreement between calculated and measured $\bar{\nu}_p$ means that $\langle E_r \rangle$ and/or $\langle S_n \rangle$ are in error. Therefore, a good calculation of $\bar{\nu}_p$ imposes a constraint on the corresponding N(E) calculation in that only the level-density parameter a is free to be adjusted.

Our calculations of $\bar{\nu}_p$ corresponding to the use of the seven-point approximation to calculate $\langle E_r \rangle$ are summarized in the first four lines of Table I. In these calculations, $\langle S_n \rangle$ is also calculated using the seven-point approximation. We calculate $\langle S_n \rangle$ as one-half of the average two-neutron separation energy $\langle S_{2n} \rangle$ in order to average over pairing effects and use the value $\langle E_\gamma^{tot} \rangle$ = 6.95 MeV given by Hoffman and Hoffman[15]. The calculated values of $\bar{\nu}_p$ are to be compared with the experimental value of 3.757 ± 0.009 obtained from the measurements of Amiel[16] and Smith[17], or the value of 3.773 ± 0.007 measured by Spencer et al.[18].

The first two calculations of $\bar{\nu}_p$ agree with experiment to within approximately 1%, whereas the second two are more than 5% low. On the other hand, the second two calculations of N(E) are closer to experiment than are the first two. From the four $\bar{\nu}_p$ calculations, their corresponding values of $\langle E_r \rangle$ and $\langle S_n \rangle$, and experiment, we conclude that the calculated values of $\langle E_r \rangle$ are probably not excessively high. From the four N(E) calculations, their corresponding values of $\langle E_r \rangle$ and a, and experiment, we conclude that $\langle E_r \rangle$ is somewhat high and/or a is somewhat low. Taking these conclusions together, we infer that the level-density parameter a is somewhat low.

However, before acting on this inference, we test it by repeating the third and fourth calculations of $\bar{\nu}_p$ contained in Table I, except that we again perform the full integration for $\langle E_r \rangle$, given by Eqs. (8) and (9), instead of using the seven-point approximation. Similarly, we perform the full integration to obtain $\langle S_n \rangle$. Indeed, with the full integration technique we are able to calculate the average prompt neutron multiplicity as a function of mass division, $\bar{\nu}_p(A_H)$, and integrate this quantity over the fragment mass-yield distribution to obtain $\bar{\nu}_p$ with greater accuracy.

Thus,

$$\bar{\nu}_p(A_H) = \frac{\langle E_r(A_H)\rangle - \langle E_f^{tot}(A_H)\rangle - \langle E_\gamma^{tot}(A_H)\rangle}{\langle S_n(A_H)\rangle - \langle \varepsilon(A_H)\rangle} \quad (13)$$

and

$$\bar{\nu}_p = \frac{\sum_{A_H} Y(A_H)\, \bar{\nu}_p(A_H)}{\sum_{A_H} Y(A_H)} \quad (14)$$

In these equations, we use the experimental results of Weber et al.[13] for $\langle E_f^{tot}(A_H)\rangle$ and $Y(A_H)$, except that the $\langle E_f^{tot}(A_H)\rangle$ are renormalized to the value $\langle E_f^{tot}\rangle$ = 185.9 MeV. We use the experimental results of Pleasonton et al.[19] for $\langle E_\gamma^{tot}(A_H)\rangle$ and we calculate $\langle \varepsilon(A_H)\rangle$ in the constant cross-section approximation, namely $(4/3)T_m(A_H)$.

Our calculations using the full integration for $\langle E_r\rangle$, $\langle S_n\rangle$, and $\bar{\nu}_p$ are summarized in the fifth and sixth lines of Table I. As in the case of $\langle E_r\rangle$, discussed earlier, the new value of $\langle S_n\rangle$ lies between those obtained in the two previous sets of calculations. The two values of $\bar{\nu}_p$ calculated using Eq. (12) are approximately 1% smaller than experiment. The more accurate calculation of $\bar{\nu}_p$, given by Eq. (14), is 1.1% less than the experimental value of 3.757. As discussed earlier, the corresponding spectra are somewhat hard in the tail region and somewhat soft in the peak region compared to experiment #7 of Boldeman et al.[11,12]. We conclude that with full integrations to obtain the average energy release $\langle E_r\rangle$ and the average fragment neutron separation energy $\langle S_n\rangle$, the nuclear level-density parameter a is still somewhat low. We therefore perform a least-squares adjustment to the spectrum of experiment #7 of Boldeman et al.[11,12], with respect to the value of the nuclear level-density parameter, and recalculate N(E) and $\bar{\nu}_p$. Using these results, we also calculate $\bar{\nu}_p(A_H)$.

Least-Squares Adjustment to Spectrum

We perform two least-squares adjustments to the experimental spectrum with respect to the level-density parameter. The first is performed using the energy-dependent cross-section calculation, Eqs. (1) and (2), with the level-density parameter given initially by Eq. (6), and the second is performed using the constant cross-section calculation, Eqs. (1) and (3), with the level-density parameter given initially by Eq. (7). To obtain an absolute value of χ^2, the normalization of the experiment is recalculated for each iteration in the value of the level-density parameter. In these calculations an iteration is taken as an increment of 0.1 MeV in the denominator of Eq. (6) or Eq. (7).

For the energy-dependent cross-section case, χ^2_{min} (per degree of freedom) occurs for a = A/(9.6 MeV) and has a value of 4.37. The range of the level-density parameter corresponding to $\chi^2 \leqq 1.5\chi^2_{min}$ is

$$A/(10.2\text{ MeV}) \leqq a \leqq A/(8.9\text{ MeV}).$$

For the constant cross-section case, χ^2_{min} occurs for a_{eff} = A/(8.4 MeV) and has a value of 7.49. In this case, the range of the level-density parameter corresponding to $\chi^2 \leqq 1.5\chi^2_{min}$ is

$$A/(9.2\text{ MeV}) \leqq a_{eff} \leqq A/(7.6\text{ MeV}).$$

In both least-squares adjustments, the level-density parameter corresponding to χ^2_{min} has increased somewhat relative to its initial value given by Eq. (6) or Eq. (7).

Our calculations of N(E) and $\bar{\nu}_p$ using the least-squares adjusted level-density parameters are summarized in the seventh and eighth lines of Table I and are illustrated in Figs. 3 and 4. The energy-dependent cross-section calculation clearly agrees better with experiment than does the constant cross-section calculation, as expected from the ratio 1.7 in the values of χ^2_{min} for the two cases. Despite the clear preference of the energy-dependent cross-section calculation, the average energies for the two cases are almost identical, as are the corresponding values of $\bar{\nu}_p$ calculated using Eq. (12). The average energies are, however, approximately 30 keV larger than the average value 2.136 MeV obtained by Boldeman et al.[11] in a Maxwellian fit to several experiments, whereas the values of $\bar{\nu}_p$ are well within 1% of the experimental value 3.757 ± 0.009 due to Smith[17] and Amiel[16] and the experimental value 3.773 ± 0.007 due to Spencer et al.[18].

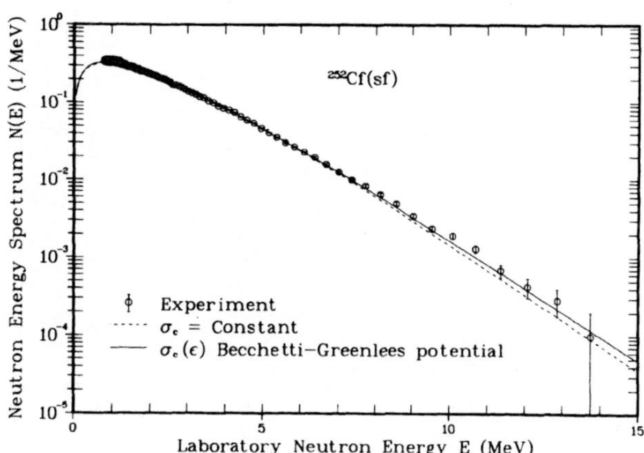

Fig. 3. Prompt fission neutron spectrum for the spontaneous fission of ^{252}Cf. The dashed curve gives the spectrum calculated with Eqs. (1) and (3) for a constant cross section, using T_m = 1.049 MeV resulting from the least-squares adjustment with a_{eff} = A/(8.4 MeV), whereas the solid curve gives the spectrum calculated with Eqs. (1) and (2) for an energy-dependent cross section, using T_m = 1.121 MeV resulting from the least-squares adjustment with a = A/(9.6 MeV). For both calculated spectra the values of E_f^L and E_f^H are 0.984 MeV and 0.553 MeV, respectively. The experimental data are from experiment #7 of Boldeman et al. (Refs. 11, 12).

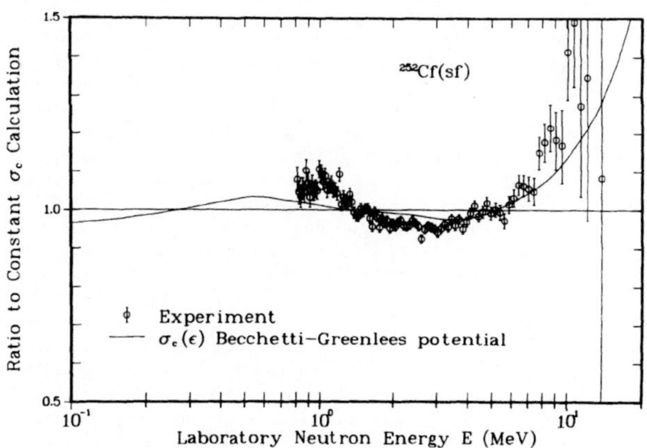

Fig. 4. Ratio of the spectrum calculated using energy-dependent cross sections and the experimental spectrum to the spectrum calculated using a constant cross section, corresponding to the curves shown in Fig. 3.

Our most accurate calculation of $\bar{\nu}_p$, using Eq. (14), yields a value of 3.783 that differs from the former experimental result by 0.7% and differs from the latter experimental result by 0.3%. The decomposition of this calculated value into $\bar{\nu}_p(A_H)$, by use of Eq. (13), is shown in Fig. 5, where the calculated values are compared to the experimental data of Walsh and Boldeman[20], which have been renormalized to the value 3.757 for $\bar{\nu}_p$. The calculation and experiment agree very well in the peaks and wings of the fragment mass distribution, but discrepancies as large as 15% occur near A_H = 138, where the descent into the valley is well under way. It is clear from the figure that even better agreement of the integral of $\bar{\nu}_p(A_H)$ with experiment can be achieved if refinements to the calculation of $\bar{\nu}_p(A_H)$ are made.

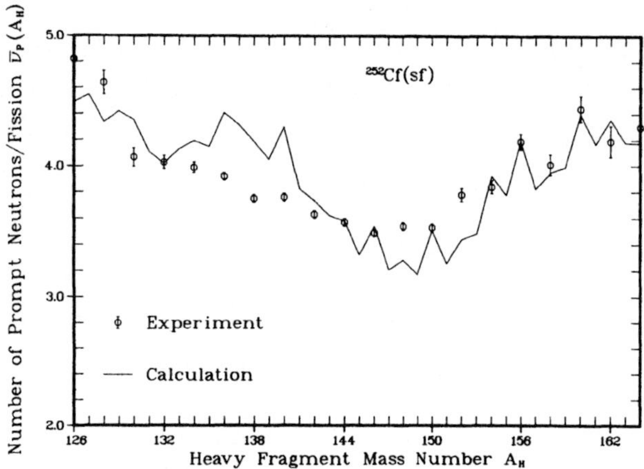

Fig. 5. Average prompt neutron multiplicity as a function of the heavy fragment mass for the spontaneous fission of ^{252}Cf. The calculation is performed with Eq. (13) using a_{eff} = A/(8.4 MeV) resulting from the least-squares adjustment to the spectrum calculated for a constant cross section. The experimental data are those of Walsh and Boldeman (Ref. 20). Note the suppressed zero of the vertical scale.

Conclusions

We have calculated the prompt fission neutron spectrum N(E) and average prompt neutron multiplicity $\bar{\nu}_p$ for the spontaneous fission of ^{252}Cf with greater accuracy than in our earlier work. Whereas our calculations of $\bar{\nu}_p$ agree very well with current measured values, our calculations of N(E) yield a harder spectrum than that measured by Boldeman et al.[11,12]. From a study of these results we have inferred that our initial values of the nuclear level-density parameters a and a_{eff} are too low for the ^{252}Cf(sf) reaction. Least-squares adjustments performed with respect to the level-density parameters have raised their values somewhat, resulting in better agreement not only with the experimental N(E), but also with the experimental $\bar{\nu}_p$. Although both the energy-dependent and constant cross-section calculations satisfactorily reproduce the experimental results, the former calculation does so somewhat better. In conclusion, we have provided a unified energy-dependent cross-section calculation that reproduces both N(E) and $\bar{\nu}_p$ with high accuracy.

References

1. D. G. Madland and J. R. Nix, Nucl. Sci. Eng. 81, 213 (1982).

2. J. Terrell, Phys. Rev. 113, 527 (1959).

3. A. H. Wapstra and K. Bos, At. Data. Nucl. Data Tables 19, 175 (1977).

4. W. D. Myers, Droplet Model of Atomic Nuclei (IFI/Plenum Data Co., New York, 1977).

5. A. H. Wapstra and K. Bos, private communication to the Nat. Nuclear Data Center, Brookhaven Nat. Laboratory (March 1982).

6. P. Möller and J. R. Nix, At. Data Nucl. Data Tables 26, 165 (1981).

7. D. G. Madland, Trans. Am. Nucl. Soc. 38, 649 (1981).

8. M. Abramowitz and I. A. Stegun, Eds., Handbook of Mathematical Functions (U. S. National Bureau of Standards, Washington, D.C., 1964).

9. J. P. Unik, J. E. Gindler, L. E. Glendenin, K. F. Flynn, A. Gorski, and R. K. Sjoblom, Proc. Third IAEA Symp. on Physics and Chemistry of Fission, Rochester, New York, 1973 (International Atomic Energy Agency, Vienna, 1974), Vol. II, p. 19.

10. F. D. Becchetti, Jr. and G. W. Greenless, Phys. Rev. 182, 1190 (1969).

11. J. W. Boldeman, D. Culley, and R. J. Cawley, Trans. Am. Nucl. Soc. 32, 733 (1979).

12. J. W. Boldeman, private communication (July 1979).

13. J. Weber, H. C. Britt, and J. B. Wilhelmy, Phys. Rev. C 23, 2100 (1981).

14. A. C. Wahl, J. Radioanalytical Chem. 55, 111 (1980).

15. D. C. Hoffman and M. M. Hoffman, Ann. Rev. Nucl. Sci. 24, 151 (1974).

16. S. Amiel, Proc. Second IAEA Symp. on Physics and Chemistry of Fission, Vienna, Austria, 1969 (International Atomic Energy Agency, Vienna, 1969), p. 569.

17. J. R. Smith, Proc. Symp. Nuclear Data Problems for Thermal Reactor Applications, Brookhaven National Laboratory, 1978, EPRI-NP-1093, Electric Power Research Institue (1979), p. 5-1.

18. R. R. Spencer, R. Gwin, and R. Ingle, Nucl. Sci. Eng. 80, 603 (1982).

19. F. Pleasonton, R. L. Ferguson, and H. W. Schmitt, private communication (April 1982).

20. R. L. Walsh and J. W. Boldeman, Nucl. Phys. A276, 189 (1977).

NEW EXPERIMENTAL DATA ON THE ENERGY SPECTRUM OF ^{252}Cf SPONTANEOUS FISSION PROMPT NEUTRONS *

M. V. Blinov, G. S. Boykov, V. A. Vitenko

V. G. Khlopin Radium Institute,
197022 Leningrad, USSR

The spectrum of ^{252}Cf spontaneous fission prompt neutrons has been measured in the neutron energy range 10 keV-7 MeV by the time-of-flight method by means of a fast ionization chamber with ^{235}U layers as a neutron detector. The results of the measurements are described by the expression $N(E) \sim E^{0.50 \pm 0.01} . \exp(E/1.418 \pm 0.024)$.

/^{252}Cf, standard, fission neutron spectrum, energy range 10 keV-7 MeV, neutron detector, ^{235}U fission cross-section/.

Introduction

The demands in application of a standard energy spectrum of neutrons are great enough both in science and technology. The IAEA has recommended to use as such a standard the spectrum of ^{252}Cf spontaneous fission prompt neutrons which is very convenient by many characteristics. It is already widely used though the accuracy with which it is known is not yet satisfactory for many tasks. One of the main requirements for refinement of this standard is the determination of a more exact shape of the spectrum in the low energy range (less than 1 MeV).

Reported in this paper new experimental data about the spectrum of ^{252}Cf fission neutrons have been obtained for the energy range 10 keV-7 MeV by the time-of-flight method by use of ^{235}U (n, f) reaction for detection of neutrons. The application of this thresholdless reaction with quite smooth and well-known dependence of the cross-section on the energy of neutrons enables to obtain in one experiment and with one detector information in a very broad energy range including the low energy region. Our preliminary data about the spectrum obtained by an analagous but less perfect than in the present work method have been reported earlier [1]. The characteristics of the spectrometer, of neutron and fragment detectors have been substantially improved compared to the ones applied in work [1]. The time resolution, the speed and the efficiency of registration have been increased, the random coincidence background has been substantially lowered and considerably decreased the influence of scattered neutrons.

Experimental method

An ionization chamber with ^{235}U layers (the assembly 100 mm in diameter, height - 12 mm) was made as a fast detector for neutron registration. Uranium containing 99.9 % of ^{235}U was deposited on both sides of thin aluminium foils (thickness 0.05 mm). The homogeneity of the layers was within the limits $\pm$ 5 % (with average density 1 mg/cm^2) and their average weights within the limits $\pm$ 1 %. The full weight of uranium in the chamber was 0.8 g. The walls of the chamber were made of cadmium foil 0.2 mm thick. During the experiment methane at atmospheric pressure was blown through the chamber. Separation of fragments pulses from alpha-particles in the chamder was good enough. But fragments registration efficiency was only 85 % because the threshold was set to suppress the background from superposition of alpha-particles. The own background of the chamber under working conditions was $2 \cdot 10^{-3}$ pulses/second.

A miniature fast ionization chamber was applied as californium fission fragments detector. Californium layers were 4 mm in diameter and 0.75 μg ($5.1 \cdot 10^5$ fiss/s) and 0.3 μg ($2.0 \cdot 10^5$ fiss/s) in weight. The contribution of spontaneous fissions of other californium isotopes and of curium isotopes

*) This work was supported by International Atomic Energy Agency (research contract 2048/RB and agreement 2791/CF).

K. H. Böckhoff (ed.), Nuclear Data for Science and Technology, 479–483.

was less than 0.2 %. Fragments registration efficiency was more than 99 %. The full time resolution of the spectrometer was 1.5 ns; the main contribution to the uncertainty being from the uranium chamber. Calibration of the time scale was done with an accuracy 0.1 % with a quarz generator. The time "zero" was determined by three methods - by the spectra coincidence on different flight bases, by simultaneous application of pulses from the generator to the current collector electrodes of the californium and uranium chambers and by replacement of uranium layers in the working chamber for one layer of a smaller diameter. All the three methods showed a good agreement and the precision of the time "zero" determination was ± 0.2 ns in consequence.

The dependence of neutron detector efficiency on the energy of neutrons was determined by the dependence of ^{235}U fission cross-sections which had the uncertainty of value not more than 4 % in the investigated energy range. This uranium ionization chamber had an important abvantage in comparison with other types of detectors due to impossibility of neutron registration by any other reactions on uranium or other elements of construction materials. The dependence of anisotropy of fission fragments emission from the uranium layer on the energy of neutrons resulted in the efficiency dependence less than by 1.5 %. The data on the cross-section values of this reaction were taken from the file ENDF/B-V.

Measurements. Data analysis

Neutron spectra were measured on three flight distances - 25, 50 and 100 cm. In order to decrease the random coincidence background the measurements were carried out in an experimental room of large volume and the walls of the neutron detectors were made of thin cadmium foil. The values of the random coincidence background and of the spectrum intensity were equal at the energy 10 keV. The intensity of the apparatus spectrum for the 25 cm base was 8000 events/channel at a maximum while the value of the random coincidence background was 3 events/channel.

In connection with high intensity of spontaneous fissions of the applied californium layers the background depending on true-random coincidences was rather high. For example, at the energy 500 keV it amounted about 30 % of the value of the effect. For suppression of this background the pile up controller was used, which analyzed time intervals between fission fragments and excluded the events for which the interval was less than 200 ns. In consequence the background of true-random coincidences was reduced to 1 % without distortion of the spectrum. The residue background is due to the dead time of the californium chamber, that equals 25 ns.

Good agreement of the spectra obtained at different flight distances was an important criterion of correctness of the measurements. On each flight base the spectrum was measured several times. The experimental data obtained at one and the same flight distance were compared and, since no systematic differences exceeding statistical uncertainty were found, the results of the series were summed up.

During the whole time of measurements in every 5-10 hours "zero" drift was controlled by the reference mark peak and, if necessary, fine adjustment was done. Before and after each series the time resolution of the chambers was controlled, which stayed practically unchangeable throughout the whole time of measurements.

When processing the experimental results corrections were made for neutron scattering by the fragment and neutron detectors and by the environment. As the ionization chambers both of the source and of the neutron detector were designed taking into consideration the decrease of their masses (the weights of the chambers are 0.5 g and 65 g, respectively), the corrections for neutron scattering by the chambers were small. They were easily accountable by calculations using single interaction approximation. The values of these corrections changed from 3 % at the energy 10 keV to 0.7 % at 1 MeV and slightly depended on the distance. The correction functions for neutron scattering by the air medium were found by calculations using program [2] which was checked experimentally. The correction for scattering by air changed for the 25 cm base from 30 % at E = 10 keV to 0.5 % at E = 1 MeV. At higher energies the value of the corrections was very small. The correction functions for neutron scattering are presented in fig. 1. Corrections for time

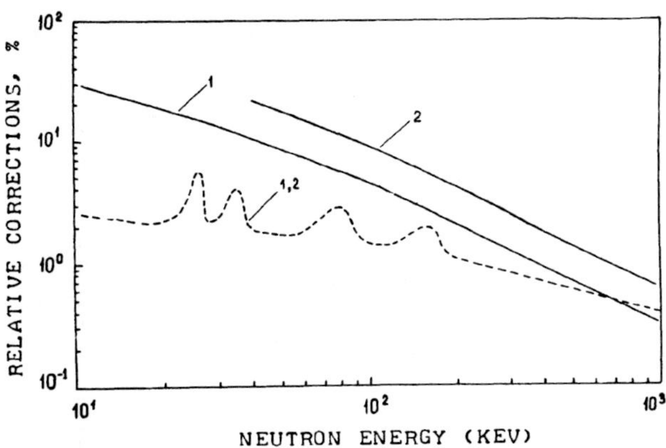

Fig. 1. Correction functions for neutron scattering by the air medium (solid lines) and by detectors (dotted curves) for flight distances: 1 - 25 cm and 2 - 50 cm.

resolution were introduced in the results. In analysis of the errors an account was taken of the statistical errors, errors in determination of the flight distance, of the time scale calibration, of the fission cross-section values, of the corrections for scattering. The errors of the corrections for scattering were mostly connected with uncertainty in using of nuclear data and with approximate character of the calculation. The main contribution in the error of spectrum intensity was due to statistical uncertainty at low energies and at high energies it was also due to uncertainty in the determination of the time "zero". At medium energies the error was explained in the main by uncertainty of the knowledge of the cross- section values. The relative errors in determination of the spectrum intensity are presented in fig. 2. The error in determination of the energy was connected with time scale calibration error, with time resolution, error in determination of the flight distance. The energy resolution changed according to neutron energy and flight base and it was less than 10 %.

Results and discussion

The results of the measurements are presented in fig. 3. The total error in intensity determination changed from ± 16 % (1σ) at the energy 10 keV to ± 6 % (1σ) at the energy 7 MeV.

In fig. 4 experimental data are presented in termes of a ratio to Maxwellian distribution with T = 1.42 MeV. As it is seen from fig. 4, the one-parameter Maxwellian expression desc-

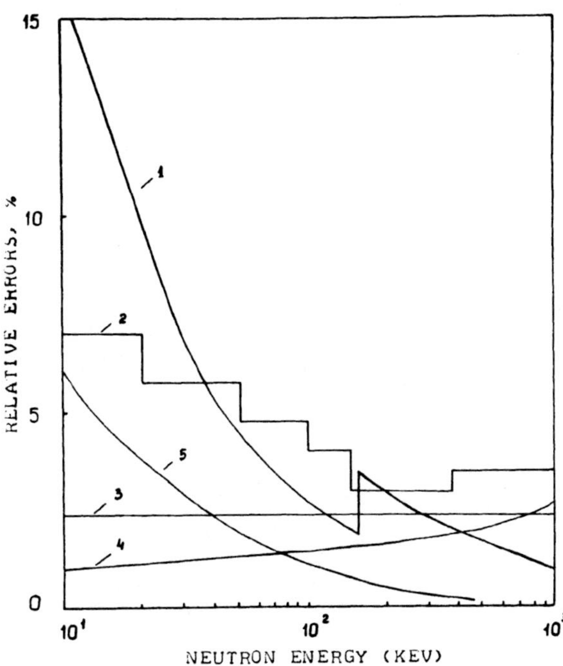

Fig. 2. The relative errors of spectrum intensity: 1 - statistical error, 2 - ^{235}U (n, f) reaction cross-section error, 3 - uncertainty of the flight base, 4 - error in determination of the time of flight, and 5 - uncertainty of determination of the correction for scattering in the air.

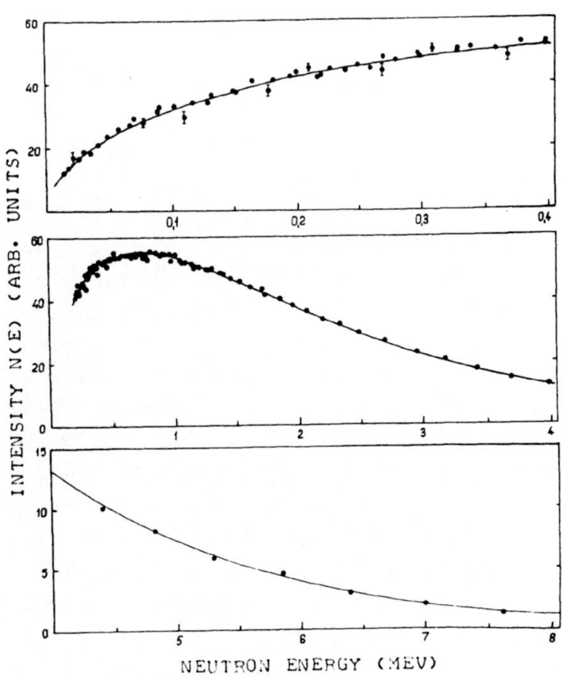

Fig. 3. Spectrum of ^{252}Cf spontaneous fission neutrons: points - experimental results, solid curve - Maxwellian distribution with T = 1.42 MeV.

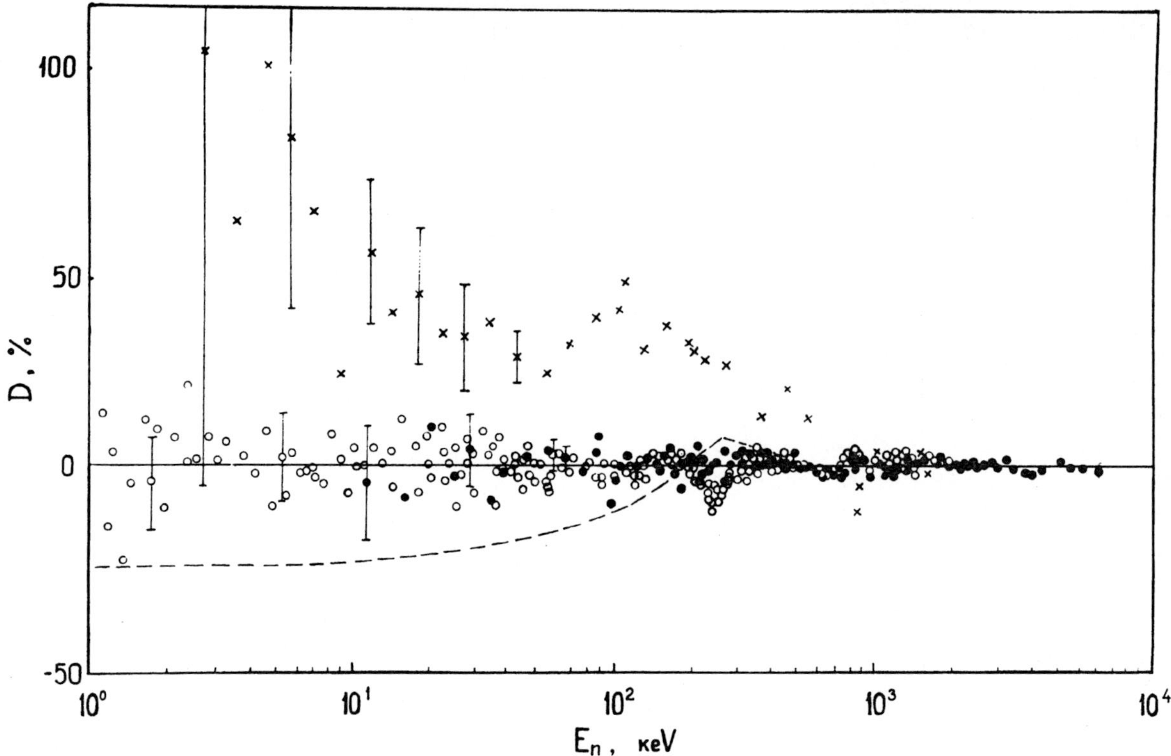

Fig. 4. Ratio to Maxwellian distribution (T = 1.42 MeV) of the experimental results (D): present measurements (•), data of the work [3] (O) and paper [4] (X). Dashed curve - results of evaluation.

ribes spectrum shape well enough not only in the energy range 1 ÷ 7 MeV but in the range of low energies too. Determination of the parameter T for our data by the method of least squares gave the value 1.418 ± 0.024 MeV. In fig. 4 there are also data (1 keV - 1 MeV) which have been obtained earlier using a low-background spectrometer on the base of a crystal ^{6}LiI (Eu) [3]. The full time resolution of the spectrometer in this work was 1 ns. Thin crystals with the thickness 2 and 4 mm were used. Multiple scattering in the crystals and in adjacent parts of the photomultiplier were accounted experimentally and by calculations using the Monte-Carlo method. Corrections were introduced for neutron scattering in the environment - the air and the walls of the rooms. A characteristic feature of the spectrometer was the presence of a fast path of analysis of pulses' amplitudes and durations in the neutron channel, that enabled to decrease by an order the gamma-quanta pulses superposition background and the random coincidences background. It is seen from fig. 4 that the data obtained by detectors of two types differ little from each other and in generel are close to Maxwellian distribution. A considerable deviation in the region of 250 keV obtained in the measurements with a lithium detector is not proved by the results of measurements with the uranium chamber. It seems that this deviation is rather connected with inaccuracy of the applied value of the reaction ^{6}Li (n, α) cross-section in the resonance region.

The results of the measurements by means of two different neutron detectors in the low energy region show that the spectrum can be approximated by the expression

$$N(E) \sim E^{0.50\pm0.01} \cdot \exp(E/1.418\pm0.24)$$

In fig. 4 there are also the results of the work [4] done by means of lithium glasses, which considerably differs from the results of this work and of work [3]. It can be attributed evidently to an important influence of the scattered neutrons in the experimental conditions of work [4].

If we compare the results of the given work

with the evalution [5], they coincide well enough by the value of T, but in the low energy region there is an essential disagreement.

The authors express their acknowledgement to V. I. Yurevich for the calculation of corrections for neutron scattering, to A. S. Krivokhatskiy, B. M. Alexandrov, P. S. Soloshenkov for preparation of californium and uranium layers and to A. S. Veshchikov for help at measurements.

References

1. M. V. Blinov, V. A. Vitenko, V. T. Touse, Neutron Standards and Applications (Proc. Intern. Symposium. Gaithersburg, 1977) NBS Special Publ. - 493 194 (1977).
2. M. V. Blinov, V. A. Vitenko, V. N. Dushin, V. I. Yurevich, Nucl. Instr. and Meth. (in press).
3. M. V. Blinov, V. A. Vitenko, V. I. Yurevich, Neutron Physics (Proc. Fifth All-Union Conference on Neutron Physics, Kiev, 1980) Moscow, 1980, part 3, p. 109 (in Russian).
4. T. W. Meadows, Phys. Rev. 157 N 4 1076 (1967).
5. L. Stewart, C. M. Eisenhauer, Neutron Standarts and Applications (Proc. Intern. Symposium, Gaithersburg, 1977) NBS Special Publ. - 493 198 (1977).

THE NEUTRON ENERGY SPECTRUM FROM THE SPONTANEOUS FISSION OF CF-252 IN THE ENERGY RANGE 2 MEV $\leq E_n \leq$ 14 MEV

R. Böttger, H. Klein
Physikalisch-Technische Bundesanstalt (PTB)
Bundesallee 100, D-3300 Braunschweig

A. Chalupka, B. Strohmaier
Institut für Radiumforschung und Kernphysik (IRK)
Boltzmanngasse 3, A-1090 Wien, Austria

Discrepancies in the description of the neutron energy spectrum of Cf-252 above 5 MeV have given rise to an investigation of this part of the spectrum. The new PTB multi-angle neutron time-of-flight spectrometer and an improved version of the IRK low mass ionization chamber were combined for the measurements. The influence of the detection efficiency for fission fragments on the neutron energy spectrum measured in coincidence with a TOF-detector was carefully studied. The measured spectra were corrected for background due to uncorrelated stops and for the efficiency of the fission chamber. The shape of the spectrum is best described in the energy range 3 MeV $\leq E_n \leq$ 13 MeV by a Maxwellian distribution with the temperature parameter E_o = 1.355 MeV.

[^{252}Cf, prompt fission neutron spectrum, 2 - 14 MeV, TOF-technique, ionization chamber, background analysis, Maxwell distribution]

Introduction

The Cf-252 neutron source is used as a reference standard in a large number of applications such as the calibration of the efficiency of neutron detectors /1/ and the energy distribution of prompt fission neutrons /2/ and the investigation of activation reactions for reactor dosimetry /3/. Existing discrepancies in the literature /4/ concerning the parametrization of the neutron energy distribution from the spontaneous fission of Cf-252 have given rise to further investigations of the spectrum. The multi-angle spectrometer for precise neutron time-of-flight (TOF) measurements at the PTB /5/ is well suited for the investigation especially of the high energy part of the spectrum. An improved "golden" version of the Vienna low mass, fast ionization chamber /6/ has been used to detect the fission fragments and to generate the fast timing signal.

First the influence of the structural material and the efficiency for the detection of fission products on the neutron energy spectrum had to be studied. Then an improved analysis of the measured TOF-spectra including the renormalization of a considerable number of events with uncorrelated stops had to be applied.

Finally, we compare the analyzed distributions with the NBS evaluation /7/ and measurements recently published /8/. A more detailed description of the experimental method and the analysis of the TOF-spectra will be published elsewhere.

Experimental Method

The neutron energy spectrum from the Cf-252 source was simultaneously measured with four large volume NE213 scintillation detectors (5.08 cm x 25.4 cm Ø) of the multi-angle time-of-flight spectrometer. The angle between two neighbouring detectors was 12.5 deg; the flight path of 12 m was the same for all detectors in all TOF experiments. The collimator /9/ consists of water tanks with polyethylene liners. An efficient n-γ- discrimination was installed to separate the few n-events (2 neutrons/s) from the approximately 250 γ- events/s resulting mainly from the surroundings of the unshielded neutron detectors. A table-cut could be set in the two-dimensional display of the light output L(E) versus the n-γ separation spectrum in an on-line or off-line analysis /10/. In addition for each event the time-of-flight (TOF) and the pulse height of the energy loss signal of one of the two fission fragments ΔE were recorded and stored on tape.

About 10^5 fissions/s from the Cf-252 source inside a small parallel-plate ionization chamber were registered. As commonly used in TOF experiments, the higher detector rate serves as STOP-input for the time-to-amplitude converters (TAC) after being properly delayed. For all detectors, the time range for the TACs was chosen as 1 μs with a channel width of about 1 ns. The integral nonlinearity of the TAC-ADC-system was checked with two different time calibration systems and found to be less than ± 0.25 ns.

The whole running time of the experiment was 324 h, regularly interrupted by measurements to check the stability of the system i.e. thresholds in both the neutron and fission fragment detector, the quality of the n-γ- discrimination and the time calibration.

The Fission Fragment Detector

Before the TOF-spectra could be recorded at the spectrometer, the properties of the fission chamber had to be carefully investigated. The Cf-252 source is deposited onto a highly polished golden backing, the radius of the active area is about 5 mm. The source emitted about 4 x 10^5 neutrons/s isotropically into the laboratory system. The effect of neutron absorption and production in the golden chamber walls and the backing was calculated using experimental total and differential elastic cross sections and neutron production spectra evaluated with the code

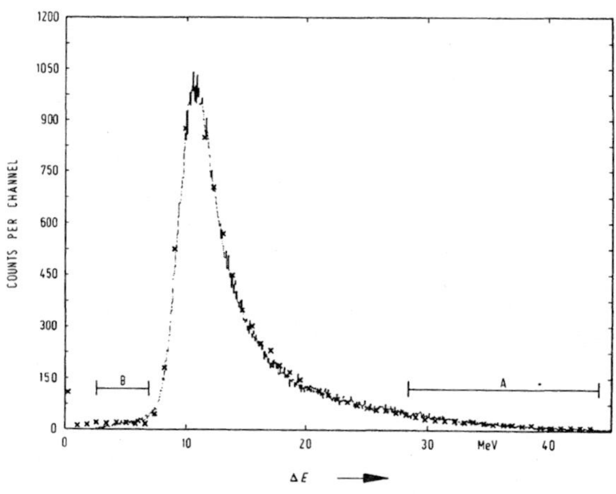

Fig. 1: Energy loss spectrum ΔE of fission fragments in the parallel-plate ionization chamber (histogram) compared with a calculated spectrum (x).

K. H. Böckhoff (ed.), Nuclear Data for Science and Technology, 484–487.

STAPRE /11/. This contribution was found to be negligible for energies of $E_n > 2$ MeV, but due to its strong increase has to be taken into account for energies below 1 MeV. The pulses after the preamplifier had rise times of about 3 ns and a length of 15 ns. ΔE-signals were derived after charge integration in a linear gate and stretcher with a gate time of about 20 ns. The gate was allowed to trigger only if a neutron was measured in one of the scintillation detectors. A typical ΔE-spectrum measured in coincidence with neutrons is shown in fig.1. A time resolution of 1.5 ns is achieved for the γ-peak if measured with a large scintillation detector with a bias level equivalent to 470 keV (± 5 %) electron energy which is used in the TOF experiment.

Both the efficiency for fragment detection and the dead time loss must be known for an absolute measurement of the neutron energy spectrum. The non-extended dead time was determined by means of the time interval distribution method /12/ as being 530 ± 10ns. In a time-of-flight arrangement with a flight path of 230 cm and a small, heavily shielded scintillation detector (3.81 cm x 3.81 cm Ø) the efficiency for fission fragment detection $\varepsilon_{cF} = N_{NF}/N_N$ was determined as a function of the angle between the flight path and the chambers axis. The number of neutron-fission coincidences N_{NF} must be corrected for dead time loss and random coincidences. For both N_{NF} and N_N (number of detected neutrons), contributions from background have to be taken into account which were measured in a separate run with the n-detector well shielded against the Cf-source. This angle dependent efficiency ε_{cF} is shown in fig. 2 together with the results of a computer simulation, the aim of which was to calculate ΔE-pulse height spectra and chamber efficiencies for various experimental conditions. The code is based on experimental data (e.g. the distribution of the fragments with respect to mass M_{fr} and energy E_{fr} /13/, the average number of neutrons per fragment $\bar{\nu}(M_{fr})$ and the average neutron energy $\bar{\eta}(M_{fr})$ /14/, $dE_{fr}/dx(M_{fr}, E_{fr})$ values in Au and CH_4 /15/, scission neutron data /13/) assuming isotropic Maxwellian neutron emission spectra in the fragments' frame. The only variable parameter is the average roughness $\bar{R}$ of the surface onto which the CF-252 source is deposited. Data in fig. 1 and fig. 2 were obtained with $\bar{R} = 0.7$ µm which also gives a total efficiency of $\varepsilon_F = 95.4 \pm 0.3$ % which is used for the normalization of the theoretical reference neutron energy distribution. As the maximum angle between the large area detectors and the Cf-252 source is 20 deg in the multi-detector experiment, no individual efficiency correction had to be applied for the detectors because $\varepsilon_{cF}(\vartheta)$ is almost constant up to 30 deg. The angle dependence of ε_{cF} and the results of our calculations support the assumption /6/ that the fission fragments not detected are predominantly emitted into an angle between 85 deg and 90 deg and would have caused a contribution in the high energy tail (region A in fig. 1) of the ΔE-pulse height spectrum. From the correlated neutron TOF-spectra, it can also be deduced that the low energy tail of the pulse height spectrum (region B in fig. 1) can be associated with fission tracks parallel to the backing surface losing their energy mainly within the source or backing. From correlation studies between neutron TOF-spectra and ΔE-fission pulse heights it can be concluded that there is no contribution from α-events in the ΔE- spectrum with a threshold set three times higher than the α-pulse height.

The Large Volume Neutron NE213 Detectors

Within the limits of the statistical accuracy the detectors used to measure the neutron spectra did not deviate from one another as far as neutron detection efficiency or time resolution of the γ-peak are concerned.

The threshold was determined using γ-sources /16/, the absolute efficiency ε_d for this threshold was measured with monoenergetic neutrons via the reaction D(d,n)He-3 for various energies relative to the fluence measurement with a proton-recoil telescope, which was also used in an international intercomparison /17/. Data of calibration measurements /5/ were available up to 14 MeV, thus limiting the investigation of the high energy part of the Cf-252 neutron spectrum.

The air attenuation of 10 % on an average with several resonances and the disturbances due to the collimator, were taken into account chosing the same geometry for the calibration and the measurement.

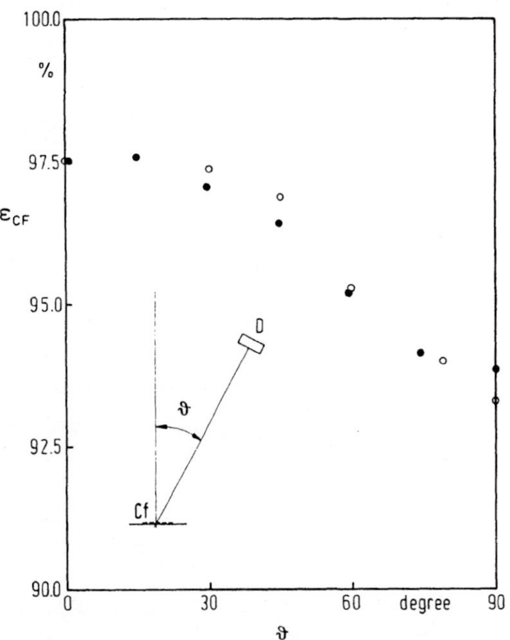

Fig. 2: Efficiency ε_{cF} for fragment detection in dependence on the position of the coincident neutron detector for a threshold of $E_0 = 0.5$ MeV. Filled symbols indicate calculations.

Analysis of TOF-Spectra

Fig. 3 shows a neutron TOF-spectrum in a time inverse scale accumulated in a running time of 324 h for each detector with a bias level equivalent to 470 keV electron energy. Two different corrections have to be taken into account /18/:

1.) Contributions from random background $I_R(K)$ such as γ-events for which the n-γ- discrimination had failed or from neutrons the associated fission fragment of which could not be detected. Assuming that the detected fission rate A' can be characterized by the Poisson statistics, the shape of this random time distribution is fully determined by the undisturbed rate A' and the dead time τ of the fission chamber electronics.

To obtain the absolute scale it is fitted to the measured distribution between the γ-peak and the end of the TAC-range T_R and subtracted afterwards (fraction a in fig. 3).

2.) After this correction, another type of background is left which is considerably larger. It results in a typical, horizontal time distribution left of the properly correlated neutron TOF distribution and is strongly channel dependent. It is caused by valid neutron start signals with an uncorrelated stop signal. This background must be calculated and subtracted and can exceed the pure dead time loss from the fission chamber in the case of $\tau < T_R$.

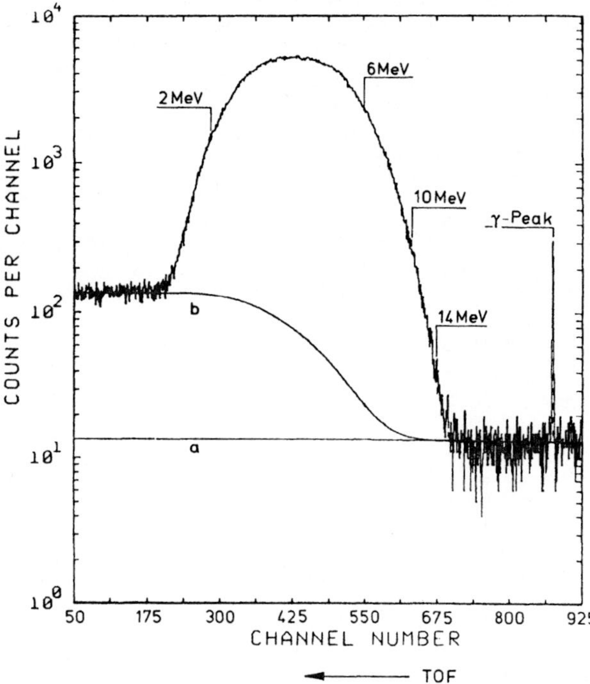

Fig. 3: Neutron TOF-spectrum as measured for Cf-252 for a 12 m flight path (histogram). The random background (a) and the background due to uncorrelated stops (b - a) are calculated (channel width 0.972 ns).

The corrected intensity distribution $I_c(K)$ can in the first order be calculated from the measured distribution I(K) diminished by the uncorrelated background (fraction b - a in fig. 3) for $t(K) \leq \tau$ according to

$$I_c(K) = (1+A'\tau)\cdot\left\{I'(K)-A'\Delta t \sum_{\nu > K}^{K_{max}} I'(\nu)\right\}$$

with the channel width Δt and

$$I'(K) = I(K) - I_R(K).$$

For $\tau \leq t(K) \leq T_R$ an additional renormalization factor $\exp\{A'(t(K)-\tau)\}$ has to be applied and changes the time-of-flight distribution, especially in the high energy tail.

As can be seen from fig. 3 there is no background left after applying the corrections.

In principle it is also possible to renormalize the fraction of neutrons $(1-\varepsilon_{CF})$ without associated fission signals from the Cf-252 chamber which are partially subtracted in the random background. On the assumption of a specific angle-dependent sensitivity to fission fragments (see sect. on fission detector) it is necessary to correct the neutron energy distribution to full 4π fission detection geometry by generating in an off-line analysis that neutron energy spectrum which is correlated to the high energy part of the ΔE-fission spectrum. The normalization of this spectrum to be added is $(1-\varepsilon_{CF})/\varepsilon_{CF}$.

This correction considerably changes the spectrum for higher efficiency losses due to the strong anisotropic emission from fully accelerated fragments in the laboratory system.

Analysis of the Neutron Energy Distribution

The corrected n-TOF distribution is converted into an energy distribution $I(E_n)$ by means of the position of the γ-peak, time channel width Δt and length of flight path as input data. A Cf-252 neutron energy distribution $I(E_n)/\varepsilon_d(E_n)$ covering the energy range between 2 MeV and 14 MeV is analyzed according to fig. 4 assuming a pure Maxwellian distribution $N(E) = c\ \sqrt{E}\ \exp(E/E_o)$.
The best fit between 3 MeV and 13 MeV yields a temperature parameter E = 1.355 ± 0.015 MeV. For comparison, a Maxwellian distribution with E_o = 1.425 MeV, normalized between 3 MeV and 4 MeV, is shown and obviously does not fit the data.

In fig. 5 the deviations between the experimentally determined distribution and the following three reference spectra

1.) NBS segment fit
2.) Maxw. distribution: E_o = 1.355 MeV
3.) Maxw. distribution: E_o = 1.425 MeV

are displayed in the energy range
$3\ \text{MeV} \leq E_n \leq 13\ \text{MeV}$.

For the absolute normalization of the reference spectra, the neutron emission from the Cf-source is assumed to be isotropical in the laboratory system with the rate $A \cdot \bar{\nu}$. (A corrected for dead time and finite detection efficiency; $\bar{\nu}$ = 3.77 /19/).

The shape of the neutron energy distribution is described by the NBS segment fit as well as the pure Maxwellian with E_o = 1.355 MeV between 6 MeV and 13 MeV, but the scaling factors differ by about 15 %.

There is no justification for describing our data by a pure Maxwellian with E_o = 1.425 MeV due to deviations from +10 % to -33 % in the energy range investigated. With a scaling factor of 1.08 applied to the reference spectrum with E_o = 1.355 MeV which is extracted from the shape analysis, we find the best agreement in shape as well as in the absolute scale with deviations not exceeding 5 % between 3 MeV and 12 MeV.

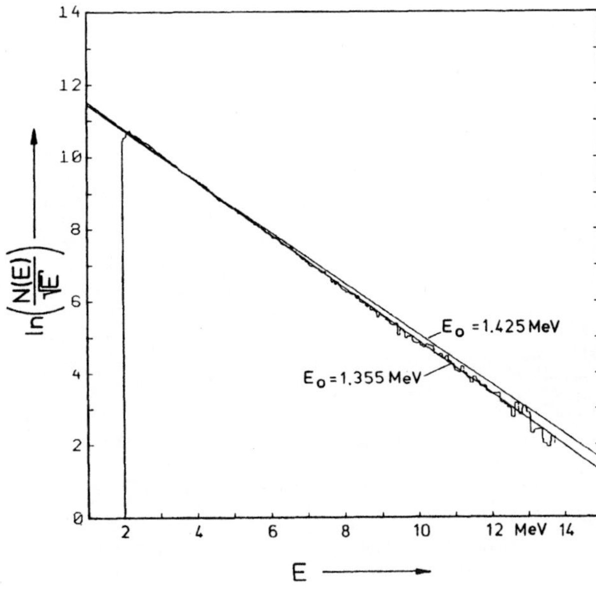

Fig. 4: The neutron energy spectrum of Cf-252 (histogram) in comparison with two Maxwellian distributions.

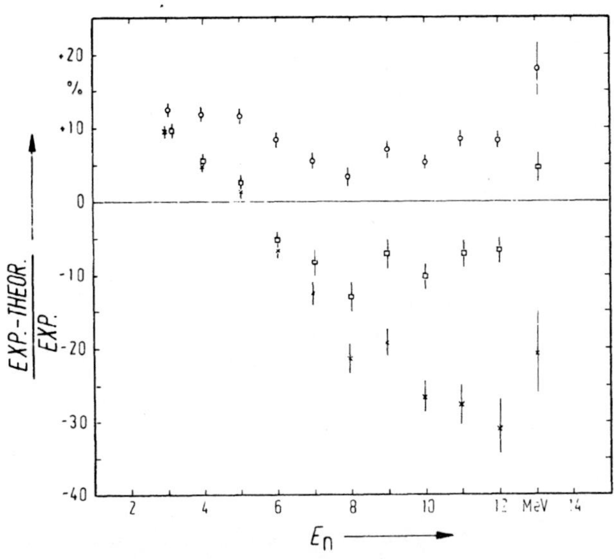

Fig. 5: The relative deviation of the normalized neutron yield (EXP) to the NBS segment fit (▫) and Maxwellian-distributions with temperature parameters E_o = 1.355 MeV (o) or 1.425 MeV (x) (THEOR). The error bars indicate the full uncertainties due to statistics and normalization.

Conclusions

The widespread use of Cf-252 requires an exact knowledge of the neutron energy distribution up to high energies. Optimal experimental conditions are fulfilled if the TOF technique is applied with very long flight paths. The efficiency of the neutron detector as well as the fission detector should be carefully investigated. Due to the strong anisotropic neutron emission from the fission fragments, the influence of the loss in fission detection on the neutron energy spectrum must be considered.

Another reason for discrepancies between various experiments may be the determination of the uncorrelated and random background in the TOF-spectra, expecially its time-dependent renormalization factor in the case of high fission rates.

The measured neutron energy distribution can suitably be described by a pure Maxwellian distribution with a temperature parameter E_o = 1.355 MeV. The NBS segment fit is confirmed although there are problems with the absolute scale as mentioned above, however, our results contradict a pure Maxwellian distribution with E_o = 1.425 MeV.

References

/ 1/ A. Smith, P. Günther, R. Sjoblom; NIM 140 (1977) 397-450

/ 2/ A.B. Smith, P. Günther, G. Winkler, R. McKnight; ANL/NDM-50 (1979)

/ 3/ W. Mannhart; contribution to this conference

/ 4/ M.V. Blinov; INDC (NDS) - 114/GT (1980) 79-106

/ 5/ R. Böttger, H.J. Brede, M. Cosack, G. Dietze, R. Jahr, H. Klein, H. Schölermann, B.R.L. Siebert; contribution to this conference

/ 6/ A. Chalupka; NIM 164 (1979) 105-112

/ 7/ J.A. Grundl, C.M. Eisenhauer; NBS - 425 (1975) 250-253

/ 8/ J.W. Boldemann, D. Culley, J.J. Cawley; Trans. Am. Nucl. Soc. 32 (1979) 733-735

/ 9/ D. Schlegel-Bickmann, G. Dietze, H. Schölermann; NIM 169 (1980) 517-526

/10/ H. Klein, D. Dietze, B.R.L. Siebert, W. Bretfeld; NIM 169 (1980) 359-367

/11/ B. Strohmaier, M. Uhl; Proc. Winter Course on Nuclear Theory for Applications, Trieste, 1978, IAEA-SMR-43 (1980) 313-326

/12/ J.W. Müller; NIM 112 (1973) 47-57

/13/ M. Forte; Phys. Rev. B 14 (1976) 956-968

/14/ V.M. Piksaikin, P.P. D'yachenko, L.S. Kutsaeva; Sov. J. Nucl. Phys. 25 (1977) 385-389

/15/ L.C. Northcliffe, R.F. Schilling, Nucl. Data Tables A7 (1970) 233-463

/16/ G. Dietze, H. Klein; NIM 193 (1982) 549-556

/17/ V.D. Huynh; Metrologia 16 (1980) 31-49

/18/ A. Chalupka; NIM 165 (1979) 103-108

/19/ R.R. Spencer; NBS - 594 (1980) 728-732

THE HIGH-ENERGETIC PART OF THE NEUTRON SPECTRUM FROM SPONTANEOUS FISSION OF ^{252}Cf

H. Märten, D. Seeliger, and B. Stobinski

Technische Universität Dresden
Sektion Physik
Mommsenstr. 13, DDR-8027 Dresden
German Democratic Republic

The high-energy end of the neutron spectrum from spontaneous fission of ^{252}Cf has been measured by the use of a very sensitive neutron spectrometer. The experimental data which were corrected for different apparatus effects indicate a hard emission component predominant at energies higher than 20 MeV. In this energy range, the measured spectrum cannot be described in the framework of a complex cascade evaporation model, whereas the calculated data agree with experimental ones up to 20 MeV satisfactorily. Non-equilibrium emission of fission neutrons is considered in interpretation.

[$^{252}Cf(sf)$, prompt fission neutron spectrum, E_n = 10-30 MeV]

Introduction

Properly fission neutron spectra extend to relatively high emission energies. Considering the energy balance in fission, on principle, energies up to about 40 MeV are conceivable. Because of the very low emission cross sections in the high-energy range, fission neutron spectra are measurable up to about 15 MeV commonly. The physical interest in this matter stimulated our effort to determine the high-energy end of fission neutron spectra experimentally as high as possible [29, 30]. The importance of such measurements is confirmed in the case of the neutron emission spectrum from spontaneous fission of ^{252}Cf, because it was recommended as a standard [1]. Hitherto, its validity at high emission energies is not founded due to the high experimental errors above 10 MeV. Data of different authors diverge substantially [2]. The result of the NBS evaluation [3] was stated by a correction function $\mu(E)$ with reference to the Maxwellian distribution with kT = 1,420 MeV:

$$N(E) = \mu(E)\cdot 0,6672\cdot E^{1/2}\cdot \exp(-E/1,420). \qquad (1)$$

Some recent measurements [4] and evaluations [5] are consistent with that of Grundl and Eisenhauer (NBS) [3].

The main mechanism of neutron emission in low-energy fission reactions is the evaporation from fully accelerated fragments. Experimental results on the prompt neutron anisotropy have led to the conclusion that a small fraction (about 10 %) of the total number of fission neutrons is emitted isotropically in the compound nucleus frame [6]. Further experimental data on scission neutron emission are poor and partially contradictory [8, 9, 10, 11]. Corresponding theoretical investigations have been based on typical rapid changes of nuclear potential in fission (move of the fissioning nucleus from saddle to scission point [12], transition of strongly deformed fragments into the equilibrium state [13]). Hitherto, the partial spectra of the different eventual kinds of scission neutrons are not founded theoretically. One way to obtain more informations about neutron emission in fission is the measurement of the high-energy spectrum parts, because scission neutrons which should be emitted due to strong one-particle excitations in fission may influence the neutron spectrum at high energies especially.

Experimental method

The employed high-sensitive neutron spectrometer, its coupling to a minicomputer as well as the analysis of experimental data were already described in detail [14, 15]. Therefore, only a brief summary is given here. A high-efficient neutron detector with a voluminous NE 213 scintillator is located in a heavy shielding. The electronic system for particle discrimination by the charge comparison method is used to suppress the background counts of the detector caused by γ-rays and penetrating components of the cosmic rays. Especially cosmic myons with energies around 1 GeV give rise to a background part with about 3,5 s^{-1} event rate and an average pulse of about 25 MeV with reference to proton recoil energy (PRE). The n/μ-discrimination method enables the suppression of the cosmic background to less than 0,2 % in the region of the myon hump of the pulse height background spectrum (see Fig. 1).

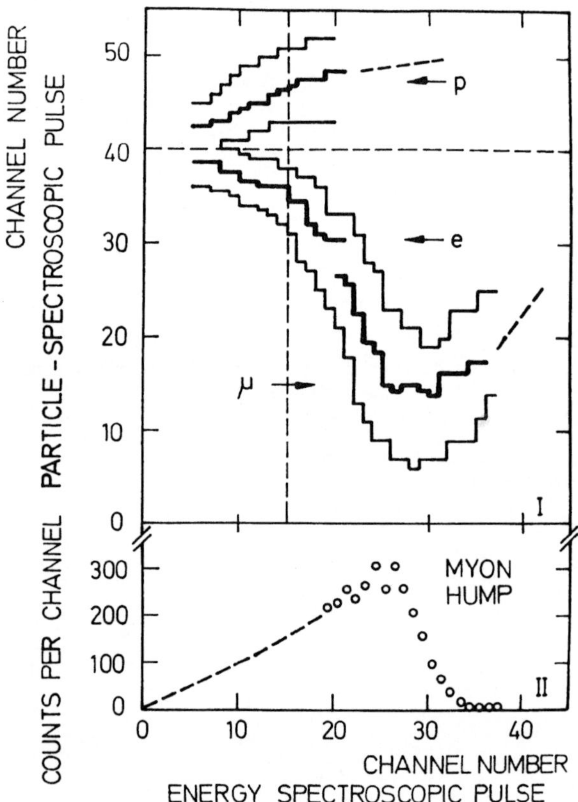

Fig. 1/I. Two-dimensional representation of the performance of the particle discrimination system (p, e, μ - particle branches of recoil protons, Compton electrons and cosmic myons respectively).

Fig. 1/II. Pulse height spectrum of cosmic myons.

K. H. Böckhoff (ed.), Nuclear Data for Science and Technology, 488–491.

Besides the use of the electronic n/γ-discrimination method, i. e. effective background suppression, the high sensitivity of the spectrometer is based on the two-dimensional measurement of neutron time-of-flight (TOF) and PRE. In this way, one is able to select the optimum (regarding background conditions) PRE-range for a given TOF-channel or channel range in analysis which is carried out cyclicly in connection with PRE-interval variation. Fig. 2 shows typical neutron-TOF-distributions which were obtained from a two-dimensional (TOF,PRE)-measurement for selected threshold energies.

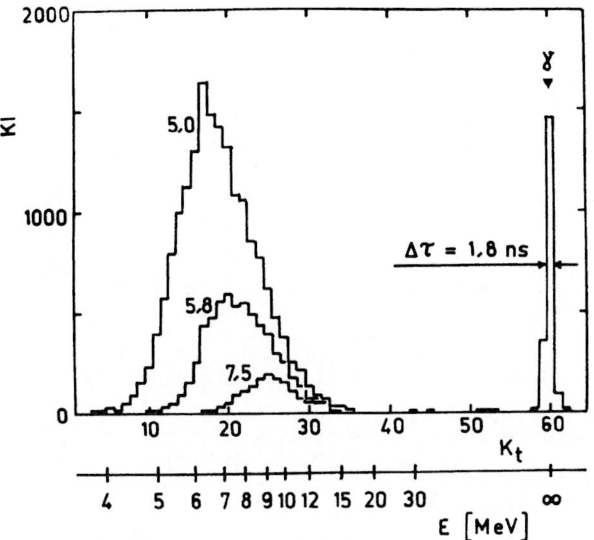

Fig. 2. Typical neutron-TOF-spectra from spontaneous fission of ^{252}Cf deduced from the two-dimensional (TOF,PRE)-measurement for selected PRE-threshold energies (4 m flight path, 62,0 h measuring time; KI - counts per channel; K_t - TOF-channel number; E - neutron energy).

The background level per TOF-unit, PRE-unit and measuring time respectively was found to be smaller than $1{,}5 \cdot 10^{-4}$ ns^{-1} MeV^{-1} h^{-1} in the PRE-ranges which were used in the analysis of the long-time measurement [25]. The stated value illustrates the sensitivity of the experiment.

The zero-time signal is obtained employing a fast ionization chamber [16] for direct fission fragment detection. It is characterized by a very light construction. The fission event rate amounted to $3{,}40 \cdot 10^4$ s^{-1} at the beginning of the measurement. The whole time resolution of the experimental arrangement is 1,8 ns regarding FWHM of the γ-peak. It is somewhat higher for neutrons due to the dimension of the scintillator and, hence, neutron energy dependent. To guarantee a sufficiently good energy resolution for high neutron energies a relatively high flight path is required (more than 4 m).

The spectrometer is coupled to a minicomputer which arranges the control of the two-dimensionally working multi-channel analyser for data acquisition, for the check (regarding TOF-peak and PRE-edge positions) and correction of the spectra as well as their analysis. The corresponding programme system was elaborated by the use of the high-level language FORTRAN 4000/4200 including CAMAC and display application subroutines [15].

The calibration of the time coordinate is carried out by additional measurements using a defined delay device and considering the γ-peak as a fixed TOF-point. The PRE-edge position (point of inflexion) for a given TOF-channel or neutron energy EN corresponds to EN but a systematic deviation because of the distortion of the PRE-response function by multiple detection processes. This effect was studied by the use of the Monte Carlo code NEUCEF [17] after determining an effective parameter which characterizes the finite pulse height resolution of the detector by a fit of calculated PRE-response functions to experimental ones. Considering the corresponding correction factor the calibration of the PRE-coordinate is possible by the use of the measured continuous spectrum itselves due to the energy selection by the TOF-measurement.

Generally the background is a function of TOF. This effect doesn't appear at sufficiently high PRE, i. e. above about 5 MeV in the present case. Therefore, it is possible to reduce the common conception of the alternating measurements with and without sample on the sole measurement with sample. In this case, the background is determined from a defined region of the (TOF,PRE)-plane, where no effect events appear for physical reasons.

The detector efficiency was calculated by the use of the code NEUCEF [17] accepting the light output data of Verbinski et al.[18] and realistic values of the mentioned pulse height resolution parameter and geometric factors. A first measurement of the ^{252}Cf(sf) neutron spectrum, which is known with an uncertainty of less than 2,5 % between 0,4 and 7 MeV [5], was aimed at the comparison of the calculated efficiency data with the measured ones for relatively high bias energies and EN up to 10 MeV. We assumed the NBS evaluated spectrum for efficiency determination. The deduced efficiency functions depending on the bias energy confirm the NEUCEF data absolutely within an error which is EN and PRE-threshold dependent. It amounts to about 5 % in the ranges of best statistics (bias around 4 MeV, EN around 7 MeV) [25]. It is emphasized that the description of the experimental efficiency data is rather good in the PRE-threshold region due to the realistic consideration of resolution effects.

The long-time experiment (1218,5 h measuring time, 4,5 m path of flight) was subdivided in single runs. It is described in Ref.[25] in detail. The deduced energy distribution, i. e. the sum of the spectra from the single runs, which were obtained for PRE-thresholds around 8 MeV, was corrected for TOF-channel width and time resolution. The latter influences the measured spectrum in the high-energy region especially. Dead-time corrections were neglegible because of the low event rate. The results of the experiment are summarized below. Here it should be mentioned that the measured spectrum extends to about 30 MeV. Therefore, the interpretation of the experiment has to be based on an enlarged spectrum calculation which has been carried out in the framework of a complex cascade evaporation model presuming suitable approximations to guarantee a sufficient accuracy at high emission energies. This implies that the first theoretical analysis is founded on the main mechanism of fission neutron emission, i. e. the evaporation from fully accelerated fragments.

Cascade evaporation model

Because of the availability of necessary experimental data, the calculation was performed for different fragment mass numbers A by averaging the initial parameters over corresponding pairs of proton and neutron numbers. To consider the excitation energy distribution $P_o(E^x)$ and the cascade evaporation of fission neutrons by steps i as well as the energy balance, i. e. introduction of the spectrum dependence on the total kinetic energy TKE, one has to generalize the equation of the standard evaporation theory for the description of the emission spectrum f(e) in the center-of-mass frame and obtains

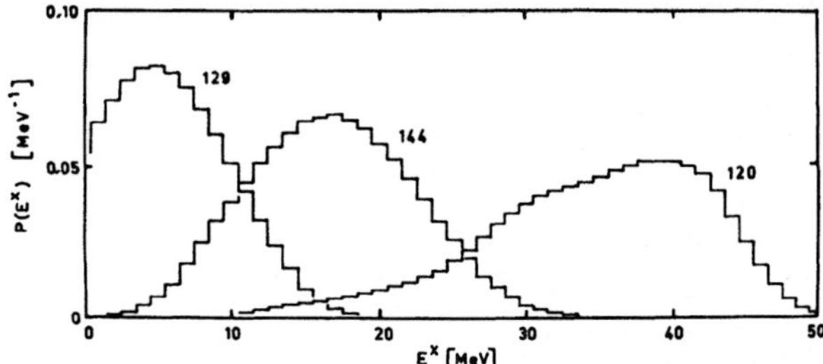

Fig. 3. Initial distributions of excitation energy for typical fragment mass numbers

$$f(e:A,TKE) = \sum_{i=0}^{i_{max}} \int_{B_{ni}}^{\infty} dE^x \cdot f(e,E^x:A-i) \cdot P_i(E^x:A,TKE) \cdot P(A,TKE), \quad (2)$$

where B_n is the neutron separation energy. $P(A,TKE)$ is the occurance probability of fission events with the stated characteristics. The analysis was based on the nuclear-level density description by Ignatyuk et al.[20] including the excitation energy dependence of shell effects. The transformation of eq. (2) into the laboratory system, which was carried out considering a small emission anisotropy in the center-of-mass frame [22], results in the spectrum $F(E:A,TKE)$. The integral energy spectrum is given by

$$N(E) = \sum_{A} \int_{TKE} dTKE \cdot F(E:A,TKE) \cdot P(A,TKE). \quad (3)$$

In the final calculation, we neglected the TKE dependence in the expressions (2) and (3), because this approximation influences the integral spectrum weakly [19]. Consequently, the laboratory system spectrum as a function of A has to be calculated by the use of the average kinetic energy of the fragments with given A. The initial distributions $P_o(E^x:A)$ have been deduced on the base of experimental data on neutron and γ-emission as a function of both A and TKE of the fragments [21] by

$$P_o(E^x:A) = \int_{TKE} dTKE \cdot P(E^x:A,TKE) \cdot P(TKE:A). \quad (4)$$

The $P_o(E^x:A,TKE)$ distributions were assumed to be Gaussian. Fig. 3 shows obtained $P_o(E^x:A)$ for typical A. The dependence of the average emission energy in the center-of-mass frame on A is illustrated in Fig. 4 in comparison with experimental data. The discrepancy between evaporation theory

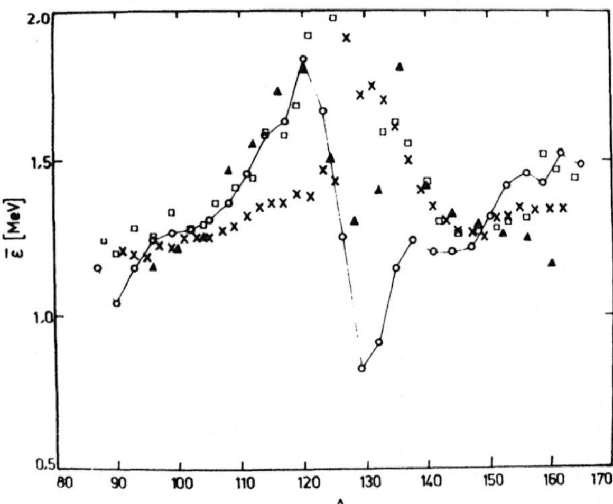

Fig. 4. The calculated average emission energies in the center-of-mass frame (o) in comparison with results deduced from experimental data (□ - Ref.[7], x - Ref.[23], ▲ - Ref.[9]).

and experiment regarding the $\bar{e}(A)$ curve around A = 132 may be explained qualitatively considering the data on the scission neutron yield as a function of both A and TKE obtained by Samyatnin et al.[10] and the fact that the average emission energy of such neutrons is somewhat higher than the corresponding value for evaporated fission neutrons. Further details of the used model, the method of determining the excitation energy distributions and obtained results are described in Ref.[19]

Results and discussion

Our experimental results on the high-energy end of the neutron spectrum from the spontaneous fission of ^{252}Cf are represented in Fig. 5 and may be summarized as follows:

i) Within the experimental errors, the NBS evaluated spectrum was confirmed up to 20 MeV (in a qualified sense for the range from 16 to 20 MeV).

ii) For the energy interval from 20 to 28 MeV, the correction function

$$\mu(E) = \exp(+0{,}65 \cdot (E-20{,}65)) \quad (5)$$

with reference to the Maxwellian distribution with kT = 1,42 MeV (eq. (1)) was determined. The integral over N(E) from 21,5 to 26,7 MeV amounts to $(6{,}0 \pm 3{,}4) \cdot 10^{-6}$. This neutron yield is much more higher as expected [25].

Furtheron, we present a comparison of our experimental data with those of other groups (Fig. 6). It illustrates the discrepancies of experimental data determined by different authors in the energy range above 8 MeV.

The integral fission neutron spectrum obtained by weighted concentration of the l. s. spectra of the fragments with A between 87 and 165 by steps 3 is

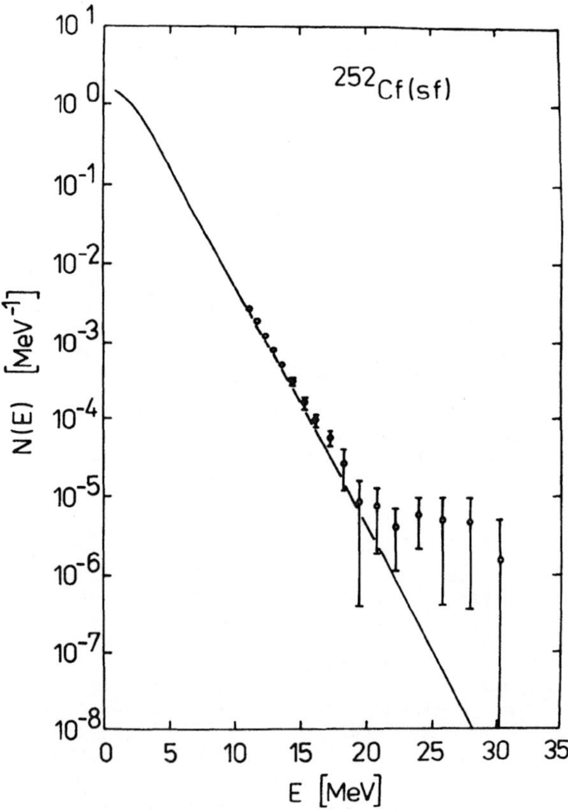

Fig. 5. The experimental data on the high-energy end of the neutron spectrum from spontaneous fission of ^{252}Cf compared with the result of the complex cascade evaporation calculation.

shown in Fig. 5 in comparison with the results of the described experiment. We were able to obtain good agreement with the experimental data on the neutron spectrum from ^{252}Cf up to the energy of about 20 MeV using the cascade evaporation model and realistic initial distributions of excitation energy. No arbitrary normalizations or free parameters were introduced. The experimental data indicate the existence of a hard emission component of fission neutrons. A similar result was already found in a measurement of the neutron spectrum from 14,5 MeV-neutron induced fission of Uranium by the use of the same experimental method [29, 30]. The high-energy component of fission neutron emission is predominant at extremely high emission energies (above 20 MeV) and cannot be explained assuming neutron evaporation from fully accelerated fragments. Hence, one should take into account non-equilibrium neutron emission which may be attributed to the typical rapid changes of nuclear potential in fission [24]. Further conclusions and outlooks are described in Ref.[19]

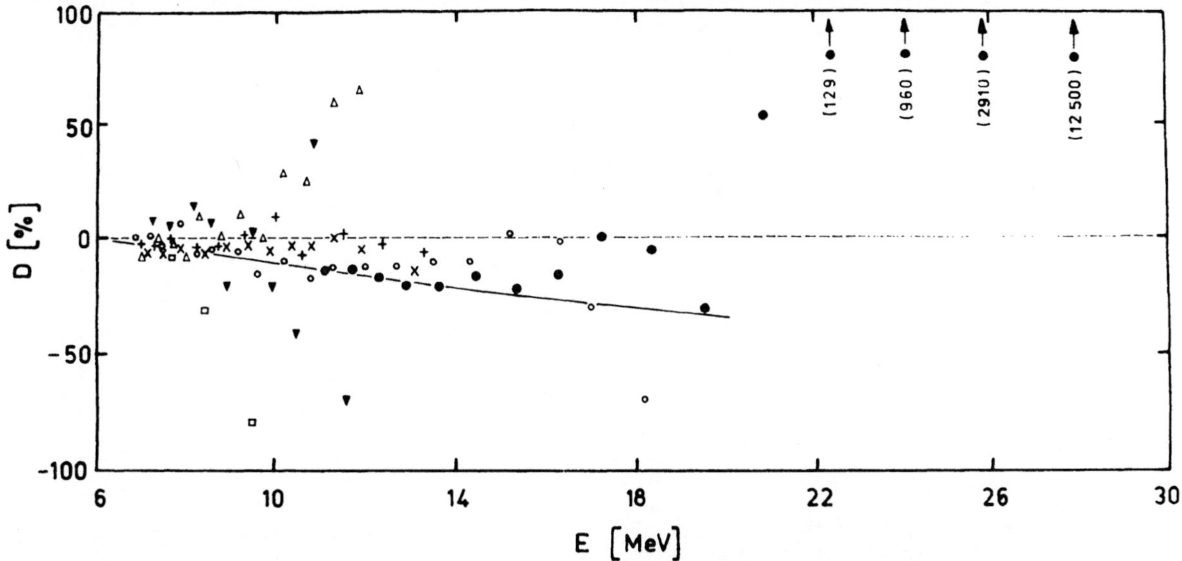

Fig. 6. Percentage departure of our data (o - 1st experiment with 62,0 h measuring time; • - 2nd experiment with 1218,5 h measuring time) from the Maxwellian distribution with kT = 1,42 MeV in comparison with the results of other groups (+ - Ref.[4]; x - Ref.[26]; ▼△ - Ref.[27]; □ - Ref.[28]) as well as with the NBS evaluated spectrum (continuous line). The representation is a supplemented one of Ref.[2]

References

1. Report INDC-36/LN, INDC/NEANDC, Nuclear Standards File, 1980 version, p. B-38.
2. M. V. Blinov, Proc. IAEA Consultants Meeting on Neutron Source Properties, Debrecen, 1980, INDC(NDS)-114/GT (1980).
3. J. Grundl, C. Eisenhauer, Natl. Bur. Stds. Publ., NBS-493 (1977).
4. J. Boldeman et al., Trans. Am. Nucl. Soc. 32, 733 (1979).
5. B. I. Starostov, A. F. Semenov, W. N. Nefedov, Yad. Konst. 2(37), 3 (1980).
6. H. R. Bowman, S. G. Thompson, J. C. D. Milton, W. J. Swiatecki, Phys. Rev. 126, 2120 (1962).
7. H. R. Bowman, J. C. D. Milton, S. G. Thompson, W. J. Swiatecki, Phys. Rev. 129, 2133 (1963).
8. M. V. Blinov, N. M. Kazarinov, I. T. Krisyuk, Yad. Fiz. 16, 1155 (1972).
9. V. M. Piksaikin, P. P. Dyatchenko, L. S. Kazaeva, Yad. Fiz. 25, 723 (1977).
10. Ju. S. Samyatnin, D. K. Ryasanov, B. G. Basova, V. A. Rabinovitch, V. A. Kostilev, Yad. Fiz. 29, 595 (1979).
11. P. Riehs, AIAU 81201 (1981).
12. Y. Boneh, Z. Fraenkel, Phys. Rev. C 10, 893 (1974).
13. V. A. Rubtchenya, Report RI-28 (Leningrad, 1974).
14. W. Grimm, H. Märten, D. Seeliger, Nejtronnaya Fizika (Proc. Vth All Union Conf., Kiev, 1980) 3, 3 (Moscow, 1980).
15. W. Grimm, H. Märten, D. Seeliger, B. Stobinski, Proc. XIth Int. Conf. on the Interaction of Fast Neutrons with Nuclei, Rathen, 1981, to be published.
16. M. Adel-Fawzy, H. Förtsch, S. Mittag, W. Pilz, D. Schmidt, D. Seeliger, T. Streil, Kernenergie 24, 107 (1981).
17. N. R. Stanton, COO-1545-92 (1971); D. Hermsdorf, ZfK-315, 192 (1977).
18. V. V. Verbinski et al., Nucl. Instr. Meth. 65, 8 (1968).
19. H. Märten, D. Seeliger, submitted to J. of Phys. G.
20. A. V. Ignatyuk, K. K. Istekov, G. N. Smirenkin, Nejtronnaya Fizika (Proc. IVth All Union Conf., Kiev, 1977) 1, 60 (1977).
21. H. Nifenecker et al., Proc. 3rd IAEA Symp. on the Physics and Chemistry of Fission, Rochester, 1973, vol. II, 117 (Vienna, IAEA).
22. A. Gavron, Phys. Rev. C 13, 2561 (1976).
23. A. Gavron, Z. Fraenkel, Phys. Rev. C 9, 632 (1974).
24. H. Märten, thesis, TU Dresden (1981).
25. H. Märten et al., INDC(GDR)-17/L (1982).
26. Z. A. Alexandrova et al., Atomnaya Energiya 36, 282 (1974).
27. N. N. Knitter et al., Atomkernenergie 22, 84 (1974).
28. H. Werle, H. Bluhm, Prompt Fission Neutron Spectra (Proc. IAEA Consultants Meeting, Vienna, 1971) p. 65 (1972).
29. H. Märten, D. Seeliger, ZfK-459, 98 (1981).
30. H. Märten, D. Seeliger, B. Stobinski, Proc. Europhysics Topical Conf. on Neutron Induced Reactions, Smolenice, 1982, to be published.

NEUTRON SPECTRA FROM Am/F AND Am/Li (α,n) SOURCES

J.G. Owen, D.R. Weaver and J. Walker

Birmingham Radiation Centre
University of Birmingham
Birmingham, UK.

Spectral measurements on neutrons from an Am/F (α,n) source have been made with a ^{3}He spectrometer; the unfolded spectrum is presented together with covariance error analysis. The results of measurements on two Am/Li sources are also reported and compared with previous published data. Am/Li has been proposed by an IAEA Consultants' meeting as a suitable basis for comparing the spectrometers used in different laboratories for delayed neutron spectral measurements.

Introduction

Radioactive neutron sources are frequently used as standards in applications such as the determination of neutron source strength, and the calibration of neutron detectors. In many cases an accurate knowledge of the source spectrum is desirable, but unfortunately large discrepancies exist in published data. Moreover, the information provided on errors is often incomplete so that a full assessment of accuracy and reliability cannot be made. Recent work at Birmingham has therefore concentrated on measuring the neutron spectra from two Am/Li (α,n) sources using high-resolution ^{3}He spectrometers; an Am/F source was also been measured. Complete covariance error information on the unfolded data has been produced, including contributions from both counting statistics and errors in the detector response function.

An additional incentive to measure the spectra of neutrons from radioactive sources exists because of the discrepancies which have appeared in measurements of delayed neutron spectra made by a number of laboratories. In particular, large differences exist between results obtained with ^{3}He spectrometers and those from proton recoil counters. This led an IAEA Consultants' meeting on delayed neutron properties[1] to recommend that all laboratories engaged in this work should measure the neutron spectrum from an Am/Li source, preferably the same one, because it is similar to that of delayed neutrons.

Spectrum Measurements

Measurements of both Am/Li sources were made with a Shalev-type ^{3}He spectrometer filled to a total pressure of 6.6 atmospheres. The upper energy limit of this spectrometer is only 2 MeV, however, and therefore for measurements on the Am/F source a second ^{3}He spectrometer filled to 10 atmospheres was employed. Both spectrometers were used in conjunction with a rise-time analysis system[2]. In all spectrum measurements the source was positioned at the centre of the (9m x 9m x 8m) Low Scatter Cell in the Birmingham Radiation Centre to minimise the effects of wall-scattered neutrons.

Similar counting geometries were used in all cases. The neutron source was placed approximately 7 cm in front of the spectrometer, with a 5 cm thick lead wall between the two to reduce the gamma flux incident upon the detector. Although pile-up of neutron and gamma pulses was unlikely in these measurements, both spectrometers are usually used to measure delayed neutron spectra where high gamma fluxes can degrade the resolution of the spectrometers. Consequently, all measurements of spectrometer efficiency and response functions have been made with the lead in position. Details of the spectrometer calibration procedure have been discussed elsewhere[2].

The Am/F Source

This source produces neutrons from the reaction

$$^{19}F + \alpha \rightarrow {}^{22}Na + n - 1.923 \text{ MeV}$$

A 1Ci source was lent to us by the National Physical Laboratory, Teddington, England. It was manufactured by Amersham International[3] and encapsulated in a standard X3 stainless steel container.

As in the case of the larger of the two Am/Li sources[4], the spectrum from Am/F was measured at three different orientations and the total counting period was approximately 25 days; the background count-rate proved to be negligible. In each of the three orientations the spectra were very similar, and therefore the pulse height distributions were summed and unfolded by an iterative procedure described previously[2] to produce the average spectrum as shown in Figure 1. Errors are indicated at the ± 1 standard deviation level and were determined from the covariance error matrix, calculated as in reference 2.

The unfolding of the results is dependent on a full knowledge of the response of the detector, and although most of the energy range has been covered in measurements with monoenergetic neutrons from the Radiation Centre's Dynamitron accelerator, more data are required for neutrons with energies above 2.2 MeV. The unfolded spectrum from the Am/F source should therefore be treated as preliminary, although the actual shape is not expected to change greatly.

The Am/Li Sources

Neutrons are produced by the reaction

$$^{7}Li + \alpha \rightarrow {}^{10}B + n - 2.79 \text{ MeV}$$

With α particles from a ^{241}Am source, the maximum neutron energy is about 1.5 MeV.

Figure 2 shows the normalised spectrum of the larger source (5 Ci, lent to us by the National Physical Laboratory, Teddington, England); once again errors represent ± 1 standard deviation, and include contributions from both counting statistics and uncertainties in the detector response function. Figure 3 compares this spectrum with that from a 100 mCi source lent to us by AERE Harwell, and in Figure 4 the spectrum of this smaller source is compared with data obtained by Werle[5] using a proton recoil counter. On the whole, the agreement between the measurements is good, although the proton recoil data contains slightly

K. H. Böckhoff (ed.), Nuclear Data for Science and Technology, 492–494.

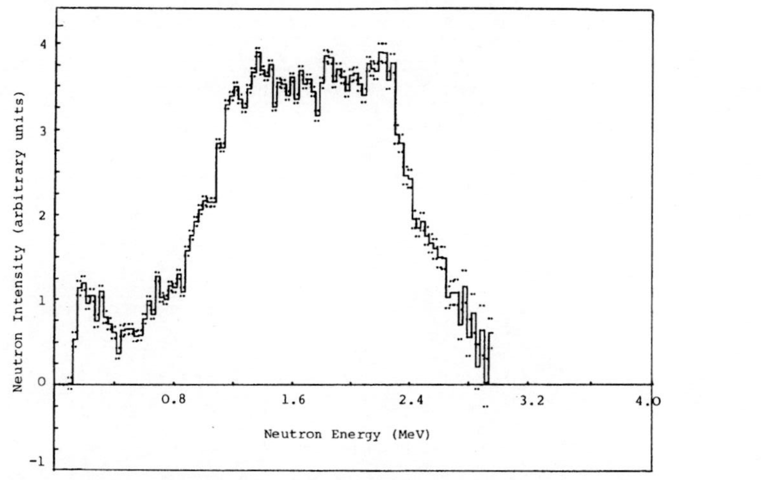

Figure 1: The neutron spectrum from the Am/F source. Each energy bin is 29.1 keV wide. Errors are indicated as $\pm$ 1 standard deviation, but include ONLY the contributions from counting statistics. The assessment of the errors due to uncertainties in the spectrometer's response function is in progress.

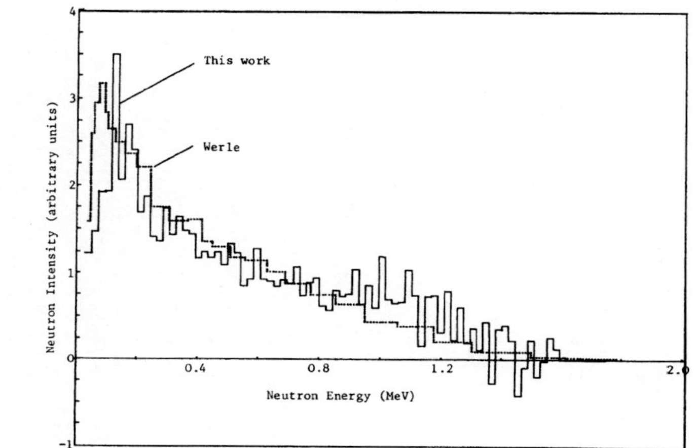

Figure 2: The neutron spectrum from the 5 Ci Am/Li source. Each energy bin is 22.5 keV wide, and the errors represent $\pm$ 1 standard deviation.

Figure 3: The neutron spectra from the 5 Ci and 100 mCi Am/Li sources.

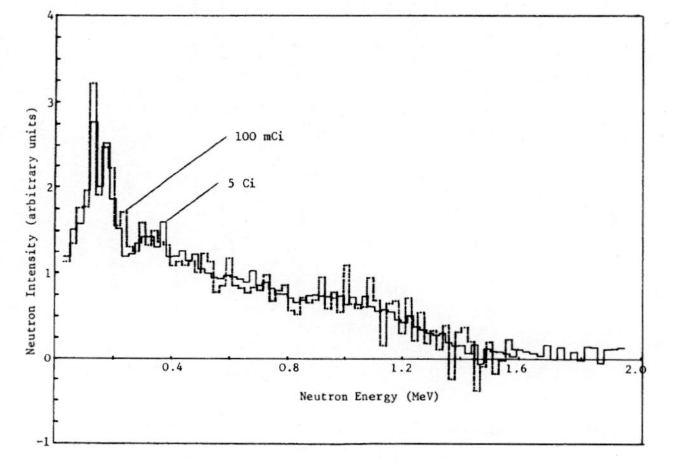

Figure 4: The neutron spectrum from the 100 mCi Am/Li source, and that obtained by Werle[5].

more neutrons at low energies.

The agreement between these measurements and those made previously confirms that an Am/Li source would prove useful as a means of intercomparison of the response of the various detectors used in delayed neutron spectroscopy, as suggested by the IAEA Consultants' meeting. For such comparisons to be meaningful full covariance error data should, of course, be supplied.

Acknowledgements

We would like to express our thanks to the National Physical Laboratory and AERE Harwell for the loan of their sources. We also thank the Science and Engineering Research Council for financial support.

References

1. Proc. Consultants' Meeting on Delayed Neutron Properties, Vienna (March 1979), International Nuclear Data Committee Report INDC(NDS)-107G + Special (1979).
2. J.G. Owen, D.R. Weaver and J. Walker, Nucl. Instr. and Meth. 188, 579 (1981).
3. Amersham International PLC, Amersham, Buckinghamshire HP7 9LL, England.
4. D.R. Weaver, J.G. Owen and J. Walker, Nucl. Instr. and Meth. 198, 599 (1982).
5. H. Werle, Karlsruhe Nuclear Research Centre Report KFK-INR-4/70-25, Karlsruhe, West Germany (1970); in English translation as ORNL-tr-2415.

The Understanding of Nuclear Reactions and Nuclear Model Calculations

BEYOND THE STATISTICAL MODEL:
RECENT PROGRESS IN NEUTRON NUCLEAR REACTION THEORY

H.A. Weidenmüller

Max-Planck-Institut für Kernphysik
P.O.B. 103980, D-6900 Heidelberg
Federal Republic of Germany

A brief exposition of the central results of the Statistical Model is followed by that of a generalization to include direct reactions and by a survey of tests of the statistical model without and with such reactions being present. The last section is devoted to recent progress in precompound processes, another generalization of the Statistical Model.

Introduction

Since the fundamental papers by Bohr, Bethe, and Wigner on the subject, the Statistical Model and, somewhat more generally, random-matrix models have played an ever-growing role in the formulation of nuclear reaction theories, and in the analyses of resonance data and reaction cross sections. It is the purpose of this review to summarize recent progress in this field, to identify open problems, and to point towards areas of probable future growth. A review of the standard Statistical Model and its implications for compound-nucleus reactions is followed by a discussion of the generalizations including direct reactions and precompound processes and their experimental tests.

1. The Statistical Model

This model describes properties of nuclear energy levels in statistical terms. Quantities of prime interest are (i) the spacings of energy levels E_λ of equal spin and parity (λ is a running index); (ii) the partial width amplitudes $\gamma_{\lambda c}$ (or, equivalently, the spectroscopic amplitudes) describing the coupling of level λ to the physical channel c ; (iii) the reduced partial widths $\Gamma^o_{\lambda c} \propto \gamma^2_{\lambda c}$. Only a few energy intervals of limited size ("windows") are experimentally available for sampling levels. Therefore, the statistical model is primarily concerned with the statistical fluctuations of the above-mentioned observables about their local mean values.

Predictions of the Statistical Model are based on the results of random-matrix theory (recently reviewed in ref. 1)). Statistical features of the nuclear Hamiltonian are modelled in terms of a variety of random-matrix ensembles. With one exception cited below, all random-matrix ensembles investigated up to date yield coinciding results for the fluctuations of the above-mentioned observables about their local mean values. Some ensembles (like the two-body random ensemble) can so far only be investigated numerically. For such models, the statement just made does not carry the same generality and weight as for the analytically tractable models. Among the latter, the Gaussian Orthogonal Ensemble (GOE) is the most thoroughly investigated; unless otherwise stated, results cited below are derived from the GOE.

The Statistical Model predicts that

(i) The level spacings and the $\gamma_{\lambda c}$ are uncorrelated random variables;

(ii) For $\lambda \neq \lambda'$ or $c \neq c'$, $\gamma_{\lambda c}$ and $\gamma_{\lambda' c'}$ are uncorrelated;

(iii) $\gamma_{\lambda c}$ is Gaussian distributed with mean value zero, the distribution law is the same for all λ . (This implies the Porter-Thomas distribution for the reduced partial widths.) Note that the distribution of the $\gamma_{\lambda c}$ is completely specified in terms of the second moment $A_c = \overline{\gamma^2_{\lambda c}}$. Numerical results obtained [2] by sampling the distribution of eigenvectors in a large-scale sd shell-model calculation disagree with the Gaussian distribution and show that the two-body random ensemble gives a result different from the one just cited. It appears that the distribution of the eigenvector components can be approximated by a sum over Gaussians, see Fig.1.1. This would suggest that

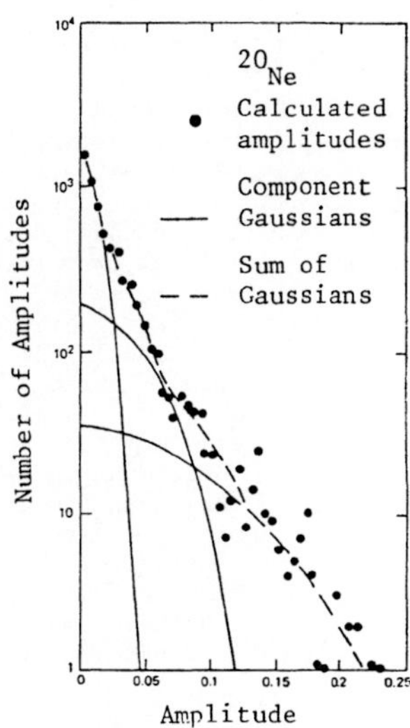

Fig.1.1 The distribution of eigenvector components in ^{20}Ne calculated with the Chung-Wildenthal interaction, and its (fitted) decomposition into a superposition of three Gaussian functions (from the last of refs. 2)) .

deviations from the Porter-Thomas distribution would be hard to detect experimentally. The numerical results [2] are confined to sd shell nuclei; it is not clear how they would have to be modified for heavy nuclei where most of the neutron data have been taken.

K. H. Böckhoff (ed.), Nuclear Data for Science and Technology, 497–505.

The level spacing distribution is known in principle ("Wishart distribution"). Two special results are of interest:

(iv) The probability distribution of nearest neighbour spacings is, for levels of the same spin and parity, given to a very good approximation by the Wigner formula $\frac{\pi}{2}\cdot x\cdot \exp(-\frac{\pi}{4}x^2)$ with x the ratio of actual level spacing over the mean level spacing d.

(v) The probability distribution for k^{th} nearest neighbours (that is, two levels with k levels between them, all of the same spin and parity) is, for $k >> 1$, well approximated by a Gaussian. The width of this Gaussian increases with k like $\ln(k+1)$. [Note that if the Wigner distribution were used to generate k uncorrelated nearest-neighbour spacings, the probability distribution for k^{th} nearest neighbours would, for $k >> 1$, be a Gaussian with a width growing like $(k+1)^{1/2}$, that is much faster. This shows that longe-range correlations exist in the spectra of random matrices. This property is referred to as the "stiffness" of the spectra.]

Experimentally, distributions are obtained by sampling a running sequence of levels of the same spin and parity. Theoretically, one calculates the distribution over the ensemble. The equality of the two distributions is assured by ergodicity theorems [1]. The same remark applies to average compound-nucleus cross sections.

The Statistical Model can also be used to model the nuclear compound scattering in terms of an ensemble of stochastic S-matrices[3]. For fixed spin and parity, each member of the ensemble has the form

$$S_{cc'} = \left[(1+iK)/(1-iK) \right]_{cc'} \qquad (1a)$$

where the statistical K-matrix

$$K_{cc'} = K_c^{(o)}\delta_{cc'} + \sum_\lambda \frac{\gamma_{\lambda c}\gamma_{\lambda c'}}{E-E_\lambda} \qquad (1b)$$

contains a diagonal constant non-random background term $K^{(o)}$ and a sum over resonance terms. The energy of the system is E, and the resonance parameters E_λ and $\gamma_{\lambda c}$ are random variables with properties described above.

In the domain of isolated resonances (where the mean level spacing d is much larger than the mean level width Γ), the evaluation of compound-nucleus cross sections constitutes a straightforward application of the Statistical Model. With increasing excitation energy of the compound nucleus, Γ becomes comparable with and, ultimately, much larger than d, and a direct analysis of the statistical properties of the underlying compound-nuc... resonances is therefore impossible. In this domain, the application of the statistical model is based on the belief that the statistical properties of nuclear levels do not change with excitation energy. This view is indirectly supported by empirical evidence: Statistical properties deduced from levels sampled near neutron threshold in medium-weight and heavy nuclei, near the ground state by sampling all mass numbers, and several MeV above proton threshold for nuclei in the f-p shell, do not differ, except for some little-understood findings reported in section 3 below. Nonetheless, direct experimental tests of the results obtained for $\Gamma \gtrsim d$ are obviously desirable.

Cross sections are calculated by writing $S_{cc'}$ in the form

$$S_{cc'} = \overline{S_{cc'}} + S^{fl}_{cc'}$$

where the bar denotes the average. The (average) compound-nucleus cross section is defined as

$$\sigma^{CN}_{cc'} = \overline{|S^{fl}_{cc'}|^2} \; .$$

(Kinematical factors are omitted.) There is also interest in the fluctuations of the cross section about this mean value.

In its full generality this problem has not been solved: We cannot calculate $\sigma^{CN}_{cc'}$ analytically, let alone the distribution function of the cross section. Analytical answers based on series expansions exist in two limits, $\Gamma << d$ and $\Gamma >> d$. They serve as anchor points for fit formulae based on numerical calculations and applicable to the domain of Γ/d values between these two limits.

A limitation of the model (to be removed in the next section) is the restriction to the absence of direct reactions. Indeed, using $\overline{\gamma_{\lambda c}\gamma_{\lambda c'}} = \delta_{cc'}\overline{\gamma^2_{\lambda c}}$ and $K^{(o)}_{cc'} = \delta_{cc'}K^{(o)}_c$ we see immediately that $\overline{S_{cc'}} = \overline{S_c}\delta_{cc'}$: The average S-matrix is diagonal, and direct reactions are not included. This limited problem only is discussed in the remainder of the present section. In applications, $\overline{S_c}$ is given in terms of an optical-model calculation. It is then the purpose of the theory to express $\sigma^{CN}_{cc'}$ in terms of $\overline{S_c}$. This is accomplished with the help of the Statistical Model.

For $\Gamma << d$, $\sigma^{CN}_{cc'}$ has been calculated long ago [4]. The formulas are not reproduced here. For $\Gamma >> d$, the elements $S^{fl}_{cc'}$ can be shown to be Gaussian distributed random variables with mean value zero and a second moment given by [3]

$$\overline{S^{fl}_{cc'}(E)\, S^{fl}_{dd'}(E')} = 0 \; ,$$

$$\overline{S^{fl}_{cc'}(E)\, S^{fl*}_{dd'}(E')} = \frac{T_c T_{c'}}{\sum_{c''} T_{c''}} \left(\delta_{cd}\delta_{c'd'} + \delta_{cd'}\delta_{c'd}\right) \frac{\Gamma}{\Gamma + i(E'-E)} \qquad (2)$$

Here, $T_c = 1 - |\overline{S_c}|^2$ are the transmission coefficients, and $\Gamma = \frac{d}{2\pi}\sum_c T_c$. With the distribution of $S^{fl}_{cc'}(E)$ completely known, average cross sections, Ericson fluctuations etc. can be worked out. We confine ourselves here to $\sigma^{CN}_{cc'}$ which is given by the content of the square bracket on the right-hand side of equation (2) with (d,d') pairwise equal to (c,c'). Note that for $c = c'$ (and $d = d' = c$), the Kronecker symbols add up to a factor 2, while they give unity for $c \neq c'$. This factor 2 is referred to as the "elastic enhancement" factor.

Moldauer [5] has pointed out that the analytical technique [6] used to establish the results for $\Gamma >> d$ uses a picket-fence model (constant spacings) for the eigenvalues E_λ instead of the full Wishart distribution. He suggested that for the latter, the elastic enhancement factor might differ from 2. Detailed numerical calculations [5,7] have led to the conclusion that for a spectrum of E_λ values which has the "stiffness" typical of random-matrix ensembles (see point (v) above), the elastic enhancement factor is 2 within the statistical accuracy of the calculations.

Fit formulae for the intermediate domain of Γ/d values are constructed as follows: Analytical expressions with the correct asymptotic limits, dependent only on the transmission coefficients and embodying overall properties established analytically, are fitted to numerical values of $\sigma^{CN}_{cc'}$. These are generated by drawing E_λ and $\gamma_{\lambda c}$ from their respective probability distributions with the help of a random-number generator, by calculating $S_{cc'}$ from the values so obtained, and by repeating the procedure many times over to sample $|S^{fl}_{cc'}|^2$. Fit formulae obtained in this fashion have been given by several authors [8]. Typically, they are based on calculations utilizing up to ten or so channels. This is perhaps not quite enough to connect to the limit $\Gamma >> d$ in a realistic fashion. Moreover, in practical calculations one encounters the situation that many weakly populated (gamma) channels are present. A special numerical study [9] was recently undertaken to include this case; improved-fit formulae have been reported.

A new analytical approach to the problem has been devised by Mello and Seligman [10]. These authors attempt to determine, for arbitrary and fixed values of the transmission coefficients (and, hence, of Γ/d), the distribution of S-matrix elements directly in terms of the available "phase space" and some constraints based on symmetry, unitarity, and some further simple and general properties of $S_{cc'}$. For $\Gamma >> d$, this approach yields results which coincide with the ones cited above. For intermediate values of Γ/d, the theory has not yet been worked out.

In many cases of practical importance, compound-nucleus cross sections cannot be measured, and use of the statistical model is the only strategy available to determine their values. A deeper theoretical understanding of, and more universal formulae for, the intermediate domain of Γ/d values would therefore be of very considerable interest. In view of its basic algebraic simplicity, the problem is intellectually challenging.

2. Inclusion of Direct Reactions

Direct Reactions, explicitly excluded in section 1, are often very important experimentally, especially for transitions between collective states, and must be included in the theory. Generalizing the procedure of section 1, we assume that the matrix elements $\overline{S}_{cc'}$ are given in terms of a coupled channels or DWBA model and aim at using the statistical model to calculate $\sigma^{CN}_{cc'}$. Again, we restrict ourselves to states of fixed spin and parity.

$\overline{S}_{cc'}$ is nondiagonal (and because of absorption, non-unitary) if and only if $\gamma_{\lambda d}\gamma_{\lambda d'} \neq 0$ for $d \neq d'$. This follows easily from eqs.(1). The existence of direct reactions implies correlations between partial width amplitudes in different channels, and vice versa. The statistical model must be generalized to account for this situation. At the same time, one wishes to retain as many as possible of the general results of random-matrix theory. This leads to the following statistical model: The distribution of the energy levels E_λ in eq. (1) is the same as before, the $\gamma_{\lambda c}$ have a Gaussian distribution independent of λ, have mean value zero, $\gamma_{\lambda c}$ and $\gamma_{\lambda' d}$ are uncorrelated for $\lambda \neq \lambda'$, but the second moments $A_{cc'} = \overline{\gamma_{\lambda c}\gamma_{\lambda c'}}$ are no longer diagonal in the channels. In fact, using eqs. (1) one finds the values of $A_{cc'}$ (and of the nondiagonal background matrix $K^{(o)}_{cc'}$) easily in terms of $\overline{S}_{cc'}$. The input of the statistical model is thus determined.

For isolated resonances, the statistical model with direct reactions can be directly tested by observing correlations between partial width amplitudes $\gamma_{\lambda c}$ and $\gamma_{\lambda c'}$ in different channels or, should this not be feasible, correlations between reduced partial widths $\Gamma_{\lambda c}$ and $\Gamma_{\lambda c'}$, and by comparing them with calculated values based on some direct reaction model. When direct reactions are present, the calculation of $\sigma^{CN}_{cc'}$ sketched in section 1, is modified as follows. The unitarity deficit of the average S-matrix $\overline{S}_{cc'}$ is expressed in terms of Satchler's Hermitean transmission matrix

$$P_{cc'} = \delta_{cc'} - \sum_d \overline{S}_{cd}\,\overline{S^*_{c'd}} .$$

We note that $P_{cc'}$ is the generalization of the transmission coefficients T_c introduced in section 1. Since $P_{cc'}$ is Hermitean, it can be diagonalized by a unitary matrix,

$$(UPU^+)_{cc'} = \delta_{cc'}P_c . \qquad (3)$$

Matrix algebra can be used to show that with $(UPU^+)_{cc'}$, also $\overline{\tilde{S}}_{cc'} = (U\overline{S}U^T)_{cc'}$ is diagonal. This suggests the idea [11] that $\tilde{S}^{fl}_{cc'} = (U\tilde{S}^{fl}U^T)_{cc'}$ does not differ in its statistical properties from a statistical S-matrix without direct reactions. It was possible to show [8] that this idea is correct, although the dynamical role of the U-transformation is not yet completely understood. Through the unitarity transformation defined in eq. (3), the problem with direct reactions can be reduced to the problem without. The latter was dealt with in section 1. In other words, we have [8,12]

$$\overline{S^{fl}_{ab}\,S^{fl*}_{cd}} = \sum_{efgh} U^*_{ea}\,U^*_{fb}\,U_{gc}\,U_{hd}\,\overline{\tilde{S}^{fl}_{ef}\,\tilde{S}^{fl*}_{gh}} \qquad (4)$$

The quantity $\overline{\tilde{S}^{fl}_{ef}\,\tilde{S}^{fl*}_{gh}}$ appearing on the r.h.s. of eq.(4) vanishes unless the four indices are pairwise equal. For $e = g$ and $f = h$ (or for $e = h$ and $f = g$) it has the form of σ^{CN}_{ef} discussed in section 1. An additional term not mentioned before is $\overline{\tilde{S}^{fl}_{ee}\,\tilde{S}^{fl}_{gg}}$ which, for $g \neq e$, does not signify a cross section, but has to be included in the calculation. Fit formulas for this term, generated along the same lines as described in section 1 for $\sigma^{CN}_{cc'}$, have also been given [8].

In summary, the case with direct reactions can be reduced quantitatively to the one without, and is seen to be completely determined in terms of the matrix elements of $\overline{S}_{cc'}$. Eq. (4) has the interesting property that $\overline{S^{fl}_{ab}\,S^{fl*}_{cd}}$ does not vanish, in general, when all four indices differ from each other. This is the most distinctive feature of the case with direct reactions, and the obvious place for experimental tests of the theory.

Isolated doorway states (isobaric analogue resonances are the best-studied example) lead to energy-dependent modulations of the average S-matrix $\overline{S}_{cc'}$. In many cases, the average compound nuclear level spacing d is very small compared to the width Γ_d of this modulation. Then, it is justified to consider $\overline{S}_{cc'}$ as energy-independent over an energy interval I with $\Gamma_d >> I >> d$, and to calculate the compound-nucleus contribution to the cross section as in the case of direct reactions.

3. Experimental Tests of the Statistical Model

In 1975, Lane [13] gave a review on direct neutron capture and the ensuing correlations between γ-widths $\Gamma_{\lambda\gamma}$ and reduced neutron widths $\Gamma^o_{\lambda n}$. This review was updated [14] in 1979. Little seems to have happened since that would go beyond the detailed tests of the Statistical Model performed earlier with neutron time-of-flight spectroscopy, except for several papers which report on strength functions, mean level spacings, optical model calculations, and Γ_γ/Γ^o_n correlations in domains of mass number and/or excitation energy not investigated previously [15,16,17]. Neutron reactions on light nuclei have recently been used to investigate properties of analogue states and isospin mixing [18,19]. Competing theoretical models employed in the analysis of $\Gamma^o_n - \Gamma_\gamma$ correlations have been analysed, and shown to be essentially consistent [14,20].

In recent years, statistical properties of nuclear levels have been studied also with ultra-high-resolution proton scattering experiments [21]. A new "window", containing nuclei with masses A<70 and accessible with proton energies of several MeV, has thereby been opened. The advantages of the method are: (i) Interference with Rutherford scattering produces characteristic, strongly ℓ-dependent resonance shapes making the ℓ-assignment from elastic data easy; (ii) Inelastic proton scattering, possibly in coincidence with deexcitation γ-rays, can be easily investigated. Novel information has thereby been obtained on amplitude correlations in several inelastic channels.(iii) Fine structure properties of analogue resonances can be investigated, and compared with the statistical model. These advantages have to be balanced against the drawback of a limited range of mass numbers and the need to take account of the exponential increase of the level density in evaluating mean values and distributions of level spacings. From a considerable body of material [21-30], only some highlights can be reported.

Spacing distributions of levels have been investigated [29,30] for 56 levels with spin $1/2^+$ in ^{57}Co, populated in the reaction $^{56}Fe(p,p)$ at bombarding energies ranging from 3.1 to 3.6 MeV. Fig.3.1 shows the distribution of kth nearest neighbour spacings for $k \leq 5$; the smooth curves are the GOE predictions. Fig.3.2 shows the widths of these distributions versus k for $k \leq 10$. The solid lines give the predictions of the GOE (labelled OE) and of the uncorrelated Wigner (UW) models. In the latter, nearest neighbour spacings

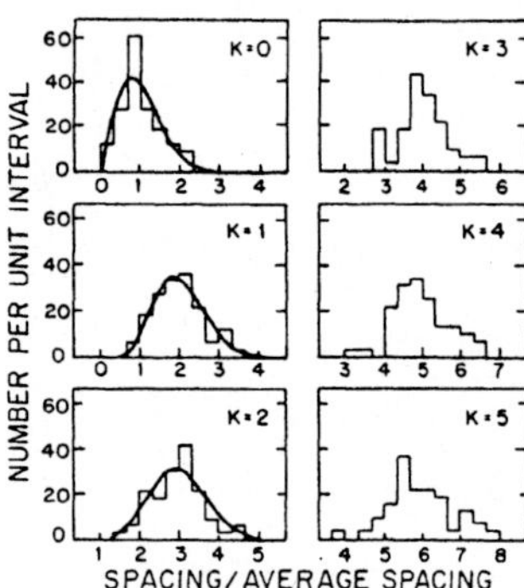

Fig. 3.1 k^{th} nearest neighbour distribution and prediction (solid line) of the GOE for ^{57}Co (from refs. 29,30)).

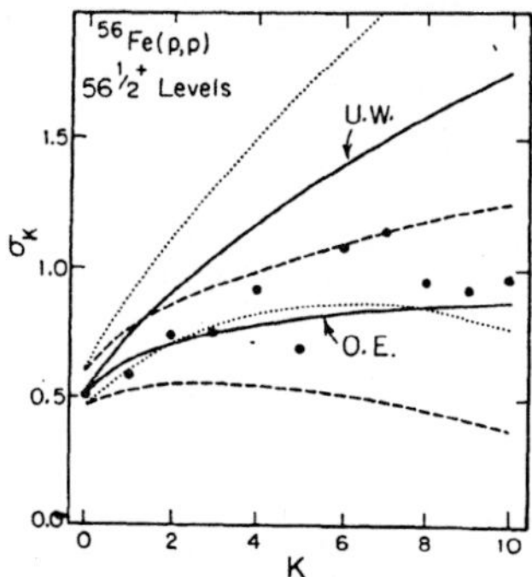

Fig. 3.2 Widths of kth nearest neighbour distributions in ^{57}Co versus k in comparison with theoretical predictions (from refs. 29,30)).

are drawn at random from a Wigner distribution, thus leading to a distribution which lacks the "stiffness" of the GOE. The dashed (dotted) curves are the 10% and 90% limits of the statistical distribution for a sample of 56 levels following the GOE (UW), respectively. The agreement with the GOE is quite good, and the data points are scattered above and below the OE prediction in a manner consistent with the statistical analysis. A summary of similar data on three other nuclei is shown [21] in Fig. 3.3 .

Fig.3.4 shows [21] how inelastic proton scattering on a 0^+ target leading to a 2^+ final state and followed by E2 deexcitation γ-rays, can be used to distinguish $1/2^+$ and $3/2^+$ intermediate compound-nucleus resonances from one another: In the former case, the angular distributions of inelastically scattered protons and of the deexcitation gamma rays are isotopic, while in the latter case, these two angular distributions together can be used to determine unambiguously the absolute values and the relative phase of the two inelastic proton amplitudes $\gamma_{\lambda \ell s}$ where $\ell = 1$ is the proton angular momentum, and $s = 3/2$ or $s = 5/2$ the channel spin. A weak point of the method is that inelastic proton scattering with $\ell \geqslant 3$ must be assumed to be negligible; this hypothesis can be ascertained by consistency tests only.

With this method correlation between partial width amplitudes can be measured, and novel tests of the Statistical Model thereby be performed. Fig.3.5 shows refs.22,23) the example of the "off-diagonal strength function" $s_{35} = \overline{\gamma_{\lambda 1\ 3/2}\ \gamma_{\lambda 1\ 5/2}}/d$ measured in the vicinity of an isobaric analogue resonance in the reaction ^{44}Ca(p,p'), where a total of 32 fine-structure resonances has been resolved. The dots are

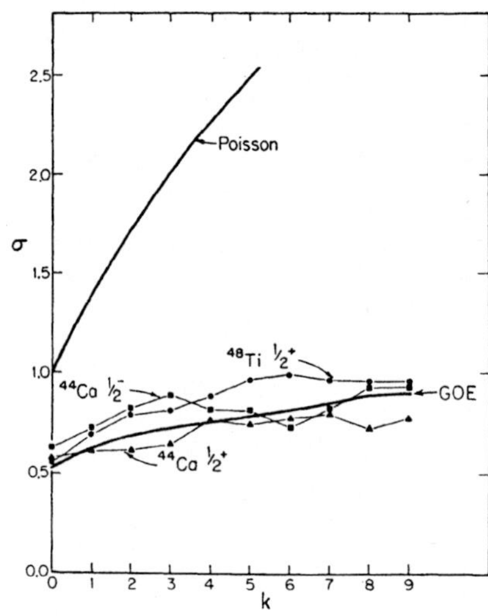

Fig. 3.3 Widths of kth nearest neighbour spacing distributions versus k in comparison with the GOE and a Poisson distribution for three other nuclei (from ref.21)) .

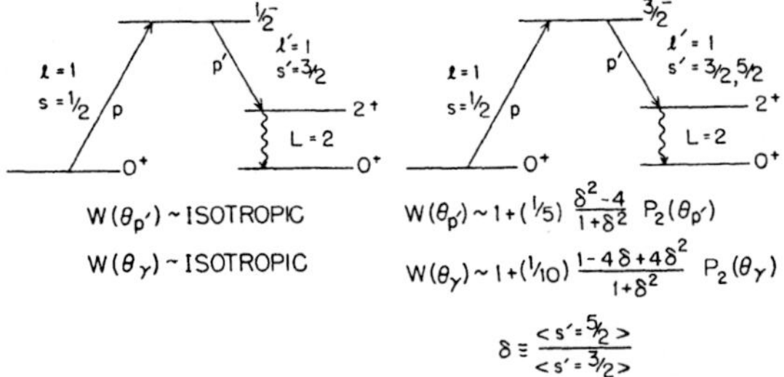

Fig. 3.4 Determination of spin values and reduced width amplitudes from inelastic resonant proton scattering (from ref. 21)).

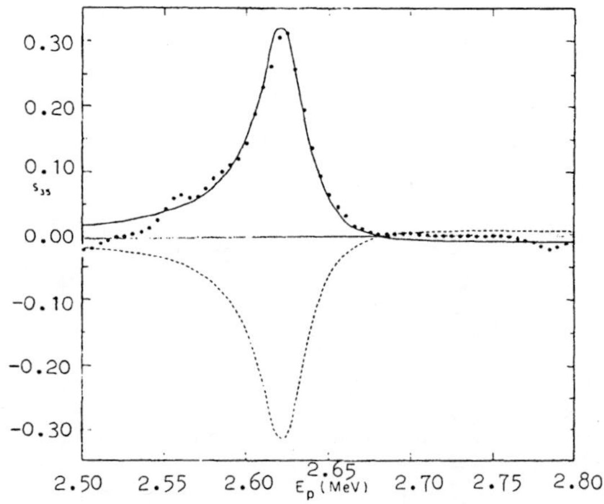

Fig. 3.5 Comparison of experimental (dots) and theoretical (full and dotted curves) results on the "off-diagonal strength function" s_{35} in ^{45}Sc in the vicinity of an analogue resonance (from refs. 22,23)) .

obtained by averaging the data points for $\gamma_{\lambda 1\ 3/2}\gamma_{\lambda 1\ 5/2}$ with a Lorentzian envelope of width 20 keV.
The theory of analogue resonances, when fitted to the diagonal strength functions s_{33} and s_{55} (also determined experimentally), predicts the shape and magnitude of s_{35} except for a sign (full and dotted curves). This very detailed test of amplitude correlations

(including the resonance enhancement and the famous Robson asymmetry) is seen to lead to a very satisfactory agreement between theory and experiment.

Certainly the most unexpected and potentially the most interesting results have been obtained by applying this method to sequences of levels void of analogue resonances [24,26-28]. The first few of these papers reported large correlations between inelastic proton amplitudes pertaining to different channel spins and suggested deviations from a Gaussian distribution of the $\gamma_{\lambda c}$'s. The latest results [28] strongly point to a violation of this central feature of random-matrix theory. A total of 53 d-wave $5/2^+$ resonances in the reaction $^{44}Ca(p,p'\gamma)^{44}Ca$ was analysed in a way similar to the one described above, and values for $|\gamma_{\lambda\,0\,1/2}|$, $|\gamma_{\lambda\,2\,3/2}|$, $|\gamma_{\lambda\,2\,5/2}|$ and their relative phase angles extracted (all $\ell \geq 4$ contributions were neglected). The distribution of reduced widths and the cumulative sums of reduced widths versus energy are completely consistent with the Statistical Model, and show no anomalies. The products of pairs of amplitudes show strong correlations. These alone, however, are not inconsistent with the Gaussian (Krieger-Porter) distribution; the correlations might in principle be accounted for in terms of some direct reaction between the inelastic channels - a process very difficult to test experimentally. The striking observation comes from the evaluation of the correlation coefficients ρ. For a pair of random variables x and y, the coefficient $\rho(x,y)$ is defined as

$$\rho(x,y) = \frac{\overline{(x-\bar{x})\,(y-\bar{y})}}{\overline{(x-\bar{x})^2}^{1/2}\,\overline{(y-\bar{y})^2}^{1/2}} \quad ;$$

it is easy to show that if the $\gamma_{\lambda c}$'s have a Gaussian distribution centered at zero (albeit with non-diagonal correlations), we must have

$$\rho(\gamma^2_{\lambda c}, \gamma^2_{\lambda c'}) = \rho^2(\gamma_{\lambda c},\gamma_{\lambda c'}) \ .$$

A comparison between these two quantities as obtained from the data is given in Table 1 which also contains the significance level (in percent) for $\rho^2(\gamma_{\lambda c},\gamma_{\lambda c'})$.

Table 1. Amplitude and width correlations [28] for levels in ^{45}Sc

Channels	$\rho^2(\gamma_{\lambda\ell s},\gamma_{\lambda\ell' s'})$	$\rho(\gamma^2_{\lambda\ell s},\gamma^2_{\lambda\ell' s'})$	Sign.Level
$0\,\frac{5}{2}\,\vert\,2\,\frac{3}{2}$	.05	.67	> 99.9 %
$0\,\frac{5}{2}\,\vert\,2\,\frac{5}{2}$	.52	.27	95 %
$2\,\frac{3}{2}\,\vert\,2\,\frac{5}{2}$	.004	-.08	40 %

Monte Carlo calculations [28] show that the discrepancy between the values in the first two rows in Table 1 is statistically highly significant. Its origin and implications are not understood, nor is it known whether the anomaly is confined to a narrow mass or energy region. It definitely deserves further study. Speculations about a connection with the theoretical findings of refs. [2] appear premature. It appears that the comparison of two correlation coefficients - not available so far for neutron data - provides a very sensitive test of the underlying distribution law.

Investigations aimed at testing modifications of average compound-nuclear scattering cross sections due to direct reactions (see eq. (4)) have commonly compared the prediction of two theoretical models with the data. One model is the one described in section 2 . The other combines incoherently a direct reaction model for the average S-matrix $\overline{S_{cc'}}$, with the standard expression for the compound-nucleus cross section in the absence of direct reactions as outlined in section 1 . We refer to the first of these models as the unified and to the second as the standard model, respectively [31-33]. In either model, some parameters determining $\overline{S_{cc'}}$, are usually fitted to the data; such fits are commonly made separately and independently for the standard model and the unified model, thereby rendering a direct comparison of the compound-nuclear contributions in the two models meaningless.

As mentioned at the end of section 2 , analogue resonances produce average S-matrix elements which can often be treated like those of a direct reaction. Early tests of the models used analogue states in nuclei near mass number 90, i.e. close to a closed neutron shell as testing grounds (for reference, see ref. [3]), because here the compound-nucleus contribution in the proton channels is particularly prominent, because the theory of analogue resonances largely fixes the form of $\overline{S_{cc'}}$, and because the analysis of the data is simplified since the analogue resonance has fixed spin and parity. The majority of the cases investigated favours the unified model.

A much more ambitious program - comparing the standard and the unified model for inelastic neutron scattering on ^{232}Th and ^{238}U - has been carried through recently by Chan, Sheldon and collaborators [31-32] and is reported at this meeting [33]. The work is presented in the next session. This enables me to be brief, and only give a few results. The two nuclei have low-lying collective states which are strongly excited via inelastic neutron scattering. This process is modelled in terms of a coupled-channels direct reaction model with suitable optical-model parameters and coupling strengths. For each spin J , the associated elements of $\overline{S_{cc'}}$, generated in this way are then used to apply the formalism of section 3 and to generate angle-integrated and energy-averaged cross sections by combining the direct and the compound-nucleus contributions as prescribed by the unified model; the results for the standard model are generated similarly, but with different values for the inelastic coupling strengths of the coupled-channels programme. Fig.3.6 shows several inelastic angle-integrated cross sections versus neutron energy for the unified (solid lines) and the standard (broken lines) models in comparison with the experimental data (dots) which have a typical error of 10%.

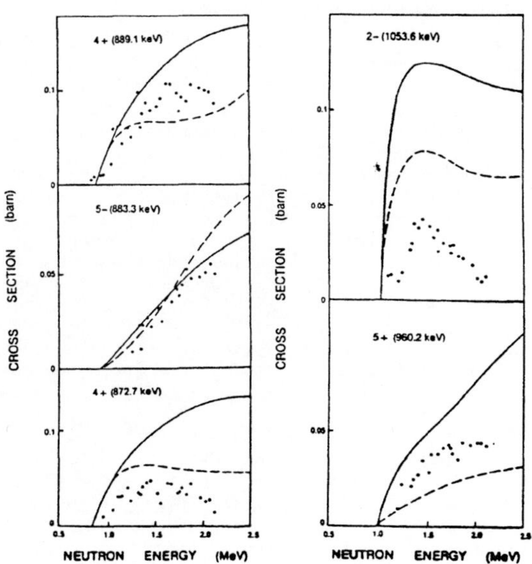

Fig. 3.6 Calculated and experimental angle-integrated inelastic neutron cross sections versus neutron energy populating final states in ^{232}Th as indicated (from ref. [32]) .

From this and similar calculations, the authors conclude [33]: "Regarded overall, the unified theory has a better record of success in fitting the experimental data than the standard theoretical approach..."

One of the difficulties in tests such as the one described lies in the determination of accurate optical-model parameters. Recent studies [34,35] put severe limits on the possible difference between neutron and proton optical potentials, thereby narrowing the choice for the former.

Deviations from a Gaussian probability distribution for the amplitudes $\gamma_{\lambda c}$ (as indicated in the Duke data described above) would also influence the form of the compound-nucleus cross section. In the limit $\Gamma >> d$, such deviations are expected to modify the value 2 of the elastic enhancement factor [5,7], although the modification appears to be a very slight one. A precice measurement of W in this region, and in the absence of direct reactions, could yield interesting information on this problem. The best available value $W = 2.09 \pm 0.14$ at present has an uncertainty of about 7 percent [36]. A reduction of the error to one or two percent would, in the light of the recent results of refs. [2,5,7], be very welcome.

4. Precompound Reactions, Isospin Mixing in the Compound Nucleus, and Relaxation Processes in Nuclear Reactions Induced by Light Particles

From a thermodynamic point of view, (applicable if $\Gamma >> d$,) the Statistical Model is a model of an equilibrated nucleus: It assigns equal a priori probability to the partial width amplitude of every compound-nuclear level, i.e., to the available phase space. In a time-dependent formulation, this postulate implies instantaneous equilibration. Hence, for $\Gamma >> d$ the Statistical Model predicts independence of formation and decay of the compound nucleus, evaporation spectra determined by the available phase space, and symmetry of the angular distributions about 90°c.m. The model is expected to work as long as the nuclear equilibration time τ_{eq} is much smaller than the nuclear lifetime $\hbar/\Gamma$. Rough estimates [6,37] of τ_{eq} yield values around 10^{-21} sec or so. The average decay width Γ increases roughly like $\exp(2\sqrt{aE})$ with excitation energy E. Taking a value for Γ at neutron threshold of $\Gamma = 1$ eV and choosing $a = 20$ MeV^{-1} (typical values for heavy nuclei), we find that $\hbar/\Gamma$ becomes comparable to τ_{eq} for neutron bombarding energies around 10 MeV.

Experimentally, deviations from the Statistical Model are indeed observed at such energies: The "evaporation peak" acquires a high-energy tail. In this tail, angular distributions are peaked in the forward (incident) direction. These features are interpreted as being due to particle emission while the compound nucleus equilibrates. The physical picture behind all models for this equilibration process is the following. Through a series of two-body collisions with the nucleons in the target, the incident particle gradually loses its memory of the incident direction; its energy becomes degraded. Particles emitted before equilibrium is reached preserve some memory of the incident direction (forward peaking), and carry more energy than when emitted after equilibration (the energy is shared by fewer degrees of freedom). This picture can be implemented using either classical ("cascade") or quantum-statistical models of various degrees of complexity and refinement; for reviews see refs. [38,39]. The quantal models use a hierarchy of states of increasing complexity with occupation probabilities that obey a Master equation. This very active field of nuclear reaction theory cannot be reviewed here: There exist numerous and detailed studies of model- and parameter dependences [40-43,48], of simplified analytical and computational procedures [44,45], of the emission of complex clusters [39,46,47], of prequilibrium processes in heavy-ion collisions [42,49,50] (of which I can give no more than the most recent references I am aware of), not to speak of various computer programs embodying further physical assumptions, and the numerous associated fits to the data. I confine myself in the sequel to a few theoretical aspects of the microscopic foundations of such models which have received attention lately.

Ref. [6] contains the first microscopic statistical model which connects the phenomenological theory of precompound reactions with a random-matrix model of the nuclear Hamiltonian. It is based on the postulate of a hierarchy of states - nowadays also referred to as "chaining hypothesis". The model, originally intended to justify the use of compound and precompound models in the calculation of angle-integrated spectra, was later extended [51] to include angular distributions. This was possible through the concept of the "fast particle" and a modified set of statistical assumptions on matrix elements containing the wave vector of the fast particle as argument. Both these modifications are, however, somewhat ad hoc and lack a substantiated justification and derivation in the framework of a microscopic random-matrix model. The results obtained - an angle-dependent Master equation - have, on the other hand, been widely used (albeit sometimes in modified form) to calculate angular distributions and spectra of preequilibrium reaction products.

Parallel to these developments, the MIT group developed an approach [52] which, although fundamentally similar, is considerably different in detail. Particularly attractive is the decomposition of Hilbert space into parts which give separately rise to "multistep direct" and "multistep compound" reactions, in my opinion an improvement over the concept of the "fast particle". Equally attractive is the use of a DWBA-type prescription for calculating the multistep-direct process, intuitively a more convincing (although at the same time much more elaborate) procedure than the use [51] of nucleon-nucleon cross sections without reference to refractive and absorptive effects. On the other hand, the statistical assumptions used in ref. [52] are made on quantities derived in the course of the formal development, not on the basic matrix elements characterizing the Hamiltonian. In another context, an analogous procedure has led to difficulties [53].

It was pointed out by Yoshida [50] and Tang [54] that the two approaches bear a considerable formal similarity which can be displayed by using a series expansion [44]. This enhances confidence in the general result. The basic quantum-statistical assumptions underlying precompound models have been critically examined in refs. [37,55]. A theory of precompound processes developed by Bunakov et al. [56] uses quantum-statistical procedures familiar from solid state physics. Precompound processes have also been analysed in the framework of a pole-expansion of the nuclear S-matrix [57]. Yet another approach to the problem [58] establishes contact with nuclear relaxation processes as described by adding a collision term to a time-dependent Hartree-Fock description of the nuclear mean field. It raises the interesting question whether the main contributions to the observed precompound processes, i.e. particles produced in the first two collisions of the incident projectile, can be described statistically at all. This question is open. Finally, classical equations have been shown to give good results, too [59].

This listing shows that several theoretical questions remain open. The microscopic connection between and the ranges of validity of random-matrix models, phenomenological models like the exciton model, models using a Boltzmann-type collision term, and classical models are not clear. Direct processes have so far not been included. Statistical assumptions are in need of a better justification.

Preequilibrium models are intuitively expected to contain the Hauser-Feshbach cross section of standard compound-nucleus theory as derived from the Statistical Model, as their equilibrium component. A nice corroboration [45] of this expectation is shown in Fig. 4.1 which also shows (on a logarithmic scale) the differences between the experimental spectrum and an equilibrium calculation. This difference is well accounted

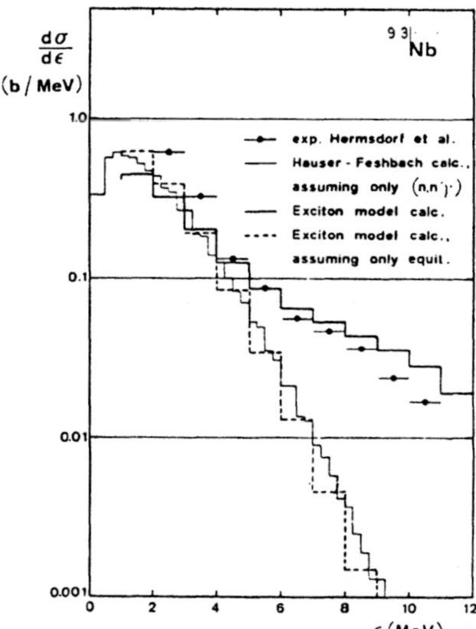

Fig. 4.1 Results of exciton-model and Hauser-Feshbach calculations (assuming no secondary particle emission which is important only for small secondary neutron energies) for the reaction ^{93}Nb(n,n') at 14.6 MeV. Also plotted is the experimental cross section versus neutron energy [60]. The data have been integrated over $\Delta\epsilon$ = 1 MeV and over the full solid angle (from ref. [45]) .

for by the preequilibrium component. The calculated curves are free of ad-hoc fit parameters; two constants in the calculations have been determined from a universal fit to many data. Angular distributions obtained in the same calculation [45] for a different nucleus are shown in Fig. 4.2 . The experimental data were summed over the energy intervals as indicated, mainly to smooth out some structures perhaps due to level-density effects. The increased forward-peaking at higher energies is clearly visible. The disagreement at the backward angles is probably due to an over-simplification of the model - diffraction and refraction effects were completely disregarded. Comparisons as the ones just shown establish the exciton model as a very useful tool in interpreting data and predicting cross sections.

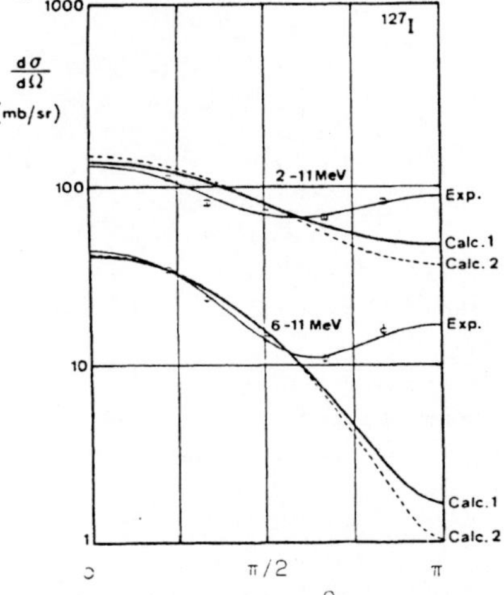

Fig. 4.2 Energy-integrated experimental [60] and calculated angular distribution of neutrons emitted in the reaction ^{127}In+n at 14.6 MeV. The curves labelled Exp. are three-term Legendre fits. Calculation 1 was obtained in terms of the usual global parameters. In calculation 2, a closed form approximation was used (from ref. [45]) .

By way of comparison, Fig. 4.3 shows the result of a multistep DWBA calculation [61] for the reaction Pb(p,p') at a proton energy of 62 MeV in comparison with the data which again are summed over bins of proton energy as indicated. Note the very much higher energy than in the previous figure. The dashed curve is the result of a one-step calculation, the full curve includes protons emitted in the second collision.

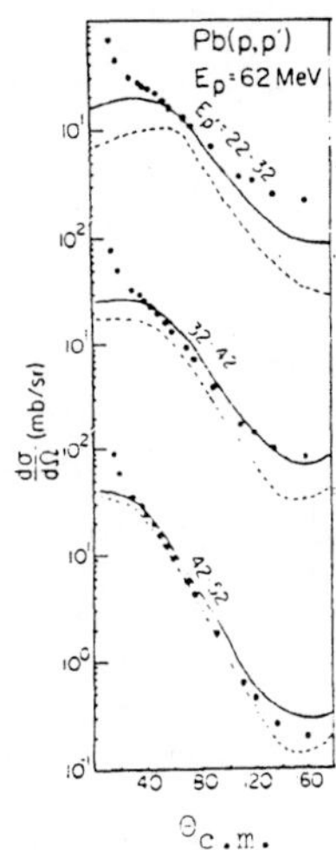

Fig. 4.3 Comparison of a one- and two-step DWBA calculation (dotted and full lines, respectively) with the experimental data (dots) (summed over energy bins as indicated) for the reaction Pb(p,p') at 62 MeV proton energy (from ref. [61]) .

Higher steps are expected to yield very little contribution. The spirit of the calculations is very much akin to ref. [52]. The agreement with the data (dots) is satisfactory and extends to large scattering angles; no arbitrary normalization factor is involved. The discrepancy at small angles is again blamed on insufficiencies of the basic DWBA cross sections.

Exciton-model calculations are much easier and faster to perform than multistep calculations using DWBA matrix elements. On the other hand, the latter contain physical ingredients - notably refraction and diffraction - missing in the former. Perhaps some compromise formulation can be found which unites the advantages of both approaches.

Reactions populating channels with different isotopic spins can be used to test isospin mixing in compound-nucleus reactions. Such mixing, in turn, can be viewed as the simplest example of a preequilibrium reaction: Two classes of strongly overlapping resonances with isospin quantum numbers T_1 and T_2 are mixed through isospin breaking forces. If the mixing time is short compared to the nuclear lifetime, only standard compound-nucleus decay is expected. In the converse case, the situation is analogous to precompound decay, with only two classes of states occurring in the hierarchy, while in real precompound processes this number is much bigger and typically ten or so. Because of this simplification, isospin mixing compound-nuclear reactions provide a very fine test on the underlying theory and on the mixing matrix elements. The autocorrelation function of the nuclear S-matrix is expected to show two correlation lengths corresponding to the two classes, and there is experimental evidence supporting this prediction. Phenomena such as the existence of several correlation lengths, or equivalently, several nuclear lifetimes, are not expected ever to be observable in standard precompound processes. Lack of space prevents me from going into details. A recent reference is ref. [62] where earlier work on the subject may be found.

5. Summary

During the last decade, the theory of nuclear reactions has made very considerable progress towards a unified theory treating direct, precompound and compound-nucleus reactions on the same footing. Aside from this - yet unattained - goal, many questions remain open. Some of them were formulated above. Others range from the foundations of random-matrix models in nuclear physics to detailed parameter studies for applications. The connection between the microscopic statistical theories sketched above and classical ("cascade") models successful at higher energies is also unexplored. Even in the domain of fairly low energies, the theory of nuclear reactions is yet far from being completed.

References

1) T.A. Brody, J. Flores, J.B. French, P.A. Mello, A. Pandey, S.S.M. Wong, Rev.Mod.Phys.53,385(1981)

2) R.R. Whitehead, A. Watt, D. Kelvin and A. Conkie, Phys.Lett. 76B, 149 (1978); J.J.M. Verbaarschot and P.J. Brussaard, ibid. 86B, 155 (1979); S.M. Grimes and S.D. Bloom, Phys.Rev.C23,1259 (1981)

3) C. Mahaux and H.A. Weidenmüller, Ann.Rev.Nucl. Part.Sci 29, 1 (1979)

4) A.M. Lane and J.E. Lynn, Proc.Phys.Soc. London LXX 8-A, 557 (1957); L. Dresner, Proc.Int.Conf. on Neutron Interactions with the Nucleus, Columbia University 1957. Report CU 175, p.71, New York: Columbia Univ. (1957); P.A. Moldauer, Phys.Rev. 123, 968 (1961) and Phys.Rev. C14, 764 (1976); G. Reffo, F. Fabbri, H. Gruppelaar, Lett. Nuovo Cimento 17, 1 (1976); H. Gruppelaar, G.Reffo, Reprint Technical Notes, Nucl.Sci.Eng. 62. 756 (1977). An interesting recent contribution to this question is T.H. Seligman, Z. Phys. A305, 69 (1982)

5) P.A. Moldauer, Nucl.Phys. A344, 185 (1980)

6) D. Agassi, H.A. Weidenmüller and G. Mantzouranis Phys. Lett. C22, 145 (1975)

7) H.M. Hofmann, T. Mertelmeier and H.A. Weidenmüller Phys. Rev. C24, 1884 (1981)

8) H.M. Hofmann, J. Richert, J.W. Tepel and H.A. Weidenmüller, Ann.Phys. (N.Y.) 90, 403 (1975); P.A. Moldauer, Phys. Rev. C11, 426 (1975); ibid. C12, 744 (1975)

9) H.M. Hofmann, T. Mertelmeier, M. Herman and J.W. Tepel, Z. Phys. A297, 153 (1980)

10) P.A. Mello, Phys.Lett. 81B, 103 (1979); P.A. Mello and T.H. Seligman, Notas de Fisica 2, 182 (1979); Nucl.Phys. A344, 489 (1980); J. de los Reyes, P.A. Mello and T.H. Seligman, Z.Phys. A259, 247 (1980); G.P. Lopez, P.A. Mello and T.H. Seligman, Z.Phys. A302, 351 (1981)

11) C.A. Engelbrecht and H.A. Weidenmüller, Phys.Rev. C8, 859 (1973)

12) For the case $\Gamma >> d$, this formula was first suggested by Z. Vager, Phys.Lett. B36, 269 (1971) and derived by M. Kawai, A.K. Kerman and K.W. McVoy, Ann.Phys. (N.Y.) 75, 156 (1973)

13) A.M. Lane, in: Neutron-capture γ-ray spectroscopy, RCN Petten 1975, p.31

14) A.M. Lane, Invited paper at the Europhysics Study conference at Hvar 1979 (unpublished)

15) J.B. Garg, V.K. Tikku, J. Halperin and R. Macklin, Phys.Rev. C23, 683 (1981)

16) J.B. Garg, V.K. Tikku, J.A. Harvey, R.L. Macklin and J. Halperin, Phys.Rev. C24, 1922 (1981)

17) D.J. Horen, J.A. Harvey and N.W. Hill, Phys.Rev. C24, 1961 (1981)

18) F. Hinterberger, P. von Rossen, S. Cierjacks, G. Schmalz, D. Erbe and B. Leugers, Nucl.Phys. A352, 93 (1981)

19) C.R. Jungmann, H. Weigmann, L. Mewissen, F. Poortmans, E. Cornelis and J.P. Theobald, preprint (1982) and contribution to this Conference

20) M. Potokar, Phys. Lett. B92, 1 (1980)

21) G.E. Mitchell, in: "Theory and Applications of Moment Methods in Many-Fermion Systems", edited by B.J. Dalton et al., Plenum Publ.Corp. p.33 (1980)

22) A.M. Lane, T.R. Dittrich, G.E. Mitchell and E.G. Bilpuch, Phys.Rev.Lett. 41, 454 (1978)

23) G.E. Mitchell, T.R. Dittrich and E.G. Bilpuch, Z. Phys. A289, 211 (1979)

24) W.K. Wells, E.G. Bilpuch and G.E. Mitchell, Z. Phys. A297, 215 (1980)

25) W.A. Watson III, E.G. Bilpuch and G.E. Mitchell, Z. Phys. A294, 153 (1980)

26) B.H. Chou, G.E. Mitchell, E.G. Bilpuch and C.R. Westerfeld, Phys.Rev.Lett. 45, 1235 (1980)

27) B.H. Chou, G.E. Mitchell, E.G. Bilpuch and C.R. Westerfeld, Z. Phys. A300, 157 (1981)

28) J.F. Shiner, Jr., E.G. Bilpuch, C.R. Westerfeld and G.E. Mitchell, Z. Phys. A305, 307 (1982)

29) W.A. Watson III, E.G. Bilpuch and G.E. Mitchell, Z. Phys. A300, 89 (1981)

30) W.A. Watson III, E.G. Bilpuch and G.E. Mitchell, Nucl. Instr. Meth 188, 571(1981)

31) D.W.S. Chan, Ph.D. Thesis, University of Lowell, August 1981 (unpublished)

32) D.W.S. Chan, J.J. Egan, A. Mittler and E. Sheldon, Phys. Rev. C (in publication)

33) D.W.S. Chan and E. Sheldon, submitted to Phys.Rev. (1982) and E. Sheldon and D.W.S. Chan, contribution to this Conference

34) R.P. DeVito, S.M. Austin, W. Sterrenburg and V.E.P. Berg, Phys. Rev. Lett.47, 628 (1981)

35) C.E. Floyd, P.P. Guss, K. Murphy, R.C. Byrd, G. Tungate, S.A. Wender, R.L. Walter and T.B.Clegg, Phys. Rev. Lett. 47, 1042 (1981)

36) W. Kretschmer and M. Wangler, Phys.Rev.Lett. 41, 1224 (1978)

37) H.Machner, Z. Phys. A302, 125 (1981)

38) M. Blann, Ann. Rev. Nucl. Sci. 25, 123 (1975)

39) E. Gadioli and E. Gadioli-Erba, preprint SMR/68/I-1 , International Centre for Theoretical Physics, Trieste, 1980, unpublished

40) M. Blann, Phys. Rev. C17, 1871(1978)

41) E. Gadioli, E. Gadioli-Erba, G. Tagilaferri, Phys. Rev. C17, 2238 (1978)

42) J.M. Akkermans and H. Gruppelaar, Z.Phys. A300, 345 (1981)

43) H. Machner, Z. Phys. A302, 125 (1981)

44) J.M. Akkermans, Z. Phys. A292, 57 (1979)

45) J.M. Akkermans, H. Gruppelaar and G. Reffo, Phys. Rev. C22, 73 (1980)

46) E. Gadioli and E. Gadioli-Erba, J.Phys. G : Nucl. Phys. 8, 83 (1981)

47) H. Machner, Phys. Lett. 86B, 129 (1979)

48) J. Ernst and J. Rama Rao, Z. Phys. A281, 129(1977)

49) M. Blann, Phys. Rev. C23, 205 (1981)

50) S. Yoshida, preprint 1982, Tohoku University

51) G. Mantzouranis, D. Agassi and H.A. Weidenmüller, Phys. Lett. 57B, 220 (1975)
G. Mantzouranis, H.A. Weidenmüller and D. Agassi, Z. Phys. A276, 145 (1976)

52) H. Feshbach, A. Kerman and S. Koonin, Ann. Phys. (N.Y.) 125, 429 (1980)

53) H.A. Weidenmüller, Phys. Rev. C9, 1202 (1974)
P.A. Moldauer, Phys. Rev. C11, 426 (1975)

54) Tang Xue-tian, preprint 1981

55) V.E. Bunakov, Z. Phys. A297, 323 (1980)

56) V. Bunakov and N. Nesterov, Phys.Lett. 60B, 417 (1976); V. Bunakov, Sov.J.Nucl.Phys. 25, 271(1977) V. Bunakov, Proceedings of the Trieste Course on Nuclear Theory for Applications, Vienna, IAEA 1980; V. Bunakov, Fiz. Elem. Chastits At. Yadra II, no. 6 (1980)

57) W.A. Friedman, M.S. Hussein, K.W. McVoy and P.A. Mello, Phys. Lett. C77, 47 (1981)

58) P. Mädler and R. Reif, Nucl. Phys. A373, 27(1982)

59) H.C. Chiang and J. Hüfner, Nucl.Phys. A349, 466 (1980)

60) D. Hermsdorf et al., Zentralinstitut für Kernforschung, Rossendorf bei Dresden, Report No. ZfK-277 (1974) unpublished

61) T. Tamura and T. Udagawa, Phys. Lett. 78B, 189 (1978)

62) H.L. Harney, H.A. Weidenmüller and A. Richter, submitted to Nucl. Phys.

Acknowledgements

The author is grateful to Dr. E. Gadioli, Dr. A.M.Lane, Dr. G. Mitchell and to Dr. E. Sheldon for reprints and preprints, and for communications, relating to the topic of this survey.

APPLICATION OF NUCLEAR MODELS TO NEUTRON NUCLEAR CROSS SECTION CALCULATIONS

Phillip G. Young

Theoretical Division, Los Alamos National Laboratory
Los Alamos, New Mexico 87545 U.S.A.

Nuclear theory is used increasingly to supplement and extend the nuclear data base that is available for applied studies. Areas where theoretical calculations are most important include the determination of neutron cross sections for unstable fission products and transactinide nuclei in fission reactor or nuclear waste calculations and for meeting the extensive dosimetry, activation, and neutronic data needs associated with fusion reactor development, especially for neutron energies above 14 MeV. Considerable progress has been made in the use of nuclear models for data evaluation and, particularly, in the methods used to derive physically meaningful parameters for model calculations. Theoretical studies frequently involve use of spherical and deformed optical models, Hauser-Feshbach statistical theory, preequilibrium theory, direct-reaction theory, and often make use of gamma-ray strength function models and phenomenological (or microscopic) level density prescriptions. The development, application, and limitations of nuclear models for data evaluation are discussed in this paper, with emphasis on the 0.1 to 50 MeV energy range.

[Nuclear reaction theory, nuclear model codes, nuclear data evaluation]

Introduction

Requirements for nuclear data are sufficiently broad that even with our present body of experimental data many areas remain where the application of nuclear theory is important. The purposes for applying theory range from providing simple interpolation tools in regions where measurements are abundant and consistent to actually predicting nuclear data for nuclei or energy regions inaccessible to experiment. The most common situation involves both these extremes in that one usually builds a theoretical parameter base from the available experimental data and then uses theory to extrapolate that information into unknown regions. The uncertainty in the final result depends, of course, upon how "far" the extrapolation extends in a physical sense.

The most stringent predictive requirements for theory involve such applications as neutron absorption and scattering by reactor fission products; production, depletion, and absorption calculations for actinides produced in reactors; dosimetry and activation calculations for unstable nuclides that will be produced in fusion reactors; and extension of the data base from the 15 to 50 MeV energy range for facilities that utilize higher energy neutrons, for example, d + Li neutron sources. It should be emphasized, however, that the application of theory for evaluation purposes remains important even for the more common materials where measurements are abundant. The reason is simply that discrepancies occur in the experimental data base, and nuclear theory can provide hints both as to whether in fact discrepancies exist in given situations and what the resolution of the discrepancies might be.

The aim of this paper is to briefly review the main nuclear reaction models that are being used to calculate data for applications and to convey an idea of their capabilities and deficiencies. To avoid redundancy with other papers, only applications of theory above $E_n \sim 100$ keV will be discussed and fission will not be considered. Emphasis will be given to the general features of the theories, as several excellent papers are available that detail the mathematics (for example, see Refs. 1-7).

We will begin with a brief discussion of applications of theory for light elements, but most of the paper will focus on analyses of neutron-induced reactions on intermediate and heavy mass materials involving spherical or deformed optical models, preequilibrium theory, and Hauser-Feshbach statistical theory. Some of the nuclear theory computer codes in common use will be summarized, and examples from several recent analyses will be described. We will conclude by briefly discussing some of the directions being pursued that offer promise for improved predictions in the future.

Applications of Theory for Light Elements

Because of the individual character of most light elements, the use of nuclear theory in developing applied data has been mainly limited to short extrapolations of experimental data using fairly simple models. An important exception to this occurs for coupled-channel R-matrix theory, which has been extensively applied in several light element systems, particularly the A = 7 and 11 systems that include the ^{6}Li(n,α) and ^{10}B(n,α) standard reactions.[8,9] Other compound systems where R-matrix methods have been used are A = 2, 3, 4, 5, 13, 16, and 17.[10]

In conjunction with the R-matrix studies, a new resonance model has recently been developed by Hale[11] to describe energy spectra of particles in reactions involving three-body final states. Typically, such spectra consist of relatively narrow peaks on top of broad, underlying structures commonly attributed to "three body phase space" contributions. However, such structures can also come from kinematically broadened resonance effects. In the new model, an expression for the transition amplitude was derived from the three-body Schröedinger equation, assuming that the relative wave functions for pairs of final-state particles are dominated by single resonances. This assumption allows the three-body spectra to be calculated in terms of known parameters for the two-body resonances, with full account being taken of interference between direct and exchange amplitudes.

Calculations of neutron emission spectra from n + d reactions with 14-MeV neutrons are compared in Fig. 1 to measurements[12] at 10 and 60° and to other calculations.[13] There is reasonable agreement in shape at both angles (note that the calculated curves have not been broadened for experimental resolution), but the 60° calculation overpredicts the data somewhat. This model is still under development, but it offers promise to broaden the scope of several R-matrix-based evaluations for light elements.

Other areas where nuclear theory is utilized for light element applications include use of optical, statistical, and intranuclear-cascade models to extrapolate data to unmeasured regions. For example, a spherical optical model fit of elastic angular distributions below 15 MeV and total cross sections from 10 to 20 MeV was recently used[14] to extrapolate a ^{7}Li evaluation to 20 MeV. Similarly, intranuclear-cascade calculations[15] are being combined with new measure-

K. H. Böckhoff (ed.), Nuclear Data for Science and Technology, 506–517.

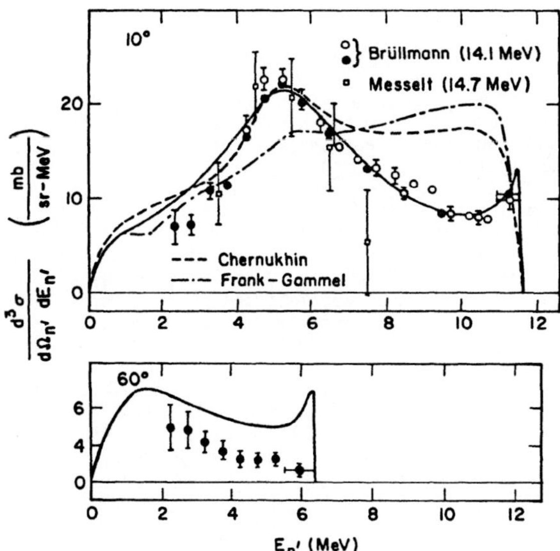

Fig. 1. Resonance model calculations (solid curves) compared to measured[12] n + d neutron spectra at θ_{lab} = 10 and 60° for 14.1-MeV incident neutrons. The dashed curves represent other calculations.[13]

ments of hydrogen and helium emission spectra from 27 to 61 MeV neutron reactions on carbon to develop a data base for several applications in this region.

Intermediate- and Heavy-Mass Nuclei

The sequence of steps followed in applying nuclear theory for data evaluation of intermediate- or heavy-mass nuclei can vary greatly depending on the nuclei involved, the energy range and reaction types required, and the accuracy needed in the evaluation. Typically, an analysis involves determination of optical model potentials for both neutrons and charged particles; development of a model for calculating gamma-ray transmission coefficients; use of a level density formulation in combination with the available experimental data on discrete states; estimation of direct and preequilibrium reaction effects; use of a fission model when appropriate; and specification of a framework for combining the above components, usually Hauser-Feshbach statistical theory. More advanced unified theories[16-18] that combine compound and direct reaction effects in a realistic manner are being explored in nuclear data calculations[18-20] but have thus far not seen wide use in applied data calculations. This approach is the subject of other papers[21,22] at this conference and will not be discussed here.

Optical Model Analyses

Most modern theoretical data evaluations are built around an optical-model analysis using either a spherical or deformed potential, depending upon the particular mass region being studied. The importance of this component to an analysis is obvious since it provides not only the total, shape elastic, and reaction cross sections but also the neutron and charged-particle transmission coefficients that are used in Hauser-Feshbach statistical theory calculations. An important and demanding requirement of such analyses is that they usually must cover a very wide energy range; typically, 1 keV to 20 MeV or higher. The low-energy transmission coefficients continue to be important even for high incident energies in order to correctly calculate particle emission in the various reaction chains.

Except for general scoping calculations or studies in regions completely devoid of data, the modern trend is to focus such analyses on the mass region of immediate interest rather than to use global optical model parameters. The SPRT method developed by Lagrange[23] and coworkers has been widely used to determine optical model parameters. Basically, this method involves fitting experimental values of s- and p-wave neutron strengths, potential scattering radii, total cross sections, and elastic and inelastic scattering data to determine the optical model parameters. Automated fitting techniques are generally not required with this method but have frequently been used[24-26] in determining spherical potentials.

Computations with deformed optical potentials are much more time consuming, of course, and one of the advantages of the SPRT method has been that automatic searching is not required. It has been observed in several analyses[24-26] that calculated (n,xn) cross sections near threshold are very sensitive to low energy transmission coefficients, and comparisons with experiment have been used to test or further optimize parameters determined by the SPRT approach.

An important development that significantly reduces computation time in deformed optical model analyses for odd-A nuclei is described in a recent paper by Lagrange, Bersillon, and Madland.[27] Using a strong coupling rotational model, it is shown that coupled-channel calculations for an odd-A nucleus can be approximated by performing the same calculation with a suitably chosen (fictitious) K = 0 rotational band and appropriately combining the results. For example, calculations for the ground-state rotational band of ^{239}Pu coupling 5-states (J = 1/2, 3/2, ..., 9/2) can be accurately approximated by a 3-state calculation (J = 0, 2, 4) with a reduction by a factor of 27 in computing time. Similarly, replacement of a ^{241}Pu calculation coupling 5-states having J = 5/2, 7/2, ..., 13/2 by an appropriate J = 0, 2, 4 calculation reduces computation time by a factor of ∿ 54, although the approximation is poorer.

A comparison of cross sections calculated at E_n = 4 MeV for the above cases is given in Table I. The approximation is nearly exact for ^{239}Pu (K = 1/2) at this energy. The fictitious values are less precise for ^{241}Pu (K = 5/2) but note that the integrated cross sections are still quite accurate. Although not shown, similar accuracies are achieved for transmission coefficients after suitable collapsing.

Even using such approximate methods, the complexity and expense involved in performing coupled-channel calculations when many levels are involved quickly becomes prohibitive. Hodgson[28] recently proposed an alternative method for calculating inelastic cross sections. In particular, he determined that, if the coupling between excited states is small, inelastic scattering can be calculated for deformed nuclei with standard DWBA theory if a deformed potential is used to determine the exit channel wave function.

Increasing emphasis has been placed in recent years on linking analyses of (n,n), (p,p), and (n,p) data by means of the Lane model.[29] Basically, this model relates the nuclear potentials for the three different reaction types through isospin considerations and permits, for example, the deduction of neutron potentials from the analysis of proton measurements. Recent reviews discussing and applying this model have been given by Rapaport[30] and by Hansen.[31] The latter review also addresses the importance of including coupling effects in calculations of deformed nuclei, and both topics are illustrated in Fig. 2, taken from that paper.

Figure 2 compares calculations of (n,n) scattering angular distributions with measurements[32] for Ta, Au,

Table I Comparison of ^{239}Pu (K=1/2) and ^{241}Pu (K=5/2) Cross Sections Using Real and Fictitious Levels in Coupled-Channel Calculations for 4-MeV Neutrons.
(σ_i refers to the cross sections of the first four excited states.)

	^{239}Pu		^{241}Pu	
	Real	Fictitious	Real	Fictitious
σ_{TOT}	7.797 b	7.796 b	7.821 b	7.831 b
σ_{CN}	3.124	3.124	3.171	3.120
σ_{EL}	4.249	4.247	4.343	4.398
σ_1	0.1233	0.1232	0.1533	0.1660
σ_2	0.1845	0.1847	0.0864	0.0751
σ_3	0.0522	0.0517	0.0418	0.0674
σ_4	0.0644	0.0646	0.0268	0.0117

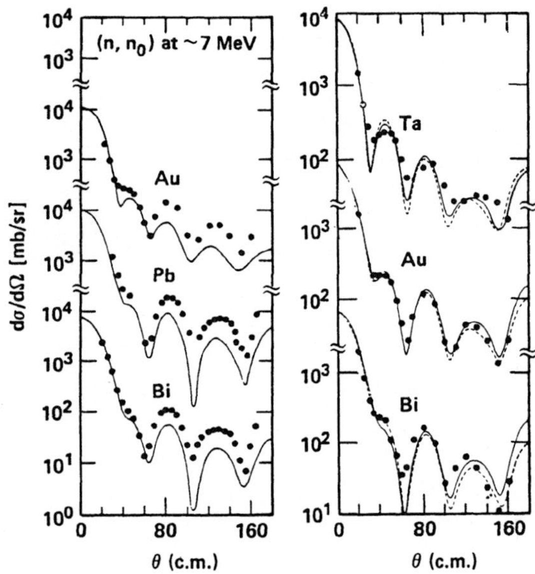

Fig. 2. Calculated and measured neutron elastic measurements near 7 MeV as presented in Ref. 31. See text for details.

Pb, and Bi near 7 MeV. The neutron potential used to calculate the solid curves on both sides of the figure were determined using the Lane model from analyses of (p,p) and (p,n) data. The curves on the left were obtained in a distorted-wave Born approximation (DWBA) calculation,[33] whereas the ones on the right result from a coupled-channel calculation.[34] The coupled-channel calculations using the Lane formalism agree about as well with the (n,n) experiments as do calculations using neutron global parameters[35] (optimized to fit neutron data), shown by the dashed curves. The agreement with experiment is much poorer for the DWBA calculations.

Other developments that hold promise for improved predictive capabilities are the efforts at several laboratories to integrate microscopic model calculations into determination of optical potentials. Starting from basic calculations by Jeukenne, Lejeune, and Mahaux[36] of the optical potential in nuclear matter, Lagrange and Brient[37] have performed microscopic calculations of elastic and inelastic scattering from ^{208}Pb in the 8.5-61 MeV energy range. Similarly, Dietrich et al[38] found reasonable agreement with 24-MeV neutron scattering data for ^{208}Pb in a microscopic folding model calculation. Microscopic calculations have also been used to enhance or refine conventional global optical model potentials.[35]

Gamma-Ray Transmission Coefficients

In a recent review of fast neutron capture calculations, Gardner[3] summarized the status of statistical, direct, and semidirect theories used in calculations. A qualitative view of the relative importance of these contributions to (n,γ) cross sections is given in Fig. 3. For orientation, the rapid falloff of the statistical contribution typically occurs near E_n = 1 MeV, and the peak in the semidirect contribution is in the neighborhood of 14 MeV, where the (n,γ) cross section is ∿ 1 mb. For most applications the statistical contribution is clearly the most important of the three.

Two models are commonly used to determine gamma-ray transmission coefficients for statistical calculations, the Weisskopf single-particle model[39] and the giant dipole resonance (GDR) model.[40] Of these, the GDR model has been most successful in reproducing gamma-ray strength functions inferred from experimental data. Normalization of the gamma-ray strength function $f(E_\gamma)$ is usually accomplished from experimental information on $<\Gamma_\gamma>$ and $<D_0>$, the average gamma-ray width and spacing for s-wave resonances, through the relation

$$\frac{2\pi<\Gamma_\gamma>}{<D_0>} = \int_0^{S_n} f(E_\gamma)E_\gamma^3\ \rho\ (S_n-E_\gamma)dE \quad , \qquad (1)$$

where S_n is the neutron binding energy and ρ is the level density of the compound system.

The strength function for electric dipole radiation, which is the dominant transition, is usually taken as Lorentzian in shape (or as a sum of 2 Lorentzians for deformed nuclei). More recently Gardner et al[3,41] have investigated the use of Breit-Wigner shapes and have developed expressions for f_{E1} based on systematics covering the mass range A > 40. A comparison of strength functions calculated with both representations is given in Fig. 4 with points inferred from measurements[42] on ^{89}Y. While the normalizations of both curves are somewhat high in this case, the shape of the dashed curve calculated using the Breit-Wigner form more nearly follows the measured points. In most applied problems where f_{E1} is used to compute capture cross sections or gamma-ray competition with particle emission, the results are not highly sensitive to this difference in shape, and the Lorentzian is still commonly used.

The preferred method[43] for performing gamma-ray calculations in regions where $<\Gamma_\gamma>$ or $<D_0>$ in Eq. (1) are unmeasured is to extrapolate the strength function f_{E1} rather than $<\Gamma_\gamma>$ and $<D_0>$. The latter quantities can vary by orders of magnitude in nearby nuclei, making reliable extrapolation difficult, whereas f_{E1} changes much more slowly.

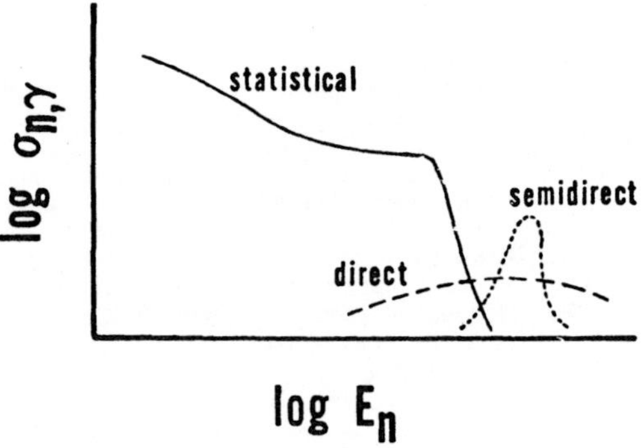

Fig. 3. Schematic view of the relative importance of different reaction mechanisms to neutron capture in a medium-weight nucleus.

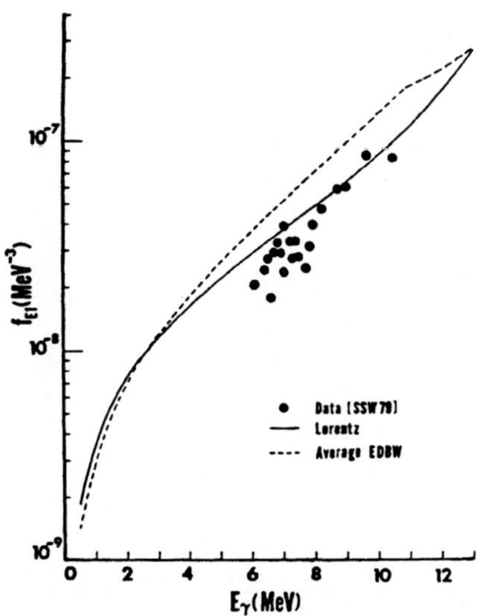

Fig. 4. Comparison of experimental values[35] of $f_{E1}(E_\gamma)$ from n + ^{89}Y with calculations using the Lorentz form and the Breit-Wigner (EDBW) form.

Nuclear Level Densities

The Hauser-Feshbach expression for the cross section from the initial state C to the final state C' through the compound nucleus spin and parity $J\pi$ is[44]

$$\sigma^{J\pi}_{CC''} = \frac{\langle\theta^{J\pi}_{C}\rangle \langle\theta_{C'}\rangle^{J\pi}}{\langle\theta^{J\pi}\rangle} W^{J\pi}_{CC'}, \qquad (2)$$

where $W^{J\pi}_{CC'}$ is a width-fluctuation correction[4] important at lower energies but which approaches one above a few MeV. The partial widths are obtained by summing the particle or gamma-ray transmission coefficients over the possible transitions to levels in the final state. Because level spacings rapidly become very small as excitation energy is increased, a continuum of levels must be introduced and the density of such levels specified.

For reasons of convenience, most calculations of applied data employ phenomenological level density models. The most commonly used are the Gilbert and Cameron[45] and back-shifted Fermi-gas models,[46] as well as a model by Ignatuyk[47] that has seen extensive use in (n,2n) calculations.[48] The Gilbert and Cameron model consists of a constant temperature form at low excitation energies, which is smoothly joined to a Fermi-gas shape at higher excitation energies. The level density parameters (a and T) are determined from empirical s-wave level spacings at the neutron binding energy ($E_x \sim$ 6 MeV) and from matching with the available discrete level data at low excitation energies. The back-shifted Fermi-gas model is a little simpler, consisting of a pure Fermi gas form. Its variables involve a level density parameter, a, and a ground-state energy-shift parameter, Δ, which are determined from the same data described above. The Ignatuyk expressions incorporate an excitation-energy dependent level-density parameter. Recent improvements to these models include a more accurate specification of spin cut-off parameters by Reffo[49] and updated fits by Cook[50] of other parameters.

Phenomenological level-density parameters are determined mainly near the neutron binding energy, and the energy dependence of shell and pairing effects is not necessarily well represented. Microscopic calculations of the state density, on the other hand, incorporate shell effects naturally because they use realistic shell-model single-particle levels, and the BCS superfluid model correctly includes pairing effects.[51] More recently, improved formalisms have been developed to handle unpaired nucleons in odd-A systems, that is, to include the blocking effect of single-particle levels due to the unpaired nucleon.[52] While such microscopic models are not necessarily more accurate at present than the phenomenological ones, they do include improved physics and are expected to better predict the energy dependence of level densities away from regions of experimental data.

A comparison of the Gilbert and Cameron and the back-shifted Fermi-gas phenomenological models with a microscopic thermodynamic model[53] was presented by Arthur[7] and is expanded in Fig. 5. In the upper half of the figure the state densities for ^{238}U calculated with the microscopic model are plotted versus excitation energy. In the lower half, the ratios of the Gilbert and Cameron and the back-shifted Fermi-gas models to the microscopic model are shown, with all the calculations normalized to experiment at the neutron binding energy ($\sim$ 6.1 MeV). No attempt was made to optimize the phenomological model parameters to represent the microscopic calculation; they were simply taken from the literature.[46,54]

The state densities from the phenomenological level densities differ from the microscopic calculation by as much as a factor of 2 between E_x = 0 and the neutron binding energy and by even greater factors at higher excitation energies. The region between E_x = 1 and 5 MeV is particularly important for calculating (n,n'), (n,xn), and (n,f) reactions. Although some cancellation of errors in level densities can occur in calculating competing reactions, this area is probably the one most in need of improvements for applied calculations.

Statistical-Preequilibrium Theory

For incident neutron energies above about 10 MeV, statistical model calculations of neutron cross sections and spectra must be corrected for nonequilibrium effects. The master equation exciton model[55] has been widely used in evaluations to calculate preequilibrium particle emission, as has the geometry-dependent hybrid model.[56] The basic idea of the master equation exciton model is that a given reaction is assumed to proceed through a series of particle-hole configurations, starting with simple ones and proceeding through more complicated ones, until equilibrium is reached. At each stage during the process, particle

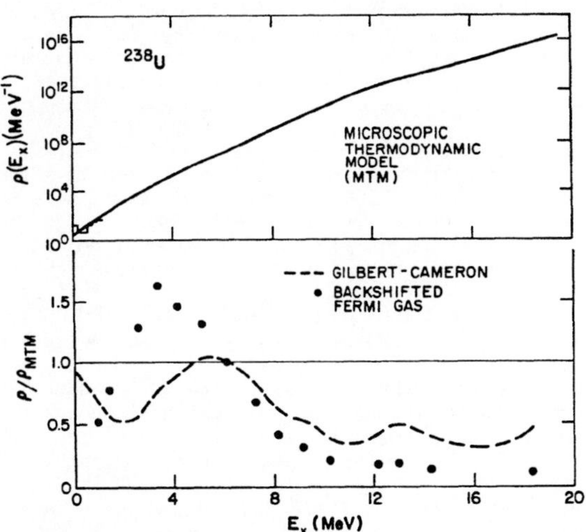

Fig. 5. Comparison of state densities calculated using a microscopic thermodynamic model[53] with values from the Gilbert and Cameron[45] and back-shifted Fermi-gas[46] models.

emission can occur with some probability, and a series of coupled equations must be solved to obtain cross sections and spectra.

Several recent reviews address preequilibrium theory in some detail. In his review, Jahn[6] considers the merits of both the master equation and geometry-dependent hybrid approaches and includes a number of example calculations. He points out that the latter model depends only on optical model parameters and is the only one that takes into account the diffuseness of the nuclear surface.

In other developments, Akkermans et al[57] have developed a unified model of equilibrium and preequilibrium emission, still based on the master equation, that permits calculation of angular distribution effects. The results are found to agree reasonably up to an outgoing energy of about 30 MeV with the semiempirical formulation of Kalbach and Mann,[58] which is commonly used in data evaluations. Applying Monte Carlo techniques, Akkermans and Gruppelaar[59] have used this model to calculate preequilibrium effects for the second and third particles in a reaction chain. Their results indicate that inclusion of preequilibrium in the tertiary steps is unnecessary below 50 MeV but is important for second particle emission above 25 MeV.

A procedure commonly followed in statistical preequilibrium calculations that carry angular momentum effects is to simply correct or scale the energy dependence of cross sections for emitted particles to account for preequilibrium effects. The statistical spin distribution of states in the final nucleus is then still maintained and does not properly reflect the preequilibrium process. Fu[60] has developed a procedure for incorporating preequilibrium effects into the angular momentum distribution of final states. Such considerations might be important, for example, in calculating production cross sections for isomers created in (n,2n) reactions. Fu has recently used this procedure to calculate cross sections for specific gamma rays created in the ^{57}Fe(n,2nγ)^{56}Fe reaction.[61] Comparisons with a measurement at 15 MeV are shown in Table II.

Table II Comparisons of Calculated and Experimental ^{57}Fe(n,2nγ)^{56}Fe Gamma-Ray Cross Sections for $E_n \sim 15$ MeV

Gamma-Ray Energy (keV)	Production Cross Section (mb) Predicted	Experiment
847	980	1071 ± 59
1238	425	451 ± 36
1811	39	33 ± 17
2113	36	41 ± 17
1038	46	61 ± 14
1303	73	117 ± 16
367	8	17 ± 6
1670	27	53 ± 11

Statistical-Preequilibrium Codes

A number of computer codes have been developed over the past few years that combine statistical and preequilibrium theory for the purpose of data evaluation. A selection of these are described here. See references 62 and 63 for more complete summaries.

The multireaction Hauser-Feshbach statistical preequilibrium codes GNASH,[64] HAUSER,[65] STAPRE,[66] and TNG[67] have been used extensively over the past few years. All four codes include full allowance for angular momentum effects and can calculate particle spectra as well as cross sections. All except HAUSER output gamma-ray spectra, and all except GNASH include width fluctuation corrections. The TNG code calculates angular distributions including preequilibrium effects, whereas GNASH, HAUSER, and STAPRE depend upon external codes for angular effects. GNASH, HAUSER, and STAPRE include fission channels with double-humped barriers and a similar capability is under development for TNG. GNASH and STAPRE are usually used in combination with the reaction theory code COMNUC[68] at lower energies. All four codes have been employed up to incident neutron and/or proton energies in the 30-50 MeV range. In addition to these codes, a more advanced multireaction Hauser-Feshbach code is under development by Uhl and Strohmaier that will automate much of the code setup and will be better adapted for evaluation work.[69]

The MSPQ[70] and ALICE[71] codes use evaporation theory for the statistical portion of the calculation and preequilibrium emission based on the master equation exciton and geometry-dependent hybrid models, respectively. Both codes calculate particle emission spectra, and MSPQ contains a fission channel as well.

The AMALTHEE[72] and PREANG[73] codes both use matrix methods to solve exactly the master equations of the exciton model without artificial division between preequilibrium and equilibrium components. PREANG has recently been modified to utilize a random walk model that simplifies and compacts calculations of multiparticle emission.[59] Both codes calculate particle emission spectra, and PREANG also calculates particle angular distributions.

Hauser-Feshbach Statistical-Preequilibrium Calculational Examples

There are a number of recent studies in which rather complete theoretical analyses have been performed in association with data evaluations. To illustrate the use of multireaction Hauser-Feshbach statistical preequilibrium calculations, some of the details of recently completed analyses of neutron reactions on ^{165}Ho, ^{169}Tm, and ^{182,183,184,186}W will be described. These analyses are linked through use of very similar deformed optical model parameterizations. Emphasis will be on the W-isotope analysis, as it preceded the Ho and Tm work, and the latter analysis is described in detail in another paper at this conference.[74] Several other recent analyses will also be briefly summarized.

Ho-Tm-W Analysis

Earlier calculations for W-isotopes using a spherical optical potential are described in a paper at the 1979 Knoxville conference.[75] The difficulty and ambiguity associated with deriving an equivalent spherical optical potential to represent deformed nuclei over extended energy ranges motivated us to revise the analysis using a deformed optical potential.[76] This approach has the advantage that a single neutron potential is used to calculate total, shape elastic, and direct inelastic cross sections as well as the neutron transmission coefficients used in compound elastic, (n,γ), (n,n'), and (n,xn) reaction calculations. To illustrate the inadequacy of spherical potentials in this mass region, a comparison is given in Fig. 6 of total and nonelastic cross sections from our coupled-channel (CC) analysis and values calculated with the spherical potential of Moldauer,[77] which gives good agreement with data for A < 140.

We used a symmetric rotational model with coupling of the ground state band members in our analysis, even though there is evidence of mixing between the two lowest band members for ^{183}W and ^{186}W. For the even-even isotopes, we chose a 0^+, 2^+, 4^+ coupling basis, while for ^{183}W the equivalent $1/2^-$, $3/2^-$, $5/2^-$, $7/2^-$, 9/2 basis was used.

The deformed optical potential was obtained by modifying the potential of Delaroche et al.[78] to give reasonable agreement with (n,2n) measurements on the four major W-isotopes near threshold. Care was taken to maintain the good agreement with W experimental data for s- and p-wave neutron strengths, potential scattering radii, neutron total cross sections to 15 MeV, (n,n) and (n,n') angular distributions at 3.4 MeV, and (p,p') angular distributions at 16 MeV that Delaroche had established in his SPRT analysis. Subsequently, the parameterization (with some modification) was found to give reasonable agreement with experimental data for ^{165}Ho and ^{169}Tm, as described in Ref. 74 and shown in Fig. 6 for $\sigma_{TOTAL}(^{165}\text{Ho})$. The calculation of the 3.4-MeV elastic and inelastic ^{184}W angular distribution and experimental data[78] are shown in Fig. 7, together with the ENDF/B-V evaluation (dashed curve).

The deformed optical model parameterization for W-isotopes from this analysis is given in Table III. The notation and form of the potential are the same as in Refs. 74 and 78. Slight modifications to the tabulated values of V and W_D were used in the actual evaluations to optimize agreement with data for the individual isotopes.

We obtained our gamma-ray transmission coefficients from an empirically determined gamma-ray strength function. A sum of two Lorentzians was used to represent f_{E1} [see Eq. (1)], with parameters taken from photonuclear measurements. The overall normalization of $f(E_\gamma)$ was achieved by comparison with (n,γ) cross-section measurements below 1 MeV for the various isotopes. Standard parameters were used for the exciton model preequilibrium calculation in the GNASH code, and level density parameters for the Gilbert and Cameron[45] formulation were obtained from the Cook tables.[54]

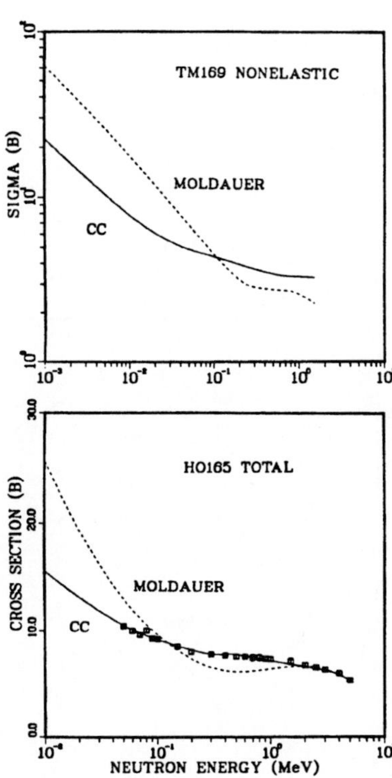

Fig. 6. Comparison of ^{169}Tm nonelastic and ^{165}Ho total cross sections calculated using a coupled-channel analysis (Parameter Set 1 in Ref. 74) with values calculated from a spherical optical potential.[77] The squares in the lower half are experimental data.[32]

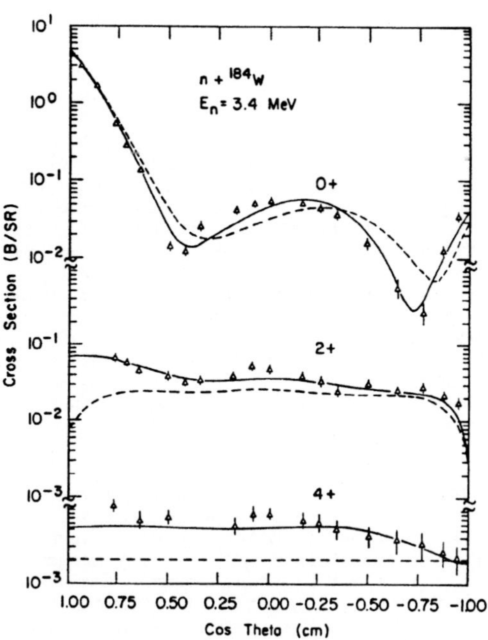

Fig. 7. Calculated[76] and measured[78] elastic and inelastic angular distributions for states in ^{184}W with 3.4-MeV incident neutrons. The dashed curves are ENDF/B-V.

Table III. Optical Parameters for Tungsten Isotopes

$$V(^n_p) = 49.8(^-_+)16\,\frac{N-Z}{A} + \Delta V_c - 0.25E$$

$$\Delta V_c = 0.4\,\frac{Z}{A^{1/3}} \quad \text{for incident protons}$$

$$= 0 \quad \text{for incident neutrons}$$

$$W_D\,(^n_p) = 5.1\,(^-_+)8\,\frac{N-Z}{A} + 0.6E \quad [E \leq 6.5]$$

$$= 9.0\,(^-_+)8\,\frac{N-Z}{A} - 0.1(E-6.5) \quad [E > 6.5]$$

$$W_V = -1.8 + 0.2\,E \quad [E > 9.0]$$

$$V_{SO} = 7.5$$

$$r_V = r_{SO} = 1.26f;\ r_D = 1.24f$$

$$a_V = a_{SO} = 0.61f;\ a_D = 0.45f$$

Isotope	β_2	β_4
^{182}W	0.223	-0.054
^{183}W	0.220	-0.055
^{184}W	0.209	-0.056
^{186}W	0.195	-0.057

Comparisons of calculated values with a few of the experimental results that were not included in the analysis are given in Figs. 8-10. Figure 8 compares the calculations (solid curves) with measurements by Smith et al[79] of inelastic scattering excitation functions for four levels in ^{184}W (the 1.125-MeV "level" is actually a cluster of three levels). The upper curves are for the 2^+ and 4^+ levels, which contain substantial direct reaction contributions, whereas the lower curves are entirely compound-nucleus calculations. The dashed curves represent ENDF/B-V.

Figure 9 compares the composite neutron emission spectrum from calculation of the four isotopes with 14-MeV measurements[34] for natural tungsten. There is some disagreement among the various measurements, but the calculation seems to represent the mean rather well above ~ 1.5 MeV and agrees with Vonach's data at lower energies.

Figure 10 compares calculated gamma-ray emission spectra for natural tungsten with three measurements[32] near 7.4 MeV. There is significant disagreement for E_γ > 2 MeV with the Dickens data, but generally reasonable agreement with the other measurements. This trend is observed in similar comparisons at other energies and could indicate an experimental problem.

As an additional illustration of the predictive capability of such analyses, a comparison is shown in Fig. 11 of preliminary experimental data by Haouat and Patin[80] with results from the ^{165}Ho and ^{169}Tm analysis described in Ref. 74. Neutron scattering

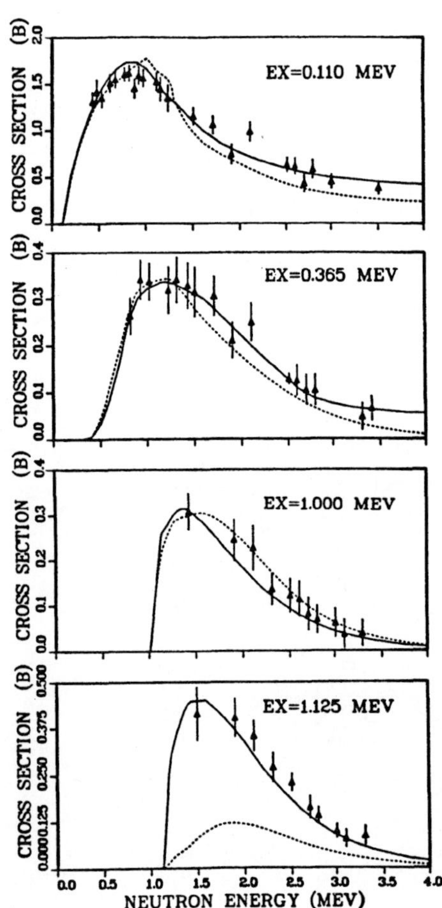

Fig. 8. Comparison of calculated[76] and measured[79] excitation functions for $^{184}W(n,n')$ reactions to four excited states. The dashed curves are ENDF/B-V.

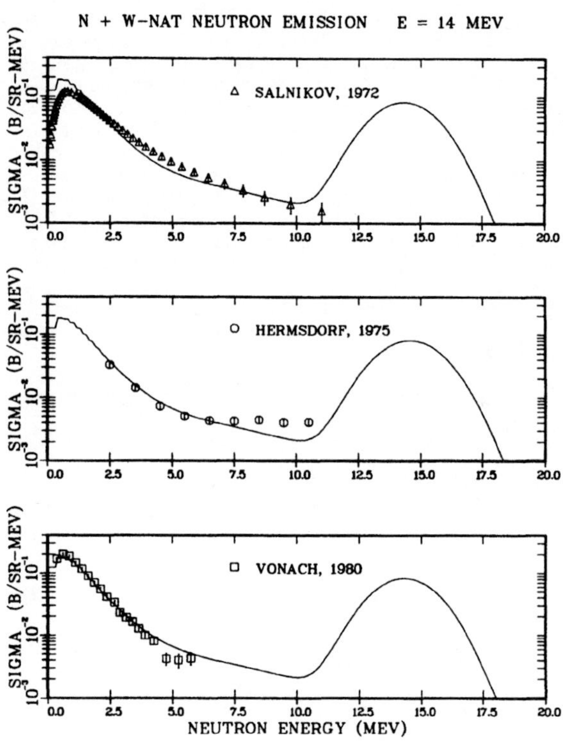

Fig. 9. Calculated[76] and measured[32] neutron emission spectra from 14-MeV neutron bombardment of natural W.

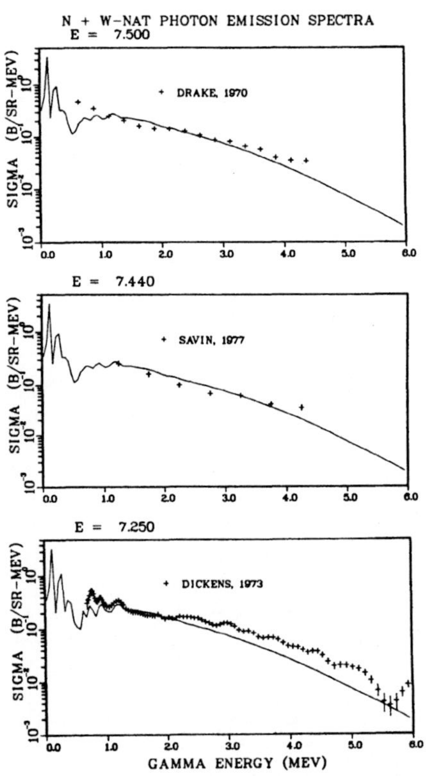

Fig. 10. Calculated[76] and measured[32] gamma-ray emission spectra from ~ 7-MeV neutron interactions on natural W.

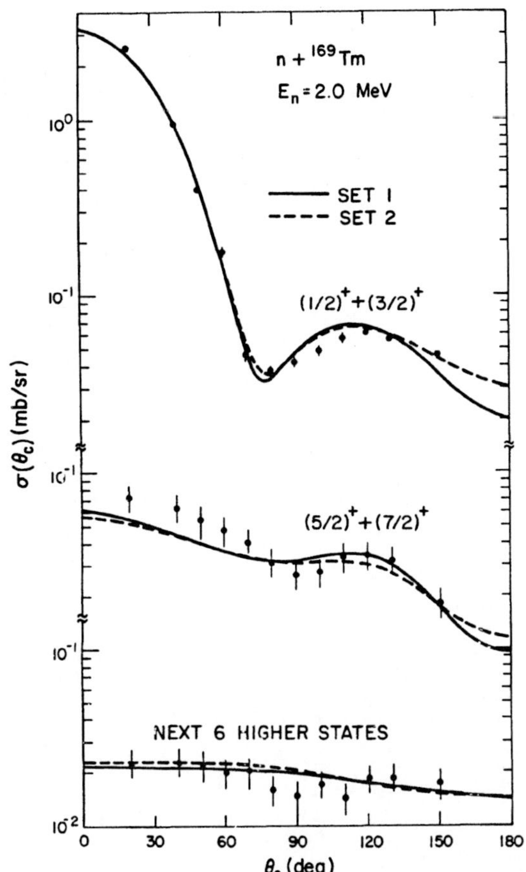

Fig. 11. Comparison of calculated elastic and inelastic neutron angular distributions at 2 MeV with the preliminary measurements of Haouat and Patin.[80] The solid and dashed curves represent coupled-channel calculations with Parameter Set 1 and Set 2 of Ref. 74.

data for ^{169}Tm were not available for that analysis, which mainly involved small modifications to the parameters of Table III to improve agreement with total cross-section data. The dashed curves represent the final results of that analysis.

Other Statistical-Preequilibrium Analyses

Other good examples of multireaction Hauser-Feshbach statistical-preequilibrium analyses include the papers at this conference by Strohmaier et al[81] describing an analysis of ^{52}Cr, ^{55}Mn, ^{56}Fe, and ^{58,60}Ni cross sections to 30 MeV with the STAPRE code; and an analysis using the COMNUC and GNASH codes of neutron data to 20 MeV for ^{209}Bi by Bersillon et al.[82] Both analyses utilize spherical optical potentials and use techniques similar to those described above.

The TNG code has recently been used by Fu[61] to update ENDF/B data for Fe and Cu, and by Hetrick et al[83] to calculate neutron-induced reactions on ^{40}Ca from 20 to 40 MeV. During the Cu analysis, a factor of 5 error was discovered in the ENDF/B-IV ^{63}Cu(n,p) cross section due to a misinterpretation of experimental data by the ENDF/B-IV evaluator. A comparison of calculated and experimental[84] proton emission spectra for 14-MeV neutrons on ^{63}Cu is shown in Fig. 12 with the individual reaction components separated. The error in ENDF/B-IV resulted because the (n,pn) reaction component of Fig. 12 was erroneously included in the (n,p) cross section. Fu's analysis[61] of Fe also supported an earlier observation by Young et al,[85] based on nuclear model calculations, that an error is likely in a 14-MeV measurement[86] of the gamma-ray emission spectrum from Fe.

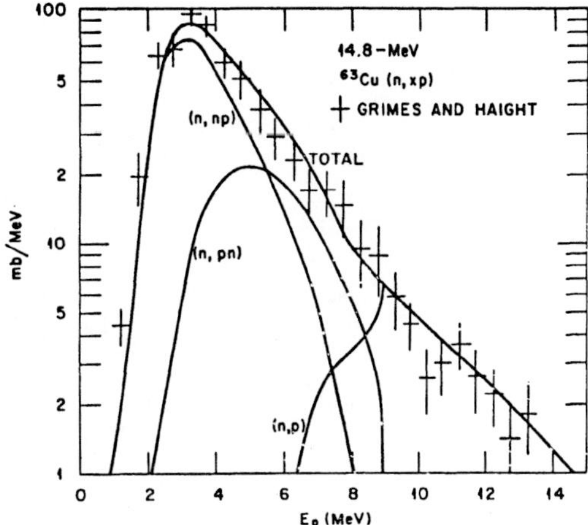

Fig. 12. Calculated[61] and measured[84] proton emission spectra from 14.8-MeV neutron bombardment of ^{63}Cu.

In the ^{40}Ca analysis by Hetrick et al,[83] experimental total cross sections and elastic angular distributions for neutron energies between 4 and 40 MeV were fit to determine the (spherical) neutron potential. A comparison of calculated and measured[32] total, elastic, and nonelastic cross sections from 12 to 80 MeV is given in Fig. 13. Agreement is seen to be very good below 40 MeV, where the fitting was done. Neutron transmission coefficients for this analysis, together with proton and alpha transmission coefficients, gamma-ray strength functions, level density parameters, and preequilibrium parameters, were used to calculate all significant neutron, proton, alpha, and gamma-ray production cross sections to 40 MeV.

Analyses similar to the ^{40}Ca study have been performed to 40 MeV for ^{54,56}Fe (Ref. 25) and to 50 MeV for ^{59}Co (Ref. 26) using the GNASH code. As was the case in the analyses discussed here, simple forms were found for the neutron and charged-particle potentials that described the reactions from very low energies to the maximum energies of the analyses.

New Developments In Spectrum Calculations

Two other developments in the application of theory for data evaluation should be mentioned. The first of these involves calculations of beta decay properties, specifically, decay spectra and half-lives. Mann et al[87] have found that by multiplying the level density parameter, a, by the ratio N/(N+Z), where N and Z are the numbers of neutrons and protons in the daughter nucleus, simple statistical theory can be used to calculate average beta decay spectra and half-lives. Using a microscopic approach, Klapdor et al[88] have reproduced measured structure in more detailed calculations of beta spectra. Both methods appear more promising than the gross theory of beta decay,[89] as is illustrated by the comparison of measured and calculated half-lives for Rb isotopes in Fig. 14.

A second development in spectral calculations is the recent work of Madland and Nix,[90] which uses standard nuclear evaporation theory to calculate both the average number of neutrons (ν_p) and neutron spectra [N(E)] from prompt fission. The calculations include the effects of first-, second-, and third-chance fission. It is shown that, using certain well-measured fission-related quantities, ν_p and N(E) can be reliably predicted. Improvements in this technique and its application to spontaneous fission of ^{252}Cf are the subject of another paper at this conference.[91]

Conclusions

It is evident that the present generation of nuclear theory and model codes used for data evaluation has

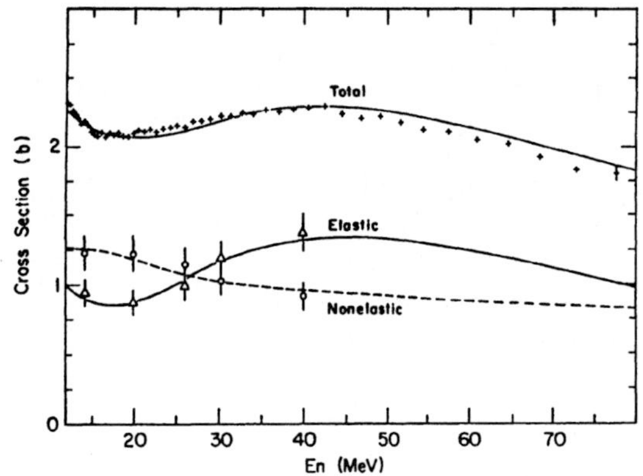

Fig. 13. The optical model analysis of Ref. 83 compared to measurements[32] of n + ^{40}Ca total, elastic, and non-elastic cross sections.

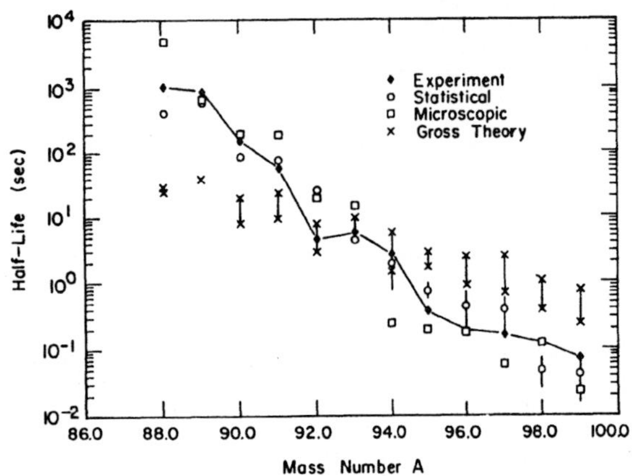

Fig. 14. Comparison of measured and calculated half-lives for Rb isotopes over the range A = 88-99.

been quite successful in describing a variety of nuclear reaction data. Substantial progress has been made in several areas of applied theory, particularly in developing techniques for determining nuclear model parameters. There remains, however, a number of areas where improvements are needed in the models, particularly if reliable extrapolations to regions away from measured data are to be made.

In several of the analyses that were described, calculations were performed to energies considerably above 20 MeV. For example, a composite of reaction cross sections from the n + ^{59}Co analysis[26] to 50 MeV is shown in Fig. 15. For this analysis, no experimental data on these reactions were available above 24 MeV and only limited data from 8-13 and 15-24 MeV. The division of the nonelastic cross section into the various reaction channels, together with calculation of emission energy spectra and angular distributions, was accomplished entirely with the simple models described above. Here we are not only depending on reliable estimates of energy dependence in the models, we are also assuming their accuracy as we drift off the line of stability. Clearly, improved methods are required for confidence in calculations such as these.

Of the topics covered in this review, level density formulation probably constitutes the area most in need of improvement. A good deal of the theoretical basis for such improvements already exists, but implementation of more detailed microscopic theories without overly complicating applied calculations has been an obstacle.

Reliable optical model analyses are obviously essential for applied calculations and continued improvement in methods and actual parameterizations is important. While significant progress has been made in developing neutron and proton potentials, relatively little advance has occurred for alpha particle potentials, and improvement is needed for reliable calculations of helium production. From the point of view of data prediction, greater use of microscopic optical model calculations should facilitate more meaningful extrapolations into unmeasured regions.

Preequilibrium models have been highly successful in calculating particle emission spectra near 14 MeV. How well such models do in describing the dependence of spectra on incident energy is less well established

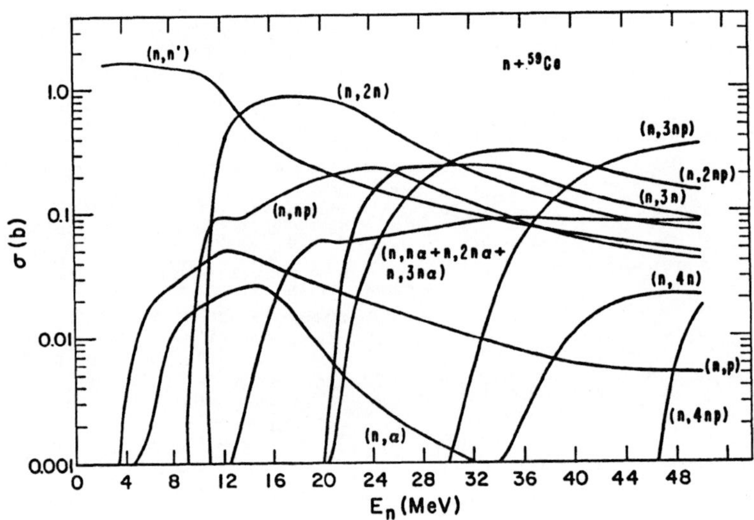

Fig. 15. Calculated[26] reaction cross sections for n + ^{59}Co interactions to 50 MeV.

and further development is certainly required for angular distribution effects. Continued advance of unified reaction theories is particularly important for higher energies and should put the entire calculational framework on a sounder theoretical footing.

Finally, although not covered in this review, fission theory remains an area much in need of improvement if reliable predictions of data are to be realized in the actinide region.

Acknowledgments

A number of individuals have contributed to this review by providing results and comments prior to publication. In particular, I wish to thank E. D. Arthur, J. P. Delaroche, R. W. Finlay, C. Y. Fu, D. G. Gardner, H. Gruppelaar, G. M. Hale, L. F. Hansen, G. Haouat, P. E. Hodgson, H. Jahn, D. Larson, J. E. Lynn, D. G. Madland, F. M. Mann, E. Menapace, C. Philis, R. E. Schenter, B. Strohmaier, and M. Uhl. I also wish to thank L. Stewart for helping review the final manuscript.

References

1. E. D. Arthur, "Calculational Methods Used to Obtain Evaluated Data above 3 MeV," Proc. Conf. on Nuclear Data Evaluation Methods and Procedures, Brookhaven, Sept. 22-25, 1980 (BNL-NCS-51363, 1981) p. 655.

2. Ch. Lagrange, "Comments on Some Aspects of the Use of Optical Statistical Models for Evaluations," ibid, p. 599.

3. D. G. Gardner, "Current Status of Fast Neutron Capture Calculations," to be published in Proc. NEANDC/NEACRP Specialists Meeting on Fast-Neutron Capture Cross Sections, Argonne, April 20-23, 1982.

4. P. A. Moldauer, "Statistical Applications of Neutron Nuclear Reactions," Proc. of Course on Nuclear Theory for Applications, Trieste, Jan. 17-Feb. 10, 1978 (IAEA-SMR-43, 1980) p. 165.

5. C. Mahaux, "Theoretical Aspects of the Optical Model," ibid, p. 97.

6. For example, H. Jahn, "Absolute values of Inelastic Neutron Scattering with Account Taken of the Preequilibrium Mechanism, ibid, p. 293; F. Gadioli and E. Gadioli Erba, "Recent Results in the Theoretical Description of Preequilibrium Processes," Proc. Course on Nuclear Theory for Applications, Trieste, Jan. 28-Feb. 22, 1980 (IAEA-SMR-68/I) p. 3.

7. E. D. Arthur, "Use of the Statistical Model for the Calculation of Compound Nucleus Contributions to Inelastic Scattering on Actinide Nuclei," Proc. Specialists Meeting on Fast Neutron Scattering on Actinide Nuclei, Paris, Nov. 23-25, 1981 (NEANDC-158U, 1982) p. 145.

8. G. M. Hale, "R-Matrix Analysis of the ^{7}Li System," Proc. Conf. on Neutron Standards and Applications, Wash., D. C., Mar. 28-31, 1977 (NBS Spec. Publ. 493, (1977) p. 30.

9. G. M. Hale, "R-Matrix Analysis of the Light Element Standards," Proc. Conf. Nuclear Cross Sections and Tech., Wash., D.C., Mar. 3-7, 1975 (NBS Spec. Publ. 425, 1975) p. 302.

10. P. G. Young, "Summary Documentation of Nuclear Data Evaluations for ENDF/B-V," LA-7663-MS (1979).

11. G. M. Hale, "Calculations of Neutron Spectra from the n + d Reaction," in "Applied Nuclear Data Research and Development, July 1 - Sept. 30, 1981," LA-9262-PR (1982) p. 1.

12. M. Brüllmann et al, Nucl. Phys. A117, 419 (1968); S. Messelt, ibid, 48, 512 (1963).

23. R. M. Frank and J. L. Gammel, Phys. Rev. 93, 463 (1954); Yu. I. Chernuklin and R. S. Shuvalov, J. Nucl. Phys. (USSR) 4, 197 (1967).

14. P. G. Young, Trans. Am. Nucl. Soc. 39, 272 (1981).

15. R. E. Prael, Trans. Am. Nucl. Soc. 41, 480 (1982).

16. C. A. Engelbrecht and H. A. Weidenmüller, Phys. Rev. C8, 859 (1973).

17. H. Feshbach et al, Ann. Phys. (NY) 125, 429 (1980).

18. T. Tamura et al, Phys. Rev. C23, 2769 (1981).

19. E. Sheldon and D. W. S. Chan, "Evaluation of (n,n') Scattering Cross Sections from 0.8 to 2.5 MeV for Higher Collective Bands of ^{232}Th and ^{238}U in 'Standard' (CN + DI) and 'Unified' (Weidenmuller S-Matrix) Formalisms," Proc. Specialists Mtg. on Fast Neutron Scattering on Actinide Nuclei, Paris, Nov. 23-25, 1981 (NEANDC-158U, 1982) p. 169.

20. R. Bonetti et al, Phys. Rev. C24, 71 (1981).

21. H. A. Weidenmüller, "Beyond the Statistical Model: Recent Progress in Neutron Nuclear Reaction Theory," Proc. Conf. Nuclear Data for Science and Tech., Antwerp, Sept. 6-10, 1982.

22. E. Sheldon, "Level Excitation Function Data for Fast Neutron Scattering on Actinide Nuclei Calculated with the Unified Statistical S-Matrix Formalism," ibid.

23. J. P. Delaroche et al, "The Optical Model with Particular Considerations of the Coupled-Channel Optical Model," IAEA-190 (1976), p. 251.

24. E. D. Arthur, Nucl. Sci. Eng. 76, 137 (1980).

25. E. D. Arthur and P. G. Young, "Evaluation of Neutron Cross Sections to 40 MeV for ^{54,56}Fe," Proc. Symp. Neutron Cross Sections from 10-50 MeV, Brookhaven, May 12-14, 1980 (BNL-NCS-51245, 1980), p. 731.

26. E. D. Arthur et al, "Calculation of ^{59}Co Neutron Cross Sections Between 3 and 50 MeV," ibid, p. 751.

27. Ch. Lagrange et al, to be published in Nucl. Sci. Eng. (1982).

28. P. E. Hodgson, "The Neutron Optical Model in the Actinide Region," Proc. Specialists Mtg. on Fast Neutron Scattering on Actinide Nuclei, Paris, Nov. 23-25, 1981 (NEANDC-158U, 1982) p. 69.

29. A. M. Lane, Phys. Rev. Lett. 8, 171 (1962); A. M. Lane, Nucl. Phys. 35, 676 (1962).

30. J. Rapaport, Phys. Reports, in press (1982).

31. L. F. Hansen, "Study of Proton-Induced Reactions and Correlation with Fast Neutron Scattering," Proc. Specialists Mtg. on Fast Neutron Scattering on Actinide Nuclei, Paris, Nov. 23-25, 1981 (NEANDC-158U, 1982), p. 116.

32. D. I. Garber and R. R. Kinsey, "Neutron Cross Sections, Volume II, Curves," Brookhaven Natl. Lab. report BNL 325, 3rd Ed., Vol. 2 (1976), and personal communication from NNDC (1981).

33. S. D. Schery et al, Nucl. Phys. A234, 109 (1974).

34. L. F. Hansen et al, Bull. Am. Phys. Soc. 25, 728 (1980); L. F. Hansen et al, UCID-18987 (1981).

35. J. Rapaport et al., Nucl. Phys. A330, 15 (1979).

36. J. P. Jeukenne et al, Phys. Rev C16, 80 (1977).

37. Ch. Lagrange and J. C. Brient, "Interpretation Semi Microscopic de la Diffusion Elastique et Inelastique de Nucleons par ^{208}Pb," submitted to J. de Physique, Paris (1982).

38. F. S. Dietrich et al, Bull. Am. Phys. Soc. 27, 543 (1982).

39. J. M. Blatt and V. F. Weisskopf, Theoretical Nuclear Physics (John Wiley, New York, 1952) p. 627.

40. D. M. Brink, Thesis, Oxford University (1955) unpublished; P. Axel, Phys. Rev. 126, 671 (1962).

41. D. G. Gardner and F. S. Dietrich, "A New Parameterization of the E1 Gamma-Ray Strength Function," Conf. Nuclear Cross Sections for Tech., Knoxville, Oct. 22-26, 1979 (NBS Spec. Publ. 594, 1980) p 770; M. A. Gardner and D. G. Gardner, "Continued Study of the Parameterization of the E1 Gamma-Ray Strength Function," Symp. Neutron-Capture Gamma-Ray Spectroscopy and Related Topics, Grenoble, Sept. 7-11, 1981 (UCRL-86265).

42. G. Szeflinska et al, Nucl. Phys. A323, 253 (1979).

43. D. G. Gardner et al, "A Study of Gamma-Ray Strength Functions," UCID-18759 (1980).

44. W. Hauser and H. Feshbach, Phys. Rev. 87, 366 (1952).

45. A. Gilbert and A. G. W. Cameron, Can. J. Phys. 43, 1446 (1965).

46. W. Dilg et al, Nucl. Phys. A217, 269 (1973).

47. A. V. Ignatuyk et al, Sov. J. Nucl. Phys. 21, 255 (1975).

48. J. Jary and J. Frehaut, "Level Density Dependence of (n,γ), (n,n'), and (n,2n) Reaction Cross Sections," in Progress Report of the Neutron and Nuclear Physics Division for the year 1979, CEA-N-2134, p. 185 (1980).

49. G. Reffo, "Parameter Systematics for Statistical Theory Calculations of Neutron Reaction Cross Sections," CNEN-RT/FI-80 (1980).

50. J. L. Cook and E. K. Rose, "An Evaluation of the Gilbert-Cameron Level Density Parameters," AAEC/E419 (1977).

51. For example, J. R. Huizenga and L. G. Moretto, Ann. Rev. Nucl. Sci. 22, 427 (1972).

52. V. Benzi et al, Nuovo Cimento A66, 1 (1981); G. Maino et al, Nuovo Cimento A57, 427 (1980).

53. S. M. Grimes et al, Phys. Rev. C10, 2373 (1974).

54. J. L. Cook et al, Aust. J. Phys. 20, 477 (1967).

55. C. Kalbach, Acta. Phys. Slov. 25, 100 (1975).

56. M. Blann, Ann. Rev. Nucl. Sci. 25, 123 (1975).

57. J. M. Akkermans et al, Phys. Rev. C22, 73 (1980).

58. C. Kalbach and F. M. Mann, Phys. Rev. C23, 112 (1981).

59. J. M. Akkermans and H. Gruppelaar, Z. Phys. A300, 345 (1981).

60. C. Y. Fu, "A Constant Nuclear Model for Compound and Precompound Reactions with Conservation of Angular Momentum," Oak Ridge report ORNL/TM 7042 (1980).

61. C. Y. Fu, "Summary of ENDF/B-V Evaluations for Carbon, Calcium, Iron, Copper, and Lead and ENDF/B-V Revision 2 for Calcium and Iron," Oak Ridge report ORNL/TM-8283 (1982).

62. P. G. Young, "Nuclear Model Codes and Data Evaluation," Proc. Symp. Neutron Cross Sections from 10 to 50 MeV, Brookhaven, May 12-14, 1980 (BNL-NCS-51245, 1980) p. 43.

63. A. Prince, "Analysis of High-Energy Neutron Cross Sections for Fissile and Fertile Isotopes," Proc. Intl. Conf. Nuclear Data for Reactors, IAEA (1970) p. 825.

64. P. G. Young and E. D. Arthur, "GNASH: A Preequilibrium Statistical Nuclear Model Code for Calculations of Cross Sections and Emission Spectra," Los Alamos report LA-6947 (1977).

65. F. M. Mann, "HAUSER-4: A Computer Code to Calculate Nuclear Cross Sections," Hanford report HEDL-TME-76-80 (1976).

66. B. Strohmaier and M. Uhl, "STAPRE - A Statistical Model Code with Consideration of Pre-Equilibrium Decay," Proc. Nuclear Theory for Applications, IAEA-SMR-43, p. 313 (1980).

67. C. Y. Fu, "Development of a Two-Step Hauser-Feshbach Code with Precompound Decays and Gamma-Ray Cascades," Proc. Nuclear Cross Sections and Technology Conf., NBS Spec. Publ. 425 (1975), p. 328 .

68. C. L. Dunford, "A Unified Model for Analysis of Compound Nucleus Reactions," Atomics International report AI-AEC-12931 (1970).

69. D. G. Gardner, "Recent Developments in Nuclear Reaction Theories and Calculations," Proc. Symp. Neutron Cross Sections from 10 to 50 MeV, Brookhaven, May 12-14, 1980 (BNL-NCS-51245, 1981) p. 641.

70. J. Jary, "MSPQ: A FORTRAN Code for Cross Section Calculations Using a Statistical Model with Preequilibrium Effects," INDC(FR)10L (1977).

71. M. Blann, "Overlaid ALICE,"University of Rochester report COO-3494-29 (1975).

72. O. Bersillon and L. Faugere, "AMALTHEE: A Code for Spectra and Cross Section Calculations within the Exciton Model," NEANDC(D)191L (1977).

73. J. M. Akkermans and H. Gruppelaar, "Calculation of Preequilibrium Angular Distributions with the Exciton Model Code PRANG," ECN-60 (1979).

74. P. G. Young et al, "Analysis of n + ^{165}Ho and ^{169}Tm Reactions," Proc. Conf. Nuclear Data for Science and Technology, Antwerp, Sept. 6-10, 1982

75. E. D. Arthur and C. A. Philis, "Calculations of Neutron Cross Sections for Tungsten Isotopes," Proc. Conf. Nuclear Cross Sections for Technology, Knoxville, Oct. 22-26, 1979 (NBS Spec. Publ. 594, 1980), p. 333.

76. E. D. Arthur et al, Trans. Am. Nucl. Soc. 39, 793 (1981).

77. P. A. Moldauer, Nucl. Phys. 47, 65 (1963).

78. J. P. Delaroche et al, Phys. Rev. C23, 136 (1981).

79. P. T. Guenther, A. B. Smith, and J. F. Whalen, "Fast Neutron Interactions with ^{182}W, ^{184}W, and ^{186}W," Argonne report ANL/NDM-56 (1981).

80. G. Haouat and Y. Patin, personal communication (1982).

81. B. Strohmaier and M. Uhl, "Nuclear Model Calculations of Neutron-Induced Cross Sections for ^{52}Cr, ^{55}Mn, ^{56}Fe, and ^{58,60}Ni for Incident Energies Up To 30 MeV," Proc. Conf. Nuclear Data for Science and Technology, Antwerp, Sept. 6-10, 1982.

82. O. Bersillon et al, "A New Evaluation of Neutron Data for ^{209}Bi Between 10^{-5} eV and 20 Mev, ibid.

83 D. M. Hetrick et al, "Evaluated Neutron-Induced Cross Sections for ^{40}Ca from 20 to 40 MeV," Oak Ridge report ORNL/TM-8290 (1982).

84. R. C. Haight and S. M. Grimes, UCRL-80235 (1977); S. M. Grimes, Phys. Rev. C19, 2127 (1979).

85. P. G. Young et al, "Application of Nuclear Models," Proc. Conf. Nuclear Cross Sections for Technology, Knoxville, Oct. 22-26, 1979 (NBS Spec. Publ. 594, 1980) p. 639.

86. G. T. Chapman et al, "A Re-Measurement of the Neutron-Induced Gamma-Ray Production Cross Sections for Iron in the Energy Range 850 keV $\leq E_n \leq$ 20.0 MeV," Oak Ridge report ORNL/TM-5416 (1976).

87. F. M. Mann et al, Phys. Rev. C25, 524 (1982).

88. H. V. Klapdor et al, Z. Phys. A299, 213 (1981).

89. K. Takahashi et al, At. Nucl. Data Tables 12, 101 (1973).

90. D. G. Madland and J. R. Nix, Nucl. Sci. Eng. 81, 213 (1982).

91. D. G. Madland and J. R. Nix, "Calculation of the Prompt Neutron Spectrum and Average Prompt Neuton Multiplicity for the Spontaneous Fission of ^{252}Cf," Proc. Conf. Nuclear Data for Science and Technology, Antwerp, Sept. 6-10, 1982.

LEVEL EXCITATION FUNCTION DATA FOR FAST NEUTRON SCATTERING ON ACTINIDE NUCLEI CALCULATED WITH THE UNIFIED STATISTICAL S-MATRIX FORMALISM

E. Sheldon

Department of Physics and Applied Physics
University of Lowell
Lowell, Massachusetts 01854
U.S.A.

Results from analyses of the most recent ^{232}Th, ^{238}U, ^{240}Pu and ^{242}Pu (n,n') level cross-section data from threshold to 2.5 MeV for inelastic neutron scattering to collective (rotational and quadrupole or octupole vibrational) levels of these deformed actinide nuclei are presented, as calculated with the unified statistical S-matrix formalism using the Bruyères optical potential and deformation parameters. The computations with the programme "NANCY" allowed for competition from other neutron exit channels, and have been corrected for fission and radiative-capture competition. The treatment took account of strong coupling (with strongly or weakly absorbing channels) in a comprehensive, consistent, cohesive approach that involved an intimate combination of direct interaction and fluctuation contributions to the cross section. The good fits obtained for (n,n') level excitation functions to quite high lying levels, especially when (n,n') measured data were combined with results from (n,n'γ) studies, prompt the application of such calculations to cases in which experimental data are unavailable or unobtainable.

[^{232}Th(n,n'), ^{238}U(n,n'), ^{240}Pu(n,n'), ^{242}Pu (n,n') scattering cross sections, E_n = 0.8 - 3.5 MeV; experimental $\sigma(E_n)$ compared with calculated unified data for individual collective levels or groups]

Introduction

Over the past decade, gratifying progress has been made in the acquisition of experimental and theoretical data for neutron interactions with actinide nuclei, essential for the efficient design and operation of nuclear power reactors. These advances, as detailed in successive nuclear data conferences[1-10] and embodied in successive nuclear data file evaluations culminating in ENDF/B-V (Refs.[11-14]) and recent publications,[15-17] have provided significant improvements in the precision, reliability and consistency of the data for the principal fissile and fertile actinide nuclei. Much of the needed fine detail in information and interpretation has become available, and substantial insights have been gained into the quantitative behaviour of neutron interactions over an appreciable energy range with the major (and, to an ever-increasing extent, several of the subsidiary) actinide isotopes. In the field of fast neutron inelastic scattering on such even-A fertile nuclei as ^{232}Th, ^{238}U, ^{240}Pu and ^{242}Pu, to which the present paper is devoted, several recent unpublished reports [18-21] supplement the information contained in the above References. Experimental data are contained in these References and in Refs.[22-47]; theoretical results are tabulated in these References and in Refs.[47-56] Latterly, the Lowell group in particular has concentrated on measuring ^{232}Th and ^{238}U (n,n') cross sections directly at high resolution, supplementing the results derived from (n,n'γ) yield measurements; a selection of our latest findings is presented in this paper and in the contribution to this Conference by Couchell et al.[57] For the theoretical analysis, we have progressed to the employment of the unified statistical S-matrix formalism, as developed by Weidenmüller et al.,[58-61] finding that even without fine adjustment of parameters satisfactorily good fits to the experimental data could be attained. In this paper, a survey of our findings for each of the above four isotopes is presented, substantiating the validity of the current approach and encouraging its application to further studies.

Theoretical Treatment

The inherent physics of the (n,n') scattering interaction is entirely contained within the transition amplitudes that make up the grand ensemble of elements in the full S-matrix array.

The present theoretical approach makes full use of this aggregate, whereas in the past only a portion was drawn into consideration as an incoherent sum was built of compound-nucleus (CN) and direct-interaction (DI) contributions to the net cross section. Among the refinements introduced into this "standard" approach were corrections of the CN component, calculated from Hauser-Feshbach theory[62] with such programs as "CINDY"[63] or "NRLY",[64] which contain provision for competition from discrete or continuum levels, radiative capture, Moldauer level-width fluctuations, and fission channels. The evidently deformed, collective nature of the actinide nuclei called for a distorted-wave coupled-channel approach to be used in the computation of the DI component with codes such as "JUPITOR"[65] or "KARJUP".[66] The Bruyères group established the optical potential and deformation parameters,[56] as well as the requisite fission-channel parameters.[19,20] The generally adopted optical model to describe the interaction is of the derivative Woods-Saxon type with a real spin-orbit term having the numerical parameters

$$V = 49.82 - 17\left(\frac{N-Z}{A}\right) - 0.3\,E_n \text{ MeV,} \tag{1}$$

$$W = 5.52 - 9\left(\frac{N-Z}{A}\right) + 0.4\,E_n \text{ MeV } [E_n \le 10 \text{ MeV}]$$

$$W = 9.52 - 9\left(\frac{N-Z}{A}\right) \text{ MeV } [E_n > 10 \text{ MeV}], \tag{2}$$

$$V_{so} = 6.2 \text{ MeV}, (r_0)_{so} = 1.12 \text{ fm}, a_{so} = 0.47 \text{ fm}, \tag{3}$$

$$a = 0.63 \text{ fm}, a' = 0.52 \text{ fm}, r_0 = r_0' = 1.26 \text{ fm}. \tag{4}$$

A listing of the deformation parameters and the numerical values of V and W used in past and present computations (as a function of E_n, the incident neutron [lab] energy in MeV) is given for each nuclide in Table I. In this Table are also specified the number of neutron exit channels included in the computations at each energy (outgoing charged-particle channels were precluded by their high Q-values) for each incident energy, and the numerical correction factors applied to the calculated cross sections to take (approximate) account of fission and radiative-capture channels. For the derivation of these reduction factors, the quotient was built of the evaluated total inelastic neutron cross section to the sum of the inelastic-neutron plus fission plus radiative-capture cross sections, using values obtained from various compilations.[48,67,68,56]

The past and present DI coupled-channels computations had to be based upon collective-band level schemes for each nuclide. For ^{232}Th and ^{238}U, these schemes have been presented in Refs.[10,16,17] and are in conformity with those given, e.g., in Nuclear Data Sheets.[69,70] As the compiler's evaluation[71] of the collective schemes for ^{240}Pu and ^{242}Pu is rather sparse and ambiguous, a mutually consistent set of band schemes has been prepared for the present calculations. The band parameters have been tabulated in Tables II and III, and the schemes have been displayed explicitly in Figs. 1(a,b).

K. H. Böckhoff (ed.), Nuclear Data for Science and Technology, 518–527.

TABLE I. Parameters featured in the evaluation of (n,n') scattering cross sections for ^{232}Th, ^{238}U, ^{240}Pu, ^{242}Pu.

Nucleus	Parameter	E_n = 0.2	0.4	0.6	0.8	1.0	1.2	1.5	2.0	2.5
^{232}Th	Deformation parameters: $\beta_2 = 0.190$, $\beta_4 = 0.071$									
	Real optical potential parameter V:	46.34	46.28	46.22	46.16	46.10	46.04	45.95	45.80	45.65
	Imaginary optical potential parameter W:	3.68	3.76	3.84	3.92	4.00	4.08	4.20	4.40	4.60
	Number of exit channels:	3 + 0	3 + 1	3 + 6	3 + 11	3 + 22	3 + 24	3 + 24	3 + 24	3 + 24
	Correction factor $[\sigma_{inel}/(\sigma_{inel}+\sigma_f+\sigma_\gamma)]$:		0.8796	0.8852	0.9122	0.9421	0.9568	0.9494	0.9444	0.9627
^{238}U	Deformation parameters: $\beta_2 = 0.198$, $\beta_4 = 0.057$									
	Real optical potential parameter V:	46.14	46.08	46.02	45.96	45.90	45.84	45.75	45.60	45.45
	Imaginary optical potential parameter W:	3.68	3.76	3.84	3.92	4.00	4.08	4.20	4.40	4.60
	Number of exit channels:	3 + 0	3 + 1	3 + 1	3 + 3	3 + 11	3 + 21	3 + 28	3 + 28	3 + 28
	Correction factor $[\sigma_{inel}/(\sigma_{inel}+\sigma_f+\sigma_\gamma)]$:	0.8703	0.9204	0.9285	0.9271	0.9201	0.9222	0.8480	0.8066	0.8190
^{240}Pu	Deformation parameters: $\beta_2 = 0.200$, $\beta_4 = 0.062$									
	Real optical potential parameter V:	46.08	46.02	45.96	45.90	45.84	45.78	45.69	45.54	45.39
	Imaginary optical potential parameter W:	3.65	3.73	3.81	3.89	3.97	4.05	4.17	4.37	4.57
	Number of exit channels:	3 + 0	3 + 1	3 + 2	3 + 6	3 + 11	3 + 21	3 + 31	3 + 33	3 + 33
	Correction factor $[\sigma_{inel}/(\sigma_{inel}+\sigma_f+\sigma_\gamma)]$:	0.7093	0.7360	0.5613	0.4988	0.4683	0.4929	0.5297	0.5440	0.5431
^{242}Pu	Deformation parameters: $\beta_2 = 0.204$, $\beta_4 = 0.051$									
	Real optical potential parameter V:	45.97	45.91	45.85	45.79	45.73	45.67	45.58	45.43	45.28
	Imaginary optical potential parameter W:	3.59	3.67	3.75	3.83	3.91	3.99	4.11	4.31	4.51
	Number of exit channels:	3 + 0	3 + 1	3 + 2	3 + 4	3 + 9	3 + 16	3 + 18	3 + 28	3 + 30
	Correction factor $[\sigma_{inel}/(\sigma_{inel}+\sigma_f+\sigma_\gamma)]$:	0.8210	0.8165	0.7207	0.6182	0.5369	0.5477	0.5631	0.5892	0.6128

TABLE II. Fitted collective band parameters for ^{240}Pu in the formula $E_J^* = E_0 + A[J(J+1)] + B[J(J+1)]^2$.

Band $K\pi$	Multiplet type	Parameters E_0 [keV]/ A [keV]/ B [eV]	Spin $J\pi$	Excitation Energy E^*_{calc} [keV]	E^*_{exp} [keV]
0+	g.s. rot.	0.00	0+	0.00	0.000
		7.16	2+	42.83	42.825
		-3.80	4+	141.69	141.686
			6+	294.03	294.314
			8+	495.84	497.6
			10+	741.65	751.4
			12+	1024.53	
0+	β vibr.	860.7	0+	860.7	860.7
		6.602	2+	900.3	900.3
		-0.357	4+	992.6	992.6
			6+	1137.4	
0+	2γ vibr.	1089.7	0+	1089.7	1089.7
		7.967	2+	1137.5	1137.5
		0.00	4+	1249.0	
0-	octupole	587.14	1-	597.4	597.4
		5.125	3-	648.9	648.9
		1.786	5-	742.5	742.5
			7-	879.7	
1-	octupole	929.99	(1-)	938.1	938.07
		3.654	2-	958.9	958.9
		194.25	3-	1001.8	1001.8
			4-	1080.8	1076.4
			5-	1214.4	
			6-	1426.1	
0	2-phonon octupole ?	1410.8	0	1410.8	1410.8
		4.617	2	1438.5	1438.5
		0.00	4	1466.2	
0+	higher g.s. in 2nd min.	2800.0	0+	2800	2800
		3.326	2+	2820	2820
		1.190	4+	2867	2867
			6+	2942	2940
			8+	3046	3039
			10+	3180	

TABLE III. Fitted collective band parameters for ^{242}Pu in the formula $E_J^* = E_0 + A[J(J+1)] + B[J(J+1)]^2$.

Band $K\pi$	Multiplet type	Parameters E_0 [keV] / A [keV] / B [eV]	Spin $J\pi$	Excitation Energy E^*_{calc} [keV]	E^*_{exp} keV]
0+	g.s. rot.	0.000	0+	0.00	0.0
		7.441	2+	44.50	44.5
		-4.048	4+	147.20	147.2
			6+	305.38	305.9
			8+	514.76	517.6
			10+	769.52	778.7
			12+	1062.28	1086.7
			14+	1384.09	
0+	β vibr.	956.50	0+	956.50	956.5
		6.50	2+	995.50	995.5
		0.00	4+	1034.50	1039.6
			6+	1229.50	
2+	2γ vibr.		2+		1102.4
0-	octupole	769.95	1-	780.30	780.3
		5.169	3-	832.29	832.3
		2.183	5-	926.98	927
			7-	1066.26	
1-	octupole	951.20	1-	963.0	[963 not seen]
		5.95	2-	986.0	986
		-25.0	3-	1019.0	1019†
			4-	1060.2	1064†
			5-	1107.2	1122†
			6-	1157.0	
2-	octupole		2-		1152
3-	octupole	955.50	3-	1019.4	1019.4†
		5.175	4-	1064.0	1064†
		12.50	5-	1122.0	1122†
			6-	1194.9	1204
			7-	1284.5	

† Note: These states may either be higher members of the K = 1- band or lower levels of the K = 3- octupole family. In the computations, they were assumed to belong to the K = 1- band.

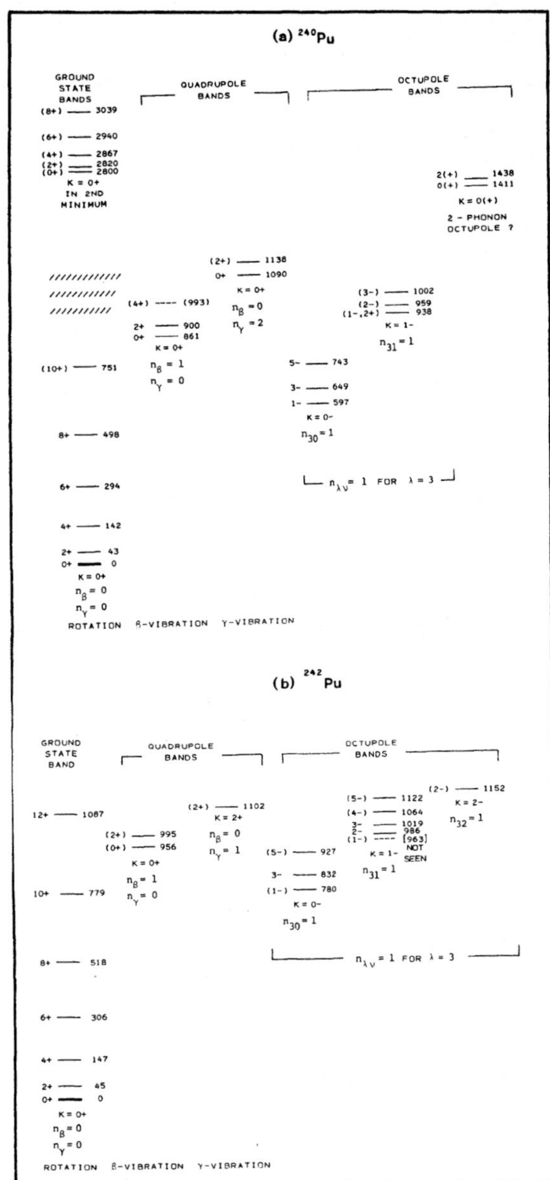

Fig. 1. Collective (rotational and vibrational) band level schemes of (a) ^{240}Pu and (b) ^{242}Pu.

For given level schemes, interaction potentials and deformation parameters, the only adjustable variables that remain are the relative band coupling strengths. By first undertaking a complete set of analyses for ^{232}Th and ^{238}U using the "standard" approach,[16-18] combining CN + DI independently, it was possible to obtain a close indication of their values by successive adjustment to obtain a best fit to the experimental level excitation functions. For the appreciably more complicated, lengthy and time-consuming "unified" analyses,[17-72] no such fine adjustment was feasible: instead, reasonable values in conformity with those deduced from the "standard" approach were employed. These values are listed in Table IV (overleaf) together with the "unified" cross sections derived therefrom. It is entirely possible that by further "fine-tuning" of the coupling strengths, still closer fits to measured data could ensue.

The "unified" cross sections were computed from the comprehensive statistical S-matrix program "NANCY", as compiled at Lowell by Chan.[17-73] This employed the formalism of Weidenmüller et al.[58-61] founded upon the manipulation of energy-averaged (statistical) S-matrix elements. The energy averaging was accomplished automatically through the inclusion of the absorptive energy-dependent imaginary W-term within the complex optical potential (ancillary studies confirmed ensemble averaging to be unnecessary). Following Moldauer,[74] a procedure was devised to extend the S-matrix subroutines in the coupled-equations program "JUPITOR" to allow for interchannel coupling of each collective state with each other (and not, as in the original version, merely with the common ground state). In practice, to conserve core storage requirements and running schedules, the coupling was limited to 3 (or at most 4) such states at any given time, the transmission coefficients for the competing extra channels being generated with the subsidiary "SCAT" procedure.[75] Thus, a full aggregate of the requisite S-matrix elements was built up, taking cognizance of the influence of (compound) fluctuation amplitudes upon the direct amplitudes. Incorporating the unitary Engelbrecht-Weidenmüller transformation[58] in matrix diagonalization procedures, and employing the unrestricted unified formalism (with inclusion of DI),[59-61] the angle-integrated scattering cross section was derived at each energy for each state as a sum of a (compound) fluctuation component and a direct component.

In an independent check of the results furnished by the code "NANCY", a separate set of normalized S-matrix elements was, with assistance from Raynal,[76] generated for several cases from his extensive coupled-equations code "ECIS" and the ensuing cross sections were compared with those furnished by "NANCY". Our preliminary findings indicate no perceptible difference. The program "NANCY" has latterly been modified to feature a more recent formula for the "elastic enhancement factor" W_a (not to be confused with the W-term in the optical potential): in the place of the expression in the intrinsic formalism[60]

$$W_a = 1 + [2/(1+p_a^{0.3+1.5\psi})] + 2[\psi - (\bar{p}/\Sigma_c p_c)]^2 \quad (5)$$

in which $\psi \equiv p_a/\Sigma_c p_c$ for an incident channel a and all open outgoing channels c, the p_i being elements of the Satchler[77] penetrability matrix, the newer expression, designed to take better cognizance of weakly absorptive channels,[61] is now used:

$$W_a = 1 + [2/(1+p_a^F)] + 87[\psi - (\bar{p}/\Sigma_c p_c)]^2 \psi^5, \quad (6)$$

where the exponent F is given by the formula

$$F = 4(\bar{p}/\Sigma_c p_c)\{[1+(p_a/\Sigma_d p_d)]/[1+3(\bar{p}/\Sigma_c p_c)]\}. \quad (7)$$

In the present applications, we have discerned hardly any difference with the new elastic enhancement factor (6), as can be seen from the specimen entry in Table IV for the 3- (731.9 keV) K = 0- octupole state in ^{238}U.

Up till now, the program "NANCY" has not contained provision for fission or radiative-capture competing channels, or for continuum competition. This is currently being incorporated; pending the completion of these new procedures we have been obliged (i) to make a rough correction for the effects of fission and radiative capture as detailed in Table I, multiplying each calculated cross section by a fractional (cross-section) correction factor, and (ii) to regard the computed cross sections beyond an incident energy of about $E_n \approx 1.5$ MeV as probably too high because the discrete levels that were entered as competition fall appreciably short of the "drainage" that continuum states would provide. The marked effect that a continuum routine, such as that in "CINDY", can provide is illustrated in Ref.[78] Accordingly, some theoretical excitation functions beyond $E_n = 1.5$ MeV in the present paper have been indicated by broken curves to suggest this likelihood of enhanced magnitude, especially in the case of Pu data.

The effect of varying the relative band-coupling strengths in the collective multiplets of each nuclide is indicated by some of the entries in Table IV: for example, there has been a consistent finding that a coupling strength of 0.01 is appropriate for the K = 0+ β-vibrational states and in certain of the negative-parity octupole states. In the remaining portions of this paper, results for each nuclide in turn are provided in graphical form for intercomparison (Figs. 2 - 9).

TABLE IV. Inelastic Neutron Scattering Cross Sections for Actinide Nuclei, ^{232}Th, ^{238}U, ^{240}Pu, ^{242}Pu, calculated from the unified statistical S-matrix formalism, using the Bruyères optical potential and deformation parameters, and corrected for fission and radiative-capture competition [σ_{inel} in mb units].

Nucleus	No.	Jπ	E* (keV)	Kπ	Band	C.S.	E_n = 0.2	0.4	0.6	0.8	1.0	1.2	1.5	2.0	2.5 MeV
^{232}Th	0	0+	0.00	0+	g.s.rot.										
	1	2+	49.37	0+	g.s.rot.			1171	1267	1268	997	868	772	764	764
	2	4+	162.1	0+	g.s.rot.			85	225	386	374	353	311	280	255
	3	6+	333.1	0+	g.s.rot.			0	2	12	26	39	53	67	72
	4	8+	556.9	0+	g.s.rot.										
	5	1-	714.3	0-	octupole	0.30				223	364	324	248	222	211
	6	0+	730.4	0+	β vib.	0.01				71	132	114	76	64	55
	7	2+	774.1	0+	β vib.	0.10				67	261	270	238	243	249
	8	3-	774.4	0-	octupole	0.30				49	256	255	224	226	229
	9	2+	785.2	2+	γ vib.	0.08					283	298	262	267	278
	10	10+	827.4	0+	g.s.rot.										
	11	3+	829.6	2+	γ vib.	0.08					127	166	157	172	182
	12	4+	873.0	0+	β vib.	0.01					44	78	95	121	131
	13	5-	883.3	0-	octupole	0.10					7	16	29	55	68
	14	4+	890.1	2+	γ vib.	0.08					34	79	103	143	156
	15	5+	960.2	2+	γ vib.	0.08					6	34	47	77	96
	16	2-	1053.6	2-	octupole	0.01						107	116	106	105
	17	2+	1072.9	2+	higher										
	18	1-	1077.5	1-	octupole	0.01						101	125	100	88
	19	0+	1078.7	0+	2γ vib.							49	71	58	54
	20	2-	1094.4	1-	octupole	0.01						79	126	118	116
	21	3-	1105.7	2-	octupole	0.01						50	92	103	115
	22	2+	1122.8	0+	2γ vib.							78	217	194	181
	23	12+	1138.1	0+	g.s.rot.										
	24	4-	1143.3	2-	octupole	0.01						15	47	71	93
	25	4+	1147.9	0+	2γ vib.							20	125	139	154
	26	3-	1182.5	3-	octupole	0.01						6	82	100	112
	27	5-	1208.9	2-	octupole	0.01							10	33	56
	28	4-	1218.1	3-	octupole	0.01							43	66	85
^{238}U	0	0+	0.0	0+	g.s.rot.										
	1	2+	44.9	0+	g.s.rot.		905	1200	1300	1223	1108	842	692	669	671
	2	4+	145.4	0+	g.s.rot.		10	103	262	416	511	387	278	237	214
	3	6+	307.2	0+	g.s.rot.			0	3	14	37	45	45	51	54
	4	8+	517.8	0+	g.s.rot.										
	5	1-	680.0	0-	octupole	0.32				247	328	256	198	183	179
	6	3-	731.9	0-	octupole	0.30	(old enhancement factor)			117	246	226	190	188	196
	6'	"	"		"	0.30	(new enhancement factor)			116	243	225	190	188	196
	6"	"	"		"	0.10				117	222	181	131	122	120
	7	5-	827.2	0-	octupole	0.01					9	18	25	39	49
	8	0+	927.2	0+	2γ vib.						73	122	79	60	48
	9	1-	930.8	1-	octupole	0.15					126	193	132	108	104
	10	2-	950.0	1-	octupole	0.10					74	167	117	98	93
	11	2+	966.3	0+	2γ vib.						64	337	259	197	163
	12	0+	993.0	0+	β vib.	0.01						72	56	46	42
	13	3-	997.5	1-	octupole	0.10						123	107	112	117
	14	2+	1037.3	0+	β vib.	0.01						124	128	119	114
	15	4+	1055.0	0+	2γ vib.							122	150	140	138
	16	2+	1060.9	2+	γ vib.	0.01						101	112	110	109
	17	3+	1105.3	2+	γ vib.	0.10						57	108	120	131
	18	4+	1127.0	0+	β vib.	0.01						28	74	97	111
	19	2-	1128.9	2-	octupole	0.01						49	80	77	79
	20	4+	1167.4	2+	γ vib.							7	66	93	110
	21	3-	1169.4	2-	octupole	0.01						13	67	75	86
	22	4-	1243	2-	octupole	0.01							32	50	67
	23	6+	1269.4	0+	β vib.								2	16	32
^{240}Pu	0	0+	0.000	0+	g.s.rot.										
	1	2+	42.825	0+	g.s.rot.		787	930	764	616	527	464	445	455	452
	2	4+	141.686	0+	g.s.rot.		10	88	165	218	243	199	171	154	138
	3	6+	294.314	0+	g.s.rot.			0	2	8	18	25	28	33	34
	4	8+	497.6	0+	g.s.rot.										
	5	1-	597.36	0-	octupole	0.10				169	162	125	96	71	62
	6	3-	648.89	0-	octupole	0.10				97	123	106	98	88	83
	7	5-	742.5	0-	octupole	0.10				1	7	13	23	34	39
	8	10+	751.4	0+	g.s.rot.										
	9	0+	860.70	0+	β vib.	0.10					68	94	80	70	65
	9'	"	"		"	0.01					56	65	43	31	26
	10	2+	900.32	0+	β vib.	0.10					81	129	128	130	131
	10'	"	"		"	0.01					73	112	96	81	61
	11	1-	938.07	1-	octupole	0.10					51	101	87	65	59
	11'	"	"		"	0.01					49	95	80	56	47
	12	2-	958.87	1-	octupole	0.10					30	93	87	70	64
	12'	"	"		"	0.01					30	93	85	67	61
	13	4+	992.6	0+	β vib.	0.10						43	54	68	73
	13'	"	"		"	0.01						32	45	59	61
	14	3-	1001.8	1-	octupole	0.10						66	76	78	79
	14'	"	"		"	0.01						56	63	64	62

(Table IV continued on next page.)

TABLE IV. *(Continuation)* : *Inelastic Neutron Scattering Cross Sections (in mb units) for ^{240}Pu and ^{242}Pu.*

Nucleus	No.	Jπ	E* (keV)	Kπ	Band	C.S.	E_n = 0.2	0.4	0.6	0.8	1.0	1.2	1.5	2.0	2.5 MeV
^{240}Pu	15	3+	1030.65	?	unassigned										
	16	4+	1037.9	?	unassigned										
	17	4-	1076.4	1-	octupole	0.10						21	30	41	46
	17'	"	"		"	0.01						21	30	41	46
	18	0+	1089.71	0+	2γ vib.										
	19	5-	1115.7	?	unassigned										
	20	2+	1137.52	0+	2γ vib.										
^{242}Pu	0	0+	0.00	0+	g.s.rot.										
	1	2+	44.54	0+	g.s.rot.		822	1020	972	868	682	570	545	542	548
	2	4+	147.2	0+	g.s.rot.		10	96	211	298	316	281	249	202	181
	3	6+	305.9	0+	g.s.rot.			0	2	10	22	31	40	43	45
	4	8+	517.6	0+	g.s.rot.										
	5	10+	778.7	0+	g.s.rot.										
	6	1-	780.3	0-	octupole	0.10				35	170	147	122	95	83
	7	3-	832.3	0-	octupole	0.10					109	112	114	111	108
	8	5-	927	0-	octupole	0.10					3	9	21	39	49
	9	0+	956	0+	β vib.	0.10					25	96	99	86	81
	9'	"	"		"	0.01					22	71	61	46	37
	10	2-	986	1-	octupole	0.10						121	118	100	90
	11	2+	995	0+	β vib.	0.10						128	159	161	162
	11'	"	"		"	0.01						118	130	110	99
	12	3-	1019.4	1-	octupole	0.10						82	100	106	108
	13	4+	1039.6	0+	β vib.	0.10						40	70	83	87
	13'	"	"		"	0.01						40	70	83	89
	14	4-	1064.0	1-	octupole	0.10						30	44	59	67
	15	12+	1086.7	0+	g.s.rot.										
	16	2+	1102	2+	γ vib.	0.10						78	166	180	196
	17	5-	1122	1-	octupole	0.10						2	12	32	45
	18	2-	1152.2	2-	octupole	0.01						23	77	75	74
	19	6-	1204	3-	octupole	0.01							4	13	24

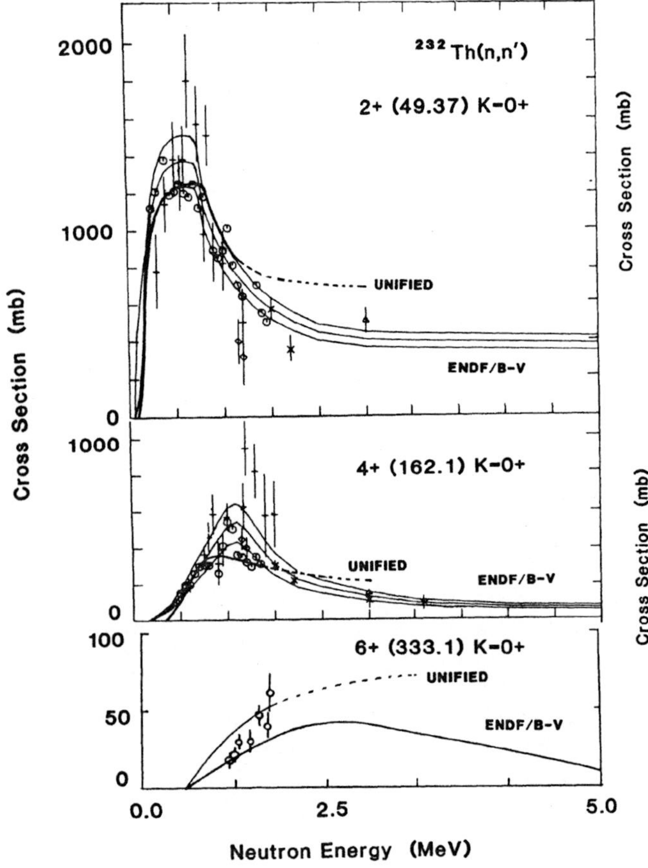

Fig. 2. Comparison of the unified level excitation functions for inelastic neutron scattering to the first three levels (2+/4+/6+) in the K = 0+ ground-state rotational band of ^{232}Th with experimental data[22-29] (identified in Ref.[11] and with the (bounded) ENDF/B-V curve.[11]

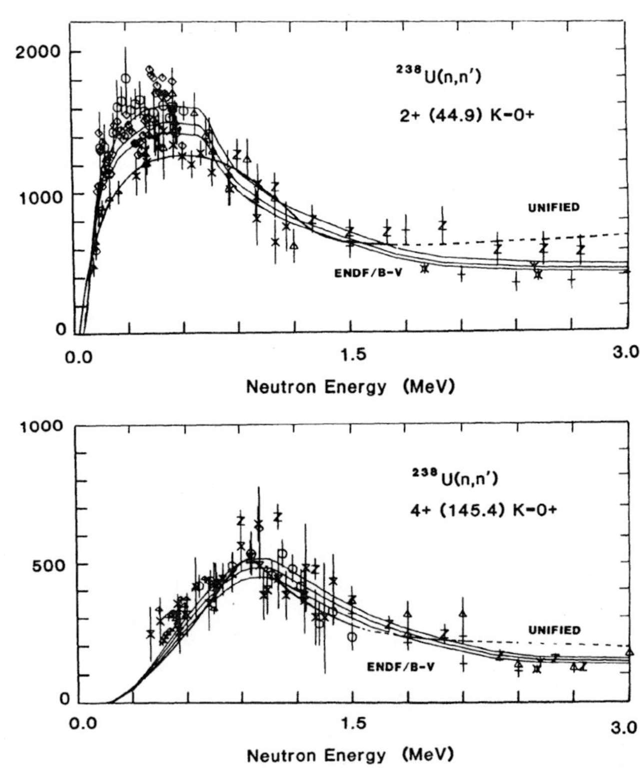

Fig. 3. Comparison of the unified level excitation functions for inelastic neutron scattering to the first two levels (2+/4+) in the K = 0+ ground-state rotational band of ^{238}U with experimental data[25,30-40] (identified in Ref.[13]) and with the evaluated ENDF/B-V curve, shown together with its upper and lower bounding limits.[13] Because of inadequate provision for continuum competition in the unified computations, the unified cross section is probably too high beyond $E_n \simeq 1.5$ MeV; the high-energy tail of the curve is therefore shown broken.

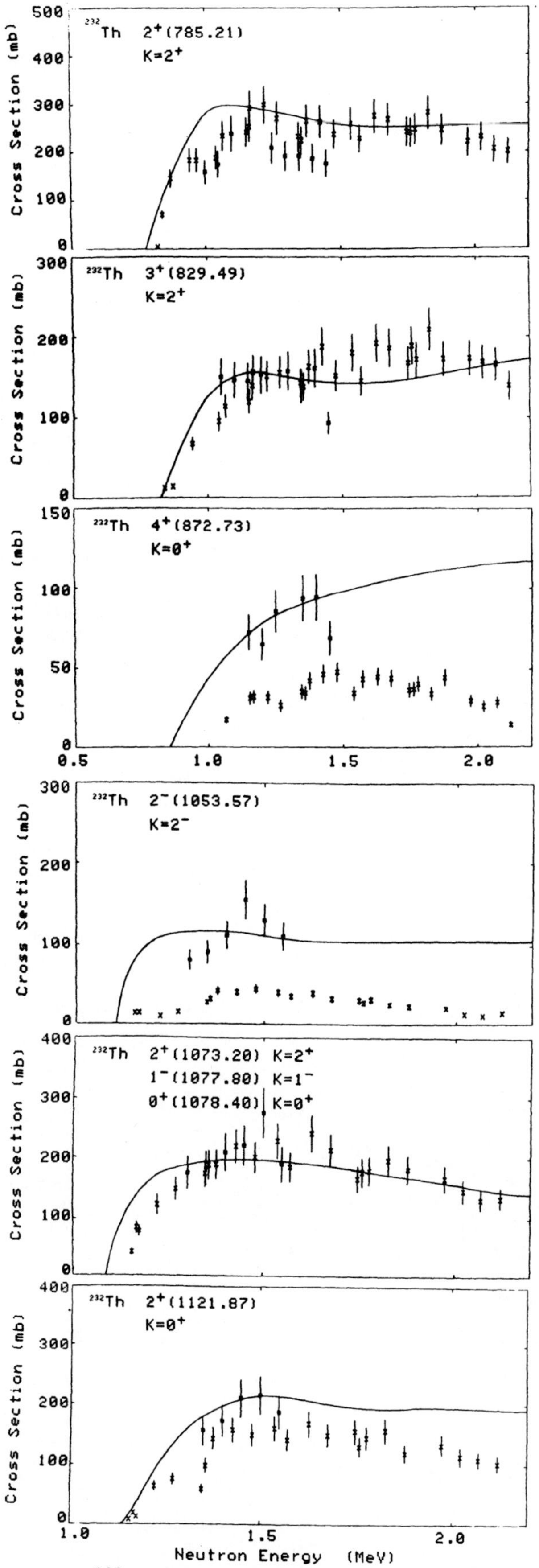

Fig. 4. ^{232}Th(n,n') excitation functions for vibrational levels; experimental data from (n,n') measurements (solid dots with error bars) and from (n,n'γ) studies (crosses, with error bars) are contrasted with unified theoretical curves.

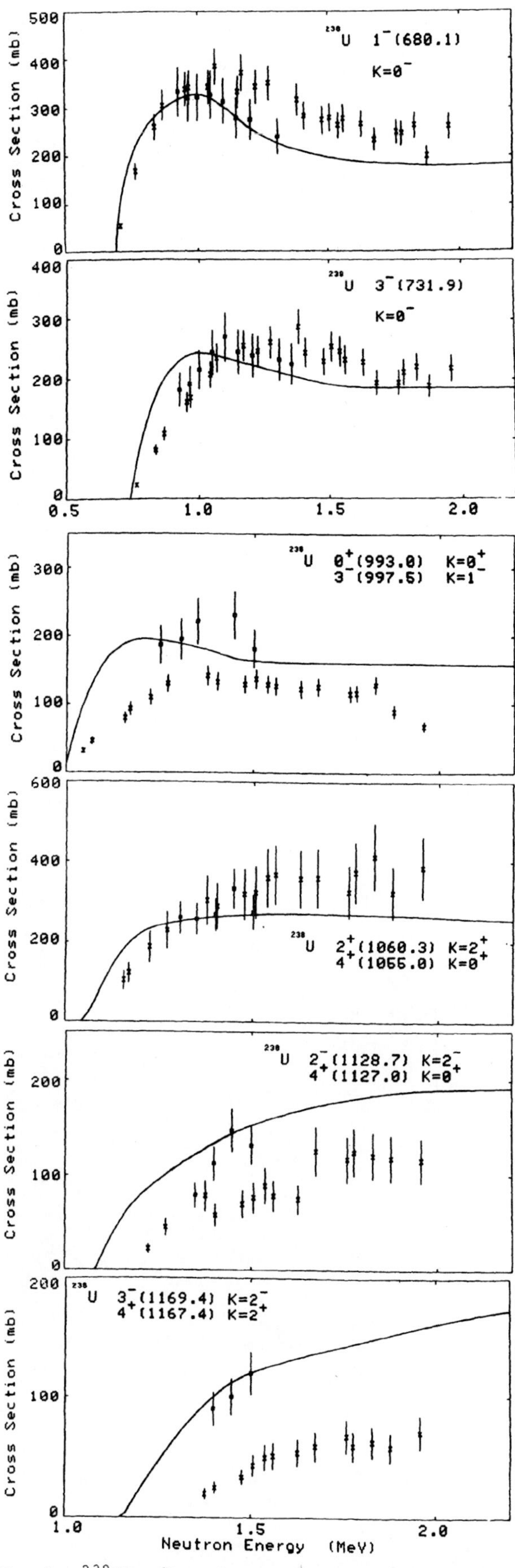

Fig. 5. ^{238}U(n,n') excitation functions for vibrational levels; Lowell experimental data from (n,n') measurements (solid dots with error bars) and from (n,n'γ) studies (crosses, with error bars) are contrasted with unified theoretical angle-integrated cross sections.

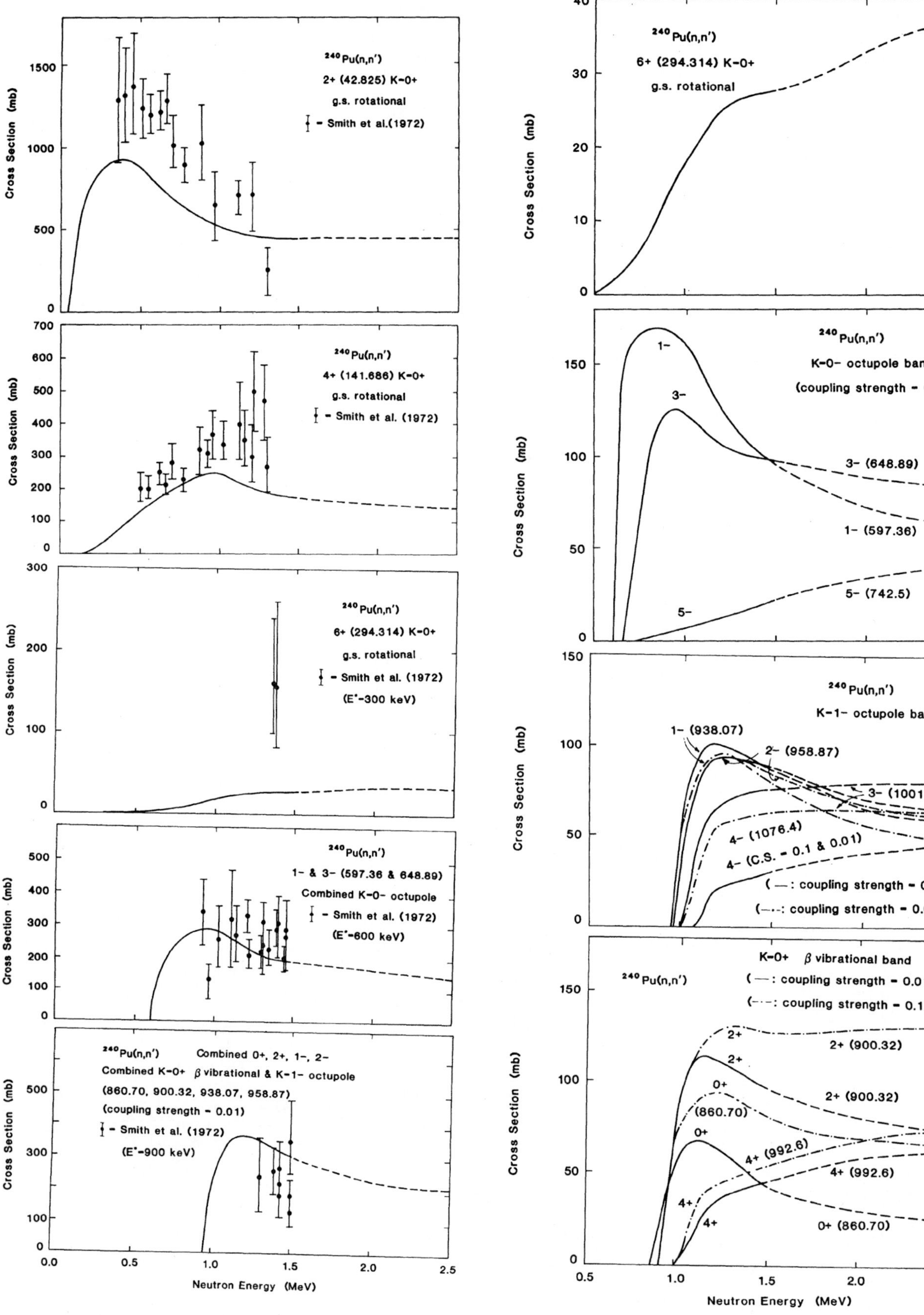

Fig. 6. ^{240}Pu(n,n') level excitation functions, contrasting experimental data[44] for single or combined groups of states with unified theoretical curves (dashed beyond 1.5 MeV to indicate probable augmentation).

Fig. 7. ^{240}Pu(n,n') unified theoretical excitation functions, calculated from the code "NANCY"[17,72,73] with a uniform coupling strength of 0.1 (and, in two instances, of 0.01). Note the suppressed energy zero.

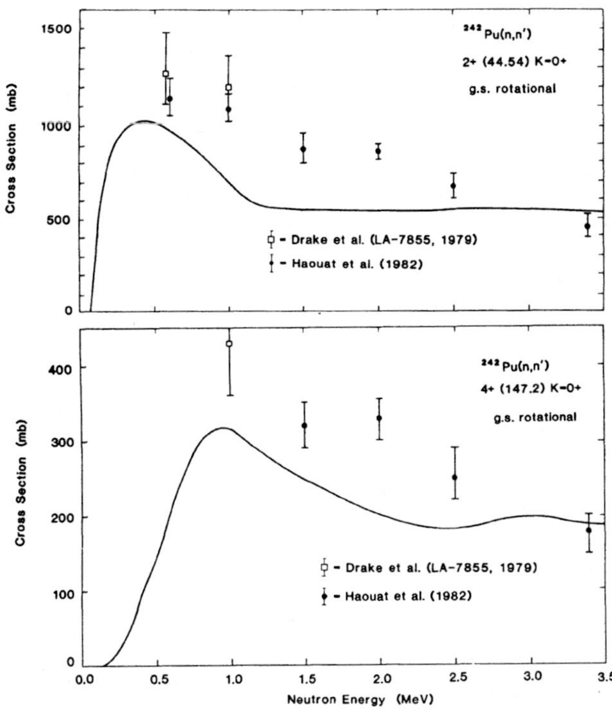

Fig. 8. ^{242}Pu(n,n') rotational level excitation functions, comparing recent experimental data[45,46,15] with unified theoretical curves up to E_n = 3.5 MeV.

Concluding Remarks

Given the lack of fine adjustment of calculational parameters, the extent of agreement between experiment and unified theory, as manifested in Figs. 2 - 6 & 8, is definitely encouraging. In particular, the cross sections derived from (n,n') measurements by the Lowell group[57] support the theoretical predictions even better than the prior (n,n'γ) data, as can be seen from Figs. 4 and 5. Under the present conditions, it appears that the new elastic enhancement factor[61] inserted in the unified formalism[60] effects little change. Substituting routines for generating S-matrix elements from Raynal's program "ECIS"[76] (kindly made available by the compiler, and modified by him for our application) for the procedures in "JUPITOR"[65] did not appreciably change the ensuing cross sections from those furnished by the unified code "NANCY".[73] The latter is in the process of being extended to make provision for fission, radiative capture, and continuum competition. When these various extensions have been incorporated and tested, it is intended to expand the investigations to other even-A and odd-A fertile and fissile actinide nuclei, for which the unified formalism now appears to be well substantiated. Although running times set a pragmatic limit upon the amount of interchannel coupling, it is hoped that (e.g., with the "ECIS" routines) the extent of coupling can be augmented still further to take advantage of the full possibilities of this propitious treatment.

References

1. R.A. Schrack, C.D. Bowman (eds.), *Nuclear Cross Sections and Technology* (Proceedings of a Conference, Washington, D.C., March 3 - 7, 1975), National Bureau of Standards Special Publication NBS SP-425, Vols. I & II (Washington, 1975).

2. E. Sheldon (ed.), *Proceedings of the International Conference on the Interactions of Neutrons with Nuclei, Lowell, 6 - 9 July, 1976*, CONF-760715-P1/2, Vols. I & II (USERDA, TIC, Oak Ridge, 1976).

3. *Nuclear Theory in Neutron Nuclear Data Evaluation*, IAEA-190, Vols. I & II (IAEA, Vienna, 1976).

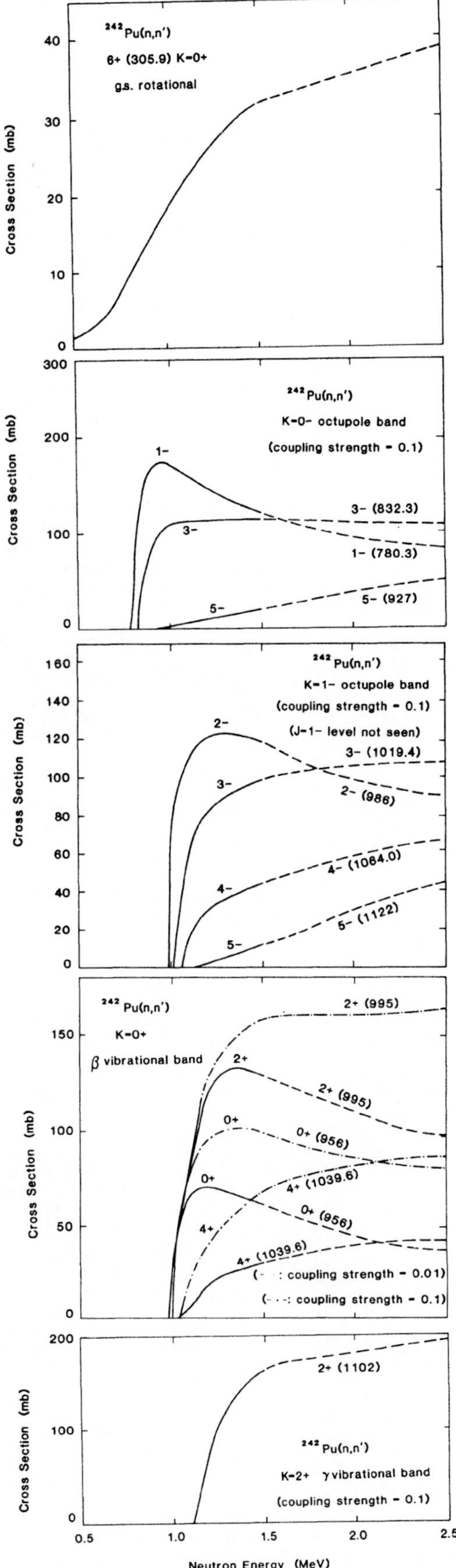

Fig. 9. ^{242}Pu(n,n') unified theoretical level excitation functions, computed with the program "NANCY" with a coupling strength of 0.1 (and, for the K = 0+ β-vibrational band, of 0.01 as the more likely value). Note the suppressed zero on the neutron energy scale.

4. *Proceedings of the Conference on Neutron Physics and Nuclear Data for Reactors and Other Applied Purposes, Harwell, Sept. 25 - 29, 1978* (Organization for Economic Cooperation and Development, OECD, Paris, 1978).

5. *Nuclear Theory for Applications,* Proceedings of the Course on Nuclear Theory for Applications, Trieste, 1978 (International Centre for Theoretical Physics, Trieste, 1980).

6. *Proceedings of the Meeting on Nuclear Data of Higher Pu and Am Isotopes for Reactor Applications, Upton, New York, Nov. 20 - 22, 1978,* BNL-50991, Brookhaven National Laboratory (1979).

7. J.L. Fowler, G.H. Johnson, C.D. Bowman (eds.), *Nuclear Cross Sections for Technology: Proceedings of the International Conference, University of Tennessee, Knoxville, Oct. 22 - 26, 1979,* National Bureau of Standards Special Publication NBS SP-594 (1980).

8. *Proceedings of the Symposium on Neutron Cross Sections from 10 - 50 MeV, Brookhaven National Laboratory, Upton, New York, May 1980.*

9. *Nuclear Theory for Applications - 1980,* Proceedings of the Interregional Advanced Training Course on Applications of Nuclear Theory to Nuclear Data Calculations for Reactor Design,Trieste, 1980 (International Centre for Theoretical Physics, Trieste, 1981).

10. *Fast Neutron Scattering on Actinide Nuclei, Proceedings of a Specialists' Meeting, OECD(NEANDC), Paris, Nov. 23 - 25, 1981* (OECD, Paris, 1982).

11. J. Meadows, W. Poenitz, A. Smith, D. Smith, J. Whalen, R. Howerton, ANL/NDM-35, Argonne National Laboratory (February, 1978).

12. J.W. Meadows, W.P. Poenitz, A.B. Smith, D.L. Smith, J.F. Whalen, R.J. Howerton, B.R. Leonard, G. De-Saussure, R.L. Macklin, G. Gwin, M.R. Bhat, *Evaluated Nuclear Data File-B, Version V (ENDF/B-V),* MAT 1390, Nat. Neutron Cross Section Center, 1979.

13. W. Poenitz, E. Pennington, A.B. Smith, R. Howerton, ANL/NDM-32, Argonne National Laboratory (Oct.,1977)

14. E. Pennington, A. Smith, W. Poenitz, M.R. Bhat, *Evaluated Nuclear Data File-B, Version V (ENDF/B-V)* MAT 1395, NNCSC, BNL (Upton, New York, 1977 & 1980).

15. G. Haouat, J. Lachkar, Ch. Lagrange, J. Jary, J. Sigaud, Y. Patin, Nucl. Sci. Eng. 81, 491 (1982).

16. D.W.S. Chan, J.J. Egan, A. Mittler, E. Sheldon, Phys. Rev. C 26(3), 1982.

17. D.W.S. Chan, E. Sheldon, Phys. Rev. C 26(3), 1982.

18. E. Sheldon, L.E. Beghian, D.W.S. Chan, J.H. Chang, J.J. Egan, G.H.R. Kegel, J. Menachery-Davé, A. Mittler, A.T.Y. Wang, poster presentation at the XIII. Polish Summer School in Nuclear Physics, Mikolajki, Sep. 1 - 12, 1980 (unpublished).

19. *Progress Report of Recent Works on Actinide Nuclear Data at Bruyères-le-Châtel,* NEANDC (E) 211 "L", INDC (FR) 41/L (Bruyères, 1981).

20. *Status of Activities on Actinide Nuclear Data at Bruyères-le-Châtel,* NEANDC (E) 227 "L", INDC (FR) 54/L (Bruyères, 1982).

21. *Compte Rendu d'Activité du Service de Physique Neutronique et Nucléaire pour l'Année 1981, Bruyères-le-Châtel,* CEA-N-2284, NEANDC (E) 232 "L", INDC (FR) 55/L (Bruyères, 1982).

22. A. Smith, Phys. Rev. 126, 718 (1962).

23. W.R. McMurray, I.J. van Heerden, E. Barnard, D.T. L. Jones, Southern Universities Nuclear Institute Annual Report SUNI-41 (1975).

24. W.R. McMurray, E. Barnard, I.J. van Heerden, D.T. L. Jones, Southern Universities Nuclear Institute Annual Report SUNI-45 (1976).

25. G. Haouat, J. Sigaud, J. Lachkar, Ch. Lagrange, B. Duchemin, Y. Patin, in Ref.[2], p. 1330 (1976).

26. R. Batchelor, J. Towle, Nucl. Phys. 65, 236 (1965).

27. R. Batchelor, J. Towle, Proc. Phys. Soc. (London) 73, 193 (1959).

28. A. Smith et al. (unpublished 1970, 1977) in Ref.[11].

29. A. Smith, J. Whalen, BNL-NCS 24273 (Brookhaven National Laboratory, 1978), p.1; see also A. Smith, P. Guenther, G. Winkler, BNL-NCS 26133 (Brookhaven National Laboratory, 1979), p.3.

30. A. Smith et al., ANL/NDM-22 (Argonne National Laboratory, 1976).

31. E. Barnard, A. Ferguson, W.R. McMurray, I.J. van Heerden, Nucl. Phys. 80, 46 (1966).

32. A. Smith, Nucl. Phys. 47, 633 (1963).

33. P. Guenther, D. Havel, A. Smith, ANL/NDM-16 (Argonne National Laboratory, 1975).

34. L. Cranberg, J. Levin, Phys. Rev. 109, 2063 (1956).

35. J.J. Egan, T.V. Marcella, B.K. Barnes, G.P. Couchell, G.H.R. Kegel, A. Mittler, D.J. Pullen, W.A. Schier, E. Sheldon, N.B. Sullivan, in Ref.[2]

36. P. Guenther et al. (private communication cited in Ref.[13]); also P.T. Guenther, D.G. Havel, A.B. Smith, Nucl. Sci. Eng. 65, 174 (1978).

37. E. Barnard, J.A.M. de Villiers, D. Reitman, in *Proceedings of the Second International Conference on Nuclear Data for Reactors, Helsinki,* CONF-700606, Vol. II, p. 103 (International Atomic Energy Agency, IAEA, Vienna, 1970).

38. L. Stromberg, S. Schwartz, Nucl. Phys. 71, 511 (1965).

39. P. Guenther, A. Smith, in Ref.[1], p. 862 (1975).

40. J.J. Egan, G.H.R. Kegel, G.P. Couchell, A. Mittler, B.K. Barnes, W.A. Schier, D.J. Pullen, P. Harihar, T.V. Marcella, N.B. Sullivan, E. Sheldon, A. Prince, in Ref.[1], p. 950; also J.J. Egan, J.D. Menachery, G.H.R. Kegel, D.J. Pullen, in Ref.[7], p. 685, and L.E. Beghian, G.H.R. Kegel, T.V. Marcella, B.K. Barnes, G.P. Couchell, J.J. Egan, A. Mittler, D. J. Pullen, W.A. Schier, Nucl. Sci. Eng. 69, 191 (1979).

41. G. Haouat, in *Neutron Induced Reactions: Proceedings of the Second International Symposium, Smolenice, June 25 - 29, 1979,* edited by I. Ribanský and E. Běták (VEDA Publishing House of the Slovak Academy of Sciences, Bratislava, 1980), p.333.

42. L.E. Beghian, J. Egan, G. Kegel, A. Mittler, in *Neutron Induced Reactions: Proceedings of the Second International Symposium, Smolenice, June 25 - 29, 1979,* edited by I. Ribanský and E. Běták (VEDA Publishing House of the Slovak Academy of Sciences, Bratislava, 1980), p. 405.

43. A.B. Smith, P.T. Guenther, ANL/NDM-63 (Argonne National Laboratory, 1982).

44. A.B. Smith, P. Lambropoulos, J.P. Whalen, Nucl. Sci. Eng. 47, 19 (1972).

45. D.M. Drake, M. Drosg, P. Lisowski, L. Veeser, LA-7855-MS (Los Alamos Scientific Laboratory,1979).

46. G. Haouat et al., in Ref.[7], p. 672.

47. *CINDA 82 (1977 - 1982),* Index to Literature and Computer Files on Microscopic Neutron Data (IAEA, International Atomic Energy Agency, Vienna, 1982).

48. H. Abou Yehia, J. Jary, J. Trochon, NEANDC (E) 204 "L", INDC (FR) 34/L (Bruyères, 1979).

49. J. Garg, BARC-990 19 (1978).

50. P.T. Guenther, D.G. Havel, A.B. Smith, Nucl. Sci. Eng. 65, 174 (1978).

51. D.G. Madland, P.G. Young, LA-7533-MS & LA-7596 (Los Alamos Scientific Laboratory, 1978).

52. V.A. Konshin, INDC(CCP)-135 (1979).

53. E.D. Arthur, LA-UR-81-3497 (Los Alamos Scientific Laboratory, 1981).

54. Ch. Lagrange, CEA-N-1970 (Bruyères, 1977).

55. J. Jary, Ch. Lagrange, C. Philis, in Ref.[6], p. 83.

56. Ch. Lagrange, J. Jary, NEANDC (E) 198 "L", INDC (FR) 30/L (Bruyères, 1978).

57. G.P. Couchell, C. Ciarcia, J.J. Egan, G.H.R. Kegel, A. Mittler, D.J. Pullen, W.A. Schier, J. Shao, this conference.

58. C.A. Engelbrecht, H.A. Weidenmüller, Phys. Rev. C 8, 859 (1973).

59. J.W. Tepel, H.M. Hofmann, H.A. Weidenmüller, Phys. Lett. 49B, 1 (1974).

60. H.M. Hofmann, J. Richert, J.W. Tepel, H.A. Weidenmüller, Ann. of Phys. (N.Y.) 90, 391 & 403 (1975).

61. H.M. Hofmann, T. Mertelmeier, M. Herman, J.W. Tepel, Z. Phys. A 297, 153 (1980).

62. W. Hauser, H. Feshbach, Phys. Rev. 87, 366 (1952).

63. E. Sheldon, V.C. Rogers, Computer Phys. Commun. 6, 99 (1973).

64. J. Jary, in Ref.[20], C-III-1 on p. 146 (1982).

65. T. Tamura, ORNL-4152 (Oak Ridge National Laboratory, 1967); Rev. Mod. Phys. 37, 679 (1965).

66. H. Rebel, G.W. Schweimer, KFK-1333 (Kernforschungszentrum Karlsruhe, 1971).

67. G. Haouat, J. Sigaud, J. Lachkar, Ch. Lagrange, B. Duchemin, Y. Patin, NEANDC (E) 180 "L", INDC (FR) 13/L (Bruyères, 1977).

68. G. Haouat, J. Lachkar, Ch. Lagrange, Y. Patin, J. Sigaud, R.E. Shamu, NEANDC (E) 196 "L", INDC (FR) 29/L (Bruyères, 1978).

69. M.R. Schmorak, Nucl. Data Sheets 20(2), 165 (1977).

70. Y.A. Ellis, Nucl. Data Sheets 21(4), 549 (1977).

71. M.R. Schmorak, Nucl. Data Sheets 20(2), 218 (1977).

72. E. Sheldon, D.W.S. Chan, in Ref.[10], p. 169.

73. D.W.S. Chan, Ph.D. Thesis, University of Lowell (1981) (unpublished); see also Ref.[17].

74. P.A. Moldauer (Argonne National Laboratory), private communications and discussions (1981).

75. W.R. Smith, Computer Phys. Commun. 1, 106 (1969).

76. J. Raynal, IAEA-SMR-918, p. 281 (Saclay, 1972), and private communication.

77. G.R. Satchler, Phys. Lett. 7, 55 (1963).

78. J.H. Dave, J.J. Egan, G.P. Couchell, G.H.R. Kegel, A. Mittler, D.J. Pullen, W.A. Schier, E. Sheldon, submitted to Phys. Rev. C (1982).

ACKNOWLEDGEMENTS

The author appreciatively acknowledges the provision of additional, hitherto unpublished, experimental data by his colleagues in the Lowell group, which is supported in part by research grants from the U.S. National Science Foundation and the Department of Energy.
To Dr. Peter A. Moldauer of Argonne National Laboratory we are indebted for much assistance in the compilation of the unified computer program "NANCY" and for valuable discussions, as also to Dr. Alan B. Smith (Argonne National Laboratory), who kindly provided details of data measurement and evaluation in advance of publication. The assistance and interest of the University of Lowell Computation Center (Suresh C. Mathur, Director) has enabled the lengthy computations to be performed successfully and expeditiously. Dr. Desmond W.S. Chan (General Physics Corporation) kindly provided help with draughtsmanship and computation. The author's participation in this and the Cambridge Meeting has been made possible by the award of an International Travel Grant (No. CPE-8207973, Dr. Royal E. Rostenbach, Engineering Energetics Division) by the U.S. National Science Foundation, and by a supplementary award from the President's Fund of The University of Lowell.

A MICROSCOPIC DESCRIPTION OF THE ELASTIC SCATTERING CHANNEL
I. THE ROLE OF THE NUCLEAR REACTION MECHANISMS

V.G. Pronyaev

Nuclear Data Section
International Atomic Energy Agency
P.O.B. 100, A-1400 Vienna
Austria

In this paper the nuclear structure approach to nuclear rections is used to evaluate the contributions of the multi-step compound and multi-step direct reaction mechanisms in the reaction cross-section. It is demonstrated that for neutrons and protons with an initial energy higher than several MeV elastic scattering from ^{40}Ca nucleus will be dominated by the multi-step direct reaction mechanism.

Introduction

The optical model plays an important role in the description of the nucleon cross-sections of nuclei. The parameters of this model are usually chosen in order to obtain the best agreement with the experimental data for total, elastic and reaction cross-sections, as well as the angular distributions and polarization from elastic scattering. Therefore the question of the justification for the choice of parameters and their calculation in the consistent microscopic approach is always of great interest [1-5].

The potentials determined in the microscopic models are essential non-local and orbital momentum dependent. Therefore it would be useful to perform a direct comparison of the microscopic model results to the available experimental data, without using those approximations which are performed when the local equivalent potential is constructed as an intermediate step of calculations.

The question of the microscopic description of the imaginary part of the optical potential (or experimentally observed total, elastic and reaction cross-sections) is closely related to the question of which reaction mechanisms make a major contribution to the imaginary part of the potential (or optical reaction cross-section) at different energies. Below we shall use the reaction mechanism definitions given by H. Feshbach [6] and generalized in ref. [7] for multi-step direct and multi-step compound reactions.

The aim of this paper is to determine, in the framework of the rather self-consistent microscopic approach, the contributions of the simplest intermediate configurations of the open and closed type; i.e. the contributions of the multi-step direct and multi-step compound nuclear reaction mechanisms to the optical reaction cross-section. This approach is similar to that used in other papers [3,4] for optical potential calculations, but differs in the following respects,

i) all calculations were done directly for transition matrix elements and cross-sections;

ii) the wide energy region, in which the contributions of different reaction mechanisms are changed noticeably, is considered;

iii) the spectroscopical characteristics of the excited states to be used in calculations were obtained in a rather self-consistent approach of the nuclear structure theory.

The microscopic model: basic relations

The elements of the transition or T-matrix describing the elastic scattering process and taking into account the definitions of the nuclear reaction mechanisms given in ref. [6,7] may be presented in the second order Born approximation as sum of three terms,

$$T_{ii} = T_{ii}^{sel} + T_{ii}^{dw} + T_{ii}^{dir} \qquad (1)$$

T_{ii}^{sel} - is the contribution of the shape-elastic term including the Coulomb component for charged particle.
T_{ii}^{dw} - is the contribution from formation and decay of the intermediate states - the simplest bound states in the continuum, or the door-way states.
T_{ii}^{dir} - is the contribution of the simplest unbound states (or open configurations) in the continuum. It has been demonstrated in many papers [3,4] that the main contribution to the imaginary part of the optical potential for initial energies up to several tens of MeVs is connected with the formation of intermediate states, such as nucleon in bound or unbound state plus some collective excitation of the target nucleus. Below we shall consider spherical nuclei with collective excitations of the vibrational type.

Using the usual designations (as for example in paper [8]) and neglecting the spin of nucleon, we can present the separate terms of the T-matrix as

$$\text{a)}\quad T_{ii}^{sel} = T_{ii}^{coul} + \frac{(4\pi)^{3/2}}{k_i^2} \sum_{\ell=0}^{\infty} e^{2i\sigma_\ell} \sqrt{2\ell+1} \int_0^{\infty} dr\, f_\ell^2(k_i r) V(r) Y_{\ell,0}(\theta) \qquad (2)$$

where T_{ii}^{coul} is the Coulomb part of the transition amplitude and $f_\ell(k_i r)$ is the radial wave function for single-particle Schrödinger equation with real potential V(r);

$$\text{b)}\quad T_{ii}^{dw} = \frac{\sqrt{4\pi}}{k_i^2} \sum_{\ell\lambda\ell''} e^{2i\sigma_\ell} \sqrt{2\ell+1}\, \beta_\lambda^2 [(\ell 0 \lambda 0 | \ell'' 0)]^2 \frac{R_{\ell\ell''}^2}{(E_i - E_d) + i/2\, \Gamma_d(E_i)} Y_{\ell,0}(\theta) \qquad (3)$$

where ℓ'' the orbital momentum of the nucleon occupying the discrete (or quasi-discrete) single-particle level with energy $E_{n\ell''}$ and λ is the multipolarity of the one-phonon excitations with energy E_λ and mean square-root parameter of deformation β_λ, forming the door-way state with energy

$$E_d = E_\lambda - E_{n\ell''}, \qquad R_{\ell\ell''} = \int_0^{\infty} f_\ell(k_i r)\, r \frac{dV(r)}{dr} \chi_{n\ell''}(r)\, dr$$

is the radial integral where $\chi_{n\ell''}(r)$ describes the radial wave function of the nucleon at the single-particle level $E_{n\ell''}$. One problem that is encountering in trying to use this last equation is the difficulty of precisely defining $\Gamma_d(E_i)$ which appears in the denominator of the expression (3) for T_{ii}^{dw}. We shall use the approximation that $\Gamma_d(E_i)$ is a sum $\Gamma_d(E_i) = \Gamma_d^{\uparrow}(E_i) + 2I$, where

$$\Gamma_d^{\uparrow}(E_i) = (k_i/2\pi)\beta_\lambda^2 [(\ell 0 \lambda 0 | \ell'' 0)]^2 |R_{\ell\ell''}|^2$$

is the width of the door-way state decay to the continuum and I is the averaging interval. The S-matrix corresponding the sum of (2) and (3) will be unitary only for I=0 and if it is possible to neglect the interference between door-way states;

$$\text{c)}\quad T_{ii}^{dir} = -\frac{\sqrt{4\pi}}{k_i^2} \sum_{\ell\lambda n} e^{2i\sigma_\ell} \sqrt{2\ell+1}\, \beta_\lambda^2 [(\ell 0 \lambda 0 | n 0)]^2 R_{\ell n\ell} Y_{\ell 0}(\theta) \qquad (4)$$

where the radial integral $R_{\ell n\ell}$ has the form

$$R_{\ell n\ell} = \frac{2m}{(k_f \hbar^2)} \int_0^{\infty} dr \int_0^{\infty} dr'\, r \frac{dV(r)}{dr} f_\ell(k_i r) f_n(k_f r_<) \qquad (5)$$

$$\times h_n(k_f r_>) f_\ell(k_i r') r' \frac{dV(r')}{dr'}$$

K. H. Böckhoff (ed.), Nuclear Data for Science and Technology, 528–533.

and $f_n(k_f r_<)$ and $h_n(k_f r_>)$ are the regular and irregular solutions of the single-particle Schrödinger equation. To take into account phenomenologically the absorption which exists for the nucleon in the intermediate states of the system we added the imaginary part to the real potential when calculating f_n and h_n.

In summary the main approximations used in the microscopic approach presented here are,

1. The solution of the problem was limited to the second order Born approximation. Only transitions between the elastic scattering channel and those intermediate states of the system which are formed by the one-phonon excitation of the target nucleus and nucleons in discrete or continuum spectrum, were considered. The influence on the elastic scattering channel of such reactions as charge-exchange reactions were omitted.

2. The spin of the nucleon was omitted and therefore as a consequence all effects connected with the spin-orbit splitting were omitted.

3. The effective forces both for nuclear structure and nucleon scattering problem were described by the separable approximation with self-consistency conditions discussed in ref. [9] and it led to the simplest form of the transitional potentials ($\sim r\frac{dV(r)}{dr}$) for one-phonon excitations.

Model Parameters and Calculation Results for ^{40}Ca

Calculations were performed for ^{40}Ca nucleus that allowed us to use the states of the harmonical oscillator model, as a basis of the single-particle states, for determination of the characteristics of the one-phonon excitations in the random phase approximation method (RPA). The results of this calculation are given in Table 1 and discussed in more detail elsewhere [10]

Table I. Spectroscopical characteristics of the one-phonon excitations in ^{40}Ca nucleus

E_λ MeV	3.30	4.51	6.06	9.90	10.5	14.0	14.5
λ^π	3^-	5^-	3^-	5^-	3^-	3^-	4^+
β_λ	0.42	0.27	0.10	0.12	0.10	0.08	0.31

E_λ MeV	15.1	15.2	15.3	17.0	19.0	20.5	25.0
λ^π	6^+	5^-	2^+	4^+	4^+	6^+	4^+
β_λ	0.22	0.18	0.26	0.104	0.09	0.11	0.20

E_λ MeV	25.6	25.8	27.2	27.6	28.2	31.0	35.0
λ^π	7^-	5^-	6^+	5^-	3^-	7^-	3^-
β_λ	0.20	0.18	0.19	0.12	0.21	0.10	0.09

E_λ MeV	36.1	38.0	38.1	39.0	39.4	40.0
λ^π	8^+	5^-	6^+	8^+	4^+	7^-
β_λ	0.17	0.161	0.24	0.11	0.18	0.19

In some cases the values of E_λ present the average-weighted value of energy and β_λ - the total intensity for several excited states of the same multipolarity. It is important to stress that the intensities of the one-phonon excitations given in Table 1 differ appreciably from those evaluated from energy-weighted sum rules [3] and are more realistic. The energies of the discrete and quasi-discrete single-particle states were taken from [11] and the parameters of the real part of the optical potential from [12].

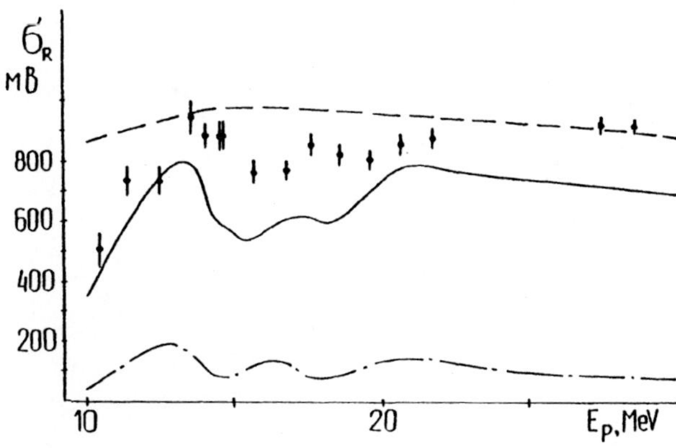

Fig. 1 Comparison of the experimental data on reaction cross-section for protons on ^{40}Ca nucleus (points) with the results of phenomenological optical model (dashed line) and microscopic description (solid line).

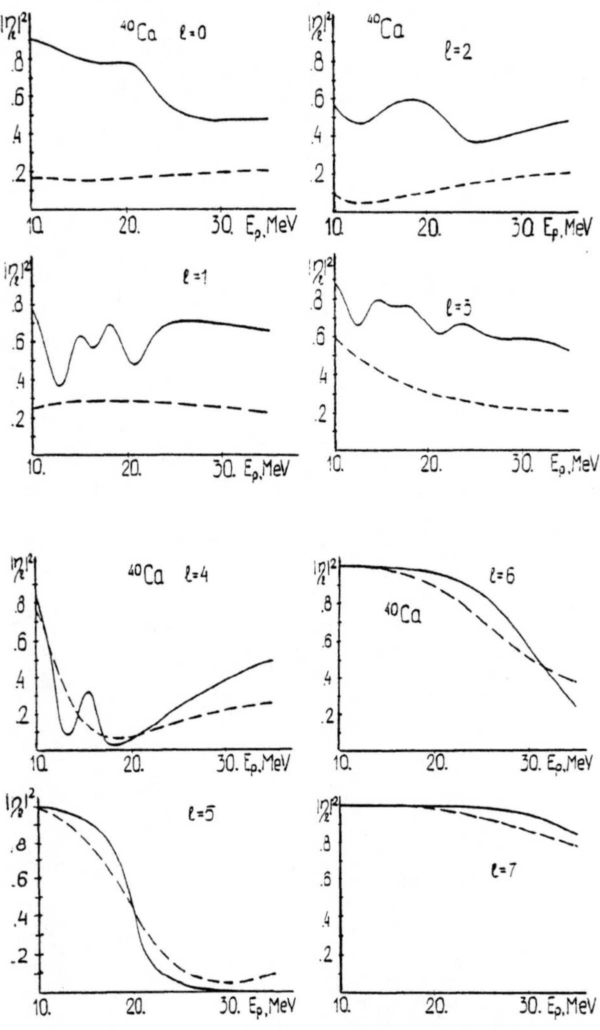

Fig. 2 The energy dependence of the square of the module of the S-matrix element for different orbital waves from phenomenological and microscopic models.

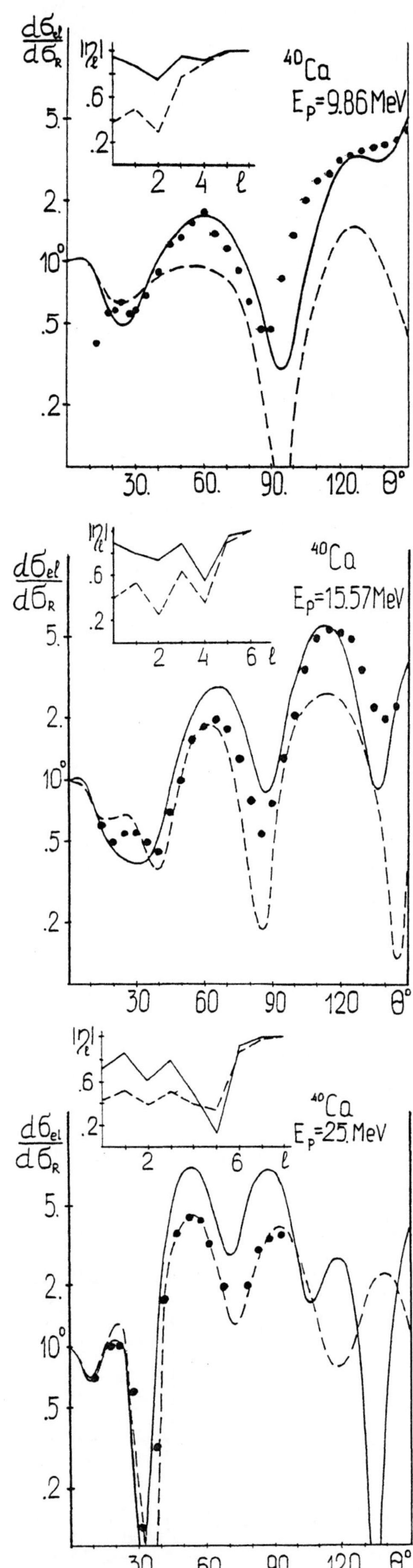

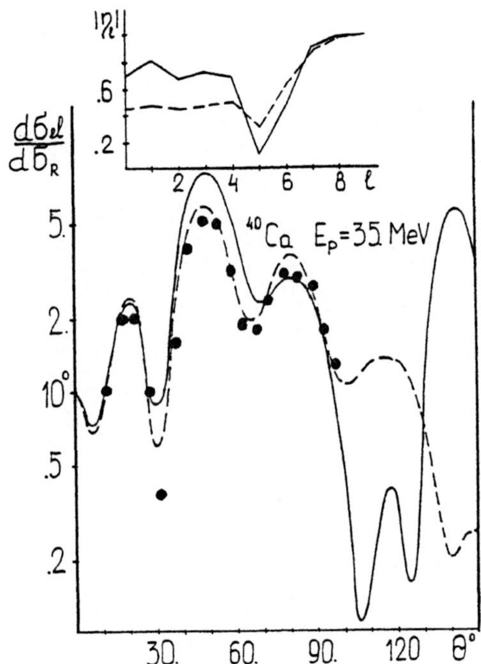

Fig. 3 Comparison of the experimentally observed elastic scattering angular distributions (points) with the results of their description in the phenomenological optical model (dashed line) and in the microscopic approach (solid line).

Fig.1 presents a comparison of the available experimental data of the reaction cross-sections for protons with an incident of energy from 10 to 30 MeV (points) with the results of the microscopic calculations taking into account both the closed and open intermediate configurations (solid line); the contribution of the door-way intermediate states for I=0.5 MeV averaging interval is shown by dot-dashed line. The dashed line presents the results of the phenomenological optical model. Because of the high threshold of the (p,n) reaction, this channel does not contribute to the imaginary part of the optical potential below an energy of 16 MeV. Fig.2 presents a comparison of the square module of the S-matrix elements of the results obtained in microscopic and phenomenological approaches. The elastic scattering angular distributions for protons with incident energy from 10 to 35 MeV is shown in Fig.3; the same designations as in Fig.1 were used.

Similar calculations were done for the elastic scattering of neutrons with incident energies from 0.1 MeV to 35 MeV on ^{40}Ca nuclei. A detailed discussion of low energy neutron scattering is given in another paper [13].

Discussion of Calculation Results

The analysis of the results of these calculations lead to the following conclusions:

1. As noted earlier [3,4], the major contribution to the imaginary part of the optical potential (or to the optical reaction cross-section as considered here) is connected with the process of the formation and decay of the open type intermediate configurations; i.e. the so-called multi-step direct reaction mechanism.

2. The contribution of the multi-step compound nuclear reaction mechanism to the reaction cross-section is important only below the inelastic scattering threshold and then diminishes rapidly at higher energies. The magnitude of this component is dependent on I, the averaging parameter.

If a non-zero I value is used the S-matrix will no longer be unitary, but this is justified since we wish to have a smooth dependence of the cross-section with energy and to take into account to some extent the effect of coupling of the doorway with more complicated states.

3. From Fig.2 we see that at low energies the phenomenological optical model predicts too strong absorption for the waves with low orbital moment; in contrast the microscopic model predicts more reasonable absorption. The discrepancy in the lower energy absorption predicted by the phenomenological optical model is due to the parameters taken from ref. [12]. For higher energies, the phenomenological optical model predicts absorption which agrees well with experimental data. In contrast the microscopic model under predicts absorption. The underprediction of absorption by the microscopic model at higher energies is due to the fact that the model does not consider charge-exchange channels.

4. Neglecting spin-orbit coupling in our calculations leads to the situation where the single-particle resonances are not split; it is possible that the result of this is some "oversaturation" of the absorption and the appearance of divergence near the position of the single-particle resonances. The peak in $|R_l|^2$ dependence for g-wave near the 16 MeV energy (Fig.2) presents an example of such divergence. It is possible such singularities will disappear when the spin-orbital splitting of the single-particle resonances are taken into account. However, it is still not obvious that this divergence will not occur for any Born type approximation [14].

Conclusion

The results obtained in this paper have shown that the multi-step direct mechanism of the nuclear reaction is the dominant mechanism which determines the cross-sections of nucleons with energy higher than several MeV.

References

[1] H. Feshbach, A.K. Kerman and R.H. Lemmer, Ann. Phys. 41 (1967) 230

[2] N. Vinh Mau, in Theory of Nuclear Structure, IAEA, Vienna (1970) p.931
N. Vinh Mau, A. Bouyssy, Nucl. Phys. A257 (1976) 189
A. Bouyssy, H. Ngo and N. Vinh Mau, Nucl. Phys. A371 (1981) 173

[3] C.L. Rao, M. Reeves, G.R. Satchler, Nucl. Phys. A207 (1973) 182
P.W. Coulter, G.R. Satchler, Nucl. Phys. A293 (1977) 269

[4] A. Lev and W.P. Beres, Phys. Rev. C9 (1974) 2416

[5] J.P. Jeukenne, A. Lejeune, C. Mahaux, Phys. Rev. C16 (1977) 80
A. Lejeune, P.E. Hodgson, Nucl. Phys. A295 (1978) 301
F.A. Brieva and J.R. Rook, Nucl. Phys. A307 (1978) 493

[6] H. Feshbach, Ann. Phys. 5 (1958) 357; 19 (1962) 287

[7] H. Feshbach, A. Kerman, S. Koonin, Ann. Phys. 125 (1980) 429

[8] A.I. Blokhin, V.G. Pronyaev, Sov. Jour. Nucl. Phys. 30 (1979) 655

[9] D.J. Rowe, Phy. Rev. 162 (1967) 866
V.P. Lunev, Izv. ANSSSR, ser. fiz. 44 (1977) 199 (in Russian)

[10] A.I. Blochin, V.G. Pronyaev, Izv. AN KazSSR, ser. fiz.-mat. 4 (1979) 26 (in Russian)

[11] Yu.L. Dobrinin, S.V. Tolokonnikov, S.A. Fayans, IAE-2593 (1975), preprint of the I.V. Kurchatov Institute of Atomic Energy (in Russian)

[12] F.D. Becchetti, Jr., G.W. Greenlees, Phys. Rev. 182 (1969) 1190

[13] V.G. Pronyaev, "A Microscopic Description of the Elastic Scattering Channel. II. The Strength Functions for the Low Energy Neutrons", paper presented at this conference.

[14] D.S. MacMillan, E.F. Redish, Phys. Rev. Lett. 48 (1982) 391

A MICROSCOPIC DESCRIPTION OF THE ELASTIC SCATTERING CHANNEL
II. THE STRENGTH FUNCTIONS FOR THE LOW ENERGY NEUTRONS

V.G. Pronyaev

Nuclear Data Section
International Atomic Energy Agency
P.O.B. 100, A-1400 Vienna
Austria

The nuclear structure approach to nuclear reactions described in detail in Part I of this paper, is applied to the description of the cross-sections and angular distributions of the elastic scattering of the low-energy neutrons incident on nuclei.

Introduction

This paper is a natural continuation of using the microscopic approach described in Part I [1] for low energy neutron scattering. A similar approach has been used in other papers [2-4]. As in Part I of this paper [1], the ^{40}Ca nucleus, where excited states have been extensively studied, is used as an example to calculate cross-sections and strength functions in the energy region from several keV up to the inelastic scattering threshold. The objective of this paper is to determine if it is possible to simultaneously describe, using the same set of initial parameters, the s-, p- and d-wave strength functions and the cross-sections for low-energy neutrons - this cannot be done using the phenomenological optical model.

Parameters and results of calculations

The basic equations used in this model are given in paper [1]. Since we will only consider the case of the low-energy neutrons, only the shape-elastic term and the term describing the elastic scattering leading to the formation of door-way states (i.e. closed configurations in the continuum) as intermediate states will contribute to the transition matrix. We will use the approximation [1] that the width $\Gamma_d^\uparrow(E_i)$ describing the decay into the continuum of the door-way state formed by neutron on the single-particle level and one-phonon excitations of the target nucleus, can be written as [1]

$$\Gamma_d^\uparrow(E_i) = \frac{k_i}{2\pi}\beta_\lambda^2\left[(l0\lambda0|l'0)\right]^2 |R_{ll'}|^2 \qquad (1)$$

From ref [1] equations (2) and (3) the important feature to note for the T-matrix in ref.[1] is that the corresponding S-matrix will be unitary only in case when the averaging interval I is equal to zero and it is possible to neglect the interference between different door-way states.

In contrast to the approach of paper [1], in order to obtain the best description of the strength functions and cross-sections, we will use as parameters of our model the latest experimental data on single-particle levels in ^{41}Ca [5] and strong low energy one-phonon excitations in ^{40}Ca [6]. The positions of the neutron single-particle levels together with their C^2 - spectroscopical factors, which have been prescribed by us from analysis of data in [5], are given in table I.

As was mentioned earlier the spectroscopical characteristics of the strong one-phonon excitations in the ^{40}Ca target nucleus for energies up to 10 MeV energy were taken from paper [6]; for higher energies those from table I of the paper [1] were used.

It was found that the results of the calculations of the strength functions and cross-sections are strongly dependent on the parameters of the real part of the optical potential. The following Wood-Saxon type parameters were used to describe simultaneously the s-, p- and d-wave strength functions and

Table I. The positions and the C^2 coefficients for neutron single-particle levels in ^{41}Ca used in the calculations

nl_j	$1f_{7/2}$	$2p_{3/2}$	$2p_{3/2}$	$2p_{1/2}$	$2p_{1/2}$	$2p_{3/2}$
E_{nlj},MeV	-8.36	-6.42	-5.90	-4.75	-4.42	-3.76
C^2	1.00	0.71	0.24	0.10	0.70	0.05

nl_j	$2p_{1/2}$	$1f_{5/2}$	$1f_{5/2}$	$1f_{5/2}$
E_{nlj},MeV	-3.61	-3.48	-2.71	-2.56
C^2	0.20	0.24	0.52	0.24

cross-sections

$$V_o = (60.3 - 0.32\ E_n)\text{MeV}$$
$$r_o = 1.21\ \text{fm}$$
$$a_o = 0.50\ \text{fm}.$$

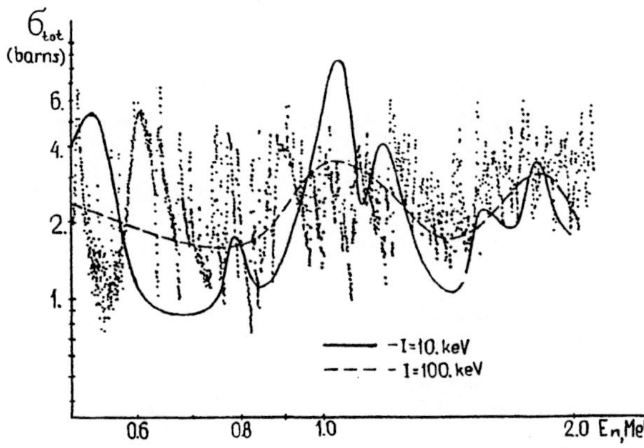

Fig. 1 Calculated total cross-sections for two different averaging intervals (solid and dashed lines) compared with experimental data (points)

Fig.1 illustrates the results obtained for the ^{40}Ca total cross-section [7]. The solid and dashed lines present results corresponding to calculations for two different averaging interval I = 10 keV and I = 100 keV. Fig.2 illustrates the results of the microscopic description of the "experimentally" observed neutron widths for s-, p- and d-waves presented in a form of a cumulative sum of the

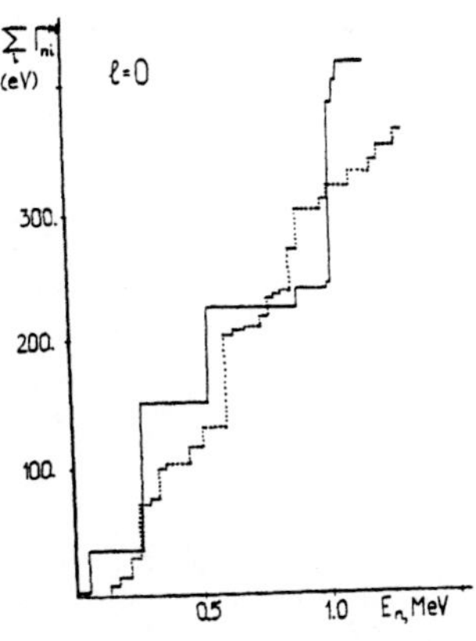

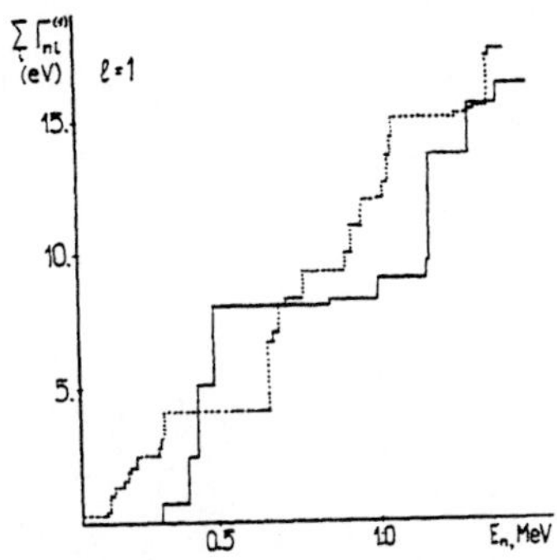

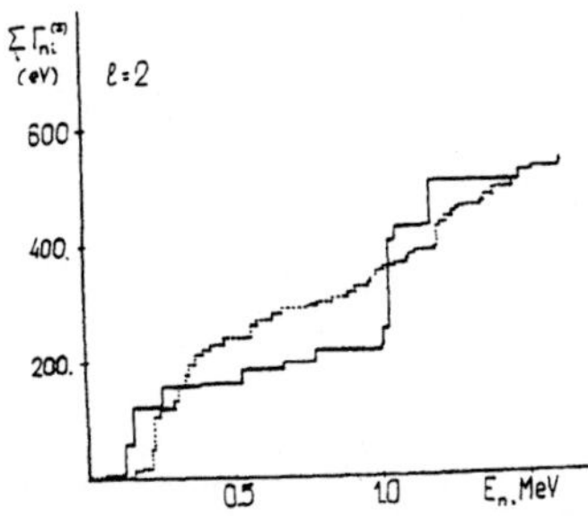

Fig. 2 Comparison of the results of the microscopic calculations and experimental data on reduced neutron widths for s-, p- and d-wave

reduced neutron widths. The word "experimentally" was placed in quotation marks because the data were taken from ref.[8] as follows,

a) the neutron widths for s-wave were taken exactly as given in ref [8];

b) all the levels which were identified as $\ell > 0$ levels (but without explicit assignment for ℓ) and having I = 3/2, 5/2 have been assumed to be d-wave.

c) Since for ^{40}Ca the p-wave strength function has a strong minimum while the s- and d-wave strength functions are close to maximum, it was assumed that all levels which were identified in [8] as strong p-wave levels with $g\Gamma_n^{(1)} > 1$eV could really be s- or d-wave levels. The I value was used to determine whether these levels were s or d-wave.

It is possible that the last assumption is too stringent and that by taking into account the spin-orbit splitting of the real part of the potential we can obtain a good description of the experimental data for the p-wave strength function without this assumption. From Fig.2 we may see, that although the experimentally measured strength

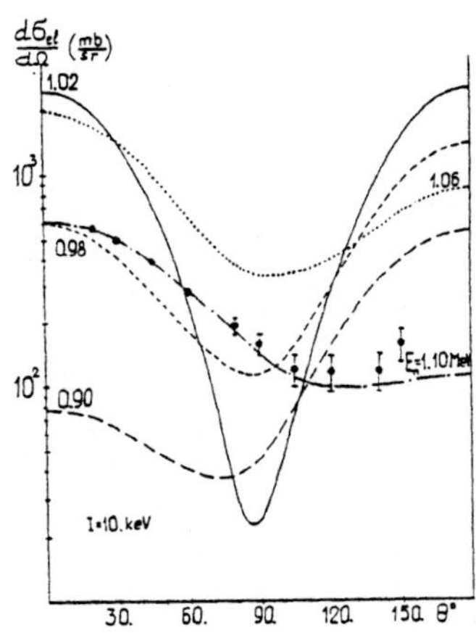

Fig. 3 Results of the microscopic calculations of the elastic scattering angular distributions for different energies of the initial neutrons near the energy of the door-way state

functions have structure which is similar to the calculated one, the number of experimentally observed levels is much higher. The reason for this effect is that we have limited our model to one-phonon excitations in target nucleus.

For the elastic scattering angular distributions Fig.3 shows the effects of the interference of the potential scattering with the scattering through formation of the strong intermediate door-way state ($1f_{5/2} \otimes 3_1$, E_d = 1.026 MeV, $\Gamma^{\uparrow}_{d,\ell=0}$ = 143 keV and $\Gamma^{\uparrow}_{d,\ell=2}$ = 11 keV). It is important to note that the angular distribution changes considerably as the energy of the incident neutrons passes through the resonance energy; only after averaging near the resonance energy does the angular distributions become symmetrical relative to the 90° scattering angle.

Conclusions

In summary, we may conclude that the nuclear structure approach to the nuclear reactions is a powerful tool for studying the multi-step compound reaction mechanism. Some effects, such as local dependence of the strength functions and elastic scattering angular distributions, which are not possible to describe in the phenomenological optical model, can be natural explained in the frame-work of this microscopic approach.

References

[1] V.G. Pronyaev, "A Microscopic Description of the Elastic Scattering Channel. I. The Role of the Nuclear Reaction Mechanisms", paper presented at this conference

[2] I. Lovas, Nucl. Phys. 81 (1966) 353

[3] G.L. Payne, Phys. Rev. 174 (1968) 1227

[4] A. Lev, W.P. Beres, M. Divadeenam, Phys. Rev. C9 (1974) 2416

[5] P.M. Endt, C. van der Leun, Nucl Phys. A310 (1978) 588

[6] C.R. Gruhn, T.Y.T. Kuo, C.J. Maggiore et al., Phys. Rev. C6 (1978) 915

[7] S. Cierjacks, P. Forti, D. Kopsch et al., KFK-1000 (1968)

[8] S.F. Mughabghab, M. Divadeenam, N.E. Holden, Neutron Cross Sections, Vol.1 Part 4 (1981)

A NEW APPROACH TO CALCULATE NUCLEAR LEVEL DENSITIES ON THE BASIS OF RECENT NUMBER THEORETICAL DEVELOPMENTS

A.M.Anzaldo Meneses*

Kernforschungszentrum Karlsruhe
Institut für Neutronenphysik und Reaktortechnik
P.O.Box 3640, D-75 Karlsruhe
Federal Republic of Germany

*DAAD-Stipendiat. On leave from the UNAM(Universidad Nacional Autonomà de Mexico).

The nuclear level density problem has been treated using modern methods of analytical number theory. In particular the asymptotical calculation of the partition function leads to more rigorous and more general results than the usual Bethe-formula. The partition function is expressed with the help of a Dirichlet series. The parameters of this series determine the obtained level density. For the case of constant single particle level density the resulting parameters yield the usual Bethe-formula. In general the parameters lead to another energy and nucleon number dependende of the level densities. They correspond to more general single particle level densities which include the shell structure. Also the BCS type of paring has been taken into account.

Introduction

As it is well known[1),2),3)] the relations actually in use to compute nuclear level densities are not able to reproduce quantitatively the experimental results over a wide energy range. Moreover the mathematical methods normally used make very restrictive physical assumptions in order to obtain simple analytical expressions.

At the very beginning of the investigations on nuclear level densities[4),5)] it was clear that for large excitation energies, there was a strong relation between the methods of number theory (computation of the so called "partitions of integer numbers") and those needed to calculate the level densities. After these very early investigations till now there has been little progress in this sense, although the mathematical methods of number theory have experienced several important advances[6)].

In this paper the recent methods of modern number theory will be applied to obtain an asymptotical expression for the partition function which leads to a more rigorous and more general result than the usual level density formulae.

Derivation of the new level density formula

We start from the usual relation for the nuclear level density including pairing effects and obtained with help of the saddle point method[7)]. The nuclear level density reads:

$$\rho(N_n,N_p,\varepsilon) = \frac{\exp S}{(2\pi)^{3/2}\sqrt{\det|\partial^2 \mu_i\mu_j S|}} \tag{1}$$

where ε is the total energy, $N_{n,p}$ is the number of neutrons (protons), $\mu_{n,p}$ are the Lagrange parameters for neutrons (protons), to be determined, S is the entropy and $\det|\partial^2\mu_i\mu_j S|$ is a certain determinant.
The entropy is given by:

$$S = \ln Z(\mu_n,\mu_p,\beta) + \beta\varepsilon - \mu_n N_n - \mu_p N_p \tag{2}$$

where $Z(\mu_n,\mu_p,\beta)$ is the partition function given by:

$$\ln Z = -\beta \sum_k (\varepsilon_k-\lambda-E_k) - \beta\frac{\Delta^2}{G} + 2\sum_k \ln\left[1+\exp(-\beta E_k)\right] \tag{3}$$

The quasiparticle energies E_k are given in terms of the single particle energies, ε_k, the chemical potential λ and the energy gap Δ by:

$$E_k = \sqrt{(\varepsilon_k-\lambda)^2+\Delta^2} \tag{4}$$

G is the related to Δ by the equation:

$$\frac{2}{G} = \sum_k \frac{1}{E_k} \operatorname{tgh}\left(\frac{1}{2}\beta E_k\right) \tag{5}$$

The saddle point equations are:

$$\partial_{\mu_n} \ln Z_n = N_n,\ \partial_{\mu_p} \ln Z_p = N_p,\ -\partial_\beta \ln Z = \varepsilon \tag{6}$$

if $\ln Z = \ln Z_n + \ln Z_p$, from which is possible to obtain:

$$S_{n,p} = 2\left[1-\beta\partial_\beta \sum_k \ln 1+\exp(-\beta E_k)\right]_{n,p} \tag{7a}$$

$$\varepsilon_{n,p} = \sum_k \varepsilon_k \left[1-\frac{\varepsilon_k-\lambda}{E_k}\operatorname{tgh}\left(\frac{1}{2}\beta E_k\right)\right]_{n,p} - \frac{\Delta^2_{n,p}}{G_{n,p}} \tag{7b}$$

$$N_{n,p} = \sum_k \left[1-\frac{\varepsilon_k-\lambda}{}\operatorname{tgh}\left(\frac{1}{2}\beta E_k\right)\right]_{n,p} \tag{7c}$$

where the subscripts n,p denote neutrons and protons respectively and $S = S_n + S_p, \varepsilon = \varepsilon_n + \varepsilon_p$.

To compute these expressions asymptotically it is useful to define the "quasiparticle partition function" $\hat{Z}(\beta)$:

$$\hat{Z}(\beta) = \prod_k (1+\exp(-\beta E_k)) \tag{8}$$

and rewrite:

$$S = 2\left[1-\beta\partial_\beta\right] \ln Z(\beta) \tag{9}$$

The estimation of $\hat{Z}(\beta)$ goes as follows:
First $|\varepsilon_k-\lambda| = \hat{\varepsilon}_k$ are substituted by integral multiples of some energy unit $1/g$:

$$\hat{\varepsilon}_k = e_k/g, \text{ for } e_k \text{ positive integers} \tag{10}$$

thus:

$$\ln\hat{Z} = \sum_n a_n \ln\left[1+\exp(-\beta\sqrt{(n/g)^2+\Delta^2}\right] \tag{11}$$

K. H. Böckhoff (ed.), Nuclear Data for Science and Technology, 534–536.

where

a_n = degeneracy of $\hat{\varepsilon}_k$, if $n=e_k$ for some e_k

and

$a_n = 0$ otherwise

We aplly now the Mellin transformation [8]:

$$\exp\{-\beta m\sqrt{(n/g)^2+\Delta^2}\} = \frac{1}{2\pi i}\int_{c-i\infty}^{c+i\infty} dz\left(\frac{n}{g}\right)^{-z}\frac{\Delta}{\sqrt{\pi}}\left(\frac{2\Delta}{\beta m}\right)^{\frac{1}{2}z-\frac{1}{2}}$$

$$\times\,\Gamma\left(\frac{z}{2}\right)K_{\frac{1}{2}z+\frac{1}{2}}(\Delta\beta m) \qquad (12)$$

where $K_\nu(t)$ is a modified Bessel function of the second kind.

$$D(t) = \sum_n \frac{a_n}{n^t} \qquad (13)$$

which will be assumed to be analytic except for simple poles at the points $\alpha>\alpha_1>\dots\dots\dots>\alpha_N>0$ with the residues A_0, A_1, etc. Thus $Z(\beta)$ will be given by:

$$\ln Z = \frac{1}{2\pi i}\int_{c-i\infty}^{c+i\infty} dz g^z D(z)\frac{\Delta}{\sqrt{\pi}}\Gamma\left(\frac{z}{2}\right)\sum_m\left(\frac{2\Delta}{\beta m}\right)^{\frac{1}{2}z-\frac{1}{2}}$$

$$\frac{(-)^{m+1}}{m}K_{\frac{1}{2}z+\frac{1}{2}}(\Delta\beta m) \qquad (14)$$

Using the representation:

$$t^\nu K_\nu(t) = 2^\nu\Gamma(\nu+1)e^{-t}\sum_{m=0}^{\infty}\frac{(\frac{1}{2}-\nu)_m}{(\frac{1}{2}+\nu)_{m+1}}L_m(2t) \qquad (15)$$

where L_m denotes the Lagrange polynomials,

$$\ln\hat{Z} = \frac{1}{2\pi i}\int_{c-i\infty}^{c+i\infty} dz g^z D(z)\{\beta^{-z}\Gamma(z)(1-2^{-z})\zeta(1+z)+O(\beta^{-z+1})\}$$

$$= \sum_i A_i(1-2^{-\alpha_i})\zeta(\alpha_i+1)\Gamma(\alpha_i)(\beta/g)^{-\alpha_i}+D(0)\ln Z +$$

$$+\,O(\beta^{-\alpha_{N+1}}) \qquad (16)$$

where $\zeta(x)$ denotes the Riemann ζ-function.
And for the entropy at the saddle-point:

$$S = 2D(0)\ln 2 + \sum_i(1-2^{-\alpha_i})\Gamma(\alpha_i+1)\zeta(1+\alpha_i)2A_i(\beta/g)^{-\alpha_i}$$

$$\left(1+\frac{1}{\alpha_i}\right) + O(\beta^{-\alpha_{N+1}}) \qquad (17)$$

which leads after some algebra to:

$$S = \sum_j M_j(\alpha_i,A_i)E^{m_j(\alpha_i)}, \quad m_j > m_{j+1} \qquad (18)$$

This is a descending series of the excitation energy $E = \varepsilon-\varepsilon_0-\delta P$ (ε_0 is the ground state energy and δP is a pairing correction). The coefficients $M_j(\alpha_i,A_i)$ and the exponents $m_j(\alpha_i)$ are given by Lagrange equations for ε and N. The leading term is of the form:

$$2^{\frac{1}{1+\alpha_0}}\left(1+\frac{1}{\alpha_0}\right)\{(1-2^{-\alpha_0})\zeta(1+\alpha_0)\Gamma(1+\alpha_0)A_0\}^{\frac{1}{1+\alpha_0}}E^{\frac{\alpha_0}{1+\alpha_0}} \qquad (19)$$

With equation (18) we have thus arrived at an explicit analytic expression for the entropy which shows how the relevant characteristics of the discrete single particle spectrum determines the nuclear level density for large excitation energies.

For the simple case of a constant single particle level density without pairing forces, it is easy to see that:

$$D(s) = \frac{1}{2}\{\zeta(s,x)+\zeta(s,1-x), \; 0<x\leq 1/2 \qquad (20)$$

(where $\zeta(s,x)$ denotes the generalized Riemann ζ-function) and to find $D(0) = 0$, $A = 1$, $\alpha = 1$ which leads

$$\rho(E) = g\left(\frac{g^2}{g_n g_p}\right)^{1/2}\frac{6^{1/4}}{12}\frac{e^{\pi\sqrt{\frac{2}{3}gE}}}{(gE)^{5/4}} \qquad (21)$$

which is the well known [1] Bethe-formula for the nuclear level density with $g = g_n + g_p$.
It is also possible to compute analytic expression for other single particle spectra. For example for an harmonic oscillator with energies $\varepsilon_k = (k+3/2)\hbar\omega$ we obtain:

$$D_{osc.}(S) = \frac{2}{(\hbar\omega)^s}\left\{\zeta(s-2,1/2)+\left(\frac{\lambda^2}{(\hbar\omega)^2}-\frac{1}{4}\right)\zeta(s-1/2)\right\} \qquad (22)$$

and for the partition function $\hat{Z}$:

$$\ln\hat{Z}_{osc.} = \frac{\pi^2}{6\hbar\omega\beta}\left[\frac{\lambda^2}{(\hbar\omega)^2}-\frac{1}{4}\right]+\frac{7\pi^2}{180(\hbar\omega\beta)^3}+O(\beta) \qquad (23)$$

The shell correction method

The mathematical treatment applied in the preceding section provides also a method to study the closely related Strutinsky calculations for ground state shell corrections [9].
The effect of the shell structure on the potential energy surface is expressed usually in the form

$$\delta U = U - \tilde{U} \qquad (24)$$

where U is the total single particle energy sum given by:

$$U = \int_{-\infty}^{\lambda} d\varepsilon\varepsilon g(\varepsilon), \quad N = \int_{-\infty}^{\lambda} d\varepsilon\, g(\varepsilon) \qquad (25)$$

with

$$g(\varepsilon) = \frac{1}{2\pi i}\int_{-i\infty}^{i\infty} d\beta\; e^{\beta\varepsilon}Z_0(\beta) \qquad (26)$$

using the partition function:

$$Z_0(\beta) = \sum_k e^{-\beta\varepsilon_k} \qquad (27)$$

The smoothed energy $\tilde{U}$ is given by the system:

$$\tilde{U} = \int_{-\infty}^{\tilde{\lambda}} d\varepsilon\varepsilon\tilde{g}(\varepsilon), \quad N = \int_{-\infty}^{\tilde{\lambda}} d\varepsilon\tilde{g}(\varepsilon) \qquad (28)$$

where $\tilde{g}(\varepsilon)$ is the smoothed level density function [10]:

$$\tilde{g}(\varepsilon) = \frac{1}{\gamma}\sum_k \zeta_M\frac{1}{\gamma}(\varepsilon-\varepsilon_k) \qquad (29)$$

defined by the energy-smoothing parameter γ and by the smearing functions $\zeta_M(x)$:

$$\zeta_M(x) = P_M(x)\omega(x) \qquad (30)$$

where $P_M(x)$ is a so-called curvature correction polynomical of Mth degree and $\omega(x)$ is a weight function.
To compute $\tilde{g}(\varepsilon)$ we proceed first with an asymptotical computation of $Z_0(\beta)$ using the Mellin transformation

$$e^{-\beta\varepsilon} = \frac{1}{2\pi i}\int_{c-i\infty}^{c+i\infty} ds(\beta\varepsilon)^{-s}\Gamma(s), \quad \mathrm{Re}\beta\varepsilon>0, \; c>0 \qquad (31)$$

to find

$$Z_o(\beta) = \sum_{j=1}^{n} B_j \Gamma(\lambda_j)\beta^{-\lambda_j} + D_o(0) + O(\beta^{-\lambda n+1}) \quad (32)$$

where

$$D_o(s) = \sum_k \frac{b_k}{\varepsilon_k^s}, \quad \varepsilon_k \neq \varepsilon_m \text{ for } k \neq m \quad (33)$$

b_k is the degeneracy of the level ε_k and we assume that $D_o(s)$ has only simple poles at $s = \lambda_j > \lambda_{j+1} > 0$ with residues B_j. We obtain in this way:

$$g(\varepsilon) = \sum_j B_j \varepsilon^{\lambda_j - 1} + D_o(0)\delta(\varepsilon) \quad (34)$$

without the introduction of any parameters in addition to those defining the single particle spectrum.
As a simple example it is easy to find for the single particle level density of a cubic box potential of side L, the smoothed expression:

$$g(\varepsilon) = \frac{1}{4\pi^2}\left(\frac{2mL^2}{\hbar^2}\right)^{3/2}\sqrt{\varepsilon} - \frac{3}{8\pi}\left(\frac{2mL}{\hbar^2}\right)^{1/2}\frac{1}{\sqrt{\varepsilon}}$$

$$-\frac{1}{8}\delta(\varepsilon) \quad (35)$$

which agrees with the result of reference [11] obtained using a semiclassical procedure.
Another example very easy to compute is the isotropic harmonic osciallator with a constant spin-orbit interaction with hamiltonian:

$$H = \frac{-\hbar^2}{2M}\nabla^2 + \frac{1}{2}M\omega^2 r^2 - k\hbar\omega \mathbf{l}\cdot\sigma \quad (36)$$

leading to

$$g(\varepsilon) = \frac{(1-k^2)\varepsilon^2}{3(\hbar\omega)^3(1-k^2)^2} - \frac{k^3\varepsilon}{(\hbar\omega)^2(1-k^2)^2} + \frac{10k^4 - 9K^2 - 3}{12\hbar\omega(1-k^2)^2}$$

$$+ \frac{5k^3 - 2k^5}{12(1-k^2)^2}\delta(\varepsilon) \quad (37)$$

which is also in agreement with the semiclassical calculation of the same quantity[12].

Conclusions

It is easy to see that the obtained asymptotic expression for the entropy (eq.(18)) will lead to more general relations than for the case of only one pole. But even for only one pole we obtain a more general expression than the usual Bethe-formula.
It must be stressed that the analytic properties of the Dirichlet series D(t) depend only on the structure of the quasiparticle spectra and that the knowledge of this would determine entirely the nuclear level density through the parameters appearing in eq.(18).
The usual assumption of a continous single particle level density where not required, representing thus another advantage of this method. Of special interest is the possibility to study through the analytic behaviour of D(t) very important contributions as those arising from the shell structure of the spectra and their expected disappearance for very high excitation energies.
The so-called a-parameter defined by

$$a = S^2/4E \quad (38)$$

can also be calculated in terms of the general characteristics of the single particle spectrum. Preliminary calculations usind deformed single particle potentials of the Nilsson type reproduce qualitatively the energy dependence of the a-parameter reported in reference [13] [14] and [15]. Figure 1 shows the results of reference [13] computed numerically for a single-component system having particle numbers 40 and 50 with a Nilsson spectrum. Figure 2 shows the parametrized energy dependence of the a-parameter of reference [14] for ^{210}Po and ^{232}U. Figure 3 shows the energy dependence of the a-parameter of the neutron channel (a_n) derived from an analysis of cross-section for the reaction ^{206}Pb (α,f) from ref.[15]. Additional calculations as well as a study of the effect of deformation on the level densities are actually in progress.

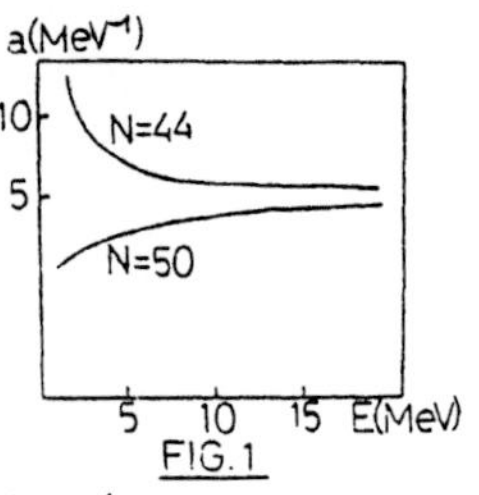

FIG. 1

Energy dependence of the a-parameter for a single component system.

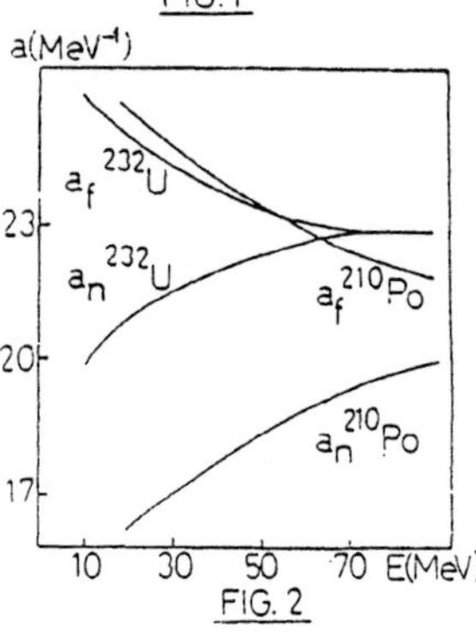

FIG. 2

In this fig. a_f is the a-parameter at the fission saddle point deformation and a_n is the a-parameter at equilibrium.

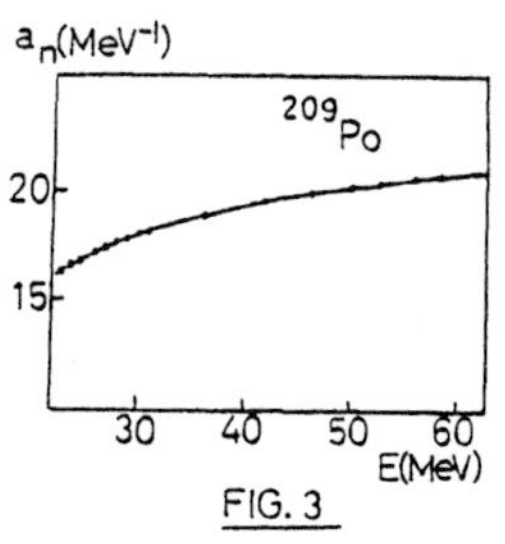

FIG. 3

Energy dependence of the a-parameter of the neutron channel (a_n) derived from an analysis of cross-sections for the reaction ^{206}Pb(α,f) from ref [15].

References

1. T. Ericson, Adv. Physics 9, (1960) 425.
2. V.S.Stavinskii, Sov. J. Part. Nucl. 3,(1973)417.
3. V.S.Rammamurthy, Nuclear Theory of Applications IAEA-SMR-43, Vienna 1980, p.187.
4. H.A.Bethe, Phys. Rev. 50,(1936)332.
5. C. van Lier, G.E. Uhlenbeck, Physica 4 (1937)531.
6. C.E.Andrews, The Theory of Partitions, Addison Wesley Publishing Co. Reading, Massachusetts (1976).
7. M.Sano and S.Yamasaki, Progr. Theor. Phys. 29, (1963) 397.
8. F. Oberhettinger, Tables of Mellin Transforms, Springer-Verlag Berlin, (1974).
9. P.A.Gottschalk, T. Lederberger, Nucl.Phys. A278, (1977)16
   G.G.Bunatyan, Yad. Fiz. 29, (1979)38
   W.Z. Reisdorf, Z.Phys. A300 (1981)227
10. V.M. Strutinsky, F.A.Ivantjuk, Nucl.Phys. A255, (1975) 405
11. R.K.Bhaduri, C.K.Ross, Phys. Rev. Lett. 27,(1971) 606
12. B.K.Jennings, R.K.Bhaduri, M.Brack, Nucl. Phys. A253, (1975)29
13. A.V.Ignatyuk,, Y.N.Shubin, Sov. J. Nucl. Phy. 8, (1969)660
14. M.Ploszajczak and M.E.Faber, Phys. Rev. C25,(1982) 1538
15. M.G.Itkis et al., Yad. Fiz.16,(1972)1150

CALCULATION AND PROCESSING OF CONTINUUM PARTICLE-EMISSION SPECTRA AND ANGULAR DISTRIBUTIONS

H. Gruppelaar, C. Costa, D. Nierop and J.M. Akkermans

Netherlands Energy Research Foundation (ECN)
P.O. Box 1, 1755 ZG Petten, The Netherlands

This paper describes some recent improvements in the calculation and processing of double-differential cross-sections using the unified exciton model. First, the model is reconsidered with respect to the state-density problem. Next, it is shown that refraction effects are quite important to describe the angular distributions. The introduction of the Kikuchi-Kawai expressions for the scattering kernel leads to further improvements. A significant refinement of this model is obtained by accounting explicitly for the energy-angle correlation of the initial emission. Comparison with experimental 14.6 and 25.7 MeV neutron-emission data and with the systematics of Kalbach and Mann shows quite good agreement. Finally, some comments are made on the processing of these data, using the energy-angle distribution format MF6 of ENDF/B.

Introduction

In the unified exciton model of pre-equilibrium and equilibrium emission [1] the double-differential cross sections are calculated from the equation

$$\frac{d^2\sigma}{d\varepsilon d\Omega}(a,b) = \sigma_a \sum_n w_b(n,\varepsilon)\tau(n,\Omega) \ , \quad (1)$$

where a and b denote projectile and ejectile nucleons, respectively; σ_a is the composite formation cross-section; $w_b(n,\varepsilon)$ is the emission rate of particle b from exciton state n with emission energy ε, $\tau(n,\Omega)$ is the mean lifetime of exciton state n at emission angle Ω. The mean life times are expressed as a Legendre polynomial series:

$$\tau(n,\Omega) = \sum_\ell \zeta_\ell(n) P_\ell(\cos\Theta); \quad \tau(n) = 4\pi\zeta_o(n) \ , \quad (2)$$

where the coefficients follow from the time-integrated master equation:

$$-\eta_\ell{}^o(n) = \mu_\ell\lambda^+(n-2)\zeta_\ell(n-2) + \mu_\ell\lambda^-(n+2)\zeta_\ell(n+2) +$$

$$-[w(n) + \lambda^+(n) + \lambda^-(n) + (1-\mu_\ell)\lambda^o(n)]\zeta_\ell(n). \quad (3)$$

In this equation λ^+, λ^- and λ^o are the intranuclear transition rates and w is the total emission rate. The eigenvalues μ_ℓ follow from the equation

$$\int d\Omega' G(\Omega,\Omega')P_\ell(\cos\Theta') = \mu_\ell P_\ell(\cos\Omega) \ , \quad (4)$$

where $G(\Omega,\Omega')$ is the scattering kernel normalized to $\mu_o = 1$. The initial condition $\eta_\ell{}^o(n)$ refers to the Legendre coefficient of the initial (t=0) occupation probability:

$$q^o(n,\Omega) = \sum_\ell \eta_\ell{}^o(n) P_\ell(\cos\Theta);$$

$$q^o(n) = 4\pi\eta_o^o(n). \quad (5)$$

For ℓ=0 Eq. (3) reduces to the angle-integrated result. In our previous work we have assumed that $G(\Omega,\Omega')$ is obtained from the assumption of free nucleon-nucleon scattering inside the nucleus:

$$G^f(\Omega,\Omega') = \frac{1}{\pi}\cos\Theta \ H(\pi/2-\Theta). \quad (6)$$

Furthermore, we have taken

$$\eta_\ell{}^o(n) = \delta_{n1}\rho_\ell\frac{(2\ell+1)}{4\pi} \quad (7)$$

with ρ_ℓ the eigenvalue following from the equation

$$\int d\Omega' R(\Omega,\Omega')P_\ell(\cos\Theta') = \rho_\ell P_\ell(\cos\Theta) \ , \quad (8)$$

where $R(\Omega,\Omega')$ is the refraction kernel normalized to ρ_o=1. In Ref. 1 the limit of an infinite refraction index has been employed. In this case it is easy to show that $R(\Omega,\Omega') = G^f(\Omega,\Omega')$ and therefore $\rho_\ell=\mu_\ell^f$.

In Refs. 1 and 2 we have studied the energy and angular distributions of neutrons scattered at 14.6 MeV incident neutron energy for a large number of target elements, using the above-mentioned theory. In this work we will extend the model towards higher incoming neutron energies, with emphasis on neutron scattering. A summary of recent model improvements is given in the following sections, together with a comparison with available experimental neutron data and systematics. Some discussion is also devoted to the processing of these data for fusion-reactor oriented applications.

State density in precompound decay

In the unified model of pre-equilibrium and equilibrium decay it is a natural requirement that the adopted state densities[3] ω_{ph}, when summed over all possible particle-hole combinations, are consistent with available experimental data. Thus, the level density

$$\rho(U,J,\pi) = \frac{2J+1}{4\sqrt{2\pi}\ \sigma^3}\exp\left[\frac{-(J+1/2)^2}{2\sigma^2}\right]\sum_{p,h}\omega_{ph}(U) \quad (8)$$

should agree with experimental information from low-lying levels and from neutron-resonance spacings. Eq. (8) contains the spin cut-off parameter σ of the compound state. When the spin and parity are explicitly considered in the exciton model, the spin distribution of each (p,h)-state needs to be accounted for, but this is irrelevant in the present model. This is not so for the (pairing) energy shifts which occur in Eq. (8). However, it is not easy to find

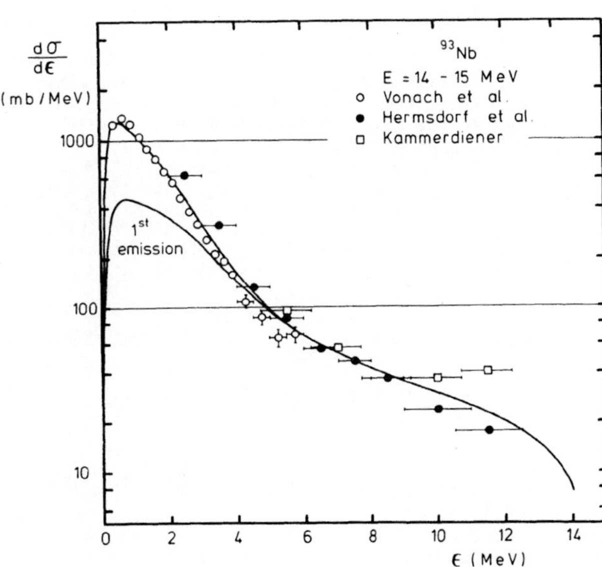

Fig. 1. Angle-integrated neutron emission spectrum for neutron-induced reactions on ^{93}Nb. The calculation has been performed with realistic level-density parameters. A fine, non-equidistant energy mesh was used to represent the curve.

K. H. Böckhoff (ed.), Nuclear Data for Science and Technology, 537–542.

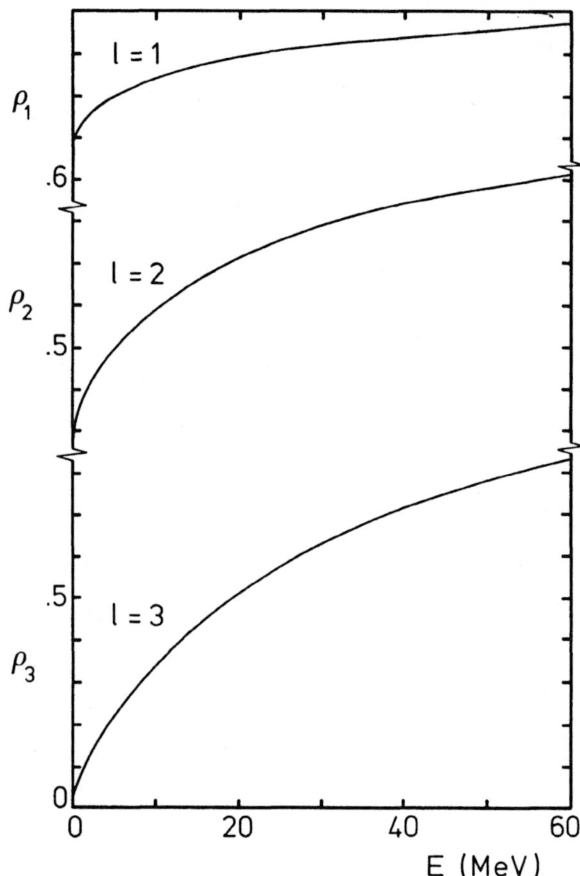

Fig. 2. Refraction coefficients ρ_ℓ as a function of incident energy for neutrons interacting with ^{93}Nb.

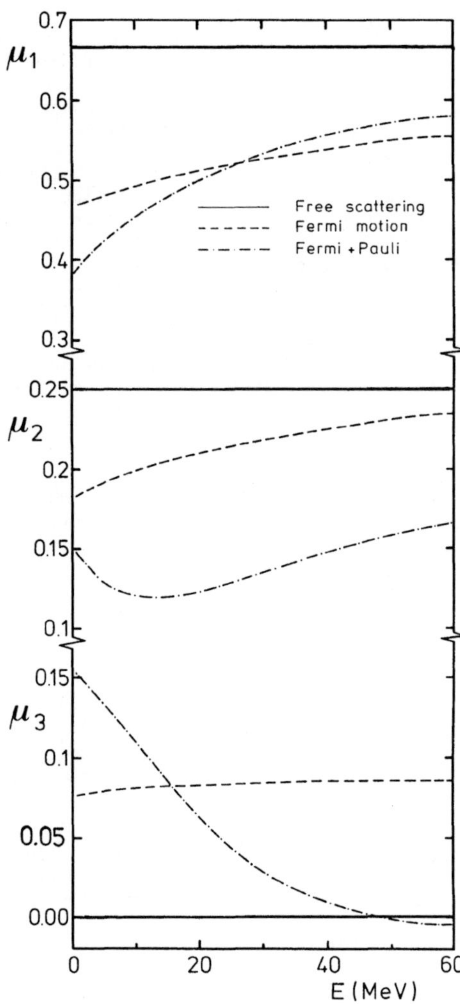

Fig. 3. Scattering coefficients μ_ℓ as a function of incident energy for nucleons in ^{93}Nb. The curves have been calculated for three different kernels, averaged over outgoing energy E'.

reliable expressions for these shifts. Moreover, the adopted formulae [3] for ω_{ph} give only rough approximations of the actual values which could be calculated from realistic models. For this reason we have decided to adopt a constant energy shift Δ, which is adjusted to fit the experimental data. In fact, there are two adjustable parameters $g = 6a/\pi^2$ and $\Delta = E-U$, like in the back-shifted Fermi-gas formula ρ_D of Dilg et al.[4]. Equating (8) to ρ_D at high energies, where the sum of $\omega_{ph}(U)$ approaches[3] $\exp(2\sqrt{aU})/(\sqrt{48}\ U)$, we may renormalize ω_{ph} with a factor

$$f(U) = \sqrt{\frac{\pi}{3}}\ \frac{U}{a^{1/4}(U+t)^{5/4}}\ , \tag{9}$$

where t is the thermodynamic temperature defined in Ref. 4.

Multiplication of Eq. (8) with f(U) has the advantage that ρf is asymptotically equal to the back-shifted Fermi-gas formula, which is frequently used in equilibrium model calculations. Since f is a simple factor, its introduction in the exciton model hardly complicates the computations. Another advantage of this renormalization is that in first instance the systematics of Dilg et al. for a and Δ could be used, although it is preferred to fit ρf to the recent data. For the calculations reported in this paper we have adopted: $g = 6.72\ MeV^{-1}$, $\Delta = -0.592$ MeV for ^{93}Nb and $g = 5.36\ MeV^{-1}$, $\Delta = -1.83$ MeV for ^{92}Nb. The negative sign of the shifts assures that the level density does not vanish at the lowest excitation energies as occurred in previous calculations[2], where the (positive) pairing-energy corrections were adopted from Ref. 5.

The above-mentioned modification was inserted into the emission rates only; the internal (λ^+) transition rates should be quite independent of the nucleus. For the same reason the value of the average transition matrix element $\langle M^2\rangle$ was assumed to be proportional[6] to g^{-3} rather than A^{-3}. This was necessary because shell effects may lead to values of g quite different from the previously assumed value $g = A/13\ MeV^{-1}$. Another modification to the emission rates was the replacement of $R_b(n)$ by $Q_b(n)$, for reasons discussed by Kalbach[7]. The finally adopted value of $\langle M^2\rangle$ which fits the 14.6 MeV neutron emission data[8-10] for ^{93}Nb is $\langle M^2\rangle = 500/(13g)^3E$. The result of this fit is shown in Fig. 1. The same parameters were used in calculations at higher incident neutron energies, resulting in excellent agreement with the recent[11] emission cross sections at 25.7 MeV, see upper part of Fig. 6b.

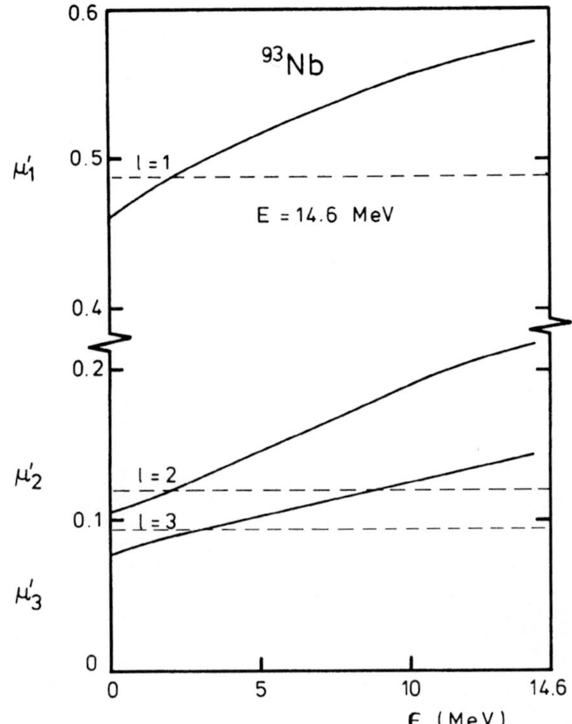

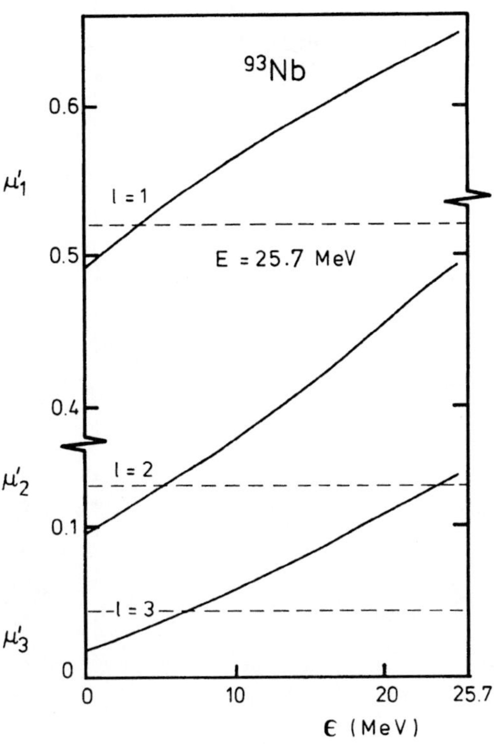

Fig. 4. Scattering coefficients $\mu'_\ell(\varepsilon)$ as a function of outgoing energy ε for two incident energies (E = 14.6 and 25.7 MeV), according to the Kikuchi-Kawai expressions, taking into account Fermi motion and Pauli principle for nucleons in ^{93}Nb. The dashed curves correspond to μ_ℓ averaged over E' (Fig. 3).

Refraction kernel

For the neutron refraction kernel we have adopted the classical expression:

$$R(\Omega,\Omega') = R(\Theta) =$$

$$\frac{n^2}{\pi}\,\frac{[(n^2+1)\cos\Theta-n(\cos^2\Theta+1)]}{(n^2-2n\cos\Theta+1)^2} * H(\Theta_n-\Theta)\ , \qquad (10)$$

where H is the Heaviside function and Θ_n is the maximum refraction angle $\Theta_n = \pi/2-\arcsin 1/n$. The refraction index n is defined as

$$n = \sqrt{\frac{E+V}{E}}\ , \qquad (11)$$

where V is the real potential, taken from Becchetti and Greenlees[12]. Eq. (10) differs from the expression given by Gadioli and Gadioli Erba[13], who have omitted the Heaviside function.

The calculated eigenvalues ρ_ℓ (Eq. 6) have been plotted as a function of energy in Fig. 2 for ℓ=1, 2 and 3. For high energies n approaches 1. Then R becomes a δ-function and ρ_ℓ approaches 1. On the other hand, maximum refraction is obtained for $n\to\infty$ leading to R equal to υ^f in the case of free nucleon-nucleon scattering or: $\rho_\ell = \mu_\ell^f$.

It has been shown in Ref. 1 that the solution of the linear equation (3) for ζ_ℓ leads to a proportionality with η_ℓ^o and thus with ρ_ℓ (see Eq. 7). This means that the refraction of the incoming beam leads to a reduction factor $\rho_\ell(E)$. Recently, it has been shown by Akkermans[14] that Eq. (10) can also be used to describe the refraction of the outgoing beam, replacing E by ε in Eq. (11). This leads to the conclusion that $\zeta_\ell(n)$ should be multiplied with $\rho_\ell(\varepsilon)$ to include the effect of refraction of the outgoing beam. Hence, in the calculations (see Eq. 2) $\zeta_\ell(n)$ should be replaced by

$$\zeta'_\ell(n,\varepsilon) = \zeta_\ell(n)\rho_\ell(\varepsilon). \qquad (12)$$

This coefficient is proportional to $\rho_\ell(E)\rho_\ell(\varepsilon)$, indicating that there is a strong dependence on incoming energy at the highest emission energy $\varepsilon\to E$ (cf. Fig. 5a).

Scattering kernel

The assumption of free scattering[15] is not very realistic, since we are dealing with collisions inside nuclear matter at relatively low energies. Referring to Sun Ziyang et al.[16] the Kikuchi-Kawai expressions[17] are used, as follows:

$$G(\Omega,\Omega') = \int \frac{d^2\sigma^{KK}}{dE'd\Omega''}\,dE' / \iint \frac{d^2\sigma^{KK}}{dE'd\Omega''}\,dE'd\Omega''\ , \qquad (13)$$

where $\Omega'' = \Omega-\Omega'$. The integration over outgoing energy is performed to obtain an average angular distribution, since in the exciton model one does not follow the individual particles and their corresponding energies, but only the total number of excited particles. Eq. (13) takes into account the Fermi motion of the target nucleons. It is required that the momenta of the projectile and target nucleons are above and below the Fermi momentum, respectively.

Since in the initial stage Δn=+2 transitions are favoured as a result of the Pauli principle we add the further restriction that the momenta of the

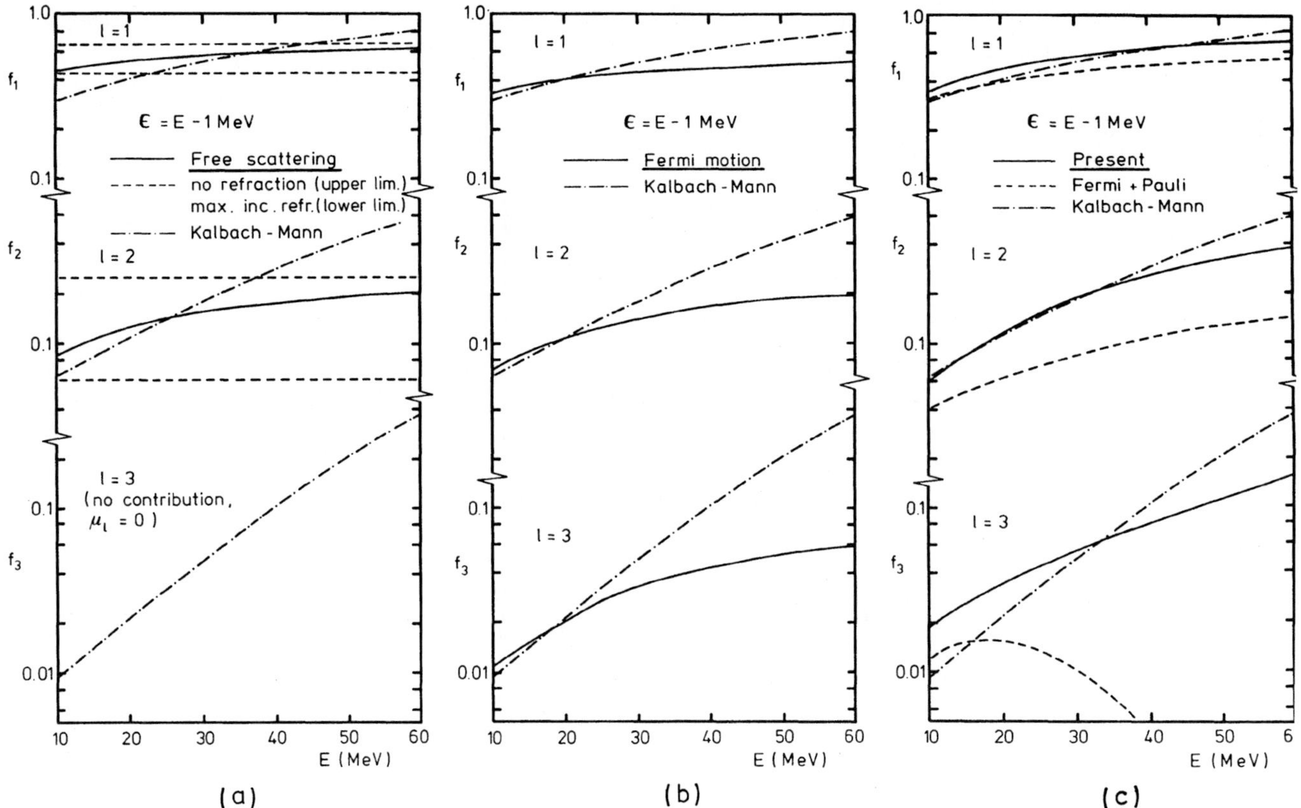

Fig. 5. Comparison of reduced Legendre coefficients f_ℓ at the highest emission energies ($\varepsilon \to E$) with the systematics of Kalbach and Mann for various scattering kernels. All curves, except the dashed lines in Fig. 5a, represent results with refraction of incoming and outgoing beam. The calculations have been performed for inelastic neutron scattering on ^{93}Nb at $\varepsilon = E - 1$ MeV.

(a) Free scattering : the dashed curves correspond with no refraction (upper limits) or maximum refraction of incoming beam only (lower limits), see previous work (Ref. 1).

(b) Fermi-motion : averaged Kikuchi-Kawai kernel without taking into account the Pauli principle.

(c) Fermi+Pauli : Kikuchi-Kawai kernel, taking into account the Pauli principle; the dashed curves correspond with the averaged kernel; the full curves correspond with the present "correlated emission model".

two scattered particles should be above the Fermi momentum. In Fig. 3 the Legendre coefficients μ_ℓ for these kernels are intercompared. We have performed these calculations with a two-dimensional integration routine as described in Ref. 18.

We note that μ_ℓ significantly depends upon the incoming particle energy. This is easily accounted for at the first collision (n_o=3). At the following collisions the average energy of the "fast particle" E_f is gradually reduced. This could approximately be described by introducing n-dependent eigenvalues $\mu_\ell(n)$, assuming $E_f(n) = E_F+(E+B)/p$ (E_F = Fermi energy; B = binding energy). One could also approximately account for differences between Δn=2,0 or -2 processes, by distinguishing between $\mu_\ell^+(n)$, $\mu_\ell^o(n)$, $\mu_\ell^-(n)$. However, from numerical calculations of angular distributions we found that the effects of these refinements are quite small and could be neglected[18].

The leading term in the angular distribution, certainly at high emission energies, comes from the initial emission. Inspection leads to the conclusion that the neglect of the energy-angle correlation by averaging over the outgoing energies in Eq. (13) is certainly not justified in the case of first emission. Therefore, we propose to use the average kernel only for the description of the relaxation of the system towards isotropy, and to adopt the energy-angular distribution of Kikuchi and Kawai for the initial emission from n_o=3. This is approximately performed by identifying E' with ε and by the replacement

$$\zeta''_\ell(n_o,\varepsilon) = \zeta'_\ell(n_o,\varepsilon)\,\mu'(\varepsilon)/\mu_\ell \;, \qquad (14)$$

where $\mu'_\ell(\varepsilon)$ and μ_ℓ are the reduced Legendre coefficients of

$\frac{d^2\sigma^{KK}}{d\varepsilon d\Omega''}$ and $\int \frac{d^2\sigma^{KK}}{dE' d\Omega''} dE'$, respectively. These coefficients have been plotted in Fig. 4 as a function of ε for nucleon scattering in ^{93}Nb at two incident energies (E = 14.6 and 25.7 MeV). From Fig. 4 it follows that the replacement of Eq. (14) significantly increases the angular distribution coefficients at high energies.
This "correlated emission model", combined with refraction effects, gives quite satisfactory results[18],

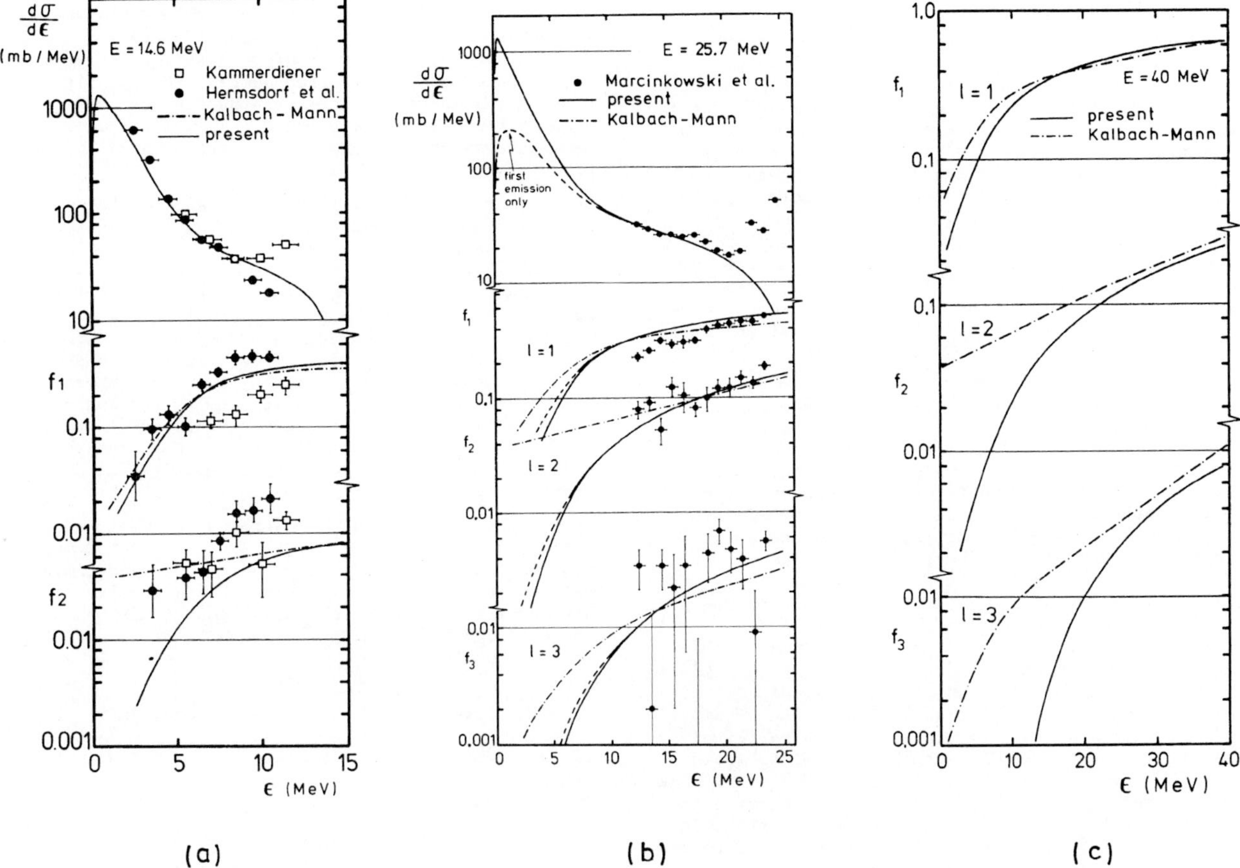

Fig. 6. Comparison of calculated Legendre coefficients with those from experiments and systematics, as a function of outgoing energy ε for neutron emission from ^{93}Nb at E = 14.6, 25.7 and 40 MeV. The full lines result from calculations with the present "correlated emission model", taking into account multiple scattering effects; the dashed-dotted curves correspond to the systematics. At low values of ε the systematics contains an additional symmetric term, leading to increased values of f_2.

(a) E = 14.6 MeV : there is a discrepancy between the two experimental data sets; note that f_2 is underpredicted by theory and systematics.

(b) E = 25.7 MeV : there is good agreement between theory and experimental data for f_1, f_2 and f_3.

(c) E = 40 MeV : there is reasonably good agreement between theory and systematics at ε > 20 MeV; at lower energies we note that the uncertainties in the systematics could be quite large.

as is illustrated in the next section.

We note that similar replacements could be made for emission from higher n-values, e.g. by assuming $E_f = E_F + B + E$ at all values of n. However, from calculations it was found [18] that the Legendre coefficients are not sensitive to the introduction of ε-dependent eigenvalues at values of $n > n_o$.

Comparison with systematics and experimental data

Calculations have been performed for total neutron emission cross sections of neutron-induced reactions in ^{93}Nb. Comparisons have been made with the systematics of Kalbach and Mann[19] and with experimental data[8,9,11] at 14.6 and 25.7 MeV.

We first show the comparison with the systematics at outgoing energies close to the incoming energy (ε→E). At these energies the n=1 (refraction) and n_o=3 components contribute most and the reduced angular distribution coefficients f_ℓ are approximately proportional to $\rho_\ell^2(E)\ \mu_\ell'(E)$. Fig. 5a shows the large improvement which is obtained when refraction is added to (free) scattering). A further improvement is obtained when the average Kikuchi-Kawai expressions are used without or with Pauli principle (full curves in Fig. 5b and dashed curves in Fig. 5c, respectively). Still, the energy dependence is too weak. The results are significantly improved by introducing $\mu_\ell'(\varepsilon)$ rather than μ_ℓ at n_o=3 (full lines in Fig. 5c). In view of the uncertainties of the Kalbach-Mann systematics (particularly at E > 40 MeV) we believe that our present model and the systematics are equally good.

The spectral dependence of the Legendre coefficients at E = 14.6 MeV, 25.7 MeV and 40 MeV is shown in Fig. 6. We see that the present theory gives about the same results as the systematics at ε > 0.5 E. At lower outgoing energies f_2 is underpredicted, possibly because our theory does not account for quantum-mechanical angular-momentum effects[20,21]. In the systematics an additional term for (multi-step) compound reactions was added[19]. Also for f_3 the systematics gives higher values than the theory at low emission energies. However, here the underlying experimental data are quite uncertain.

We note that there is excellent agreement with the measured data[11] at 25.7 MeV (f_1 is perhaps somewhat too large). At 14.6 MeV there is a serious discrepancy between the two sets of experimental data[8,9]. Still, it is clear that all theories as well as the systematics give systematically too low values for f_2 at E = 14.6 MeV. We think that this is at least partly due to the neglect of angular-momentum effects which would enhance the symmetric components. The relative enhancement is expected to be more important at low energies than at high energies. Also, these effects may be more important for neutrons than for protons, of which the maximum values of the orbital angular momenta are limited due to Coulomb effects. For practical calculations we suggest to add a corrective constant term to μ_2', fitted to 14.6 MeV data.

Multi-particle emission and processing

Multi-particle emission has been introduced in the PRANG programme library in a quite natural way, as already described in Ref. 22. Meanwhile, we have inserted a variable, non-equidistant energy mesh (triangle function representation[23]) and have improved the integration and sorting of the cross-section components. These modifications allow us to calculate emission spectra at the lowest outgoing energies, where most of the multi-particle emission occurs. As an example we refer again to Fig. 1. About 45 energy points from ε = 1 eV to 15 MeV were used (constant-lethargy intervals up to 100 eV) with a fine mesh near the evaporation peak and the (n,2n)-threshold (ε = 6.17 MeV). We note that for the calculation of multi-particle emission this is a large number of points, which can only be processed with relatively fast codes, such as PRANG. Using less points would lead to a poor representation of the evaporation peak and the multi-group transfer matrices. The effect of multi-particle emission on the angular distributions is shown in Ref. 22 and Fig. 6b (dashed curves correspond with first emission only).

Following the recommendations of the "Working Group on Neutron Transport and Gamma Ray Production" at the IAEA Advisory Group Meeting for Fusion Reactors Technology (Vienna 1978) we calculate the Legendre expansion coefficients of the double differential neutron-emission cross section $\sigma_{n,em}(E,\varepsilon,\Theta)$, rather than processing its various components $\sigma_{nn'}$, σ_{n2n}, σ_{npn}, etc. (the elastic scattering component still has to be added). The Legendre coefficients of $\sigma_{n,em}$ are obtained with the sorting module MCFRLN and are stored on a file in ENDF/B format. For the energy-angle distributions the new MF6-format, proposed by Macfarlane[24] is used. An option to transform the data to laboratory coordinates is being prepared. Since there are as yet no multi-group constant processing codes available for the calculation of transfer matrices stored in the MF6-format, we have developed a new module called GROUPXS. This code calculates multi-group constants and Legendre coefficients of total scattering and transfer matrices of continuum reactions only. It will be used to supplement multi-group constants obtained with a conventional processing code. The results are to be used to study the relevance of correlated angle-energy distributions and pre-equilibrium effects in calculations for fusion-reactor applications.

Conclusions

A step towards further unification of exciton and equilibrium models is to require that the adopted level densities agree with available experimental data. Our simple renormalization approach seems to be useful in this respect.

Two important improvements with regard to the description of angular distributions are the introduction of incoming and outgoing refraction and the use of the Kikuchi-Kawai expressions[17] for intranuclear scattering. We have shown that averaging over the emission energies[16] is not justified at the initial emission, where the energy-angle correlation of the Kikuchi-Kawai expressions should be used. Our "correlated emission model", combined with refraction effects, gives good agreement with experimental data at 14.6 and 25.7 MeV and with the systematics of Kalbach and Mann[19] (up to at least 40 MeV). The incident-energy dependence of results for $\varepsilon \rightarrow E$ of our model and of the systematics seems to be equally good. A remaining problem concerns the neglect of angular momentum effects which are expected to enhance the symmetric components, particularly at low energies.

The results of the model are to be applied in fusion-reactor studies. The recently introduced description of multi-particle emission in the framework of the exciton model[22] allows for a consistent calculation of cross sections at the lowest emission energies, provided that a fine, non-equidistant energy mesh is introduced. For the calculation of Legendre coefficients of multi-group transfer matrices a new processing module has been written, reading the data in the new MF6 format[24] of ENDF/B. The processing of total neutron emission cross sections, rather than their various components, leads to a fast and reliable method for the production of multi-group constants of continuum cross sections.

We are grateful to Drs. Zhuo Yizhong and Wang Shunuan (Beijing), C.Y. Fu (Oak Ridge) and R.E. Macfarlane (Los Alamos) for valuable communications and conversations and to Drs. R.W. Finlay (Ohio) and A. Marcinkowski (Warsaw) for their kind permission to use the 25.7 MeV data, prior to publication.

References

1. J.M. Akkermans, H. Gruppelaar and G. Reffo, Phys. Rev. C22, 73 (1980)
2. H. Gruppelaar and J.M. Akkermans, ECN-84, Netherlands Energy Research Foundation (1980).
3. F.C. Williams, Nucl. Phys. A166, 231 (1971).
4. W. Dilg, W. Schantl, H. Vonach and M. Uhl, Nucl. Phys. A217, 269 (1973).
5. A. Gilbert and A.G.W. Cameron, Can. J. Phys. 43, 1446 (1965).
6. C.Y. Fu, ORNL/TM-7042, Oak Ridge National Laboratory (1980), to be published.
7. C. Kalbach, Z. Phys. A283, 401 (1977).
8. D. Hermsdorf et al., ZfK-277, Zentralinstitut für Kernforschung, Rossendorf bei Dresden (1974).
9. J.L. Kammerdiener, UCRL-51232, Lawrence Livermore Laboratory (1972).
10. H. Vonach, A. Chalupka, F. Wenninger and G. Staffel, Proc. Symp. Neutron cross section from 10 to 50 MeV, BNL-NCS-51245, vol. 1, 343, Brookhaven National Laboratory (1980).
11. A. Marcinkowski et al., to be published in Nucl. Sci. Eng.
12. F.D. Becchetti and G.W. Greenlees, Phys. Rev. 182, 1190 (1969).
13. E. Gadioli and E. Gadioli Erba, Proc. Course on Nuclear theory and applications-1980, Trieste, IAEA-SMR-68/1, 3, International Atomic Energy Agency, Vienna (1981).
14. J.M. Akkermans, thesis, State University of Groningen (1982).
15. G. Mantzouranis, H.A. Weidenmüller and D. Agassi, Z. Phys. A276, 145 (1976).
16. Sun Ziyang, Wang Shunuan, Zhang Jingshang and Zhuo Yizhong, Z. Phys. A305, 61 (1982); see also W. Shunuan et al., this conference.
17. K. Kikuchi and M. Kawai, Nuclear matter and nuclear reactions, p. 41, North-Holland (1968).
18. C. Costa, H. Gruppelaar and J.M. Akkermans, to be published.
19. C. Kalbach and F. Mann, Phys. Rev. C23, 112 (1981).
20. A.V. Ignatyuk, V.P. Lunev and G. Pronyaev, Izv. Akad. Nauk. SSSR, Ser. Fiz. 39, 2144 (1975).
21. C.Y. Fu, Proc. Symp. Neutron cross sections from 10 to 50 MeV, BNL-NCS-51245, vol. 2, 675 Brookhaven National Laboratory (1980).
22. J.M. Akkermans and H. Gruppelaar, Z. Phys. A300, 345 (1981).
23. F. Schmittroth and R.E. Schenter, Nucl. Sci. Eng. 74, 168 (1980).
24. R.E. Macfarlane, private communication (1980).

A MODEL FOR ANGULAR DISTRIBUTIONS IN PREEQUILIBRIUM REACTIONS

S.K. Gupta and A. Chatterjee

Nuclear Physics Division
Bhabha Atomic Research Centre
Bombay 400 085

A new model is proposed for calculating angular distributions in preequilibrium reactions. In this model, as in the model of Feshbach et al the system consisting of target plus projectile initially branches into two sets of states with either no particle in the continuum (multistep compound states) or with at least one particle in the continuum (multistep direct states). In the present model, these two chains of states are treated independently by solving two sets of master equations. The multistep compound emission is assumed to be isotropic while the angular distribution of the multistep direct emission is described using the fast particle model of Mantzouranis et al. The angular distributions for 14.6 MeV neutrons calculated using this model are found to be in better agreement with the data than the fast particle model.

$^{31}P(n,n')$, $^{93}Nb(n,n')$, $^{197}Au(n,n')$, angle dependent spectra, new model, fast particle model

Introduction

Preequilibrium models[1] have been used with a good degree of success in the description of angle-integrated spectra of particles emitted from nuclear reactions initiated by particles in the energy range of several tens of MeV.

Mantzouranis et al[2] proposed a generalization of the usual exciton model master equation in order to describe angular distributions. In this model the exciton states are characterized by a direction Ω in addition to the exciton number n. The direction Ω corresponds to the direction of the projectile for the initial exciton state and it is assumed that successive binary collisions with the target nucleons lead to a gradual loss of correlation with the initial direction. However, a crucial assumption is that at each stage there is an identifiable leading particle whose direction Ω determines the direction of any emitted particle. This assumption breaks down for larger exciton numbers, but the forward-peaked angular distribution is determined largely by the first few collisions.

The mathematical formulation of Akkermans[3] reducing the integro-differential equation of Mantzouranis et al[2] to a system of linear differential equations enabled an extensive application[4] to the calculation of angular distributions in the (n,n') reaction at 14.6 MeV. These calculations show a systematic underestimation of the emission at backward angles.

In view of the backward angle discrepancy several calculations of preequilibrium angular distributions have been made using different approaches[5]. In the present work, the fast particle model[2] is modified. The exciton states are further classified as multistep direct (MSD) and multistep compound (MSC) in the spirit of the model of Feshbach, Kerman and Koonin[6]. The MSC states include only bound excited states while the MSD states have at least one unbound particle. The forward peaking of the angular distribution is determined by the MSD emission while the MSC emission is assumed isotropic. The initial target-projectile interaction is assumed to cause a branching into an MSC or an MSD chain of states with a branching ratio related to the relative density of states. The successive interactions are described using an exact solution[7] of the exciton model master equations including its generalization for angular distributions[2] for the MSD branch and neglecting transitions between the MSD and MSC states. Since the master equations are solved exactly, the preequilibrium and equilibrium components of the reaction are included in an unified manner without any arbitrary separation of the two phases. The present model leads to a better agreement of the calculated emission at backward angles with the experiment.

Generalised Master Equations

The generalised master equations[2] are given by,

$$\frac{d}{dt} q(n,\Omega,t) = \sum_m \int d\Omega' \, q(m,\Omega',t) W_{mn}(\Omega',\Omega) - q(n,\Omega,t) \sum_m \int d\Omega' W_{nm}(\Omega,\Omega'), \qquad (1)$$

where $q(n,\Omega,t)$ is the probability of finding the system in a class characterized by (n,Ω) at time t and $W_{nm}(\Omega,\Omega')$ is the transition rate from class (n,Ω) to (m,Ω') The latter is a product of the usual transition rates λ^+ , λ^- or λ^0 and an angle dependent factor,

$$G(\Omega,\Omega') = G(\Omega',\Omega) = \left[\int d\Omega \frac{d\sigma}{d\Omega}\right]^{-1} \frac{d\sigma}{d\Omega}(\Omega,\Omega') \qquad (2)$$

K. H. Böckhoff (ed.), Nuclear Data for Science and Technology, 543–546.

where $d\sigma/d\Omega$ is the differential free nucleon-nucleon scattering cross section.

The integro-differential equation (1) was reduced to the following set of equations[3],

$$\frac{d}{dt}\eta_\ell(n,t) = \mu_\ell \lambda^+_{n-2}\eta_\ell(n-2,t) + \mu_\ell\lambda^-_{n+2}\eta_\ell(n+2,t) - \eta_\ell(n,t)\left[W_n + \lambda^+_n + (1-\mu_\ell)\lambda^0_n + \lambda^-_n\right], \quad (3)$$

where $\eta_\ell(n,t)$ are the expansion coefficients of the occupation probabilities $q(n,\Omega,t)$ in a Legendre Polynomial series,

$$q(n,\Omega,t) = \sum_\ell \eta_\ell(n,t)\, P_\ell(\cos\theta), \quad (4)$$

and W_n are the total emission rates. The μ_ℓ are eigenvalues of the differential operator occuring in (1) and are determined in a closed form[4] from the expression,

$$\mu_\ell = 2\int_0^1 x\, P_\ell(x)\, dx \quad (5)$$

The initial condition is taken as[2],

$$q(n,\Omega,0) = \Theta(\pi/2-\theta)\cdot\frac{\cos\theta}{\pi}\cdot\delta_{nn_0} \quad (6)$$

and the Legendre series coefficients of the initial condition are $\eta_\ell\,\delta_{nn_0}$ where,

$$\eta_\ell = (2\ell+1)\,\mu_\ell/4\pi \quad (7)$$

The system of equations (3) can be expressed in the time integrated form,

$$-\eta_\ell\delta_{nn_0} = \Lambda^+_\ell(n-2)\zeta_\ell(n-2) - \zeta_\ell(n)/\tau_n + \Lambda^-_\ell(n+2)\zeta_\ell(n+2), \quad (8)$$

where $\Lambda^+_\ell(n) = \mu_\ell\lambda^+_n$, $\Lambda^-_\ell(n) = \mu_\ell\lambda^-_n$

$$\tau_\ell(n) = \left[W_n + \lambda^+_n + (1-\mu_\ell)\lambda^0_n + \lambda^-_n\right]$$

and $$\zeta_\ell(n) = \int_0^\infty \eta_\ell(n,t)\,dt \quad (9)$$

For calculating particle spectra it is sufficient to solve the system of algebraic equations (8). The double differential cross section for a reaction (α,β) is,

$$\frac{d^2\sigma}{d\epsilon\, d\Omega} = \sigma_\alpha \sum_{n\ (\Delta n=2)} W_\beta(n,\epsilon)\, t_n(\theta), \quad (10)$$

where the partial lifetimes of the exciton states for particle emission at an angle θ are given by,

$$t_n(\theta) = \sum_\ell \zeta_\ell(n)\, P_\ell(\cos\theta). \quad (11)$$

An exact solution[7,8] of the tridiagonal system of equations (8) has been given which enables the determination of the Legendre coefficients of the lifetimes as,

$$\zeta_\ell(n) = \mathcal{J}_{n_0}\tau_\ell(n)\prod_{\substack{k=n_0\\(\Delta k=2)}}^{n-2}\Lambda^+_\ell(k)\tau_\ell(k)\mathcal{J}_{k+2}, \quad (12)$$

where the product is unity for $n = n_0$ and the $\mathcal{J}_n$ are continued fractions,

$$\mathcal{J}_n = \frac{1}{1-}\,\frac{F_n}{1-}\,\frac{F_{n+2}}{1-}\cdots\frac{F_N}{1} \quad (13)$$

Here N is the maximum exciton number taken as $2\sqrt{g_0 E}$ where g_0 is the single particle level density, E is the composite system excitation energy and,

$$F_n = \Lambda^+_\ell(n)\,\tau_\ell(n)\,\Lambda^-_\ell(n+2)\,\tau_\ell(n+2) \quad (14)$$

These continued fractions can be generated recursively from the identity,

$$\mathcal{J}_n = 1/(1-F_n\mathcal{J}_{n+2}) \quad (15)$$

starting with $\mathcal{J}_N = 1$.

The angle dependent spectra for the (n,n') reaction at 14.6 MeV for several targets were calculated[4] using the approach described above. In comparison to the measured spectra[9] the calculated emission was consistently underpredicted at backward angles and high emission energies. Also in ref.4, the usual value $n_0 = 3$ for neutron induced reactions was replaced by $n_0 = 1$ with no emission from the first exciton state. Although this procedure led to some improvement, the spectra were still underestimated at backward angles and high emission energies. Moreover there does not appear to be any justification for taking $n_0 = 1$.

New Model

A quantum mechanical treatment of statistical multistep direct and statistical multistep compound processes has been given by Feshbach, Kerman and Koonin[6]. The definition of MSD and MSC states used in the present model is that given by Kalbach[10]. The MSC states are taken to mean both bound and unbound states, but for which the system has previously passed through a bound state. Once the system encounters a bound state all correlation effects are assumed to be lost and the process is MSC. For MSC the leading particle approximation 2 does not hold. The MSD states are unbound at every stage and here the leading particle model is assumed to be valid.

In order to keep the model simple, transitions between the two chains of states is neglected. This makes the two MSC and MSD branches independent of each other so that the usual master equation can be used separately for each branch. The generalised master equation is used for MSD and the angle independent master equation is used for MSC. This approximation is justified from the transition rate calculations in ref 10 where it is seen that the forward transition rate in the MSD branch, λ^+_{uu} is much larger than the forward transition rate from MSD to MSC states, λ^+_{ub} and similarly $\lambda^+_{bb} \gg \lambda^+_{bu}$ for the lower exciton states. Since the forward peaking of the angular distribution is determined by the MSD emission in the initial stages and the subsequent emission is increasingly isotropic the effect of interbranch transitions at the later stages on the angular distribution is negligible. We also assume that the transition and particle emission rates for the two identical and given by the usual expressions[11] for the exciton model. For the MSC this is justified because both bound and unbound states are included with the only requirement that the system should have passed through a bound state. For the initial MSD states this is justified if the density of unbound states is much larger than that of bound states. This is true at sufficiently high excitation energies.

In order to specify the model completely the only other quantities required are the probabilities R_u and R_b of the initial branching into MSD and MSC ($R_u+R_b=1$). The initial state with 3 excitons for a neutron induced reaction can have either $\pi=1$, $\nu=1$, $\bar{\pi}=1$, $\bar{\nu}=0$ or $\pi=0$, $\nu=2$, $\bar{\pi}=0$, $\bar{\nu}=1$ where π and ν are the number of excited protons and neutrons respectively and $\bar{\pi}$ and $\bar{\nu}$ are the corresponding number of holes. The former case arises as a result of an n-p interaction while the latter arises as a result of an n-n interaction. The probability for the MSD branch can be expressed as,

$$R_u = \frac{Z\sigma_{np}}{Z\sigma_{np}+N\sigma_{nn}} \frac{\omega^u(1,1,1,0,E)}{\omega(1,1,1,0,E)} + \frac{N\sigma_{nn}}{Z\sigma_{np}+N\sigma_{nn}} \frac{\omega^u(0,2,0,1,E)}{\omega(0,2,0,1,E)} \quad (16)$$

where Z and N are the proton and neutron numbers of the target, σ_{nn} and σ_{np} are nucleon-nucleon cross-sections and $\omega^u(\pi,\nu,\bar{\pi},\bar{\nu},E)$ and $\omega(\pi,\nu,\bar{\pi},\bar{\nu},E)$ are the unbound and total state densities. The state densities taking account of proton-neutron distinction are given by,

$$\omega(\pi,\nu,\bar{\pi},\bar{\nu},E) = g^n E^{n-1}\left[\pi!\,\nu!\,\bar{\pi}!\,\bar{\nu}!\,(n-1)!\right]^{-1}, \quad (17)$$

where g is the single particle level density for neutrons or protons ($g \approx g_o/2$). The unbound state density has been given by Kalbach[12]. Neglecting Pauli correction and including proton-neutron distinction, the neutron unbound state density is,

$$\omega^u(\pi,\nu,\bar{\pi},\bar{\nu},E) = g^n(E-S)^{n-1}\left[\pi!\,\bar{\pi}!\,\bar{\nu}!\,(\nu-1)!\,(n-1)!\right]^{-1}F(\nu), \quad (18)$$

where S is the neutron binding energy. The factor $F(\nu)$ corrects for multiply unbound states,

$$F(\nu) = \frac{1}{\nu}\sum_{i=1}^{\nu}(-1)^{i+1}\binom{\nu}{i}\Theta(E-iS)\left(\frac{E-iS}{E-S}\right)^{n-1}. \quad (19)$$

Assuming that $E > 2S$,

$$R_u = \left(1-\frac{S}{E}\right)^2\left[\frac{Z\sigma_{np}}{Z\sigma_{np}+N\sigma_{nn}} + \frac{2N\sigma_{nn}}{Z\sigma_{np}+N\sigma_{nn}}\left\{1-\frac{1}{2}\left(\frac{E-2S}{E-S}\right)^2\right\}\right] \quad (20)$$

Taking $\sigma_{nn}/\sigma_{np} \approx 0.3$, the value of R_u can be calculated from (20).

Since the MSD and MSC chains are assumed to have identical transition rates the calculated angle integrated spectra are identical for the two chains. The angle dependent spectra in the new model is given by,

$$d\sigma/(d\epsilon\, d\Omega) = (1-R_u)\sum_n W_\beta(n,\epsilon)\,\mathcal{T}_o(n) + R_u\sum_\ell P_\ell(\cos\theta)\sum_n W_\beta(n,\epsilon)\,\mathcal{T}_\ell(n), \quad (21)$$

where the Legendre coefficients of the lifetimes, $\mathcal{T}_\ell(n)$ are obtained from the solution of the generalised master equation in the same manner as described in the earlier section. Equation (21) predicts identical angle integrated spectra as eq(10) but an angular distribution which is less forward peaked.

Calculations

The angle dependent neutron spectra for neutron induced reactions at 14.6 MeV have been calculated using the model described above for ^{31}P, ^{93}Nb, and ^{197}Au and compared with the calculation using the usual fast particle model and the experimental data of Hermsdorf et al[9]. In both the calculations the initial exciton state is taken to be $n_o=3$, the single particle level densities are taken as A/13 MeV^{-1}, pairing energies are taken from Gilbert and Cameron[13], optical reaction cross-sections are calculated from empirical expressions[14] the matrix element squared is taken from Kalbach[15] and (n,2n) contributions have been calculated using a constant temperature statistical model. The results of the calculations are shown in Figs 1-3.

Discussion

The calculated emission using the new model is generally less than that for the fast particle model at the forward angles and larger at the backward angles especially at higher emission energies. Thus there is

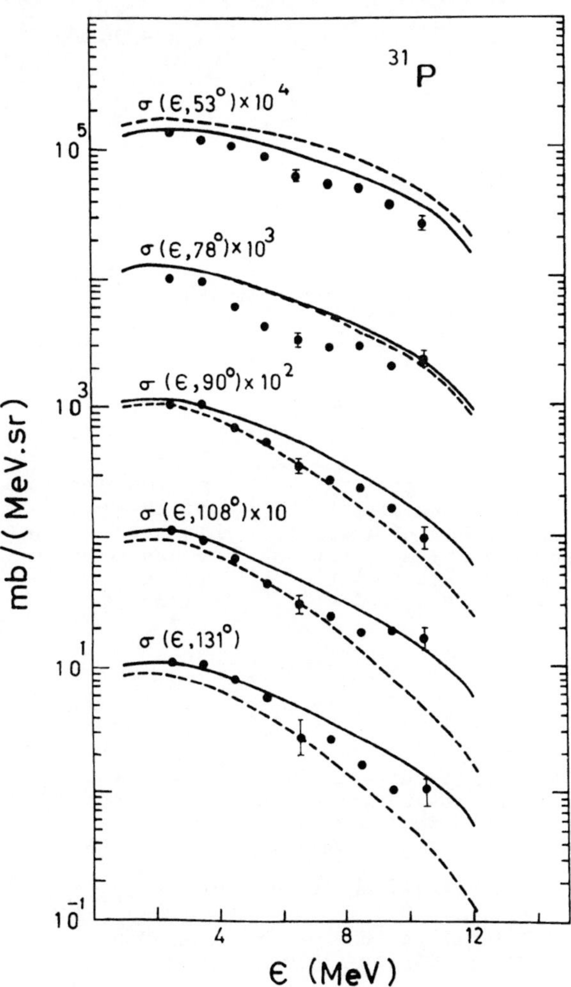

Fig. 1

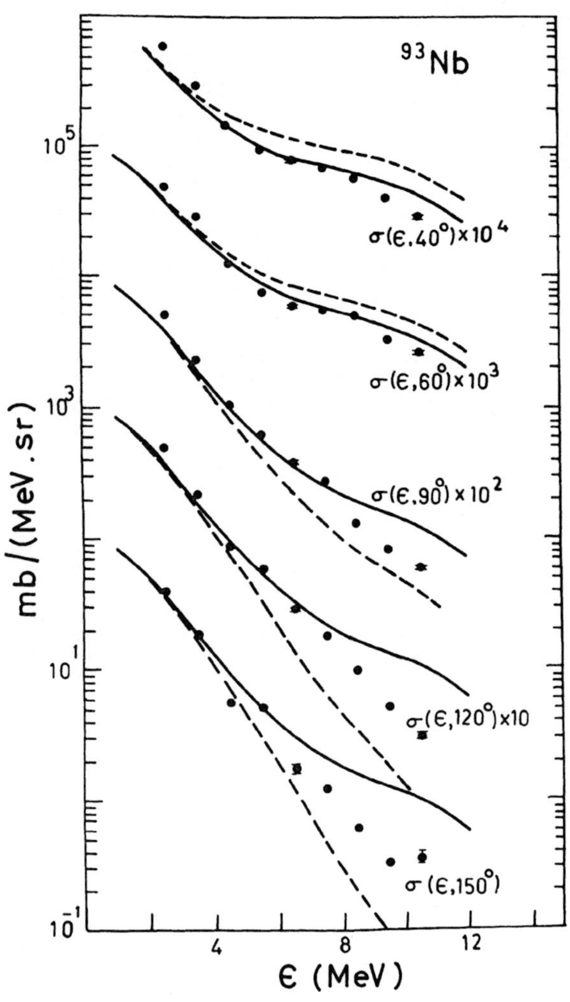

Fig. 2

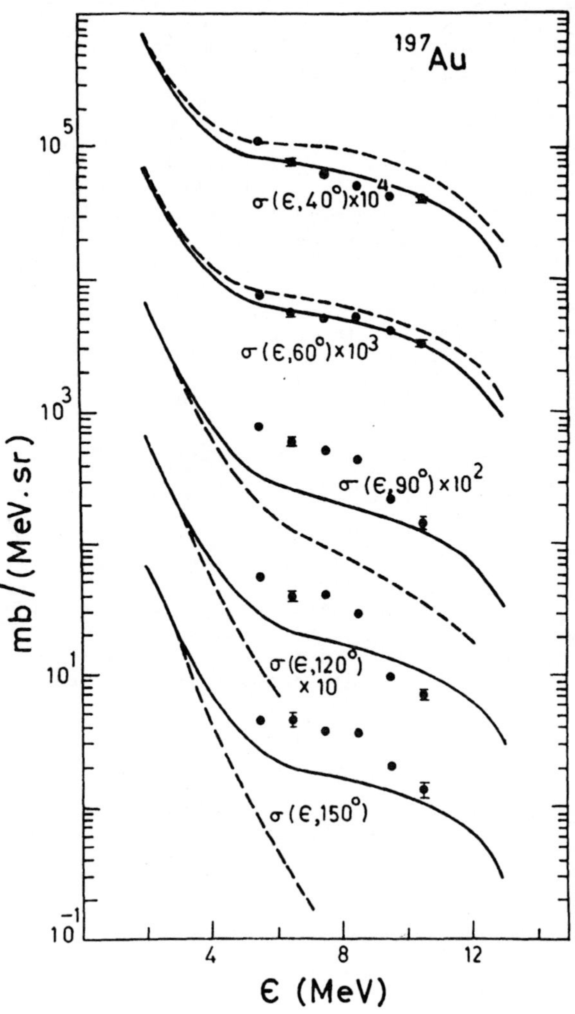

Fig. 3

an improved agreement with the experiment in all the cases and the best case is that of ^{197}Au. Some of the differences between the theory and experiment could be due to structure effects in the level densities which have not been included in the present calculations.

References

1. M. Blann, Annu. Rev. Nucl. Sci. 25, 123 (1975)

2. G. Mantzouranis, D. Agassi and H.A. Weidenmuller, Phys. Lett. 57B, 220 (1975); G. Mantzouranis, H.A. Weidenmuller and D. Agassi, Z. Physik A276, 145 (1976)

3. J.M. Akkermans, Phys. Lett. 82B, 20(1979); J.M. Akkermans and H. Gruppelaar, ECN-60, Netherlands Energy Research Foundation Report Series (1979)

4. J.M. Akkermans, H. Gruppelaar and G. Reffo, Phys. Rev. C22, 73 (1980) ; H. Gruppelaar and J.M. Akkermans, ECN-84, Netherlands Energy Research Foundation Report Series (1980).

5. P. Mädler and R. Reif, Nucl. Phys. A273, 27 (1982), and A337, 445 (1980); R. Bonetti, M. Camnasio, L. Colli Milazzo and P.E. Hodgson, Phys. Rev. C24, 71 (1981)

6. H. Feshbach, A. Kerman and S. Koonin, Ann. Phys.(N.Y.) 125, 429 (1979)

7. A. Chatterjee and S.K. Gupta, Z. Phys. A301, 275 (1981); S.K. Gupta and A. Chatterjee, Int. Conf. on Nuclear Cross Sections for Technology, Knoxville (October 1979), Contributed Paper.

8. J.M. Akkermans, Z.Phys.A 292, 57 (1979)

9. D. Hermsdorf et al, rep.ZfK-277 (1974)

10. C. Kalbach, Phys.Rev. C23, 124 (1981)

11. C. Kalbach, Z.Phys. A 283, 401 (1977)

12. C. Kalbach, Phys.Rev. C24, 819 (1981)

13. A. Gilbert and A.G.W. Cameron, Can. J. Phys. 43, 1446 (1965)

14. A. Chatterjee, K.H.N. Murthy and S.K. Gupta, Pramana 16, 391 (1981)

15. C. Kalbach, Z. Phys. A287, 319 (1978).

ROLE OF PREEQUILIBRIUM EMISSION ON (n, xn) CROSS SECTIONS

M.L. Jhingan

Tata Institute of Fundamental Research
Colaba, Bombay 400 005

and

R.P. Anand, S.K. Gupta, and M.K. Mehta

Nuclear Physics Division
Bhabha Atomic Research Centre
Bombay 400 085

Our previous method for calculating (n,2n) and (n,3n) cross sections based on the statistical equilibrium model, has been improved further by incorporating preequilibrium emission at the first stage of the calculation. The effect of gamma emission is taken into account by using the apparent level density of Pearlstein. The matrix element of the preequilibrium component is the one recently given by Kalbach. Calculated cross sections for (n,2n) and (n,3n) processes agree with the measured ones within 10% over the mass range 89 to 238 without using any new parameter. Cross sections for (n,4n) reaction agree within a factor of two with the data wherever available.

(n,2n), (n,3,), (n,4n) cross section calculated threshold to 28 MeV. For ^{89}Y, ^{169}Tm, ^{197}Au, ^{248}U.

Introduction

The knowledge of (n,2n) and (n,3n) cross sections is quite essential in the reactor technology. Recently hybrid fusion-fission reactors have gained considerable importance. The energy of the neutrons from the fusion reaction D-T is about 14 MeV, above the threshold of (n,2n) and (n,3n) reactions in most of the reactor materials. These cross sections are needed in shielding and breeding calculations also. Many of the nuclides produced in the reactor have short half lives and it is not possible to measure their cross sections directly. Also it is interesting to see the role played by the preequilibrium emission in the case of (n,xn) reactions. Here a simple method to calculate (n,xn) cross sections has been developed. In our earlier calculations[1,2] the preequilibrium mode of decay was not taken explicitly but all nonequilibrium effects were taken care by an empirical factor obtained by Kondaiah[3] by the analysis of a large number of (n,2n) cross sections measured at 14 MeV. This empirical factor is valid only around 14 MeV and does not hold good at higher energy. As a result in our earlier calculations upto about 16 MeV there was agreement with the measured cross sections but at higher energy systematically calculated (n,2n) cross sections were lower than measured ones and the reverse was true in case of (n,3n). In the present work the preequilibrium mode of decay along with the equilibrium mode is considered for the first particle emission. Subsequent emissions are considered to be due to the equilibrium mode only. In the preequilibrium decay both proton and neutron channels are considered while in the equilibrium decay proton channel is ignored as in this case proton is likely to have low energy insufficient to cross the Coulomb barrier. Gamma deexcitation competes with neutron emission near the threshold as the neutron having low energy is angular momentum forbidden and further the level density is low at that energy. This effect is indirectly compensated by using the level density parameters of Pearlstein[4] which are lower by about a factor of 2.7 as compared to those of Gilbert and Cameron[5].

Method of Calculation

Cross sections for (n,xn) reactions are calculated on the following asumptions:

(1) Neutrons are emitted in a statistical manner from the composite nucleus formed after the capture of incident neutron.

(2) The emission of the first particle is considered to be due to both preequilibrium and equilibrium processes while subsequent emissions are due to the equilibrium process only.

(3) In the case of the preequilibrium emission both proton and neutron channels are considered while in the case of the equilibrium emission the proton channel is neglected.

(4) The competition due to the gamma deexcitation is taken care indirectly by using level density parameters given by Pearlstein[4] which are less by a factor of 2.7 as compared to those given by Gilbert and Cameron[5].

(5) Equilibrium emission is calculated according to the Weisskopf model neglecting angular momentum.

When a target nucleus of mass A captures a neutron of energy E_n a composite nucleus of mass A+1 is formed at an excitation energy E_n above the ground state of nucleus A. The composite nucleus deexcites by an emission of neutron of energy ϵ_1 and the residual nucleus is left with an excitation energy $E_n - \epsilon_1$. If this energy is more than the binding energy of one neutron in the nucleus

K. H. Böckhoff (ed.), Nuclear Data for Science and Technology, 547–550.

A, a second neutron is emitted with energy otherwise further neutron emission cannot take place and the event results in (n,n'). Similarly after the emission of a second neutron if the residual nucleus possess sufficient energy to emit a third neutron the same is emitted. Otherwise the event results in (n,2n). In this work maximum four neutron emission is considered. The first neutron emission consists of two component preequilibrium and equilibrium and may be written as follows as used by Chatterjee and Gupta[6)]

$$\frac{df(\epsilon_1)}{d\epsilon_1} = \left\{ \frac{df_{PE}(\epsilon_1)}{d\epsilon_1} + (1-\delta)\frac{df_{EQ}(\epsilon_1)}{d\epsilon_1} \right\} \qquad (1)$$

where $f_{PE}(\epsilon_1)$ is the fraction of neutrons emitted with energy ϵ_1 due to the preequilibrium process and $f_{EQ}(\epsilon_1)$ is a similar term due to the equilibrium process. δ is the sum of neutron and proton fractions emitted due to the preequilibrium process.

Preequilibrium Component

The emission of particles of type ν, of kinetic energy ϵ_ν from an n-exciton state at an excitation energy E is proportional to the emission rate $w_\nu(n, E, \epsilon_\nu)$ multiplied by the time $\tau(n,E)$ the system spends in a particular state and accordingly $\frac{df_{PE}}{d\nu}$ is given by

$$\frac{df_{PE}}{d\epsilon_\nu} = \sum_{\substack{n=3 \\ \Delta n=2}} w_\nu(n,E,\epsilon_\nu)\cdot\tau(n,E)\cdot D_n \qquad (2)$$

The emission rate $w_\nu(n,E,\epsilon_\nu)$ of particle ν from an n-exciton state composed of p particles and h holes $(n=p+h)$ is given by

$$w_\nu(n,E,\epsilon_\nu) = \frac{2s_\nu+1}{\pi^2\hbar^3}\mu_\nu\,\epsilon_\nu\,(inv)\cdot R_\nu \times \frac{1}{gE}\left(\frac{U}{E}\right)^{n-2}\frac{(n+1)(n-1)}{2} \quad \ldots(3)$$

R_ν is a factor which arises due to proton neutron distinguishability. g is the single particle level density and is taken to be A/13. μ_ν is the reduced mass of ν particle and s_ν is its spin. τ is the life time of n exciton state and is given by

$$\tau(n,E) = \frac{1}{\lambda_n^+ + \gamma_n} \qquad (4)$$

where λ_n^+ is the transition rate to the next higher exciton state and is given by

$$\lambda_n^+ = \frac{g^3E^3\,\overline{|M|^2}}{2(n+1)}\cdot\frac{2\pi}{\hbar} \qquad (5)$$

$\overline{|M|^2}$ is the average squared matrix element and is taken to be dependent on average energy per exciton as given by Kalbach[1)]. γ_n in equation (4) is given as follows,

$$\gamma_n = \sum_{\nu=n,p}\int_0^{\epsilon_{max}} w_\nu(n,\epsilon_\nu)\,d\epsilon_\nu \qquad (6)$$

Dn in equation (2) is the term responsible for depletion and is given by

$$D_{n+2} = D_n(1-\gamma_n\tau_n) \text{ and } D_3=1 \qquad (7)$$

δ in equation (1) is related with Dn as follows

$$\delta = 1 - D_{\bar{n}+2} \qquad (8)$$

Inverse reaction cross sections have been obtained by the parametrization developed by Chatterjee et al[8)] to get Wilmore-Hodgson[9)] optical model reaction cross sections. At energy lower than 2 MeV their parametrization is not accurate. Hence at these energies an expression given by Blett and Weisskopf[10)] is used by matching the cross section at 2 MeV.

Equilibrium Component

The equilibrium component is calculated according to the Weisskopf model neglecting angular momentum. The probability of neutron emission with energy ϵ is given as follows

$$P \sim \epsilon\,\sigma\,\rho(U) \qquad (9)$$

σ is the inverse reaction cross section and is calculated as explained in section III. $\rho(U)$ is the level density at the excitation energy U and is given by

$$\rho(U) = C\,e^{2\sqrt{aU}} \qquad (10)$$

C is taken to be constant, any slow variation in C is neglected. a is level density parameter given by Pearlstein. Accordingly, the expression for (n,2n) cross-section is given as follows,

$$\sigma_{n,2n} = \sigma_m \int_0^{E_n-EB1} \left\{ \frac{df_{PE}}{d\epsilon_1} + \frac{(1-\delta)\,\epsilon_1\sigma_1\rho_1}{\int_0^{E_n}\epsilon_1\sigma_1\rho_1\,d\epsilon_1} \right\} \cdot P_2\,d\epsilon_1 \quad \ldots(11)$$

σ_m is taken to be σ_R for non-fissionable nuclei and for fissionable nuclei it is taken as $\sigma_R - \sigma_f$. σ_f is the fission cross section and is taken from the measured data if available otherwise it may be estimated from the empirical formula given by Jhingan et al [11)]. P_2 is the probability that after the emission of the first neutron the second neutron is emitted with such an energy that the third neutron cannot be emitted and is given as follows

$$P_2 = \frac{\int_{E_n-\epsilon_1-EB2}^{E_n-\epsilon_1-EB1} \epsilon_2\sigma_2\rho_2\,d\epsilon_2}{\int_0^{E_n-\epsilon_1-EB1} \epsilon_2\sigma_2\rho_2\,d\epsilon_2} \qquad (12)$$

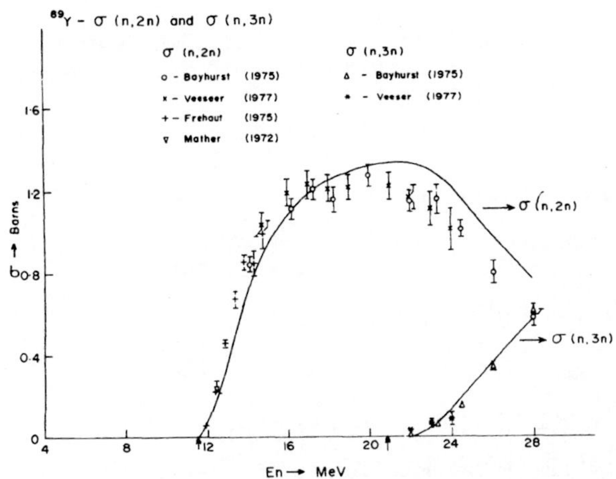

Fig.1

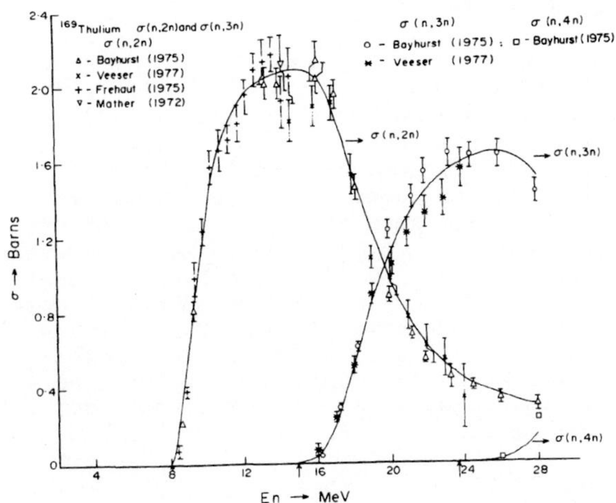

Fig.2

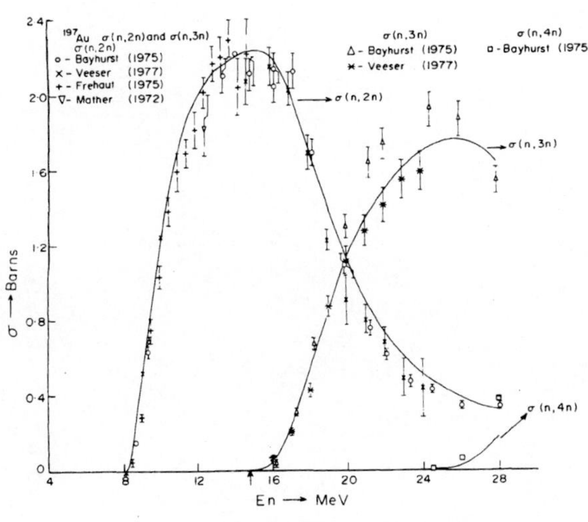

Fig.3

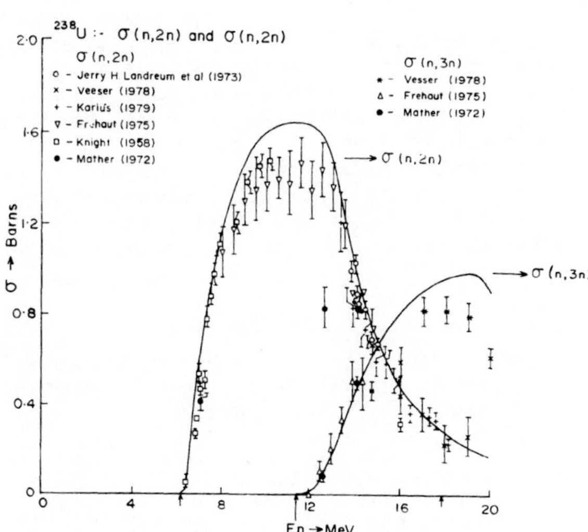

Fig.4

EB1 and EB2 are the binding energy of one and two neutrons in the nucleus A respectively. If $E_n - \epsilon_1$ -EB2 happens to be negative it is taken to be zero. In the numerator limits of integration ensure that second neutron is emitted with an energy so that third neutron cannot be emitted. In a similar way expressions for (n,3n) and (n,4n) are written.

In this work a maximum of four neutron emissions have been considered. Accordingly a computer code has been developed. All the integrals have been evaluated using Simpson's rule. Cross sections have been calculated and compared for a number of nuclei in the mass region 89 to 238 at incident neutron energies upto 28 MeV. However cross sections for ^{238}U have been calculated upto 20 MeV only as the fission cross sections have been taken from ENDF-B-IV where data are given upto 20 MeV only. In some typical cases the comparisons between calculated and measured cross sections are shown in figures 1 to 4.

Conclusion

The agreement in case of (n,2n) and (n,3n) cross section is within 10%. Data on (n,4n) are scanty and wherever measured data are available they have been compared with the calculated ones and the agreement is within a factor of two. The preequilibrium component is found to be considerable at energies higher than about 16 MeV and its inclusion is essential. Average squared matrix element given by Kalbach[7] can be used without any adjustment. It is not very sensitive to (n,xn) cross sections. In the present work there is no fitting parameter. All the parameters have been taken from the literature. The level density parameters given by Pearlstein give good agreement in the case of (n,2n) and (n,3n) cross sections. It appears to compensate the effect of neglecting gamma deexcitation and makes the calculation easier. The (n,xn) cross section for unstable nuclei can be predicted by this method where direct measurement is not possible.

References

1. M.L. Jhingan, R.P. Anand, S.K. Gupta and M.K. Mehta, Proc. Nucl. Physics & Solid State Phys. Symp. 20B, 141 Pune (1977)

2. M.L. Jhingan, R.P. Anand, S.K. Gupta and M.K. Mehta, Proc. Int. Conf. on Neutron and Nucl. Data Harwell (1978)

3. E. Kondaiah, J. Phys. A. Math. Nucl. Gen. 7, 1457 (1974)

4. S. Pearlstein, Nucl. Sci. Engg. 23, 233 (1965)

5. A. Gilbert and A.G.W. Cameron, Can J. Phys. 34, 804 (1965)

6. A. Chatterjee and S.K. Gupta, Phys. Rev C 18, 2118

7. C. Kalbach, Z. Physik A 287, 319 (1978)

8. A. Chatterjee, K.H.N. Murthy and S.K. Gupta, Pramana 16, 391 (1981)

9. D. Wilmore and P.E. Hodgson, Nucl. Phys. 55, 673 (1964)

10. John M. Blatt and Victor F. Weisskopf, Theoretical Nucl. Phys., John Wiley and Sons New York (1952)

11. M.L. Jhingan, R.P. Anand, S.K. Gupta and M.K. Mehta, Annals of Nucl. Energy, 6, 495 (1979).

References for Figures

Bayhurst (1975) Phys. Rev. C12, 451

Frehaut (1975) Proc. National Soviet Conf. on Neutron Physics, Kiev Vol.4, 303

Karius (1979) J. Phys. G 5, 715

Knight (1958) Phys. Rev. 112, 259

Landrum (1973) Phys. Rev. C8, 1938

Mather (1972) AWRE Rept. No.072/72

Veeser (1977) Phys. Rev. C16, 1792

Veeser (1978) Conf. on Neutron Phys. and Nucl. Data, Harwell, 1054.

PARAMETRIZATION PROBLEMS OF PRECOMPOUND MODELS

Helmut Jahn

Kernforschungszentrum Karlsruhe
Institut für Neutronenphysik
und Reaktortechnik

Postfach 3640
D-75 Karlsruhe 1
Federal Republic of Germany

With most of the precompound models to calculate nuclear reaction cross-sections new parametrization problems have been introduced in addition to those already connected with the Hauser-Feshbach-formalism.
For the different versions of precompound models these parametrization problems are investigated and compared with each other. It is shown how far a certain arbitrariness arising from the parametrization problems can be reduced and whether the degree of uniqueness can be reached to obtain the predictive capability which is necessary to close gaps of experimental information.

Paper not available.

K. H. Böckhoff (ed.), Nuclear Data for Science and Technology, 551.

NUCLEAR MODEL CALCULATIONS OF NEUTRON INDUCED CROSS SECTIONS FOR ^{52}Cr, ^{55}Mn, ^{56}Fe AND ^{58,60}Ni FOR INCIDENT ENERGIES UP TO 30 MeV

B. Strohmaier and M. Uhl

Institut für Radiumforschung und Kernphysik der Universität Wien
Boltzmanngasse 3, A-1090 Vienna, Austria

For neutrons incident on ^{52}Cr, ^{55}Mn, ^{56}Fe and ^{58,60}Ni, total cross sections, differential elastic cross sections, production spectra of emitted particles and photons integrated over solid angle, production cross sections of solid and gaseous transmutation products were calculated for neutron energies up to 30 MeV. For the calculations, the optical model, the compound nucleus model and the exciton model were employed, considering in the precompound stage the emission of neutrons, protons and α-particles, in the compound nucleus stage the emission of these particles and of photons. The dependence of the calculated cross sections on the model parameters was studied. A consistent set of parameters was chosen so as to achieve an overall best fit to all experimental cross section data which, however, exist mainly for incident energies below 20 MeV.

[^{52}Cr, ^{55}Mn, ^{56}Fe, ^{58,60}Ni(n,a), (n,ab), (n,abc), (n,abcd) cross sections (a,b,c..n,p,α,γ;d..γ). Neutron induced reactions. Cross section evaluation. E_n = threshold - 30 MeV.]

Introduction

For the most abundant isotopes of Cr, Mn, Fe and Ni we calculated the following neutron induced cross sections to 30 MeV incident neutron energy:

- total
- differential elastic
- activation cross sections for particular reaction paths, as e.g. (n,pn) and (n,np)
- production spectra of emitted particles and photons integrated over the full solid angle
- production cross sections of all residual nuclei summed over all considered reaction paths
- H and He production cross sections

The above elements are structural materials, and this work was performed in connection with a planned target construction at the CCR Ispra for an intense neutron source for fusion related material radiation damage studies. A large amount of experimental data for several of the above cross sections exists extending, however, to energies not higher than 20 MeV. Therefore, for supplying these cross sections in an energy region exceeding this limit and for predicting unmeasured cross sections, model calculations are required. The abundance of available experimental cross sections can be exploited for the adjustment and verification of model parameters.

Procedure

For an incident neutron energy of 30 MeV, reactions with up to 3 or 4 emitted particles are allowed by energy conservation. As emitted particles we considered neutrons, protons and α-particles. This means, that for a given target nucleus, 27 to 81 different reaction paths have to be considered.

The following nuclear reaction models were applied for the cross section calculations: compound nucleus evaporation model for the treatment of multiple emission of particles and γ-rays, exciton model for 1st chance precompound emission, optical model for the total and the elastic cross sections, the direct inelastic cross sections and for the creation of transmission coefficients. For the computations, we used the code STAPRE[1] which incorporates the former two models, the spherical optical model code ABACUS II[2] and the DWBA code DWUCK[3].

The results of the calculations critically depend on a set of model parameters which for the particular target nuclei, the considered emitted particles and the energy range of interest, are required for about 80 nuclei. In general all these model parameters are not known very accurately and often one has to resort to systematics.

We attempted to find a consistent set of model parameters which in the mass region of interest reproduces simultaneously total and differential elastic neutron cross sections, average resonance data as spacings and strength functions, cross sections for competing reactions of the type (n,nx), (n,px) and (n,αx), cross sections for (p,nx) and (α,nx) reactions. The advantages of such a consistent set of model parameters are:

- confirmation of the applicability of the models
- possibility of determining level density parameters for nuclei with no resonance data
- extension of the calculations to less abundant isotopes possible

For a simultaneous fit to many different experimental cross section data one often has to compromise: in general, a particular cross section will not be reproduced as well as if the model parameters were adjusted to fit this very cross section.

Determination of the model parameters

The choice of the model parameters and their adjustment by comparison to experimental data was performed as follows:

Transmission coefficients

i. Neutrons

We utilized the optical potential of Arthur and Young[4], who made very extensive calculations of neutron induced cross sections for ^{54,56}Fe. This potential reproduces differential elastic, total and reaction cross sections of ^{56}Fe very well and has been verified also for the calculations of reaction cross sections of ^{52}Cr, ^{55}Mn, ^{58,60}Ni by comparison with experimental data. Our most recent studies, however, which concern an improvement of the reproduction of differential elastic cross sections of these 4 nuclei, the inclusion of direct reaction contributions to inelastic neutron scattering as well as the calculation of reaction cross sections of less abundant isotopes in this mass region, showed that the potential of Ref.[4] is less suitable for these nuclei for which it had not been determined. Therefore, we tried to extract global parameters from the optical potentials determined by Prince[5] for the isotopes of Cr, Fe and Ni. These investigations are in progress.

ii. Protons

For protons, we chose the optical potential by Mani et al.[6] which besides (n,p) cross sections also reproduces experimental (p,n) cross section data (Fig.1); the calculated cross sections are corrected for isospin effects[7].

iii. Alpha-particles

For α-particles we used a potential which was derived from that of McFadden et al.[8] To test the potential again via the inverse reactions, calculated (α,n) cross sections were confronted to experimental ones (Fig. 2). In fact, this potential was our 2nd choice after we had realized that with the potential of Huizenga et al.[9] the reproduction of the experimental (n,α) cross sections required level density parameters which are inconsistent with resonance data.

K. H. Böckhoff (ed.), Nuclear Data for Science and Technology, 552–555.

iv. Photons

For E1 radiation we used the Brink-Axel model[10] which relates the strength function to the photo-absorption cross section. In spite of the severe fragmentation of the giant dipole resonance for A < 60 we approximated it by a single Lorentz curve with resonance energy $75/A^{1/3}$ MeV and width 5.5 MeV consistent with Ref.[11] For improving the reproduction of the experimental 14 MeV photon production spectra, the E1 strength function was reduced by a factor of o.55 below 6 MeV. For the strength functions of the other multipole types, the Weisskopf model[12] was used. They were normalized relative to the E1 strength function at the neutron binding energy according to the systematics of McCullagh et al.[13] in the case of M1 radiation and according to Weisskopf's estimate for E2, M2 and E3. An overall normalization factor was determined so as to reproduce experimental capture cross sections for some nuclei, however, without consideration of valence capture. A weighted average over the thus obtained normalization factor was taken for those nuclei for which no experimental capture cross sections are available.

Exciton model parameters

The particle-hole state densities were calculated with the Pauli principle corrected formula of Williams[14] with $g=(6/\pi^2)A/8$ and with a simple pairing correction. The emission rates for nucleons account for the type of the projectile[15], those for α-particles for cluster preformation[16]. For the internal transition rates we used Williams' [17] expressions. The quantity FM which via the relation[18] $|M|^2=FM/(A^3.U)$ defines the squared transition matrix element for given excitation energy U, was adjusted to reproduce the high energy portion of (n,pγ) excitation functions (Fig.3) and the high energy tail of proton production spectra (Fig.4) at 15 MeV incident neutron energy. As a further verification of the model parameters we compared neutron production spectra with measurements for outgoing energies for which preequilibrium emission dominates.

Levels and level density parameters

The level density is calculated within the framework of the back-shifted Fermi gas model[19] . Using the same procedure as described in Ref.[19] we redetermined the parameters a and Δ for those nuclei for which more recent data on levels[20] and resonance spacings (mostly from Refs.[21,22]) were available; the rigid body value with r_o = 1.25 fm was assumed for the effective moment of inertia. If required for the reproduction of experimental cross sections we varied the level density parameters within their uncertainties. If no resonance information was available we used systematics. Also for the explicit consideration of levels, the information was taken from Ref.[20].

Deformation parameters

The deformation parameters of low lying collective states required for the DWBA calculations were in general taken from (p,p') and (α,α') data, quoted in Nuclear Data Sheets. For the first quadrupole and octupole states, we used the (n,n') data of Refs.[23,24,25]

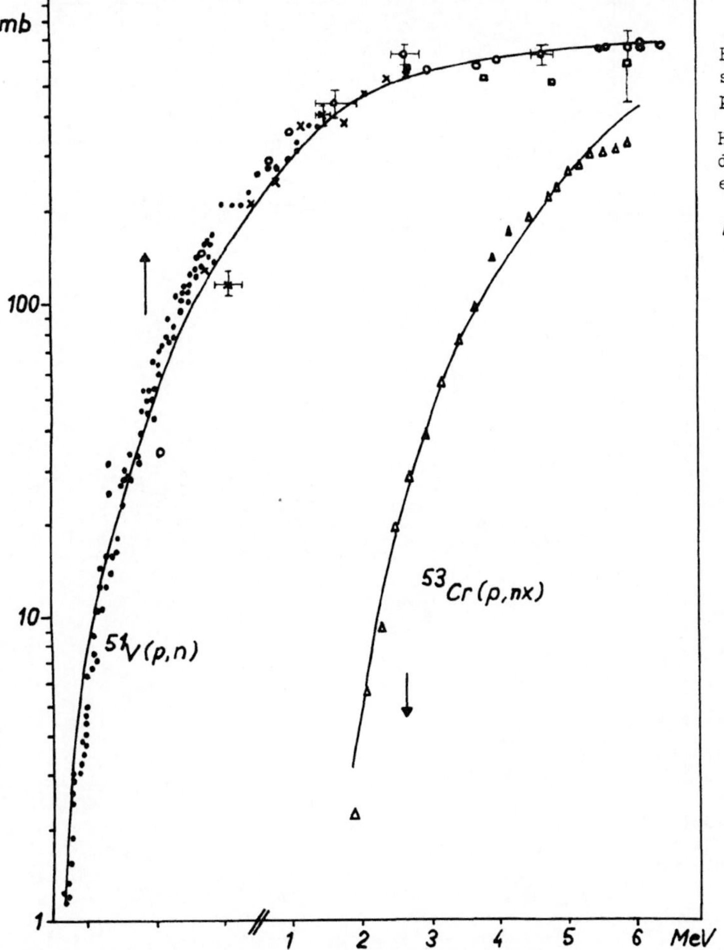

Fig. 1. (p,nx) and (p,nγ) cross sections for ^{51}V and ^{53}Cr. Energy in lab.system. Solid lines: calculation.
□ (p,nγ) S.Tanaka et al., J.Phys.Soc.Jap.15(1960)1547
•,Δ (p,nx) C.H.Johnson et al., Rept.ORNL-2910(1960) and Rept.ORNL-2501(1958)
o (p,nγ) J. Wing et al., Phys.Rev.128(1962)280
× (p,nγ) G.F.Dell et al., Nucl.Phys.64(1965)513

Results and discussion

Reaction paths with up to 3 steps were considered. Further, reaction paths with charged particles in each step were neglected.

How sensitively the results depend on the considered reaction paths, is illustrated by way of example of the He and H production cross sec-

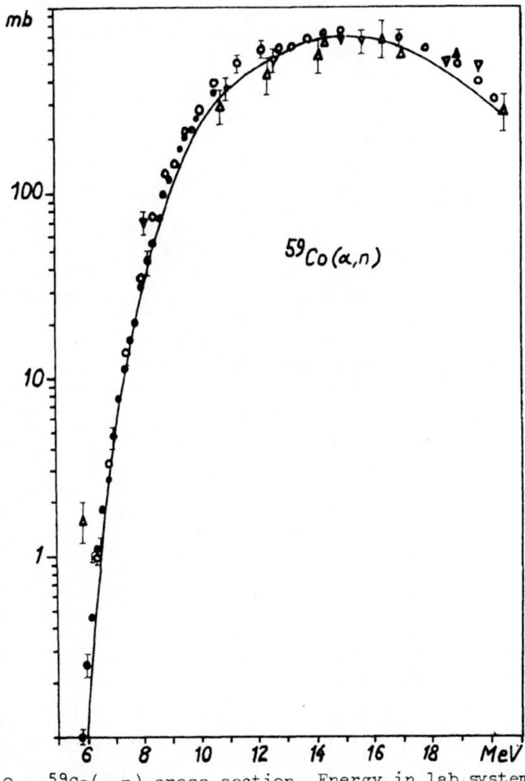

Fig. 2. ^{59}Co(α,n) cross section. Energy in lab.system. Solid line: calculation.
• (α,nx) P.H.Stelson et al., Phys.Rev.133(1964)B911
∇ (α,nγ) J.M.D'Auria et al., Phys.Rev.168(1968)1224
Δ (α,nγ) O.A.Zhukova et al., Sov.J.Nucl.Phys.16(1973)134
o (α,nγ) F.M.Mann et al., Nucl.Phys.A255(1975)287

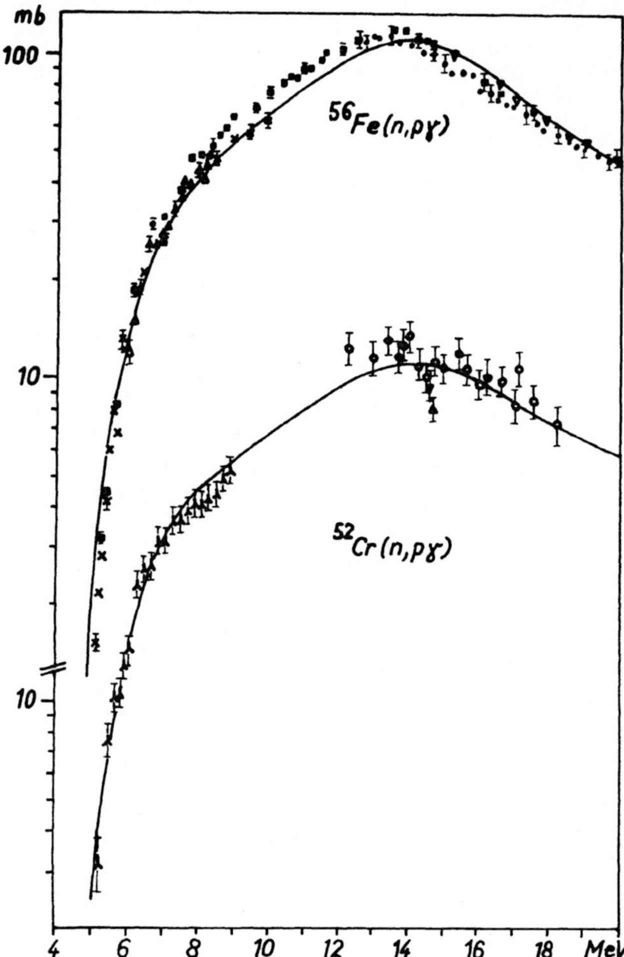

Fig. 3. (n,pγ) cross sections for ^{56}Fe and ^{52}Cr. Energy in lab. system. Solid lines: calculation.
■ D.C.Santry et al., Can.J.Phys.42(1964)1030
● H.Liskien et al., J.Nucl.En.A/B19(1965)73
Δ H.Liskien et al., Nukleonik 8(1966)315
+ H.K.Vonach et al., Proc.Conf.Nucl.Cross Sections and Technology, Washington, 1968, p. 885
× D.L.Smith et al., Nucl.Sci.Eng.58(1975)314
▽ T.B.Ryves et al., Metrologia 14(1978)127
○ B.D.Kern et al., Nucl.Phys.10(1959)226
▲ N.I.Molla et al., Nucl.Phys.A283(1977)269
⅄ D.L.Smith et al., Nucl.Sci.Eng.76(1980)43
▼ P.Holmberg et al., J.Inorg.Nucl.Chem.36(1974)715

tions. The results obtained under consideration of reaction paths with only 1 and up to 2 charged particles were compared. Whereas for ^{58}Ni the reaction paths with 2 charged particles increase the hydrogen production by 45% and the He production by about 10% at 30 MeV neutron energy, in the case of ^{56}Fe the contribution of such paths is negligible throughout the energy region.

To convey an insight into the accuracy of the calculated cross sections, we present some further comparisons with experimental data:

i. (n,2nγ) cross sections (Fig. 5):

In the case of ^{58}Ni, the use of the Prince[5] potential has improved the reproduction compared to previous results[26]. For ^{52}Cr, the high energy part of the excitation function disagrees with the only measurement.

ii. (n,pγ) cross sections and proton production spectra:

For ^{52}Cr, ^{56}Fe and ^{58}Ni the (n,pγ) data are well reproduced; the same is true for all proton production spectra (examples Figs. 3 and 4).

iii. (n,αγ) cross sections and α-particle production spectra:

The calculated ^{55}Mn (n,αγ) cross section is in agreement with the trend of the experimental data (Fig. 6).

The α-spectra (for ^{60}Ni: Fig.7) measured at 15 MeV neutron energy by Grimes et al. are well reproduced except the high energy end. Here, inclusion of a direct reaction contribution populating low excited levels of the residual nucleus might improve the agreement.

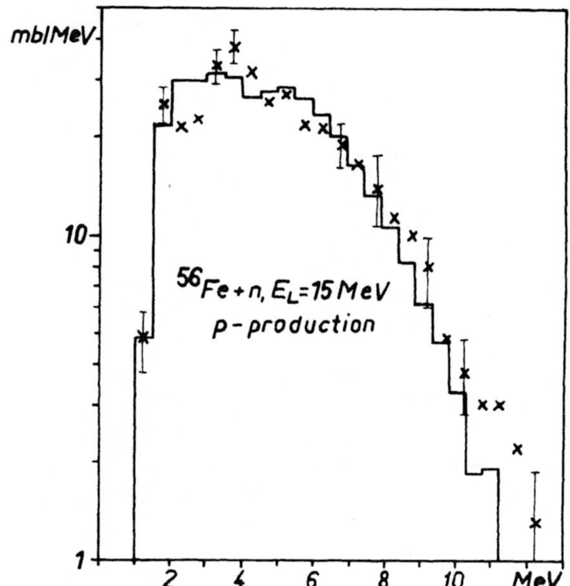

Fig. 4. Proton production spectrum for ^{56}Fe at 15 MeV neutron energy (lab.system). Proton energy lab.system. Histogram: calculation. × S.M.Grimes et al., Phys.Rev. C19(1979)2127

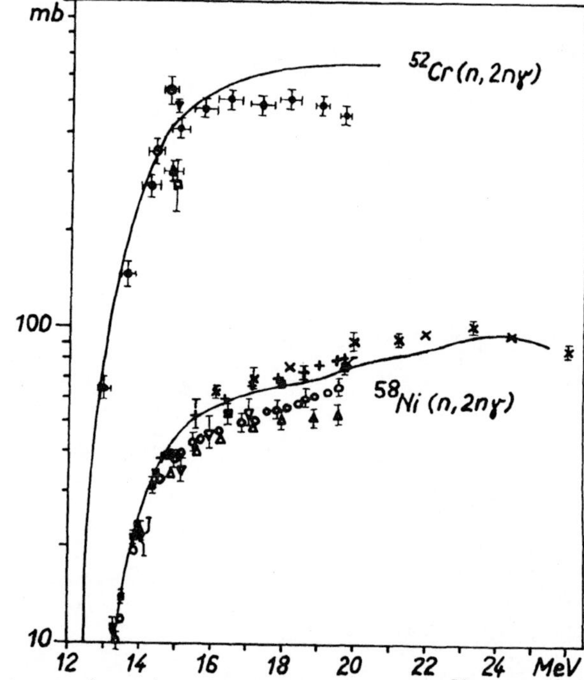

Fig. 5. (n,2nγ) excitation functions for ^{52}Cr and ^{58}Ni Energy in lab.system. Solid lines: calculation.
□ R.Wenusch et al., Anz.ÖAW, Math.-Nat.Kl.99(1962)1
● M.Bormann et al., Nucl.Phys.A115(1968)309
▲ S.M.Qaim, Nucl.Phys.A185(1972)614
⊙ G.N.Maslov et al., Rept.YK-9(1972)50
▼ K.Sailer et al., Proc.All Union Conf.Neutron Phys., Kiev, 1977, p.246
■ R.J.Prestwood et al., Phys.Rev.121(1961)1438
○ A.Paulsen et al., Nukleonik 7(1965)115
Δ M.Bormann et al., Rept.EANDC(E)66(1966)42
× B.P.Bayhurst et al., Phys.Rev.C12(1975)451
▽ C.G.Hudson et al., Ann.Nucl.En.5(1978)589
+ G.Winkler et al., this conference

iv. neutron production spectra (Fig. 8):

Calculations for ^{56}Fe are compared to a measurement for natural iron. The direct contributions obtained with DWBA were grouped in 0.5 MeV bins. Equally good reproduction was achieved for ^{52}Cr and ^{58}Ni.

The calculations reproduce the experimental data in general within 20%, in many cases better. We estimate errors of 20-30% for extrapolated cross sections which at some energy are confirmed by data. Only slightly less accurate are the predicted unknown cross sections which involve nuclei whose level density is confirmed by other data. For reactions which populate nuclei far from the line of β-stability, the errors may be much larger. For the gas production cross sections, we estimate an accuracy of 20-30% as well as for the particle production spectra except the high energy end.

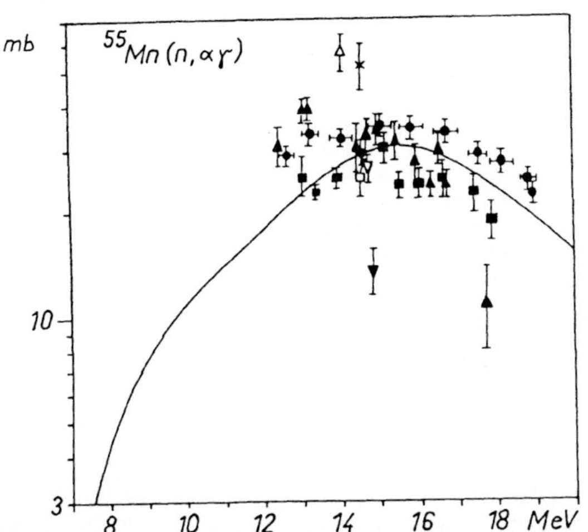

Fig. 6. $^{55}Mn(n,\alpha\gamma)$ excitation function. Energy in lab. system. Solid line: calculation.
× E.B.Paul et al., Can.J.Phys.31(1953)267
△ C.S.Khurana et al., Nucl.Phys.13(1959)88
▲ F.Gabbard et al., Phys.Rev.128(1962)1276
● M.Bormann et al., Nucl.Phys.63(1965)438
▼ E.Frevert, Acta Phys.Austriaca20(1965)304
□ A.Peil, Nucl.Phys.66(1965)419
▽ B.Minetti et al., Z.Physik 199(1967)275
■ J.Turkiewicz, Proc.Symp.Fast Neutron Interactions.., Debrecen, 1975, paper 26PM5

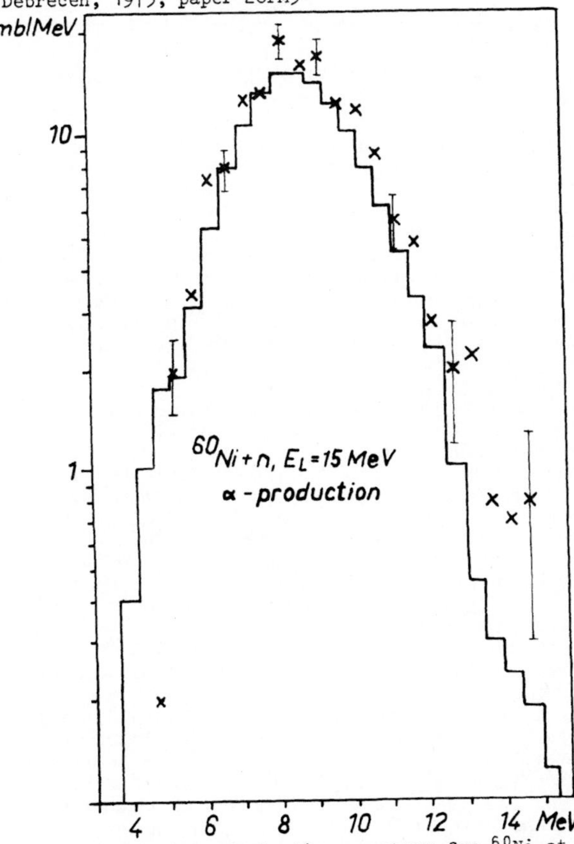

Fig. 7. α-particle production spectrum for ^{60}Ni at 15 MeV neutron energy (lab.system). α-particle energy lab. system. Histogram: calculation. × S.M.Grimes et al., Phys.Rev.C19(1979)2127

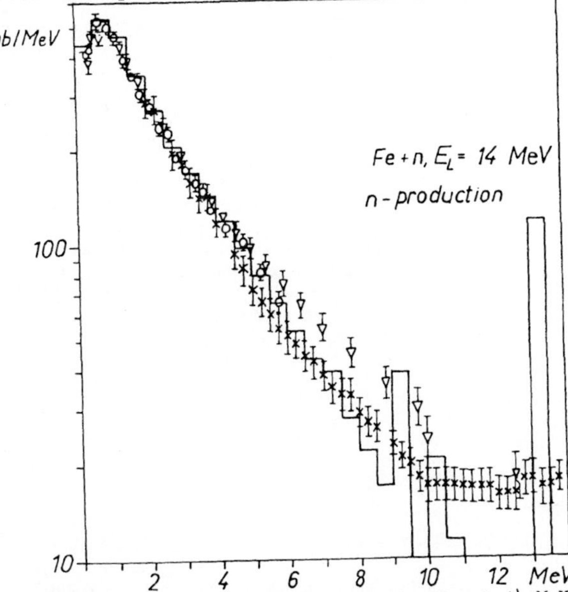

Fig. 8. Neutron production spectrum for Fe at 14 MeV incident neutron energy (lab.system). Outgoing neutron energy C.M.system. Histogram: calculation for ^{56}Fe. Measurements for natural iron.
o H.Vonach et al., Proc.Int.Symp.Neutron Cross Sections from 10-50 MeV, Brookhaven, 1980, p.343
× G.Seeliger, priv.comm. to H.Vonch,1980
▽ O.A.Salnikov et al., Sov.J.Nucl.Phys.12(1971)620

References

1. B. Strohmaier and M. Uhl, Proc. Winter Course Nucl. Theory for Applications, Trieste, 1978, IAEA-SMR-43 (1980), p. 313
2. E.H. Auerbach, Rept. BNL-6562 (1962)
3. P.D. Kunz, Univ. of Colorado, unpublished
4. E.D. Arthur and P.G. Young, Rept. LA-8626-MS (ENDF-304) and Proc. Int. Symp. Neutron Cross Sections from 10 to 50 MeV, Brookhaven, 1980, p. 731
5. A. Prince, priv. comm. and this conference
6. G.S. Mani et al., Rept. C.E.A.-2379 (1963)
7. S.M. Grimes et al., Phys.Rev. C5 (1972) 85
8. L. McFadden and G.R. Satchler, Nucl.Phys.84(1966)177
9. J.R.Huizenga and G.J.Igo, Rept.ANL-6373 (1961)
10. D.M.Brink, Thesis, Oxford Univ.(1955) and P.Axel, Phys.Rev. 126 (1962) 671
11. B.L. Berman and S.C.Fultz, Rev.Mod.Phys.47(1975)713
12. J.M.Blatt and V.F.Weisskopf, "Theoretical Nuclear Physics", J.Wiley & Sons Inc., (1952)
13. C.M.McCullagh et al., Phys.Rev.C23(1981)1394
14. F.C.Williams, jr., Nucl.Phys.A166(1971)231
15. E .Gadioli et al., Nucl.Phys.A217(1973)589
16. L.Millazzo-Colli et al., Nucl.Phys.A210(1973)297
17. F.C.Williams, jr., Phys.Lett.31B(1970)184
18. C.Kalbach-Cline, Nucl.Phys. A210(1973)590
19. W.Dilg et al., Nucl.Phys. A217(1973)269
20. C.M.Lederer and V.S.Shirley (eds.), "Table of Isotopes", 7th ed., J.Wiley & Sons Inc., N.Y. (1978)
21. G.Rohr, Proc. Specialists Meeting Neutron Data of Structural Materials for Fast Reactors, Geel, 1977 Pergamon Press (1979) p.614
22. F.H.Fröhner, Proc.Int.Conf.Neutron Physics and Nucl. Data for Reactors ..., Harwell, 1978, p.268
23. Y.Yamanouti, Proc.Int.Conf. Nucl.Cross Sections for Technology, Knoxville, 1979, NBS Spec.Publ., p. 146
24. P.H.Stelson et al., Nucl.Phys.68(1965)97
25. M.Hyukatake et al., J.Phys.Soc.Jap. 38(1975)606
26. B.Strohmaier et al. , Proc.Adv.Group Meeting Nucl.Data for Radiation Damage Assessment .., Vienna, 1981, IAEA-TECDOC-263, p.135

CALCULATION OF ^{239}Pu NEUTRON INELASTIC CROSS SECTIONS

Edward D. Arthur

Theoretical Division, Los Alamos National Laboratory
Los Alamos, New Mexico 87545 U.S.A.

We have calculated cross sections for neutron-induced reactions on ^{239}Pu between 0.001 and 5 MeV, with particular emphasis on inelastic scattering. Coupled-channel and Hauser-Feshbach statistical models were used. Within the coupled-channel calculations we employed neutron optical parameters derived from simultaneous fits to total, elastic, inelastic, and resonance data. The resulting transmission coefficients were used in Hauser-Feshbach statistical calculations having a fission channel based on a double-humped barrier representation. Barrier parameters and transition state enhancements needed to reproduce well the (n,f) cross sections between 0.001 and 5 MeV were in general agreement with those from other published analyses. Calculated compound-nucleus and direct-reaction components for inelastic scattering were combined incoherently, and the resultant cross sections agreed well with the Bruyères-le-Châtel measurements for scattering from levels occupying the ground state rotational band. Our results are in substantial disagreement with ENDF/B-V values for these levels. We are presently performing DWBA calculations to determine direct-reaction components for states occupying higher-lying vibrational bands.

[^{239}Pu(n,n), ^{239}Pu(n,n'), ^{239}Pu(n,f), coupled channel optical and Hauser-Feshbach statistical model calculations, E_n = 0.001 to 5 MeV]

Introduction

As a step towards improvement of the ^{239}Pu evaluation appearing in the ENDF/B evaluated data library, we have performed neutron cross-section calculations in the incident energy range between 0.001 and 5 MeV. We placed particular emphasis on the realistic determination of inelastic scattering cross sections since relevant experimental data are sparse, and since such information is important for fast reactor applications. To accomplish this we used sophisticated nuclear models employing parameters constrained to reproduce concurrently a variety of available experimental data. This technique has proved successful in the extrapolation or prediction[1] of cross-section data where experimental measurements are not complete or do not exist.

Models and Parameters

Our ^{239}Pu calculations involved application of two main reaction models, the first being the coupled-channel optical model to describe direct-reaction contributions to inelastic scattering from collective states. The ECIS[2] program was used for this purpose as well as to provide neutron transmission coefficients for the second portion of the calculation involving application of the Hauser-Feshbach statistical model code COMNUC.[3] This technique ensures consistency between the direct-reaction and compound-nucleus facets of the calculation. Cross-section contributions obtained separately from these two models were then combined incoherently to produce the final results.

ECIS calculations were made using the first six states of the ground state rotational band ($1/2^+$, $3/2^+$, $5/2^+$, $7/2^+$, $9/2^+$, and $11/2^+$) as the coupling basis. The optical potential was represented in a standard manner (see Ref. 4 for details), while the coupling form factors needed in the expansion of the optical potential were assumed complex. We used neutron optical parameters based on the Bruyères-le-Châtel results[5] that were obtained primarily from fits to actinide total, elastic, and inelastic cross sections, as well as s- and p-wave strength functions. We did modify them slightly to produce better agreement to ^{239}Pu total cross sections measured by Poenitz,[6] particularly around 1 MeV. Our resulting optical and deformation parameters appear in Table I.

The Hauser-Feshbach portion of the calculation utilized the phenomenological level density model of Gilbert and Cameron[7] along with the parameters of Cook.[8] A maximum amount of discrete level information was included for each nucleus appearing in the calculation. Such data were used to adjust the constant temperature level density parameters so as to reproduce the cumulative number of levels while joining smoothly to the Fermi-gas form at higher excitation energies. Gamma-ray transmission coefficients were calculated using a Brink-Axel expression[9] that utilized two Lorenztian forms to represent the split giant dipole resonance. These gamma-ray transmission coefficients were normalized to reproduce measured $2\pi\langle\Gamma_\gamma\rangle/\langle D\rangle$ data[10] available for s-wave resonances near the neutron binding energy.

Table I Optical Model and Deformation Parameters Used in the Coupled-Channel Calculations (Depths in MeV, geometrical parameters in fm)

		r	a
V	= 46.2 - 0.3E	1.26	0.615
W_{SD}	= 3.6 + 0.4E	1.24	0.50
V_{SO}	= 6.2	1.12	0.47
β_2	= 0.21	β_4 = 0.065	

The ability to fit measured (n,f) cross sections accurately (to within approximately 5%) over a wide energy range introduces important constraints on the Hauser-Feshbach calculation of cross sections for other competing channels. Thus, in order to reproduce (n,f) experimental data reasonably well using realistic fission channel parameters, we incorporated a double-humped fission model into the COMNUC Hauser-Feshbach code (Ref.11 provides a complete description). In this application, two uncoupled oscillators were used for the barrier representation, and penetrabilities through each barrier were calculated from a Hill-Wheeler expression.[12] The spectrum of transition states occurring at each barrier was constructed utilizing bandhead information[13] available for the ^{240}Pu compound system. At higher excitations, we assumed a continuum of transition states that we calculated using the Gilbert-Cameron level density expressions and parameters applicable for the ground state deformation case. To these calculated densities we applied enhancements directly to account for deviations from symmetry present at each barrier. In this mass region the inner barrier has associated with it a triaxial shape while the outer one is mass

K. H. Böckhoff (ed.), Nuclear Data for Science and Technology, 556–559.

asymmetric and axially symmetric.[14] Theoretical enhancements[15] associated with these barrier shapes are $\sigma\sqrt{8\pi}$ (σ is the level density spin cutoff parameter) for the inner barrier and two for the outer one.

Our fission model also included corrections for Class II fluctuations based on the picket-fence approximation of Lynn et al.[16] These corrections are important primarily at low energies and were applied in addition to the width-fluctuation corrections[17] utilized throughout our Hauser-Feshbach calculations. Our ^{239}Pu(n,f) data fits yielded the barrier parameters and density enhancement factors shown in Table II. These are in general agreement with other published values[13,18] and in particular the density enhancements agree well with results[13] utilizing microscopic level-density expressions.

TABLE II Barrier Parameters and Density Enhancements Used to Calculate ^{239}Pu(n,f) Cross Sections

	Barrier Height (MeV)	$\hbar\omega$ (MeV)	Density Enhancement
Barrier A	5.80	0.8	16
Barrier B	5.45	0.6	2

Results

Calculated scattering cross sections for 0.7 MeV neutrons incident on ground band members of ^{239}Pu are compared in Fig. 1 to recent Bruyères-le-Châtel measurements.[5] At this energy, compound contributions can be significant so that direct and compound nucleus calculations can be tested in such a comparison. Figure 2 presents a comparison at 2.5 MeV to data of Smith et al[19] that includes elastic scattering as well as contributions from states having excitation energies up to 0.20 MeV.

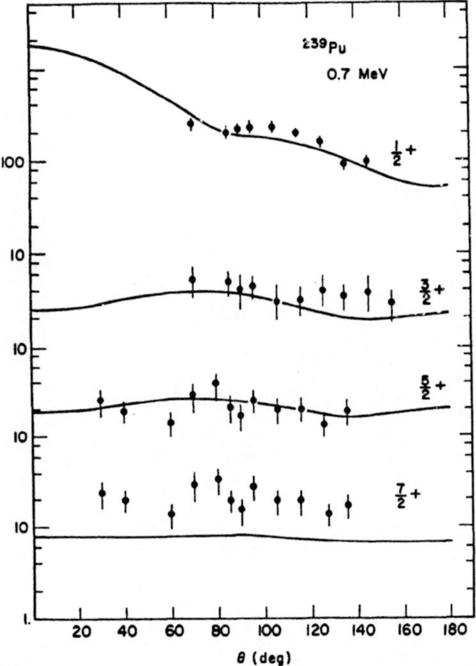

Fig. 1. Our calculated angular distributions are compared to recent measurements[5] of elastic and inelastic scattering on ^{239}Pu at a neutron energy of 0.7 MeV.

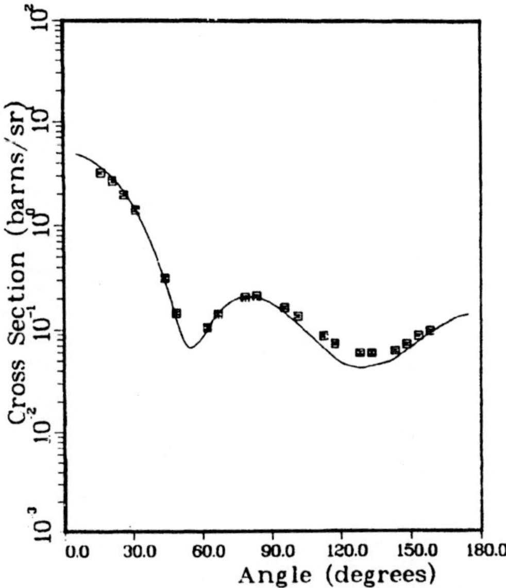

Fig. 2. Our calculations of cross sections for scattering reactions that excite ^{239}Pu states having energies $\leq$ 200 keV are compared to the data of Smith et al[19] for 2.5 MeV incident neutrons.

Figure 3 illustrates the large differences that exist between our calculated results and data appearing in the present ENDF/B-V evaluation. The comparison is made for excitation functions resulting from inelastic scattering on the 0.057 MeV ($5/2^+$) state and the 0.164 MeV ($9/2^+$) state. In the first example, the difference occurs because direct-reaction contributions were included in our calculations but not in ENDF/B. In the second example, our calculated excitation function shape occurs because of the state's relatively high spin (9/2) and because of direct-reaction components. The ENDF/B-V evaluation, on the other hand, probably employed a shape similar to those acsumed for scattering from states with lower spin values.

Inelastic scattering reactions that leave the ^{239}Pu nucleus with 1 to 2 MeV of excitation are important for fast breeder reactor systems because of the energy transfers involved. This excitation energy region lies well above that encompassing the ground state rotational band so that experimental data are sparse and theoretical efforts are generally restricted to application (as in our present effort) of the Hauser-Feshbach statistical model. The experimental situation has been improved by recent measurements made by Smith et al[19] whereby the total inelastic cross section to levels lying above a given excitation energy can be inferred. Such thresholds range from 0.08 to 0.3 MeV for these measurements. The comparison of our calculations, as is done in Fig. 4, to these data provides indirect evidence regarding the behavior of inelastic scattering to higher levels for which we assumed only compound nucleus contributions. The solid cure represents our results, which agree reasonably with the experimental data but which lie significantly lower than the dashed curve representing ENDF/B-V.

Recently we began calculations of direct-reaction contributions to scattering cross sections from states occupying higher-lying collective bands, particularly the K^{π} = 1/2- octupole vibrational band. These efforts were prompted by (n,n') cross sections inferred from (n,n'γ) measurements[20,21] on ^{232}Th and ^{238}U that indicate sizeable inelastic cross sections for similar states in the neutron energy range between 2 to 4 MeV. Initially we performed Distorted Wave Born Approximation (DWBA) calculations for scattering from the 0.555 MeV 7/2- state assuming an ℓ = 3 angular momentum transfer and using the spherical iteration of the

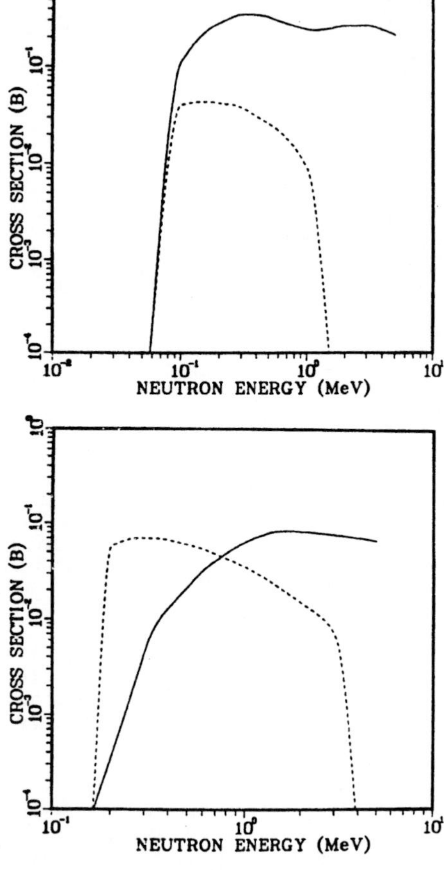

Fig. 3. Two calculated excitation functions (solid curves) for inelastic scattering from states in ^{239}Pu are compared to the ENDF/B-V evaluation (dashed curve). (a) 0.057 MeV $5/2^+$ state; (b) 0.164 MeV $9/2^+$ state.

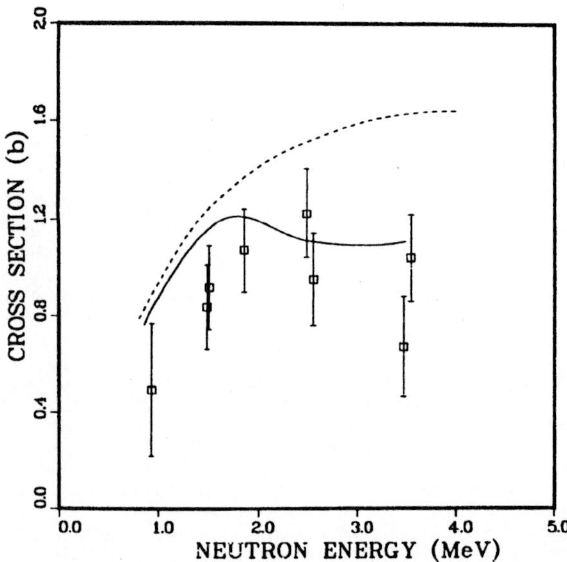

Fig. 4. Comparisons of our calculated inelastic cross sections (solid curve) for production of ^{239}Pu states above a given excitation energy (0.08-0.3 MeV over this neutron energy range) to cross sections inferred from measurements by Smith et al.[19] The dashed curve is ENDF/B-V.

Madland-Young neutron optical parameters.[22] To determine the β_3 deformation parameters necessary for normalization of the calculated DWBA results, we employed B(E3) values extracted from charged-particle reactions on similar octupole states in even plutonium nuclei[22] in the following expression

$$B(E\lambda) = (\frac{3}{4}\pi\ Ze\ R^{\lambda}\ A^{1/3\lambda})^2 \beta_{\lambda}^{\ 2} . \qquad (1)$$

Here R_0 and B(E3) values of 1.24 fm and 0.4 e^2b^3 were used, respectively. The β_3 values determined in this manner yielded DWBA contributions on the order of 5-10 mb for neutron energies between 2 and 4 MeV. These results lie factors of 5-10 lower than values one would infer from the (n,n'γ) measurements mentioned previously.

As a further test we analyzed (p,p') measurements[23,24] for scattering from collective states in ^{238}U, with particular emphasis on data available for the 0.731 MeV 3- octupole state. Although such data are sparse, we were able to compare the relative strength for excitation of this state to reasonably known cross sections for ground-state rotational band members. This provided an independent normalization method applicable to our $\ell = 3$ DWBA calculations, and and when inserted into Eq. (1) above produced a B(E3) value of 0.35 e^2b^3, which agrees with published values.[22,25] Figure 5 shows the results obtained when this normalization is applied to the calculated DWBA direct reaction contribution which is then added to the compound nucleus contribution (sum given by the solid line) for inelastic neutron scattering exciting the 0.731-MeV state in ^{238}U. The data are those of Olsen et al,[20] which were again inferred from (n,n'γ) cross sections. In the energy range where direct reaction contributions dominate, our calculations are in disagreement with these measurements as well as theoretical analyses[26] performed for similar data[21] where the relative coupling between bands was treated as an adjustable parameter. We therefore see a possible discrepancy between cross sections extracted from such (n,n'γ) measurements and cross sections obtained using information inferred from charged-particle data.

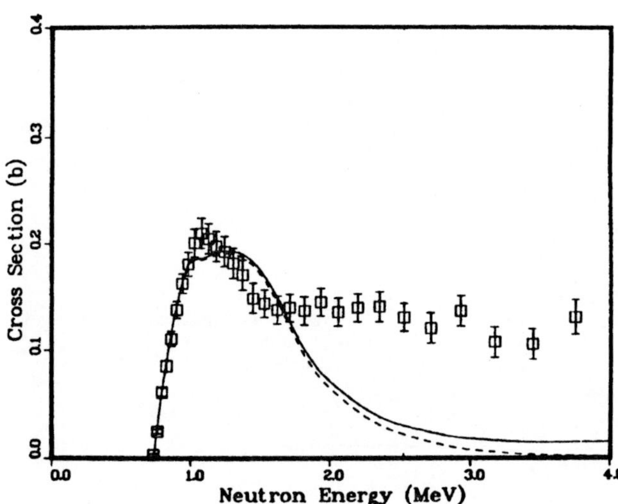

Fig. 5. Comparison of the calculated (n,n'γ) cross section for excitation of the 0.731 MeV 3- octupole state in ^{238}U to data of Olsen et al[20] as deduced from (n,n'γ) measurements. The dashed curve is the compound-nucleus (CN) contribution; the solid curve contains both CN and direct-reaction (DWBA) components.

Conclusion

Using nuclear-model calculations that incorporate realistic parameter sets, we are able to reproduce neutron inelastic scattering data for the ground state rotational band members of ^{239}Pu at energies less than 5 MeV. Similarly, our calculations of integrated inelastic cross sections agree well with such information inferred from recent experimental measurements. There do appear to be discrepancies between cross sections extracted from (n,n'γ) measurements for levels occupying higher collective bands and results obtained from charged-particle measurements.

References

1. E. D. Arthur, Nucl. Sci. Eng. 76, 137 (1980).

2. J. Raynal, "Optical Model and Coupled-Channel Calculations in Nuclear Physics," Int. Atomic Energy Agency report IAEA-SMR-9/8 (1970).

3. C. L. Dunford, "A Unified Model for Analysis of Compound-Nucleus Reactions," Atomics International report AI-AEC-12931 (1970).

4. C. M. Perey and F. G. Perey, Atomic Data and Nuclear Data Tables 17, 1 (1976).

5. G. Haouat et al, Nucl. Sci. Eng. 81, 491 (1982).

6. W. P. Poenitz et al, "Total Neutron Cross Sections of Heavy Nuclei," Proc. Intl. Conf. Nuclear Cross Sections for Technology, NBS Spec. Publ. 594, 698 (1980).

7. A. Gilbert and A.G.W. Cameron, Can. J. Phys. 43, 1446 (1965).

8. J. L. Cook et al, Aust. J. Phys. 20, 477 (1967).

9. P. M. Brink, thesis, Oxford University (1955) unpublished; P. Axel, Phys. Rev. 126, 671 (1962).

10. S. F. Mughabghab and D. I. Garber, "Neutron Cross Sections, Resonance Parameters," Brookhaven National Lab report BNL 325, 3rd Ed., V. 1 (1973).

11. E. D. Arthur, "Use of the Statistical Model for the Calculation of Compound Nucleus Contributions to Inelastic Scattering on Actinide Nuclei," Proc. Specialists Mtg. on Fast Neutron Scattering on Actinide Nuclei, NEANDC-158U, 145 (1982).

12. D. L. Hill and J. A. Wheeler, Phys. Rev. 89, 1102 (1953).

13. B. B. Back et al, Phys. Rev. C9, 1924 (1974).

14. H. C. Britt, "Experimental Survey of the Potential Energy Surfaces Associated with Fission," Proc. Symp. Physics and Chemistry of Fission, Julich, 1979, Int. Atomic Energy Agency report IAEA-SM-241/A1, Vol. 1, p 3 (1980).

15. S. Bjornholm et al, "Role of Symmetry of the Nuclear Shape in Rotational Contributions to Nuclear Level Densities," Proc. Symp. Physics and Chemistry of Fission, Rochester, 1973, Intl. Atomic Energy Agency report IAEA-SM-174/205, Vol. 1, p 367 (1973).

16. J. E. Lynn and B. B. Back, J. of Phys. A71, 395 (1974).

17. P. A. Moldauer, Phys. Rev. C11, 426 (1975).

18. J. E. Lynn, "Systematics for Neutron Reactions on Actinide Nuclei," Harwell report AERE-R-7468 (1974).

19. A. B. Smith and P. T. Guenther, "On the Neutron Inelastic-Scattering Cross Sections of ^{232}Th, ^{233}U, ^{235}U, ^{238}U, ^{239}Pu, and ^{240}Pu," Argonne National Laboratory report ANL/NDM-63 (1982).

20. D. K. Olsen et al, "Measurements of ^{238}U(n,n'γ) Cross Sections," Proc. Intl. Conf. Nuclear Cross Sections for Technology, NBS Spec. Publ. 594, p. 677 (1980).

21. A. Mittler et al, "Neutron Inelastic Scattering Cross Sections of ^{238}U via (n,n'γ)," Proc. Intl. Conf. Nuclear Cross Sections for Technology, NBS Spec. Publ. 594, p 680 (1980).

22. R. C. Thompson et al, Phys. Rev. C12, 1227 (1975).

23. C. H. King et al, Phys. Rev. C20,2084 (1979).

24. L. F. Hansen et al, Rev. C25, 189 (1982).

25. J. S. Boyno et al, Nucl. Phys. A209, 125 (1973).

26. E. Sheldon and D. W. S. Chan, "Evaluation of (n,n') Scattering Cross Sectons from 0.8 to 2.5 MeV for Higher Collective Bands of ^{232}Th and ^{238}U," Proc. Specialists Meeting on Fast Neutron Scattering on Actinide Nuclei, NEANDC-158U, p 169 (1982).

THE NUMBER OF DEGREES OF FREEDOM FOR STATISTICAL DISTRIBUTION OF S WAVE REDUCED NEUTRON WIDTH FOR SEVERAL NUCLEI

Zhao Zhixiang

Institute of Atomic Energy, Academia Sinica,
P.O.Box 275, Beijing, China

The least squares fit has been performed using chi-squared distribution function for all available evaluated data for s wave reduced neutron width of several nuclei. The number of degrees of freedom and average value have been obtained. The missing levels of weak s wave resonances and extra p wave levels have been taken into account, if any. For ^{75}As and ^{103}Rh, s wave population has been separated by Bayes' theorem before making fit. The results thus obatained are consistent with Porter-Thomas distribution, i.e., chi-squared distribution with $\nu=1$, as one would expect. It has not been found in this work that the number of degrees of freedom for the distribution of s wave reduced neutron width might greater than one as reported by H.C.Sharma et al..

[least squared fit, Γ_n^0 , $\bar{\Gamma}_n^0$, degrees of freedom]

The knowledge of statistical distribution of reduced neutron widths is of importance to the calculation of neutron cross section in unresolved resonance region. It also provides important experimental informations which are helpful improve the resonance theory. It is well known that the reduced neutron widths obey chi-squared distribution with degrees of freedom $\nu=1$, i.e., Porter-Thomas distribution [1]. However, the analysis of H.C.Sharma et al. [2] showed that the number of degrees of freedom are indeed 1 for most nuclei which have been analysed by them, but $\nu=2$ for ^{195}Pt, ^{155}Gd, ^{179}Hf, ^{239,240}Pu and ^{182}W and $\nu>2$ for ^{75}As, ^{103}Rh and ^{241}Pu.

The least squares fit has been performed by using chi-squared distribution function with parameters ν and $\bar{\Gamma}_n^0$ (or $2g\bar{\Gamma}_n^0$) for s wave reduced neutron widths of several nuclei, especially nuclei with $\nu\geqslant 2$ reported by Sharma et al.. Resonance parameters recommended by CNDC(Chinese Nuclear Data Center)[3] and ref.4 have been used in analysis. To eliminate in-fluence of unobserved weak s wave resonances, at first histogram of resonance number vs neutron energy was plotted as usual way. From histogram, chi-squared distribution fit was made in an energy region which corresponding to the linear region of no missing levels. When sample is too small for least squares fit due to the rejection of data above liear energy region, all data were used but fit is only performed above a given width (i.e., so called detection threshold). To take into account the possible extra p wave resonances, the detection threshold was never specified to zero for all nuclei analyed. For ^{75}As and ^{103}Rh, s wave populations have been separated by Bayes' theorem before making fit because l assignment has not given by ref.4 for many resonances. The results of these analyses are given in the table below.

From the table it may be seen that for most nuclei the number of degrees of freedom ν approximately equal to 1. For some nuclei, ν is approching to 2, however, test of goodness of fit shows that either fit with $\nu=1$ or 2 are good enough and the difference is statistically insignificant. The results obtained by this work are consistent with Porter-Thomas distribution, as expected.

The $\bar{\Gamma}_n^0$ (or $2g\bar{\Gamma}_n^0$) obtained from the fit and sample average $\bar{\Gamma}_n^0 = \frac{1}{m}\sum_{i=1}^{m}\Gamma_{ni}^0$ were also given in the table. The fitting value and sample average value are consistent within one standard error. It shows that the least squares fit method for obtaining average parameters is as feasible as the sample average method and the maximum likelihood method. In fact, the average parameter obtained by the least squares fit is sometimes more reliable than that obtained by other methods because it is insensitive to uncertainty of resonance widths and missing resonances.

K. H. Böckhoff (ed.), Nuclear Data for Science and Technology, 560–561.

Table Results of Analyses

Nuclei	J	N	ν*(meV)	$\overline{\Gamma}_n^0$*(meV)	Reduced χ^2	$\overline{\Gamma}_n^0=\frac{1}{m}\sum_{i=1}^{m}\Gamma_{ni}^0$	m
^{56}Fe	½	30	0.4	5000			
			1	7900	0.97	6760±2470	15
^{75}As	1,2	81	1.1	35			
			1	36	0.22	34.8±5.4	81
^{103}Rh	0,1	133	0.8	3.4			
			1	3.5	0.67	3.2±0.4	129
^{110}Cd	½	51	1.2	8.2			
			1	7.6	0.43	9.5±2.3	40
^{111}Cd	0,1	118	0.5	1.52			
			1	1.49	0.70	1.7±0.3	96
^{112}Cd	½	68	0.8	14.3			
			1	13.2	0.64	12.3±4.5	15
^{149}Sm	3,4	89	1.1	2.8			
			1	2.7	0.87	3.4±0.6	40
^{151}Sm	3,4	120	1.2	1.5			
			1	1.3	0.96	1.1±0.1	60
^{155}Gd	2,3	92	1.1	0.65			
			1	0.64	0.14	0.66±0.12	59
^{179}Hf	4	40	0.7	1.0			
			1	1.3	0.72	2.0±0.4	40
^{179}Hf	5	42	1.5	2.1			
			1	1.9	0.10	2.3±0.4	42
^{182}W	½	138	1.9	20.8			
			1	16.8	0.36	16.1±2.7	70
			2	21.3	0.26		
^{195}Pt	0	11	0.9	14.0			
			1	14.1	0.26	15.2±6.5	11
^{195}Pt	1	27	2.0	4.5	0.27		
			1	4.0	0.59	4.5±1.2	27
^{239}Pu	0	44	1.0	0.48	0.30	0.53±0.11	44
^{239}Pu	1	76	0.8	0.76			
			1	0.82	0.57	0.81±0.13	76
^{240}Pu	½	105	0.6	0.94			
			1	1.20	0.4	1.67±0.22	105
^{241}Pu	2	59	0.8	0.15			
			1	0.17	0.67	0.25±0.05	44
^{241}Pu	3	64	1.67	0.31			
			1	0.25	0.10	0.28±0.08	24
			2	0.33	0.15		

* The best fit values are given in first row for every nuclide.

References

1. C.E.Porter & R.G.Thomas, Phys. Rev., 104, 483 (1956)
2. H.C.Sharma et al., Inter. CONF. on Interactions of Neutrons with Nuclei, Lowell, p.1428 (1976)
3. CNDC, Neutron Resonance Theory and Resonance Parameter (1982)
4. S.F.Mughabghab & D.I.Garber, BNL-325, 3rd ed. (1973)

GENERALIZED OPTICAL MODEL AT LOW NEUTRON ENERGIES

M.B. FEDOROV

Institute for Nuclear Research, Academy of Sciences of the Ukrainian SSR
Kiev UkrSSR

The consideration of the role of collective states leads to the grounding of one phonon generalized optical model version with a preferential absorption in the channels corresponding to the excitation of 2+ states. The proposed model satisfactorily describes the mass dependence of neutron strength functions and the energy dependence of total cross sections without introducing any formal dependence of the absorption potential parameters on energy.

[optical model, neutrons, strength functions, absorption, vibrational states, reaction mechanism, E= 0 - 10 MeV]

Introduction

The progress in the nuclear physics and technology fields produces the increased requirements both to the accuracy of experimental neutron cross sections and to the improvement of their theoretical description. The necessity of the more detailed description of neutron-nucleus interactions naturally leads to the search of improved model versions because the predicting possibilities of generally used versions are insufficient. The typical example of optical model insufficiency is the lack of reliable systematics of mass and energy absorption potential dependence that reflects our imperfect knowledge of compound nucleus formation. So in the exceedingly important energy field of reactor spectrum 0-15 MeV it is necessary to choose different strengths of the imaginary part of potential for low and high energies, the introduction of individual energy dependence of potential depth parameter for different nuclides being insufficiently grounded from the physical point of view. It may be also said about the sensitivity of individual parameter values to shell structure of particular nuclei.

The difficulties of description of strength function mass dependence are well known. The necessity to perfect our understanding the mechanism of a neutron leaving the entrance channel is closely associated with the problem of investigation of the effect of compound system states directly coupled with this channel on the formation of imaginary part of optical potential. First of all this statement concerns the channels corresponding to the excitations of collective nucleus target states.

Theory

On the basis of a number of approachs (for instance [1]) it is possible to examine the potential properties of spherical optical model (SOM) which is equivalent to the generalized optical model (GOM) with the channel coupling.

Take for simplicity the two channel model

$$
\begin{aligned}
(H_a - E)u_a(r) + V_{ba}(r)u_b(r) &= 0, \\
(H_b - E)u_b(r) + V_{ab}(r)u_a(r) &= 0,
\end{aligned} \qquad (1)
$$

where H_i is a diagonal Hamiltonian with a complex optical model potential in the channel i=a,b, $u_a(r)$ - neutron radial wave function, $V_{ab}(r) = V_{ba}(r) = V(r)$ - real nondiagonal coupling potential. It is easy to show that in the entrance channel a the energy shift caused by the channel coupling near the resonance energy E_b is small, the absorption width of the equivalent SOM may be presented as

$$\tilde{\Gamma}_a^a = \Gamma_a^a + v^2 \Gamma_b / \left[(E_b - E)^2 + \Gamma_b^2/4 \right] \qquad (2)$$

where

$$v = \int_0^{\mathcal{R}} q_a(r) V(r) q_b(r) dr,$$

$q_i(r)$ are neutron radial wave functions in the interior of the nucleus $0 \leqslant r \leqslant \mathcal{R}$ which are normalized in this region, $\Gamma_i = \Gamma_i^a + \Gamma_i^p$; Γ_i^p - one particle width; Γ_i^a - absorption width caused by the imaginary potential part in channel i=a,b. In the case of channel b corresponding to the excitation of 2+ collective level matrix element v of nondiagonal coupling potential is easy to calculate [3]; in average its value is ~3 MeV.

The appearence of an additional resonance width in the entrance channel in consequence of channel coupling may be presented as the result of adding a nonlocal absorption potential

$$W_a(r,r') = -V_{ab}(r) q_b(r) \frac{i\Gamma_b/2}{(E_b - E)^2 + \Gamma_b^2/4} q_b(r') V_{ba}(r') \qquad (2a)$$

to the potential in the channel a. Obviously the second term in (2) and the potential (2a) are conditioned by the possibility of direct inelastic scattering with the exciting of level b if the corresponding channel is open ($\Gamma_b^p \neq 0$) and by the possibility of compound nucleus formation through the intermediate channel b.

It is evident from (2) that the contribution of the channel b to the absorption width $\tilde{\Gamma}_a^a$ is maximum near resonance energy E_b. With decreasing energy this contribution increases due to decrease in Γ_b^p. This conclusion correlates with the numerical calculation practice (for instance [4]).

In the traditional approach $\Gamma_a^a = \Gamma_b^a = \Gamma^a$ near E=0 ($\Gamma_b^p = 0$) formula (2) may be presented in a form

$$\tilde{\Gamma}_a^a = \Gamma^a \left\{ 1 + v^2 \left[(E_b - E)^2 + \Gamma^{a2}/4 \right]^{-1} \right\}, \qquad (3)$$

which is received in [5] on the basis of shell approach. From formula (3) there can be received the condition of the essential role of collective 2_1+ state in the process of compound nucleus formation [5]. For s-wave in the entrance channel this condition may be written as

K. H. Böckhoff (ed.), Nuclear Data for Science and Technology, 562–564.

$$\xi = (22/\Gamma^{a2}) \gtrsim 1 . \qquad (4)$$

In a traditional GOM version with the parameters selected at moderate and high energies Γ^a equals $\sim$10 MeV and $\xi \sim 0.2$ so that collective channel coupling does not play an essential role in the imaginary potential part formation. In fact the numerical calculations show that both the GOM and the SOM provide the same total cross sections values if the GOM absorption potential strength is reduced by no more than 15-20% versus SOM parameter [6].

It has been shown in [2] that the consideration of collective state effects on the imaginary optical potential part leads to different absorption potentials in the different directly considered channels. Thus in the simple vibrational approximation there is no direct coupling of the elastic scattering and multiphonon channels. Therefore the consideration of averaged effects of these permits to justify the one phonon GOM version with a preferential absorption in the one phonon channels directly coupled with multiphonon ones.

In the case of $\Gamma_b^a > \Gamma_a^a$ the condition (3) must be generalized as

$$(\Gamma_b^a / \Gamma_a^a)(22/\Gamma_b^{a2}) \gtrsim 1 . \qquad (3a)$$

It is seen from (3a) that if Γ_b^a / Γ_a^a is sufficiently large the condition of the essential role of collective 2_1+ states is easily fulfilled.

The choice of small absorption potential strength $W_s^{(0)}$ for the entrance channel and corresponding strength $W_s^{(1)} > W_s^{(0)}$ for one phonon channels leads to some obvious consequences:

far from one particle resonances the equivalent SOM absorption potential strength is lagely determined by the small absorption potential strength for the GOM entrance channel resulting in the absorption cross section minimum deepening;

the small values $W_s^{(0)}$ have the most essential influence on cross sections at low energies when the one particle width is small and the total width is almost exclusively determined by the absorption width;

if the proposed GOM version is correct the predicted absorption neutron cross sections should reveal noticeable individual variations caused by the absorption potential dependence on individual collective nucleus properties.

Calculation results

The possibility of obtaining deepened strength function minima in the frame of proposed GOM version follows from (2a) accounting the fundamental property of grouping the one particle levels with the same parity in a relatively small energy range. Therefore the collective channel contribution to the SOM imaginary potential part is made primarily near the entrance channel shape resonance. The typical experimental intervals of energy averaging are not so large to provide the same absorption potential near and far from one particle resonances, therefore the usually used SOM version with W_s=const is a highly crude approximation.

The choice of GOM surface absorption potential parameters shows that at low energies the averaged mass dependence of neutron strength functions can be described satisfactirily if $W_s^{(0)}$=2 MeV and $W_s^{(1)}$=13 MeV (fig. 1 and 2). The coupling potentials were calculated on the basis of the simple vibra-

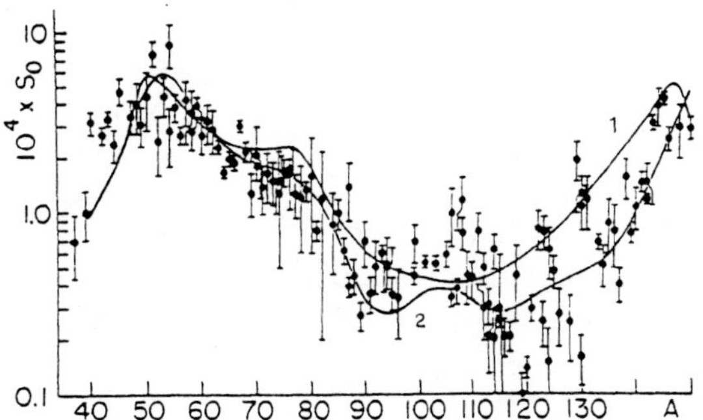

Fig. 1 Comparison of theoretical and experimental values of the s-wave neutron strength function. The curve 1 represents calculations of ref. 7, the curve 2 is based on GOM calculations of the present work.

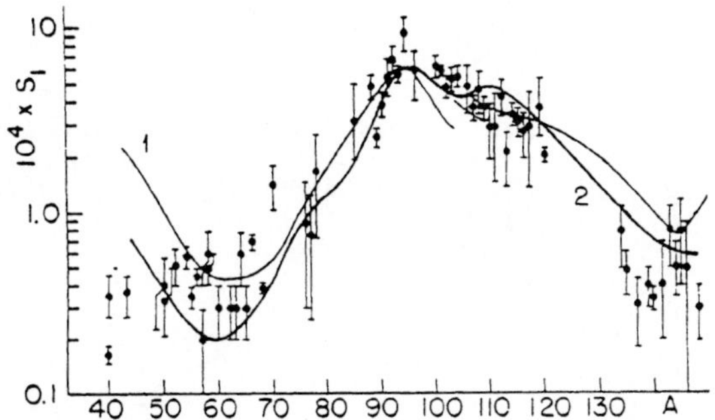

Fig. 2 Comparison of theoretical and experimental values of the p-wave neutron strength function. The symbols are the same as in fig. 1

tion model [3] and were chosen to be real (this choice is grounded in [1]). Another parameters are: the strength of Wood-Saxon real potential V=52.47-29.2(N-Z)/A MeV, its diffuseness a=0.46+0.8(N-Z)/A [8], the radius $R=1.25A^{1/3}$ Fm (the same for both the real and imaginary potential parts), the diffuseness of the imaginary part of Wood-Saxon derivative form a_W=0.47 Fm, the depth of spin-orbit potential of Thomas form V_{so}=7.5 MeV. The averaged mass dependence of quadrupole deformation parameters was determined basing on the data [9].

Figures 1 and 2 show that GOM with a preferential absorbtion in the one phonon channels enables to reflect some peculiarities of neutron strength functions in particular the depth of minima, the presence of minimum near A=120 and small maximum near A=110 for s-wave functions.

The GOM parameters fixed at the strength function analysis lead to satisfactory agreement with the data on total cross sections and elastic differential cross sections at MeV energies [10] and also improve describing the mass dependence of total cross sections at low energies near 3p resonance [11].

Fig.3 illustrates the possibilities of GOM with a small absorption in the entrance channel representing as an example the energy dependence of total neutron cross section σ_t for natural nickel. The GOM parameters $W_s^{(0)}$=3 MeV and $W_s^{(1)}$=13 MeV were used leading to the satisfactory agreement of the

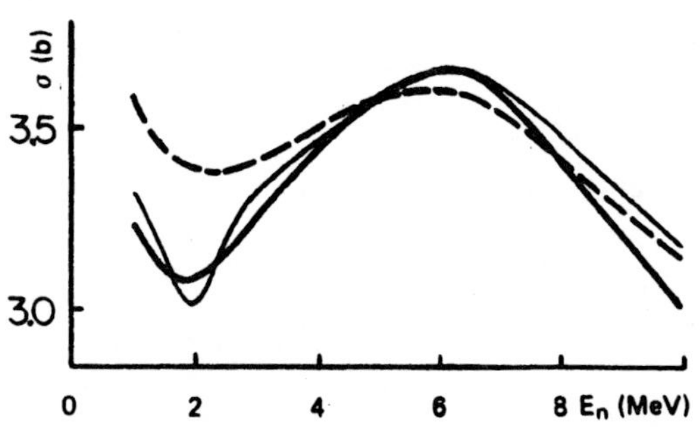

Fig. 3 Dependence of total neutron cross section on the energy for natural nickel. The thin solid curve shows the recomended behavior of ref. 12, the dashed curve represents the SOM calculations of ref. 12, the thick solid curve represents the GOM calculations of the present work.

theoretical s-wave strength functions for ^{58}Ni, ^{60}Ni with the experimental data. So fixing the absorption parameters it is possible to describe satisfactorily the averaged energy behavior of σ_t for ^{58}Ni, ^{60}Ni, ^{Nat}Ni from 0.4 MeV to 7 MeV reflecting for all this the σ_t values in the minimum near 2 MeV. Disagreements appear at higher energies indicating apparently to the necessity of introducing an additional volume absorpting potential. The same result has been obtained also for titanium and chromium. In general the application of generally used model versions for these nuclei requires the introduction of artificial potential parameter dependence on the energy.

The choice of a small absorption potential strength in the GOM entrance channel leads to the reduction in SOM absorption potential strength and to corresponding decrease in absorption cross section for nearly magic nuclei. Such a decrease is caused by the small values of quadrupole deformation parameters for these nuclei and is revealed at the phenomenological data analysis.

Conclusions

The choice of a preferential absorption in the channels corresponding to the excitation of 2_1+ target nucleus states suggests a great probability of compound nucleus formation through the intermediate stage of one particle - collective nature. Such a mechanism leads to the great relative contribution of the one particle - phonon states to the total absorption optical potential. This contribution is greater than it is directly or indirectly supposed in the generally used optical model.

The numerical calculations compared with the experimental data show that for spherical nuclei the suggested approach leads on the whole to more satisfactory results as compared to the traditional optical model versions.

References

1. М.Б.Фёдоров,Спектрометрия нейтронов средних энергий, Наукова думка, Киев, 1979.
2. М.Б.Фёдоров,Ядерная физика 32,392(1980).
3. T.Tamura, Rev.Mod.Phys. 37, 679 (1965).
4. E.Zijp,C.C.Jonker,Nucl.Phys.A 222,93 (1974).
5. Б.В.Адамчук,В.К.Сироткин,Препр.ИАЭ2560,1975
6. В.М.Бычков, В.В.Возяков, А.Г.Довбенко, А.В.Игнатюк, В.П.Лунёв, В.Г.Проняев, В.С.Шорин - В кн.: Нейтронная физика (Материалы 2-й Всесоюзной конференции по нейтронной физике, Киев, 1973), Обнинск, 1974, часть I, стр. 316.
7. S.F.Mughabghab,M.Divadeenam,N.E.Holden, Neutron cross sections, v.1, Neutron resonance parameters and thermal cross sections, Part A Z=1-60, Academic Press, 1981.
8. М.В.Пасечник, М.Б.Фёдоров, Т.И.Яковенко, Укр.физ.журнал 20, 388 (1975).
9. P.H.Stelson, L.Grodzins, Nucl.Data A 1, 21 (1965).
10. М.Б.Фёдоров - В кн.: Нейтронная физика (Материалы 5-й Всесоюзной конференции по нейтронной физике, Киев, 1980), Москва, ЦНИИатоминформ, 1980, часть I, стр. 299.
11. М.В.Пасечник, М.Б.Фёдоров, В.Д.Овдиенко, Г.А.Сметанин, Т.И.Яковенко - В кн.: Нейтронная физика (Материалы 5-й Всесоюзной конференции по нейтронной физике, Киев, 1980), Москва, ЦНИИатоминформ, 1980, часть I, стр. 304.
12. В.М.Бычков, В.Н.Манохин, В.Г.Проняев, В.И.Попов, А.Б.Пащенко - В кн.: Нейтронная физика (Материалы 3-й Всесоюзной конференции по нейтронной физике, Киев, 1975), Москва, ЦНИИатоминформ, 1976, часть I, стр. 160.

MICROSCOPIC OPTICAL POTENTIAL CALCULATION WITH SKYRME FORCES

Shen Qingbiao, Tian Ye, Zhang Jingshang, and Zhuo Yizhong
Institute of Atomic Energy, Academia Sinica, Beijing, China

A calculation method of the microscopic optical potential based on various kinds of Skyrme forces has been introduced. The first- and second-order mass operators in asymmetric nuclear matter have been derived and the real and imaginary parts of the optical potential for finite nuclei have been obtained by applying a local density approximation. The spin-orbit potential has also been obtained approximately. In most cases the expressions are analytical. The calculated results for some sets of Skyrme forces are in good agreement with the empirical values and experiments.

[microscopic optical potential, Skyrme forces, nuclear matter approach]

1. Introduction

The optical model is one of the most fundamental theoretical tools in analysis of nuclear reaction data. The phenomenological optical model with many adjustable parameters can reproduce the experimental data quite well, but cannot, however, predict the unknown data with certainty. Thus, the derivation of the optical potential from the more basic theory is one of the most important problems in nuclear theory which is of both theoretical and practical interest.

As early as 1959 Bell and Squires[1] showed that from the point of view of the many-body theory the optical potential can be identified with the mass operator of the one-particle Green function. This identification has made it possible to utilize the many-body theory technique to obtain a microscopic optical potential (MOP) without any free parameters. In practical application, however, the mass operator cannot be calculated exactly since it would mean to solve the many-body problem itself. Therefore this theory remained as a formal one for a long time in the past. Only in recent years after some appropriate approximations have been developed the actual calculations of the MOP have become possible. Basically two different approaches have been formulated. One is the "nuclear matter approach"[2-4] in which one starts from the realistic nuclear force to calculate the mass operator by Brueckner-Hartree-Fock (BHF) approximation in nuclear matter, then the MOP for finite nuclei is obtained by making a local density approximation (LDA). Thus the MOP for all target nuclei can be obtained in principle. Another approach is the "nuclear structure approach"[5,6] in which one only applies the two-body effective nuclear force by the random phase approximation (RPA) to calculate the MOP for some specific nuclei. Thus this approach includes specific features of the nuclear structure of the target nucleus, while the "nuclear matter approach" includes the effects of the nuclear structure only in an average way (via the LDA).

Both approaches have their merits and linitations. Since we are more interested in the global properties of the optical potential, encouraged by the success of the nuclear matter approach and the success of the phenomenological Skyrme interactions, we adopt a simpler and more economical way for MOP calculation. We calculate the mass operator of the one-particle Green function only up to the second order in nuclear matter by using the Skyrme interactions. The optical potential for the finite nuclei is then obtained as usual by LDA[7-9]. Since the Skyrme interactions can be viewed as an effective G matrices in the Hartree-Fock (HF) calculations[10,11], we let only the first-order mass operator $M^{(1)}$ represent the real part of the optical potential and consider the imaginary part of the second-order mass operator $M^{(2)}$ as the imaginary part of the optical potential[12]. The spin-orbit potential has also been obtained approximately. Of course, this approach is not as basic as the MOP in certain sense, but by using the Skyrme interaction we can get very simple, in most cases analytical, expressions for the optical potential, which is physically more transparent and easily to be applied. In addition, at present there exist many sets of Skyrme interactions, most of which are quite successful in the calculations of the average ground state properties of nuclei. As is known, the different Skyrme interactions give the same good agreement for HF calculations but yield quite different results for excited states[13]. Thus from another point of view our work might provide a good testing ground for different Skyrme interactions in reproducing not only the nuclear ground state properties but also the excited state properties of nucleus. So we have performed the calculations and comparisons with nearly all kinds of the Skyrme interactions availabele. Our calculated results show that there indeed exist some sets of Skyrme interactions with which we can obtain MOP which are in good agreement with empirical data and the phenomenological optical potential POP and comparable with the MOP based on realistic nucleon-nucleon interactions.

In Sect.2 we present the different Skyrme forces and their parameters. The expressions of the first-order and second-order mass operators including two-body and three-body interactions are given in Sect.3. In Sect.4 we deal with the optical potential for finite nuclei by applying LDA. The calculation results and analysis are given in Sect.5. Finally, some tentative conclusions are presented in Sect.6.

2. The Skyrme Forces

At present there exist many sets of Skyrme interactions. We summarize them as follows.

2.1 The Conventional Skyrme Force

The two-body interactions of the conventional Skyrme force are as follows[10]:

$$V_{12}(\vec{R},\vec{r})=t_0(1+x_0P_\sigma)\delta(\vec{r})+\frac{1}{2}t_1(\vec{k}'^2\delta(\vec{r})+\delta(\vec{r})\vec{k}^2)+t_2\vec{k}'\cdot\delta(\vec{r})\vec{k}+iW_0(\vec{\sigma}_1+\vec{\sigma}_2)\cdot\vec{k}'\times\delta(\vec{r})\vec{k} \quad (1)$$

where

$$\vec{r}=\vec{r}_1-\vec{r}_2\ , \qquad \vec{R}=\frac{1}{2}(\vec{r}_1+\vec{r}_2). \quad (2)$$

The relative momentum operators are

$$\vec{k}=\frac{1}{2i}(\vec{\nabla}_1-\vec{\nabla}_2) \quad \text{acting on the right}$$
$$\vec{k}'=-\frac{1}{2i}(\vec{\nabla}_1-\vec{\nabla}_2) \quad \text{acting on the left}\ . \quad (3)$$

P_σ is the spin exchange operator and $\vec{\sigma}_i$ are the Pauli spin matrices. The three-body interaction has the following form:

$$W_{123}(\vec{r}_1,\vec{r}_2,\vec{r}_3)=t_3\delta(\vec{r}_1-\vec{r}_2)\delta(\vec{r}_2-\vec{r}_3). \quad (4)$$

In (1) and (4) t_0, t_1, t_2, t_3, x_0 and W_0 are

K. H. Böckhoff (ed.), Nuclear Data for Science and Technology, 565–573.

Table 1. The parameters of the conventional Skyrme force

	t_0 MeV.fm^3	t_1 MeV.fm^5	t_2 MeV.fm^5	t_3 MeV.fm^6	x_0	W_0 MeV.fm^5
SII	−1,169.9	585.6	−27.1	9,331.1	0.34	105
SIII	−1,128.75	395.0	−95.0	14,000	0.45	120
SIV	−1,205.6	765.0	35.0	5,000	0.05	150
SV	−1,248.29	970.56	107.22	0	−0.17	150
SVI	−1,101.81	271.67	−138.33	17,000	0.583	115

Table 2. The parameters of the extended Skyrme force

	t_0 MeV.fm^3	t_1 MeV.fm^5	t_2 MeV.fm^5	t_3	t_4 MeV.fm^8	x_0	x_1	x_2	x_3	x_4	W_0 MeV.fm^5	α	Ref.
GS1	−1268	887	−77.3	14485	−1853	0.150	0	0	1	1	105	1	13
GS2	−1177	670	−49.7	11054	−775	0.124	0	0	1	1	105	1	13
GS3	−1037	336	−7.3	5774	883	0.074	0	0	1	1	105	1	13
GS4	−1242	760	−146.2	19362	−2157	0.206	0	0	1	1	105	1	13
GS5	−1152	543	−118.2	15989	−1079	0.182	0	0	1	1	105	1	13
GS6	−1012	209	−76.3	10619	579	0.139	0	0	1	1	105	1	13
SGOI	−1089	558.8	−83.7	8272	0	0.412	0	0	0	0	130	1	16
SGOII	−2248	558.8	−83.7	11224	0	0.715	0	0	0	0	130	1/6	16
Ska	−1602.78	570.88	−67.70	8000	0	−0.02	0	0	−0.286	0	125	1/3	17
Skb	−1602.78	570.88	−67.70	8000	0	−0.165	0	0	−0.286	0	125	1/3	17
SKM	−2645	385	−120	15595	0	0.09	0	0	0	0	130	1/6	18
SGI	−1603	515.9	84.5	8000	0	−0.02	−0.5	−1.731	0.1381	0	115	1/3	19
SGII	−2645	340	−41.9	15595	0	0.09	−0.0588	1.425	0.06044	0	105	1/6	19

t_5(MeV.fm^8)=0 $\quad x_5$=0

the parameters[10,14] (see Table 1).

2.2 The Generalized Skyrme Force

The generalized Skyrme force is only a two-body effective interaction described by a density- and momentum-dependent delta function, such as[13]

$$V_{12}(\vec{R},\vec{r})=t_0(1+x_0P_\sigma)\delta(\vec{r})+\frac{1}{6}t_3(1+x_3P_\sigma)\rho(\vec{R})\delta(\vec{r}) +\frac{1}{2}t_1(1+x_1P_\sigma)\,(\vec{k}'^2\delta(\vec{r})+\delta(\vec{r})\vec{k}^2) +\frac{1}{2}t_4(1+x_4P_\sigma)\,(\vec{k}'^2\rho(\vec{R})\delta(\vec{r})+\delta(\vec{r})\rho(\vec{R})\vec{k}^2) +t_2(1+x_2P_\sigma)\vec{k}'\cdot\delta(\vec{r})\vec{k}+t_5(1+x_5P_\sigma)\vec{k}'\cdot\rho(\vec{R})\delta(\vec{r})\vec{k} +iW_0(\vec{\sigma}_1+\vec{\sigma}_2)\cdot\vec{k}'\times\delta(\vec{r})\vec{k}\,. \tag{5}$$

The conventional Skyrme interactions include a three-body term to simulate the density dependence. This term is equivalent to a two-body density-dependent interaction

$$V_{12}=\frac{1}{6}t_3(1+P_\sigma)\rho(\vec{R})\delta(\vec{r}) \tag{6}$$

when used in HF calculations of doubly even nuclei, as shown by Vautherin and Brink[10]. The three-body term is, however, not desirable, because this term produces unstable spin-saturated HF ground states[15]. Thus a density-dependent version with two-body forces is only used.

2.3 The Modified Skyrme Force

In order to overcome the shortcomings of the conventional Skyrme forces, in recent years another form of Skyrme force called as modified Skyrme interaction has been introduced[16-19]. The form of the modified Skyrme force is

$$V_{12}(\vec{R},\vec{r})=t_0(1+x_0P_\sigma)\delta(\vec{r})+\frac{1}{6}t_3(1+x_3P_\sigma)\rho^\alpha(\vec{R})\delta(\vec{r}) +\frac{1}{2}t_1(1+x_1P_\sigma)\,(\vec{k}'^2\delta(\vec{r})+\delta(\vec{r})\vec{k}^2)+t_2(1+x_2P_\sigma)\vec{k}' \cdot\delta(\vec{r})\vec{k}+iW_0(\vec{\sigma}_1+\vec{\sigma}_2)\cdot\vec{k}'\times\delta(\vec{r})\vec{k}\,. \tag{7}$$

Both the generalized Skyrme force (5) and the modified Skyrme force (7) can be written in a unified form as follows:

$$V_{12}(\vec{R},\vec{r})=t_0(1+x_0P_\sigma)\delta(\vec{r})+\frac{1}{6}t_3(1+x_3P_\sigma)\rho^\alpha(\vec{R})\delta(\vec{r}) +\frac{1}{2}t_1(1+x_1P_\sigma)\,(\vec{k}'^2\delta(\vec{r})+\delta(\vec{r})\vec{k}^2) +\frac{1}{2}t_4(1+x_4P_\sigma)\,(\vec{k}'^2\rho(\vec{R})\delta(\vec{r})+\delta(\vec{r})\rho(\vec{R})\vec{k}^2) +t_2(1+x_2P_\sigma)\vec{k}'\cdot\delta(\vec{r})\vec{k}+t_5(1+x_5P_\sigma)\vec{k}'\cdot\rho(\vec{R})\delta(\vec{r})\vec{k} +iW_0(\vec{\sigma}_1+\vec{\sigma}_2)\cdot\vec{k}'\times\delta(\vec{r})\vec{k}\,. \tag{8}$$

We call (8) the extended Skyrme force. The sets of its parameters known to us so far are listed in Table 2.

3. Mass Operator

The Hamiltonian simultaneously composed of the two-body and three-body interactions can be written as

$$H=H_0+H_1 \tag{9}$$

where

$$H_0=\sum_i(t_i+U_i) \tag{10}$$

$$H_1=\frac{1}{2!}\sum_{i\neq j}V_{ij}+\frac{1}{3!}\sum_{i\neq j\neq k}W_{ijk}-\sum_i U_i\,. \tag{11}$$

Here H_0 is the single-particle Hamiltonian, H_1 is the residual interaction, and U_i is the mean field of the single particle.

Let us write the single-particle Green function $G_{\alpha\beta}(t_1-t_2)$ and expand it into the perturbation series, then perform the Fourier transformation and obtain

$$G_{\alpha\beta}(\omega)=\delta_{\alpha\beta}G^{(0)}_{\alpha}(\omega)+G^{(1)}_{\alpha\beta}(\omega)+G^{(2)}_{\alpha\beta}(\omega)+\cdots \tag{12}$$

which satisfies the Dyson equation

$$G_{\alpha\beta}(\omega)=\delta_{\alpha\beta}G^{(0)}_{\alpha}(\omega)+G^{(0)}_{\alpha}(\omega)\sum_\gamma\left[U_{\alpha\gamma}-M_{\alpha\gamma}(\omega)\right]G_{\gamma\beta}(\omega) \tag{13}$$

where $U_{\alpha\gamma}$ is the mean field, $M_{\alpha\gamma}(\omega)$ is the mass operator, and

$$M_{\alpha\gamma}(\omega)=M^{(1)}_{\alpha\gamma}+M^{(2)}_{\alpha\gamma}(\omega)+\cdots\,. \tag{14}$$

The mass operator $M_{\alpha\alpha}(E)$ can be identified with the optical model potential for the scattering process with the energy of E.

The HF mean field U is expressed as

$$U_{\alpha\alpha}=M^{(1)}_{\alpha\alpha}=\sum_\rho V_{\alpha\rho,\alpha\rho}\,n_\rho+\frac{1}{2}\sum_{\rho\sigma}W_{\alpha\rho\sigma,\alpha\rho\sigma}\,n_\rho n_\sigma \tag{15}$$

where

$$n_\rho=\begin{cases}1 & \text{below the Fermi surface}\\ 0 & \text{above the Fermi surface.}\end{cases} \tag{16}$$

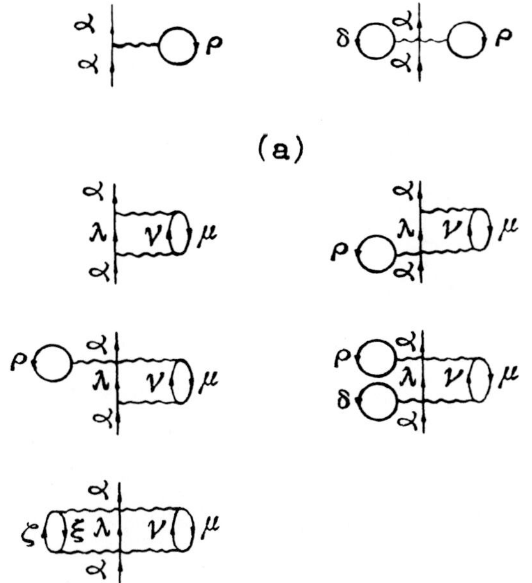

Fig.1. Feynman diagrams of the one-particle Green functions discussed in the text. (a) First-order diagrams. (b) Second-order diagrams.

In the right-hand side of (15) the first term comes from the contribution of the two-body interactions and the second term is that of the three-body interactions (Fig.1 (a)). Their matrix elements are, respectively,

$$V_{\alpha\rho,\alpha\rho} = \langle \alpha\rho | V | \alpha\rho \rangle_A \tag{17}$$

$$W_{\alpha\rho\delta,\alpha\rho\delta} = \langle \alpha\rho\delta | W | \alpha\rho\delta \rangle_A \tag{18}$$

where A denotes the antisymmetrization. The HF mean field gives the real part of the MOP and the second-order diagrams are the lowest order diagrams to contribute to the imaginary part of the MOP. We only need to calculate the five second-order diagrams (Fig.1 (b)). Their mass operators (retarded part) are as follows:

$$M^{(2)}_{\alpha\alpha}(E) = \sum_{\mu\lambda\nu} \frac{A_{\alpha\mu\lambda\nu}}{E+\varepsilon_\mu-\varepsilon_\lambda-\varepsilon_\nu+i\eta} n_\mu (1-n_\lambda)(1-n_\nu) + \sum_{\mu\xi\lambda\nu\zeta} \frac{B_{\alpha\mu\xi\lambda\nu\zeta}}{E+\varepsilon_\mu+\varepsilon_\xi-\varepsilon_\lambda-\varepsilon_\nu-\varepsilon_\zeta+i\eta} n_\mu n_\xi \cdot (1-n_\lambda)(1-n_\nu)(1-n_\zeta) \tag{19}$$

where

$$A_{\alpha\mu\lambda\nu} = \frac{1}{2} V_{\alpha\mu,\lambda\nu} V_{\lambda\nu,\alpha\mu} + \frac{1}{2}\Big[V_{\alpha\mu,\lambda\nu}\big(\sum_\rho W_{\lambda\nu\rho,\alpha\mu\rho} n_\rho\big) + \big(\sum_\rho W_{\alpha\mu\rho,\lambda\nu\rho} n_\rho\big) V_{\lambda\nu,\alpha\mu}\Big] + \frac{1}{2}\big(\sum_\rho W_{\alpha\mu\rho,\lambda\nu\rho} n_\rho\big)\big(\sum_\delta W_{\lambda\nu\delta,\alpha\mu\delta} n_\delta\big) \tag{20}$$

$$B_{\alpha\mu\xi\lambda\nu\zeta} = \frac{1}{12} W_{\alpha\mu\xi,\lambda\nu\zeta} W_{\lambda\nu\zeta,\alpha\mu\xi} \; . \tag{21}$$

In (19) the first term is the contribution of the diagrams with 2p-1h in the intermediate processes and the second term is that of the fifth diagram with 3p-2h in the intermediate process.

4. Formulation of the Microscopic Optical Potential

In the Fermi gas model for symmetric nuclear matter which has equal numbers of neutrons and protons, the density ρ of the nuclear matter is related to the Fermi momentum K_F by

$$\rho = \frac{2}{3\pi^2} K_F^3 \; . \tag{22}$$

But for asymmetric nuclear matter where the neutrons and protons have different Fermi momentum K_n and K_p, the neutron density ρ_n and proton density ρ_p are expressed by

$$\rho_n = \frac{1}{3\pi^2} K_n^3 \; , \quad \rho_p = \frac{1}{3\pi^2} K_p^3 \; . \tag{23}$$

Let us define the asymmetric parameter α_0 as

$$\alpha_0 = (\rho_n - \rho_p)/\rho \; . \tag{24}$$

Then (23) can be written as

$$\rho_n = \frac{1}{2}(1+\alpha_0) \; , \quad \rho_p = \frac{1}{2}(1-\alpha_0) \tag{25}$$

and

$$K_n = (1+\alpha_0)^{1/3} K_F \; , \quad K_p = (1-\alpha_0)^{1/3} K_F \; . \tag{26}$$

In order to simplify the calculation we use the nuclear matter approach, and the wave funtion of nucleon α is just the plane wave function

$$\psi_\alpha(\vec{r}) = \frac{1}{\sqrt{\Omega}} e^{i\vec{k}_\alpha \cdot \vec{r}} \chi_{\sigma_\alpha} \chi_{\tau_\alpha} \tag{27}$$

where χ_{σ_α} and χ_{τ_α} are the spin and isospin wave functions, respectively, and Ω is the volume.

Our purpose is to investigate the MOP for even-even nuclei with the asymmetric nuclear matter approximation. Since the expressions of the MOP for both conventional and extended forces are different, they are described in 4.1 and 4.2, respectively.

4.1 Case of Conventional Skyrme Force

The real part of the MOP for a nucleon has the form of

$$V_{\tau_\alpha} = \frac{m^*_{\tau_\alpha}}{m_{\tau_\alpha}} \Big\{ t_0 \big[(1+\frac{x_0}{2})\rho - (x_0+\frac{1}{2})\rho_{\tau_\alpha}\big] + \big[\frac{1}{4}(t_1+t_2)\rho - \frac{1}{8}(t_1-t_2)\rho_{\tau_\alpha}\big] \frac{2m_{\tau_\alpha}}{\hbar^2}\big(\frac{M}{M+m_{\tau_\alpha}} E_L - V_c\big) + \frac{1}{40\pi^2}(t_1+3t_2) K^5_{\tau_\alpha} + \frac{1}{20\pi^2}(t_1+t_2)(2K_F^3 - K^3_{\tau_\alpha})^{5/3} + \frac{t_3}{4}(\rho^2 - \rho^2_{\tau_\alpha}) \Big\} \tag{28}$$

where the effective mass is

$$\frac{m^*_{\tau_\alpha}}{m_{\tau_\alpha}} = \frac{1}{1+\frac{2m_{\tau_\alpha}}{\hbar^2}\big[\frac{1}{4}(t_1+t_2)\rho - \frac{1}{8}(t_1-t_2)\rho_{\tau_\alpha}\big]} \tag{29}$$

τ_α refers to the neutron or proton. m_{τ_α} and M are the mass of the incident particle and the target nucleus, respectively, E_L is the energy of the incident nucleon in the laboratory frame, V_c is the Coulomb potential and $V_c=0$ for the neutron. (28) shows clearly that the real part of the MOP is linearly dependent on the incident energy.

According to the formula of the principal value integral

$$\frac{1}{x+i\eta} = P\frac{1}{x} - i\pi\delta(x) \tag{30}$$

one can derive the imaginary part of the MOP from the second-order mass operator $M^{(2)}_{\alpha\alpha}(E)$ from (19) . The contributions of the first four second-order diagrams are

$$W_A = -\frac{1}{64\pi^5} \sum_{i=1}^{7} W_i \tag{31}$$

where

$$W_1 = 2\big[(1+x_0+x_0^2)t_0^2 + (1+\frac{1}{2}x_0)t_0 t_3 \rho + \frac{1}{4}t_3^2\rho^2\big] \cdot \big[I_1(\tau_\alpha) + I_1(\tau_\alpha,-\tau_\alpha)\big] - \big[(1+4x_0+x_0^2)t_0^2 + (1+2x_0)t_0 t_3\rho + \frac{1}{4}t_3^2\rho^2\big] I_1(\tau_\alpha)$$

$$W_2 = \frac{t_1}{2}\Big\{2\big[(1+\frac{1}{2}x_0)t_0 + \frac{1}{2}t_3\rho\big]\big[I_2(\tau_\alpha) + I_2(\tau_\alpha,-\tau_\alpha)\big] - \big[(1+2x_0)t_0 + \frac{1}{2}t_3\rho\big] I_2(\tau_\alpha)\Big\}$$

$$W_3 = \frac{t_1^2}{16}\big[I_3(\tau_\alpha) + 2I_3(\tau_\alpha,-\tau_\alpha)\big]$$

$$W_4 = t_2\big[(1+\frac{x_0}{2})t_0 + \frac{1}{2}t_3\rho\big]\big[I_4(\tau_\alpha) + I_4(\tau_\alpha,-\tau_\alpha)\big]$$

$$W_5 = \frac{t_1 t_2}{4}\big[I_5(\tau_\alpha) + I_5(\tau_\alpha,-\tau_\alpha)\big]$$

$$W_6=\frac{t_2^2}{16}\left[3I_6(\tau_\alpha)+2I_6(\tau_\alpha,-\tau_\alpha)\right]$$

$$W_7=\frac{W_0^2}{4}\left[2I_7(\tau_\alpha)+I_7(\tau_\alpha,-\tau_\alpha)\right] \quad . \tag{32}$$

The explicit expressions of $I_i(\tau_\alpha)$ and $I_i(\tau_\alpha,-\tau_\alpha)$ can be found in appendix 1 and 3 of reference 9.

In the calculation of the symmetric nuclear matter approximation it is found that the contribution of the last diagram is very small as compared with those of the other diagrams. It is known that the asymmetic parameter α_0 is a small quantity for most of the stable nuclei, so the correction of asymmetric nuclar matter to the symmetric nuclei is also small. Therefore, it is believed that such correction in the last diagram can be ignored and treated just as the symmetric nuclear matter which is

$$W_B=-3t_2^2\frac{\pi}{3(2\pi)^{12}}\int d\vec{K}_\mu d\vec{K}_\xi d\vec{K}_\lambda d\vec{K}_\nu d\vec{K}_\zeta$$
$$\cdot\delta(E+\varepsilon_\mu+\varepsilon_\xi-\varepsilon_\lambda-\varepsilon_\nu-\varepsilon_\zeta)\delta(\vec{K}_\alpha+\vec{K}_\mu+\vec{K}_\xi-\vec{K}_\lambda-\vec{K}_\nu-\vec{K}_\zeta) \tag{33}$$

where the ranges of integration are confined by $K_\mu, K_\xi \leq K_F$ and $K_\lambda, K_\nu, K_\zeta \geq K_F$. After coordinate transformation the integral can be reduced to a fivefold integral and it is carried out numerically.

The total imaginary part of the MOP should be

$$W_{\tau_\alpha}=W_A+W_B \quad . \tag{34}$$

4.2 Case of Extended Skyrme Force

The real part of MOP obtained from (8) is as follows:

$$V_{\tau_\alpha}=\frac{m^*_{\tau_\alpha}}{m_{\tau_\alpha}}\Big\{t_0\left[(1+\frac{x_0}{2})\rho-(x_0+\frac{1}{2})\rho_{\tau_\alpha}\right]+\frac{t_3}{6}\rho^\alpha\Big[(1+\frac{x_3}{2})\rho$$
$$-(x_3+\frac{1}{2})\rho_{\tau_\alpha}\Big]+\frac{1}{4}\Big\{t_1\left[(1+\frac{x_1}{2})\rho-(x_1+\frac{1}{2})\rho_{\tau_\alpha}\right]+t_4\rho$$
$$\cdot\left[(1+\frac{x_4}{2})\rho-(x_4+\frac{1}{2})\rho_{\tau_\alpha}\right]+t_2\left[(1+\frac{x_2}{2})\rho+(x_2+\frac{1}{2})\rho_{\tau_\alpha}\right]$$
$$+t_5\rho\left[(1+\frac{x_5}{2})\rho+(x_5+\frac{1}{2})\rho_{\tau_\alpha}\right]\Big\}\frac{2m_{\tau_\alpha}}{\hbar^2}\left(\frac{M}{M+m_{\tau_\alpha}}E_L-V_c\right)$$
$$+\frac{1}{40\pi^2}\left[t_1(1-x_1)+t_4\rho(1-x_4)+3t_2(1+x_2)\right.$$
$$\left.+3t_5\rho(1+x_5)\right]K^5_{\tau_\alpha}+\frac{1}{20\pi^2}\left[t_1(1+\frac{x_1}{2})+t_4\rho(1+\frac{x_4}{2})\right.$$
$$\left.+t_2(1+\frac{x_2}{2})+t_5\rho(1+\frac{x_5}{2})\right](2K_F^3-K^3_{\tau_\alpha})^{5/3}\Big\} \tag{35}$$

where the effective mass is

$$\frac{m^*_{\tau_\alpha}}{m_{\tau_\alpha}}=\Big\{1+\frac{2m_{\tau_\alpha}}{\hbar^2}\,\frac{1}{4}\Big\{t_1\left[(1+\frac{x_1}{2})\rho-(x_1+\frac{1}{2})\rho_{\tau_\alpha}\right]$$
$$+t_4\rho\left[(1+\frac{x_4}{2})\rho-(x_4+\frac{1}{2})\rho_{\tau_\alpha}\right]+t_2\left[(1+\frac{x_2}{2})\rho+(x_2+\frac{1}{2})\rho_{\tau_\alpha}\right]$$
$$+t_5\rho\left[(1+\frac{x_5}{2})\rho+(x_5+\frac{1}{2})\rho_{\tau_\alpha}\right]\Big\}\Big\}^{-1} \quad . \tag{36}$$

The imaginary part of MOP obtained from (19) is

$$W_{\tau_\alpha}=-\frac{1}{64\pi^5}\sum_{i=1}^{7}W_i \tag{37}$$

where

$$W_1=2\left[(1+x_0+x_0^2)t_0^2+\frac{1}{6}(2+x_0+x_3+2x_0x_3)t_0t_3\rho^\alpha\right.$$
$$\left.+\frac{1}{36}(1+x_3+x_3^2)t_3^2\rho^{2\alpha}\right]\left[I_1(\tau_\alpha)+I_1(\tau_\alpha,-\tau_\alpha)\right]$$
$$-\left[(1+4x_0+x_0^2)t_0^2+\frac{1}{3}(1+2x_0+2x_3+x_0x_3)t_0t_3\rho^\alpha\right.$$
$$\left.+\frac{1}{36}(1+4x_3+x_3^2)t_3^2\rho^{2\alpha}\right]I_1(\tau_\alpha)$$

$$W_2=\frac{1}{2}\Big\{\left[(2+x_0+x_1+2x_0x_1)t_0t_1+(2+x_0+x_4+2x_0x_4)t_0t_4\rho\right.$$
$$+\frac{1}{6}(2+x_3+x_1+2x_3x_1)t_3t_1\rho^\alpha+\frac{1}{6}(2+x_3+x_4+2x_3x_4)$$
$$\left.\cdot t_3t_4\rho^{1+\alpha}\right]\left[I_2(\tau_\alpha)+I_2(\tau_\alpha,-\tau_\alpha)\right]$$
$$-\left[(1+2x_0+2x_1+x_0x_1)t_0t_1+(1+2x_0+2x_4+x_0x_4)t_0t_4\rho\right.$$
$$+\frac{1}{6}(1+2x_3+2x_1+x_3x_1)t_3t_1\rho^\alpha+\frac{1}{6}(1+2x_3+2x_4+x_3x_4)$$
$$\left.\cdot t_3t_4\rho^{1+\alpha}\right]I_2(\tau_\alpha)\Big\}$$

$$W_3=\frac{1}{16}\Big\{2\left[(1+x_1+x_1^2)t_1^2+(2+x_1+x_4+2x_1x_4)t_1t_4\rho\right.$$
$$\left.+(1+x_4+x_4^2)t_4^2\rho^2\right]\left[I_3(\tau_\alpha)+I_3(\tau_\alpha,-\tau_\alpha)\right]$$
$$-\left[(1+4x_1+x_1^2)t_1^2+2(1+2x_1+2x_4+x_1x_4)t_1t_4\rho\right.$$
$$\left.+(1+4x_4+x_4^2)t_4^2\rho^2\right]I_3(\tau_\alpha)\Big\}$$

$$W_4=\frac{1}{2}\left[(2+x_0+x_2+2x_0x_2)t_0t_2+(2+x_0+x_5+2x_0x_5)t_0t_5\rho\right.$$
$$+\frac{1}{6}(2+x_3+x_2+2x_3x_2)t_3t_2\rho^\alpha+\frac{1}{6}(2+x_3+x_5+2x_3x_5)$$
$$\left.\cdot t_3t_5\rho^{1+\alpha}\right]I_4(\tau_\alpha,-\tau_\alpha)$$

$$W_5=\frac{1}{8}\left[(2+x_1+x_2+2x_1x_2)t_1t_2+(2+x_1+x_5+2x_1x_5)t_1t_5\rho\right.$$
$$+(2+x_4+x_2+2x_4x_2)t_4t_2\rho+(2+x_4+x_5+2x_4x_5)$$
$$\left.\cdot t_4t_5\rho^2\right]I_5(\tau_\alpha,-\tau_\alpha)$$

$$W_6=\frac{1}{16}\Big\{2\left[(1+x_2+x_2^2)t_2^2+(2+x_2+x_5+2x_2x_5)t_2t_5\rho\right.$$
$$\left.+(1+x_5+x_5^2)t_5^2\rho^2\right]\left[I_6(\tau_\alpha)+I_6(\tau_\alpha,-\tau_\alpha)\right]$$
$$+\left[(1+4x_2+x_2^2)t_2^2+2(1+2x_2+2x_5+x_2x_5)t_2t_5\rho\right.$$
$$\left.+(1+4x_5+x_5^2)t_5^2\rho^2\right]I_6(\tau_\alpha)\Big\}$$

$$W_7=\frac{1}{4}W_0^2\left[2I_7(\tau_\alpha)+I_7(\tau_\alpha,-\tau_\alpha)\right] \quad . \tag{38}$$

The simplest way to obtain MOP for a finite nucleus is to use the LDA[20]. We assume that the densities of the neutrons and protons in a spherical nucleus have the same geometrical distributions and are expressed by Negele's empirical formula

$$\rho_K(r)=\frac{\rho_{0K}}{1+\exp[(r-c)/a]} \qquad K=N \text{ or } Z \tag{39}$$

where

$$\rho_{0K}=\frac{3K}{4\pi c^3(1+\pi^2a^2/c^2)} \qquad K=N \text{ or } Z \tag{40}$$

$$c=(0.978+0.0206A^{1/3})A^{1/3} \quad , \quad a=0.54 \quad . \tag{41}$$

In this case the asymmetric parameter can be deduced as

$$\alpha_0=(N-Z)/A \quad . \tag{42}$$

In general the optical potential has the following form:

$$U_{\tau_\alpha}(r)=V_{\tau_\alpha}(r)+iW_{\tau_\alpha}(r)+\left[V^{\tau_\alpha}_{so}(r)+iW^{\tau_\alpha}_{so}(r)\right](\vec{\sigma}\cdot\vec{l}) \tag{43}$$

However, we have only got the $V_{\tau_\alpha}(r)$ and $W_{\tau_\alpha}(r)$ without spin-orbit parts $V^{\tau_\alpha}_{so}$ and $W^{\tau_\alpha}_{so}$ which, as is known, vanish in nuclear matter. In order to keep the spin-orbit term, let us start from the HF calculation for the finite nuclei. In the HF calculation of spherical nuclei, for the conventional Skyrme force, the spin-orbit term of the real part is read[10]

$$V^{\tau_\alpha}_{so}(r)=\frac{1}{2}W_0\frac{1}{r}\,\frac{d}{dr}\left[\rho(r)+\rho_{\tau_\alpha}(r)\right]+\frac{1}{8r}(t_1-t_2)J_{\tau_\alpha}(r) \tag{44}$$

and for the extended Skyrme force is

$$V^{\tau_\alpha}_{so}(r)=\frac{1}{2}W_0\frac{1}{r}\,\frac{d}{dr}\left[\rho(r)+\rho_{\tau_\alpha}(r)\right]$$
$$-\frac{1}{8r}\left[t_1x_1+t_2x_2+t_4x_4\rho(r)+t_5x_5\rho(r)\right]J(r)$$
$$+\frac{1}{8r}\left[t_1-t_2+t_4\rho(r)-t_5\rho(r)\right]J_{\tau_\alpha}(r) \tag{45}$$

where $J_{\tau_\alpha}(r)$ is the spin density. The numerical results show that the contribution of the term produced by the central force involveng $J_{\tau_\alpha}(r)$ is much smaller than the first term directly arising from two-body spin-orbit force [10,21]. Thus we only keep the first term and (44) and (45) can be simultaneously reduced to

$$V_{so}^{\tau_\alpha}(r)=\frac{1}{2}W_0\frac{1}{r}\frac{d}{dr}\left[\rho(r)+\rho_{\tau_\alpha}(r)\right] \qquad (46)$$

where $\rho(r)$ and $\rho_{\tau_\alpha}(r)$ are described by Negele's empirical formula (39) as before, then (46) is an analytical expression.

The imaginary part of the spin-orbit potential $W_{so}^{\tau_\alpha}(r)$ is usually very small, which is often omitted even in POP[22]. We also omit its contribution to MOP in our work.

5. Results and Analyses

From the point of view of fitting the expirimental data the optical potential is not unique. However, the calculated results are mainly sensitive to the volume integrals per nucleon, namely

$$J_V=-\frac{1}{A}\int V(r)d\vec{r}\ , \quad J_W=-\frac{1}{A}\int W(r)d\vec{r} \qquad (47)$$

and the root mean square (rms) radii of the real and imaginary parts of the MOP

$$\langle R_V^2\rangle^{1/2}=\left[\int V(r)r^2d\vec{r}/\int V(r)d\vec{r}\right]^{1/2}$$

$$\langle R_W^2\rangle^{1/2}=\left[\int W(r)r^2d\vec{r}/\int W(r)d\vec{r}\right]^{1/2} . \qquad (48)$$

Firstly, we use the conventional Skyrme forces SII-SVI to calculate the MOP and their volume integrals. In certain energy regions the calculated MOP are in reasonable agreement with the

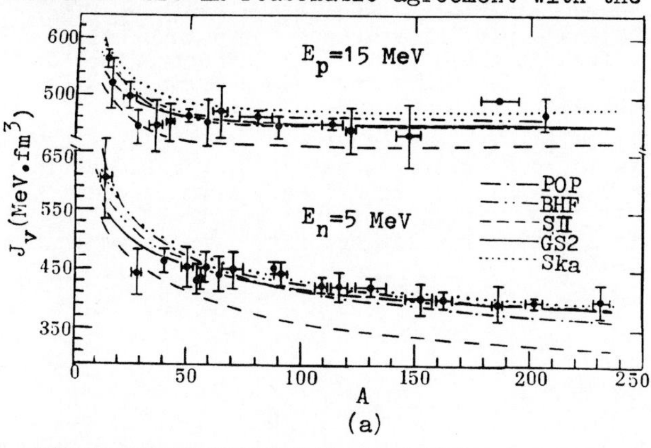

(a)

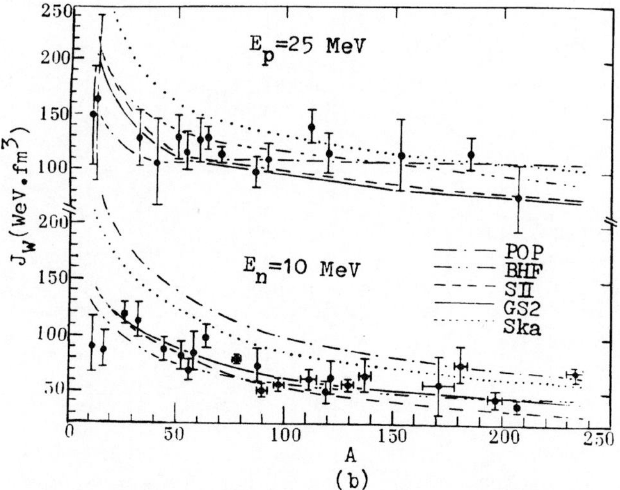

(b)

Fig.2. The volume integral per nucleon of the MOP against mass number A, for neutrons at the bottom, for protons on the top. The dots and crosses represent the empirical values[23]. (a) The real parts. (b) The imaginary parts.

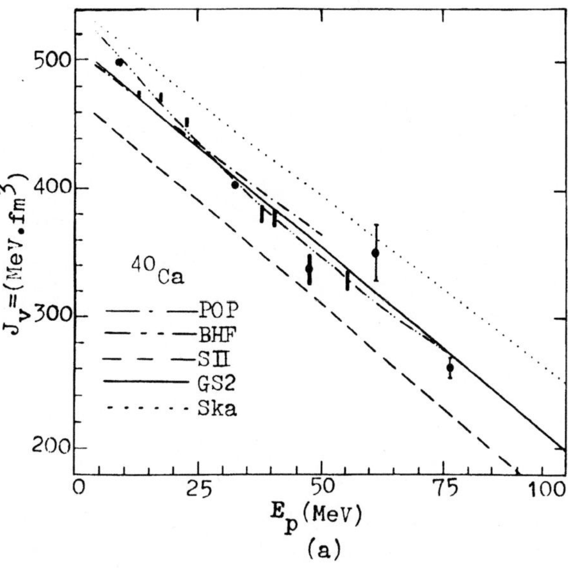

(a)

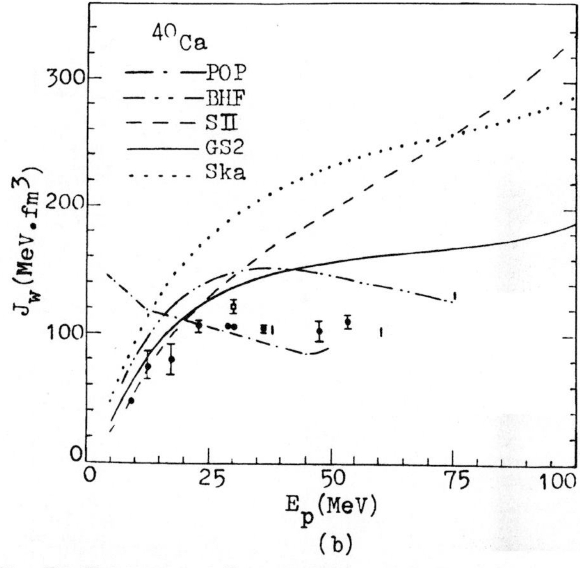

(b)

Fig.3. The energy E_p variation of the volume integral per nucleon of the proton MOP with different Skyrme forces for ^{40}Ca. The dots are empirical values [3,23]. (a) The real parts. (b) The imaginary parts.

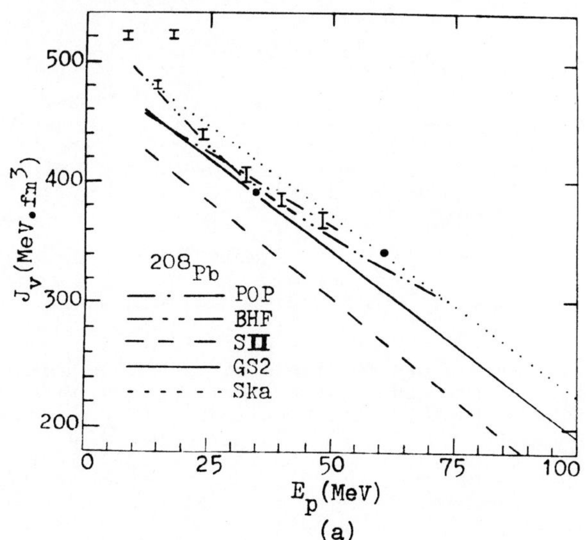

(a)

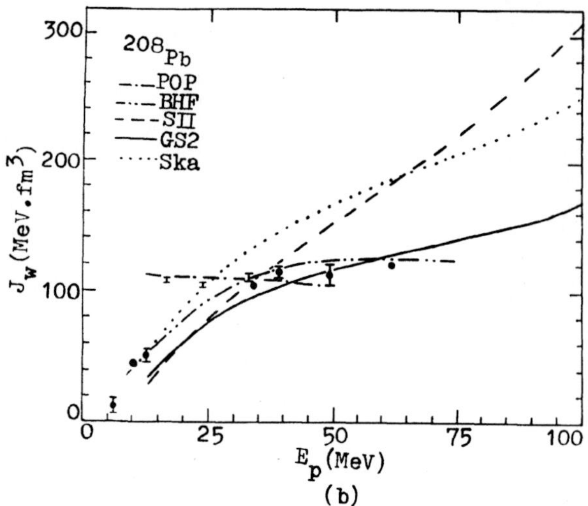

Fig.4. Same as Fig.3, for ^{208}Pb.

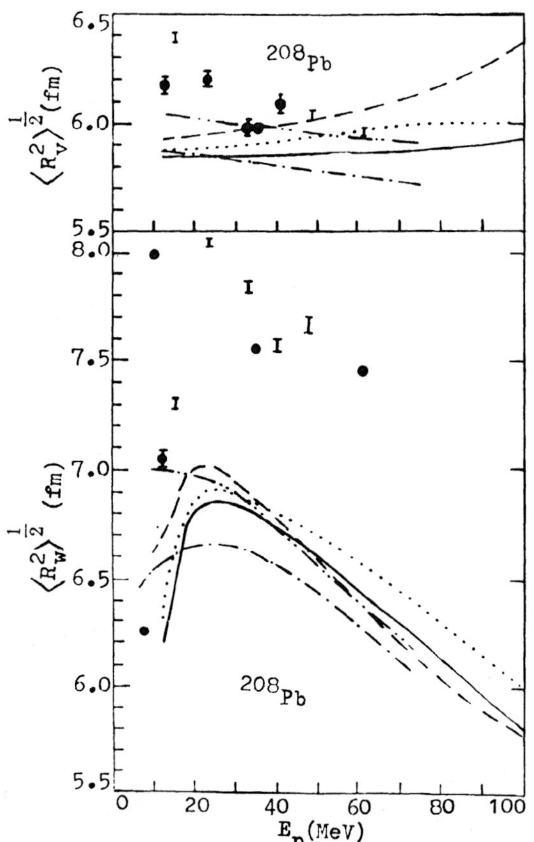

Fig.5. The root mean square radii of the real parts (top) and the imaginary parts (bottom) of the MOP for proton scattering by ^{208}Pb against proton energy E_p. The solid curve are the theoretical values for GS2, the dots line correspond to Ska, the long dashes to SII, the dash-dots to the simple LDA and the dash-dot-dots to the improved LDA in the BHF approximation[3,23]. The dots are empirical values [3,23].

POP[22] and those MOP based on the BHF approximation[3]. Compared the results calculated with five sets of parameters SII-SVI, we have found that the real part of the MOP calculated with SIII is the best, but the imaginary parts of that is too bad, because its energy dependence is rising too steep as compared with the empirical one [3,23] . Considering both the real and the imaginary parts as a whole it seems that the MOP with SII is the best one. In addition, there also appear some shortcomings in the MOP calculated with the conventional Skyrme forces. For example, the surface region of the real parts has the unnecessary peak, the values of both the volume integral per nucleon and the depth of the real parts are smaller than the POP and the empirical ones, the energy variation of the imaginary parts is too fast and so on. Then, we come to use the generalized Skyrme forces to calculate MOP to test whether this kind of Skyrme forces can improve the results. The calculations show that indeed with the generalized Skyrme forces many shortcomings due to the conventional Skyrme forces can be overcome, and the agreements with the empirical data have obviously been improved. The results by GS1-GS3 are all quite good, among which the GS2 is the best one. We have also tried the modified Skyrme forces, in which the Ska and Skb are better than the others. In the following we just show the calculated results with SII, GS2 and Ska as the representatives, and compare them with those based on the BHF approximation, and with the POP, empirical and experimental values.

The dependences on the mass number of the volume integrals per nucleon of the real and imaginary parts of the computed optical potential and the comparison with the empirical values [24,25] are shown in Fig.2 (a) and (b), in which a number of even-even nuclei ranging from A=12 to 238 near β-stable line are chosen. Fig.2 shows that the volume integrals of the computed MOP with GS2 and Ska are very close to the empirical values and the theoretical values obtained by POP and BHF approximation[23], but J_v with SII are somewhat underestimated.

Fig.3 (a) shows that the J_V of ^{40}Ca with GS2 is in good agreement with POP, BHF and empirical values for $E_p < 75$ MeV, but the J_V with SII is somewhat too small and the J_V with Ska is a little bit larger. Fig.3 (b) shows that

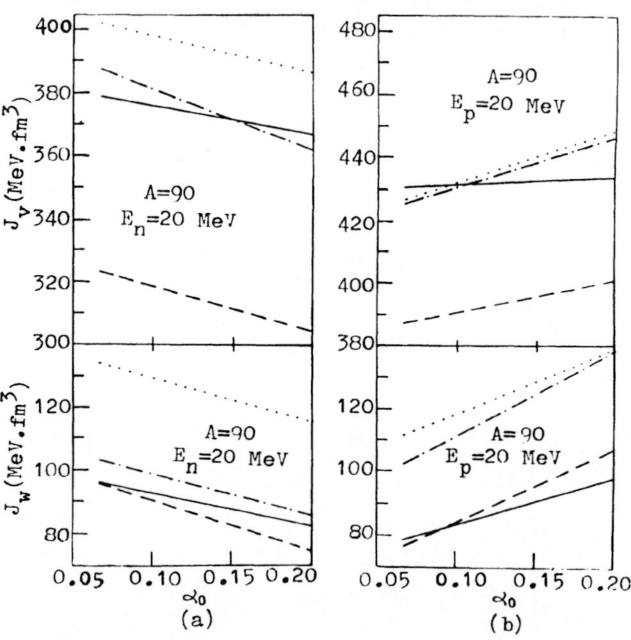

Fig.6. The volume integral per nucleon of the real parts (top) and the imaginary parts (bottom) of the MOP against asymmetry factor α_0 for A=90 nuclei at E_L=20 MeV. The dash-dots correspond to POP, the long dashes to SII, the solid line to the GS2 and the dots line to the Ska. (a) For neutrons. (b) For protons.

the energy dependence of the J_W for ^{40}Ca with SII, GS2 and Ska are all larger than empirical values, among which the GS2 is the best and very close to the one with BHF. In Fig.4 it is the same as in Fig.3 but for ^{208}Pb. It is noted that the whole situation is better for this heavier nucleus than for the lighter one. It is also noted that GS2 is still the best one.

The computed rms radii for ^{208}Pb with SII, GS2 and Ska are compared with the empirical ones and those BHF calculations. It seems our results are close to the one obtained by BHF (see Fig.5). Because of the surface peak in the real part of the MOP computed with SII in high energy region, so the $\langle R_V^2 \rangle^{1/2}$ for this case is rising with the increase of energy in contrast to the empirical one. Fig.5 shows that all theoretical values of $\langle R_W^2 \rangle^{1/2}$ (including the BHF one) are less than empirical values. Perhaps this is because the surface absorption of all the theretical optical potential within the framework of nuclear matter approach is too weak.

We have also investigated the dependence of J_V and J_W on the asymmetric parameter α_0. The four nuclei $^{90}_{36}$Kr, $^{90}_{38}$Sr, $^{90}_{40}$Zr, and $^{90}_{42}$Mo with the same atomic number A=90 are chosen to compare the volume integral J_V and J_W of the neutron or proton with the incident energy E_L=20 MeV in Fig.6. The linear dependence on the asymmetric parameter α_0 is found. The volume integrals decrease for neutrons and

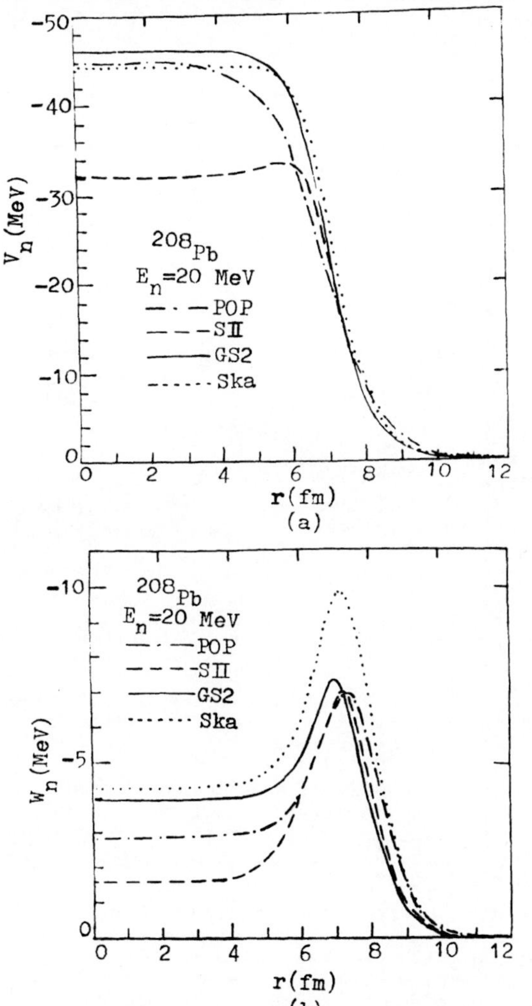

Fig.7. Radial dependence of the neutron MOP for ^{208}Pb at energy E_n=20 MeV. (a) The real parts. (b) The imaginary parts.

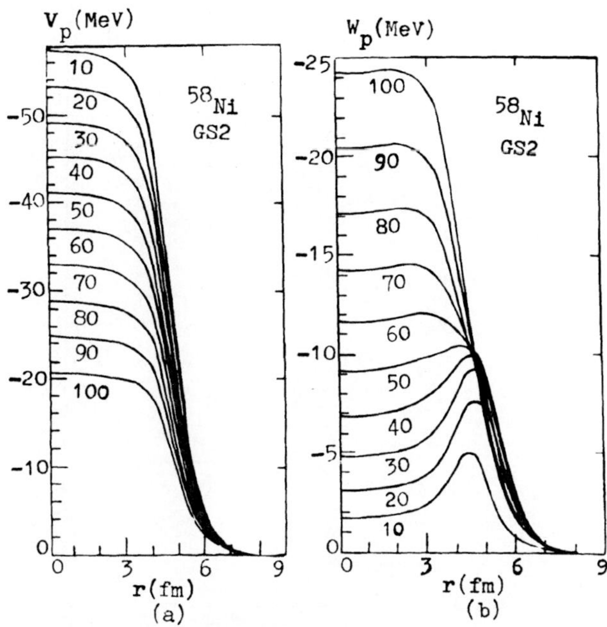

Fig.8. Radial dependence of the proton MOP for ^{58}Ni at energies 10-100 MeV. The curves are the theoretical values computed with GS2. (a) The real parts. (b) The imaginary parts.

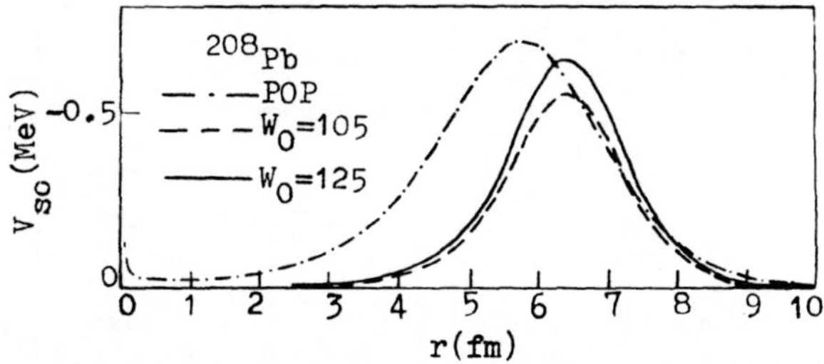

Fig.9. Radial dependence of the real parts of the neutron spin-orbit potential for ^{208}Pb.

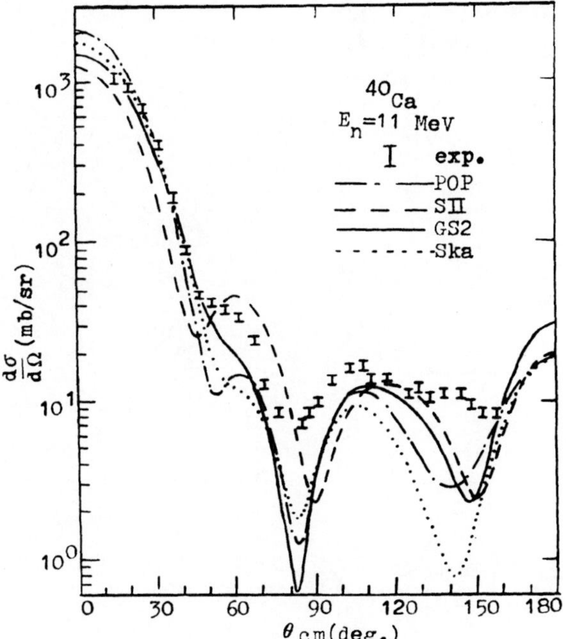

Fig.10. Comparison of the theoretical values of the neutron elastic scattering angular distribution for ^{40}Ca at E_n=11 MeV with experiments[26].

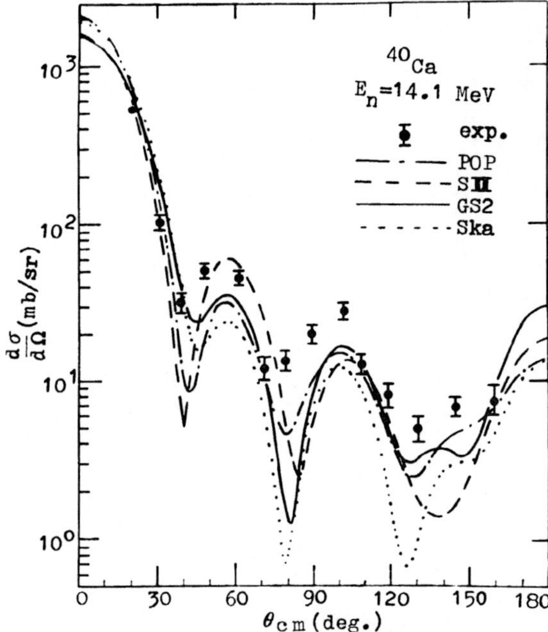

Fig.11. Same as Fig.10, for E_n=14.1 MeV. The experimental points are taken from reference 27.

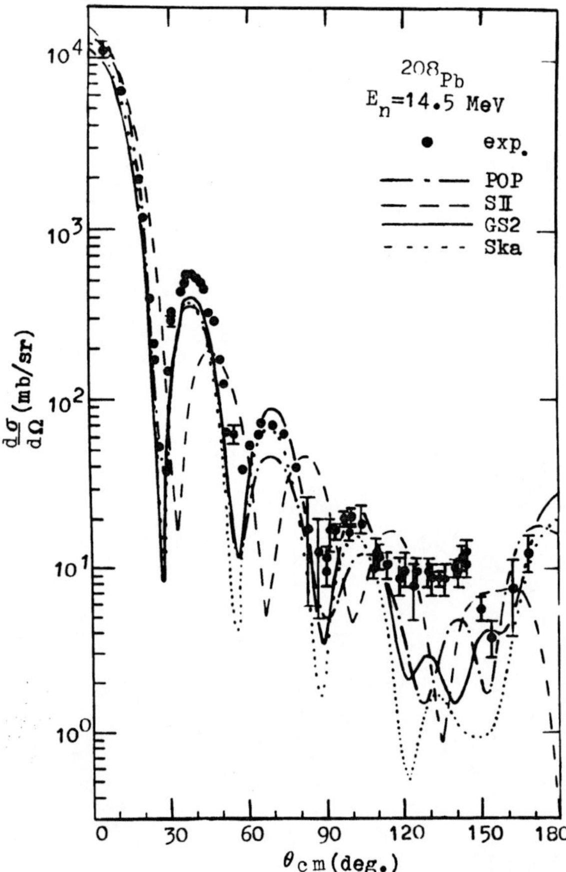

Fig.12. Same as Fig.10, for ^{208}Pb at E_n=14.5 MeV. The experimental points are taken from reference 28.

increase for protons as α_0 increases, which is the same tendency as with the POP. But the variation rates of J_V and J_W on α_0 with GS2 are less than others.

Fig.7 shows that the shape of the real parts of the neutron MOP for ^{208}Pb at E_n=20 MeV

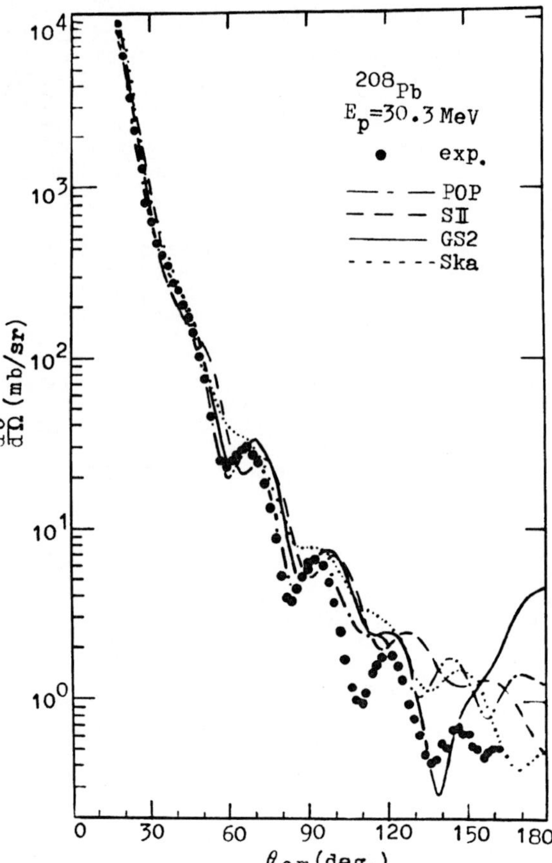

Fig.13. Same as Fig.10, for proton elastic scattering of ^{208}Pb at E_p=30.3 MeV. The experimental points are taken from reference 29.

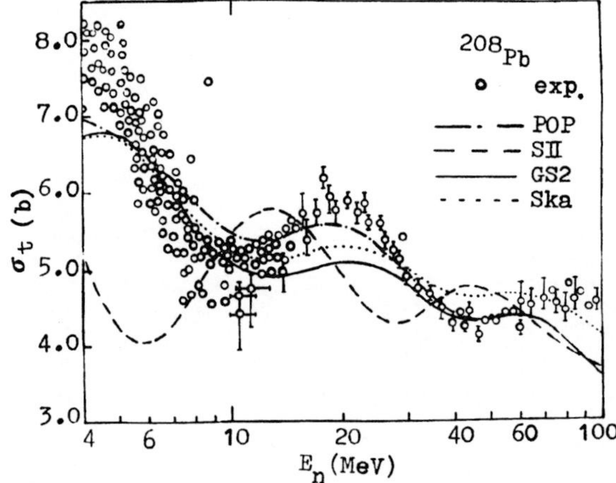

Fig.14. Comparison of the theoretical values of the neutron total cross sections for ^{208}Pb with experiments[30].

computed with GS2 and Ska are very close to the POP, but the one with SII is too shallow and has a surface peak, whereas the imaginary parts of the MOP all have the evident surface absorption peaks.

In Fig.8 the real and imaginary parts of the proton MOP for ^{58}Ni at energy E_p=10-100 MeV computed with GS2 are illustrated. The imaginary part of the MOP changes from the dominant surface absorption into the volume absorption as energy increases as expected from the POP.

Fig.9 shows the real parts of the neutron spin-orbit potential for ^{208}Pb calculated with (46). Their shape and value are close to the POP.

We have calculated the angular distributions of the elastic scattering for ^{40}Ca and ^{208}Pb, which are shown in Figs.10-13, by using the MOP with SII, GS2, Ska as well as the POP, and compared them with the experimental data [26-29]. They all roughly agree with the experimental data, in which the one with GS2 is better than the others. We have also calculated the neutron total cross sections for ^{208}Pb for energy region of E_n=4-100 MeV and compared them with the experimental data[30]. Fig.14 shows that the computed results by GS2 and Ska are in agreement with the experiments, but the one by SII fits the experiments quite poorly.

6. Conclusions

We have presented a wide variety of results and comparisons which show that for some sets of Skyrme forces in certain energy regions the potential depth, shape, relative contributions of the surface and volume parts, as well as the energy and asymmetric parameter dependences of our MOP are in reasonable agreement with the POP and those based on the BHF approximation. The volume integrals of the MOP is in good agreement with the empirical values and other theoretical approaches. The calculated total cross sections and angular distributions of the elastic scattering with our MOP without adjusting the parameters fit the experiments pretty well. Thus, our MOP can be regarded as a parameter-free one. Although our approach is not as fundamental as those based on the realistic nucleon-nucleon interactions,however, it is physically more transparent and simple enough to be applied to the nuclear data calculations.

References

1. J.S.Bell, E.J.Squires, Phys. Rev. Lett. 3, 96 (1959)
2. J.P.Jeukenne, A.Lejeune, C.Mahaux, Phys. Rep. 25C, 83 (1976)
3. J.P.Jeukenne, A.Lejeune, C.Mahaux, Phys. Rev. C16, 80 (1977)
4. F.A.Brieva, J.R.Rook, Nucl. Phys. A291, 299, 317 (1977)
5. N.Vinh Mau, A.Bouyssy, Nucl. Phys. A257, 189 (1976)
6. V.Bernard, Nguyen Van Giai, Nucl. Phys. A327, 397 (1979); A348, 75 (1980)
7. Shen Qingbiao, Zhang Jingshang, Tian Ye, Zhuo Yizhong, Phys. Energ. Fortis Phys. Nucl. Vol.6(1), 91 (1982)
8. Shen Qingbiao, Tian Ye, Ma Zhongyu, Zhang Jingshang, Zhuo Yizhong, Phys. Energ. Fortis Phys. Nucl. Vol.6(2), 185 (1982)
9. Shen Qingbiao, Zhang Jingshang, Tian Ye, Ma Zhongyu, Zhuo Yizhong, Z.Phys. A—Atoms and Nuclei, 303, 69 (1981)
10. D.Vautherin, D.M.Brink, Phys. Rev. C5, 626 (1972)
11. J.W.Negele, D.Vautherin, Phys. Rev, C5, 1472 (1972)
12. Tian Ye, Shen Qingbiao, Zhuo Yizhong, Phys. Energ. Fortis Phys. Nucl. (to be Published)
13. S.Krewald, V.Klemt, J.Speth, A.Faessler, Nucl. Phys. A281, 166 (1977)
14. M.Beiner, H.Flocard, Nguyen Van Giai, P.Quentin, Nucl. Phys. A238, 29 (1975)
15. B.D.Chang, Phys. Lett. 56B, 205 (1975)
16. J.Treiner, H.Krivine, J.Phys. G2, 285 (1976)
17. H.S.Köhler, Nucl. Phys. A258, 301 (1976)
18. H.Krivine, J.Treiner, O.Bohigas, Nucl. Phys. A336, 155 (1980)
19. Nguyen Van Giai, H.Sagawa, Phys. Lett. 106B, 379 (1981)
20. J.W.Negele, Phys. Rev, C1, 1260 (1970)
21. Shen Qingbiao, Tian Ye, Liu Ruizhe, Chin. J.Nucl. Phys. (to be published)
22. F.D.Becchetti, G.W.Greenlees, Phys. Rev. 182, 1190 (1969)
23. C.Mahaux, Microscopic optical potentials, Lecture Notes in Physics, Vol. 89, p.1, Berlin, Heidelberg, New York: Springer 1979
24. S.Kailas, S.K.Gupta, Phys. Rev. C17, 2236 (1978)
25. S.Kailas, S.K.Gupta, Phys. Lett. 71B, 271 (1977)
26. J.C.Ferrer, et al., Nucl. Phys. A275, 325 (1977)
27. W.J.McDonald, et al., Nucl. Phys. 59, 321 (1946)
28. F.Perey, B.Buck, Nucl. Phys. 32, 353 (1962)
29. B.W.Ridley, J.F.Turner, Nucl. Phys. 58, 497 (1964)
30. D.I.Garber, R.R.Kinsey, Neutron Cross Sections, Vol.2, BNL-325, 3rd ed., 1976

OPTICAL MODEL CALCULATIONS FOR THE CHROMIUM, IRON AND NICKLE ISOTOPES IN THE ENERGY RANGE OF 0.5 TO 100.0 MeV

A. Prince

Brookhaven National Laboratory
Upton, New York 11973
U. S. A.

Optical model (O. M.) calculations have been carried out for the Cr, Fe and Ni isotopes at energies just above the resolved resonance region (0.5 to 0.7 MeV) to the high energy (100.0 MeV) region. A consistent set of O. M. parameters that covered this region produced in addition to the S- and P-wave strength functions the total elastic and differential angular distributions that was in very good agreement with the experimental data. Reproduction of the broad maxima and minima (Nuclear Ramsauer Effect) in addition to S_o, S_1 and R were the established constraints for this analysis.

[$^{50,52,53,54}Cr$, $^{54,56,57,58}Fe$ and $^{58,60,61,62,64}Ni$ Total and Elastic scattering cross sections E=0.5 to 100.0 MeV O. M. Calcs.]

Introduction

During the past few years the interests in neutron cross sections in the high energy region (E >20 MeV) has been steadily growing. This interest was stimulated by the data needs in dosimetry reactions, damage studies for the Fusion Materials Irradiation Test (FMIT) Facility, Hybrid Fusion-Fission Reactors, Shielding Studies, etc.

At two symposia held in 1977 and 1980[1,2] it was stressed that more model calculations would be necessary to supplement the sparse amount of experimental data at such high energies.

One of the major observations resulting from these two symposia was the lack of agreement resulting from the use of ''global''parameter sets in the analysis of the total cross section at these high energies.

At BNL an effort was undertaken to establish a remedy for this situation. The following is a brief outline of the method of analysis used in this study.

Procedural Analysis

The area of concentration was started with the analysis of the cross sections for the main constituents in the structural materials namely Fe, Ni and Cr.

As is usual in carrying out an evaluation, an effort was undertaken to first produce a set of data for the total, elastic and non-elastic neutron cross sections. The customary procedure of using a ''global'' set of optical metal parameters for calculating these cross sections did not prove fruitful.

An indication of the inadequacy of two well known O. M. parameters Refs. 3 and 4 is shown in Figures 1 and 2.

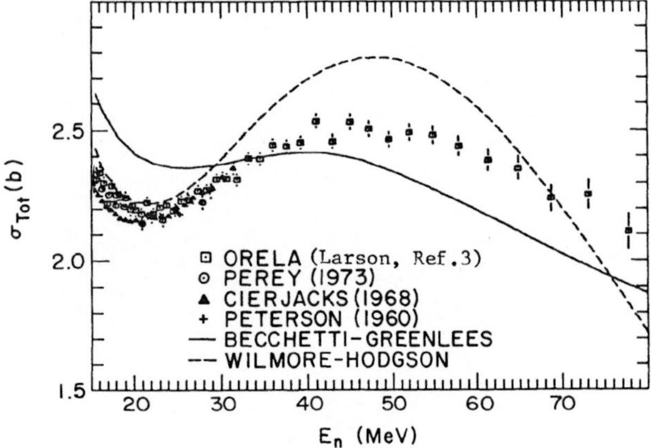

Fig. 1 Calculation of σ_{Tot} for Cr (N a t) Using Global O.M. Parameters.

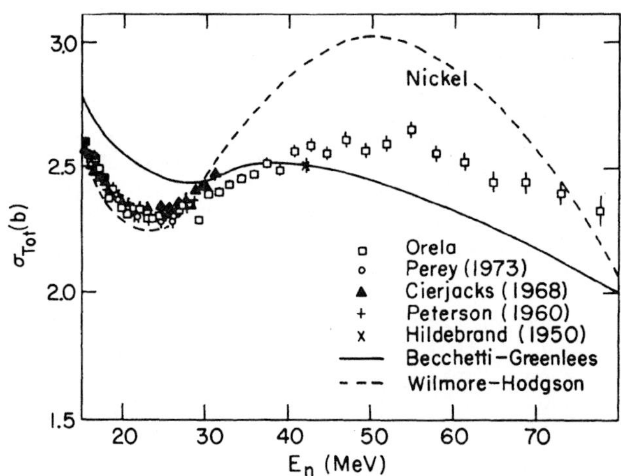

Fig. 2 Calculation of σ_{Tot} for Ni (N a t) Using Global O.M. Parameters.

It should be pointed out, however, that these global fits were derived from X^2 fitting of experimental cross sections at energies E_n <25 MeV and thus are not necessarily capable of being extended to fit data beyond this range.

Faced with this problem, a survey of O. M. parameter sets that spanned the low energy (E_n 1.0 MeV) to high energy (E_n - 100 MeV) was undertaken. This search produced only two such sets namely Englebrecht and Fiedeldey[5] and Aver'yanov and Purtseladze.[6]

Engelbrecht and Fiedeldey (EF) fitted the experimental data with a non-local potential (NLP) and derived an equivalent local potential (ELP) up to E - 200 MeV. Aver'yanov and Purtseladze (AP) fitted experimental cross sections up to about E = 100 MeV with an energy dependent potential as well as one that depended on the target mass and isospin.

Continuing the trend of including an isospin dependence in the O. M. parameters, the choice was made to investigate and hopefully extend the treatment of (AP).

The real and imaginary potentials of (AP) were Woods-Saxon and Gaussian respectively. The depths of the wells of these potentials are written in the form shown below

K. H. Böckhoff (ed.), Nuclear Data for Science and Technology, 574–579.

Real Well Depth

$$V_1(E, A, N, Z) = V_{10}(E) - V_{11}(E)\left(\frac{N-Z}{A}\right) \quad (1)$$

Imaginary Well Depth

$$V_2(E,A) = V_{20}(E) + V_{21}(E)\, A^{1/3} \quad (2)$$

(AP) carried out a least squares fit based on experimental data for six nuclei spanning the mass region of A = 12 to 238 which consisted of C, Al, Cu, Cd, Pb and U. The experimental data was from 2.5 to 96 MeV which produced values for V_{ij} of Eqs. 1 and 2. These results are given in Table I.

Table I

Neutron Optical Model Parameters (Ref. 6)

E_n	V_{10}	V_{11}	V_{20}	V_{21}	E_n	V_{10}	V_{11}	V_{20}	V_{21}
2.5	53.0	35.0	5.0	0.0	37.0	37.2	36.0	3.0	2.0
4.1	51.0	29.0	6.5	0.0	45.0	33.2	126.0	1.3	2.2
7.0	47.0	9.0	8.0	0.0	60.0	30.5	98.0	0.0	2.5
14.5	43.2	0.0	9.0	0.0	72.0	28.8	60.0	0.0	2.3
24.0	40.0	0.0	9.8	0.0	84.0	26.4	53.0	1.0	2.1
30.0	38.0	0.0	7.0	0.8	96.0	24.0	38.0	3.5	1.5

Figure 3 shows the calculated values for σ_{tot} for U, Cd and Cu obtained by (AP) using the parameter set of Table I.

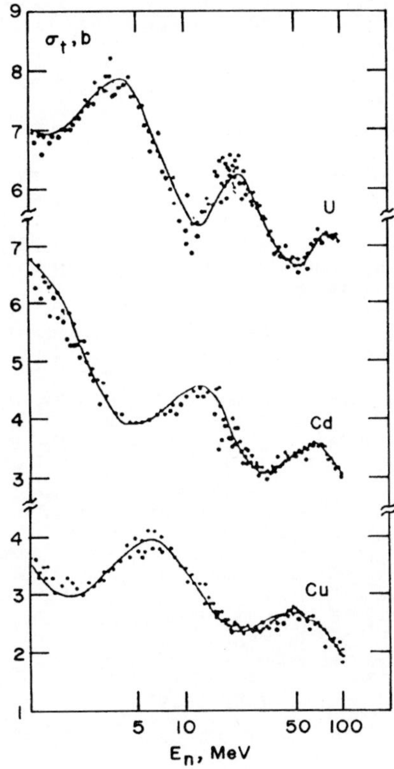

Fig. 3 Total Cross Section, for U, Cd & Cu
. EXPT'L
- CALC

Despite the apparent ambiguities for the various V_{ij} in Table I, it was found that an apparent trend in V_1 and V_2 as a function of energy did exist. Upon using the O. M. parameters from Table I a least squares fit was made for V_1 and V_2 which led to an energy dependence for both the real and imaginary potentials of the form

$$V_{Pot} = V_o + \alpha_1 E + \alpha_2 E^2 \quad (3)$$

Unfortunately these parameters failed to reproduce the experimental data for Cr, Fe and Ni especially at the higher energies.

The parameters α_i in Eq. 3 were close in magnitude to the various global real potentials but the imaginary potential exhibited an energy dependence having maximum in three distinctive energy regions.

The most striking feature of these maxima was that they appeared to coincide with the maxima of the total cross sections of many nuclides.

Now ordinarily one usually envisions the magnitude of V_{IM} to vary linearly with energy as depicted in the Wilmore-Hodgson and Becchetti-Greenlees global parameters. However, as mentioned earlier, these global sets are based on data for E < 25 MeV. If one looks at the shape variation of V_{IM} at high energies as reported by some investigators,[7,8] it is seen that indeed, both empirical and theoretical forms of V_{IM} have shapes that indicate structure (i.e. minima and maxima).

These three maxima in V_{IM} which coincided approximately with the maxima in σ_{tot} immediately brought to mind the ''Ramsauer'' effect interpretation of Peterson.[9] Peterson showed that the relationships among the three families of broad maxima and minima appearing in σ_{tot} can be explained in terms of a semi-classical treatment wherein the oscillations in σ_{tot} result from the interference between the neutron waves traversing the nucleus and those going around it.

This is described in terms of the average phase difference Δ between the two waves, and the condition for maximum (minimum) cross sections is given by
$\Delta = n\pi$ n (odd) = max.
n (even) = min.

The loci of maxima-minima for different values of n are shown in Fig. 4.

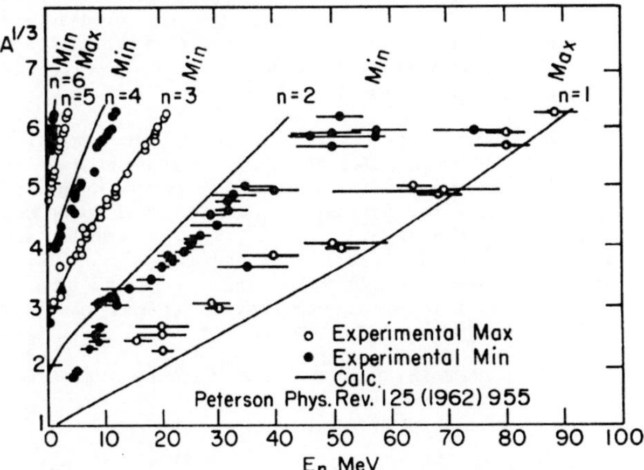

Fig. 4 LOCI of Maximum & Minimum Neutron Cross Sections

This ''Ramsauer Effect'' is plainly evident for U in Fig. 3 where from Fig. 4 it is seen that a maximum occurs at approximately E = 90, 70 and 5 MeV and minima at 50, 15 and 2 Mev.

Since the amount of experimental data has tremendously increased since Peterson's analysis and also in an attempt to improve the fit for n = 1, 2 and 4 in Fig. 4, it was decided to carry out a similar type investigation using this additional data in order to refine the loci of minimum and maximum in the total neutron cross sections.

Following Peterson's procedure but allowing for a more physical meaning to the parameter r_o used in defining the nuclear radius a much improved fit was established as can be seen in Fig. 5.

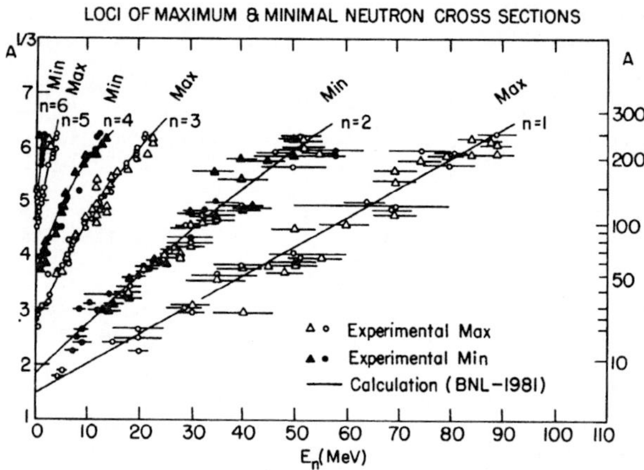

Fig. 5 Improved Fit to LOCI of Maximum & Minimum Neutron Cross Sections.

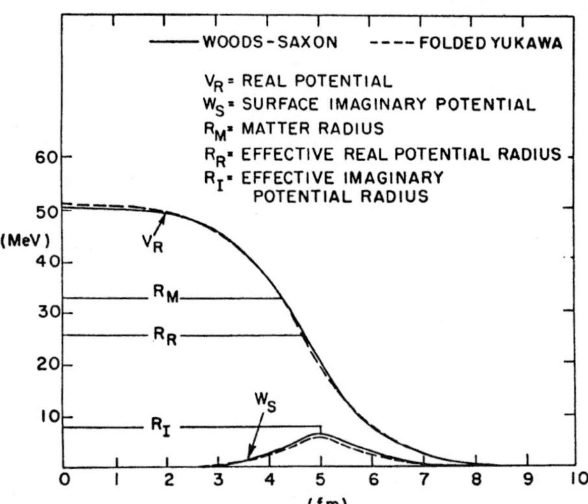

Fig. 6 Comparison of Woods-Saxon & Folded Yukawa Potentials for ^{52}Cr.

Having established a reasonable method for estimating the local maximum and minimum in the total cross section, it was decided to employ this as an additional constraint in the least squares fit for finding O. M. parameters.

An effort was first made to fit the minimum (~20 MeV) and maximum (~45 MeV) which occurs in σ_{tot} for the Cr isotopes by modifying the (AP) LSQ fit. Some progress resulted in the fits - but only at the expense of non-agreement in σ_{non-el} and/or $d\sigma/d\Omega$. It was noticed, however, that shifting the peak (maximum) of V_{IM} caused a shifting in the minima/ maxima of σ_{tot}. This led to the idea that while the energy variation for the real and imaginary potentials might follow the form given by Eq. 3, the magnitude along with the associated geometrical factors (radii and diffusivities) might have to be modified to reflect the intrinsic characteristics of each isotope (similar to the isospin dependence).

One method suggesting itself was to use the simulated Woods-Saxon potential derived from an assumed nuclear matter distribution folded with a Yukawa type of nuclear-nuclei interactions. Such an analysis was presented at the Lowell Conf.[10] by the author and was shown to be moderately successful in predicting cross sections for E< 20 MeV. This procedure may be likened to that used by Greenlees et al[4,11] in their Reformulated Optical Model (ROM).

The equation for the folded Yukawa potential is taken from Ref. 10 and is given by

$$V(\bar{r}) = -\frac{V_o\mu^3}{4\pi}\int \frac{e^{-|\bar{r}-\bar{r}'|\mu}}{|\bar{r}-\bar{r}'|}\,\rho(\bar{r}')\,d^3\bar{r}' \qquad (4)$$

where V_o is the well depth experienced by the neutron; μ the range of the Yukawa folding function and a constant matter density ρ is assumed.

As pointed out in Ref. 10, V(r) in Eq. (4) is shown to depend on both (N-Z)/A and the binding energy, thus yielding a unique potential for every isotope.

A typical potential of this type is compared with the conventional Woods-Saxon potential for ^{52}Cr in Fig. 6 where it is assumed that V_o and R_o are the same for both potentials.

The next step was to renormalize the potential given in Eq. (3) by V(r) at R = 0 in Eq. (4).

A similar procedure for the imaginary potential when V_{IM} (r) = dV(r)/dr (5) produced an imaginary potential similar to the usual surface form (i.e. Gaussian or surface-derivative) as is seen in Fig. 6.

The magnitude of the imaginary potentials derived from the least squares fit of (AP) were subsequently renormalized to this potential.

These simulated potentials along with the resulting radii and diffusivities were used in O. M. calculations for Cr, Fe and Ni for energies up to 200 MeV.

While the results were favorable, it was found that in order to fit the minima and maxima in the total cross sections over the energy range of 0.5 to 100 MeV, it was necessary to assume an energy dependence for both the real and imaginary radii.

The maxima and minima as predicted from the earlier analysis used in establishing the loci in Fig. 5, provided the empirical device leading to this energy dependence.

Having derived the O. M. parameters using this recipe, calculations were carried out for the Cr, Fe and Ni isotopes. A typical calculation of σ_{tot} is seen in Fig. 7 for ^{52}Cr where the fit to experimental data is very good from E - 1.5 to 100 MeV. The dashed curve in Fig. 7 shows this fit below 1.5 MeV where the calculated σ_{tot} is seen to be somewhat higher than the experimental values.

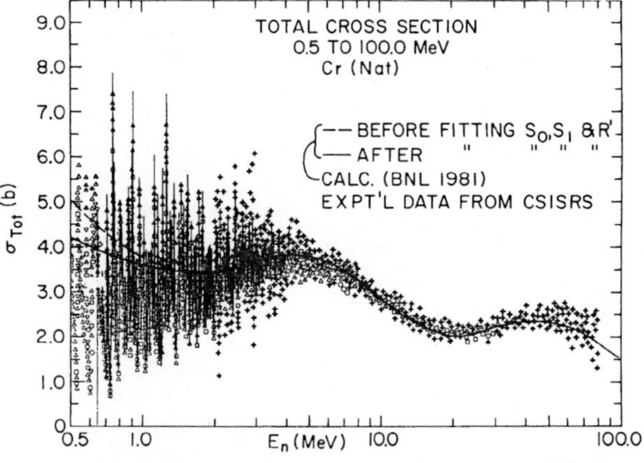

Fig. 7 Comparison of O.M. Calcs of σ_{Tot} with Experimental Data for Cr (N a t).

Due to the extreme structure in this energy region, an averaging process was employed which yielded an average set of values over selected energy bins. This too indicated that the calculated values were too high.

In order to overcome this difference, a ''SPRT'' type analysis was undertaken, however, rather than using the X^2 procedure of Delaroche et al,[12] a much simpler approach was taken.

Porter[13] showed that the S-wave strength function is proportional to the imaginary potential and is given by

$$S_o \; \alpha \int_0^{\infty} dr \, |U_0(r)|^2 \, W(\bar{r}) \qquad (6)$$

where

$U_0(\bar{r})$ is the gross neutron wave function at zero energy: Moldauer[14] observed this relationship also when he fitted the S-wave strength functions for A~100. Thus merely shifting the maximum of the imaginary potential obtained from the modified Avery'anov parameters, a value of V_{IM} was found that produced not only a lower σ_{tot} that agreed with the average σ_{tot} for E < 1.5 MeV (see Fig. 7) but also excellent agreement with S_0, S_1, R.

The resulting O. M. potential and radii are of the following forms

Real Potential (Woods - Saxon Form)

$$V_{Real} = V_0 + \alpha_1 E + \alpha_2 E^2$$

Imaginary Potential (Gaussian Form)

$V_{IM}(I) = V_1 + \beta_1 E + \beta_2 E^2$ E < 25 MeV

$V_{IM}(II) = V_2 + \beta_3 E + \beta_4 E^2$ $25 \leq E \leq 45$ MeV

$V_{IM}(III) = V_3 + \beta_5 E + \beta_6 E^2$ E > 45 MeV

Nuclear Potential Radii

$r_{Real} = r_o + \gamma_1 E$, $r_{IM} = r_1 + \gamma_2 E$ E < 15 MeV

$r_{Real} = r_{IM} = r_2 + \gamma_3 E$ $15 \leq E < 100$ MeV

Due to space limitations the O. M. parameters are given below only for the main isotopes of Cr, Fe and Ni, however, the remaining O. M. parameters for the other isotopes are available upon request.

All energies in MeV (Lab) geometrical parameters in fm.

^{52}Cr

$V_{RE} = 48.37 - 0472E + 0.0020E^2$

$V_{IM}(I) = 12.0 + 0.358E - 0.007E^2$ E<25MeV

$V_{IM}(II) = 0.625 + 0.888E - 0.01E^2$ $25 \leq E \leq 45$

$V_{IM}(III) = 3.23 + 0.53E - 0.00333E^2$ E>45

$V_{SO} = 6.44$ MeV

$r_{RE} = 1.2649 + 0.00252E$, $r_{IM} = 1.5329 - 0.01533E$ E<15MeV

$r_{RE} = r_{IM} = 1.3353 - 0.00218E$ $15 \leq E \leq 100$MeV

$r_{SO} = r_{RE}$

$a_{RE} = 0.7692$ fm $a_{SO} = a_{RE}$

$a_{IM} = 0.6203$ fm

^{56}Fe

$V_{RE} = 48.06 - 0.469E + 0.0022E^2$

$V_{IM}(I) = 12.0 + 0.358E - 0.007E^2$ E<25MeV

$V_{IM}(II) = 0.543 + 0.90E - 0.010E^2$ $25 \leq E < 45$MeV

$V_{IM}(III) = 2.52 + 0.543E - 0.00342E^2$ E>45MeV

$V_{SO} = 6.7$ MeV

$r_{RE} = 1.2605 + 0.00149E$, $r_{IM} = 1.49 - 0.0149E$ E<14 MeV

$r_{RE} = r_{IM} = 1.308 - 0.0191E$ $14 \leq E \leq 100$

$a_{RE} = 0.7776$ $r_{SO} = r_{RE}$

$a_{IM} = 0.6271$ $a_{SO} = a_{RE}$

^{58}Ni

$V_{RE} = 49.33 - 0.48E + 0.0024E^2$

$V_{IM}(I) = 12.0 + 0.358E - 0.007E^2$ E<25

$V_{IM}(II) = 0.445 + 0.908E - 0.011E^2$ $25 \leq E < 45$MeV

$V_{IM}(III) = 2.30 + 0.55E - 0.00346E^2$ E>45MeV

$V_{SO} = 6.75$ MeV

$r_{RE} = 1.2583 + 0.00258E$, $r_{IM} = 1.4645 - 0.0146E$ E<12MeV

$r_{RE} = r_{IM} = 1.3128 - 0.00196E$ $12 \leq E \leq 100$MeV

$r_{SO} = r_{RE}$

$a_{RE} = 0.7813$ $a_{SO} = a_{RE}$

$a_{IM} = 0.6300$

Figures 8 and 9 show the calculations for the natural elements of Fe and Ni based on the O. M. parameters discussed herein.

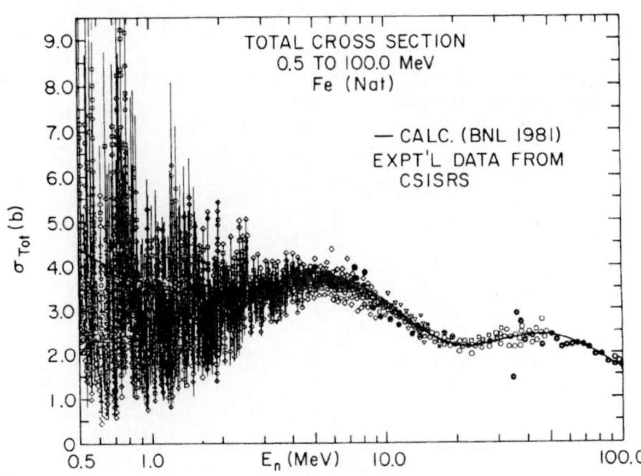

Fig. 8 Comparison of O.M. Calcs. of σ_{Tot} with Experimental Data for Fe (N a t).

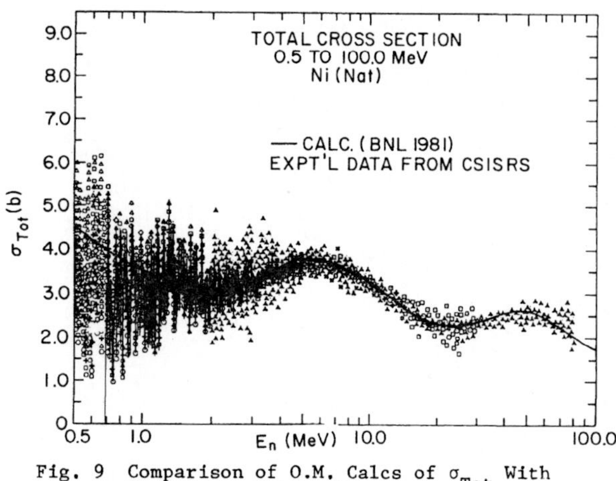

Fig. 9 Comparison of O.M. Calcs of σ_{Tot} With Experimental Data for Ni (Nat).

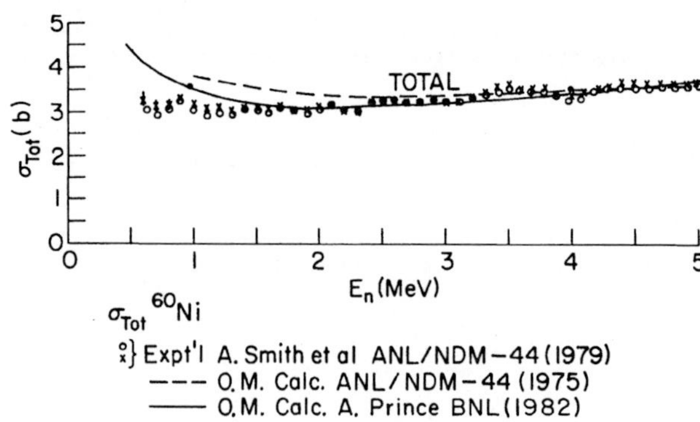

Fig. 11 Comparison of O.M. Calcs With Experimental Data for ^{60}Ni.

Table II shows a comparison of the calculated values of S_o, S_1 and R for the isotopes of Cr, Fe and Ni with other values quoted from experimental data.

Table II

	S_o(Calc)x10⁴	S_o(Expt'l)x10⁴	S_1(Calc)x10⁴	S_1(Expt'l)x10⁴	R´(Calc) fm	R´(Expt'l) fm
^{50}Cr	2.90	3.6 ± 0.8	0.41	0.33 ± 0.12	5.67	5.0 ± 0.30
^{52}Cr	2.68	2.5 ± 0.9	0.40	0.52 ± 0.12	5.20	5.2 ± 0.4
^{53}Cr	3.49	4.5 ± 1.1	0.65	--	5.43	5.4 ± 0.3
^{54}Cr	2.98	2.8 ± 1.0	0.43	--	5.59	5.3 ± 0.3
^{54}Fe	2.40	8.7 ± 2.4	0.58	0.58 ± 0.11	6.10	5.0 ± 0.3
^{56}Fe	2.70	2.6 ± 0.6	0.46	0.45 ± 0.05	6.06	6.1 ± 0.3
^{57}Fe	3.33	4.2 ± 1.2	0.45	0.2 ± 0.1	5.99	5.9 ± 0.3
^{58}Fe	2.58	3.6 ± 1.2	0.49	0.6 ± 0.2	6.00	6.1 ± 0.7
^{58}Ni	2.72	2.8 ± 0.6	0.58	0.5 ± 0.1	6.97	8.0 ± 0.5
^{60}Ni	2.30	2.7 ± 0.6	0.63	0.3 ± 0.1	6.30	6.7 ± 0.3
^{61}Ni	3.27	3.2 ± 0.8	0.47	--	6.54	6.5 ± 0.3
^{62}Ni	2.40	2.8 ± 0.7	0.70	0.3 ± 0.1	6.70	6.2 ± 0.3
^{64}Ni	2.64	2.9 ± 0.8	0.66	0.6 ± 0.2	7.0	7.55 ± 0.3

Experimental values from Ref. 15.

Figures 10 and 11 display the results for the isotopes ^{58,60}Ni in the low energy region.

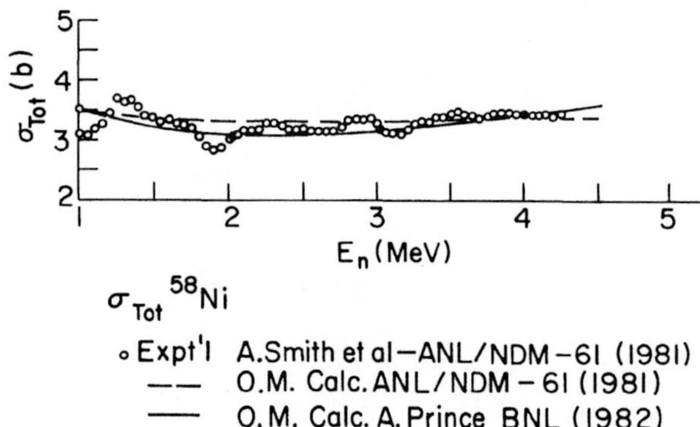

Fig. 10 Comparison of O.M. Calcs. With Experimental Data for ^{58}Ni.

In Figures 12 and 13 the calculated differential neutron scattering at 14.0 and 24.8 MeV for Cr and Fe respectively compare favorably with the experimental data. Agreement with the nonelastic in the high energy region, while not shown, was also achieved.

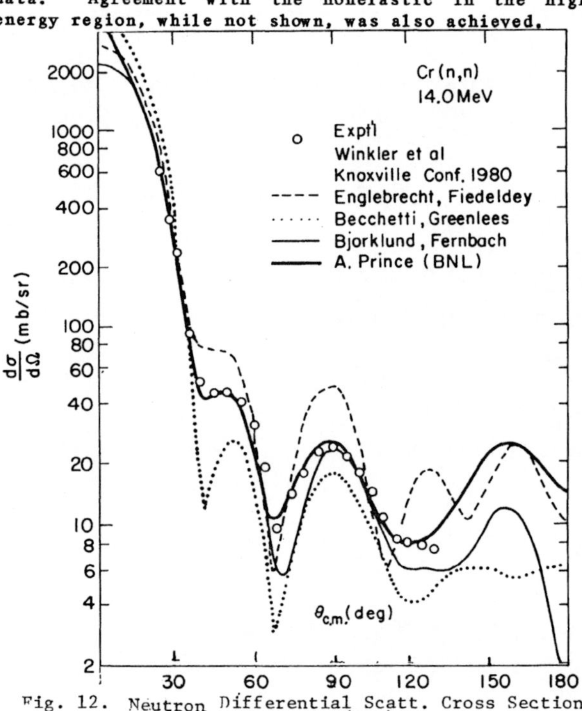

Fig. 12. Neutron Differential Scatt. Cross Section at 14.6 MeV for Cr (Nat).

Calculations for all the isotopes of Cr, Fe and Ni have been carried out up to 100 MeV.

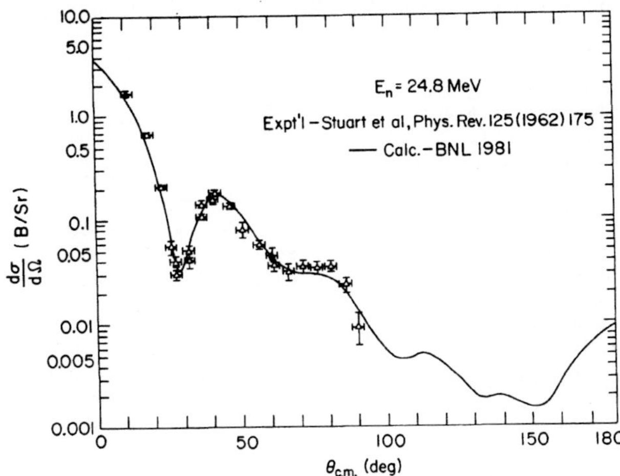

Fig. 13 Neutron Differential Scatt. Cross Section for ^{56}Fe.

Conclusion

Using a rather simple prescript, it has been possible to obtain a set of O. M. parameters which yield satisfactory agreement with experimental data on the structure materials for σ_{tot}, $(d\sigma_{el})/dr$ and σ_{non}, from the low to high (100 MeV) regions.

The emphasis thus far has been devoted only to this mass region. The next logical step is to see how well this procedure works in other mass regions. Of course, it is anticipated that deformation effects will play a role in such mass regions as the fission products, rare earth and fissile and fertile isotopes.

A remark concerning the three region structure for the imaginary potential (see Fig. 13) should be made here, namely, it is felt that although the approach is phenomenological the coincidental shifting of the maxima/minima of V_{IM} to reproduce the ''Ramsauer'' effect in σ_{tot} might serve as a paradigm worthy of further investigation.

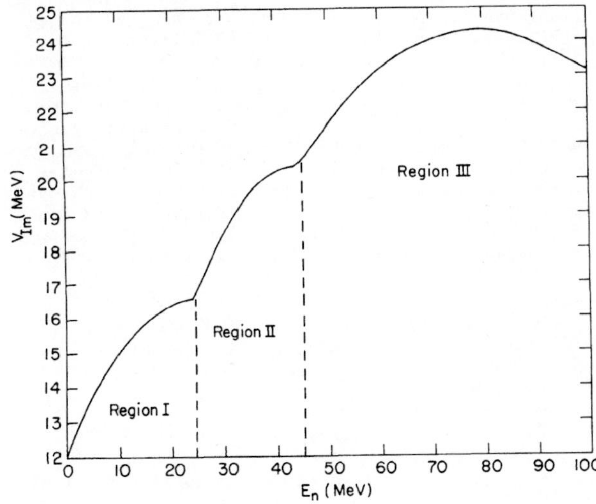

Fig. 14 Imaginary Potential V_{IM} For ^{52}Cr $V_{IM}(r)$ Gaussian Form.

It should also be noted that the use of only a Gaussian form to span the low to high energy region is unusual, since a volume term is also considered in many global sets, however, since the starting point in this investigation was with the work of (AP) it was more convenient to adhere to this approach. It might also be mentioned that in this aforementioned work of (EF), that the imaginary volume term vanished at low energies and only a Gaussian potential remained.

Preliminary investigation to obtain a O. M. set of potentials for protons has indicated that a volume term will be necessary to obtain agreement with the experimental proton reaction cross sections.

It is hoped that such an additional analysis will provide sufficient means to produce the necessary transmission coefficients used in the statistical model calculations of the various reaction cross sections.

References

1. ''Symposium on Neutron Cross Sections from 10 to 40 MeV,'' BNL-NCS-50681 (1977).

2. ''Symposium on Neutron Cross Sections from 10 to 50 MeV,'' BNL-NCS-51245 (1980).

3. D. Wilmore and P. E. Hodgson Nucl. Phys. 55, 673 (1964).

4. G. W. Greenlees et al., Phys Rev. 171, 1115 (1968).

5. C. A. Engelbrecht and H. Fiedeldey, Ann. of Phys. 42, 262 (1967).

6. I. K. Aver'yanov and Z. Z. Purtseladze, Sov. J.Nucl. Phys. 6 (2), 212 (1968).

7. R. H. Lemmer et al., Nucl. Phys. 12, 619 (1956).

8. A. M. Lane and C. F. Wandal, Phys. Rev, 98, 1524 (1955).

9. J. M. Peterson, Phys. Rev. 125, 955 (1962).

10. A. Prince, Int'l Conf on Interaction of Neutrons with Nuclei, (1976), Lowell, Mass.

11. G. J. Pyle and G. W. Greenlees, Phys Rev. 181, 1444 (1969).

12. J. P. Delaroche et al., IAEA Consultants Mtg. Trieste (1975).

13. C. E. Porter, Phys Rev. 100, 935 (1955).

14. P. A. Moldauer, Nucl. Phys., 47, 65 (1963).

15. ''Neutron Resonance Parameters and Thermal Cross Sections,'' Vol. I, Part A (Z=1 to 60), S. F. Mughabghab et al., Academic Press, N.Y. 1981.

INTERACTION DIRECTE DES NEUTRONS RAPIDES DIFFUSES PAR LES TETES DE BANDES DE VIBRATION β ET γ POUR LES ISOTOPES PAIRS DU TUNGSTENE

J.P. Delaroche

Service de Physique Neutronique et Nucléaire
Centre d'Etudes de Bruyères-le-Châtel
B.P. n° 561
92542 Montrouge Cedex, France

As an extension of a previous report on fast neutron scattering from the first and second excited states of $^{182,184,186}W$, we have investigated the question of the importance of direct interaction of neutrons with β- and γ- band levels of these nuclei. For that purpose we have performed new coupled channel calculations using extended coupling bases and collective wave functions from the Rotation-Vibration and Kumar and Baranger Models. The predictions are compared with recent scattering data obtained at incident energies below 4 MeV.

[$^{182}W(n,n')$, $^{184}W(n,n')$, $^{186}W(n,n')$, differential and integrated cross sections, vibrational states, E = 0.9-4 MeV, coupled channels calculations]

Introduction

Si l'excitation des états de vibration β et γ dans les noyaux déformables et déformés par diffusion inélastique de particules chargées est un phénomène relativement bien connu, il n'est pas encore fermement établi que les neutrons rapides excitent ces mêmes états aux énergies incidentes (E) de quelques MeV. En fait, quelques résultats expérimentaux relatifs à des noyaux des terres rares ($^{184,186}W$) et des actinides (^{232}Th, ^{238}U) semblent indiquer qu'une telle interaction existe pour les têtes de bandes de vibration γ [1,2]. Malheureusement, aux energies incidentes inférieures à quelques MeV, la diffusion inélastique par formation du noyau composé est encore importante, ce qui rend délicate toute estimation précise de la fraction des sections efficaces issue du mécanisme d'interaction directe pour ces états de vibration. Dans le présent travail nous montrons comment décrire l'interaction directe des neutrons avec les états de la bande du fondamental, de la bande β, et de la bande γ dans le cadre du modèle optique en voies couplées, d'une façon simple qui implique l'usage d'un nombre très réduit de paramètres. Des illustrations sont données pour les noyaux ^{182}W, ^{184}W et ^{186}W.

Modèle optique en voies couplées

Bien qu'elles n'autorisent que le couplage des états de la bande de rotation construite sur l'état fondamental (gB), les équations couplées telles qu'elles sont écrites [3] pour le modèle rotationnel constituent le point de départ des études présentées dans ce rapport. L'extension du formalisme en voies couplées aux états de vibration β(βB) et γ(γB) est faite en insérant dans ces équations des éléments de matrice réduits $<I||M(E\lambda)||I'>$ de l'opérateur de transition M(Eλ). Ici l'état physique |IM> est un état collectif appartenant soit à gB, soit à βB ou γB. L'approximation [4] qui est introduite consiste à utiliser les facteurs de forme du potentiel optique liés au modèle rotationnel "simple". Cette approximation s'est déjà révélée fructueuse [5] dans l'analyse de la diffusion inélastique de protons par des isotopes du platine. Les états |IM> introduits dans les calculs faits à l'aide du code ECIS79 [6] sont issus de modèles de structure nucléaire censés décrire de façon convenable les propriétés collectives des noyaux considérés présentement. Pour les noyaux situés dans la région des terres rares et des actinides, le modèle Rotation-Vibration (RVM) proposé par Faessler et al.[7] semble être approprié ; il en va de même pour le modèle de Kumar et Baranger (KB) qui a été utilisé de façon systématique dans la région des terres rares [8]. Des éléments de matrice réduits $<I||M(E2)||I'>$ pour les terres rares ont été calculés à l'aide du modèle KB et présentés dans une publication ancienne [9]. Les éléments de matrice réduits du modèle RVM peuvent être calculés à l'aide d'expressions analytiques et de tables publiées [7].

Un exemple de calculs en voies couplées utilisant les fonctions d'onde KB pour le noyau ^{184}W est présenté en Fig. 1. Dans celle-ci la contribution de l'interaction directe à la section efficace $\sigma(2^+_\gamma)$ est représen-

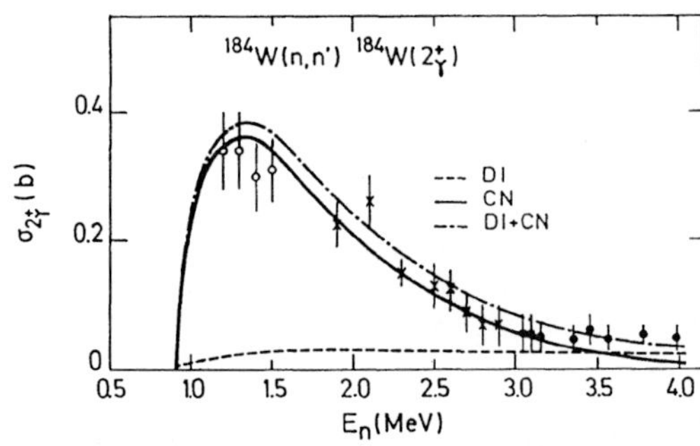

Fig. 1. Fonction d'excitation pour la section efficace $\sigma(2^+_\gamma)$ du noyau ^{184}W. Les points expérimentaux sont issus de la Réf. 1. Les courbes représentent des calculs expliqués dans le texte.

tée par la ligne en tirets. Ces calculs ont été effectués avec la base de couplage (0^+, 2^+, 4^+, 2^+_γ) et à l'aide de paramètres de potentiel et de déformation publiés récemment [10]. On peut voir que la section efficace DI croît depuis le seuil (E_X = 0.905 MeV) et atteint rapidement un plateau ($\sigma_{DI} \sim 30$ mb) qui se prolonge au moins jusqu'à 4 MeV. A cette énergie la contribution [10] σ_{CN} du noyau composé (ligne en trait plein) est plus faible que σ_{DI} de moitié. Combinées de façon incohérente (ligne en points-tirets), les composantes σ_{DI} et σ_{CN} représentent assez bien les valeurs expérimentales de $\sigma(2^+_\gamma)$ obtenues entre 3 et 4 MeV. Des résultats semblables à ceux-ci ont également été obtenus pour l'isotope ^{186}W. Ces calculs en voies couplées avec des bases de couplage incluant des états de vibration 2^+_γ ou 0^+_β prédisent aussi des distributions angulaires $d\sigma/d\Omega(2^+_\gamma)$ ou $d\sigma/d\Omega(0^+_\beta)$. Quand elles sont combinées avec celles produites [10] par le modèle statistique, un bon accord entre les prédictions et les résultats expérimentaux est réalisé. C'est le cas pour les distributions angulaires $d\sigma/d\Omega(2^+_\gamma)$ et $d\sigma/d\Omega(0^+_\beta)$ mesurées [1], respectivement, pour les noyaux : ^{186}W(E = 3 MeV) et ^{182}W(E = 2.7 MeV).

Discussion

Une extension des calculs en voies couplées aux états de vibration β et γ a été présentée pour des isotopes pair-pair du tungstène. La méthode utilisée pour y parvenir est simple et, malgré une hypothèse restrictive faite sur les potentiels de transition, conduit à des prédictions raisonnables pour la forme et la grandeur des sections efficaces d'interaction directe

K. H. Böckhoff (ed.), Nuclear Data for Science and Technology, 580–581.

induites par des neutrons rapides pour ces états. Ce type de calcul peut être étendu à d'autres noyaux, en particulier à ceux situés dans la région des actinides qui ont des bandes de vibrations β et γ localisées assez bas en énergie d'excitation. Cependant, l'adéquation de ces calculs en voies couplées nécessite un plus grand nombre de données expérimentales aux énergies incidentes pour lesquelles le mécanisme d'interaction directe est prépondérant.

Les calculs présentés ici pour des isotopes du tungstène sont jusqu'à un certain point incomplets car les couplages aux états 2^+_γ et 0^+_β ont été réalisés en prenant uniquement en compte les transitions E2. L'insertion de transitions multipolaires d'ordre plus élevé dans les calculs a montré dans le cadre du modèle RVM que l'on pouvait sous-estimer la grandeur des sections efficaces et ainsi altérer le bon accord théorie-expérience mentionné ci-dessus. De plus, les paramètres du modèle optique définis et optimisés en Réf. 10 par rapport à la base de couplage (0^+, 2^+, 4^+) ne conduisent plus au meilleur accord possible entre les calculs et les sections efficaces de diffusion inélastique mesurées pour les premier et second états excités de $^{184,186}W$ quand la base de couplage utilisée est (0^+, 2^+, 4^+, 2^+_γ) ou (0^+, 2^+, 4^+, 0^+_β), ou encore (0^+, 2^+, 4^+, 2^+_γ, 0^+_β). Ces résultats confirment de façon indirecte que les différentes bandes de rotation interagissent assez fortement entre elles. Ultérieurement il y aurait donc lieu de réoptimiser l'ensemble des paramètres du modèle optique adapté aux isotopes du tungstène.

Références

1. P.T. Guenther, A.B. Smith, J.F. Whalen, ANL/NDM-56 (1981)
2. E. Sheldon, D.W.S. Chan, Proceedings of a Specialists'Meeting on Fast Neutron Scattering on Actinide Nuclei, NEANDC-158 "U" (1982).
3. T. Tamura, Rev. Mod. Phys. 37, 679 (1965)
4. F.T. Baker et al., Nucl. Phys. A258, 43 (1976)
5. P.T. Deason et al., Phys. Rev. C23, 1414 (1981)
6. J. Raynal, Code ECIS79 (non publié)
7. A. Faessler, W. Greiner, R.K. Sheline, Nucl. Phys. 70, 33 (1965)
8. K. Kumar et M. Baranger, Nucl. Phys. A122, 273 (1968)
9. K. Kumar, Phys. Lett. 29B, 25 (1969)
10. J.P. Delaroche et al., Phys. Rev. C23, 136 (1981)

ON THE INELASTIC SCATTERING OF FAST NEUTRONS IN STRONGLY DEFORMED NUCLEI

J.R. Fernández Díaz and R. Cabezas Solórzano

National Institute for Nuclear Research
Academy of Sciences of Cuba
P.O.B. 6795, Havana
Cuba

A study of elastic and inelastic neutron scattering by the nucleus ^{186}W with incoming energies of 1.8 and 2.75 Mev using the coupled channel method has been made. The 2^+ (0.122 Mev), 4^+ (0.3966 Mev), 2^+ (0.7375 Mev), 3^+ (0.8618 Mev) and 4^+ (1.031 Mev) excited states are considered. It is shown that in this range of energies, the referred process can be satisfactorily described by the Davydov-Filippov model, considering ^{186}W as a deformed nucleus with nonaxial symmetry, given the quadrupole and the hexadecapole deformations. The scattering process through the compound nucleus is calculated according to the Hauser-Feshbach formula.
It is shown that the presence of direct excitation processes can be partly due to the consideration of the nonaxiality of ^{186}W.

[^{186}W (n,n'), inelastic cross sections, E= 1.8 and 2.75 Mev, Davydov-Filippov model, coupled channel calculation, Hauser-Feshbach formula, hexadecapole deformation].

Introduction

Currently, there is great interest in studying the inelastic scattering at low energy neutrons in deformed transitional nuclei such as the ^{186}W. A number of theoretical predictions have been made suggesting that, in these nuclei, a deviation from the Hauser-Feshbach statistical theory is observable and manifests itself in an increase of the inelastic cross sections for certain excited states of those nuclei. Sit'ko et al.[1-3] investigated the neutron inelastic scattering in the ^{152}Sm at 2.47 Mev confirming a deviation from the Hauser-Feshbach theory for the 4_1^+ state. This deviation in Ref. 1 is atributed to the appearance of direct excitation effects on several levels of this nuclei.

In the case of the ^{186}W isotope which is a strongly deformed nucleus with nonaxial symmetry, it has been observed that the excitation of its ground-state rotational band is weakened, due to a more intensive excitation of the γ-vibrational band. For this reason, it is interesting to analize the deviation of the Hauser-Feshbach theory in this energy range, and to calculate the inelastic cross sections,taking into account more accurately the structure of the excited levels of this nucleus by considering the nucleus' nonaxiality and the inclusion of radius parametrization up to hexadecapole deformation. In Ref. 3 the consideration of the hexadecapole deformation is proposed as a possible way to explain the large value of the cross sections for the 4_1^+ level in transitional nuclei. In several investigations [4,5] large negative hexadecapole moments and deformations are predicted for W isotopes.

In this paper, an analysis of the elastic and inelastic scattering for neutrons at energies of 1.8 Mev and 2.75 Mev in the ^{186}W has been made. The direct and the compound nucleus processes were calculated considering the ^{186}W as a nonaxial nucleus and including the hexadecapole deformation up to the term $(Y_{42} + Y_{4-2})$.

Analysis

Since the ^{186}W is a deformed nucleus with nonaxial symmetry, it is possible to use the Davydov-Filippov model[6], with the asymmetry parameter equal to 35°.

To calculate the direct cross section the coupled channel method has been considered, using the ECIS code[7].

In our analysis the optical potential has three parts : a real one, a surface imaginary deformed one and a spin-orbital one, as shown in the following expression :

$$V_{opt}(r,\theta',\phi') = -V\frac{1}{1+\exp[(r-R_v(\theta',\phi'))/a_v]} + 4iW_d\, a_w \frac{d}{dr}\frac{1}{1+\exp[(r-R_w(\theta',\phi')/a_w]} + (\frac{\hbar}{M_\pi c})^2 V_{so}(\hat{\sigma}.\hat{\ell})\frac{1}{r}\frac{d}{dr}\frac{1}{1+\exp[(r-R_{o_{so}})/a_{so}]} \quad (1)$$

This potential was parametrized taking into account that

$$R_i(\theta',\phi') = R_{oi}\,[1+\alpha_{20}Y_{20}+\alpha_{22}(Y_{22}+Y_{2-2}) + \alpha_{40}Y_{40}+\alpha_{42}(Y_{42}+Y_{4-2})] \quad (2)$$

with $i= v$ and w. The angle θ'and ϕ'refers to the body-fixed system. For the quadrupole deformation, the Bohr's parametrization was used[8]

$$\alpha_{20} = \beta_2 \cos\gamma \quad ; \quad \alpha_{22} = \alpha_{2-2} = \frac{1}{\sqrt{2}}\beta_2 \,\mathrm{sen}\,\gamma$$

and for the hexadecapole deformation the Raynal's parametrization was used[7]

$$\alpha_{40} = \beta_4 \cos\gamma \qquad \alpha_{42} = \alpha_{4-2} = \beta_4 \,\mathrm{sen}\, 2\gamma$$

α_{44} term is omitted in our calculation since no $K= 4^+$ vibrational band is considered here. Equation (2) allows the 4^+ states at 0.3966 and the 4^+ at 1.031 Mev to be included simultaneously in a consistent way.

For the parameters β_2 and β_4, values 0.23 and -0.193 were taken from Ref. 4 respectively.

The initial set of parameters was taken from Ref. 10 and it was fitted to our calculation using the χ^2 method with the following results :

K. H. Böckhoff (ed.), Nuclear Data for Science and Technology, 582–584.

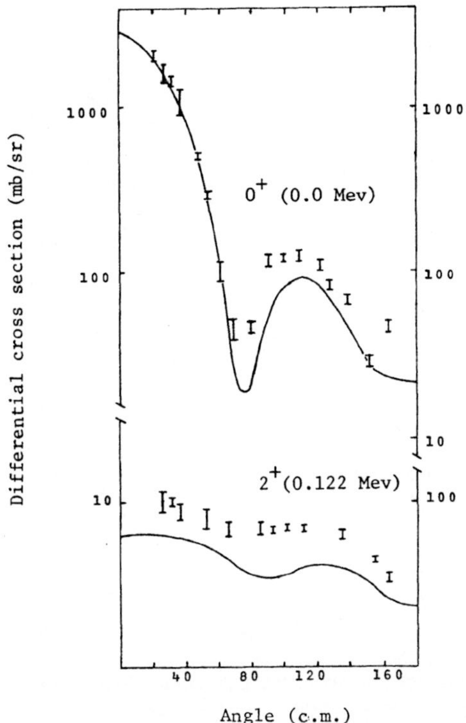

Fig. 1. Comparison of the calculated differential cross section (full line) with the experiment for ^{186}W (n,n') at E = 1.8 Mev.

V = 45.05 - 0.3 E Mev $\quad W_d$ = 6.0 Mev $\quad V_{so}$ = 6.2 Mev

R_v= 1.25 $A^{1/3}$ f $\quad R_w$ = 1.25 $A^{1/3}$f R_{so} =1.25 $A^{1/3}$f

a_v = 0.65 f $\quad a_w$ = 0.47 f $\quad a_{so}$= 0.65 f

The compound-nucleus cross section was determined by the Hauser-Feshbach formalism using the ECIS code. The required paramaters such as the level density and the spin cutoff were taken from Ref. 11.

Discussion

First, considering the coupling scheme $0^+ - 2_1^+ - 2_2^+$, the differential cross sections for the 0^+ and 2^+ (0.122 Mev) states were calculated at an incidental neutron energy of 1.8 Mev. Fig 1 shows a comparison of the calculations with the results of the experiments[12].

Table I. Comparison of symmetrical and asymmetrical rotor models at En = 2.75 Mev.

I^π	σ_{dir}(mb)	Symmetric rotor σ_{HF}(mb)	σ_{tot}(mb)
0^+	3699.32	134.75	3834.07
2_1^+	498.00	198.73	696.73
2_2^+	0.00	177.63	177.63
		Asymmetric rotor	
0^+	3524.61	124.00	3648.61
2_1^+	434.79	186.39	621.18
2_2^+	68.74	159.65	228.39

The calculations of total cross sections made by means of the symmetric rotor model and the asymmetric rotor model with γ= 35° are shown in Table I. In the same way, it is seen that the calculations made in the frame of the asymmetric rotor model closely agrees with the experimental results for the 2^+ state of the γ-vibrational band. It can also be observed that the consideration of the nucleus' asymmetry causes the direct excitation processes in the 2^+ state of the γ-vibrational band.

The calculations of total cross sections for the 0^+, 2^+ (0.122 Mev), 4^+ (0.3996 Mev), 2^+ (0.7375 Mev), 3^+ (0.8618 Mev) and 4^+ (1.031 Mev) states from the inelastic scattering at 2.75 Mev were made and compared with the results obtained in Ref. 2. As Table II shows, the results are somewhat under the experimental ones, but in general fairly close to them. This slight difference may be due to the effects of direct excitation on the different excited levels, as mentioned in Ref. 1,2 for ^{186}W and which were also observed in Ref. 1-3 for ^{152}Sm.

Table II. Calculated cross sections in the ^{186}W (n.n') process at En= 2.75 Mev considered only up to the quadrupole deformation.

I^π	σ_{dir}(mb)	σ_{HF}(mb)	σ_{tot}(mb)	σ_{exp}(mb)[1]
0^+	3565.34	109.76	3675.11	-
2_1^+	357.96	174.88	532.84	-
4_1^+	34.46	110.44	144.90	$250 \pm ^{50}_{80}$
2_2^+	50.27	143.53	193.80	260 ± 55
3_2^+	0.87	125.42	126.29	200 ± 60
4_2^+	8.59	89.02	97.61	65 ± 20

Throughout our calculations, we observed that the differential cross sections for the low-lying states were very sensitive, not only to the parameter used but also to the number of higher excited states considered in the calculations. To illustrate this, in Fig 2 the differential cross sections for the 0^+, 2^+ (0.122 Mev) and 2^+ (0.7375 Mev) states were calculated with different coupling schemes : $0^+ -2_1^+ -2_2^+$ and $0^+ -2_1^+ -2_2^+ -4_1^+ -3_2^+ -4_2^+$. The same sets of parameters were used in both cases. It may be noted that the inelastic differential cross section is decreased when a larger number of excited states is considered. The total cross sections show a similar behavior (see Table I and II).

Table III. Calculated cross sections in the ^{186}W (n,n') process at 2.75 Mev, including : a) up to the α_{40} Y_{40} additional term b) up to the α_{42} (Y_{42} + Y_{4-2}) vibrational mode.

I^π		σ_{dir}(mb)	σ_{HF}(mb)	σ_{tot}(mb)	σ_{exp}(mb)
4_1^+	a	61.67	107.82	169.49	$250 \pm ^{50}_{80}$
	b	114.28	114.10	228.38	
2_2^+	a	50.12	135.16	185.28	260 ±55
	b	29.74	143.18	172.92	
3_2^+	a	2.57	116.02	118.59	200 ±60
	b	2.47	123.83	126.30	
4_2^+	a	18.75	80.45	99.20	65 ±20
	b	5.65	84.02	89.67	

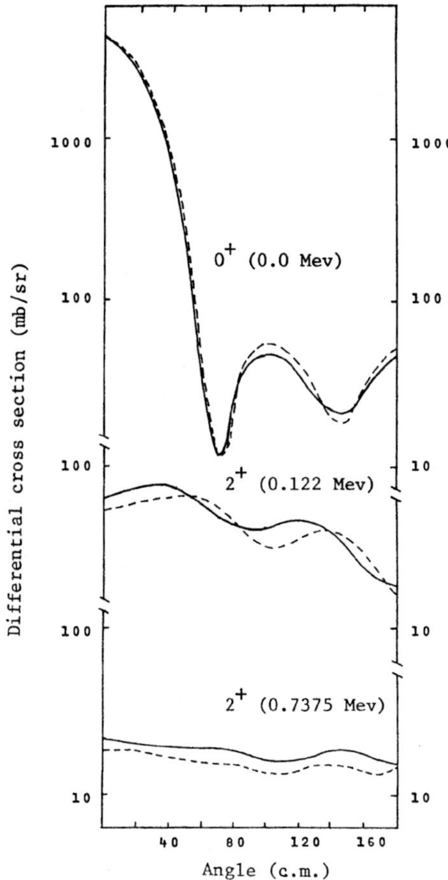

Fig.2. Angular distribution for the $0^+,2^+$ (0.122 Mev) and 2^+ (0.7375 Mev) states with different coupling schemes : $0^+ -2_1^+ -2_2^+$ (full line) and $0^+ -2_1^+ -2_2^+ -4_1^+ -3_2^+ -4_2^+$ (dash line).

Then, calculations were made again, now considering up to the hexadecapole deformation, the results of which can be seen in Table III. It is shown that when considering $\alpha_{40}\ Y_{40}$ additional term, values of the cross section for 4_1^+ state increase. This is due to the inclusion of the direct transition from the state to the 4_1^+ state. In order to consider direct transition from the ground state to the 4_2^+ state ,it is necessary to consider up to α_{42} ($Y_{42} + Y_{4-2}$) vibrational mode. This consideration, nevertheless, while increasing the value of the cross section for 4_2^+ state over the experimental variation range,lowers the cross sections for the rest of the states, as compared with the results obtained with the inclusion of Y_{40}.

Conclusions

The scattering process of ^{186}W (n,n') at energy ranges of 2 to 3 Mev can be satisfactorily described using the Davydov-Filippov model, considering ^{186}W as a deformed nucleus with nonaxial symmetry, with a 35° asymmetry parameter.

Due to the fact that ^{186}W is a nucleus with nonaxial symmetry, the proyection of the quantum number K on the symmetry axis is not conserved. This causes the equations for the coupled channels to vary considerably, resulting in a higher contribution of the direct processes to the states of the γ-vibrational band observed in the experiment. Therefore, we can state that the processes of direct excitation are partly due to the nonaxial status of ^{186}W

Consideration of hexadecapole deformation by means of ($Y_{42} + Y_{4-2}$) terms makes it possible to assess more thoroughly the transitions that may take place among the different nuclei states, but this deformation is not accountable for the high values of the inelastic cross sections obtained in the experiment.

The inclusion of the hexadecapole deformation rises the contribution of direct processes in the excitation of 4_1^+ and 4_2^+ levels, but the increase is not significant enough for the rest of the excited states, therefore we can affirm that such deformation, while having influence, is not the only determinant of the low cross section values obtained, in comparison with the experiments.

These differences could be due to nonconsideration of additional terms of the $\alpha_{\lambda 2}$ ($Y_{\lambda 2} + Y_{\lambda -2}$) form, coupling to other bands, or even inadequacies of the simple macroscopic collective rotational model.

It is, thus, interesting to make these calculations by means of a semiphenomenological model such as the folding model, that make it possible to consider, more realistically, the deformation of nuclear surface through parameters determining the form of the nucleus instead of those determining the form of potential.

The authors are grateful to Dr. J. Raynal for his information about the ECIS code and to Francisco Fernández for his help in processing the programs.

References

1. E.A. Andreev, V.K. Basenko,S.P. Sit'ko, Proc. of the Int. Conf. of the Interaction of Neutrons with Nuclei. Lowell, Mass., July, 1976,p.1334.
2. S.P. Sit'ko, Izvestiya AN SSSR. Ser. Fiz.42, 1809 (1978).
3. S.P. Sit'ko, E.A. Andreev, V.K. Basenko, Yad. Fiz. 25, 1119 (1977).
4. S.G. Rohozinski, A. Sobiczewski, Act. Phys.Pol. 812, 1001 (1981).
5. W. Brückner, D. Husar, D.Pelte, K. Traxel, M. Samuel, U. Smilansky, Nucl. Phys. A231, 159 (1974).
6. A.S. Davydov, G. E. Filippov, Nucl. Phys.8,237 (1958).
7. J. Raynal, CEN-SACLAY, Dph- T/71-48 (1971)
8. A. Bohr, Dgl. Danske Videnskab. Selskab. Mat.-Fys. Medd. Bd26, No. 14 (1952).
9. W. Nazarewcicz, P. Rozmej, Nucl . Phys. A369,396 (1981).
10. D.F. Coope, S.N. Tripathi, Phys. Rev. C 16,2223 (1977).
11. A. Gilbert, A.G.W. Cameron, Can . J. Phys. 43, 1446 (1965).
12. A. B. Kulkarni, 20th Nucl. Phys, and Sol. State Phys. Symposium, India 2, 142 (1979).

AN EFFECTIVE INTERACTION FOR THE NUCLEON-NUCLEAUS OPTICAL POTENTIAL

S.K. Gupta and D.R. Chakrabarty

Nuclear Physics Division
Bhabha Atomic Research Centre
Bombay 400 085, India

and

K.H.N. Murthy

Department of Physics
University of Mysore
Mysore 570 006, India

An effective two-body interaction in the nuclear matter has been defined using the t-matrices calculated by Brieva, Rook and Geramb. This interaction is expressed as a sum of two Yukawa form factors, one attractive and the other repulsive. While the range parameters are found independent of the density and the energy E, the interaction-strengths are parametrized as polynomials in ρ and E upto 60 MeV. This is accomplished for the real and the imaginary parts of both the isoscalar and the isovector potentials. Four prescription for folding the interaction to obtain the optical potentials have been discussed. Volume integrals for $^{40}Ca+n/p$ and $^{208}Pb+n/p$ are calculated and compared with phenomenology and other calculations.

Effective interaction as a function of energy, density, folding model; optical potential real, imaginary parts for $^{40}Ca+n/p$, $^{208}Pb+n/p$

Introduction

The phenomenological analyses of the nucleon-nucleus scattering employing the optical model potential have been quite successful. An understanding of the optical potential has been obtained through the microscopic calculation[1,2] exploiting the nuclear matter approach. This approach is based on the calculation of the inter-nucleon effective t-matrix in infinite nuclear matter within the framework of the Brueckner theory. The infinite nuclear matter results are to be extended to the finite nuclei to obtain the optical potential. However, the transition from the infinite to the finite is not quite clear and is ambiguous. Two alternatives have been followed in literature. Jeukenne, Lejeune and Mahaux (JLM)[1] evaluate the optical potential in finite nuclei using the local density approximation. However the r.m.s radii of the potential are underpredicted and are corrected further, using a phenomenological range parameter. In the other method, developed by Brieva and Rook[2], an effective interaction is defined in the coordinate space. The optical potential is obtained by folding the matter density of the finite nucleus with this effective interaction. Though this approach does not use free parameters, it is quite cumbersome in practice due to the availability of the interaction only in the form of quite large tables. Thus inspite of the tabulation of the interaction, by Geramb[3] in terms of five Gaussian form factors as well as five Yukawa form factors, the application of these results is tedious. Further, a comparison of the GBR (Geramb, Brieva and Rook) approach with that of JLM is not easy.

In the present work, we describe a new approach to calculate an effective interaction in nuclear matter and to obtain the optical potential for finite nuclei by folding the interaction over the nuclear density distribution. Our proposed interaction is based on the GBR tabulation but it is simpler and is a sum of two Yukawa form factors, The ranges of the form factors are independent of the energy E and the density ρ and the strength have been parametrized as polynomial functions of ρ and E. The ambiguity in connecting the infinite matter to the finite nuclei arises because of the density variation in finite nuclei. It is not clear what density should be used in calculating the effective interaction between two nucleons located at two points of different densities. We discuss four prescriptions of obtaining the optical potential in finite nuclei and the differences in their predictions. The potentials for proton and neutron scattering from ^{40}Ca and ^{208}Pb are discussed.

Effective Two Body Interaction

Brieva and Rook[2] define the optical potential in nuclear matter as

$$U_{NM} = \int \frac{1}{8}\left[\sum_{ST}(2S+1)(2T+1)\,t^{ST}(r,\rho_{NM},E)\left\{\frac{1}{2}(\rho_{NM} \pm \rho_{NM}^{EX} j_0(kr))\right\}\right] \times d^3r \quad (1)$$

Here t^{ST} are the t-matrices of the effective interaction between two nucleons with total spin S and total isospin T. These are functions of the separation r, the matter density ρ_{NM} and also the energy E. ρ_{NM}^{EX} is defined as

$$\rho_{NM}^{EX} = \rho_{NM} \frac{3}{rk_F} j_1(rk_F) \quad (2)$$

where k_F is related to ρ_{NM} as

$$\rho_{NM} = \frac{2}{3\pi^2} k_F^3 . \quad (3)$$

j_0 and j_1 are the spherical Bessel functions. The momentum k is related to E

$$E = \frac{\hbar^2 k^2}{2m} + Re\, U_{NM} \quad (4)$$

K. H. Böckhoff (ed.), Nuclear Data for Science and Technology, 585–588.

where m is the nucleon mass and Re U_{NM} is the real part of the optical potential. The + and - signs appearing in eqn (1) are for the odd and the even combinations of (S+T) respectively. Summing explicitely over S and T, one can write

$$U_{NM} = \rho_{NM} \int [T_D + \frac{3}{r k_F} j_1(r k_F) j_0(kr) T_{Ex}] \times d^3 r \qquad (5)$$

where the direct and the exchange parts of the interactions are defined as

$$T_D = (t^{00} + 3t^{01} + 3t^{10} + 9t^{11})/16 \qquad (6)$$

and $$T_{Ex} = (-t^{00} + 3t^{01} + 3t^{10} - 9t^{11})/16 . \qquad (7)$$

In the nuclear matter we define an effective interaction T_{eff} which gives the same optical potential and so from eqn (5) we write

$$T_{eff} = T_D + \frac{3}{r k_F} j_1(r k_F) j_0(kr) T_{Ex} . \qquad (8)$$

Defining the volume integral of the interaction as

$$J_0 = \int T_{eff} \, d^3 r . \qquad (9)$$

We thus write

$$U_{NM} = \rho_{NM} J_0 . \qquad (10)$$

The effective interaction defined by us thus have similar meaning as defined by JLM.

In the finite nuclei, we propose to obtain the optical potential by folding this T_{eff} over the matter density distribution. We thus combine the direct and the exchange interactions of Brieva and Rook into a single effective interaction. This makes the folding procedure also much simpler. We shall discuss in the next section the various procedures followed in obtaining the nuclear potential.

We have also separated the effective interaction into the isoscalar and the isovector parts, which give rise to the isoscalar U_0 and the isovector U_1 parts of the nuclear optical potential defined through the equations,

$$U_p = U_0 - \frac{N-Z}{A} U_1 \qquad (11)$$

and $$U_n = U_0 + \frac{N-Z}{A} U_1 \qquad (12)$$

where p and n refer to proton and neutron respectively. N,Z,A are the neutron, the proton and the mass numbers of the target nucleus. By writing down the effective interactions for the pp and pn systems in terms of t^{ST} - matrices, we deduce that for the isoscalar part, T_D & T_{EX} will be as defined in eqn (6) and (7) and for the isovector part we write

$$T_D^{I.V.} = (-t^{00} + t^{01} - 3t^{01} + 3t^{11})/16 \qquad (13)$$

and $$T_{Ex}^{I.V} = (t^{00} + t^{01} - 3t^{10} - 3t^{11})/16 \qquad (14)$$

An effective interaction for obtaining the isovector potential is then constructed by combining (13) and (14) with eqn (8).

These four components of the effective interaction (the isoscalar and the isovector components, both for the real and the imaginary parts) have been expressed as a sum of two Yukawa form factors for their r-dependence for various densities and energy values. The parameters of the form factors have been calculated so as to conserve the volume integral J_0 and the second moment J_2 of the interaction. We have chosen eight energy values, viz. E=2,6,10,20,30,42,51 and 60 MeV and eight density values corresponding to k_F = 0.6,0.8,0.9,1.0,1.1,1.2,1.3 & 1.4 fm^{-1}. We calculated the J_0 and J_2 values for T_{eff} at these energy and density combinations from the parametrized t-matrices in the GBR tables using the five Yukawa parametrization. Defining our interaction as,

$$T_{eff}(r) = V_c \frac{e^{-\mu_c r}}{\mu_c r} - V_0 \frac{e^{-\mu_0 r}}{\mu_0 r} . \qquad (15)$$

We calculate V_c, V_0, μ_c and μ_0 so as to conserve J_0 and J_2 as referred to earlier. We find that the values of μ_c and μ_0 can be taken independent of ρ and E for all the four components of the effective interaction they are given by

$$\mu_c = 4.81 \, fm^{-1} \qquad (16)$$

and $$\mu_0 = 2.46 \, fm^{-1} .$$

The first part in (15), which is of shorter range, is called the core part and the second part, which is of longer range is called the outer part by us. With these choices of μ_c and μ_0 V_c and V_0 have been calculated for the above choices of and E. These strength parameters have been then parametrized as polynomials in ρ and E. In this work E is in MeV and ρ in fm^{-3}.

Real Part

For both the isoscalar and the isovector parts, we expressed the strength V_c and V_0 as

$$V(\rho, E) = \sum_{\substack{i=1,3 \\ j=1,3}} a_{ij} \rho^{i-1} E^{j-1} . \qquad (17)$$

The values of a_{ij} for the core part and the outer part for both the isoscalar and the isovector components are tabulated in Table I. The isoscalar part is fitted on an average within 5% while isovector part is fitted within 10%.

Imaginary Part

The imaginary part could not be fitted with the expression (17) with only nine parameters and we used the relation,

$$V(\rho, E) = \sum_{\substack{i=1,3 \\ j=1,3}} a_{ij} \rho^{i-1} E^{j-1} + K(\rho - \xi)^2 \sum_{\substack{i=1,2 \\ j=1,3}} b_{ij} \rho^{i-1} E^{j-1} \qquad (18)$$

Here K = 1 if $\rho > \xi$

$= 0$ if $\rho \le \xi$

and $\xi = 0.067548\ fm^{-3}$ corresponding to $k_F = 1\ fm^{-1}$. However the above fit did not contain the effective interaction calculated from the GBR table for $k_F = 0.6\ fm^{-1}$ and we used semi-classical values for this density at various energies. The reason for this is that we find that the volume integral of the effective interaction oscillates as a function of energy E at $k_F = 0.6\ fm^{-1}$, when calculated from the GBR table. This looks unphysical and completely spoils the fit. We therefore used semiclassical values (which exhibit a smooth behaviour as a function of energy) calculated as described below.

The imaginary potential W is defined as

$$W = J_W \rho \qquad (19)$$

where J_W is the volume integral of the interaction, W is calculated from the relation[4])

$$W = \frac{1}{2}\hbar c \beta \sigma_{NN} \rho P(\zeta) \qquad (20)$$

Here ζ is a parameter defined as

$$\zeta = E_F / E_K \qquad (21)$$

where E_F is the Fermi energy,

$$E_F = \hbar^2 k_F^2 / 2m \qquad (22)$$

and $$E_K = E - Re\, U^{OMP} \qquad (23)$$

The function $P(\zeta)$ is defined as

$$P(\zeta) = 1 - \frac{7}{5}\zeta \qquad (\zeta \le \tfrac{1}{2})$$
$$= 1 - \frac{7}{5}\zeta + \frac{2\zeta}{5}\left(2 - \frac{1}{\zeta}\right)^{5/2} \quad (\zeta > \tfrac{1}{2}) \qquad (24)$$

β is defined as $$\beta = \sqrt{2E_K/mc^2} \qquad (25)$$

σ_{NN}, which is the nucleon-nucleon scattering cross section, was expressed as a sum of isoscalar and isovector parts. Incorporating these J_W values for $k_F = 0.6\ fm^{-1}$, we fit eqn(18) to the strengths of the imaginary interaction. Parameters a_{ij} and b_{ij} of eqn (18) are tabulated in Table II. Both the isoscalar and the isovector parts are fitted on an average within 10%.

Table I a_{ij} values for Real Part

Type	i \ j	1	2	3
Isoscalar Core Part	1	33317	-702.1	7.0965
	2	-247770	11169	-117.8
	3	850540	-39420	414.66
Isoscalar Outer Part	1	-5563.7	117.2	-1.1533
	2	41522	-1743.3	18.104
	3	-137560	6018.6	-62.161
Isovector Core Part	1	-14812	530.31	-5.5051
	2	183400	-9036.4	97.927
	3	-700440	35657	-393.41
Isovector Outer Part	1	2563.6	-87.538	0.87985
	2	-30294	1419.1	-15.084
	3	110690	-5468.2	59.298

Table II a_{ij} and b_{ij} values for Imaginary Part

Type		i \ j	1	2	3
Isoscalar Core	a_{ij}	1	9330	202.1	-3.305
		2	-165200	-4458	76.16
		3	856300	32760	-507.9
	b_{ij}	1	-6685	-40590	333.1
		2	-3087000	36120	694.3
Isoscalar Outer	a_{ij}	1	-1670	35.87	0.6232
		2	31420	817.6	-15.32
		3	175000	-6050	105.8
	b_{ij}	1	45700	7382	-81.29
		2	468200	-6024	-99.41
Isovector Core	a_{ij}	1	-8261.1	-27.096	1.6589
		2	194520	461.01	-46.1
		3	-1296600	-4760.6	348.03
	b_{ij}	1	1033400	12736	-385.85
		2	913220	-33497	149.06
Isovector Outer	a_{ij}	1	1260.3	9.2207	-0.31592
		2	-27427	-224.29	8.7291
		3	170920	1968	-65.348
	b_{ij}	1	-98902	-3676	70.979
		2	-259180	7094	-22.605

Before proceeding to describe the calculation of the optical potential, we note a few points. The GBR parametrizations of the t-matrices in terms of the Gaussian form factors and the Yukawa form factors give the same value of the volume integrals of the interaction but the second moments differ by a factor of two. We have preferred the Yukawa parametrization.

The r-dependence of the calculated $T_{eff}(r)$ is reproduced quite well by our proposed two Yukawa form factors, demonstrating that the conservation of the zeroeth and the second moment of the interaction is a good criterion for the parametrization. We also note that the contribution to the interaction from the exchange part is quite substantial specifically in the case of the isoscalar part ranging upto 70%.

Calculation of the Optical Potential

We calculate the optical potential in the finite nuclei, by folding the effective interaction over the density distribution of the nucleus, viz.

$$U(r) = \int \rho(\vec{r}')\, T_{eff}(|\vec{r}-\vec{r}'|, \rho_x, E_x)\, d^3r' \qquad (26)$$

$\rho(\vec{r}')$ is the density of the target nucleus at $\vec{r}'$. We have chosen a two-parameter Fermi-distribution for $\rho(r)$ with parameters as given in ref.5,

$$\rho(r') = \frac{\rho_0}{1 + \exp\frac{r'-c}{a}} \qquad (27)$$

where $\rho_0 = \frac{3A}{4\pi c^3}\left(1 + \frac{\pi^2 a^2}{c^2}\right)^{-1}$

$$c\,(fm) = 1.116 A^{1/3} - 0.813/A^{1/3}$$
$$\text{and} \quad a\,(fm) = 0.525 \qquad (28)$$

T_{eff} in eqn(26) is a function of $|\vec{r}-\vec{r}'|$, the effective energy Ex and the density ρ_x. The effective energy Ex is defined as

$$E_x = E - V_c(r') \qquad (29)$$

where $V_c(r')$ is the Coulomb potential. We have chosen four prescriptions for the choice of T_{eff} calculation, viz.

(1) $\rho_x = \rho(r)$ (2) $\rho_x = \rho(r')$

(3) $T_{eff} = \frac{1}{2}[T_{eff}(\rho(r)) + T_{eff}(\rho(r'))]$ and $\rho_x = \frac{\rho(r)+\rho(r')}{2}$.

The prescriptions (1) and (2) yield the limiting choices of the density

Results and Discussion

We have calculated the optical potentials, their volume integrals per nucleon J and r.m.s. radii for ^{40}Ca+n/p and ^{208}Pb+n/p. The effect of using various choices of ρ_x is exemplified in Table III where we tabulate the calculated J and r.m.s. radii for ^{208}Pb+n system.

Table III Volume Integrals/nucleon in MeV fm^3 and the rms radii in fm with prescriptions 1-4

	E_n(lab) MeV		Prescription 1	2	3	4
REAL PART	8.05	J	345.2	316.2	330.7	323.5
		$\langle r^2\rangle^{1/2}$	6.38	6.21	6.30	6.28
	30.3	J	255.8	247.6	250.4	250.4
		$\langle r^2\rangle^{1/2}$	6.27	6.20	6.23	6.23
IMAGINARY PART	8.05	J	64.8	43.7	54.3	47.6
		$\langle r^2\rangle^{1/2}$	7.4	7.03	7.26	7.25
	30.3	J	83.5	59.9	71.7	64.5
		$\langle r^2\rangle^{1/2}$	7.17	6.75	7.0	6.96

It is seen with increasing energy, the various prescriptions give similar results for the real part. At low energies they differ, the difference in the real volume integral is maximum 15%. The rms radii agree within 4%. However the imaginary volume integral varies as much as by 60%. The prescriptions (1) and (2) predict the extreme values and (3) and (4) give intermediate values. Henceforth we shall concentrate on the prescription (4) which looks to be the most reasonable one.

Comparing our results with those of Brieva and Rook, we find that our volume integrals for the real part are consistently less than those of these authors by about 10 to 12%. For the imaginary part on the other hand, the agreement is better than the real part. The predicted values differ mainly due to the different approach we have adopted for the finite nuclei, because in the nuclear matter the volume integrals of the effective interaction are as same those of Brieva and Rook. The r.m.s. radii agree within 5%.

Our volume integrals for the real part are systematically less than the phenomenological values with increasing energy. We have found a normalisation factor N_R to fit the phenomenological values, given by

$$N_R = 1.14 + 0.0047\,(E - Z_P Z_T / A_T^{1/3}). \qquad (30)$$

For the imaginary part our volume integrals agree well with the phenomenological values within 10%, except for the case of ^{40}Ca+n at 7.91 MeV and ^{208}Pb+p at 30.3 and 40 MeV. In Table IV we compare the volume integrals of our calculation (normalised for the real part) with the phenomenological values as well as with the JLM calculation. For the real part the JLM values agree quite well with the phenomenological values although for the imaginary part they are high. It is worthwhile to point out that the volume integral of the real part in the nuclear matter, is differently predicted by JLM and the GBR calculations by as much as 25% in some cases.

Table IV Comparison of the volume integrals/nucleon (MeV fm^3) from the present work with the phenomenological and the JLM values. The real part is normalised using eqn (30)

Reaction	E_{lab} (MeV)	Present work Re	Im	Pheno. Re	Im	Ref	JLM Re	Im
^{40}Ca	7.91	486	85.0	482	60.5	6	476	106
+n	30.3	386	106	413	113	7	397	152
	40.0	356	109	348	97.2	7	365	151
^{40}Ca	30.3	415	101	424	103	8	421	148
+p	40.0	382	106	379	102	9	388	151
	8.05	381	47.6	388	43.8	10	386	54.8
^{208}Pb	30.3	321	64.5	320	67.3	7	330	91.3
+n	40.0	300	68.4	296	74.6	7	307	94.5
^{208}Pb	30.3	427	62.5	413	104	11	417	106
+p	40.0	393	71.2	375	105	11	388	120

Conclusion

We have proposed a simplified effective two-body interaction based on the microscopic calculation by Brieva and Rook. The real part of the optical potential calculated by us is low compared to the phenomenological values and we deduce a normalisation factor. The imaginary part agrees in general.

References

1. J.P. Jeukenne, A. Lejeune and C.Mahaux, Phys. Rev. C 16, (1977) 80
2. F.A. Brieva and J.R. Rook, Nucl. Phys. 291 (1977) 299 ; F.A. Brieva and J.R. Rook, Nucl. Phys. A291 (1977) 317
3. H.V. Geramb, (1979) University of Hamburg tables (Private Communication)
4. K. Kikuchi and M. Kawai, Nuclear Matter and Nuclear Reactions, 1968 (North-Holland) chapter 2
5. S.K. Gupta and S. Kailas, Phys. Rev. C23 (1981) 928
6. J.D. Reber and J.D. Brandenberger, Phys. Rev. 163 (1967) 1077
7. R.P. Devito, Ph.D. Thesis (1979) Michigan State University
8. B.W. Ridley and J.F. Turner, Nucl. Phys. 58 (1964) 509
9. L.N. Blumberg et al, Phys. Rev. 147 (1966) 812
10. B. Holmquist and T. Wiedling, A.E.C. Sweden AE-430 (1971)
11. W.T.H. Van Oers et al, Phys. Rev. C10 (1974) 307.

THE NEUTRON OPTICAL MODEL PARAMETERS FOR $^{235,238}U$ and $^{239,240}Pu$

Shen Qingbiao, Chao Xiaolin, and Gu Yingqi

Institute of Atomic Energy, Academia Sinica, Beijing, China

The neutron optical model parameters for $^{235,238}U$ and $^{239,240}Pu$ have been determined by fitting the experimental data of the cross sections and elastic scattering angular distributions for incident energy from 1 keV to 20 MeV.

[Neutron optical model parameters, ^{235}U, ^{238}U, ^{239}Pu, ^{240}Pu, E_n=1 keV-20 MeV]

1. Introduction

^{235}U, ^{238}U, ^{239}Pu and ^{240}Pu are the most important nuclear fuels in thermal and fast neutron reactors. Although for these nuclei a lot of data have been measured, it is still necessary to calculate some nuclear data with the help of nuclear theory to fill the gap of experimental data.

The optical model is one of the most important theories in nuclear data calculations. In addition, the Hauser-Feshbach theory and the evaporation model, describing the compound nucleus process, as well as the exciton model, describing the pre-equilibrium emission process, all depend on the transmission coefficient T_{lj} or the compound nucleus formation cross section σ_a which again should be calculated by optical model. The calculated results of optical model are mainly decided by its parameters. Thus choosing and adjusting the optical model parameters are the crucial steps in nuclear data calculations.

In this note we describe the neutron optical model parameters for $^{235,238}U$ and $^{239,240}Pu$ determined by fitting the experimental data of the cross sections and elastic scattering angular distributions for incident energy from 1 keV to 20 MeV. In section 2 the form of the optical model potentials and the method of automatically searching the parameters by computer are described. Finally, the cross sections and elastic scattering angular distributions calculated with obtained optical model parameters and the comparison with the experimental data are given in section 3.

2. The Adjustment of the Optical Model Potential

The optical model potential V(r) is defined as Woods-Saxon form:

$$V(r)=-Uf_1(r)-iW_vf_2(r)+4iW_sa_2\frac{df_2(r)}{dr}+U_{so}\frac{\lambda_\pi^2}{r}\frac{df_1(r)}{dr}(\vec{\sigma}\cdot\vec{l}) \qquad (1)$$

where

$$f_i(r)=\{1+\exp[(r-R_i)/a_i]\}^{-1} \qquad i=1,2 \qquad (2)$$

$$R_i=r_iA^{1/3} \qquad i=1,2 \qquad (3)$$

λ_π is the compton wavelength of pion, $\lambda_\pi^2=2.00\ fm^2$.

Our purpose is to find out a set of potential parameters, which can reproduce the experimental data well for each fissile nucleus in the energy region from 1 keV to 20 MeV. So we assume that U, W_s and W_v are dependent upon the incident neutron energy E_n in laborotory system as follows:

$$U=U_0-U_1E_n+U_2E_n^2-24\ (N-Z)/A$$

$$W_s=S_0+S_1E_n-12\ (N-Z)/A \text{ or zero, whichever is greater} \qquad (4)$$

$$W_v=W_0+W_1E_n \text{ or zero, whichever is greater}$$

Because the spin-orbit potential has less effects on the computed cross sections and elastic scattering angular distributions, we fix U_{so}=6.2 MeV. The rest 11 parameters, U_0, U_1, U_2, S_0, S_1, W_0, W_1, r_1, r_2, a_1 and a_2, are treated as free parameters to be adjusted.

The adjustment of the optical model potential are performed automatically with computer codes to minimize a quantity called χ^2. The χ^2 is defined as follows:

$$\chi^2=\sum_{j=1}^{N_1}\left(\frac{\sigma_t^T(j)-\sigma_t^E(j)}{\Delta\sigma_t^E(j)}\right)^2+W_{non}\sum_{j=1}^{N_2}\left(\frac{\sigma_a^T(j)-\sigma_{ce}^T(j)-\sigma_{non}^E(j)}{\Delta\sigma_{non}^E(j)}\right)^2+W_e\sum_{j=1}^{N_3}\frac{1}{n_j}\sum_{i=1}^{n_j}\left(\frac{\sigma_{se}^T(\theta_i^j)+\sigma_{ce}^T(\theta_i^j)-\sigma_{el}^E(\theta_i^j)}{\sigma_{el}^E(\theta_i^j)}\right)^2 \qquad (5)$$

K. H. Böckhoff (ed.), Nuclear Data for Science and Technology, 589–592.

These three terms represent total cross sections, nonelastic scattering cross sections and elastic scattering angular distributions, respectively. W_{non} and W_e are weight factors. N_1, N_2 and N_3 are the numbers of the energy points. n_j is the number of the angles of the angular distribution for certain energy. The superscripts T and E represent the theoretical and experimental values, respectively. The compound elastic scattering cross sections σ_{ce}^T and angular distributions $\sigma_{ce}^T(\theta)$ are taken from the statistical theory calculation[1]. We regard χ^2 as the function of the adjustable parameters. In order to search for the minimum of the χ^2, the adjusted parameters are constantly changing along the direction in which the χ^2 decreases fastest. Usually, in this way the solution is not unique. Namely there exist many minimum values for χ^2. Therefore the choice of the initial values of the potential parameters is very important. Our work starts from the Becchetti-Greenlees potential[2]. This potential can't give good results for fissile nuclei over the whole energy range from 1 keV to 20 MeV. In order to find out a set of better initial parameter values, we have studied the influence of each potential parameter on the computed cross sections. In general the real part of the central potential U and the nuclear radius parameters r_i have the strongest influence on the cross sections, furthermore the values of computed cross sections fluctuate as they are changing. The compound formation cross sections σ_a increase as the depths of the imaginary parts of potential W_s and W_v increase. When a_1 and a_2 increase, all kinds of cross sections will increase simultaneously, and the peaks of elastic scattering angular distributions will move toward the small angle region. According to above-mentioned general rules, we make some changes to the Becchetti-Greenlees potential so as to reproduce the experimental data as well as possible. Then we take this set of parameters as the initial values in the program of the automatically searching optical model parameters. In order to make the calculation self-consistent, we use the obtained optical model parameters to perform the statistical theory calculation, from which we get the σ_{ce}^T and $\sigma_{ce}^T(\theta)$ which are put back into the optical model parameters searching program to obtain the new optical model parameters. Such procedure is repeated for several times until the satisfactory results are obtained. The final results of the 11 parameters thus obtained for above-mentioned four fissile nuclei are listed in Table 1.

Table 1 Optical Model Parameters

	^{235}U	^{238}U	^{239}Pu	^{240}Pu
U_o	48.87	49.72	48.92	49.50
U_1	0.465	0.439	0.398	0.440
U_2	0.0235	0.0142	0.0193	0.0156
S_o	6.39	6.51	5.96	6.41
S_1	0.128	0.152	0.116	0.153
W_o	0.291	1.450	0.846	1.480
W_1	0.0059	-0.0351	0.0103	-0.0325
r_1	1.265	1.268	1.260	1.263
r_2	1.350	1.351	1.374	1.348
a_1	0.567	0.602	0.563	0.580
a_2	0.642	0.505	0.603	0.501

3. The Comparison of Theoretical and Experimental Values

The optical model calculations of the cross sections and elastic scattering angular distributions have been carried out for ^{235}U, ^{238}U, ^{239}Pu and ^{240}Pu for energy ranging from 1 keV to 20 MeV. The comparison between the theoretical and the recomended experimental values for total cross sections σ_t, elastic scattering cross sections σ_{el} and nonelastic cross

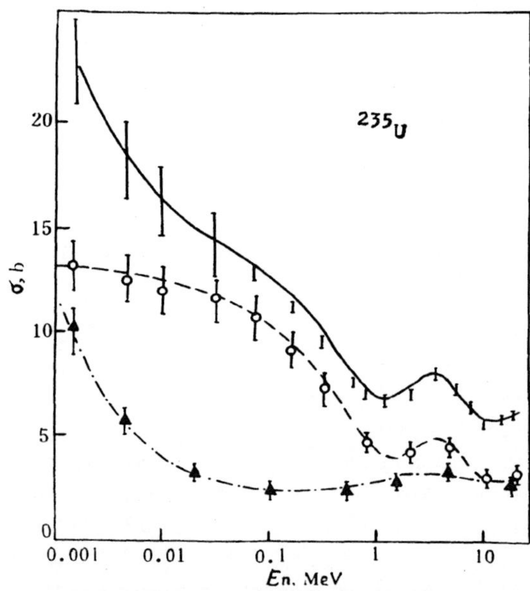

Fig.1. The comparison of theoretical and experimental values of cross sections for ^{235}U.

—— theoretical values of σ_t,
- - - - theoretical values of σ_{el},
-·-·- theoretical values of σ_{non},
I experimental values of σ_t,
ǿ experimental values of σ_{el},
Ɨ experimental values of σ_{non}.

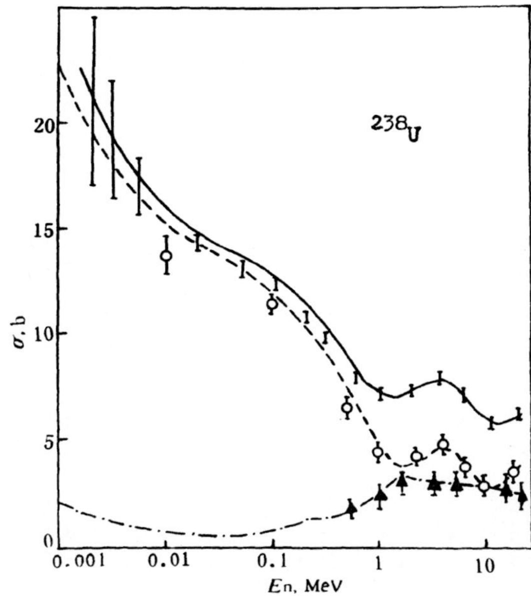

Fig.2. Same as Fig.1, for ^{238}U.

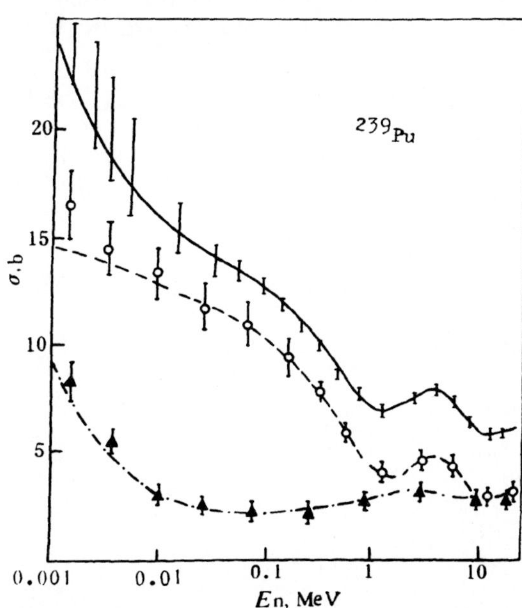

Fig.3. Same as Fig.1, for ^{239}Pu.

sections σ_{non} are shown in Fig.1-4, where

$$\sigma_{el}=\sigma_{se}+\sigma_{ce}$$
$$\sigma_{non}=\sigma_{a}-\sigma_{ce} \quad (6)$$

σ_{ce} is not zero only below 3 MeV. These figures show the total cross sections for ^{235}U, ^{238}U and ^{239}Pu are in good agreement with experiments. Most of the theoretical values go through or near the error bars of experiments. But the experimental values of total cross sections for ^{240}Pu are smaller than those for the other three nuclei, especially at 1.5 MeV. The computed values of total cross sections for ^{240}Pu are a little larger than experiments, otherwise the nonelastic cross sections will be too small as compared with the experiments.

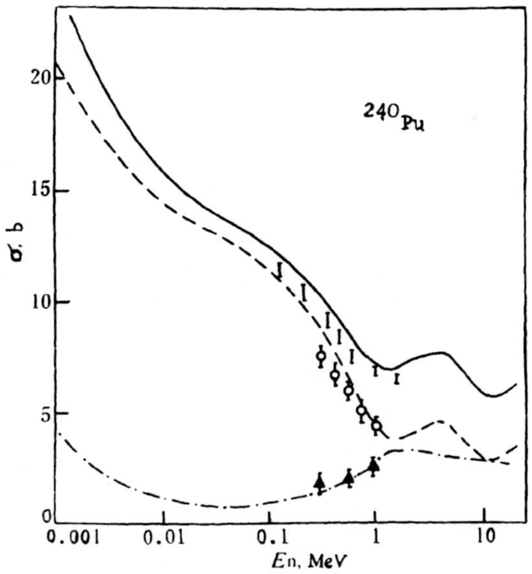

Fig.4. Same as Dig.1, for ^{240}Pu.

ments. The figures also show that the theoretical values of the nonelastic and elastic cross sections are basically in agreement with the experiments.

We have also compared the theoretical values of elastic scattering angular distributions with experimental ones. In general they are in good agreement with each other (see Fig.5 and 6). But in some cases the theoretical curves of elastic scattering angular distributions exhibit some valleys which are too deep as compared with experiments. This might be due to the facts that the experimental data of elastic scattering cross sections and angular

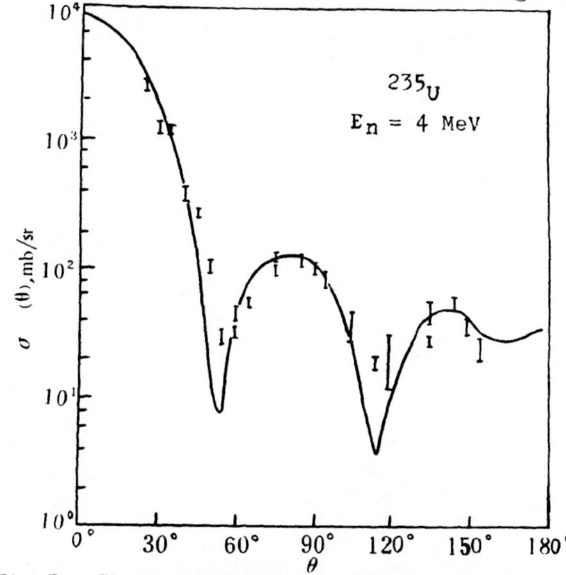

Fig.5. The comparison of the theoretical and experimental values of elastic scattering angular distributions for ^{235}U at energy E_n=4 MeV.

—— theoretical values,
I experimental values.

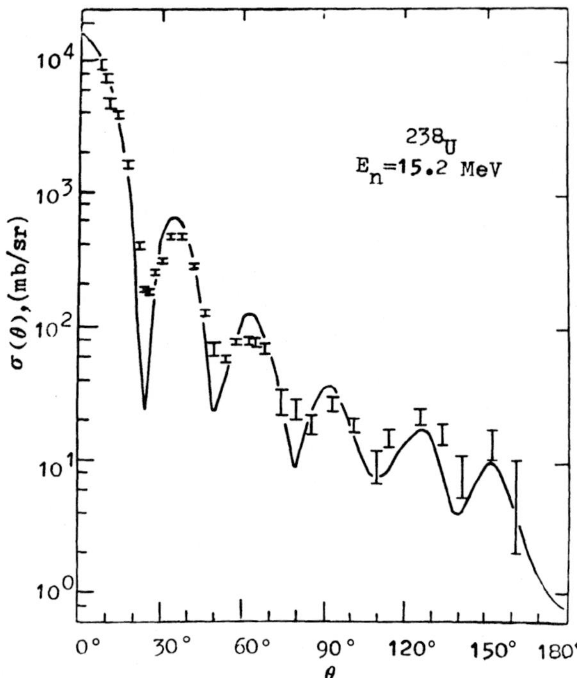

Fig.6. Same as Fig.5, for ^{238}U at energy E_n=15.2 MeV.

distributions include the contribution arising from the inelastic scattering of the low-lying energy states. The elastic scattering angular distributions can also be expressed in the form with the Legendre expansion coefficients. Define

$$\sigma_{el}(\theta)=\frac{\sigma_{el}}{4\pi}\sum_L(2l+1)f_LP_L(\cos\theta) \qquad (7)$$

The formula for the calculation of f_L is given

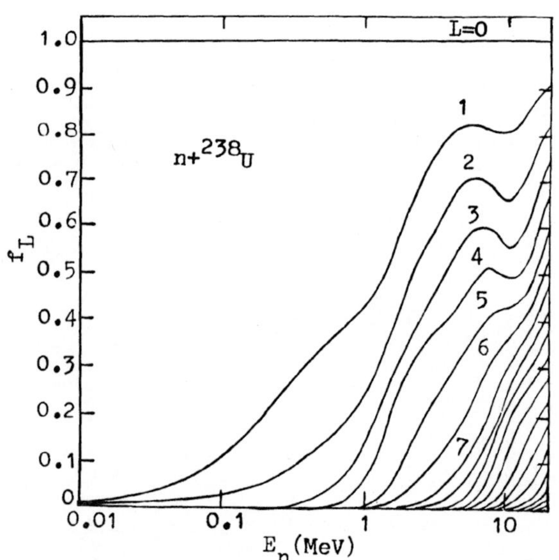

Fig.7. The energy dependence of the Legendre expansion coefficients f_L of elastic scattering angular distributions for ^{238}U.

in appendix. Fig.7 show the energy dependence of f_L for ^{238}U. It is clear that the number of expansion terms increases apparently as the incident energy increases.

Appendix

The Formula for the Calculation of the Legendre Expansion Coefficient of the Elastic Scattering Angular Distribution

The elastic scattering angular distribution $\sigma_{el}(\theta)$ is composed of the shape elastic scattering and the compound elastic scattering. That is

$$\sigma_{el}(\theta)=\sigma_{se}(\theta)+\sigma_{ce}(\theta) \qquad (A1)$$

In the optical model the shape elastic scattering angular distribution can be written as

$$\sigma_{el}(\theta)=\frac{1}{4\pi}\sum_L(2l+1)S_LP_L(\cos\theta) \qquad (A2)$$

In the statistical theory the compound elastic scattering angular distribution can be written as

$$\sigma_{ce}(\theta)=\frac{1}{4\pi}\sum_L(2l+1)C_LP_L(\cos\theta) \qquad (A3)$$

Compared with (7) we get

$$f_L=(S_L+C_L)/\sigma_{el} \qquad (A4)$$

S_L is expressed as follows:

$$S_L=\frac{\pi}{(2l+1)k^2}\sum_{l_1l_2}\Big[(f^{AR}_{l_1}f^{AR}_{l_2}+f^{AI}_{l_1}f^{AI}_{l_2})(C^{L\,0}_{l_1 0 l_2 0})^2 - (f^{BR}_{l_1}f^{BR}_{l_2}+f^{BI}_{l_1}f^{BI}_{l_2})\sqrt{l_1(l_1+1)l_2(l_2+1)}\; C^{L\,0}_{l_1-1 l_2 1}C^{L\,0}_{l_1 0 l_2 0}\Big] \qquad (A5)$$

where

$$f^{AR}_l=(l+1)(1-\mathrm{Re}S^{l+½}_l)+l(1-\mathrm{Re}S^{l-½}_l)$$
$$f^{AI}_l=(l+1)\mathrm{Im}S^{l+½}_l+l\ \mathrm{Im}S^{l-½}_l$$
$$f^{BR}_l=\mathrm{Re}S^{l+½}_l-\mathrm{Re}S^{l-½}_l$$
$$f^{BI}_l=\mathrm{Im}S^{l+½}_l-\mathrm{Im}S^{l-½}_l \qquad (A6)$$

$S^{l\pm½}_l$ are the S—matrix elements obtained from the optical model calculations.

References

1. Su Zongdi et al., Chin. J.Nucl. Phys. 1, 81 (1979)
2. F.D,Becchetti, G.W.Greenlees, Phys. Rev. 182, 1190 (1969)

POPULATION OF DELAYED NEUTRON GRANDDAUGHTER STATES AND THE OPTICAL POTENTIAL

R. E. Schenter and F. M. Mann

Westinghouse Hanford Company
Hanford Engineering Development Laboratory
Richland, Washington 99352, U.S.A.

R. A. Warner and P. L. Reeder

Pacific Northwest Laboratory
Richland, Washington 99352, U.S.A.

Using a statistical treatment of beta decay and the Hauser-Feshbach model of nuclear reactions, calculations were made and compared to recent experimental measurements of the population of granddaughter states of several delayed neutron precursors (^{144,145,147}Cs and ^{96}Rb). Emphasis of this paper is on the sensitivity and interpretation of experimental results to various standard low energy neutron optical model potentials and variations in their forms and parameters. Results for these precursors show qualitative agreement with experiment for all the optical potential models used and good quantitative agreement for two ("Moldauer" and "Becchetti-Greenlees"). Questions such as (N-Z) terms, deformation and nonlocality dependence will be presented.

[^{144}Cs, ^{145}Cs, ^{147}Cs, ^{96}Rb, beta decay, delayed neutron emission, optical potentials]

Introduction

Recent measurements have been made which have a potential of providing results that would give significant insight into the study of problems describing beta decay and nucleon-nucleus reaction theories for nuclei far from the line of stability. These experiments involve measuring gamma spectra in coincidence with neutrons following the beta decay of the delayed-neutron precursors $^{95-99}$Rb and $^{143-147}$Cs. Results from these measurements can be used to test and provide parameters for statistical approaches to beta decay and optical model potentials for scattering from target nuclei in their excited states. One specific output from the experimental analysis is the population of "granddaughter" states following the beta decay and neutron emission of the rubidium and cesium precursors.

In this paper results are presented from an initial experimental analysis for the precursors ^{144,145,147}Cs and ^{96}Rb. A statistical treatment of beta decay and the Hauser-Feshbach model of nuclear reactions[1,2] were used to calculate the granddaughter state population. These provided tests of optical models, beta strength functions, Q values and spin-parity assignments for the "parent," "daughter" and "granddaughter" nuclei. Experimental methods, calculations and results, and conclusions are given in the following sections.

Experimental Methods

The on-line isotope separator TRISTAN[3] was used to produce Rb and Cs delayed neutron precursors from thermal neutron fission of ^{235}U. As depicted in Figure 1, the mass-separated ion beam is deposited on an aluminized mylar tape in the center of the SOLAR neutron counter (SNC). The SNC consists of forty ^{3}He-filled proportional counters embedded in polyethylene in three concentric rings around the beam tube. Its efficiency in this configuration ranged from 58 to 42% for neutron energies from thermal to 1 MeV respectively. More detail may be found in References 4 and 5. Gamma rays were detected in an HPGe diode abutting the end of the beam tube. Its absolute efficiency was a maximum of 4% at 122 keV and fell off to 0.6% at 1 MeV.

Pulses derived from the gammas started a time-to-amplitude converter (TAC) which was stopped by pulses from the SNC, within which the average neutron's residence before detection was 38 μs. For each TAC interval less than 100 μs, three addresses were recorded in list mode for off-line sorting: the gamma energy, the TAC output, and a pulse identifying the ring of proportional counters in which the neutron was detected. A gamma energy spectrum with no coincidence requirement was accumulated simultaneously.

NEUTRON-GAMMA COINCIDENCE SYSTEM

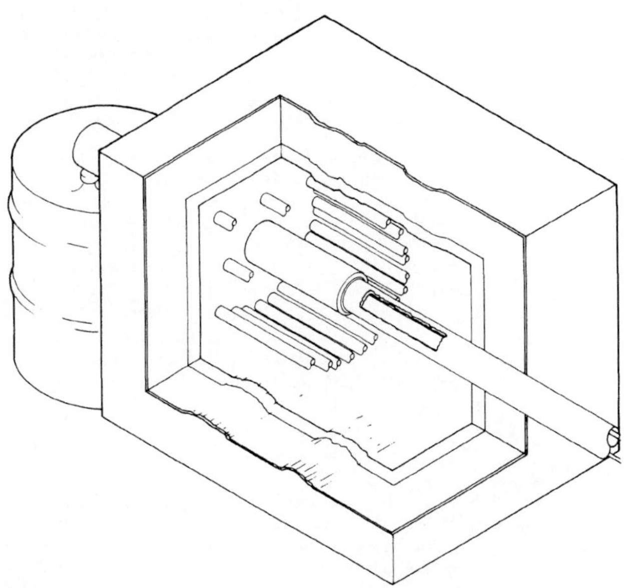

FIG. 1

Neutron gated gamma spectra for the delayed-neutron precursors $^{95-99}$Rb and $^{143-147}$Cs have been accumulated as described. Analysis includes corrections for chance coincidences and for the presence of neighboring masses in the ion beam. Neutron feeding to states in the final nucleus were then calculated. The results presented here are restricted to those decay for which we have completed the analysis on one set of data and for which adequate level schemes are available.

Calculations and Results

Calculations were made with the BETA computer code which uses a statistical treatment of beta decay and the Hauser-Feshbach model of nuclear reactions.[1,2] A special output option for the code was written to provide calculated results for this particular problem. Figure 2 illustrates the physical process being studied, where population of states of ^{144}Ba occurs following β decay and neutron emission from ^{145}Cs. An important parameter for the problem is the energy available for delayed neutron emission or "Energy Window" (EW) and as shown in Figure 2 is the difference

K. H. Böckhoff (ed.), Nuclear Data for Science and Technology, 593–596.

POPULATION OF βa-144 FROM β DECAY OF Cs-145

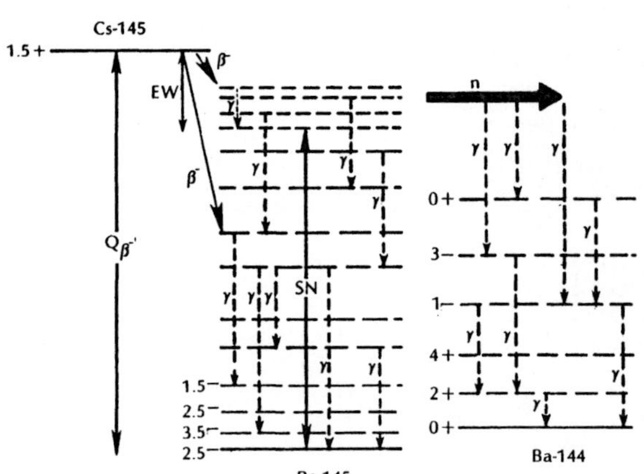

FIG. 2

Cs-145 BETA-DELAYED NEUTRON DECAY, EFFECT OF OPTICAL POTENTIALS, J•Pi=+1.5

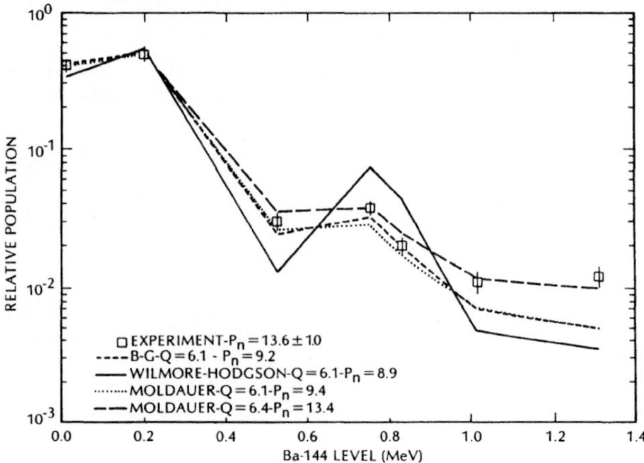

FIG. 3

between the β-decay energy (Q_{β^-}) of the "parent nucleus" (^{145}Cs) and the neutron separation energy (SN) of the "daughter nucleus" (^{145}Ba). Values of spin and parity of the "parent" and "granddaughter" (^{144}Ba) nuclei states also have quantitative effects on the final results.

Competition between neutron and γ emission in the "daughter nucleus" was calculated according to the Hauser Feshbach model, where the optical potential is used to calculate neutron transmission coefficients. Because of the reciprocity theorem these potentials are the same that describe neutron scattering from the ground and excited states of the "granddaughter nucleus." For calculations of this type it is usually assumed that the same optical potential is used for target nuclei in all states of excitation. For the calculations performed in this paper these same assumptions were made.

Parameters for three different local phenomenological neutron optical model potentials were input to the BETA code for these calculations on rubidium and cesium isotopes. The authors of these well known potentials are Becchetti and Greenlees,[6] Moldauer,[7] and Wilmore and Hodgson.[8] The real radial dependence of all three of these potentials is basically the same, given by the usual Woods-Saxon form. Well depth and radius parameters are A(N+Z) and energy dependent. The real well depth (MeV), for example, is 56.3-0.32E-24(N-Z)/A, 36.0 and 37.01-0247E-0.0018E^2, respectively for the above three potentials. Note that the Becchetti-Greenless potential has an "N-Z" term which is about 10% of the real well depth for all the nuclides considered in this study.

The main results from these calculations are shown in Figures 3 through 9, where BETA generated values of relative population of "granddaughter" states versus excitation energy are plotted with the experimental results. Sensitivity to the three optical potential types is shown in Figures 3 through 5 for ^{145}Cs, ^{147}Cs and ^{96}Rb. For the cesium comparisons, especially, the Becchetti-Greenlees (B-G) and Moldauer results give a good quantitative description of the population fall-off with excitation. This is significant since this decrease is very rapid and with structure occurring due to various spin-parity assignments for the "parent" and "granddaughter" states. Figures 6 through 9 show the sensitivity to "Energy Available" (Q_{β}-, SN, EW). The higher energy excitation results are especially sensitive to this. Figure 3 also illustrates this point. The variation of Q_{β}- presented in these figures is not large compared to the uncertainty in

Cs-147 BETA-DELAYED NEUTRON DECAY, EFFECT OF OPTICAL POTENTIALS, J•Pi=+1.5, Q=7.27 MeV

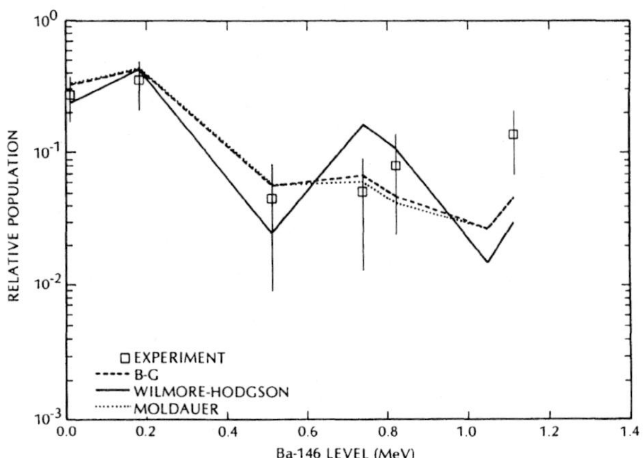

FIG. 4

Rb-96 BETA-DELAYED NEUTRON DECAY, EFFECT OF OPTICAL POTENTIALS

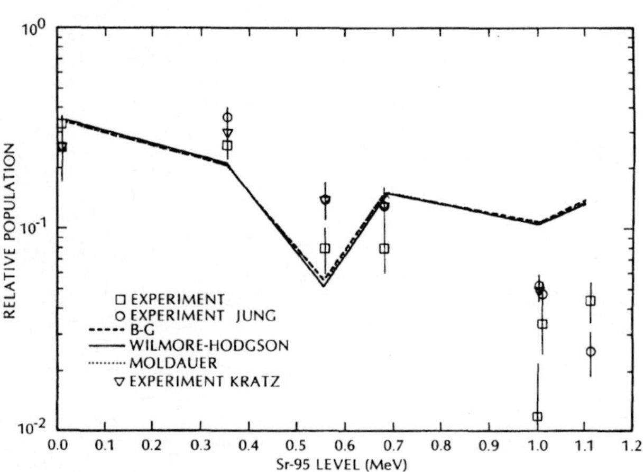

FIG. 5

Cs-144 BETA-DELAYED NEUTRON DECAY,
EFFECT OF ENERGY AVAILABLE, J*Pi=+0.0,
WILMORE-HODGSON POTENTIAL

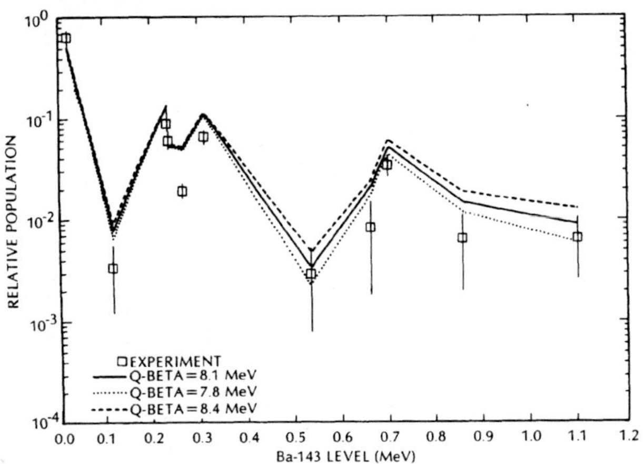

FIG. 6

Cs-145 BETA-DELAYED NEUTRON DECAY,
EFFECT OF Q VALUE CHANGES,
MOLDAUER POTENTIAL

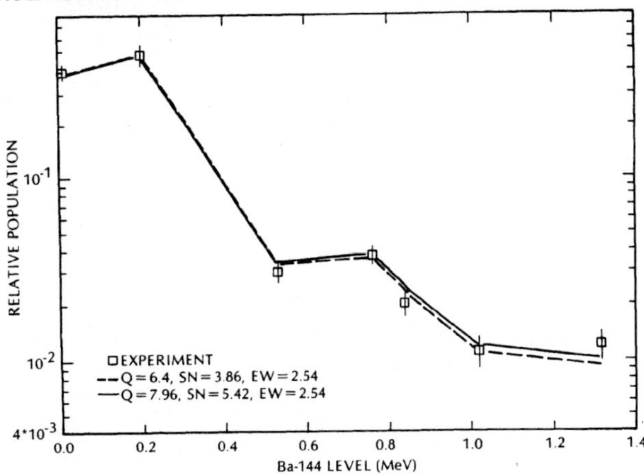

FIG. 9

Cs-145 BETA-DELAYED NEUTRON DECAY,
EFFECT OF ENERGY AVAILABLE, J*Pi = +1.5
WILMORE-HODGSON POTENTIAL

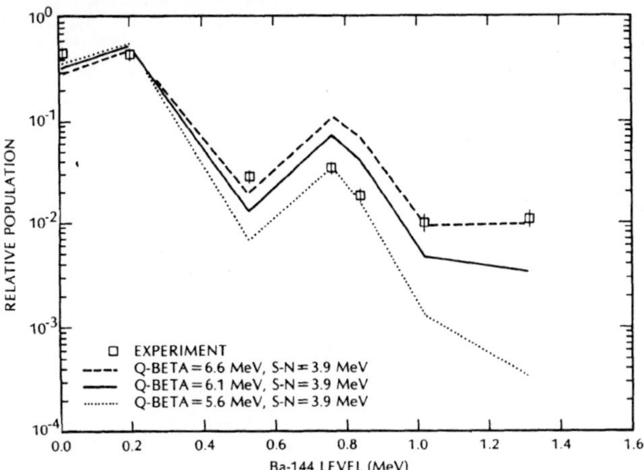

FIG. 7

Rb-96 BETA-DELAYED NEUTRON DECAY,
EFFECT OF ENERGY AVAILABLE,
WILMORE-HODGSON POTENTIAL

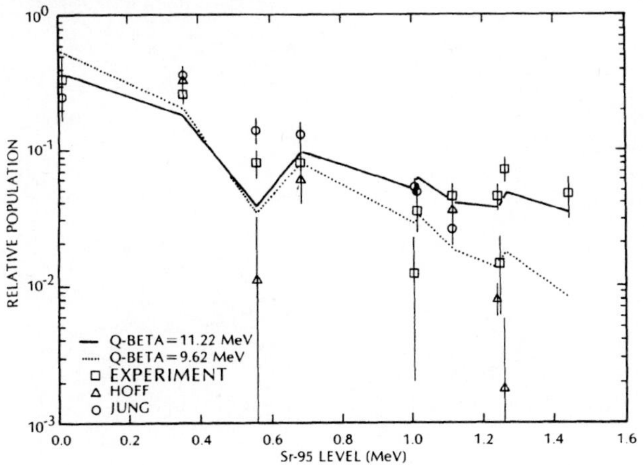

FIG. 8

present mass law descriptions and experimental results for strontium and cesium[9] and hence indicates a utility for these results which may help to reduce that uncertainty.

Returning to the figures for a more detailed look shows that in Figure 3 are also given values of "P_n" the delayed neutron probability factor (%) from the experimental results and from the BETA calculations. This mainly tests the beta-decay description of the problem and the excellent agreement for the ^{145}Cs result gives additional confidence for this procedure. In Figure 4 the large uncertainties in the experimental results indicate a need for additional work on ^{147}Cs to reduce those values. Experimental results from other authors ("Kratz et al.,"[10] "Hoff,"[11] "Jung"[12]) are also given in Figures 5 and 8. The large differences between them needs to be resolved. In Figures 6 through 8 the Wilmore-Hodson potential was used to generate the theoretical results and Q_β was varied with a fixed separation energy (SN). For these cases the main effect of energy variation is due directly to a change in the "Energy Window" (EW). This is shown explicitely in Figure 9 for ^{145}Cs where a large Q_β value change with a constant EW produce small changes in the calculated result.

Conclusions

Results given in this paper represent only a start in the analyses, both experimental and theoretical. However, the good quantitative agreement for the cesium and rubidum nuclides analyzed in this paper using reasonable optical model potential descriptions, Q values and spin-parity assignments suggests this study provides an excellent guide to future work. Planned experimental improvements involve complete analysis of data not used here, and extension of measurements to other precursors. For many nuclides, complete analysis will require the determination of level schemes by conventional gamma spectroscopy. Finally on the theoretical side a more extensive selection of optical model potentials should be tried inclduing phenomenological and theoretical non local potentials[13-15] and recent studies of local phenomenological potentials.[16] Mass laws and recent experimental results[9] to determine Q values and separation energies should be considered.

Acknowledgements

The authors are grateful to A. Wolf and T. R. Yeh for help with the counting system and to the entire

TRISTAN staff at Brookhaven National Laboratory for reliable ion beams. The experimental work presented here was supported by the U.S. Department of Energy, Basic Energy Sciences, under contract number DE-AC06-76RLO-1830.

References

1. F. M. Mann, C. Dunn and R. E. Schenter, Phys. Rev., C25 (1982) 524.

2. F. M. Mann, Hanford Engineering Development Laboratory Report, HEDL-TME 78-83, 1979.

3. R. E. Chrien, M. L. Stelts, V. Manzella, R. L. Gill, F. K. Wohn, and J. C. Hill, in Nuclear Spectrometry of Fission Products, The Institute of Physics Conference Proceedings No. 51, edited by Till von Egidy (The Institute of Physics, Bristol and London, 1980), p. 44.

4. P. L. Reeder, J. F. Wright and L. J. Alquist, Phys. Rev., C15 (1977) 2098.

5. P. L. Reeder and R. A. Warner, Nucl. Instrum. Meth., 180 (1981) 173.

6. F. D. Becchetti, Jr. and G. W. Greenlees, Phys. Rev., 182 (1969) 1190.

7. P. A. Moldauer, Nucl. Phys., 47 (1963) 65.

8. D. Wilmore and P. E. Hodgson, Nucl. Phys., 55 (1964) 673.

9. M. Epherre, et al., Phys. Rev., C19 (1979) 1504.

10. K. L. Kratz, A. Schröder, H. Ohm. M. Zendel, H. Gabelmann, W. Ziegert, P. Peuser, G. Jung, B. Pfeiffer, K. D. Wünsch, H. Wollnik, C. Ristori and J. Crancon, Z. Phys. A - Atoms and Nuclei 306, 239-257 (1982).

11. P. Hoff, Nucl. Phys., A359 (1981) 9.

12. G. Jung, PhD Thesis, University of Giessen, Giessen Germany, 1981.

13. F. G. Perey and B. Buck, Nucl. Phys., 32 (1962) 353.

14. Y. Yamaguchi, Phys. Rev., 95 (1954) 1628.

15. A. D. MacKellar and R. E. Schenter, Hanford Engineering Development Laboratory Report, HEDL-TME 71-154, 1972.

16. D. G. Foster, Jr. and E. D. Arthur, Los Alamos National Laboratory Report, LA-9168-MS, 1982.

INELASTIC SCATTERING CROSS SECTIONS IN THE ENERGY RANGE 2.0 TO 4.5 MeV CALCULATED WITH DIFFERENT FORMALISMS FOR LEVEL WIDTH FLUCTUATION CORRECTIONS

E. Ramström

The Studsvik Science Research Laboratory,
S-611 82 Nyköping, Sweden

Two different methods, i.e. the Moldauer formalism and the Tepel formalism, have been used for the calculation of level width fluctuation correction factors to average compound-nucleus cross sections with special emphasis on neutron inelastic scattering cross sections. Thus corrected inelastic scattering cross sections have been calculated with the two formalisms for ten elements covering the atomic mass region 27 to 209 for incident neutrons in the energy range 2.0 to 4.5 MeV. The calculated inelastic scattering cross sections have been compared to the experimental ones reported in a previous work of Almén-Ramström.

[^{27}Al, ^{51}V, ^{56}Fe, ^{59}Co, ^{63}Cu, ^{65}Cu, ^{89}Y, ^{206}Pb, ^{207}Pb, ^{209}Bi(n,n'), inelastic scattering cross sections, E_n = 2.0 - 4.5 MeV].

Introduction

The basic expression for the average compound-nucleus cross section σ_{ab} for an initial channel a and a decay channel b can be written as

$$\sigma_{ab} = \sigma_{ab}^{HF} W_{ab}$$

where σ_{ab}^{HF} is the cross section given by the well known Hauser-Feshbach formula[1] and W_{ab} is a factor which corrects for the fact that the level widths fluctuate in the resonances. The factor W_{ab} can be calculated by using different statistical model theories[2-6]. Some properties of this factor have also been reviewed by Gruppelaar et al.[7]. In the present work two different formalisms, i.e. the Tepel formalism[5,6] and the Moldauer "Q" parameter (or overlap parameter) formalism[2], have been used for the calculation of the correction factor W_{ab} with special emphasis on neutron inelastic scattering cross sections. In order to investigate the usefulness of the two formalisms, the calculated cross sections have also been compared with experimental ones.

Theory

In the Moldauer "Q" parameter formalism[2] the corrections for level width fluctuations are obtained with the classical integration method assuming that the fluctuations of the level widths in a resonance follow a Porter-Thomas distribution (i.e. a χ^2 distribution with one degree of freedom). It has been shown by Moldauer[3] that the derivation of the expressions for the level width fluctuations in the Moldauer formalism is not correct if the level separation is not large compared to the level width, because channel-channel and level-level correlations may become important. Moldauer has also shown[3], that the formula can still be used provided that the number of degrees of freedom for the distribution of level widths is adjustable between 1 and 2 for each neutron channel. In the "Q" parameter formalism used in this work the number of degrees of freedom was equal to 1 for each channel.

In the Tepel formalism[5,6] the values of W_{ab} are algebraically evaluated. The assumptions used in this formalism were based on a random-matrix model for the S matrix. This model includes a semi-empirical relation for the elastic enhancement parameter.

Calculations

In order to make a systematic comparison between the inelastic scattering cross sections obtained with the two formalisms mentioned above a relatively large set of calculations is necessary. Thus calculations have been performed for ten elements covering the atomic mass region 27 to 209 for incident neutrons in the energy range 2.0 to 4.5 MeV.

For the performance of the statistical model calculations transmission coefficients are necessary. These have been calculated by using a local spherical optical model potential of the form

$$-V(r) = Uf(r) + iWg(r) + U_{SO}\left(\frac{\hbar}{\mu_\pi c}\right)^2 \frac{1}{r}\frac{d}{dr}|f(r)|\ \overline{\sigma}\cdot\overline{\ell}$$

The radial dependences of the real and the imaginary parts of the potential are described by the Saxon-Woods and derivative Saxon-Woods form factors, respectively. The sets of parameters were taken from an earlier investigation at our laboratory on a generalized optical model potential based on experimental neutron elastic scattering data[8]. However, for computational reasons the spin-orbit potential was set equal to zero in the present work instead of 8 MeV as was used in the previous work. This change of the potential does not affect the values of the inelastic cross sections with more than 5 per cent. For the calculations with the Moldauer "Q" parameter formalism the ABACUS-NEARREX computer code[9] was used while the calculations with the Tepel formalism were performed with the computer code HAUSER 5[10].

Results

Comparisons between the inelastic scattering cross sections calculated with the Moldauer "Q" parameter formalism and those calculated with the Tepel formalism show agreements within 10 per cent for most of the levels in the ten elements studied. However, disagreements of up to 20 per cent were found for several elements and for several incident neutron energies for such channels where the energies of the outgoing neutrons were less than about 0.6 MeV. This is in agreement with the statement of Gruppelaar et al.[7] that the Tepel formalism is useful only for high values of the transmission coefficients.

There are also some other exceptions from the agreement within 10 per cent. In these cases the deviations between the calculated cross sections are not due to low values of the transmission coefficients. Thus, as shown in Fig. 1, there are discrepancies of the order of 20 per cent for the two first excited states in ^{209}Bi for incident neutrons in the energy range 2.0 to 3.0 MeV. Discrepancies of the same order are also found for the first excited state in ^{207}Pb in the same energy region.

The calculated inelastic scattering cross sections have also been compared to experimental ones reported in a previous work of Almén-Ramström[11,12]. The purpose of that work was to investigate the usefulness of the "Q" parameter formalism in describing the experimental data. In that work it was found that the inelastic scattering cross sections calculated in this way in most cases describe the experimental data well within 20 per cent. However, there are some exceptions. Thus, disagreements of the order of a factor of 2 between theory and experiment were obtained for the two first excited states in ^{209}Bi at primary neutron energies below 3.5 MeV. The same order of disagreement was also obtained for the 0.894 MeV $3/2^-$ level (i.e. the second excited state) in ^{207}Pb at the lowest primary neutron energies studied, while for the other levels in ^{207}Pb rather good agreements (within 25 per cent) were obtained.

K. H. Böckhoff (ed.), Nuclear Data for Science and Technology, 597–598.

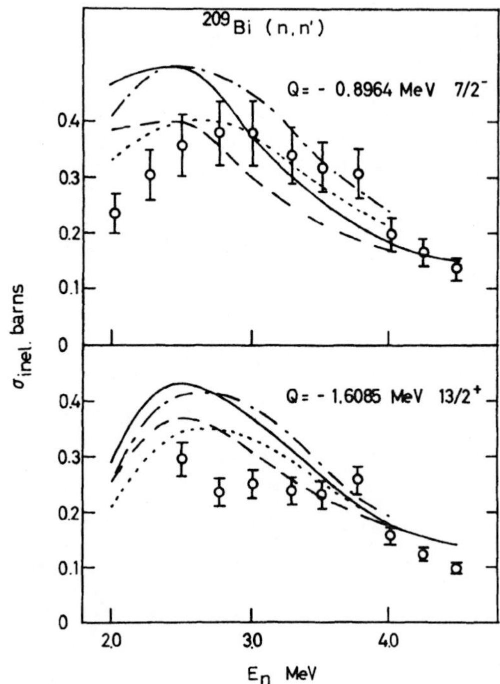

Fig. 1. The excitation functions for inelastic scattering from ^{209}Bi. Experimental results (circles) are taken from Refs.[11,12]. Moldauer "Q" parameter calculations with optical model parameters from Ref.[8] (solid curves) and from Ref.[13] (point dashed curves). Tepel calculations with parameters from Ref.[8] (dashed curves) and from Ref.[13] (dotted curves).

As mentioned above there are discrepancies between the inelastic scattering cross sections calculated with the "Q" parameter formalism and those calculated with the Tepel formalism for the two first excited states in ^{209}Bi and for the first excited state in ^{207}Pb in the energy range 2.0 to 3.0 MeV. Generally better descriptions of the experiments are obtained with the Tepel formalism than with the Moldauer "Q" parameter formalism. This is shown for ^{209}Bi in Fig. 1. However, for the levels in these two elements there are deviations in the shapes of the experimental and theoretical curves when applying both the formalisms. These deviations are decreased considerably as shown for ^{209}Bi in Fig. 1 by using the optical model parameters derived in a recent work on elastic scattering from radiogenic lead and bismuth in the energy region 1.5 to 4.0 MeV[13] instead of the generalized parameters from Ref.[8] .

References

1. W. Hauser and H. Feshbach, Phys. Rev. 87, 366 (1952).
2. P.A. Moldauer, Phys. Rev. 135B, 642 (1964).
3. P.A. Moldauer, Phys. Rev. C11, 426 (1975).
4. P.A. Moldauer, Phys. Rev. C14, 764 (1976).
5. J.W. Tepel, H.M. Hofmann and H.A. Weidenmüller, Phys. Lett. 49B, 1 (1974).
6. J.W. Tepel, H.M. Hofmann and M. Herman, Nucl. Cross Sections for Technology, Proc. Conf. Oct. 22-26, 1979, Knoxville, Tennessee (ed. Sept. 1980).
7. H. Gruppelaar and G. Reffo, Nucl. Sc. Eng. 62, 756 (1977).
8. B. Holmqvist and T. Wiedling, J. Nucl. Energy 27, 543 (1973).
9. P.A. Moldauer, C.A. Engelbrecht and G.J. Diffy, Argonne National Laboratory, Report ANL-6978 (1964).
10. F.M. Mann, Hanford Engineering Development Laboratory, Report HEDL-TME 78-83 (1978).
11. E. Almén-Ramström, Atomic Energy Company, Sweden, Report AE-503 (Dissertation, University of Lund).
12. E. Ramström, Nucl. Phys. A315, 143 (1979).
13. N. Olsson, B. Holmqvist and E. Ramström, Nucl. Phys. A385, 285 (1982).

MEASUREMENT AND HYBRID MODEL ANALYSIS OF INTEGRAL EXCITATION FUNCTIONS FOR α-INDUCED REACTIONS ON VANADIUM AND MANGANESE

R. Michel, G. Brinkmann, and R. Stück
Institut für Kernchemie der Universität zu Köln
Zülpicher Str.47, D 5000 Köln - 1
Federal Republic of Germany

α-induced reactions on vanadium and manganese have been investigated for energies between 12 and 172 MeV. Using the stacked-foil technique 28 excitation functions were measured for the production of ^{42}K, ^{43}K, ^{47}Ca, ^{43}Sc, ^{44g}Sc, ^{44m}Sc, $^{46m+g}Sc$, ^{47}Sc, ^{48}Sc, ^{48}V, ^{48}Cr, ^{51}Cr, ^{52}Mn, ^{54}Mn from vanadium; and of ^{42}K, ^{43}K, $^{46m+g}Sc$, ^{48}V, ^{48}Cr, ^{51}Cr, ^{52}Mn, ^{54}Mn, ^{56}Mn, ^{52}Fe, ^{55}Co, ^{56}Co, ^{57}Co, $^{58m+g}Co$ from manganese. The experimental data are compared with a priori calculations considering equilibrium as well as preequilibrium reactions. Severe discrepancies are to be seen between theory and experiment originating from a variety of particular reaction types, such as incomplete break-up of the α-particles, direct reactions, as well as preequilibrium emission of complex particles, none of which is adequately described by theory.

Introduction

In the course of a systematic study of integral excitation functions of α-induced reactions with target elements $22 \leq Z \leq 28$ we earlier investigated the elements Co [1] as well as Ti, Fe and Ni [2]. The measurements were performed with a twofold aim:
1) to provide the necessary cross section data for the interpretation of the production of cosmogenic nuclides in extraterrestrial matter by solar α-particles (see also the respective contribution to this conference by Michel and Stück) and 2) to test the applicability of actual theories of nuclear reactions to the a priori calculation of integral excitation functions. The second viewpoint is of particular interest for the various fields of applications of integral cross section data.

Such calculations applying the hybrid model of Blann [3] and the concept of the compound nucleus in statistical equilibrium according to Weisskopf and Ewing [4] had proved to be very successful for p-induced reactions up to E_p = 45 MeV (e.g.[5]), but the tests on α-induced reactions [1,2] showed severe shortcomings of the theoretical approach. Since, however, the investigation of multi-isotope elements, such as Ti, Fe and Ni [2] hardly can give clear-cut evidence about the systematic trends of discrepancies between theory and experimental data, we present in this work the measurement and hybrid model analysis of integral excitation functions of α-induced reactions on vanadium and manganese,which are, or at least can be regarded as, single isotope targets.

Experimental

The excitation functions were measured using the stacked-foil technique. For each target element 2 types of stacks were used in order to avoid too large errors in the particle energies due to the effect of range straggling. The first one had to degrade the α-energy from 172.5 MeV to about 80 MeV, the second one from 90 to about 15 MeV. Each stack consisted of a number of target foils, alternately followed by a high purity 25 µm Al catcher foil and by commercial grade Al degrader foils. For manganese 25 µm foils of a Mn-Ni alloy, containing 88 % Mn and 12 % Ni, supplied by Goodfellow Metals Ltd., U.K., were used. For the Mn-irradiation the contribution of Ni to the production of radionuclides had to be corrected for using the earlier determined excitations for α-induced reactions on Ni [2]. However, only for the excitation functions of the (α,xn)-reactions, x = 1-4, leading to the product nuclides $^{55,56,57,58}Co$, the Ni-contribution became a limiting factor for the high energy parts of these excitation functions.

The α-irradiations were performed at the isochronous cyclotron JULIC of the KFA Juelich, using the irradiation facility BK 1. Since the experimental procedure was identical to that described earlier [1,2] and also the nuclear data used for the evaluation of cross sections, such as half-lives, γ-energies and intensities, were the same as used earlier [1,2,5] we prefer to avoid a more detailed description of the experimental procedure.

Experimental Results

For the target element vanadium 14 excitation functions were measured for α-energies between 25.2 and 171 MeV, describing (α,n) to (α,6p7n)-reactions (Table I). The new measurements extend the earlier cross section data [6-10] with regard to the particle energy covered as well as to the product nuclides investigated.With the exception of the work of Bowman and Blann[9] who used α-energies up to 120 MeV all the earlier authors only applied energies below 50 MeV. For a detailed comparison of our new data with those of other authors see Ref.[11].

With respect to manganese, 14 excitation functions for α-energies between 12.7 and 171.2 MeV were determined (Table II), describing reactions of the types (α,n) to (α,8p9n). Here, only two groups of authors have reported some earlier results. Matsuo et al. [12] investigated the reaction $^{55}Mn(\alpha,n)^{58m+g}Co$ and Tanaka et al. [13] (α,xn)- and (α,2pxn)-reactions on Mn. However, in both references α-energies were below 40 MeV. Our results for Mn are in excellent agreement with the data of the former authors.

Comparison with Theoretical Predictions

Theoretical excitation functions were calculated using the program OVERLAID ALICE [14] considering equilibrium reactions according to Weisskopf and Ewing [4] and preequilibrium reactions according to the hybrid model of Blann [3].

The same set of parameters was used as in our earlier calculations of α-induced reactions [1,2] in order to preserve the "a priori" character of our calculations, and no adjustment of parameters was done.

The initial exciton configuration n_o = sum of particle and hole exciton states is of particular importance for the preequilibrium reactions. From our analysis of α-induced reactions on Co [1] we derived n_o to lie between 4(2n-2p-oh) and 5(3n-2p-oh) or (2n-3p-oh) in agreement with the work of other authors. Generally, n_o = 4 gives a more pronounced preequilibrium contribution. Since a decision about an unambiguous choice of n_o was not possible up to now, we decided to present theoretical excitation functions for both initial exciton configurations. However, we did not distinguish between the (3n-2p-oh) and (2n-3p-oh) initial configurations, since the respective results differed only slightly.

A comparison of experimental cross sections and theoretical excitation functions is given in Figs.1 and 2 for some selected reactions which, however, provide a systematic survey on the types of agreement and, even more important, of the discrepancies observed between theory and experiment for α-induced reactions.

Generally, (α,xn)-reactions are well described by theory, as it is shown in Fig.1 for the reactions $^{51}V(\alpha,3n)^{52}Mn$ and $^{51}V(\alpha,n)^{54}Mn$. The initial configurations again are lying between n_o = 4 and n_o = 5.

K. H. Böckhoff (ed.), Nuclear Data for Science and Technology, 599–602.

Table I Cross Sections [mb] for α-Induced Reactions on Vanadium

E_α [MeV]	^{54}Mn	^{52}Mn	^{51}Cr	^{48}Cr	^{48}V	^{48}Sc	^{47}Sc	$^{46m+g}Sc$	^{44m}Sc	^{44g}Sc	^{43}Sc	^{47}Ca	^{43}K	^{42}K
170.97 ± 0.43	0.119 ± 0.016	1.11 ± 0.14	59.5 ± 6.0	0.627 ±0.056	72.1 ± 6.5	7.42 ±0.82	29.1 ± 2.0	67.7 ± 4.7	35.7 ± 2.5	17.8 ± 2.0	5.16 ±0.52	0.378 ±0.038	6.35 ±0.44	12.1 ± 1.5
161.59 ± 0.61	0.179 ± 0.023	1.50 ± 0.15	69.0 ± 6.2	0.673 ±0.074	74.9 ± 6.8	7.36 ±0.74	28.8 ± 2.0	68.3 ± 5.5	35.3 ± 2.5	14.0 ± 1.5	4.98 ±0.50	0.349 ±0.038	5.93 ±0.47	11.3 ± 1.6
151.91 ± 0.72	0.229 ± 0.030	2.00 ± 0.22	78.2 ± 6.3	0.683 ±0.055	77.3 ± 5.5	7.05 ±0.78	28.2 ± 2.0	66.6 ± 5.3	34.2 ± 2.4	13.4 ± 1.3	4.88 ±0.49	0.306 ±0.040	5.33 ±0.37	10.4 ± 1.2
143.07 ± 0.81	0.305 ± 0.049	2.63 ± 0.26	89.5 ± 9.0	0.723 ±0.065	81.4 ± 6.7	6.94 ±0.76	27.8 ± 1.9	69.2 ± 4.8	33.2 ± 2.3	13.1 ± 2.1	4.15 ±0.42	0.269 ±0.032	4.86 ±0.39	10.1 ± 1.2
133.72 ± 0.89	0.323 ± 0.036	3.44 ± 0.34	102. ± 10.	0.735 ±0.059	82.2 ± 7.5	6.68 ±0.80	26.8 ± 1.9	66.5 ± 5.3	29.8 ± 2.1	11.4 ± 2.4	3.63 ±0.36	0.246 ±0.027	4.45 ±0.36	9.03 ± 0.99
123.87 ± 0.96	0.399 ± 0.044	4.75 ± 0.43	118. ± 8.	0.736 ±0.081	82.9 ± 6.7	6.27 ±0.69	25.7 ± 1.8	65.0 ± 5.9	23.8 ± 1.7	10.5 ± 2.0	2.56 ±0.26	0.202 ±0.014	4.09 ±0.29	7.62 ± 0.91
114.88 ± 1.02	0.437 ± 0.039	6.44 ± 0.58	137. ± 11.	0.670 ±0.053	84.3 ± 7.6	6.04 ±0.73	25.6 ± 1.8	66.9 ± 6.0	18.6 ± 1.3	7.61 ± 0.99	3.30 ±0.33	0.156 ±0.011	3.80 ±0.27	5.28 ± 0.69
105.32 ± 1.10	0.539 ± 0.049	8.79 ± 0.79	159. ± 11.	0.469 ±0.037	76.6 ± 6.9	5.33 ±0.48	24.8 ± 1.7	64.7 ± 4.5	15.3 ± 1.1	5.26 ± 0.95	3.17 ±0.32	0.144 ±0.022	2.79 ±0.28	2.77 ± 0.42
95.06 ± 1.17	0.828 ± 0.075	12.8 ± 1.2	192. ± 15.	0.214 ±0.021	61.6 ± 5.6	4.92 ±0.54	25.4 ± 1.8	50.2 ± 5.5	17.5 ± 1.2	7.21 ± 1.00	1.94 ±0.19	0.135 ±0.020	1.22 ±0.09	1.74 ± 0.26
87.65 ± 0.64	0.78 ± 0.12	14.9 ± 1.2	207. ± 19.	0.092 ±0.016	57.4 ± 4.0	4.51 ±0.36	22.7 ± 1.8	38.1 ± 4.6	17.6 ± 1.2	6.35 ± 0.89	0.933 ±0.112	0.091 ±0.012	0.585 ±0.047	1.55 ± 0.22
83.87 ± 1.25	1.24 ± 0.11	19.1 ± 1.7	240. ± 17.	0.0409 ±0.0061	61.1 ± 5.5	4.69 ±0.52	21.4 ± 1.5	24.0 ± 1.9	16.0 ± 1.1	5.14 ± 0.51	0.343 0.051	0.060 ±0.009	0.381 ±0.031	1.74 ± 0.26
81.77 ± 0.74	1.11 ± 0.14	18.7 ± 1.7	239. ± 31.	0.0293 ±0.0053	59.8 ± 4.8	4.50 ±0.36	20.0 ± 1.6	25.0 ± 2.5	15.6 ± 1.1	4.98 ± 0.55	0.161 ±0.024	0.054 ±0.021	0.340 ±0.027	1.77 ± 0.37
75.53 ± 0.84	1.50 ± 0.15	24.0 ± 1.9	278. ± 33.	0.0075 ±0.0034	73.8 ± 5.2	4.54 ±0.36	14.0 ± 1.3	16.2 ± 1.5	9.22 ± 0.65	2.28 ± 0.46	0.0118 ±0.0024		0.271 ±0.024	1.64 ± 0.23
68.88 ± 0.92	2.07 ± 0.19	31.3 ± 2.5	365. ± 51.		96.4 ± 6.8	4.07 ±0.33	7.15 ± 0.57	15.8 ± 1.6	2.28 ± 0.18	0.65 ± 0.13			0.323 ±0.032	0.92 ± 0.11
61.69 ± 1.02	2.95 ± 0.29	44.5 ± 4.0	476. ± 67.		104. ± 7.	2.35 ±0.21	3.31 ± 0.26	20.8 ± 1.9	0.118 ± 0.024	0.0291 ± 0.0064			0.319 ±0.028	0.221 ± 0.031
53.74 ± 1.13	4.74 ± 0.43	80.7 ± 6.5	510. ± 71.		55.2 ± 3.9	0.56 ±0.06	2.67 ± 0.24	24.7 ± 3.0					0.109 ±0.010	0.061 ± 0.019
47.96 ± 1.22	6.10 ± 0.55	142. ± 13.	358. ± 43.		11.7 ± 0.9	0.126 ±0.023	3.30 ± 0.26	15.6 ± 1.9					0.023 ±0.009	
41.24 ± 1.34	9.00 ± 0.63	232. ± 26.	93. ± 15.		0.354 ± 0.057	0.039 ±0.019	3.91 ± 0.35	3.51 ± 0.28						
33.95 ± 1.50	14.9 ± 1.2	233. ± 26.	8.0 ± 0.8				1.99 ± 0.20	0.077 ± 0.013						
25.15 ± 1.79	36.8 ± 3.7	19.4 ± 1.7	1.82 ± 0.16				0.175 ± 0.023							

The trend to a higher n_o with increasing energy which may be read off the $^{51}V(\alpha,3n)^{52}Mn$-reaction cannot be seen in the (α,n)-reaction. However, it has to be pointed out that these initial configurations do not describe (α,pxn)-reactions in the same way [2]. (α,pxn)-reactions are partially strongly underestimated by theory and show e.g. influences of incomplete break-up of the α-particle. The latter effect is demonstrated in Fig.2 for (α,2pxn)-reactions. The production of ^{56}Mn from ^{55}Mn is at higher energies underestimated by theory by one order of magnitude. Here, the incomplete break-up of the α-particle or direct reactions, which both are not considered by theory, may be taken as an explanation. (See [1] for a detailed discussion). But also the reaction $^{55}Mn(\alpha,2p3n)^{54}Mn$ shows severe differences between theory and experiment. The discrepancies can be divided into two classes which are observed for all {(α,xpyn), x,y ≥ 2)}-reactions. The first one is to be found for relatively low α-energies, here between 40 and 60 MeV, where the experimental data show the preequilibrium emission of a complex α-particle. This reaction mechanism is not described by the present theory. The second type of discrepancy is observed in the region of higher energies. Generally, the experimental excitation functions show a strong levelling off or even an increase with energy for α-energies > 100 MeV. This effect is not explained by theory. For $^{55}Mn(\alpha,2p3n)^{54}Mn$ the respective difference is nearly one order of magnitude. However, with an increasing number of neutrons the agreement between theory and experiment for (α,2pxn)-reactions becomes better, as it is demonstrated by the $^{55}Mn(\alpha,2p5n)^{52}Mn$-reaction. Obviously, this is due to the fact that due to the increasing threshold energies the excitation functions are more and more dominated by a compound nucleus reaction and preequilibrium effects scarcely can be observed. Also in the case of the (α,4pxn)-reactions the calculations do not adequately describe the experimental data (Fig.1). For example, at 60 MeV the production of ^{47}Sc from ^{55}Mn shows a strong contribution of preequilibrium emission of α-particles. The high energy tail of this excitation function, similar to that one of the $^{51}V(\alpha,4p3n)^{48}Sc$-reaction, exhibits rather strong contributions of direct reactions and/or incomplete particle break-up. But both reaction types cannot be described in the framework of the hybrid model. It should be pointed out that also for $^{51}V(\alpha,5p3n)^{47}Ca$ the high energy cross sections are completely misinterpreted by the theory which demonstrates that the initial phases of α-induced reactions are partially dominated by direct reactions at high energies and that they are more complex than assumed by the hybrid model.

For (α,6pxn)-reactions in the high energy region (E > 100 MeV) the shape of the excitation functions is fairly reproduced by theory. But also for these reaction channels, having 12 and 13 nucleons in the exit channel, either preequilibrium emission of α-particles or evaporation of more complex particles are to be

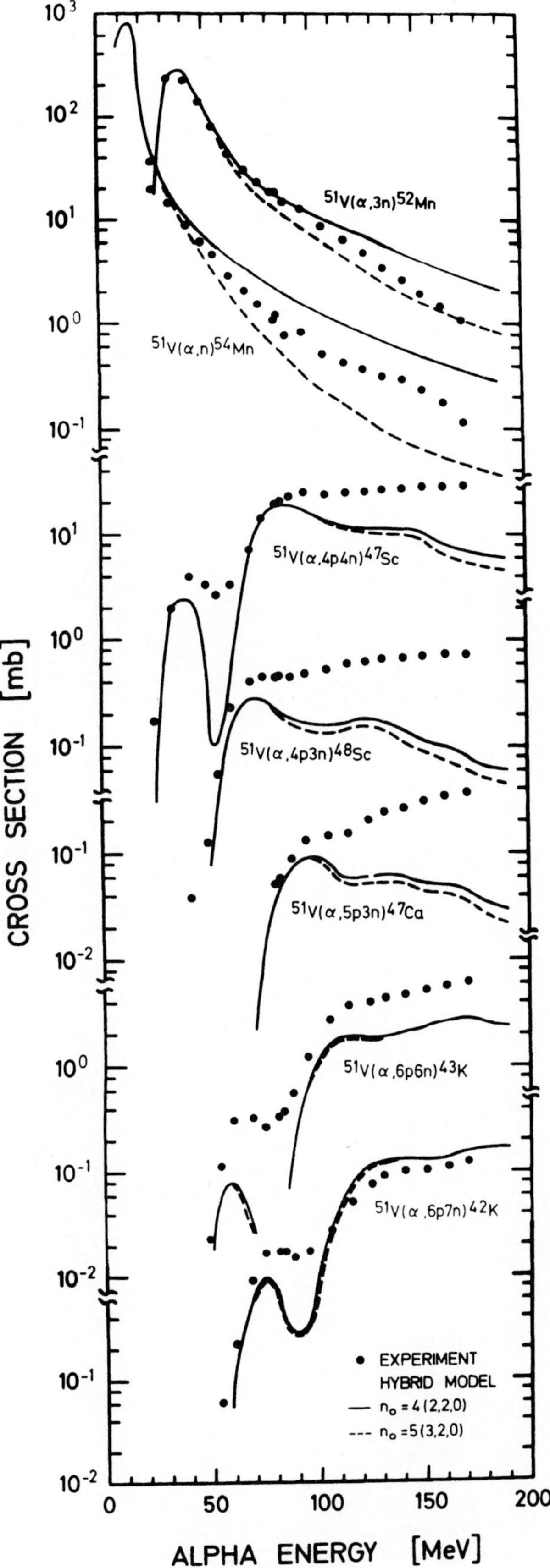

Fig. 1. Experimental and Theoretical Excitation Functions for α-Induced Reactions on Vanadium.

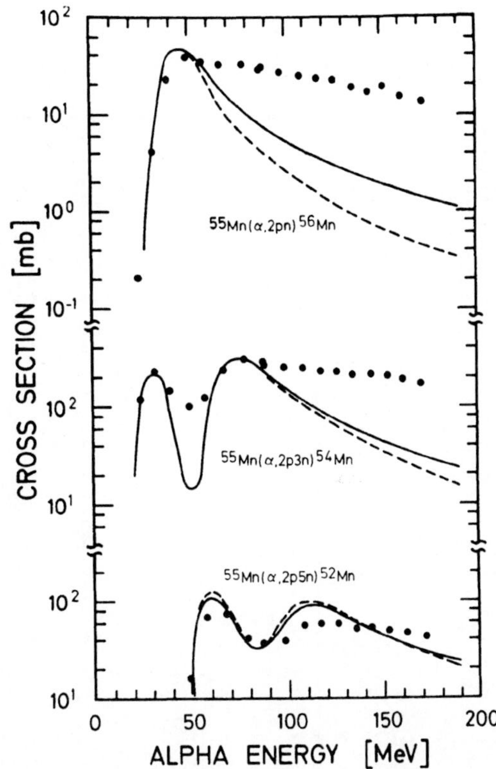

Fig. 2. Experimental and Theoretical Excitation Functions for {(α,2pxn), x = 1,3,5}-Reactions.

considered at lower energies between 50 and 90 MeV.

Summarizing, it can be said that it is not possible to describe quantitatively integral excitation functions of α-induced reactions for energies up to 200 MeV on the basis of the hybrid model of preequilibrium reactions in the form of OVERLAID ALICE. This is not only due to the neglection of the preequilibrium emission of complex particles, but also due to the increasing influence of direct reactions and of the break-up of the α-particle.

Acknowledgments

Our thanks are due to Prof.Dr.W.Herr for his continuous interest and many helpful discussions. We are also grateful to the directors and to the members of the IKP of the KFA Juelich for the hospitality and to the staff of the cyclotron for the assistance during the irradiations. This work was supported by the Bundesminister für Forschung und Technologie, Bonn.

References

1. R. Michel, G. Brinkmann, Nucl.Phys.A 338, 167 (1980)
2. R. Michel, G. Brinkmann, R. Stück, submitted to the Radiochim.Acta (1982)
3. M.Blann, Phys.Rev.Lett.27, 337, errata ibid. 27, 1550 (1971)
4. V.F.Weisskopf, D.H. Ewing, Phys.Rev. 57, 472 (1940)
5. R. Michel, G. Brinkmann, H. Weigel, W. Herr, Nucl.Phys.A 322, 40 (1979)
6. A. Iguchi, H. Amano, S. Tanaka, J.At.Energ.Soc.Jap.2, 682 (1960)
7. E.F. Neuzil, R.H. Lindsay, Phys.Rev. 131, 1697 (1963)

Table II Cross Sections [mb] for α-Induced Reactions on Manganese

E_α [MeV]	$^{58m+g}Co$	^{57}Co	^{56}Co	^{55}Co	^{52}Fe	^{56}Mn	^{54}Mn	^{52}Mn	^{51}Cr	^{48}Cr	^{48}V	$^{46m+g}Sc$	^{43}K	^{42}K
171.15 ± 0.36				0.24 ± 0.44	0.158 ±0.094	13.9 ± 1.3	177. ± 19.	42.9 ± 3.7	137. ± 13.	0.572 ±0.069	43.2 ± 3.5	19.6 ± 1.9	1.060 ±0.098	2.14 ±0.29
162.24 ± 0.53				0.45 ± 0.62	0.205 ±0.070	15.5 ± 1.4	194. ± 18.	46.9 ± 4.2	147. ± 17.	0.567 ±0.072	44.1 ± 3.4	19.2 ± 1.7	0.960 ±0.098	1.94 ±0.28
152.88 ± 0.64				0.39 ± 0.59	0.220 ±0.070	19.7 ± 3.8	201. ± 19.	48.7 ± 4.3	149. ± 22.	0.518 ±0.060	42.6 ± 3.4	17.9 ± 1.9	0.823 ±0.088	1.70 ±0.21
144.37 ± 0.72				0.64 ± 0.59	0.241 ±0.088	16.8 ± 3.2	214. ± 20.	52.6 ± 4.5	154. ± 13.	0.479 ±0.049	42.2 ± 3.2	17.1 ± 1.3	0.69 ±0.10	1.42 ±0.16
135.32 ± 0.79				0.52 ± 0.58	0.234 ±0.062	18.7 ± 2.1	205. ± 17.	50.6 ± 4.4	147. ± 15.	0.320 ±0.036	34.9 ± 2.6	14.4 ± 1.1	0.544 ±0.059	0.74 ±0.19
125.91 ± 0.83			1.6 ± 2.7	1.14 ± 0.62	0.260 ±0.061	22.8 ± 2.1	226. ± 21.	56.8 ± 4.5	153. ± 13.	0.237 ±0.028	29.1 ± 2.4	13.1 ± 1.2	0.363 ±0.031	0.56 ±0.14
117.40 ± 0.92			1.6 ± 2.9	1.17 ± 0.64	0.215 ±0.063	23.6 ± 2.4	223. ± 18.	56.1 ± 4.4	142. ± 10.	0.093 ±0.018	20.6 ± 1.4	8.98 ± 0.70	0.174 ±0.026	0.310 ±0.062
108.43 ± 0.98			5.1 ± 3.0	1.95 ± 0.69	0.183 ±0.046	25.1 ± 2.5	253. ± 18.	54.9 ± 4.4	151. ± 14.	0.0130 ±0.0082	17.9 ± 1.4	5.00 ± 0.29	0.103 ±0.019	0.315 ±0.050
98.8 ± 1.0			4.2 ± 3.2	1.96 ± 0.69	0.050 ±0.018	26.7 ± 3.5	244. ± 20.	39.6 ± 3.1	154. ± 14.		17.1 ± 1.2	2.06 ± 0.17	0.039 ±0.012	0.176 ±0.029
88.70 ± 0.50			7.9 ± 3.4	3.53 ± 0.67	0.032 ±0.015	30.3 ± 2.3	264. ± 16.	36.8 ± 3.4	186. ± 14.		19.9 ± 1.5	2.22 ± 0.20	0.031 ±0.013	
88.4 ± 1.1			8.8 ± 3.5	3.17 ± 0.78	0.012 ±0.016	29.6 ± 3.3	283. ± 23.	37.0 ± 2.7	192. ± 15.		17.1 ± 1.2	2.06 ± 0.17	0.038 ±0.012	
78.68 ± 0.68		6.5 ± 7.1	14.3 ± 3.1	5.10 ± 0.77		32.2 ± 2.6	300. ± 21.	41.8 ± 3.9	206. ± 18.		14.2 ± 1.2	2.52 ± 0.18		
67.53 ± 0.79		11.7 ± 6.3	22.9 ± 3.5	8.8 ± 1.3		32.4 ± 3.2	240. ± 14.	72.4 ± 6.1	89.7 ± 7.7		1.20 ± 0.19	1.015 ± 0.071		
57.77 ± 0.90		21.3 ± 4.5	39.4 ± 5.5	11.6 ± 1.1		34.9 ± 2.7	126. ± 8.	68.6 ± 5.7	7.00 ± 0.46			0.162 ± 0.049		
49.4 ± 1.0		39.2 ± 4.6	87.6 ± 9.1	6.74 ± 0.52		39.6 ± 2.8	101. ± 7.	15.8 ± 1.3	1.01 ± 0.22					
39.8 ± 1.1	5.5 ± 1.5	106.9 ± 9.0	154. ± 12.	0.225 ± 0.067		22.8 ± 1.6	146. ± 10.							
32.8 ± 1.3	16.7 ± 1.4	336. ± 27.	93.3 ± 7.3			4.33 ± 0.42	225. ± 15.							
24.4 ± 1.5	70.2 ± 4.8	646. ± 47.	0.284 ± 0.043			0.211 ± 0.033	120. ± 8.							
12.7 ± 2.1	582. ± 52.	74.7 ± 5.2												

8. P.P. Dimitriev, I.O. Konstantinov, N.N. Krasnov, Atom.Energ.USSR 26, 467 (1969)

9. W.W. Bowman, M. Blann, Nucl.Phys.A 131,513 (1969)

10. A.E. Vlieks, J.F. Morgan, S.L. Blatt, Nucl.Phys.A 224, 492 (1974)

11. G. Brinkmann, Thesis, University of Cologne(1979)

12. T. Matsuo, J.M. Matuszek, N.D. Dudey, T.T. Sugihara, Phys.Rev. 139, B 886 (1965)

13. S. Tanaka, M. Furukawa, T. Mikumo, S. Iwata, M. Tagi, H. Amano, J.Phys.Soc.Japan 15, 545 (1960)

14. M. Blann, Computer-Code COO-3494-29, COO-3494-32 and UR-NSRL-181 (1978)

Further Study of Several Physical Effects on the Calculation of Angular Distribution Based on Exciton Model

Sun Ziyang, Wang Shunuan, Zhang Jingshang, Zhuo Yizhong
Institute of Atomic Energy, Academia Sinica, P.O.Box 275, Beijing

Abstract

In order to understand angular distribution deviation between the calculation based on conventional exciton model and the experimental data of the preequilibrium and equilibrium decays, especially the underestimate of angular distribution by the theory at backward angles, an exact closed form solution to the time-integrated master equation of the exciton model by including the effects of the Fermi motion, the Pauli principle, the finite nuclear size and the refraction of the incident wave at the nuclear surface is applied to the calculation of the angular distribution. In this paper the neutron inelastic scattering cross section for 34 elements at 14.6 MeV have been calculated and compared with experimental data measured by Hermsdorf et al.[5] The results show that because of taking into account the influence of several physical effects mentioned above, the theory can explain the presently existing experimental data, especially solve the problem of the double differential cross section for emitted neutrons with higher energy at backward angles.

1. Introduction

Preequilibrium statistical theory has been proved to be very useful for the description of light projectiles induced reaction in the energy range of several tens of MeV. An important problem in preequilibrium theory is how to describe the angular distribution of emitted particles. For the analysis of preequilibrium angular distribution, Mantzouranis et al[2]. have proposed a generalized master equation of exciton model and derived the distribution probability of two-nucleon collision from one direction to the other based on the approximation of free nucleon-nucleon scattering. As we know, when the incident energy is much higher than Fermi energy Ef, this approximation might be adequate. However, when the incident energy is in the range of several tens of MeV, which is comparable with or less than the Fermi energy, one would expect the Fermi motion and the Pauli principle to play a role. Thus in our previous paper(1), on the basis of the approaches of Mantzouranis et al[2]. and Akkermans [3] we have also derived an exact closed form solution for the time-integrated master equation of exciton model but with the emphasis of the importance by taking into account the influence of both the Fermi motion and the Pauli principle. The main difference lies in the calculation of the distribution probability of two-nucleon collision from one direction to the other $G(\Omega,\Omega')$. In this respect we use the Fermi gas model to describe the nucleon-nucleon interaction inside the nucleus, in this way the Pauli principle are properly taken into account. The solution is then applied to calculate the double differential cross section and the angle-integrated energy spectra for both the preequilibrium and equilibrium decays. In our previous paper(1), we have analysed the variation of the shape of the angular distribution due to Pauli principle and Fermi motion in detail and as an example to compare with the experiments we have calculated the neutron-induced reaction ^{93}Nb(n,n') at En=15 MeV. It seems that the most significant impoprovement of our approach is the rise of the backward direction of the double differential cross section for higher energy emitted neutrons.

Akkermans et al[4]. have calculated the neutron inelastic scattering cross section for 34 elements at 14.6 MeV by using of the conventional exciton model by taking into account the effect of refraction of the incident particle at the nuclear surface based on the approximation of free nucleon-nucleon scattering. Their results show that the angular distribution at backward angles is much underestimated. Thus, they have reached the conclusion that further study with regard to the physics of the model is required.

In order to understand the deviation between the calculation based on conventional exciton model and experimental data of the preequilibrium and equilibrium decays, our model[1] which contains some new physical ingredients such as the effects of the Fermi motion, the Pauli principle, the finite nuclear size and the refraction of the incident wave at the nuclear surface is applied to calculate the neutron inelastic scattering for 34 elements at 14.6 MeV and compared with the experimental data measured by Hermsdorf et al[5].. The results show that because of taking into account the influence of several physical effects mentioned above, the theory can explain the presently existing experimental data, especilly solve the problem of the rise at the backward angles of the double differential cross section for the higher energy emitted neutrons.

In Sec.2 a general formalism of our theory develeped in(1) is briefly described. The main results and discussions are given in Sec.3.

2. General Formalism of the Model

The generalized master equation takes the form of

K. H. Böckhoff (ed.), Nuclear Data for Science and Technology, 603–605.

$$\frac{d}{dt}q(n,\Omega,t)=\sum_m\int q(m,\Omega',t)\,W_{m\to n}(\Omega',\Omega)\,d\Omega$$

$$-\sum_m\int q(n,\Omega,t)\,W_{n\to m}(\Omega,\Omega')\,d\Omega'. \quad (1)$$

Where $G(\Omega,\Omega')$ is the distribution probability of two-nucleon collision from direction Ω to direction Ω' inside the nuclear matter.

The exact closed form solution to the time-integrated master equation in the form of a partial wave reads

$$\zeta_l(n_0)=\tau_l(n_0)\eta_l\mathcal{F}_{n_0} \quad (n=n_0)$$

$$\zeta_l(n)=\tau_l\eta_l\mathcal{F}_{n_0}\prod_{\substack{k=n_0\\ \Delta k=2}}^{n-2}\mu_l\lambda_+(k)\tau_l(k)\mathcal{F}_{k+2}. \quad (n>n_0) \quad (2)$$

According to the definition the quantity $G(\Omega,\Omega')$ is expressed as

$$G(\Omega,\Omega')=\frac{d\sigma(\Omega,\Omega')}{d\Omega'}\Big/\int\frac{d\sigma(\Omega,\Omega')}{d\Omega'}d\Omega'. \quad (3)$$

In the papers of Mantzouranis et al. (2) and Akkermans (4) the $G(\Omega,\Omega')$ is derived by means of free nucleon-nucleon scattering.

$$G(\Omega,\Omega')=\frac{\cos\theta}{\pi}\Theta\left(\frac{\pi}{2}-\theta\right). \quad (4)$$

Then, they obtain μ_l based on formula(4)

In the present paper in order to consider the influence of the Fermi motion and the Pauli principle we have derived $G(\Omega,\Omega')$ and μ_l based on the Fermi gas model instead of free nucleon-nucleon scattering.

Let P_{L_1}, P_{L_2}, $P_{L'_1}$, $P_{L'_2}$ be the momenta of a pair of nucleons before and after collision, respectively. Because of the restriction of the Pauli principle the following conditions must be satisfied for the $\Delta n=2$ transition process:

$$|P_{L_1}|>|P_f|, \quad |P_{L_2}|\leq|P_f|,$$
$$|P_{L'_1}|>|P_f|, \quad |P_{L'_2}|>|P_f| \quad (5)$$

where P_f is the Fermi momentum. If we consider only the Fermi motion without the Pauli principle, then only conditions $|P_{L_1}|>|P_f|$ and $|P_{L_2}|\leq|P_f|$ are needed.

Thus $G(\Omega,\Omega')$ and μ_l can be obtained, which are expressed as

$$G(\Omega,\Omega')=\int\frac{d^2\sigma}{dE_{L'_1}d\Omega_{L'_1}}dE_{L'_1}/\bar{\sigma}. \quad (6)$$

$$\mu_l=\frac{1}{\bar{\sigma}}\iint\frac{d^2\sigma}{dE_{L'_1}d\Omega_{L'_1}}P_l(\cos\theta_{L'_1})dE_{L'_1}d\Omega_{L'_1}. \quad (7)$$

where

$$\bar{\sigma}=\iint\frac{d^2\sigma}{dE_{L'_1}d\Omega_{L'_1}}dE_{L'_1}d\Omega_{L'_1}.$$

$d^2\sigma/dE_{L'_1}d\Omega_{L'_1}$ is the double differential cross section of effective nucleon-nucleon scattering in the nuclear matter.

Compared with the $G(\Omega,\Omega')$ and μ_l derived from the free nucleon-nucleon scattering mentioned above where the target nucleons are supposed to sit still, in the present paper these quantities are derived from the double differential cross section of the effective nucleon-nucleon scattering in the nuclear matter as expressed by (6) and (7) where $G(\Omega,\Omega')$ and μ_l are dependent on total excitation energy E of the system and exciton number n. The struck target nucleons have non-zero momenta but less than Fermi momentum P_f and because of the restriction of the Pauli principle $|P_{L'_1}|$ and $|P_{L'_2}|$ must be greater than P_f after collision in nuclear matter.

In the calculations of μ_l, E_f =36 MeV and $E_{L_1}=E_f+E/n$ are used. Here E is the excitation energy of the whole system.

We have proved that when $E_f=0$, $G(\Omega,\Omega')$ can be exactly reduced to free nucleon-nucleon scattering. That is

$$G(\Omega,\Omega')\to\frac{\cos\theta}{\pi}\Theta\left(\frac{\pi}{2}-\theta\right).$$

The influence of geometry effect of the finite size of the nucleus on the shape of angular distribution is studied in detail in the paper of ref(1). As a rough estimate for the finite nuclear size we just cut off the summation over partial wave number l up to a certain value l_{max}. In the present paper we take l_{max} =2 for all of the 34 elements because the experimental data were also analysed with only l=0,1,2 three terms.

The shape of the angular distribution is largely determined by the emission from the lowest exciton states (i.e. from the first few collisions). Thus, the effect of the refraction of the incident wave at the nuclear surface should be taken into account. Qualitatively speaking, the effect always leads to a smoothing of the angular distribution. Similar to the previous paper (4), we have also used the crude approach to include the refraction of the incident wave based on the simple quasi-classical consideration, in which the initial exciton number n_0=1 with the total particle emission rate $W_{n_0=1}$=0 has been taken.

3. Numerical Results and Discussions

We have calculated the neutron inelastic scattering cross section for 34 elements in a large mass range at 14.6 MeV by including the effects of the Fermi motion, the Pauli principle, the finite nuclear size and the refraction of the incident wave. The composite nucleus formation cross section and the inversion cross section are calculated by means of an empirical formula taken from (6). The free parameter of the square of the average two-body interaction matrix element K=190 MeV^3 is used. The theoretical results both of Akkermans et al. and ours together with the experimental data over the energy range from 6 to 11 MeV for 34 elements have been plotted in Figs. 1-4. These 34 elements are ^{9}Be, ^{12}C, ^{23}Na, ^{24}Mg, ^{27}Al, ^{28}Si, ^{31}P, ^{32}S, ^{40}Ca, ^{48}Ti, ^{51}V, ^{52}Cr, ^{55}Mn, ^{56}Fe, ^{59}Co, ^{58}Ni, ^{63}Cu, ^{64}Zn, ^{69}Ga, ^{80}Se, ^{79}Br, ^{90}Zr, ^{93}Nb, ^{114}Cd, ^{115}In, ^{120}Sn, ^{121}Sb, ^{127}I, ^{181}Ta, ^{184}W, ^{197}Au, ^{202}Hg, ^{208}Pb, ^{209}Bi. From the comparison between the calculated results and the data one can clearly see that the theory of Akkermens et al. is seriously underestimated and our model can fit experimental data quite well. Our model can explain both the shape and the magnitude of the angular distribution cross section very well for the ^{209}Bi, ^{208}Pb, ^{202}Hg, ^{127}I, ^{184}W, ^{120}Sn, ^{115}In, ^{24}Mg, eight elements; for ^{56}Fe, ^{55}Mn, ^{12}C, ^{58}Ni four elements the fitting is still rather good at backward-angles, but the calculated values are underestimated at forward-angles; in contrast, for ^{23}Na, ^{29}Br, ^{69}Ga, ^{80}Se four elements the agreement between the theory and the experimental data is better at the forward angles than at the backward angles, where the calculated values are underestimated. In general, except ^{48}Ti and ^{32}S our results are much better than the ones of Akkermens et al.. According to the experimental data, for these two nuclei the behavior of the cross section at backward-angles decreases with the increase of the angle, which is not the same as other 32 elements. It is not clear to us why they behave so irregularly. If the data are reliable, further study on this point is necessery.

In summary in this paper the model proposed by us in(1) has been further developed and applied to compare with a number of experimental data. It seems that our model works quite well, especially greatly improves the agreement with the experimental data of the cross section in the backward angle range for higher energy emitted particles. However, our calculation is restricted by the situation that up to now the existing experimental data have only been analysed with three partial waves, because of lack of data. Thus, in order to have a deeper understanding of the preequilibrium reaction mechanism, further experimental measurement on double differential cross section is absolutely needed.

As in (1), in the present paper the Pauli principle for the density of the exciton states has not been considered, which might be

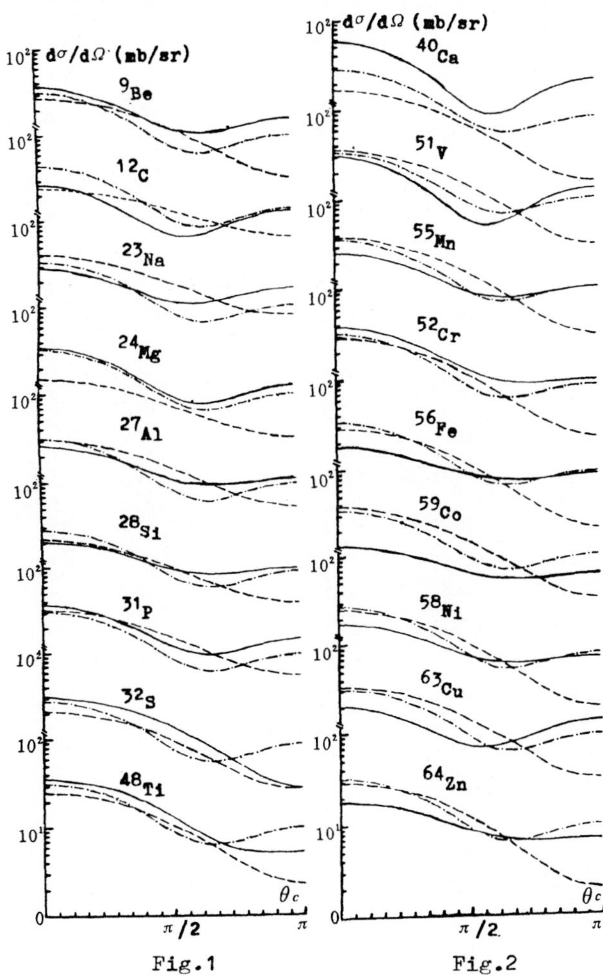

Fig.1 Fig.2

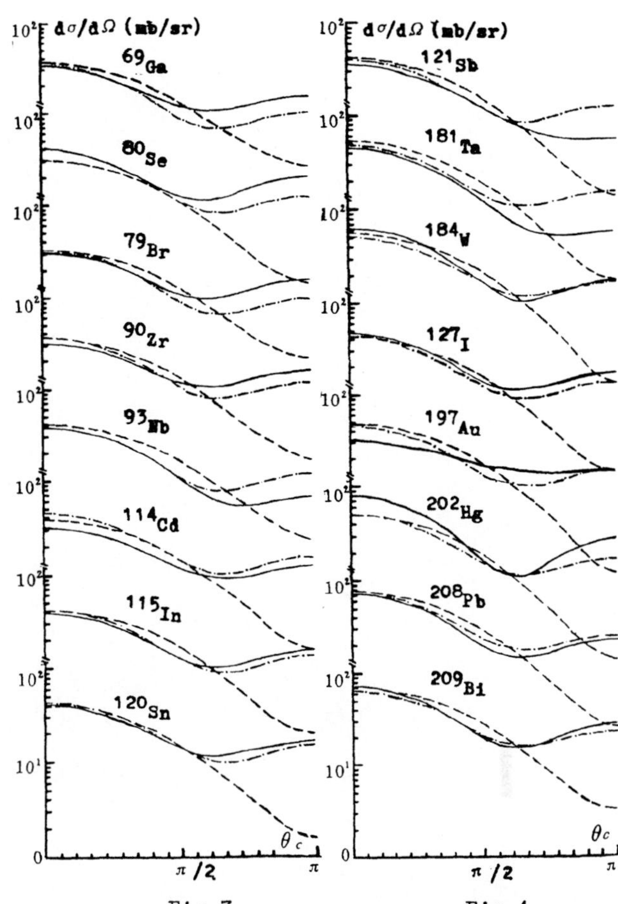

Fig.3 Fig.4

Fig.1-4 The comparison of angular distribution calculation with experimental data of neutron inelastic scattering cross section for 34 elements at 14.6 MeV for energy range of 6-11 MeV.
——— for the experimental data
------ for the calculation performed by Akkermens et al.
-.-.-. for the present calculation

important for equilibrium decay. Our description for equilibrium decay has not been improved over the previous ones. In addition, in the present paper the effect of finite nuclear size and the refraction of the incident wave at the nuclear surface are treated with simple quasi-classical consideration. How to deal with them in a more consistent way is remained to be further studied.

References

1. Sun Ziyang et al., Z.Phys. A 305, 61-68 (1982).
2. G.Mantzouranis, H.A.Weidenmüller, D.Agassi, Z.Phys. A 276, 145 (1976).
3. J.M.Akkermans, Phys. Lett. B 82, 29(1979).
4. J.M.Akkermans, H.Gruppelaar, Phys. Rev. C.Vol. 22, number 1, 73(1980).
5. D.Hermsdorf et al., Zentralinstitut für Kernforschung, Rossendorf bei Dresden Report No.ZfK-277(1974).
6. I.Dostrovsky et al., Phys. Rev., 116, 683(1959).

Neutron Data Analysis and Evaluations

NUCLEAR DATA EVALUATION

Ju.L. Kalachev, V.M. Kolobashkin, M.V. Lepeshkin, T.M. Televinova and V.P. Chist'yakov
Moscow Engineering Physics Institute
Kashirskoe ch. 31, Moscow 115409
Union of Soviet Socialist Republics

Some statistical problems relevant to combining a number of estimates are considered. The effect of allowance for correlations between estimates as well as that of substitution of theoretical weights by approximate ones are studied. The approximate formulae of calculating the covariance matrix of combined estimate vector are given. Problems of Combining Estimates subject to Functional relationships between data are investigated. It is shown, that the account of functional relationships increase the estimates accurace.

[Nuclear data evaluation, variance, estimate, common random value, functional dependence, decay scheme, yield]

There is a demand for accuracy and reliability of nuclear data used in nuclear technology that makes it necessary to elaborate mathematical methods for experimental data evaluation. Evaluation process may be presented as three stages:

1. The expert treatment, a preliminary analysis of articles by a "physicist-evaluator" to select author's values which have not obvious systematic components.
2. The statistic processing, testing of values selected at the first stage for homogeneity and formation of homogeneous samples.
3. The construction of unified estimates by unifying sample values obtained at the second stage.

The report concerns statistical problems that appear in the constructing of unified estimates [1-2].

Let $y_1, \dots y_k$ and $s_1^2, \dots s_k^2$ be estimates of a quantity θ and estimates of variances of $y_1, \dots y_k$. Suppose these estimates to be unbiased, the covariance matrix $c = \mathrm{cov}(y_i y_j) = \rho_{ij}\sigma_i\sigma_j$ ($\sigma_i^2 = Es_i^2 = Dy_i$) to be known and its determinant not to vanish. The least squares method gives the following estimate of the quantity θ:

$$\hat{\theta} = W^{-1} \cdot \sum_{i=1}^{k} W_i y_i \tag{1}$$

where

$$W_i = \sum_{j=1}^{k} C_{ij}^{-1}, \quad W = \sum_{i=1}^{k} W_i = \sum_{i,j=1}^{k} C_{ij}^{-1}.$$

The variance of this estimate $D\theta = W^{-1}$. In case of zero correlation the matrix C becomes diagonal, the calculation of $\hat{\theta}$ being greatly simplified.

Consider what results from neglecting of correlation provided the correlation to be small: $\rho_{i\ell} = \epsilon\rho_{i\ell}^{(0)}$, $\epsilon \ll 1$.

Then we have

$$\frac{D\hat{\theta}_0 - D\hat{\theta}}{D\hat{\theta}} \cong \frac{\epsilon^2}{2} W_0^2 \cdot \sum_{i,j=1}^{k} \frac{(\Delta_i - \Delta_j)^2}{\sigma_i^2 - \sigma_j^2} \tag{2}$$

where

$$\hat{\theta}_0 = W_0^{-1} \cdot \sum_{i=1}^{k} W_i^{(0)} \cdot y_i \; ; \; D\hat{\theta}_0 = W_0^{-1} \; ; \; \Delta_i = \sum_{\ell \neq i} \frac{\rho_{i\ell}^{(0)} \cdot \sigma_i}{\sigma_\ell},$$

W_0, $W_i^{(0)}$ are weights calculated with vanishing correlation. The discrepancy $D\hat{\theta}$ from $D\theta_0$ may be small in many cases, the estimate $\hat{\theta}_0$ being preferable in view of construction simplicity. Note that weights $W_i \cdot W^{-1}$ and $W_i^{(0)} \cdot W_0^{-1}$ depend on theoretical covariances σ_i^2. When theoretical values σ_i^2 are replaced by their estimates s_i^2, we obtain "random weights". Influence of such a substitution is investigated below.

Unification with Random Weights

Let $\overline{W}_0$, $\overline{W}_i^{(0)}$ be weights in which the substitution $\sigma_i^2 \to s_i^2$ is made; they give rise to a new estimate

$$\overline{\theta} = \overline{W}_0^{-1} \cdot \sum_{i=1}^{k} \overline{W}_i^{(0)} \cdot y_i .$$

Consider an approximate formula for variance of $\overline{\theta}$ provided the sample size n_i of i-th author to be large enough. Let n_i may be written in the form $n_i = d_i n$ where d_i = const $< \infty$ and $n \to \infty$. Using expansion $D\overline{\theta}$ in powers of $1/n$ we obtain:

$$D\overline{\theta} = \frac{1}{W_0} [1 + 2 \cdot \sum_{i=1}^{k} \frac{\eta_i(1-\eta_i)}{n_i}] + O(n^{-5/2}) \tag{3}$$

where $\eta_i = W_0^{-1} \cdot (1/\sigma_i^2)$ are theoretical normalized weights. Note that the paper [3] is concerned with the formula (3) but the order of the remainder is not pointed out.

An estimate of $D\overline{\theta}$ may be naturally used

$$\hat{D\overline{\theta}} \cong \frac{1}{\overline{W}_0} [1 + 2 \cdot \sum_{i=1}^{k} \frac{h_i(1-h_i)}{n_i}]$$

where $h_i = \overline{W}_0^{-1} \cdot (1/s_i^2)$ are random weights. The main term $\overline{B}^2 = \overline{W}_0^{-1}$ may differ significantly from $D\overline{\theta}$ when we have the scanty sample. In such a case the correction terms must be taken into account.

Covariance Matrix Calculation for Combined Estimates Vector

Examine a more general case when $\mathrm{cov}(y_i y_j) \neq 0$ for $i \neq j$. The correlation between estimates usually appears as a result of using common random parameters. Such parameters may be coefficients of graduated curve of a spectrometer for example.

Let combined vector

$$\vec{y} = [f_1(\vec{X}_1, \vec{\varphi}), \dots, f_k(\vec{X}_k, \vec{\varphi})]$$

be a function of independent estimates $\vec{X}_1, \dots, \vec{X}_k$ and common random vector $\vec{\varphi}$, vector $\vec{X}_i$ having m_i components $X_{i_1}, \dots, X_{im_i}$ and vector $\vec{\varphi}$ having N components $\varphi_1, \dots, \varphi_N$. Then in linear approximation we obtain

$$\mathrm{cov}(y_i y_j) \cong \sum_{r,s=1}^{N} \frac{\partial f_i}{\partial \varphi_r} \mathrm{cov}(\varphi_r \varphi_s) \frac{\partial f_j}{\partial \varphi_s} + \sum_{r,s=1}^{m_i} \frac{\partial f_i}{\partial X_{ir}} \mathrm{cov}(X_{ir}, X_{is}) \frac{\partial f_i}{\partial X_{is}} \cdot \delta_{ij}, \tag{4}$$

δ_{ij} is the Kronecker symbol. Thus $\mathrm{cov}(y_i y_j)$ for $i \neq j$ is proportional to the sum of parameter covariances.

K. H. Böckhoff (ed.), Nuclear Data for Science and Technology, 609–611.

Unification of Estimates in Case of Functional Relationships Between Data

Let us have M quantities a_j ($j = 1, 2, \ldots M$) and L quantities b_k ($k = 1, 2, \ldots L$) to be estimated. These quantities are assumed to be connected by the functional dependence

$$a_j = F_j(b_1, \ldots, b_L) . \tag{5}$$

Suppose quantities a_j, b_k be respectively evaluated by I_j, I_k authors, their estimates being X_{ij} for a_j ($i = 1, 2, \ldots, I_j$) and y_{ik} for b_k ($i = 1,2,\ldots, I_k$). All the estimates are supposed to be independent. Thus there are $\sum_{j=1}^{M} I_j$ estimates of a_j and $\sum_{k=1}^{L} I_k$ ones of b_k. In such a case the logarithmic function of likelihood is

$$L = \sum_{j=1}^{M} \sum_{i=1}^{I_j} \frac{[X_{ij} - F_j(b_1, \ldots, b_L)]^2}{\sigma^2_{X_{ij}}} + \sum_{k=1}^{L} \sum_{i=1}^{I_k} \frac{(y_{ik} - b_k)^2}{\sigma^2_{y_{ik}}} \tag{6}$$

where $\sigma^2_{X_{ij}}$ and $\sigma^2_{y_{ik}}$ are estimates of variances of X_{ij} and y_{ik} given by i-th author. Minimizing (6) gives estimates of b_k which coincide with estimates obtained by minimizing

$$L' = \sum_{j=1}^{M} \frac{[\overline{X}_j - F_j(b_1, \ldots, b_L)]^2}{\sigma^2_{\overline{X}_j}} + \sum_{k=1}^{L} \frac{(\overline{y}_k - b_k)^2}{\sigma^2_{\overline{y}_k}} \tag{7}$$

where

$$\overline{X}_j = \sum_{i=1}^{I_j} h_{X_{ji}} \cdot X_{ji}; \quad h_{X_{ji}} = [\sigma^2_{X_{ji}} \sum_{i=1}^{I_j} \frac{1}{\sigma^2_{X_{ji}}}]^{-1};$$

$$\sigma^2_{\overline{X}_j} = [\sum_{i=1}^{I_j} \frac{1}{\sigma^2_{X_{ji}}}]^{-1}$$

and means $\overline{y}_k$ and $\sigma^2_{\overline{y}_k}$ are given in a similar manner.

Thus the estimation of quantities connected by the functional dependence may be made in two stages:

1. First of all by ignoring dependence (5) and obtaining initial estimates, "author's weighted means" $\overline{X}_j$; $\overline{y}_j$;
2. next by taking into account (5) and obtaining improved estimates.

The possibility of such a "consequent estimation" is very important for data on decay as the functional dependences between data are revealed only after the construction of decay scheme and the "author's means" are used to construct the decay scheme.

The Case of Linear Functional Relationship

In case of linear functions (5) the formulae were obtained for the improved estimates, the variances of improved estimates being smaller than those of primary estimates. It may be seen in the following simple example. Let the estimates a, b, c and estimates of their variances Da, Db, Dc of some parameters $a^\star$, $b^\star$, $c^\star$ be given and there is a linear relationship $c^\star = a^\star + b^\star$. Then by minimizing the function

$$L = \frac{(a-a^\star)^2}{Da} + \frac{(b-b^\star)^2}{Db} + \frac{[c-(a^\star+b^\star)]^2}{Dc}$$

we shall obtain

$$Da^\star = Da \cdot \frac{Db + Dc}{Da + Db + Dc} < Da$$

$$Db^\star = Db \cdot \frac{Da + Dc}{Da + Db + Dc} < Db$$

$$Dc^\star = Dc \cdot \frac{Db + Da}{Da + Db + Dc} < Dc$$

The mentioned decrease of variance estimates exists actually for the non-linear relationship as well. This property may be used in processing of nuclear data, for example, when one deals with the estimation of energy characteristics in decay scheme.

Suppose that the following data on β-decay are known:

a) the authors' weighted means of electromagnetic energy (e.-m.) transitions with estimates of their variances - E_i^{em}; D_i^{em}; i refers to i-th transition ($i = 1, 2, \ldots, I$);

b) the authors' weighted means of β-transition energy E_k^{β} with estimates of their variances D_K^{β} (K refers to level populated by β-transition);

c) estimate of mass difference Q between the parent's and daughter's nuclei ground states and estimate of its variance DQ.

The functional relationships may be shown as following:

$$\overset{\star}{E}{}_i^{em} = \sum_{j=1}^{J} \lambda_{ij} \cdot \overset{\star}{E}{}_j^{\ell ev} \quad (i = 1, 2, \ldots, I) \tag{8}$$

where I is the total number of electromagnetic transition,

$$\lambda_{ij} = \begin{cases} -1, & \text{if i-th e.-m. transition populates j-th level;} \\ +1, & \text{if i-th e.-m. transition discharges j-th level;} \\ 0, & \text{for all other cases;} \end{cases}$$

$$\overset{\star}{Q} = \overset{\star}{E}{}_k^{\ell ev} + \overset{\star}{E}{}_k^{\beta} \tag{9}$$

Subscript k refers to level populated by β-transition; $E_k^{\ell ev}$ is the energy of the daughter's nucleus. The unknown parameters are marked by stars ($\overset{\star}{E}, \overset{\star}{Q}$).

To solve the problem it is necessary to minimize the function

$$L = \sum_{i=1}^{I} \frac{(E_i^{em} - \sum_{j=1}^{J} \lambda_{ij} \cdot \overset{\star}{E}{}_j^{\ell ev})^2}{D_i^{em}} + \sum_{k=1}^{J} \frac{(E_k^{\beta} - \overset{\star}{Q} + \overset{\star}{E}{}_k^{\ell ev})^2 \cdot d_k}{D_k^{\beta}} + \frac{(Q - \overset{\star}{Q})^2}{DQ} \tag{10}$$

$$\text{where } d_k = \begin{cases} 1, & \text{if there is a } \beta\text{-transition on k-th level;} \\ 0, & \text{if a } \beta\text{-transition on k=th level is forbidden.} \end{cases}$$

The function (10) may be minimized analytically.

An example of Non-linear Functional Relationships

The data evaluation in case of non-linear functional relationships may be shown on processing fission product yields as an example. The experimental data are usually obtained by the relative method, the value of nuclide "X" yield being determined with respect to the value of nuclide "R" yield, i.e. the ratio $N_{X,R} = Y_X/Y_R$ is measured in this case. The value Y_R is a common random parameter of yields $\{Y_X\}$ to be determined. The existence of this parameter results in a correlation of yields $\{Y_X\}$.

One of the methods for treatment of relative yields is used in the ENDF/B format file [2] for the recommended fission product yields preparing. The method in which instead of authors' value of reper nuclide yield is used recommended yield has two main drawbacks. One of them is that the correlation between averaged renormalized data is not taken into account. The second is that the renormalization process is unidirectional, in other words, the renormalization procedure must foresee the possibility of the yield change for both reper and measured yield.

If we suppose the statistical independence of experimental values of normalization coefficients and normality for distributions both relatively and absolutely measured yields, then the likelihood function is

$$L = \sum_i \sigma^{-2}_{N_i} [N_{X_i,R_i} - \frac{Y_{X_i}}{Y_{R_i}}]^2 + \sum_j \sigma^{-2}_{\hat{Y}_j} [\hat{Y}_{X_j} - Y_{X_j}]^2. \quad (11)$$

Here:

N_{X_i,R_i}	is i-th experimental value of the nuclide "X_i" normalizing coefficient when we use reper "R_i";
$Y_{X_j}, Y_{X_i}, Y_{R_i}$	are nuclide "X_j", "X_i", "R_i" yields (parameters of likelihood function);
$\sigma^2_{N_i}$	is a normalizing coefficient variance;
$\hat{Y}_{X_j}$	is an experimental value of nuclide "X_j" absolutely measured yield;
$\sigma^2_{X_j}$	is a variance of nuclide "X_j" yield.

Summation is taken over all experimental points, only the normalizing coefficients for relatively measured yield being used for the first group components and the authors' yield values of Y_{X_j} for the absolute measured yields being used for the second group of components. The yield estimates Y_X are obtained as a result of minimizing (11).

References

1. Горожанкин В.М., Колобашкин В.М., Лепешкин М.В., Покровский В.Н., Телевинова Т.М., Чистяков В.П. "К вопросу объединения оценок". Препринт Р-80-579 ОИЯИ, Дубна, 1980.

2. B.F. Rider, M.E. Meek, "Compilation of fission product yields", Rpt. NEDO-12154-2(E) (1978)

3. P. Meier, "Variance of a weighted mean", Biometrics, 1953, v.9, p. 59

THE EVALUATION OF NEUTRON RESONANCE PARAMETERS

Zhou Delin, Zhuang Youxiang, Zhao Zhixiang, Yu Baosheng,
Gu Fuhua, Liu Jicai, Zhou Enchen, Zhou Chunmei

Institute of Atomic Energy, Academia Sinica
P.O.B.275(41), Beijing, China

Yan Yiming, Liu Yili

Beijing Normal University
Beijing Normal University, Beijing, China

Zhang Zhuoliang, He Jinchang, Li Xiangcheng, Li Mianfeng

Zhongshan University
Zhongshan University, Guangzhou, China

The neutron resonance parameters as well as the thermal neutron cross sections of more than fifty nuclides have been evaluated for CENDL. From the recommended resolved resonance parameters, the average resonance parameters have been estimated and recommended. Some characteristics of the average parameters have been studied. Efforts are being made to compare the cross sections calculated on the basis of the evaluated parameters with the measured ones and to obtain the residual cross sections.

(resonance parameters, average resonance parameters, thermal cross sections, evaluation)

Introduction

Neutron resonance parameters are important for nuclear technology and other applications. Interpretation and prediction of reactor properties such as resonance absorption, resonance escape probability, resonance self-shielding and temperature dependent reactivity (Doppler coefficient) require both a detailed understanding of resonance cross sections and comprehensive machine-readable resonance data files.

The evaluations of resonance parameters and thermal cross sections for more than fifty nuclides have been made in last three years in order to provide the evaluated resonance parameters for Chinese Evaluated Nuclear Data Library (CENDL). These include fissile, fertile and transplutonium nuclides, fission products, structural and control materials, which are useful for nuclear engineering and other applications.

At the same time the method of evaluating resonance parameters, the calculation of resonance cross sections on the basis of evaluated resonance parameters and the estimation of average resonance parameters and its confidence interval have been investigated.

Evaluation Method

Experimental data of resonance parameters were surveyed through CINDA.

As usual, the process of evaluation was separated into two phases. At first, the parameters were evaluated on the basis of measured parameters those were collected as complete as possible (including the existing evaluation reports and compilations of evaluated parameters), and the resonance parameters given by different laboratories were analyzed one by one. For a particular nuclide, one best set of parameters was selected and recommended or several sets of parameters were selected and a suitable combination was made from which the recommeded parameter set was obtained. In a similar way the thermal neutron cross sections (especially for transplutonium nuclides) have also been evaluated. Secondly, the cross sections are calculated with evaluated parameters and compared with the measured ones. Then the parameters are slightly modified, if necessary, and the residual cross sections are obtained. This step is in progress.

Resonance Energy   Usually, results from transmission measurements are recommended and those from the yield measurements are taken for reference. But if the transmission data

K. H. Böckhoff (ed.), Nuclear Data for Science and Technology, 612–614.

are not available, the results of yield measurements are adopted as recommended data.

Sometimes, a problem of evaluation of resonance energy is how to align their positions one by one correctly. In order to see if there is a systematic deviation among the various measurements and to combine these results and give the recommended ones reasonably, it is necessary to find out those measured resonances from different laboratories which are actually the same ones, although the measured energy positions of that resonance are different from each other. In some cases it is rather difficult to do this. After the resonance energy are corrected and aligned, a suitable combination is made for the recommended value.

Neutron and Reaction Widths It is suitable that the neutron widths obtained from transmission measurements were taken as the basic data. Those from yield measurements were adopted if there were no transmission data, but only for reference. Because the reaction yield measurement is more sensitive than the transmission ones, some weak resonances can easily be found in a reaction yield experiment, so in some cases (for example, $\Gamma_n << \Gamma_\gamma$) this is the only way to obtain the neutron width.

In order to get more exact values of widths, both transmission and reaction yield measurements are usually needed. In general, however, the reaction yield measurement is more difficult than the transmission one and the discrepances often exist between different measurements. We have paid more attention to the evaluation of reaction width. It is well known, resonance area is independent on the resolution of the spectrometer and it is much better to examine and combine such quantities. From combined resonance area and other given parameters reaction width were obtained. Certainly, it would be better to derive a complete set of the resonance parameters from analyzing the reaction yield or scattering and transmission measurements simultaneously. Such a complete set of parameters would be self-consistent, and it could represent the various cross sections better. This is what we want to do, and yet we haven't done so far.

Orbital Angular Momentum l and Spin J Spin and parity of neutron resonance were evaluated on the basis of the reported measured data. The reliability of such data to a great extent depends on the measuring method used. If more than one measurements are available, the value which was measured by most reliable method was selected and recommended. The first priority is polarization method. And the method such as primary gamma-rays detected by Ge-Li detectors, (n, gamma) and (gamma, n) angular distribution, low-energy gamma-rays detected by Ge-Li detectors, the 2-step cascade method, scattering neutron angular distribution and the gamma-ray multiplicity are considered as better methods than the others. If we can't say anything about which one of the measurements is the best, then the results which have the same J value and obtained by majority of measurements were considered as the most reliable one and selected as the recommeded value. Some of the methods described above may shed light, in some cases, on the l value of the incident neutron. Furthermore, if necessary, Bayes' theorem has been used to distinguish the different l values.

Average Resonance Parameters From the evaluated resolved resonance parameters the average resonance parameters were estimated by means of a proper method and compared the cross cross sections calculated based on such parameters with measured ones over unresolved energy region, if necessary and possible. In order to recommend average resonance parameters reasonably, we have examined the number of degrees of freedom nu of statistical distribution of the s wave reduced neutron width with a method of least squares fit by using chi-squared distribution function for about twenty nuclides, and have not found any resolved resonance parameter set with the number of degrees of freedom differ (in a specific statistical significance) from nu=1.

The dependence of the s wave strength function on spin have also been studied. In the early seventies, a couple of papers on this subject were published and quite contradictory conclusions were presented. Since then, we have not found any further study on this subject. Based on the existing evaluated resolved parameters, our investigation demonstrates that in a specific confidence interval, the s wave neutron strength function is independent on spin J. We have studied the confidence interval of average resonance parameters, i.e., strength function, mean reduced neutron width and level spacing of s wave neutron estimated from small sample. The probability density function for various estimator were derived and the confidence interval of those estimators have been given.

Conclusion

Most of the neutron resonance parameters as well as thermal cross sections evaluated by us have been collected in a compilation entitled "Neutron Resonance Theory and Resonance Parameters" which will be published in the near future. (In this book, works on neutron resonance theory and other related subjects finished in recent years are also included.) Among those evaluations one may find out that we have paid more attention to the evaluation for nuclides which are more useful in nuclear engineering and other applications. We also have paid more attention to the evaluation for transplutonium nuclides, especially its thermal cross sections. Efforts for checking the evaluated resonance parameters and obtaining the residual cross section are now being made. That is, by means of comparing the cross sections calculated on the basis of evaluated resonance parameters with the measured ones, the recommended resonance parameters (maybe it is adjusted more or less) as well as the residual cross sections will be obtained and will be adopted as part of CENDL. For the convenience of the further evaluation work one interactive evaluation system is being planned.

Acknowledge

The authers wish to thank Drs. Liu Deshun, Li Jieping, Zhong Wenfa, Su Zongdi and Mr. Qiu Cheng, they compiled and provided the codes of cross section calculation with resolved resonance parameters in different forms and the code of average cross section calculation with avrage resonance parameters for us; and sincerely appreciate Drs. Qiu Guochun and Lu Guoxiong for their helpful work on resonance theory concerned.

Present Status and Benchmark Tests of JENDL-2

Yasuyuki KIKUCHI and Members of JNDC

Japanese Nuclear Data Committee
Japan Atomic Energy Research Institute
Tokai-mura, Ibaraki 319-11, Japan

The second version of Japanese Evaluated Nuclear Data Library (JENDL-2) consists of the evaluated data from 10^{-5} eV to 20 MeV for 176 nuclides including 99 fission product nuclei. Complete reevaluation has been made to heavy actinide, fission product and main structural material nuclides. Benchmark tests have been made on JENDL-2 for fast reactor application. Various characteristics in core center have been tested with one-dimensional model for total of 27 assemblies, and more sophisticated problems have been examined for MOZART and ZPPR-3. Furthermore analyses of JUPITER project give useful information. Satisfactory results have been obtained as a whole. However, the spectrum is a little underestimated above a few hundred keV and below a few keV. The positive sodium void reactivity worth is much overestimated. As to the latter, the sensitivity analysis with the generalized perturbation method suggests that the fission cross section of ^{239}Pu below a few keV has an important role.

[JENDL-2, evaluation, benchmark tests, k_{eff}, reaction rate ratio, reactivity worth, Doppler coefficients, control rod worth, sodium void reactivity, sensitivity analysis]

Introduction

Japanese Evaluated Nuclear Data Library (JENDL) has been developed as the standard domestic library of microscopic cross sections by JAERI Nuclear Data Center with cooperation of Japanese Nuclear Data Committee (JNDC). Its first version[1,2] (JENDL-1), which was released in 1976, mainly aimed to provide data for fast reactor calculations. The second version (JENDL-2) has aimed at wider applications such as thermal reactor, radiation shielding, fusion neutronics, nuclear fuel cycle etc. Hence the number of nuclides to be evaluated was enlarged from 72 (including 28 fission product (FP) nuclides) to 176 (including 99 FP nuclides). The maximum energy was extended from 15 MeV to 20 MeV. The compilation work of JENDL-2 is nearly finished. The general purpose file containing 77 non-FP nuclides is going to be released, and the FP nuclide file will be also released at the end of the fiscal year 1982.

At the first step of the compilation, the highest priority was put to evaluation of the most important nuclides for fast reactors: ^{235}U, ^{238}U, ^{239}Pu, ^{240}Pu, ^{241}Pu, Cr, Fe and Ni*, whose cross sections mostly determine characteristics of a typical fast reactor. This decision was made responding to an urgent request to use JENDL-2 for analyses in the JUPITER project[3], joint USA-Japan mock-up experiments of large fast reactors using the ZPPR facility. The evaluation of the eight nuclides was completed in November 1979.

Benchmark tests have been made on these evaluated data by JNDC. The data of JENDL-1 were used for other nuclides. This combined library of JENDL-2 for the eight nuclides and JENDL-1 for the others is called JENDL-2B library. The first stage of the benchmark tests was made on the core center characteristics with one-dimensional model in 1979 and satisfactory results were obtained[4]. As the second stage, applicability of JENDL-2B was tested to more sophisticated problems with two-dimensional model in 1980. On the other hand, JENDL-2B library has been used for the analyses in the JUPITER project and satisfactory results have been also observed[5]. Some problems encountered in the benchmark tests have been further studied.

In this paper, a brief description is given on the evaluation for JENDL-2. The results of the benchmark tests are discussed as well as the results of analyses of the JUPITER program. The problems encountered in the benchmark tests are discussed. These experiences will be reflected on the evaluation work for the third version of JENDL(JENDL-3).

Nuclear Data Evaluation

Status of JENDL-2

The evaluation for JENDL-2 was made on the basis of the microscopic nuclear data, but various problems pointed out[6,7] through the benchmark tests on JENDL-1 were taken into account.

Table I shows nuclides contained in the general purpose file. Evaluated are all the neutron induced cross sections including the threshold reactions, the resonance parameters, the angular and/or energy distributions of emitted neutrons and some quantities pertaining to the fission properties such as ν and χ. Table II gives nuclides contained in the fission product file. For the FP file, the threshold reactions such as (n,2n), (n,3n), (n,p), (n,α) etc. are not included.

Though JENDL-2 aimed at the fusion neutronics application too, the direct reactions were not sufficiently considered. This problem is left for JENDL-3.

Light Nuclides

New evaluation was made for D, ^{7}Li, ^{9}Be and ^{19}F. The evaluation was mainly based on the measured data. For the other nuclides, the data of JENDL-1 were adopted with some slight modification.

Medium Heavy Nuclides

Newly added are ^{40}Ca, ^{45}Sc, ^{51}V and ^{59}Co. As to the main structural materials, however, it was pointed out through the benchmark tests[6] of JENDL-1 that the total and elastic scattering cross sections of Cr, Fe and Ni in JENDL-1 were overestimated in the energy region above resolved resonances up to several MeV. This is caused by our adopting the calculated total cross sections with the optical model instead of following the resonance structure appeared in the measured data. The optical potential adopted in the JENDL-1 evaluation overestimates the total cross section below 1 MeV. Moreover, the resonance self-shielding effects were ignored by smoothing away the remaining resonance structure. It was proved by benchmark tests[7] for shielding problems that the resonance self-shielding effects in this energy range had considerable effects on the neutron spectrum and penetrability.

* There exist other important nuclides such as ^{10}B, C, O, Na, and Al. It was found through the benchmark tests of JENDL-1[6] that the evaluated or recommended data of JENDL-1 were satisfactory for these nuclides. Hence little change has been made for JENDL-2.

K. H. Böckhoff (ed.), Nuclear Data for Science and Technology, 615–622.

Table I Nuclide List of JENDL-2 General Purpose File

H, D, ^{6}Li, ^{7}Li, ^{9}Be, ^{10}B, ^{12}C, ^{19}F, ^{23}Na, ^{27}Al, Si, ^{40}Ca, ^{45}Sc, ^{51}V, Cr, ^{50}Cr, ^{52}Cr, ^{53}Cr, ^{54}Cr, ^{55}Mn, Fe, ^{54}Fe, ^{56}Fe, ^{57}Fe, ^{58}Fe, ^{59}Co, Ni, ^{58}Ni, ^{60}Ni, ^{61}Ni, ^{62}Ni, ^{64}Ni, Cu, ^{63}Cu, ^{65}Cu, ^{93}Nb, Mo, Hf, ^{174}Hf, ^{176}Hf, ^{177}Hf, ^{178}Hf, ^{179}Hf, ^{180}Hf, ^{181}Ta, Pb, ^{204}Pb, ^{206}Pb, ^{207}Pb, ^{208}Pb, ^{228}Th, ^{230}Th, ^{232}Th, ^{233}Th, ^{234}Th, ^{233}Pa, ^{233}U, ^{234}U, ^{235}U, ^{236}U, ^{238}U, ^{237}Np, ^{239}Np, ^{236}Pu, ^{238}Pu, ^{239}Pu, ^{240}Pu, ^{241}Pu, ^{242}Pu, ^{241}Am, ^{242}Am, ^{242m}Am, ^{243}Am, ^{242}Cm, ^{243}Cm, ^{244}Cm, ^{245}Cm.

Table II Nuclide List of JENDL-2 FP File

^{85}Rb, ^{87}Rb, ^{86}Sr, ^{87}Sr, ^{88}Sr, ^{90}Sr, ^{89}Y, Zr, ^{90}Zr, ^{91}Zr, ^{92}Zr, ^{93}Zr, ^{94}Zr, ^{95}Zr, ^{96}Zr, ^{92}Mo, ^{94}Mo, ^{95}Mo, ^{96}Mo, ^{97}Mo, ^{98}Mo, ^{100}Mo, ^{99}Tc, ^{100}Ru, ^{101}Ru, ^{102}Ru, ^{103}Ru, ^{104}Ru, ^{106}Ru, ^{103}Rh, ^{104}Pd, ^{105}Pd, ^{106}Pd, ^{107}Pd, ^{108}Pd, ^{110}Pd, Ag, ^{107}Ag, ^{109}Ag, Cd, ^{110}Cd, ^{111}Cd, ^{112}Cd, ^{113}Cd, ^{114}Cd, ^{116}Cd, ^{115}In, Sb, ^{121}Sb, ^{123}Sb, ^{124}Sb, ^{127}I, ^{129}I, ^{131}Xe, ^{132}Xe, ^{133}Xe, ^{134}Xe, ^{135}Xe, ^{136}Xe, ^{133}Cs, ^{135}Cs, ^{137}Cs, ^{134}Ba, ^{135}Ba, ^{136}Ba, ^{137}Ba, ^{138}Ba, ^{139}La, ^{140}Ce, ^{142}Ce, ^{144}Ce, ^{141}Pr, ^{142}Nd, ^{143}Nd, ^{144}Nd, ^{145}Nd, ^{146}Nd, ^{148}Nd, ^{150}Nd, ^{147}Pm, ^{147}Sm, ^{148}Sm, ^{149}Sm, ^{150}Sm, ^{151}Sm, ^{152}Sm, ^{154}Sm, Eu, ^{151}Eu, ^{152}Eu, ^{153}Eu, ^{154}Eu, ^{155}Eu, Gd, ^{155}Gd, ^{156}Gd, ^{157}Gd, ^{158}Gd, ^{160}Gd.

Hence complete reevaluation was made for JENDL-2. The resonance parameters were evaluated on the basis of recently measured data. The resonance truncation effect, i.e., the effect of tails from resonances located outside the considered resonance region, were carefully examined and corrected. The effects of the energy dependence of effective scattering radius were also taken into consideration. The resonance structure of the total cross section observed in high resolution measurements was traced up to several MeV. This work was made with Neutron Data Evaluation System, NDES[8].

In the nuclear model calculation, the optical potential parameters and the level density parameters were determined[9,10] by carefully examining the systematic trends over the neighboring nuclides. Furthermore, the calculated cross sections were carefully normalized to the measured data particularly in the capture and inelastic scattering cross sections.

Fission Products

In addition to the 28 FP nuclides contained in JENDL-1, other 34 nuclides were evaluated[11] in 1977 with the same manner as JENDL-1[2]. At the same period, however, considerable progress was reported in both experimental work and evaluation methodology in the advisory group meeting held at ECN Petten in 1977. Hence it was decided to make a reevaluation for JENDL-2 by taking account of this progress.

As to the resonance parameters, newly measurements have been reported by JAERI, ORNL and AAEC. Many of these measurements gave capture areas of p-wave resonances for which the experimental information had been very scarce at the time of the JENDL-1 evaluation. Taking account of these new measurements, the quality and quantity of the evaluated resonance parameters are much improved.

The optical potential parameters were determined by the SPRT method, taking account of the systematics among neighboring nuclei. The level density parameters were determined from the measured resonance spacing and the low lying level structures by taking account of the systematics among neighboring nuclei. Special care was paid for smooth connection between the Fermi gas and constant temperature models. The γ-ray strength function was so determined as to reproduce the available experimental capture data. As lots of new measurements have been reported, the reliability of the γ-ray strength function is much improved.

Though the present evaluation was mainly based on the microscopic data, we took into consideration the results of the integral test[12] of the previously evaluated cross sections with the sample worth data in STEK[13] and activation data in CFRMF[14]. The outline of the present evaluation is given in Ref. 12.

Heavy Nuclides

The simultaneous evaluation method was applied to the evaluation for the main fissile and fertile materials. For the nuclides other than ^{235}U, the measured data are often reported as the ratio to the fission cross section of ^{235}U. The reliable fission cross section of ^{235}U is required so as to deduce the cross section values from the ratio data. As to the status of the ^{235}U fission cross section data, however, considerable discrepancies exist even among recent measurements. Hence the following simultaneous evaluation method was adopted to keep the consistency among different nuclides:

1. The fission cross section of ^{235}U is evaluated on the basis of the recently measured data.
2. The cross sections of the other nuclides are deduced from the ratio data by using the fission cross section of ^{235}U.
3. The deduced cross section data are compared with the absolutely measured data. If there exist some systematic discrepancies between the deduced and absolute cross sections, the way to diminish the discrepancies is suggested on the fission cross section of ^{235}U.
4. The fission cross section of ^{235}U is reevaluated by taking account of the suggestions from the other nuclides.
5. Procedures 2∿4 are repeated until the consistency is obtained among all the nuclides.

The theoretical calculations, based on the optical, statistical and evaporation models, were used for the evaluation of the quantities whose measured data were not sufficient. Various parameters used in the calculation were determined by taking account of their systematic trends in this mass region. Particularly the optical potential parameters were determined so that the total cross sections and strength functions might be well reproduced for all the nuclides with a simple systematics of the parameters.

After preliminary results were obtained, simple benchmark tests were made and slight modification was applied. Details of the evaluation for the heavy nuclides are given in Ref. 15.

Benchmark Tests with One-dimensional Model

The benchmark tests were made with the same manner[6] as used in the tests of JENDL-1. The group constants of JAERI-Fast II type[16] with 70 group structure were used in the analyses. The adopted integral data are those of 18 assemblies selected by Hardie et al.[17], of two MOZART cores (MZA and MZB) and of 7 FCA cores. As to Doppler coefficients, we adopted small sample Doppler coefficient measurements as well as whole core Doppler measurements. The same calculations were made with JENDL-1, JAERI-Fast-II (JFS-2) and ENDF/B-IV, and the present results were compared with them. The calculated results are given in the form of C/E values, and they are arranged in the form of average and standard deviation.

Effective Multiplication Factor: Table III

JENDL-2B gives the average C/E value of 0.998 with standard deviation of 0.65%. The discrepancy between the Pu and the U cores is 0.2% and is smaller than those for JENDL-1 and ENDF/B-IV. The C/E values are plotted against the fertile to fissile ratio in Fig. 1. The C/E value shows little dependence on the ratio for JENDL-2B, while the value for JENDL-1 decreases with increase of the fertile to fissile ratio.

The C/E difference between ZPR-3-48 and ZPR-3-49 is interesting. The core of ZPR-3-49 was made by removing sodium from the core of ZPR-3-48. JENDL-2B predicts the k_{eff} value of 0.5% higher for ZPR-3-49 than for ZPR-3-48, while JENDL-1 and JFS-2 give nearly same values for both assemblies. This tendency of JENDL-2B may be related to the overestimate of the sodium void reactivity worth as will be discussed later.

Table III Effective Multiplication Factor

No of Cores	*	JENDL-2B	JENDL-1	JFS-2	ENDF/B-IV
Pu-Cores	A	0.9971	0.9978	1.0017	0.9859
16	B	0.0049	0.0074	0.0044	0.0057
U-Cores	A	0.9992	1.0067	1.0033	0.9960
10	B	0.0082	0.0077	0.0100	0.0104
All-Cores	A	0.9979	1.0012	1.0023	0.9898
26	B	0.0065	0.0087	0.0071	0.0093
ZPR-3-48		1.0005	1.0005	1.0031	0.9885
ZPR-3-49		1.0055	1.0001	1.0042	0.9918

* A: Average of C/E B: Standard deviation of C/E

Central Reaction Rate Ratio: Table IV

The ratio of ^{238}U fission to ^{235}U fission is overestimated by 6% on an average. It can be seen from Fig. 1, however, that the C/E values of some assemblies much deviate from unity with every library set. The C/E values of ^{240}Pu fission to ^{235}U fission are also extreme for the same assemblies. The experimental condition should be checked more carefully for such cases, since the fission rate of a fertile material such as ^{238}U is very sensitive to the detector position in the cell. We believe that the C/E values of JENDL-2B are not so much overestimated, eliminating such extreme cases. The ratio of ^{239}Pu fission to ^{235}U fission is underestimated by 1.2% but is improved by comparison with JENDL-1. The C/E values are satisfactory for the ratio of ^{238}U capture to ^{235}U fission. On the other hand, the ratio of ^{238}U capture to ^{239}Pu fission is overestimated by a few % for the Pu cores.

Table IV Central Reaction Rate Ratios

Quantities*	Cores	**	JENDL-2B	JENDL-1	JFS-2	ENDF/B-IV
$F(^{238}U)/F(^{235}U)$	Pu	A	1.09	1.01	1.02	1.03
		B	0.08	0.08	0.07	0.07
	U	A	1.02	0.95	1.02	1.02
		B	0.07	0.07	0.07	0.08
	All	A	1.06	0.99	1.01	1.03
		B	0.09	0.08	0.07	0.07
$F(^{239}Pu)/F(^{235}U)$	Pu	A	0.98	0.96	0.97	0.98
		B	0.03	0.03	0.03	0.03
	U	A	0.99	0.97	0.99	0.99
		B	0.04	0.04	0.04	0.04
	All	A	0.98	0.97	0.98	0.98
		B	0.04	0.04	0.03	0.03
$F(^{240}Pu)/F(^{235}U)$	Pu	A	1.07	1.01	1.06	1.06
		B	0.14	0.13	0.12	0.13
	U	A	1.04	1.01	1.06	1.08
		B	0.08	0.08	0.07	0.08
	All	A	1.06	1.01	1.06	1.07
		B	0.13	0.12	0.11	0.11
$C(^{238}U)/F(^{235}U)$	Pu	A	1.01	1.00	1.00	1.02
		B	0.03	0.03	0.03	0.03
	U	A	0.99	0.98	0.98	0.98
		B	0.02	0.01	0.01	0.02
	All	A	1.00	0.99	0.99	1.00
		B	0.03	0.02	0.02	0.03
$C(^{238}U)/F(^{239}Pu)$	Pu	A	1.03	1.04	1.03	1.05
		B	0.03	0.03	0.03	0.03
	U	A	0.97	0.99	0.97	0.97
		B	0.04	0.04	0.04	0.03
	All	A	1.01	1.02	1.00	1.01
		B	0.05	0.05	0.04	0.05

* F: Fission, C: Capture
** A: Average of C/E, B: Standard Deviation of C/E.

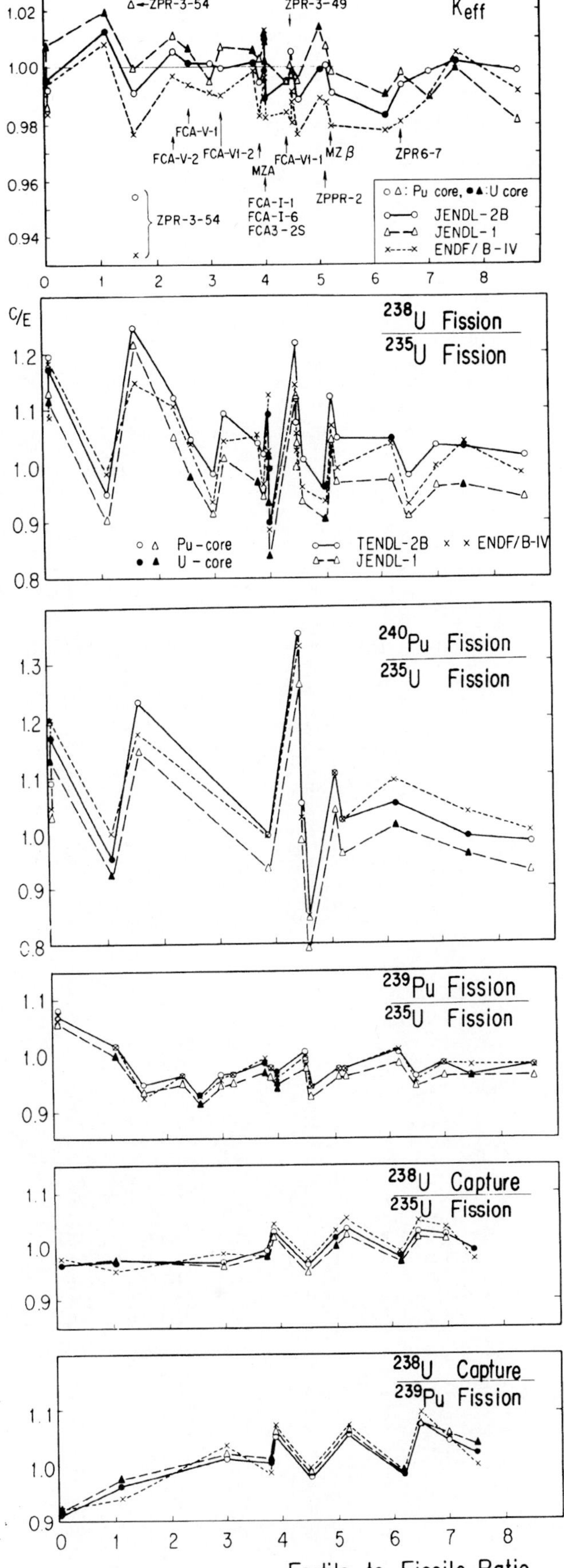

Fig. 1 C/E of k_{eff} and reaction rate ratios vs. fertile to fissile ratio of assemblies.

Central Reactivity Worth: Table V
The central reactivity worths were calculated with the first order perturbation. The conversion factor from inhour/kg to %Δk/k was calculated from the delayed neutron yield and spectrum recommended by Tuttle[17] and Saphier et al.[18], respectively.

All the sets overestimate the worths of ^{239}Pu and ^{235}U by 15∿20%. Moreover, discrepancies of 10% are observed between the Pu and U cores. For ^{238}U, however, little discrepancies exist between the Pu and U cores for all the sets. JENDL-2B predicts the worths of ^{10}B well, while JENDL-1 overestimates them by 5%. This difference must be caused by the difference in the real and adjoint neutron spectrum, because the same ^{10}B cross sections were used in both JENDL-1 and JENDL-2B. The unbalance of worths between ^{239}Pu and ^{10}B may cause the underestimate of the control rod worths. The worths are overestimated and the core dependences are observed for Cr and Ni, while the overestimate is modest and no core dependence is observed for Fe.

Thus the overestimate of the fissile worths and the discrepancies between the Pu and U cores were not resolved even by adopting the recent delayed neutron data[17,18].

Table V Central Reactivity Worths

Sample	Cores	*	JENDL-2B	JENDL-1	JFS-2	ENDF/B-IV
^{239}Pu	Pu	A	1.17	1.19	1.17	1.19
		B	0.09	0.08	0.09	0.09
	U	A	1.06	1.05	1.07	1.07
		B	0.04	0.04	0.05	0.05
	All	A	1.14	1.15	1.14	1.15
		B	0.09	0.10	0.09	0.10
^{235}U	Pu	A	1.18	1.23	1.17	1.22
		B	0.08	0.08	0.07	0.09
	U	A	1.09	1.11	1.09	1.10
		B	0.05	0.05	0.05	0.05
	All	A	1.15	1.19	1.14	1.18
		B	0.09	0.09	0.07	0.10
^{238}U	Pu	A	1.08	1.19	1.12	1.10
		B	0.09	0.09	0.08	0.08
	U	A	1.08	1.16	1.07	1.08
		B	0.05	0.05	0.04	0.06
	All	A	1.08	1.18	1.10	1.09
		B	0.08	0.08	0.08	0.08
^{10}B	Pu	A	1.01	1.07	1.06	0.99
		B	0.09	0.10	0.09	0.08
	U	A	0.96	0.99	0.94	0.93
		B	0.06	0.06	0.06	0.06
	All	A	1.00	1.05	1.03	0.97
		B	0.09	0.10	0.10	0.08
Cr	Pu	A	1.29	1.22	1.57	1.67
		B	0.12	0.11	0.13	0.16
	U	A	1.08	1.07	1.41	1.47
		B	0.08	0.06	0.17	0.15
	All	A	1.24	1.19	1.53	1.62
		B	0.14	0.12	0.16	0.18
Fe	Pu	A	1.13	1.05	1.12	1.26
		B	0.23	0.17	0.16	0.29
	U	A	1.10	1.01	1.29	1.30
		B	0.05	0.06	0.06	0.03
	All	A	1.12	1.04	1.15	1.27
		B	0.20	0.15	0.15	0.25
Ni	Pu	A	1.42	1.32	1.29	1.33
		B	0.16	0.16	0.09	0.10
	U	A	1.21	1.12	1.26	1.22
		B	0.08	0.10	0.09	0.14
	All	A	1.35	1.26	1.28	1.30
		B	0.17	0.17	0.09	0.13

* A: Average of C/E, B: Standard deviation of C/E

Doppler Coefficient: Table VI
Doppler coefficients are very satisfactorily predicted with JENDL-2B, while JENDL-1 overestimates them by 10% and JFS-2 and ENDF/B-IV underestimate them by 10%. As will be discussed later, however, these results become lower by 10% uniformly, when an advanced type of group constants is applied.

Table VI Doppler Reactivity Coefficients (C/E)

		JENDL-2B	JENDL-1	JFS-2	ENDF/B-IV
FCA	V -1	0.96	1.09	0.78	0.91
	V -2	0.86	0.98	0.74	0.78
	VI-1	1.00	1.13	0.94	0.93
	VI-2	0.93	1.03	0.90	0.87
ZPPR-2	(Normal)	1.13	1.25	1.08	0.93
	(Na voided)	0.85	0.96	0.81	0.81
ZPR-3-47		0.97	1.04	0.95	0.92
SEFOR		1.05	1.12	1.05	1.04
Average of C/E		0.97	1.08	0.91	0.90
Standard Deviation		0.09	0.09	0.12	0.08

Benchmark Tests with Two-dimensional Model

After completion of the benchmark tests with one-dimensional model, applicability of JENDL-2B was tested to more sophisticated problems such as reaction rate distribution, control rod worth and sodium void reactivity in MOZART and ZPPR-3 cores with two-dimensinal model.

Reaction Rate Distribution: Fig. 2 and Table VII
The axial and radial reaction rate traverses in MZB were analyzed. The radial fission rate traverses are shown in Fig. 2 as the form of C/E value. The C/E values, normalized to unity at the core center, stay near unity within 3% in the outer core with JENDL-2B and are better predicted than with JENDL-1. In the radial blanket, however, the C/E values have strong position dependence and deviate from unity by 10% at maximum. Moreover, obvious discontinuities are often observed in the C/E values on the boder of core and blanket. This anomalous C/E behavior in the radial blanket appears with every set and suggests that the spectrum is not well predicted in the blanket region. This probably comes from some drawbacks in analysis method.

^{235}U fission rate distribution in ZPPR-3 was calculated for the following three control rod patterns:
(1) Three rods are inserted in the inner ring and three in the outer ring (Reference),
(2) all the six rods are inserted in the inner ring (AIR), and
(3) six rods are inserted at the even positions of the outer ring (EOR).
The C/E values were normalized to unity near the core center. The C/E values are near unity within 3% in the inner core for any control rod patterns. The minimum, maximum and average values of C/E in the outer core are given in Table VII. With JENDL-2B the fission rates are underestimated by 3%, but little dependence is observed on the inserted control rod pattern, while the C/E values with JENDL-1 are decreased down to 0.88 with insertion of the control rods in the outer ring. This suggests that the k_∞ value is well predicted with JENDL-2B and that it is underestimated with JENDL-1.

As a whole, the reaction rate distributions in the inner and outer cores are better predicted with JENDL-2B than with JENDL-1.

Control Rod Worth: Tables VIII and IX
We analyzed the control rod worths measured at the core center of MZC with changing the ^{10}B enrichment. The C/E values with JENDL-2B are about 0.97 and 2% lower than those with JENDL-1. The C/E values depend little on the ^{10}B enrichment with both JENDL-2B and JENDL-1.

The worths of multi control rods in ZPPR-3 were analyzed. The maximum inserted worth is $45 with 18 control rods. JENDL-2B gives average C/E value of 0.94, whcih is 3% lower than that obtained with JENDL-1. No apparent rod pattern dependence is observed for both sets.

For both the cases, the difference between JENDL-2B and JENDL-1 is consistent with that observed in the ^{10}B worth.

Sodium Void Reactivity: Fig. 3
The sodium void traverse experiments in MZB were analyzed. JENDL-2B seems to overestimate the positive reactivity of sodium void worth particularly in the core center. The C/E value reaches 1.4 for the center void. On the other hand, the overestimation disappears for the void in the outer core and blanket regions. The prediction of JENDL-1 is satisfactory for all the regions.

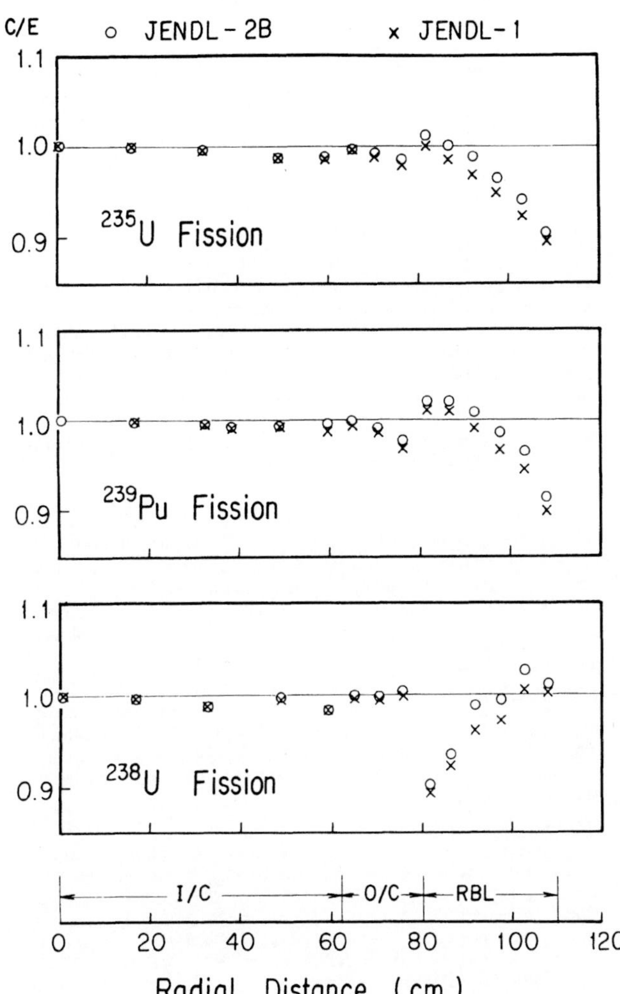

Fig. 2 C/E of reaction rate distribution in MZB.

Table VII C/E Value of ^{235}U Fission Rates in the Outer Core of ZPPR-3

Case	JENDL-2B Max	Min	Average	JENDL-1 Max	Min	Average
Ref.	1.00	0.94	0.97±0.01	0.99	0.91	0.95±0.02
AIR	0.99	0.93	0.96±0.01	1.00	0.92	0.96±0.02
EOR	1.00	0.94	0.97±0.02	0.97	0.88	0.92±0.02

Table VIII Central Control Rod Worth of MZC

^{10}B Enrichment %	Experimental Worth %Δk/kk'	C/E-Value JENDL-2B	JENDL-1
Natural	0.808±1.3%	0.97	1.00
30	0.993±1.4%	0.97	0.99
80	1.525±1.8%	0.97	0.99
90	1.629±1.9%	0.96	0.98

Table IX Control Rod Worths of ZPPR-3 Phase 1B Core

No. of Control Rods: Center	Inner Ring	Outer Ring	Measured Worth ($)	C/E-Value JENDL-2B	JENDL-1
		1	1.94	0.93	0.94
	1		2.02	0.96	0.99
	1	1	3.58	0.96	0.98
1			4.14	0.93	0.96
		2	4.24	0.93	0.95
	2		4.26	0.95	0.99
	3		6.51	0.96	1.00
	3	3	14.33	0.95	0.98
		6	14.88	0.94	0.96
	6	3	20.75	0.92	0.95
	3	6	22.65	0.94	0.97
		12	28.96	0.95	0.97
	3	9	28.99	0.94	0.97
	6	6	30.12	0.92	0.96
	6	12	44.76	0.92	0.96
Average of C/E				0.94	0.97
Standard Deviation of C/E				0.01	0.02

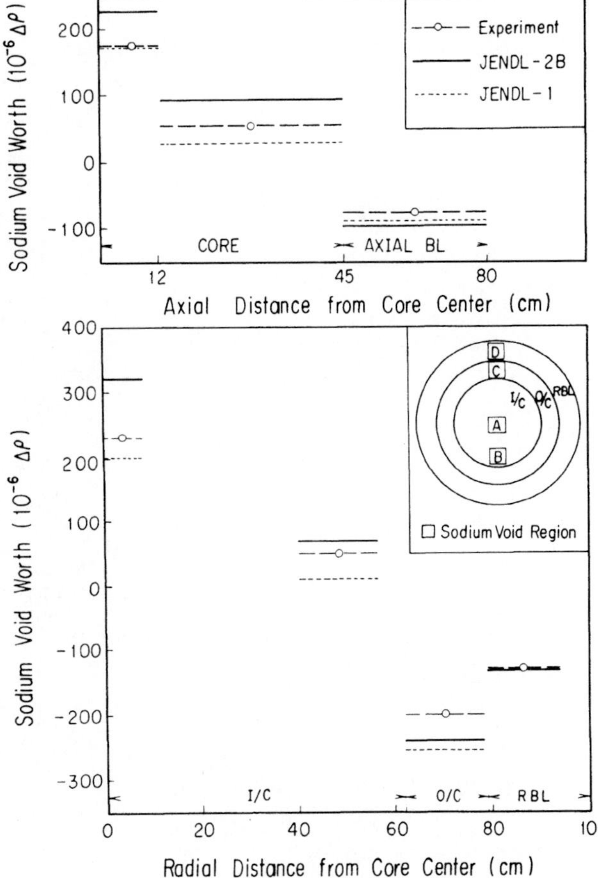

Fig. 3 Sodium void worth in MZB.

JUPITER Analysis

JUPITER experiments were analyzed with JENDL-2B by Japanese makers under contracts with Power Reactor and Nuclear Fuel Development Corporation. The analysis method is equivalent to three-dimensional transport model with 15 group structure, by applying various corrections on the calculations based on three-dimensional diffusion model with 7 groups. The results averaged over ZPPR-9, -10A∿-10D are summarized in Table X with the results analyzed by ANL with ENDF/B-IV.

The k_{eff} values are underestimated by 8%. This underestimate seems larger than the results of our benchmark tests. The fission rate ratio of ^{235}U to ^{239}Pu is by 4% overestimated as is expected from the results of benchmark tests. The ratio of ^{238}U capture to ^{239}Pu fission is overestimated by 7%, and this overestimate is a little larger than the results of the benchmark tests.

As to the sample worth, the C/E values for fissile materials are 1.06∿1.08 and much better than the results by ANL and those of our benchmark tests. The balance of the worths between ^{239}Pu and ^{10}B is also satisfactory, while the C/E value of ^{10}B worth is 15% lower than that of ^{239}Pu in the benchmark tests. The Doppler reactivity coefficients are consistent with the results of the benchmark tests.

The positive sodium void reactivities are overestimated by 25∿40%. This tendency is consistent with that in MZB. The control rod worth is fairly well predicted as a whole. However, the average C/E values depend considerably on the assembly. Furthermore, significant dependences are observed on the control rod position and its size. These problems may be partly caused by drawbacks of the analytical model and method.

The reaction rate distributions show good C/E values in the inner and outer cores. However, discontinuities are observed on the boundary between the core and blanket, as is seen in MZB.

Table X Summary of JUPITER Analyses (C/E)

	JENDL-2B	ENDF/B-IV*
K_{eff}	0.9915±0.0014	0.9844±0.0012
Reaction rate ratio**		
$F(^{235}U)/F(^{239}Pu)$	1.04	1.04
$C(^{238}U)/F(^{239}Pu)$	1.07	1.09
$F(^{238}U)/F(^{239}Pu)$	0.99	0.93
Sample worth		
^{235}U	1.08	1.20
^{238}U	1.12	1.15
^{239}Pu	1.06	1.16
^{240}Pu	1.09	1.18
Fe	1.18	1.37
Ni	1.25	1.36
Cr	1.16	1.43
^{10}B	1.09	1.05
UO_2 Doppler reactivity	1.02±0.03	0.93±0.02
Large zone cumulative Na void reactivities		
ZPPR-9	1.27±0.10	1.08±0.05
ZPPR-10A	1.40±0.02	1.20±0.02
ZPPR-10B	1.25±0.11	1.19±0.06
ZPPR-10D	1.32±0.02	1.16±0.02
Control rod worths		
ZPPR-9	0.974±0.018	
ZPPR-10A	1.011±0.012	
ZPPR-10B	1.036±0.016	
ZPPR-10C	1.018±0.018	
ZPPR-10D	1.040±0.040	

* Performed by ANL

** F: fission, C: capture

Discussion

The evaluation of JENDL-2 was performed by taking account of the problems encountered through experiences of JENDL-1. Consequently JENDL-2B predicts various characteristics of fast reactors better than JENDL-1 as a whole. Summarizing the results observed through the benchmark tests, the following could be pointed out on JENDL-2B.

Effective Multiplication Factor

The average C/E value is 0.998±0.006. No apparent dependence is observed either on the fertile to fissile ratio or on the core volume. It should be noted, however, that the k_{eff} values are underestimated by 0.8% in the most precise analyses of ZPPR-9 and 10's.

Reaction Rate Ratio

The fission rate ratio of ^{239}Pu to ^{235}U is underestimated by a few % for any case. This must come from some misprediction of the spectrum above 100 keV, since the microscopic fission cross section ratio hardly seems to be underestimated. The ratio of ^{238}U capture to ^{239}Pu fission is overestimated by 4% for the Pu benchmark cores and by 7% for JUPITER cores. This tendency is also observed with the other sets. As this reaction rate ratio is very important for prediction of the breeding ratio, further study is required if this overestimate comes from the nuclear data or from the spectrum.

Central Reactivity Worth

The central reactivity worths are overestimated in the benchmark tests for fissile materials. Moreover, considerable discrepancies are observed between the Pu and U cores. On the other hand, overestimate is more moderate in the JUPITER analysis. The balance between ^{10}B and ^{239}Pu worths is also inconsistent between the benchmark tests and JUPITER analysis. These problems partly come from different prediction of the real and adjoint spectra.

Doppler Reactivity

The Doppler coefficients seem to be well predicted. It is found, however, that the C/E values become lower by 10% when more advanced treatment is applied on the elastic removal cross section (REMO correction: See Appendix). Hence it is concluded that the Doppler coefficients are underestimated by 10% with JENDL-2B.

Reaction Rate Distribution

The C/E values are improved in the outer core with JENDL-2B. The improvement is significant when control rods are inserted in the outer core. The C/E discontinuity on the boundary between the core and radial blanket might be caused by some reasons other than nuclear data.

Control Rod Worth

The control rod worths predicted with JENDL-2B are 2∿3% lower than those with JENDL-1. This tendency is consistent with prediction of the ^{10}B worth, and must come from the difference of the real and adjoint spectra. However, the C/E values themselves are discrepant with one another among MZC, ZPPR-3, ZPPR-9 and ZPPR-10's. This discrepancy might come from other reasons than nuclear data.

Sodium Void Reactivity

The positive sodium void worths are much overestimated with JENDL-2B, while JENDL-1 predicts the worths satisfactorily. This is consistent with the C/E difference of k_{eff} between ZPR-3-48 and ZPR-3-49.

By comparing the perturbation components, it is found that the positive moderation term is larger with JENDL-2B than with JENDL-1. Table XI gives the components for the core center void and the outer core void in MZB.

The difference of the moderation term is caused by the difference of the adjoint spectrum. The energy gradient of adjoint spectrum is larger with JENDL-2B than with JENDL-1 as seen in Fig. 4.

Table XI Perturbation Components of Sodium Void Worth

($\times 10^{-6}\Delta\rho$)

Component	Core Center Void		Outer Core Void	
	JENDL-2B	JENDL-1	JENDL-2B	JENDL-1
Fission	-108	-148	-15	-32
Absorption	+189	+216	+37	+48
Moderation	+550	+414	+131	+94
Leakage	-311	-283	-396	-364
Total	+321	+198	-242	-255
Measured	+231±5		-201±5	

First order perturbation.

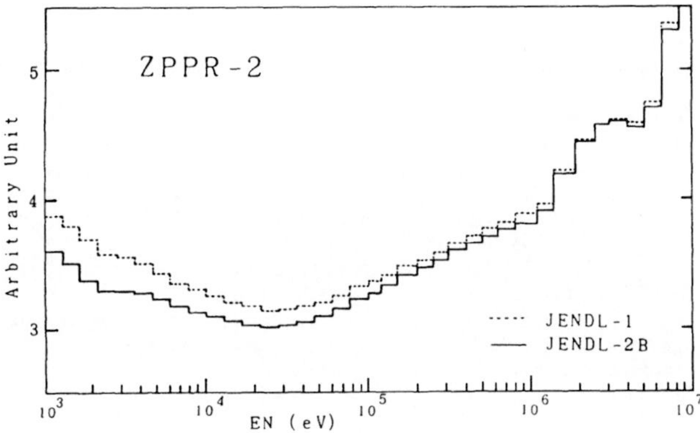

Fig. 4 Core center adjoint spectrum in ZPPR-2.

Sensitivity Analysis with Generalized Perturbation

In order to know which cross sections cause the overestimate of sodium void worth, a sensitivity analysis was performed by using the generalized perturbation theory. Table XII gives the sensitivity coefficients of important reaction cross sections for ZPPR-9. In addition to these, the fission cross section of ^{240}Pu also has considerable sensitivity coefficients.

Significant are contributions of the fission and capture cross sections of ^{239}Pu below the sodium resonance of 2.85 keV. The ^{239}Pu fission cross section of JENDL-2 is lower by 20∿30% than that of JENDL-1 in the energy region between 600 eV and 2 keV as seen in Fig. 5. Replacing the ^{239}Pu fission cross sections of JENDL-2 by that of JENDL-1 in this energy range, the adjoint flux increases below 10 keV and the energy gradient decreases above 50 keV. This improves not only the sodium void coefficients but also the Doppler coefficients and the balance of the sample worths between ^{239}Pu and ^{10}B, with little affecting the k_{eff} and the reaction rate ratios.

Generally the energy range below a few keV is the unresolved resonance region for fissile nuclei, where the measured data depend on experimental conditions and are discrepant with one another. Hence the uncertainty is considerably large in this energy region. Further experimental works are much required.

On the other hand, the integral data should be further considered in the evaluation of the cross sections of fissile and fertile materials in the energy region below the sodium resonance so that the sodium void and Doppler reactivities might be well predicted. This problem is now further studied. At least, the present sensitivity analysis reveals the importance of the cross sections below a few keV for fast reactor calculations.

Table XII Sensitivity Coefficients for ZPPR-9

(%$\Delta\rho/\rho/\Delta\sigma/\sigma$)

E_{mim}*	E_{max}*	^{239}Pu fission	^{239}Pu capture	^{238}U capture	^{238}U scattering
6.5 +6	1.05+7	0.0	0.0	-0.0	-1.6
4.0 +6	6.5 +6	0.6	0.0	-0.1	-4.5
2.5 +6	4.0 +6	0.6	0.0	-0.2	-8.3
1.4 +6	2.5 +6	-2.3	0.0	-0.4	-9.8
8.0 +5	1.4 +6	3.3	-0.1	-2.5	-5.6
4.0 +5	8.0 +5	2.5	-0.2	-3.7	-1.5
2.0 +5	4.0 +5	18.0	1.0	-7.8	-4.2
1.0 +5	2.0 +5	3.5	-0.3	-3.5	-2.2
4.65+4	1.0 +5	15.7	-0.8	-9.0	-3.9
2.15+4	4.65+4	0.9	0.5	1.1	0.4
1.0 +4	2.15+4	-7.4	2.2	9.8	0.7
4.65+3	1.0 +4	1.0	0.9	1.9	0.4
2.15+3	4.65+3	2.8	-0.3	-1.7	0.3
1.0 +3	2.15+3	-43.0	15.2	31.0	0.3
4.65+2	1.0 +3	-42.9	16.8	26.1	0.3
THR	4.65+2	-39.0	15.4	18.4	0.1

* 6.5+6 indicates 6.5×10^{6}

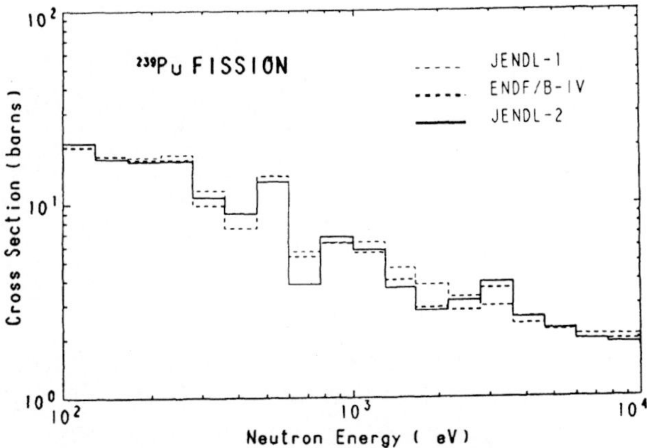

Fig. 5 Average fission cross sections of ^{239}Pu.

Concluding Remarks

Applicability of JENDL-2 to fast reactor calculations is proved through the benchmark tests. Though JENDL-2 predicts various characterestics very satisfactorily as a whole, the following two problems are pointed out:

1) The underestimate in the fission rate ratio of ^{239}Pu to ^{235}U suggests that the spectrum must be too soft in the energy region above 100 keV. The underestimate of Doppler coefficients and ^{10}B worths indicate that the flux or adjoint flux is underestimated in the low energy region. To improve the prediction of high energy spectrum, further study is required on the fission spectrum and the inelastic scattering cross sections of ^{238}U and of the structural materials.
2) The overestimate of positive sodium void worths is mainly caused by the error in the adjoint spectrum. The sensitivity analysis with the generalized perturbation theory suggests that the fission cross section of ^{239}Pu below the sodium resonance is underestimated. The other cross sections such as ^{239}Pu capture, ^{238}U capture and inelastic scattering, and ^{240}Pu fission should be further checked.

It is expected that the Doppler reactivity and ^{10}B worth will be improved, if the adjoint spectrum is changed so as to improve the sodium void reactivity.

In addition to these problems, one of the severest drawbacks of JENDL-2 is its insufficient treatment of the direct reaction process in the evaluation. This reflects the fact that most of requirements to JENDL-2 came from the fission reactor calculations. Recently, however, nuclear data become required from the fusion neutronics and shielding calculations.

Responding to these requests, we started the program for the third version of JENDL. As JENDL-2 contains enough number of nuclides, we add minimum number of nuclides to JENDL-3. The main efforts are devoted to improve the quality of the data particularly in the high energy region above a few MeV. The direct and preequilibrium processes will be considered, and the measured data of the double-differential cross sections at 14 MeV will be utilized in the evaluation. The γ-ray production cross sections will be included for important nuclides. Compilation of JENDL-3 is scheduled to be completed at the end of 1984.

References

1. S. Igarasi, T. Nakagawa, Y. Kikuchi, T. Asami, T. Narita, JAERI-1261 (1979).
2. Y. Kikuchi, T. Nakagawa, H. Matsunobu, M. Kawai, S. Igarasi, S. Iijima, JAERI-1268 (1981).
3. T. Inoue, K. Shirakata, K. Kinjo, T. Ikegami, M. Yamamoto, J. At. Energy Soc. Jpn., 23, 310 (1981) [in Japanese].
4. Y. Kikuchi, T. Narita, H. Takano, J. Nucl. Sci. Technol., 17, 567 (1980).
5. K. Shirakata, Y. Kato, Y. Kaise, Proc. 1981 Seminar on Nuclear Data, JAERI-M 9999, p.161 (1982) [in Japanese].
6. Y. Kikuchi, A. Hasegawa, H. Takano, T. Kamei, T. Hojuyama, M. Sasaki, Y. Seki, A. Zukeran, I. Otake, JAERI-1275 (1982).
7. M. Kawai, N. Yamano, K. Koyama, Proc. 1979 Knoxville Conf., p.586, NBS Special Publication 594 (1980).
8. T. Nakagawa, J. At. Energy Soc. Jpn., 22, 559 (1980) [in Japanese].
9. M. Kawai, JNDC Progress Report, p.18, NEANDC(J)-61/U, INDC(JAP)-47/U (1979).
10. T. Yoshida, ibid, p.16.
11. S. Iijima, Fission Product Nuclear Data, Petten, 1977, p.279, IAEA-213 (1977).
12. S. Iijima, T. Watanabe, T. Yoshida, Y. Kikuchi, H. Nishimura, Proc. Specialists' Meeting on Neutron Cross Section of FP Nuclei, Bologna, 1979, p.317, RIT/FIS-LDN(80) 1.
13. J. J. Veenema, A. J. Janssen, ECN-10 (1976).
14. Y. D. Harker, J. W. Rogers, D. A. Millsap, TREE-1259 (1978).
15. H. Matsunobu, Y. Kanda, M. Kawai, T. Murata, Y. Kikuchi, Proc. 1979 Knoxville Conf., p.715, NBS Special Publication 594 (1980).
16. H. Takano, A. Hasegawa, M. Nakagawa, Y. Ishiguro, S. Katsuragi, JAERI-1255 (1978).
17. R. W. Hardie, R. E. Schenter, R. E. Wilson, Nucl. Sci. Eng., 57, 222 (1975).
18. R. J. Tuttle, Nucl. Sci. Eng., 56, 37 (1975).
19. D. Saphier, D. Ilberg, S. Shalev, S. Yiftah, Nucl. Sci. Eng., 62, 660 (1977).

Appendix 1: REMO Correction

The group constants of JAERI-Fast-II type are produced by averaging the cross sections with the weight of the fission spectrum above 1 MeV and the 1/E spectrum below 1 MeV. This gives enough precision for most of the group cross sections, when the number of groups is as many as 70. The only exception is the elastic removal cross section for which the 1/E spectrum weighting results considerable overstimate in the energy region below a few hundred keV.

Recently JAERI developed the conception of JAERI-Fast-III, in which the weighting spectrum was changed to the spectrum in the core center of the Japanese demonstration fast reactor. The elastic removal cross section is much improved by this new weighting, which is called REMO correction. The effect of the REMO correction is significant on the Doppler reactivities, ^{10}B and Mo sample worths, and reaction rates in the sodium follower of CRP.

Appendix 2: Comments on Other Sets

The benchmark calculations were made with the other sets such as JENDL-1, JFS-2 and ENDF/B-IV. Here some comments are given on the results obtained with these sets.

JENDL-1

The average k_{eff} value is 1.001 with the standard deviation of 0.9%. There exists a difference of 0.9% between the Pu and U cores. Half of this difference comes from the fact that different ν_p-values of ^{252}Cf were used to normalize the ν_p-values for ^{235}U and ^{239}Pu. JENDL-1 has a tendency to underestimate the k_{eff} value for the assemblies with high fertile to fissile ratio, as seen in Fig. 1.

The C/E difference between ZPR-3-53 and ZPR-3-54 is interesting. These two assemblies have the same core but have different blanket; a natural uranium blanket for ZPR-3-53 and an iron reflector for ZPR-3-54. All the sets except JENDL-1 give very low k_{eff}-values for ZPR-3-54 (C/E=0.93∿0.95). This is believed to come from drawbacks of one-dimensional diffusion model for the assembly having a thin iron reflector. On the other hand, JENDL-1 gives higher k_{eff} value for ZPR-3-54 than for ZPR-3-53. This suggests that JENDL-1 has the Fe cross sections which give much less leakage. Investigating the microscopic cross sections, it was found that the total and elastic scattering cross sections of Cr, Fe and Ni were much overestimated in the energy region from a few hundred keV to a few MeV.

The reaction rates are underestimated a little in the outer core region. This underestimation is particularly enhanced, when the outer core is decoupled from the inner core by inserting control rods on the boundary between inner and outer cores. This suggests that the k_∞ value is underestimated in the outer core. From the observation above mentioned, it could be concluded that JENDL-1 underestimated the k_∞ values and this underestimate be compensated with underestimate of the leakage due to overestimate of the elastic scattering cross sections for the structural materials.

On the other hand, the sodium void and Doppler reactivities are predicted satisfactorily with JENDL-1.

ENDF/B-IV

The k_{eff} values are much underestimated with ENDF/B-IV, and a difference of 1% exists between the Pu and U cores. On the other hand, Hardie et al.[17] reported much satisfactory values in their benchmark tests: the average of C/E is 0.9939 for the Pu cores and 1.0038 for the U cores. The reason of the inconsistency is not clear, because we adopted the same models and corrections for the 18 assemblies.

JAERI-Fast-II(JFS-2)

JAERI-Fast-II is an adjusted group constants set. The cross sections of the main fissile and fertile materials were adjusted by using the same benchmark cores as the present tests. For the other materials, the ENDF/B-IV data were mainly adopted. The k_{eff} values and reaction rate ratios are predicted very well naturally as a result of the adjustment.

REVIEW OF METHODS FOR LEVEL DENSITY ESTIMATION FROM RESONANCE PARAMETERS

F.H. Fröhner

Kernforschungszentrum Karlsruhe
Institut für Neutronenphysik und Reaktortechnik
Postfach 3640, D-7500 Karlsruhe
West Germany

A number of methods are available for statistical analysis of resonance parameter sets, i. e. for estimation of level densities and average widths with account of missing levels. The main categories are (i) methods based on theories of level spacings (orthogonal-ensemble theory, Dyson-Mehta statistics), (ii) methods based on comparison with simulated cross section curves (Monte Carlo simulation, Garrison's autocorrelation method), (iii) methods exploiting the observed neutron width distribution by means of Bayesian or more approximate procedures such as maximum-likelihood, least-squares or moment methods, with various recipes for the treatment of detection thresholds and resolution effects. The present review will concentrate on (iii) with the aim of clarifying the basic mathematical concepts and the relationship between the various techniques. Recent theoretical progress in the treatment of resolution effects, detectability thresholds and p-wave admixture is described.

1. Introduction

Accurate information on level densities, apart from providing test material for level density theories[1],[2], is indispensable for level-statistical calculations of average partial cross sections. Neutron capture cross sections in the unresolved resonance region, for instance, are nearly proportional to the level density. At excitation energies just above the neutron separation energy, in the resolved resonance region, one can obtain level densities most directly by counting the neutron resonances observed in a given energy interval. Due to the preponderance of small neutron widths in the Porter-Thomas distribution[3] there is, however, always a large fraction of levels missing for which the count must be corrected. Even the best high-resolution resonance data, for instance those on ^{238}U s-wave levels[4], are affected by about 20 % missing levels, and 30-40 % are quite common. Level densities uncorrected for missing levels are therefore useless for most purposes. Nor does it help to take only the low-energy portion of the cumulative level count N(E) with its typical nicely linear behaviour. This linearity is occasionally mistaken as an indication that no levels are missed. Of course it indicates merely that the missing fraction does not depend on energy.

The whole problem of level density estimation is truely an evaluator's item - replete with missing data, shaky statistical models suggested by extremely formal nuclear theories, rigorous equations which are so intractable that approximations must be invoked, logical and numerical traps etc. Even benchmark calculations have recently made their appearance in this field[5], showing that (and why[6]) impeccably conceived and elaborately tested programs (the author's, among others) can produce rather unsatisfactory results.

Liou[7] has reviewed a number of level density estimation techniques, giving a short functional description of each one. In the present paper the emphasis will be on the mathematical aspects which do not seem to be coherently treated in the literature so far. Since the whole concept of level density is statistical it is appropriate to use the tools of mathematical statistics to develop and to compare methods for the estimation of level densities and missing levels. It will be seen that many of the seemingly quite different techniques that exist are mere variants of the same basic approach.

2. Theory of Level Statistics

Strictly speaking there is nothing random or statistical about resonance energies or widths. They are determined as eigenvalues and by the eigenfunctions of a Schrödinger equation with suitable boundary conditions. The justification for use of a statistical description lies only in the complexity of the spectra, which in turn is caused by the complicated interaction between many nucleons in the nuclides which we consider here. The square roots of the reduced neutron widths, for instance, are essentially surface integrals over rapidly oscillating functions in the 3A-dimensional configuration space associated with the A nucleons of a given compound nucleus. Not knowing much more one may therefore expect them to have a normal distribution around zero. This hypothesis leads immediately to the Porter-Thomas distribution[3] for the reduced neutron widths,

$$p(\Gamma|\langle\Gamma\rangle)d\Gamma = \frac{e^{-x}}{\sqrt{\pi x}}\,dx\ , \quad 0 < x \equiv \frac{\Gamma}{2\langle\Gamma\rangle} < \infty\ . \qquad (1)$$

where $\langle\Gamma\rangle$ is the ensemble average (for brevity we omit the usual sub- and superscripts for reduced widths)

The level spacing distribution is much more difficult to find. The level energies are the eigenvalues of a Hamiltonian matrix H which can be taken as real and symmetric since the nuclear interaction is invariant under time reversal. Furthermore, the probability density function p(H) must be invariant under rotations in Hilbert space because all representations of H, including the diagonal form, are equally valid. The additional requirement that the matrix elements be independent of each other yields the Gaussian orthogonal ensemble[8]. The last requirement is, however, unfounded and leads to an unrealistic semi-circular dependence of level density on energy. Dyson (see [8]) introduced the circular orthogonal ensemble by assuming that some unitary (otherwise unspecified) function S of H has its eigenvalues distributed uniformly around the unit circle. He showed that with this very general assumption one can reproduce any reasonable energy dependence of the level density. Mello et al.[9] studied the more physical statistical shell model where not the elements of H but only those of the residual (pair) interaction are considered as random variables. Both the orthogonal ensembles and the statistical shell model (or two-body random ensemble) yield a level spacing distribution that is very close to Wigner's famous surmise (see [8])

$$p(D|\langle D\rangle)dD = 2xe^{-x^2}dx\ , \quad 0 < x \equiv \frac{\sqrt{\pi}}{2}\frac{D}{\langle D\rangle} < \infty\ , \qquad (2)$$

where $\langle D\rangle$ is the ensemble average. In addition to the level repulsion (improbability of small spacings) implied by (2) all random-matrix models predict hat nuclear level sequences possess "nearly crystalline" regularity in the sense that the cumulative level count N(E) follows closely a straight line with slope $\rho = 1/\langle D\rangle$, excursions by more than one unit being extremely unlikely. This is possible only if spacings are correlated in such a way that a large spacing is followed by a short one more often than not and vice versa. The mean-square deviation from a best-fit

K. H. Böckhoff (ed.), Nuclear Data for Science and Technology, 623-627.

straight line in an interval containing N levels, called the Δ_3 statistic by Dyson and Mehta[10], has the expectation value

$$\langle\Delta_3\rangle = \frac{1}{\pi^2}\{\ln(2\pi N) + \gamma - \frac{\pi^2}{8} - \frac{5}{4}\} \quad (3)$$

(γ = 0.5772.. is Euler's constant) and the variance

$$\text{var}\,\Delta_3 = \frac{1}{\pi^4}\left(\frac{4\pi^2}{45} + \frac{7}{45}\right). \quad (4)$$

Absence of levels or presence of spurious levels from other sequences obviously increases Δ_3. One has therefore tried to use it as a test statistic for the purity of level sequences. According to Dyson the best test statistic for the presence of spurious or missing levels in an almost pure and complete level sequence is (see [11])

$$F_\lambda = \sum_{\mu\neq\lambda} \text{arcosh}\frac{I/2}{|E_\mu - E_\lambda|}, \quad (5)$$

where μ runs through all levels between E_λ-I/2 and E_λ+I/2 and I is an arbitrary test interval (for instance 20 times $\langle D\rangle$). Expectation value and variance are, with $m \equiv \pi I/(2\langle D\rangle)$,

$$\langle F_\lambda\rangle = m - \ln m - \gamma + 2, \quad (6)$$

$$\text{var}\,F_\lambda = \ln m, \quad (7)$$

if E_λ is a true member of the sequence. If it is the energy of a spurious level in an otherwise pure sequence one gets

$$\langle F_\lambda\rangle = m \quad (8)$$

so that a spurious or missing level produces, on average, a peak or a dip of magnitude ~ln m in an almost constant trend. The catch, however, lies in the words "on average" (see [12]). In practice one finds that the Δ_3 and F_λ test criteria for purity and completeness are often satisfied for samples that are known to be neither pure nor complete.

3. Estimation Based on Level Positions and Widths

The seemingly straightforward approach of fitting the Wigner distribution (2) to the observed distribution of level spacings is ruled out by the bad distortion of the latter if 20 % or more of all levels are missing. Moreover, it must be stressed that none of the tests described so far permits unambiguous identification of spurious or missed levels. Nevertheless, as the Columbia group demonstrated, one can purify almost pure level sequences further by a combination of all available tests[12]. Such an ambitious program involves much judgement and is therefore not easily cast into the form of a computer code. Moreover, as already mentioned the tests based on orthogonal-ensemble theory are not as sensitive as one might expect ([13],[14]). In fact they are quite useless for resonance data afflicted by severe level overlap and unknown level spins such as those for ^{233}U, ^{239}Pu and other fissile nuclei with nonzero target spin.

In this situation one may use Monte Carlo techniques. One generates artificial cross sections from resonance parameters sampled from the relevant distributions. By varying the mean widths and spacings one tries to make the artificial, Doppler and resolution broadened cross section curves statistically as similar as possible to the measured data (see e. g. [15],[16]).

The difficult judgement of the statistical similarity between experimental and artificial cross sections was put on a quantitative basis by Garrison[17]. From each of the two cross cross section sets he generates a bivariate distribution in matrix form by considering all pairs of data points that are separated by the same energy difference ΔE (which is to be chosen as comparable to the mean level spacing). If the two cross sections of such a pair fall into the i-th and the k-th cross section bin the value one is added to the (i,k) matrix element. The matrices thus created from the experimental and the artificial cross section are then compared by means of either a maximum-likelihood or a chi-square criterion to determine the degree of statistical similarity between both. The method is able to cope with data that are quite badly affected by missing levels and unresolved doublets. Garrison's estimate for ^{235}U, $\langle D\rangle$ = 0.38 ± 0.04 eV, deduced from spin-merged high-resolution data, is consistent with the value $\langle D\rangle$ = 0.44 ± 0.04 eV found later by Moore et al. from spin-separated data measured with polarised beam and sample[13]. Previous, less quantitative comparisons with Monte-Carlo-generated cross sections had given much higher values[15].

It is obvious that simpler techniques are required for routine extraction of level densities from the vast body of modern resonance parameter data.

4. Estimation Based on Neutron Widths only

In contrast to the spacing distribution the neutron width distribution is only slightly affected by missing levels. The upper part of the Porter-Thomas distribution, corresponding to the strong levels, can usually be regarded as virtually unperturbed. It is then possible to estimate the mean width from this part. The expected number of missing levels and the level density can be calculated once the mean width is known. This is the basis of most presently used level density estimation techniques. We shall begin with the simplest case.

4.1 Unperturbed Porter-Thomas Distribution

Suppose we have a sample of reduced neutron widths $\Gamma_1, \Gamma_2, \ldots \Gamma_N$ from an unperturbed Porter-Thomas distribution. The joint probability that in a random sample of size N, drawn from the distribution (3), the sample values lie in the infinitesimal intervals $d\Gamma_1$ at Γ_1, $d\Gamma_2$ at $\Gamma_2, \ldots$ $d\Gamma_N$ at Γ_N is

$$L(\Gamma_1, \ldots \Gamma_N|\langle\Gamma\rangle)d\Gamma_1 \ldots d\Gamma_N = \prod_{i=1}^{N} p(\Gamma_i|\langle\Gamma\rangle)d\Gamma_i. \quad (9)$$

The probability density function L here is called the likelihood function. It specifies the relative probabilities for different samples if the parent distribution and its parameter(s) are given. Our problem, however, is just the reverse. We want the probability density function not for the sample (that is given) but for the parameter $\langle\Gamma\rangle$ of the parent distribution. The recipe for the necessary "inversion" of conditional probabilities is provided by Bayes' theorem (see e. g. [18]) which thus constitutes the very basis for all scientific inference from experimental (uncertainty-affected) data. It states that the required (a-posteriori) probability density function is the product of the likelihood function and the a-priori probability $p_0(\langle\Gamma\rangle)d\langle\Gamma\rangle$,

$$p(\langle\Gamma\rangle|\Gamma_1, \ldots \Gamma_N)d\langle\Gamma\rangle \propto L(\Gamma_1, \ldots \Gamma_N|\langle\Gamma\rangle)p_0(\langle\Gamma\rangle)d\langle\Gamma\rangle. \quad (10)$$

Now $\langle\Gamma\rangle$ is a scale factor in our problem. Jaynes[19], using invariance arguments, showed that for such scale parameters the appropriate prior probability is $d\langle\Gamma\rangle/\langle\Gamma\rangle = d\ln\langle\Gamma\rangle$, thus proving a conjecture due to Jeffreys[18]. The distribution (10) constitutes the complete information about $\langle\Gamma\rangle$ which we get from the sample. Now the question arises which value of $\langle\Gamma\rangle$ should be quoted as the estimate, the maximum (mode) of p or the expectation value or some other value? The corresponding Porter-Thomas distribution should clearly be the same whether we estimate $\langle\Gamma\rangle$ or $1/\langle\Gamma\rangle$ (both must be equally possible for a scale parameter). The Jeffreys-Jaynes prior probability suggests that we consider $\ln\langle\Gamma\rangle$ as the basic parameter, hence L as its probability density function. The maximum of L with respect to $\ln\langle\Gamma\rangle$ is then determined by

$$\frac{dL}{d\ln\langle\Gamma\rangle} = \langle\Gamma\rangle\frac{dL}{d\langle\Gamma\rangle} = \frac{1}{\langle\Gamma\rangle}\frac{dL}{d(1/\langle\Gamma\rangle)} = 0, \quad (11)$$

which shows that in each case we have to maximise the likelihood function, and that the solution is unique.

We could have avoided Bayes' theorem and the Jeffreys-Jaynes prior probability by using the conventional maximum-likelihood technique (see e. g. [18]). Writing down L explicitly for the Porter-Thomas distribution one sees that L is a product of one factor containing the sample values and a second factor which depends only on the true mean $\langle\Gamma\rangle$ and the sample average

$$\bar{\Gamma} \equiv \frac{1}{N}\sum_{i=1}^{N} \Gamma_i \; . \tag{12}$$

The factorisation shows that $\bar{\Gamma}$ is a minimal sufficient statistic, i. e. it is a number that can be calulated from the sample, contains all information about $\langle\Gamma\rangle$ that the sample contains and has the smallest scatter around its expectation value among all possible sufficient statistics. Small scatter is one property which a useful estimator must have. The second property is that it should be unbiased which means that its expectation value should be equal to the estimated true value. Our sample average $\bar{\Gamma}$ has both properties. Its distribution is a χ^2-distribution with N degrees of freedom[20],

$$p(\bar{\Gamma}|\langle\Gamma\rangle)d\bar{\Gamma} = \Gamma(N/2)^{-1}e^{-y}y^{N/2-1}dy \; ,$$

$$0 < y \equiv \frac{N\,\bar{\Gamma}}{2\langle\Gamma\rangle} < \infty \tag{13}$$

(where $\Gamma(N/2)$ is a gamma function, not a width) as follows upon substitution $\xi_i \equiv \sqrt{\Gamma_i/(2\langle\Gamma\rangle)}$ and integration in the space of the ξ_i over all angles, for fixed radius. Now (13) is seen to be basically the distribution of the ratio $\bar{\Gamma}/\langle\Gamma\rangle$. It can be interpreted either as the distribution of $\bar{\Gamma}$ for given $\langle\Gamma\rangle$or, equally well, as that of $\langle\Gamma\rangle$ for given $\bar{\Gamma}$,

$$p(\langle\Gamma\rangle|\bar{\Gamma})d\langle\Gamma\rangle = p(\bar{\Gamma}|\langle\Gamma\rangle)d\bar{\Gamma} \; . \tag{14}$$

Exactly the same estimate and the same distribution are obtained by insertion of the Porter-Thomas distribution and the Jeffreys-Jaynes prior in Bayes' theorem, Eq. (10), and determination of the most probable value of $\ln\langle\Gamma\rangle$. In fact, whenever a scale parameter is estimated the Bayesian solution coincides with the maximum-likelihood solution [6], although in more general cases the maximum-likelihood solution only approximates the rigorous Bayesian solution.

4.2. Porter-Thomas Distribution with Known Threshold

Let us now consider a less academic case. We assume that the sample contains only reduced widths that exceed a detection threshold Γ_c. If the threshold is energy dependent we must start from the bivariate distribution (properly normalised to unity)

$$p(\Gamma_i,E_i|\langle\Gamma\rangle,\Gamma_c)d\Gamma_i dE_i = \frac{1}{\overline{\mathrm{erfc}\sqrt{x_c}}}\,\frac{e^{-x_i}}{\sqrt{\pi x_i}}\,dx_i\,\frac{dE_i}{E_b-E_a} \; ,$$

$$x_c \equiv \frac{\Gamma_c(E)}{2\langle\Gamma\rangle} < x_i \equiv \frac{\Gamma_i}{2\langle\Gamma\rangle} < \infty \; , \; E_a < E < E_b \tag{15}$$

where the bar over the complementary error function denotes the energy average in the interval $(E_a \ldots E_b)$. The joint probability for the widths to be observed in intervals $d\Gamma_i$ at Γ_i and for the level energies in intervals dE_i at E_i ($i = 1,\ldots N$) is

$$L\, d\Gamma_1 \ldots d\Gamma_N dE_1 \ldots dE_N = \Big(\frac{2}{\sqrt{\pi}\;\overline{\mathrm{erfc}\sqrt{x_c}}}\Big)^N e^{-\xi^2} d^N\xi \frac{dE_1 \ldots dE_2}{(E_b-E_a)^N} \tag{16}$$

where $\xi^2 = (N/2)(\bar{\Gamma}/\langle\Gamma\rangle)$ is the squared radius, $d^N\xi$ the volume element in the space of the ξ_i defined before. Factorising L one sees again that the sample average $\bar{\Gamma}$ is a minimal sufficient statistic. As before its distribution can be obtained by integration over all angles in the space of the ξ_i for given level energies E_i. The resulting solid angle factor must then be averaged over level energies. The final distribution can again be interpreted either as that for $\bar{\Gamma}$ or as that for $\langle\Gamma\rangle$. Since the energy-averaged solid-angle factor does not depend on $\langle\Gamma\rangle$ one finds in the latter case

$$p(\langle\Gamma\rangle|\bar{\Gamma},\Gamma_c)d\langle\Gamma\rangle \propto (\overline{\mathrm{erfc}\sqrt{yz}})^{-N}e^{-y}y^{N/2-1}dy \; ,$$

$$0 < y \equiv \frac{N\,\bar{\Gamma}}{2\langle\Gamma\rangle} < \infty \; , \qquad z \equiv \frac{\Gamma_c(E)}{N\bar{\Gamma}} \; . \tag{17}$$

Since we estimate a scale factor, the maximum-likelihood solution coincides again with the Bayesian solution. Maximising L with respect to $\langle\Gamma\rangle$ one obtains

$$\bar{\Gamma} = \langle\Gamma\rangle\Big(1 + \frac{2}{\sqrt{\pi}}\,\frac{\overline{e^{-x_c}\sqrt{x_c}}}{\overline{\mathrm{erfc}\sqrt{x_c}}}\Big) \; . \tag{18}$$

This equation can be solved for $\langle\Gamma\rangle$ for example with the Newton-Raphson method.

Although it may not be readily apparent from the available documentation this is the approach of

- Fuketa and Harvey[21] (with $\Gamma_c/\bar{\Gamma} = c \cdot E^b$, c and b given constants)
- Ribon et al.[22] (ESTIMA code, Γ_c = const chosen so as to exclude practically all p-wave levels)
- Rohr et al.[23] (modified Fuketa-Harvey code with threshold chosen so as to restrict p-wave admixture to a given small value)

4.3. Porter-Thomas Distribution with Unknown Threshold

If the threshold is unknown one can extend the estimation procedure as follows. We introduce the notation

$$u_i \equiv \mathrm{erfc}\sqrt{x_i} \quad , \qquad u_c \equiv \mathrm{erfc}\sqrt{x_c} \quad , \tag{19}$$

so that u_c is the fraction of observed levels in the interval dE at E. In this representation the bivariate distribution of level energies and reduced neutron widths (normalised to unity) assumes the simple form

$$p(\Gamma_i,E_i|\langle\Gamma\rangle,\Gamma_c)d\Gamma_i dE_i = \frac{du_i}{\bar{u}_c}\,\frac{dE_i}{E_b-E_a} \; ,$$

$$0 < u_i < u_c(E_i) \; , \quad E_a < E < E_b \; . \tag{20}$$

This means that the sample points (E_i,u_i) are uniformly distributed in the (E,u) plane below the threshold $u = u_c(E)$ as shown schematically below.

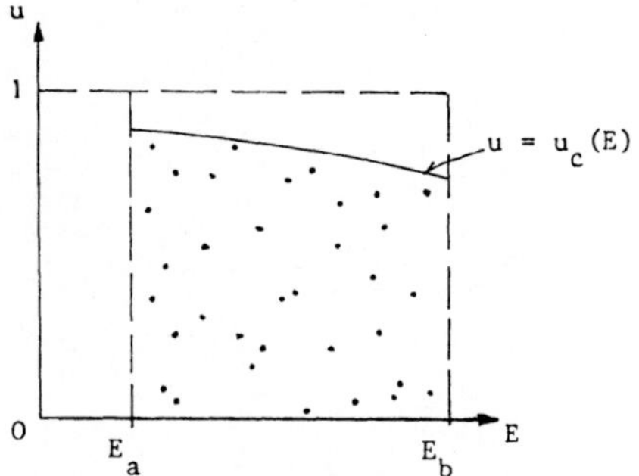

Fig. 1 - Threshold and distribution of sample points in (E,u) representation

Next we write

$$u_c(E) = \bar{u}_c\, f(E) \tag{21}$$

and assume the energy dependence $f(E)$ as known. This function can easily be obtained with adequate precision by least-squares fitting of a suitable test function to the cumulative level number $N(E)$. A parabolic fit, corresponding to a linear energy dependence of u_c, is usually quite adequate[24].

Clearly $\langle\Gamma\rangle$ is again a scale parameter but the role of u_c is less clear. Not knowing the prior probability we cannot invoke Bayes' theorem. Instead we try to find sufficient statistics by factorisation of L and then their probability distribution by integration over as many widths and energies as possible. We start with a constant threshold, $f(E) = 1$. The joint probability for the whole sample is

$$p(\Gamma_1,\ldots\Gamma_N|\langle\Gamma\rangle,\Gamma_c)d\Gamma_1\ldots d\Gamma_N$$

$$= \frac{1}{\bar{u}_c^N}\frac{e^{-N\bar{\Gamma}/(2\langle\Gamma\rangle)}}{\langle\Gamma\rangle^{N/2}}\prod_{i=1}^{N}\frac{d\Gamma_i}{\sqrt{2\pi\Gamma_i}}H(\Gamma_1-\Gamma_c) \tag{22}$$

where H is the Heaviside function and Γ_1 the smallest width. This shows that $\bar{\Gamma}$ and Γ_1 are jointly sufficient statistics. Integration over all angles in the (n-1)-dimensional space spanned by $\xi_2,\ldots\xi_N$ results in the distribution

$$p(\bar{\Gamma},\Gamma_1|\langle\Gamma\rangle,\Gamma_c)d\Gamma_1 d\bar{\Gamma} \propto \frac{du_1}{\bar{u}_c^N}e^{-y}y^{(N-1)/2-1}dy\ ,$$

$$y \equiv \frac{N\bar{\Gamma}-\Gamma_1}{2\langle\Gamma\rangle}\ , \tag{23}$$

which is the product of the joint probability that Γ_1 lies in $d\Gamma_1$ and all other widths are larger, viz.

$$p(u_1>u_2\ldots>u_N|\bar{u}_c)du_1 = N\left(\frac{u_1}{\bar{u}_c}\right)^{N-1}\frac{du_1}{\bar{u}_c}\ , \tag{24}$$

and the probability that the sum of these other widths is $N\bar{\Gamma}-\Gamma_1$ provided Γ_1 is the lower threshold, viz.

$$p(\bar{\Gamma}|\langle\Gamma\rangle,\Gamma_1)d\bar{\Gamma} \propto \frac{1}{\bar{u}_c^{N-1}}e^{-y}y^{(N-1)/2-1}dy\ . \tag{25}$$

Going to $\langle\Gamma\rangle$ and u_c as variables one finds from (23) the joint distribution of the estimated parameters,

$$p(\langle\Gamma\rangle,u_c|\bar{\Gamma},\Gamma_1)d\langle\Gamma\rangle du_c \propto \frac{u_1}{\bar{u}_c^N}e^{-y}y^{(N-1)/2-1}dy\frac{du_c}{\bar{u}_c}$$

$$0 < y \equiv \frac{N\bar{\Gamma}-\Gamma_1}{2\langle\Gamma\rangle} < \infty\ ,\qquad 0 < u_c < 1\ . \tag{26}$$

For given $\langle\Gamma\rangle$ the probability becomes maximal if $\bar{u}_c$ is minimal, i. e. $\bar{u}_c = u_1$. This estimate is biased, being always low. From (24) we can calculate the expectation value of $u_1/\bar{u}_c$ with the result that

$$\bar{u}_c = \frac{N+1}{N}u_1 \tag{27}$$

is an unbiased estimator of the observable fraction of levels. Differentiation of L with respect to $\langle\Gamma\rangle$ yields

$$\frac{1}{N-1}\sum_{i=2}^{N}\Gamma_i = \langle\Gamma\rangle\left(1+\frac{2}{\sqrt{\pi}}\frac{e^{-x_1}\sqrt{x_1}}{\operatorname{erfc}\sqrt{x_1}}\right)\ . \tag{28}$$

Thus $\langle\Gamma\rangle$ can be found from (28) whereupon $\bar{u}_c$ and the estimated true number of levels $N/\bar{u}_c$ follow from (27). If the threshold depends on energy (28) remains valid but instead of (27) one finds

$$\bar{u}_c = \frac{N+1}{N}\frac{\operatorname{erfc}\sqrt{x_1}}{f(E_1)} \tag{29}$$

In both equations the subscript 1 refers now to the sample point which relatively speaking is closest to the threshold, i. e. which has the highest ratio $u_i/f(E_i)$ (but not necessarily the smallest Γ_i).

So far we assumed the threshold to be sharp. In reality, however, thresholds are diffuse. It is then better to base the estimation not on all members of the sample, but to discard the points in the region of the diffuse threshold. It is not difficult to derive the corresponding equations. If the sample members are enumerated in descending order of $u_i/f(E_i)$ and the estimation is based on the members k to N only one finds as generalisation of Eqs. 28, 29

$$\frac{1}{N-k}\sum_{i=k+1}^{N}\Gamma_i = \langle\Gamma\rangle\left(1+\frac{2}{\sqrt{\pi}}\frac{e^{-x_k}\sqrt{x_k}}{\operatorname{erfc}\sqrt{x_k}}\right)\ , \tag{30}$$

$$\bar{u}_c = \frac{N+1}{N+1-k}\frac{\operatorname{erfc}\sqrt{x_k}}{f(E_k)} \tag{31}$$

One can begin the estimation with the outermost point (k=1) and then move inward point by point to check the stability of the results against threshold variations. It should be pointed out that the rigorous result presented here differs from the maximum-likelihood result given in [6], especially for small samples.

4.4 Porter-Thomas Distribution Distorted by Unresolved Multiplets

Let us now consider the problem that levels are missed not because of a detection threshold but because finite instrumental resolution causes pairs, triples etc. of closely spaced spaced levels to be mistaken for single peaks. We assume that this happens whenever spacings are smaller than some limiting separation D_c which of course must be of the order of the instrumental resolution. The fraction of missing levels is then

$$q = \int_0^{D_c} p(D)dD \tag{32}$$

where $p(D)dD$ is the level spacing distribution (usually for a mixture of level sequences). If one assumes that the apparent neutron width extracted from an unresolved multiplet peak is equal to the sum of the true component widths one can show that the observed width distribution is given by[6]

$$p(\Gamma|\langle\Gamma\rangle,q)d\Gamma = (1-q)(1+v)\frac{e^{-x}}{\sqrt{\pi x}}dx \tag{33}$$

with

$$v = \sqrt{\pi}ze^{z^2}(1+\operatorname{erf} z)\ ,\qquad z = q\sqrt{x}\ . \tag{34}$$

The distortion v reduces the relative frequency of small widths and increases that of large ones. This analytic treatment of resolution effects is more convenient than the usual Monte Carlo simulations. It is used in a new version of the STARA code for statistical resonance analysis[6]. This feature was added to the code after the recent NEA benchmark[5] had shown that all the participating codes underestimated the level densities by several percent because they did not account for unresolved multiplets (nor did the various Monte Carlo tests with which the authors had tested them).

4.5. Mixtures of Level Sequences

So far we treated only a single s-wave level sequence as occurs for target nuclei with spin 0. For target nuclei with nonzero spin one has two s-wave level sequences. It is well known that the quantities $g\Gamma$ of the mixed sample (g being the spin factor) are again members of a Porter-Thomas distribution provided that the strength function is the same for both sequences and their level densities can be taken as proportional to g. This means that the methods discussed so far are applicable to all isotopically pure s-wave samples, from both even and odd nuclei.

The problem of p-wave admixture (with unidentified level parities) is more difficult. One can use artificial thresholds to discard the p-wave levels (which usually have very small widths) as is done in ESTIMA [22], in Rohr's code[23] and in the original version of STARA[24], or one can estimate the p-wave strength function together with the s-wave strength function and the level density, as is done in Stefanon's CAVE code[25]. In this case it has not been possible so far to derive rigorous (Bayesian) equations. Therefore one uses the maximum-likelihood approach. This is also necessary with such concepts as analytical treatment of resolution effects (see Sect. 4.4) or diffuse threshold which are used in the new version of the STARA code[6]. It seems that in their present form these elaborate methods are not much better than the very simple one of finding $\langle\Gamma\rangle$ for the s-wave levels and their density from the upper part of the width distribution, then subtract the extrapolated s-level distribution from the lower part and fit the remainder with a p-wave strength function. Another unorthodox but well working estimator has been introduced by Moore[13]. He equates the sample averages of Γ and $\sqrt{\Gamma}$ for a truncated Porter-Thomas distribution with the corresponding expectation values:

$$\overline{\Gamma} = \frac{1}{N}\sum_{i=1}^{N}\Gamma_i = \int_{\Gamma_c}^{\infty}\Gamma\, p(\Gamma)d\Gamma \Big/ \int_{\Gamma_c}^{\infty} p(\Gamma)d\Gamma\,, \tag{35}$$

$$\overline{\sqrt{\Gamma}} \equiv \frac{1}{N}\sum_{i=1}^{N}\sqrt{\Gamma_i} = \int_{\Gamma_c}^{\infty}\sqrt{\Gamma}\, p(\Gamma)d\Gamma \Big/ \int_{\Gamma_c}^{\infty} p(\Gamma)d\Gamma\,. \tag{36}$$

Inserting (1) one gets

$$\frac{\overline{\Gamma}}{(\overline{\sqrt{\Gamma}})^2} = \sqrt{\frac{\pi}{2}}e^{x_c}\operatorname{erfc}\sqrt{x_c}\left(\sqrt{\frac{\pi}{2}}\, e^{x_c}\operatorname{erfc}\sqrt{x_c}+\sqrt{2x_c}\right)\,. \tag{37}$$

For given sample averages this is an equation for x_c. It can be solved by iteration. If the threshold Γ_c is known one obtains $\langle\Gamma\rangle = \Gamma_c/(2x_c)$. Moore, on the other hand, prescribes a value for x_c (he takes e. g. 1/8 for ^{238}U) which determines the right-hand side, and then varies the threshold and thus N, beginning with a high threshold and lowering it gradually so that smaller and smaller widths are included, until the left-hand side equals the right-hand side. The true number of levels is then estimated as $N/\operatorname{erfc}\sqrt{x_c}$.

Recently Moore extended the procedure in a very elegant way so that it can cope with admixture of unidentified p-wave levels also: for each member of the sample he determines the a-posteriori probabilities that it belongs to one of the possible level sequences. These values are then added to the two sums (35), (36) for each sequence separately, i.e. each member of the sample is split in fractions equal to the a-posteriori probabilities and distributed accordingly over the various sequences. A similar, conceptually slightly simpler, procedure could be based on Eq. 18. Assuming with Moore that the threshold does not depend on E, on can rewrite (18) as

$$\frac{\overline{\Gamma}}{2\Gamma_c} = \frac{1}{x_c}\left(1 + \frac{2}{\sqrt{\pi}}\frac{e^{-x_c}\sqrt{x_c}}{\operatorname{erfc}\sqrt{x_c}}\right)\,. \tag{38}$$

For given right-hand side one can again vary N (via the threshold) until equality is achieved between left- and right-hand side, with the same result for for the true level density as above, $\rho = N/\operatorname{erfc}\sqrt{x_c}$.

Unresolved multiplets could be accounted for easily by using the distorted Porter-Thomas distribution, Eqs. 33, 34, instead of Eq. 1.

5. Summary

Level density estimation methods have been reviewed with emphasis on the mathematical and statistical aspects. A new rigorous solution is given for the problem of simultaneous estimation of mean width and level density (or true number of levels) for a Porter-Thomas distribution affected by a detection threshold with known energy dependence but unknown height. Furthermore, a new analytical treatment of resolution effects for s-wave levels is presented which can replace Monte Carlo simulation in many cases. More work is needed to derive similarly convenient prescriptions for mixed (s- and p-wave) resonance samples.

Acknowledgement

It is a pleasure to acknowledge stimulating discussions with M. Stefanon, G. Rohr, E. Fort, M. Moore and F. Perey.

References

1. A.M. Anzaldo Meneses, this conference
2. T.A. Brody, J. Flores, J.B. French, P.A. Mello, A. Pandey and S.S.M. Wong, Rev. Mod. Phys. 53 (1981)385; O. Bohigas, R.U. Haq and A. Pandey, this conference
3. C.E. Porter and R.G. Thomas, Phys. Rev. 104(1959)483 vol. I, p. 463
4. D.K. Olsen, G. de Saussure, R.B. Perez, E.G. Silver, F.C. Difilippo, R.W. Ingle and H. Weaver, Nucl. Sci. Eng. 62(1977)479; G. de Saussure, this conference
5. P. Ribon and A. Thompson, this conference
6. F.H. Fröhner, IAEA Meeting on Uranium and Plutonium Isotope Resonance Parameters, INDC(NDS)-129/GJ, Vienna (1981), p. 103
7. H.I. Liou, Conf. on Nucl. Data Eval. Meth. and Proced., Brookhaven 1980, BNL-NCS-51363(1981), Vienna (1981), p. 103
8. C.E. Porter (ed.), Statistical Theories of Spectra, New York and London (1965)
9. P.A. Mello, J. Flores, T.A. Brody, J.B. French, S.S.M. Wong, Proc. Conf.on Interact. of Neutrons with Nuclei, Lowell (1976), vol. I, p. 496
10. F.J. Dyson and M.L. Mehta, J. Math. Phys. 4(1963)701
11. M.L. Mehta, Statistical Properties of Nuclei (ed. J.B. Garg), New York and London (1972), p. 179
12. H.I. Liou, H.S. Camarda and F. Rahn, Phys. Rev. C5(1972)450
13. M.S. Moore, J.D. Moses, G.A. Keyworth, J.W.T. Dabbs and N.W. Hill, Phys. Rev. C18(1978)1328
14. E.Fort, H. Derrien and D. Lafond, Meeting on Neutron Cross Sections of Fission Product Nuclei, Bologna 1979, RIT/FIS-LDN(80)1, p. 121
15. A. Michaudon, H. Derrien, P. Ribon, M. Sanche, Nucl. Phys. 69(1965)545
16. H. Derrien and B. Lucas, Nucl. Cross Sections and Technol., NBS Spec. Publ. 425, Washington D.C. (1975) vol. II, p. 637
17. J.D. Garrison, Phys. Rev. Letters 29(1972)1189
18. H. Jeffreys, Theory of Probability, Oxford (1939)
19. E.T. Jaynes, Trans. Systems Sci. and Cybern. 4(1968)227
20. D.D. Slavinskas and Kennett, Nucl. Phys. 85(1966)641
21. T. Fuketa and J.A. Harvey, Nucl. Inst. Meth. 33(1965)107
22. P. Ribon, E. Fort, J. Krebs, Tran Quoc Thuong CEA-N-1832 167L (1975)
23. G. Rohr, L. Maisano and R. Shelley, Meeting on Neutron Cross Sections of Fission Product Nuclei, Bologna 1979, RIT/FIS-LDN(80)1, p. 197
24. F.H. Fröhner, Meeting on Neutron Cross Sections of Fission Product Nuclei, Bologna 1979, RIT/FIS-LDN(80)1, p. 145
25. C. Coceva and M. Stefanon, Nucl. Phys. A315(1979)

A BENCHMARK TEST OF COMPUTER CODES FOR CALCULATING AVERAGE RESONANCE PARAMETERS

P. Ribon

CEN Saclay
91191 Gif-sur-Yvette CEDEX, France

A. Thompson

NEA Data Bank
91191 Gif-sur-Yvette CEDEX, France

Sets of resonance parameters, designed to simulate data derived from experimental measurements, were generated. They were used to compare the performance of a group of computer codes for calculating average parameters. Some of these codes have been implemented and are now available for general distribution.

Introduction

At both the 1977 and 1979 Specialists' Meetings on Fission Product Nuclear Data held at Petten, the Netherlands, and Bologna, Italy, respectively, E. Fort pointed out significant discrepancies in reported average nuclear resonance parameters calculated from the same experimental data by different evaluators using different methods (IAEA213(1978) and NEANDC(E)-209-L). The latter meeting accepted a proposal by P. Ribon to organise an international benchmark test to compare the various calculation techniques in use. A set of resonance parameters would be generated from known, but secret, average values; the parameters would then be adjusted to mimic experimental data by including the effects of Doppler broadening, resolution broadening and statistical fluctuations. Average parameters calculated from the dataset by various computer codes could then be compared with each other, and also with the true values.

The benchmark test is fully described in the report NEANDC160-U (NEA Data Bank Newsletter No. 27, July 1982); the present paper is a summary of this document.

The Problem Considered by the Codes

The central problem investigated by this benchmark test is the calculation of the mean reduced width, level spacing and strength function of a given set of s-wave neutron width data. Such a set of widths is expected to obey a Porter-Thomas distribution. However, any experimental distribution will be distorted due to instrumental minimum detection and resolution thresholds, and by the contamination of the s-wave data with p-wave widths.

In general the codes calculated s-wave average parameters either by a maximum likelihood fit to an appropriate truncated and distorted Porter-Thomas distribution, or by using information derived from moments of this distrbution. Expressions derived from Bayes' Theorem were often used to calculate the probability that any particular width was s- or p-wave, and estimates of p-wave average parameters could frequently be obtained.

The First Stage

Pseudo-experimental datasets were generated representing two different fission product nuclei, one with spin 5/2 in the mass region A=150, the other with spin 1/2 near A=100. In order to differentiate systematic and random errors, three datasets were generated for each case. The spin 5/2 nucleus was expected to be an easier problem - the s-wave strength function is large in this mass region and the p-wave contamination is correspondingly small; the other nucleus is perhaps representative of the most difficult problem to be encountered.

These datasets, together with a report describing their origin (NEANDC-213-AL, Part 1) were distributed to all physicists wishing to take part. Nine sets of results were obtained from the following: M. Caner and Y.S. Gur (SOREQ, Israel), H. Derrien and E. Fort (CEN Cadarache, France), F.H. Fröhner (KFK, Karlsruhe, Fed. Rep. of Germany), G. Rohr and H. Weigmann (CBNM, Geel, Belgium). Although the sample of codes tested was not large it did represent most of the sophisticated methods currently in use.

The results can be summarized as follows:

i) Spin 5/2 nucleus (easier case)

The s-wave function was calculated to within 1-3 percent and showed no systematic deviation; the spacings were calculated with a dispersion of 1-4 percent, and were systematically about 7 percent too large. The systematic error was found to be due to levels missed because of resonance overlap (F.H. Fröhner, "Level Density Estimation with Account of Unrecognized Multiplets Applied to Uranium and Plutonium Resonance Data" - contribution to the IAEA Specialists' Meeting on Uranium and Plutonium Resonance Parameters, Vienna, 28th September to 2nd October, 1981).

ii) Spin 1/2 nucleus (more difficult case)

The average parameters differed systematically by up to a factor of two, and the discrepancies were generally well outside the uncertainties assigned by the participants.

The Second Stage

The origin of the discrepancies discovered in the first stage was not clear, and it was decided to organise a second stage designed to identify the sources of error. This also gave participants an opportunity to test any modification they wished to make to their codes.

Six new sets of resonance parameters were generated, each representing the more difficult nucleus of the first stage: two consisted merely of an undistorted set of s-wave widths, two were a mixture of s- and p-waves without any simulation of experimental effects, and two simulated experimental data. The second stage was open to all who had participated in the first stage; in fact seven of the original nine took part. Results were discussed at a workshop held on 15th and 16th October, 1981, at the NEA Data Bank, Saclay, France. The workshop was also attended by J.S. Story (AEE, Winfrith, United Kingdom).

The results were, unfortunately, still not completely satisfactory. While the very large discrepancies observed in the first stage appeared to have been eliminated, even the most elaborate methods occasionally seemed to show unexplained disagreements with the true values of 20 to 30 percent for s-wave parameters, and up to a factor of two in p-wave estimates.

The Third Stage

It was decided to devote attention to three or four codes which would be fully documented and sent to the NEA Data Bank for distribution. In order to test how well this documentation enabled a user to make the subjective judgements inherent in many of the methods, a "blind test" of the codes by a physicist unfamiliar with the problem was organised.

This test was carried out at the NEA Data Bank by A.

K. H. Böckhoff (ed.), Nuclear Data for Science and Technology, 628–629.

Thompson, using three stage-two datasets. In the case of both codes available from NEADB, the calculated s-wave average parameters appear to be independent of the user - the results obtained by A. Thompson and the respective authors are 95 percent correlated. A third code, not currently available from NEADB, showed an anticorrelation between the errors in the calculated results. In this case, A. Thompson had developed a method of using the code which was rather different from that used by the author.

The Future

As a result of this exercise, two codes, using rather different methods, are available from NEA Data Bank. They are M. Moore's BAYESZ and E. Fort's ESTIMA. A benchmark test such as this can never be said to have completely finished. Many questions remain to be satisfactorily answered and it is to be hoped that the discussion on the behaviour of the codes will continue, not only with the authors themselves but also involving other users.

Conclusions

The major conclusions drawn from the exercise are:

1. The two codes available from NEA Data Bank, BAYESZ and ESTIMA, can be used equally well by an unfamiliar user as by their authors to obtain average resonance parameters from experimental neutron width data with a standard deviation of about 8 percent on the s-wave spacing and about 2 per cent on the s-wave strength function in the difficult cases.
2. Only two codes, those of M. Moore and H. Weigmann, provide good estimates of p-wave strength functions.
3. There still remain unexplained discrepancies between calculated results, even for codes using similar methods.

There was an initial scepticism about the usefulness of such an exercise and a fear that it could develop into a self-perpetuating project. It was, however, concluded that this benchmark exercise had three important results:

i) The physicists involved had to increase the sophistication of their mathematical and numerical techniques in order to correct for physical effects which were initially assumed to be unimportant.

ii) The exercise enabled the accuracy of the codes to be more realistically assessed

iii) In order to participate in the third stage, authors made an effort to increase the portability and extend the documentation of their codes.

Finally, it is hoped that the techniques developed to compare computer codes for calculating average resonance parameters will be usefully applied to other topics.

Figures

The following figures are examples of the calculated values for the s-wave average level spacing D^0 and the strength function S^0 (ratio of mean reduced width to level spacing). The different codes are labelled A to H; error bars indicate the range of values computed by each code; crosses are used when no error was calculated.

The value labelled "theoretical" was used to generate the data. Their average spacing and strength is labelled "true"; the difference between these two values is a measure of the statistical spread in the sample.

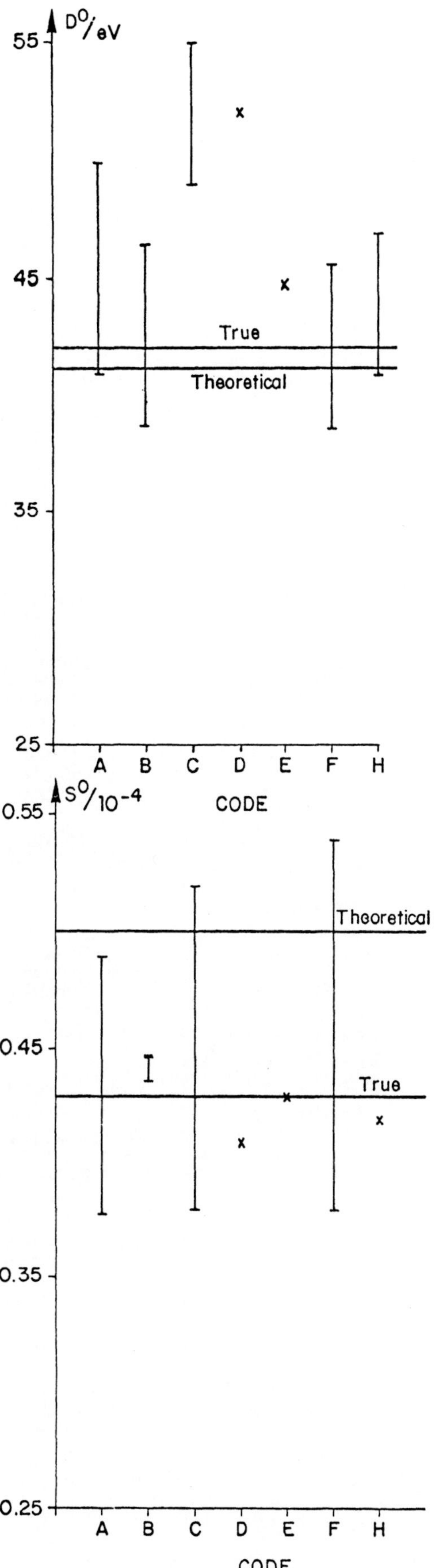

Fig. 1. A difficult case from stage 1.

ANALYSIS OF AVERAGE RADIATION WIDTHS OF NEUTRON RESONANCES

V.I.Bondarenko, M.G.Urin

Moscow Physics Engineering Institute, Moscow, USSR

H.Malecki, A.B.Popov, K.Trzeciak

Laboratory of Neutron Physics, Joint Institute for Nuclear Research, Dubna, USSR

The average radiation width estimates made using the known experimental data are compared with different theoretical calculations, which give the dependence of Γ_γ on atomic weight A, excitation energy U, level density parameter a in the form $\Gamma_\gamma = cA^\alpha U^\beta a^\gamma$.

The report communicates the results of the systematization of average radiation widths and of their comparison with different theoretical estimates from Ref.[1], which were analysed in Ref.[2]. An emphasis was made on the the quantitative comparison of theoretical and semi-empirical estimates of Γ_γ with experimental data.

Usually the theoretical analysis of Γ_γ is carried out under the following assumptions. 1. The Brink's hypothesis holds and thus one can describe in a similar way the gamma-quanta absorption by the nucleus in the ground state and in the excited state

$$\bar{\sigma}_{cE1}(E_\gamma, U^*) = \bar{\sigma}_{cE1}(E_\gamma, U^*=0) \equiv \bar{\sigma}_{cE1}(E_\gamma). \quad (1)$$

Here $\bar{\sigma}_{cE1}$ is the average cross section of dipole photoabsorption, U^* is the excitation energy of the nucleus. 2. Mainly the E1-transition contributes to the Γ_γ width. 3. The Γ_γ width is formed mainly by the transitions between the compound states.

The expression for the average radiation width can be written as follows:

$$\Gamma_\gamma = \frac{1}{\pi^2}\rho_0^{-1}(U)\int_0^U k_\gamma^2 \bar{\sigma}_{cE1}(E_\gamma)\rho_0(U^*)dE_\gamma, \quad (2)$$

where $U^* = U - E_\gamma$, $(2J+1)\rho_0(U) \sim (2J+1)\exp(2\sqrt{aU})$ is the level density of compound nuclei for the excitation energy U amd moment J , a is the level density parameter.

Different theoretical approaches use different parametrization descriptions of the low energy part of $\bar{\sigma}_{cE1}$. The simplest estimate for this cross section is the Weiskopff's single particle estimate (see Ref.[3]):

$$\bar{\sigma}_{cE1}(E_\gamma) = \frac{9\pi^2}{4}\frac{e^2}{\hbar c}\frac{R^2}{D_0}E_\gamma, \quad (3)$$

here $R = r_0 A^{1/3}$, $D_0 = E_0 A^{-1/3}$

The substitution of (3) into (2) and the subsequent integration gives:

$$\Gamma_\gamma = c_1 A\left(\frac{U}{a}\right)^2, \quad c_1 = \frac{13.5}{E_0}\frac{e^2}{\hbar c}\left(\frac{r_0}{\hbar c}\right)^2. \quad (4)$$

One may take into consideration the collective nature of photoabsorption and the dissipation of simple excitation modes to many particles ones. Then by the extrapolation to the energy region of the Lorentz's similar $\bar{\sigma}_{cE1}(E_\gamma)$ dependence with parameters $E_G = E_0' A^{1/3}$ and Γ_G , found near the maximum of the giant electric dipole resonance (GEDR) (see Ref[4]) normalized over classical sum rule one may obtain

$$\Gamma_\gamma = c_2 A^{7/3}\left(\frac{U}{a}\right)^{5/2}, \quad c_2 = \frac{4!}{\pi}\frac{e^2}{\hbar c}\frac{1}{Mc^2}\frac{\Gamma_G}{(E_0')^4}. \quad (5)$$

It is more correct to use the relaxation width $\Gamma^\downarrow$ instead of the total GEDR width of Γ_G . The empirical dependence of $\Gamma^\downarrow$ on E_G was given by the formula $\Gamma^\downarrow = 0.026 E_G^2$ (Ref.5). In this case the expression (5) is transformed into

$$\Gamma_\gamma = c_3 A^{5/3}\left(\frac{U}{a}\right)^{5/2}, \quad c_3 = \frac{4!}{\pi}\frac{e^2}{\hbar c}\frac{1}{Mc^2}\frac{0.026}{(E_0')^2} \quad (6)$$

The low energy part of $\bar{\sigma}_{cE1}$ can be also described in the frame of the shell or optical model. This approach allows one to refuse from the assumption (1) and obtain under some simplification the analitical expressions for $\bar{\sigma}_{cE1}$ and Γ_γ , respectively, (Refs.[2,6]).

$$\Gamma_\gamma = c_4 A^{7/3}\left(\frac{U}{a}\right)^{7/2}, \quad c_4 = \frac{2}{3}\frac{6!}{\pi}\left(1+\frac{\pi^2}{10}\right)\frac{e^2}{\hbar c}\frac{\alpha}{Mc^2(E_0')^4} \quad (7)$$

Here α gives the average dependence of the imaginary part of optical potential on excitation energy: $W(U^*) = \alpha(U^*)^2$.

Formulas (4)-(7) can be given in the form

$$\Gamma_\gamma = cA^\alpha U^\beta a^\gamma. \quad (8)$$

The latter together with exps. (4)-(7) were used in the analysis of experimental values of Γ_γ . According to the statistical model a=kA, k=0.125 MeV^{-1} (Ref.[7]). Under this assumption Γ_γ in formulas (4)-(8) should not depend on the level density parameter a . The dependence on A changes also. The C_i coefficients are replaced by C_i' . Let's mark the modified expressions as (4')-(8').

The analysis of experimental data was made on the basis of expressions (4)-(8) and (4')-(8'). The following assumptions were made: 1) the formulas (4)-(7) (4')-(7') give the correct dependence of Γ_γ on A, U, a. The coefficients C_i are chosen to give the best description of experimental data; 2) the Γ_γ width is described by formula (8, 8'), the parameters of C, α, β, γ are chosen to describe in the best way the experimental data.

The parameters were determined using the program of the minimizing functional

$$\chi^2 = \sum_{i=1}^{m}\left(\frac{\Gamma_{\gamma i}^{calc} - \Gamma_{\gamma i}^{exp}}{\Delta\Gamma_{\gamma i}^{exp}}\right)^2. \quad (9)$$

Here m is the number of nuclei for analysis, $\Gamma_{\gamma i}^{calc}$ and $\Gamma_{\gamma i}^{exp}$ are the theoretical and experimental values of $\Gamma_{\gamma i}$, $\Delta\Gamma_{\gamma i}^{exp}$ are the experimental errors.

The average relative deviations of Γ_γ^{calc} from Γ_γ^{exp} were also calculated

$$\delta = \frac{1}{m}\sum_{i=1}^{m}\frac{|\Gamma_{\gamma i}^{calc} - \Gamma_{\gamma i}^{exp}|}{\Gamma_{\gamma i}^{exp}}. \quad (10)$$

This value is a more real measure of the confirmity of the theory with experiment, since χ^2 depends both on differences of $\Gamma_{\gamma i}^{calc}$ and $\Gamma_{\gamma i}^{exp}$ and on estimated errors $\Delta\Gamma_\gamma^{exp}$.

The analysis was carried out for 131 nuclei in the interval $50 < A < 250$. The parameters U and a were taken from Ref.8. The results are summarized in the Table. One may conclude that formulas (4')-(8') give a better agreement on χ^2/m and δ values of

K. H. Böckhoff (ed.), Nuclear Data for Science and Technology, 630–631.

Table

Ver-sion	For-mulae	Formulae with empirical C_i	χ^2/m	δ	C_i^{th}
1	(4)	$\Gamma_\gamma = 2.86\ A(\frac{U}{a})^2$	15.9	0.28	115 MeV^{-3}
2	(5)	$\Gamma_\gamma = 0.0044A^{7/3}(\frac{U}{a})^{5/2}$	34.6	0.44	0.0072 MeV^{-4}
3	(7)	$\Gamma_\gamma = 0.0122A^{7/3}(\frac{U}{a})^{7/2}$	32.6	0.37	0.0006 MeV^{-6}
4	(6)	$\Gamma_\gamma = 0.15A^{5/3}(\frac{U}{a})^{5/2}$	21.4	0.32	0.25 MeV^{-4}
5	(8)	$\Gamma_\gamma = 38.7\frac{U^{2.4}}{A^{0.63}a^{0.31}}$	8.5	0.23	-

Ver-sion	For mulae	Formulae with empirical C_i	χ^2/m	δ	C_i^{th}
1	(4')	$\Gamma_\gamma = 235\frac{U^2}{A}$	9.0	0.23	7360 MeV^{-1}
2	(5')	$\Gamma_\gamma = 1.19\ \frac{U^{5/2}}{A^{1/6}}$	15.3	0.32	1.31 $MeV^{-3/2}$
3	(7')	$\Gamma_\gamma = 27.2\ \frac{U^{7/2}}{A^{7/6}}$	17.8	0.33	0.93 $MeV^{-5/2}$
4	(6')	$\Gamma_\gamma = 37.8\ \frac{U^{5/2}}{A^{5/6}}$	8.6	0.26	45.9 $MeV^{-3/2}$
5	(8')	$\Gamma_\gamma = 32.1\frac{U^{2.56}}{A^{0.823}}$	8.6	0.24	-

Γ_γ^{calc} results with experimental data than those calculated with formulas (4)-(8). The Fig. shows the ratios $\Gamma_{\gamma i}^{calc}/\ \Gamma_{\gamma i}^{exp}$ for formulas (4')-(8'). The digitals in Fig. indicate the version in the Table. The results of calculation demonstrate that one cannot give preference to this or that theoretical version because χ^2/m and δ in different formulas have about the same values. Nevertheless, one can affirm that the versions 1,4 and 5 are in better agreement with experimental data that the 2nd and 3rd ones. It is interesting to compare the empirically selected C_i and C_i' values with theoretical ones. The values of C_i^{th} and $C_i'^{th}$ were estimated using the following parameters $\Gamma_G = 5$ MeV E_o=48 MeV, E_o' =78 MeV (Ref.[9]), α =0.06MeV^{-1} (Ref.[2]) We obtain (see Table) that $C_i'^{th}$ for the 4th version is more like the empirical value of C_i', moreover, the powers of the 4th version (formula (6')) coincide with the empirical α' and β' parameters in the 5th version.

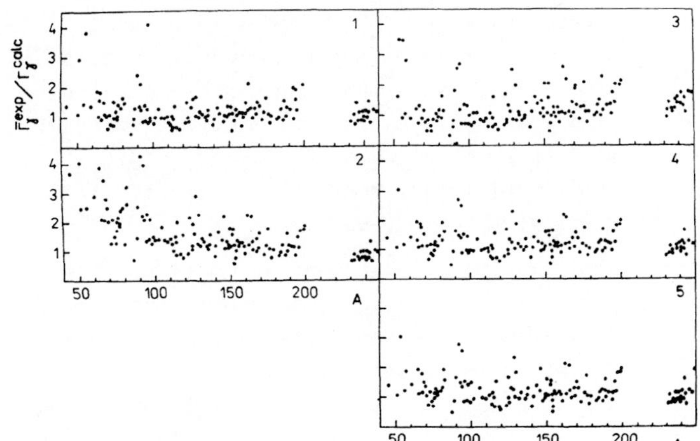

Fig. Dependence on A of $\Gamma_\gamma^{exp}/\ \Gamma_\gamma^{calc}$ ratios in the case of a=kA. Digitals 1 - 5 correspond to the version number of the lower half of the Table.

Note, in all the variants an irregular scattering of the $\Gamma_{\gamma i}^{calc}/\ \Gamma_{\gamma i}^{exp}$ values for the neighbouring nuclei is observed. Thus no available theoretical approach can describe in detail the experimental data on Γ_γ. Apparently, one should admit that the description of Γ_γ with A,U,a parameters is a rough approximation.

For a more detailed description of radiation widths it is necessary to draw a joint analysis of physical values conditioned by the relaxation of GEDR, i.e. the radiation strength functions of transitions between compound states, the partial radiation strength function for the transition to the ground state, the total width of GEDR, etc. Then it will be possible to choose a more suitable theoretical model and to determine more exactly its parameters.

References

1. H.Malecki, A.B.Popov, K.Trzeciak, JINR,P3-82-11, Dubna, 1982.
2. V.I.Bondarenko, M.G.Urin, Yad.Fiz.35,675, 1982.
3. J.Blatt, V.Weisskopf. Theoretical nuclear physics New-York-London, 1952.
4. C.H.Johnson. Phys.Rev.,C16, 2238, 1977.
5. P.Carlos, H.Beil, R.Bergere et al., Nucl.Phys., A219, 39, 1974.
6. D.F.Zaretski, V.K.Sirotkin. Jad.Fiz., 27,1534,1978.
7. A.A.Lukianov. Struktura nejtronnykh sechenij, Moskva, Atomizdat, 1978.
8. W.Dilg, W.Shantl, H.Vonach, M.Unl. Nucl.Phys.A217 269, 1973.
9. F.E.Bertrand. Nucl.Phys., A354, 129, 1981.

SYSTEMATICS OF AVERAGE TOTAL RADIATIVE WIDTHS FOR S- AND P-WAVE RESONANCES

Zhuang Youxiang, Wang Shunuan, Zhou Delin

Institute of Atomic Energy, Academia Sinica
P. O. B. 275(41), Beijing, China

Jia Zhize

Research Institute of Science and Technology
Information, the Ministry of Nuclear Industry,
Beijing, China
P. O. B. 2103, Beijing, China

The experimental values of total radiative widths, $\langle\Gamma_\gamma\rangle$, for 208 nuclides (Z=11-98, A=23-249) are compiled and analysed. Two empirical formulas are presented for the radiative widths, which are dependent on the atomic number Z, the mass number A and the neutron separation energy Sn of the compound nucleus: $\langle\Gamma_\gamma^0\rangle = 31\ Z^{-2.5}\ A^{0.56}\ Sn^{0.83}$ (in eV) and $\langle\Gamma_\gamma^1\rangle = 24\ Z^{1.3}\ A^{-2.6}\ Sn^{0.95}$ (in eV). Using the strong-coupling dipole model, the coupling coefficients Cr which are varied smooth with the mass numbers A are obtained. Some maxima of Cr around the magic numbers are obtained. This is an evidence of the shell effect for Cr. It is found that Cr of the odd neutron nuclides generally are larger than those of the even neutron nuclides. This is due to the pairing effect.

Introduction

Average total radiative widths of neutron resonances are one of the resonance parameters which are necessary for application in astrophysics, reactor physics and fission physics. But there are quite a number of nuclides which lack the experimental values $\langle\Gamma_\gamma\rangle$. Thus it is necessary to study systematically average radiative widths to derive $\langle\Gamma_\gamma\rangle$ of those nuclides. Moreover, the determination of the dependence of radiative widths on neutron spacing, spin and parity of the initial state, excitation energy, nuclear size and nuclear structure effects is of particular interest from a theoretical and evaluation point of view.

In 1956 J.S.Levin and D.J.Hughes(1) carried out the first attempt to study systematically average radiative widths and to interpret them in terms of Blatt-Weisskopf's estimates of partial radiative widths. In 1957 by a similar study A.Stolovy and J.A.Harvey(2) arrived at the following empirical expression for $\langle\Gamma_\gamma\rangle$:

$$\langle\Gamma_\gamma\rangle = 5.3\cdot10^{-4}A^{2/3}[D(U)]^{0.25}U^{4.3} \text{ (in meV)}, \quad (1)$$

where U and D(U) are the effective excitation energy in MeV and the spacing of levels with the same spin and parity as the radiating levels around the excitation energy U, respectively.

In 1971 H.Malecky et al(3). noted that the previously derived relations of total radiative widths can be expressed in terms of a combination of the parameters A, a, U and T. A least squares fit of the experimental data for 108 nuclides (Z=27-96, A=59-246) was carried out. It has the following form

$$\langle\Gamma_\gamma\rangle = 9U^{0.9}A^{-0.9}a^{-0.57}(1-0.011\ I^2) \text{ (in eV)}, \quad (2)$$

where I is the spin of the target nuclides, a is the level density parameter, and U and a^{-1} are expressed in MeV. As pointed out by H.Malecky et al(3)., there is a general agreement between the experimental values and those calculated on the basis of Eq.(2) with a few exceptions. But these $\langle\Gamma_\gamma\rangle$ compiled by H.Malecky et al. are published in 1966-1969, and in addition, there are no average radiative widths for p-wave resonance.

The purpose of this work is to extend the range of the nuclides, to compile more recent experimental values $\langle\Gamma_\gamma\rangle$ as much as possible, to fit them using a empirical expression for $\langle\Gamma_\gamma\rangle$, to carry out theoretical analysis and calculation of average radiative widths and from which some regularity may be found out for setting up a more complete systematics of average radiative widths.

The Experimental Systematics of Average Total Radiative Widths

The experimental values of average radiative

K. H. Böckhoff (ed.), Nuclear Data for Science and Technology, 632–638.

widths for neutron resonance are collected in terms of CINDA 1982, and average radiative widths of S- and P-wave neutron resonances for 208 nuclides (Z=11-98, A=23-249) are selected, in which 207 nuclides for S-wave and 56 nucledes for P-wave. These experimental or evaluated data are published in 1973-1981, most of them are taken from BNL-325 (the fourth edition, 1981)[4] and our evaluation (1981)[5]. They are given in appendix.

The variation of the average S- wave radiative widths with mass number A is illustrated in Fig.1. Two interesting features emerge from this plot: (a) the general monotonic decrease of $\langle\Gamma_\gamma^0\rangle$ with A in the mass region 60-190 and (b) the maxima at about the mass numbers A=50 and 208, which are related to the magic numbers Z=28 and 82. This is an evidence of shell effect for $\langle\Gamma_\gamma^0\rangle$.

The variation of P-wave radiative widths with mass number is shown in Fig.2, which demonstrates the same qualitative features as for S-wave radiative widths. A detailed comparison of S- and P-wave radiative widths shows the following results:

$$\begin{array}{ll} \langle\Gamma_\gamma\rangle_S > \langle\Gamma_\gamma\rangle_P & \text{at } A\approx 50, \\ \langle\Gamma_\gamma\rangle_P > \langle\Gamma_\gamma\rangle_S & \text{at } A\approx 100. \end{array} \quad (3)$$

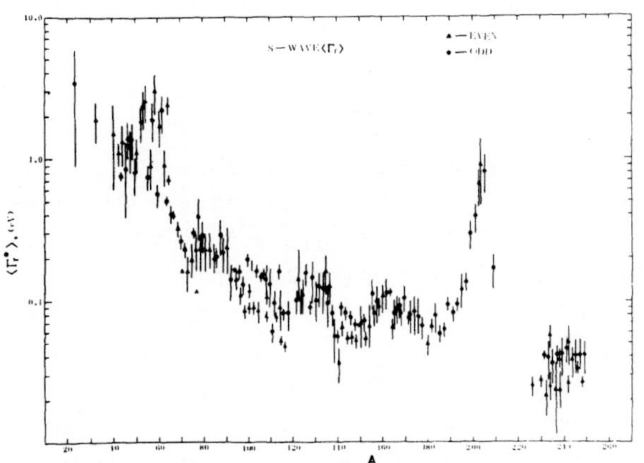

Figure 1. The average S-wave radiative widths plotted versus mass number A of target nucleus.

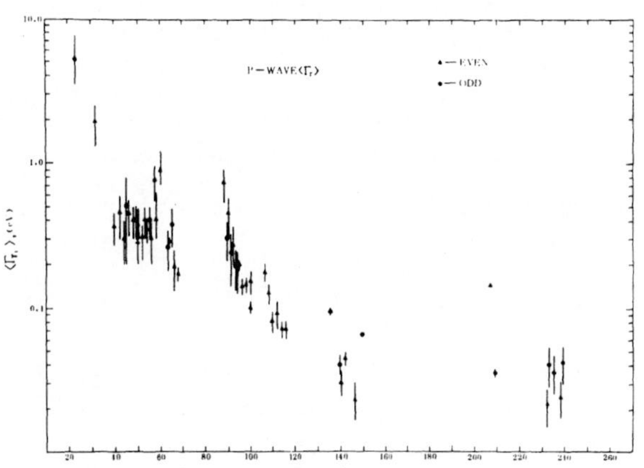

Figure 2. Variation of the average P-wave radiative widths with mass number A of target nucleus.

By means of an empirical expression $\langle\Gamma_\gamma\rangle=K\ Z^a A^b Sn^c$, which is similar to reference (3), and the least squares analysis the average radiative widths are fitted with the following forms:

$$\langle\Gamma_\gamma^0\rangle=31\ Z^{-2.5}A^{0.56}Sn^{0.83} \quad \text{(in eV)}, \quad (4)$$

$$\langle\Gamma_\gamma^1\rangle=24\ Z^{1.3}A^{-2.6}Sn^{0.95} \quad \text{(in eV)}. \quad (5)$$

Where the neutron separation energy Sn is expressed in MeV. Generally, the experimental values are in resonable agreement with those calculated on the basis of Eq.4 and 5, but not including the nuclides which are around A=50 and 208 for S-wave radiative widths and $^{23}_{11}Na$, $^{32}_{16}S$, $^{60}_{28}Ni$, $^{88}_{38}Sr$, $^{140}_{58}Ce$, $^{146}_{60}Nd$ and $^{207}_{82}Pb$ for P-wave.

Because ref.(3) didn't show the maxima at about the mass numbers A=50 and 208, it is obvious to use the same expression is difficult for describing all the experimental values $\langle\Gamma_\gamma\rangle$ which we compiled.

Theoretical Analysis and Calculation for Average Total Radiative Widths

Various theoretical formulas are available to estimate the average total radiative widths $\langle\Gamma_\gamma\rangle$ of neutron resonances. There are two methods which are used often: the Blatt-Weisskopf strong-coupling dipole model and the giant-dipole resonance model. The former is adopted because it has a relatively complete set of parameters which can be utilized to analyse and calculate $\langle\Gamma_\gamma\rangle$ in a wide range. We have calculated 192 nuclides among 207 ones. However there are still 15 nuclides have not been analyzed for lack of the parameters.

For dipole emission, which usually gives the largest contribution to the radiative decay process, an estimate of average total radiative width in terms of the strong-coupling dipole model, as described by J.E.Lynn[6], can be written in

$$\langle\Gamma_\gamma(E)\rangle=\frac{C_\gamma}{\rho(E)}\int_0^E \varepsilon^3\rho(E-\varepsilon)\,d\varepsilon. \quad (6)$$

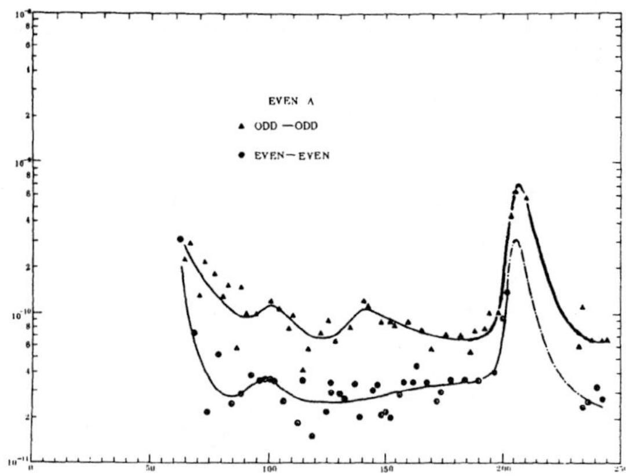

Figure 3. The coupling coefficients Cr for even A nucleus plotted versus mass numbers A of compound nucleus.

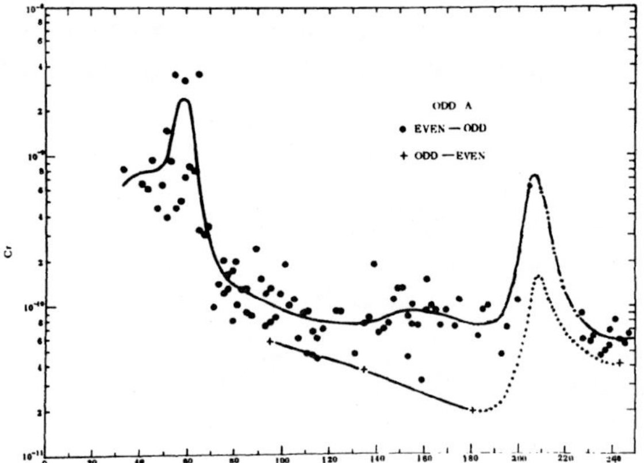

Figure 4. Variation of the coupling coefficients Cr for odd A nucleus with mass numbers A of compound nucleus.

Where Cr is the coupling coefficient, and ρ is the level density of compound nucleus.

Thus the radiative width at the neutron binding energy Bn is

$$\langle\Gamma_\gamma(B_n)\rangle = \frac{C_r}{\rho(B_n)}\int_0^{B_n}\varepsilon^3\rho(B_n-\varepsilon)\,d\varepsilon\ . \qquad (7)$$

The level density formula used here was given by A.Gilbert and A.G.W.Cameron[7]. At high excitation energies it is the regular Fermi gas formula and at low excitation energies it is a constant nuclear temperature representation. The boundary between them is at energy E_T. They are connected smooth at E_T. That is

$$\rho(E)=\begin{cases}\dfrac{\sqrt{\pi}\,\exp(2\sqrt{a(E_T-\Delta)})}{12\sqrt{a}\,(E-\Delta)^{3/2}}\,, & E>E_T\\[2ex] \dfrac{\sqrt{\pi}\,\exp(2\sqrt{a(E_T-\Delta)})}{12\sqrt{a}\,(E-\Delta)^{3/2}}\exp((E-E_T)/T)\,, & E\le E_T.\end{cases} \qquad (8)$$

E_T is determined by the following form[8]

$$E_T=2.5+150/A+\Delta, \qquad (9)$$

where Δ is pairing correction. The values Δ are taken from J.L.Cook's work.[9] The constant temperature factor has the following expression:

$$T=\frac{E_T-\Delta}{\sqrt{a(E_T-\Delta)}-1.5}\ , \qquad (10)$$

where a is the level density parameter.

In the practical calculation of $\langle\Gamma_\gamma\rangle$, the following expression[10] is used at $E\le E_T$:

$$\langle\Gamma_\gamma(E)\rangle = C_r T^4\left[6-e^{-E/T}\cdot F(E/T)\right]\ , \qquad (11)$$

where $F(x)=5+3x+(1+x)^3$. (12)

At $E/T<0.1$,

$$\langle\Gamma_\gamma(E)\rangle = C_r E^4/4\cdot(1-0.8x+\tfrac{1}{3}x-\tfrac{2}{21}x^3)\,. \qquad (13)$$

For $E>E_T$, $\langle\Gamma_\gamma(E)\rangle$ is calculated with the numerical integral method.

Firstly, the known experimental values $\langle\Gamma_\gamma^0\rangle$ are utilized to determine a sets of the coupling coefficients $C_r(\exp)$, and it is found that for even-even, even-odd, odd-even and odd-odd nucleus, the variation of the coupling coefficients C_r with mass number A is smooth (see Fig.3 and 4). Then the C_r from the smoothed curves are used to calculate the radiative widths $\langle\Gamma_\gamma^0\rangle$ in turn. There is an overall agreement between the experimental values and those calculated on the basis of Eq.(7) (see appendix), where the fluctuation for even-odd nucleus is rather large. Consequantly, the radiative widths of the nuclides which are deficient in the experimental data could be calculated from the smoothed values of C_r.

Two interesting results emerge from Fig.3 and Fig.4: (a) The maxima of C_r are presented at about the magic numbers, specially Z=20, 28 and 82. This is an evidence of the shell effect for C_r; (b) the coupling coefficients C_r of odd neutron nucleus are generally larger than those of even ones, which should be closely related to the nuclear structure effects.

Acknowledgment

The authours wish to thank Dr. Sun Zuxun for his kindly help.

References

(1) J.S.Liven and D.J.Hughes, Phys. Rev. 101, 1328 (1956).

(2) A.Stolovy and J.A.Harvey, Phys. Rev. 108, 353 (1957).

(3) H.Malecky, et al., Yad. Phys. 13, 240 (1971).

(4) S.F.Mughabghab, M.Divadeenam and N.E.Holden, BNL-325, Vol.1, (the Fourth Edition, 1981).

(5) The Centre of Nuclear Data of China, Neutron Resonance Theory and Neutron Resonance Parameters, (1981).

(6) J.E.Lynn, Theory of Neutron Resonance Reaction, (1968).

(7) A.Gilbert and A.G.W.Cameron, Can. J.Phys. 43, 1446 (1965).

(8) C.L.Dunford, AI-AEC-12931 (1970).

(9) J.L.Cook, Aust. J.Phys. 20, 477 (1967).

(10) Zhang Jingsang and Wang Shunuan, hsj-78227 (lljs) (1978).

Appendix

Table 1 Average S-wave radiative widths

Target nucleus	$\langle\Gamma_\gamma^0\rangle$(exp) (meV)	$\Delta\langle\Gamma_\gamma^0\rangle$(exp) (meV)	$\langle\Gamma_\gamma^0\rangle$(fit) (meV)	$\langle\Gamma_\gamma^0\rangle$(calc) (meV)	Cr(exp) $10^{-11}MeV^{-3}$	Cr(smooth) $10^{-11}MeV^{-3}$	References
11-Na-23	3400	1500	2200				(1)
16-S -32	1900	600	1200	1500	81.6	64.0	(2)
20-Ca-40	1500	900	800	1800	66.7	79.0	(2)
20-Ca-42	1100	200	800	1500	60.1	80.0	(2)
20-Ca-43	750	40	1010				(2)
20-Ca-44	1300	400	700	1100	93.2	81.0	(2)
21-Sc-45	840	460	750				(2)
22-Ti-46	1400	400	690	2470	45.3	80.0	(2)
22-Ti-47	1200	400	900				(2)
22-Ti-48	1400	400	700	1800	64.4	86.0	(2)
22-Ti-49	810	240	850				(2)
22-Ti-50	1100	300	700	700	145	95	(2)
24-Cr-50	1500	500	600	3600	39.2	95.0	(2)
24-Cr-52	1850	550	540	2820	91.7	140	(2)
24-Cr-53	2300	700	600				(2)
24-Cr-54	2530	700	450	1290	353	180	(1)
25-Mn-55	750	150	470				(2)
26-Fe-54	1800	500	500	720	44.9	180	(2)
26-Fe-56	900	300	400	420	50.9	24.0	(2)
26-Fe-57	1900	600	600				(2)
26-Fe-58	3000	900	400	2200	322	240	(2)
27-Co-59	560	100	410				(2)
28-Ni-58	2600	800	400	8600	72.4	24.0	(2)
28-Ni-60	1700	500	400	4000	84.9	200	(2)
28-Ni-61	2200	700	500	1800	30.5	25.0	(2)
28-Ni-62	910	270	360	1360	80.3	120	(2)
28-Ni-64	2400	700	300	420	347	61.0	(2)
29-Cu-63	500	150	371	570	23.3	27.0	(2)
29-Cu-65	395	120	344	319	29.7	24.0	(2)
30-Zn-64	726	60	346	1363	32.4	61.0	(2)
30-Zn-66	400	20	318	518	30.1	39.0	(2)
30-Zn-67	460	140	430	434	7.24	6.80	(2)
30-Zn-68	320	40	300	280	34.5	30.0	(2)
31-Ga-69	262	20	321	356	13.6	18.5	(2)
31-Ga-71	237	15	285	164	24.5	17.0	(2)
32-Ge-70	165	50	291	388	9.77	23.0	(2)
32-Ge-72	162	50	274	208	14.0	18.0	(2)
32-Ge-73	145	45	387	250	2.22	3.75	(2)
32-Ge-74	195	60	269	156	20.0	16.0	(2)
32-Ge-76	115	35	258	109	15.8	15.0	(2)
33-As-75	300	90	277	230	18.2	14.0	(2)
34-Se-74	280	85	275	387	11.6	16.0	(2)
34-Se-76	230	70	261	261	13.2	15.0	(2)
34-Se-77	390	120	351	226	5.22	3.03	(2)
34-Se-78	230	70	251	187	17.3	14.0	(2)
34-Se-80	230	70	250	150	20.2	13.5	(2)
35-Br-79	293	90	261	264	12.7	11.5	(2)
35-Br-81	300	100	300	200	15.3	11.0	(2)
36-Kr-78	230	70	250	410	7.87	14.0	(2)
36-Kr-80	230	70	240	310	10.0	13.5	(2)
36-Kr-82	230	70	240	240	12.5	13.0	(2)

Table 1 Average S-wave radiative widths (continue)

Target nucleus	$\langle\Gamma_\gamma^0\rangle$(exp) (meV)	$\Delta\langle\Gamma_\gamma^0\rangle$(exp) (meV)	$\langle\Gamma_\gamma^0\rangle$(fit) (meV)	$\langle\Gamma_\gamma^0\rangle$(calc) (meV)	Cr(exp) $10^{-11}MeV^{-3}$	Cr(smooth) $10^{-11}MeV^{-3}$	References
36-Kr-83	200	60	320	220	2.59	2.80	(2)
36-Kr-84	200	60	230	180	13.3	12.0	(2)
37-Rb-85	205	35	255	346	5.91	10.0	(2)
37-Rb-87	115	20	193	74	15.1	9.70	(2)
38-Sr-84	290	90	230	390	8.93	12.0	(2)
38-Sr-86	265	80	235	359	8.49	11.5	(2)
38-Sr-87	290	90	300	280	2.97	2.90	(2)
38-Sr-88	225	70	189	103	23.9	11.0	(2)
39-Y -89	130	40	189	123	9.95	9.40	(2)
40-Zr-90	240	70	190	180	14.9	11.0	(2)
40-Zr-91	140	40	217	115	3.99	3.30	(2)
40-Zr-92	140	40	178	128	11.5	10.5	(2)
40-Zr-94	130	40	174	103	12.6	10.0	(2)
41-Nb-93	165	50	179	163	9.92	9.80	(2)
41-Nb-94	188	60	205	330	5.70	10.0	(2)
42-Mo-92	190	60	183	278	7.18	10.5	(2)
42-Mo-94	140	20	172	181	7.72	10.0	(2)
42-Mo-95	160	20	207	158	3.53	3.50	(2)
42-Mo-96	110	15	163	119	8.42	9.10	(2)
42-Mo-97	130	20	199	129	3.62	3.60	(2)
42-Mo-98	85	10	147	68	12.0	9.50	(2)
42-Mo-100	90	10	137	45	19.0	9.30	(2)
43-Tc-99	160	50	160	140	12.6	11.0	(2)
44-Ru-99	195	20	197	207	3.38	3.60	(2)
44-Ru-100	120	15	148	120	9.18	9.30	(2)
44-Ru-101	180	15	191	166	3.79	3.50	(2)
44-Ru-102	90	10	139	82	10.0	9.10	(2)
44-Ru-104	85	15	127	64	11.9	8.90	(2)
45-Rh-103	160	15	146	157	10.7	10.5	(2)
46-Pd-105	145	8	180	172	2.60	3.10	(2)
46-Pd-108	77	5	127	73	8.91	8.40	(2)
46-Pd-110	60	10	120	50	9.24	8.20	(2)
47-Ag-107	140	40	137	167	7.89	9.40	(2)
47-Ag-109	130	40	132	129	9.67	8.40	(2)
48-Cd-106	155	15	139	217	6.05	8.50	(2)
48-Cd-108	105	10	132	154	5.60	8.40	(2)
48-Cd-110	71	6	128	122	4.73	8.20	(2)
48-Cd-111	96	20	164	142	1.83	2.70	(2)
48-Cd-112	77	5	123	93	6.74	8.10	(2)
48-Cd-113	160	20	160	120	3.55	2.60	(2)
48-Cd-114	53	4	118	70	6.07	8.05	(2)
48-Cd-116	47	4	113	53	7.07	8.00	(2)
49-In-113	75	5	128	141	4.15	7.80	(2)
49-In-115	77	4	122	102	5.72	7.60	(2)
50-Sn-112	110	40	127	191	4.67	8.10	(2)
50-Sn-114	90	30	126	171	4.25	8.05	(2)
50-Sn-117	80	20	152	134	1.55	2.60	(2)
51-Sb-121	100	20	110	90	7.23	6.80	(2)
51-Sb-123	100	20	110	80	8.98	6.80	(2)
52-Te-122	140	40	110	118	9.33	7.90	(2)
52-Te-123	124	20	143	142	2.27	2.60	(2)
52-Te-124	110	30	106	93	9.20	7.80	(2)
52-Te-125	157	30	140	125	3.28	2.60	(2)
53-I- 127	90	10	100	110	6.35	7.50	(2)
54-Xe-129	144	45	131	138	2.81	2.70	(2)
54-Xe-131	124	34	129	120	2.79	2.70	(2)
54-Xe-135	95	30	119	76	3.38	2.70	(2)
55-Cs-133	120	40	100	140	8.03	9.10	(2)
55-Cs-134	160	50	123	328	3.76	7.70	(2)
56-Ba-130	100	30	101	166	4.70	7.80	(2)
56-Ba-134	120	20	97	121	7.61	7.70	(2)
56-Ba-135	150	20	121	131	3.09	2.70	(2)
56-Ba-136	125	30	97	118	8.28	7.80	(2)
56-Ba-137	80	15	117	104	2.07	2.70	(2)
56-Ba-138	55	20	71	23	19.3	8.00	(2)
57-La-139	55	20	74	47	12.9	11.0	(2)
58-Ce-140	35	9	74	43	6.57	8.10	(2)
59-Pr-141	88	9	76	85	11.9	11.5	(3)
60-Nd-142	64	8	76	77	7.04	8.50	(2)
60-Nd-143	80	9	93	70	3.08	2.70	(2)
60-Nd-144	54	5	72	61	7.76	8.80	(2)
60-Nd-145	75	9	91	64	3.30	2.80	(2)
60-Nd-146	54	6	68	43	11.2	9.00	(2)
60-Nd-148	51	6	66	35	13.2	9.10	(2)
60-Nd-150	67	25	70	47	13.3	9.40	(2)
61-Pm-147	68	6	72	76	8.72	9.80	(4), (5)
62-Sm-147	65	20	90	84	2.10	2.70	(4)

Table 1 Average S-wave radiative widths (continue)

Target nucleus	$<\Gamma_\gamma^0>$(exp) (meV)	$\Delta<\Gamma_\gamma^0>$(exp) (meV)	$<\Gamma_\gamma^0>$(fit) (meV)	$<\Gamma_\gamma^0>$(calc) (meV)	Cr(exp) $10^{-11}MeV^{-3}$	Cr(smooth) $10^{-11}MeV^{-3}$	References
62-Sm-149	65	2	89	83	2.27	2.90	(1)
62-Sm-151	69	10	92	99	2.08	3.00	(1),(6)
62-Sm-152	65	20	70	73	8.48	9.50	(7)
62-Sm-154	79	16	70	70	10.7	9.50	(6)
63-Eu-151	92	12	71	93	8.68	9.00	(1),(6)
63-Eu-153	95	12	73	101	8.25	8.80	(1),(6)
64-Gd-152	54	7	70	110	4.54	9.50	(6)
64-Gd-154	65	21	70	110	5.61	9.50	(6)
64-Gd-155	108	28	89	118	2.83	3.10	(6)
64-Gd-156	82	12	70	100	7.48	9.50	(6)
64-Gd-157	91	22	84	88	3.47	3.10	(6)
64-Gd-158	90	13	84	250	3.19	9.20	(6)
64-Gd-160	98	15	64	58	15.1	9.00	(6)
65-Tb-159	97.0	7.5	68.1	90.5	8.67	8.10	(8)
66-Dy-160	108	30	66	106	9.20	9.00	(1)
66-Dy-161	112	35	81	101	3.45	3.10	(1)
66-Dy-162	112	35	65	93	10.8	8.90	(1)
66-Dy-163	109	6	77	76	4.45	3.10	(6)
66-Dy-164	65	10	61	60	9.35	8.60	(6)
67-Ho-165	80	23	63	80	7.81	7.81	(6)
68-Er-166	88	20	63	99	7.51	8.50	(6)
68-Er-167	91	12	73	84	3.47	3.20	(6)
68-Er-168	81	15	60	70	9.34	8.10	(6)
69-Tm-169	77	13	62	94	5.90	7.20	(9)
70-Yb-171	79	10	71	99	2.63	3.30	(6)
70-Yb-172	80	20	59	90	7.10	8.00	(6)
70-Yb-173	68	8	67	74	3.05	3.30	(6)
70-Yb-174	83	25	55	61	10.8	8.00	(7)
71-Lu-175	77	23	57	77	7.17	7.20	(7)
72-Hf-177	66	13	62	66	3.31	3.30	(4)
73-Ta-180	48.5	8.8	63.2	47.1	1.96	1.90	(10)
73-Ta-181	64.6	8.2	52.4	65.3	7.02	7.10	(11)
74-W -182	62	13	52	71	6.19	7.30	(4)
74-W -183	77	19	60	74	3.53	3.40	(4)
74-W- 184	69	15	49	54	9.36	7.20	(4)
74-W- 186	61	8	47	43	10.3	7.20	(4)
75-Re-185	57	7	50	68	5.68	6.80	(4)
75-Re-187	61	9	49	56	7.57	7.00	(4)
76-Os-189	95	9	60	95	3.50	3.50	(4)
77-Ir-191	81	11	48	75	7.91	7.30	(4)
77-Ir-193	94	11	47	75	10.1	8.00	(4)
78-Pt-192	50	15	47	81	4.72	8.00	(4)
78-Pt-194	69	30	46	87	7.08	9.00	(4)
78-Pt-195	119	33	58	117	4.07	4.00	(4)
79-Au-197	135	20	48	135	10.6	10.6	(4)
80-Hg-198	150	45	47	86	10.8	6.20	(4)
80-Hg-199	293	65	55	282	9.44	9.30	(4)
80-Hg-201	384	90	54	376	14.3	14.0	(4)
81-Tl-203	650	200	46	650	46.2	46.2	(12)
81-Tl-205	800	240	46	800	62.7	62.7	(12)
82-Pb-204	920	470	46	920	62.1	62.1	(4)
83-Bi-209	164	45	33	164	59.2	59.2	(13)
88-Ra-226	24.5	3.8	29.3	19.6	8.99	7.20	(14)
90-Th-230	27	3	31	30	5.86	6.00	(6)
90-Th-232	21.2	6.4	29.2	21.2	6.26	6.00	(6)
91-Pa-231	40	2	32	47	6.06	7.20	(15)
91-Pa-233	56	10	31	35	11.4	7.00	(4)
92-U- 233	39.4	12.0	37.2	45.4	2.43	2.80	(1),(4)
92-U- 234	25	6	30	32	4.69	6.00	(4)
92-U- 235	35	10	36	35	2.61	2.61	(1)
92-U- 236	23.0	15.0	29.5	27.7	4.99	6.00	(1)
92-U- 238	24	7	28	21	6.92	6.10	(1)
93-Np-237	41	3	30	39	6.57	6.20	(16)
94-Pu-238	38.6	6.0	30.5	43.5	5.41	6.10	(16)
94-Pu-239	41.6	12.0	34.4	33.4	3.11	2.50	(1)
94-Pu-240	40.8	4.5	28.8	30.9	7.99	6.05	(1),(16)
94-Pu-241	30.8	2.6	33.6	28.8	2.68	2.50	(1),(16)
94-Pu-242	25.4	3.7	28.0	25.9	5.92	6.05	(1),(16)
95-Am-241	44	9	29	43	6.80	6.60	(1)
95-Am-242	50	15	33	49	4.15	4.10	(1)
95-Am-243	39.3	4.8	28.7	37.8	6.94	6.60	(16)
96-Cm-243	40	34					(1)
96-Cm-244	36.2	8.4	28.7	38.6	5.02	6.00	(16)
96-Cm-245	40	12	33				(4)
96-Cm-246	31	6	27	29	6.49	6.00	(16)
96-Cm-247	40	12	32				(1)
96-Cm-248	26	2	25				(16)

Table 1 Average S-wave radiative widths (continue)

Target nucleus	$\langle\Gamma_\gamma^0\rangle$(exp) (meV)	$\Delta\langle\Gamma_\gamma^0\rangle$(exp) (meV)	$\langle\Gamma_\gamma^0\rangle$(fit) (meV)	$\langle\Gamma_\gamma^0\rangle$(calc) (meV)	Cr(exp) $10^{-11}MeV^{-3}$	Cr(smooth) $10^{-11}MeV^{-3}$	References
97-Bk-249	35.7	2.0	39.1				(16)
98-Cf-249	40	12	32				(1)

Table 2 Average P-wave radiative widths

Target nucleus	$\langle\Gamma_\gamma^1\rangle$(exp) (meV)	$\Delta\langle\Gamma_\gamma^1\rangle$(exp) (meV)	$\langle\Gamma_\gamma^1\rangle$(fit) (meV)	References
11-Na-23	5100	2600	900	(1)
16-S -32	1900	600	800	(2)
20-Ca-40	360	90	580	(2)
20-Ca-42	450	150	490	(2)
20-Ca-44	300	100	410	(2)
21-Sc-45	500	300	500	(2)
22-Ti-46	440	130	480	(2)
22-Ti-48	400	100	400	(2)
22-Ti-49	400	120	510	(2)
22-Ti-50	280	80	290	(2)
24-Cr-50	370	110	460	(2)
24-Cr-52	310	90	360	(2)
24-Cr-54	330	80	260	(1)
25-Mn-55	400	100	300	(2)
26-Fe-54	400	100	420	(2)
26-Fe-56	300	100	300	(2)
26-Fe-57	750	230	400	(2)
26-Fe-58	400	100	300	(2)
28-Ni-60	900	300	300	(2)
29-Cu-63	260	80	282	(2)
29-Cu-65	370	110	234	(2)
30-Zn-64	272	30	285	(2)
30-Zn-66	190	60	235	(2)
30-Zn-68	170	20	200	(2)
38-Sr-88	721	210	139	(2)
39-Y -89	300	90	150	(2)
40-Zr-90	440	150	160	(2)
40-Zr-91	240	70	182	(2)
40-Zr-92	360	110	140	(2)
40-Zr-94	185	60	127	(2)
41-Nb-93	190	60	150	(2)
42-Mo-92	275	85	177	(2)
42-Mo-94	200	30	154	(2)
42-Mo-96	140	20	135	(2)
42-Mo-98	145	20	112	(2)
42-Mo-100	100	10	98	(2)
44-Ru-100	150	30	129	(2)
48-Cd-106	175	25	144	(2)
48-Cd-108	125	20	127	(2)
48-Cd-110	80	15	116	(2)
48-Cd-112	90	20	100	(2)
48-Cd-114	70	10	90	(2)
48-Cd-116	70	10	80	(2)
54-Xe-135	95	30	91	(2)
57-La-139	40	7	60	(2)
58-Ce-140	30	6	63	(2)
60-Nd-142	44	5	71	(2)
60-Nd-146	23	7	58	(2)
62-Sm-149	65	2	84	(1)
82-Pb-207	145	2	48	(17)
83-Bi-209	34	3	31	(13)
90-Th-232	21.2	6.4	26.9	(6)
92-U -233	39.4	12.0	38.5	(1)(4)
92-U -235	35	10	36	(1)
92-U -238	24	7	26	(1)
94-Pu-239	41.6	12.0	35.5	(1)

References

(1) The Centre of Nuclear Data of China, Neutron Resonance Theory and Neutron Resonance Parameters, (1981).

(2) S.F.Mughabghab, M.Divadeenam and N.E.Holden, BNL-325, Vol.1, (the Fourth Edition, 1981).

(3) A.R.deL.Musgrove, et al., Neutron Phys. and Nuclear Data for Reactors and Other Applied Purposes, P.449, Harwell, 1978.

(4) S.F.Mughabghab and D.I.Garber, BNL-325 (the Third Edition, 1973).

(5) G.J.Kirouac and H.M.Eiland, Nucl. Sci. Eng., 52, 310 (1973).

(6) V.Benzi, Neutron Phys. and Nuclear Data for Reactors and Other Applied Purposes, P.288, Harwell, 1978.

(7) G.Hacken and H.I.Liout, NBS-SP-425 (1975).

(8) M.Mizumoto, et al., Phys. Rev.,C17, 522 (1978).

(9) S.Joly and D.M.Drake, Phys. Rev.,C20, 2072 (1979).

(10) J.A.Harvey and N.W.Hill, ORNL-5025 (1974).

(11) T.S.Belanova, et al., Atom. Energ.,38, 430 (1975).

(12) H.L.Liou, et al., Phys. Rev., C12, 102 (1975).

(13) R.Macklin and J.Malperin, Phys. Rev., C14, 1389 (1976).

(14) S.M.Kalebin, et al., Yad. Phys.,14, 22 (1971).

(15) A.R.Z.Hussein, J.A.Harvey, et al., Nucl. Sci. Eng., 78, 370 (1981).

(16) M.S.Moore, Neutron Phys. and Nuclear Data for Reactors and Other Applied Purposes, P.313, Harwell, 1978.

(17) S.Raman, et al., Phys. Rev. Lett.,40, 1306 (1978).

SIMULTANEOUS EVALUATION OF NEUTRON CROSS SECTION AND THEIR COVARIANCES FOR SOME REACTION OF HEAVY NUCLEI

Y. Uenohara and Y. Kanda

Department of Energy Conversion
Graduate School of Engineering Sciences
Kyushu University
33 Sakamoto Kashuga-shi, Fukuoka-Ken
Japan

We have developed the simultaneous evaluation method and been applied to evaluate the cross sections and their covariances of $^{235}U(n,f)$, $^{238}U(n,f)$, $^{238}U(n,\gamma)$, $^{249}Pu(n,f)$ and $^{197}Au(n,\gamma)$.

In this method, the excitation functions of cross sections and cross section ratios have been represented by B-spline functions and the logarithms of data have been taken in order to linearlize them and to use the linear least square formulae including a weight matrix deduced from experimental informations.

We have shown that the need for a large memory can be eliminated in the calculation if a sufficiency which is a concept defined in Statistics is applied.

[$^{235}U(n,f)$, $^{238}U(n,f)$, $^{238}U(n,\gamma)$, $^{239}Pu(n,f)$, $^{197}Au(n,\gamma)$, covariance, simultaneous evaluation]

Introduction

Cross sections used in applications are in most cases absolute values. The majority of neutron cross sections, however, have been measured relative to ones for orher reactions. These relative values have to be fully utilized in evaluation works. We distinguish the experiments into three types, absolute, ratio, and shape measurements. The absolute measurements include the ratio measurements taken relatively by using well-known standard cross sections for H(n,n), $^{7}Li(n,\alpha)$, and $^{10}B(n,\alpha)$, which have been measured and evaluated more reliably than others. The ratio measurements are the experiments performed relative to other reaction cross sections. The shape measurements mean that in a excitation function measurements the ratios are taken relative to the value at a given energy point on itself. We call hereafter some times both ratio and shape measurements as relative measurements.

In usual evaluations of neutron cross sections, the data of the relative measurements have been renormalized with the values which evaluators adopt as standards in their works. It is possible that even if the evaluated cross sections for individual reactions agree well with experimental data their mutual ratios disagree with the corresponding relative measurements. The cross section ratios can be measured more acculately than absolute ones. It is reasonable that the evaluations are performed by using the original values obtained in the experiments. A simultaneous evaluation developed in the present work make it possible. The covariances for evaluated cross sections are recently interested both in evaluation and application fields. We have obtained them as well as the cross sections and their ratios, in the present work.

B-spline functions are used to present the excitation function and the logarithms of the data are taken to linearize the relations between the absolute and relative values. The linear least square method is used to search for the optimum solutions which are fitted to all kinds of experimental data.

We demonstrate to be able to evaluate the cross section of $\sigma_f(^{235}U)$, $\sigma_f(^{238}U)$, $\sigma_\gamma(^{238}U)$, $\sigma_f(^{239}Pu)$ and $\sigma_\gamma(^{197}Au)$ and their covariances using the absolute data[1-18] for them and the relative data[19-30] of $\sigma_f(^{238}U)/\sigma_f(^{235}U)$, $\sigma_\gamma(^{238}U)/\sigma_f(^{235}U)$, $\sigma_\gamma(^{238}U)/\sigma_\gamma(^{197}Au)$, $\sigma_f(^{239}Pu)/\sigma_f(^{235}U)$ and $\sigma_\gamma(^{197}Au)/\sigma_f(^{235}U)$. Non diagonal components of a weight matrix used in the least square method, the covariances of the experimental data, are estimated mainly from the errors of refered values used in the shape measurements.

Calculation methods

1. Numerical representation of the excitation function

The excitation function for fast neutrons cannot be represented with the simple functions or the formula based on available nuclear models. Spline functions are useful to represent them. They have been used for the nuclear data evaluations in some works. The well-known spline functions are cubic-spline functions, which were applied to evaluate the H(n,n) cross sections by Horsely.[31] The cubic-spline functions are not always appliciable to all kinds of reaction cross sections, since they oscilate possibly in case of evaluations of complicated excitation functions, when the spline functions are used in the nuclear data evaluations, it is desirable that the order of them can be easily selected because of avoiding abnormal oscilations. The B-spline functions are suitable for it.

They can be written by the following recurence formulae

$$M_{rj}(x)=\frac{(x-\xi_{j-r})M_{r-1,j-1}(x)+(\xi_j-x)M_{r-1,j}(x)}{(\xi_j-\xi_{j-r})} \quad (1)$$

$$M_{1j}(x)=(\xi_j-\xi_{j-1})^{-1} \quad (2)$$

$$N_{rj}(x)=M_{rj}(x)(\xi_j-\xi_{j-r}) \quad (3)$$

$M_{rj}(x)$; the j-th B-spline function of (r-1)-th order.

ξ_j; the j-th node

$N_{rj}(x)$; the normalized B-spline function

When r is 2, the normalized B-spline functions are "roof functions" introduced by Schmitroth in his evaluation works.[32]

The excitation function f(x) can be represented by using the normalized B-spline function.

$$f(x)=\sum_j N_{fj}(x)\theta_j \quad (4)$$

The spline functions of r-th order can be differentiated in r-1 times. The derivatives of the "roof functions", are not continuous. The excitation functions, however, are practically used to be integrated. Therefore, the B-spline functions of the first order are enough to express the evaluated excitation functions.

This is one of the reasons that we use them in the present work. The computer program can be easily changed to calculate with the higher order B-spline functions. We have tried to apply 2-nd order of them and obtained more smoothed curves as they are expected.

K. H. Böckhoff (ed.), Nuclear Data for Science and Technology, 639–642.

2. Simultaneous evaluation

2-1 Handling of the ratio measurement data

We took logarithms of the cross sections and cross section ratios in order to apply a linear liast square method and to evaluate them and their covariances. They are very useful since they are not negative. Their relative errors reduced from the experimental data are not so large in the present work that logarithms of the excitation functions are represented by

$$\ln(\sigma_i(E)) = B_k(E)\theta_k \quad (5)$$

$$\ln(\sigma_i(E)/\sigma_j(E)) = B_k(E)\theta_k^i - B_{k'}(E)\theta_{k'}^j \quad (6)$$

$\sigma_i(E)$; the excitation function for the reaction i

$B_k(E)$; the k-th B-spline function of the first order

θ_k^i; the parameter which decides the absolute value of the excitation function σ_i

Using the above equations, the relations between experimental data and parameters can be written by

$$\mathbf{y} = \begin{pmatrix} \mathbf{y}_1 \\ \mathbf{y}_2 \\ \mathbf{y}_3 \\ \mathbf{y}_4 \\ \mathbf{y}_5 \\ \mathbf{y}_6 \\ \mathbf{y}_7 \\ \mathbf{y}_8 \\ \mathbf{y}_9 \\ \mathbf{y}_{10} \end{pmatrix} = \begin{pmatrix} \Phi_{11} & 0 & 0 & 0 & 0 \\ 0 & \Phi_{22} & 0 & 0 & 0 \\ -\Phi_{31} & \Phi_{32} & 0 & 0 & 0 \\ 0 & 0 & \Phi_{43} & 0 & 0 \\ -\Phi_{51} & 0 & \Phi_{53} & 0 & 0 \\ 0 & 0 & \Phi_{63} & 0 & -\Phi_{65} \\ 0 & 0 & 0 & \Phi_{74} & 0 \\ -\Phi_{81} & 0 & 0 & \Phi_{84} & 0 \\ 0 & 0 & 0 & 0 & \Phi_{95} \\ -\Phi_{101} & 0 & 0 & 0 & \Phi_{105} \end{pmatrix} \begin{pmatrix} \theta_1 \\ \theta_2 \\ \theta_3 \\ \theta_4 \\ \theta_5 \end{pmatrix}$$

$$= \Phi\,\theta \quad (7)$$

$\mathbf{y}$; the experimental data vector

$\mathbf{y}_i$; the sub-experimental data vector

Φ_i; the sub-design matrix

θ; the parameter vector for the set of the cross sections to be evaluated simultaneously

θ_j; the sub parameter vector

i; the experimental data identification number given in Table I.

j; the evaluated cross section identification number given in Table II.

The least square estimators of the cross sections become

$$\theta = (\Phi^T W \Phi)^{-1}(\Phi^T W \mathbf{y})) \quad (8)$$

and estimated covariance matrix is

$$M = (\Phi^T W \Phi)^{-1} \hat{s}^2 \quad (9)$$

$$\hat{s}^2 = \frac{(\mathbf{y} - \Phi\theta)^T W (\mathbf{y} - \Phi\theta)}{m - p} \quad (10)$$

$$W = V^{-1} \quad (11)$$

V; the covariance matrix of the used experimental data

2.2 Handling of the shape measurement data

The shape measurement data are probably more accurate than the other kinds of data with respect to the shape of the excitation function.

The shape measurement data σ_i and σ_j, correlate strongly. Their data can be represented by

$$\sigma_i = R_i\,\sigma_0 \quad (12)$$

$$\sigma_j = R_j\,\sigma_0 \quad (13)$$

σ_i; the data measured at the i-th energy point

R_i; the shape data at the i-th energy point

σ_0; the standard value

The variances and covariances of the shape measurement data are written by

$$\frac{\langle \Delta^2\sigma_i \rangle}{\sigma_i^2} = \frac{\langle \Delta^2 R_i \rangle}{R_i^2} + \frac{\langle \Delta^2\sigma_0 \rangle}{\sigma_0^2} \quad (14)$$

$$\frac{\langle \Delta\sigma_i\,\Delta\sigma_j \rangle}{\sigma_i\,\sigma_j} = \frac{\langle \Delta^2\sigma_0 \rangle}{\sigma_0^2} \quad (15)$$

We assume that the shape data are not correlated.

$\langle \Delta^2\sigma_0 \rangle$; the variance of the standard value: square of normalization error

2.3. Sufficiency[33]

The cross sections repesented by Eq. (8) are estimator in linear least square equation. Therefore they satisfy with a sufficiency which is a concept defined in Statistics. It is practicaly useful character. When least square method are applied to the problems including correlations, executions are limited with the size of memories of a computer. If the estimator is sufficient, it eliminate the need for a large memory. Eq.(8) can be rewritten as following equations, if a group of the data sets is not correlating with another group mutually. The group is identified by a suffix i.

$$\hat{\theta} = \left(\sum_i \Phi_i^T W_i \Phi_i\right)^{-1}\left(\sum_i \Phi_i^T W_i \mathbf{y}_i\right) \quad (16)$$

$$M = \left(\sum_i \Phi_i^T W_i \Phi_i\right)^{-1} \hat{s}^2 \quad (17)$$

Table II. The reaction identification number

j	1	2	3	4	5
Type	$\sigma_f(^{235}U)$	$\sigma_f(^{238}U)$	$\sigma_\gamma(^{238}U)$	$\sigma_f(^{239}Pu)$	$\sigma_\gamma(^{197}Au)$

Table I. The identification numbers for the kinds of used experimental data

i	1	2	3	4	5	6	7	8	9	10
Type	$\sigma_f(^{235}U)$	$\sigma_f(^{238}U)$	$\frac{\sigma_f(^{238}U)}{\sigma_f(^{235}U)}$	$\sigma_\gamma(^{238}U)$	$\frac{\sigma_\gamma(^{238}U)}{\sigma_f(^{235}U)}$	$\frac{\sigma_\gamma(^{238}U)}{\sigma_\gamma(^{197}Au)}$	$\sigma_f(^{239}Pu)$	$\frac{\sigma_f(^{239}Pu)}{\sigma_f(^{235}U)}$	$\sigma_\gamma(^{197}Au)$	$\frac{\sigma_\gamma(^{197}Au)}{\sigma_f(^{235}U)}$

$$W = \begin{pmatrix} W_1 & & & & \\ & \cdot & & 0 & \\ & & W_i & & \\ & 0 & & \cdot & \\ & & & & W_n \end{pmatrix} \tag{18}$$

$$\mathbf{y} = \begin{pmatrix} \mathbf{y}_1 \\ \cdot \\ \cdot \\ \mathbf{y}_i \\ \cdot \\ \cdot \\ \mathbf{y}_n \end{pmatrix} = \begin{pmatrix} \Phi_1 \\ \cdot \\ \cdot \\ \Phi_i \\ \cdot \\ \cdot \\ \Phi_n \end{pmatrix} \boldsymbol{\theta} \tag{19}$$

The sufficiency is defined by Eqs.(16) and (17). It can be found in these equations that calculations are able to be proceeded independently at every data set. Therefore, the size of the memory demanded in the calculations are reduced drasticaly. In the present work, amount of experimental data input is about 10^3. If the calculations are performed straightforwards the memories of about 10^6 are occupied by a W-matrix. The maximum number of the correlated data, however, is about 2×10^2 and the memories needed for the matrix are 4×10^4.

Results and discussions

We have demonstraited applicability of the simultaneous evaluation method to determinning the cross sections for five reactions, ^{235}U(n,f), ^{238}U(n,f), ^{238}U(n, γ), ^{239}Pu(n,f) and ^{197}Au(n,γ) consistently with their ratio data as well. The typical results are shown in Fig. 1 through Fig. 3 . The results compairing with the values of JENDL-2 and ENDF/B-V are also shown in Fig. 4 .

A part of the evaluated covariances is listed in Table III . The presented values are not final ones because of lack of examinations for experimental data.

Applications of the method are restricted to the cross sections varring moderately and being many available data unbiasedly over the energy range. One of key points in these calculations is a setting of nodes for the spline function. A structure at a broad peak near 1 MeV in ^{238}U(n,γ) shown in Fig. (2) comes probably from interference between the setting of nodes and numbers of data points in this energy range. If a trial curve is assumed as an initial guess this difficulty can be eliminated. It can be executted in the present work with minor changes of the program.

In the present work, the correlations in the shape measurements are assumed to be strong. It is found from the present studies that the variances and covariances for the evaluated results are sensitive to the correlations in experimental data. If correlations in the experimental data are not taken into account the variances decrease gradually as the number of data points increase. It comes from the law of large numbers in Statistics. The experimental data measured by Time-of-Flight method have a large number of data points. They are, however, very strongly correlated. We have to estimate reasonable values of non-diagonal components in the weight matrix.

Table III. A part of correlation matrix for ^{238}U(n,f)

E (MeV)									
1.80E+00	100								
1.70E+00	37	100							
1.60E+00	49	38	100						
1.50E+00	34	48	40	100					
1.40E+00	33	38	50	30	100				
1.30E+00	24	29	30	35	10	100			
1.20E+00	21	23	24	21	27	-1	100		
1.10E+00	18	18	18	16	16	21	-7	100	
1.00E+00	12	12	10	7	11	9	19	-11	100

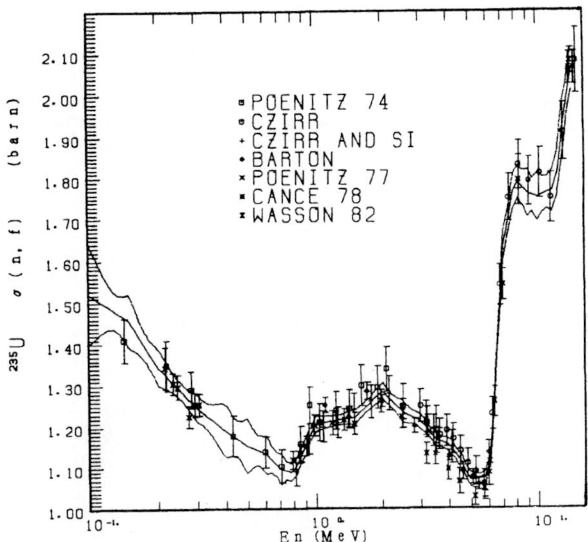

Fig. 1. The evaluated ^{235}U(n,f) cross sections and their standard deviations over the energy ranges from 100 keV to 15 MeV. A part of the used experimental data are shown to avoid cluttering.

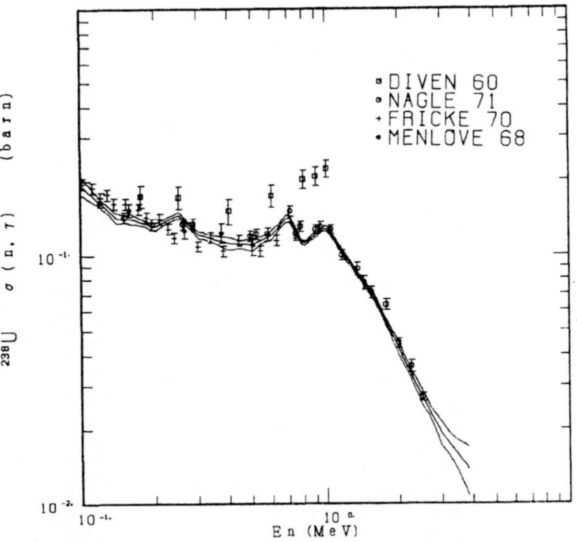

Fig. 2. The evaluated ^{238}U(n,γ) cross sections and their standard deviations over the energy ranges from 100 keV to 4 MeV.

In the evaluated covariance matrix, negative values are frequently obtained mainly near the diagonal line, it is proved mathematically that they are resulted from the application of the first order B-spline function to represent the data. There are not any physical meening. These negative values disappear as increasing correlations in the experimental data.

In the present method, uncertainties for neutron energies are not taken into account. It is possible that they are reduced to the errors for cross sections. We do notagree with this method because it is not valid to give correctly the errors for the neutron energies of peaks. We must develop an useful method in which the errors for both ordinates and abscissa can be built in. Otherwise, the systematic errors in the measurements of neutron energies by Time-of-Flight method can not be introduced in the evaluations.

When new experiments are reported after evaluating the cross sections based on the available data at the time, we can easily revise the previous evaluated results using the new ones. Sufficiency described in the subsections 2. 3. assures that the results obtained from both previous evaluated values and new data are equivalent to ones from all available data.

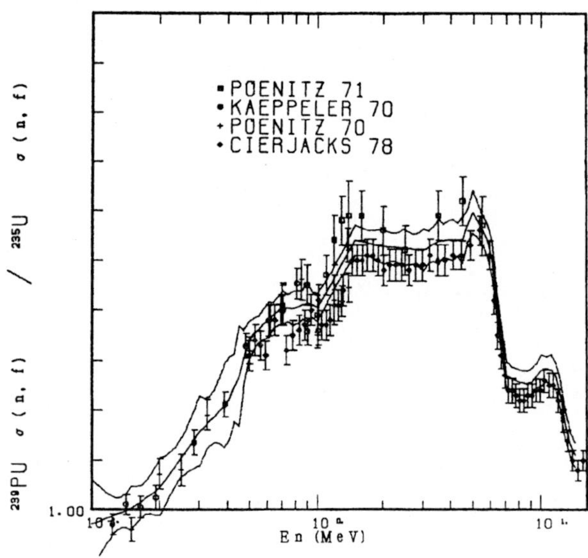

Fig. 3. The evaluated cross section ratios $^{239}Pu(n,f)$ to $^{235}U(n,f)$ and their standard deviations over the energy ranges from 100 keV to 15 Mev. A part of the used data are shown to avoid cluttering.

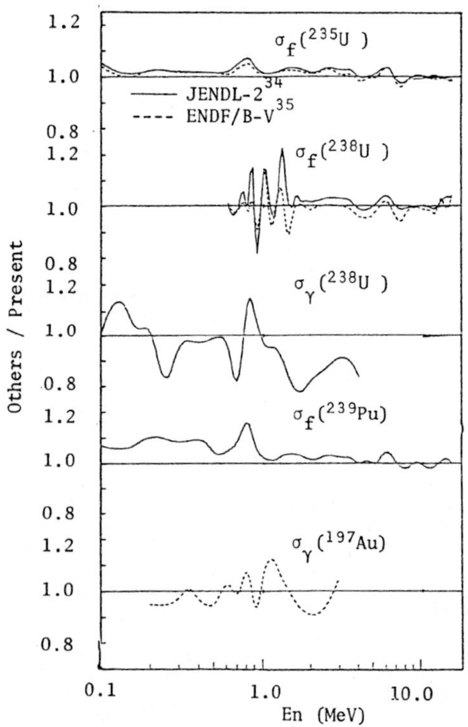

Fig. 4. The present results are compaired with JENDL-2 and ENDF/B-V.

References

1, W. P. Poenitz, Nucl. Sci. Eng. , 53 370 (1974)

2, J. B. Czirr and G. S. Sidhu, Nucl. Sci. Eng. , 57 18 (1975)

3, J. B. Czirr and G. S. Sidhu, Nucl. Sci. Eng. , 58 371 (1975)

4, D. H. Barton et al., Nucl. Sci. Eng. , 60 369 (1976)

5, W. P. Poenitz, Nucl. Sci. Eng. , 64 894 (1977)

6, M. Cance and G. Grenier, Nucl. Sci. Eng. , 68 197 (1978)

7, R. Leugers et al., "The U-235 and U-238 Neutron Induced Fission Cross Sections Relative to the H(n,n) Cross Sections", ANL-76-90 246 (1976)

8, R. J. Nagel et al., Proc. Knoxville Conf. , 259 (1971)

9, B. C. Diven et al., Phys. Rev. , 120 556 (1960)

10, M. P. Fricke, HELSI-70 15 (1970)

11, H. O. Menlove and W. P. Poenitz, Nucl. Sci. Eng. , 33 24 (1968)

12, M. C. Davis et al., "Absolute Measurements of U-235 and Pu-235 Fission Cross Sections with Photoneutron Sources", ANL-76-90 225 (1976)

13, I. Szabo and G. P. Marquette, "Measurement of the Neutron Induced Fisson Cross Section of Uranium-235 and Plutonium-239 in the MeV Energy Range", ANL-76-90 208 (1976)

14, J. B. Czirr, Nucl. Sci. Eng. , 52 299 (1973)

15, B. C. Diven et al., Phys. Rev. , 120 556 (1960)

16, W. P. Poenitz, Nucl. Sci. Eng. , 57 300 (1975)

17, Robertson, J. Nucl. Energ. , 23 305 (1968)

18, Poeto, J. Nucl. Energ. , 21 797 (1967)

19, J. W. Meadows, Nucl. Sci. Eng. , 58 255 (1975)

20, J. W. Behrens and G. W. Carlson, Nucl. Sci. Eng. , 63 250 (1977)

21, F. G. Difilippo, Nucl. Sci. Eng. , 68 197 (1978)

22, M. Lindner, Nucl. Sci. Eng. , 59 381 (1976)

23, R. R. Spencer and F. Kaeppeler, WASH-75 620 (1975)

24, Yu. G. Panitkin and V. A. Tolstikov, HELSI-70 57 (1970)

25, Yu. G. Panitkin and V. A. Tolstikov, Atomnaya Energiya 33 825 (1972)

26, W. P. Poenitz et al., Nucl. Sci. Eng. , 78 239 (1981)

27, W. P. Poenitz, Nucl. Sci. Eng. , 57 300 (1975)

28, F. Kaeppeler, Nucl. Sci. Eng. , 40 375 (1976)

29, W. P. Poenitz, Nucl. Sci. Eng. , 47 228 (1972)

30, S. Cierjacks et al., NEANDC(E) 182 5 (1977)

31, A. Horsley et al., Nucl. Inst. Methd. , 62 29 (1968)

32, F. Schmitroth and R. E. Schenter, Nucl. Sci. Eng., 74 168 (1980)

33, B. R. Martine, "Statistics for Physicists", Academic Press Inc. , (1971)

34, Private Comunications, Japanese Nuclear Data Committee , (1982)

35, INDC, Nuclear Standard File , NEANDC , (1981)

EVALUATION OF ACTIVATION CROSS SECTIONS OF CORROSION PRODUCTS, COVER-GAS NUCLIDES AND OTHER NUCLIDES IN THE PRIMARY COOLING CIRCUIT OF A FAST POWER REACTOR

H. Gruppelaar and H.A.J. van der Kamp

Netherlands Energy Research Foundation (ECN)
P.O. Box 1, 1755 ZG Petten
The Netherlands

Neutron activation and transmutation cross sections of nuclides in the primary cooling circuit of a fast breeder reactor have been evaluated for the purpose of radioactive-inventory calculations of the Na coolant. These evaluations were performed to supplement existing data files obtained from KEDAK-3 and ENDF/B-IV. The work has been completed for the following nuclides: ^{22}Na (produced from the ^{23}Na(n,2n) reaction), ^{36}Ar, ^{38}Ar, ^{40}Ar (cover-gas), ^{50}Cr, ^{54}Fe, ^{58g}Co, ^{58m}Co, ^{58}Ni, ^{62}Ni, ^{64}Ni (corrosion products) and ^{64}Zn, ^{112}Sn (contaminants). For these nuclides all important neutron cross sections have been evaluated in the energy range from 1 meV to 20 MeV (RCN-2 evaluation).

Introduction

In the primary cooling circuit of a fast-power reactor the sodium is contaminated with various nuclides, that are activated in the reactor core and vessel. In order to be able to estimate the radioactive inventory of the coolant one needs to know the activation cross sections of all the nuclides in the sodium. The following categories of materials may be distinguished: (a) sodium, (b) argon cover-gas, (c) stainless-steel corrosion products, (d) impurities and contaminants. The activation cross sections of some of these materials are available from well-known evaluated data libraries, e.g. from the ENDF/B-V dosimetry file. However, a number of other reactions have been studied less extensively. This work fulfills a request (Table I) to supplement the KEDAK-3 and ENDF/B-IV data files with evaluations for the above-mentioned purpose. Table I also contains requests for some radioactive nuclides (^{22}Na, ^{58g}Co, ^{58m}Co) of which the absorption cross sections are expected to be very high at low neutron energies. By neutron capture in the reactor vessel part of these nuclides will transmute into stable isotopes, thus reducing the originally produced radioactivity.

Table I Evaluation requests [a)]

Nuclide [b)]	Important reaction [c)]	Reaction product
^{22}Na (2.6a)	absorption	stable
^{38}Ar	(n,γ)	^{39}Ar (269a)
^{40}Ar	(n,γ)	^{41}Ar (1.83h)
^{50}Cr	(n,γ)	^{51}Cr (27.7d)
	(n,np)	^{49}V (331d)
^{54}Fe	(n,γ)	^{55}Fe (2.7a)
^{58m}Co (9.1h)	absorption	stable
^{58}Ni	(n,np)	^{57}Co (271d)
^{62}Ni	(n,γ)	^{63}Ni (100d)
^{64}Ni	(n,γ)	^{65}Ni (2.52h)
^{64}Zn	(n,γ)	^{65}Zn (244d)
^{112}Sn	(n,γ)	^{113}Sn (115d)

a) DeBeNe cooperation on fast power-reactor development.

b) We have also considered ^{36}Ar and ^{58g}Co (71.3d) cross sections.

c) The evaluation has been performed for *all* possible neutron cross sections at 1 meV to 20 MeV.

This paper describes the evaluation of all possible neutron cross sections for the target nuclides mentioned in Table I at energies from 1 meV to 20 MeV, with emphasis on the important activation and absorption reactions. The general evaluation procedure is outlined in the next sections, followed by more specific comments on the corrosion products, the cover-gas isotopes and the radioactive nuclides. The evaluation procedure for the contaminants ^{64}Zn and ^{112}Sn is the same as the one followed for the corrosion products and the fission products (Ref. 1), respectively. The last section contains a summary and some conclusions.

Evaluation Method

In this section a short survey is given of the adopted evaluation methods. Many of these methods are essentially the same as those used for the evaluation of fission-product cross sections [1]. However, since we are dealing with an other mass range some modifications had to be introduced, which are discussed below.

Resolved Resonance Range

For most nuclides considered in this paper the resonance structure in the cross sections extends to quite high energies. Therefore the resolved resonance range requires special attention in the evaluation. For the corrosion products of the important structural materials resonance parameters have been recently evaluated by Fröhner [2], who has also calculated pointwise given cross sections for σ_{el}, $\sigma_{n\gamma}$ and σ_t up to 300 keV using the multi-level Reich-Moore formula for the description of s-wave resonances. The data were Doppler broadened (by Fröhner) to represent cross sections at T=900 K. We have adopted these data without modifications for ^{50}Cr, ^{54}Fe, ^{58}Ni, ^{62}Ni and ^{64}Ni.

For the nuclides ^{22}Na, ^{36}Ar, ^{38}Ar and ^{58}Co no resolved resonance parameters are available. In these cases we have fitted one positive resonance and, if necessary, a negative resonance, to obtain agreement with experimental thermal cross sections and the resonance absorption integral I_a. In this process the contribution of distant resonances to the thermal absorption cross section σ_a and resonance integral was also included [3]. In most cases the capture widths of the fictitious resonances were obtained from an average value predicted by systematics, whereas the neutron width and the resonance energy were varied to obtain a good fit with the available data. For the radioactive nuclides ^{22}Na, ^{58g}Co and ^{58m}Co hypothetical resonances with energies above 0.5 eV were fitted to reach the observed high values of σ_a and I_a.

For the remaining nuclides (^{40}Ar, ^{64}Zn and ^{112}Sn) some resonance parameters were available [4] and the multilevel Breit-Wigner formula was used to calculate the cross sections. In the most recent BNL-compilation [5] many new parameters for ^{64}Zn have been included, which were not available at the time this evaluation was performed.

Optical model

At higher energies spherical optical-model calculations were performed using a set of model parameters, fitting the available data of the natural elements according to the "SPRT" approach [6]. In this process various recommended "local" and "global" sets of parameters were utilized, occasionally with slightly different values. In most cases satisfactory fits with σ_t and σ_{el} were obtained over a wide energy range. For ^{22}Na the optical model used by Larson [7] for natural Na was selected. For most of the corrosion products (except for ^{50}Cr) the potential of Fort and Lafond [8], fitted to the nickel isotopes, was used. The generalized optical model of Wiedling et al. [9] was adopted for

K. H. Böckhoff (ed.), Nuclear Data for Science and Technology, 643–650.

the Ar isotopes, ^{50}Cr and ^{64}Zn. The cross sections for ^{112}Sn were calculated using the same optical model as used to fit the Pd data [10]. For the Ar-isotopes, ^{64}Zn and ^{112}Sn the s- and p-wave neutron transmission coefficients calculated at low energies were replaced by those derived from s- and p-wave strength functions obtained from experimental data or systematics [1]. The proton and alpha potentials were taken from the global sets of Becchetti and Greenlees [11] and Igo and Huizenga [12], respectively.

Statistical Model

The calculated neutron transmission coefficients were used in Hauser-Feshbach model calculations, taking into account width-fluctuation corrections. For these calculations a quite flexible modular code system was used. The fast main model-code [1] calculates σ_t, σ_{el}, $\sigma_{nn'}$, σ_{n2n} and $\sigma_{n\gamma}$ by taking into account the competition of a given "residual" cross section σ_r, which may consist of σ_{np}, $\sigma_{n\alpha}$, σ_{npn}, σ_{n3n}, etc. This competition is effectuated by multiplying the calculated reaction cross sections with the factor $(\sigma_{cf}-\sigma_r)/\sigma_{cf}$, where σ_{cf} is the compound-formation cross section. This procedure works quite well, as was checked by performing calculations with the more rigorous codes CERBERO [13] and ERINNI [14] for neutron-induced reactions on ^{59}Co.

An advantage of this method is that we may adjust the charged-particle emission cross sections individually and still obtain a consistent set of reaction cross sections. Therefore we adopted as standard procedure to calculate first of all cross sections with the CERBERO and ERINNI codes and subsequently to adjust or to renormalize the individual charged-particle emission cross sections using evaluated experimental data, if necessary. In this process the experimental 14 MeV data and the recommended values of Qaim [15] were very useful. Checks were also made with 14 MeV systematics [16-18]. The (n,d) cross sections were supplied by a THRES-2 calculation [19], renormalized to 14 MeV data.[16,17] Next, the adjusted values of σ_r were introduced into our main model code to calculate the neutron scattering and radiative capture cross sections. If necessary, the ratio between $\sigma_{nn'}$ and σ_{n2n} was modified to fit available experimental data. Finally, the value of the average total radiation width at the neutron binding energy $\langle\Gamma_\gamma\rangle$ was adjusted to obtain the best possible agreement with radiative capture data. Since the neutron optical-model parameters were already fitted to agree with σ_t and σ_{el} this procedure usually led to satisfactory agreement with all available data in a consistent way.

The above-mentioned method allows for a large flexibility in the evaluation by combining model calculations with experimental data in a consistent way with the possibility to correct for deficiencies of the statistical model which often has to be used beyond its limits of validity, i.e.for very light nuclides or at too low energies. Direct and precompound effects were not considered explicitly, though the above-mentioned renormalizations may partly compensate for this neglect. In one case (for ^{64}Zn) we have added a direct component to $\sigma_{nn'}$ for scattering to the first-excited state, based upon experimental evidence.

We finally add some notes on the evaluation of the important (n,γ) activation cross sections, for which we have followed the methods as described in Ref. 1. The important parameters, characterizing $\sigma_{n\gamma}$ are the mean s-wave level spacing D_o, and the average total radiative widths $\langle\Gamma_\gamma\rangle_\ell$ at the neutron binding energy. As input to our code we may give different normalizing values for $\langle\Gamma_\gamma\rangle_\ell$ at even and odd ℓ values; in addition a valency component is added. In the mass range of the corrosion products s-wave valency effects are large, complicating the experimental estimate of the statistical s-wave component, which should be comparable to the d-wave component. Since the experimental uncertainties in $\langle\Gamma_\gamma\rangle$ for s-, p- and d-wave components are usually quite large, adjustments of $\langle\Gamma_\gamma\rangle$ were made to fit available (n,γ) cross sections in the MeV range. In absence of these data occasionally we have fitted $\langle\Gamma_\gamma\rangle$ to obtain agreement between the calculated and averaged experimental cross sections in the resolved resonance range. We finally note that the valency component for s-waves has a marginal influence on the calculated cross section at energies in the MeV range.

Corrosion products

Above the resolved resonance range, which extends to about 300 keV for ^{50}Cr, ^{54}Fe, ^{58}Ni, ^{62}Ni and ^{64}Ni, optical-model [8,9] and statistical-model calculations were performed. In Tables II and III the important parameters D_o and $\langle\Gamma_\gamma\rangle$ are listed.

Throughout this work we have used for the level density $\rho(J,\pi,E)$ the composite Gilbert-Cameron [20] formula with a modified expression for the spin cut-off parameter [1]. The evaluated experimental s-wave level spacings <D> of Fröhner [2] were used to derive the level-density parameters a, according to the expression

$$\frac{1}{\langle D\rangle} = \frac{1}{\Delta E} \sum_{J\pi} \int_{B}^{B+\Delta E} \rho(J,\pi,E)dE,$$

where ρ is a function of a. It has been assumed that <D> is obtained from resonances measured in an interval ΔE above the neutron binding energy B. The integration over energy is needed to account for the energy dependence of ρ over the interval ΔE. The importance of this effect follows from Table II, where we have calculated the (reduced) values of the level spacings at the neutron binding energy from the derived a-parameters.

Table II S-wave level spacings for the corrosion products (in keV).

Nuclide	<D> a)	D_o b)
^{50}Cr	15.4±1.5	17.1
^{54}Fe	17.0±2.0	19.2
^{58}Ni	16.7±2.1	19.1
^{62}Ni	18.6±1.9	23.3
^{64}Ni	19.9±2.0	26.1

a) Experimental average s-wave level spacing, evaluated by Fröhner [2].

b) Reduced values at the neutron binding energy.

The systematics of a is given in Fig. 1 from which it follows that the data for the odd-mass compound nuclides are linear functions of N for values larger than the magic number 28. The lowest values of a occur for values of N (or Z) close to 28. Similar "local" systematics of a versus N have been found [1] at N > 50 or N > 82. For the even-mass compound nuclides the a-values are usually lower. This odd-even effect has also been observed [1] for the Pd-isotopes [10], and the Sm and Nd isotopes.

The large experimental s-wave radiative widths are difficult to evaluate, partly because of uncertainties in experimental corrections and partly because of non-statistical effects. Therefore there are large fluctuations in these data, complicating the assessment of average total radiation widths. The results of the evaluation of s- p- and d-wave widths by Fröhner [2] are given in Table III. The values adopted in our model calculations (also listed in Table III) were obtained in the following way. First, an s-wave valency component was calculated [21] or guessed (from the difference between experimental s- and d-wave widths). Next, first estimates of the statistical even-ℓ and odd-ℓ components were made from the experimental data. These data were used in the initial model-calculations of which the results were compared with experimental data in the resolved resonance range (averaged cross sections [22-24]) or in the MeV range [25] (only for ^{64}Ni). Finally, the values of the statistical components of $\langle\Gamma_\gamma\rangle$ were adjusted to fit these data, if necessary.

Table III Average radiation width for the corrosion products (in eV).

Nuclide	Type[a]	ℓ	Experimental[b]	Adopted
^{50}Cr	v	0	1.5±1.0	0.56[c]
	s	0		0.74[d]
	s	1	0.8	0.55[c]
	s	2	-	0.74[d]
^{54}Fe	v	0	1.38±1.0	2.1[c]
	s	0		0.5[d]
	s	1	0.5±0.25	0.5[d]
	s	2	1.0	0.5[d]
^{58}Ni	v	0	2.3	0.9[c]
	s	0		1.4
	s	1	0.5	0.5
	s	2	1.5	1.4
^{62}Ni	v	0	0.5	≈0.1
	s	0		0.23[d]
	s	1	0.3	0.17[d]
	s	2	0.4	0.23[d]
^{64}Ni	v	0	1.0	≈0.3
	s	0		0.2[f]
	s	1	0.4	0.2[f]
	s	2	0.7	0.2[f]

a) v = valency; s = statistical.

b) Evaluated by Fröhner [2].

c) Calculated according to Ref. 21.

d) Fitted to data in resolved resonance range or to group constants; see Figs. 2-4.

e) Our own analysis.

f) Fitted to experimental data in MeV-range [25]; see Fig. 5.

The results, which have been obtained with the capture widths given in the last column of Table III are shown in Figs. 2 to 5. The histograms represent averaged cross sections, taken from recent measurements [22,23] or group constants derived from Fröhner's evaluation [2]. The calculated cross sections are indicated by full curves above 300 keV and by dotted or dotted-dashed curves at lower energies. We have also made a comparison with the JENDL-1 evaluations (dashed curves) which appear to be in good agreement with the present calculations, except for ^{64}Ni. Above 10 MeV the JENDL-1 cross sections are quite low, due to the neglect of direct- and collective effects.

We conclude this section with some comments on the evaluated (n,n'p)+(n,d) activation cross sections for ^{50}Cr and ^{58}Ni, that lead to the long-lived radioactive nuclides ^{49}V (331 d) and ^{57}Co (271 d), respectively. These cross sections increase very rapidly at energies above about 10 MeV and become quite large at about 14.5 MeV. The calculated values for these activation cross sections are 430 and 533 mb, respectively, in quite good agreement with the recently measured values of Qaim [17]: 405±65 and 520±50 mb, respectively. The evaluated cross section for the ^{50}Cr(n,n'p) reaction was normalized to the previously known experimental value of 121±8.5 mb [26]. At present this value appears to be much too low; accordingly, a correction to the ^{50}Cr data file will be applied, returning to the originally calculated values. For the ^{58}Ni(n,n'p)+(n,d) activation cross section our adopted experimental value was 541±22 mb at 14.5 MeV (weighted average of many experimental data), in good agreement with the model calculations. These two examples show that the model calculations of the (n,n'p) cross sections are apparently quite good. Probably model calculations should be preferred, unless recent data of good quality are available, e.g. those compiled in Refs. 15, 17.

Cover-gas Isotopes

Natural argon mainly consists of ^{40}Ar for which a set of resolved resonance parameters is known upto about 600 keV [4,5]. No resolved resonances have been measured for the other two stable isotopes ^{36}Ar and ^{38}Ar.

For ^{36}Ar there is strong evidence that σ_t upto 5 keV is determined by one negative resonance with parameters fitted by Mughabghab [27] to the total cross section measured by Chrien et al. [28]. This resonance leads to the high thermal scattering cross-section of 73.7 b, which significantly contributes to σ_t and σ_{el} of the natural element, see Fig. 6. The recommended [4] thermal capture cross section of ^{36}Ar, $\sigma_\gamma = 5\pm1$ b, is also reproduced. We have assumed a second, fictitious s-wave resonance with average parameters ($\Gamma_n^o = 2.7$ eV, $\Gamma_\gamma = 0.38$ eV) at a value of +20 keV, which is 30 keV (D_o) above the negative resonance energy. The calculated capture cross-section upto 1 MeV is shown in Fig. 7, together with the statistical-model prediction of the unresolved distant-resonances contribution. Note that at energies just above the resonance the calculated cross section is reduced due to the level-repulsion effect, which has been taken into account according to the method described in Ref. 3.

A similar procedure was followed for ^{38}Ar, for which no other information than the thermal cross section, $\sigma_\gamma = 0.8\pm0.2$ b, is available [5]. We have assumed average parameters ($\Gamma_n^o = 12.7$ eV; $\Gamma_\gamma = 0.36$ eV) for two hypothetical resonances of which the energies are -5 keV (fitting σ_γ) and +135 keV, i.e. one level spacing higher.

Also for ^{40}Ar a negative resonance was needed to fit the thermal capture cross section to 0.660±0.010 b [5]. Unfortunately, it was not possible to get agreement with the measured resonance integral [5]: I_γ = 0.41±0.03 b, of which the calculated value is only 0.29 b. The thermal and resolved resonance ranges of the evaluated radiative capture cross-section are shown in Figs. 8a and b. The resonances are based upon the parameters compiled in Refs. 4,5, supplemented with theoretical values of 0.6 eV and 0.3 eV for the s- and p-wave capture widths (see below). The calculated averaged cross sections (histogram in Fig. 8b) are below the broad-resolution data of Bostrom et al. [31]

Above the resonance range for each of the three isotopes the optical model of Wiedling et al. [9] was used, with modified transmission coefficients upto 1.2 MeV, based upon experimental s- and p-wave neutron strength functions of Liou et al. [29]. In Table IV all adopted average resonance parameters are summarized. The value of D_o for ^{40}Ar was obtained from Liou et al. For ^{36}Ar and ^{38}Ar this information was deduced from charged-particle induced resonances, from the data compiled by Beckermann [30] for nuclides with mass numbers from A = 23 to 40. From this compilation we derived experimental level-density parameters for all nuclides involved in the statistical-model calculations. There is no clear systematics for these data.

Since no capture widths are known for the argon isotopes these data were calculated [1] with a normalization to fit available experimental information on the radiative capture cross section of ^{40}Ar: the data point at 1 MeV with $\sigma_{n\gamma} = 0.9\pm0.3$ mb [31] and the fission-spectrum average [32] $\langle\sigma_{n\gamma}\rangle_f = 0.93$ mb. The adjusted calculated data for ^{40}Ar are: 1.3 mb (see Fig. 8c) and 0.84 mb, respectively. The same normalization was used to obtain values for $\langle\Gamma_\gamma\rangle$ of the other two isotopes. The resulting values of $\sigma_{n\gamma}$ at 1 MeV are 4.0 and 0.79 mb for ^{38}Ar and ^{40}Ar, respectively; the shape of the calculated cross section is similar to that of ^{40}Ar (Fig. 8c). The reason why the value for ^{36}Ar is much higher than that of ^{38}Ar is mainly due to the different level spacings (Table IV). This, in turn, is related to the quite different neutron-binding energies.

As a conclusion we note that the available experimental information is clearly insufficient to obtain a reliable data file for these nuclides. Moreover, the adopted models are near the limit of validity.

Table IV Average resonance parameters for the cover-gas isotopes.

Quantity	^{36}Ar	^{38}Ar	^{40}Ar
D_0 (keV)	30±10[a]	141±35[a]	113±20[b]
S_0 (x10^{+4})	$0.91^{+0.77}_{-0.37}$	$0.91^{+0.77}_{-0.37}$	$0.91^{+0.77}_{-0.37}$ [c]
S_1 (x10^{+4})	$0.33^{+0.14}_{-0.09}$	$0.33^{+0.14}_{-0.09}$	$0.33^{+0.14}_{-0.09}$ [c]
$<\Gamma^{v}_{\gamma,\ell=0}>$ (eV)	0.12 [d]	0	0
$<\Gamma^{s}_{\gamma,\ell=0}>$ (eV)	0.26 [e]	0.36 [e]	0.6 [e]
$<\Gamma^{s}_{\gamma,\ell=1}>$ (eV)	0.64 [e]	0.31 [e]	0.3 [e]

a) From charged-particle induced reactions [30].

b) Deduced from value of Liou et al. [29]: $<D> = 87^{+13}_{-11}$ keV.

c) Experimental data from Liou et al. [29]

d) Calculated according to Ref. 21.

e) Theoretical values, normalized to fit $\sigma_{n\gamma}$ of ^{40}Ar to experimental data [31,32].

Radioactive nuclides

Common characteristics of the radioactive nuclides ^{22}Na, ^{58g}Co and ^{58m}Co are their very large thermal absorption cross section σ_a and resonance integral I_a and the absence of much further information (see Table V). The fact that σ_a is so large suggests that one resonance dominates the absorption. Furthermore, since $I_a >> \sigma_a$ the resonance is most likely located just above the Cd cut-off energy at about 0.5 eV. This means that it is very difficult to analyse the measured data, particularly the resonance integrals. There are quite large discrepancies in the reported values for ^{22}Na and ^{58m}Co (see Table V).

From measurements on ^{22}Na it has recently been found [34] that most of the absorption is due to the (n,p) reaction, rather than radiative capture. This is confirmed by our present model calculations, which predict a large fraction of (n,p) at thermal energies for all three nuclides. For ^{22}Na the (n,α) reaction may also contribute. Fortunately, all possible absorption reactions lead to stable nuclides reducing the produced activity of the coolant.

In order to estimate the parameters of the resonance, which is responsible for the high values of σ_a and I_a, we have varied the resonance parameters (E_r, Γ_n, Γ_a) of a fictitious resonance. This process is illustrated in Fig. 9 for ^{58g}Co, which shows the range of possible values for E_r and Γ_n ($\Gamma_a = 0.5$ eV), leading to the experimental values $\sigma_a = 1.88 \pm 0.12$ kb and $I_a = 7 \pm 1$ kb. The adopted resonance parameters (bottom part of Table V) are certainly not unique. For ^{58m}Co the same procedure failed: no solutions were obtained, unless it was assumed that I_a could be lower than reported [37-39] or when Γ_a was lowered. A similar situation was found for ^{22}Na.

It turned out that the combination of very high σ_a and I_a with the condition $I_a >> \sigma_a$ was only possible for E_r close to 0.5 eV and $\Gamma = \Gamma_n + \Gamma_a$ less than $E_r - 0.5$ eV. For instance, in the case of ^{22}Na the recommended value of σ_a is 29±1 kb (Ref. 4), whereas the measured values of I_a vary from 100 to 200 kb (Refs. 35, 36). From all the possible solutions that fit $\sigma_a = 29$ kb it was tried to obtain the maximum value of I_a. We found about 140 kb at $E_r = 0.65$ eV, $\Gamma_n = 0.08$ eV and $\Gamma_a = 0.05$ eV ($g = 1/2$). However, such a low absorption width seems

Table V Thermal data and fictitious resonances for radioactive absorbers.

Quantity	Unit	^{22}Na	^{58g}Co	^{58m}Co
σ_a(exp)	kb	29±1[a]	1.88±0.12[a]	136±10[a]
σ_a(fit)	kb	29	1.88	136
σ_γ(guess)	kb	2.9	0.17	101
σ_p(guess)	kb	25.2[b]	1.7	35
σ_α(guess)	kb	0.9[e]	-	-
σ_{el}(calc.)	kb	0.85	0.055	6.38
σ_{in}(guess)	kb	-	-	1.65
I_a (exp)	kb	100-200[c]	6.89[a]	250-550[d]
I_a (calc)	kb	50	6.69	182
E_r	eV	4.25	9.0	1.09
g	-	0.571	0.5	0.5
Γ_a	eV	1.0	0.5	0.5
Γ_n	eV	0.423	0.412	0.161
R'	fm	5.7	6.5	6.5

a) Recommended value of Ref. 4; Ref. 5 gives (almost) the same value (these values are denoted by σ_γ in Refs. 4,5).

b) Measured values: 40±20 kb (Ref. 33) and 30.6±0.6 kb (Ref. 34).

c) Refs. 35, 36; recommended in Ref. 5: 200±50 kb.

d) Refs. 37-39; recommended values not given in Refs. 4,5.

e) Experimental evidence [34] gives an upper limit of only 0.06 kb.

to be quite unrealistic, since it must consist mainly of Γ_p, which is estimated [34] to be much larger than 1 eV, whereas Γ_γ should be in the order of 1 eV (Table VI). As a compromise we have adopted $\Gamma_a = 1$ eV for ^{22}Na, leading to a too low value of I_a (= 50 kb). It was tentatively assumed that 10% of this value is due to radiative capture (i.e. $\Gamma_\gamma = 0.1$ eV); the division between proton and alpha emission is based upon a statistical-model estimate. In this way most of the capture is due to the (n,p) reaction [34].

Similar arguments hold for ^{58m}Co, where we have adopted $\Gamma_a = 0.5$ eV, leading to $\sigma_a = 136$ kb and $I_a = 182$ kb (measured values range from 250 to 550 kb; Refs. 37-39). For ^{58g}Co and ^{58m}Co the division between radiative capture and the (n,p) process is based upon a statistical-model estimate.

The calculations performed at higher energies are based upon the optical [7,8] and statistical models, with parameters mostly obtained from theory or systematics (Table VI). These calculations showed the importance of (n,p) reactions at the lowest energies. For ^{22}Na also the (n,α) reaction contributes at low energies. The assumed (n,α) contribution at thermal

Table VI Average resonance parameters for radioactive nuclides.

Quantity	^{22}Na	^{58g}Co	^{58m}Co
D_o (keV)	30.8 [a]	1.17 [b]	1.00 [b]
$<\Gamma_{\gamma,\ell=0}>$ (eV)	1.1 [c]	0.5 [d]	0.5 [d]
$<\Gamma_{\gamma,\ell=1}>$ (eV)	5.4 [c]	0.5 [d]	0.5 [d]

a) From charged-particle induced reactions [30].

b) From $a = 7.4$ MeV^{-1}, see systematics of Fig. 1.

c) Theoretical estimates (cf. Ref. 1).

d) Guessed value, valid for ^{59}Co.

energies is too high compared with the experimental upper limit of $\sigma_{np}/500$ (Ref. 34). In Figs. 10 and 11 the results of some of the calculated cross sections for ^{22}Na and ^{58m}Co are shown. The model calculations are performed above 20 and 0.5 keV, respectively.

For the calculation of the cross sections of metastable targets we have slightly modified our codes. One interesting feature is that neutron scattering to the ground state is possible at the lowest thermal neutron energies. These inelastically scattered neutrons gain an energy of 25 keV. A rough estimate leads to a thermal inelastic scattering cross section of 1.7 kb. This is shown in Fig. 11b. The sharp threshold at 28 keV refers to scattering from the metastable state at 24.9 keV to the second-excited state at 52.8 keV. All inelastic scattering below 28 keV goes exclusively to ^{58g}Co.

Finally we note that these evaluations are rather speculative and that the applied models are near or beyond their validity limits.

Summary and conclusions

For the purpose of radioactive-inventory calculations of the coolant of a fast power-reactor we have made evaluations of cross sections for neutron activation reactions of corrosion products, cover-gas isotopes and contaminants. Also the absorption cross sections of some radioactive nuclides were considered, see Table I.

The evaluation methods have been reviewed. We have followed a flexible method to combine model calculations with experimental data in a consistent way to correct for deficiencies of the statistical model, which often had to be used near or beyond its limits of validity.

For the *corrosion products* ^{50}Cr, ^{54}Fe, ^{58}Ni, ^{62}Ni and ^{64}Ni the resolved resonance part of the cross sections upto 300 keV have been adopted from Fröhner [2]. At higher energies model calculations have been performed with average parameters given in Tables II and III. The adopted level-density parameters a for the corrosion products show clear systematic trends (Fig.1), including a significant odd-even effect, also observed in other mass regions [1]. Only for ^{64}Ni there are experimental data [25] of $\sigma_{n\gamma}$ in the MeV range. We have fitted the calculated values of $\sigma_{n\gamma}$ mainly to averaged data in the resolved range by adjusting the average capture widths, see Figs. 2 to 5. The model calculations for the (n,n'p) cross sections of ^{50}Cr and ^{58}Ni resulted in very good agreement with the recently measured high values [17] at 14.5 MeV.

For the *cover-gas isotopes* ^{36}Ar, ^{38}Ar and ^{40}Ar negative resonances were fitted to obtain agreement with the measured thermal capture cross-sections [5]. The high thermal scattering cross-section of ^{36}Ar significantly contributes to σ_t and σ_{el} of natural argon (Fig.6). In the absence of experimental information for ^{36}Ar and ^{38}Ar at higher energies, a second (positive) resonance was assumed at one s-wave level spacing above the negative resonance. In the statistical-model calculations the repulsion effect of this fictitious resonance was accounted for (Fig.7). The average resonance parameters (Table IV) are partly based upon level spacings of resonances observed in charged-particle induced reactions [30]. For ^{40}Ar the available resonance parameters [5] were utilized, supplemented with average capture widths, which were adjusted to fit the calculated capture cross section to the scarce experimental information in the MeV range [31,32]. The evaluated capture cross section for ^{40}Ar (Fig.8) is below the experimental points of Bostrom et al.[31] at 0.1 to 0.5 MeV. Also the calculated resonance integral is lower than measured [5]. Clearly, more experimental data are needed for these isotopes.

The *radioactive nuclides* ^{22}Na, ^{58g}Co and ^{58m}Co have very large thermal absorption cross sections and resonance integrals. We have argued that this is probably due to just one resonance above 0.5 eV which contains a significant width for proton decay, in particular for ^{22}Na (Ref.34) and ^{58g}Co. We have fitted a fictitious resonance reproducing the recommended values [4,5] of σ_a. For ^{58g}Co it was also possible to fit the recommended [4] absorption integral (Fig.9). For ^{22}Na and ^{58m}Co we could not meet the very high experimental values [35-39] of I_a, without assuming unrealistic values of the absorption widths. In view of the large uncertainties in the resonance integrals, we therefore have decided to adopt lower values for I_a(Table V). The calculated cross sections (Figs. 10,11) are rather speculative. An interesting feature is that inelastic scattering of ^{58m}Co to ^{58g}Co may occur at thermal energies with a guessed cross section of 1.7 kb, see Fig. 11b.

Evaluations for the *contaminants* ^{64}Zn and ^{112}Sn have not been discussed in detail, as the evaluation for ^{64}Zn needs to be revised, using new experimental data [40], and the evaluation of ^{112}Sn still needs some final checks.

We are grateful to Dr. F.H. Fröhner (KfK) for providing us with resolved resonance data on the Cr, Fe and Ni isotopes.

References

1. H.Gruppelaar, Contr.to the NEANDC/NEACRP Specialists' Mtg. on Fast-neutron capture cross sections, Argonne, 20-23 April, 1982 (ECN-82-045).
2. F.Fröhner, Proc.Spec.Mtg. on Neutron data of structural materials for fast reactors, Geel, Dec. 1977, p.138, Pergamon, Oxford and private communication (1977-1980).
3. H.Gruppelaar, Proc.Int.Symp.on Neutron capture gamma-ray spectroscopy and related topics, Petten, 1974, p.760, Reactor Centrum Nederland, Petten (1975).
4. S.F.Mughabghab and D.I.Garber, BNL-325, third edition, Vol.1, 1973.
5. S.F.Mughabghab, M.Divadeenam and N.E.Holden, Neutron cross sections, 1A, 1981, Academic Press, New York.
6. Ch.Lagrange, Proc.Spec.Mtg. on Neutron data of structural materials for fast reactors, Geel, Dec. 1977, p.756, Pergamon, Oxford.
7. D.C.Larson, ORNL-5662, Oak Ridge National Laboratory (1980).
8. E.Fort and D.Lafond, private communication 1979, CEA, Cadarache.
9. T.Wiedling, E.Ramström and B.Holmqvist, Proc. Consultants Mtg. on the use of nuclear theory in Neutron nuclear data evaluation, Trieste 1975, IAEA-190, 2, 205 (1976).
10. H.A.J. van der Kamp and H. Gruppelaar, ECN-108 Netherlands Energy Research Foundation (1981).
11. F.D.Becchetti and G.W.Greenlees, Phys.Rev. 182, 1190 (1969).
12. G.Igo and J.R.Huizenga, Nucl.Phys. 29, 462 (1962).
13. F.Fabbri, G.Fratamico and G.Reffo, RT/FI(77) 6, Comitato Nazionale Energia Nucleare (1977).
14. F.Fabbri and G.Reffo, RT/FI(77) 4, Comitato Nazionale Energia Nucleare (1977).
15. S.M. Qaim, Handbook of Spectroscopy, 3, 141 (1981), CRC Press, Boca Raton, Florida.
16. S.M. Qaim, Proc.of the Int.Conf. on Neutron physics and nuclear data for reactors and other applied purposes, Harwell, 1958, p.1088, OECD.
17. S.M. Qaim, A systematic study of (n,d),(n,n'p) and (n,pn) reactions, to be publ. in Nucl. Phys. A.
18. H.Gruppelaar and B.P.J. van den Bos, ECN-78 and references quoted therein, Netherlands Energy Research Foundation (1979).
19. S.Pearlstein, Program THRES-2, December 1975, code received from NEA Data Bank, Saclay.
20. A.Gilbert and A.G.W.Cameron, Can.J.Phys.43, 1446 (1965).
21. B.J.Allen and A.R.de L.Musgrove in: Advances in Nuclear Physics, Plenum Press, N.Y. (1977).
22. B.J.Allen and A.R.de L.Musgrove, Proc.Specialists' Mtg. on Neutron data of structural materials for fast reactors, 1977, p.447, Pergamon Press (1978).
23. B.J.Allen, AAEC/E403 (1977). Australian Atomic Energy Commission.

24. A.Ernst and F.H.Fröhner, KfK-1231 (1970), Kernforschungszentrum Karlsruhe.
25. H.A. Grench, Phys.Rev. 140B, 1277, 1965.
26. D.L.Allen, Nucl.Phys. 24, 274 (1961).
27. S.F.Mughabghab and B.A.Magurno, Proc.Conf. on Nuclear cross sections and technology, Washington DC, 1975, NBS-SP425, vol.2, p.774, National Bureau of Standards, Washington.
28. R.E.Chrien, A.P.Jain and H.Palevsky, Phys.Rev.125, 275 (1962).
29. H.I.Liou et al., Phys.Rev. C11, 457 (1975).
30. M.Beckermann, Nucl.Phys. A278, 333 (1977).
31. N.A.Bostrom et al., WADC-TN-59, 107 (1959), quoted in BNL-325, third edition, vol.II (1976).
32. D.J.Hughes, R.C.Garth and J.S.Levin, Phys.Rev. 91, 1423 (1953).
33. R.Ehehalt, H.Moringa und Y.Shida, Z.Naturforschung 26A, 590, 1971.
34. J.Kvitek et al., Z.Phys. A299, 187 (1981).
35. G.H.E.Sims, J.Inorg.Chem. 29, 593 (1967).
36. M.Elgart et al., Nucl.Sc.Eng. 58, 291 (1975).
37. C.H.Hogg, L.D.Weber, E.C.Yates, IDO-16744 (1962).
38. J.Halperin et al., ORNL-3679, 16(1964) Oak Ridge National Laboratory.
39. I.A.Konderov et al., Energie Atomique 74, 38 (1968).
40. J.B.Garg, V.K.Tikku and J.A.Harvey, Phys.Rev. C23, 671 (1981).

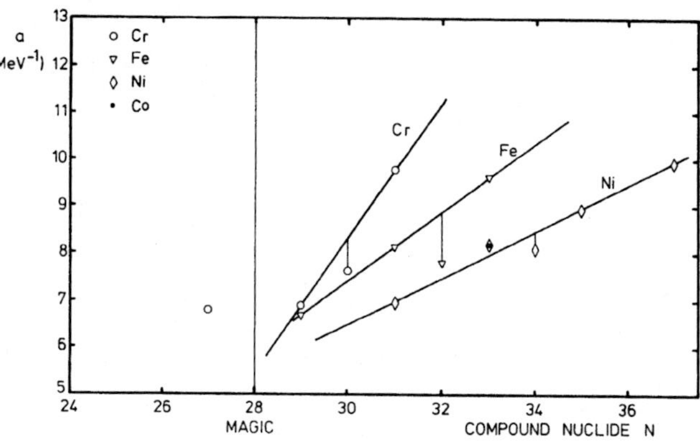

Fig. 1. Systematics of the level-density parameter a as a function of N based upon level spacings given in Table II. The data for odd values of N are located as straight lines for each Z; the data for even values of N are below these lines.

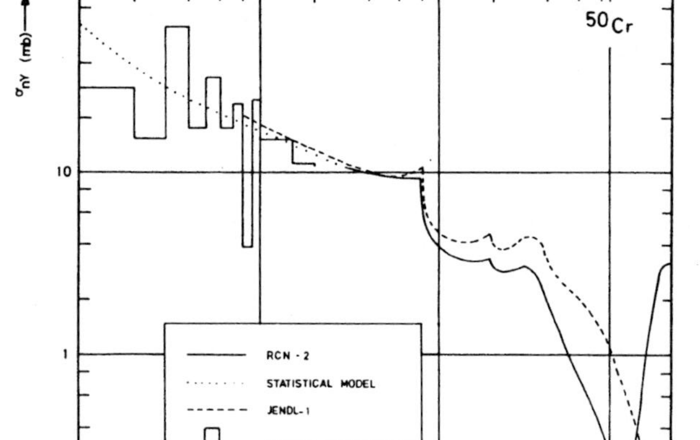

Fig. 2.

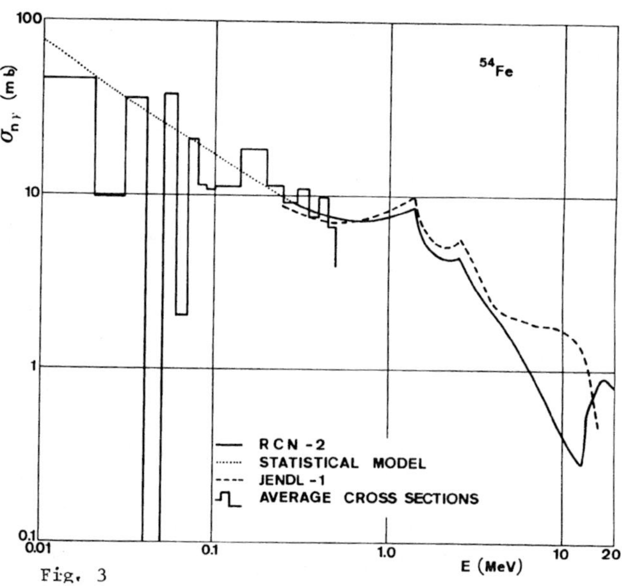

Fig. 3

Figs. 2,3. Results from the present (RCN-2) and JENDL-1 evaluations for $\sigma_{n\gamma}$ of ^{50}Cr and ^{54}Fe. Below 300 keV the statistical-model estimates (dotted curves) are compared with experimental averaged data [22,23] (histograms).

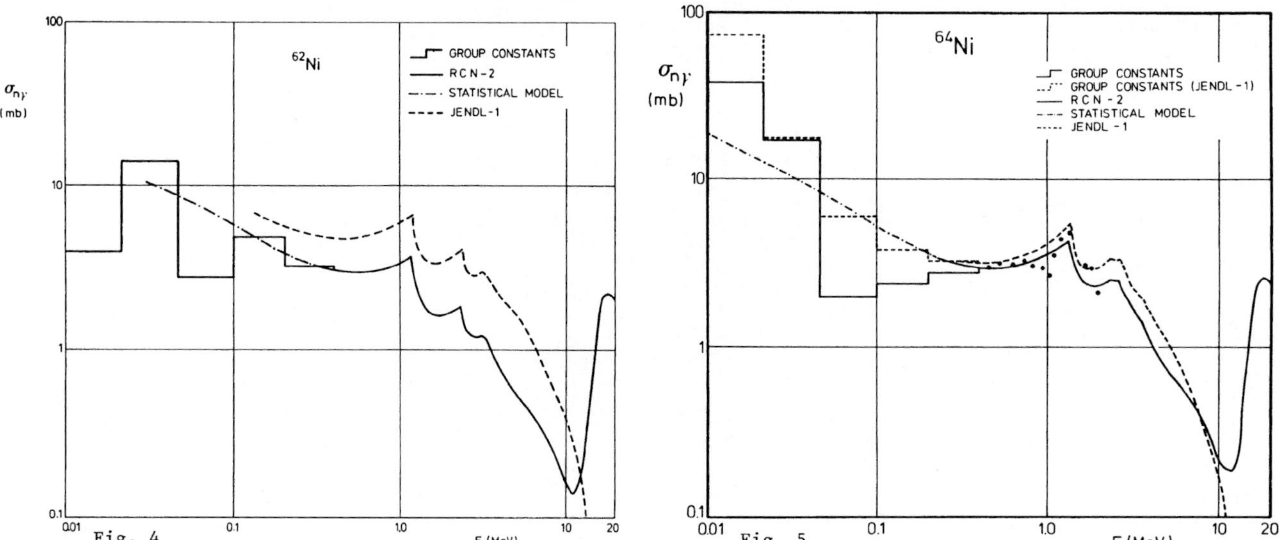

Fig. 4 Fig. 5

Figs. 4,5 Results from the present (RCN-2) and JENDL-1 evaluations for $\sigma_{n\gamma}$ of ^{62}Ni and ^{64}Ni. Below 300 keV the statistical-model estimates (dotted-dashed curves) are compared with group cross sections (histograms). The experimental data for ^{64}Ni are from Ref. 25.

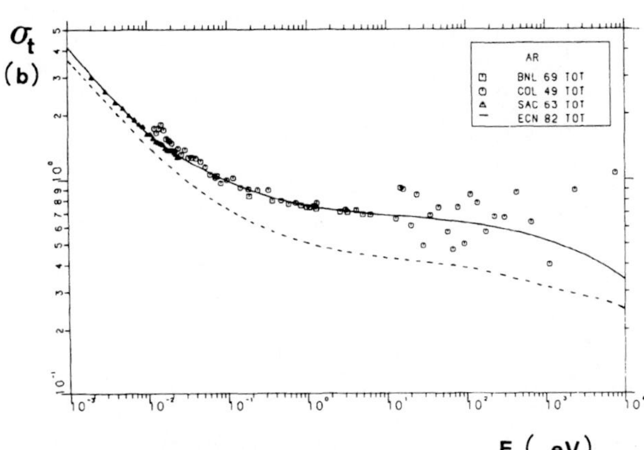

Fig. 6. Calculated total cross section of natural Ar compared with experimental data points (see CINDA data index) in the range 1 meV to 10 keV. The dashed curve represents the total cross section of ^{40}Ar (abundance: 99.6%); the difference is mainly due to ^{36}Ar (abundance: 0.337%), which has a thermal scattering cross section of 73.7 b.

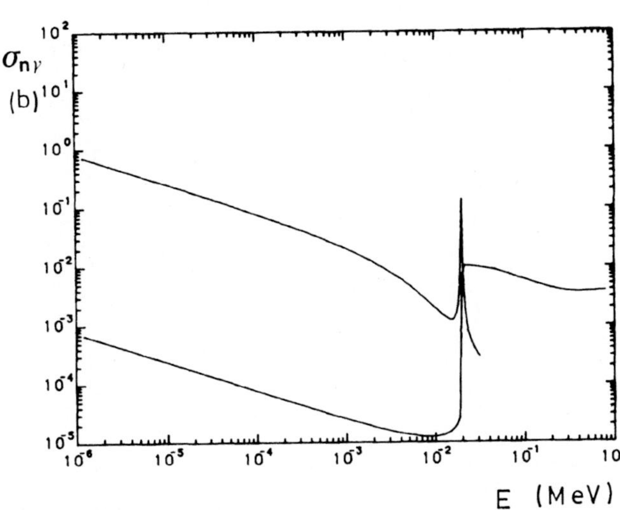

Fig. 7. The calculated radiative capture cross section of ^{36}Ar consists of two components: a Breit-Wigner component calculated from a negative and a positive (fictitious) resonance and a statistical-model component to account for "unresolved" distant resonances. The last-mentioned component [3] reduces the cross section upto about one level spacing (≈ 30 keV) above the resonance at 20 keV, to account for its level-repulsion effect.

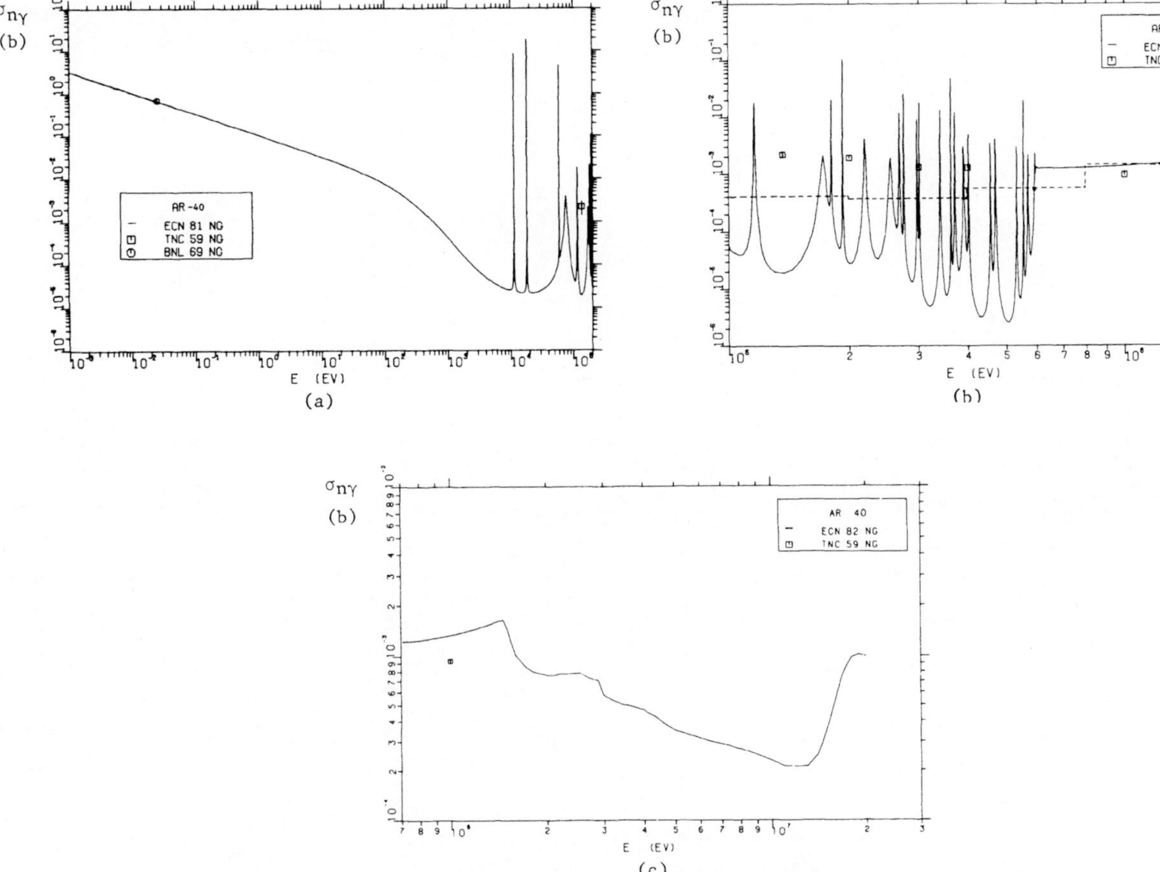

Fig. 8. Calculated radiative capture cross section of ^{40}Ar at 1 meV to 20 MeV. In the thermal range (a) a negative resonance was assumed to fit σ_γ. The dashed-line histogram represents group cross sections calculated from the present evaluation (b). The calculated curve is below the experimental data [31] in the resolved range; at 1 MeV the calculation is above the measured point (c). The average capture cross section in a fission spectrum is calculated to be 0.84 mb, below the measured value of 0.93 mb [32].

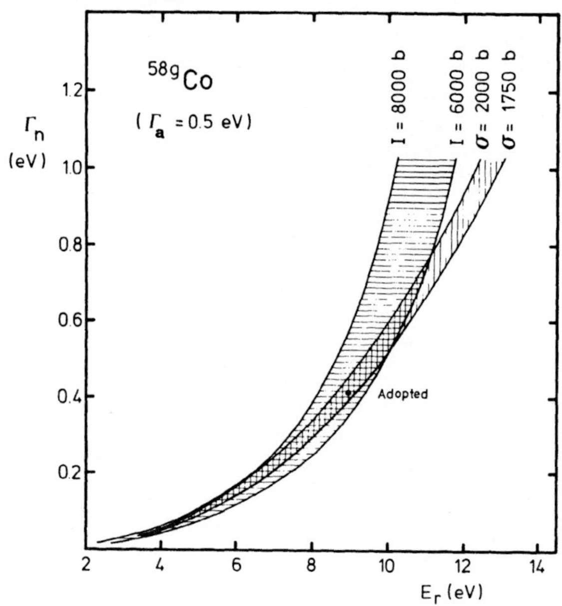

Fig. 9. Range of possible parameters E_r and Γ_n (Γ_a = 0.5 eV and g = 1/2) of a resonance in ^{58g}Co, fitting the experimental absorption cross section σ = 1.88±0.12 kb and the resonance integral I = 7±1 kb.

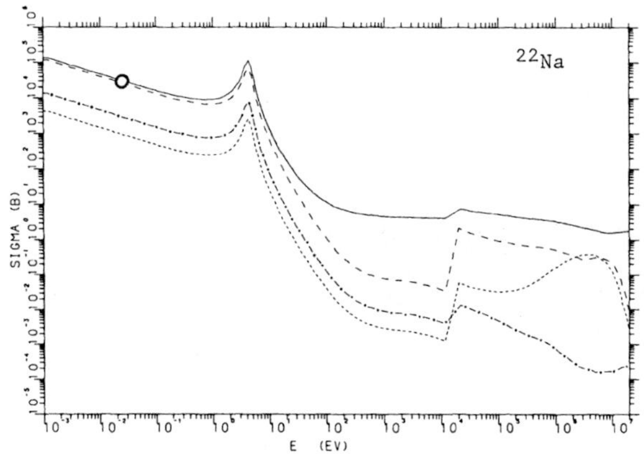

Fig. 10. Calculated cross sections for ^{22}Na based upon one fictitious resonance at 4.25 eV and model calculations above about 15 keV. The total, (n,p), (n,γ) and (n,α) cross sections are denoted by ——, ----, -·-·-, and .... respectively. The main absorption comes from the (n,p) reaction for which the experimental thermal cross section is about 30 kb.

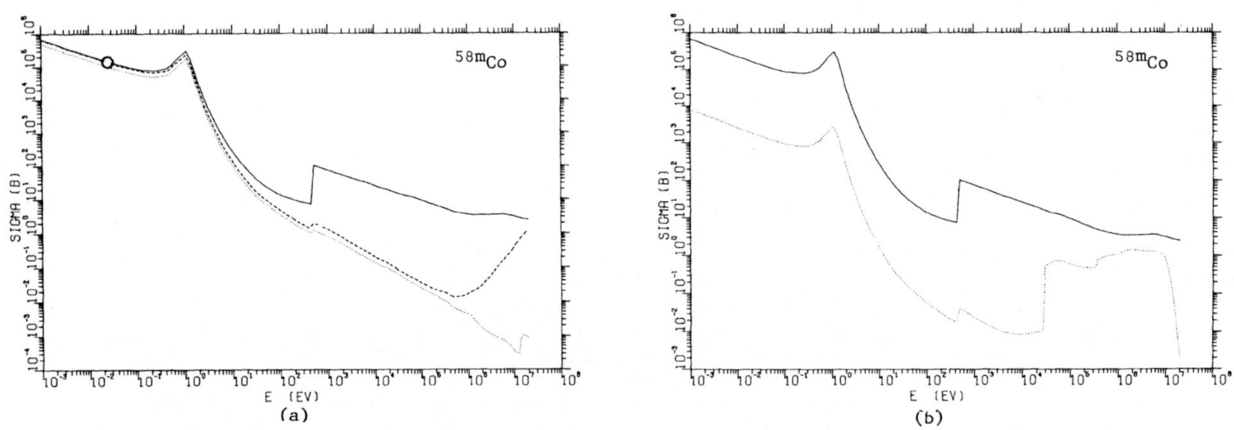

Fig. 11. Calculated cross sections for ^{58m}Co based upon one fictitious resonance at 1.09 eV and model calculations above about 0.5 keV. The total cross section is indicated by a full line. In graph (a) the dashed and dotted curves represent the absorption cross section and the (n,γ) cross section, respectively. At thermal energies the absorption cross section is about 140 kb. In graph (b) the dotted curve represent the inelastic scattering cross section, which contains a significant contribution at thermal energies, see text.

REDUCED R-MATRIX FORMULAE FOR NON S-WAVE NEUTRON RESONANCES

M. F. James

United Kingdom Atomic Energy Authority
Atomic Energy Establishment, Winfrith
Dorchester, Dorset, U.K.

The reduced R-matrix or Reich-Moore formalism is as simple to apply to neutron cross-sections for non-fissile, zero spin nuclides as the more usual multi-level Breit-Wigner formalism, and is less approximate. One disadvantage is the shift in the positions of non s-wave resonances away from the eigenvalues. This paper shows how the shifts arise, and how they may be reduced by a suitable choice of boundary condition. The residual shifts due to the interaction of resonances may be readily calculated by an iterative method, involving little extra computing time.

Introduction.

For neutrons incident on a non-fissile nuclide of spin zero, the reduced R-matrix (Reich-Moore) resonance formalism requires no more parameters than the Breit-Wigner formalism, and needs slightly less computing time.

Thus the reduced R-matrix approximation is to be preferred, provided the basic assumption is justified, that there are many capture channels with random phases. However, for medium weight nuclides with many resolved p- and d-wave resonances there is a serious disadvantage: non s-wave resonances may be appreciably shifted in energy. This note shows how the shift can be calculated and allowed for. The basic formulae may be found in, for example, Ref. 1, but it is believed that many evaluators and experimenters may not be aware of the details of the problem.

The Form of the Collision Function near a Resonance.

For one particle channel the neutron elastic element of the collision matrix is:-

$$U^J_{nn} = e^{-2i\phi}\left[1 + 2i(1-RL^0)^{-1}RP\right] \qquad (1)$$

where

$$L^0 = S^0 + iP \qquad (2)$$

ϕ is the phase shift,

P the penetration factor,

$S^0 = S - B$, where S is the shift factor and

B the logarithmic boundary condition on the eigenfunctions, and the reduced R function is:

$$R = \sum_\lambda \frac{\gamma^2_{\lambda n}}{\widetilde{E}_\lambda - E - \frac{i}{2}\Gamma_{\lambda\gamma}} \qquad (3)$$

where

$\widetilde{E}_\lambda$ are the eigenvalues, $\gamma^2_{\lambda n} = \left(\frac{\Gamma_{\lambda n}}{2P}\right)$ is the λth reduced neutron width,

$\Gamma_{\lambda\gamma}$ is the total capture width for resonance λ,

and E is the neutron energy.

The sum in (3) is over all resonances of a particular spin J and parity π.

If we write

$$Q = RP = \tfrac{1}{2}\sum_\lambda \Gamma_{\lambda n}/(\widetilde{E}_\lambda - E - \tfrac{i}{2}\Gamma_{\lambda\gamma}) \qquad (4)$$

and

$$\widetilde{S} = S^0/P,$$

$$U^J_{nn} = e^{-2i\phi}\,W$$

where

$$W = 1 + 2iQ/\left[1-(i+\widetilde{S})Q\right] \qquad (5)$$

Suppose the neutron energy E is close to resonance μ, and that the other resonances of the same J and π are at distances $|E_\lambda - E_\mu|$ large compared with $\Gamma_{\lambda tot} = \Gamma_{\lambda n} + \Gamma_{\lambda\gamma}$.

Then we can write

$$Q = Q_\mu + q \qquad (6)$$

where the contribution from the nearby resonance μ is

$$Q_\mu = \tfrac{1}{2}\Gamma_{\mu n}/\varepsilon \qquad (7)$$

with

$$\varepsilon = \widetilde{E}_\mu - E - \tfrac{1}{2}i\Gamma_{\mu\gamma} \qquad (8)$$

and where the contribution from all other resonances of the same J and π is

$$q = \tfrac{1}{2}\sum_{\lambda\neq\mu}\Gamma_{\lambda n}/(\widetilde{E}_\lambda - E - \tfrac{1}{2}i\Gamma_{\lambda\gamma}) \qquad (9)$$

Assuming well-spaced resonances, $|Q_\mu| >> |q|$ for $E \approx \widetilde{E}_\mu$; and for $\Gamma_{\lambda\gamma} << |\widetilde{E}_\lambda - E|$,

$Im\{q\} << Re\{q\}$, so that q may be considered real.

With these substitutions, Eq. (5) becomes:

$$W = 1 + 2i\left[(\tfrac{1}{2}\Gamma_{\mu n}/\varepsilon) + q\right]/\left[1-(i+\widetilde{S})(\tfrac{1}{2}\Gamma_{\mu n}/\varepsilon + q)\right]$$

$$W = 1 + i\left[\frac{\Gamma_{\mu n} + 2q\varepsilon}{1-(i+\widetilde{S})q}\right]/\left\{\varepsilon - \tfrac{1}{2}\frac{\Gamma_{\mu n}(i+\widetilde{S})}{1-(i+\widetilde{S})q}\right\} \qquad (10)$$

K. H. Böckhoff (ed.), Nuclear Data for Science and Technology, 651–653.

The term in curly brackets determines the resonant nature of the cross-section, which will be a maximum at the value of E for which the term is a minimum. Considering the term in more detail:

$$\varepsilon - \frac{1}{2}\frac{\Gamma_{\mu n}(i+\tilde{S})}{1-(i+\tilde{S})q} = \tilde{E}_\mu - E - \tfrac{1}{2}i\Gamma_{\mu\gamma}$$

$$- \tfrac{1}{2}\Gamma_{\mu n}\left[\frac{\tilde{S}(1-\tilde{S}q) - q + i}{(1-\tilde{S}q)^2 + q^2}\right]$$

$$= E'_\mu - E - \tfrac{1}{2}i\Gamma'_\mu \quad (11)$$

where

$$E'_\mu = \tilde{E}_\mu - \tfrac{1}{2}\Gamma_{\mu n}\left[\tilde{S}(1-\tilde{S}q) - q\right] / \left[(1-\tilde{S}q)^2 + q^2\right] \quad (12)$$

$$\approx \tilde{E}_\mu - \tfrac{1}{2}\Gamma_{\mu n}\tilde{S} - \tfrac{1}{2}\Gamma_{\mu n}\tilde{S}q/(1-\tilde{S}q) \quad \text{for } |q| << 1$$

is the effective resonance energy,

and

$$\Gamma'_\mu = \Gamma_{\mu\gamma} + \Gamma_{\mu n}/\left[(1-\tilde{S}q)^2 + q^2\right] \quad (13)$$

$$\approx \Gamma_{\mu\gamma} + \frac{\Gamma_{\mu n}}{(1-\tilde{S}q)^2} \quad \text{for } |q| << 1$$

is the effective total resonance width.

From Eq. (13), the effective neútron width is

$$\Gamma'_{\mu n} = \Gamma'_\mu - \Gamma_{\mu\gamma} \approx \Gamma_{\mu n}/(1-\tilde{S}q)^2 \quad (14)$$

Thus, resonance μ is shifted in energy by an amount given by Eq. (12), and has an effective neutron width differing from the true value by the factor given by Eq. (14).

In Eq. (12), the second term, $-\frac{1}{2}\Gamma_{\mu n}\tilde{S}$, gives a shift that is independent of the other resonances. This shift is assumed also in the multi-level Breit-Wigner approximation, and computer programmes such as SIGAR (Ref. 2) allow for it. The last term however arises from the more accurate treatment of the reduced R-matrix formalism, and it is worth considering in more detail.

The Shift due to other Resonances.

From Eq. (12) this is:

$$\Delta_\mu = -\tfrac{1}{2}\Gamma_{\mu n}\tilde{S}q/(1-\tilde{S}q) \quad (15)$$

where

$$\tilde{S}q = \frac{S^0}{P}q$$

$$= S^0 \sum_{\lambda \neq \mu} \frac{\gamma^2_{\lambda n}}{(\tilde{E}_\lambda - E)} \quad (16)$$

For $E \approx \tilde{E}_\mu$, $|\tilde{E}_\lambda - E| \sim D$, the mean neutron spacing, for the nearest neighbours to μ. Thus the sum is of the order of the neutron strength function, $\langle\gamma_n^2\rangle/D$. Note that the usually quoted strength function is $\langle\Gamma_n^{(1)}\rangle/D$ where $\Gamma_n^{(1)}$ is the normalised reduced width (Ref. 3, p.289) given by

$$\Gamma_n^{(1)} = \Gamma_n/(\nu_\ell(E)\sqrt{E})$$

where

$$\nu_\ell = \frac{P}{ka}$$

Hence,

$$\Gamma_n^{(1)} = 2\gamma_n^2 k_o a$$

when k_o is the wave number of a 1 eV neutron, and a is the nuclear radius. For $a \sim 5$fm, $k_o a \sim 10^{-3}$. Thus, for conventional strength functions $\langle\Gamma_n^{(1)}\rangle/D \sim 10^{-4}$, $\langle\gamma_n^2\rangle/D \sim 0.1$ and consequently $|q| \sim 0.1$; but it may be appreciably greater if the nearest resonances are large and close.

For neutrons with orbital angular momentum quantum number ℓ,

$$S^0 = S - B$$

where $S \quad - \ell + (ka)^2/(2\ell - 1) + O((ka)^4)$

where k is the neutron wave-number and a is the nuclear radius.

Frequently, a boundary condition B = 0 is used: then $S^0 \approx -\ell$ so that $\tilde{S}q \sim 0.1\,\ell$. Indeed for some p-wave resonances in ^{58}Fe. $\tilde{S}q \sim 1$ so that the shift becomes large, and it is impossible to correlate the computed peaks with the input eigenvalues.

Clearly it is much preferable to use a boundary condition such as

$$B = -\ell;$$

then $S^0 \approx (ka)^2/(2\ell - 1) \sim 0.3$ for $E \sim 0.3$ MeV, and $a \sim 5$fm.

Thus even at high energies the effective shift is much reduced, and at low energies (~ 1 KeV) S^0 is very small.

The residual shift Δ_μ given by Eq. (15) may still be large enough in some cases for a programme such as SIGAR (Ref. 2), which distributes energy points closely around each input resonance energy but more widely elsewhere, to represent a shifted narrow peak very inadequately, and in some cases to miss it completely. Consequently, the set of Equations (12) and (13) are solved iteratively, for both resonance energies and neutron widths.

There are two different cases, each of which will be allowed for in the revised version of SIGAR.

(1) The positions E'_μ and effective widths Γ'_μ of cross-section maxima are specified (for example, from area analysis or single level Breit-Wigner fits). Then Eqs. (12) and (13) are solved iteratively (2 iterations are usually sufficient) to determine $\tilde{E}_\mu$ and $\Gamma_{\mu n}$, the true eigenvalues and neutron widths (with a boundary condition $B = -\ell$). The program uses the input maxima positions E'_μ and width Γ'_μ to select an energy mesh, and then performs a reduced R-matrix calculation with the derived $\tilde{E}_\mu$ and $\Gamma_{\mu n}$; the whole calculation takes no more computing time than a multilevel Breit-Wigner fit.

(2) Given a reduced R-matrix fit, values of the eigenvalues $\tilde{E}_\mu$ and neutron widths $\Gamma_{\mu n}$ are input (with the specified boundary condition), and the effective energies E'_μ and widths Γ'_μ are computed from Eqs. (12) and (13). These latter are then used in the point selection routine.

Conclusions.

(1) If the Reich-Moore or reduced R-matrix formalism is used for non s-wave neutrons, there will be large shifts in resonance positions unless the boundary condition $B = -\ell$.

(2) If $B = -\ell$ the shift may still not be negligible relative to Γ, and is dependent on the other resonances of the same spin and parity.

Consequently, experimenters or evaluators fitting non s-wave resonances with a reduced R-matrix formula should

(a) state the boundary condition used

and (b) give the complete set of parameters (eigenvalues and widths) for all the resonances considered.

With these provisos, the formula is as simple as the Breit-Wigner formulae, and more accurate.

References.

1. A. M. Lane and R. G. Thomas, Rev. Mod. Phys. 30, 257 (1958).

2. A. L. Pope, R. W. Smith and J. S. Story, Users' Guide for SIGAR7, A Fortran Programme for Calculating Resonance Cross-Sections with Doppler Broadening (To Be Published).

3. J. E. Lynn, The Theory of Neutron Resonance Reactions (Clarendon Press, Oxford) (1968).

INFLUENCE OF CROSS-SECTION STRUCTURE ON UNFOLDED NEUTRON SPECTRA

C. Ertek, M.F. Vlasov, B. Cross and P.M. Smith

International Atomic Energy Agency
Vienna, Austria

The influence of cross-section structure on neutron spectra unfolded by multiple foil activation technique, SAND-II case, has been studied. For three reactions with evident structure in neutron cross-section above threshold: $^{27}Al(n,\alpha)^{24}Na$, $^{31}P(n,p)^{31}Si$ and $^{32}S(n,p)^{32}P$, two remarkably different sets of evaluated data were selected from the available evaluations; one set of data was "smooth", the structure having been averaged over by a smooth curve; the other set was "sharp" with structure given in detail. These data were used in unfolding procedure together with other reactions, the same in both cases (as well as input spectra and measured reaction rates). It was found that during unfolding calculations less iteration steps were needed to unfold the neutron flux spectrum with the set of "sharp" data. In case of "smooth" data it was difficult to obtain an agreement between measured and calculated activity values even by increasing the number of iteration steps. Contrary to expectations, considerable deformation of unfolded neutron flux spectrum has been observed in the case of the "smooth" data set.

This work has been performed within the IAEA programme on standardization of reactor radiation measurements, one of the important objectives of which is assistance to laboratories in member states to implement the multiple foil activation technique for neutron spectra unfolding, an especially useful technique for in-pile neutron measurements. The importance of this method, e.g. for radiation damage studies is well recognized [1-2].

In order to unfold a neutron spectrum, the following information is required:

a) measured saturation activities of the irradiated detector foils;
b) a set of energy-dependent neutron cross-sections for each foil;
c) a computer programme for unfolding spectra using an input spectrum and data noted in (a) and (b).

The ENDF/B Dosimetry File is finding increasing use as a reference cross-section data set; however, it does not include some important dosimetry reactions. Therefore, the IAEA Nuclear Data Section has initiated an activity to evaluate these additional reactions. It is hoped that this expanded file will form a basis for an internationally recommended data set for neutron dosimetry applications.

The IAEA Seibersdorf Laboratory, with the support of the IAEA Computer Section, is currently involved in the intercomparison of available computer programmes for spectrum unfolding to recommend the best one (or a few) for general use. For the time being, the SAND-II [3] and CRYSTAL BALL [4] programmes have been implemented [5-u]. The RFSP-Jül [8] programme is under consideration and preparation have been started to implement the STAY'SL [9] unfolding code. The SAND-II programme has been compared with a new generalized least squares method by J.T. Routti [10].

As a part of these activities, we plan to investigate related problems such as the influence of input spectra on the solution spectrum, consistency of measured reaction rates, effects of structure in the energy-dependent neutron cross sections on the shape of the unfolded spectrum, etc. As the first step, we have studied the influence of cross-section structure on spectra unfolded with the SAND-II programme. The pressurized water reactor (PWR) type spectrum was chosen.

Three reactions ($^{27}Al(n,\alpha)^{24}Na$, $^{31}P(n,p)^{31}Si$ and $^{32}S(n,p)^{32}P$) with evident energy dependent cross-section structure were selected. For each reaction two sets of the evaluated data, "smooth" (structure averaged by smooth curve) and "sharp" (structure given in detail), were taken from the available evaluations and converted, where necessary, into SAND-II format. The cross sections together with the response functions in a Watt neutron spectrum are given in Fig. 1-6. In addition to these three reactions, seven other reactions, with identical evaluations in both "smooth" and "sharp" cases, were used to unfold the PWR type neutron spectrum. The saturation activities calculated with this spectrum were compared with the measured ones. This comparison is given in Table I, from which it can be seen, that for the $^{31}P(n,p)^{31}Si$ reaction, the deviation of the

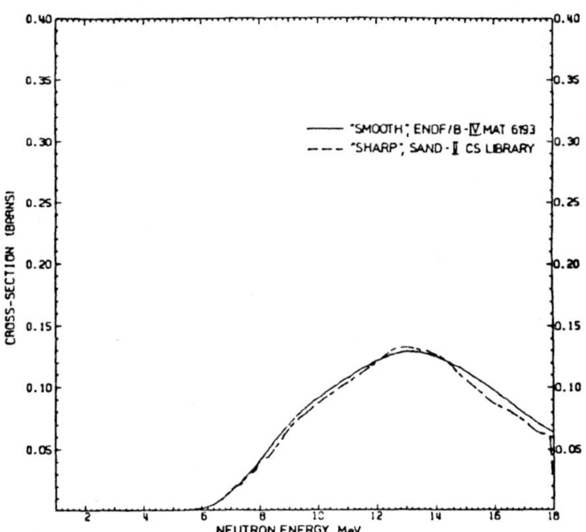

Fig. 1. $^{27}Al(n,\alpha)^{24}Na$ neutron cross section

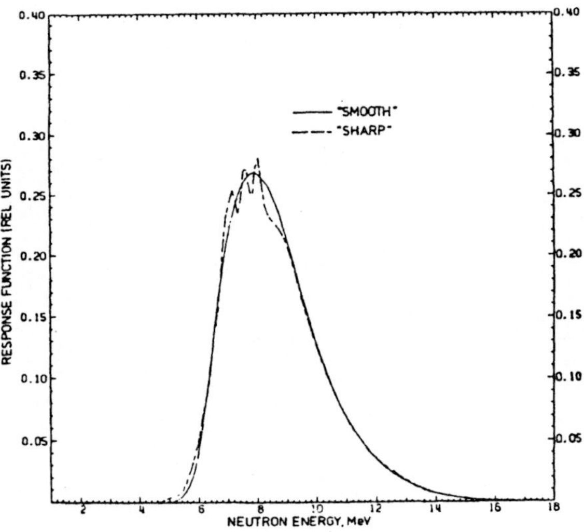

Fig. 2. $^{27}Al(n,\alpha)^{24}Na$ response in Watt spectrum

K. H. Böckhoff (ed.), Nuclear Data for Science and Technology, 654–656.

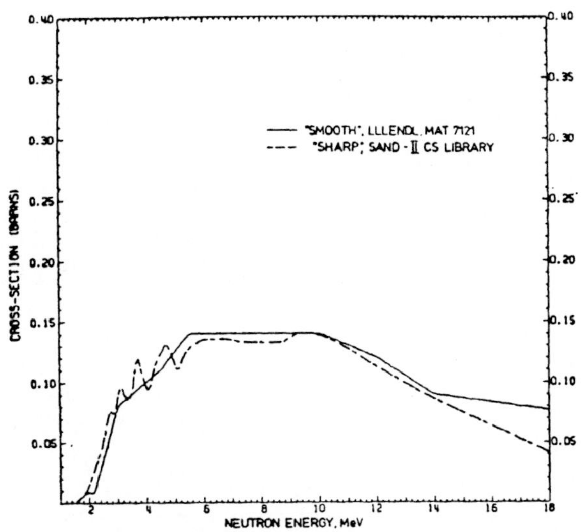

Fig. 3. $^{31}P(n,p)^{31}Si$ neutron cross section

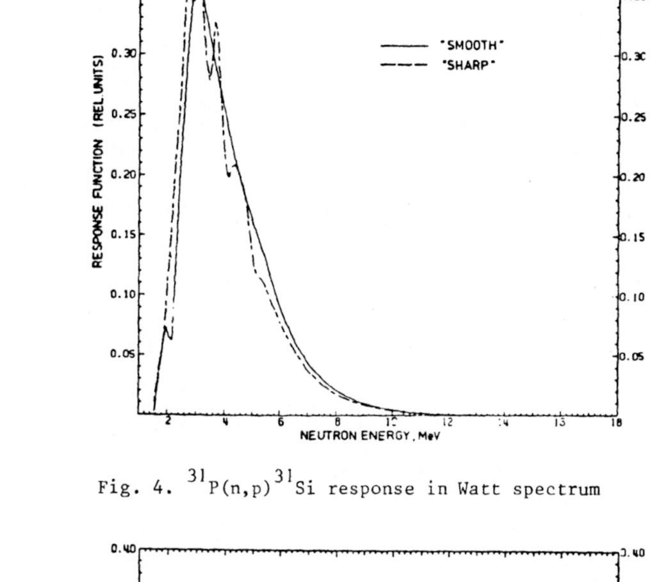

Fig. 4. $^{31}P(n,p)^{31}Si$ response in Watt spectrum

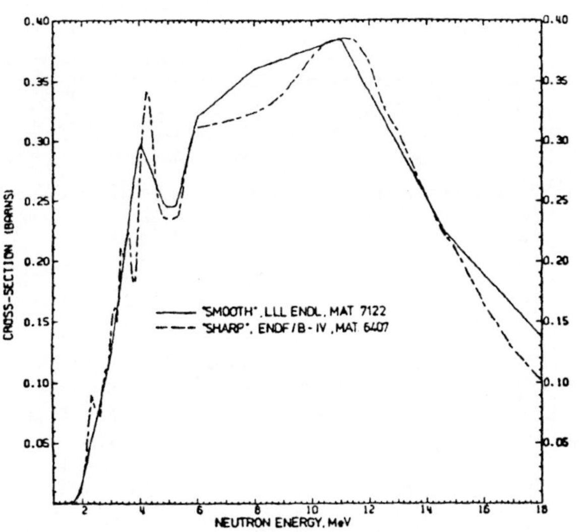

Fig. 5. $^{32}S(n,p)^{32}P$ neutron cross section

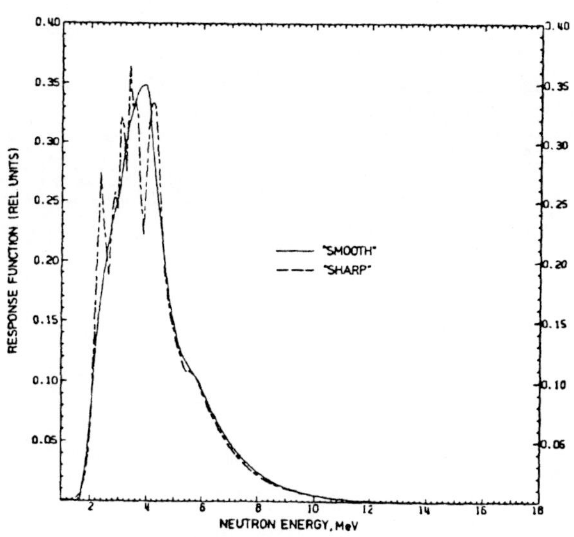

Fig. 6. $^{32}S(n,p)^{32}P$ response in Watt spectrum

Table I Results obtained after 12 iterations

Foil Reaction	Saturated Measured Activity (DPS/Nucleus)	Saturated Calculated Activities (DPS/Nucleus)	Ratio Measured to Calculated Activities	Deviation of Measured from Calculated Activity (Percent)
Results obtained after 12 iterations ("Smooth")				
$^{45}Sc(n,\gamma)^{46}Sc$	1.713E-24	1.734E-24	0.9880	- 1.20
$^{23}Na(n,\gamma)^{24}Na$	3.997E-26	3.974E-26	1.0057	0.57
$^{197}Au(n,\gamma)^{198}Au$	6.836E-23	6.839E-23	0.9996	- 0.04
$^{59}Co(n,\gamma)^{60}Co$	4.936E-24	4.936E-24	1.0001	0.01
$^{232}Th(n,f)FP$	1.727E-26	1.749E-26	0.9877	- 1.23
$^{54}Fe(n,p)^{54}Mn$	1.825E-26	1.822E-26	1.0019	0.19
$^{235}U(n,f)FP$	4.084E-23	4.065E-23	1.0047	0.47
$^{27}Al(n,\alpha)^{24}Na$	1.281E-28	1.281E-28	1.0000	0.00
$^{31}P(n,p)^{31}Si$	7.836E-27	7.324E-27	1.0699	6.99
$^{32}S(n,p)^{32}P$	1.474E-26	1.564E-26	0.9425	- 5.75
		Standard Deviation of Measured Activities (Percent)		3.08
Results obtained after 12 iterations ("Sharp")				
$^{45}Sc(n,\gamma)^{46}Sc$	1.713E-24	1.733E-24	0.9883	- 1.17
$^{23}Na(n,\gamma)^{24}Na$	3.997E-26	3.973E-26	1.0061	0.61
$^{197}Au(n,\gamma)^{198}Au$	6.836E-23	6.837E-23	0.9999	- 0.01
$^{59}Co(n,\gamma)^{60}Co$	4.936E-24	4.934E-24	1.0004	0.04
$^{232}Th(n,f)FP$	1.727E-26	1.738E-26	0.9938	- 0.62
$^{54}Fe(n,p)^{54}Mn$	1.825E-26	1.771E-26	1.0305	3.05
$^{235}U(n,f)FP$	4.084E-23	4.064E-23	1.0049	0.49
$^{32}S(n,p)^{32}P$	1.474E-26	1.515E-26	0.9727	- 2.73
$^{31}P(n,p)^{31}Si$	7.836E-27	7.796E-27	1.0052	0.52
$^{27}Al(n,\alpha)^{24}Na$	1.281E-28	1.283E-28	0.9981	- 0.19
		Standard Deviation of Measured Activities (Percent)		1.47

measured from the calculated activity is ~ 7 % in the "smooth" case, while it is only 0.5 % in the "sharp" case.

For the $^{32}S(n,p)^{32}P$ reaction this deviation is ~ -6 % and -3 % respectively. Less than 0.1 % deviation is observed in both cases for $^{27}Al(n,\alpha)^{24}Na$ reaction. The overall standard deviation of the measured activities is ~ 3 % in the "smooth" case as compared to 1.5 % in the case of "sharp" data. Increasing the number of the iteration steps in the case of "smooth" data does not decrease the final deviation of the measured from calculated activity.

At first, one might expect that sharp structure in the cross-section would perturb the smooth shape of the unfolded spectrum; however, contrary to expectations, considerable deformation of the unfolded neutron flux spectrum has been observed in the case of the "smooth" data set in the energy range from 2.2 MeV up to ~ 6 MeV as shown in Fig. 7-8. It is difficult to explain in detail the observed results. However, it is clear that the "sharp" data, which more accurately represent the measured neutron cross section, are preferable. At the same time these results suggest, that smoothing of the input cross-section data should be done very carefully in order to avoid introducing distortions of the type seen in this work.

Acknowledgements

The authors would like to thank Dr. D.W. Muir for useful discussions and Mr. K. McLaughlin for assistance with computer plotting programmes.

References

1. 1st ASTM-EURATOM Symposium on Reactor Dosimetry, Petten, Netherlands, EUR 5667 e/f parts I and II (September 1975)

2. 2nd ASTM-EURATOM Symposium on Reactor Dosimetry, Palo Alto, Calif., USA (October 1977) to be published

3. W.N. McElroy et al., "SAND-II, Neutron Flux Spectra Determinations by Multiple Foil Activation Iterative Method", RSIC Computer Code Collection CCC112, ORNL, (May 1969)

4. F.B.K. Kam, F.W. Stallman, CRYSTAL BALL, Report ORNL-TM-4601, (June 1974)

5. C. Ertek, "On the Penetration of Mono-energetic Neutrons Inside the Detector Foils and Related Problems", IAEA/RL/44 (February 1977)

6. C. Ertek et al., "Progress Report on the IAEA Activity on Neutron Spectra Unfolding by Activation Technique", paper presented at the Second ASTM-EURATOM Symposium on Reactor Dosimetry, Palo Alto, Calif., USA (October 1977)

7. C. Ertek, "Report on Some Interlaboratory IAEA Activities on Neutron Spectra Unfolding by Activation Technique", IAEA/RL/47 (May 1978)

8. A. Fischer, "RESP-JUL: A Programme for Unfolding Neutron Spectra from Activation Data", Jül-1475 (December 1977)

9. F.G. Perey, "Least-Squares Dosimetry Unfolding: The Programme STAY'SL",ORNL/TM-6062(October 1977)

10. J.T. Routti et al., "Threshold Detector Measurements of Fast Neutron Spectra in TRIGA Reactor with Application of a New Unfolding Method", Atomkernenergie (ATKE) Bd. 26 (1975), Lfg. 2

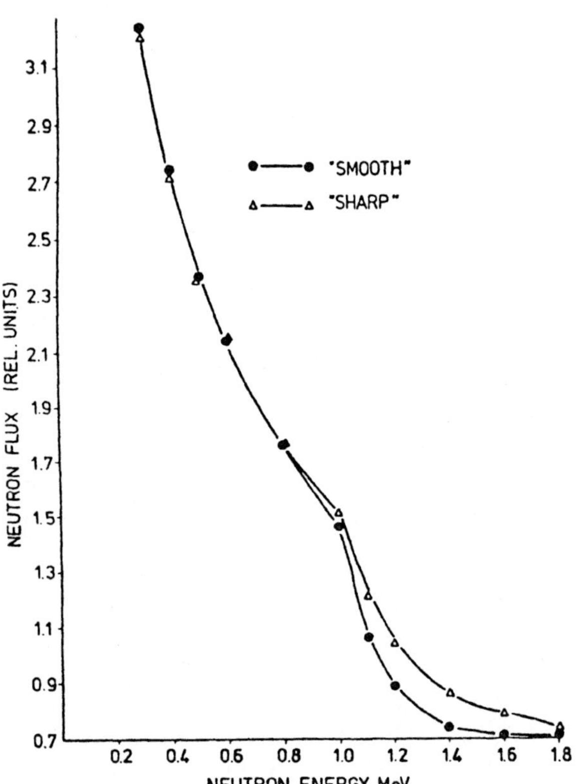

Fig. 7. PWR unfolded flux spectrum

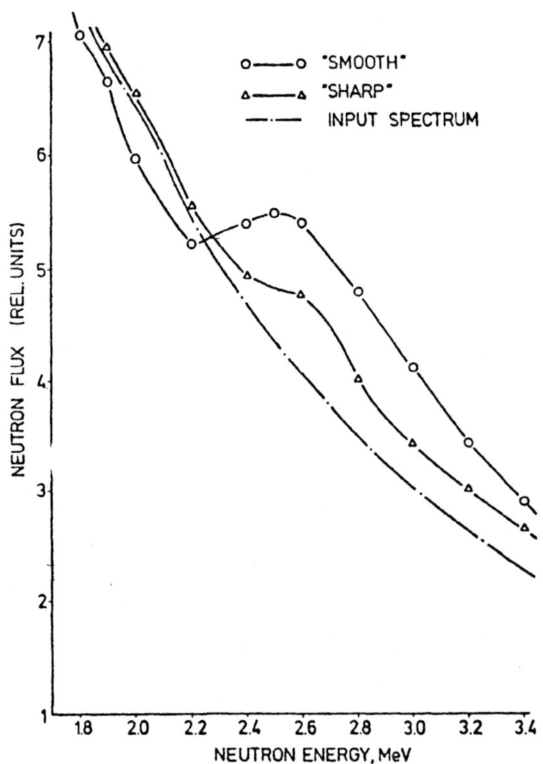

Fig. 8. PWR unfolded flux spectrum

EVALUATION OF THE ^{232}Th NEUTRON CAPTURE AND FISSION CROSS SECTION ABOVE 50 keV

H.M. Jain and M.K. Mehta

Bhabha Atomic Research Centre, Bombay 400 085, India

This paper describes evaluations of measured differential data for ^{232}Th (n,γ) and (n,f) cross section above 50 keV which is undertaken at BARC. Most of the existing differential measurements are reviewed. Data sets are renormalized to ENDF/B-V values of the standard cross sections. The input data base, the evaluation procedures and the judgements are outlined. The evaluated cross sections above 50 keV are smooth SPLINE fitted curves which follow the trend of the most recent measurements. The end results of both these evaluations are sets of "recommended" values at energy intervals close enough to enable linear interpolation. The present evaluations are compared with some recent ones. Regions of larger uncertainties are pointed out where new measurements would be of value.

[^{232}Th(n,γ), ^{232}Th(n,f) cross section evaluation E_n = 50 keV - 20 MeV SPLINE fits.]

I. Introduction

During the last few years there has been a renewed interest in neutron induced capture and fission cross section for ^{232}Th, mainly because of the feasibility of the alternate ^{232}Th - ^{233}U nuclear reactor fuel cycle. In addition to superior fuel cycle performance, ^{233}U also offers the possibility for isotopic denaturing with ^{238}U which may improve proliferation and diversion resistance.

As a part of the IAEA-NDS sponsored Coordinated Research Programme on the Intercomparison of Evaluation of Actinide Neutron Nuclear Data, a programme has been undertaken at BARC to carry out the evaluation of nuclear data of the isotopes of importance for Th-U fuel cycle. This paper describes the evaluation, of the measured differential neutron fission and capture cross sections of ^{232}Th.

II. Methods of Data Evaluation

Two techniques are generally used for capture cross section measurements namely; prompt capture gamma ray measurements and activation method. Similarly measurement of fission cross section also use two common techniques namely, ratio measurement relative to ^{235}U or ^{238}U standard fission cross section and absolute measurement. The available measured data sets were scrutinised for following points:

(a) Measurement technique used.
(b) Standard used or method used in determination of the absolute neutron flux.
(c) Corrections and normalization applied.
(d) Errors reported, and
(e) For fission cross section measurement only, the method used in determination of the number of ^{235}U and ^{232}Th nuclei or their ratio.

In determining the degree of reliability and validity of measured data, the initial consideration was whether the literature accessible contained a detailed description of the experiment with full particulars and analysis of the corrections made, along with an estimate of errors. Wherever some information was lacking, their weightage was reduced in SPLINE fitting by increasing the reported error appropriately. In cases where even this could not be done, the data sets were not included in this work. In a number of cases the sets were renormalized to ENDF/B-V standard values.

III. Data Base

A survey of all available experimental data and discrepancy among them has already been reported in Ref. 1 and 2. Recently these cross sections were evaluated by Meadows et al.[3] and Vasiliu et al[4]. Since that time two new capture measurements by Poenitz et al.[5] and Kobayashi et al.[6] have been reported and Macklin et al.[7] have revised their measured data. Similarly two new fission measurements have been reported i.e. Nordborg et al.[8] and Meadows[9]. Blons et al[10] have repeated their measurements with better energy resolution. All these new data sets are included in this work, most of them supplied by IAEA-NDS from EXFOR Library. The experimental information has been complemented to a certain extent using CINDA 81 and scrutinizing the published work referred to in this Index for obtaining experimental details. Only ten capture (Refs. 5-7, 11-17) and seven fission (Refs. 8-10, 18-21) measurements considered to be most reliable and valid on the basis of criteria mentioned in II were chosen for these evaluations.

Among the ten capture measurements, the data of Lindner et al[11] and Poenitz et al[5] seem to be consistent with each other within their respective errors. Lindner's set of activation measurements is found to be complete in all aspects and considered suitable for comparing other similar data. Macklin et al.[7] revised their values for ^{6}Li(n,α) cross section using ^{235}U(n,f) ENDF/B-V and corrected for computer coding error (multiplied by a factor of 1.1131) have been used in this work. The three points of Jain et al.[12] do not agree fully with either Lindner or Poenitz data and are therefore inconclusive. Kobayashi et al.[6] data are preliminary and found to be lower than Lindner and Poenitz data. Basically Miskel et al.[13] measurements are identical to Lindner and differ only in accuracy and therefore the reported error (10%) has been increased to (20%) for reducing its weightage. Data of Tolstikov et al.[14] measured by activation technique, have been renormalized at 24 keV to cross section value of 520 mb and the reported errors (15-20%) have been used in this work. Data of Stavisski et al[15] have been renormalized to the ^{127}I (n,γ)ENDF/B-IV cross section values, and measured values of Barry et al.[16] are renormalized to ^{238}U(n,γ) ENDF/B-IV cross section values at 600 keV. Reported errors in both these sets have been multiplied by a factor of 3. Similarly, after rejecting a few inconsistent data points from the old measurement of Hanna et al.[17], remaining data points

K. H. Böckhoff (ed.), Nuclear Data for Science and Technology, 657–660.

are included with the reported error multiplied by a factor of three.

Among these chosen seven fission data sets, the recent ratio measurements of Behrens et al[18], Nordborg et al.[8] and Meadows[9] are found to be consistent with each other within their respective errors. Behrens set is found to be most acceptable from the evaluator's point of view, as it covers the widest range in energy, reports consistency checks on energy scale, simultaneous measurement at all energies assuring uniform efficiency and more accurate mass deposition measurements for the Th/U mass ratio determination. These ratios are converted to ^{232}Th(n,f) cross section by using ^{235}U(n,f) ENDF/B-V values for this evaluation. The data sets from Blons et al[10] is also a ratio measurement converted to ^{232}Th(n,f) cross section by the author using ^{235}U(n,f) ENDF/B-IV values. Most of the ^{232}Th(n,f)/ ^{238}U(n,f) ratio measurements are clustered around 14.2 MeV neutron energy and used for normalization. The weighted average value of 382 $\pm$ 21.6 mb at 14.6 MeV corrected for ^{238}U(n,f) ENDF/B-V given in Meadows evaluation is also adopted in the present work. In addition ^{232}Th(n,f)/^{238}U(n,f) ratio measurements by Rago et al[19] are converted to ^{232}Th(n,f) cross section using ^{238}U(n,f) ENDF/B-V values and renormalized. After reassigning errors, these data are included in this evaluation. Pankratov et al.[20] is the only absolute measurement considered in this work.

IV. Result of Evaluation

In all 176 experimental data points from the chosen ten capture measurements in the energy range 50 keV to 4 MeV were obtained after the processing described above. These data are SPLINE fitted to get a smooth curve representing the present evaluation. This curve along with the experimental points and other recent evaluation are shown in Fig. 1.

Similarly, 1062 experimental data points from the chosen seven fission measurements were processed and SPLINE fitted in three segments of energy i.e. from threshold to 2.45 MeV, 2.38 to 5.85 MeV and 5.38 to 20.4 MeV with 563, 421 and 142 experimental data points in each region respectively. This smooth curve, representing the present evaluation is shown in Fig. 2 and 3 along with the data points. The evaluated data in detailed tabulated form at energy interval suitable for linear interpolation are given in Ref. 22 and 23. In fig. 4 the present evaluation is compared with Meadows et al[3] and Vasiliu[4] evaluations. The figure also shows the measured ratio data.

V. Discussion and Conclusion

From 50 to 580 keV, present capture evaluation follows the measurements of Lindner, Poenitz and Macklin (corrected data); from 600 keV to 2.8 MeV it is the same as Lindner and then upto 4 MeV it follows Miskel. On this basis, the evaluated errors estimated for the fitted values are $\approx \pm$ 10% for 50 to 100 keV range $\pm$ 5% for 100 keV to 2.8 MeV range and then $\pm$ 10% upto 4 MeV. Above 4 MeV, the only measurement is of Perkin et al al.[24] at 14.5 MeV. When the Spline fitted curve is extrapolated beyond 4 MeV to 14 MeV, a value of $\approx$ 1 mb is obtained for 14 MeV. This value is also suggested in ORNL/TM-6161 (1977) and is probably more correct than the value of 5.2 $\pm$ 0.8 mb reported by Perkin which is a very old measurement. Considering the scatter of data which is slightly more than the quoted errors on recent measurements between 400 keV to 1 MeV, one can conclude that more measurements are required to improve evaluation. Between 1 and 3 MeV there are three sets of measurements with accuracies better than 10% which agree within their quoted errors. Beyond this upto 4 MeV only one set of Miskel et al exists; which is an old measurement. More measurement in this range are required. No measurements exist between 4 and 14 MeV except only one very old measurement at 14.5 MeV. Again measurements in this region are required to produce a more reliable evaluation upto 15 MeV. However, in the high energy region the cross section decreases rapidly and the sensitivity of most reactor calculation to this cross section above 4 MeV is expected to be small.

It can be seen from Fig. 2 and 3 that for fission cross section the difference between the values given by the SPLINE fitted smooth curve and actual measured ones varies from less than 1% to as much as 20%. On careful examination of the various data it becomes clear that this is mainly due to systematic differences between the various sets especially so between 2.4 and 6.0 MeV where Blons data set is systematically lower by as much as 20% as compared to Behren's set. Above 6 MeV when Blons data set terminates, the variation is relatively small. Even below 6 MeV if Blons data are not included, the "evaluated" curve gives values much closer to the measured ones reported by Behrens, Meadows and Nordborg, which are ratio measurements. Even amongst these three sets Meadows values are systematically higher than Behren's data. In the present evaluation the errors estimated for fitted values are $\approx \pm$ 10%, which are much larger than the required accuracy, around 5% as per WRENDA 81/82. It is clear that the required accuracies can be met once the systematic differences between the data sets are removed by proper normalisation. However, the actual normalisation between the sets could not have been done before the present evaluation was carried out. The reevaluation is now underway and it is expected that after that is over the required accuracies for the ^{232}Th(n,f) cross section would be satisfied and no new measurements would be necessary.

References

1. M.K.Mehta, H.M. Jain, IAEA-TECDOC-232, page 287 to 336 (1980)
2. M.K.Mehta, H.M.Jain, INDC Discrepancy File - 1979, INDC-32/L page 2 to 4 (1980)
3. J. Meadows et al., ANL/NDM-35(1978)
4. G. Vasiliu et al., INDC(RUM)-10 (1980)
5. W.P. Poenitz, O.L. Smith, ANL/NDM-42(1978)
6. K.Kobayashi et al., Conf. Neutron Physics and Nuclear Data, Harwell, 432 (1978)
7. R.L. Macklin, J. Halperin, Nucl. Sci. Eng. 64 849 (1977)
8. C. Nordborg et al., Harwell Conf. 910 (1978)
9. J.W. Meadows, Paper EB4 presented at Knoxville Conf., Tennessee (1979)
10. J. Blons et al., Phys. Rev. Lett. 35 1749 (1975); Phys. Rev. Lett. 41 1289 (1978)
11. M. Lindner et al., Nucl. Sci. Eng. 59 381(1976)
12. H.M.Jain et al., Conf. Neutron Physics and Nuclear Data, Harwell, 1108(1978).

13. J.A.Miskel et al., Phys. Rev.,128 2717 (1962)
14. V.A. Tolstikov et al.,At. Energ.,15 414 (1963)
15. Y.Y. Stavisski et al., At. Energ.,10 508 (1961)
16. J.F. Barry et al., Proc. Phys. Soc. 74 685 (1959)
17. R.C. Hanna, B. Rose, J. Nucl. Energy 8 197 (1959)
18. J.W. Behrens et al., UCID-17442 (1977)
19. P.F. Rago; N. Goldstein, Health Phys., 13 654 (1967).
20. V.M. Pankratov et al., At. Energ. 14 177 (1963).
21. H.M. Jain, M.K. Mehta, NDE(BARC)-9 (1981)
22. H.M. Jain. M.K. Mehta, NDE (BARC) - 10 (1982)
23. D'Hondt P. et al., Annals of Nuclear Energy 7 367 (1980)
24. J.L. Perkin et al., Proc. Phys. Soc. 72 505 (1958)

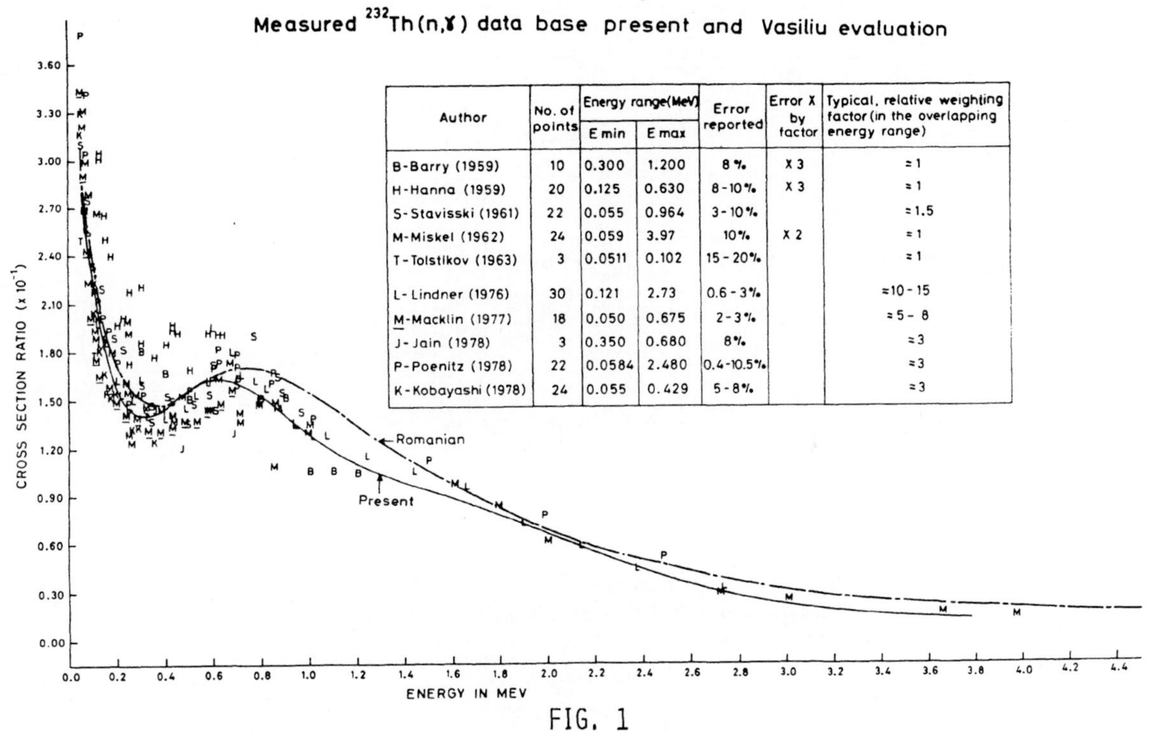

Author	No. of points	Energy range(MeV)		Error reported	Error X by factor	Typical, relative weighting factor (in the overlapping energy range)
		E min	E max			
B-Barry (1959)	10	0.300	1.200	8 %	X 3	≈ 1
H-Hanna (1959)	20	0.125	0.630	8-10 %	X 3	≈ 1
S-Stavisski (1961)	22	0.055	0.964	3-10 %		≈ 1.5
M-Miskel (1962)	24	0.059	3.97	10 %	X 2	≈ 1
T-Tolstikov (1963)	3	0.0511	0.102	15-20 %		≈ 1
L-Lindner (1976)	30	0.121	2.73	0.6-3 %		≈10-15
M-Macklin (1977)	18	0.050	0.675	2-3 %		≈ 5- 8
J-Jain (1978)	3	0.350	0.680	8 %		≈ 3
P-Poenitz (1978)	22	0.0584	2.480	0.4-10.5 %		≈ 3
K-Kobayashi (1978)	24	0.055	0.429	5-8 %		≈ 3

FIG. 1

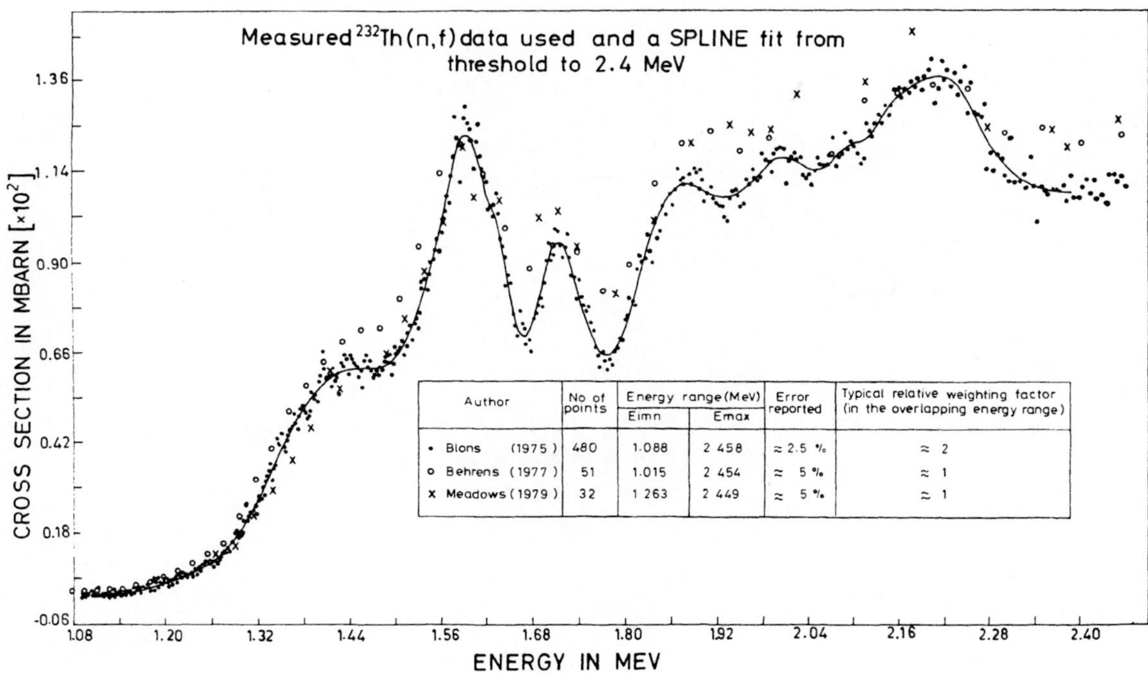

Author	No of points	Energy range(MeV)		Error reported	Typical relative weighting factor (in the overlapping energy range)
		Eimn	Emax		
• Blons (1975)	480	1.088	2 458	≈ 2.5 %	≈ 2
o Behrens (1977)	51	1.015	2 454	≈ 5 %	≈ 1
x Meadows (1979)	32	1 263	2 449	≈ 5 %	≈ 1

FIG. 2

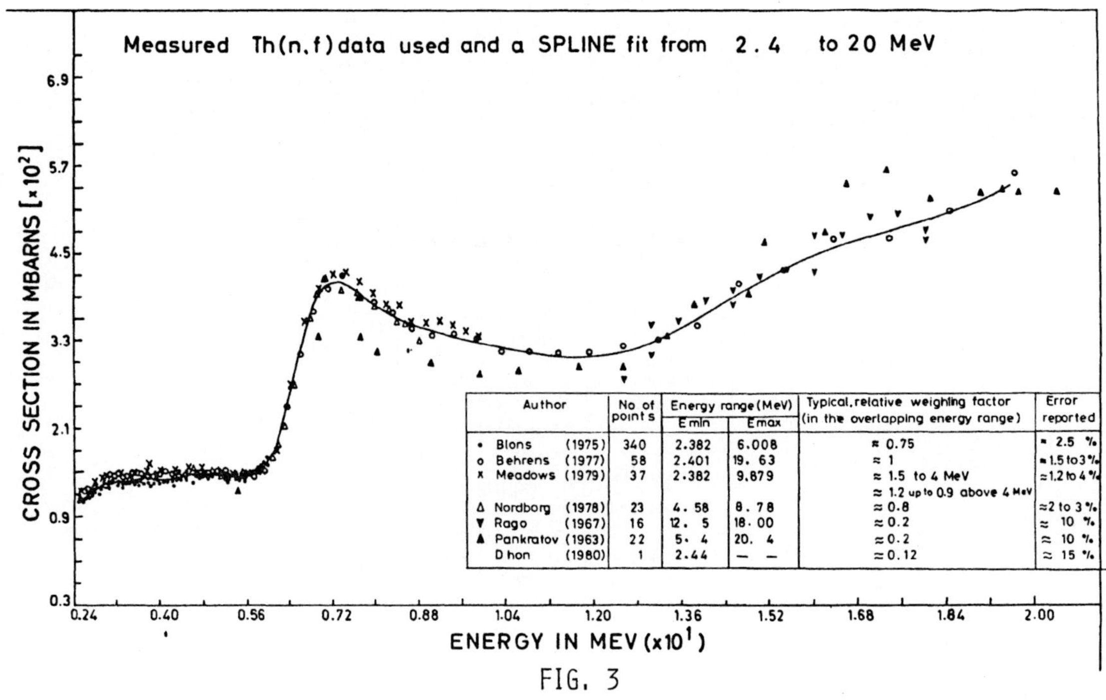

Author	No of points	Energy range (MeV) Emin	Emax	Typical, relative weighing factor (in the overlapping energy range)	Error reported
• Blons (1975)	340	2.382	6.008	≈ 0.75	≈ 2.5 %
o Behrens (1977)	58	2.401	19. 63	≈ 1	≈ 1.5 to 3 %
x Meadows (1979)	37	2.382	9.879	≈ 1.5 to 4 MeV ≈ 1.2 up to 0.9 above 4 MeV	≈ 1.2 to 4 %
Δ Nordborg (1978)	23	4. 58	8. 78	≈ 0.8	≈ 2 to 3 %
▼ Rago (1967)	16	12. 5	18. 00	≈ 0.2	≈ 10 %
▲ Pankratov (1963)	22	5. 4	20. 4	≈ 0.2	≈ 10 %
D hon (1980)	1	2.44	— —	≈ 0.12	≈ 15 %

FIG. 3

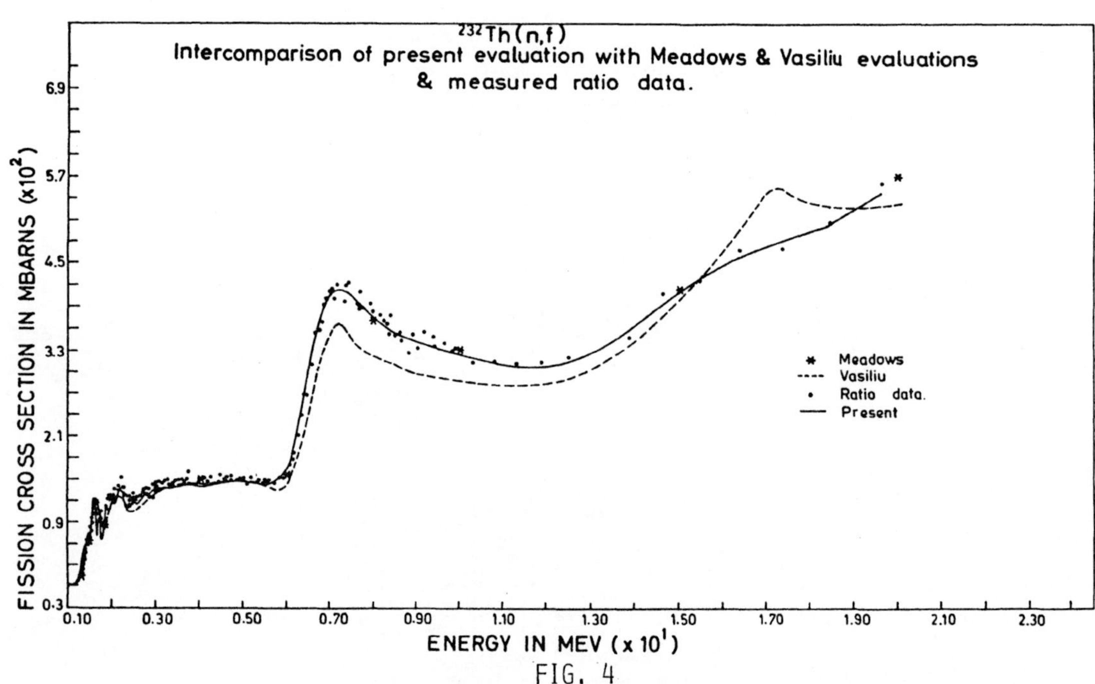

FIG. 4

MULTIGROUP CONSTANTS CALCULATED FROM EVALUATED DATA LIBRARIES FOR REACTOR CALCULATIONS

A. Trkov, M. Ravnik, M. Budnar, A. Perdan

J. Stefan Institute
E. Kardelj University
P.O.B. 53, 61001 Ljubljana
Yugoslavia

Multigroup constants for Cd(nat), Ag-107, Ag-109 and In-115 had been calculated using FEDGROUP-C package. Cross-section data were retrieved from ENDF/B IV and V files. Multigroup constants were inserted into the WIMS library where they were used in lattice cell calculations by S-WIMS code. Control rod worths were estimated and compared against the project specifications and experimental results showing good agreement with experiments.

Introduction

The lattice cell code S-WIMS (1,2) has been chosen as part of the power reactor core calculations program package. The S-WIMS code has been proven for its flexibility and reliability, however a number of users (4,5) have noted problems in using the nuclear data library (3) supplied with the package. The library is based on fairly old evaluations but modified to reproduce the measurements from some integral experiments. Thus it may be unreliable in certain cases. It also lacks some materials which are important as structural or control components.

The capability to update and extend the multigroup constants library offers the possibility to test the dependence of reactor parameters on the evaluated data source (6). Also, the library can be extended to include some important structural materials not yet included in the library.

The program package FEDGROUP-C (10) has been used because it is available with all the auxilliary programs for data manipulation and because it was found sufficiently accurate for the purpose.

Multigroup constants preparation for Cd, Ag and In

The standard version of the S-WIMS cross-section library was used for criticality and burn-up calculations of the NPP Krško reactor (PWR Westinghouse 2-loop type 16 x 16 grid). The results are in good agreement with measured values. For the control rod worth calculations the WIMS library had to be extended for certain absorber materials. The control rods of the NPP Krško are of the In-Cd-Ag type. Apart from cadmium these materials do not exist in the WIMS library as structural materials (they are treated as fission products). They had been recalculated from ENDF/B IV and V evaluated data files (7,8,9) using multigroup package FEDGROUP-C.

From the available data all the required group constants were calculated including the resonance integrals (except for Indium). Upscattering and temperature dependence in the thermal region were neglected. The group constants were inserted into the WIMS library.

From Fig. 1 and 2 it can be seen that the thermal and epithermal cross-sections for cadmium from ENDF/B are slightly lower than the ones in the original WIMS library and that they exhibit a structure in the resonance region. The WIMS supercell calculations show that the differences in these cross-sections alone were not insignificant. The absorption cross-section data for Ag-109 and In-115 treated in the WIMS library as fission products are compared to the ENDF/B estimates in Fig. 3 and 4. Also, for these two materials a number of group constant types are missing. Because of differences found in the fast and the epithermal region for absorption cross section data and lack of important group constant types, the data originally in the WIMS library may not be taken for structural components (3). The group constants calculated from ENDF/B files were used instead for the control rod calculations.

The control rod calculations were performed in 3-D diffusion 2-group approximation. The unit and super-cell calculations were performed by S-WIMS for different types of the cells (fuel, fuel & control rod, fuel & burnable poison, fuel & instrumentation channel, etc.) in 18-group (integral) transport approximation. The (super) cells were homogenized in fuel elements with the program ELCON (12). 3-D calculations were performed by programmes FASVER and AXIAL (13), developed at the institute. The results were compared with the measurements at hot zero power conditions. They are in excellent agreement with experiment, as can be seen from Fig. 5.

Discussion of Results

The calculations of the control rods for NPP Krško had been performed using multigroup constants for Cd, Ag and In from ENDF/B IV and V which were calculated with FEDGROUP-C. Excellent agreement with the experimental results was achieved. The improvement in the calculated results can be partly attributed to the more recent cross-section used and partly to the calculational methodology. The agreement with the experiment inspires confidence into the extended WIMS library which forms the basis for the safety and control analysis of the reactor. The extended WIMS library is available from the authors at Jožef Stefan Institute.

K. H. Böckhoff (ed.), Nuclear Data for Science and Technology, 661–664.

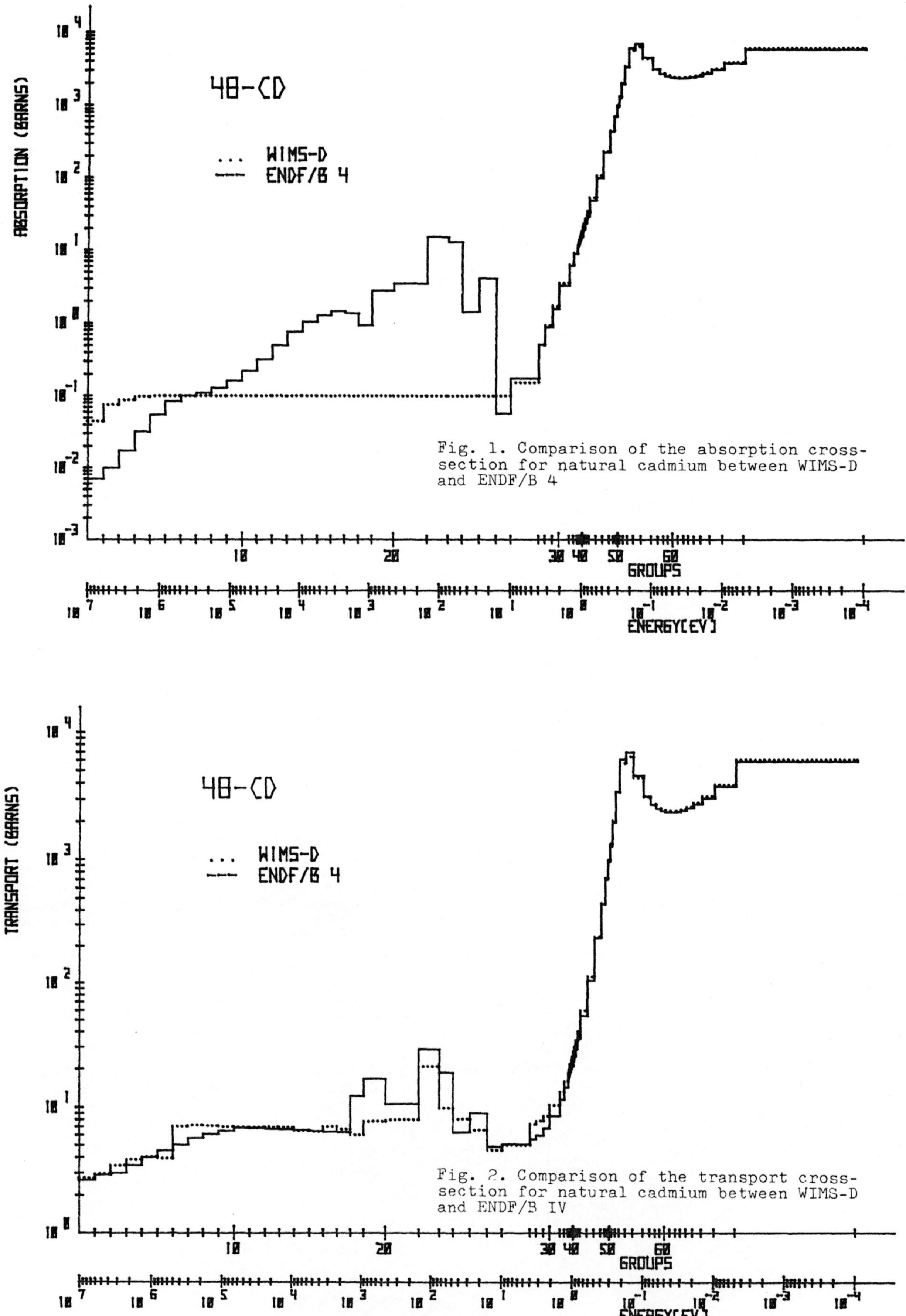

Fig. 1. Comparison of the absorption cross-section for natural cadmium between WIMS-D and ENDF/B 4

Fig. 2. Comparison of the transport cross-section for natural cadmium between WIMS-D and ENDF/B IV

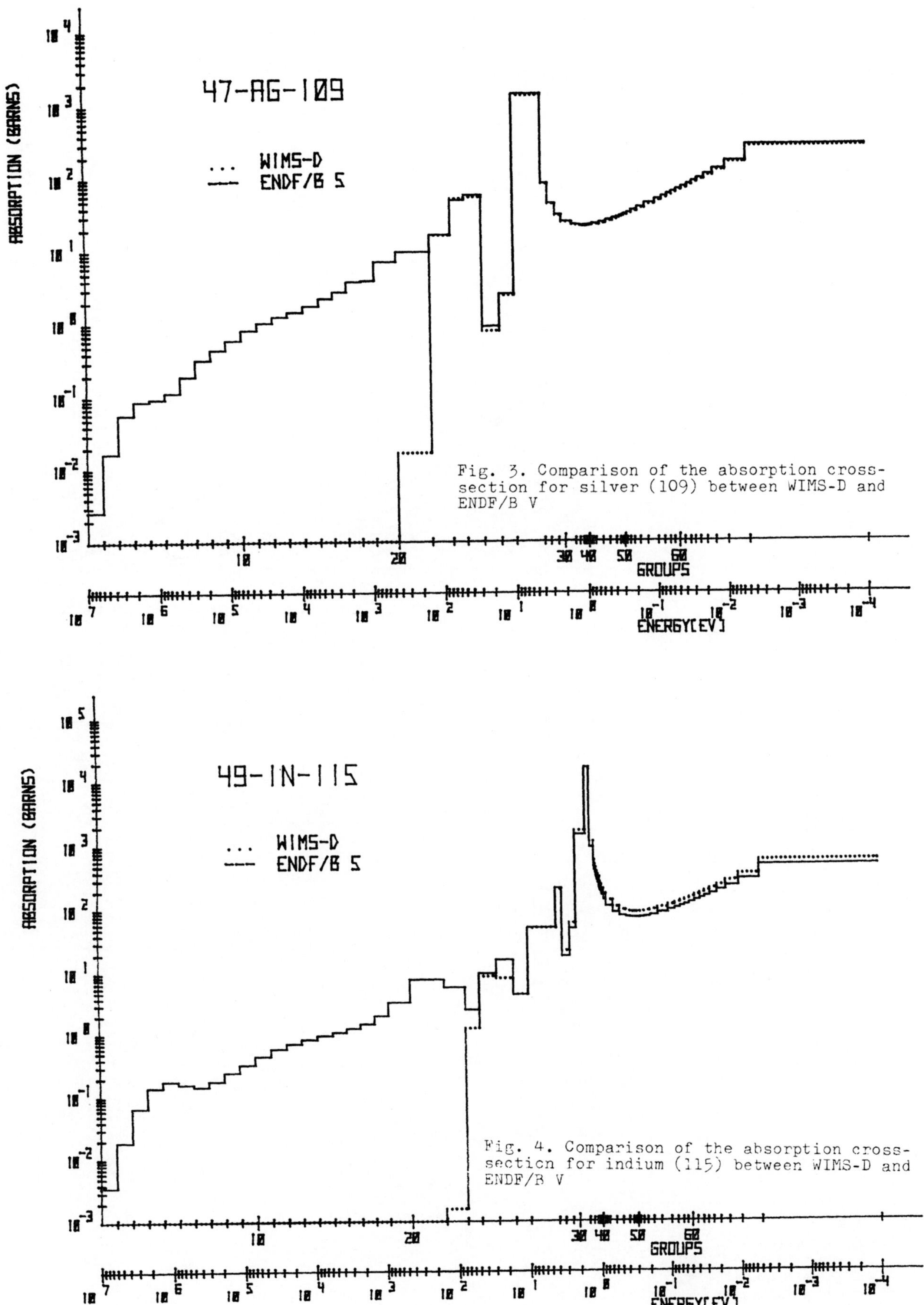

Fig. 3. Comparison of the absorption cross-section for silver (109) between WIMS-D and ENDF/B V

Fig. 4. Comparison of the absorption cross-section for indium (115) between WIMS-D and ENDF/B V

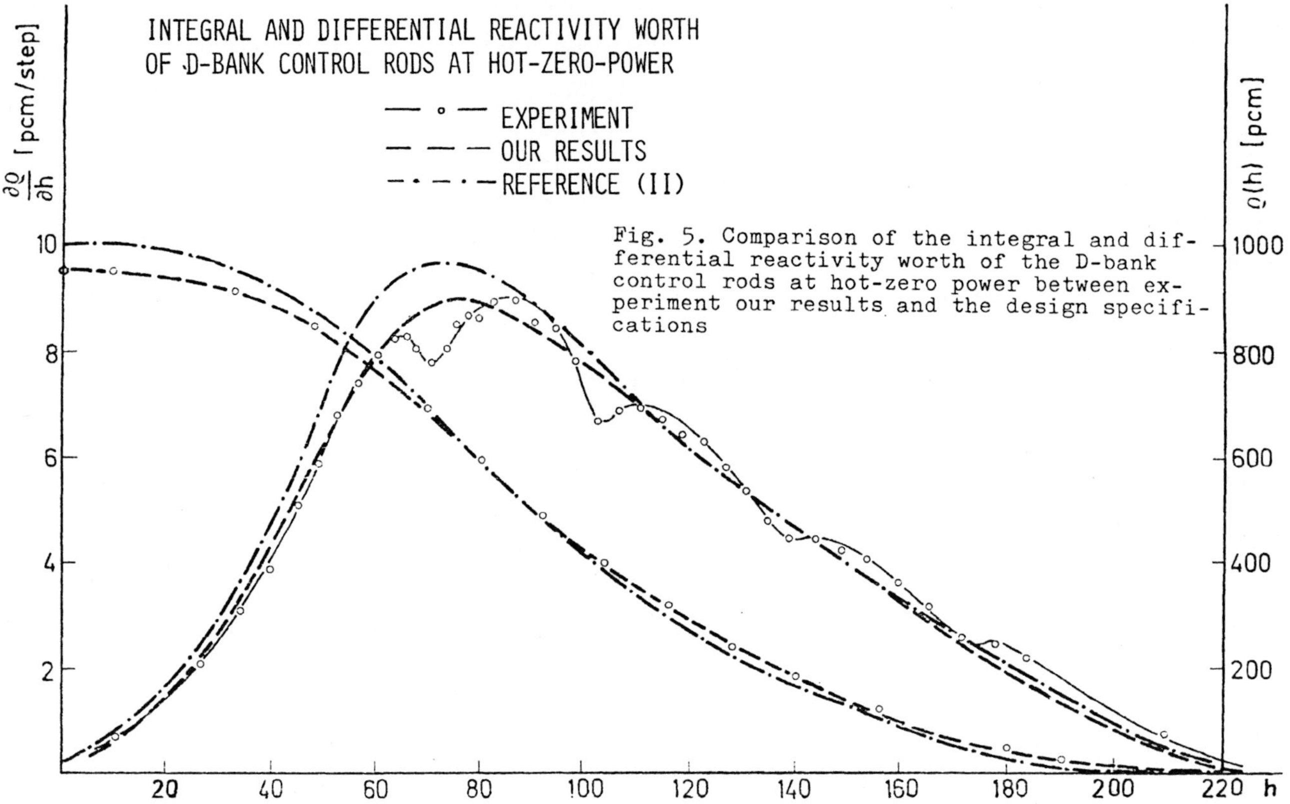

Fig. 5. Comparison of the integral and differential reactivity worth of the D-bank control rods at hot-zero power between experiment our results and the design specifications

References

1. A General Description of the Lattice Code WIMS, J.R. Askew, F.J. Fayers, P.B.Kemsharle, J.B.N.E.S., 564 (1968)
2. The S-WIMS Code for the CYBER-72 Computer, K. Kowalska, NEA-CPL, Saclay, Gif-sur-Yvette, France
3. The WIMS 69 Group Library, C.J. Toubman, AEEW-M-1342, July 1975
4. An Assessment of Methods and Data for Predicting Integral Properties for Thermal Reactor Physics Experiments, R. Chawla, AEEW-R-797, May 1972
5. WIMS-D Performance in Cell Parameters Calculations for UO_2-D_2O Systems, R. Paviotti, Corcuera, A.V. Galia, R.E. Pezzoni, Newsletter of the NEA Data Bank 23, Nov. 1979
6. Calculation of Multigroup Constants in WIMS format with Programs FEDGROUP-C and FLANGE and Comparison of the Results Obtained Using Different Evaluated Libraries, A. Trkov, M. Budnar, A. Perdan, M. Ravnik, INDC(YUG)-7, Jan. 1982
7. ENDF/B IV General Purpose Library Cd, MAT 1281, BNL-NCS-17541 (ENDF-201), 2nd Edition (Oct. 1975), D. Garber, Brookhaven National Laboratory, Upton, N.Y.
8. ENDF/B V Fission Products Library AG-107 MAT 1371, Ag-109 MAT 1373, BNL-NCS-17541 (ENDF-201), 3rd Edition (July 1979), R. Kinsley, Brookhaven National Laboratory, Upton, N.Y.
9. ENDF/B V Dosimetry Library, MOD-2 In-115 MAT 6437, BNL-NCS-17541 (ENDF-201) 3rd Edition (July 1979), R. Kinsley, Brookhaven National Laboratory, Upton, N.Y.
10. FEDGROUP-C, An Improved and Modified CDC Version of the Program Package for Processing Evaluated Nuclear Data in KEDAK, UKNDL and ENDF/B Format, A. Trkov, A. Perdan, M. Budnar, J. Stefan Institute, E. Kardelj University of Ljubljana, Yugoslavia, (to be published)
11. Krško NPP Final Safety Analysis Report, Westinghouse 1978
12. Use of Small Computers in Research Reactor Operations, IAEA Report of Consultants Meeting, Vienna, Austria, December 7-9, 1981
13. M. Ravnik, M. Čopič, Testing of a Two Group Twodimensional Diffusion Code FASVER on the IAEA benchmark Problem, Seminar/Workshop on Thermal Reactor Benchmark Calculations, May 17-18, 1982, Brookhaven National Laboratory

A NEW EVALUATION OF NEUTRON DATA FOR THE ^{209}Bi BETWEEN 10^{-5} eV AND 20 MeV

O. Bersillon, B. Caput, C.A. Philis

Service de Physique Neutronique et Nucléaire
Centre d'Etudes de Bruyères-le-Châtel
B.P. n° 561, 92542 Montrouge Cedex, France

A new evaluation of neutron induced cross-sections on ^{209}Bi has been completed within the full energy range 10^{-5} eV - 20 MeV and put under ENDF format. A careful study of the resonance region led to a consistent set of resonance parameters. On this basis, the tabulated cross-sections (total, elastic, capture) have been calculated using the Reich-Moore formalism. At higher energies a consistent set of optical model parameters has been obtained by fitting mainly the total cross-section between 0.7 and 150 MeV and elastic scattering angular distributions from 4 to 24 MeV. The so obtained neutron penetrabilities have been used for Hauser-Feshbach statistical model calculations which have been completed with pre-equilibrium and direct interaction components to get elastic and inelastic cross sections, angular distributions, secondary neutron spectra and gamma production. All the results are generally in good agreement with the available experimental data.

[^{209}Bi, nuclear data evaluation, neutron cross-sections, neutron-production, gamma-production, resonances, nuclear models]

Introduction

Une connaissance précise et complète des sections efficaces neutroniques du ^{209}Bi est indispensable à certaines études de réacteurs pour lesquelles on a envisagé son emploi comme matériau constitutif (en particulier les réacteurs hybrides fusion-fission). Ces études faisant appel à des calculs de transport, il est nécessaire de disposer d'un fichier de données évaluées cohérent et suffisamment complet présenté dans un format standard. La présente contribution résume dans ses grandes lignes l'évaluation complète des interactions neutroniques du ^{209}Bi dans la gamme d'énergie 10^{-5}eV - 20 MeV (BR).

Evaluation

Du fait de sa structure particulière, voisine du noyau doublement magique ^{208}Pb, le ^{209}Bi a fait l'objet de nombreuses mesures expérimentales ($\sim$ 15000 valeurs), mais la répartition en énergie et selon les différentes réactions possibles n'est pas suffisante pour construire une évaluation complète et cohérente. Différents modèles de réactions nucléaires ont donc été utilisés, les paramètres étant ajustés sur un grand nombre de données expérimentales. La gamme d'énergie envisagée est divisée ici en deux parties : la zone des résonances résolues et infra, et la zone d'énergies supérieures.

La zone des résonances résolues s'étend, pour ce noyau, jusqu'à 265 keV du fait du grand espacement entre les résonances, en particulier celles induites par les ondes s. Après compilation, vérification et sélection des données expérimentales, on a constitué un jeu de paramètres de résonances comportant 50 résonances ℓ = 0 jusqu'à 265 keV et 72 résonances ℓ = 1 jusqu'à 77 keV (pour ces dernières, certaines largeurs neutroniques non déterminées expérimentalement ont été estimées à partir de la fonction force S_1). Certains de ces paramètres ayant été déterminés par l'analyse des données expérimentales soit à l'aide du formalisme de Reich-Moore, soit de celui de Breit-Wigner à un niveau, nous avons comparé les sections efficaces calculées avec l'un ou l'autre formalisme. Il apparaît que sans effet Doppler des différences notables n'apparaissent que dans certaines zones d'interférence négative. Il est à noter que ces différences disparaissent dès que l'on applique, à des fins de comparaison avec les résultats expérimentaux les corrections d'effet Doppler et de résolution, ce qui rend alors les deux formalismes quasiment indiscernables. La figure 1 montre un exemple de l'accord obtenu pour la section efficace totale dans la gamme d'énergie 8-32 keV entre les données évaluées (BR) corrigées (voir ci-dessus) et les données expérimentales (GA64,SI76). La section efficace totale extraite de l'évaluation de SMITH (SM) également reportée sur cette figure décrit moins en détail les résonances observées.

A très basse énergie (< 0,1 eV), la section efficace de diffusion élastique est fortement affectée par la structure cristalline rhomboédrique de ^{209}Bi, comme l'illustre la figure 2 où les discontinuités correspondent aux différents plans réticulaires du cristal. Dans le fichier de données évaluées où ces effets ne sont pas pris en compte, la section efficace est constante ($\sim$ 9,1 b).

Au-delà de la zone des résonances, les différents modèles de réactions nucléaires utilisés font appel aux paramètres ou aux quantités déduites du modèle optique sphérique. Les paramètres de celui-ci ont donc été déterminés avec soin à l'aide du code SCAT2 (BE81) de façon à reproduire au mieux en particulier les grandeurs suivantes :

- la section efficace totale entre 0,4 et 30 MeV,
- les distributions angulaires élastiques entre 4 et 24 MeV,
- les fonctions force S_0 et S_1 et le rayon de diffusion R' issus de la zone des résonances.

Le jeu de paramètres du modèle optique sphérique ainsi obtenu est indiqué Table I. La figure 3 montre la comparaison calcul-expérience des distributions angulaires élastiques pour différentes énergies incidentes.

Table I Paramètres du modèle optique sphérique de ^{209}Bi

	r(fm)	a(fm)
V(MeV) = 45,45 - 0,22E	1,265	0,65
W(MeV) = 2,2 + 0,47E (E⩽10 MeV) 6,9 - 0,045E (E>10 MeV)	1,235	0,5
W_V(MeV) = 0,08E	1,235	0,5
V_{SO}(MeV) = 5,5	1,08	0,6

Outre les pénétrabilités fournies par le modèle optique sphérique, le modèle statistique nécessite la connaissance des niveaux excités et des densités de niveaux des différents noyaux résiduels. Du fait de la proximité de la double couche fermée de ce noyau, les niveaux discrets présentent une répartition particulière et les densités de niveaux peuvent s'écarter singulièrement d'une répartition statistique. Nous avons néanmoins utilisé des densités de niveaux statistiques de la forme Gilbert et Cameron (GI65)

K. H. Böckhoff (ed.), Nuclear Data for Science and Technology, 665–668.

mais en ajustant certains paramètres sur les niveaux discrets et sur l'espacement moyen des niveaux à l'énergie de liaison pour le ^{210}Bi. Par ailleurs, pour traduire l'écart local (observé expérimentalement et théoriquement) existant vers 4,5 MeV par rapport à une loi en température de la densité de niveaux du ^{209}Bi, nous avons introduit un facteur correctif de la forme :

$$\rho(E) = \rho(E)[1 + 1,2 \exp(-(E - 4,5)^2)]$$

Sur la base de ces paramètres, et dans une gamme d'énergie approximative 0,2 - 10 MeV, les sections efficaces de capture, élastique et inélastiques ainsi que les distributions angulaires correspondantes ont été calculées avec le code COMNUC (DU70). La section efficace de capture, pour laquelle la largeur radiative a été légèrement ajustée de façon à bien reproduire le plateau situé entre 0,5 et 1 MeV (MA76), est donnée à titre d'exemple figure 4 entre 10 keV et 20 MeV. La forme de la composante de section efficace correspondant aux mécanismes direct et semi-direct à haute énergie a été prise identique à celle du ^{208}Pb et normalisée à 14 MeV. D'une façon générale on observe un bon accord avec les résultats expérimentaux, notamment les résultats récents de VOIGNIER (VO82).

Toujours sur la base des paramètres établis, la multiplicité des sections efficaces à plus hautes énergies et la possibilité de calculer les spectres neutrons et gamma émis pour chaque réaction nous ont conduits à utiliser le code GNASH (YO77). La figure 5 montre l'accord calcul-expérience obtenu pour les sections efficaces (n,2n) et (n,3n). Il faut noter que les effets du mécanisme de pré-équilibre devenant importants aux environs de 9 MeV, une correction a été introduite dans les calculs. Les sections efficaces (n,p) et (n,α) ainsi que les spectres des particules émises ont été calculés avec le code AMALTHEE (BE77).

Le dernier mécanisme pris en compte est le mécanisme d'intéraction directe utilisé pour décrire la section efficace inélastique du groupe de 7 niveaux excités du ^{209}Bi centrés vers 2,6 MeV. Cette composante a été calculée avec le code DWUCK (KU) en utilisant le jeu de paramètres optiques définis plus haut. Les résultats de la composition des mécanismes statistique et d'intéraction directe sommée sur les 7 niveaux à des fins de comparaison sont indiqués figure 6 et sont en bon accord avec les résultats expérimentaux.

Les figures 7 et 8 concernent les spectres de particules émises, calculées avec le code GNASH (YO77). La figure 7 montre l'accord calcul-expérience obtenu pour une énergie incidente de 14,5 MeV, accord qu'on ne peut améliorer au-dessus de 6 MeV sans détériorer celui que l'on observe aux énergies inférieures. La figure 8 montre le spectre de rayonnements gamma émis à 14,5 MeV. Aux énergies inférieures à 4 MeV, on observe un bon accord avec les résultats expérimentaux, tandis que pour les énergies supérieures il n'en est pas de même, mais l'intensité dans cette zone est faible.

Finalement, l'ensemble des résultats obtenus a fait l'objet de la création d'un fichier de format ENDF ($\sim$ 30000 cartes) dont toutes les sections sont tabulées et qui est actuellement en cours d'exploitation. Il est à noter que tout au long de cette évaluation, le système SYNOPSIS (CO80) a facilité considérablement la manipulation du grand nombre de données aussi bien expérimentales qu'évaluées.

Références

AL75 E. Almen-Ramstrom, AE-503 (1975)

AS58 V.J. Ashby, H.C. Catron, L.L. Newkirk, C.J. Taylor, Phys. Rev. 111 (1958) 616

BE65 I. Bergqvist, B. Lundberg, Antwerp (1965) p. 550

BE77 O. Bersillon, L. Faugère, NEANDC(E)191"L"

BE80 V. Bezotosnyi, V. Gorbachev, M. Shvetson, L. Iurov, Atomnaya Energ. 49 (1980) 239.

BE81 O. Bersillon, CEA-N-2227 (1981)

BO58 R. Booth, W.P. Ball, M.H. McGregor, Phys. Rev. 112 (1958) 226.

BR O. Bersillon, B. Caput, C.A. Philis, Rapport à paraître

BU79 M. Budnar, F. Cvelbar, E. Hodgson et al., INDC(YUG)-6

CO80 M. Collin, A. Schett, C.A. Philis, NEANDC(E)207"L", INDC(FR)37/L (1980)

CS67 J. Csikai, G. Petö, M. Buczko, Z. Miligy, N.A. Eissa, Nucl. Phys. A95 (1967) 229

DE67 V.G. Degtyarev, V.N. Protopopov, Energ. Atom. 22 (1967) 118

DE75 F. Deaf, S. Gueth, J. Inczédy, A. Kiss Acta Physica Scient. Hung. 38 (1975) 209

DI60 B.C. Diven, J. Terrell, A. Hemmendiger, Phys. Rev. 120 (1960) 556

DI70 M. Diksic, P. Strohal, G. Petö, P. Bornemisza-Pauspertl et al., Acta Physica Scient. Hung. 28 (1970) 257

DU70 C.L. Dunford, AI-AEC-12931 (1970)

EG53 P.A. Egelstaff, AERE-N/R-1147 (1953)

FO71 D. Foster, D. Glasgow, Phys. Rev. C3 (1971) 576

FR80 J. Fréhaut, Brookhaven (1980) p. 399

GA64 J.B. Garg, J. Rainwater, W.W. Havens Jr., CR-1860 (1964)

GI61 J.H. Gibbons, R.L. Macklin, P.D. Miller, J.H. Neiler, Phys. Rev. 122 (1961) 182

GI65 A. Gilbert, A.G.W. Cameron, Can. J. Phys. 43 (1965) 1446

GO64 G.V. Gorlov, N.S. Lebedeva, V.M. Morozov, Doklad. Akad. Nayk. 158 (1964) 574

HA48 W.W. Havens Jr., E.J. Rainwater, C.S. Wu, J.R. Dunning, Phys. Rev. 73 (1948) 963

HA80 G. Haouat,Comm.privée(1980)et CEA-N-2200(1981)

HE74 D. Hermsdorf, A. Meister, S. Sassonof, D. Seeliger, K. Seidel, ZfK 283 (1974)

HU53 D.R. Hughes, H. Palevsky, Phys. Rev. 92 (1953) 1206

KI57 Kimball, Monahan, Mooring, WASH 745 (1957) 7

KO76 L. Koester, W. Nistler, W. Maschkowski, Phys. Rev. Lett. 36 (1976) 1021

KU P.D. Kuntz, The Program DWUCK, Boulder, Unpublished

KU72 P. Kuijper, J.C. Veefkind, C.C. Jonker, Nucl. Phys. A181 (1972) 545

MA73 M. Matoba, M. Hyakutake, H. Tawara et al., Nucl. Phys. A204 (1973) 129

MA76 R. Macklin, J. Halperin, Phys. Rev. C14 (1976) 1389

PE58 J.L. Perkin, L.P. O'Connor, R.F. Coleman, Proc. Phys. Soc. 72 (1958) 505

PO59 A. Poularikas, R.W. Fink, Phys. Rev. 115 (1959) 989

RO57 L. Rosen, L. Stewart, Phys. Rev. 107 (1957) 824

SI76 U.N. Singh, J. Rainwater, H.I. Liou, G. Hacken, J.B. Garg, Phys. Rev. C13 (1976) 124

SM A.B. Smith, J. Whalen, E. Barnard, J. Devilliers, D. Reitman, Nucl. Scien. Eng. 41 (1970) 63

SM A.B. Smith, P. Guenther, D. Smith, J. Whalen, R. Howerton, ANL/NDM-51 (1980)

A.B. Smith, P. Guenther, J. Whalen, Nucl. Scien. Eng. 75 (1980) 69

ST65 P.H. Stelson, R.L. Robinson, H.J. Kim, J. Rapoport, G.R. Satchler, Nucl. Phys. 68 (1965) 97

TA72 S. Tanaka, Y. Tomita, K. Ideno, S. Kikuchi, Nucl. Phys. A179 (1972) 593

VE77 L.R. Veeser, E.D. Arthur, P.G. Young, Phys. Rev. 16 (1977) 1792

VO82 J. Voignier, S. Joly, G. Grenier, This Conference

WI77 L. Wilde, H. Mennekes, V. Schröder, W. Scobel, J. Phys. G3 n° 5 (1977) L99

YO77 P.G. Young, E.D. Arthur, LA-6947 (1977)

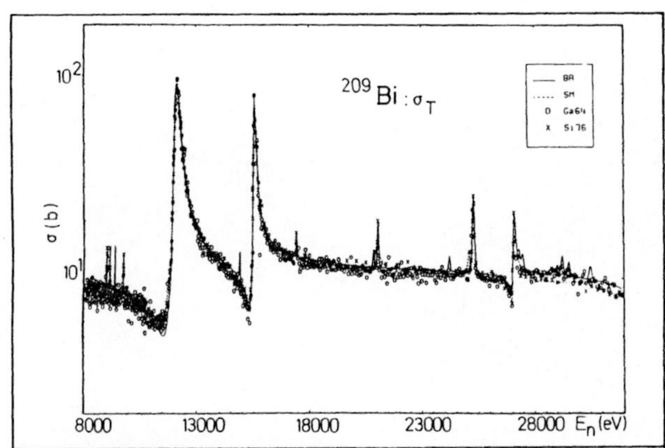

Fig. 1. Section efficace totale : données évaluées et expérimentales dans la gamme d'énergie 8 keV - 32 keV. Les données évaluées (BR) ont été convoluées par une Gaussienne pour pouvoir être comparées aux données expérimentales.

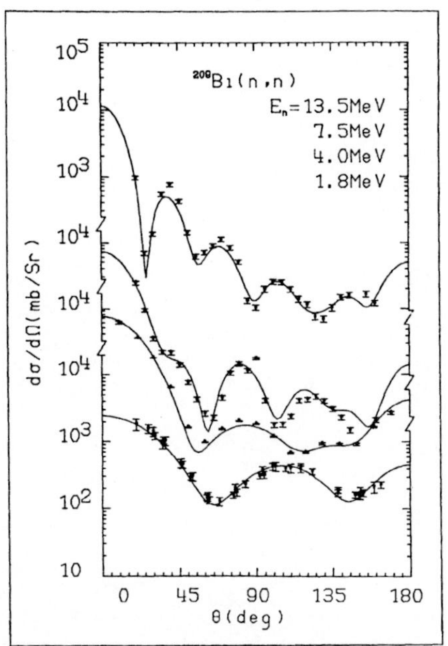

Fig. 3. Section efficace différentielle de diffusion élastique de ^{209}Bi : comparaison des données évaluées (courbes continues) aux données expérimentales pour des énergies incidentes de 1,8 MeV (SM), 4 MeV (GO64), 7,5 et 13,5 MeV (HA80).

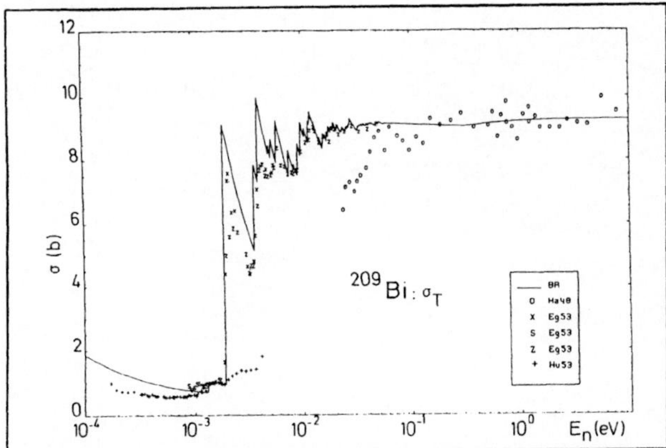

Fig. 2. Section efficace totale : les données expérimentales dans la gamme d'énergie 10^{-4} eV - 10 eV sont comparées aux données évaluées pour lesquelles les effets cristallins ont été introduits.

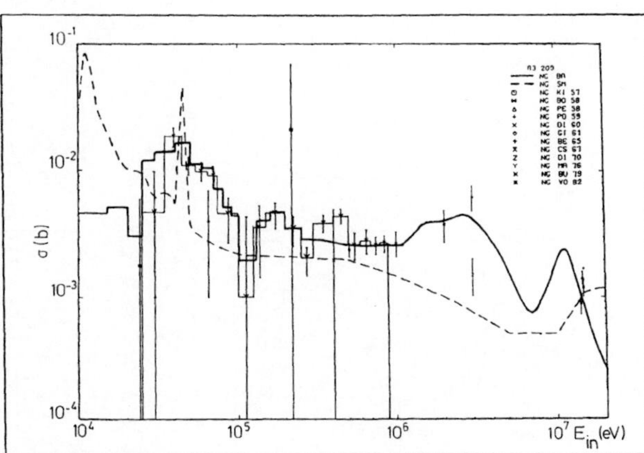

Fig. 4. Section efficace de capture de ^{209}Bi de 10^4 eV à 20 MeV. Au-dessous de 0,3 MeV, nos données ont été mises en groupes pour faciliter la comparaison avec les autres données évaluées (SM) et les données expérimentales.

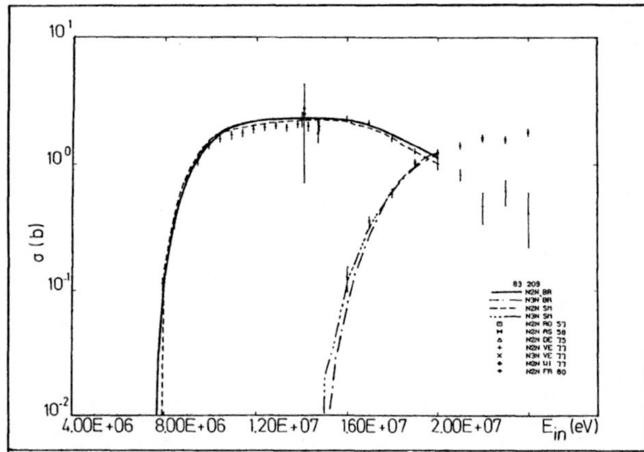

Fig. 5. Sections efficaces (n,2n) et (n,3n) de ^{209}Bi : comparaison entre les différentes données évaluées et les résultats expérimentaux.

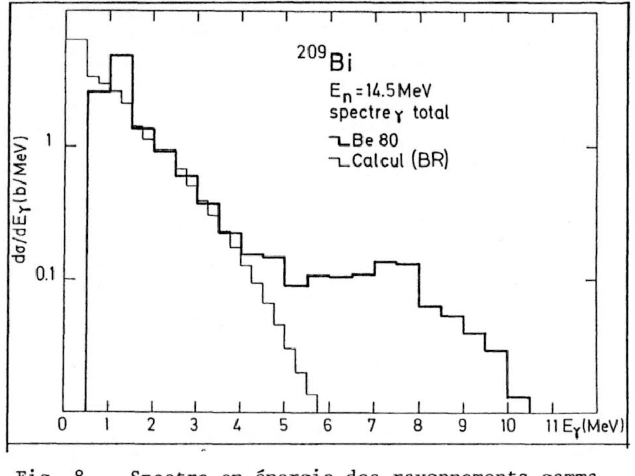

Fig. 8. Spectre en énergie des rayonnements gamma émis au cours de l'intéraction d'un neutron de 14,5 MeV avec ^{209}Bi : comparaison du résultat expérimental (BE80) avec les données évaluées (BR).

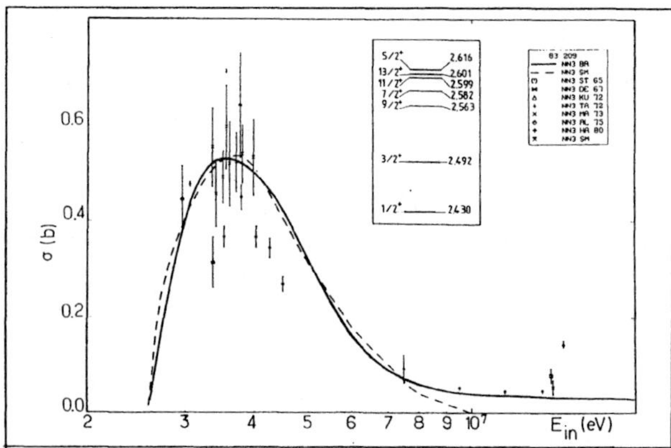

Fig. 6. Section efficace de diffusion inélastique de ^{209}Bi pour le groupe de 7 niveaux d'énergies 2,43 à 2,62 MeV (voir encarté). Comparaison entre les résultats expérimentaux, les données évaluées de A. Smith (SM) et la somme des 7 sections efficaces partielles que nous avons évaluées (BR).

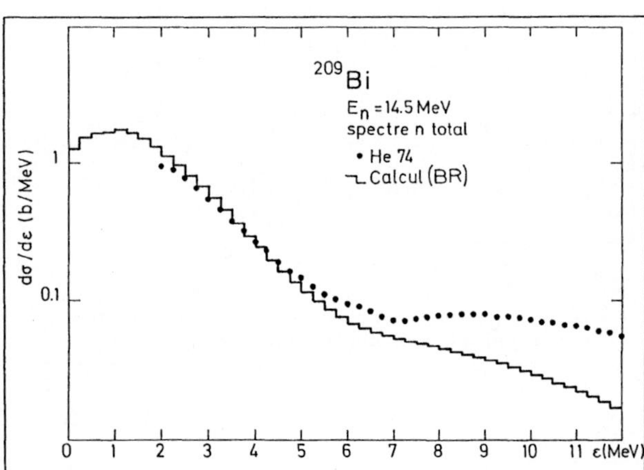

Fig. 7. Spectre en énergie des neutrons secondaires émis au cours de l'intéraction d'un neutron d'énergie 14,5 MeV avec ^{209}Bi : comparaison du résultat expérimental (HE74) avec les données évaluées (BR).

EVALUATION DES SECTIONS EFFICACES NEUTRONIQUES DE ^{238}Pu de 10^{-5} ev A 16 Mev

H. Derrien

Centre d'Etudes Nucléaires de Cadarache
B.P. n° 1, 13115 Saint Paul Lez Durance France

Abstract

The ^{238}Pu neutron cross-sections have been evaluated in the energy range 10^{-5} ev to 16 Mev. The anomalies observed in the measured total cross-sections (very high value of the potential scattering cross-section) and the measured capture cross-sections (high value of the non resonant part in the resolved and unresolved region) have been examined and considered as being due to some unknown experimental effect. More realistic cross-sections have been proposed from statistical and optical model calculations based on average parameters deduced from the resonance analysis and the experimental fission cross-sections which seem to be reasonably well known on the entire energy range.

Résumé

Les sections efficaces de ^{238}Pu ont été évaluées dans le domaine d'énergie de 10^{-5} ev à 16 Mev. Les anomalies observées dans les sections efficaces totales mesurées (valeur très haute de la section efficace de diffusion potentielle) et dans les sections efficaces de capture (valeur haute de la partie non résonante dans le domaine résolu et non résolu) ont été examinées et considérées comme étant dues à des effets expérimentaux non connus. Des sections efficaces plus réalistes ont été proposées à partir de calculs par modèle optique et statistique basés sur des paramètres moyens obtenus à partir de l'analyse des résonances et sur la section efficace de fission expérimentale qui semble être assez bien connue dans tout le domaine d'énergie.

Région thermique et des résonances résolues

Les données expérimentales. Il y a relativement peu de données expérimentales disponibles pour l'évaluation des sections efficaces de ^{238}Pu dans le domaine des résonances résolues. On trouve les 4 ensembles suivants dans le fichier international EXFOR :

1/ - les aires de fission de 5 résonances obtenues par GERASIMOV et al.[1] à partir d'une mesure de section efficace de fission de 0.02 ev à 400 ev ;

2/ - les aires de fission de 16 résonances obtenues par STUBBINS et al.[2] également à partir d'une mesure de section efficace de fission de 2 ev à 300 ev ;

3/ - les largeurs neutroniques de 14 résonances obtenues par YOUNG et al.[3] par une analyse de forme de transmissions expérimentales dans l'intervalle d'énergie de 0.01 ev à 200 ev ;

4/ - les largeurs neutroniques et de fission de 49 résonances obtenues par SILBERT et al.[4,5] par une analyse d'aires des sections efficaces de fission et de capture mesurées simultanément dans l'intervalle d'énergie de 20 ev à 500 ev (explosion nucléaire PERSIMMON).

Les paramètres de résonances. Aucune de ces 4 séries d'expériences n'avait une résolution suffisante pour permettre une bonne définition et une bonne séparation des résonances. Les résultats obtenus par SILBERT et al. sont les plus complets et les plus précis. Au-dessous de 140 ev où les résultats de YOUNG et al. et de SILBERT et al. peuvent être comparés, il y a un bon accord d'ensemble sur les largeurs neutroniques Γn. Quant aux largeurs de fission, il y a également un bon accord entre les résultats de STUBBINS et al. et de SILBERT et al. du moins au-dessous de 100 ev. On peut obtenir la largeur de capture radiative pour trois résonances seulement en combinant les résultats de YOUNG et al. et de STUBBINS et al. : 36,8 mev ; 30,3 mev 35,2 mev respectivement pour les résonances à 2,88 ev; 9,98 ev et 18,56 ev. Une valeur moyenne de capture radiative de 34 mev avait été retenue par SILBERT et al. pour obtenir les largeurs neutroniques et les largeurs de fission à partir des aires expérimentales de capture et de fission.

L'ensemble des paramètres de résonance recommandé a été établi de la façon suivante jusqu'à l'énergie de 500 ev :

1/ - les 2 résonances négatives sont celles utilisées par YOUNG et al. pour reproduire la forme de la section efficace totale mesurée à basse énergie à partir de l'énergie de 0,008 ev ;

2/ - les paramètres de résonances à 2,885 ev ; 9,976 ev et 18,562 ev ont été obtenus en combinant les résultats de YOUNG et al. et de STUBBINS et al. ;

3/ - au-dessus de 20 ev les résultats de SILBERT et al. ont été retenus dans leur ensemble, sauf la valeur de $\Gamma n°$ de la résonance à 320 ev qui a été prise égale à (7 ± 5) mev au lieu de la valeur (10 ± 10) mev donnée par SILBERT et al.

Les sections efficaces calculées. Les paramètres de résonance permettent de calculer les sections efficaces de 10^{-5} ev à 400 ev. Le formalisme BREIT-WIGNER à un niveau est suffisant pour le calcul des sections efficaces de fission ; la précision supplémentaire qui serait obtenue en utilisant un formalisme multiniveaux est négligeable comparée à l'imprécision qui existe sur les données expérimentales.

Deux remarques sont à faire concernant les sections efficaces calculées :

1/ - YOUNG et al. ont utilisé une section efficace potentielle de 17 barns pour décrire leurs résultats expérimentaux entre 0,008 ev et 200 ev. Ils n'ont pas réussi à expliquer cette valeur anormale par des effets expérimentaux tels que la contamination par l'eau des échantillons de PuO_2 utilisés dans les expériences de transmission. Il suffirait, par ailleurs, d'admettre 1 % d'erreur systématique sur les transmissions mesurées (erreur de bruit de fond ou erreur de normalisation) pour expliquer près de 5 barns d'erreur absolue sur les sections efficaces totales, l'épaisseur maximum des échantillons utilisés n'étant que de 0,0023 at/b. Nous avons retenu une valeur plus réaliste de (11 ± 1) barns pour la section efficace potentielle, ce qui correspond à $R' = 9,36$ fm ;

2/ - les paramètres de résonances ne reproduisent pas les valeurs élevées de la section efficace de capture observées entre les résonances dans les résultats expérimentaux de SILBERT et al. La différence

K. H. Böckhoff (ed.), Nuclear Data for Science and Technology, 669–672.

entre la section efficace calculée et la section efficace mesurée est en moyenne de 4,6 barns entre 20 ev et 300 ev. Cette différence est trop importante pour être expliquée par la contribution de résonances négatives ou tout autre effet tel que la capture directe. On doit donc l'attribuer à un effet expérimental non signalé par les auteurs et ne pas en tenir compte dans les résultats de l'évaluation.

Les sections efficaces calculées à 0,0253 ev sont données dans le tableau I et comparées aux valeurs que l'on peut trouver dans ENDF/B-V et KEDAK.

TABLEAU I - Sections efficaces à 0,0253 ev en barns

	Ce Travail	ENDF/B-V	KEDAK
σ_T	583,75	590,0	575,5
$\sigma n, \gamma$	546,74	552,4	541,5
σn, f	17,15	17,0	17,0
σn, n	19,86	20,6	17,0

Valeurs moyennes des paramètres de résonances

Espacement moyen et fonction densité. Toutes les résonances visibles dans les sections efficaces expérimentales ont été considérées comme résonances s. L'excès de petites valeurs observées par SILBERT et al. dans la distribution des largeurs neutroniques réduites a été considéré par les auteurs comme pouvant être dû à une sous-estimation des petites valeurs, ce qui ne serait pas surprenant dans une méthode d'analyse où les largeurs neutroniques ont été évaluées à partir des aires de capture et de fission en attribuant à toutes les résonances une valeur constante pour la largeur de capture radiative. D'autre part, étant donné la mauvaise qualité de résolution des expériences de SILBERT et al. ou de YOUNG et al., il est très probable qu'un nombre important de grandes valeurs de $\Gamma n°$ corresponde à des multiplets non résolus. Il n'est donc pas surprenant que la distribution des largeurs neutroniques réduites s'adapte mal à une loi de PORTER-THOMAS. En fait, toute tentative pour déterminer les paramètres moyens par les méthodes usuelles appliquées à la distribution des largeurs neutroniques réduites donne des résultats très peu précis.

L'espacement moyen des résonances dans l'intervalle d'énergie 0 ev à 250 ev est de 8,3 ev. Nous avons adopté la valeur $\langle D\rangle_{\ell=0}$ = (7,7 ± 0,6) ev en estimant que 10 % des niveaux n'étaient pas observés dans cet intervalle d'énergie.

La fonction densité a été évaluée à :

$$So = (\sum_0^E \Gamma n°)/E = (1,17 \pm 0,20) \times 10^{-4}$$

dans le domaine d'énergie analysé.

Les voies de fission. La principale caractéristique des largeurs de fission apparaît sur la figure 1 due à SILBERT et al. La structure intermédiaire due à une barrière de fission à deux bosses est évidente.

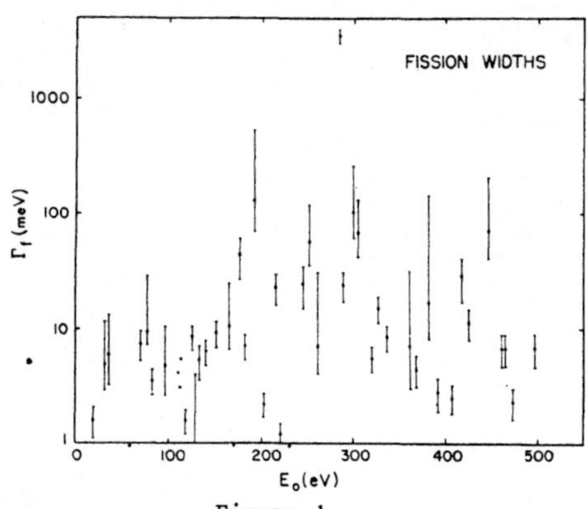

Figure 1

Il semblerait que toutes les largeurs de fission mesurées dans l'intervalle d'énergie de neutrons de 0 à 500 ev subissent l'influence d'un état de classe II qui serait situé aux environs de 285 ev. Dans ce cas il est très difficile d'interpréter la distribution des largeurs de fission (fig. 2), dont la forme suggère l'existence de deux familles de résonances à largeurs moyennes très différentes, mais non compatibles avec la présence d'un seul état de spin (I=0, ℓ=0, J=1/2). La séparation en deux familles faite sur la figure 2 n'est donnée qu'à titre indicatif. SILBERT et al. d'une part et BOWMAN et al.[6] d'autre part ont pensé que la composante à faible valeur moyenne pourrait être due au processus (n, γf). Mais le nombre de degré de liberté faible (ν=3) et la valeur moyenne (7 mev) ne sont pas compatibles avec ce processus, comme l'a d'ailleurs souligné LYNN[7].

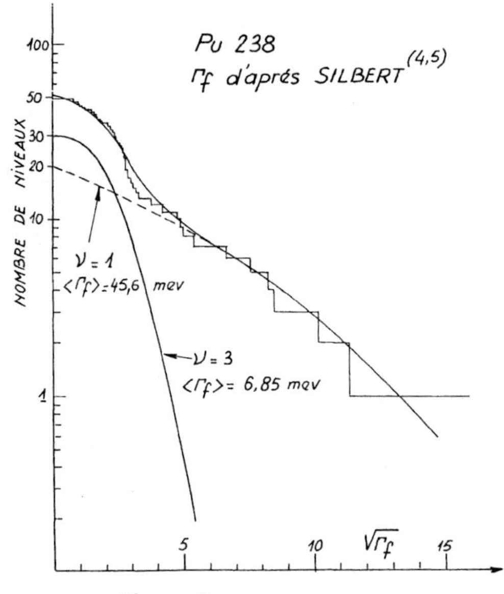

Figure 2

Dans notre évaluation nous tenons compte implicitement de la structure intermédiaire en moyennant les sections efficaces expérimentales de fission de SILBERT et al. dans la région non résolue.

Largeur moyenne de capture radiative. La valeur retenue est égale à (34 ± 6) mev. C'est celle qui a été utilisée par SILBERT et al. Elle représente la moyenne arithmétique de trois valeurs expérimentales. Elle est en accord avec la valeur de (35,7 ± 2,2) mev relevée dans la systématique de MOORE[8].

Les paramètres du modèle optique

Il n'y a pas de données expérimentales de sections efficaces totales, de diffusion élastique ou inélastique au-dessus de quelques kev. L'évaluation doit donc être basée sur des résultats de calcul par modèle optique et statistique. Un ensemble de paramètres de modèle optique a été proposé par C. LAGRANGE[9] pour les isotopes pairs du Pu. Il a utilisé les paramètres de déformation de MOLLER[10] et un potentiel optique basé sur les données expérimentales de ^{232}Th et ^{238}U. Ces paramètres reproduisent à basse énergie les fonctions densité So = 0,944 x 10^{-4} et S1 = 1,90 x 10^{-4}, et un rayon effectif de diffusion R' = 9,26 fm. La fonction densité So de ce modèle optique est 20 % inférieure à celle obtenue à partir des paramètres de résonances. Un meilleur accord est obtenu en modifiant légèrement les paramètres de LAGRANGE, c'est à dire en utilisant les paramètres du tableau II. On obtient dans ce cas : So = 1,12 x 10^{-4}, S1 = 1,97 x 10^{-4} et R' = 0,931. Ce potentiel optique a été utilisé pour les calculs dans le domaine non résolu et à haute énergie.

Tableau II - Paramètres du potentiel optique utilisé pour ^{238}Pu

Potentiel	Profondeur (Mev)	rayon (fm)	Diffusivité (fm)
réel	47,2-0,3E	1.244	0.62
Imaginaire de surface	2,7+0,4E (E<10Mev) 6,7 (E>10Mev)	1.280	0.58
Spin orbite	7,5	1.240	0.62
Paramètres de déformation : β_2=0,219 β_4=0,089			

Les sections efficaces dans le domaine non résolu (0.4 kev à 40 kev).

Dans ce domaine d'énergie les données expérimentales disponibles sont les résultats des mesures des sections efficaces de capture et de fission de SILBERT et al., et quelques points de fission de ERMAGAMBETOV et al.[12]. La structure intermédiaire qui est visible dans la région des résonances résolues dans la section efficace de fission est encore apparente jusqu'à vers 10 kev, l'espacement moyen des états de classe II étant à peu près égal à 1 kev. Quant à la composante non résonnante anormale qui existe à basse énergie dans la section efficace de capture de SILBERT, il n'y a pas de raison qu'on ne la retrouve pas à plus haute énergie dans le domaine non résolu. Pour évaluer son importance, les sections efficaces partielles de réaction ont été calculées par le code de modèle statistique FISINGA[11], en imposant à la section efficace de fission calculée d'être égale aux moyennes expérimentales dans 5 intervalles d'énergie choisis. Les résultats sont montrés dans le tableau III. Dans chaque intervalle d'énergie la section efficace de capture calculée ne représente qu'à peu près 60 % de la section moyenne mesurée. Cette différence a été interprétée comme étant due à une erreur systématique dans les sections efficaces de capture mesurées.

Tableau III - Comparaison entre la section efficace de capture mesurée (réf.5) et la valeur calculée (barns)

Energie (kev)	σ (n, f) Exp. (réf.5)	σ (n, γ) Exp. (réf.5)	σ (n, γ) calculée
0,06-1	3,98	12,11	7,75
1 - 4	1,220	3,970	2,670
4 - 10	0,980	2,250	1,350
10 - 20	0,752	1,744	0,950
20 - 40	0,803	1,152	0,675

Par suite, l'évaluation a été conduite de la façon suivante :

1/ - entre 400 ev et 4 kev les sections efficaces totales de YOUNG et al. et les sections efficaces de capture et de fission de SILBERT et al. ont été moyennées par intervalle d'énergie de 50 ev entre 400 ev et 1000 ev et par intervalle d'énergie de 200 ev entre 1 kev et 4 kev. La section efficace de YOUNG a été normalisée sur les résultats des calculs par le code FISINGA. Les sections efficaces de capture de SILBERT ont été corrigées de 3.10 b à 400 ev et de 1.15 b à 4 kev et, entre ces deux énergies, d'une valeur obtenue par interpolation log-log ;

2/ - entre 4 kev et 40 kev, les données de SILBERT ont été moyennées par intervalle d'énergie de 1 kev entre 4 kev et 10 kev et par intervalle d'énergie de 2 kev entre 10 kev et 40 kev. La correction à la capture de SILBERT était de 1.15 b à 4 kev et de 0.45 b à 40 kev. Les sections efficaces totales et de formation du noyau composé ont été obtenues par le code FISINGA, et la section efficace de diffusion élastique par différence entre la section efficace totale calculée et la somme des sections efficaces de capture et de fission.

Les résultats de l'évaluation dans le domaine non résolus sont reproduits dans la figure 3.

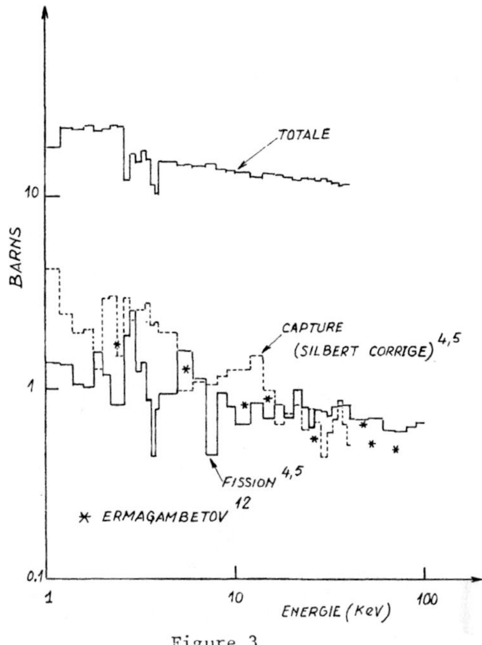

Figure 3

La région des hautes énergies (40 kev à 14 Mev)

Les données expérimentales. A part la section efficace de capture qui a été mesurée jusqu'à 200 kev par SILBERT et al.[5] seules des données de fission existent à haute énergie[4,5,12-18]. Les résultats de VOROTNIKOV et al.[14] ont été normalisés relativement à une mesure absolue à 720 kev avec une précision de 20 %. Les résultats de BUTLER et al.[13] proviennent également de mesures absolues et sont en accord avec ceux de VOROTNIKOV. Les autres mesures ont été faites relativement à la fission de l'^{235}U et ont utilisé le standard de DAVEY[19] pour la normalisation. Malheureusement, au dessus de 0.5 Mev, les résultats des mesures absolues sont à environ 30 % au-dessus de ceux obtenus par des mesures relatives à l'^{235}U. Par ailleurs, les calculs basés sur la systématique de BEHRENS et al.[20] ou de LYNN[21] donnent raison aux mesures relatives à l'^{235}U. Pour cette raison nous n'avons pas tenu compte des résultats de VOROTNIKOV et al. et de BUTLER et al.

Une évaluation de la section efficace de fission a été faite jusqu'à 5 Mev en moyennant l'ensemble choisi des données expérimentales. Au-dessus de 5 Mev, nous n'avons connaissance que des valeurs préliminaires de KNITTER et al.[22] et de quelques valeurs d'ERMAGAMBETOV et al.[17] aux environs de 15 Mev.

Les sections efficaces calculées. Elles ont été obtenues par le code de modèle statistique FISINGA. On a utilisé les coefficients de transmission neutroniques calculés par le code de modèle optique en voies couplées ECIS[23] avec le potentiel optique du tableau II. Les paramètres des barrières de fission, la densité des voies de fission, la densité des voies inélastiques dans le continuum sont ceux déduits de la systématique de LYNN[24]. Au-dessus de 5 Mev, le code FISINGA a été couplé au code SI2N[25] pour calculer la section efficace de fission de 2ème chance et la section efficace (n, 2n). Quelques valeurs des sections efficaces de fission calculées sont comparées aux résultats de l'évaluation des données expérimentales dans le tableau IV. L'accord est satisfaisant. On a retenu pour l'évaluation définitive les valeurs déduites des sections efficaces expérimentales ; les autres sections efficaces partielles calculées ont été ajustées pour conserver la valeur de la section efficace totale. Toutefois, au-dessus de 5 Mev, on a conservé les sections efficaces de fission calculées qui sont d'ailleurs assez voisines des valeurs préliminaires de KNITTER et de celles d'ERMAGAMBETOV.

Tableau IV Comparaison entre l'évaluation expérimentale et quelques valeurs calculées

Energie (Mev)	σ (n, f) barns expérimentale	σ (n, f) barns calculée
0,040	0,708	0,669
0,070	0,620	0,588
0,100	0,613	0,559
0,400	1,184	1,184
0,700	1,905	1,944
1,000	2,128	2,098
1,500	2,184	2,141
3,000	2,195	2,113

Comparaison avec d'autres évaluations et avec les mesures intégrales

Les figures 4, 5, et 6 permettent de comparer la présente évaluation à ENDF/B-V, KEDAK et aux résultats des calculs de P. THOMET[26], pour les sections efficaces de diffusion, de capture et de fission.

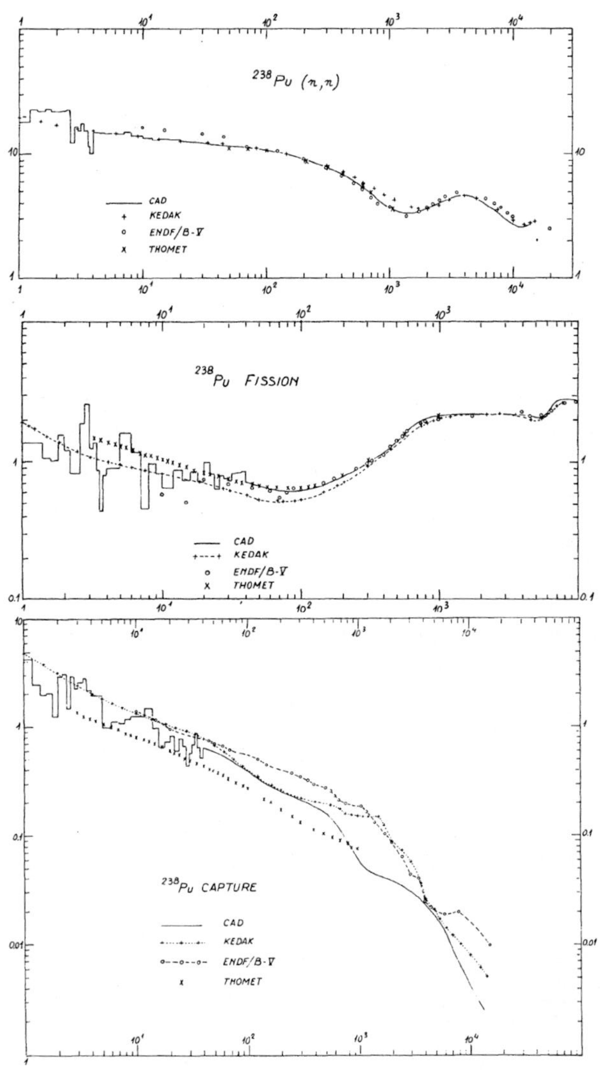

Energie en Kev

Figures 4, 5, 6

En ce qui concerne les mesures intégrales, une comparaison a été faite avec les résultats des mesures de taux de fission et de capture dans ERMINE et PHENIX, qui ont permis d'ajuster le jeu CARNAVAL IV à partir du jeu CARNAVAL III[27]. En fait, la section efficace de capture à un groupe calculée à partir de la présente évaluation est en accord à moins de 4 % avec celle que l'on peut calculer à partir du jeu ajusté CARNAVAL IV ; par contre, si la comparaison est faite groupe par groupe dans le jeu à 25 groupes, on note des écarts de - 30 % à + 30 % dans les régions significatives du spectre. Quant à la section efficace de fission, la valeur à un groupe obtenue à partir de l'évaluation est supérieure de 26 % à celle obtenue à partir du jeu CARNAVAL IV. Ceci est évidemment incompatible avec la précision de l'ajustement CARNAVAL IV (moins de 3 %) et la précision de l'évaluation ou des mesures des données microscopiques de fission (6 à 7 % dans la région significative du spectre).
Des mesures ou des évaluations supplémentaires sont donc nécessaires pour résoudre ce problème de désaccord.

Références

1. GERASIMOV, Nuclear Data for Reactors Vol.2, p.129 (Paris 1966)
2. W.F. STUBBINS, C.D. BOWMAN, G.F. AUCHAMPAUGH, M.S. COOPS, Phys. Rev. 154, 1111 (1967)
3 T.E. YOUNG, F.B. SIMPSON, J.R. BERRETH, M.S. COOPS Nucl. Scien. Eng. 30, 355 (1967)
4 M.G. SILBERT, A. MOAT, T.E. YOUNG, Nucl. Scien. Eng. 52, 176 (1973)
5. M.G. SILBERT, J.R. BERRETH, Nucl. Scien. Eng. 52, 187 (1973)
6. C.D. BOWMAN, G.F. AUCHAMPAUGH, W.E. STUBBINS, T.E. YOUNG, F.B. SIMPSON, M.S. MOORE, Phys. Rev. Let. 18, 15 (1967)
7. J.E. LYNN, The theory of neutron resonance reactions, p. 108, Clarendon Press, OXFORD (1968)
8. M.S. MOORE, Neutron Physics and Nuclear Data, p. 313 (HARWELL 1978)
9. Ch. LAGRANGE, CEA-N-1970 (1977)
10. P. MOLLER, Nucl. Phys. A192, 529 (1972)
11. E. FORT, D. LAFOND, F. RIBAUD, à paraitre
12. S.B. ERMAGAMBETOV, G.N. SMIRENKIN, Sov. At. En. 25, 1364 (1968)
13. D.K. BUTLER and R.K. SJOBLOM, Bul. Am. Phys. 8, 369 (1963)
14. P.E. VOROTNIKOV, S.M. DUBROVINA, G.A. OTROSHCHENKO V.A. SHIGIN, Sov. J. Nucl. Phys. 3, 348 (1966)
15. D.M. BARTON, P.G. KOONTZ, Phys.Rev. 162, 1070(1967)
16. E.F. FOMUSHKIN, E.K. GUTNIKOVA, Sov. J. Nucl. Phys. 10, 529 (1970)
17. S.B. ERMAGAMBETOV, G.N. SMIRENKIN, JETP letters 9, 309 (1969)
18. D.M. DRAKE, C.D. BOWMAN, M.S. COOPS, R.W. HOFF, LA-4420, p. 101 (1970)
19. W.G. DAVEY, Nucl. Scien. Eng. 26, 149 (1966), Nucl. Scien. Eng. 32, 35 (1968)
20. J.W. BEHRENS, R.J. HOWERTON, Nucl. Scien. Eng. 65, 464 (1978)
21. J.E. LYNN, The theory of neutron resonance reactions, Clarendon Press, OXFORD (1968)
22. H. KNITTER, communication privée
23. R. RAYNAL, Rapport LYCEN-6804
24. J.E. LYNN, AERE-R-7468 (1974)
25. E. FORT et J.P. DOAT, DRNR/SPNR 79/5703
26. P. THOMET, rapport CEA-R-4631
27. A. GIACOMETTI, M. DARROUZET, P. HAMMER, M. LUCAS, F. PROST-MARECHAL, M. ROBIN, Fast Reactor Physics, Vol.I, p.521, Aix-en-Provence Sept. 1979.

EVALUATION DES SECTIONS EFFICACES NEUTRONIQUES DE ^{237}Np ENTRE 5 MeV et 16 MeV. ETUDE PARTICULIERE DE LA REACTION (n,2n) POUR L'APPLICATION AUX CALCULS DE PRODUCTION DE ^{236}Pu

E. FORT, H. DERRIEN, J.P. DOAT

DRNR/SEDC/SPNR
Centre d'Etudes Nucléaires de Cadarache
Boîte Postale n°1
F.13115 Saint-Paul-Lez-Durance
France

In the energy range 5MeV-16MeV the ^{237}Np neutron cross sections have been evaluated using theoretical models. More specifically the (n,2n) cross section evaluation is based on the fission probabilities of ^{236}Np and ^{235}Np as functions of energy. Concerning ^{236}Pu production the calculated microscopic values are consistent with integral data through some assumptions.

I - Introduction

L'importance de ^{237}Np résulte de la réaction de fission utilisée dans les opérations de dosimétrie et de la réaction (n,2n) qui conduit à ^{236}Np puis à ^{236}Pu. Ce dernier noyau est le point de départ d'une chaîne de décroissance qui conduit à ^{208}Te émetteur d'une raie γ à 2.614 MeV. Le dimensionnement optimisé des protections nécessaires lors du retraitement et du transport du combustible demande un calcul aussi exact que possible de la production de ^{236}Pu lors de l'irradiation.

II - Méthode d'évaluation et formalisme

II.1 - Calcul de la section efficace de la réaction $^{237}Np(n,2n)$ ^{236}Np

Aux énergies supérieures à 6.8 MeV l'émission de particules multiples devient possible :(n,n'f), (n,2n), (n,2nf), (n,3n), ...).

Dans le cadre du formalisme choisi, la section efficace d'une réaction sur un noyau A, fissile ou fertile, qui donne lieu à x neutrons pour 1 neutron incident, s'exprime par :

$$\sigma_{n,xn_i} = \sigma_c^{A+1}(E^*_{A+1}).P_{ne}(E^*_{A+1}).P(E^*_{A+1},xn_i) \qquad (1)$$

i = γ ou fission
avec des notations évidentes.

$$P_{ne} = \left(\frac{\Gamma n}{\Gamma_T}\right)^{A+1} = \frac{\sigma_c^{A+1} - (\sigma_{n,\gamma} + \sigma^1_{n,f} + \sigma^{\text{composé}}_{n,n})}{\sigma_c^{A+1} - \sigma^{\text{composé}}_{n,n}}$$

$$P(E^*_{A+1}, xn_i) =$$

$$\frac{\int_o^{E^*_{A+1} - B_n^{A+1} - B_n^A - \dots B_{n_A}^{A-(x-2)}} [\varepsilon\, \sigma^{A+1}(\varepsilon)\rho^A(E^*_A - \varepsilon)]^n \left(\frac{\Gamma n}{\Gamma T} E^*\right).P(E^*_{A.}(x-1)n_i)d\varepsilon}{\int_o^{E^*_{A+1}} [\varepsilon_1 \sigma_c^{A+1}(\varepsilon_1)\rho^A(E^*_A - \varepsilon_1)]\, d\varepsilon_1} \qquad (3)$$

Dans les quantités entre crochets ne figure pas la composante de prééquilibre. On peut tenir compte de celle-ci d'une façon phénoménologique en suivant l'approche suggérée par CLINE et BLANN /1/, adoptée par JARY /2/ et transformer la quantité

$\sigma_c^{A+1}(\varepsilon)\ \rho^A(E^*_A - \varepsilon)$ en la suivante :

$$[(1-R(E)\ \varepsilon\ \sigma_c^{A+1}(\varepsilon)\ \rho^A(E^*_A - \varepsilon)] + \rho^{A+1}(E^*_{A+1})\ \frac{8g_r}{g_c^5 E^3}\ \frac{1}{2\Gamma c}\ \sigma_c(\varepsilon)\ x\ \frac{A^3}{K^2}$$

$$x \sum_{h=1}^{6} h(h+1)\ \frac{g_r.(E_A^* - \varepsilon)^{2h-1}}{g_c.E^*_{A+1}}$$

où les diverses variables ont la signification donnée en /1/.

Les largeurs partielles sont calculées de la façon suivante :

$$\Gamma_n^A = C_n \frac{\int_o^{E^*_{A-1}} \varepsilon\sigma_c(\varepsilon)\ \rho_{A-1}(E^*_{A-1} - \varepsilon)d\varepsilon}{\rho_A(E^*_A)}$$

$$\Gamma_\gamma = C_\gamma \frac{\int_o^{E^*_A} E^*_A\ \emptyset\ (\varepsilon)\ \rho_A(E^*_A - \varepsilon)d\varepsilon}{\rho_A(E^*_A)}$$

$$\Gamma_f = C_f \frac{\Sigma_i\ T_i^{equi}(E^*_A) + \int_o^{E^*_A} E_{cf}\ \rho_F\ T_{fo}^{equi}\ (E^*_A - \varepsilon)d\varepsilon}{\rho_A(E^*_A)}$$

$\emptyset(\varepsilon)$ représente le facteur de forme de BRINCK-AXEL utilisé dans le cadre du modèle dipolaire électrique.

T_f^{equi} représente la penétrabilité d'une barrière unique équivalente à 2 barrières traitées dans le cadre de l'amortissement total du couplage entre états I et états II.

En s'inspirant des travaux de LYNN /3/ qui montrent que les densités de voies de fission concernant ce type de noyaux pour la barrière "intérieure" 1 sont le double de celles de la barrière "extérieure" 2, on a :

$$T_{fo}^{equi} = \frac{3}{2}\ \frac{T_1 T_2}{T_1 + \frac{1}{2} T_2}\ , \quad T_{1,2}(u) = \frac{1}{1+\exp \frac{2\pi}{hw}(V_{1,2}-U)}$$

Les différentes constantes permettent de normaliser les largeurs partielles à leurs valeurs à "basse énergie", obtenues par la mesure ou la systématique.

K. H. Böckhoff (ed.), Nuclear Data for Science and Technology, 673–676.

Les sections efficaces composées sont calculées par modèle optique déformé à l'aide du code ECIS /4/. Les autres sections efficaces dans P_{ne} sont calculées par modèle statistique de HAUSER-FESBACH à l'aide du code FISINGA /5/.

Il est à remarquer que les quantités intervenant dans P (E^{*}_{A+1},xni) sont prises indépendantes du Spin et de la parité. Une comparaison avec des calculs effectués à l'aide de codes tels que GNASH ou STAPRE permettrait de chiffrer l'importance de l'effet "Jπ" qui est à comparer avec l'économie de temps de calcul (probablement considérable) que l'approximation effectuée permet de réaliser.

Il est facile de montrer que les vrais paramètres de l'évaluation sont les probabilités de fission.

En effet, à condition de négliger les largeurs γ (ce qui est justifié lorsque l'énergie d'excitation du noyau est supérieure à l'énergie de liaison du dernier neutron), les sections efficaces de fission, de réactions (n,2n) et(n,3n) s'expriment en fonction des probabilités de fission P_f^{A+1}, P_f^{A}, P_f^{A-1}, la sensibilité à ces paramètres étant importante.

Quand aux autres quantités intervenant dans le calcul des sections efficaces mentionnées plus haut, elles sont ou bien connues ou à l'origine de faibles sensibilités. C'est le cas des sections efficaces composées et des densités de niveaux.

Pour nos calculs nous nous sommes reférés aux valeurs expérimentales de probabilité de fission de ^{236}Np et ^{235}Np obtenues par WILHELMY et al /6/ à LOS ALAMOS par réaction (^{3}He,d) sur ^{236}U et ^{235}U respectivement (Fig. 1 et 2).

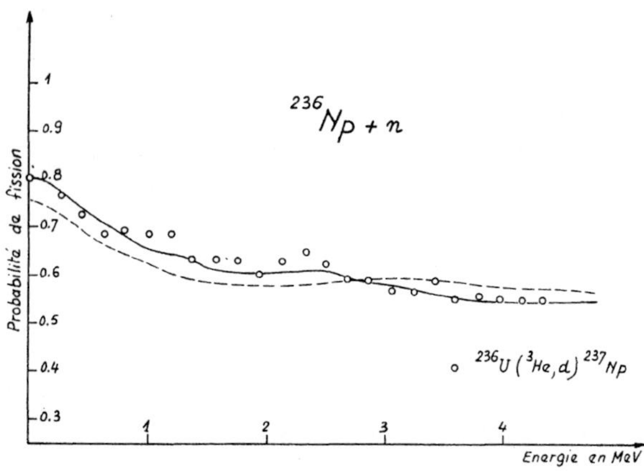

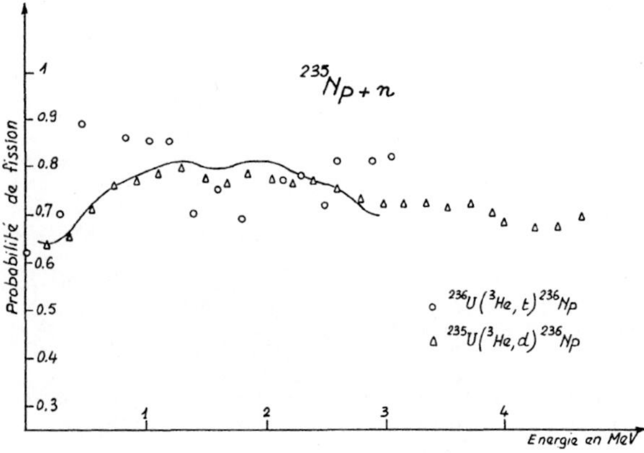

Fig.1. et Fig.2. Valeurs expérimentales de probabilité de fission, de ^{236}Np et de ^{235}Np comparés aux valeurs calculées. La courbe (-) représente une interprétation théorique des valeurs expérimentales tandis que l'interprétation des valeurs de section efficace de fission de ^{237}Np jugées les meilleures conduit à la courbe (--).

Dans le cadre de la méthodologie utilisée la comparaison de la section efficace calculée avec les valeurs expérimentales de section efficace de réaction de fission est un test de cohérence des diverses sections efficaces entre elles pour ^{237}Np. (Fig.3).

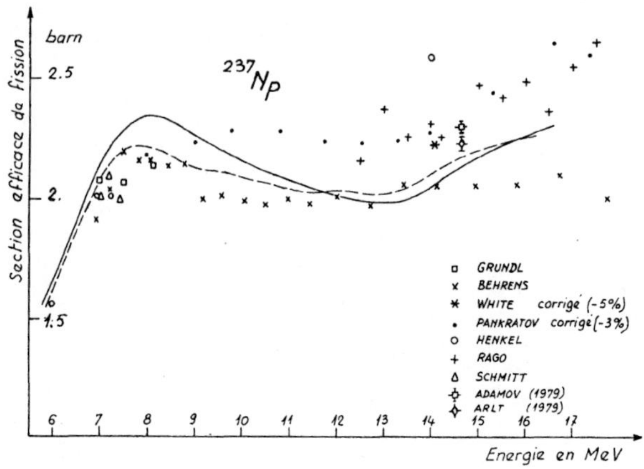

Fig.3. La courbe (-) représente la section efficace de fission de ^{237}Np calculée suivant la méthode proposée. La courbe (--) correspond aux valeurs jugées les meilleures.

La figure 5 montre la section efficace calculée pour la réaction (n,2n) ainsi que, en tirets, celle qui est obtenue en imposant à la section efficace de fission calculée d'être strictement égale aux valeurs expérimentales. La comparaison de ces deux courbes illustre bien la sensibilité importante de $\sigma_{n.2n}$ de ^{237}Np à la probabilité de fission de ^{236}Np.

II.2 - Calcul de la section efficace de production de ^{236}Pu

Dans la production de ^{236}Pu par décroissance β^- de ^{236}Np deux états sont concernés : l'état fondamental (6^-) à vie longue (1.15.10^5 a) (contribution 8.9%) et un état (1^-) à vie courte (22H5) (contribution 48%) à une énergie d'excitation de 0.0446 MeV selon la version récente de ENSDF. La section efficace de production de ^{236}Pu, ^{237}Np(n,2n) ^{236}Np → ^{236}Pu, et que nous noterons $\sigma^{236}Pu^{*}$ est une fraction de la section efficace (n,2n). Soient R(E) cette fraction et r(E) le rapport isomérique de ^{236}Np pour la réaction (n,2n).

$$\frac{\sigma^{236}Pu^{*}}{\sigma n.2n} = R(E) = \frac{0.48 \times r(E)}{1 + r(E)}$$, où E est l'énergie d'excitation de ^{236}Np liée à l'énergie incidente E_{inc} par : $E = E_{inc} - 6.8$.

En principe R(E) peut être évalué :

- par le calcul théorique de r(E), très imprécis compte-tenu du manque d'informations sur la densité et la spectroscopie des niveaux des noyaux de la famille des Np.

- à partir de données expérimentales. Celles-ci sont rares, puisque seule une valeur à E_{inc} = 14,8 MeV à été mesurée par LANDRUM et al /8/.

$$R(E{=}8) = \frac{1}{2.76} = 0.362.$$

Face à la pauvreté des informations disponibles, il n'est possible que d'estimer à l'oeil une courbe R(E) à partir de principes de continuité et de monotomie et des considérations supplémentaires suivantes :

- au seuil de la réaction n,2n,
  $6.8 < E_{inc} < 6.8 + 0.0446$, R = 0

Pour des énergies E comprises entre 0.0446 et 0.1 MeV (énergie prévisible pour le niveau suivant), le rapport isomérique varie comme le rapport de la section efficace inélastique du 1er niveau à la section efficace élastique. Le calcul effectué à l'aide de FISINGA montre que R(E) reste faible sur cette plage d'énergie.

- La donnée de NISHI /9/ à E_{inc} = 9.6 MeV (E= 2.8 MeV) permet d'évaluer R en supposant que la valeur calculée pour $\sigma_{n,2n}$ est correcte. R(2.8) = 0.312.

Cette valeur est confortée par les résultats obtenus en 1972 par LECOQ et VEYSSIERE /11/ qui, par des mesures de sections efficaces photonucléaires sur ^{237}Np ont montré que pour des énergies de photons supérieures à 9 MeV la section efficace de formation de l'état (1^-) était deux fois plus grande que celle de l'état (6^-) r(2.7) = 2.33, R(2.7) = 0.336).

- Le rapport isomérique et par suite R(E) (fig.4) doit vraisemblablement varier lentement pour des énergies incidentes comprises entre 10 et 15 MeV.

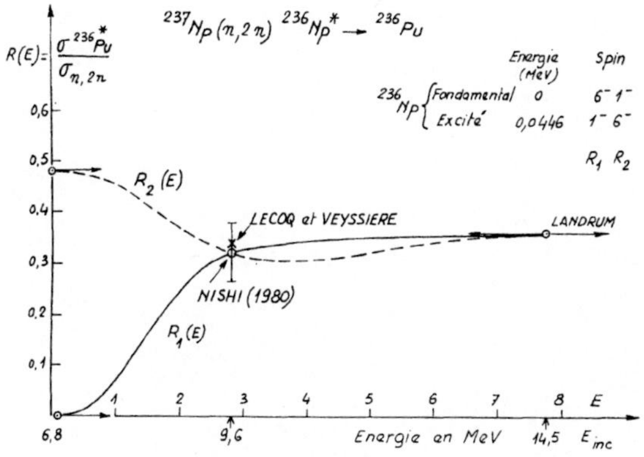

Fig.4. Evolution avec l'énergie du rapport de la section efficace productrice de ^{236}Pu à la section efficace totale (n,2n) de ^{237}Np.

II.3 - Estimation des incertitudes

Compte tenu des incertitudes expérimentales (≈10%) et de la procédure utilisée pour l'évaluation, les incertitudes estimées sur les valeurs évaluées sont les suivantes :

- Section efficace (n,2n): ±10% sur tout le domaine
- Section efficace de production de ^{236}Pu

± 10 %,	intervalle d'énergie		6.8 MeV - 7 MeV
± 80 %,	"	"	7 MeV - 9 MeV
± 30 %,	"	"	9 MeV - 11 MeV
± 20 %,	"	"	11 MeV - 13 MeV
± 15 %,	"	"	13 MeV - 16 MeV

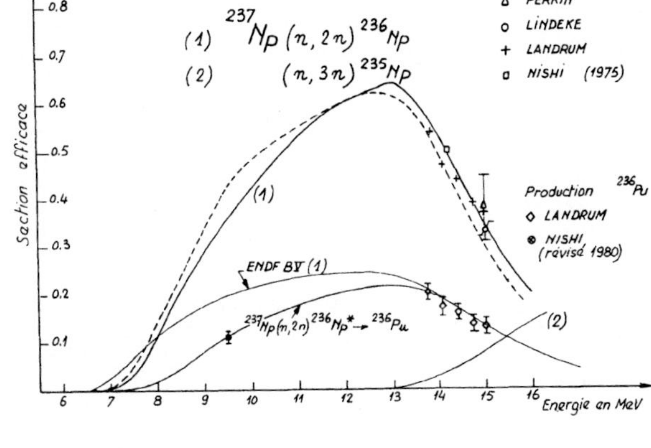

Fig.5. La section efficace de production de ^{236}Pu résulte de l'application de R(E) à la section efficace totale de réaction (n,2n) recommandée (-). A noter pour cette dernière l'influence de l'utilisation de probabilités de fission différentes pour ^{236}Np(--).

III - Cohérence avec les données intégrales

L'expérience de PAULSON et HENNELY /10/ est une expérience d'irradiation de ^{235}U pour laquelle ^{237}Np est un sous produit. La quantité de ^{236}Pu formée et analysée provient des réactions $^{237}Np(\gamma,n)\ ^{236}Np \xrightarrow{\beta^-}\ ^{236}Pu$ et $^{237}Np(n,2n)\ ^{236}Np \xrightarrow{\beta^-}\ ^{236}Pu$. Compte tenu des faibles concentrations d'isotopes autres que ^{235}U (<1 %) le spectre énergétique de la population neutronique a été assimilé à un spectre de fission de ^{235}U. Dans un tel spectre la section efficace ^{237}Np a été mesurée égale à 0.063 barn et la section efficace de production de ^{236}Pu à 0.0315 ± 0.03.

Pour les mêmes conditions spectrales, la section efficace de production de ^{236}Pu calculée est égale à :

$$< \sigma^{236}Pu^* > = \frac{\int_{6.8}^{20} R(E)\ \times\ \sigma_{n.2n}(E)\ \times\ ^{235}\chi(E)\ dE}{\int_{6.8}^{20} {}^{235}\chi(E)dE} = 0.0345\ b$$

IV - Conclusion

L'accord acceptable entre valeur calculée et valeur expérimentale montre que la cohérence des données intégrales et microscopiques peut être assurée aux conditions suivantes :

1) Le niveau a vie courte est le niveau (1^-) à une énergie d'excitation de 0.0446 MeV.

2) Le rapport de la section efficace productrice de ^{236}Pu et de la réaction (n,2n) totale ainsi que le rapport isomérique sont des fonctions appropriées de l'énergie.

Il convient de souligner le caractère spéculatif attaché à la fonction R(E). Pour le faire disparaître de nouvelles mesures sont nécessaires, intégrales et microscopiques mais en remarquant pour ces dernières que le domaine de plus grande sensibilité correspond aux énergies inférieures à 10-11 MeV.

Références

1. C.K. Cline, M. Blann
Nucl. Phys. A 172 (1971), 225.

2. J. Jary
INDC(FR) IOL, NEANDC(E) 175 "L".

3. J.E. Lynn
AERE - R 7468.

4. J. Raynal
Equations couplées et DWBA, Aussois (1968), Report Lycen -6804.

5. E. Fort, D. Lafond
Le code de modèle statistique FISINGA, non publié.

6. Wiehelmy
L.A.S.L. Communication privée.

7. Nuclear data sheets 20, 192.

8. T. Landrum et al
Phys. Rev. 8, 5, 1938 (1973)

9. T. Nishi et al
NEANDC (J) 42 L (1975), valeur révisée d'après INDC discrepancy file. 19 juin 1980.

10. C.K. Paulson et E.J. Hennely
N.S.E 55, 24-27 (1974)

11. G. Lecoq et A. Veyssiere
Communication privée.

REMARK ON THE ENDF/B-V EVALUATIONS OF THE NEUTRON CAPTURE CROSS SECTIONS IN THE REGION AROUND 10 MeV NEUTRON ENERGY

F. Cvelbar, R. Martinčič and A. Likar

J. Stefan Institute and Faculty of Natural Sciences and Technology
E. Kardelj University
P.O.B. 53, 61001 Ljubljana
Yugoslavia

From the description of the ENDF/B-V General Purpose Library it follows that in the evaluations of the neutron radiative capture excitation function around 10 MeV neutron energy, the direct-semi-direct capture model (DSD) in some cases has not been considered et all. In some cases too rough calculations have been performed. Some examples of fitting (n,γ) excitation functions are presented to show the predictive power of DSD model.

$^{89}Y(n,\gamma)^{90}Y$, $^{140}Ce(n,\gamma)^{141}Ce$ and $^{208}Pb(n,\gamma)^{209}Pb$ calculated excitation functions E_n = 2 - 20 MeV.

Fusion energy programs induced the interest for neutron cross sections in the region around 10 MeV neutron energy. Except at 14 MeV, neutron data are scarce in this region. That is especially true for capture cross sections. Systematic measurements (excitation functions and angular distribution for isolated γ-ray transitions) were performed only in few cases (1,2) (Si, Ca, As, Y, Pb, Bi). Excitation function are characterised by the broad peak in the region of giant dipole resonance (GDR). Integrated cross sections were measured at 14 MeV neutron energy for all easily available natural targets. Data are scattered around the value of 1 mb, except in the region of $A < 50$, where around $A = 30$ reach the value of 0.3 mb.

The capture process in the region of GDR treats the so called direct-semi-direct (DSD) model, which was intensivelly studied during the last decade (1,2). Calculations based on the recent approaches to the model, reproduce satisfactorily well the experimental excitation functions and angular distributions. From these analyses it follows, that the predictive power of the model is strong enough so that the (n,γ) data prepared for libraries should be based on it. For $A>50$ nuclei the averall accuracy should be about $\pm$ 25 %.

The main remark, concerning the recent fast neutron (n,γ) evaluations (ENDF/B-V, KEDAK III) is that the DSD model has not been used intensively enough.

The ENDF/B-V general purpose library is not yet generally accesible. Some nuclei/elements are available in Dosimetry, Fission products and Standards Libraries. The others we know only from their description (3).

The (n,γ) excitation functions for some nuclei/elements (Si, Ca, Fe, Co-59, Ag-107, Pb) were in the ENDF/B-V Library obtained by the rough interpolation/extrapolation, based on scarce experimental points. DSD model in these cases was not considered. GDR enhancement is usually missing.

In some other cases (Rh, Ta, W) the results of DSD model were obtained in a very crude way - without calculating the matrix elements (3). There are only few cases of possibly more exact use of DSD model (e.g. Pu-242).

General conclusion is, that the neutron capture ENDF/B-V data above 5 MeV neutron energy cannot be used safely without the additional analysis, based on proper DSD calculations in which recently available data for free parameters (4) (see Fig. 1) are used.

To show how appropriate is the DSD model for the description of the radiative capture of energetic neutrons, in Figs. 2, 3 in 4 the experimental data are compared with the results of DSD calculations for Y-89, Ce-140 and Pb-208 respectively. The differences in calculated results, if the experimental or calculated single particle levels are used, are pointed out.

References

1. L. Nilsson, Fast Neutron Radiative Capture, 4th (n,γ) Int. Symp., Grenoble 1981, Inst. Phys. Conf. Ser. No. 62, p. 465, and references therein.

2. M. Potokar, The Role of Simple Direct Models in Fast-Nucleon Radiative Capture, 4th (n,γ) Int. Symp., Grenoble 1981, Inst. Phys. Conf. Ser. No. 62, p. 477, and references therein.

3. R. Kinsley, ENDF/B-V General Purpose Library BNL-NCS-17541 (ENDF-201), 3rd edition (July 1979), Brookhaven National Laboratory, Upton, N.Y.

4. V.Benzi, R.D'Orazi, G. Reffo, and M. Vaccari, Fast Neutron Radiative Capture Cross Sections of Stable Nuclei with 29 Z 79, Centro di calcolo di Bologna DOC.CEC(79)9, 1971

5. R. Martinčič, A. Likar, F. Cvelbar, and M. Mikuž, Mass Dependence of the Strength of the Imaginary Dipole Form Factor in Fast Neutron Capture, 4th (n,γ) Int. Symp., Grenoble 1981, Inst. Phys. Conf. Ser. No. 62, p. 526, and references therein.

K. H. Böckhoff (ed.), Nuclear Data for Science and Technology, 677–678.

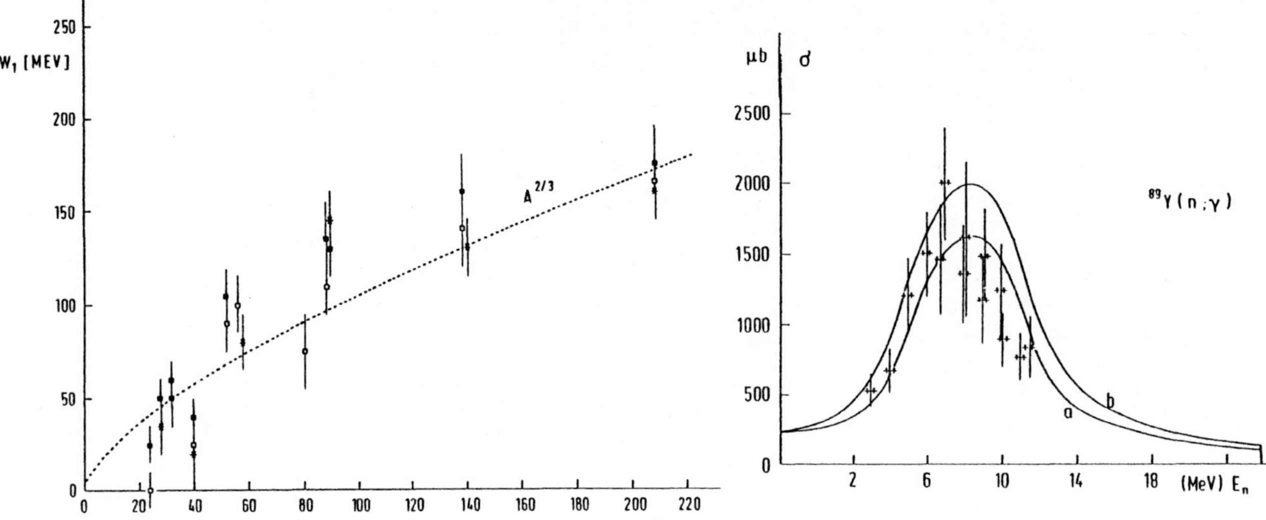

Fig. 1 Mass dependence of the imaginary part of the strenght of the particle-vibration coupling, which can be treated as free parameter in the DSD model. The strength of the real part was taken as constant V_1 = 75 MeV (5).

Fig. 2 Comparison of experimental and DSD model data for Y-89 (n, γ) reaction: a) calculation based on experimental single particle levels, b) calculation based on calculated single particle levels. V_1 = 75 MeV, W_1 = 145 MeV.

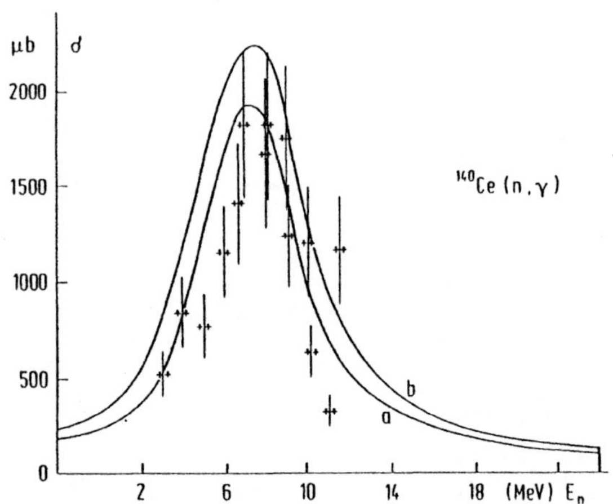

Fig. 3 Comparison of experimental and DSD model data for Ce-140 (n, γ) reaction: a) calculation based on experimental single particle levels, b) calculation based on calculated single particle levels. V_1 = 75 MeV, W_1 = 145 MeV.

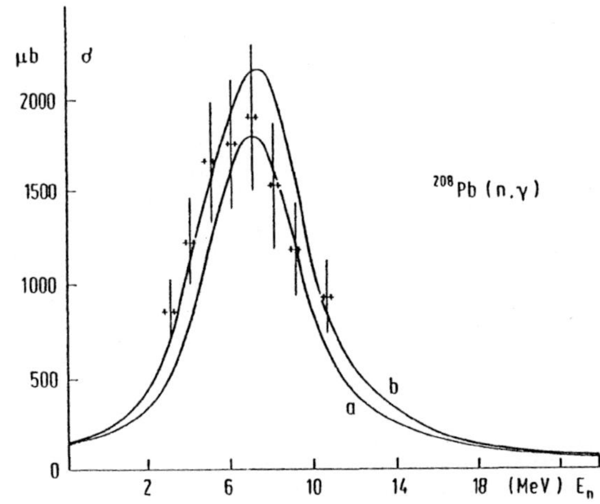

Fig. 4 Comparison of experimental and DSD model data for Pb-208 (n, γ) reaction: a) calculation based on experimental single particle levels, b) calculation based on calculated single particle levels. V_1 = 75 MeV, W_1 = 165 MeV.

EVALUATION OF THE EXCITATION FUNCTION OF THE $^{238}U(n,2n)^{237}U$ REACTION FOR NEUTRON ENERGIES FROM THRESHOLD TO 19 MeV*

N.V. Kornilov, V.N. Vinogradov, E.V. Gay, N.S. Rabotnov, O.A. Salnikov
Fiziko-energeticheskiy Institut, Obninsk, USSR

P. Raics, S. Daróczy, S. Nagy, J. Csikai
Institute of Experimental Physics, Kossuth L. University,
Debrecen, Bem tér 18/A, H-4026, Hungary,
P.O.B. 105.

Experimental results for the $^{238}U(n,2n)$ reaction were collected from the literature and evaluated. The normalisation of the measured cross sections was carried out using recent values for the cross sections of standard monitor reactions as well as new nuclear decay data. The evaluated excitation function was then obtained by the Padé-approximation.

Measurement techniques, data evaluation

The nuclear reaction $^{238}U(n,2n)^{237}U$ plays an important role in the nuclear fuel cycle [1]. That is why there are many works on the excitation function from threshold up to 19 MeV.

The determination of the cross section has generally been carried out by two methods: (i) detection of the radioactivity of the final product nuclei (activation), and (ii) detection of the (n,2n) events by a large liquide scintillator tank (neutron multiplicity). The neutron flux density has been measured by standard reactions such as $^{238}U(n,f)$, $^{27}Al(n,alpha)$, $^{65}Cu(n,2n)$, $^{58}Fe(n,p)$, H(n,n), etc. $D(d,n)^3He$ and $T(d,n)^4He$ reactions as neutron sources were applied in the experiments.

There were sixteen works containing 86 data points available from the literature which have been collected, and critically analysed with respect to the measurement technique, the applied corrections (neutron field, counting losses, etc.) and nuclear data used for calculations. The quoted errors of the measured cross sections were in some cases increased: (i) an error of 15% was added quadratically to the original ones when the data were available from figures, only: (ii) if the authors have not mentioned clearly the nuclear data used for evaluation, their errors were increased by an additional 5-10%. Recent cross section values for the monitor reactions as well as new decay data were then used for the normalisation of the experimental values. The details of the experiments and the normalisation are described elsewhere [2].

The recommended excitation function of the $^{238}U(n,2n)$ reaction

The normalised experimental results were fitted using the second order Padé-approximation [3] which gives the recommended excitation function in the form

$$\sigma(E) = \sum_{i=1}^{2} \frac{\alpha_i \cdot (E-\xi_i)+\beta_i}{(E-\xi_i)^2+\gamma_i^2} + \frac{a_1}{E-P_1} \qquad (1)$$

for $6.55 \leq E \leq 19$ MeV.

(σ is in $10^{-31}m^2$ or mbarns if E is given in MeV.) The best values for the parameters of this empirical function were found from the minimisation of the chi-square. The goodness of fit can be seen from the value $\chi^2 = 52.2$ which should be compared to the number of freedom, 76. The coefficients are listed in Table I.

Table I Coefficients of the function (1)

parameters	coefficients i=1	i=2
α_i	3520.727	-1179.224
β_i	-1178.012	5676.409
ξ_i	6.69328	12.5233
γ_i	1.9349	2.55837
a_i	172.638	-
p_i	5.80306	-

The excitation function can not be described appropriately by the function (1) close to the reaction threshold. It seemed a simple 3rd order polynomial to be useful in the form

$$\sigma(E) = 1,625.(E-6.169)^3 \quad \text{for} \quad 6.169 \leq E \leq 6.550 \text{ MeV} \qquad (2)$$

Here $E_{th} = 6.169$ MeV is the threshold energy ($Q_{2n} = -6.143$ MeV).

The recommended excitation function together with the normalised data points are shown in Fig.1. Great many experiments have been carried out around 14 MeV. Two-three sets of measurements can be found in the 6.5-11 and 15-19 MeV regions. The excitation function is not so well defined by experiments from 11 to 13.5 MeV.

It is quite difficult to give acceptable uncertainties to the recommended excitation function. The function (1) is well defined by the weighted data: 82.56 % of the normalised cross sections lie within $\pm 1\sigma$ of their accepted errors around the curve and 96.51 % of them are within the $\pm 2\sigma$ level. However, it should be taken into account the accepted errors of the individual experiments as well as the technical difficulties of the

*Work supported by the Ministry of Education and the Academy of Sciences, Hungary

K. H. Böckhoff (ed.), Nuclear Data for Science and Technology, 679–680.

measurements (determination of the flux density, energy distribution of the source

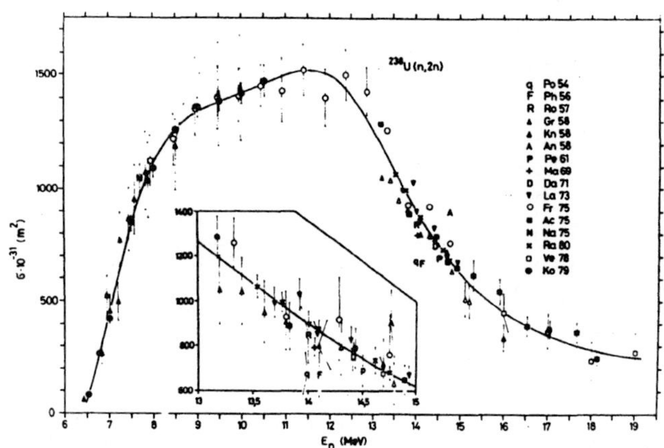

Fig. 1. The excitation function of the $^{238}U(n,2n)$ reaction. Points: normalised experimental cross sections of the quoted authors (Refs. [4-19]); curve:fitted empirical function (1) as recommended values. The region around 14 MeV is enlarged in the insert.

neutrons, etc.). All of these facts together with the evaluators' subjective feelings result in, for orientation,error values $\delta\sigma$, of the cross sections in different neutron energy regions (ΔE_n) which are listed in Table II.

Table II Proposed uncertainties (1σ) of the recommended excitation function

ΔE_n, MeV	6.5-7.5	7.0-8.0	8.0-11.0	11.0-12.5	12.5-13.5	13.5-15.0	15.0-19.0
$\delta\sigma$,%	7	5	4	6	9	3	8

It can be pointed out that the WRENDA requirements [20] are generally fulfilled by the recent status of the excitation function for the $^{238}U(n,2n)$ reaction from 6.5 up to 19 MeV.

References

1. T.S. Zaritzkaya, C.M. Zaritzkiy, A.K. Kruglov et al, Atomnaya Energiya 48, 67 (1980)
2. N.V. Kornilov, V.N. Vinogradov, E.V. Gay, N.S. Rabotnov, O.A. Salnikov, P. Raics, S. Daróczy, S. Nagy, J.Csikai, Voprosy Atomnoy Nauky i Tehniki, ser. Yadernye Konstanty 1(45) 33 (1982)
3. V.N. Vinogradov, E.V. Gay, N.S. Rabotnov, Preprint FEI-434, Obninsk, 1974; Neytronnaya Fizika vol.I. 44 (Mat. 3. Vsesoyuznoy Konferentziy po Neytronnoy Fizike, Kiev, 9-13 iyunya 1975)
4. M.J. Pool, 1954; data from Ref. 10.
5. J.A. Phillips, AERE NP/R-2033, 1956; data from Ref. 10.
6. L. Rosen, L. Stewart, Report LA-2111, Los Alamos, 1957.
7. E.R. Graves, J.P. Coner et al., 1958, cited in Ref. 8.
8. J.D. Knight, R.K. Smith, B. Warren, Phys. Rev. 112, 259 (1958)
9. G.P. Antropov, Yu.A. Zysin, A.A. Kovrizhnykh, A.A. Ldov, Atomnaya Energiya 5, 456 (1958)
10. J.L. Perkin, R.F. Coleman, Reactor Sci. Techn. 14, 69 (1961)
11. D.S. Mather, L.P. Pain, AWRE-047/69, 1969.
12. S. Daróczy, E. Germán, P. Raics, S. Nagy, J. Csikai, Neytronnaya Fizika vol.III. 323 (Mat. 2. Vsesoyuznoy Konf. Kiev 28 maya - 1 iyunya 1973; Obninsk, 1974)
13. J.H. Landrum, R.J. Nagle, M. Lindner, Phys. Rev. C8, 1938 (1973)
14. J. Fréhaut, G. Mosinski, CEA-R-4627, CEN-Saclay, Gif-sur-Yvette, France, 1974, and Nuclear Cross Sections and Technology, vol.II. 855 (NBS Spec. Publ. 425; Proc. Conf. 3-7 March 1975, Washington)
15. A. Ackermann, B. Anders, M. Borman, W. Scobel, Nuclear Cross Sections and Technology, vol.II. 819. (NBS Spec. Publ. 425, Proc. Conf. 3-7 March 1975, Washington)
16. S. Nagy, K.F. Flynn, J.E. Gindler, J.W. Meadows, L.E. Glendenin, Phys. Rev. C17, 163 (1978)
17. Raics P., Thesis, Institute of Experimental Physics, Kossuth University, Debrecen, 1978 (in Hungarian)
18. L.R. Veeser, E.D. Arthur, Neutron Physics and Nuclear Data, 1054 (Proc. Int. Conf. Sept. 1978, Harwell; OECD Nuclear Energy Agency, Paris-UKAEA, Harwell)
19. N.V. Kornilov, B.V. Zhuravlev, O.A. Salnikov, P. Raics, S. Daróczy, S. Nagy, K. Sailer, J. Csikai, Atomnaya Energiya 49, 283 (1980)
20. WRENDA 78/77, INDC/SEC/-55/URES, Vienna, 1976.

ANALYSIS AND EVALUATION OF THERMAL AND RESONANCE NEUTRON ACTIVATION DATA

S.M. Jefferies, T.D. Mac Mahon, J.G. Williams, A. Ahmad

Reactor Centre, Imperial College at Silwood Park,
Ascot, Berkshire, SL5 7PY, England.

T.B. Ryves

National Physical Laboratory,
Teddington, Middlesex,
TW11 OLW, England.

The paper describes the measurement and evaluation of thermal and epithermal neutron activation data for fission reactors, including the 2200 ms^{-1} activation cross sections, the resonance activation integrals and decay data for the neutron capture products. These are the data required for routine neutron metrology, activation analysis and calculations of activation in reactor materials. The greatest practical problems which arise in this type of work for most isotopes are in the treatment of resonance self-shielding and in the method of accounting for the effects of departures from a (1/E) slowing down spectrum in resonance integral data. New methods of treating these problems are described.

Introduction

In the characterization of thermal and epithermal neutron fields the assumption has often been made that the neutron flux distribution can be represented by a Maxwellian thermal component and an epithermal slowing down spectrum proportional to 1/E. The well known conventions of Westcott[1], Stoughton and Halperin[2], and Hogdahl[3] all make use of this assumption. In these conventions the reaction rate for a radiative capture reaction, in which the fast neutron contribution is negligible, depends on the effective thermal cross-section and on the resonance integral defined here as

$$I_o = \int_{E_{Cd}}^{\infty} \sigma(E)\, E^{-1}\, dE \qquad (1)$$

where E_{Cd} is the effective cut-off energy of a cadmium box, assumed here to be 0.55 eV for a 1 mm thick box[4].

The 1/E slowing down spectrum arises in systems in which the slowing down density is constant, which can only be expected in the absence of the effects of leakage and absorption. It has been shown by Williams[5] that in the case of energy independent buckling and absorption cross section the flux per unit energy is approximately proportional to $1/E^{1+\alpha}$. Functions containing this type of deviation from the 1/E spectrum have also been proposed on empirical grounds by many authors[6-13].

In this paper an adaptation of the older flux conventions[1-3] is used in which the epithermal spectrum is assumed to be proportional to $1/E^{1+\alpha}$. The validity of this assumption has been tested in a consistency analysis using reaction rates and cadmium ratios measured in three research reactors[14].

We assume here that the flux can be represented by means of three empirical parameters. These are the conventional thermal flux Φ_{th}, defined to be the ratio of the thermal neutron induced reaction rate to the 2200 m/s cross-section of a 1/v detector, the epithermal flux parameter ϕ_e, equal to the flux per unit lnE at 1 eV, and α, the parameter describing the deviation from the 1/E spectrum.

The reaction rate for a radiative capture reaction is given by[14]

$$A = g\sigma_o\{\Phi_{th} + \phi_e\left[f_1(\alpha) + \frac{W'}{g} + f_2(\alpha) + \left(\frac{E_1}{\bar{E}_r}\right)^{\alpha}\left\{\frac{I_o}{g\sigma_o} - f_2(o)\right\}\right]\} \qquad [E_1 \text{ is } 1 \text{ eV}] \qquad (2)$$

where g is the Westcott factor[15] accounting for the deviation of the cross-section from the 1/v law in the thermal region, σ_o is the 2200 m/s cross-section, W' is a correction, as tabulated by Ryves and Zieba[16], accounting for deviation of the cross-section from the 1/v law in the energy range μkT to E_{Cd}, and $\bar{E}_r$ is the effective resonance energy as defined and tabulated by Moens et al[17]. The functions $f_1(\alpha)$ and $f_2(\alpha)$ account for the epithermal activation of a 1/v detector and are defined as follows:

$$f_1(\alpha) = \int_{\mu kT}^{E_{Cd}} \frac{v_o}{v}\frac{E_1^{\alpha}}{E^{1+\alpha}}\, dE \qquad (3)$$

$$f_2(\alpha) = \int_{E_{Cd}}^{\infty} \frac{v_o}{v}\frac{E_1^{\alpha}}{E^{1+\alpha}}\, dE \qquad (4)$$

The saturated gamma-ray emission rate per unit mass of target element is given by

$$N_s = N_{Av}\, \eta\, A/g\sigma_o, \qquad (5)$$

where A is calculated using equation (4), N_{Av} is Avogadro's number and η is a compound nuclear data parameter defined by

$$\eta = \theta P_\gamma g\sigma_o M^{-1} \qquad (6)$$

where θ is the mole fraction of the target isotope, P_γ is the emission probability of the relevant gamma-ray and M is the element atomic mass.

Self-Shielding Factors

In the previous section no account was taken of self-shielding. In neutron fluxes greater than $10^{15}m^{-2}s^{-1}$ it is usually possible to use foil detectors which are sufficiently dilute that any self-shielding effects are negligible. It is frequently necessary, however, to make measurements in much lower fluxes, for example in standard flux facilities. It is then necessary to use relatively thick foils in which self-shielding factors are important. The self-shielding factor, G, for a detector is defined as the ratio of the experimental activation to the theoretically expected activation without self-shielding. It has been found convenient to consider separate self-shielding factors, G_{th} and G_r, for thermal and epithermal neutrons respectively.

The resonance, or epithermal self-shielding factor is usually the one which causes greatest difficulty.

K. H. Böckhoff (ed.), Nuclear Data for Science and Technology, 681–684.

Although these factors have been studied previously by several authors[19-23] there are still many detector materials for which the literature contains no precise data. Even when literature values are available it is not always evident how they should be applied, because of two conflicting definitions which are in use. For example, the experimental self-shielding factors, G_r, of Baumann[22] and of Jacks[23] should be used as factors premultiplying the resonance integral, I_o, defined in equation (1). Axton's[21] values, however, should premultiply a reduced resonance integral, I_o', given by the normal resonance integral minus the 1/v contribution:

$$I_o' = I_o - g\sigma_o f_2(o) \; . \tag{7}$$

We will denote this type of resonance self-shielding factor by G_r'.

In calculations of resonance self-shielding factors values have been published[19,20] which relate to individual resonances. If I_i' is taken to be the contribution to the reduced resonance integral of an individual resonance and G_i is the self-shielding factor relating to this resonance then one can calculate G_r' as follows:

$$G_r' = \sum_i I_i' G_i / \sum_i I_i' \; , \tag{8}$$

where the summations are taken over all resonances.

Because of the lack of sufficient reliable self-shielding data for the materials used in this work, a Monte-Carlo code has been written to calculate them for plane foils and the results have been compared with the available experimental and theoretical values found in the literature already cited.

Table I shows the results of the Monte-Carlo calculations for G_i for the 337 eV resonance in ^{55}Mn, compared with values calculated by Selander[20]. As can be seen, the agreement is better than 2%. Similar agreement has been found for the 1098 eV resonance in the same isotope. The resonance parameters used were the same as those used by Selander[20].

Table I G_i factors for the ^{55}Mn 337 eV resonance

Foil Thickness (mm)	Monte-Carlo Calculation	Selander[20]
6.125×10^{-3}	0.972 ± 1.7%	0.979
1.715×10^{-2}	0.939 ± 1.3%	0.944
5.145×10^{-2}	0.862 ± 1.6%	0.862
1.016×10^{-1}	0.772 ± 0.9%	0.778
1.524×10^{-1}	0.709 ± 0.9%	0.716
2.032×10^{-1}	0.658 ± 0.7%	0.671

Using equation 8 above, the self-shielding factor, G_r', for use with the reduced resonance integral can be deduced from the individual G_i.

Table II shows the Monte-Carlo results for G_r' for ^{197}Au taking account of the resonances at 4.9 eV, 60 eV and 107 eV compared with experimental G_r' values determined by Axton[21]. Again the agreement is very good.

Self-shielding factors, G_r, for the total resonance integral (i.e. including 1/v contribution) are determined as follows:

$$G_r = \frac{\sum_i G_i I_i' + g\sigma_o f_2(o)}{I_o} \tag{9}$$

Table II G_r' factors for ^{197}Au

Foil Thickness (mm)	Monte-Carlo Calculation	Axton[21]
3.105×10^{-3}	0.727 ± 1.2%	0.710
1.552×10^{-2}	0.440 ± 1.1%	0.441
2.277×10^{-2}	0.378 ± 1.3%	0.380

G_r factors for the important nuclei ^{55}Mn, ^{59}Co and ^{197}Au are shown in Table III.

Table III G_r factors for ^{55}Mn, ^{59}Co and ^{197}Au

Thickness mg cm^{-2}	G_r		
	^{55}Mn	^{59}Co	^{197}Au
2	0.982±0.014	0.947±0.009	0.877±0.007
15	0.948±0.010	0.802±0.006	0.587±0.008
30	0.916±0.007	0.718±0.005	0.461±0.011
70	0.853±0.007	0.602±0.005	0.327±0.012
90	0.832±0.005	0.572±0.005	0.294±0.013
120	0.805±0.005	0.538±0.005	0.263±0.013
140	0.790±0.005	0.521±0.005	0.247±0.013

In principle the resonance self-shielding factors must depend on the shape of the epithermal spectrum in which irradiation takes place. To investigate this dependence Monte-Carlo calculations have been performed for ^{55}Mn and ^{197}Au resonance self-shielding factors for epithermal spectra with α values up to 0.2. The calculations showed that the effect of α on G_r was negligible.

Consistency Analysis

The model described by equations 2 and 5, together with analogous formulae for cadmium covered activities and cadmium ratios, has been tested by means of experimental activation data from nine irradiation positions in three research reactors[14]. The measured quantities can be written as a vector $[E]$, with its associated variance-covariance $[V_e]$, where non-diagonal terms arise because of correlated uncertainties in detector efficiency[25].

The data in $[E]$ are to be compared with the predictions of the model. These predictions, which form a vector $[C]$, depend on the parameters, $[P_o]$, used in the model, including both nuclear data and flux parameters. We need to find best estimates of these and their associated variance-covariance matrix. The method chosen to estimate $[P_o]$ was the generalised least-squares method in which we seek the values $[P_o]$ which minimize chi-squared:

$$\chi^2 = [P_1 - P_o]^T [V_p]^{-1} [P_1 - P_o] + [E-C]^T [V_e]^{-1} [E-C] \tag{10}$$

in which the symbol T represents the transpose of a matrix and -1 the inverse. $[P_1]$ is a vector containing prior information about the parameters and $[V_p]$ is the variance-covariance matrix for $[P_1]$. Information on the data used in $[P_1]$ is given in the next section.

The iterative search procedure to find the values of $[P_o]$ which minimize chi-squared was carried out using the CERN library code MINUIT developed by James and Roos[24]. The code also supplies a variance-covariance matrix for the solution. The details are given by Ahmad[25].

Results of the consistency analysis which have been reported in detail in an earlier paper[14], showed that the model used provided an adequate and useful representation of the neutron flux in the irradiation positions studied. In the following section we describe the application of the consistency analysis to a new set of activation data.

Input Data and Results

Foils of Na (as NaCl), Mn, Co, Cu, In, W and Au were irradiated in the standard thermal neutron facility of the National Physical Laboratory. This facility consists of a graphite stack driven by a deuteron beam providing neutrons by the $^{9}Be(d,n)^{10}B$ reaction, and has been described by Ryves and Paul[10]. Foils of Mn and Au were also irradiated under cadmium covers. After irradiation the saturated gamma ray emission rates (N_s in equation 5) were determined by gamma counting using a calibrated hyper-pure germanium detector with resolution 1.73 keV FWHM at 1.33 MeV.

In order to carry out the consistency analysis on this data it is necessary to construct the vector $[P_1]$ containing prior knowledge of the three flux parameters (ϕ_{th}, ϕ_e and α) and of the nuclear data appearing in equations 2 and 5 above (η, $I_o/g\sigma_o$, $\bar{E}_r$, W'/g and the decay constant λ). The last three of these do not contribute significantly to the total uncertainty of the activities considered here, nor do the values of θ and M which appear in η (equation 6). All of these have therefore been treated as known constants and the values were taken from the literature[16,17,26,27,29]. Uncertainties in P_γ, I_o and σ_o, however, contribute significantly to uncertainties in the values of η and $I_o/g\sigma_o$. Values of P_γ and their standard deviations were taken from the Table of Isotopes[26]. In the case of ^{187}W, P_γ values were taken from Zijp and Baard[30]. I_o and σ_o values have been evaluated using experimental data included in the compilation of Gryntakis and Kim[28]. Some measurements were excluded from the evaluation, either because no errors were given or because they were discrepant from the weighted mean of the remaining values by more than three standard deviations. Table IV lists the evaluated input data used for the consistency analysis. E_γ is not used as input data but is listed to identify the gamma rays for which η is determined. The Westcott g factor was unity for all isotopes except indium (1.019) and gold (1.0052). The cross-sections for cobalt refer to the production of ^{60m}Co, and for indium, $^{116m_1}In$.

When the consistency analysis was applied to the experimental data and the input nuclear data described above the value of χ^2 obtained (see equation 10) failed a two tailed test even at a 99.95% confidence level indicating that there were severe inconsistencies in the data. It was observed that the major contributions to χ^2 arose from the η values for In and W, which implicates the P_γ values as the probable source of the inconsistencies. Other sets of P_γ data were then tested, and the set of data shown in Table V, when included in the consistency analysis, gave a value of χ^2 of 43 for 36 degrees of freedom which passed the two-tailed test at the 95% confidence level. In Table V the P_γ values for ^{116}In were taken again from the Table of Isotopes[26] but were the third set given therein (with reference code RRL-16-241(74)) rather than the first, recommended, set. In the case of ^{187}W the P_γ data were taken directly from Brenner and Meyer[31]. A comparison of the precision of the input and output values of η in Table V show clearly the value of the consistency analysis and its ability to improve knowledge of nuclear data. These output values are highly correlated with one another. Their variance-covariance matrix is available on request. The output of the consistency analysis also contains the three flux parameters for the irradiation position:

$$\phi_{th} = (1.259 \pm 0.009) \times 10^7 \text{ n cm}^2\text{s}^{-1}$$
$$\phi_e = (1.42 \pm 0.05) \times 10^5 \text{ n cm}^2\text{s}^{-1}$$
$$\alpha = 0.048 \pm 0.010$$

These flux parameters agree well with the previous measurements in the same position reported by Ryves and Paul[10].

The output of the consistency analysis also provides improved information of the $I_o/g\sigma_o$ values. Examples of this improvement were given in an earlier paper[14]. This work is continuing and more comprehensive lists of output nuclear data will be published in the future.

References

1. C.H. Westcott, J. Nucl. Energy 2, 59 (1955).
2. R.W. Stoughton and J. Halperin, Nucl. Sci. Eng., 6, 100 (1959).
3. O.T. Hogdahl, Radiochemical Methods of Analysis, IAEA Vienna, 1, 23 (1965).
4. H. Goldstein, J.A. Harvey, J.S. Story and C.H. Westcott, EANDC-12, (1961).
5. M.M.R. Williams, The Slowing Down and Thermalization of Neutrons, North Holland (1966).
6. J.W. Connoly, A. Rose and T. Wall, Integral Reaction Rates and Neutron Energy Spectra in Well Moderated Reactors, AAEC/TM191 (1963).
7. L.C. Schmid, J.H. Lauby, W.P. Stinson and V.O. Uotinen, Summary of Results of EBWR Critical Experiments, Phys. Res. Quart. Rept., HW-84608 (1965).
8. P. Schumann and D. Albert, Kernenergie, 8, 88 (1965).
9. K.W. Geiger and L. Van Der Zwan, Metrologia, 2, 1 (1966).
10. T.B. Ryves and E.B. Paul, J. Nucl. Energy, 22, 759 (1968).
11. T.B. Ryves, Metrologia, 5, 119 (1969).
12. T. Bereznai and T.D. Mac Mahon, J. Radioanal. Chem. 45, 423 (1978).
13. F. De Corte, L. Moens, A. Simonits, A. De Wispelaere and J. Hoste, J. Radioanal. Chem. 52, 295, (1979).
14. A. Ahmad, S.M. Jefferies, T.D. Mac Mahon J.G. Williams and T.B. Ryves, Proc. 4th ASTM-EURATOM Symp. on Reactor Dosimetry, Maryland, U.S.A. March 1982, CONF-820321/V2 p.745-753.
15. C.H. Westcott, Effective Cross-Section Values for Well Moderated Thermal Reactor Spectra, AECL-1101 (1960).
16. T.B. Ryves and K.J. Zieba, J. Phys. A : Math. 7, 2318 (1974).
17. L. Moens, F. De Corte, A. Simonits, A. De Wispelaere and J. Hoste, J. Radioanal. Chem. 52, 295 (1979).
18. W.L. Zijp, "Intermediate Neutrons" in Neutron Fluence Measurements, Tech. Rept. 107, IAEA Vienna (1974).
19. G.M. Roe, The Absorption of Neutrons in Doppler Broadened Resonances, Rept. KAPL-1241 (1954).
20. W.N. Selander, Theoretical Evaluation of Self-Shielding Factors Due to Scattering Resonances in Foils, Rept. AECL-1077 (1960).
21. E.J. Axton, J. Nucl. Energy, Parts A/B, 17, 125 (1963).
22. N.P. Baumann, Resonance Integrals and Self-Shielding Factors for Detector Foils, Rept. DP-817 (1963).
23. C.M. Jacks, A Study of Thermal and Resonance Flux Detectors, Rept. DP-608 (1961).
24. F. James and M. Roos, Computer Physics Communications, 10, 343 (1975).
25. A. Ahmad, Ph.D. Thesis, (February 1982).
26. C.M. Lederer and V.S. Shirley, Table of Isotopes, 7th Edition, John Wiley and Sons, 1978.
27. Chart of Nuclides, Kernforschungszentrum Karlsruhe, 1974.
28. E.M. Gryntakis and J.I. Kim, NEA Data Bank, ND1540, 1980.
29. S.F. Mughabghab and D.I. Garber, Neutron Cross-Sections BNL-325, 3rd Edition, Vol. 1, New York, 1973.
30 W.L. Zijp and J.H. Baard, ECN-70, Netherlands Energy Research Foundation (1979).
31. D.S. Brenner and R.A. Meyer, Phys. Rev. C, 13, 1288-1294 (1976).

Table IV Initial Input Nuclear Data

Target	σ_o (barns)	I_o (barns)	$I_o/g\sigma_o$	E_γ keV	P_γ	η barn. mol. kg^{-1}
^{23}Na	0.530 ± 0.004	0.306 ± 0.005	0.577 ± 0.010	1368	1.00	23.05 ± 0.18
^{55}Mn	13.3 ± 0.2	13.97 ± 0.08	1.05 ± 0.02	847	0.9887 ± 0.0004	239 ± 4
^{59}Co	18.8 ± 1.5	39.7 ± 4.3	2.1 ± 0.3	59	0.020 ± 0.001	6.4 ± 0.6
^{65}Cu	2.17 ± 0.03	2.27 ± 0.07	1.05 ± 0.04	1039	0.08 ± 0.01	0.84 ± 0.11
^{115}In	161 ± 3	2595 ± 56	15.8 ± 0.5	138	0.033 ± 0.002	45.6 ± 2.9
				417	0.324 ± 0.015	443 ± 22
				819	0.116 ± 0.006	159 ± 9
				1097	0.557 ± 0.015	762 ± 25
				1294	0.85 ± 0.02	1163 ± 35
				1507	0.102 ± 0.005	140 ± 7
^{186}W	37.4 ± 1.1	470 ± 10	12.56 ± 0.45	480	0.234 ± 0.010	13.6 ± 0.7
				512	0.0069 ± 0.0006	0.40 ± 0.04
				619	0.067 ± 0.003	3.9 ± 0.2
				626	0.0117 ± 0.0005	0.68 ± 0.04
				686	0.293 ± 0.012	17.0 ± 0.9
				773	0.0442 ± 0.0018	2.57 ± 0.13
^{197}Au	98.8 ± 0.3	1549 ± 27	15.60 ± 0.27	412	0.955 ± 0.001	481.5 ± 1.6

Table V Revised Input Data and Output η Values

Target	E_γ keV	P_γ	Input η	Output η
^{23}Na	1368	1.00	23.05 ± 0.18	22.83 ± 0.16
^{55}Mn	847	0.9887 ± 0.0004	239 ± 4	233 ± 2
^{59}Co	59	0.020 ± 0.001	6.4 ± 0.6	6.86 ± 0.10
^{65}Cu	1039	0.08 ± 0.01	0.84 ± 0.11	0.87 ± 0.03
^{115}In	138	0.0330 ± 0.0014	45 ± 2	50.8 ± 0.6
	417	0.278 ± 0.012	380 ± 18	370 ± 4
	819	0.116 ± 0.005	159 ± 7	164.6 ± 2.0
	1097	0.573 ± 0.022	784 ± 34	786 ± 8
	1294	0.846 ± 0.020	1157 ± 35	1139 ± 14
	1507	0.107 ± 0.005	146 ± 7	137 ± 3
^{186}W	480	0.253 ± 0.005	14.7 ± 0.5	15.4 ± 0.2
	512	0.00747± 0.00017	0.435 ± 0.016	0.468 ± 0.006
	619	0.0727 ± 0.0014	4.23 ± 0.15	4.36 ± 0.04
	626	0.0126 ± 0.0003	0.733 ± 0.027	0.755 ± 0.008
	686	0.316 ± 0.007	18.4 ± 0.7	18.9 ± 0.2
	773	0.0477 ± 0.0010	2.78 ± 0.10	2.85 ± 0.03
^{197}Au	412	0.955 ± 0.001	481.5 ± 1.6	481.7 ± 1.6

EVALUATION OF THE THERMAL NEUTRON CONSTANTS FOR ^{233}U, ^{235}U, ^{239}Pu AND ^{241}Pu*

J.R. Stehn, M. Divadeenam, and N.E. Holden

National Nuclear Data Center
Brookhaven National Laboratory
Upton, New York, 11973, U.S.A.

A consistent set of best values of the 2200 meter/second neutron cross sections, Westcott g-factors, and fission neutron yields for ^{233}U, ^{235}U, ^{239}Pu and ^{241}Pu are presented.

A least squares fitting program, LSF, is used to obtain the best fit and to estimate the sensitivity of these fissile parameters to the quoted uncertainties in experimental data.

The half-lives of the uranium and plutonium nuclides have been evaluated and these have been used to reassess the significant experimental data. The latest revision of the spontaneous fission neutron yield $\bar{\nu}$, of ^{252}Cf and the foil thickness corrections to the fission neutron yield ratios of fissile nuclei to ^{252}Cf are included. These lead to greater consistency in the data used for $\bar{\nu}$ (^{252}Cf). Similarly, the ^{234}U half-life as revised leads to improved consistency in the ^{235}U fission cross section.

Comparison is made with the values from ENDF/B-V and other evaluations.

Introduction

The present work, like the efforts that preceded it in 1965,[1] 1969,[2] 1974,[3] 1975,[4] and 1977[5] analyzes and combines all the relevant experimental measurements that lead to a knowledge of the thermal neutron constants for the four principal fissile isotopes: the absorption, fission, and scattering cross sections, and the prompt and total nubar values for thermal neutron fission. Measurements of nubar for ^{252}Cf are included as well. The results of both absolute and relative measurements are combined by the same iterative least-squares fitting program LSF[6] that was used in previous efforts.

The analysis involves adjusting older data to be consistent with current values of the standard neutron cross sections and of the half-lives of the fissile materials shown in Table 5.

The Westcott g-factors have been updated by using the values proposed by Leonard[7,8] for ^{235}U and ^{239}Pu, and those of ENDF/B-V for ^{233}U and ^{241}Pu. Westcott's[9] original uncertainty values are used except for fission in ^{235}U and ^{239}Pu, where Leonard's uncertainty values are quoted. We simplify the interpretation of the old scattering and total cross section measurements by ignoring the small differences in scattering that used to be attributed to crystal structure effects (e.g., metal vs. liquid). We also simplify the calculation of corrections for neutron detector sensitivity in measurements of prompt nubar ratios by adopting the mean fission neutron spectrum energies given by A. B. Smith[10], rather than treating these means as parameters to be adjusted in the LSF fitting (we find the results of the fit to be insensitive to the set of mean energies chosen). We accept J. R. Smith's evaluation[11,12,13] of the manganese bath measurements of ^{252}Cf nubar total, and we accept Boldeman's[14] evaluation of ^{252}Cf prompt nubar and nubar ratio measurements.

New measurements and analyses, made since the previous efforts, result in a more consistent fit to the data than were obtained in past efforts.

Recent Data and Analyses

Since Lemmel's paper[4] at the Washington Conference in 1975, there have been significant changes in some auxiliary data used to determine the parameters. The uranium half-lives have been evaluated here at NNDC[15] and the ^{234}U value has increased by about 0.4% compared to the most accurate values quoted in the literature. The latest experimental data on the half-life of ^{239}Pu have cleared up the discrepancy between calorimetric and alpha-counting half-lives and resulted in a decrease of 1.2%. These changes produced corresponding changes in the cross sections for the ^{235}U and ^{239}Pu whenever the amount of fissile material was determined by alpha counting, and this affected some of the most precise measurements in the input data set. Monte Carlo studies have refined the interpretations of certain measurements of alpha[16] (neutron capture to neutron fission cross-section ratio) and eta[17,18] (number of neutrons released per neutron absorbed).

*This work was supported by the U.S. Department of Energy

However, the most extensive evaluation work has dealt with the determination of nubar. ^{252}Cf is used as a standard and the various fissile nuclide nubar values are measured as ratios to ^{252}Cf. Boldeman[14] and Smith[11,12,13] have evaluated various ^{252}Cf nubar measurements. Boldeman[19] has estimated a thickness correction for the foil samples in his earlier experiments on nubar ratios of the fissile nuclides to ^{252}Cf.

Following Smith's last review[13] in 1980, there has been little activity on the ^{252}Cf nubar problem although the results from liquid scintillator measurements tend to disagree with the results for manganese bath experiments. However, recently Axton[20] has commented on the Smith evaluation of Axton's experiment and Smith has replied.[21] Spencer[22] has published final results on his liquid scintillator measurement at Oak Ridge, and Edwards[23] has reported on a new measurement at Harwell.

Treatment of Uncertainty (Errors)

In general, the measurer's estimates of the uncertainties in their results are used - although we do not hesitate to follow evaluators' recommendations when they have shown good reason to change the original. We are indebted particularly to H.D. Lemmel and to B. R. Leonard, Jr. for their labors in this regard.

The LSF calculation scheme is such that the uncertainties (standard deviations) calculated for the output values are realistic if the uncertainties assigned to the input values are realistic. That is, if each experimental input datum is drawn from a population of data whose standard deviation is known

K. H. Böckhoff (ed.), Nuclear Data for Science and Technology, 685–688.

to be equal to the experimental error assigned to it, then the LSF output errors are correct. We assume this contrary-to-fact situation to be the case. Even though there are good reasons for questioning the validity of this assumption, our results suggest that experimenters on the whole are realistic, even pessimistic, in assigning errors to their results. We find that not one of the individual data used differs from the LSF output value by more than twice the standard deviation of the difference between the two. The input data as a whole tend to be slightly more consistent with each other, judged by the Chi-squared test, than would be expected.

Results

The results of our fits for ^{233}U, ^{235}U, ^{239}Pu and ^{241}Pu are presented in Tables 1, 2, 3, and 4. The standards used in this work are shown in Table 5.

Table 1

^{233}U Results at 2200 m/s

Quantity	Lemmel(75)	Steen(72)	ENDF/B-V(79)	NNDC(8/82)
σ_{abs}	575.2±1.3	571.0±2.5	574.2	576.1±1.3
g_a	1.001±0.002	0.9990	0.9990	0.999±0.001
σ_{fiss}	529.9±1.4	525.1±2.4	528.4	530.2±1.2
g_f	0.9967±0.002	0.9966	0.9966	0.996±0.001
σ_γ	45.3±0.9	45.9±0.2	45.8	45.9±0.6
η	2.283±0.006	2.297±0.007	2.296	2.293±0.004
α	0.086±0.002	0.0874±0.0005	0.0866	0.0866±0.0011
ν_T	2.479±0.006	2.498±0.008	2.495	2.492±0.004
$^{252}Cf\ \nu_T$	3.746±0.009	3.783±0.014	3.766	3.767±0.004
$T_{1/2}(^{233}U)$	159000±200	159300±2300		159100±200

In Table 1, column 2 lists Lemmel's recommended values,[4] Column 3 lists Steen's evaluation[24], Column 4 lists the ENDF/B-V values, and Column 5 our most recent result.

Although Lemmel had later presented a talk at the 1977 Standards Meeting[5] at NBS, his conclusion was that the discrepancy between the fit to the 2200 m/s data and the fit to the 20°C Maxwellian data would not allow him to recommend either result. Accordingly, we have used his 1975 recommendations[4] in Column 2 to provide some comparison with our fit.

Our nubar and eta values are larger than Lemmel's due to our 0.6% larger value for ^{252}Cf nubar, which is primarily due to Spencer's measurement[22] performed after Lemmel's evaluation.

Table 2

^{235}U Results at 2200 m/s

Quantity	Lemmel(75)	Leonard(76)	ENDF/B-V(79)	NNDC(8/82)
σ_{abs}	680.9±1.7	681.9±1.9	681.9	681.5±1.2
g_a	0.980±0.003	0.9782	0.9781	0.9781±0.0009
σ_{fiss}	583.5±1.3	583.5±1.7	583.5	582.9±1.1
g_f	0.9758±0.002	0.9775±0.0011	0.9775	0.9771±0.0010
σ_γ	97.4±1.6	98.38±0.76	98.38	98.7±0.9
η	2.071±0.006	2.071±0.003	2.085	2.078±0.003
α	0.167±0.003	0.1686±0.0014	0.1686	0.1692±0.0017
ν_T	2.416±0.005	2.4205±0.012	2.437	2.430±0.004
$^{252}Cf\ \nu_T$	3.746±0.009		3.766	3.767±0.004
$T_{1/2}(^{234}U)$	244700±200			245700±500

In Table 2, we have listed Leonard's evaluation [7] in Column 3 and the other Columns remain the same. Our ^{235}U nubar value is larger than the corresponding result of Lemmel by 0.6% due primarily to our higher ^{252}Cf nubar value. The increased half-life for ^{234}U reduces the ^{235}U fission cross section in Column 5 of Table 2. The ENDF/B fission cross-section value is based on Leonard's evaluation and as such is not an independent evaluation.

Beer's analysis[16] of Lounsbury's Chalk River alpha measurements results in a lower value for alpha (Maxwellian). However, Lemmel's g-factors for capture and alpha are much larger than ours and results in his larger Maxwellian alpha being reduced to a lower 2200 m/s alpha value compared to us.

Table 3

^{239}Pu Results at 2200 m/s

Quantity	Lemmel(75)	Leonard(81)	ENDF/B-V(79)	NNDC(8/82)
σ_{abs}	1011.2±4.1	1028.6±5.1	1011.9	1016.3±3.0
g_a	1.081±0.004	1.0762	1.0764	1.078±0.003
σ_{fiss}	744.0±2.5	754.8±4.5	741.7	748.3±1.7
g_f	1.0555±0.0024	1.0535±0.0015	1.0562	1.054±0.001
σ_γ	267.2±3.3	273.75±2.7	270.2	268.0±2.5
η	2.106±0.007	2.111±0.008	2.119	2.121±0.006
α	0.359±0.005	0.363±0.004	0.3643	0.358±0.003
ν_T	2.862±0.008	2.877±0.013	2.891	2.881±0.006
$^{252}Cf\ \nu_T$	3.746±0.009		3.766	3.767±0.004
$T_{1/2}(^{239}Pu)$	24290±70	24110		24100±12

Table 3 lists the results for ^{239}Pu. Once again, the half-life change has directly affected the fission cross section. Our value is 0.6% larger than Lemmel's value and 0.9% larger than the ENDF/B fission cross section. Our nubar value is again larger than Lemmel's and is due almost entirely to the increase in the ^{252}Cf standard. It can be noted that our larger nubar and fission cross section values produce an eta value which is 0.7% larger than Lemmel's value.

In the case of ^{239}Pu, the most recent shape measurements by Deruytter[25] ($g_f = 1.0553 \pm 0.0013$) and Gwin[26] ($g_f = 1.055 \pm 0.002$) agree better with Lemmel than with Leonard or ourselves, but the differences are minor. The most precise fission cross section measurement in our input set is also by Deruytter.[25] The half-life correction adjusts his reported value of 741.9 barns to an input value of 751.6 barns. Since his measurement carries a considerable weight in this fit, our fission cross section of 748.3 barns is significantly larger than both Lemmel's value and the ENDF/B-V value. Since Leonard's evaluation was performed much later than Lemmel's or the ENDF/B-V evaluation, he had the lower half-life value available and obtained a larger cross section as a result.

Table 4

^{241}Pu Results at 2200 m/s

Quantity	Lemmel(75)	Leonard(81)	ENDF/B-V(79)	NNDC(8/82)
σ_{abs}	1378.±9.	1368.5	1376.4	1373.7±10.5
g_a	1.039±0.003		1.043	1.044±0.002
σ_{fiss}	1015.±7.	1003.8	1015.0	1011.6±8.1
g_f	1.044±0.005		1.0452	1.046±0.005
σ_γ	362.±6.	364.7	361.4	362.1±5.4
η	2.155±0.010	2.166	2.178	2.167±0.007
α	0.357±0.007	0.363	0.3560	0.358±0.006
ν_T	2.924±0.010	2.953	2.953	2.943±0.009
$^{252}Cf\ \nu_T$	3.746±0.009		3.766	3.767±0.004

Most of our effort to date has dealt with the ^{235}U and ^{239}Pu data. However in order to complete the picture, we provide Table 4 which lists the present results for ^{241}Pu.

In Table 4, the various results listed for ^{241}Pu are generally consistent. Our eta value is 0.6% larger than Lemmel's value due entirely to the change in the ^{252}Cf nubar value.

In general, for all four fissile materials, we find no significant discrepancies between 2200 m/s and Maxwellian data. Like Lemmel, we have compared the fits obtained separately for each set of data. They are different, it is true; but the differences are neither large nor systematic. In neither set was the fitted value of any quantity different by more than twice its error from the value fitting the combined input data set.

We suggest that the recent evaluations and the different g-factors which we use have affected the Maxwellian data such that Lemmel's conclusion that these differ significantly from 2200 m/s data is no longer valid.

Table 5
Standards

Quantity	Value	Unit	Reference
$T_{1/2}(^{233}U)$	1.592±0.002	10^5 years	BNL-NCS-51320
$T_{1/2}(^{234}U)$	2.454±0.006	10^5 years	BNL-NCS-51320
$\sigma_\gamma(^{197}Au)$	98.65±0.09	barns	BNL-NCS-51388
$\sigma_\gamma(^{59}Co)$	37.18±0.06	barns	BNL-NCS-51388
$\sigma_\gamma(^{55}Mn)$	13.3±0.2	barns	BNL-NCS-51388
$\sigma_a(^{10}B)$	3838.±6.	barns	BNL-NCS-51388
$\sigma_a(^{6}Li)$	941.±3.	barns	BNL-NCS-51388
g-factors	ENDF/B-V, Leonard		EPRI NP-167, EPRI NP-1763
g-error	Westcott, Leonard		AECL-3255, ERPI NP-167,-1763

Conclusion

Although we have not yet examined all the input data and revised them with the latest values of the auxiliary data, we have reevaluated those data sets which carry the largest weight in the fit, e.g. Deruytter's ^{235}U and ^{239}Pu fission cross section measurements,[25,27] the Rumanian ^{235}U and ^{239}Pu fission measurments,[28,29] Okazaki's alpha measurement,[30,31] and Bigham's and Keith's fission ratio measurements.[32,33,34]

We presented a preliminary version of this work at a BNL Conference.[34] We intend to complete the evaluation of the older (lower weight) measurements but we expect these changes to have a minor affect on the final recommendations. Contrary to Lemmel's 1977 paper,[5] we believe that the present situation is satisfactory in regard to choosing a best-fitting set of values for the 2200 m/s parameters of the fissile materials.

Acknowledgement

The authors appreciate the help of H. D. Lemmel (IAEA) in providing information in advance of publication of the report of his complete analysis of all experimental input data used in his 1977 evaluation.[5]

References

1. C. H. Westcott, et al., Atom. Energy Rev., 3,3 (1965).

2. G. C. Hanna, et al., Atom. Energy Rev., 7, 3 (1969).

3. J. R. Stehn, Trans. Am. Nucl. Soc. 18 (1), 351 (1974).

4. H. D. Lemmel, Proc. Conf. Nuclear Cross Sections and Technology, Wash. D.C. 3-7 Mar. 1975, NBS Sp. Pub. 425, 1, 286 (1975).

5. H. D. Lemmel, Proc. Int. Specialists Symp. Neutron Standards and Applications, Gaithersburg, Maryland, March 28-31, 1977, NBS Spec. Pub. 493, 170 (1977).

6. D. McPherson, J. H. Johnson, "LSF:an Interpreter for a Class of Least Squares Fitting Problems," AECL-3415, Chalk River Laboratory (1972).

7. B. R. Leonard, Jr., D. A. Kottwitz, J. K. Thompson, "Evaluation of the Neutron Cross Sections of ^{235}U in the Thermal Energy Regions," EPRI-NP-167, Electric Power Research Institute (1976).

8. B. R. Leonard, Jr., J. K. Thompson, "Evaluation of the Thermal Cross Sections of ^{239}Pu and ^{241}Pu," EPRI-NP-1763, Electric Power Research Institue (1981).

9. C. H. Westcott, "A Study of the Accuracy of g-Factors for Room Temperature Maxwellian Spectra for U and Pu Isotopes,"AECL-3255, Chalk River Laboratory (1969).

10. A. B. Smith, "Prompt Fission Neutron Spectra", Proc. Consultants Meeting on Prompt Neutron Spectra; Vienna, Aug. 25-27, 1971, p.3, IAEA, Vienna, (1972).

11. J. R. Smith, Symp. Proc. Nuclear Data Problems for Thermal Reactor Applications, Brookhaven National Laboratory, May 22-24, 1978, EPRI-NP-1078, 5-1, Electric Power Research Institute (1979).

12. J. R. Smith, "Status of the Quest for ^{252}Cf $\overline{\nu}$," EPRI-NP-1258, Electric Power Research Institute, (1979).

13. J. R. Smith, informal report "Residual Differences in Manganese Bath Measurements and a New Measurement of $\overline{\nu}$ ^{252}Cf" in $\overline{\nu}$ Workshop Summary, NBS Nov. 20-21, 1980 (unpublished).

14. J. W. Boldeman, Proc. Int. Specialists Symp. Neutron Standards and Applications, Gaithersburg, Maryland, March 28-31, 1977, NBS Spec. Pub. 493, 182 (1977).

15. N. E. Holden, "The Uranium Half-lives: a Critical Review," BNL-NCS-51320, Brookhaven National Laboratory (1981).

16. M. Beer, et al., "A Monte Carlo Analysis of a Chalk River Experiment on Cross Sections of Fissile Nuclides," EPRI-NP-163, Electric Power Research Institute (1975).

17. J. J. Ullo, M. Goldsmith, Nucl. Sci. Eng., 60, 239 (1976).

18. M. Goldsmith, J. J. Ullo, Nucl. Sci. Eng., 60, 251 (1976).

19. J. W. Boldeman, J. Frehaut, Nucl. Sci. Eng., 76, 49 (1980).

20. E. J. Axton, A. Bardell, priv. comm. March 19, 1982.

21. J. R. Smith, priv. comm. April 26, 1982.

22. R. R. Spencer, R. Gwin, R. Ingle, Nucl. Sci. Eng., 80, 603 (1982).

23. G. Edwards, D. J. S. Findlay, E. W. Lees, Ann. Nucl. Energy, 9, 127 (1982).

24. N. M. Steen, "Revision of the Thermal Parameters of U-233," WAPD-TM-1052, Bettis Atomic Power Laboratory (1972).

25. A. J. Deruytter, W. Becker, Ann. Nucl. Energy, 1, 311 (1974).

26. R. Gwin, Nucl. Sci. Eng., 61, 116 (1976).

27. A. J. Deruytter, J. Spaepen, and P. Pelfer, J. Nucl. Energy 27, 645 (1973).

28. I. Berceanu et al., Rev. Roum. Phys. 23, 867 (1978).

29. I. Berceanu et al., Rev. Roum. Phys. 24, 437 (1979).

30. A. Okazaki, M. Lounsbury, R. W. Durham, "A Determination of the Ratio of Capture to Fission Cross Section of ^{233}U", AECL-2148 (1964).

31. A. Okazaki, M. Lounsbury, R. W. Durham, I. H. Crocker, (1964) "A Determination of the Ratio of Capture to Fission Cross Sections of ^{235}U," AECL-1965.

32. C. B. Bigham, et al., Proc. 2nd Int. Conf. Peaceful Uses Atomic Energy, Geneva 16, 125 (1959).

33. C. B. Bigham, Nucl. Sci. Eng., 59, 50 (1976).

34. R. L. G. Keith, A. McNair, A. L. Rodgers, J. Nucl. Energy 22, 477 (1968).

35. N. E. Holden, M. Divadeenam, J. R. Stehn, "Least-Squares Fit of the Thermal Neutron Data for the Fissile Nuclei ^{233}U, ^{235}U, ^{239}Pu and ^{241}Pu" in Seminar-Workshop on "Thermal Reactor Benchmark Calculation Techniques, Results, and Applications, Brookhaven National Laboratory May 17-18, 1982, EPRI report (to be published).

EVALUATION OF THERMAL NEUTRON CAPTURE CROSS SECTION OF GOLD

Yuan Hanrong

Institute of Atomic Energy, Academia Sinica
P.O.Box 275, Beijing
People's Republic of China

The thermal neutron capture cross section of gold has been evaluated. The relevant experimental data published both at home and abroad have been collected as complete as possible. The experimental methods concerned (i.e. activation method, transmission method, pile oscillator method, neutron life time method and 4π detector-transmission method) are discussed. Based on the analysis of available data 14 experimental results have been selected, and some of them have been renormalized and corrected. After proper treatment the recommended value of the thermal neutron capture cross section of gold is obtained. It is

$$\sigma_r = 98.67 \pm 0.09 \text{ barns.}$$

[$^{197}Au(n,r)$, capture cross section, E_n=0.0253 eV]

I. Introduction

In slow neutron energy region the slight departure from 1/v in absorption cross section of gold is existing, but because of its monoisotopic nature, its chemical purity, its large thermal neutron capture cross section which can be measured by different methods accurately, and the simple decay scheme and suitable half life for measurement of the product nucleus formed by neutron capture, the thermal neutron capture cross section of gold has become one of the most important thermal neutron standard cross sections and has been widely used to determine other capture cross sections by relative measurements, which can be performed with much greater accuracies than absolute determinations. This standard cross section has been also applied to measure the neutron flux and by means of flux standarization to calibrate the neutron source strength.

In view of the importance of this neutron cross section as a standard, the experimental measurements and evaluations on this cross section were considered seriously. Since the fourties a large number of measurements have been performed by the measurers of different countries [1-27].

Owing to the limitation of experimental conditions and the lack of sufficient knowlege of the factors which could affect the result of measurement, the results of the early measurements were not satisfactory and the accuracies were rather poor. Later with the improvement of the experimental conditions and technique of measuring, the accuracies were significantly improved and the experimental results agreed one another within 1% in general.

In respect of the evaluations, since the publication of second edition of BNL-325 published in 1958, in the course of about 20 years the recommended value 98.8±0.3 barns of this cross section had been adopted by many compilers [28-33]. Generally speaking this recommended value is still useful as a reference value for some ordinary measurements, but it is not accurate enough for the high precision measurements. For example, it is desired that the accuracy of the thermal neutron capture cross section of gold is with an accuracy of 0.1% for standarization of thermal neutron flux density measurements. At the same time the recommended value mentioned above seems to be out of keeping with the present experimental data situation. So it is necessary to revise this recommended value. In recent years two new evaluations of this cross section were made by S.F.Mughabghab [34] and N.Holden [35]. And the Holden's result has been adopted as the recommended value of this cross section of BNL-325, IV edition [36].
Taking the special importance of this cross section in application into account, to carry out a new evaluation on this cross section is worthy.

II. General Review of Experimental Methods

In the thermal neutron energy region the capture cross section of gold is just the same

K. H. Böckhoff (ed.), Nuclear Data for Science and Technology, 689–693.

as its absorption cross section. To put it briefly, this cross section could be measured by following methods: activation method, transmission method, pile oscillator method, neutron lifetime method (pulsed source method) and 4π detector-transmission method.

1. Activation Method

Because in this method one needs to determine absolutely both the radioactivity strength and neutron flux, the experimental technique is rather complicated. And then the accuracy of the measurement is rather poor. Thus it is not a good method for measuring accurately the thermal neutron capture cross section of gold.

2. Transmission Method

This method is suitable for measuring the absorption cross section when the absorption cross section is much larger than its scattering cross section. In the case of gold the scattering cross section is sufficiently large as to affect the final accuracy seriously. Especially, the scattering cross section is difficult to measure accurately in a separate experiment, and its calculation is uncertain because the coherent scattering is reduced by extinction effects in the microcrystals, and is modified by preferred orientation of the crystal grains of samples usually used. Therefore the ordinary transmission method is not quite good for the high precision determination of this cross section.

However, the effect of scattering can be avoided by measuring the cross section at the long-wavelength region. This is because beyond the wavelengths longer than the crystal cut off the coherent scattering is not observable, and the incoherent scattering and inelastic scattering in this region are negligible compared to absorption. So using the transmission method with long-wavelength neutrons one should be able to obtain a perfect result. The accuracies of measurement made since the sixties by this method have reached 0.3% or even better. So we may conclude that this is the most important and most accurate method for measuring the thermal neutron capture cross section of gold at the present time.

3. Pile Oscillator Method

It is a common method for relative measurements of thermal neutron absorption cross section of non-fissile materials but not good enough for standard cross section measurement. Because it is a relative measurement method after all, its accuracy depends on the accuracy of the cross section which was used as a standard. Besides, some problems related to spectrum transform have yet to be solved. In general, the accuracy of measurement carried out by this method is difficult to go beyond the limit of 1%.

4. Neutron Lifetime Method

This is a method of direct measurement of thermal neutron absorption cross section by the pulsed neutron source. It is based on the measurement of the decay constant of the thermal neutrons in the solutions of different concentrations. Because the used solutions are diluted, the energy spectrum and space distribution of the neutrons in these solutions are roughly the same as that in the pure water. And the diffusion characteristics of the thermal neutrons in solutions are not very different from that in water. Thus the variation of decay constant of the thermal neutrons in water and in solutions mainly depends on the absorption characteristics of the solute materials. In this case the correction is rather small. So using this method people could obtain a quite good result. In the case of gold, its accuracy is about 0.5%.

5. 4π Detector-Transmission Method

This is a method combined the common transmission measurement with the measurement of transmission-scattering rate measured by 4π detector. In deriving the value of absorption cross section the knowledge of exact value of scattering cross section is no longer needed. It seems to be a pretty good method of this measurement. However, in the case of using the $MnSO_4$ solution as the 4π detector [23], the total efficiency of detection was rather low, because the determination of the transmission-scattering rate was realized by sampling measurements of the activated $MnSO_4$ solution. This could not but impar the accuracy of the measurement. So it is not an ideal method for this measurement too.

Table I shows the brief introductions and results of the measurement of thermal neutron capture cross section of gold.

III. Selection, Adjustment and Recommendation of Data

1. Selection of Experimental Data

The selection of experimental data was per-

Table I Brief introductions and results of the measurements of thermal neutron capture cross section of gold

Author	Lab	Date	Experimental method	Observed σ_r (barns)	Ref.
B.D.McDaniel et al.	LASL	1947	Transmission; SC	94.6	1
Leo Seren et al.	ANL	1947	Activation	96.4	2
A.Wattenberg	ANL	1948	Transmission; SC	93.0	3
L.W.Cochran	ORNL	1949	Transmission; CS	95.9	4
R.G.Allen et al.	ORNL	1951	Transmission; CS	93.9	5
R.G.Allen et al.	ORNL	1953	Transmission; CS	94.55±0.78	6
R.S.Carter et al.	BNL	1953	Transmission at long-wavelengths; SC	98.7±0.6	7
R.G.Allen et al.	ORNL	1954	Transmission; CS	96.5±0.7	8
P.A.Egelstaff	AERE	1954	Transmission at long-wavelengths; SC	98.4±0.9	9
R.E.Wood et al.	BNL	1955	Transmission, B-W formula; CS	93.7±2.3	10
F.T.Gould et al.	COL BNL	1955	Transmission at long-wavelengths; CS	99.0±2.0	11
F.T.Gould et al.	COL BNL	1957	Transmission at long-wavelengths; CS	98.8±0.3	12
J.W.Wade	SRL	1957	Pile oscillator	97±6	13
J.D.Cummins	AERE	1957	Pile oscillator	98.5±2.0	14
Ignatev et al.		1958	Transmission	99.2±1.0	15
D.Jowitt et al.	AERE	1959	Pile oscillator	99.3±1.5	16
F.T.Gould et al.	COL BNL	1960	Transmission at long-wavelengths: CS	98.8±0.3	17
J.W.Meadows et al.	ANL	1961	Neutron life time	98.2±0.5	18
N.Teutsch	BUC	1962	Transmission at long-wavelengths; SC	98.9±0.3	19
G.Huettel	ROS	1963	Pile oscillator	98.5±0.6	20
J.Als-Nielsen et al.	RIS	1963	Transmission at long-wavelengths: CS	98.61±0.18	21
J.Als-Nielsen et al.	RIS	1964	Transmission at long-wavelengths; CS	98.6±0.2	22
Lu Hanlin et al.	IAP	1965	4π detector-Transmission; CS	98.8±1.7	23
S.J.Friesenbahn et al.	GA	1968	Activation; TOF	98.5±2	24
A.Steyerl et al.	MUN	1972	Transmission at long-wavelengths; SC	99.3±0.5	25
W.Dilg et al.	MUN	1973	Transmission at long-wavelengths; SC	98.68±0.12	26
D.Antonini et al.	CAS	1978	Activation, Intermittent irradiation	98.9±7.3	27

formed according to the following principles: (1) The data possessed large errors were discarded; (2) The data obtained in the way except direct measurement were discarded; (3) Only the data published last were considered when the same work was published in different places; (4) The remained experimental results were considered, so long as these data were reliable. The data measured with different methods were given consideration as far as possible.

2. Adjustment for Reported Values

(1) An adjustment was carried out for the data obtained by the method of measuring absorption cross section with long-wavelength neutrons. Namely, a correction for the non-$^1/_v$ portion of the gold cross section due to the 4.9 eV resonance was made by using the data calculated from the parameters of this resonance given by H.Tellier et al..[37] This portion contributes 0.90±0.03 barns to the thermal neutron cross section.

(2) A correction for the value of standard cross sections was made for the data measured by pile oscillator method. The new standard cross sections $\sigma_r(Ag)=63.3\pm0.4$ barns [36] and $\sigma_a(B)=761\pm3$ barns [38] were used to correct the data of references 14 and 20 respectively.

(3) Based on the review of the experimental methods discussed in the preceding section some adjustments of the experimental errors were made. The errors given in references 8 and 20 were enlarged.

3. Recommended Value

According to the selection principles mentioned above, 14 data were selected and then the recommended value was calculated. The reported and adjusted values are shown in table II. χ^2 test of this set of experimental data has been made, the result of which indicates that these data are statistically consistent. Hence in data processing it is suitable to take the

Table II Selected data of thermal neutron capture cross section of gold

Reference	Reported value (barns)	Adjusted value (barns)	Comments
Carter[7] (1953)	98.7±0.6	98.7±0.6	
Allen[8] (1954)	96.5±0.7	96.5±2.0	Reported error was enlarged
Egelstaff[9] (1954-1957)	98.4±0.9	98.6±0.9	Correction for non-1/v portion was made
Cummins[14] (1957)	98.5±2.0	98.8±2.0	Correction for standard cross section was made
Ignatev[15] (1958)	99.2±1.0	99.2±1.0	
Gould[17] (1960)	98.8±0.3	98.7±0.3	Correction for non-1/v portion was made
Meadows[18] (1961)	98.2±0.5	98.2±0.5	
Teutsch[19] (1962)	98.9±0.3	98.9±0.3	
Huettel[20] (1963)	98.5±0.6	99.3±1.0	Correction for standard cross section was made and reported error was enlarged
Als-Nielsen[22] (1964)	98.6±0.2	98.6±0.2	
Lu[23] (1965)	98.8±1.7	98.8±1.7	
Friesenbahn[24] (1968)	98.5±2.0	98.5±2.0	
Steyerl[25] (1972)	99.3±0.5	99.3±0.5	
Dilg[26] (1973)	98.68±0.12	98.63±0.12	Correction for non-1/v portion was made

reciprocal square of their errors as the weights.

The weighted average of these results is

$\sigma_r(Au)$=98.667±0.053 (external error) barns

and

$\sigma_r(Au)$=98.667±0.087 (internal error) barns.

We take the internal error as the error of the recommended value. Thus the recommended value of the thermal neutron capture cross section of gold is

$\sigma_r(Au)$=98.67±0.09 barns.

IV. Conclusion

1. From the experimental data listed in table II we can see that almost all of these experimental data are in agreement with the recommended value within the quoted error. Among them several high precision results [17,19,22,26] are in agreement with the recommended value within 0.2%. Owing to the fact that the accuracy of the latest measurement has reached an reasonable value of about 0.14%, it seems to be believable that the recommended value has an accuracy of 0.1%.

2. The new recommended values (98.7±0.2 barns [39], 98.71 barns [34] and 98.65±0.09 barns [35,36]) are highly accordant with ours. It means, in spite of the fact that the collected data and the data processing methods were not the same for various evaluators, the important data concerned would not be omitted and the evaluations on these data would not be diverged obviously.

3. All things considered, the recommended values of this work and references [35,36] are rather resonable. In a sense they can reflect the present situation of related data.

References

1. B.D.McDaniel et al., Phys. Rev., 72, 729 (1947).
2. Leo Seren et al., Phys. Rev., 72, 888 (1947).
3. A.Wattenberg, ANL-4174 (1948).
4. L.W.Cochran, ORNL-481 (1949).
5. R.G.Allen et al., ORNL-1164 (1951).
6. R.G.Allen et al., TID-14868 (1953).
7. R.S.Carter et al., Phys. Rev., 92, 716 (1953).
8. R.G.Allen et al., Phys. Rev., 96, 1297 (1954).
9. P.A.Egelstaff, J.Nucl. Energy, 1, 57 (1954); J.Nucl. Energy, 5, 41 (1957).
10. R.E.Wood et al., Phys. Rev., 98, 639 (1955).
11. F.T.Gould et al., Phys. Rev., 100, 1248 (1955).
12. F.T.Gould et al., Bull. Am. Phys. Soc., Ser. II, 2, 42 (1957).
13. J.W.Wade, DP-207 (1957).
14. J.D.Cummins, AERE R/R 2333 (1957).
15. Ignatev et al., Proc. 2nd Intern. Conf. of

Peaceful Uses of Atomic Energy, Geneva, Vol. 16, p.209, 1958.

16. D.Jowitt et al., AERE R/R 2516 (1959).
17. F.T.Gould et al., Nucl. Sci. Eng., 8, 453 (1960).
18. J.W.Meadows et al., Nucl. Sci. Eng., 9, 132 (1961).
19. N.Teutsch, Nucleonik, 4, 165 (1962).
20. G.Huettel, Kernenergie, 6, 336 (1963).
21. J.Als-Nielsen et al., ENDC(OR) 21"U" (1963).
22. J.Als-Nielsen et al., Phys. Rev., 133, B925 (1964).
23. Lu Hanlin et al. Atomic Energy Science and Technology 10, 848 (1965).
24. S.J.Friesenbahn et al., J.Nucl. Energy, 22, 191 (1968).
25. A.Steyerl et al., Z.Physik, 250, 166 (1972).
26. W.Dilg et al., Z.Physik, 264, 427 (1973).
27. D.Antonini et al., Nucl. Instr. Meth., 151, 567 (1978).
28. D.J.Hughes et al., BNL-325, 2nd Edition, 1958.
29. M.D.Goldberg et al., BNL-325, 2nd Edition, Supplement No.2, 1965.
30. S.F.Mughabghab et al., BNL-325, 3rd Edition, Vol. 1, 1973.
31. R.Sher, Technical Reports Series No.156, IAEA, Vienna, 1974.
32. S.F.Mughabghab et al., BNL-NCS-50439 (1974).
33. C.Michael Ledere et al., Table of Isotopes, Seventh Edition, Lawrence Berkeley Laboratory, University of California, Berkeley, 1978.
34. S.F.Mughabghab, ENDF/B Material No. 1379, 1979.
35. N.Holden, Proc. of the Intern. Conf. on Nuclear Cross Sections for Technology, Knoxville, Tennessee, 22-26 Oct. 1979, GA5.
36. S.F.Mughabghab et al., Neutron Cross Sections. Vol. 1, Part A, p.47-1, Academic Press, 1981.
37. H.Tellier et al., Saclay CEA Report 1230 N (1969).
38. Yuan Hanrong, Evaluation of thermal neutron absorption cross section for ^{10}B and natural boron, to be published.
39. C.B.Bigham, Nucl. Sci. Eng., 59, 50 (1976).

MAJOR QUESTIONS ABOUT DERIVATION OF VARIANCE-COVARIANCE INFORMATION FOR NUCLEAR DATA EVALUATIONS*

R. W. Peelle

Oak Ridge National Laboratory
Oak Ridge, Tennessee 37830
United States of America

The uncertainties in and correlations among some nuclear data are now evaluated to permit estimation of data-related uncertainties in the outputs of neutronic calculations and to focus data improvement efforts. Questions are discussed that arise in trying to obtain adequate numerical files of variance-covariance uncertainty information. These involve 1) discrepant data, 2) experimental data with incompletely reported uncertainties, 3) uncertainties in nuclear model results, 4) uncertainty data for the resonance regions and for angle and energy distributions, and 5) the role of integral data in nuclear data evaluation. The question also arises whether files of uncertainty data designed for technological applications can suffice to represent past knowledge in an evaluation that includes new data. Directions are indicated toward resolving these questions.

Introduction

In developing a file of nuclear data for engineering applications, initial priority is given to sufficient completeness and to obtaining results from neutronics calculations that are comparable to those from experiments. At a later stage, the need is felt to quantify the quality of the codified data to permit estimation of the data-related uncertainties in computed results and to help identify candidates for measurement programs. Without quantified statements of uncertainty and correlation there is no guide to the quality of a data file other than comparisons with integral experiments that are also used to judge methods and are sometimes used to adjust cross-section data.

Efforts toward adequately complete characterization of uncertainty in evaluated nuclear data have been seriously undertaken at a modest level for a decade. These efforts are bearing fruit in development and use of improved evaluation techniques,[1] inclusion of extensive uncertainty files in practical evaluated data files,[2] and in the publication of many uncertainty analyses for neutronics calculations.[3]

Some definitions are required to clarify what is meant by evaluated nuclear data uncertainty information. Cross section data elements of interest here, s_1, s_2, ..., have exact true values that remain unknown to us. A multivariate density function $f(s_1', s_2', \ldots)$ characterizes our knowledge of these values based on the research that has been performed, such that we believe the probability that the true value of the evaluated data vector lies in the element $\underline{ds}' = ds_1' \, ds_2' \ldots$ at $\underline{s}'$ is just $f(\underline{s}')\underline{ds}'$. Then the evaluated cross section

$$s_i \equiv \int \cdots \int s_i' \, f(\underline{s}') \, \underline{ds}'$$

is the expected value (mean) of this density function, the variance or mean square "error" or uncertainty is

$$\mathrm{var}(s_i) \equiv \sigma_i^2 \equiv \int \cdots \int (s_i' - s_i)^2 \, f(\underline{s}') \, \underline{ds}' \quad ,$$

and the covariance between two data elements is

$$\mathrm{cov}(s_i, s_j) \equiv \int \cdots \int (s_i' - s_i)(s_j' - s_j) \, f(\underline{s}') \, \underline{ds}' \quad .$$

The effort up to now has been to parameterize the complex multivariate density function by its moments of first and second degree; we do not assume the density function is normal in form. This focus on the second moment of the density function is convenient because of the easy rules for combining variances when a quantity can be considered a linear function of the variables of the density function. Also, knowledge of the second moment allows the Chebyshev inequality[4] to be invoked to limit the probability of large deviations from the mean.

Note that if the evaluated cross sections must be represented directly as continuous functions of energy, some approximation must be employed to prevent codified uncertainty data from being intractably voluminous; in any case we know very little about the value of the cross section at a "point" in energy. For energy dependent "smooth" cross sections, piecewise constant covariance quantities are therefore employed in the ENDF/B system.

If the evaluated data are in multigroup form, then uncertainty (or "covariance") data can be written as the matrix of variance and covariance elements:

$$V_{ij} = \begin{cases} \mathrm{var}(s_i), & i = j \; ; \\ \mathrm{cov}(s_i, s_j), & i \neq j \; . \end{cases}$$

Based on first order theory, the variance of a computed response R induced by the nuclear data $\underline{s}$ from which R is computed is given by

$$\mathrm{var}(R) \cong \sum_{ij} \frac{dR}{ds_i} \frac{dR}{ds_j} V_{ij} \quad .$$

If the terms in this sum are of comparable magnitude and if the density function goes to zero not too far from the evaluation $\underline{s}$, the central limit theorem assures that computed results depending on many cross sections will have nearly normal density functions.

Certain issues are sidestepped by development of nuclear data uncertainty files along the idealized lines implied above. Systematic evaluation failures or biases are not treated, yet the formats chosen to represent the evaluated cross sections themselves mandate both bias and methods-related uncertainty. The nonuniqueness of evaluated cross sections occurs because the various evaluation efforts have not utilized _exactly_ the same knowledge bases even if the same experiments and nuclear models were referenced. Finally, it is apparent that judicious approximations are required to identify, obtain, and codify the significant portion of the defined variance-covariance information for an entire cross-section set.

The sections below discuss some of the major questions that now arise in developing and codifying nuclear data uncertainty information; the initial intractable problem of estimating uncertainty data retrospectively for existing evaluated data is set aside here because evaluated data and their uncertainties are intrinsically the results of a single process.

Incompletely Reported Experimental Data

Nuclear data evaluators seek to combine the world's knowledge of the relevant cross sections, yet the experimental results with which evaluators must work are infrequently reported with enough attention to their needs. (In turn, evaluations seem even more rarely to be sufficiently documented for scholarly uses.) Information must be included to enable update of the results themselves to compensate for revised standards etc. More relevant here, the published variance-covariance

K. H. Böckhoff (ed.), Nuclear Data for Science and Technology, 694–697.

information must be sufficiently explicit that the corresponding matrix intended by the author for the presented results can be generated without ambiguity. When only a few results are given, the matrix itself can be reproduced,[6] while in more complex cases the publication of sufficiently complete information has not posed insuperable difficulties.[7] Note that it is insufficient in general to just quote "systematic uncertainties" for the various experimental points -- the covariance elements are rarely well defined that way. The recognized difficulty of estimating systematic uncertainties does not justify partial treatment of the information that is available.

Almost as useless to an evaluator as ambiguously specified variance and covariance elements is the unjustified assertion by the experimenter of very small systematic uncertainties. Taken at face value, the claimed uncertainties might require the evaluator to neglect all other experiments though they appear to have been conducted with comparable care. Evaluators need to be bold in revising claimed uncertainties under such circumstances, even though severe criticism is inevitable. The next section covers the case where experiments appear to be discrepant because the assigned uncertainties are too small.

Conflicting Data

We intuitively abhor ambiguity, so tend to be concerned about differences even when the range of experimental results is narrower than the claimed uncertainties. Data points appear to differ when their "error bars" (one standard deviation) no longer overlap, yet in the comparison of any two large consistent data sets one should observe very many points that differ by twice the combined uncorrelated uncertainties. Real discrepancy is identified by differences three or more times as large as the uncertainties in those differences; where more than two sets are being combined, chi-square tests are helpful indicators of discrepancy though one must remember that the tabulated probabilities are not quantitative unless the density functions of the values being considered are normal in form. Since the scatter among experimental results can scarcely be ignored in the assignment of uncertainty, methods like SUR[8] or SURP[9] based entirely on data scatter may sometimes be justified.

The relatively rigorous methods of data combination based on various forms of least-square theory[1] can aid combination of basically consistent data sets and perform the propagation of uncertainties. However, such methods of data combination are even more helpless than our intuition in the face of real data discrepancies. Figure 1 lists methods used by evaluators to deal with discrepant inputs. Normative symbols indicate this author's judgement about their typical suitability. Those with the plus (+) symbol are recommended, those in parentheses are discouraged, and the remainder are suggested to be used with caution. The recommended

+ WORK WITH AUTHORS TO LOCATE PROBLEMS AND ADJUST UNCERTAINTIES
+ USE THEORY OR SYSTEMATICS TO WEED OUT BADLY DISCREPANT RESULTS
DISCARD DATA OBTAINED USING DISCREDITED TECHNIQUES
(DISCARD ALL OLD DATA)
+ DE-WEIGHT POORLY DOCUMENTED EXPERIMENTS (i.e., ASCRIBE LARGER EXPERIMENTAL UNCERTAINTY THAN CLAIMED BY THE EXPERIMENTER)
SCALE UP ALL UNCERTAINTIES TO OBTAIN χ^2/df = 1
SCALE UP UNCERTAINTIES FOR REGION WHERE MISFIT OCCURS

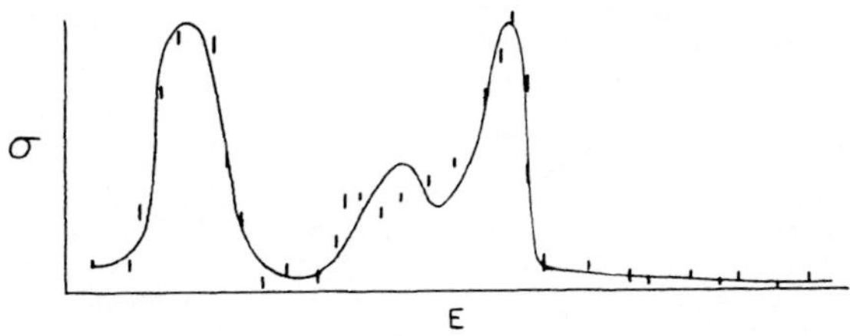

FIG. 1. DATA MODIFICATION TECHNIQUES USED IN THE EVALUATION OF INCONSISTENT DATA

procedures are time consuming but are much preferable to arbitrary selection or weighting of data, or to propagation of input uncertainties without concern for consistency. In this figure "poor documentation" includes failure to give complete variance/covariance information as discussed in the previous section.

The schematic resonance parameter fit at the bottom of Fig. 1 was chosen to illustrate that the misfit or modeling error near the center of the figure should primarily affect the uncertainties of those parameters that represent the center peak. The same principle applies when the basis chosen for the parameterization is the cross-section set itself. Where discrepancy is general between data sets rather than concentrated as in Fig. 1, one may without fear of serious error expand the uncertainties from least-squares procedures to "cover" the discrepancies. (However, a rigorous justification of this common technique cannot be given.)

In the presence of real data discrepancy, the role of the evaluator is here seen as the strong one of updating input quantities given by the experimenter. This uncomfortable role requires consultation with the providers of data and documentation of actions taken and the reasons for them.

Uncertainty in the Results From Nuclear Models

Particularly in the energy range above several MeV, evaluators often adopt the outputs of nuclear reaction models after some parameter adjustment or verification by comparison with sparse direct experimental results. The parameters of these models have generally been obtained from fitting of experimental results for theoretically related quantities, but for a variety of reasons there is little quantified information available on the uncertainties in and correlations among the model parameters. Moreover, even for cases where the general class of nuclear models to be used is well agreed upon, the models are approximate representations of the physics and results are expected to deviate from true cross sections even when there is enough direct data to adjust parameters for a particular case.

A few authors have addressed the uncertainty of nuclear model outputs and have achieved some success.[10] To this author, it appears that a general treatment of uncertainty in model calculations would require 1) obtaining a variance-covariance matrix of the model parameters from analysis of the experiments that led to the parameter selection, including whatever specific data was applied to adjust the model for the case of concern, 2) propagation of the parameter variance-covariance matrix to give the uncertainties and correlations for all the cross sections computed with that model. Since bias caused by systematic inadequacy of a model cannot be handled by present uncertainty treatments the first step may include supplementing the "pure" physics model with additional parameters needed to represent known systematic model failures. Experience concerning the applicability of a global model to a specific nuclide must be considered so that available data for that particular nuclide will carry appropriately high weight. More examples are needed of techniques that allow one to estimate and codify the patterns of uncertainty in model outputs.

The Resonance Region

In the resonance cross-section region the evaluator is challenged with detailed and complex cross-section behavior not always clearly seen in experiments, but is aided by some knowledge of the formulae that should be applicable to represent this behavior.

For cases in which only a few resonances are important, the existing data can in principle be fit and the variance-covariance matrix of all the resonance parameters can be entered directly into an evaluated file to allow whatever uncertainty propagation is required.[11] This direct approach is rarely exercised because the resonance parameter uncertainties usually obtained from fitting codes are unrealistically small. This effect is believed to occur because resonance analysis codes up to now have typically not permitted the user to include the systematic covariances among the input data that should dominate the parameter uncertainty when many data points combine to determine the value of the parameter itself. (The theorem on which rests the calculation of the variance-covariance matrix of least-squares fitting parameters requires that the complete variance-covariance matrix of the input data be included.) Artificially small parameter variances may also be obtained when values of poorly determined parameters are "frozen" to obtain computational stability. This simplest case is solved in principle; it remains to obtain experience with more advanced resonance analysis codes that accept input covariance data.[12]

For nuclides that have many resonances that are of likely importance, such as ^{56}Fe or ^{238}U, an almost prohibitively large variance-covariance matrix for resonance parameters would have to be developed if the above solution were to be applied, and obtaining the needed sensitivity coefficients might be uneconomic. In any case, since the long-range uncertainties important to any broad-group cross section set must depend on the correlated uncertainties in the underlying experimental data, it appears that the covariances among values of corresponding parameters of the various resonances could be evaluated and codified by energy region just as are smooth cross sections. This hypothesis has not been verified, so what is needed is derivation of analytical results for simple cases, verification using numerical examples, and finally application of the resulting technique. Through such a development we could hope to obtain the uncertainties in shielded cross sections throughout the resolved resonance regions.

In evaluations where a portion of the resonance region is represented with "statistical" resonance parameters, additional problems impede progress toward achieving the ability to obtain appropriately complete uncertainty representations.[13] So little experience is available and there is so much question about the evaluation techniques used that it must suffice here to ask whether the local cross section fluctuations expected within any unresolved resonance treatment should be represented in the uncertainty file, and to note that the techniques to be developed must take into account the evaluation method used to obtain the statistical parameters and the uncertainties in those methods as well as in the data employed.

Angle and Energy Distributions

There is little experience known to the author in parameterizing the uncertainties in angle, energy, or joint angle-energy distributions. Since these distributions are a function of neutron energy, the large quantity of data that appear to be required has inhibited trials and format definitions.

In the case of secondary angle distributions evaluated in terms of their Legendre coefficients, the natural solution will be to tabulate piecewise constant uncertainties in and among these coefficients as a function of incident neutron energy for a few low-order coefficients. If the evaluated file instead contains tabulated angular distributions, similarly defined quantities based on symmetric and assymetric components of the anisotropy may be appropriate parameters. The main research required before implementation of such representations is a demonstration that covariance information in these forms can readily be processed to obtain the quantities needed for uncertainty propagation in transport calculations.

The problem in representing the uncertainties in secondary energy distributions such as fission spectra is that not only are the parameters uncertain but also the forms of the spectra are not very well known. Therefore, the variance and covariance elements that could readily be obtained as a function of incident neutron energy would likely underestimate the uncertainties at very high secondary neutron energies, for instance. Nevertheless, codifying the uncertainties in the parameters would be a valuable step. Lucius and Marable[14] have computed sensitivity coefficients of some integral experiment results to the mean and the variance of the secondary energy distribution; the uncertainties in these quantities could be sought and tabulated. Further progress in this area awaits additional practical demonstrations.

The Role of Integral Data

Numerous examples have been given[15] of techniques for inclusion of integral with differential data to allow refinement of cross sections. One can at least improve the ability to calculate the results of integral experiments similar to those used in the data "adjustment." Such refinement is sound in principle provided that full covariance data for the integral experiments and calculations are included, the cross section set being adjusted is sufficiently complete, and calculational methods errors ("biases") are negligible. In any such technique, care must be exercised that the integral data used to adjust the cross section set have not already been used to guide the evaluation of the input differential data.

Examples for which cross-section adjustments might readily be attempted using a basis of cross-section variables too incomplete to allow extensive application of the results include (1) multigroup sets for shielding applications with broad energy groups and low-order angular expansions, and (2) multigroup representations in the resonance regions for nuclides having high number densities. In the first case the apparent solution is to perform adjustments on "tailored" sets designed to recognize important cross-section features, and in the second case variables other than the group cross sections must reflect the fact that resonance self-protection factors are independent quantities.

Covariance Representation for Incremental Evaluation

In principle it is possible to perform cross-section evaluations to include new data by considering an earlier evaluation, together with its variance-covariance file, as a Bayesian prior.[16] (The same equations are solved in the complete adjustment codes if the new differential data are utilized in a manner similar to the integral data often employed.) This technique has been successfully employed, but with two difficulties of interest here.

Existing covariance evaluations often carefully represent the uncertainties that are correlated over broad energy ranges but do not represent the uncorrelated uncertainties that exist between cross sections at nearby energies. This approach is adequate for uncertainty propagation for engineering applications, but for incremental evaluation has the effect that the implied correlations among the prior evaluated cross sections at the energies of new measurements are far too close to unity. The new evaluated cross sections from the code sometimes include apparently unphysical results, results that would be correct if the covariance data supplied to the code had been correct.[17] How may this problem be overcome? It appears that the covariance data required for incremental updating of evaluations is more detailed than the representation needed for propagation of uncertainty in engineering applications. An obvious solution is for the evaluator to derive uncorrelated variance information on a finer grid than that needed by general users of the evaluation. Another approach has been to parameterize the covariance information in terms of correlations effective over small and large energy ranges.[18] This approach is convenient but may not correspond to the actual information content of the underlying experiments and might sometimes correspond to variance-covariance matrices that are not positive definite.

Another problem is that the users of "Bayesian" evaluation codes obtain computing economy by assuming that the "new" data being combined with the prior evaluation are uncorrelated with the data base of the prior evaluation. In some cases this assumption is not valid, and then the idea of incremental evaluation is helpful only as a concept.

Final Comments

Reasonably complete representations of the variance-covariance information corresponding to evaluated cross-section sets are not available after a decade of experience developing such data for the simple yet important cases. Approaches to more thorough treatments are apparent and further development appears to be warranted for those cross sections most important for practical applications of nuclear data. Until covariance data and uncertainty analysis techniques become somewhat more comprehensive, one cannot be sure about the full extent of the covariance data that will be required in a fully adequate evaluated nuclear data file.

Acknowledgments

The author acknowledges numerous conversations with experimenters and with evaluators of nuclear data active in the Covariance Data Subcommittee of the U.S. Cross Section Evaluation Working Group. Interactions with Francis Perey and Jim Marable have been particularly stimulating.

References

*Research sponsored by the U. S. Department of Energy, under contract W-7405-eng-26 with the Union Carbide Corporation.

1. As examples: F. Schmittroth, Nucl. Sci. Eng. 72, 19 (1979); D. M. Hetrick and C. Y. Fu, ORNL/TM-7341 (October 1980); W. Poenitz, Argonne National Laboratory, private commuication (1982).

2. R. Kinsey, compiler, ENDF-201, BNL-NCS-17541, third edition (ENDF/B-V), 1979; D. W. Muir and R. J. LaBauve, LA-8733-M (March 1981); J. D. Smith and B. L. Broadhead, ORNL/TM-7389 (1981).

3. Many examples are referenced in Sections 5 and 6 of R. W. Peelle and T. W. Burrows, BNL Report, ENDF-328, (to be published).

4. W. Feller, An Introduction to Probability Theory and Its Applications, John Wiley and Sons, vol. 1, p. 219.

5. F. G. Perey, ORNL/TM-5938 (1979). Reprinted in R. Kinsey, ENDF-102, BNL-NCS-50496 (1979).

6. G. Winkler, D. L. Smith, and J. W. Meadows, Nucl. Sci. Eng. 76, 30 (1980). J. K. Dickens *et al.* Nucl. Sci. Eng. 74, 106 (1980).

7. L. W. Weston and J. H. Todd, Nucl. Sci. Eng. 79, 184 (1981). D. Larson, ORNL, private communication (1982).

8. F. C. Difilippo, ORNL/TM-5223 (March 1976).

9. R. W. Peelle, in "Sensitivity and Uncertainty Analysis of Reactor Performance Parameters," vol. 14 of Advances in Nuclear Science and Technology, M. Becker, ed., Chapter 2-C3, Plenum Press, New York, 1982.

10. J. B. Dragt *et al*. Nucl. Sci. Eng. 62, 119 (1977). B. Strohmaier, S. Tageson, and H. Vonach, Physics Data Nr. 13-2, ISSN 0344-8401 (1980).

11. A. Gandini, CNEN-RP/FI(74)3; E. Greenspan et al. ORNL/RSIC-42, p. 231, 1977; B. Broadhead, Trans. Am. Nucl. Soc. 39, 929 (1981).

12. N. M. Larson and F. G. Perey, ORNL/TM-7485, Nov. 1980.

13. G. de Saussure and R. B. Perez, Annals of Nuclear Energy 9, 79 (1982).

14. J. L. Lucius and J. H. Marable, Trans. Am. Nucl. Soc. 32. 731 (June 1979).

15. D. R. Harris *et al.* LA-5987 (1975); A. Gandini, CNEN-RT/FI(73)5, (1973); J. B. Dragt *et al.*, Nucl. Sci. Eng. 62, 119 (1977); J. H. Marable in "Sensitivity and Uncertainty Analysis of Reactor Performance Parameters," vol. 14, of Advances in Nuclear Science and Technology, M. Becker, ed., Plenum Press, New York, 1982; F. G. Perey, ORNL/TM-6267, 1978.

16. R. W. Peelle, *op. cit.*, Chapter 2-C1. See also Reference 1.

17. D. Hetrick and C. Y. Fu, in LANL informal report T-2-IR(82)-1, E. D. Arthur, ed. (1982).

18. F. Schmittroth and D. W. Wootan, HEDL report TC-1588 (1979).

ANALYSIS OF NEUTRON DATA ON GROUNDS OF PROTON DATA IN LIGHT NUCLEAR SYSTEMS

H. Zankel

Institut für Theoretische Physik
Karl-Franzens-Universität Graz
A-8010 Graz, Austria

Coulomb corrections including the internal Coulomb distortion are calculated for p-d and p-^{4}He elastic scattering by an effective two-body method. Measured differences of the nucleon analysing powers and the differential cross sections in the charge symmetric N-d and N-^{4}He scattering can be qualitatively explained by these Coulomb effects. The new neutron data thus do not provide evidence for the need of substantial deviation from nuclear charge symmetry.

[^{2}H(N,N), E_N = 5,10 and 14 MeV; ^{4}He(N,N), E_N = 17,30 and 50 MeV, Coulomb corrections to nucleon analysing powers and differential cross sections.]

Introduction

Neutron data in both the three- and five-nucleon elastic scattering systems are becoming available now with an accuracy that approaches the standard that is usually achieved with charged particle scattering only. These new neutron data can be used for more restrictive tests of dynamical calculations without having the problem of how to treat the Coulomb force. The recently measured neutron analysing power and the differential cross sections[2] in neutron-deuteron (n-d) elastic scattering will be a challenge for the Faddeev type calculations, but one has to keep in mind that due to the complicated spin structure of the n-d system these observables are by far not enough to uniquely determine the scattering amplitudes[3]. Further progress in understanding the nucleon-nucleon interaction through studying the three-nucleon system can presumably be made only in the p-d system, where data are quite abundant.[4] In the N-^{4}He system the spin structure is much simpler and at lower energies neutron data, not as accurate as the proton ones,are at hand that allow a full parameterization of the scattering matrix. However, unlike the Faddeev equation in the nonrelativistic three-body-system, the theoretical models to be tested with the data are not exact and an analysis of improved neutron data would hardly yield more reliable information on the underlying nuclear interaction models.

Another way to look at the new neutron data is to compare them with the better known proton data and to question eventual differences. Such an approach requires, in the first place, a proper treatment of the Coulomb force. Not before the Coulomb contributions to the charged branch of the reaction have been removed differences with the neutral reaction can be discussed in the context of possible deviations from nuclear charge symmetry. However, an exact calculation of the Coulomb corrections can be performed in a model dependent way only, which is already complicated and at present not fully solved in the three-body system[5]. Consequently on approximate description of the Coulomb effects should be employed for a rough qualitative interpretation of the neutron data. One approximation has been proposed within R-matrix theory and has been successfully applied for charge symmetric predictions in light nuclear systems[6,7]. Here we present another approximation and we mention at the onset - without going into detail - that this approximation bears resemblance to the R-matrix Coulomb corrections although it has been derived within the complete different frame of integral equations (originally from Lippmann-Schwinger equations in the N-N system[8] and later from the Faddeev equations[9]). An advantage of our approximation is its close connection with the - in principle - exact solvable Faddeev equation, whereas a disadvantage is that the method as it stands now fails when narrow resonances appear[10]. This restricts for the moment a wider application of our method.

The Coulomb Correction

Invoking the channel distortion approximation[11] and using methods developed in the two-body system[8] expressions can be obtained from the Faddeev equations that describe the Coulomb distortion in the three-body system in an effective two-body manner[9,12]. The Coulomb modified nuclear three-body reaction matrix then reads for each J

$$Q^{sc}_{mn}(q) = Q^{s}_{mn}(q) + \frac{\mu e^2}{\pi q}\left[\frac{1}{\tilde{C}_m(q)} + \frac{1}{\tilde{C}_n(q)}\right] \left[Q^{s}_{mn}(q) + q\frac{d}{dq} Q^{s}_{mn}(q)\right] - \delta_{mn}\frac{\pi}{2}\tilde{V}_{mn}(q) \qquad (1)$$

where

$$\tilde{C}_l(q) = \mathcal{P}\!\!\int_0^\infty \frac{x^2 dx}{1-x^2}\int_{-1}^{1} \frac{dy P_l(y) f\{q^2(1+x^2-2xy)\}}{1+x^2-2xy} < 0 \qquad (2)$$

The labels m and n denote the possible three-body states, q is the on-shell center-of-mass momentum,μ the reduced mass, f the form factor of the finite charge distribution of the deuteron and $\tilde{V}(q)$ is the first order correction of the Coulomb potential due to the finite extension of the deuteron charge. The second expression on the right hand side of Equ.(1) represents the approximate (short range) Coulomb distortion whereas the third term is solely due to the finite charge distribution of the composite scattering particle. A stri-

K. H. Böckhoff (ed.), Nuclear Data for Science and Technology, 698–704.

king feature of Eq.(1) is that the strong reaction matrix Q^s occurs only on the energy shell. Thus in the calculation of the approximate Coulomb distortion solely on-shell information, i.e. dataare required. We can either use Q^s as input or in an inverse procedure use Q^{sc} as input and solve for Q^s in an iterative way. In the actual calculation of the Coulomb corrections to observables we have to further add the asymptotic Coulomb phases which represent the "trivial" long range part of the Coulomb corrections. Equ. (1) and (2) can be extended to elastic scattering in other few-body systems by simply adapting the charge distribution of the composite particle. The N-^{4}He system with its tightly bound composite particle is certainly a good candidate for an effective two-body treatment, but it should be mentioned that the same method was also successsful in explaining observed differences in $\pi^{\pm}$- ^{16}O differential cross sections[13].

In comparing our calculation with the data we will have to keep in mind that our Coulomb corrections are approximate ones and that their quality can be judged not till exact calculations will be available. There are however, some hints indicating that this effective two-body on-shell approximation could be a sensible one.

i) Merkuriev who developed a configuration space Faddeev-type method to exactly describe the three-body system including the Coulomb force has published first numerical results[14]. He quotes Coulomb corrections to the unsplitted S-waves between 2 and 14 MeV. These corrections remarkably coincide in trend with the results as given by the effective two-body approximation[4]. Up to 7 MeV even quantitatively the results are in good accord.

ii) For higher partial waves no exact calculations are available. On the other hand, the effective two-body method should improve in higher partial waves. Given its quality in the lowest partial wave one should expect a reasonable description of the true Coulomb effect in the three-nucleon system. Certainly the approximation should also be better the more compact the composite particles are, i.e., it should be better in p-^{4}He than in p-d scattering.

Applications

On-shell collision matrix elements respectively reaction matrix elements are necessary to compute the Coulomb distortion through Equ. (1). Preferably one would like to take a collision matrix that is parameterized by the data i.e. as given by a phase shift analysis. Typically such a phase shift analysis is available rather for the charged branch of a charge symmetric system than for the neutral one. In the three-nucleon system, however, even for p-d scattering not enough data are at hand to allow a phase shift analysis over a wider range of energy (above 6 MeV). Instead we can take the on-shell collision matrix elements as e.g. provided by the solution of the Faddeev equation using local or separable potentials as nucleon-nucleon interaction models. Here we employ the n-d reaction matrix as given by the calculation using the separable potential[15] but it should be mentioned that it has been shown that e.g. the Coulomb distortion to the nucleon analysing powers is not significantly different when a local potential is being used[16]. From the Coulomb distorted Q^{sc} charge symmetric predictions of p-d observables can be obtained.

Thereby we make use of the relation

$$T = \frac{1}{2i}(U-1) = Q(1-iQ)^{-1} \qquad (3)$$

where U is the collision operator and T the transition operator. The spin-space matrix elements of the transition operator read

$$\tau = \sum_{JMS'l'Sl} |JMS'l'\rangle T^J_{S'l'Sl} \langle JMSl| \qquad (4)$$

when $|JMSl\rangle$ are the spin-angle eigenfunctions of total angular momentum and parity. τ is related to Wolfenstein's M-matrix by

$$M_{S'\mu',S\mu} = (S'\mu'|\tau|S\mu) \qquad (5)$$

From the M-matrix elements the observables can be obtained in the usual way. To actually obtain the observables of the charged branch of the reaction (p-d or p-^{4}He) the asymptotic Coulomb phase shifts have to be added to the Coulomb distorted nuclear T-matrix in each J^π

$$T_{S'l',Sl} \rightarrow e^{i(\sigma_{l'}+\sigma_l)} T^{sc}_{s'l',Sl} \qquad (6)$$

and the Coulomb amplitude has to be added to the diagonal elements of the M-matrix. It is clear that, the predictions, will contain the shortcomings of the model n-d collision matrix used as input in the calculation. Therefore in comparing with the data we will look at the calculated difference between p-d and n-d rather than at the absolute value of the prediction. In the five-nucleon-system, on the other hand, we are able to make predictions for n-^{4}He observables on the basis of numerous p-^{4}He data.

We apply an iterative inversion of Equ.(1), i.e. we use the on-shell Q^{sc} that can be calculated with the help of Equ.(3) from the collision matrix elements of a phase-shift analysis (after pure Coulomb corrections like in Equ.(6) and the pure diagonal Coulomb amplitudes have been removed) as input information and solve for Q^s. Below 20 MeV we take the phase shift solution of Schwandt et al.[17], between 20 and 30 MeV the one of Plattner et al.[18] and above 30 MeV the solution I of Saito[19]. The charge symmetric n-^{4}He predictions can then be compared directly with the new neutron data.

Three Nucleons - Results

In Figure 1 the neutron analysing power at 10.85 MeV as given by the collision matrix element of the solution ACS4-M5 P of Ref.15 (states with L>4 were neglected) is displayed together with the charge symmetric p-d prediction. In addition a p-d prediction (p-d 1) is shown where the Coulomb distortion to the S-wave in the $J^\pi = 3/2^+$ state was slightly enlarged to be about equal to the Coulomb correction of the uncoupled ^{4}S-wave in the exact calculation of Merkuriev[14]. The n-d[1] and p-d[20] data measured at 10 MeV (Figure 2) with similar accuracy have to be compared with the results in Figure 1. As mentioned before the absolute value of the p-d prediction cannot be better than the n-d input and indeed the prediction comes out somewhat low. Looking at the difference between p-d and n-d data we find reasonably good agreement with the data in the very forward direction which is governed by pure Coulomb contributions (i.e. asymptotic Coulomb phases) and around the maximum. There the difference between the data can be explained by the approximate Coulomb distortion. The slight angular shift of the neutron curve towards smaller angles is not confirmed by the data. The slight adjustment of the Coulomb correction of the ^{4}S-wave in the $3/2^+$ state (p-d 1 in Figure 1) makes this angular shift almost vanish without disturbing the good agreement at other angles.

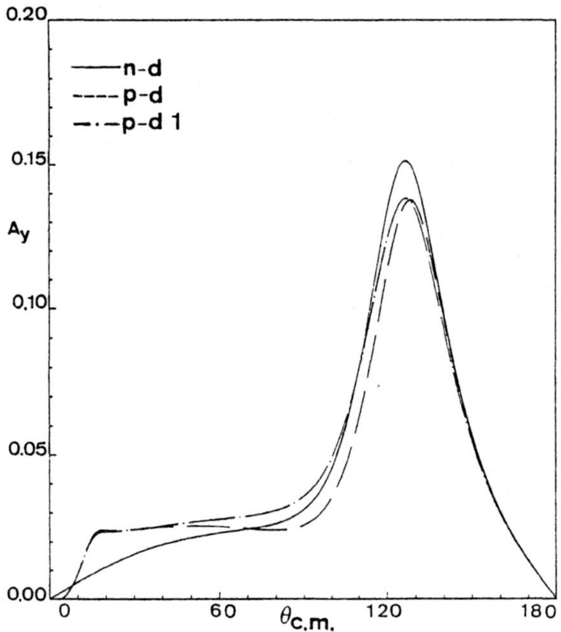

Fig. 1. Calculated nucleon analysing powers at 10.85 MeV nucleon lab. energy.

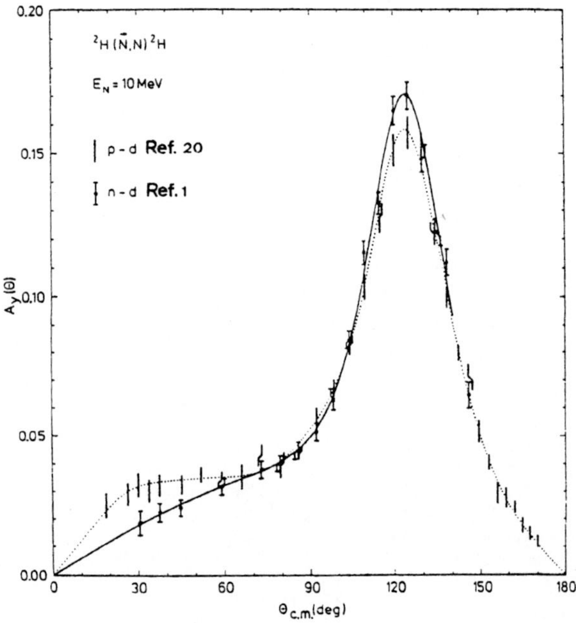

Fig.2. Nucleon analysing power data at 10 MeV. Curves are polynomial fits through the data (taken from Ref.1)

The curve through the new neutron data was found by a polynomial fit of the product $Ay(\Theta)\sigma(\Theta)$ whereby $\sigma(\Theta)$ was taken from p-d experiments. It is not yet clear to what extent the $\sigma(\Theta)$ recently measured in n-d scattering[2] might affect this fit.

Calculated nucleon analysing powers at 14.95 MeV are shown in Figure 3 and experiments at 14 MeV are displayed in Figure 4.

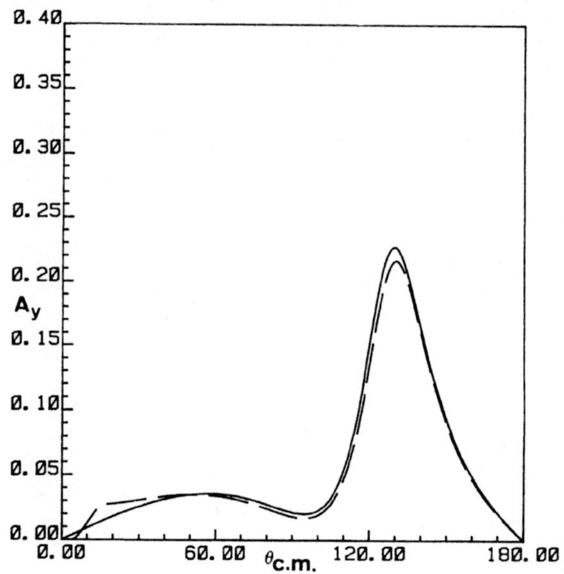

Fig. 3. Calculated nucleon analysing powers at 14.95 MeV. Solid curve is the n-d result and the dashed curve the p-d prediction.

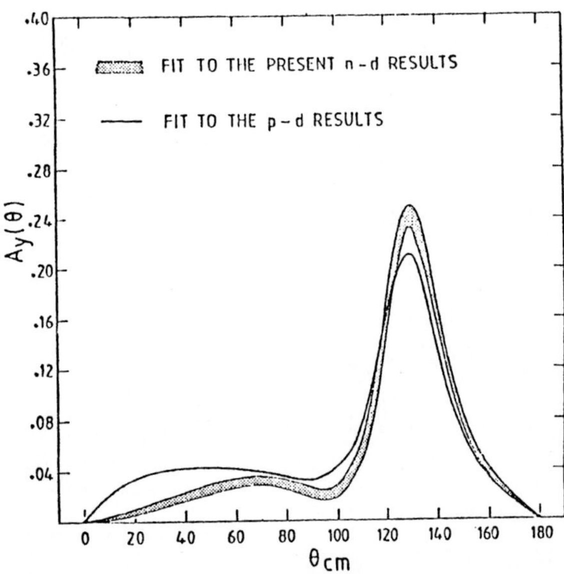

Fig. 4. Fit to the nucleon analysing power data at 14 MeV (taken from Ref. 21).

The experimental situation of the neutron data is not as good as at 10 MeV. The n-d curve in Figure 4 was obtained[21] by fitting three different data sets where in particular the one set of the Auckland group led to the enhanced difference of p-d[22] and n-d data around $\Theta_{c.m.}=120^o$. This relatively big difference is not predicted by the calculation and was similarly not found by experiments at 12 MeV[23]. If Coulomb effects should be responsible for the difference in the measured analysing powers one would not expect the difference to increase strongly with energy. Another independent measurement under way at TUNL laboratory[1] will hopefully contribute to resolve this question. At backward and very forward direction the calculation is compatible with the experiments.

The new n-d differential cross sections measured at Karlsruhe[2] are compared with available p-d[20,24,25] data and with our calculations in the Figures 5-7.

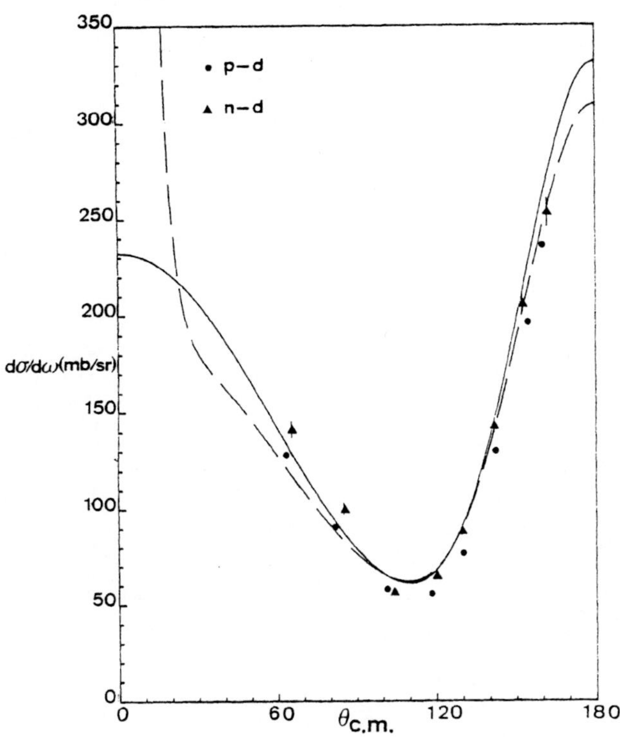

Fig. 5. N-d differential cross sections. Solid curve is n-d and dashed curve predicted p-d at 4.5 MeV. Proton data at 5 MeV are from Ref. 24 and neutron data from Ref. 2.

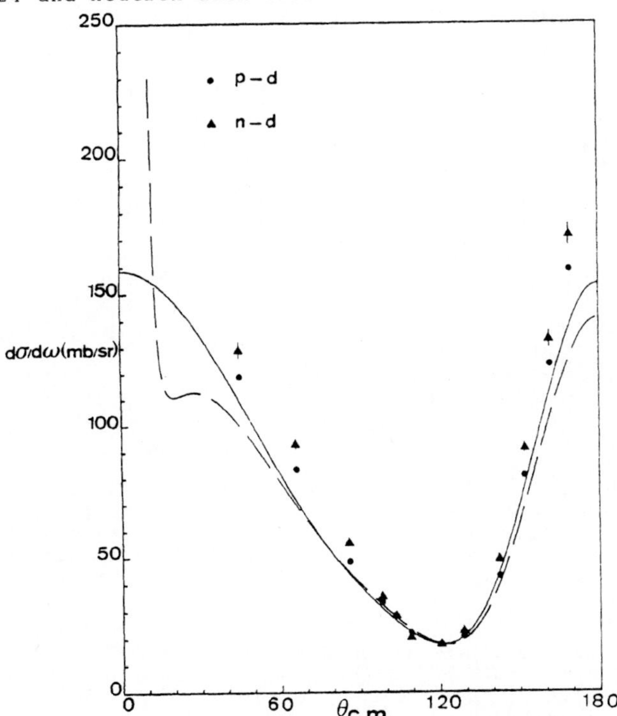

Fig. 6. N-d differential cross sections. Curves are the same as in Fig. 5, but calculated at 10.85 MeV. Proton data at 10 MeV are taken from Ref. 20 and neutron data at 10.25 MeV from Ref. 2.

Throughout the calculated difference of p-d and n-d is in semi-quantitative agreement with the measured difference. Except in the very forward direction the Coulomb distortion largely accounts for the observed difference, whereby this difference becomes smaller with increasing energy (note that the scales are different in Figs. 5-7). Taking pure Coulomb corrections only as proposed by Doleschall[26] would even qualitatively not explain the experimental situation at a number of angles. The deficiency of the p-d prediction at very forward and backward angles is partly due to the fact that the n-d collision matrix has been truncated at three-nucleon orbital angular momentum $\ell=4$. To what extent nucleon-nucleon off-shell effects play an additional role, particular at the backward angles, is not yet clear.

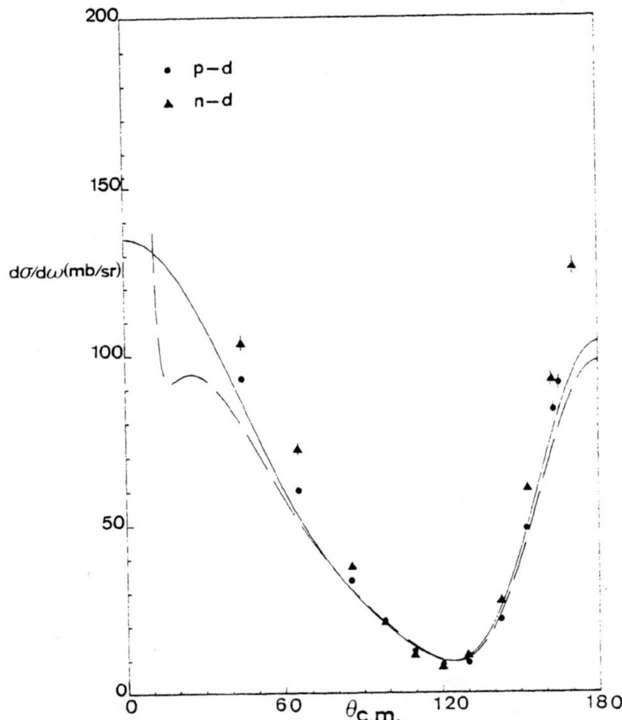

Fig. 7. N-d differential cross sections. Curves are the same as in Fig. 5, but calculated at 14.95 MeV. Proton data at 14 MeV are taken from Ref. 25 and neutron data from Ref. 2.

Five Nucleons - Results

The Karlsruhe experiments[27] to measure the neutron analysing power cover energies from 15 to 50 MeV (neutron lab. energy). Below 30 MeV they have to be compared with older data by Broste et al.[28] and at 50 MeV with a recent experiment by York et al.[29]. Between 30 and 50 MeV they will provide first experimental information on neutron polarization. The Karlsruhe data are still preliminary ones with systematical errors not yet taken into account. We have chosen to show neutron predictions at 27.5, 30.43 and 50.4 MeV together with data from Refs. 27-29. The calculated neutron analysing powers are charge symmetric predictions based on proton data i.e. phase shifts. The phase shifts of Plattner et al.[18] match the phases of Saito[19] at 30 MeV reasonably well. Problems with the phase shifts arise at energies above 45 MeV. It is there that a smooth energy behaviour of the phases is very hard to obtain without spoiling the fit to the data above 45 MeV. Since in Equ.(1) the on-shell momentum derivative of the reaction matrix is contained we, in fact, need a smooth energy dependence of said quantity. Therefore we have tried to fit the energy independent phase shifts of Saito in two ways.

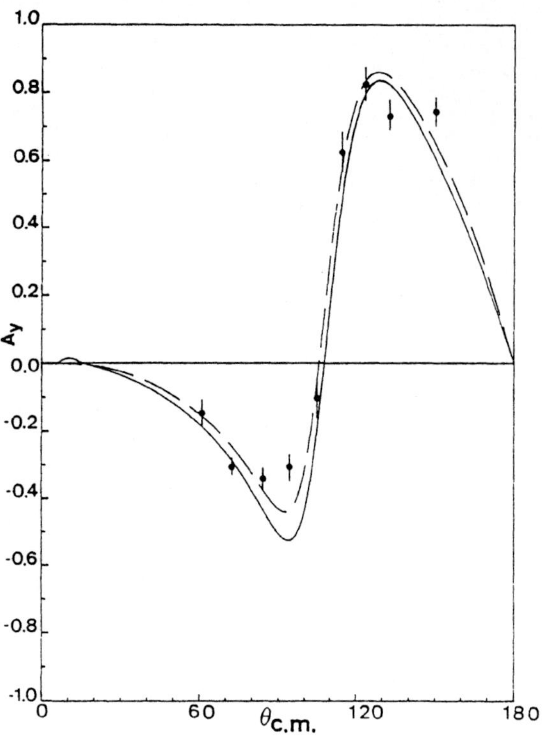

Fig. 8. Nucleon analysing powers. Solid curve is the interpolated result at 27.5 MeV (proton lab. energy) form p-^{4}He phase shift analysis (Ref. 18) and the dashed curve is the n-^{4}He charge symmetric prediction. The neutron data at 27.3 MeV are taken from Ref. 28.

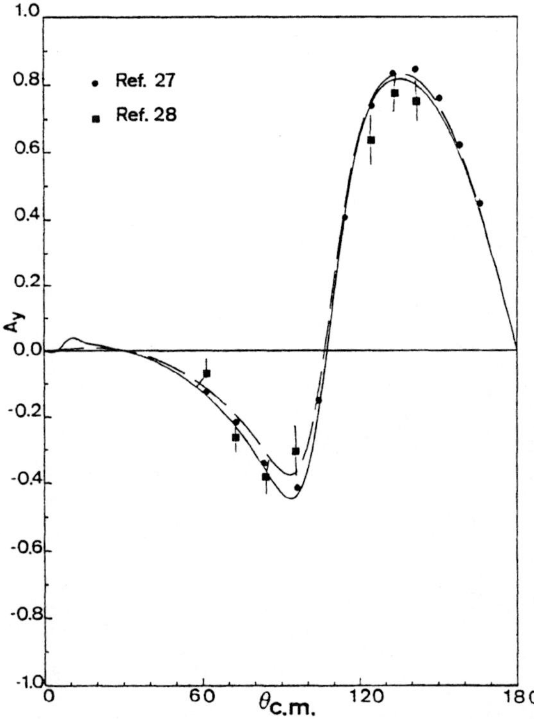

Fig. 9. Nucleon analysing powers. Solid curve is the p-^{4}He phase shift analysis result at 30.4 MeV taken from Ref. 19 and dashed line is the neutron prediction. The neutron data are from Refs. 27 and 28.

One set was found by putting weight on the data below 45 MeV (set II) and one set by emphasizing the data above 45 MeV (set I). In this way we have found error bounds for our prediction above 45 MeV. Simultaneously we have observed that the $^2P_{1/2}$ and $^2D_{5/2}$ phase shifts are very sensitive and ought to be treated with special care in a future phase shift analysis of both the p-^{4}He and the n-^{4}He system. In Figure 8 we show the interpolated proton analysing power[18] at 27.5 MeV together with the neutron prediction and the neutron data[28] at 27.3 MeV. The trend of the prediction to yield a smaller minimum and a higher maximum relative to the proton curve seems to agree with the data (respectively with a single phase shift analysis of the data[28]). The difference in the calculated curves originates from pure Coulomb as well as from Coulomb distortion effects whereby the latter must not be neglected. The proton analysing power[19] at 30.43 MeV, the neutron prediction, preliminary neutron data from Karlsruhe[27] at 30 MeV and older data from Broste et al.[28] at 30.3 MeV are displayed in Figure 9. The prediction seems to coincide in trend with the Karlsruhe data with the absolute value in the minimum, however, being quite far apart. The second data set turns out to be too small in both the minimum and the maximum which seems to be reflected in it's strange single energy phase shift solution[28].

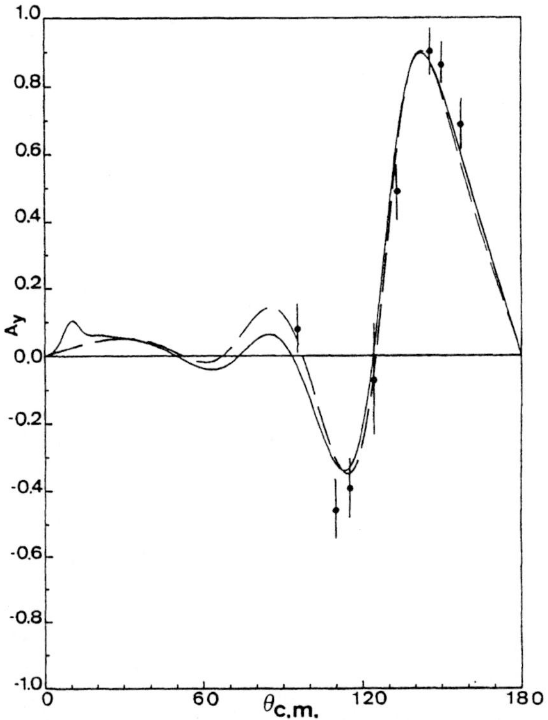

Fig. 10. Nucleon analysing powers. The curves are the same as in Fig. 9 with set I of the fitted phase shifts of Saito[19] used as proton input at 50.4 MeV. The neutron data are taken from Ref. 29.

Interpolated proton analysing power[19] at 50.4 MeV, neutron prediction and neutron data[29] at 50.4 MeV are shown in Figure 10. The neutron prediction is based on the fit I of the proton data. The prediction with fit II would be different by a few percent in the second maximum around $\Theta_{c.m.}=70^o$, but similar to solution I elsewhere. Again agreement in the trend of the data and the calculation is apparent but the quantitative difference around 120° is remarkable. Preliminary data from Karlsruhe[30] show a closer agreement with the prediction in particular for angles smaller than 140° whereby expecially the data points of the second minimum are now compatible with the prediction. There are no measurements of n-^{4}He differential cross sections above 30 MeV available. Since it is not clear whether the difficulties in calibrating the detectors will be overcome soon we just quote a prediction at 17 MeV where experiments are at hand to demonstrate that the prediction yields acceptable results at energies where the p-d data base[17] is well established. In Table I the predicted n-^{4}He differential cross sections at 17 MeV and data[31] at 17.6 MeV are shown.

Table I. n-^{4}He Differential Cross Sections (units are mb/sr)

$\theta_{c.m.}$ (deg)	experiment of Niller et al. (Ref. 31) E_n = 17.6 MeV	predictions from p-^{4}He phase shifts (Ref. 17) E_n = 17 MeV
37.2	198.±7.	194.0
49.3	153.±5.	148.4
61.1	101.±3.	105.0
72.6	69.4±2.2	68.0
83.7	37.6±1.3	39.5
94.4	18.7±0.9	20.3
104.6	9.5±0.4	10.27
114.4	7.9±0.4	7.79
123.7	9.8±0.5	10.94
132.6	15.8±0.7	17.70
141.1	24.6±1.0	26.16
149.3	29.2±1.1	34.78

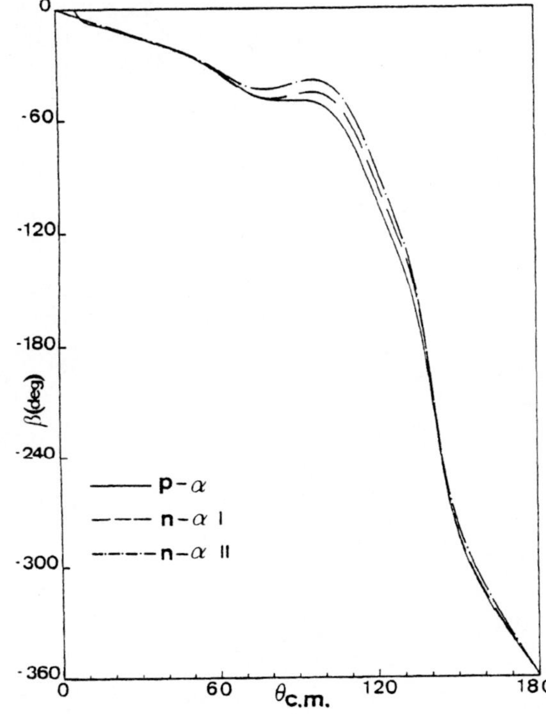

Fig. 11. N-^{4}He rotation parameter. Solid curve is set I of the fitted p-^{4}He phase shifts of Saito[19] at 50.4 MeV, dashed curve is the n-^{4}He prediction and the dashed-dotted curve is another prediction using set II of the fitted p-^{4}He phase shifts as input for the calculation of the Coulomb distortion.

The calculated cross sections are about compatible with the data except at the very backward direction. Here the measured neutron cross section is much smaller than the measured proton one which remains hard to interpret. Pure Coulomb and approximate Coulomb distortion effects cannot cover this difference in particular since their contribution acts in the opposite direction. Apart from this angular region the Coulomb distortion is essential in explaining the good agreement of the calculation and the data. Finally a p-^{4}He rotation parameter at 50.4 MeV and two sets of n-^{4}He predictions (set I and II of fits to Saito's[19] phase shifts) are given in Figure 11. In the advent of the new neutron analysing power data measurements of polarization transfer parameters become feasible. However, it is quite unlikely that an accuracy will be achieved that would be better than the differences between the theoretical curves.

Summary

Recent neutron data with improved accuracy measured in both the three- and five-nucleon elastic scattering provide the possibility of testing dynamical calculations that are not hampered by the difficulties inherent in the few-body Coulomb problem. On the other hand, it is interesting to analyse the neutron data on grounds of existing proton data thus trying to extract some information on nuclear charge symmetry which still is a contraversial issue. For a direct comparison of data in charge symmetric reactions like e.g. p-d and n-d scattering the Coulomb contributions have to be taken into account. Only then the question of charge symmetry can be discussed. An approximate treatment of the few-body Coulomb problem was proposed within R-matrix theory[6,7] yielding no strong evidence for the need of nuclear charge asymmetry to explain data of various charge symmetric reactions in light nuclear systems. Alternatively,charge symmetric experiments in the four-nucleon system have been discussed by Grüebler[4] leaving room for deviations from charge symmetry. However, no attempts have been made there to include the effect of the Coulomb distortion which on the other hand is, at least partly, contained in the above mentioned R-matrix approach though not in a very transparent way. The starting point for our approximation was originally the two-potential formalism which leads to a transition matrix that is decomposed into a pure Coulomb part and a nuclear distorted part. This second part which represents an effect screened by the nuclear interaction (here we should note the similarity with the R-matrix concept; there the interior wave function is matched to the Coulomb wave function at a certain radius and additional "internal" Coulomb corrections are applied) was subject of our approximation. In nucleon-nucleon scattering an on-shell formula was derived[8] in analogy to nucleon-nucleon Bremstrahlung[32] and was successfully used in phase shift analysis[33]. For three-body systems a similar simple on-shell formula was found[9] from three-body scattering theory where the Coulomb distortion had been described in an effective two-body manner but with the finite charge distribution of the composite particle taken into account. This concept was extended to other light nuclear systems with the restriction that narrow resonances are not well described. An exact calculation of the Coulomb distortion in p-d scattering[14] has shown that the on-shell approximation provides a reasonable description of the true Coulomb effect in this system. Given this success we can expect that our Coulomb corrections though approximate are sensible ones.

The differences in these data of charge symmetric three- and five-nucleon elastic scattering considered here can overall be explained, at least qualitatively, by Coulomb corrections. The Coulomb distortion thereby plays an important role and must not be neglected. It seems that no substantial deviation from nuclear charge symmetry can be deduced unless a rather strange cancellation of neglected Coulomb few-body aspects and nuclear charge asymmetry have occurred. A quantitative discussion of the symmetry problem, however, awaits exact few-body calculations that include the full Coulomb problem. Therefore, a more promising candidate for quantitative considerations should be looked for. The neutron-proton scattering system seems to be the best choice at the moment[34]. There the Coulomb force is absent and direct electromagnetic contributions are small. Simultaneously the accuracy of the experiments[35] is expected to be about 10^{-3} which should be good enough to provide a clean test of nuclear charge symmetry.

References

1. W.Tornow, C.R.Howell, R.C.Byrd, R.S.Pedroni and R.L.Walter, Phys. Rev. Lett. 49, 312 (1982).
2. H.Klages, private communication.
3. M.Simonius, Lecture Notes in Physics, Vol. 30, ed. D.Fick (Springer, Berlin, 1974) p. 38.
4. W.Grüebler, in "The Few Body Problem", (Eugene, 1980), ed. F.S.Levin, Nucl. Phys. A353, 75c (1981).
5. C.Chandler, in Ref. 4, p. 31c.
6. G.M.Hale and D.C.Dodder, in "Proceedings of the Knoxville Conference on Nuclear Cross Sections for Technology", 1979, p.650.
7. G.M.Hale and D.C.Dodder, in "Proceedings of the 9th Int. Conf. on the Few Body Problem", Eugene, Vol. I (1980), p. 43.
8. J.Fröhlich, L.Streit, H.Zankel and H.Zingl, J.Phys. G6, 841 (1980).
9. M.I.Haftel and H.Zankel, Phys. Rev. C24, 1322 (1981).
10. H.Kriesche and H.Zankel, J. Phys. G6, 853 (1980).
11. G.Bencze, Nucl. Phys. A196, 135 (1972).
12. H.Zankel and G.M.Hale, Phys. Rev. C24, 1384 (1981).
13. J.Fröhlich, G.Schlaile, L.Streit and H. Zingl, Z. Phys. A302, 89 (1981).
14. S.P.Merkuriev, Acta Phys. Austriaca, Suppl. XXIII, 65 (1981).
15. C.Fayard, G.Lamot and E.Elbaz, in "Few Body Problems in Nuclear and Particle Physics, ed. by R.J.Slobodrian, B.Cujec and K. Ramavatram (Les Presses de l' Universite Laval, Quebec, Canada, 1975), p. 503.
16. H.Zankel and G.M.Hale, submitted for publication.
17. P.Schwandt, T.B.Clegg and W.Haeberli, Nucl. Phys. A163, 432 (1971).
18. G.R.Plattner, A.D.Bacher and H.E.Conzett, Phys. Rev. C5, 1158 (1972).
19. T.Saito, Nucl. Phys. A331, 447 (1979).
20. W.Grüebler, V.König, P.A.Schmelzbach, B. Jenny, H.R.Bürgi, P.Doleschall, G.Heidenreich, F.Seiler and W.Reichart, Phys. Lett. 74B, 173 (1978).
21. A.Chisholm, J.C.Duder and R.Garret, in "Polarization Phenomena in Nuclear Physics - 1980", ed. by G.G.Ohlsen et al., AIP Conf. Proc. No. 69, 1488 (1981), J.E.Brock, A.Chisholm, J.C.Duder and R.Garret, submitted for publication.
22. J.C.Duder, M.Sosnowski and D.Melnick, Phys. Lett. 85B, 206 (1979).
23. W.Tornow, P.W.Lisowski, R.C.Byrd and R.L. Walter, Nucl. Phys. A296, 23 (1978).
24. D.C.Kocher and T.B.Clegg, Nucl. Phys. A132, 455 (1969).
25. S.Kikuchi, J.Sanada, S.Suwa, I.Hayaski, K.Nisimura and K.Fukunaga, J. Phys. Soc. Japan 15, 9 (1960).
26. P.Doleschall, W.Grüebler, V.König, P.A. Schmelzbach, F.Sperisen and B.Jenny, Nucl. Phys. A380, 72 (1982).
27. H.Klages, private communication.
28. W.B.Broste, G.S.Mutchler, J.E.Simmons, R.A.Arndt and L.D.Roper, Phys. Rev. C5, 761 (1972).
29. R.L.York, J.C.Hiebert, L.C.Northcliffe and H.L.Woolverton, in "Polarization Phenomena in Nuclear Physics - 1980", ed. by G.G. Ohlsen, AIP Conf. Proc. No. 69, 1281 (1981), R.L.York, J.C.Hiebert and L.C. Northcliffe, submitted for publication.
30. J.C.Hiebert, private communication.
31. A.Niller, M.Drosg, J.C.Hopkins, J.D. Seagrave and E.C.Kerr Phys. Rev. C4, 36 (1971).
32. L.Heller, Phys. Rev. 174, 1580 (1965).
33. R.Dubois, D.Axen, R.Keeler, M.Comyn, G.A. Ludgate, J.R.Richardson, N.M.Stewart, A. S.Clough, D.V.Bugg and J.A.Edgington, Nucl.Phys. A377, 354 (1982).
34. W.T.H. van Oers, Comments Nucl. Part. Phys. 10, 251 (1982).
35. S.E.Vigdor, A.D.Bacher, D.Duplantis, W.S. Jakobs, H.O.Meyer, G.L.Moake, P.Schwandt, E.J.Stephenson, L.D.Kuntson, P.A.Quin, J.Sowinski, B.P.Hichwa and P.L.Jolivette, in "Polarization Phenomena in Nuclear Physics - 1980", ed. by G.G.Ohlsen et al., AIP Conf. Proc. No. 69, 1455 (1981).

Underlying Physics and Supporting Data

THE NEUTRON CAPTURE PROCESS

B.J. Allen

Australian Atomic Energy Commission Research Establishment,
Lucas Heights Research Laboratories,
Sutherland, NSW 2232,
Australia.

The nature of the neutron capture process depends on the coupling of the incident neutron with target nucleons. Weak, intermediate and strong coupling assumptions are discussed which lead to the application of single particle, doorway and statistical models. While the Bohr statistical model has widespread application, there is abundant evidence for the intermediate coupling assumption in which the distribution of single particle strength is limited to local resonances only. External capture is shown to be important in both potential and resonant reactions. Width correlations for resonances and final states are predicted by the valence model and are observed near the magic neutron numbers. However quasiparticle-phonon interactions can be more dominant, and evidence for the relationship between quasiparticle hierarchy and level density is presented. The capture mechanism in ^{54}Fe is highlighted.

[neutron-nucleus coupling, channel interaction, non-resonant and resonant neutron capture, E1 and M1 strength functions; statistical, valence and quasiparticle-phonon models in ^{54}Fe]

Introduction

Neutron capture research has been an important part of neutron physics since the discovery of the neutron 50 years ago. Fermi and collaborators[1] conducted a series of neutron activation experiments in 1934, discovering many new isotopes. Neutron capture γ-rays from hydrogen were detected by Lea[2] in 1934, and Moon and Tillman[3] in 1936 were able to show that neutrons could reach equilibrium with the thermal vibrational energies of paraffin molecules. Insights into resonance structure came from the observation of neutron absorption bands in 1935 by Fermi and Amaldi[4] and Szilard[5]. The development of a mechanical velocity selector by Dunning et al.[6] enabled the study of these bands by the neutron time of flight method. Within a year, Bohr[7] and Breit and Wigner[8] were able to interpret these absorption bands as resonances or highly excited states of the compound nucleus. Strong absorption of the incident nucleon by the target resulted in an immediate coalescence of the neutron-target system into a many body 'compound state'. Bohr[9] also pointed out in 1938 that a direct transition from the entrance to the exit channel could also occur.

Neutron reactions near threshold excite levels at 4 to 10 MeV in the compound nucleus. At these energies Bohr's compound nucleus model was initially considered to provide an adequate description of the neutron interaction which exhibits narrow resonances in the scattering and capture cross sections. Since the wave functions of these resonances are highly complicated, their properties were expected to be described by statistical theory.

However, the observation[10,11] in thermal capture spectra of correlations between the reduced γ-ray intensities and the spectroscopic factors of final states, led to the proposal of a direct capture mechanism. Lane and Wilkinson[12] pointed out that the matrix elements of the (n,γ) and (d,p) reactions contain a common or parentage overlap factor which would yield final state correlations in the extreme case of a unique parent. For the above reactions this is the target ground state.

Lane and Lynn[13] in 1960 recognised three components of the resonance capture cross section: the compound nucleus (or resonance internal), channel (or resonance external), and direct capture (hard sphere potential and distant resonances). The last two components preferentially feed single particle final states. Furthermore, the resonance external part could give rise to enhanced transitions to single particle states if the resonance reduced width is large.

The foregoing theoretical developments were applied to s-wave capture, mostly at thermal energies. With the advent of improved reactor choppers and pulsed Van de Graaff and linear accelerators, γ-ray spectra could be measured with Ge(Li) detectors from both s- and p-wave resonances. These measurements also showed final state correlations in many cases[14] and when sufficient resonances could be studied, initial state correlations between the reduced neutron widths and partial or total radiation widths were observed[15].

For neutron energies in the giant dipole resonance region, the direct and compound nucleus models could not reproduce the measured capture cross sections, and a collective semi-direct capture mechanism[16,17] was introduced to account for the data. This theory assumes that the incident nucleon is captured into a lower orbit exciting the target nucleus into its giant dipole resonance. This intermediate state subsequently decays by enhanced γ-ray emission. Interference between the direct and semi-direct mechanism[18] and the introduction of a complex coupling function[19] are required for the description of high energy capture.

Developments in the 1960's resulted in a renaissance in neutron capture research. Some twenty years later, it is time to take stock and to determine whether the theories and data have stood the test of time.

An affirmative answer must be given, and this review is therefore as much a tribute to the Lane and Lynn theory as it is a statement of the status quo of the neutron capture process.

Neutron-Nucleus Coupling

Prior to the 1960's, the interaction of the neutron with the target nucleus was described using either weak or strong coupling assumptions, giving rise to the independent particle model and the Bohr statistical model respectively.

(a) Weak Coupling

The independent motion of neutrons in the nucleus is a consequence of the weakness of the nuclear long range attraction and the Pauli exclusion principle. Nucleons are therefore assumed to move in a single particle potential U(i) which depends on their spatial, spin and charge coordinates. In addition a perturbation arises from the effective residual interaction between nucleons. The nuclear Hamiltonian is written as

$$H = H_o + \sum_{i<j=1}^{A} V(ij)$$

where H_o is the unperturbed Hamiltonian summed over all particles with relative kinetic energy T, i.e.

$$H_o = \sum_{i=1}^{A} \{T(i) + U(i)\}.$$

The residual interaction is just the difference between the exact and model Hamiltonian.

In the independent particle model the residual interaction is assumed to be weak. Nucleons move nearly independently and two particles interact only in close proximity (~ $1/k_F$ where k_F is the Fermi momentum) with no memory of preceding interactions. Excited states of the compound nucleus are just single particle states of the form $\psi_\mu = \phi_i U_p$ where ϕ_i is the core wave function and U_p the radial single particle wave function.

Shell model studies of the residual interaction have led to its definition in terms of Bartlett (spin), Heisenberg (isospin), Majorana (space) and Wigner (ordinary) exchange operators. The interaction of valence

K. H. Böckhoff (ed.), Nuclear Data for Science and Technology, 707–718.

particles is included in shell model calculations which have had outstanding success in predicting nuclear properties, but only in rather limited mass regions near the magic numbers.

The weak coupling assumption is of limited application to the highly excited states above the neutron threshold. It is most relevant in very light nuclides where the reduced neutron widths approach the Wigner limit and level spacings are in the MeV range (^{6}Li, ^{12}C). Ground and low lying states near closed shells are often strongly single particle in nature. These states have (d,p) spectroscopic factors close to unity ($\Theta_\mu^2 \sim 1$) and can be represented by $\psi_\mu = \phi_o U_p$ where ϕ_o is the wave function of the unperturbed core.

(b) Strong Coupling

In the strong coupling assumption many collisions occur in the compound nucleus (the residual interaction is large) and the strength of the single particle state is fragmented over many fine structure states. There is no appreciable free motion of the incident neutron inside the compound nucleus and the entrance and exit channels are uncorrelated. Many single particle states contribute with random amplitudes to the fine structure states, and the distribution of amplitudes is assumed to be gaussian with zero mean. The distribution of reduced neutron widths and partial radiation widths is therefore that of a chi-squared distribution with one degree of freedom[20].

(c) Intermediate Coupling

The Bohr compound nucleus or statistical model has found widespread application, but could not account for the observation in thermal neutron capture of correlations between reduced γ-ray intensities and (d,p) spectroscopic strengths[10,11]. Lane et al.[27] evolved the concept of intermediate coupling wherein the residual interaction is not strong enough to invalidate the notion of single particle motion in the nucleus. Instead the single particle strength is fragmented over compound states of the system about the energy of the single particle state $E_{n\ell}$ over a region 2w where w is the absorption potential. In the limiting case for w large, the single particle states are completely smeared out (i.e. the strong interaction assumption).

The impact of these assumptions on the neutron capture mechanism is best understood by expanding the level functions (X) for resonances (λ) and final states (μ) into a complete set of basis functions .

$$\text{i.e. } X_\lambda = \Sigma_{cp} C_{cp} X_{cp}$$

where $X_{cp} = \phi_c U_p(a_c)$ and $U_p(a_c)$ are the radial wave functions of the set of single particle states p at the channel radius a_c in a real potential field, and ϕ_c are the channel surface functions. The coefficients C_{cp} depend on the assumed model for the neutron interaction as shown in fig. 1.

(d) Single Particle Model

For weak coupling the neutron interacts only with the central potential field and there is no residual interaction between nucleons. The single particle strength therefore remains intact. When the resonance energy corresponds to the single particle energy, i.e. $E_\lambda = E_{cp}$, then $C^\lambda_{cp} = 1$. Otherwise all other terms are zero.

(e) Statistical Model

In the strong coupling assumption, the strengths of single particle states are fragmented rather uniformly over the compound nucleus states , as a result of strong residual nuclear forces between nucleons.

The coefficients C^λ_{cp} are therefore assumed to be small, with random sign, and are independent of the proximity of E_λ to E_{cp}.

(f) Valence Model

If the dissolution of single particle strength is assumed to occur mainly in the vicinity of the single particle state, the coefficients C^λ_{cp} are appreciable (but with random sign) for $E_\lambda \sim E_{cp}$ but are small elsewhere. It is in this context that the valence model of resonance neutron capture is derived.

General expressions are given by Lynn[21] for the reduced width amplitude in channel $c(\gamma_{\lambda c})$ and for the valence radiation amplitude$(\Gamma^{\frac{1}{2}}_{\lambda\mu})$ in terms of the above expansion.

$$\gamma_{\lambda c} = \left(\frac{\hbar^2}{2M_c a_c}\right)^{\frac{1}{2}} \sum_p C^\lambda_{cp}\, U_p(a_c)$$

$$\Gamma^{\frac{1}{2}}_{\lambda\mu} = \frac{(8\pi)^{\frac{1}{2}}\, K_\gamma^{\ell+\frac{1}{2}}\, (\ell+1)^{\frac{1}{2}}}{\ell^{\frac{1}{2}}(2\ell+1)!!\,(2J_\lambda+1)^{\frac{1}{2}}} \sum_{cp',p''} C^\lambda_{cp'}\, C^\mu_{cp''} \langle U_{p'}\phi_{cj'} || H_T^{(\ell)} || U_{p''}\phi_{cj''}\rangle$$

where M_c is the reduced mass, K_γ the photon wave number $(E_\gamma/\hbar c)$, single and double prime quantities denote the resonance and final single particle states (p) and channel spins (j).

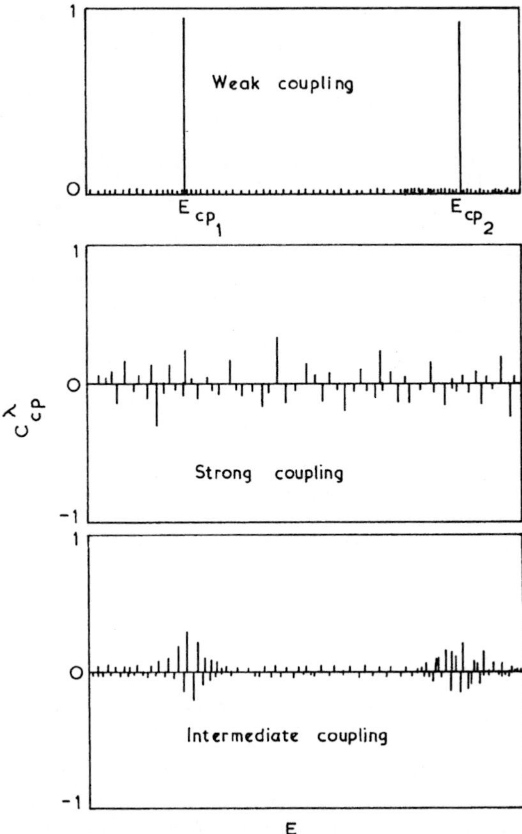

Fig. 1. Model dependence of expansion coefficients of internal eigen functions X_λ for channel c. (ref. 21)

(g) Doorway State Model

The resonance capture reaction can be pictured as a series of two-body interactions[22] beginning with the entrance channel or one particle-zero hole state (i.e. 1p-0h with respect to the target nucleus), exciting 2p-1h or collective states and through a succession of more complex p-h interactions, leading ultimately to the statistical interaction involving many nucleons. Because the 2p-1h state (which is not an eigen state of the system) provides the link between the entrance channel and more complex states, it is called a doorway state. If a number of such states are excited coherently, then a collective vibrational state may result which is characterized as a one phonon state[23].

The resonance wave function therefore contains many configurations, any of which can undergo a radiative de-excitation when overlap with the final state wave function occurs. Radiative decay which occurs in the first two stages of the interaction (Fig. 2a) can be non-resonant in character as a consequence of the uncertainty principle.

Resonant processes can be divided into valence,

doorway and statistical mechanisms (Fig. 2b). When the resonance is in the entrance channel configuration, the valence neutron can undergo a radiative transition without perturbing the core. Radiative decay can also occur from the doorway configurations of a resonance, either by a particle-hole annihilation or by particle transition in the presence of an excited core (i.e. phonon excitation). All other decay modes are grouped together under the heading of statistical interactions.

a NON-RESONANT INTERACTION

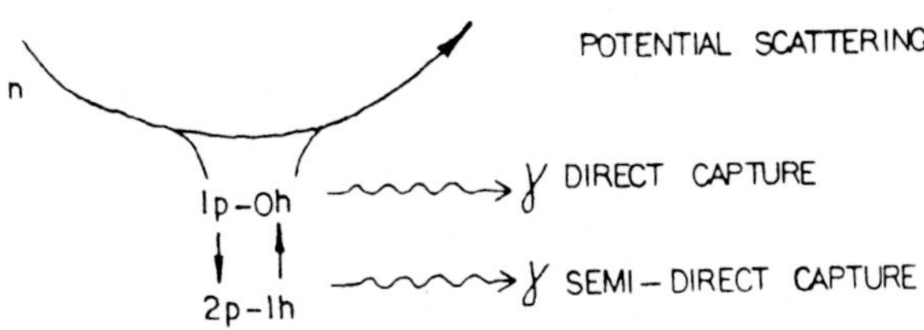

b RESONANT INTERACTION

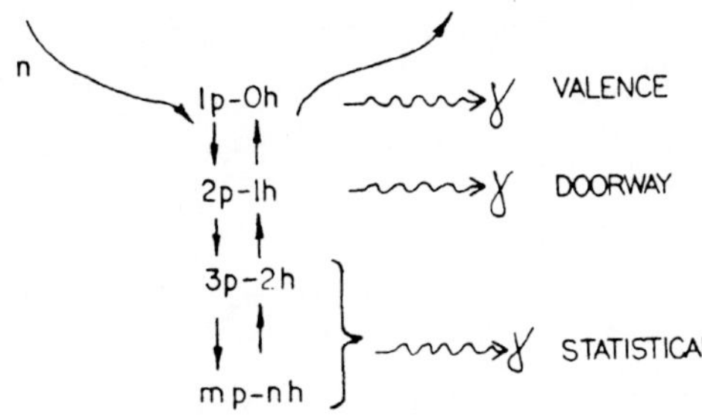

Fig. 2. Schematic representation of (a) non-resonant and (b) resonant capture (ref. 15).

The p-h excitations, however, are unlikely to exist at their unperturbed (i.e. IPM) energies. Rather, according to the Brown-Bolsterli schematic model[24], the residual interaction causes a coherent mixing of the IPM states, such that coherent states are formed at lower excitation energies (i.e. collective states) and one or more states are found at much higher energies (~80 $A^{1/3}$ MeV), taking most of the dipole strength (GDR). This energy shift depends on the strength of the residual p-h interaction.

More recently, Brown and Speth[25] have suggested that the unperturbed energies of the particles (and holes) are too low because coupling with vibrations rearranges the particle (and hole) states. This would have the effect of reducing the nucleon effective mass and consequently pushing up the calculated GDR in Pb to the observed value of 14 MeV.

The weak coupling assumption implies that only a limited interaction occurs, whereas strong coupling would give effect to the complex mp-nh configurations. Intermediate coupling implies a more limited excitation of nucleons with the single particle strength fragmented over 2p-1h or perhaps 3p-2h fine structure resonances. The width of the single particle state is ~MeV with lifetime ~10^{-22}s, and would be considered a non-resonant reaction. For doorway states, widths of tens to hundreds of keV are envisaged, whereas the strong coupling assumption gives rise to long lived resonances (10^{-15} s) with narrow widths in the eV range. The question to be asked is does the width and spacing of resonances tell us something about the p-h hierarchy? This is discussed in a later section.

The Channel Interaction

The importance of the external part (r>R where R is the interaction radius) of the nuclear wave function was first recognised by Thomas[26]. Both Moringa and Ishii[28] and Lane & Lynn[13] included the external part in their calculations of neutron capture cross sections.

The capture cross section can be expressed in terms of the diagonal element of the collision matrix which Lane & Lynn divided into resonance and hard sphere (HS) parts

$$U = U(\mathrm{res}) + U(\mathrm{HS}).$$

The resonance term can be divided into internal and external (or channel) parts

$$U(\mathrm{res}) = U^{\mathrm{int}}(\mathrm{res}) + U^{\mathrm{ch}}(\mathrm{res})$$
$$= -i\, e^{-i\phi_n} \sum_\lambda (\Gamma^{\frac{1}{2}}_{\lambda\mu} + \delta\Gamma^{\frac{1}{2}}_{\lambda\mu})\Gamma_{\lambda n}\, [E_\lambda - E - \tfrac{1}{2} i \Gamma_\lambda]^{-1}$$

The resonance term can also be separated into local and distant resonances, the latter then contributing to the potential (direct or non-resonant) capture cross section.

$$U = U^{\mathrm{int}}(\mathrm{loc.res}) + U^{\mathrm{int}}(\mathrm{dist.res}) + U^{\mathrm{ch}}(\mathrm{loc.res}) + U^{\mathrm{ch}}(\mathrm{dist.res}) + U(\mathrm{HS})$$

In the strong coupling model the random signs of the channel amplitudes cause the distant resonance contribution to be zero. Then the potential part of the collision matrix element, excluding local resonance contributions, is just

$$U(\mathrm{pot}) = U(\mathrm{HS})$$

This is not the case in the intermediate coupling model where the single particle motion persists and is described by the distant resonance terms in the dispersion sum. Neglecting the local resonance contribution

$$U(\mathrm{pot}) = U(\mathrm{HS}) + U^{\mathrm{int}}(\mathrm{dist.res}) + U^{\mathrm{ch}}(\mathrm{dist.res})$$

Under appropriate conditions, the external or channel part can be an order of magnitude larger than the internal part[13]. Potential capture is therefore dominated by channel capture because the major contribution to the dipole overlap integral comes from the channel region.

The importance of the external region is evident for the 3s→2p transition in ^{40}Ca. The transition probability is proportional to the electric dipole radial matrix element

$$\left| \int \psi_{2p}\, r\, \psi_{3s}\, r^2 dr \right|^2,$$

and a factor of twenty enhancement occurs for calculations with a diffuse rather than a square well, achieving improved agreement with experiment[28]. This sensitivity to the shape arises from the fact that the main contribution to the matrix element comes from the channel region (r>R) as shown in fig. 3. Note that this is a mathematical statement and does not imply that the physical process of γ-ray emission occurs mainly outside the nucleus. Such processes cannot be localised in the quantal description of Lane & Lynn[29].

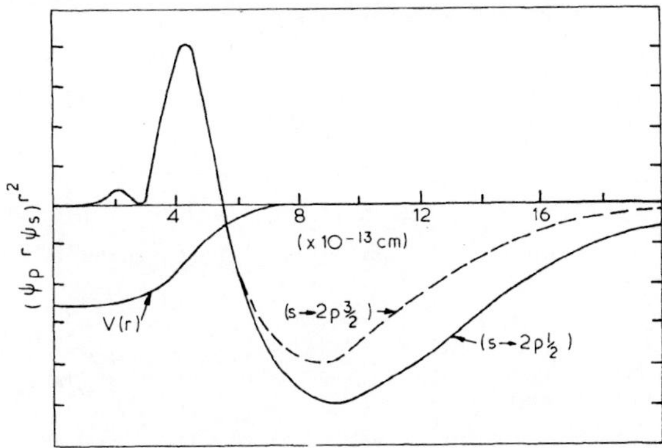

Fig. 3. Integrands of radial matrix elements for E1 transitions from s-wave capture to p-wave final states in ^{40}Ca. The Woods-Saxon potential V(r) is shown (ref. 28).

In the intermediate coupling model, the single particle strength is localised in an energy region so that, in effect, only one single particle state is dominant in resonance reactions. The reduced neutron width $(\Gamma^{\ell}_{\lambda n})$ is a measure of the strength of that single particle state and Lane & Lynn[13] showed that channel resonance capture will exhibit initial state correlations between $\Gamma_{\lambda\mu}$ and $\Gamma^{\ell}_{\lambda n}$. Both channel resonance and hard sphere capture select single particle final states, and will show final state correlations between $(2J_{\mu}+1)\Theta^2_{\mu}$ and $\Gamma_{\lambda\mu}/E^3_{\gamma}$. The detailed theory[13] allows quantitative calculations of both potential and resonance capture in the external region.

Thermal and Non-Resonant Neutron Capture

(a) Resonant Capture at Thermal Energy

The mechanism of thermal neutron capture depends on the proximity of an s-wave excited state to the neutron separation energy. The thermal capture cross section (barn) ascribed to a single resonance at energy E_{λ} (eV) is

$$\sigma_{\gamma}(\text{th}) = \tfrac{1}{4}\,\frac{2\cdot 608\ 10^6}{E_{\lambda}}\left(\frac{A+1}{A}\right)^2 g\ \sqrt{\frac{E_{\lambda}}{E}}\ \frac{\Gamma_{\lambda n}(E_{\lambda})\Gamma_{\lambda\gamma}}{E^2_{\lambda}}$$

where $E = 0.0253$ eV, $g = \dfrac{2J_{\lambda}+1}{2(2I+1)}$ and $\Gamma_{\lambda n}(E_{\lambda})$ is the energy dependent neutron width evaluated at the resonance energy.

When thermal capture is dominated by a local resonance, the γ-ray spectrum at thermal is similar to that of the resonance, and a resonance capture mechanism is operative. These mechanisms are discussed in detail later. However a good case for thermal valence capture is ^{54}Fe[30], whereas the large correlations observed in ^{43}Ca would be attributed to a doorway process[31], the valence part being estimated at 2.5%. Statistical spectra are observed for ^{113}Cd, ^{149}Sm, ^{155}Cd and ^{157}Gd[32].

The great majority of nuclides have a number of resonances contributing to thermal capture. The coherent sum of partial radiation width amplitudes results in random γ-ray intensities, with a Porter-Thomas (ν=1) distribution. However for these nuclides an exponential (ν=2) distribution was found but this can result from the incoherent summation of intensities of the two s-wave spin states ($J = I \pm \tfrac{1}{2}$) for odd A target nuclides[33].

(b) Potential Capture - strong coupling

There is a class of nuclides which have very small (<<1b) thermal capture cross sections. When neither positive nor negative energy resonances contribute significantly to thermal capture then both U^{int}(loc.res) and U^{ch}(loc.res) are negligible. In the strong coupling model, U(dist.res) is also zero because of the random sign and magnitude of the partial radiation width amplitudes. The potential or non-resonant capture cross-section is then given by the hard sphere cross-section[13,34] to final state μ (for a square potential well)

$$\sigma_{\gamma\mu}(\text{Pot}) = \sigma_{\gamma\mu}(\text{HS}) = \frac{0.062}{R\sqrt{E_n}}\left(\frac{Z}{A}\right)^2 M\ \frac{2J_{\mu}+1}{6(2I+1)}\,\Theta^2_{\mu}\ \frac{Y_{\mu}+3}{Y_{\mu}+1}^2\,Y_{\mu}^{\ 2}$$

where E_n (ev) is the neutron energy,

Θ^2_{μ} is the (d,p) spectroscopic factor,

I, J_{μ} are the target and final state spins,

M is the channel spin multiplicity,

($M = 1$ for $I = 0$, $J_{\mu} = I \pm {}^3/_2$; $M = 2$ for $J_{\mu} = I \pm \tfrac{1}{2}$)

$Y^2_{\mu} = R^2k^2$ and $k = (2mE_{\gamma}/\hbar^2)^{\frac{1}{2}} = 0.219\sqrt{E_{\gamma}}$(Mev) fm^{-1}, and

the interaction or effective hard sphere radius is $R = 1.35\ A^{1/3}$ fm.

Hard sphere capture is a consequence of the scattering of the incident neutron by the nuclear potential into a bound, single particle state. The γ-ray energy dependence is just y^2 or $E_{\gamma}^{\ 1}$.

The potential capture cross section will be modified by resonance channel contributions (i.e. U^{ch}(loc.res) > 0) from nearby resonances. This is accounted[13,34] for by the introduction of the coherent scattering length. Then

$$\sigma_{\gamma\mu}(\text{pot}) = \sigma_{\gamma\mu}(\text{HS})\left(1 + \frac{R - a_{coh}}{R}\ Y_{\mu}\ \frac{Y_{\mu}+2}{Y_{\mu}+3}\right)^2$$

For resonance energies not too close to thermal, the coherent scattering length is given by

$$a_{coh} = R' - 2.277.10^3\ \frac{A+1}{A}\ \sum_{\lambda}\frac{\Gamma^{\circ}_{\lambda n}}{E_{\lambda}}$$

and is the difference between the potential scattering amplitude $R'(\sigma = 4\pi R'^2)$ and the sum of resonance terms with the same J^{π}.

When $\Gamma^{\circ}_{\lambda n}/E_{\lambda} << 1$, then $a_{coh} \approx R'$. Maximum values of $\sigma_{\gamma\mu}$(Pot) occur at A ~40, 140 when R>R' for many nuclides, as shown in Fig. 4.

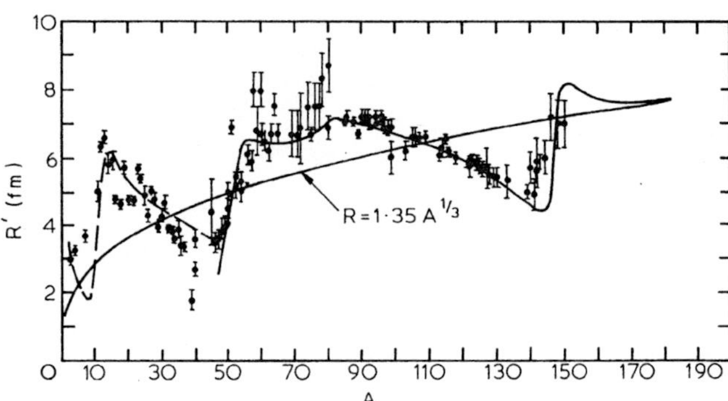

Fig. 4. Mass number dependence of potential scattering length R' (ref. 39).

The expected energy dependence for potential capture (E^1_{γ}) was observed by Spitz and Akkerman[35] when the energy dependence exponent (n) was varied to maximise the final state correlation coefficient

$$\rho_F\left((2J_{\mu}+1)\Theta^2_{\mu},\ I_{\gamma\mu}E_{\gamma}^{-n}\right)$$

However the interpretation of this effect is ambiguous because an energy dependent correlation may arise randomly, or more probably, from the energy dependence of the (d,p) spectroscopic strength[36]. In many cases the (d,p) strength is concentrated in the ground and low lying states, causing ρ_F to maximise for a small exponent.

Mughabghab[34,37] applied the Lane & Lynn theory with considerable success in these mass regions to calculate the magnitude of the potential partial and total capture cross sections in ^{42}Ca (fig. 5), ^{136}Xe, ^{138}Ba, ^{144}Sm. These results, together with those of other authors, are listed in table I. This body of data represents a remarkable, quantitative verification of potential capture theory calculated in the framework of the strong coupling model. Many additional direct capture cross sections have been calculated[38] for A≤70 and A~140 which represent a large fraction of the measured thermal capture cross sections.

An essential part of this theory is the reliance on the external or channel contribution to the radial integral. However the use of a square well rather than a diffuse well reduces the external contribution[28], and realistic detailed estimates are expected only when a diffuse potential is used (ref. 13, p.578): it is therefore somewhat surprising that agreement is so good.

(c) Intermediate Coupling (complex potential model)

In the intermediate coupling model, the neutron strength function shows strong peaking close to the single particle neutron levels corresponding to a real potential. The projection of the neutron-nucleus wave function on the channel function (describing the coupling of the neutron and the ground state of the target), gives rise to a wave function whose average value corresponds to a particle in a complex well. The real part of this wave function gives the potential scattering cross section and its radial overlap integral with a final single particle wave function in the real part of the well leads to the potential capture cross

Table I Partial and Total Potential Capture Cross Sections

Target[(1)]	R' fm	a_{coh} fm	σ_{th} mb	σ^{P}_{γ} mb	$\sigma_F^{(3)}$	n	$\sigma^{P}_{\gamma\mu}$ mb	μ=0	1	2	3	4	5	6	7	8	ref.
^{12}C	6.3	6.13	3.53 ±7	3.53			cal exp	2.28 2.38	1.11 1.14								84
^{13}C	6.6	5.47+ 6.59-	1.37 ±4	1.37			cal exp	1.19 1.15									84
^{32}S	3.92	2.72	530 ±40	401			cal exp	228 300	28 25	10 15	60 88	23 30	15 11	6 2.5			54
^{34}S	3.6	3.38	227 ±5	224	1.0	0.8	cal exp	133 162	18 18	16 16	30 44	32 33					85
^{37}Cl	3.4	2.93+ 3.10-	443 ±6	400	0.98	1.1	good agreement with spectroscopic factors										86
^{40}Ca	3.6	4.87	410	230			good relative agreement but $\sigma_p \sim 2\sigma_{exp}$										87
^{42}Ca		3.08	680 ±70	680			cal exp	71 54	18 19	13 16	21 23	21 16	29 27	300 344	13 6	29 37	37
^{49}Ti	4.0	0.7	2200 ±300	2010	0.85	1.3	cal exp	240 550	(p state)								88
^{128}Te	5.5	5.6	215 ±8	127	1.0		cal exp	3.4 4.3	4.6 5.1	42.1 43.2	22.0 22.8	22.8 21.9	3.4 2.8	2.6 11.6	9.1 7.5		89
^{130}Te	5.4	5.5	272 ±6	125	1.0	1.0	cal[(2)] exp	13.2 21.8	44.9 96.1	1.6 -	17.0 44.2	2.1 -	1.7 -	3.8 4.6			90
^{136}Xe	6.9		260 ±20		0.98	1.2	cal exp	104 102	30.6 33.3	23.9 26.9	12.7 11.7	3.3 3.4	14.1 14.8	13.5 2.2			34
^{138}Ba			360 ±36	375			good agreement with selected spectroscopic factors										34
^{144}Sm	7.1		~700		0.99		cal exp%	490 49.4	243 24.5	18 1.8	93 9.4	24 2.4	26 2.6	31 3.1	33 3.3		34

(1) R', $a_{coh} = \frac{A}{A+1} b, \sigma_{th}, \sigma^P_\gamma$ taken from ref. 38 and are not necessarily the same as used in $\sigma^P_{\gamma\mu}$ calculations.

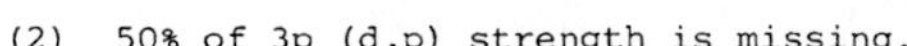

(2) 50% of 3p (d,p) strength is missing. (3) ρ_F is maximum correlation for E_γ^{-n}.

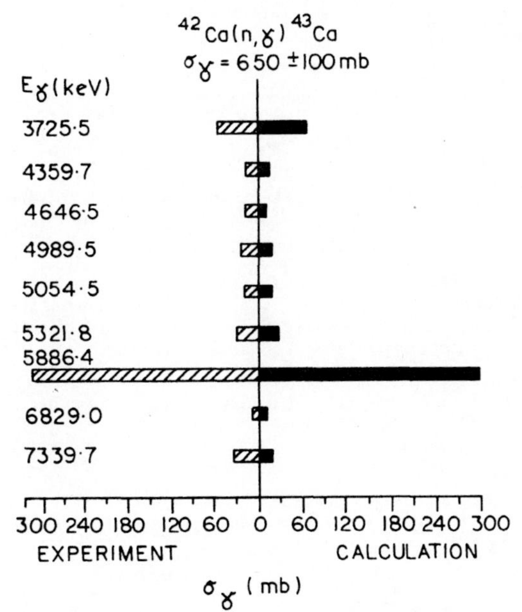

Fig. 5. Comparison of calculated and measured partial capture cross sections in ^{42}Ca(n,γ) at thermal energy (ref.34).

section. Lane and Lynn[13] used a diffuse, complex well which gave a broad reproduction of the mass dependence of the neutron strength functions, shown in fig. 6 for the 3s region.

More recently Cugnon & Mahaux[39] used optical model and shell model formulations to obtain mass dependent non-resonant capture cross sections.

A consequence of the intermediate coupling model is the existence of a correlation between $\gamma_{\lambda n}$ and $\gamma_{\lambda\mu}$. This correlation can be related to the imaginary part of the R-matrix i.e. $R_{n(\gamma\mu)} = \Sigma_\lambda \gamma_{\lambda n}\gamma_{\lambda\mu}(E_\lambda - E)^{-1}$ at the complex energy $E^+ = E + i\varepsilon$, and

$$\mathrm{Im}R(E^+) = \pi\ \overline{\gamma_{\lambda n}\gamma_{\lambda\mu}}\ D_\lambda^{-1}$$

when Porter-Thomas statistics are assumed (i.e. $\overline{\gamma_{\lambda n}} = \overline{\gamma_{\lambda\mu}} = 0$). For the linear correlation $\gamma_{\lambda\mu} = a\gamma_{\lambda n} + b$, $\overline{\gamma_{\lambda n}\gamma_{\lambda\mu}} > 0$, and the expression[40] relating the distant resonance contribution (DR) in the channel to the resonance width correlation coefficient $\rho_I(\Gamma_{\lambda n}, \Gamma_{\lambda\mu})$ is

$$\sigma_{\gamma\mu}(DR) = 2\pi^2 ƛ^2 g \frac{\langle\Gamma_{\lambda n}\rangle}{\langle D\rangle}\frac{\langle\Gamma_{\lambda\mu}\rangle}{\langle D\rangle}\rho_I\left(\Gamma^{\ell}_{\lambda n}, \Gamma_{\lambda\mu}\right)\frac{\pi}{2}\left(\frac{\mathrm{Re}\ R_{n(\gamma\mu)}}{\mathrm{Im}\ R_{n(\gamma\mu)}}\right)^2$$

where the real and imaginary parts of the R-matrix, are closely related and comparable in magnitude for a common doorway between photon and neutron channels.

The hard sphere and distant resonance amplitudes add coherently to yield the observed non-resonant capture cross section. These components can only be resolved under appropriate conditions - e.g. when R < R' (so as to minimise the hard sphere and local resonance part) and ρ is large; or conversely.

Measurements of the non-resonant capture cross section can also be made by observing interference effects in partial capture γ-ray channels at and in between resonances. Many measurements have been made using Ge(Li) detectors at Brookhaven in the eV range. Complications arose from count rate and resonance-resonance interference effects[41] which tend to lead to ambiguous results. A study of eV capture in the 4S region[42] supports the presence of non-resonant cross sections at thermal energy which are comparable with the predicted values:

$$\sigma_{\gamma\mu}(\text{HS}) = 80\ \Theta_\mu^{\ 2}\ \text{mb}\ (J_\mu = \tfrac{1}{2})$$
$$= 160\ \Theta_\mu^{\ 2}\ \text{mb}\ (J_\mu = \tfrac{3}{2}$$
$$\sigma_{\gamma\mu}(\text{DR}) = F\rho_I(\Gamma_{\lambda n}^{\ 1}, \Gamma_{\lambda\mu})\,\text{mb}$$ where F varies from 13 for ^{164}Dy to 175 for ^{162}Dy.

The partial capture cross section is given by

$$\sigma_{\lambda\mu}(E) = \pi \lambda\!\!\!^{-} \lambda\!\!\!^{-0} \sum_J g(J) \left| D_\mu^0(J) + \frac{\Sigma_\lambda \Gamma_{\lambda n}^0 \Gamma_{\lambda\mu}^{\frac{1}{2}}}{(E-E_\lambda) + i\Gamma_\lambda/2} \right|^2$$

where D^0 is the energy independent amplitude for the non-resonant capture cross section. Measured values of a few hundred mb in ^{162,164}Dy and ^{152}Sm were found for the stronger (d,p) states which are comparable with either or both Hard Sphere and Distant Resonance cross sections. However for the above nuclides, initial state correlations have not been reported, and the observed results should be close to the Hard Sphere capture cross section.

An alternative method to the Ge(Li) measurements of capture γ-ray spectra is provided by the inverse threshold photoneutron reaction. Observations of asymmetries in the ground state channel (γ,n_o) for ^{28}Si, ^{52}Cr, ^{207}Pb final nuclides have been interpreted to result from interference effects[43]. Again experimental problems arose because of possible, unresolved $\ell>0$ contributions. An asymmetry (low energy tail) in the 90 keV resonance of ^{52}Cr corresponds to a thermal potential cross section of 0.38b, while ^{60}Ni shows little interference at the 12 keV resonance and σ_{th} (Pot) = 0.008b.

These results are consistent with the expected sign of the resonance - potential interference term below (constructive) and above (destructive) the 3s single particle resonance[21] (see fig. 6).

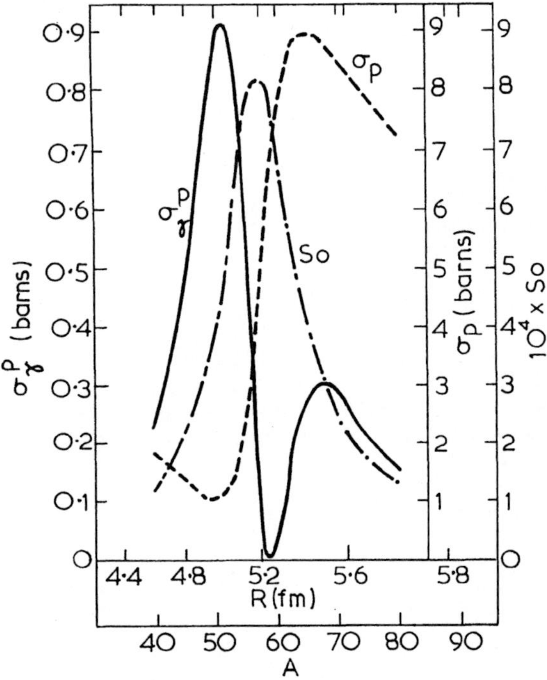

Fig. 6. Mass number dependence of the potential capture cross section (σ_γ^P), the potential scattering cross section (σ_p) and the s-wave neutron strength function (10^4 So) (ref. 13).

At the peak of this size resonance is ^{54}Fe, but total capture measurements of the 7.7 keV resonance do not show asymmetry and an upper limit of σ_{th}(pot) < 0.1b has been set[44]. Optical model calculations using the Moldauer potential[39] are consistent with this upper limit but other calculations which predict a peak at ^{54}Fe are not substantiated[13]. More recent measurements on ^{54}Fe confirm the symmetry of the 7.7 keV resonance[45] and the dominance of valence transitions to the ground state[30].

Further, the large initial state correlation observed for ^{54}Fe ($\rho_I \approx 0.4 - 0.9$) gives σ_{th}(DR) = 1-2b at thermal energy. Because the 7.7 keV resonance contributes 1.74b to the observed thermal cross section of 2.25b, this calculation is 2-4 times too large, and is inconsistent with the shape analysis of this 7.7 keV resonance.

Measurements of interference in the 41 keV s-wave resonance in ^{207}Pb have also yielded conflicting results, as described in ref. 15. Early (γ,n) measurements showed a strong asymmetry which was explained by postulating a large background cross section derived from the GDR. However the (n,γ_o) resonance capture cross section showed no indication of asymmetry. Later (γ,n) measurements gave σ_{th}(Pot) = 0.59b, a much reduced value that was at least consistent with resonance-resonance interference and the potential cross section, without recourse to anomalous non-resonant processes. Direct measurements at 2 and 24 keV were also consistent with a small, non-resonant cross section.

Perhaps one of the most convincing demonstrations of the Lane & Lynn theory is the $^{17}O(\gamma,n_o)\,^{16}O$ differential photoneutron cross section of Holt et al.[46]. The internal, channel and potential capture mechanisms are explicitly observed in this work (fig. 7). A multilevel R-matrix formalism was used with the Lane & Lynn collision matrix

$$U_{\gamma n} = U_{n\gamma} = ie^{-i\phi_\ell}\left\{\Sigma_{\lambda\mu} A_{\lambda\mu} \Gamma_{\lambda n}^{\frac{1}{2}} [\Gamma_{\lambda\mu}^{\frac{1}{2}} - \delta\Gamma_{\lambda\mu}^{\frac{1}{2}}] + D_\mu\right\}$$

where $\Gamma_{\gamma\mu}^{\frac{1}{2}}$ and $\delta\Gamma_{\lambda\mu}^{\frac{1}{2}}$ are reduced radiation width amplitudes for internal and channel capture, $A_{\lambda\mu}$ is an element of the level matrix and D the hard sphere capture amplitude.

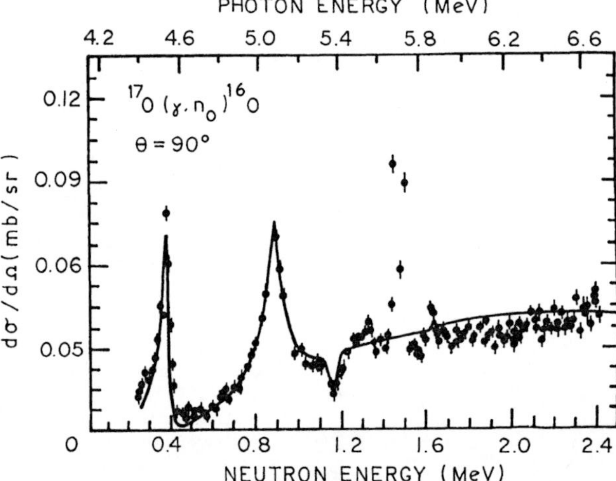

Fig. 7. Differential cross section for $^{17}O(\gamma,n_o)\,^{16}O$ is well reproduced by a multilevel R-matrix formalism. The minimum (5.38 Mev) results from interference between channel and potential capture amplitudes (ref. 46).

The symmetric minimum observed at E = 5.38 MeV implies that $\Gamma_{\lambda\mu}^{\frac{1}{2}}$ (a real quantity because of capture within the channel radius) is equal to the real part of $\delta\Gamma_{\lambda\mu}^{\frac{1}{2}}$, which is a complex quantity because it is related to the outgoing part of the scattering wave function.

The large non-resonant cross-section is seen to be well reproduced by the theory, and was ascribed to the potential capture mechanism.

Resonance Neutron Capture

(a) Statistical Model

The basis of the statistical (strong coupling) model is the Bohr condition that the lifetime of an excited state is much longer than the time required to traverse the nucleus. The incoming particle experiences many collisions and the probability of forming an excited state λ in channel c is independent of the probability of decay in channel c'. The overlap of resonance (λ) and channel (c) wave functions is small and $\langle\lambda|c\rangle^2 \approx 10^{-4} - 10^{-7} = \gamma_{\lambda c}^2/\gamma_{sp}^2$ where γ_{sp}^2 is the single-particle width.

The basic tenets of the statistical model are:

The reduced widths $\gamma_{\lambda c}$ and $\gamma_{\lambda c'}$ have random signs such that

$$\sum_{\lambda} \gamma_{\lambda c}\gamma_{\lambda c'} = 0$$

over many levels in the range ΔE. As a consequence, the amplitude correlation coefficient is also zero.

i.e. $\rho(\gamma_{\lambda c}, \gamma_{\lambda c'}) = 0$

The strength function is independent of energy

$$\frac{1}{\Delta E} \sum_{\lambda} \gamma^2_{\lambda c} = \text{constant.}$$

(i) Distribution of partial radiation widths

The partial radiation width amplitude $\Gamma^{\frac{1}{2}}_{\lambda\mu}$ is expected to be a highly complicated but essentially real quantity. The expression for $\Gamma^{\frac{1}{2}}_{\lambda\mu}$ shows a dependence on C^{λ}_{cp} which is related to $\langle\lambda|c\rangle$ and the reduced neutron width $(\gamma^2_{\lambda n})$. Consequently the radiation amplitude is expected to obey the same statistics as $\gamma_{\lambda n}$ which are normally and independently distributed. The partial radiation width will therefore be described by a chi-squared distribution with one degree of freedom i.e. the Porter-Thomas distribution[20].

Many attempts have been made to measure these distributions for nuclides with A>100 and level spacings less than 50 eV. Overall, the results are now consistent with one degree of freedom (ν~1)[14] but are quite sensitive to the quality of the measurement. Some exceptions to ν=1 persist, for example, certain transitions in ^{238}U.

While the observation of a Porter-Thomas distribution of reduced widths is indicative of a statistical interaction, non-statistical effects may still be observed for the reduced width amplitudes. In inelastic proton scattering on ^{48}Ti, Chou et al.[47] observed amplitude correlations between the ($\ell'=0$, $s'=^5/_2$) and ($\ell'=2, s'=^5/_2$) decay channels although the width distributions were in excellent agreement with the P-T distribution.

By averaging γ-ray intensities over many (N) resonances, the Porter-Thomas fluctuations can be reduced to a relative variance 2/N. In this way the energy dependence of γ-ray transitions can be investigated. Results for E1 transition in Cu at keV energies exhibited on E^3_γ dependence[48], but thermal capture in ^{45}Sc,[49] and averaged eV measurement in Pt[50] and other heavy nuclides[51] support an E^5_γ dependence. This latter result was consistent with the Axel prediction[52] of the influence of the Lorentzian tail of the E1 GDR.

In agreement with statistical theory, Bollinger[51] showed that the average γ-ray intensities are independent of the structure of the low lying states in heavy nuclides.

(ii) γ-ray functions

(a) Electric dipole

The E1 radiation width for an s→ p single particle transition is[21]

$$\Gamma_\gamma = \frac{4}{3} k^3_\gamma \left\{ \bar{e} \int_0^\infty dr\ r\ U_0(r)\ U_1(r) \right\}^2$$

where $\bar{e} = -\frac{Z}{A}e$ is the effective charge of the neutron and $u_\ell(r) = r\psi(r)$ is the radial wave function for the initial and final single particle states. Lynn[21] used the Eckart potential to include numerically the dipole integrals and found the empirical relation.

$$\left| \int_0^\infty dr\ r\ u_0(r)\ u_1(r) \right|^2 \simeq 0.012\ A^{\frac{1}{2}}\ 10^{-24}\ \text{cm}^2.$$

The single particle level spacing for states with the same ℓ value (e.g. 2s,3s,4s) is $2\hbar w \sim 82A^{-1/3} = 15\text{-}30$ MeV. Using these values and $\bar{e} = 0.45e$ (for A~100), the γ-ray strength function is

$$k^V_L(E1) = \langle\Gamma_{\lambda\mu}\rangle[E^3_\gamma\ A^{\frac{1}{2}}\ D_\lambda]^{-1} = (2.5\text{-}5)10^{-9}\ \text{MeV}^{-3}$$

for E_γ MeV, $\Gamma_{\gamma\mu}$ and D_λ in eV, which is comparable to the equivalent Blatt and Weisskopf approximation[21]

$$k(E1) = \langle\Gamma_{\lambda\mu}\rangle\ [E^3_\gamma\ A^{2/3}\ D_\lambda]^{-1} = (2\text{-}4)\ 10^{-9}\ \text{MeV}^{-3}$$

This latter strength function is in current use in the literature, but the Lynn equation is the more soundly based. Lynn also obtained an expression derived from the strong coupling version of the valence model

$$k^V_L(E1) = \langle\Gamma_{\lambda\mu}\rangle\ [E^3_\gamma\ A^{5/6}\ D]^{-1} = 0.16\ 10^{-9}\ \text{MeV}^{-3}$$

In a generalized extrapolation of the GDR Lorentzian to the threshold region, Axel[52] derived the strength function

$$S(E1) = \langle\Gamma_{\lambda\mu}\rangle\ [E^5_\gamma\ A^{8/3}\ D_\lambda]^{-1} = 6.1\ 10^{-15}\ \text{MeV}^{-5}$$

for A ~ 100, E_γ ~ 7 MeV and Γ ~ 5 MeV).

In the most recent survey of E1 and M1 γ-ray strength functions[53], the E1 global average for the single particle model (SPM) is $k(E1) = (2.9 \pm 0.3)\ 10^{-9}$ MeV^{-3} which is comparable to the single particle average.

The Axel formulation gives $S(E1) = (4.2 \pm 0.4)\ 10^{-15}$ MeV^{-5} and is also in good agreement with the predicted value. A more accurate method is to extrapolate the Lorentzian using individual photo-absorption parameters. McCullagh et al.[53] find the average ratio of experiment to theory to be 0.70 ± 0.06, and Raman[54] observes a similar result for energy dependent strength functions for a number of nuclides. To obtain better agreement, an empirical energy dependence of the width of the GDR would be required i.e. $\Gamma(E) = \Gamma_R(E/E_0)^n$ where $n \simeq 0.5$. The square root dependence would give $E_\gamma^{5.5}$ which is more acceptable than the E^7_γ resulting from the inclusion of an energy dependence of the damping of the GDR[53].

At this point it should be emphasized that the presence of intermediate structure in the low energy tail of the GDR could be grounds for the rejection of the Axel hypothesis. This is often the case for A<100 and for nuclides near closed shells. Further, for very light nuclides (e.g. ^{13}C, ^{15}N, ^{17}O) it is apparent that the (γ,n_0) channel does not participate in the GDR, but accounts for much of the low energy strength. Isotopic spin selection rules for photon transitions apply ($\Delta T = \pm1$, 0 and $\Delta T = \pm1$ for N = Z), but do not adequately account for the data (J.Jury,Priv.Com.). In ^{32}Si(n,γ), the ground and first excited state channels show evidence of single particle strength below the very weak GDR[55]. Elastic and inelastic photo scattering (γ,γ_i) on ^{28}Si and ^{52}Cr exhibit markedly different giant resonance structure, demonstrating a pronounced dependence on the final state configuration[56]. There is little (γ,n_0) data available for the photo-reaction, but (n,γ_0) results for ^{40}Ca, ^{89}Y, ^{140}Ce and ^{208}Pb all exhibit a GDR in the ground state channel[57].

The situation for heavy nuclides (A>181) is also not favourable. The GDR extrapolation does not reproduce the structure observed in the E1 strength function (Bartholomew et al. 1973) in the threshold region .

(b) Magnetic dipole

For magnetic dipole transitions, the Blatt & Weisskopf estimate is

$$k(M1) = \langle\Gamma_{\lambda\mu}\rangle\ [E^3_\gamma\ D_\lambda]^{-1} = (0.7\text{-}1.4)\ 10^{-9}\ \text{MeV}^{-3}$$

Lynn[21] has obtained a mass dependent expression with a strong coupling version of the valency nucleon model.

$$k^V_L(M1) = \langle\Gamma_{\lambda\mu}\rangle\ [E^3_\gamma\ A^{1/3}\ D_\lambda]^{-1} = 0.66\ 10^{-9}\ \text{MeV}^{-3}$$

The global average of McCullagh et al.[53] is

$$k(M1) = (30 \pm 4)10^{-9}\ \text{MeV}^{-3}$$

which is somewhat higher than earlier surveys, and indicates that M1 transitions are strongly enhanced over the Weisskopf value. A great deal of mass dependent structure is also apparent in the McCullagh survey, suggesting the strong influence of the M1 Giant Dipole Resonance in the threshold region.

(c) Ratio of E1 and M1 transition strengths

From eV averaging measurements[51] for A>100, the ratio of E1 and M1 transition strengths is

$$R = \langle\Gamma_{\gamma\mu}(E1)\rangle E^{-3}_\gamma\ /\ \langle\Gamma_{\gamma\mu}(M1)\rangle E^{-3}_\gamma = 7 \pm 1$$

and is in good agreement with R = 6.1 ± 0.6 obtained from the discrete resonance averaged data[59]. However for 15<A<90, this ratio fluctuates around R~1.6 (fig. 8). The Weisskopf mass dependence is $R = 3.2\ A^{2/3}$ which is an order of magnitude up on the overall fit to the data (15<A<240) of $R = 0.1\ A^{2/3}$, because of the enhanced M1 strengths. The best fit to the data is found for $R = 0.03A^1$. On the other hand the GDR mass dependence of $A^{8/3}$ greatly exceeds the best fit to the data when an E_γ^5 dependence is assumed.

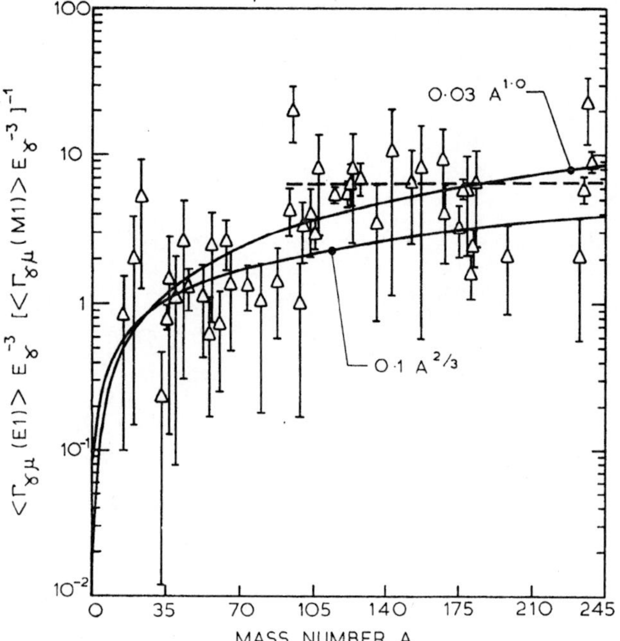

Fig. 8. Ratio of reduced E1 and M1 radiation widths as a function of mass number. The dashed line is the 'eV averaging' value and the curves represent the single particle and overall best fits to the data (ref. 59).

(iii) Radiation widths

Adopting the generalized definition of the γ-ray strength function[60]

$$f_{E1}(E_\gamma) = \langle \Gamma_{\lambda\mu}^{J} \rangle \left[E_\gamma^3\, D_\lambda \right]^{-1}$$

the average total radiation width is

$$\langle \Gamma_{\lambda\gamma}{}^{J} \rangle = \int_0^{E_\lambda} E_\gamma{}^3\ f(E1) \sum_{I=J-1}^{J+1} \frac{\rho_I(E_\lambda - E_\gamma)}{\rho_J(E_\lambda)}\ dE_\gamma$$

If the energy dependence of the photoeffect is assumed to be independent of the structure of the initial state, the same γ-ray strength function can be applied to excited states as to the ground state, i.e. the Brink-Axel hypothesis. The strength function can then be obtained from the E1 photoabsorption cross section (σ_γ mb)

$$f_{E1}(E_\gamma) = \frac{26\ 10^{-8}\ \sigma_\gamma(E_\gamma)}{g_J E_\gamma}\ \text{MeV}^{-3} \quad \text{where } g_J = \frac{2J+1}{2I+1},$$

and $\sigma_\gamma(E_\gamma) = E_\gamma^2 \Gamma_G^2 \sigma_m \left[(E_\gamma^2 - E_G^2)^2 + \Gamma_G^2 E^2 \right]^{-1}$

The parameters of the giant dipole resonance[61] are

$$\sigma_m \Gamma_G = 46(NZ/A)\,\text{MeV mb},\quad E_G = 31.2\ A^{-1/3} + 20.6\ A^{-1/6}\,\text{MeV},$$
$$\text{and } \Gamma_G = 88\ A/NZ\ \text{MeV}$$

Bearing in mind the limitation on the use of the GDR strength function, it remains to determine the appropriate level density.

While experiment determines $\rho(E_\lambda)$, we require $\rho(E)$ for all energies up to E_λ. Gilbert and Cameron[62] used the constant nuclear temperature representation at low energies and a Fermi gas formula at high excitations. Hardy[61] has recently calculated average radiation widths using updated parameters for the Gilbert & Cameron formalism and also the back shifted fermi gas model for the mass range 60<A<150, and finds a markedly better fit to the data for the latter level density method. The average ratio of $\langle\Gamma_\gamma\rangle_{exp}/\langle\Gamma_\gamma\rangle_{cal} = 1.08$, with standard deviation of 32% (fig. 9) is a markedly superior result to that obtained with the Gilbert-Cameron formulae (1.92, 40% SD).

A concise review of the theory of radiation widths is given in ref. 38.

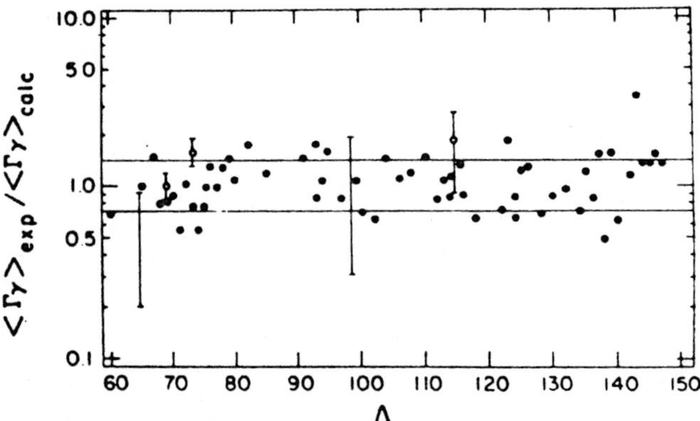

Fig. 9. Ratio of experimental to calculated total radiation widths using the Brink-Axel hypothesis and the back shifted fermi gas level density. Error bars denote exotic nuclides (ref. 61).

(b) Valence Model

First formulated by Lane & Lynn[13] and Lynn[21] the Valence model was successfully applied to resonance spectra in ^{92,98}Mo by Mughabghab et al[63]. The optical model formalism was developed by Lane & Mughabghab and further investigated by Barrett & Terrasawa, Cugnon, and Cugnon & Mahaux who evaluated the valence model and the mutual relationships between the S-matrix, optical model and shell model approaches (see ref. 15 for references). The optical model approach was favoured by Halderson and Castel[64] because of the possible contributions from the external region and the lack of sensitivity to the potential used. Note that the radial integration in the valence model is from zero to infinity and therefore includes both resonance internal and channel capture.

Valence model calculations have been made by a number of authors in the 3s region[65,66] and across the periodic table[15]. The systematic properties of partial and total radiation widths and correlations were also reviewed by Allen & Musgrove[15], with particular emphasis on the 2p, 3s, 3p and 4s neutron strength function regions. The most recent results for magic neutron nuclides are given in ref. 67.

The valence model, specialized to exclude core excitations, describes the change of state of the entrance channel configuration by the emission of dipole radiation in the field of a spectator core. The partial width for an E1 transition from resonance λ to final state μ is[21]

$$\Gamma_{\lambda\mu} = \frac{16\pi\ k_\gamma^3}{9}\ \frac{\left| \langle X_\lambda(J_\lambda) || H_E^{(1)} || X_\mu(J_\mu) \rangle \right|^2}{(2J_\lambda + 1)}$$

where $H_E^{(1)}$ is the irreducible tensor operator specialized to E1 radiation. This equation is evaluated by expanding the eigen functions (X) into radial wave functions for a set of single particle states (P). In the intermediate coupling assumption only one single particle state is assumed to be significant and the reduced neutron width is a measure of its amplitude squared in the resonance wave function. The success of the valence model is therefore rather unexpected when $\langle \lambda | \theta_0\ \mu_p \rangle^2 \sim 10^{-4}$ in typical cases. The single particle strength of the final state is given by the (d,p) spectroscopic factor (θ_μ^2).

A simplified expression for the valence partial radiation width[15] is

$$\Gamma_{\lambda\mu}^{v} = q_{\lambda\mu}(E_\lambda)\ E_\gamma^3\, \theta_\mu^2\, (Z/A)^2\ \Gamma_{\lambda n}^{\ell}$$

where $\Gamma^{\ell}_{\lambda n} = \Gamma_{\lambda n}(P_{\ell}E^{\frac{1}{2}})^{-1}$ and $q_{\lambda\mu}$ is an energy dependent optical model parameter describing the radial overlap integral and spin factors. The valence width is directly proportional to the single particle strength of the resonance and final state and, as a consequence, the valence model is manifested by the width correlations

$$\rho_I[\Gamma\lambda^{\ell}_n, \Gamma_{\lambda\gamma(\mu)}] \text{ and } \rho_F[(2J_{\mu}+1)\Theta^2_{\mu}, \Gamma_{\lambda\mu}/E^3_{\gamma}].$$

In the valence model, an effective charge for the neutron is taken as $e_n = -\frac{Z}{A}e$, and represents the recoil of the nuclear charge when the neutron changes its state. There may, of course, be components in the nuclear wave function corresponding to an excited core. The annihilation of proton p-h components requires the effective proton charge $e_p = \frac{N}{A}e$. Further polarization of the core can occur, because of the non-spherical field generated by the valence nucleon. A vibrational state, which in the first approximation is a constructively interfering linear combination of 1p-1h states, may result. (If the core is deformed, a rotational state may be formed). Then the core can contribute to the radiative transition probability. The consequent enhanced transition probability can be approximated by assigning the valence nucleons an effective charge. Of course, if a complete set of wave functions are used to describe the resonance and final state, and these are chosen so as to diagonalize the Hamiltonian, there is no need for the effective charge concept. This is not the case for the valence model, and the success of this model depends in the first instance on the assumption that core excitations are not significant. As such the effective charge used is valid.

Johnston and Castel[68] have derived an energy dependent effective charge for ^{28}Si by coupling to the GDR. Good agreement with excited state energies and observed E1 transition intensities is found, using an effective charge which varies from $e_n = -0.13e$ at 5 Mev to $e_n = -0.08e$ at 7.8 Mev, depending on the single particle energy gap. These values are much lower than the normal value of -0.5e. Nevertheless, the radiation widths of the $p^{3}/_{2}$ resonances at 565 and 813 Mev are best reproduced by the valence model using the standard effective charge[69].

There is a great deal of evidence for non-statistical behaviour in resonance neutron capture and the valence model appears to account for much of the data. As reviewed in Allen & Musgrove[15], initial and final state correlations have maximum values in the 3s and 3p regions, in agreement with valence expectations (fig. 10). However, initial state correlations continue into the 4s region when valence effects are negligible. In the 3s and 3p region, the largest radiation widths are well accounted for by the model, but many other large widths are observed for which the valence contribution is small. Generally, the calculated total valence widths ($\Gamma^{V}_{\lambda\gamma} = \Sigma\,\Gamma^{V}_{\lambda\mu}$) are consistent with observed widths, and rarely exceed them, suggesting a qualitative validity to the calculations, and the isolation of valence strength from the GDR.

For all but one of the best cases in the 3s and 3p regions (table II), the calculated valence strength function, expressed in terms of the Axel model, is found to be substantially in excess of the Lorentzian extrapolation in the same mass region, i.e. $10^{15}\langle s\rangle = 7.7\ Mev^{-5}$. Consequently the valence model predicts a marked enhancement in the γ-ray strength function over the GDR value, and indicates effective decoupling of E1 transitions from the GDR.

The observation of resonance asymmetries[44,45] in the total capture cross section in ^{54}Fe (fig. 11) has been interpreted to result from the fragmentation of the 3s state and the coherent summing of partial radiation amplitudes in the intermediate coupling model[70]. The conclusion is fully consistent with the valence model predictions, substantiated to a high degree in the 7.7 kev resonance[30], (as shown in fig. 12).

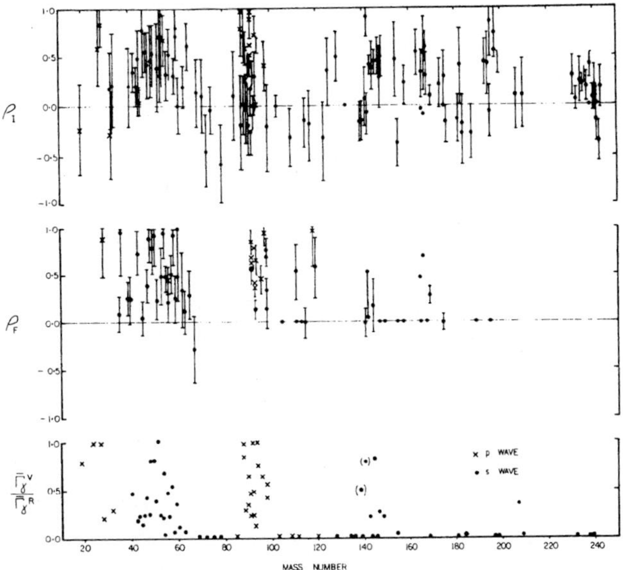

Fig. 10. Initial (ρ_I) and final (ρ_F) state correlation coefficients peak at the magic numbers N = 28 for s-wave (•) and N = 14, 50 (x) for p-wave capture, and correspond to the largest valence contributions relative to the non-statistical width (ref. 15).

Table II. Valence Strength Functions

Target	E_λ keV	$(\ell,J)_\lambda$	$(\ell,J)_\mu$	$10^{15}.s^V$ Mev^{-5}	$10^{9}.k^V$ Mev^{-3}
^{54}Fe	<200	$s\frac{1}{2}$	$p^{3}/_{2}$	23.1	5.8
			$p\frac{1}{2}$	11.2	2.6
^{58}Fe	<200	$s\frac{1}{2}$	$p^{3}/_{2}$	12.5	1.8
			$p\frac{1}{2}$	17.2	1.7
^{88}Sr	<350	$p^{3}/_{2}$	$d^{5}/_{2}$	21.5	6.7
			$s\frac{1}{2}$	47.8	10.5
^{90}Zr	<225	$p^{3}/_{2}$	$d^{5}/_{2}$	7.4	3.1

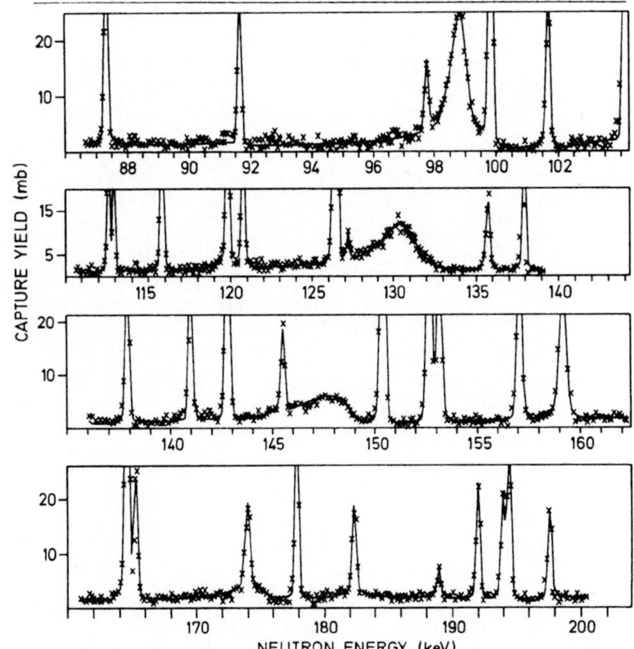

Fig. 11. Neutron capture γ-ray yield for ^{54}Fe (n,γ). Narrow ℓ>o resonances are superimposed on asymmetric s-wave resonances which are well fitted in a two channel multilevel formalism (ref. 45).

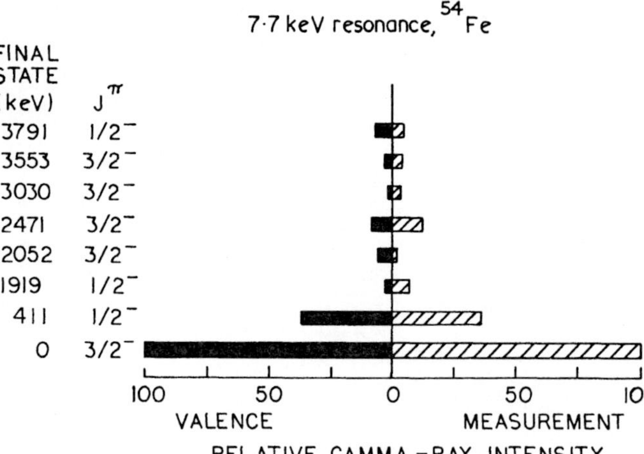

Fig. 12. Comparison of relative valence and measured partial radiation widths for the 7.7 keV s-wave resonance in ^{54}Fe. The total radiation widths agree within errors (ref. 30).

There are, nevertheless, some glaring inconsistencies. Notably:

(a) Observation of asymmetries in ^{54}Fe total capture at 99 and 137 kev whereas the estimated valence contributions are only 22 and 40% resp.[45]

(b) Weak resonance correlations in ^{96}Zr when valence effects are expected to be very strong.[71]

(c) Absence of valence transitions[72] in ^{100}Mo.

Such difficulties are explained by postulating interference between valence and doorway amplitudes[15]. While this ubiquitous argument is difficult to prove, it is substantiated in detailed calculations by Soloviev & Stoyanov[73] with the quasiparticle-phonon model.

At best the valence model provides a ready means of investigating single particle effects in resonance neutron capture. However, quantitative successes of the model are restricted to certain magic neutron nuclides in the 2p,3s and 3p regions, and shortcomings in its application are becoming more apparent.

(c) Quasiparticle-Phonon Model (QPM)

The fragmentation of simple configurations was first investigated by Brown et al.[74]. These authors showed for ^{16}O and ^{40}Ca that the coupling of the one quasiparticle state (i.e. 1p-0h) with the nuclear surface shifted an appreciable part of the single particle spectroscopic strength several MeV with respect to the unperturbed energy. In the QPM[73], the fragmentation of the one quasiparticle (1q) strength in odd mass nuclides is due to the interaction of the 1q state with the one phonon excitations of the core (i.e 2q and vibrational states). It is these few q components of highly excited states that largely determine the spectroscopic factors, neutron and γ-ray strength functions and photoexcitation cross sections.

The coupling of the single particle and collective motions is described by the quasiparticle-phonon interaction. There are no free parameters in this interaction and satisfactory agreement with experiment has been achieved in doubly even nuclides for calculation of the widths of the GDR and the neutron and radiative strength functions[75].

Calculations of the fragmentation of quasiparticle and quasiparticle plus phonon (q+ph) states have been made for spherical nuclides[73] and these serve as a basis for estimating resonance γ-ray strength functions to 1q final states in odd A nuclides.

The Hamiltonian of the QPM includes the average nuclear field potential, the superconducting pairing interaction and separable isoscalar and isovector multipole and spin multipole forces. The model Hamiltonian is expressed in terms of the creation and annihilation operators for the quasiparticles ($\alpha^+_{jm},\alpha_{jm}$) and phonons ($Q^+_t,Q_t$) where (nl)jm are the single particle quantum numbers; $t\equiv\lambda\mu i$ and λ is the multipole order of the phonon with z component μ and phonon state number i. The wave function of the highly excited state (ν) of an odd A spherical nuclide is

$$\psi_\nu(JM) = C_{J\nu}\Big\{\alpha^+_{JM} + \Sigma_{\lambda ij} D^{\lambda i}_j(J\nu)[\alpha^+_{jm}Q^+_t]_{JM} + \Sigma_{jI\lambda_1 i_1\lambda_2 i_2} P^{\lambda_1 i_1\lambda_2 i_2}_{jI}(J\nu)\Big[\alpha^+_{jm}[Q^+_{t_1}Q^+_{t_2}]_{IM'}\Big]_{JM}\Big\}\psi_o$$

where ψ_o is the g.s. wave function of the doubly even nucleus, and the coefficients C, D and P determine the contribution of q, q+ph and q+2ph components to the normalisation condition of the wave equation.

The matrix element

$$<\psi^*_o\,\alpha_{JM}|H_{qph}|\Big[\alpha^+_{jm}\,Q^+_t\Big]_{JM}\,\psi_o>$$

determines the strength distribution of the 1q state over the more complex 'quasiparticle + phonon' states. Matrix elements of this type describe the fragmentation of the 1q component of the wave function, as in the valence model. The QPM also accounts for the matrix elements coupling the states α^+Q^+ with the more complex (q + 2h) configurations $\alpha^+Q^+Q^+$. The fragmentation of phonon states arises from the matrix element

$$<\psi^*_o[Q_t]_{JM}|H_{qph}|[Q^+_{t_1}Q^+_{t_2'}]_{\lambda\mu}\,\Psi_o>$$

where H_{qph} describes the q-ph interaction.

Now the radiative width amplitude can be expressed as the sum of amplitudes[76], where the first term is proportional to the 1q part of the resonance and final state (i.e. $1q_i \rightarrow 1\,q_f$); the second term describes the phonon annihilation $(q + ph)_i \rightarrow q_f$; the third and fourth terms represent transitions to $(q + ph)_f$ components; and the last term includes transitions between more complex configurations. Only the first two terms contribute to the 1q component of the final state wave functions.

$$\psi_{\nu_f}(j_f m_f) = C_{j_f\nu_f}\,\alpha^+_{j_f m_f}\,\psi_o$$

It is these terms that are calculated by Soloviev and Stoyanov[73] for ^{54,56}Fe and ^{58,60}Ni targets, but only results for ^{54}Fe are discussed here.

The reduced Eλ transition probability to final state f is expressed as

$$B(E\lambda;J\nu\rightarrow j_f\nu_f) = c^2_{j_f\nu_f}\,c^2_{J\nu}|(\text{valence} + (q + ph)|^2(2J+1)^{-1}$$

where the valence term contains the Eλ operator and the (q + ph) term the one phonon operator. Formula for Mλ transitions were also derived by changing the single particle matrix element of the electric operator for that of the magnetic operator.

Radiative strength functions for E1 and M1 transitions were calculated using parameters which fit the GDR and $B[E\lambda;0^+_{g.s}\rightarrow 2^+_1\ (3^-_1)]$ values. A large phonon space is required for the QPM calculation, and phonons with $I^\pi = 1\pm$ to $7\pm$ and energy up to 15 MeV were included. Several dozens of one phonon states were calculated for each value of I^π.

The QPM reproduces for ^{54}Fe the spectroscopic factors of the $3/2^-$ and $\frac{1}{2}$- ground and first excited states (in ^{55}Fe) and the s and p-wave neutron strength functions. However the valence part of the calculation gives a strength function in the threshold region which is substantially lower than the valence model estimate for E_n = 0 - 200 keV (table III). While this discrepancy is unresolved, the interesting feature of the QPM is the shift in energy of the (q + ph) states and the threefold increase in strength function for the $p^{3/2}$ transition when the (q + ph) interaction is included (fig. 13a,b). The main contribution is from $[2p^{3/2}\otimes 1^-_5]_{\frac{1}{2}^+}$ near threshold whereas the $[2p\frac{1}{2}\otimes 1^-_1]_{\frac{1}{2}}^+$ strength occurs at higher energies (fig. 13c). Fig. 13 also shows a marked influence of the (q + ph) interaction on the M1 γ-ray strength function in the threshold region.

Note that interference terms between the q and q+ph components were excluded from the calculations on the grounds that their contributions were not large.

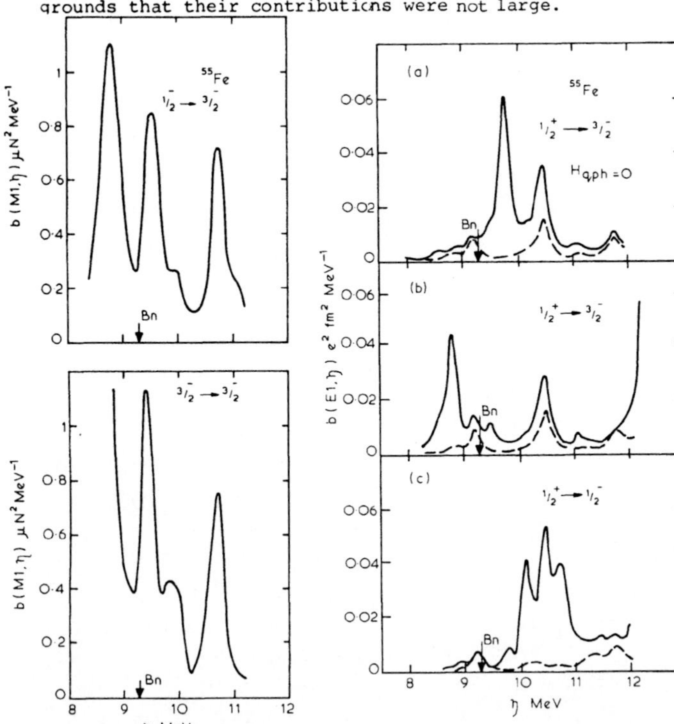

Fig. 13. QPM estimates of E1 and M1 reduced width distributions for transitions to the ground ($^3/_2$ -) and first excited ($\frac{1}{2}$-) states of ^{55}Fe. The E1 valence component is the broken line. The quasiparticle-phonon interaction is set at zero in figure (a), but when this interaction is turned on, a redistribution of q + ph strength occurs, as shown in figure (b) (ref. 73).

TABLE III Photon Strength Functions in ^{54}Fe (n,γ)

$<k(M1)>\times 10^9$ MeV^{-3}		$<k(E1)>\times 10^9$ MeV^{-3}		Mechanism
$p^{3/2} \rightarrow p^{3/2}$	$p^{\frac{1}{2}} \rightarrow p^{3/2}$	$s^{\frac{1}{2}} \rightarrow p^{3/2}$	$s^{\frac{1}{2}} \rightarrow p^{\frac{1}{2}}$	
-	-	5.8	2.6	Valence(tableII)
0.1	0.1	0.6	0.25	Valence[73]
6.6	13	1.7	0.3	Valence+q-ph[73]

The observation of initial state correlations in nuclides for which valence effects are negligible can also be a manifestation of the QPM. Becvar et al.[77] have made measurements of partial radiation widths for resolved neutron resonances in ^{154}Gd, ^{167}Er, $^{171,173}Yb$ and ^{185}Re, using the fast pulsed reactor at Dubna. Statistically significant correlations were observed in most of these nuclei, the strongest correlations being found for transitions to rotational bands built on 2q states. These results suggest a dominant role of a small number of few quasiparticle components in the radiative decay of resonances for nuclides in the deformed region A = 150-190.

The QPM is seen to predict the existence of intermediate structure with widths of a few hundred keV in Fe and Ni. This structure results from the q+ph interaction which would occur while the compound nucleus is in a doorway state. According to Bloch & Feshbach[78] the width of the doorway state is $\Gamma_D = \Gamma\uparrow + \Gamma\downarrow$ where the decay width $\Gamma\uparrow$ is given by

$$\Gamma\uparrow = 2\pi <\phi_1|V_R|\psi_E>^2,$$

the spreading width $\Gamma\downarrow$ is

$$\Gamma\downarrow = 2\pi <\phi_{n+1}|V_R|\phi_n>^2 D_{n+1}^{-1},$$

and n is the particle-hole hierarchy with level spacing D_n. The intermediate structure is described by

$$<\sigma>_E = \pi\lambda\!\!\!^{-2} \frac{\Gamma\uparrow(\Gamma\uparrow + \Gamma\downarrow)}{(E-E_D)^2 + \frac{1}{4}(\Gamma\uparrow + \Gamma\downarrow)^2}$$

and is most pronounced for $\Gamma\uparrow \sim \Gamma\downarrow$.

The overlapping of doorway states and configuration mixing caused by the residual interaction smears out the intermediate structure and results in the statistical properties of fine structure resonances. The best examples of intermediate structure[79,80] are found for ^{56}Fe (n,n)^{56}Fe and ^{206}Pb (n,n)^{206}Pb. Such examples are difficult to find and Rohr[81] has turned to the level density systematics of fine structure resonances above the neutron threshold to gain an insight into the neutron capture mechanism.

An evaluation of s-wave level densities at the neutron separation energy provides the data base. The level densities are reduced to the level density parameter 'a' at the binding energy by means of the Fermi gas level density expression of Gilbert and Cameron[62]. The level density parameter for compound nucleus states 'a_C' in the Fermi gas model is proportional to the level density of single particle states at the Fermi surface energy and is expected to be proportional to the mass number.

The experimental data are compared to the level density parameter a_D obtained from a calculation of the density of doorway states at the binding energy (fig. 14). The calculation is based on Nilsson single particle states (without deformation) and the pairing force is taken into account in a quasi-particle formalism. The energy, spin and parity of these states, corresponding to the first collision, are calculated in this independent quasiparticle model. Whereas the mass dependence of a_D is monotonic, that for a_C shows clearly the influence of the closed shells. When $a_C \simeq a_D$ for A<38, the first hierarchy (doorway states) adequately accounts for the observed level density and implies that the compound resonances are in fact doorway resonances. Three steps in a_C are observed at A = 38,69,94 and Rohr[81] interprets these changes in the compound level density as the addition of two quasiparticles participating in the excitation process of the compound nucleus.

In the region of closed shells, the hierarchial level tends to reduce to the 3 quasiparticle doorway state (i.e. $a_C \rightarrow a_D$). For ^{56}Fe, the ratio a_C/a_D corresponds to a level density ratio of 11, which is in good agreement with the 13 fine structure resonances observed in the 359 keV doorway structure[79].

Further calculations of the density of hierarchies with more than 3 quasiparticles are needed to confirm this interpretation of the level systematics.

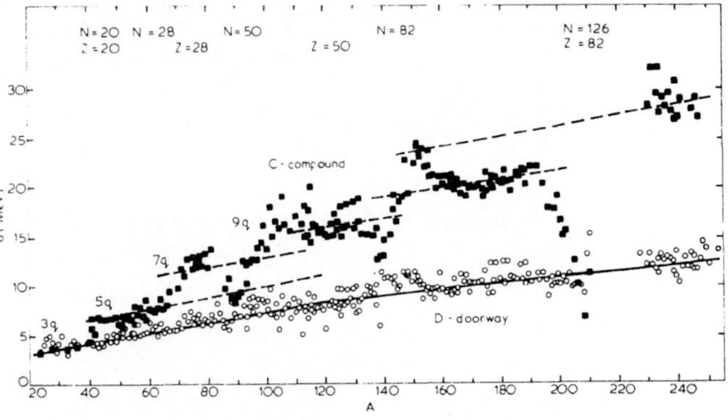

Fig. 14. Systematics of the level density parameter 'a' show ledge structures which could be indicative of quasiparticle hierarchy (q). The calculated values for 3q doorway states are also shown (ref. 81).

For the lighter nuclides, there are many examples of resonance capture γ-ray spectra which support the dominance of a limited p-h hierarchy. The 7.7 keV resonance in ^{54}Fe appears to be predominantly valence in character[30] - where is the statistical part?

s and p wave resonances in ^{35}Cl, ^{52}Cr, $^{56,58}Fe$ exhibit strong correlations between partial widths[54,82] and large final state correlations are observed in ^{28}Si when valence effects are negligible[83]. These observations are readily explained in terms of a 2p-1h hierarchy in the capture process. Support for the hierarchy concept also occurs in the mass range 181 to 205. In this region an apparent inverse correlation exists between the quasiparticle hierarchy and the magnitude of the 'anomalous 5 MeV bump' observed in (n,γ) and(d,pγ) experiments[58]. The implication here is that decoupling of E1 strength from the GDR increases for less than 8 quasi-particles, reaching a maximum effect at 3 quasiparticles.

References

1. E. Fermi et al.,Proc.Roy.Soc. A146, 438 (1934)
2. D.F. Lea, Nature 133, 24 (1934)
3. P.B. Moon,R.Tillman, Proc.Roy.Soc.153, 421 (1936)
4. E. Fermi,E. Amaldi, La Recerciv Scientifica A6, 544 (1935)
5. L. Szilard, Nature 136, 849 (1935)
6. J.R. Dunning et al., Phys.Rev. 48, 704 (1935)
7. N. Bohr, Nature 137, 344 (1936)
8. G. Breit , E.P. Wigner, Phys.Rev. 49, 519 (1936)
9. N. Bohr, Nature 141, 326 (1938)
10. B.B. Kinsey et al., Phys.Rev., 83, 519 (1951)
11. L.V. Groshev et al., Int.Conf.Peaceful Uses of Atomic Energy, 2nd Geneva Conf. 15, 138 (1958)
12. A.M. Lane, D.H.Wilkinson, Phys.Rev.97, 119 (1955)
13. A.M. Lane, J.E. Lynn, Nucl.Phys. 17, 553 (1960)
14. R.E. Chrien, Int.Symp.Neutron Capture γ-ray Spectroscopy, Studsvik, 627 (1969)
15. B.J. Allen,A.R.Musgrove,Adv.Nucl.Phys.10,129(1978)
16. G.E. Brown, Nucl.Phys. 57, 339 (1964)
17. C.F. Clement et al., Nucl.Phys. 66, 273 (1965)
18. G. Longo,F.Saporetti, Nuova Cim. 56B, 264 (1968)
19. M. Potokar, Phys.Lett. 46B, 346 (1976)
20. C.E.Porter, R.G.Thomas, Phys.Rev., 104, 483 (1950)
21. J.E.Lynn, Theory of Neutron Resonance Reactions Clarendon Press (1968)
22. H.Feshbach et al., Ann.Phys. 41 230 (1967)
23. A. Lande, G.E.Brown, Nucl.Phys. 75, 344 (1966)
24. G.E.Brown, M.Bolsterli, Phys.Rev.Lett.3, 472 (1959)
25. G.E.Brown, J.Speth, Neutron Capture γ-ray Spectroscopy, Brookhaven, Ed.R.E.Chrien, & W.R. Kane, Plenum Press, 181 (1979)
26. R.G. Thomas, Phys.Rev. 84, 1061 (1951)
27. A.M. Lane,R.G.Thomas,E.P.Wigner,Phys.Rev.98,693 (1955)
28. H.Moringaga,C.Ishii, Prog.Theor.Phys.23,161 (1960)
29. A.M. Lane,J.E.Lynn, Nucl.Phys. 11,646 (1959)
30. S. Raman et al., Phys.Rev. 22, 328 (1980)
31. A.R.Musgrove etal.,Nucl.Phys.A279, 317 (1977)
32. L.V.Groshev et al.,Nucl.Phys. 16, 645 (1960)
33. L.M.Bollinger, Phys. Rev. C3, 2071 (1971)
34. S.F. Mughabghab, Phys.Lett. 35B, 469 (1971)
35. A.M.J.Spits,J.A.Akkermans,Nucl.Phys.A215,260(1973)
36. S.Mughabghab ,R.E.Chrien, see ref. 25, p.266
37. S. Mughabghab,Proc.Spec.Mtg.Neutron Cross Sections of Fission Product Nuclei, Bologna, Ed. C.Coceva, G.C.Panini (1979) NEANDC(E) 209L
38. S.F.Mughabghab,M.Divadeenam,N.E.Holden, Neutron Cross Sections, vol.1A,Academic Press (1981)
39. J.Cugnon,C.Mahaux, Ann.Phys. 94, 128 (1975)
40. A.M.Lane, Ann.Phys. 63, 171 (1971)
41. R.E.Chrien, Proc.Int.Symp. Neutron Capture γ-ray Spectroscopy, Petten 247 (1974)
42. G.W.Cole, R.E. Chrien - ibid 271 (1974)
43. H.E. Jackson - ibid 437 (1974)
44. B.J. Allen, A.R. Musgrove, W.K. Bertram, Phys. Lett. 72B, 323 (1978)
45. A. Brusegan et al., Neutron Capture γ-ray Spectrotroscopy,Grenoble,Ed.T.von Egidy et al., Inst. Phys.Conf.Ser. 62, 408 (1982)
46. R.J.Holt et al.,Phys.Rev. C18, 1962 (1978)
47. B.H.Chou et al.,Phys.Rev.Lett. 45, 1235 (1980)
48. B.J. Allen, Nucl.Phys. A111, 1 (1968)
49. T. Tielens et al.,Nucl.Phys. A376, 421 (1982)
50. L.Bollinger,G.Thomas, Phys.Rev.Lett.18,1143(1967)
51. L.Bollinger,G.Thomas,Phys.Rev.C2,1951 (1970)
52. P.Axel, Phys.Rev. 126, 671 (1962)
53. C.M.McCullagh, et al. Phys.Rev. C23, 1394 (1981)
54. S.Raman, see ref. 45, 357 (1982)
55. M.Micklinghoff,B.Castel, Ann.Phys.114,452 (1978)
56. B.Ishakhanov et al.,Sov.J.Nucl.Phys.32,757(1980)
57. I.Bergqvist,M.Potokar, see ref. 25, 299 (1979)
58. G.Bartholomew et al.,Adv.Nucl.Phys.7,229 (1974)
59. J.Kopecky, ECN 81-04 (1981)
60. A.M.Lone, see ref. 25, 161 (1979)
61. J.C. Hardy, Phys.Lett. 109B, 242 (1982)
62. A.Gilbert, A.Cameron, Can.J.Phys.43, 1446 (1965)
63. S.F.Mughabghab,Phys.Lett.35B, 469 (1971)
64. D.Halderson,B.Castel, Phys.Rev. C18, 1542 (1978)
65. O.Harouna,J.Cugnon, J.Phys. 37, 1377 (1976)
66. H.Beer, J.Phys. 40, 339 (1979)
67. B.J. Allen, see ref. 45, 395 (1982)
68. I.Johnston,B.Castel,Nucl.Phys.A213,341 (1973)
69. S.Joly et al., Nucl.Phys.A334, 269 (1980)
70. B.J. Allen, see ref. 45, 393 (1982)
71. A. Brusegan et al. see ref. 45, 406 (1982)
72. H. Weigmann et al., Phys.Rev. C20, 115 (1979)
73. V.Soloviev,Ch.Stoyanov, Nucl.Phys.A382,206 (1982)
74. G.E.Brown et al., Nucl.Phys. 45, 164 (1963)
75. V.G. Soloviev et al., Nucl.Phys. A342,261 (1980)
76. M. Beer, Ann.Phys. 65, 181 (1971)
77. F. Becvar et al.,Sov.J.Nucl.Phys.33,1 (1981)
78. B.Bloch,H.Feshbach, Ann.Phys. 23, 47 (1963)
79. A.Elwyn,J.Monohan, Nucl.Phys.A123,33 (1969)
80. J.A.Farrell et al.,Phys.Lett. 17, 286 (1965)
81. G.Rohr, see ref. 45, 322 (1982)
82. L.K. Peker, R.E.Chrien,see ref. 25, 714 (1979)
83. M.J.Kenny et al., Nucl.Phys. A170, 164 (1976)
84. S.F.Mughabghab et al., BAPS 26, 1138 (1981), also M.A. Lone see ref. 45, 391 (1982)
85. R.F. Carlton et al. see ref. 46 (1982)
86. J.Kopecky, A.Spits, Interactions of Neutrons with Nuclei, Lowell, Ed.E.Sheldon,conf.760715,1286(1976)
87. G. Longo et al., see ref. 45, 413 (1982)
88. J.F.A.G. Ruyl, see ref. 45, 417 (1982)
89. J.Honzatko et al., Z.Phys. A299, 183 (1981)
90. J.Honzatko et al., Czech J.Phys. B30, 763 (1980).

NUCLEAR FISSION: FROM SADDLE TO SCISSION

J.P. Theobald

Institut für Kernphysik der Technischen Hochschule Darmstadt
D-6100 Darmstadt, W. Germany

After the origin of mass asymmetry in fission is understood today in the framework of static macroscopic-microscopic models, present experiments are focussed on correlations between various fragment characteristics and on fine structures in fragment parameter distributions. While the gross properties of the fragments, like mass asymmetry and excitation energy, are essentially determined at the (second) saddle of the fission barrier, fine structures on the distribution attributed to fragment shells and nucleon pairing are generated during the descent of the fissioning nucleus from saddle to scission. At the last stages of fission the emission of light charged particles although occuring very rarely provide valuable information on nuclear configurations close to scission. Some recent nuclear data to these topics will be presented and discussed.

Introduction

Nuclear deformation is illustrated by potential energy surfaces (PES), i.e. the dependency of the deformation energy on nuclear shape parameters, e.g. the elongation, the neck constriction and the mass asymmetry. Fission is characterized by a configuration point following a certain path on the PES, fission of actinides by essentially three stationary points on the path distance:

- the first saddle at an energy E_A ,
- the second potential energy minimum E_{II} , and
- the second saddle at E_B

This situation is displayed in Fig. 1[1]. There is some experimental indication for an even stronger modulation of the fission barrier for some thorium isotopes[2] and a theoretical argument for an additional barrier near the scission point[3]. The hills and valleys on the PES are caused by the interplay between liquid-drop model energies, pairing energies and compound-nucleus as well as fragment shell effects[4]. After for about one decade experiments were favouring nuclear spectroscopy of low lying nuclear states above the second potential minimum E_{II}, the so called excited shape isomeric states[5], there is now an increasing interest in the late stages of fission, mainly induced by the advent of heavy ion physics[6]. From the results of the static fission model displayed in Fig. 1 it is obvious that symmetric fission is hindered by a potential well of about 8 MeV height, for a neck radius of about 5 fm. For mass asymmetric fragmentation, the barrier is lower by 2.3 MeV thus favouring the asymmetric fragment mass ratio A_L/A_H = 91/145. From saddle to scission the most probable mass ratio of the light to heavy fragment A_L/A_H adjusts to about 96/140, in agreement with experiments. The excitation energy of the fragments is mainly determined by the nuclear configuration at the (second) saddle. If this shape is close to the scission configuration, excitation is low, otherwise high. The

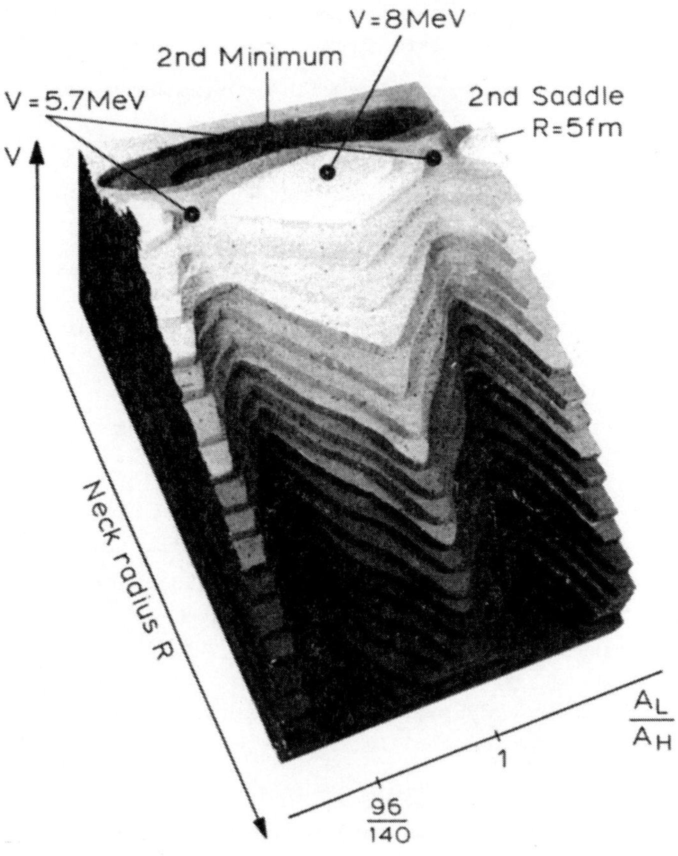

Fig. 1. The two dimensional energy surface for ^{236}U. The energy has been minimized with respect to the elongation of one of the fragments and the maximum dimensions of the fragments perpendicular to the fission direction (from ref.[1]).

K. H. Böckhoff (ed.), Nuclear Data for Science and Technology, 719–729.

number of broken proton (neutron) pairs is correlated with the fragment excitation energy. Thus, the gross properties of the fragments: mass asymmetry and excitation are prearranged at the saddle. The total kinetic energy of the fragments has two components, the prescission kinetic energy E_O obtained by the fission mode during the descent from saddle to scission and the Coulomb repulsion energy E_∞ acquired at scission. The latter part is distributed according to the dispersion of the fragment deformations. Consequently the total energy release Q_∞ at scission is given by

$$Q_\infty = E_\infty + E_O + \sum_{LH} E_{def} + \sum_{LH} E^* \quad (1)$$

where $\sum_{LH} E_{def} + \sum_{LH} E^*$ are the sums of the deformation and (intrinsic) excitation energies of the light and heavy fragments. In this review we are particularly interested in single particle effects, i.e. shell and pairing effects on the fission fragment parameters. In order to avoid divergency, we restrict ourselves not only to neutron induced fission, but to a few topics:

i) Fragment mass and nuclear charge distributions as function of light fragment kinetic energy[7,8],

ii) fragment energy and mass distributions for spontaneous and neutron induced fission of the same compound system[9],

iii) the total kinetic energy (TKE) as function of compound nucleus excitation energy (EX)[10], and

iv) charged particle associated fission[11].

Only very recent investigations will be considered, which are of co rse based on numerous previous studies on these subjects. Recent work on fission-barrier parameter systematics is not included, since it has recently been reviewed in ref.[12].

Experimental Techniques to Measure Fragment Mass, Nuclear-charge and Energy Distrubtions.

Besides γ- or x-ray spectroscopic and radiochemical methods to measure fragment mass distributions (FMD) and fragment nuclear charge distributions (FCD), there are essentially two types of techniques to determine simultaneously FMD and fragment kinetic energy distributions FKED making use of

a) the conservation of linear momentum in fission or

b) ion separation in combined magnetic and electric fields together with energy resolving particle detection.

The technique under a) is either a simultaneous measurement of the velocities of both fragments or of their kinetic energies. It is applied in the measurements of topics ii) and iii). The technique under b) is of topics i, in which the fission product spectrometer "LOHENGRIN" at the high flux reactor of the Institute Max von Laue - Paul Langevin (ILL) at Grenoble (France) has been used[13].

Direct methods for FCD measurements are

c) ΔE/E detection,

d) absorber techniques, and recently

e) Bragg-curve spectroscopy[14], in which the ionization charge density along the fragment track is recorded and analysed.

Absorber techniques are used at "LOHENGRIN". In all FCD measurements, ionization chambers are applied.

Experimental Data

i) FMD and FCD of light fission fragments have been determined for neutron induced fission of ^{233}U, ^{235}U and ^{239}Pu at "LOHENGRIN" for light fragment kinetic energies between 83 and 118 MeV, (from 83 - 108 MeV in steps of 5 and from 109 MeV - 118 MeV in steps of 1 MeV), at present ^{239}Pu data are available up to 111 MeV. FCD are measured between 83 and 108 MeV in steps of 5 MeV for the Uranium isotopes. The recorded nuclides are displayed in nuclide chart cuts in Fig. 2 and 3 together with the energy integrated yields for ^{233}U and ^{239}Pu, resp.. The data on ^{235}U are already published[7]. For the elements in Fig. 2a, an proton odd-even effect is obvious, which is less prominent in Fig. 3. Nuclear structure effects on FMD are more pronounced at higher fragment kinetic energies (Fig. 4); the same holds for proton and neutron pairing effects (Fig. 5 and Fig. 6).

ii) FMD and FKED of spontaneous and neutron induced fission of the ^{242}Pu system were recently investigated by E. Allaert et al.[9]. The mass distributions are shown in Fig. 7. The main results of this experiment can be summarized as follows:

a) The light and heavy fragment mass peaks are higher, their widths smaller in spontaneous compared to induced fission

b) The ^{242}Pu spontaneous fission total kinetic energy (TKE) is shifted by about 2.8 MeV (in the maximum of the distribution) towards higher energies compared to induced fission

c) The TKE distribution is Gaussian in the induced, but asymmetric with greater yields at low energy in the spontaneous fission case.

iii) From the very elaborate and detailed work of Y. Patin et al.[10], a study of the ^{233}U(d,pf) reaction, only one of their interesting results is considered here. The dependency of the average fragment kinetic energy $\bar{E}_k$ on the compound

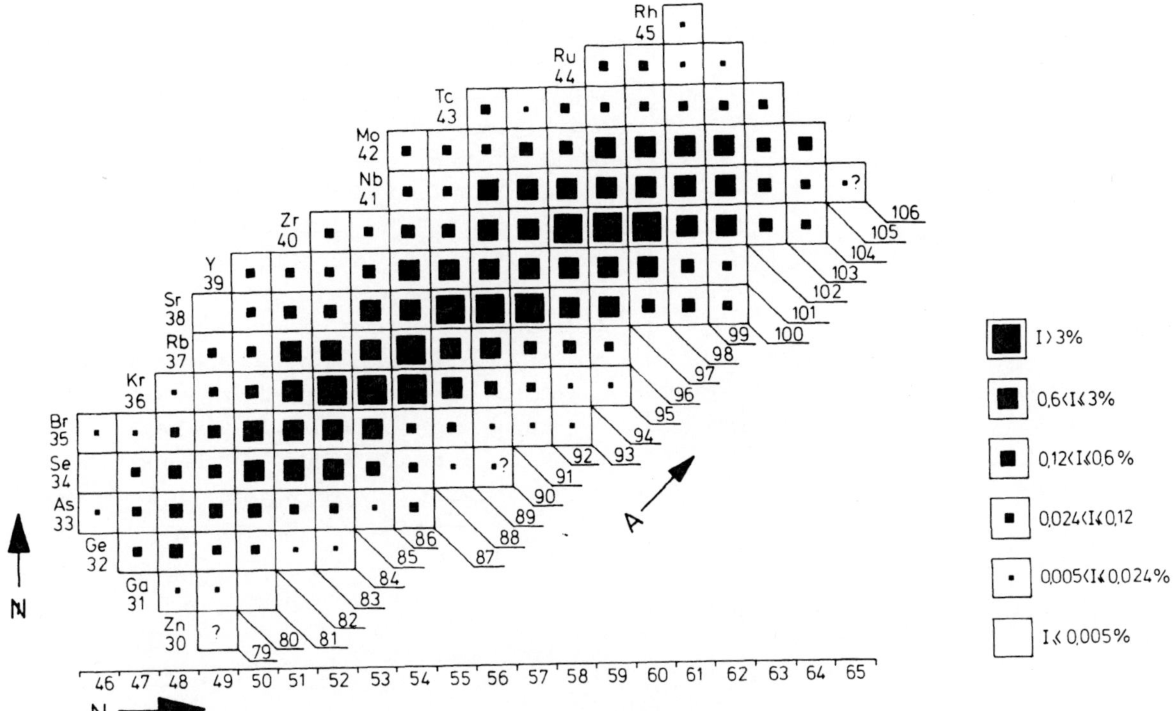

Fig. 2. Nuclide yields summed over the kinetic energy in ^{233}U (n,f)

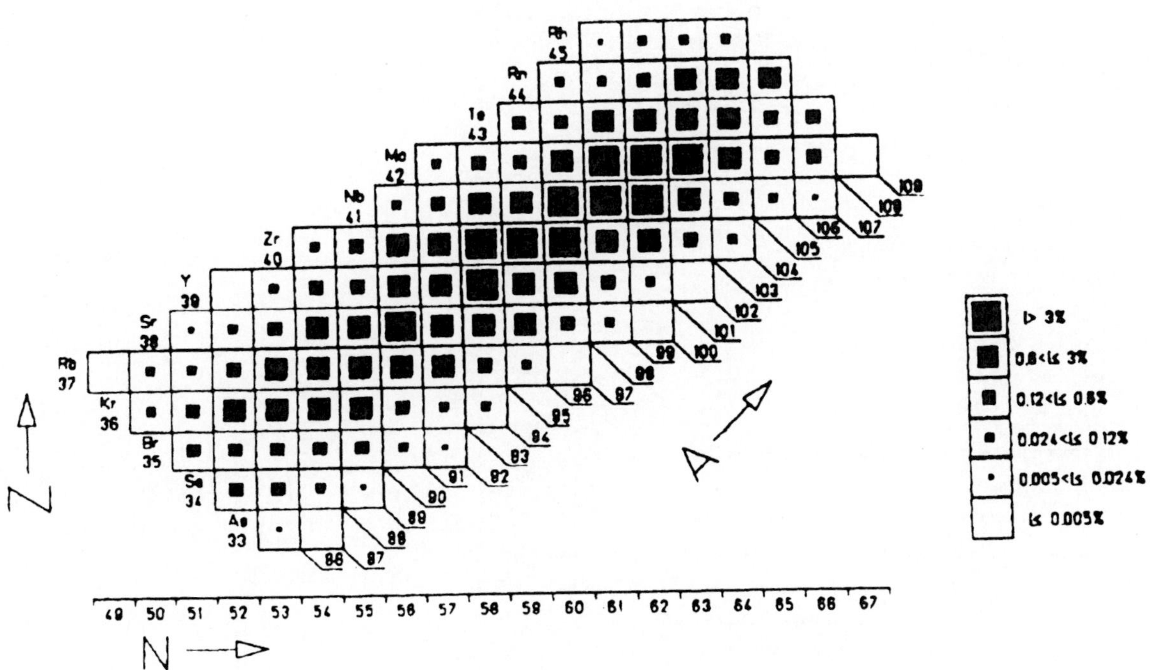

Fig. 3. Nuclide yields summed over the kinetic energy in ^{239}Pu (n,f)

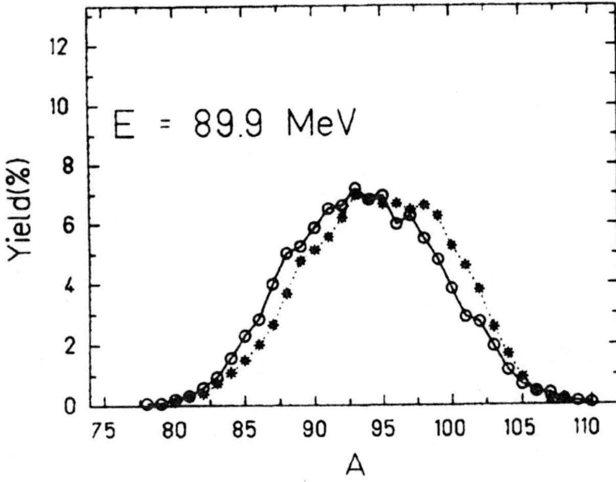

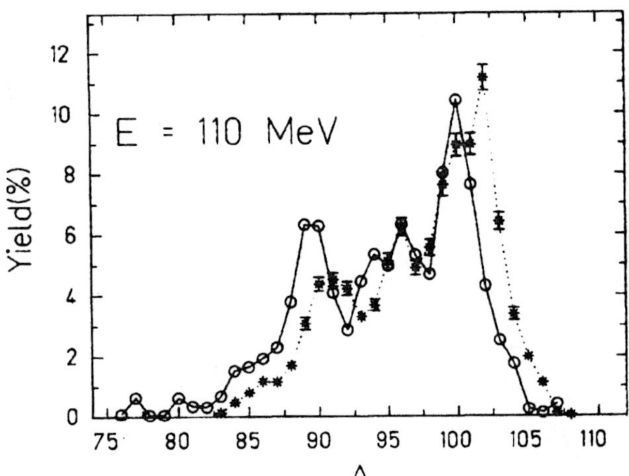

Fig. 4. Fragment mass distributions
a) at 89,9 MeV
b) at 110 MeV
light fragment kinetic energies:
full lines for $^{233}U + n$,
dotted lines for $^{235}U + n$.

nucleus excitation energy (EX) between 5.7 and 9.7 MeV is in good approximation linear with slopes $d(\bar{E}_k)/d(EX)$ as shown in Fig. 8. While these derivatives are generally negative or zero, there is a significant jump to positive values for mass symmetric fission. Similar results have been reported by E. Jacobs et al. for photofission of ^{235}U and ^{238}U[15].

iv) Of particular interest for our understanding of the late stages of the fission process is the message, which is provided by the light charged particles (p, α, ... , O) emitted by a small fraction of fission events. Close to scission the α particle formation is predicted preferentially in the neck region[16]. The typical fragment characteristics, discussed in this paper, are besides trivial modifications similar for binary and particle accompagnied fission. In connection with the experimental results presented here a measurement of C.R. Guet et al. has to find our attention[11]. From the set of their data only the width of the angular distribution (FWHM) as function of TKE in given mass intervals is displayed here in Fig. 9, essentially a decreasing function with increasing TKE. In the following chapter the data summarized above are discussed on a level on which we understand to-day the asymmetric mass fragmentation in actinide fission, on the level of static models of B.D. Wilkins et al.[17]. It is a synthesis of P. Fong's statistical[18] and W. Nörenberg's thermodyna-

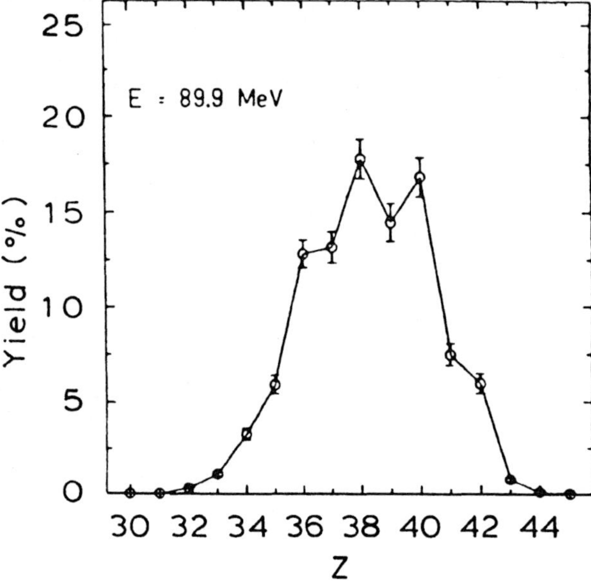

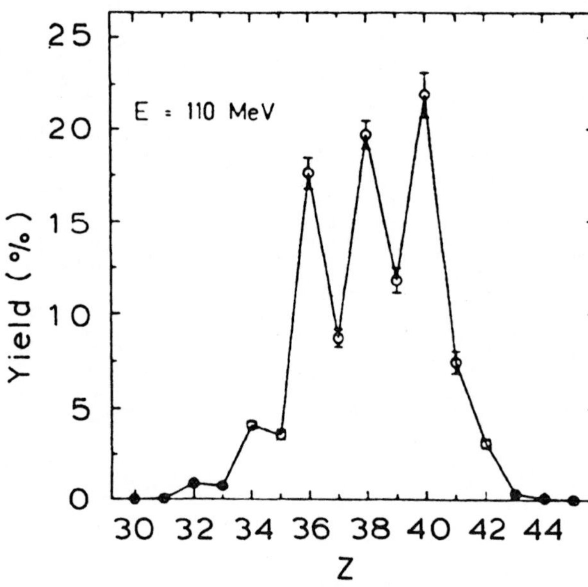

Fig. 5. Element yield distributions for light fission fragments at two kinetic energies in $^{233}U + n$

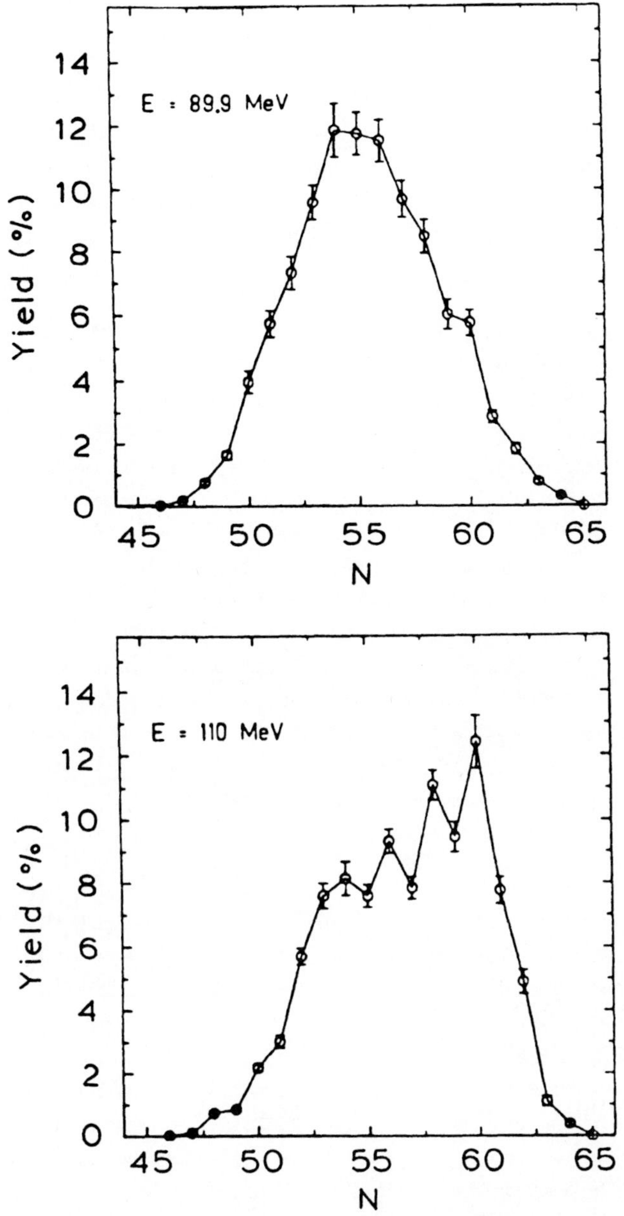

Fig. 6. Isotonic yield distributions at two light fragment kinetic energies in $^{233}U + n$

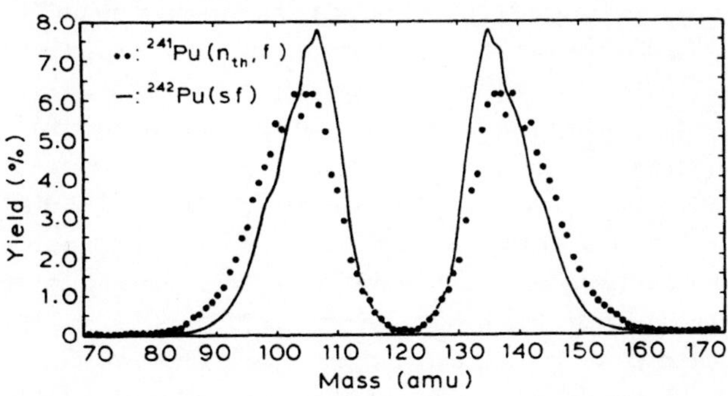

Fig. 7. Comparison of pre-neutron mass yields for thermal neutron induced fission of ^{241}Pu (dots) and the spontaneous fission of ^{242}Pu (full line) ref.[9]

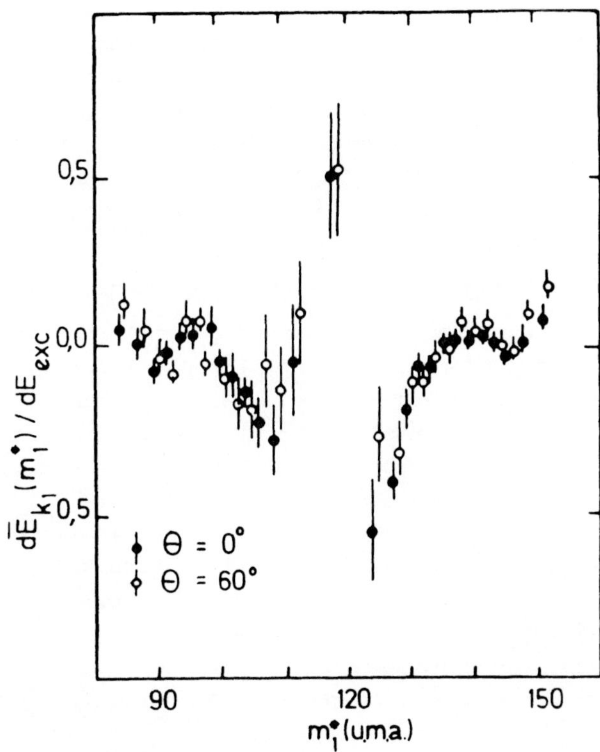

Fig. 8. Slopes d $\bar{E}_k$/d(CEX) as functions of primary fragment masses (ref.[10])

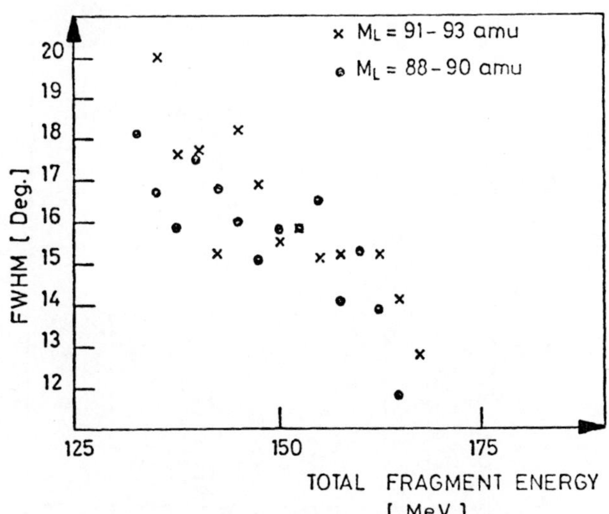

Fig. 9. Width of the angular distribution of long range α-particles as function of total fragment kinetic energy under the conditions that 14 MeV ≤ E_α ≤ 16 MeV and A_L = 88 - 90 amu (circles) or A_L = 91-93 amu (crosses) ref. [11]

mical model[3] which makes the assumption that on descent from saddle to scission the collective degrees of freedom are strongly coupled among each other and to the fission degree of freedom, but only weakly to intrinsic excitation, which is essentially determined at the (second) saddle. One of Wilkin's result is a potential energy surface for the fission fragments at the scission point.

Interpretation of the Experimental Data

We are particularly interested in shell and pairing effects on fission fragment parameter distributions: FMD, FCD and FKED. They cause fine structures and modulations on the smooth distribution functions. After the descent from saddle to scission through the fission valley of the potential surface on Fig. 1, the configuration point, describing the nuclear deformation, is trapped in one of the potential minima on the potential landscape of Wilkin's scission point model. Fig. 10 shows a contour map deduced from this model; only minima lower than - 1 MeV below the liquid drop potential energy are outlined. They correspond to peaks in the mass yield distribution.

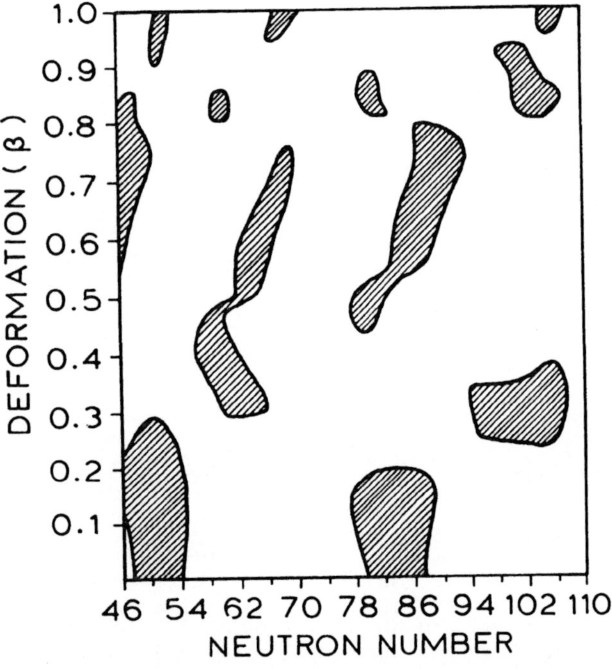

Fig. 10. Neutron-shell corrections as function of deformation (β) and fragment neutron number. The contours represent only the strongest shell corrections containing all values lower than - 1 MeV (ref.[17])

Shell Effects

a) on fragment mass distribution (FMD)

Particularly effective is the spherical neutron shell around N = 82. It stabilizes the position of the heavy fragment mass peak when going from one fissioning actinide nucleus to another and generates in the light FMD a fine structure at A = 100 for the compound nucleus ^{234}U, at A = 102 for ^{236}U and at A = 106 for ^{240}Pu (see Fig. 4). These mass yields are also favoured by the N $\geq$ 60 deformed shell in the light fragment. The corresponding fragments are:

$^{134}Te^{52}_{82}$ + $^{100}Zr^{40}_{60}$ for ^{234}U

$^{134}Te^{52}_{82}$ + $^{102}Zr^{40}_{62}$ for ^{236}U and

$^{134}Te^{52}_{82}$ + $^{106}Mo^{42}_{64}$ for ^{240}Pu

Minor shell effects are attributed to the deformed N $\approx$ 88 shell, favouring $^{144}Ba^{56}_{88}$ for ^{234}U and $^{146}Ba^{56}_{90}$ for ^{236}U together with the rigid light fragment $^{90}Kr^{36}_{54}$, and the N = 50 shell, leading in both fissioning Uranium systems to $^{84}Se^{34}_{50}$. The Z = 38 deformed shell is weakly effective in the generation of $^{96}Sr^{38}_{58}$. Minor shell effects are however masked by a simultaneous proton pairing effect in the Z = 34, 36, 38, 40 and 42 elements.
The difference in the FMD between ^{241}Pu(n,f) and ^{242}Pu(s.f.) is a result of an interplay between the spherical N = 82 and deformed N $\approx$ 88 shells. The strong shell effect of the N = 82 shell is reduced in the neutron induced reaction, with an compound nucleus excitation energy EX = 6.3 MeV and the N $\approx$ 88 shell becomes effective.

b) on fragment nuclear charge distribution (FCD)

Proton shell effects on the FMD are in general weak. There is however a significant effect of the Z = 50 shell on the deviation ΔZ of the average nuclear charge $\bar{Z}$ integrated over the fragment kinetic energy from the unchanged charge distribution Z_{UCD}. ΔZ is plotted in Fig. 11 as function of light fragment pre-neutron mass. The Z = 50 shell effect gives rise to the strong variation of ΔZ from the average behaviour on the left of Fig. 11. The modulation of ΔZ in this figure is a proton pairing effect.

c) on fragment kinetic energy distribution (FKED)

The data in Fig. 9 tell us, that the width of the angular distribution of long range α-particles emitted in fission decreases with increasing fragment kinetic energy. This result is incompatible with trajectory calculations, which demonstrate that the faster the fragments seperate the less the α-particles are focussed, thus causing a broadening of the angular distribution as the fragment kinetic energy increases. This discrepancy can be avoided by the thermodynamical model of W. Nörenberg[3]. In this model prescission kinetic energy is small and constant at about 8 MeV. FKED is mainly determined by the distribution of scission shapes (stretching modes). This is our starting point for the discussion of shell effects on FKED. Compact scission configurations are correlated with high, elongated

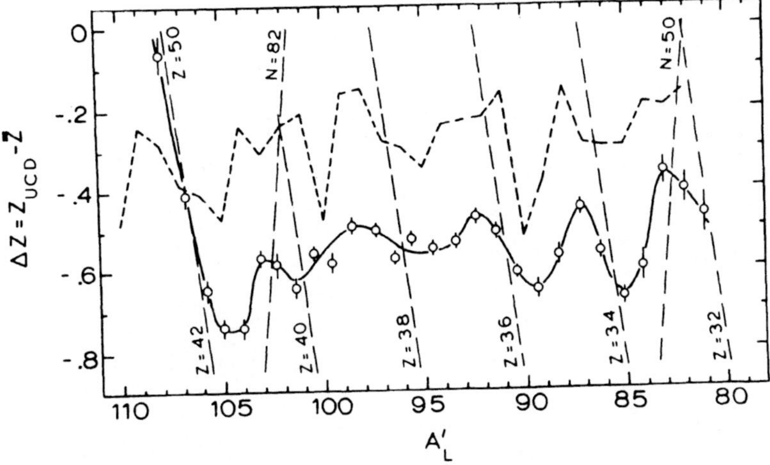

Fig. 11. Deviation of the isobaric average nuclear charge summed over the kinetic energy as function of pre-neutron light fragment mass (ref. [7]). The dashed line was calculated by Wilkins et al. (ref.[17])

with low fragment kinetic energies. In Fig. 12 isobaric fragment kinetic energies are plotted as function of mass number. The kinetic energy of the fine structure fragment masses at A_L = 100/102 for $^{234/236}$U is obviously enhanced due to the compact N = 82 fragments. The higher kinetic energy in ^{242}Pu (s.f.) compared to the ^{241}Pu (n,f) case has the same reason (see end of section a). The influence of the N = 82 on mass fragmentation is stronger than in the neutron induced reaction, where the deformed N ≈ 88 shell becomes effective. The asymmetric KED in ^{242}Pu (s.f.) is explained in ref.[9] . The fast drop of the FKED for more symmetric fragment masses is also a shell effect. Deformed symmetric fragments are weakly shell stabilized at relatively large deformations[17] $\beta \approx 0.8$. Small excitation energies of the fissioning nucleus relax the shell stabilization and the fragmentation is controlled by the liquid drop potential energy, which has a minimum at the smaller deformation $\beta = 0.6$. The fast increase from negative to positive values of the derivatives d $\bar{E}_K$/d(EX) measured by Y. Patin et al.[10] falls into the same mass intervall, in which the FKED drops. Positive derivatives are typical for subthreshold fission as has been shown by I. Lachkar et al.[19]. This is the so called superfluid regime of fission in contrast to the viscous one above the fission barrier. It seems that the jump in d E_K/d(EX) is associated with shell rather than with pairing effects provided that the picture of the dynamical deformation antipairing effect[20] is not applicable. The transition from shell stabilized deformed to less deformed liquid drop fragments leads to higher fragment kinetic energies with increasing compound nucleus excitation energy. The probability of a nucleus to get excited instead of staying in the (dynamic) ground state increases with the slope of those levels, which cross the Fermi energy. It seems that fast changing level energies near the Fermi surface characterize shell stabilized deformed symmetric fission, which at small compound nucleus excitation energy goes over into less damped liquid drop fission with less deformed fragments. Finally it should be mentioned that W. Lang et al.[7] found a correlation between fragment kinetic energy and nuclear charge distribution predicted by W. Nörenberg already in 1966[21].

Pairing Effects

a) on fragment mass and nuclear charge distributions (FMD and FCD)

While shell effects are dominated by the neutrons, pairing effects are preferably brought about by the protons (see Fig.5 and 6). Consequently are the pairing effects on FMD caused by odd-even effects in

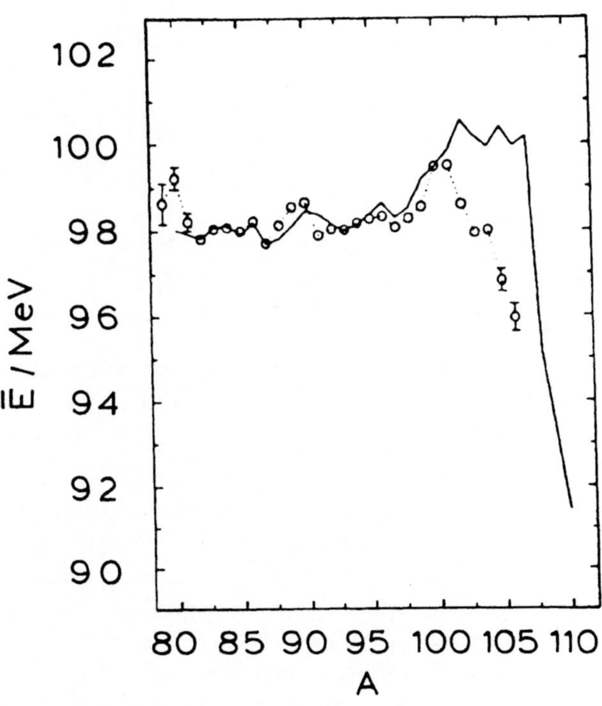

Fig. 12. Isobaric energy distributions of light fission products as function of fragment mass dotted line for ^{233}U + n, full line for ^{235}U + n (ref.[7]).

the charge distribution, which give rise to mass yield modulations with a period of about 5. Fig. 11 shows a proton odd-even effect.

The proton odd-even effect δ is defined as the difference between the yields of even Z and odd Z elements in percent.

$$\delta = \frac{Y_e - Y_o}{Y_e + Y_o}$$

Recently an excellent review on odd-even effects in fission has been given by H. Nifenecker[22]. We shall summarize the characteristics of this effect as follows:

1. Even Z elements are preferentially formed.
2. The odd-even effect modulates the widths of the isobaric nuclear charge distributions (Fig. 13). These widths increase with decreasing fragment kinetic energies,
3. grows with the fragment kinetic energies (Fig. 14),
4. falls from 35 % for Thorium to about 14 % for Plutonium (eventually 4 % in ^{241}Pu + n).

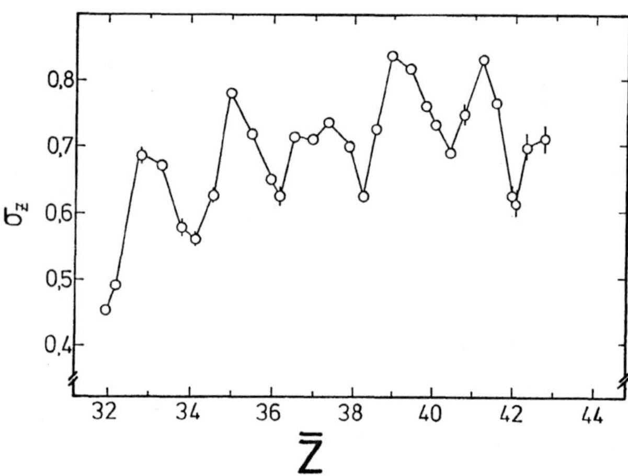

Fig. 13. Standard deviations of post-neutron isobaric charge distributions as function of the average nuclear charge

As the odd-even effect is a consequence of level crossings (Zener effect) on the descent from saddle to scission, a process, which exceeds the possibilities of static fission models, we shall renounce an interpretation in this review. From our investigation of "cold fragmentation[23,24] which will be shortly presented in the last chapter, we know that during fission about a factor of two more neutron pairs are broken than proton pairs. As neutron density in the neck of the fissioning nucleus is higher than proton density, as protons prefer to cluster in the nascent fragments, neutron stay in levels described by complicated wave functions with amplitudes in the neck between the seperating fragments. Neutrons are more excited than protons and consequently more neutron pairs will be broken than proton pairs during fission. Probably

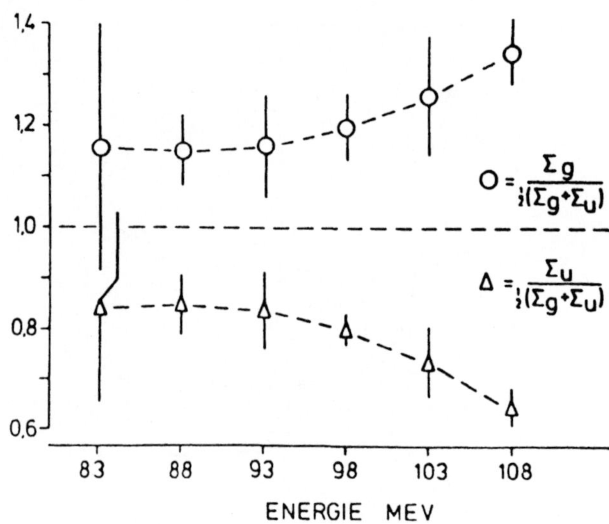

Fig. 14. Kinetic energy dependency of proton odd-even effect for ^{233}U + n, u: odd, g: even fragment yield

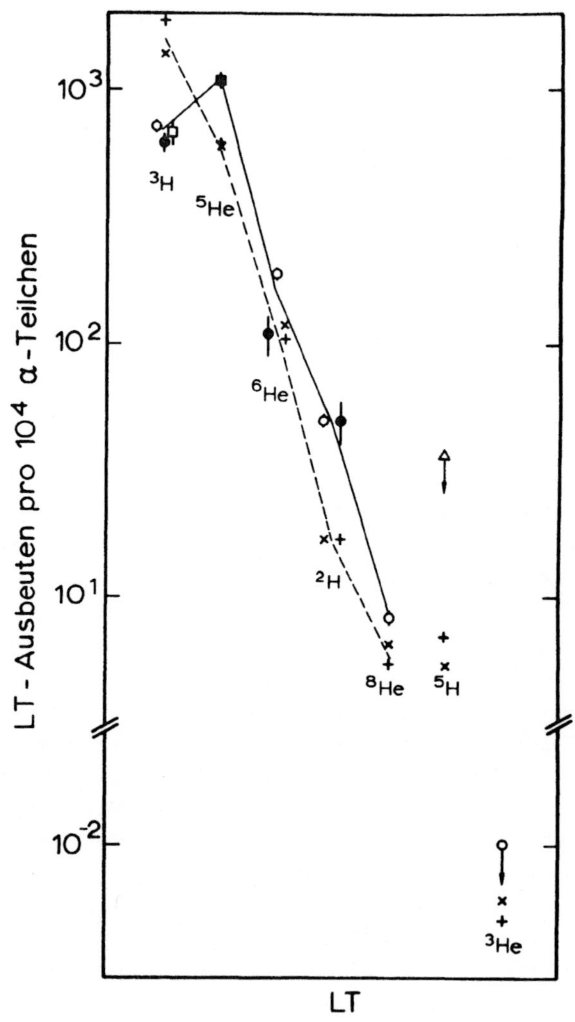

Fig. 15. Relative yields of H and He nuclei emitted in neutron induced fission of ^{235}U calculated with a neutron-neck model (ref. [25]). + calculated, other symbols experimental values

particle associated fission supports this picture of the "neutron neck model" A. Koldobski[25] has predicted light charged particle yields with an surprisingly good accuracy (Fig. 15).

b) on fragment kinetic energy distribution (FKED)

W. Lang et al.[7] report that the kinetic energy distributions of even-Z elements are shifted by about 0.4 MeV to higher kinetic energies with respect to their odd-Z neighbours. They explain this effect quantitatively by assuming that two components contribute to the even-Z yield: one with one broken proton pair and another superfluid component (all protons paired).
Paired nucleon systems favour spherical shapes, and thus compact scission configurations with higher fragment kinetic energies.

Cold Fragmentation

We have seen that nuclear structure effects generate fine structures on the FMD, which are more articulate at higher fragment kinetic energies (see Fig. 4). At very high kinetic energies exhausting the Q-value of the corresponding mass fragmentation up to about 1 MeV, energy conservation alone governs the fission process. Fig. 16 shows the attenuation of the mass split with light fragment mass $A_L = 90$. The yield decreases by more than one order of magnitude per MeV.

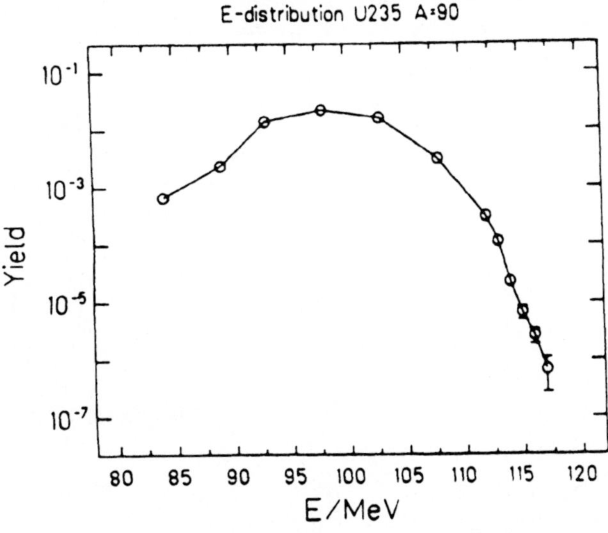

Fig. 16. Yield of fragment mass A = 90 as function of light fragment kinetic energy for $^{235}U + n$

By raising the kinetic energy barrier the dying out of the last energetically possible mass splits can be observed as shown in Fig. 17.
At given yield levels, eg. 10^{-5}/fission and 10^{-6}/fission the fragment TKE can be compared to the corresponding nuclear charge averaged Q-values. This has been done for the systems $^{233}U + n$ and $^{235}U + n$ in Fig. 18. There is obviously a proton odd-even effect modulating the experimental TKE as a function of light fragment masses.
The following conclusion can be drawn from the present data

i) Cold fragmentation is a general phenomenon for all mass splittings with light fragment masses in the range $80 \leq A_L \leq 104$
ii) Variations of 1 MeV in the cold fragment kinetic energy change drastically the mass yields. The slope averages of the light fragment yields at the 1 ppm level are (1.40 ± 0.05) orders of magnitude per MeV.
iii) Even fragments are (0.90 ± 0.10) MeV higher excited than odd ones. This fact is seen directly in Fig. 18.
iv) While in nearly all fission events neutron pairs are broken, in about half the cases proton pairs resist during fission. A direct determination of the odd-even effect $(Y_e - Y_o)/(Y_e + Y_o)$ from independent yield measurements at 112 MeV, where Y_e and Y_o are yields of even and odd fragments, resp., provides a value of 11 % for neutron and 46 % for protons as an average over the mass range.

Cold fragmentation is a mode of fission characterized by practically non-excited fragments near their ground states. The descent from saddle to scission is accompanied by a constriction of the neck without considerable nuclear stretching. From the resulting compact scission shapes fragments with high kinetic energies are released. The corresponding fission path is indicated in Fig. 19.
The production of (super) heavy elements by heavy ion fusion has to be the inverse process of cold fragmentation, as target and projectile nuclei in the entrance channel are in their ground states.
Cold fragmentation is also a process which allows to test mass predictions for nuclei far from β-stability, although the present fission product spectrometers do not allow to seperate fragments below their first excited states, because of limitations in their kinetic energy resolution (ΔE of "LOHENGRIN" is about 1 MeV).

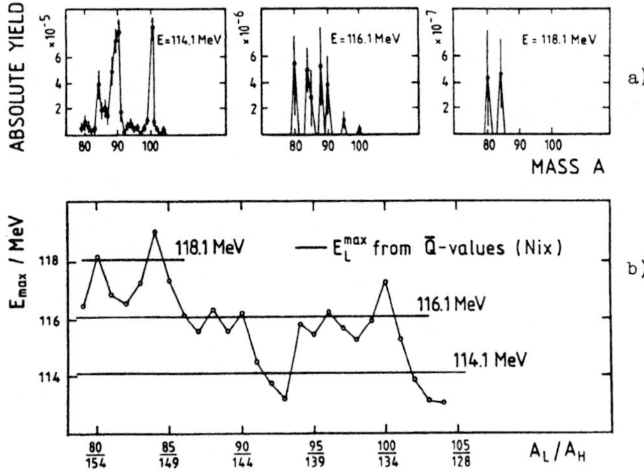

Fig. 17 a) Mass yield for cold fragmentation of ^{234}U at 3 light fragment kinetic energies (LFKE)
b) Maximum light fragment kinetic energy calculated from Q-values together with the corresponding mass yields

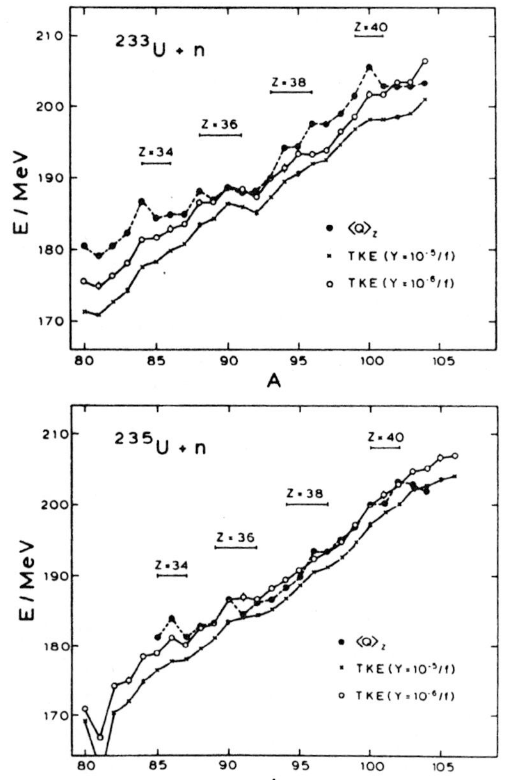

Fig. 18 Experimental total kinetic fragment energy at 10^{-6} and $10^{-5}/f$ yield level together with Z-weighted average Q-values as function of light fragment mass. a (^{233}U + n) b (^{235}U + n)

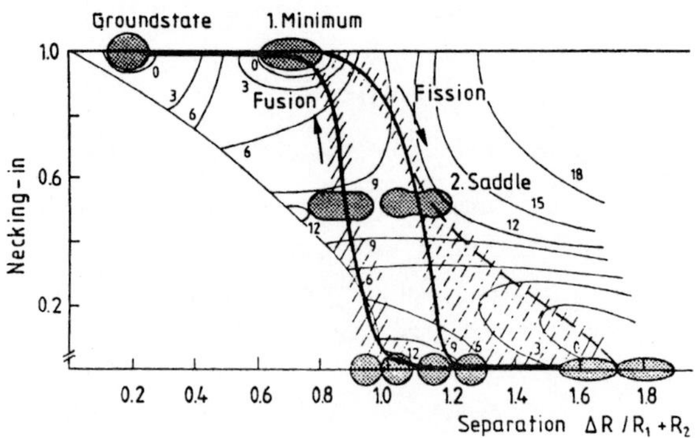

Fig. 19 Scheme of a two-dimensional energy surface for symmetric fragmentation with different fragment elongation (ref. 26)

Conclusion

We have presented recent data on fission fragment parameters (mass, nuclear charge and kinetic energy) and discussed shell and pairing effects on their distributions (fine structures, modulations) in the framework of a static scission point model. To a certain extent we were able to explain also experimental data on the dependency of fragment kinetic energies on the fissioning nucleus excitation energy. However, we felt that a dynamical fission model would be more appropriate to understand the sharing between fragment kinetic and axcitation energy at all stages of the fission process. Several groups are working on this subject[26].

Acknowledgement

I thank P. Armbruster, H.-G. Clerc, M. Mutterer, and C. Signarbieux for fruitful discussions and U. Quade and C. Schmitt for their appreciable help during the data evaluation.

References

1. M.G. Mustafa, U. Mosel, and H.W. Schmitt, Phys. Rev. C 7 (1973) 1519
2. J. Trochon, J. Fréhaut, J.W. Boldeman, G. Simon, and Y. Pranal, this conference
3. W. Nörenberg, Physics and Chemistry of Fission, proceedings IAEA Wien (1969), p. 51
4. V.M. Strutinsky, Nucl. Phys. A 122 (1968) 1
5. V. Metag, Physics and Chemistry of fission, proceedings (1980) IAEA Wien, p. 153
6. W.J. Swiatecki and S. Bjørnholm, Phys. Rep. 4 C (1972) 326
7. W. Lang, H.-G. Clerc, H. Wohlfarth, H. Schrader, and K.H. Schmidt, Nucl. Phys. A 345 (1980) 34

8. P. Armbruster, J.P. Bocquet, R. Brissot, H.-G. Clerc, F. Gönnenwein, A. Guessous, M. Mutterer, H. Nifenecker, U. Quade, K. Rudolph, C. Schmitt, and J.P. Theobald, to be published

9. E. Allaert, C. Wagemans, G. Wegener-Penning, A.J. Deruytter, and R. Barthélémy, Nucl. Phys. A 380 (1982) 61

10. Y. Patin, F. Cocu, J. Lachkar, J. Sigaud, G. Haouat, and S. Cierjacks, Nucl. Phys. A 382 (1982) 31

11. C. Guet, C. Signarbieux, P. Perrin, H. Nifenecker, M. Asghar, F. Caitucolli, and B. Leroux, Nucl. Phys. A 314 (1979) 1

12. M. Dahlinger, D. Vermeulen, and K.H. Schmidt, Nucl. Phys. A 376 (1982) 94

13. E. Moll, H. Schrader, G. Siegert, M. Asghar, J.P. Bocquet, G. Bailleul, J.P. Gautheron, J. Greif, G.I. Crawford, C. Chauvin, E. Ewald, H. Wollnik, P. Armbruster, G. Fiebig, H. Lawin, and K. Sistemich, Nucl. Inst. and Meth. 123 (1975) 615

14. C.R. Gruhn, M. Binimi, R. Legrain, R. Loveman, W. Pang, M. Roach, D.K. Scott, A. Shotter, T.J. Symons, J. Wouters, M. Zisman, R. Devries, Y.C. Peng, and W. Sondheim, Nucl. Inst. and Meth. 196 (1982) 33

15. E. Jacobs, A. De Clercq, H. Thierens, D. De Frenne, P. D'hondt, P. De Gelder, and A. Deruytter, Phys. Rev. C 20 (1979) 2249

16. N. Cârjan, A. Sandulescu, and V.V. Pashkevich, Phys. Rev. C 11 (1975) 782

17. B.D. Wilkins, E.P. Steinberg, and R.R. Chasman, Phys. Rev. C 14 (1976) 1832

18. P. Fong, Physics and Chemistry of Fission, proceedings IAEA Wien (1980) p 373 and: "Statistical Theory of Nuclear Fission", Gordon and Breach Science Publishers, New York (1969)

19. J. Lachkar, Y. Patin, and J. Sigaud J. de Phys. 36 (1975) L 79, Supplément

20. F. Dickmann, Physics and Chemistry of Fission proceedings IAEA Wien (1980) p. 439

21. W. Nörenberg, Z. f. Phys. 197 (1966) 246

22. H. Nifenecker, International Symposium on "Nuclear Fission and Related Collective Phenomena and Properties of Heavy Nuclei" Bad Honnef, W.-Germany (1981), proceedings in "Lecture Notes in Physics" W.-Berlin, to be published

23. U. Quade, K. Rudolph, P. Armbruster, H.-G. Clerc, W. Lang, M. Mutterer, J. Pannicke, C. Schmitt, J.P. Theobald, F. Gönnenwein, and H. Schrader, dto.

24. M. Montoya, Thèse â l'université de Paris XI report ORSAY No d'ordre 2510

25. A. Koldobski, Moscow, priv. comm.

26. Phys. and Chemistry of Fission, proceedings IAEA Wien, p. 387 ff.

MEASUREMENT OF ALPHA VALUE ON URANIUM-235 RESONANCES

Yu.V. Adamchuk, M.A. Voskanyan, G.V. Muradyan, P.Yu. Simonov, Yu.G. Shchepkin

I.V. Kurchatov Atomic Energy Institute
Moscow, USSR

An absolute measurement of the alpha value at the Uranium-235 resonances has been performed. For most resonances the accuracy is better than 2 % which considerably exceeds that of the available alpha values.

[Alpha value, multiplicity spectrum, multisectional detector, U-235]

Problem Statement

The ratio of the neutron capture to neutron fission cross-section (alpha) of U-235 is one of the most important quantities for reactor design. Knowledge of this ratio is required over a wide neutron energy range and with a high accuracy. However, information available on the alpha value does not satisfy the required accuracy because of experimental difficulties in exactly classifying capture and fission events, in finding their detecting efficiency, in reducing scattered and fission neutrons background and in determining the said background level. From the point of view of obtaining high accuracy throughout the required energy region it is not optimal to solve the above-mentioned problems simultaneously. To measure the alpha value over the wide energy range, it is expedient to do it in two stages. At the first stage the measurements of absolute values in the resonant range are taken and at the second one, the measurements of relative energy variation over the wide energy range and normalization on the measured resonance values are performed.

In this paper the absolute alpha value measurements have been taken on the U-235 resonances employing the method of multiplicity spectrometry of excited nuclei radiation [1/2/3]. With the help of this method the above-mentioned difficulties are successfully overcome.

When taking measurements in the resonant range ($E_n \lesssim 50$ eV) it is sufficient to employ thin (n,γ)-converter made of B-10 to shield NaI(Tl) crystals from neutrons scattered from the sample. Mind that in the range of tens of keV use is made of thick converters. The transfer to a thin converter makes it possible to decrease the absorption of γ-quanta arising from the capture and, in particular, from the fission and, thus, to significantly improve the quality of separating the capture and fission events and to increase their detection efficiency practically up to 100 %. Due to the high efficiency of the fission events detection it is possible to simultaneously eliminate the background caused by the fission products detection, i.e. the neutrons of fission and radiation of fission fragments, later on. To eliminate this background, the detector electronics is locked by each detected pulse. In order to eliminate all the other events which are essentially the background for the given resonance peak, one should perform the background subtraction. In so doing, a part of the effect may be different from the true alpha value corresponding to the total effect in the resonance range. However, let us stress that the aim of this paper is to obtain data for normalization of measurements in the high energy range. Therefore, if in both the measurements the "effect" range and the "background" one are taken equal, the correct normalization will be obtained. Mind that the above-mentioned resonance alpha values are of interest in themselves, since they represent Γ_γ-to-Γ_f widths ratio of the resonances and make it possible to compare various procedures and results of the measurements unambiguously.

Procedure and Measurements

The measurements have been taken on the LEA of the I.V. Kurchatov AEI on the 26-m path length using the 48-section scintillation detector with 4π-geometry (48SSD) made on the basis of NaI(Tl) crystals employing the converter made of B-10 of ~ 0.3 g/cm^2 thickness. The detector sections are made of individual scintiblocks in the form of rectangular parallelepipeds. The detector central part employs 300 x 132 x 132 mm ones. The scintillator minimal thickness as to the path length of γ-quanta emitted by the sample is ~ 130 mm, the detector geometrical efficiency is ~ 98 %. The scintillator total volume is ~ 200 l. The detector inner cavity is 400 x 400 x 450 mm overall.

The U-235 sample under study is arranged at the converter centre and consists of six identical disks of 29 mm diameter and 0.06-g/cm^2 thickness, each. The disks are spaced 10 mm apart from each other so as to decrease γ-quanta absorption in the sample.

As a result of the measurements for each time channel (interval) i, corresponding to some definite neutron energy, 4 groups (j = 1 - 4) of coincidence multiplicity spectra from 1 to 15 and above 15 have been stored by a computer provided that the energy release over the detector total volume at each time exceeds 0.6, 0.8, 1.1 and 1.5 MeV. In each section the energy discrimination level is set to be as low as possible and equals ~ 25 keV. All the four spectrum groups are measured simultaneously. The time coder channel minimal width is 0.16 μs. After each detected event the detector electronics is locked for a time period of 50 μs. In so doing, the dead time correction is measured with the aid of a generator producing random pulses which pass throughout the detector electronics along with the effective pulses.

Data Treatment

The spectra obtained as a result of the experiment in the form of coincidence counting as a function of multiplicity N_j (i, K) for each group j = 1 - 4 have the region of fission "proper" (coincidence multiplicities K > 9), where only the fission events are recorded, and the overlap region K'_γ, where the capture events as well as the fission ones are recorded ($K \leqslant 9$). To determine the alpha value, one should find the fission and capture fractions in the overlap region (i.e. to separate the capture and fission events). The lower is the contribution of fission in the said region, the less stringent are the requirements for the separation procedure accuracy. The problem of separation could have been easily solved if there existed the capture region "proper" along with the fission region "proper". Then, we could have taken two resonances with different alpha values to define the capture cross-sections ratio ν_j (for example, that of the first level (i = 1) to that of the second one (i = 2)) over the capture region "proper" (K_γ). This done, we could have multiplied the spectrum $N_j(2,K)$ by the above-mentioned

K. H. Böckhoff (ed.), Nuclear Data for Science and Technology, 730–732.

ratio and then subtracted the spectrum $N_j(1,K)$. In so doing, at all values of K the capture is subtracted and the effect proportional to the fission turns out to be the difference:

$$\nu_j = \frac{\Gamma_j(1,K_1)}{\Gamma_j(2,K_1)} = \frac{\Gamma_j(1,K_2)}{\Gamma_j(2,K_2)} = \frac{\Gamma_j(1,K_3)}{\Gamma_j(2,K_3)} = \ldots\ldots =$$

$$\sum_{K_\gamma} \Gamma_j(1,K) / \sum_{K_\gamma} \Gamma_j(2,K) \qquad (1)$$

$$N_j(2,K) = \Gamma_j(2,K) + F_j(2,K)$$

$$N_j(1,K) = \Gamma_j(1,K) + F_j(1,K)$$

$$\nu_j \cdot N_j(2,K) - N_j(1,K) = \nu_j \cdot F_j(2,K) - F_j(1,K) = \Delta F_j(K)$$

Here, $\Gamma_j(i,K)$ and $F_j(i,K)$ are contributions made by the capture and fission to $N_j(i,K)$. Then the form of the fission multiplicity spectrum $f_j(K)$ is defined:

$$f_j(K) = \Delta F_j(K) / \sum_{K=1}^{16} \Delta F_j(K)$$

and, respectively, the fission contribution in the range K'_γ (or its part from K'_{γ_1} to K'_{γ_2}):

$$\sum_{K'_{\gamma_1}}^{K'_{\gamma_2}} F_j(i,K) = \sum_{K'_{\gamma_1}}^{K'_{\gamma_2}} f_j(K) / \sum_{K=10}^{16} f_j(K) \cdot \sum_{K=10}^{16} N_j(i,K)$$

i.e. one of the basic questions, that is the separation of the capture and fission processes, could be solved. But in reality there is no capture region "proper". It is illustrated by Table I which presents the Cf-252 fission multiplicity spectrum form being other than zero at all values of the multiplicity K. To exclude the fragment radioactivity detection, the curve has been obtained taking into account the count coincidence of 48SSD and fragment semiconductor detector (SD) arranged together with a layer of Cf-252 at the centre of 48SSD. Refer to the same table for the multiplicity spectrum of the U-235 fission and the capture $\gamma_j(K)$ of neutron in U-235 for the discrimination level of 0.6 MeV.

Owing to the fact that the fission region "proper" ($K > 9$) is available, $\gamma_j(K)$ can easily be defined on the analogy of the procedure [1].

However, inspite of the fact that there is no capture region "proper", and due to a high quality of the equipment separation, it is sufficient to approximately know the fission form in the overlap region. Even if we neglect the fission at K = 3, that is when K = 3 is assumed to be the capture region "proper", it results in the error of α being equal to merely 7 % at $\alpha = 1$. Actually, the fission contribution is taken into account using the measurements with Cf-252. The value $\mu_j = f_j(K_1)/f_j(K_2)$ has been introduced knowing which one can easily define ν_j:

$$\nu_j = \left\{\mu_j N_j(1,K_2) - N_j(1,K_1)\right\} / \left\{\mu_j N_j(2,K_2) - N_j(2,K_1)\right\}$$

and consequently, the fission form. The multiplicity K = 5 is taken as K_1 and K_2 is essentially the sum of multiplicities 3 and 4. It turned out that the value μ_j was extremely stable at any preset value of the fission spectrum peak K_m and weakly depended upon variation of $f_j(K)$ on the "wings". By changing the measurement conditions (the number of the detector sections, coincidence time duration, the discrimination level of a fragment energy at the SD), we obtained $f_j(K)$ having the maximum at various "K".

Extrapolating the values μ_j obtained at various K_m and discrimination levels j, the values μ_j have been selected for the value K_m which corresponds to the U-235 multiplicity spectrum peak. In so doing, the drift of K_m by a value of ~ 0.5 does not result in an intolerable drift of μ_j. As for the value K_m for U-235, one can define the fomer accurately enough by assuming $\mu_j = 0$, since K_m depends weakly upon μ_j. For example, for the level at 14.02 eV with $\alpha \approx 0.08$ the capture contribution to the region $K_m = 8.5$ amounts to as low as ~ 1 % of the fission. It is obvious that such an insignificant capture ingredient does not affect the accuracy of defining K_m.

Having separated the capture and fission events and defined the total (over "K") detected events number $\Gamma_j = \sum_K \Gamma_j(k)/\sum_K \gamma_j(k)$ and $F_j = \sum_K F_j(k)/\sum_K f_j(k)$, the detection efficiency of the said processes were determined. To this end, the detected numbers Γ_j and F_j have been plotted as function of the energy discrimination level. The extrapolation to the zero level yields the true number of capture $\Gamma_{(o)}$ and fission $F_{(o)}$ events which have happened during the measurements, while the efficiencies are as follows:

$$\epsilon_{\gamma j} = \Gamma_j/\Gamma_{(o)}; \quad \epsilon_{fj} = F_j/F_{(o)}.$$

To prove the fact, the measurements of dependence of the total count of 48SSD, with the ^{137}Cs and ^{60}Co sources having known intensities, upon the discrimination level have been taken, as well as the analogous measurements with the ^{252}Cf source in coincidence with fragments count of the SD. The intensity of the ^{137}Cs and ^{60}Co sources has been defined using an assemblage made of the same scintiblocks. Provided in the assemblage is only a channel being 10 x 5 mm used to intake the source. The minimum crystal widths for the γ-quanta is ~ 26 cm. For ^{60}Co the extrapolation to the zero level of the count dependence upon the discrimination level obtained with the 48SSD detector yields count values coinciding with the source intensity obtained at the assemblage, thus confirming the correctness of $\epsilon_{\gamma j}$ and ϵ_{fj} defining procedure for U-235, since in case of neutron capture in ^{235}U the multiplicity of the γ-radiation and its total energy is higher than those of ^{60}Co. The obtained efficiency

Table I

K	1	2	3	4	5	6	7	8	9	10	11	12	13	14	15	16
Cf-252 $f_{0.6}(K)$ (%)	0.026	0.17	0.57	1.30	2.43	3.95	5.95	7.38	9.10	10.06	10.20	10.24	8.95	7.77	6.21	14.45
U-235 $f_{0.6}(K)$ (%)	0.021	0.53	1.47	3.30	5.63	8.02	9.92	11.03	11.20	10.53	9.26	7.75	6.16	4.69	3.47	7.00
U-235 $\gamma_{0.6}(K)$ (%)	2.68	10.95	21.28	25.00	20.24	11.89	5.37	1.95	0.64	0.15						

values for ^{235}U at the discrimination level of 0.6 MeV constitues ϵ_γ = (96.8 ± 0.6)% and ϵ_f = (99.76 ± 0.1)%. Such high values of ϵ_γ and ϵ_f allows us to define them to a high accuracy and are also essential for excluding the fission products detection.

To conclude the section, presented are the basic parameters of the installation:

1. The fission contribution in the region K ⩽ 6 where ~ 92 % of the capture events are concentrated amounts to as low as ~ 19 %.
2. The efficiency of the neutron capture event detection at the discrimination level of 0.6 MeV is ϵ_γ = (96.8 ± 0.6)%, while that of the fission event, ϵ_f = (99.76 ± 0.1)%.
3. The influence of the fission products after effect on the alpha value being defined has been practically eliminated.

Due to the above-mentioned, as well as to the elimination of all backgrounds by means of subtracting the resonance background, one succeeds in obtaining the absolute alpha values on the resonances to a high accuracy.

Results

The obtained alpha values for 19 resonances of U-235 are presented in Table II. Shown in the same table are the energy intervals against which the mean values of the "effect" and "background" are defined taking into account one time channel. Note that when employing the obtained alpha values for the relative measurements normalization, one should take the "effect" and "background" within the same limits.

The alpha value error includes the systematic and statistical errors. The systematic error consists of two parts. The first part $(\Delta\, a/a)\mu_j$ is due to the fact that we do not know the value μ_j exactly. The error $\Delta\,\mu_j$ can be estimated by varying the measurement conditions at the stable value of the fission multiplicity spectrum peak (K_m) when ^{252}Cf is employed (see above). The said variation is approximately 4.2 %.

The second part is due to the error of values ϵ_γ and ϵ_f, being 0.6 and 0.1 %, respectively. The total errors $(\Delta\, a/a)$ allowing for the systematic and statistical errors are presented in Table II.

As is evident from the table, for most of the levels the alpha value has been obtained with a high accuracy (~ 2 %) which considerably exceeds the accuracy of the data available.

Presented in the last column of the table are the alpha values calculated employing the numerical data of ref. [4]. In so doing, the particular resonances have been selected proceeding from the requirements of the numerical data adequate minuteness, so as to exclude the possibility of additional errors due to the discrepancy between the averaging intervals of the present paper and those of ref. [4]. One can see that the given alpha values are systematically below those presented in ref. [4]. Note that the effect of the continuous background inaccurate account on the alpha value is eliminated in the present comparison.

References

1. G.V. Muradyan, Yu.G. Shchepkin, Yu.V. Adamchuk, G.I. Ustroev, Study of neutron cross-sections and quantum characteristics of nuclear levels on the basis of excited nucleus radiation multiplicity spectrometry - Preprint IAE-2634, 1976.
2. G.V. Muradyan, Multiplicity spectrometry, Atomnaya Energya (Sov. J. Atom. Phys.), 1981, Vol. 50, 6, pp. 394-398.
3. G.V. Muradyan, Yu.G. Shchepkin, Yu.V. Adamchuk, M.A. Voskanyan, A measurement of U-235 absolute alpha value in the neutron energy range from 0.1 to 30 keV, in proc.: Nuclear Cross Sections for Technology (International Conf., USA, Knoxville, Oct. 22-26, 1979), 1980, pp. 488-490.
4. T. Gwin, E.G. Silver, R.W. Ingle, H. Weaver, Measurement ^{239}Pu of the Neutron Capture and Fission Cross Sections of ^{239}Pu and ^{235}U, 0.02 eV to 200 keV, the Neutron Capture Cross Sections of ^{197}Au, 10 to 50 keV, and Neutron Fission Cross Sections of ^{233}U, 5 to 200 keV, Nucl. Sci. and Eng., 1976, Vol. 59, pp. 79-105.

Table II

E_o(eV)	"Effect" range (eV)	"Background" 1 range (eV)	"Background" 2 range (eV)	a	$\Delta\, a/a$	$(\Delta\, a/a)\mu_j$	a [4]
2.028	1.980 - 2.103	2.291 - 2.445		3.415	1.9	0.84	
3.145	3.047 - 3.242	3.311 - 3.417		0.307	5.1	2.80	
3.61	3.509 - 3.720	4.144 - 4.274	2.299 - 2.353	0.732	1.9	1.50	
4.85	4.754 - 4.945	4.242 - 4.403		8.262	1.4	0.73	
6.39	6.287 - 6.507	7.857 - 7.928	6.689 - 6.802	3.033	1.2	0.87	4.03
7.08	6.957 - 7.213	7.857 - 7.928		1.188	1.5	1.20	1.44
8.78	8.609 - 8.982	7.618 - 7.928		0.397	2.4	2.30	
9.28	9.178 - 9.381	9.432 - 9.643		0.595	2.3	1.80	
10.18	10.053 - 10.285	10.686 - 10.941		0.646	3.6	1.70	
11.66	11.520 - 11.820	10.674 - 10.941		7.691	1.2	0.74	8.68
12.39	12.192 - 12.599	10.612 - 10.941		1.466	1.3	1.10	1.66
14.02	13.617 - 14.270	14.765 - 15.166		0.078	10.1	9.00	
15.40	15.243 - 15.568	14.765 - 15.074	17.307 - 17.489	0.843	2.0	1.40	
16.08	15.904 - 16.250	17.307 - 17.489		1.882	1.4	1.00	2.14
16.68	16.487 - 16.853	17.307 - 17.489		0.381	3.1	2.40	
18.05	17.835 - 18.274	18.387 - 18.586	17.282 - 17.489	0.355	3.2	2.50	
19.30	19.084 - 19.539	18.387 - 18.586	17.230 - 17.621	0.681	1.8	1.60	
21.07	20.87 - 21.25	21.83 - 22.23		1.345	1.4	1.10	1.65
32.07	31.80 - 32.35	32.68 - 32.95	31.23 - 31.48	0.658	1.9	1.60	

The errors are presented as %.

EFFETS DYNAMIQUES DANS LA FISSION DE ^{230}Th ET ^{232}Th INDUITE PAR NEUTRONS

J.Trochon, J.Fréhaut, Y.Pranal, G.Simon

Service de Physique Neutronique et Nucléaire
Centre d'Etudes de Bruyères-le-Châtel
B.P. N° 561
92542-MONTROUGE CEDEX, France

J.W.Boldeman

AAEC, Lucas-Heights, NSW (Australie)

The fission fragment caracteristics of the two odd thorium isotopes ^{231}Th* and ^{233}Th* have been measured in an attempt to study the evolution of the fissioning nucleus from saddle point to scission. In an other hand, the partial fission channel at the saddle point have been deduced from a fission fragment angular distribution and fission cross section analysis.

Dramatic changes with energy in the number of neutron $\bar{\nu}_p$ emitted per fission and the total fragment kinetic energy $\overline{TKE}$ have been observed in the fission threshold region. A rather good fit of $\bar{\nu}_p$ and $\overline{TKE}$ values has been obtained on the basis of a correlation of these quantities and the partial fission channel ratios.

This leads to expect for these isotopes a passage from saddle point to scission sufficiently rapid for the coupling between collective and intrinsec excitation to be very weak.

L'évolution du noyau entre le point selle et la scission est la phase la plus mal connue du processus de fission. Nous avons essayé dans ce travail d'en faire une approche expérimentale dans le cas de la fission induite par neutrons dans les isotopes du thorium ^{230}Th et ^{232}Th. Pour cela, nous avons tout d'abord mesuré les propriétés des fragments de fission, puis déterminé les configurations des noyaux au point selle et enfin nous avons recherché une corrélation entre ces différentes grandeurs. La gamme d'énergie étudiée comprend le seuil de fission où de grandes résonances apparaissent dans la section efficace de fission, attribuées à des états vibrationnels dans un troisième puits de la barrière de fission [1].

Les caractéristiques des fragments de fission mesurées pour ^{232}Th sont l'énergie cinétique moyenne totale $\overline{TKE}$, la distribution en masse, le nombre moyen de neutrons prompts $\bar{\nu}_p$ et l'énergie totale moyenne $\bar{E}_\gamma$ du rayonnement γ prompt émis par fission. Seules les deux premières grandeurs ont été étudiées pour ^{230}Th. Ces mesures ont été effectuées auprès du Van de Graaff de 4 MV de BRC, utilisé comme source monoénergétique de neutrons avec une résolution en énergie de ± 20 KeV.

Mesure de l'énergie cinétique totale moyenne $\overline{TKE}$ et de la distribution en masse des fragments de fission.

Cette expérience a déjà été décrite [2]. L'énergie $\overline{TKE}$ et la distribution en masse sont déduites, selon la méthode de Schmitt, des hauteurs d'impulsions délivrées par deux jonctions à barrière de surface situées de part et d'autre du dépôt fissile. La résolution sur la masse, propre à la méthode, n'est que de 4 uma environ. Par contre les très grandes précautions prises pour contrôler la stabilité du système au cours de l'expérience ont permis d'obtenir une résolution de l'ordre de 200 keV sur TKE.

Les résultats sur l'énergie $\overline{TKE}$ de ^{232}Th en fonction de l'énergie E_n des neutrons incidents sont portés sur la figure 1b. Ils présentent d'abord des valeurs basses jusqu'à E_n = 2,2 MeV, avec une structure vers E_n = 1,7 MeV. Les valeurs croissent ensuite linéairement avec une pente de 0,40 MeV/MeV jusqu'à E_n = 4 MeV.

Une comparaison avec les résultats de Dyachenko et coll. [4] et de Budtz-Jørgensen et coll. [12] est présentée sur la figure 2b. Les mesures sont en assez bon accord, sauf au dessous de E_n = 1,3 MeV.

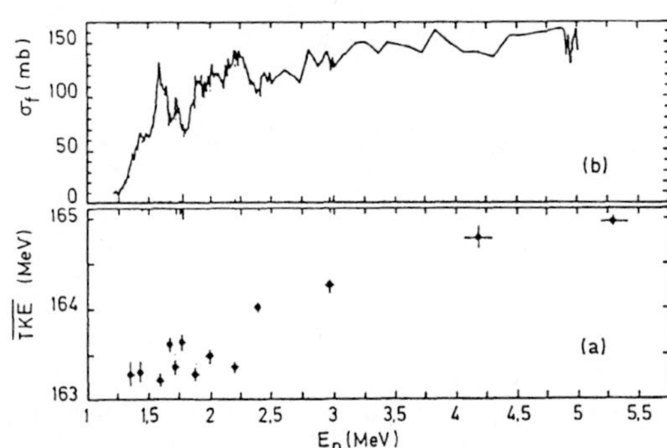

Fig. 1a. Section efficace de fission de ^{232}Th + n (d'après [3]) en fonction de l'énergie des neutrons incidents E_n.

Fig. 1b. Valeurs expérimentales de l'énergie $\overline{TKE}$.

Les distributions en masse associées présentent deux caractéristiques : la masse moyenne lourde reste à peu près constante jusqu'à E_n = 2 MeV, puis décroît rapidement pour être 1,5 uma plus faible à E_n = 4 MeV. Parallèlement le rendement en masse dans la région de la symétrie croît rapidement à partir de E_n = 2,5 MeV. Cet effet, déjà observé par ailleurs [5], est généralement attribué à l'ouverture d'une voie de fission symétrique. Il est habituellement accompagné d'un léger changement de la pente $\overline{TKE}/dE_n$ mais cet effet est ici masqué par l'augmentation brutale de $\overline{TKE}$ à E_n = 2,4 MeV.

Les valeurs de l'énergie $\overline{TKE}$ pour ^{230}Th sont portés sur la figure 3c. On observe une augmentation importante de $\overline{TKE}$ entre E_n = 1,1 et 2,2 MeV, avec une pente de 1,6 MeV/MeV entre 1,1 $\leqslant E_n \leqslant$ 1,6 MeV. A plus haute énergie la pente s'infléchit et se stabilise à 0,35 MeV/MeV entre 2,5 $\leqslant E_n \leqslant$ 5 MeV. Par ailleurs, on observe une décroissance monotone de la valeur moyenne de la masse lourde ainsi que l'apparition d'un mode plus symétrique vers E_n = 3 MeV. Ces résultats sont

K. H. Böckhoff (ed.), Nuclear Data for Science and Technology, 733–736.

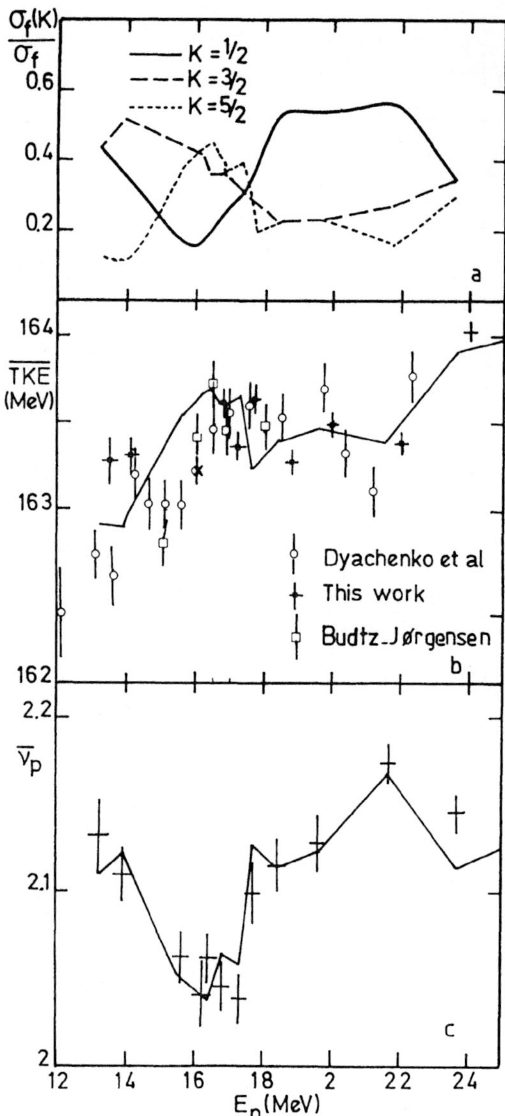

Fig. 2. Fission de $^{232}Th + n$
a) contribution relative des différentes sections efficaces de fission partielles à la section efficace de fission totale.
b) comparaison des valeurs de l'énergie $\overline{TKE}$ obtenues dans ce travail, celui de Dyachenko [4] et de Budtz-Jørgensen [12]. Le trait plein correspond au calcul effectué avec les hypothèses formulées dans le texte.
c) valeurs expérimentales du nombre $\bar{\nu}_p$ de neutrons émis par fission en fonction de l'énergie des neutrons incidents. La ligne en trait plein correspond à un ajustement à l'aide des contributions relatives σ_f^K de a).

en accord remarquable avec ceux obtenus par Sicre et Coll. à partir de la réaction (d,pf) [6]. La probabilité de former le noyau composé dans un état de spin donné étant sensiblement différente pour les deux réactions, ce bon accord semble indiquer que les propriétés de la fission dépendent peu de l'état de spin du noyau lui-même.

Mesure du nombre moyen $\bar{\nu}_p$ de neutrons prompts et de l'énergie $\bar{E}_\gamma$ du rayonnement γ prompt émis par fission.

Une chambre à fission contenant 1 g de ^{232}Th et un dépôt de ^{252}Cf pour la normalisation sont placés à l'intérieur d'un gros scintillateur liquide chargé au gadolinium. Des détails sur la méthode expérimentale sont donnés dans une communication séparée à ce congrès [8].

Les résultats sont présentés sur la figure 2c. On observe une importante décroissance de $\bar{\nu}_p$ jusqu'à E_n = 1,75 MeV, suivie d'une remontée brutale puis un accroissement monotone jusqu'à E_n = 2,2 MeV. Après une diminution de $\bar{\nu}_p$ entre 2,2 $\leqslant E_n \leqslant$ 2,4 MeV (corrélée à une augmentation de l'énergie $\overline{TKE}$), on observe au-delà une croissance linéaire de pente $d\bar{\nu}_p/dE_n$ = 0,0915 neutrons/MeV jusqu'à E_n = 4 MeV.

D'autre part, les résultats sur l'énergie $\bar{E}_\gamma$ montrent que cette quantité est proprotionnelle au nombre $\bar{\nu}_p$, avec la relation :

$$\bar{E}_\gamma = (0{,}636 \pm 0{,}045)\ \bar{\nu}_p + (4{,}48 \pm 0{,}10)\ \text{MeV}.$$

Détermination des états de transition au point selle.

^{232}Th : Une analyse complète des différentes sections efficaces - totale, capture, élastique et inélastique et fission - ainsi que de la distribution angulaire des fragments de fission a été effectuée pour ce noyau [9]. Ces différentes sections efficaces ont été calculées à l'aide du formalisme de Hauser-Feshbach. Les coefficients de transmission utilisés dérivent d'un calcul avec un modèle optique en voies couplées effectué par Lagrange [10].

La section efficace de fission σ_f et les distributions angulaires $W(\theta)$ résultent d'une sommation sur les différentes voies de sortie de fission caractérisées par le nombre quantique K, projection du spin J sur l'axe de symétrie du noyau, tel que :

$$\sigma_f = \sum_K \sum_{J=K}^{Jmax} \sigma_f^{K,J}$$

où $\sigma_f^{K,J}$ est la section efficace de fission partielle de l'état K de spin J. Bien que les résonances observées dans la section σ_f proviennent probablement d'états dans un troisième puits, une barrière de fission à deux bosses a été utilisée car les calculs montrent que le sommet de la première bosse est inférieure à l'énergie de liaison d'un neutron [11]. Cependant les résultats d'une mesure récente indiquent une contribution sensiblement différente des voies de sortie partielles de fission dans la gamme 1,5 $\leqslant E_n \leqslant$ 1,8 MeV [12] qui, comme nous le verrons par la suite, permet un meilleur ajustement des résultats expérimentaux. Ces résultats, complétés par notre calcul, sont portés sur la figure 2a.

^{230}Th : Nous nous sommes limités pour l'instant à la détermination de la contribution relative des différentes voies de sortie de fission à partir des distributions angulaires seulement.

La mesure de ces distributions a été effectué en collaboration à l'Université de Bordeaux [13]. A partir des courbes théoriques élémentaires $W_K^J(\theta)$ de la distribution angulaire des fragments de fission d'un état K de spin J à un angle θ, nous avons recherché par une méthode de moindres carrés quelle est la combinaison des coefficients $B_K^J(E_n)$ qui donne le meilleur ajustement des distributions angulaires expérimentales, tel que :

$$W^\theta_{(E_n)} = \sum_K \sum_{J=K}^{Jmax} B_K^J(E_n)\ W_K^J(\theta).$$

Un test effectué sur les distributions angulaires de ^{232}Th avait donné satisfaction [14]. Les résultats sont donnés sur la figure 3c.

Interprétation des résultats

Cas de ^{232}Th : Au-delà de E_n = 2,4 MeV, le nombre $\bar{\nu}_p$ et l'énergie TKE varient linéairement avec E_n. Si on suppose constant le Q de la réaction, on obtient à partir des valeurs expérimentales la relation :

$$(d\overline{TKE}/dE_n) + 5{,}92 + (d\bar{E}_\gamma/d\bar{\nu}_p)\ (d\bar{\nu}_p/dE_n) = 1$$

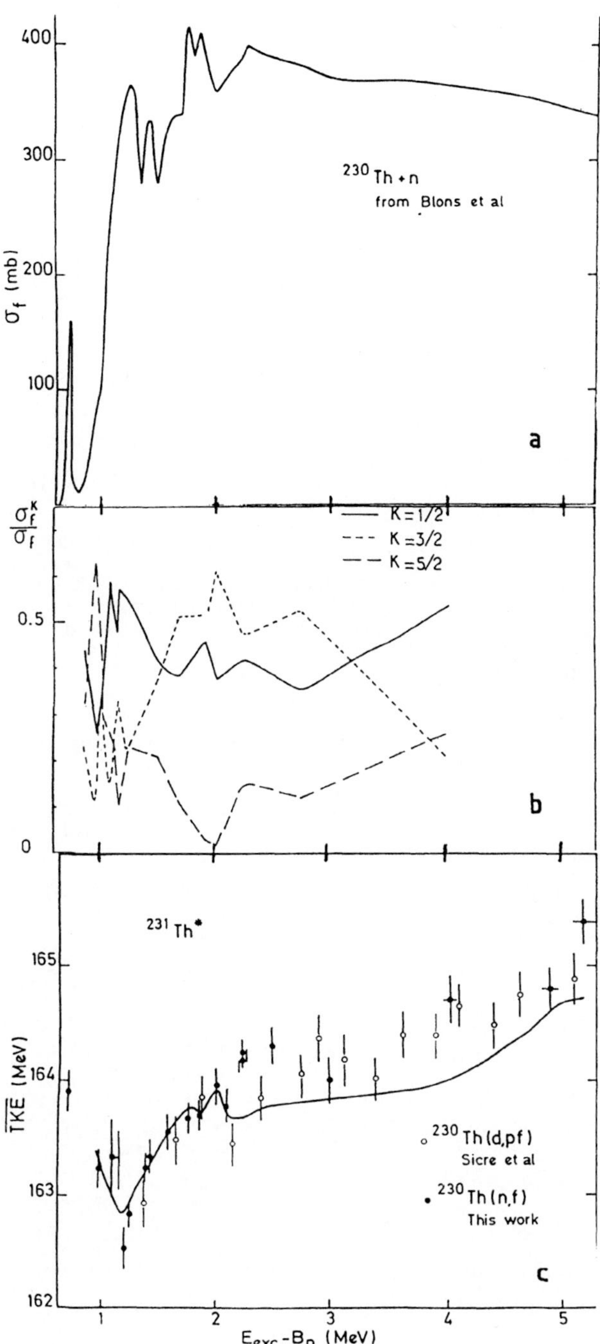

Fig. 3. Fission de $^{231}Th*$
a) section efficace de fission $^{230}Th + n$ (d'après [7])
b) contribution relative des différentes sections efficaces de fission partielles
c) valeurs expérimentales de l'énergie $\overline{TKE}$. La courbe en trait plein correspond à un calcul effectué avec les équations du texte.

où la valeur 5,92 MeV/neutron représente l'énergie moyenne nécessaire au fragment pour émettre un neutron. Cette valeur conduit à une énergie moyenne de liaison d'un neutron d'environ 4 MeV dans les fragments, ce qui est cohérent avec l'hypothèse du Q de la réaction constant.

Au dessous de E_n = 2,4 MeV la comparaison des valeurs expérimentales de $\bar{\nu}_p$, de $\overline{TKE}$ et des sections efficaces partielles de fission σ_f^K semble indiquer une certaine corrélation, au moins dans la région 1,2 ⩽ E_n ⩽ 2,2 MeV (Voir figure 3).

Afin de préciser cette corrélation, nous avons cherché à reproduire simultanément les valeurs expérimentales de $\bar{\nu}_p$ et de $\overline{TKE}$ à partir des sections σ_f^K, en faisant certaines hypothèses :

i) A chaque voie de sortie de fission caractérisé par le nombre quantique K correspond des propriétés particulières des fragments. Nous supposons que les propriétés sont les mêmes pour les voies de nombre K identique. Nous n'avons considéré que K = 1/2, 3/2 et 5/2.

ii) Pour des voies de fission de nombre K donné, nous avons supposé que le nombre $\bar{\nu}_{p(K)}$ et l'énergie $\overline{TKE}_{(K)}$ variaient linéairement. Cependant la pente de cette variation a été supposée identique pour toutes les voies et égale à celle observée pour 2,5 ⩽ E_n ⩽ 4,0 MeV.

Dans ces conditions, le nombre $\bar{\nu}_p$ et l'énergie $\overline{TKE}$ s'écrivent :

$$\bar{\nu}_p = \frac{\sum_K \bar{\nu}_K(E)\ .\ \sigma_f^K(E)}{\sum_K \sigma_f^K(E)} = \sum_K (a\ E_n + \bar{\nu}_{oK})\ \sigma_f^K(E)/\sigma_f(E)$$

avec a = 0,0915 neutron/MeV.

$$\overline{TKE} = \sum_K (bE_n + E_{oK})\ \sigma_f^K(E)/\sigma_f(E)$$

avec b = 0,40 MeV/MeV.

Ces hypothèses sont réalistes :

i) Pour ces noyaux, les voies de sortie de fission avec K > 5/2 ont une faible contribution au dessous de E_n = 2,5 MeV.

ii)Une variation non linéaire de $\overline{TKE}$ et $\bar{\nu}_p$ semble peu compatible avec le comportement à plus haute énergie.

iii) Enfin les pentes utilisées sont en accord raisonnable avec l'énergie de liaison d'un neutron dans les fragments ; elles peuvent différer légèrement d'une voie à l'autre car des variations sur les distributions en masses ont été observées [16], mais pas suffisamment pour influer de façon importante sur les ajustements.

Dans un premier temps, nous avons ajusté les paramètres $\bar{\nu}_{oK}$ pour obtenir un bon accord avec les données expérimentales de $\bar{\nu}_p$. Si l'ajustement à l'aide de notre jeu de σ_f^K est moyen, il devient extrêmement bon avec celui de Budtz-Jørgensen [12] qui introduit essentiellement une composante K = 5/2 à la résonance à 1,732 MeV (Voir Figure 2c). La brutale décroissance observée à 2,4 MeV (et anti-corrélée à $\overline{TKE}$) peut être reproduite en admettant une contribution de 30% pour K = 5/2 à cette énergie, ce qui est compatible avec les mesures de Auchampaugh et coll. [15]. Pour les trois valeurs de K considérées, les équations sont alors :

$$\bar{\nu}_p 1/2 = 0,0915\ E_n + 2$$
$$\bar{\nu}_p 3/2 = 0,0915\ E_n + 2,06$$
$$\bar{\nu}_p 5/2 = 0,0915\ E_n + 1,7$$

Dans un second temps, nous avons essayé d'ajuster les valeurs expérimentales de $\overline{TKE}$ en supposant une valeur unique de Q pour les trois états K considérés. L'ajustement obtenu avec les valeurs :

$$\bar{E}_K 1/2 = 0{,}4\ E_n + 163{,}23$$
$$\bar{E}_K 3/2 = 0{,}4\ E_n + 168{,}92$$
$$\bar{E}_K 5/2 = 0{,}4\ E_n + 164{,}3$$

est assez mauvais (figure 2b). D'autres tentatives faites en supposant une valeur de Q différente pour chaque bande ce qui introduit des variations du Q total dans la limite des variations relatives de sections σ_f^K n'ont pas donné de meilleurs résultats.

Ces difficultés appellent les remarques suivantes :

- au-dessous de 1,3 MeV, les calculs sont plutôt compatibles avec des valeurs basses de $\overline{TKE}$ comme celles publiées par Dyachenko [4].

- il est impossible de reproduire correctement la structure fine entre E_n = 1,6 et 1,7 MeV.

- entre 1,8 et 2,4 MeV, l'ajustement est assez correct mais au-delà il faut augmenter la valeur de Q d'environ 400 keV pour obtenir une bonne reproduction. En d'autres termes, Q ne peut pas être conservé.

Cas de ^{230}Th : Un travail similaire a été effectué pour ce noyau sur l'énergie $\overline{TKE}$, avec les mêmes hypothèses, en utilisant une pente de 0,35 MeV/MeV (Figure 3). Les équations utilisées sont :

$$\overline{TKE}\ (1/2) = 0{,}35\ E_n + 161{,}45$$
$$\overline{TKE}\ (3/2) = 0{,}35\ E_n + 164{,}40$$
$$\overline{TKE}\ (5/2) = 0{,}35\ E_n + 162{,}85$$

L'ajustement est correct jusqu'à 2,1 MeV, à l'exception de la valeur à 0,732 MeV, compte tenu des incertitudes sur la détermination des σ_f^K. Au-delà, il faut aussi augmenter la valeur du Q de la réaction d'environ 400 keV pour retrouver les valeurs expérimentales.

Conclusion

Pour les deux isotopes du thorium étudiés, l'évolution de l'énergie $\overline{TKE}$ avec l'énergie d'excitation du noyau dans la région du seuil de fission présente des structures importantes. De même des variations sont obbservées sur le nombre $\bar{\nu}_p$ pour ^{232}Th et on s'attend à des variations encore plus prononcées pour ^{230}Th.

Ces structures ne sont pas corrélées directement aux structures de la section efficace de fission mais semblent l'être plutôt aux différentes voies partielles de fission σ_f^K. Ainsi nous avons pu reproduire dans leur ensemble les valeurs expérimentales de $\overline{TKE}$ et $\bar{\nu}_p$ au dessous de E_n = 2,5 MeV avec des hypothèses simples, malgré l'incertitude sur notre connaissance des σ_f^K qui peuvent expliquer certaines difficultés rencontrées avec $\overline{TKE}$ de ^{232}Th. Au-delà de cette énergie, l'énergie $\overline{TKE}$ et le nombre $\bar{\nu}_p$ évoluent de façon monotone probablement à cause du très grand nombre de voies de sortie mis en jeu. Cette évolution est plus ou moins linéaire, de pente $d\overline{TKE}/dE_n \sim 0{,}35$ MeV/MeV et $d\bar{\nu}_p/dE_n \sim 0{,}1$ neutron/MeV. Dans cette même région apparaît une composante de fission symétrique, tandis que la valeur du Q de la réaction est en moyenne 400 keV supérieure à celle au dessous de E_n = 2,5 MeV.

Ceci est confirmé par un calcul de Q à partir des distributions en masses expérimentales.

La corrélation observée suggère que le passage du point selle au point de scission est extrêmement rapide pour ces noyaux, de sorte qu'un couplage fort entre les états collectifs et les états intrinsèques ne peut avoir lieu. Cependant cette propriété ne peut expliquer la répartition de l'énergie disponible entre $\overline{TKE}$ et $\bar{\nu}_p$ qui dépendent de la déformation relative des deux fragments [17].

Références

1. J. Blons, C. Mazur, D. Paya, M. Ribrag et H. Weigmann, Phys. Rev. Lett. 41 (1978) 1282.

2. J. Trochon, H. Abou Yehia, F. Brisard, Y. Pranal, Nucl. Phys. A318 (1979) 63.

3. J. Blons, C. Mazur, D. Paya, Phys. Rev. Lett. 35 (1975) 1749.

4. N. Dyachenko, B.O. Kuzminov, V.F. Mitrofanof et A.I. Sergachef, Sov. J. Nucl. Phys. 26 (1977) 365.

5. E. Koneckny, H.J. Specht et J. Weber, "Physics and Chemistry of Fission", C.R. Conf. Rochester (AIEA Vienne) II (1973) 3.

6. A. Sicre, G. Auchampaugh, H.C. Britt, A. Gavron, Y. Patin, J. Van der Plicht et J.B. Wilhelmy, à paraître.

7. J. Blons, C. Mazur, D. Paya, Communication privée.

8. J. Fréhaut, G. Mosinski, R. Bois et M. Soleilhac, Rapport CEA-R-4626 (1974).
   J. Fréhaut, R. Bois, Cette Conférence.

9. H. Abou Yehia, J. Jary, J. Trochon, J.W. Boldeman, A.R. de L. Musgrove, "Nuclear Cross Sections for Technology", C.R. Conf. Knoxville (1980) 469.

10. Ch. Lagrange, Communication privée.

11. P. Möller et J.R. Nix, "Physics and Chemistry of Fission" C.R. Conf. Rochester (AIEA Vienne) I (1973) 103.

12. C. Budtz-Jørgensen et J. Meadow, Proceedings of Symposium Nuclear Fission and Related Collective Phenomena and Properties of Heavy Nuclei, Bad Honnef, Oct. 1981
    Lecture Notes in Physics, 158, 111 (82) Springer Verlag Ed. Prof. Dr. P. David

13. T. Benfoughal, T.P. Doan, B. Leroux, A. Sicre, B. Bruneau, H. Abou Yehia, J. Trochon, Note CEA-N-2134 (1979).

14. G. Simon, J. Trochon, Note CEA-N-2284 (1981).

15. G.F. Auchampaugh, S. Plattard, N.W. Hill, G. de Saussure, R.B. Perez et J.A. Harvey, Phys. Rev. Letters (1981) Vol. 46 n° 10, p. 633-636.

16. S.K. Lisin, L.N. Morozov, V.A. Pchelin, L.V. Chistyakov, V.A. Shigin et V.M. Shubko, Sov. J. Nucl. Phys. 24 (1976) 570.

17. B.D. Wilkins, E.P. Steinberg et R.R. Chasman, Phys. Rev. C14 (1976) 1832.

ENERGY AND MASS DISTRIBUTIONS FOR $^{241}Pu(n_{th},f)$, $^{242}Pu(s.f.)$ AND $^{244}Pu(s.f.)$-FRAGMENTS

E. Allaert*, C. Wagemans**, G. Wegener-Penning

SCK-CEN, B-2400 Mol, Belgium

and

Nuclear Physics Lab., B-9000 Gent, Belgium

A.J. Deruytter, R. Barthélémy

CEC - JRC, Geel Establishment

Central Bureau for Nuclear Measurements, B-2440 Geel, Belgium

Abstract

The energy and mass distributions and their correlations have been studied for the spontaneous fission of ^{240}Pu, ^{242}Pu and ^{244}Pu and for the thermal neutron induced fission of ^{239}Pu and ^{241}Pu. A comparison of the $^{242}Pu(s.f.)$ and the $^{241}Pu(n_{th},f)$ results shows a narrower mass distribution, a much higher peak yield and a more pronounced fine structure for the spontaneous fission than for the neutron induced fission. Also the average masses differ by 2 mass units. The average total kinetic energy is higher for $^{242}Pu(s.f.)$ than for $^{241}Pu(n_{th},f)$, and also the energy-mass correlations behave differently in both cases. Preliminary results seem to indicate that the fissioning system ^{240}Pu behaves in a very similar way as ^{242}Pu. Also the $^{244}Pu(s.f.)$ data are very similar to these for $^{242}Pu(s.f.)$. All these results are discussed and interpreted in the frame of the scission point model of Wilkins et al.

Introduction

The characteristics of the energy and mass distributions of the $^{240}Pu(s.f.)$-fragments remain a controversial problem in the fission physics field. Indeed, since 1958 several comparative measurements of the $^{239}Pu(n_{th},f)$ and the $^{240}Pu(s.f.)$ reactions have been performed yielding quite discrepant results [1-7] for both the energy and the mass characteristics. The most striking fact was that three of the experimental groups reported a higher total fission fragment kinetic energy for $^{240}Pu(s.f.)$ than for $^{239}Pu(n_{th},f)$ despite the 6.5 MeV more excitation energy available in the last case. The opposite behaviour was observed by four experimental groups. Also for the $^{240}Pu(s.f.)$ fragments mass distribution results strongly different from that of $^{239}Pu(n_{th},f)$ were reported by some of the authors mentioned above.

A very similar comparison can be made when considering the $^{242}Pu(s.f.)$ and the $^{241}Pu(n_{th},f)$ reactions. Here only one experiment has been reported up to now [8] yielding a higher total kinetic energy for the $^{242}Pu(s.f.)$ than for the $^{241}Pu(n_{th},f)$-fragments. Also strongly different mass distributions were obtained for both reactions.

In an effort to clarify this situation, a series of experiments was started at the Central Bureau for Nuclear Measurements, Geel (Belgium) in which the $^{239}Pu(n_{th},f)$, $^{240}Pu(s.f.)$, $^{241}Pu(n_{th},f)$, $^{242}Pu(s.f.)$ and $^{244}Pu(s.f.)$ reactions are studied. Final results obtained for ^{241}Pu, ^{242}Pu and ^{244}Pu are reported in the present paper, together with very recent partial results for ^{239}Pu and ^{240}Pu. They provide a consistent data base for the study of the variation of the fission fragments' energy and mass characteristics for fissioning systems with the same Z but with different A-values.

Experimental procedure

The fission measurements were performed at an 8 m flight path of the linear accelerator GELINA of the Central Bureau for Nuclear Measurements at Geel. For the neutron induced fission measurements the well-known time-of-flight technique was used to select thermal neutrons out of the broad spectrum of moderated neutrons produced by GELINA. The spontaneous fission measurements on the other hand were carried out when GELINA was stopped. The experimental set-up for determining mass and energy of the fission fragments was based on the double - energy method, using two surface barrier detectors which were calibrated by means of the Schmitt-Neiler calibration procedure.

The targets used were made by the CBNM Sample Preparation Group. For the comparative study of the $^{241}Pu(n_{th},f)$ and $^{242}Pu(s.f.)$ reactions a combined ^{241}Pu - ^{242}Pu target was used. It consisted of nearly equal amounts of ^{241}Pu(26 $\mu g/cm^2$) and ^{242}Pu (20 $\mu g/cm^2$) sandwiched between two 30 $\mu g/cm^2$ thick polyimide backings. By rotating such a target over π radians when switching over from a spontaneous to a neutron induced fission measurement, one greatly eliminates the systematic errors in the comparison of both measurements. In the ^{244}Pu case, 30 $\mu g/cm^2$ of plutonium was also sandwiched between two 30 $\mu g/cm^2$ polyimide foils, which permitted us to conserve the energy calibration obtained via the $^{241}Pu(n_{th},f)$ measurement [9]. The ^{239}Pu-^{240}Pu target was a 58 $\mu g/cm^2$ thick mixed target, containing about 25% of ^{239}Pu and 75% ^{240}Pu. It was prepared by evaporation of plutonium-fluoride, while all other layers were obtained by electrospraying of plutoniumacetate. The isotopic enrichment of the target materials were 91.0% for ^{241}Pu, 99.8% for ^{242}Pu and 87.6% for ^{244}Pu.

The storage and the analysis of the data were done with a Hewlett Packard 1000E computer. Pre-neutron emission fission fragment mass-and energy results for the $^{241}Pu(n_{th},f)$, $^{242}Pu(s.f.)$and $^{244}Pu(s.f.)$ experiments were acquired using the $^{241}Pu(n_{th},f)$ neutron emission data of Caïtucoli et al.[10] normalized to the appropriate $\bar{\nu}$-value. Similarly, for the $^{239}Pu(n_{th},f)$ and $^{240}Pu(s.f.)$ measurements, the $^{239}Pu(n_{th},f)$ neutron emission data of Milton and Fraser [11] were used.

Since the counting rate in the spontaneous fission experiments is low, the long term stability of the measuring chains is very important. This stability was mainly realised by cooling the detectors at a very constant temperature. Moreover, the peak position(s) of the natural α-particles emitted by the plutonium isotopes enables a permanent and intrinsic control of the stability.

Results

Table 1 summarizes the main characteristics of the $^{241}Pu(n_{th},f)$, the $^{242}Pu(s.f.)$ and the $^{244}(s.f.)$ fragment energy and mass distributions obtained in the present work. They will be discussed in detail below. All results quoted are pre-neutron emission data.

* IWONL

** NFWO

K. H. Böckhoff (ed.), Nuclear Data for Science and Technology, 737–739.

Table 1. Main characteristics of the $^{241}Pu(n_{th},f)$, ^{242}Pu(s.f.) and ^{244}Pu(s.f.) fragment mass and energy distributions; the errors are only statistical

	$^{241}Pu(n_{th},f)$	^{242}Pu(s.f.)	^{244}Pu(s.f.)
$\overline{E}_K$ (MeV)	176.56±0.07	180.03±0.09	182±1 [a)]
$\overline{E}_K^*$ (MeV)	178.97±0.07	181.78±0.09	184±1 [a)]
σ_{E_K} (MeV)	12.0	11.9	12.1
$\overline{E}_L^*$ (MeV)	103.10±0.05	103.21±0.06	103.7±0.7 [a)]
$\overline{E}_H^*$ (MeV)	75.87±0.05	78.57±0.06	80.3±0.7 [a)]
$\overline{m}_L^*$ (amu)	102.45±0.04	104.50±0.05	106.31±0.08
σ_{m^*L}(amu)	6.72	5.82	6.45
$\overline{m}_H^*$ (amu)	139.55±0.04	137.50±0.05	137.69±0.08
σ_{m^*H}(amu)	6.72	5.82	6.45
$\overline{m}_L$ (amu)	100.64±0.04	103.12±0.05	104.76±0.08
σ_{mL} (amu)	6.47	5.79	6.25
$\overline{m}_H$ (amu)	138.45±0.04	136.77±0.05	136.92±0.08
σ_{mH} (amu)	6.41	5.78	6.25
$\Delta\overline{E}_K^*$ (MeV)	25.0±3.2	18.3±4.7	19.3±6.1
$\overline{\nu}_t$	2.927	2.10	2.30
N	$3x10^4$	$1.8x10^4$	$6x10^3$

[a)] Statistical error+errors due to energy loss corrections.

a) Mass distributions and mass-energy correlations

The mass yield distributions for $^{241}Pu(n_{th},f)$ and ^{242}Pu(s.f.) fragments are compared in fig. 1. In fig.2 similar curves are shown for ^{242}Pu(s.f.) and ^{244}Pu(s.f.) and compared with our preliminary ^{240}Pu(s.f.) data. The $^{241}Pu(n_{th},f)$ fragments' mass characteristics agree (within the accuracy of the Schmitt-Neiter calibration method) with those reported in the literature [9,10,12]. Fig. 1 reveals a striking difference between the fragments' mass distribution of the $^{241}Pu(n_{th},f)$ and ^{242}Pu(s.f.) reactions: the peak yield is much higher in the spontaneous fission case, resulting in a narrower distribution. Moreover, there is a difference of about 2 amu between the average light (and heavy) fragment mass, as also observed by Dyachenko et al.[8] and Trochon et al. [7].

Fig. 2 illustrates the similar behaviour of ^{240}Pu, ^{242}Pu and ^{244}Pu for spontaneous fission decay. The heavy fragment mass distributions are practically superposed, apart from fine structures at masses 132 - 134 (^{244}Pu) and 142 - 144 (^{240}Pu). The $\overline{m}_L^*$ and $\overline{m}_H^*$ values reported in table 1 confirm that both additional neutrons in the ^{244}Pu nucleus compared to ^{242}Pu are going to the light fragment.

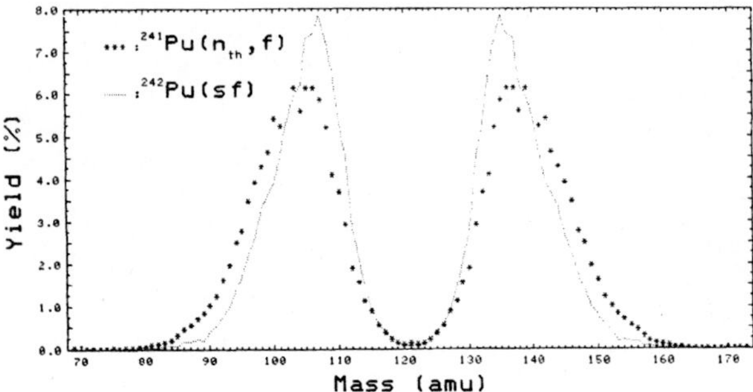

Fig. 1. Comparison of the (pre-neutron emission)mass yield distribution for the thermal-neutron induced fission of ^{241}Pu(stars) and the spontaneous fission of ^{242}Pu (full line).

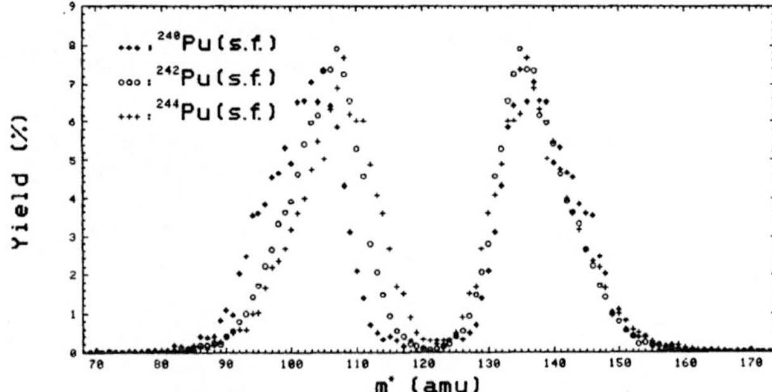

Fig. 2. Comparison of the (pre-neutron emission) mass yield distributions for the spontaneous fission of ^{244}Pu, ^{242}Pu and ^{240}Pu.

b) Kinetic energies and energy-mass correlations

Average kinetic energy values are listed in table 1. The symbol $\Delta\overline{E}_K^*$ represents the difference between the maximum total kinetic energy value and the value for symmetric mass divisions (cf. fig. 5). For $^{241}Pu(n_{th},f)$ we obtain a value of 25.0±3.2 MeV compared to 22.6±1.3 MeV as reported by Caïtucoli et al.[10]. For spontaneously fissioning isotopes, no $\Delta\overline{E}_K^*$ values have been reported up to now.

The total kinetic energy distributions of the $^{241}Pu(n_{th},f)$ and ^{242}Pu(s.f.) fragments are compared in fig. 3. Again differences between both distributions are obvious. First of all the $^{241}Pu(n_{th},f)$ kinetic energy distribution seems to be shifted towards lower energies, resulting in a lower $\overline{E}_K^*$ value (cf. table 1),despite the 6.3 MeV excitation energy of the fissioning nucleus. Moreover this distribution is fairly symmetric as it supports a gaussian fit (point line), which is clearly not the case for the ^{242}Pu(s.f.) data. Preliminary results indicate that the same situation occurs in the $^{239}Pu(n_{th},f)$ - ^{240}Pu(s.f.) case. Also the ^{244}Pu(s.f.) fragments'total kinetic energy distribution is similar to that of ^{242}Pu(s.f.).

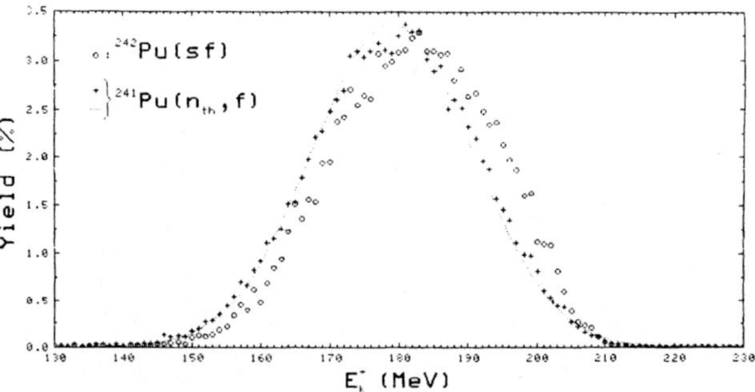

Fig. 3 Comparison of the total kinetic energy distribution for the spontaneous fission of ^{242}Pu (circles) and the thermal-neutron induced fission of ^{241}Pu (crosses). The point line represents a gaussian fit through the thermal data.

In an attempt to explain the different behaviour of the spontaneous and the thermal neutron induced fission with the help of the doubly magic shell N=82, Z=50 the total kinetic energy distributions were split up into two parts [6], according to the mass of the heavy fragment : 130 ≤ m_H^* (amu) ≤ 135(containing the N=82, Z=50 region) and 121 ≤ m_H^* (amu) < 130 plus 135 < m_H^* (amu) ≤ 174. The result is shown in fig. 4. The average energies in both mass-intervals are very different : 185.3 MeV for $^{241}Pu(n_{th},f)$, 187.0 MeV for ^{242}Pu(s.f.) and 189.1 MeV for ^{244}Pu(s.f.) in the interval 130 - 135, compared to respectively 177.0 MeV, 179.2 MeV and 181.9 MeV in the other interval .

Fig. 5 shows the $\overline{E}_K^*(m_H^*)$ curves for $^{241}Pu(n_{th},f)$ and ^{242}Pu(s.f.). They both show the typical maximum at

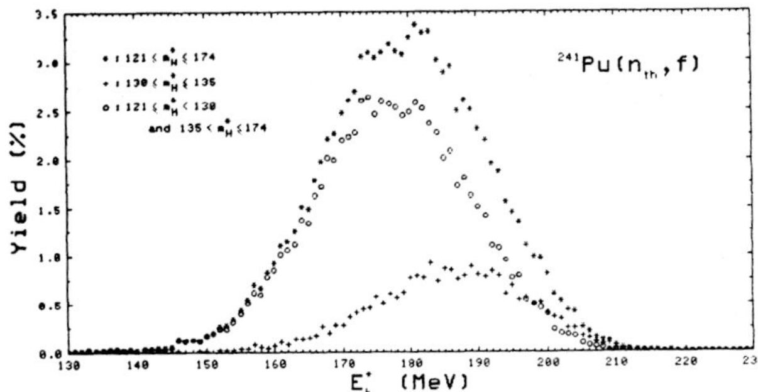

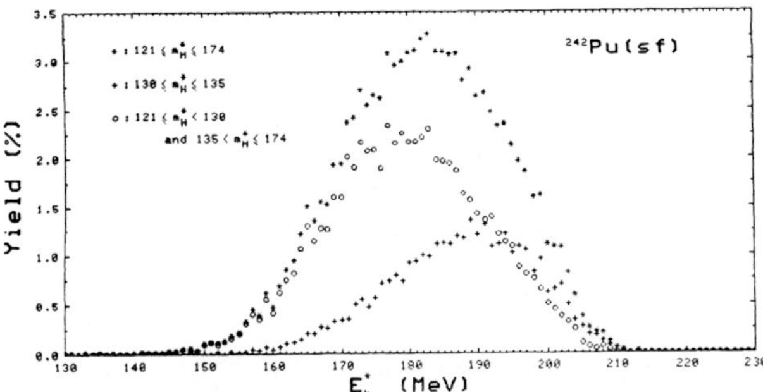

Fig. 4. Total kinetic energy distributions for several mass-intervals for the thermal neutron-induced fission of ^{241}Pu (upper part) and the spontaneous fission of ^{242}Pu (lower part).

mass 132 - 134. Moreover, they intersect at $m_H^* \cong 145$ amu. This is not the case for the ^{242}Pu(s.f.) and the ^{244}Pu(s.f.) curves, which are about parallel.

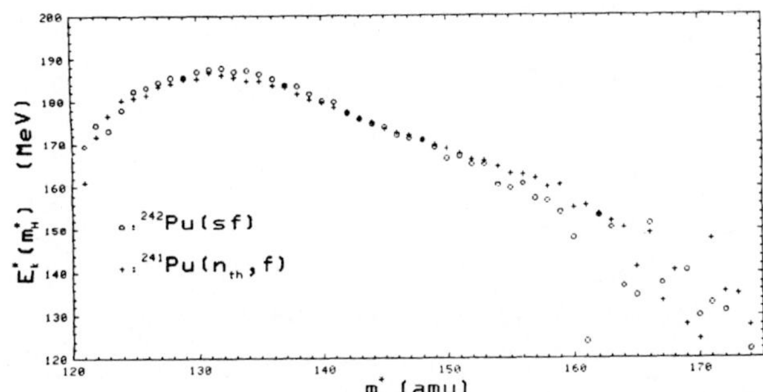

Fig. 5. Average total fission fragment kinetic energy for ^{242}Pu(s.f.) and ^{241}Pu(n_{th},f) as a function of the heavy fragment mass.

Discussion

These results can be interpreted in terms of the nuclear shell effects suggested in the scission point model of Wilkins et al. [13]. The asymmetry of the total kinetic energy distributions for ^{242}Pu(s.f.) and, partly, also its higher $\overline{E}_K^*$ value can be accounted for by the shell effects at N=82. This shell will have its maximum influence for heavy fragments with $m_H^* \cong 132$ - 134 amu. As can be seen in fig. 4, the asymmetry in the total kinetic energy distribution for ^{242}Pu(s.f.) is mainly due to this mass region. On the other hand the corresponding partial kinetic energy distribution for ^{241}Pu(n_{th},f) is rather symmetric and can be approximated by a gaussian distribution. The observed asymmetry for ^{242}Pu(s.f.) can be ascribed to the preferential formation of a shell-stabilized configuration with small deformation ($\beta_1 + \beta_2 \approx 0.9$) for mass splits with the heavy mass in the region of the closed neutron shell N=82, compared to a second stable - but very deformed ($\beta_1 + \beta_2 \cong 1.5$) - configuration in this same region [15], favoured by the liquid drop behaviour. The first configuration becomes less important for the ^{241}Pu(n_{th},f) reaction, as the 6.3 MeV excitation energy decreases shell corrections and enhances the liquid drop behaviour. This is accompanied with a much smaller heavy mass yield in the region 130 amu $\leq m_H^* \leq 135$. The same shell-stabilized configuration with $\beta_1 + \beta_2 \cong 0.9$ is also responsible for the higher average kinetic energy values observed for ^{242}Pu(s.f.) in the $E_K^*(m_H^*)$ curves for masses around 132 - 134(fig. 5). Fig. 2 demonstrates first of all the similarity and stability of the heavy fragment peaks for the mass yields of the three spontaneously fissioning Pu-isotopes studied. Such a stability was already observed previously for other fissioning systems. It can be explained by the combined influence of the N=82 and the deformed N=88 neutron shells. A second observation is the difference in fine structure, although one should be cautious with the preliminary ^{240}Pu data.

Another striking result are the average values of the total kinetic energy. For ^{244}Pu(s.f.) we obtain $\overline{E}_K^*$ = 184 MeV, dropping first to 181.8 MeV for ^{242}Pu(s.f.) and further on to 177.5 MeV for ^{240}Pu(s.f.). So these isotopes show an increase in $\overline{E}_K^*$ with increasing mass, which seems to be in contradiction with the general expectation of a linear $Z^2/A^{1/3}$ dependence [14]. However, if we consider the detailed calculations of Wilkins et al.[13] for the most probable deformation of the light fragment (A-142) for various A values, we observe a slight decrease of this deformation as A increases from 235 to 244. This should be reflected in an increase for $\overline{E}_K^*$, which is in agreement with our results.

When the excitation energy of the fissioning system increases, the tendency of a given mass split to change its total deformation will only be small if this deformation approximates the liquid-drop value $\beta_1 + \beta_2 \cong 1.25$. Therefore, one could consider mass splits with $\beta_1 + \beta_2 \cong 1.25$ to decide on whether additional excitation energy of the parent nucleus appears as kinetic energy or as excitation energy of the fragments. The data of Wilkins et al.[13] reveal that for ^{242}Pu this condition is only met for $144 \leq m_H^*$ (amu) ≤ 150. Fig. 5 demonstrates that for such mass splits the addition of 6.3 MeV excitation energy in the ^{241}Pu(n_{th},f) experiment yields no higher kinetic energies. This points out a strong damping of the fission mode below the barrier for the fissioning system ^{242}Pu, quite opposite to the thesis of Lachkar et al. [15], who suggested a superfluid motion based on results for the fissioning system ^{240}Pu. However, their conclusion was strongly influenced by the $\overline{E}_K^*$ value for ^{240}Pu(s.f.), which is rather controversial.

References

1) T. Mostovaya, Int.Conf.on Peaceful Uses of Atomic Energy, Geneva (1958) p.2031.
2) J. Toraskar & E. Melkonian, Phys.Rev.C4(1971) 1391.
3) A. Deruytter & G. Wegener-Penning, Physics and Chemistry of Fission, Vol.2 (IAEA Vienna,1974) p.51.
4) B. Basova, INDC Report 1977.
5) C. Wagemans, G. Wegener-Penning, H. Weigmann, R. Barthélémy, Physics and Chemistry of Fission, Vol.2 (IAEA Vienna, 1980) p.143.
6) Thierens, A. De Clercq, E. Jacobs, D. De Frenne, P. D'hondt, P. De Gelder, A. Deruytter, Phys. Rev. C23 (1981) 2104.
7) J. Trochon, J. Boldeman, F. Brisard, Y. Franal, this conference.
8) N. Dyachenko, V. Kabenin, N. Kolosov, B. Kuzminov, A. Sergachev, Sov. J. Nucl. Phys. 17 (1973) 362.
9) J. Neiler, F. Walter, H. Schmitt, Phys.Rev.149 (1966) 894.
10) F. Caïtucoli, C. Wagemans, P. Perrin, E. Allaert, P. D'hondt, M. Asghar, Nucl.Phys. A369 (1981) 15.
11) J. Milton & J. Fraser, Ann.Rev.Nucl.Sci.16(1966) 894.
12) V. Vorobeva, N. Dyachenko, N. Kolosov, B. Kuzminov, A. Sergachev, Sov. J. Nucl. Phys. 19 (1974) 621.
13) B. Wilkins, E. Steinberg, R. Chasman, Phys. Rev. 14 (1976) 1832.
14) V. Viola & T. Sikkeland, Phys.Rev. 130 (1963) 2044.
15) J. Lachkar, Y. Patin, J. Sigaud, J. de Phys. L36 (1975) 79.

FISSION FRAGMENT ANGULAR DISTRIBUTIONS AND TOTAL KINETIC ENERGIES FOR $^{235}U(n,f)$ FROM 0.18 TO 8.83 MeV

J.W. Meadows

Applied Physics Division
Argonne National Laboratory
Argonne, Illinois 60439, U.S.A.

C. Budtz-Jørgensen

Commission of the European Communities
Joint Research Centre
Central Bureau for Nuclear Measurements
Geel, Belgium

A gridded ion chamber was used to measure the fission fragment angular distribution and total kinetic energy for the $^{235}U(n,f)$ reaction from 0.18 to 8.83 MeV neutron energy. The anisotropies are generally in good agreement with earlier measurements. The average total kinetic energy is ~ 0.2 MeV greater than the thermal value at neutron energies < 2 MeV and shows a sudden decrease of ~ 0.8 MeV between 4 and 5 MeV neutron energy, well below the (n,n'f) threshold.

[nuclear reaction $^{235}U(n,f)$, $W(\theta)$, $\overline{TKE}$ 0.18 MeV < E_n < 8.83 MeV]

Introduction

It has recently been demonstrated [1] that gridded ion chambers can be used for measurements of angular distributions as well as total kinetic energies of fission fragments. The large angular efficiency (~ 4π) and the radiation resistant performance of such detectors are important advantages when neutron induced fission is investigated.

A detector was assembled at the Argonne National Laboratory Fast Neutron Generator (FNG) and used to measure the fragment angular distribution and average kinetic energy for $^{235}U(n,f)$ from thermal energies to ~ 9 MeV.

The angular distributions are important for the application of the $^{235}U(n,f)$ standard cross section and have been investigated in previous measurements using several techniques. Leachman and Blumberg [2] and Simmons and Henkel [3] have covered this energy range, while several investigators [4-9] have reported results at lower energies with generally good agreement.

Less is known concerning the dependence of the average fragment kinetic energy on neutron energy although some measurements have been reported [10-15]. The only measurements at substantially higher energies were made by D'yachenko et al.[15] who made measurements up to 7 MeV with an isolated point at 15 MeV. They observed little change in the kinetic energy below 3.5 MeV but their results do indicate a substantial decline somewhere between 3.5 and 5 MeV.

Experimental Apparatus

A detailed description of the experimental set-up has been reported elsewhere [16] and only the essentials will be given below.
Neutrons with energies < 5 MeV were produced by the $^7Li(p,n)^7Be$ reaction, whereas neutrons with higher energies were produced by the $D(d,n)^3He$ reaction. The neutron energy resolution varied from ~ 0.02 MeV below 1 MeV to ~ 0.1 MeV above 5 MeV. The background produced by (d,n) reactions in the gas target structure was measured by bombarding the empty gas target at each energy.

The detector was a double-gridded ion chamber with a common cathode, mounted so that the chamber axis coincided with the FNG beam line. The ^{235}U deposits were mounted on both sides of the cathode. These were prepared by molecular deposition of highly enriched (99.86%) ^{235}U, were 2.5 cm in diameter and had thicknesses of 76 and 54 $\mu g/cm^2$ respectively. The chamber was constructed of stainless steel with polytetrafluoroethylene insulators. It was operated as a flow chamber, and a constant absolute pressure was maintained by means of a pressure regulator of the cartesian diver type. The chamber design and dimensions were chosen with the following requirements in mind : Fission fragments must stop within the volume defined by the cathode and grid. The grid must effectively shield the anode from ions in the active volume. The detector structure should be as light as practical to minimize neutron scattering. The experimental set-up is shown in fig. 1. Each fission produced a cathode and an anode signal (q_c,q_a) in one of the detectors. The pairs of cathode and anode pulses were digitized to 1024 channels each, tagged to indicate the detector of origin, and sent to an on-line computer system which stored them sequentially on a magnetic disk. All processing was done off-line.

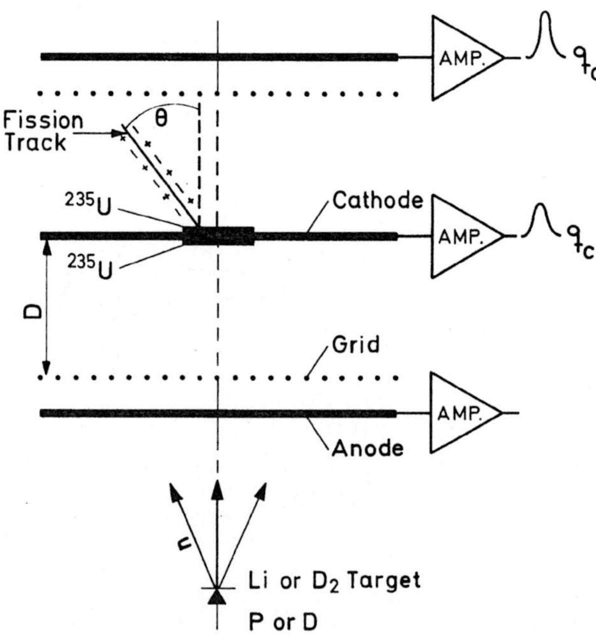

Fig. 1. A schematic diagram of the experimental arrangement.

The Angular Distribution Measurement

The method used for obtaining angular distribution information from the present detector is described in detail in refs.[1,16] and only some main characteristics will be given below. Consider a fast-parallel-plate ion chamber with a grid inserted between the anode and cathode at a distance D from the cathode. It is

K. H. Böckhoff (ed.), Nuclear Data for Science and Technology, 740–743.

assumed that electron capture is negligible and that the amplifier time constants are long compared to the electron transit time but very short compared to the positive ion transit time. The electrons formed along the track of a fission fragment will then induce a signal at the cathode given by

$$q_c = C \cdot E(1 - \bar{X}(E)/D \cdot \cos\theta)$$

where C is a constant proportional to the amplifier gain, E is the initial energy of the fragment, $\bar{X}(E)$ is the distance from the origin of the track to the center of gravity of the ionization charge and θ is the angle of the track with respect to the normal of the electrodes.

The anode signal is, due to the inserted grid, proportional to the number of ion-electron pairs so that

$$q_a = C \cdot E.$$

Hence the ratio $q_c/q_a = (1 - \bar{X}(E)/D \cdot \cos\theta)$ depends only on $\cos\theta$ and $\bar{X}(E)/D$, and will be distributed between $1 - \bar{X}(E)/D$ and 1. For each fission fragment energy the distribution of q_c/q_a was established and $\bar{X}(E)/D$ was readily found as the difference between the end and start point of the q_c/q_a distribution belonging to the fragment energy E. In order to obtain the full angular distribution integrated over all fragments the quantity :

$$(\cos\theta)_m \equiv (1 - q_c/q_a)/(\bar{X}(E)/D))$$

was formed. Fig. 2A displays the $(\cos\theta)_m$ distribution for fission fragments stemming from a self transferred ^{252}Cf source. The angular isotropy of the emitted fragments is reflected by the rectangular distribution. Fig. 2B shows the similar distribution for fission fragments produced in the 76 $\mu g/cm^2$ thick ^{235}U deposit by thermal neutrons. This distribution is also flat but shows some loss of events near $\cos\theta = 0$, which is to be expected for a sample of finite thickness.

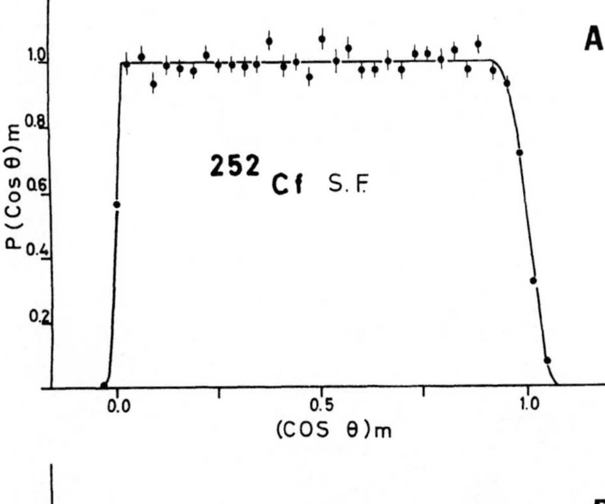

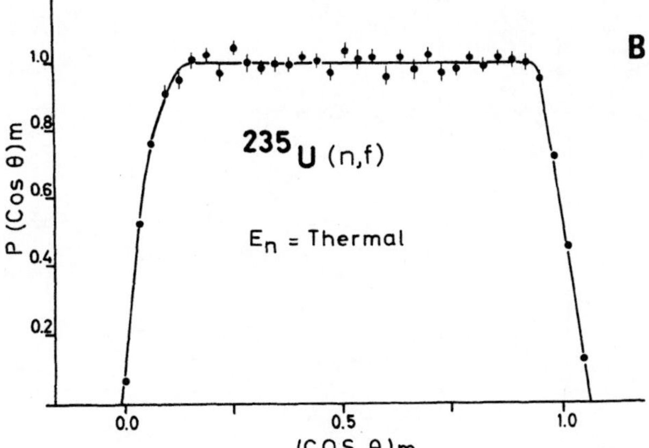

Fig. 2. The distribution of $(\cos\theta)$m obtained from
A) ^{252}Cf spontaneous fission fragments
B) thermal neutron induced fission of ^{235}U.

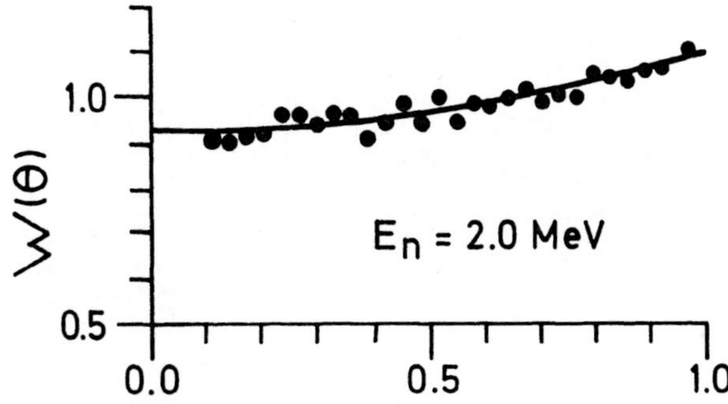

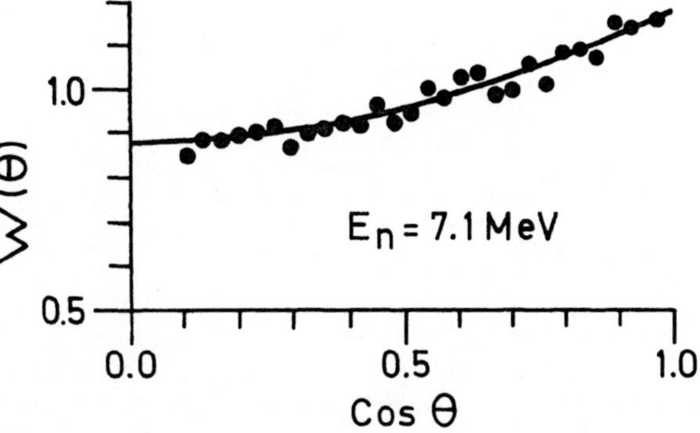

Fig. 3. Two angular distributions for fast neutron induced fission of ^{235}U obtained at 2.0 and 7.1 MeV

In order to correct for distortions induced by the uranium deposit and by the angular resolution, the final fast neutron angular distributions were obtained by dividing the $(\cos\theta)_m$ distributions by the thermal distribution.

For each incident neutron energy the distributions for the forward and backward detector were averaged. This effectively corrected for the center of mass motion in the investigated neutron energy range. All the angular distributions were fitted with a polynomial of the type

$$W(\theta) = \sum_{k=0}^{n} A_{2k} \cdot \cos^{2k}\theta$$

The best fit for about 2/3 of the measurements was obtained with n = 1. In most of the remaining cases higher order polynomials gave only slight improvement so all anisotropies were based on n = 1 fits. Fig. 3 shows two representative angular distributions obtained

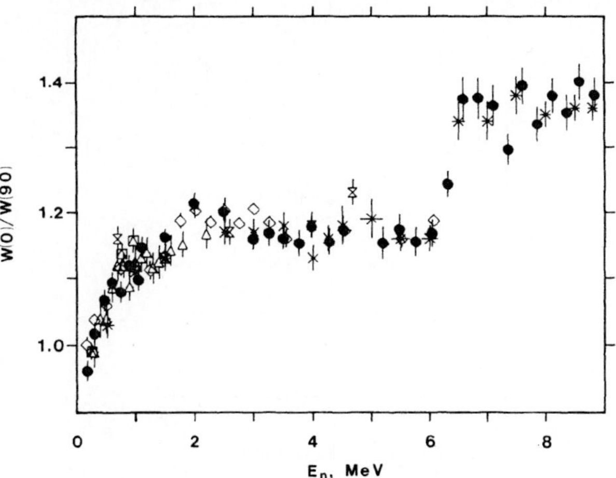

Fig. 4. W(0)/W(90) versus neutron energy for ^{235}U(n,f) • - the present work, ⌶- ref. 2, ✕- ref. 3, ◇- ref. 4, ⊠ - ref. 8, Δ - ref. 9.

at 2.0 and 7.1 MeV respectively. The anisotropies were determined from these fits with errors between 0.015 and 0.025. Corrections were made for lower energy neutrons from room scattering and from the secondary source reactions. The W(0)/W(90)-ratios are plotted in fig. 4 where they are compared with several other measurements [2,4,8,9]. The agreement is fairly good at all energies. The increase in the anisotropy beginning at ~ 6 MeV is very well correlated with the increase in fission cross section due to second chance fission. The change of the sign of the anisotropy below ~ 300 MeV as observed by Hsue et al.[8] and Ahmed et al.[9] is confirmed.

The Average Kinetic Energy Measurements

The kinetic energy measurements were based on the anode signal recorded simultaneously with the angular information. An absolute calibration procedure for the detector was not available, therefore the measurements were made relative to the average total kinetic energy for thermal neutron induced fission of ^{235}U.

Energy loss in the uranium deposits made the average anode signal dependent on the fragment angle and consequently on the angular distribution. This effect was eliminated by dividing the cos θ-fragment-energy distribution into 8 equal intervals in cos θ. Then the average kinetic energy $\bar{E}$, relative to the value for thermal neutron induced fission $\bar{E}(E_{th})$, is

$$\bar{E}(E_n) = \bar{E}(E_{th}) \sum_i^k \bar{P}_j(E_n) / \sum_i^k \bar{P}_j(E_{th})$$

where E_n is the incident neutron energy, i and k specify the range of cos θ intervals, and $\bar{P}_j$ is the average pulse height for the jth interval. The thermal fission value for $\bar{E}(E_{th})$ was taken as 170 MeV. Only intervals 5-8 (0-60deg.) were used to calculate $\bar{E}$ since the energy spectra in the first intervals were badly distorted due to energy loss in the deposits and to fragment scattering. Averaging the results for the forward and backward detectors effectively corrected for the center of mass motion. Corrections for fissions induced by lower energy neutrons from secondary source reactions and room scattering were made by an iteration procedure. These were generally small, less than 0.01 MeV at most energies.

The correction for neutron emission was made using the values for the average number of prompt neutrons $\bar{\nu}_p$ as given by ref.[17] It was assumed that all neutrons were emitted isotropically from the fully accelerated fragments.

The consistency of the energy measurements was tested by determining the effect of the incident neutron momentum on the average fragment energy and comparing it with the expected value. $\Delta\bar{E}_m$ was defined as the change in the average fragment energy produced by the incident neutron momentum, and was found experimentally as the difference between the average energies obtained for the forward and backward detector respectively. The deviations between the measured $\Delta\bar{E}_m$ values and the calculated $\Delta\bar{E}_m \propto E_n^{1/2}$ where within the statistical accuracy of ~ 0.15 MeV as demonstrated in fig. 5.

The results of the average kinetic energy measurements are compared with other measurements [10,14,15] in fig. 6 where

$$\Delta\bar{E}(E_n) = \bar{E}(E_n) - \bar{E}(E_{th}).$$

An average of the present data below 2 MeV give 0.22± 0.06 MeV. There is as slow decline to ~ 4 MeV then a precipitous decrease of ~ 0.8 MeV between 4 and 5 MeV. Above 5 MeV $\Delta\bar{E}$ is constant with an average value of -0.80±0.06 MeV. There is no significant effect in the region of the (n,n'f) threshold. Blyumkina et al. [10] have reported a step of ~ 0.7 MeV near 0.4 MeV neutron energy with a corresponding change in the number of prompt neutrons. Our data show no such step and measurements by Ajitanand and Boldeman [14] find $\Delta\bar{E} \approx 0$ from 0.2 to 0.9 MeV. D'yachenko et al.[12,13,15] have reported a number of measurements. They find $\Delta\bar{E}$ to be ~ + 0.2 MeV below 2 MeV neutron energy and ~ -0.4 MeV above 5 MeV.

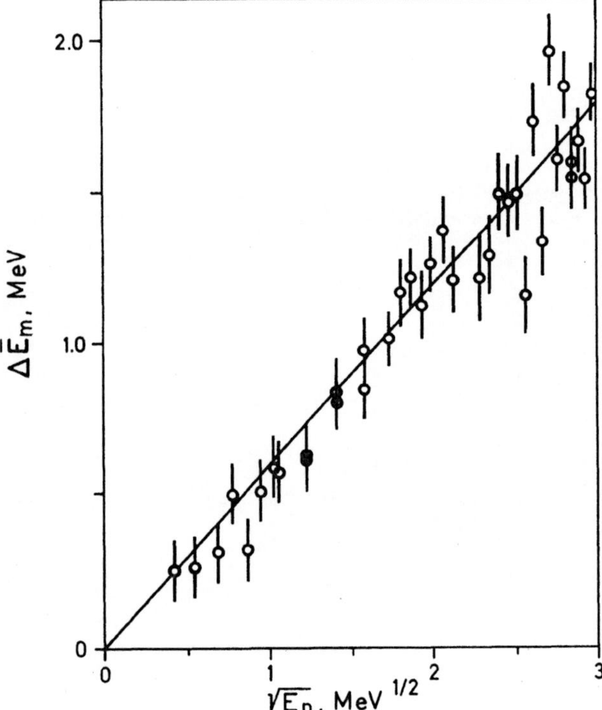

Fig. 5. The change in the average fragment energy produced by the incident neutron momentum. The angular range is 0 - 60 deg. The expected values are given by the full line.

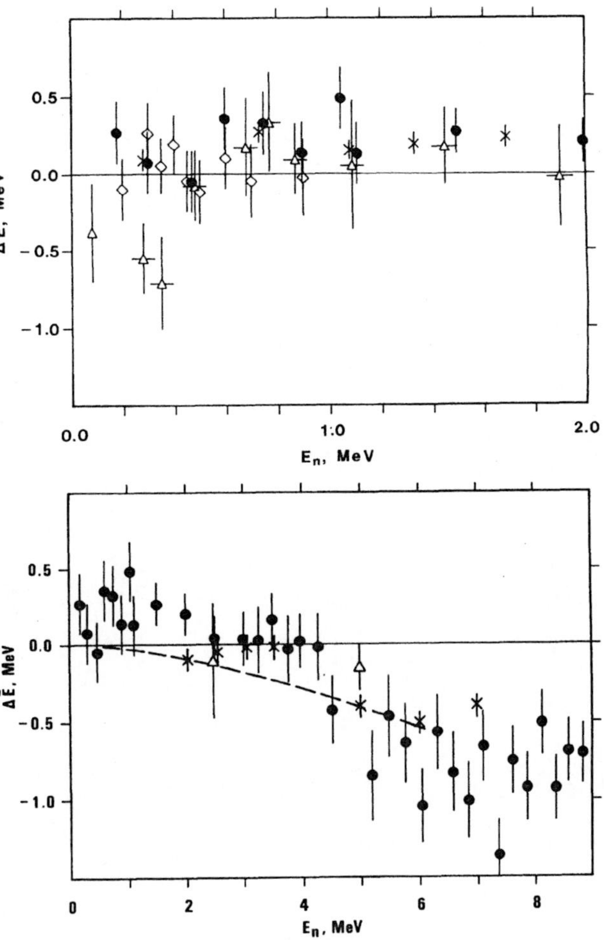

Fig. 6. The change in the average total kinetic energy relative to the thermal value.● - the present work, △ - ref. 10, ◇ - ref. 14, ✱ - ref. 15. The dashed line was calculated from the change in the mass distribution.

The most prominent feature of the present work is the sudden decrease of $\Delta\bar{E}$ near 4.5 MeV neutron energy. At this energy the number of fission channels is large and the opening of a few additional channels should have little effect on $\bar{E}$. Other quantities associated with ^{235}U neutron induced fission were inspected for anomalies near this energy. Krick and Evans [18] have observed that the delayed neutron yield decreases rapidly between 4 and 5 MeV although it shows no significant change below that energy. This indicates a change in the yield of the delayed neutron precursors and suggests that the decrease in $\Delta\bar{E}$ is associated with a change in the mass and charge distributions.

It is known that the kinetic energy is a function of fragment mass ratio so factors such as a change in the amount of symmetric fission and in the width of the mass distribution may affect $\Delta\bar{E}$. Thus the total kinetic energy for symmetric fission of ^{236}U is ~ 20 MeV lower [19] than for the most probable mass division. Glendenin et al.[20] measured the $^{235}U(n,f)$ mass distribution for incident neutron energies of 0.17 to 8.81 MeV and found that the peak-to-valley ratio decreased exponentially with neutron energy up to the (n,n'f) threshold then remained nearly constant. From 0.17 to 6 MeV the peak-to-valley ratio is

$$P/V \approx 501 \exp(-0.55\, E_n)$$

For asymetric fission the variance of the mass distribution increases from ~34 amu^2 at thermal energies to ~ 46 amu^2 at 6 MeV.[19].

An approximate calculation of $\Delta\bar{E}$ based on these factors was made assuming that the symmetric and asymmetric mass distributions were represented by Gaussians. The mass dependent total kinetic energy was taken from ref.[19]. The results are shown in fig. 5 as a dotted line and are in quantitative sense in surprisingly good agreement with the experimental data although the precipitous decline near 4.5 MeV is not reproduced.

An explanation for this dramatic decrease might be found within the frame of the static scission-point model of Wilkins et al.[21], where the total potential energy at scission is the sum of the liquid-drop energies of the nascent fragments with shell and pairing correction terms plus coulomb and nuclear interaction terms. A decrease in shell effects with increasing excitation energy permits a more elongated configuration at the scission point and thus a decrease in coulomb repulsion with a corresponding decrease in $\bar{E}$. Normally, this is a gradual process but for some mass divisions the deformation can change quite rapidly. For the 134/102 mass division the deepest potential minimum at low excitation corresponds to a nearly spherical heavy fragment. However there is a secondary minimum at ~ 30% larger deformation. As the shell corrections decrease with increasing excitation energy the depth of the first minimum decreases most rapidly and, at some point, the secondary minimum becomes the deepest one. When this occurs there is a rapid increase in its relative contribution to the fission yield and its larger deformation produces a decrease in the total kinetic energy.

References

1. H.-H. Knitter and C. Budtz-Jørgensen, Proc. Int. Conf. Nuclear Cross Sections for Technology, Knoxville, Tenn., October 22-26, 1979, NBS Special Publication 594, U.S. National Bureau of Standards (1980).
2. R.B. Leachman and L. Blumberg, Phys. Rev., 137B, 814 (1965).
3. J.E. Simmons and R.L. Henkel, Phys. Rev., 120, 198 (1960).
4. V.G. Nesterov, G.N. Smirenkin and D.L.Shpak, Sov. J. Nucl. Phys., 4, 173 (1967).
5. D.M. Nadkarni, S.S. Kapoor and P.N. Rama Rao, Proceedings of a Symposium on Nuclear Physics and Solid State Physics, vol.II, p. 133, Bombay(1968).
6. J. Caruana, J.N. Mathur, J.W. Boldeman and R.L. Walsh, Proceedings of a Symposium on Nuclear Physics and Solid State Physics, Trombay (1974).
7. G.N. Smirenkin, D.L. Shpak, Yu. B. Ostapenko and B.I. Furson, JETP Lett., 11, 133 (1970).
8. S.T. Hsue, G.F. Knoll and J.W. Meadows, Nucl. Sci. Eng., 66, 24(1978).
9. S. Ahmad, M.M. Islam, A.H. Khan, M. Khaliquazzaman, M. Husain and M.A. Rahman, Nucl. Sci. Eng. 71, 208, (1979).
10. Yu. A. Blyumkina, I.I. Bondarenko, V.F. Kuznetsov, V.G. Nesterov, V.N. Okolovitch, G.N. Smirenkin and L.N. Vsachev, Nucl. Phys., 52, 648 (1964).
11. J.W. Boldemann, W.K. Bertram and R.L. Walsh, Nucl. Phys., A265, 337 (1976).
12. P.P. D'yachenko, B.D. Kuz'menov and M.Z. Tarasko, Sov. J. Nucl. Phys., 8, 165 (1968).
13. P.P. D'yachenko, B.D. Kuz'menov and A. Lajatai, Phys. Letters, 31B, 122 (1970).
14. N.N. Ajitanand, J.W. Boldeman, Nucl. Phys., A144, 1 (1970).
15. P.P. D'yachenko, B.D. Kuz'menov, L.S. Kutsaeva and V.M. Piksaikin, Sov. J. Nucl. Phys., 14, 629 (1972).
16. J.W. Meadows and C. Budtz-Jørgensen ANL/NDM-64 (1982).
17. ENDF/B Summary Documentation, BNL-NCS-17451(ENDF-201) 3rd ed. (ENDF/B-V), R. Kinsey, ed., available from the National Nuclear Data Center, Brookhaven National Laboratory (July , 1979).
18. M.S. Krick and A.E. Evans, Nucl. Sci. Eng., 47, 311 (1972).
19. J.W. Meadows, Phys. Rev. 177, 1817 (1968).
20. L.E. Glendenin, J.E. Gindler, D.J. Henderson and J.W. Meadows, Phys. Rev. 24C, 2600 (1981).
21. B.D. Wilkins, E.P. Steinberg and R.R. Chasman, Phys. Rev. 14C, 1832 (1976).

BARRIER HEIGHTS OF PLUTONIUM ISOTOPES FROM (n,n'f) - THRESHOLDS

H.-H. Knitter and C. Budtz-Jørgensen

Commission of the European Communities
Joint Research Centre
Central Bureau for Nuclear Measurements
Geel, Belgium

The neutron induced second chance fission cross section for the isotopes ^{238}Pu, ^{239}Pu, ^{240}Pu, ^{241}Pu, ^{242}Pu and ^{244}Pu are studied in the region of the threshold using a simple model. Numerical values are obtained for the inner fission barrier heights of the mentioned isotopes and of the nuclear temperatures governing the neutron evaporation process at incident neutron energies around the second chance fission threshold. The comparisons of the present parameters with those obtained by other methods give hints to possible insufficiencies of experimental cross section data in the region of the second chance fission threshold.

[second chance fission cross section $^{238-242}$Pu, ^{244}Pu, fission barrier height, nuclear temperature]

Introduction

The fission barrier parameters for the odd plutonium isotopes have been obtained usually from the behaviour of the fission cross section in the region of the (n,f)-thresholds of the corresponding even isotopes and also from resonance parameters measured below the fission thresholds. Since the neutron binding energies for the odd plutonium isotopes are larger than the fission barrier heights, one cannot measure directly fission barrier parameters for the even plutonium isotopes by the neutron induced fission process. The fission barrier parameters of the even plutonium isotopes have been deduced mainly from charged particle induced fission processes [1]. The analysis of the neutron induced second chance fission cross section in the threshold region gives in principle however the opportunity to obtain information on fission barrier parameters of the target nucleus itself and therefore for all the plutonium isotopes. The second chance fission threshold is well pronounced in the experimental neutron induced fission cross sections of the plutonium isotopes ^{238}Pu [2], ^{239}Pu [3,4], ^{240}Pu [5,6], ^{241}Pu [7], ^{242}Pu [4,7] and ^{244}Pu [7,8], as may be seen from the experimental data sets of the mentioned references. These data sets were used in the present study to obtain barrier parameters of these six isotopes from the (n,n'f)-process.

Contrary to the shapes of the fission cross sections in the regions of the (n,f)-thresholds of the even plutonium isotopes, the cross section shapes in the regions of the (n,n'f)-thresholds are not only determined by the barrier characteristics, but are also broadened by the neutron inelastic scattering process. One expects then a decreased sensitivity of the fission barrier parameters to the second chance fission cross section shape, compared with the situation at the (n,f)-thresholds. Nevertheless, from an analysis of the second chance fission cross section one might get some insight at least in the difference in compatibility of experimental data with model calculations for the different isotopes.

Having these limiting conditions in mind, a very simple model was set up, which is described in the next section.

Representation of the (n,n'f)-Cross Section

The (n,n'f)-cross section as function of the incident neutron energy can in general terms be written as

$$\sigma_{n,n'f)}(E_n) = \sigma_{(n,n')}(E_n) \cdot \int_0^{E_n} N(\epsilon) \cdot \frac{T_f(E_n-\epsilon)}{T_f(E_n-\epsilon) + \sum_c T_c(E_n-\epsilon)} \, d\epsilon \qquad (1)$$

where $\sigma_{(n,n')}(E_n)$ is the inelastic neutron scattering cross section as function of the incident neutron energy. $T_f(E_n-\epsilon)$ is the transmission coefficient for fission of the target nucleus, which is excited by the neutron inelastic scattering process. The $T_c(E_n-\epsilon)$ are the transmission coefficients for the decay of the

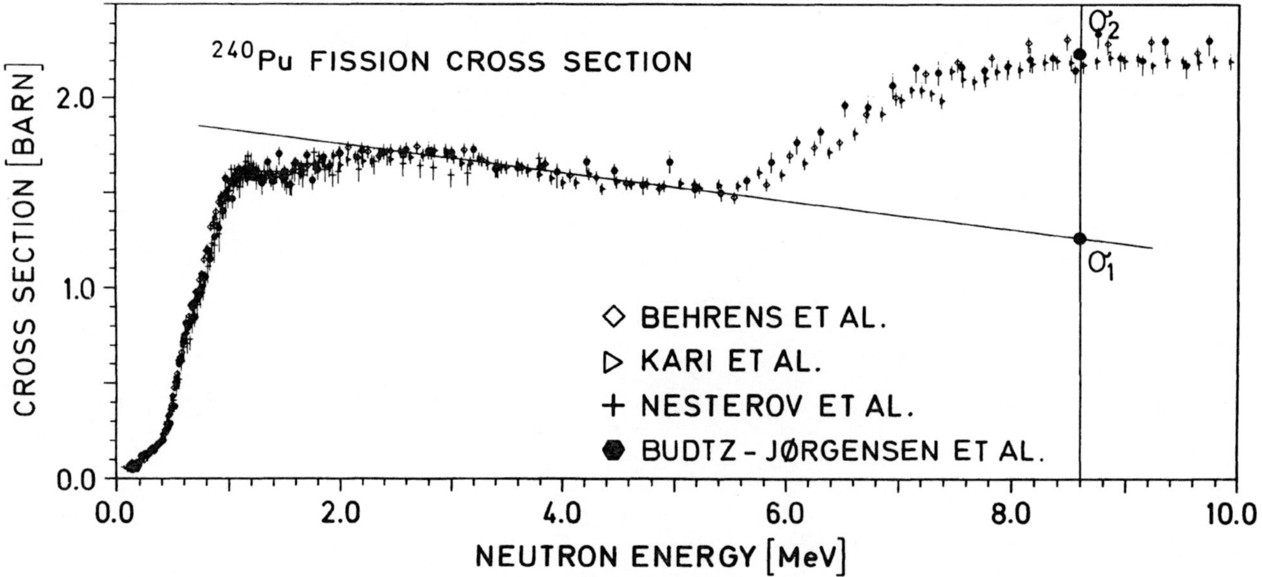

Fig. 1. Fission cross section of ^{240}Pu versus the incident neutron energy covering first and second chance fission threshold.

K. H. Böckhoff (ed.), Nuclear Data for Science and Technology, 744–747.

excited target nucleus into all other open channels than fission. $N(\epsilon)$ is the energy spectrum of the inelastically scattered neutrons :

$$N(\epsilon) = \frac{1}{T^2} \cdot \epsilon \cdot e^{-\epsilon/T} \tag{2}$$

$$\text{with } \int_0^\infty \frac{1}{T^2} \cdot \epsilon \cdot e^{-\epsilon/T} d\epsilon = 1 \approx \int_0^{E_n} \frac{1}{T^2} \cdot \epsilon \cdot e^{-\epsilon/T} d\epsilon \tag{3}$$

where T is the nuclear temperature of the nucleus after neutron evaporation and ϵ is the energy of the evaporated neutron.

Fig. 1 shows as an example experimental neutron induced fission cross sections of ^{240}Pu, as published in ref. [6], covering the (n,f)- and (n,n'f)-thresholds. In the region below the (n,n'f)-threshold the (n,f)-cross section can be approximated by a straight line as shown in fig. 1. This approximation shown in fig. 1 for the case of ^{240}Pu is possible for all the mentioned plutonium isotopes. It is now assumed that the (n,f)-cross section continues this behaviour at higher incident neutron energies , and that the(n,n'f)-cross section can be obtained as the difference between the total fission cross section and the indicated straight line.

As the essential change of the cross section in the

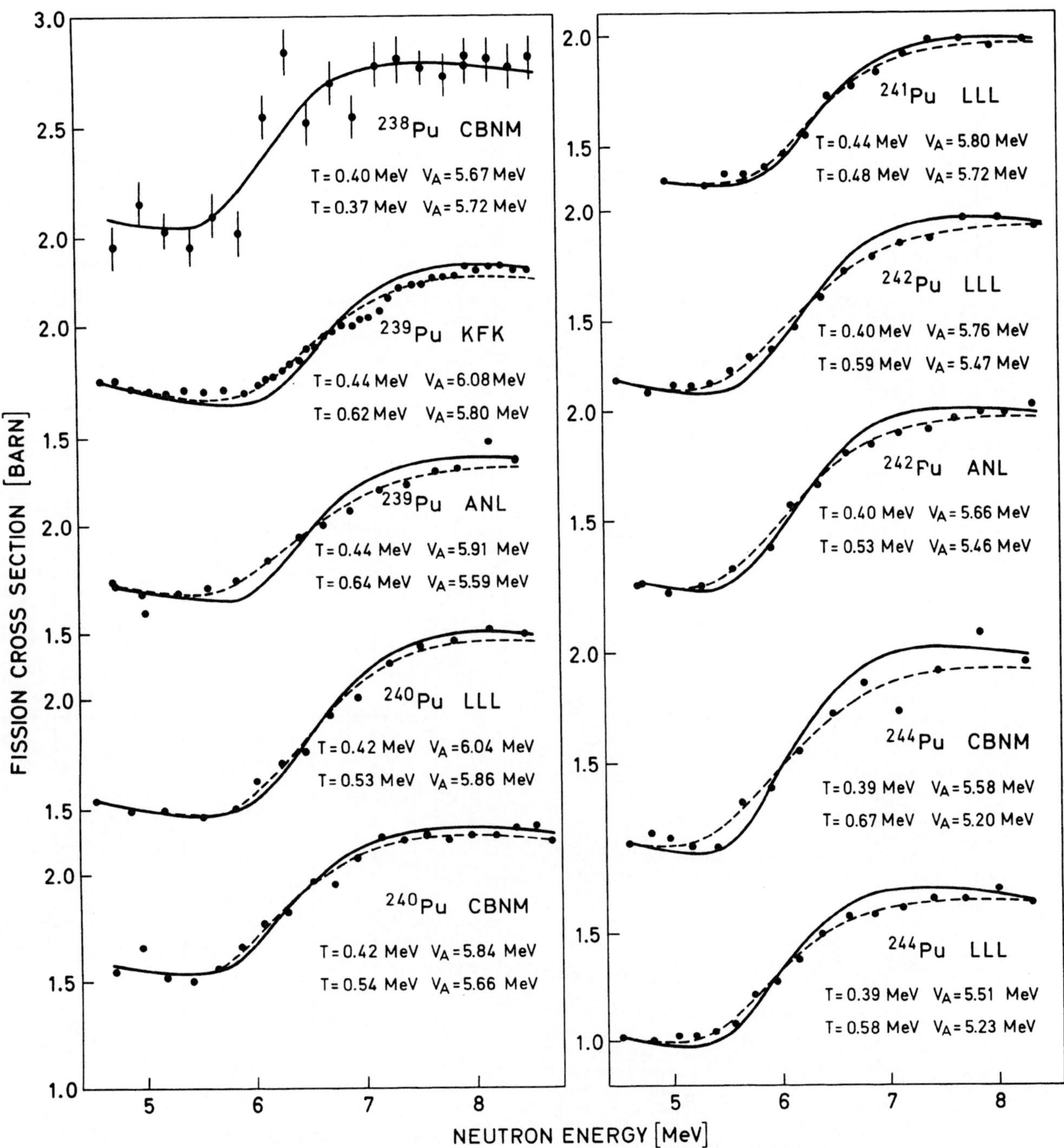

Fig. 2. The experimental fission cross sections of several plutonium isotopes are plotted versus the incident neutron energy in the region of the second chance fission thresholds. Dashed lines are results of fits with the free parameters V_A and T, and $T_{\acute{c}}$ = 1; full lines are results of fits with the free parameter V_A and T equal to the Fermi-gas model value. The lower and upper lines give the numerical values of the first and second fit respectively.

region of the second chance fission threshold occurs for all Pu-isotopes within a relatively small incident neutron energy range of 1.5 MeV to 2 MeV, it is assumed that $\sigma_{(n,n')}(E_n)$ is constant, as well as the sum of the transmission coefficients for the decay into all other channels than fission. With these simplifying assumptions equation (1) reduces to :

$$\sigma_{(n,n'f)}(E_n) = \sigma_{(n,n')} \cdot \int_0^{E_n} N(\epsilon) \cdot \frac{T_f(E_n-\epsilon)}{T_f(E_n-\epsilon) + T_c} \, d\epsilon \quad (4)$$

For incident neutron energies above the barrier, where $T_f \approx 1$, one obtaines the condition

$$(\sigma_2 - \sigma_1) \approx \sigma_{(n,n')} \cdot \frac{1}{1 + T_c} \quad (5)$$

The difference $(\sigma_2 - \sigma_1)$ is taken well above the threshold as indicated in fig. 1. This condition was used to suppress the parameter $\sigma_{(n,n')}$ in equation (4).

All the plutonium isotopes from 238 to 244 have a double humped fission barrier, the inner barrier between 400 keV and 600 keV higher than the outer one, as e.g. evaluated by Bjørnholm and Lynn [1]. Because of this large energy difference the influence of the outer barrier can be neglected in the calculation of the transmission coefficient $T_f(E_n - \epsilon)$ for excitation energies of the target nucleus near to the second chance fission threshold or near to and above the inner barrier V_A. In this situation the expression for the transmission through an inverted parabolic barrier as given by Hill and Wheeler [9] can be used :

$$T_f(E_n - \epsilon) = [\,1 + \exp\,(2\,\pi\,(V_A - E_n + \epsilon)/\hbar\omega_A)]^{-1} \quad (6)$$

where V_A is the barrier height, and $\hbar\omega_A$ is describing the curvature of the parabola. In doing so, only one transmission state, namely the ground state at the barrier saddle point, has been considered.

If equations (5) and (6) are inserted in equation (4) an expression with the four free parameters V_A, $\hbar\omega_A$, T and T_c is obtained, which can be used to fit the cross section data in the region of the second chance fission threshold.

Fitting to the (n,n'f)-Cross Section

The information contained in the experimental cross section data for the (n,n'f)-threshold region is not sufficient to determine all four parameters. The step of the cross section at the (n,n'f)-threshold is smeared out due to neutron evaporation and hence the shape is not strongly dependent on the curvature parameter $\hbar\omega_A$. This parameter had to be fixed to 0.8 MeV and 1.04 MeV for the odd and even plutonium isotopes respectively, according to values given in the review by Bjørnholm and Lynn [1].

For the parameter T_c an estimate with the help of equation (5) is reached if numerical values for $\sigma_{(n,n')}$ from cross section evaluations of plutonium isotopes are used. T_c should get values between about 0.5 and 3.

Two parameters V_A and T were determined by fitting equation (4) for the (n,n'f)-cross section to the experimental data in the neutron energy range between 4.6 MeV and 8.6 MeV for the isotopes ^{238}Pu, ^{239}Pu, ^{240}Pu, ^{241}Pu, ^{242}Pu and ^{244}Pu. In these fits the parameter T_c was fixed to values of 0.5, 1 and 2. Fig. 2 shows the resulting fits with T_c= 1 as dashed lines. The data are all fairly well represented by the dashed lines. A change of T_c by a factor of two gave changes in V_A and T of less than 50 keV and 2 keV respectively. The results for the parameters T and V_A from these fits with T_c= 1 are given in fig. 2 in the lower lines and they are plotted in fig. 3 with open symbols. The symbols ◇, △, ▽, and ○ represent the results using the experimental data sets from the laboratories KFK [3], ANL [4], LLL [5,7] and CBNM [2,6,8] respectively.

In the lower part of fig. 3 comparison is made between the present nuclear temperatures and the values obtained from the Fermi-gas model as given e.g. by Gilbert and Cameron [11]. The temperatures, as plotted in fig. 3 with the ✦ -symbols, were calculated corresponding to $T = \sqrt{U/a}$, with U = E-P(Z)-P(N), where for the excitation energy E an energy equal to the corresponding barrier height was used; P(Z) and P(N) are the proton and neutron pairing energies which were also taken from ref. [11], as well as the numerical values for the level density parameter a. Considering the crudeness of the present physical picture the agreement for the temperatures is good for the isotopes ^{238}Pu, ^{240}Pu, ^{241}Pu, less good for ^{242}Pu and worse for ^{239}Pu and ^{244}Pu.

In a second fitting attempt also the parameter T was kept fixed and equal to the corresponding Fermi-gas model value. The full lines in fig. 2 represent the fitted curves. The barrier heights V_A obtained by this procedure are plotted in fig. 3, as above mentioned, but with full symbols. The numerical values for V_A and for the Fermi-gas model temperatures T are given in fig. 2 in the upper lines.

In the upper part of fig. 3 comparison is made between the barrier heights from the present fits to the (n,n'f)-thresholds and values obtained by other methods. The ■-symbols represent the values for the inner barrier heights for the different isotopes as evaluated by Bjørnholm and Lynn [1]. For the four even isotopes the barrier heights obtained by the present rough picture are in agreement with the corresponding values of ref. [1]. One exception is the barrier height from the fit with one free parameter to the data set of ref. [5], which is too high compared to the evaluated value of ref. [1].

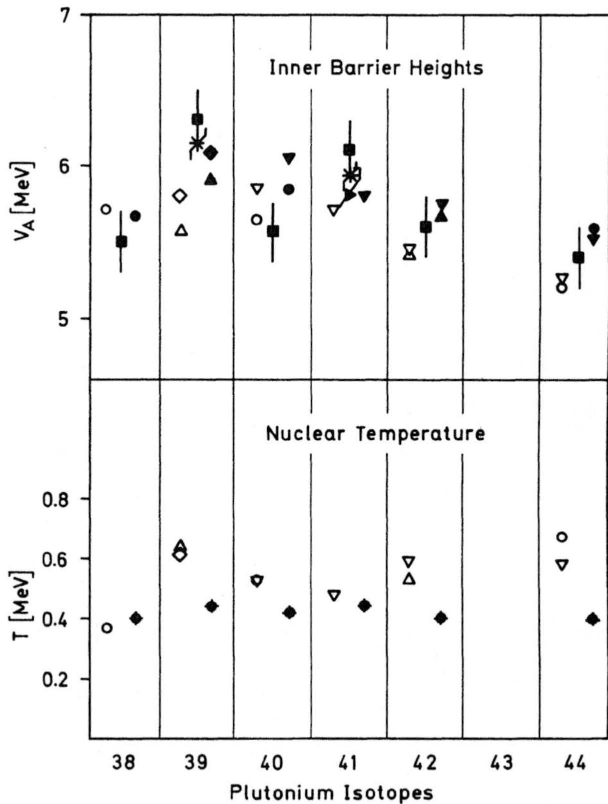

Fig. 3. The full and open symbols ◇, △, ▽ and ○ represent values from fits with one and two free parameters respectively. The ■- and ▶ -symbols correspond to values for V_A from ref. [1] and [10] respectively. The ✱-symbols represent presently evaluated barrier heights from own (n,f)-threshold measurements [2,6]. The ✦ -symbols are Fermi-gas model nuclear temperatures.

In the case of ^{239}Pu comparison can be made with the value of ref. [1] and with V_A = (6.14 ± 0.15) MeV, which we obtain from the position of the (n,f)-threshold in our measurements on ^{238}Pu and which is plotted with the ✱-symbol in fig. 3. Between these two barrier heights and the results from the fits with two free parameters no agreement is found. The results from the fits with one free parameter and T equal to the Fermi-gas model value are closer to the value of ref.[1] and to the ^{238}Pu(n,f)-threshold value, however, the cross section shape is then badly described.

In the case of ^{241}Pu comparison for the barrier heights can be made with the value of ref. [1], with a value obtained from data given by Auchampaugh and Weston [10] and with a value of V_A = (5.94 ± 0.10) MeV which we obtain from the position of the (n,f)-threshold in our measurement on ^{240}Pu [6]. The present value from the (n,f)-threshold is plotted with the ✱-symbol in fig. 3 and agreement is found with the results from the fits within the uncertainty limits. Auchampaugh and Weston [10] obtained from the analysis of sub-barrier resonance parameters of the ^{240}Pu(n,f)-reaction a result $(V_A - S_n)/\hbar\omega_A = (0.71^{+0.31}_{-0.10})$ where S_n = 5.24 MeV is the neutron separation energy for ^{241}Pu. Using also here $\hbar\omega_A$ = 0.8 MeV one obtains V_A plotted with the ▶-symbol in fig. 3, which is within 90 keV in agreement with our present result from the fit with two free parameters and in even better agreement with the result from the fit with V_A as free parameter and T equal to the Fermi-gas model value. Compared to the values we obtain from the fits, the value of ref. [1] is 300 to 400 keV too high. The cross section shape for the (n,n'f)-threshold of ^{241}Pu is well described by the two kinds of fits as may be seen in fig. 2.

Conclusion

The simple physical picture presented here is able to describe the behaviour of the experimental second chance fission cross sections of the isotopes ^{238}Pu, ^{240}Pu, ^{241}Pu and ^{242}Pu with values for the nuclear temperatures near to those of the Fermi-gas model and with inner fission barrier heights which compare well with those obtained by other methods. For the fit of ^{239}Pu, where T and V_A is allowed to vary, T becomes too large compared with theoretical and also with experimental values as measured by Coppola and Knitter [12]. Moreover, V_A is by 400 keV to 700 keV too low with respect to data from other sources. When T is kept to the theoretical value, V_A rises to values near to others, but the cross section shape is then badly described. For ^{244}Pu the situation is much better compared with ^{239}Pu, but not as good as one would expect from the situations for the four other plutonium isotopes treated in the present study.

Ackowledgement : The authors like to thank Prof. Dr. A.J. Deruytter for the careful reading of the manuscript.

References

1. S. Bjørnholm and J.E. Lynn, Rev. Mod. Phys. 52, 725 (1980).
2. C. Budtz-Jørgensen, H.-H. Knitter and D.L. Smith "Neutron induced fission cross section of ^{238}Pu in the energy range from 5 eV to 10 MeV", this conference.
3. K. Kari and S. Cierjacks, Rept. KFK-2673 (1978).
4. J.W. Meadows, Nucl. Sci. Eng. 68, 360 (1978).
5. J.W. Behrens, R.S. Newbury and J.W. Magana, Nucl. Sci. Eng. 66, 433 (1978).
6. C. Budtz-Jørgensen and H.-H. Knitter, Nucl. Sci. Eng. 79, 380 (1981).
7. G.W. Carlson and J.W. Behrens, Nucl. Sci. Eng. 68, 128 (1978).
8. M.S. Moore, C. Budtz-Jørgensen, H.-H. Knitter, C.E. Olsen, J.A. Wartena, and H. Weigmann, "Neutron induced fission cross section of ^{244}Pu", this conference.
9. D.L. Hill and J.A. Wheeler, Phys. Rev. 89, 1102 (1953).
10. G.F. Auchampaugh and L.W. Weston, Phys. Rev. C12 1850 (1975).
11. A. Gilbert and A.G.W. Cameron, Can. J. Phys. 43 1446 (1965).
12. M. Coppola and H.-H. Knitter, Z. Physik 232, 286 (1970).

ISOTOPIC DISTRIBUTIONS AND ELEMENT YIELDS IN THE PHOTOFISSION OF ^{235}U AND ^{238}U WITH 12-, 15-, 20- AND 30-MeV BREMSSTRAHLUNG

D. De Frenne, H. Thierens, B. Proot, E. Jacobs, P. De Gelder and A. De Clercq

Laboratorium voor Kernfysika
42, Proeftuinstraat, B 9000 Gent
Belgium

Using chemical separation techniques, γ-spectrometry of fission product catcherfoils and direct γ-spectrometry of irradiated uraniumsamples, the fractional independent or cumulative chain yields of about 50 fission products were measured for the photofission of ^{235}U and ^{238}U with 12-, 15-, 20- and 30-MeV bremsstrahlung. In addiation a number of fractional independent chain yields were calculated using the charge distribution curves. This enables us to study the isotopic distributions for Kr, Rb, Sr, Sn, Sb, Te, I, Xe, Cs and Ba isotopes for the photofission of ^{235}U and for Rb, Sn, Sb, Te, I and Xe isotopes for the photofission of ^{238}U. The corresponding element yields for all investigated isotopic distributions were calculated. As well for the photofission of ^{235}U as for ^{238}U at all bremsstrahlung end point energies a very low yield of ^{136}I and ^{134}Sb was obtained.

[$^{235}U(\gamma,F)$, $^{238}U(\gamma,F)$, $E\gamma_{max}$: 12-, 15-, 20- and 30-MeV, measured fission product γ-ray spectra, deduced isotopic distributions and element yields]

Introduction

The extremely high (n_{th},f) cross section for the fissile materials as e.g. ^{235}U and ^{239}Pu, allows the use of sophisticated methods, as high-resolution mass separation, for the study of the fission products. As a consequence very detailed information about the isotopic distributions and element yields is available for the thermal neutron induced fission of these fissile materials. However only very few experimental results for these fission characteristics are known for almost all other fissioning systems. Especially for photofission no isotopic distributions or element yields are reported in the literature. The lower value of the photofission cross section, e.g. the $^{235}U(\gamma,F)$ cross section shows a maximum value of about 200 mb in the giant resonance region, limits the experimental techniques which can be used for the measurement of yields of long-lived fission products ($T_{1/2}$ >3.0 min) to chemical separation techniques or γ-spectrometry of fission product catcherfoils. For the determination of the yields of short-lived fission products ($T_{1/2}$ >2 s) we developed a method based on the direct γ-spectrometry of natural uranium samples and uranium samples enriched in ^{235}U. Irradiations were performed at 4 different bremsstrahlung end point energies ranging from 12- to 30 MeV. Using those three techniques we measured the fractional independent yields of about 50 fission products for the photofission of ^{235}U and ^{238}U with 12-, 15-, 20- and 30-MeV bremsstrahlung and studied the charge-, isotopic- and isotonic distributions for respectively different mass-, charge- and neutron numbers. In this paper we will report the results obtained for the isotopic distributions for different elements.

Experimental Procedure

The fractional independent or cumulative yields of about 40 short-lived fission products with half-lives as short as 2s were determined by direct γ-spectrometry of irradiated uranium samples. For the study of $^{235}U(\gamma,F)$ 0.2mm thick metallic disks, enriched up to 97% in ^{235}U were used, while for the $^{238}U(\gamma,F)$ experiments the tickness of the natural uranium disks was 0.1mm. For as well the ^{235}U as the ^{238}U photofission experiments, 13 samples were prepared at the Central Bureau for Nuclear Measurements (Euratom, Geel) and enclosed into very pure nickel capsules (99.99%). The capsules were irradiated for 30s with 12-, 15-, 20- or 30-MeV bremsstrahlung beams, produced in a 0.1mm thick gold bremsstrahlung converter. After irradiation the capsules were transported from the irradiation site to the Ge(Li)-detector set-up with a pneumatic transport system. The measurement of 60 consecutive γ-ray spectra, each with a measuring time of 10s, started 5s after the end of irradiation. These γ-ray spectra were registered as one single file on a RK05 disk using a PDP11/10 computersystem with a CA11C CAMAC interface. The time loss between two successive γ-ray spectra was less than 150μsec. The variation of the dead time during the measurements was followed with the pulser technique. For the analysis of the data, a number of successive γ-ray spectra, depending on the half-life of the fission products to be studied, was added up.

The independent yields of ^{96}Nb, $^{124}Sb^{g}$, $^{132}I^{g}$, $^{132}I^{m}$, $^{134}I^{g}$, $^{134}I^{m}$, ^{134}Cs and ^{136}Cs for $^{238}U(\gamma,F)$ were obtained by chemical separation on 1g $UO_2(NO_3)_2.6H_2O$ (natural uranium) irradiated for appropriate times. The chemical separation of the niobium, antimony, iodine and cesium fractions was based on the procedures given by Morris et al.[1)], Dropesky and Orth[2)], Troutner et al.[3)] and Cuninghame et al.[4)] respectively. As fission yield monitors respectively ^{97}Nb, ^{127}Sb, ^{131}Sb, ^{135}I and ^{137}Cs were used. The independent yields of $^{130}I^{g+m}$, $^{132}I^{g+m}$, ^{133}I, $^{134}I^{g}$, $^{134}I^{m}$ in the photofission of ^{235}U were obtained by chemical separation of the iodine fraction from the fission product catcherfoils. Several successive γ-ray spectra were taken using a 50 cm^3 ORTEC Ge(Li)-detector and a conventional measuring chain.

The fractional independent yields of $^{126}Sb^{m}$, $^{128}Sb^{m}$, $^{131}Te^{g}$, $^{131}Te^{m}$, $^{133}Te^{g}$, $^{133}Te^{m}$, ^{132}I, ^{134}I and ^{135}Xe are determined by γ-spectrometry of fission product catcherfoils irradiated for appropriate times. The experimental set-up for the measurement of the γ-ray spectra was the same as for the chemical separation experiments.

Data Analysis

The identification of all observed γ-rays was mainly based on the γ-ray catalogs of Reus et al.[5)] and Blachot and Fiche[6)]. The peak areas in the γ-ray spectra were calculated using a modified version of the programs MARKER and CAOS of W. Westmeier[7)]. The half-life of the γ-rays used in the calculation of the fractional independent or cumulative yields of the short-lived fission products was checked carefully and the corresponding yields calculated using a modified version of the CLSQ fitting program of Cumming et al.[8)] in which successive β-decay is taken into account.

The contribution of the slow neutron induced fission in our photofission experiments on the ^{235}U samples was estimated to be at most 2% for 30-MeV bremsstrahlung. This number was obtained by replacing the uranium targets by scandium and indium samples.

From the measured fractional independent or cumulative yields we deduced simultaneously the width parameter c and the most probable charge Z_p for the mass chains A = 130-136 and 140 for the photofission of ^{235}U. For the photofission of ^{238}U we obtained both parameters of the charge distribution for the mass chains A = 131-136.

By averaging the experimentally obtained c-values for each bremsstrahlung end point energy for the photofission of ^{235}U and ^{238}U, Z_p values for another 20 mass-chains were calculated together with the fractional independent yields of the different members of the

K. H. Böckhoff (ed.), Nuclear Data for Science and Technology, 748–750.

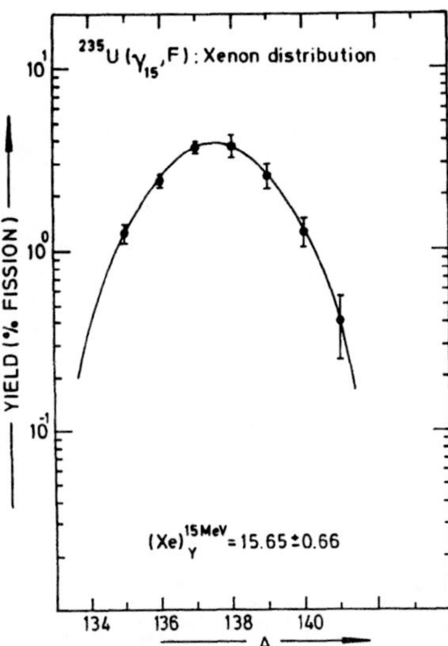

Fig. 1a: Xenon distribution for the photofission of ^{235}U with 15-MeV bremsstrahlung. The full line represents a gaussian fit through the data.

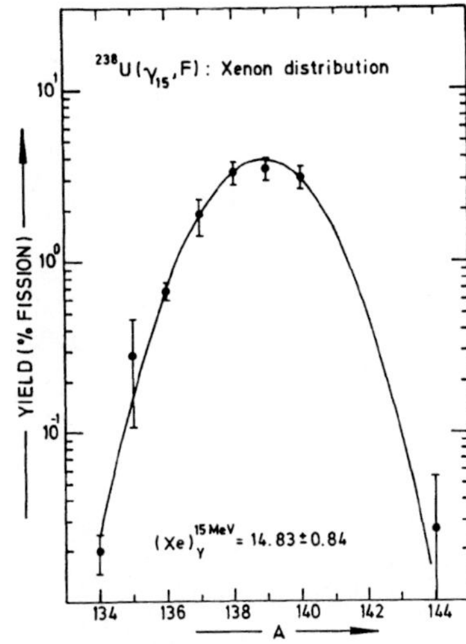

Fig. 1b: Xenon distribution for the photofission of ^{238}U with 15-MeV bremsstrahlung. The full line represents a gaussian fit through the data.

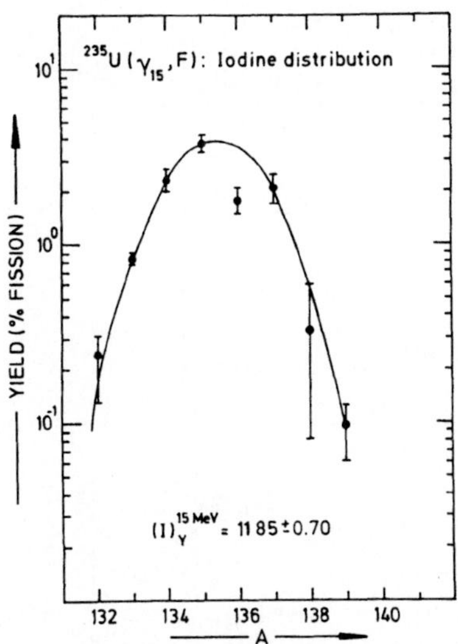

Fig. 2a: Iodine distribution for the photofission of ^{235}U with 15-MeV bremsstrahlung. The full line represents a gaussian fit through the data (^{136}I excluded).

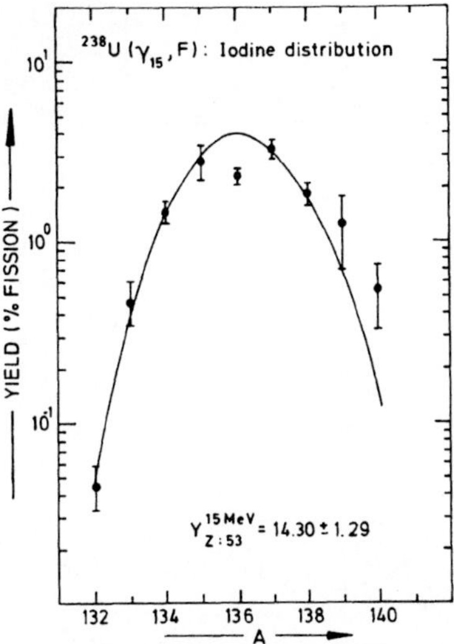

Fig. 2b: Iodine distribution for the photofission of ^{238}U with 15-MeV bremsstrahlung. The full line represents a gaussian fit through the data (^{136}I excluded).

considered A-chains. Combining directly measured fractional independent or cumulative yields and those calculated from the charge distribution curves, the isotopic distributions of Kr, Rb, Sr, Sn, Sb, Te, I, Xe, Cs and Ba were studied for the photofission of ^{235}U with 12-, 15-, 20- and 30-MeV bremsstrahlung while for $^{238}U(\gamma,F)$ at the same bremsstrahlung end point energies the isotopic distributions of Rb, Sn, Sb, Te, I and Xe were investigated.

Experimental results and discussion

As an example, the isotopic distributions for the xenon- and iodine isotopes for the photofission of ^{235}U and ^{238}U with 15-MeV bremsstrahlung are given in fig.1a,b and fig.2a,b.These figures show clearly that the yield of $^{136}I(Z=53)$ is about 50% smaller than the expected value obtained by fitting a gaussian through the experimental data. This effect is observed systematically for all bremsstrahlung end point energies. A similar effect is present for another odd Z,N=83 nucleus : ^{134}Sb. However it seems to be absent for the even Z,N=83 nuclei ^{135}Te and ^{137}Xe. This different behaviour for the odd and even Z,N=83 nuclei cannot be explained by a difference in the very low neutron separation energy of those nuclei because the situation of the odd or even Z,N=83 nuclei is very similar. A possible reason for the abnormally low yields of ^{134}Sb and ^{136}I could be either wrong or incomplete spectroscopic data for the β^--decay of both isotopes. The element yields, which are obtained by adding up the individual yields of the different members of the corresponding isotopic distributions are given in table I for the photofission of ^{235}U and in table II for the photofission of ^{238}U. The bracketed values of tables I and II represent the percentages of the element yields which could not be measured directly or calculated indirectly using the charge distribution curves. They are only given in the cases where they exceed 1% of the total element yield. The missing yields are obtained by interpolation or extrapolation,

Table I Element yields $^{235}U(\gamma,F)$ and $^{235}U(n,f)$ in % fission

Element	$E_{\gamma max}$ (MeV) 12	15	20	30	$^{235}U(n_{th},f)$ [9]	$^{235}U(n_{14},f)$ [10]
Kr	11.88±0.66(17)	12.31±0.81(17)	11.20±0.71(17)	10.90±0.64(23)	15.7 ± 0.6	9.8 ± 1.0
Rb	13.78±0.88(26)	14.00±0.96(27)	13.12±0.83(25)	-		
Sr	15.87±0.81(28)	15.27±0.82(28)	15.04±0.86(28)	-		
Sn	-	3.89±0.38(22)	3.73±0.17(17)	4.17±0.13(15)		
Sb	8.54±0.56(4)	8.06±0.28(2)	8.01±0.36(2)	7.79±0.30(2)		
Te	13.52±0.75	13.41±0.60	13.18±0.68	13.68±0.89		
I	12.38±0.75	11.85±0.70	11.74±0.69	11.66±0.74		
Xe	16.35±0.64	15.65±0.66	14.67±0.98	14.41±0.85	19.1 ± 1.0	13.5 ± 1.3
Cs	14.36±1.00(15)	14.40±1.20	14.37±0.99	14.15±1.09		
Ba	-	12.47±0.87(27)	12.46±0.94(26)	11.60±0.83(23)		

Table II Element yields $^{238}U(\gamma,F)$ and $^{238}U(n_{14},f)$ in % fission

Element	$E_{\gamma max}$ (MeV) 12	15	20	30	$^{238}U(n_{14},f)$ [10]
Rb	10.62 ± 1.52(21)	11.19 ± 1.22(22)	11.32 ± 1.15(18)	-	
Sr	4.85 ± 0.31(2)	5.28 ± 0.22(5)	4.98 ± 0.75(7)	5.55 ± 0.27(8)	
Sb	10.10 ± 0.54	10.57 ± 0.54	10.34 ± 0.42	9.68 ± 0.51	
Te	15.28 ± 0.91	14.61 ± 0.86	15.21 ± 0.61	13.43 ± 0.64	
I	13.61 ± 1.09	14.30 ± 1.29	14.25 ± 0.96	12.51 ± 0.62	
Xe	14.92 ± 1.45(25)	14.83 ± 0.84(13)	14.47 ± 0.66(14)	14.38 ± 0.62(9)	13.1 ± 1.0

fitting a gaussian through all experimental data. For comparison the element yields of krypton and xenon for $^{235}U(n_{th},f)$ [9], $^{235}U(n_{14},f)$ [10] and $^{238}U(n_{14},f)$ [10] are also given in tables I and II.

As could be expected from the experimentally determined peak-to-valley ratios and the pronounced proton odd-even effect in the case of $^{235}U(n_{th},f)$, the values for the element yields of krypton and xenon in the photofission of ^{235}U with 12-, 15-, 20- and 30-MeV bremsstrahlung lie between those for thermal and 14-MeV neutron induced fission of ^{235}U and are for the photofission of ^{238}U somewhat higher than for the 14-MeV neutron induced fission of ^{238}U.

Acknowledgements

This research was supported by the National Fund for Scientific Research, Interuniversity Institute for Nuclear Sciences. We are very grateful to Prof.Dr. A.J.Deruytter for his continuous interest in this work. Ing.J.Buysse and Mr.L.Schepens are acknowledged for the careful chemical separations. We also thank the Linac team of our laboratory.

References

1. D.F. Morris and D. Scargill, Anal. Chem. Acta 14, 57 (1956)
2. B.J. Dropesky and C.J. Orth, J. Inorg. Nucl. Chem. 24, 1301 (1962)
3. D.E. Troutner, M. Eichor and C. Pace, Phys. Rev. C1, 1044 (1970)
4. J.G. Cuninghame, J.A.B. Goodall, G.P. Kitt, C.B. Webster and H.H. Willis, Report n° AERE-R-5587 (unpublished)
5. U. Reus, W. Westmeier and J. Warnecke, G.S.I. Report 79-2 (unpublished)
6. J. Blachot and C. Fiche, At. Data Nucl. Data Tables 20, 241 (1977)
7. W. Westmeier, private communication
8. J.B. Cumming, National Academy of Sciences, National Series Report N.A.S. NS-3107 (1962) (unpublished)
9. R. Brissot, J. Crançon, Ch. Ristori, J.P. Bocquet and A. Moussa, Nucl. Phys. A255, 461 (1975)
10. J.P. Bocquet, R. Brissot, J. Crançon and A. Moussa, Nucl. Phys. A189, 556 (1972)

A PREDICTED DIRECTIONAL BIAS OF THE MASS ASYMMETRY IN $^{230}Th(n,f)$

D.W. Lang

Australian Atomic Energy Commission Research Establishment
Lucas Heights Research Laboratories,
Private Mail Bag, Sutherland, NSW 2232, Australia

Heavy and light fragments in fission usually share a common angular distribution, associated with a defined parity for an isolated saddle-point state. In ^{230}Th (n,f) at 708 keV, the triple-humped fission barrier model suggests an overlap of two saddle-point states of opposite parity based on the same asymmetric configuration.

The model then predicts fore-and-aft asymmetry for, say, the distribution of heavy fragments at a given neutron energy. For neutrons polarized perpendicular to the beam there is asymmetry in the third perpendicular direction. Experimental verification may depend critically on energy resolution.

At lower energies the effect may be more general in fissionable nuclei and there is a possibility of detecting and identifying p-wave resonances at much lower energies.

Many general features of the fission process have been described in terms of "double-humped" models in which a broad barrier is replaced by two separated narrower barriers[1]. Parameters describing these barriers can be predicted from shell model considerations[2]. In detailed fits to data, more specific prediction becomes important. Study has been concentrated on a broad resonance near 715 keV neutron energy in the ^{230}Th (n,f) cross section[3].

A recent discussion[4] of the experimental data for ^{230}Th (n,f) has shown that both the fission cross section and the fission fragment angular distribution can be explained in terms a model with a triple-humped fission barrier and asymmetric deformations. Earlier calculations with symmetric deformations[5] described a double-humped barrier with the inner barrier 3 MeV lower than the outer barrier and considerably displaced in deformation. The observed resonance in the fission cross section was then not simply explained. Moller and Nix[6] showed that if asymmetric deformations are included the outer hump is split into two peaks separated by a narrow valley. The nucleus has axial symmetry but is most easily visualised as pear shaped. It does not have reflection symmetry in a plane perpendicular to the symmetry axis.

The single-particle states of nucleons in a nucleus constrained to lie in the shallow valley lead to an unpaired neutron in a state with angular momentum projection ½ on the symmetry axis. The intrinsic state is then parent to rotational states of both parities and all spins consistent with the angular momentum carried by the incoming neutrons.

It is assumed[4] that the axis of the fission event is the intrinsic axis of symmetry of the nucleus. The description of the cross section for ^{230}Th (n,f) and of the angular distribution of fission fragments then follows.

As a further consequence of the model, described here, an experiment can be set up with appropriately sited detectors so that fission events have a preferred sense, along the defined axis of detected events, for the mass asymmetry. The prediction concerns the existence of the asymmetry but not, as yet, the sign.

The axis of the fission event is taken here as the axis of symmetry near the saddle point configuration of the fissioning nucleus. States which in the laboratory have a definite spin and parity include components of several intrinsic configurations with the given axis of symmetry. For an intrinsic state with spin ½ and considerable octupole and quadrupole deformation, the value K of the projection of angular momentum on the symmetry axis can be ± ½ and the octupole deformation can independently have positive or negative sign when referred to the axis. These four configurations are written $\phi^{\frac{1}{2}}_{pp}$, $\phi^{\frac{1}{2}}_{pn}$, $\phi^{\frac{1}{2}}_{np}$ and $\phi^{\frac{1}{2}}_{nn}$ where the first subscript refers to the sign of K and the second to the sign of the octupole deformation.

The state of positive spin projection and defined parity π is then

$$\frac{1}{\sqrt{2}} \left(\phi^{\frac{1}{2}}_{pp} + (-1)^{\pi} \phi^{\frac{1}{2}}_{pn}\right).$$

By convention $(-1)^{\pi}=+1$ for positive parity, -1 for negative parity. The component, with a specified intrinsic axis, of a state of parity π spin J and laboratory projection M of that spin is

$$\left(\frac{2J+1}{32\pi^2}\right)^{\frac{1}{2}} \left\{\left(\phi^{\frac{1}{2}}_{pp} + (-)^{\pi}\phi^{\frac{1}{2}}_{pn}\right) D^J_{M\frac{1}{2}} + (-)^{J+\pi+\frac{1}{2}} \left(\phi^{\frac{1}{2}}_{np} + (-)^{\pi}\phi^{\frac{1}{2}}_{nn}\right) D^J_{M-\frac{1}{2}}\right\}$$

where $D^J_{M\frac{1}{2}}$ is the appropriate element of the rotation group. The present discussion concentrates on J = ½ since the model used[4] suggests that, for fission at 708 keV, there is a considerable overlap of $\frac{1}{2}^+$ and $\frac{1}{2}^-$ laboratory states in the saddle point configuration. If the intrinsic axis associated with K makes an angle Θ with the laboratory axis associated with M, the underlying dependence of $D^{\frac{1}{2}}_{MK}$ on Θ is

$$\left(\cos \frac{\Theta}{2}\right)^{|K+M|} \left(\sin \frac{\Theta}{2}\right)^{|K-M|}.$$

For isolated resonances in the neutron channel, the parity of the compound nucleus is settled by the resonance involved. In the neighbourhood of 708 keV there is already considerable overlap of resonances and a mixture of parities may occur in the compound nucleus. In what follows, interference of levels of opposite parities is assumed at all energies of interest.

Similarly, the intrinsic deformed state $\phi^{\frac{1}{2}}$ and the rotational states to which it is parent have substantial energy widths due to leakage in either direction through the appropriate barriers. Among the states defined in the first well, there is a broad state for each spin and parity corresponding to maximum overlap with a state formed by tunnelling from the corresponding state in the shallow second barrier well. Next, coupling of these states with neutron doorway states is assumed to be coherent so that the phase relations of opposite parity components in the neutron beam are reflected in the relationship (with a possible slow energy dependence) of opposite parity components in the shallow well.

If the intrinsic axis of the nucleus at its saddle point is assumed to become the axis of the fission event, the sign of the octupole deformation can be expected to govern the sense of the mass asymmetry. The decay of the component $\Theta^{\frac{1}{2}}_{pp}$, say, would thus give a preferred sense of the mass asymmetry along the intrinsic axis.

Consider then decay along the beam direction. The orbital angular momentum of the neutrons has zero projection on the beam axis. If the neutron spin has projection + ½ the wave function at the saddle-point is

$$\alpha_0 \left\{\psi^{\frac{1}{2}+}_{+\frac{1}{2}} + \alpha_1 e^{i\Theta} \psi^{\frac{1}{2}-}_{+\frac{1}{2}}\right\}$$

where the wave function $\psi^{J\pi}_M$ in the laboratory frame has spin J with projection M on the chosen axis and has parity π. The quantity $\alpha_1 e^{i\Theta}$ contains the information about relative strengths of S + P waves in a plane wave, a Clebsch Gordan coefficient (½1½0/½½),

K. H. Böckhoff (ed.), Nuclear Data for Science and Technology, 751 – 752.

and the relative strength of the nuclear coupling between the neutron doorway state and the fission channel. With the detectors aligned along the beam axis, $\phi^{\frac{1}{2}}_{pp}$ has relative intensity $|1 + \alpha_1 e^{i\Theta}|^2$ and results in a positive asymmetry, while $\phi^{\frac{1}{2}}_{pn}$ has intensity $|1 - \alpha_1 e^{i\Theta}|^2$ and results in a negative asymmetry.

Similarly for the neutron spin projection $-\frac{1}{2}$, the wave function at the saddle-point is

$$\alpha_o \left\{ \Psi^{\frac{1}{2}+}_{-\frac{1}{2}} - \alpha_1 e^{i\Theta} \Psi^{\frac{1}{2}-}_{-\frac{1}{2}} \right\}$$

The difference from the case of positive spin projection is a negative sign attached to α_1, since the Clebsch Gordan coefficient $(\frac{1}{2}\ 1\ \frac{1}{2}\ 0\ /\ \frac{1}{2}\ \frac{1}{2}) = 1/\sqrt{3}$ and $(\frac{1}{2}\ 1 - \frac{1}{2}\ 0\ /\ \frac{1}{2}\ -\frac{1}{2}) = -1/\sqrt{3}$

Then $\phi^{\frac{1}{2}}_{np}$ has relative intensity $|1 + \alpha_1 e^{io}|^2$

and $\phi^{\frac{1}{2}}_{nn}$ has relative intensity $|1 - \alpha_1 e^{io}|^2$

The relative intensities are the same as those for positive spin projection for the same asymmetry. The asymmetry divided by the total yield is then

$$\frac{2\cos\Theta}{\alpha_1 + \alpha_1^{-1}}$$

Consideration can also be given to neutrons completely polarized in a direction perpendicular to the beam. Polarized neutrons in the energy range around 700 keV can be generated using the Li (p,n) reaction[7]. The neutrons have approximately 30 per cent polarization at a laboratory angle 50° over a considerable range of energies. The polarization direction is perpendicular to both the proton and neutron directions.

In this direction, neutrons with $M_s = +\frac{1}{2}$ give rise to positive parity states of $(J,M) = (\frac{1}{2},\frac{1}{2})$ but, because the p waves have no component with orbital angular projection M = 0, the negative parity state becomes $(J,M) = (\frac{1}{2}, -\frac{1}{2})$. There is no interference, and hence no asymmetry, along the axis. If the polarization direction is taken as the y axis, the asymmetry occurs in the x,z plane. Initially, the axis of quantization is taken, as is usual, as the z axis and aligned with the beam direction. Complete polarization in the y direction then gives components[8] in the z direction

$$\frac{1}{\sqrt{2}} e^{\frac{-i\pi}{4}} \text{ with } M = \tfrac{1}{2} \text{ and } \frac{1}{\sqrt{2}} e^{\frac{+i\pi}{4}} \text{ with } M = -\tfrac{1}{2}.$$

The saddle point wave function is

$$\frac{e^{\frac{-i\pi}{4}}}{\sqrt{2}} \alpha_o \left\{ \psi^{\frac{1}{2}+}_{+\frac{1}{2}} + i\psi^{\frac{1}{2}+}_{-\frac{1}{2}} + \alpha_1 e^{i\Theta} (\psi^{\frac{1}{2}-}_{\frac{1}{2}} - i\psi^{\frac{1}{2}-}_{-\frac{1}{2}}) \right\}$$

where the negative sign of the fourth amplitude again results from the Clebsch Gordan coefficient $\frac{1}{2}1-\frac{1}{2}0/\frac{1}{2}-\frac{1}{2})$. The relevant ratios of decay magnitudes are given by

$\phi^{\frac{1}{2}}_{pp}$, $|1 + \alpha_1 e^{i\Theta}|^2$; $\phi^{\frac{1}{2}}_{pn}$, $|1 - \alpha_1 e^{i\Theta}|^2$; $\phi^{\frac{1}{2}}_{np}$, $|1 + \alpha_1 e^{i\Theta}|^2$ and $\phi^{\frac{1}{2}}_{nn}$, $|1 - \alpha_1 e^{i\Theta}|^2$, so that the asymmetry ratio is

$\frac{2\cos\Theta}{\alpha_1 + \alpha_1^{-1}}$ as before.

The wave function at the saddle-point can be rewritten with the axis of quantization in the x-direction. The appropriate Euler angles are in turn $(0, -\frac{\pi}{2}, 0)$. Then

$$d^{\frac{1}{2}}_{\frac{1}{2}\frac{1}{2}}(-\frac{\pi}{2}) = d^{\frac{1}{2}}_{-\frac{1}{2}-\frac{1}{2}}(-\frac{\pi}{2}) = \frac{1}{\sqrt{2}}$$

$$d^{\frac{1}{2}}_{-\frac{1}{2}\frac{1}{2}}(-\frac{\pi}{2}) = -d^{\frac{1}{2}}_{\frac{1}{2}-\frac{1}{2}}(-\frac{\pi}{2}) = -\frac{1}{\sqrt{2}}$$

Hence $\phi^{\frac{1}{2}}_{pp}$ and $\phi^{\frac{1}{2}}_{np}$ both have relative decay magnitude $|1 - i\alpha_1 e^{io}|^2$, while $\phi^{\frac{1}{2}}_{pn}$ and $\phi^{\frac{1}{2}}_{nn}$ both have $|1 + i\alpha_1 e^{io}|^2$ along the x-axis. In the x direction the asymmetry ratio has magnitude $\frac{2\sin\Theta}{\alpha_1 + \alpha_1^{-1}}$.

In order to use the y-direction as the axis of quantization there is a further non-zero Euler rotation angle. The components of opposite parity acquire opposite spins as already discussed, and there is no asymmetry.

Measurements in the x and z directions using a known polarization in the y direction allows calculation of Θ. Measurements, using a completely polarised beam, made with the detector axis at an angle Θ to the z axis in the direction of the x axis give a maximum value of the asymmetry. Measurements at right angles to this axis give zero asymmetry. It is obviously crucial for the quantities α_1 and Θ to vary slowly within the energy resolution of the experiment.

While this work was in preparation it was noticed that some experimental evidence of mass asymmetry at much lower energies had been reported[9]. These experiments have been interpreted in terms of a parity-violating weak interaction[10] component. It is worth noting that the p-wave amplitude at one eV for a fissionable nuclide is about 10^{-4} of the s-wave amplitude. It should also be noted that the design reported for the experiment would be expected to eliminate interference from p-waves. As mentioned above, positive parity orbital angular momentum gives rise to even projections on an axis perpendicular to the beam. Odd parity orbital angular momentum similarly has odd integer projections. Interference of saddle-point states with differing parity and the same projected spin is only expected perpendicular to the beam if it is also perpendicular to the direction of polarization.

In a direction perpendicular both to the beam and an axis of polarization, interference can lead to mass asymmetry which can be interpreted without a need to invoke weak interactions or parity violation. The experiments reported need especial care to ensure that the polarization is indeed in the planned direction.

The investigation reported here was undertaken because one case appeared especially favourable. In addition, low energy experiments now seem possible. Fission experiments with an unpolarised beam, but well resolved in energy, should display fore-and-aft mass asymmetry, changing sign as a p-wave resonance is passed. A similar effect would appear with a beam polarized in a direction perpendicular to the line of flight. The asymmetry then is detected in the third perpendicular direction.

Thanks are due to Drs. J.W. Boldeman and R.L. Walsh for helpful suggestions concerning presentation.

References

1. V.M. Strutinsky, Nucl. Phys. A95 (1967) 420.
   J.E. Lynn, AERE-R5891 (Harwell 1968)
   Int.Symp. Nuclear Structure, Dubna 1961, (IAEA, Vienna, 1968) p. 463.
   H. Weigmann, Z. Phys. 214 (1968) 7.
2. V.M. Strutinsky, Nucl. Phys., A122 (1968) 1.
3. G. Yuen, G.T. Rizzo, A.N. Behkami and J.R. Huizenga, Nucl. Phys. 171 (1971) 614.
   G.D. James, J.E. Lynn and L.G. Earwaker, Nucl. Phys. 189 (1972) 225
   J. Blons, C. Mazur, D. Paya, M. Riburg and H. Weigmann, Phys. Rev. Lett. 41 (1978) 1282.
4. J.W. Boldeman, D. Gogny, A.R. de L. Musgrove and R.L. Walsh, Phys. Rev. C 22 (1980) 627.
5. H.C. Pauli and T. Ledergerber, Nucl. Phys. 175 545.
6. P. Moller and R. Nix, IAEA Symp. Physics and Chemistry of Fission, Rochester (1973), Vol. 1, p. 103.
7. W. Haeberli, Fast Neutron Physics Part II, Marion and Fowler (ed) Interscience Publishers Division of John Wiley, New York and London (1963), Ch, VG. p. 1379.
8. T.A. Welton, ibid, Ch. VF, p. 1317-1377.
9. V.V. Vladimirskii and V.M. Andreev J.E.T.P. (USSR) 41, p. 663-665, G.V. Dunilyan et al., Symposium in Physics and Chemistry of Fission, Jülich 1979, p. 111.
10. A.P. Budnik and M.S. Rabotnoy, Phys. Lett. 46B, 155, V.V. Flambaum and O.P. Sushkoy, Phys. Lett. 94B, 277.

SYSTEMATICS OF FISSION BARRIER HEIGHTS AND THE NUCLEAR SURFACE TENSION CONSTANT

S.K. Kataria, S.S. Kapoor and V.S. Ramamurthy
Bhabha Atomic Research Centre, Trombay
Bombay 400 085, India

A global analysis of available fission barrier data without any explicit assumption on the functional form of the surface energy coefficient has been carried out. The deduced values are also consistent with the available angular distribution data. The present work brings out clearly the density dependence of the nuclear surface tension (σ) which is of significance in the accurate predictions of fission barrier heights of nuclei which are far off beta stability.

Introduction

In the framework of the Liquid Drop Model (LDM), the macroscopic part of the fission barrier heights of nuclei are sensitive to their fissility parameters, defined as the ratio of the Coulomb energy to twice the surface energy, both taken for the spherical configuration of the nucleus. In the same model, the shape of the nucleus at the fission transition state which decides the fragment angular distribution is also determined by the fissility parameter. Theoretical estimates of the fissility parameters are based on the Coulomb radii of the nuclei and the nuclear surface energies, both obtained from the semi-empirical nuclear mass formulae. While the theoretically estimated fissility parameters are reasonably consistent with the available experimental data on macroscopic fission barrier heights and fragment angular distributions, a number of systematic deviations have been pointed out in the past [1-2], which can not be resolved by a simple readjustment of the mass formula coefficients. In the present work, we have analysed the experimental data on macroscopic fission barrier heights and effective moments of inertia of the saddle point shapes deduced from angular distributions for a large number of nuclei in the mass region $100 < A < 250$. It was found that while the two sets of data are mutually consistent, the fissility parameters themselves do not follow the Liquid drop or the droplet model predictions. Possible reasons for these deviations are discussed.

Determination of fissility parameters from macroscopic fission barrier heights and fragment angular distributions

The fissility parameter x of a nucleus is defined as the ratio $E_c^\circ/2E_s^\circ$, where E_c° and E_s° are the Coulomb and the surface energies of the nucleus in its spherical nuclei. For a given fissility, the shape and the energy in units of E_s° of the saddle point have been numerically computed by Cohen and Swiatecki [3]. Making use of these results and the experimental macroscopic fission barrier heights, one can deduce the fissility parameters if one knows either the Coulomb energy E_c° or the surface energy E_s°. For the present analysis, the experimental macroscopic fission barrier heights for nuclei in the mass region $100 < A < 250$ were taken from a number of compilations in literature [4-6]. Starting from different prescriptions for the Coulomb energy E_c°, the fissility parameters of nuclei, the effective surface tension constants, and the saddle point effective moments of inertia were computed. Fig. 1 shows a plot of the deduced effective moments of inertia with the conventional prescription for the Coulomb energy $E_c^\circ = 0.72\ Z^2/A$. Also shown in figure are the effective moments of inertia parameters obtained from the angular distribution data. It can be seen that the two sets of data are in quite good agreement over the entire mass region. This agreement can be interpreted to imply that the concept of a deforming liquid drop provides a consistent picture of the fission process both in terms of the barrier height and the barrier shape. However the deduced surface tension constants exhibit the same discrepancies noticed earlier in literature, namely, the data does not follow the universal form $A_s = C_s(1 - K_sI^2)$ over the entire mass region. In particular, medium

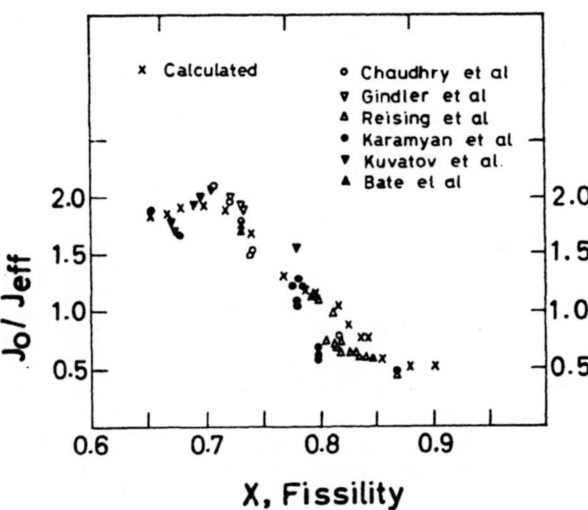

Fig. 1. A plot of reciprocal moment of inertia at saddle point in normalized used (J_o/J_{eff}) obtained from fission fragment angular distribution data taken from Ref. [1]. The crosses (x) represent the values deduced by fitting the macroscopic fission barrier heights.

mass nuclei have a substantially lower surface tension than what is predicted by the liquid drop or droplet model.

As was mentioned earlier, a calculation of the fissility parameter and the surface energy requires a knowledge of the Coulomb energy while it is customary to start with the assumption that the radius R of a nucleus is given by $R = r_o A^{1/3}$, leading to the Coulomb energy expression $E_c^o = a_c Z^2/A^{1/3}$ experimental Coulomb radii exhibit considerable deviations from this simple relation [7]. In particular strong isospin dependence and shell effects are known to be present [7]. Since any isospin dependence of the Coulomb energy is expected to affect the deduced isospin dependence of the surface tension constant, we have examined this aspect in more detail. Fig. 2 shows a plot of the deviation of the experimental radii from the predictions of simple LDM relations $R = r_o A^{1/3}$ and $R = r'_o Z^{1/3}$ are shown for the different isotopes of the nucleus. It can be seen that both those forms represent rather poorly the isospin dependence of the nuclei. On the basis of a global analysis of a large number of experimental Coulomb radii, we have arrived at the following empirical relation.

$$R = \alpha(ZA)^{1/6} + \beta(ZA)^{1/3} + c$$

for the Coulomb radii. As shown in Fig. 2, this relation provides a much better description of the isospin dependence of the Coulomb radii. The smooth form also ensures that the shell effects are smoothed out on the spirit of the liquid drop model.

K. H. Böckhoff (ed.), Nuclear Data for Science and Technology, 753–754.

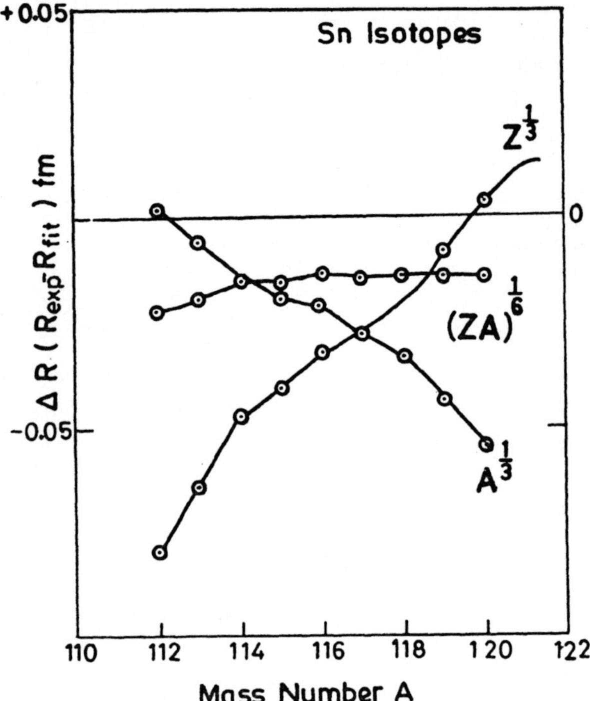

Fig. 2.

With the nuclear radii calculated using this empirical relation and the fissility parameters X fitted to the experimental fission barrier heights, we have obtained the nuclear surface tension constand σ per fm^2. These values of $\sigma(I^2)$ do not show isospin dependence for any isotopic chain in the preactinide and actinide mass region [4-8]. This observation raises serious doubts whether the expression $a_s' = C_s(1 - K_s I^2)$ for the surface tension constant is not a simple consequence of the constant radius parameter r_o approximation. Further the maximum deviations of σ from average trends have been observed for those nuclei which have smaller central densities. Fig. 3 shows a plot of σ versus the effective radius parameter r_o' related to central density ($n_c = 3/4\,\pi\, r_o'^3$). It can be seen that there is an unambiguous correlation between deduced σ values and r_o' parameter, and the much reduced values of σ for medium nuclei are not due to any new physical feature but can be correlated to the central density dependence of the nuclear surface tension. Recently such correlations have been the

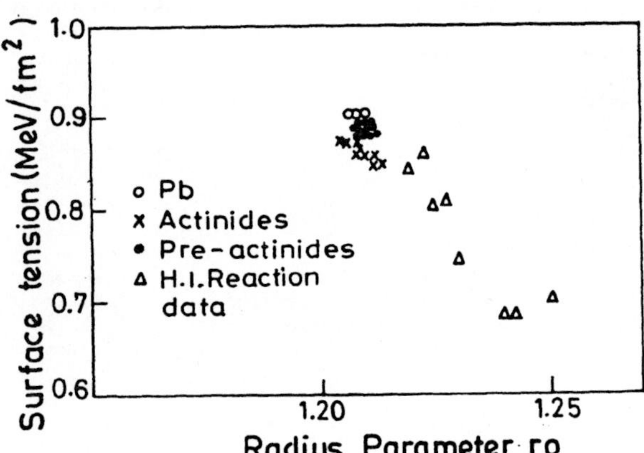

Fig. 3. A plot of deduced values of surface tension constant per fm^2 (σ) versus effective radius parameter r_o' obtained from eqn. (1)

subject of a number of theoretical investigations of nuclear surface tension [9-10], and the present work provides a basis for evaluating these models.

Summary

A global analysis of macroscopic fission barrier heights over the periodic table has been carried out. Unlike earlier analysis, the nuclear radii have been taken from a smoothed fit to the experimental radii using an expression involving both Z and A. It was found that the usual linear dependence of surface tension on I^2 may be a consequence of the constant density approximation. We have also shown that the data including those of medium mass nuclei becomes consistent if one takes into account the natural density dependence of surface tension.

References

1. W.D. Myers, "Droplet Model of Atomic Nuclei" Plenum Press, New York (1977)

2. R.W. Hasse, Ann. of Phys. 68 (1971) 377

3. S. Cohen, W.J. Swiatecki, Ann. of Phys. 22 (1963) 406

4. H.C. Pauli, T. Ledergerber, Nucl. Phys. A175 (1971) 545

5. A.V. Ignatuk, M.G. Itkis, V.N. Oklovich, G.N. Smirenkin, A.S. Tishin, Sov. J. Nucl. Phys. 21 (1975) 612

6. M. Beckermann, M. Blann, Phys. Rev. C17 (1978) 1615

7. I. Angeli, M. Csatlos, Nucl. Phys. A288 (1977) 480

8. S.S. Kapoor, V.S. Ramamurthy, Pramana 5 (1975) 129

9. W. Stocker, Nucl. Phys. A342 (1980) 293

10. C. Campi, S. Stringari, Nucl. Phys. A337 (1980) 313

ON THE VALENCE CONTRIBUTION TO THE RADIATIVE WIDTH OF s-WAVE RESONANCES IN Fe AND Ni ISOTOPES

A. Mengoni

Kernforschungszentrum Karlsruhe GmbH
Institut für Angewandte Kernphysik
P.O.B. 3640, D-7500 Karlsruhe
Federal Republic of Germany

G. Reffo, F. Fabbri

E.N.E.A.
Centro di Calcolo
Via Mazzini, 2
I - 40131 Bologna, Italy

Recent experiments and evaluations in the mass region A = 60 provide an improved data basis for a reinvestigation of the valence mechanism. For the test cases $^{56,58}Fe$ and $^{58,60}Ni$ we calculated the valence contribution to the radiative widths of s-wave resonances. The total width was derived as a sum of compound and valence contribution including the interference term. From a statistical analysis of the results there appears to be no necessity for the inclusion of a valence contribution in the investigated isotopes.

Introduction

In recent years the necessity for a valence contribution to the radiative width of the neutron resonances in some structural materials like Iron and Nickel was illustrated in several publications[1,2,3]. With the availability of recent, more precise experiments it was realized that the preceding work was based on resonance data affected by systematic errors[4]. In the light of these observations we have reinvestigated the valence hypothesis. To this end we have considered recent measurements of the s-wave resonances in $^{56,58}Fe$[5,6] and an evaluation for $^{58,60}Ni$[7] which provided values for the total and partial radiative widths.

A rigorous application of the Brink-Axel model and the optical model approach were used for the theoretical estimates of compound nucleus and valence contributions, respectively. In addition, we introduced an interference term between the amplitudes of partial transitions.

Calculations

Valence Model. The necessity of including a channel contribution to the radiative width was first recognized by Lane and Lynn[8]. This contribution introduces a component in the gamma transition probability related to the incident channel quantity $\Gamma_{\lambda n}$ (neutron width).

For the derivation of the fundamental relation between partial gamma width $\Gamma_{\lambda\gamma\mu}$ and the neutron width we follow Lynn[9]: From the basic assumption of a Lorentzian distribution for the coefficients in the initial state expansion (intermediate coupling model) it follows that the gamma amplitude for a single "valence" transition is:

$$(\Gamma_{\lambda\gamma\mu}^{1/2})^{VAL} = \left(\frac{16\pi}{9}\right)^{1/2} K_\gamma^{3/2} \bar{e}\, C_\lambda C_\mu I_{\lambda\mu}(E_\lambda) A_{\lambda\mu} \qquad (1)$$

where K_γ is the gamma-wave number and $\bar{e}$ the effective charge. $|C_\lambda|^2 = \Gamma_{\lambda n}/2 P_\ell \gamma^2_{sp}$, where γ^2_{sp} is the single particle reduced width for the initial state λ and P_ℓ is the penetration factor for incident ℓ-wave neutrons, $|C_\mu|^2$ is the spectroscopic factor for the final state μ and $A_{\lambda\mu}$ is a geometrical factor involving only angular momentum coupling coefficients.

Different from Lynn, and according with ref. 10 the initial single particle state is a scattering state. This assumption has the advantage to avoid the large uncertainties in the overlap integral $I_{\lambda\mu}(E)$ coming from the uncertainty in the energy of the initial bound state as adopted in the model of Lynn (up to a factor 5 for 3 s1/2 → 2 p3/2 transitions in the A=60 mass region and for initial bound energies ranging from 0.05 to 0.75 MeV). The radial wave functions $\psi_o^E\lambda(r)$ and $u_1(r)$ in

$$I_{\lambda\mu}(E_\lambda) = \int_0^\infty \psi_o^E\lambda(r)\, r\, u_1(r)\, dr \qquad (2)$$

were obtained from the solution of the Schrödinger equation for the neutron scattering in an optical model potential and for the wave function of a single particle state (s.p.s.) bound in a Saxon-Wood well, respectively.

The optical model parameters were fitted to the strength functions and the scattering radius and the parameters for the Saxon-Wood potential to the quantum characteristics E,n,l,j of the low-lying levels.

The total valence contribution to the widths of single resonances can be obtained by summing over all final states μ, $\Gamma_\lambda^{VAL} = \sum_\mu \Gamma_{\lambda\gamma\mu}^{VAL}$, whereas the average valence contribution over all the resonances is in this case:

$$\langle\Gamma_{\lambda\gamma}^{VAL}\rangle_\lambda \propto \langle D\rangle_\lambda S_o$$

where $\langle D\rangle_\lambda$ is the level spacing of the s-wave resonances and S_o is the s-wave neutron strength function.

The calculations of average total valence widths, the total valence widths for single resonances and the partial valence widths for single transitions to discrete levels, (for given resonances), allow for a comparison to as many different types of experimental values.

Statistical Model. For the calculation of the radiative width contribution via the compound nucleus mechanism, the Brink-Axel model was adopted as applied in refs. 11,12.

The statistical fluctuations of the total radiative widths Γ_γ^{CN} were considered. To do this, according to ref. 14, an effective number of degrees of freedom was estimated from the variance of the lumped distribution of all partial transitions:

$$\nu_{eff} = \frac{(\bar{\Gamma}_\gamma^{CN})^2}{\sum_\mu (\Gamma_{\gamma\mu}^{CN})^2} \qquad (3)$$

Interference Term. In spite of the completely different mechanisms (compound and valence) leading to the capture of a neutron in a resonance state, we have that the amplitude for a gamma transition $\lambda \to \mu$ is:

$$(\Gamma_{\lambda\gamma\mu}^{1/2})^T = (\bar{\Gamma}_{\lambda\mu}^{1/2})^{CN} + (\Gamma_{\lambda\gamma\mu}^{1/2})^{VAL} \qquad (4)$$

and therefore, under the limit assumption that the

K. H. Böckhoff (ed.), Nuclear Data for Science and Technology, 755–757.

Table I Comparison between experimental radiative widths and the results of our calculations. Details are explained in the text

Isotope	E_λ (keV)	$\Gamma^{o}_{\lambda n}$ (eV)	$\Gamma^{exp}_{\lambda\gamma}$ (eV)	Ref.	$\bar{\Gamma}^{CN}_{\gamma} \pm$ S.D. (eV)	$\Gamma^{VAL}_{\lambda\gamma}$ (eV)	$\bar{\Gamma}^{CN}_{\gamma}+\Gamma^{VAL}_{\lambda\gamma}$ (eV)	$\Gamma^{T}_{\lambda\gamma}$ (eV)
^{56}Fe	27.66	8.72	1.04	A		0.145	1.215	0.646
	73.9	1.97	0.885			0.055	1.125	0.756
	83.4	4.16	0.497			0.125	1.195	0.641
	129.7	1.32	(1.0)	B	1.07 ± 0.47	0.050	1.120	0.769
	140.2	7.12	(1.4)		(ν_{eff}=11.4)	0.282	1.352	2.183
	169.2	2.43	(1.0)			0.105	1.175	1.683
	187.8	8.36	(1.6)			0.381	1.451	2.417
		ρ=0.57	$\bar{\Gamma}^{exp}_{\lambda\gamma}$ =1.06					$<\Gamma^{T}_{\lambda\gamma}>_\lambda$ =1.59±0.47
^{58}Fe	10.42	2.81	0.790	A		0.025	0.928	0.753
	43.30	24.99	1.247		0.903±0.44	0.387	1.290	1.986
	66.70	3.55	0.427		(ν_{eff}=9.4)	0.073	0.976	0.674
	93.0	26.53	1.0			0.574	1.477	2.325
		ρ=0.83	$\bar{\Gamma}^{exp}_{\lambda\gamma}$ =0.866					$<\Gamma^{T}_{\lambda\gamma}>_\lambda$ =1.38±0.44
^{58}Ni	15.4	9.19	1.46±0.22	C		0.062	2.242	1.745
	63.0	14.34	2.3 ±0.3			0.303	2.483	3.573
	107.7	4.57	3.8 ±0.8			0.144	2.324	3.079
	124.0	2.13	3.5 ±0.6		2.18±0.896	0.075	2.255	2.798
	136.8	4.76	2.2 ±0.4		(ν_{eff}=11.8)	0.179	2.359	1.520
	139.7	9.26	2.2 ±0.5			0.353	2.533	1.353
	146.5	15.68	3.9 ±1.9			0.616	2.796	4.356
		ρ=-0.35	$\bar{\Gamma}^{exp}_{\lambda\gamma}$=2.637					$<\Gamma^{T}_{\lambda\gamma}>_\lambda$=2.73±.90
^{60}Ni	12.30	23.98	2.73±0.5	C		0.127	1.177	1.668
	28.60	4.73	0.6 ±0.15			0.044	1.094	0.805
	42.90	0.58	0.98±0.3			0.007	1.057	0.941
	65.12	2.35	1.9 ±0.3		1.05±0.43	0.037	1.087	1.355
	86.35	1.12	1.5 ±0.3		(ν_{eff}=12.2)	0.022	1.072	1.276
	97.79	2.78	1.2 ±0.25			0.058	1.108	1.441
	107.77	1.86	1.35±0.25			0.041	1.091	1.372
		ρ=0.76	$\bar{\Gamma}^{exp}_{\lambda\gamma}$=1.466					$<\Gamma^{T}_{\lambda\gamma}>_\lambda$=1.38±0.43
$\bar{\xi}_\lambda$±S.D.	(see eq. 6)				-0.16±0.47		-0.03±0.43	-0.00±0.37

A) Values from ref. 5; B) From ref. 1 with an empirical correction of the values obtained in that measurement; C) Values from ref. 7.

Table II Comparison of experimental values for partial radiative widths and the results of our calculations. For details see text.

Isotope	E_λ (keV)	E_μ (MeV)	J_μ^π	$\bar{\Gamma}^{CN}_{\gamma\mu}$ (eV)	$\Gamma^{VAL}_{\lambda\gamma\mu}$ (eV)	$\Gamma^{T}_{\gamma\mu}$ (eV)	$\Gamma^{exp}_{\gamma\mu}$ (eV)
^{56}Fe	27.66	0.0	$1/2^-$	0.183	0.012	0.103	0.145 ± 0.025
		0.014	$3/2^-$	0.182	0.071	0.026	0.035 ± 0.013
^{58}Ni	15.4	0.0	$3/2^-$	0.339	0.035	0.157	0.124 ± 0.017
		0.465*	$1/2^-$	0.305	0.018	0.176	0.11 ± 0.019
	63.0	0.0	$3/2^-$	0.339	0.184	1.022	0.258 ± 0.033
		0.465*	$1/2^-$	0.305	0.076	0.685	0.114 ± 0.016
		0.878	$1/5^-$	0.237	0.015	0.372	0.078 ± 0.012
^{60}Ni	12.30	0.0	$3/5^-$	0.178	0.06	0.444	0.514 ± 0.072
		0.283	$1/2^-$	0.149	0.057	0.38	0.289 ± 0.046

* The measured energy (Ref. 10) was 0.471 MeV, but we adopted the value from ref. 13.

phase coefficient is 1:

$$\Gamma^{T}_{\lambda\gamma\mu} = \bar{\Gamma}^{CN}_{\gamma\mu} + \Gamma^{VAL}_{\lambda\gamma\mu} + 2(\bar{\Gamma}^{1/2}_{\gamma\mu})^{CN}(\Gamma^{1/2}_{\tau\gamma\mu})^{VAL} \tag{5}$$

The sign of the third term in this equation is not known a priori because the unknown signs of the coefficients C_λ and C_μ in eq. 1. In our calculations we we choose the appropriate sign by comparison with the experimental values of the gamma widths.

Comparison with Experiment

Recent experimental information[5,7] for total and partial[5] radiative widths of s-wave resonances in ^{56,58}Fe and ^{58,60}Ni were used to test the predictions of the illustrated model.

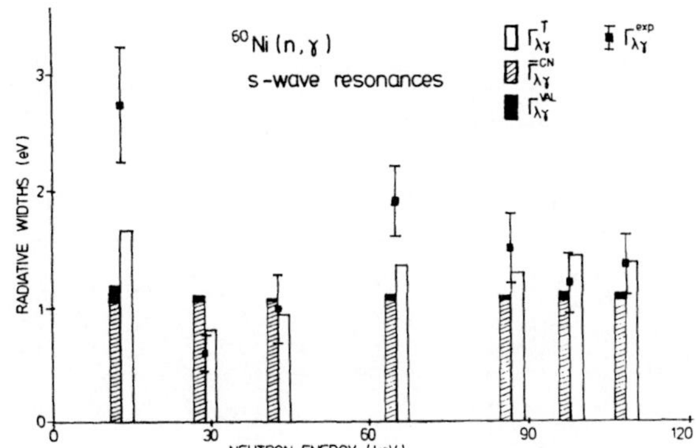

Fig. 1 Experimental and calculated radiative widths for some resonances in ^{60}Ni.

Total radiative widths $\Gamma^{exp}_{\lambda\gamma}$ are given in Table I with the mean values for each isotope.

In column 5 the Brink-Axel estimate for the total radiative width $\bar{\Gamma}^{CN}_{\gamma}$ is given with ν_{eff} and the standard deviation (S.D.) of the corresponding statistical distribution.

From the comparison of various quantities in columns 4 and 5 one sees that the fluctuations observed in $\Gamma^{exp}_{\lambda\gamma}$ over the resonances are in any case within the large variances characterizing $\bar{\Gamma}^{CN}_{\gamma}$.

However, some degree of correlation was found between neutron width and radiative width as one can see from the correlation coefficient $\rho(\Gamma^{o}_{\lambda n},\Gamma^{exp}_{\lambda\gamma})$ given in column 2, and we attempted to verify whether such a correlation may be explained in terms of the valence mechanism. The results of our calculations for $\Gamma^{VAL}_{\lambda\gamma}$ are listed in column 6.

The simple sum $\bar{\Gamma}^{CN}_{\gamma}+\Gamma^{VAL}_{\lambda\gamma}$ is given in column 7 while the total radiative width of eq. (5) is given in the last column 8 assuming that for any given resonance the sign of the interference term (the third term in eq. 5) is the same for all the final states μ.

The respective average radiative width:

$$<\Gamma^{T}_{\lambda\gamma}> = \bar{\Gamma}^{CN}_{\gamma} + <\Gamma^{VAL}_{\lambda\gamma}>_{\lambda}$$

over all the resonances λ are given for any isotope in column 8 with its standard deviation. In this case the interference drops because of the random distribution of the sign over many resonances. Although ν_{eff} for the compound contribution is much larger than for valence transitions (∿11 and ∿2, respectively), one has that

$$\text{S.D.}(CN) \gtrsim \Gamma^{VAL}_{\lambda\gamma}$$

because the valence width is at most 30% of the total width. As a first result this illustrates the role of statistical fluctuations in the radiative width.

In order to enlarge the size of our sample we also performed a statistical analysis over all isotopes and characterized it with global parameters like the mean of the quantity

$$\xi_\lambda = (1 - \frac{\Gamma^{exp}_{\lambda\gamma}}{\Gamma^{calc}_{\lambda\gamma}}) \tag{6}$$

and the standard deviation of the corresponding distribution (bottom of columns 5,7,8). In general, we do not see any appreciable difference by comparing ξ from column 5 with the respective values from columns 7,8. This means, that in spite of the observed correlations which seem to suggest the existence of non-statistical effects, these are in any case smaller than the statistical fluctuations of the compound nucleus contribution.

A similar situation also holds for the comparison of experimental and calculated partial widths for single transitions as showed in Table II. In fact, in spite of the excellent agreement obtained introducing the interference term, the situation for the fluctuations of $\Gamma^{CN}_{\lambda\mu}$ is even worse because in this case the number of degrees of freedom is 1 (Porter-Thomas distribution) with a standard deviation of $\sqrt{2}\cdot\Gamma_{\lambda\mu}$).

From the present analysis one may say that the experimental data could be explained in terms of the compound nucleus contribution and its statistical fluctuations, only. In addition, it is not possible to draw any conclusion on the validity of the valence hypothesis because if such an effect exists, it is hidden by the statistical fluctuations of the dominant compound nucleus contribution.

Our investigation appears to contradict the results of similar work[1,2,3]. This is due mostly because in the recent measurements the large background from scattered neutrons has been taken into account, leading to a considerable loss of correlation between neutron and radiative widths.

In our opinion a decisive conclusion on the valence hypothesis cannot be made until experimental values of partial gamma widths for a statistically significant sample of resonance have been determined. Such a sample could reduce the drawbacks of the statistical fluctuations of the compound nucleus contribution. Test cases where the correlations are more pronounced and the level densities are high enough to reduce statistical fluctuations of $\bar{\Gamma}^{CN}_{\gamma}$ would be of particular interest.

References

1 B.J. Allen, A.R. de L. Musgrove, J.W. Boldeman, M.J. Kenny, R.L. Macklin, Nucl. Phys. A269 (1976)408.

2 B.J. Allen, R.L. Macklin, J. of Physics, G6 (1980) 381.

3 B.J. Allen, A.R. de L. Musgrove, Advances in Nuclear Physics, (New York: Plenum Press), Vol. 10, p. 129.

4 B.J. Allen, A.R. de L. Musgrove and W.K. Betram, Proc. Conf. Neutron Data of Structural Materials for Fast Reactors, Geel (1978) p. 497.

5 F. Käppeler, K. Wisshak, L.D. Hong, private communication.

6 H. Beer, R.R. Spencer, F. Käppeler, Z. Physik A294 (1978) 173.

7 F. Fröhner, Proc. Conf. Neutron Data of Structural Materials for Fast Reactors, Geel (1978) p. 138.

8 A.M. Lane, J.E. Lynn, Nucl. Phys. 17 (1960) 563.

9 J.E. Lynn, Theory of Neutron Resonance Reactions (Oxford, Clarendon Press) 1968.

10 A.M. Lane, S.F. Mughabghab, Phys. Rev. C10 (1974) 412.

11 G. Reffo, Winter Course in Nuclear Physics and Reactors, Trieste, 1978, published by IAEA as IAEA-SMR-43 (1980).

12 V. Benzi, G. Reffo, M. Vaccari, Proc. Conf. "Fission Product Nuclear Data", Bologna (1973), published by IAEA as IAEA-169 (1974) 123.

13 C.M. Lederer et al, Table of Isotopes, New York: Wiley and Sons, 1978.

14 H.Gruppelaar,G.Reffo,Nucl.Sci.Eng.62(1977)756.

RADIATION WIDTHS OF IODINE, CESIUM AND IRIDIUM NEUTRON RESONANCES

Zo In Ok, A.B.Popov, K.Trzeciak

Laboratory of Neutron Physics, Joint Institute for Nuclear Research, Dubna, USSR

The neutron transmission through natural samples of iodine, cesium and iridium were measured in the energy interval up to 200 eV at the JINR neutron spectrometer with a resolution of about 10 ns/m. In addition to neutron widths the new data about radiation widths were obtained from the analysis of the resonance shape. The average radiation widths were estimated for the investigated nuclei.

The determination of radiation widths is an old problem of neutron spectroscopy. The area method applied to transmission data does not provide high precision for Γ values from which one may than obtain $\Gamma - g\Gamma_n/g = \Gamma_\gamma$. The use of radiative capture data improves the precision of Γ_γ determination, but leads to the necessity of calibrating the gamma-detector efficiency and neutron beam. In last years in Dubna the shape method was applied for the estimation of radiation widths using the measured with good resolution transmission data (Ref.[1]). Our "shape method" program reported in Ref.[2] describes the transmission by the sum of Breit-Wigner isolated resonances and accounts for the Doppler broadening and for the resolution function of the neutron spectrometer. The following fitting parameters were used in the calculation: E_{oi}, $g\Gamma_{ni}$, $\Gamma_{\gamma i}$, the potential scattering radius R' and the constant introduced to make correction for the transmission normalization between resonances. The resonance spins were assumed known. As a rule the processing of the transmission curves was made simultaneously for the two sample thicknesses. The statistical accuracy of $2g\Gamma_n$ and Γ_γ for the resonances below 50 eV was better than 5%. However, we took into account also the uncertainty of the resolution function by variation of its parameters within ±10% limits, corresponding to the deviations of $2g\Gamma_n$ and Γ_γ being considered as errors.

Our choice of samples for the investigation is explained by the absence of radiation width data for iodine at the time we started the measurements and by very poor data on cesium and iridium (Ref.[3]). If one compares our results on iodine summarized in Ref.[4] with data reported recently in Ref.[5] it can be noted that most of $2g\Gamma_n$ values are in agreement within errors, but for a large discrepancy observed for the 31.2 and 45.3 eV resonances. The small 145.7 eV resonance has also a visibly different $2g\Gamma_n$ value. The value $2g\Gamma_n$ =974 meV from Ref.[5] is obviously a mistake for the 168.8 eV resonance. Let us note that the Γ_γ values from Ref.[5] are systematically lower than ours. We obtained $\bar{\Gamma}_\gamma = 120 \pm 9$, while Ref.[5] gave =90±10 meV. Such difference in data obtained by different research groups is not very rare (Ref.[6]). As always one finds it difficult to show the reason for the discrepancy. The iodine data in Ref.[5] were given with a reference to the Rohr's private communication about the new radiative capture data which was apparently used for the Γ_γ estimation. In this case the correctness of experimental data callibration is questionable.

Our data on cesium resonance parameters reported in Ref.[4] are in good agreement with the data from Refs.[3,5] our errors in determination of the $2g\Gamma_n$ and Γ_γ values being smaller.

The Table contains iridium data. Our $2g\Gamma_n$ values are in satisfactory agreement with the data from Ref.[3] but for the $2g\Gamma_n$ values for the 25.2; 26.1 and 41.3 eV resonances.

Using our results given in the Table and Ref.[4] one can obtain values of average radiation widths (we made average arithmetic estimations): $\bar{\Gamma}_\gamma$ =120±9, 118±7, 86±4 and 103±8 meV for ^{127}I, ^{133}Cs, ^{191}Ir and ^{193}Ir, respectively.

References

1. A.B.Popov, K.Trzeciak, Hvan Cher Gu. Yad.Fiz., 32, 603, 1980.
2. A.B.Popov, I.I.Shelontsev, N.Yu.Shirikova. JINR, 3-9742, Dubna, 1976.
3. S.F.Mughabghab, D.I.Garber. BNL-325, 3-d Ed., v.1, 1973.
4. A.B.Popov, K.Trzeciak, JINR, P3-81-721, Dubna, 1981.
5. S.F.Mughabghab, M.Divadeenam, N.E.Holden. Neutron Cross Sections, v.1, Part A, Academic Press, 1981.
6. A.B.Popov, K.Trzeciak, JINR, P3-81-19, Dubna, 1981.

Table Neutron resonance parameters of Iridium

A	E_o eV	Γ_γ meV present paper	Γ_γ meV Ref.[3]	$2g\Gamma_n$ meV Present paper	$2g\Gamma_n$ meV Ref.[3]
191	10.4	120±5		0.26±0.01	0.33±0.03
191	19.2	72±1		3.10±0.02	2.6 ±0.5
191	20.2	82±2		2.30±0.02	1.8 ±0.2
191	24.2			0.58±0.02	
193	24.5			4.4 ±0.4	4.3 ±0.4
193	25.0			5.2 ±0.8	4.2 ±0.4
191	25.2			7.4 ±1.2	14.0 ±2.5
193	26.1	103±4		4.84±0.08	1.7 ±0.3
191	26.9			0.12±0.02	
191	29.9	80±4		21.6 ±0.6	16.8 ±0.9
191	31.6	79±4		7.00±0.14	5.5 ±1.5
191	36.5	82±4		5.28±0.08	3.5 ±0.8
191	40.4	83±11		6.38±0.20	5.9 ±0.9
191	41.5			1.0 ±0.2	2.0 ±0.4
193	41.7			2.72±0.20	2.8 ±0.4
193	43.2		102±38	56 ± 2	50 ±6
191	44.6	66±17		1.0 ±0.4	1.2 ±0.3
191	45.7	108±16		2.8 ±0.2	2.3 ±0.5
191	51.1	74±7	87±20	48 ±2	49 ±3
193	52.6			4.0 ±1.0	5.3 ±0.5
191	52.8			7.8 ±1.6	8.2 ±1.6
193	54.1	93±11		17 ±1	14.3 ±1.8
191	56.9	104±22		2.26±0.10	2.6 ±0.7
191	62.9	95±12		6.8 ±0.4	5.3 ±1.3
191	64.9	87±12		8.0 ±0.2	7.2 ±1.8
191	65.8	71±10		8.8 ±0.4	11.4 ±2.5
191	68.1	72±20		9.8 ±0.6	11.6 ±2.7
193	69.4			70 ±10	84 ±5
193	71.7	143±32		2.36±0.12	2.2 ±0.3
191	76.6	88±23		7.0 ±0.4	7.4 ±1.8
193	77.9	94±20		32 ±6	42 ±4
191	78.9	113±20		18 ±2	20 ±4
191	80.6			10.8 ±1.0	11.2 ±2.2
193	80.9			3.0 ±0.6	2.7 ±0.9
191	82.0	102±50		3.0 ±0.2	3.0 ±0.7

K. H. Böckhoff (ed.), Nuclear Data for Science and Technology, 758.

MESURE DE LA SECTION EFFICACE DE CAPTURE RADIATIVE DU LANTHANE, DU BISMUTH, DU CUIVRE NATUREL ET DE SES ISOTOPES POUR DES NEUTRONS D'ENERGIE COMPRISE ENTRE 0,5 ET 3 MeV

J. Voignier, S. Joly, G. Grenier

Service de Physique Neutronique et Nucléaire
Centre d'Etudes de Bruyères-le-Châtel
B.P. N° 561
92542-MONTROUGE-CEDEX, France

Absolute neutron capture cross sections of ^{63}Cu, ^{65}Cu, Cu, La and ^{209}Bi have been measured in the 0.5 - 3.0 MeV energy range with a pulsed 4 MV Van de Graaff accelerator. Prompt capture gamma rays emitted by the sample were detected by a NaI spectrometer. The number of captures was determined from the ratio of the observed capture gamma ray spectrum to the gamma ray energy distribution calculated for one capture reaction.

[^{63}Cu(n,γ), ^{65}Cu(n,γ), Cu(n,γ), La(n,γ), ^{209}Bi(n,γ), capture cross sections, E_n = 0.5 - 3.0 MeV].

Introduction

La section efficace absolue de capture radiative de neutrons a été mesurée dans la gamme d'énergie 0,5 - 3,0 MeV pour les éléments ^{63}Cu, ^{65}Cu, Cu, La et ^{209}Bi. Cette mesure est à la limite des possibilités expérimentales actuelles. En effet, la section efficace de capture du bismuth et du lanthane est particulièrement faible dans ce domaine d'énergie, où elle représente respectivement 0,03% et 0,08% de la section efficace totale.

Deux méthodes sont utilisées habituellement pour mesurer les sections efficaces de capture : la méthode d'activation et la méthode de mesure directe des rayonnements γ émis par la capture des neutrons. La méthode d'activation est limitée aux cas où le noyau final peut se désintégrer avec émission d'un rayonnement mesurable avec une période pas trop courte pour pouvoir être mesuré en différé ; il faut aussi que les schémas de désintégration soient bien connus. D'autre part, l'influence des neutrons secondaires doit être correctement corrigée.

La méthode de mesure directe des rayonnements γ émis par la capture est plus générale. Ces rayonnements sont détectés à l'aide de gros scintillateurs liquides, de détecteurs de Moxon-Rae ou de Maier-Leibnitz (technique de pondération des amplitudes) ou de scintillateurs NaI (méthode du spectre intégré ou mesure du spectre d'énergie). C'est cette dernière méthode que nous avons utilisée.

La section efficace de capture est déterminée à partir du spectre des rayonnements γ en supposant que la distribution angulaire des rayonnements γ de capture est isotrope. L'angle solide de détection du scintillateur NaI étant petit (3.10^{-3}sr), la probabilité de détecter plus d'un rayonnement γ de la cascade est faible. On mesure effectivement la distribution en énergie des rayonnements γ de capture, ce que ne permettent pas les autres techniques de détection. Le nombre total de captures produites dans l'échantillon est donné par le rapport du spectre expérimental et du spectre calculé pour une capture. Au-dessus de l'énergie $E_\gamma = E_n$, la forme du spectre expérimental est utilisée pour calculer la distribution en énergie des rayonnements γ émis dans une capture. La partie basse énergie du spectre ($E_\gamma < E_n$) qui contient essentiellement des rayonnements γ des réactions (n,n'γ) et (n,γn') doit être extrapolée. L'erreur commise sur cette extrapolation est faible (1 à 2%) et elle est bien inférieure à l'erreur statistique (3 à 20%).

Conditions expérimentales

Le dispositif expérimental utilisé pour la mesure des sections efficaces de capture est représenté schématiquement sur la figure 1. Le faisceau de protons délivré par l'accélérateur Van de Graaff 4 MV de Bruyères-le-Châtel est d'abord pulsé à une fréquence de répétition de 2,5 MHz, puis regroupé à l'aide d'un aimant Mobley. La largeur de la bouffée de protons est d'environ 1 ns et le courant moyen de 10 μA. Les neutrons de 0,5 MeV sont produits par la réaction ^{7}Li(p,n) et ceux de 1-2-3 MeV à l'aide de la réaction T(p,n). La cible de lithium métallique est évaporée sur un support de tantale de 1 mm d'épaisseur et la cible de tritium est constituée de tritium adsorbé dans du titane sur un support d'or. Les cibles sont refroidies par un jet d'air froid comprimé et leur épaisseur correspond à une perte d'énergie des protons comprise entre 40 et 80 keV.

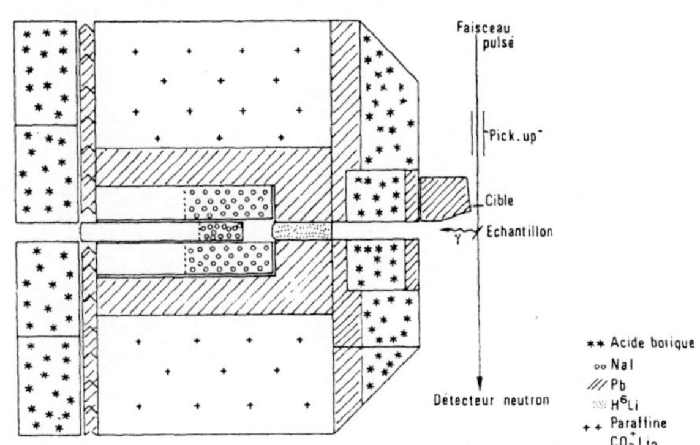

Fig.1. Dispositif expérimental utilisé pour les mesures de sections efficaces de capture radiative de neutrons.

Le flux de neutrons est mesuré à l'aide d'un long compteur directionnel (BF_3) par transmission à travers l'échantillon. L'efficacité de ce moniteur est connue avec une bonne précision (± 2 % à 1 MeV).

La technique du temps de vol est utilisée pour discriminer les rayonnements γ provenant de l'échantillon de ceux produits par diffusion inélastique ou par capture autour du détecteur ou dans la salle de mesure.

K. H. Böckhoff (ed.), Nuclear Data for Science and Technology, 759–761.

Les échantillons de ^{63}Cu et ^{65}Cu sont cylindriques (25 x 20) mm et leur masse correspond à 1 mole. L'enrichissement isotopique de l'échantillon ^{63}Cu est de 99,7% et celui de ^{65}Cu est 99,2%. Les échantillons de cuivre naturel (60 x 3) mm, de lanthane (52 x 4) mm et de bismuth (60 x 12) mm sont utilisés sous forme de disque.

Les rayonnements γ sont détectés par un cristal NaI de 7,6 cm de diamètre et de 15,2 cm de longueur entouré d'un cristal annulaire qui permet de diminuer le bruit de fond et de simplifier la fonction de réponse du cristal central. Les spectres d'amplitude sont déconvolués par la fonction de réponse du spectromètre puis corrigés de l'auto-absorption des rayonnements γ dans l'échantillon et de l'efficacité de détection.

Résultats et discussion

La section efficace absolue de capture est déterminée par la méthode du spectre d'énergie décrite en détails par ailleurs [1] [2] [3]. Les sections efficaces obtenues sont corrigées de l'anisotropie de la source de neutrons, de l'auto-absorption et de la diffusion multiple des neutrons dans l'échantillon. Ces corrections sont calculées par la méthode de Monte Carlo à l'aide du programme GAMSCT écrit initialement par D.L. Smith [4] pour des échantillons cylindriques. Nous avons modifié ce programme pour tenir compte des échantillons sous forme de disque et des neutrons d'énergie supérieure à l'énergie du dernier niveau excité pour lequel la section efficace de diffusion inélastique a été mesurée. La correction d'auto-absorption des neutrons dans l'échantillon annulant partiellement l'effet de la diffusion multiple, la correction finale n'est pas très importante. Elle atteint 14% dans le cas le plus défavorable (échantillon cylindrique des isotopes du cuivre à E_n = 3 MeV).

Les sections efficaces de capture de Cu, ^{63}Cu, ^{65}Cu, La et ^{209}Bi sont reportées sur les figures 2-3-4-5 et 6.

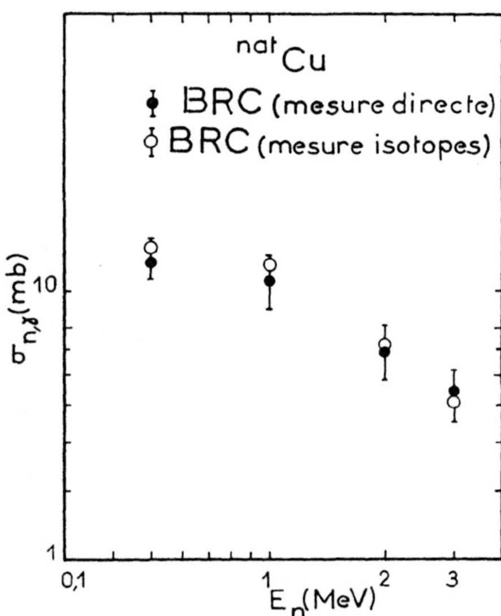

Fig. 2. Section efficace de capture du cuivre naturel (mesuré et déduit).

Nous avons porté sur la figure 2 la section efficace de capture du cuivre naturel. Il y a un bon accord entre notre mesure directe et les valeurs déduites de nos mesures faites sur les isotopes séparés, en tenant compte de l'abondance isotopique de l'élément naturel.

En ce qui concerne les isotopes du cuivre (figures 3 et 4) il existe un bon accord entre nos mesures et les valeurs de Tolstikov [5], Zaikin [6] et Goldberg [7].

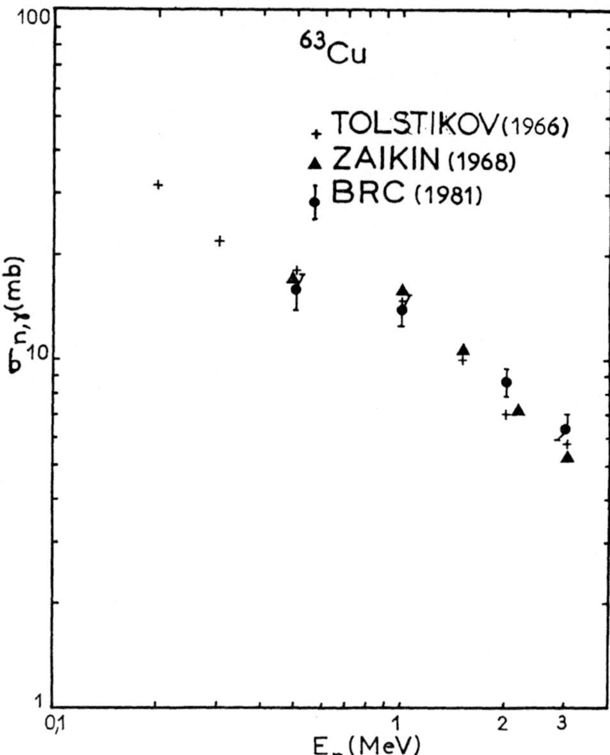

Fig. 3. Section efficace de capture de ^{63}Cu.

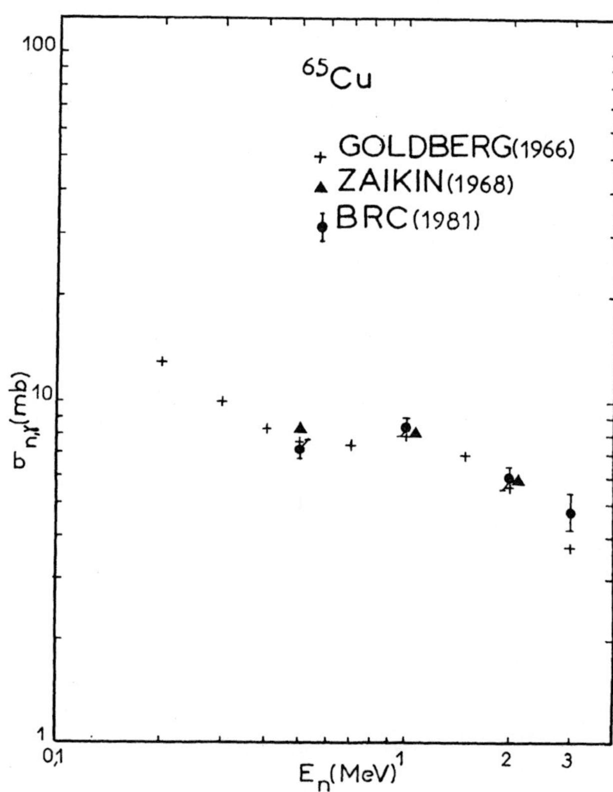

Fig. 4. Section efficace de capture de ^{65}Cu.

L'élément ^{139}La existe à l'état naturel avec une abondance isotopique de 99,9%. La figure 5 rassemble diverses mesures faites par activation sur ^{139}La dans le domaine d'énergie 100 keV-4 MeV [8 à 12]. Nous y

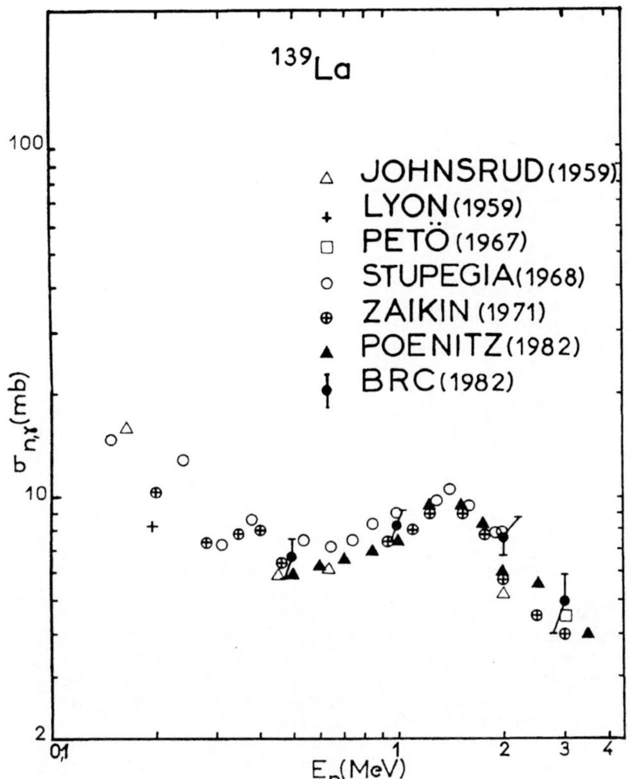

Fig. 5. Section efficace de capture de ^{139}La.

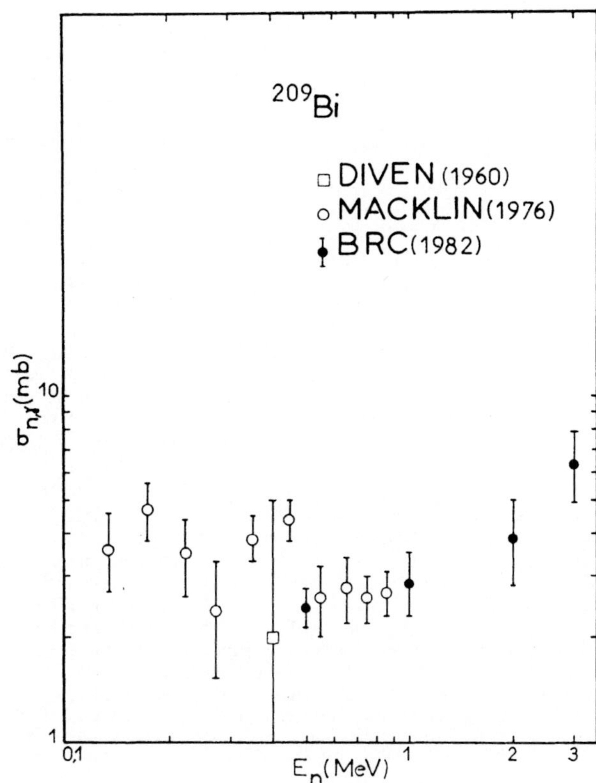

Fig. 6. Section efficace de capture de ^{209}Bi.

avons ajouté les résultats de Poenitz [13] et les nôtres obtenus avec le lanthane naturel. L'accord est généralement bon ; cependant les valeurs de Stupegia [11] et les nôtres sont 15 à 20% supérieures à celles de Zaikin [12] et de Poenitz [13].

En ce qui concerne le bismuth, nos résultats sont en bon accord avec la mesure de Macklin [14] effectuée jusqu'à 900 keV.

Références

1. S. Joly et Coll., Rapport CEA-R-4884 (1977).
2. S. Joly et Coll., Nucl. Sci. Eng. 70 53 (1979).
3. J. Voignier, S. Joly et G. Grenier, Rapport CEA-R-5089 (1981).
4. D.L. Smith, Rapport ANL/NDM-17 (1975).
5. V.A. Tolstikov et Coll., Atomnaya Energiya 21 45 (1966).
6. G.G. Zaikin et Coll., Atomnaya Energiya 25 256 (1968).
7. M.D. Goldberg et Coll., BNL 325, Suppl. 2, Vol. II A (1966).
8. A. Johnsrud, M. Silbert et H. Barshall, Phys. Rev. 116 927 (1959).
9. W. Lyon et R. Macklin, Phys. Rev. 114 1619 (1959).
10. G. Peto, Z. Miligy et I. Hunyadi, J. of Nucl. Energ. 21 797 (1967).
11. D. Stupegia, Marcia Schmidt, C. Keedy et A. Madson, J. of Nucl. Energ. 22 267 (1968).
12. G. Zaikin et Coll., Ukrainskij Fizicesky Zurnal 16 1205 (1970) et INDC (CCP) - 15U (1971).
13. W.P. Poenitz, NEANDC/NEA CRP Specialist's Meeting on Fast Neutron Capture Cross Sections Argonne Nat. Lab., 20-23 April 1982.
14. R. Macklin et J. Halperin, Phys. Rev. C14 1389 (1976).

AVERAGE RADIATION WIDTHS AND THE GIANT DIPOLE RESONANCE WIDTH

F.-K. Thielemann*

Max-Planck-Institut für Physik und Astrophysik
Garching b. München, FRG

and

M. Arnould†

Institut d'Astronomie et d'Astrophysique
Université Libre de Bruxelles, Belgium

The average E1 radiation width can be calculated in terms of the energy E_G and width Γ_G of the Giant Dipole Resonance (GDR). While various models can predict E_G quite reliably, the theoretical situation regarding Γ_G is much less satisfactory. We propose a simple phenomenological model which is able to provide Γ_G values in good agreement with experimental data for spherical or deformed intermediate and heavy nuclei. In particular, this model can account for shell effects in Γ_G, and can be used in conjunction with the droplet model. The Γ_G values derived in such a way are used to compute average E1 radiation widths which are quite close to the experimental values. The method proposed for the calculation of Γ_G also appears to be well suited when the GDR characteristics of extended sets of nuclei are required, as is namely the case in nuclear astrophysics.

1. Introduction

The total photon transmission function from a compound nucleus excited state is one of the key ingredients for statistical cross section evaluations. This function is normally dominated by the electric-dipole (E1) contribution, which can be calculated on grounds of the Lorentzian representation of the Giant Dipole Resonance (GDR), at least for nuclei with atomic mass $A \gtrsim 40$. In this model, the E1 transmission function for a photon of energy ε_γ in a nucleus with N neutrons and Z protons (A=N+Z) is given by[1,2]

$$T_{E1}(\varepsilon_\gamma) = \frac{8}{3}\frac{NZ}{A}\frac{e^2}{\hbar c}\frac{(1+\chi)}{mc^2}\sum_{i=1}^{2}\frac{i}{3}\frac{\Gamma_{Gi}\varepsilon_\gamma^4}{(\varepsilon_\gamma^2-E_{Gi}^2)^2+\Gamma_{Gi}^2\varepsilon_\gamma^2}, \quad (1)$$

where χ accounts for the neutron-proton exchange contribution, and the summation allows the application of the formula to statically deformed spheroidal nuclei. In these cases, the GDR indeed splits into an oscillation along (i=1) and one perpendicular (i=2) to the axis of rotational symmetry, and can be represented by the superposition of two noninterfering Lorentz lines with energy at maximum E_{Gi} and full width at half-maximum Γ_{Gi}. In spherical nuclei $E_{G1}(\Gamma_{G1})$ is to be set equal to $E_{G2}(\Gamma_{G2})$ in Eq.(1).

Many microscopic or more macroscopic models have been devoted to the calculation of the GDR energies and widths [e.g. refs.[3-5]], including in particular a droplet-model approach[6]. This model gives excellent results for the GDR energy. As far as the width is concerned, microscopic models can achieve some success in reproducing the experimental data [e.g. ref.[7]], but are not free from difficulties. On the other hand, more macroscopic models of the hydrodynamical type[8,9] can explain more or less correctly the average trends of the Γ_G variations with A, but cannot account for the pronounced observed shell effects.

The aim of this paper is to propose a phenomenological model for the GDR width which satisfactorily reproduces the experimental data for spherical and deformed $A \gtrsim 70$ nuclei, and which appears well suited namely for the calculation of total E1 transmission functions.

2. A Phenomenological Model for the GDR Width of Spherical Nuclei

We assume that the direct decay contribution to Γ_G is negligible. This is expected to be a generally acceptable approximation for the $A \gtrsim 70$ nuclei under consideration. In such conditions, Γ_G is expressed as

$$\Gamma_G = \Gamma_d + \Delta\Gamma. \quad (2)$$

In this expression, the damping width Γ_d results from the spreading of the simple dipole states due to the coupling to many more complicated intrinsic states of the nucleus. On the other hand, the broadening term $\Delta\Gamma$ translates the splitting up of the dipole strength over a certain energy range.

The damping width Γ_d has been calculated in many microscopic models [e.g. refs.[4,7]], in macroscopic approaches[6,11] making use of the one-body damping mechanism[10], as well as in hydrodynamical treatments where viscous damping leads approximately to $\Gamma_d \propto E_G^\delta$, with $1 \lesssim \delta \lesssim 2$, depending upon the model[8,9] (note that $\delta \simeq 1.8$-1.9 is derived from a fit of the experimental widths to a E_G^δ form). In view of the uncertainties still involved in these models, as well as in the light of certain other studies[12,13], δ values outside the above mentioned range do not appear to be absolutely excluded. We therefore explore the consequences of adopting δ in the whole $0 \leq \delta \leq 1.8$ range.

As far as the broadening term $\Delta\Gamma$ is concerned, it can be evaluated in the framework of the Dynamic Collective Model. Considering only the coupling between dipole oscillations and quadrupole surface vibrations, $\Delta\Gamma$ can be approximated, when relating the second moment of the strength function to the width, by[7,14] $\Delta\Gamma \simeq 2.35\sqrt{5/8\pi}\,E_G\beta_2$, where β_2 is the root-mean-square amplitude of the quadrupole surface vibration.

The above considerations allow rewriting Eq.(2) as

$$\Gamma_G = \alpha E_G^\delta + 2.35\sqrt{5/8\pi}\,E_G\beta_2, \quad (3)$$

where α is a free parameter. For each selected δ in the $0 \leq \delta \leq 1.8$ range mentioned above, α is derived from a least-squares fit to the experimental GDR widths of spherical nuclei for which E_G and β_2 are also known[15-19] [β_2 is related to the reduced electric quadrupole transition rate $B(E2,0^+\to2^+)$ by $\beta_2 = 4\pi\sqrt{B(E2,0^+\to2^+)}/(2ZeR^2)$].

Fig. 1 demonstrates that the Γ_G values derived in such a way from Eq.(3) can reproduce the experimental data very nicely for δ values in the whole considered range, even if the agreement is slightly better in the least squares sense for $\delta \simeq 0$. It is especially gratifying that the marked shell effects observed in Γ_G can be very satisfactorily accounted for.

3. A Droplet Model Evaluation of E_G and β_2

Having demonstrated the virtues of Eq.(3) when E_G and β_2 are known experimentally, the question now arises of its applicability when these quantities are not measured. We show that droplet-model estimates of E_G and β_2 can be adequately used in Eq.(3).

The E_G droplet model values

$$E_G = \hbar\left[\frac{8J}{R^2 m^*}\left(1+\mu-\frac{1+\varepsilon+3\mu}{1+\varepsilon+\mu}\right)^{-1}\right]^{1/2}, \quad (4)$$

where $\mu \equiv 3Jr_o/(QR)$, are in excellent agreement with experiment if the droplet model parameters J=36.8 MeV, Q=17 MeV, r_o=1.18 and ε=0.0768 are adopted, along with an effective mass m^*= 0.7 m.

* supported in part by NSF-grant (PHY79-23638) during a stay at CalTech
† Chercheur Qualifié F.N.R.S. (Belgium)

K. H. Böckhoff (ed.), Nuclear Data for Science and Technology, 762–765.

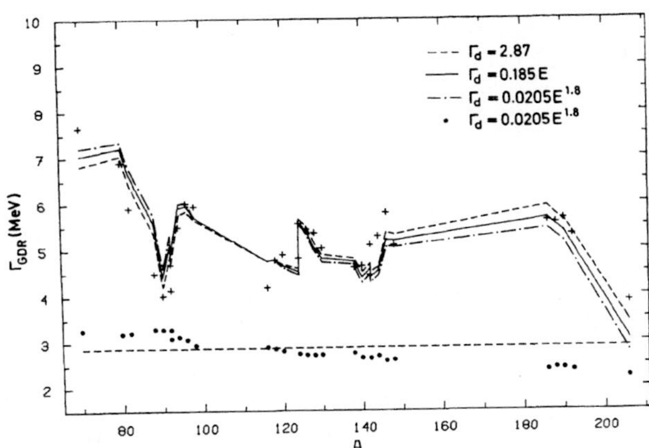

Fig.1. Comparison between Γ_G derived from Eq.(3) and from experiment (+) for various spherical nuclei whose E_G and β_2 values are also known. The various lines connect results calculated with the different displayed δ values. The Γ_d widths obtained with the extreme δ=0 and 1.8 values considered here are also indicated for comparison in the lower part. The experimental β_2 values are from ref.[15], while the experimental GDR characteristics are adopted from refs.[16-19]. Note that the displayed Os data correspond to single-Lorentz line fits. More recent data suggest that two-Lorentz curve fits characteristic of deformed nuclei (see Sec.4) might be more appropriate[20].

The droplet-model evaluation of β_2 is based on the relation[21]

$$\beta_2{}^2 = \frac{5}{2}\hbar\,(C_2D_2)^{-1/2}\ , \tag{5}$$

where C_2 and D_2 are the restoring force parameter and mass parameter, respectively. These quantities are related to the kinetic and potential energies associated to small β-deformation surface vibrations by[21]

$$E_{pot} \simeq \frac{1}{2}C_2\beta^2 \quad ; \quad E_{kin} \simeq \frac{1}{2}D_2\dot{\beta}^2 \tag{6}$$

to leading order in β.

The liquid-drop model can provide an estimate of C_2. However, the resulting C_{2LD} values do not agree very well with the experimentally derived

$$C_{2exp} = \frac{5}{2}E(2^+)(\frac{3}{4\pi}ZeR^2)^2/B(E2,0^+\rightarrow 2^+)\ , \tag{7}$$

where $E(2^+)$ is the energy of the first 2^+ excited state in even-even nuclei. As shown in Fig.2a, C_{2exp} may differ from C_{2LD} by more than a factor of 10. These discrepancies reveal the profound importance of the shell structure for the low-frequency quadrupole mode, and might be cured by a droplet model containing deformation-dependent shell correction terms[22]. Noting that E_{pot} [Eq.(6)] is just the difference in binding energy B due to deformation, C_2 is approximately given by

$$C_2 \simeq -\frac{\partial^2 B}{\partial\beta^2}\Big|_{\beta=0} \tag{8}$$

The C_{2DM} values derived from Eq.(8) and the droplet model of ref.[22] are in substantially better agreement than C_{2LD} with C_{2exp}, as shown in Fig.2b. The general trends, including shell structures, are in fact reproduced in a reasonable way.

Adopting these C_{2DM} values, β_2 can be estimated from Eq.(5) if D_2 is known. This latter quantity can be easily evaluated from $D_2 = \hbar^2C_2/E^2(2^+)$ in the case of even-even nuclei with known $E(2^+)$. If $E(2^+)$ is not measured, the simplest model for evaluating D_2 relies on the assumption of the irrotational flow of a constant density fluid. As can be seen in Fig.3a, the resulting values $D_{2irr} = (3/8\pi)AmR^2$ are, however, systematically lower than the experimentally deduced

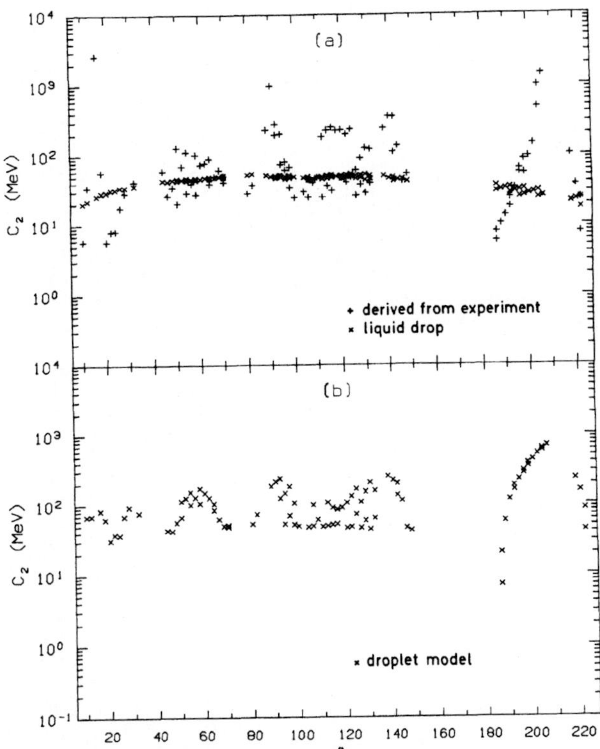

Fig.2.(a) Experimental (+) and liquid-drop (x) values of C_2; (b) C_{2DM} values based on the droplet model of ref.[22].

D_{2exp} data, and of course, do not exhibit the required shell structure. Those two shortcomings can be avoided by the purely phenomenological fit

$$D_2 = D_{2irr}\exp(2.42-0.195\ E_{shell}), \tag{9}$$

where E_{shell} is the shell correction term (in MeV) of the droplet model of ref.[22].

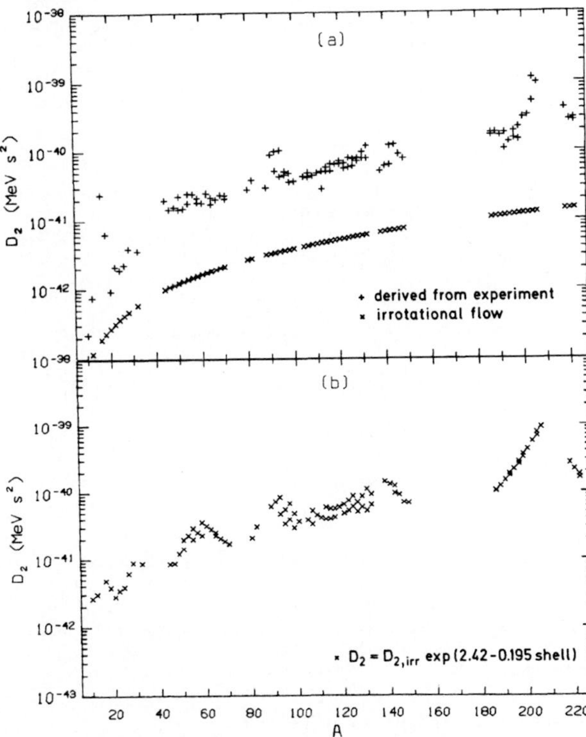

Fig.3.(a) Experimental (+) and irrotational flow (x) values of D_2; (b) D_2 values derived from Eq.(9) with the droplet model shell corrections of ref.[22]

Combining the predictions for C_{2DM} and D_2 [Eq.(9)] to determine β_2 [Eq.(5)], and adopting the droplet model predictions for E_G[Eq.(4)], Γ_G can be estimated from Eq.(3) in absence of experimental informations. Fig.4 compares the Γ_G values derived in such a way with all the experimental data for spherical nuclei made available in refs.[16-19]. The agreement is seen to be almost as good as in Fig.1. Certain discrepancies mainly arise from some inadequacies in the predicted C_2 values. As seen in Fig.2, this quantity is underestimated in the $A \approx 90$ region (N=50), while its rather large measured scatter in the Sn isotopes cannot be fully accounted for. Some defect in the Os isotopes could be traced back to an overestimate of C_2 or, as already discussed in connection with Fig.1, a two-Lorentz line representation might be more appropriate. In addition, the adopted droplet model is probably not very reliable in that mass region. Finally, in the $A \simeq 208$ region, the slight discrepancy already observed in Fig.1 is seen not to be significantly enhanced in Fig.4.

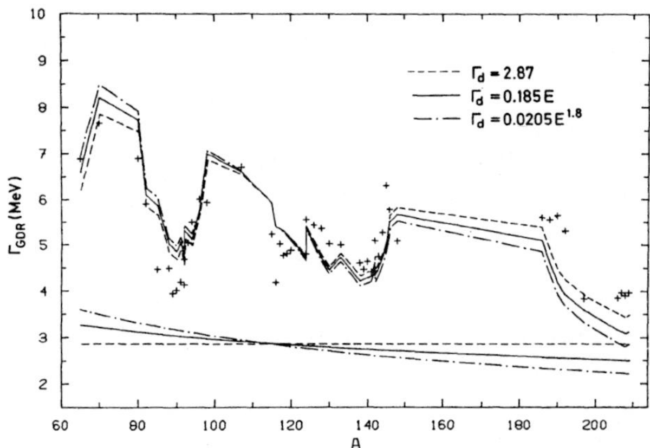

Fig.4. Experimental Γ_G values (+) compared with the results of Eq.(3) in which use is made of the droplet-model predictions for E_G[Eq.(4)] and β_2[Eq.(5)]. As in Fig.1, the various curves show the effect of different assumptions for $\Gamma_d(\delta=0,1,$ or $1.8)$.

4. Extension to Deformed Nuclei

No expression equivalent to Eq.(3) is available for deformed nuclei. In these conditions, our evaluation of $\Delta\Gamma$ for such nuclei merely relies on some analogy with the spherical case, in which Eq.(3) results from the adoption of an interaction hamiltonian proportional to first order to βE_G, where β is the actual amplitude of the quadrupole surface vibrations.

Our trial treatment for the deformed nuclei assumes that the vibrations are purely of the β-type, so that the interaction hamiltonian takes the simple form

$$H_{int} = H_1 + H_2; \quad H_i \propto E_{Gi} G_{i-1} \xi \quad (i=1,2);$$
$$G_\mu = (-1)^\mu \left(\frac{4\pi}{5}\right)^{-\frac{1}{2}} \left[\frac{-1}{1+(-\frac{1}{2})^\mu (\frac{4}{5}\pi)^{-\frac{1}{2}} \beta_o} + \frac{0.08}{1+0.08(-\frac{1}{2})^\mu (\frac{4\pi}{5})^{-\frac{1}{2}} \beta_o} \right], \quad (10)$$

where $\xi = \beta - \beta_o$ describes the displacement of the surface with respect to the static deformation β_o. As in Eq.(1), the subscripts 1 and 2 indicate vibrations along and perpendicular to the axis of rotational symmetry, respectively. In analogy with the spherical case, the broadening terms $\Delta\Gamma_{1,2}$ are then assumed to have the simple form

$$\Delta\Gamma_i = a_i |G_{i-1}| E_{Gi} \xi_2 \quad (i=1,2), \quad (11)$$

where ξ_2 is the root-mean-square value of ξ, and a_i are constants to be determined by comparison with experiment.

The kinetic and potential energies for the considered β-vibrations are still given by Eq.(6) in which ξ has to replace β. In addition, C_2 is given by Eq.(8) evaluated at $\beta=\beta_o$, while Eq.(9) is assumed to remain valid in deformed nuclei. In such conditions, ξ_2 can be calculated from Eq.(5) in which $\xi_2{}^2$ replaces $\beta_2{}^2$. Finally, E_{Gi} are evaluated from[24]

$$E_{G1} + 2E_{G2} = 3\,\overline{E}_G; \quad E_{G2}/E_{G1} = 0.911\,\eta + 0.089, \quad (12)$$

where $\overline{E}_G$ calculated from Eq.(4) is interpreted as an average peak energy, while the ratio η (diameter along the nuclear symmetry axis/diameter perpendicular to it) can be derived from a droplet model[22].

Using such a prescription for the evaluation of E_{Gi} and ξ_2, and generalizing Eq.(3) into $\Gamma_{Gi} = \alpha E^{\delta}_{Gi} + \Delta\Gamma_i$, where $\Delta\Gamma_i$ is given by Eq.(11), it is found that the values

$$\Gamma_{Gi} = 0.185\,E_{Gi} + 0.57\,i|G_{i-1}|E_{Gi}\xi_2 \quad (13)$$

are in good agreement with the experimental data for deformed nuclei ($A \simeq 75$; $150 < A < 180$; $A > 220$). This can be seen in Fig.5, which also displays the spherical results of Fig.4.

Note that, in the $150 < A < 180$ region, the Γ_{G1} values obtained from total gamma-absorption cross section measurements[19] are generally higher than those derived from photoneutron cross section experiments[16-18]. An explanation for this discrepancy which favors the former determinations is proposed in ref.[19]. Around $A \simeq 75$, only photoneutron data are available, so that the disagreement with our calculations may be artificial. The problems encountered in spherical nuclei and the question of the Os isotopes have already been discussed previously.

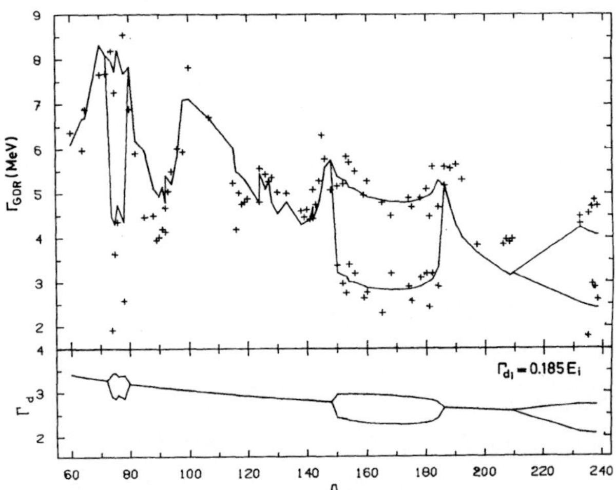

Fig.5. Comparison between experimental GDR widths (+) and theoretical predictions for spherical and deformed nuclei based on Eqs.(3) and (13), respectively. The splitting of the GDR for deformed nuclei is taken into account. In all cases, the droplet model estimates for $E_{G(i)}$ and $\beta_2(\xi_2)$ are used for evaluating the broadening term, while the damping width shown in the lower part of the figure is obtained with the choice $\delta=1$ and $\alpha=0.185$.

5. The Total Radiation Width in Neutron-Capture Reactions

The GDR characteristics evaluated as described in Secs. 2-4 are now used to calculate average radiation widths. The total photon transmission function for E1 radiation from a compound nucleus state with energy E_c, spin J_c and parity π_c is given by

$$T_\gamma(E_c,J_c,\pi_c) = \sum_{\mu=o}^{\omega c} T_\gamma(E_c,J_c,\pi_c,E_\mu,J_\mu,\pi_\mu) + \int_{E_\omega}^{E_c} \sum_{J,\pi} T_\gamma(E_c,J_c,\pi_c,E,J,\pi)\rho(E,J,\pi)dE\,, \quad (14)$$

where the first term represents a summation over the known low-lying states μ up to an excitation energy E_ω, while an integration involving the level density ρ is performed at higher excitation. In Eq.(14), T_γ is either zero if the E1 selection rules are violated, or $T_\gamma(E_c,J_c,\pi_c,E_\mu,J_\mu,\pi_\mu) = T_{E1}(\varepsilon_\gamma = E_c-E_\mu)$ otherwise, T_{E1} being given by Eq.(1). The average radiation width $\langle\Gamma_\gamma(E_c,J_c,\pi_c)\rangle$ is related to the total photon transmission function by

$$\langle\Gamma_\gamma(E_c,J_c,\pi_c)\rangle = T_\gamma(E_c,J_c,\pi_c)/2\pi\rho(E_c,J_c,\pi_c). \qquad (15)$$

The most extended available sets of measured average radiation widths relate to thermal s-neutron capture[25]. In these conditions, the relevant theoretical quantity to be compared with the experimental data is

$$\langle\Gamma_\gamma\rangle_o = \frac{J_i+1}{2J_i+1}\langle\Gamma_\gamma(S_n,J_i+\tfrac{1}{2},\pi_i)\rangle + \frac{J_i}{2J_i+1}\langle\Gamma_\gamma(S_n,J_i-\tfrac{1}{2},\pi_i)\rangle, \qquad (16)$$

where S_n is the compound nucleus neutron separation energy, while $J_i(\pi_i)$ is the spin(parity) of the target nucleus.

The $\langle\Gamma_\gamma\rangle_o$ calculations are performed with the use of χ=0.2 in Eq.(1) [cf. ref.[17]]. The departure of the photoabsorption cross sections from a Lorentz curve at low energy is also taken into account in Eq.(1) by the introduction of the energy-dependent width[26] $\Gamma_G(\varepsilon_\gamma) \simeq \Gamma_G(\varepsilon_\gamma/E_G)^{1/2}$. Finally, a back-shifted Fermi gas model is adopted for the level densities, the involved parameters being obtained by a fit to experiment[27] (a rough estimate of these parameters can be obtained from ref.[28] in absence of experimental data).

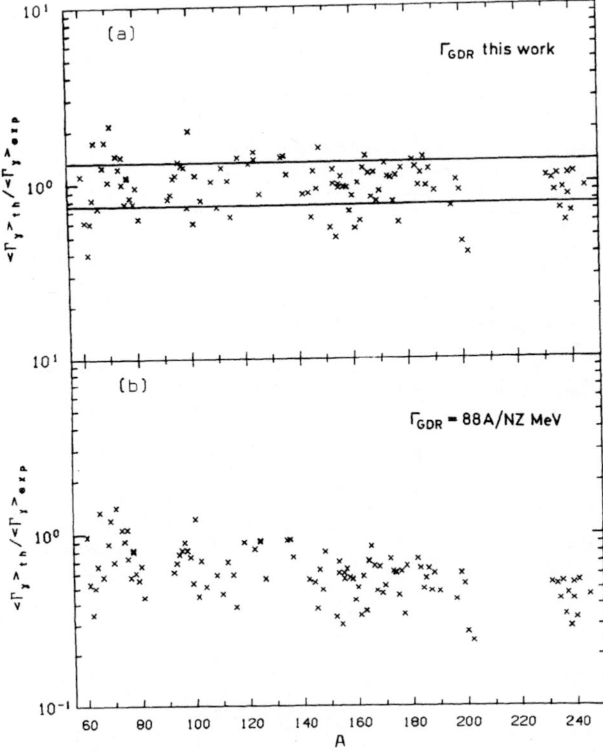

Fig.6.(a) Ratios of theoretical-to-experimental $\langle\Gamma_\gamma\rangle_o$ versus A. The experimental data are from ref.[25], while the calculations are based on Eq.(16), the droplet-model E_G and Γ_G values being used. 2/3 of the values lie within a factor of 1.33 (0.75-1.33); (b) same as in (a), but with the adoption of Γ_G = 88A/NZ

The overall agreement between theory and experiment for $\langle\Gamma_\gamma\rangle_o$ is seen to be very satisfactory [Fig.6a], indicating that the droplet-model estimates of E_G and Γ_G of this work are well suited for the calculation of total E1 transmission functions, while the use of the approximation formula of Γ_G = 88A/NZ MeV, adopted in recent studies[29,30] underestimates $\langle\Gamma_\gamma\rangle_o$ for large A.

6. Conclusions

A phenomenological model for Γ_G is presented. This quantity is expressed as the sum of a damping part Γ_d due to viscous damping, and of a broadening term $\Delta\Gamma$ due to the coupling of the dipole mode to low-frequency quadrupole surface vibrations. This latter term is evaluated in the framework of the droplet model and can account for the shell effects in Γ_G. In fact a very good overall agreement with experiment is achieved for spherical as well as deformed nuclei.

These droplet-model estimates of Γ_G are used in conjunction with the E_G droplet-model values to calculate average E1 radiation widths. The agreement with experiment is very satisfactory.

In addition to its reliability, our model is computationally so simple that it can be easily applied to large sets of nuclei, as is namely required in nuclear astrophysics.

Of course, it remains to be seen to which extent the method needs to be amended for dealing with very exotic nuclei (see e.g. ref.[31] for a preliminary discussion of the level density aspect of such a problem).

References

1. D.M. Brink, D.Phil thesis, Oxford (1955)
2. J.E. Lynn, Inst.Phys.Conf.Ser. No.62 p.267 (1981)
3. C.Y. Wong, Phys.Rev. C17, 1832 (1978)
4. J. Speth and A. van der Woude, Rep.Progr.Phys. 44, 719 (1981)
5. T. Suzuki, Prog.Theor.Phys. 64, 1627 (1980)
6. W.D. Myers, W.J. Swiatecki, T. Kodama, L.J. El-Jaick and E.R. Hilf, Phys.Rev. C15, 2032 (1977)
7. P.F. Bortignon and R.A. Broglia, Nucl.Phys. A376, 405 (1981)
8. N. Auerbach and A. Yeverechyahu, Ann.Phys. 95, 35(1975)
9. R.W. Hasse and P. Nerud, J.Phys.G: Nucl.Phys. 2, L101 (1976)
10. J. Blocki, Y. Boneh, J.R. Nix, J. Randrup, M. Robel, A.J. Sierk and W.J. Swiatecki, Ann.Phys. 113, 330(1978)
11. F.-K. Thielemann and M. Arnould, Caltech report OAP-617 (1981)
12. D.H. Wilkinson, Physica 22, 1039 (1956)
13. C.B. Dover, R.H. Lemmer and F.J.W. Hahne, Ann.Phys. 70, 458 (1972)
14. J. Le Tourneux, Mat.Fys.Medd.Dan.Vid. Selsk. 34, 11 (1965)
15. P.H. Stelson and L. Grodzins, Nucl. Data A1, 21(1965)
16. B.L. Berman, At. Data Nucl. Data Tables 15, 319(1975)
17. B.L. Berman and S.C. Fultz, Rev.Mod.Phys. 47, 713(1975)
18. P. Carlos, R. Bergère, H. Beil, A. Lepêtre and A. Veyssière, Nucl. Phys. A219,61 (1974)
19. G.M. Gurevich, L.E. Lazareva, V.M. Mazur, S.Yu. Merkulov, G.V. Solodukhov and V.A. Tutin, Nucl.Phys. A351, 257 (1981)
20. B.L. Berman, D.D. Faul, R.A. Alvarez, P. Meyer and D.L. Olson, Phys.Rev. C19, 1205 (1979)
21. A. Bohr and B.R. Mottelson, in Nuclear Structure, vol.II (W.A. Benjamin, Reading, MA, 1975)
22. E.R. Hilf, H. v.Groote and K. Takahashi, CERN-Report 76-13, p.142 (1976)
23. M. Eisenberg and W. Greiner, in Nuclear Theory, vol.I (North Holland, Amsterdam and American Elsevier, N.Y. 1970)
24. M. Danos, Nucl.Phys. 5, 23 (1958)
25. H. Weigmann and G. Rohr, Reactor Centrum, Nederland Report 203, p.194 (1973)
26. C.M. McCullagh, M.L. Stelts and R.E. Chrien, Phys.Rev. C23, 1394 (1981)
27. W. Dilg, W. Schantl, H. Vonach and M. Uhl, Nucl. Phys. A217, 269 (1973)
28. J.A. Holmes, S.E. Woosley, W.A. Fowler and B.A. Zimmerman, At.Data Nucl. Data Tables 18, 306 (1976)
29. C.H. Johnson, Phys.Rev. C16, 2238 (1977)
30. J.C. Hardy, Phys. Lett. 109B, 242 (1982)
31. M. Arnould and F. Tondeur, CERN-report 81-08 p.229 (1981)

SCATTERING OF 7-TO 15-MEV NEUTRONS FROM 1-P SHELL NUCLEI

C.R. Gould and J. Dave

Physics Department, North Carolina State University
Raleigh, NC 27650 USA and
Triangle Universities Nuclear Laboratory
Duke Station, Durham, NC 27706 USA

and

R.L. Walter

Physics Department, Duke University and TUNL
Duke Station, Durham, NC 27706 USA

A program of measurements of cross sections for neutron elastic and discrete inelastic scattering from the 1-p shell nuclei is described. Experiments have been performed at an FN Tandem van de Graaff facility using the D(d,n) reaction as a neutron source. In our new data acquisition system, time-of-flight data are stored in a VAX-11/780 computer interfaced to CAMAC via a micro programmed branch driver, the MBD-11. We review our results for neutron scattering from ^{6}Li, ^{7}Li, ^{9}Be, ^{10}B, ^{11}B, ^{12}C, ^{13}C, and ^{16}O. Data have generally been accumulated in 1-MeV steps from 7 to 15 MeV at angles from 25° to 160°. Cross sections are compared to previous work. Much of the neutron scattering data is well described by the spherical optical model (SOM). However, SOM parameters for heavier nuclei do not reproduce the data well. Parameter sets for the individual nuclei are presented along with the results of a global SOM search over 45 neutron scattering angular distributions for all 1-p shell nuclei. Volume integrals for the real and imaginary wells are compared to recent theoretical predictions. The SOM predictions for proton elastic scattering are also compared to the available data.

[^{6}Li, ^{7}Li, ^{9}Be, ^{10}B, ^{11}B, ^{12}C, ^{13}C, ^{16}O(n,n), differential elastic cross sections, E_n=7 to 15 MeV, spherical optical model parameters]

Introduction

A systematic program of measurements of cross sections for neutron elastic and discrete inelastic scattering from the 1-p shell nuclei has been carried out at TUNL over the last six years. The 1-p shell nuclei comprise many elements of interest to fusion reactor design studies. The lithium isotopes are components of the cooling and tritium breeding systems, beryllium is a neutron multiplier, and boron, carbon and oxygen are components of shielding or structural materials. Cross section measurements are important for light nuclei because reaction model parameters which work very well for heavier nuclei are usually much less reliable for light nuclei. Measurements of discrete scattering angular distributions are therefore important for improving the overall nuclear data base as well as for input to neutron transport calculations.

Energies up to 14 MeV are particularly relevant to the first generation of fusion reactors, based on the d-t reaction cycle. The TUNL program has concentrated on measurements using the D(d,n) reaction with neutrons from 7 to 15 MeV. This is an energy region in which there have been few previous studies. The energies are below those generally accessible to d-t neutron generator sources, but are above those accessible to single ended Van de Graaff accelerators. The elements studied include ^{6}Li and ^{7}Li (1), ^{9}Be (2), ^{10}B and ^{11}B (3), ^{12}C (4), ^{13}C (5), and ^{16}O (6). In general measurements have been made at 34 angles between 25° and 160° in 1- or 2-MeV steps. Smaller energy steps were taken in regions where resonance structure was prominent in the total neutron cross section of the nucleus being studied.

Fig. 1 illustrates the point about predicting cross sections for A<16 nuclei based on reaction model parameters for heavier nuclei. It shows the measured ^{10}B differential elastic cross sections for 12-MeV neutrons and compares them to spherical optical model (SOM) predictions based on well known global parameter sets from the literature. The calculations were made with the code GENOA of F. Perey (7). The Becchetti and Greenlees parameters (8) are based primarily on proton data. The Wilmore and Hodgson (9) and Rapaport (10) parameters are specifically for neutron data. They are all parameter sets which reproduce scattering cross sections well for medium mass and heavy nuclei. The agreement with the measured ^{10}B data is not good however. The Watson parameters (11), although based mostly on fits to proton data, were specifically for light nuclei and do in fact reproduce the data somewhat better than the other sets. There is still room for improvement though, and a further motivation for the present work, was to obtain improved global SOM parameter sets based specifically on neutron scattering data.

We are able to reproduce the bulk of the 1-p shell proton and neutron scattering data reasonable well with the spherical optical model, particularly where resonance structure is not prominent. We have obtained energy dependent parameter sets for the individual nuclei, as well as an A dependent parameter set for all nuclei. The fits are generally not as good as those seen for heavier nuclei however the parameter sets show the same qualitative trends as those seen for heavier nuclei, and indicate that the spherical optical model is a physically reasonable way of describing the main features of elastic

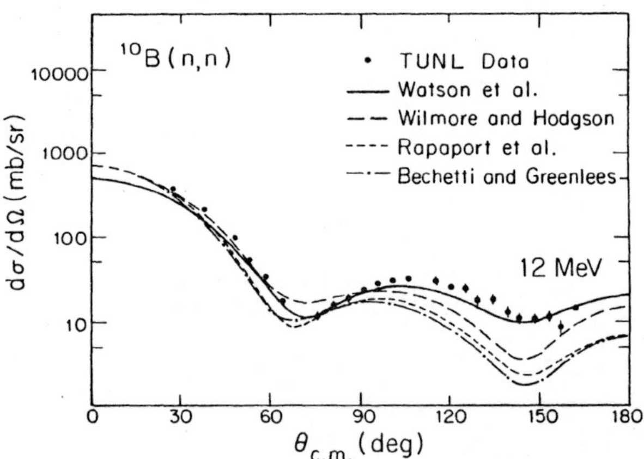

Fig. 1. Measured angular distribution for elastic scattering of 12-MeV neutrons from ^{10}B, and spherical optical model predictions using the parameters from Refs. 8-11. Disagreements indicate that optical model parameter sets based on fits to proton data, or fits to data for heavier nuclei, do not in general reproduce neutron scattering data from light nuclei.

K. H. Böckhoff (ed.), Nuclear Data for Science and Technology, 766–772.

scattering from light nuclei.

Experimental Procedures

Our data acquisition procedures have recently been summarized by El-Kadi et al. (12). The neutron beam is produced by deuteron bombardment of a 3-cm long deuterium gas cell filled to a pressure of 2 atm. The beam current is typically 3μA, pulsed at a 2 MHz repetition rate and bunched to a width of about 2 ns. The scattering targets are in the form of right circular cylinders suspended vertically from a stainless steel wire a distance of 9.25 cm from the center of the gas cell. The samples were typically 1" high and 1/2" to 3/4" in diameter. The ^{6}Li, ^{7}Li, ^{10}B, ^{11}B and ^{13}C samples were monoisotopic and were enclosed in thin walled aluminum cans. The ^{9}Be and ^{12}C samples were pure and the ^{16}O sample was in the form of BeO, enclosed between two aluminum endcaps. The samples and energies studied are listed in Table I. Details specific to each of the 1-p shell scattering experiments are discussed in Ref. 1-6.

Table I. Sample masses and energies studied

SAMPLE	MASS (gms)	PURITY	ENERGY RANGE (MeV)
^{6}Li	3.32	94.9% ^{6}Li	7.5-14
^{7}Li	3.80	99.8% ^{7}Li	7-14
^{9}Be	14.08	99.9% ^{9}Be	7-15
^{10}B	8.64	92.4% ^{10}B	8-14
^{11}B	8.87	97.1%	8-14
^{12}C	12.12	natural C	9-15
^{13}C	8.52	99% ^{13}C	10-18
^{16}O	12.28	natural BeO	9.2-15

The scattered neutrons are detected by the time of flight method in two heavily shielded liquid scintillators, one a 3 1/2" x 2" NE218 scintillator at a flight path of 4m, and the other a 5" x 2" NE213 scintillator at a flight path of 6m. Each detector can be placed at angles from 0° to 160° (on opposite sides of the beam) and is shielded by a tungsten metal shadow bar from the direct flux of neutrons from the gas cell. The detectors are operated at ^{137}Cs edge bias, corresponding to a 1.9-MeV neutron cutoff. Pulse shape discrimination using the Canberra 2160 PSD module in the n-gamma mode is used to distinguish neutron events from gamma-ray events in the scintillators. A monitor NE213 detector, mounted above the scattering plane, views the gas cell directly and is used to provide a relative normalization of all the scattering data taken at a given energy. Sample out spectra were accumulated at each angle to correct for backgrounds due to scattering from the sample containers and suspension wire. For the case of oxygen the 'sample-out' spectrum was taken with a small Be sample containing the same numbers of beryllium atoms as the BeO sample.

Spectra obtained with 14 MeV neutrons and the BeO and Be samples are shown in Fig. 2. The large continuum below channel 400 in the BeO and Be spectra comes from scattering of deuterium break up neutrons and from (n,2n) reactions on Be. The monoenergetic peaks in the BeO spectrum are associated with Be and O elastic scattering (unresolved), inelastic scattering to the 2.43 MeV state in Be and inelastic scattering to the 6 MeV states in O. After subtraction only the oxygen scattering peaks remain, as shown in the bottom of the figure. The inelastic groups were not well separated from the break up neutrons here, and no inelastic scattering data were obtained for oxygen. In general we have been able to extract inelastic scattering cross sections for well isolated levels at excitation energies below about 5 MeV. These cross sections are not considered further here, however.

The absolute normalization at each energy is obtained from hydrogen scattering from a polyethylene sample. The normalization runs are performed at angles near 30° where the hydrogen scattering peak is well separated from the carbon elastic and inelastic scattering peaks. At higher energies the separation of the hydrogen and carbon peaks is less good and a small carbon sample is used to obtain a 'sample-out' spectrum to subtract from the polyethylene 'sample-in' spectrum. Usually these normalization runs are taken three times in the course of an experiment, each accumulating about 10^{4} counts in the hydrogen scattering peak. A complete angular distribution, including normalization runs, takes about 18-24 hours.

The most recent data at TUNL have been taken using the new DEC VAX-11/780 computer system (13) shown in fig. 3. The VAX is a powerful computer for

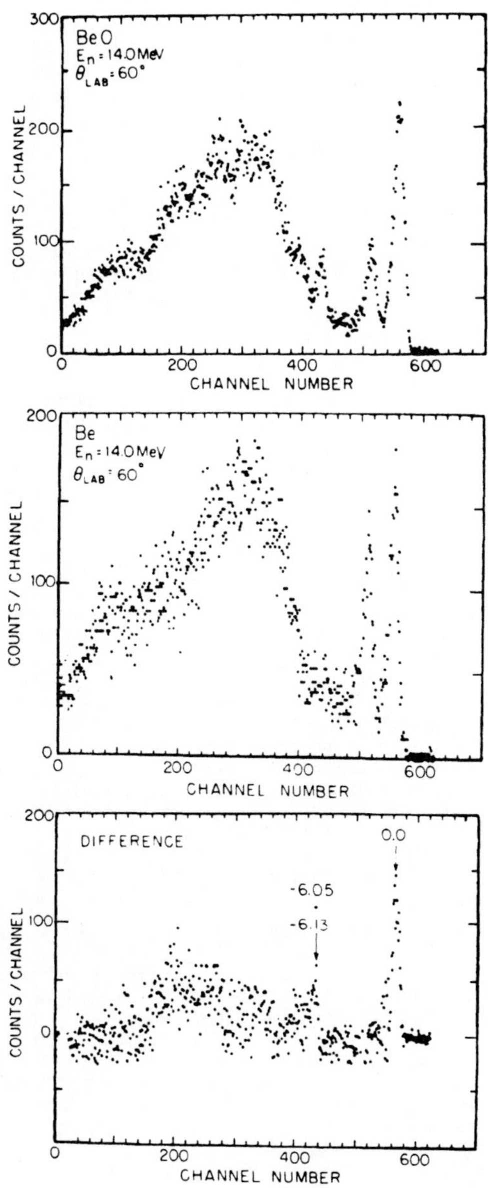

Fig. 2. Neutron time-of-flight spectra for 14 MeV neutrons scattered at 60° from BeO and Be samples. The flight path is 6m. The difference spectrum, resulting from subtraction of the normalized Be spectrum from the BeO spectrum, shows only oxygen scattering.

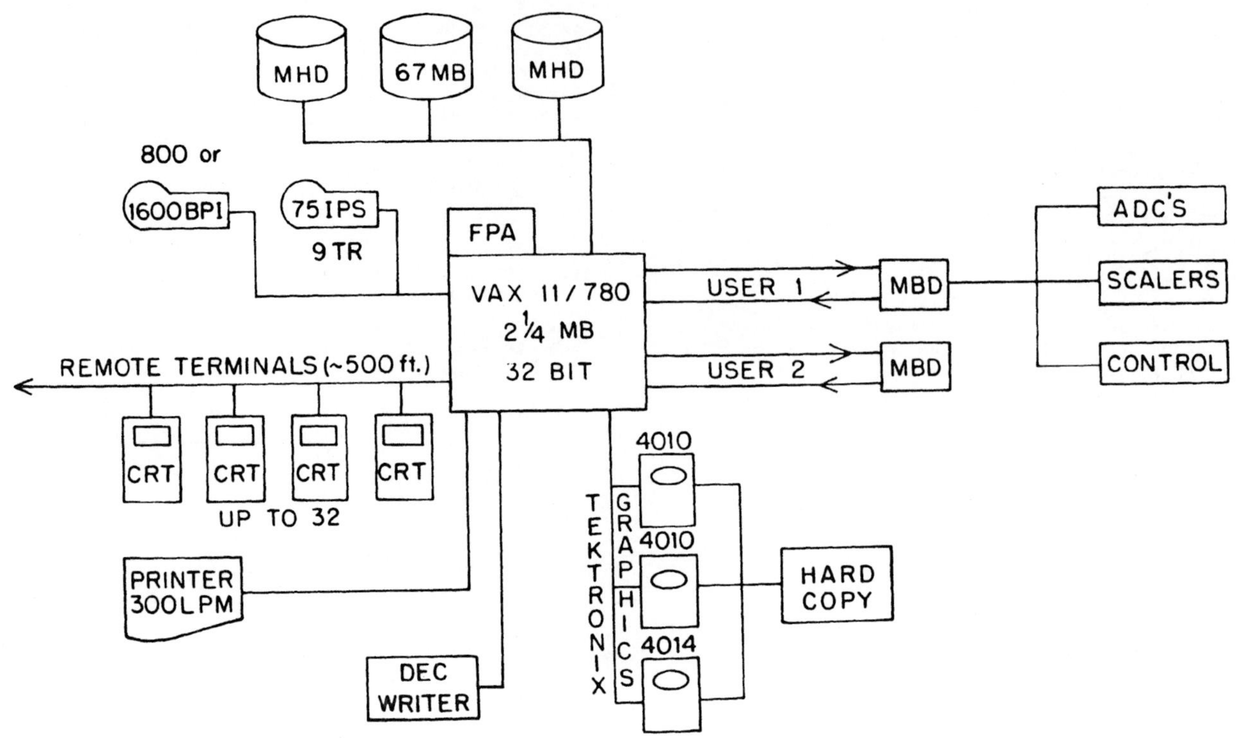

Fig. 3. Hardware configuration of the TUNL VAX-11/780 computer system, used for both on-line data acquisition as well as for general off-line computing.

calculations and is also well suited for real time data acquisition functions. The computer serves both the on-line and off-line needs of the laboratory and runs under the DEC operating system VMS. We have about 60 users with account codes and about 30 processes supported in interactive or batch mode at peak times. VMS and systems programs occupy one disk, user programs occupy the second disk and the third disk is for general use, primarily for storage of spectra for the on-line and off-line users. A typical job mix consists of: one data acquisition user (up to 10 processes and sub-processes), three data analysis users (interactive peak fitting and spectra replays), four batch processes (Monte Carlo calculations of neutron multiple scattering corrections, optical model or coupled channels calculations (GENOA, ECIS)) and about ten interactive processes (text editing and word processing).

The VAX is interfaced to CAMAC via the microprogrammed branch driver developed at LAMPF, the MBD-11. This device is a 390-ns cycle time microprocessor and functions as an intelligent eight channel DMA controller, linking CAMAC with the VAX UNIBUS. Data can be read from up to six Northern Scientific NIM ADC's as well as from three 12 input Lecroy CAMAC ADC's. Scaler read in and output control of external devices are also supported through CAMAC. A single word can be read from CAMAC into the VAX in about 14 μs. For multiparameter events the first word is read in 14 μs, and each subsequent word is read in 6 μs. The words are stored in large buffers in the VAX memory (up to 10240 words long) and the MBD interrupts the VAX only when the buffer is filled. In this way the millisecond long overhead times associated with interrupts on a multi user system are spread over many events and do not contribute significantly to the dead time for an individual event.

The TUNL data acquisition and analysis system, XSYS supports one data acquisition user and any number of data analysis users (the number is limited by physical resources such as memory, display terminals, bulk storage). The large address space of 32-bit computers is extremely valuable for histogramming applications. Maximum spectra allocation per user is presently 1/2 MB which can be split into up to 256 different one or two dimensional spectra. Single or multiparameter events (up to 30 parameters per event) are sorted into spectra by the event analysis language EVAL (14). EVAL allows the user to specify efficient and flexible sorting algorithms for the events in the input data stream. The algorithms are specified in a simple higher level language and are easily modified by the user during an experiment if the need arises. Single parameter event rates of 50 KHz take about 40% of the CPU time. Typical count rates in neutron scattering experiment are much lower and the experimenter normally uses less than 10-15% of the CPU time. Capabilities exist in XSYS for Gaussian fitting, background subtraction, spectrum subtraction and normalization. This means that much of the data analysis can be completed on-line during the actual experiment.

Experimental Results and Comparisons to Other Work

The peak yields in the time-of-flight spectra are converted to uncorrected angular distributions by normalizing to hydrogen elastic scattering cross sections. The angular distributions are then corrected for effects due to the extended neutron source, the finite sample size and the finite detector size. These corrections are carried out via Monte Carlo simulation using the code EFFIGY written at TUNL. The code calculates a neutron time-of-flight spectrum at each angle, based on the experimental geometry and input libraries of cross sections versus energy for all elements in the sample. EFFIGY can handle one or two element samples, as in the case of the oxygen scattering data. The code can take energy loss

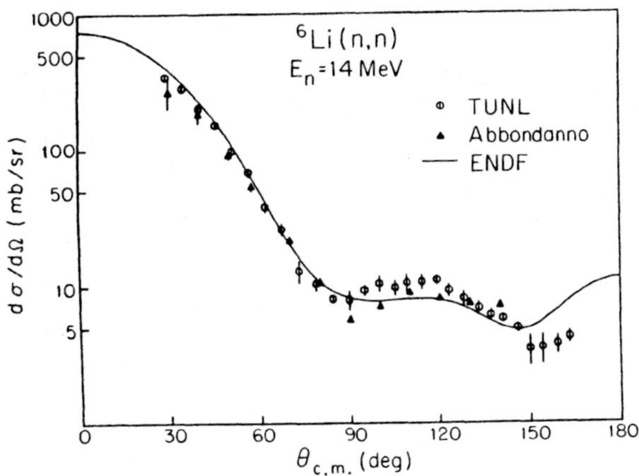

Fig. 4. TUNL differential cross sections for ^{6}Li(n,n) compared to the results of Ref. (15) and the current ENDF evaluation (16).

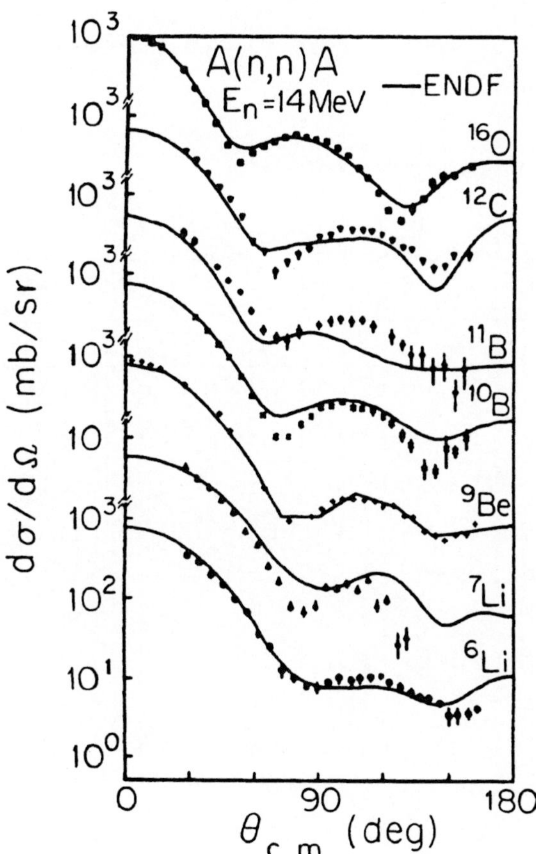

Fig. 5. TUNL differential cross sections for seven 1-p shell neutron scattering distributions compared to ENDF values (17).

following elastic scattering into account. This proves to be an important correction in multiple scattering of neutrons from light elements because the energy loss can have the effect of removing the scattered neutrons from the time-of-flight peak of interest. One or two iterations are sufficient to reach agreement between the calculated and measured peak yields in the TOF spectra.

Below 14 MeV there are relatively few experiments with which we can compare our results. An exception is the carbon data set from 8.0 to 14.5 MeV of Haouat et al. (28). Above 14 MeV there have been a number of studies of neutron scattering using the d-t reaction as a neutron source and the ENDF evaluations rely heavily on these measurements. As an example, fig. 4 shows our ^{6}Li results compared to the measurements of Abbonadanno et al. (15) and the current ENDF evaluation (16). The agreement is quite good here. In other cases, discrepancies exist. This can be seen in Fig. 5, which compares all our 14 MeV data to the ENDF evaluations (17). Based on cases where cross section results are readily accessible in the literature, we see reasonable agreement with other studies at forward angles, and possible systematic discrepancies at back angles, and in deep diffraction minima. We estimate our total elastic cross sections to be good to about 5% at the higher energies, with somewhat smaller errors at the lower energies.

Optical Model Fits to Neutron Data

Spherical optical model fits have been made to the elastic scattering angular distributions using the code GENOA. Many of the 1-p shell nuclei are non spherical but it is known that deformation effects can be masked by suitable choices of SOM parameters (18). Resonance structure cannot of course be fit by the optical model and we have focussed on fitting the general trends of the data rather than seeking the best fit at each energy for every nucleus. The potential is of the standard form, having a Woods-Saxon real volume term and surface derivative imaginary and spin orbit terms. The form of the potential is listed in Table II. The data are from the present work except those for ^{14}N, which are from Bauer et al. (19). We do not include compound nucleus contributions to the elastic scattering and have excluded the lower energy distributions from the searches. Polarization data were not considered in the present work and the spin orbit parameters were fixed early in the search at close to standard values (11, 18). The value of Wick's limit was included as a data point at zero degrees. It was assigned an error of 10%. Although Wick's limit is only a lower limit on the zero degree cross section it is known to predict the cross section fairly accurately for heavier nuclei, and for light nuclei where resonant structure is absent (20).

Searches were first made for individual nuclei, taking into account the known ambiguities in Vr^2 and Wa. We allowed the real and imaginary well depths to vary linearly with energy, but did not include energy dependence in the geometric parameters. The parameters obtained in these fits are listed in Table II, together with the minimised values of χ^2 per degree of freedom. Even though the fits are made over a relatively restricted range of energies, the energy dependences of the well depths are physically reasonable. The real well decreases with increasing energy while the imaginary well increases with increasing energy. The fits are shown in Fig. 6 as the dashed lines. The parameters for the different nuclei are similar except for r_o. A second search was therefore attempted over all the nuclei except nitrogen, using as starting parameters the average values from the individual searches. For this search, called global, a dependence of the radius parameter r_o on the mass number A was used. Such a dependence has been used by Rapaport et al. (10) for heavy nuclei. Unlike these authors we find r_o has to be decreasing linearly with A for light nuclei. The fits obtained with this global parameter set are shown as the solid line in Fig. 6. The parameters are listed under global in

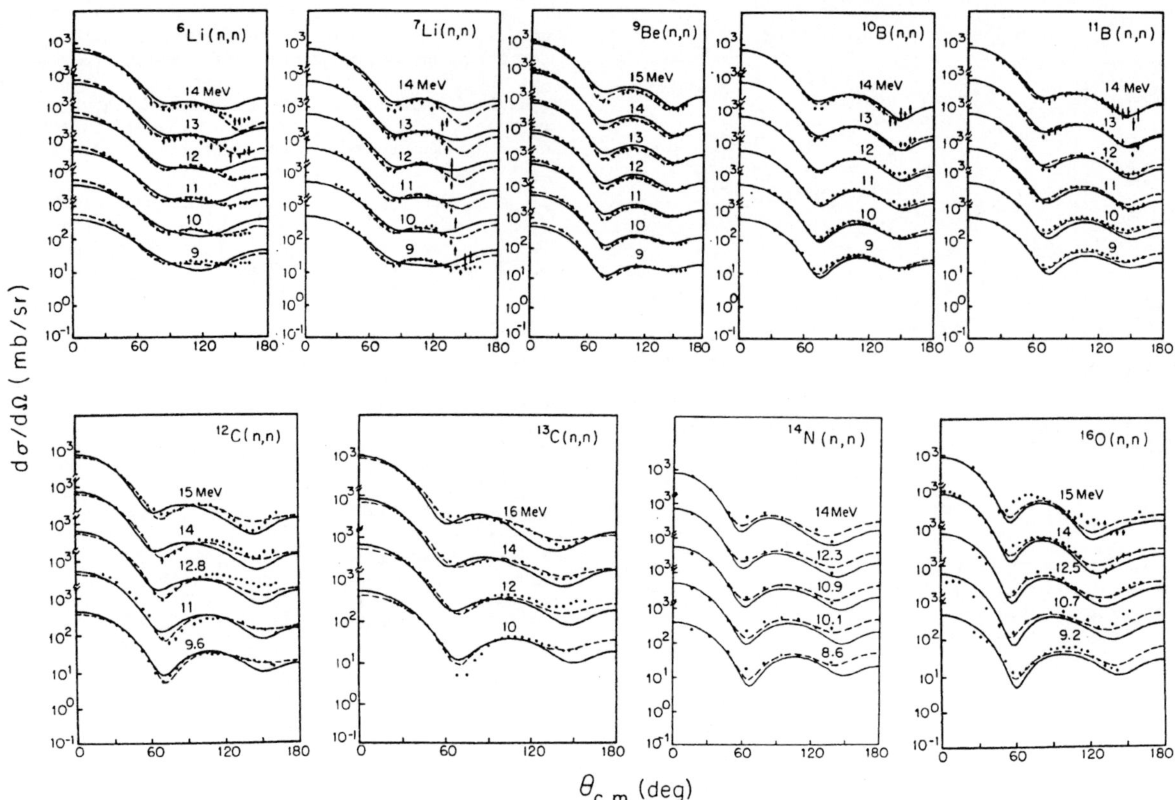

Fig. 6. Neutron elastic scattering angular distributions compared to SOM predictions based on the parameters of Table II. Dotted lines are for the individual parameter sets, solid lines are for the global parameter set.

Table II. Spherical Optical Model parameters for scattering from 1-p shell nuclei (Potentials in MeV, Radii in fm and center-of-mass energy E, in MeV.)

	V_o	r_o	a_o	W_d	r_i	a_i	V_{so}	r_{so}	a_{so}	x^2/N
^{6}Li	39.62-0.027E	1.507	0.663	9.30+0.646E	1.616	0.196	5.5	1.15	0.5	6.2
^{7}Li	37.62-0.013E	1.593	0.473	6.68+1.916E	1.396	0.111	5.5	1.15	0.5	8.4
^{9}Be	38.50-0.145E	1.447	0.387	1.67+0.365E	1.368	0.424	5.5	1.15	0.5	4.0
^{10}B	47.91-0.346E	1.387	0.464	0.66+0.810E	1.336	0.278	5.5	1.15	0.5	10.1
^{11}B	45.66-0.004E	1.337	0.548	0.003+1.256E	1.438	0.182	5.5	1.15	0.5	17.8
^{12}C	58.86-0.663E	1.211	0.434	12.65+0.045E	1.387	0.163	5.5	1.15	0.5	18.8
^{13}C	61.23-0.765E	1.131	0.561	16.40+0.136E	1.412	0.112	5.5	1.15	0.5	22.5
^{14}N	50.54-0.006E	1.209	0.573	12.07+0.703E	1.415	0.105	5.5	1.15	0.5	10.1
^{16}O	48.25-0.053E	1.255	0.536	4.418+0.556E	1.352	0.205	5.5	1.15	0.5	30.4
Global	$45.14-0.020E-23.48\frac{(N-Z)}{A}$	1.508-0.0133A	0.5	$11.32+0.237E-16.08\frac{(N-Z)}{A}$	1.353	0.200	5.5	1.15	0.5	

The potential is of the form

$$U_{OPT} = -V_o\, f(r,r_o,a_o) + 4\, i\, a_i W_d \frac{df(r,r_i,a_i)}{dr} + 2V_{so} \frac{\sigma\cdot\ell}{r} \frac{df(r,r_{so},a_{so})}{dr}$$

where f is the usual Woods-Saxon form factor

$$f(r,r_x,a_x) = [1 + \exp\,(r-r_x A^{1/3})/a_x]^{-1}$$

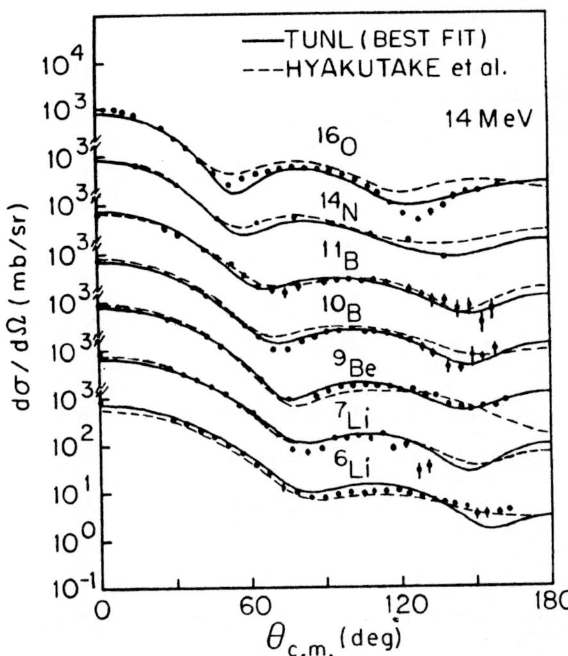

Fig. 7. TUNL 14-MeV data and individual SOM predictions compared to the SOM predictions of Hyakutake et al. (21).

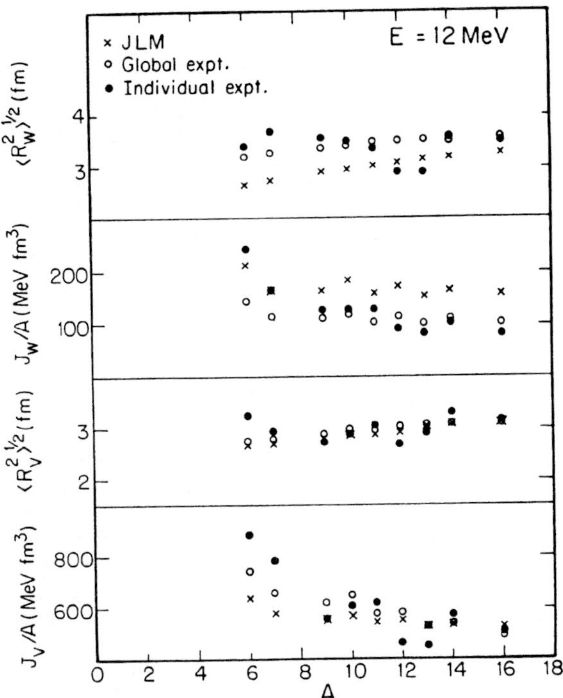

Fig. 8. Volume integrals and root mean square radii for the real and imaginary wells of the optical potentials found in the present work for 12-MeV neutrons. The values are compared to the theoretical values of Jeukenne et al. (22) labelled JLM.

Table II. The best fits are obtained for the lighter nuclei (A<11). For carbon, nitrogen and oxygen, resonance structure is much more prominent in the total cross section, and the fits are, as expected, somewhat poorer.

The only other systematic SOM search over 1p shell neutron scattering data has been that of Hyakutake et al. (21). Their work was confined to the single energy 14 MeV and they used a Gaussian imaginary well form factor as opposed to a surface derivative term. Their parameters reproduce our data quite well, however. This is seen in fig. 7 which shows our data and individual SOM predictions at 14 MeV, along with their individual SOM predictions (parameter set S in ref. 21). They also found a global parameter set in which r_o decreased with increasing A, as in the present work.

Volume integrals are a convenient way of comparing different optical model parameter sets. Jeukenne et al. (22) have recently made extensive calculations of root mean square radii and volume integrals of optical potentials. While these would not necessarily be expected to be appropriate for light nuclei, it is interesting to compare their predictions with the values obtained from the parameter sets found in the present work. Fig. 8 shows this comparison for 12 MeV neutrons for the individual and global parameter sets as a function of mass number A. The calculated values from Ref. 22 are marked with crosses. The J/A are the volume integrals per nucleon for the real (V) and imaginary (W) wells. The other values are the root mean square radii, also for the real and imaginary wells. Overall trends with A are reproduced, however the integrals and radii for the imaginary well are systematically off. Of course the local density approximation is potentially a much more drastic simplification for light nuclei than for heavy nuclei. Disagreements here more reflect the fact that light nuclei, while exhibiting scattering behavior similar to heavier nuclei need to be considered separately as regards the determination of general trends in SOM parameter sets.

Optical Model Fits to Proton Data

We can assess the usefulness of the neutron SOM parameter sets obtained here by seeing how well they predict proton scattering data for the same nuclei. For the T=0 nuclei (^{6}Li, ^{10}B, ^{12}C, ^{14}N and ^{16}O) the comparison is particularly straight forward. The only change in the parameters should arise from Coulomb correction terms in the real and imaginary well depths. Isospin dependent terms will necessarily be zero. Coulomb correction terms arise from the energy dependences of the well depths. Since V and W are energy dependent, we expect correction terms ΔV and ΔW in both the real and imaginary well depths respectively.

The SOM searches have been made for the proton data of Harrison and Whitehead (23) for ^{6}Li, Watson et al. (11) for ^{10}B, Daehnick and Sherr (24) for ^{12}C, Hansen et al. (25) for ^{14}N and Hiddleston et al. (26) for ^{16}O. In the first search the individual parameters of Table II were used but V and W were allowed to vary. (The Coulomb radius was set equal to the real well radius). The results of this search are shown as the dotted lines in fig. 9. The fits are quite good. The resulting differences in the proton and neutron potentials ΔV and ΔW are plotted in fig. 10. These are the Coulomb correction terms for light nuclei. The dotted lines are predictions of Rapaport (27) based on a study of scattering data from heavier T=0 nuclei. Rapaport's imaginary well correction term includes a slight energy dependence, and the values in fig. 9 are calculated for 12 MeV neutrons. A clear pattern in our values is not discernible because of the large errors. However the values found for the light nuclei are not necessarily inconsistent with those found for heavier nuclei.

In the second search we used our global neutron parameter set and fixed the Coulomb correction terms at standard values ΔV=0.4*Z/A**(1/3) and ΔW=0. The results of this fit are shown as the solid lines in fig. 9. The agreement here is quite poor, particularly for ^{14}N and ^{16}O. In looking at fig. 10, it is clear that the discrepancies are associated with setting ΔW=0. Based on these results we see that we are able

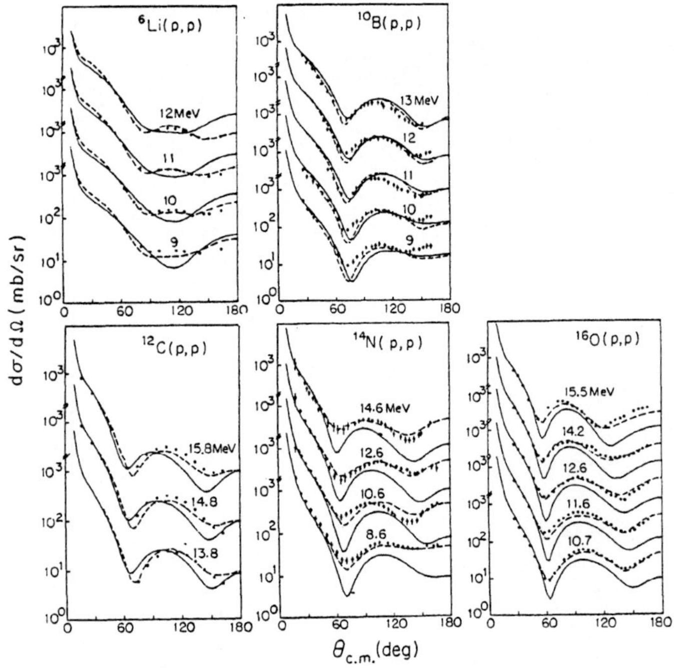

Fig. 9 Proton elastic scattering angular distributions for T=0 nuclei compared to SOM predictions based on the neutron parameter sets. Dotted lines are for the individual parameter sets allowing V and W to vary. The solid line is for the global parameter set including a fixed Coulomb correction term for the real well and a correction term of zero for the imaginary well (see text).

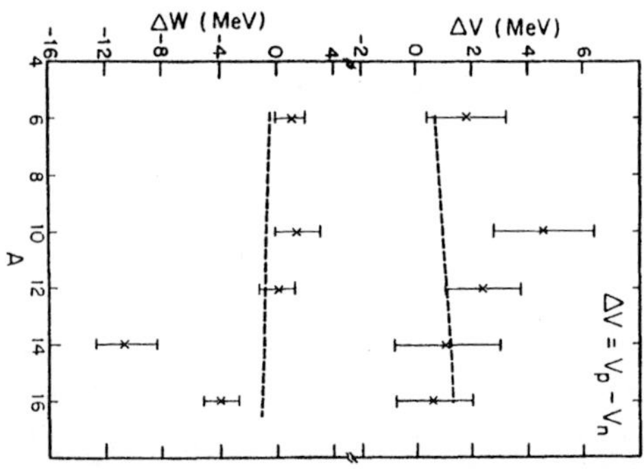

Fig. 10 Coulomb correction terms for light nuclei based on fits to the proton data of fig. 9. The dotted lines are the predicted values of Rapaport (27).

to fit neutron and proton scattering data from light nuclei quite well as long as Coulomb correction terms are included for both the real and imaginary well depths.

Summary

We have reviewed a program of measurements of neutron elastic scattering cross sections for 1-p shell nuclei. Data have been accumulated for neutrons in the energy range 7- to 15-MeV and have been compared to spherical optical model predictions. Energy dependent parameter sets have been obtained for individual nuclei and provide a good description of the main features of the scattering distributions. A global parameter set containing an A dependent real well radius parameter has also been found. The best fits are obtained for the lithium and boron isotopes, and beryllium. We have compared our volume integrals and root mean square radii to recent theoretical predictions. Qualitative trends are reproduced but the imaginary well parameters are systematically off. By including Coulomb correction terms with the individual neutron parameters, we can fit proton scattering data for T=0 nuclei. A non zero imaginary Coulomb correction term appears to be important in obtaining acceptable fits to the proton data. Overall the spherical optical model describes the bulk of neutron and proton scattering data quite satisfactorily.

Acknowledgements

We would like to thank the past and present members of the TUNL neutron scattering group for their invaluable help and contributions to the experiments discussed here. This work was supported in part by the U.S. Department of Energy.

References

1. H.H. Hogue et al., Nucl. Sci. Eng. 69 (1979) 22.
2. H.H. Hogue et al., Nucl. Sci. Eng. 68 (1978) 38.
3. S.G. Glendinning et al., Nucl. Sci. Eng. 80 (1982) 256.
4. D.G. Glasgow et al., Nucl. Sci. Eng. 61 (1976) 521.
5. J.H. Dave et al., Nucl. Sci. Eng. 80 (1982) 388.
6. S.G. Glendinning et al., Nucl. Sci. Eng. (to be published)
7. F. Perey (private communication).
8. F.D. Becchetti and G.W. Greenlees, Phys. Rev. 182 (1969) 1190.
9. D. Wilmore and P.E. Hodgson, Nucl. Phys. 55 (1964) 673.
10. J. Rapaport, V. Kulkarni and R.W. Finlay, Nucl. Phys. A330 (1979) 15.
11. B.A. Watson, P.P. Singh, R.E. Segel Phys. Rev. 182 (1969) 977.
12. S.M. El-Kadi et al., Nucl. Phys. (to be published).
13. C.R. Gould et al., IEEE Trans. Nucl. Sci. NS-28 (1981) 3708.
14. L.G. Holzsweig and R.V. Poore IEEE Trans. Nucl. Sci. NS-28 (1981) 3815, and references therein.
15. U. Abbondanno et al., Nuovo. Cim. 66 (1970) 139.
16. ENDF/B-V data file for ^{6}Li (MAT 1303), evaluation by G. Hale, L. Stewart, P.G. Young (LANL), LA-7663-MS (1978).
17. ENDF/B Summary Documentation, BNL-NCS-17541 (ENDF-201) 3rd edition (ENDF/B-V) edited by R. Kinsey, available from NNDC. Brookhaven National Laboratory, Upton, NY (July 1979).
18. H.J. Votava et al., Nucl. Phys. A204 (1973) 529.
19. R.W. Bauer et al., Nucl. Phys. A93 (1967) 673.
20. W. P. Bucher et al., Nucl. Sci. Eng. 54 (1974) 416.
21. M. Hyakutake et al., J. of Nucl. Sci. and Tech. 11 (1974) 407.
22. J.P. Jeukenne, A. Lejeune and C. Mahaux, Phys. Rev. C16 (1977) 80.
23. W.D. Harrison and B.A. Whitehead, Nucl. Phys. A330 (1979) 15.
24. W.W. Daehnick and R. Sherr, Phys. Rev. 133 (1964) B934.
25. L.F. Hansen et al., Phys. Rev. C8 (1973) 2078.
26. H.R. Hiddleston et al., Nucl. Phys. A242 (1975) 323.
27. J. Rapaport Phys. Letts. 92B (1980) 233.
28. G. Haouat et al. Nucl. Sci. Eng. 65 (1978) 331.

RESONANCE ENHANCEMENT OF PARITY VIOLATION EFFECTS IN NEUTRON-NUCLEAR INTERACTION

V.P.Alfimenkov, S.B.Borzakov, Vo Van Thuan, Yu.D.Mareev, L.B.Pikelner, A.S.Khrykin and E.I.Sharapov

Laboratory of Neutron Physics, Joint Institute for Nuclear Research, Dubna, USSR

The results on parity-nonconserving (PNC) effects in p-wave neutron resonances are reported. The experiments were carried out at the IBR-30 pulsed reactor with the neutron beam polarized by transmission through a polarized proton target. The dependence of total cross section on longitudinal polarization was found for the resonances at 0.88 eV ^{81}Br, 5.53 eV ^{111}Cd, 1.33 eV ^{117}Sn and 0.75 eV ^{139}La. The PNC effect has the resonance shape. These results were analysed together with thermal neutron data in the frame of the model of compound states mixed due to weak nucleon interaction.

[^{81}Br, ^{111}Cd, ^{117}Sn, ^{139}La, polarized neutrons, transmission, p-resonances, E_0, $g\Gamma_n$, Γ parameters, parity violation effect.]

Introduction

In the middle of sixties after the experimental observation of P-odd effects in gamma-transitions (see review Ref.[1]) it became clear that the excited states of nuclei have mixed parities. In Refs.[2-6] some new experiments on parity-nonconserving (PNC) effects in neutron-nuclear interaction were proposed. The analogues of the optical activity of neutrons such as rotation (by an angle ϕ) of neutron polarization about the direction of beam propogation through the nonpolarized target as well as the helicity dependence of the neutron total cross section (asymetry A) were in particular discussed. But, nowhere the mixing of compound states was taken into account. Only the influence of the PNC single particle potential was considered. Therefore, the estimates of the proposed effects (e.g. for a rotating power of 10^{-7}-10^{-6} radian/cm) were on the werge of experimental possibilities of their observation. In Ref.[4] it was also pointed out that the PNC effects should be enhanced near a single-particle potential p-resonance.

In Ref.[7], published in 1980, it was stated that the parity mixing of compound states should enhance the PNC effects which must be especially large near the compound nucleus p-resonances. The theoretical predictions in Ref.[7] were : ϕ $=10^{-2}$ radian/cm for the rotating power near the resonance and A $=10^{-2}$-10^{-1} for the relative change of target transmission upon the reverse of neutron longitudinal polarization.

Simultaneously in 1980 the experiment performed at the Laue-Langevin Institute, Grenoble, was reported (Ref.[8]). There the large PNC effects $\phi=(3.7\pm0.3).10^{-5}$ rad/cm and A $=(-9.78\pm4.01).10^{-6}$ for the ^{117}Sn target and cold neutrons were found. Next year the measurement of the helicity dependence of the thermal neutron cross section for ^{117}Sn was made at the Nuclear Physics Institute, Leningrad, with a much better accuracy and the analogous PNC effect was found in the ^{139}La target. It was also shown that the transmission effect is due to the (n,γ) cross section PNC effect. This indicates the importance of the compound states mechanism of parity violation. Really, these nuclei have very weak neutron resonances near the thermal neutron energy region (E_0=1.3 eV for the ^{117}Sn- and E_0=o.75 eV for the ^{139}La-target. They are p-resonances with a high probability, i.e. their parity is the opposite one to that of neutron s-resonances. Therefore, the theoretical papers (Refs.[10-13]), which followed the experimental discoveries explain the new PNC effects by parity mixing of compound states having the same spin and different parity values.

To have a direct evidence of the correctness of the PNC effects explanation first given in Ref.[7], one should show experimentally that the new effects enhance considerably with neutron energy approaching the p-levels in question. This was made in our experiment with polarized resonance neutrons performed in the Laboratory of Neutron Physics of the Joint Institute for Nuclear Research, Dubna.

The PNC Effect in Neutron Cross Sections

Let us following Ref.[7] consider the neutron capture into the p-resonance with account for parity violation. On neutron capture the nucleus is in the compound state with energy E_i and spin J, which is a superposition of states with parities $\mathfrak{h}$ and $\bar{\mathfrak{h}}$. The wave function of such states in the first approximation of the perturbation theory may be presented in the form:

$$|i\rangle = |J^{\mathfrak{h}}\rangle + \sum_j \alpha_{ij}(E)\,|J_j^{\bar{\mathfrak{h}}}\rangle, \tag{1}$$

$$\alpha_{ij} = \frac{\langle J_j^{\bar{\mathfrak{h}}}\,|H_W|\,J^{\mathfrak{h}}\rangle}{E - E_j - i\Gamma_j/2}, \tag{1a}$$

where H_W is the parity nonconserving Hamiltonian due to nucleon weak interaction. The matrix element $\langle\,|H_W|\,\rangle$ is imaginary in the case of T-invariant interactions and with the Condon and Shortly phase choice for the spherical functions $Y_{\ell m}$.

The incident plane neutron wave with momentum $\vec{k}$ and helicity $\pm1/2$ can be expanded in a series of states $|\ell, j, j_z\rangle$ with a definite angular momentum j and orbital momentum ℓ =0,1 (the axis z is along $\vec{k}$):

$$e^{i\vec{k}\vec{r}}\chi^{\pm} \approx \sqrt{4\pi}\left[Y_{00} + \frac{ikr}{\sqrt{3}}Y_{10}\right]\chi^{\pm} = \tag{2}$$

$$= \sqrt{4\pi}\left(|0,\tfrac{1}{2},\pm\tfrac{1}{2}\rangle \mp i\frac{kr}{3}|1,\tfrac{1}{2},\pm\tfrac{1}{2}\rangle + i\frac{kr\sqrt{2}}{3}|1,\tfrac{3}{2},\pm\tfrac{1}{2}\rangle\right).$$

The capture of the neutron into the p-resonance described by the wave function (1) may as usual go from the p-wave (2nd and 3rd terms in (2)), and besides, from s-wave (1st term in (2)) because of parity violation. The capture amplitudes caused by the 1st and 2nd terms in (2) experience interference and the interference term changes its sign with the change of neutron helicity. This is the reason for the PNC effect in neutron cross sections.

The total angular momentum J of a resonance is conserved. The values $J=I\pm1/2$, $I\pm3/2$ are possible, in general, for p- and $J=I\pm1/2$ for s-resonances (here I is the target angular momentum). Evidently, because of J-value conservation only p-resonances with $J=I\pm1/2$ can mix with s-resonances. According to Ref.[7] the expression for the capture cross section $\sigma_p^{\pm}$ near the p-resonance for neutrons with $\pm$ helicity may be written as

$$\sigma_p^{\pm} = \sigma_p\left(1 \pm \mathcal{P}(k)\right), \tag{3}$$

K. H. Böckhoff (ed.), Nuclear Data for Science and Technology, 773–776.

$$\sigma_p = \frac{\pi}{k^2} \frac{g \Gamma_p^n(k) \Gamma_p}{(E - E_p)^2 + \Gamma_p^2/4} , \qquad (3a)$$

$$\mathcal{P} = 2\alpha \sqrt{\frac{\Gamma_s^n(k)}{\Gamma_p^n(k)} \frac{\Gamma_{p1/2}^n(k)}{\Gamma_p^n(k)}} \simeq 2\alpha \sqrt{\frac{\Gamma_s^n(k)}{\Gamma_p^n(k)}} . \qquad (4)$$

Here σ_p is the usual capture cross section in the absence of neutron polarization, E_p is the resoance energy, $\Gamma_p^n = \Gamma_{p1/2}^n + \Gamma_{p3/2}^n$ is the neutron width of p-resonace, Γ_p is the total width. The latter expression in (4) is obtained under approximation that $\Gamma_{p1/2}^n (j=\frac{1}{2}) \simeq \Gamma_p^n$. The neutron widths have the usual energy dependence $\Gamma_s^n(k) = [\Gamma_s^n]_{res} (k/k_s)$ and $\Gamma_p^n(k) = [\Gamma_p^n]_{res} (k/k_p)^3$ for p- and s-resonances, respectively.

The matrix elements for the different admixt levels in (1a) are believed to be of the same order of magnitude. This leads to the assumption of the strongest mixing between the nearest s- and p-levels and to the so-called two-resonance approximation used to obtain expressions (3) and (4). The estimate of the mixing coefficient α, made in Ref.[7], is 10^{-4}-10^{-5}, the corresponding prediction for the PNC effect in the total cross section near the centre of resonance is $\mathcal{P}(E_p) = 10^{-1}$-$10^{-2}$. The stronger the admixt s-wave resonance (and the weaker p-resonance) is, the higher effect $\mathcal{P}(E_p)$ is expected (kinematic enhancement). The effect $\mathcal{P}(E_s)$ in an s-resonance, on the contrary, is $2\alpha\sqrt{\Gamma_p^n/\Gamma_s^n}$, i.e. by 10^4 times smaller than $\mathcal{P}(E_p)$.

In the experimental study of the PNC effect it is convenient to measure the transmission effect $\varepsilon = \frac{N_p - N_a}{f_n(N_p + N_a)}$ upon the reverse (p-parallel, a-antiparallel) of the sign of the longitudinal neutron polarization f_n. For the transmission effect ε one can obtain the following simple expression

$$\varepsilon = -\frac{n}{2} (\sigma_t^+ - \sigma_t^-) , \qquad (5)$$

where n is the target thickness in nuclei per cm^2. The σ_t is the total cross section for two helicity states and consists of the sum $\sigma_s^\pm + \sigma_p^\pm + \sigma_{pot}^\pm$ of s-wave resonance cross section σ_s, p-wave resonance cross section σ_p and potential scattering cross section σ_{pot}. In the frame of the considered model the contribution from σ_{pot} is small everywhere and the contribution from σ_s is small near the p-resonance studied. Therefore, from (5) and (3) it follows that

$$\varepsilon(E) = -n \mathcal{P}(E_p) \sigma_p(E). \qquad (6)$$

To compare the PNC effects in the resonance region and the thermal neutron region one may use the relation

$$\mathcal{P}_{therm} \simeq \mathcal{P}(E_p) \frac{\sigma_p(E_p)}{\sigma_t(therm)} \left(\frac{\Gamma_p}{2E_p}\right)^2 \left(1 - \frac{E_p}{E_s}\right). \qquad (7)$$

Here σ_t(therm) is the total neutron cross section for neutrons in the thermal energy region. The eq.(7) follows from Refs.[7-11] as well as from eq.(3) if one accepts that the thermal point is close enough to the p-resonance to us eq.(3).

Experimental Method

The measurements were performed by the timeof-flight method on the beam of polarized resonance neutrons from the IBR-30 pulsed reactor, Ref.[14]. The neutron pulse duration was 70 μs and the reactor mean power -- 20 kW. At neutron energies above 1.5 eV the booster mode of the IBR-30 operation with the electron accelerator was used to improve resolution. In this case the pulse duration was 4 μs and the mean power of the booster 5 kW. The resonance neutrons were polarized in a modernized apparatus described earlier in Ref. [15] by their transmission through the polarized proton target.

The experimental arrangement is shown in Fig. 1. The

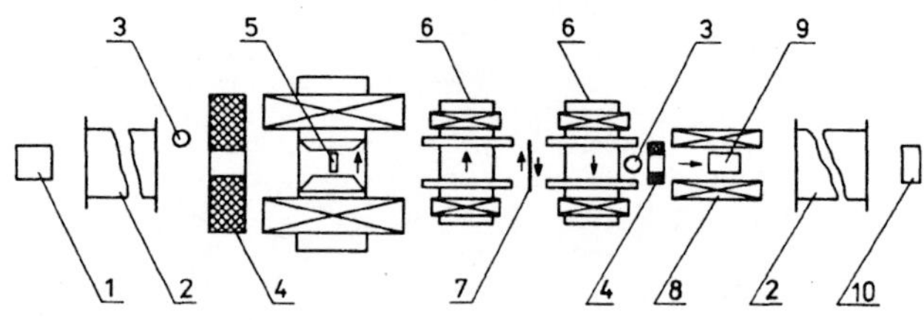

Fig. 1. Experimental arrangement. 1-reactor, 2-evacuated neutron beam tubes, 3-monitors, 4-collimators, 5-polarized proton target, 6-electromagnets of the guide field, 7-current sheet, 8-solenoid, 9-target, 10-neutron detector. Arrows along the neutron beam show the direction of the magnetic fields.

polarizer 4 placed at a 32 m distance from the reactor core produced the neutron polarization f_n 0.6, the intensity loss being a factor of 10. The neutron beam with a transverse horizontal polarization passed the gaps of the two electromagnets 6 having the guide fields of 200 Oe. It was possible to change the direction of the guide field in the second electromagnet by switching its current. The current sheet 7 was placed between electromagnets to to create the nonadiabatic conditions of neutron transmission in this region. So the polarization was reversed relative to the magnetic field of the second magnet after switching its current. Then the beam polarization followed adiabatically the turn of the field into the longitudinal magnetic field of the solenoid 8 where the sample 9 was placed. The adiabatical conditions were fulfilled up to 100 eV neutron energy. In this way the longitudinally polarized neutron beam was obtained. Its helicity was changed by switching the current in the second magnet 6 .

The neutron detector was placed at a 58 m distance. The acquisition of the time-of-flight spectra and the control after polarization and experiment were performed by the automatic system on the basis of the minicomputer. The polarization was reversed every 40 sec. Every 48 hours the sign of the proton polarization and therefore, the sign of the PNC effect was changed to control the false effects. The total measuring time amounted to 200 hrs for one target. The transmission effect was calculated from the spectra N_p and N_p corresponding to two helicity states of the beam.

Results

The helicity dependence of total neutron cross sections was investigated for 11 weak resonances in the nuclei ^{81}Br, ^{111}Cd, ^{117}Sn, ^{127}I, ^{129}La, ^{238}U. Some results were partly reported earlier in Ref.[16-18]. In the case of ^{111}Cd and ^{117}Sn the isotopic targets were used. The isotopic assignment for the 0.88 eV resonance discovered by us in the Bromine was made in the special measurements of low-energy gamma-rays from the Br(n,γ) reaction in the resonance region. The PNC effect was found in 4 cases which are shown in Figs.2-5. There the time-of-flight spectra N and the transmission effect ε are given in dependence on the time-of-flight channel t, neutron energy in eV. The arrow with a digital above is the position and energy of the p-resonance.

As is shown above the helicity dependence of the neutron cross section near the p-wave resonance is characterized by the quantity $\mathcal{P}$(E), which is nearly constant and equal to $\mathcal{P}(E_p)$ in the energy interval $\sim \Gamma_p$ around the p-resonance. The values of $\mathcal{P}(E_p)$ were obtained from ε according to (6) in case of good resolution. In the opposite case the corrections for the resolution function and the Doppler effect were introduced. The samples thicknesses n

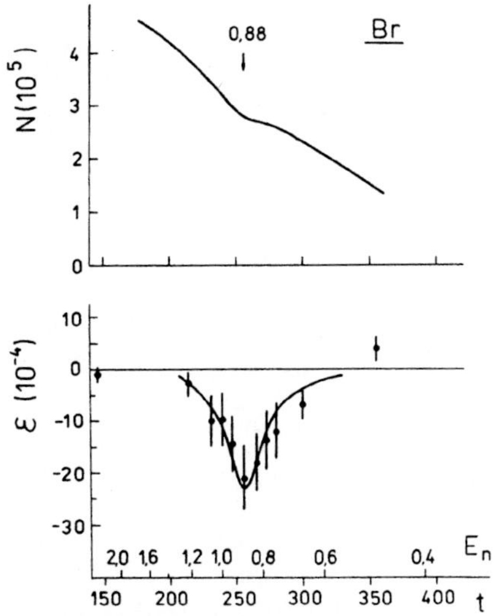

Fig. 2 The time-of-flight spectrum N and the transmission effect ϵ for Bromine.

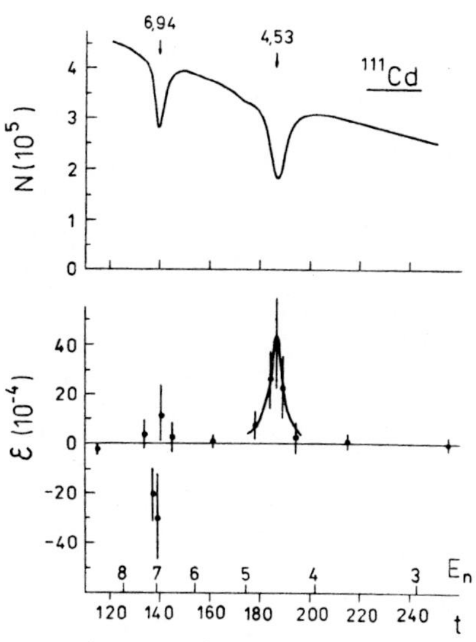

Fig. 4 The time-of-flight spectrum N and the transmission effect ϵ for Cadmium-111

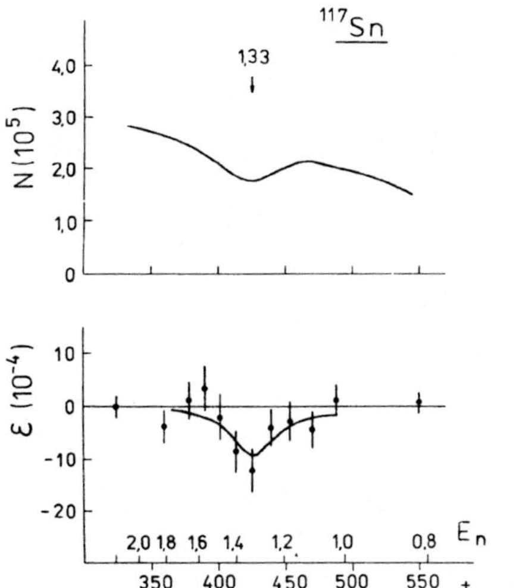

Fig. 3 The time-of-flight spectrum N and the transmission effect ϵ for Tin-117.

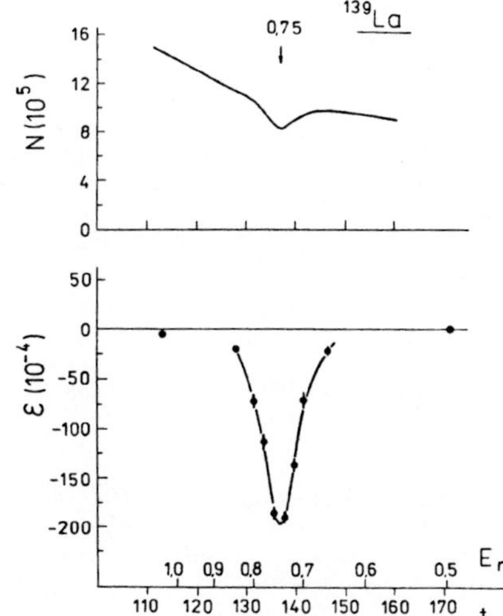

Fig. 5 The time-of-flight spectrum N and the transmission effect ϵ for Lanthanum.

Table 1 Parity violation data and related parameters of neutron p-resonances

Nucleus	E_p eV	Γ_p 10^{-3} eV	$g\Gamma_p^n$ 10^{-8} eV	E_s eV	$g\Gamma_s^n$ 10^{-3} eV	$P(E_p)$ 10^{-3}	$\|\alpha\|$ 10^{-5}	$\|H_W\|$ meV
^{81}Br	0.88±0.01	190±20	5.8±0.3	101.0±0.1	9.7±0.7	24±4	3.0±0.5	3.0±0.5
^{111}Cd	4.53±0.03	163±10	107±5	-4	0.95	-8.2±2.2	9.4±2.5	0.80±0.22
	6.94±0.07	143±13	108±8			4.1±3.3	3.8±3.0	0.42±0.33
^{117}Sn	1.33±0.01	230±20	19±1.5	-29	5.5	4.5±1.3	1.25±0.35	0.38±0.10
^{127}I	7.6±0.1	90±10	13±2	37.7±0.1	4.3±0.4	11.2±8.0	1.8±1.3	0.5±0.4
	10.4±0.1	90±10	320±40			0.3±0.4	0.3±0.4	0.1±0.1
	14.0±0.2	90±10	150±20			1.3±1.0	0.6±0.5	0.15±0.11
^{139}La	0.75±0.01	45±5	3.6±0.3	-48.6	84	73±5	2.6±0.2	1.28±0.12
^{238}U	4.41±0.01	(25)	11.1±0.2	6.67±0.02	0.59±0.01	3.7±3.7	1.7±1.7	0.04±0.04
	11.32±0.02	(25)	35±6	20.9±0.1	1.9±0.5	-2.5±2.5	1.7±1.7	0.08±0.08
	19.50±0.2	(25)	140±70			0.±1	0±0.7	0±0.01

in our measurements were 1.1(^{81}Br), 2.05(^{111}Cd), 3.9 (^{127}I), 1.25 (^{129}La) and 1.5 (^{238}U) in 10^{23} nucl/cm^2. Table 1 contains information on $\mathcal{P}(E_p)$ values and neutron resonance parameters E_0, Γ, $g\Gamma^n$. In all cases, but for ^{238}U, the resonance parameters of p-levels were obtained from our measurements by the shape analysis of spectra. The spins of all the p-resonances in Table 1 are unknown. They may be equal to $I\pm 3/2$ in some cases and then no P-odd effect is observed. Table 1 contains also the parameters from Ref.[19] of s-resonances, which are believed to mix with the p-resonances in question.

If is known which of the s-resonances mix with the p-resonances, then from the $\mathcal{P}(E_p)$ value through eq.(4) and eq.(1a) one may derive the corresponding values of modules of mixing coefficients $|\alpha|$ and of matrix elements $|H_W|$. The last columns in Table 1 contain the $|\alpha|$ and $|H_W|$ values obtained under assumption that only s-resonance having the maximum value of $\Gamma_s^n/(E_p-E_s)^2$ mixes with the p-resonance. In this way one obtains the lower estimate of $|H_W|$ for the resonances where within available accuracy $\mathcal{P}(E_p)\neq 0$.

Discussion

One can see from Table 1 that the experimental values of $\mathcal{P}(E_p)$ are within the 10^{-1}-10^{-3} interval, i.e. they are in excellent agreement with theoretical prediction made in Ref.[7]. The corresponding $|\alpha|$ values lie within 10^{-4}-10^{-5} and $|H_W|$ values -- (0.3-3) meV. No evident effect was observed in the resonance at 6.9 eV in ^{111}Cd and in the resonance at 7.6 eV in ^{127}I, though there are indications to the matrix element of the same order. No effect was found in the 2 resonances in ^{127}I and 3 resonances in ^{238}U. Their matrix elements are either less than o.2 meV or the resonances have the unfavourable for mixing spin values $I\pm 3/2$.

It seems interesting to compare the PNC effects in the resonance and thermal energy region in order to proove the energy dependence of PNC-effects given by the formula (7). Thermal region data are available for ^{117}Sn and ^{139}La from Ref.[9] and for Br from Ref.[20]. The results of comparison are given in Table 2, where $\mathcal{P}^*$(therm) is the expected value for the thermal neutron region,calculated using the resonance $\mathcal{P}(E_p)$ data by formula (7), and the $\mathcal{P}$ (therm) value is the measured value. The agreement between $\mathcal{P}^*$(therm) and $\mathcal{P}$ (therm) is quite good having in mind that $\mathcal{P}$(therm) and $\mathcal{P}(E_p)$ differ by 3 orders of magnitude. One can also compare the PNC transmission data with the neutron spin rotation result (the revised value is $\phi = -(3.7\pm 0.27)\cdot 10^{-5}$ rad/cm according to Ref.[21]). The equation one should use in this case is $\phi = \frac{\varepsilon(\text{therm})}{t}(E-E_p)\ell/\Gamma_p$ which follows from Ref.[7] (here t is the length of the target). The expected value ϕ lies in the interval $-(1.1-2.5)\cdot 10^{-5}$ depending on various values ε(therm) (directly measured or calculated from resonance region). Therefore, we have a perfect agreement in sign and quantitative agreement in the value of $|\phi|$.

Table 2 Comparison of PNC effects in thermal neutron region and related data

Nucl.	E_p eV	$\sigma_p(E_p)$ b	σ_t (therm) b	$\mathcal{P}$(therm) 10^{-6}	$\mathcal{P}^*$(therm) 10^{-6}
^{117}Sn	1.33	1.6 ± 0.2	3.7 ± 0.4	6.2 ± 0.7	14.5 ± 5.5
^{139}La	0.75	2.8 ± 0.4	19.6 ± 2.0	9.0 ± 1.4	9.3 ± 2.9
^{81}Br	0.88	0.9 ± 0.1	15.5 ± 1.5	19.6 ± 2.0 1)	16.2 ± 5.1
^{111}Cd	4.53	3.8 ± 0.5	29 ± 4	-	0.4 2)

1) Data from Ref.[20] recalculated per isotope ^{81}Br.

2) Upper limit of $\mathcal{P}^*$(therm) under assumption that the formalism of Ref.[11] is applicable for the description of the PNC effect in a wide energy range.

Conclusions

The present work has shown that the P-odd helicity dependence of the total neutron cross section has the resonance behaviour. All experimental data on PNC-effects in neutron-nuclear interaction are explained by the model of mixed compound states from Ref.[7]. There are indications to the universal character of the considered effect, the corresponding matrix elements determining the parity mixing of s- and p-resonances are approximately the same.

The further deepening of our understanding of the phenomenon requires the extension of the study to a wider range of nuclei, the improvement of experimental accuracy, the study in the resonances of the P-odd assymetry of gamma-emission investigated earlier in the thermal energy range only (Refs.[1,22,23]) and determination of spins of p-resonances. The parity violation experiments with resonance neutrons may have a bright future.

References

1. Yu.G.Abov, P.A.Krupchitsky. Usp.Fiz.Nauk, 118, 141 (1976).
2. F.C.Michel. Phys.Rev. 133, B329 (1964).
3. L.Stodolsky. Phys.Lett. 50B, 352 (1974).
4. M.Forte. In: Fundamental Physics with Reactor Neutrons and Neutrino: Paper from the Workshop..., Grenoble 1977, p.86. Bristol and London: Institute of Physics, 1978.
5. G.Karl, D.Tadic. Phys.Rev.C16, 1726 (1977).
6. A.Barroso, F.Margasa. J.Phys.G: Nucl.Phys. 6, 657 (1980).
7. O.P.Sushkov, V.V.Flambaum. Pisma ZhETF, 32,377 (1980).
8. M.Forte et al. Phys.Rev.Lett.45,2088 (1980).
9. E.A.Kolomensky. Phys.Lett. 107B, 272 (1981).
10. O.P.Sushkov, V.V.Flambaum.Usp.Fiz Nauk, 136,2 (1982).
11. L.Stodolsky. Phys.Lett., 96B, 127 (1980).
12. V.E.Bunakov, V.P.Gudkov. Z.Phys.A, 303, 285 (1981).
13. G.A.Lobov. Jad.Fiz. 35, 1408 (1982).
14. I.M.Frank. Particles and nucleus, 2, 805 (1972).
15. F.L.Shapiro. In: Nuclear Structure Study with Neutrons: Proc.of Int.Conf. 9-23 July 1965, Antwerp, p.223. Amsterdam: North Holland Publishing Company, 1966.
16. V.P.Alfimenkov et al. Pisma ZhETP, 34, 308 (1981).
17. V.P.Alfimenkov et al. Pisma ZhETP, 35, 42 (1982).
18. V.P.Alfimenkov et al. JINR Preprint P3-82-66, Dubna (1982).
19. S.F.Mughabghab, D.I.Garber. Neutron Cross Section BNL-325, 3rd ed.,1(1973).
20. V.A.Vesna et al. Pisma ZhETP, 35, 351 (1982).
21. B.Heckel. Private Communication, ILL, Grenoble (1982).
22. G.V.Danilyan et al. Pisma ZhETP, 24, 380 (1976).
23. H.Benkoula et al. Phys.Lett. 71B, 287 (1977).

RESONANCE STRUCTURE OF $^{32}S + n$ FROM TRANSMISSION AND DIFFERENTIAL ELASTIC SCATTERING EXPERIMENTS

C.R. Jungmann*, E. Cornelis**, L. Mewissen***, F. Poortmans***, J.P. Theobald****, and H. Weigmann*

Commission of the European Communities
Joint Research Centre
Central Bureau for Nuclear Measurements
Geel, Belgium

High resolution transmission and differential elastic scattering measurements were performed on ^{32}S between 0.18 MeV and 19 MeV neutron energy. The resonance parameters, including spin and parity, were determined for most of the 88 levels observed up to 1.7 MeV.

The distribution of neutron strength was determined for s1/2,p1/2, p3/2, d3/2, d5/2 and f-wave levels, and is compared to recent model calculations. One T = 3/2 level has been identified in ^{33}S and candidates for a second one have been found. Their neutron widths have been determined ,thus adding to the systematic knowledge of isospin impurities of T = 3/2 levels in low mass T_Z = 1/2 nuclei.

Introduction

This paper reports on a study of the resonance structure of the reaction ^{32}S+n by means of high resolution transmission and differential elastic scattering measurements.

The resonances of ^{32}S+n have previously been studied by Halperin et al.[1] and Johnson et al.[2] with neutron transmission and radiative capture experiments.

The very good energy resolution of the present measurements allowed a resonance parameter analysis up to 1.7 MeV neutron energy which is a substantial improvement compared to the data of ref.[1-2]. Moreover, from our scattering data, we were able to determine the spin and parity of a large number of resonances.

Experimental Method

The experimental details will be described elsewhere [3], and only a short summary is given here:

The measurements were performed at the neutron time-of-flight spectrometer of the 150 MeV electron linac of CBNM, Geel. The accelerator was operated with a burst width of 4.5 nsec (FWHM) and a repetition rate of 800 Hz. The direct, unmoderated neutron beam from the neutron producing U-target was used in the present measurements.

For the transmission measurements a 388 m flight path was used. The neutrons were detected in a 15.24 x 15.24 x 2.54 cm NE 110 plastic scintillator. Four 1.9 cm diameter photomultiplier tubes were mounted side-on to the scintillator. Typical figures for the over-all energy resolution (FWHM) are 46 eV, 210 eV, and 1.2 keV at 200 keV, 600 keV, and 2 MeV neutron energy respectively.

For the elastic scattering measurements six NE 110 plastic scintillator detectors with 5.08cm diameter and 2.54 cm thickness were placed around the sample at laboratory angles of 30°, 45°, 77°, 90°, 130° and 150°. Flight path length, accelerator parameters and time-of-flight resolution were the same as for the transmission measurements.

Resonance Parameter Analysis

The transmission data have been analyzed with the multi-level Breit-Wigner (MLBW) code SIOB [4]. The analysis yields the product $g\Gamma n$ for each resonance. In the case of isolated resonances with a total width larger than the resolution width, it also yields the statistical weight factor g and hence the resonance spin. For a limited number of resonances, the orbital angular momentum could be determined from the resonance shape : S-wave resonances are identified by their strong interference with potential scattering; also well isolated and sufficiently strong p-wave resonances could be assigned from their asymmetric shape due to interference with p-wave potential scattering. Most of the resonance spins and parities however were determined from the scattering data as explained below.

An example of a fit to the transmission data is shown on fig. 1 for the energy range from 450 keV up to 1000 keV.

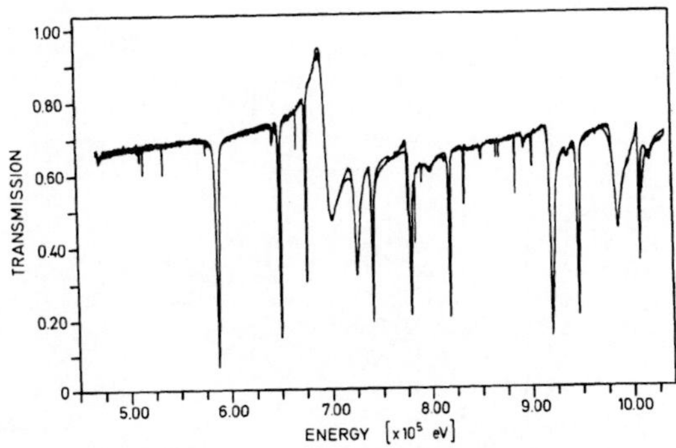

Fig. 1. Example of a fit of the transmission data by the multi-level Breit-Wigner fitting routine. The fitted curve is distinguishable from the data only at a few interference minima of the cross section (maxima of the transmission).

The purpose of the elastic scattering measurements was not to obtain an independent quantitative measure of the neutron widths, but to determine spin and parity of the resonances from the angular distribution of the scattered neutrons. Therefore, the analysis of the scattering data did not consist in a quantitative fit of the measured angular distributions. We rather limited ourselves to the theoretical calculation of the cross section expected for different assumptions on the various resonance spins, and a qualitative comparison of the calculated cross section with the experimental one.

The program used for the theoretical calculation is based on the Blatt-Biedenharn formalism [5]. It calculates the differential elastic scattering cross

* CEC, JRC, CBNM, B-2440 Geel, Belgium
** RUCA, B-2020, Antwerp, Belgium
*** SCK-CEN, B-2400 Mol, Belgium
**** TH Darmstadt. D-6100 Darmstadt, W-Germany

K. H. Böckhoff (ed.), Nuclear Data for Science and Technology, 777–780.

section as a function of energy for a number of preselected c.m. scattering angles,corresponding to the positions of the detectors in the experiment. Resonance-resonance interference may be taken into account in the MLBW-approximation. Doppler broadening and experimental resolution are taken into account in an approximate fashion : both effects are lumped together and treated as a single resolution function with rectangular shape.

Examples of scattering cross sections calculated this way are shown and compared to the experimental data in the energy range from 640 keV to 760 keV in fig. 2. It can be clearly seen that the calculations reproduce the different behaviour of resonances with different orbital angular momentum very well. Also the influence of different compound spins for resonances with the same ℓ-value can be clearly observed for the two p-wave resonances with spin 1/2 and 3/2 at 726.0 keV and 741.6 keV neutron energy. The assignment of the spin of d-wave resonances required a quantitative comparison of calculated and experimental resonance areas. It was possible to deduce the orbital angular momentum for 79 and the compound spin for 59 out of totally 88 observed resonances.

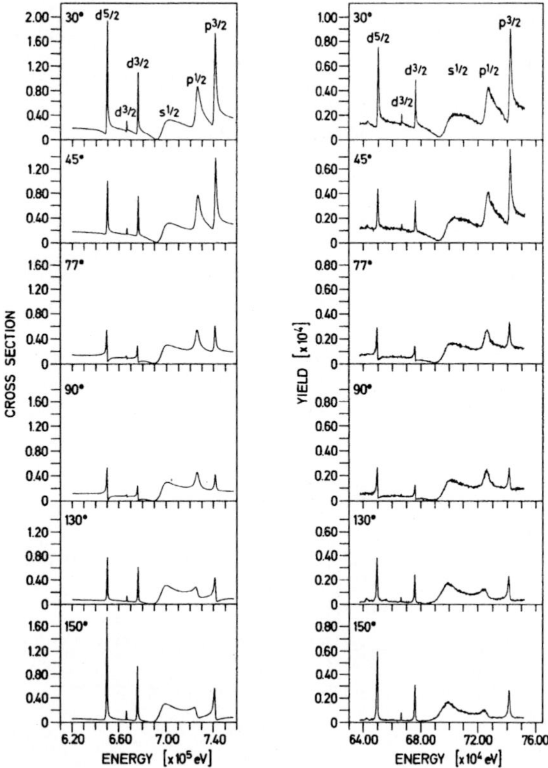

Fig. 2. Comparison of the calculated elastic scattering cross section (left) to the experimental yield spectra (right) for 6 different scattering angles in the energy range from 620 to 750 keV.
The resonance spins assumed for the calculations are indicated in the 30° spectrum.

Results and Discussion

A complete list of resonance parameters as obtained from the analysis of the transmission and differential elastic scattering data is given in ref. [3].

For further discussion of resonance properties, we include the parameters of resonances below 200 keV neutron energy as given by Halperin et al. [1]

Distribution of Strength

The level density in a light nucleus like ^{33}S being sufficiently low, it may be hoped that shell model or similar calculations extended into the continuum might be able to reproduce some of the gross features of the measured cross section. Such calculations (based on a particle + vibration model) and a comparison to experimental data have indeed been done with considerable success for the case of ^{28} Si + n by Halderson et al.[6]

For ^{32}S + n calculations within a pure quasi-particle model have been done as early as 1968 by Payne [7]. These calculations yield the position and neutron widths of 3-quasi-particle states above neutron separation energy in ^{33}S. They also indicate that the density of 5-quasi-particle states should be very small in the energy range of interest here.

Later, Halderson et al.[8] have performed similar calculations as those for ^{29}Si mentioned above also for ^{33}S. They use a particle + vibration model for the negative parity states and a particle-hole picture for the positive parity ones. The results of their calculation are compared to the present experimental data in fig.3. The figure shows the cumulative sums of experimental and theoretical reduced widths for each compound spin separately; both, the experimental and the theoretical widths of ref.[8] have been converted to reduced widths using an interaction radius of 4.7 fm.

A comparison of the number of calculated and experimental levels shows no strong discrepancies for the positive parity levels, whereas the number of negative parity levels is strongly underpredicted by the pure particle + vibration model. Nevertheless it is worthwile to compare the calculated strength of predicted simple states to the experimental one in the same energy region : the doorway character of the simple configurations insures that their strength is merely distributed over a larger number of levels. Indeed, for the p-wave as well as the s-wave resonances the total strengths in the energy range investigated coincide to within 50% on the average. In detail, however, there is little similarity between the staircase-functions representing experiment and theory in fig. 3. For the d-wave resonances even the total strength is very different. In the d 5/2 case, we have left out the calculated level at 17 keV. Since that ℓ = 2 level is at very low energy, after conversion to reduced widths, it would contribute dominantly to the cumulative sum for this spin and the theoretical strength would exceed the experimental one by a large factor.

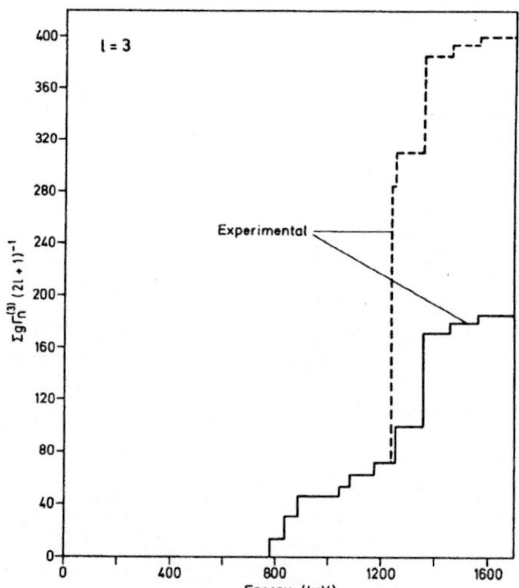

Fig. 4. Cumulative sum of experimental $g\Gamma_n^{(3)}$-values for f-wave resonances. The broken line results if the resonance at 1242.2 keV is included.

Finally, fig.4 shows the cumulative sum of experimental $g\Gamma_n^{(3)}$-values for f-wave resonances. No spin assignments have been attempted for ℓ = 3.

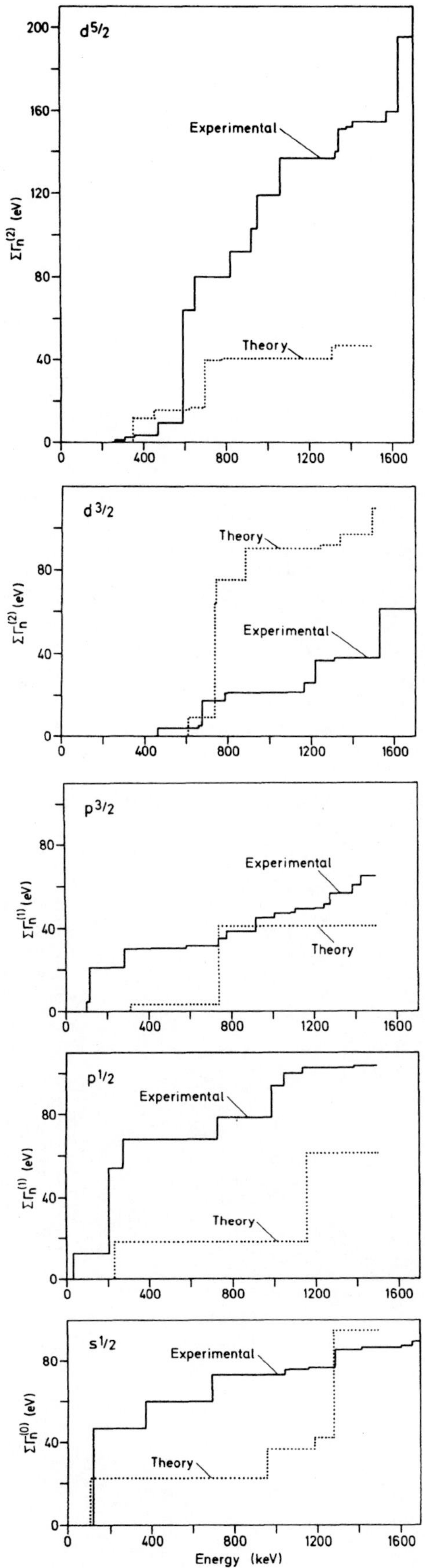

Fig. 3.: Comparison of cumulative sums of calculated[8] and experimental reduced widths.

From the cumulative sums of experimental reduced widths we have calculated strength functions for each compound spin according to

$$S(J^{\pi}) = \langle \Gamma_n^{(\ell)} (J^{\pi}) \rangle / D (J^{\pi}) \qquad (1)$$

The resulting values are given in column 2 of Table I. For f-wave resonances only the overall ℓ = 3 strength is given; the entry in brackets results if the resonance at 1242.2 keV neutron energy, the f-wave assignment of which is uncertain, is included in the set of f-wave resonances. For comparison, in column 3 of the table we also list strength functions obtained for 1 MeV neutron energy from an optical model calculation with the code CERBERO-3 of Fabbri et al.[9], using the standard optical model parameter set of Igarasi et al.[10]. The agreement is surprisingly good for positive parity, whereas the p-wave strength functions disagree by a factor of about 2.5.

Table I : Strength Functions

J^{π}	$S(J^{\pi})$ [10^{-4}] experim.	optical model
$1/2^+$	$0.56 \begin{smallmatrix} +0.37 \\ -0.20 \end{smallmatrix}$	0.53
$1/2^-$	$0.72 \begin{smallmatrix} +0.56 \\ -0.36 \end{smallmatrix}$	1.88
$3/2^-$	$0.48 \begin{smallmatrix} +0.21 \\ -0.13 \end{smallmatrix}$	1.17
$3/2^+$	$0.55 \begin{smallmatrix} +0.36 \\ -0.19 \end{smallmatrix}$	0.68
$5/2^+$	$1.40 \begin{smallmatrix} +0.63 \\ -0.39 \end{smallmatrix}$	1.44
$\ell = 3$	$2.29 \begin{smallmatrix} +1.51 \\ -0.81 \end{smallmatrix}$ ($4.97 \begin{smallmatrix} +3.12 \\ -1.67 \end{smallmatrix}$)	2.18

Isobaric Analog States

As is well known, the neutron decay of isobaric analog states is isospin forbidden. Nevertheless, in recent high resolution work isobaric analog states have been observed as neutron resonances [11-12-13]. The interesting point in this observation is that, by the measurement of the neutron width of the resonance, direct information on the isospin purity of the analog state can be obtained.

In first order perturbation theory the reduced neutron width of the analog state is given by

$$\Gamma_n^{(\ell)}(T = 3/2) = \left| \sum_i \frac{\langle T = 3/2 \mid V \mid i \rangle}{E_i - E(T = 3/2)} \left(\Gamma_n^{(\ell)}(i)\right)^{1/2} \right|^2 \qquad (2)$$

where the sum extends over all T = 1/2 states of the same spin and parity.

From this expression two approximations may be derived for a 'ZEROTH ORDER' guess and a lower limit of the average isospin mixing matrix element $\overline{\langle T = 3/2 | V | i \rangle}$. The expression for the 'ZEROTH ORDER' value is :

$$\Gamma_n^{(\ell)}(T = 3/2) = \left| \overline{\langle T = 3/2 \mid V \mid i \rangle}_0 \right|^2 \sum_i \frac{\Gamma_n^{(\ell)}(i)}{(E_i - E(T=3/2))^2} \qquad (3)$$

whereas the lower limit is given by :

$$\Gamma_n^{(\ell)}(T=3/2) = \left| \overline{\langle T=3/2 \mid V \mid i \rangle}_{min} \right|^2 \left| \sum_i \frac{\pm \sqrt{\Gamma_n^{(\ell)}(i)}}{| E_i - E(T=3/2) |} \right|^2 \qquad (4)$$

T = 3/2 states in ^{33}S are known experimentally only below neutron separation energy. In the energy range covered by the present experiment one expects the T = 3/2 analogs to levels [14] in ^{33}P at 3.490 MeV ($5/2^+$) 3.628 MeV ($7/2^+$), 4.048 MeV ($3/2^+$, $5/2^+$), 4.194 MeV ($5/2^+$) and 4.226 MeV ($7/2^-$). In order to estimate the positions of the analog states in ^{33}S, we simply assume that the coulomb displacement energies are the same as those observed for the T = 3/2 multipletts below neutron separation energy. According to a recent compilation by Benenson and Kashy [15] this assumption should be correct to within 20 keV, where because of a possible contribution from the magnetic spin-orbit term, it is preferable to compare multipletts with the same spin and parity.

The lowest T = 3/2 level in the energy range of the present experiment is expected to be the analog of the 5th excited state (3.490 MeV) in ^{33}P. Since its spin is 5/2+, we adopt the coulomb displacement energy as deduced from the excitation energy (7.337 MeV) of the analog to the 2nd excited state (1.848 MeV 5/2+) in ^{33}P. With this, we obtain for the neutron energy at which the analog of the 3.490 MeV level is expected, E_n = 347.5 keV. The only 5/2+ resonance observed within 20 keV of this energy is the one at E_n = 353.57 keV. Assuming that this resonance is indeed the T = 3/2 state expected, we deduce 'GUESS' and 'MINIMUM' values for the isospin mixing matrix element from equ.(4) and (5), respectively, as given in line 1 of Table II.

We omit the analogue to the 3.628 Mev (7/2+)state, because we do not observe resonances with spin 7/2+, and consider next the analogue of the 7th excited state (4.048 MeV) in ^{33}P. The spin of the parent state is given as either 3/2+ or 5/2+ in ref.[14]. Considering first the possibility of J = 3/2+ and using the coulomb displacement energy as obtained from the excitation energy (6.905 MeV) of the analog to the 1st excited state (1.432 MeV, 3/2+) in ^{33}P, a neutron energy of E_n = 906.5 keV is obtained for the analog to the 4.048 MeV level. The only candidate would be the resonance at 901.85 keV. The isospin mixing matrix elements obtained for this case are given in line 2 of Table II.

However, Wagner et al.[16] give arguments for a preference of the J = 5/2+ assignment for the 4.048 MeV level in ^{33}P. With this assignment, using again the coulomb displacement energy as obtained from the analog to the 1.848 MeV state, we obtain a neutron energy of E_n = 923.0 keV for the analog to the 4.048 MeV level. There are two possible candidates in this case: one of them is again the 901.85 keV resonance. The isospin mixing matrix elements which would result are given in line 3 of Table II.

The second candidate is the 5/2+ resonance at 920.54 keV neutron energy. At first sight this resonance seems to be rather strong for a T = 3/2 level. However, this strength may be due to the fact that it is rather close in energy to the 947.88 keV resonance. The similar magnitude of the neutron widths of these two resonances would indeed tell that it is not allowed to treat their mixing in first order perturbation theory. On the other hand, the contributions to the neutron width of 920.54 keV resonance from levels other than the 947.88 keV one, would be sufficiently small to be neglected, such that the case could be treated in a two-level approximation. This is discussed in more detail in ref.[3]. The resulting isospin mixing matrix element is given in the bottom line of Table II.

Table II : Average mixing matrix elements for possible isobaric analog resonances in ^{33}S.

E_{ex}(parent) [MeV]	J^π	E_{res} [keV]	$\langle T=3/2 \vert V \vert i \rangle_0$ equ.(3) [keV]	$\langle T=3/2 \vert V \vert i \rangle_{min}$ equ.(4) [keV]
3.490	$5/2^+$	353.57	18	6.6
4.048	$3/2^+$	901.85	35	13
	$5/2^+$	901.85	4.0	2.0
		920.54	13.5	

Finally the analogs to the 8th and 9th excited states at 4.194 and 4.226 MeV in ^{33}P are expected at neutron energies of approximately 1074 and 1106 keV, respectively. There are at least 3 candidates for each of them. Because of this uncertainty, we did not attempt to deduce information on mixing matrix elements for these levels.

The isospin mixing matrix elements of Table II are generally somewhat smaller than those obtained in ref.[11] for ^{25}Mg and in ref.[12] for ^{29}Si.

Acknowledgements

The authors wish to thank Mr. J.M. Salomé and the linac operation group of CBNM as well as Mr.C. Cervini and the computer operation team for their continuous effort.

References

1. J. Halperin, C.H. Johnson, R.R. Winters and R.L. Macklin, Phys. Rev. C21 (1980) 545.
2. C.H. Johnson and R.R. Winters,Phys. Rev. C21 (1980) 2190.
3. C.R. Jungmann, H. Weigmann, L. Mewissen, F. Poortmans, E. Cornelis, and J.P. Theobald to be published.
4. G. De Saussure, D.K. Olsen and R.B. Perez, Report ORNL/TM-6286, 1978.
5. J.M. Blatt and L.C. Biedenharn, Rev. Mod. Phys. 24 (1952) 258.
6. D. Halderson, B. Castel, M. Divadeenam and H.W. Newson, Annals of Phys. 103 (1977) 133.
7. G.L. Payne,Phys. Rev. 174 (1968) 1227.
8. D. Halderson, B. Castel and G. Aizer, J. Phys. G 6 (1980) 59, and private communication.
9. F. Fabbri, G. Fratamico and G. Reffo, CNEN-Report RT/FI (77) 6, 1977.
10. S.G. Igarasi et al., Report JAERI 1228 (1974) 41.
11. H. Weigmann, R.L. Macklin and J.A. Harvey, Phys. Rev. C14 (1976) 1328.
12. S. Cierjacks, S.K. Gupta and I. Schouky, Phys. Rev. C17 (1978) 12.
13. F. Hinterberger, P. Von Rossen, S. Cierjacks, G. Schmalz, D. Erbe and B. Leugers, Nucl. Phys. A352 (1981) 93.
14. P.M. Endt and C. van der Leun, Nucl. Phys. A310 (1978) 1.
15. W. Benenson and E. Kashy, Rev. Mod. Phys. 51 (1979) 527.
16. P. Wagner, J.P. Coffin, M.A. Ali, D.E. Alburger and A. Gallman, Phys. Rev. C7 (1973) 2418.

DIFFERENTIAL NEUTRON SCATTERING CROSS SECTIONS AND AVERAGE NEUTRON PARAMETERS OF TIN ISOTOPES

V.G.Nikolenko, A.B.Popov, G.S.Samosvat

Laboratory of Neutron Physics, Joint Institute for Nuclear Research, Dubna, USSR

Abstract: The differential cross sections of elastically scattered neutrons were measured for tin isotopes at the pulsed reactor IBR-30 within the 1-200 keV energy region. The s- and p- strength functions S^0, $S^1_{1/2}$, $S^1_{3/2}$ and radii R'_0, R'_1 of even-even tin isotopes were obtained.

As early as 1964 [1] a possibility was suggested to extract the s- and p-neutron average parameters from the differential neutron scattering cross section. The present paper is apparently the first attempt made to estimate S^0, $S^1_{1/2}$, $S^1_{3/2}$, R'_0 and R'_1 on the basis of measured $\sigma(\vartheta)$.

The intensities of neutrons scattered by tin samples were measured on the 250 m flight path of the pulsed reactor IBR-30 under angles of 45°, 90° and 135° using the equipment described briefly in [2]. We used the metallic tin samples enriched to 97-98 %. Their weights varied from 120 to 250 g. The measurements were performed both in the reactor and booster mode of the reactor. The parameters of the differential scattering cross section described by the formula $\sigma(\vartheta)=\frac{\sigma_s}{4\pi}\left[1+\omega_1 P_1(\cos\vartheta)+\omega_2 P_2(\cos\vartheta)\right]$ were found by processing the experimental spectra measured at above mentioned angles with and without the sample. The cross section parameters σ_s, ω_1, ω_2 were calculated for 11 energy intervals up to 60 keV in the reactor mode and for 18 energy intervals up to 200 keV in the booster mode. The results of measurements on carbon sample were calibrative for the determination of σ_s. Under assumption that in the investigated energy region only s- and p-wave neutrons are scattered on tin isotopes and by averaging over resonances of the differential cross section expression given in [3] one may obtain for the even-even targets the following expressions for σ_s, ω_1 and ω_2:

$$\sigma_s=4\pi B_0 \quad \omega_1=B_1/B_0 \quad \omega_2=B_2/B_0 \qquad (1)$$

$$B_0=\frac{1}{k^2}\Big\{\sin^2\delta_0+3\sin^2\delta_1+\pi\sqrt{E}\Big[\frac{1}{2}S^0F_0+\frac{1}{2}(S^1_{1/2}F_{1/2}+2S^1_{3/2}F_{3/2})v-\sin^2\delta_0\cdot S^0-\sin^2\delta_1(S^1_{1/2}+2S^1_{3/2})v\Big]\Big\} \qquad (2)$$

$$B_1=\frac{1}{k^2}\Big\{6\sin\delta_0\sin\delta_1\cos(\delta_0-\delta_1)-\pi\sqrt{E}\big[3\sin\delta_1\sin(2\delta_0-\delta_1)\cdot S^0+\sin\delta_0\sin(2\delta_1-\delta_0)(S^1_{1/2}+2S^1_{3/2})v\big]\Big\} \qquad (3)$$

$$B_2=\frac{2}{k^2}\Big\{3\sin^2\delta_1+\pi\sqrt{E}\cdot v\Big[\frac{1}{2}S^1_{3/2}F_{3/2}-\sin^2\delta_1\cdot(S^1_{1/2}+2S^1_{3/2})\Big]\Big\} \qquad (4)$$

Here

$$\delta_0=-\kappa R+\operatorname{arctg}(\kappa R R_0^\infty) \qquad (5)$$

$$\delta_1=-\kappa R+\operatorname{arctg}(\kappa R)+\operatorname{arctg}\left[\frac{(\kappa R)^3R_1^\infty}{1+(\kappa R)^2+R_1^\infty}\right] \qquad (6)$$

In the expressions (1) - (6) the generally accepted symbols are used, other symbols are defined below:

$$v=\frac{(\kappa R)^2}{1+(\kappa R)^2} \qquad (7)$$

(We assumed $R = 1.4\,A^{1/3}$ in fermi);

$$F_\mu=\frac{1}{\sqrt{2\pi}}\int_0^\infty\frac{x\sqrt{x}}{x+a_\mu}e^{-x/2}dx \qquad (8)$$

This is the term which accounts for the neutron widths distribution in the averaging of the resonance contribution and here

$$a_0=\frac{S^0_\gamma}{S^0\sqrt{E}} \qquad a_{1/2}=\frac{S^1_\gamma}{3S^1_{1/2}v\sqrt{E}} \qquad (9)$$

$$a_{3/2}=\frac{2S^1_\gamma}{3S^1_{3/2}v\sqrt{E}}$$

S^0 is the s-neutron strength function, $S^1_{1/2}$ and $S^1_{3/2}$ are the p-neutron strength functions for the spins 1/2 and 3/2, S^0_γ, S^1_γ are the s-and p-radiation strength functions, respectively.

Formulas (1) - (9) have permitted us to carry out the analysis of the experimentally obtained values of $\sigma_s(E)$, $\omega_1(E)$ and $\omega_2(E)$ in order to determine S^0, $S^1_{1/2}$, $S^1_{3/2}$, R_0^∞ and R_1^∞. The least square program FUMILI [4] was used in the analysis. The obtained results are shown in the table. The table gives also $R'_0 = R(1-R_0^\infty)$ values and R'_1 values (both in fermi) found from $R'_1 = R(1-\frac{3R_1^\infty}{1+R_1^\infty})$ [5].

During calculation the values of S^0_γ and S^1_γ were assumed to known and varied within $0.5.10^{-4}$ - -10.10^{-4}. The variations of S^0_γ and S^1_γ within this interval did not affect the values and errors of other parameters. For illustration the ^{124}Sn results are shown in fig. The points stand for experimental data, solid lines are the calculated σ_s, ω_1, ω_2 values corresponding to the parameters given in the table.

The errors indicated in the table are the statistical ones given by the program FUMILI and corrected for the factor $[\chi^2/(m-n)]^{1/2}$, where m is the number of experimental points, n is the number of parameters. Our S^0 values agree within errors with data from [6]. The R_0^∞ values are also not in contradiction with estimates of R'_0 made in [6] for the low energy region. As far as the $S^1_{1/2}$, $S^1_{3/2}$ and R_1^∞ parameters are concerned, they were obtained for the first time.

Note that the calculation made with fixed values of R_0^∞ in the interval from 0.08 to 0.16 displaid

K. H. Böckhoff (ed.), Nuclear Data for Science and Technology, 781–782.

Table Results

Isotope	$\chi^2/m-n$	S^0	$S^1_{1/2}$	$S^1_{3/2}$	R_0^∞/R_0'	R_1^∞/R_1'
^{116}Sn	1.8	0.18±0.04	7.0±1.3	2.14±0.15	0.17±0.03 5.67±0.20	-0.16±0.02 10.8 ±0.6
^{118}Sn	2.1	0.16±0.05	5.7±1.7	1.96±0.22	0.20±0.04 5.50±0.27	-0.18±0.03 11.4 ±0.9
^{120}Sn	1.6	0.06±0.04	2.4±1.5	2.02±0.18	0.12±0.04 6.07±0.27	-0.14±0.02 10.4±0.4
^{122}Sn	2.0	0.17±0.05	4.9±1.5	2.07±0.21	0.20±0.04 5.55±0.21	-0.19±0.02 12.0 ±0.8
^{124}Sn	1.7	0.19±0.03	10.5±1.0	1.35±0.18	0.29±0.03 5.0 ±0.3	-0.23±0.02 13.4 ±0.9

the correlated variations of S^0 and $S^1_{1/2}$ outside error limits without an essential increase of χ^2, the values of $S^1_{3/2}$ and R_1^∞ being stable.

The authors would like to thank Dr. Zo In Ok for his help in the preparation of the report.

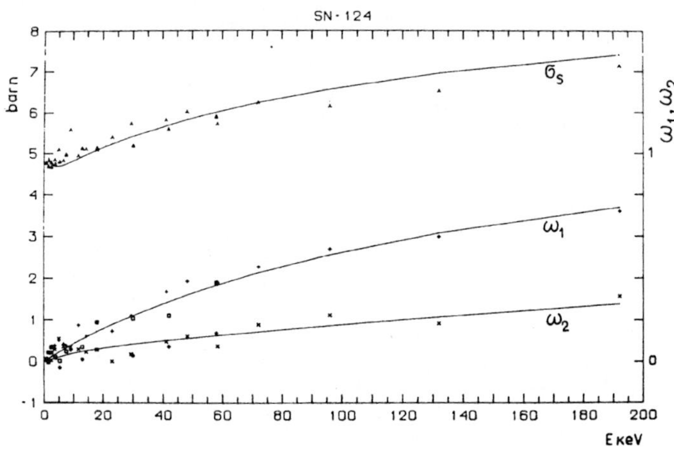

Fig. 1. Dependence of σ_s , ω_1 , ω_2 on neutron energy for ^{124}Sn: points - experimentally obtained values, solid lines - calculated results corresponding to the parameters given in the Table.

References

1. Yu.P.Popov, Yu.I.Fenin. Materialy rabochego soveschanija po vzaimodejstviju nejtronov s jadrami. JINR-1845, 89, 1964.
2. V.G.Nikolenko, G.S.Samosvat. Jad.Fiz. 23, 1159, 1976.
3. J.M.Blatt, L.C.Biedenharn. Rev.Mod.Phys. 24, 258, 1952.
4. I.N.Silin. In: Statisticheskie metody v experimentalnoj fizike. p.319, Moscow, Atomizdat, 1976.
5. Yu.A.Alexandrov et al. Materialy 5-oj vsesojuznoj konferentsii po nejtronnoj fizike. Kiev, 15-19, sept. 1980.
6. S.F.Mughabghab, M.Divadeenam, N.E.Holden. Neutron cross sections, v.1, part A, Academic Press, 1981.

NEUTRON ANALYZING POWER AND ELASTIC DIFFERENTIAL CROSS SECTION FOR CARBON-12 AND SOME MEDIUM AND HEAVY ELEMENTS (Si, S, Ca, Cu, La, Pb, Bi, and U)

G. Bulski, W. Grum, J. W. Hammer, H. Postner, G. Schleussner, E. Speller

Universität Stuttgart
Institut für Strahlenphysik
Allmandring 3
D-7000 Stuttgart 80
Federal Republic of Germany

The neutron analyzing power and differential cross sections of ^{12}C, natural Si, S, Ca, Cu, La, Pb, ^{209}Bi, ^{238}U have been determined using a scattering experiment for optical model determination purposes. The experimental method employed is briefly described.

{Analyzing power $A_y(\Theta)$ and differential cross section $d\sigma/d\Omega$ for ^{12}C (E_n=7.55, 7.65, 7.85, 8.05, 8.73MeV), natural Si, S, Ca, Cu, La, Pb, and ^{209}Bi, ^{238}U (E_n=7.75MeV), optical model, unfolding of proton recoil spectra, response matrix of NE213 detector, high current Dynamitron accelerator}

Introduction

Elastic scattering data of polarized neutrons are necessary for the determination of the optical model of the nucleus, especially for providing the spin-orbit-term. There exist very few data for neutron energies above 4MeV, often proton data are used for the determination of the optical model parameters.

Employing a special experimental technique it has become possible to measure analyzing power $A_y(\Theta)$ and differential cross section $\sigma(\Omega)$ in a relative small laboratory simultaneously.

Using transverse polarized neutrons generally differential cross sections for neutrons with 'spin up' and 'spin down' are obtained, $\sigma\uparrow(\Theta)$ and $\sigma\downarrow(\Theta)$. The analyzing power of neutrons with a polarization degree P is defined as

$$A_y(\Theta) = \frac{1}{P}\,\frac{\sigma\uparrow(\Theta) - \sigma\downarrow(\Theta)}{\sigma\uparrow(\Theta) + \sigma\downarrow(\Theta)}$$

Experimental Method

Production of the Used Polarized Neutrons

For producing polarized neutrons the $^{9}Be(\alpha,n)^{12}C$-reaction is used. In an α-energy-range of E_α=2...5MeV this reaction shows a rather good figure of merit and delivers neutron polarisation degrees of up to 60%. Due to the high Q-value of 5.7MeV one obtains neutrons in the energy range of E_n=7...9MeV with usable yield. Indeed the cross section of about 10mb/sr in the average is rather small. This disadvantage has to be compensated with high particle currents. At Stuttgart we use a high current Dynamitron accelerator capable of delivering α-particle currents (He^{1+}) of up to 1mA at energies up to 4MeV. A special target technique had to be developed for dissipating power of 4kW (or 10kW/cm^2). In total we obtain typically 10^6 neutrons/s on the scattering sample section.

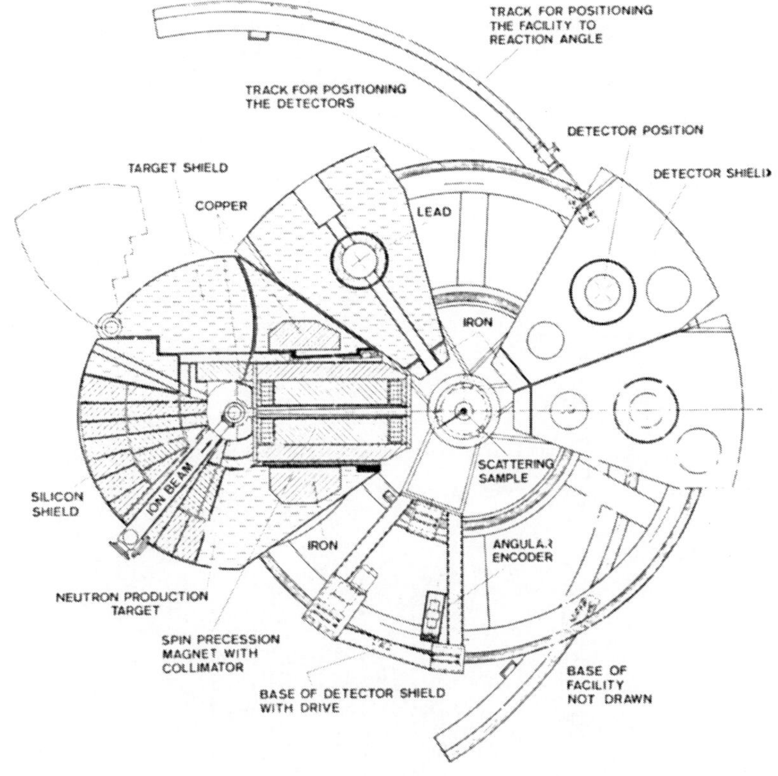

Fig. 1a The Stuttgart neutron scattering facility, top view (Fig. 1b see next page)

Scattering Experiment

The scattering experiment consists essentially of a neutron producing target, spin-precession magnet, scattering sample, 4 neutron detectors, and 3 monitors. Due to the limited experimental area the distances neutron producing target - scattering sample - detectors are relatively small (1.25m and 1.00m, respectively) thus target and detectors had to be

K. H. Böckhoff (ed.), Nuclear Data for Science and Technology, 783–791.

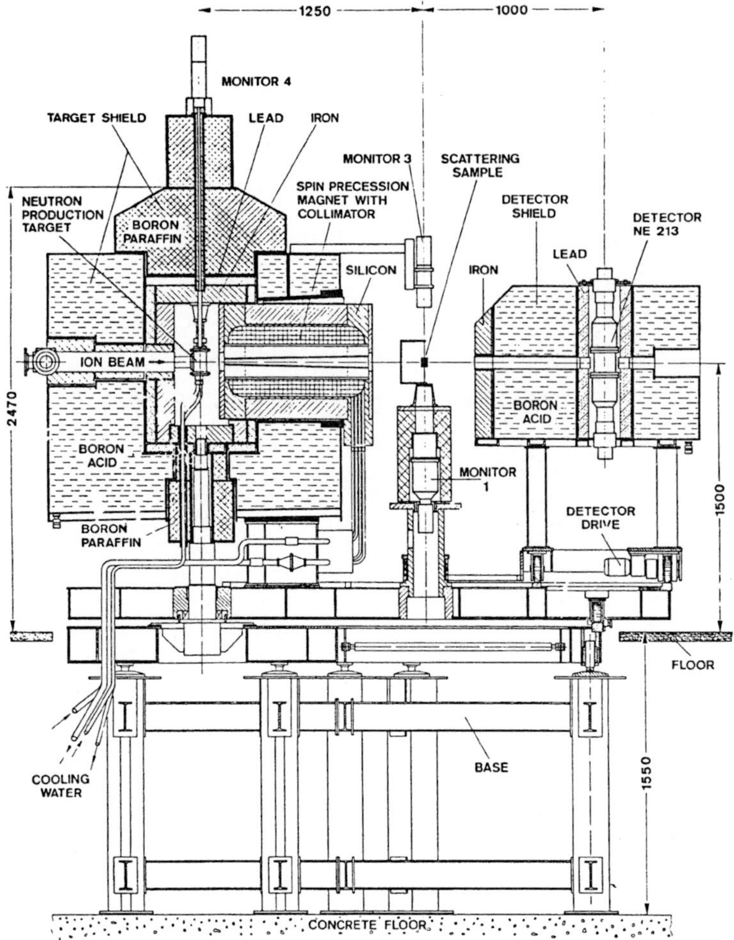

Fig. 1b The Stuttgart neutron scattering facility, side elevation

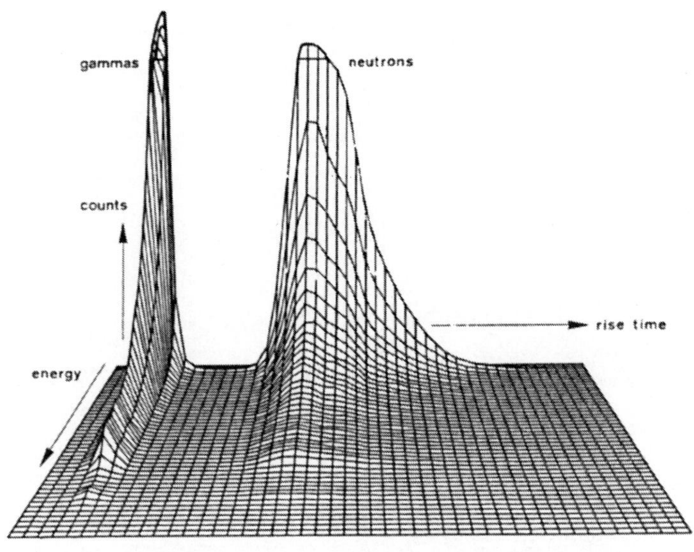

Fig. 2 n-γ-discrimination, 3-dimensional spectrum

shielded very properly. This shielding consists of several layers of iron, copper, lead, boric acid or boric paraffin. The reaction angle ϕ between α-particles and neutrons can easily be changed between -10° and +70°, the scattering angles Θ however can be varied from -124° to +124°. Fig. 1a shows the experimental setup in a top view, Fig. 1b shows a side elevation.

Detection of Neutrons

Due to our narrow geometry we have to use pulse height spectroscopy instead of the common time-of-flight technique. The available energy resolution is better than 10% and thus is comparable to a TOF-experiment with flight paths of some meters. Our resolution has been optimized using a method developed by Schölermann and Klein[1] at PTB. In order to eliminate apparative asymmetries and drifts the spin of the neutrons is flipped by 180° every 5min. Thus spectra for both spin states can be recorded quasi-simultaneously. The detectors are built up from NE213 scintillation counters of size 3"x3" with optical contact to two 5" photomultipliers. The n-γ-discrimination which has been constructed in our laboratory following well-known principles had to be very efficient, Fig. 2 shows a typical n-γ-spectrum: the separation occurs at a bias of about 100keV electron energy respectively about 500keV proton energy and is better than 1°/oo. Special efforts are necessary for measuring and evaluating neutron pulse height spectra.

Evaluation of Spectra

The continuous proton recoil spectra have to be transformed to line spectra by an unfolding procedure. For that purposes we use modified versions of the codes FERDOR respectively FORIST [2,14]. These matrix inversion procedures need a response matrix (i. e. a catalog of typical spectrometer response functions) with narrow energetic steps. We use energy steps of 125keV, which is sufficient enough because our neutron energy width is about 150keV. Formerly we used a response matrix computed with the code 05S [3], recently we use the code NRESP4 developed by G. Dietze [4]. Fig. 3 shows our response matrix computed with NRESP4 using the latest cross section data.

As an example for the quality of the computed response functions Fig. 4 shows the neutron spectrum of the $^{9}Be(\alpha,n)^{12}C$-reaction with the two neutron groups n_0 and n_1 at the energies 3.4MeV and 7.75MeV, one spectrum is determined experimentally, the other one is computed using NRESP4. The remaining small differences (which are hardly to be seen) are due to collimators and shieldings which are not considered by our calculations.

Figs. 5a-c show the results of the unfolding procedure of a pure $^{9}Be(\alpha,n)^{12}C$-spectrum in 3 steps:

1. pure matrix inversion with remaining more or less strong oscillations due to the statistics of the spectrum
2. consideration of the detector resolution in a smoothing procedure
3. optimization of smoothing.

The method of digital filtering will be applied if proton recoil spectra with bad statistics are to be evaluated [5,6]. Thus physical meaningless fluctuations are suppressed.

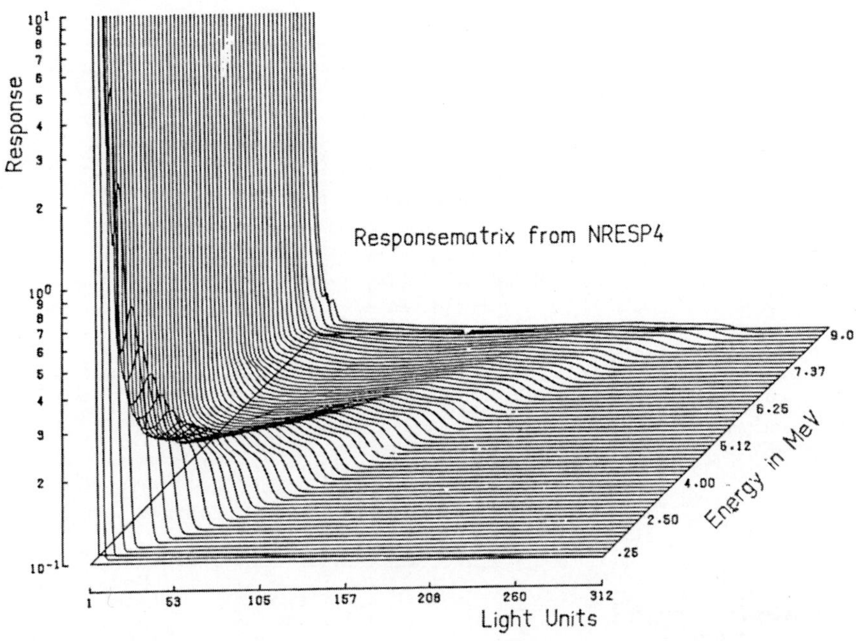

Fig. 3 Response matrix for a 3"x3"-NE213-n-detector from NRESP4

From proton recoil spectra unfolded with the NRESP4 based response matrix we receive absolute values of the neutron fluence with an accuracy of about 5%. Data received by using our former response matrix based on O5S were only relative ones.

Fig. 6 shows a real unfolded spectrum of neutrons scattered by Silicon at $\Theta=110^{o}$. The peaks of elastic and inelastic scattering can be well separated.

Precise energy calibration of the neutron detectors is an important condition for correct unfolding of spectra [7]. For this calibration we use the Compton-edges of two γ-sources. Neutron background is determined separately for each angle position and is subtracted from the scattering spectra. For determination of the background spectra the collimators of the detectors are closed by long stoppers made from copper and polyethylene, the scattering sample remains at its place in the center.

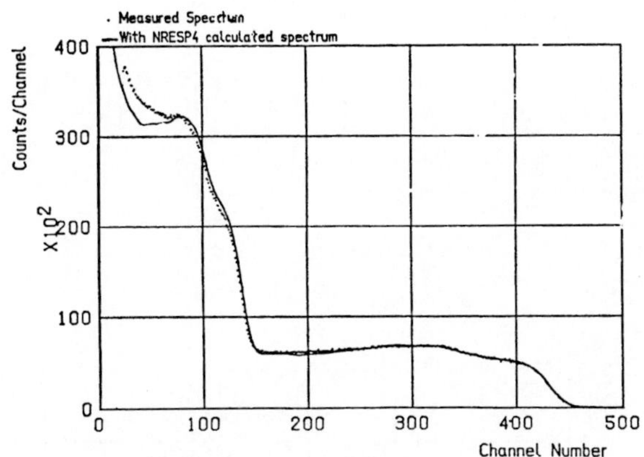

Fig. 4 Comparison between measured and computed neutron spectrum

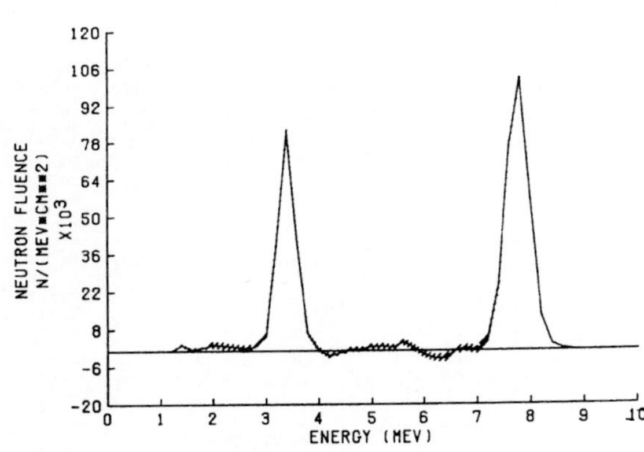

Fig. 5b first smoothing step

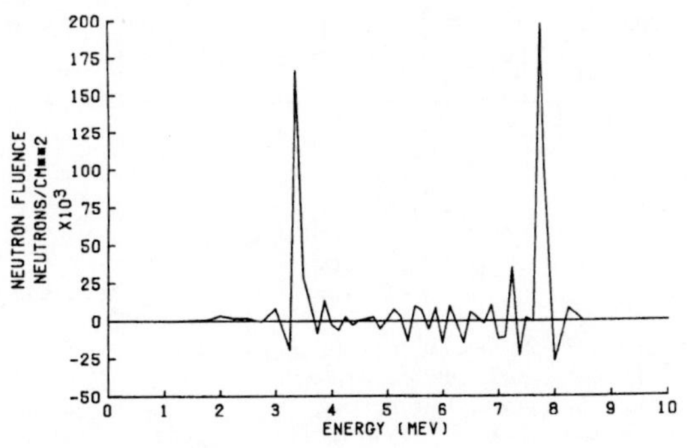

Fig. 5a pure matrix inversion

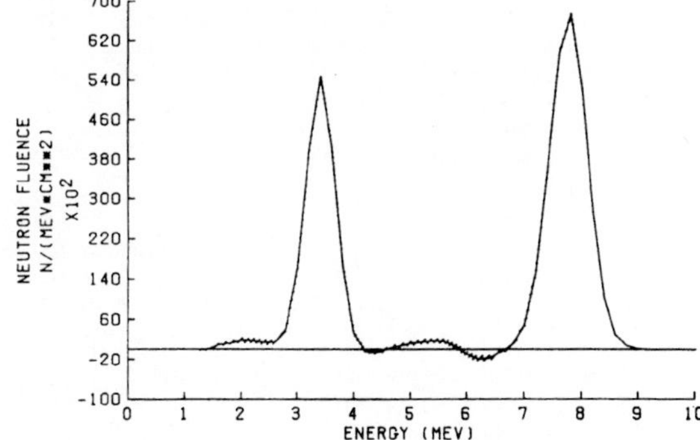

Fig. 5c optimization of smoothing

Fig. 5 Unfolding of a neutron spectrum of the $^{9}Be(\alpha,n)^{12}C$-reaction in 3 steps (a-c)

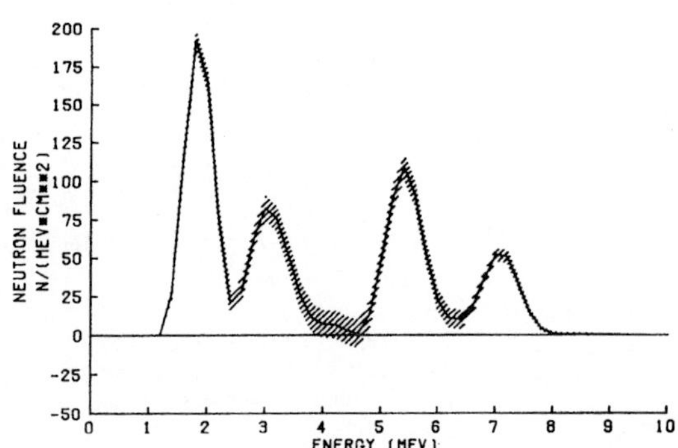

Fig. 6 Unfolded spectrum of neutrons scattered by Silicon at $\Theta=110^{o}$. Elastic and inelastic scattered neutrons can be well separated.

Experimental Results

Investigations of Carbon

Due to finite geometry effects of the scattering setup several corrections have to be carried out. In order to study scattering corrections investigations at the well-suited nucleus Carbon-12 have been done. Using this nucleus one has no problems with inelastic scattering and one is able to compare computations and experiment in an ideal way. The analyzing power of 3 sample sizes has been determined by an experiment and was corrected with the Monte-Carlo-code JANE by E. Woye [8]. This code is up to now the only one capable of correcting the analyzing power for effects of the polarization of the multiple scattered neutrons. As it is shown in Figs 7a-c the corrected angular distributions show good agreement within the experimental errors. Neutron energy of the spectra shown in Figs. 7 is 7.65MeV.

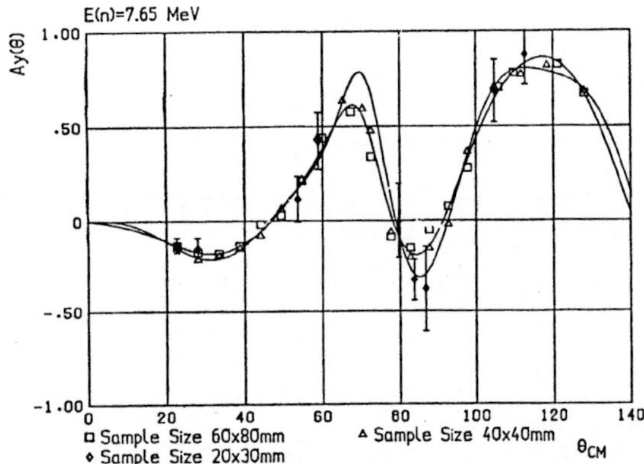

Fig. 7a Analyzing power of Carbon-12 for 3 sample sizes without correction for finite geometry

Moreover angular distributions of ^{12}C at neutron energies E_n=7.55, 7.65, 7.85, 8.05, and 8.73MeV have been determined. Fig. 7c shows the effect of finite geometry corrections for the neutron energy 7.55MeV.

Because Carbon shows resonances in this energy region which can be seen in the energy dependence of the total cross section (Fig. 8), the shape of analyzing power and differential cross section changes drastically with energy. The energy dependance of analyzing power and differential cross section is shown in Figs. 9a-e and 10a-e. It has to be mentioned that the cross section data are uncorrected experimental data. The analyzing power data have been corrected.

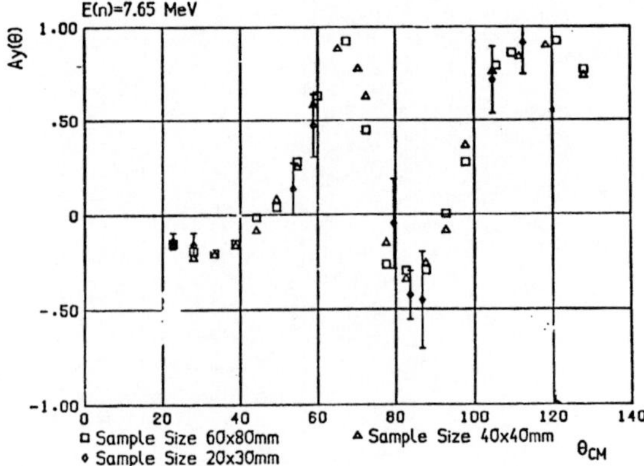

Fig. 7b Analyzing power of Carbon-12 for 3 sample sizes with corrections for finite geometry.

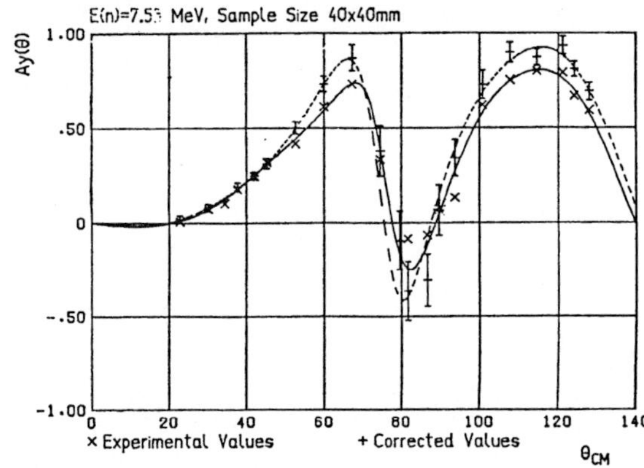

Fig. 7c Effect of finite geometry corrections for E_n=7.55MeV

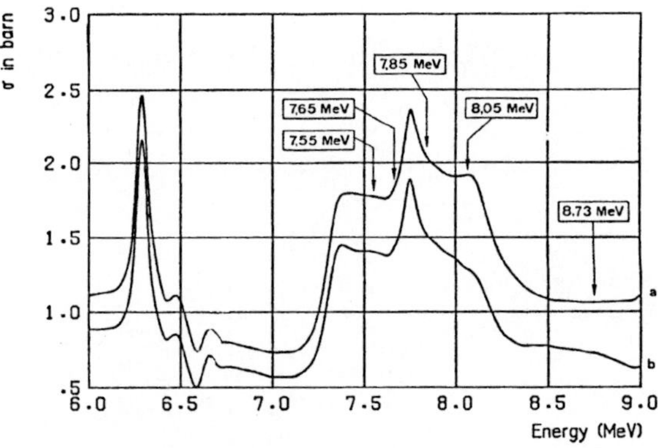

Fig. 8 Total cross section of Carbon-12 (from ENDF/B-V), curve a shows total cross section, curve b shows elastic cross section

The averaging of the cross section due to the energy width of 160keV of our polarized neutrons has to be taken into account. The results of our measurements fit quite well into a phase shift analysis of Carbon-12 worked out by W. Tornow [9] for a wider energy range.

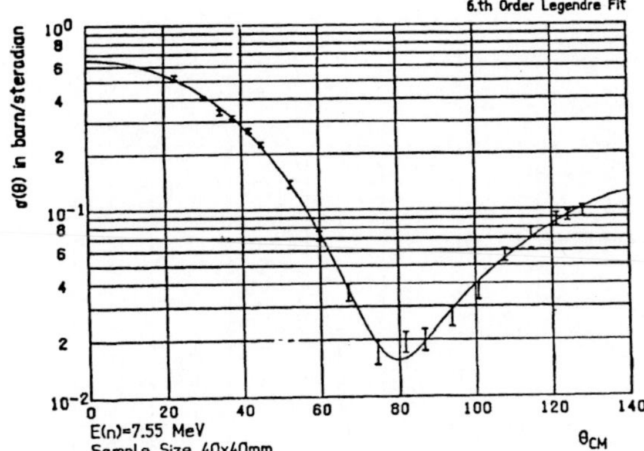

Fig. 9a

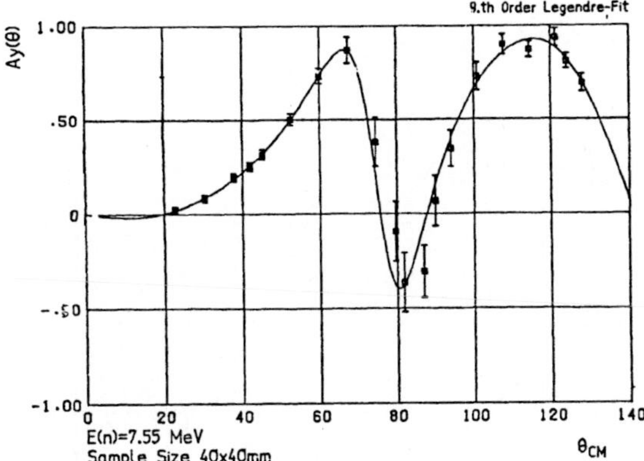

Fig. 10a

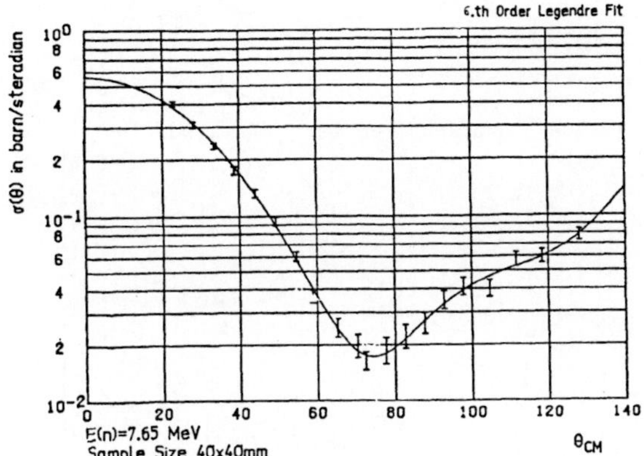

Fig. 9b

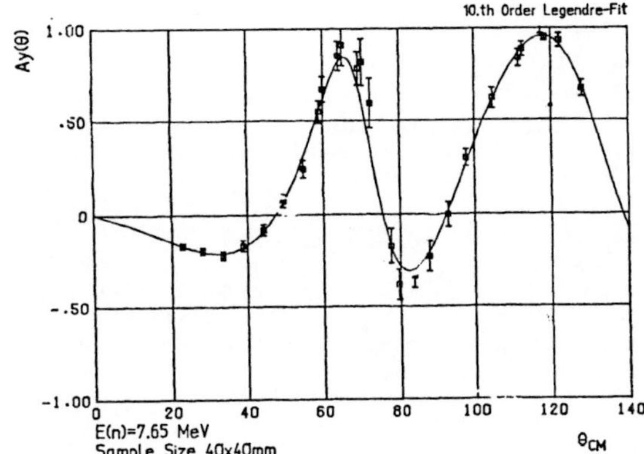

Fig. 10b

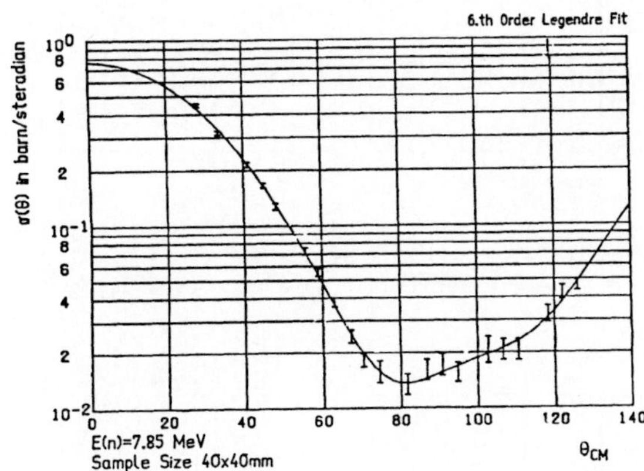

Fig. 9c

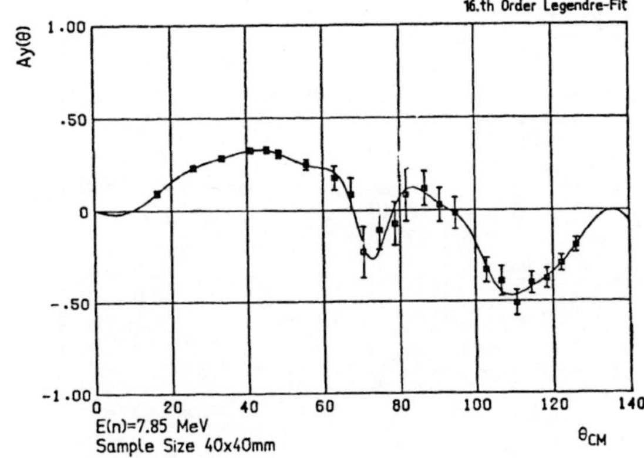

Fig. 10c

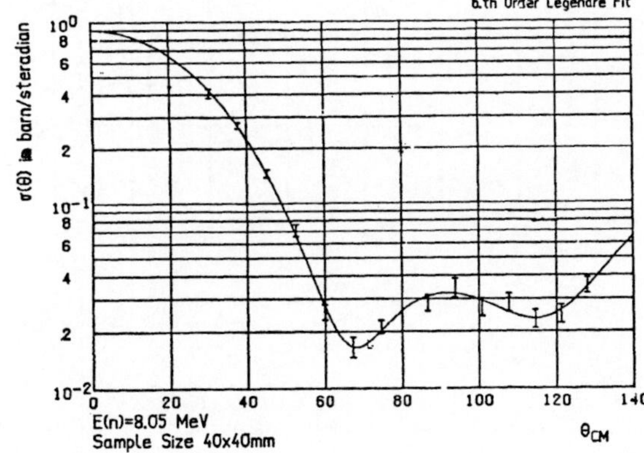

Fig. 9d
explanations see text

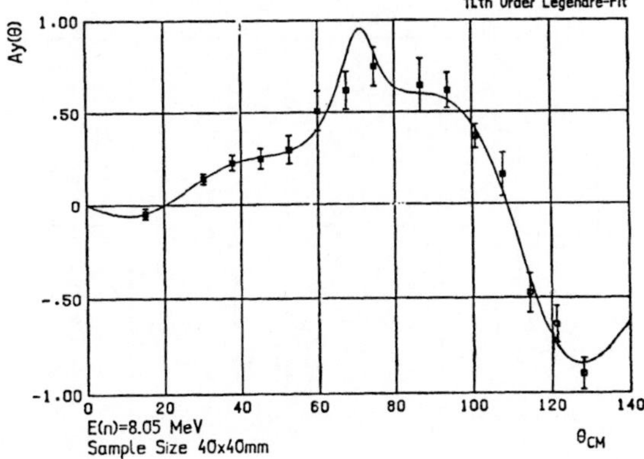

Fig. 10d
explanations see text

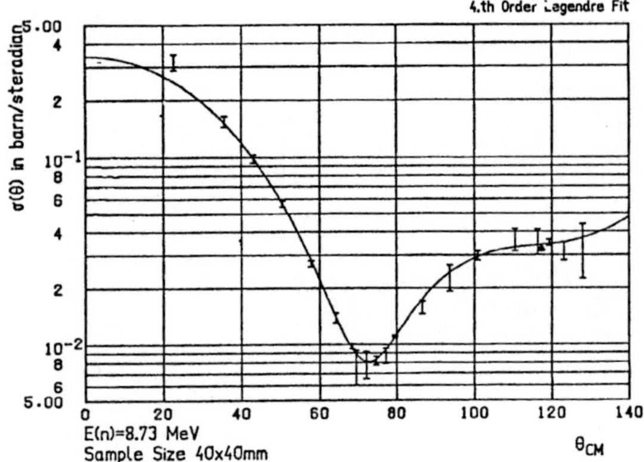

Fig. 9e

Fig. 9 explanations see text

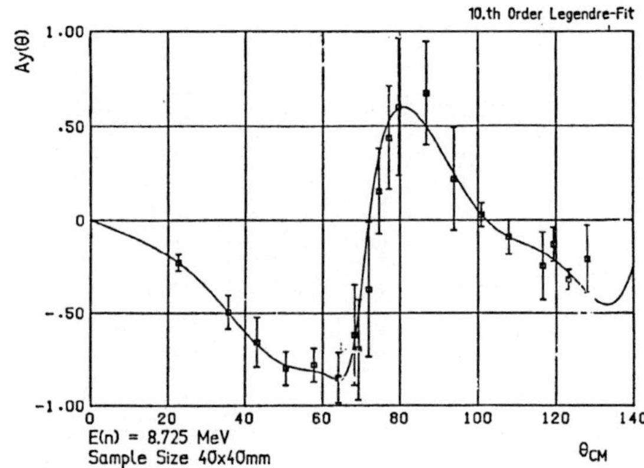

Fig. 10e

Fig. 10 explanations see text

Investigations of Some Medium Mass Nuclei: Si, S, Ca, Cu, and La

In the same manner as already described analyzing power and differential cross section of elastic scattered neutrons have been determined for a neutron energy of 7.75MeV. For Silicon inelastic scattered neutrons could be evaluated too. Figs. 11-15 show the analyzing power of the elements Silicon, Sulfur, Calcium, Copper, and Lanthanum, respectively.

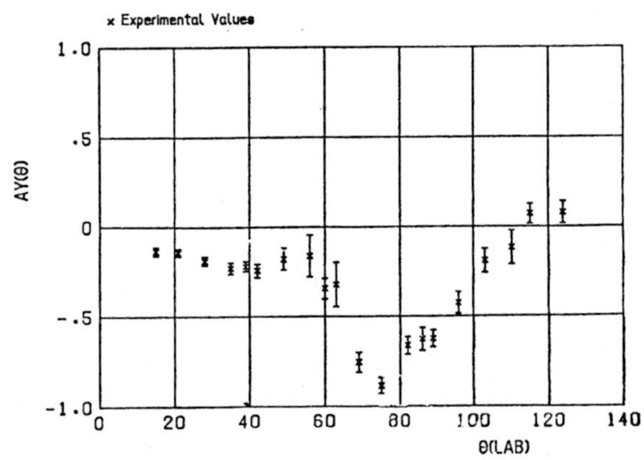

Fig. 11 Analyzing power of natural Silicon, E_n=7.75 MeV, experimental data without corrections

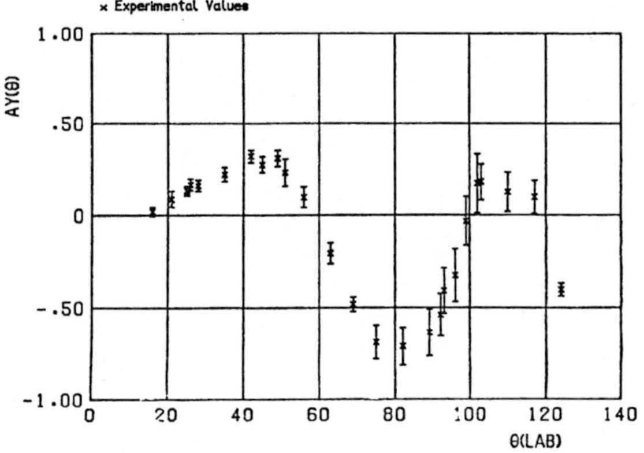

Fig. 12 Analyzing power of natural Sulfur, E_n=7.75 MeV, experimental data without corrections

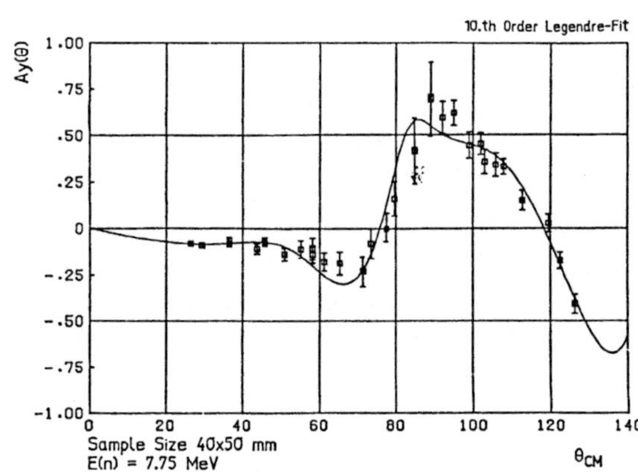

Fig. 13 Analyzing power of natural Calcium, E_n=7.75 MeV, experimental data corrected using JANE

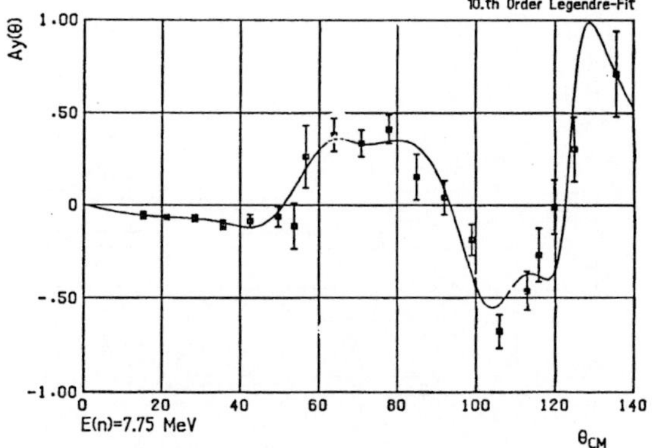

Fig. 14 Analyzing power of natural Copper, E_n=7.75 MeV, experimental data corrected using JANE

As an example the experimental differential cross section of natural Lanthanum is plotted in Fig. 16, still without finite geometry corrections.

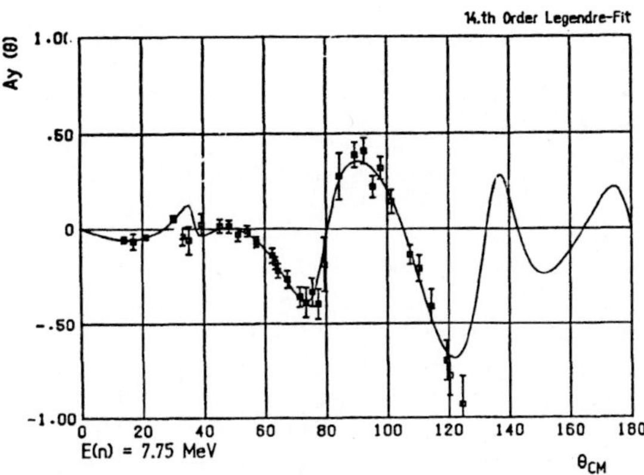

Fig. 15 Analyzing power of natural Lanthanum, E_n=7.75MeV, experimental data corrected using JANE

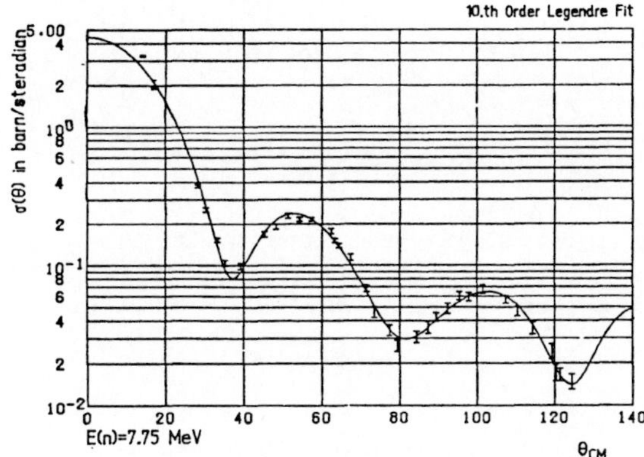

Fig. 16 Differential cross section of natural Lanthanum without corrections, E_n=7.75MeV

The analyzing power of Silicon for inelastic scattered neutrons is shown in Fig. 17.

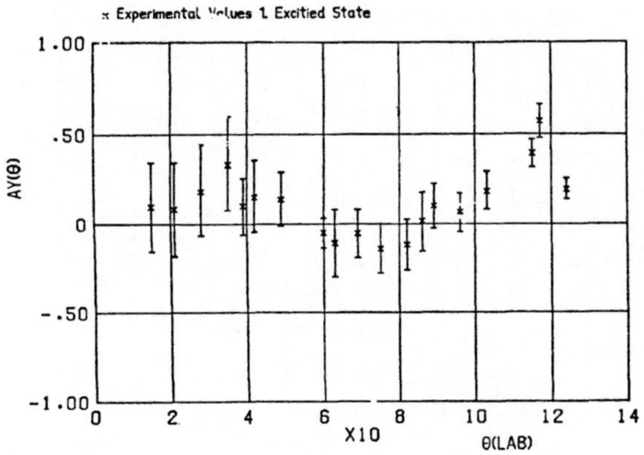

Fig. 17 Analyzing power of Silicon for inelastic scattered neutrons

Investigations of Some Heavy Nuclei: ^{209}Bi, Natural Pb, and ^{238}U

For reasons of optical model calculations analyzing power and differential cross section of some heavy nuclei have been determined. For deformed nuclei one gets also the deformation parameters. One more reason for the investigations of Bismuth is the fact that it is going to be a secondary scattering standard for neutrons. Natural Bismuth consists of one isotope (A=209), its first excited state is rather high with 897keV. This makes separation of elastic scattering easy. At Bismuth we employed analytical scattering correction methods which fail at lighter nuclei. Fig. 18 shows the differential cross section of Bismuth at E_n=7.75MeV; Fig. 19 shows the same but corrected with a method suggested by Kinney [10].

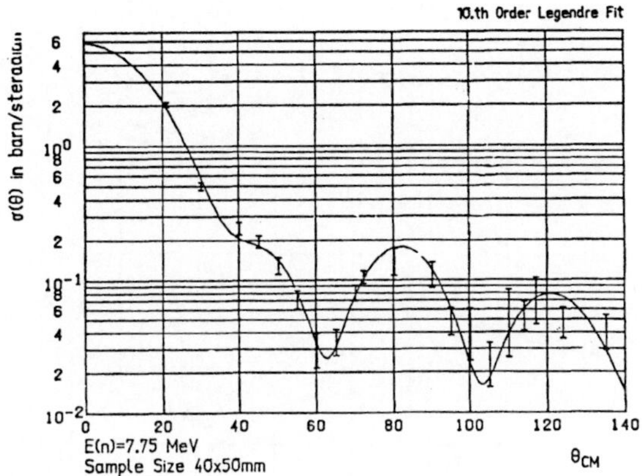

Fig. 18 Differential cross section of ^{209}Bi at E_n=7.75MeV

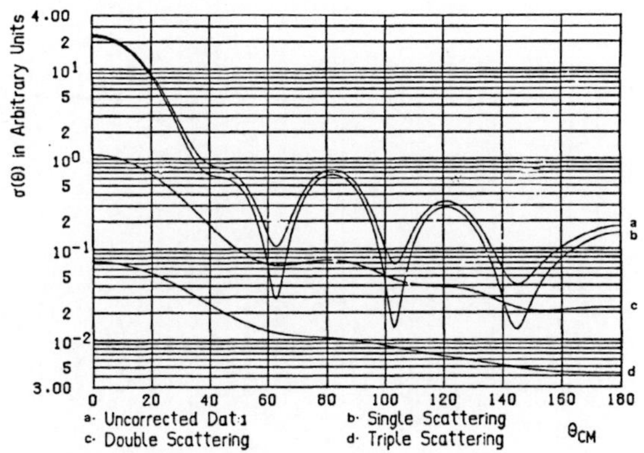

Fig. 19 Differential cross section of ^{209}Bi at E_n=7.75MeV corrected with the method of Kinney

The analyzing power of ^{209}Bi is plotted in Fig. 20.

Figs. 21 and 22 show the analyzing power of natural Lead and ^{238}U, both corrected with JANE. The solid lines are least square fits using associated Legendre polynomials.

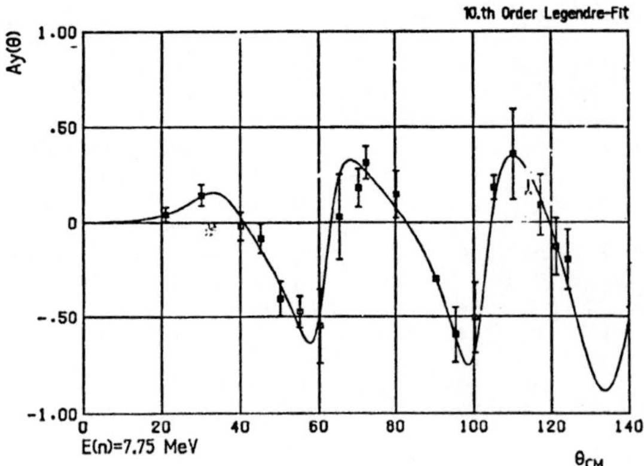

Fig. 20 Analyzing power of ^{209}Bi at E_n=7.75MeV

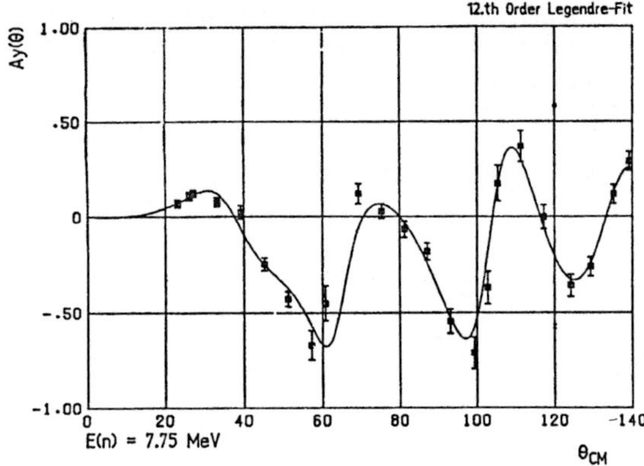

Fig. 21 Analyzing power of nat. Lead at E_n=7.75MeV

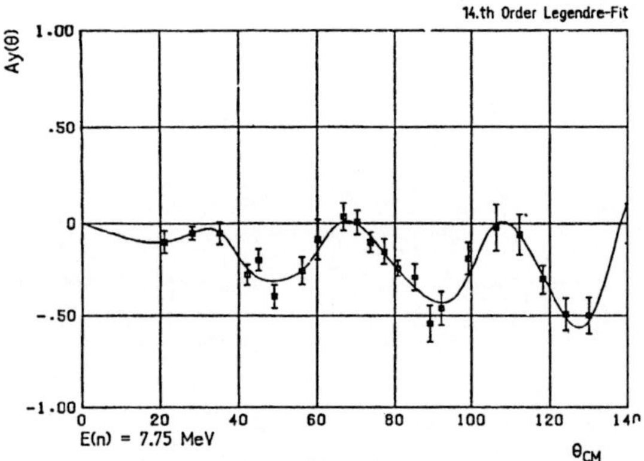

Fig. 22 Analyzing power of ^{238}U at E_n=7.75MeV

Optical Model Analysis

Optical model analysis is still in progress, some technical and other problems are to be solved. We have carried out some calculations with sets of standard parameters, for natural Lanthanum (0.089% ^{138}La, 99.911% ^{139}La) we made an analysis using the code ECIS by J. Raynal [11]. In Fig. 23 the experimentally determined analyzing power of Lanthanum is plotted together with preliminary results obtained from ECIS.

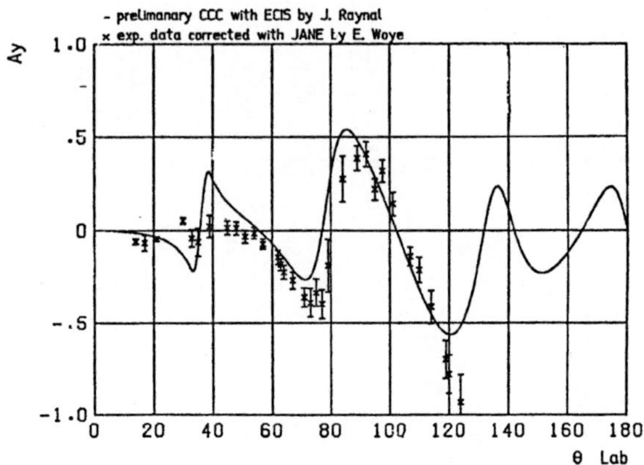

Fig. 23 Optical model calculations for La and experimental data, E_n=7.75MeV

Table 1 Optical model parameters for La

V_R	= 45.5MeV	r_R	= $1.24A^{1/3}$fm	a_R	= 0.62fm
W_D	= 6.0MeV	r_D	= $1.26A^{1/3}$fm	a_D	= 0.58fm
V_{SO}	= 7.0MeV	r_{SO}	= $0.85A^{1/3}$fm	a_{SO}	= 0.20fm
Deformation		β_2	= 0.26		

For Bismuth we have made an analysis using the code JUPITOR by T. Tamura [12]. Fig. 24 shows the results of these computations carried out with a standard set of parameters by G. Haouat [13].

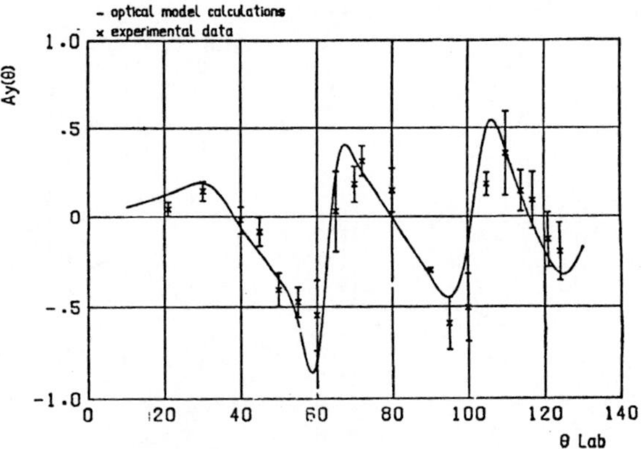

Fig. 24 Optical model calculations for Bi and experimental data, E_n=7.75MeV

The analyzing power of natural Lead can actually only be compared with an optical model calculation for the isotope ^{208}Pb, which is only 52.3% of natural Lead, this is plotted in Fig. 25. There is already good qualitative description of the experimental data. The optical model calculations for Pb have been carried out using ECIS and a standard parameter set by G. Haouat [13].

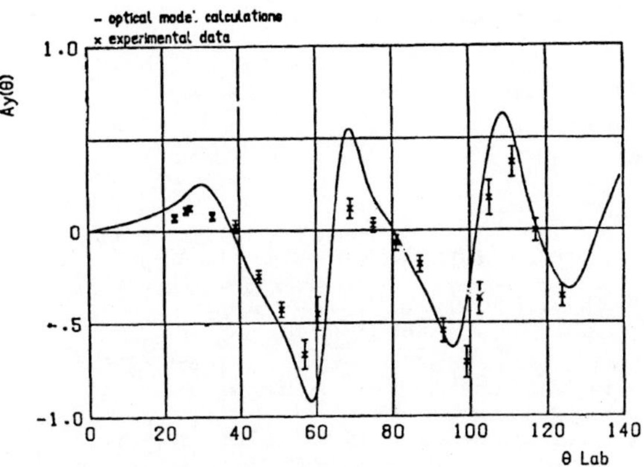

Fig. 25 Optical model calculations for ^{208}Pb and experimental data, parameter set by Haouat [13]

This work has been supported in part by the Volkswagen Foundation.

References

1 Schölermann, H. and H. Klein, Nucl. Instr. Meth. 169 (1980) 25

2 FERDOR: Verbinski, V. V. and W. Burrus, Nucl. Instr. Meth. 65 (1968) 8
FORIST: RSIC ORNL, PSR-92

3 Textor, R. E. and V. V. Verbinski, ORNL-4160 (1968)

4 Dietze, G. and H. Klein, PTB-report ND22

5 Inouye, Y. et al., Nucl. Instr. Meth. 67 (1969) 125

6 Sloan, W. R. and G. L. Woodruff, Nucl. Instr. Meth. 150 (1978) 253

7 Dietze, G., Nucl. Instr. Meth. 193 (1982) 549

8 Woye, E., Thesis Tübingen (1982)

9 Tornow, W., Univ. Tübingen, private communication

10 Kinney, W. E., Nucl. Instr. Meth. 83 (1970) 15

11 Raynal, J., in: "Computing as a language of Physics", Int. Atomic Energy Agency, Vienna 1972

12 Tamura, T., Rev. Mod. Phys. 37 (1965) 679

13 Haouat, G., NEANDC(E) 194 "L"

14 Bulski, G., Diplomarbeit Stuttgart (1982)

ANALYSIS OF $n+^{165}Ho$ AND ^{169}Tm REACTIONS

P. G. Young and E. D. Arthur

Theoretical Division, Los Alamos National Laboratory
Los Alamos, New Mexico 87545 U.S.A.

C. Philis, P. Nagel, and M. Collin

Centre d'Études de Bruyères-le-Châtel
92542 Montrouge CEDEX FRANCE

Experimental data for neutron-induced reactions on ^{165}Ho and ^{169}Tm have been theoretically analyzed in preparation for calculations on the unstable isotopes of Tm. A set of deformed optical model parameters was determined from measurements of s- and p-wave neutron strength functions, total cross sections, elastic angular distributions, and 16-MeV proton scattering to the ^{165}Ho ground and first excited states. The parameters for the ^{165}Ho and ^{169}Tm nuclei were linked by means of an isospin term in the real and imaginary well depths, together with adjustment of the β_2 and β_4 deformation parameters based on systematics in this mass region. Transmission coefficients from this analysis were used in Hauser-Feshbach statistical model calculations of the $^{169}Tm(n,\gamma)$ cross section as well as the $^{169}Tm(n,2n)$ and (n,3n) cross sections to 23 MeV, after application of suitable preequilibrium corrections. The results of these calculations are in good agreement with most of the available experimental data on ^{165}Ho and ^{169}Tm.

[$^{165}Ho(n,n)$, $^{169}Tm(n,\gamma)$, $^{169}Tm(n,n)$, $^{169}Tm(n,n')$, $^{169}Tm(n,2n)$, $^{169}Tm(n,3n)$, ^{169}Tm total cross sections, nuclear models, E_n = 0.01-20 MeV]

I. Introduction

Information on nuclear cross sections is required for the element thulium because of its application in radiochemical diagnostic studies. The primary aim of the present work was to derive a set of nuclear model parameters that could be used to reliably calculate neutron-induced cross sections for ^{169}Tm (natural thulium) and several of its unstable isotopes. The general approach followed in the analysis was to optimize parameters using combined optical model and statistical theory calculations to reproduce the available experimental data for ^{165}Ho and ^{169}Tm. Because both nuclei are strongly deformed, we used a deformed optical model as the basis for the analysis rather than attempting to determine an equivalent spherical optical potential that might be physically unrealistic. Consistency between the statistical and optical models was obtained by using transmission coefficients from the deformed optical model analysis in the statistical theory calculations.

II. Deformed Optical Model Analysis

Because of the sparsity of pertinent nuclear data for ^{169}Tm, we chose to first analyze the more abundant data for the neighboring nucleus ^{165}Ho. The deformed optical model parameters obtained for ^{165}Ho were adapted for ^{169}Tm through use of an isospin term in the real and imaginary well depths along with adjustment of the β_2 and β_4 deformation parameters from systematics in this mass region. These parameters were then used to calculate the available ^{169}Tm data, and a final adjustment was made to optimize the agreement.

The coupled-channel code ECIS[1] was employed for all the deformed optical model calculations. A potential of the following form was used:

$$U = -Vf(r,a_V,R_V) + 4i\, a_D W_D \frac{d}{dr} f(r,a_D,R_D) \qquad (1)$$

$$-iW_V f(r,a_V,R_V) + 2\lambda_\pi^2 V_{so} \vec{\ell}\cdot\vec{s} \frac{1}{r}\frac{d}{dr} f(r,a_{so},R_{so}) ,$$

where $f(r,a_i,R_i) = [1 + \exp(r-R_i)/a_i]^{-1}$, and

$$R_i = r_i A^{1/3}[1 + \beta_2 Y_2^0 (\Omega') + \beta_4 Y_4^0 (\Omega')].$$

The notation of Ref. 2 is followed in Eq. 1, and the usual real, imaginary surface derivative, imaginary volume, and spin-orbit terms are included.

Because of the high spin of the ground-state rotational band for ^{165}Ho, we only coupled the first three states in those calculations, whereas we were able to include the first five states for ^{169}Tm. The states used in the calculations are listed in Table I. In the case of ^{169}Tm, it was possible to approximate closely the 5-state calculations using fictitious 0+, 2+, 4+ states (see Table 1) according to the scheme of Lagrange, Bersillon, and Madland[3]. We empirically verified that our approximate calculations reproduce the more precise 5-state ones to better than ~ 1% except near thresholds for the (n,n') cross sections.

Table I States Included in the Coupled-Channel Calculations

^{169}Ho		^{169}Tm		^{169}Tm (fictitious states)	
E_x (keV)	J^π	E_x (keV)	J^π	E_x (keV)	J^π
0	$7/2^-$	0	$1/2^+$	0	0^+
94.7	$9/2^-$	8.4	$3/2^+$	72	2^+
209.8	$11/2^-$	118.2	$5/2^+$		
		138.9	$7/2^+$	240	4^+
		331.9	$9/2^+$		

A. Parameter Set 1

The initial set of parameters for our analysis was obtained (after slight modification) from a recent study of neutron-induced reactions on tungsten isotopes[4]. The parameters for the nuclear potential, written in the Lane[5] form, are as follows:

$$V\binom{n}{p} = 49.8 \binom{-}{+} 16 \frac{N-Z}{A} + \Delta V_c - 0.25E$$

$$\Delta V_c = 0.4 \frac{Z}{A^{1/3}} \text{ for incident protons}$$

$$= 0 \text{ for incident neutrons.} \qquad (2)$$

$$W_D\binom{n}{p} = 5.1 \binom{-}{+} 8 \frac{N-Z}{A} + 0.6E \qquad [E \leq 6.5]$$

$$= 9.0 \binom{-}{+} 8 \frac{N-Z}{A} - 0.1(E-6.5) \quad [E > 6.5]$$

K. H. Böckhoff (ed.), Nuclear Data for Science and Technology, 792–795.

$W_v = -1.8 + 0.2E$ [E > 9.0]

$V_{so} = 6$

$r_V = r_D = r_{so} = 1.26f$

$a_V = a_{so} = 0.63f;\ a_D = 0.48f,$

where all energies and well depths are given in MeV. This parameter set, together with the specific β_2 and β_4 values for ^{165}Ho and ^{169}Tm discussed below, is referred to as Set 1 in this paper.

For the ^{165}Ho calculations, deformation parameters of $\beta_2 = 0.30$ and $\beta_4 = -0.02$ were estimated from systematics in this mass region[6]. These values were used with the above potential in the ECIS code to calculate the available experimental data for ^{165}Ho. Values of the s- and p-wave neutron strength functions ($S_0 = 2.0 \times 10^{-4}$; $S_1 = 2.0 \times 10^{-4}$) were inferred from the ECIS calculations of transmission coefficients at 10 keV. These values, as well as the potential scattering radius (R' = 7.2f), agree reasonably with experiment[7] ($S_0 = 1.8 \pm 0.2 \times 10^{-4}$; $S_1 = 1.6 \pm 0.3 \times 10^{-4}$; R' = 7.6 ± 0.5f). Additionally, the energy dependence of measured total cross sections[8] for ^{165}Ho between 0.1 and 20 MeV was well reproduced by these parameters. Comparisons were then made to neutron scattering angular distribution measurements between 3 and 7.5 MeV by Fasoli et al.[9] and at 11 MeV by Ferrer et al.[10] In both cases the agreement was reasonably good, and the latter comparison is shown in Fig. 1. Finally, measured ^{165}Ho(p,p) and (p,p') angular distributions for the $9/2^-$ ground state and the $11/2^-$ first-excited state were calculated at E_p = 16 MeV and verified by comparison with the experimental data of Kruse et al.[11]

Parameter Set 1 was also used to calculate the available experimental data for neutron-induced reactions on ^{169}Tm, changing only the deformation parameters to reflect systematic trends in this mass region[6]. Values of $\beta_2 = 0.29$ and $\beta_4 = -0.01$ were used for ^{169}Tm. The s- and p-wave neutron strength functions and the potential scattering radius for ^{169}Tm were calculated as before with these parameters, as well as the total cross section between 0.01 and 20 MeV. The results, labeled Set 1, are compared to the experimental data in Table II and Fig. 2. While the overall agreement with experiment is certainly reasonable, the calculated total cross section between 8 and 12 MeV is systematically lower than the data by about 4%. Because of this discrepancy, an effort was made to further optimize the parameters for ^{169}Tm.

B. Parameter Set 2

The Coulomb excitation measurements of Olesen and Elbek[12] yielded a value for the intrinsic ^{169}Tm quadrupole moment of $Q_0 = 7.52$. Using the potential from Parameter Set 1, β_2 was varied until this quadrupole moment was reproduced by numerical integration* of the expression

$$Q_\lambda = Z \frac{\int V(r) Y_\lambda^0(\theta,\phi) r^\lambda d\vec{r}}{\int V(r) d\vec{r}},$$

where Z is the charge of the nucleus and $Y_\lambda^0(\theta,\phi)$ is a spherical harmonic of order λ. This calculation resulted in a value of $\beta_2{}^c = 0.35$ for the Coulomb deformation, which implies a value of $\beta_2 \sim 0.31$ for

*We are indebted to Ch. Lagrange for suggesting this approach and providing the computer code for this calculation.

Table II Summary of n + ^{169}Tm S_0, S_1, R' Results

	Exp.	Set 1	Set 2
S_0 (x 10^{-4})	1.5 ± 0.2	1.97	1.65
S_1 (x 10^{-4})	0.5 - 1.5*	2.38	3.60
R' (f)	7.7 ± 0.5	7.77	7.55

*Inferred from systematics

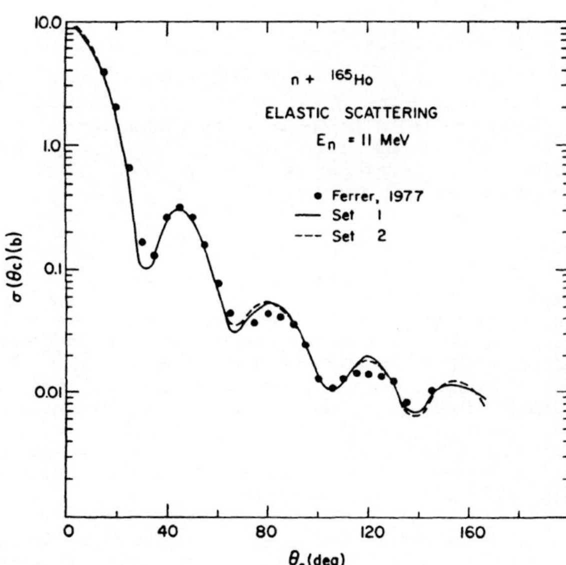

Fig. 1 Comparison of calculated ^{165}Ho(n,n)^{165}Ho + ^{165}Ho(n,n')^{165}Ho*(95 keV) angular distributions with the measurement of Ferrer et al.[10].

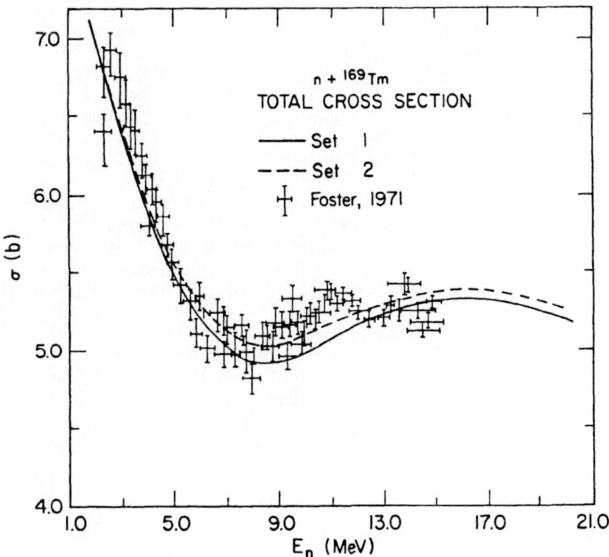

Fig. 2. Calculated and measured[8] values of the n + ^{169}Tm total cross section.

the nuclear deformation. Using this β_2 deformation, we carried out sensitivity calculations aimed at bringing the calculated total cross section into better agreement with experiment, while simultaneously maintaining reasonable agreement with S_o,S_1, and R' results.

From this analysis a new set of parameters, labeled Set 2 in the tables and figures, was obtained. These parameters are identical to Set 1 except

$$r_V = r_D = r_{so} = 1.27f$$

$$W_D = 8.7 - 8\,\frac{N-Z}{A} - 0.03(E-6) \qquad E \geq 6 \text{ MeV} \qquad (3)$$

$$\beta_2(^{169}\text{Tm}) = 0.31 \; .$$

Calculated values from Parameter Set 2 for the ^{169}Tm neutron total cross section are included in the experimental comparisons of Fig. 2. Similarly, values of S_o, S_1, and R' from these parameters are given for ^{169}Tm in Table II, and a calculated ^{165}Ho scattering angular distribution for E_n = 11 MeV is included in Fig. 1. Except for S_1, somewhat improved agreement with experiment is obtained with Parameter Set 2, especially in the ^{169}Tm neutron total cross section.

III. Statistical Theory Calculations

To further test both parameter sets against available experimental data, ^{169}Tm(n,γ) cross sections were calculated with the statistical theory code COMNUC[13], and ^{169}Tm(n,2n) and (n,3n) cross sections were calculated with the statistical-preequilibrium code GNASH[14], using neutron transmission coefficients from the deformed optical model calculations. Different codes were used for the (n,γ) and (n,xn) reactions because COMNUC includes width-fluctuation corrections and is better suited for low-energy calculations, whereas GNASH accounts for preequilibrium effects, which are important at incident neutron energies above ~ 10 MeV. Both codes use the Gilbert and Cameron[15] level density formulation, the Cook[16] tabulation of level density parameters, and all the available information on discrete states in the various residual nuclei. The level density parameters for the constant temperature Gilbert-Cameron expressions are, in turn, appropriately matched to the discrete level data. The two codes have been shown[17] to closely agree when identical model assumptions are used in each.

The COMNUC calculations of the ^{169}Tm(n,γ) cross section are compared to the available experimental data[8] in Fig. 3. The calculations include width fluctuation corrections[18] and rely on a giant dipole resonance model[19] (GDR) to compute gamma-ray transmission coefficients. The GDR calculations are normalized to the ratio of the experimental values for the average gamma-ray width ($\langle\Gamma_\gamma\rangle$ = 0.084 eV) and the s-wave resonance spacing ($\langle D_o\rangle$ = 7.3 eV) at the neutron binding energy, increased by 10% to improve agreement with the experimental data. Two Lorentzian curves centered at 12.1 and 15.5 MeV with widths of 2.9 and 4.5 MeV were used to describe the GDR shape appropriate for a deformed nucleus. The resulting gamma-ray strength function for ^{170}Tm is in reasonable agreement with that deduced by Joly et al.[20] from measurements of gamma-ray spectra and is illustrated in Fig. 4. We did not include the resonance structure at ε_γ = 3.5 MeV in the theoretical shape, because the calculation of the integrated (n,γ) cross section is not very sensitive to such detail.

The GNASH calculations of the ^{169}Tm(n,2n) and (n,3n) cross sections are compared to experiment[8] in Fig. 5. Consistent gamma-ray strength functions from the (n,γ) analysis were used in these calculations following the method outlined by Arthur[21]. The preequilibrium calculations utilized the Kalbach exciton model[22] and were performed with standard parameters[21] of g_o = A/13 and k = 130 MeV^3. The ^{168}Tm level density parameter given in the Gilbert and Cameron formulation was increased by 10% to optimize agreement at energies above the (n,3n) threshold. The calculated results in Fig. 5 are seen to reasonably reproduce the measurements, except for the (n,3n) results near threshold.

As a final check, the (n,2n) cross section for the unstable isotope ^{168}Tm was calculated for comparison with a recent measurement at 14.5 MeV[23], using the transmission coefficients and model parameters from the ^{169}Tm analysis (Set 2). The calculated value (2.08 b) agrees well with the experimental result (2.1 ± .1 b).

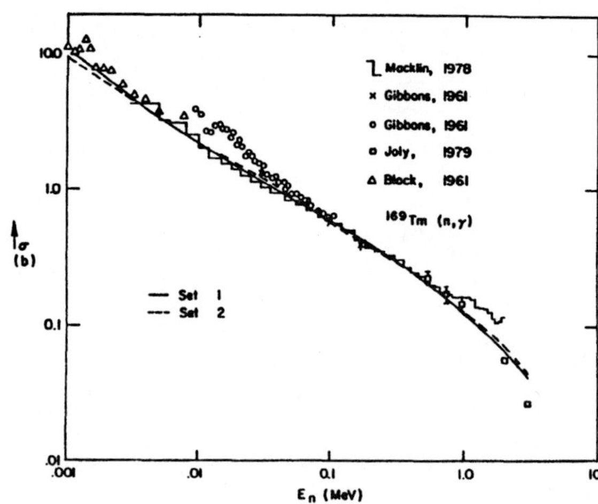

Fig. 3. Comparison of the calculated ^{169}Tm(n,γ) cross section with experimental data[8].

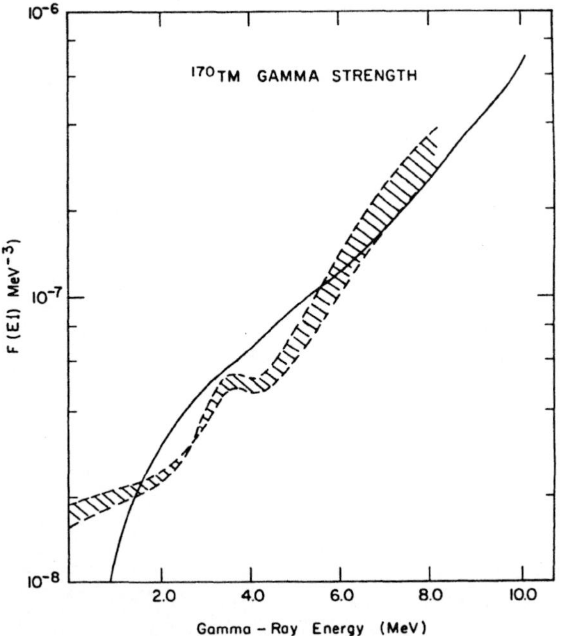

Fig. 4. The ^{170}Tm gamma-ray strength function from the present calculation (solid curve) compared to that extracted from spectral ^{169}Tm(n,γ) measurements[20] (cross hatched).

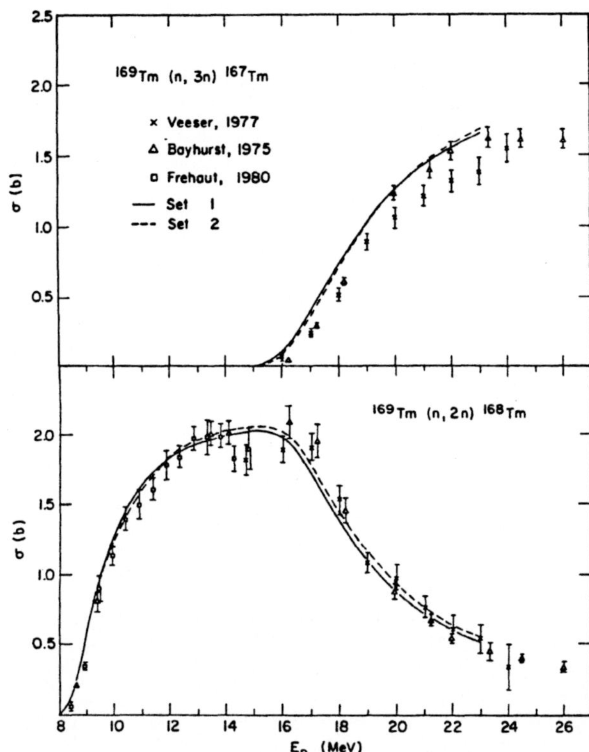

Fig. 5. Comparison of calculated and measured[8] $^{169}Tm(n,2n)$ and $^{169}Tm(n,3n)$ cross sections.

IV. Conclusions

There is good overall agreement between a variety of experimental data and our deformed optical model and statistical theory calculations using both sets of parameters described in this paper. There is some preference for Set 2, however, on the basis of the total cross section, s-wave strength function, and (n,2n) cross-section comparisons. It is expected that this parameterization will permit reliable calculation of nuclear data for other Tm isotopes and neighboring nuclei when appropriate nuclear deformations are used.

References

1. J. Raynal, "Optical Model and Coupled-Channel Calculations in Nuclear Physics," Int. Atomic Energy Agency Report, IAEA-SMR-9/8 (1970).

2. J. P. Delaroche, G. Haouat, J. Lachkar, Y. Patin, J. Sigaud, and J. Hardine, Phys. Rev. C23, 136 (1981).

3. Ch. Lagrange, O. Bersillon, and D. G. Madland, to be published in Nuc. Sci. Eng. (1982).

4. E. D. Arthur, P. G. Young, A. B. Smith, and C. A. Philis, Trans. Am. Nucl. Soc. 39, 793 (1981).

5. A. M. Lane, Phys. Rev. Lett. 8, 171 (1962); Nuc. Phys. 35, 676 (1962).

6. K. E. G. Lobner, M. Vetter, and V. Honig., Nucl. Data Tables A7, 495 (1970).

7. S. F. Mughabghab and D. I. Garber, "Neutron Cross Sections Volume I, Resonance Parameters," Brookhaven Nat. Lab. report BNL 325, Third Ed., V. 1 (1973).

8. D. I. Garber and R. R. Kinsey, "Neutron Cross Sections Volume II, Curves," Brookhaven Nat. Lab. report BNL 325, Third Edition, V. 2 (1976); and personal communication from National Nuclear Data Center, Brookhaven Nat. Lab. (1981).

9. U. Fasoli, P. Sambo, D. Toniolo, and G. Zago, Nuc. Phys. A133, 572 (1969).

10. J. C. Ferrer, J. D. Carlson, and J. Rapaport, Nuc. Phys. A275, 325 (1977).

11. T. Kruse, W. Makofske, H. Ogata, W. Savin, M. Slagowitz, M. Williams, and P. Stoler, Nucl. Phys. A169, 177 (1971).

12. M. C. Olesen and B. Elbek, Nucl. Phys. 15, 134 (1960).

13. C. L. Dunford, "A Unified Model for Analysis of Compound Nucleus Reactions," Atomics International report AI-AEC-12931 (1970).

14. P. G. Young and E. D. Arthur, "GNASH: A Preequilibrium Statistical Nuclear Model Code for Calculation of Cross Sections and Emission Spectra," Los Alamos Sci. Lab. report LA-6947 (1977).

15. A. Gilbert and A. G. W. Cameron, Can. J. Phys. 43, 1446 (1965).

16. J. L. Cook, H. Ferguson, and A. R. L. Musgrove, Aust. J. Phys. 20, 477 (1967).

17. E. D. Arthur, in "Applied Nuclear Data Research and Development, July 1-Sept. 30, 1978," C. I. Baxman and P. G. Young, Eds., LA-7596-PR, p. 5 (1978).

18. P. A. Moldauer, Phys. Rev. C11, 426 (1975).

19. P. M. Brink, thesis, Oxford University (1955), unpublished; P. Axel, Phys. Rev. 126, 671 (1962).

20. S. Joly, D. M. Drake, and L. Nilsson, Phys. Rev. C20, 2072 (1979).

21. E. D. Arthur, Nucl. Sci. Eng. 76, 137 (1980).

22. C. Kalbach, Z. Phys. A283, 401 (1977); ibid A287, 319 (1978).

23. D. R. Nethaway, Lawrence Livermore National Lab., personal communication (1982).

ETUDE DE LA DEFORMATION DES NOYAUX DE LA COUCHE S-D A L'AIDE DE LA DIFFUSION DE NEUTRONS RAPIDES

G. Haouat, Ch. Lagrange, Y. Patin
Service PNN, C.E. de Bruyères-le-Châtel
92542 Montrouge Cedex, France

R. De Swiniarski
Institut des Sciences Nucleaires
Avenue des Martyrs, 38026 Grenoble, France

F. Dietrich
Lawrence Livermore National Laboratory
Livermore, California, U. S. A.

A. Virdis
Institut de Physique Nucleaire
BPl 91406, Orsay, France

Nuclear deformation of ^{24}Mg, ^{28}Si and ^{32}S has been investigated by means of fast neutron scattering. These nuclei lie in a mass region of the s-d shell where nuclear shape changes very rapidly from one nucleus to the other. Differential cross sections were measured at 9.76 and 14.83 MeV incident neutron energies over the angular range from 15 to 160 deg. Angular distributions were obtained for the elastic scattering and the inelastic scattering to levels of up to 6 MeV excitation energy. An analysis of the data using the deformed and the statistical models is presented and discussed. Quadrupole and hexadecapole deformations of the nuclear potential, deduced from the analysis of the 0^+, 2^+, 4^+ states of the ground state rotational band, are compared to those obtained by other excitation processes.

[Nuclear Reactions. ^{24}Mg, ^{28}Si, ^{32}S (n,n),(n,n'); E_n=9.76-14.83 MeV; measured σ (E_n', θ), deduced optical model and deformation parameters. Natural targets.]

Introduction

Les noyaux de la couche s-d ont été abondamment étudies du point de vue de la forme nucléaire. On observe dans cette région de masse une trés rapide transition tant dans la forme des noyaux que dans l´amplitude des déformations nucléaires. Ainsi, le noyau ^{24}Mg est considéré comme prolate dans son état fondamental [1], tandis que l´on admet que l´isotope ^{28}Si est de forme oblate [2-6]; quant à l´isotope ^{32}S on me peut s´accorder sur sa nature rotationnelle ou vibrationnelle [2-4,7].

Les paramètres de déformation de ces noyaux ont été mesurés en utilisant des techniques basées soit sur l´interaction électromagnétique, soit sur la diffusion inélastique de hadrons. On a cependant constaté que les valeurs des déformations déduites des mesures semblent dépendre du processus utilisé et du type de la particule employée comme sonde nucléaire. Plusieurs explications de ce fait expérimental ont été proposées parmi lesquelles la différence entre les distributions de charge et de masse à l´intérieur du noyau [8], les dimensions différentes des projectiles dans le cas de la diffusion de hadrons [9], ou bien les pénétrations différentes des sondes dans le noyau, dont les déformations varient sensiblement au voisinage de la surface nucléaire [10]. Le choix du modèle d´interaction utilisé dans l´analyse des données de diffusion inélastique joue aussi un rôle trés important dans la détermination des déformations nucléaires.

Parmi les hadrons les neutrons sont une sonde intéressante car à la différence des particules chargées ils sont sensibles uniquement au champ nucléaire, et contribuent d´une façon particulière à notre connaissance des déformations nucléaires. Nous avons entrepris l´étude de la diffusion élastique et inélastique de neutrons sur les noyaux ^{24}Mg, ^{28}Si et ^{32}S aux énergies de 9,76 et 14,83 MeV dans le but de déterminer les paramètres de déformation de ces noyaux et de les comparer à ceux obtenus par d´autres méthodes d´investigation.

Nous présentons ici la technique de mesure des sections efficaces différentielles et l´analyse des données expérimentales basée sur les formalismes de l´interaction directe et du noyau composé statistique.

Méthode Expérimentale

Les sections efficaces différentielles de diffusion de neutrons ont été mesurées avec le spectromètre de neutrons par temps de vol du Centre d´Etudes de Bruyerès-le-Chatel [11]. Les neutrons primaires de 9,76 et 14,83 MeV étaient produits à l´aide de la réaction $^2H(d,n)^3He$ en utilisant le faisceau pulsé de deutérons délivré par L´Accélérateur Tandem Van de Graaff et une cible gazeuse de deutérium. Les mesures ont été effectuées sur des échantillons cylindriques de Mg, Si et S naturels; ils avaient un diamètre de ≃ 2cm et une masse de 23,9 g, 32,2 g et 25,4 g respectivement pour Mg, Si et S. Ils étaient placés à environ 12cm du centre de la cible gazeuse et à 0 deg par rapport à l´axe du faisceau des déuterons incidents. Les neutrons diffusés étaient détectés par les cinq chaines de mesure constitutant le spectromètre [11]. Chaque détecteur était placé à 8m de l´axe de l´échantillon. Les conditions expérimentales ont été choisies pour satisfaire les exigences suivantes: avoir une bonne statistique de comptage et pouvoir séparer aisément, dans les spectres de temps de vol, les groupes de neutrons de diffusion élastique et de diffusion inélastique sur les premiers états excités de ^{24}Mg, ^{28}Si et ^{32}S, respectivement l´isotope le plus abondant de chaque échantillon.

Les sections efficaces différentielles ont été mesurées, pour chacune des deux energies, à 28 angles répartis entre 15 et 160 deg. La valeur absolue des sections efficaces à été obtenue en mesurant avec le même détecteur le flux des neutrons diffusés et le flux des neutrons primaires, et en tenant compte de la dépendence en enérgie de l´efficacité du détecteur. Les sections efficaces mesurées ont été ensuite corrigées des effets de l´atténuation des flux de neutrons incident et diffusé sur l´échantillon, des effets de diffusions multiples dans ce dernier et des effets d´ouverture angulaire du faisceau des neutrons incidents. Le calcul des corrections à été fait à l´aide d'un code analytique et d'un code basé sur la méthode de Monte Carlo; les résultats obtenus par les deux méthodes sont très voisins, les différences les plus significatives ne dépassant pas 2%. Les incertitudes sur les données expérimentales proviennent de la statistique de comptage, de la mesure des distances, de la mesure indirecte du flux des neutrons incidents,

K. H. Böckhoff (ed.), Nuclear Data for Science and Technology, 796–799.

de l'estimation de l'efficacité relative des détecteurs et enfin du calcul des corrections d'échantillon. On trouvera dans la référence 11 une description plus détaillée du dispositif expérimental et de la méthode de mesure.

Résultats Expérimentaux et Interprétation

Les sections efficaces de diffusion de neutrons sur ^{24}Mg, ^{28}Si et ^{32}S ont été mesurées aux énergies de 9,76 et 14,83 MeV. Le choix de ces énergies s'appuie sur des calculs préliminaires qui indiquaient que la contribution du mécanisme d'interaction directe, sensible aux effets de la déformation nucléaire, est prépondérante dans les sections efficaces de diffusion sur les premiers états collectifs des trois noyaux étudiés. Les sections efficaces différentielles ont été obtenues pour la diffusion élastique et la diffusion inélastique sur les niveaux d'énergie d'excitation inférieure ou égale à 6 MeV [11]. La présence dans le spectre des neutrons primaires du continuum des neutrons provenant de la cassure du deutéron rendait difficile l'extraction des données pour les niveaux d'énergie d'excitation plus élevée. Les figures 1 et 2 illustrent les résultats obtenus sur le noyau ^{28}Si respectivement à 9,76 et 14,83 MeV, pour l'état fondamental 0^+ et les premiers états excités 2^+ (1,779 MeV) et 4^+ (4,618 MeV); les tracés continus sont le résultat de calculs qui sont présentés et discutés ci-dessous.

Dans l'analyse des données les sections efficaces sont décrites par une superposition incohérente des contributions des mécanismes de l'interaction directe et du noyau composé. Les calculs ont été effectués en adoptant, pour chaque noyau, un potentiel optique

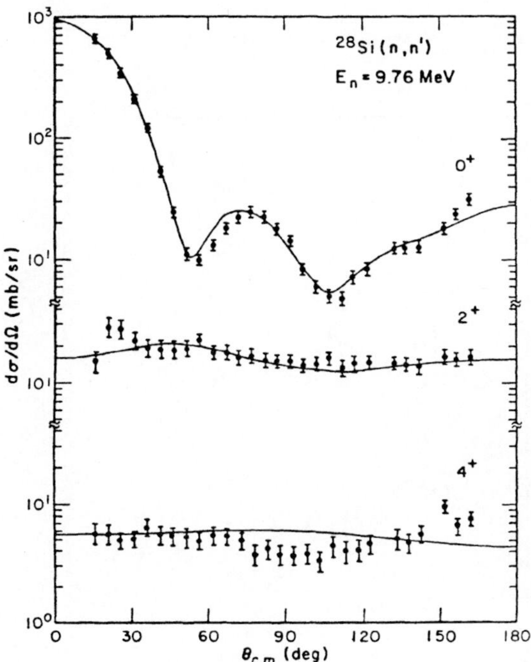

Fig. 1. Sections efficaces de diffusion de neutrons de 9,76 MeV sur ^{28}Si.

déformé avec un terme réel et un terme imaginaire de surface dépendants de l'énergie, le terme de spin-orbite étant constant et le terme imaginaire de volume étant nul. Pour le calcul des sections efficaces nous avons procédé en trois étapes.

Nous avons analysé dans une première phase, les présentes données de diffusion élastique à 9,76 et 14,83 MeV ainsi que la partie non fluctuante de la section efficace totale entre 0,5 et 15 MeV [12]. De cette analyse, effectuée en supposant que les noyaux sont des rotors symétriques avec déformations quadrupolaires et héxadécapolaires, nous avons extrait, pour chaque noyau, un jeu de paramètres du potentiel optique avec les dépendances en énergies des termes réel et imaginaire de surface. Comme paramètres de départ nous avons utilisé ceux des références 2,6,7.

Ce jeu de paramètres, assez voisin du jeu final (tableau I), a été introduit dans le code de modèle statistique HELMAG [13] pour le calcul des sections efficaces différentielles de diffusion élastique et inélastique composée. Ce code prend en compte environ 40 niveaux discrets; les états excités d'énergie plus élevée sont traités sous forme d'une densité de niveaux. Les sections efficaces de noyau composé, issues du calcul, ont été normalisées pour tenir compte des voies de sortie (n,p) et (n,α).

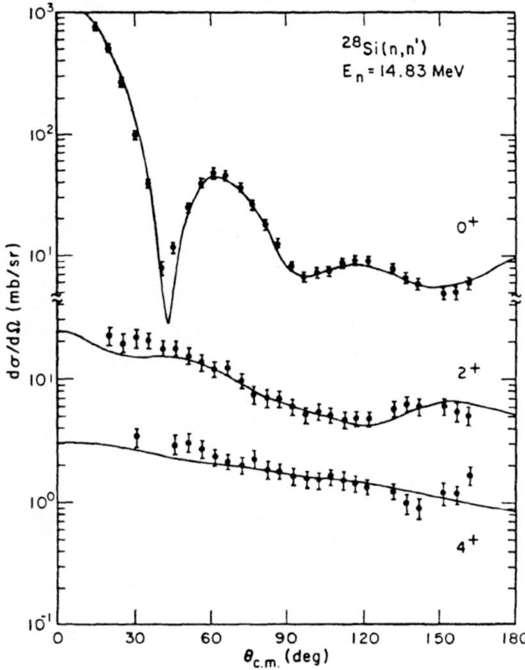

Fig. 2. Sections efficaces de diffusion de neutrons de 14,83 MeV sur ^{28}Si.

Une fois soustraite la contribution du noyau composé des sections efficaces expérimentales, l'analyse des distributions angulaires de diffusion élastique et inélastique "directe" nous a permis d'ajuster avec précision les paramètres du potentiel optique déformé. Les sections efficaces d'interaction directe ont été calculées à partir du modèle collectif en utilisant la méthode des équations couplées. Les calculs ont été effectués avec une version modifiée [14] du code JUPITOR I [15].

Les calculs sur le noyau ^{28}Si ont été effectués dans le cadre du modèle rotationnel symétrique en couplant les états 0^+ (fondamental), 2^+ (1,779 MeV), 4^+ (4,618 MeV) de la bande fondamentale K=0. Nous avons supposé que les potentiels réel et imaginaire de surface ont des déformations quadrupolaire et héxadécapolaire. Les valeurs des paramètres du potentiel optique et des déformations nucléaires sont rassemblées dans le tableau I. Les calculs présentés dans les figures 1 et 2 comprennent les contributions de l'interaction directe et du noyau composé. L'accord avec les données à 14,83 MeV est satisfaisant pour les trois niveaux; le léger désaccord entre les valeurs calculées et mesurées pour le niveau 4^+ à 9,76 MeV pourrait être attribué à une estimation imparfaite des sections efficaces d'interaction directe.

Pour le noyau ^{24}Mg, l'analyse à tout d'abord été effectuée avec le modèle rotationnel symètrique et en couplant les états 0_0^+ (fondamental), 2_0^+ (1,370 MeV), 4_0^+ (4,120 MeV) de la bande fondamentale K=0. Les paramètres obtenus sont donnés dans le tableau I et libellés avec la lettre S (modèle rotationnel symètrique).

Tableau I: Paramètres du potentiel optique et déformations nucléaires.

Les potentiels V, W, $V_{S.O}$ et l'énergie E sont exprimés en MeV, les rayons et diffusivités en fm. Les lettres S et T refèrent respectivement aux modèles du rotor symètrique et du rotor asymètrique. B. C. signifie base de couplage.

	Noyau	^{24}Mg	^{28}Si	^{32}S
Pot. Réel	V	54,60 - 0,24E	55 - 0,30E	52,70 - 0,18E
	R	1,15	1,15	1,16
	a	0,61	0,61	0,70
Pot. Imag. Surf	$W_{(S)}$	5,10 + 0,10E	4,90 + 0,12E	4,20 + 0,20E
	$W_{(T)}$	4,50 + 0,10E	---	---
	R	1,15	1,15	1,22
	a	0,58	0,58	0,64
Pot. S.O.	$V_{(S.O.)}$	6,00	6,00	6,00
	R	1,15	1,15	1,16
	a	0,58	0,58	0,70
Modèle S	B.C.	$0_0^+,2_0^+,4_0^+$	$0_0^+,2_0^+,4_0^+$	$0_0^+,2_0^+,4_0^+$
	β_2	0,50	-0,42	0,36
	β_4	0,00	0,20	-0,10
Modèle T	B.C.	$0_0^+,2_0^+,4_0^+,2_2^+,3_2^+$	--	--
	β_2	0,53	--	--
	γ	21°	--	--

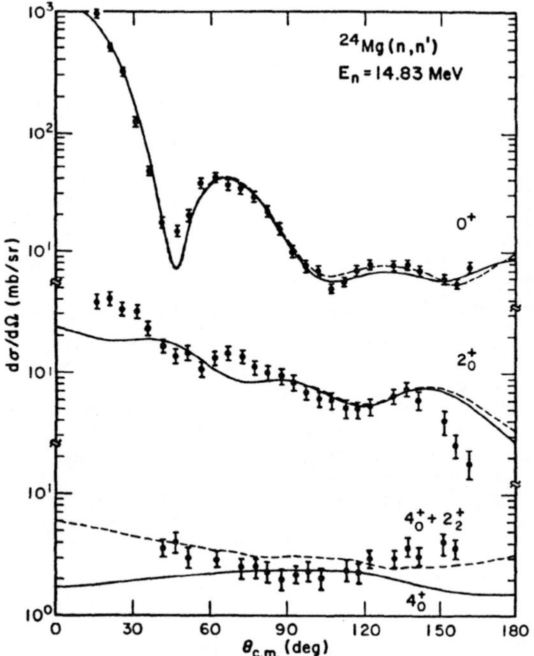

Fig. 3. Sections efficaces de diffusion de neutrons de 14,83 MeV sur ^{24}Mg.

Les valeurs calculées à 14,83 MeV sont présentées dans la figure 3 sous forme de tracés continus et comparées aux données expérimentales. L'accord entre le calcul et les mesures est satisfaisant pour la diffusion élastique; par contre, nous observons un net déphasage entre les distributions angulaires calculée et mesurée pour l'état 2_0^+, et les valeurs calculées pour l'état 4_0^+ sont nettement inférieures, vers les angles avant et arrière, aux valeurs mesurées pour le groupe d'états non résolus 4_0^+ (4,120 MeV), 2_2^+ (4,240 MeV). Nous avons alors entrepris un calcul dans lequel le noyau ^{24}Mg est supposé être un rotor triaxial et la base de couplage comprend les états 0_0^+, 2_0^+, 4_0^+ de la bande fondamentale K=0 et les états 2_2^+ (4,240 MeV), 3_2^+ (5,240 MeV) de la bande K=2, sans mélange de bandes. Les paramètres obtenus sont donnés dans le tableau I et libellés avec la lettre T (modèle triaxial). Les résultats du calcul sont montrés dans la figure 3 sous forme de tracés discontinus. Nous observons un meilleur accord pour le groupe de niveaux (4_0^+, 2_2^+), mais pour l'état 2_0^+ les deux calculs donnent des résultats pratiquement indentiques.

Nous avons analysé les données de diffusion sur l'état fondamental 0^+ et le premier état excite 2^+ (2,230 MeV) de ^{32}S en supposant que ce noyau est un rotor symètrique avec des déformations quadrupolaire et héxadécapolaire. Les calcul à 14,83 MeV sont présentés dans la figure 4 avec les mesures. L'accord est satisfaisant pour la diffusion élastique, mais pour le niveau 2^+ nous observons un désaccord autour de 90 deg. Des calculs précédents [16], effectués en supposant que le noyau ^{32}S est soit vibrationnel soit rotationnel de forme prolate ou oblate, donnent des résultats très similaires, indiquant ainsi que les données de diffusion sur les états 0^+ et 2^+ sont peu sensibles à la forme de ce noyau.

Les valeurs du paramètre de déformation quadrupolaire β_2, et de la longueur de déformation, $\delta = \beta_2.R$

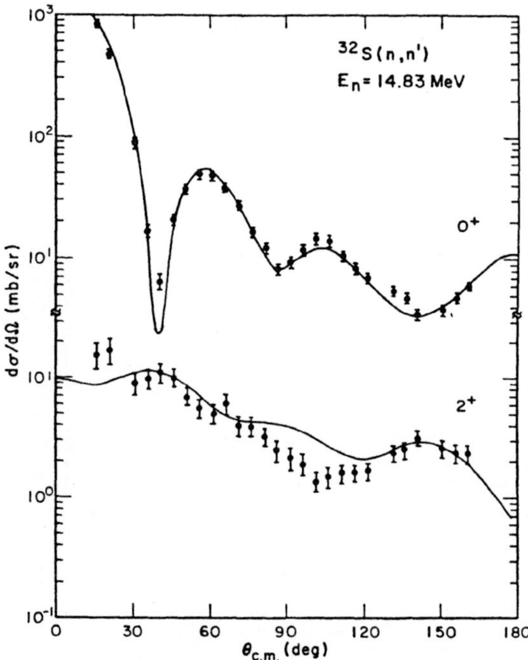

Fig. 4. Sections efficaces de diffusion de neutrons de 14,83 MeV sur ^{32}S.

(R étant le rayon du puits de potentiel réel), obtenues dans différents travaux et avec différentes sondes nucléaires sont présentées dans le tableau II. On constate, à l'examen de ce tableau, que les valeurs de β_2 varient notablement suivant le type de sonde utilisé.

Tableau II: Déformations quadrupolaires et longueurs de déformation des noyaux ^{24}Mg, ^{28}Si et ^{32}S.

Réaction	Ref.	^{24}Mg		^{28}Si		^{32}S	
		β_2	δ(fm)	β_2	δ(fm)	β_2	δ(fm)
nn'	C.T.	0,50	1,66	-0,42	1,47	0,36	1,33
nn'	2			-0,48	1,67	0,33	1,26
nn'	17			0,40	1,54		
nn'	18			0,39	1,50		
pp'	19	0,47	1,56				
pp'	20			-0,40	1,35	0,30	1,14
pp'	21					0,28	1,04
pp'	22			0,41	1,68		
$\alpha\alpha'$	23	0,34	1,76				
$\alpha\alpha'$	24			-0,42	1,47		
$^{16}O,^{16}O'$	25	0,47	1,60				
$^{16}O,^{16}O'$	26			-0,34	1,35		
e,e'	27	0,45	1,46	-0,39	1,31		
E.C.	28			0,41	1,49	0,32	1,20

Par contre, les différences sont moins marquées pour le paramètre δ; ce dernier paramètre est en fait une meilleure mesure de la déformation du potentiel d'interaction [29].

Conclusion

Nous avons entrepris l'étude des déformations nucléaires dans la région de la couche s-d en mesurant les sections efficaces de diffusion élastique et inélastique de neutrons de 9,76 et 14,83 MeV sur les noyaux ^{24}Mg, ^{28}Si et ^{32}S. L'analyse des données sur les premiers états de chaque noyau dans le cadre du modèle collectif, en utilisant la méthode des équations couplées et en tenant compte de la contribution du noyau composé statistique, nous a fourni les paramètres de déformation de ces noyaux. Cette étude va se poursuivre avec l'analyse des niveaux excités d'énergie plus élevée afin d'avoir plus d'informations sur la forme de ces noyaux.

References

(1) K. Van der Borg et al, Nucl. Phys. A325 (1979) 32.

(2) A. W. Obst et J. L. Weil, Phys. Rev. C7 (1973) 1076.

(3) M. Mermaz et al., Phys. Rev. 187 (1969) 1466.

(4) R. De Swiniarski et al., Nucl. Phys. A261 (1976) 111.

(5) F. Huang et D. McDaniels, Phys. Rev. C2 (1970) 1342.

(6) W. Pilz et al., Proc. of the IX° Intern. Symp. on the Interaction of Fast Neutrons with Nuclei (Nov. 1979). ZFK.410 Gaussig, DDR (1980).

(7) M. Abdel-Fawzi et al. ibid.

(8) A. Kurepin et N. Topil'skaya, Sov. J. Nucl. Phys. 20 (1975) 585.

(9) D. Hendrie, Phys. Rev. Lett. 31 (1973) 478.

(10) R. De Swiniarski et al., Zeit. für Phys. 298 (1980) 37.

(11) A. Virdis, Rapport CEA-R-5144 (1981).

(12) M. Goldberg et al., Neutron Cross Sections, BNL 325 2nd Edition, Suppl. No.2 Phys. TID 4500 (1966).

(13) B. Duchemin et Ch. Lagrange, Code de modèle statistique HELMAG, Centre d'Etude de Bruyères-le-Châtel (1974) non publié.

(14) Ch. Lagrange et N. Mondon, Rapport Interne, Centre d'Etudes de Limeil (1973) non publié.

(15) T. Tamura, Computer Code JUPITOR-I for Coupled Channel Calculations. ORNL-4152 (1967).

(16) G. Haouat et al. Conférence sur "Nuclear Structure and Particle Physics" Oxford U. K. 6-8 Avril 1981.

(17) J. Höhn et al., Nucl. Phys. A134 (1969) 289.

(18) S. Kliczewski et Z. Lewandowski, Nucl. Phys. A304 (1978) 269.

(19) R. Lombard et al., Phys. Rev. C18 (1978) 42.

(20) R. DeSwiniarski et al., Nucl Phys. A261 (1976) 111.

(21) R. De Leo et al., Nuovo Cem. 59A (1980) 101.

(22) P. Cole et al., Nucl. Phys. 75 (1966) 241.

(23) T. Tamura, Nucl. Phys. 73 (1965) 241.

(24) C. Fulmer et al. Phys. Rev. C18 (1978) 621.

(25) W. Mittig et al., Nucl. Phys. A233 (1974) 48.

(26) A. Dudek-Ellis et al., Phys. Rev. C18 (1978) 1039.

(27) Y. Horikawa et al., Phys. Lett. 36B (1971) 9.

(28) G. Ball et al., Nucl. Phys. A349 (1980) 241.

(29) J. Blair, Direct Interactions and Nuclear Reaction Mechanisms, edited by E. Clementel and C. Villi, New York, Gordon and Breach (1963) p. 669.

ABSOLUTE MEASUREMENT OF THE DIFFERENTIAL CROSS SECTION OF NEUTRONS ON H, C, Fe, Pb AT FORWARD ANGLES BETWEEN 45 AND 75 MEV

A. Bol, P. Leleux, P. Lipnik, P. Macq

Université Catholique de Louvain
Institut de Physique
Chemin du Cyclotron 2
B - 1348 Louvain-la-Neuve, Belgium

The differential cross section for elastic scattering of 45 to 75 MeV neutrons on hydrogen, carbon, iron and lead (nautral targets) has been measured at forward angles (3°-10° in the lab system).

Principle of the measurement

The rate of neutrons scattered at a given angle θ by a target of thickness x, made of N nuclei/cm³, is given by :

$$N(\theta) = I_0 Nx \frac{d\sigma}{d\Omega}(\theta)\ \Delta\Omega\varepsilon(\theta)$$

where I_0 is the incident neutron flux on the target
$\Delta\Omega$ is the detection solid angle
$\frac{d\sigma}{d\Omega}(\theta)$ is the differential cross section
and $\varepsilon(\theta)$ is the efficiency of the neutron detector.

To become independent of a difficult incident neutron flux absolute measurement, one can normalize the cross section measurement to the neutron flux detected at 0° :

$$N(0°) = I_0\varepsilon(0°)$$

Since we limit ourselves to the diffusion at small angles, we can consider the detection efficiency as a constant so that : $\varepsilon(\theta) = \varepsilon(0°)$; the differential cross section is then given by :

$$\frac{d\sigma}{d\Omega}(\theta) = \frac{N(\theta)}{N(0°)\Delta\Omega Nx} \qquad (1)$$

Experimental set-up (fig. 1)

A variable energy proton beam, from the isochronous cyclotron of Louvain-la-Neuve, bombards a 6 mm thick lithium target ; the neutrons produced at 0° by the $^7Li(p,n)^7Be$ reaction are collimated on the target (liquid hydrogen, C, Fe or Pb) at about 4 m from the lithium.
The neutron detectors are plastic scintillators, 10 cm thick and 6.3 cm diameter, shielded by a 15 cm long iron collimator.

Measurements and analysis

The scattered neutrons rates (N(θ)) are measured simultaneously at 8 angles (from 3 to 10° in the lab. system with 1° step), alternately with and without target (or empty target for LH_2).
The neutron rates at 0° (N(0°)) are obtained by placing each of the 8 detectors one after another at 0°.
The informations stored on magnetic tape via a CAMAC interface are the pulse heights of the 8 detectors and the time-of-flight signal of the neutrons from the production target to the detectors.
The neutron rates are obtained by integration of the TOF spectrum monokinetic peak for a given selection on the pulse height. The numbers N(θ) result from the substraction :

$$N(\theta) = N_{in}(\theta) - \alpha N_{out}(\theta)$$

where $N_{in}(\theta)$ and $N_{out}(\theta)$ are the rates measured at θ respectively with and without target.
α takes into account the background attenuation through the target.

The main correction to the results obtained with the formula (1) is the following : the neutrons coming from the target can be scattered by the detector shielding iron collimator into the detector. Such an effect leads to an effective detection solid angle greater than the geometric one. This correction, evaluated by a Monte-Carlo method and experimentally tested, comes to about 15 %.

Results

With this method, we measured the differential cross section of neutron on H at 45 - 50 - 55 - 60 - 65 - 70 and 75 MeV as well as on C, Fe and Pb at 45 and 75 MeV.
Some of the results are reproduced on the figures 2 to 5.

Remark

We also performed a measurement of the differential cross section of neutron on H at backward angles (between 168 and 180° in the c.m. systemp between 45 and 75 MeV (with 5 MeV step) ; that absolute measurement has been normalized to the total cross section of n-p radiative capture.

K. H. Böckhoff (ed.), Nuclear Data for Science and Technology, 800–801.

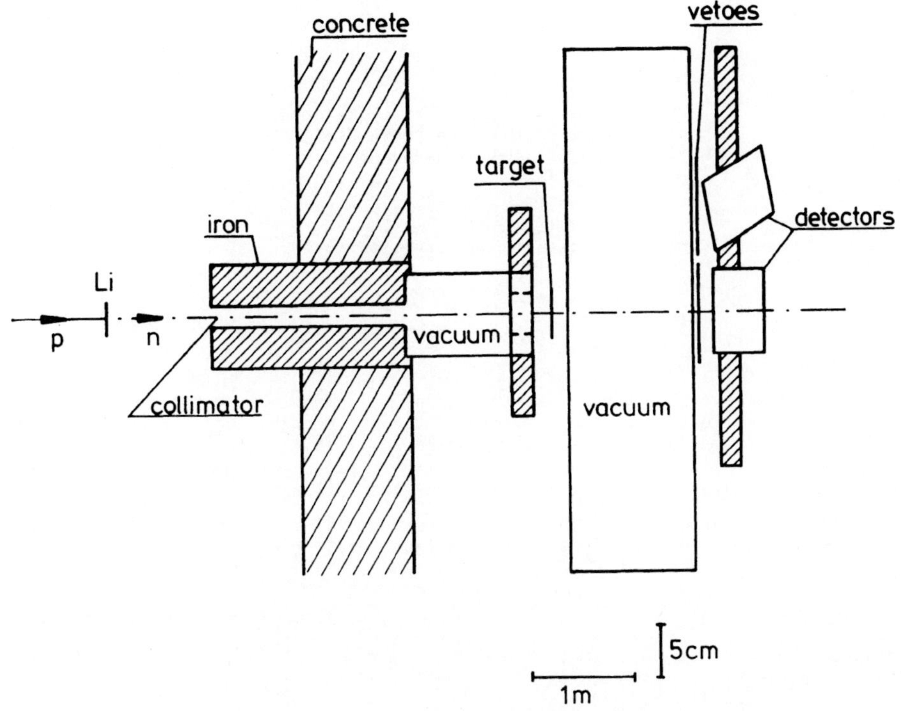

Fig. 1 : Experimental set-up

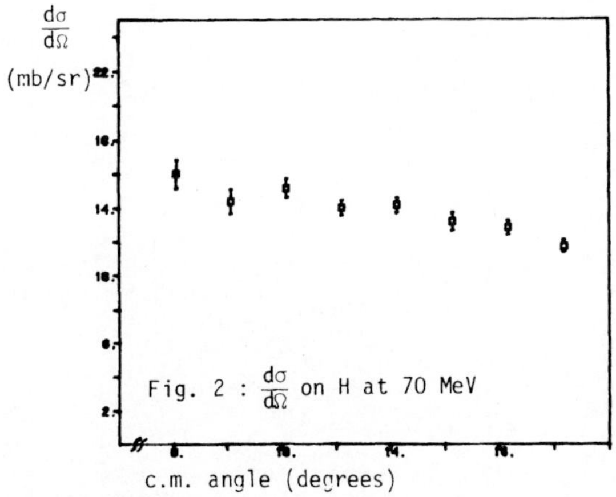

$\frac{d\sigma}{d\Omega}$ (mb/sr)

Fig. 2 : $\frac{d\sigma}{d\Omega}$ on H at 70 MeV

c.m. angle (degrees)

$\frac{d\sigma}{d\Omega}$ (b/sr)

Fig. 4 : $\frac{d\sigma}{d\Omega}$ on Fe at 75 MeV

Lab. angle (degrees)

$\frac{d\sigma}{d\Omega}$ (mb/sr)

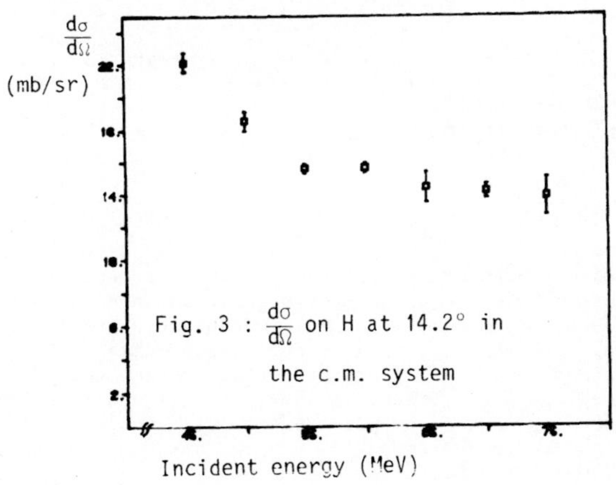

Fig. 3 : $\frac{d\sigma}{d\Omega}$ on H at 14.2° in the c.m. system

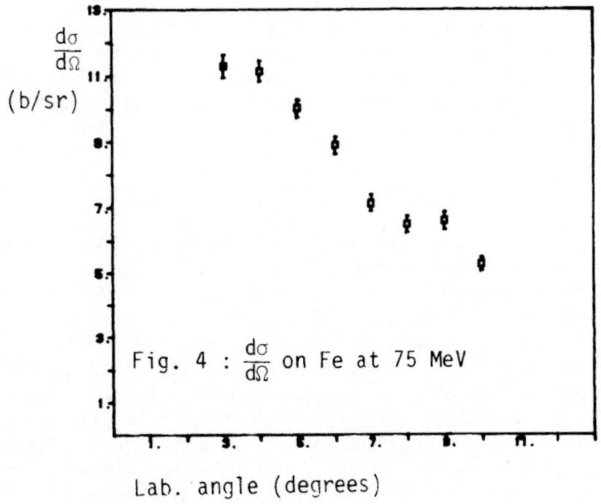

Incident energy (MeV)

$\frac{d\sigma}{d\Omega}$ (b/sr)

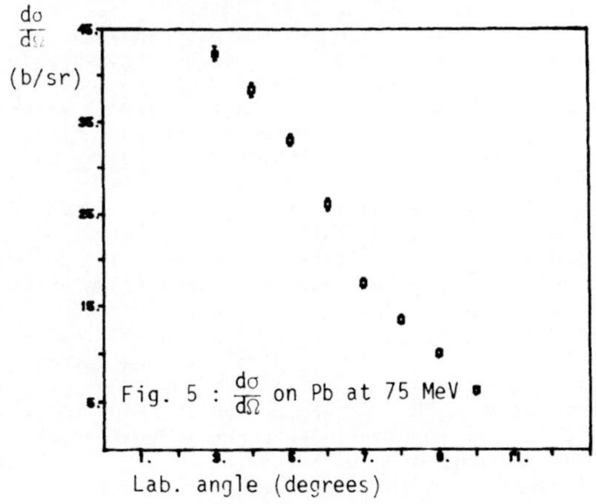

Fig. 5 : $\frac{d\sigma}{d\Omega}$ on Pb at 75 MeV

Lab. angle (degrees)

EFFETS DU COUPLAGE AUX RESONANCES GEANTES SUR LA DIFFUSION ELASTIQUE ET INELASTIQUE DE NEUTRONS RAPIDES

J.P. Delaroche

Service de Physique Neutronique et Nucléaire
Centre d'Etudes de Bruyères-le-Châtel
B.P. n° 561, 92542 Montrouge Cedex, France

P. Guss, C.E. Floyd et R.L. Walter

Duke University et Triangle Universities Nuclear Laboratory (USA)

W. Tornow

Université de Tübingen (RFA)

While the inelastic scattering of high energy hadrons is commonly used for the study of giant resonances in nuclei, it is just recently that one has thought to take into account these states in the analysis of proton scattering at low incident energies ($E \lesssim 50$ MeV). Feedback of coupling to giant resonances on optical model predictions are investigated for fast neutron elastic and inelastic scattering from vibrational nuclei and for neutron strength functions S_0 and S_1.

[Coupled channels calculations, coupling to giant resonances, vibrational nuclei, neutron elastic and inelastic scattering cross sections, strength functions]

Introduction

Les résonances géantes isoscalaires ont fait l'objet d'importantes études expérimentales tout au long de ces dernières années. Des travaux ont ainsi permis de mieux connaître les propriétés (énergies d'excitation, intensités,...) de ces états collectifs pour de nombreux noyaux vibrationels [1]. L'insertion de tels états dans des calculs en voies couplées pour des énergies incidentes inférieures à 50 MeV a permis de montrer que ces couplages avaient un effet assez important sur la diffusion élastique de protons par des noyaux vibrationels [2]. Dans la présente contribution, nous avons étudié l'influence de ces couplages sur le mécanisme de réaction des neutrons rapides pour de tels noyaux.

Calculs en voies couplées

Des mesures de diffusion élastique et inélastique de neutrons ont été faites récemment au Triangle Universities Nuclear Laboratory (TUNL) pour des noyaux choisis entre ^{40}Ca et ^{208}Pb, à des énergies incidentes (E_n) comprises entre 8 et 17 MeV. L'analyse des résultats expérimentaux a été faite à l'aide du modèle optique en voies couplées où sont exclusivement pris en compte, dans un premier temps, les états collectifs de basse énergie. Un exemple de ces calculs est présenté en Fig. 1 (courbes en tirets) pour le noyau ^{120}Sn à 10 MeV. La base de couplage utilisée comprend l'état fondamental (0^+) ainsi que les premiers états excités : 2^+, 3^-, 4^+ et 5^-. Les calculs reproduisent assez bien les distributions angulaires mesurées. L'insertion de la résonance géante octupolaire à basse énergie (LEOR) et de la résonance géante quadrupolaire (GQR) dans les calculs a été faite en conservant les paramètres du potentiel optique ajustés précédemment. Seule l'intensité du potentiel d'absorption a été réduite de $\sim 0{,}2$ MeV pour tenir compte qualitativement du changement de base de couplage. Les calculs où sont inclus de façon séquentielle les états LEOR et GQR sont représentés en Fig. 1 par les courbes en pointillés et en traits pleins. A l'énergie incidente considérée, le couplage aux résonances géantes produit un effet important sur la diffusion élastique aux grands angles. Pour la diffusion inélastique, l'effet observé est moins important et se traduit, pour l'essentiel, par une renormalisation des prévisions. On peut aussi remarquer que les effets du couplage à l'état LEOR sont plus importants que ceux dus à l'état GQR. Des calculs semblables effectués pour les noyaux ^{40}Ca, ^{116}Sn et ^{208}Pb conduisent à des effets dont l'ampleur est comparable à celle qui est montrée en Fig. 1.

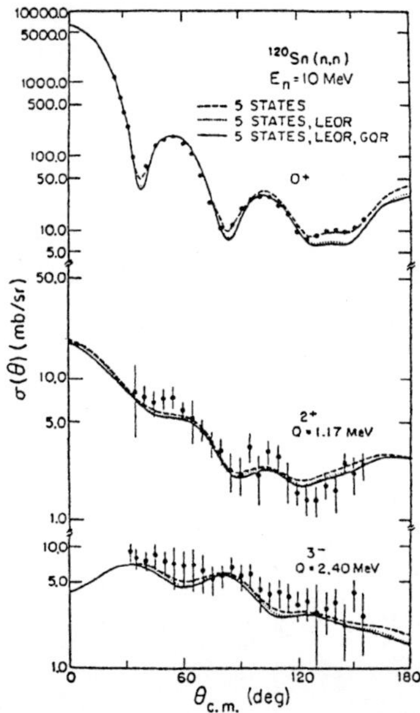

Fig. 1. Diffusion élastique et inélastique de neutrons par ^{120}Sn. Les valeurs expérimentales ont été obtenues à TUNL. Les courbes représentent des calculs décrits dans le texte.

L'effet du couplage aux résonances géantes sur les prédictions du modèle optique a également été étudié pour les fonctions densité de neutrons s et p (S_0 et S_1). Dans le cas du noyau ^{40}Ca, les prédictions pour S_0/S_1 varient de $\sim 20\%$ quand les bases de couplage $(0^+, 3^-, 5^-)$ et $(0^+, 3^-, 5^-, \text{GQR})$ sont successivement utilisées.

Conclusion

Les quelques études de sensibilité mentionnées ci-dessus indiquent clairement que les résonances géantes jouent un rôle actif dans le mécanisme de réaction des neutrons rapides avec des noyaux collectifs. Vu le caractère spécifique des effets occasionnés par le couplage à ces états, il pourrait être opportun, dans certains cas, de ne plus ignorer systématiquement l'existence de résonances géantes au moment du choix d'une base de couplage.

Références

1. F. Bertrand, Nucl. Phys. A354, 129C (1981)
2. M. Pignanelli et al;, Phys. Rev. C24, 369 (1981)

K. H. Böckhoff (ed.), Nuclear Data for Science and Technology, 802.

SURFACE CONTRIBUTIONS TO NUCLEAR LEVEL DENSITIES

V.S. Ramamurthy and S.K. Kataria

Bhabha Atomic Research Centre, Trombay
Bombay 400 085, India

M. Asghar

Centre of Nuclear Science and Technology
Alger-Gave, Algerie

A general leptodermous expansion for the density of single particle levels in thin skinned potential wells is derived and is used to study the finite size corrections to the macroscopic level density parameters of nuclei. With droplet model values for the potential parameters, the calculated level density parameters for nuclei along the beta stability line show systematic deviations from the experimental estimates.

Introduction

In the last few years, a number of investigations [1-4] have shown that the shell independent part of the level density parameters of nuclei as derived from the experimental neutron resonance spacings data exhibit characteristic deviations from a strict proportionality to the mass number of the nuclei. These deviations arise both as a result of a surface correction to the single particle level density in the shell model single particle potential well and as a result of the complicated mass and charge dependence of the parameters of the shell model potential wells themselves. We have examined both these aspects with a view to understand the mass and charge dependence of the macroscopic part of the level density parameters of nuclei.

Mean distribution of single particle levels in potential wells

Considering a nucleus as a portion of infinite nuclear matter the density of single particle levels is given by the usual Thomas-Fermi relation

$$\frac{dN}{dE} = \frac{2m}{\hbar^2}\frac{Deg}{4\pi^2} kV$$

where V is the volume of the nucleus, Deg is the degeneracy of the single particle levels and $k = \sqrt{2mE/\hbar^2}$.

A number of attempts [5-8] have been made in the past to introduce finite size corrections to the Thomas-Fermi relation for the density of single particle levels dN/dE. Starting from simple model potentials, correction terms proportional to the surface area and integrated curvature are obtained. We present below a general leptodermous expansion applicable for realistic potential shapes with a finite depth and finite diffuseness. The starting point of the present procedure is the single particle wave function $\psi(\underline{\nu})$ inside the nucleus. In the interior of the potential well, ψ is a plane wave with a momentum k and energy $E = \hbar^2 k^2/2m - V_o$. The corresponding particle density in the interior is given by

$$\rho(\underline{\nu}) = \int_o^{k_f} d^3k\, \psi^\star(\underline{\nu})\, \psi(\underline{\nu})\, \frac{Deg}{(2\pi)^3}$$

If A is the total number of nucleons in the nucleus,

$$A = \int d^3\gamma \int d^3k\, \psi^\star(\nu)\psi(\nu)\, \frac{Deg}{(2\pi)^3}$$

Since the finite size corrections arise in the surface region, where the wave function deviates from simple plane waves, we choose the following approximate representation for $\psi(\underline{\nu})$.

$$\psi(\underline{\nu}) = \exp(k_n \cdot \gamma_n)\, \psi_\perp(s)$$

where s is the distance of γ from the potential surface along the normal and k_n is the component of the momentum in a direction parallel to the potential surface. $\psi_\perp$ is the nomalized wave function in the direction perpendicular to the potential surface. In terms of this wave function,

$$\rho(\underline{\nu}) = \frac{Deg}{8\pi^3}\int_o^{k_f} d^3k\, \psi_\perp^\star(s)\, \psi_\perp(s)$$

$$= \frac{Deg}{4\pi^2}\left[\int k_f^2\, dk_\perp\, |\psi_\perp|^2 - \int dk_\perp k_\perp^2 |\psi_\perp|^2\right]$$

By definition

$$\left.\frac{dN}{dE}\right|_{E=E_f} = \frac{d\rho}{dk_f}$$

Therefore one obtains

$$\frac{dN}{dE} = \frac{2m}{\hbar^2}\frac{Deg}{4\pi^2}\int d^3\gamma \int_o^k dk_\perp\, |\psi_\perp|^2$$

Separating out the volume term, it is easy to obtain

$$\frac{dN}{dE} = f_oV + f_1S + f_2C + f_3G$$

where

$$f_o = \frac{2m}{\hbar^2}\frac{Deg}{4\pi^2}k$$

$$f_1 = \frac{2m}{\hbar^2}\frac{Deg}{4\pi^2}\left[\int_{-\infty}^{\infty} ds \int_o^k dk_\perp\left(\psi_\perp^2 - \frac{\Theta(s)}{2}\right)\right]$$

$$f_2 = \frac{2m}{\hbar^2}\frac{Deg}{4\pi^2}\left[\int_{-\infty}^{\infty} ds \int_o^k dk_\perp\left(\psi_\perp^2 - \frac{\Theta(s)}{2}\right) s\right]$$

$$f_3 = \frac{2m}{\hbar^2}\frac{Deg}{4\pi^2}\left[\int_{-\infty}^{\infty} ds \int_o^k dk_\perp\left(\psi_\perp^2 - \frac{\Theta(s)}{2}\right) s^2\right]$$

$$\Theta(s) = 1; \quad s \leqslant 0, \quad = 0 \quad s \geqslant 0$$

It is thus possible to write down a formal leptodermous expansion for the density of single particle states in a potential well, by simply obtaining the one dimensional wave function in a direction perpendicular to the surface. It should be mentioned here that the above representation of the wave function is strictly valid only for a parallelepiped box and the resulting curvature correction is only approximate for a potential well with a curved surface. To test the accuracy of the method, we show in Fig. 1, the density of levels at the Fermi level for a system with 126 particles in a potential well of radius R = 8.0 fm, diffuseness a = 0.67 fm and depth ranging from 30 to 200 MeV. The same quantity was also obtained by first computing the single particle levels in the potential well exactly and then smearing the levels by the Strutinsky smearing procedure. The two results can not be distinguished in the figure. It can therefore be concluded that the present expansion based on the one dimensional wave

K. H. Böckhoff (ed.), Nuclear Data for Science and Technology, 803–804.

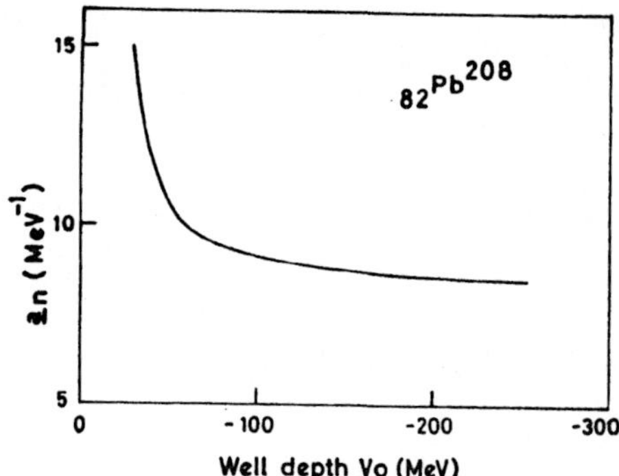

Fig. 1. Density of levels at the Fermi level for 126 particles in a potential well
Radius = 8.0 fm; Diffuseness = 0.67 fm;
Depth = V_o MeV

function provides an accurate estimate of the density of levels in any potential well with a thin surface.

Level density parameters of nuclei

Having obtained the density of single particle levels, a calculation of the level density parameter requires a determination of the Fermi energy. The particle number equation

$$A = \int \frac{dN}{dE} \, dE$$

provides the relation between the Fermi energy and the potential parameters. For the present calculations, we have used the droplet model values of the potential parameters [9]. An approximate correction for the decrease of the potential depth for protons due to the Coulomb field inside the nucleus and for the presence of the Coulomb barrier outside has also been introduced. Fig. 2 shows a plot of the calculated a values along with the "experimental" values [2].

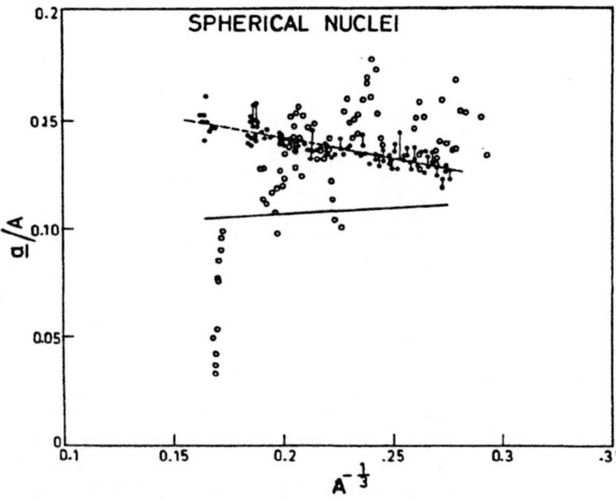

Fig. 2. Calculated level density parameters for nuclei along the beta stability line using the present procedure and droplet model parameters (continuous line). Also shown in figure for comparison are the experimental values with (closed circles) and without (open circles) shell corrections.

It can be seen that, in addition to a difference of 20-30 % in the magnitude of the predicted and the experimental values, an important difference in the sign of the $A^{1/3}$ dependence also persists. Further investigations are necessary to explain these deviations.

Conclusion

A general leptodermous expansion for the density of single particle levels in thin-skinned potential wells is presented. With droplet model parameters for the potential wells, the calculated values not only deviate from the experimental estimates by about 20-30 % but also fail to reproduce the $A^{-1/3}$ dependence.

References

1. A.V. Ignatyuk, M.G. Itkis, V.N. Okolovich, G.N. Smirenkin, A.S. Tishin, Yad. Fiz 21 (1975) 1185 (Sov. J. Nucl. Phys. 21 (1975) 612)
2. S.K. Kataria, V.S. Ramamurthy, S.S. Kapoor, Phys. Rev. 18C (1978) 549.
3. A.S. Jensen, J. Sandberg, Physica Scripta 17 (1978) 107
4. W. Reisdorf, Z. Phys. A300 (1981) 227
5. D.L. Hill, J.A. Wheeler, Phys. Rev. 89 (1953) 1102
6. R. Balian, C. Bloch, Ann. Phys. 60 (1970) 401
7. Ph.J. Siemens, A. Sobicjewski, Phys. Letters 41B (1972) 16
8. J. Toke, W.J. Swiatecki, Nucl. Phys. A 372 (1981) 141
9. W.D. Myers, Nucl. Phys. A145 (1970) 387

VALENCY EFFECTS IN COMPOUND NUCLEUS LEVEL SPACINGS

J.L. Cook and E.K. Rose

Lucas Heights Research Laboratories
Australian Atomic Energy Commission
Research Establishment
Sutherland 2232, NSW, Australia

It is shown that nuclides whose proton or neutron numbers lie within three units of a magic number have a level density parameter that is very strongly correlated with the Myers-Swiatecki shell correction to the mass formula. Using this correlation, 93 level densities are calculated from only two adjustable constants, in a semi-empirical fashion.

It is shown that since weaker correlations exist in five regions of the periodic table, intermediate and heavy nuclides which lie between the strong correlation ranges also give satisfactory fits, thus making a twelve-parameter fit overall.

Introduction

Knowledge of the average level spacing between resonances of a compound nucleus has important applications in astrophysics, reactor physics and fission physics. Until now the most suitable formula has come from the evaluation of Gilbert and Cameron[1] shell and pairing corrections carried out by Cook, Ferguson and Musgrove[2], and modified by Rose and Cook[3]. Wapstra and Gove[4] published a thorough evaluation of neutron binding energies which was used by Rose and Cook, together with experimental values of the level spacing tabulated by Gyulassy and Perkins[5], Mughabghab and Garber[6] and Musgrove[7], to re-evaluate the Gilbert-Cameron parameters.

Unfortunately, the Gilbert-Cameron theory requires many parameters in the form of shell and pairing corrections which, for about 300 values, requires about 200 constants to be fitted. In this report it is shown that there is a strong correlation between the Myers and Swiatecki[8] shell correction to the semi-empirical mass formula and the level density mass parameter, which permits accurate calculation of level densities for 93 spherical nuclides in terms of only two adjustable parameters. The physical significance of the level density parameters is discussed fully in Lang[9].

Gilbert-Cameron Theory

The densities of states of spin J at an energy E above the ground state were derived by Gilbert and Cameron as

$$\rho(E,J) = \frac{\sqrt{\pi}\,\exp\{2(aU)^{\frac{1}{2}}\}\,(2J+1)\exp\{-(J+\frac{1}{2})^2/2\sigma^2\}}{24\,a^{\frac{1}{4}}\,U^{\frac{5}{4}}\,(2\pi)^{\frac{1}{2}}\sigma^3} \tag{1}$$

U is the effective excitation energy, given as

$$U = E - \Delta E \quad ,$$

where ΔE is the nucleon pairing energy which is fixed at the Green and Edwards[10] value:

$$\left.\begin{aligned} \Delta E\ (\text{odd-odd}) &= 0 \\ \Delta E\ (\text{even-odd}) &= 11\,A^{-\frac{1}{2}} \\ \Delta E\ (\text{even-even}) &= 22\,A^{-\frac{1}{2}} \end{aligned}\right\} \tag{2}$$

where A is the compound nucleus mass number.

The spin cutoff parameter σ was determined by Gilbert and Cameron to be

$$\sigma^2 = 0.0888(aU)^{\frac{1}{2}}\,A^{\frac{2}{3}} \ . \tag{3}$$

More recent estimates[11] give the constant coefficient of 0.146. This agrees with the result of Lang. The quantity a is the level density parameter, which Gilbert and Cameron assumed to be

$$a/A = \alpha\,S(Z,N)+\beta \quad , \tag{4}$$

where α and β are constant and S(Z,N) is the shell correction to the semi-empirical mass formula. Cameron[12] used the relationships

$$\begin{aligned} S(Z,N) &= S(Z)+S(N) \\ \Delta E &= P(Z)+P(N) \end{aligned} \tag{5}$$

and worked out tables of S(Z), S(N), P(Z) and P(N) which fitted the measured masses. This treatment gives many adjustable parameters, but interpolation to unmeasured values is hazardous. Figure 1 shows the scatter of experimental a/A values with compound nucleus mass number A.

The Myers-Swiatecki Shell Corrections

A theoretical derivation of the shell correction was given by Myers and Swiatecki. They found the expressions

$$\text{(i)}\quad S(Z,N) = C\left[\frac{F(N)+F(Z)}{(\frac{1}{2}A)^{\frac{2}{3}}}\right] - cA^{\frac{1}{3}} \quad , \tag{6}$$

$$\text{(ii)}\quad F(X) = \int_0^X [q(N)-n^{\frac{2}{3}}]\,dn \quad ,$$

$$\text{(iii)}\quad q(n) = \frac{3}{5}\ \frac{M_i^{\frac{5}{3}} - M_{i-1}^{\frac{5}{3}}}{M_i - M_{i-1}} \text{ for } M_{i-1} < n < M_i \ .$$

The M_i are the magic numbers 14,28,50,82,126,184 and 258 for both Z and N. The values of constants C and c are

$$C = 4.8\text{ MeV} \quad , \quad c = 0.26 \ . \tag{7}$$

For deformed nuclides, one replaces S by $S_o[1+\ell nS/S_o]$, where S_o is the Myers-Swiatecki spherical limit.

On examining the data compiled by Rose and Cook at the neutron binding energy, a correlation was naturally found between the calculated a/A from the experimental values of $\langle D\rangle = 1/\rho$ and the Myers-Swiatecki shell correction for those nuclides with either Z or N within three units of any of the magic numbers M_i. The correlation coefficient was 0.903 between a/A and S(Z,N) for these valency nuclides, 93 of which have been measured. The Myers-Swiatecki shell corrections were then applied to S(Z,N) and a correlation coefficient of 0.365 was calculated for the remaining 107 nuclides. Green's pairing correction for ΔE was used. The relationship (Equation 4)

K. H. Böckhoff (ed.), Nuclear Data for Science and Technology, 805–808.

for these others, most of which were deformed nuclei, was therefore rejected.

For the valency nuclides, a linear fit gives

$$a/A = (0.01018\pm0.00036)S(Z,N)+(0.12746\pm0.00050). \quad (8)$$

The experimental values for a/A and the fitted values are shown in Appendix A. Errors obtained by including the experimental errors for <D> are also presented. Figures 2 to 7 show the variation of experimental a/A with S(Z,N) for each group together with the fitted line $\alpha S(Z,) + \beta$.

For neutron reactions, the proper excitation energy is given by

$$E = E_n + B \quad , \qquad (9)$$

where E_n is the kinetic energy of the neutron and B is the neutron binding energy. The kinetic energy was assumed to be about one half of the last resolved resonance energy and the binding energies were obtained from Wapstra and Gove. Since formula (1) applied to both parities, only one parity prevails at low energies, so for s-waves

$$\langle\rho\rangle = \frac{1}{\langle D\rangle} = \frac{1}{2} \sum_{J=I-\frac{1}{2}}^{J=I+\frac{1}{2}} \rho(U,J) \quad , \qquad (10)$$

where I = the target nucleus spin. The recalculated values of <D> are given in Appendix B together with the experimental value.

One can perceive from the coefficients in Table 1 that for group 5, a constant value of a/A is quite acceptable as a fit. This happens to be the range for the most strongly deformed nuclei; when S(Z,N) assumes values well away from magic numbers, the correlation is lost.

The overall situation regarding the possibility of using a broader group structure to reduce the number of parameters is presented in Table 2. Here we postulate that in the weaker correlation ranges, a satisfactory fit is achieved by replacing the linear dependence with a constant average value of a/A. The value of χ^2/n is given at each stage and it is apparent from the table that the new scheme is the optimum one for satisfactory predictions of <D> . The group structures are summarised in Table 3.

With regard to the calcium isotopes, which make up the second group, it was found that these light isotopes departed from the strong correlation expected near the semi-magic number 20; in reality it should be expected that the Fermi gas model would be unreliable in this range.

In the case of fission product data, which is the ultimate purpose of this study, there is no need to be concerned about isotopes in this range, so the fit to group 2 would never be needed. Our rigorous statistical analysis reveals that five semi-empirical constants are required to fit about two hundred for intermediate and heavy nuclides. This is a satisfactory result for the prediction of unmeasured values of the level spacing.

Conclusion

A satisfactory overall fit to measured values of the level spacing is obtained with twelve adjustable parameters. The 93 nuclides with valency 3 or less are very well fitted with just two adjustable constants. The purpose of carrying out these fits was to reduce the number of degrees of freedom from the large number required for a Gilbert and Cameron type of theory. Extrapolations and interpolations to unmeasured values of $\bar{D}$, such as are required in astrophysics and reactor physics, can serve as a check on Gilbert and Cameron values and probably provide more reliable results.

References

1. A. Gilbert and A.G.W. Cameron, Can. J. Phys. 43, 1446 (1965).
2. J.L. Cook, H. Ferguson, and A.R. de L. Musgrove, Aus. J. Phys. 20, 477 (1967).
3. E.K. Rose and J.L. Cook, AAEC/E419 (1977).
4. A.H. Wapstra and N.B. Gove, Nuc. Data Tables 9 (1971).
5. M. Gylassy and S.T. Perkins, UCRL-50400, Vol. 13 (1972).
6. S.F. Mughabghab and D.I. Garber, BNL325 Vol. 1, (1973).
7. A.R. de L. Musgrove, (1976) Private communication
8. W.P. Myers and W.J. Swiatecki, Nucl. Phys. 81, 1, (1966).
9. D.W. Lang, Nucl. Phys. 77, 545, (1966)
10. A.E.S. Green and D.F. Edwards, Phys. Rev. 91, 46, (1953).
11. D.G. Gardner (1980) - Private communication.
12. A.G.W. Cameron, Can. J. Phys. 36, 1040, (1958).

TABLE 1

COEFFICIENTS FOR $a/A = \alpha S(Z,N)+\beta$

Group	Number of nuclides (n)	α	$\Delta\alpha$	β	$\Delta\beta$	χ^2/n	C_R
1	93	0.01018	0.00036	0.12746	0.00050	0.92	0.89
2	3	-0.01028	0.02135	0.18438	0.03641	0.04	-0.52
3	9	0.02976	0.01703	0.08391	0.02873	0.09	0.91
4	14	0.05209	0.01513	0.01041	0.02568	0.31	0.78
5	56	0.00631	0.00288	0.11319	0.00496	1.04	0.18
6	22	0.07947	0.04179	-0.03514	0.06008	0.45	0.53

Appendices are available from authors.

TABLE 2

χ^2/n VALUES AS A FUNCTION OF GROUP STRUCTURE

Group	Nuclear Range	n	αS+β Fit	a/A Fit	5 Groups	4 Groups		3 Groups		2 Groups
1	all spherical 4-181	93	0.92	0.92	0.92	0.92	0.92	0.92	0.92	0.92
2	1-3	3	0.04	0.12	0.12	0.12				
3	35-46	9	0.08	0.42			0.89		0.89	
					0.95			2.30		
4	56-71	14	0.31	1.16		2.34				80.0
5	111-167	56	1.03	1.12	1.12		1.12			
									1.0	
6	182-204	22	0.45	0.62	0.62	0.62	0.62	0.62		
		197								
7										

7 omitted owing to high χ^2 in a/A fit.

Groups 5 and 6 only marginally worse for a/A fit.

αS+β fits to 6 regions still best.

For deformed nuclides 6-group structure is such that each group lies between magic numbers in either Z or N or both.

TABLE 3

DEFORMED GROUP STRUCTURE

Group	Magic Number Range for Z	Magic Number Range for N' = N+1
2	14 < Z ≦ 28	14 < N' ≦ 28
3	28 < Z ≦ 50	28 < N' ≦ 50
4	28 < Z ≦ 50	50 < N' ≦ 82
5	50 < Z ≦ 82	82 < N' ≦ 128

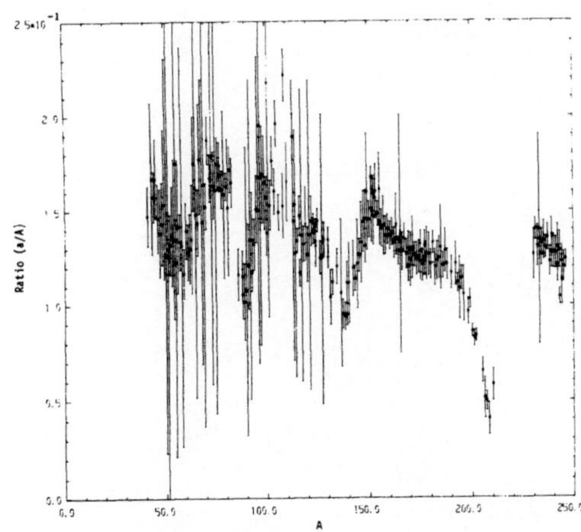

FIGURE 1. SCATTER OF EXPERIMENTAL a/A VALUES WITH COMPOUND NUCLEUS MASS NUMBER A

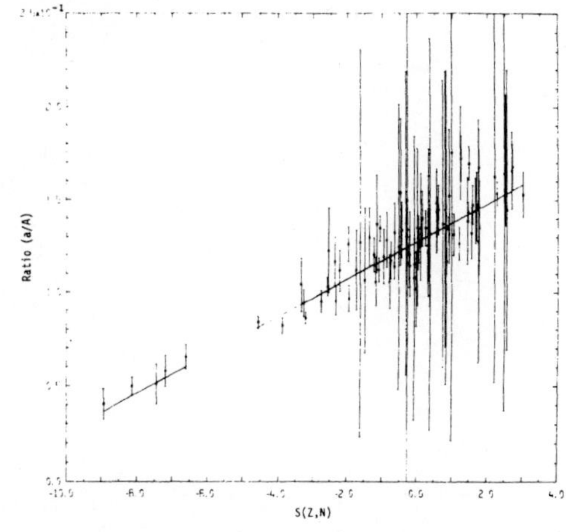

FIGURE 2. VARIATION OF EXPERIMENTAL a/A WITH S(Z,N) AND FITTED LINE α S(Z,N) + β, GROUP I

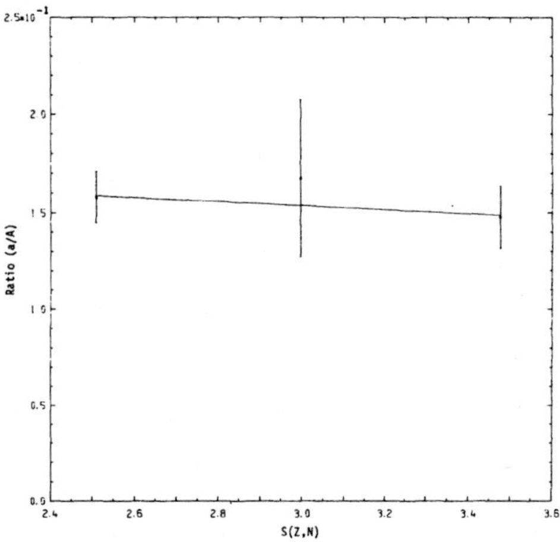

FIGURE 3. VARIATION OF EXPERIMENTAL a/A WITH S(Z,N) AND FITTED LINE α S(Z,N) + β, GROUP 2

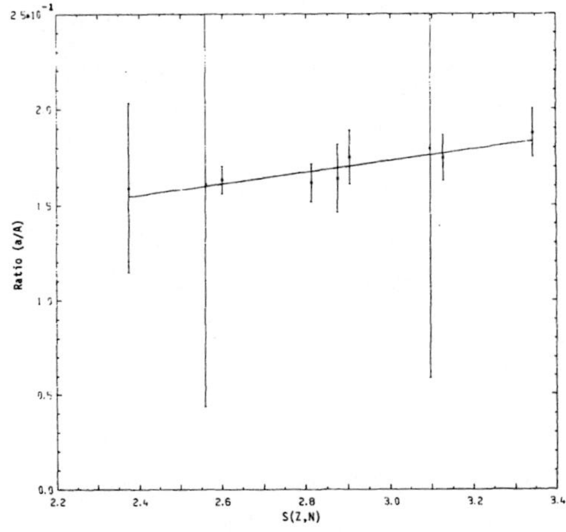

FIGURE 4 . VARIATION OF EXPERIMENTAL a/A WITH S(Z,N) AND FITTED LINE α S(Z,N) + β, GROUP 3

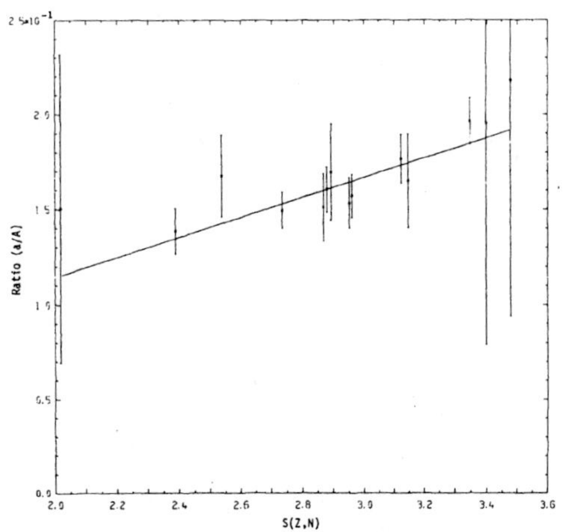

FIGURE 5. VARIATION OF EXPERIMENTAL a/A WITH S(Z,N) AND FITTED LINE α S(Z,N) + β, GROUP 4

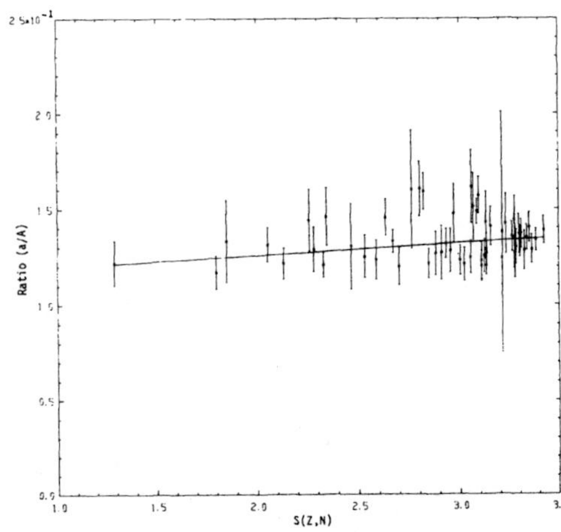

FIGURE 6. VARIATION OF EXPERIMENTAL a/A WITH S(Z,N) AND FITTED LINE α S(Z,N) + β, GROUP 5

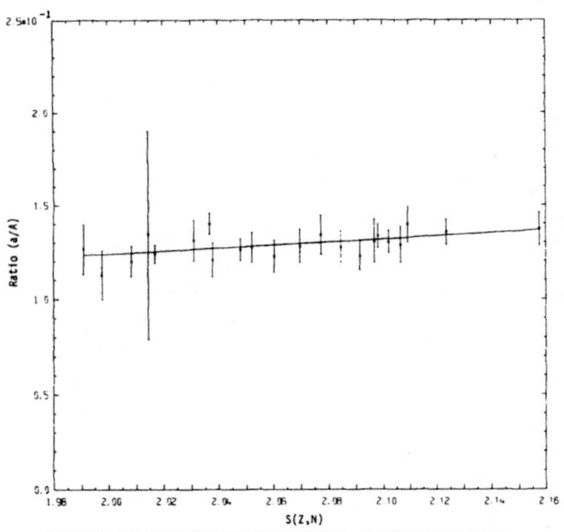

FIGURE 7. VARIATION OF EXPERIMENTAL a/A WITH S(Z,N) AND FITTED LINE α S(Z,N) + β, GROUP 6

FLUCTUATION PROPERTIES OF NUCLEAR ENERGY LEVELS AND WIDTHS : COMPARISON OF THEORY WITH EXPERIMENT

O. Bohigas
Division de Physique Théorique*, Institut de Physique Nucléaire
91406 Orsay Cedex, France

R.U. Haq
Department of Physics, University of Toronto, Toronto
Ontario M5S 1A7, Canada

A. Pandey
Service de Physique Théorique, CEN-Saclay
91191 Gif-sur-Yvette Cedex, France

*Laboratoire associé au CNRS.

We analyze the fluctuation properties of nuclear energy levels and widths with new spectrally averaged measures. A remarkably close agreement between the predictions of random-matrix theories and experiment is found.

[Fluctuation properties of energy levels, width fluctuations, level repulsion, random matrices, neutron resonances]

1. Introduction

This paper deals with the current status of the random matrix theories with regard mainly to the nuclear energy level fluctuations i.e. departures from spectral uniformity, and also to width fluctuations. Specifically the question is how well the ensemble results agree with experiment. And if the agreement is good, then : what is the source of this agreement and what limits can one impose on the mechanisms which affect the fluctuations. We are mainly concerned here with the first question for which we rely for the most part on our recent work[1]. We refer only briefly to the latter two questions for which also the answers are emerging.

The nuclear data relevant for fluctuation studies consists of slow-neutron resonances of medium and heavy nuclei and also of proton resonances of some light nuclei. Most of the neutron data come from systematic measurements of total neutron cross-sections performed at Columbia during the sixties and early seventies. There are also measurements in which the spin of the levels are measured directly. These include mainly (n,γ) measurements performed at Geel, Livermore and Oak Ridge. Besides the neutron measurements, detailed nuclear resonance parameters have been extracted from high resolution inelastic proton scattering experiments performed at Duke.

The principal random matrix model, valid for space rotation and time reversal invariant systems, is Wigner's Gaussian Orthogonal Ensemble (GOE), consisting of asymptotically large real symmetric matrices, in which the distinct matrix elements are distributed independently with essentially the same zero-centered Gaussian law. For energy levels the main predictions are due to Dyson and Mehta[2]. The close-lying levels behave as if they are repelling each other and the distant ones display a long range order, the resultant spectral rigidity being then akin to a crystalline structure. For widths the model predicts a χ_1^2 -distribution known after the names of Porter and Thomas[3]. The early theoretical work in the subject is reviewed in Refs.[4-6]. For later developments as well as uptodate references see Ref.[7].

The GOE predictions are very sensitive to spurious and missing levels, especially for energy level fluctuations, imposing stringent conditions on the quality of data. Ideally the observed level sequence should be pure (all levels should have the same spin and parity) and complete (no missing level). Because of this, while the main GOE predictions have been known for almost two decades, good quality data required to test the model have come relatively recently. Earlier analyses of individual level sequences indicate "overall good agreement" between theory and experiment ; see Ch.VI of Ref.[7] for a review.

The present analysis makes much sharper comparisons between theory and experiment than the previous ones. The main reason for this improvement is not due to the additional data of the last few years, but rather to making better use of the theory when compared with experiment. Indeed, there are two disconcerting features in the previous analyses : the predictions of the theory have not been tested in a systematic way nor much attention has been paid to sample size effects. We resolve these difficulties here. We ask : what features of the theory are observed in the data and with what accuracy. We attempt to test directly the (fundamental) k-level correlation functions which are essentially the joint probability functions and of which the two level properties are the most significant. We go on to other fluctuation measures (which in fact derive from the correlation functions) so as to improve the statistical significance of the testing procedure. Moreover, since the theory is parameter-free, we combine all the data into a nuclear data ensemble (NDE) of approximately 1700 levels, this improving the statistics even more dramatically. Our main conclusion is that the agreement between theory and experiment is remarkably good.

Some comparison between theory and experiment is also made for widths, although in this case conclusions are less significant because the theoretical distribution itself has been used to a large extent to make spin and parity assignment of the levels.

In the next section we review the GOE predictions for the energy levels and in section 3 we present the results of comparison with the NDE. Section 4 deals with width fluctuations. The concluding section discusses the implications of the agreement.

2. GOE Predictions for the Energy Levels

Dyson's[8] k-level cluster functions* $Y_k (k>1; \bar{Y}_1 \equiv 1)$ provide a natural hierarchy for energy level fluctuation studies. Essentially these are the k-level correlation functions, viz the joint-probability densities of observing k levels at given positions, from which the lower order correlation effects have been subtracted out. In particular $(1-Y_2(r))dr$ gives the probability of observing a level in an infinitesimal interval dr

---

*For a random variable W, we denote its ensemble average by $\overline{W}$ and its ensemble variance $\overline{(W-\overline{W})^2}$ by Var W. Note, moreover, that we are considering spectra normalized to unit local average spacing.

K. H. Böckhoff (ed.), Nuclear Data for Science and Technology, 809–813.

at a distance r from a given level. These functions are designed in such a way that for the Poisson ensemble, in which the energy levels are chosen randomly in a given interval, the Y_k are all zero for $k>1$.

For GOE[5,8]

$$Y_2(r) = \left(\frac{\sin\pi r}{\pi r}\right)^2 + \left(\int_r^\infty \frac{\sin\pi r'}{\pi r'} dr'\right)\frac{d}{dr}\left(\frac{\sin\pi r}{\pi r}\right), \quad (1)$$

which is normalized to a total unit integral. It has the value unity at r=0 implying level repulsion and goes to zero as $(\pi r)^{-2}$ for large $r(\gtrsim 1)$ implying spectral rigidity. The latter aspect is more apparent from the number variance[2,7] $\Sigma^2(\bar{n})$, namely the variance of the number statistic n, the number of levels in a given interval of length $\bar{n}$:

$$\Sigma^2(\bar{n}) = \bar{n} - 2\int_0^{\bar{n}} (\bar{n}-r)Y_2(r)dr$$

$$= \frac{2}{\pi^2}\left\{\ln(2\pi\bar{n})+\gamma+1+\frac{1}{2}[Si(\pi\bar{n})]^2 - \frac{\pi}{2}Si(\pi\bar{n})-\cos(2\pi\bar{n}) - Ci(2\pi\bar{n}) + \pi^2\bar{n}\left[1-\frac{2}{\pi}Si(2\pi\bar{n})\right]\right\}$$

$$\xrightarrow{\bar{n}\gtrsim 1} \frac{2}{\pi^2}\left\{\ln(2\pi\bar{n}) + \gamma + 1 - \frac{\pi^2}{8}\right\}. \quad (2)$$

Note that its value is $\sim$ unity for $\bar{n}$=100 and even for $\bar{n} \sim 10^6$ the fluctuation is not more than a couple of levels. This should be compared with $\Sigma^2(\bar{n})=\bar{n}$ for the Poisson ensemble.

Another measure of the spectral rigidity is the Δ_3-statistic of Dyson and Mehta[2]. It measures, for a fixed interval $[x,x+\bar{n}]$, the least square deviation of the staircase N(E), the number of levels with energy less than or equal to E, from the best straight line fitting it :

$$\Delta_3(\bar{n};x) = (1/\bar{n}) \min_{A,B} \int_x^{x+\bar{n}} (N(E)-AE-B)^2 dE \quad . \quad (3)$$

Its ensemble average is related to Σ^2, and hence to Y_2, by[9]

$$\bar{\Delta}_3(\bar{n}) = (2/\bar{n}^4) \int_0^{\bar{n}} (\bar{n}^3-2\bar{n}^2r+r^3)\, \Sigma^2(r)\, dr \quad , \quad (4)$$

which may be integrated numerically, or for $\bar{n}\gtrsim 15$ one may use[2] $\bar{\Delta}_3 \simeq \pi^{-2} \ln \bar{n} - 0.007$ with good accuracy. Again, in comparison with the Poisson value $\bar{\Delta}_3=\bar{n}/15$, the spectrum is seen to exhibit a long range order. (Note that because of stationarity[10] the ensemble properties are independent of x, the position of the interval).

Usually the most one can hope for is to test the theory for the two level functions and its integrals such as Σ^2 and $\bar{\Delta}_3$. However from a very long sequence of levels or from a collection of many short sequences as is the case for nuclear spectra, it may be possible to observe some higher-than-two level effects as well. Moreover the higher order functions are needed for the calculation of the sample errors of the testing procedure ; thus for example for testing two-point measures one needs to know the k-level functions for $k\leq 4$. Although all cluster functions have been evaluated for GOE[11], very few analytic results are available for higher-than-two-point measures. For properties related to Y_k with $k>2$, we rely mostly on Monte Carlo results that we have obtained from a sample, adequate for our purpose, of 100 GOE matrices of dimensionality 300.

Some examples of higher-than-two-level measures are provided by the distributions and correlation properties of the Δ_3 as well as the number (n) statistics. In this paper some results for Δ_3 are considered (see Fig.2c for Var Δ_3 and Figs.3a,3b for the Δ_3 distribution for $\bar{n}$=10,20). For very large $\bar{n}$, Var Δ_3 approaches[2] the value .011 which should be compared with the Poisson result Var $\Delta_3 = (1+11\bar{n}/30)\bar{n}/210$. Once again it is seen that the fluctuations are small for GOE.

In order to make optimal use of an observed spectrum, we introduce spectral averaged quantities. Consider a sequence containing (p+1) levels. For a statistic $W(\bar{n};x)$ defined in an interval of length $\bar{n}$ at x, the spectral average

$$\langle W(\bar{n})\rangle_p = (p-\bar{n})^{-1} \int_x^{x+(p-\bar{n})} W(\bar{n},y)\, dy \quad (5)$$

is an estimator of $\bar{W}(\bar{n})$. The variance of $\langle W\rangle_p$, which is the square of the sample error, is also given by the theory :

$$\mathrm{Var}\, \langle W(\bar{n})\rangle_p \equiv \overline{\langle W(\bar{n})\rangle_p^2} - \overline{\langle W(\bar{n})\rangle_p}^2$$

$$= \frac{2\,\mathrm{Var}\, W(\bar{n})}{(p-\bar{n})^2} \int_0^{(p-\bar{n})} (p-\bar{n}-r)C(r;\bar{n})dr$$

$$\simeq \frac{2\,\mathrm{Var}\, W(\bar{n})}{(p/\bar{n})} \int_0^1 C(\lambda\bar{n};\bar{n})d\lambda \ , \quad (6)$$

where $C(r;\bar{n})$ is the autocorrelation function of W :

$$C(r;\bar{n}) = (\mathrm{Var}\, W(\bar{n}))^{-1}\left\{\overline{W(\bar{n};x)W(\bar{n};x+r)} - (\bar{W}(\bar{n}))^2\right\} . \quad (7)$$

The last step of Eq.(6), valid for $p\gg\bar{n}$, follows from the fact that the autocorrelation functions for measures discussed in this paper fall off rapidly, being close to zero for $r\gtrsim\bar{n}$ for $\bar{n}\gtrsim 1$. Notice that as $p\to\infty$, Var $\langle W\rangle_p$ goes to zero, which is nothing but the ergodic property[10] : the spectral average coincides with the ensemble average.

It is seen from (6) that for the sample errors we need Var W as well as the integral $\int C(\lambda\bar{n};\bar{n})d\lambda$ both of which we estimate from our Monte Carlo sample. The details will be published elsewhere. In practice instead of (5) we use a less smooth average in which the integral is replaced by a summation.

3. Comparison with Experiment for Energy Levels

In the conventional analysis one treats the data from each nucleus, containing about 50-100 levels, separately and consequently the statistical significance is poor. As seen from (5) the testing procedure improves as the sample size increases. Moreover since the correlation functions, Eq.(7), fall off rapidly it is not necessary to have one long sequence but a collection of many relatively short sequences could also be used.

Since one expects that the resonance energies of every compound nucleus will share the same fluctuation properties, it seems natural, and in the spirit of random matrix theory, to treat the set (NDE) of nuclear resonance energy data of different nuclei as a sampling of eigenvalues of GOE matrices ; see Ref.[12] for a similar notion for complex atomic spectra. As an example of this procedure, the nearest neighbor spacing distribution is shown in Fig.1. It is clear that a histogram containing $\sim$ 1700 spacings corresponding to the NDE (Fig.1b) is statistically much more significant, when compared to a theoretical prediction, than a histogram containing only $\sim$ 100 spacings corresponding to a single nucleus (Fig.1a).

Our NDE consists of 1762 resonance energies corresponding to 36 sequences of 32 different nuclei :

i) Slow-neutron resonance data on ^{64,66,68}Zn (103,65,45 levels[13]), ^{114}Cd (17 levels[14]), ^{152,154}Sm (70, 27 levels[15]), ^{154,156,158,160}Gd (19,47,21,54 levels[16]), ^{160,162,164}Dy (18,46,20 levels[17]), ^{166,168,170}Er (109,48,31 levels[18]), ^{172,174,176}Yb (55,19,23 levels[19]), ^{182,184,186}W(41,30,14 levels[20]), ^{186,190}Os (56,17 levels[21]), ^{232}Th (178 levels[22]) and ^{238}U (146 levels[22]),

ii) (n,γ) reaction data on ^{177}Hf (34'J=3' levels, 23'J=4' levels[23]), ^{179}Hf(25'J=4' levels, 22'J=5' levels[23]) and ^{235}U(58'J=3' levels, 68'J=4' levels[24]),

iii) proton resonance data on ^{44}Ca(52'1/2$^-$' levels, 39'1/2$^+$' levels[25]), ^{48}Ti(66'1/2$^+$' levels[25]) and ^{56}Fe (56'1/2$^+$'levels[26]).

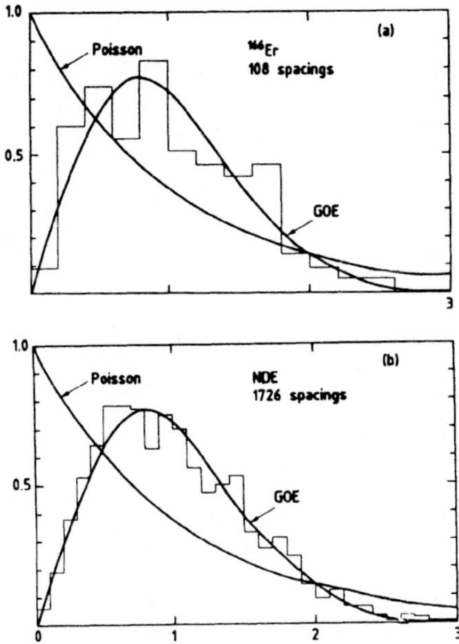

Fig.1. Nearest neighbor spacing histogram for : (a) ^{166}Er, (b) Nuclear Data Ensemble (NDE).

The criterion* for inclusion in the NDE is that the individual sequences be in general agreement with GOE.

We have made an extensive test of the theory by studying the aforementioned functions $Y_2(r)$, $\Sigma^2(\bar{n})$, $\Delta_3(\bar{n})$ as well as others. A complete account will be given elsewhere. Here we restrict ourselves to a few typical examples for the NDE.

Consider first the two-level measures $\Sigma^2(\bar{n})$ and $\bar{\Delta}_3(\bar{n})$. The procedure for calculations is to evaluate for each of the 36 sequences the spectral-averaged measure, say $<\Delta_3(\bar{n})>$ for $\bar{\Delta}_3(\bar{n})$, and then take their average, weighted according to the size of each sequence. We have considered the range $0\leq\bar{n}\leq 100$. The results for $\bar{n}<25$ are shown in Figs.2a, 2b. They are in a remarkably good agreement with GOE. We emphasize that typically the "figure of merit", e.g. $(\mathrm{Var}<\Delta_3>)^{1/2}/\bar{\Delta}_3$ for $\bar{\Delta}_3$, is of the order of a few percent in the present analysis, whereas in the conventional analysis it is $\sim$ 20% or more. The values corresponding to the Poisson case as well as the Gaussian Unitarity Ensemble (GUE ; see section 5 below) are also shown for comparison.

To explore higher-than-two level effects we consider the ensemble variance of $\Delta_3(\bar{n})$, which is sensitive to Y_k upto k=4, and the distribution function of Δ_3 for various values of $\bar{n}$ (which involves in principle all order effects). The results are presented in Fig.2c for Var Δ_3 for $\bar{n}<25$ and two examples ($\bar{n}$=10,20) of the distribution function of Δ_3 are shown in Fig.3. The agreement between the GOE predictions and data is again very good.

----

*In the previous paper[1] our NDE consisted of 1407 resonance energies corresponding to 30 sequences of 28 different nuclei. In the present NDE we have added 8 new sequences. Moreover, we have not included the modifications suggested by the authors of the Columbia data to account for some missing levels. Instead, we have considered shorter sequences in a few cases and have dropped from the list the ^{110,112}Cd-spectra which involved large modifications. Our conclusions, as exemplified by Figs.2,3, are still the same as in Ref.[1].

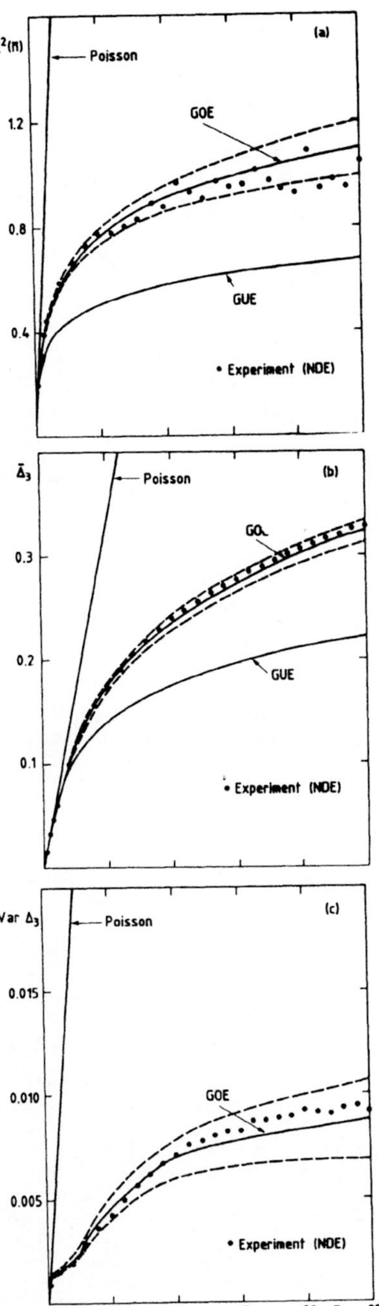

Fig.2. (a) Σ^2,(b) $\bar{\Delta}_3$,(c) Var Δ_3 as functions of $\bar{n}$. Dashed lines correspond, for GOE, to one standard deviation from the average.

4. Width Fluctuations

The main question is whether the transition widths (via which the resonance levels are detected and identified) have a χ_1^2 (Porter-Thomas[3]) distribution :

$$\rho(x) = (2\pi x)^{-1/2} \exp(-x/2) \; ; \; x = \Gamma_{ic}|\bar{\Gamma}_{ic} , \qquad (8)$$

where the width Γ_{ic} measures the rate of transition from an initial (statistical) state i to a channel state c. Or, equivalently the question is whether the transition amplitudes, whose squares (to within a penetrability factor) are the widths, are Gaussian random variables. This prediction follows directly from the orthogonal invariance of the random matrix ensembles as is the case with GOE ; see Ch.VII of Ref.[7].

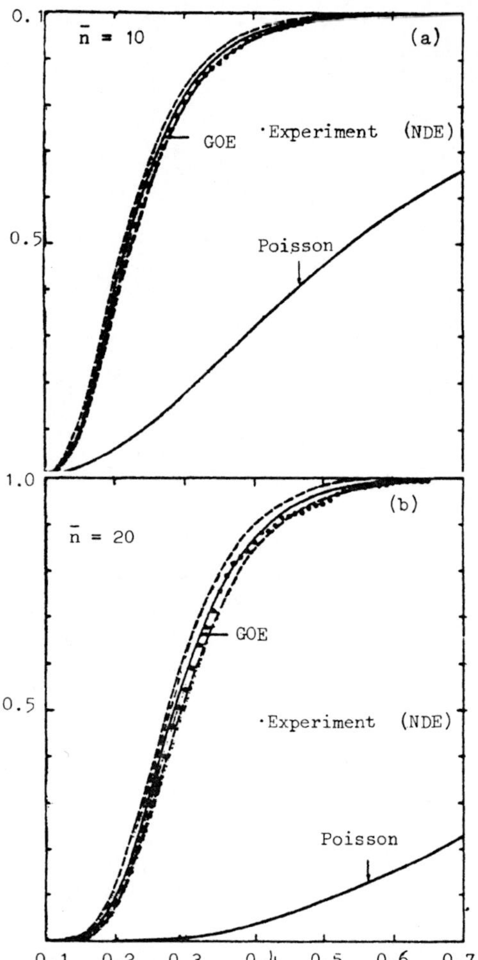

Fig.3. Distribution function of Δ_3 : (a) $\bar{n}$=10, (b) $\bar{n}$=20. Dashed lines correspond, for GOE, to one standard deviation from the average.

The amount of data that can be used to test the theory is much larger than in the case of energy levels. This is because even if some of the resonances are not detected due to very small widths, Eq.(8) can be tested for $x \geq \alpha$, an experimental threshold. In this section we present the results for some of the neutron-widths. A more detailed analysis will be taken up later. For a similar analysis of much older data, see Ref.[27].

We consider an NDE containing 1182 widths corresponding to 21 level sequences* of the previous section : ^{64,66,68}Zn, ^{114}Cd, ^{152,154}Sm, ^{154,156,158,160}Gd, ^{162}Dy, ^{166,168,170}Er, ^{172,174,176}Yb, ^{182,184}W, ^{232}Th and ^{238}U. The transition amplitude distributions are shown in Fig.4 for an individual sequence as well as the NDE. The agreement between theory and data is seen to be very good.

For further tests we calculate ν, the number of degrees of freedom, assuming that the width distribution itself is χ^2_ν with ν not necessarily an integer. A statistically significant departure from ν=1 would then be an indication of a failure of the Porter-Thomas model. A commonly-used procedure is to construct the likelihood function which is the product of the χ^2_ν - densities of the observed (renormalized) widths and then to maximize this function with respect to ν. The sample errors associated with this estimation procedure are obtained by studying the deviations from the maximum. The result would however show departure if some small widths are missing. For a better procedure we may assume a (relative) cut-off α below which the widths are likely to be missed and then do the above analysis for all widths of values $\geq \alpha$ with a truncated

----

* The widths in each sequence are renormalized to unit average.

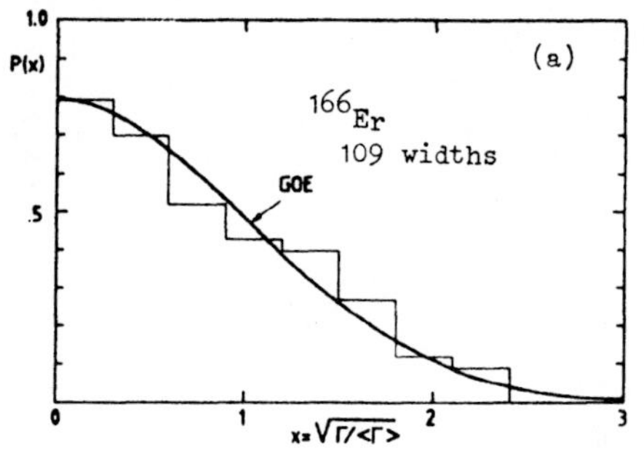

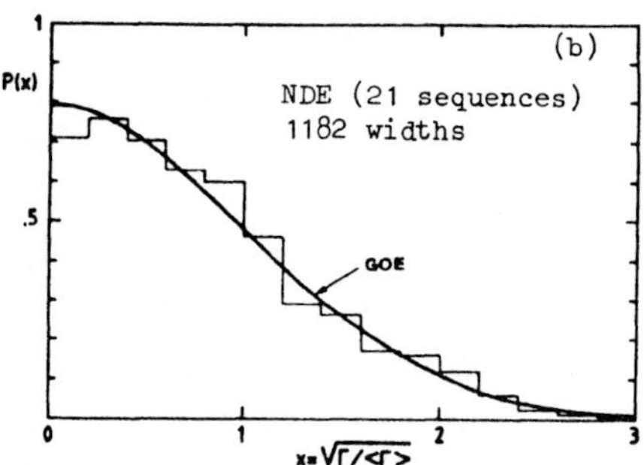

Fig.4. Histogram for the absolute values of transition amplitudes : (a) ^{166}Er, (b) NDE. The theory predicts $P(x)=(2/\pi)^{1/2}\exp(-x^2/2)$.

χ^2_ν-distribution;for details see Refs.[3,27]. We have studied ν as a function of α and a plot is given in Fig.5. The departures from theory for $\alpha \lesssim 10^{-3}$ may be due to the missing small widths. For larger values of α the agreement is again very good.

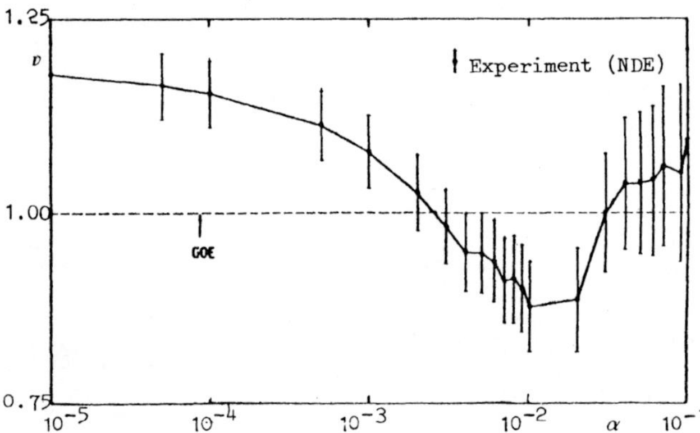

Fig.5. The degree of freedom ν as a function of cut-off α. The error bars correspond to one standard deviation from the estimated ν.

Finally we turn to the question of correlations between the width and energy level fluctuations. The theory predicts the two to be statistically independent. To test this we calculate the correlation coefficient

$$r = \frac{1}{N}\sum_{i=1}^{N}\left(\frac{\Gamma_{ic}-\langle\Gamma\rangle}{\sigma_\Gamma}\right)\left(\frac{E_i-\langle E_i\rangle}{\sigma_E}\right) , \qquad (9)$$

for each sequence. Here N is the number of levels in the sequence and Γ_{ic} the width associated with the level E_i. $\langle\Gamma\rangle$ and σ_Γ are the mean and standard deviation of the Γ_{ic}, and

$$\langle E_i\rangle = Ai+B,\ \sigma_E^2 = N^{-1}\sum_{i=1}^{N}(E_i-\langle E_i\rangle)^2 , \qquad (10)$$

where A and B are fixed by minimizing* the expression for σ_E^2. For the NDE we take the average of the correlation coefficients for each sequence, weighted according to the size of the sequence. We obtain r(NDE)= 0.017 as compared to the theoretical value 0±0.029 confirming again the predictions.

5. Conclusion

We have established an astonishingly good agreement between a parameter-free theory (GOE) and the data. We emphasize that, apart from rotation and time-reversal invariance resulting in the real symmetric nature of the matrices, GOE takes no account of the specific properties of the nuclear Hamiltonian, e.g. its (1+2)-body nature, its large pairing and quadrupole components, etc . We also recall that the eigenvalues and eigenvector components of realistic (non-random) nuclear shell model matrices show the same fluctuation patterns[7,28]. How can this be understood theoretically ?

To answer this we recall first that there are other classical ensembles in which, like GOE, the orthogonal invariance plays a key role and which also give the same fluctuations[11,29]. But more significantly for the nuclear Hamiltonian, there are Monte Carlo results[28,30] which indicate that ensembles of two-body operators acting in many-particle (shell model) spaces also yield fluctuations characteristic of GOE. More recently[31] it has been demonstrated that adding a GOE matrix H to any real symmetric matrix K,$(K+\alpha H)$, leads very quickly, as α increases, to the same fluctuations. Intermediate fluctuation patterns are to be expected only when the random matrix elements are not much larger than the local average spacing of the given (non-random) matrix K. The good agreement with experiment, coupled with the theoretical understanding which is now emerging, reinforces the belief that GOE fluctuations are to be found in nature under very general conditions.

We mention finally that one can use the close agreement to impose restrictions on mechanisms which would change the fluctuations. In particular if time reversal is not an exact symmetry, the appropriate model would be an ensemble of complex Hermitian matrices in which the real and imaginary parts are sampled independently. The Gaussian Unitary Ensemble (GUE) in which the two parts have the same norm is one such example. It gives very different fluctuation properties than GOE[4-7]. For example, $Y_2(r)=(\sin\pi r/\pi r)^2$ implying stronger level repulsion than GOE (see Figs. 2a,2b for Σ^2 and $\overline{\Delta}_3$) and widths follow a χ_2^2-distribution. Analytic results for ensembles in which the real and imaginary parts have unequal weights have recently been given[31-33]. It is shown for energy level fluctuations (and is expected to be valid for width fluctuations as well) that even a small magnitude of the imaginary part induces major changes in fluctuation properties, leading very rapidly from GOE to GUE type fluctuations. The results are being used[32] to derive an upper bound on the time reversal non-invariant part of the nuclear Hamiltonian.

References

1. R.U. Haq, A. Pandey and O. Bohigas, Phys. Rev. Lett. 48, 1086 (1982)
2. F.J. Dyson and M.L. Mehta, J. Math. Phys. 4, 701 (1963),reprinted in Ref.[4]
3. C.E. Porter and R.G. Thomas, Phys. Rev. 104, 483 (1956), reprinted in Ref.[4]
4. C.E. Porter (Ed.), Statistical Theories of Spectra: Fluctuations, Academic Press, New York, 1965

----

*We remark parenthetically that the expression for A thus obtained is a good estimator of the average spacing ; see section X.B of Ref.[7]. The spectra used in section 3 were normalized by this procedure.

5. M.L. Mehta, Random Matrices and the Statistical Theory of Energy Levels, Academic Press, New York 1967
6. E.P. Wigner, SIAM Rev. 9, 1 (1967)
7. T.A. Brody, J. Flores, J.B. French, P.A. Mello, A. Pandey and S.S.M. Wong, Rev. Mod. Phys. 53, 385 (1981)
8. F.J. Dyson, J. Math. Phys. 3, 166 (1962), reprinted in Ref.[4]
9. A. Pandey, Ann. Phys. (N.Y.) 119, 170 (1979), Appendix II
10. A. Pandey, Ref.[9]
11. F.J. Dyson, Commun. Math. Phys. 19, 235 (1970) ; M.L. Mehta, Commun. Math. Phys. 20, 245 (1971)
12. N. Rosenzweig and C.E. Porter, Phys. Rev. 120, 1698 (1960), reprinted in Ref.[4]
13. J.B. Garg, V.K. Tikku and J.A. Harvey, Phys. Rev. C23, 671 (1981); J.B. Garg, V.K. Tikku, J.A. Harvey, R.L. Macklin and J. Halperine, Phys. Rev. C24, 1922 (1981) ; J.B. Garg, private communication
14. H.I. Liou, G. Hacken, F. Rahn, J. Rainwater, M. Slagowitz and M. Makofske, Phys. Rev. C10, 709 (1974)
15. F. Rahn, H.S. Camarda, G. Hacken, W.W. Havens,Jr., H.I. Liou, J. Rainwater, M. Slagowitz and S. Wynchank, Phys. Rev. C6, 251 (1972)
16. F. Rahn, H.S. Camarda, G. Hacken, W.W. Havens, Jr., H.I. Liou and J. Rainwater, Phys. Rev. C10, 1904 (1974) ; for ^{156}Gd, C. Coceva and M. Stefanon, Nucl. Phys. A315, 1 (1979) and private communication
17. H.I. Liou, G. Hacken, J. Rainwater and U.N. Singh, Phys. Rev. C11, 462 (1975)
18. H.I. Liou, H.S. Camarda, S. Wynshank, M. Slagowitz, G. Hacken, F. Rahn and J. Rainwater, Phys. Rev. C5, 974 (1972)
19. H.I. Liou, H.S. Camarda, G. Hacken, F. Rahn, J. Rainwater, M. Slagowitz and S. Wynchank, Phys. Rev. C7, 823 (1973)
20. H.S. Camarda, H.I. Liou, G. Hacken, F. Rahn, W. Makofske, M. Slagowitz, S. Wynchank and J. Rainwater, Phys. Rev. C8, 1813 (1973)
21. J.C. Browne and B.L. Berman, Phys. Rev. C23, 1434 (1981) and private communication
22. F. Rahn, H.S. Camarda, G. Hacken, W.W. Havens,Jr., H.I. Liou, J. Rainwater, M. Slagowitz and S. Wynchank, Phys. Rev. C6, 1854 (1972)
23. C. Coceva, F. Corvi, P. Giacobbe and M. Stefanon, in Statistical Properties of Nuclei, J.B. Garg (Ed.), Plenum Press, New York, 1972
24. M.S. Moore, J.D. Moses, G.A. Keyworth, J.W.T. Dabbs and N.W. Hill, Phys. Rev. C18, 1328 (1978)
25. W.M. Wilson, E.G. Bilpuch and G.E. Mitchell, Nucl. Phys. A245, 285 (1975)
26. W.A. Watson, E.G. Bilpuch and G.E. Mitchell, Z. Phys. A300, 89 (1981)
27. J.D. Garrison, Ann. Phys. (N.Y.) 30, 269 (1964)
28. S.S.M. Wong and J.B. French, Nucl. Phys. A198, 188 (1972)
29. M.L. Mehta, J. Math. Phys. 17, 2198 (1976)
30. O. Bohigas and M.J. Giannoni, Ann. Phys. (N.Y.) 89, 393 (1975)
31. A. Pandey, Ann. Phys. (N.Y.) 134, 110 (1981)
32. J.B. French, V.K.B. Kota and A. Pandey, to be published
33. A. Pandey and M.L. Mehta, Commun. Math. Phys., to be published.

NEUTRON INDUCED CHARGED PARTICLE REACTIONS ON ^{23}Na

H. Weigmann*, G.F. Auchampaugh, P.W. Lisowski,
M.S. Moore, and G.L. Morgan.

Los Alamos National Laboratory, Los Alamos, NM, U.S.A.

High resolution measurements of neutron induced charged particle reactions on ^{23}Na have been performed. A NaI(Tl) detector served as both target and detector, with pulse shape discrimination being applied for the separation of protons and alpha-particles from each other and from events involving gamma-ray detection. The neutron energy was measured by time-of-flight, using an 80 m flight path at the Los Alamos National Laboratory WNR facility.

Relative cross sections for ^{23}Na(n,p) and ^{23}Na(n,α) have been determined in the neutron energy range between 5 and 11 MeV. Up to 9 MeV it has also been possible to separate the (n,p_o) reaction leading to the ^{23}Ne ground state from the (n,p') reaction feeding the first five excited states of ^{23}Ne.

All observed cross sections show strong fluctuations in most of the energy intervals investigated, especially below 7 MeV, where marked peaks are observed in the (n,p_o) cross section. In the range between 7 and 9 MeV neutron energy, a significant correlation is found between the fluctuations in the (n,p_o) and (n,p') cross sections, whereas the fluctuations in the (n,α) cross section are not correlated with either of the former two. This observation is indicative of doorway states common to the proton channels, possibly due to states in ^{24}Na with isospin $T = T_Z + 1$.

Introduction

Earlier measurements of the ^{23}Na(n,p) cross section done by the activation method [1,2] showed strong fluctuations in the excitation function, particularly in the neutron energy range from 5 to 8 MeV. The ^{23}Na(n,α) cross section had not been studied with sufficiently small energy steps to be conclusive with respect to the presence or absence of fluctuations [1].

In order to study the structure of these cross sections in more detail, and possibly to obtain some information on the nature of the observed fluctuations, high resolution measurements of the relative (n,p) and (n,α) cross sections of ^{23}Na have been performed over the neutron energy range from 5 MeV to 11 MeV.

Experimental Method

The neutron time-of-flight method was used for the present measurements with the spallation source of the Los Alamos National Laboratory WNR facility serving as the pulsed neutron source. The effective time resolution was about 1.2 ns (FWHM). With the 80 m flight path, typical values of the energy resolution are 6.1 keV at 6 MeV, and 13.1 keV at 10 MeV neutron energy. A 2" diameter and 2" thick NaI(Tl) scintillation crystal served as both target and detector. Pulse shape discrimination (PSD) [3] was used for the separation of protons and alpha particles from each other and from events involving gamma-ray detection. The shape of the neutron spectrum was determined with a plastic scintillation detector with well known response function.

Two examples of PSD spectra for two pulse-height intervals comprising most of the data used in the analysis are shown in fig. 1. As can be seen from the figure, a rather clean separation of protons and alpha particles by the PSD information is possible. The cross contamination is of the order of 10%. Also, the positions of the "p" and "α" peaks are sufficiently independent of pulse height, such that fixed PSD windows may be used for the further analysis.

In fig. 2, three examples of pulse height spectra for different neutron energy intervals are shown. They were obtained when selecting events from the "proton" group in the PSD spectra as indicated in fig. 1.

The pulse height groups labelled "p_o" and "p'" refer to reactions leaving the final nucleus ^{23}Ne in its ground state and in excited states, respectively.

* Visiting scientist from CEC, JRC-CBNM, Geel, Belgium.

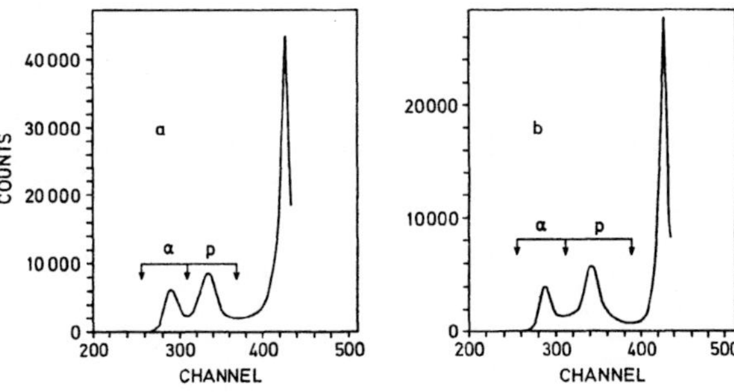

Fig. 1. PSD spectra for two pulse height (PH)intervals:
a) 25 < PH < 50 (arbitrary units)
b) 50 < PH < 75 (arbitrary units)

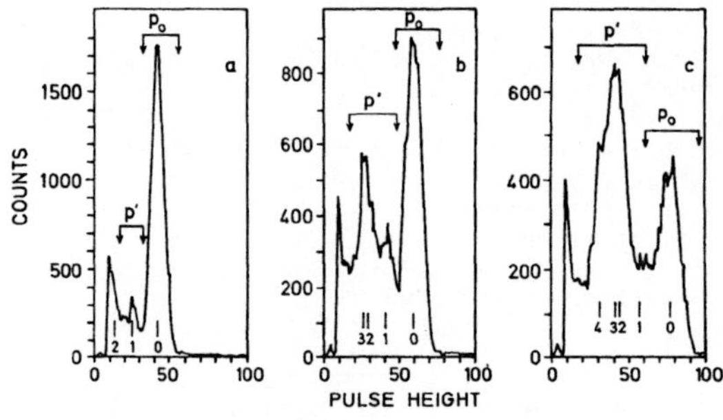

Fig. 2. Pulse height spectra for three neutron energy intervals :
a) 6 MeV $< E_n <$ 6.5 MeV
b) 7 MeV $< E_n <$ 7.5 MeV
c) 8 MeV $< E_n <$ 8.5 MeV

At low neutron energies (6 MeV $< E_n <$ 6.5 MeV)the "p'" group cointains mainly transitions to the first excited state in ^{23}Ne. Above 7 MeV, the "p'" group mainly comprises transitions to the second, third and fourth excited states; the positions of the first five ^{23}Ne levels are roughly indicated in fig. 2 by the vertical bars.

A reasonable separation of the "p_o" group from the "p'" group was possible up to E_n = 9 MeV. Since in the discussion below we will be interested in correla-

K. H. Böckhoff (ed.), Nuclear Data for Science and Technology, 814–817.

tions between the p_o and p' yields, the question of cross contamination is important. We estimate that in the 7- to 9-MeV range both the contamination of the "p'" group by p_o, and the contamination of the "p_o" group by p' do not exceed 5%. Transitions to the first excited level are not separated from the "p_o" group above 8 MeV, but their contribution is obviously very small.

Results

The relative cross sections for ^{23}Na(n,p) and ^{23}Na(n,α) as obtained in the present experiment are shown in fig. 3; the data have been normalized to the respective cross sections given in ENDF/B-IV in the 8- to 9-MeV range. Typical statistical error bars are indicated in the figure at regular energy intervals.

Fig. 4 shows the relative cross sections obtained for the (n,p_o) and (n,p') contributions up to 9 MeV neutron energy.

In the following discussion we are interested in the local structure of the cross sections rather than their gross energy dependence. Therefore, we divide the experimental cross sections by the smooth curves indicated in figs. 3b, 4a and 4b. The smooth cross sections have been obtained from a routine statistical model calculation. The transmission coefficients used in this calculation were first obtained from the optical model routine CERBERO of Fabbri et al.[4] with standard optical model parameter sets as given by Perey and Buck [5], Perey [6], and Makowska et al.[7]. These transmission coefficients were then adjusted, however, to better reproduce the gross shape of the experimental spectra. The ratios of the experimental cross sections and the smooth curves of figs. 3b and 4 will be referred to as fluctuation spectra f(n,x) in the following discussion. In order to reduce somewhat the structure due to counting statistics in the fluctuation spectra we have collapsed every two channels into one.

Discussion

All observed cross sections show strong fluctuations. Particularly noteworthy are the strong peaks in the (n,p_o) cross section at 5.9 MeV, 6.25 MeV, 6.4 MeV and 7.2 MeV neutron energy. Some of them, especially those at 6.4 MeV and 7.2 MeV neutron energy, do have the appearance of almost isolated narrow resonances with widths of the order of 50 keV, although already at E_n = 7 MeV neutron energy one expects the average width Γ to be about twice the average spacing for given spin, D_j, thus calling for strong interference affects : our statistical model treatment yields typical figures of Γ = 100 keV and D_j = 50 keV at E_n = 7 MeV. Unfortunately, the (n,p') and (n,α) reactions are still too weak below E_n = 6.5 MeV to be investigated for the presence of similar peaks at the same positions. We therefore compare the (n,p_o) fluctuation spectrum above E_n = 5.5 MeV to the total neutron cross section of ^{23}Na as measured by Larson et al.[8] in fig. 5. Obviously the structures in both spectra are not correlated, showing that the strong peaks in the (n,p_o) spectrum are related to the exit- rather than the entrance channel. A possible mechanism for enhanced

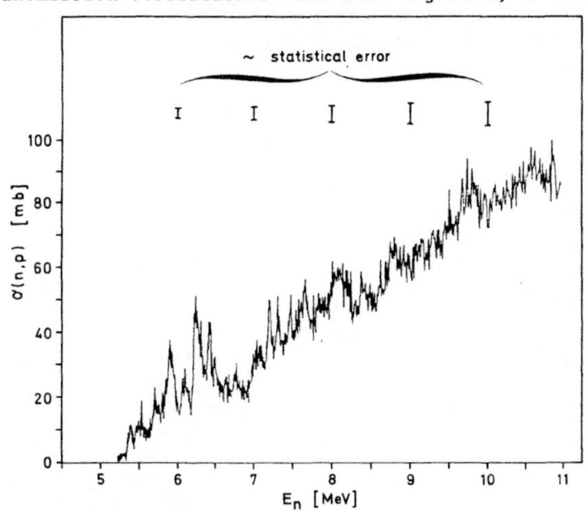

Fig. 3a. Measured relative cross section for ^{23}Na(n,p)^{23}Ne.

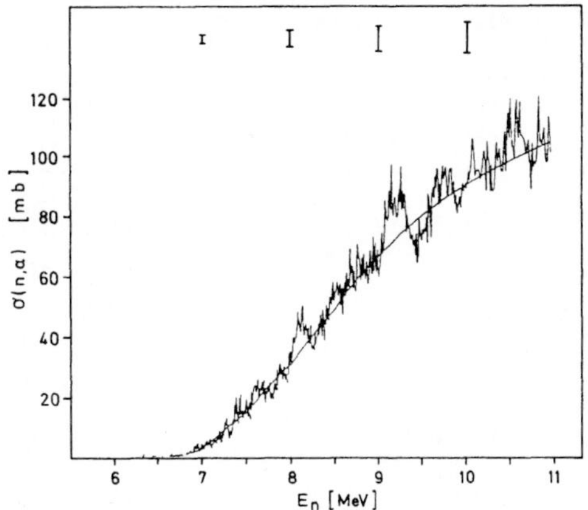

Fig. 3b. Measured relative cross section for ^{23}Na(n,α)^{20}F; smooth curve: statistical model.

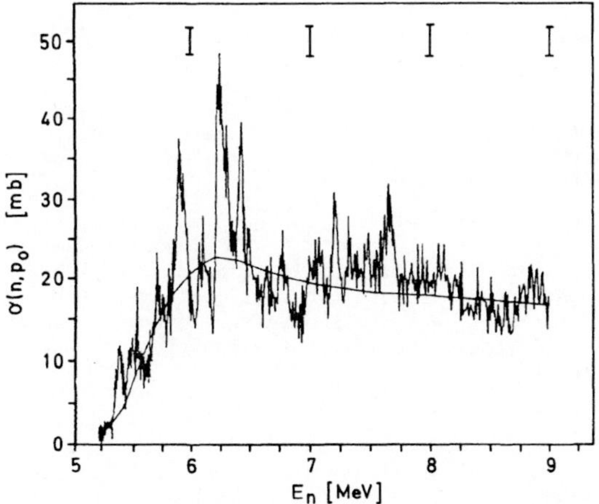

Fig. 4a. Measured relative cross section for ^{23}Na(n,p_o) leading to the ground state of ^{23}Ne; smooth curve : statistical model calculation.

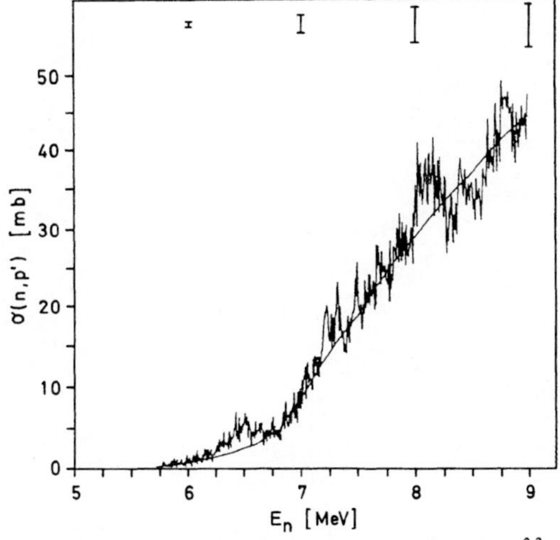

Fig. 4b. Measured relative cross section for ^{23}Na(n,p') feeding the first five excited states of ^{23}Ne; smooth curve : statistical model calculation.

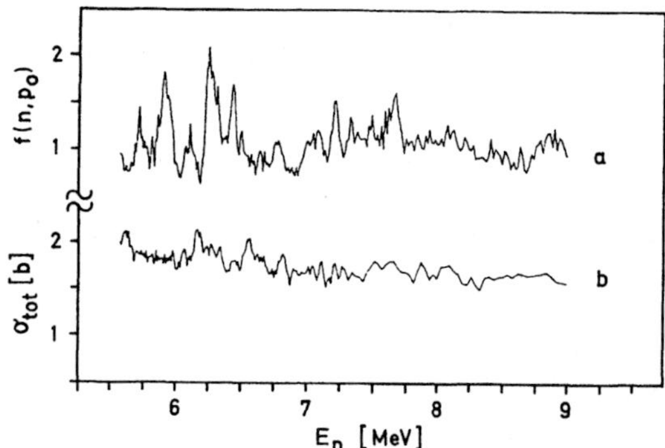

Fig. 5. Comparison of the (n,p_o) fluctuation spectrum (a) to the total neutron cross section [8] (b) of ^{23}Na.

structure in the proton channels is the presence of isobaric analogue states in ^{24}Na acting as doorway states for the proton decay of the compound nucleus. Unfortunately, nothing is known about parent states in ^{24}Ne at the relevant excitation energies.

In the neutron energy range from 7 MeV to 9 MeV the three partial spectra measured in the present experiment may be directly compared to each other. For this purpose, in fig. 6, we have plotted the fluctuation spectra for the (n,p_o), (n,p'), and (n,α) reactions on a common energy scale. As may be seen from the figure, many structures are correlated in the (n,p_o) and (n,p') fluctuation spectra, whereas most of the structures in the (n,α) fluctuation spectrum do not line up with structures in the other two spectra.

In order to arrive at a more quantitative description of the situation, we have calculated correlation coefficients between the three spectra of fig. 6. The correlation coefficient is defined here as

$$r(a,b)=\frac{\sum_{i=1}^{n}(f_i(n,a)-\overline{f(n,a)})(f_i(n,b)-\overline{f(n,b)})}{[\sum_{i=1}^{n}(f_i(n,a)-\overline{f(n,a)})^2\sum_{i=1}^{n}(f_i(n,b)-\overline{f(n,b)})^2]^{1/2}} \qquad (1)$$

with a standard error of

$$\Delta r = (1 - r^2)/\sqrt{n-1} \qquad (2)$$

The sums in equ.(1) extend over all data points in the neutron energy range from 7 MeV to 9 MeV. Their number, n=129, is sufficiently large for equ.(2) to be a good approximation to the uncertainty or r.

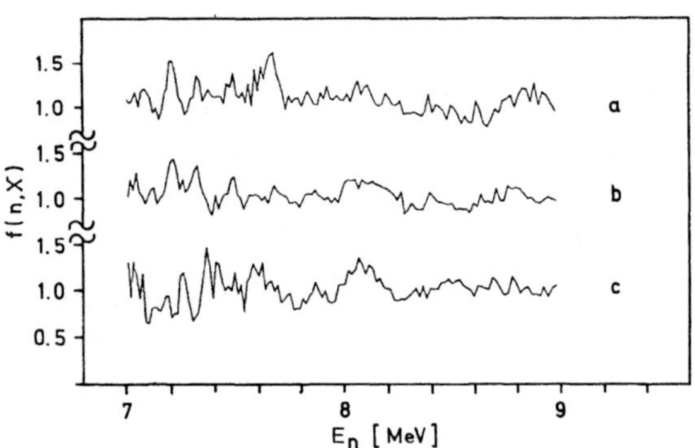

Fig. 6. Comparison of fluctuation spectra for : a) $^{23}Na(n,p_o)$ b) $^{23}Na(n,p')$ c) $^{23}Na(n,\alpha)$

The resulting correlation coefficients and uncertainties are :
$r(p_o,p')=0.49\pm0.07$ $r(p_o,\alpha)=0.05\pm0.09$ $r(p',\alpha)=-0.03\pm0.09$

An important question here is the one of spurious correlations due to experimental cross contaminations. Cross contamination can be expected between the (n,p_o) and (n,p') spectra due to insufficient pulse-height resolution, and between the (n,p') and (n,α) spectra (no correlation observed, however) due to insufficient PSD resolution. (n,p_o) and (n,α) events are separated by both PSD information and pulse height, thus very little cross contamination is expected between these two spectra.

The possible amount of correlation due to cross contamination is investigated in the following way: Starting from the two uncorrelated fluctuation spectra $f(n,p_o)$ and $f(n,\alpha)$ we construct two artificially contaminated spectra according to

$$f_1 = (1-c)\ f(n,p_o) + c\ f(n,\alpha)$$

$$f_2 = c\ f(n,p_o) + (1-c)\ f(n,\alpha)$$

Subsequently the effect of independent counting statistics is simulated by modifying the spectra f_1 and f_2 channel by channel according to their statistical uncertainty in a random fashion. Finally the correlation coefficient between f_1 and f_2 is calculated. The whole procedure is repeated many times and thereby the average correlation coefficient and its standard deviation are obtained. The result for c = 0.05 which represents the maximum possible experimental cross contamination between the (n,p_o) and (n,p') spectra (see above) is :
$r(1,2) = 0.12 \pm 0.06$ $(c = 0.05)$
For c = 0.025, a more realistic estimate of the average cross contamination between (n,p_o) and (n,p') we obtain :
$r(1,2) = 0.08 \pm 0.06$ $(c = 0.025)$
Thus the above correlation between the (n,p_o) and (n,p') fluctuation spectra of $r(p_o,p') = 0.49 \pm 0.07$ cannot be explained by any reasonable amount of cross contamination and is thus considered physically real.

In the neutron energy region between 7 MeV and 9 MeV we are dealing with overlapping levels, with Γ/D_j ranging from about 2 to 5, respectively, according to the above-mentioned statistical model calculation. Therefore, no correlation between the fluctuations in different partial reaction cross sections is expected, unless a special mechanism producing such correlations is operative. The observed correlation between the (n,p_o) and (n,p') fluctuation spectra with no correlation between any of these and the (n,α) spectrum, is indicative of doorway states common to the proton channels, but not effective in the (n,α) reaction. Isobaric analogue states in ^{24}Na with $T_> = T_z + 1$ are possible candidates.

In order to demonstrate that the order of magnitude of the observed correlation between the (n,p_o) and (n,p') fluctuation spectra can be explained by the effect of isobaric analogue states, we again have performed a numerical cross section simulation : We again start from the spectra $f(n,p_o)$ and $f(n,\alpha)$ as examples of uncorrelated fluctuation spectra. We then superimpose the effect of isobaric analogue states as local enhancements in the way described below. and calculate the resulting correlation coefficient.

Summing over all relevant compound nuclear spins, i.e. those spins for which neutron and proton transmission coefficients are sufficiently large to contribute significantly to the cross sections, we expect totally 10 $T_>$-states in the neutron energy interval between 7 MeV and 9 MeV. The mean spreading width of the $T_>$-states is

$$\Gamma^{\downarrow} = \frac{2\pi}{D_j} V_{12}^2 \qquad (3)$$

where $D_j \approx 35$ keV is the spacing of $T_<$-states of given spin, and the isospin mixing element is estimated as V_{12} = 25 keV [9,10]. Thus the $T_>$-states are treated as isolated resonances with a width of $\Gamma^{\downarrow} \approx 110$ keV.

The relative contributions of these resonances to the (n,p_o) and (n,p') fluctuation spectra vary according to the proton widths of the individual analogue states.

We thus have sampled resonance areas for the $T_>$-states randomly from the relevant distribution laws, i.e. χ^2-distributions with effective numbers of degrees of freedom between 1 and 3 (for (n,p') feeding effectively three final states).

The expectation values for the $T_>$-contributions to the (n,p_0) and (n,p') fluctuation spectra have been determined in the following way: According to Harney, Weidenmüller and Richter [11] the cross section from channel a to channel b, suppressing spin statistical and kinematical factors, is

$$\sigma_{a,b} = \frac{T_{a1}T_{b1}}{T_1}\left(1 - \frac{T_2}{T_1}\nu\right) + \frac{T_{a1}T_{b2}}{T_1}\nu + \frac{T_{a2}T_{b1}}{T_1}\nu + \frac{T_{a2}T_{b2}}{T_2}(1-\nu) \quad (4)$$

Here, the T_{a1},... are transmission coefficients, the indices 1 and 2 referring to $T_<$- and $T_>$-states, respectively, and

$$\nu = \left[\frac{\Gamma_2^{\uparrow}}{\Gamma_2^{\downarrow}} + \frac{T_2}{T_1} + 1\right]^{-1} \quad (5)$$

is a parameter describing the degree of isospin mixing ($\Gamma_2^{\uparrow}$ is the mean escape width of the analogue states). Equ.(4) was originally derived for the case of overlapping $T_>$-levels, but it has been shown by Lane [12] to be valid also for isolated $T_>$-states. For the present case of a neutron induced reaction $T_{a2} = 0$, thus

$$\sigma_{a,b} = \frac{T_{a1}T_{b1}}{T_1}\left(1 - \frac{T_2}{T_1}\nu\right) + \frac{T_{a1}T_{b2}}{T_1}\nu = \frac{T_{a1}T_{b1}}{T_1}\left(1 + \frac{T_{b2}}{T_{b1}}\nu\right) \quad (6)$$

where to obtain the second expression we have also neglected the term with (T_2/T_1), because only the $T_<$-states are allowed to decay by neutron emission, and we therefore expect $T_2 \ll T_1$. The ratio T_{b2}/T_{b1} is simply estimated from the ratio of the relevant isovector coupling coefficients as $T_{b2}/T_{b1} = 1/3$. Thus the final condition imposed on the expectation values for the $T_>$-resonance areas is that their relative total contribution to the (n,p_0) and (n,p') fluctuation spectra is $\nu/3$.

With $T_2 \ll T_1$, the above estimate for $\Gamma_2^{\downarrow}$, and a rough estimate of $\Gamma_2^{\uparrow} \approx 40$ keV, based on $T_{b2} = T_{b1}/3$ and the assumed spacing of $T_>$-states, we expect for the mixing parameter ν a value between 0.5 and 1.

The simulation procedure described above has been repeated a sufficient number of times such that a meaningful average correlation coefficient $r(\nu)$ and its standard deviation (scatter of individual correlation coefficients around $r(\nu)$) could be determined. The result

$r(\nu=1) = 0.31 \pm 0.21 \qquad r(\nu=0.5) = 0.25 \pm 0.17$

shows that the experimentally observed correlation between the (n,p_0) and (n,p') fluctuation spectra is indeed consistent with the expected effect of the isobaric analogue states in ^{24}Na.

Conclusion

All of the three measured partial reaction cross sections of ^{23}Na show considerable fluctuations. The observed correlation between the fluctuations in the (n,p_0) and (n,p') cross sections with absence of correlation between either of these and the (n,α) cross section, can be understood in terms of isobaric analogue states acting as doorway states for the proton channels.

References

1. C.F. Williamson, Phys. Rev. 122 (1961) 1877.
2. R. Bass, F. Saleh, and B. Staginnus, EANDC(E) 66U (1966) 64.
3. C.M. Bartle, Nucl. Instr. Meth. 124 (1975) 547.
4. F. Fabbri, G. Fratamico, and G. Reffo, CNEN-Report RT/FI (77) 6, (1977).
5. F.G. Perey and B. Buck, Nucl. Phys. 32 (1962) 462.
6. F.G. Perey, Phys. Rev. 131 (1963) 745.
7. M. Makowska et al., Inst. Nucl. Phys., Cracow, Rep. No. 735/PL (1970)
8. D.C. Larson, J.A. Harvey, and N.W. Hill, Oak Ridge Nat. Lab. report ORNL/TM-5614 (1976).
9. H. Weigmann, R.L. Macklin, and J.A. Harvey, Phys. Rev. C14 (1976) 1328.
10. C.R. Jungmann, E. Cornelis, L. Mewissen, F. Poortmans, J.P. Theobald, and H. Weigmann, these proceedings
11. H.L. Harney, H.A. Weidenmüller, and A. Richter, Phys. Rev. C16 (1977) 1774.
12. A.M. Lane, Phys. Rev. C18 (1978) 1525.

IBA DESCRIPTION OF COLLECTIVE STATES IN NEODYMIUM ISOTOPES

G. Maino, E. Menapace and A. Ventura

ENEA, via G. Mazzini 2,
I-40138 Bologna, Italy

The nuclear structure properties of Neodymium isotopes in the range A=136-154 are analyzed in the frame of the Interacting Boson Approximation (IBA) Model. Collective levels up to $E^* \leq 2.5$ MeV are reproduced with satisfactory accuracy, suggestions are made in a few cases of uncertain spin-parity attribution and the existence of undetected levels is predicted. E2 and E3 transitions are analyzed. All the symmetry limits of the Interacting Boson Model, corresponding to a) γ-unstable rotors, b) vibrators, c) axially symmetric rotors, together with intermediate shape transitions, are observed in the Nd isotope chain.

[Interacting Boson Model, $^{138-148-152}$Nd, collective levels, $E^* \leq 2.5$ MeV, E2 and E3 transitions]

Introduction

Nuclear data of Neodymium isotopes, in particular $^{143-145}$Nd, play an important role in the estimate of long-term reactivity effects due to fission products in fast reactors [1]. An accurate description of the low-lying levels of target and compound nuclei is very important for calculations of neutron capture and inelastic cross sections. This fact is confirmed by the discrepancies of recent evaluations (ENDF/B, JENDL, RCN, CNEN) of Nd capture cross sections in the fast neutron region: such discrepancies are related to the different level schemes adopted by evaluators. Moreover, heavier even-even nuclides, like $^{152-154}$Nd, are detected as short-lived products in the spontaneous fission of ^{252}Cf. [2]

The present analysis refers to the collective spectra of even-even isotopes and may be considered as the necessary introduction to the study of odd-mass isotopes, planned for the near future.

The Model

The interacting boson model [3], developed in recent years by Arima and Iachello, has proved a very flexible mathematical tool to describe the collective properties of medium-heavy nuclei: its simple Hamiltonian is built up with one- and two-body operators for s- and d-bosons, which simulate pairs of like nucleons coupled to $L^\pi=0^+$ and 2^+, respectively. These operators are expressed in terms of the generators of a Lie group, U(6): with a particular choice of parameters, the general U(6) Hamiltonian is reduced to simpler symmetries corresponding to the subgroups U(5), SU(3), O(6): the diagonalization of a reduced Hamiltonian with an underlying dynamical symmetry generates collective level spectra characteristic of a) vibrators (U(5)), b) axially symmetric rotors (SU(3)), c) γ-unstable rotors (O(6)). Neodymium spectra are an interesting application of the interacting boson model, since all three limit symmetries appear along the isotope chain.

The explicit form of the s-d boson Hamiltonian and the parameters used to fit experimental positive-parity spectra from ^{136}Nd to ^{154}Nd have been published in a preliminary work [4], focused mainly on the shape transitions between the limit symmetries.

In order to describe negative-parity states and parity changing (E3) transitions, we have adopted in the present work a more general Hamiltonian containing also f-bosons, which simulate like nucleons coupled to $L^\pi=3^-$. The form adopted for the complete s-d-f Hamiltonian is the following:

$$\hat{H} = \hat{H}_{sd} + \hat{H}_{df} \qquad (1)$$

Here, $\hat{H}_{sd}$ is the s-d part of the Hamiltonian:

$$\hat{H}_{sd} = \varepsilon \hat{n}_d + a_o \hat{P}^+ \cdot \hat{P} + a_1 \hat{L}\cdot\hat{L} + a_2 \hat{Q}\cdot\hat{Q} + a_3 \hat{T}_3\cdot\hat{T}_3 + a_4 \hat{T}_4\cdot\hat{T}_4 \qquad (2)$$

The meaning of the terms at r.h.s. of formula (2) has already been illustrated in ref.[4]. H_{df} is the d-f part of the Hamiltonian, responsible for negative-parity states:

$$\begin{aligned}\hat{H}_{df} = {} & \varepsilon_f f^+ \cdot \tilde{f} - 2\sqrt{70}\, f_1 \left[(d^+\tilde{d})^{(1)} \cdot (f^+\tilde{f})^{(1)}\right] \\ & -2\sqrt{7}\, f_2 \left[(s^+\tilde{d}+d^+\tilde{s})^{(2)} - (\sqrt{7}/2)(d^+\tilde{d})^{(2)}\right] \cdot \left[(f^+\tilde{f})\right]^{(2)} \\ & + f_3 (d^+\tilde{f})^{(3)} \cdot (f^+\tilde{d})^{(3)}\end{aligned} \qquad (3)$$

Here, s^+, d^+_μ ($\mu=-2,\dots,+2$), f^+_μ ($\mu=-3\dots,+3$) create s-,d-, and f-bosons, respectively, and $\tilde{f}_\mu=(-1)^\mu f_{-\mu}$ are phase-modified annihilators. The angular momentum couplings can be understood from the following example:

$$(f^+\tilde{f})^{(2)}_\nu = \sum_{-3^{\mu,\mu'}}^{+3} (3,\mu;3,\mu'|2,\nu) f^+_\mu \tilde{f}_{\mu'} \qquad (4)$$

The numerical diagonalization of $\hat{H}$ has been perfomed by means of the PHINT code [5].

Collective spectra: three examples

Since it is not possible to give here a detailed description of the level schemes for all the isotopes of the chain, we have chosen examples of three different symmetries: ^{138}Nd(O(6) symmetry), ^{148}Nd(U(5)-SU(3) transition) and ^{152}Nd(SU(3) symmetry). The parameters used in the IBA Hamiltonian (1) are listed in Table I.

Table I IBA Parameters for $^{138-148-152}$Nd

	^{138}Nd	^{148}Nd	^{152}Nd
N(bos.numb)	4	5	10
f_3(KeV)	0.	0.	0.
ε_d (KeV)	0.	10.	0.
a_o (KeV)	420.	10.4	29.0
a_1 (KeV)	35.	27.1	6.0
a_2 (KeV)	0.	-33.	-16.7
a_3 (KeV)	155.	0.	0.
a_4 (KeV)	-75.	0.	0.
ε_f (KeV)	2330.	1670.	0.
f_1 (KeV)	17.	9.5	0.
f_2 (KeV)	100.	107.	0.

K. H. Böckhoff (ed.), Nuclear Data for Science and Technology, 818–821.

a) ^{138}Nd: the collective spectrum is generated by the interaction of 2 hole-like protonic bosons, counted from the subshell closure at Z=64, and 2 hole-like neutronic bosons counted from the shell closure at N=82. The influence of the Z=64 subshell on the structure of lighter Nd isotopes has already been

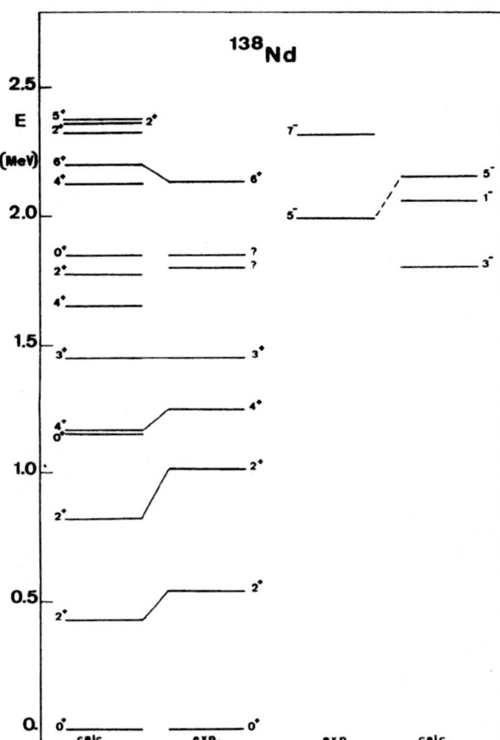

Fig. 1. ^{138}Nd, experimental and calculated levels.

discussed in ref. [4]. Fig. 1 shows a comparison between experimental [6] and calculated levels up to E*~2.5 MeV. The spectrum is consistent with a weakly broken O(6) symmetry. The quasi-ground state band is satisfactorily reproduced up to the 6^+ state, around 2.1 MeV. The quasi-gamma band is also seen, starting around 1 MeV, with 2_2^+ and 3_1^+ numbers, experimentally identified, and a 4_1^+ member, whose experimental counterpart could be the level at 1843 KeV: in fact, it decays to 2_2^+ (1014 KeV) through an E2-M1 transition whose intensity is consistent with the calculated B(E2) coefficient. However, another candidate to 4^+ attribution might be the level at 1801 KeV. Both these levels are close in energy to a calculated (4_2^+, 2_3^+, 0_2^+) triplet, and only poor experimental information prevents us from formulating a precise spin-parity attribution. Negative-parity levels are poorly known, with the exception of the 5_1^- and 7_1^- states, at 1990 and 2320 KeV, respectively, interpreted in ref. [7] as having two-neutron-hole configurations. The calculated odd-parity states have been obtained through a simple extrapolation of the f-boson parameters, adjusted in the case of ^{140}Nd, where the experimental situation is better defined.

b) ^{148}Nd: the experimental [8] levels of both parities are fairly well reproduced up to about E*=1.7 MeV by means of IBA parameters which define a strongly broken U(5) symmetry, already close to the SU(3) limit: the interacting bosons are in this case 2 hole-like protonic bosons, counted from Z=64, and 3 particle-like neutronic bosons, counted from N=82. The transitional character of the spectrum

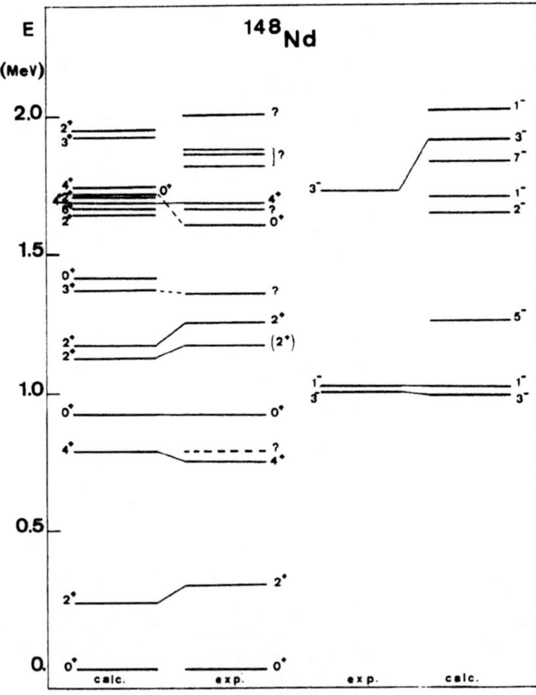

Fig. 2. ^{148}Nd, experimental and calculated levels.

in Fig. 2 appears, for instance, in the energy ratio of the 4_1^+ to the 2_1^+ state, of the order 2.5, still lower than the rotational limit, 10/3. Very close to the 4_1^+ state, a level at 791 KeV has been tentatively proposed [9], without spin-parity assignment, on the basis of energy-sum relationships and γ-γ coincidence results in an experiment on ^{148}Pr decay. This level has not been observed in other experiments and is not reproduced by our fit, whose parameters are intermediate between those of the adjacent even-even isotopes. Another level, at 1356 KeV, detected in Coulomb excitation experiments and decaying to the 4_1^+ state [8], could be identified with the theoretical 3_1^+ level. The experimental information about odd parity states is even poorer. Also in this case, transitional features appear: the 1_1^- level is slightly higher than 3_1^-, while it is expected to become lower than the latter in the SU(3) limit. A number of theoretical odd-parity levels below 2 MeV are not observed in experiments.

c) ^{152}Nd: this is an example of well-deformed nucleus, with sharp rotational spectrum. Following the suggestion of other Authors [10], we assume that the influence of the Z=64 closure disappears for N>90; therefore, we have in this case 5 particle-like protonic bosons, counted from the Z=50 closure, and 5 particle-like neutronic bosons counted from the N=82 closure. As Fig. 3 shows, the positive-parity levels [11] are well reproduced up to E*~1.3 MeV. Other levels up to 2.5 MeV, excited in the ^{150}Nd(t,p)^{152}Nd reaction [12], appear to be at the level of calculated 0^+ states: in any case, the experimental information is too poor to allow certain identification. Our choice of IBA parameters is consistent with an almost pure SU(3) symmetry. No experimental data have been found for negative-parity states.

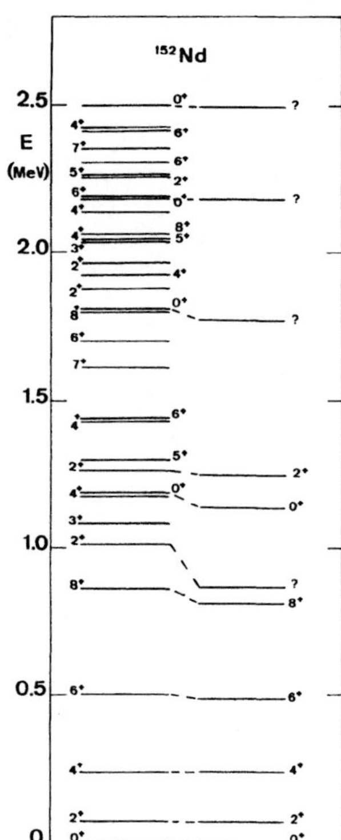

Fig. 3. ^{152}Nd, experimental and calculated levels.

the E2 operator of formula (5):

$$e_2=0.140 \text{ e.b} \; ; \; \chi_2= -0.894$$

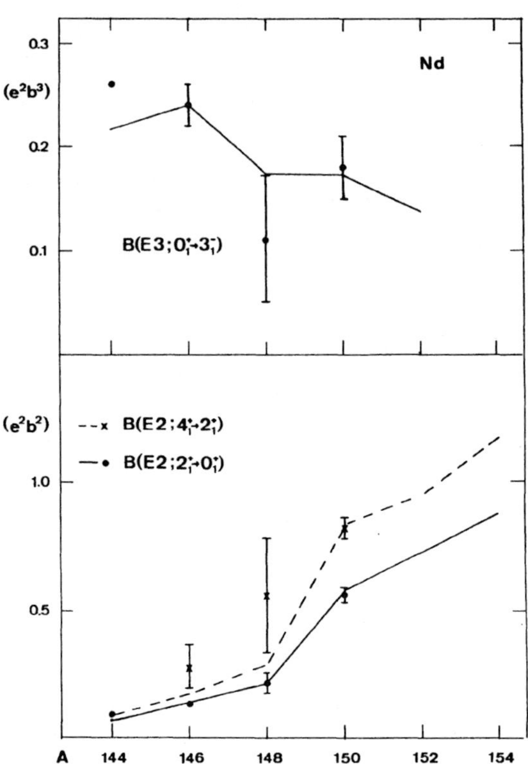

Fig. 4. a) Calculated and experimental E2 strengths; crosses and dots: experimental data from refs. [8-14-15-18]. b) Calculated and experimental E3 strengths; dots: experimental data from refs. [8-16-17-18].

Electromagnetic Transitions

The interacting boson model permits evaluation of the electromagnetic transitions between collective levels, by means of electric and magnetic multipole operators built up with creators and annihilators of s-,d- and f- bosons. Each of these operators is characterized by one or more effective charges, to be adjusted on experimental data.

In this section we limit ourselves to an analysis of E2 and E3 transitions; the relevant multipole operators are the following:

$$\hat{T}(E2)=e_2\left\{(s^+\tilde{d}+d^+\tilde{s})^{(2)}+\chi_2(d^+\tilde{d})^{(2)}\right\} \tag{5}$$

$$\hat{T}(E3)=e_{3f}\left\{s^+\tilde{f}+f^+\tilde{s}\right\}^{(3)}+e_3\left\{\left[(s^+\tilde{d}+d^+\tilde{s})^{(2)}+\chi_3(d^+\tilde{d})^{(2)}\right]\cdot\left[s^+\tilde{f}+f^+\tilde{s}\right]^{(3)}\right\}^{(3)} \tag{6}$$

The reduced strengths, $B(E\lambda)$, to be compared with experimental data, are given by the formula:

$$B(E\lambda;J_i\to J_f)=\frac{1}{2J_i+1}|\langle J_f\|\hat{T}(E\lambda)\|J_i\rangle|^2 \tag{7}$$

Here, the brackets denote a reduced matrix element between an initial state with spin J_i and a final state with spin J_f.

The parameters e_2, χ_2 for E2 transitions, and e_{3f}, e_3, χ_3, for E3 transitions have to be adjusted on experimental data.

As for B(E2), the experimental information is relatively accurate, at least for $^{146-148-150}$Nd and may be reasonably reproduced by means of the following parameters in the E2 operator of formula (5) (see above).

Fig. 4a shows a comparison between experimental and calculated values for $B(E2;\ 2_1^+\to0_1^+)$; as expected, the B(E2) values rapidly increase towards the rotational region, A>150.

Coming now to E3 transitions, the only experimental data which have been found refer to $(0_1^+\to3_1^-)$: they have been reproduced by means of the E3 operator of formula (6), with the following parameters:

$$e_{3f}=0.158 \text{ e.b}^{3/2};\ e_3=0.012 \text{ eb}^{3/2};\ \chi_3=-0.894$$

A comparison between experimental and calculated $B(E3;0_1^+\to3_1^-)$ is shown in Fig. 4b. As expected, the B(E3)'s decrease with increasing mass number towards the rotational region.

All the calculations of E2 and E3 transitions have been performed by means of the FBEM code [5].

Conclusions

The phenomenological analysis of collective levels and electromagnetic transitions in even Nd isotopes, only sketched out in the present work, will be described in detail in a more extensive paper. Our results indicate that the Interacting Boson Model [3] can reproduce a considerable quantity of experimental data and gives useful indications where data are lacking. There is, however, an important exception in the isotope chain, the semimagic nucleus ^{142}Nd, where low-lying levels have strong single-particle components which cannot be described by the present version of the IBA model. Microscopic analysis of the spectroscopic properties of semimagic nuclei might be carried out with the help of other models, like the Broken Pair Approximation [13], which also makes it possible to derive, microscopically, the IBA parameters near a shell closure. Such a microscopic study as well as extension of the present analysis to odd-mass isotopes might be interesting subjects for future work.

References

1. E. Fort, J. Krebs, P. Ribon, Tran Quoc Thuong, E. Menapace, M. Motta and G. Reffo, Proc. of the 4th Conference on Neutron Physics, Kiev, v.4,p.3 (1977).
2. E. Cheifetz, R.C. Jared, S.G. Thompson and J.B. Wilhelmy, Proc. of an Int. Conf. on the Properties of Nuclei far from the region of β-stability, Leysin, CERN 70-30, v.2, p.883 (1970).
3. For a recent review, see: A. Arima and F. Iachello, Ann. Rev. of Nucl. and Part. Sci. 31, 75 (1981).
4. G. Maino and A. Ventura, Lett. Nuovo Cimento 34, 79 (1982).
5. O. Scholten, report KVI-63, Groningen (1979).
6. L.K. Peker, Nucl. Data Sheets 26, 473 (1979).
7. M. Müller-Veggian, H. Beuscher, D.R. Haenni, R.M. Lieder, A. Neskakis and C. Mayer-Böricke, Nucl. Phys. A344, 89 (1980).
8. B. Harmatz and J.R. Shepard, Nucl. Data Sheets, 20, 373 (1977).
9. H. Yamamoto, C. Ichihara, K. Kawade and T. Katoh, J. of Phys. Soc. of Japan 41, 729 (1976).
10. R.F. Casten, D.D. Warner, D.S. Brenner and R.L. Gill, Phys. Rev. Lett. 47, 1433 (1981).
11. C.M. Baglin, Nucl. Data Sheets 30, 1 (1980).
12. R. Chapman, W. McLatchie and J.E. Kitching, Nucl. Phys. A186, 603 (1972).
13. See, for instance: J.N.L. Akkermans, Thesis, Free University of Amsterdam (1981).
14. P.A. Crowley, J.R. Kerns and J.X. Saladin, Phys. Rev. C3, 2049 (1971).
15. S.W. Yates, N.R. Johnson, L.L. Riedinger, A.C. Kahler, Phys. Rev. C17, 634 (1978).
16. J.K. Tuli, Nucl. Data Sheets 27, 97 (1979).
17. T.W. Burrows, Nucl. Data Sheets 14, 413 (1975).
18. A. Ahmed, R.M. Ronningen, A.V. Ramayya, R.B. Piercey, H. Kawakami, C.F. Maguire, J.H. Hamilton and P.H. Stelson, Bull. Am. Phys. Soc. 22, 614 (1977).

NEUTRON PICK-UP $(n,{}^{2}n^{\star})$ REACTION

Mile Zadro and Đ. Miljanić

"Ruđer Bošković Institute"
41001 Zagreb, P.O.B.1016
YUGOSLAVIA

An effort is made to estimate direct ${}^{2}n^{\star}$ contributions to the (n,2n) reaction cross sections (${}^{2}n^{\star}$ being singlet neutron-neutron final state interaction). Standard zero-range DWBA together with experimentally obtained spectroscopic factors and l assignments (e.g.from (p,d) reactions) are used in these neutron pick-up $(n,{}^{2}n^{\star})$ reaction cross section calculations. These calculations show that the ${}^{2}n^{\star}$ contributions could be large enough to be discriminated experimentally against other 2n contributions.

Introduction

In recent years many different nuclear reactions with unbound ejectiles were discovered and used for nuclear structure studies. Among them the most prominent ones are the (a,đ) and $(b,{}^{2}He)$ reactions (see e.g. Ref.[1-5]. đ and ${}^{2}He$ stand for the singlet neutron-proton and proton-proton final state interaction (FSI), respectively, and a and b are different projectiles. As far as we know, there are no experimental or theoretical studies on analogous $(c,{}^{2}n^{\star})$ recations - ${}^{2}n^{\star}$ being singlet neutron-neutron final state interaction. However, it would be important to establish, whether the ${}^{2}n^{\star}$ contributions to the (c,xn) reaction cross sections are significant. As it is well known there are the discrepancies between the statistical theory predictions and experimental (n,2n) reaction cross sections, which are attributed to the presence of preequilibrium processes. In this short note an estimate is made, whether the neutron pick-up $(n,{}^{2}n^{\star})$ reaction could also be a cause of these discrepancies.

Calculations

van Driel et al.[3] have found that the neutron stripping $({}^{3}He,{}^{2}He)$ reaction can be well described by the standard zero-range DWBA, assuming that in outgoing channel one can use an optical potential acting on the centrum-of-mass of two protons without affecting their relative motion. Using a similar procedure one can calculate the contributions of the neutron pick-up $(n,{}^{2}n^{\star})$ reaction. Ingredients into these calculations are :

a) DWBA cross section, calculated with the code DWUCK[6];
b) Normalization factor which depends on the neutron-neutron relative energy and is connected with neutron-neutron effective range parameters, a_{nn} and r_{ef};
c) Spectroscopic factors,S, and l assignments obtained from the analysis of other neutron pick-up reactions (e.g. (p,d) reactions).

These calculations are performed for the ${}^{106}Pd(n,{}^{2}n^{\star}){}^{105}Pd$ reaction. Optical potential parameters for neutrons are taken from Ref.[7], while for the ${}^{2}n^{\star}$ the deuteron parameters from Ref.[8] are used. Spectroscopic factors and l assignments for the transitions below 1 MeV in excitation of ${}^{105}Pd$ are taken from the study of the ${}^{106}Pd(p,d){}^{105}Pd$ reaction at E_p= 22.9 MeV.[9]

Results and Discussion

Fig. 1 shows calculated angular distributions for the transition to the ground state of ${}^{105}Pd$ at E_{o}= 22.9 MeV. Full line represents the angular distributions of neutrons from the $(n,{}^{2}n^{\star})$ reaction. As expected angular distribution is forward peaked. Differential cross sections for neutrons from ${}^{2}n^{\star}$ are an order of magnitude lower than (p,d) cross sections.

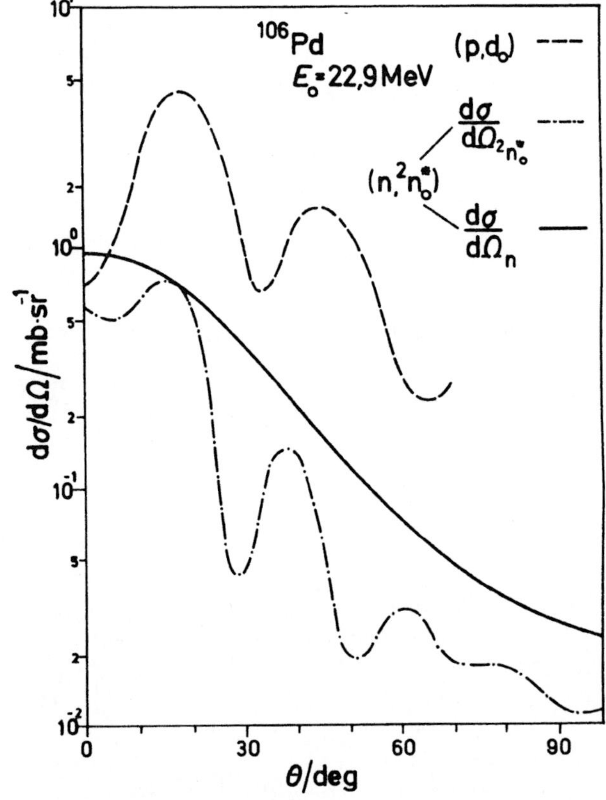

Fig. 1. Calculated angular distributions of deuterons, ${}^{2}n^{\star}$ and neutrons from (p,d_{o}) and $(n,{}^{2}n^{\star}_{o})$ reactions, respectively, on ${}^{106}Pd$ at E_{o}= 22.9 MeV.

Fig. 2 shows the contributions of the neutron pick-up $(n,{}^{2}n^{\star})$ reaction to the energy spectrum of neutrons from (n,2n) reaction at $\theta = 40°$. ${}^{105}Pd$ states below 1 MeV in excitation are taken into account only. Two curves correspond to different maximum relative energies of two neutrons.

It could be seen that these contributions are affecting mainly central part of the (n,2n) reaction energy spectra (i.e. lower part of the single neutron spectra). From the comparison of different experimental data on (n,2n) reactions with these calculations one can conclude that these $(n,{}^{2}n^{\star})$ reaction contributions to the neutron spectra could amount up to few percent for a given energy region of the spectrum and for the forward emission angles. Although

K. H. Böckhoff (ed.), Nuclear Data for Science and Technology, 822–823.

these contributions are not large they could be experimentally observed anyway due to the energy and space correlation of two neutrons from $^2n^{\star}$.

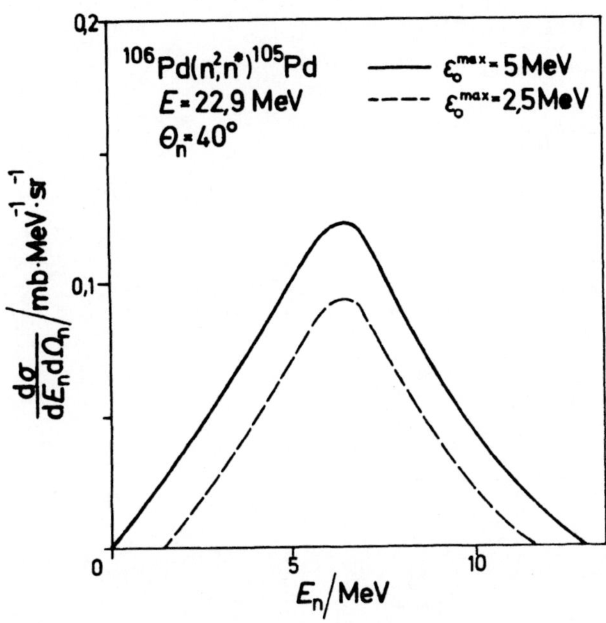

Fig. 2. Contributions of the $(n,^2n^{\star})$ reaction to the energy spectrum of neutrons at $\theta=40°$.

References

1. B.L. Cohen, E.C. May, T.M. O'Keefe and C.L. Fink, Phys. Rev. 179, 962 (1969).
2. R. Jahn, D.P. Stahel, G.J. Wozniak, R.J. de Meijer and Joseph Cerny, Phys. Rev. C18, 9 (1978).
3. J. van Driel, R. Kamermans, R.J. de Meijer and A.E.L. Dieperink, Nucl. Phys. A342, 1 (1980).
4. J. van Driel, R. Kamermans and R.J. de Meijer, Nucl. Phys. A350, 109 (1980).
5. D.P. Stahel, R. Jahn, G.J. Wozniak and Joseph Cerny, Phys. Rev. C20, 1680 (1979).
6. P.D. Kunz, University of Colorado, unpublished.
7. F.D. Becchetti, Jr. and G.W. Greenless, Phys. Rev. 182, 1190 (1969).
8. G.E. Brient, P.J. Brient, P.S. Riley, H. Seitz and S. Sen, Phys. Rev. C6, 1837 (1972).
9. R.E. Anderson, R.L. Bunting, J.D. Burch, S.R. Chinn J.J. Kraushaar, R.J. Peterson, D.E. Prull, B.W. Ridley, R.A. Ristinen, Nucl. Phys. A242, 75 (1975).

A COMPILATION OF EVALUATED NUCLEAR ACTIVATION AND DECAY DATA FOR USE IN (n,γ) REACTOR NEUTRON ACTIVATION ANALYSIS

F. De Corte, L. Moens, A. De Wispelaere, J.Hoste

Institute for Nuclear Sciences, State University
Proeftuinstraat 86, B-9000 Gent (Belgium)

A. Simonits

Central Research Institute for Physics
H-1525 Budapest, P.O.Box 49 (Hungary)

A survey is given of the Gent-Budapest cooperative work on the evaluation of nuclear activation and decay data for use in absolute (n,γ) reactor neutron activation analysis. Special attention is paid to the following parameters : isotopic abundance, 2200 $m.s^{-1}$ cross-section, absolute gamma-intensity , resonance integral to thermal cross-section ratio, effective resonance energy, half-life, and k_0-factor - a composite nuclear constant nowadays available with high accuracy. The present state of the art, including data for 80 isotopes, is summarized in a comprehensive report, which is available on request.

[evaluation ; nuclear data library ; reactor neutron activation analysis]

Introduction

(n,γ) Reactor neutron activation analysis (RNAA) is an important tool for the quantitative determination of major, minor and trace elements or impurities in a large variety of materials. For instrumental, multi-element routine analysis it is most convenient to use "absolute" RNAA (preferably combined with high-resolution Ge(Li) or HPGe gamma-spectrometry), where calculation of the elemental concentrations is based on the fundamental equations for (n,γ) activation and decay :

$$\rho(\text{ppm}) = \frac{\dfrac{A_p}{SDCW}}{\left(\dfrac{A_p}{SDCw}\right)^{*}} \; \frac{M\theta^{*}\sigma_0^{*}\gamma^{*}}{M^{*}\theta\sigma_0\,\gamma} \; \frac{f + Q_0^{*}(\alpha)}{f + Q_0\,(\alpha)} \; \frac{\varepsilon_p^{*}}{\varepsilon_p} \qquad (1)$$

with

$$Q_0(\alpha) = [(Q_0 - 0.429)\overline{E}_r^{-\alpha} + 0.429/(2\alpha+1)(0.55)^{\alpha}].(1\text{eV})^{\alpha} \qquad (2)$$

In eqs.(1) and (2) :

$*$ = flux monitor, coirradiated with the sample ; w = weight in microgram ;

A_p = average count rate ; = N_p/t_m, with N_p = number of counts (corrected for pulse losses)collected under the full-energy peak during measuring time t_m ;

W = sample weight in gram ;

f = conventional thermal (subcadmium) to epithermal flux ratio ;

ε_p = full-energy peak detection efficiency ;

α = factor denoting the non-ideality of the $1/E^{1+\alpha}$ epithermal neutron flux distribution ;

S = $1-\exp(-\lambda t_{irr})$; $\lambda = (\ln 2)/T^1/_2$; t_{irr} = irradiation time ;

D = $\exp(-\lambda t_d)$; t_d = decay time ;

C = $(1-\exp(-\lambda t_m)) / \lambda t_m$.

In the above, the following nuclear data have to be introuced :

M = atomic weight ;

θ = isotopic abundance ;

σ_0 = 2200 $m.s^{-1}$(n,γ) cross-section ;

γ = absolute gamma-intensity ;

$Q_0 = I_0/\sigma_0$, with I_0 = (n,γ) resonance integral =

$$\int_{0.55}^{\infty} \frac{\sigma_{n,\gamma}(E)dE}{E} \quad ;$$

$\overline{E}_r$ = effective resonance energy[1,2],defined as :

$$\ln \overline{E}_r = \frac{\sum_i \dfrac{g_i\Gamma_{n,i}\Gamma_{\gamma,i}}{\Gamma_i} \dfrac{\ln E_{r,i}}{E_{r,i}^2}}{\sum_i \dfrac{g_i\Gamma_{n,i}\Gamma_{\gamma,i}}{\Gamma_i} \dfrac{1}{E_{r,i}^2}} \qquad (3)$$

with g = statistical weight factor ; E_r = resonance energy ; Γ_n = neutron width ; Γ_γ = radiative width ; Γ = total width.

$T^1/_2$ = half- life.

Although for a well-chosen flux monitor(e.g.^{197}Au (n,γ)^{198}Au ; E_γ = 411.8 keV) these nuclear data are known with sufficient accuracy, this is not always the case for most of the analytically interesting isotopes. It has been shown[3-5] that inaccuracies of randomly selected literature nuclear data can cause intolerable systematic errors(sometimes up to several tens of percents) on the analytical results of absolute RNAA. This is a serious drawback, e.g. compared to "classical" relative RNAA (accuracy ∿ 1-2%), where a standard is coirradiated for each element to be determined.

Since 1975, a cooperation exists between the Institute for Nuclear Sciences(INW,Gent/Belgium)and the Central Research Institute for Physics(KFKI,Budapest/Hungary) with the aim to optimize absolute RNAA and to make its accuracy comparable to relative standardization. One of the goals is the accurate experimental determination, critical selection or calculation of the nuclear activation and decay data for use in eqs.(1) and (2). This paper gives a condensed survey of the present state of the art.

Requested accuracy and status of nuclear data for use in RNAA

M - It needs no further comment that a requested accuracy on M-data of e.g. better than 0.1% is met for all elements which are analytically important in RNAA.

θ - Since any uncertainty on θ is directly transferred to the analytical results (see eq.(1)), it seems reasonable to request an accuracy of better than 1% on this parameter. Although this source of errors in absolute RNAA is usually disregarded, the above requirement is far from being generally

K. H. Böckhoff (ed.), Nuclear Data for Science and Technology, 824–827.

fulfilled, not only for minor[6] but also for major isotopes. For instance, when scanning the latest SAIC-evaluation (1981)[7], it appears that for the analytically interesting target isotopes ^{68}Zn, ^{116}Sn, ^{121}Sb, ^{123}Sb, ^{174}Yb, ^{179}Hf and ^{202}Hg an uncertainty of more than 1% is specified on the isotopic abundance values.

It might be instructive to have a closer look at the remarkable history of the ^{58}Fe θ-value[22]. The few selected examples in Fig.1 show that former compilations[8-12] report values ranging from 0.29-0.33%, mainly based on the results of Valley et al. (1941;1947)[13,14] and Hibbs (1949)[15]. Although an experimental θ as low as 0.28% had been reported earlier (Nier (1939)[16] ; Chenouard (1966)[17]), it is obviously only under the influence of the most recent findings of Schmidt et al.(1979)[18] and James et al.(1980)[19], confirming the 0.28-value, that the latest SAIC (1981)[7] evaluation mentions $\theta=(0.28 \pm 0.01)\%$, also quoted (without error assignment) by the 1981 BNL edition[20]. Note that this SAIC-recommendation still reveals an uncertainty of 3.6% (relative), which exceeds the 1% accuracy requested in RNAA. In view of the above, it is regrettable that the 1982 NNDC Computope Chart[21] persists in reporting a value of 0.3%, which is thus probably in error with 7%.

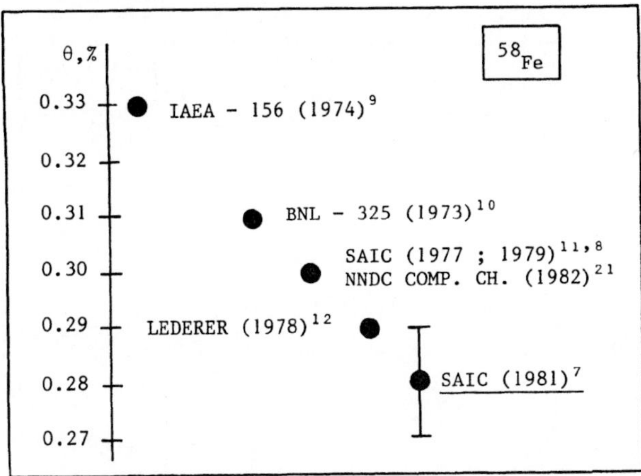

Fig. 1. Compiled and evaluated data for $\theta(^{58}Fe)$

It is hoped that spectroscopists will find interest in the accurate remeasurement of hitherto poorly established isotopic abundance values.

σ_0- For the same reason as mentioned above, an accuracy of better than 1% is requested on this parameter. Intercomparative scanning of some frequently cited compilations[9,10,20,21,23,24] reveals that the above requirement is far from being met for most cases. From the quoted values in the recent BNL(1981) edition[20] it turns out that the average uncertainty on σ_0 amounts to ∿6%. Besides, occasionally large systematic errors seem to exist, and conformity between different compilations is by no means an indication of accuracy. This is illustrated in Table I, where for some striking examples frequently listed σ_0-values are compared to "recommended" ones. The latter values are obtained as an evaluation result from comparison of theoretically calculated and experimentally determined k_0-factors (see later) or are determined from the well known "activation method"; both methods yield $(\theta.\sigma_0.\gamma)$ as a primary result and can give reliable σ_0-results only when θ and γ are accurately known.

It might be interesting to consider again the case of $^{58}Fe(n,\gamma)^{59}Fe$. The above mentioned re-evaluation of the ^{58}Fe isotopic abundance value evidently necessitates a corresponding adjustment of the cross-section. This has been done in a former paper[22] by normalizing - whenever possible - all previously reported experimental data (obtained by the activation method) to $\theta(^{58}Fe)=0.28\%$; this yielded as an average $\sigma_0=(1.34 \pm 0.03)$b. From the activation method applied at the INW/KFKI, a value of (1.31 ± 0.02)b was found[22]. This is in good agreement with the new BNL (1981)[20] evaluation of $\sigma_0=(1.28 \pm 0.05)$b, which is thus consistent with the quoted value of θ=0.28%. On the other hand, it is confusing that the NNDC Computope Chart (1982)[21] evaluates a combination of σ_0=1.28b and θ=0.3%, which is certainly incorrect.

Table I. Comparison of σ_0-values [(a) INW/KFKI : from activation method ; (b) INW/KFKI : evaluated]

Reaction	σ_0, barn	
	Frequently listed	Recommended
$^{58}Fe(n,\gamma)^{59}Fe$	1.14-1.16	1.31±0.02[22] (a)
$^{112}Sn(n,\gamma)^{113m}Sn$	0.3-0.35	0.155[25] (b)
$^{174}Yb(n,\gamma)^{175}Yb$	65	141[26] (b)
$^{179}Hf(n,\gamma)^{180m}Hf$	0.34	0.433[27] (b)

γ - On this parameter an accuracy of better than 1% should be requested as well. Careful scanning of the available large scale compilation and evaluation works[12,28,29,30], including the recent editions of the Nuclear Data Sheets, shows that this requirement is only fulfilled for the most intense gamma-lines and for efficiency calibration sources. Typical examples for the latter can for instance be found in NBS-626 (1981)[31]. In general, it can be stated that the average uncertainty on γ-values for analytically interesting lines amounts to ∿4%. This is not satisfactory, but it should not be denied that during the past five years the quality of reported data improved considerably[3,5].

Nevertheless, the scattering of the literature data e.g. for the 165.9 keV line of ^{139}Ba(γ=18.8-22.05%), as shown in Fig.2, reveals that occasionally large systematic errors can be introduced in RNAA by randomly selecting a γ-value. A critical INW/KFKI evaluation, based on comparison of calculated and determined k_0-factors[5] (see later), leads to the selection of $\gamma=(23.76 \pm 0.25)\%$ (Gehrke, 1980)[33].

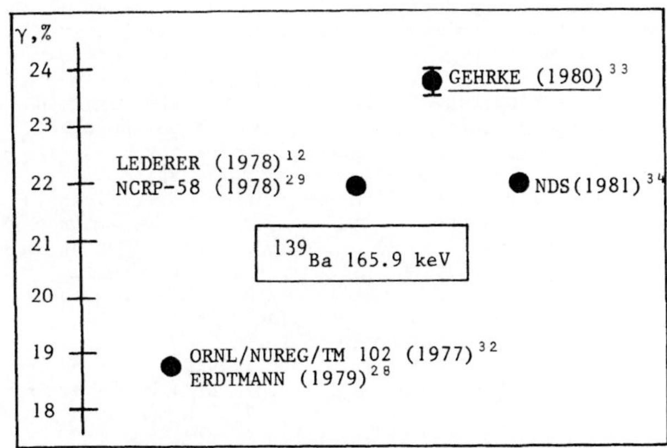

Fig. 2. Compiled and evaluated data for γ(165.9 keV ^{139}Ba)

Q_0- Due to the $(f+Q_0(\alpha))$-summing terms in eq.(1), the error propagation factor on $Q_0(=I_0/\sigma_0)$ is smaller than unity, and - of course - depending on the values of Q_0 and f. For medium Q_0 and f, an accuracy of better than 4% on Q_0 is requested to obtain an uncertainty contribution of less than 1% on the analysis result. This requirement is far from being fulfilled for most (n,γ) reactions. The observed large literature scattering probably is mainly caused by not correcting the results of Cd-ratio experiments for non-ideality (non-1/E) of the epithermal flux distribution[2]. In large compilations [9,10,20,24], Q_0 is not given as such, but has to be calculated from the quoted σ_0 and I_0-values, the latter however being not available e.g. in the Karslruher Nuklidkarte (1981)[23] and in the NNDC Computope Chart (1981)[21]. It is regretted that often no separate information on σ_0 and especially I_0-values for (n,γ) formation of g and m-states is available;however scanning of the new BNL(1981)[20] reveals an improvement in this respect.

An example of the sometimes huge scattering of literature Q_0-values is given in Fig. 3. for the reaction $^{94}Zr(n,\gamma)^{95}Zr$, showing a range from 4.61 to 6.0. It should be remarked that at the INW/KFKI systematic remeasurements of Q_0-values are performed[35,36], based on the Cd-ratio method and including careful correction for non-1/E epithermal neutron flux distributions. The result of our common effort for $^{94}Zr(n,\gamma)^{95}Zr$, as shown in Fig.3 , yielded Q_0= 5.88±0.07.

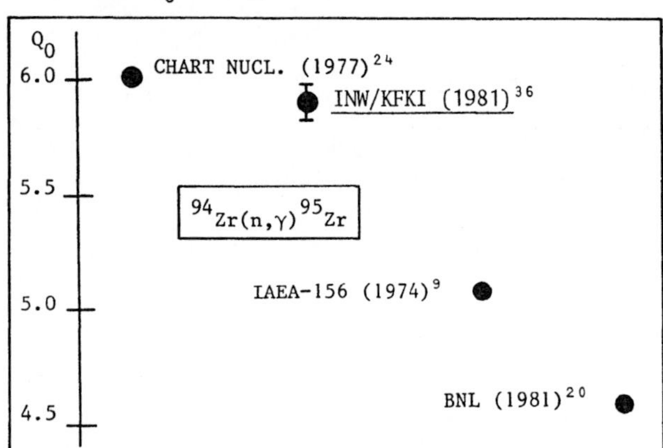

Fig. 3. Compiled and evaluated data for $Q_0(^{94}Zr(n,\gamma)^{95}Zr)$

$\bar{E}_r$- This parameter, recently introduced in RNAA[2], should be entered in eq.(2) to convert the resonance integral I_0(or the $I_0/\sigma_0(=Q_0)$ratio), defined in an ideal 1/E epithermal spectrum, to $I_0(\alpha)$(or $Q_0(\alpha)$)for use in a deviating spectrum, approximated by a $1/E^{1+\alpha}$ epithermal neutron flux distribution.

From error propagation theory applied to eqs.(1) and (2) it follows that an accuracy of 10-20% on this parameter is sufficient to keep the uncertainty contribution on the analysis results below 1%. A cooperative effort of the INW/KFKI has led to the calculation of $\bar{E}_r$-values for 96 isotopes[2], mainly based on the resonance parameter data (see eq.(3)) from BNL-325(1973)[10], with additional data from the NEA Data Bank. These former results will be updated, taking profit of the numerous new data from BNL (1981)[20]. It is appreciated that in the latter edition in many cases $g\Gamma_n\Gamma_\gamma/\Gamma$-values are listed as such. Still, lack of data for some isotopes prevents the calculation of $\bar{E}_r$-values ; this is for instance the case for ^{96}Ru and ^{180}Hf. In this context, it should be remarked that at the INW/KFKI a procedure for experimental $\bar{E}_r$-determination has been developed[4,37,38]. As shown in Table II, good correspondence is obtained between experimental and calculated results for isotopes with well known resonance parameters (^{64}Zn ; ^{68}Zn), and the data for ^{96}Ru and ^{180}Hf can thus be considered as the first reliable $\bar{E}_r$-values available for these isotopes.

Table II. Calculated and measured $\bar{E}_r$-values for some Isotopes

Reaction	$\bar{E}_r$,eV	
	calculated	measured [37,38]
$^{64}Zn(n,\gamma)^{65}Zn$	853 ± 111*+	832 ± 141
$^{68}Zn(n,\gamma)^{69m}Zn$	515 ± 1+	521 ± 99
$^{96}Ru(n,\gamma)^{97}Ru$	no data	776 ± 124
$^{180}Hf(n,\gamma)^{181}Hf$	no data	64.7± 10.4

* assuming J=1/2 for the 0.2818 keV resonance (cfr.ref.[10])
+ assuming Γ_γ constant for all resonances

$T^1/_2$-Taking into account that $T^1/_2$ appears in the S,D and C factors of eq.(1), it follows from error propagation theory that in normal working conditions an accuracy better than 0.5-1% is required on this parameter. Although this requirement is fulfilled for the majority of the analytical interesting radioisotopes, it turns out that occasionally - especially for short-lived isotopes - more accurate data are needed. This is for instance the case for ^{125m}Sn, where repeated measurements at the KFKI[40] yielded a $T^1/_2$-value of (9.414±0.010) min, in contrast with the usually reported data scattering between 9.5 and 9.7 min (see Fig.4.).It should be noted that half-life remeasurement for other problematic cases is in progress at the KFKI.

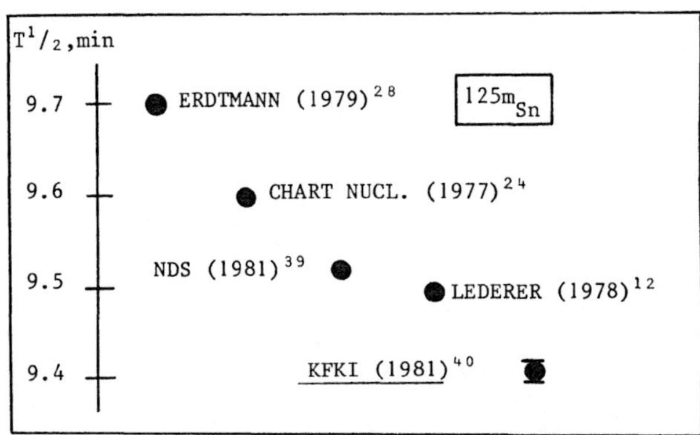

Fig. 4. Compiled and evaluated data for $T^1/_2(^{125m}Sn)$

k_0- Although not explicitly mentioned yet,the cooperative INW/KFKI group developed a standardization methodology for absolute RNAA,wherein the individual absolute nuclear data (M,θ,σ_0,γ-see eq.(1)) are replaced by a so called k_0-factor, defined as :

$$k_0 = \frac{M^* \theta \sigma_0 \gamma}{M \theta^* \sigma_0^* \gamma^*} \qquad (4)$$

(e.g.$k_{0,Au}$ with * = $^{197}Au(n,\gamma)^{198}Au$; E_γ = 411.8 keV)

k_0-factors can be determined experimentally[3,5], thus avoiding possible inaccuracies on the arbitrary chosen individual nuclear data. Up to now, thus measured k_0's with an accuracy of 1-2% are available for 70 isotopes (including for each isotope the analytically important gamma-lines)and extension is in progress.

Although individual nuclear data (M,θ,σ_0 andγ)are thus no longer needed in absolute RNAA, comparison of experimentally determined and calculated k_0-factors (eq.(4)) can give information about the quality of the literature data introduced in eq.(4).

A comparison of calculated and measured $k_{0,Au}$-factors is shown in Table III. Note that for $^{59}Co(n,\gamma)^{60}Co$ (E_γ= 1173.2 keV), with well-known nuclear data, excellent agreement is obtained. For $^{127}I(n,\gamma)^{128}I$(E_γ= 442.9 keV), on the other hand, no nuclear data are available which can lead to better correspondence than shown in Table III. This means in fact that absolute RNAA, introducing nuclear data in eq.(1), will inevitably lead to a systematic error of 45% in the analysis results of iodine.

Table III. Comparison of calculated and measured (recommended) k_0-factors

Formed isotope	E_γ, keV	k_0(* : ^{198}Au, E_γ = 411.8 keV)	
		calculated(eq.4) (best possible)	measured(rel. err.,%) (recommended)
^{60}Co	1173.2	1.31	1.32 (0.4)
^{128}I	442.9	1.63 E - 02	1.12E-02(1.7)

INW/KFKI compilation of evaluated nuclear data for use in RNAA

The present status of the above described evaluation work is layed down in a INW/KFKI Interim Report(Sept. 1982)[41],which is available on request. This report is in fact a user-oriented compilation of nuclear data for isotopes of interest in RNAA. Listed data are : name of the element ; atomic weight ; thermal neutron absorption cross-section and absorption resonance integral ; target isotope ; isotopic abundance (θ) ; 2200 $m.s^{-1}(n,\gamma)$ cross-section (σ_0) ; infinite dilution resonance integral (I_0) ; Q_0(= I_0/σ_0)-ratio ; Cd-transmission factor for epithermal neutrons(F_{Cd}) ; effective resonance energy ($\bar{E}_r$) ; formed isotope together with schemes, data and formulae for branched activation, daughter formation, etc. ; half-life($T^1/_2$) ; analytically interesting gamma-lines (E_γ) ; absolute gamma-intensities (γ) ; recommended (measured)$k_{0,Au}$-factors ; and for comparison, theoretically calculated $k_{0,Au}$-factors.

Acknowledgments

The financial support of the "Nationaal Fonds voor Wetenschappelijk Onderzoek"(F.D.C.)is highly appreciated.

References

1. T.B.Ryves, Metrologia 5, 119 (1969)
2. L.Moens, F.De Corte, A.Simonits, A.De Wispelaere, J.Hoste, J.Radioanal.Chem. 52, 379 (1979)
3. A.Simonits, L.Moens, F.De Corte, A.De Wispelaere, A.Elek, J.Hoste, J.Radioanal.Chem. 60, 461 (1980)
4. A.Simonits, F.De Corte, L.Moens, J.Hoste, J.Radioanal.Chem. (in press ; proceedings MTAA Toronto 1981)
5. L.Moens, F.De Corte, A.Simonits, A.De Wispelaere, A.Elek, E.Szabó, J.Hoste, J.Radioanal.Chem(in press)
6. P.De Bièvre, F.De Corte, L.Moens, A.Simonits, J.Hoste, Intern.J.Mass Spectr. (in press)
7. Table of Isotopic Compositions as determined by Mass Spectrometry, SAIC (1981) ; P.De Bièvre, private communication
8. N.E.Holden, Atomic Weights of the Elements 1979, Pure & Appl.Chem. 52, 2349 (1980)
9. Handbook of Nuclear Activation Cross-Sections, IAEA Techn.Rept.Ser. 156 (1974)
10. S.F.Mughabghab, D.I.Garber, Neutron Cross Sections, Vol.I, Resonance Parameters, BNL-325, 3rd ed. (June, 1973)
11. N.E.Holden, Atomic Weights of the Elements 1977, Pure & Appl.Chem. 51, 405 (1979)
12. C.M.Lederer, V.S.Shirley(eds.), Table of Isotopes, 7th ed., J.Wiley & Sons, Inc., N.Y. (1978)
13. G.E. Valley, H.H.Anderson, Phys.Rev. 59, 113(1941)
14. G.E. Valley, H.H.Anderson, J.Am.Chem.Soc. 69, 1871 (1947)
15. R.F.Hibbs, Rept. AECU - 556 (1949)
16. A.O.Nier, Phys.Rev. 55, 1143 (1939)
17. J.Chenouard, Adv.Mass Spectrom. 3, 583 (1966)
18. P.F.Schmidt, J.E.Riley, Jr.,Anal.Chem. 51, 306 (1979)
19. W.D.James, J.J.Carni, J.Radioanal.Chem. 57, 223 (1980)
20. S.F.Mughabghab, M.Divadeenam, N.E.Holden, Neutron Cross Sections, Vol 1, Part A (Z=1-60), Acad. Press, N.Y. (1981)
21. NNDC Computope Chart (March 1982)
22. A.Simonits, F.De Corte, L.Moens, J.Hoste, J.Radioanal.Chem (in press)
23. W.Seelmann-Eggebert, G.Pfennig, H.Münzel, H.Klewe-Nebenius, KFK Nuklidkarte (November 1981)
24. F.W. Walker, G.J.Kirouac, F.M.Rourke, Chart of the Nuclides, 12th ed., Gen.Electr.Co., Schenectady N.Y. (1977)
25. W.Maenhaut, F.Adams, J.Hoste, J.Radioanal.Chem. 16, 39 (1973)
26. G.H.E.Sims, D.G.Juhnke, J.Inorg.Nucl.Chem. 32, 2839(1970)
27. W.Mannhart, Z.Phys. A 272, 273 (1975)
28. G.Erdtmann, W.Soyka, The Gamma Rays of the Radionuclides : Tables for Applied Gamma Ray Spectrometry, Verlag Chemie Weinheim/N.Y. (1979)
29. A Handbook of Radioactivity Measurements Procedures, NCRP Rept. No 58 (1978)
30. J.Legrand, J.P.Perolat, F.Lagoutine, Y.Le Gallic, Tables des Radionucléides, CEA/BNM/LMRI (1975)
31. NBS Special Publication 626 (in press, October 15, 1981)
32. D.C.Kocher, Nuclear Data for Radionuclides occuring in Routine Releases from Nuclear Fuel Cycle Facilities, ORNL/NUREG/TM-102 (Aug. 1977)
33. R.J.Gehrke, Intern.J.Appl. Radiation and Isot.31, 37 (1980)
34. L.K.Peker , Nuclear Data Sheets 32, 1 (1981)
35. L.Moens, A.Simonits, F.De Corte, J.Hoste, J.Radioanal.Chem. 54, 377 (1979)
36. A.Simonits, F.De Corte, T.El Nimr, L.Moens, J.Hoste, J.Radioanal.Chem. (in press)
37. F.De Corte, L.Moens, A.Simonits, K.Sordo-El Hammami, A.De Wispelaere, J.Hoste, J.Radioanal.Chem. (in press ; proceed. MTAA Toronto 1981)
38. A.Simonits, S.Jovanović, F.De Corte, L.Moens, J.Hoste, J.Radioanal.Chem. (in press)
39. T.Tamura, Z.Matumoto, M.Ohshima, Nuclear Data Sheets 32, 497 (1981)
40. A.Simonits, unpublished work (1981)
41. F.De Corte, A.Simonits, L.Moens, A.De Wispelaere, J.Hoste, INW/KFKI Report (Sept. 1982)

Facilities, Instruments and Methods for Neutron Data Measurements

CHARACTERISTICS OF THE WNR - A PULSED SPALLATION NEUTRON SOURCE

G. J. Russell, P. W. Lisowski, S. D. Howe, N. S. P. King and M. M. Meier

University of California
Los Alamos National Laboratory
Los Alamos, New Mexico 87545 USA

The Weapons Neutron Research facility (WNR) is a pulsed spallation neutron source in operation at the Los Alamos National Laboratory. The WNR uses part of the 800-MeV proton beam from the Clinton P. Anderson Meson Physics Facility accelerator. By choosing different target and moderator configurations and varying the proton pulse structure, the WNR can provide a white neutron source spanning the energy range from a few meV to 800 MeV. The neutron spectrum from a bare target has been measured and is compared with predictions using an Intranuclear Cascade model coupled to a Monte Carlo transport code. Calculations and measurements of the neutronics of WNR target-moderator assemblies are presented.

[800-MeV protons, spallation, neutron spectra, 1 meV-800 MeV neutrons, Intranuclear Cascade Predictions]

Introduction

The Weapons Neutron Research facility[1,2] is an operational pulsed spallation neutron source at the Los Alamos National Laboratory. At the WNR, a portion of the intense 800-MeV proton beam produced by the Clinton P. Anderson Meson Physics Facility (LAMPF)[3] bombards a target to produce an intense white-neutron source. Variable proton pulse widths are available at the WNR giving time-of-flight (TOF) experimental capability for neutrons from a few meV to 800 MeV. The flexibility and capability of the WNR will be enhanced by the addition of a Proton Storage Ring (PSR)[4]. The PSR, presently under construction and scheduled for operation in 1986, will be able to alter the intensity, time structure, and repetition rate of the WNR proton pulse. We present here a description of the WNR, its operating characteristics, and some neutron spectra measurements and calculations.

Facility Description

The WNR is one of several experimental facilities located at LAMPF; the layout of the WNR is illustrated in Fig. 1. At present, the high-current area (target 1) is capable of accepting up to 20 μA of proton beam. Target 1 has a vertical target, is surrounded by a 2-m diam by 2-m high cylindrical void, and is shielded by a 3.7-m-thick laminated iron-concrete structure. Target 1 has a very flexible target-moderator —reflector handling scheme; all three configurations shown in Fig. 2 are employed. The orientation of the reflected 'T'-shape moderator in target 1 is illustrated in Fig. 3. For use with the PSR, the shielding and handling scheme for targets, moderators, and reflectors is being upgraded for operation with 100 μA of proton current. Neutron beams are extracted at 90° to the target axis; flight paths from 5 to several hundred meters are available. Basic research in nuclear physics and materials science research is done in the high-current target area.

The low-current target area (target 2) can accept up to 0.1 μA of proton beam or be used for measurements with neutrons from target 1. The horizontal proton beam in target 2 strikes targets which can be viewed at a variety of angles (from 7.5° to 165°) to the proton beam. We have implemented 30-m flight paths at 7.5°, 15°, and 30° to study (p,xn)- and (p,xp)-type reactions, and are considering longer flight paths and other angles. In target 2, we have also developed the capability to measure (for 800-MeV-proton spallation reactions): 1) spatial distributions of (absolute) 'thermal' neutron surface fluxes from moderators, 2) absolute neutron spectra for energies < 10 eV, and 3) neutron pulse shapes for energies < 0.172 eV. We use this unique experimental capability (combined with excellent computational support) to optimize the neutronics of the high-current production target system in terms of neutron beam fluxes and pulse widths, and minimizing high-energy neutron beam contamination and associated backgrounds[5,6,7].

Using targets placed in the WNR beam channel, we have constructed a 200-m flight path at 0° to the proton beam. This flight path combined with a unique WNR proton pulse structure* allows high resolution experiments to be performed at neutron energies up to 800 MeV[8].

Operating Characteristics

With or without the PSR proton beam, the WNR can operate in two modes. These modes are as follows:

- without PSR
  - micropulse mode
  - microsecond mode
- with PSR
  - short bunch high frequency mode
  - long bunch high frequency mode

A summary of the WNR proton pulse characteristics is given in Table 1. At present, the width of the WNR proton pulse can be varied from 200 ps to 8 μs. The PSR will provide additional pulse-width and repetition-rate combinations with higher instantaneous intensities, and more average proton currents.

Neutronic Characteristics

The various designs of target, target-moderator, and target-moderator -reflector configurations for target 1 (see Fig. 2) are strongly influenced by the need for neutrons with energies ranging from a few meV to 800 MeV having pulse widths as narrow as practical. We use a bare target to produce neutrons with energies $\gtrsim 100$ keV[1], an unreflected 'fast' moderator for neutrons with energies 1 ev $\lesssim E \lesssim 100$ keV[9], and a reflected system for neutrons with energies $\lesssim 1$ eV.

The high-resolution bare target presently used in target 1 consists of a Ta cylinder (2.5-cm diam by 15-cm long) placed in a water cooled Al canister. We measured the neutron spectrum (at 90° to the proton beam) from such a target; the results are shown in Fig. 4 and compared with calculated predictions. The bare-target data were measured

* The minimum proton pulse width at the WNR was measured to be approximately 200 ps FWHM. This corresponds to a single LAMPF micropulse.

K. H. Böckhoff (ed.), Nuclear Data for Science and Technology, 831–835.

at a 29.4-m flight path using TOF techniques and a scintillator of known efficiency. For these measurements, we used 200-ps-wide proton pulses spaced 11 μs apart at 12 Hz. Charged particle contamination of the neutron beam was observed to contribute significantly to the detector count rate at energies > 100 MeV and was eliminated using sweep magnets. The experimental data have been normalized to calculated results at 10 MeV.

In target 2, we measured the neutron beam flux from a reflected 'T'-shape moderator using a BF_3 detector, a 5.6-m flight path, and 35 ns proton pulses at 120 Hz[7]. The measured data are shown in Fig. 5 for a 100 cm^2 field-of-view at the moderator surface. The target was a 4.5-cm-diam by 25-cm-long W rod. The CH_2 moderator had a 0.0025-cm-thick Gd poison sheet located 1.91 cm from the viewed surface. The moderator was isolated from the Be reflector by 0.076 cm of Cd. The Be reflector was a cube with a 46-cm side. Measured neutron pulses from the Cd-decoupled, Gd-poisoned CH_2 moderator are illustrated in Fig. 6.

State-of-the-art Monte Carlo codes are operational on the Los Alamos computing system. These codes include the ORNL High Energy Nucleon Meson Transport Code (HETC)[10], and the sophisticated Los Alamos coupled neutron-photon transport code MCNP[11]. We used these codes in the calculation shown in Fig. 4. We have also computed the neutronics of moderated systems. In particular, we have calculated the outward neutron current at a moderator surface for the two configurations shown in Fig. 7; the results are given in Fig. 8. We used a 2.5-cm-diam by 15-cm-long Ta target in the unreflected slab-moderator computations. A 5-cm-diam by 25-cm-long W Target was used in the reflected wing-moderator calculations. The reflector was a Be cube (50 cm on a side), and the moderator was 3.5 cm from target center. The Cd-decoupler was 0.076-cm thick, and the Gd poison sheet was 0.0025-cm thick placed 1.5 cm from the viewed surface. Note in Fig. 8 the decrease in thermal neutron current when a reflected system is decoupled and poisoned; the thermal neutron pulse characteristics are, however, significantly better for the decoupled/poisoned system[7]. In Fig. 8, the neutron current from the unreflected slab-moderator is averaged over 400 cm^2 whereas the neutron current from the wing-moderators were averaged over 100 cm^2. The presence of a high-energy neutron component, beginning at ~1 MeV, in a center-looking field-of-view from a slab moderator (compared to the offset field-of-view in wing geometry) can be seen in Fig. 8. This high-energy neutron component can extend up to a few hundred MeV[9].

The characterization of the WNR neutronic capability will continue in an effort to intercompare calculations with experiments on an absolute basis.

Acknowledgements

This work was performed under the auspices of the U.S. Department of Energy. We acknowledge the help of E.R. Whitaker in preparing the figures and R.E. Prael in our computing effort. Our thanks to Anselma Martinez for typing the paper.

References

1. G. J. Russell, LA-6020, Los Alamos National Laboratory (1977).

2. G. J. Russell, P. W. Lisowski and N. S. P. King, Int. Conf. on Neutron Phys. and Nucl. Data for Reactors and other Appl. Purposes, Harwell, England (1978).

3. M. S. Livingston, LA-6878-MS, UC-28 and UC-34, Los Alamos National Laboratory (1977).

4. G. P. Lawrence, et. al., XIth Intl. Conf. on High Energy Accel., Geneva, Switzerland (1980).

5. G. J. Russell, J. S. Gilmore, R. E. Prael, H. Robinson and S. D. Howe, Symp. Neut. Cross Sec. from 10-50 MeV, Brookhaven Natl. Lab. (1980).

6. G. J. Russell, M. M. Meier, J. S. Gilmore, R. E. Prael, H. Robinson and A. D. Taylor, KENS Report II, KEK, Tsukuba, Japan (1980).

7. G. J. Russell, M. M. Meier, H. Robinson and A. D. Taylor, Jülich-Conf-45, ISSN-0344-5789, KFA, Jülich, West Germany (1981).

8. P. W. Lisowski, et. al., Phys. Rev. Letters, 49, 255 (1982).

9. G. J. Russell, Trans. Am. Nucl. Soc., 27, 861 (1977).

10. T. W. Armstrong and K. C. Chandler, ORNL-4744, Oak Ridge National Laboratory (1972).

11. Los Alamos Monte Carlo Group, LA-7396-M, Los Alamos National Laboratory (1981).

TABLE I
LAMPF AND WNR PROTON BEAM CHARACTERISTICS

		PRE PSR		POST PSR			
		WNR		WNR		WNR/PSR	
	LAMPF[a]	MICROPULSE MODE	MICROSECOND MODE	MICROPULSE MODE	MICROSECOND MODE	SBHF[b] MODE	LBLF[c] MODE
PROTONS/PULSE	5.2×10^{13}	2.5×10^{8}	5×10^{11}	3.1×10^{8}	$\leqslant 5.2\times10^{12}$	1.04×10^{11}	5.2×10^{13}
PULSE WIDTH	833 μs	200 ps	8 μs	200 ps	⩽ 83 μs	1 ns	270 ns
REPETITION RATE	120 Hz	1 Hz – 6 kHz	120 Hz	1 Hz – 166.6 kHz	120 Hz	720 Hz	12 Hz
PROTONS/s AVERAGE	6.25×10^{15}	$2.5\times10^{8} - 1.5\times10^{12}$	6×10^{13}	$3.1\times10^{8} - 5.2\times10^{13}$	$\leqslant 6.25\times10^{14}$	7.5×10^{13}	6.25×10^{14}
AVERAGE PROTON CURRENT	1 mA	40 pA – 240 nA	9.6 μA	50 pA – 8.3 μA	⩽ 100 μA	12 μA	100 μA

[a]The LAMPF beam structure contains micropulses separated by 5 ns intervals and grouped in 833-μs-long macropulses at 120 Hz for an assumed 10% duty factor. Present LAMPF current is 700 μA at a 9% duty factor.

[b]Short bunch high frequency.

[c]Long bunch low frequency.

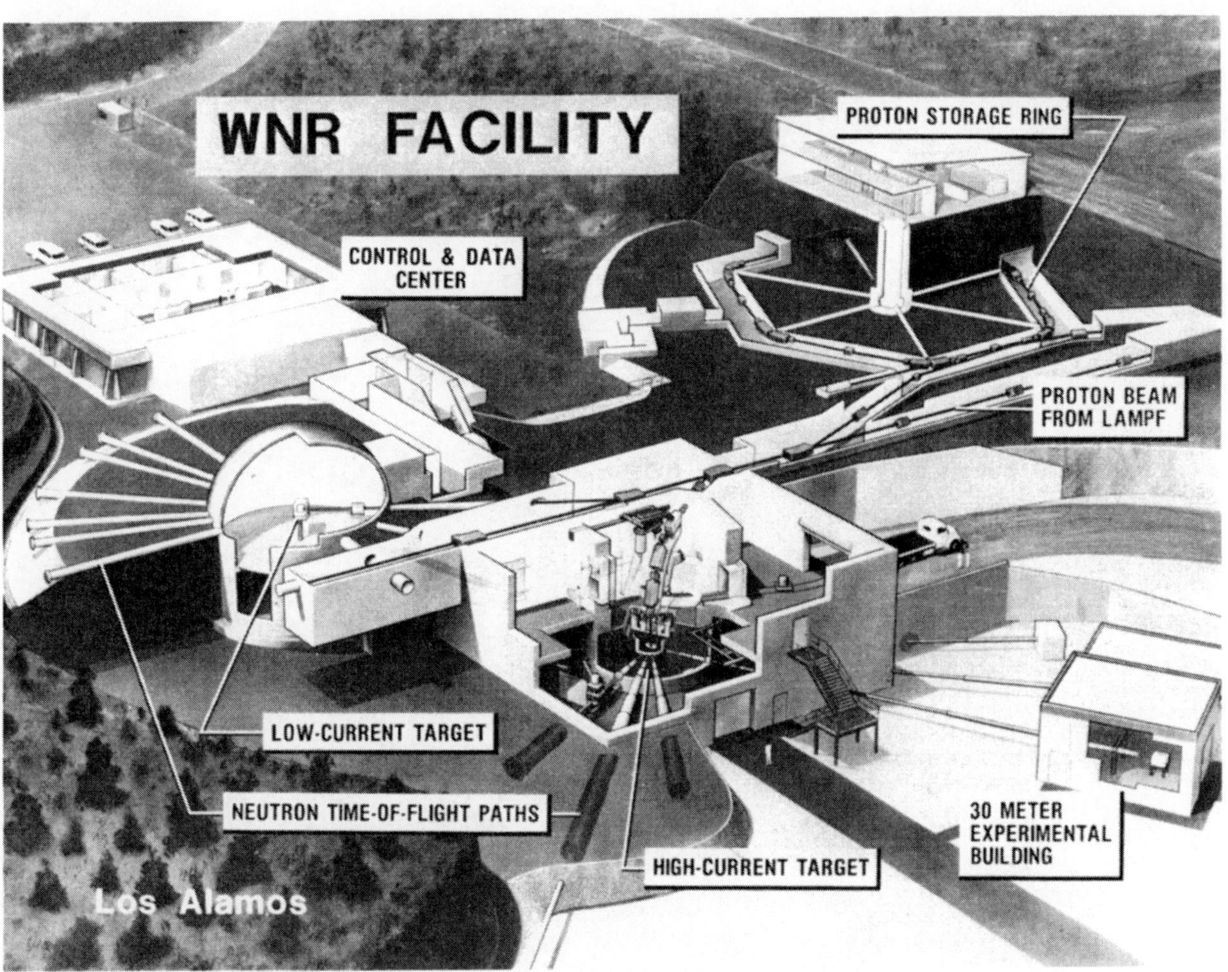

Fig. 1 This figure shows a layout of the WNR and the location of the proton storage ring. The high-current target is located in a vertical proton beam and is viewed by 11 horizontal flight paths. The low-current target is located in a horizontal proton beam and viewed by 11 horizontal flight paths and one vertical flight path. The 0° flight path extends out the end of the proton beam channel between the high- and low-current target areas.

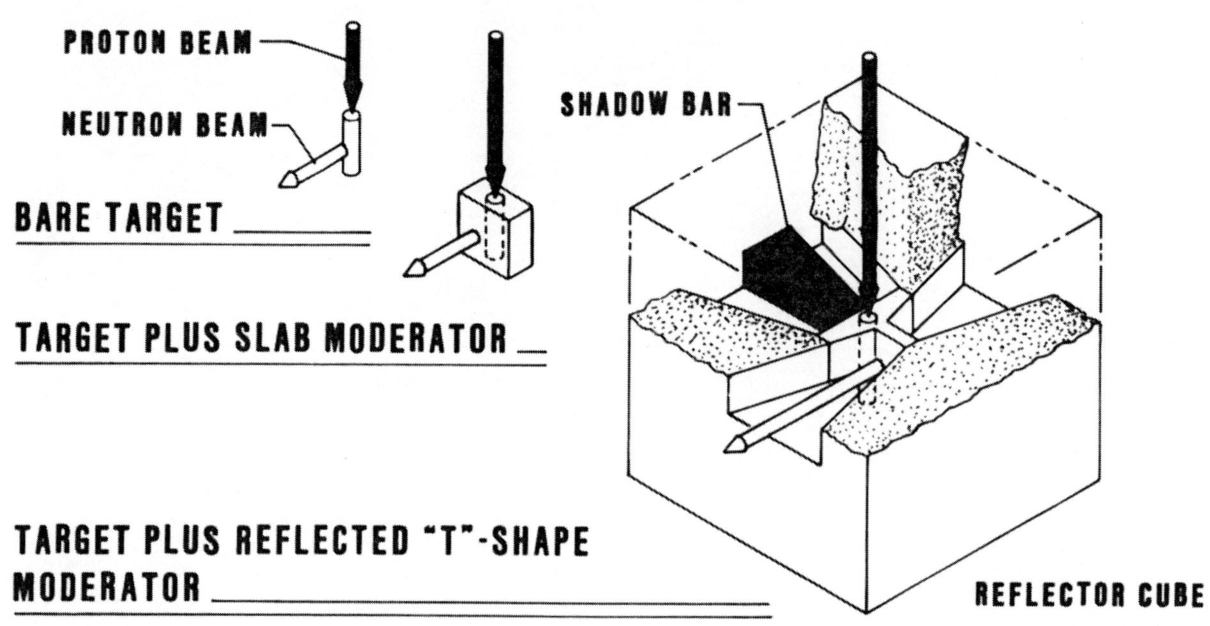

Fig. 2 Neutron production configurations presently in use in the high-current target area.

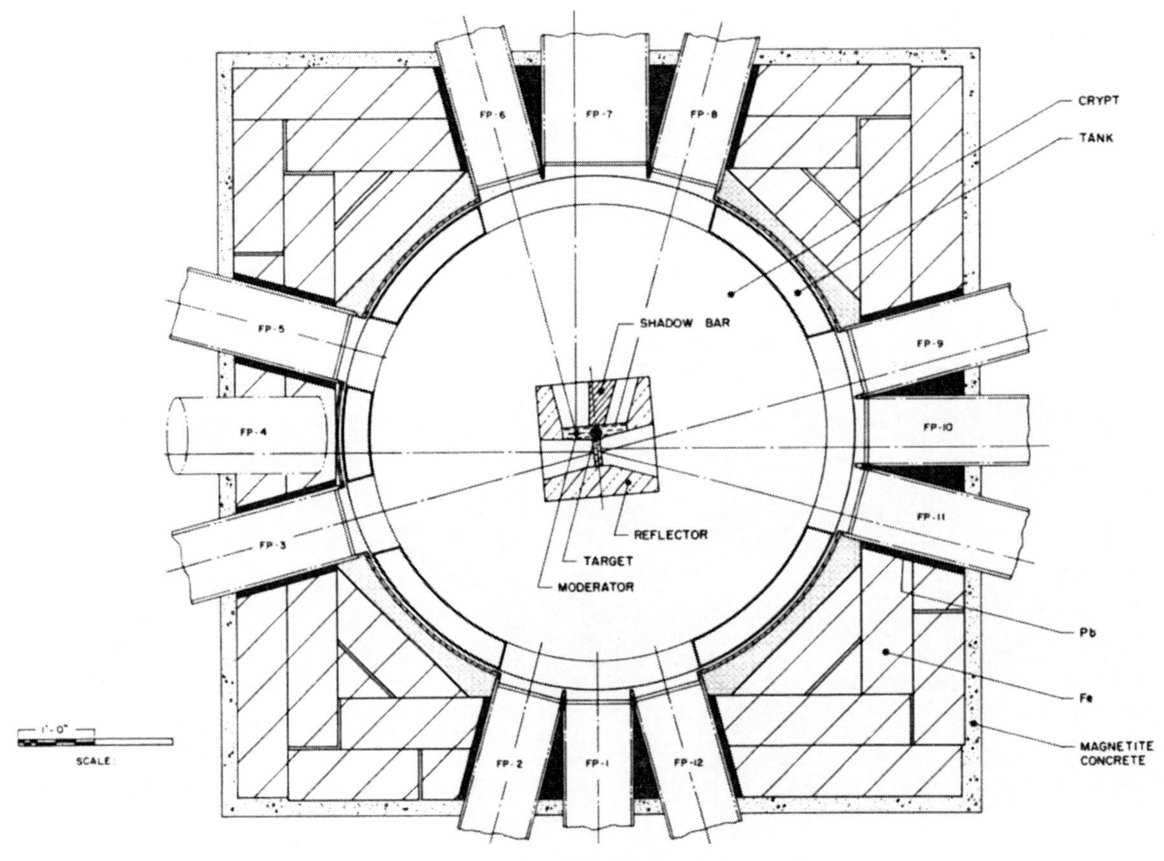

Fig. 3 Present orientation of the reflected 'T'-shape moderator in the high-current target area.

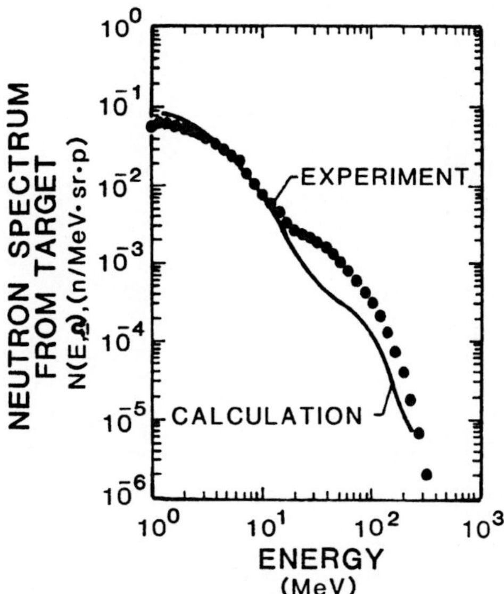

Fig. 4 Bare target spectrum (at 90° to the proton beam axis) emitted from the cylindrical surface of the WNR Ta target. The field-of-view at the target surface encompassed the full target diameter but limited the height seen to 4.4 cm centered 3.8 cm from the top of the target.

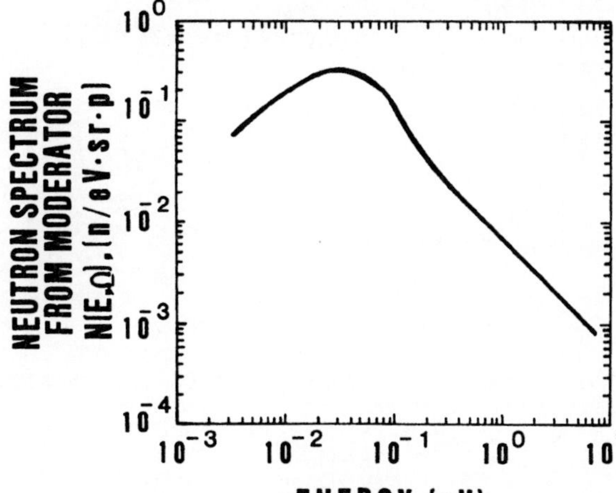

Fig. 5 Measured neutron spectrum emitted (at 90° to the moderator surface) from a reflected wing-moderator.

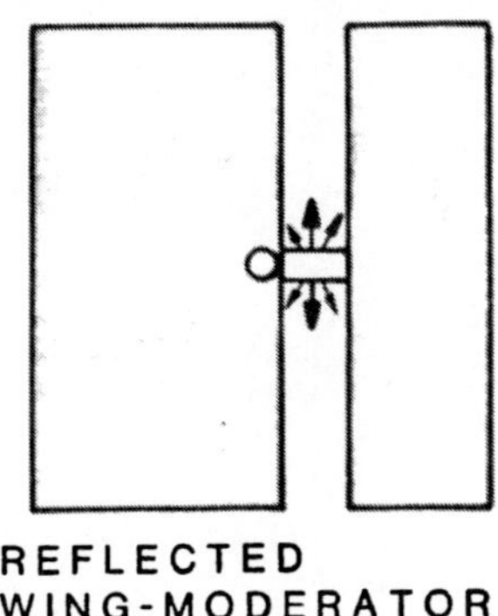

Fig. 7 Illustration of geometries used in calculating neutron leakage currents.

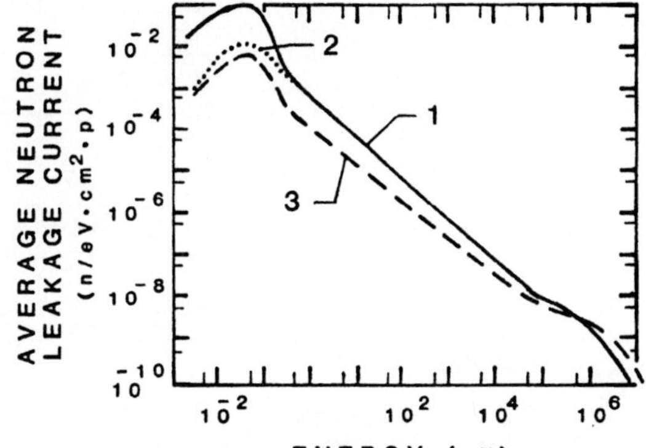

Fig. 8 Calculated neutron leakage currents averaged over the entire moderator surface for three configurations: 1) coupled reflected wing-moderator (5 cm by 10 cm by 10 cm H_2O), 2) Cd-decoupled Gd-poisoned reflected wing-moderator (5 cm by 10 cm by 10 cm H_2O), and 3) unreflected slab-moderator (5 cm by 20 cm by 20 cm H_2O).

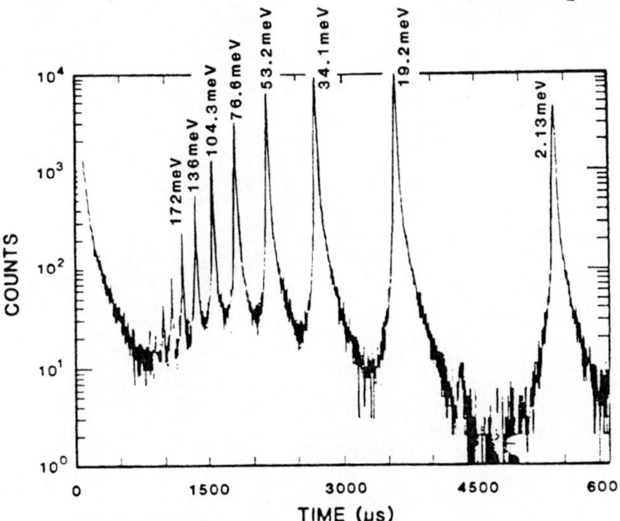

Fig. 6 Measured neutron time distributions from a Cd-decoupled Gd-poisoned reflected wing-moderator.

A MULTI-ANGLE TIME-OF-FLIGHT SPECTROMETER FOR FAST NEUTRON SCATTERING EXPERIMENTS

R. Böttger, H.J. Brede, M. Cosack, G. Dietze, R. Jahr, H. Klein, H. Schölermann, B.R.L. Siebert
Physikalisch-Technische Bundesanstalt (PTB)
Bundesallee 100, D-3300 Braunschweig
Fed. Rep. of Germany

A new TOF-spectrometer for the precise determination of fast neutron spectra is described, which shows some advantages compared with existing multi-angle facilities. As the system was mainly designed and optimized for scattering experiments in the energy range of 6 MeV ≤ E_n ≤ 14 MeV, the properties of the various components such as the gas target, collimator and detector assembly were at first investigated with respect to this particular energy region. The spectrometer has recently been successfully applied in neutron energy spectroscopy and cross section measurements.

[neutron TOF-spectrometer; multi-angle array; detector efficiency $\varepsilon(E_n;E_{thres})$; E_n ≤ 14 MeV]

Introduction

Precise fast neutron cross section data are required for fusion reactor design /1/ as well as for neutron dosimetry and therapy /2/. While in the energy region up to 4 MeV precise reference cross sections are well established /3/ there is a lack of sufficient data in the energy region below 14 MeV. Investigations in this energy region are difficult because there is practically no reaction available to produce a pure monoenergetic neutron field. As primary charged particle beams up to 12 MeV are needed for the reaction D(d,n)He-3 only a few facilities with neutron TOF spectrometers /4,5,6/ are able to investigate scattering experiments in this energy region.

At the PTB a multi-angle neutron TOF spectrometer has recently been completed especially designed for precise scattering cross section measurements. The general setup and the properties are described. Applications are discussed in separate contributions to this conference /7,8/.

Design Criteria and General Setup

The neutron source, the collimator and the TOF detectors had to be carefully designed to meet the following main requirements:

(1) energy resolution $\Delta E_n/E_n \leq$ 1 % and angular resolution $\Delta \vartheta_n \leq$ 1 deg, if the neutron producing reactions are directly investigated. These conditions chiefly determine the size of the detectors with respect to the flight path considering that the beam pulse width is practically limited to $\Delta t \geq$ 1 ns.

(2) precise absolute determination of the scattering cross sections $\Delta\sigma/\sigma \leq$ 5 % in a wide angular range. These requirements influence the design of the collimator and necessitate careful investigation of the neutron detection efficiency.

(3) stability for long time runs to produce reliable data. This condition determines the data acquisition and control system as well as methods for analysing the data, i.e. Monte Carlo simulation instead of analytical approximations, if necessary.

The general setup of our particular spectrometer is shown in fig. 1. A compact cyclotron /9/ is mounted on a swivel arm. A long collimator system /10/ consisting mainly of water tanks could therefore be in a fixed position in front of a 1.7 m concrete wall. Five liquid TOF detectors can be positioned at flight paths between 10 m and 30 m. To realize different angles for the neutron production or scattering, and to normalize subsequent measurements the accelerator and a heavily shielded TOF-

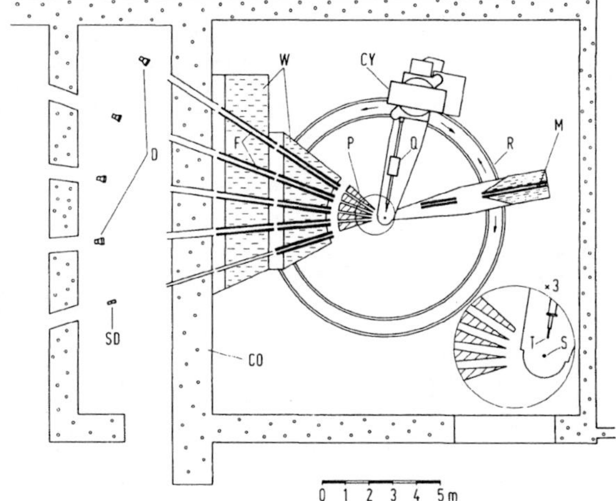

Fig. 1: The fast neutron scattering facility (cyclotron (CY); rail (R); quadrupole (Q); gas target (T); scatterering sample (S); TOF monitor (M); polyethylene (P); water (W); concrete (CO); small (SD) and large area (D) TOF detectors)

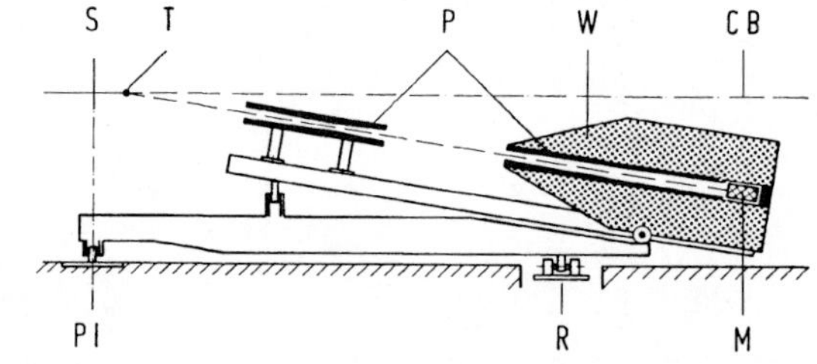

Fig. 2: Setup of the TOF monitor (notation see fig. 1; pivot (PI); cyclotron beam (CB))

K. H. Böckhoff (ed.), Nuclear Data for Science and Technology, 836–839.

monitor can be synchronously moved. The water tank and the preshield of the latter are situated below the reaction or scattering plane (fig. 2) to avoid any distortion of the scattering process. The five TOF detectors as well as the TOF monitor are simultaneously analysed in a multi-parameter computer analyser /11/. Some of the details will be discussed in the following sections.

The Cyclotron

The conventional compact cyclotron CV28* covers the energy range necessary to produce fast neutrons up to 14 MeV via the reaction D(d,n)He-3. Our particular accelerator has been modified by an internal pulsing system to select the repetition rate as required and to shorten the pulse duration time into the subnanosecond region, i.e. at a repetition rate of about 1 MHz and a pulse width of about 1 ns average currents up to 1 μA are available /9/. Thus the beam serves as an ideal system to produce a pulsed neutron source for TOF spectroscopy.

The Gas Target

The deuterium gas target (fig. 3) shows some special features. Although a low mass construction, the entrance foil as well as the beam stop can easily be changed. If a small angle straggling of the charged particle beam /12/ is required, i.e. if the primary neutrons are directly used for calibration measurements, Ni-foils may be used. Mo-foils are preferable if the background between the main neutron peak due to the reaction D(d,n)He-3 and the breakup continuum from the reaction D(d,np)D has to be minimized for scattering experiments. In addition the source strength can be increased due to the better mechanical and thermal stability of Mo-foils. The Au-backing is regularly changed after a series of measurements, started at the highest energy, to avoid additional neutrons from self-targets as these contributions cannot be correctly subtracted by means of 'gas out'-runs.

The net spectra obtained as the difference from 'gas in' and 'gas out' runs are shown in fig. 4 for a mean energy $\langle E_d \rangle$ = 6.8 MeV within the gas target, 3 cm in length at 1.1 bar pressure. The 0-deg spectrum (fig. 4a) as seen by the scatterer is practically free of background within the gap. In the peak region of the monitor spectrum (fig. 4b) about 10 to 15 % of the intensity had to be subtracted in the 'gas out'-run.

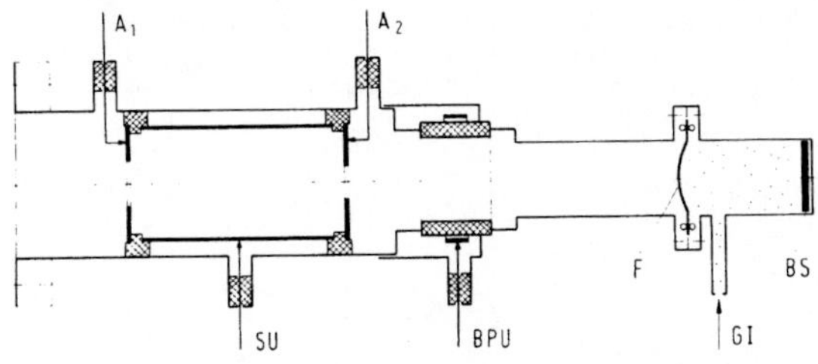

Fig. 3: Gas target assembly and the beam-pulse pickup (beam stop (BS); foil (F); gas inlet (GI); apertures (A); suppressor (SU); beam pickup (BPU))

* (delivered by TCC, Berkeley, USA)

The Collimator

As a long distance was available to shield the detectors against the neutron producing target, large water tanks could be used as moderator and absorber in front of a 1.7 m thick concrete wall. Only the edges of the polyethylene preshield are of iron /10/. The five channels start with a rectangular-shaped profile as cylindrical scatterers are commonly positioned perpendicularly to the scattering plane and end with circular tubes in front of the cylindrical detectors. The channels are lined with polyethylene to minimize the disturbance of discrete neutron spectra. The inscattering of neutrons with less than 10 % energy loss is estimated to contribute less than 0.5 % of the peak intensity /10/.

For a distance of 17.5 cm between the centers of the gas target and the scatterer as used in first scattering experiments /8/; the collimator system allows an angular range between 12.5 deg and 160 deg. While the forward angle may be decreased to about 7.5 deg by increasing the distance up to 30 cm, the backward angles are limited due to construction details of the collimator, the cyclotron and the beam transport system.

The Time-of-Flight Detectors

Five liquid scintillation detectors are arranged behind the neutron channels equidistantly separated by $\Delta\vartheta$ = 12.5 deg. Flight paths between 10 m and 35 m can easily be realized as all detectors are mounted free without any additional shielding. In general, flight paths of 12 m are used to keep the flux attenuation in air sufficiently small.

Large area detectors have been carefully designed to increase the solid angle covered without having any detrimental effect on the time resolution. Four liquid scintillators, 25.4 cm in diameter and 5.08 cm in length (D_ν), are fitted via 7.5 cm long, conical light guides to phototubes, 11.4 cm in diameter. The light pipes are partially coated with diffuse reflecting paint in such a way that the light transmission varies with the locus of interaction by only about 5 % /13/. Thus independent Monte Carlo calculations of the response functions for monoenergetic photons /14/ and neutrons /15/ can be applied to determine thresholds and efficiencies. For the same reason the small detectors, 2.54 cm or 5.08 cm in length and 10.16 cm or 5.08 cm in diameter respectively (SD1 or SD2), as well as the monitor scintillator (M: 3.8 cm x 3.8 cm Ø) are coupled via partially coated cylindrical light pipes to the proper phototubes.

Any detector event is analysed three times for

(a) the integrated pulse height L. The signal is decoupled at the 11.th dynode for the better linearity and used to determine the response spectrum for selected TOF regions. These spectra can be compared to calibrated response spectra for absolute scaling.
(b) the fraction of the long tail component of the fast anode current pulse PSD /16/. A pulse height dependent table-cut condition PSD(L) is used to separate photon and neutron induced events optimally.
(c) a fast timing signal from the anode pulse by means of a constant fraction

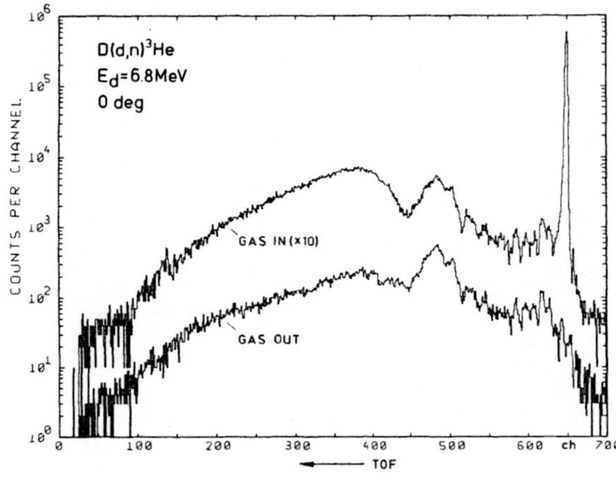

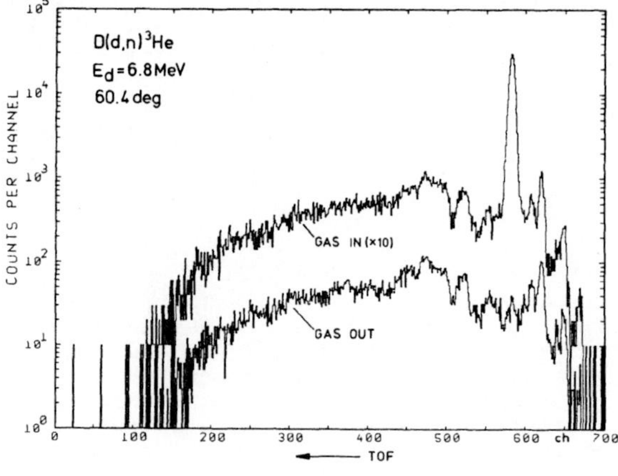

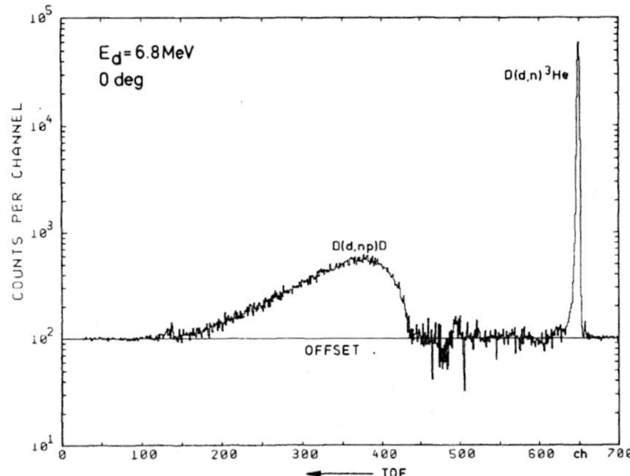

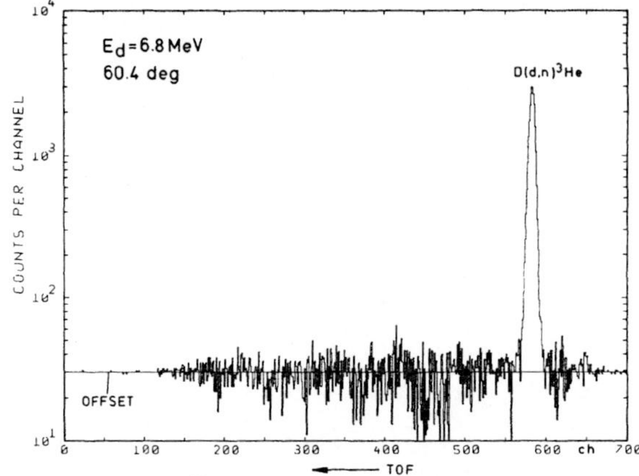

Fig.4a: Neutron TOF spectra for 'gas in' and 'gas out' runs (upper part) as well as the difference spectrum (lower part) at 0 deg

Fig 4b: Neutron TOF spectra for 'gas in' and gas out' runs (upper part) as well as the difference spectrum (lower part) at 60.4 deg (monitor)

discriminator /17/ which serves as the START signal for photon or neutron TOF measurements in an inverse scale to the properly delayed beam pickup timing.

All three analogue signals are gated by means of a slow coincidence event into a multi-parameter pulse height analyser /11/ which simultaneously serves up to 8 independent detector systems. The originally analysed data are dumped on tape for detailed off-line inspection, but the most important information such as n-TOF spectra and selected response spectra may immediately be stored using various cut conditions. The monitor detector is commonly used at an angle of about 60 deg to the beam direction because the angular distribution of the neutron producing reaction shows a second maximum with respect to the yield /18/. In addition the gradient $dE_n/d\vartheta_n$ and the energy resolution at a 6.5 m flight path are suitable to control the mean beam direction on-line. However, the main scope of the monitor system is to survey the source strength and to transfer the neutron fluence as measured with a proton recoil telescope to the scattering sample for absolute scaling.

Fluence Measurement and Detector Calibration

The neutron fluence near the target is measured by means of a proton recoil telescope of the Los Alamos type /19/. Two pill-box proportional counters and a thick surface barrier detector trigger in coincidence if a fast neutron knocks a recoil proton out of the 10.82 mg/cm^2 tristearine radiator. The spectrum of the semiconductor detector (fig. 5) which corresponds to the energy spectrum measured with a TOF detector behind the telescope (see fig. 4a) is analysed by means of the code SINENA /20/. This Monte Carlo simulation includes all geometrical details of the beam, the gas target and the telescope itself. The uncertainty in determining the neutron fluence at 20 - 30 cm in front of the target is estimated to be less than 3 %, as has also been confirmed at selected energies in an international intercomparison /21/. By means of the same SINENA code, the neutron fluence can be transferred to the detectors at any angle on the basis of recent evaluations of the differential D(d,n)- cross sections /18/. In this way the various detectors have been calibrated with respect to the detection efficiency for fast neutrons in the range of 2 MeV up to 14 MeV (fig. 6). The experimental data of various runs performed for the large area detectors within two years agree within a ± 5 % uncertainty. The efficiency curve calculated for this particular threshold by means of a Monte Carlo code NEFF4 /15/ shows significant deviations of up to 10 % in the region of 10 MeV neutrons. These discrepancies are not yet understood and are the object of further investigations. In general the experimentally calibrated efficiencies properly interpolated are used for the absolute scaling of TOF spectra.

Conclusion

A powerful fast neutron TOF spectrometer has been completed, particurlarly suitable in the energy range of 2 MeV ≤ E_n ≤ 14 MeV. An extension up to 30 MeV is possible as soon as a T-gas target is available. Precise neutron energy spectroscopy compensates the absence of an analysing magnet in front of the target.
The spectrometer has been successfully applied for the determination of the neutron spectrum from the spontaneous fission of Cf-252 /7/, the investigation of the detector properties for neutron energies below 14 MeV and for first scattering experiments on carbon samples /8/.

Acknowledgement: The authors wish to acknowledge the work of Dr. Schlegel-Bickmann in optimizing the shape of the collimator and of Mr. M. Bittner in mechanically designing the collimator.

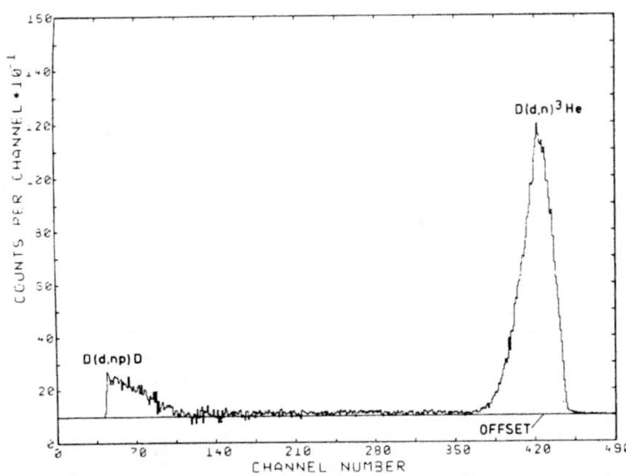

Fig. 5: Proton recoil spectrum for $\langle E_n \rangle$ = 9.92 MeV

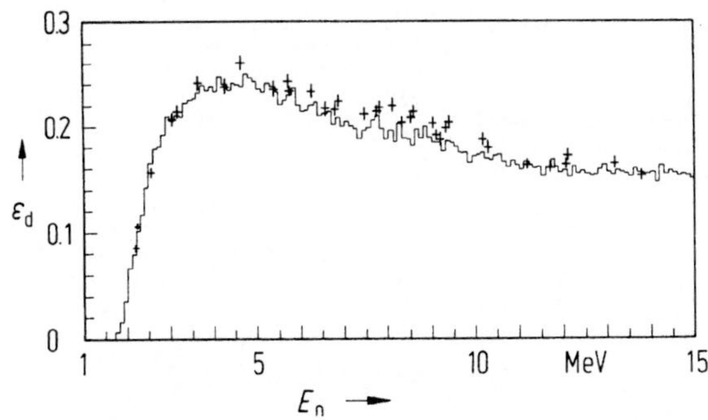

Fig. 6: Absolute detection efficiency for the large area detectors (D_ν) and a threshold energy E_e = 0.5 MeV. The experimental data are compared with MC calculations (histogram).

References

/ 1/ World Request List of Nuclear Data WRENDA 81/82; INDC (SEC)-78/URSF (1981)

/ 2/ R. Jahr, H.J. Brede; Phys. Med. Biol. 25. (1980) 923-926 and ext. PTB-report ND-15 Braunschweig (1979)

/ 3/ A. Smith, R. Holt, J. Whalen; NSE 70 (1979) 281-293

/ 4/ D.W. Glasgow, F.O. Purser, H. Hogue, J.C. Clement, K. Stelzer, G. Mack, J.R. Boyce, D.H. Epperson, S.G. Buccino, P.W. Lisowski, S.G. Glendinning, E.G. Bilpuch, H.W. Newson; NSE 61 (1976) 521-533

/ 5/ G. Haouat, J. Lachkar, J. Sigaud, Y. Patin, F. Cocu; NSE 65 (1978) 331-346

/ 6/ V. Eckstein, H. Helfer, D. Kätzner, J. Kayser, D. Lehmann, W. Pilz, J. Rumpf, D. Schmidt, D. Seeliger, T. Streil; Kernenergie 20 (1977) 161

/ 7/ R. Böttger, H. Klein, A. Chalupka, B. Strohmeier; contribution to this conference

/ 8/ H. Klein, B.R.L. Siebert, R. Böttger, H.J. Brede, H. Schölermann; contribution to this conference

/ 9/ H.J. Brede, M. Cosack, G. Dietze, H. Gumpert, S. Guldbakke, R. Jahr, M. Kutscha, D. Schlegel-Bickmann, H. Schölermann; NIM 169 (1980) 349-358

/10/ D. Schlegel-Bickmann, G. Dietze, H. Schölermann; NIM 169 (1980) 517-526

/11/ H. Klein, H.J. Barrenscheen, G. Dietze, B.R.L. Siebert, W. Bretfeld; NIM 169 (1980) 359-367

/12/ H. Klein, H.J. Brede, B.R.L. Siebert; NIM 193 (1982) 635-644

/13/ H. Schölermann, H. Klein; NIM 169 (1980) 25-31

/14/ G. Dietze, H. Klein; NIM 193 (1982) 549-556

/15/ G. Dietze, H. Klein; ext. PTB-report ND-22, Braunschweig (1982)

/16/ P. Sperr, H. Spieler, M.R. Maier, D. Evers; NIM 116 (1974) 55-59

/17/ M.R. Maier, P. Sperr; NIM 87 (1970) 13-18

/18/ M. Drosg; NSE 67 (1978) 190-220

/19/ J.H. Johnson in: Fast Neutron Physics ed. by J.B. Marion, J.C. Fowler Vol. I, 247-295 Interscience Publ. New York (1960)

/20/ B.R.L. Siebert, H.J. Brede, H. Lesiecki; ext. PTB-report ND-23, Braunschweig (1982)

/21/ V.D. Huynh; Metrologia 16 (1980) 31-49,

A NEW APPROACH TO A FAST NEUTRON FACILITY

P. Leleux, P. Lipnik, P. Macq, A. Ninane

Université Catholique de Louvain
Institut de Physique
2, Chemin du Cyclotron
B - 1348 Louvain-la-Neuve (Belgium)

The facility is aimed at the production of a fast (40 to 75 MeV) monokinetic neutron beam, through the $^7Li(p,n)$ reaction at 0°. Contrary to our previous set-up, the incident charged beam is not bent by a magnet behind the Li target, but dumped and integrated in a Carbon Faraday cup. The target and the Faraday cup are enclosed in the center of a large iron shielding surrounded by a boron-loaded paraffin jacket.

Introduction

Fast neutrons are generated by a primary accelerated beam reacting with a well chosen target material. It is difficult to build an all-purpose neutron facility and so each set-up is more or less dedicated to a particular set of experiments. In all cases, two questions have to be answered before starting :

1) How to produce the neutrons ? (what reaction to use ?)
2) What to do with the primary beam behind the production target ?

As we request a high-flux monoenergetic neutron beam, the answer to question 1) is the $^7Li(p,n)$ reaction with a Q-value of - 1.6 MeV. The Li target is easy to handle, and it has good thermal properties. This choice is not something new[1].
For what concerns the second question, the usual answer was to bend the primary beam with a dipole magnet and to focus the bended beam into a heavily shielded Faraday cup. In this paper we report on a new answer to this question, namely to set immediately behind the production target, a beam stopper made of a material with a large negative Q-value for neutron production.

The charged beam stopper

Large negative Q-values for the (p,n) reaction are found only for light target nuclei. Two of the most interesting materials, i.e. graphite and water, were checked as a function of the proton beam energy. A Li target thickness of 6 mm and a stopper thickness of 5 cm were used. Neutrons were detected at 0° by a 10 cm long and 6.3 cm diameter plastic scintillator, at 4 m from the Li target. Neutron time-of-flight spectra were recorded. In each spectrum the high energy monoenergetic peak and the low energy tail ("continuum") were integrated and normalized to the incident beam current. Results are presented at table I. One can remark the attenuation of the monoenergetic peak in the beam stopper as well as the increase of the tail due to low energy neutron production in the stopper. Although water turned out to be slightly better, carbon was chosen owing to its handling properties. The target and stopper region of our new set-up is showed at figure 1. The Li target and the graphite block are water-cooled. The graphite block closes the beam line vacuum and is electrically isolated. The Li target can be removed from its copper support.

Table I

ENERGY	MONOKINETIC			CONTINUUM		
	Li	Li+C	Li+H_2O	Li	Li+C	Li+H_2O
48	2.12	1.31	1.57	1.15	0.78	0.96
65	2.67	2.04		1.75	2.51	
78	3.0	3.0	2.85	2.1	6.63	3.94

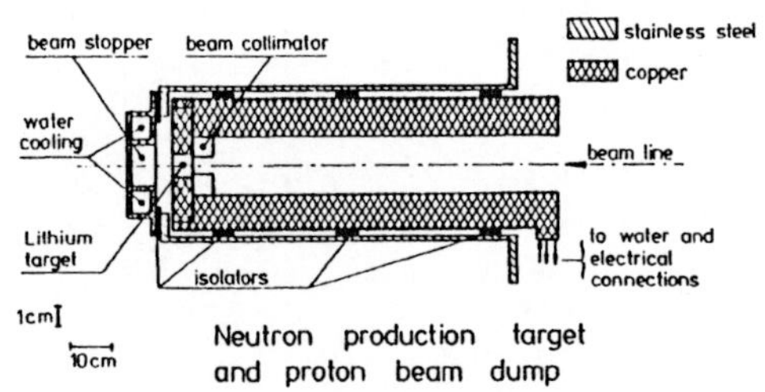

Figure 1. Target (Li) and Stopper (C) region of our new set-up

Description of the new facility

The old and the new facility are presented at figure 2. Both were builded on the same charged beam line and the Li target position was nearly conserved. Due to space requirements, the unused primary beam could only be bent by 20° in the old facility. The same reason limited the source shielding to 3 x 3 x 3 m^3 in the new facility. As our present experimental program only uses unpolarized neutrons, the 20° beam line was not kept. A qualitative comparison of both facilities is presented in table 2. The new source is described in what follows (see figure 2) :

K. H. Böckhoff (ed.), Nuclear Data for Science and Technology, 840–842.

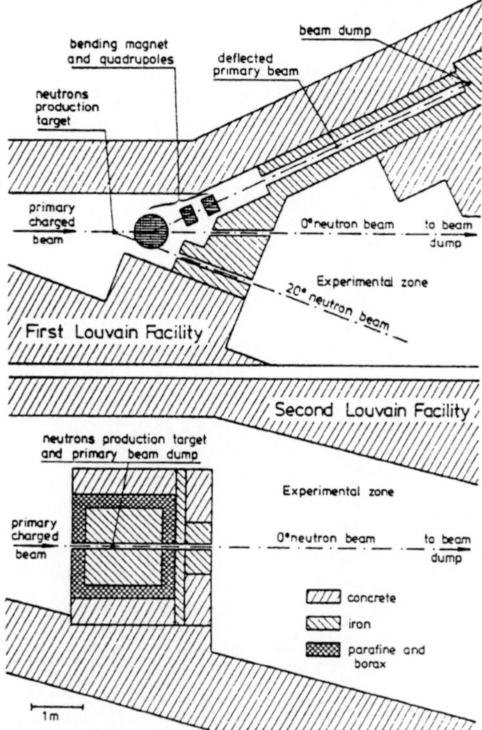

Figure 2. Old and new Louvain neutron facility

At our energies not only the neutron production but also the neutron scattering are strongly forward peaked and this explains why the Li-C assembly is not centered in the shielding, but put more backward. The neutron beam is defined at 0° with respect to the charged beam line, by a 7.6 cm diameter stainless steel tube. Iron, being the less expensive high density moderator for fast neutron, is used as a shielding core. This core is surrounded by a 20 cm-thick jacket of boron-loaded paraffine, to shield against the neutron transmission windows of iron. In front of the experimental area, a 20 cm-thick iron wall followed by 50 cm concrete ends up the shielding, round a sweep-out magnet (figure 3).

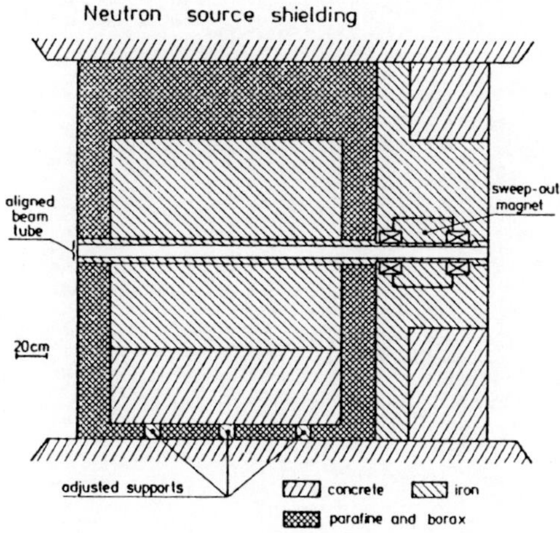

Figure 3. Side-view of the shielding of the new facility

Table 2. Qualitative comparison of the new and old facility

New

In favor :
- . localized shielding makes transmission calculations possible
- . efficient shielding allows primary beam up to 50 μA
- . easy beam handling

Against :
- . complex neutron spectrum
- . higher γ-ray contamination of the neutron beam

Old

In favor :
- . clean neutron spectrum $^7Li(p,n)$
- . low in-beam gamma over-neutron ratio
- . easy target access

Against :
- . dispersed shielding elements make calculations difficult and dispersed background sources
- . primary beam limited to 5 μA
- . more difficult charged beam transportation, beyond the production target

Properties of the neutron beam

i) Intensity

The highest proton energy attainable with high intensity at our cyclotron is 78 MeV. Such a 20 μA-beam was used to bombard a 1 cm-thick natural Li target. The neutron monokinetic flux was then about 1.5 10^7 per sec at 2 m from the Li target, in a 2 cm diameter collimator. A neutron spectrum at this distance is shown in figure 4.

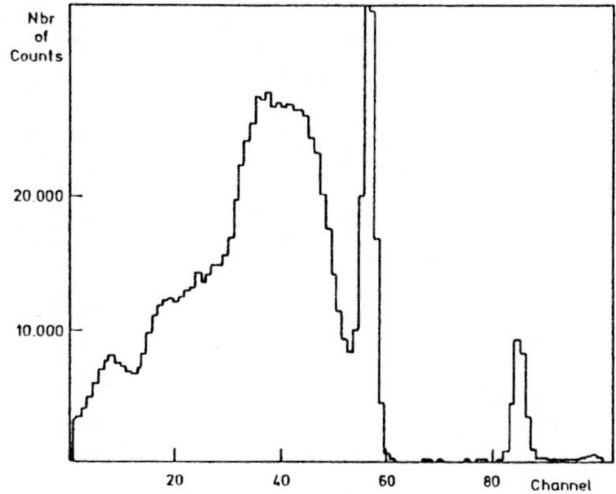

Figure 4. Neutron spectrum at 2 m from the Li target ; the small peak at the right of the figure is due to gamma-rays

ii) Spatial extension of the beam

The monokinetic neutron beam profile was scanned in the horizontal and vertical direction with a 5 mm diameter, 3 cm thick plastic scintillator, at 5.8 m and 9.6 m from the Li target. An example of measurement is shown in figure 5. Three regions are visible : first a uniformely distributed direct neutron flux in the central region, then a sudden drop coming from the eclipse of the source spot by the collimator exit and third a tail region, or beam halo, due to neutrons scattered in the collimator walls.
Calculations are underway to reproduce those measurements.

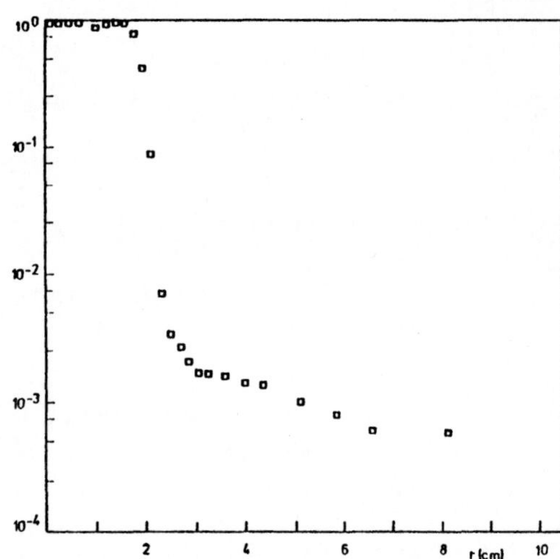

Figure 5. Neutron beam profile

1. J.A. Jungerman, F.P. Brady, W.J. Know, T. Montgomery, M.R. Mc Gie, J.L. Romero and Y. Ishizaki, Nucl. Instr. Meth. 94 (1971) 421 ; M. Bosman, P. Leleux, P. Lipnik, P. Macq, J.P. Meulders, R. Petit, C. Pirart et G. Valenduc, Nucl. Instr. Meth. 148 (1978) 363

A FACILITY FOR FAST NEUTRON DATA MEASUREMENTS IN BRAZIL

A.G. da Silva and L.T. Auler

Instituto de Engenharia Nuclear - CNEN
Caixa Postal 2186
20001 Rio de Janeiro
Brasil

A facility is described that features a CV-28 variable energy cyclotron, at the Instituto de Engenharia Nuclear in Rio de Janeiro, Brazil, for among other purposes, the acquisition of neutron nuclear data for the Brazilian fast reactor programme. Acceptance tests, which resulted in measurements of beam emittance, energy width and time width, to assess its potential, will be presented. The initial part of the fast neutron data measurement programme, supported by the IAEA, will be concerned with the production of fast monoenergetic neutrons, in the several MeV region, to measure excitation functions of (n, 2n) reactions by the activation method. Installation of a beam transport system to allow cleaner experiments is under way. Neutron detectors are being built. Possibilities of pulsing the machine by simple suppression of some of the microstructure pulses will be discussed.

Introduction

Nuclear data, in all their aspects, are required in any programme of energy production using fission or fusion reactors as can be seen in the topics covered at this conference and former ones[1] of similar scope. Fast neutron data in particular, are also required for activation analysis[2], radiation damage[3], astrophysics[4] and many other fields[5]. Besides, a programme on fast neutron data acquisition will have an enormous value in the training of personnel in experimental methods and techniques[6].

Several facilities throughout the world[7] have been used to collect most of the required neutron data, but still the need exists for more and sometimes better data[8].

With the purpose, among others, of collecting neutron nuclear data for the fast reactor programme going on in Brazil, a CV-28* variable energy cyclotron has been installed and tested to assess its possibilities as a tool in a fast neutron data acquisition programme to be supported by the International Atomic Energy Agency as a part of the Inter-regional Project on nuclear data techniques and instrumentation (INT/1/018).

The facility, occupying 1800 m^2 of floor space, was built between 1972 and 1975. The relevant part of the lay-out of the cyclotron building is shown in Fig. 1. The cyclotron has been installed in a cave with 1.8m walls adjacent to an experimental area to which the

*Supplied by The Cyclotron Corporation - Berkeley, Cal, U.S.A.

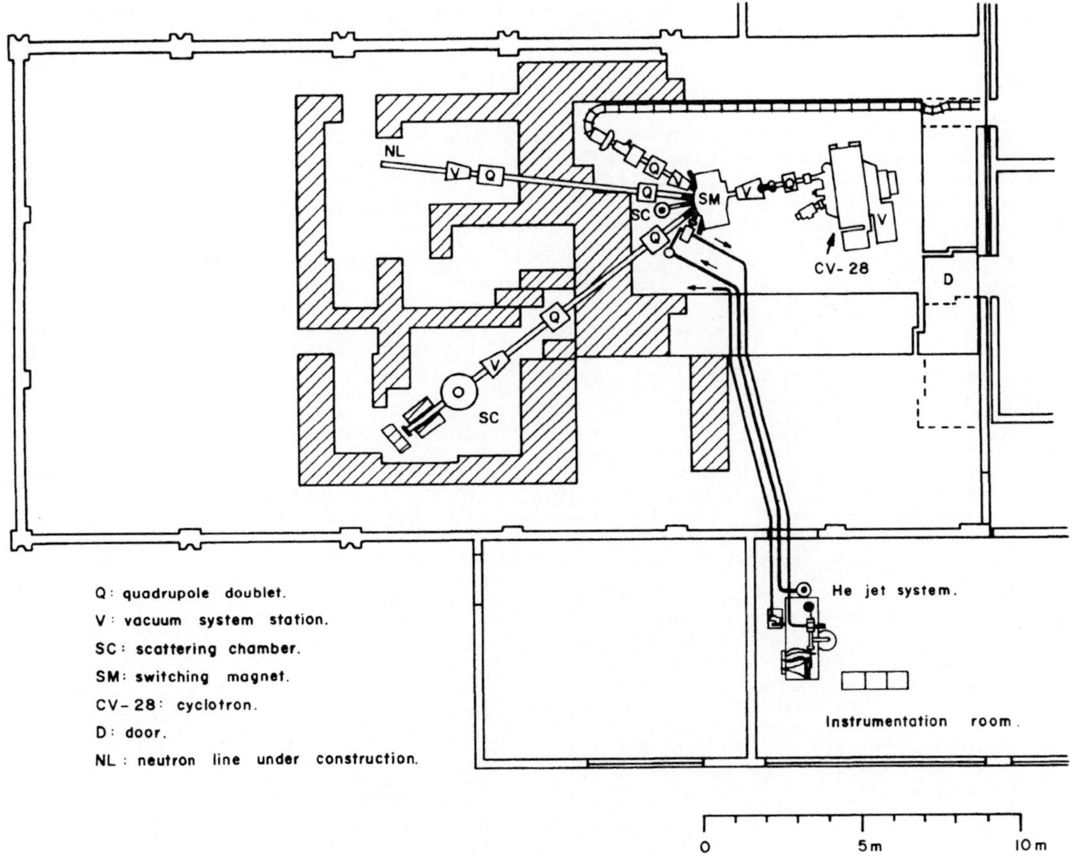

Fig. 1. Relevant part of the lay-out of the cyclotron facility.

K. H. Böckhoff (ed.), Nuclear Data for Science and Technology, 843–846.

beam will be transported. Removable concrete blocks provide the necessary shielding and allow a flexible geometry for the experiments. The beam transport system for the fast neutron work, consisting of beam pipes, vacuum system and magnetic quadrupole doublets, is being built and is expected to be operational by the end of the year. This transport system is a copy of the one already operational leading to the southern cave which was designed and built by the nuclear physics group of the Physics Department of the Universidade Federal do Rio de Janeiro[9]. The CV-28 cyclotron

Table I CV-28 Specification Values

Particle	Energy (MeV)	Ext. current (μA)
H^+	2 - 24	70*
D^+	3 - 14	100*
$^3He^{++}$	5 - 36*	15 - 70*
$^4He^{++}$	6 - 28*	10 - 50*

*Actually measured

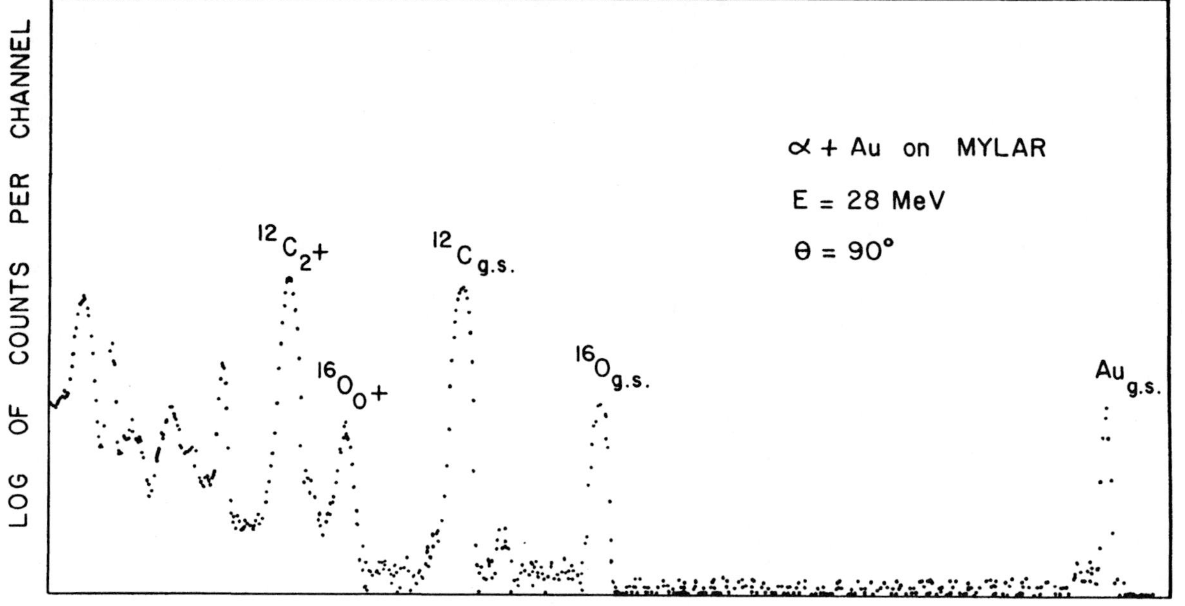

Fig. 2. Energy spectrum of α particles scattered at 90° from Au deposited onto MYLAR. Peaks used in obtaining the incident 28 MeV α particle energy are identified.

has been described in a number of papers[10]; however this model had special power supplies installed so that the magnetic field current could be stabilized to 1 part in 10^5 together with a set of remote-controlled internal beam defining slits for phase selection. Table I shows the energy range and current available for each particle. Using thin targets and the appropriate angle, low intensity "monochromatic" neutron beams of any energy between 0.5 and 22 MeV can be obtained[11]. Tritium and deuterium targets (absorbed in titanium) are already available. A neutron spectrometer that has two Au surface barrier detectors sandwiching a ^{6}Li foil has been built[12] and NE213 liquid scintillater has been ordered.

A helium-jet nuclear reaction recoil transport system[13] has been installed, connecting the 60° port of the switching magnet to the instrumentation room. This He-jet facility may be used to study excitation functions of short-lived reaction products of (n, x) reactions using monochromatic neutrons.

The experimental area is ten meters high and will have enough open space to allow time of flight experiments. These experiments will take some time to initiate because the cyclotron has no provision yet for supression of some of the microstructure pulses. Recently, adjacent to the counter room, a computer terminal has been installed for off-line data processing. At the present it is mainly used for processing gamma-ray spectra[14] and statistical model calculation programmes like ALICE[15].

Tests

In order to assess the possibilities of neutron production, machine characteristics such as maximum beam current, maximum energy, beam emittance, energy width and time width of the beam for each particle have been measured.

Currents were measured in a deep Faraday cup with electron supression. Changing supression voltages from zero to a few hundred volts, did not affect the current measurements. The beam currents were compared with current standards from a calibrated electrometer. These measurements were also checked against activities induced by nuclear reactions of known cross-sections[16] using a current integrator[17]. Results are shown in Table I.

The energy of the particles was measured by inelastic scattering and nuclear reaction Q-values. The energy axis of particle spectra (obtained with silicon surface barrier detectors) was calibrated using reaction kinematics and the well known Q-values[18] of some reactions on carbon and oxygen. The scattering chamber was carefully aligned for these experiments and the particles elastically and inelastically scattered from thin targets (normally formvar or polyethilene onto which Au was deposited) were measured at 90° to the beam. A spectrum is shown in Fig. 2 and the results of several runs, are shown in Table II. The energy width of the beam was obtained from the half-width of the peaks, when very thin targets were used, and amounted to 100 KeV for 28 MeV alpha particles, for example.

Two methods were used to measure beam emittance. One is the method used by Grunder et al[19] in which the

Table II Results of Energy Measurements Against Cyclotron Radio-frequency

Particle	Frequency (MHz)	Energy (MeV)
^{3}He	18.460	36.22
^{3}He	17.406	32.65
^{3}He	15.568	26.54
^{3}He	13.891	21.60
^{4}He	14.000	27.75
^{4}He	14.066	28.00

beam is allowed to pass through a carbon grating and the beam current through each slit is measured by a wire located at a distance ℓ downstream. To each slit located at a position x corresponds two angular apertures θ_1 and θ_2. When plotted in an x vs θ diagram these points will form a figure whose area will be the emittance in the x direction. The grating was located at the exit port of the cyclotron and the beam scanner (ℓ = 185 cm for horizontal plane and ℓ = 194 cm for vertical plane) was used to measure the shape and position of the beam current through each slit.

The other method, used by Hartwig et al[20] is based on the assumption that the square of the full width at half maximum H of the beam in each plane at a fixed distance from a magnetic quadrupole lens is given by:

$$H^2 = C\left(\frac{S_2}{f} - A\right)^2 + B,$$

where S_2 is the distance from the center of the lens to the point of measurement and f is the focal length of the lens. Fig. 3 shows a curve obtained with alpha particles of 28 MeV and Table III results of some of the measurements for p, d and α particles of maximum energies, by both methods.

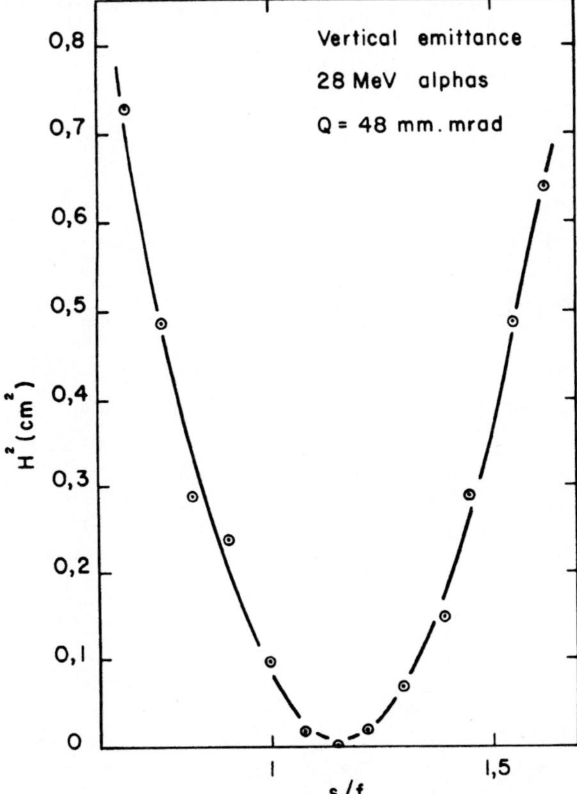

Fig. 3. Vertical emittance (Q) for 28 MeV α particles by the method of Hartwig et al[20]. The line through the experimental points is the best fit parabola from which Q of 90% of the beam is calculated:
$Q\ (90\%) = 260\ \sqrt{BC}/S_2$ (mm).

An upper limit of the time width of the beam was obtained with a calibrated time to amplitude converter. The start pulse was provided by the alpha particles scattered at 90^0 from a thin carbon foil and the stop pulses by the radio-frequency of the accelerating voltage. Fig. 4 shows a time spectrum for 28 MeV alpha particles for which the radio-frequency is about 14 MHz. The FWHM is around 10 nsec.

Table III Results of Beam Emittance Measurements

Particle	Energy (MeV)	Q_H mm.mrad	Q_v mm.mrad	Method (ref.)
p	24	81	58	(20)
d	14	90	63	(20)
α	28	41	71	(20)
p	24	80	-	(19)
d	14	80	-	(19)
α	28	60	54	(19)

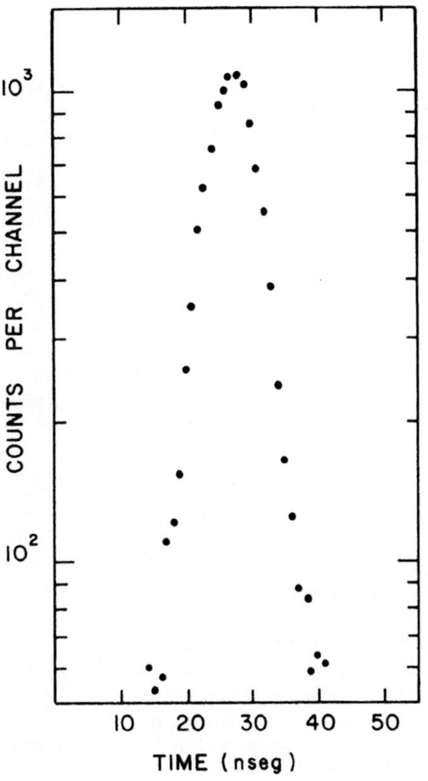

Fig. 4. Time spectrum of 28 MeV incident α particles scattered at 90^0 from C obtained with a time to amplitude converter: start pulse from the α particle, stop pulse from the radio-frequency of the cyclotron.

Conclusions

From the above measurements it's apparent that the cyclotron can be used for many purposes to gather neutron data. The currents and maximum energies are reasonably high, and the energy width is low enough to allow experiments with around 1% energy resolution with 10 MeV neutrons. The beam emittance is adequate to allow transport to the experimental area, without much scattering from, and activation of, the beam pipes and slits.

The time width of about 14% of a full cycle however (54^0 of phase angle, FWHM) is not satisfactory for high resolution time of flight work. Using phase selection slits[21] it would be possible to cut down the time width but at the expense of beam current. Supressing some of the pulses using a gated ion source[22] would of course decrease the effective current and also would demand single turn extraction which is possible judging from achievements with other machines. Except for very low intensity beam work, a pulsing system is essential; this would deliver several pico-coulombs in pulses of 1 nsec width at variable pulsing rates by bunching each microstructure pulse and skipping a fixed number of pulses[23].

Acknowledgements

It's a pleasure to thank the cyclotron crew for their help in running the machine. Thanks are also due to Dr R.H. Töpke who took part in some of the measurements. We thank Genice Nascimento for the typing and Prof. G.W.A. Newton for critical reading of the manuscript. We thank A.M. Reis for the drawings.

References

1. "Nuclear Data in Science and Technology", Proc. of a Symposium, Paris, 12-16 March 1973, IAEA, Vienna (1973);

"Nuclear Cross Sections for Technology", J.L. Fowler, C.H. Johnson and C.D. Bowman (eds.), NBS special publication 594, NBS, Washington, DC (1980).

2. "Modern Trends in Activation Analysis", J.R. DeVoe (ed.) NBS special publication 312, NBS, Washington, DC (1969).

3. "Nuclear Data for Radiation Damage Assessment and related Safety Aspects", IAEA-TECDOC-263 (1982).

4. R.A. Smith, in "Proc. Int. Conf. on the Interactions of Neutrons with Nuclei", E. Sheldon (ed.) p. 597 (1976) ERDA Conf. 760715.

5. Proc. Int. Conf. Neutron Physics and Nuclear Data for Reactors and other Applied Purposes, Harwell, September (1978).

6. IAEA Technical Assistance Project INT/1/018.

7. S.W. Cierjacks, in "Proc. Int. Conf. on the Interactions of Neutrons with Nuclei", E. Sheldon (ed.) p. 893 (1976) ERDA Conf. 760715.

8. World Request List of Nuclear Data 81/82, report INDC (SEC) 78/URSF.

9. H.E. Wolf, J. Eichler, S. de Barros and L.F.V. Paiva, report 001/1982 IF/FIN, Universidade Federal do Rio de Janeiro, Brasil (1982).

10. G.O. Hendry, in "Cyclotrons - 1972", J. Bugerjon and A. Strathdee (eds.) p. 616, American Institute of Physics, New York (1972);

H.J. Brede, M. Cosack, G. Dietze, H. Gumpert, S. Guldbakke, R. Jahr, M. Kutscha, D. Schegel-Bickmann and H. Schölermann, Nucl. Inst. Meth. 169, 349 (1980).

11. M. Drosg, Nucl. Sci. Eng. 67, 190 (1978).

12. M.A. Fabro, M.Sc. thesis, COOPE, UFRJ, Brasil (1973).

13. S.C. Cabral, A.G. da Silva, O.F. Lemos Jr. and L.T. Auler, Rev. Bras. Fis., Vol. Especial, p. 107, Sept. (1981).

14. L.T. Auler, report DEAT/SF/SFN-2/77, Instituto de Engenharia Nuclear, Rio de Janeiro, Brasil (1977).

15. M. Blann, ALICE code, University of Rochester, New York, (1976).

16. E.A. Bryand, D.F. Cochran and J.D. Knight, Phys. Rev. 130, 1512 (1963);

E. Lebowitz and M.W. Greene, Int. J. Appl. Radiat. Isotopes, 21, 625 (1970).

17. J.A.D. Furlanetto, unpublished.

18. A.H. Wapstra and K. Bos, Atomic Data and Nuclear Data Tables 19, 215 (1977).

19. H.A. Grunder, F.B. Selph and H. Atterling, Proc. Int. Conf. on Sector Focused Cyclotrons, p. 59, CERN 63-19, (1963).

20. D. Hartwig, W. Linder, M.E. Lösel, G. Schatz and H. Schweickert, KFK - 754 (1966).

21. H.G. Blosser, in "Fifth Int. Cyclotron Conf.", R.W. Mcllroy (ed.), p. 257, Butterworths, London (1971).

22. R.F. Bentley, L.A. Erb, D.A. Lind, C.D. Zapiratos and C.S. Zaidius, Nucl. Inst. Meth. 83, 245 (1970).

23. G.O. Hendry, TCC report 3005B, January (1972).

A 0.2 ns BEAM PULSE FOR THE 6 MV VAN DE GRAAFF ACCELERATOR

J.J. Kritzinger, W.R. McMurray, C.V. Wikner,
T. Swart and H. Schmitt

Southern Universities Nuclear Institute,
P.O. Box 17, Faure, South Africa.

The 1.5 ns pulsed beam of the SUNI Van de Graaff has been used for neutron time-of-flight studies. To provide sufficient resolution for neutron scattering measurements at 22 MeV, a post acceleration bunching system has been installed. Bunching is achieved in a simple quarter-wave coaxial resonator chamber designed for high Q and low power. The bunched pulse has a 0.2 ns FWHM.

[Post acceleration bunching system, 0.2 ns (FWHM) pulses, 2-6 MeV p,d and He-3 beams.]

Introduction

As is wellknown, the limiting resolution in fast neutron time-of-flight spectroscopy is usually set by the length of the beam pulse derived from the accelerator. The SUNI 6 MV single ended Van de Graaff accelerator has been provided with top terminal sweep (2 MHz) and bunching systems[1] which routinely achieve better than 1.5 ns pulse widths (FWHM) at the neutron producing target for pulsed proton, deuteron and helium-3 beams. For emitted neutron energies around 20 MeV, a timing resolution of 1.5 ns FWHM over a neutron flight path of 3 m corresponds to an energy resolution of 1.2 MeV. Better energy resolution can be obtained by using much larger flight paths, but this is usually not a realistic option for Van de Graaff accelerators. The alternative is a shorter beam pulse width[2]. This too has limitations imposed by the physical size of targets and detectors and the intrinsic timing resolution of neutron detectors such as organic scintillators. In practice a beam pulse width of about 0.2 ns is sufficiently small to make the other contributions to timing uncertainties more important. A relatively simple post acceleration bunching system which achieves 0.2 ns pulses (FWHM) has been installed on the SUNI Van de Graaff.

The bunching system

Bunching is achieved by beam energy modulation within the width of the beam pulse. The energy modulation is given by a synchronised RF voltage across gaps in a buncher tube through which the beam is directed. The distance between the buncher gaps has been made adjustable (100 to 140 mm) to make possible the optimum bunching of p, d and He-3 beams with energy between 2 and 6 MeV. The required gap separation is also related to the RF frequence (90 MHz) selected for convenience of frequency multiplication and to enable the use of a standard 1 kW FM transmitter as power amplifier.

Calculations showed that a peak voltage of 33.5 kV is required for bunching of 6 MeV deuterons. The buncher is placed 7.9 m from the beam target with a quadrupole doublet situated 5 m from the target. The design of the high-Q quarter-wave coaxial resonator was optimised to enable the required voltage to be reached without exceeding the 1 kW power level. The diam of the outer conductor is 200 mm, the diam and length of the inner conductor are 28 and 600 mm respectively. The buncher tube is made of brass, the inner conductor of polished copper tubing, and the outer cylinder of rolled and welded aluminium plate. These materials were selected for convenience of manufacture. The inner conductor is water cooled, the outer conductor is air cooled.

Power is coupled into the resonator through an adjustable coupling capacitor which allows the impedance to be matched to the 60 Ω cable from the power amplifier. Coarse tuning of the resonance frequency to 90 MHz is done by adjustment of the gap widths. Fine tuning is accomplished by the remotely controlled rotation of a tuning loop mounted on the short circuiting plate of the resonator. The resonator chamber is vacuum pumped to 10^{-5} torr through the buncher gaps using the existing pumps on the beam line.

Synchronisation of the buncher voltage to the basic pulses from the accelerator is achieved by using a capacitative beam pick-off 1.5 m ahead of the buncher. The 2 MHz pick-off pulses are shaped, and multiplied to 90 MHz to provide a drive for the 1 kW amplifier, with variable level and phase. The final output goes to the resonator which is tuned by minimising the observed reflected power. Initial conditioning (4 hours) was required to overcome electron multipacting caused by the surface conditions in the resonator. If vacuum is maintained, subsequent start-up conditioning is achieved in 5 minutes. After about 20 minutes both phase and amplitude of the bunching voltage show sufficient stability. Automatic control loops have not been provided.

Buncher Power

The measured Q-value of the resonator is 4900. For this Q-value, the buncher power required for optimum can be calculated and is given in Table I for different particles and beam energies.

Table I Optimum Power Requirements

Beam Energy (MeV)	Buncher Power (Watts)		
	Protons	Deuterons	He-3
3	250	125	85
4	600	300	200
5	1160	580	390
6	2000	1000	670

If the commercial grade copper used for the construction of the inner conductor of the resonator is silver plated, the expected improvement in surface conductivity will increase the Q to 6500. This will reduce the power levels required for optimum bunching to 75% of the tabulated values (viz., 750 Watts for 6 MeV deuterons).

Bunched Pulses

The bunched beam at the target can be observed on a 1 GHz sampling oscilloscope connected to beam pick-offs near the target. This observation is sufficient for setting up the bunching system. The details of the pulse can only be observed by using a nuclear time-of-flight detection system of sufficient resolution. We observed the γ-peak from a ^{19}F target using a Pilot U scintillator on a fast photomultiplier. (Comparison with an identical detector showed that these detectors had intrinsic timing resolution of 0.15 ns FWHM.)

K. H. Böckhoff (ed.), Nuclear Data for Science and Technology, 847–848.

The γ-peaks measured before and after bunching of 3 MeV proton pulses are given in Fig. 1. The bunched pulse with a FWHM of about 0.2 ns is shown to be without distortion or tails.

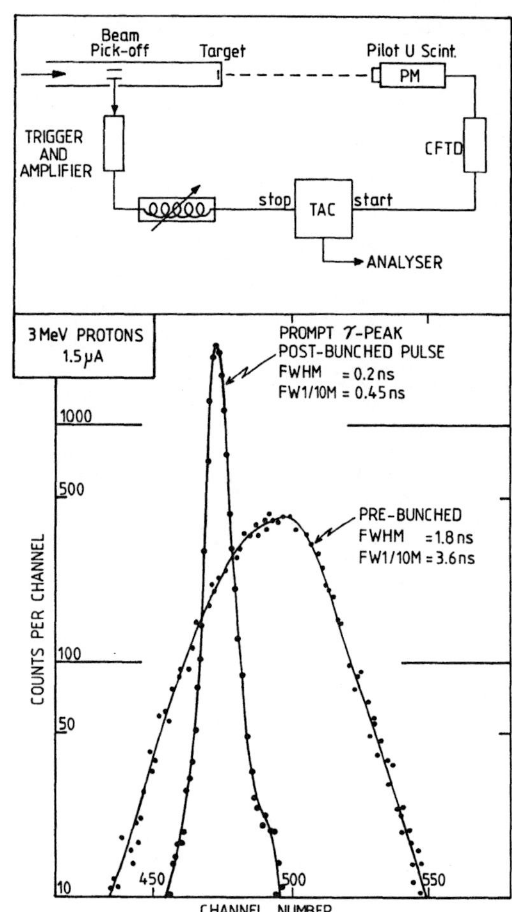

Fig. 1. Experimental time-of-flight spectra for the prompt γ-peak from 3 MeV protons on fluorine, with and without post acceleration bunching.

Conclusion

A post acceleration bunching system has been successfully installed on the SUNI Van de Graaff. It will be used to provide adequate timing (hence energy) resolution for studies of neutron scattering and polarization at 22 MeV. The price one has to pay for post acceleration bunching is an increased energy spread in the accelerated beam (about 35 keV for 6 MeV deuterons). This increase in energy spread is, however, unimportant for high energy neutron time-of-flight.

References

1. J.J. Kritzinger, W.R. McMurray, R.J.N. Nuspliger, N.I.M. 101 (1972) 573.

2. N. Olsson, N.I.M. 187 (1981) 341.

A HYBRID CHARGED-PARTICLE GUIDE FOR STUDYING (n, CHARGED PARTICLE) REACTIONS*

R. C. Haight and R. M. White

Lawrence Livermore National Laboratory
Livermore, California 94550

S. J. Zinkle+

University of Wisconsin
Madison, WI 53706

Charged-particle transport systems consisting of magnetic quadrupole lenses have been employed in recent years in the study of (n, charged particle) reactions. We have completed a new transport system that is based both on magnetic lenses as well as electrostatic fields. The magnetic focusing of this charged-particle guide is provided by six magnetic quadrupole lenses arranged in a CDCCDC sequence (in the vertical plane). The electrostatic field is produced by a wire at high voltage which stretches the length of the guide and is physically at the center of the magnetic axis. The magnetic lenses are used for charged particles above 5 MeV; the electrostatic guide is used for lower energies. This hybrid system possesses the excellent focusing and background rejection properties of other magnetic systems. For low energy charged-particles, the electrostatic transport avoids the narrow band-passes in charged-particle energy which are a problem with purely magnetic transport systems. This system is installed at the LLNL Cyclograaff facility for the study of (n, charged particle) reactions at neutron energies up to 35 MeV.

[Charged-particle spectrometer, magnetic quadrupoles, electrostatic guide, neutron-induced reactions.]

Introduction

In recent years, magnetic quadrupole spectrometers have been used extensively in the study of neutron-induced reactions that produce charged particles.[1-3] These spectrometers consist of a magnetic quadrupole lens multiplet that focuses the charged-particle reaction products onto detectors that are a few meters from the neutron source and the target foil. The detectors subtend a small solid angle for background events due to neutrons and secondary gamma rays and thus these backgrounds can be reduced significantly. The magnetic quadrupole lenses restore a large solid angle (several msr) for detecting the charged particles. The net result is a large increase in the signal-to-background ratio.

One disadvantage of the magnetic quadrupole spectrometer is that the magnetic lens acts as a band-pass filter; that is, it accepts only a finite range of energies for a given magnetic field gradient setting. Typical band-widths are in the range of 25 to 35% in $\Delta E/E$. To measure continuum spectra of protons produced by 14-MeV neutron bombardment of target nuclides below A=100, we have found that about 10 magnet settings are required for reliable data, and about half of these are for charged particles below 4 MeV. For targets of light nuclides even more magnet settings are required to investigate the charged particles below 1 MeV. Thus the measurement of the lower energy charged particles can consume a major fraction of the experimental effort.

To measure low energy charged particles more efficiently we have developed a hybrid charged-particle guide that retains the advantages of the magnetic quadrupole spectrometer at the higher energies. For a good signal-to-noise ratio at all energies, the charged particles are transported from the target to the detectors that are a few meters away. The essence of this hybrid spectrometer then is the transportation of the low energy charged particles by a method different from that of using magnetic lenses. Previous work on electrostatic guiding fields[4] led us to the current design.

Spectrometer Design

Design Considerations

The spectrometer consists of the hybrid charged-particle guide and a particle detector. It is designed for use at the LLNL Cyclograaff facility which can be used as a neutron source by accelerating protons or deuterons and directing them to suitable neutron production targets. The source reactions of most interest are the $^2H(d,n)$ and $^3H(d,n)$ reactions. The Cyclograaff can accelerate deuterons to energies from low energies (around 1 MeV) to 20 MeV and thereby produce neutrons up to 35 MeV. The neutron flux on target is much less than that available in previous experiments at the LLNL Rotating Target Neutron Source--I (RTNS-I), however, so that a new magnetic lens configuration has been constructed for a larger solid angle of acceptance. The electrostatic transport of low energy particles is all the more necessary to achieve reasonable experimental running times with the reduced flux. Since the angular distribution of the charged particles is to be studied, the spectrometer must in addition be rotatable around an axis passing through the target.

Quadrupole Lens Multiplet

A symmetric magnetic quadrupole lens sextuplet (see Fig. 1) was chosen for the magnetic transport. The two outer lenses have 10 cm diameter apertures and the inner four lenses have 15 cm diameter apertures. The first lens is placed as close as possible to the target for maximum solid angle of acceptance. The order of the lenses is CDCCDC (C = converging, D = diverging) in the vertical plane. This symmetric configuration was chosen so that the lens system could be lengthened by separating the first three lenses from the last three. This option will reduce further the background of neutron-induced events in the detectors.

To achieve rotation of the system about an axis passing through the target, the entire spectrometer is placed on a gun mount that allows rotation from 0 to 160 degrees with respect to the incident neutron direction.

Electrostatic Particle Guide

Electrostatic fields have been used previously to guide charged particles produced by the bombardment of materials with neutrons from a reactor.[4] These studies quantified the efficiency of such a system and furthermore showed that it could be used in a high radiation environment. This latter result is important to the implementation of the present charged-particle guide in the neutron flux from the Cyclograaff and especially so if the concept is to be used in the more intense, 14-MeV neutron flux from the RTNS-I.

K. H. Böckhoff (ed.), Nuclear Data for Science and Technology, 849–850.

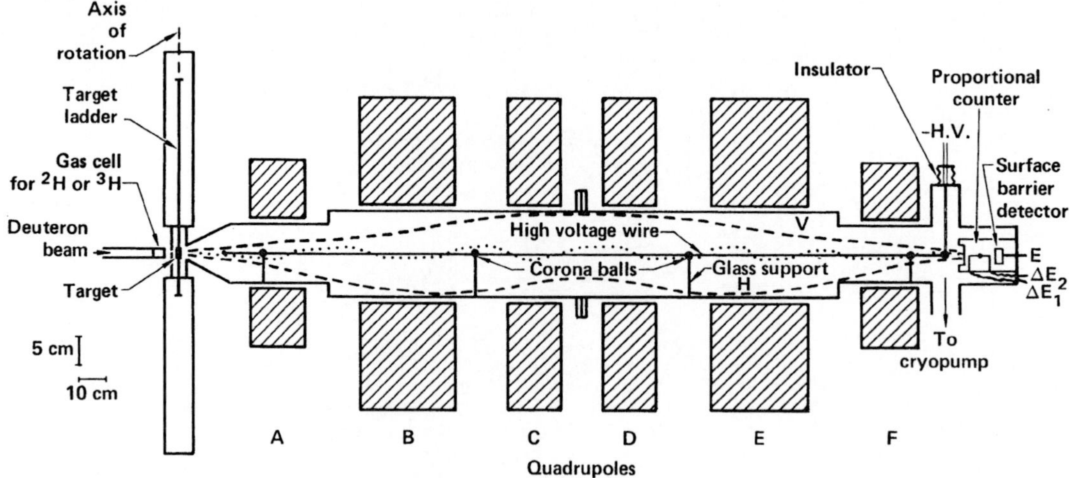

Fig. 1. Hybrid charged-particle guide for studying (n, charged particle) reactions. For low charged-particle energies the central high voltage wire acts as a guide for the charged particles which follow trajectories typified by the dotted curve around the wire. For particles of higher energy, the magnetic quadrupole lenses focus the particles which follow trajectories such as shown by the dashed lines for the horizontal (H) and vertical (V) planes.

The electrostatic guide chosen here is similar in principle, but differs in detail from that of Ref. 4. The wire is 1.5 mm diameter nichrome supported by 3 mm glass stalks that are drawn to approximately 1 mm near the wire. The junction between the wire and the stalks is provided by a corona ball, 4.8 mm diameter, through which the wire passes. The diameter of this ball was chosen to improve the voltage-holding properties of the system by reducing the electrical field where the glass stalk meets the wire. The vacuum in the spectrometer is maintained by a cryopump to be better than 1.3×10^{-4} Pa. Voltages of -45 kV can be maintained reliably in this manner.

Test Results

The hybrid charge-particle guide has been tested with a 6 mm diameter ^{244}Cm alpha-particle source (5.7 MeV) in both the magnetic and electrostatic modes. The detector was a silicon surface barrier diode collimated to 7 mm diameter, which is approximately the size of the alpha-particle source. The high-voltage wire was approximately 14 cm from the source on one end and from the detector on the other.

In the magnetic mode the solid angle accepted by the spectrometer for a source of this size was measured to be 6.9 msr or somewhat larger than the 3.1 msr of the magnetic quadrupole triplet used previously at this Laboratory. (These solid angles depend on the size of the source of charged particles and hence are larger than the values given in Refs. 1 and 2 for sources of larger size.) The present system places the detector 2.9 m from the neutron source, more than the 2.65 m of the triplet system of Refs. 1 and 2, so that the ratio of solid angle accepted to the square of the distance to the source is better by a factor of 2.7. This value is a measure of the improved signal-to-background ratio.

In the electrostatic mode the acceptance of the spectrometer for 5.7 MeV alpha particles is 0.22 msr or considerably less than that in the magnetic mode of operation. One reason is that the particles diverge after leaving the guide field and therefore many will miss the detector. This effect can be compensated for by moving the detector closer to the end of the wire or by using a larger detector. At lower charged-particle energies, the efficiency is expected to be larger as shown in Ref. 4. If collisions with the wire or its supports are neglected, the efficiency should vary as 1/E for $E \gg q V$ where E and q are the particle energy and charge, and V is the voltage on the wire.

A prototype of the electrostatic guide was tested in the neutron flux of RTNS-I to see if the voltage-holding characteristics deteriorated under bombardment by 14-MeV neutrons. In this case the voltage on the wire was 50 kV and the neutron flux was 6×10^7 neutrons/cm^{-2} sec^{-1}. No deterioration was seen after bombardment for several hours. Thus we are confident that the system will perform well in the less intense radiation environment at the Cyclograaff facility.

Conclusion

We have built and tested a hybrid charged-particle spectrometer for investigating neutron-induced reactions that produce charged particles. This spectrometer uses magnetic focusing for charged particles above a few MeV and an electrostatic guide field for particles of lower energy. The efficiency of the spectrometer in the magnetic mode is larger than that of its predecessor at this Laboratory and the background is expected to be lower. In the electrostatic mode, the new spectrometer is used to measure charged particles of low energy without changing the field. The reduction in the number of runs is approximately a factor of 5 for these low energies.

References

1. K. R. Alvar, H. H. Barschall, R. R. Borchers, S. M. Grimes, and R. C. Haight, Nucl. Instr. and Meth. 148, 303 (1978).

2. S. M. Grimes, R. C. Haight, K. R. Alvar, H. H. Barschall, and R. R. Borchers, Phys. Rev. C 19, 2127 (1979).

3. G. Randers-Pehrson, R. W. Finlay, P. Grabmayr, V. Kulkarni, R. O. Lane, and J. Rapaport, Proc. Symp. on Neutron Cross Sections from 10 to 50 MeV, Brookhaven Natl. Lab. Report BNL-NCS-51245 (1980) 389.

4. N. S. Oakey and R. D. MacFarlane, Nucl. Instr. and Meth. 49, 220 (1967).

---

*Work performed under the auspices of the U.S. Department of Energy by Lawrence Livermore National Laboratory under contract #W-7405-Eng-48.

+US DOE Magnetic Fusion Energy Technology Fellow.

THE DOUBLE ROTOR NEUTRON MONOCHROMATOR FACILITY AT THE ET-RR-1 REACTOR

M. Adib, R.M.A. Maayouf, A. Abdel-Kawy, S.E. Gwaily and I. Hamouda

Reactor and Neutron Physics Department, Nuclear Research Centre
Atomic Energy Establishment, Cairo, Egypt

A double rotor neutron monochromator recently installed in front of one of the ET-RR-1 reactor horizontal channels is described. The system consists of two rotors, suspended in magnetic field, spinning at speeds up to 16000 rpm with a constant phase angle relative to each producing bursts of monochromatic neutrons at the sample. Each of the rotors, 32 cm in diameter and 27 Kg in weight, has two slits to produce two neutron bursts per revolution. The slits are with radius of curvature 65.65 cm and 7 x 10 sq.mm cross-sectional area. The jitters of the phase between the rotors were measured at different rotation rates and was found not to exceed $\pm$ 1.5 μsec.
The transmission function of one rotor system was measured and found to be in agreement with that theoretically predicted.

Introduction

Valuable information, concerning the dynamic properties of solids and liquids, can be obtained from neutron inelastic experiments. Most of the neutron inelastic scattering experiments are performed with either a triple-axis crystal spectrometer or a pulsed time-of-flight spectrometer.

Egelstaff [1] and Krebs [2] have reported the parameters of their pulsed TOF spectrometers based on a phased rotor system which applies wire suspension. However, as reported by Mook et al. [3], such high speed rotating equipment is expensive to construct, quite often needs a fair amount of maintenance and requires special safety precautions.

Recently Kalebin et al. [4] designed a phased rotor system using magnetic suspension instead of the wire one. The present work deals with both the description and main parameters of a double rotor neutron monochromator system constructed using Kalebins method.

Experimental Arrangement

The horizontal view of the general arrangement of the double rotor neutron monochromator is represented in Fig. 1, for scattering angles 0 and 90°. Neutrons are emitted from one of the ET-RR-1 reactor (2 MW) horizontal channels (100 mm in diameter). An inpile collimator, (60 cm long and with beam hole 2.5 cm in diameter) made from lead, paraffin and boric acid is placed inside the reactor channel used.

The collimator and the rotors are surrounded by shielding from boron carbide and paraffin. The distance between the rotors could be varied from 1-3 meters. The sample is located at distance 40 cm from the centre of the second rotor. Two helium-3 neutron detector batteries are mounted at the ends of two mobile flight tubes (each 1.6 meters in length), presently situated at 0° and 90° with respect to the incident neutron beam direction.

Construction of the Double Rotor Facility

The double rotor facility consists of two similar rotor systems, each of them is mounted on its mobile platform. The general view of the rotor system is given in Fig. 2. The rotor (1) is installed in evacuated chamber, in order to minimize its friction during rotation. Each of the rotors, 32 cm in diameter and 27 Kg in weight, is made from nickel alloy. The rotor has two slits. The slits are with radius of curvature 65.65 cm and 7 x 10 sq.mm cross-sectional area. The magnetic pole (2) is manufactured from magnetic iron material, this helps its suspension in the magnetic field. The rotor tail (3), duralaluminium, is for rotating the rotor when a rotating magnetic field is induced in the motor's coil (4). The steel ball (5) is used for the rotor's landing, where the position of the landing point is controlled by the indicator (6). The transmitter (7) feeds the current into the magnet's coil (8) in a negative feedback loop. Consequently it controlls the height level of the suspended rotor in space within an accuracy better than 5 μm.

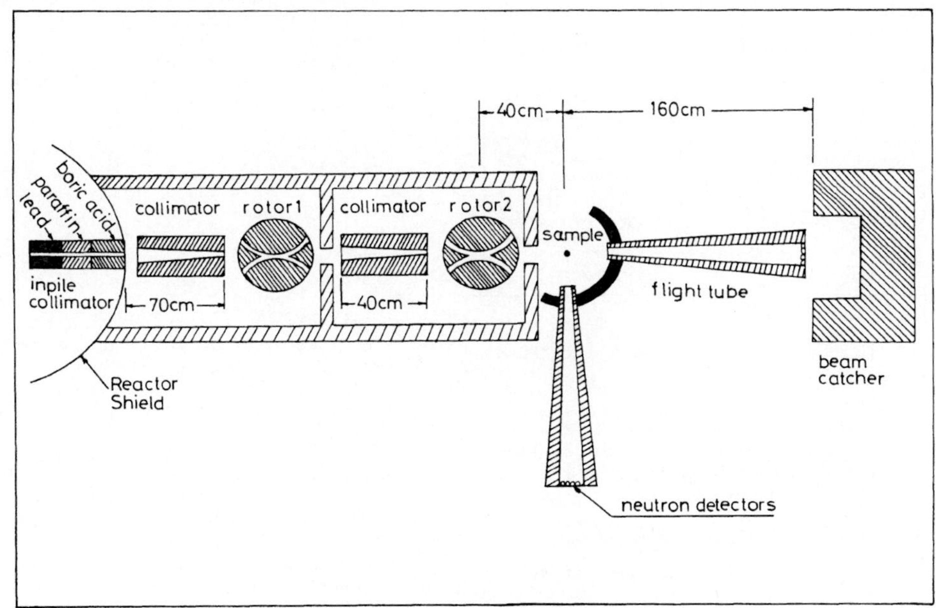

Fig. 1. A schematic of the geometrical arrangement of the double rotor system

K. H. Böckhoff (ed.), Nuclear Data for Science and Technology, 851–853.

1- The rotor. 2- Magnetic pole.
3- Duralaluminium cylinder.
4- Motor's coil. 5- Landing ball.
6- Indicator. 7- Vertical transmitter. 8- Magnet's coil.
9- Horizontal transmitter.
10- Damping coils.
11- Emergency contact.
12- Duralaluminium disc.
13- Rubber ring. 14- Landing cone.
15- Brass ring inside bearing.

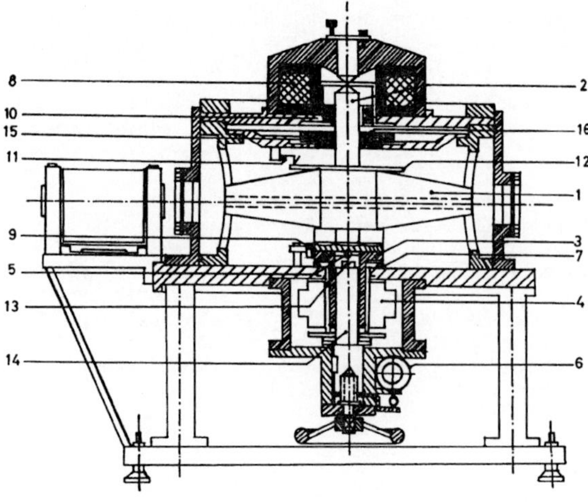

Fig. 2. Schematic design of a rotor system

The transmitter (9) is used, along with the damping coils (10), for controlling the horizontal position of the rotor.

The angular velocity of the suspended rotor, as well as the start of the time analyzers, is controlled by light signals reflected by a mirror fixed on the rotor's body, then received by two diametrically opposite, optical systems.

In order to avoid the collision of the rotor with the magnet's pole, in case of emergency due to unexpected failures of electronic devices, a mechanical contact (11) along with a light duralaluminium disc (12) (fixed on the rotor's body) are used. The cone (14), made from brunze and surrounded by a rubber ring (13), is used for damping the vibrations due to emergency landing of the rotor. The horizontal vibrations are limited by a brass ring (15) which is installed in a bearing. The safety of rotation is ensured while the rotor is rotating up to a maximum velocity 16000 rpm and during long times of operation. The neutron spectra, transmitted through one rotor system, observed at different rotations are represented in Fig. 3.

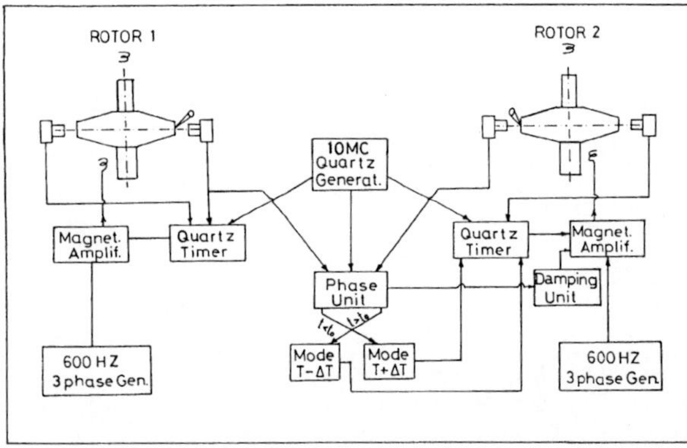

Fig. 3. The neutron spectra transmitted through one rotor

The dashed line is the thermal neutron reactor spectrum, theoretically calculated for the rotor's parameters. It is noticeable that both measured and calculated spectra are in good agreement.

The Synchronization of the Rotating Rotors

The method of synchronizing rotors is based on the use of the advantages of the rotor's suspension in magnetic field, such as the small friction and the absence of uncontrolled mechanical effects upon the angular velocity. The simplified block-diagram of the synchronized rotors is represented in Fig. 4.

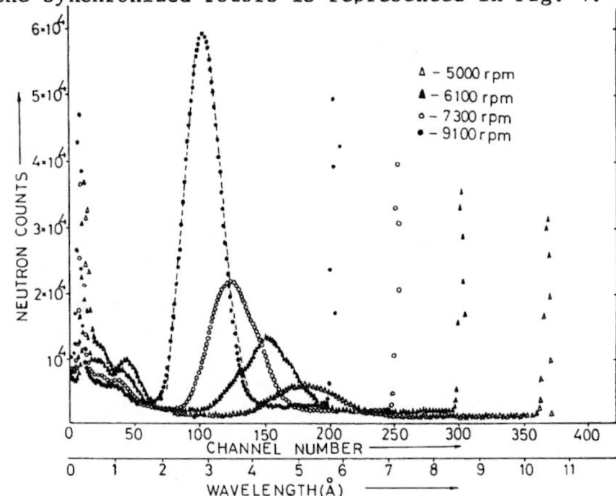

Fig. 4. A simplified block diagram of the rotor's synchronization system

The phase and angular velocity control are realized using signals from the optical devices and comparing them with a fixed period of a 10 MC quartz generator. The rotor (1) rotates with a constant period T and is derived by a quartz timer unit, while the phase unit ensures the rotation of rotor (2) with a period T-ΔT (mode A) or with period T+ΔT (mode B). The pulses from rotor (2) (in the time scale) are regarded relative to the signals from rotor (1) which are considered as starting ones. In such case the phase of rotor (2) t_o ($0 \leqslant t_o \leqslant T$) can be selected using the phase unit. The phase unit changes the operation of the second rotor's quartz timer from mode A to mode B and vice versa. For example if the phase unit is operating in mode A, the second rotor then rotates with a velocity higher than that of the first rotor. Thus continuously decreasing the time elapsed between the signals of the first and second rotors from the value T and down to zero. A moment t_o, within the time interval, occurs and then the phase unit operates in mode B. Thus the second rotor will have a period of rotation equal to T+ΔT and the operation of the phase unit will be opposite to that discussed above. Therefore the fluctuations of the phase shift between the two rotors will be only within Δt. A damping unit is designed specially for decreasing such fluctuations. The value of ΔT can be varied between 0.1 - 2 μsec. The jitters of the phase between the two rotors, when rotating at different rotations, are represented in Fig. 5. The double peak observed, in all of them, around the zero phase is due to the effect of the damping unit.

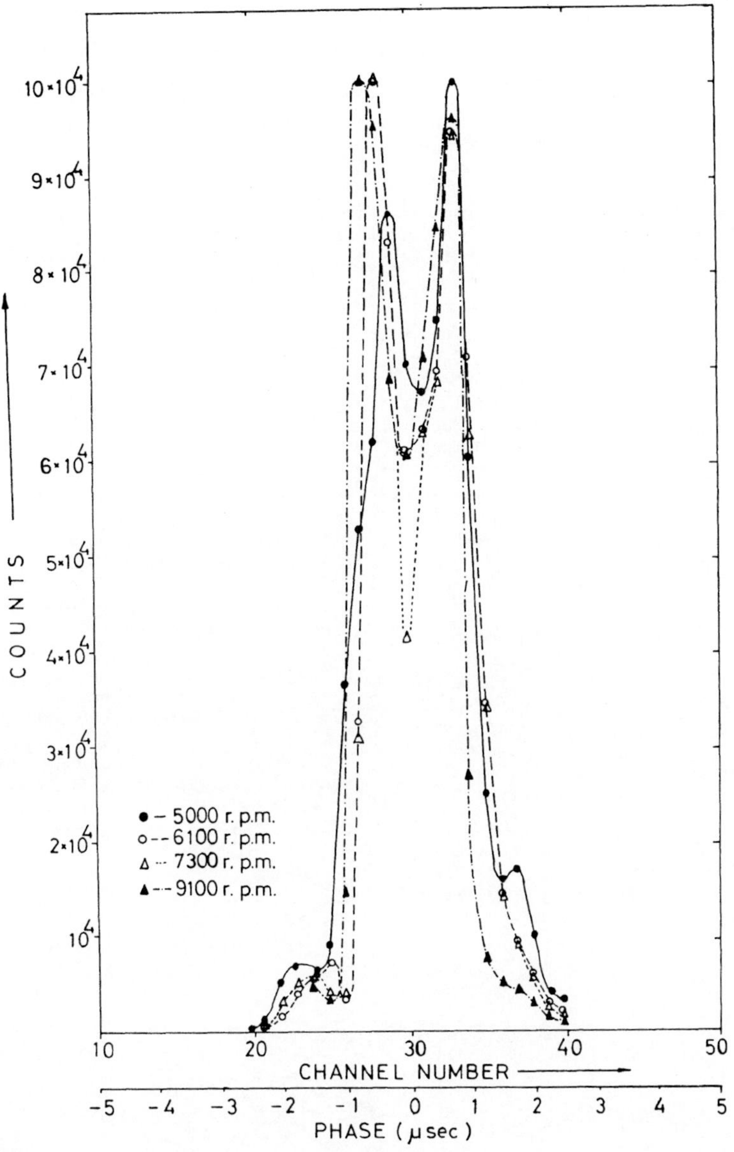

Fig. 5 The jitters of the phase between the two rotors

From Fig. 5 one can notice that the jitters of the phase do not exceed the value $\pm$ 1.5 μsec, which corresponds to an accuracy 0.04°, of the phase between both rotors, at 5000 rpm.

Acknowledgement

The authors would like to express their deepest gratitude to the division of technical assistance of the IAEA for the financial support of the present work.

References

1. P.A. Egelstaff et al., Inst Atomic Energy Agency, Proc. Symp. on Inelastic Scattering of Neutrons in Solids & Liquids, Vienna (1960) p. 165

2. K. Krebs, Proc. Symp. on Neutron Inelastic Scattering, Vol. 2, (1968)

3. H.A. Mook et al., Nucl. Inst. & Meth. 116, 205 (1974)

4. S.M. Kalebin, G.V. Rukalino, M. Adib, ITEPh, Rep. 698, Moscow, (1969)

A FACILITY FOR THE INVESTIGATION OF (n,CHARGED PARTICLE) REACTIONS

G. Vourvopoulos, E. Kossionides and T. Paradellis

Tandem Accelerator Laboratory
N.R.C. Demokritos, Aghia Paraskevi Attikis, Greece

A system comprised of a gas cell and a magnetic quadrupole triplet was developed for measuring (n, charged particle) reactions. The large (2.68m) distance between target and detector results in appreciable reduction of the background while the solid angle (about 15 msr) ensures measurement of small cross sections.

Introduction

The study of neutron induced reactions in the energy range E_n=5-25 MeV has been extremely limited till now, especially for the cases where the reaction products are charged particles. Studies of nuclear structure or reaction mechanisms require adequate energy resolution of the neutron beam as well as a detection system capable of subtending a large solid angle and having good resolution and large efficiency. Furthermore, the neutron beam generated has to have the largest possible intensity, which means the utilization of intense primary charged particle beams.

The above requirements have been incorporated in the facility described below and which is composed of a a) high intensity electrostatic accelerator for the primary charged particle beam b) a gas cell for the production of monochromatic neutrons and c) a quadrupole triplet system focusing the charged particles emanating from the (n, charged particle) reactions on a detector located at a large distance from the target.

Quadrupole doublets and triplets have been used frequently in the past[1] for focusing charged particles at some distance from the target. This characteristic with the subsequent large reduction of the background has also been used previously[2] for the study of (n, charged particle) reactions.

Neutron Production

For the production of the primary charged particle beam the T11/25 Tandem Van de Graaff accelerator of NRC Demokritos is utilized. This is a high intensity electrostatic accelerator with a maximum terminal voltage of 5.5 MV. Figure 1 shows the detailed structure of the gas cell and the target ladder. Analyzed deuteron beams of up to 50 μA can be delivered on target. The neutrons are produced through the reaction D(d,n) taking place in a gas cell. (A). The neutrons emerge from the gas cell with energies between 5 and 13 MeV for deuteron energies between 2 and 10 MeV. A gas target is used because it can give approximately a factor of 10 more neutrons for the same energy loss as do standard solid targets such as ZrD. The gas cell is made of stainless steel with a wall thickness of 0.2 mm. length 5 cm and external diameter 1 cm. There is a 7 μm Mo entrance foil (B) to the gas cell. A number of materials were considered as entrance foils (Havar, Ni) but Mo was chosen for the combination of beam intensity, cell pressure and thickness. Because of the entrance foil, the beam intensity has to be kept at a lower level than the one the accelerator is capable of delivering . The beam is kept at a current of less than 10 μA since foils tend to have a short lifetime otherwise. Several tests have indicated that if the beam is controlled, runs of between 24-36 hrs of continuous

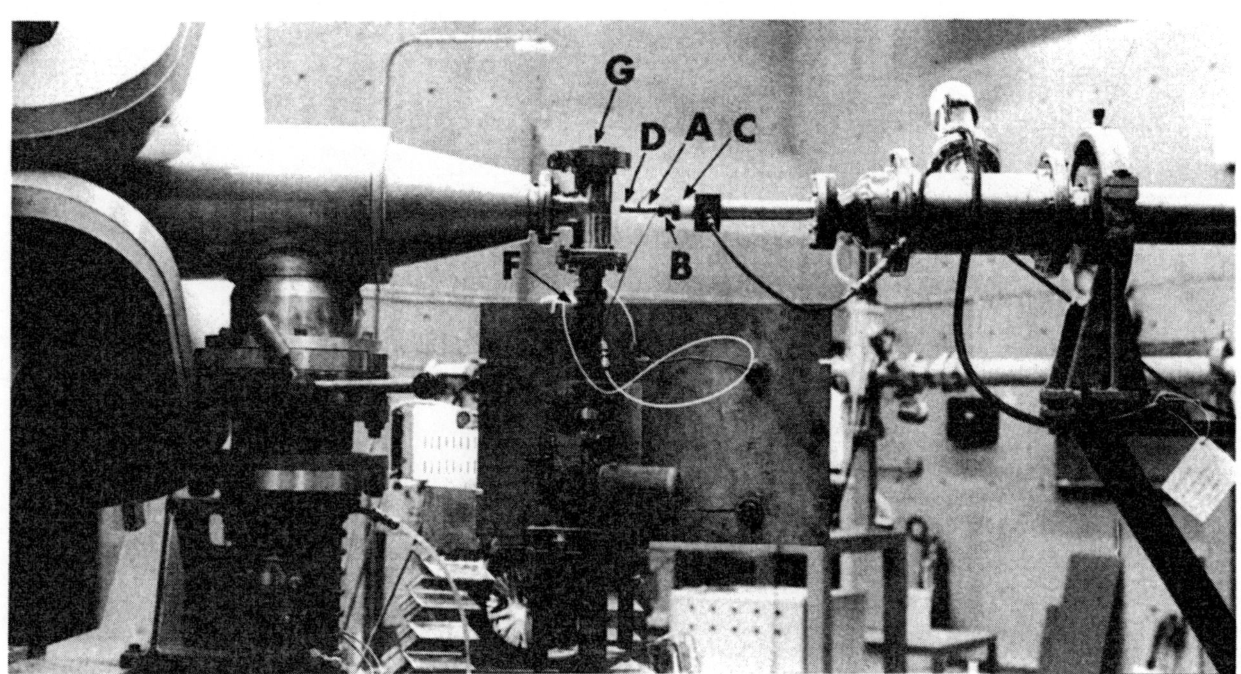

Fig.1 A close up view of the gas cell and the target ladder. A: gas cell, B: Mo entrance foil, C: Ta Collimator, D: Pt beam stop. F: target rod. G: target chamber.

K. H. Böckhoff (ed.), Nuclear Data for Science and Technology, 854–855.

Fig.2 An overview of the facility

bombardment can be achieved without foil rupture. There is a 3 mm Ta collimator (C) in front of the foil and the beam stop (D) is Pt 0.9 mm thick, both appropriately cooled. The cell operating pressure is 1000 Torr.

Target Ladder

The targets are positioned approximately 3.5 cm from the end of the gas cell. The target assembly is mounted rigidly on the floor (F) while the target chamber (G) rotates with the quadrupole triplet. The target chamber is made of stainless steel with a thickness of 1 mm at the position of the neutron entry into the chamber. The target ladder can be servomechanically controlled from the accelerator control room.

Magnet Quadrupole Triplet

The background from neutron induced reactions in the detector material and its surrounding has beset many experimental set ups used for (n, charged particle) measurements[3]. By moving further away from the target in order to reduce the background there is a severe drop in the solid angle subtended by the detector. The use of a quadrupole triplet as a means of focusing the charged particles at some distance from the target offers the following advantages

a) large reduction of the background
b) appreciable solid angle
c) good energy resolution

Due to the non-dispersivity however of the quadrupole spectrometer, since it acts as an energy band bass filter there is a drawback on the range of energies that it can accept. For a given quadrupole current setting one obtains $\Delta(B\rho)/B\rho(FWHM)=0.2$ to 0.3. The Demokritos quadrupole spectrometer is composed of three identical quadrupoles that have an internal diameter of 20 cm (Figure 2). The pole length is 22 cm and the maximum induction is 0.99 T at 10 cm from the axis. They are mounted on a rotating platform which can cover a range of angles from -30° to $+140^{\circ}$. The quadrupole triplet can also be moved radially on two cylindrical tracks allowing thus their placement closer or further away from the target as well as a variation of the distance between them.

The magnetic field calculations and the ray tracing for the system were performed with a computer code written at Orsay[4] which takes into account all the optical aberrations. These calculations show that for a total distance of 268 cm between target and detector, at 17 msr solid angle is obtained with a magnification of 1, a horizontal acceptance angle of 2.5° and vertical acceptance angle of 9°.

At the present , the detector for the charged particles is a solid state detector. The overall energy resolution of the facility (excluding target thickness effects) is less than 300 keV.

The authors wish to thank F. Trouposkiadis for designing the quadrupole carriage and J.P. Schapira (Orsay) for his help in bringing to existence this facility.

References

1. H. Laurent and J.P. Schapira, Nucl. Instr. and Meth. 162 (1979) 181

2. K.R. Alvar, H.H. Barschall, R.R. Borchers, S.M. Grimes, R.C. Haight, Nucl. Instr. and Meth. 148 (1978) 303

3. H. Vonach, Proc. 2nd Symp. on Neutron Induced Reactions, I. Ribansky and E. Betak editors, (Slovak Academy of Sciences Bratislava 1980)

4. H. Laurent, J.P. Schapira, S. Fortier, S. Gales and J.M. Maison, Nucl. Instr. and Meth. 117 (1974) 17

NEUTRON FILTERS FOR PRODUCING MONOENERGETIC NEUTRON BEAMS

J. A. Harvey and N. W. Hill
Oak Ridge National Laboratory, Oak Ridge, Tennessee 37830 USA

J. R. Harvey
CEGB Berkeley Nuclear Laboratories, Berkeley, England

Neutron transmission measurements have been made on high-purity, highly-enriched samples of ^{58}Ni (99.9%), ^{60}Ni (99.7%), ^{64}Zn (97.9%) and ^{184}W (94.5%) to measure their neutron "windows" and to assess their potential usefulness for producing monoenergetic beams of intermediate energies from a reactor. Transmission measurements on the Los Alamos Sc filter (44.26 cm Sc and 1.0 cm Ti) have been made to determine the characteristics of the transmitted neutron beam and to measure the total cross section of Sc at the 2.0 keV minimum. When corrected for the Ti and impurities, a value of 0.35 ± 0.03 b was obtained for this minimum.

[^{58}Ni, ^{60}Ni, ^{64}Zn, ^{184}W, ^{45}Sc, σ_T, 0.05-5000 keV, neutron filters]

Introduction

Neutron filters of thick samples (~ 50 cm) of iron, (also ^{56}Fe), scandium, silicon and ^{238}U have been used for many years at several reactors to produce monoenergetic neutron beams with energies of 24, 2.0, 140 and 0.186 keV, respectively. Often it is necessary to add a secondary material in the beam (such as Al with an iron filter) to reduce the intensity of other energy neutrons from secondary windows. These monoenergetic neutron beams have been used for basic nuclear physics research, such as average resonance capture gamma ray spectroscopy and for accurate bench-mark type measurements at specific energies. The neutron filter technique has been combined with a pulsed neutron source such as ORELA to measure the inelastic neutron scattering of ^{238}U[1] and to utilize less intense neutron groups from higher energy windows in the filter for detector development.

An international filter beam project[2,3] has been proposed to produce beams of intermediate energies which could be used for many applications. These include calibration and development of instruments, fundamental radiobiology, and boron neutron capture radiotherapy for brain tumors. Earlier studies[2,4] proposed several isotopes which might be useful as filters. Transmission measurements on some of these isotopes were undertaken at ORELA and are reported in this paper.

Knowledge of the cross section at the 2.0-keV minimum in ^{45}Sc is important in order to determine the optimum thickness of Sc to be used as the filter. Early measurements indicated that this cross section was very small (<0.1 b); hence, very thick filters could be used with little loss of 2.0 keV neutrons. However, recent measurements have reported values from 0.71 ± 0.03,[5] 0.26 ± 0.07,[6] and 0.24 ± 0.07[7] b for this minimum. It was hoped that a measurement on the Los Alamos Sc filter would contribute to the solution of the discrepancy.

Experimental Method

The transmission measurements were made using a 12 mm thick, 111 mm dia. ^{6}Li glass scintillator located 78.203 m from the ORELA water moderated Ta neutron target. The linac was operated at 800 Hz with a pulse width of 30 nsec. The samples were located at 9 m and were alternated with a blank holder and background samples on a 30 minute cycle. The metallic samples of ^{58}Ni, ^{60}Ni, ^{64}Zn, and ^{184}W, were carefully prepared by P.R. Kuehn, Target Center, ORNL, to minimize the oxide and water content of the samples. The Sc sample (44.26 cm Sc and 1.0 cm Ta) was loaned by E.T. Jurney, LANL. The Sc_2O_3 was heated to remove water and the sample was prepared in an inert atmosphere to prevent absorption of water.

Data Analysis

The time-of-flight spectra were corrected for the dead-time of the time digitizer, a time dependent background due to neutron capture in the water moderator (17.6 µsec capture decay time), and a constant room background. The contributions of the 1 cm of Ti and the air displaced by the Sc sample were applied to the measured transmission of the Los Alamos Sc filter in order to obtain the total cross section of Sc.

Results and Discussion

The transmissions of the ^{58}Ni (N = 0.60 a/b, 99.9%) and ^{60}Ni (N = 0.65 a/b, 99.7%) filters are shown in Figures 1 and 2. The transmissions for the lowest energy windows are in agreement with predictions[8] based on resonance parameters. When such thick samples are used, small amounts of oxygen or water can reduce the transmission significantly. For a 1/E reactor flux, the relative importance of each window can be estimated from the plot of transmission vs log energy.

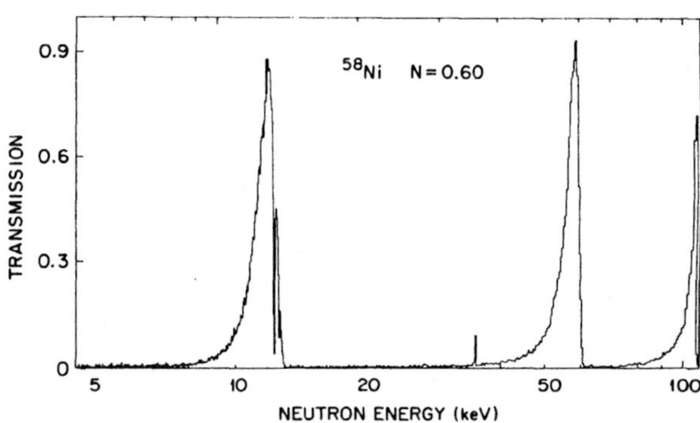

Fig. 1. Transmission of ^{58}Ni (N = 0.60 a/b, 99.9%).

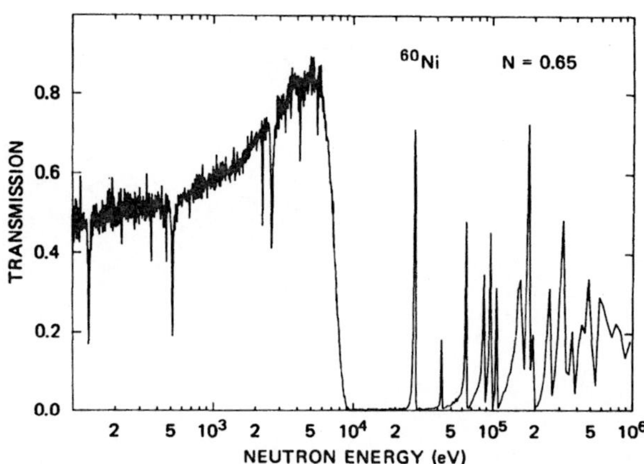

Fig. 2. Transmission of ^{60}Ni (N = 0.65 a/b, 99.7%).

K. H. Böckhoff (ed.), Nuclear Data for Science and Technology, 856–858.

The transmissions of the ^{64}Zn (N = 0.35 a/b, 97.9%) and ^{184}W (N = 0.41 a/b, 94.5%) filters are shown in Figures 3 and 4. These measurements are useful for confirming the theoretical calculations and giving information on the higher energy windows. The measured transmissions of the lowest energy windows for both these filters are considerably lower than those calculated from the resonance parameters and the enrichments of the samples. An analysis for oxygen in the ^{64}Zn metal, showed that the oxygen content of 0.8% would only account for about 15% of the discrepancy. Hence, the discrepancy is most likely due to hydrogen in the form of a hydroxide. The presence of hydrogen in the ^{184}W sample is also suspected, since the metal was plated out of an aqueous solution.

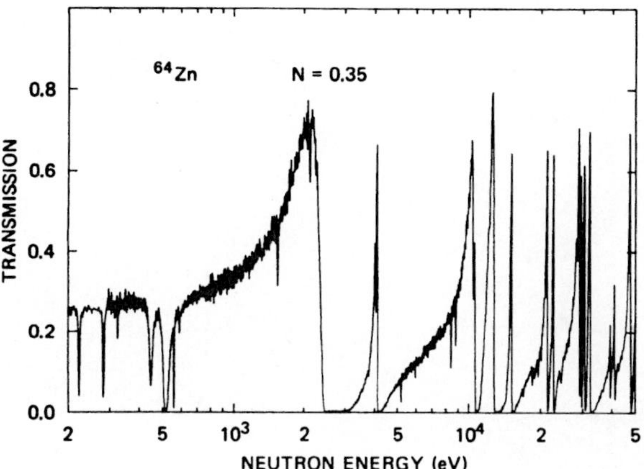

Fig. 3. Transmission of ^{64}Zn (N = 0.35 a/b, 97.9%).

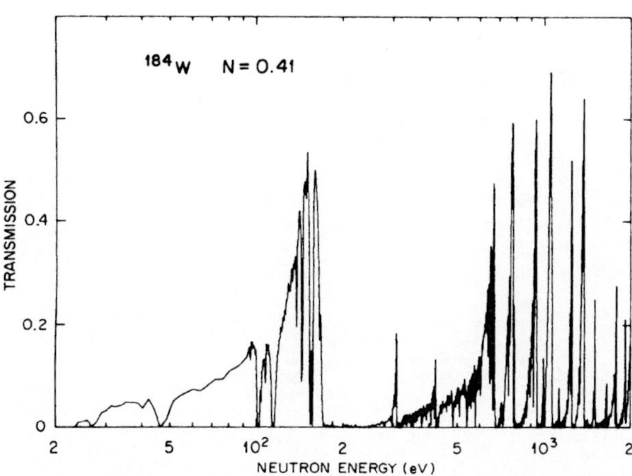

Fig. 4. Transmission of ^{184}W (N = 0.41 a/b, 94.5%).

The transmission of the Los Alamos Sc + Ti filter is shown in Figure 5. The cross section of Sc in the region of 2 keV compared to that reported by BNL/RPI is shown in Fig. 6. The cross section of Sc derived from the Los Alamos filter at the 2-keV minimum is 0.41 ± 0.02 b. Spectroscopic analysis of the Los Alamos sample showed the presence of Mn (105 ppm), Ta (70 ppm), V (40 ppm) and Er (20 ppm) impurities. After correcting for the Mn impurity (but not for the possible presence of hydrogen or oxygen) a value of 0.36 ± 0.03 b was obtained. This is considerably lower than the BNL/RPI value of 0.71 ± 0.03 b which had been corrected for hydrogen, oxygen, and tantalum impurities.

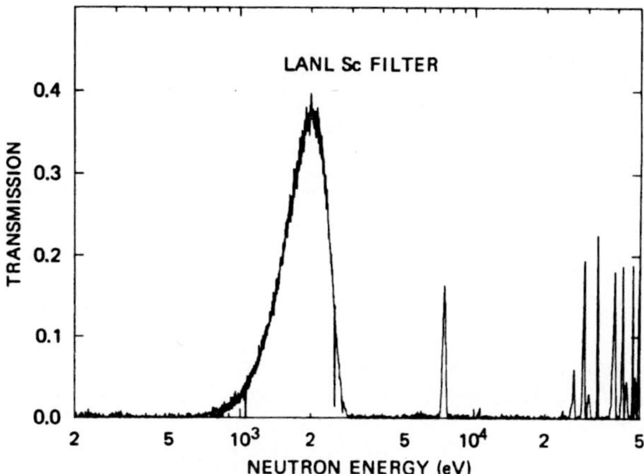

Fig. 5. Transmission of Los Alamos Sc Filter (44.26 cm Sc + 1.0 cm Ti, (N_{Sc} = 1.736, N_{Ti} = 0.057).

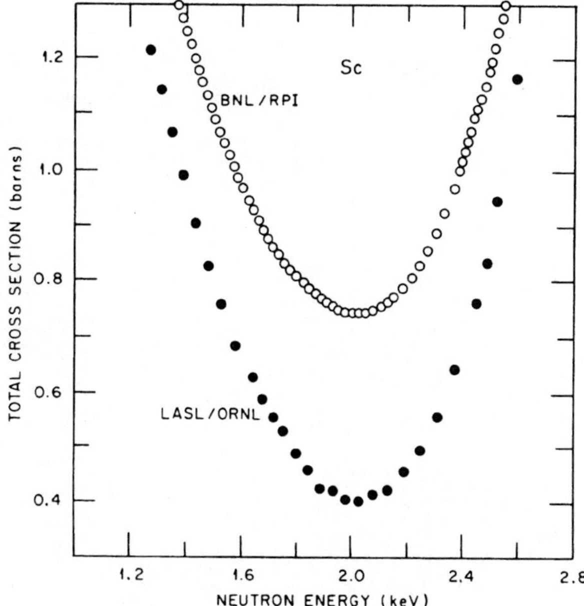

Fig. 6. Total cross section of "impure" Sc.

Transmission measurements on the Sc_2O_3 sample when corrected for the oxygen contribution (measured by using samples of BeO and O), gave a value of 0.47 ± 0.04 b, uncorrected for any impurities in the sample. Our value of 0.36 ± 0.03 b for the LANL Sc is somewhat higher than the recently reported values of 0.26 ± 0.07[6] and 0.24 ± 0.07[7] b but significantly lower than the value obtained for the BNL Sc.

Conclusions

Neutron filters at reactors for producing monoenergetic neutron beams of intermediate energies have been useful in the past and will continue to be in the future. Neutron filters of isotopes such as ^{58}Ni, ^{60}Ni, ^{64}Zn and ^{184}W have unique characteristics, and will probably be used for special applications. Neutron filters combined with a pulsed neutron source is also a valuable technique.

There is still some uncertainty in the minimum of the total cross section of Sc at 2.0 keV because of the possibility of impurities in the sample. Experimenters planning to use thick filters should obtain highly pure samples since thick samples are needed, and the transmission of the filter they plan to use should be measured.

Research was sponsored by the Division of Nuclear Sciences, U. S. Department of Energy, under Contract No. W-7405-eng-26 with the Union Carbide Corporation.

References

1. R.R. Winters, N.W. Hill, R.L. Macklin, J.A. Harvey, G.L. Morgan, Nucl. Sci. Eng. 78, 147 (1981).
2. A.J. Mills and J.R. Harvey, 1978, Intermediate Energy Neutron Production: A Survey of Existing Techniques, a Proposed Service and its Applications, Euratom Report EUR 6107 EN, UK CEGB Report RD/B/N4324.
3. A. Mill and J.R. Harvey, this conference.
4. A.J. Mill and J.R. Harvey, 1980, Reactor- and Accelerator-Based Filter Beams, UK CEGB Report RD/B/N4776.
5. H.I. Liou, R.E. Chrien, R.C. Block, and K. Kobayashi, Nucl. Sci. Eng. 67, 327 (1978).
6. V.F. Razbudey, A.V. Muravitsky, V.P. Vertebnyi, A.L. Kiriluk, Nuclear Cross Sections for Technology, ed. J.L. Fowler, C.H. Johnson and C.D. Bowman, NBS Special Publication 594, p. 890.
7. Y. Fujita and K. Kobayashi, private communication 1981.
8. A.J. Mill, 1979, A Computer Code for Calculating Neutron Cross-Sections from Resonance Parameter Data, CEGB Report RD/B/N4632.

APPLICATION OF A NOVEL 14-MEV NEUTRON ACTIVATION ANALYSIS SYSTEM FOR CROSS-SECTION MEASUREMENTS WITH SHORT-LIVED NUCLIDES

B. Anders*, R. Pepelnik and H.-U. Fanger
Institut für Physik, GKSS Forschungszentrum Geesthacht
Federal Republic of Germany

* I. Institut für Experimentalphysik, Universität Hamburg
Federal Republic of Germany

Nuclear data measurements are reported that have been performed by means of a new intense 14-MeV neutron generator, named KORONA, with cylindrical acceleration structure. The precise neutron energy as well as the flux distribution have been calculated and experimentally investigated. Those experiments were performed with the aid of two different methods: a) the activity-ratio method and b) the reaction-threshold technique. The mean neutron energy was found to be 14.6 ± 0.2 MeV.

With a new electronic system, designed by Westphal, several unknown or uncertain cross-sections for (n,n'γ)-reactions leading to isomeric states with halflives below 12 s have been determined: ^{79}Br(n,n'γ), ^{167}Er(n,n'γ), ^{168}Er(n,2n), ^{176}Yb(n,n'γ), ^{183}W(n,n'γ), ^{184}W(n,2n), ^{191}Ir(n,n'γ), ^{192}Os(n,n'γ), ^{207}Pb(n,n'γ) + ^{208}Pb(n,2n). In addition the activation cross-sections of ^{51}V(n,p), ^{174}Yb(n,p) and ^{190}Os(n,n'γ)^{190m}Os have been measured.

Introduction

Cross-section data for (n,n'γ)-reactions leading to short-lived isomeric states are rather scarce [1] but of considerable theorectical and practical interest. First, 14-MeV neutrons can excite compound states with high spin and thus are a useful tool to investigate the so-called Yrast-traps. Second, as the (n,n'γ) cross-sections are often very high in fast-neutron reactions, a good elemental sensitivity in neutron activation analysis is obtained, especially when the dominant (n,2n)-reaction leads to a comparatively long-lived radionuclide.

Apparatus

At the GKSS Research Center Geesthacht a new intense 14-MeV neutron generator has been put in operation in March 1981. The main components of the facility, named KORONA, are a sealed neutron tube and an integrated fast pneumatic rabbit system with sample transfer times less than 140 ms.

A schematic view of KORONA is presented in Fig. 1. Powdered as well as liquid samples in polyethylene (PE) containers (Fig. 1,d) are irradiated on the axis of a cylindrical closed-end hole of the neutron tube.

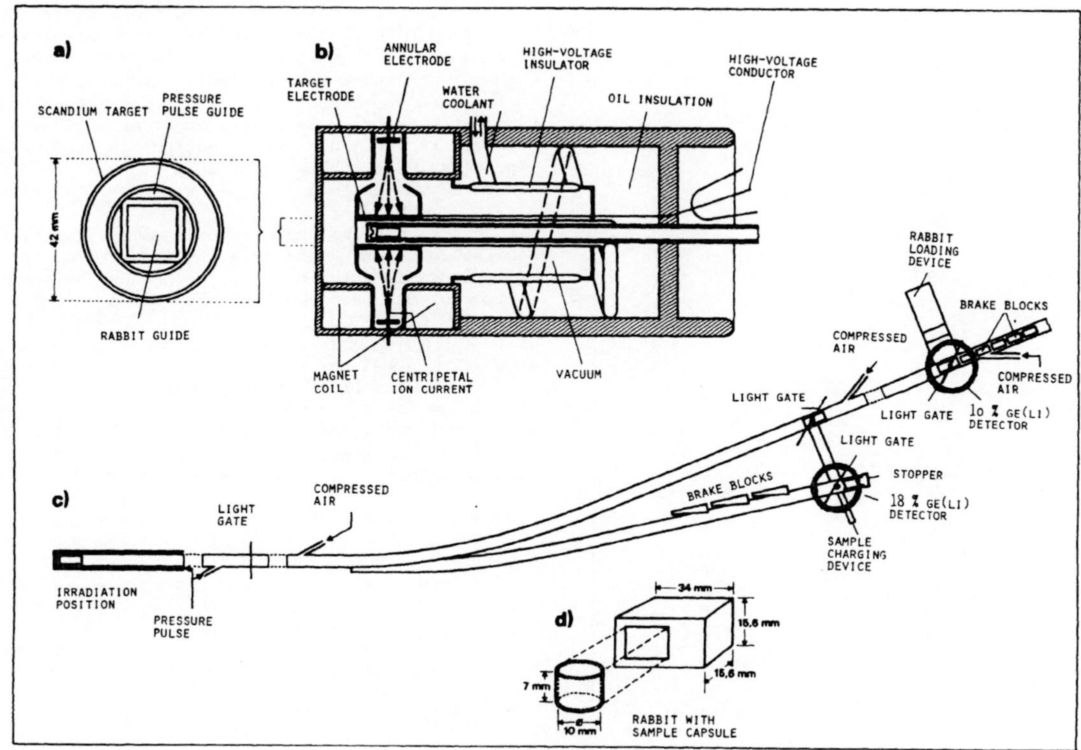

Fig. 1: Schematic display of KORONA. a) Cross-sectional view of the cylindrical target
b) Schematic longitudinal section of the neutron tube
c) Top view of the pneumatic transport system
d) Dimensions of rabbit and capsule

K. H. Böckhoff (ed.), Nuclear Data for Science and Technology, 859–862.

The neutrons are produced by centripetal ion bombardement of a thin scandium layer deposited on a cylinder of 42 mm diameter (s. cross-sectional views a) and b) of Fig. 1) where the T(d,n) reaction takes place. The activated samples carried by a rugged polyamide rabbit are pneumatically transferred to a 16 m distant detector station in a separate building where, in a final phase of flight before stopping, rabbit and sample container are separated by means of centrifugal force in a curved branch of the guiding tube. Details have been already described elsewhere [1, 2].

The γ-ray spectroscopy system consists of a Canberra-2001 preamplifier, a Canberra-2020 main amplifier, a Laben-8215 ADC and a Nuclear Data-66 MCA. It has been improved for counting rates from originally 200000 cps to about 700 000 cps by means of a DC-level controlled charge restoration for the preamplifier and a novel real-time correction of counting losses. This was necessary to cope with the high initial activity of short-lived isotopes produced by an intense neutron flux of more than $3 \cdot 10^{10}$ n/cm^2 s. The new electronic components were designed by Westphal [3].

Neutron flux and energy distribution

For determining the important neutron flux and neutron-energy distribution, several experimental and theoretical efforts have been made. A special rabbit for the simultaneous transport of several small-sized wires was used to measure the flux distribution over the rabbit volume. The activity of the irradiated wires was counted (each wire separately)with the 10 % Ge(Li)-detector at the rabbit loading position (Fig. 1c).. The result confirms earlier theoretical calculations [4] of the flux distribution of KORONA, predicting rather constant values within the rabbit.

The neutron energy was determined with the aid of two different experimental methods: a) the activity-ratio method [5] using the (n,2n)-reactions on Zr and Nb as well as on Zr and U and b) the reaction-threshold technique with 22 different element samples [6].

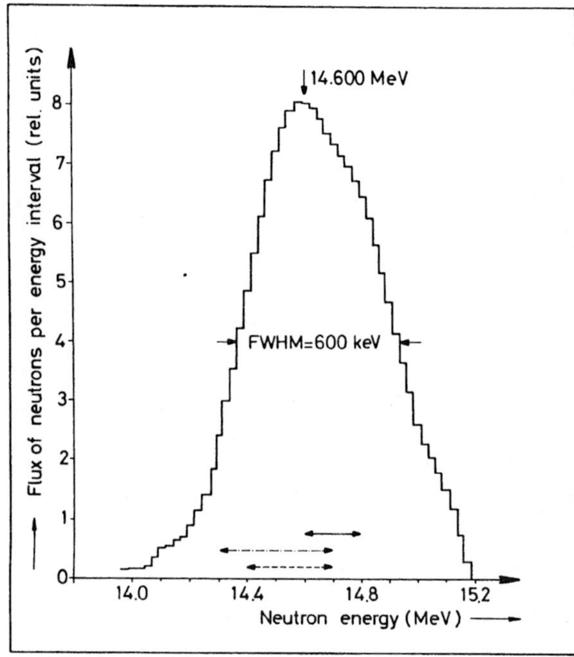

Fig. 2: Experimentally determined average neutron energy (-·- SANDC; -- Nb/Zr-method;—Zr/U-method, for explanation s. text) and theoretical neutron spectrum for KORONA [10].

With the activity-ratio method the average neutron energy is easily determined, because the well-known ^{90}Zr(n,2n) and the ^{238}U(n,2n) cross-sections vary rapidly (but with opposite slope) in the 14 to 15 MeV region while the ^{93}Nb(n,2n) cross-section keeps constant. The reaction-threshold technique requires a computer program for unfolding the energy distribution of neutrons from the measured activities of various radioisotopes. As already mentioned with the flux distribution measurements, we used several high-purity wires irradiated in a special transport rabbit. The induced γ-ray spectra were recorded and analyzed to get the activation cross-sections. These data were used as input for the computer program SANDC [6 - 8] run at our laboratory computer ND 6620. The convergence of the iteration procedure was very slow in the 14 to 15 MeV region; we think, that further improvement could be achieved, e.g., with the recently published computer program LOUHI 78 [9]. The FWHM of the neutron energy distribution (calculated with SANDC) is 600 keV.

The average neutron energy from all three experimental methods was found to be 14.6 ± 0.2 MeV, as can be seen from Fig. 2. The results are in very good agreement with an elaborate theoretical calculation of the neutron spectrum of KORONA [10], performed in our institute at Geesthacht.

With the help of several (n,γ)-reactions the contribution of thermal neutrons to the fast total neutron flux could be estimated to be less than $5 \cdot 10^{-3}$ [6].

Sample irradiation

For Br, Yb, Os, Ir and Pb natural samples of high purity (s. first column of Table I) were used. For Er and W enriched isotopes were available, so that their (n,n'γ) cross-sections could be determined with negligible contribution of the neighbouring isotope that populates the same metastable state via the (n,2n)-reaction. The amount of interfering (n,γ)-activities was less than 0.3 % for the nuclides investigated and could be neglected. During activation the neutron source strength of about $3 \cdot 10^{12}$ n/s was monitored by means of a long counter and a multi-channel scaler. For absolute flux calibration high-purity vanadium powder was added to the sample material. The ^{51}V(n,p) ^{51}Ti reaction with $T_{1/2}$ = 5.76 m was chosen because the cross-section is nearly constant in the 14 to 15 MeV region. This cross-section had been measured previously at KORONA to be 28 ± 1.5 mb, based on the well-known (n,α)- and (n,p)-reaction of aluminum.

For all reactions under investigation, ten cycles were sufficient for a good counting statistics. The irradiation and measuring times were about twice the half-lives of the isotopes given in column 3 of Table I.

Analysis

The absolute efficiency calibration of the 18 % Ge(Li)-detector for our counting geometry was performed using a homogeneously mixed solution of radioactive sources, commercially available. All spectra have been recorded on the multi-channel analyzer ND 66 using 4096 channels.

Because of the close distance of the activated capsule bottom to the detector cap (26.5 mm) a coincidence summing correction had to be performed in order to take into account cascading photons. These corrections amounted to about 20 % for the ^{176m}Yb- and the ^{190m}Os-lines, where up to 6 γ-quanta can follow each other. Also for ^{183m}W and ^{174}Tm summing corrections (about 10 %) were necessary, although for the tungsten measurements the source to detector distance was enhanced to 56.5 mm.

For all γ-lines with energies below 300 keV (with the exception of those from the low Z bromine) absorption correction factors were taken into account. Using the absorption coefficients from Veigele [11], these corrections were around 10 % with the exception of the tungsten isotope, were the 108 keV-line must be corrected by 100 %, the 160 keV-line by 29 % and the 210 keV-line by 14 %.

Results and Discussion

For the reactions considered in this work only the metastable cross-sections were measured. The results are given in Table I, together with relevant spectroscopic data. The indicated errors are composed as follows:

a) the error of the peak area estimated by the peak-analysis program, ranging from 0.3 to 3 %;

b) summing correction errors, ranging from 1.5 to 2.5 %;

c) absorption correction errors, taken as 5 % for all cases [11];

d) the error in flux and γ-ray efficiency calibration, estimated ± 3 %;

e) γ-intensity errors taken into account individually, as given by ref. [13], ranging from 0.2 to 9 %. For the γ-lines of the tungsten and the iridium isotope, an error of 9 % was assumed, because of lack of other information.

Errors due to halflives ($\leq$ 1 %), mass determinations ($\leq$ 0.1 %) and isotopic abundances ($\leq$ 0.5 %) can be neglected, compared to other errors. All contributions to the total error were thought to be independent of each other and thus added geometrically.

For bromine, an accurate result for the (n,n'γ) cross-section can be given now. The previously published values differ by a factor of two.

The literature cross-section data for ytterbium are scarce or uncertain. For the (n,n'γ)-reaction we only found a confirmation for the old (but not inaccurate) value from ref. [15]. The (n,p) cross-section was also redetermined as a by-product.

For iridium, we found no literature value at all. This is astonishing because the cross-section is quite high and thus is one of the most sensitive reactions for 14-MeV neutron activation analysis.

Only one published value was found for the 6.1 s isomer in osmium, which is little smaller than our result. The long-lived isomer was also investigated from the same activation. The cross-section is in excellent agreement with the accurate measurement of the corresponding excitation curve [25].

In the case of lead, it was necessary to reinvestigate the 14-MeV neutron activation cross-section for the ^{207m}Pb-isomer, because of the great uncertainties in the already published values. It turns out that only the values from refs. [27, 31] agree with our result. For this comparison, all cross-sections were normalized with respect to the abundance of the ^{208}Pb-isotope.

For erbium, a similar measurement with enriched isotopes is reported by Rurarz et al. [19]; the order of magnitude of their results is confirmed by our analysis. For the other cross-sections, found in the literature, only the summed contribution of both nuclear reactions can be compared. It can be seen that our result corresponds roughly to the mean value of the already published data.

In the case of tungsten, a potential structural material in fusion technology, the same intercomparison is made. The agreement is surprisingly good, in spite of the big summing and absorption corrections (see above) which had to be applied for the rather low-energetic γ-rays of the tungsten isotope.

In summary, it has been proved that the KORONA-facility together with the high-rate γ-spectroscopy system can be successfully used for the determination of 14-MeV neutron activation cross-sections involving short-lived nuclides ($T_{1/2} \leq 0.5$ s), for which cyclic activation must be applied to get accurate results.

Table I: Neutron reaction cross-section measurement

Target	Reaction a)	$T_{1/2}$ b) (s)	E_γ b) (I_γ b)) (keV)	Cross-section (mb)	
				Present study (14.6 ± 0.2 MeV)	Literature
NaBr (99.5 %)	^{79}Br(n,n'γ)^{79m}Br	4.9	207 (0.76)	294. ± 16.	266 ± 35 [15]; 230 ± 30 [16] 450 ± 80 [17]; 653 ± 65 [18]
$^{167}Er_2O_3$ (95.6 %)	^{167}Er(n,n'γ)^{167m}Er			252. ± 18.	343 ± 34 [19]
$^{168}Er_2O_3$ (98.3 %)	^{168}Er(n,2n)^{167m}Er	2.3	208 (0.417)	581. ± 43.	403 ± 40 [19]
	^{167}Er(n,n'γ)^{167m}Er + ^{168}Er(n,2n)^{167m}Er			795.$^+$± 59.	1053$^+$± 222[17]; 967$^+$± 96 [18] 694$^+$± 69 [19]; 690$^+$± 110[20]
Yb (99.9 %)	^{174}Yb(n,p)^{174}Tm	324.0	273 (0.887)	3.0 ± 0.22	2.1 ± 0.3[21]; 3.5 ± 0.7[22]
	^{176}Yb(n,n'γ)^{176m}Yb	11.7	293 (0.93) 389 (0.908)	19.7 ± 1.7	16.7 ± 3.5 [15]
^{183}W (77.4 %)	^{183}W(n,n'γ)^{183m}W	5.3	108 (0.18 c))	127. ± 14.	
^{184}W (96.3 %)	^{184}W(n,2n)^{183m}W		160 (0.049c))	656. ± 74.	
	^{183}W(n,n'γ)^{183m}W + ^{184}W(n,2n)^{183m}W		210 (0.009c))	715.$^+$± 81.	762$^+$ ± 94 [17] 790$^+$ ± 90 [20]
Os (99.9 %)	^{190}Os(n,n'γ)^{190m}Os	594.0	361 (0.94); 502 (0.93) 616 (0.986)	14.0 ± 1.1	15 ± 1.5 [23,24]; 13.0 ± 1.8 [25] 11 ± 1.5 [26]
$IrCl_3$ (99 %)	^{191}Ir(n,n'γ)^{191m}Ir	4.9	129 (0.24 c))	221. ± 22	
	^{192}Os(n,n'γ)^{192m}Os	6.1	453 (0.59) 569 (0.70)	2.6 ± 0.27	2.1 ± 0.3 [23]
PbS (99 %)	^{207}Pb(n,n'γ)^{207m}Pb + ^{208}Pb(n,2n)^{207m}Pb	0.81	570 (0.979) 1064 (0.90)	1365$^+$ ± 68.	1340$^+$± 174[27]; 990$^+$± 120 [20] 1700$^+$± 300[28]; 682$^+$± 99 [17] 1310$^+$± 105[29]

+ Cross-section value is taken with respect to the abundance of the heavier isotope

a) Isotopic abundances taken from ref. [12]; b) $T_{1/2}$, E_γ, I_γ taken from ref. [13], c) I_γ taken from ref. [14]

Acknowledgment

We would like to thank Mr. E. Bössow and Mr. H. Krüger for operating the neutron generator. One of us (B.A.) is grateful to Prof. Dr. W. Michaelis, GKSS-Research Center Geesthacht, and to Prof. Dr. H. Brückmann, University of Hamburg, for a research appointment.

References

1) H.-U. Fanger, R. Pepelnik, W. Michaelis;
J. Radioanal. Chem. 61, (1981) 147 (GKSS 81/E/30)

2) R. Pepelnik, H.-U. Fanger, W. Michaelis, B. Anders;
J. Radioanal. Chem. 72, (1982) 393

3) G.P. Westphal;
J. Radioanal. Chem. 70, (1982) 387

4) P. Rybaczok, H.-U. Fanger;
GKSS-report 80/E/2

5) V.E. Lewis, K.J. Zieba;
Nucl. Instr. & Meth. 174, (1980) 141

6) B. Anders;
GKSS-report 82/E.. (in preparation)

7) R. Schmidt;
Dissertation, Hamburg 1978

8) R. Dierckx, M.L. Nimis, V. Sangiust, M. Terrani;
Nucl. Instr. & Meth. 105, (1972) 1

E.L. Draper; Nucl. Science & Eng. 46, (1971) 22

9) M. Hyvönen-Dabek, P. Nikkinen-Vilkki;
Nucl. Instr. & Meth. 178, (1980) 451

J.T. Routti, J.V. Sandberg;
Comp. Phys. Comm. 21, (1980) 119

10) B. Bahal , H.-U. Fanger;
GKSS-report 82/E/22

11) W.J. Veigele;
Atomic Data Tables 5, (1973) 51

12) W. Seelmann-Eggebert, G. Pfennig, H. Münzel;
Chart of Nuclides, KFK, Karlsruhe 1974

13) C.M. Lederer, V.S. Shirley;
Table of Isotopes, 7th ed., Wiley & Sons, New York 1978

14) G. Erdtmann, W. Soyka;
The Gamma Rays of the Radionuclides,
Verlag Chemie, Weinheim 1979

15) P. Bornemisza-Pauspertl, P. Hille;
Sitzungsbericht d. österr. Adad. d. Wiss.,
Math.-Naturwiss. Klasse (Abteilg. II),
176, (1967) 227

16) E. Rurarz, Z. Haratym, T. Kozlowski, J. Wojtkowska;
Acta Phys. Polon. B1, (1970) 97

17) J. Janczyszyn, L. Gorski;
J. of Radioanal. Chem. 14, (1973) 201

18) S.L. Sothras, G.N. Salaita;
Int. Conf. on the Interactions of Neutrons with Nuclei, Vol. II,
Lowell 1976, pg. 1391

19) E. Rurarz, Z. Haratym, M. Pietrzykowski, A. Sulik;
Acta Phys. Polon. B1, (1970) 415

20) R. Prasad, D.C. Sarkar, C.S. Khurana;
Nucl. Phys. 88, (1966) 349

21) T. Tuurnala, V. Pursiheimo;
Physica Scripta 5, (1972) 183

22) S.C. Gujrathi, S.K. Mukherjee;
Indian J. Phys. 41, (1967) 633, 667

23) A. Pakkanen, D.W. Heikkinen;
Physica Fennica 8, (1973) 345

24) E. Rurarz, J. Chwaszczewska, Z. Haratym,
M. Pietrzykowski, A. Sulik;
Acta Phys. Polon. B2, (1971) 553

25) M. Herman, A. Marcinkowski, G. Bielewicz;
Nucl. Phys. A 297, (1978) 335

26) P. Bornemisza-Pauspertl, P. Hille;
Radiochimica Acta 27, (1980) 71

27) J. Karolyi, J. Csikai, G. Petö;
Nucl. Phys. A 122, (1968) 234

28) V.L. Glagolev, P.A. Yampolskii;
Sov. Phys. JETP 13, (1961) 520

29) S.L. Sothras, G.H. Salaita;
J. Inorg. Nucl. Chem. 40, (1978) 585

THE MEASUREMENT OF SHORT-LIVED RADIONUCLIDES USING A CYCLIC ACTIVATION SYSTEM

C.A. Adesanmi and N.M. Spyrou

Medical and Environmental Physics Group
Department of Physics
University of Surrey
Guildford, Surrey GU2 5XH
England

The method of cyclic activation analysis is selective in enhancing the signal-to-noise ratio of short-lived radionuclides of interest and suppresses the longer lived nuclides in the matrix when compared with the conventional one-shot irradiation and counting sequence. The important features of the cyclic activation system available at the University of London Reactor Centre, are briefly outlined. The method can be used to determine in the same experiment, the half-life of the radionuclides of interest in order to confirm identification in complex γ-ray spectra and thus making the determination of the yield of short-lived fission products possible. Results of experiments undertaken in the detection and measurement of short-lived fission products from natural uranium and thorium are discussed.

Introduction

The method of cyclic activation analysis has been developed primarily for the determination of elemental concentrations in biological and environmental samples through the measurement of short-lived radionuclides as a result of neutron capture (1). A number of irradiation facilities have been designed or modified in the last decade in order to incorporate the method of cyclic neutron activation analysis (CNAA) including reactors, neutron generators and isotopic neutron sources (2). The method is selective in enhancing the signal-to-noise ratio of shorter-lived radionuclides of interest and suppresses the longer lived nuclides in the matrix, when compared with the conventional one-shot irradiation and counting sequence. It is believed that this method can be extended to applications in industrial process analysis and control particularly in nuclear fuels fabrication (3). Given a total experimental time in which identification of material composition must be carried out, optimum timing conditions can be calculated for maximum detector response.

The method is used for the analysis of complex spectra obtained from short-lived fission products produced on neutron irradiation of natural uranium and thorium in both a mixed reactor flux and under a cadmium sleeve in the cyclic activation system (CAS) at the University of London Reactor Centre (ULRC), Ascot, England.

Theory

The cumulative detector response for all n repeated irradiate-wait-count-return cycles has been derived elsewhere (1).

$$D_c = \frac{N_0 fm}{A} R\varepsilon(E_\gamma)I(E_\gamma)\,\frac{(1-e^{-\lambda t_i})(e^{-\lambda t_w})(1-e^{-\lambda t_c})}{\lambda} \times \left|\frac{n}{1-e^{-\lambda T}} - \frac{e^{-\lambda T}(1-e^{-n\lambda T})}{(1-e^{-\lambda T})^2}\right| \qquad (1)$$

where N_0 is Avogadro's number, f the fractional abundance of target isotope, m the mass of target element, A the atomic weight of target element, R the reaction rate per target nucleus, $\varepsilon(E_\gamma)$ the efficiency (energy dependent) of the detector, $I(E_\gamma)$ the intensity of radiation of interest, and λ the decay constant of the isotope of interest.

T is the cycle period and it is defined as: $T = t_i + t_w + t_c + t_r$ where t_i is the irradiation time, t_w is the waiting time, t_c the counting time, and t_r the return time.

The maximum D_c value occurs when $t_w = t_r = 0$ and $t_i = t_c = T/2$ (1). The former is an unrealistic situation because it is practically unattainable; but t_w can be made equal to t_r for optimum D_c and therefore $t_i = (T/2 - t_w)$.

The detector response for three total experimental times (72s, 162s, 312s) used in the analysis has been calculated from equation (1), where $(N_0 fm/A)R\varepsilon(E_\gamma)I(E_\gamma)$ is taken to be constant and plotted versus a range of half-lives. The cycling conditions for each case were $t_i = t_c = 2s$, $t_w = t_r = 0.4s$; $t_i = t_c = 5s, t_w = t_r = 0.4s$; $t_i = t_c = 10s$, $t_w = t_r = 0.4s$ respectively Figure 1 indicates the range of half-lives best detected under the three conditions and shows the relative signal suppression particularly of the longer-lived isotopes.

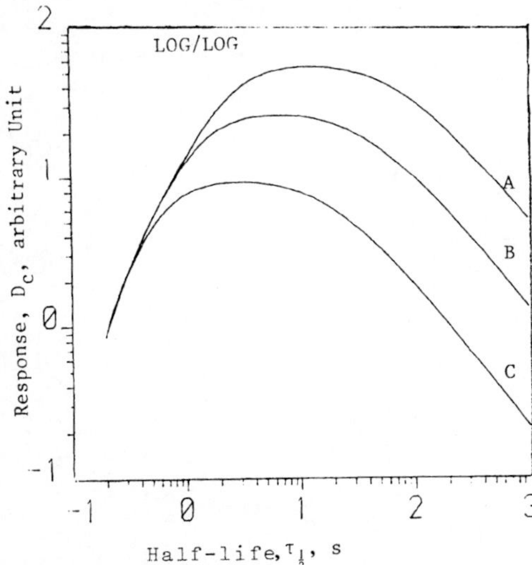

Fig. 1. Variation of detector response (signal), D_c, with half-life, $\tau_{\frac{1}{2}}$, of cyclic activation for three different irradiation conditions: $t_w=t_r=0.4s$, n = 15 cycles, $A(t_i=t_c=10s), B(t_i=t_c=5s), C(t_i=t_c=2s)$.

K. H. Böckhoff (ed.), Nuclear Data for Science and Technology, 863–865.

The Cyclic Activation System (CAS)

In a meeting in 1980 at the workshop on short-lived radionuclides (2) in activation analysis the consensus of opinion was that there are still challenges to be met and that emphasis should be placed on their detection and accurate measurement. There is a demand for rapid and reliable analytical techniques with good sensitivity for trace elemental determinations and multi-elemental capabilities. The development of a system to handle a large number of samples in routine analysis which is necessary for environmental and industrial process samples has taken place with respect to the irradiation facility at ULRC (2). Figure 2 shows the system diagramatically.

CAS has been designed so that irradiation can be carried out in a position in the core of the Consort-II reactor in an aluminium tube. In addition, an alternative flight path allows irradiation to be carried out under cadmium. This unique feature has therefore also made it possible to measure improvement factors for radionuclides with relatively large resonance integrals and thus provide selective activation under cadmium (4). The technique has been termed epicadmium cyclic neutron activation analysis (ECNAA).

Another important feature is the 'operational modes'. The system has 4 modes of operation i.e. A, B, C and D. Mode A was designed to utilise reactor time to the full. In this mode, 3 samples can be in the system with one sample each in the core, waiting and counting positions. In mode B, the waiting position which is the direction control valve is dispensed with thus only 2 samples are in the system. In mode C, only one sample is in the system throughout the irradiate-wait-count-return cycle. The fourth and the last mode D is for automation of cyclic activation. One sample at a time, goes through a preset number of cycles, ejected and replaced by another sample. This is repeated until all the samples, a maximum of 30, have been cycled.

Nitrogen gas pressure of 2 atmospheres has been used to achieve a transfer (waiting) time between core and counting position of 400ms. The counting system consists of a coaxial Ge(Li) detector of $30{,}000 mm^3$ active volume with a resolution of 1.69 keV (FWHM) at 1.332 MeV. The source-to-detector distance is about 18mm. The detector is connected through a preamplifier and an amplifier to the Laben 800 multichannel analyser. The data collected are stored on a magnetic tape and analysed using the modified SAMPO program (5).

Experiment and Results

Between 5-40μg of natural uranium and 1-5mg of thorium were irradiated in both the bare tube and cadmium sleeve. The main timing parameter used is $t_i=t_c$ = 10s, $t_w=t_r$ = 0.4s. Other timing parameters were tried ($t_i=t_c$ = 2s, $t_w=t_r$ = 0.4s; $t_i=t_c$ = 5s, $t_w=t_r$ = 0.4s). Fig. (3) shows the cumulative detector counts for natural uranium using condition $t_i=t_c$ = 10s, $t_w=t_r$=0.4s for n = 5, 10 and 15. Table 1 lists the energy of the photopeak and likely candidates.

The data from each individual cycle was stored separately on a magnetic tape. This enables the dead time correction to be carried out on each individual spectrum before summing. In addition it allows the estimation of the half-life of the radionuclide of interest from the photopeak from each spectrum in order to confirm identification in complex γ-ray spectrum (6).

The cumulative detector response, D_c, in equation (1) is reduced, for large n to

$$D_c = D_1 \left| \frac{n}{1-e^{-\lambda T}} - \frac{e^{-\lambda T}}{(1-e^{-\lambda T})^2} \right|$$

i.e.

$$D_c = \frac{D_1}{1-e^{-\lambda T}} \cdot n - \frac{D_1 e^{-\lambda T}}{(1-e^{-\lambda T})^2}$$

This is a linear equation which can take the form

$$D_c \; a \cdot n + b$$

where $a = \frac{D_1}{1-e^{-\lambda T}}$ and $b = -\frac{D_1 e^{-\lambda T}}{(1-e^{-\lambda T})^2}$

a plot of D_c against n gives a straight line at large n with slope a and intercept b. The half-life of the radionuclide of interest can be expressed as

$$\tau_{\frac{1}{2}} = \frac{T \ln 2}{\ln(1-\frac{a}{b})} \qquad (2)$$

This estimation of half-life has been made simply to identify or confirm the presence of radionuclides eg in the case where the estimated half-life of radionuclide of interest was found to be 4.9 ± 0.6s for ^{87}Se at the photopeak energy 469.1 keV. The estimated half-life agrees with the literature value within experimental

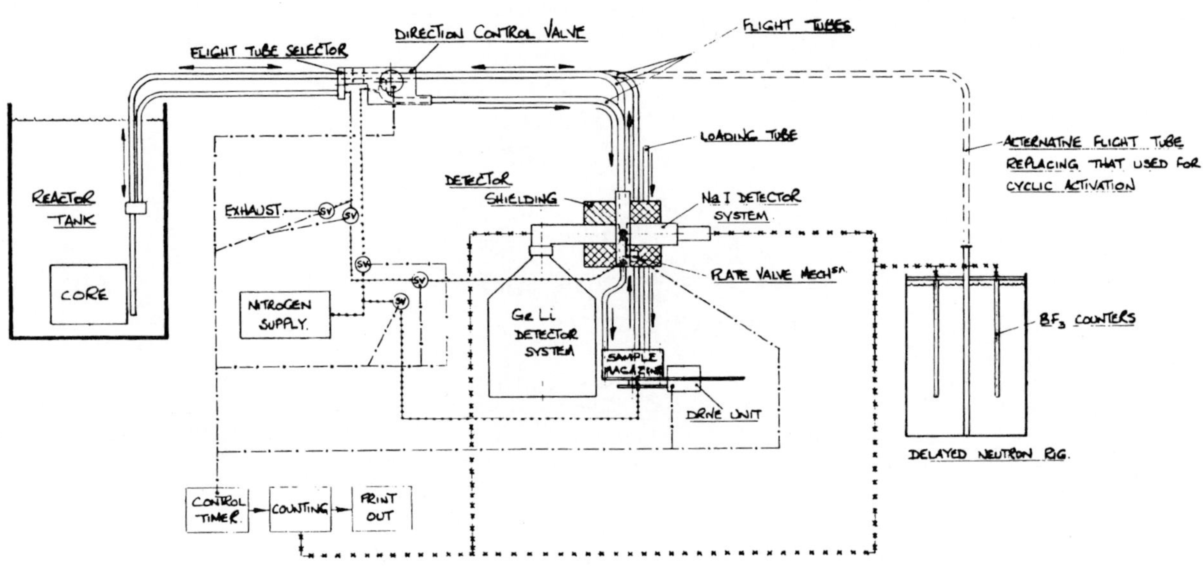

Fig. 2. SCHEMATIC DIAGRAM of Cyclic Activation System

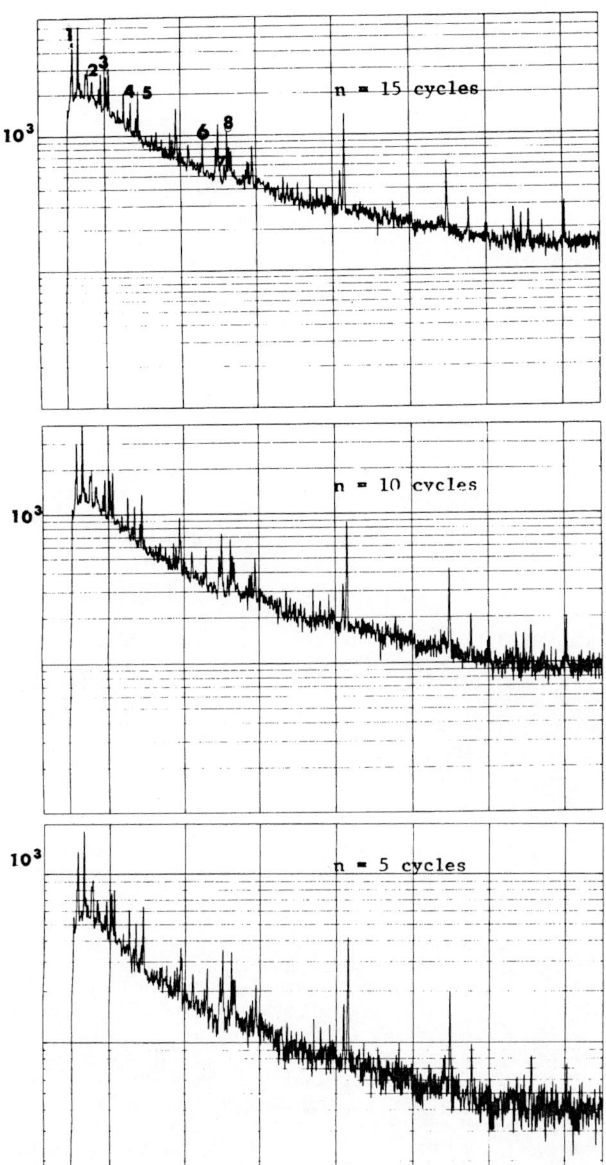

Fig. 3. The cumulative spectra of fission products of natural uranium for 5, 10, 15 cycles illustrating the growth of individual photopeaks depending on the half-life of the isotope of interest, under condition $t_i = t_c = 10s$, $t_w = t_r = 0.4s$, n = 15 cycles.

error. On the other hand there are cases where the estimated half-life could not do much to ascertain the identity of the radionuclide except to indicate multiple contributions.

Cyclic activation therefore can be a useful method in determining the yield of short lived fission products and confirming their identities where interference may occur through estimate of the half-life of the isotope of interest. If the latter still proves that more than one radionuclide has contributed to the peak then the timing parameters can be changed and the relative contributions of the isotopes in the photopeak change. This can therefore lead not only to confirmation of identity but also to respective yields.

It should be noted that because $\lambda t \gtrsim 1$ Poisson statistics do not apply, therefore data for the determination of half-life and yield must be treated by Ruark-Devol statistics (7).

Acknowledgement

Our thanks are extended to Maurice Kerridge, Director of the University of London Reactor Centre and his staff particularly E.A. Caesar, for their help.

C.A. Adesanmi would like to acknowledge the financial support of the University of Ife, Ile-Ife, Nigeria.

Table I Measured half-lives to identify fission products detected in natural uranium using CNAA.

Peak	Energy (keV)	Estimated Half-life (s)	Likely Candidates *
1.	119.4	2.5 ± 0.3	^{101}Zr(2.0s); ^{148}Ce(48.0s)
2.	172.8	10.2 ± 1.2	^{144}Ba(10.7s)
3.	210.3	19.3 ± 2.3	^{143}Ba(12.0s); ^{103}Tc(54.2s)
4.	276.4	7.8 ± 0.9	^{101}Nb(7.1s)
5.	297.0	7.3 ± 0.8	$^{134}Sb^F$(10.3s)
6.	469.1	4.9 ± 0.6	^{87}Se(5.8s)
7.	528.2	7.9 ± 0.9	$^{118}In^F$(5.0s)
8.	534.6	8.9 ± 1.1	^{88}Se(1.5s); $^{100}Nb^F$(1.5s)

* Likely candidates are obtained from compilation of gamma-ray and half-life data for the fission products from Ref. 8.

References

1. N.M. Spyrou, S.A. Kerr, J. Radioanal. Chem. 48 (1979) 169.
2. N.M. Spyrou, J. Radioanal. Chem. 61 No. 1-2 (1981) 211.
3. S.J. Gage, G.D. Atkinson, Jr. and G.D. Bouchey Nucl. Tech. 17 (1973) 247.
4. S.J. Parry, J. Radioanal. Chem.(1982) (in press).
5. G.D. Burholt, E.A. Caesar and T.C. Jones (to be published in Nucl. Instr. and Method).
6. N.M. Spyrou, C.A. Adesanmi, P.M. Kidd, L.G. Stephens-Newsham, A.Z. Ortaova, F. Ozek, J. Radioanal. Chem. (1982) (in press).
7. I.P. Matthews, K. Kouris, M.C. Jones, N.M. Spyrou Nucl. Instr. and Meth. 171 (1979) 369.
8. J. Blachot and C. Fiche, Atomic and Nucl. Data Tables, 20 (1977) 241-311.

ON NEUTRON CAPTURE CROSS SECTION MEASUREMENTS WITH THE ACTIVATION TECHNIQUE IN THE MeV REGION

P. Andersson, R. Zorro and I. Bergqvist

Department of Physics, University of Lund
Sölvegatan 14, S-223 62 Lund
Sweden

Sources of systematic errors in cross-section measurements with the activation technique have been investigated with the aid of the reaction $^{115}In(n,\gamma)^{116m}In$.

The main problems are related to low-energy background neutrons produced by charged-particle reactions (e.g. (p,n) and (d,n) reactions) in the target material and by nonelastic neutron reactions (e.g. (n,n´) and (n,np) reactions) in the sample and surrounding materials. Methods to correct for the background neutrons have been developed and capture cross-sections for ^{115}In have been determined in the energy region 2.0-4.5 MeV.

| Neutron capture cross-sections in the MeV region, activation technique|

Introduction

Capture cross-section measurements using the activation technique in the MeV region appear to be difficult primarily due to accompanying background neutron sources. On the average, the background neutrons are of a much lower energy than the primary neutrons. Since the (n,γ) cross-section increases rapidly with decreasing neutron energy it means that even a small contamination by background neutrons can seriously disturb the measurements.

Studies (1-3) with 14-15 MeV neutrons indicate that the main sources of background neutrons are (n,2n) and (n,n´) reactions in the target backing and the sample itself in addition to the uniformly distributed flux of room-scattered neutrons. The corrections due to the secondary neutrons may be determined by systematically varying the experimental parameters. These studies show that the results of early activation measurements at 14-15 MeV are in error; in some cases bv more than a factor of ten.

Numerical estimates suggest that secondary neutrons mav significantly disturb the (n,γ) activation measurements also in the MeV region. It is then important to study these effects. Rather few (n,γ) cross-sections have been reported for energies above 3 MeV and those below 3 MeV often show large discrepancies. The differences in the results, even for well-studied nuclei like ^{197}Au and ^{238}U, are several times greater than the quoted uncertainties which indicates the precens of serious systematic errors.

This report describes activation measurements for $^{115}In(n,\gamma)^{116m}In$ in the energy range 2.0-4.5 MeV with the primary objective being to investigate the contributions of background neutrons and to find methods which can be used to correct for these contributions. A second objective is to apply the methods to determine the capture cross section for ^{115}In.

The reaction $^{115}In(n,\gamma)^{116m}In$ offers favorable experimental conditions. The ^{116m}In decay has a suitable half life ($T_{1/2}$=54.2 min.) and yields a γ-ray, E_γ=1293 keV, which is easy to detect. Furthermore, the capture cross section for low-energy neutrons is rather large so the activation yield is sensitive to low-energy background neutrons. It is also an advantage that inelastic neutron scattering of ^{115}In exites a metastable state at E_γ=336 keV with $T_{1/2}$=4.5 h. The cross-section for this reaction is known to increase with energy above the threshold up to about 2.4 MeV and then to become almost constant with energy (4). Because it is a threshold reaction it is not sensitive to low-energy background neutrons, and hence the reaction can be used as a reference reaction in the activation studies.

Experimental arrangement and procedure

Monoenergetic neutrons were produced either by the reaction $T(p,n)^3He$ or by $D(d,n)^3He$ with the protons or the deuterons being accelerated by a 3 MV tandem accelerator. The beam was foccused onto the target with a beam spot of a diameter less than 3 mm. The beam conditions were checked during irradiation with the aid of the two apertures along the beam line. We required that the current on these apertures be less than 1% of that on the target.

Solid target were used for the measurements. The targets consisted of a thin titanium layer (about 2 mg/cm^2 thickness), in which the hydrogen was absorbed, evaporated onto a 0.5 mm thick backing. In order to determine the background from neutrons produced by (p,n) or (d,n) reactions in the backing and absorption materials and in the apertures along the beam line, measurements were also made with targets with natural hydrogen. These targets, which we call "empty targets", were prepared in the same way and at the same time as the T(Ti) and D(Ti) targets.

The irradiation of the samples took place in a target chamber which was originally made for the experiments with 14-15 MeV neutrons (3). The chamber was designed to reduce the amount of material around the target-sample region and thereby to reduce the production of secondary neutrons. A great number of samples, all circular disks, were used in the measurements. The samples were placed at distances such that the solid angle subtended by the samples at the target was the same in all cases. For the studies of the thickness dependence of the activation yield, indium samples were used with a diameter of 10 mm and thicknesses ranging from 0.18 to 1.66 g/cm^2. The target-sample distance dependence was determined using 0.6 g/cm^2 samples with diameters ranging from 5 to 20 mm.

The neutron flux was monitored by a long counter placed in the forward direction about 1.2 m from the target (5). The long counter served both as a detector of neutrons from various source reactions and as a monitor to ensure that the time variation of the neutron flux during the sample irradiations was kept below about 10%.

The γ-rays from the induced activities (Table I) were detected by an 80 cm^3 open end Ge(Li) detector. The relative detector efficiency was determined by means of a ^{152}Eu source.

Table I: Decay properties

REACTION	ENERGY (keV)	HALF-LIFE (s)	INTENSITY (%)	REF.
$^{115}In(n,\gamma)^{116m}In$	1293.54	3247±3	85±2	6,7
$^{115}In(n,n')^{115m}In$	336.0	16150±14	45.9±0.3	8

The capture cross-sections were determined relative to the cross-section of the $^{115}In(n,n')^{115m}In$ reaction, which means that neither the neutron flux nor the detector efficiency have to be measured in an absolute manner. Furthermore, it is advantageous to utilize a reference reaction which is built-in in the sample in such a way that the irradiation and analysis geometries will be the same for the sample and the reference.

K. H. Böckhoff (ed.), Nuclear Data for Science and Technology, 866–868.

Background of primary neutrons

We have found that for experiments in the MeV region primary reactions other than $T(p,n)^3He$ and $D(d,n)^3He$ cause at least as many problems as the secondary neutrons. The main sources are (p,n) or (d,n) reactions in the target deposition and backing materials.

In the studies of the primary neutron background we started with measurements of the neutron yield by the long counter from the T(Ti) target and from the corresponding empty target. The results show that, because of the primary neutron background, the determination of the (n,γ) activation cross-section becomes increasingly uncertain with T(Ti) targets as the neutron energy exeeds 4.2 MeV (i.e. for E_p=5.0 MeV). In practice, we have set the limit to E_p=5.3 MeV corresponding to E_n=4.5 MeV. For experiments at higher energies we have to look for other target materials.

For the study of the background yield in connection with the $D(d,n)^3He$ source reaction we considered a solid target with deuterium absorbed in titanium and a deuterium gas target with a 4.2 mg/cm^2 thick molybdenum foil as the entrance window and gold as beam stop. The results show that above E_n=3 MeV deuterium targets are to be preferred to the T(Ti) targets. With solid targets the energy range can be extended by about 1 MeV and with gas targets by several MeV.

Gas targets would be advantageous for (n,γ) activation measurements provided they could be made short enough and with sufficiently little material in the target-sample vicinity. The provisos are due to the importance of room-scattered neutrons, which necessitates short distances between target and sample, and of secondary neutrons produced in surrounding material. These effects will be discussed in the next section.

Preliminary measurements have been made with new targets which have been prepared specifically for (n,γ) experiments in the MeV region. Measurements were made with a set of $T(^{90}Zr)$ targets as well as targets with deuterium absorbed in natural zirconium. In all cases, gold was used as the backing material. We observe a ignificant reduction of primary background neutrons compared with our previous targets. It seem possible to extend the (n,γ) activation experiment up to about 6 MeV with solid targets. We are also working on a gas target cell that fulfills the requirements dictated by the need to reduce the production of secondary neutrons to a minimum.

Influence of secondary neutrons

As described in considerable detail in a previous report (3) the contribution of secondary neutrons from various sources can be determined by varying the experimental parameters. One important source is (n,n′) reactions within the sample itself. It can easily be shown that for thin slabs the correction due to secondary neutrons is approximately proportional to the thickness of the slab. Hence, this contribution may be studied by varying the sample thickness. This is illustrated in Fig. 1 which shows the influence on the $^{115}In(n,\gamma)^{116m}In$ activation yield at 3.4 MeV. At this energy there is an observable contribution of secondary neutrons but it should be pointed out that the samples are somewhat thicker than normally used in activation experiments. The dashed line represents a linear fit to the data which indicates a correction of about 1 mb for a sample thickness of 1 g/cm^2. The slope of the line agrees quite well with the contribution of secondary neutrons which is obtained from model calculations (solid line) (9).

The calculations indicate that the thickness dependence should not be precisely linear and that the contribution at 1 g/cm^2 should be nearly 2 mb. According to the calculations the corrections for secondary neutrons in the sample should be approximately the same in the energy interval 2.0-4.5 MeV. We made an attemt to varify these results at 2.5 and 4.5 MeV but the effects were too small to be observable with the accuracy of the present measurements.

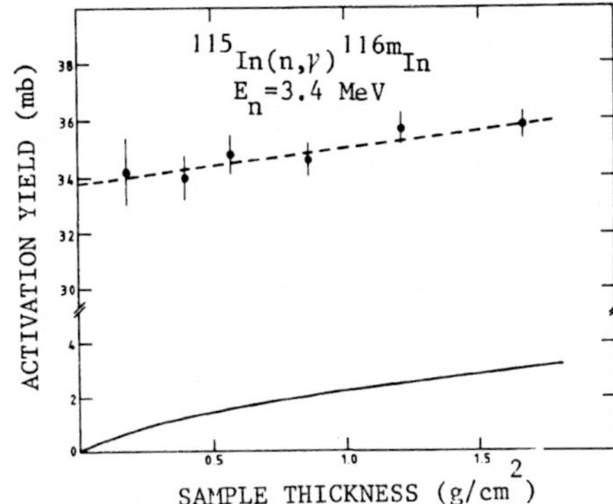

Fig. 1. Dependence of the activation yield on the sample thickness. The dashed line is a linear fit to the data points. The solid line shows the contribution obtained from model calculations. Statistical uncertainties.

Other important sources of secondary neutrons are (n,n′) reactions in structural materials near the target-sample assembly and neutrons scattered from the walls of the room. The room-scattered neutrons should be rather uniformly distributed in the vicinity of the target. The (n,γ) activity in the sample induced by these neutrons is then constant with target-sample distance. The activity due to the primary source, assumed to be a point source, varies as $1/r^2$ whereas the contribution from neutrons scattered in the structual materials near the target and the sample should exhibit a distance dependence between constant and $1/r^2$. Hence, the activity, I(r), should be given by

$$I(r) \sim \sigma_0\Phi_0(r) + \langle\sigma_{room}\rangle\Phi_{room} + \langle\sigma_{sec}\rangle\Phi_{sec}$$

where Φ_0 is the primary neutron flux, Φ_{room} and Φ_{sec} are the room-scattered and secondary fluxes and σ_0, σ_{room} and σ_{sec} are (n,γ) activation cross-sections corresponding to primary, room-scattered and secondary neutrons, respectively. The activation yield, Y(r), is obtained by normalizing to unit incident flux

$$Y(r) = \sigma_0 + c_1r^2 + c_2f(r)$$

By studying the dependence of the activation yield on the target-sample distance one may determine the contribution of room-scattered and secondary neutrons from sources outside the sample itself. Extrapolation to zero distance gives the corrected cross-section.

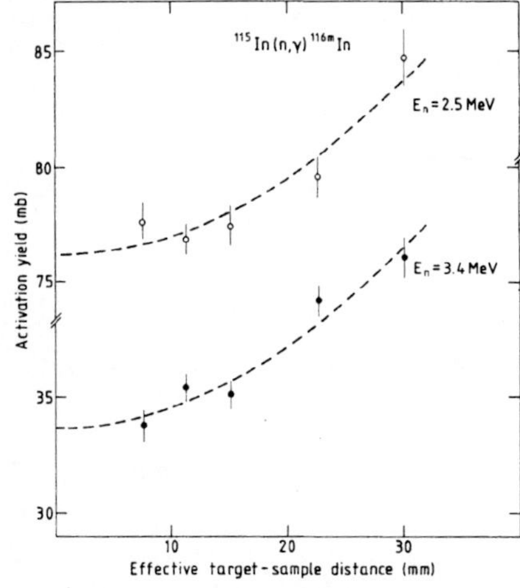

Fig. 2. Dependence of the activation yield on the effective target-sample distance. This distance is calculated as the average value weighted over the beam spot on the target and over the sample. The dashed curves are r^2 least squares fits to the data points. Statistical uncertainties.

The results of measurements of the activation yield at various sample distances between 8 and 30 mm at 2.5 and 3.4 MeV are shown in Fig. 2. The distance dependence is in all cases rather strong. The room-scattered background seems to dominate at large distances and the experimental points can be reasonably well fitted by an r^2 function (dashed curves). There is no obvious contribution of a linear term as was the case in the 14-15 MeV experiments (1,3). This contribution should be weak with the target-sample arrangement used in this work. In order to study it, one has to perform measurements at distances well below 8 mm. A noticeable result is that the magnitude of the room-scattered background is the same at 2.5 and 3.4 MeV, e.g. it is about 3.5 mb at a distance of 20 mm in both cases. The measurements at 4.5 MeV indicate roughly the same result although the uncertainty of the experimental points is rather large. This lead us to believe that the intensity and energy distribution of the room-scattered neutrons are approximately independent of the primary neutron energy in this energy range.

In conclusion, we have found that the contributions from secondary neutrons, from the sample itself and from surrounding materials, may be determined by varying the sample thickness and the target-sample distance. The results indicates that the magnitude of the contributions are approximately constant with primary neutron energy between 2.5 and 4.5 MeV. More data are needed to show how the contributions vary over a wider energy region. For reasonable thick samples, about 0.5 g/cm^2, we observe a fairly weak effect due to secondary neutrons produced in the sample. The influence of room-scattered neutrons is greater. We have noted that this influence can not be significantly reduced by surrounding the target chamber with cadmium (0.25 mm thick) which means the room-scattered neutrons are epi-thermal rather than thermal. However, our resent measurements show that a considerable improvement can be achieved by increasing the distance between the target chamber and the closest wall.

Capture cross sections

Cross-sections were measured for the $^{115}In(n,\gamma)^{116m}In$ reaction in the energy range 2.0-4.5 MeV in intervals of about 200 keV. The cross-sections were determined relative to the $^{115}In(n,n')$ cross-section which was taken from ref. 4. The results are summarized in Table II and compared in Fig. 3 with the results of previous experiments (11-14).

Table II: The present cross section results together with data for the reference reaction used (4)

	$^{115}In(n,\gamma)^{116m}In$		$^{115}In(n,n')^{115m}In$	
E_n (keV)	σ (mb)	Δσ/σ (%)	σ (mb)	Δσ/σ (%)
1.96	132	6.8	262	6.1
2.06	132	6.9	287	6.1
2.16	126	6.8	313	6.0
2.21	127	6.8	319	6.0
2.26	119	6.8	323	6.0
2.35	102	6.8	330	6.0
2.45	75.4	6.7	336	6.0
2.68	60.0	6.9	343	6.0
2.93	47.0	6.9	348	6.0
3.17	39.0	8.2	348	7.3
3.42	32.9	8.1	341	7.3
3.68	26.3	7.7	336	6.0
3.86	22.4	7.7	328	6.0
4.06	20.4	8.0	327	6.0
4.21	17.3	8.0	325	6.0
4.36	15.5	7.9	321	6.0
4.46	13.2	15.2	321	6.0

The cross-sections have been corrected for contributions of secondary neutrons by using the measured dependencies of the target-sample distance and the sample thickness. The corrections were obtained by extrapolating the yield curves (Figs. 1-2) to zero effective distance and zero thickness. As discussed in the previous section the magnitude of the contribution of secondary neutrons for a given sample at a given distance is approximately independent of the neutron energy in this range. The corrections amount to 2.0 mb. The error bars reflect statistical uncertainties (4-5%) and uncertainties in the cross-section for the reference reaction (6%). Some contribution comes from uncertainties in the relative detection efficiency (2%), in the γ-ray correction factors (2%) and in the decay and branching ratios (2%).

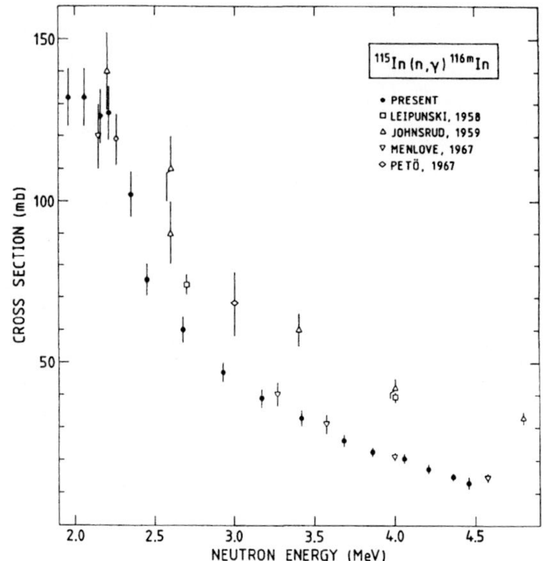

Fig. 3. Capture cross section of ^{115}In as a function of neutron energy.

The present results agree very well with those of Menlove et al. (11) but are significantly lower than earlier results (12-14). Menlove et al. corrected for room-scattered neutrons and for neutrons scattered from the target vicinity. The influence of room-scattered neutrons was determined from measurements with long target-sample distances (0.5 and 1.0 m). The effects of neutron scattering from materials in the target vicinity (gas cell assembly, sample packet and the fission chamber) were measured in a way that may not properly account for the influence of secondary neutrons at higher energies, say above 5 MeV (see ref. 3). However, the corrections may be adequate in the range displayed in Fig. 3. It seems reasonable to assume that sufficient corrections were not applied in earlier work (11).

References

1. M. Valkonen and J. Kantele, Nucl. Instr. Meth. 103 (1972)549.
2. K. Ponnert, G. Magnusson and I. Bergqvist, Physica Scripta 10(1974)35.
3. G. Magnusson and I. Bergqvist, Nucl. Technol. 34 (1977)114.
4. D. Smith, ANL/NDM-26 Argonne National Laboratory 1976
5. A.O. Hanson and J.L. McKibben, Phys. Rev. 72(1947)673
6. K.H. Beckurts, M. Brose, M. Knoche, G. Kruger, W. Poenitz and H. Schmidt, Nucl. Sci. Eng. 17(1963)329.
7. D. Rabenstein, Z. Phys. 240(1970)244.
8. INDC/NEANDC Nucl. Standard File 1980 Version B-51.
9. G. Magnusson, P. Andersson and I. Bergqvist, Physica Scripta 21(1980)21.
10. G. Magnusson and P. Andersson, University of Lund Report, LUNFD6/(NFFR-3030)/1-37/1979.
11. H.O. Menlove, K.L. Coop and H.A. Grench, Phys. Rev. 163(1967)1299.
12. G. Petö, Z. Miligy and I. Hunyadi, J.Nucl.Energy 21(1967)
13. A.E. Johnsrud, N.G. Silbert and H.H. Barschalt, Phys. Rev. 116(1959)927.
14. A.I. Leipunsky et al. Proc. of the Second UN Conf. on Peacful Uses of Atomic Energy, Geneva, 1958.

SIMPLE MEASUREMENT OF NEUTRON CROSS SECTION RATIO FOR REACTIONS LEADING TO THE SAME RADIONUCLIDE

J. Janczyszyn

University of Mining and Metallurgy
Institute of Physics and Nuclear Techniques
Al. Mickiewicza 30, 30-059 Kraków, Poland

The theoretical background of the activation method of cross section determination, which uses as a reference reaction the one leading to the same radionuclide as the investigated reaction, is shown. Based upon this method, cross sections of 26 Mg(n, α) 23 Ne, 30 Si(n, α) 27 Mg and 31 P(n, α) 28 Al reactions for 14.9 MeV neutrons were determined.

[Activation method, cross section ratio, radionuclide, reference reaction, neutron flux density monitoring, activation monitor, error]

Introduction

The increasing interest in projects of modern fission and fusion reactors and recently of spallation breeders has resulted in the need for precise nuclear cross section data for their computation. In particular, the neutron reaction cross sections for expected construction materials in the neutron energy range from hundreds of keV to hundreds of MeV are needed. The existing cross section libraries are far from complete and frequently the provided values are so scattered that it is impossible to recommend a reliable one.

The absolute cross section measurement is difficult and frequently inaccurate. Most of the measurements and especially those performed with the activation method are relative ones. They are based on the reference between the investigated reaction and a reaction with a well known cross section value. The choice of such a reference reaction has an important influence on the accuracy of the obtained results. The measurement error can be greatly reduced, even in a poorly equipped laboratory, when both reactions, the investigated and the reference ones, lead to the same radionuclide.

This work shows the theoretical background of such a version of the activation method, some results of cross section measurements based upon it, discussion of the results and evaluation of their errors.

Theory

The activation method of cross section measurement is based on the well-known formula:

$$\sigma = \frac{\lambda N_c A \exp(\lambda t_d)}{m f \Phi N_A \varepsilon P G [1-\exp(-\lambda t_i)][1-\exp(-\lambda t_c)]} \quad (1)$$

where:

- σ - cross section
- N_c - number of counts
- λ - decay constant
- t_d - decay time
- t_i - irradiation time
- t_c - counting time
- m - mass of the element
- A - atomic mass
- f - isotopic abundance
- Φ - neutron flux density
- N_A - Avogadro constant
- ε - detection efficiency
- P - peak to total ratio
- G - number of gamma-rays per disintegration

The cross sections ratio of the investigated (denoted "x") and reference (denoted "s") reactions is then:

$$\frac{\sigma_x}{\sigma_s} = \frac{N_{cx}\lambda_x A_x m_s f_s \Phi_s \varepsilon_s P_s G_s [1-\exp(-\lambda_s t_{is})]}{N_{cs}\lambda_s A_s m_x f_x \Phi_x \varepsilon_x P_x G_x [1-\exp(-\lambda_x t_{ix})]} \cdot \frac{[1-\exp(-\lambda_s t_{cs})]}{[1-\exp(-\lambda_x t_{cx})]} \exp(\lambda_x t_{dx} - \lambda_s t_{ds}) \quad (2)$$

The version of the activation method discussed here assumes that both reactions produce the same radionuclide. Thus the following relations hold:

$$\lambda_x = \lambda_s, \quad \varepsilon_x = \varepsilon_s, \quad P_x = P_s, \quad G_x = G_s$$

And when it is possible to perform measurements in the same time schedule:

$$t_{ix} = t_{is}, \quad t_{dx} = t_{ds}, \quad t_{cx} = t_{cs}$$

the above formula can be reduced to the simplest form:

$$\frac{\sigma_x}{\sigma_s} = \frac{N_{cx} A_x m_s f_s \Phi_s}{N_{cs} A_s m_x f_x \Phi_x} \quad (3)$$

Values of the atomic masses, element masses and isotopic abundances either can be measured or are known very accurately. Thus the most important sources of error are the measurements of numbers of counts and neutron flux densities ratio. The first depends mainly on the available neutron flux density. The second can be greatly reduced by use of the proper measuring method. One can distinguish three such methods:

1. Neutron counter as the source yield monitor.

In this method the number of counts measured during an activation of a sample is the measure of neutron flux density. This number is proportional to the average, during the activation time, neutron yield - Y. Assuming then that neutron flux density is proportional to yield we obtain:

$$N_{\Phi x} : N_{\Phi s} = Y_x : Y_s = \Phi_x : \Phi_s$$

The last assumption is true when the shape of spatial neutron distribution is constant and when the investigated sample and reference sample are activated in the same position with respect to the neutron source. After taking into account the above relations the equation (3) changes to:

$$\frac{\sigma_x}{\sigma_s} = \frac{N_{cx} A_x m_s f_s N_{\Phi s}}{N_{cs} A_s m_x f_x N_{\Phi x}} \quad (4)$$

K. H. Böckhoff (ed.), Nuclear Data for Science and Technology, 869–872.

Serious errors are commited when the above adopted assumptions and also the assumption hidden in the formula (1), i.e. that:

$$\lambda \int_0^{t_i} \Phi(t) \exp(-\lambda t)\, dt = \bar{\Phi}\, [1-\exp(-\lambda t_i)] ,$$

are not strictly fulfilled in measurements. Those errors were thoroughly discussed by S.Vass, B.Vorsatz and B.Keszei [1].

The second and third method of neutron flux density monitoring consist of the simultaneous irradiation of a sample and a monitor-sample placed in the nearest possible positions. The monitor-sample, called the activation monitor, should be made of the element leading to the same radioisotope as the investigated sample element or to the radioisotope with almost eaquel half-life. The number of counts of the activation monitor is given by the transformed formula (1):

$$N_\Phi = m_\Phi f_\Phi N_A \Phi \varepsilon_\Phi P_\Phi G_\Phi [1-\exp(-\lambda_\Phi t_i)] \cdot \frac{[1-\exp(-\lambda_\Phi t_{c\Phi})]}{\lambda_\Phi A_\Phi \exp(\lambda_\Phi t_{d\Phi})} \qquad (5)$$

The ratio of the activation monitors number of counts of the reference and the investigated samples is then:

$$\frac{N_{\Phi s}}{N_{\Phi x}} = \frac{m_{\Phi s}\Phi_s [1-\exp(-\lambda_\Phi t_{is})][1-\exp(-\lambda_\Phi t_{c\Phi s})]}{m_{\Phi x}\Phi_x [1-\exp(-\lambda_\Phi t_{ix})][1-\exp(-\lambda_\Phi t_{c\Phi x})]} \cdot \exp[\lambda_\Phi (t_{d\Phi x} - t_{d\Phi s})] \qquad (6)$$

And for similar time schedules:

$$t_{is} = t_{ix},\ t_{c\Phi s} = t_{c\Phi x},\ t_{d\Phi s} = t_{d\Phi x}$$

it becomes:

$$\frac{\Phi_s}{\Phi_x} = \frac{N_{\Phi s}\, m_{\Phi x}}{N_{\Phi x}\, m_{\Phi s}} \qquad (7)$$

Formula (3) changes then to:

$$\frac{\sigma_x}{\sigma_s} = \frac{N_{cx} A_x m_s f_s N_{\Phi s} m_{\Phi x}}{N_{cs} A_s m_x f_x N_{\Phi x} m_{\Phi s}} \qquad (8)$$

The activation monitor allows for a serious reduction of the neutron flux density ratio error. When the same radionuclide is applied as in the investigated reaction, the error caused by instabilities of the neutron field during activation is eliminated. On the other hand, the monitor leading to another radionuclide, but mixed with the sample, reduces errors caused by spatial changes of the neutron field with respect to the samples. Those two conditions lead to two versions of activation monitor techniques.

2. Outer activation monitor.
This monitor is not placed in the same activation position as the sample. An example of such a case is shown in Fig. 1. In such geometry the monitor can be made of the same element as the investigated sample because after simultaneous irradiation their gamma spectra can be counted separately.

3. Inner activation monitor.
In this case the element chosen for the monitor is homogeneously mixed with the sample. It should lead to a radionuclide with the nearest possible half-life to the investigated one and its gamma spectrum cannot disturb the sample gamma lines.

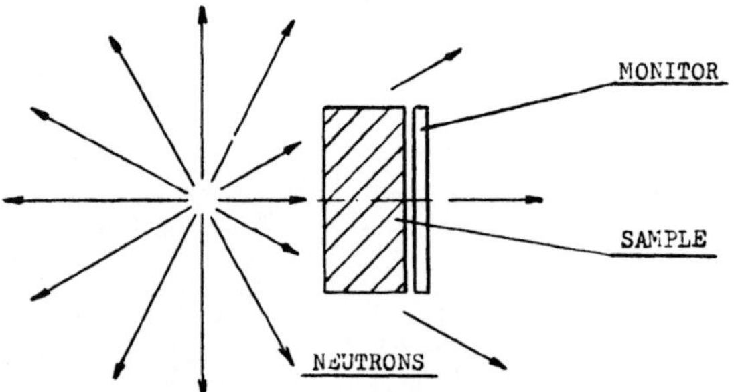

Fig. 1. An example of the outer activation monitor geometry.

Experimental

Using the method discussed here, cross section ratios of fast neutron induced reactions have been measured. Neutrons from the D-T reaction have been used. The applied deuteron accelerating voltage was 120 kV. The equipment consisted of: neutron generator yielding ~ 10^9 n/s, pneumatic sample transport system, large semiconductor Ge-Li detector, scintillation NaI (Tl) detector sized 75 x 75 mm, scintillation neutron monitor of the proton recoil type, 4000-channel pulse height analyser (ICA-70) and 16 kbyte minicomputer (TPA-70/25).

Samples were packed in polyethylene containers (Fig. 2). Metallic samples or powder pressed tablets dimentions were: Φ17x1.5 mm or Φ17x7.5 mm.

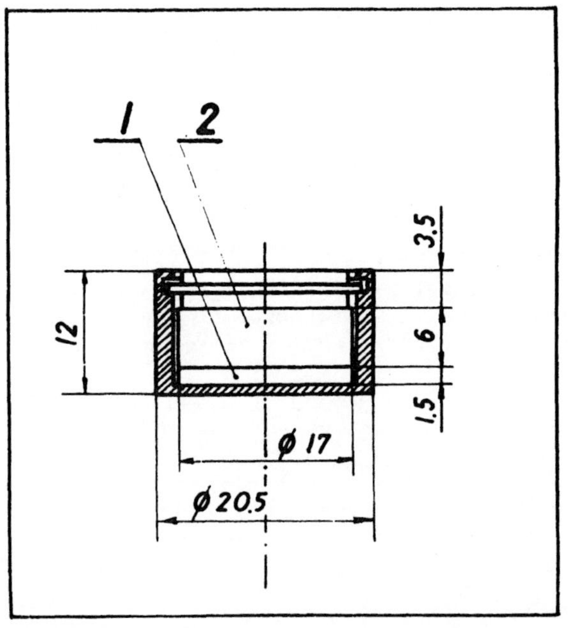

Fig. 2. Polyethylene container with a sample (1) and a perspex filling (2).

Activation and counting geometries are shown in Fig. 3 and experimental parameters and sample characteristics are given in Table I.

To minimize the apparatus error, both the investigated and the reference samples were measured in quick succesion. A minicomputer was used for processing the obtained gamma spectra and the signal was calculated ac-

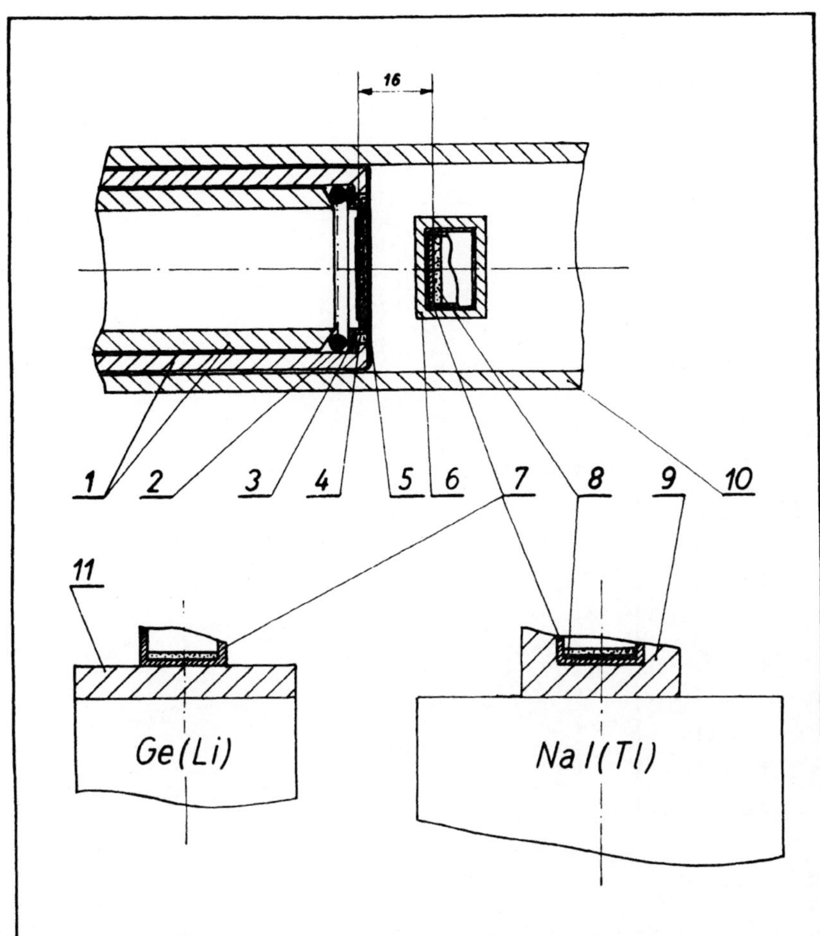

Fig. 3. Activation and counting geometries.

1 - stainless steel construction elements,
2 - vacuum seal,
3 - T/Ti target backing, 1mm copper plate,
4 - cooling water layer 0.5 mm thick,
5 - 0.3 mm stainless steel sheet,
6 - pneumatic tube, 2 mm Cu,
7 - polyethylene container bottom, 1 mm,
8 - sample,
9 - perspex end of the pneumatic tube, 8 mm,
10 - "tarnamide" holder of the pneumatic tube,
11 - perspex absorber, 10 mm.

Table I Applied samples and experimental conditions.

Sample material *	Sample mass [mg]	Applied detector	Product radionuclide	Half-life	γ - ray energy [keV]	Times [min] t_i	t_d	t_c
NaF	318	Ge(Li)	^{23}Ne	37.2 s	439.0	0.67	0.50	0.67
Mg	423							
Al	755.5	Ge(Li)	^{27}Mg	9.46 m	843.76 1014.4	5.0	23.5	10.0
SiO_2	120.5					3.3	18.0	10.0
Al_2O_3 + celulose **	1734.2					1.0	1.0	10.0
Si **	3930					3.3	7.5	10.0
Al **	4277					0.5	6.5	8.3
Si **	3930					3.3	12.7	8.3
SiO_2	120.5	Ge(Li) and NaI(Tl)	^{28}Al	2.24 m	1778.8	3.3 1.7	7.5 1.0	2.5 2.0
P + celulose	574.2					3.3 1.7	10.5 1.0	5.0 2.0

*/ Target element underlined. **/ 7.5 mm thick sample.

cording to Sterliński's formula. Neutron energy was calculated on the basis of the neutron source - sample geometrical configuration and deuteron accelerating voltage. The evaluated value was:

$$E_n = 14.9 \pm {}^{0.0}_{0.3} \text{ MeV.}$$

Results and discussion

In Table II are gathered the results of the measurements.

Table II Determined and literature values of cross sections.

Reaction	$\frac{\sigma_s}{\sigma_x}$	σ_s* [mb]	σ_x [mb] this work	σ_x [mb] lite-* rature
x $^{26}Mg(n,\alpha)$ ^{23}Ne s $^{23}Na(n,p)$	0.724 ± 0.026	43 ± 5	59 ± 7	72 ± 10
x $^{30}Si(n,\alpha)$ ^{27}Mg s $^{27}Al(n,p)$	0.880 ± 0.035	75 ± 7	85 ± 9	≈ 150
x $^{31}P(n,\alpha)$ ^{28}Al s $^{28}Si(n,p)$	2.018 ± 0.058	210 ± 20	104 ± 10	115 ± 12

*/ Reference [2]

The cross section ratios are determined with a relative error of 3 - 5%. This error, caused mainly by inaccurate measurement of neutron flux density ratio, can be reduced by applying of the activation monitor. A systematic error can take place when different reactions starting from two or more isotopes of the same element lead to the same radionuclide e.g.:

$^{30}Si(n,\alpha)^{27}Mg$ and $^{29}Si(n,^{3}He)^{27}Mg$ or

$^{28}Si(n,p)^{28}Al$, $^{29}Si(n,d)^{28}Al$, $^{30}Si(n,t)^{28}Al$.

The (n,t) and $(n,^{3}He)$ reactions have cross sections smaller by three or more orders of magnitude than the (n,α) and (n,p) ones and do not cause serious errors. More dangerous are the (n,d) reactions, cross sections of which are rarely known and are of the order of (n,p) reaction cross sections. In the above case this can cause up to a 5% underestimation of the 31 P(n,α) 28 Al reaction cross section.

References

1. S. Vass, B. Vorsatz, B. Keszei, Nucl. Instr. Meth. 130, 271 (1975).
2. M. Bormann, H. Neuert, W. Scobel, Tables of Cross Sections in Handbook on Nuclear Activation Cross Section, IAEA Technical Reports , Series No. 156, Vienna, 1974.

MEASUREMENT OF REACTION CROSS-SECTION RATIOS OF SOME NEUTRON REACTIONS USING GAMMA AND X-RAY SPECTROMETRY

A. REGGOUG, G. PAIĆ† and A. CHIADLI
Laboratoire de Physique Nucléaire, Faculté des Sciences,
B.P. 1014, Rabat, Morocco

Natural In samples have been irradiated with 14,7 MeV neutrons and the ratios of (n,2n) reactions cross-section measured by both γ and x-ray methods.
The conversion coefficients extracted from tables served to calculate the corresponding γ and x-ray intensities for each case.
The results show a good agreement between γ and x-ray data and we believe that such measurements could in some cases contribute to increase the accuracy of cross-section measurements.
[NUCLEAR REACTIONS: $^{115}In(n,2n)^{114m}In$, $^{113}In(n,2n)^{112m}In$ and $^{113}In(n,2n)^{112g}In$
E_n = 14,7 ± 0,15 MeV, measured σ ratios by activation, γ and x-ray spectrometry HPGe, GeLi]

Introduction

One of the important source of uncertainty in the nuclear reaction cross-sections measurement using the activation method is the uncertainty in the branching ratios of the emitted gamma and or x-rays.

A cursory glance at the latest edition of the Table of Isotopes[1)] indicates that the usual uncertainty in branching ratios is 4 - 10 percent.

In cases when part of the decay goes through emission of gamma and the other part through emission of an x-ray or when electron capture is followed by a gamma there is an interesting possibility offered by the comparison of the cross-section or ratios of cross--sections obtained by both methods, i.e. gamma and x-ray detection.

We have started a programme of measurement of that type. We present here the results obtained so far for the reactions:
$^{115}In(n,2n)^{114m}In$, $^{113}In(n,2n)^{112m}In$ and
$^{113}In(n,2n)^{112g}In$.

Experimental method

Two natural indium samples were irradiated with the neutron generator of the Nuclear Physics Laboratory in Rabat. The irradiation was done at 7 mm from the tritium target with a mean neutron energy of 14,7 ± 0,15 MeV. The samples were foils of 1 cm diameter and 0,1 mm thick. They were chosen thin for several reasons:

a) to minimize the x-ray absorption in the sample.

b) to avoid characteristic x-rays induced not by internal conversion or electron capture but rather by the ionization of the K-shell by either β or γ rays emitted by the sample.

This source of interference is important to be checked upon whenever the x-rays of the isotope of the sample matrix are measured.

The x-rays have been measured using a hyperpure Ge detector of 7 mm active depth and 16 mm diameter with a 125 μm Be window. The counting rate was chosen so that the dead time correction was less than 3 %.

The γ rays were detected with a GeLi detector of 67 cm^3. The x-rays were counted in subsequent time intervals using a Canberra 80 MCA in a programmed mode.

Data Analysis

The areas under the peaks were analyzed off line for the different decay components present in the observed Cd and In x-rays.

The conversion coefficients were extracted from tables[2)] assuming single multipole transitions and using quadratic interpolation with mean relative error of 1 %. These conversion coefficients were used to calculate the corresponding γ intensities with a mean relative error of 2 %. In the majority of cases the agreement between our values and the ones quoted in ref. 1 are good as visible from Table I. With the exception of:

I_γ (%) and e_k/γ values for ^{114m}In and ^{112m}In

Results

The results for the raio of cross-sections obtained by both detection methods is shown in Table II.
In Table III we quote the absolute values of the cross-sections as measured by γ spectrometry relative to the Al(n,α) monitor $\sigma_{n,\alpha}$=112,3 mb[3] compared to literature values.

Discussion

The present results indicate a remarkable agreement in the case of the transitions to the isomeric states for data obtained by both methods. The ratio of the

†IAEA expert. Present address: Ruđer Bošković Institute, POB 1016, 41001 Zagreb, YU

K. H. Böckhoff (ed.), Nuclear Data for Science and Technology, 873–874.

cross sections for the transitions to the ground and isomeric state of ^{113}In measured by gamma rays is much smaller than the one obtained by x-rays. This discrepancy seems to be due to the higher yield observed for the $^{113}In(n,2n)^{112g}In$ detected by x-rays.

We do not have an explanation for that feature so far. Measurements are in course to elucidate that question.

Table I: γ and x-ray data used in this work compared with literature values

	$^{115}In(n,2n)^{114m}In$	$^{113}In(n,2n)^{112m}In$	$^{113}In(n,2n)^{112g}In$
Half-life	49,51 d	20,9 m	14,4 m
E_γ (keV)	190,29	155	617
I_γ (%) used	15,64	12,66	6
I_γ (%)^1	15,64 17,82	13_1	6
e_k/γ used	2,57	5,27	-
e_k/γ ^{1}	$2{,}37_{31}$ $1{,}95_{29}$	$5{,}8_{12}$	
EC(%)^1 used	$3{,}3_2$	-	34
$\frac{EC(K)}{EC}^1$ used	0,88	-	0,87
In x-ray observed	yes	yes	no
Cd x-ray observed	yes	no	yes

E_γ(keV): Energy of the γ used, I_γ(%): γ intensity in percent, e_k/γ : Internal conversion coefficient, EC(%): branching ratio of electron capture, EC(K)/EC: K-electron capture ratio

Table II: Cross-sections ratios obtained by both detection methods

	$\frac{\sigma[^{115}In(n,2n)^{114m}In]}{\sigma[^{113}In(n,2n)^{112m}In]}$	$\frac{\sigma[^{113}In(n,2n)^{112g}In]}{\sigma[^{113}In(n,2n)^{112m}In]}$
by x-ray method	1,16 ± 0,08	0,82 ± 0,08
by γ method	1,16 ± 0,09	0,23 ± 0,05

Table III: Measured cross-sections compared with literature values

	This work at E_n = 14,7 MeV	Ref. 3) at E_n = 14 MeV	Ref. 4) at E_n = 14,7MeV	Ref. 5) at E_n = 14,9MeV
$^{115}In(n,2n)^{114m}In$	1540±55	1530	1470±120 1580±80	1515±100
$^{113}In(n,2n)^{112m}In$	1330±48	1290	1490±150	1450±100
$^{113}In(n,2n)^{112g}In$	310±30	1240 360	300±30	320±25

References

1) C.M. Lederer and al., Table of Isotopes, Seventh edition, Wiley, New York, 1978.

2) Rösel and al., Atomic Data and Nuclear Data Tables, Vol. 21, N° 2-3, Feb-Mar. 1978.

3) J. Csikai, Proceedings of the interregional advanced training course on applications of nuclear theory to nuclear data calculations for reactor design, Held at Trieste, 28 January - 22 February 1980.

4) H. Rötzer, Nucl. Phys. A109 (1968) 694-696.

5) Handbook on Nuclear Activation Cross-Sections, Technical reports series n° 156 I.A.E.A. 1974.

AN ABSOLUTE NEUTRON FLUX DETECTOR FOR THE 1-20 MeV ENERGY REGION

M.S. Dias*, R.G. Johnson, and O.A. Wasson

National Bureau of Standards
Washington, D.C. 20234, USA

A neutron flux detector for use in the 1-20 MeV energy range is described. A dual thin scintillator configuration yields a proton recoil spectrum which approaches the ideal thin scintillator response. Corrections due to multiple neutron scattering and escape of recoil protons from the scintillator are small. The detector efficiency was experimentally calibrated at 2.47 and 14.1 MeV using the associated particle technique and was extended to other energies by means of a Monte Carlo calculation. The efficiency uncertainty is estimated to be 1-2% in the 1-20 MeV region.

[associated particle, detector calibration, flux detector, Monte Carlo, neutrons, scintillators, Van de Graaff]

Introduction

For fast neutron cross section measurements there is a need for absolute neutron flux detectors having a fast timing and a simple response with neutron energy. A detector with these characteristics has been built at the National Bureau of Standards to be used for cross section experiments with the NBS 150 MeV Electron Linac.

Organic scintillators have been extensively used as neutron flux monitors employing the proton recoil technique. In order to keep multiple scattering corrections small, thin scintillators are necessary. However, at neutron energy of several MeV, the escape of protons produces a large distortion in the proton recoil spectrum. For example, in a 2.5 mm thick scintillator and 14 MeV neutrons, the escape of protons is around 30%.

In the present work a detector is described where the escape of protons is eliminated experimentally, and the multiple scattering correction is low. The detector consists of two thin plastic scintillators optically separated from each other and independently coupled to phototubes. The protons which escape from the first scintillator are detected by the second scintillator which is placed behind the first one.

Because of the low multiple scattering and the spectrum discrimination, there is relatively little dependence on the carbon cross sections or angular distributions. Therefore the detector efficiency is essentially dependent on the hydrogen cross section, which is known with accuracies $\lesssim$ 1%, as well as on light tables and on the hydrogen areal density, which are parameters that can be checked experimentally.

The detector has been calibrated at 2.47 MeV and 14.1 MeV, using the associated particle technique with the NBS Positive Ion Van de Graaff as the neutron source. Details of the calibration are described later.

Theoretical calculations of the detector efficiency and pulse height distributions were performed using a Monte Carlo code in order to extend the detector efficiency to other energies between 1 and 20 MeV.

Detector Design

A schematic diagram of the detector is shown in Fig. 1. It consists of two cylindrical pieces of NE110 plastic scintillators 0.254 cm thick, and 4.70 and 4.90 cm in diameter, respectively. Each scintillator was coupled to a pair of RCA8850 phototubes in head-on geometry. Perspex light guides are used, with the same thickness as the scintillators. As a result, most of the light that reaches the photocathode is by total internal reflection. A small air gap was allowed between light guide and phototube because it was shown to result in a better light collection uniformity at different points on the scintillator.

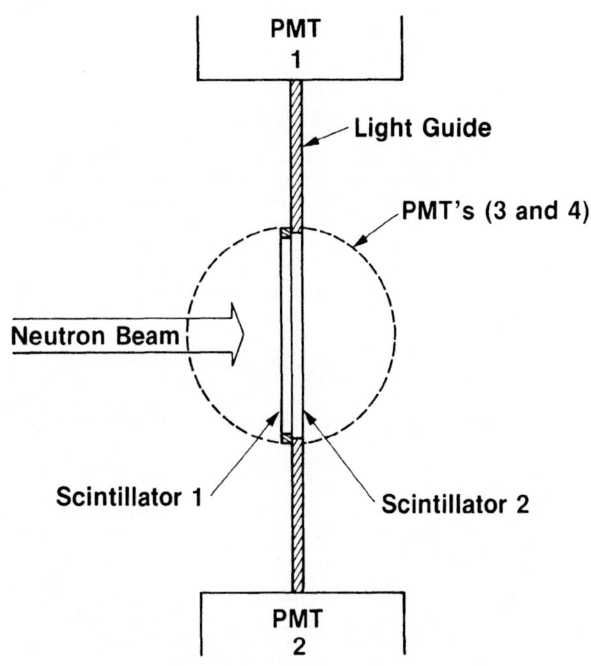

Fig. 1. Neutron flux detector.

Each scintillator was wrapped in a 6.6 μm thick aluminum foil, except at the interface between the two scintillators, where a 0.66 μm thick aluminum foil was used. The phototubes and scintillators were held together inside of a box which has a 54 μm thick aluminum window on each side, in order to make the system light tight.

Electronics

Figure 2 shows a simplified block diagram of the electronics for the flux detector. For each photomultiplier, both a time and linear signal is derived. The linear signals from the pair of phototubes of each scintillator are added and fed into a preamplifier and amplifier system. After adjusting the proper delays, the signals from each amplifier are joined into a sum amplifier. A linear gate placed in the second scintillator leg is triggered by the timing signal from the first scintillator, therefore

*Instituto de Pesquisas Energeticas e Nucleares - São Paulo - Brazil

K. H. Böckhoff (ed.), Nuclear Data for Science and Technology, 875–877.

all single events in the second scintillator are rejected. The pulses from the sum amplifier are single events from the scintillator 1 plus sum pulses produced by coincident events between the two scintillators.

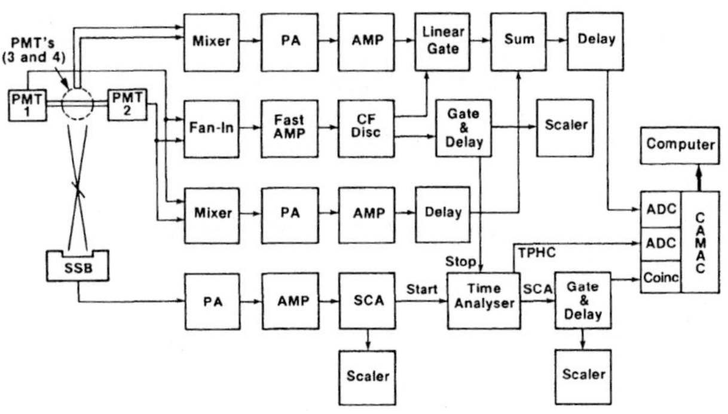

Fig. 2. Electronics for detector calibration.

Detector Calibration

Checks of the absolute efficiency of the detector have been performed by means of the associated particle technique at the NBS 3 MV Positive Ion Van de Graaff. For this purpose, a monoenergetic 500 keV molecular deuterium beam was utilized. The molecular deuterium dissociates upon impact with the target, allowing a 250 keV atomic beam to interact with the target nuclei. The $T(d,n)^4$ He reaction on a 3 mg/cm^2 copper backed TiT target was used to generate 14.1 MeV neutrons[1]. The associated ^{4}He particles were detected by a Silicon Surface Barrier detector located at 85 degrees with respect to the deuterium beam axis. The neutrons were detected at 90 degrees and 99% of the associated neutron flux was contained in a cone of 6.0 degrees half angle. The cone subtended by the neutron flux detector was 11.0 degrees half angle, which was large enough to account for variations in the beam profile due to deterioration of the tritiated titanium target.

The $D(d,n)^3$ He reaction on a 0.62 mg/cm^2 copper backed TiT target was used to generate 2.47 MeV neutrons. In this case, the associated ^{3}He particles were detected at 70 degrees with respect to the deuterium beam. The neutrons were again detected at 90 degrees and 99% of the associated neutron flux was contained in 10 degrees half angle. The profile was measured by a monitor consisting of a 2.5 cm diameter, 7.5 cm thick NE110 plastic scintillator. Corrections to the measured profile were applied due to target deterioration, and to the finite sizes of the monitor and target beam spot. The cone subtended by the neutron flux detector was 13.0 degrees half angle, and also considered sufficiently large to account for variations in the beam profile.

The electronics for the associated particle are also included in Fig. 2. The time distribution between the associated particles and proton recoil events was measured with a time analyzer. The analog signal from the sum amplifier of the neutron flux detector was fed into an analog to digital converter (ADC). Another ADC was used to receive the signals from the time analyzer. The data from both ADC's were then transferred through a CAMAC interface into a 32 x 128 work array in the computer for storage.

The experimental neutron detector efficiency is given by:

$$\varepsilon = \frac{Y_{coinc}}{Y_a} \cdot f$$

where ε is the efficiency

Y_{coinc} is the number of coincidences between the associated particle and the proton recoils.

Y_a is the number of associated particles detected.

f is the total correction factor including: neutron beam attenuation through the materials between the target and the neutron detector; dead-time losses; divergence of the neutron beam; associated particle background.

Calculated Efficiency

In order to extend the detector efficiency to broader energies, a Monte Carlo code is being developed to calculate the neutron detector efficiency and the expected pulse height distribution of proton recoils as a function of neutron energy. The computer code CARLO BLACK[2], originally written for use with the "Black Detector," has been modified to be compatible with the characteristics of the present detector. The main modifications are: a) dual scintillator configuration; b) up-dated cross sections using ENDF/B-V extended up to 20 MeV neutron energy range; c) use of forced first collision technique; d) effect of escape of protons; e) use of proton light table obtained experimentally.

The calculated pulse height distribution is fitted to the experimental spectrum by means of a code[3] which takes into account the Poisson statistics in the number of photoelectrons detected.

Results and Discussion

Figure 3 shows proton recoil spectra obtained with 14.1 MeV neutrons in an associated particle measurement. Spectrum A is the proton recoil spectrum for energy deposited in scintillator 1. Spectrum B is the proton recoil spectrum due to protons that lose a fraction of their energy inside each scintillator. Spectrum C is the observed spectrum including protons totally absorbed in scintillator 1 and pulses due to protons that lose a fraction of their energy inside each scintillator. Therefore, spectrum C approximates the thin scintillator response and it is the one used to calculate the efficiency. The solid curves are the theoretical spectra obtained with the Monte Carlo code.

Figure 4 shows the experimental efficiencies at 2.47 and 14.1 MeV. The calculated efficiency curve is also plotted as a function of the neutron energy and normalized at 2.47 MeV. The 14.1 MeV point agrees with the theory within 1.8%. The behavior of the calculated curve follows essentially the H(n,n)H cross section with some small irregularities due to C(n,n)C cross section. Therefore, the interpolation in the efficiency curve for the intermediate energy region may be obtained with good accuracy. Our present detector is optimized for energies up to 15 MeV. To measure above that, it is only necessary to replace the second scintillator by another one 4.5 mm thick. The slight increase in the efficiency is small and can be calculated.

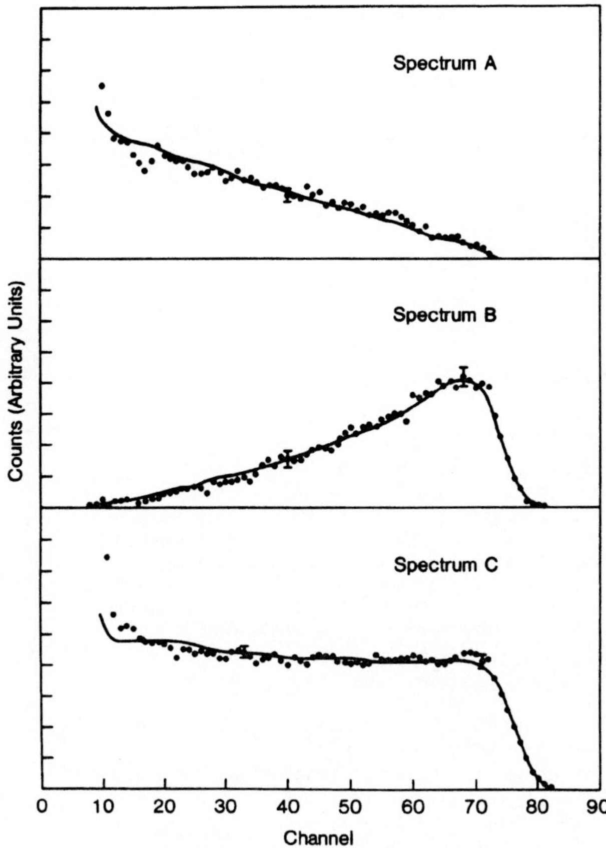

Fig. 3. Experimental proton recoil spectra. Solid curve is a Monte Carlo calculation.

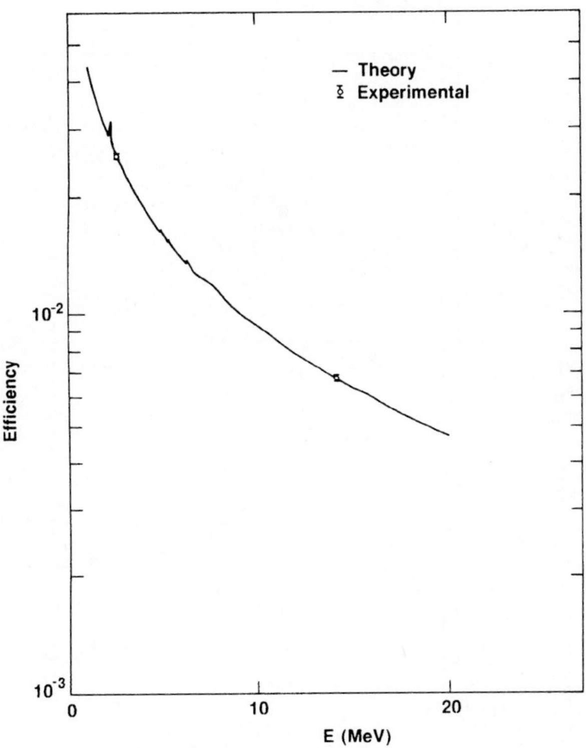

Fig. 4. Detector efficiency curve. The points at 2.47 and 14.1 MeV were obtained experimentally. The calculated curve is normalized at 2.47 MeV.

From these results it can be concluded that the detector can be used as a flux monitor in the 1-20 MeV range. The present accuracy at 2.47 and 14.1 MeV is 1.3%. Because the corrections involved in the calculation are relatively small and the proton recoil spectrum approximates the ideal scintillator response, it is expected that the estimated accuracy for the 1-20 MeV range is around 1-2%.

This detector is presently being used as a flux monitor to measure the $^{235}U(n,f)$ cross section in the 1-6 MeV range with the NBS Linac as the source of neutrons.

References

1. K.C. Duvall and O.A. Wasson, IEEE Trans., Nucl. Sci. Eng., Vol. NS-28, No. 2, April 1981.
2. W.P. Poenitz, Report ANL-7915 (1972).
3. M.M. Meier, NBS Internal Report, February 1978.

PERFORMANCE AND EFFICIENCY OF A LARGE LIQUID SCINTILLATION COUNTER FOR THE DETECTION OF FAST NEUTRONS

J. Marton, M. Cargnelli, H. Fuhrmann, P. Kammel, P. Pawlek

Austrian Academy of Sciences
Boltzmanngasse 3, A-1090 Vienna
Austria

W. H. Breunlich

University of Vienna
Boltzmanngasse 3, A-1090 Vienna
Austria

W. H. Bertl

ETH Zürich, c/o SIN
CH - 5234 Villigen
Switzerland

This neutron detector of high efficiency consists of a large cell (26 x 14 x 10 cm) filled with liquid scintillator NE 213. Two photomultipliers (XP 2041) viewing the volume from both ends allow a compensation for geometrical effects on uniformity and time resolution. The n-γ pulse shape discrimination (PSD) has been tested using the sum of the anode signals of the photomultipliers. The applied PSD method is based upon the comparison of two weighted time integrals of the detector signal. Its perfomance in the low energy range (> 0.06 MeV in equivalent electron energy) is represented by the figure of merit and an alternative definition. The energy- and pulse height-dependent efficiency (response matrix) has been calibrated in a time of flight arrangement with a 252-Cf-fission chamber.

[Large NE 213 neutron counter, homogenity of light collection,pulse shape discrimination, efficiency]

Introduction

For the measurements of fast neutrons originating from muon induced processes in hydrogen isotopes /1/ a neutron counter was developed providing a large solid angle and high intrinsic efficiency.
Liquid scintillator NE-213 was used as the active medium due to its excellent pulse shape discrimination (PSD) capability and its fast response time.
To suppress in neutron measurements the rather large gamma background, special attention has to be given to the PSD performance. Since the neutron energy spectrum of the neutrons following muon capture in deuterium is continous and has its highest intensity at about 1 MeV, extensive studies were made to achieve a good PSD in the low energy range.
Another part of this work was devoted to the improvement and test of the homogenity of light collection. For the measurements of total neutron fluxes the efficiency of the neutron counter has to be known. A Monte Carlo program /2/ was developed, which permits the calculation of this quantity. The theoretical results were compared with the experimental values, measured in an calibration experiment by using a Cf-252 fission chamber.

Detector Design

The scintillator cell is made of 2 mm aluminum (thickness of the neutron entrance window: 1 mm) in a prismatic shape (28 x 14 x 10 cm). The inside of this box is coated with diffuse reflecting paint. The light guides are mounted at the glass windows on both ends, matching the rectangular shape of the cell to the circular photomultipliers (XP 2041).

Discrimination of Neutrons and Gammas

For neutron measurements in our current experiments at SIN the discrimination of the rather large gamma background is very important, especially in the low energy range.
To study the detector's PSD performance two neutron detectors with equal design were irradiated by a Cf 252 fission source. As input signal for the pulse shape discriminator (Link 5010)the actively added anode outputs of the detector's two photomultipliers were used. Since this PSD unit provides energy and PSD information, a two dimensional distribution with a neutron and a gamma branch can be displayed. The quality of n-γ discrimination was determined by making cuts along the energy axis and calculating the following parameters:

$$M = \Delta P/(W_n + W_\gamma). \qquad S = \sqrt{(\sigma_n^2 + \sigma_\gamma^2)}/\Delta P.$$

ΔP..... Distance between n and γ peak
W_n, W_γ... FWHM of the peaks
σ_n, σ_γ... Standard deviation assuming Gaussian shape of the n and γ peak

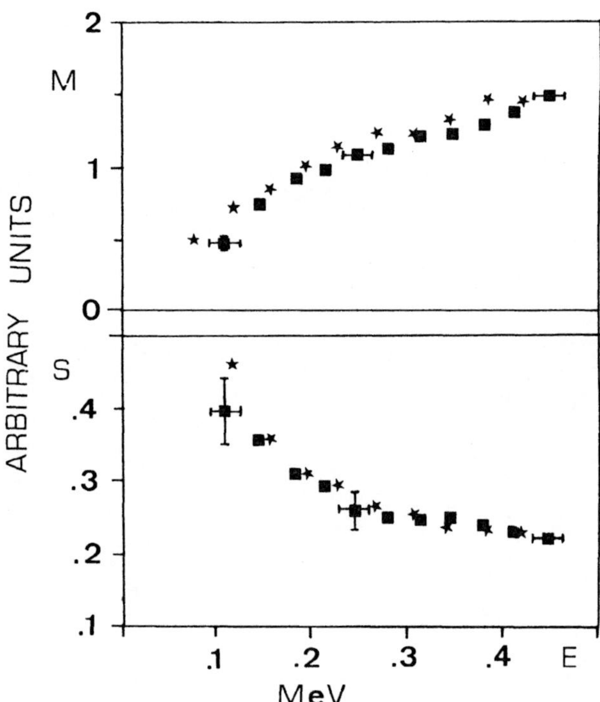

Fig. 1: M and S as functions of the electron energy E (MeV) for the two neutron detectors.

K. H. Böckhoff (ed.), Nuclear Data for Science and Technology, 878–879.

Homogenity of Light Collection

Optimization of energy and timing resolution is provided by maximizing the amount of light transported to the photocathodes and making it independent of its point of origin within the scintillator.
Collimated gamma ray sources (e.g. Cs137) were used to study the "homogenity" of light collection. A measure for this homogenity is the variation of the location of the Compton half height in the electron energy spectrum. This energy distribution was also studied by Monte Carlo simulation of the gamma ray interaction process /3/.
With these calculations systematical errors due to the source-position dependent edge effects could be eliminated.

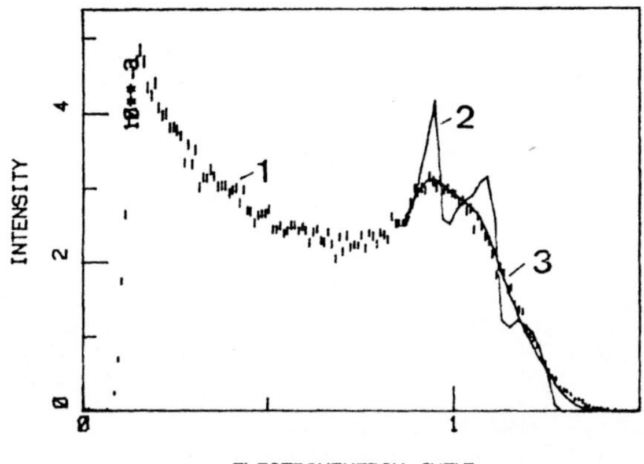

Fig. 2: Measured (1) and calculated (2, 3) energy spectrum of a Co60 source. The summed anode outpute were used for this measurement. Curve 2 shows the Compton distribution, neglecting the instrumental resolution. Folding with the resolution function gives 3.

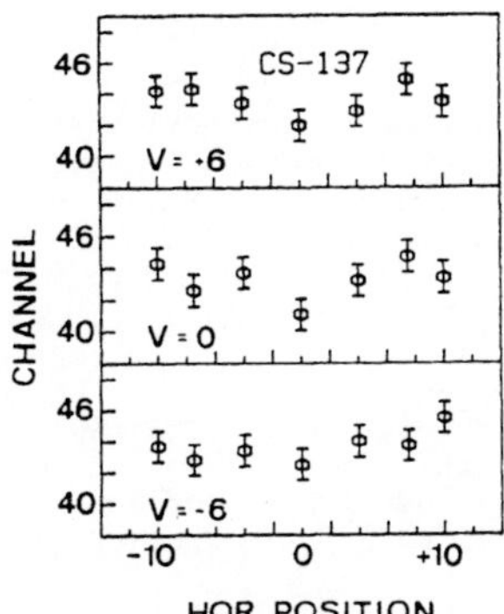

Fig. 3: Location of the Compton half height $L_{1/2}$ as function of the horizontal position of a collimated Cs137 source for three line scans at different vertical positions. Reference point is the center of the detector's entrance window (0/0).

Efficiency

The knowledge of the neutron detection efficiency is essentialfor any absolute neutron yield measurements. The efficiency depends on the size of the scintillator cell of the peculiar detector, the lower energy threshold and the neutron energy. There are two possible ways to obtain this efficiency: in an experiment or by computer simulation of the detection process.

For the experimental determination, a Cf252 fission chamber as previously described /4/ was used in a time of flight arrangement. From the continous neutron energy spectrum 'monoenergetic' neutrons could be selected by setting an appropriate time window, providing recoil spectra for various neutron energy bins. In connection with the known fission rate, the neutron multiplicity per fission, the neutron energy spectrum and the solid angle, absolute efficiency calibration was possible within an accuracy of about ± 4 %.

On the theoretical side a Monte Carlo program was developed to simulate the neutron interaction with the scintillator medium. Therefore recoil spectra could be generated at any incidence energy of the neutrons (Fig. 5). The comparison of the calculated spectrum with the measured shows a good agreement, supporting the validness of the used light output function and the instrumental resolution. Furthermore we computed the intrinsic efficiency as function of the neutron energy for different energy thresholds and compared these values with the experimental data and also a good agreement could be obtained.

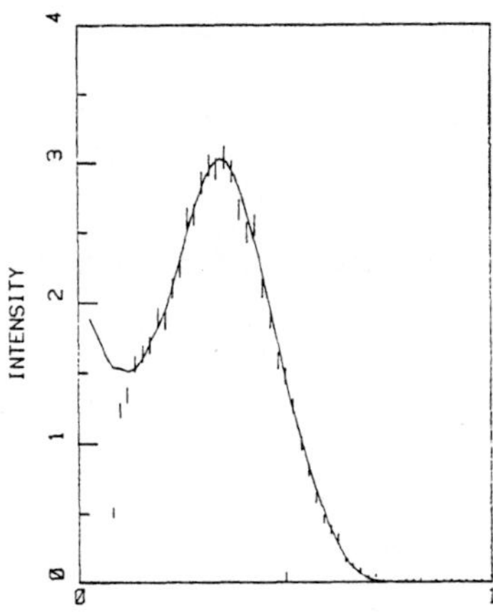

Fig. 4: Comparison between a measured and a calculated recoil spectrum (neutron energy window 1.75 - 2.29 MeV).

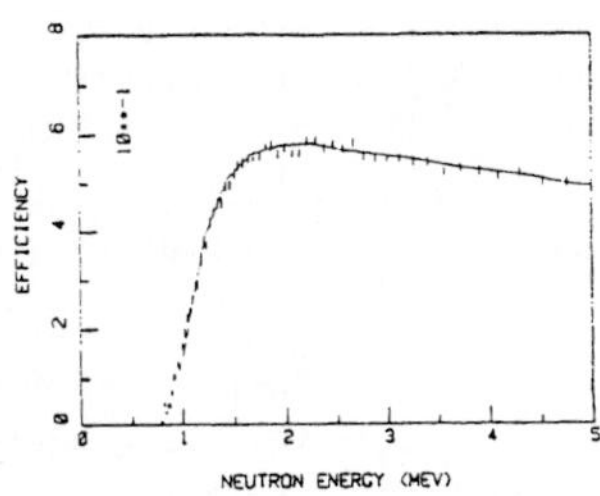

Fig. 5: Measured and calculated efficiency for an energy threshold at 165 keV electron energy.

References

/1/ W. H. Breunlich et al., SIN - proposal R-73-04.1

/2/ M. Cargnelli, to be published.

/3/ M. Cargnelli, to be published.

/4/ A. Chalupka, Nucl. Instr. Meth. 164 (1979) 105

COMPARATIVE MEASUREMENTS BETWEEN A Li-6 GLASS AND A HE-3 HIGH-PRESSURE GAS SCINTILLATOR

H.G. Priesmeyer, P. Fischer, U. Harz, B. Soldner
Institut für Reine und Angewandte Kernphysik der Universität Kiel
D-2054 Geesthacht, Reaktorstation
Federal Republic of Germany

The He-3 high-pressure gas scintillation neutron detector, described by LEE AAMODT et al. (Rev. Sc. Instr. 37 1338 (1966)), commercially available as LND 800, has been compared to a Li-6 glass - scintillator type NE 912. (n,γ) pulse height discrimination capabilities and neutron detection efficiencies have been determined. The objective of these measurements was to try to improve the Kiel Fast-Chopper TOF detector system by using a gas scintillator which could cover the neutron beam geometry and by which gamma ray background contributions could be reduced. The time response always meets the requirements of a chopper experiment, but the neutron detection efficiency of the Li-6 glasses now used had to be maintained.

Introduction

The neutron detector bank at the Kiel University Fast-Chopper consists of 11 Li-6 glass scintillators (2" Ø x 1" thick). The glasses have been fixed to XP 2030 photomultipliers using NE 508 optical cement and have been reflector-painted. These integral-line mountings operate without loss of gain and gain stability since more than seven years.

One disadvantage is the fact that for constructional purposes only about 70 % of the beam area are covered by the glasses. This could be overcome with a gas-scintillation detector, whose outlines could match the geometry of the neutron beam. Furthermore a detector with lower gamma ray sensitivity than the glasses would be necessary to reduce the background of gammas present around experimental facilities at a reactor (e.g. leakage through the chopper rotor, neutron capture gamma rays or activation products like Ar-41 [E_γ = 1.29 MeV], which cannot be completely excluded by pulse height discrimination).

In their paper [1], the authors mention the high neutron detection efficiency and low gamma sensitivity of their detector, but for the decision of whether or not to change the detector system at our chopper, their findings had to be more quantified. Especially a loss of neutron counts due to lower detection efficiency could not be tolerated.

Therefore the following topics have been investigated:

a. the comparison of pulse-height spectra of neutrons and gamma rays
b. the background-to-signal ratios under true chopper conditions
c. the neutron detection efficiencies.

Experimental details

The He-3 high-pressure gas scintillation detector is commercially available *). The gas pressure is 246 kg/cm^2, the He-3 density is 0.03 g/cm^3. The containment material is INCONEL 718, closed to one side by a high-pressure sapphire window. Because of the low light output of the scintillator, a current selection of the photomultiplier is necessary.

The measurements were performed with one of the detector mountings used at Geel **), since our own detector had lost its gas filling for reasons unknown. An RCA 4516 type multiplier selected to have an anode blue current of > 9 µA and an anode dark current of < 0.1 nA is used in this assembly.

For the background-to-signal ratios to be determined by the black-resonance method at the chopper TOF spectrometer it was necessary to compare two detectors of almost equal active volume. Therefore a 10 mm Ø x 1" Li-6 glass NE 912 was mounted on a PM 1910 multiplier, the resolution of which had been determined in the usual way (NaJ/Cs-137) to be ≈ 10 %. One of the detectors now in use at the chopper (NE 912, 2" Ø x 1" thick, mounted on an XP 2030) was also included into these comparative measurements. The time-of-flight spectra were taken at a flight-path of 5 m with the Fast-Chopper. Black resonances of Au, U, Sm, W, and Co were used. The efficiency comparisons were done in the calibrated neutron fields of the Sc- and Fe- filtered beams at the PTB Braunschweig ***) with fluxes of 2000, resp. 2200 n/cm^2 s [2]. These figures were used for the efficiency calculation.

Results

Figures 1 to 3 show linear-scale, preset-counts pulse height spectra of neutrons (Cf-252, moderated by 10 cm of polyethylene) and Co-60 gamma rays of the "small" and "big" glass scintillator and the He-3 detector. It has been checked that the Co-60 gammas do represent the average reactor gamma rays very well. It can be well understood that the Compton edge of scattered gammas is more pronounced in the "big" glass, but no answer can yet be given why the neutron pulse height resolution of the "small" glass is 44 %, while it normally is in the vicinity of 20 %. This question is being investigated by the manufacturer of the glass.

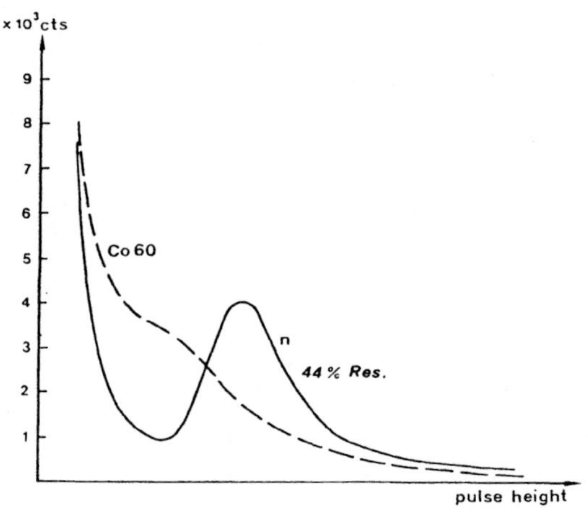

Fig. 1 Pulse height spectrum of NE 912, 10 mm Ø x 1"

K. H. Böckhoff (ed.), Nuclear Data for Science and Technology, 880–882.

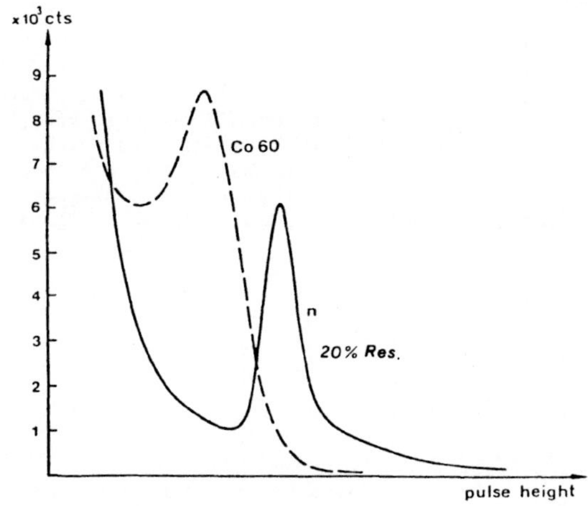

Fig. 2 Pulse height spectrum of NE 912, 2" Ø x 1"

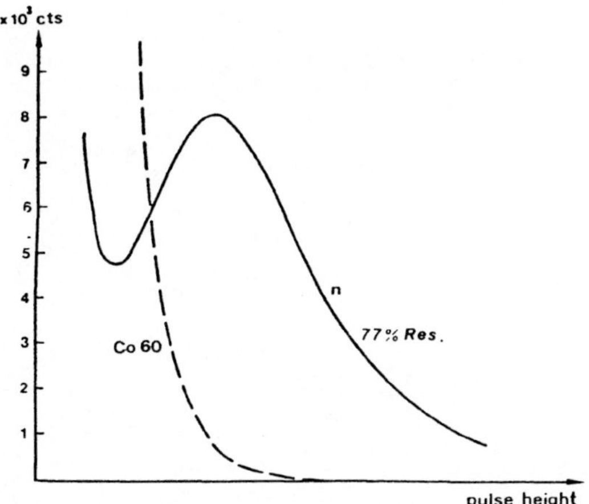

Fig. 3 Pulse height spectrum of LND 800

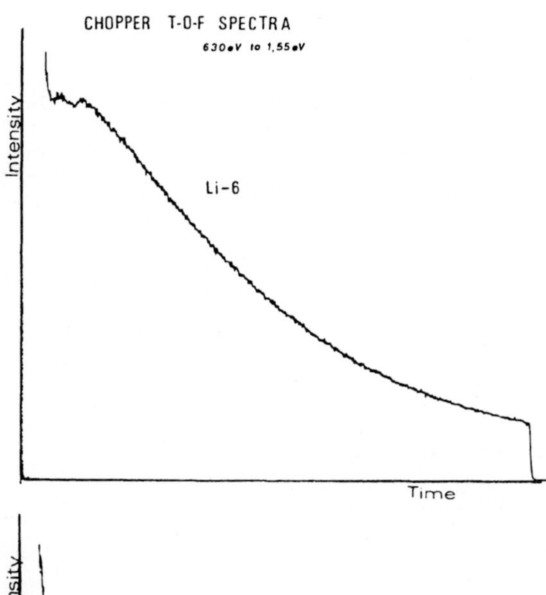

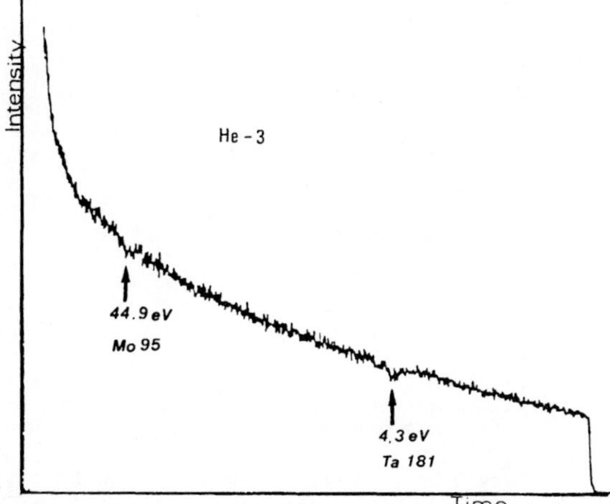

Figures 4 and 5 Open-beam chopper time-of-flight spectra (cf. text)

As can be seen from Fig. 3, (n,γ) pulse-height discrimination for the gas scintillator can be performed efficiently despite the 77 % resolution, found in this case.

Figures 4 and 5 compare the open beam time-of-flight spectra with very low resolution in energy. The selective absorption of the He-3 detector at 4.3 eV and 44.9 eV is due to the alloy constituents Mo (≈ 3.05 %) and Ta (Nb + Ta: 5.15 %) (in INCONEL 718). This fact may also be important at higher energies.

The background-to-signal ratio under our special conditions seems to be slightly better for the He-3 detector. The comparison is not unique, since neither detector at 5 m is optimized to the chopper beam geometry. (For the 4.9 eV - gold resonance, which as black over > 100 channels, the "big" glass had ≈ 36 % and the He-3 detector ≈ 18 % background-to-signal ratio.)

The time-of-flight independent background counts per channel and minute measuring time, normalized to the detector volumes, are $5.4 \cdot 10^{-3}$ n/ch min cm^3 for the "big" glass and $1.5 \cdot 10^{-4}$ n/ch min cm^3 for the He-3 detector.

The efficiency measurement results are given in Table I:

Table I

	"big" Li-6 glass	He-3 gas scintillator
2 keV Sc-filter	30 %	17 %
24 keV Fe (+Al+S+Ti) filter	9 %	4 %

As detector efficiency we define the neutron capture probability $\varepsilon = 1 - \exp(-d/\lambda)$, where d is the detector thickness in cm.

The efficiencies of the Li glass yield a mean free path $\lambda = (0.17 \pm 0.01) \cdot \sqrt{E}$ [cm] (E in [eV]) and fit well with the 1/v-cross-section of the Li-6 (n,α) reaction. (This value is a factor of two lower than the value given in the manufacturers data sheet for NE 912 !)

For the He-3 detector we find $\lambda \approx (0.18 \pm 0.2) \cdot \sqrt{E}$ at 2 keV and $\lambda \approx (0.23 \pm 0.2) \cdot \sqrt{E}$ at 24 keV, assuming the average detector thickness to be 1.5 cm. The difference can be explained by the deviation of the He-3 (n,p) cross-section from the 1/v-behaviour in the high energy range.

As can be seen from Figure 6, the He gas-scintillator emission spectrum is in the UV, peaking at about 800 Å. It was no more tried to see the direct emission photons, which are out of the range of the usual UV sensitive photomultipliers.

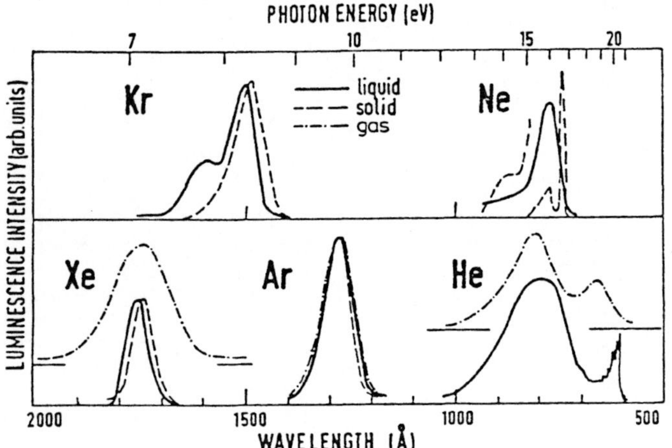

Fig. 6 Noble gas emission spectra (from [5])

*) LND 800 from LND, Inc.OCEANSIDE, L.I., N.Y.11578

**) Special thanks are due to Dr. F. POORTMANS of Geel, who helped us with one of his He-3 detector assemblies and with his experimental experience.

***) We thank Dr. W.G. ALBERTS and his group for the possibility of doing the measurements at the PTB tailored beams.

More detailed information about the measurements will be published [4].

Conclusion

The investigation did not show a convincing advantage of a high-pressure He-3 gas scintillator over the Li-6 glasses now in use, especially when the gas price is also taken into consideration. Furthermore the scattering in the performances of individual He-3 detectors may be considerably larger than for the NE 912 glasses.

References

1. Lee Aamodt, Brown, Smith, Rev. Sc. Instr. 37 1338 (1966)
2. W.G. Alberts PTB Report PTB-FMRB-70 (1978)
3. W.G. Alberts, H. Lisiecki, priv. comm., to appear in Rad. Prot. Dosimetry, 1982
4. B. Soldner, Staatsexamensarbeit Universität Kiel, to be published as GKSS-E report
5. N. Schwentner, Internal Report DESY F 41, HASYLAB 80/11, October 1980.

A NEW TYPE OF NEUTRON DETECTOR USING MULTIWIRE CHAMBERS

F.P. Brady, T.D. Ford, C.M. Castaneda M.L. Johnson
G.A. Needham,* and J.L. Romero

Department of Physics and Crocker Nuclear Laboratory
University of California, Davis
Davis, California 95616

*Now at Rockwell International, Canoga Park, CA, U.S.A.

A new type of neutron flux and energy-measuring detector, which uses large-area ΔE and E detectors, and multiwire chambers (MWC) is being developed. The purpose is to allow one to measure neutron elastic and inelastic scattering and in particular to measure (n,xn) neutron continuum over a wide energy range. The detector offers many distinct advantages over the conventional neutron time-of-flight (TOF) method. The system was tested using a large area CH_2 converter to produce recoil protons, which are tracked in the two MWC's and registered in the ΔE and E detectors. A ^{7}Li(p,n) spectrum from 62 MeV protons was measured and a ^{40}Ca(n,n') spectrum from 65 MeV neutrons was also measured.

Introduction

Currently we are investigating the feasibility of measuring neutron elastic, inelastic scattering (for example to giant resonances[1]), and (n,xn) neutron continuum energy spectra from nuclei in the neutron beam energy range 25-65 MeV. There exist little published data in this energy range, though there are a good quantity of quality proton data including that for isotopic targets. Most of the neutron data are at E_n = 14 MeV, although recent work at Ohio University has obtained some good data up to 25 MeV (using neutron TOF) with ±5% measurements of elastic and some inelastic scattering with good energy resolution (≲1 MeV).[2] At MSU similar results have been obtained at 40 MeV.[3]

Our recent measurements of "removal" cross sections,[4] leading to total non-elastic cross sections,[5] have provided good evidence that optical model (OM) potentials for 1p shell nuclei are not very accurate (±20%) in predicting the magnitude of differential cross sections for neutrons. Elastic measurements of neutron scattering are needed.

The excitation of collective giant resonances states is currently a topic of great interest in nuclear physics. These giant resonances are one of the outstanding manifestations of collective phenomena in nuclei. They have been investigated with a number of probes and reactions such as inelastic electron and proton scattering, but there exist no data on their excitation via inelastic neutron scattering.

There is also considerable interest in neutron scattering from research efforts in other fields: For light nuclei resident in tissue there is a contribution to KERMA (kinetic energy released) from neutron elastic and inelastic scattering which is of importance in neutron therapy and neutron radiobiology for energies up to 65 MeV.[6] Elastic and inelastic scattering of neutrons and neutron continua play an important role in neutron material damage studies.[7] The Fusion Materials Irradiation Tests (FMIT) facility being constructed at Hanford will have neutron energies up to 50 MeV from ^{7}Li + d (35 MeV). (See ref. 7) for recent reviews on neutron data needs in our energy range).

Conventional neutron detectors in our energy range consist of NE213 scintillators, typically 20 cm diameter by 7.5 cm thick where neutron energy is determined by time-of-flight (TOF). At our energies (up to 65 MeV) TOF measurements require relatively long flight paths (at least 10 m), which reduce solid angle and overall efficiency. One major problem with TOF is that there are large backgrounds of neutrons (and gamma rays) associated with the production of a neutron beam via (intense) proton or deuteron beams. Elastic cross sections away from forward angles as well as inelastic and continuum cross sections are relatively small and difficult (and very time consuming) to measure against the backgrounds present. To get energy resolutions of the order of 1 MeV, longer flight paths are needed. Also the efficiency of these detectors is only known to within ≅10% at 50 MeV.[9] Because of the wrap-around from earlier cyclotron beam bursts, it is difficult to measure continuum spectra using TOF. Beam sweepers reduce average beam intensities one to two orders of magnitude. The technique used to measure the continuum neutron spectra at lower eneriges is to numerically unfold the pulse height spectra from the scintillator.[10] This is in general difficult to achieve at our energies and requires very good statistics and confidence in the efficiency calculations.

The Detector

The system we have tested uses the arrangement shown in Fig. 1. The CH_2 acts as a converter so that a recoil proton produced by a neutron is tracked in the multiwire chambers (MWC) and registered in the ΔE (0.10 cm thick NE102 plastic scintillator) and E (12.7 cm diameter NaI crystal, 1.9 cm thick) detectors. The X and Y positions from each chamber plus ΔE and E signals are recorded event-by-event on magnetic tape. For neutrons coming from a small target area, the measurement of the angle between the path of the scattered neutron and the path of the recoil proton allows one to determine the neutron energy.

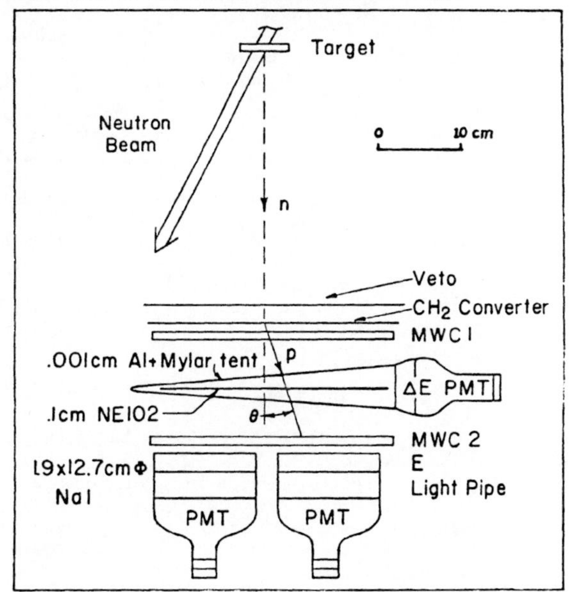

Figure 1. Multiwire Chamber Neutron Detection System.

K. H. Böckhoff (ed.), Nuclear Data for Science and Technology, 883–885.

The closeness of the detector system to the target allows for a large solid angle, measurement of a wide range of neutron scattering angles simultaneously and compensates for the low efficiency (compared to large plastic scintillators many meters away). Also one expects small target associated backgrounds, which are easily measured by removal of the target.

This new technique is particularly useful when one wants to measure a neutron spectrum over a wide range of energies as in (n,xn) reactions and when searching for giant resonance structures. Because TOF is not used there is no long flight path and no wrap-around due to previous beam bursts.

The Multiwire Chambers

The two MWC's are used as position detectors to track recoil protons from the converter. Because pulse height analysis is not used, the chambers are operated above the proportional region (+3500 V on the anode plane) for optimal detection efficiency. The MWC's have the mixture "Magic Gas" flowing through so that they are just above atmospheric pressure (Magic gas is: 23.5% Isobutane, 4.0% Methylal, 0.5% Freon 13B1, and 72.0% Argon). For details on multiwire chamber theory and operation, see reference 11.

The chambers are constructed of four layers of Aluminum frames sandwiched together with "O" rings to prevent gas leakage. The first layer is a 4.76 mm sheet 48 cm square with an offset 30 cm square window epoxied over with .025 mm thick aluminized mylar. This layer also has the gas inlet and outlet near the top. The second layer is the same size frame with .0064 mm aluminized mylar epoxied, which acts as an inactive cathode. This frame is slotted to allow easy gas flow around the window (gas flows on both sides to prevent bulging as the inactive cathode must be parallel to the wire planes for field uniformity). The third layer is the same size again but with inlaid epoxy-fiberglass boards to which the anode sense wires are epoxied. On one side the fiberglass board has etched on its surface the 2.5 ns/cm delay line to which the wires are soldered. The fourth frame is 9.5 mm thick and holds the cathode sense wires and a matched delay line. Epoxied on the back is a .025 mm aluminized mylar exit window. Also on the back are the BNC type electrical feedthroughs for the four signals coming from each chamber.

When assembled the anode wire plane is 4 mm from both cathode planes, and the active area is about 28x28 cm The wires are 2 mm apart and made from 20 μm gold plated tungsten. The high voltage is applied through a 3 MΩ resistor, and signals are taken from each side of the delay line across capacitors. The cathode plane is at ground, with wires 1 mm apart and made from 80 μm copper clad aluminum. Signals are taken directly from each end of the delay line.

There are eight signals in all from the two chambers, that are delayed, discriminated, and fed into an Ortec TD811 time digitizer. We can then manipulate these signals with our software. The time difference between the signals from the ends of a given delay line is related to the position along that delay line. The chambers are calibrated using a collimated ^{55}Fe source and making a linear fit for X and Y on each chamber. The cathode plane gives the X position within 1 mm, and the anode plane gives the Y position to about 2 mm.

Experimental Data

The system is presently well debugged, for both hardware and software, having been used extensively for (n,p) and (n,d) reactions on a wide variety of targets. In a preliminary test of the system using the (n,p) reaction on a .031 g/cm^2 CH_2 target acting as a converter we were able to reproduce the primary neutron beam spectrum. The effect of the carbon in the CH_2 was directly measured and found to be $\lesssim$5% at the highest neutron energies (50-60 MeV) and smaller at lower energies.

A second test was performed on the system with a .075 g/cm^2, large area CH_2 converter directly before the front MWC. The detector system was placed at zero degrees, in the primary neutron beam (at low intensity) produced from a $^7Li(p,n)$ reaction. Figure 2 shows the spectrum we obtained. The energy resolution is approximately 2.2 MeV at 60 MeV or 3.5%. One has the option of using a thicker converter when greater counting efficiency is desired. This, of course, results in poorer resolution. Data were also obtained for $^{12}C(p,n)$ and Fe(p,n) with very good results. The ground state (and 1 MeV excited state), the 4 MeV state, and the giant resonance at 7 MeV were all clearly seen in the carbon data, and the isobaric analog state at 3.5 MeV was seen in the iron data.[12]

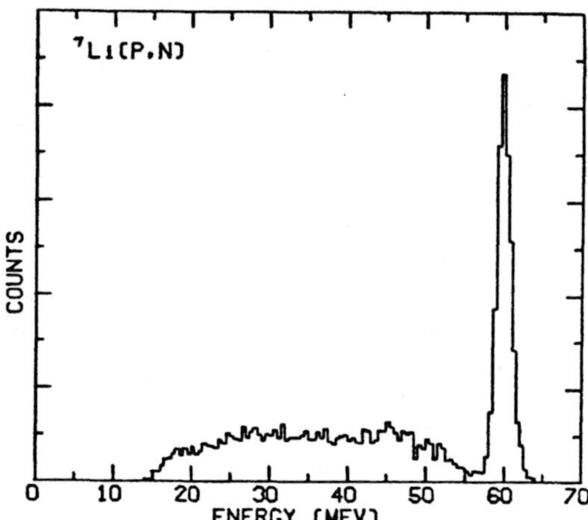

Figure 2. Spectrum from $^7Li(p,n)$ at 61.8 MeV, 0°.

Most recently we have collected data on ^{40}Ca using the (n,xn) reaction at E_n=65 MeV. The detector was configured as shown in Fig. 1, with the detector at 16 degrees relative to the neutron beam. Data were collected in the angular range 4 to 36 degrees from a 1 cm thick calcium target about 30 cm from the converter. The spectrum shown in Fig. 3 represents the neutrons scattered between 7 and 9 degrees for approximately 7.5 hours of counting (including background). The full width at half maximum of the elastic peak is 3.9 MeV, or about 6%. We expected poorer resolution than in the case of (p,n) as we used a thick target and a large beamspot (2x4 cm) to increase the counting rate. Also this data must still be corrected for small gain shifts of the detector.

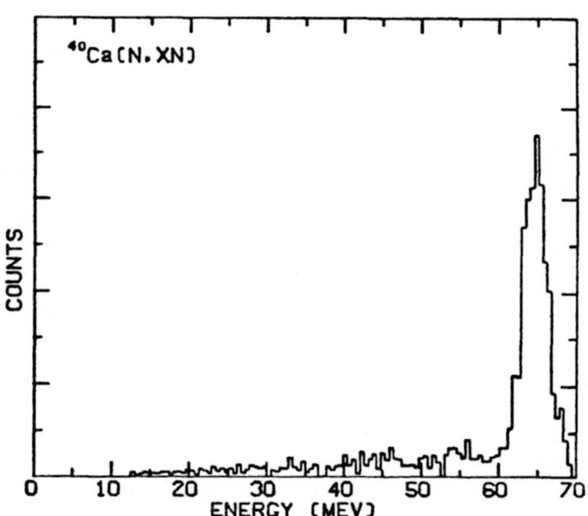

Figure 3. Spectrum from $^{40}Ca(n,n')$ at 65 MeV, 8°.

Data Analysis

Data are recorded event by event through our CAMAC interface with a PDP 15/40 onto a 7 track magnetic tape. For analysis the data are read from the tape, one event at a time. Each event contains position information from the MWC's, ΔE and E and TOF (only for selecting a window around the neutron beam peak). The ΔE and E detectors, due to their large areas suffer from rather varied response relative to the particle position (especially the ΔE). These detectors have been mapped for response across the surface and are corrected for nonuniform response. The proton energy is corrected for energy loss in the components of the detector. The (n-p) angle is calculated and used to determine the scattered neutron energy. The position where (n-p) conversion occurred allows separation of spectra into angular bins.

The analysis program then calculates the differential (n-p) cross section from the Binstock parametrization[13] and calculates a weighting factor which corrects for the efficiency of the detector as a function of energy. A record is kept of total real counts and total weighted counts, then the next event is read. The real counts are saved so that the error can be calculated for the spectra produced.

Discussion

This detection system has demonstrated its value as a tool in measuring neutron spectra. The large solid angle, compactness, versatility, and elegance of the system make it desirable over the conventional TOF method. We are working to upgrade the system to increase counting efficiency and measure more forward angles (to zero degrees). We plan to collect more (n,xn) spectra by the end of this year.

Acknowledgment

This work was supported by the National Science Foundation, grant No. PHY-8121003.

References

1. F.P. Brady and G.A. Needham, "The (n,p) Reaction on Light N=Z Nuclei at 60 MeV", in The (p,n) Reaction and the Nucleon-Nucleon Force, C.A. Goodman, et al., editors (1980).

2. J. Rapaport, et al., "Neutron Elastic Scattering on T=0 Nuclei", Nucl. Phys. A286, 232 (1977).

3. R.P. DeVito, S.M. Austin, W. Sterrenburg, U.E.P. Berg "Limit on Charge Symmetry Breaking in the Optical Model and the Coulomb-Energy Anomaly". Phys. Rev. Lett. Vol. 47, No. 9, p. 628 (1981).

4. F.P. Brady, J.L. Romero, C.I. Zanelli, M.L. Johnson, G.A. Needham, J.L. Ullmann, P.P. Urone, D.L. Johnson, "A Method for Measuring Neutron Total Nonelastic Cross Sections", Nucl. Inst. and Meth. 178, 427 (1980).

5. C.I. Zanelli, P.P. Urone, J.L. Romero, F.P. Brady, M.L. Johnson, G.A. Needham, J.L. Ullmann, and D.L. Johnson, "Total Nonelastic Cross Section Measurements of Neutron on C, O, Ca, and Fe at 40.3 and 50.4 MeV", Phys. Rev. C, Vol. 23, no. 3, p. 1051 (1981).

6. See, for example, R.S. Caswell, J.J. Coyne, M.L. Randolph, "Kerma Factors for Neutron Energies Below 30 MeV", Radiat. Res. Vol. 83, no. 2, p. 217 (1980).

7. C.W. Maynard, "Neutrons and Fusion", in "Proceedings of the International Conference on the Interaction of Neutrons with Nuclei", E. Sheldon, ed., CONF-760715-P1, p. 790 (1976).

8. M.R. Bhat and S. Pearlstein, eds., "Symposium on Neutron Cross Sections from 10 to 50 MeV", BNL-NC5-51245, 2 volumes, 1980.

9. R.A. Cecil, B.D. Anderson, and R. Madey, "Improved Predictions on Neutron Detection Efficiency for Hydrocarbon Scintillators from 1 MeV to about 300 MeV", Nucl. Inst. and Meth. 61, 439 (1979).

10. R.H. Johnson, D.T. Ingersoll, B.W. Wehring, and J.J. Dorning, "NE-213 Neutron Spectrometry System for Measurements from 1.0 to 20 MeV", Nucl. Inst. and Meth. 145, 337 (1977).

11. F. Sauli, Principles of Operation of Multiwire Proportional and Drift Chambers, Lectures given in the Academic Training Programme of CERN 1975-1976, Geneva, 1977.

12. H.W. Fielding, L.D. Rickertsen, P.D. Kuns, D.A. Lind, C.D. Zafiratos, "Study of the Reaction $^{56}Fe(p,n)^{56}Co$ to the Antianalog State", Phys. Rev. Lett., vol. 33, no. 4, p. 226 (1974).

13. J. Binstock, "Parametrization of σ_{tot}, $\sigma(\theta)$, and $P(\theta)$ for 25-100 MeV np elastic scattering", Phys. Rev. C, Vol., 10, no. 1, p. 19 (1974).

BACKGROUND AND RESOLUTION FUNCTIONS IN NEUTRON TIME-OF-FLIGHT SPECTROMETERS

D.B. Syme, B. Thom, and A.D. Gadd

Nuclear Physics Division
A.E.R.E. Harwell,
Oxfordshire,
England, U.K.

In the context of the more complete characterisation of time-of-flight spectrometers based on pulsed white sources, neutron scattering effects in the source and detector areas are likely sources of resolution function asymmetry and background generation. A detailed monte carlo simulation has been made of the effects of source room return of scattered neutrons on pulsed and moderated white sources. It reveals significant resolution broadening effects only for compact sources (cyclotrons), but not for larger source enclosures (linacs). Source room return is unlikely to be a significant source of background for either cyclotrons or linac sources.

The preliminary findings of a set of complementary calculations for scintillation detectors in practical shields are that both background generation and resolution broadening are likely to be significant following neutron scattering in this region.

Introduction

There is an established procedure for analysis of neutron time-of-flight data from white sources. After making dead time corrections and subtracting the known time independent backgrounds, the experimenter subtracts a time dependent background defined by notch filter measurements. The residual spectra are then considered to be background free and analysable in terms of a simple resolution function characteristic of the source and detector combination.

In general this procedure does not lead to a completely successful conclusion. For example, few of the published data in the resonance region have peak cross-sections reaching $4\pi\lambda^2$. Modern shape analysis codes can and do indicate systematic errors in data consistent with erroneous values of backgrounds and resolution functions [1,2].

There are several reasons for this:

1) The resolution function is in general much more complicated and asymmetric than that given by simple calculations.

2) The source and nature of the background events is just not well understood in any general way and any correlation it may have with the foreground is neglected.

3) It has recently been shown that the notch filter method under its usual analysis gives systematically erroneous results except under special and unusual conditions [1].

Little has been done in the way of a general attack on this area of limited understanding although several individual improvements have been made, for example in measuring separately gamma ray backgrounds and calculating detector response in detail [3] and in improving the notch filter method [1]. This paper describes some work on the contributions of neutron scattering effects near the source and near the detector to the resolution function and to the generation of backgrounds in the neutron time-of-flight spectrometers based on white sources. In this context, the background is the part of the tail on the resolution function observable by notch filter methods and the two are closely related. Notch filters use strong resonances with $\Gamma/E_o \approx 2\%$ to 20% to suppress events due to neutrons near energy E_o. Events observed in the corresponding part of the time-of-flight spectrum are defined as background. Scattered neutrons can be included only if their energy is different from E_o by $dE>\Gamma$. Scattered neutrons with $dE\lessdot\Gamma$ must be considered part of the resolution function and their effects calculated or otherwise measured.

Some of the many possible sources of background generation and resolution function skewing are noted in Table I. Neutron scattering effects cover the majority of these. It is important to attempt to model these effects, firstly to determine their relative importance and help in their reduction and secondly for predictions of their energy and time dependences to help interpret the behaviour of notch filter measurements.

Table I Some contributions to Resolution Broadening and Background Generation in White Source Based Neutron Time-Of-Flight Spectrometers

Source	Potential contribution to: Resolution Function	Background
A) PRIMARY BEAM		
1) Delayed source events	?	Y
2) Source/moderator (coupling)	Y	?
3) Neutron scattering near source		
a) Supporting structure	Y	
b) Air	?	?
c) Room return	Y	?
B) COLLIMATOR SCATTERING (negligible for long flight paths)		
C) DETECTOR REGION		
1) Detector and electronics		
a) Primary neutron interaction	Y	
b) Neutron multiple scattering	Y	?
c) Delayed events after a) (luminescence, after pulsing)	?	Y
d) Electronic	Y	Y
2) Neutron Scattering near detector		
a) Light guides	Y	
b) Photomultiplier	Y	
c) Supports	Y	
d) Air	?	?
e) Shielding	?	Y
f) Room		Y

In this paper is described a detailed monte carlo simulation of neutron scattering in and near a pulsed moderated source. This calculation is carried through to the prediction of time-of-flight spectra and the effect of notch filters in these spectra. It is

K. H. Böckhoff (ed.), Nuclear Data for Science and Technology, 886–890.

found that significant resolution broadening arises from neutron scattering in the surroundings of compact sources (cyclotrons etc) but not in accelerators with larger source enclosures (linacs). Background generation is not significant. Some preliminary results are given for neutron scattering effects due to detector surroundings and these indicate that resolution broadening and background generation are more significant in this case.

Generation of resolution and background effects by neutron scattering near a moderated source

1) Simple estimates

Source room return can generate neutrons different in time from the source ones by $dT \gtrsim V.2r$, where V is the initial neutron velocity and r is the radius of the enclosure. If the neutron time-of-flight is measured over distance L then the corresponding energy difference for neutrons observed at the same time is

$$dE/E = 2dT/T \sim 4r/L \quad (1)$$

The order of magnitude of the returned flux can naively be estimated as

$$I \propto \frac{S}{4\pi r^2} . Pn \quad (2)$$

where S is the surface area of the target (moderator) and $Pn \lesssim 1$ is the probability of re-emission from the target room walls. As cyclotrons typically have r=15cm and linacs have r=2m, I is of the order of $^1/_9$ and 6×10^{-4} respectively for these accelerators. For L=50m, dE/E is 1.2% and 16% respectively.

As notch filter resonances are ⩾ 2% wide, neutrons with dE/E less than this from the 'true' energy are attenuated by the filter and scattering effects from compact cyclotrons are not observed as background. They constitute a tail on the resolution function which must be calculated and included in e.g. a proper shape analysis of resonance data.

For linacs a larger fraction of the room-returned neutrons are potentially observable as background by the notch filter method. Although this fraction will vary with the width of the observing resonance this is unimportant as the returned intensity is small.

2) Detailed calculation and results

The monte carlo neutronics code MORSE[4] used to model in detail the required geometry and track neutrons originating in the tungsten target of the Harwell Synchrocyclotron to their exit at the face of the adjacent water moderator (Fig.1). The geometry has been described previously[5]. The tungsten target lies 7mm below the midway plane between the iron magnet poles which have a 30cm gap and a 127cm radius.

The programme was operated with a notional 'point detector' 1m from the moderator face and sampling neutrons emitted along the normal to the moderator. In fact neutrons interacting in the water were flagged according to whether they had come directly from the tungsten source or via one or other of the magnet poles. The statistics accumulated were probabilities of reaching the point detector following an interaction in the water, for each of the flag conditions. For a given input neutron energy (or energy spectrum) the results are in the form of bi-dimensional arrays of exit neutron intensity versus energy and time for each flag condition.

Confidence in the input scattering data and the program itself was increased by preliminary calculations of spectra from isolated moderators and comparison of these with results of a simpler monte carlo code written by one of the authors, and with published results.

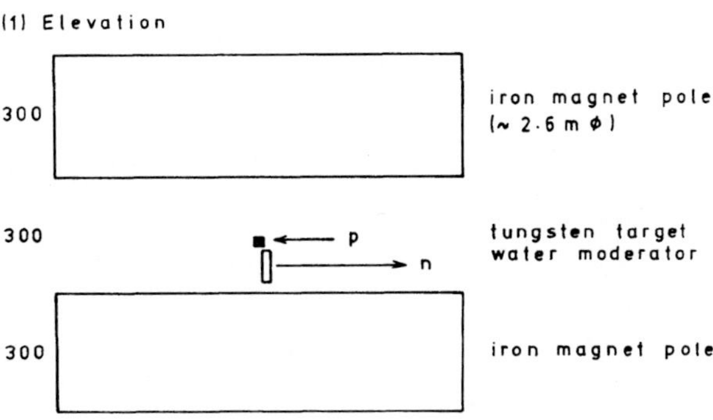

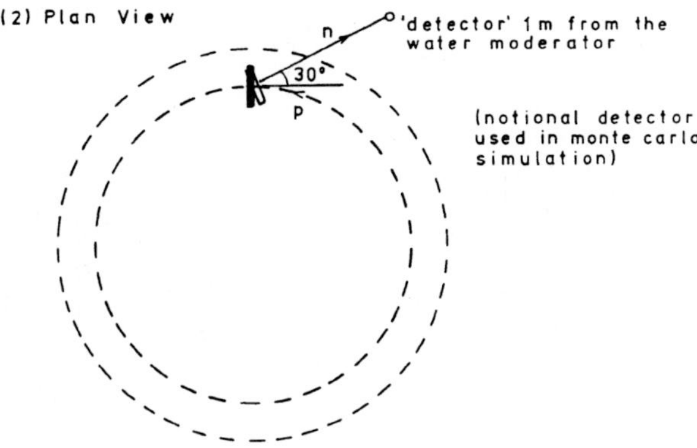

Fig.1. Geometry of tungsten target, water moderator and magnet poles for a monte carlo simulation of the Harwell synchrocyclotron pulsed and moderated neutron source.

Some results from the monte carlo calculations are given in Fig.2. These are time spectra for neutrons emitted from the moderator face in the energy range 111-224 keV following a delta function burst of 1 MeV neutrons in the tungsten target. The prompt spectrum is well fitted by a function with the usual time dependence of moderator spectra.

$$\phi(X) \sim \tfrac{1}{2}.x^2.e^{-x} \quad ; \quad x = t/Tm \quad Tm = 1/\Sigma.V \quad (3)$$

Here t is the moderator time delay, Tm is the mean time between collisions in the moderator, Σ is the macroscopic cross-section and V is the neutron velocity. The 'background' moderator spectrum is for neutrons which have entered the moderator via one or other of the adjacent magnet poles. This is decreased in intensity and begins later than the prompt spectrum because of the pole-moderator distance. It is represented well by the sum of two functions of type ϕ(X) of different delay and intensity. These two components arise from neutrons which have or have not suffered inelastic scattering in their history of scatterings in the iron magnet pole. The iron acts as an intermediate moderator, with its own characteristic Tm, delaying and reducing in energy the incident neutron pulse from the source and re-emitting it towards the moderator. The early component arises from neutrons undergoing inelastic scattering which instantaneously reduces their energy by 845 keV, whereas many collisions and considerable time are required for this to occur by elastic scattering (later component).

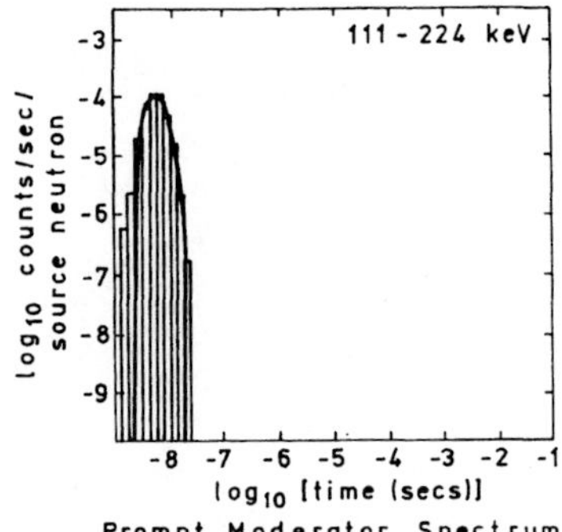

Note:

Curve is moderator shape $\frac{1}{2}\ x^2\ e^{-x}$, $x = \frac{t}{T_m}$

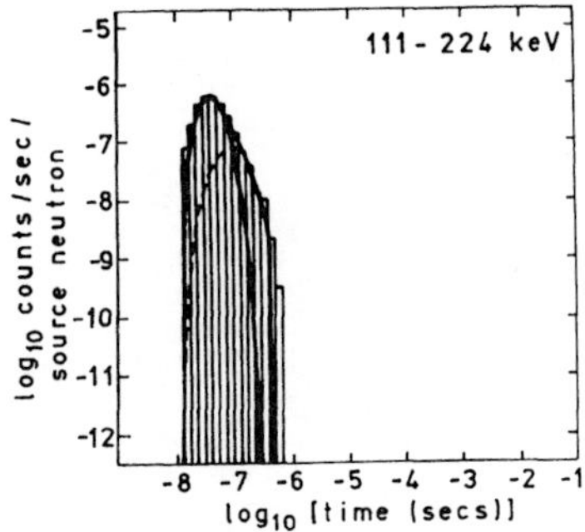

Background Moderator Spectrum (via poles)

Note:

1. Decreased intensity
2. Extra time delay
3. Two components - (pole elastic scattering + pole inelastic scattering)

Fig.2. Comparison of prompt and pole-scattered neutron spectra emitted by the water moderator of the pulsed neutron source of the Harwell synchrocyclotron - results of a monte carlo simulation.

The final monte carlo runs used an evaporation source spectrum. The time/energy spectra were examined by an interactive computer program and at each energy the spectra were characterised in terms of functions of type $\phi(X)$. More precisely the functions were

$$\phi(t) = \tfrac{1}{2}.(I/Tm).((t-tz)/Tm).\exp(-(t-tz)/Tm) \qquad (4)$$

and one found the intensity factor I, the zero time tz and the mean time between collisions Tm. The results are given for the prompt spectra as a function of energy in Fig.3. Note that $I\sim 1/E$ and $Tm\sim 1/V\sim 1/\sqrt{E}$ as expected. At lower energies $tz\sim 1/V$ also, but the transit time between tungsten source and water moderator becomes a limit at high energies. At low energies the trends for the two components of the pole-returned spectra are similar to the prompt trends, but above 100eV the delaying effects of the iron pole moderation modify these trends (Fig.4).

The overlap of the prompt and pole-returned spectra increases at lower energies as shown in Fig.5. At any energy a vertical slice through Fig.5 would give the resolution function of the source as a function of time. A further interactive program allowed the generation and storage of the resolution function as a function of energy for each time channel in a time-of-flight spectrum from this simulated source at any prescribed flight path.

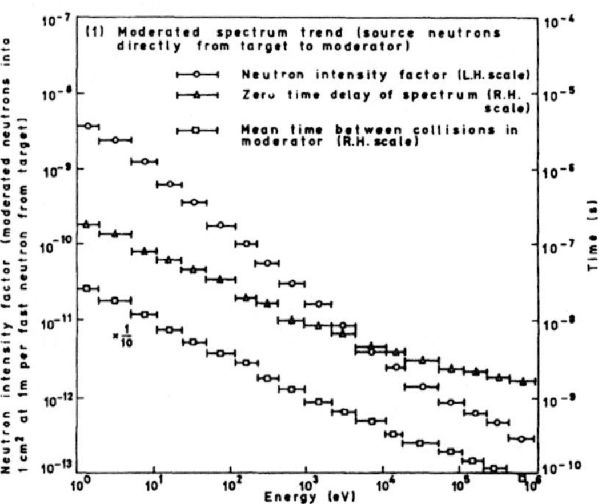

Fig.3. Energy dependence of coefficients describing characteristic moderator spectra from the monte carlo simulation of the Harwell synchrocyclotron source. These data are for neutrons which have gone directly from the tungsten source to the water moderator.

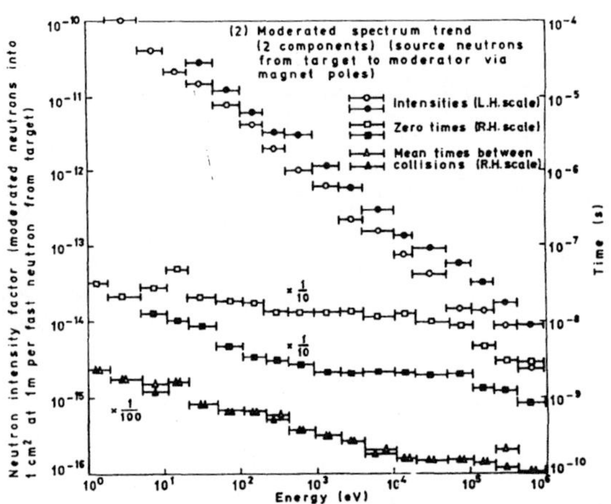

Fig.4. As Fig.3 but now for the neutrons which have reached the moderator via the magnet poles, i.e. the walls of the source enclosure. There are two components, as described in the text.

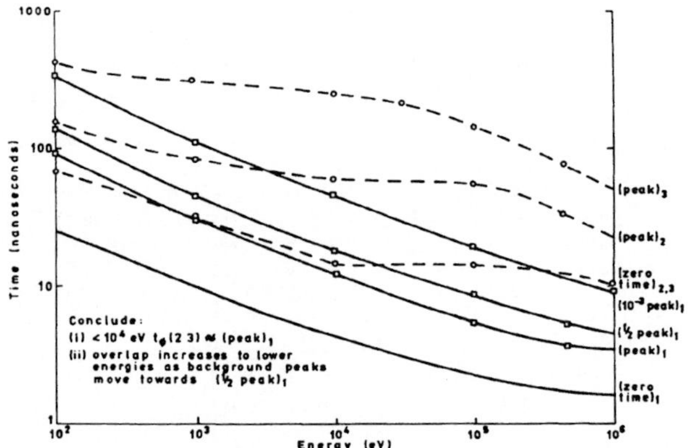

Fig.5. Trend with energy of the prompt and pole-scattered neutron spectra.

In Fig.6 are shown such resolution functions for prompt and pole-scattered neutrons for a time channel 2.5ns wide in a 100m time-of-flight spectrum corresponding to a nominal energy of 1308eV. The scattered component is nearly 10% of the prompt one and is essentially all found within a fractional energy difference of dE/E=1.6%. Therefore it is not expected to be observable as a background by the notch filter method but must be considered an important part of the resolution function lineshape. This is confirmed in Fig.7 where such a time-of-flight spectrum has been mathematically passed through a notch filter exhibiting a Breit-Wigner resonance of Γ/E_o=2%. It is clear that not only the prompt spectrum but also the pole-scattered component is severely attenuated. As this resonance is similar to the narrowest one now commonly used as a notch filter to indicate backgrounds, that method clearly reveals nothing about the pole-scattered components. Thus the simple calculations are confirmed.

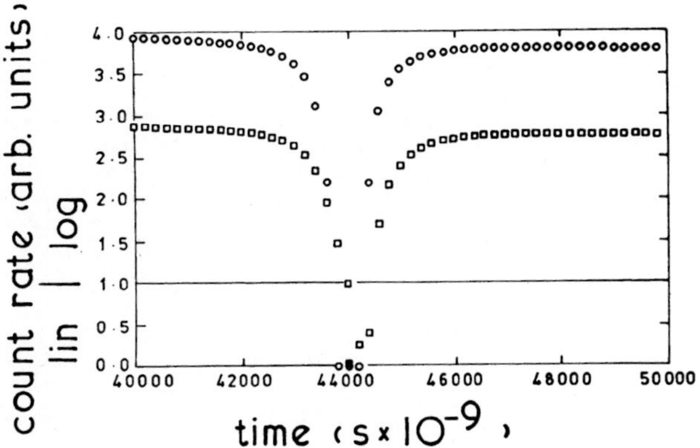

Fig.7. Calculated transmission of a notch filter with a Breit-Wigner resonance having Γ/E_o=2% for the prompt and pole-scattered components of the spectra calculated in this simulation of the Harwell synchrocyclotron.

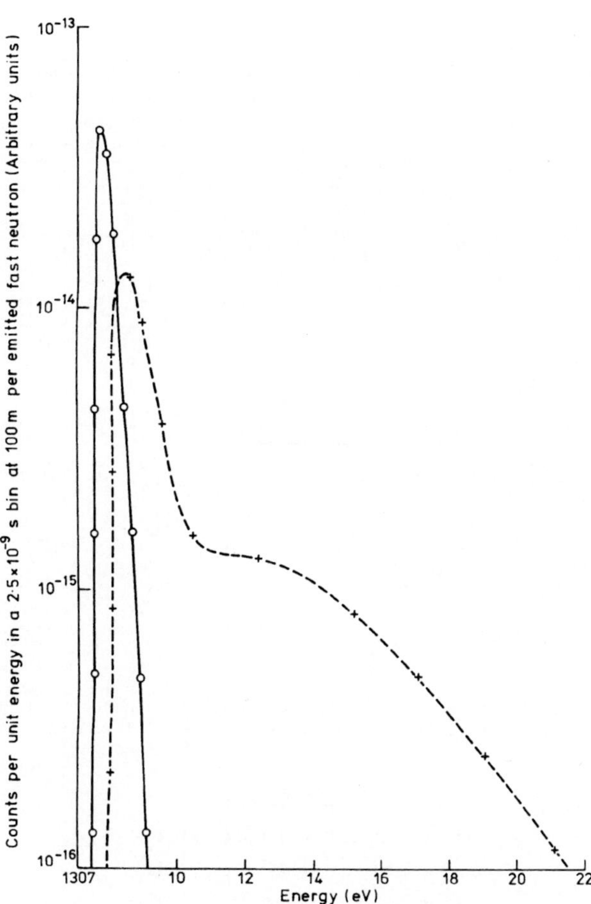

Fig.6. Calculated effective resolution function components for a single 2.5 x 10^{-9}s time channel of a time-of-flight spectrum taken 100m from the Harwell synchrocyclotron source, as simulated in the text.

The results for monte carlo runs with different sizes of enclosure are compared in Fig.8. The background intensity is reduced and the delay increased as expected when the dimensions of the enclosure are increased. This leads to the conclusion that the room return for linac sources will be of negligible intensity.

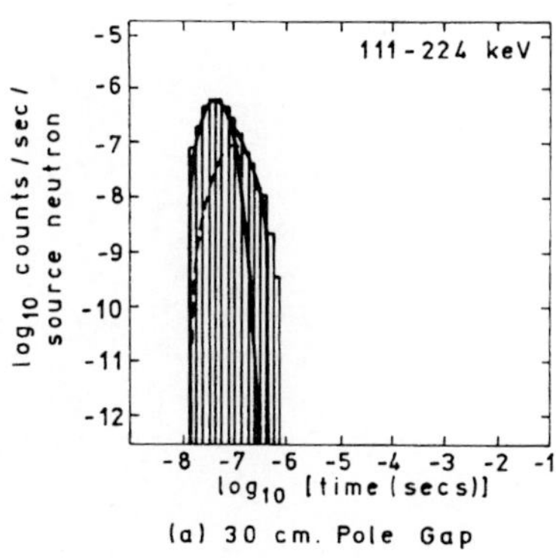

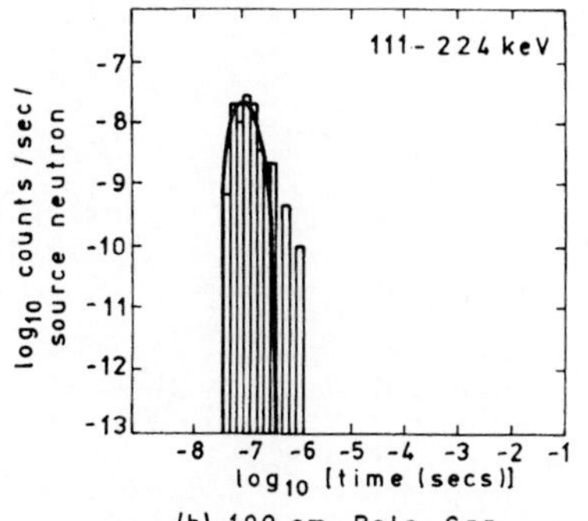

Note:
Peak reduced ×40
Delay increased
Statistics worse

Fig.8. Results of a monte carlo calculation of source room return of scattered neutrons for some enclosures of different sizes.

Response functions of detectors in shields

Quite detailed calculations of response functions of isolated detectors are common, for the importance of multiple scattering effects in the detector itself are well known. The inclusion of scattering effects from adjacent materials which may even be in the neutron beam (light guides, photomultiplier cathodes) is unusual, although the effects on the resolution function are significant and their neglect will be a source of systematic error in resonance shape analysis[1,2].

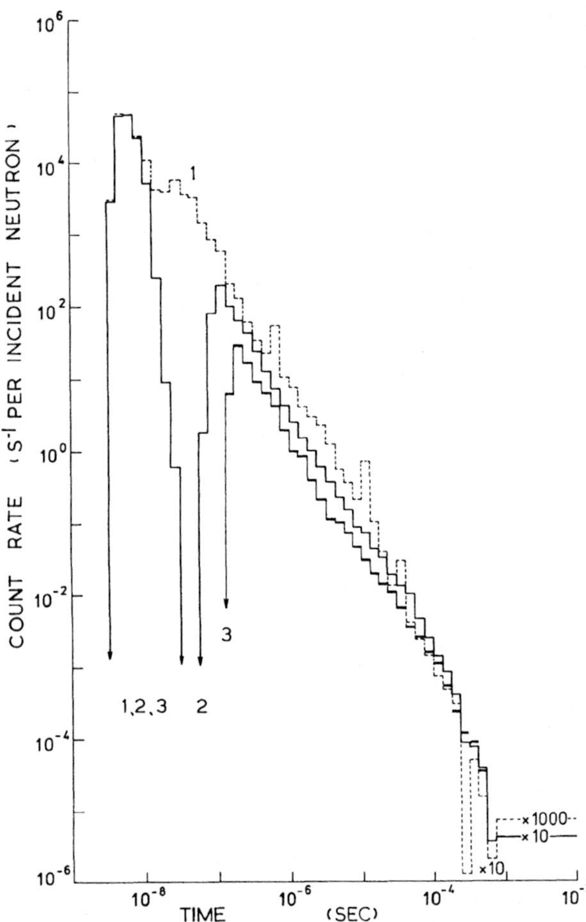

Fig.9. Calculated neutron fluxes caused by scattering of a 1MeV parallel beam by a 10cm diameter, 1.27cm thick ^{6}Li glass detector at the centre of shields of various sizes. The shields 1,2,3 are of 10cm lead followed by 30cm of borated resin in the form of cubes of inner dimensions of 0.2, 1.0 and 2.0 metres respectively. The neutron fluxes are those passing through a notional point detector 5cm outside the edge of the ^{6}Li glass detector.

In a search for significant sources of background it is interesting to calculate the return to the detector from surrounding shielding etc., of neutrons originally scattered out of the incident beam by the detector itself. The range of sizes of detector shields makes it likely that backgrounds observable by notch filters can be generated in this way. Most pulsed white sources have an energy spectrum composed of an intense high energy evaporation component centred around 1MeV and a moderated (1/E) tail of considerably lower intensity extending to lower energies. The high energy component will preferentially scatter out of the detector and may be returned from the shield, delayed and at lower energy. Returned neutrons are likely to be captured with increased efficiency in a detector with a 1/V response and may constitute a significant background effect. This has been confirmed, e.g. by Gayther, who improved the signal-to-background of low energy capture measurements by insertion of a helium filter which preferentially removed high energy neutrons near the 1.15MeV resonance[6].

It is difficult to register a significant number of returned events in the obvious monte carlo simulation because of the combination of low probability scatters etc., involved and it is likely that the process may need to be split into parts. At present the neutron flux scattered out from a parallel beam by a detector and returning towards the detector has been examined with the detector at the centre of cubical shields of lead and borated resin. Examples with a 10cm diameter and 1.27cm thick ^{6}Li glass detector are given in Fig.9. The early peak is the outgoing flux and is the same for all the enclosure sizes shown. The returned flux is the remainder and is separated from the outgoing flux by the expected times for cubical shields of 1 and 2m inside dimensions. The outgoing and incoming fluxes overlap for the small shield (20cm inside cube) partially because of the position of the notional point detector for flux estimation. The integrated returned flux has the expected inverse square variation with enclosure size. This has been confirmed experimentally with capture detectors by Moxon[7]. The time delays of the returned flux will make at least some of it observable with notch filters, depending also on the flight path length. The energy spectra of the returned neutrons will effect their detection efficiency and this needs to be included and related to the incident flux energy spectrum in a more complete calculation. Gamma ray generation in the shield also needs to be included.

Conclusions

There are two main sets of conclusions of interest for users of pulse white neutron sources:

1) Experimenters using sources with compact source enclosures need to model source room return as an important contribution to the resolution function asymmetry. This should not be important for large source rooms. In either case the generation of observable background by this mechanism should be negligible.

2) Detector room return of scattered neutrons deserves serious study as a potentially more important source of observable background.

References

1. D.B.Syme, N.I.M RD9 198(2,3) 357-364 (1982)
2. H. Derrien, BNL-NCS-50451, 156 (1975)
3. D.K. Olsen, G. de Saussure, R.B. Perez, F.C. Difilippo, R.W. Ingle and H. Weaver, Nucl.Sci.and Eng.69, 202 (1979)
4. M.B.Emmett, The MORSE Monte Carlo Radiation Transport Code System, ORNL-4972 (1975), also RSIC CCC-203 (1978)
5. G.D. James, D.B. Syme, P.H. Bowen, A.D. Gadd and I.L. Watkins in 'Neutron data of structural materials for fast reactors', proceedings of a specialists' meeting held at CBNM, Geel, 1977, K.H. Bockoff, Editor, 571 (1979)
6. D.B. Gayther, (NEANDC Topical Conference on neutron capture cross-section measurements, Harwell, 1975), Report AERE-R8082, 4 (1975) (Edited by G.D.James)
7. M.C. Moxon, private communication

SAMPLE SIZE CORRECTIONS OF NEUTRON SCATTERING DATA AND THE ANALYSIS OF ANGULAR DISTRIBUTIONS

H. Klein, B.R.L. Siebert, R. Böttger, H.J. Brede, H. Schölermann
Physikalisch-Technische Bundesanstalt (PTB)
Bundesallee 100, D-3300 Braunschweig
Fed. Rep. of Germany

The new multi-angle neutron time-of-flight spectrometer at the PTB and the Monte Carlo code STREUER II which was developed in order to correct for sample size effects have been tested by means of the scattering of 9.9 MeV neutrons on various carbon samples. A full simulation of the TOF-spectra is needed for all samples to describe the position and the shape of the TOF-peaks due to elastic and inelastic scattering on carbon as well as the underlying background originating from multiple neutron scattering.
A new matrix inversion method has been applied which with only one interation produced a set of Legendre coefficients to describe the experimental data sufficiently for all samples. The differential cross sections extracted confirmed neither the ENDF/B-V data nor the TUNL-data set.
[natural C; polyethylene; $\sigma_{el}(\vartheta_n)$; $\sigma_{inel}(\vartheta_n)$; $\langle E_o \rangle = 9.9$ MeV; $12.5 \leq \vartheta_n \leq 160$; MC-code STREUER II; matrix inversion method]

Introduction

The finite size of target, sample and detector as well as their relative distances necessitate corrections in fast neutron scattering experiments which can be obtained by comparing with Monte Carlo simulations. Simple disc approximations for the neutron flux attenuation /1/ and recursion formulas for the multiple scattering correction /2/ cannot be applied if light elements are involved, i.e. polyethylene (reference) and air (background).

This paper is concerned with fast neutron scattering experiments using a recently completed multi-angle neutron time-of-flight (TOF) spectrometer /3/. The facility was mainly designed and optimized for the investigation of neutron scattering on tissue materials in the energy range of 6 MeV $\leq E_n \leq$ 14 MeV.

As the cross sections on carbon seemed to be well known in this energy region the experimental method was checked by means of the scattering of 9.9 MeV neutrons on various carbon samples. It will be shown that a careful Monte Carlo simulation of the TOF-spectra is necessary to precisely determine the energy of the scattered neutrons and the differential cross sections.

Experimental Method

The spectrometer is described in detail in ref /3/. A ns-pulsed deuteron beam of a movable compact cyclotron produces the primary neutrons via the reaction D(d,n)He-3. The energy gap between the mean monoenergetic neutrons and the continuous neutron distribution due to the breakup reaction D(d,np)D allows an investigation of the inelastic scattering into the first excited state (E_x = 4.439 MeV) of carbon, too.

Two carbon samples with a mass ratio of 4.013 and a polyethylene sample of the same size as the small carbon sample, but only 0.455 of the content of carbon atoms, have been subsequently used (see Table I).

Behind a fixed collimator at a 12 m flight path five liquid scintillators are installed as TOF-detectors. Simultaneously three parameters are analysed for each detector - the integrated pulse height L, a signal PSD analysing the pulse shape and the time-of-flight signal TOF in an inversed time scale with respect to a beam-pickup signal. Neutron - and photon-induced events are on-line separated by means of carefully chosen table cut conditions PSD(L). Besides the neutron TOF-spectra response spectra $L(E_n)$ are on-line recorded for time windows corresponding to the elastic and inelastic scattering on carbon to be compared with calibrated response spectra. In addition the original data are dumped on magnetic tape to allow further off-line analysis.

Table I: Properties of the Scattering Samples and Related Parameters

	C - I	C - II	$(CH_2)n$
distance target to sample center (cm)		17.5 ± 0.1	
diameter (cm)	3.89 ±0.01	2.45 ±0.01	2.46 ±0.01
height (cm)	7.90 ±0.01	5.00 ±0.01	4.98 ±0.01
mass (g)	167.4 ±0.3	41.72 ±0.01	22.43 ±0.01
mean angle target to sample (deg)	6.95	4.7	4.7
mean scattering energy $\langle E_o \rangle$ (MeV)	9.904	9.944	9.944
spread of scatt. angle FWMH (deg)	10.0	6.2	6.2

For any position of the movable cyclotron three separate runs are performed. The background due to neutrons from the structural materials of the target (gas out, sample in) and the scattering of the surrounding air as well as the incomplete flux attenuation of the source neutrons by means of the collimator (gas in, sample out) can be eliminated from the full spectra (gas in, sample in). All different runs in the "gas in" mode were relatively normalized by means of a TOF-monitor movable simultaneously with the cyclotron /3/.

Monte Carlo Simulation

The scattering experiment is fully simulated including all details of
(a) the neutron source as the width of the beam in energy and position, the angle and energy straggling in the entrance foil of the gas target, the energy loss in the gas filling and the strongly anisotropic neutron emission
(b) the size, mass and atomic density of all constituents of the scattering sample and
(c) the geometrical position within the scattering plane and the size of the TOF-detectors.

Besides various important mean values and their variances some distributions are recorded simultaneously:
(α) the energy distribution $f(E_n)$ before the scattering

K. H. Böckhoff (ed.), Nuclear Data for Science and Technology, 891–894.

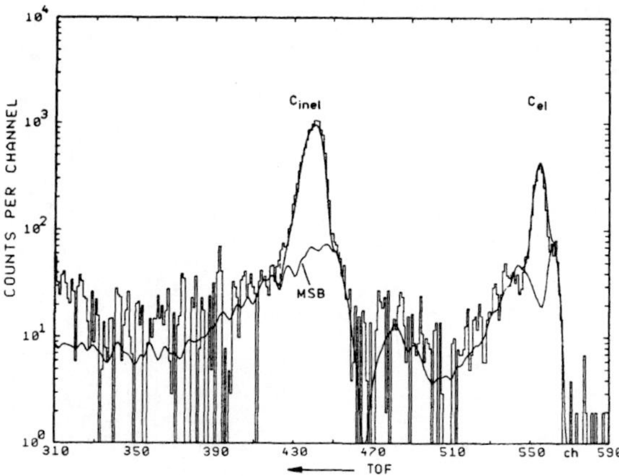

Fig. 1: Neutron time-of-flight spectrum at 72.5 deg for the scattering of 9.9 MeV neutrons on the large carbon sample compared with the normalized MC-simulation (channel width 0.964 ns). The background due to multiple scattered neutrons is indicated (MSB).

(β) the TOF or energy spectra of the scattered neutrons $g(E_n, \vartheta_n)$. The background originating from multiple scattering is recorded separately (fig. 1). The simulation in general uses cross sections of the evaluated data file ENDF/B-V /4/, but alternative data sets can be inserted, i.e. the angular distributions extracted from this measurement.

In the case of the primary neutron distribution it seems to be noticeable in further comparisons that the mean energy $\langle E_0 \rangle$ differs by about 40 keV for the two sample sizes.

As demonstrated with a typical TOF-spectrum measured with the large carbon sample (fig. 1), the position of the peaks corresponding to the elastic and inelastic scattering as well as the underlying multiple scattering background are well described within the statistical uncertainties of the measured and calculated spectra.

Energy of Scattered Neutrons

In this particular scattering experiment an ideal time reference is available from photons produced within the sample by inelastic neutron scattering. As the Compton spectrum of 4.4 MeV photons shows a pulse height spectrum very similar to a spectrum induced by 8 MeV neutrons in the liquid scintillators, the contribution of the time walk and time jitter of the fast timing electronics as well as the beam pulse distribution in time are inherently considered in the analysis of the n-γ-TOF difference.

The mean energies $\langle E^1_{MC}(\vartheta_n) \rangle$ calculated from first collision event TOF or energy spectra deviate up to 50 keV from the energies obtained from relativistic kinematics for the adjusted geometrical angles (broken lines in fig. 2). These energy shifts disappear if the mean angles $\langle \vartheta_n \rangle$ instead of the center angles ϑ_n are used. Taking into account the background due to multiple neutron scattering the energy $\langle E^{tot}_{MC} \rangle$ associated to the mean TOF-difference is partially shifted up to 50 keV (full lines in fig. 2) and is in better agreement with the experimental data. The error bars correspond to an uncertainty of ±0.5 ns estimated for the absolute determination of the TOF-differences.

In fact a simultaneous analysis of the elastic and inelastic scattering by comparing it with the MC-simulated TOF-spectra allows the mean neutron energy $\langle E_0 \rangle$ before scattering to be determined with an uncertainty of 20 keV, even for the large carbon sample. This uncertainty finally results from the integral nonlinearity of the time-to-amplitude converters which was checked with two independent time calibrators and found to be ± 0.25 ns within 95 % of the 1 µs time range used in all experiments. The geometrical adjustment of the detector center (± 2 mm) and the energy dependent shift of the effective center due to the flux attenuation (≤ 5 mm) may introduce an additional uncertainty of 10 keV at the most.

Ratio of Scattered Intensities

Next, the MC simulation may be tested without any absolute scaling of the experimental data but only taking the ratio of the net amount of scattered neutrons including the statistical uncertainties which result from the background subtraction and the relative normalization via the monitor (fig. 3). Obviously the ratio calculated on the basis of the ENDF/B-V cross section disagrees at almost all angles (dashed-dotted lines in fig. 3) for both elastic and inelastic scattering. As the data file is mainly based on the French evaluation /5/, we alternatively inserted a second data set /6/ available in this energy region. The agreement of the calculated and the measured ratio is sufficient for forward and backward angles, but the deep minimum of the elastic scattering at about 65° is not reproduced (broken lines in fig. 3). In order to detect the origin of these discrepancies we had to determine the absolutely scaled cross sections.

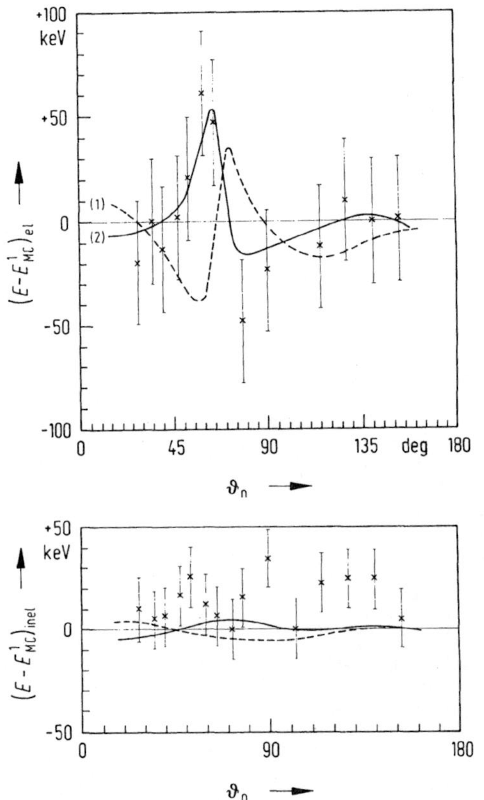

Fig. 2: Energy shift of the MC-simulated mean energy of first collision events $\langle E^1_{MC} \rangle$ with respect to the mean energy $E_{kin}(\vartheta_n)$ (1) as well as the full MC-simulated energy $\langle E^{tot}_{MC} \rangle$ (2), compared with the experimental data for the elastic (a) and inelastic (b) scattering of 9.9 MeV neutrons on the large carbon sample.

Absolute Scaling

For an absolute scaling, the fluence of the primary neutrons hitting the scatterer was measured in a separate run by means of a proton recoil spectrometer which had been successfully used in an international intercomparison of fast neutron fluence meters /7/. The code SINENA /8/, which includes the same detailed description of the target properties, is used to transfer the fluence measured in a 10.85 mg/cm^2 tristearine radiator foil to the scatterer which cover a similar solid angle.

In addition this code was used to calibrate the neutron detection efficiency $\varepsilon_{d}(E_n, E_{thres})$ of the liquid scintillators with respect to the telescope (see fig. 6 in ref. 3).

Comparing the absolutely scaled experimental data with the calculations on the basis of ENDF/B-V cross sections, we find a diffraction pattern for the elastic scattering (fig. 4) which would fully disappear if an angle offset of $\Delta\vartheta \sim 1.2$ deg is assumed. Thus the angle alignment had to be checked experimentally.

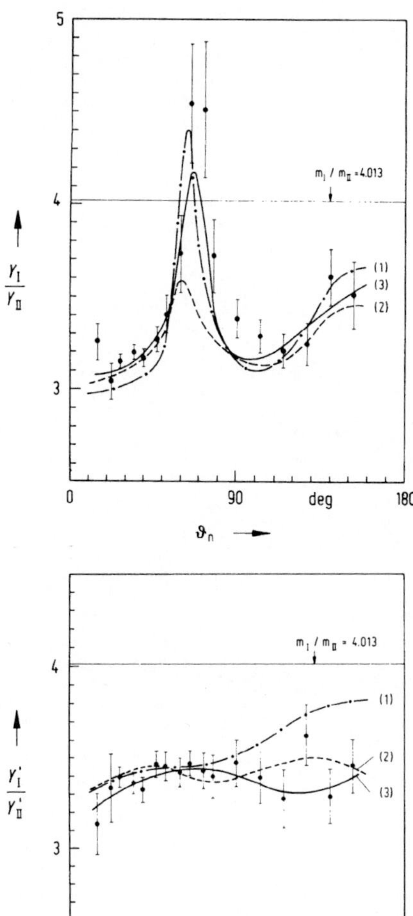

Fig. 3: Ratio of the yields measured for the elastic (a) and inelastic (b) scattering of about 9.9 MeV neutrons on two different carbon samples (see table I). The experimental data are compared to MC-simulations on the basis of the ENDF/-V /4/ data file (1), the TUNL /6/-data set (2) or differential cross sections fitted to our data (3).

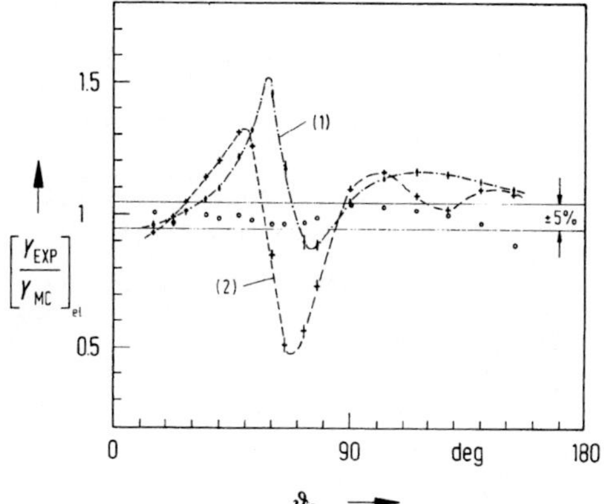

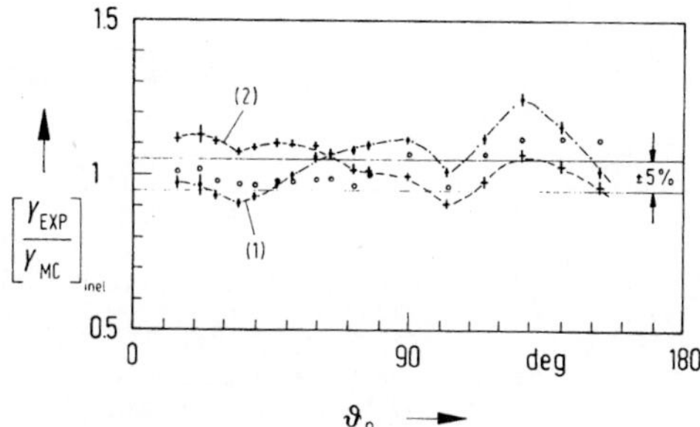

Fig. 4: Ratio of the absolutely normalized experimental yield Y_{exp} for the elastic (a) and inelastic (b) scattering of about 9.9 MeV neutrons on the large carbon sample to the expected yield Y_{MC} as calculated on the basis of ENDF/B-V (1), TUNL (2), or PTB (o) data sets.

Scattering Angle

An angle offset of 1 deg would correspond to a displacement of 3 mm of either the center of the beam within the gas-target or the center of the scatterer with respect to the reference axis optically aligned. Although being improbable the hypothesis was checked using a polyethylene sample which was adjusted in the same way above the pivot of the cyclotron by means of two theodolite arranged perpendicularly. In this

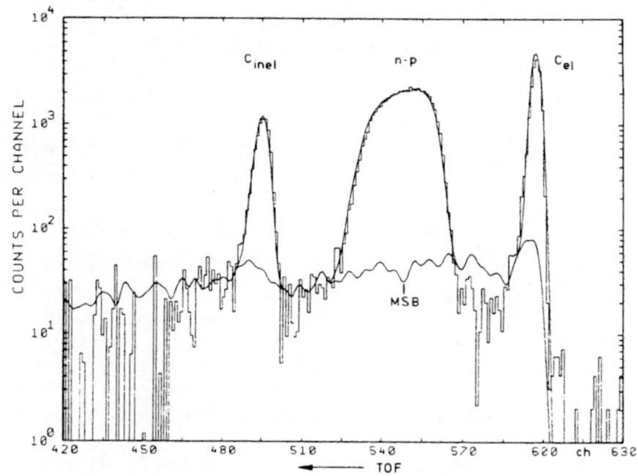

Fig. 5: Experimental neutron TOF spectrum at 27.5 deg for the scattering of 9.94 MeV neutrons on the polyethylene sample (channel width 0.972 ns).

method we made use of the fact, that the gradient $dE_n/d\vartheta_n$ is only 20 keV/deg at maximum at 90 deg for the scattering on carbon, while the neutron energy changes with 170 keV/deg at maximum at 45 deg for n-p-scattering. As the MC-simulation sufficiently describes the completely different multiple scattering background (fig. 5) a slight shift of the TOF distribution originating from n-p-scattering with respect to the two TOF-peaks from elastic and inelastic scattering on carbon may be attributed to an angle offset $\Delta\vartheta$. Three of five detectors in the range $15 \leq \vartheta_n \leq 57.5°$ were simultaneously analysed, if the peaks were separated. From two independent runs we obtained a mean offset of $\Delta\vartheta = 0.09 \pm 0.1$ deg where the uncertainty includes the reproducibility for independent geometrical adjustments.

In addition the absolute scaling is reproduced within ±3 % with respect to the n-p scattering.

Legendre Expansion Coefficients

As the MC simulations disagree with the experimental data in a different way for the two data sets available, it seemed warranted to extract a third angular distribution from our data. A new matrix inversion method is applied to calculate the Legendre expansion coefficients from the background corrected, absolutely scaled experimental data. The full geometry including the flux attenuation within the sample is considered in matrix coefficients which can be calculated in the first MC run on the basis of total cross sections only. Finally the multiple scattering correction has to be determined iteratively starting with the measured distribution or differential cross sections interpolated from evaluations or model calculations.

Table II: Legendre polynomial coefficients (ENDF expansion format), as effectively used for the MC simulation of the scattering on the large sample C-II

	ENDF/B-V	TUNL	this work
σ_{el} (b)	0.632	0.635	0.633
a1	0.5387	0.5130	0.5192
a2	0.4192	0.4074	0.4012
a3	0.3209	0.3065	0.2962
a4	0.1509	0.1568	0.1193
a5	0.0303	0.0432	0.0082
a6	0.0041	0.0209	--
σ_{inel} (b)	0.330	0.314	0.303
a1	0.3378	0.2775	0.3342
a2	0.1758	0.1373	0.1486
a3	0.0297	0.0265	0.0322
a4	0.0182	0.0198	0.0155
a5	0.0023	0.0064	0.0095

For this analysis we started with ENDF/B-V predictions and found the final parameters listed in table II with one iteration only. The new parameters differ significantly from both data sets but describe the experimental data consistently for all samples used.

Conclusion

From precise neutron TOF measurements we conclude that the cross sections of the evaluated data file ENDF/B-V are inadequate to describe the scattering of neutrons with a mean energy of about 9.9 MeV. In the case of two different carbon samples and a polyethylene sample it is demonstrated that a full Monte-Carlo simulation of the experiment allows the energy of the scattered neutrons to be determined by means of TOF measurements with a 20 keV uncertainty. In particular, the completely different shape of the multiple scattering background if pure carbon samples or a polyethylene scatterer are used clearly prove that the sample size corrections can only be determined by comparing the measured TOF-spectra to complete MC-simulations.

The scattering on polyethylene may not only be used for absolute scaling, but is suitable for fixing the effective scattering angle to within ±0.1 deg. Finally, a new matrix inversion method could be successfully applied to extract Legendre expansion coefficients.

This work has been supported by the Commission of the European Communities (Contract No. BIO-A-284-80-D).

References

/1/ W.E. Kinney; NIM 83 (1970) 15-18

/2/ S.A. Cox; NIM 56 (1967) 245-253

/3/ R. Böttger, H.J. Brede, M. Cosack, G. Dietze, R. Jahr, H.Klein, H. Schölermann, B.R.L. Siebert; contribution to this conference

/4/ ENDF/B-V; Brookhaven (1979)

/5/ G. Haouat, J. Lachkar, J. Sigoud, Y. Patin, F. Cocu; NSE 65 (1978) 331-346
J.C. Lachkar, F. Cocu, G. Haouat, P. LeFloch, Y. Patin, J. Sigoud; NEANDC(E) "L" - INDC(Fe) - 7/L (1975)

/6/ D.W. Glasgow, F.O. Purser, H. Hogue, J.C. Clement, K. Stelzer, G. Mack, J.R. Boyce, D.H. Epperson, S.G. Buccino, P.W. Lisowski, S.G. Glendinning, E.G. Bilpuch, H.W. Newson, G.R. Gould; NSE 61 (1976) 521-533

/7/ V.D. Huynh; Metrologia 16 (1980) 31-49

/8/ B.R.L. Siebert, H.J. Brede, H. Lesiecki; ext. PTB-report ND-23, Braunschweig (1982)

A NEW UNFOLDING METHOD APPLIED TO RECOIL SPECTRA FROM A LIQUID SCINTILLATOR

M. Cargnelli

Austrian Academy of Sciences
Boltzmanngasse 3, A-1090 Vienna
Austria

A recently developed numerical procedure for the reconstruction of the neutron- and gamma-energy-spectra (by unfolding the recoils) is presented together with test results using a response function from a large liquid scintillator NE-213 and various experimental and simulated recoil spectra. The basic problem in spectra-unfolding is the instability of the backward transformation, which leads to useless pseudo-solutions unless special care is taken.
The essential improved performance of the new method is due to the suppression of the pseudo-components by two measures: The energy distribution is approximated by a sum of either harmonic- or Legendre-polynomial basis functions (thus preventing high frequency spurious oszillations) and this spectrum is constrained to be nonnegative. The basis functions are folded with the response function and then used as components for a constrained least-square fit to the pulse-height data. So the essence of the procedure is a sequence of multi-parameter minimizations subject to a set of unequality constraints.

[Unfolding of recoil spectra, resolution-improvement, response of a large NE-213 detector]

The Problem

Liquid scintillators are widely used to detect fast-neutrons via their charged recoil particles (protons). However the analysis of these data is not so easy: to begin with, it is rather difficult to determine the detector efficiency accurately. Furthermore,to determine the number of incident neutrons it is necessary to know the shape of the neutron energy spectrum. That problem arises from the characteristic of recoil spectra: for neutron energy E the pulse-heights P are distributed between zero and P-max. Therefore each pulse-height channel contains contributions from the whole neutron spectrum above the threshold.
The relation between neutron flux X(E) and pulse-height spectrum Y(P) is described by the 'folding equation' (1).

(1) $Y(P_i) = \int dE\ A(P_i,E)\ X(E) \quad + R_i \quad i=1,\ldots N$

A = Response function (efficiency for each pulse-height channel), R = statistical error component.

In symbolic writing: $Y = A \cdot X \ + R$

If the neutron energy spectrum is a priori unknown the 'unfolding problem' (to determine X from (1)) can not be avoided. The serious problem in the backward transformation is the instability, which can amplify errors R by many orders of magnitude and thus obscure the solution completely ('pseudo-solutions', 'spurious oscillations', 'artefacts').

The Algorithm

Since the error component in the unfolded spectrum only reflects the statistical errors of the measurement, it can be suppressed dramatically by recognizing its random structure.
Approximating the original energy spectrum by a polynomial or harmonic sum does a good deal, but still, for peak shaped spectra, the true solution will not be visible.
The additional utilization of the positivity constraint ($X(E) \geq 0$) is essential. The suppression of oscillations around the zero line also effectively suppresses the error amplitude at parts of the spectrum where the intensity is concentrated.
In other words: Legendre-polynomial-, or sine-, cosine- basisfunctions covering a certain energy range are folded to give fit components for the measured spectrum. The fit is constrained in the way that the superposition of the original basis-functions, which is just the unfolded energy spectrum, is positive.
The neutron-energy-, and the pulse-height- variable (E, P) are substituted by log(E), log (P) (or $\sqrt{E}$, $\sqrt{P}$) to have an approximately constant number of data points in the response function's edge-width.

Now, with $B_k(E)$ the k-th basisfunction, the ansatz is

(2) $$X = \Sigma\ B_k\ V_k \qquad k=1,\ldots M$$
$$X\ (E) \geq 0 \qquad \text{for all } E$$

The folding of this neutron flux is the superposition of the folded basisfunctions ($A \cdot B_k = C_k$).

(3) $Z = \Sigma\ C_k\ V_k \qquad k=1,\ldots M$

The coefficients V (and thus the unfolded spectrum X) are determined from minimizing the deviation of the fitspectrum Z from the measured spectrum Y under the constraint $X \geq 0$.

(4) $\text{Min}\ |\ Y - C\ V\ | \qquad \text{subject to } B\ V \geq 0$

For the norm $|\ |$ the sum of weighted quadratic differences is used (least squares fit).

The Response Function

The neutron response of a large NE-213 detector was measured in a time-of-flight arrangement with a Cf-252 source in a fission chamber as described in the poster of J. Marton et.al.
To check by an independant method, to expand the energy region and to get rid of the smearing due to the widths of the neutron energy bins, the response function was also calculated. A Monte Carlo code was written to treat multiple scattering events on both hydrogen and carbon. It uses a light-output-curve matched to the experimental data.
For the pulse-height calibration using gamma sources a second Monte Carlo program was written to deal with multiple compton scattering and thus to calculate the total electron energy deposited in the detector by gamma sources. This program was also used to construct a gamma response function, for unfolding gamma spectra.

Test Results

The performance of the algorithm may be illustrated by a few figures.

Fig.1: Measured pulse-height spectrum: Neutrons from the muon catalyzed fusion process $d + d \rightarrow$ 3-He + n

Fig.2: Unfolded spectrum of fig.1. The dominating peak of fusion neutrons at 2.45 MeV is accompanied by two peaks at lower energies due to inelastic scattering on Fe and Cu.

Fig.3: Measured pulse-height spectrum: Gammas from a Co-60 source.

Fig.4: Unfolded Co-spectrum displaying the two gamma lines at 1.17 and 1.33 MeV.

K. H. Böckhoff (ed.), Nuclear Data for Science and Technology, 895–896.

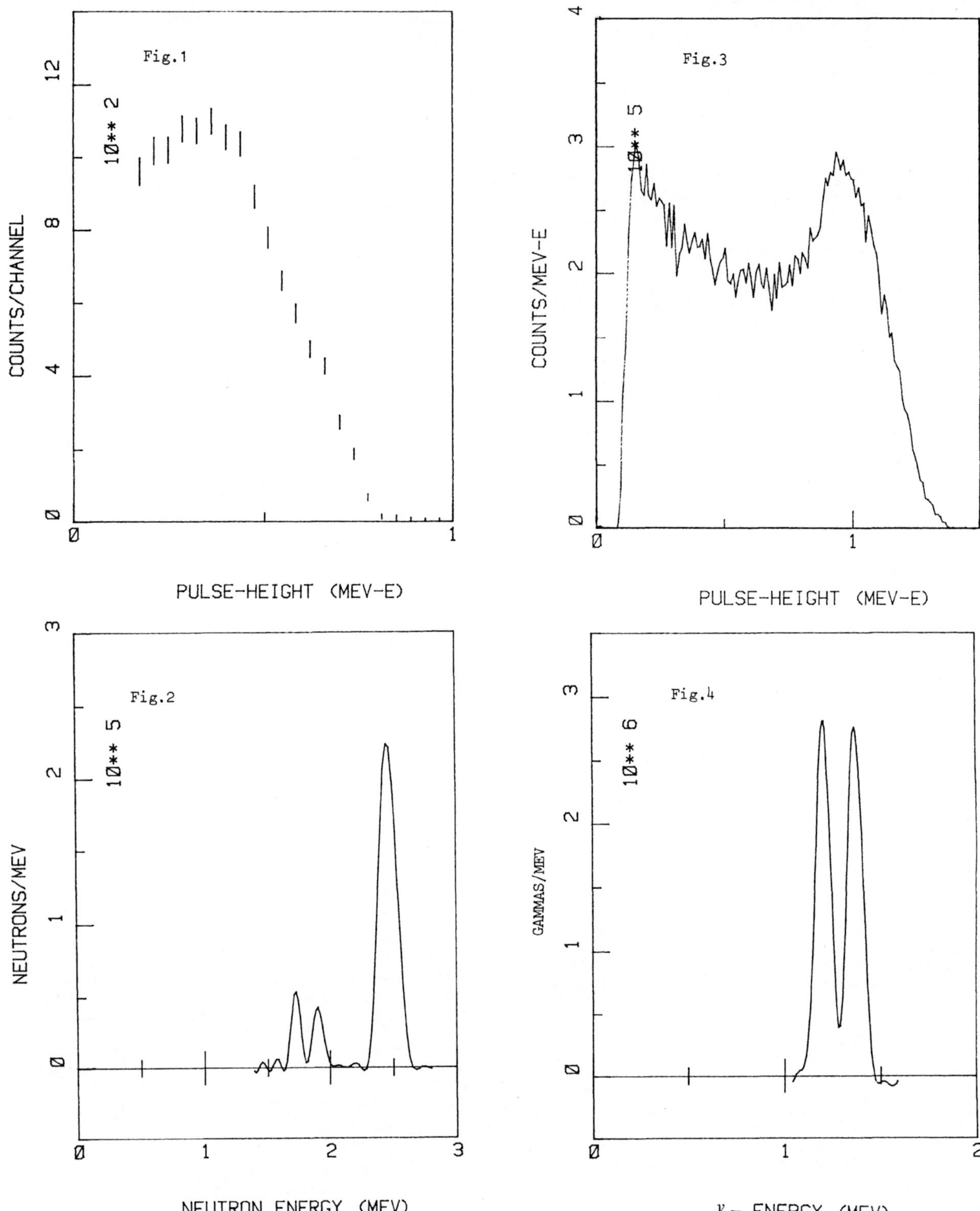
Fig.1
COUNTS/CHANNEL
10** 2
PULSE-HEIGHT (MEV-E)
Fig.3
COUNTS/MEV-E
10** 5
PULSE-HEIGHT (MEV-E)
Fig.2
NEUTRONS/MEV
10** 5
NEUTRON ENERGY (MEV)
Fig.4
GAMMAS/MEV
10** 6
γ – ENERGY (MEV)

USE OF THE RESPONSE FUNCTION IN THE ANALYSIS OF COMPLEX NEUTRON SPECTRA

G.H.R. Kegel, C. Ciarcia, G.P. Couchell and J. Shao

Physics Department, University of Lowell
Lowell, Massachusetts 01854, U.S.A.

Neutron time-of-flight spectra with overlapping peaks must be unfolded to yield contributions of individual neutron groups. This requires an accurate knowledge of the resolution profile of each group. It is also desirable to know the shape of the spectra of neutrons which were scattered more than once in the scatterer, so that corrections for multiple interactions can be made. These resolution profiles and spectra shapes are not readily available. We have developed a series of measures to account for these effects in our work. We monitor the neutron target thickness during target preparation with a separate time-of-flight spectrometer; we measure detector and accelerator time resolutions for different neutron energies using a thin target and we use computer codes to simulate those factors not amenable to direct measurement.

Introduction

The University of Lowell Group has found the response function to be a useful tool in the design of neutron time-of-flight (TOF) experiments and in the analysis of TOF spectra. This function simulates the flight time distribution of a single neutron group in a scattering experiment. The present paper outlines procedures to obtain this function and gives examples of its application.

The Response Function

Consider a typical time-of-flight experiment. Neutrons are produced by irradiating a metallic lithium target with a pulsed proton beam; they are scattered by nuclei of the sample under investigation into the detector. Several factors determine the time spread of the recorded TOF spectrum; it is convenient to separate them into three types. The first type consists of factors which are amenable to direct evaluation: the Li(p,n) angular distribution and kinematics, the sample geometry, attenuation and scattering cross sections, the scattering kinematics, the scattering angle and the sample-detector flight path. Code IMBUI[1], written in our laboratory, determines G(t), the time-of-flight spectrum, taking these factors into consideration. (In this paper we shall ignore the fact that the G and experimental TOF spectra are functions of a discreet variable and we shall use the continuous flight time t instead). A companion code, GAVEA, as yet unpublished, computes similar TOF spectra for twice scattered neutrons.

The second type of factors includes those which are empirically obtained, but which remain constant during a series of experiments. They include the proton beam energy spread, the time distribution of the protons in an accelerator proton burst, the time resolution of the detector electronics, and the scintillator thickness. The contribution of these terms can be obtained in a convenient way by placing the neutron detector at $\Theta=0°$ and by measuring the TOF spectrum of the primary neutrons. A very thin lithium target should be used in these measurements. We shall refer to these spectra as prompt resolution profiles. Usually their shape changes with neutron energy, E_n, so that a series of measurements should be made. We use the notation $P(t,E_n)$ for these profiles.

Often $P(t,E_n)$ is needed for an energy for which no measurements were made. To obtain these functions we approximate the measured profiles by a linear combination of Laguerre functions,

$$P(t,E_n) = \sum_{m=0}^{N} A_m(E_n)\, x^{\alpha/2} e^{-x/2} L_m^{(\alpha)}(x) \qquad (1)$$

Here x = Ht, with H a constant, $L_m^{(\alpha)}(x)$ is a Laguerre polynomial and N<10. Laguerre functions are quite convenient in this expansion because the zero-order function already reflects the assymmetry often found in resolution profiles. A similar function has also been encountered by Post and Schiff[2] in their analysis of the timing of nuclear events with scintillation detectors.

Linear interpolation can now be used to obtain the $A_m(E_n)$ and hence $P(t,E_n)$ for an energy for which no measurements are available. A computer program, LAGUE, is used to obtain the coefficients, A_m and the synthetic profiles, $P(t,E_n)$.

The last type consists of a single factor, the target thickness. In a study of highly excited states in ^{232}Th and ^{238}U[3] good energy resolution was required to separate neutron groups from adjacent nuclear states, which differ in excitation energy by typically 20 keV. The target thickness should not exceed 10 keV in this case. We prepare targets of this thickness by evaporating lithium metal onto a tantalum backing mounted in the accelerator vacuum system. The evaporation takes place with the proton beam on target with E_p=2.25 MeV(E_n=517 keV). A separate "target making" TOF spectrometer with a 5-m flight path is used to measure the primary neutron energy spread during lithium deposition. Evaporation is terminated when the time spread of these neutrons corresponds to an energy spread of 8 to 10 keV.

We have found that over time targets deteriorate under proton bombardment and therefore new targets are prepared when appropriate. While these targets usually are less than 10 keV thick, some variability in thickness is expected and is found. Thus the distribution of primary neutron energies may change from measurement to measurement. In the present paper we use $F(E_n)$ to designate the primary neutron energy spectrum; the corresponding TOF spectrum is designated $F'(t)$. The two distributions are related by

$$F'(t) = \frac{dE_n}{dt} F(E_n) \qquad (2)$$

The response function R(t) can now be computed as the convolution of G(t), $P(t,E_n)$ and $F'(t)$:

$$R(t) = \iint G(t-t_0) P(t_0-t_1) F'(t_1)\, dt_0\, dt_1 \qquad (3)$$

We use our code LAPA to obtain R(t). This code accepts $F(E_n)$ and other parameters as input data, determines G(t) and $P(t,E_n)$ and computes R(t).

Applications

The use of R(t) is illustrated in Figure 1. This is a TOF spectrum of neutrons inelastically scattered by a Th scatterer at E_n=1.45 MeV. Different neutron groups are identified by the excitation energy of the scattering state. Response functions were calculated for each neutron group using LAPA; the entire spectrum was then approximated, in a least squares form, by a linear combination of these response functions using our code TINA. This approximation is shown as s solid line in Figure 1. A detailed account of these measurements will be presented at this conference[3].

On occasion prompt resolution profiles have a long, trailing exponential "tail" with time constant, τ, specially if the detection efficiency is good for low energy neutrons. A typical tail of this kind is shown in Figure 2.

K. H. Böckhoff (ed.), Nuclear Data for Science and Technology, 897–899.

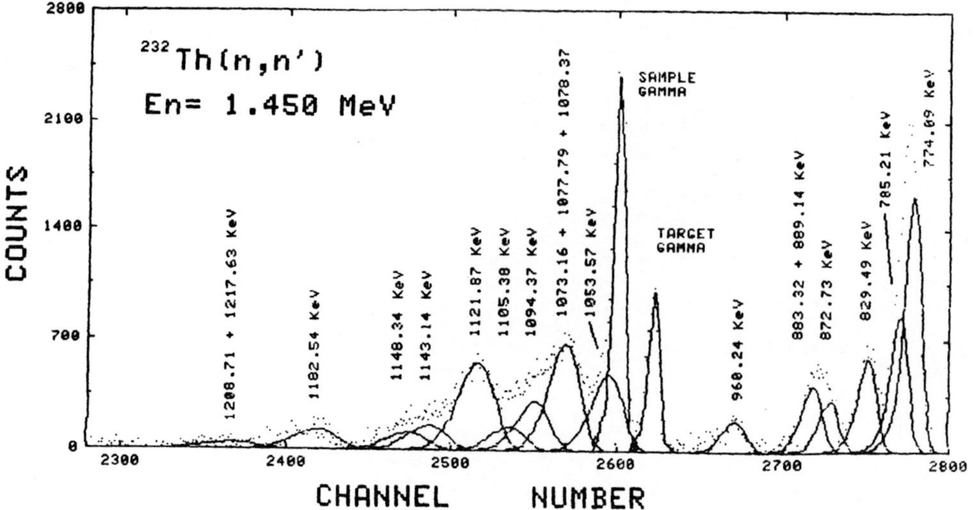

Figure 1. A neutron time-of-flight spectrum.

Tails such as this never appear appended to gamma ray peaks and hence they are not experimental artifacts caused by improper operation of the accelerator bunching system or by faulty operation of the timing electronics. We assume that they are caused by neutrons which have undergone multiple interactions in the detector shield.

If $P(t,E_n)$ includes a tail, the expansion (1) shows poor convergence. In this case we remove the tail, expand the remaining peak, compute the reduced response function r(t), and reappend the tail to obtain R(t). To do this in a form which is independent of resolution and subjective bias, we write

$$P(t,E_n)=p(t,E_n)+a\int p(t-u)\exp(-u/\tau)\,du \qquad (4)$$

The parameter a is small, a ≈ 0.15 hence eq. (4) is readily solved for $p(t,E_n)$. With these p distributions we generate reduced response functions r(t) and obtain R(t) by a convolution

$$R(t)=r(t)+a\int r(t-u)\exp(-u/\tau)\,du \qquad (5)$$

Response functions with tails are encountered in the next example.

A series of low energy (200 keV $\leq E_n \leq$ 900 keV) measurements of neutron scattering cross sections of ^{232}Th and ^{238}U has been initiated in our laboratory. A typical TOF spectrum is shown in Figure 3.

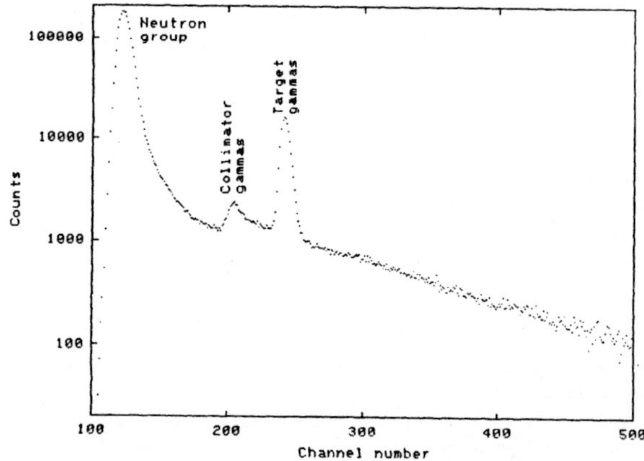

Figure 2. A prompt resolution profile.

Lithium targets were prepared as previously described, with target thicknesses close to 10 keV. To obtain $F(E_n)$ and R(t) we use a trial-and-error approach: The R(t) for the elastic group was calculated with an assumed $F(E_n)$; this distribution was modified until good agreement between R(t) and the measured data resulted. $R(E_n)$ obtained in this way will be called the solution distribution. This procedure was repeated for each run.

Solution spectra all exhibited the same general features. They had approximately trapezoidal shape, with a rapid drop-off at the high energy side, a slower drop-off at the low energy side and with a width at half height of about 10 keV.

Response functions for the inelastic group were obtained using the solution distribution and were multiplied with appropriate constants to match the data. Good fits were obtained in most cases, see e.g. Figures 4-7. In these figures, the simulations are shown as dotted lines, solid lines indicate background-subtracted, measured spectra.

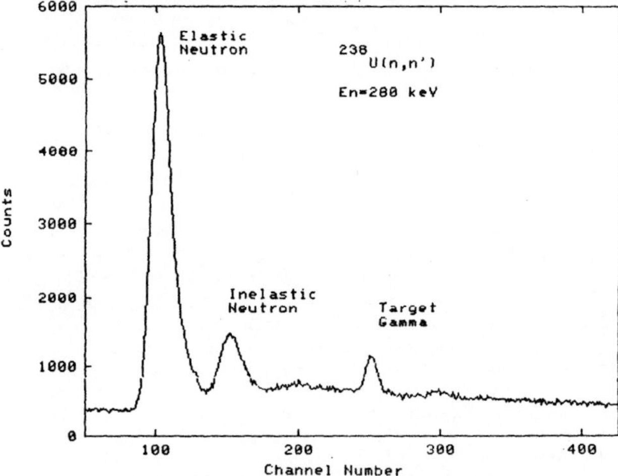

Figure 3. A TOF spectrum, uncorrected.

There is poor agreement of simulation and measurement for ^{232}Th, E_n=240 keV. Kinematic considerations suggest that an oxygen contamination of the scatterer might cause the observed discrepancy. A similar contamination in a Th sample has been reported by McMurray et al.[4]. In fact, if an oxygen elastic scattering response function is added to the thorium functions, a good fit to the data can be obtained. The contamination is not seen at E_n=280 keV because the oxygen peak virtually coincides with the inelastic one.

The U data of Figures 4 and 5 were corrected for neutron attenuation and for neutron multiple scattering and cross section ratios, $\sigma_{inel.}/\sigma_{elastic}$, were obtained. The results are shown in Table I; they are in good agreement with the ENDF/B-V evaluation.

Table I $\sigma_{inelastic}/\sigma_{elastic}$, ^{238}U

	240 keV	280 keV
ENDF-V	0.150	0.160
This Paper	0.145	0.166

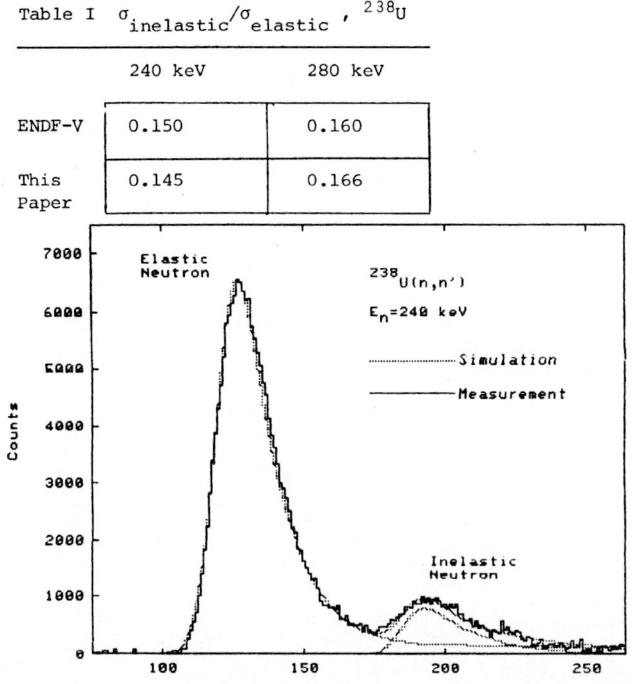

Figure 4. A TOF spectrum.

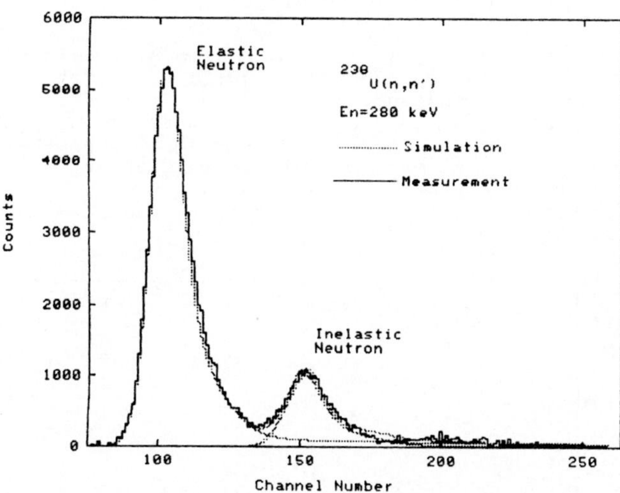

Figure 5. A TOF spectrum.

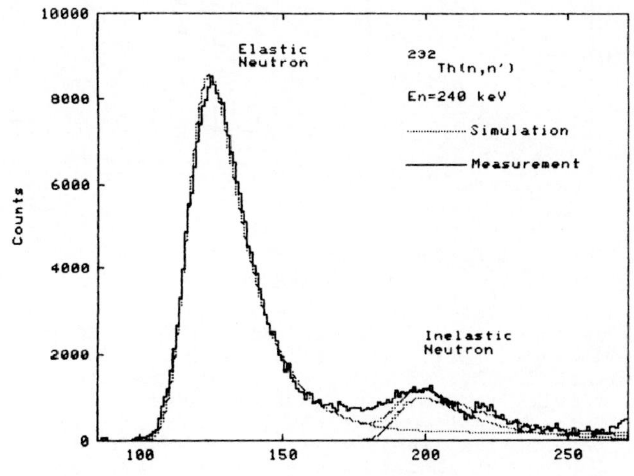

Figure 6. A TOF spectrum.

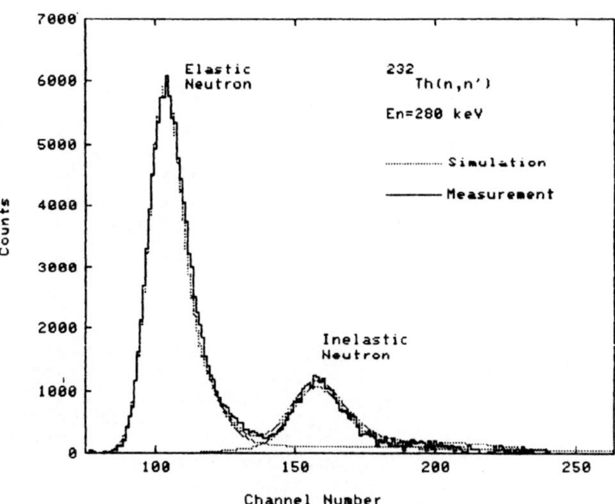

Figure 7. A TOF spectrum.

Acknowledgements

The authors appreciate the help received from their colleagues. This work has been supported by the U.S. Department of Energy.

References

1. G.H.R. Kegel, Computer Physics Communications 24, 205 (1981).
2. R.F. Post and L.J. Schiff, Phys. Rev. 80, 1113 (1950).
3. G.P. Couchell, C. Ciarcia, J.J. Egan, G.H.R. Kegel, A. Mittler, D.J. Pullen, W.A. Schier and J. Shao, "Recent Results of Neutron Inelastic Scattering to Higher-Excited States in ^{232}Th and ^{238}U", this conference.
4. W.R. McMurray, E. Barnard, I.J. van Heerden, D.T.L. Jones, section 1.1.4. of the 1979 Annual Research Report, Southern University Nuclear Institute, Faure, South Africa.

NEW CONCEPT OF MEASURING THE NEUTRON ABSORPTION PROPERTIES OF BULK MEDIA

A. Kreft, P. Maloszewski, K. Morstin*

Institute of Physics and Nuclear Techniques
Akademia Gorniczo-Hutnicza
al. Mickiewicza 30, 30-059 Krakow, Poland

*at present: Institut für Reaktorentwicklung, KFA Jülich GmbH
D-5170 Jülich, W-Germany

A simple method for Σ_a determination is proposed. Neither sophisticated instrumentation nor a complex geometric set-up are required. It makes use of changes in the spectrum of registered neutrons caused by shielding the detector with polyethylene. Relative "1/v" signal variations univocally depend on moderating ratio of the surrounding medium. Possible applications include evaluation of strong neutron absorbers (B, Cd, Hg, rare earths, etc.) in ore deposits and on-line testing PWR coolant for boron content.

Although the numerous ways of determining thermal-absorption properties are known, additional methods of cheap and rapid in-situ measurements in bulk media seem to be desirable. Evaluation of strong neutron absorbers (B, Cd, Hg, rare earths, etc.) in ore deposits is still a difficult problem, since the routine neutron well-logging methods are sensitive to frequent changes in probe positioning, rock moisture, bore-hole diameter and saturation, etc. The on-line control of boron content in the primary loop of a modern PWR would be an attractive alternative for reactor engineers, especially if reactor neutrons would serve for this purpose, instead of forming the interferring background, as it is in the "neutron-through-pipe" transmission method.

Since the shape of the thermal neutron spectrum is mainly governed by the absorption properties of the investigated medium, evaluation of the macroscopic absorption cross section Σ_a can be performed by determining the so-called Maxwellian temperature of neutrons. Such a measurement should be in principle insensitive to geometric conditions around the detector and to the neutron source applied. For this purpose Guberman and Yakubson[1] propose to shield a "1/v" detector with some absorption filters of varying thickness. Resulting hardening of the spectrum of registered neutrons enables the Maxwellian temperature, and then the moderating ratio $\Sigma_a/\xi\Sigma_s$, to be determined. However, due to the low detection efficiency caused by the absorption filters, such measurements are very time-consuming.

Instead, we propose to shield the detector with a polyethylene filter. One may expect the opposite effect, the relative variations of the "1/v" detector signal being roughly proportional to the moderating ratio. In order to check the feasibility of the above proposal, we performed some measurements in a large cylindrical tank (60 cm x ∅ 50 cm), filled with water and intersected by an aluminium pipe (∅ 4 cm) in which the Pu-Be source and the BF_3 detector were located. Changes in absorption properties of water were realized by adding H_3BO_3 (up to 2.4% w). The detector was shielded with polyethylene filters of varying thickness (up to 12 mm). Fig. 1 presents the results interpreted in the form of dependence $R(\Delta)$ for given shield thicknesses d, Δ denoting the moderation ratio, while R being defined as

$$R = \frac{I_{sh}/I_b}{I^o_{sh}/I^o_b} ,$$

where subscripts "sh" and "b" mean "shielded" and "bare", respectively, while index "o" denotes the signal in the case of absence of neutron absorber. Normalization to the signal ratio for "non-absorbing" medium compensates for the influence of polyethylene on the space distribution of thermal neutrons. It is evident that the investigated relation is almost linear.

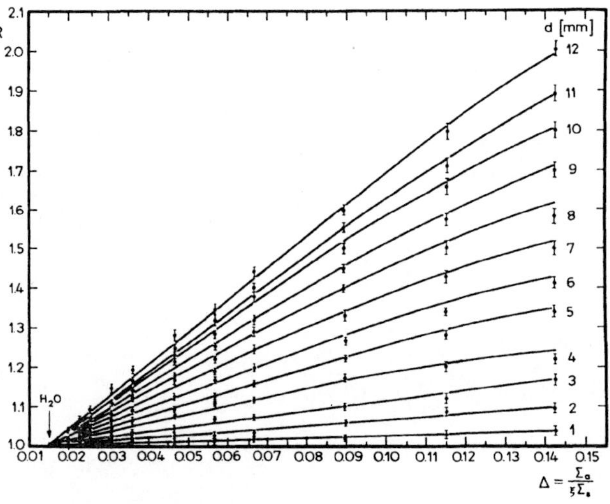

Fig.1. Relative signal ratio R vs. moderation ratio $\Sigma_a/\xi\Sigma_s$ in borated water.

Similar measurements were carried out in the second tank of different dimensions (110 cm x ∅ 60 cm), filled with water and sand, alternatively, in order to verify suggested insensitivity to geometry and matrix composition. This time only the 3.7 mm thick filter was applied. The results are compared in Fig. 2 with interpolation curves from Fig. 1 for 3 mm and 4 mm of polyethylene. It may be concluded that in the frames of achieved accuracy the hypothesis under verification has

K. H. Böckhoff (ed.), Nuclear Data for Science and Technology, 900–901.

been proved. Sensitivity of such determinations improves with increasing PE thickness, since the R(Δ) curves become steeper then.

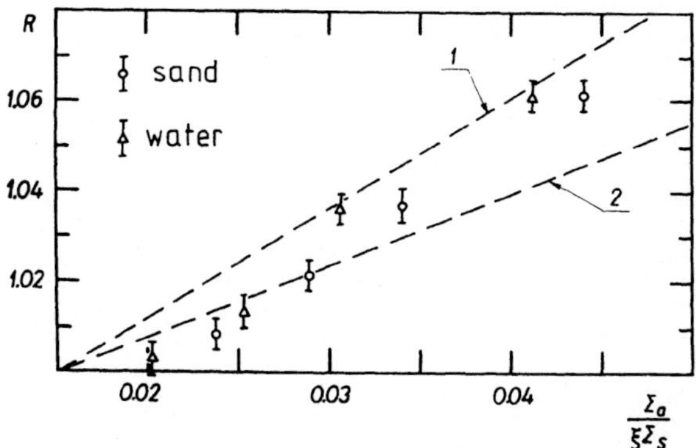

Fig. 2. R vs. $\Sigma_a/\xi\Sigma_s$ for borated sand (circles) and water (triangles) measured with 3.7 mm PE filter. Curves 1 and 2 are taken from Fig. 1 for 4 mm and 3 mm filters, respectively.

It gives rise for designing simple probes consisting of two detectors equally spaced from the source, one of them being shielded with polyethylene. In fact, in the case of the neutron-bearing environment (e.g. LWR) the presence of an external source is even not necessary, since interpreted results are not influenced by source distribution. Such probes could be easily calibrated when assuming linearity of R(Δ) relations, which can be accepted for moderating ratio ranging, say, up to 0.1 (i.e. ca. 3000 ppm of boron in light water). In the case of varying moderation properties (e.g. in rocks) the changes in $\xi\Sigma_s$ could be simultaneously detected by introducing an additional epithermal detector, its signals being interpreted such as proposed e.g. by Wozniak[2].

References

1. Sh.A. Guberman and K.I. Yakubson, Vozmozhnost' ispolzovaniya temperatury teplowykh neytronov dla issledovaniya gornykh porod, in "Problemy Yadernoy Geofiziki", Nedra, Moskva 1964

2. J. Wozniak, Calibration of neutron gauges for measuring the soil moisture content in the 4 π geometry, Rep. INT, 27/I, Krakow 1973

THICK-TARGET $^9Be(d,n)^{10}B$ NEUTRON SPECTRUM AT E_d = 7 MeV

A. Crametz, H.-H. Knitter, and D.L. Smith*

Commission of the European Communities, J.R.C.
Central Bureau for Nuclear Measurements
B - 2440 Geel, Belgium

The CBNM Van de Graaff will be equipped with a post buncher to reduce the pulse width from 1.5 ns down to about 250 ps. Therefore the $^9Be(d,n)^{10}B$ thick-target yield obtained by a pulsed 7 MeV deuteron beam was studied, especially in view of using this source for neutron cross section measurements in the energy range up to 11.3 MeV. For this purpose the spectrum shape was measured in this neutron energy range and the integral was normalized to values from literature. An absolute neutron fluence comparison was made between this source and the GELINA spectrum normalized to the same inverse velocity resolution of 0.1 ns.m^{-1}. In order to show the great potential of this source, the total cross sections of carbon and silicon were determined in an energy range from a few 100 keV to 11 MeV in a measuring time of a few hours with a statistical accuracy, e.g. at 6 MeV, better than 2 %, and with an energy resolution of 60 keV. Comparison of the experimental results was made with the evaluated data sets of ENDF/B-V.

[$^9Be(d,n)^{10}B$ thick-target neutron spectrum, total neutron cross section, C, Si]

Introduction

The energy spectrum of this neutron source reaction has been studied in more detail by Weaver et al.[1] and by Lone et al.[2] covering the incident deuteron energy range up to the 20 MeV. The aim of the present work was more specific. We wanted to study the possibilities for using this neutron source reaction in conjunction with the CBNM's 7 MV pulsed Van de Graaff accelerator for neutron cross section measurements. The neutron spectrum of this source reaction extends for a deuteron energy of 7.0 MeV up to 11.3 MeV. This accelerator, which can deliver proton, deuteron and helium beams, pulsed with a maximum repetition frequency of 2.5 MHz, a pulse width of 1.5 ns and an average beam current of 6 μA, is being equipped with an external klystron buncher. The post bunching system is been realized in collaboration with the Max Planck Institut für Kernphysik in Heidelberg[3]. It is expected to reduce the presently available pulse width of 1.5 ns to less than 250 ps without major loss in the average beam current. Due to the narrow pulse width which will be available in future, it seemed worthwhile to study the pulsed thick-target neutron source of a deuteron energy of E_d = 7.0 MeV, especially with regard to neutron cross section measurements using the time-of-flight technique.

For this purpose, the neutron spectrum emitted at zero degree with respect to the incident deuteron beam was measured and compared to the neutron flux available from the Geel electron linear accelerator (GELINA), normalized to the same time resolution. Furthermore, in order to judge in a realistic manner the experimental possibilities which will be available with the 250 ps pulsing system, the total cross sections of carbon and silicon were measured with the presently available resolution of 1.5 ns covering a neutron energy range from several 100 keV to 11 MeV.

The Spectrum Shape

The neutron detector used for the spectrum shape measurements consisted of a NE 224 liquid scintillator of 5 cm diameter and 3.8 cm length, viewed directly by a photomultiplier of type RCA-8850 mounted on a constant fraction timing base, model Ortec-270. No provision was made for neutron-gamma discrimination in the measurements other than the natural selection provided by time-of-flight. This detector has been used in earlier spectrum shape measurements of fission neutrons and therefore its relative neutron detection efficiency was known[4]. The geometrical arrangement of the experimental apparatus is shown schematically in fig.1. The detector was positioned in a collimating shield at a distance of (11.30 ± 0.02 m) from the beryllium target and at zero degree with respect to incident deuteron beam direction. The beryllium target consisted of a metallic disc, 40 mm in diameter and 0.5 mm thick. Between the detector and the target, a second collimating shield was positioned to reduce the background at the detector position. Also the target was surrounded by a shield made of a mixture of paraffin and Li_2CO_3.

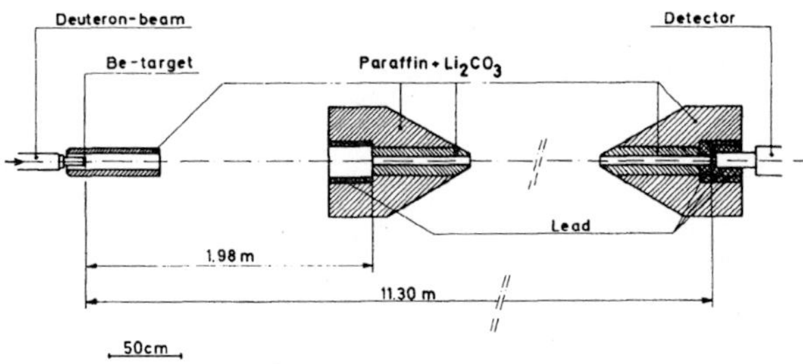

Fig. 1. Schematical drawing of experimental set-up.

The accelerator was operated at E_d = 7.0 MeV with a pulse repetition frequency of 0.625 MHz, the maximum frequency reduced by a factor of 4. The average beam current was 1.5 μA. A neutron time-of-flight spectrum was recorded using the detector signal and the signal obtained from a beam pulse pick-up loop as start and stop signals for a time-to-pulse height converter respectively. The converter output signals were sorted according to their pulse heights by a Nuclear Data Inc. ND 6660 data acquisition system. Neutrons with flight-times t longer than T = 1600 ns, the time between two successive deuteron pulses, were recorded at a nominal time (t-T). This occurred for neutrons with energies smaller than 260 keV. Due to the decreasing detector efficiency and decreasing neutron energy width per channel for these overlap neutrons, this disturbing effect was less than 1 % at the high energy side of the spectrum and decreasingly smaller for lower neutron energies. For this relatively small effect, no correction was applied.
A constant background, as measured in the valley between the neutron spectrum, ending at about 11.3MeV, and the γ-peak was subtracted from the time-of-flight spectrum. Its relative magnitude with respect to the spectrum counts amounted to 5.5 %, 1 % and 7 % at neutron energies of 1 MeV, 4 MeV and 10 MeV respectively. The time-of-flight spectrum was then converted to a neutron energy spectrum using relativistic kinematics. The spectrum is presented in form of a histogramme in fig. 2. Below 5 MeV neutron energy, the measured data points were put together in 100 keV

* Visiting scientist from Argonne National Laboratory Argonne, Illinois, U.S.A.

K. H. Böckhoff (ed.), Nuclear Data for Science and Technology, 902–905.

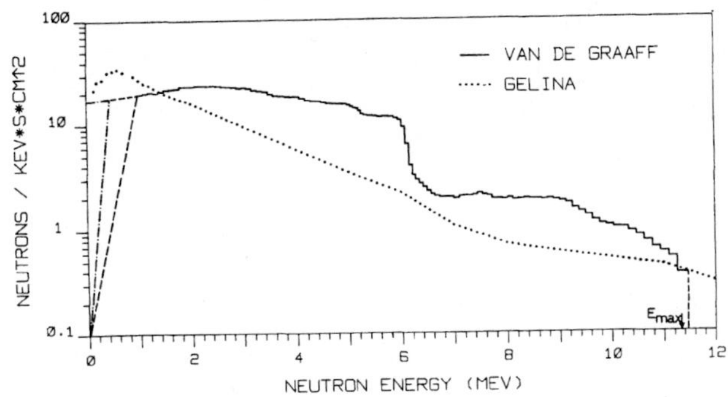

Fig. 2. The histogramme gives the measured number of neutrons per cm^2, s, and keV, for the $^9Be(d,n)^{10}B$ thick-target yield at 2.5 m distance from the target in the zero degree direction for a deuteron current of 1.5 μA at an energy of 7.0 MeV. The spectrum plotted with a dotted line represents the spectrum of GELINA with a beam power of 4 kW on a bare uranium target, a repetition frequency of 800 Hz, a burst width of 4 ns and for the same resolution of 0.1 $ns \cdot m^{-1}$.

intervals. Above 5 MeV, the histogramme intervals have a width corresponding to 1.76 ns, which is about the timing resolution of the experiment. To obtain an absolute scale for the neutron fluence per unit time, the integral of our spectrum

$$I = \int_{0\ \mathrm{MeV}}^{11.3\ \mathrm{MeV}} \Phi(E)\,dE$$

was normalized to the integrated thick-target zero degree neutron yield of $0.5 \cdot 10^{10}\ s^{-1} \cdot \mu C^{-1}\ sr^{-1}$ as read from an appropriate figure of the paper by Weaver et al.[1]. For the determination of the integral, the spectrum was extrapolated from 1 MeV towards zero neutron energy and zero fluence per unit time as indicated in fig. 2. The uncertainty of this integral was estimated to $\pm$ 3.5 %, including the error due to the extrapolation towards zero neutron energy. A confirmation of this normalization was obtained by comparing the measured average cross section of the $^{27}Al(n,\alpha)^{24}Na$ reaction for the present spectrum with the calculated average cross section using the shown shape of the spectrum and the known ENDF/B-V $^{27}Al(n,\alpha)$ cross section. This neutron activation measurement was performed by Liskien et al.[5]. The agreement between the two normalizations was within 3 %, very well within the expected limits.

Besides the statistical errors, there are errors originating from the neutron energy determination and from the relative detection efficiency of the detector, which should be considered. The relative detection efficiency as given in ref.[4] was measured in the four energy ranges, 0.5 MeV $\leq E_n \leq$ 2.5 MeV, 2.5 MeV $\leq E_n \leq$ 6.3 MeV, 6.3 MeV $\leq E_n \leq$ 8.3 MeV and 8.3 MeV $\leq E_n \leq$ 20 MeV, in independent experiments, and it was normalized at the corresponding endpoints of these ranges.

$$\epsilon(E_n) = \epsilon_1(E_n) \qquad 0.5 \leq E_n \leq 2.5\ \mathrm{MeV}$$
$$\epsilon(E_n) = n_2 \cdot \epsilon_2(E_n) \qquad 2.5 \leq E_n \leq 6.3\ \mathrm{MeV}$$
$$\epsilon(E_n) = n_2 \cdot n_3 \cdot \epsilon_3(E_n) \qquad 6.3 \leq E_n \leq 8.3\ \mathrm{MeV}$$
$$\epsilon(E_n) = n_2 \cdot n_3 \cdot n_4 \cdot \epsilon_4(E_n) \qquad 8.3 \leq E_n \leq 20\ \mathrm{MeV}$$

Table 1 Columns 1, 4 and 7 give the centre neutron energies of the intervals. Columns 2, 5 and 8 give the absolute thick target yield per cm^2, s, keV and per 1.5 μA for a target distance of 2.5 m under zero degree for the $^9Be(d,n)^{10}B$ reaction at E_d = 7 MeV. The statistical errors are given in columns 3, 6 and 9.

Energy [MeV]	Φ $[cm^{-2}.s^{-1}.keV^{-1}]$	$\Delta\Phi$ [%]	Energy [MeV]	Φ $[cm^{-2}.s^{-1}.keV^{-1}]$	$\Delta\Phi$ [%]	Energy [MeV]	Φ $[cm^{-2}.s^{-1}.keV^{-1}]$	$\Delta\Phi$ [%]
1.05	19.41	1.09	4.56	16.09	1.61	7.35	2.02	5.13
1.15	19.87	1.05	4.65	15.70	1.61	7.44	2.06	5.05
1.25	20.45	1.07	4.74	15.63	1.60	7.53	2.17	4.91
1.35	20.00	1.06	4.85	15.52	1.28	7.62	2.09	4.96
1.45	20.73	1.07	4.97	15.45	1.57	7.71	2.02	5.01
1.55	21.54	1.05	5.05	15.16	1.21	7.81	1.89	5.15
1.65	21.89	1.06	5.10	14.65	2.25	7.90	1.90	5.11
1.76	22.08	1.06	5.15	14.21	1.26	8.00	1.88	5.12
1.86	23.10	1.05	5.20	13.98	2.28	8.10	1.92	5.01
1.95	23.14	1.08	5.25	13.33	2.32	8.20	1.84	5.06
2.05	23.19	1.06	5.30	12.44	2.39	8.31	1.87	4.99
2.16	23.47	1.01	5.36	12.13	2.40	8.41	1.92	4.90
2.26	23.47	1.13	5.44	11.88	1.70	8.52	1.88	4.89
2.35	23.36	1.02	5.55	11.78	1.71	8.63	1.88	4.85
2.46	23.45	1.08	5.64	11.59	2.40	8.74	1.86	4.86
2.55	23.28	1.17	5.73	11.67	1.69	8.86	1.89	4.76
2.66	23.01	1.05	5.85	11.62	1.68	8.97	1.84	4.77
2.76	22.86	1.11	5.94	11.25	2.40	9.09	1.80	4.81
2.86	22.66	1.09	6.00	10.84	2.43	9.21	1.72	4.85
2.96	22.10	1.21	6.07	9.31	2.61	9.34	1.53	5.11
3.05	22.41	1.19	6.13	6.33	3.15	9.46	1.44	5.23
3.15	22.00	1.19	6.20	3.99	3.93	9.59	1.31	5.43
3.25	21.20	1.19	6.27	3.24	4.36	9.72	1.18	5.68
3.35	20.86	1.18	6.34	2.90	4.58	9.86	1.08	5.86
3.46	20.44	1.17	6.41	2.69	4.74	9.99	1.05	5.90
3.56	19.30	1.37	6.48	2.48	4.91	10.13	1.02	5.96
3.65	18.68	1.37	6.55	2.29	5.09	10.28	1.00	5.94
3.76	18.51	1.17	6.63	2.12	5.26	10.42	0.93	6.14
3.87	18.45	1.34	6.70	2.05	5.32	10.57	0.84	6.34
3.95	18.33	1.63	6.82	1.97	3.81	10.72	0.76	6.67
4.04	18.38	1.30	6.97	1.98	3.76	10.88	0.66	7.12
4.15	17.62	1.32	7.10	1.93	5.37	11.04	0.60	7.40
4.26	16.85	1.33	7.18	2.01	5.21	11.20	0.53	7.79
4.36	16.59	1.61	7.26	2.04	5.14	11.37	0.36	9.23
4.46	16.52	1.30						

The errors of the normalization constants n_2, n_3 and n_4 were estimated to be 1 %, 2 % and 4 % respectively. The estimated errors for the relative efficiencies ϵ_1, ϵ_2, ϵ_3 and ϵ_4 in the 4 energy groups are 2 %, 3 %, 2 % and 2 % respectively. Due to these normalizations a correlation is introduced which should be considered in the error analysis.

The neutron energy is entirely determined by the flight path length of L = (11.30 ± 0.02)m, the time per channel of τ = (0.888 ± 0.01)ns/channel, and the channel number difference between the γ-peak channel and the channel corresponding to a particular neutron energy $(x_\gamma - x_i)$. This difference was estimated to be known within ± 1 channel. All these error sources are common for all neutron energies and therefore correlations exist. In order to be able to calculate the proper error for each spectrum point in the case of correlated errors, the covariance matrix $M_{\Phi_{ij}}$ for the spectrum must be known. The i and j are indices numbering the neutron energy intervals in table 1. Therefore the covariance matrix for the present case has, in principle, 103 x 103 elements and is given by

$$\begin{aligned} M_{\Phi_{ij}} &= \delta_{ij}\, \Phi_i\, \Phi_j\, (e_{st})_i \cdot (e_{st})_j \\ &+ \Phi_i\, \Phi_j \cdot M_{\epsilon_{ij}} / (\epsilon_i \cdot \epsilon_j) \\ &+ \left(\frac{\partial \Phi}{\partial E}\right)_i \cdot \left(\frac{\partial \Phi}{\partial E}\right)_j \cdot M_{E_{ij}} \end{aligned}$$

where Φ_i and Φ_j represent the spectrum values of table 1, $(e_{st})_i$ and $(e_{st})_j$ are the corresponding fractional statistical errors, ϵ_i and ϵ_j are the efficiencies, and M_ϵ and M_E are the covariance matrices for the efficiency and neutron energy determination respectively. A very detailed description of how to obtain the covariance matrices in practical cases is given by Smith[6] and the present problem is analyzed explicitly by Liskien et al.[5] and will therefore not be repeated here. However the information given in the present paper is complete for the construction of M_Φ.

In order to compare different pulsed neutron sources the same time resolution should be used. For this purpose, we have chosen an inverse velocity resolution of 0.1 ns·m^{-1}. This resolution should be feasible after installation of the pulse compression system at a target-detector distance of 2.5 m. Therefore, in fig. 2 the number of neutrons is given per cm^2, s, and keV, for a target-detector distance of 2.5 m, a deuteron energy of E_d = 7.0 MeV, for an average target current of 1.5 μA and a repetition frequency of 0.625 MHz.

Fig. 2 shows as a dotted line the same quantity for GELINA plotted versus the incident neutron energy. This spectrum, obtained by private communication[7], is valid for the following GELINA parameters: 4 kW beam power dissipated on a bare uranium target, 800 Hz repetition frequency and 4 ns pulse width. The comparison is made for a target distance of 40 m in order to also have 0.1 ns·m^{-1} resolution. From the two curves in fig. 2, one can see that in the 1.5 MeV to 11 MeV range the Van de Graaff, when equipped with the pulse compression system will give more neutrons than GELINA for time-of-flight experiments where a resolution of 250 ps can be fully exploited. If a repetition frequency of 2.5 MHz can be used in an experiment, the average target current increases from 1.5 to 6 μA and therefore the neutron output rate of the Van de Graaff spectrum increases by a factor 4 compared to fig. 2. Table 1 gives the numerical values of the middle neutron energies of the histogramme intervals, the normalized number of neutrons per cm^2, s, and keV, as well as the statistical errors.

Total Cross Section Measurements

In order to demonstrate the usefulness of this pulsed white spectrum neutron source, especially when the pulsing is improved to 250 ps, the already well-known total cross sections of carbon and silicon were measured in the neutron energy range from several 100 keV to 11 MeV with the presently available accelerator pulsing conditions. The total cross sections were measured in a transmission experiment with the same flight path length, however with a detector threshold increased to about 300 keV. The total cross section is obtained by:

$$\sigma_T(E) = \frac{1}{n \cdot t}\, \ell n\, \frac{N_o(E)}{N_i(E)}$$

where n is the number of nuclei per cm^3, t the sample thickness in cm and N_i and N_o are the number of counts with and without sample corresponding to a neutron energy E after background subtraction and normalization with respect to a relative neutron monitor. Fig. 3 shows three time-of-flight spectra, measured without transmission sample, with carbon sample and with the collimator plugged with a 1 m long nylon cylinder. All spectra are normalized

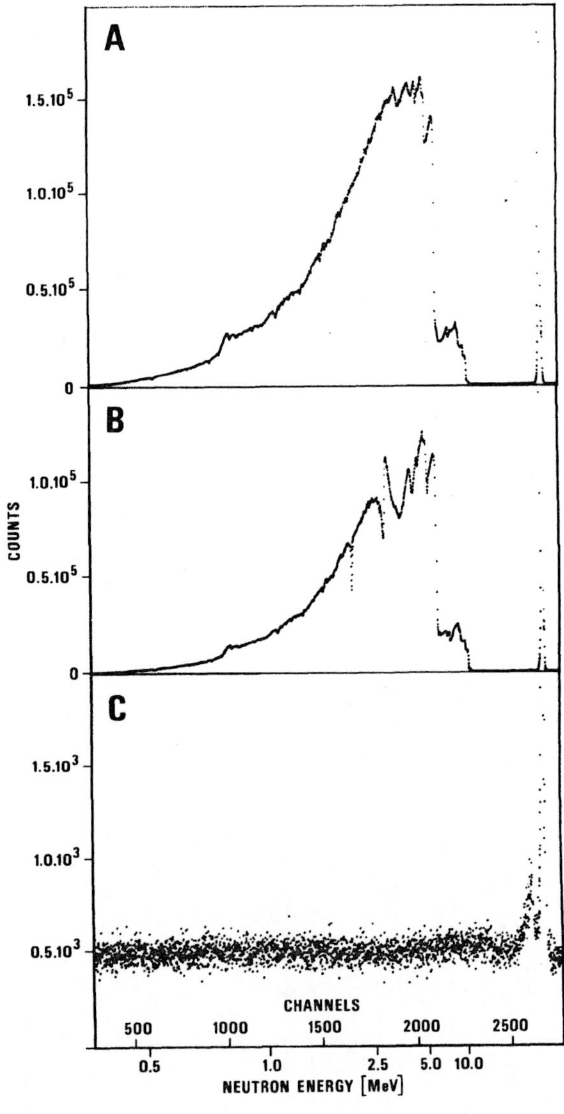

Fig. 3. The time-of-flight spectra measured without transmission sample, with carbon and with the collimator plugged with a 1m long nylon cylinder are shown in part A, B, and C respectively. Part A and B are taken for the same number of counts of the relative neutron monitor detector, whereas the "background" spectrum in part C was normalized to the same number of counts by a factor 4.839.

with respect to the same number of counts of the relative neutron monitor. Carbon and silicon samples of 3 cm thickness and with densities of 1.665 and 2.339 g/cm^3, respectively were used. The product n·t had an uncertainty of 0.3 %. Errors introduced by deadtime effects and in-scattering were negligible. A comparison is made in figs. 4 and 5 with the most recent and complete evaluations of these cross sections. Each cross section was obtained in a measuring

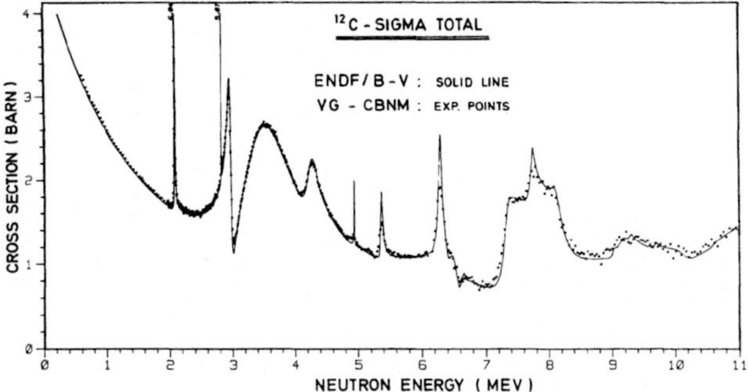

Fig. 4. The total cross section of carbon is plotted versus the incident neutron energy and compared to the ENDF/B-V values.

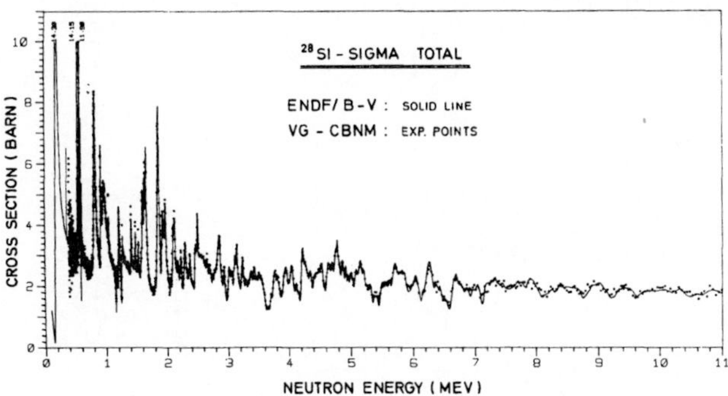

Fig. 5. The total cross section of silicon is plotted versus the incident neutron energy and compared to the ENDF/B-V values.

time of about 8 hours, and a statistical accuracy of less than 2 % at 6 MeV and 5 % at 10 MeV was reached in both cases. At 6 MeV and 10 MeV the neutron energy resolutions were 80 keV and 160 keV FWHM respectively. Both measured cross sections show good agreement in the absolute values compared to the ENDF/B-V curve, and also all the structures appear at exactly coinciding neutron energies. However, the measured values of the carbon total cross section do not follow sharply the evaluated curve in some peaks and valleys. This is a sign of insufficient energy resolution in order to fully resolve the true structure of the cross section, and leads even to a wrong average total cross section $< \sigma_T(E) >_{\Delta E}$. Indeed:

$$< \sigma_T(E) >_{\Delta E} = < \frac{1}{n \cdot t} \cdot \ell n \frac{N_o(E)}{N_i(E)} >_{\Delta E}$$

$$\neq \frac{1}{n \cdot t} \cdot \ell n \frac{< N_o(E) >_{\Delta E}}{< N_i(E) >_{\Delta E}}$$

And the second equation is not valid if σ_T varies within the interval ΔE [8]. Therefore it is indispensable for total cross section measurements to work with neutron energy resolutions smaller than the expected structures in the cross section, since otherwise one does not even measure a quantity which corresponds to the cross section averaged over the resolution interval. In fig. 5, where the total cross section of silicon is given, one can see, that the present experimental results below about 2.2 MeV give, for some peaks, a larger total cross section than the evaluation of ENDF/B-V. That implies, that the energy resolution of the present experiment is probably better than those used to obtain the experimental data base for the evaluation.

Conclusions

The $^9Be(d,n)^{10}B$ white spectrum pulsed neutron source, which will be available at the Van de Graaff accelerator at a deuteron energy of 7 MeV and with a pulse width of 250 ps, will be a powerful source for neutron data measurements in the energy range from 1 to 11 MeV. When this narrow pulsing can be fully exploited this neutron source will deliver, in the indicated neutron energy range, more neutrons than GELINA for the same inverse velocity resolution. The $^9Be(d,n)^{10}B$ source has now also been used successfully in an activation experiment to determine the cross section ratio of $^{27}Al(n,\alpha)^{24}Na$ to $^{63}Cu(n,\alpha)^{60}Co$ [5].

In a programme for total cross section measurements of isotopes relevant to fusion reactor design, an energy range up to 20 MeV should certainly be covered. For this purpose a hybrid target containing ^7Li and ^9Be would be well suited for the production of a neutron spectrum ranging from zero to about 20 MeV, with an integral neutron intensity very similar to that for the $^9Be(d,n)$ neutron source reported here. Many of the total cross sections between 2 and 20 MeV are insufficiently well known not only for fusion applications, but also for calculations of nuclear model parameters from total cross sections, because e.g. resonance self-shielding has only seldom been considered.

The 7 MV Van de Graaff with the klystron compression system in conjunction with a $^7Li/^9Be$-target would give to the CBNM a powerful neutron source covering the whole neutron energy range of interest in the fusion field.

References

1. K.A. Weaver, J.D. Anderson, H.H. Barschall and J.C. Davis, Nucl. Sci. Eng. 52, 35 (1973)
2. M.A. Loan, A.J. Ferguson and B.C. Robertson, Nucl. Instr. Meth. 189, 515 (1981)
3. E. Jaeschke, R. Repnow, Th. Walcher, H. Ingwersen, G. Ihmels, B. Kolb, H. Schwarz and F. Gamp, I.E.E.E. Trans. Nucl. Sci. 24, 1136 (1977)
4. H.-H. Knitter, A. Paulsen, H. Liskien and M.M. Islam, ATKE 22, 84 (1973)
5. H. Liskien, D.L. Smith and R. Widera, this conference
6. D.L. Smith, Report ANL/NDM-62 (1981)
7. K.H. Böckhoff, private communication
8. C. Budtz-Jørgensen, P.T. Guenther, A.B. Smith and J.F. Whalen, ANL/NDM-61 (1981)

Nuclear Data for Biomedical Applications

RECENT NUCLEAR DATA MEASUREMENTS RELEVANT TO THE CYCLOTRON PRODUCTION OF SOME MEDICALLY IMPORTANT SHORT-LIVED RADIOHALOGENS

S.M. Qaim, G. Stöcklin, R. Weinreich, H. Backhausen, Z.B. Alfassi, and J.H. Zaidi

Institut für Chemie 1 (Nuklearchemie), Kernforschungsanlage Jülich GmbH, D-5170 Jülich, Federal Republic of Germany

Besides organic elements (C, H, N, O, P etc.) members of the family of halogens (F, Cl, Br, I etc.) form strong covalent bonds with C. Some of the short-lived neutron deficient radioisotopes of those elements are therefore very useful for labelling biomolecules or their analogues which eventually find applications in regional functional studies in-vivo using either single photon emission computed tomography (e.g. ^{123}I) or positron emission computed tomography (e.g. ^{18}F and ^{75}Br). We recently performed nuclear data measurements in the following cases: i) ^{3}He- and d-induced reactions on ^{20}Ne with a view to determining the contribution of the $^{18}Ne \xrightarrow[1.67\ sec]{} {}^{18}F$ precursor process to the total ^{18}F formation ii) (^{3}He,xn)- and (α,xn)-reactions on arsenic as well as (d,xn)-reactions on natural bromine up to 90 MeV with the aim of determining the optimum conditions for the production of positron emitter ^{75}Br iii) (d,xn)-reactions on enriched ^{122}Te up to incident deuteron energy of 40 MeV with the objective of investigating the (d,n)-reaction as a potential route for ^{123}I production.

Some considerations relevant to the large scale production of radioisotopes are discussed.

[$^{20}Ne(d,p3n)^{18}Ne$, E_d=35-75 MeV; $^{20}Ne(^{3}He,\alpha n)^{18}Ne$, $E_{^3He}$=18-36 MeV; $^{75}As(^{3}He,xn)^{74m-77}Br$, $E_{^3He}$=10-70 MeV; $^{75}As(\alpha,xn)^{74m-77}Br$, E_α=15-130 MeV; $^{nat}Br(d,xn)^{75-77}Kr$, E_d=20-90 MeV; $^{122}Te(d,xn)^{121-123}I$, E_d=7-40 MeV; excitation functions; thick target yields; radionuclidic impurities].

Introduction

The criteria for the choice of a radioisotope for medical applications include short half-life, decay by electron capture or β^+ emission, γ-rays of suitable energy and intensity, and strong chemical bond formation between C and the radioelement (cf. 1). The "organic" β^+ emitters $^{11}C(T_{1/2}$=20.3 min), $^{13}N(T_{1/2}$=10 min) and $^{15}O(T_{1/2}$=2.0 min) are very suitable for labelling biomolecules containing these elements without any biochemical changes. The short half-lives of those radioisotopes, however, limit their application to the immediate vicinity of a cyclotron. Since members of the family of halogens (F, Cl, Br, I etc.) also form strong covalent bonds with C they are also regarded as "organic" elements. In recent years some of the neutron deficient radioisotopes of those elements having suitable decay properties have found increasing applications in diagnostic nuclear medicine, particularly with the increasing application of single photon emission computed tomography (SPECT) or positron emission computed tomography (PECT)[2]. The decay data[3] of the three main radioisotopes used so far are given in Table I.

For the production of those radioisotopes several nuclear reactions can be used (cf. 4). Some of the processes have been thoroughly investigated; the information available on the others was scanty. The importance of nuclear reaction cross section data in optimizing the yield and purity of the desired radioisotope cannot be overemphasized (cf. 1,4). We describe below some of our recent measurements relevant to the production of the three radioisotopes mentioned. Special attention is paid to ^{75}Br which was proposed and applied for the first time in diagnostic nuclear medicine by us[5].

Table I. Some Important Radiohalogens for in-vivo Function Studies with Analogue Tracers.

Radio-isotope	$T_{1/2}$	Mode of decay (%)	E_{β^+} (KeV)	Main γ-rays keV (%)	Imaging device
^{18}F	1.83h	β^+(97) EC(3)	635		Positron tomograph
^{75}Br	1.63h	β^+(75) EC (25)	1740	286(92) 141(7)	Positron tomograph
^{123}I	13.2h	EC(100)		159(83)	Conventional γ-camera (SPECT)

Nuclear Data for Production

^{18}F

In recent years ^{18}F-labelled radiopharmaceuticals have gained increasing interest for in-vivo studies using PECT. Some of the compounds such as ^{18}F-fluorodesoxyglucose[6,7] have already found wide application for studies of brain functions and others such as long-chain ^{18}F-fluorofatty acids (cf. 8) are presently being tested as agents for measuring regional heart metabolism.

^{18}F is generally produced via $^{16}O(^{3}He,p)^{18}F$, $^{20}Ne(d,\alpha)^{18}F$ or $^{20}Ne(^{3}He,\alpha p)^{18}F$ reaction. Recently the $^{18}O(p,n)^{18}F$ reaction has also been used. At the commencement of this work the excitation functions of the first three reactions were known up to about 40 MeV[9,10]. No information was available at energies above 40 MeV. Furthermore, the contribution of ^{18}F formed via the decay of its 1.67 sec ^{18}Ne precursor was not known so that no

K. H. Böckhoff (ed.), Nuclear Data for Science and Technology, 909–915.

distinction could be made between the yields of primary and decay-produced ^{18}F. We measured this contribution for the deuteron and ^{3}He-particle induced nuclear reactions on ^{20}Ne for the first time[11].

In the case of deuterons on neon, the thresholds of the $^{20}Ne(d,\alpha)^{18}F$ and $^{20}Ne(d,p3n)^{18}Ne$ reactions are quite different. A measurement of the ^{18}F formed in both the direct and the indirect way should allow a separation of the two excitation functions. In the case of ^{3}He on neon, however, both the $^{20}Ne(^{3}He,\alpha p)^{18}F$ and $^{20}Ne(^{3}He,\alpha n)^{18}Ne$ reactions occur in an almost identical energy region; here a rapid separation of ^{18}Ne from the directly formed ^{18}F was absolutely necessary.

The target system consisted of 4 single cylinders filled with neon at normal pressure. They were arranged and adjusted similar to a stacked foil system. Irradiations were carried out using both the cyclotrons at Jülich (CV28 and JULIC). The $^{18}Ne(T_{1/2}=1.67$ sec) itself could not be measured because of the low abundance (9%) of its 1041 KeV γ-line; instead the ^{18}F daughter was measured. In the case of deuterons on neon, the ^{18}F activity produced was completely removed from the target wall by washing the target 6 times with a 2 ml aqueous NaF solution. In ^{3}He-induced reactions on neon, however, a flow system was used to separate the indirectly formed ^{18}F. The neon gas from the target containing ^{18}Ne and ^{18}F was allowed to pass through four Al_2O_3 columns: the primary ^{18}F was quantitatively removed in the first column, the secondary ^{18}F from the decay of ^{18}Ne was absorbed upon formation during the passage of ^{18}Ne through the columns. The details of the flow system have been described elsewhere[11]. After irradiation the ^{18}F content in each Al_2O_3 column was determined. The ^{18}Ne activity was then obtained by a summation of the decay corrected ^{18}F activities in all the columns.

The excitation functions for deuteron induced nuclear reactions on neon are given in Fig. 1. Our measurements in the high-energy region are combined with the low-energy cross section values of Nozaki et al[9]. In order to ascertain the contribution of the low-energy (d,α) reaction to the total cross section in the high energy region, the excitation function of the (d,α) reaction was extrapolated to higher energies using a simple compound nucleus model. The contribution of direct processes to the (d,α) reaction is unknown and would cause some extra uncertainty in the (d,p3n) cross sections. The measured $^{20}Ne(d,p3n)^{18}Ne$ excitation function does not exhibit any maximum. Different from the low energy (d,α) part the curve seems to increase to saturation, which is indicative of a high contribution from direct reactions, as well as of the complexity of the process.

The theoretical yield of ^{18}F formed via the ^{18}Ne precursor was calculated from the excitation function and lies at 5 mCi/μAh for a deuteron energy degradation of 65 to 60 MeV. With high-energy deuteron beam currents of 10 μA one should obtain a ^{18}F activity of about 100 mCi after an irradiation of 2 hrs. A comparison with the yield obtained via direct ^{18}F production shows that the latter process gives rise to higher yields. The advantage of the ^{18}Ne-produced fluorine thus lies in the possibility to produce anhydrous ^{18}F species in a relatively easy way.

The measured excitation function for the sum of $^{20}Ne(^{3}He,\alpha p)^{18}F$ and $^{20}Ne(^{3}He,\alpha n)^{18}Ne \xrightarrow[1.67s]{\beta^+} {}^{18}F$ is shown in Fig. 2; it is more or less in agreement with the earlier results[9]. The excitation function of the $^{20}Ne(^{3}He,\alpha n)^{18}Ne$ reaction alone, measured by us, is also shown in Fig. 2. It can be seen that the cross sections for the formation of ^{18}Ne in ^{3}He-particle induced reactions on neon are smaller by one to two orders of magnitude than the deuteron induced reactions (see above) and by two orders of magnitude than those of the $(^{3}He,\alpha p)$ reaction.

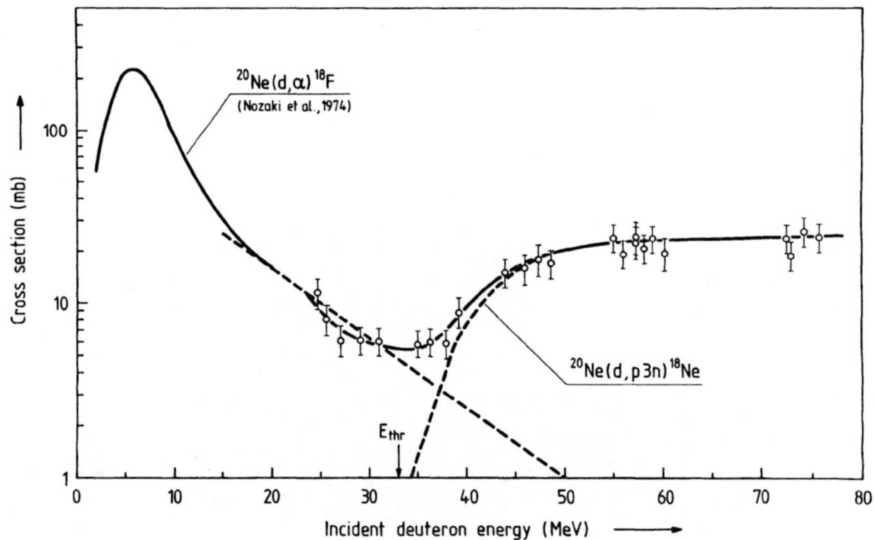

Fig. 1 Excitation functions for some deuteron induced nuclear reactions on neon (cf. refs. 9,11). The (d,3pn)-reaction product contains some contribution from the (d,d2n) and (d,tn) processes.

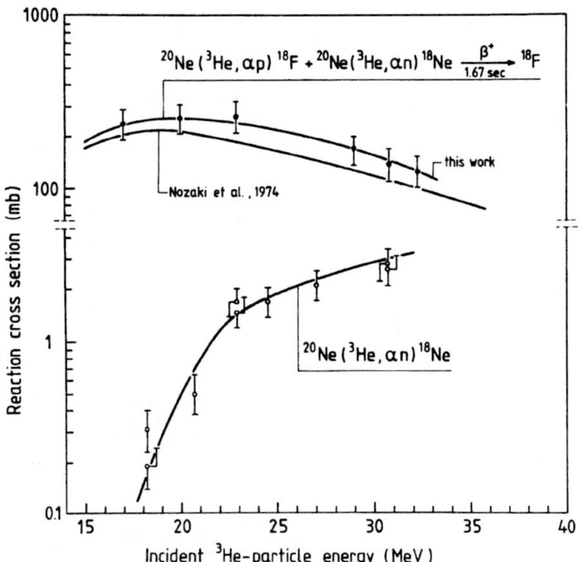

Fig. 2 Excitation functions of some ^3He-particle induced nuclear reactions on neon (cf. refs. 9,11).

The theoretical yield of ^{18}F formed via the $^{20}Ne(^3He,\alpha n)^{18}Ne \xrightarrow[1.67s]{\beta^+} {}^{18}F$ reaction was calculated to be 0.1 mCi/μAh for ^3He-particle energy degradation of 33 to 17 MeV. In comparison to the deuteron induced reaction with significantly higher yields it can be stated that this reaction will not be useful for production of ^{18}F radiopharmaceuticals. The process might, however, be of interest in hot atom chemistry for studying the reactions of decay produced ^{18}F species.

^{75}Br

The radioisotopes of bromine have great potential for applications in nuclear medicine (cf. 12). In comparison to the mostly used iodine isotopes, bromine has the advantages of a stronger bond to the carbon atom in biomolecules and of non-accumulation in the thyroid. Among the various radioisotpes of bromine, ^{75}Br is of particular interest since it has a high positron branching and is especially suited for PECT (cf. 2,5). The use of this radioisotope in nuclear medicine was first suggested by our laboratory[5]. In the meantime its large scale production has been developed[13] and the radioisotope has already been used in labelling ^{75}Br-(p-bromophenyl)-pentadecanoic acid[14] for heart tomography and L-3-^{75}Br-α-methyltyrosine[15] for investigations on pancreas. Both the radiopharmaceuticals have found applications in man.

^{75}Br can be generally produced via the following four ways:

i) ^3He or α-particle induced reactions on arsenic
ii) p and d induced reactions on isotopically enriched selenium isotopes
iii) decay of its precursor (4.5 min ^{75}Kr)
iv) (p,α) reaction on ^{78}Kr

We investigated the ^3He- and α-particle activation of natural arsenic in detail. Excitation functions were measured by the conventional stacked foil technique. Each foil consisted of metallic arsenic suspended in a self-supporting polystyrene film. A thin Al foil was used as backing material. The amount of arsenic in each sample was determined after the completion of the experiment by neutron activation analysis. Further experimental details are given elsewhere[16].

The excitation functions for the formation of $^{74m-77}Br$ in ^3He-particle induced nuclear reactions on arsenic are given in Fig. 3. Results of some other measurements[17] up to 45 MeV are also shown. The $^{75}As(^3He,n)^{77}Br$ reaction has a very low cross section and is similar to other (^3He,n) reactions in this mass region (cf. 18). The (^3He,3n) reaction leading to the formation of ^{75}Br occurs with the highest cross section. The long tails of the various reactions suggest that at high energies direct processes make significant contributions. The optimum energy range for the production of ^{75}Br is between 36 and 25 MeV, with an expected ^{76}Br-impurity level of about 2%.

The excitation functions of the α-particle induced nuclear reactions on arsenic measured by us and those reported by others[17,19,20] are given in Fig. 4. The maximum of the excitation function seems to decrease as the number of emitted neutrons increases (n=2 to 5). Here also some contributions from direct processes are apparent. The (α,4n) reaction leading to the formation of ^{75}Br occurs at $E_\alpha > 45$ MeV, demanding thereby a high energy cyclotron. Furthermore, the expected level of ^{76}Br-impurity is ∿ 6%. This reaction is therefore not as suitable as the (^3He,3n) reaction for the production of ^{75}Br.

The proton and deuteron induced nuclear reactions on enriched ^{76}Se were investigated by Paans et al[17]. The cross sections of the reactions leading to the formation of ^{75}Br are higher than those of the ^3He- and α-particle induced reactions on arsenic. Construction of a high current target using selenium is, however, very difficult (see below) and limits the use of those reactions for large scale production of ^{75}Br.

In order to investigate the production of ^{75}Br via its 4.5 min ^{75}Kr precursor, excitation functions of the deuteron induced reactions on bromine were measured[21,22] and the results on the (d,xn) reactions are sketched out in Fig. 5. Evidently the formation of ^{75}Kr occurs only at high energies and the cross section is very low. Even in the optimum deuteron energy range of 90 → 68 MeV the expected yield of ^{75}Br is quite low.

As far as the $^{78}Kr(p,\alpha)^{75}Br$ reaction is concerned, so far only some preliminary results have been reported[23]. The practicability of this method using high current irradiations has not yet been demonstrated.

A summary of the various methods investigated for the production of ^{75}Br is given in Table II. From the viewpoint of purity and theoretically expected yields the $^{76}Se(p,2n)^{75}Br$ reaction appears to be the most suitable. However, use of enriched material, need of rather high proton energies and, above all, difficulties associated with the construction of a high current target and subsequent isolation of ^{75}Br formed have precluded the use of this reaction for large scale production of ^{75}Br. In the case of arsenic as target material, on the other hand, a high current Cu_3As alloy target has been developed[13] which is capable of withstanding up to 2100 Wcm^{-2} power density.

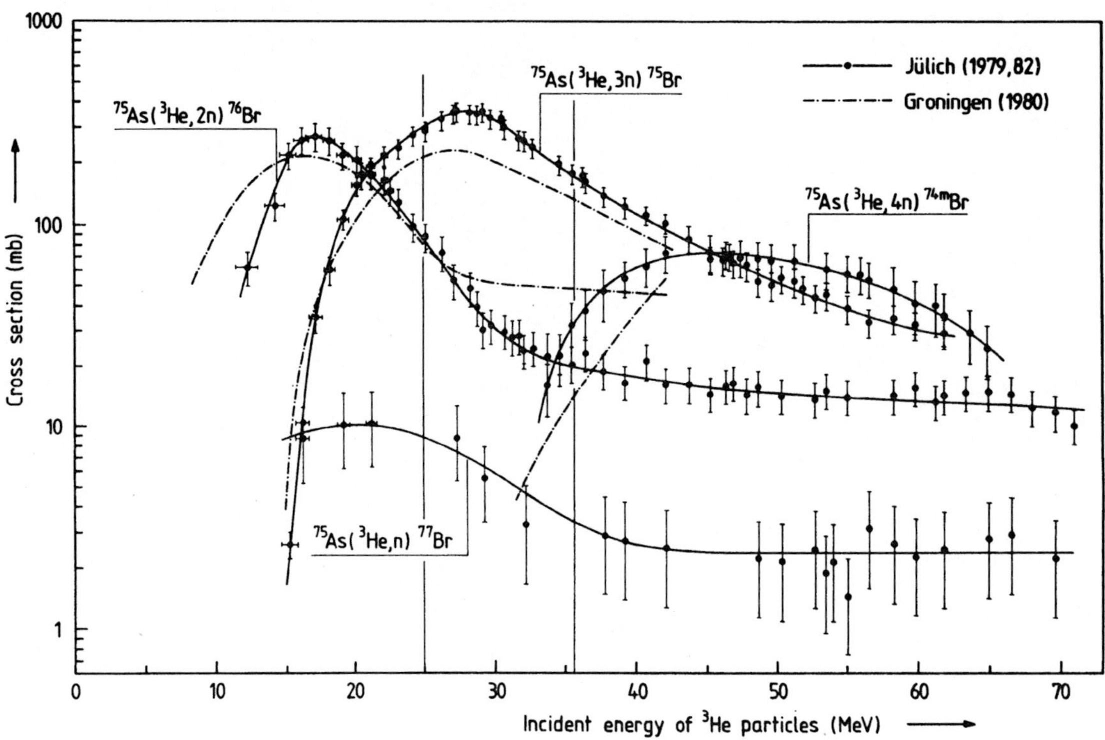

Fig. 3 Excitation functions for the reactions $^{75}As(^{3}He,xn)^{74m-77}Br$. (cf. refs. 5,16 (Jülich), 17 (Groningen)).

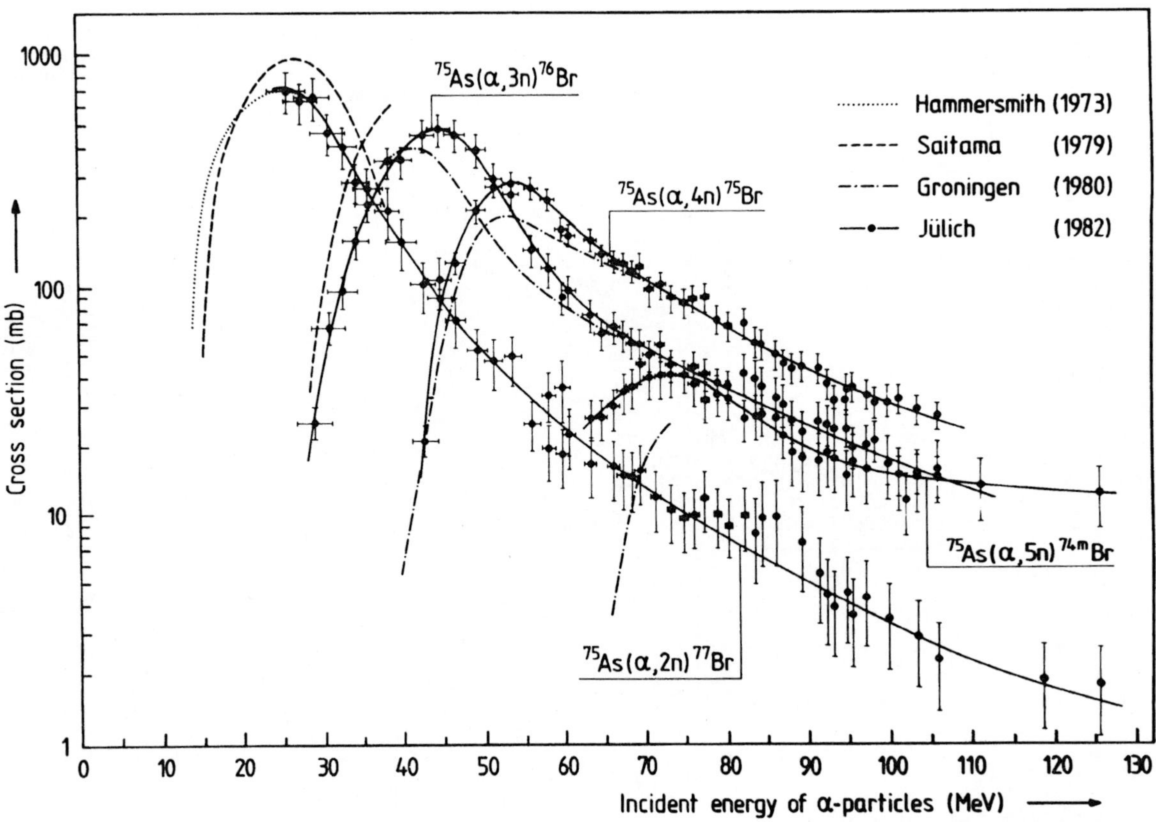

Fig. 4 Excitation functions of α-particle induced nuclear reactions on arsenic. (cf. refs. 19 (Hammersmith), 20 (Saitama), 17 (Groningen), 16 (Jülich)).

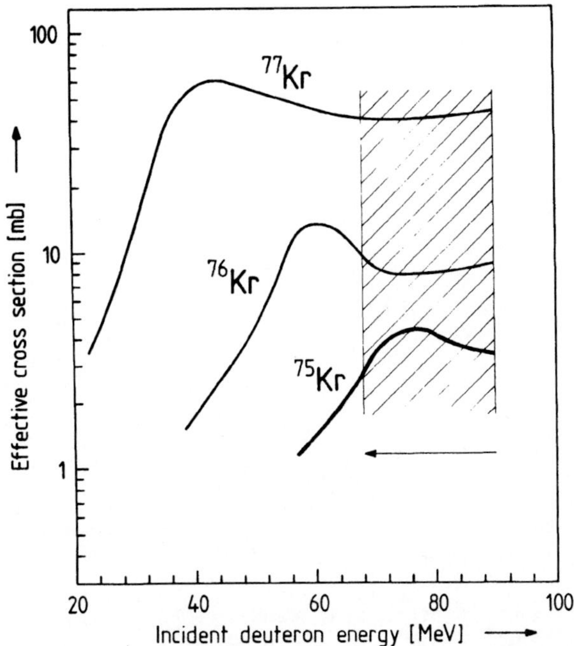

Fig. 5 Excitation functions of $^{nat}Br(d,xn)$ $^{75-77}Kr$ reactions (cf. refs. 21,22).

Furthermore, radiobromine can be distilled rather easily at 850 °C from that target. The $^{75}As(^{3}He,3n)^{75}Br$ process developed by us is therefore the most suited method for the routine large scale production of ^{75}Br.

^{123}I

^{123}I is one of the more commonly used radioisotopes and has almost ideal properties for single photon emission computed tomography. A large number of biomolecules labelled with ^{123}I have been synthesized and several of them are already finding routine applications in diagnostic nuclear medicine.

For the production of ^{123}I about twenty-five nuclear processes have been suggested (cf. 12, 24). All those processes, however, can be grouped under two general headings:

i) Indirect methods, i.e. those which make use of the $^{123}Xe \xrightarrow[2.0h]{EC,\beta^+} {}^{123}I$ precursor system

ii) Direct methods.

The energy ranges involved in the various processes are given in Table III. The ^{123}Xe-^{123}I precursor method demands the use of medium and high energy machines and gives rise to high-purity ^{123}I, the major impurity being 60d ^{125}I (0.2-0.4%). For routine and large scale production of ^{123}I the $^{127}I(p,5n)^{123}Xe$ (cf. 25-27) and $^{127}I(d,6n)^{123}Xe$ (ref. 28) reactions are commonly used. In recent years the spallation of Cs and La has also been increasingly used (cf. 29,30).

The direct methods of production of ^{123}I require low to medium energy cyclotrons and include proton and deuteron induced nuclear reactions on enriched tellurium isotopes. The three major reactions considered so far are also given in Table III.

Extensive measurements on the cross section data of the $^{124}Te(p,2n)^{123}I$ reaction have been reported (cf. 31). This reaction leads to very high yields of ^{123}I. The ^{124}I impurity formed via the (p,n) reaction on ^{124}Te, however, is the major limiting factor in the widespread use of this reaction. The level of this impurity depends upon the enrichment of the ^{124}Te target material and the energy range of the incident protons. In general this impurity corresponds to ∿ 0.8% at the end of bombardment (EOB).

The (p,n) and (d,n) reactions appeared to be promising but so far no detailed studies on nuclear data measurements have been reported. Some investigations on the $^{123}Te(p,n)^{123}I$ reaction using a 91.5% enriched ^{123}Te target have shown[32] that the level of ^{124}I impurity is reduced to 0.35%. However, no information on the other impurities is available. We investigated the (d,n) reaction in detail.

Excitation functions of the (d,xn) and (d,pxn) reactions on ^{122}Te were measured by the stacked foil technique mentioned above. Thin foils were obtained by electroplating 96.45% enriched ^{122}Te on 25 or 10 μm thick Ti-backing, the thickness of the ^{122}Te film in each case amounting to about 7 μm. After low current irradiations the various reaction products were determined via high-resolution γ- or x-ray spectroscopy. The excitation functions of the (d,xn) reactions on ^{122}Te are shown in Fig. 6. The optimum energy for the production of ^{123}I via the (d,n) reaction at a compact cyclotron is between 14 and 8 MeV. Integration of the excitation function over this energy region gives a theoretical yield value of 1.7 mCi/μAh. This yield is only about 25% of the ^{123}I yield

Table II. Summary of Nuclear Processes for the Production of ^{75}Br

Nuclear reaction	Energy range (MeV)	Theoretical thick target yield (mCi/μAh)	Major impurity at EOB (%)	References
$^{75}As(^{3}He,3n)^{75}Br$	36 → 25	8	^{76}Br(2.0)	5,16,17
$^{75}As(^{4}He,4n)^{75}Br$	64 → 54	7.5	^{76}Br(6.0)	16,17
$^{76}Se(p,2n)^{75}Br$	28 → 22	118	^{76}Br(1.5)*	17
$^{76}Se(d,3n)^{75}Br$	35 → 29	82	^{76}Br(4.5)*	17
$^{nat}Br(d,xn)^{75}Kr \xrightarrow[4.2\ min]{EC,\beta^+} {}^{75}Br$	90 → 68	0.21	^{77}Br(5.7)	22
$^{78}Kr(p,\alpha)^{75}Br$	15 → 12	2**	$^{76,77}Br$(0.1)	23

* With 92.4% enriched ^{76}Se

** With 100% enriched ^{78}Kr, 31 cm target length, 2 bar (practical yield)

Table III. Energy Ranges of Important Nuclear Processes for the Production of ^{123}I

Indirect Methods:		
$^{127}I(p,5n)^{123}Xe$	$E_p = 70 \rightarrow 50$ MeV	$^{123}Xe \xrightarrow[2.0\ h]{EC,\beta^+} {}^{123}I$
$^{127}I(d,6n)^{123}Xe$	$E_d = 78 \rightarrow 64$ MeV	
Cs, La(p,spall)^{123}Xe	$E_p = 660 \rightarrow 200$ MeV	
Direct Methods:		
$^{124}Te(p,2n)^{123}I$	$E_p = 30 \rightarrow 20$ MeV	
$^{123}Te(p,n)^{123}I$	$E_p = 17 \rightarrow 10$ MeV	
$^{122}Te(d,n)^{123}I$	$E_d = 14 \rightarrow 8$ MeV	

expected from the (p,2n) reaction under optimum conditions.

The impurities associated with the (d,n) reaction could not be determined via excitation function measurements and an integral determination was necessary. A thick $^{122}TeO_2$ target was therefore irradiated with 14 MeV deuterons, radioiodine was separated by dry distillation (cf. 33) and the levels of various impurities determined. The levels of ^{124}I, ^{126}I and ^{131}I impurities were rather small (0.08, 0.06 and 0.07%, respectively); the major impurity, however, being the 12.4 h ^{130}I (1.5%). Due to the comparable half-lives of ^{130}I and ^{123}I the level of ^{130}I impurity remains constant over the whole period of application.

The above discussion suggests that high purity ^{123}I is produced via the $^{123}Xe \rightarrow {}^{123}I$ precursor method. Among the low energy processes the $^{122}Te(d,n)^{123}I$ reaction offers an alternative method of production since the ^{124}I impurity level is lower than in the case of the $^{124}Te(p,2n)^{123}I$ process. The (p,2n) reaction, however, is still to be preferred because of higher yields.

Conclusions

The above discussion shows that the number of nuclear processes suggested for the production of radiohalogens is rather large and, in recent years, considerable progress has been made in nuclear data measurements. It should, however, be emphasized that an accurate knowledge of nuclear data is not the only criterion for the choice of a process for production purposes. Suitability of the target material for constructing a high current target and ease of subsequent chemical separation of the desired radioisotope play important roles in the choice of a production method.

The major reactions for the production of ^{18}F are $^{16}O(^{3}He,p)^{18}F$, $^{20}Ne(d,\alpha)^{18}F$, $^{20}Ne(^{3}He,\alpha p)^{18}F$ and $^{18}O(p,n)^{18}F$. In the deuteron and ^{3}He-particle induced reactions on neon the contribution of ^{18}F formed via the decay of its 1.67 sec ^{18}Ne precursor has been found to be small.

Among the various processes suggested for the production of ^{75}Br, the $^{75}As(^{3}He,3n)^{75}Br$ reaction is the most suitable and large scale

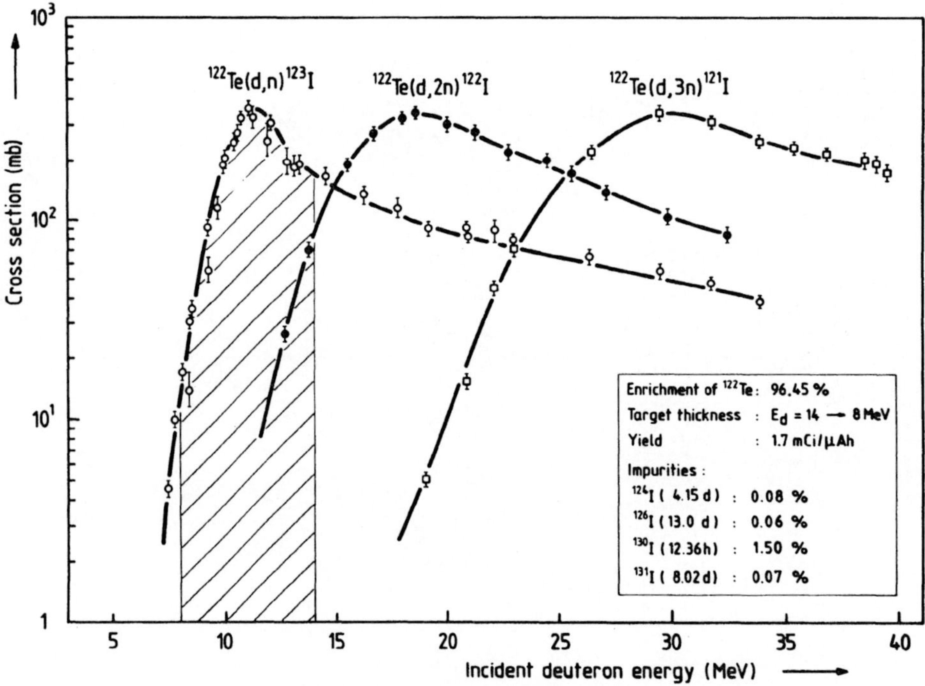

Fig. 6 Excitation functions of deuteron induced nuclear reactions on 96.45% enriched ^{122}Te.

production of ^{75}Br using that reaction has been developed. This radioisotope has great potential in PECT.

As far as ^{123}I production is concerned, high-purity product is achieved through the $^{123}Xe \rightarrow ^{123}I$ precursor method. The low-energy processes, e.g. $^{124}Te(p,2n)^{123}I$, $^{123}Te(p,n)^{123}I$ and $^{122}Te(d,n)^{123}I$, though suitable at low energy machines lead to higher impurities.

References

1. G. Stöcklin and S.M. Qaim, Proc. Int. Conf. on Neutron Physics and Nuclear Data for Reactors and other Applied Purposes, Harwell, September 1978 (NEA, Paris, 1979) p. 667
2. G. Stöcklin and G. Kloster in "Computed Emission Tomography" (Edited by P.J. Ell and B.L. Holman), Oxford University Press, Oxford (1982), p. 299
3. Table of Isotopes, 7th Edition (Eds. C.M. Lederer and V.S. Shirley), John Wiley & Sons, Inc., 1978
4. S.M. Qaim, Radiochimica Acta 30, 147 (1982)
5. R. Weinreich, Z.B. Alfassi, G. Blessing and G. Stöcklin, Proc. 17th Intern. Ann. Meeting Soc. Nucl. Med., Innsbruck, September 1979, cf. Nuklearmedizin: Suppl. 17, 202 (1980)
6. B.M. Gallagher, A. Ansari, H. Atkins, V. Casella, D.R. Christman, J.S. Fowler, T. Ido, R.R. MacGregor, P. Som, C.N. Wan, A.P. Wolf, D.E. Kuhl and M. Reivich, J. Nucl. Med. 18, 990 (1977)
7. M.E. Phelps, S.C. Huang, E.J. Hoffman, C. Selin, L. Sokoloff and D.E. Kuhl, Ann. Neurol. 6, 371 (1979)
8. E.J. Knust, Ch. Kupfernagel and G. Stöcklin, J. Nucl. Med. 20, 1170 (1979)
9. T. Nozaki, M. Iwamoto and T. Ido, Int. J. Appl. Radiat. Isotopes 25, 393 (1974)
10. J. Fitschen, R. Beckmann, U. Holm and H. Neuert, Int. J. Appl. Radiat. Isotopes 28, 781 (1977)
11. H. Backhausen, G. Stöcklin and R. Weinreich, Radiochimica Acta 29, 1 (1981)
12. G. Stöcklin, Int. J. Appl. Radiat. Isotopes 28, 131 (1977)
13. G. Blessing, R. Weinreich, S.M. Qaim and G. Stöcklin, Int. J. Appl. Radiat. Isotopes 33, 333 (1982)
14. H.H. Coenen, M.-F. Harmand, G. Kloster and G. Stöcklin, J. Nucl. Med. 22, 891 (1981)
15. F. Ritzl, G. Kloster, H.H. Coenen, U. Tisljar and G. Stöcklin, J. Nucl. Med. 22, P87 (1981)
16. Z.B. Alfassi and R. Weinreich, Radiochimica Acta 30, 67 (1982)
17. A.M.J. Paans, J. Welleweerd, W. Vaalburg, S. Reiffers and M.G. Woldring, Int. J. Appl. Radiat. Isotopes 31, 267 (1980)
18. He Youfeng, S.M. Qaim and G. Stöcklin, Int. J. Appl. Radiat. Isotopes 33, 13 (1982)
19. S.L. Waters, A.D. Nunn and M.L. Thakur, J. Inorg. Nucl. Chem. 35, 3413 (1973)
20. T. Nozaki, M. Iwamoto and Y. Itoh, Int. J. Appl. Radiat. Isotopes 30, 79 (1979)
21. S.M. Qaim, G. Stöcklin and R. Weinreich, Int. J. Appl. Radiat. Isotopes 28, 947 (1977)
22. S.M. Qaim and R. Weinreich, Int. J. Appl. Radiat. Isotopes 32, 823 (1981)
23. A.M. Friedman, O.J. DeJesus, P. Harper and C. Armstrong, 4th Int. Symp. on Radiopharmaceutical Chemistry, Jülich, August 1982, Abstract p. 157
24. Proc. Panel Discussion on Iodine-123 in Western Europe, KFA Jülich, Federal Republic of Germany, 13 February 1976 (Edited by S.M. Qaim, G. Stöcklin and R. Weinreich) KFA Report JÜL-CONF-20 (1976)
25. S.R. Wilkins, S.T. Shimose, H.H. Hines, J.A. Jungerman, F. Hegedüs and G.L. De Nardo, Int. J. Appl. Radiat. Isotopes 26, 279 (1975)
26. A.M.J. Paans, W. Vaalburg, G. Van Herk and M.G. Woldring, Int. J. Appl. Radiat. Isotopes 27, 465 (1976)
27. D.B. Syme, E. Wood, I.M. Blair, S. Kew, M. Perry and P. Cooper, Int. J. Appl. Radiat. Isotopes 29, 29 (1978)
28. R. Weinreich, O. Schult and G. Stöcklin, Int. J. Appl. Radiat. Isotopes 25, 535 (1974)
29. N.F. Peek and F. Hegedüs, Int. J. Appl. Radiat. Isotopes 30, 631 (1979)
30. M. Adilbish, C.G. Chumin, V.A. Khalkin, O. Knotek, M.Ja. Kuznetsova, Ju.V. Norseev, V.I. Fominykh and N.G. Zaitseva, Int. J. Appl. Radiat. Isotopes 31, 163 (1980)
31. K. Kondo, R.M. Lambrecht and A.P. Wolf, Int. J. Appl. Radiat. Isotopes 28, 395 (1977)
32. R.C. Barrall, J.F. Beaver, H.B. Hupf and F.F. Rubio, Eur. J. Nucl. Med. 6, 411 (1981)
33. H. Michael, H. Rosezin, H. Apelt, G. Blessing, J. Knieper and S.M. Qaim, Int. J. Appl. Radiat. Isotopes 32, 581 (1981)

201-Tl PRODUCTION FOR MEDICAL USE BY (p,xn) NUCLEAR REACTIONS ON Tl AND Hg NATURAL AND ENRICHED TARGETS

M. Bonardi, C. Birattari, and A. Salomone

Istituto di Scienze Fisiche
Cyclotron Laboratory
Università di Milano
via Celoria 16, 20133 Milano, Italy

High-specific activity 201-Tl for biomedical uses might be produced by proton activation on: a) Hg targets by direct (p,xn) nuclear reactions, b) Tl targets via its parent radionuclide 201-Pb, c) Pb targets via its parent radionuclides 201-Bi and 201-Pb. With the Milan University AVF Cyclotron (maximum proton energy 45 MeV) the only nuclear reactions, which can be used for its production are the 203-Tl(p,3n) 201-Pb → 201-Tl and the Hg(p,n/p,2n/p,4n)201-Tl ones. So, thin-target excitation function measurements and thick-target-yield calculations have been carried out in the proton energy range available, for the (p,xn) nuclear reactions on nat-Tl, 81% enriched 203-Tl, nat-Hg and 98.6% enriched 202-Hg targets. The yields and contaminations obtainable with the various methods are discussed in some details.

(Nuclear medicine, 201-Tl radiopharmaceutical, Tl(p,xn), Hg(p,xn), 203-Tl(p,3n), 202-Hg(p,2n), thin-target excitation functions, thick-target-yields, contamination)

Introduction

The 201-Tl (t1/2 = 74 h; Hg-K-X-rays, 98%; 135 keV, 2 %, 167 keV, 8%) is widely employed in nuclear medicine for the imaging of the myocardium (1).

It is generally produced by (p,3n/p,5n) nuclear reactions on thallium targets of natural isotopic composition (203-Tl, 29.5%; 205-Tl, 70.5%), as well as on 203-Tl or 205-Tl enriched targets by (p,3n) and (p,5n) nuclear reactions respectively, in any case via the decay of its parent radionuclide 201-Pb (see Tables I and II).

Table I Produced Radionuclides and Decay Parameters

Nuclide	t 1/2	E_γ (keV)	(%)
199-Tl	7.4 h	455	14
200-Tl	26.1 h	368	89
201-Tl	74. h	167	8
202-Tl	12. d	439	92
200-Pb	21.5 h	147	28
201m-Pb	61. s	628	55
201-Pb	9.4 h	331	81
202m-Pb	3.6 h	422	85
202-Pb	long	no	--
203-Pb	52.1 h	279	81
201m-Bi	59. m	846	?
201-Bi	1.8 h	628	?
201m-Po	8.9 m	967	43
201-Po	15.2 m	890	53

Table II Tl(p,xn) nuclear reactions and Q values

Target	Reaction	Q(MeV)	Ref
nat-Tl	(p,xn)201-Pb → 201-Tl		2,3,4,5
203-Tl	(p,3n) " "	-17.4	6,7
205-Tl	(p,5n) " "	-31.6	8

With this method the total contamination level due to the other thallium-radionuclides 200-Tl and 202-Tl can be reduced to a value lower than 1% at the end of the target chemical processing (EOP). This contamination level can be obtained only following optimization of the parameters :

---incident proton energy on the Tl target,
---proton energy loss (MeV) in the Tl target itself,
--waiting time between the end of bombardment (EOB) and the first chemical separation (Tl target/201-Pb),
--waiting time between the first chemical separation and the second chemical separation (201-Pb/201-Tl).

A second production method needing a highest proton energy (about 70 MeV) has been studied by Lagunas-Solar and coworkers (9), being based on the nuclear reactions reported in Table III.

Table III Pb(p,xn/p,pxn) nuclear reactions

Target	Reaction	Q(MeV)
nat-Pb	(p,xn)201-Bi→201-Pb→201-Tl	
204-Pb	(p,4n)201-Bi→201-Pb→201-Tl	-29.0
206-Pb	(p,6n)201-Bi→201-Pb→201-Tl	-43.8
207-Pb	(p,7n)201-Bi→201-Pb→201-Tl	-50.6
208-Pb	(p,8n)201-Bi→201-Pb→201-Tl	-58.0
nat-Pb	(p,pxn)201-Pb→201-Tl	
204-Pb	(p,p3n)201-Pb→201-Tl	-24.0
206-Pb	(p,p5n)201-Pb→201-Tl	-38.9
207-Pb	(p,p6n)201-Pb→201-Tl	-45.6
208-Pb	(p,p7n)201-Pb→201-Tl	-53.0

The possibility to induce 209-Bi(p,9n)201-Po→201-Bi →201-Pb→201-Tl and 209-Bi(p,p8n)201-Bi→201-Pb→ 201-Tl nuclear reactions, whose Q values are -67.8 MeV and -61.8 MeV respectively, has not yet been considered by any researcher for production purposes.

Lastly, 201-Tl can be produced directly by (p,xn) nuclear reactions on mercury targets (see Table IV).

Table IV Hg(p,xn) nuclear reactions and Q values

Target	Reaction	Q(MeV)	Ref
nat-Hg	(p,xn)201-Tl		10,11
201-Hg	(p,n)201-Tl	- 1.19	
202-Hg	(p,2n)201-Tl	- 8.95	
204-Hg	(p,4n)201-Tl	-22.44	

This method, which in principle, needs a short waiting

K. H. Böckhoff (ed.), Nuclear Data for Science and Technology, 916–918.

time from the EOB to deliver the 201-Tl, is promising for the very simple chemical separation involved (12), but is not commonly employed and poor data are available in the literature (10,11).

The Milan University AVF Cyclotron accelerates external proton beams, with energy variable up to 45 MeV, so the only production methods of 201-Tl which can be considered are based on:

A) the 203-Tl(p,3n)201-Pb→201-Tl scheme, employing either nat-Tl or 203-Tl targets (2-7,12),
B) the Hg(p,n/p,2n/p,4n)201-Tl nuclear reactions employing either nat-Hg or 201,202,204-Hg targets. About this production methods poor data are available in the literature (10-12).

Thin-Target Excitation Functions

In order to start a routine production of 201-Tl in Milan, the thin-target excitation functions (μCi/μAh MeV = 10.28 MBq/C MeV) for 201-Tl production by nat-Tl (p,xn), 81% enriched 203-Tl(p,xn), nat-Hg(p,xn) and 98.6% enriched 202-Hg(p,xn) nuclear reactions have been measured in the proton energy range available, with the single target activation method, followed by delayed measurement of the radionuclides produced, by Ge(Li) spectrometry. The following formula was employed for thin-target-yield (MBq/C MeV) calculations:

$$Y = \left[\frac{C\,\lambda^2 \tau \exp(\lambda t)}{\varepsilon\,\alpha\,S\,(dE/dX)\,Q}\right] (1 - \exp(-\lambda\tau))^{-1} (1 - \exp(-\lambda\Delta))^{-1} \qquad (1)$$

where:

- C = integrated photopeak counts
- S = target thickness (g/cm^2)
- Q = integrated proton charge (μC)
- dE/dX = stopping power of the target (MeV cm^2/g)
- λ = decay constant (s-1)
- τ = irradiation time (s)
- t = waiting time (s)
- ε = photopeak efficiency
- α = abundance of the gamma emission employed
- Δ = measuring time (s).

In the Figs. 1,2,3 and 4, the thin-target-yields at the end of an "instantaneous bombardment (EOIB), calculated from the eq. 1 for the main nuclear reactions mentioned above are shown.

Thick-Target-Yields and Contamination Calculations

From the integration of the excitation functions the thick-target-yield (μCi/μAh = 10.28 MBq/C) for the various radionuclides produced can be obtained as a function of the target thickness (MeV) and of the incident proton energy on the target itself.

A) 203-Tl(p,3n)201-Pb→201-Tl

With this method the optimum yield for the 201-Tl is 4.9 GBq/C from natural thallium and 14.3 GBq/C from 81% enriched 203-Tl targets, for a 8 MeV target thickness, at a bombarding proton energy of 27 MeV. The total contamination level from 200,202-Tl would be 0.28% and 0.29% for the natural and the enriched targets respectively at the end of target processing (EOP). These yields are calculated from Figs. 1 and 2 after two chemical separations: the first one being performed at the end of an instantaneous bombardment (EOIB), while the second one is carried out 32 hours later, in order to reach saturation for the 201-Pb→201-Tl generator. These results are in good agrrement with those found in the literature, whitin the experimental errors (4,5).

B) 202-Hg(p,2n)201-Tl

Thick-target-yield calculations on the excitation functions for nat-Hg targets (see Fig. 3) show that a very high contamination Level has to be expected from other thallium radionuclides (199,200,202-Tl) of the order of 15% under the most favourable irradiation conditions. This contamination can be reduced using a 202-Hg enriched target. For example, as a comparison to the method A), the contamination for a 6 MeV target thickness, at the optimal bombarding proton energy of 19 MeV, is 3.4% for a 98.6% enriched 202-Hg target, after the optimum waiting time from the EOIB of 60 hours. Under the irradiation and waiting conditions mentioned above the 201-Tl yield is 14.2 GBq/C.

Fig. 1.

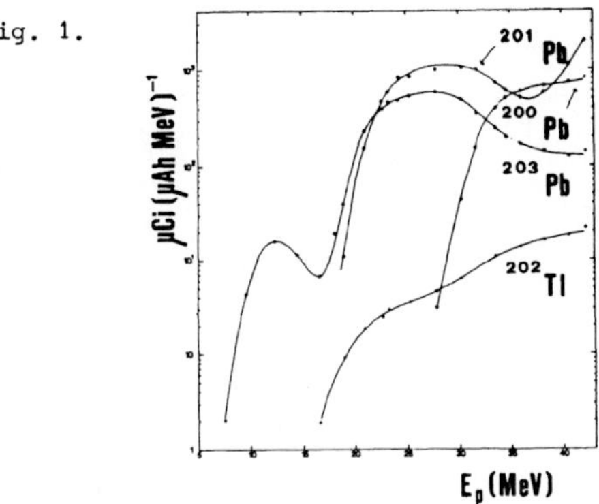

Fig. 2.

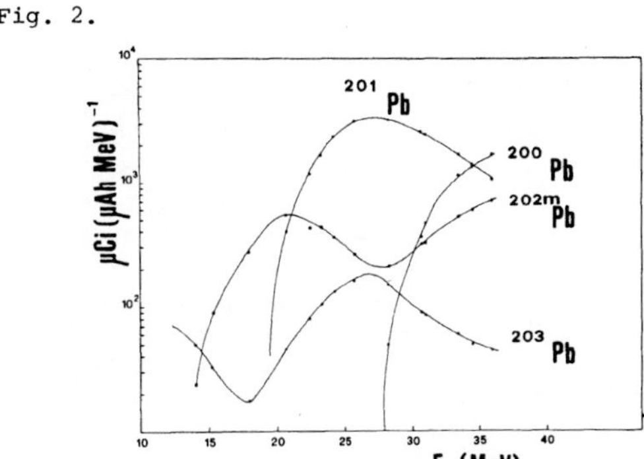

Fig. 3.

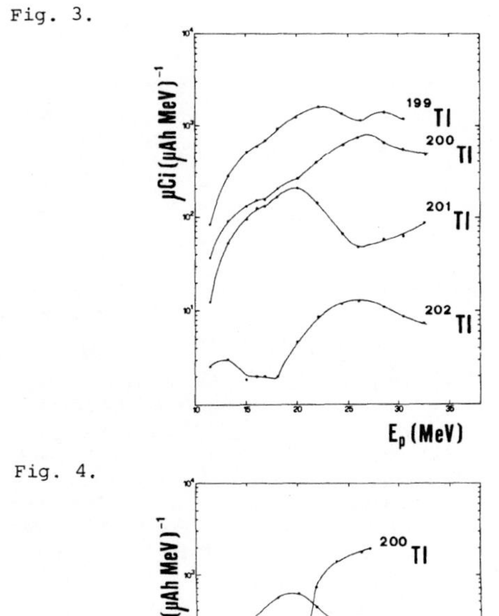

Fig. 4.

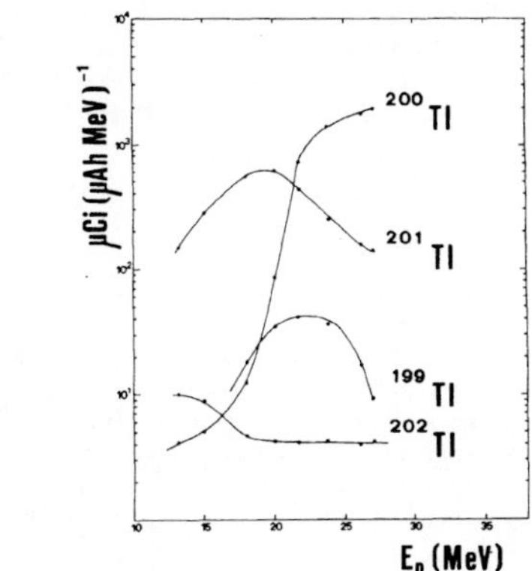

Figs. 1,2,3,4. Thin-target excitation functions for the main (p,xn) nuclear reactions on nat-Tl, 81% enriched 203-Tl, nat-Hg and 98.6% enriched 202-Hg targets. In Fig. 1 the 202-Tl yield is due to the nat-Tl (p,pxn) nuclear reactions plus the charging by the short-lived 202m-Pb.

In Table V the 201-Tl thick-target yields and total contamination levels obtainable for different incident proton energies, with total proton energy absorption are reported.

Table V 201-Tl Thick-Target-Yields, for different incident proton energies, at the waiting times tw from the EOIB when the contamination by 199-200-202-Tl is minimum

E_p (MeV)	tw(h)	Y(MBq/C)	Contamination(%)
16	20	9.3	7.2
18	30	16.2	5.0
19	50	17.6	4.5
20	70	18.0	4.7
22	160	9.6	7.7
24	200	6.7	10.0
26	220	6.5	12.0

In Table VI the 201-Tl thick-target-yields and total contamination levels obtainable, are reported as a function of the target thickness at the optimal incident proton energy of 19 MeV.

Table VI 201-Tl Thick-Target-Yields, for different target thicknesses, at the optimal incident proton energy of 19 MeV, at the waiting time tw from the EOIB when the contamination by 199,200,202-Tl is minimum

ΔE_p (MeV)	tw(h)	Y(GBq/C)	Contamination(%)
9	50	17.6	4.5
6	60	14.2	3.4
3	90	7.1	2.3

Discussion

Despite the higher contamination level in respect to the use of thallium targets, the 202-Hg(p,2n) method seems to be promising for the simplicity of the chemical separation involved (12). The employ of a 202-Hg target of very high isotopic purity (99.9%) could not change the situation from the point of view of contamination, which is mainly due to the 202-Hg(p,n)202-Tl nuclear reaction, which cannot be eliminated at all.

The employ of the (p,n) nuclear reaction on 201-Hg targets of enriched isotopic composition, was not considered at present for two reasons:
-- the higher cost of the enriched 201-Hg in respect to the enriched 202-Hg,
-- the lower cross-section of the 201-Hg(p,n) nuclear reaction in respect to the 202-Hg(p,2n) one.

Nevertheless, being the results of the employ of 202-Hg targets not completely satisfactory from the point of view of the contamination, seems to be interesting to start a study of production of the 201-Tl by the 201-Hg (p,n) nuclear reaction, for which, in principle, no contamination has to be expected.

Finally, the use of the 204-Hg(p,4n) nuclear reaction was not considered at all, being the 204-Hg enriched target very expensive, while the contamination due to the 204-Hg(p,3n)202-Tl nuclear reaction would be probably similar to that obtained by the use of 202-Hg enriched targets.

References

1. G. Subramanian , et al., Radiopharmaceuticals, The Soc. Nucl. Med., Inc., New York (1975).
2. F. Girardi, et al., Int. J. Appl. Radiat. Isotopes 26, 267 (1975).
3. E. Acerbi, et al., Status report on the program of radioisotope production at the Milan AVF Cyclotron, Report of the Istituto Nazionale di Fisica Nucleare, INFN/TC-77/9, Frascati, Rome (1977).
4. M. Lagunas-Solar, J. A. Jungerman, N. F. Peak, R. M. Theus, Int. J. Appl. Radiat. Isotopes 29, 159 (1978).
5. S. M. Qaim, R. Weinreich, H. Ollig, Int. J. Appl. Radiat. Isotopes 30, 85 (1979).
6. M. Lagunas-Solar, et al., Ref. 4.
7. M. Bonardi, Radiochem. Radioanal. Letters 42, 35 (1980).
8. M. Lagunas-Solar, J. A. Jungerman, D. W. Paulson, Cyclotron Production of 201-Tl via the 205-Tl(p, 5n)201-Pb Reaction, Proc. 2nd Int. Symph. on Radioph., pages 779-789, Seattle, Washington (1979).
9. M. Lagunas-Solar, F. E. Little, S. L. Waters, J. A. Jungerman, Cyclotron production of carrier-free 201-Tl via the 207-Pb(p,7n) reaction, Proc. 3rd Int. Symph. on Radiopharm. Chem., St.Louis, Missouri (1980).
10. D. Comar, C. Crouzel, Radiochem. Radioanal. Letters 23, 131 (1975).
11. L. Goetz, E. Sabbioni, E. Marafante, J. Edel-Rade, C. Birattari, M. Bonardi, J. Radioanal. Chem. 67, 183 (1981).
12. C. Birattari, M. Bonardi, A. Salomone, 201-Tl production studies by 203-Tl(p,3n)201-Pb and 202-Hg (p,2n) nuclear reactions, Proc. 4th Int. Symph. on Radiopharm. Chem., Jülich, FRG (1982).

This has been carried out in the frame of "Progetti Finalizzati" biomedical tecnology of Research National Council (C.N.R.) Italy.

KERMA FROM NEUTRON-INDUCED SPALLATION IN TISSUE

George H. Harrison

Department of Radiation Oncology
University of Maryland
Baltimore, Maryland 21201 USA

Alice C. Mignerey, Farideh Moghadami and Ali Gökmen

Department of Chemistry
University of Maryland
College Park, Maryland 20742 USA

Spallation kerma factors were calculated for ^{12}C and ^{16}O irradiated with neutrons between 40 and 100 MeV. Spallation kerma arises from secondary charged particles 6≤A≤15 following n-^{12}C and n-^{16}O reactions. The method of calculation was based on existing experimental data for proton-induced reactions, which were used to represent neutron-induced reactions. The most complete set of experimental data was for ^{16}O irradiated at 50, 55, 65 and 75 MeV giving energy spectra for spallation fragments. A Fermi break-up model with a preceeding Monte Carlo pre-equilibrium decay step was used in conjunction with the available experimental data. Kerma from densely ionizing ^{16}O spallation (A≥6) was calculated to be 28 to 48% of the ^{1}H kerma for tissue irradiated with neutrons between 40 and 100 MeV, respectively. The corresponding ^{12}C spallation kerma was calculated to be approximately 3% of the ^{1}H kerma in this energy range.

Introduction

In recent years there has been an increasing number of clinical trials at neutron therapy facilities world-wide. Many of the radiation beams used at these centers have neutron energies greater than 30 MeV, ranging in some instances up to 80 MeV. Since increased energies permit more efficient dose delivery to deep-seated tumors, high-energy beams are being recognized as desirable and are being sought in the United States, the United Kingdom and the Republic of South Africa. Neutron reaction cross section data serve as necessary input to many types of calculations, relating not only to cancer patient dosimetry, but also to the design of neutron shields and collimators and to the questions of radiological protection for occupationally exposed personnel. Specifically, these cross sections are required for the conversion of neutron spectra into microdosimetric distributions and the calculation of kerma per unit fluence in biological materials, hydrocarbon or water shielding and materials used in fabricating tissue-similar dosimeters. In the interpretation of ion chamber response, kerma factors are used to supply the conversion from dose measured in chambers not precisely tissue equivalent to the expected dose for tissue (if the neutron energy spectrum is known) or to the direct calculation of dose (if the energy spectrum and fluence are known).

In microdosimetry, the knowledge of the number and energy spectrum of every possible secondary charged particle consitutes the most fundamental and complete specification of radiation quality. In this paper we have used available experimental data for proton-induced reactions with ^{12}C and ^{16}O and, using the assumption that the products with A≥6 from these reactions will not differ significantly from those obtained for neutron-induced reactions, calculated the kerma factor k_f (kerma per unit fluence) in tissue resident carbon or oxygen due to secondary charged particles of masses 6≤A≤16 from neutron irradiations between 40 and 100 MeV. These kerma contributions were calculated separately for each secondary particle type and summed to yield k_f due to all ions heavier than alpha particles.

In a previous report the densely ionizing kerma contributions from n-^{16}O elastic scattering and alpha particle production were obtained for high incident neutron energies[1]. As shown in Fig. 1, it was also observed that the n-^{16}O spallation products (6≤A≤11) do not exist with significant cross section below neutron energies of 20 MeV[2]. However, these products may contribute significantly to tissue kerma above 40 MeV. This is confirmed by the new results presented in this paper.

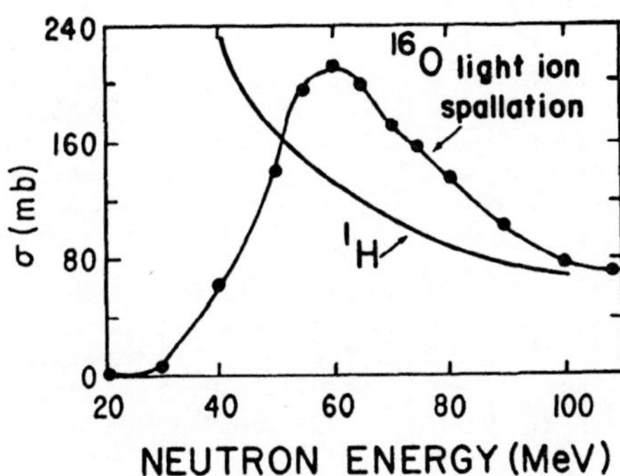

Fig. 1. Nuclear reaction cross sections (σ) for n-^{1}H scattering and n-^{16}O spallation production (6≤A≤11).

Method and Data Base

The basic approach is to use an appropriate theoretical model, with results checked against experimental data, to calculate the energy spectra of spallation products from reactions of 40-100 MeV neutrons with ^{12}C and ^{16}O. For the neutron energies considered, experimental proton-induced reaction cross sections were used to represent neutron induced results. This is justified because the incident energies far exceed the Coulomb barrier and because of the N=Z symmetry of ^{12}C and ^{16}O.

Very limited data on ^{12}C and ^{16}O spallation are available in published form[3-9]. Much of the data are incomplete giving total cross sections, angular distributions and/or energy spectra of only a few products. These data serve as a check of theoretical predictions. The main experimental results used and those presented in this paper represent a partial analysis of unpublished data obtained at the University of Maryland Cyclotron by the Nuclear Chemistry

K. H. Böckhoff (ed.), Nuclear Data for Science and Technology, 919–921.

group in connection with other work[3,4,7,10]. These data include complete sets of energy spectra for spallation products with A≥6 at several incident energies, making possible the direct calculation of k_f.

Theoretically immediate energy (i.e., 40-100 MeV/nucleon) nucleon-induced reactions on light target nuclei (e.g., ^{12}C, ^{14}N and ^{16}O) may be treated by a two-step model. The initial stage of interaction involves the collision of the projectile with the nucleons in the target nucleus. The reaction evolves as a cascade process of successive nucleon-nucleon collisions[11]. The energetic particles resulting from this fast process may escape from the nucleus if they have enough energy and leave a residual nucleus at high excitation energy. The system will eventually attain equilibrium with the internal energy distributed over all degrees of freedom. The equilibrized and high internal energy of this relatively small system (i.e., A~10) is dissipated by spallation into several products following the Fermi break-up statistical law[12].

In the cascade stage of interaction the track of each cascade particle is followed and the recoil momentum and the emission angle of the residual nucleus is determined. In the Fermi breakup de-excitation process the highly excited recoiling nucleus can decay to any particle stable state of fragments according to their statistical weight. The relative distribution of final fragments is determined by the available levels of all possible products and their formation Q-values. In contrast to the more widely used approach of particle evaporation[13] which must sequentially evaporate particles with A≤4, the Fermi break-up model counts all possibilities simultaneously. Any possible break up from 2-body to 5-body has been included in these calculations. The emission angle and energy of each product is calculated conserving particle number, energy and momentum throughout.

The end products of the two-step calculations are energy spectra $d^2\sigma/d\Omega dE$ for all possible isobars resulting from the reaction. Integrating the spectra over emission angle, the energy distribution $d\sigma/dE$ is obtained for a given product mass A, which is the salient quantity used in calculating the kerma factor,

$$k_f(\text{cGy-cm}^2) = 1.6\times10^{-8}(N_A/A)\int E(d\sigma/dE)dE(\text{MeV})$$

where N_A is Avogadro's number, E is in MeV and $d\sigma/dE$ is in units of cm^2/MeV. The available experimental data included only mass identification of the reaction products, therefore, the calculations were performed in terms of A. However, the probable fragment distribution among charges is as follows[10]: (A = 6: Li); (A = 7: 50% Be + 50% Li); (A = 9: Be); (A = 10: Be); (A = 11: 50% B + 50% C); (A = 12: C); (A = 13: 50% C + 50% N); (A = 14: N); (A = 15: 50% N + 50% O); (A = 16: O).

Results

As a primary check on the theoretical model, the angle-integrated cross sections for fragment emission were compared with experimental results. Figure 2 shows such a comparison for both ^{12}C and ^{16}O bombarded by neutrons at 75 MeV, as a function of spallation fragment mass. Agreement is fairly good, with the fit to the data better for the lighter fragments. Theoretical and experimental results were also compared as a function of incident neutron energy for both ^{12}C and ^{16}O. The results for A = 7, 9 and 11 are shown in Fig. 3. Here the fits to the experimental data seem fair above 40 MeV, with ^{16}O being modelled better than was ^{12}C. Figure 3 also shows the total cross section for the production of spallation products with A≥6. Here the theoretical estimate is in excellent agreement with experiment. These results seemed encouraging, especially since no adjustment to the model was performed in order to improve the agreement with experimental results.

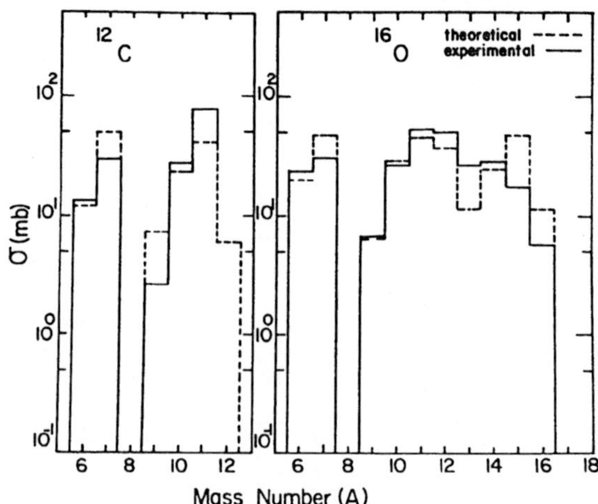

Fig. 2. Experimental and theoretical reaction cross sections for spallation reactions of 75 MeV neutrons bombarding ^{12}C and ^{16}O, as a function of spallation fragment mass. Experimental data from p + ^{16}O reactions.

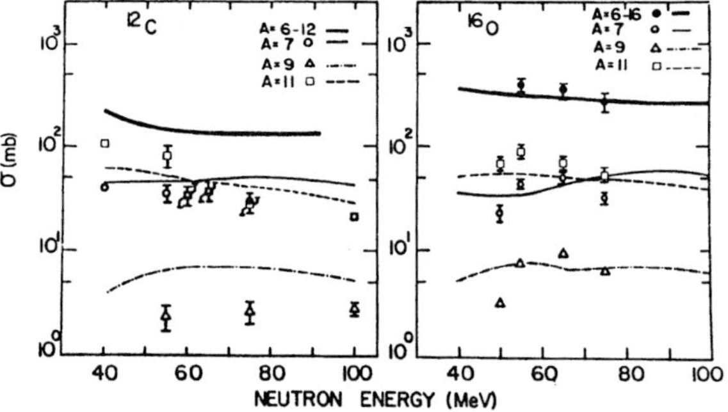

Fig. 3. Theoretical and experimental reaction cross sections for spallation products with A = 7, 9 and 11, as a function of the energy of neutrons bombarding ^{12}C and ^{16}O. Also shown is the total cross section for all spallation products with A≥6. Experimental data from p + ^{16}O reactions.

The k_f values were tabulated for numerous neutron energies and fragment masses. For example, the same theoretical results in Fig. 2, along with associated k_f values, are plotted in Fig. 4 for 75 MeV neutrons bombarding ^{12}C and ^{16}O. The average energy of emitted fragments can be calculated as a function of fragment mass from Fig. 4. Thus, for 75 MeV incident neutrons, the average energies of spallation fragments range from 3.4 MeV (for A = 15 fragments from ^{16}O) to 18 MeV (for A = 6 fragments from ^{12}C). The final computation consisted of summing theoretical k_f values for all spallation fragments with A≥6, at a given neutron energy. In this case, complete experimental data were available only for ^{16}O bombarded at 55 and 65 MeV. The results are shown in Fig. 5, with k_f for ^1H (as tabulated by Bassel and Herling[14] included for comparison). Also shown is a previous theoretical estimate of the total n + ^{16}O k_f, calculated with a particle evaporation code[13]. Very good agreement was obtained between our theoretical model and the available experimental data for the summed spallation k_f. To illustrate the significance of the magnitude of the spallation kerma in tissue, the ^{16}O spallation kerma is plotted relative to ^1H tissue kerma in Fig. 6, assuming a tissue composition $C_5H_{40}O_{18}N$.

Discussion

Figure 6 demonstrates the significant contribution of ^{16}O spallation products to the total tissue kerma at

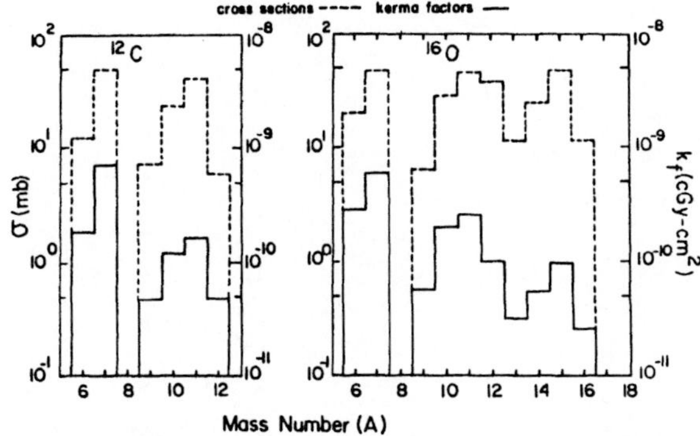

Fig. 4. Theoretical reaction cross sections and k_f values for spallation reactions of 75 MeV neutrons bombarding ^{12}C and ^{16}O, as a function of spallation fragment mass.

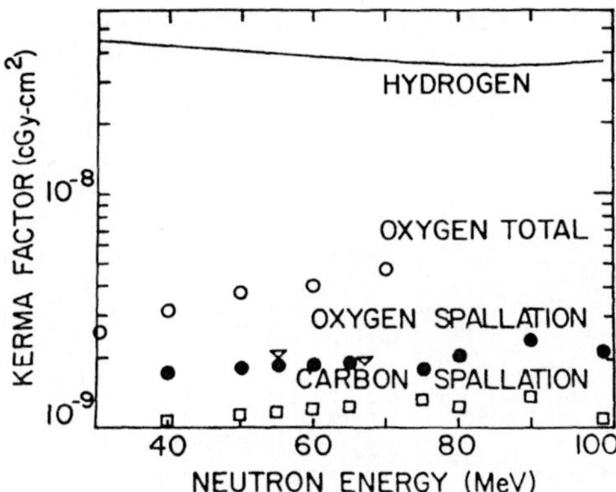

Fig. 5. Theoretically calculated k_f for all spallation products with A≥6 from ^{12}C and ^{16}O as a function of incident neutron energy. The triangles for ^{16}O at 55 and 65 MeV represent the available experimental p + ^{16}O data. For comparison, k_f for ^1H is also shown as well as the k_f for all n + ^{16}O reactions as calculated by Alsmiller and Barish using particle evaporation model[13].

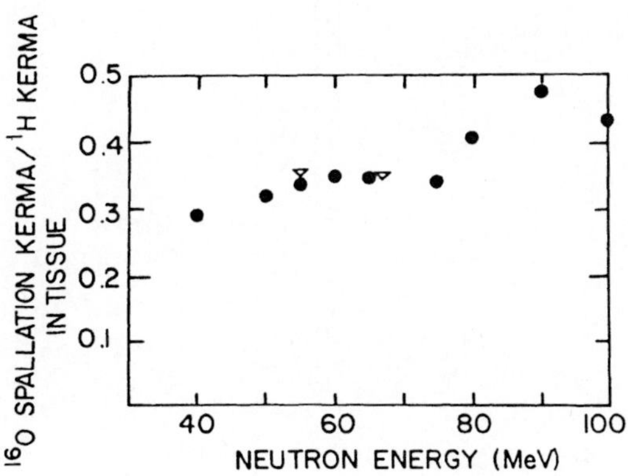

Fig. 6. Kerma for all spallation products of ^{16}O with A≥6 as a proportion of ^1H kerma in tissue approximated by $C_5H_{40}O_{18}N$, as a function of incident neutron energy. As in Fig. 5, the triangles represent the experimental data.

higher neutron energies. Note that the kerma due to light fragments including α-particles is not included in this comparison. The present results reinforce the idea that the character of the densely-ionizing dose component in tissue irradiated with neutrons changes dramatically with increasing neutron energy. Below 20 MeV, most of the densely-ionizing dose arises from slow proton recoil energy from the n-^{1}H reaction. Above 50 MeV, the proton recoil energy rises so that most of the associated energy deposition is lightly ionizing. The densely ionizing dose component is contributed mainly by ^{16}O reaction products with ^{12}C playing a relatively minor role.

To analyze the ^{16}O kerma, it is convenient to identify 3 groups of spallation products: light ions (A≥4), intermediate ions (6≤A≤11) and heavy ions (A>11). Figure 1 demonstrates definite structure of intermediate-ion cross-sections (and thus k_f) as a function of incident neutron energy, with a threshold below 30 MeV and a maximum near 60 MeV. This structure is not reproduced by the limited amount of theoretical modelling we have performed to date. It should be noted that the removal of the Coulomb barrier experienced by the protons used to obtain the data may result in a less distinct threshold for the neutrons we are trying to simulate. Nevertheless, other available data and calculations confirm the view that very little intermediate ion kerma exists in tissue irradiated with neutrons below 30 MeV[15,16]. We hope soon to calculate the light-ion component of n + ^{16}O kerma, so that the total non-elastic ^{16}O kerma can be calculated, combined with the small elastic scatter recoil kerma component, and compared to ^1H kerma as a function of neutron energy. This information should be directly applicable to problems in high-energy neutron dosimetry.

Even if there emerges a reasonably accurate description of all the secondary charged particle interactions induced by high-energy neutrons in tissue, it will remain a challenging task to correlate this physical description with biological response, since the bioeffect additivity of such a wide spectrum of radiation species has yet to be demonstrated.

References

1. G.H. Harrison, C.R. Cox, E.B. Kubiczek, EUR 5848, pp. 471-480 (1978)
2. G.H. Harrison, E.B. Kubiczek, Radiat. Res. 83, 90 (1980)
3. C.T. Roche, R.G. Clark, G.J. Mathews, C.E. Viola, Jr., Phys. Rev. C14, 410 (1976)
4. R.A. Moyle, B.G. Glagola, G.J. Mathews, V.E. Viola, JR., Phys. Rev. C19, 631 (1979)
5. S.M. Jung, C. Jaquot, C. Baixeras-Aiguabella, R. Schmitt, H. Braun, Phys. Res. C1, 435 (1970)
6. C.N. Davis, H. Laumer, S.M. Austin, Phys. Rev. C1, 270 (1970)
7. C.T. Roche, R.G. Clark, G.J. Mathews, V.E. Viola, Jr., NBS SP-425, pp. 504-507 (1975)
8. H. Laumer, S.M. Austin, L.M. Panggabean, Phys. Rev. C10, 1045 (1974)
9. M.G. Albouy, J.-P. Cohne, M. Gusakow, N. Paffe, H. Sergolle, L. Valentin, Phys. Lett. 2, 306 (1962)
10. V.E. Viola, Jr., private communication (1980)
11. G.H. Mathews, B.G. Glagola, R.A. Moyle, V.E. Viola, Jr., Phys. Rev. C25, 2181 (1982)
12. A. Gökmen, G.J. Mathews, V.E. Viola, Jr., Phys. Rev. (to be published)
13. R.G. Alsmiller, JR., J. Barish, Health Phys. 33, 98 (1977)
14. R.H. Bassell, G.H. Herling, Radiat. Res. 69, 210 (1977)
15. R.S. Caswell, J.J. Coyne, M.L. Randolph, Radiat. Res. 83, 217 (1980)
16. P.J. Dimbylow, Phys. Med. Biol. 25, 637 (1980)

CHARGED PARTICLE SPECTRA AND CALCULATED KERMA FROM 25 TO 60 MEV NEUTRONS

F.P. Brady, J.L. Romero, T.S. Subramanian*

Crocker Nuclear Laboratory and Department of Physics
University of California, Davis, California, U.S.A.

*Present address: Swedish American Hospital, Section of Medical Physics, Rockford, Illinois, U.S.A.

Charged particle proton, deuteron, triton, helium-3, and alpha particle spectra have been measured over a wide angular range for neutron energies of 27.4, 39.7, and 60.7 MeV, incident on Carbon, Nitrogen, and Oxygen. The main purpose of these measurements is to provide data for neutron dosimetry and for tests of nuclear models and their interpolation and extrapolation to other energies, particles, nuclei, etc. Good agreement is found in comparisons with charge symmetric proton-induced energy spectra, such as in (p,d) vs (n,d). Comparisons with models are being carried out. Kerma have been calculated at the three energies for each nucleus including elastic recoil contributions. The contributions from other reaction channels (^{6}Li, etc.) have been estimated using sum rules and, where available, proton induced data. The total Kerma one obtains for C, N, and O appear to be considerably less than recent predictions of several authors.

Introduction

Differential cross-section measurements for charged particles arising from neutron-induced reactions at neutron energies above the d-t generator energies are virtually nonexistent. These higher energy ranges (up to about 50 MeV) have recently drawn closer attention because neutrons of these energies are realized in configurations such as the d-Li irradiation facilities and the d-Be medical therapy facilities. These cross sections, where available, help to evaluate the performance of a given element as a constituent of a reactor structure or human tissue. In addition the charged-particle cross sections for carbon for instance, are relevant to the evaluation of neutron detection efficiencies of plastic scintillator, or for use in neutron spectrum tailoring. Similarly, the cross sections for oxygen are relevant to shielding calculations. Combining the cross sections for hydrogen (elastic proton), oxygen, carbon, and nitrogen, one can obtain quantities relevant to neutron dosimetry, such as KERMA in tissue. Here we report on cross sections measured at the Crocker Nuclear Laboratory in Davis, and KERMA calculations for carbon, nitrogen, and oxygen.

Experimental

The unpolarized neutron beam line at the Crocker Nuclear Laboratory is designed to provide nearly monoenergetic neutron beams in the 20-60 MeV range via the ^{7}Li(p,n)^{7}Be reaction[1-4] (Fig. 1). The neutron spectrum consists of a monochromatic peak containing 60% of the neutrons; the remaining 40% spread out in the flat low energy tail region. For a typical 40 MeV, 10 μA proton beam incident on a Li target, the neutron flux at 3 m is 5×10^5 (cm^2/sec) in a 1 MeV FWHM peak. A scattering chamber housing telescopes under vacuum has been incorporated in the beam line. See ref. 5 for further details.

Polysteyrene (CH) was chosen as a target for carbon instead of pure carbon sheets. The overwhelming reason for such a choice is the ease of direct normalization to the concurrently measured n-p elastic scattering, whose cross section is well known. Since solid targets help minimize low energy cut-offs in the measured charged-particle spectra, particularly those of alphas, foils containing N and O were used for their respective targets: Melamine ($C_3H_6N_6$) for nitrogen and Kodacel plastic ($C_3H_4O_2$) for oxygen. The N and O cross sections were obtained by subtracting out the contributions from carbon in the respective targets. Again, the absolute normalization was obtained by referring to the concurrently measured n-p elastic peaks. The targets so used were typically 5 mg/cm^2 thick. The cross sections presented here

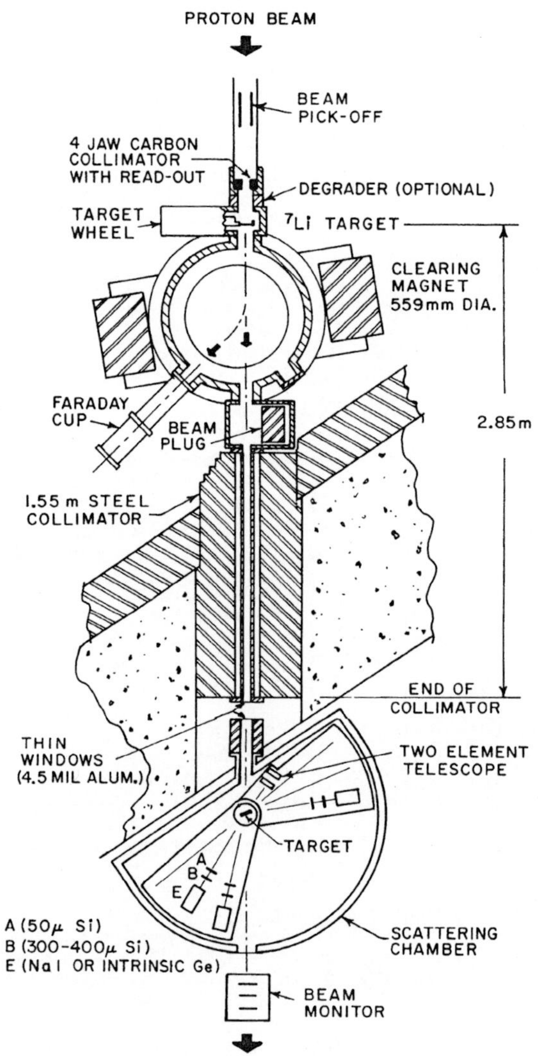

Fig. 1. Experimental set-up showing the neutron facility, scattering chambers, and detectors.

K. H. Böckhoff (ed.), Nuclear Data for Science and Technology, 922–925.

have been corrected for (alpha) particle and energy loss in targets.[5,6]

Table I gives the targets used and the particles detected.

Table I

Target	Composition	Thickness (mg/cm^2)	Particles Detected
Polystyrene	$(CH)_n$	4.1	p,d,t,^{3}He,α
Kodacel	$C_3H_4O_2$	5.2	"
Melamine	CH_2N_2	4.4	"
No target (background)		0	"

Double Differential Cross Sections

We have carried out measurements of the light hydrogen and helium ion production cross sections for neutron energies of 27.4, 39.7, and 60.7 MeV. These are energies for which there exists a considerable amount of data for proton-induced measurements of hydrogen (proton, deuteron, and triton) and helium (helium-3 and alpha particle) ion production. Thus proton and neutron-induced reactions which are charge-symmetric can be compared to check how well the data allow this approximate symmetry. This charge symmetry is true in large measure for the hadronic component of the nucleon-nucleon force but is violated by the Coulomb force. Exact charge-symmetry says that the p-p = n-n force. Typical spectra of deuterons, tritons, ^{3}He, and alphas measured from carbon for an incident neutron beam of 60.7 MeV are shown in Fig. 2. The error bars in our data do not include normalization errors; they are only indicative of the statistics. Charge symmetric proton-induced cross sections of Bertrand and Peelle[7] are shown in the same figure. In comparing for example ^{12}C(n,dx) with ^{12}C(p,dx) and ^{12}C(p,^{3}He) with ^{12}C(n,^{3}H) we find good agreement. Since these tissue resident nuclei have N(# neutrons) = Z(# protons) and are light (small total Coulomb effects) the agreement between charge-symmetric reactions is reasonable. Figure 3 shows similar results for Oxygen at 60.7 MeV.

Angle Integrated Spectra

The differential cross sections have been obtained at nine different angles in the range 15° and 150°. For those sets of data with all nine angles covered, the integration is obtained via Gaussian quadrature by assuming that the cross-section levels off on either end of the measured angular range. For sets of data deficient in larger angles, an extrapolation is done first to obtain cross sections for the next missing angle, and then the integration is performed. The errors shown in the plot are purely statistical. The additional uncertainties associated with the angle integrals are estimated to be 5% for the cases where 9 angles were measured and 20% when only 6 angles were measured. Such angle-integrated spectra are presented in the bottom row of each of the figures 2-3. Again, charge symmetric (p,xz) cross sections of Bertrand, et al.[7] are compared as histograms.

Predictions from calculations, using the code TNG with pre-equilibrium effects included, have been compared for protons and alphas by Herling and Bassel.[8] Predictions for Oxygen at 60.7 MeV, for Carbon at 39.7 MeV and for Nitrogen at 27.4 MeV are presented as solid curves in Figures 4,5, and 6 respectively. Figure 5 for Carbon at 39.7 MeV shows also a recent prediction by Bremmer and Prael.[9] They use an intranuclear cascade code which takes into account specific nuclear properties of Carbon, in particular, alpha clustering. It also allows excited nuclei to deexcite by a Fermi break up mechanism.

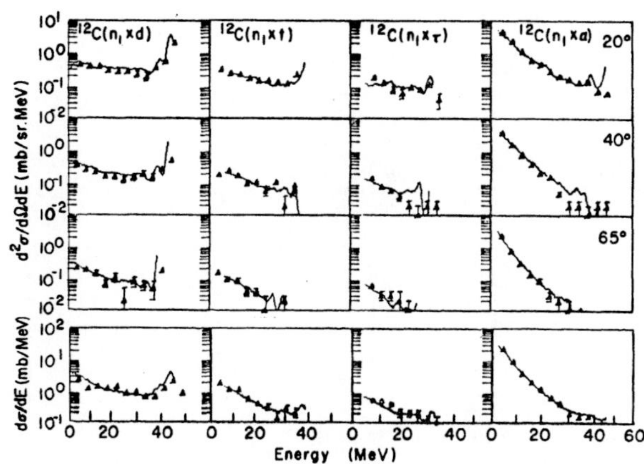

Fig. 2. Double differential cross sections for charged particles obtained from Carbon at 60.7 MeV. Data points are the present neutron-induced measurements. Curves are the corresponding symmetric proton-induced cross sections.[7]

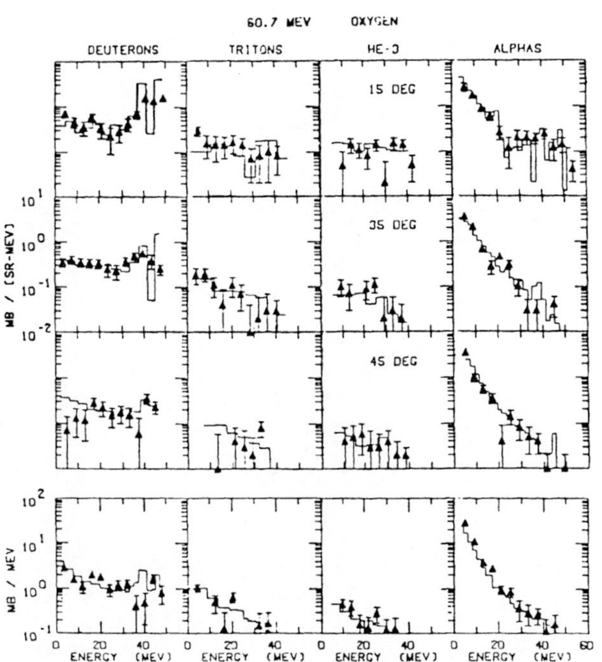

Fig. 3. Typical spectra of differential cross sections for charged particles obtained from Oxygen at 60.7 MeV. See legend for Fig. 2.

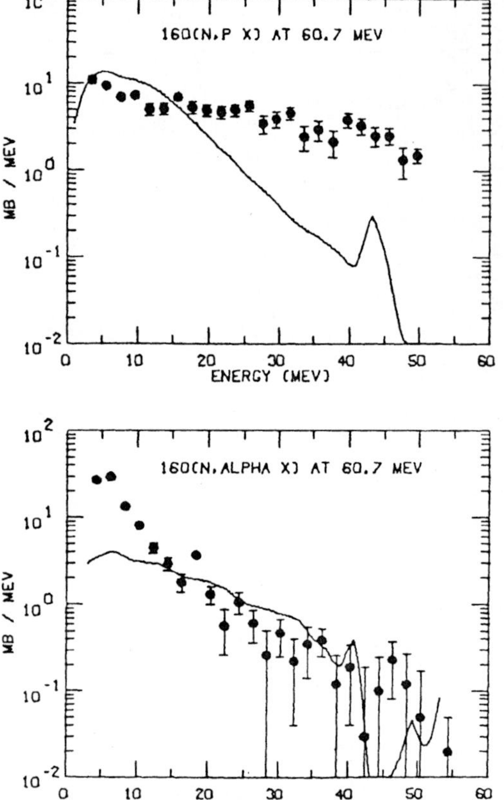

Fig. 4. Comparison with model predictions for angle integrated cross sections for protons and alphas from Oxygen at 60.7 MeV. Solid curves are from ref. 8.

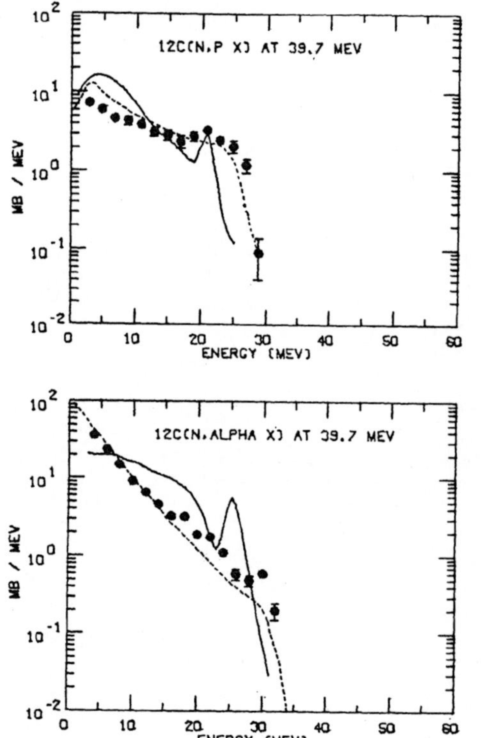

Fig. 5. Comparison with model predictions for angle integrated cross sections for protons and alphas from Carbon at E_n = 39.7 MeV. Solid curves are from ref. 8, dotted curve from ref. 9.

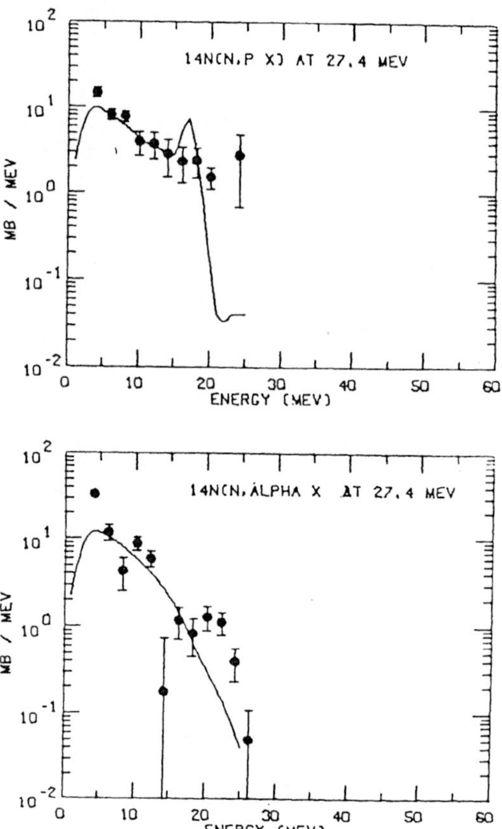

Fig. 6. Comparison with model predictions for angle integrated cross sections for protons and alphas for Nitrogen at E_n = 27.4 MeV. Solid curves are from ref. 8.

Kerma

For a given incident neutron energy on a particular element (say Carbon) the energy transfer per neutron per unit area (KERMA factor K) can be written as

$$\frac{1}{N} K = \sum_j \int \sigma_j(E)\, E dE + \bar{E}_{el} \cdot \sigma_{el},$$

where $\sigma_j(E)$ [$\equiv d\sigma_j(E)/dE$] is the (inclusive) cross section (in the laboratory) to detect particle j with kinetic energy between E and E+dE (i.e., our angle-integrated cross sections). The sum is over all charged particles (j = p,d,t,τ,α, ... up to ^{13}C in the case of a C target). $\bar{E}_{e\ell}$ is the average kinetic energy transfered elastically to the recoil nuclei. $\sigma_{e\ell}$ is the total elastic cross section and N is the number of nuclei of the element per gram.

Figure 7 shows KERMA factors obtained for Carbon at E_n = 27.4, 39.7, and 60.7 MeV. Our KERMA values were obtained using above expression, but include only j = p,d,t,τ, and α. A small correction due to finite energy thresholds (∿4 MeV) in the detectors has been included. The elastic recoil contribution was obtained from the optical model values of Herling and Bassel.[8]

Total KERMA have been calculated by a number of authors. See fig. 7. Recent values of Caswell et al.[10] extend up to 29 MeV neutron energy and those of Dimbylow[11] up to 50 MeV. These values are signi-

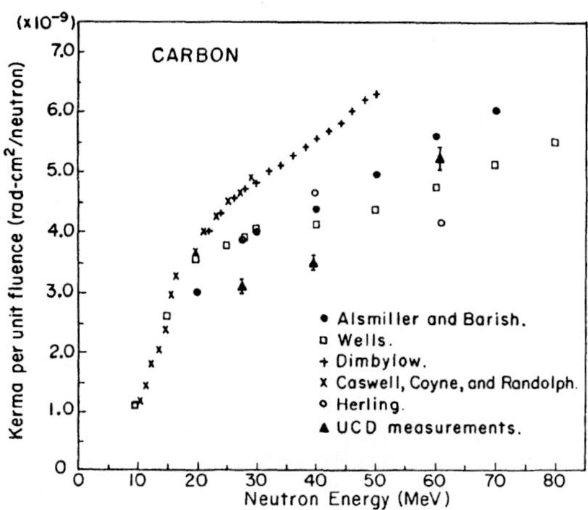

Fig. 7. Measured KERMA factors for Carbon as a function of incident neutron energy compared with various calculations. Our measurements do not include corrections due to detector threshold and contributions due to non-elastic heavy particles.

ficantly higher than the KERMA we obtain from the measured light ion spectra. Internuclear cascade model calculations,[12,13] which do not do very well in predicting individual proton, alpha particle, and other spectra, nevertheless do a reasonable job, it appears, of predicting the trend of total KERMA. The calculations of Herling, et al.[8] do reasonably well at 27.4 and 39.7 MeV but at 60.7 MeV are much too low. Similar results are obtained for O and N.

The data of fig. 7 do not include the non-elastic heavy recoil contribution for nuclei such as ^{11}B, ^{10}B, ^{7}Li, etc., which were not detected. Based on energy and baryon number conservation sum rules[14] and on proton-induced data,[7,15,16] we estimate the above contribution to be about 5-10% for carbon at 60 MeV, increasing up to about 20% at 27 MeV. On the basis of these estimates our KERMA values will still be ≅20% (at 27.4) and ≅30% (at 39.7 MeV) below the KERMA estimates of Caswell et al.[10] and of Dimbylow.[11]

Acknowledgments

This work was supported by the National Cancer Institute, grant PHS CA-16261. We thank also the National Science Foundation for additional support through grant PHY 77-05301.

References

1. J.A. Jungerman and F.P. Brady, Nucl. Inst. and Meth. 89, 167 (1970).
2. J.A. Jungerman, F.P. Brady, W.J. Knox, T. Montgomery, M.R. McGie, J.L. Romero, and Y. Ishizaki, Nucl. Inst. and Meth. 94, 421 (1971); J.L. Romero, F.P. Brady, and J.A. Jungerman, Nucl. Inst. and Meth. 134, 537 (1976).
3. F.P. Brady, W.J. Knox, and S.W. Johnsen, Nucl. Inst. and Meth. 89, 309 (1970).
4. J.L. Romero, T.S. Subramanian, F.P. Brady, N.S.P. King, and J.F. Harrison, BNL-NCS-50681, 247 (1977).
5. T.S. Subramanian, J.L. Romero, and F.P. Brady, Nucl. Inst. and Meth. 174, 475 (1980).
6. M.L. Johnson, J.L. Romero, T.S. Subramanian, and F.P. Brady, Nucl. Inst. and Meth. 169, 179 (1980).
7. F.E. Bertrand, and R.W. Peelle. ORNL Report No. 4799 (1973) unpublished.
8. G.H. Herling and R.H. Bassel, private communication.
9. D.J. Brenner and R.E. Prael, private communication.
10. R.S. Caswell, J.J. Coyne, and M.L. Randolph, Rad. Res, in press.
11. P.J. Dimbylow, Phys. Med. Biol., June 1980.
12. R.G. Alsmiller, Jr., and J. Barish, Health Phys. 33, 981 (1977).
13. A.H. Wells, Rad. Res. 80, 1 (1980).
14. C.E. DeTar, D.Z. Freedman, and G. Veneziano, Phys. Rev. D4, 906 (1971).
15. C.T. Roche, R.G. Clark, G.J. Mathews, and V.E. Viola, Jr., Phys. Rev. C14, 410 (1976).
16. E.M. Rimmer and P.S. Fisher, Nucl. Phys. A108, 561 (1968).

STUDY OF THE REACTION $^{12}C(n,3\alpha)n$ FROM THRESHOLD TO E_n=35 MeV

B.Antolković, I.Šlaus and D.Plenković

"Ruđer Bošković" Institute, 41001 Zagreb, Yugoslavia

and

P.Macq and J.P.Meulders

Universite de Louvain, Louvain la Neuve, Belgium

Kinematically complete measurement of the reaction $^{12}C(n,3\alpha)n$ has been performed using nuclear emulsions exposed to neutrons of continuous energy. Alpha particle energy spectra and cross sections for the reaction $(n,3\alpha)$ as well as for specific channels $^{12}C(n,^{8}Be_{gs})n\alpha$ and $^{12}C(n,^{12}C_{9.63})n$ have been measured up to 35 MeV incident energy with an accuracy of about 20%.

[Nuclear reaction $^{12}C(n,3\alpha)n$, E_n=10-35 MeV, natural target, measured alpha energy spectra, cross sections for $^{12}C(n,3\alpha)n$ and partial cross sections for $^{12}C(n,^{8}Be_{gs})n$ and $^{12}C(n,^{12}C_{9.63})n$]

Introduction

Accurate nuclear data for $n+^{12}C$ interaction are necessary[1] for dosimetry, transport in shielding materials and for medical use of fast neutrons in radiotherapy. The calculation of the kerma depends on reaction cross section and mean energies of outgoing charged particle energy spectra.

The $n+^{12}C$ interaction at energies above 7.887 MeV leads to a breakup of ^{12}C into three alphas. In the energy interval of 12-20 MeV the cross section for $^{12}C(n,3\alpha)n$ reaction amounts to more than 20% of the total $n+^{12}C$ cross section[2]. The cross sections of the reaction $^{12}C(n,3\alpha)n$ have been measured up to 19 MeV[3,4]. The recommended cross sections up to 20 MeV have been obtained[2] from available direct measurements[3,4] and from the difference between the total cross section and the sum of the elastic scattering and all other reactions.

We have performed kinematically complete measurements up to 35 MeV neutron energy determining the cross sections for the reaction $^{12}C(n,3\alpha)n$ as well as for channels $^{12}C(n,^{8}Be_{gs})n\alpha$ and $^{12}C(n,^{12}C_{9.63})n$.

Experimental Method

Nuclear emulsion exposed to neutrons provide a 4π detector which kinematically overdetermines the neutron induced reaction whenever the final state does not contain more than one neutral particle. The reaction $^{12}C(n,3\alpha)n$ is a three prong star event and it can be discriminated by kinematics from three prong stars representing other neutron induced reaction on the constituents of the nuclear emulsion. In multiparticle reactions outgoing charged particles have energy spectra which extend to zero energy. Though nuclear emulsions allow a very low energy cutoff (typically considerably lower than other counter techniques), the correction due to this low energy cutoff is the leading one. It depends on the reaction mechanisms. An extensive study of the mechanism of the reaction $^{12}C(n,3\alpha)n$ up to 35 MeV demonstrates[5] that sequential decays involving ground and 2.9 MeV states of ^{8}Be and 9.63 MeV and higher excited states of ^{12}C dominate, quasifree processes are negligible, and three $(n+\alpha+^{8}Be)$ and four body statistical simultaneous decays could account as much as 50% of the reaction cross section. The correction C at a given incident energy is obtained from the corrections for specific reaction channels weighted by their corresponding contribution to the total cross section. These corrections are obtained as ratios between the calculated number of all three alpha events over the whole phase space and the number of those events for which none of the alpha particles has an energy less than 0.6 MeV, which is the low energy cutoff. Corrections C are listed in Table I. Comparison between corrections in columns two, four and five shows the dependence on reaction mechanisms.

In the neutron energy region above about 20 MeV one cannot produce monoenergetic neutrons. Thus, we have used neutrons of continuous energy (Fig.1b) produced by the reaction $^{9}Be(d,n)$ induced with the 55 MeV deuteron beam of the Louvain la Neuve cyclotron. Neutron spectrum has been measured by the time-of-flight at E_d=50 MeV in the conditions identical to those of the present experiment. The actual neutron spectrum at 55 MeV has been obtained by performing the extrapolation of the measured spectrum at 50 MeV to 55 MeV. This extrapolation is based on experimentally determined dependence of the neutron flux on the incident deuteron energy (Fig.1a), and the assumption that the reaction mechanism of the reaction $^{9}Be(d,n)$ is the same at both energies. The shape of the extrapolated spect-

K. H. Böckhoff (ed.), Nuclear Data for Science and Technology, 926–929.

rum has been compared with the measurement at 53.8 MeV[6] (Fig.1b). The overall uncertainty in the incident neutron flux determination is 10% except at energies larger than 26 MeV where it could be as large as 17% reflecting a possible error in extrapolation.

Table I
Corrections due to the loss of events with one alpha energy less than 0.6 MeV

E_0(MeV)	C	ΔC(%)	C($^8Be_{gs}$)	C($^{12}C_{9.63}$)
11	2.61	28	2.61	2.61
12	2.24	25	2.30	2.53
13	1.90	23	2.06	2.44
14	1.69	21	1.87	2.36
15	1.52	18	1.74	2.28
16	1.46	15	1.64	2.21
17	1.40	13	1.57	2.13
18	1.35	12	1.52	2.07
19	1.32	11	1.48	2.01
20	1.29	10	1.45	1.94
21	1.27	10	1.43	1.89
22	1.26	9	1.41	1.83
23	1.24	9	1.39	1.78
24	1.23	8	1.37	1.74
25	1.22	7	1.36	1.70
26	1.21	7	1.35	1.66
27	1.20	6	1.34	1.62
28	1.19	6	1.33	1.59
29	1.18	5	1.32	1.56
30	1.18	5	1.31	1.53
31	1.17	4	1.30	1.50
32	1.17	4	1.29	1.48
33	1.16	4	1.28	1.45
34	1.16	4	1.27	1.43
35	1.15	4	1.26	1.42

Three additional corrections have to be applied:

1. Since the emulsions have a finite size, some tracks do not terminate in the emulsion. The loss occurs near the surface of the emulsion and it depends on the range of alpha particle and consequently on the incident neutron energy. To calculate this loss the number of $^{12}C(n,3\alpha)n$ events has been measured as a function of the depth in the nuclear emulsion as shown on Fig.2a. Obviously, the loss of such events is larger the nearer they are to the surface of the emulsion. The lost part of events is represented by the shaded area on Fig.2a. For the events generated in the central part of the emulsion all the tracks end in the emulsion. From this region the average number of events N_{ave} per unit depth layer of nuclear emulsion was extracted. The total number of events is then: $N_{tot}=N_{ave}\times$ depth. The ratio of N_{tot} and the actually measured number of $^{12}C(n,3\alpha)n$ events in a specified interval of incoming neutron energies yields the corrections plotted in Fig.2b as a percentage by which the data have to be increased.

2. Even a long track can have a short projection to the plane of scanning if it is almost perpendicular. This loss has been estimated studying the angular distributions of two processes in which particles are isotropically distributed in space. Two such processes are: a) collinear $\alpha+t$ tracks induced by thermal neutrons on ^6Li (total length of $\alpha+t$ tracks is about 45 μm simulating an alpha energy of 8.5 MeV) and b) almost collinear two alphas emitted from $^8Be_{2.9}$ decaying at rest in the hammer tracks (two alphas average energy is 1.5 MeV and their range is 4.8 μm) from the reaction $^{11}B(n,\alpha)^8Li(\beta^-)^8Be(2\alpha)$. Fig.3a shows the angular distribution of collinear two-alpha tracks as a function of the angle θ between this two-alpha tracks and the direction coinciding with the longer edge of the horizontal emulsion plane (θ =0°).

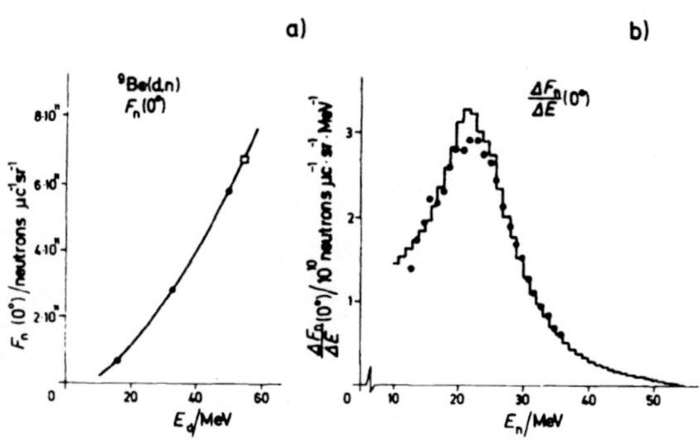

Fig.1 a) Dependence of the neutron flux on deuteron energy for the reaction $^9Be(d,n)$ at $\theta_n=0°$.

b) Neutron energy spectra at $\theta_n=0°$: extrapolated E_d=55 MeV spectrum (hystogram) and E_d=53.8 MeV spectrum from data of Ref.6 multiplied by a factor 1.12 (points).

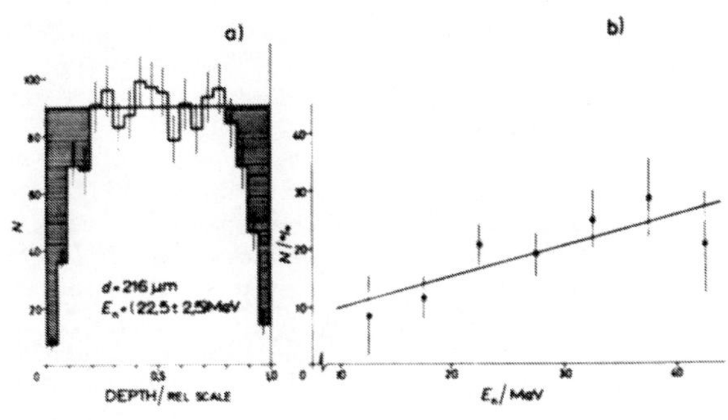

Fig.2 a) Number of three alpha events as a function of the depth in the emulsion.

b) Percentage correction as a function of the incident neutron energy.

Table II
Cross sections for the reaction $^{12}C(n,3\alpha)n$

E_o (MeV)	ref.3	ref.4	present	ref.2	ref.8
11			33±20	130	
12			76±40	195	
13	190±50		130±49	185	
14	230±50	180±100	301±89	235	
15	316±73	200± 80	312±84	294	
16		310±110	429±97	395	
17		400±130	349±76		
18		350±110	367±74	360	
19	283±59	270±110	367±70		
20			320±59	300	388
21			323±59		
22			356±62		399
23			320±57		
24			328±56		386
25			287±50		
26			336±57		366
27			291±57		
28			287±60		344
29			283±63		
30			244±60		318
31			313±56		
32			280±66		293
33			289±71		
34			310±73		274
35			256±62		263

Those tracks which are perpendicular to the plane of emulsion are emitted in a zone of a sphere, $\sin\theta d\theta$, corresponding to $\theta=90^{o}$. Since these events are missing, the measured angular distribution shows a deficiency in the region near 90° with regard to the isotropic angular distribution $\sin\theta d\theta$. The ratio between the area under the isotropic angular distribution (dashed curve in Fig.3a) and the measured angular distribution presented by a hystogram gives the correction coresponding to 1.5 MeV alphas. The solid curve on Fig.3b presents the correction as a function of neutron energy. It is obtained by the following model calculation. Assuming the equipartition of the available energy among the four emitted particles the average energy of alpha particles as a function of the incident neutron energy is obtained. The corresponding average alpha particle range is then determined via the range-energy relations for particles in nuclear emulsion. The shortest horizontal projection which can still be measured and the average alpha particle range define the cone of the sphere. Tracks emitted inside this cone are lost during the measurement. Thus, the ratio of the surface of the cone and the surface of the hemisphere yield the percentage of the lost events.

3. Since the measurements of three alphas provide 9 data and the four particle final state is a function of 8 independent kinematical variables one can determine the incident neutron energy corresponding to each event providing the direction of the beam is known. Due to experimental errors the obtained incident energies are subjects to an uncertainty. Studies at 18.2 MeV[7] and 45 MeV have shown that the incident energy spectrum determined from

Table III
Cross sections for $^{12}C(n,^{8}Be_{gs})n\alpha$ and $^{12}C(n,^{12}C_{9.63})n$ processes

E_o (MeV)	$^{12}C(n,^{8}Be_{gs})n$ present	$^{12}C(n,^{8}Be_{gs})n$ ref.3	$^{12}C(n,^{12}C_{9.63})n$ present	$^{12}C(n,^{12}C_{9.63})n$ ref.2	$^{12}C(n,^{12}C_{9.63})n$ ref.3
11	27±17		25±20	40	
12	56±24		49±23	70	
13	76±33	166	84±37	64	85
14	187±64	181	118±84	65	76
15	147±46	209	76±29	83	76
16	155±42		86±29	100	
17	146±39		65±23		
18	114±29		70±22	70	
19	87±25	104	40±15		35
20	80±20		29±11	45	
21	98±23		52±16		
22	104±22		53±16		
23	44±13		9± 6		
24	66±17		16± 8		
25	44±13		11± 6		
26	73±18		30±11		
27	45±14		26±10		
28	31±11		11± 7		
29	55±18		8± 6		
30	46±17		18±10		
31	55±19		28±10		
32	38±16		16±10		
33	38±17		6± 6		
34	50±20		14±10		
35	40±20		7± 7		

measured alpha energies is smeared and has an asymmetric shape. Using these results we construct a procedure to unfold the measured neutron spectrum. This unfolding procedure introduces an error of less than 10% in the absolute cross section.

Results

Tables II and III summarize our cross section data and compare them with the previous measurements[2-4] and with the predictions[8] of the statistical model calculations for various channels normalized so that the total reaction cross section agrees with the optical model fit to the available data. The energy spectra of alpha particles are given in Fig.4 for two incident neutron energy intervals. In these spectra every event enters through 3 alpha energies. The experimental spectra are presented by histograms and the error bars indicate typical statistical errors. The dashed curves are the calculated cutoff spectra S_{co} normalized to the experimental spectra. In both spectra the events in which at least one alpha particle has the energy below the cutoff energy of 0.6 MeV have been omitted from the analysis. The cutoff energy of 0.6 MeV has been found to be the optimum. With the higher cutoff energy the end result would be equal: the number of events in the cutoff spectrum would be smaller but the correction would be larger. However, with a higher cutoff energy too much of the experimental data are cut off yielding a larger statistical uncertainty. With lower cutoff energy a part of the experimental spectrum at low energies

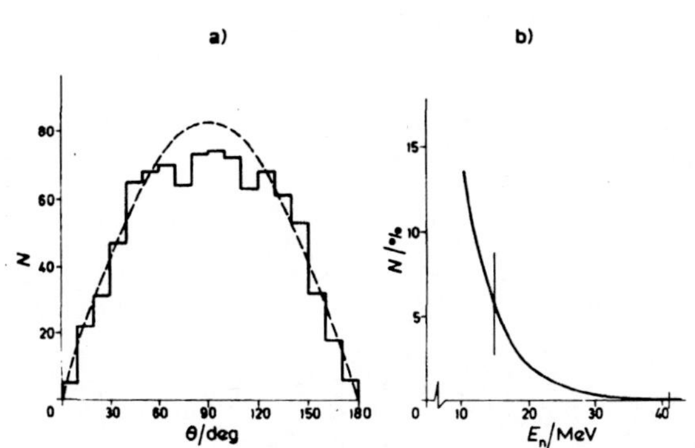

Fig.3 a) Angular distribution of collinear alpha tracks from $^{8}Be_{2.9}$ decay at rest in the reaction $^{11}B(n,\alpha)^{8}Li(\beta^{-})^{8}Be(2\alpha)$: hystogram-data, dashed curve-theoretical angular distribution.

b) Percentage correction as a function of incident neutron energy. Solid curve is a model calculation (see text). Data are from the two measured reactions (see text).

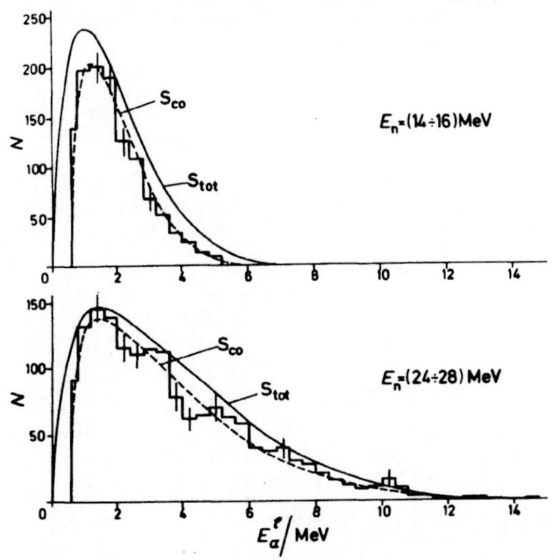

Fig.4 Energy spectra of alpha particles for two incident neutron energy intervals. Data are given as hystograms with representative statistical errors. Curves are calculated spectra: cutoff spectra-S_{co} (dashed) and total spectra-S_{tot}(solid).

---

would be left in which lost of events is already present and the fit with the calculated cutoff spectrum would not be satisfactory. Solid curves S_{tot} are the corresponding alpha particle spectra which involve three alpha correlated energies of all events emitted in the whole space.

Discussion

Our data agree quite well with the existing data in the region up to 19 MeV. The prediction of Ref.8 tends to be consistently higher. The cross sections and mean alpha energies are crucial ingredients to determine the kerma. The determination of both of these quantities depends on the understanding of the reaction mechanism and on the reliability of the extrapolation of charged particle spectra from the low energy cutoff to zero energy. It follows from Fig.4 that large values for the low energy cutoff prevent the accurate determination of the mean energy. The mean alpha energy determined from our Fig.4 for the E_n=14-16 MeV data is 1.89±0.10 MeV. It is instructive to compare this value with those obtained[9] at E_n=14.1 MeV assuming various reactions mechanisms: E_{mean}=1.64 MeV for four body simultaneous statistical decay and 2.2 MeV for $^{12}C(n,^{12}C_{12.76})n$. This is a considerable difference appreciably influencing the calculated kerma.

References

1. WRENDA 79/80, D.W. Muir, IAEA
2. J.Lachkar, F.Cocu, G.Haout, P.Le Floch, Y.Patin, J.Sigaud, NEANDC(E) 168 "L", INDC(FR) -7/L (1975)
3. G.M.Frye, L.Rosen and L.Stewart, Phys.Rev. 99, 1375 (1955)
4. S.S.Vasilev, V.V.Komarov and A.M.Popova, JETP 6 1016 (1958)
5. B.Antolković, I.Šlaus, D.Plenković, P.Macq and J.P.Meulders to be published
6. G.W.Schweimer, Nucl.Phys. A100 537 (1969)
7. B.Antolković, Fizika 8, 163 (1976)
8. P.J.Dimbylow, NRPB-R78 (1978)
9. R.S.Caswell and J.J.Coyne, preprint

DIFFERENTIAL CROSS SECTIONS OF THE $^{12}C(n,\alpha)^{9}$BE REACTION IN THE ENERGY RANGE FROM 8 TO 10 MEV EXTRACTED FROM NE 213 SCINTILLATION DETECTOR RESPONSE FUNCTIONS

G. Dietze, H.J. Brede, H. Klein and H. Schölermann
Physikalisch-Technische Bundesanstalt (PTB)
Bundesallee 100, D-3300 Braunschweig
Fed. Rep. of Germany

The differential cross section of the reaction $^{12}C(n,\alpha)^{9}Be$ has been measured in the neutron energy range from 8 to 10 MeV. An NE 213 scintillation detector has been simultaneously used as a carbon target, an α-particle detector and a neutron fluence monitor. Measured and calculated response spectra have been compared in order to separate the fraction of the spectra induced by the reaction $^{12}C(n,\alpha)^{9}Be$. The differential cross section could then be obtained by analyzing the pulse height distribution on the basis of an accurately determined light output function for α-particles in an NE 213 scintillator. The angular distributions vary significantly with the neutron energy and are consistent with data from the inverse reaction $^{9}Be(\alpha,n_0)^{12}C$.

[$^{12}C(n,\alpha)^{9}Be$, differential cross sections, E_n = 8-10 MeV, NE 213 scintillation detector]

Introduction

An NE 213* liquid scintillation detector is often used for the detection of fast neutrons, mainly because of its excellent n-γ discrimination properties, its high neutron response and its good timing properties. For neutron spectroscopy with a time-of-flight method or by the unfolding of a pulse height spectrum, the neutron response functions must be known for the neutron energy range investigated. Neutron response functions for an NE 213 scintillation detector for incident monoenergetic neutrons may be determined either by measurements or by calculations /1,2,8/ using the Monte Carlo method. The neutron response and the shape of a response function obtained from calculations depend strongly on the neutron cross sections for hydrogen and carbon, the main components of an NE 213 liquid scintillator (ρ = 0.874 g/cm^3; mole fraction H 54.8 % C 45.2 %). On the other hand, a comparison of calculated and accurately-measured response functions can give information about the cross sections of neutron reactions in the scintillator, mainly carbon cross sections in this case.

For neutron energies from 8.0 to 10 MeV, the fraction of a pulse height spectrum corresponding to the reaction $^{12}C(n,\alpha_0)^{9}Be$ can be separated. From the analysis of this part in relation to the upper part of the spectrum, which is mainly determined by recoil protons from the n-p scattering, the cross section /3/ and the differential cross section can be obtained in an absolute scale.

Experiments

By means of the ns-pulsed deuteron beam of a compact cyclotron /4/ focussed into a deuterium gas target, "monoenergetic" neutrons in the energy range from 8.0 to 10.0 MeV were produced. An NE 213 scintillation detector (5.07 cm in height, 5.06 cm in diameter) encapsulated by BA1-type* aluminum housing and coupled via a light pipe to a fast photomultiplier XP2020** was positioned at a distance of about 12 m, behind a collimator system /5/. A time-of-flight signal, an integrated dynode signal and a signal derived from the anode current for n-γ discrimination were simultaneously analyzed in a multiparameter pulse height analyzer /6/. A small time window was set to separate the neutron energy of interest.

The associated pulse height spectra were measured down to electron energies of about 12 keV equivalent to proton energies of about 100 keV. At very low pulses, difficulties arise due to an incomplete n-γ discrimination on the basis of a pulse shape analysis /7/. In order to avoid any loss of charged particle events the n-γ discrimination was only applied for electron energies above 100 keV. These correctly separated electron spectra were then extrapolated to the lower pulse height region by using the shape of the spectra independently measured for random background photons. In the region of interest, only a small γ-contribution had to be subtracted.

Response Functions

The neutron response functions were calculated by the Monte Carlo code NRESP4 /8/ which simulated the response of an organic scintillator to neutrons from 50 keV to 20 MeV.

The main features of the code are:

1. The detector geometry includes an aluminum cylinder and a lucite light pipe in addition to the NE 213 scintillator volume.
2. Neutron scattering and neutron induced reactions are taken into account for the nuclide ^{1}H, ^{12}C and ^{27}Al (s. table I). Integral and differential cross sections are taken from ENDF/B-IV and V data files /9/. Relativistic kinematics are used.
3. The relation between the output pulse height and the energy of a secondary charged particle totally stopped in the scintillator (subsequently referred to as light output function L(E)), is accurately determined for protons, α-particles and ^{12}C-ions from pulse height spectra measured for monoenergetic neutrons with NE 213 detectors of different sizes.

* Nuclear Enterprise Ltd. Edinburgh, Scotland

** Valvo GmbH, Hamburg

K. H. Böckhoff (ed.), Nuclear Data for Science and Technology, 930–933.

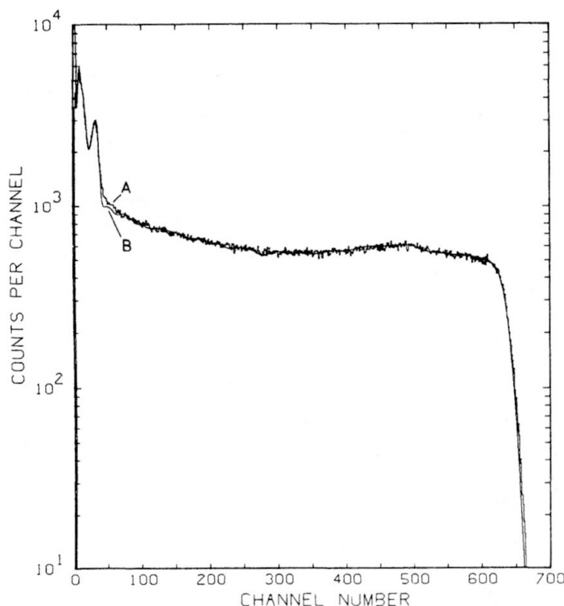

Fig. 1: Pulse height spectrum of an NE 213 scintillation detector for neutrons with E = 9.83 MeV
A measured spectrum
B calculated spectrum

Table I: Nuclear reactions considered in NRESP4

Reaction	Q-value MeV	Angular Distribution
$^{1}H(n,n)^{1}H$	-	non-isotropic
$^{12}C(n,n)^{12}C$	-	non-isotropic
$^{12}C(n,n')^{12}C^{*}$	-4.439	non-isotropic
$^{12}C(n,\alpha)^{9}Be$	-5.695	non-isotropic
$^{12}C(n,\alpha')^{9}Be^{*} \rightarrow n'+2\alpha$	-8.13	isotropic
$^{12}C(n,n')^{12}C^{*} \rightarrow \alpha + {}^{8}Be \rightarrow 2\alpha$	-7.65	isotropic
	-9.63	"
	-10.80	"
	-11.80	"
	-12.70	"
$^{12}C(n,p)^{12}B$	-12.59	isotropic
$^{12}C(n,d)^{11}B$	-13.73	isotropic
$^{27}Al(n,n)^{27}Al$	-	non-isotropic
$^{27}Al(n,n')^{27}Al$	$-E_n - 0$	isotropic
$^{27}Al(n,x)$	absorption	-

4. Wall effects are considered only for protons.

A calculated response spectrum L(E) must be folded with a Poisson or Gaussian distribution taking into account the pulse height dependent resolution function $\delta L/L$ of the particular scintillation detector /10/.

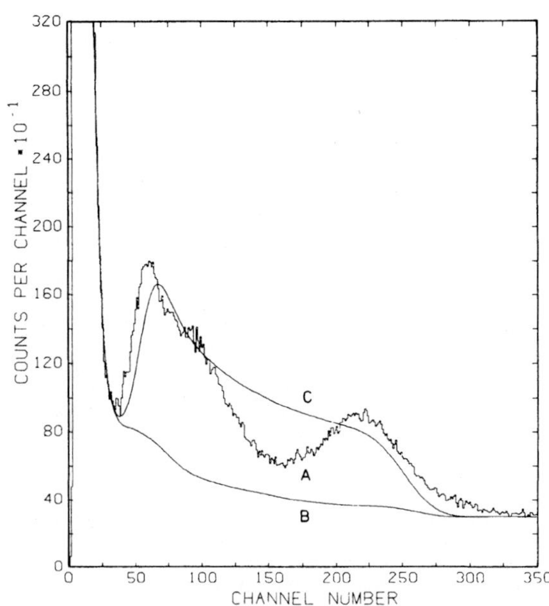

Fig. 2: Detail of a pulse height spectrum from an NE 213 scintillation detector for neutrons with E = 9.83 MeV
A measured spectrum
B calculated spectrum without $^{12}C(n,\alpha_0)^{9}Be$ induced events
C calculated spectrum with data from ENDF/B-V (isotropic angular distribution for (n,α))

Data Analysis

For monoenergetic neutrons the measured and the calculated response spectra were compared at the upper part of the spectra, which mainly corresponds to recoil protons from n-p collisions (see fig. 1). By this method, a calibration factor with respect to the neutron fluence at the detector position was obtained. This calibration was compared with another calibration taken from fluence measurements with a proton recoil telescope. For neutron energies from 8 to 13 MeV the differences were within 6 %.

The fraction of the spectrum induced by α-particles of reactions on carbon was separated. Taking into account only the contributions due to proton and carbon recoils as well as multiple scattering effects, the calculated response spectrum was fitted to the measured spectrum above channel 300 and then subtracted (see curve B in fig. 2). The residual spectrum at low pulse heights originates from α-particles produced by the reaction $^{12}C(n,\alpha_0)^{9}Be$. Integrating this spectrum the (n,α) cross section can be given in relation to the n-p cross section /3/. In addition the pulse height is correlated to the α-particle energy and to the angle of emission of the reaction $^{12}C(n,\alpha_0)^{9}Be$. For this reason the differential cross section can be extracted from the pulse height distribution by using an unfolding method and taking into account the relation between the α-particle energy and its angle of emission, the light output functions $L_\alpha(E)$ and $L_{Be}(E)$ for α-particles and the associated Be-ions in an NE 213 scintillator, and the detector resolution, which smears the pulse height distribution.

In this work a slightly different method has been used. In the center-of-mass system, the differential cross section can be described by a sum of Legendre polynomials

$$\frac{d\sigma}{d\Omega}(\Theta) = \sum_{i=0}^{n} A_i \cdot P_i(\Theta)$$

and the integral cross section for this reaction results in

$$\sigma_{n,\alpha} = 4 \cdot \pi \cdot A_0$$

The angular distribution corresponding to each single Legendre polynomial was separately transferred into a pulse height distribution of an NE 213 scintillator detector. The calculation was carried out numerically in three steps:

1. Calculation of an α-particle energy distribution in the laboratory system assuming an angular distribution $P_i(\Theta)$ in the center-of-mass system.

2. Calculation of a pulse height distribution on the basis of the experimentally-determined light output function $L_\alpha(E)$ for α-particles in an NE 213 scintillator and a linear function of $L_{Be}(E)$.

3. Folding of the pulse height spectrum by a Gaussian distribution with a channel-dependent width equal to the detector resolution function

$$(\delta L/L)_{FWHM} = (\alpha^2 + \beta^2/L + \gamma^2/L^2)^{1/2}$$

(FWHM full width half maximum). The parameters $\alpha = 1.2$ %, $\beta = 10.0$ % and $\gamma = 0.2$ % were determined separately /10/.

An example is given in fig. 3a and b. For a neutron energy of 9.83 MeV, 7 pulse height spectra had to be considered corresponding to the Legendre polynomials P_0 to P_6 (fig. 3a). A linear superposition of these spectra was fitted to the measured pulse height distribution (see fig. 3b). A least-squares method was used to determine the coefficients a_i for the superposition. The Legendre coefficients A_i of the angular distribution function in the center-of-mass system which describes the measured pulse height distribution was obtained by normalizing a_i with the factor $\sigma_{n,\alpha}/4 \cdot \pi \cdot a_0$.

The discrepancy at the upper part of the spectrum (fig. 3b) also appears at other energies and is not yet understood.

Results and Discussion

Neutron cross sections and differential cross sections have been determined for several neutron energies from 8.00 to 9.83 MeV. In table II the cross section data are compared with evaluated data from the ENDF/B-V data file.

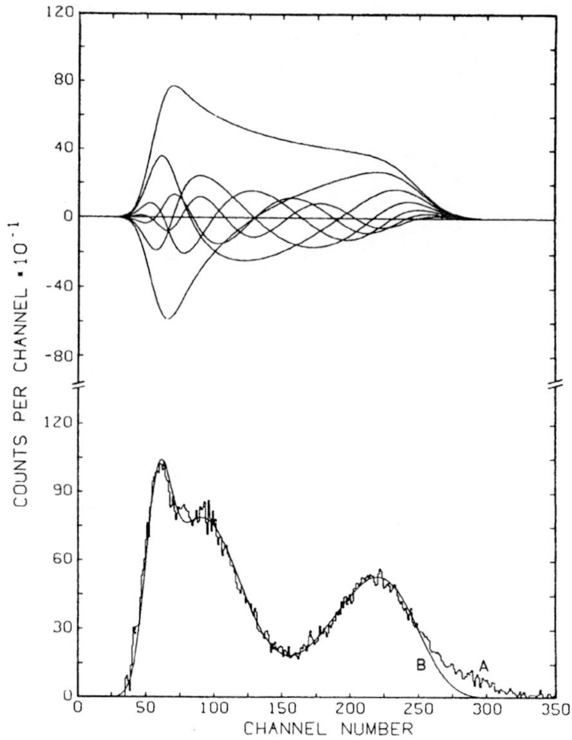

Fig. 3a: Pulse height spectra as calculated for the Legendre polynomials P_0 to P_6.
b: Pulse height spectra of α-particles from $^{12}C(n,\alpha_0)^9Be$
A obtained from measurements
B weighted sum of the spectra from single Legendre polynomials in order to fit the measured spectrum (A)

Table II: Cross sections of the reaction $^{12}C(n,\alpha_0)^9Be$

E	$\sigma_{n,\alpha}$ in mb		
MeV	ENDF/B /9/	Geiger /11/	this work
8.00	171	156	131.3
8.64	59.5	50.4	65.0
8.99	178	202	186.6
9.22	322	305	285.5
9.41	249	265	254.9
9.83	184	164	174.2

In addition data from an evalution of $^9Be(\alpha,n)^{12}C$ cross sections by Geiger and Van der Zwan /11/ have been considered by calculating the cross section $\sigma_{n,\alpha}$ on the basis of the detailed-balance theorem /12/ from

$$\sigma_{n,\alpha} = 2 \cdot \frac{M_\alpha \cdot E_\alpha}{M_n \cdot E_n} \cdot \sigma_{\alpha,n}$$

M_α, M_n, E_α, E_n are the reduced masses and kinetic energies of the incident α-particles or neutrons in the laboratory system. The same relation has been applied for the calculation of differential cross sections from the data of the inverse reaction. Differences of up to 25 % exist between the three data sets listed in table II.

Table III: Legendre expansion coefficients for the differential cross section of $^{12}C(n,\alpha_0)^9Be$ in the center of mass system

E MeV	coefficients in mb						
	A_0	A_1	A_2	A_3	A_4	A_5	A_6
8.00	10.45	7.67	3.87	3.00	2.22	-	-
8.64	4.72	3.20	-3.78	-2.99	0.85	1.05	-
8.99	14.85	-6.85	4.46	0.24	11.73	-	-
9.22	22.72	-10.80	20.42	4.61	26.85	-4.40	7.45
9.41	20.28	-10.39	18.59	4.21	23.05	-3.55	8.92
9.83	13.86	-1.83	11.63	1.70	6.08	-16.01	5.28

The coefficients of the Legendre polynomials describing the differential cross section of the $^{12}C(n,\alpha_0)^9Be$ reaction in the center-of-mass system are given in table III. For each energy, the same number of coefficients has been used as given by Geiger and Van der Zwan /11/ for the inverse reaction.

The differential cross sections for four energies calculated with these coefficients from 0 to 180 degrees are shown in figure 4. In addition, these distributions are compared with angular distributions obtained from the inverse reaction. No normalization has been applied to the data. The angular distributions vary significantly with the incident neutron energy. The agreement between the data from the inverse reaction and our data is sufficient for all energies.

The results of this work will be used to improve the response function calculations for NE 213 scintillation detectors, particularly with regard to the shape in the low pulse height region.

This work has been supported by the Commission of the European Communities (Contract No. BIO-A-284-80-D).

References

1. V.V. Verbinski, W.R. Burrus, T.A. Love, W. Zobel, N.W. Hill, R. Textor; Nucl. Instr. and Meth. 65, 8 (1969)
2. J.A. Lockwood, C. Chen, L.A. Friling D. Swartz, R.N.St. Onge; Nucl. Instr. and Meth. 138, 353 (1976)
3. G. Dietze, H.J. Brede, H. Klein, H. Schölermann; Proceedings of Fourth Symp. on Neutron Dosimetry, EUR 7448 Vol. I, 373 (1981)
4. H.J. Brede et al.; Nucl Instr. and Meth. 169, 349 (1980)
5. D. Schlegel-Bickmann, G. Dietze, H. Schölermann; Nucl. Instr. and Meth. 169, 517 (1980)
6. H. Klein, G. Dietze, B.R.L. Siebert, H.J. Barrenscheen; Nucl. Instr. and Meth. 169, 359 (1980)
7. P. Sperr, H. Spieler, M.R. Maier, D. Evers; Nucl. Instr. and Meth. 116, 55 (1974)
8. G. Dietze, H. Klein; PTB-Bericht ND-22 (1982)
9. NNCSC, Evaluated Nuclear Data File ENDF/B-V, BNL, Upton, New York (1979)
10. G. Dietze, H. Klein; Nucl. Instr. and Meth. 193, 549 (1982)
11. K.W. Geiger, L. Van der Zwan; NRCC 15303, Ottawa (1976)
12. P. Marmier, E. Sheldon; Physics of Nuclei and Particles, Vol. I, Academic Press, New York (1971)

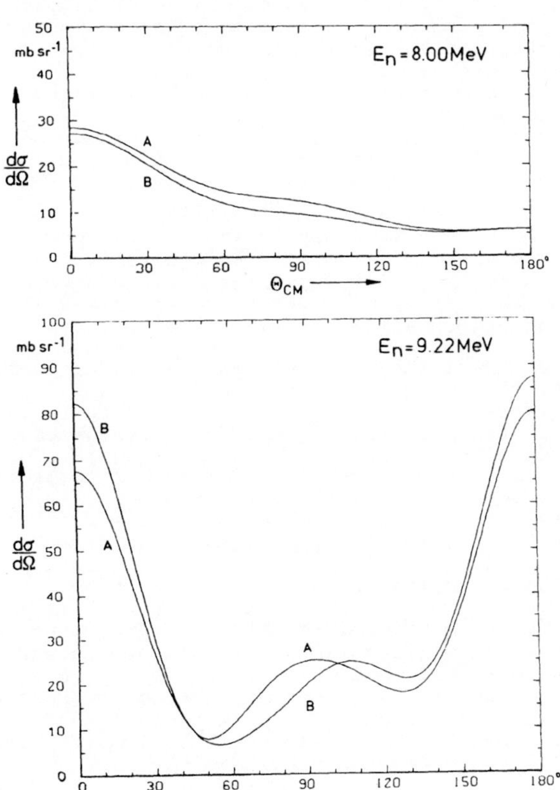

Fig. 4a - d:
Differential cross section of $^{12}C(n,\alpha_0)^9Be$ for various neutron energies

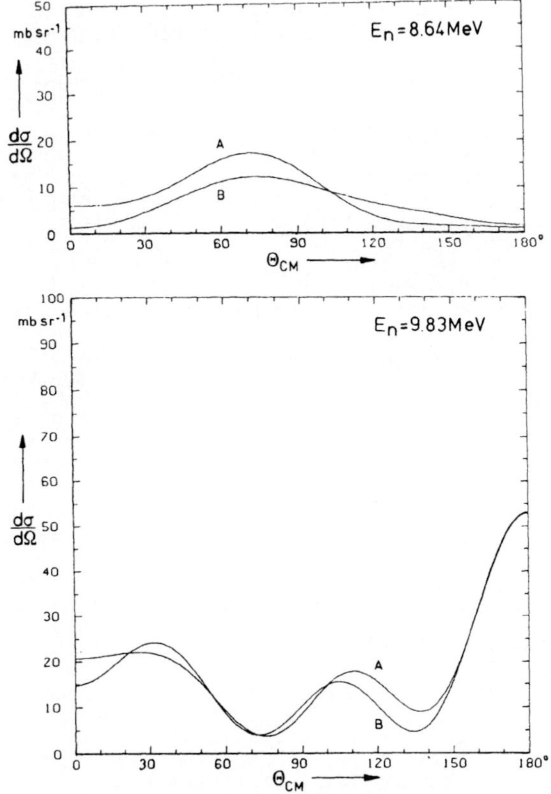

A obtained from this work
B calculated from of the inverse reaction $^9Be(\alpha,n_0)^{12}C$ with data evaluated by Geiger and Van der Zwan /11/

USE OF NUCLEAR DATA AND TRANSPORT CALCULATIONS FOR EVALUATION OF BIOEFFECTS IN MEDIA IRRADIATED WITH FAST NEUTRONS

K. Morstin* and B. Kawecka

Institute of Physics and Nuclear Techniques
Adademia Gorniczo-Hutnicza
al.Mickiewicza 30, 30-059 Krakow, Poland

*at present: Institut für Reaktorentwicklung, KFA Jülich GmbH
D-5170 Jülich, W.Germany

Numerical treatment of the transport equation enables the space-energy distributions of neutrons and neutron-induced gamma radiation to be computed for media of arbitrary geometry and composition. The S_N calculations of radiation fields and resulting energy deposition are presented here for a radiotherapeutical phantom irradiated with fast neutrons from different sources. Nuclear data up to 20 MeV were applied, results being compared with dosimetric measurements.

A simple algorithmic description of cell mutations and lethality was applied. It enables some biological effects of a neutron-induced mixed radiation field to be predicted. Thus calculated bioeffect vs. depth relations were obtained for several cell communities positioned in the phantom exposed to different neutron beams. They are compared with similar results obtained with the use of the Katz radiobiological model.

The data needs for extension of such calculations towards higher neutron energies are also discussed.

Introduction

Since fast neutrons are more and more used for biomedical purposes (e.g. cancer therapy, neutron radiography and computer-assisted tomography[1], bioengineering, etc.) the need for methods of evaluation of resulting bioeffects in bulk media seems to be obvious. Medical treatment should be performed with respect to internal dose distributions as well as to expected cellular radiation effects. But one can hardly imagine performance of radiobiological experiments in vivo.

Advanced simulation techniques are sometimes used[2,3] for analysing the secondary charged-particle radiation fields in continuous media irradiated with neutrons, thus enabling the local variations in so-called microdosimetric spectra to be determined, provided the related stopping-power data are known with sufficient accuracy. In principle, the biological response of the investigated system could be then predicted with the use of some sophisticated radiobiological models[4-8], but the huge amount of biochemical, molecular, and biological data is still required. These data, which characterize each individual bioeffect, vary from one secondary to another depending on their energy yet, and - for most cases - are simply not evaluated till now.

So it seems that at present the only realistic way of calculating the space distributions of bioeffects in bulk media exposed to neutrons is to make use of some final effect vs. dose dependences already accumulated in many laboratories[9-13] from in-air irradiations of separated biological material with neutrons of different energies. The amount of physical data required for such a semi-empirical approach is reduced to multigroup cross-sections and kerma factors such as applied for standard radiation-shielding calculations.

Depth dose distributions

First, the space-energy distributions of neutrons and neutron-induced gamma radiation have to be determined, thus enabling the absorbed dose vs. depth to be estimated (as good as kerma approaches the absorbed dose). The standard radiation-shielding transport codes solving the Boltzmann equation in the multigroup approximation can be applied for this purpose. The widely distributed and easy-to-operate one-dimensional programs, such as ANISN or DTF-IV, are usually sufficient, unless the dose distributions outside the incidence field of a collimated neutron beam are of primary interest.

All ANISN calculations for this work were performed in S_8-P_3 approximation for a cubic (30 cm x 30 cm x 30 cm) radiotherapeutical phantom exposed to collimated neutron beams. The transverse buckling corrections were introduced to allow for finite lateral dimensions. The coupled DLC-23 (based on ENDF/B-III) and DLC-31 (based mainly on DNA evaluations) multigroup cross-section sets were applied, some of the results being compared in Table I. The standard-tissue group kerma responses were elaborated from the well-known fluence-to-kerma conversion evaluations for neutrons[14] and gamma rays[15]. The resulting dose rate distributions for d+Be neutron beam from the Krakow cyclotron (E_d=12.5 MeV) are compared in Fig. 1 with indications of TE ionization-chamber dosimeters[16]. All calculations were made for an incident neutron flux of $4 \cdot 10^7$ $n \cdot cm^{-2}s^{-1}$ uniformly distributed on a phantom surface, while dosimetric measurements were in fact performed for varying incidence fields, the results being then normalized to the calculated curve at 1 cm. Slow-neutron (below 0.1 MeV) and gamma-ray contributions to the total dose are also shown in Fig. 1.

K. H. Böckhoff (ed.), Nuclear Data for Science and Technology, 934–937.

Table I. Results of depth dose calculations for the phantom exposed to the Krakow beam ($\bar{E}_n \sim 6$ MeV, $\phi_o \sim 4 \cdot 10^7$ n/cm^2/s)

	DLC-23	DLC-31
inlet dose rate	10.9 rad/min	10.9 rad/min
max.dose rate(at2.5mm)	12.7 rad/min	12.8 rad/min
dose rate at 5 cm	8.9 rad/min	9.0 rad/min
γ-contribution at 5 cm	5.2 %	5.4 %
half-dose distance (ref.max.dose)	8.2 cm	8.2 cm

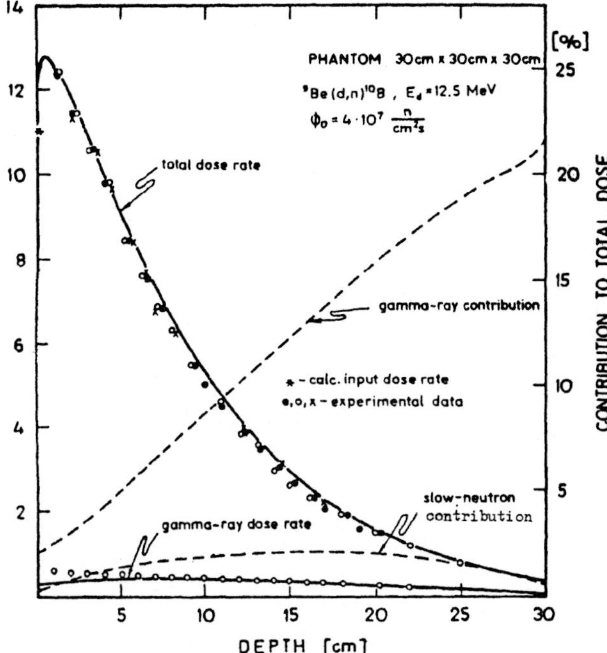

Fig. 1. Absorbed dose distributions in the phantom irradiated with cyclotron neutrons (experimental data normalized at 1 cm).

When introducing the dose-averaged LET values for given neutron energies and material composition the space distributions of the effective LET can be estimated by simple averaging over previously calculated spectra. Results of such a procedure for different neutron beams are presented in Fig. 2. It reflects the changes in radiation quality along the beam centerline in the investigated phantom. The most favourable results occur for fission neutrons close to the front wall. With hardening the incoming neutron spectrum the in-phantom radiosensitivity decreases but significantly flattens. As it derives from Table I and Fig. 2, as long as hydrogeneous moderators are considered, the energy deposition calculations are not very sensitive to the cross-section libraries applied.

Cellular radiation effects

Taking into account the reported radiobiological experiments[10-12] in which separated cells and/or tissues were exposed to monoenergetic neutrons and reference x-radiation, the following may be assumed:

- the frequency of somatic cell mutations and/or lethal events caused by neutrons is proportional to the incident neutron dose, the proportionality factor being dependent on neutron energy as well as on the biological end-point under consideration;
- the dependence of mutation frequency on x-ray dose can be described, except for very low and very high doses, by the power relationship, the exponent being dependent on the dose rate rather than on biological cirumstances.

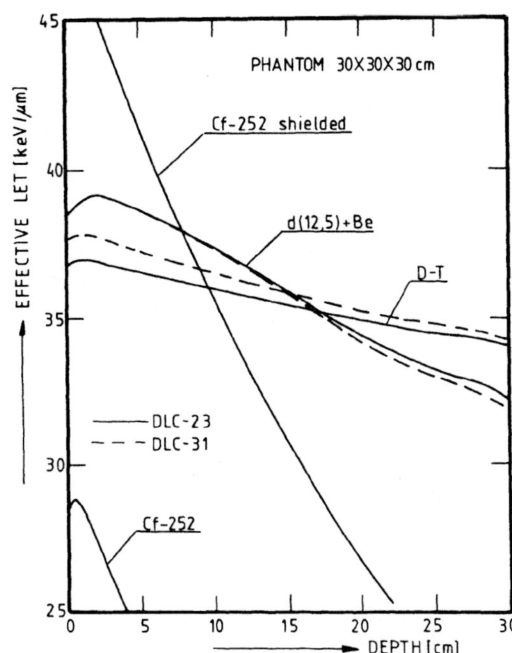

Fig.2. Effective LET vs. depth calculated for different neutron beams

Hence, the surviving (non-mutant) fraction of the cell community positioned in the mixed neutron-gamma radiation field can be estimated as

$$S = \exp\left[-\frac{D(1-\gamma)}{D_o} - \left(\frac{D\cdot\gamma}{D_G}\right)^k\right],$$

where γ stands for relative gamma-ray contribution to the total dose D. The constants D_o and D_G can be determined by fitting in-air data to the above-mentioned dose-effect relations and then averaged over in-body spectra previously calculated. In the present calculations the exponent k was taken as

$$k = a \ln \dot{D} + b, \quad k \geq 1,$$

with the parameters a and b obtained by fitting to some experimental dependences[12] on dose rate $\dot{D}$. This approach was first proposed in Ref.17 and does not include any "philosophy" - it is nothing but an algorithmic description of experimental observations. Since it is also applicable for reference x-radiation, the RBE values can be then easily determined for any mixed radiation field. In Fig. 3 thus calculated RBE-dose relations for pink mutations of Tradescantia stamen hairs are compared with experimental data from ^{252}Cf irradiations[13].

Another approach that does not require the detailed knowledge of secondary distributions is that proposed by Katz et.al.[18]. It makes use of a concept of an "equivalent" secondary and results in a simple formula

$$S = \exp\left[-\sigma_o\frac{(1-\gamma)P}{L}D\right]\{1-\left[1-\exp\left(-\frac{1-P+P\gamma}{E_o}D\right)\right]\},$$

where σ_o, E_o and m are pure cellular parameters, while P and L depend also on neutron energy but can be easily averaged over the spectrum considered. However, as it was pointed out by Günther[19], the way in which the authors evaluate the parameters P and L is wrong when applied to neutrons, especially of energies around 1 MeV, and leads to an overestimated RBE. Indeed, when applying Katz's original parameters for some cultured mammalian cells we obtained rather poor agreement with experiment.

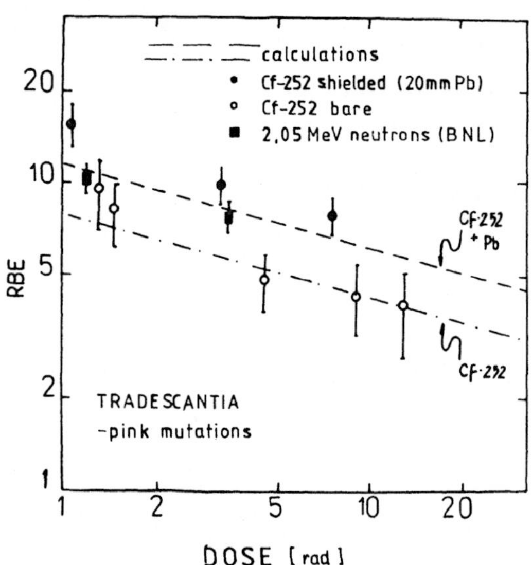

Fig. 3. RBE-dose relations measured and calculated for pink mutations of Tradescantia stamen hairs due to ^{252}Cf in-air irradiation

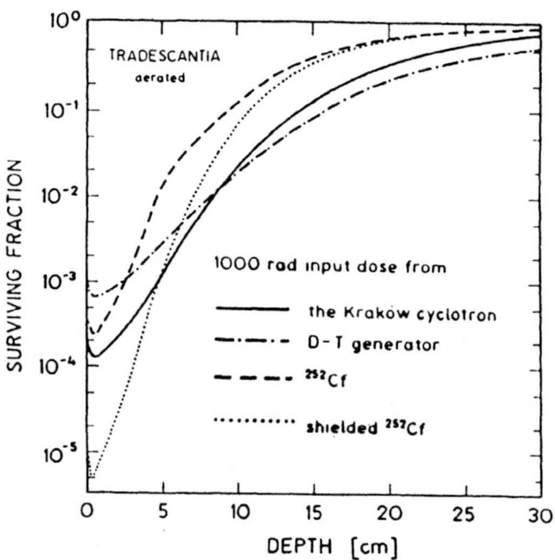

Fig. 4. Dependence of surviving fraction on depth for aerated Tradescantia stamen hairs, following a 1000 rad input dose from various sources.

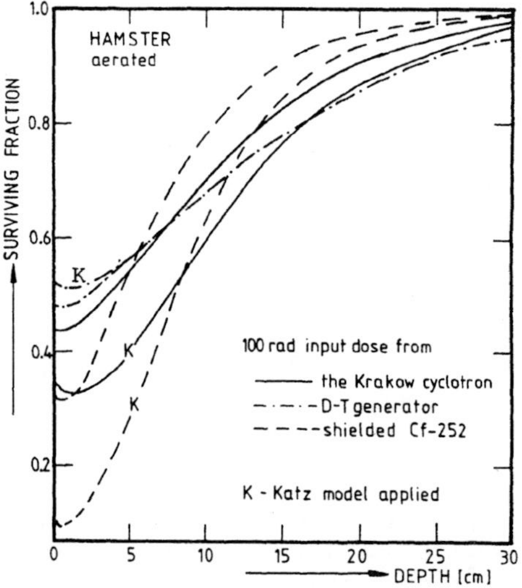

Fig. 5. Surviving fraction vs. depth for V-79 Chinese hamster cells in the phantom irradiated with 100 rad input dose from various sources.

Spatial bioeffect distributions

Once having the depth dose calculations and the algorithmic description of cellular radiation effects experimentally verified, the space distributions of bioeffects in the irradiated medium can be reliably predicted. An extensive analysis was carried out for several cell cultures positioned along the beam axis in the radiotherapeutical phantom exposed to different neutron beams, some of the results being presented here. Both aerobic and hypoxic conditions were considered, thus enabling the local variations in the oxygen enhancement ratio (OER) to be estimated. All calculations were performed with both DLC-23 and DLC-31 multigroup data sets but no significant differences were distinguished. Most of the presented results were obtained with the use of above mentioned phenomenological description of cellular radiation effects.

Fig. 4 presents the depth increment of aerobic Tradescantia survivals in the phantom exposed to 1000 rad inlet dose (free-in-air kerma) from different neutron sources. Similar results for V-79 Chinese hamster cells (this time for 100 rad inlet dose) are compared in Fig. 5 with those obtained with the use of the Katz model. As it has been already noticed, Katz's approach seems to overestimate cellular radiation effects, especially for fission neutrons. Regardless of displayed discrepancies, it is evident that the optimum choice of the therapy beam strongly depends on localization of the point of interest (e.g. tumour) in the irradiated body.

In Fig. 6 and 7 the depth distributions of RBE and OER are given. Again the fission neutron source exhibits the most advantageous features, provided its primary gamma radiation is shielded off and only shallow penetration is considered.

Discussion

It seems that the numerical approach as presented here provides a useful tool for planning neutron treatment in biomedical applications. It is based on well verified radiation transport calculations and makes direct use of some final results of radiobiological experiments in vitro. So its reliability depends mainly on uncertainties of nuclear data applied as well as on adequacy of information achieved from laboratory irradiations of cultured cells for cancer therapy problems. Although main attention was devoted here to external neutron beams, gamma and x-ray beams, as well as internal sources, can be considered in the same manner. All the calculations presented referred to the one-field exposure. In the case of cross irradiations, which are mostly applied in practice, the problem should be obviously treated at least in two dimensions.

Due to the present trends in neutron therapy extension of such calculations towards higher neutron energies would be reasonable. At present the main limitation is the lack of data. Even though there exist some multigroup cross-section libraries exceeding 20 MeV[20,21] based on nuclear model calculations, they are neither properly verified nor widely distributed. Furthermore, there is still no applicable radiobiological information in this neutron energy range. It is not only demagogy to say that clinicians and patients wait for further cross-section evaluations, such as recently performed by Dimbylow[22], and for results from radiobiological experiments in the larger electrostatic machines.

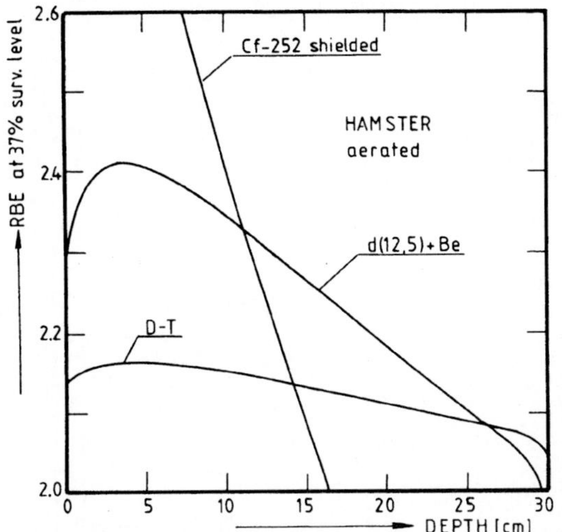

Fig.6. Dependence of RBE on depth, calculated at the 37% survival level for V-79 Chinese hamster cells in aerobic conditions for various neutron sources.

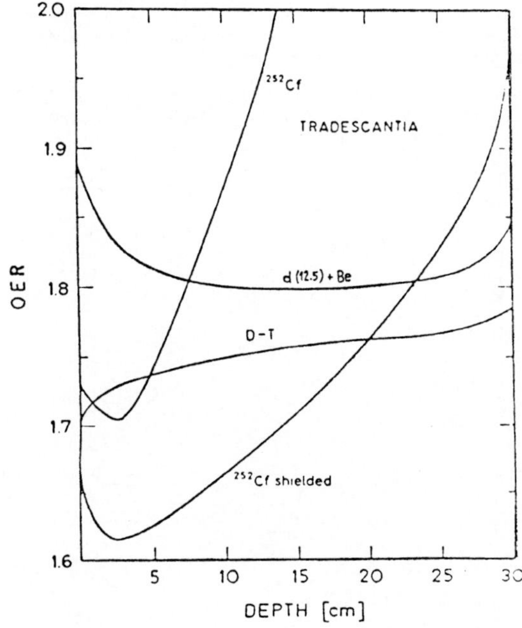

Fig. 7 Dependence of OER on depth, calculated for Tradescantia stamen hairs for various neutron sources.

References

1. D.F. Jackson, Nuclear Physics A354, 237c (1981)
2. R.G. Alsmiller et.al., Radiat. Res. 60, 369 (1974)
3. G. Pfister et.al., in Proc. 4th Symp. on Neutron Dosimetry (Neuherberg, June 1981) EUR7448EN, Vol.II, p.91, CEC, Luxemburg 1981
4. A.M. Kellerer and H.H.Rossi, Current Topics in Radiat. Res. Q. 8, 85 (1972) and Radiat. Res. 75, 471 (1978)
5. R. Katz et.al., Radiat. Res. 47, 402 (1971) and in Topics in Radiation Dosimetry, Suppl. 1, p. 317, Academic Press, New York 1972
6. H.P. Leenhouts and K.H. Chadwick, Adv. Radiat.Biol.7, 58 (1978)
7. C.A. Tobias et.al., in Radiation Biology in Cancer Research, Raven Press, New York 1980
8. W. Pohlit, in Proc. 4th Symp. on Neutron Dosimetry (Neuherberg, June 1981), EUR7448EN, Vol.I, p. 153, CEC, Luxemburg 1981
9. G.W. Barendsen and J.J. Broerse, in Biophysical Aspects of Radiation Quality, p. 55, IAEA, Vienna 1968
10. A.M. Kellerer and E.J. Hall, in Ann. Rep. on Res.Proj., p. 278, COO-3243-2, Columbia University, New York1973
11. E.J. Hall et.al., Radiat. Res. 64, 245 (1975) and Radiat.Res.65, 172 (1976)
12. A.G. Underbrink and A.H. Sparrow,in Proc. Symp. Biological Effects of Neutron Irradiation, IAEA-SM-179/31, p.185, IAEA, Vienna 1974
13. K. Morstin et.al., to be presented at 50th Anniv. Discovery of Neutron (Cambridge, Sept. 1982)
14. R.S. Caswell et.al., Radiat. Res. 83, 217 (1980)
15. B.J. Henderson, General Electric Co. Rep. XDC-59-8-179 (1959)
16. J. Skolyszewski et.al., Nowotwory 29, 169 (1979)
17. K. Morstin and B. Kawecka, in Proc. 4th Symp. on Neutron Dosimetry (Neuherberg, June 1981), EUR7448EN, Vol.II, p.115, CEC, Luxemburg 1981
18. R. Katz et.al., Health Physics 30, 148 (1976)
19. K. Günther, Int.J. Radiat.Biol. 30, 495 (1976)
20. R.G. Alsmiller and J. Barish, Nucl. Sci. Eng. 69, 378 (1979) and Nucl. Sci.Eng. 80, 448 (1982)
21. G. Pfister et.al., Int. Symp. on Biomedical Dosimetry (Paris 1980)
22. P.J. Dimbylow, Phys. Med. Biol. 25, 637 (1980) and in Proc. 4th Symp. on Neutron Dosimetry (Neuherberg, June 1981), EUR7448EN, Vol.I, p.341, CEC, Luxemburg 1981

COMPARISON OF THE NEUTRON FIELDS PRODUCED BY 50 MEV DEUTERONS AND 65 MEV PROTONS ON ^{9}Be AT THE LOUVAIN-LA-NEUVE CYCLOTRON

J.P. Meulders,
Institut de Physique, Université Catholique de Louvain
Chemin du Cyclotron, 2 - 1348 Louvain-la-Neuve

and

S. Vynckier, P. Pihet and A. Wambersie,
Neutronthérapie Expérimentale,
Université Catholique de Louvain
Avenue Hippocrate, 54 - 1200 Bruxelles

Intense neutron beams are produced for therapy purposes at the cyclotron of the University of Louvain. The paper deals with the improvement of the target system which allows the use of a 65 MeV proton beam instead of the previously used 50 MeV deuteron beam on Be target. The optimum composition of the target assembly is described. The performance of the new neutron beam is compared with the previous one : dose rate, γ contribution, depth doses and induced radioactivity.

Introduction

Until recently, most of the neutron beams used for radiotherapy have been produced by accelerating deuterons on a thick Be target, which appeared the most efficient way to obtain intense neutron fluxes with the existing cyclotrons. The low energy cyclotrons, accelerating deuterons of 16 MeV or less like in London (Hammersmith Hospital), Edinburgh or Essen, are mostly hospital based machines but the poor penetration of the neutron beam in tissue does not allow to treat deep lying tumours at those energies. On the contrary, the higher energy cyclotrons accelerating deuterons of more than 35 MeV, like at Texas A & M or at Louvain-la-Neuve, produce sufficiently penetrating neutron beams but are located in research laboratories located far from hospitals.
Most of the existing neutrontherapy facilities have other limitations like a fixed beam direction or a fixed collimator with a limited number of field sizes.

A new generation of hospital based cyclotrons are being installed or are planned ; they will solve, if not all, at least some of the previously mentioned problems ; they will accelerate proton beams at a fixed energy : around 45 MeV for the three american cyclotrons, 50 MeV at Nice (France) and 60 MeV at Liverpool (G.B.).

At the Louvain-la-Neuve cyclotron, neutrontherapy has been routinely performed since March 1978 by accelerating 50 MeV deuterons on a 10 mm thick Be target and 327 patients have been treated at this energy. During 1981 the production of neutrons with a 65 MeV proton beam on Be has been considered for the following reasons : the penetration in tissue could be improved in a significant way - the possibility of a "time sharing" use of the beam could be considered between therapy and isotope production, like ^{123}I - the energy is comparable to the high energy neutrontherapy facility at Fermilab (Batavia, Ill.).

A movable target system has been installed and is used in routine for radiotherapy since March 1982.

The new target configuration

The neutrontherapy beam at Louvain-la-Neuve has a fixed vertical direction, the treatment room and the related medical facilities being located one level below the main experimental hall at the cyclotron vault[1]. The difficult access to the target - which implies the dismantling of the major part of the collimator - and the request not to interrupt the clinical treatments limited the possibility of experiments on the target assembly.

The energy spectra of the neutrons produced by p on a thick Be target are very different of those produced by deuterons.

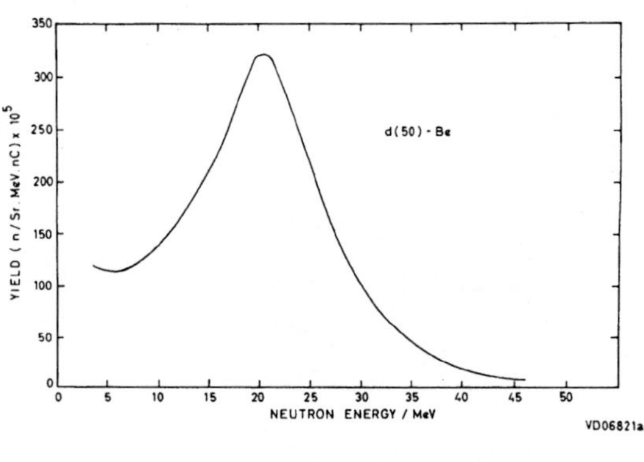

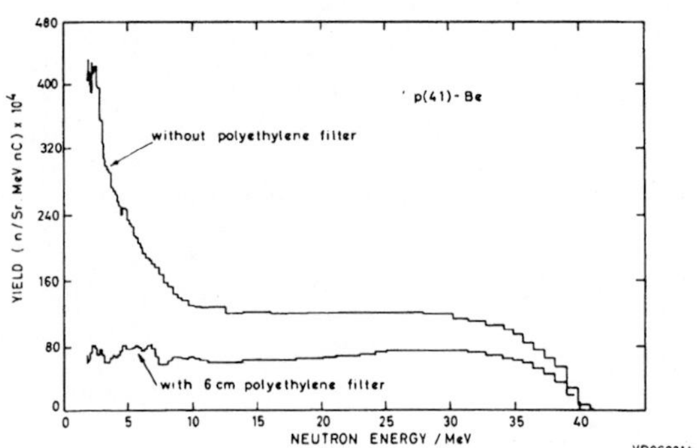

Fig. 1. Energy spectra of the neutrons produced by 50 MeV deuterons[3] and 41 MeV protons[2] on thick Be targets.

K. H. Böckhoff (ed.), Nuclear Data for Science and Technology, 938–940.

The Fig. 1 compares the spectra obtained with 41 MeV p by the Houston group[2] with that obtained previous at Louvain with 50 MeV d[3] with the time of flight method. In the case of p on Be, an important low energy neutron component shows up ; removal of this component should harden the energy spectrum and improve the penetration in tissue.
This can be obtained in two ways :

- by filtering the beam with a hydrogenous material placed between the target and the patient.
- by avoiding the production of those lower energy neutrons with a thinner Be target and an appropriate backstop material.

1. The effect of filtering the neutron beam with polythene, already observed by accelerating p of different energies on a thick Be target at our cyclotron[4], has been measured in our actual treatment room by accelerating 65 MeV p on the 10 mm thick Be target (ΔE_p = 17.5 MeV) inserted in a cooled brass support, 7 mm thick. Table I shows some of the results obtained with a tissue equivalent (T.E.) ionization chamber flushed with T.E. gas. The increase in depth of the 50 % isodose correlates to the hardening of the energy spectrum but at the expense of the dose rate. As a compromise, a 2 cm thick polythene filter has been chosen.

Table I. Depth of the 50 % isodose and dose rate for p(65)-Be neutron beams (16 cm x 20 cm field) as a function of the polythene filter thickness. The dose rate without filter is taken as reference.

filter thickness/cm	50 % isodose depth/cm H_2O	relative dose rate
0 (no filter)	18.1	1
2 cm	18.8	0.87
5 cm	19.2	0.74

2. The choice of target thickness and backstop material. The dose rate obtained with 65 MeV p on the 10 mm thick Be target with brass backing is 6.6 lower than the dose rate obtained with 50 MeV d on the same target (Table II). It seemed thus advisable to try a thicker target in order to gain with this respect. The actual Be target is 17.3 mm thick and ΔE_p is 34.8 MeV. In order to stop the remaining 30.2 MeV protons, Carbon is an interesting backstop material, due to the large negative Q value of the (p,n) reaction : - 18.1 MeV.

Table II. Neutron dose rate from d(50)-Be and p(65)-Be.
10 x 10 cm² field - SSD = 162.5 cm.

Reaction and target	DOSE RATE at maximum	DOSE RATE at maximum (Intense beams)	DOSE RATE at 10 cm depth
d(50) → 10 mm Be + 7 mm brass	9.3 rad/min per µA	47 rad/min for 5 µA	30 rad/min for 5 µA
p(65) → 10 mm Be + 7 mm brass + 2 cm filter	1.40 rad/min per µA	21 rad/min for 15 µA	15 rad/min for 15 µA
p(65) → 17.3 mm Be + 8.5 mm C + 4 mm brass + 2 cm filter	1.67 rad/min per µA	25 rad/min for 15 µA	18 rad/min for 15 µA

Measurements performed in collaboration with D. Bewley[5] have shown that a 30 MeV p beam stopped in different materials produce a neutron dose, measured with a N_p fission counter, 20 times lower for C than for Be and 5 times lower for C than for copper.
The final target consists of 17.3 mm Be with 8.5 mm Carbon to stop the proton beam ; the whole target is inserted in brass (4 mm thick) and surrounded by a water cooling system. The additional 2 cm filter is located between the target and the collimator.
Figure 2 shows an exploded view of the new target system. The upper target is the Be-C target used with the 65 MeV p beam, the lower target is the 10 mm thick Be target which can still be used for the 50 MeV d beam. The hole in the middle of the block is designed for future proton therapy. The whole system fits in an aluminium box which closes the beam line. An 100 µm havar foils holds the system under vacuum.
Finally the Be-C target is automatically removed from the central axis of the beam at the end of each patient irradiation in order to reduce the residual activity.

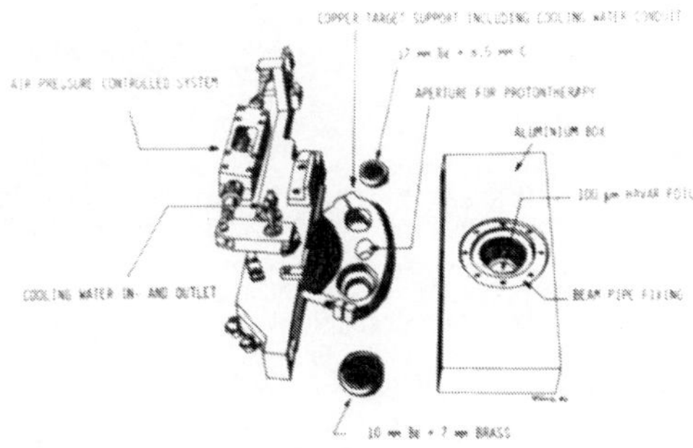

Fig. 2. Exploded view of the removable target system.

Results

Dose rate

By increasing the Be target thickness, a gain of 20 % in dose rate has been measured (Table II). The cyclotron accelerates routinely 15 µA of 65 MeV protons. This intensity combined with the improved depth dose, allows to work at a 10 cm depth with a dose rate of 60 % of the previous d(50)-Be neutron beam, still sufficient for patient irradiation, usually 65 rad/session. Evaluation of the neutron flux can be obtained by refering to the 50 MeV d on Be flux : 5.8 10^{11}n/µC. sr.[2]

Depth dose

The figure 3 compares the depth doses in water for a standard 10 cm x 10 cm field. The 50 % isodose increases from 13.6 cm to 16.6 cm by bombarding the 10 mm thick Be target respectively with 50 MeV d and 65 MeV p. With the new Be-C target, the depth is even increased with 1 cm which gives an overall improvement of 4 cm for the 50 % isodose.
By combining the depth dose with the dose in the build-up region, a penetration comparable to a 6 MV electron accelerator is obtained.

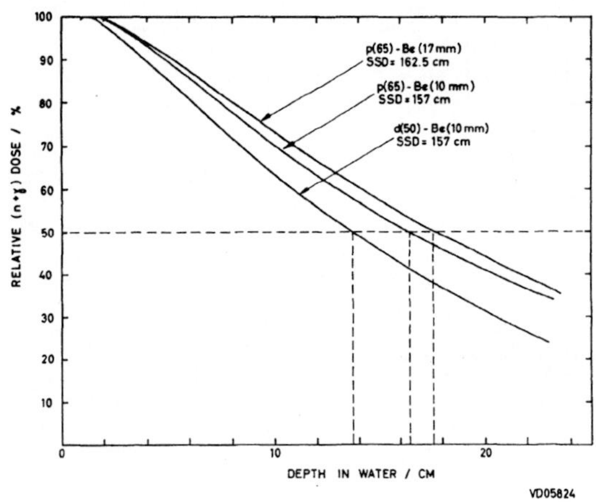

Fig. 3. Depth dose curves in water for different neutron beams in a 10 cm x 10 cm field.

Table III. Residual activation after 4 hours of therapy treatment

	ACTIVATION AFTER SEVERAL TREATMENTS	
reaction and target	1 m from collimator at 1.5 m height	in beam axis at patient position
d(50) → 10 mm Be + 7 mm brass	13 mRem/h	± 200 mRem/h
p(65) → 10 mm Be + 7 mm brass + 2 cm filter	20 mRem/h	± 2500 mRem/h
p(65) → 17.3 mm Be + 8.5 mm C + 4 mm bras + 2 cm filter	target removed 10 mRem/h	target in axis ± 800 mRem/h target removed ± 15 mRem/h

γ Contribution

There is a definite increase of the γ contribution by accelerating p instead of d on the different Be targets. The γ contribution to the total dose has been measured in a polystyrene phantom at different depths and for different field sizes by means of the 2 detector method (GM and TE chamber). For a small field (6 cm x 8 cm), the 0.8 % γ contribution at 50 MeV d increases to 6.4 % with 65 MeV on the same Be target and reduces to 2.8 % with the Be-C target. This higher γ contribution is thus much dependent of the target composition and has to be considered at the neutrontherapy facilities using the Be(p,n) reaction.

Ambient activity

One of the main reasons to change the target to a movable target system has been the activation measured with the proton beams on Be. The table III give some figures observed at the end of a therapy session for the different neutron beams at 2 positions, one is representative of the medical staff, the other is at the irradiation area of the patient. With the 65 MeV protons on the Be-brass target, unacceptable high activation rates have been measured, up to 2500 mRem/h in the beam axis. Replacing brass by Carbon lowers the dose in the axis to 800 mRem/h. Definite improvement is obtained by removing the target away from the central axis of the collimator. The final situation is better than with the previous 50 MeV deuteron beam. The solution of a movable target should be seriously considered at the new hospital based proton cyclotrons, particularly if an isocentric head is provided. On the other hand, this target assembly allows to use directly the proton beam in view of protontherapy applications.

References

1. A. Wambersie, J.P. Meulders, Proc. of 3d Symposium on Neutron Dosimetry in Biology and Medicine - München (1977) 315
2. J.P. Meulders, P. Leleux, P.C. Macq, C. Pirart, Phys. in Med. and Biol. 20 (1975) 235
3. R.S. Graves, J.B. Smothers, P.R. Almond, W. Grant, V.A. Otte, Med. Phys. 6 (1979) 123
4. D.K. Bewley, J.P. Meulders, M. Octave-Prignot, B.C. Page, Phys. Med. Biol. 25 (1980) 887
5. B.D. Bewley, B.C. Page, J.P. Meulders, Cambridge Meeting, Sept. 1982, The neutron and its applications.

Nuclear Data for the Understanding of Stellar Nucleo-Synthesis

^{187}Os + n RESONANCE PARAMETERS IN THE INTERVAL 27-500 eV NEUTRON ENERGIES*

R. R. Winters

Denison University
Granville, Ohio 43023 U.S.A.

R. F. Carlton

Middle Tennessee State University
Murfreesboro, Tennessee 37130 U.S.A.

J. A. Harvey and N. W. Hill

Oak Ridge National Laboratory
Oak Ridge, Tennessee 37830 U.S.A.

The neutron total cross section for ^{187}Os, in the energy range, 27 eV to 500 eV, has been measured at the ORELA facility by the neutron time-of-flight technique, utilizing a 2.0 gm Osmium sample ($n = 0.008401 \frac{\text{Os-nuclei}}{\text{barn}}$) enriched to 70.38% ^{187}Os. Measurements were performed at a 80 m flight station with an energy resolution, $\Delta E/E$, of 0.1% using a ^{6}Li glass scintillator. Resolved resonances have been analyzed by a Reich-Moore multilevel code (SAMMY) to obtain parameters for 85 resonances up to 500 eV. Preliminary determinations of the level spacing (5 eV) and s-wave strength function (3.9×10^{-4}) for ^{187}Os are in agreement with recent analyses of the Osmium isotopes, made in connection with the use of the Re/Os chronometer for estimating the duration of stellar nucleosynthesis.

[^{187}Os, total cross sections, E_n = 0-500 eV, nucleosynthesis, Re/Os - cosmochronometer]

Introduction

The s-wave neutron strength functions and average level spacings for ^{187}Os + n are important input parameters for the use of the ^{187}Re/^{187}Os beta decay as a cosmochronometer. For example, values for the s-wave strength function (S_0) were assumed in the statistical model analysis of the ^{187}Os average capture cross sections performed by Winters and Macklin[1] on their measurements of the total neutron capture cross sections $\sigma_\gamma(187)$. Their analysis yielded estimates for the p-wave strength function (S_1), gamma ray strength function (S_γ) and for the inelastic scattering cross section near 30 keV $\langle\sigma(n,n')\rangle$. The value for $\langle\sigma(n,n')\rangle$ is a critically important prameter in estimating the amount of ^{187}Os made by s-process stellar nucleosynthesis from ^{186}Os. In fact a variation in this parameter from ~0.5 to ~1 bn, results in estimates based on Re/Os for the age of the universe ranging from ~13 to ~17 billion years. The latter estimate is close to the Hubble time and, if correct, might have implications about closure of the universe[2].

Experimental Details

The neutron transmission measurements were made using ≈ 2 gm isotopically enriched samples of 186, 187 and 188. The samples' isotopic compositions are shown in Table Ia. The ORELA operating conditions are shown in Table Ib. Since only the average total cross sections are of interest above ~2 keV, adequate energy resolution for resonance analysis below 2 keV was achieved using the 80 m flight path and a 30 ns beam pulse width.

Table Ia ISOTOPIC ABUNDANCE (ATOM-%)

SAMPLE	^{186}Os	^{187}Os	^{188}Os	^{189}Os	^{190}Os	^{192}Os
^{186}Os	78.4	1.62	5.07	4.00	5.15	5.67
^{187}Os	0.93	70.38	12.79	5.28	5.41	5.26
^{188}Os	0.13	0.17	94.47	2.77	1.42	1.04

Table Ib
EXPERIMENTAL CONDITIONS

Neutron Source	ORELA
Repetition Rate	800Hz
Pulse Width	33ns
Flight Path	80m
Detector	^{6}Li glass
Energy Resolution	0.1%

In figure 1 is shown a typical energy region 200 eV to 230 eV showing predominately resonances in ^{187}Os. Some of the smaller transmission dips are due to other Osmium isotopes. The other two samples, 186 and 188, plus the work of Browne and Berman[2] and Stolovy[3] et al. served to allow identification of all ^{187}Os resonances up to about 500 eV. Moreover, the isotopic impurity resonances in each sample could be analyzed as a "thin" sample to provide internal consistency checks on the derived resonance parameters.

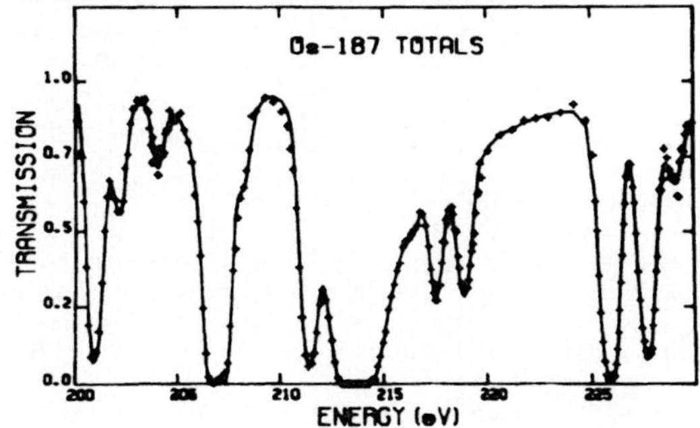

Figure 1. ^{187}Os Transmission vs. Incident Neutron Energy

Data Analysis

The data were analyzed using the Reich-Moore multi-level code (SAMMY[4,5]). A typical fit is shown by the the solid curve in Figure 1. Note that the fit also includes several isotopic impurity resonances which "black out" in the thicker sample in which they correspond to the major isotope. The spin (J) assignments up to 140 eV are taken from Stolovy[3] et al. The resulting resonance parameters will be published elsewhere, but the average properties S_o, D, and $\langle\gamma_\lambda^2\rangle$ are given in Table 2.

TABLE 2. SUMMARY OF DEDUCED NEUTRON PARAMETERS FOR ^{187}Os.

J-value	no. resonances	$\langle\Gamma_n^0\rangle$ (meV)	$S_0 x10^4$	$\langle D\rangle$ (eV)
J=0[a]	8	9.5 ± 7.1	9.0 ± 5.0	10.6 ± 2.1
J=1[a]	18	2.1 ± 1.1	3.2 ± 1.6	6.5 ± 0.8
J=0,1[b]	85	5.3 ± 1.1	3.6 ± 0.8	5.6 ± 0.3

[a] Energy range - 25 to 140eV; J-assignments from A. Stolovy, et.al., Phys. Rev. *C14*, 965 (1976).
[b] Energy range - 25 to 500eV.

Comparisons of the reduced neutron width distributions with the expected Porter-Thomas distributions reveal not more than one $J^\pi = 0^-$ missed level up to 140 eV and not more than two $J^\pi = 1^-$ missed levels. Over the region 25-500 eV, we estimate not more than ten s-wave levels are missed in this work. The level spacings and strength functions in Table 2 are corrected for missing levels.

Age of the Universe

In their statistical model analysis of the Osmium capture cross sections, Winters and Macklin[1] used $S_o = 2.2x10^{-4}$, a value characteristic of the Osmium mass region. They then proceeded to analyze their $^{187}Os(n,\gamma)$ cross sections above 2.5 keV for S_1, S_γ, and $\langle\sigma(n,n')\rangle$. Of particular importance in the context of the Re/Os cosmochronometer is Winters and Macklin's value for $\langle\sigma(n,n')\rangle \approx 0.30$ bn near 30 keV. This result is consistent with an upper bound $\langle\sigma(n,n')\rangle \lesssim 0.7$ bn near 30 keV reported by Winters[6] et al. which placed critical constraint on a calulation by Woosley and Fowler[7] of the correction factor F to be applied to the laboratory ratio of capture cross sections $\langle\sigma_\gamma(186)\rangle/\langle\sigma_\gamma(187)\rangle$ to account for the effects of excited states at stellar temperatures. In order to obtain such small values of $\langle\sigma(n,n')\rangle$, Woosley and Fowler arbitrarily reduced the strength function for the ^{187}Os excited states by a factor of two relative to the assumed ground state value $S_o = 2.2x10^{-4}$. This resulted in $F \approx 1.0$ and an estimate[1] for the age of the universe ≈ 14.5 billion years. This is consistent with, but larger (by ≈ 1-2 billion years) than, estimates derived from observational astronomy and stellar modeling techniques. As shown by their[7] Table I, the calculation of F by Woosley and Fowler is rather sensitive to the relative values of S_o and S_1, and also depends on the value of S_γ.

Preliminary statistical model re-analysis of the Winters and Macklin ^{187}Os capture cross sections using the value for S_o from this work leaves the value for $\langle\sigma(n,n')\rangle$ unchanged. However, S_γ is decreased by 25% and the value for S_1 is decreased by 30%. These results and Woosley and Fowler's Table I[7] suggest that F may be less than unity which would <u>increase</u> the already high estimate for the age of the universe derived from Re/Os.

*Research supported in part by the U.S. Department of Energy under contracts No. DE-AC02-76-ER02696 and DE-AS05-80ER10710.

References

1. R. R. Winters and R. L. Macklin, Phys. Rev. C 25, 208 (1982).
2. J. C. Browne and B. L. Berman, Phys. Rev. C 23, 1434 (1981).
3. A. Stolovy et al., Phys. Rev. C 14, 965 (1976).
4. George F. Auchampaugh, LA-5473-MS, Los Alamos Report (1974).
5. N. M Larson and F. G. Perey, ORNL/TM-7485, ENDF-297 (1980).
6. R. R. Winters, F. Käppeler, K. Wisshak, B. L. Berman and J. C. Browne, BAPS 24, 854 (1979).
7. S. E. Woosley and W. A. Fowler, Astrophys. J 233, 411 (1979).

THE NEUTRON CAPTURE CROSS SECTIONS OF $^{178,179,180}Hf$ AND THE ORIGIN OF NATURE'S RAREST STABLE ISOTOPE ^{180}Ta

H. Beer

Kernforschungszentrum Karlsruhe GmbH
Institut fur Angewandte Kernphysik
P.O.B. 3640, D-7500 Karlsruhe
Federal Republic of Germany

R. L. Macklin

Oak Ridge National Laboratory,
Oak Ridge, Tennessee 37830 USA

The neutron capture cross sections of $^{178,179,180}Hf$ were measured in the energy range 2.6 keV to 2 MeV. The average capture cross sections were derived and fitted in terms of strength functions. Resonance parameters for the observed resonances below 10 keV were determined by shape analysis. Maxwellian averaged capture cross sections were computed for thermal energies with kT between 5 and 100 keV. The cross sections for kT = 30 keV were used to determine the population probability of the 8^- isomeric level in ^{180}Hf by neutron capture as $(1.24 \pm 0.06)\%$ and the r-process abundance of ^{180}Hf as 0.0290 (Si $\equiv 10^6$). These quantities served to analyze s- and r-process nucleosynthesis of ^{180}Ta, nature's rarest stable isotope.

[$^{178,179,180}Hf(n,\gamma)$, capture cross sections, E_n = 2.6 keV-2 MeV, nucleosynthesis]

Introduction

The neutron capture cross sections of ^{179}Hf and ^{180}Hf in the keV neutron energy range are of particular importance to the stellar nucleosynthesis of the rare odd-odd isotope ^{180}Ta, actually an isomeric state[1] with the impressively long half life of $>3\times10^{14}$ yr.[2] The buildup of $^{180}Ta^m$ is thought to occur via a small branching in the main flow of the common s- and/or r-process nucleosynthesis where the branching is mediated by an allowed Gamow-Teller beta transition from an 8^- isomeric state in ^{180}Hf.[3]

In order to quantitatively assess the possible s- and r-process contributions to the solar abundance of $^{180}Ta^m$ a variety of quantities must be determined experimentally. This concerns not only fractional beta transitions ($^{180}Hf^m \rightarrow {}^{180}Ta^m$ for s- and r-processes and $^{180}Lu \rightarrow {}^{180}Hf^m$ for the r-process only) but also Maxwellian averaged capture cross sections. It is the aim of the present investigation to completely satisfy this second requirement by the measurement of the neutron capture cross sections of three hafnium isotopes ($^{178,179,180}Hf$) and by a careful statistical model calculation of the 30 keV neutron capture cross section of $^{180}Ta^m$. The measurements on $^{178,179,180}Hf$ covered the energy range from 2.6 keV to 2 MeV. The average isotopic cross sections were described in terms of strength functions.

The data allowed us to determine accurately 1) the population probability of the 8^- isomeric state in ^{180}Hf by s-process nucleosynthesis and 2) the r-process abundance of ^{180}Hf which is important for the calculation of the r-process abundance of $^{180}Ta^m$.

Experimental Method

The measurements were carried out at the Oak Ridge Electron Linear Accelerator (ORELA) in the energy range 2.6 keV to 2 MeV using the time-of-flight technique with a flight path of 40.12 m. The accelerator was operated at a repetition rate of 800 pulses per second with an electron burst width of 15 to 18 ns full width half maximum. The neutron beam was collimated to provide for an approximately 2.6 x 5.2 cm rectangular beam profile at the sample position. A 0.05 cm thick ^6Li glass detector 43 cm in front of the sample was used to monitor the neutron flux.[4] The neutron capture events in the sample were counted via the prompt emitted capture gamma-radiation with a pair of fluorocarbon based liquid scintillation detectors symmetrically placed outside the neutron beam at the position of the sample. The Hf-samples which consisted of HfO_2 powder + 10% sulfur binder were pressed to thin 2.6x2.6 cm squares and exposed to the neutron beam in a 6.4 µm thin Mylar foil bag. The amount and composition of the samples are summarized in Table I.

The capture events were accumulated into 128 pulse height and 18 K time-of-flight channels. For the pulse height a sharp digital threshold was set at 153 keV. The energy calibration of the time-of-flight channels was made by well-known resonances in ^{27}Al at 5.903 keV and 1.091 MeV. The pulse height scale was frequently checked with the Compton edge of the 4.43 MeV gamma line of a PuBe source. More details of the experimental technique are found in ref. 5,6.

Table I

Sample Characteristics

Weight[a] HfO_2 (g)	Thickness (mm)	Isotopic Fractions[b] (%)				
		^{176}Hf	^{177}Hf	^{178}Hf	^{179}Hf	^{180}Hf
3.988 4.746	1.4 1.6	0.022	1.54	94.72	1.84	1.69
4.3785 4.3795	1.5 1.5	0.56	3.42	5.47	81.85	8.74
4.394 4.395	1.5 1.5	0.23	1.0	2.22	2.66	93.89

a. These weights include 10% by weight of sulfur which was added as a binder.
b. The content of ^{174}Hf was <0.05%.

Data Reduction

For the accumulated capture events, on line pulse height weighting was performed. The data were corrected for dead time and the various backgrounds were subtracted. The samples require a correction for gamma ray absorption (10%) and a correction for neutron multiple scattering and self protection (at 30 keV this correction factor lies between 1.018 and 1.032 for the individual isotopes). In order to get the final cross sections for the enriched Hf samples the sulfur and oxygen contributions were taken out. The sample cross sections were unscrambled to derive the pure isotope cross sections. The minor ^{176}Hf and ^{177}Hf contributions were approximated by ^{178}Hf and ^{179}Hf, respectively.

K. H. Böckhoff (ed.), Nuclear Data for Science and Technology, 945–947.

Data Analysis and Results

For the determination of the effective $^{178,179,180}Hf$ cross sections in the whole energy region from 2.6 keV to 2 MeV the sample yield data were averaged in >250 eV bins. This procedure smears out individual resonance fluctuations because more than 10 resonances are combined and represents an adequate basis for a parametrization of the cross section in terms of strength functions. The strength function analysis was limited to the energy range below 93 keV for $^{178,180}Hf$ because this energy value is the threshold where the first inelastic channel opens. In Figs. 1 and 2 our average neutron capture cross sections for $^{178,179,180}Hf$ are shown and the strength function fits are indicated by the solid lines. The level density for the even isotopes was too low to allow fitting the s-wave strength function so the value fitted to the ^{179}Hf data was used for the even isotopes without adjustment. The final individual strength functions of our analysis are included in Figs. 1 and 2.

Systematic Uncertainties

Systematic uncertainties in the cross sections (Table II) are dominated by the saturated resonance calibration and the energy dependence of the $^6Li(n,\alpha)$ cross section. The saturated resonance technique which was applied to the 4.9 eV resonance in ^{197}Au serves to normalize the detection efficiency. The systematic uncertainties are summarized in Table II.

Table II

Systematic Uncertainties in Average Capture Cross Section

Saturated resonance calibration		2%
Shape of the $^6Li(n,\alpha)$ cross section	at 50 keV	1%
	at 250 keV	2%
	>500 keV	3%
Pulse height weighting technique		1%
Neutron sensitivity of detection system (sample scattered neutrons)		<1.6%
Gamma ray self absorption in the sample		0.9%
Multiple scattering and self-protection		0.5%
Detector bias extrapolation (E_{bias}=153 keV)		0.9%
Misalignment of sample or neutron beam		<0.2%
Uncertainty in detector efficiency by gain drifts of electronics		<0.4%[a]

a. 1.9% for the ^{179}Hf measured during a time when the gain of one detector decreased significantly and probably not uniformly.

Maxwellian Averaged Capture Cross Sections

Maxwellian averaged capture cross sections (represented by the symbol σ) were computed from the differential data by numerical integration according to the following formula:

$$\sigma = \frac{2}{\sqrt{\pi}} \int_0^\infty \sigma(E)E \exp(-E/kT)dE / \int_0^\infty E \exp(-E/kT)dE$$

where kT designates the thermal energy which is conventionally listed at 30 keV although cross sections for a broad range of values can also be derived from the data.

In Table III the required cross sections are summarized.

With the capture cross sections of ^{179}Hf from Table III the population probability by neutron capture of the 8- ^{180}Hf isomeric state was calculated to be $\sigma(^{179}Hf^*)/\sigma(^{179}Hf) = (1.24 \pm 0.06)\%$.

Table III

Maxwellian Averaged Cross Sections of Relevant Nuclides at kT = 30 keV

A_Z	σ(mb)	A_Z	σ(mb)
^{179}Hf	991 ± 30	^{180}Hf	179 ± 5
^{179}Hf*	12.2 ± 0.6	$^{180}Ta^m$	1800 ± 200

*The ^{179}Hf cross section to the $^{180}Hf^m$ isomeric state from ref. 3.

The $^{180}Ta^m$ Capture Cross Section

The unmeasured $^{180}Ta^m$ capture cross section was determined via a statistical model calculation with average $^{180}Ta^m$ resonance parameters from Harvey et al.[7] derived from a measurement in the energy range 0.3 to 300 eV. The quoted values are: $S_{\ell=0} = (2.4 \pm 0.4) \times 10^{-4}$, $\bar{\Gamma}_\gamma$ = 51 ± 1 meV and $\bar{D}$ = 1.1 ± 0.1 eV. The p- and d-wave strength functions were estimated using the systematic behavior of the neighboring nuclei Hf and W. Fig. 3 shows the analysis. The individual contributions of s-, p- and d-wave capture are also indicated. A 30 keV $^{180}Ta^m$ cross section of 1800 ± 200 mb is obtained.

s- and Post r-Process Nucleosynthesis of $^{180}Ta^m$

The capture cross sections in Table III and our result on $^{180}Ta^m$ have a straightforward bearing on the possible neutron capture nucleosynthesis of nature's rarest stable isotope $^{180}Ta^m$. According to ref. 3 we can write for the s- and post r-process generated abundances $N_s(^{180}Ta^m)$ and $N_r(^{180}Ta^m)$, respectively:

$$N_s(^{180}Ta^m) = \frac{\sigma(^{179}Hf^*)}{\sigma(^{179}Hf)} f^m_{\beta^-}(^{180}Hf) \frac{\sigma N_s(A=180)}{\sigma(^{180}Ta^m)} \quad (1)$$

$$N_r(^{180}Ta^m) = f^m_{\beta^-}(^{180}Hf)\, f^{180}_m(^{180}Lu) \times \left[N_\odot(^{180}Hf) - \frac{\sigma N_s(A=180)}{\sigma(^{180}Hf)}\right] \quad (2)$$

$f^m_{\beta^-}(^{180}Hf)$ and $f^{180}_m(^{180}Lu)$ are fractional β transitions. $f^m_{\beta^-}(^{180}Hf)$ is the fractional β decay from $^{180}Hf^m$ to $^{180}Ta^m$ and $f^{180}_m(^{180}Lu)$ stands for the population probability of the 8- ^{180}Hf isomer via the ^{180}Lu β decay. $f^m_{\beta^-}(^{180}Hf)$ was estimated to be 0.7%.[8] It should be noted that this estimate includes thermal effects to be expected in the hot stellar environment of the s-process. The value based on laboratory half lives amounts to 0.34%. $N_\odot(^{180}Hf)$ represents the solar abundance of ^{180}Hf as given by Cameron[9] and $\sigma N_s(A=180)$ = 5.53(mb x Si ≡ 10^6) is the capture cross section times s-process abundance at mass number 180 taken from Käppeler et al.[10]

Inserting the relevant quantities into eq. (1) shows that the s-process can produce only ~ 11% of the solar $^{180}Ta^m$ abundance. In eq. (2) the expression in brackets represents the r-process abundance of ^{180}Hf. This quantity was determined to be $N_r(^{180}Hf)$ = 0.0290 (Si ≡ 10^6). As the factor $f^{180}_m(^{180}Lu)$ is unknown $N_r(^{180}Ta^m)$ cannot yet be determined. But a β branching of $f^{180}_m(^{180}Lu)$ ≤2.5% in the ^{180}Lu decay to $^{180}Hf^m$ would be sufficient already to reproduce the whole solar abundance of $^{180}Ta^m$ by the r-process.

Research was sponsored by the Division of Nuclear Sciences, U.S. Department of Energy, under Contract No. W-7405-eng-26 with the Union Carbide Corporation.

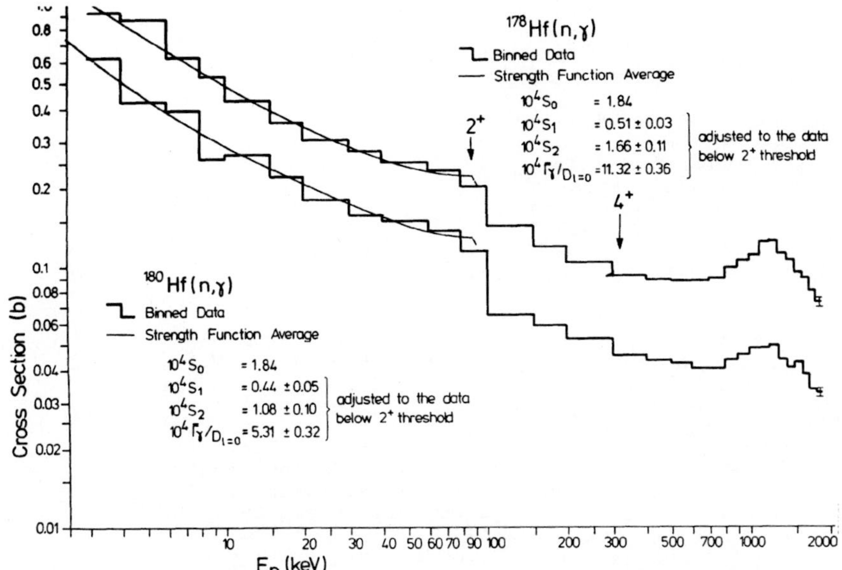

Fig. 1 Effective cross sections for ^{178,180}Hf(n,γ). The curves are a statistical model fit to the data below the inelastic threshold. The arrows mark the locations of the first few excited levels.

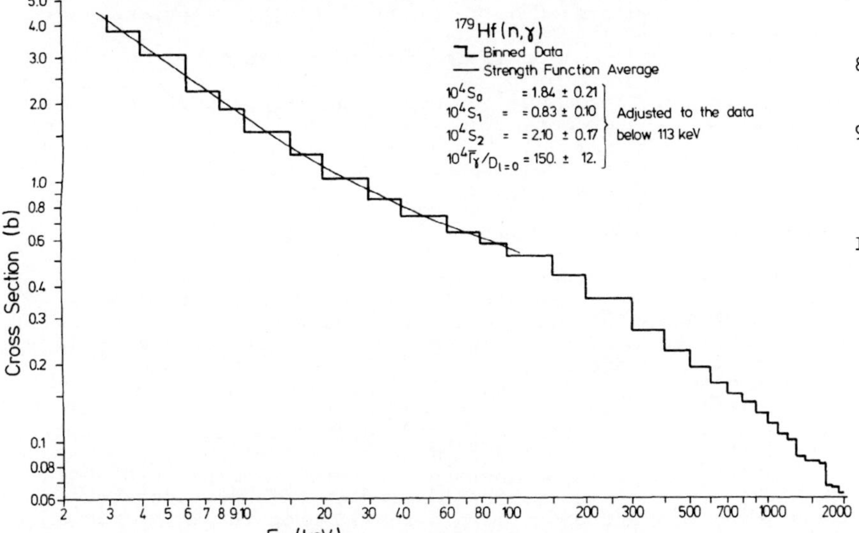

Fig. 2 Effective cross section for ^{179}Hf(n,γ). The curve is a statistical model to fit to the data below the inelastic threshold.

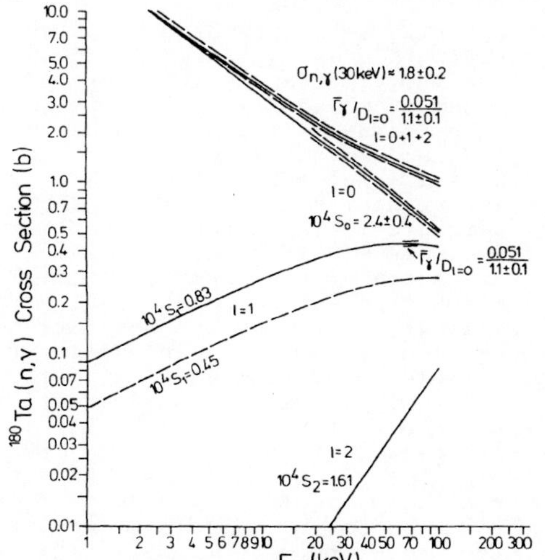

Fig. 3 The average cross section of ^{180}Ta as a function of energy calculated from the statistical model. The various contributions from s-, p-, and d-waves are shown separately. The uncertainties are indicated by dashed lines.

References

1. K.S. Sharma, R.J. Ellis, V.P. Derenchuck, R.C. Barber, H.E. Duckworth, Phys. Lett. 91B, 211 (1980).
2. E.B. Norman, private communication
3. H.Beer and R.A. Ward, Nature 291, 308 (1981)
4. R.L. Macklin, N.W. Hill, B.J. Allen, Nucl. Instrum. Method 96, 509 (1971)
5. B.J. Allen, R.L. Macklin, R.R. Winters, and C.Y. Fu, Phys. Rev. C 8, 1508 (1973)
6. R.L. Macklin and B.J. Allen, Nucl. Instrum. Method 91, 565 (1971)
7. J.A. Harvey, N.W. Hill and E.R. Mapoles, ORNL Report 5025 (1975)
8. H. Beer and R.L. Macklin, Phys. Rev. C (in press)
9. A.G.W. Cameron, in Nucl. Astrophysics (ed. C.A. Barnes, D.D. Clayton, D.N. Schramm), Cambridge Univ. Press (1981)
10. F. Käppeler, H. Beer, K. Wisshak, D.D. Clayton, R.L. Macklin, and R.A. Ward, Ap. J. 257, 821 (1982)

NEUTRON CAPTURE CROSS SECTIONS OF ^{20,21,22}Ne BETWEEN 5 AND 400 keV AND THE NEUTRON BALANCE IN s-PROCESS NUCLEOSYNTHESIS

J. Almeida[+)] and F. Käppeler
Kernforschungszentrum Karlsruhe GmbH
Institut für Angewandte Kernphysik
P.O.B. 3640, D-7500 Karlsruhe
Federal Republic of Germany

+) Present address: LNETI-DEEN, Est. Nac. 10, 2685 Sacavem, Portugal

The neutron capture cross sections of the three stable neon isotopes have been measured by the time-of-flight method in the energy range from 5 to 400 keV with an accuracy of better than 1 mb. The neutron source of the s-process is believed to be the ^{22}Ne(α,n) reaction, taking place in the He burning shell of a pulsating red giant. From the discussion of the s-process neutron balance it was found that neutron capture in the neon isotopes is a minor effect thus not affecting the main (α,n)-reaction on ^{22}Ne. As a result it was proved that the neutron balance is satisfied. In addition, the ratio of the (α,n) and (α,γ) reaction rates on ^{22}Ne could be used to derive a lower limit for the s-process temperature which is consistent with the assumption of the He-burning shell as site for the s-process.

[^{20}Ne(n,γ), ^{21}Ne(n,γ), ^{22}Ne(n,γ), capture cross sections, E_n=5-400 keV, nuclear astrophysics, nucleosynthesis]

Introduction

Since the discovery of Technetium lines in S stars by Merrill[1] it is known that the s-process must be occurring in red giants. The analysis of the observed solar system abundances also shows that the s-process must take place under temperature and density conditions that are characteristic of the He burning shell of red giants. The most successful s-process model is unquestionably that of Iben[2] assuming an intermediate mass red giant undergoing thermal pulses. The neutron source in this model is the ^{22}Ne(α,n) reaction taking place in the He burning shell. There, the ^{22}Ne is continuously supplied by two α-captures on ^{14}N which is the main leftover from hydrogen burning in the CNO-cycle and which is brought into the He-burning zone by the thermal pulses. In order to calculate the fate of the released neutrons, one needs the cross sections of the isotopes present in the shell, foremost those of the most abundant species. The ^{22}Ne has as progenitors the C,N and O involved in the CNO cycle and is therefore, apart from ^{4}He and ^{12}C, the most abundant isotope, the second most abundant being ^{20}Ne.

It has been shown recently by Käppeler et al. [3] that the traditional s-process model which assumes a steady neutron flux and an exponential distribution of neutron irradiations (Seeger, Fowler and Clayton[4]) represents a quantitative basis for s-process analyses. Within this model the requirement of neutron balance can be used to test the particular stellar s-process site mentioned above, provided that the relevant cross sections are known.

Although ^{20}Ne and ^{22}Ne are the most abundant neutron absorbers in this model, they had not yet been experimentally investigated, because their cross sections are very small (∿1 mb). Theoretical estimates for such light nuclei are even less reliable than in the mass region above A∿60, where they exhibit systematic uncertainties of 50% on average. In the following section we describe our measurements of the total and capture cross sections of the three stable neon isotopes ^{20,21,22}Ne. The calculation of the neutron balance is presented in the last section. Beside this result at which we aimed primarily, we found from our discussion of the neutron balance that we were also able to derive a lower limit for the temperature during the s-process. This further result is obtained by comparing the rates of the competing ^{22}Ne(α,n) and ^{22}Ne(α,γ) reactions.

Cross Section Measurements

In the stellar model under consideration, the s-process occurs at a temperature T∿3.5 10^8 K, or kT ∿ 30 keV. Therefore, one needs the 30 keV Maxwellian averaged capture cross sections, $\langle\sigma\rangle_{kT=30\ keV}$ as defined by [5]

$$\langle\sigma\rangle = \frac{2}{\sqrt{\pi}} (kT)^{-2} \int_0^\infty dE\ E\ \sigma(E)\ e^{-E/kT} \qquad (1)$$

In order to evaluate this expression, we measured σ(E) in the energy range from 5 to 400 keV. The total cross sections, which are needed for the analysis of our capture data, were also measured in the energy range 5-800 keV. All the data were taken at the Karlsruhe 3 MV pulsed Van de Graaff accelerator, the neutrons were produced by the ^{7}Li(p,n) reaction, and their energy determined by the time-of-flight method. A short description of these measurements is given in the next subsections, and a detailed discussion of the experiments has been reported elsewhere[6].

a) Total Cross Sections

In this experiment two series of runs were made, covering the energy ranges 5-350 keV and 150-800 keV. The samples were placed in an automatic sample changer with 5 positions, 3 samples with the neon isotopes contained in thin walled, high pressure gas cells, an empty container, and a position with no container. The transmission spectra of these 5 samples were cyclically measured. Samples were changed every 10 min in order to average over smooth variations of the neutron yield. The measuring position was immediately after a ^{6}LiCo$_3$ collimator, the neutron beam at the sample position having a smaller diameter than the sample to prevent it from touching the 10 cm long cylindrical container walls. The transmitted neutrons were detected at 3 m flight path by a 10 x 10 cm ^{10}B slab, through the ^{10}B(n,α)^{7}Li* reaction. The 478 keV gamma rays were detected in two C_6D_6 liquid scintillator detectors. The combined time resolution of the detectors, the accelerator pulsing system and the electronics was better than 1.2 ns, and the total time-of-flight (TOF) resolution was 0.4 ns/m. By carefully shielding the detector, all sources of time-dependent backgrounds were eliminated.

K. H. Böckhoff (ed.), Nuclear Data for Science and Technology, 948–951.

With respect to the analysis of the capture cross sections, it is important to note that none of the neon isotopes has strong scattering resonances in the relevant energy region below 150 keV. This and the small magnitude of the total cross sections means that the effect of sample scattered neutrons on the capture cross section measurement will be small and can be simulated by a graphite scatterer.

b) Capture Cross Sections

The capture cross sections of the neon isotopes were measured relative to the standard gold cross section using the set-up of Fig.1.

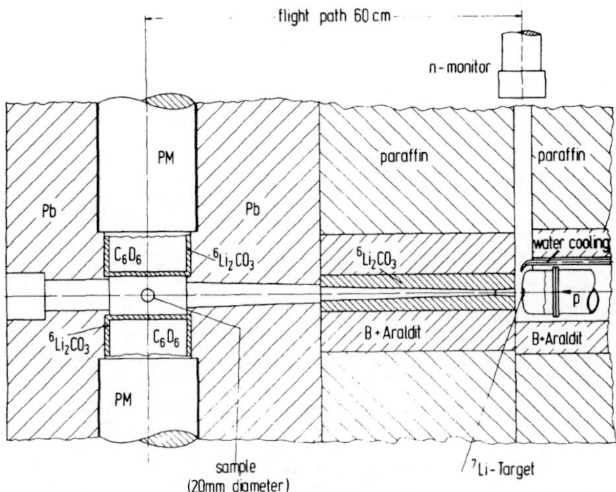

Fig. 1 Schematic experimental set-up

The capture events were counted by detecting the gamma-ray cascades following neutron capture, with the same two C_6D_6 scintillators used in the transmission measurements. Six spherical stainless steel containers with 20 mm diameter and 0.5 mm wall thickness were placed on an automatic sample changer: 3 with the neon isotopes, under pressures of up to 300 atmospheres, 1 empty container, 1 with the gold standard, 1 with a graphite sphere, to simulate the effect of sample scattered neutrons. The flight path was 60 cm and the total time resolution achieved 1.2 ns. At 30 keV neutron energy this led to an energy resolution of 0.7 keV, of which 0.5 keV are due to the flight path uncertainty caused by the 20 mm diameter of the sample.

For each detected event the pulse height response of the detector was recorded in addition to the timing signal allowing for off-line application of the pulse height weighting technique. The neon cross section for any TOF channel, labelled by its corresponding neutron energy, E_n, is simply

$$\sigma(E_n) = \sigma_{Au}(E_n) \frac{N_{Au}\ B_{Au}}{N_{Ne}\ B_{Ne}} \frac{NEON(E_n)}{GOLD(E_n)} \qquad (2)$$

where the B_i are the neutron separation energies and the N_i the number of nuclei per unit surface seen by the neutron beam, and NEON and GOLD are the numbers of detected events, after applying the weighting procedure and after substraction of all backgrounds. The main experimental difficulty was to obtain a large enough signal from neutron captures in the neon. Neutron collimator, gas containers and sample changer were specifically designed to reduce the background, while allowing the detectors to be placed at a distance of only 2 cm from the sample. By coating the detectors with a layer of $^6LiCO_3$, it was possible to reduce considerably the background due to the detection of scattered neutrons which are thermalized in the liquid scintillator, and subsequently captured by its residual ^{1}He. The remaining background from scattered neutrons has a time-dependent component, due to fast neutrons captured in the lead shielding and in the aluminium walls of the scintillator containers. Its shape was given by the graphite TOF spectrum, and its magnitude was calculated using our measured total cross sections.

About 95 % of the net count rate of the neon samples was due to the steel container. For this reason, extreme care was taken in the summing of the short 10 min runs: all of them were individually inspected, and the ratios (count rate/neutron fluence) computed, in order to detect any variations of these ratios, which, for each sample, should not vary with time. Only series of runs showing no variation over several hours were summed and normalized to the same neutron flux.

Additional small corrections were made to account for the following effects: (i) Pile-up events: Given the large solid angle of the detectors seen by the sample, there is a non-negligible probability for two (or more) gamma rays of the same cascade to be simultaneously detected in the same detector (ii) Neutron multiple scattering and self shielding and gamma-ray self-absorption in the sample. (iii) Lower threshold of the electronics: Events with a pulse height lower than this threshold, which corresponds to E_γ=90 keV, are lost.

The energy ranges 5-200 keV and 170-400 keV were covered by independent runs. The cross sections between 5 and 200 keV are shown in Fig. 2. Obviously, there is almost no resonant capture in the neon isotopes in the energy range considered.

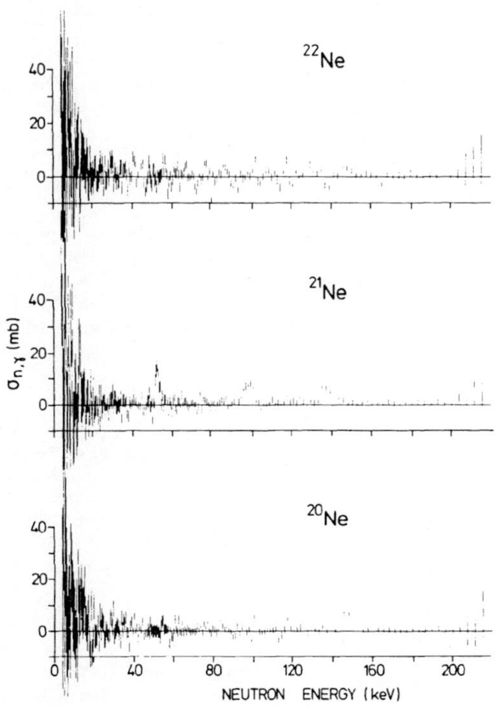

Fig. 2 The measured neutron capture cross section for the stable neon isotopes. Error bars represent the statistical uncertainty only.

Table I Corrections applied in the analysis of the capture cross sections and related uncertainties of $\langle\sigma\rangle$

Correction	Uncertainty (%)
Background from scattered neutrons	40-60
Background from sample container	6
Background from neutron self shielding and multiple scattering	2
Pile-up events	2
Gamma-ray self absorption	1
Effect of electronic threshold	1
Pulse height weighting	0.5
Statistics	25-40

While from 5 to 200 keV data with sufficient statistics could be obtained, the results of the runs at higher energies suffer from somewhat reduced statistical accuracy. An upper limit of about 1 mb was found for the average cross sections between 200 and 400 keV.

The estimated systematic uncertainties in $\langle\sigma\rangle$ associated with the various corrections are listed in Table I. The error from the correction for scattered neutrons dominates all other experimental uncertainties because this systematic error becomes very large for $E_n<20$ keV, and neutrons in this low energy range make up about half the total area of the Maxwell energy distribution. For $E_n>30$ keV, however, this error is greatly reduced as can be seen from the correspondingly smaller uncertainties for kT=50 keV in Table II.

That this uncertainty is so much larger than the others is due to the fact that, together with the statistical uncertainty, it is directly related to the sensitivity of the experiment and hence depends on the cross section magnitude. The other uncertainties are due to sample effects or to the experimental technique and reflect the state of the art for this type of experiments. It should be noted that in spite of the large fractional uncertainty, the precision achieved in the present measurement for a small, non-resonant cross section as for ^{22}Ne, is quite remarkable.

In Table II we have summarized the resulting Maxwellian average cross sections for various thermal energies. From here on σ denotes always these average cross sections.

The s-Process Neutron Balance

In this section we try to verify whether the neutron balance condition is satisfied under the assumptions of Iben's stellar model[2].

During the s-process, the abundance N_A, of any isotope of mass A along the synthesis path is given by

$$\frac{dN_A}{dt} = \phi(t)\sigma_{A-1}N_{A-1}(t) - (\phi(t)\ \sigma_A+\lambda_\beta)N_A \qquad (3)$$

where ϕ and σ_i are the neutron flux and the neutron capture cross sections. The β-decay rate λ_β has to be considered only if A is unstable. This system of coupled differential equations for all isotopes in the s-process chain can be solved by means of a few simplifying assumptions. It turnes out that the observed solar σN curve cannot be reproduced by subjecting all the seed nuclei to the same neutron exposure τ, τ being the time integrated neutron flux $\int_0^\infty \phi(t)dt$. Rather, a continuous distribution of neutron exposures $\rho(\tau)$ is called for, $N_{56}\rho(\tau)d\tau$ being the number of ^{56}Fe seed nuclei that had an exposure between τ and $\tau+d\tau$.

The present best fit to the solar σN-curve was found for the distribution

Table II Maxwellian averaged capture cross sections of the neon isotopes

Thermal Energy (keV)	σ+Δσ (mb) ^{20}Ne	^{21}Ne	^{22}Ne
20	1.8	1.8	1.7
25	1.6	1.7	1.2
30	1.5±0.7	1.6±0.9	0.9±0.7
35	1.5	1.5	0.8
40	1.4	1.4	0.8
50	1.4±0.3	1.3±0.3	0.7±0.3

$$\rho(\tau)=\rho_1(\tau)+\rho_2(\tau)=s_1N^\odot_{56}\exp(-\tau/\tau_{01}) + s_2N^\odot_{56}\exp(-\tau/\tau_{02}) \qquad (4)$$

where $sN^\odot_{56}$ represents that fraction of the solar ^{56}Fe abundance which is required as seed. The parameters are $s_1=(2.7\pm0.2)\%$, $\tau_{01}=(0.056\pm0.005)\ \mathrm{mb}^{-1}$, and $s_2=(0.092\pm0.015)\%$, $\tau_{02}=(0.24\pm0.01)\mathrm{mb}^{-1}$ (Ref. 3).

The weaker irradiation term, ρ_1, is needed to account for the higher σN values below A∿80. Possible production sites for these isotopes seem to be the He burning cores of massive stars[7]. The main term, ρ_2, reproduces the solar σN curve precisely in the mass range in which Iben's stellar model is found to synthesize the s-only isotopes in solar proportions. This type of pulsating red giant seems therefore a very promising s-process site.

In our approach we assume that the operation of the s-process in this environment does indeed produce the observed solar σN curve and, using the traditional treatment of the s-process, we determine what conditions should then be satisfied by the stellar model. We assume that the s-nuclides with A>80 were produced by the second term of eq. (4). The resulting σN values can be evaluated analytically

$$\sigma_A N_A = \frac{s_2\ N^\odot_{56}}{\tau_o} \prod_{N=56}^{A} \frac{\tau_o\sigma_i}{1+\tau_o\sigma_i} \qquad (5)$$

At each pulse, along with ^{56}Fe, other light elements are introduced in the He burning region, hence, they are exposed to the same neutron flux as ^{56}Fe. Although their contribution to the synthesis of heavy isotopes (A>56) is very small, they nevertheless absorb a considerable fraction of the available neutrons. Each of these light elements is also an s-process seed, building heavier isotopes, A, to abundances given by

$$\sigma_A N_A = \frac{s_2 N^\odot_{A_{seed}}}{\tau_o} \prod_{i=A_{seed}}^{A} \frac{\tau_o\sigma_i}{1+\tau_o\sigma_i} \qquad (6)$$

For each heavier isotope, A, thus produced, $(A-A_{seed})$ neutrons are captured by its progenitors following the seed. Therefore, the s-process progeny of a species "A_{seed}" captures a total number of neutrons

$$n_c(A_{seed}) = \sum_{A=A_{seed}}^{200} (A-A_{seed})N_A(A_{seed}) \qquad (7)$$

The total number of neutrons captured during the s-process is $\Sigma\ n_c(i)$, where i runs over all seeds. We assume that the elements from C to Fe initially present had the same relative abundances as in the solar system. Since we know the total amount of ^{56}Fe processed, $N_{56} = s^2N^\odot_{56}$, we know also that fraction of the light isotopes, $A_A=s_2N^\odot_A$, and with (6) and (7) we can compute $\Sigma n_c(i)$. The neutron source is the ^{22}Ne(α,n) reaction. The initial abundance of ^{22}Ne, N_{22}, prior to the onset of

the $^{22}Ne(\alpha,n)Mg$ reaction is, $(N_C+N_N+N_O+N_{22})^o_{solar}$, or N_{CNO} for short since all the C,N and O end up as ^{22}Ne.
A fraction, f, of the ^{22}Ne undergoes the (α,n) reaction. The number of neutrons produced is fN_{CNO}, an amount $(1-f)N_{CNO}$ of ^{22}Ne and fN_{CNO} of ^{25}Mg are left to act as s-process seeds.

We can now write the neutron balance condition

(number of neutrons produced) = (number of neutrons captured)

$$\text{or } f\,N_{22} = f\,s_2 N^{\odot}_{CNO} = n_c(^{56}Fe) + n_c(^{22}Ne) + n_c(^{25}Mg) + n_c(\text{others}) \qquad (8)$$

from which we can extract f, since it is the only unknown quantity. A condition for the validity of this s-process model, where ^{22}Ne is the only neutron source, is that this equation should yield a value $f<1$. A value $f>1$ would mean that the $^{22}Ne(\alpha,n)$ reaction cannot provide enough neutrons to reproduce the abundances corresponding to the second term of eq. (4).

Our analysis of eq. (8) yields f=0.95: the neutron balance is satisfied if, during each pulse, 95 % of the ^{22}Ne present in the He shell produce neutrons by the (α,n) reaction. The uncertainty of f is dominated by two quantities: (i) The 4% uncertainty of the neutron exposure parameter τ_{02} propagates almost linearly into Δf. (ii) Because practically all of ^{22}Ne is converted to ^{25}Mg by (α,n) reactions, this isotope accumulates to a high abundance and is by far the largest neutron absorber. The 10% uncertainty of its cross section contributes another 4 % to Δf. The very large uncertainty $\Delta\sigma/\sigma$ of most other isotopes results in only a small additional error in Σn_c. If in the range $20<A<56$ every cross section except σ_{25} and σ_{56} is increased by 50%, the total neutron absorption increases by less than 10%. As an additive error propagation is unlikely we conservatively obtain $f=(95\pm10)\%$.

The s-Process Temperature

We now try to determine whether the condition f=95% can be fulfilled under the conditions prevailing in the He shell. In the extreme case in which all the ^{22}Ne is burned by (α,n) and (α,γ), none being left to capture neutrons, one has $f^n+f^\gamma = 1$, f^n and f^γ being the fractions consumed by (α,n) and (α,γ) respectively, and of course $f^n/f^\gamma = \lambda^n/\lambda^\gamma$, the reaction rate ratio. Fig. 3 shows this ratio as a function of temperature, calculated according to Fowler, Caughlan and Zimmerman[11]. For the value f^n=95 %, deduced from the neutron balance equation, $f^\gamma = 1-f^n = 5\%$, or $f^n/f^\gamma=19$. Since it is unreasonable to expect that no neutrons at all are captured by ^{22}Ne f^γ must be <5 %, and therefore $\lambda^n/\lambda^\gamma>19$, a condition satisfied only if kT > 35 keV.

In our derivation of the value of f we assumed the abundance of ^{22}Ne to be $N_{CNO}+N_{22}$, thus neglecting the ^{22}Ne produced in the s-process itself by neutron capture on lighter isotopes, especially on ^{20}Ne. Taking into account this additional amount of fuel for the $^{22}Ne(\alpha,n)$ neutron source causes the value of f to decrease by 5 % to f^n=90%. If we allow for the 10% uncertainty in f^n, its lower limit is f=80%, and therefore $f^n/f^\gamma>4$. The lower limit for the temperature is then $T_s>3.2\ 10^8$ K, or $kT_s>27$ keV and this result is completely consistent with the temperature range of the stellar model calculation of Cosner, Iben and Truran[12].

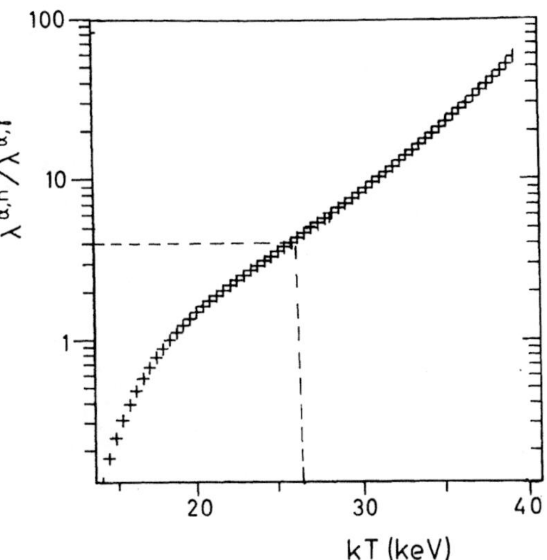

Fig. 3 Ratio of the (α,n) and (α,γ) reaction rates as a function of thermal energy (Fowler, Caughlan and Zimmerman 1975).

Conclusions

With the measured neon cross sections and the empirical neutron fluences derived from the traditional s-process model[3] it has been shown from the discussion of the neutron balance condition that the $^{22}Ne(\alpha,n)$ reaction provides sufficient neutrons for the s-process in the He burning shell of a pulsating red giant. Neutron absorption in the light isotopes from ^{20}Ne to ^{56}Fe is dominated by ^{25}Mg which is abundantly produced by the $^{22}Ne(\alpha,n)$ reaction. The ratio of (α,n) and (α,γ) reaction rates that follows from the neutron balance provides an estimate for the s-process temperature: $T_s>3.2\ 10^8$K or kT>27 keV. These results are based exclusively on experimental solar abundances and cross sections but not on the model dependent physical conditions in the He burning shell.

In summary, we conclude that the traditional s-process treatment of the neutron balance yields strong evidence in favor of the $^{22}Ne(\alpha,n)$ reaction as the effective neutron source to build most s-process abundances.

References

1 P.W. Merrill, Science 115, 484 (1952).
2 I. Iben Jr., Ap. J. 196, 525 (1975)
3 F. Käppeler, H. Beer, K. Wisshak, D.D. Clayton, R.L. Macklin, R.A. Ward, Ap. J. 257, 821 (1982).
4 P.A. Seeger, W.A. Fowler, D.D. Clayton, Ap. J. Suppl. 11, 121 (1965).
5 B.J. Allen, R.L. Macklin, J.H. Gibbons, Adv. in Nucl. Phys. 4, 205 (1971).
6 J. Almeida, KfK-3347, Kernforschungszentrum Karlsruhe (1982).
7 S.A. Lamb, W.M. Howard, J.W. Truran, I. Iben Jr., Ap. J. 217, 213 (1977)
8 D.D. Clayton, R.A. Ward, Ap. J. 193 397 (1974).
9 H. Weigmann, R.L. Macklin, J. Harvey, Phys. Rev. C14, 1328 (1976).
10 R.L. Macklin, R.R. Winters, Nucl. Sci. Eng. 78, 110 (1981).
11 W.A. Fowler, G.R. Caughlan, B.A. Zimmerman, Ann. Rev. Astr. and Astrophys. 13, 69 (1975).
12 K. Cosner, I. Iben Jr., J.W. Truran, Ap. J. 238, L91 (1980)

INTEGRAL EXCITATION FUNCTIONS AS A TOOL TO DECIPHER THE SOLAR RECORD IN EXTRATERRESTRIAL MATTER

R. Michel and R. Stück

Institut für Kernchemie der Universität zu Köln
Zülpicher Str.47, D 5000 Köln - 1
Federal Republic of Germany

A large variety of stable and radioactive nuclides is produced by the interaction of cosmic ray particles with extraterrestrial matter. In order to interpret these cosmogenic nuclides in lunar samples, in meteorites and cosmic dust, one has to differentiate between p- and α-induced reactions with solar and galactic cosmic rays. Particularly, for a satisfactory description of the interaction of solar particles with dense matter, a detailed knowledge of the cross section of the respective nuclear reactions is necessary. On the basis of integral excitation functions, model calculations of the depth dependent radionuclide productions by solar protons and α-particles in the lunar surface and in meteorites are presented. Theoretical profiles are compared with experimental data and several cosmochemical applications are discussed, concerning questions about the constancy of the cosmic radiation in time and space.

Introduction

The investigation of cosmogenic nuclides in samples of extraterrestrial origin yields knowledge about the fluxes, the spectral distribution and the constancy of the cosmic radiation, as well as about particular events like the break-up of meteorite parent bodies and impact events on the lunar surface. In order to interpret cosmogenic nuclides one has to differentiate between proton and α-induced reactions with solar (SCR) and galactic (GCR) cosmic rays (Fig.1). The galactic cosmic radiation which has its origin outside the solar system consists of 90 % protons, 9 % α-particles and 1 % heavier nuclides. Its energy distribution has a maximum at about 500 MeV and extends to extreme energies, having an energy distribution proportional to $E^{-2,5}$. Below 1 GeV the flux of galactic particles is modulated by the solar magnetic field The long term averaged 4 π integral flux at 1 A.U. for protons with energies > 1 GeV is ~ 1.7 $cm^{-2} \cdot s^{-1}$.

The solar cosmic rays have considerably lower energies. SCR-particles are emitted from the sun during short term eruptions, the solar flares. Also the SCR consist mainly of protons (~90%) with a contribution of α-particles and heavier particles which are varying from flare to flare. Considering large time intervals, the average spectra of solar flare particles can be described as a function of the rigidity R by

$$J(R)\ dR = k \cdot \exp(-R/R_O)\ dR \qquad (1)$$

k is a normalization constant and R_O is the characteristic rigidity describing the ha-dness of the spectrum.

For the interpretation of cosmogenic nuclides particular models have been derived which describe their depth-dependent production of cosmogenic nuclides. For the irradiation conditions of the lunar surface, there exist theoretical models for both components of the cosmic radiation [1,2] which allow to satisfactorily describe the measured concentrations of cosmogenic nuclides in lunar samples. These theories are, moreover, compatible with terrestrial simulation experiments of the interaction of cosmic rays with dense matter[3].

In this work we concentrate on the interactions of solar particles with extraterrestrial material. For SCR-particles the production of secondaries which again might undergo nuclear reactions can be neglected. So, the production rate $P_j(d)$([atoms/min • kg rock] or [dpm/kg rock]) of a cosmogenic nuclide j at a depth d in an extraterrestrial body can be described as

$$P_j = \sum_{i=1}^{n} N_i \int_{o}^{\infty} \sigma_{ij}(E) \cdot J(E,d)\ dE \qquad (2)$$

with N_i being the number of target atoms of element i per kg material. Since the depth-dependent fluxes J(E,d) can be calculated with adequate accuracy from the stopping of SCR particles considering the special geometry of the particular body and N_i can be determined from the chemical analysis of the object, the accuracy of the calculated SCR production rate depth profiles is exclusively determined by the accuracy of the excitation function $\sigma_{ij}(E)$.

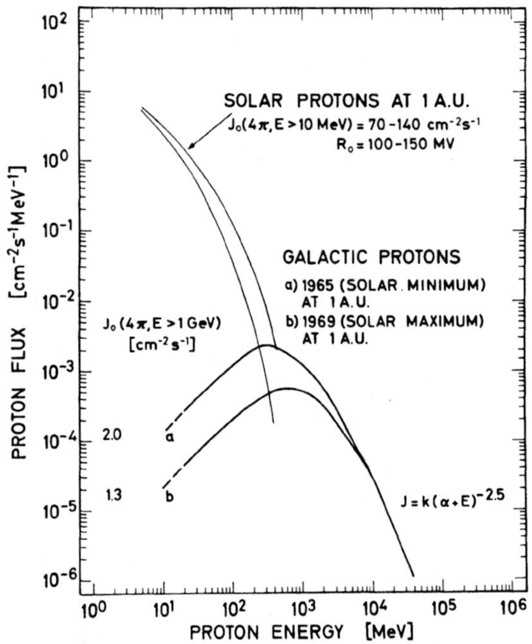

Fig. 1. Differential energy spectra of solar and galactic cosmic rays at 1 A.U..

In the last years we have established a consistent set of integral excitation functions of p- and α-induced reactions for particle energies up to 200 MeV [4,5] describing the production of cosmogenic radionuclides 40 ≤ A ≤ 60 from target elements 22 ≤ Z ≤ 28. On the basis of such experimental cross section data it is now possible to decipher the solar record in extraterrestrial matter by analyzing the cosmogenic nuclides. In the course of this study we shall describe some well-established applications considering the interpretation of SCR-interactions with lunar surface material, as well as some new developments with respect to the understanding of SCR-effects in meteorites.

SCR-Produced Nuclides in Lunar Surface Material

In lunar surface materials ~ 20 cosmogenic radionuclides have been observed and for all the rare gases isotopic abundance anomalies have been measured which are due to SCR- and GCR-interactions. For a detailed review of the radionuclide production in lunar surface material including experimental results as well as model calculations of production rate depth profiles we refer to one of our earlier publications[4].

By the interpretation of experimental data on cosmo-

K. H. Böckhoff (ed.), Nuclear Data for Science and Technology, 952–955.

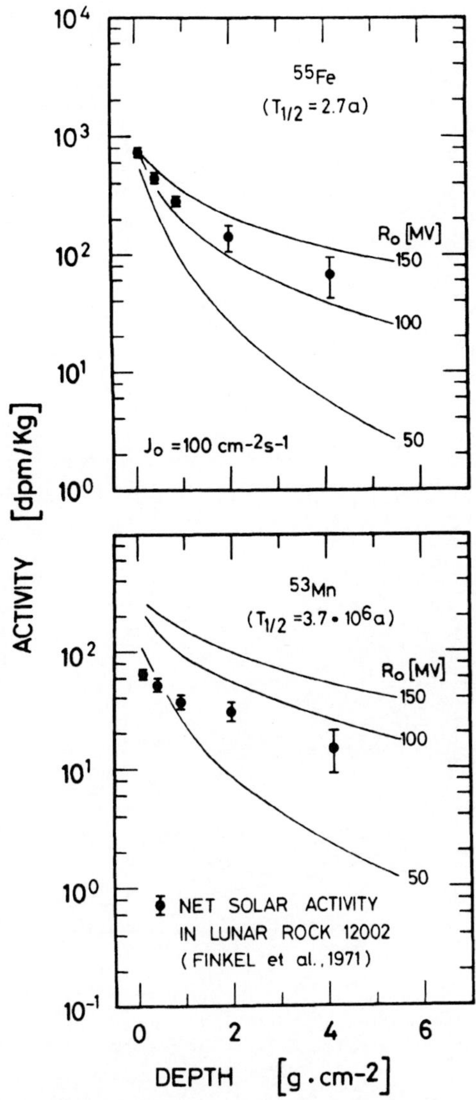

Fig. 2. Observed net solar saturation activities [6] and theoretical production depth profiles of ^{55}Fe and ^{53}Mn in lunar rock 12002.

genic nuclides it is possible to detect variations of the solar activity on a time-scale of several million years by comparing the production rates of short-lived with long-lived radionuclides. The first ones are representing the status of the present and the latter ones are recording the ancient sun.

Besides, satellite data of individual solar flare events, the radionuclides ^{56}Co, ^{54}Mn, ^{55}Fe and ^{44}Ti are well suited to describe the today's sun. Due to their particular lifetimes these nuclides reveal influences of particular flare events or give time-averaged SCR-fluxes which cover the solar eleven year's activity cycle. On the other hand, long-lived nuclides as ^{26}Al, ^{10}Be and ^{53}Mn provide information about the sun up to several million years ago. In Fig.2 experimental SCR-production rates for ^{55}Fe and ^{53}Mn are compared with the theoretical depth profiles. The ^{55}Fe-data in lunar rock 12002 are fairly well described by a 4π integral flux, $J_o(4\pi, E_p > 10 \text{ MeV}) = 100 \text{ cm}^{-2} \cdot s^{-1}$ and a characteristic rigidity between R_o = 100 and 150 MeV.

The ^{53}Mn-data of rock 12002 seem on the first glance to reveal a lower J_o and a higher characteristic rigidity. But from a crude comparison it can already be seen that the solar flux parameters could not have changed by more than a factor of 2 with regard to R_o and a factor of 3 with regard to J_o.

The presentation of the ^{53}Mn depth profiles in Fig.2, however, is rather simplified in one essential point, namely it neglects the phenomenon of the space erosion, which for rock 12002 is about 0.5 mm/10^6 a ref.[6]. This neglect of space erosion lowers the apparent J_o- and increases the apparent R_o-value. A detailed analysis also accounting for space erosion, which is beyond the topic of this paper, changes the theoretical production rate depth profiles, so that the experimental ^{53}Mn data are best fitted by J_o=70 $cm^{-2}s^{-1}$ and R_o=100 MV [7].

Taking into consideration all the investigations dealing with this question, there is no evidence for the solar proton flux to have differed from the recent flux during the last 10^7 a. The uncertainties of the flux parameters are those given in Fig.1 for the SCR-spectra. Several investigations yielded ranges of $J_o(4\pi, E_p > 10 \text{ MeV}) = 70 - 140 \text{ cm}^{-2-1}$ and R_o=100 - 150 MV (see [8] for detailed references).

In spite of the fact that the SCR contains about 90 % protons, also the influence of the solar α-particles cannot be neglected for the production of some particular cosmogenic radionuclides. Contributions of solar α-particles are to be seen in the nuclides ^{57}Co, ^{58}Co, and ^{59}Ni with the magnitude of the α-contribution increasing in this order. In Fig.3 we have plotted theoretical depth profiles for the p- and α-induced production of ^{59}Ni in lunar rock 12002 demonstrating that up to depth of about 3 g/cm^2 the α-induced production exceeds that due to solar protons.

From the analysis of lunar rock 12002 Lanzarotti et al. [9] derived that the ratio p/α = 10 remained constant over the last 10^5 years within a factor of approximately 4. The main uncertainty of this result was lying in the scarcely known excitation functions. On the basis of our newly established excitation functions which also were used for the depth profiles of Fig.3 this constancy may be claimed even within a factor of 2.

Besides these applications of SCR-produced nuclides in lunar surface materials there are several investigations regarding the irradiation history of individual stones and the dating of particular impact events. Even special irradiation conditions due to single

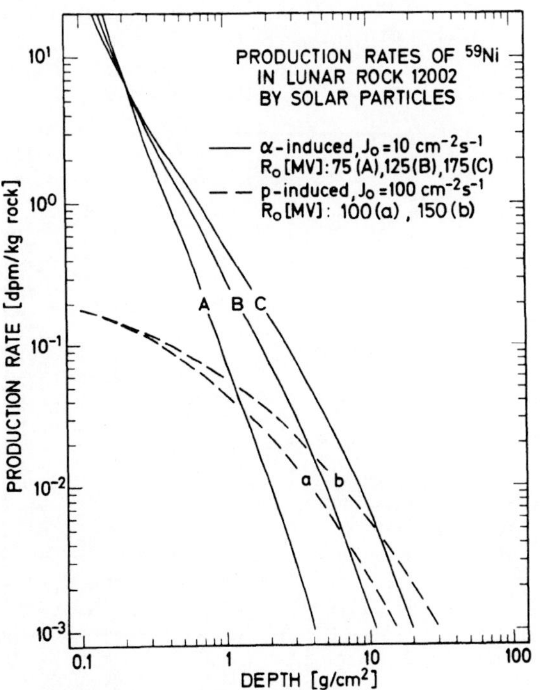

Fig. 3. Depth dependent production of ^{59}Ni (T = 7.5 x 10^4 a) in lunar rock 12002 by solar protons and α-particles.

flares were reported. A survey may be found elsewhere [4].

Effects of Solar Protons in Meteorites

For the production of cosmogenic nuclides in meteorites several model calculations exist [10-12], which, however, are not as satisfactory as for lunar samples. This is mainly due to the fact that the geometry dependent built-up of the field of GCR-secondaries is still not known. Besides, for meteorite calculations it is always adopted that the effects of solar particles are wiped out by the ablation of surface material during the atmospheric transit of the meteorite. However, the availability of particular meteorite pieces having suffered few ablation and of an increasing number of small meteorites from the Antarctic meteorite search make it possible to detect SCR-effects also in meteorites and to investigate the solar particle fluxes not only at 1 A.U. but also at the meteorite orbits of 3-5 A.U.. Recently, we published comprehensive model calculations of the SCR-interactions in meteorites [8] based on our new set of excitation functions. These calculations give evidence that for a successful interpretation of cosmogenic nuclides in surface samples and in particularly small meteorites one has to consider also the interaction of solar particles with meteoritic material.

In order to demonstrate up to which meteorite sizes SCR-effects should be dominant, in Fig.4 we have compared the SCR-production of ^{53}Mn in the centre of spherical stone meteorites adopting $J_o(4\pi, E>10$ MeV$)$ = 100 $cm^{-2}s^{-1}$ and R_o=100 and 150 MV with estimates of the centre production rates due to galactic cosmic rays. The latter estimates (together with a 10 % uncertainty range) are based on experimental data for high meteorite radii (R>15 cm) and on the theoretical GCR-production rate of 80 dpm ^{53}Mn/kg Fe at R = 0, representing the production due to primary GCR-protons with $J_o(4\pi, E>1$ GeV$)$ = 1.7 $cm^{-2}s^{-1}$. The cross sections were the same as discussed in our earlier work [8].

Since in the centre points of meteorites the SCR-contributions are minimal and the GCR-production rates are maximal, the data of Fig.4 are well suited to forecast that up to meteorite radii about 5 cm SCR-effects are not to be neglected even if one considers the expected radial gradient of solar particle fluxes between the earth and meteorite orbits [8].

Besides investigations of the constancy of solar particle fluxes in time and space, the study of

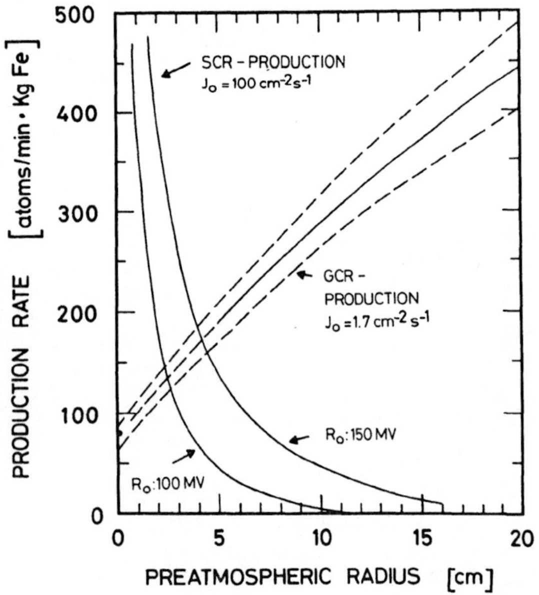

Fig. 4. Production of ^{53}Mn in the centre of spherical H-chondrites (ρ = 3.66 $g \cdot cm^{-3}$).

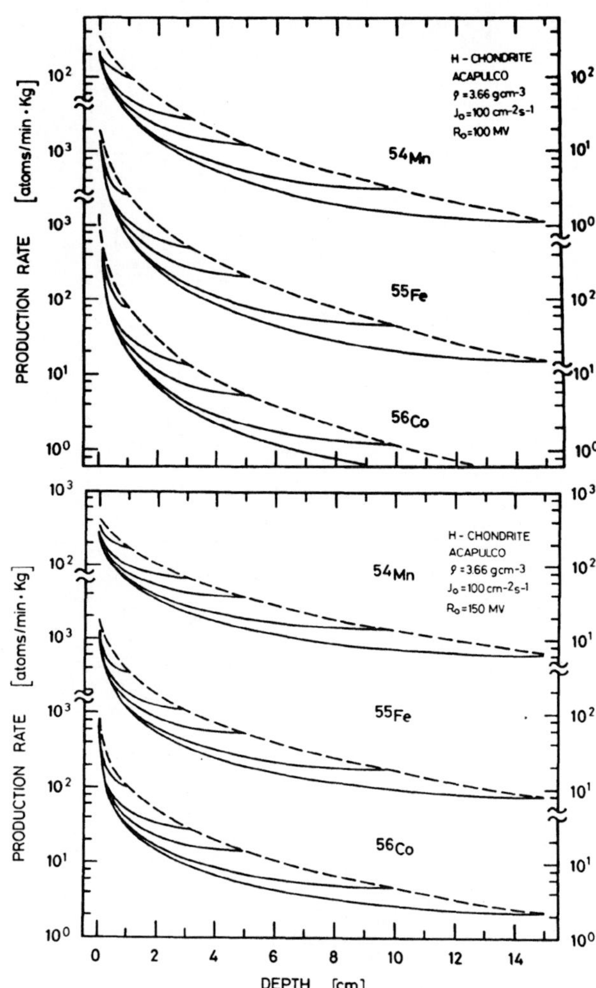

Fig. 5. Depth profiles for the SCR-production of ^{54}Mn, ^{55}Fe and ^{56}Co in H-chondrites for preatmospheric radii of 1, 3, 5, 10 and 15 cm.

SCR-effects of shortlived radionuclides, such as ^{54}Mn, ^{55}Fe and ^{56}Co (Fig.5) in freshly fallen meteorites will allow for the reconstruction of the final earth-crossing orbits of these extraterrestrial bodies. This will offer the unique opportunity to untackle the "last moment" irradiation history of these objects. Besides these depth profiles demonstrate that even in larger objects the GCR-effects are not to be neglected in surface samples. So for instance in the case of the H6-chondrite San Juan Capistrano which fell on March 15,1973 and which had a preatmospheric radius of 8 to 10 cm. Finkel et al.[13] reported for a sample 6 cm below the surface saturation activities of 60±8 dpm ^{54}Mn/kg rock, 45±18 dpm ^{55}Fe/kg rock and 16±7 dpm ^{56}Co/kg rock. These activities are by far exceeded by the production rates in the outmost 2 cm of a H-chondrite with 8 to 10 cm radius.

Up to now we only dealt with radioactive isotopes and also our former cross section work was directed to radioactive product nuclides. But with respect to the investigation of cosmogenic nuclides in extraterrestrial matter the production of stable cosmogenic nuclides, namely of rare gas isotopes, is of particular interest. On the one hand, this is because of the larger time interval that is recorded by the stable isotopes and, on the other hand, because of the analytical power of rare gas measurements in extraterrestrial matter.

Recently, Bhandari and Potdar [14] reported on the GCR-production of Ne isotopes in meteorites, giving absolute depth profiles for the production of Ne-isotopes as well as results for the ^{22}Ne/^{21}Ne ratio which is regarded as a reliable depth indicator in meteorites.

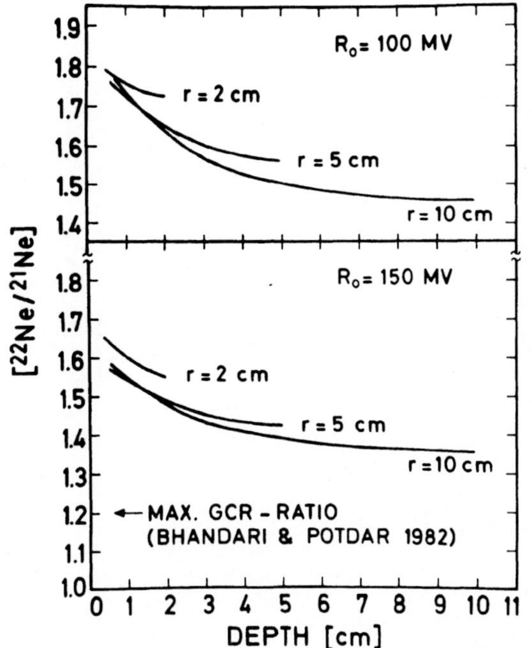

Fig. 6. $^{22}Ne/^{21}Ne$ ratios of solar proton produced Ne in meteorites. The maximum GCR-ratios were calculated by Bhandari and Potdar[14].

In spite of the fact that the depth dependent GCR-fluxes adopted by these authors may not be the final solution of the GCR problem, it is of interest to compare their results with those for pure SCR-production rates.

For stable rare gas production rate calculations still a considerable lack of high precision integral excitation functions exist. Such data are urgently needed for more reliable model calculations. For the Ne-isotopes Bhandari and Potdar [13] reviewed the published cross section data. Without discussing the quality and the inherent discrepancies of these cross section data, we have calculated the depth dependence of the $^{22}Ne/^{21}Ne$ ratio as well as the ^{21}Ne depth profiles due to pure SCR-production (Figs.6 and 7). It is to be seen that the SCR $^{22}Ne/^{21}Ne$ ratios exceed by far the maximum GCR-ratio of $^{22}Ne/^{21}Ne$ = 1.2 and also the measured ^{21}Ne-GCR-production rates in meteorites which are in the range of 0.3 to $0.5 \cdot 10^{-8} cm^3 STP/10^6$ a • g.

At the moment, surely the interpretation of cosmogenic nuclides in meteorites is still not satisfactory, but at least the description of the SCR-effects in meteorite surfaces and in small meteorites can be tackled with sufficient accuracy, since these calculations can be done with a good a priori accuracy which is only determined by the accuracy of the cross section data used. Here high precision measurements of integral excitation functions provide the basis for a successful investigation of the irradiation history of extraterrestrial matter.

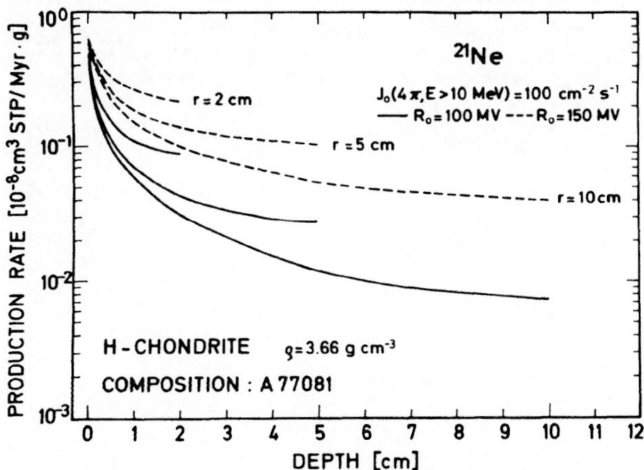

Fig. 7. Depth profiles of the SCR-production of ^{21}Ne in meteorites.

Acknowledgment

This work was supported by the Bundesminister für Forschung und Technologie, Bonn.

References

1. R.C. Reedy, J.R. Arnold, J.Geophys.Res.77,537 (1972)
2. Y. Yokoyama, R. Auger, R. Bibron, R.Chesselet, F. Guichard, C. Leger, H. Mabuchi, J.L. Reyss, J. Sato, Geochim.Cosmochim.Acta, Suppl.3,2 1733 (1972)
3. H. Weigel, R. Michel, U. Herpers, W. Herr, Radiochimica Acta 21, 179 (1974)
4. R. Michel, G. Brinkmann, J.Radioanal.Chem. 59, 467 (1980)
5. R. Michel, G. Brinkmann, R. Stück, Radiochimica Acta; submitted (1982)
6. R.C. Finkel, J.R. Arnold, M. Imamura, R.C. Reedy, J.S. Fruchter, H.H. Loosli, J.C. Evans, A.C. Delany, J.P. Shedlovsky, Proc.Sec.Lun.Sci. Conf., Geochim.Cosmochim.Acta, Suppl.2,2, 1773 (1971)
7. C.P. Kohl, M.T. Murrell, G.P. Russ III, J.R. Arnold, Proc.Lun.Sci.Conf.9th, Geochim.Cosmochim. Acta, Suppl.9,2, 2299 (1978)
8. R. Michel, G. Brinkmann, R. Stück, Earth Planet. Sci.Letters 59, 33 (1982)
9. L.J. Lanzerotti, R.C. Reedy, J.R. Arnold, Science 179, 1232 (1973)
10. A.K. Lavrukhina, G.K. Ustinova, Earth Planet. Sci.Letters 15, 347 (1972)
11. R.C. Reedy, G.F.Herzog, E.K. Jessberger, Earth Planet.Sci.Letters 44, 341 (1979)
12. S.K. Bhattacharya, M. Imamura, N. Sinha, H. Bhandari, Earth Planet.Sci.Letters 51, 45(1980)
13. R.C. Finkel, C.P. Kohl, K. Marti and B. Martinet, Geochim.Cosmochim.Acta 42, 241 (1978)
14. N. Bhandari and M.B. Potdar, Earth Planet.Sci. Letters 58, 116 (1982)

CHARGED PARTICLE REACTION CROSS SECTIONS RELEVANT FOR NUCLEAR ASTROPHYSICS

J. Szabó, M. Várnagy, Z.T. Bődy, J. Csikai

Kossuth University
Institute of Experimental Physics
P.O.B. 105, H-4001 Debrecen
Hungary

Cross section and cross-section factor data have been determined for (p,α), (d,α), (d,p) and (d,n) reactions on ^{6}Li, ^{7}Li, ^{10}B between $E_{p,d}$= 60-180 keV using activation or/and direct detection methods. Different methods of extrapolation for cross section data have been employed and compared down to the energy region of astrophysical interest.

[^{10}B(p,α), ^{6,7}Li(p,α), ^{6}Li(d,α), ^{6}Li(d,p), ^{6}Li(d,n), cross sections, cross-section factors, $E_{p,d}$= 60-180 keV, deduced astrophysical cross sections]

Introduction

The investigation of nuclear reactions induced by low-energy charged particles on light nuclei is important from the point of view of nuclear astrophysics as well as for planning controlled thermonuclear reactors (CTR). The cross sections of these nuclear processes are very small at the relevant stellar energies and they cannot be measured directly in laboratories. The usual procedure is to study nuclear reactions over a wide interval, starting with the lowest practical energy, and then to extrapolate the data to the stellar energy region. The solid state nuclear track detectors (SSNTD), as they have a lot of advantages [1,2], proved to be useful for the investigation of charged particle reactions having small cross sections.

Experimental Technique

For the measurements a Cockcroft-Walton generator of 180 kV was used. The energy calibration of the accelerator was performed using the ^{11}B(p,γ)^{12}C resonance at E_p = = 163.1 keV and the current of the analyser magnet was calibrated by measuring the high voltage in a direct way. The uncertainty in the bombarding energy was estimaged to be less than 4 keV. In every case thick targets were bombarded with analyzed and collimated proton or deuteron beams with an intensity of 1-10 μA. To determine the cross section and cross-section factor data,two detection methods were used:i) direct method by SSNTD, ii) activation method by NaI(Tl) or Ge(Li) detectprs. The details of the measurements and evaluations can be found in our pa - pers [3-7].

Cross section determination and other results

^{10}B(p,α)^{7}Be reaction [3]

This reaction is an important destructive process and may have considerable effect on the extremely low boron abundance in the universe. The measurements were made with target of ^{10}B, the enrichment being to 93% in ^{10}B and the intensity of proton beam was about 10 μA. To measure the cross sections, activation and direct α detection methods were used in the energy range of E_p = = 100-180 keV and E_p= 60-180 keV, respectively. In the case of activation method, the 477.6 keV γ-rays coming from the first excited state of ^{7}Li following the decay of ^{7}Be were detected by a NaI(Tl) detector of 7.6x5.0 cm. At the direct method for the detection of α-particles SSNTD (T-Cellit, cellulose acetate) were used. Cross sections values were calculated by the well-known formula

$$\sigma(E_0) = \frac{1}{N}\left(\frac{dY}{dE}\right)_{E_0}\left(\frac{dE}{dx}\right)_{E_0} \qquad (1)$$

which is valid for a thick target, are summarized in Table I.

Table I The measured cross section data for the ^{10}B(p,α)^{7}Be reaction

E_p (keV)	σ (μb) activation method	σ (μb) direct method
180	882 ± 125	1277 ± 164
160	560 ± 88	744 ± 96
140	332 ± 38	320 ± 38
120	148 ± 17	202 ± 17
100	51.0 ± 8.5	60.9 ± 6.9
80		16.2 ± 1.3
60		1.41 ± 0.04

This arrangement also rendered it possible to determine the branching ratio ^{7}Be(ε)^{7}Li, which was found to be R = 0.104 ± 0.003. Using the experimental cross section data, the values of the cross-section factor S(E) were calculated [8]. The S=S(E) function obtained by both activation and direct methods was found not to be constant, which can be explained by the presence of the 8.694 MeV state in the compound nucleus ^{11}C having a width of $\Gamma \approx$ 300 keV. From astrophysical point of view the cross section value is especially important at the Gamow energy (E_0). For a temperature of T_6=13 E_0 is 19.1 keV. Using the S(E) values extrapolated to the Gamow energy a cross section $\sigma(E_0) \approx 1.0\times10^{-36}$ cm^2 was estimated.

^{6,7}Li(p,α)^{3,4}He reaction [4]

The purpose of this investigation was i) to clear the disagreements between the values obtained for the cross section and cross -

K. H. Böckhoff (ed.), Nuclear Data for Science and Technology, 956–957.

-section factor data in the case of $^6Li(p,\alpha)^3He$ reaction below E_p= 200 keV, ii) to study the possibility of mass discrimination for ^3He and ^4He particles having about the same energy. For the measurement natural lithium targets were used. The intensity of proton beam on the target was about 10 μA. To detect ^3He and ^4He SSNTD (cellulose-acetate, T-Cellit) has been applied so that the SSNTD sheets covered an angular interval of 45-165° for the outgoing particles. The alternate-etch method, developed in our laboratory, proved to be useful to obtain well visible and distinguishable tracks of the particles from the $^6Li(p,\alpha)^3He$ reaction on one surface, and the ^4He tracks from the reaction $^7Li(p,\alpha)^4He$ on the other surface of the same sheet simultaneously. Using this method, angular distribution and cross sections can easily be determined. Cross section values obtained for the $^6Li(p,\alpha)^3He$ reaction in the energy of E_p= 100-180 keV are shown in Table II.

Table II Cross section values for the $^6Li(p,\alpha)^3He$ reaction

E_p (keV)	σ (mb)
100	2.65
120	4.37
140	7.16 (± 15 % for all values)
160	11.78
180	19.49

The cross section factor was calculated and found to be constant with a value $S(E) \approx$ 2.7 MeV.barn.

$^6Li(d,\alpha)^4He$, $^6Li(d,p)^7Li$, $^6Li(d,n)^7Be$ reactions[5-7]

In the progress of investigations on CTR, exotic reactions are becoming more and more interesting. Among these the ^6Li+d process has the highest priority and cross section data are needed with a precision less than 25% over a wide energy range (E_d=0.1-5 MeV). The measurements were performed with ^6LiF target enriched to 87.9 % in ^6Li. The deuteron beam intensity was 1-3 μA. In the case of (d,α) and (d,p) reactions direct detection method was used. Two 100 μm thick SSNTD sheets (cellulose nitrate, CA 80-15) covered an angular interval of 45-165°, and made it possible to registrate both alpha-particles and protons, i.e. to determine the angular distributions simultaneously. For the (d,n) reaction activation method was used with a Ge(Li) detector of 40 cm^3. The process to get the cross section values was similar to the one used in ref.[3] as the residual nucleus ^7Be is the same. Cross section data for all the three reactions were calculated by formula (1) and are shown in Table III.

Table III Measured and extrapolated (*) cross section data for $^6Li(d,\alpha)^4He$, $^6Li(d,p)^7Li$, $^6Li(d,n)^7Be$ reactions

E_d(keV)	(d,α)(mb)	(d,p) (mb)	(d,n) (mb)
100	0.55±0.05	0.76±0.07	0.63±0.16
120	0.91±0.09	1.30±0.17	1.08±0.22
140	1.52±0.13	2.23±0.23	1.86±0.33
160	2.52±0.22	3.81±0.43	3.19±0.62
180	4.16±0.32	6.49±0.55	5.44±1.28
16*	14 pb	18 pb	15 pb

Extrapolation of cross sections down to low energy region

For astrophysical purpose the cross sections have to be extrapolated down to the thermal energy region. It is common to factorize the cross section in terms of a rapidly changing penetration factor (P) and the cross-section factor (S) varying slowly with energy (E):

$$\sigma(E) = \frac{S}{E} P \quad (2)$$

There are several suggestions [8,9] for the E-dependence of S:

S= constant, (3a)

S depends linearly on energy (3b)

$$S = \frac{t}{(1-\gamma D)^2 + (\gamma P)^2} \quad (3c)$$

Formula (3c) was deduced from R-matrix theory, here t and γ are adjustable parameters, D and P can be expressed in terms of Coulomb wave functions and their derivatives taken at the channel radius as a parameter. Formulae (3) were investigated for the ^6Li+d reactions. It seems that the use of formula (3c) is the best one. However, due to inaccuracies in experimental data, sophisticated methods having more parameters do not offer any extra advantage For all the three ^6Li+d reactions extrapolated cross sections were calculated at E_d= 16 keV. Results obtained are indicated in Table III under the mark * .

References

1. M. Várnagy, Thesis (Kossuth University, Debrecen, 1970).
2. G. Somogyi, I. Hunyadi, E. Koltay and L. Zolnai, Nucl. Instr. and Meth. 147 (1977) 287
3. J. Szabó, J. Csikai and M. Várnagy, Nucl. Phys. A195 (1972) 527
4. M. Várnagy, J. Csikai, J. Szabó, S. Szegedi and J. Bánhalmi, Nucl. Instr. Meth. 119 (1974) 451
5. M. Várnagy, J. Szabó and S. Szegedi, Nucl. Instr. Meth. 154 (1978) 557.
6. Z.T. Bődy, J. Szabó and M. Várnagy, Nucl. Phys. A330 (1979) 495.
7. J. Szabó, Z.T. Bődy, S. Szegedi and M. Várnagy, Nucl. Phys. A289(1977) 526
8. D.D. Clayton, Principles of stellar evolution and nucleosynthesis (McGraw-Hill, New York, 1968)
9. J.E. Monahan, A.J. Elwyn and F.J.D. Serduke, Nucl. Phys. A269 (1976) 61

Solid State, Molecular and Atomic Effects on Nuclear Data

SOLID STATE EFFECTS ON THERMAL NEUTRON CROSS SECTIONS AND ON LOW ENERGY RESONANCES

J. A. Harvey, H. A. Mook, N. W. Hill and O. Shahal†
Oak Ridge National Laboratory, Oak Ridge, Tennessee 37830 USA

The neutron total cross sections of several single crystals (Si, Cu, sapphire), several polycrystalline samples (Cu, Fe, Be, C, Bi, Ta), and a fine powder copper sample have been measured from 0.002 to 5 eV. The Cu powder and polycrystalline Fe, Be and C data exhibit the expected abrupt changes in cross section. The cross section of the single crystal of Si is smooth with only small broad fluctuations. The data on two "single" Cu crystals, the sapphire crystal, cast Bi, and rolled samples of Ta and Cu have many narrow peaks ~ 10^{-3} eV wide. High resolution (0.3%) transmission measurements were made on the 1.057-eV resonance in ^{240}Pu and the 0.433-eV resonance in ^{180}Ta, both at room and low temperatures to study the effects of crystal binding. Although the changes in Doppler broadening with temperature were apparent, no asymmetries due to a recoilless contribution were observed.

[Si, Cu, Be, C, Bi, Ta, Sapphire, σ_T, E_n = 0.002-5 eV, 1.057-eV resonance ^{240}Pu, solid state effects, Doppler broadening]

Introduction

In the thermal energy region the measured total cross section often depends on the physical form of the sample. The total cross sections of polycrystalline materials such as Be, C and Fe show abrupt changes of ~ 2 barns or ~ 30% at energies corresponding to scattering at 180° from lattice planes in the crystal as shown in Figure 1. At higher energies, $\gtrsim$ 0.2 eV, so many planes are involved that the cross section is quite smooth (providing there are no nuclear resonances). At energies $\gtrsim$1 eV, the recoil energy is sufficient that the atoms are knocked out of the lattice and the free atom cross section is observed. Severe self-shielding or extinction effects can occur in the thermal energy region if the crystal grains are not very fine (~ 10^{-4} cm); the observed cross section of a polycrystalline material can be reduced by several barns. The cross section of a single crystal at thermal energies can be very small at room temperature and can be reduced even further by cooling the crystal. Large single crystals are in use both at room and low temperatures as neutron band pass filters[1] to pass thermal neutrons while excluding fast neutrons and gamma rays.

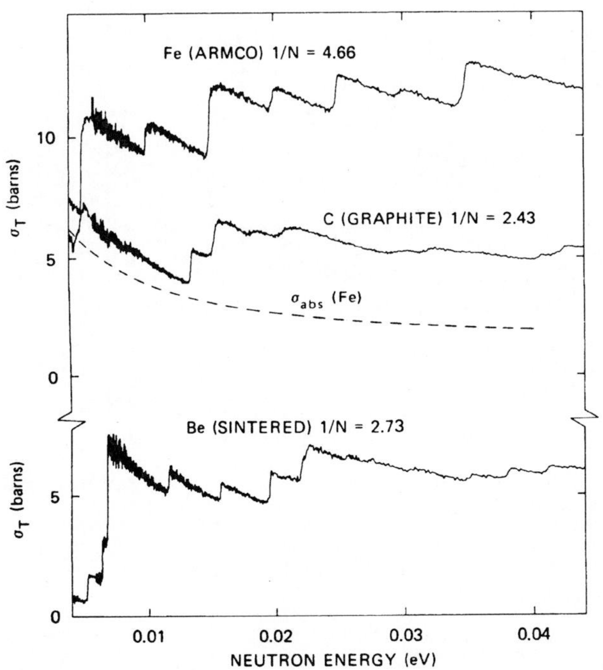

Fig. 1. Total cross sections of polycrystalline Fe, C and Be.

The thermal motion of the atoms in the sample also influences the cross section of a nuclear resonance. In 1937, Bethe and Placzek[2] calculated the Doppler broadening assuming a Maxwellian velocity distribution of gas atoms. In 1939, Lamb[3] calculated the resonance shape for nuclei bound in a Debye crystal. In the weak binding limit where the Doppler width, Δ, plus the natural width, Γ, is >> $2\theta_D$ (θ_D = Debye temperature), the observed resonance shape would be the same as for a gas, but an effective temperature which is somewhat greater than the temperature of the crystal must be used. Lamb also predicted that if an atom were more tightly bound in the lattice, i.e., medium binding where $(\Delta + \Gamma) \sim 2\theta_D$, some fine structure should appear in the resonance shape. Several experiments have been performed to observe the structure of resonances in both solids[4,5,6] and gases.[6,7] For a crystal at a temperature $< \theta_D$ and for a low energy resonance with a small Γ, Lamb showed that a "recoilless" Breit Wigner peak could be expected, as well as structure or general broadening at higher energies. Trammell[8] has pointed out that a detailed study of the 1.057-eV resonance in ^{240}Pu might be able to give detailed information concerning the crystal vibrational spectrum. Recently, Liou and Chrien[9] studied this resonance using both metal and oxide samples.

Experimental Method

The neutron total cross sections of several single crystals (silicon, copper, sapphire), several polycrystalline samples (rolled Cu, armco Fe, sintered Be, graphite, cast Bi and rolled Ta) and a fine powder copper sample were measured from 0.002 to 5 eV. The measurements were made at ORELA using neutrons from the water moderated Ta target with the linac producing ~ 30 nsec pulses at 25 Hz. A 1 mm thick 111 mm dia. ^{6}Li glass scintillator was located at 17.870 m resulting in a neutron energy resolution $\Delta E/E$ = 0.3%. A few measurements were made on the "single" copper crystal with a 25 mm dia. ^{6}Li glass scintillator to reduce the angular divergence of the detected neutrons. The samples were placed at 9 m and the effect of a small rotation of the Cu "single" crystal from the direction of the neutron beam was studied. Samples were alternated with an open beam and background samples every ~ 30 minutes; runs were of 1-2 days duration.

Transmission measurements were also made on ^{240}Pu sample using a 17.858-m flight path at 400 Hz. A Cd filter was used to eliminate overlap neutrons and 2 cm of Pb to reduce the intensity of gamma rays from the target. Measurements were made at room temperature and at 95 K which is much lower than the Debye temperature (~ 175 K). In the 1 eV energy region ~ 40000 counts per channel were collected for a channel width of 0.5 x 10^{-3} eV. The sample was an Al-Pu alloy containing only 0.726% ^{240}Pu with a thickness of ^{240}Pu of 2.12 x 10^{-6} a/b. The uniformity of the sample was ~ 2% over the area used. The Ta metal sample consisted of 15 1.5 mm thick rolled sheets and was measured at 11 K and room temperature using a repetition rate of 25 Hz. The

†Visiting scientist from Nuclear Research Centre-Negev, Beer-Sheva, Israel.

K. H. Böckhoff (ed.), Nuclear Data for Science and Technology, 961–964.

$^{180}Ta_2O_5$ sample was a fine powder and was measured at room temperature and 175 K. Each sample was measured for several days at each temperature.

Data Analysis

The time-of-flight spectra were corrected for the dead-time of the time digitizer, 1.1 μsec, and a constant room background which was ≲1%. Transmissions were computed using the monitor counts from a fission chamber; the total cross sections were computed from the average thicknesses of the samples.

The resonance parameters for the 1.057-eV resonance in ^{240}Pu were obtained by R. F. Carlton using the code SAMMY[10] and by R. R. Spencer using the code SIOB.[11]

Results and Discussion

The total cross sections of polycrystalline Fe, C and Be samples are shown in Figure 1. The scattering cross section for this Armco Fe sample is ~ 9 barns in the thermal energy region and drops to ~ 1 barn at 0.0050 eV, which corresponds to twice the lattice spacing of the 1,1,0 planes for Fe; Fe is body centered cubic with a = 2.86645 A. At lower energies the scattering arises from isotope incoherent scattering (0.4 b) and thermal inelastic scattering; the latter can be reduced by cooling the sample. The other breaks arise from planes with higher Miller indices. Recently, Johnson and Bowman[12] have observed a large number of these breaks at higher energies in samples of rolled Fe.

The scattering cross sections of sintered Be and C (graphite) are ~ 5 b except at low energies. (The absorption cross sections of Be and C are only 10 and 3 mbs.) The breaks in Be at 5.3 and 6.9 meV are well known, but the intermediate one at 6.5 meV has not been reported previously, probably because of insufficient energy resolution. The size of the step depends on the number of atoms in the reflecting plane and the Debye-Waller factor. Above each break the coherent scattering cross section has a 1/E energy dependence. At low energies the thermal inelastic scattering can be reduced to <0.1 b by cooling the sample. Thick filters of polycrystalline materials have been used at reactors to produce beams of "cold" neutrons.

The total cross sections for the copper powder sample and two different measurements on a "single" copper crystal are shown in Figure 2. The scattering is ~ 9 barns for the copper powder but only ~ 2 barns for the "single" crystal. This reduction is consistent with the coherent cross section of 7.5 b. The remainder arises from isotope and spin incoherent scattering (~ 0.5 b), and thermal inelastic scattering. In addition, there are many narrow peaks associated with lattice planes in the single crystal. The widths of the peaks is a measure of the mosiac spread of the crystal, the angular resolution, and the neutron energy resolution. The two lower curves are for the same Cu crystal but oriented differently with respect to the direction of the neutron beam. Additional measurements have been made rotating the crystal by ~ 0.2°, and the peaks shifted appreciably in energy (~ 1 meV) with this small change in angle.

The data for cast Bi are shown in Figure 3. The scattering is ~ 40% of that expected from the free atom scattering of 9.3 barns. The extinction effects are due to the large size of the crystallites in this cast sample. The two low energy peaks at 3.38 and 4.22 meV are only ~ 0.04 meV wide. Since the absorption and the spin incoherent cross sections of Bi are very small, 9 and ~50 mbs respectively, a filter of a cold large single crystal of Bi in a reactor neutron beam would be a valuable technique for eliminating gamma rays and greatly reducing resonance and high energy neutrons from the neutron beam.

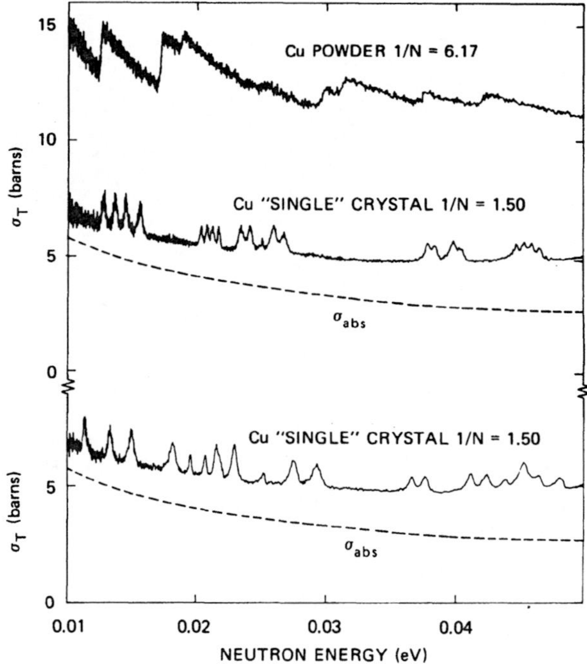

Fig. 2. Total cross sections of Cu powder and two measurements on a "single" Cu crystal.

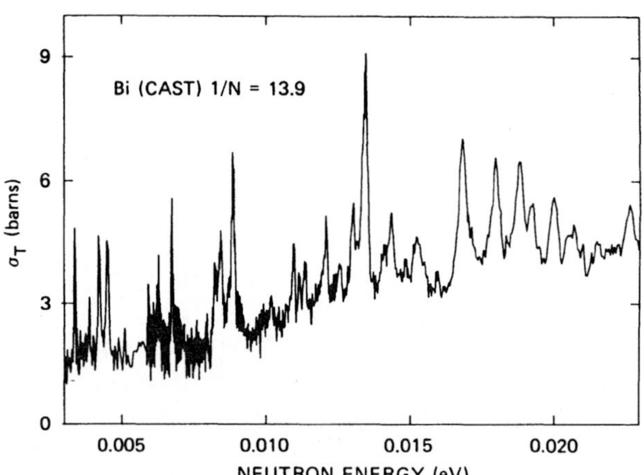

Fig. 3. Total cross section of Cast Bi.

Figure 4 shows the data for single crystals of Si and sapphire. The scattering of Si at thermal energies is only ~ 10% of its value above a few eV energy and arises from incoherent scattering (9 mb) and thermal inelastic scattering. This "perfect" single crystal has a very small mosiac spread; hence, the intensities of any narrow peaks are too small to be observed. The sapphire "single" crystal shows much structure due to the mosiac spread. The gradual increase of the cross section in Figure 5 is due to the Debye-Waller factor and the small broad fluctuations with ~ 50 meV spacing may be due to crystal vibrations.

The transmission of the rolled Ta sample (consisting of 15 1.5 mm thick plates) is shown in Figure 6. The sharp break at 11.3 meV corresponds to scattering from

1,1,0 planes expected for a sample with small crystallites. The sharp peaks are probably due to coherent scattering from lattice planes similar to that observed with the Cu single crystal. The broad structures must be combinations of sharp breaks, overlapping clusters of narrow peaks and possibly phonons.

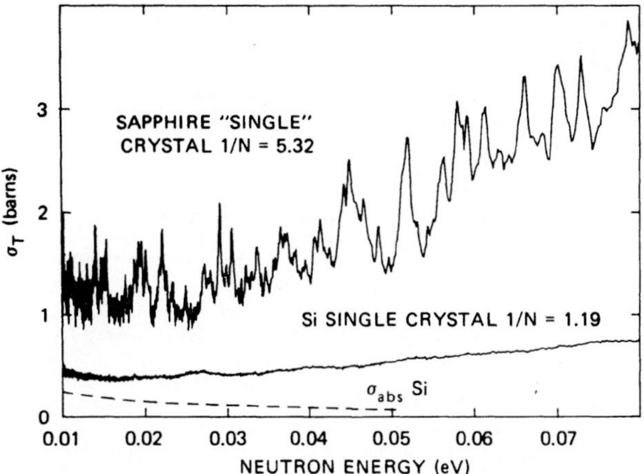

Fig. 4. Total cross sections of Si single crystal and sapphire.

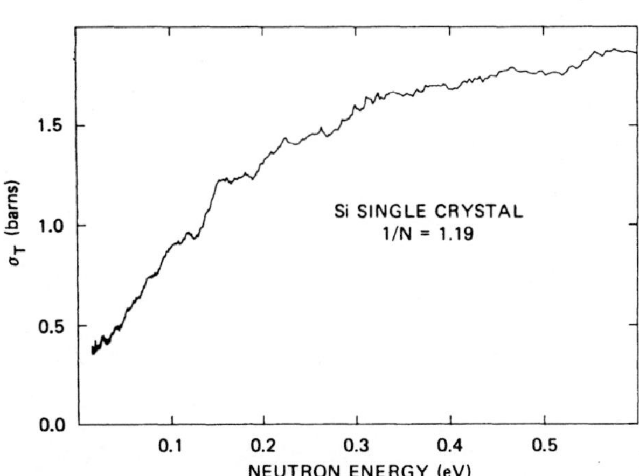

Fig. 5. Total cross section of the Si single crystal.

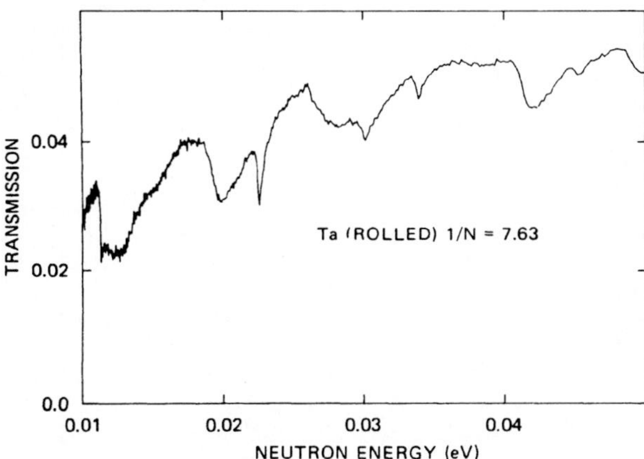

Fig. 6. Total cross section of rolled Ta.

The transmission data for the 1.057-eV resonance in ^{240}Pu are shown in Figure 7 with fits from the code SAMMY using the free gas approximation for the Doppler broadening. The effective temperatures used for the two measurements were 122 K and 300 K based on a Debye temperature of 175 K. The goodness of fit was 0.91 for the cold run and 0.93 for the room temperature run. Both runs gave exactly the same resonance energy of 1.0565 ± 0.0002 eV. Values of 30.6 ± 0.4 meV for Γ_γ and 2.35 ± 0.02 meV for Γ_n were obtained from the cold run and 30.8 ± 0.6 meV and 2.37 ± 0.02 meV from the room temperature run. The uncertainties do not include systematic uncertainties. A 9° K uncertainty of the effective temperature of the cold sample would produce a 0.6 meV uncertainty on Γ_γ. The average value of Γ_γ of 30.7 ± 0.6 meV is somewhat lower than the most recent determination.[9] Including an estimated uncertainty of 2% for the sample thickness, the average value for Γ_n of 2.36 ± 0.05 meV agrees with that obtained by Liou.[9]

Trammell[13] has estimated that at 95 K approximately 70% of the absorption by this resonance should be in the undisplaced (Mossbauer) peak and should have the nuclear width Γ. The remaining 30% should be at higher energies and should be Doppler broadened. The recoil energy after neutron absorption by this resonance is 4 meV. At room temperature only 17% would be in the

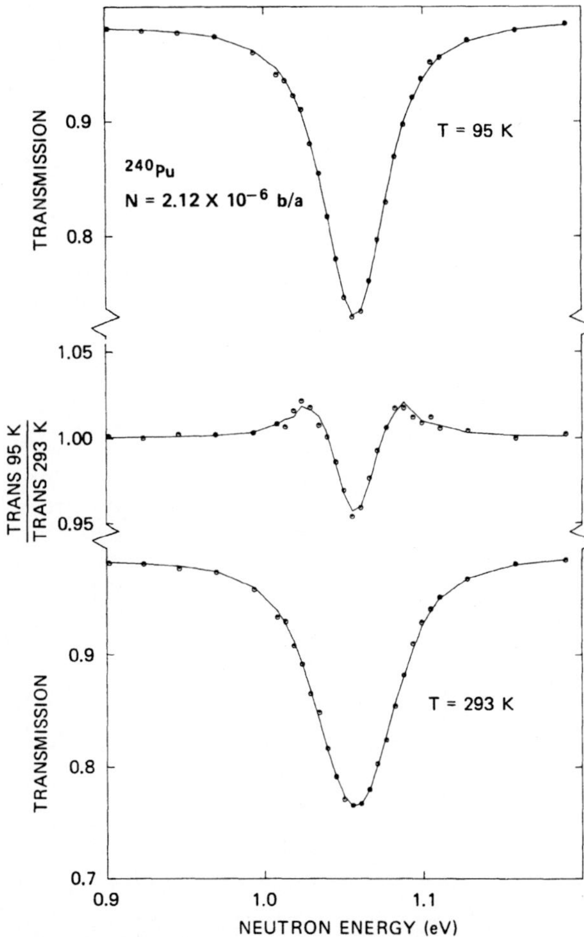

Fig. 7. Transmission of the ^{240}Pu sample at 95 K (top), transmission at 293 K (bottom) and ratio (middle). The statistical uncertainties in the region of the resonance are 0.2% for the transmissions and 0.3% for the ratio.

recoilless Mossbauer peak. The excellent agreement for both temperatures using the gas approximation and the fact that both gave the same resonance energy is not is not consistent with the predicted change in the intensity of the recoilless peak for these 2 temperatures. The middle curve in Figure 7 shows the ratio of the measured transmissions at the two temperatures compared to the ratio of calculations using the gas approximation. Any appreciable change in the intensity in the recoilless peak from the cold to room temperature should produce a noticeable asymmetry. Similar data on the 0.433-eV resonance in ^{180}Ta also did not show any asymmetry.

Conclusions

In the thermal energy region the physical form of a sample can have a large effect on the measured total cross section. Some crystalline materials and "single" crystals can have large self shielding and exhibit narrow peaks arising from the mosaic spread of the crystal. No convincing evidence was found for a recoilless peak for the 1.057-eV resonance in ^{240}Pu or the 0.433-eV resonance in ^{180}Ta.

Research was sponsored by the Division of Nuclear Sciences, U.S. Department of Energy, under Contract No. W-7405-eng-26 with the Union Carbide Corporation.

References

1. R. M. Brugger and W. Yelon, Proceedings of the Conference on Neutron Scattering, Gatlinburg, TN., June 6-10, 1976, Vol. II, p. 1117.
2. H. Bethe and G. Placzek, Phys. Rev. 51, 462 (1937).
3. W. E. Lamb, Phys. Rev. 55, 190 (1939).
4. H. H. Landon, Phys. Rev. 94, 1215 (1954).
5. H. E. Jackson and J. E. Lynn, Phys. Rev. 127, 461 (1962).
6. K. Seidel, A. Meister, D. Pabst, L. B. Pikel'neV, and W. Pilz, Sov. J. Nucl. Phys. 34(5), Nov. 1981.
7. C. D. Bowman and R. A. Schrack, Phys. Rev. 21, 58 (1980).
8. G. T. Trammell, Phys. Rev. 126, 1045 (1962).
9. H. I. Liou and R. E. Chrien, to be published.
10. N. M. Larson and F. G. Perey, Oak Ridge National Laboratory Report ORNL-7485.
11. G. deSaussure, D. K. Olsen, R. B. Perez, "SIOB: A Fortran Code for Least Squares Shape Fitting Several Neutron Transmission Measurements Using the Breit-Wigner Multilevel Formula", ORNL-TM-6286 (1978).
12. R. G. Johnson and C. D. Bowman, Symposium on Neutron Scattering, Argonne National Laboratory, August 12-14, 1981.
13. G. T. Trammell, private communication, May 1979.

MEASUREMENTS OF THE TOTAL NEUTRON CROSS-SECTIONS OF U AND UO_2 BELOW 2 eV AT DIFFERENT TEMPERATURES

M. Adib, R.M.A. Maayouf, A. Abdel-Kawy, A. Ashry, Y. Abbas, A. Abu-Zahra and I. Hamouda

Reactor and Neutron Physics Department, Nuclear Research Centre
Atomic Energy Establishment, Cairo, Egypt

The total neutron cross-sections of natural uranium and its oxide are measured using two time of flight spectrometers, installed in front of two of the ET-RR-1 reactor horizontal channels, and also by a neutron diffraction spectrometer. The measurements were carried out at room temperature in the energy range from 2 eV - 0.002 eV and at 210°C, for neutron energies below 0.005 eV. The coherent scattering cross-section of U was deduced both from the Bragg cut-offs observed in the behaviour of the total neutron cross-section of both U and UO_2 at cold neutron energies and the neutron diffraction pattern obtained at room temperature. The estimations of thermal diffuse scattering of U and UO_2 were compared with the experimental results obtained at different temperatures.

Introduction

Work has been started at the ET-RR-1 reactor on measurements of the total cross-sections of U and UO_2 at neutron energies below 1 eV. These measurements are carried out with time-of-flight technique, for micro-crystalline samples, at temperatures starting from the liquid nitrogen temperature and up to 1000°K. Such work should provide some new values for the incoherent and thermal diffuse scattering cross-sections. These are required both for the build up of the fuel element spectra, and the study of the effects of the phase transformation in the uranium crystal structure.

The total neutron cross-section measurements of U and UO_2 were carried out by Beshai [1] in the neutron energy range from 4.5 meV to 28 meV at 20°C. This work also contains some calculations, carried out both for U and UO_2, concerned with the energy positions of the Bragg edges. However Beshai reported that the measurements at 750°C were not successful due to the changes which take place in the sample's density[1].

The present work deals with total neutron cross-section measurements carried out both for U and UO_2 in the energy range from 2.5 meV - 1.0 eV at room temperature and at 483°K.

Experimental Details

The natural uranium sample

The natural uranium sample used is from U-0.05 at.% Ru alloy. The alloy was prepared under vacuum, in a Cu crucible, using an electron-beam melting furnace. The Ru powder was compressed into small pellets which were placed underneath the U-button before starting the melting operation. To obtain well homogenized alloy, the specimens were melted four times and were inverted after each melt.

Chemical analysis of 3 specimens obtained from each button showed that the alloy was perfectly homogeneous. Gamma graphic examination of every button showed that the specimens were perfectly sound and contained neither defects nor cavities. The chemical analysis of the uranium used in these experiments is given as:

(in ppm weight-average)

Cd	Cr	Cu	Fe	Mn	Ni	Si	Na	C	O_2	N_2	Al
< 0.5	6	5	50	5	21	27	5	295	53	44	< 20

As	Co	Li	Mg	Pb	K	Ag	Be	H_2
< 5	< 2	< 2	10	< 3	< 20	3	< 0.2	5

From the chemical analysis one can see that the contribution due to impurities, as well as the inserted Ru, to the total cross-section of U sample does not exceed 0.5 barns at the thermal point. One U sample of thickness 15.8 gm/cm^2 was used during the present cross-section measurements.

The lattice parameters a, b and c of the orthorombic alpha uranium lattice having addition of 0.05 at.% Ru were calculated from the x-ray diffraction charts. These charts were obtained from the U specimens using a Siemens x-ray diffractometer. Only the reflections of x-rays from the surface layers of a polycrystalline polished surface were recorded. The radiation employed was that from a cobalt target, with an iron filter to eliminate the K_β-radiation.

The diffraction lines, using a scintillation counter, were obtained over the range 2θ = 11 - 145°.

The x-ray diffraction pattern of the U-0.05 at.% Ru alloy showed that the (111), (002), (021), and (110) peaks were shifted from those of α-uranium to higher 2θ values; thus indicating contraction of the a, b and c parameters. These peaks were used for calculation of the lattice parameters by a trial and error method. The results are listed in Table I. It was pointed out that Ru is an effective grain refining element when added to uranium even in very small amounts.

The UO_2 Sample

The sample used was prepared from spec. pure fine UO_2 powder (grain size less than 20 μ). The UO_2 powder is packed in a Cu container with a thin Cu window (0.2 mm thick). One sample of thickness 8.43 gm/cm^2 was used during the measurements.

Table I Lattice Parameters of the U-0.05 at.% Ru Alloy

Ru (at.%)	2θ (deg.) for α-Uranium (h,k,l)				Lattice Parameters (Å)			Volume of Unit Cell (Å)	Average Grain size
	(110)	(021)	(002)	(111)	a	b	c	a.b.c.	μ
0	40.80	41.50	42.34	46.24	2.848	5.861	4.944	82.5259	200
0.05	40.82	41.52	42.35	46.25	2.846	5.860	4.942	82.4205	100

K. H. Böckhoff (ed.), Nuclear Data for Science and Technology, 965–967.

The neutron diffraction pattern of the UO_2 powder was measured using the crystal diffractometer, installed in front of one of the horizontal channels of the ET-RR-1 reactor. Monochromatic neutrons with wavelength λ = 1.02 A selected out of the reactor spectrum by diffraction from Zn single crystal cut along the (002) plane. The UO_2 sample was enclosed in a vanadium cylindrical container (5 cm height and 1.5 cm in diameter). The neutron diffraction pattern obtained for angles 2θ between 10°- 60° at room temperature is represented in Fig. 1. Eight well separated peaks were identified confirming the UO_2 crystal structure.

incoherently subtracted from the measured total cross-section; the residual cross-section (^{238}U) is also given in Fig. 2 (as closed triangles). From Fig. 2, one can see that the behaviour of the uranium cross-section measured at room temperature is in good agreement with the reported in the BNL-325 [5]. The coherent scattering amplitude of U was calculated from the Bragg cut-offs, observed in the total cross-section after correction both for the Debye Waller factor and the spectrometer's resolution, and found to be b_{coh} = (8.4 ± 0.2) fm. Consequently the coherent scattering cross-section of U is σ_s=(8.9 ± 0.4)

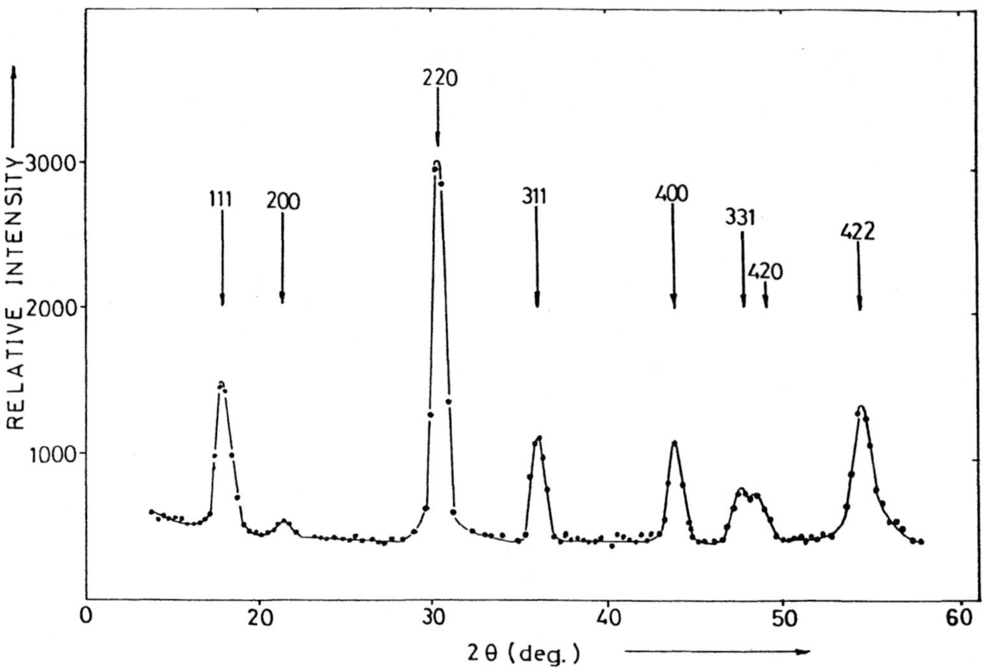

Fig. 1. The UO_2 neutron diffraction pattern obtained at RT.

The integrated intensities for the observed reflections were calculated using the area method and then corrected for background and Lorentz factor. The coherent scattering amplitude of U was deduced by fitting the corrected integrated intensities using the reliability factor method. The minimum reliability factor R was found to be 3 % for a coherent scattering amplitude value b_u = (8.5 ± 0.2) fm of uranium, with a value of the coherent scattering amplitude of oxygen b_o = 5.8 fm.

Total Neutron Cross-Section Measurements

The measurements were performed using two TOF spectrometers installed in front of two of the ET-RR-1 reactor horizontal channels. The measurements were carried out for neutrons below 5 meV using an automatically controlled heater, where the temperature could be fixed within ± 5°C in the interval between 25-500°C. The spectrometer's resolution at different intervals of the whole energy range, under consideration, could be varied from 3 μsec/m to 20 μsec/m. The spectrometers are described in Refs. [2-4].

Results and Discussion

Natural uranium

Fig. 2 shows the dependence of the total neutron cross-section (closed circles), on both wavelength and energy, as measured for natural uranium at room temperature in the energy range from 2.0 eV - 2 meV. In the same figure are also represented the total neutron cross-sections (open circles) measured at 483°K for neutron energies below 5 meV. The errors indicated in Fig. 2 are statistical ones. The contribution due to ^{235}U was

barns. This value is in good agreement with the values (8.5 ± 0.06) fm and (8.90 ± 0.16) b reported in Ref.[5].

The difference between the U total cross-section measured at 483°K and that measured at room temperature was attributed to the contribution of the one phonon annihilation process. The measured difference was found to be (1.2 ± 0.2) b at neutron energy 3 meV. This difference was calculated, under the same conditions of Ref. [6], considering a U Debye temperature 207°K and found to be 0.8 b. Such slight disagreement may be due to the contribution of the multi-phonon process.

The uranium oxide

Fig. 3 shows the dependence of the total neutron cross-section (closed circles) measured for UO_2 at room temperature in the energy range from 2 eV- 2 meV. The cross-sections measured for neutron energies below 5 meV at 483°K are also represented in the same figure (open circles). The measured cross-sections are in good agreement with those reported before [1], in the energy range from 0.028 eV - 4.5 meV. The coherent scattering amplitude of U was calculated from the Bragg cut-offs, observed in the total cross-section of UO_2 at room temperature, considering that the coherent scattering amplitudes of U and oxygen are of the same sign. The value [5] 5.8 fm was chosen for the coherent scattering amplitude of oxygen. It was found that the value of the coherent scattering amplitude is (8.6 ± 0.5) fm which is in reasonable agreement with the value (8.5 ± 0.2) fm calculated from the neutron diffraction pattern of the same sample. The contribution due to the one phonon annihilation process in the total neutron cross-section

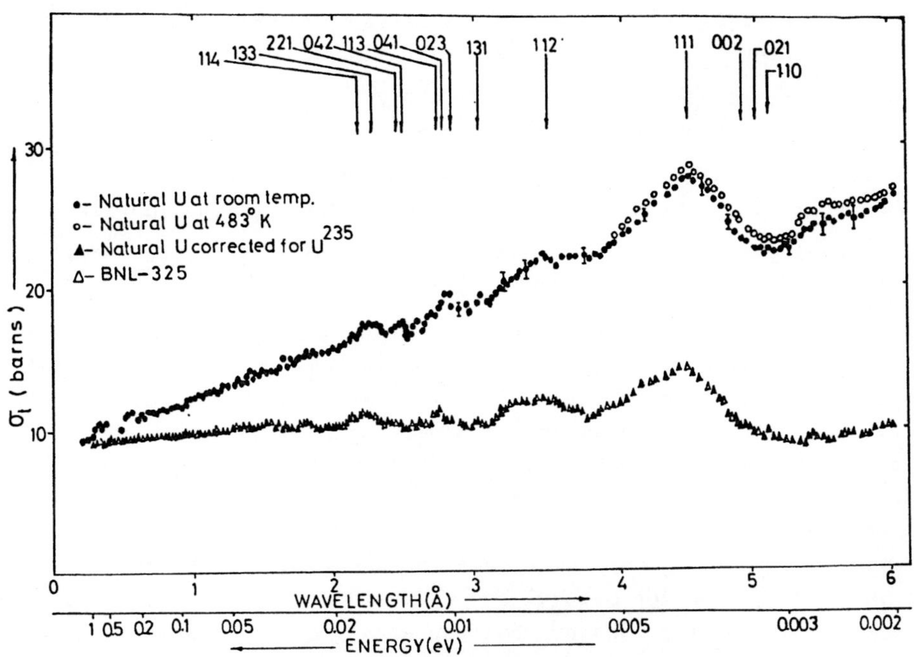

Fig. 2. The total neutron cross-section of U.

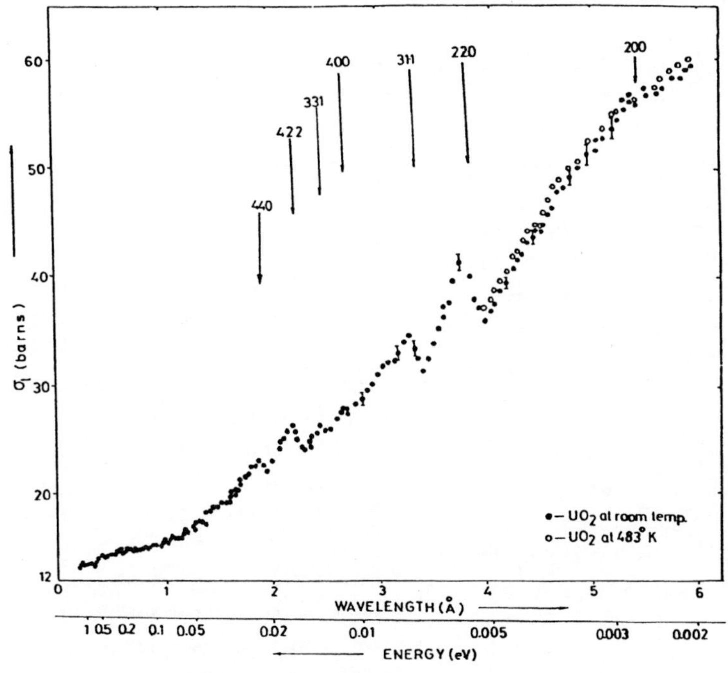

Fig. 3. The total neutron cross-section of UO_2.

was calculated for 3 meV neutron energy and the difference between the 483°K and the room temperature, and found to be 0.6 b. This value is of the same order of magnitude of the difference between the total neutron cross-sections measured at 483°K and room temperature.

References

1. S.F. Beshai, AE 222, Aktiebolaget Atomenergie, Stockholm, (1966).

2. I.F. Hamouda, R.M.A. Maayouf et al., Nucl. Inst. & Meth. 40, 153 (1966).

3. M. Adib et al., Nucl. Inst. & Meth. 116, 509 (1974).

4. M. Adib, A. Abdel-Kawy, R.M.A. Maayouf et al., Proc. Int. Conf. on Nucl. Cross-Sections for Technology, Knoxville, TN, October 22-26 (1979).

5. S.F. Mughabghab, D.I. Garber, BNL-325, Third Edition, Vol. 1, U.S. Atomic Energy Commission, June (1973).

6. M. Adib, A. Abdel-Kawy, R.M.A. Maayouf et al., INDC (EGY)-2/L, IAEA, September (1981).

DOPPLER BROADENING OF ^{238}U NEUTRON RESONANCES IN CRYSTAL LATTICES AND MOLECULAR GAS COMPARED WITH THE FREE-GAS APPROXIMATION

A. Meister, S. Mittag[+], D. Papst ([+]), W.Pilz[+], D. Seeliger, K. Seidel, H. Tschammer[+], R. Tschammer[+], D. Hermsdorf

Technische Universität Dresden - Sektion Physik - Mommsenstr. 13, DDR - 8027 Dresden, German Democratic Republik

[+]Joint Institute for Nuclear Research - Laboratory of Neutron Physics P.O.B. 79, 101000 Moscow, U.S.S.R.

For several chemical compounds of Uranium, transmission spectra have been measured with time-of-flight technique at the pulsed fast reactor IBR-30 with a procedure allowing the determination of differences between the spectra. The observed different Doppler broadening of low-lying resonances can not be interpreted in the frame of the commonly used free-gas-model. A description is possible with relatively simple weighted normal-mode frequency spectra of the crystal and molecular oscillations.
Direct comparisons of the Doppler broadening in crystal with the free-gas approximation are drawn for U-metal and UO_3 at 300 K and 600 K. The influence of intramolecular oscillations on the resonance shape is discussed for UF_6. Furthermore, the order of magnitude is indicated, as resonance parameters extracted by the shape analysis method depend on these Doppler broadening procedures.

^{238}U, neutron resonances, Doppler broadening, U-metal UO_3, UF_6

The Doppler broadening of the strong low-energetic resonances of ^{238}U has influence, expressed f. e. over the neutron capture integral, on the reactivity of a reactor core. A careful treatment of the Doppler broadening is also necessary for the extraction of resonance parameters from experimental spectra. Commonly a gas model approximation is used, where the thermal motion of Uranium atoms is described by the velocity spectrum of an ideal gas. Originated in a paper of Lamb [1], who has studied the neutron capture by crystal atoms, in the Doppler width an effective temperature T_{eff} is used instead of the real sample temperature T, connected with the mean energy of oscillation degree of freedom $\langle \epsilon_U \rangle = k \cdot T_{eff}$. Furthermore, for numerical reasons, the gas model broadening formalism is also applied to the approximated by symmetric Ψ and antisymmetric χ functions resonance cross section expression [2].
In experiments, searching for the chemical shift of neutron resonances [3], differences in the Doppler broadening between the crystal lattices of several Uranium compounds had to be studied. In the present paper the results are compared with the gas model approximation and the $\chi\Psi$-formalism. Measurements with gaseous UF_6 allow to evaluate the influence of intramolecular oscillations on the Doppler broadening, a problem, which may be of interest too at very high temperatures.

Experiment

At the pulsed reactor IBR-30 of the JINR Dubna transmissions spectra (time-of-flight technique, flight path 50...60m) have been measured for different chemical compounds of Uranium with as equal as possible experimental conditions, including multiple sample changing, stability test etc.[3, 4].
Two examples are shown in Fig. 1.
The interpretation of the differences between the spectra is possible with an extented Einstein model for the weighted normal-mode frequency spectra of the crystals, taking into account the very different masses of both atoms in the compounds [5].

$$\rho(h\nu) = a_1\delta(h\nu - h\nu_1) + a_2\delta(h\nu - h\nu_2) \quad (1)$$

with $a_1 + a_2 = 1$ and for Uranium metal $a_2 = 0$. The parameters in (1) influence the Doppler broadening in first order in their combination to

$$\langle \epsilon_U \rangle = \frac{1}{2}\int_0^\infty d(h\nu)\, h\nu\, \rho(h\nu) \coth\left(\frac{h\nu}{2kT}\right) \quad (2)$$

At T = 300 K it has been found $\langle \epsilon_U \rangle / kT$ = 1.08 for UO_3 and $UO_2(NO_3)_2$, 1.015 for

K. H. Böckhoff (ed.), Nuclear Data for Science and Technology, 968–970.

metal and values between these for UF_4 and UO_2. (In Fig. 1, left side-lower part, the dashed line corresponds to $\langle \epsilon_U \rangle$ /kT = 1.038 for Uranium oxid).

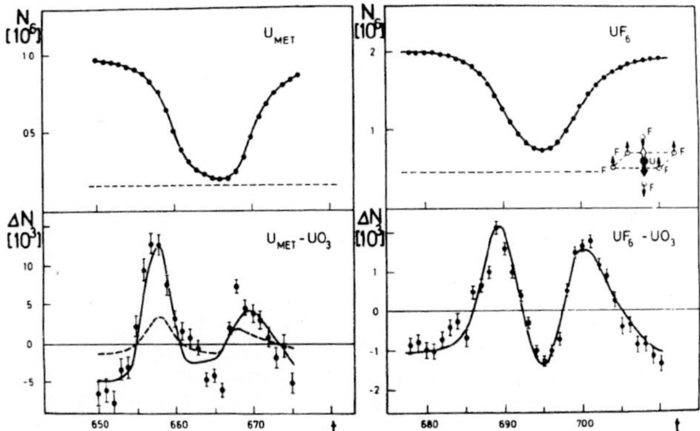

Fig. 1. Measured time-of-flight transmission spectra and their interpretation in the region of the 6.67 eV resonance (upper part) and differences between the spectra of the indicated compounds. Experimental conditions: sample at room temperature with $1.2 \cdot 10^{21}$ nuclei/cm^2, time channel width 2.5 μs, flight path 60 m (left hand side) and 373 K, $4 \cdot 10^{20}$ nuclei/cm^2, 2 μs and 58 m (right hand side) respectively.

The Doppler broadening in UF_6 is additionally to the gaseous movement of the rigid molecule strongly affected by the both in Fig. 1 indicated oscillation modes within the molecule, described by the approach of Letokhov [6] as done also by Bowmann [7], who has demonstrated the influence of intramolecular oscillations at first (For more details see Ref.[4]).

Comparisons with the gas model approximation

Fig. 2 shows, that the relatively to the resonance maximum asymmetric behaviour of the cross section differences can not be described, if the quantization of the lattice vibrations is neglected, as done in the gas model. The additional approximations of the $\chi\Psi$-formalism bring in the case of the considered differences no further remarkable deviations.

In Fig. 3 and 4, calculated Doppler broadened cross sections of U-metal, UO_3 and UF_6 are directly compared with those of the gas model predictions. The differences are about five times larger for UO_3 as in

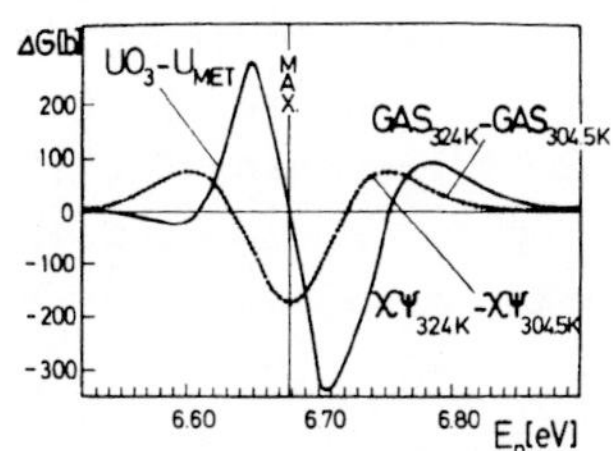

Fig. 2. Cross section differences in the region of the 6.67 eV resonance, calculated with the extented Einstein model formula (1), (——), in gas model approximation (---) and $\chi\Psi$-formalism (....), the last two with T_{eff}, corresponding to $\langle \epsilon_U \rangle$ in the UO_3 and U-metal lattice respectively.

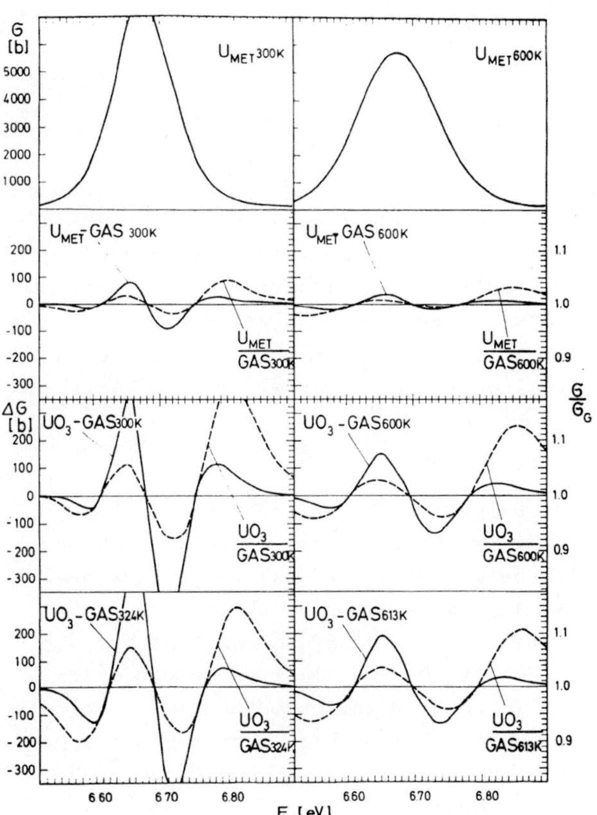

Fig. 3. Resonance cross sections, differences (full lines) and quotients (dashed lines) between crystal oscillator treatment and gas model at 300 K and 600 K sample temperature.

the case of U-metal, which has the lowest Debey temperature. The use of T_{eff} instead of T in the gas model has not a very important influence on the differences. For UF_6 an U-atomic gas is a better approximation than rigid UF_6 molecules without inner oscillations, but the remaining differences are visible already in the cross section graph. An enhancement of the temperature from 300 K to 600 K diminishes the cross section difference to about the half.

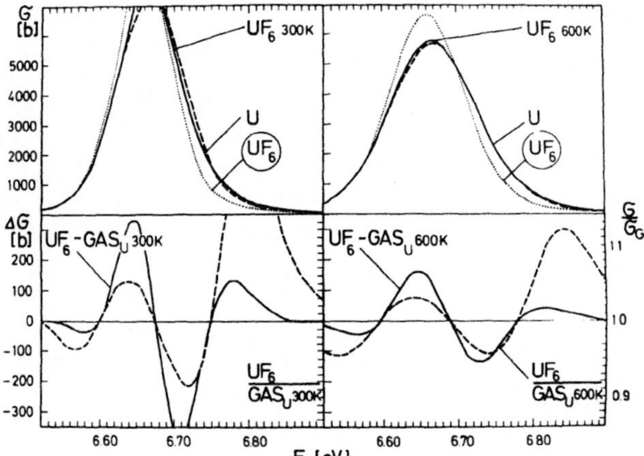

Fig 4. Resonance cross sections at 300 K and 600 K for UF_6, taking into account intramolecular oscillations (——), for an U-atomic gas (---) and rigid UF_6 gas particles (....), in the upper parts. Lower parts: Differences and quotients between UF_6 and U-atom gas cross sections.

To estimate an upper limit for the influence of the gas model approximation on the determination of resonance parameters with the commonly used shape analysis method, a Doppler broadened 6.67 eV resonance line has been fitted by variation of Γ_n and Γ to the resonance shape calculated with the gas model. The resonance parameter differ by 4 % and 0.4 % for Γ and Γ_n respectively.

References

1. W.E. Lamb, Phys. Rev. 55, 190 (1939)
2. F.H. Fröhner, IAEA - SMR - 43, 59 (1980)
3. K. Seidel et al., Dokl. Akad. Nauk SSSR, 256, 360 (1981)
   A. Meister et al., Nucl. Phys. A 362, 18 (1981)
4. K. Seidel et al., Yad. Fiz. 34, 1173 (1981)
5. H.E. Jackson, J.E. Lynn, Phys. Rev. 127, 461 (1972)
6. V. S. Letokhov, Phys. Rev. A 12, 1954 (1975)
7. C. D. Bowmann, R. A. Shrack, Phys. Rev. C 21, 58 (1980)

NEUTRON-INDUCED ATOMIC EXCITATION AND NEUTRON MODERATION

C. D. Bowman* and R. G. Johnson

National Bureau of Standards
Washington, D.C. 20234, U.S.A.

The excitation of electrons in atoms due to neutron-nucleus scattering has been examined. The cross section for neutron scattering with an accompanying excitation of a particular electron has been derived. In addition, a procedure for estimating the probability of any electron excitation in neutron scattering has been formulated. Using these probability estimates the effect of electronic excitations in neutron moderation problems has been connected to a small fractional increase (of the order of 10^{-4}) in the average logarithmic energy decrement. In special cases this small increase may lead to effects of about 1%.

[atoms, average logarithmic energy decrement, electronic excitation, neutron moderation, neutron scattering, theoretical neutron cross sections]

Introduction

In a recent paper by S. W. Lovesey and the present authors[1] the transitions between different electronic configurations in atoms (and molecules) due to neutron-nucleus scattering were discussed. Here we repeat and extend those calculations for atoms and consider the effect of such electronic excitations on neutron moderation problems.

When a neutron scatters from the nucleus of an atom the nucleus recoils under the electrons of the atom. Consequently the electrons acquire a momentum oppositly directed to the momentum transferred to the nucleus. The states of the target atom can in general be described in terms of the motion of the center-of-mass and the relative motion of the electrons and nucleus. Written in terms of the position of the scattering nucleus these coordinates are related by

$$\vec{R}_n = \vec{R} - \left(\frac{m_e}{m_n}\right)\vec{r} \tag{1}$$

where m_e and m_n are the masses of the electron and the nucleus, respectively. The mass of the atom is denoted by M.

The scattering causes the target to change from a state ψ_μ with the energy E_μ to a state $\psi_{\mu'}$ with energy $E_{\mu'}$. The μ labels refer to the atomic quantum numbers and to the momentum of the center of mass, $\vec{q}$. The associated changes in momentum and energy of the neutron are $\vec{Q}$ and ω. The partial differential cross section for scattering from the nucleus of an atom at position $\vec{R}_n$ is[2]

$$\frac{d^2\sigma}{d\Omega dE'} = \frac{k'}{k}\sum_{\mu\mu'} P_\mu |\langle\psi_{\mu'}|b\ \exp(i\vec{Q}\cdot\vec{R}_n)|\psi_\mu\rangle|^2 \cdot \delta(\omega+E_\mu-E_{\mu'}) \tag{2}$$

where P_μ is the initial state probability, $\omega = E-E' = (k^2-k'^2)/2M$, and b is the scattering length for the nucleus.

The state functions are

$$\psi_\mu = \exp(i\vec{q}\cdot\vec{R})R_{n\ell}(r)\ Y^\ell_m(\hat{r})/V^{\frac{1}{2}} \tag{3}$$

and

$$E_\mu = \varepsilon_\mu + q^2/2m \tag{4}$$

where V is the volume of the target, $R_{n\ell}(r)$ is the radial wave function of the electron, and $Y^\ell_m(\hat{r})$ is a spherical harmonic. ε_μ is the electronic energy.

The problem essentially reduces to calculating the matrix element in Eq. (2). The result is

$$\langle\psi_{\mu'}|\exp(i\vec{Q}\cdot\vec{R}_n)|\psi_\mu\rangle = \frac{(2\pi)^3}{V}\delta(\vec{Q}-\vec{q}'+\vec{q}) \cdot \int_0^\infty dr\ r^2\ R_{n'\ell'}(r)\ R_{n\ell}(r)\ G^{mm'}_{\ell\ell'}(\vec{Q},r) \tag{5}$$

where

$$G^{mm'}_{\ell\ell'}(\vec{Q},r) = (4\pi)^{\frac{1}{2}}(-1)^m\{(2\ell+1)(2\ell'+1)\}^{\frac{1}{2}} \cdot \sum_{LM}(-i)^L j_L\left(\frac{m_e}{m_n}Qr\right)\cdot Y^L_M(\vec{Q}) \cdot (2L+1)^{\frac{1}{2}}\begin{pmatrix}L & \ell & \ell'\\ 0 & 0 & 0\end{pmatrix}\begin{pmatrix}L & \ell & \ell'\\ -M & m & -m\end{pmatrix} \tag{6}$$

and j_L is a spherical Bessel function of order L. In Eq. (6) the first 3-j symbol is non-zero only if the sum is an even integer.

Application to Hydrogenic Atoms

As a simple example we consider the scattering associated with the transition from $n = 1$ to $n' = 2$ in a hydrogenic atom. In this case there are four final states with an energy ε_2. The cross section is determined by the matrix element squared, *i.e.*,

$$\sum_{\ell'm'}|\int_0^\infty dr\ r^2R_{2\ell'}(r)\ R_{10}(r)\ G^{om'}_{o\ell'}(\vec{Q},r)|^2 = 32(\beta^4+S^2\beta^2)/(\beta^2+S^2)^6 \tag{7}$$

where $\beta = Qa_om_e/Zm_n$, $S = 3/2$, and $a_o = 0.529$ Å. The first term in Eq. (7) arises from transitions to $\ell' = 0$ and the second term from transitions to $\ell' = 1$. The entire factor vanishes for both $Q \to 0$ and $Q \to \infty$; it achieves a maximum value for $Qa_o = 3Zm_n/4m_e$ at which the value is 0.102. In other words the scattering amounts to some 10% of the nuclear scattering. Note that for hydrogen this can be obtained at 28 keV for 90° center-of-mass collisions.

Probability Estimates for Atomic Excitation

A useful estimate for the probability that in a neutron scattering an electronic transition will occur can be obtained by calculating the probability amplitude $M_{\mu\mu}$ that an electron will stay in its initial state, ψ_μ, where

$$M_{\mu\mu} = \langle\psi_\mu|\exp\{-i(m_e/m_n)\ \vec{Q}\cdot\vec{r}\}|\ \psi_\mu\rangle\ . \tag{8}$$

The probability of excitation to all other states, P_e, is given by

$$P_e = 1 - |M_{\mu\mu}|^2\ . \tag{9}$$

*Present address: Los Alamos National Laboratory, PDO D-434, P.O. Box 1663, Los Alamos, NM 87545.

K. H. Böckhoff (ed.), Nuclear Data for Science and Technology, 971–973.

This method is advantageous for obtaining estimates for experiments since it gives the probability for all excitations. It is approximate in the sense that Eq. (9) allows both bound and unbound states of the electron, and the neutron energy must be high enough to allow any final state to be energetically accessible. The method is also advantageous since it requires a knowledge only of the initial wave function.

For a particular electron the desired matrix element can be written

$$M_{\mu\mu} = \int_0^\infty dr\ r^2\ R^2_{n\ell}(r)\ G^{mm}_{\ell\ell}(\vec{Q},r) \quad . \tag{10}$$

For more than one electron in the same initial state a product of matrix elements of the type given in Eq. (10) would be formed.

Using Eqs. (5) and (6) and noting that for $(m_e/m_n)(Qr) \ll 1$ only the $L = 0$ term is significant and that

$$G^{mm}_{\ell\ell}(\vec{Q},r) = j_0\left(\frac{m_e}{m_n} Qr\right) = 1 - \frac{1}{6}\left(\frac{m_e}{m_n} Qr\right)^2 + \ldots$$

we have

$$M_{\mu\mu} = \int_0^\infty dr\ r^2\ R_{n\ell}(r) \left\{1 - \frac{1}{6}\left(\frac{m_e}{m_n} Qr\right)^2\right\}$$

$$= 1 - \frac{1}{6}\left(\frac{m_e}{m_n} Qa_o\right)^2 \frac{\langle r^2\rangle}{a_o^2} \quad . \tag{11}$$

The expectation value of r^2 has been calculated by Hartree-Fock[3] and Dirac-Fock[4] methods. The values of $\langle r^2\rangle$ tabulated in these references have been used to determine the matrix elements expressed by Eq. (11).

Finally for $M_{\mu\mu} \ll 1$ we can write

$$P_e = 1 - |\pi_j M_{\mu\mu}|^2$$

$$= 1 - \left|\pi_j \left\{1 - \frac{1}{6}\left(\frac{m_e}{m_n} Qa_o\right)^2 \frac{\langle r^2\rangle}{a_o^2}\right\}\right|^2$$

$$= \frac{2j}{6}\left(\frac{m_e}{m_n} Qa_o\right)^2 \frac{\langle r^2\rangle}{a_o^2} \tag{12}$$

where the possibility of j electrons in the same state has been included.

In Table I the probability for electronic excitation for the states of helium, lithium, boron, and sodium have been calculated using the particular example of a 10 keV neutron scattering at 90° in the center-of-mass.

The highest probability listed in Table I is for the 2S electron of lithium, approximately 9%. However, that electron also has one of the lowest ionization potentials.

For the hydrogen atom the probability of electronic excitation is by far the largest, 0.203 (at 10 keV 90° center-of-mass scattering). In this case however we must recognize that hydrogen is always bound in a molecule. Although the excitation of electrons in molecules was discussed in Ref. (1), we cannot as yet extend the analysis to practical problems. Consequently in the next section we will have to exclude the interesting case of hydrogen.

Table I. Probability of Electron Excitation

State	$\langle r^2\rangle$ (a.u.)	j	P_e	I.P.[a] (eV)
Helium				
1S	1.185	2	3.55×10^{-3}	54.4
Lithium				
1S	0.447	2	4.65×10^{-3}	122.
2S	17.74	1	9.22×10^{-2}	5.39
Boron				
1S	0.143	2	7.14×10^{-4}	340.
2S	4.709	2	2.35×10^{-2}	37.9
2P	6.146	1	1.53×10^{-2}	8.30
Sodium				
1S	0.0274	2	2.70×10^{-5}	1649.
2S	0.729	2	7.19×10^{-4}	300.
2P	0.820	6	2.42×10^{-3}	209.
3S	20.65	1	1.02×10^{-2}	5.14

[a] Ionization Potential

Neutron Moderation

The effect of electronic excitation in neutron moderation problems can now be estimated, if some estimate of the energy spectrum of the excited electrons can be made. This estimate can be made by noting that the operator in Eq. (8) is within a constant the same as the operator in photon absorption. Consequently, the spectrum of electronic excitations should be similar to that in photon absorption, *i.e.*, a function which peaks at the ionization potential of the atom.

If we then use the ionization potential as the average energy loss in an excitation of a particular electron, the effect of these excitations on neutron moderation can be formulated as a correction to the usual average logarithmic energy decrement for isotropic elastic scattering. At a particular incident neutron energy the average energy loss due to electronic excitation is a product of the probability for such an excitation and the energy of excitation. The probability is on the average proportional to the incident energy while the average energy of excitation is assumed to be the ionization potential. Consequently, the average energy loss divided by the incident energy is a constant (over a wide range of energies). It can now be used as a correction to the average logarithmic energy decrement, $\bar{\xi}$.[5]

When the procedure outlined in the previous paragraph is followed we find that the fractional correction to $\bar{\xi}$ varies from 0.5×10^{-4} to 5×10^{-4} for the states listed in Table I where the largest increase is for the 2P state of Na. In practice these small increases will probably not be observable in the integral quantities of neutron moderation problems where $\bar{\xi}$ tends to enter the solution linearly. However, if the problem is sensitive to the accumulated effect of many collisions the small increase in $\bar{\xi}$ may be significant. For example, 130 collisions in Na reduces the energy by 10^5 and in this case the effect of electronic energy loss decreases the final energy by about 1%. Another more practical example is the effect on the resonance escape probability where $\bar{\xi}$ enters as an exponential quantity. Here the small increase in $\bar{\xi}$ can cause an increase in the resonance escape probability of again about 1%.

Conclusions

We have derived equations which can be used to predict the cross section for neutron scattering accompanied by a particular electronic excitation and equations which can be used to estimate the probability of electron excitation in neutron scattering. The effect of these excitations on neutron moderation problems has been estimated as a small increase, of the order of 10^{-4}, in the average logarithmic energy decrement. In most cases this small increase in $\overline{\xi}$ is expected to be negligible. However, in special cases effects on the order of 1% may be observed.

References

1. S.W. Lovesey, C.D. Bowman, and R.G. Johnson, Z. Phys. (to be published).
2. W. Marshall and S.W. Lovesey, in Theory of Thermal Neutron Scattering, Oxford University Press, London (1970).
3. C. Froese Fischer, Atomic Data 4, 301 (1972).
4. J.P. Desclaux, Atomic Data and Nucl. Data Tables 12, 311 (1973).
5. E. Amaldi, in Encyclopedia of Physics, Vol XXXVIII (Neutron and Related Gamma Ray Problems), Springer-Verlag, Berlin (1959).

General Interest

CAN WE DO MORE TO ACHIEVE ACCURATE NUCLEAR DATA?

M. S. Coates, D. B. Gayther, G. D. James,
M. C. Moxon, B. H. Patrick, M. G. Sowerby
and D. B. Syme

Nuclear Physics Division, A.E.R.E., Harwell,
Didcot, Oxon., OX11 ORA, U.K.

We ask whether more effective methods can be used to reduce the discrepancies which often persist between individual measurements of neutron nuclear data. With this in mind we trace the historical development of a few selected measurements to see what lessons can be learned from the past. Several common sources of systematic error are noted. These have been qualitatively recognised for many years but have nevertheless continued to be the most likely causes of discrepancies. Experimenters have just not been sufficiently thorough in demonstrating an adequate treatment of these systematic effects. We suggest several improved procedures for future measurements.

1. Introduction

Despite many years of effort it is a disagreeable fact that it is rare to find a set of different neutron measurements on a given parameter which are consistent within the errors assigned to the individual determinations. These discrepancies are of course signals of the presence of systematic errors and as such present major obstacles to be overcome before we can reach the accuracy goals asked for in request lists.

In this paper we are concerned primarily with the systematic effects occurring in experiments which aim to measure neutron nuclear data accurately. We are not concerned with the technological factors which determine the ultimate accuracy we can achieve such as the intensity of neutron sources or the analytical power of computer codes. We shall concentrate rather on the factors which prevent us from achieving the best accuracy at a given stage.

In looking for possible ways to improve our rate of progress in dealing with systematic problems it is difficult to fault the broad strategy used by the measurement community. An iterative process is used which is essentially the fundamental way of making an accurate determination of any physical quantity. Individual laboratories compare their measurements which often turn out to be discrepant. These problems are discussed by the body of experimenters at meetings and conferences and as a result recommendations are made which point the way ahead to better methods of measurement and analysis. Thus a feedback process is established which in principle leads eventually through re-measurement, re-analysis and evaluation to an accurate consensus. Nevertheless with most of our neutron data measurements we are still far away from this idealised conclusion although marked improvements in consistency have come about in many measurements in the course of time. To function properly this iterative method needs a wide experimental base in which both the number of measurements and the diversity of techniques are extensive enough to allow a high probability of revealing all important systematic errors. It is increasingly difficult to meet this condition in the nuclear data field as world effort continues to shrink. Even so the basic correctness of the approach cannot be profitably questioned. We must therefore look for ways to improve the efficiency of the methods used in practice within this framework. To this end we have examined in some detail the historical development of four separate neutron nuclear data measurements and ask what lessons can be learned from experience in the past. In doing so we have neglected in an arbitrary way many early experiments. Often the main purpose of these was to look for phenomenological effects and we cannot expect to gain by considering them. The measurements we have studied are:

1) The ^{6}Li(n,α)t cross-section in the vicinity of the 247 keV resonance.
2) The ^{235}U fission cross-section near 14 MeV.
3) The ^{238}U resolved resonance parameters.
4) The parameters of the 1.15 and 27.7 keV resonances in ^{56}Fe.

The sets of measurements naturally follow different courses and we have not attempted to give a unified or systematic treatment, even if this were possible. For the sake of clarity and to provide a concise foundation for the general discussion given in section 3, we have chosen to summarise the conclusions reached on each case before proceeding to a consideration of the next.

2. Investigation of Discrepancies

2.1 The ^{6}Li(n,α)t cross-section in the vicinity of the 247 keV resonance

The ^{6}Li(n,α)t reaction is important for both standards work and for fusion systems, although it is only in the first of these roles that our present interest lies. On the face of it this should be one of the easiest partial cross-sections to measure. Its magnitude is suitably large and the variation with neutron energy is smooth over a wide energy range, being characterised by a 1/v behaviour up to about 10 keV and a strong resonance peak at 247 keV. The high Q value for the reaction (about 4.8 MeV) simplifies the detection of the product charged particles and practicable detectors in the form of ^{6}Li loaded scintillation counters have been available for many years. Further, the only competing reaction, at least up to 1.7 MeV, is neutron elastic scattering, since the (n,γ) reaction cross-section is negligibly small. Yet despite these apparent advantages the direct experimental determinations of the (n,α) cross-section - that is those which measure the reaction product yield and then derive the cross-section using a subsidiary flux determination - present overall an amazingly inconsistent picture, particularly in the energy region above about 100 keV.

It is only since the mid-1970's that a reasonably satisfactory situation has emerged. It is worth quoting part of the summary made in 1977 by Derrien and Edvardson[1] on the measurements in the region of the resonance. They wrote:

> "Between 1950 and 1960, nine measurements were performed with a discrepancy of 25% at the peak of the 245 keV resonance; seven measurements between 1960 and 1970 give approximately the same discrepancy; thirteen measurements have been performed since 1970 and the inconsistency is still 25% in this period of time. If we consider all the measurements between 1950 and 1977, the difference between the lower and higher values in the peak of the resonance is about 40% relative to the mean value. Apparently no improvement has been made in 27 years, and the evident conclusion drawn by the unbiased observer would be that we do not know how to measure the ^{6}Li(n,α) cross-section with high accuracy in the vicinity of the 245 keV resonance and at higher energies."

Fortunately direct measurement is not the only route to the determination of the ^{6}Li(n,α) cross-section. The ^{7}Li system is simple enough to allow reliable

K. H. Böckhoff (ed.), Nuclear Data for Science and Technology, 977–986.

theoretical estimates of the (n,α) cross-section to be made from the analysis of the experimental data on the relatively few other reactions possible, incorporating the constraints imposed by unitarity. This approach was first used by Diment and Uttley[2] in 1969 using the parameters of total and scattering cross-sections. Over the next few years the predicted cross-section was put on a firmer footing by Hale[3,4] and by Knitter[5]. Hale's multichannel, multilevel R-matrix analysis was particularly comprehensive and by 1975 was widely considered to be reliable enough to provide the touchstone on which direct measurements should be judged. By 1980 the agreement between the most recent of the direct measurements and the calculated (n,α) cross-section had improved sufficiently for Hale[6] to comment thus on the status:

> "Measurements of the neutron cross-sections for ^{6}Li made before 1975 were inconsistent with unitary constraints relating them, particularly near the peak of the 240 keV resonance. That situation has improved significantly in the past few years in that recent measurements of the (n,α), total, and elastic cross-sections agree to the order of a few percent with each other and with calculations that impose unitary consistency."

The lack of detail in measurement reports makes it essentially impossible to establish how this improvement came about, but let us briefly trace the course of events from 1970 when Uttley et al[7] evaluated the ^{6}Li cross-sections. This was shortly after the Diment and Uttley calculated (n,α) cross-section had been published. Fig. 1 shows their prediction together with the existing direct measurements at that time. The experimental data are clearly discrepant among themselves and with the Diment and Uttley values. The general pattern is for the experimental data to lie well below (12-25%) the calculated values and to give the resonance peak at a higher energy. Only one measurement, a preliminary white spectrum measurement on a linac, is in reasonable agreement with the Diment and Uttley values although there is local disagreement at the resonance peak. All of the other experimental measurements were made on Van de Graaff accelerators.

Uttley et al concluded, correctly as it turned out, that the Diment and Uttley value was the best available representation of the (n,α) cross-section across the resonance region despite the disagreement shown with the mass of the experimental data. They were able to identify serious deficiencies in many of the Van de Graaff measurements. These related to multiple scattering corrections for the thick scintillators generally used; the way in which flux measurements had been made; the determination of neutron energy; and the estimation of energy resolution. Many, though by no means all, of the experiments could therefore be discounted, at least on a qualitative basis. Although there was some reservation about the rigour of the theoretical treatment used by Diment and Uttley their results effectively forced experimenters to examine the techniques used in the apparently good direct measurements and to search for further sources of systematic error.

During the next few years this led on the one hand to a fairly detailed examination of the factors affecting the use of ^{6}Li scintillators[8] (e.g. multiple scattering effects, ^{6}Li content and spatial distribution; angular distribution of reaction products; after-pulsing) and the establishment of more accurate energy standards[9] and on the other hand, to the reassessment of existing experimental data and to further direct measurements of the (n,α) cross-section. In parallel with experimental work the theoretical predictions were being given a sounder basis through the analysis of new measurements on other reactions in the ^{7}Li system.

By 1975[10] the best experimental data, for no clearly identifiable reason, showed considerably improved agreement with the predicted (n,α) values but still lay, as a group, some 10% lower than theory in the neighbourhood of the resonance peak. The situation was confused to some extent at this stage by the results of a direct measurement, on a linac, by Friesenhahn et al[11] which aroused considerable interest because an ion chamber containing a ^{6}Li metal foil was used rather than a scintillator for the reaction yield measurement. However these data were discrepant with both theory and all other direct measurements and unfortunately time pressures prevented any further work from being done on the experiment. However the investigations carried out on the ^{6}Li glass scintillators, although they showed up unexpected features such as the non-uniform distribution of ^{6}Li in the glass, revealed no fundamental reason why they should not be used as accurate detectors of the (n,α) reaction. The practice of mounting even thin (0.1 - 1.0 mm) glasses away from the photo-multiplier system was recommended to minimise multiple scattering effects, but this had been appreciated by at least some experimenters for many years. Several new direct measurements[12,13,14] have been made since the middle 1970's using white spectrum sources on linacs with thin ^{6}Li scintillators and it is these which give the improved consistency with the best available theoretical prediction of the (n,α) cross-section. Again it is unclear how this improvement has come about. Even so systematic discrepancies of a few percent still exist which need further investigation.

The main conclusions to emerge are:

1) There is no technical reason why the ^{6}Li(n,α) cross-section should not have been measured at least 12 years ago to its presently known accuracy. Discrepancies of a few percent still exist between the most recent direct measurements.

2) The calculated values of the (n,α) cross-section, obtained from the analysis within a unitary framework of relatively few measurements on other reactions in the ^{7}Li system, were a dominating influence on the course of progress of the direct measurements. We have not found it possible to identify clearly the factors which have led to the overall improvement in consistency between the more recent direct measurements and the calculated values. This is a puzzling mystery which is unlikely to be answered. As an aside it is interesting to speculate what the present state of the direct measurements would be had no reliable theoretical prediction been available.

3) The fact that the calculated value of the (n,α) cross-section was soundly based, yet discrepant with direct measurements, forced experimenters to examine techniques which would probably otherwise have been left unchecked.

2.2 Fission cross-section measurements near 14 MeV

The relative ease with which high intensity, monoenergetic beams of 14 MeV neutrons can be produced allows cross-sections to be measured with an accuracy which is, generally speaking, second only to that achieved in the thermal region. As long ago as 1965, it was thought that accuracies of 2% could be achieved in fission cross-section measurements and yet, if we examine the data on ^{235}U, we find that the most recent values in the region of 14 MeV are consistently about 4% lower than the earlier ones.

The early measurements of the ^{235}U fission cross-section near 14 MeV are dominated by that of White[15] in which considerable attention to detail was paid in both the performance and documentation. The associated particle technique was used with the d-t reaction, but no coincidence between alpha particles and fission events was required. This method necessitates a good knowledge of the geometrical arrangement of the detectors and the total mass of the ^{235}U sample.

In the last few years, several measurements of the ^{235}U fission cross-section in the region of 14 MeV have been published by Cancé and Grenier[16], Arlt et al[17], Adamov et al[18] and Wasson et al[19]. The results are shown in Fig. 2 along with the values of White, Uttley and Phillips[20] and Moat[21], the difference between the early and later measurements being obvious. It is almost certainly significant that all the recent measurements employed the time-correlated associated particle method in which a coincidence between an alpha particle and a fission event is required. The earlier measurements did not. As a result the early measurements have suffered from a fission background produced by scattered neutrons or neutrons arising from other reactions (see Fig. 3). Although the presence of such a background was recognised by the early experimenters, the calculation of the magnitude is not easy and the lesson to be learned is that prevention is better than estimation.

Although the time-correlated method has significant advantages, there is a price to pay for the improvement. The areal density of the fission deposit must be quite uniform and accurately known. This imposes a much more severe problem on the experimenter than when only the sample mass is required, and in the case of Wasson et al, a large fraction of the experimental work had to be devoted to the determination of the areal density.

The characterisation and long term stability of samples is a very important aspect of most measurements and in the USA much attention has been paid to the establishment of reference fissile deposits at various laboratories (Poenitz et al[22]). This has resulted, through very careful assaying and intercomparison, in deposits with an uncertainty of 0.1% in the total mass and 0.26% in the areal density. Wasson et al were able to assay their experimental foils relative to one such reference deposit and thereby achieve an areal density uncertainty of only 0.4%, this being a rather small contribution to the total experimental error. This is to be contrasted with the more usual situation in which the error on the sample assay is a significant part of the total error (e.g. in the measurement of Cancé and Grenier, the uncertainty attributed to the areal density is 1.4% out of an overall measurement error of 1.94%). These considerations lead to the conclusion that there is a need to extend the intercomparison of fission foils to laboratories outside the USA so that more measurers can take advantage of the high accuracy assays now available and it is encouraging to note that such an exercise is currently being carried out[23].

The following conclusions can be drawn from the above discussion:

1) The recent measurements of the ^{235}U fission cross-section at 14 MeV are superior because (a) the background produced by scattered neutrons (or neutrons arising from other reactions) has been eliminated and (b) fission foil assays have improved.

2) The problems in 1(a) above were foreseen and corrected for, but eliminating a correction is better than calculating it.

3) There is need for a wider intercomparison of fission foil assays.

2.3 Resolved resonance parameters of ^{238}U

For many resonances in ^{238}U there are now a number of determinations of resonance parameters. Usually there is no great disagreement on the energy and spin but the partial widths, particularly the neutron widths, are often discrepant.

Let us first consider some examples in the energy range below 1 keV where overlapping s- and p-wave levels should not be a problem. It has been noted[24] that most experiments in ^{238}U tend to give discrepant Γ_n data at their high energy limit when they are running out of energy resolution. Good resolution is always desirable in resonance analysis because it minimises the problem of overlapping resonances and it enables good background measurements to be made. Measurements for two typical resonances at 20.9 eV and 958.6 eV are shown in Figs. 4 and 5. Let us discard from these the poor resolution data (i.e. the data of Harvey et al, Lynn and Pattenden, Fluharty et al, Levin and Hughes, and Bollinger et al for the 20.9 eV resonance, and Rosen et al, Firk et al and Malecki et al for the 958.6 eV resonance). If we also disregard other data where there are grounds for rejection (Asghar et al (error in normalisation) and Garg et al (superceded by Rahn et al)), we are left with a consistent data set for the 958.6 eV resonance but an inconsistent one for the 20.9 eV case. However, this inconsistency is largely produced by the datum of Block et al where Γ_n is more than three standard deviations from the mean.

We now turn our attention to the region above 1 keV where there are fewer measurements although all of these have adequate resolution. It is noticeable that for some cases the Γ_n values are in good agreement while for neighbouring resonances there are discrepancies. A typical example is the pair of strong s-wave resonances at 2581.3 eV and 2597.8 eV whose data are shown in Fig. 6. The resonances are reasonably close but well resolved so it is unlikely that experimental errors will produce discrepancies for one and not the other. However, there is a significant chance that p-wave resonances could overlap with one or both resonances. If there is overlap then it can be shown[24] that if area analysis is performed the apparent Γ_γ will always tend to be greater than the value for a single resonance. Poortmans et al[25] have data on Γ_γ for about 90 resonances above 1 keV and approximately half of these appear to have anomalously high values of Γ_γ. One of these is the 2581 eV resonance where Γ_γ is 25.8±0.5 meV, while for the neighbouring 2598 eV resonance Γ_γ has the normal value of 22.5±1 meV. It is therefore suggested that one reason for inconsistencies in this energy range is overlapping resonances and these can be identified because they have anomalously high Γ_γ values.

It is more usual in resonance analysis to use area than shape methods because (1) it is less expensive and (2) knowledge of the resolution and Doppler broadening functions are not so important. Derrien and Ribon[26], however, have discussed the advantages of shape analysis and conclude that

a) shape analysis provides all the information resulting from area analysis - even if the resolution and Doppler broadening functions are in error

b) shape analysis can identify errors in normalisation and background. Some codes allow these quantities to be adjusted during the fitting procedure.

In recent years there have been significant improvements to shape analysis techniques. Such analysis can now be performed simultaneously on different data types (e.g. capture and transmission) and for several sample thicknesses[27]. However these codes have to be used with care because there are many parameters not varied in obtaining the final fit. Usually the variation of these parameters would produce little or no effect on the values of the other adjustable ones. For the rare instances where this is not true, fits corresponding to local minima are obtained with underestimated error values. Some examples of parameters which are unexpectedly different from the general consensus may be explained by this effect (e.g. Block et al at 20.9 eV[31] and Haste et al for the 189 eV resonance[32]). More recently Bayes' Theorem has been utilised to improve the analysis[28] and this appears to remove some of the problems encountered with conventional least squares techniques which take no account of previous information and often do not take into full account the correlated errors in the data.

Fig. 6 also shows the changes in Γ_n when Derrien[29] performed shape analysis on the data of Cararo and Kolar and Rahn et al which had previously been analysed by area analysis. It can be seen that shape analysis leads to consistent data for both resonances. This was true for all resonances analysed by Derrien. The large change in Γ_n between shape and area analysis of the data of Rahn et al is due to a background adjustment which is required to obtain a good fit to the data. We believe that shape analysis is more likely than area analysis to obtain the correct answer when there are overlapping resonances. We would like to see shape analysis applied simultaneously to high resolution ($\leq$ 0.1 ns/m) capture and transmission data to confirm this.

There is good evidence that the analysis of ^{238}U data using single and multilevel formalisms can produce different resonance parameters[30]. It follows therefore that for accuracy, multilevel formalisms must be used unless it can be shown that the use of the single level formalism leads to negligible errors.

Finally on a cautionary note we must remember that sophisticated codes of the sort under discussion are only useful if they have been validated in some acceptable way.

The following points emerge from this discussion on ^{238}U:

1) Resonance parameters obtained from poor resolution experiments are unreliable.
2) There are occasions where measurements produce parameters which are significantly different from the general consensus for no apparent reason.
3) In the energy range above 1 keV overlapping resonances are a cause of some of the discrepancies.
4) The most reliable resonance analyses are performed if
   a) a multilevel formalism is adopted
   b) shape analysis is used and
   c) a full error analysis is performed, preferably using Bayes' theorem
5) Shape analysis can identify significant background errors

2.4 The parameters of the 1.15 and 27.7 keV resonances in ^{56}Fe

We now turn our attention to two resonances of technological importance in iron, the 1.15 keV p-wave resonance and the 27.7 keV s-wave resonance, both of which occur in the principal isotope ^{56}Fe. These are of a very different nature. The former is narrow (Γ/E_r = 5.7 x 10^{-4}) and isolated from other resonances. The latter is broad (Γ/E_r = 5.2 x 10^{-2}) and the analysis of experimental data for this resonance has to take into account the presence of other resonances. An examination of published measurements for these two examples reveal some important problems which arise in the measurement and analysis of resonance data.

Measurements for the 1.15 keV resonance are presented in Fig. 7 and it can be seen that the spreads in the values of Γ_n and Γ_γ are consistent with the quoted errors. This is probably because the resonance is isolated and other resonances have a negligible effect. As a result, area analysis (which is often the only type of analysis used in the earlier work) and shape analysis give equally satisfactory parameters. An interesting discrepancy has recently emerged in the measurements of the capture area of this resonance. Rohr[33] has reported that the value of $\Gamma_n\Gamma_\gamma/\Gamma_n + \Gamma_\gamma$ obtained with a C_6D_6 total energy detector is 70 meV when the measurements are normalised to low energy Au resonances. If the measurements are normalised to the Fe capture cross-section at thermal energies, where the capture gamma-ray spectrum is similar to the 1.15 keV resonance then the value obtained is 52 meV. Macklin[34] has now also reported a high value of 73±1.5 meV obtained with a C_6D_6 total energy detector normalised to Au resonances. It can be seen from Fig. 7 that of the other available data only the inaccurate datum (72.6±11.8 meV) of Hockenbury et al[35] suggests such a value. Rohr concluded from the Geel data that total energy detectors using the pulse height weighting technique are not working as expected for hard capture gamma-ray spectra. Macklin's result does not provide enough evidence to alter this conclusion because (1) a value of about 50 meV is obtained by a whole range of transmission and capture measurements and (2) a value of 50 meV is required to obtain the correct thermal capture cross-section from the Geel data. (Since $\Gamma_\gamma >> \Gamma_n$ for this resonance, transmission measurements provide an accurate determination of $\Gamma_\gamma\Gamma_n/\Gamma_n + \Gamma_\gamma$). Brusegan et al[36] had previously reported other evidence that the pulse height weighting technique was not working satisfactorily. From his paper it is noted that the number of counts at the 1.15 keV Fe resonance and at the Au normalising resonance are multiplied by weighting factors of about 30 and 13 respectively.

A problem of a different kind arose in the initial analysis of Harwell capture cross-section data for this resonance. A preliminary value for the neutron width of the resonance was anomalously high, indicating some systematic error. A detailed examination of the raw time-of-flight data revealed a time-dependent background component at long flight times when a constant background was expected. This effect was traced to the use of an incorrect dead time value in the analysis program. Correct allowance for the dead time removed the anomaly in the background and a final value of Γ_n consistent with other measurements was obtained. The above error also explained why the simultaneous shape analysis of data from three sample thicknesses had required normalisation factors inconsistent with the experimental values. This salutary experience showed the importance of understanding every feature of a measurement (in this case the shape of the time-dependent background). Undoubtedly, incorrect allowance for, or even neglect of, dead time effects accounts for some of the unexplained differences in nuclear data.

The results of measurements on the 27.7 keV resonance are summarised in Fig. 8. The Γ_n values appear to be in reasonable agreement but there are large discrepancies in the Γ_γ values. A close examination of these measurements teaches us some important lessons.

If we look first at the transmission measurements we find that there is a surprising lack of good data. In most of the measurements the peak cross-section does not correspond to the expected value of $4\pi\lambda\!\!\!^{-2}$, indicating incorrect normalisation or background errors. Although this appears not to have drastically altered the derived values of Γ_n, this would certainly not be the case for all resonances. In the Harwell Synchrocyclotron data, for example, it was found that a good fit to the transmission could only be achieved if the background for the open beam and the sample data was increased by about a factor of two. This was not understood at the time but has subsequently been verified by Syme[37] and has led to a better understanding of background effects and resolution functions in the time-of-flight measurements.

Let us now consider the radiation width of the 27.7 keV resonance. This can only be determined from measurements with prompt capture gamma-ray detectors. Prior to the 1977 meeting at Geel on the structural materials[38], values of Γ_γ tended to cluster round 1.5 eV, but during the meeting considerable doubt was thrown on these measurements due to lack of knowledge of detector sensitivity to scattered neutrons. Since Γ_n/Γ_γ for the resonance is about 1500, the response of the detector to prompt scattered neutrons must be a factor of about 10^4 smaller than its response to gamma radiation if a substantial correction is to be avoided. Subsequent to the Geel meeting Γ_γ has fallen to its present value of about 1 eV by using detectors with low neutron sensitivity and by more careful calculation in the resonance analysis programs of the effect of this sensitivity.

We have already mentioned that because of the large neutron width of the 27.7 keV resonance, shape analysis is necessary to obtain the resonance parameters. Because of the proximity of other large resonances a multilevel formalism is strongly preferred. A transmission measurement will only yield the neutron width whereas a capture measurement will allow both the neutron width and the radiation width to be determined.

The following points are worth noting from this discussion:

(1) The discrepancy in $\Gamma_n\Gamma_\gamma/\Gamma_n + \Gamma_\gamma$ for the 1.15 keV resonance and the recent reduction in the Γ_γ of the 27.7 keV resonance both indicate the need for experimenters to understand fully the response of detector systems.

(2) Discussion of the $\Gamma_n\Gamma_\gamma/\Gamma_n + \Gamma_\gamma$ discrepancy has shown the importance of making measurements by as many techniques as possible. The importance of good total cross-section data is demonstrated in this case.

(3) Reliable background measurements are needed for all experiments. Alternative tests of background levels are vital e.g. the $4\pi\lambda\!\!\!^{-2}$ peak cross-section test for good resolution transmission measurements for resonances where $\Gamma_n >> \Gamma_\gamma$. Good knowledge of the resolution function is also important.

(4) Serious errors can be produced if the expected response of complex electronic systems is not demonstrated.

(5) Further work is required before we can be satisfied that total energy detectors are suitable for measuring accurate data for structural materials.

(6) For accurate data we ideally would like a system where the corrections to the raw experimental data are small. The pulse height weighting technique does not satisfy this requirement.

(7) In resonance analysis, shape analysis is strongly preferred to area analysis and the multilevel rather than the single level formalism should be adopted in most cases.

3. General Discussion

3.1 Weaknesses shown up by the review

We outlined in the Introduction the basic iterative approach for arriving at accurate nuclear data which has led over the years to an improved consistency between measurements. It seems to us however that the review we have made shows up serious shortcomings in the effectiveness of the detailed procedures generally used and that a more rigorous and co-ordinated approach is needed.

In support of this view we begin by emphasising two important features in the overall picture we have obtained. Firstly, it seems that the main sources of systematic error which cause us trouble now have been identified in a qualitative sense for many years. While the presence of unidentified systematic errors cannot be ruled out it seems most likely that the lack of suitably deep understanding of some of the known error sources and inadequate care in the treatment of others are both much more important factors. Secondly, but closely connected with the first point, there is a remarkable shortage of well-documented evidence on systematic errors. With few exceptions their study receives only cursory and peripheral treatment in published papers. This has made it generally difficult and sometimes impossible to trace in a reasonably precise way how consistency improvements have occurred in measurements. Only for ^{235}U of the cases we looked at is the situation satisfactorily clear-cut, and there is little doubt that this is a rare example. There are other consequencies of this unsatisfactory treatment of errors in measurement reporting. It is difficult often to judge the quality of an experiment in anything but a subjective way and, not unconnected with this aspect, many experiments are reported which appear at face value to be merely repetitive. Another more insidious effect is that pressure is not brought to bear on experimenters to question the validity of their own results. In the absence of strong evidence to the contrary every experimenter naturally tends to believe his own work is as good as that of anyone else. We can learn from the experience of others but this experience has to be available.

Viewed in the perspective given above it seems, quite bluntly, that our practices in general are not meeting the normal scientific standards expected for measurements which are intended to be accurate. The present pattern must somehow be broken to make more effective progress in the future. We need therefore to adopt a more rigorous approach in which experimenters show a more critical awareness of systematic problems both in the design and execution of experiments and demonstrate this in their measurement reports. In particular some of the long-standing sources of systematic error need to be more thoroughly investigated, and more appropriately detailed treatment needs to be given to those which are understood already.

There are important additional reasons which make the changed approach desirable. The measurement community is an ageing one and also a shrinking one. This means on the one hand we are in increasing danger of losing people of great experience. Much of this experience is of the type under discussion here and unless it is adequately documented for others to make use of a new generation will be forced into much unnecessary work of rediscovery. On the other hand with fewer laboratories now active we need to pay particular attention to reducing systematic errors in the individual determinations of nuclear data by judiciously changing elements of the experimental arrangement. Youden[39], for example, has shown how the accuracy of measurement can be increased in this way.

Although at a fundamental level such changes can only be made through the attitude of the active experimenters there is undoubtably a need for a formal mechanism of some sort to provide both a forum for the detailed discussion of systematic problems and to steer and monitor, through subsequent reports, the progress made. It seems natural that the existing specialist meetings should have a role to play here. There seems no reason why part of these meetings should not be devoted specifically to this aspect of our work.

3.2 Suggested treatment of the sources of systematic error identified in the review

We turn now to the particular sources of systematic error recognised in the review and suggest in outline how they might be treated in the future. For convenience we list the most important of these in Table I.

Table I Recognised sources of systematic error

Experimental	Analytic
1. Background	1. Inadequately validated computer programs
2. Inadequate understanding of detector response	2. Inadequately sophisticated data analysis
3. Sample characterisation	3. Inadequate application of significant corrections
4. Faulty apparatus (e.g. electronic faults)	4. Inadequate use of available statistical information in assigning errors
5. Faulty data acquisition (e.g. dead time effects)	
6. Inaccurate energy calibration	
7. Inaccurately known energy resolution function	

A main point is the apparently trite one that experimenters simply need to demonstrate that their measurements are carefully done. However some of the sources of systematic error call for collaborative types of investigation and we begin with a consideration of them. These are:

(a) Background. Uncertainty in background determination is the most universal and long standing systematic problem of all. Although its importance is more widely appreciated than any of the effects we are discussing, virtually no work until recently[40] has been done to investigate background origins. Methods of background measurement too need to be reappraised. In particular the limitations of the notch filter technique are not well enough understood despite long experience with this method.

(b) Understanding of detector response. The lack of understanding of detector response to different types of radiation is another widespread problem. With any new detector there are inevitably pressures to measure nuclear data at the expense of investigating the detector behaviour in a sufficiently comprehensive way. The dangers of yielding to such pressures have been illustrated most clearly with the detectors used to study neutron capture in the important structural materials like Fe.

(c) Comparison of complex computer codes. The wide use of large and sophisticated computer codes in nuclear data analysis makes it essential to show that different versions give consistent and accurate results. Although individual experimenters can go a long way to verify and validate their own codes, inter-laboratory comparisons of test cases are vital. Some work along these lines has already started. Recently we have made a comparison of Doppler broadened cross-sections calculated by several codes using the same resonance parameters. The results were not in agreement and errors were found in all of the codes considered. Another example which showed up discrepancies in a similar way is the recent benchmark intercomparison[41] to study methods used to determine average parameters from resolved resonance parameters. Both examples highlight the need for extensive collaborative effort to investigate large codes.

(d) Sample characterisation. Sample purity and uniformity are of fundamental importance most particularly in absolute nuclear data measurements. The encouraging experience with ^{235}U fission foil assays indicates that more attention should be given to this topic in general.

The remaining effects in Table I are best dealt with by individual experimenters. The nature of these effects are for the most part understood so the main requirement is for experimenters to spend enough time to produce convincing evidence that the various factors have been treated carefully enough.

On the experimental side subsidiary experiments are necessary to measure energy resolution functions, to measure energy calibrations, and to investigate dead time effects. Repeat measurements are needed, for example, with replacement units to demonstrate that electronic instrumentation functions properly under all the conditions of the experiment. The optimisation of effort must depend on the particular data measurement but it should be possible to formulate guide lines to help with these decisions. This aspect touches on Youden's work referred to in section 3.1 and perhaps his ideas can be adapted.

On the analytic side we have already stressed the importance of computer code validation. There are strong ties here with the choice of methods used in resonance analysis. In particular the importance of using shape analysis with a multilevel formalism has been clearly demonstrated. Computer codes are often important also in calculating correction factors to be applied to data. Ideally it should be arranged for these factors to be small, but if this cannot be done it is essential to show that significant corrections are suitably accurate.

Lastly we come to the problem of making full use of the available statistical information in the overall assignment of errors. The importance of using a more complete treatment than implied by the conventional use of standard deviations has been increasingly stressed in recent years. (Müller[42], for example, has produced an excellent paper on the assessment of errors.) Perey has been a leader in drawing our attention to correlations and the importance of the covariance matrix in the analysis of nuclear data. More recently Bayes' Theorem has been adapted effectively in data analysis and should be further exploited.

We end this section with some comments on the documentation of results. Full reporting is a key factor in the treatment of future work. This is likely to present problems when papers are submitted for publication in journals because editors are unlikely to accept such detail. Maybe the system here should be questioned. However this difficulty can be overcome by preparing comprehensive laboratory reports which could be presented and discussed at specialist meetings along the lines indicated in section 3.1.

4. Concluding Remarks and Recommendations

The main conclusion we draw is that a more rigorous and co-ordinated approach is needed in the future if we are to improve the effectiveness of our basic procedures for achieving accurate nuclear data. Not surprisingly perhaps we have not uncovered any new or unexpected obstacle to progress. The general lesson learned is rather that we have not adopted a scientifically stringent enough attitude to the treatment of the sources of systematic errors that have been recognised for years. More specifically, there are some effects which need further subsidiary experimentation to obtain a better understanding of them and there are others which are understood that need more careful attention (see section 3.2). The generally reduced effort in the nuclear data field makes it desirable for laboratories to collaborate closely so that available resources for the various kinds of nuclear data measurements needed are deployed in an optimised way. We suggest that specialist meetings should play a leading role in co-ordinating work on problems connected with systematic errors. Of course, discrepancies between data are discussed in some depth already at these meetings but usually only the broadest of guide lines are recommended for future work. We need a system which provides more specifically detailed discussion and direction and which encourages experimenters to report back and assess the results of their investigations.

We have to recognise that experiments will take longer to carry out with the changed approach we suggest. This has to be accepted as an inevitable consequence of achieving the increased accuracy we require.

To conclude we recommend that:-

1. Experimenters need to show more critical awareness of systematic errors in both the design and execution of their experiments. In particular they should demonstrate in their measurement reports that recognised sources of systematic error have been adequately treated.

2. Collaborations should be arranged to study and report on recognised sources of systematic error which are not well enough understood.

3. Specialist meetings should devote sessions specifically to the close co-ordination of work on the problems connected with the sources of systematic error.

References

[1] H. Derrien and L. Edvardson. NBS Special Publication 493, 14 (1977).

[2] K. M. Diment and C. A. Uttley. Harwell Report AERE-PR/NP 15 (1969).

[3] G. M. Hale. NBS Special Publication 425, Vol. 1, 302 (1975).

[4] G. M. Hale. NBS Special Publication 493, 30 (1977).

[5] H. H. Knitter, C. Budtz-Jorgensen, M. Mailly and R. Vogt. CBMN Report EUR 5726e (1977).

[6] G. M. Hale. Int. Nucl. Data Committee Report INDC-36/LN, B-8 (1981).

[7] C. A. Uttley, M. G. Sowerby, B. H. Patrick and E. R. Rae. CONF-70100, AEC Symp. Ser. 23, 80 (1971).

[8] G. P. Lamaze. NBS Special Publication 493, 37 (1977).

[9] G. D. James. Ibid 319 (1977).

[10] A. D. Carlson. NBS Special Publication 425, Vol. 1, 293 (1975).

[11] S. J. Friesenhahn, V. J. Orphan, A. D. Carlson, M. P. Fricke and W. M. Lopez. Ibid 232 (1975).

[12] D. B. Gayther. Harwell Report AERE-R 8856 (1977).

[13] G. P. Lamaze, R. A. Schrack and O. A. Wasson. Nucl. Sci. and Eng. 68, 183 (1978).

[14] C. Renner, J. A. Harvey, N. W. Hill, G. L. Morgan and K. Pusk. Bull. Am. Phys. Soc. 23, 526 (1978).

[15] P. H. White. J. Nucl. Energ. A/B 19, 325 (1965).

[16] M. Cancé and G. Grenier. Nucl. Sci. Eng. 68, 197 (1978).

[17] Arlt et al. Proc. Int. Conf. on Nuclear Cross Sections for Technology held at Knoxville, Tennessee and published as NBS Special Publication 594 (U.S. Department of Commerce) 990 (1980).

[18] Adamov et al. Ibid 995 (1980).

[19] O. A. Wasson, A. D. Carlson and K. C. Duvall. Nucl. Sci. Eng. 80, 282 (1982).

[20] C. A. Uttley and J. A. Phillips. Harwell Report AERE NP/R 1996 (1956).

[21] A. Moat. Unpublished Report (1958).

[22] W. P. Poenitz, J. W. Meadows and R. J. Armani. Argonne Nat. Lab. Report ANL/NDM-48 (1979).

[23] K. M. Glover. Private communication.

[24] M. C. Moxon and M. G. Sowerby. Harwell Report AERE-R 10597 (1982).

[25] F. Poortmans, L. Mewissen, G. Rohr, R. Shelley, T. Van der Veen, H. Weigmann, E. Cornelis and G. Vanpraet. Private communication (1981).

[26] H. Derrien and P. Ribon. NEANDC(E)163/U, 63 (1974).

[27] M. C. Moxon. Private communication (1982). (See also EUR 6108, 644 (1979).

[28] N. M. Larson and F. G. Perey. Oak Ridge National Laboratory Report ORNL/TM-7485 (1980).

[29] H. Derrien. Brookhaven National Laboratory Report BNL-NCS-50451, 156 (1975).

[30] D. K. Olsen, G. de Saussure, R. B. Perez, E. G. Silver, F. C. Difilippo, R. W. Ingle and H. Weaver. Nucl. Sci. and Eng. 62, 479 (1977).

[31] R. C. Block, D. R. Harris, S. H. Kim and K. Kobayashi. EPRI NP-996 (1979).

[32] T. J. Haste, M. C. Moxon and J. E. Jolly. Harwell Report AERE-R 8966 (1979).

[33] G. Rohr. Paper presented to NEANDC/NEACRP Specialists Meeting on Fast Neutron Capture Cross-sections, Argonne, April 1982.

[34] R. L. Macklin. Private communication (1982).

[35] R. W. Hockenbury, Z. M. Bartolome, J. R. Tartarczuk, W. R. Moyer and R. C. Block. Phys. Rev. 178, 1746 (1969).

[36] A. Brusegan, F. Corvi, G. Rohr, R. Shelley and T. Van der Veen. Proc. Conf. Nuclear Cross-sections and Technology, Knoxville, NBS Special Publication 594, 163 (1980).

[37] D. B. Syme. Harwell Report AERE-R 10244 (1981).

[38] "Neutron Data of Structural Materials for Fast Reactors", Editor K. H. Böckhoff, Pergamon Press (1979).

[39] W. J. Youden. Physics Today 14, 32 (1961).

[40] D. B. Syme. Paper at this Conference (1982).

[41] "The determination of average resonance parameters", Newsletter No. 27, NEA Data Bank (1982).

[42] J. W. Müller. Paper presented to Second International Conference on Precision Measurement and Fundamental Constants, Gaithersburg, June 1981.

[43] B. J. Allen. Paper presented to NEANDC/NEACRP Specialists Meeting on Fast Neutron Capture Cross-sections, Argonne, April 1982.

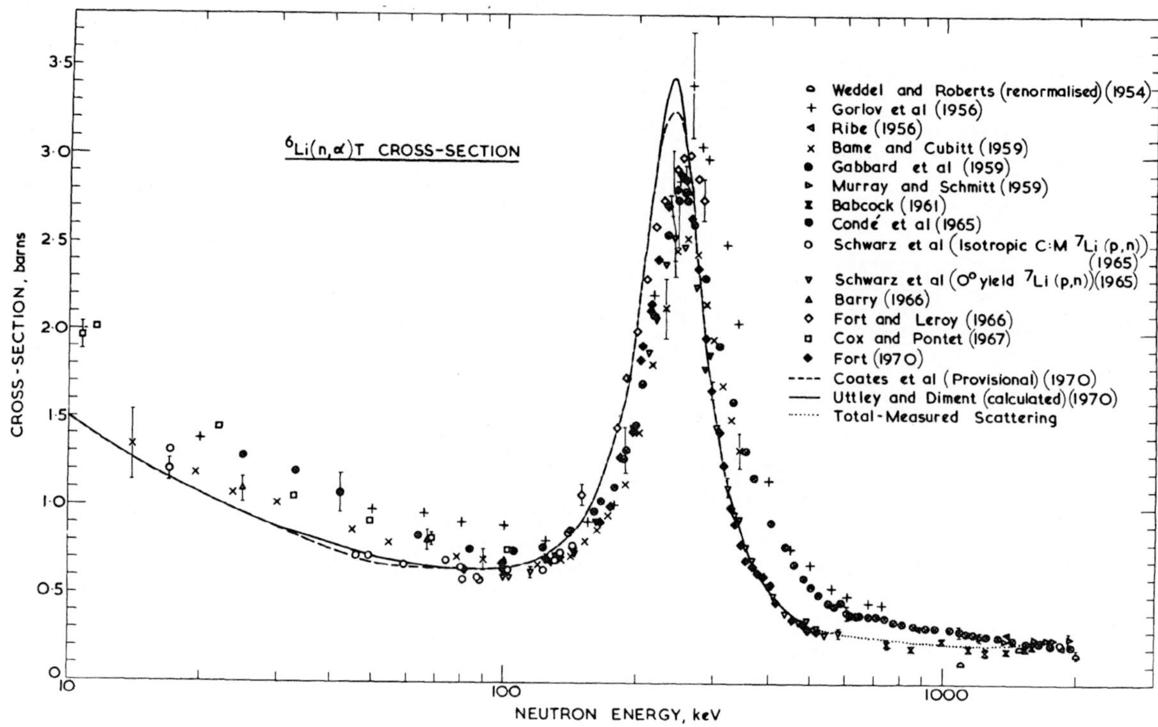

Fig. 1 The measurements of the ^{6}Li(n,α)t cross-section available in 1970.

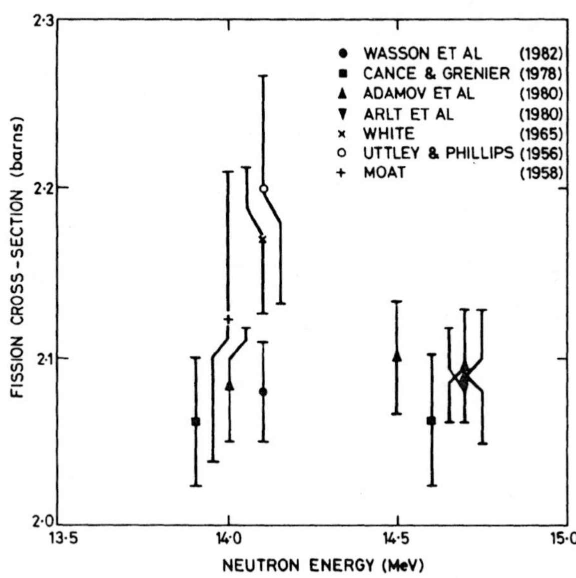

Fig. 2 Measurements of the ^{235}U fission cross-section in the region of 14 MeV.

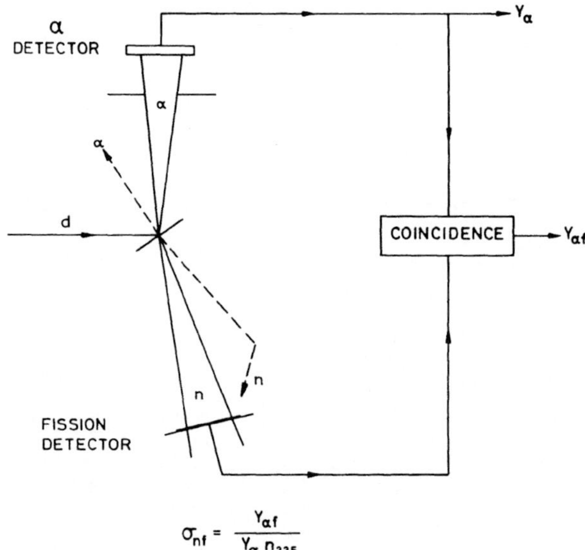

Fig. 3 Schematic representation of the time correlated associated particle technique. The dashed lines indicate an alpha-particle and its associated neutron which would not be recorded by the coincidence. However if no time correlation were required the neutron could induce a fission reaction which would be recorded, whereas the corresponding alpha-particle would not be counted.

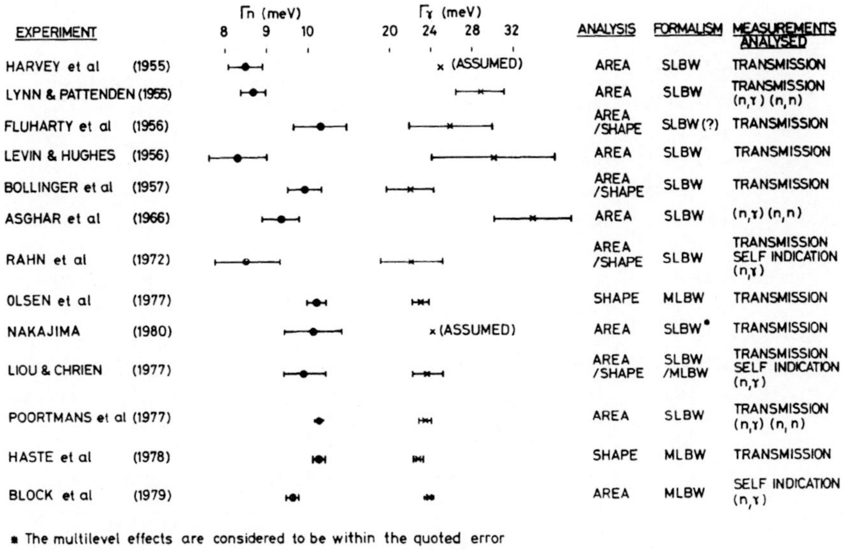

EXPERIMENT		Γ_n (meV)	Γ_γ (meV)	ANALYSIS	FORMALISM	MEASUREMENTS ANALYSED
HARVEY et al	(1955)		(ASSUMED)	AREA	SLBW	TRANSMISSION
LYNN & PATTENDEN	(1955)			AREA	SLBW	TRANSMISSION (n,γ) (n,n)
FLUHARTY et al	(1956)			AREA /SHAPE	SLBW(?)	TRANSMISSION
LEVIN & HUGHES	(1956)			AREA	SLBW	TRANSMISSION
BOLLINGER et al	(1957)			AREA /SHAPE	SLBW	TRANSMISSION
ASGHAR et al	(1966)			AREA	SLBW	(n,γ) (n,n)
RAHN et al	(1972)			AREA /SHAPE	SLBW	TRANSMISSION SELF INDICATION (n,γ)
OLSEN et al	(1977)			SHAPE	MLBW	TRANSMISSION
NAKAJIMA	(1980)		(ASSUMED)	AREA	SLBW*	TRANSMISSION
LIOU & CHRIEN	(1977)			AREA /SHAPE	SLBW /MLBW	TRANSMISSION SELF INDICATION (n,γ)
POORTMANS et al	(1977)			AREA	SLBW	TRANSMISSION (n,γ) (n,n)
HASTE et al	(1978)			SHAPE	MLBW	TRANSMISSION
BLOCK et al	(1979)			AREA	MLBW	SELF INDICATION (n,γ)

* The multilevel effects are considered to be within the quoted error

Fig. 4 Measurements of the resonance parameters of the 20.9 eV resonance in ^{238}U.

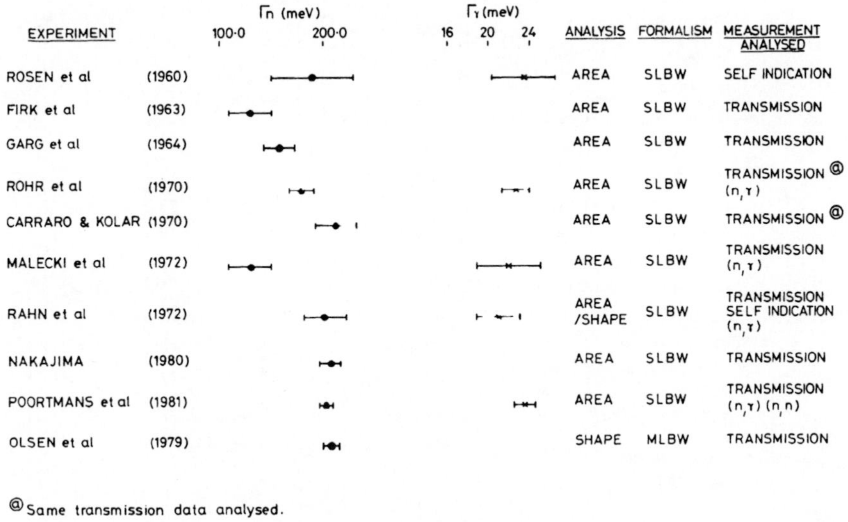

EXPERIMENT		Γ_n (meV)	Γ_γ (meV)	ANALYSIS	FORMALISM	MEASUREMENT ANALYSED
ROSEN et al	(1960)			AREA	SLBW	SELF INDICATION
FIRK et al	(1963)			AREA	SLBW	TRANSMISSION
GARG et al	(1964)			AREA	SLBW	TRANSMISSION
ROHR et al	(1970)			AREA	SLBW	TRANSMISSION @ (n,γ)
CARRARO & KOLAR	(1970)			AREA	SLBW	TRANSMISSION @
MALECKI et al	(1972)			AREA	SLBW	TRANSMISSION (n,γ)
RAHN et al	(1972)			AREA /SHAPE	SLBW	TRANSMISSION SELF INDICATION (n,γ)
NAKAJIMA	(1980)			AREA	SLBW	TRANSMISSION
POORTMANS et al	(1981)			AREA	SLBW	TRANSMISSION (n,γ) (n,n)
OLSEN et al	(1979)			SHAPE	MLBW	TRANSMISSION

@ Same transmission data analysed.

Fig. 5 Measurements of the resonance parameters of the 958.6 eV resonance in ^{238}U.

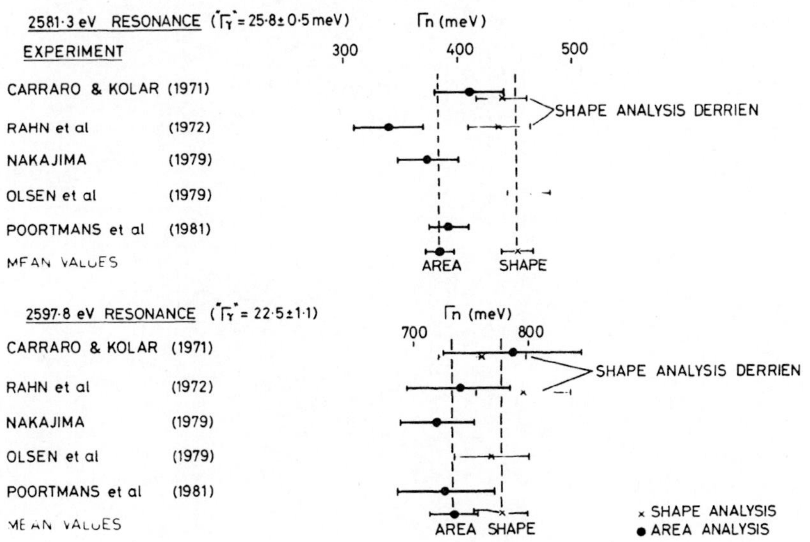

Fig. 6 The Γ_n values for the 2581.3 eV and 2597.8 eV resonances in ^{238}U.

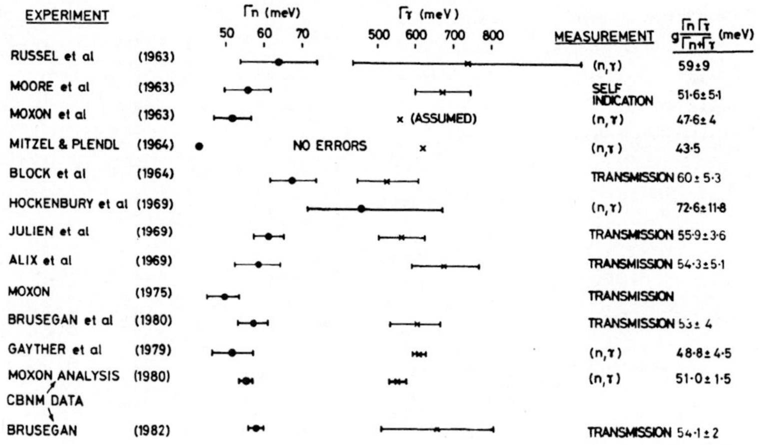

EXPERIMENT		Γn (meV) 50 60 70	Γγ (meV) 500 600 700 800	MEASUREMENT	$g\frac{\Gamma_n\Gamma_\gamma}{\Gamma_n+\Gamma_\gamma}$ (meV)
RUSSEL et al	(1963)			(n,γ)	59±9
MOORE et al	(1963)			SELF INDICATION	51·6±5·1
MOXON et al	(1963)		× (ASSUMED)	(n,γ)	47·6±4
MITZEL & PLENDL	(1964)	NO ERRORS		(n,γ)	43·5
BLOCK et al	(1964)			TRANSMISSION	60±5·3
HOCKENBURY et al	(1969)			(n,γ)	72·6±11·8
JULIEN et al	(1969)			TRANSMISSION	55·9±3·6
ALIX et al	(1969)			TRANSMISSION	54·3±5·1
MOXON	(1975)			TRANSMISSION	
BRUSEGAN et al	(1980)			TRANSMISSION	53±4
GAYTHER et al	(1979)			(n,γ)	48·8±4·5
MOXON ANALYSIS	(1980)			(n,γ)	51·0±1·5
CBNM DATA					
BRUSEGAN	(1982)			TRANSMISSION	54·1±2

Fig. 7 Measurements of the parameters of the 1.15 keV resonance in ^{56}Fe.

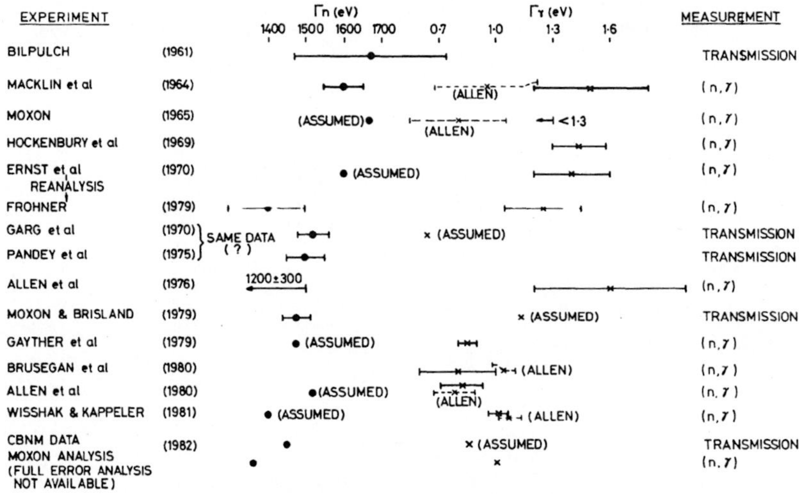

EXPERIMENT		Γn (eV) 1400 1500 1600 1700	Γγ (eV) 0·7 1·0 1·3 1·6	MEASUREMENT
BILPULCH	(1961)			TRANSMISSION
MACKLIN et al	(1964)		(ALLEN)	(n,γ)
MOXON	(1965)	(ASSUMED)	(ALLEN) <1·3	(n,γ)
HOCKENBURY et al	(1969)			(n,γ)
ERNST et al	(1970)	(ASSUMED)		(n,γ)
REANALYSIS FROHNER	(1979)			(n,γ)
GARG et al	(1970)	SAME DATA (?)	× (ASSUMED)	TRANSMISSION
PANDEY et al	(1975)			TRANSMISSION
ALLEN et al	(1976)	1200±300		(n,γ)
MOXON & BRISLAND	(1979)		× (ASSUMED)	TRANSMISSION
GAYTHER et al	(1979)	(ASSUMED)		(n,γ)
BRUSEGAN et al	(1980)		(ALLEN)	(n,γ)
ALLEN et al	(1980)	(ASSUMED)	(ALLEN)	(n,γ)
WISSHAK & KAPPELER	(1981)	(ASSUMED)	(ALLEN)	(n,γ)
CBNM DATA MOXON ANALYSIS (FULL ERROR ANALYSIS NOT AVAILABLE)	(1982)		× (ASSUMED)	TRANSMISSION (n,γ)

Fig. 8 Measurements of the parameters of the 27.7 keV resonance in ^{56}Fe. The dashed values are the revised values of Allen [43].

NUCLEAR DATA ACTIVITIES IN CHINA

Zhou Delin

Institute of Atomic Energy, Academia Sinica
P. O. B. 275 (41), Beijing, China

A summary of nuclear data activities in China, including nuclear data measurement, evaluation, theoretical calculation, etc., is given.

(CNDC, nuclear data, measurement, evaluation, calculation)

Introduction

Work on nuclear data was started quite late in China. In 1958, a heavy water research reactor and a 1.2 m cyclotron as well as some other facilities for nuclear experiment were set up in Institute of Atomic Energy (IAE). We might say that only at that time and even some later we just started our nuclear data activities. Afterwards, a Van de Graaff accelerator with terminal voltage 2.5 MV and a 450 kV Cockcroft-Walton accelerator in IAE and a few similar facilities in other laboratories in China were installed successively. Using these facilities some measurements of nuclear data have been carried out during 1960s. In 1974, a nuclear data compilation & evaluation group, and some later, the Nuclear Data Center was founded in IAE. Since then, work in various areas of nuclear data were developed succesively and more systematically. Up to now, there are four groups in our nuclear data center which take the various responsibilities for nuclear data work, namely: (1) nuclear data evaluation, (2) nuclear data calculation and theoretical research, (3) group cross section generation and nuclear data test and (4) management of the evaluated nuclear data library, respectively. Nuclear Data Center also coordinates a cooperative effort of more than 10 institutes and universities on nuclear data work, and takes the responsibilities for international information exchange and cooperation.

In IAE, several groups which belong to three different labs are working continually in the field of nuclear data measurement. Besides, a few nuclear data measurements have been carried out intermittently in some institutes else in China.

Some years later we will have a tandem accelerator of type HI-13 in our hands. The construction of an electron linear accelerator with high current and short pulses is also planned. After the operation of these facilities and some other equipments which are being established in other institutes or universities in China, more experimental work on nuclear data could be carried into effect.

Nuclear Data Measurement

The majority of measurements were accomplished at VDG, HT and variable energy cyclotron in IAE. A brief description is given as follows:

1. Fission Product Yields The yields of fission products for several fissioning nuclei were measured by means of radiochemical technique or the method of gamma-ray spectrum using Ge(Li) detector. Experiments of ^{235}U fission induced by thermal, 14 MeV, 5 MeV and 3 MeV neutron as well as ^{252}Cf spontaneous fission were accomplished in recent years. The product nuclides measured by radiochemical method includes ^{99}Mo, ^{95}Zr, ^{103}Ru, ^{106}Rh, ^{144}Ce, ^{147}Nd etc., while with the gamma-spectrum method various mass chain yields were obtained.

2. Nuclear Fission Characteristics Some measurements on mass and kinetic energy distribution and angular distribution of neutron induced fission or spontaneous fission fragments have been carried out with nuclear emulsion, solid state nuclear track detector or/ and surface barrier detector. Angular distribution of fission fragments at the $^{238}U(n,f)$ and $^{238}U(n,2n'f)$ threshold, kinetic energy distribution of fission fragments of fast neutron induced $^{235}U(n,f)$ reaction as well as mass and kinetic energy distribution of ^{252}Cf spontaneous fission fragments have been measured recently.

3. Fission Neutron Number With a Gd-loaded liquid scintillation tank 60 cm in diameter, a series experiments of precise determination $\bar{\nu}$ (the average number) and $P(\nu)$ (the multiplicity probability) of prompt neutrons

K. H. Böckhoff (ed.), Nuclear Data for Science and Technology, 987–990.

for neutron induced and spontaneous fission have been carried out. For instance, in absolute measurement of $\bar{\nu}_p$ of ^{252}Cf spontaneous fission the accuracy reached with this facility was about 0.5%. While for neutron induced fission the accuracy was about 1%.

4. Fission Cross Section Thermal and fast neutron fission cross section of some nuclides have been measured absolutely and relatively. Measurements finished in recent years are listed in table I.

5. Excitation Function of Fast Neutron Induced Activation Reaction Excitation functions of some fast neutron induced activation reactions were measured.

Table I Measurements of Fission Cross Section in China

Nuclide	Neutron Energy (MeV)	Comments
Bi-209	14.7	An upper limit of 10^{-32} cm^2 was settled
U-233	thermal	Measurement relative to U-235 with error 2%
U-233	0.03-5.6 14 - 18	Measurement relative to U-235. Absolute measurement with error 5% relative error 2.5-3%.
U-235	$(0.02-0.30)\times10^{-6}$	Absolute measurement with error 0.34%
U-233, Pu-239	$(0.02-0.30)\times10^{-6}$	Measurement relative to U-235 with error 2%
U-235, Pu-239	14	Absolute measurement with error 2.5%
U-238	3.3-5.6	Absolute measurement with error 3%
Pu-239	1.0-1.6	Absolute measurement with error 2-3%
Pu-239	1.0-1.6 2.3-5.6	Measurement relative to U-235

The gamma and beta activities were measured with a 3"x3" Na I(Tl) spectrometer, a 136 cm^3 Ge(Li) gamma-ray spectrometer and an anticoincident scintillation spectrometer, respectively. Monoenergetic neutrons were produced by T(p,n), ^{7}Li(p,n), D(d,n) and T(d,n) reactions cover different energy region for different measurements using 2.5 MV Van de Graaff and Cockcroft-Walton accelerators. Various corrections were made including corrections of multiple scattering in the target and sample, if necessary, which were calculated by Monte-Carlo method. The excitation functions of about 20 fast neutron induced reactions, such as ^{27}Al $(n,\alpha)^{24}$Na, ^{56}Fe$(n,p)^{56}$Mn, ^{197}Au$(n,2n)^{196}$Au and ^{197}Au$(n,\gamma)^{198}$Au etc. have been measured.

6. Neutron Capture and Gamma-Ray Production Cross Section Gamma-ray production cross sections from interactions of 14.9 MeV neutrons with several elements (including C, F, Al, Si, Fe and Cu etc.) have been measured. The associated particle coincidence method and the ring-geometry were employed. A 200x100 NaI(Tl) and a 42 cm^3 Ge(Li) detector gamma-ray spectrometer were used to measure the spectra and the production cross sections. Several neutron capture cross section measurements have also been performed.

7. Fast Neutron Spectroscopy Several fast neutron TOF spectrometers were installed in IAE,Fudan Univ., Sichuan Univ. at Van de Graaff, Cockcroft-Walton neutron generator or cyclotron accelerator. Some other techniques, i.e., recoil proton scintillation spectrometer with stilbene crystal, nuclear emulsion, multisphere spectrometer and technique using activation foils with different thresholds were also used to measure fast neutron spectra in different conditions. With the TOF facilities in IAE, e.g. the angular distribution and

differential cross section of elastic and inelastic scattering on C (11.6 and 14.7 MeV), ^{7}Li, ^{9}Be (14.7 MeV) have been obtained. The prompt neutron spectrum of ^{252}Cf and ^{235}U fission were also measured.

8. Decay Scheme Study The gamma-ray spectrum from beta-decay of some nuclides have been studied with a 70 cm^3 coaxil Ge(Li) spectrometer or a superpure Ge spectrometer. Energies and relative intensities of those measured gamma-rays were determined. Some data of decay scheme were derived.

Nuclear Data Compilation and Evaluation

It mainly means the compilation and evaluation of microscopic experimental neutron data and some charged particle induced reaction cross section data for light nuclides which are useful to fission and fusion reactors and other applied purposes. Evaluation work in following areas have been done or are in progress.

1. Neutron Data for General Purpose New evaluations (for all reaction channels, over the energy range from thermal to 20 MeV, including resonance parameters) for about 30 nuclides, such as fissile, fertile and fusion nuclides, structural materials, transplutonium nuclides and some standards for nuclear measurements, are in progress. For example, the experimental data of n+H (from thermal to 20 MeV) and n+^{6}Li ($E_n \leq 1.7$ MeV) reaction have been evaluated and analyzed with phase shift and R-matrix theory respectively. The calculated values reproduce the experimental data (including t+α data in the R-matrix analysis for Li-7 system) quite good and were adopted as the recommended values.

2. Excitation Function of Fast Neutron Induced Activation Reaction These include excitation functions of activation reactions induced by fast neutron for those nuclides on which measurements have been carried out in IAE.

3. Individual Fission Yield Individual fission yield for about seventy fission products of $^{233,235,238}U$, ^{239}Pu etc. have been evaluated.

4. Decay Data Decay data of about one hundred nuclides have been evaluated. Most excitation functions were fitted by orthogonal polynomial or spline. For all recommended data, the errors of which have been estimated as reasonable as possible. The accidental and fixed errors were treated separately.

Theoretical Calculation of Nuclear Data

The energy regions or reaction channels, in which the experimental data are not available, have been complemented by theoretical calculated data. Successful nuclear model theories such as optical model (including spherical and deformed), direct interaction theory, statistical theory (including H-F theory and evaporation model), pre-equilibrium emission theory, R-matrix theory, resonance group theory, three body reaction theory of light mucleus and phase shift analysis etc. have been adopted to calculate the data of specified reaction channels and energy regions, respectively.

Suitable parameters of a particular model theory were adjusted to fit the existing measured data as good as possible and then the data gaps were filled with the values calculated based on the adjusted parameters.

In order to get better agreement between calculated and measured values, some efforts are being made for improving the calculated values. For example, we have studied the influence of the Pauli exclusion principle and Fermi motion on the calculated shape of angular distribution for both the pre-equilibrium and equilibrium decay of the neutron induced reactions. Calculations for a series of elements were performed at neutron energy 14.6 MeV and the results were compared with experimental data. We have concluded that the influence of these effects on the shape of the angular distribution is rather significant for pre-equilibrium decay in the energy range of several tens of MeV, which improves the agreement with experimental data in the backward angle range for high energy emitted particles.

Group Cross Section Generation and Integral Test of Nuclear Data

We have just started to do some work on these areas. After comparing the current methods of group cross section generation, a code of calculating multigroup cross section for the thermal neutron reactor was compiled and group cross sections with 54 and 68 group structures have been calculated for some nuclei. The code of generating multigroup cross section for fast neutron reactor is now being made. Some integral quantities such as K_{eff}, center reaction rate etc. have been calculated with an one-dimensional Sn code and compared with the reported experimental results. An one-dimensional diffusion programme NDP is being made.

Nuclear Data Library

Great efforts are being made to establish the evaluated nuclear data library, it will be named CENDL (Chinese Evaluated Nuclear Data Library). The computerized evaluation system and storage and retrival system are being established.

THE EUROPEAN-JAPANESE JOINT PROGRAMME ON NEUTRON DATA EVALUATION

C.G. Campbell

AEE Winfrith, Dorchester
DT2 8DH Dorset, United Kingdom

C. Nordborg

NEA Data Bank
91191 Gif-sur-Yvette CEDEX, France

Following an initiative by the NEA Committee on Reactor Physics (NEACRP) at their meeting in September 1980, the NEA Steering Committee approved the setting up of a Joint European-Japanese Programme on Neutron Data Evaluations. A Scientific Co-ordinating Group was formed to determine the content of the Joint Evaluated File (JEF) and to co-ordinate work in national research organisations associated with the periodic revisions to this library. The NEA Data Bank is taking care of the secretariat functions of the Programme and will also, among other things, assemble and maintain the Joint Evaluated File (JEF). The plans are to release a first version of the file at the end of 1982, followed by a period of re-evaluations and benchmark testing, which will lead to the release of a second version at the end of 1985. The availability of the file will be restricted to the member countries of the NEA Data Bank.

Introduction

Following an initiative by the NEA Committee on Reactor Physics (NEACRP), a first meeting of an ad hoc working group was held in Paris in November, 1980. The main objectives of this working group were: to agree on a common format for evaluated files; to discuss common evaluation procedures; to establish a mechanism for selection of existing evaluations and co-ordination and review of new evaluations; and to establish a common Data Library.

In October 1981, the NEA Steering Committee approved the setting up of a Joint Programme on Neutron Data Evaluations. A Scientific Co-ordinating Group (SCG) was set up to continue the work of the ad hoc working group. The SCG comprises experts drawn from two sections of the nuclear data community, users and evaluators. In the first phase, the user representatives are mainly responsible for the policy, scope of the file and determination of priorities, while the function of the evaluators is to recommend the detailed composition of the common Data Library, called the Joint Evaluated File (JEF).

The NEA Data Bank coordinates the work on the Joint Evaluated File (JEF) and is responsible for the assembly and maintenance of the library. Other tasks assigned to the NEA Data Bank are: secretariat services for the Joint Programme; assistance in the preparation and publication of summary documentation for the JEF and implementation and use of computer codes for verification and consistency checking of evaluated files.

First Phase of the Joint Evaluated File

The library is being assembled in the standard ENDF/B-V format. The Scientific Coordinating Group (SCG) will continue to discuss and adopt recommendations from nominated scientists in participating countries on the evaluations to be selected for important elements and isotopes. These selected evaluations are translated, when necessary, into the ENDF/B-V format and converted to point-wise cross section representation by the staff of the NEA Data Bank, with the assistance, both scientific and if necessary computing resources, of scientists in national research organisations. The selected and re-formatted files are being assembled into a Joint Evaluated File (JEF) at the NEA Data Bank. Copies of all or part of this library will be made available on request to scientists in Member countries of the NEA Data Bank.

The aim of the first phase is to prepare a complete neutron cross section library (JEF-1) for release at the end of 1982. The process of selection and assembly of files is now largely complete and the consistency checks should be completed at the end of 1982. These consistency checks will include the comparison of 2200 m/s, thermal Maxwellian, resonance integrals and fission spectrum averages with available experimental data.

Experiencies and difficulties in the compilation work

By June 1982, the Joint Evaluated File contained complete evaluations for about 70 of the principal elements and isotopes of current importance in reactor design and shielding, notably: standard reference materials, structural materials, fuel materials, secondary actinides and fission products. The Scientific Co-ordinating Group has decided to adopt the ENDF/B-V (International) Dosimetry File and the ENDF/B-V Standard File. Modifications could be made to parts of these files, where the reactions are not considered to be an international standard, e.g. the resonance parameters of U-235 have been revised for JEF-1. Table 1 contains a list of the special selected isotopes and elements that are included in JEF-1. Most of the datasets have been checked with the ENDF/B-V processing codes CHECKER-5 and FIZCON-5 . The calculations of average cross sections have been made with the code INTER-5 .

In the case where the cross sections in the resonance region were represented by resonance parameters, the corresponding point-wise cross sections were calculated with the program RECENT . It was found that the normal six digit format output was not accurate enough to describe some resonances and that the nine digit floating point format was accurate only to 7 digits, when the program was run in single precision on an IBM-3033. The problem was in this case solved by using a UNIVAC-1110 computer, which has an accuracy of at least 9 digits in single precision mode. Another problem in connection with the calculation of point-wise cross-sections from resonance parameters was that in some cases these calculations gave rise to negative cross-sections when the resonance region was characterised by the single-level Breit-Wigner formula. This was due to incorrectly given background data and the problem was in most cases solved by employing the multi-level Breit-Wigner formula instead.

The program CRECTJ5 was used to a very large extent in the compilation work of the Joint Evaluated File at the NEA Data Bank. The program compiles evaluated data in the ENDF/B-IV or ENDF/B-V format and includes the following functions: it combines partial range and individual reaction data files in one file; it performs operations such as addition, subtraction, multiplication and division on an evaluated file; it modifies cross sections by replacing sections of data with other evaluated data; it creates an index in the general information part of the evaluated file and it outputs data in either 6-, 7- or 9-digit format for both ENDF/B-IV and ENDF/B-V formatted files.

The computer code KTOE, that translates evaluated files from the KEDAK format to the ENDF/B format, was thoroughly modified by its author. The latest ver-

K. H. Böckhoff (ed.), Nuclear Data for Science and Technology, 991–992.

sion, called KTOE-3 , has many new features like: automatic construction of the background cross section when the resonance parameters are translated and also the possibility of translating energy distributions.

The JEF evaluated library was at first decided to be a library with a pointwise representation of the cross-sections. Futher discussions within the Scientific Co-ordinating Group made it clear that the original resonance parameters should be kept. It was found impossible according to the ENDF/B-V format specifications to include these parameters in the pointwise cross section file and the only solution was therefore to create two parallel libraries, one with pointwise cross section representation and one with the original resonance parameters. The inclusion of error information has also been discussed and the conclusion was that covariance matrix information should be given in the JEF library.

Plans for a second phase of the Joint Evaluated File

In the course of the assessment of the files for inclusion in the first version of the Joint Evaluated File (JEF-1), a number of nuclei have been identified as in need of at least partial re-evaluation. Some minor improvements will be included in JEF-1, but the major improvements, which also take into account the results of new measurements, will be left for the proposed reevaluations for version 2 (JEF-2). The responsibility for these imporvements to the file is being assigned on a voluntary basis by the Scientific Co-ordinating Group (SCG) to scientists in participating countries, taking into account national research programmes and specialist expertise and national resources available for this work.

Following the release of JEF-1 at the end of 1982, a programme of benchmark testing will be carried out. This will be in two stages, Stage 1, to be completed at the end of 1983, will be to analyse simple benchmarks, like several of the benchmarks used for ENDF/B-V testing, and some selected European benchmarks. The distribution of work in this stage will be decided in September 1982. The NEA Data Bank plans to participate in this stage by, for example, preparing group cross sections and by calculating simple integral properties using codes available at the Data Bank.

In the second stage of benchmark testing, more complex assemblies will be analysed, including assemblies for which the composition data have not been published. In this case the results of the analyses will be communicated to other participants so that the performance of the library for a very wide range of applications can be understood. This second stage of testing is estimated to cover a period of about two years and the outline plan is to release a second version of the Joint Evaluated File (JEF-2) at the end of 1985.

Table 1

Specially selected isotopes and elements for JEF-1

O-16	Mo-100	Eu-151
Na-23	Tc-99	Eu-153
Ti-nat	Ru-101	Gd-nat
Ti-46	Ru-102	Gd-154
Ti-47	Ru-103	Gd-155
Ti-48	Ru-104	Gd-156
Cr-nat	Rh-103	Gd-157
Cr-50	Pd-105	Gd-158
Fe-nat	Pd-107	Gd-160
Fe-54	Ag-109	U-235
Fe-56	Cs-133	U-238
Fe-58	Cs-135	Np-237
Co-58	Ba-134	Pu-239
Co-59	Ba-135	Pu-240
Ni-nat	Ba-136	Pu-241
Ni-58	Ba-137	Pu-242
Ni-59	Ba-138	Am-241
Ni-60	Pr-141	Am-243
Ni-62	Nd-143	Cm-242
Zn-64	Nd-145	Cm-243
Zr-93	Nd-148	Cm-244
Mo-95	Sm-147	Cm-245
Mo-97	Sm-149	
Mo-98	Sm-151	

References

1. ENDF-110 (BNL-50300), ENDF/B-V Processing Programs, May 1980.
2. D.E.Cullen: "Program RECENT: Reconstuction of Energy-Dependent Neutron Cross Sections from Resonance Parameters in the ENDF/B Format", UCRL-50400, Vol. 17, Part C, 1979.
3. T. Nakagawa: To be published.
4. C.G. Panini : To be published.

THE NEA DATA BANK

G. Coddens

NEA Data Bank
91191 Gif-sur-Yvette CEDEX, France

The NEA Data Bank provides the nuclear data and computer programs necessary for reactor design and other calculations over a wide range of nuclear energy applications. The rôle which the Data Bank plays in international cooperation efforts and the procedures to follow to obtain data and programs from the Data Bank are described.

Introduction

An important aspect of the work of the Nuclear Energy Agency of the Organisation for Economic Co-operation and Development (OECD-NEA) concerns the exchange of scientific data and computer programs needed for nuclear energy applications, and this work is carried out by the NEA Data Bank.

The Data Bank, which was set up in January 1978, took over the tasks of two previously separate centres, the Computer Program Library in Ispra, Italy (CPL) and the Neutron Data Compilation Centre in Saclay, France (CCDN). It is financed by sixteen of the NEA Member countries.

The NEA Data Bank offers to scientists and technologists the nuclear physics information and computer programs necessary for reactor design and other calculations over a wide range of nuclear energy applications. One peculiarity of the Data Bank is that its services cover all stages from storing experimental data values to their use in reactor physics applications, and span the continuing feedback processes necessary for the gradual refinement of these data. In parallel the computer programs undergo a similar process of refinement through testing, feedback of user experience and benchmark calculations. Integration over this broad scope is achieved through continuing close contacts with the corresponding NEA scientific policy committees: the NEA Committee on Reactor Physics (NEACRP), the NEA Nuclear Data Committee (NEANDC), and also the Committee on the Safety of Nuclear Installations (CSNI).

The involvement of the Data Bank in the chain of acquisition and then refinement and end use of nuclear data and programs is exceptionally close.

The diagrams in Fig. 1 visualizes the technical outline of the most important data and computer program activities in which the Data Bank takes part, and contributes to the chain of refinement and improvement of these key factors for any nuclear energy programme. This figure shows only the most important parts of a similar but excessively complicated diagram representing the organisational links which are the coordinating mechanism for this thoroughly international field of work. The Data Bank itself acts as a node at which nuclear data and computer program material from many sources are concentrated, at varying stages of refinement, and from which they are distributed to users in OECD countries, in a standard form and at a modest but useful level of verification.

The interface between nuclear data and the computer program using them covers an area in which uncertainties in calculational results may be due either to measurement uncertainties in the input data or to errors in the coding and use of the programs themselves. To help clear some uncertainties of this kind, the Data Bank has joined actively in several benchmark tests over the last few years, and the accompanying paper[1] on the Ribon intercomparison exercise for computer prediction of average resonance parameters describes one such example.

Computer Program Services

The computer program services of the Data Bank extend from collecting programs in the field of nuclear energy from their authors on receipt of user requests, through compiling and testing these programs in an appropriate computer environment and ensuring that the computer program package is complete and adequately documented, to distribution of program packagesto users in OECD Member countries in Western Europe and Japan. The categories most strongly represented in the Data Bank's program collection are:

- Cross section and resonance integral calculations
- Spectrum calculation, generation of multigroup cross section sets, lattice and cell problems
- Static design studies
- Depletion, fuel management, cost analysis, reactor economics
- Space time and space independent kinetics, coupled neutronics/hydrodynamics/thermodynamics, excursion simulation
- Radiological safety, hazard and accident analysis
- Deformation and stress distribution computation, core configuration studies and composite structure analysis
- Gamma heating and shield design
- General mathematical and computer system routines.

Other important but newer categories such as fusion reactor technology are as yet represented by only a few codes. When a program is requested by a "user" establishment, the Data Bank contacts the authors (or NESC for programs developed in the U.S.), and will test the package when it is received and before distribution.

The program will be compiled, and executed with one or more test cases supplied by the authors. Where very wide interest is shown in a particular program, more detailed testing and possibly conversion for use on a different computer (normally CDC to IBM or vice versa) may be undertaken. The package sent out to requesters normally consists of a magnetic tape containing the source code for the program, any auxiliary routines or library data necessary to run it, and the input and output data for the test case. The program report, plus other documentation which the Data Bank feels may be of value to the users, completes the package. Feedback from users of these packages is collected and published as part of the SECU program (Service on Experience in Code Utilisation).

A similar service is provided to IAEA countries, not members of OECD, through an IAEA liaison officer stationed permanently at the Data Bank. Program exchange with the United States and Canada passes through the National Energy Software Centre at ANL, and the Radiation Shielding Information Centre at ORNL. Thus those U.S. and Canadian programs which are freely available for release to users in OECD countries, or to all countries, can be obtained through the Data Bank, which in turn supplies copies of nuclear energy programs received from the rest of the world to the U.S. centres.

Neutron and Other Nuclear Data

Nuclear Data Services are provided in co-operation with a number of different data centres, but most notably within a four-centre network for world exchange of neutron physics data. The other three centres are the National Nuclear Data Centre (NNDC) at Brookhaven National Laboratory (BNL), U.S.A., the Nuclear Data Section (NDS) of the International Atomic Energy Agency (IAEA) in Vienna, Austria, and the USSR Nuclear Data Centre at Obninsk. Experimental and evaluated neutron data, and bibliographic information about work in this field, are compiled and exchanged with the other centres, so that requesters from OECD countries (other than the US and Canada which are served from NNDC) can obtain neutron data compiled worldwide by a single request to the Data Bank. Particular groups of files

K. H. Böckhoff (ed.), Nuclear Data for Science and Technology, 993–996.

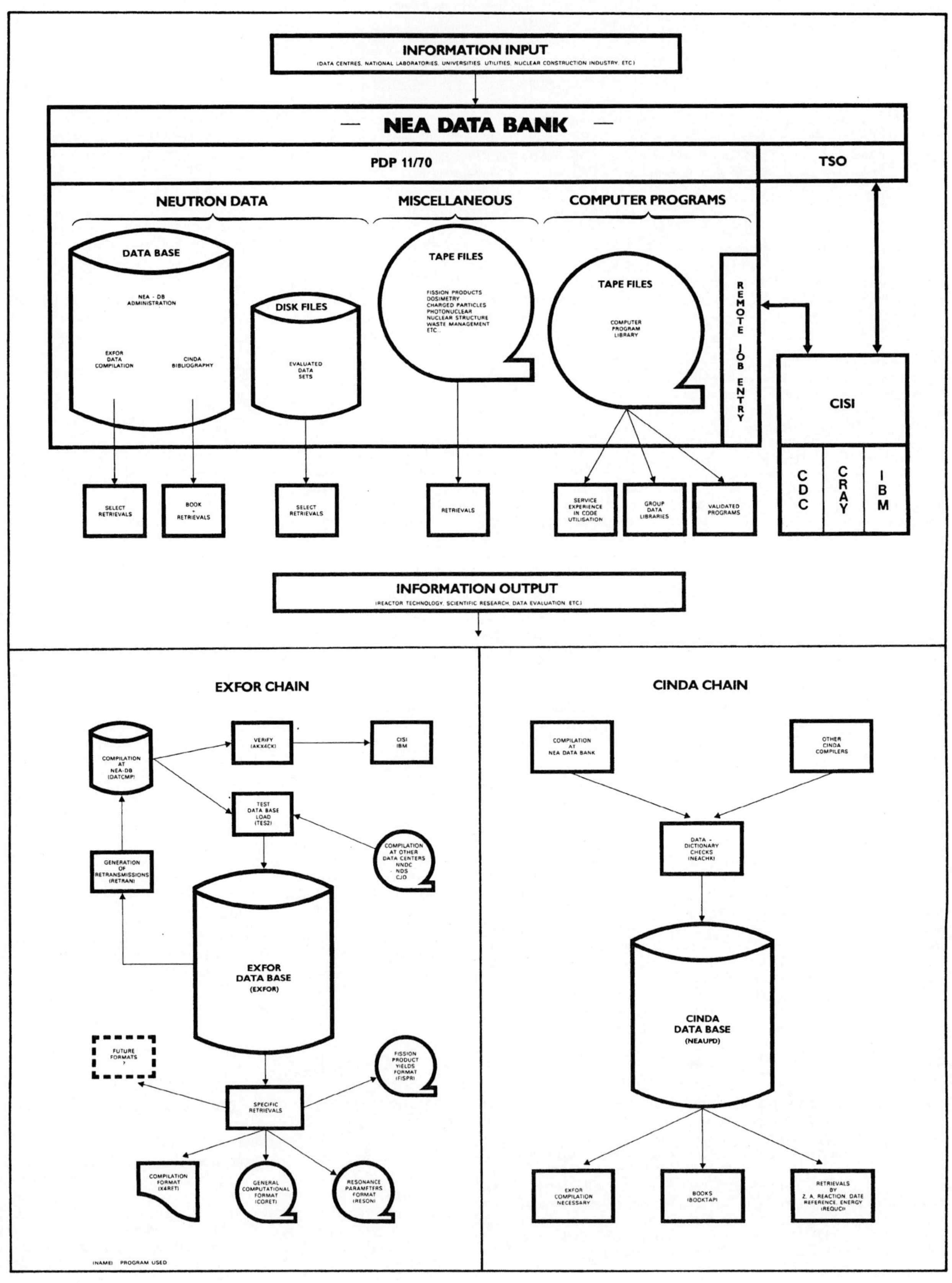

Fig. 1 - top: global organisation of the data in the data bank.
- bottom left: flow chart of the EXFOR-chain.
- bottom right: flow chart of the CINDA-chain.

maintained here and available for searching to answer user's requests are:

1. CINDA Computer Index to Neutron Data (Bibliography) (some 190,000 entries)
2. Numerical data of neutron cross section measurements in the exchange format (EXFOR): some 3 million data points in all, compiled by the four centres, plus several special files
3. Evaluated neutron data:
   - General cross section evaluations: ENDF/B, ENDL, JENDL, KEDAK, UKNDL and others
   - Dosimetry reactions
   - Fission product data
4. The JEF-1 "starter file". This file will be the vehicle for data exchange between Data Bank Member countries in the common evaluation effort involving several European specialist groups.
5. Photonuclear and charged particle nuclear data
6. Nuclear structure and decay data, especially the Evaluated Nuclear Structure Data File (ENSDF) used to produce "Nuclear Data Sheets" and its associated Nuclear Structure References (NSR) file.

The most widely used files can be searched very selectively, with output defined by a number of parameters, while others at present are given to customers only as tape copies of the whole file. One physicst works full-time on preparing answers to user requests for nuclear data: the number of data files and their volume and index complexity makes this personal service essential if users are to get the data they really need in an appropriate form.

An important part of the physics work at the Data Bank is devoted to compiling numerical data from neutron physics measurements, and to maintaining the detailed CINDA index to the neutron physics literature and to these experimental data files. The remaining files are contributed by other data centres under exchange agreements, or by individual physicsts, for distribution to users.

Data Bank cooperation with national laboratories

Various problems have been identified in connection with the data and programs handled by the Data Bank, either by physicists and programmers working there, or by the scientific committees NEACRP and NEANDC. It was never intended that the Data Bank should itself take on scientific projects, but where the scope and interest of the problem warrants it, scientists at the Data Bank can play a useful rôle in coordinating the work carried out in parallel by several laboratories in Member countries. A further encouragement for this rôle lies in the secretariat duties performed by the Data Bank for the two NEA scientific committees, the Reactor Physics Committee (NEACRP) and the Nuclear Data Committee (NEANDC). Three current activities of this kind are:

- The Joint Evaluated File Project. Work is in progress in several European Member countries on reviewing and selecting evaluations for inclusion in the first "starter" file, JEF-1, and at the Data Bank on translating data from the various evaluations selected into ENDF/B V format, merging data sets into the composite JEF files for each isotope and carrying out simple verifications, such as checks on resonance and spectrum integrals, before assembling the JEF-1 file for issue in late 1982. A second stage project, now under consideration, would lead over three years to full integral testing of JEF-1, with generation in parallel of a revised file, JEF-2, to include a number of new evluations.

- Benchmark exercises to compare programs and data. Two such exercises are at present under way at NEADB: a comparison of nuclear model codes, and the calculation of average resonance parameters. The Data Bank will join in a third exercise organised by IAEA, on the use of different data processing codes to produce group cross section sets. Here results from different codes are being compared using essentially the same input.

- The complementary approach, comparison of calculations and measurements taken in simple and very carefully controlled experiments, is used to assess the quality of data, and in particular for testing evaluation data libraries. JEF-1 will be tested in this way in the proposed second phase of the JEF project, by national laboratories with some participation from NEA-DB. The long-running NEACRP cooperative programme on reactor shielding benchmarks is another example of a benchmark exercise carried out internationally, and where the data rather than the programs are called in question.

- Extended testing of large computer programs. In cooperation between CEN Saclay and RSIC Oak Ridge, the French Monte Carlo program TRIPOLI is being extended to fulfill some shielding applications requirements. The Data Bank's contribution will be more than usually vigorous testing of the program and examination of its documentation.

Handling of user requests for programs and data

The difference in the nature and use of the computer program packages and nuclear data files held at the Data Bank has caused the evolution of rather different procedures for customer service. The Data Bank is financed by its Member countries, and services are provided free of charge to users in those countries.

Neutron physics data, charged particle data and nuclear structure and decay data

Customers in OECD Member countries outside North America should write, telephone, telex or even visit the Data Bank, specifying the data they require and the form in which it should be sent (magnetic tape or listings).

It is possible that a new requester who is not familiar with the volume of data available at the Data Bank and the various data files to be consulted, may need help on his first attempt in defining a request for data, and not all requests can be carried out without further contact between the requester and the Data Bank. Occasionally a request will in its original form result in a huge volume of data (possibly 100,000 data points): if the consumer wishes to consult these data on a listing it will be necessary to limit the volume of data, perhaps by defining his needs more precisely or by selecting data from a single file.

If the Data Bank knows how and for what purpose the data asked for will be used, it may be able to make suggestions about other data files to be accessed, or about the most suitable data for the purpose in view, and it is helpful if users will give as much information as possible when making their request. In particular, where data is to be supplied on magnetic tape, they should specify the characteristics of the tape to be sent: No label/label, 800/1600 bpi, ASCII/EBCDIC character sets, 9/7 track, parity, etc. Record formats are of course determined by the files from which the data are extracted. In case of doubt, the Data Bank will normally contact the customer to help define the request more precisely.

Multigroup cross section data are used as input to computer codes, and less often as a source of data for direct use where no appropriate point cross section data are available. The Data Bank can supply a number of the better-known multigroup libraries for either purpose.

Over the past two years, the Data Bank has acquired hardware and software for graphical presentation of neutron data, and as costs can be brought down hopes to make more use of this form in answering user requests and preparing data for review.

Graphical output has been developed in different forms.

Computer programs, associated multigroup cross section sets

The computer codes supplied by the Data Bank are often large and sometimes very large (10,000 or more Fortran statements). In order to ensure that requests from a particular institution are co-ordinated, and in order to check at the source that program requests are justified in terms of the institutes' programme of work and that the same program is not installed twice, all requests for programs pass through a Liaison Officer who is nominated by the institution to centralise contacts with the Data Bank's computer program services. Users should have no difficulty in finding their Liaison Officer, since he is also responsible for distributing information within his institution about the Data Bank's program services and the codes available. In case of doubt, or where no Liaison Officer for the laboratory or institution has been appointed, please write to the Data Bank for advice.

Where a program has been tested and is included in the Data Bank's master files, a copy of the package (source code and documentation) can be supplied very quickly. Where the program must first be obtained from the author and then tested before it can be sent out, waiting times may become long. The Data Bank may in such cases be able to suggest a code comparable to the one asked for, and which can be made available at once. Documentation on the contents of the files is published in three forms, and copies distributed free of charge through the Liaison Officers to users in participating institutions:

- A book of Program Abstracts, which is updated at least once a year, gives details of programs available from the library. The format used in this publication is the same as that used by the National Energy Software Centre at Argonne, USA.
- A KWIC Index gives brief summaries of existing computer programs and is prepared in "KWIC" (Key Words In Context) format.
- Every two months "News from NEA Data Bank" supplies update information on the current state of the programs.

In addition, a Newsletter is published at intervals containing the proceedings of specialist meetings or seminars, SECU studies on codes in a particular field, and papers contributed by Member establishments.

Users' feedback on their experience with particular codes is considered very valuable, both to the authors of the codes and to other users. Users own programs may be valuable to other requesters, and they are invited to be generous in contributing their own codes to the Data Bank.

Member Countries of the NEA Data Bank

Austria	The Netherlands
Belgium	Norway
Denmark	Portugal
Finland	Spain
France	Sweden
Fed. Rep. of Germany	Switzerland
Italy	Turkey
Japan	United Kingdom

References

1. P. Ribon and A. Thompson, NEANDC-160-U (Newsletter of the NEA Data Bank, No. 27), 1982, and Proceedings of this Conference.

ON THE DESIGN OF RELATIONAL DATA BANK FOR THE PROBLEMS OF NUCLEAR TECHNOLOGY

A.F. Grashin, V.M. Kolobashkin, T.M. Televinova
A.A. Tsyganov and B.A. Shchukin

Moscow Engineering Physics Institute
Kashirskoe ch. 31, Moscow 115409
Union of Soviet Socialist Republics

Relational data model is suggested for nuclear data base. The conceptual scheme of the data base on nuclear structure, decay and fission is proposed. The process of date base creation and querying is described.

[nuclear data base, relational model, conceptual scheme, nuclear structure, decay, fission]

The general problem of automatized data bank creation for scientific constants and specifically of the data bank on decay, nucleus structure and yields of fission products is closely connected with the question about data sources. The fact is that the corresponding constants are dissipated in the world scientific literature, therefore it is necessary for a specialist to gather, analyze and evaluate the relevant data, this tedious work being too hard for automatization. So the creation of the nuclear data banks is reasonable on the base of international exchange files ENSDF, ENDF/B, EXFOR etc.

The construction of the automized data bank as usual is defined by the directions of its application. As to the nonneutron nuclear data bank it is necessary to outline the two directions of data use. The first is connected with the problem of data refinement. In this case detailed information on literature sources, authors, experiment methods and etc. are required. All this information allows a specialist "to estimate" the quality of digital data, obtained by the different experimentators. The second direction of data use is connected with the production of both scientific and engineering computations. In this case the data should be conveniently organised and restrected by some integrity constraints [1,2,3].

Really the scientific and engineering problems may demand for their solving complex calculations, based on data from different fields of knowledge. Consequently, the data bank organization must provide the expansion of used data nomenclature. So the system of data organization for the solution of problems is advisable to construct on the basis of relational data model [4] as the most promising in theory, rather than with orientation on the concrete data exchange formats (ENSDF, etc.). In the relational data model the data base is a set of time-varying relations. Query language within the relational model is based on a special algebra (relational algebra) or calculus (relational culculus). Interest to relational model for the scientific data banks is increased [5,6].

In spite of the fact that the data base scheme determined by some relations may vary with time, it is possible to use the main two groups of relations as a conceptual model of nuclear data base:

1. The static characteristics of nuclear states (A,Z,E), where (A,Z,E) is the identifier of nucleus state.
2. Dynamic relationships that characterize the connections between different nuclear states (A,Z,E).

According to this classification the mass of ground states M or mass excess ΔM, spin-parity J^p, magnetic moment μ, quadropole electric moment Q, etc. must be included into the relation of the first type, i.e. they have the form:

$$(A,Z,E),\ J^p,\ Q, \mu,\ \ldots$$

$$(A,Z,M),\ J^p,\ Q, \mu,\ \ldots \qquad \text{if } E=0.$$

The relations of the second type must contain the transition rate (decay constant) λ_{ik} between i-th nuclear state $(A,Z,E)_i$ and k-th nuclear state $(A,Z,E)_k$ through the channel ε, where

$$\varepsilon = (\beta^{\pm}, \gamma, EC, em, n)$$

i.e. the relations of the second type have the form:

$$(A,Z,E)_i \ \lambda^{\varepsilon}_{ik} \ (A,Z,E)_k.$$

In this case the transition rate $\lambda^{\varepsilon}_{ik}$ is connected with life time τ_i of nuclear state "i" in the following way:

$$\lambda_i = \sum_{k,\varepsilon} \lambda^{\varepsilon}_{ik} = \sum_{\varepsilon} \lambda^{\varepsilon}_{i} = \frac{1}{\tau_i}$$

From the relations of these two types any secondary relation may be organized, for example, the relations reflected in exchange formats of nuclear data ENSDF, ENDF/B, EXFOR. So, the relation from ENDF/B containing the branching coefficient $(BR)^{\varepsilon}_{pm}$ at ε-transition of nucleus $(A,Z,E)_p$ into the nucleus $(A,Z,E)_m$:

$$(A,Z,E)_p \ (BR)^{\varepsilon}_{pm} \ (A,Z,E)_m$$

may be obtained from these base relations of two types. For this purpose the following relationships must be taken into account:

$BR^{\varepsilon}_i = \dfrac{\lambda^{\varepsilon}_i}{\lambda_i}$ is ε -channel branching rate for transitions from states "i";

$W^{\varepsilon}_{if} = \dfrac{\lambda^{\varepsilon}_{if}}{\lambda^{\varepsilon}_i}$ is a relative probability of transition $i \xrightarrow[\varepsilon]{} f$ $(\sum_f W_{if} = 1)$;

K_{pf} is a population coefficient of state f in the cascade transitions

$$p \xrightarrow[\varepsilon]{} f;\ p \xrightarrow[\varepsilon]{} k \xrightarrow[em]{} f;\ p \xrightarrow[\varepsilon]{} k \xrightarrow[em]{} s \xrightarrow[em]{} f$$

K. H. Böckhoff (ed.), Nuclear Data for Science and Technology, 997–998.

Here the long-lived states (ground and isometric) are indicated by the indexes P,M, L, ... (any capital letter of Roman alphabet) and short-lived states are indicated by the indexes k,s; the indexes i,f are used for any states. The population coefficients are obtained from following recurrent relationships:

$$K^{\varepsilon}_{PM} = W^{\varepsilon}_{PM} + \sum_{s>M} K^{\varepsilon}_{Ps} \cdot W^{em}_{sM} \cdot BR^{em}_{s} ,$$

where

$$K^{\varepsilon}_{Ps} = W^{\varepsilon}_{Ps} + \sum_{k>s} W^{em}_{ks} \cdot BR^{em}_{k} \cdot W^{\varepsilon}_{Pk} +$$

$$\sum_{l>k} \sum_{k>s} W^{\varepsilon}_{Pl} \cdot W^{em}_{lk} \cdot BR^{em}_{l} \cdot W^{em}_{ks} \cdot BR^{em}_{k} + \ldots ;$$

the branching coefficient

$$BR^{\varepsilon}_{PM} = K^{\varepsilon}_{PM} \cdot BR^{\varepsilon}_{P} ,$$

$\sum_{i>f}$ indicates the summation over states "i" with energies E_i greater than that of state "f"; i.e. $E_i > E_f$.

It should be noted that the solution of some physical problems connected with the participation of exited nucleus states produced as a result of the different reactions is possible only with the use of the second type relations.

The relations reflecting the fission processes, reactions, i.e. the multidimensional connections have a more complex form. For example, the relation describing fission has the following form:

$$(A,Z,E)^{fuel}_{} \quad n_f \quad Y \; (A,Z,E)^{prod}$$

where $(A,Z,E)^{fuel}$ is a fuel identifier;
n_f is a fission type identifier;
$(A,Z,E)^{prod}$ is a fission product identifier;

Y is a probability of similar fission process.

Among the relations of the second type it is convenient to put forward the secondary relations for the long-lived nucleus states ($\tau > \tau_0$). The value of the nucleus state life-time τ_0 depends on the problem type for which the relations are given, it may be taken τ_0 =0.1s.

So, our scheme of the data base consists of the above mentioned fission relation and of the relations of three types for data on decay and nucleus structure;

1. $(A,Z,E), \Delta M, T_{1/2}, J^{P}, BR^{\beta-}, BR^{\beta+}, BR^{EC}, BR^{em}, BR^{DN}$

where $BR^{DN}_{P} = \sum_{s} K^{\varepsilon}_{Ps} BR^{n}_{s}$ is the emission coefficient of delayed neutrons in the cascade transitions $P \xrightarrow{\varepsilon} s \xrightarrow{n} \ldots$.

2. $(A,Z,E)_i, W^{\varepsilon}_{if}, E_{if}, (A,Z,E)_f$.

3. $(A,Z,E)_P, BR^{\varepsilon}_{PM}, (A,Z,E)_M$.

That is the main idea of the data base for the Automatized System of Information on Nuclear Data (ASIND) [1,2].

The relation formation process of the data bank for the structure and decay calculations was divided into three stages [7].

At the first stage ENSDF file transformation is performed into the form in which it is enough to have the input/output statements for the separation of any data from selected record. At the second stage the relations containing the data directly recorded in the ENSDF file are formed. Every relation characterizes the certain object: nucleus levels, γ -transition, etc. At the third stage on the base of these relations the new two types of relations are formed. These relations contain the static characteristics of nucleus states and the dynamic relations between different nuclear states. As the ENDSF file is slowly replenished (near 10% in a year) the expenditures for transformations will be significant only for the initial version of the file.

The complex of programs that is the basis for functioning of the nuclear data bank provides facilities for the maintenance of relations using the international exchange files, the interpretation of algebraic expressions and keeping data base integrity. The experience that has been obtained nowadays in its exploitation allows to give a positive evaluation of its performance.

References

I. К вопросу о построении банков ядерных данных. В.М. Колобашкин, Б.А. Щукин, Т.М. Телевинова, С.Б. Поскакухин, А.А. Цыганов - В кн. Нейтронная физика. Материалы 4-й Всесоюзной конференции по нейтронной физике. ч.4, М.: ЦНИИАТОМИНФОРМ, I977, с.I22-I26.

2. Концептуальный подход к построению банка ядерных данных. А.Ф. Грашин, С.П. Зинькевич, В.М. Колобашкин, Т.М.Телевинова, А.А. Цыганов, Б.А. Щукин - "Экспериментальные методы ядерной физики". Под ред. В.М. Колобашкина, М.: Атомиздат, I980, вып.6, с.II9-I22.

3. Автоматизация информационного обеспечения ядерно-физического эксперимента (Проблема целостности базы данных). Е.С. Богомолова, Т.М. Телевинова, Л.Г. Тюрина, Б.А. Щукин, "Автоматизация физического эксперимента". Под ред. В.М. Колобашкина. М.: Энергоиздат, I98I, с.6I-70.

4. Codd E.F., A relational model of data for large shared data banks. - Comm. ACM, 1970, v.13, n.6, p.377-387.

5. Jones S.E., Ries D.R., A relational data bank management system for scientific data, in proceedings of the 6th International biennial CODATA conference, Oxford, Pergamon Press, 1979, p.353-357.

6. Hartwig R., Bergin M., Pistor P., Shauer U., Schmutz H., Interactive storage, retrieval, transformation and evaluation of scientific data with an experimental system, in Proceedings of the 5th International biennial CODATA conference, Oxford, Pergamon Press, 1977, p.487-492.

7. Формирование базы данных АСИЯД. Е.С. Богомолова, А.А. Цыганов, Б.А. Щукин. "Экспериментальные методы ядерной физики". Под ред. В.М. Колобашкина. М.: Атомиздат, I980, вып.7, с.43-5I.

A PROPOSED INTERNATIONAL FILTERED BEAM PROJECT

A.J. Mill and J.R. Harvey

Central Electricity Generating Board,
Berkeley Nuclear Laboratories,
Berkeley,
Gloucestershire,
GL13 9PB
UK

Intense, roughly-mono energetic sources of neutrons would find application in the fields of nuclear physics, instrument development and calibration; areas which have direct relevance to nuclear measurements. Other applications exist in radiography, radiotherapy, radiation protection and radiobiology. Appropriate sources can be produced with filters and filter-scatterer combinations in the neutron beam from a high flux reactor. At least 16 energies in the range 60 eV - 2 MeV are possible in principle and many have been realised in practice. An extensive survey of European scientific institutions has shown that a central facility giving a range of such beams would be heavily utilised. This is an ideal area for an international agency to play a coordinating role and preliminary discussions, to which interested delegates are invited to contribute, will take place in 1983.

Introduction

The neutrons produced in high flux reactors, and in accelerator-induced fission and spallation reactions, represent the most intense sources of neutrons available for research. However, the neutrons from these sources are not mono-energetic and cover the broad energy range from 10^{-3} eV up to 10^{7} eV - or even higher in the case of spallation sources. Many measurements with these sources can be made at high resolution with the time-of-flight technique using a mechanical chopper or a pulsed source. However, there are several situations where this method cannot be applied - for example when the response time of the irradiated object is comparable with practical flight times or when high intensities are required. For many measurements it is also desirable to eliminate the large accompanying gamma ray flux which may make the evaluation of the effects of neutrons or detection of neutrons and capture gamma rays difficult. Examples where such difficulties may arise are: the evaluation of the biological effect of neutrons, the evaluation of the response of radiation dosemeters and measurement of inelastic scattering cross-sections.

One method of obtaining monoenergetic neutrons from such sources is that of filtration. One type of filter particularly useful is the neutron window or interference filter. This type of filter has a large dip (the window) in its cross-section and if an appropriate length of this material is inserted in a beam tube of a reactor only neutrons of energy corresponding to the window energy will pass through. The filter is also very effective in stopping gamma radiation. The technique was first developed to produce mono-energetic neutrons at 2 keV by using scandium as a filter. Later, filters of iron and silicon were developed to give mono-energetic neutrons at 24 keV and 144 keV respectively. More recently filters of uranium, giving 186 eV neutrons and liquid oxygen, giving 2.35 MeV neutrons have been developed. In principle, mono-energetic neutron beams of a large number of different energies are possible by using other elements and enriched stable single nuclides as filters. However, perhaps the most important aspect of the neutron window filter technique is that it appears to be the only method of producing high intensity mono-energetic neutron beams in the energy range from several electronvolts to several tens of kiloelectronvolts.

As can be seen from Section 3 of this paper the possible areas in which filtered beams can be applied is very wide and many of these have direct or indirect relevance to nuclear measurements.

Neutron Windows and Neutron Window Filters

The Breit and Wigner (1936) formula gives a good description of the cross-section in the region of an isolated s-wave resonance. For such a resonance the cross-section, $\sigma(E)$, is given by:

$$\sigma(E) = 4\pi(R')^2 + \frac{\pi \lambda\!\!\!^{-2} g \Gamma_n \Gamma}{(E-E_o)^2 + (\Gamma/2)^2} + \frac{4\pi \lambda\!\!\!^{-} R' g \Gamma_n (E-E_o)}{(E-E_o)^2 + (\Gamma/2)^2}$$

where Γ = total width at resonance

E_o = resonance energy

$\lambda\!\!\!^{-}$ = reduced neutron wavelength

R' = nuclear radius for potential scattering

g = a statistical weighting factor

$= \left|\frac{2I+1\pm 1}{2(2I+1)}\right|$

I = spin of target nucleus

This simple formula is valid for $R' \ll \lambda\!\!\!^{-}$ i.e. for $E \lesssim$ 0.5 MeV. The first term of the equation describes pure resonance scattering without the formation of a compound nucleus and the second term describes pure resonance interaction. The third term describes the interference between resonance and potential scattering. For energies less than the resonance energy the interference term is negative and there can occur an energy band where potential and resonance scattering cancel each other out to produce small cross-sections, although this can only occur in s-wave interactions. These regions of low cross-section are known as windows. For neutron energies outside the range $R' \ll \lambda\!\!\!^{-}$ the interference dip will occur at higher energies with respect to the resonance energy and in fact may occur above the resonance energy.

For nuclides of odd mass number the total angular momentum of the compound nucleus can take on two possible values in s-wave scattering, however, the interference between resonance and potential scattering will occur only in the resonant spin state. Hence there will still be a contribution to the total cross-section in the window from the non-resonant spin state. Typically, for an isolated resonance the cross-section at the minimum will be about a half of the potential scattering cross-section (as $g \cong \frac{1}{2}$ for both spin states). However, for nuclides with even atomic numbers and even mass numbers there is only one possible spin state (and $g = 1$) and for these nuclides the cross-section can fall to very low values in the window. (In fact if there were no other decay modes the cross-section would fall to zero apart from contributions from other levels and from p-wave

K. H. Böckhoff (ed.), Nuclear Data for Science and Technology, 999–1004.

scattering.) The cross-section can also reach these low values in odd mass nuclides when two resonances, of opposite spin state, occur close enough together such that their interference dips overlap. Such a deep minimum occurs in scandium-45 at 2 keV.

The existence of cross-section windows suggests the means for making suitable neutron filters; and nuclides with even mass numbers and even atomic numbers are the most promising in this respect. However, all nuclides which have even numbers of protons and neutrons occur in mixtures of isotopes, the result being that, in the natural element, a window in one isotope is usually obscured by the potential or resonance scattering background of the other isotopes present. Nevertheless, useful filters have been constructed using elemental iron (91.7%, iron-56), silicon (92.2%, silicon-28), oxygen (99.8%, oxygen-16) and uranium (99.3%, uranium-238) as well as scandium (100%, scandium-45) - which is an odd mass number nuclide. Elemental sulphur (95.0%, sulphur-32) is also a promising filter material. In order to extend the range of filter materials and make use of the numerous windows present in even-even nuclides it is necessary to use enriched nuclides as filters. At present this approach has been adopted only for iron-56 (Liou, Chrien, Block and Singh, 1979) although a proposal to use other enriched nuclides has been made (Mill and Harvey, 1978); and there are proposals to use zinc-64 in conjunction with scandium to give 2 keV neutrons (Brugger, 1980).

One major drawback with neutron window filters is that they invariably have more than one isolated s-wave resonance, and hence there may be several windows present in addition to the main one. Transmission of neutrons through these subsidiary windows may be significant and it is generally unwise to rely simply on a single filter component to obtain a mono-energetic neutron beam. There are two techniques which can be adopted, either singly or in combination, which are usually effective in eliminating the effect of neutrons transmitted through secondary windows. The first method is to use another (secondary) filter which has a window (or a region of relatively low cross-section) at the same energy as the main window in the primary filter while no other regions of low cross-section coincide. For example, iron and aluminium both have windows at 25 keV. Generally this method is of limited application when using elemental materials. Alternatively materials for use as secondary filters can be chosen on the basis that they have a resonance at an energy corresponding to one of the secondary windows in the main filter. In this case only a small thickness of the secondary filter is required. An example here is the use of titanium in scandium filtered beams to suppress the transmission through the secondary windows at 7 and 24 keV in scandium.

The second method is the use of resonance scattering. Here a material having a large scattering resonance peak at the energy of interest is placed in the filtered beam and will preferentially scatter these energy neutrons out of the primary beam. Other resonance peaks in the scatterer should not coincide with any of the secondary windows in the filter. In order to minimise scattering of neutrons whose energies do not coincide with the resonance energy, the scatterer should be thin. Manganese can be used as a scatterer for a scandium filtered beam. It is also possible to use the scatterer prior to filtration to preferentially scatter neutrons corresponding to the resonance peaks into the filter. This is particularly effective when the filter is placed in the through tube of a reactor.

One major advantage with neutron window filters is that the large length of material used generally reduces the gamma-ray dose-rates to very low levels.

There are several factors to be considered when designing a filter facility. For instance, window filters act by scattering neutrons out of the beam, and not by absorption. Therefore it is important not to locate the sample too close to the end of the filter where scattered and thermalised neutrons may be present. Other factors to be considered are the effects of collimation, neutron streaming, the design of the sheilding, and, of course, the actual thickness of the filter. For example, the bandwidth of the filtered neutron beam may be made narrower by increasing the filter length. However, this occurs at the expense of decreasing the intensity of the transmitted beam; although background (both gamma-rays and fast neutrons) will also be decreased, and by a much larger factor.

Filtered Neutron Beam Facilities

There are, all told, about a dozen filtered beam facilities associated with high flux research reactors, currently in existence throughout the world. A review of these has been recently published (Mill and Harvey, 1980). The most widely utilised combination is that of iron, aluminium and sulphur, presumably because of the low material costs combined with the relative ease of obtaining a pure 25 keV neutron beam. Several scandium beams have also been developed despite some difficulties with high energy neutron contamination. Titanium has usually been used as a secondary filter. The problem with the fast neutron background arose because the cross-section of scandium was originally thought to be about 0.05 barn at 2 keV while, in fact, it is actually about ten times higher than this. However, by using a resonance scatterer of manganese it is possible to reduce the fast neutron background considerably. Silicon may be used to produce mono-energetic neutron beams at 55 keV and 144 keV and as there are no windows at energies above 144 keV there is not the problem of a fast neutron background. However, as silicon is a polycrystalline material it transmits neutrons below the Bragg cut-off at 2 meV; in addition a significant gamma ray component may be also present. These contaminating radiations may be reduced by the addition of boron-10 and lead respectively.

In addition to the scandium, iron and silicon filters, uranium and liquid oxygen filters have been developed to give filtered beams at 186 eV and 2.35 MeV respectively.

Application of Filtered Beams

Some typical applications of filtered intermediate energy neutron beams have been discussed in detail elsewhere (Mill and Harvey, 1980; Brugger and Simpson, 1973). A summary of the most important aspects of these applications is included here.

Cross-section Measurements

Filtered beams, both in experimental reactors and also in pulsed time-of-flight experimental systems, have found increasing applications in recent years for the measurement of cross-sections. It might seem, at first sight, that filtered beams would have few advantages compared with broad spectrum neutron sources combined with highly accurate time-of-flight systems since filtered beams are available at relatively few energies. However, difficulties do arise in white source measurements due to scattered neutrons which interfere with the neutrons of interest. Thus the white source techniques are ideal for measuring the relative magnitude of cross-sections at different energies but can give difficulties when absolute values of cross-section are required with high precision. The filtered beams are ideal for benchmark measurements of high accuracy to which, in principle, other measurements can be related.

Filtered beams are also useful for precision measurements of the cross-section at the minima in the cross-section of the material of the filter itself, since contaminating radiation is kept to a minimum. Such minimum values are of great importance in shielding calculations since the precision with which the flux at depth in a shield can be estimated depends crucially on the value of the cross-section at the minima. Accurate cross-section measurements of fissile and fertile materials are similarly of importance for reactor design.

In addition filtered beams may be used to measure the effect of Doppler broadening, and for accurate measurements of (n,p) and (n,γ) cross-sections.

(n,γ) Spectrometry

Filtered beam facilities lend themselves to a study of (n,γ) reactions because of the high fluxes and energy spreads in the beams. For low-Z nuclei, which have widely spaced resonances, the capture gamma-ray lines often arise from discrete resonances whereas for higher-Z targets, the capture gamma-ray spectrum is a composite resulting from many resonances. Such spectra give information on the multipolarities of the primary gamma-ray transitions and the possible range of spin values of the final states populated by the primary transitions.

Filtered beams have also been used to give information on the (n,γ) spectra of nuclides which are very difficult to study with thermal neutrons because neighbouring or impurity nuclides have very large thermal neutron cross-sections giving rise to serious contamination problems.

Shielding Studies

Filtered beams are also ideally suited for bench-mark measurements to check shielding calculations. The time-scale of moderation and absorption processes in a thick shield precludes the use of time-of-flight techniques and since 2 keV neutrons are at the limit of energies obtainable by alternative systems, a scandium beam yields valuable information if used with integrating detection systems. The iron and oxygen filtered beams give neutrons of energies at which shield penetration can be a maximum since, of course, these materials are incorporated in shields.

Instrument Calibration and Development

The development of many neutron sensitive instruments has long been bedevilled by the fact that appropriate neutron sources are not available throughout much of the intermediate energy region. The use of time-of-flight techniques is often difficult or impossible because of either the requirement to modify the detector in the instrument or the fact that the response time of the instrument is comparable with, or greater than, the flight times involved. This latter consideration is particularly relevant to all instruments which are based on large moderating assemblies. Examples are the widely used Andersson-Braun type of instrument used for environmental surveys for radiological protection purposes, the "long counter" instruments, often used as primary neutron flux transfer standards and spectrometry systems based on a range of moderating spheres.

A knowledge of the response of health physics instruments in this energy region is important for three reasons. Firstly, many surveys have shown that a significant proportion of neutron dose in many different working environments around reactors and other neutron producing installations is due to neutrons with energies less than 30-100 keV. Secondly, a widely used type of personal dosemeter, the albedo dosemeter, which utilises devices which are sensitive to neutrons reflected from the wearer's body is sensitive principally below 10 keV. Inadequacies in understanding of the characteristics of the survey instruments are therefore reflected in the overall accuracy of the personal dosimetry system. The third reason for the importance of this energy region in the context of health physics instrumentation stems from the fact that the dose equivalent-fluence relationship changes abruptly at 10 keV. The response function of a typical detector based on neutron moderation on the other hand is likely to change smoothly with neutron energy in this region. It follows that such instruments which purport to measure dose-equivalent rate are most likely to be in error at energies around 10 keV.

Proton recoil spectrometry systems can also be calibrated with the filtered beams. The routine calibration of the proportional counter is frequently based on either alpha particles from an internal source or protons from the $^{3}He(n,p)^{3}H$ reaction in added helium-3 gas. This later reaction has a Q value of 765 keV. Overall calibration and system linearities are checked by reference to presently available mono-energetic sources at energies above 10 keV or so and an extension to lower energies is valuable.

Biological and Medical Studies

It is possible to use filtered intermediate energy neutron beams to produce significant fluxes of thermal neutrons inside the human body. The advantage of using intermediate energy neutrons is that they produce much lower thermal neutron fluxes at the surface of the body for a given flux at depth than do thermal beams and also do not give a significant knock-on proton dose in the thermalisation process as do fast neutrons.

Two applications have been proposed to take advantage of this thermal neutron flux profile:- boron neutron capture therapy and "in vivo" measurement of the cadmium content of organs.

Direct irradiation of biological specimens with intermediate energy beams could give useful information about the mechanisms of the interaction of ionising radiation with cellular systems. Neutrons, although themselves non-ionising, produce densely ionising short range particles within irradiated tissue. The dimensions of the tracks of these particles may be comparable to the dimensions of important cellular components and can give valuable insights into the mechanisms of radiation injury.

Neutron Radiography

There are a number of reports of the use of filtered beams for radiography. The most obvious application is the use of an iron filtered beam for radiographing materials encased in a thick iron container because the container will in general not attenuate the neutron flux significantly. The principal difficulty is in finding a detection system which is particularly sensitive to intermediate energy neutrons. For this reason a tomographic technique is attractive since the detector can then be a thermal neutron detector inside a moderator.

Recent Developments and Future Possibilities

As discussed earlier there is considerable scope for expanding the range of energies obtainable using the filtered beam technique. Theoretical work has concentrated on three aspects:- (a) the use of enriched nuclides, (b) the utilisation of several potentially useful windows in uranium and (c) the identification of other elements which may be useful as filters.

Enriched Nuclide Filters

It has been suggested (Mill and Harvey, 1978; Harvey and Mill, 1978) that the range of energies available could be extended considerably by using filters of enriched single nuclides, which surprisingly are not prohibitively expensive. The proposed new window energies were at 60 eV (^{170}Er), 160 eV (^{184}W), 400 eV (^{68}Zn), 5 keV (^{60}Ni), 13 keV (^{58}Ni) and 2 keV (^{64}Zn). These windows were identified by theoretical means (Mill, 1979). However accurate cross-section measurements of thick samples of four of these separated nuclides were kindly undertaken at Oak Ridge National Laboratory (ORNL) by J.A. Harvey and his colleagues using well-developed time-of-flight techniques on the Oak Ridge Electron Linear Accelerator facility (ORELA). Details of the four samples used for these measurements are given in Table I.

Table I

Isotope	Isotopic abundance	Energy of principal window	Sample thickness /cm
Tungsten-184	0.9452	155 eV	6.32
Zinc-64	0.9787	2.15 keV	6.06
Nickel-60	0.9968	∿ 5 keV	7.30
Nickel-58	0.99927	12.8 keV	6.45

The results of the measurements, which are discussed in more detail elsewhere (Mill 1979) are summarized in Figures 1-4. It can be seen that for three of the nuclides the agreement between theory and experiment is very good. The agreement in the case of tungsten-184 was less good, the measured cross-section within the window being somewhat higher than predicted. It is suspected that this high cross-section is due to the presence of hydrogen in the sample, probably in the form of hydroxide since the metal was electroplated out of aqueous solution.

The measurements are valuable in giving confidence in the theoretical predictions of the energies at which windows occur and the values of the cross-section minima. They are also valuable in giving information on competing windows, the structure of which is difficult to predict theoretically. This information can be used to quantify design predictions.

Uranium-filtered Beams

Natural uranium comprises over 99% uranium-238, a nuclide having even numbers of neutrons and protons, and therefore the total neutron cross-section is expected to display neutron interference windows making it suitable for use as a filter material for producing beams of monoenergetic neutrons. In fact close inspection of transmission data reveals the presence of no less than forty windows between 2 eV and 3 keV.

One of the deeper windows in uranium has been used to produce a filtered beam at 186 eV, however in principle it should be possible to utilise several of the windows in uranium to give a series of filtered beam energies by using a main filter of uranium in conjunction with suitable secondary filters to suppress transmission through unwanted windows, together with scattering foils to preferentially scatter filtered neutrons at the required energy. The range of beam energies obtainable with reasonably pure beams lies between 100 eV and 2.5 keV (Mill, 1982) with outputs more than adequate for health physics purposes.

One of the main advantages of adding uranium to an existing filtered beam facility is the relative cheapness of the materials required.

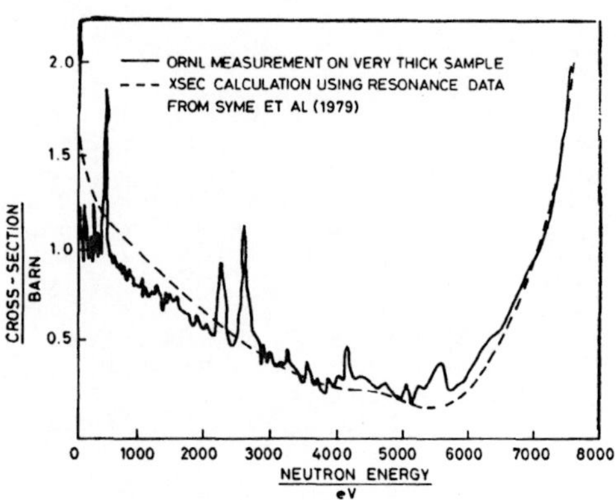

Fig. 1 Measurement and calculation of Ni 60 cross section.

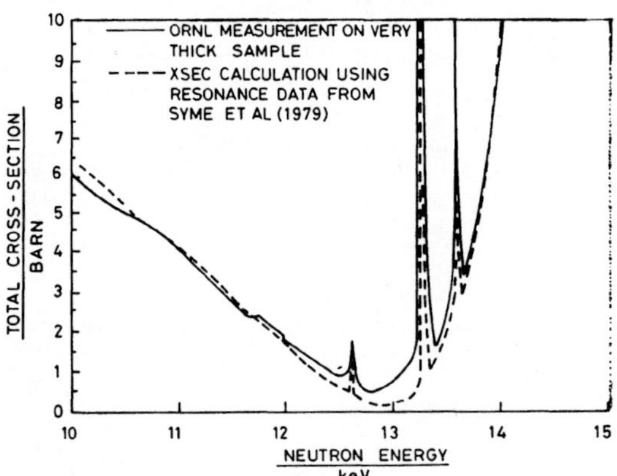

Fig. 2 Measurement and calculation of Ni 58 cross section.

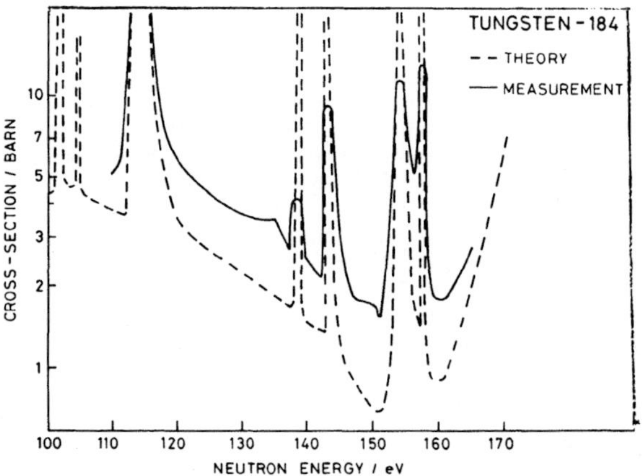

Fig. 3 Measurement of calculation of W 184 cross section.

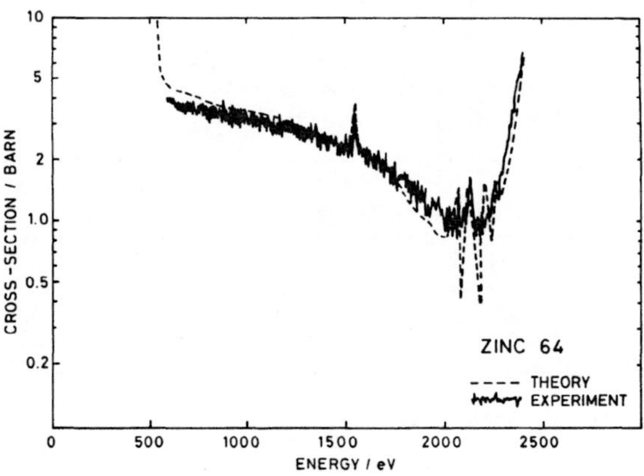

Fig. 4 Measurement and calculation of Zn 64 cross section.

Other Low-cost Filters

Other possible, but as yet unexplored, filter materials are sulphur to give 74 keV neutrons (Mill and Harvey, 1980); copper giving 7.5 keV neutrons (since few other windows exist in copper this could well be used to give an inexpensive and simple filter) and neon to give 420 keV neutrons. Another idea (Brugger, 1980) is the possibility of generating 14 MeV neutrons when a thermal neutron beam impinges on a mixture of helium-3 and deuterium gases and initiates the reaction $^3He(n,p)^3H$ which has an excess of energy of 0.77 MeV, sufficient to induce the reaction: $^3H(d,n)^4He$ to give 14 MeV neutrons.

A Proposed Facility

The considerations discussed above led us to propose the setting-up of a centralised European intermediate energy neutron facility based on a high flux reactor (Harvey and Mill, 1978; Mill and Harvey, 1978; Harvey and Mill, 1981). A notional facility based on presently available reactors and materials was designed. This notional facility assumed the location of filters in a radial beam tube directed at the core of the reactor. When a scattering foil is required this is located within the filtered beam external to the reactor.

The initial proposal was for a facility based on scandium and enriched nuclide filters with an overall cost of approximately £5 x 10^5. Such a facility would provide high purity mono-energetic beams down to 60 eV. More recently, however, more emphasis has been put on the low cost beams, such as uranium. These do not provide either the unique range of energies available with enriched nuclides nor do the beams have the high outputs or high purity obtainable with the enriched nuclides. Nevertheless they do provide a low cost option ($\sim$ £2 x 10^5) which could be adopted in the initial stages of development of a centralised facility.

Seven high flux reactors in Europe would be particularly appropriate from a scientific point of view (Mill and Harvey, 1978): BR-2, Mol, Belgium; Pluto/Dido, Harwell, England; DR-3, Risö, Denmark; Dido-Julich, Julich, Federal Republic of Germany; HFR, Petten, Holland; High Flux Reactor, Grenoble, France and the R-2, Studsvik, Sweden. It is also possible to utilise neutrons produced by high energy charged particle accelerators, many of which are being converted to neutron generation (Dore, 1978; Sourthworth, 1976). Rough calculations suggest that the time-averaged intermediate fluence in the moderator block of such a device would be one or two orders of magnitude less than in the core of a high flux reactor. This disadvantage would be offset by the fact that filters could probably be placed closer to the source than with a high flux reactor since less biological shielding would be needed. Also a pulsed source would allow very good beam spectrometry by time-of-flight methods.

The Need for Filtered Beams

In October 1978 a questionnaire was circulated to 147 laboratories or scientific institutions in Europe in order to gauge the potential utilization of a centralized facility giving a range of neutron energies. This questionnaire was circulated with the report (Mill and Harvey, 1978) giving the full details of the proposed facility, including an estimation of the cost of hiring such a facility. Analysis of replies from 47 institutions indicated that 25 would make use of the facility, some within more than one discipline, while a further 12 gave positive support to the development of such a facility. The projected utilization varied between a few hours and several months per year. An analysis of the requirements broken down in terms of time required and scientific discipline is given in Figure 5.

The total number of user institutions is 39. This exceeds the actual number of institutions since some are represented in more than one discipline. The facility would be required for 531 days per year, about twice the time which would be available on a single beam tube. A detailed analysis of the replies has been published (Harvey, 1980).

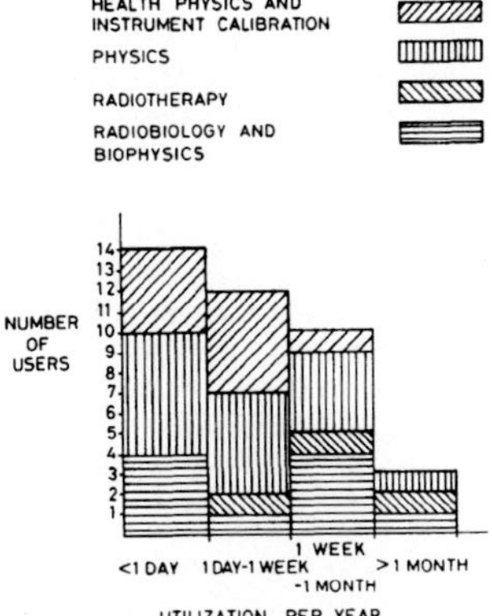

Fig. 5 Number of user institutions by discipline and utilization period.

Conclusions

It is now clear that neutron beams covering a wide range of energies can be produced with a range of relatively inexpensive filters. The range could be extended by using single isotope material. Some possibilities are summarized in Table II. If a range of filters were arranged in, say, a turret assembly so that they could be changed easily, then a wide range of energies could be provided at a single location within a short time-scale.

Table II A summary of some filtered beams

			ENERGY 10eV	100eV	1keV	10keV	100keV	1MeV	10MeV
RELATIVELY INEXPENSIVE FILTERS	BEAMS ALREADY PRACTICALLY REALIZED	URANIUM		×					
		SCANDIUM			×				
		IRON				×			
		SILICON				×			
		SILICON					×		
		OXYGEN						×	
	POSSIBLE BEAMS	URANIUM		×					
		COPPER			×				
		SULPHUR				×			
		NEON					×		
		MOCK FISSION						×	
		D-T							×
EXPENSIVE FILTERS		ERBIUM 170	×						
		TUNGSTEN 184		×					
		ZINC 68		×					
		NICKEL 60			×				
		NICKEL 58				×			

Such a facility would be very useful for certain applications where neutron beams are required within a wide energy range, such as in the development and calibration of neutron sensitive instruments. At present it is expensive and time consuming to have access to a range of neutron energies. Energies from a few keV to 14 MeV can be derived from a number of charged particle reactions with a Van de Graaff accelerator, but to cover the full range involves time consuming procedures such as measuring threshold energies for each target. Energies below a few keV are not available from charged particle reactions.

In summary, the filtered beam facilities which are available at a few reactor establishments have been shown to be useful in a wide range of scientific disciplines. There is considerable scope for extending the energies available and many scientific groups have indicated a need for a range of filtered neutron beams. It is possible that if the potential range of energies and simplicity of operation can be realised, filter beams could replace accelerators in applications where fixed energy beams are required over a wide energy range.

The considerations presented above indicate the need for a centralised facility in Europe, and this would be an ideal area in which an international agency could play a coordinating role. Preliminary discussions about possible initiatives are planned for 1983 under the auspices of CENDOS (collection and evaluation of neutron dosimetry data) and delegates at this conference who are interested in contributing are invited to contact one of the authors of this paper.

Acknowledgement

This paper is published with the permission of the Central Electricity Generating Board.

References

G. Breit and E.P. Wigner, 1936, Capture of Slow Neutrons, Physics Review, 49, 519.

R.M. Brugger, 1979, Private communication.

R.M. Brugger and O.D. Simpson, 1973, Resonance Window Filters of Neutrons for Research and Development. In: Proceedings of a Symposium on Irradiation Facilities for Research Reactors. (Teheran, November 6-10, 1972) STI/PUB/316. International Atomic Energy Agency, Vienna.

J. Dore, Pulsed Neutrons at the Rutherford, New Scientist, 78, 160.

J.R. Harvey, 1980, Analysis of Replies to a Questionnaire on the Potential Utilisation of a Proposed Intermediate-Energy Neutron Facility, Report RD/B/N4726, Central Electricity Generating Board, Berkeley, England.

J.R. Harvey and A.J. Mill, 1978, A Proposed Intermediate Energy Neutron facility and Possible Applications. In: Proceedings of the Third Symposium on Neutron Dosimetry in Biology and Medicine (Munich, 23-27, May, 1977) EUR5848 DE-EN-FR, Commission of the European Communities, Luxembourg.

J.R. Harvey and A.J. Mill, 1981, The Status of a Proposed International Filtered Beam Project. In: Proceedings of the Fourth Symposium on Neutron Dosimetry (Munich-Neuherberg, June 1-5, 1981), EUR 7448EN, 1, 431, Commission of the European Communities, Luxembourg.

H.I. Liou, R.E. Chrien, R.C. Block and U.N. Singh, 1979, The Transmission of Neutrons through Iron-56 at 24.37 keV, Nuclear Science and Engineering, 70, 150.

A.J. Mill, 1979, A Computer Code for Calculating Neutron Cross-Sections from Resonance Parameter data, Report RD/B/N4632, Central Electricity Generating Board, Berkeley, England.

A.J. Mill, 1982, Uranium-Filtered Neutron Beams at Intermediate Energies, Report TPRD/B/0004/N82, Central Electricity Generating Board, Berkeley, England.

A.J. Mill and J.R. Harvey, 1978, Intermediate Energy Neutron Production; a Survey of Existing Techniques, a Proposed Source and its Application, Report RD/B/N4324, (and Report EUR 6107EN), Central Electricity Generating Board, Berkeley, England.

A.J. Mill and J.R. Harvey, 1980, Reactor-and Accelerator-Based Filtered Beams. In: Proceedings of the IAEA Consultants' Meeting on Neutron Source Properties (Debrecen, March 17-21, 1980), INDC(NDS)-114/GT, 135, International Atomic Energy Agency, Vienna (also Report RD/B/N4776, Central Electricity Generating Board, Berkeley).

B. Southworth, 1976, Neutron Source at Rutherford, CERN Courier, 16, 170.

EVALUATION ET MESURE DE DONNEES NUCLEAIRES, PROGRAMME ET PERSPECTIVES

N. Coursol, F. Lagoutine

Laboratoire de Métrologie des Rayonnements Ionisants

CEN-Saclay - B.P. 21, 91190 Gif-sur-Yvette, France

Le Laboratoire de Métrologie des Rayonnements Ionisants, LMRI, s'est engagé depuis 1973 dans l'évaluation des paramètres de schémas de désintégration de radionucléides importants pour diverses applications : médecine nucléaire, énergie nucléaire, protection, etc... Les résultats de ces travaux figurent dans la Table de Radionucléides publiée par le LMRI.

L'effort de recherche du LMRI vise d'autre part à améliorer la précision des mesures de constantes nucléaires par une réflexion sur les méthodes employées et la réalisation de mesures précises de paramètres de certains radionucléides.

Introduction

L'utilisation pratique des radionucléides est en grande partie conditionnée par la connaissance précise des paramètres qui les caractérisent.
Les utilisateurs ont actuellement à leur disposition un grand nombre de données (tables, fichiers informatiques,...) concernant des résultats expérimentaux et des valeurs calculées des paramètres nucléaires des radionucléides. Obligés de faire un choix parmi les valeurs disponibles ils regrettent souvent de ne pas avoir de valeur recommandée à leur disposition. Etablir une table de telles valeurs constitue, depuis 1973, l'une des tâches du LMRI.

Programme et perspectives

Le Laboratoire de Métrologie des Rayonnements Ionisants a mené d'une part, depuis 1973 et de façon systématique l'évaluation approfondie des paramètres de schémas de désintégration des radionucléides importants pour les différentes applications. Des règles d'évaluation ont été élaborées de telle sorte que les incertitudes sur les paramètres évalués soient déterminées de façon comparable.

Les résultats de ces travaux figurent dans la Table des Radionucléides publiée par le LMRI. Les deux premiers volumes de cette table concernent quatre vingt onze radionucléides ; la première partie du troisième volume en comporte une trentaine (dont ^{239}Pu, ^{226}Ra et ses descendants). On en trouvera la liste détaillée en appendice. Une "Sélection médicale" de la Table de Radionucléides concerne les radionucléides couramment utilisés en médecine nucléaire.

Une réflexion a d'autre part été engagée par le LMRI sur les méthodes employées dans la détermination des paramètres nucléaires :

. comparaison des valeurs calculées et des mesures expérimentales de coefficients de conversion interne (N. Coursol 1980) ;
. détermination expérimentale précise de périodes de radionucléides (F. Lagoutine 1972, 1978, 1982) ;

. détermination expérimentale d'intensités d'émission gamma (J. Morel 1975, 1980, 1981) ;
. détermination du rendement de fluorescence K (P. Magnier 1978).

Notre programme à court et moyen terme a été planifié selon les besoins exprimés par les utilisateurs et par les actions menées en collaboration par notre laboratoire avec des organismes tels l'AIEA, l'Euratom.

Ce programme exige une confrontation permanente de la Table de Radionucléides avec les nouvelles données expérimentales, et donc des mises à jour : une mise à jour du premier volume de cette table est en cours d'édition. Ce programme met d'autre part en relief l'insuffisance des mesures expérimentales actuelles pour certains radionucléides.

Références

1. N.F. Coursol, Etude de coefficients de conversion interne pour des énergies de transition voisines du seuil, Rapport CEA-R-5052, (1980)

2. F. Lagoutine, J. Legrand, C. Perrot, J.P. Brethon, J. Morel, Int. J. Appl. Radiat. Isot. 23, (1972), 219

3. F. Lagoutine, J. Legrand, C. Bac, Int. J. Appl. Radiat. Isot. 29, (1978), 269

4. F. Lagoutine, J. Legrand, Int. J. Appl. Radiat. Isot. à paraître.

5. J. Morel, Etude des différents problèmes posés par l'étalonnage précis d'un spectromètre γ, application aux mesures des énergies et intensités absolues des raies γ de ^{152}Eu, Rapport CEA-R-4656(1975)

6. B. Chauvenet, J. Morel, J. Legrand, Intercomparison of the measurement of photon emission rates of X- and γ-rays emitted by barium 133, Report ICRM-S-6, (1980)

7. G.C. Lowenthal, C. Bac, F. Lagoutine, J. Morel, R. Vatin, Lutetium-176m : a re-measurement of its half-life and the energies and intensities of photon emissions following its decay, J. Phys. G : Nucl. Phys. 7, (1981), 1557.

8. P. Magnier, J. Bouchard, M. Blondel, J. Legrand, J.P. Pérolat, R. Vatin, Precise measurement of the X_K emission rate following the electron capture-decay of ^{54}Mn - fluorescence yield ω_K of Cr, Z. Phys. A 284, (1978), 389.

K. H. Böckhoff (ed.), Nuclear Data for Science and Technology, 1005–1006.

Appendice

Table de Radionucléides

F. Lagoutine, N. Coursol, J. Legrand

Le premier volume présente l'évaluation de 41 radionucléides :

^{3}H, ^{14}C, ^{22}Na, ^{24}Na, ^{24m}Na, ^{51}Cr, ^{54}M, ^{57}Co, ^{58}Co, ^{58m}Co
^{60}Co, ^{60m}Co, ^{75}Se, ^{85}Kr, ^{85}Sr, ^{99}Mo, ^{99m}Tc, ^{99}Tc, ^{103}Ru,
^{103m}Rh, ^{110m}Ag-^{110}Ag, ^{109}Cd, ^{111}In, ^{125}Sb, ^{125m}Te, ^{125}I,
^{133}Xe, ^{133m}Xe, ^{131}Cs, ^{134}Cs, ^{134m}Cs, ^{139}Ce, ^{144}Ce-^{144}Pr,
^{144}Pr, ^{169}Er, ^{176}Lu, ^{176m}Lu, ^{186}Re, ^{197}Hg, ^{197m}Hg,
^{203}Hg.

ainsi qu'une introduction comportant une brève description des phénomènes radioactifs et les règles d'évaluation utilisées.
Pour chaque radionucléide les meilleures valeurs et les incertitudes associées sont données pour les principaux paramètres de schéma de désintégration et les intensités des rayonnements émis ainsi qu'une table de décroissance. Des spectres γ, X et quelquefois d'électrons de conversion sont également donnés.

Le deuxième volume présente l'évaluation de 50 radionucléides :

^{32}P, ^{42}K, ^{43}k, ^{45}Ca, ^{47}Ca-^{47}Sc, ^{47}Sc, ^{52}Fe-^{52m}Mn,
^{52m}Mn, ^{52}Mn, ^{55}Fe, ^{56}Co, ^{59}Fe, ^{64}Cu, ^{65}Zn, ^{67}Ga, ^{81}Rb,
^{81m}Kr, ^{82}Br, ^{86}Rb, ^{88}Rb, ^{89}Sr, ^{90}Sr-^{90}Y, ^{90}Y, ^{91}Y,
^{95}Zr-^{95m}Nb, ^{95m}Nb, ^{95}Nb, ^{106}Ru-^{106}Rh, ^{113}Sn-^{113m}In,
^{113m}In, ^{123m}Te, ^{123}I, ^{127m}Te-^{127}Te, ^{127}Te, ^{129}I,
^{131m}Te-^{131}Te, ^{131}Te, ^{131}I, ^{131m}Xe, ^{133}Ba, ^{137}Cs-^{137m}Ba,
^{137m}Ba, ^{140}Ba-^{140}La, ^{140}La, ^{141}Ce, ^{143}Pr, ^{192}Ir, ^{198}Au,
^{201}Tl.

Mots-clés : désintégration, évaluations comparatives, précision, radio-isotopes.

COMPARISON OF EXPERIMENTAL WITH THEORETICAL PHOTON ATTENUATION CROSS SECTIONS BETWEEN 10 eV AND 100 GeV*

Henry Gerstenberg and J. H. Hubbell
Photon and Charged Particle Data Center
Center for Radiation Research
National Bureau of Standards
Washington, D.C. 20234 U.S.A.

A computerized photon attenuation data base has been developed by the NBS Photon and Charged Particle Data Center for the photon energy range 10 eV to 100 GeV and for elements, with Z = 1 to 94. An example of use of this data base in the critical evaluation of a theory-based data-set is presented.

[attenuation coefficient, critical evaluation, cross sections, data base, photons, x-rays]

1. Statement of the Problem:

Photon attenuation coefficients are required in a variety of nuclear science, technology and medical applications. The mass attenuation coefficient μ/ρ is a measure of the probability per unit path-length (in mass per unit area) for an interaction (of any kind) of a photon with an absorber material. Thus the measurable attenuation I/I_0 of a monoenergetic narrow beam of photons by an absorber of thickness t (in mass per unit area) is

$$I/I_0 = \exp\left[-(\mu/\rho)\ t\right] \qquad (1)$$

from which

$$\mu/\rho = -t^{-1} \ln (I/I_0). \qquad (2)$$

The aim of the present work is to develop computer-assisted critical evaluation procedures to aid in converting available photon attenuation information into a "best set" of μ/ρ (and component cross section) data for practical applications.

2. Sources of Data

On the one hand, there is a large body of experimental μ/ρ information [1]. Measured μ/ρ values, primarily for elemental materials, have been accumulating in the literature since 1907 [2]. As a function of photon energy E and atomic number Z this accumulation is rather uneven, with considerable redundancy in regions of readily available source energies and absorber materials, and gaps or large uncertainties in other regions.

On the other hand, theoretical predictions over wide E- and Z-regions are available for the principal component cross sections (See Fig. 1): atomic photoeffect [3], τ, coherent (Rayleigh) scattering [4,5], σ_{COH}, incoherent (Compton) scattering [4], σ_{INCOH}, nuclear-field pair production [6], κ_n, and electron-field pair production (triplet) [6], κ_e. Comparable theory is not available for predicting the photonuclear cross section $\sigma_{PH.N}$, which depends irregularly on both A and Z. Experimental information on $\sigma_{PH.N}$, considered a minor perturbation in the context of this work, is indexed in references [7] and [8]. Some indication of the relative magnitude of $\sigma_{PH.N.}$ (2% to 7% of σ_{TOT} at peak) is given in table 2.17 of ref. [9] and in table 3 of ref. [10].

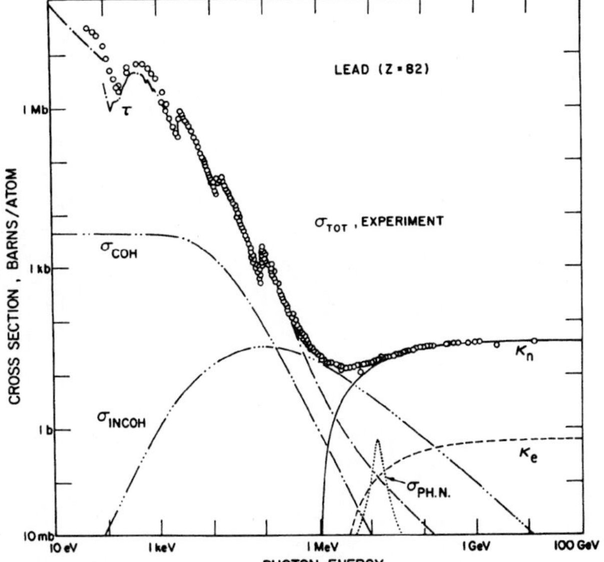

Fig. 1 Contributions of (a) atomic photo-effect, τ, (b) coherent scattering, σ_{COH}, (c) incoherent (Compton) scattering σ_{INCOH}, (d) nuclear-field pair production, κ_n, (e) electron-field pair production, κ_e, and (f) nuclear photoabsorption, $\sigma_{PH.N.}$, to the total measured cross section, σ_{TOT} (circles) in lead over the photon energy range 10 eV to 100 GeV. The measured σ_{TOT} points are not all shown in regions of high measurement density.

3. Strategies

A number of currently-used "best set" μ/ρ compilations developed at NBS and elsewhere are listed in references [11] and [12]. More recent compilations include those by Plechaty et al. [13] and by Henke et al [14]. These compilations all rely, at least partially, on theoretical predictions in addition to experimental information. Also, these compilations are rarely independent, as they frequently draw on previous compilations for one or more of the component cross sections. The NBS Photon and Charged Particle Data Center is presently developing a new "best set", and is now evaluating an interim set [10] based almost entirely on available theory [3-6].

As part of the critical evaluation and compilation process, it is useful to compare the adopted data with the raw experimental data. From this confrontation one can assess the need, if any, for further empirical adjustments of the μ/ρ "best set" compilation.

K. H. Böckhoff (ed.), Nuclear Data for Science and Technology, 1007–1009.

4. The NBS Data File

At the NBS, this comparison process has recently been facilitated by converting an extensive file of abstracted experimental μ/ρ data from a handwritten to a machine-readable form. This file consists of approximately 20,000 data points abstracted from over 500 independent literature sources. Photon energies in this file range from 10 eV to 13.5 GeV and 82 elements are included with Z from 1 to 94.

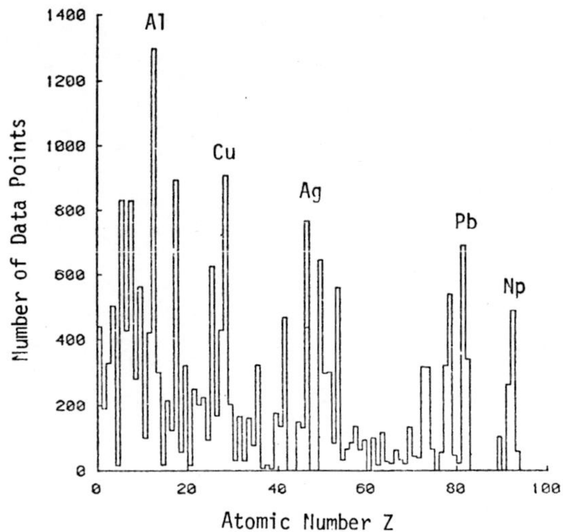

Fig. 2 Distribution of measured data points as a function of atomic number Z.

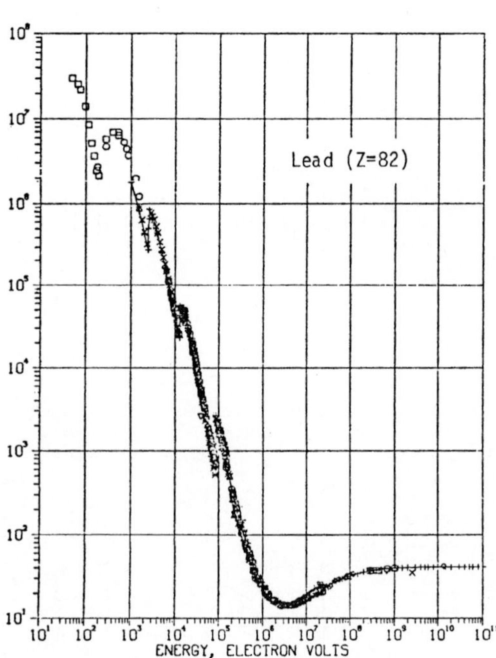

Fig. 3 Automated plot of the 653 measured data points, with different symbols distinguishing the 120 data sets for lead.

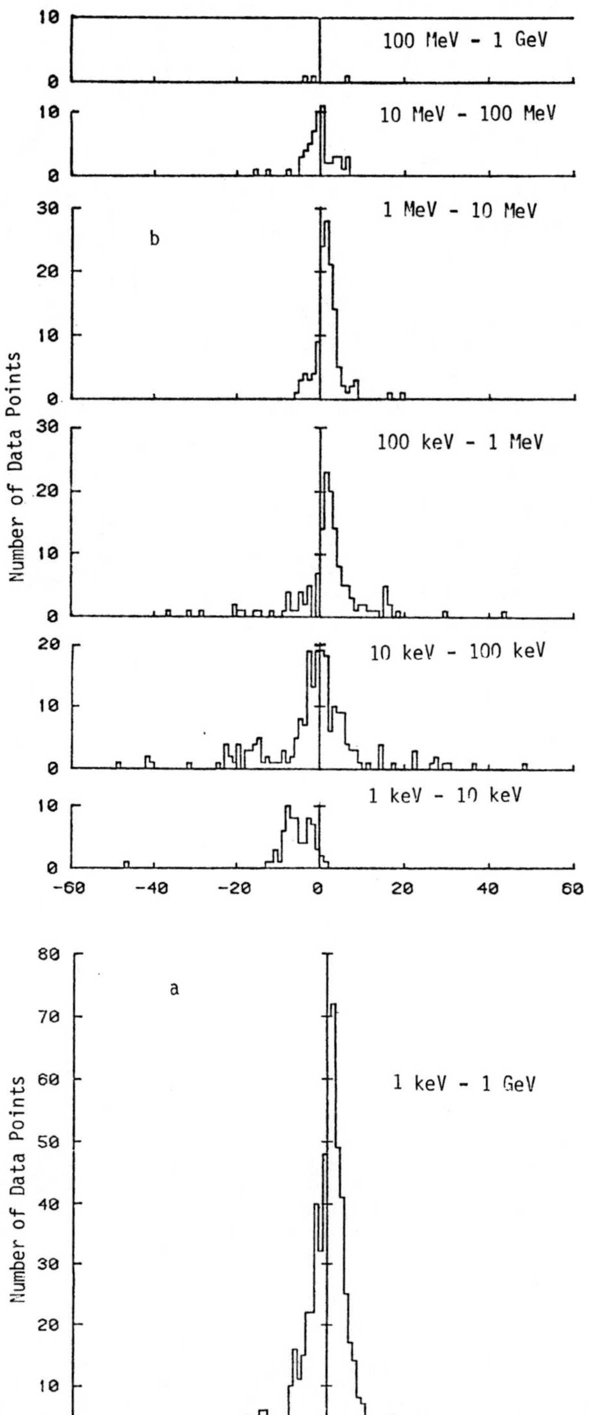

Fig. 4 Comparison of critically evaluated photon attenuation coefficients μ(theory) with raw experimental values μ(exp.) for lead (a) for photon energies 1 keV to 1 GeV and (b) for individual logarithmic decades in photon energy.

Figure 2 shows the distribution of the data points as a function of Z indicating Al as the most intensively measured. Also seen in Fig. 2 are regions where measurements are few (Z = 55 to 72) or non-existent (Z = 84 to 89, $Z \geq 95$). The data-points for lead are machine-plotted with cross section vs. photon energy in Fig. 3 on a log-log scale as a general illustration of the data-file contents.

5. Example of Use of the Data File

The automated data-base permits easier statistical analysis of the measured $\mu(exp)/\rho$ values, as well as a variety of computer-graphics aids to critical evaluation of "best set" theoretical or quasi-theoretical compiled $\mu(theor)/\rho$ data. Fig. 4a, again using lead as an example, shows the distribution of percent deviations of the current NBS $\mu(theor)/\rho$ [10] values from the corresponding $\mu(exp)/\rho$ values in the above data base for the broad energy range 1 keV to 1 GeV. In Fig. 4b this information is broken down into logarithmic photon-energy decades 1 to 10 keV, 10 to 100 keV, 100 keV to 1 MeV, 1 MeV to 10 MeV, 10 MeV to 100 MeV and 100 MeV to 1 GeV.

Information such as that displayed in Fig. 4 is useful in estimating and assigning uncertainties, both systematic and "scatter" to the compiled "best set" data. This information also suggests regions in which both theory and measurements should be examined more critically for sources of systematic error. Also clearly indicated in this sample display for lead are some "far out" measurement-sets which should be rejected in any automated use of the experimental data base.

Acknowledgements

The authors thank Dr. Martin Berger for his advice and useful criticism. A useful discussion with Dr. Richard Hennig is also acknowleged. Part of this work was performed while one of us (HG) was a participant at the Lewes Center for Physics. Their generous support, as well as the stimulating environment, were very much appreciated.

References

1. J.H. Hubbell, Survey of photon-attenuation-coefficient measurements 10 ev to 100 GeV, At. Data 3, 241 (1971).

2. C.G. Barkla and C.A. Sadler, Secondary x-rays and the atomic weight of nickel, Philosoph. Mag, 14, 408 (1907).

3. J.H. Scofield, Theoretical photoionization cross sections from 1 to 1500 keV, Lawrence Livermore Lab. Rep. UCRL-51326 (1973).

4. J.H. Hubbell, W.G. Veigele, E.A. Briggs, R.T. Brown, D.T. Cromer and R.J. Howerton, Atomic form factors, incoherent scattering functions, and photon scattering cross sections, J. Phys. Chem. Ref. Data 4, 471 (1975); errat. in 6, 615 (1977).

5. J.H. Hubbell and I. Øverbø, Relativistic atomic form factors and photon coherent scattering cross sections, J. Phys. Chem. Ref. Data 8, 69 (1979).

6. J.H. Hubbell, H.A. Gimm and I. Øverbø, Pair, triplet and total atomic cross sections (and mass attenuation coefficients) for 1 MeV to 100 GeV photons in elements Z = 1 to 100, J. Phys. Chem. Ref. Data 9, 1023 (1980).

7. E.G. Fuller, H.M. Gerstenberg, H. Vander Molen and T.C. Dunn, Photonuclear reaction data, 1973, NBS Spec. Publ. 380 (1973).

8. E.G. Fuller and H. Gerstenberg, Photonuclear data index, 1973-1981, NBSIR 82-2543 (1982).

9. J.H. Hubbell, Photon cross sections, attenuation coefficient, and energy absorption coefficients from 10 keV to 100 GeV, Report NSRDS-NBS 29 (1969).

10. J.H. Hubbell, Photon mass attenuation and energy-absorption coefficients from 1 keV to 20 MeV, Int. J. Appl. Radiation and Isotopes (in press).

11. K. Way, Atomic data related to x and XUV radiation, At. Data Nucl. Data Tables 22, 125 (1978).

12. A. Lorenz, Compilations and evaluations of data on the interaction of electromagnetic radiation with matter, IAEA Report INDC (NDS)-94/N (1978).

13. E.F. Plechaty, D.E. Cullen and R.J. Howerton, Tables and graphs of photon-interaction cross sections from 0.1 keV to 100 MeV derived from the LLL evaluated-nuclear-data library, Lawrence Livermore Nat. Lab. Report UCRL-50400, Vol. 6, Rev. 3 (1981).

14. B.L. Henke, P. Lee, J.T. Tanaka, R.L. Schimabukuro and B.K. Fuyikawa, Low-energy x-ray interaction coefficients: photoabsorption, scattering, and reflection. E = 100-2000 eV, Z = 1-94, At. Data Nucl. Data Tables 27, 1 (1982).

Closing Session

CLOSING TALK

André Michaudon

Commissariat à l'Energie Atomique
Centre d'Etudes de Limeil
Boîte postale n° 27
94190 Villeneuve-St-Georges
France

Introduction

This Conference is almost finished. Almost, because, before it is officially closed, a final talk is scheduled in the programme. Whether there should be such talks is a rather controversial subject. Some Conferences have, others have not. When they have, the final session may take several forms, including that of a panel. Then, this simplifies the task of the reviewers, each one covering one part only of the Conference. Here, one person has been asked to speak. This simplified the task of the organisers if not of the speaker. Yet, this talk is called a closing not a summary talk and this is very fortunate because obviously, it is not possible to compress a one-week Conference into thirty minutes. But half an hour is also too long for just saying good bye. Therefore this talk has the wrong duration. Similarly, the atomic nucleus which has been discussed at great length during this Conference has the wrong nucleon number : too large for an exact study using the nucleon-nucleon interaction only but too small also for treating the nucleus like infinite nuclear matter. Awkward situation : the wrong time for the wrong subject ! Moreover, many different aspects were covered at this Conference : basic nuclear theory, evaluation, microscopic measurements, integral experiments, data needs for nuclear reactors : design, safety, operation, fuel cycle, etc... In this context, it is even less possible to reach the ambitious goals set up by Mr Dinkespiler in his opening address : not only taking stock of the work accomplished but also deciding whether we need more money and why. This duty is in fact left to all of us when we get back and fight for more funding. At least, this Conference has given us more information to sharpen our arguments.

It is quite timely at this Conference to remember that the neutron which plays such a dominant role in nuclear energy, was discovered by Chadwick in 1932, just fifty years ago. A special Conference will celebrate next week in Cambridge (England) the 50th anniversary of this important event which opened the way to the discovery of fission six years later. This last discovery was immediately followed by intensive research to find out whether it could be used for the production of nuclear energy. Nuclear data though extremely scarce at that time were nevertheless crucial to answer this question of paramount importance. Microscopic measurements were barely possible and Joliot and his team (among others) had to rely, already, on what is now called integral experiments to demonstrate the feasibility of using fission energy. After fifteen years of secrecy because of its military as well as future commercial applications, the nuclear data field opened up in 1955 at the first Geneva Conference where information about nuclear energy, including nuclear data, was released. This jamboree, so called by B. Rose, started a new era for nuclear data and its evolution since then has been briefly summarized in his key-note speech.

Nuclear data for thermal- and fast-neutron reactors

I shall mention now rapidly some salient points about nuclear data for thermal- and fast-neutron reactors.

a) Thermal-neutron reactors

Nuclear data needs for pressurized-water reactors (PWR) were reviewed in an invited paper. Areas for new work have been clearly spelled out. The variation of η for ^{235}U below 1 eV should be reexamined for integral experiments now predict a different trend as a function of neutron energy. This is the renewal of an old subject. Twenty five years ago some nuclear physicists studied the low-energy shape of the fission cross-section using reactor neutrons to pin-point the properties of a postulated negative-energy resonance. This study has been stopped since then but can be reactivated with accelerator-based neutron sources operated at very low repetition frequency.

In the same manner, the low energy variation of the capture cross-section for ^{238}U should be reexamined since integral measurements predict a decrease faster than the commonly accepted $^1/_v$ dependence.

The present status of all ^{238}U cross-sections above 1 eV was very thoroughly assessed in an invited paper. As far as PWR are concerned, only the first resonances matter and their parameters are now well known with a low capture width of 23 meV.

For secondary actinides, several measurements were reported using either microscopic or integral methods. Attention was drawn to ^{244}Cm build-up because of its high spontaneous fission rate. Relevant cross-sections for its production are for ^{242}Pu and ^{243}Am resonant capture.

The fission product effect is dominated by Xe^{135} and Sm^{149} whose cross-sections are known sufficiently well.

Future trends include :

- higher burnup, hence more importance given to secondary actinides and their fission products.
- use of poisons (Hf, Gd) to compensate for greater ^{235}U depletion.
- use of reprocessed fuel having therefore more ^{236}U content. Consideration should be given to the effect of this nucleus which presents a strong resonance at 5.6 eV.

b) Liquid-Metal Fast Breeder Reactors (LMFBR)

Target accuracies of 0.2 % for the effective multiplication factor, 2 % for the breeding ratio and 10 % for the Doppler effect result in stringent requirements for the main actinides used in this type of reactor.

Despite much progress, several areas need a more thorough investigation for ^{238}U. In the resolved-resonance region, the situation looks satisfactory below 1.5 keV but above, up to 4 keV, many discrepancies are

K. H. Böckhoff (ed.), Nuclear Data for Science and Technology, 1013–1016.

observed in the determination of the parameters. They seem to depend on the sample thickness and the method of analysis. Some basic invariance seems to be lost somewhere on the way and should be clarified.

The fluctuations in the total cross-section in the unresolved-resonance region above 4 keV are not correctly described. The largest resonances which contribute to it should be better known. Also total cross-section measurements with different sample thicknesses at various temperatures should be made. The capture cross-section is not known with sufficient accuracy. Further improvements may be hampered by the technological limit of the prompt-γ detection technique. Activation may offer an alternative. Scattering is known globally but the various components of inelastic scattering should be determined more accurately. A still unresolved discrepancy persists as to the ratio of the ^{238}U capture to ^{239}Pu fission cross-sections as obtained from integral measurements.

For ^{235}U, large differences in α have been reported for the low-energy resonances using different techniques in the Soviet Union and the U.S.A. These differences are still unexplained.

This casts some doubt as to the correct values of α for ^{239}Pu and the resonance parameters for this nucleus.

It is surprising that very little work has been made for a long time about the resonances of these two nuclei. It is too great an honour given to old work which should be checked. But it is not an honour if this old work is not taken into account in the evaluated data files.

The situation for structural materials has been reviewed in great detail in an invited paper, both from the microscopic and integral points of view. Large discrepancies exist. Internal consistency is not achieved among microscopic data. For example the 1.15 keV Fe resonance, though easy to measure, is not known accurately and the large spread in the parameter values is not understood even for the neutron width. Integral measurements such as those carried out in Japan cannot be reproduced by evaluations. Also, measurements made in France with large thicknesses of sodium and steel can be explained only by modifying the cross-sections. Improvements are necessary for all aspects : evaluations, microscopic and integral measurements.

Many fission products contribute to the negative reactivity effect in contrast to PWR. The overall 10 % accuracy required for their capture cross-sections is not met. Samples are scarce and when results are available they are commonly discrepant by 15 % - 25 %, sometimes more. One has to rely on model calculations which also show large spreads in their results.

The situation is similar for secondary actinides. Their role is weak for the neutronics (except for Pu isotopes) but important for the fuel cycle. The data base is larger for fission than for capture as a consequence of the size of samples. Model calculations rely on that data base. They become less and less reliable for nuclei far from that data base.

The decay heat curve is obtained both by microscopic and integral methods. Some agreement between these methods is obtained for ^{235}U thermal-neutron fission but not so much for ^{239}Pu especially for times around 10^3 s. Moreover for ^{239}Pu, the β-radioactivity is overestimated in the summation method in contrast to the γ-radioactivity. An overall estimate from the whole data base including the ratio of ^{235}U to ^{239}Pu results leads to more conservative uncertainties of about 5 % for ^{235}U and 10 % for ^{239}Pu. These uncertainties may be greater for fast-neutron fission though it does not seem to be very much different from thermal-neutron fission.

The delayed neutrons play a role in the kinematics of the reactor and their properties (yields, spectra, precursor half-lives) are not known with sufficient precision (10 % compared to 5 % requested) for both types of reactors.

The overall problem of the fuel cycle was discussed in several papers. This is a big issue involving about 700 nuclides (actinides, fission products, activated elements) and their relevant data. One must be able to predict the mass, activity and radiation properties of all these nuclides. The γ-ray energy is a sensitive parameter for shielding. This enormous task must be tackled by defining priorities for the various steps of the fuel cycle, from extraction of the ore to waste disposal.

This brief review illustrates that despite all the progress accomplished up to now, many problems still remain unsolved.

Nuclear data for other applications

Because of the lack of time, I shall pass rapidly over other applications :

a) Fusion

The present status of data for the fusion programme was presented in a review paper. The needs for data are being defined more precisely. A broad data base is necessary to study various technical options. In addition, specific needs are requested depending on the option. For example, the breeding ratio should be known to about 2 %. The Li cross-sections are of course critical to achieve this goal. The present status of the ^{6}Li(n,α) cross-section seems satisfactory. For some systems the ^{7}Li(n,n',α) reaction needs to be known to 3 % and this is not presently achieved. Data were also reported for materials which may be used in the blanket and for charged-particle reactions.

b) Medicine

About 50 neutron-poor β^+ emitters have been identified as having a potential interest. Their production was treated in a paper which showed also that in addition to data needs many technical problems need to be solved. We had also several papers on neutron therapy and Kermas. Not mentioned at this Conference is the reevaluation of the Radiobiological effect (RBE) for neutrons as a function of the dose. This comes from a change in the data base for the neutron irradiation conditions at Hiroshima and Nagasaki.

c) Intense Neutron Sources

Two volumes on this subject will soon be published by Pergamon Press under the auspices of NEANDC and so I will not comment about this.

d) Standards

I shall not speak either about Standards though work was presented about Cf^{252} fission neutron spectrum (measurements and calculations), and ^{6}Li cross-sections.

Underlying Physics

It is impossible to have a good nuclear data situation without a sound physical basis.

a) Fission is the process which started the whole venture of nuclear energy. The knowledge of the fission barrier has greatly improved with the use of the Strutinsky prescription which resulted in the well-known double-humped barrier for actinides. An area of discrepancy has remained for a long time concerning light actinides, the so-called "Th anomaly". Though not reported at this Conference, this anomaly is now solved by very detailed measurements and analysis of the 720 keV vibrational resonance in ^{230}Th which can now be interpreted in terms of a triple-humped fission barrier. Such a barrier shape seems to apply to light actinides in general.

Fission cross-section calculations are still based on the channel theory of Bohr which emphasizes the major role played by the transition states above the saddle-point. Their spectrum is not well known especially for odd fissioning systems and cannot be determined directly by experiments. This spectrum can be deduced from fits to known fission cross-sections and, when possible, to fission fragment angular distribution data which depend on the J and K quantum numbers of these states.

The dynamics of fission from the saddle to the scission points is now actively studied. Viscosity calculations using either one-body or two-body dissipation concepts have been made but only for hot systems because they have negligible shell effects. But such systems do not appreciably fission at least for actinide nuclei because they evaporate neutrons first. When they fission after neutron emission they are cold and the dynamics are then dominated by shell effects. The influence of such effects on fragment properties was discussed in a review paper. Of interest is cold fragmentation in which the fragments are formed almost in their ground state at scission. This process appears to be the inverse process for heavy-ion fusion. Though a priori remote from the nuclear data field, these studies can help understanding the fission energy share between kinetic and excitation energies of the fission fragments, hence the properties of fission neutrons for example.

b) A thorough survey of all possible radiative capture mechanisms has been made in a review paper. These mechanisms involve various types of coupling : weak, strong, intermediate depending on the strength of the residual interactions. Different types of states can be formed in capture : one particle (direct or channel capture, valence capture), a few particles (semi-direct-doorway), many particles (compound nucleus). This field has now reached a high level of sophistication, very interesting from the fundamental point of view but also very useful for calculations. For example, several non-statistical effects occur for some structural materials.

c) More generally, the neutron-nucleus interaction presents many aspects depending on the hierarchy of states which are involved in the interaction.

The optical model is an example of the one-particle aspects and despite its phenomenological nature works extremely well for predicting many cross-sections. It becomes more and more refined. Used first with a spherical shape and a global parameterisation, it is now used with a deformed shape (when necessary) and with coupling to collective excitations. The parameters of the model are adjusted for each nucleus as in the SPRT method. In this manner, cross-sections are calculated more accurately and with less adjustable parameters. Let us also mention the use of proton data to determine the model parameters.

The few-nucleon states are met not only in radiative capture but also in preequilibrium reactions at high energy ($E > 10$ MeV) : several examples were given at this Conference.

The excitation of many nucleons corresponds to the compound nucleus and the statistical model widely used, especially for heavy nuclei.

The basic theory of these aspects and the use of this theory for evaluations were treated separately in two review papers. In addition several examples were given to illustrate these presentations.

d) Microscopic theories of the nucleus (Hartree-Fock or Hartree-Fock-Bogolyubov) play a growing role in nuclear physics but were rarely mentioned at this Conference in spite of their interest : optical model, level density, fission, etc...

e) The level density is a subject of great interest. It is as old as nuclear physics because the nuclear levels are the eigenstates of the nuclear Hamiltonian and their energy spectrum is a good test of the theory. The level density plays also a great role in calculations because of the Fermi golden role which states that the transition probability is given by the product of the absolute square of the matrix element by the density of the final states. Often physicists focus their attention on the matrix element but forget about the level density. Yet it is not well known as was pointed out in a paper even in the resonance region where a priori it could be derived very easily. In fact, even in these simple cases, elaborated professional methods must be used. It is too serious a problem to be left to amateurs.

The level spacing distributions are even more difficult to study because of the difficulty in having a good data base. Of course, if the data base is purified using some theory, the decontaminated data will be in agreement with the theory.

f) Solid-state effects were treated in several papers. This is also an old subject. Such effects were predicted by Lamb in 1939, observed first for recoiless γ-ray emission and absorption in the 1960s by Mössbauer and confirmed for neutrons shortly afterwards. Since then they did not receive much attention but they now suddenly show up again not only in microscopic but also in integral measurements.

g) Though quite far from our main stream of studies let us mention to finish this part, the role of the neutron for testing grand-unified theories of interaction between elementary particles. This is achieved for example by detailed study of its decay and search for possible neutron anti-neutron oscillations. At this Conference we heard about parity violation in the resonant capture of neutrons with some nuclei and this may help in the future to make progress in this direction.

Concluding remarks

This Conference covered a wide range of aspects as mentioned earlier and tried to achieve a balance between all these aspects, for example between detailed nuclear data and basic nuclear physics. To illustrate this let us mention the review of the nuclear data status for ^{238}U. It is the first time that a thorough survey for one single nucleus is given at a Conference of this type and this proved to be a very good idea which could be used again for other Conferences. At the other extreme, we had several papers on basic physics : Nuclear Reactions and Models, Radiative Capture, Fission. Balance could be seen also between subjects treated by reactor physicists and nuclear physicists, between integral and microscopic experiments.

Unity is difficult to achieve in these conditions because of the variety of subjects. Yet some progress has been made in this direction. Communication seems to improve between experts from different fields. Perhaps the fact that this Conference had no parallel sessions helped us to stay together and talk to each other. Nevertheless I am afraid that the nuclear data community is drifting away from the fundamentalists who are less and less interested in basic neutron physics and this is a source of concern.

The field of nuclear data seems now to enter a phase of consolidation. No major surprise is expected from improved data but they can help to understand better and predict more accurately many aspects of nuclear energy.

Several areas for more accurate work have been identified at this Conference. To make progress in these areas, we may have to reexamine our own methods of work. Many long-standing inconsistencies in the data should not be tolerated any longer. A more efficient follow-up is absolutely necessary otherwise we may very well meet again in several years and repeat the same things. This is partly the task of nuclear data committees. NEANDC will meet in two weeks and should take stock of these problems and stimulate new work. But this may not be sufficient. In addition a few important items should be selected and a task force set up to solve each of them especially if they are easy to solve.

Documentation is essential to make progress. This is a very important point which is stressed and which was mentioned again in the last talk. One has to find the right type of publication for nuclear data since most scientific journals are reluctant to accept papers with too many technical details or with too many numbers.

A more critical attitude towards our work should be encouraged. This was mentioned on several occasions. Of course, we must be nice to one another but we should have also the breadth of mind to make or accept criticisms without making enemies.

This is true in particular for the analysis of the data which should be more thorough. It was also mentioned several times that parameters derived from the same data depend on the type of analysis if not on the particular individual in charge of the analysis. Such inconsistencies should be clarified.

Nuclear theory is making progress and could be used more extensively for calculations. Several good examples were given at this Conference. The evaluators have now at their disposal very sophisticated methods and powerful tools for calculations. Yet, the parameters used in the calculations cannot be derived from pure nuclear theory but are obtained from fits to selected experimental data, hence the importance of measurements of good quality even for calculations.

A more general problem, crucial for the quality of the nuclear data work is the aging of our community, already mentioned in the key-note speech. Our activities have now levelled off to a situation which is very fragile. Expertise is essential for the quality of the work and this expertise takes a very long time to be reached. Yet it is now shared among too few people. Before not too long, these people may disappear from the stage without being replaced. The greying of the nuclear data community may not be followed by the greening of a new one like PHENIX rising up from its ashes. We may very well end up with a serious lack of scientific background on which to rely and the quality of the data will inevitably suffer from it. The future will tell us fairly shortly whether these views are too pessimistic.

The near future at least will gather us again at other similar Conferences. The cycle which has been started a few years ago will probably continue. Plans are already made to have the next Conference in the U.S.A. in 1984 then in the Soviet Union in 1986 before returning to the Europe-Japan area.

Before finishing, I would like to apologize for this oversimplified presentation. A more complete talk would have required more time and also a more thorough study of the very interesting papers given at this Conference. We are all looking forward to receiving shortly the Proceedings.

I would also like to thank our Conference Chairman K. Böckhoff and his colleagues who have worked very hard to make this Conference a real success. I thank you all also for having come and for the care taken in the presentation of your work either during the plenary sessions or in the form of posters.

Thank you very much for your attention and for the honour and the pleasure of speaking to you at the end of this Conference.

List of Participants

AUSTRALIA

Allen, B.J.
Australian Atomic Energy Commission
Lucas Heights Research Laboratories
Private Mail Bag
Sutherland, NSW 2232

Cook, J.L.
Australian Atomic Energy Commission
Lucas Heights Research Laboratories
Private Mail Bag
Sutherland, NSW 2232

AUSTRIA

Chalupka, A.
Institut für Radiumforschung und
Kernphysik der Universität Wien
Boltzmanngasse 3
A 1090 Vienna

Marton, J.
Institut für Radiumforschung und
Kernphysik der Universität Wien
Boltzmanngasse 3
A 1090 Vienna

Strohmaier, B.
Institut für Radiumforschung und
Kernphysik der Universität Wien
Boltzmanngasse 3
A 1090 Vienna

Vonach, H.K.
Institut für Radiumforschung und
Kernphysik der Universität Wien
Botzmanngasse 3
A 1090 Vienna

Winkler, G.
Institut für Radiumforschung und
Kernphysik der Universität Wien
Boltzmanngasse 3
A 1090 Vienna

Zankel, H.
Institut für Theoretische Physik
Universität Graz
Universitätsplatz 5
A 8010 Graz

BANGLADESH

Islam, M.M.
Bangladesh Atomic Energy Commission
P.O. Box 158
Ramna, Dacca-2

BELGIUM

Allaert, E.
SCK-CEN
Boeretang 200
B 2400 Mol

Arnould, M.L.L.
Institut d'Astronomie Physique
Université Libre de Bruxelles
Avenue F.D. Roosevelt 50
B 1050 Brussels

Bol, A.
Institut de Physique
Chemin du Cyclotron 2
B 1348 Louvain-la-Neuve

Belgium (continued)

Bouquegneau, C.R.
Faculté Polytechnique de Mons
Physics Department
Rue de Houdain 9
B 7000 Mons

Cornelis, E.
Rijksuniversitair Centrum Antwerpen
Groenenborgerlaan 171
B 2020 Antwerpen

De Backer, G.
Vrije Universiteit Brussel
Cyclotron Lab.
Laarbeeklaan 101
B 1090 Brussel

Debrue, J.
Département de Physique des Réacteurs
SCK-CEN
Boeretang 200
B 2400 Mol

De Corte, F.
Institute for Nuclear Sciences
Proeftuinstraat 86
B 9000 Gent

De Frenne, D.J.A.
Nuclear Physics Laboratory
Proeftuinstraat 86
B 9000 Gent

De Raedt, C.M.J.
Reactor Physics Department
SCK-CEN
Boeretang 200
B 2400 Mol

De Wispelaere, A.
Institute for Nuclear Sciences
Proeftuinstraat 86
B 9000 Gent

D'Hondt, P.J.C.H.D.
Laboratorium voor Kernphysica
Proeftuinstraat 86
B 9000 Gent

Hermanne, A.
Vrije Universiteit Brussel
Cyclotron Lab.
Laarbeeklaan 101
B 1090 Brussel

Jacobs, E.T.
Nuclear Physics Laboratory
Proeftuinstraat 86
B 9000 Gent

Leleux, P.
University of Louvain
Chemin du Cyclotron 2
1348 Louvain-la-Neuve

Lipnik, P.
University of Louvain
Chemin du Cyclotron 2
1348 Louvain-la-Neuve

Mertens, J.
Vrije Universiteit Brussel
Cyclotron Lab.
Laarbeeklaan 101
B 1090 Brussel

Belgium (continued)

Meulders, J.P.
University of Louvain
Institut de Physique
Chemin du Cyclotron 2
B 1348 Louvain-la-Neuve

Mewissen, W.L.
SCK-CEN
Boeretang 200
B 2400 Mol

Moens, L.J.M.G.
Institute for Nuclear Sciences
Proeftuinstraat 86
B 9000 Gent

Nêve de Mévergnies, M.
SCK-CEN
Boeretang 200
B 2400 Mol

Ninane, A.
University of Louvain
Chemin du Cyclotron 2
B 1348 Louvain-la-Neuve

Poortmans, F.
SCK-CEN
Boeretang 200
B 2400 Mol

Van Assche, P.
SCK-CEN
Boeretang 200
B 2400 Mol

Van de Vijver, R.E.
Laboratoria voor Kernfysika
Rijksuniversiteit Gent
Proeftuinstraat 86
B 9000 Gent

Vanpraet, G.
Rijksuniversitair Centrum Antwerpen
Groenenborgerlaan 171
B 2020 Antwerpen

Wagemans, C.
SCK-CEN
Boeretang 200
B 2400 Mol

BRAZIL

Auler, L.T.
Instituto de Engenharia Nuclear
Cidade Universita
C.P. 2186
20001 Rio de Janeiro

Martins, J.B.
Centro Brasileiro de Pesquisas Fisicas
Av. Wenceslau
22290 Rio de Janeiro

Nair, R.K.
Instituto de Estudos Avançados
CTA C.P. 6044
12200 Soa José dos Campos

CANADA

Ho, Y.U.
Theoretical Physics Branch
Atomic Energy of Canada Ltd.
Chalk River Nuclear Laboratories
KOJ IJO Chalk River, Ontario

CUBA

Cabezas Solorzano, R.
Comision de Energia Atomica de Cuba
P.O. Box 6689
La Havana

Fernandez Nodarse, F.
Comision de Energia Atomica de Cuba
P.O. Box 6689
La Havana

CYPRUS

Evangelides, G.
c/o Daresbury Laboratory
Keckwick Lane
Daresbury, Warrington WA4 4AD
United Kingdom

CZECHOSLOVAKIA

Kliment, V.
Institute of Physics
Dúbravská cesta
84228 Bratislava

DEMOCRATIC REPUBLIC OF GERMANY

Seeliger, D.
Technische Universität Dresden
Sektion Physik
Mommsenstrasse 13
8027 Dresden

EGYPT

Adib Shihata, M.
Atomic Energy Establishment
101 Kasr El-Eini Street
Academy of Scientific Research
Cairo

FEDERAL REPUBLIC OF GERMANY

Anders, B.
GKSS- Forschungszentrum
Postfach 1160
D 2054 Geesthacht

Anzaldo Meneses, A.M.
Institut für Neutronenphysik und Reaktortechnik
Kernforschungszentrum Karlsruhe
Postfach 3640
D 7500 Karlsruhe

Becker, H.W.
Institut für Kernphysik
Domackstrasse 71
D 44 Münster

Federal Republic of Germany (continued)

Beer, H.
Institut für Angewandte Kernphysik
Kernforschungszentrum Karlsruhe
Postfacht 3640
D 7500 Karlsruhe

Bosselmann, P.
Technischer Uberwachungsverein Norddeutschland E.V.
Grosse Bahnstrasse 31
D 2000 Hamburg

Böttger, R.
Physikalisch Technische Bundesanstalt
Abteilung 6.5
Bundesallee 100
D 3300 Braunschweig

Cierjacks, S.
Institut für Angewandte Kernphysik
Kernforschungszentrum Karlsruhe
Postfach 3640
D 7500 Karlsruhe

Cloth, P.
Institut für Reaktorentwicklung
Kernforschungsanlage Jülich
Postfach 1913
D 5170 Jülich

Coyne, J.J.
Physikalisch Technische Bundesanstalt
Bundesalle 100
D 3300 Braunschweig

Filges, D.
Kernforschungsanlage Jülich
Postfach 1913
D 5170 Jülich

Finckh, E.
Physikalisches Institut
Erwin-Rommelstrasse 1
D 8520 Erlangen

Fischer, P.
Institut für Reine und Andgewandte Kernphysik
der Universität Kiel
Reaktorstation
D 2054 Geesthacht

Fischer, U.
Institut für Neutronenphysik und Reaktortechnik
Kernforschungszentrum Karlsruhe
Postfach 3640
D 7500 Karlsruhe

Fröhner, F.H.
Institut für Neutronenphysik und Reaktortechnik
Kernforschungszentrum Karlsruhe
Postfach 3640
D 7500 Karlsruhe

Goel, B.P.
Kernforschungszentrum Karlsruhe
Postfach 3640
D 7500 Karlsruhe

Grum, W.
University of Stuttgart
Institut für Strahlenphysik
Allmandring 3
D 7000 Stuttgart

Hammer, J.W.
University of Stuttgart
Institut für Strahlenphysik
Allmandring 3
D 7000 Stuttgart

Heeringa, W.
Kernforschungszentrum Karlsruhe
Postfach 3640
D 7500 Karlsruhe

Henschel, H.
Fraunhofer-Institut für Naturwissenschaft
Technische Trendanalysen
Appelsgarten 2
D 5350 Euskirchen

Hofmann, K.
Institut für Kernphysik I
Kernforschungszentrum Karlsruhe
Postfach 3640
D 7500 Karlsruhe

Hino, Y.
Institut für Angewandte Kernphysik
Kernforschungszentrum Karlsruhe
Postfach 3640
D 7500 Karlsruhe

Jahn, H.
Institut für Neutronenphysik und Reaktortechnik
Postfach 3640
Kernforschungszentrum Karlsruhe
D 7500 Karlsruhe

Jungmann, C.R.
Siegfriedstrasse 12a
D 6023 Hochheim/Ts

Käppler, F.
Institut für Neutronenphysik und Reaktortechnik
Kernforschungszentrum Karlsruhe
Postfach 3640
D 7500 Karlsruhe

Klein, H.
Physikalisch Technische Bundesanstalt
Abteilung 6
Bundesallee 100
D 3300 Braunschweig

Küsters, H.
Institut für Neutronenphysik und Reaktortechnik
Kernforschungszentrum Karlsruhe
Postfach 3640
D 7500 Karlsruhe

Mannhart, W.
Physikalisch Technische Bundesanstalt
Bundesallee 100
D 3300 Braunschweig

Mattes, M.
Institut für Kernenergetik
Universität Stuttgart
Pfaffenwaldring 31
D 7000 Stuttgart

Mengoni, A.
Institut für Angewandte Kernphysik
Kernforschungszentrum Karlsruhe
Postfach 3640
D 7500 Karlsruhe

Michel, R.
Institut für Kernchemie der Universität zu Köln
Zülpicherstrasse 47
D 5000 Köln 1

Pepelnik, R.
GKSS Forschungszentrum
Institut für Physik
Max-Planck-strasse
D 2054 Geesthacht

Federal Republic of Germany (continued)

Priesmeyer, H.G.
T.K.K. Universität Kiel
D 2054 Geesthacht

Qaim, S.M.
Institut für Chemie 1
Kernforschungsanlage Jülich
D 5170 Jülich

Rosenstock, W.
Fraunhofer INT
Appelsgarten 2
D 5350 Euskirchen

Schleussner, G.
Universität Stuttgart
Institut für Strahlenphysik
Allmandring 3
D 7000 Stuttgart 80

Schölermann, H.
Physikalisch Technische Bundesanstalt
Bundesallee 100
D 3300 Braunschweig

Stöcklin, G.L.
Institut für Chemie 1
Kernforschungsanlage Jülich
Postfach 1913
D 5170 Jülich

Theobald, J.
Institut für Kernphysik der Technische Hochschule
Schlossgartenstrasse 9
D 61 Darmstadt

Thielemann, F.K.
Max Planck Institut für Kernphysik
Saupfercheckweg 1
D 6900 Heidelberg

Weidenmüller, H.A.
Max Planck Institut für Kernphysik
Saupfercheckweg 1
D 6900 Heidelberg

Wiese, H.W.
Institut für Neutronenphysik
Kernforschungszenturm Karlsruhe
Postfach 3640
D 7500 Karlsruhe

Wisshak, K.
Kernforschungszentrum Karlsruhe
Postfach 3640
D 7500 Karlsruhe

Wölfle, R.
Institut für Chemie 1
Kernforschungsanlage Jülich
Postfach 1913
D 5170 Jülich

FRANCE

Blachot, J.
DRF/LCPN
C.E.N. de Grenoble
B.P. 85
F 38041 Grenoble

France (continued)

Bouchard, J.
DRE/SEN
C.E.A. - C.E.N. de Cadarache
B.P. 1
F 13115 Saint Paul lez Durance

Communeau, F.
Commissariat à l'Energie Atomique
B.P. 27
F 94190 Villeneuve St. Georges

Costa, L.
C.E.A. - C.E.N. de Cadarache
B.P. 1
F 13115 Saint Paul lez Durance

Coursol, N.
LMRI
C.E.A. - C.E.N. de Saclay
B.P. 2
F 91190 Gif sur Yvette

Delaroche, J.P.
Centre D'Etudes de Bruyères le Châtel
B.P. 561
F 92542 Montrouge Cedex

Duchemin, B.
C.E.A. - C.E.N. de Saclay
B.P. 2
F 91190 Gif sur Yvette

Fort, E.
DRNR/SEDC/SPNR
C.E.A. - C.E.N. de Cadarache B. 230
B.P. 1
F 13115 Saint Paul lez Durance

Fréhaut, J.
Centre D'Etudes de Bruyères le Châtel
Service de Physique Nucléaire
B.P. 561
F 92542 Montrouge Cedex

Girieud, R.
DRE/SEN
C.E.A. - C.E.N. de Cadarache
B.P. 1
F 13115 Saint Paul lez Durance

Golinnelli O.
C.E.A. - C.E.N. de Cadarache
B.P. 1
F 13115 Saint Paul lez Durance

Haouat, G.
Centre D'Etudes de Bruyères le châtel
Service de Physique Neutronique et Nucléaire
B.P. 561
F 92542 Montrouge Cedex

Lachkar, J.
DS/DPG/P2N
Centre D'Etudes de Bruyères le Châtel
B.P. 561
F 92542 Montrouge Cedex

Lepage-Bertrand, F.J.
C.E.A. - C.E.N. de Saclay
B.P. 2
F 91190 Gif sur Yvette

France (continued)

Martin-Deidier, L.
DRE/SEN
C.E.A. - C.E.N. de Cadarache
B.P. 1
F 13115 Saint Paul lez Durance

Michaudon, A.
Département de Physique Générale
C.E.A. - Centre d'Etudes de Limeil
B.P. 27
F 94190 Villeneuve St. Georges

Pandey, A.
C.E.A. - C.E.N. de Saclay
B.P. 2
F 91190 Gif sur Yvette

Pannetier, R.
C.E.A. - Centre d'Etudes de Limeil
B.P. 27
F 94190 Villeneuve St. Georges

Philis, C.A.
Centre d'Etudes de Bruyères le Châtel
B.P. 561
F 92542 Montrouge Cedex

Raynal, J.
C.E.A. - C.E.N. de Saclay
Service de Physique Théorique
F 91190 Gif sur Yvette Cedex

Ribon, P.
C.E.A. - C.E.N. de Saclay
Service de Physique Théorique
F 91190 Gif sur Yvette Cedex

Trapp, J.P.
DRNR/SEDC/SPNR/LPR
C.E.A. - C.E.N. de Cadarache
B.P. 1
F 13115 Saint Paul lez Durance

Trochon, J.
Centre d'Etudes de Bruyères le Châtel
B.P. 561
F 92542 Montrouge Cedex

Voignier, J.J.
Centre d'Etudes de Bruyères le Châtel
Service de Physique Neutronique et Nucléaire
B.P. 561
F 92542 Montrouge Cedex

GREECE

Dritsa, S.
N.R.C. 'Demokritos'
Greek Atomic Energy Commission
Aghia Paraskevi Attikis

Synetos, S.
77, Third September Street
Athens 104

HUNGARY

Csikai, J.
Institute of Experimental Physics of Kossuth
Lajos University
P.O. Box 105
Bem Ter 18/A
H 4001 Debrecen

INDIA

Bahal, B.M.
c/o GKSS - Kernforschungszentrum
Postfach 1160
D 2054 Geesthacht
Federal Republic of Germany

Ganesan, S.
Reactor Research Centre Kalpakkam
P.O. 603 102
Kalpakkam
Tamil Nadu, Chengalpattu District

Gupta, S.K.
Nuclear Physics Division
Bhabha Atomic Research Centre
Trombay
400085 Bombay

Mehta, M.K.
Head of the Nuclear Physics Division
Bhabha Atomic Research Centre
Trombay
400085 Bombay

ISRAEL

Levin, P.
Nuclear Research Centre Negev
P.O. 9001
Beer Sheva

Mantel, M.
Israel Atomic Energy Commission
Nuclear Chemistry Division
70600 Yavne

Salmi, U.
Racah Institute of Physics
The Hebrew University, Jerusalem
91904 Jerusalem

Shafrir, N.H.
Department of Nuclear Engineering
Technion-Israel Institute of Technology
32000 Haifa

Wagschal, J.J.
Racah Institute of Physics
The Hebrew University
91904 Jerusalem

Yeivin, Y.
Racah Institute of Physics
The Hebrew University
91904 Jerusalem

Yiftah, S.
Atomic Energy Commission
Soreq Nuclear Research Centre
70660 Yavne

ITALY

Bonardi, M.
Istituto di Scienze Fisiche
Via Celoria 16
20133 Milano

Coceva, C.
Direttore Divisione Fisica
E.N.E.A.
Via Mazzini 2
40138 Bologna

Italy (continued)

Gabrielli, P.
E.N.E.A.
Casaccia
S.P. Anguillarese 1 + 300
00062 Roma

Menapace, E.
E.N.E.A.
Via Mazzini 2
40138 Bologna

Ventura, A.
Laboratorio Dati Nucleari
E.N.E.A.
Via Mazzini 2
40138 Bologna

JAPAN

Aizawa, O.
Atomic Energy Research Laboratory
Musashi Institute of Technology
971 Oosenji Tama-Ku
Kawasaki-Shi, Kanagawa-Ken

Akiyama, M.A.
Nuclear Engineering Research Laboratory
University of Tokyo
2-22 Shirane-Shirakata, Tokai-Mura
Naka-Gun, Ibaraki-Ken

Kadotani, H.
Century Research Corporation, Ozu-Honkan Building
Hon-Cho, Nihonbashi
Department n° 4, 3-2
103 Chou-Ku, Tokyo

Kanda, Y.
Department of Energy
Kyushu University
816 Kasunga-Shi, Fukuoka-Ken

Kikuchi, Y.
Japan Atomic Energy Research Institute
Nuclear Data Center
319-11 Tokai - Ibaraki-Ken

Kimura, I.
Research Reactor Institute
Kyoto University
Kumatori-Cho, Sennan-Gun
Osaka 590-04

Nakazawa, M.
Nuclear Engineering Research Lab.
University of Tokyo
Jaeri 319-11
Tokai Mura Ibaraki-Ken

Sugiyama, K.
Department of Nuclear Engineering
Tohoku University
Aramaki-Aoba
Sendai 980

Takahashi, A.
Department of Nuclear Engineering
Osaka University
Yamadaoka 2-1
565 Suita/Osaka

Uenohara, Y.
Department of Energy Conversion
Kyushu University
6-10-1 Hakozaki Higashi-Ku
812 Fukuoka

Japan (continued)

Umezawa, H.
Division of Chemistry
Tokai Research Establisment
319-11 Tokai-Mura
Ibaraki-Ken

Yamamuro, N.
Research Lab. for Nuclear Reactor
Tokyo Institute of Technology O-Okayama
Megro-Ku Tokyo 152

LIBYA

Abologasem, M.
Lybian Tajura Research Center
P.O. Box 30556
Tajura

Irhouma, A.
Lybian Tajura Research Center
P.O. Box 30556
Tajura

MALAYSIA

Sarmani, S.B.
Department of Nuclear Science
University Kebangsaan Malaysia
Jalan Pantai Baru
22-12 Kuala Lumpur

MOROCCO

Viennot, M.
16 rue Azegza
Rabat Agdal

NETHERLANDS

Brinkman, G.A.
Nikhef
Oosterringdijk 18
1009 AJ Amsterdam

Costa, C.
Netherlands Energy Research Foundation
Physics Department
Reactor Physics Group
1755 ZG Petten

Gruppelaar, H.
Netherlands Energy Research Foundation
Physics Department
Reactor Physics Group
1755 ZG Petten

Zijp, W.L.
Netherlands Energy Research Foundation
Radiation Metrology
1755 ZG Petten

NIGERIA

Olomo, J.B.
Nuclear Technology Unit
Physics Department University IFE
Ile-Ife

PAKISTAN

Gul, K.
Pakistan Institute of Nuclear Science and Technology
P.O. Nilore
Rawalpindi, Pakistan

PEOPLE'S REPUBLIC OF CHINA

Huang, S.
Institute of Atomic Energy
Academia Sinica
P.O. Box 275(15)
Beijing

Hu, J.
Institute of Atomic Energy
Academia Sinica
P.O. Box 275(41)
Beijing

Shen, Q.
Institute of Atomic Energy
Academia Sinica
P.O. Box 275(41)
Beijing

Zhou, D.
Institute of Atomic Energy
Academia Sinica
P.O. Box 275(41)
Beijing

POLAND

Morstin, K.
Institute of Physics and Nucl. Techn. Akadem.
Gorniczo-Hutnicza
Al. Mickiewicza 30
30-059 Krakow

REPUBLIC OF SOUTH AFRICA

Mc Murray, W.R.
Director of the Southern
Universities Nuclear Institute
P.O. Box 17
Faure 7131

Müller, E.Z.
Atomic Energy Board
Priv. Bag X 256
Pretoria 0001
Transvaal

SWEDEN

Andersson, P.
Lunds Universitet Fysika Institutionen
Solvegatan 14
223 62 Lund

Condé, H.
Forsvarets Forskningsanstalt
Box 27322
102 54 Stockholm

Olsson, N.
The Studsvik Science Research Laboratory
611 82 Nykoping

Sweden (continued)

Ramstrom, E.
The Studsvik Science Research Laboratory
611 82 Nykoping

Sjostrand, N.G.
Reactor Physics
Chalmers University of Technology
41296 Göteborg

Trostell, B.
The Studsvik Science Research Laboratory
611 82 Nykoping

SWITZERLAND

Pelloni, S.
Swiss Federal Institute for Reactor Research
5303 Wuerenlingen

Richmond, R.
Swiss Federal Institute for Reactor Research
5303 Wuerenlingen

THAILAND

Vilaithong, T.
Department of Physics
Chiangmai University
Chiangmai

UNION OF SOVIET SOCIALIST REPUBLIC

Blinov, M.
V.G. Khoplin Radium Institute
Roetgenstr. 1
197022 Leningrad

Grashin, A.F.
Moscow Engineering Physics Institute
Kashirskoie Sh. 3
115409 Moscow

Muradyan, G.
Kurchatov's Atomic Energy Institute
Kurchatov
Moscow

Paseschnyk M.
Prospeckt Nauky of Science
252028 Kiev

Popov, A.B.
Joint Institute for Nuclear Research
Haed Post Office
P.O. Box 79
Moscow 101000

Sharapov, E.I.
Joint Institute for Nuclear Research
Head Post Office
P.O. Box 79
Moscow 101000

UNITED KINGDOM

Adesanmi, C.A.
Department of Physics
University of Surrey
Surrey
GU 25XH Guilford

Campbell, C.G.
A.E.E. Winfrith
Building B 21
Dorchester, Dorset DT2 8DH

Coates, M.S.
A.E.R.E. Harwell
Nuclear Physics Division
Building 418.15
Harwell, Didcot
Oxfordshire OX11 ORA

Cox, A.J.
Physics Department
University of Aston
B4 7ET Birmingham

Davies, B.S.J.
C.E.G.B.
Berkeley Nuclear Laboratories
Berkeley, Gloucestershire GL11 5SP

Fudge, A.J.
A.E.R.E. Harwell
Chemistry Division
Building 220
Harwell, Didcot
Oxfordshire OX11 ORA

Holloway, S.P.
Rush Common House
Dorchester Crescent
Abingdon, Oxon OX14 2AJ

Jefferies, S.M.
University of London Reactor Centre
Silwood Park, Ascot
Berkshire SL5 7PY

Mac Mahon, T.D.
University of London Reactor Centre
Silwood Park, Ascot
Berkshire SL5 7PY

Moxon, M.C.
A.E.R.E. Harwell
Nuclear Physics Division
Building 418.15
Harwell, Didcot
Oxfordshire OX11 ORA

Owen, J.G.
Radiation Centre
Birmingham University
Birmingham B15 2TT

Patrick, B.H.
A.E.R.E. Harwell
Nuclear Physics Division
Building 418.15
Harwell, Didcot
Oxfordshire OX11 ORA

Powell, H.J.
DNPDE - D 1200
Dounreay
KW147 T2 Thurso
Scotland

Rose, B.
Fitzharry's Road 16
Abingdon, Berkshire OX14 1EJ

Rowlands, J.L.
A.E.E. Winfrith
Reactor Physics Division
Building B 21
Dorchester, Dorset DT2 8DH

Sowerby, M.G.
A.E.R.E. Harwell
Nuclear Physics Division
Building 418.15
Harwell, Didcot
Oxfordshire OX11 ORA

Stevenson, J.M.
A.E.E. Winfrith
Reactor Physics Division
Dorchester, Dorset DT2 8DH

Syme, D.B.
A.E.R.E. Harwell
Nuclear Physics Division
Building 418.15
Harwell, Didcot
Oxfordshire OX11 ORA

Thomas, D.J.
Division of Radiation Science and Acoustics
National Physical Laboratory
Teddington, Middlesex TW11 OLW

Todd, D.M.
Department of Physics
University of Liverpool
P.O. Box 147
Liverpool L69 3BX

Walker, J.
Birmingham Radiation Centre
University of Birmingham
Birmingham B15 2TT

Ward, N.J.
Oliver Lodge Laboratory
University of Liverpool
P.O. Box 147
Liverpool L69 3BX

Wiltshire, R.A.
A.E.R.E. Harwell
Chemistry Division
Building 220
Harwell, Didcot
Oxfordshire OX11 ORA

UNITED STATES OF AMERICA

Abdou, M.A.
Argonne National Laboratory
9700 South Cass Avenue
Argonne, Illinois 60439

Behrens, J.W.
U.S. Department of Commerce
National Bureau of Standards
RADP B-119/532
Washington, D.C. 20234

Brady, F.P.
Department of Physics
University of California
Davis, California 95616

Ciarcia, C.A.
Department of Physics
University of Lowell
1 University Avenue
Lowell, Massachussetts 01854

Clark, D.D.
Ward Laboratory
Cornell University
Ithaca, New York 14853

United States of America (continued)

Couchell, G.P.
Department of Physics
University of Lowell
1 University Avenue
Lowell, Massachussetts 01854

De Saussure, G.
Building 6010
Oak Ridge National Laboratory
P.O. Box X
Oak Ridge, Tennessee 37830

Dias, M.S.
U.S. Department of Commerce
National Bureau of Standards
RADP B-119/532
Washington, D.C. 20234

Finlay, R.W.
Physics Department
Ohio University
Athens, Ohio 45701

Gerstenberg, H.M.
U.S. Department of Commerce
National Bureau of Standards
Washington, D.C. 20234

Gould, C.R.
Physics Department
North Carolina State University
Raleigh, North Carolina 27650

Harvey, J.A.
Physics Division
Oak Ridge National Laboratory
P.O. Box X
Oak Ridge, Tennessee 37830

Howe, S.
Los Alamos National Laboratory
Los Alamos, New Mexico 87544

Jarmie, N.
Los Alamos National Laboratory
Mail Stop 456
Los Alamos, New Mexico 87544

Johnson, R.G.
U.S. Department of Commerce
National Bureau of Standards
Washington, D.C. 20234

Kegel, G.H.
Physics Department
University of Lowell
1 University Avenue
Lowell, MAssachussetts 01854

Lamaze, G.P.
U.S. Department of Commerce
National Bureau of Standards
Building 245
Washington, D.C. 20234

Mehrzad, M.
University of Michigan
2540 Londonderry
Ann Harbor, Michigan 48109

Mughabghab,
Brookhaven National Laboratory
Upton, Ney York 11973

Peelle, R.W.
Neutron Physics Division
Oak Ridge National Laboratory
P.O. Box X
Oak Ridge, Tennessee 37830

Perey, F.
Oak Ridge National Laboratory
P.O. Box X
Oak Ridge, Tennessee 37830

Poenitz, W.P.
Argonne National Laboratory
Building 316
9700 South Cass Avenue
Argonne, Illinois 60439

Prince, A.
Brookhaven National Laboratory
Upton, New York 11973

Reich, C.W.
Nuclear Physics Branch
Idaho National Engineering Laboratory
EG & G Idaho Incorporation
P.O. Box 1625
Idaho Falls, Idaho 83415

Robertson, J.C.
Department of Chemical and Nuclear Engineering
University of New Mexico
Albuquerque, New Mexico 87131

Schenter, R.E.
Westinghouse Hanford Company
P.O. Box 1970
Richland, Washington 99352

Sheldon, E.
University of Lowell
Department of Physics
1 University Avenue
Lowell, Massachussetts 01854

Smith, A.B.
Applied Physics Division
Argonne National Laboratory
9700 South Cass Avenue
Argonne, Illinois 60439

Stehn, J.R.
National Nuclear Data Centre
Brookhaven National Laboratory
Upton, New York 11973

Strachan, J.D.
Princeton University
Plasma Physics
P.O. Box 451
Princeton, New Jersey 08544

Walter, R.L.
Duke University
Department of Physics
Durham, North Carolina 27706

White, R.M.
Lawrence Livermore National Laboratory
Lab. L-280
P.O. Box 808
Livermore, California 94550

Winters, R.R.
Department of Physics
Denison University
Granville, Ohio 43023

Young, P.G.
Los Alamos National Laboratory
Group T-2, Mail Stop 243
P.O. Box 1663
Los Alamos, New Mexico 87545

YUGOSLAVIA

Cvelbar, F.
Institute 'Jozef Stefan'
Jamova 39
61000 Ljubljana

Miljanič, Đ.P.
'Ruder Boskovič' Institute
Bijenička 54
41001 Zagreb

Paič, G.
'Ruder Boskovič' Institute
Bijenička 54
41001 Zagreb

Slaus, I.
'Ruder Boskovič' Institute
Bijeničla 54
41001 Zagreb

Trkov, A.
Institute 'Jozef Stefan'
Jamova 39
61000 Ljubljana

COMMISSION OF THE EUROPEAN COMMUNITIES

Bastian, C.
C.E.C. - Joint Research Centre
Geel Establishment
Steenweg naar Retie
2440 Geel

Böckhoff, K.-H.
C.E.C. - Joint Research Centre
Geel Establishment
Steenweg naar Retie
2440 Geel

Brusegan, A.
C.E.C. - Joint Research Centre
Geel Establishment
Steenweg naar Retie
2440 Geel

Budtz-Jørgensen, C.
C.E.C. - Joint Research Centre
Geel Establishment
Steenweg naar Retie
2440 Geel

Corvi, F.
C.E.C. - Joint Research Centre
Geel Establishment
Steenweg naar Retie
2440 Geel

Crametz, A.
C.E.C. - Joint Research Centre
Geel Establishment
Steenweg naar Retie
2440 Geel

Darvas, J.
C.E.C. - D.G. XII
Brussels

De Bièvre, P.
C.E.C. - Joint Research Centre
Geel Establishment
Steenweg naar Retie
2440 Geel

Deruytter, A.J.
C.E.C. - Joint Research Centre
Geel Establishment
Steenweg naar Retie
2440 Geel

Knitter, H.-H.
C.E.C. - Joint Research Centre
Geel Establishment
Steenweg naar Retie
2440 Geel

Liskien, H.O.
C.E.C. - Joint Research Centre
Geel Establishment
Steenweg naar Retie
2440 Geel

Matthes, M.
C.E.C. - Joint Research Centre
Ispra Establishment
Ispra, Varese
Italy

Paulsen, A.
C.E.C. - Joint Research Centre
Geel Establishment
Steenweg naar Retie
2440 Geel

Rohr, G.
C.E.C. - Joint Research Centre
Geel Establishment
Steenweg naar Retie
2440 Geel

Salomé, J.M.
C.E.C. - Joint Research Centre
Geel Establishment
Steenweg naar Retie
2440 Geel

Shelley, R.
C.E.C. - Joint Research Centre
Geel Establishment
Steenweg naar Retie
2440 Geel

Wartena, H.
C.E.C. - Joint Research Centre
Geel Establishment
Steenweg naar Retie
2440 Geel

Wattecamps, E.
C.E.C. - Joint Research Centre
Geel Establishment
Steenweg naar Retie
2440 Geel

Weigmann, H.
C.E.C. - Joint Research Centre
Geel Establishment
Steenweg naar Retie
2440 Geel

Wellum, R.
C.E.C. - Transuraneninstitut
Postfach 2266
75 Karlsruhe
Federal Republic of Germany

INTERNATIONAL ATOMIC ENERGY AGENCY

Ertek, C.
I.A.E.A.
Wagramerstrasse 5
P.O. Box 100
1400 Vienna - Austria

Lemmel, H.D.
Nuclear Data Section
I.A.E.A.
Wagramerstrasse 5
P.O. Box 100
1400 Vienna - Austria

International Atomic Energy Agency (continued)

Lorenz, A.
Nuclear Data Section
I.A.E.A.
Wagramerstrasse 5
P.O. Box 100
1400 Vienna - Austria

Pronyaev, V.G.
I.A.E.A.
Wagramerstrasse 5
P.O. Box 100
1400 Vienna - Austria

Schmidt, J.J.
Nuclear Data Section
I.A.E.A. - V.I.C.
Wagramerstrasse 5
P.O. Box 100
1400 Vienna - Austria

ORGANISATION FOR ECONOMIC CO-OPERATION AND DEVELOPMENT/ NUCLEAR ENERGY AGENCY

Coddens, G.P.
NEA Data Bank
91191 Gif sur Yvette
France

Neumann, B.D.
NEA Data Bank
3.P.9.
91191 Gif sur Yvette
France

Nordborg, C.I.
NEA Data Bank
91191 Gif sur Yvette
France

Tubbs, N.
NEA Data BANK
B.P. 9 - Bâtiment 45
91191 Gif sur Yvette Cedex
France

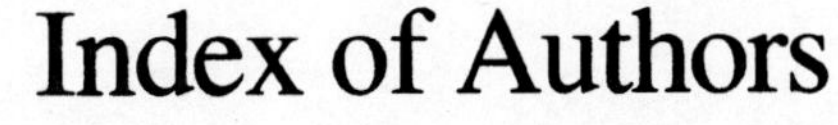
Index of Authors

CINDA Index

ELEMENT Z A	QUANTITY	ENERGY (EV) MIN	MAX	TYPE	DOCUMENTATION REF	PAGE	DATE	LAB	COMMENTS
H 001	Diff Elastic	4.5+7	7.5+7	EXPT	82ANTWER	800	8209	LVN	BOL+ 3-10 DEGR LAB-SYS, PLAST SCINT
D 002	Neut Emissn	1.4+7		EXPT	82ANTWER	360	8209	OSA	TAKAHASHI+ DIFF IN ANG,E' TOF PRELIM
LI 000	Neut Emissn	1.4+7		EXPT	82ANTWER	360	8209	OSA	TAKAHASHI+ DIFF IN ANG,E' TOF PRELIM
LI 006	Total Xsect	8.0+4	3.0+6	EXPT	82ANTWER	451	8209	GEL	KNITTER+ VDG TOF
LI 006	Elastic Scat	6.0+2	8.0+4	EXPT	82ANTWER	353	8209	DUB	ALFIMENKOV+ COMP R-MATRIX
LI 006	Diff Elastic	7.0+6	1.5+7	REVW	82ANTWER	766	8209	TNL	GOULD+ SPHERICAL OPT MDL COMP.
LI 006	Diff Elastic	8.0+4	3.0+6	EXPT	82ANTWER	451	8209	GEL	KNITTER+ VDG TOF
LI 006	Diff Inelast	7.0+6	1.5+7	REVW	82ANTWER	766	8209	TNL	GOULD+ SPHERICAL OPT MDL COMP.
LI 006	N,Triton	2.0+5	3.0+6	EXPT	82ANTWER	447	8209	TLU	CONDE+ (+UPP) + INVERSE REACTION.
LI 006	N,Triton	8.0+4	5.0+5	EXPT	82ANTWER	451	8209	GEL	KNITTER+ VDG TOF ANGDIST
LI 006	N,Alpha Reac	FAST		REVW	82ANTWER	397	8209	RCN	ZIJP+ INTEGRAL DATA CHECK,DOSIMETRY
LI 007	Diff Elastic	7.0+6	1.5+7	REVW	82ANTWER	766	8209	TNL	GOULD+ SPHERICAL OPT MDL COMP.
LI 007	Diff Inelast	7.0+6	1.5+7	REVW	82ANTWER	766	8209	TNL	GOULD+ SPHERICAL OPT MDL COMP.
LI 007	N,N Triton	4.7+6	1.6+7	EXPT	82ANTWER	349	8209	GEL	LISKIEN+ ACT METHOD
BE 009	Total Xsect	2.4+4		EXPT	82ANTWER	143	8209	JAP	AIZAWA+ (MUSASHI INST) COMP ENDFB4.
BE 009	Total Xsect	2.0-3	5.0+0	EXPT	82ANTWER	961	8209	ORL	HARVEY+ (+NEG) POLYCR SAMPLE
BE 009	Diff Elastic	7.0+6	1.5+7	REVW	82ANTWER	766	8209	TNL	GOULD+ SPHERICAL OPT MDL COMP.
BE 009	Neut Emissn	1.4+7		EXPT	82ANTWER	360	8209	OSA	TAKAHASHI+ DIFF IN ANG,E' TOF
BE 009	Diff Inelast	7.0+6	1.5+7	REVW	82ANTWER	766	8209	TNL	GOULD+ SPHERICAL OPT MDL COMP.
BE 009	Diff Inelast	1.5+7		THEO	82ANTWER	603	8209	AEP	ZIYANG+ EXCITON MDL CALC.
BE 009	N,Triton	1.4+7	1.5+7	EXPT	82ANTWER	368	8209	KOS	BOEDY+ ACTIV + T-BETA COUNTING
B 010	Diff Elastic	7.0+6	1.5+7	REVW	82ANTWER	766	8209	TNL	GOULD+ SPHERICAL OPT MDL COMP.
B 010	Diff Inelast	7.0+6	1.5+7	REVW	82ANTWER	766	8209	TNL	GOULD+ SPHERICAL OPT MDL COMP.
B 010	N,Alpha Reac	FAST		REVW	82ANTWER	397	8209	RCN	ZIJP+ INTEGRAL DATA CHECK,DOSIMETRY
B 011	Diff Elastic	7.0+6	1.5+7	REVW	82ANTWER	766	8209	TNL	GOULD+ SPHERICAL OPT MDL COMP.
B 011	Diff Inelast	7.0+6	1.5+7	REVW	82ANTWER	766	8209	TNL	GOULD+ SPHERICAL OPT MDL COMP.
C 012	Total Xsect	2.4+4		EXPT	82ANTWER	143	8209	JAP	AIZAWA+ (MUSASHI INST) COMP ENDFB4.
C 012	Total Xsect	2.0-3	5.0+0	EXPT	82ANTWER	961	8209	ORL	HARVEY+ (+NEG) POLYCR SAMPLE
C 012	Total Xsect	6.0+6	3.5+7	EXPT	82ANTWER	380	8209	TOH	TAKAKASHI+ BENCHMARK, DLC-58(ENDFB4)
C 012	Diff Elastic	4.5+7	7.5+7	EXPT	82ANTWER	800	8209	LVN	BOL+ 3-10 DEGR LAB-SYS, PLAST SCINT
C 012	Diff Elastic	7.6+6	8.7+6	EXPT	82ANTWER	783	8209	THS	BULSKI+ BE-9 N-SOURCE,FOR OPTMDL
C 012	Diff Elastic	7.0+6	1.5+7	REVW	82ANTWER	766	8209	TNL	GOULD+ SPHERICAL OPT MDL COMP.
C 012	Diff Elastic	1.0+7		EXPT	82ANTWER	891	8209	PTB	KLEIN+ SAMPLE SIZE CORR,CMP ENDFB5
C 012	Diff Elastic	6.0+6	3.5+7	EXPT	82ANTWER	380	8209	TOH	TAKAKASHI+ BENCHMARK, DLC-58(ENDFB4)
C 012	Polarization	7.6+6	8.7+6	EXPT	82ANTWER	783	8209	THS	BULSKI+ BE-9 N-SOURCE,FOR OPTMDL
C 012	Neut Emissn	1.4+7		EXPT	82ANTWER	360	8209	OSA	TAKAHASHI+ DIFF IN ANG,E' TOF
C 012	Diff Inelast	7.0+6	1.5+7	REVW	82ANTWER	766	8209	TNL	GOULD+ SPHERICAL OPT MDL COMP.
C 012	Diff Inelast	1.0+7		EXPT	82ANTWER	891	8209	PTB	KLEIN+ SAMPLE SIZE CORR,CMP ENDFB5
C 012	Diff Inelast	6.0+6	3.5+7	EXPT	82ANTWER	380	8209	TOH	TAKAKASHI+ BENCHMARK, DLC-58(ENDFB4)
C 012	Diff Inelast	1.5+7		THEO	82ANTWER	603	8209	AEP	ZIYANG+ EXCITON MDL CALC.

ELEMENT Z A	QUANTITY	ENERGY (EV) MIN	MAX	TYPE	DOCUMENTATION REF	PAGE	DATE	LAB	COMMENTS
C 012	N,Proton	2.7+7	6.1+7	EXPT	82ANTWER	922	8209	DAV	BRADY+ 3ES,ANG NT,KERMA,DOSIM APPL.
C 012	N,Deuteron	2.7+7	6.1+7	EXPT	82ANTWER	922	8209	DAV	BRADY+ 3ES,ANG NT,KERMA,DOSIM APPL.
C 012	N,Triton	2.7+7	6.1+7	EXPT	82ANTWER	922	8209	DAV	BRADY+ 3ES,ANG NT,KERMA,DOSIM APPL.
C 012	N,Alpha Reac	2.7+7	6.1+7	EXPT	82ANTWER	922	8209	DAV	BRADY+ 3ES,ANG NT,KERMA,DOSIM APPL.
C 012	N,Alpha Reac	8.0+6	1.0+7	EXPT	82ANTWER	930	8209	PTB	DIETZE+ FROM NE213 RESPONSE FCTNS.
C 012	N,N Alpha	1.0+7	3.5+7	EXPT	82ANTWER	926	8209	RBZ	ANTOLKOVIC+ (LVN) KERMA DETERMINED.
C 012	N,He3 Reactn	2.7+7	6.1+7	EXPT	82ANTWER	922	8209	DAV	BRADY+ 3ES,ANG NT,KERMA,DOSIM APPL.
C 013	Diff Elastic	7.0+6	1.5+7	REVW	82ANTWER	766	8209	TNL	GOULD+ SPHERICAL OPT MDL COMP.
C 013	Diff Inelast	7.0+6	1.5+7	REVW	82ANTWER	766	8209	TNL	GOULD+ SPHERICAL OPT MDL COMP.
N 014	N,Proton	2.7+7	6.1+7	EXPT	82ANTWER	922	8209	DAV	BRADY+ 3ES,ANG NT,KERMA,DOSIM APPL.
N 014	N,Deuteron	2.7+7	6.1+7	EXPT	82ANTWER	922	8209	DAV	BRADY+ 3ES,ANG NT,KERMA,DOSIM APPL.
N 014	N,Triton	2.7+7	6.1+7	EXPT	82ANTWER	922	8209	DAV	BRADY+ 3ES,ANG NT,KERMA,DOSIM APPL.
N 014	N,Alpha Reac	2.7+7	6.1+7	EXPT	82ANTWER	922	8209	DAV	BRADY+ 3ES,ANG NT,KERMA,DOSIM APPL.
N 014	N,He3 Reactn	2.7+7	6.1+7	EXPT	82ANTWER	922	8209	DAV	BRADY+ 3ES,ANG NT,KERMA,DOSIM APPL.
O 016	Diff Elastic	7.0+6	1.5+7	REVW	82ANTWER	766	8209	TNL	GOULD+ SPHERICAL OPT MDL COMP.
O 016	Neut Emissn	1.4+7		EXPT	82ANTWER	360	8209	OSA	TAKAHASHI+ DIFF IN ANG,E' TOF PRELIM
O 016	Diff Inelast	7.0+6	1.5+7	REVW	82ANTWER	766	8209	TNL	GOULD+ SPHERICAL OPT MDL COMP.
O 016	N,Proton	2.7+7	6.1+7	EXPT	82ANTWER	922	8209	DAV	BRADY+ 3ES,ANG NT,KERMA,DOSIM APPL.
O 016	N,Deuteron	2.7+7	6.1+7	EXPT	82ANTWER	922	8209	DAV	BRADY+ 3ES,ANG NT,KERMA,DOSIM APPL.
O 016	N,Triton	2.7+7	6.1+7	EXPT	82ANTWER	922	8209	DAV	BRADY+ 3ES,ANG NT,KERMA,DOSIM APPL.
O 016	N,Alpha Reac	2.7+7	6.1+7	EXPT	82ANTWER	922	8209	DAV	BRADY+ 3ES,ANG NT,KERMA,DOSIM APPL.
O 016	N,He3 Reactn	2.7+7	6.1+7	EXPT	82ANTWER	922	8209	DAV	BRADY+ 3ES,ANG NT,KERMA,DOSIM APPL.
F 019	N,2N	FISS		EXPT	82ANTWER	429	8209	PTB	MANNHART. CF252,CFD ENDFB-V
F 019	N,2N	FAST		REVW	82ANTWER	397	8209	RCN	ZIJP+ INTEGRAL DATA CHECK,DOSIMETRY
NE 020	N,Gamma	5.0+3	4.0+5	EXPT	82ANTWER	948	8209	KFK	ALMEIDA+ TOF N-BALANCE IN S-PROCESS.
NE 021	N,Gamma	5.0+3	4.0+5	EXPT	82ANTWER	948	8209	KFK	ALMEIDA+ TOF N-BALANCE IN S-PROCESS.
NE 022	N,Gamma	5.0+3	4.0+5	EXPT	82ANTWER	948	8209	KFK	ALMEIDA+ TOF N-BALANCE IN S-PROCESS.
NA 022	Evaluation	1.0-3	1.5+7	EVAL	82ANTWER	643	8209	RCN	GRUPPELAAR+ RCN-2 EVL,KEDAK-FORMAT
NA 022	Total Xsect	1.0-3	1.5+7	EVAL	82ANTWER	643	8209	RCN	GRUPPELAAR+ RCN-2 EVL,GRPH.
NA 022	Reson Params	1.0-2	1.0+3	EXPT	82ANTWER	150	8209	DUB	GLEDENOV+ TOF, CMPRD F19(N,A)
NA 022	N,Gamma	1.0-3	1.5+7	EVAL	82ANTWER	643	8209	RCN	GRUPPELAAR+ RCN-2 EVL,GRPH.
NA 022	N,Proton	1.0-2	1.0+3	EXPT	82ANTWER	150	8209	DUB	GLEDENOV+ TOF, CMPRD F19(N,A)
NA 022	N,Proton	1.0-3	1.5+7	EVAL	82ANTWER	643	8209	RCN	GRUPPELAAR+ RCN-2 EVL,GRPH.
NA 023	Diff Inelast	1.5+7		THEO	82ANTWER	603	8209	AEP	ZIYANG+ EXCITON MDL CALC.
NA 023	N,Gamma	MAXW		EXPT	82ANTWER	681	8209	LON	JEFFERIES+ (+NPL) ACT DATA.
NA 023	N,Gamma	FAST		REVW	82ANTWER	397	8209	RCN	ZIJP+ INTEGRAL DATA CHECK,DOSIMETRY
NA 023	N,Proton	5.0+6	1.1+7	EXPT	82ANTWER	814	8209	LAS	WEIGMANN+ TOF,WNR-FACILITY
NA 023	N,Alpha Reac	5.0+6	1.1+7	EXPT	82ANTWER	814	8209	LAS	WEIGMANN+ TOF,WNR-FACILITY
MG 000	Total Xsect	1.0-3	3.0-1	EXPT	82ANTWER	143	8209	JAP	AIZAWA+ (MUSASHI INST) COMP ENDFB4.
MG 024	Diff Elastic	1.5+7		EXPT	82ANTWER	796	8209	BRC	HAOUAT+15-160DEGR,DEFORMD+STAT MDL.

ELEMENT Z A		QUANTITY	ENERGY (EV) MIN	MAX	TYPE	DOCUMENTATION REF	PAGE	DATE	LAB	COMMENTS
MG	024	Diff Elastic	9.8+6		EXPT	82ANTWER	796	8209	BRC	HAOUAT+15-160DEGR,DEFORMD+STAT MDL.
MG	024	Diff Inelast	1.5+7		EXPT	82ANTWER	796	8209	BRC	HAOUAT+15-160DEGR,DEFORMD+STAT MDL.
MG	024	Diff Inelast	9.8+6		EXPT	82ANTWER	796	8209	BRC	HAOUAT+15-160DEGR,DEFORMD+STAT MDL.
MG	024	Diff Inelast	1.5+7		THEO	82ANTWER	603	8209	AEP	ZIYANG+ EXCITON MDL CALC.
MG	024	N,Proton	FISS		EXPT	82ANTWER	429	8209	PTB	MANNHART. CF252,CFD ENDFB-V
MG	024	N,Proton	FAST		REVW	82ANTWER	397	8209	RCN	ZIJP+ INTEGRAL DATA CHECK,DOSIMETRY
MG	026	N,Alpha Reac	1.5+7		EXPT	82ANTWER	869	8209	ITJ	JANCZYSZYN. ACT, D-T N-SOURCE.
AL	027	Neut Emissn	1.4+7		EXPT	82ANTWER	360	8209	OSA	TAKAHASHI+ DIFF IN ANG,E' TOF
AL	027	Diff Inelast	2.0+6	4.5+6	THEO	82ANTWER	597	8209	SWR	RAMSTROEM. LVL WIDTH FLUCT CALC.
AL	027	Diff Inelast	1.5+7		THEO	82ANTWER	603	8209	AEP	ZIYANG+ EXCITON MDL CALC.
AL	027	N,Proton	FISS		EXPT	82ANTWER	421	8209	KOS	BENABDALLAH+ (MOR) (CF-252)+CD111DIN
AL	027	N,Proton	1.4+7		EXPT	82ANTWER	404	8209	MOR	CHIADLI+ REL AL-27(N,A) CU-63(N,2N)
AL	027	N,Proton	1.4+7	1.5+7	EXPT	82ANTWER	414	8209	KOS	CSIKAI. ACT, GELI OR NATL DETECTOR
AL	027	N,Proton	FISS		EXPT	82ANTWER	418	8209	KOS	DEZSO+ (CF252) REL IN115(N,N')IN115M
AL	027	N,Proton	FISS		EXPT	82ANTWER	429	8209	PTB	MANNHART. CF252,CFD ENDFB-V
AL	027	N,Proton	FAST		REVW	82ANTWER	397	8209	RCN	ZIJP+ INTEGRAL DATA CHECK,DOSIMETRY
AL	027	N,Alpha Reac	1.4+7		EXPT	82ANTWER	404	8209	MOR	CHIADLI+ REL AL-27(N,P)
AL	027	N,Alpha Reac	1.4+7	1.5+7	EXPT	82ANTWER	414	8209	KOS	CSIKAI. ACT, GELI OR NATL DETECTOR
AL	027	N,Alpha Reac	FAST		REVW	82ANTWER	397	8209	RCN	ZIJP+ INTEGRAL DATA CHECK,DOSIMETRY
AL	CMP	Total Xsect	2.0-3	5.0+0	EXPT	82ANTWER	961	8209	ORL	HARVEY+ (+NEG) SAPPHIRE CRYSTAL
SI	000	Total Xsect	1.0-3	3.0-1	EXPT	82ANTWER	143	8209	JAP	AIZAWA+ (MUSASHI INST) COMP ENDFB4.
SI	000	Total Xsect	2.0-3	5.0+0	EXPT	82ANTWER	961	8209	ORL	HARVEY+ (+NEG) POLYCR SAMPLE
SI	000	Diff Elastic	7.8+6		EXPT	82ANTWER	783	8209	THS	BULSKI+ BE-9 N-SOURCE,FOR OPTMDL
SI	000	Polarization	7.8+6		EXPT	82ANTWER	783	8209	THS	BULSKI+ BE-9 N-SOURCE,FOR OPTMDL
SI	028	Diff Elastic	1.5+7		EXPT	82ANTWER	796	8209	BRC	HAOUAT+15-160DEGR,DEFORMD+STAT MDL.
SI	028	Diff Elastic	9.8+6		EXPT	82ANTWER	796	8209	BRC	HAOUAT+15-160DEGR,DEFORMD+STAT MDL.
SI	028	Diff Inelast	1.5+7		EXPT	82ANTWER	796	8209	BRC	HAOUAT+15-160DEGR,DEFORMD+STAT MDL.
SI	028	Diff Inelast	9.8+6		EXPT	82ANTWER	796	8209	BRC	HAOUAT+15-160DEGR,DEFORMD+STAT MDL.
SI	028	Diff Inelast	1.5+7		THEO	82ANTWER	603	8209	AEP	ZIYANG+ EXCITON MDL CALC.
SI	028	N,Proton	FISS		EXPT	82ANTWER	418	8209	KOS	DEZSO+ (CF252) REL IN115(N,N')IN115M
SI	030	N,Alpha Reac	1.5+7		EXPT	82ANTWER	869	8209	ITJ	JANCZYSZYN. ACT, D-T N-SOURCE.
SI	OXI	Inelst Gamma	1.4+7		EXPT	82ANTWER	377	8209	BIA	COX+ CONCRETE. 20-90 DEGR ANGDIST
P	031	Diff Inelast	1.5+7		THEO	82ANTWER	543	8209	TRM	GUPTA+ ANG DISTR. NEW PREEQUILIB MDL
P	031	Diff Inelast	1.5+7		THEO	82ANTWER	603	8209	AEP	ZIYANG+ EXCITON MDL CALC.
P	031	N,Alpha Reac	1.5+7		EXPT	82ANTWER	869	8209	ITJ	JANCZYSZYN. ACT, D-T N-SOURCE.
S	000	Diff Elastic	7.8+6		EXPT	82ANTWER	783	8209	THS	BULSKI+ BE-9 N-SOURCE,FOR OPTMDL
S	000	Polarization	7.8+6		EXPT	82ANTWER	783	8209	THS	BULSKI+ BE-9 N-SOURCE,FOR OPTMDL
S	032	Total Xsect	1.8+5	1.9+7	EXPT	82ANTWER	777	8209	GEL	JUNGMANN+ TOT+DEL XS ANALYSIS
S	032	Reson Params	1.8+5	1.9+7	EXPT	82ANTWER	777	8209	GEL	JUNGMANN+ TOT+DEL XS ANALYSIS
S	032	Strngth Func	1.8+5	1.9+7	EXPT	82ANTWER	777	8209	GEL	JUNGMANN+ TOT+DEL XS ANALYSIS

ELEMENT Z A	QUANTITY	ENERGY (EV) MIN	MAX	TYPE	DOCUMENTATION REF	PAGE	DATE	LAB	COMMENTS
S 032	Diff Elastic	9.8+6		EXPT	82ANTWER	796	8209	BRC	HAOUAT+15-160DEGR,DEFORMD+STAT MDL.
S 032	Diff Elastic	1.5+7		EXPT	82ANTWER	796	8209	BRC	HAOUAT+15-160DEGR,DEFORMD+STAT MDL.
S 032	Diff Elastic	1.8+5	1.9+7	EXPT	82ANTWER	777	8209	GEL	JUNGMANN+ TOT+DEL XS ANALYSIS
S 032	Diff Inelast	1.5+7		EXPT	82ANTWER	796	8209	BRC	HAOUAT+15-160DEGR,DEFORMD+STAT MDL.
S 032	Diff Inelast	9.8+6		EXPT	82ANTWER	796	8209	BRC	HAOUAT+15-160DEGR,DEFORMD+STAT MDL.
S 032	Diff Inelast	1.5+7		THEO	82ANTWER	603	8209	AEP	ZIYANG+ EXCITON MDL CALC.
S 032	N,Proton	FISS		EXPT	82ANTWER	418	8209	KOS	DEZSO+ (CF252) REL IN115(N,N')IN11SM
S 032	N,Proton	FAST		REVW	82ANTWER	397	8209	RCN	ZIJP+ INTEGRAL DATA CHECK,DOSIMETRY
AR 036	Evaluation	1.0-3	1.5+7	EVAL	82ANTWER	643	8209	RCN	GRUPPELAAR+ RCN-2 EVL,KEDAK-FORMAT
AR 036	N,Gamma	1.0-3	1.5+7	EVAL	82ANTWER	643	8209	RCN	GRUPPELAAR+ RCN-2 EVL,KEDAK-FORMAT
AR 036	N,Gamma	1.0-3	1.5+7	EVAL	82ANTWER	643	8209	RCN	GRUPPELAAR+ RCN-2 EVL,GRPH.
AR 038	Evaluation	1.0-3	1.5+7	EVAL	82ANTWER	643	8209	RCN	GRUPPELAAR+ RCN-2 EVL,KEDAK-FORMAT
AR 040	Evaluation	1.0-3	1.5+7	EVAL	82ANTWER	643	8209	RCN	GRUPPELAAR+ RCN-2 EVL,KEDAK-FORMAT
AR 040	N,Gamma	1.0-3	1.5+7	EVAL	82ANTWER	643	8209	RCN	GRUPPELAAR+ RCN-2 EVL,GRPH.
CA 000	Diff Elastic	7.8+6		EXPT	82ANTWER	783	8209	THS	BULSKI+ BE-9 N-SOURCE,FOR OPTMDL
CA 000	Polarization	7.8+6		EXPT	82ANTWER	783	8209	THS	BULSKI+ BE-9 N-SOURCE,FOR OPTMDL
CA 040	Strngth Func	8.0+6	1.7+7	THEO	82ANTWER	802	8209	BRC	DELAROCHE+ (TNL,TUE) CPLD CHAN CALC
CA 040	Strngth Func	5.0+5	2.0+6	THEO	82ANTWER	528	8209	IAE	PRONYAEV. MICROSC DESCR.
CA 040	Diff Elastic	8.0+6	1.7+7	THEO	82ANTWER	802	8209	BRC	DELAROCHE+ (TNL,TUE) CPLD CHAN CALC
CA 040	Diff Inelast	6.5+7		EXPT	82ANTWER	883	8209	DAV	BRADY+ TOF,NEUTRON E' SPECT,NEW DET
CA 040	Diff Inelast	8.0+6	1.7+7	THEO	82ANTWER	802	8209	BRC	DELAROCHE+ (TNL,TUE) CPLD CHAN CALC
CA 040	Diff Inelast	1.5+7		THEO	82ANTWER	603	8209	AEP	ZIYANG+ EXCITON MDL CALC.
CA 040	N,Triton	1.5+7		EXPT	82ANTWER	368	8209	KOS	BOEDY+ ACTIV + T-BETA COUNTING
CA 040	N,He3 Reactn	1.5+7		EXPT	82ANTWER	371	8209	RBZ	MILJANIC+ TELESCOPE, ANGDIST
SC 045	N,Gamma	FAST		REVW	82ANTWER	397	8209	RCN	ZIJP+ INTEGRAL DATA CHECK,DOSIMETRY
TI 000	N,Gamma	FAST		EXPT	82ANTWER	110	8209	CAD	DARROUZET+ ERMINE,COMP CARNAVAL-4
TI 046	N,2N	1.4+7	1.5+7	EXPT	82ANTWER	414	8209	KOS	CSIKAI. ACT, GELI OR NATL DETECTOR
TI 046	N,Proton	FISS		EXPT	82ANTWER	418	8209	KOS	DEZSO+ (CF252) REL IN115(N,N')IN11SM
TI 046	N,Proton	1.4+7	1.5+7	EXPT	82ANTWER	406	8209	MOR	VIENNOT+ 0-120DEGR,ACTIV,REL AL XSS
TI 046	N,Proton	FAST		REVW	82ANTWER	397	8209	RCN	ZIJP+ INTEGRAL DATA CHECK,DOSIMETRY
TI 047	N,Proton	FISS		EXPT	82ANTWER	418	8209	KOS	DEZSO+ (CF252) REL IN115(N,N')IN11SM
TI 047	N,Proton	FAST		REVW	82ANTWER	397	8209	RCN	ZIJP+ INTEGRAL DATA CHECK,DOSIMETRY
TI 048	Diff Inelast	1.5+7		THEO	82ANTWER	603	8209	AEP	ZIYANG+ EXCITON MDL CALC.
TI 048	N,Proton	1.4+7	1.5+7	EXPT	82ANTWER	406	8209	MOR	VIENNOT+0-120DEGR,ACTV,REL AL(NA/NP)
TI 048	N,Proton	FAST		REVW	82ANTWER	397	8209	RCN	ZIJP+ INTEGRAL DATA CHECK,DOSIMETRY
TI 049	N,Deuteron	1.5+7		EXPT	82ANTWER	356	8209	JUL	QAIM+ ACTIV, ND+NNP SIG
TI 049	N,N Proton	1.5+7		EXPT	82ANTWER	356	8209	JUL	QAIM+ ACTIV, ND+NNP SIG
TI 050	N,Proton	1.4+7	1.5+7	EXPT	82ANTWER	406	8209	MOR	VIENNOT+0-120DEGR,ACTV,REL AL(NA/NP)
V 050	N,Proton	MAXW		EXPT	82ANTWER	147	8209	ILL	D'HONDT+ HIGH FLUX REAC
V 051	Diff Inelast	2.0+6	4.5+6	THEO	82ANTWER	597	8209	SWR	RAMSTROEM. LVL WIDTH FLUCT CALC.

ELEMENT Z A	QUANTITY	ENERGY (EV) MIN	MAX	TYPE	DOCUMENTATION REF	PAGE	DATE	LAB	COMMENTS
V 051	Diff Inelast	1.5+7		THEO	82ANTWER	603	8209	AEP	ZIYANG+ EXCITON MDL CALC.
CR 000	Neut Emissn	1.4+7		EXPT	82ANTWER	360	8209	OSA	TAKAHASHI+ DIFF IN ANG,E' TOF
CR 000	N,Gamma	FAST		EXPT	82ANTWER	106	8209	BOL	AZZONI+ RB-2 REACT VALUES.
CR 000	N,Gamma	FAST		EXPT	82ANTWER	110	8209	CAD	DARROUZET+ ERMINE,COMP CARNAVAL-4
CR 000	N,Alpha Reac	5.0+7		EXPT	82ANTWER	156	8209	GEL	WATTECAMPS+ DIFF SIG AT 5 ANGLES.
CR 050	Evaluation	1.0-3	1.5+7	EVAL	82ANTWER	643	8209	RCN	GRUPPELAAR+ RCN-2 EVL,KEDAK-FORMAT
CR 050	Total Xsect	5.0+5	1.0+8	THEO	82ANTWER	574	8209	BNL	PRINCE. OPT MDL FIT.
CR 050	Strngth Func	5.0+5	1.0+8	THEO	82ANTWER	574	8209	BNL	PRINCE. OPT MDL FIT.
CR 050	Elastic Scat	5.0+5	1.0+8	THEO	82ANTWER	574	8209	BNL	PRINCE. OPT MDL FIT.
CR 050	Diff Elastic	5.0+5	1.0+8	THEO	82ANTWER	574	8209	BNL	PRINCE. OPT MDL FIT.
CR 050	N,Gamma	1.0-3	1.5+7	EVAL	82ANTWER	643	8209	RCN	GRUPPELAAR+ RCN-2 EVL,GRPH.
CR 050	N,Deuteron	1.5+7		EXPT	82ANTWER	356	8209	JUL	QAIM+ ACTIV, ND+NNP SIG
CR 050	N,N Proton	1.5+7		EXPT	82ANTWER	356	8209	JUL	QAIM+ ACTIV, ND+NNP SIG
CR 052	Total Xsect	5.0+5	1.0+8	THEO	82ANTWER	574	8209	BNL	PRINCE. OPT MDL FIT.
CR 052	Total Xsect	0.0+0	3.0+7	THEO	82ANTWER	552	8209	IRK	STROHMAIER+OPT,CMPD NUCL,EXCITON MDL
CR 052	Strngth Func	5.0+5	1.0+8	THEO	82ANTWER	574	8209	BNL	PRINCE. OPT MDL FIT.
CR 052	Elastic Scat	5.0+5	1.0+8	THEO	82ANTWER	574	8209	BNL	PRINCE. OPT MDL FIT.
CR 052	Diff Elastic	1.5+6	7.0+6	EXPT	82ANTWER	159	8209	IFU	KORZH+ TOF, CMP SPHERICAL OPT MDL.
CR 052	Diff Elastic	5.0+5	1.0+8	THEO	82ANTWER	574	8209	BNL	PRINCE. OPT MDL FIT.
CR 052	Diff Elastic	0.0+0	3.0+7	THEO	82ANTWER	552	8209	IRK	STROHMAIER+OPT,CMPD NUCL,EXCITON MDL
CR 052	Diff Inelast	1.5+6	7.0+6	EXPT	82ANTWER	159	8209	IFU	KORZH+ TOF, CMP SPHERICAL OPT MDL.
CR 052	Diff Inelast	1.5+7		THEO	82ANTWER	603	8209	AEP	ZIYANG+ EXCITON MDL CALC.
CR 052	N,Gamma	0.0+0	3.0+7	THEO	82ANTWER	552	8209	IRK	STROHMAIER+OPT,CMPD NUCL,EXCITON MDL
CR 052	N,Proton	0.0+0	3.0+7	THEO	82ANTWER	552	8209	IRK	STROHMAIER+OPT,CMPD NUCL,EXCITON MDL
CR 052	N,Proton	FAST		REVW	82ANTWER	397	8209	RCN	ZIJP+ INTEGRAL DATA CHECK,DOSIMETRY
CR 052	N,Alpha Reac	0.0+0	3.0+7	THEO	82ANTWER	552	8209	IRK	STROHMAIER+OPT,CMPD NUCL,EXCITON MDL
CR 053	Total Xsect	5.0+5	1.0+8	THEO	82ANTWER	574	8209	BNL	PRINCE. OPT MDL FIT.
CR 053	Strngth Func	5.0+5	1.0+8	THEO	82ANTWER	574	8209	BNL	PRINCE. OPT MDL FIT.
CR 053	Elastic Scat	5.0+5	1.0+8	THEO	82ANTWER	574	8209	BNL	PRINCE. OPT MDL FIT.
CR 053	Diff Elastic	5.0+5	1.0+8	THEO	82ANTWER	574	8209	BNL	PRINCE. OPT MDL FIT.
CR 054	Total Xsect	5.0+5	1.0+8	THEO	82ANTWER	574	8209	BNL	PRINCE. OPT MDL FIT.
CR 054	Strngth Func	5.0+5	1.0+8	THEO	82ANTWER	574	8209	BNL	PRINCE. OPT MDL FIT.
CR 054	Elastic Scat	5.0+5	1.0+8	THEO	82ANTWER	574	8209	BNL	PRINCE. OPT MDL FIT.
CR 054	Diff Elastic	5.0+5	1.0+8	THEO	82ANTWER	574	8209	BNL	PRINCE. OPT MDL FIT.
MN 055	Total Xsect	0.0+0	3.0+7	THEO	82ANTWER	552	8209	IRK	STROHMAIER+OPT,CMPD NUCL,EXCITON MDL
MN 055	Diff Elastic	0.0+0	3.0+7	THEO	82ANTWER	552	8209	IRK	STROHMAIER+OPT,CMPD NUCL,EXCITON MDL
MN 055	Diff Inelast	1.5+7		THEO	82ANTWER	603	8209	AEP	ZIYANG+ EXCITON MDL CALC.
MN 055	N,2N	FISS		EXPT	82ANTWER	429	8209	PTB	MANNHART. CF252,CFD ENDFB-V
MN 055	N,2N	FAST		REVW	82ANTWER	397	8209	RCN	ZIJP+ INTEGRAL DATA CHECK,DOSIMETRY
MN 055	N,Gamma	FAST		EXPT	82ANTWER	110	8209	CAD	DARROUZET+ ERMINE,COMP CARNAVAL-4

ELEMENT Z A	QUANTITY	ENERGY (EV) MIN	MAX	TYPE	DOCUMENTATION REF	PAGE	DATE	LAB	COMMENTS
MN 055	N,Gamma	MAXW		EXPT	82ANTWER	681	8209	LON	JEFFERIES+ (+NPL) ACT DATA.
MN 055	N,Gamma	0.0+0	3.0+7	THEO	82ANTWER	552	8209	IRK	STROHMAIER+OPT,CMPD NUCL,EXCITON MDL
MN 055	N,Gamma	FAST		REVW	82ANTWER	397	8209	RCN	ZIJP+ INTEGRAL DATA CHECK,DOSIMETRY
MN 055	N,Proton	0.0+0	3.0+7	THEO	82ANTWER	552	8209	IRK	STROHMAIER+OPT,CMPD NUCL,EXCITON MDL
MN 055	N,Alpha Reac	0.0+0	3.0+7	THEO	82ANTWER	552	8209	IRK	STROHMAIER+OPT,CMPD NUCL,EXCITON MDL
FE 000	Total Xsect	2.0-3	5.0+0	EXPT	82ANTWER	961	8209	ORL	HARVEY+ (+NEG) POLYCR SAMPLE
FE 000	Diff Elastic	4.5+7	7.5+7	EXPT	82ANTWER	800	8209	LVN	BOL+ 3-10 DEGR LAB-SYS, PLAST SCINT
FE 000	Neut Emissn	1.4+7		EXPT	82ANTWER	360	8209	OSA	TAKAHASHI+ DIFF IN ANG,E' TOF
FE 000	Inelst Gamma	1.4+7		EXPT	82ANTWER	377	8209	BIA	COX+ 20-90 DEGR ANGDIST
FE 000	Inelst Gamma	1.4+7		EXPT	82ANTWER	373	8209	AEP	XIAMIN+ GELI, G PROD XS,PRE-EQ MDL
FE 000	N,Gamma	FAST		EXPT	82ANTWER	106	8209	BOL	AZZONI+ RB-2 REACT VALUES.
FE 000	N,Gamma	FAST		EXPT	82ANTWER	110	8209	CAD	DARROUZET+ ERMINE,COMP CARNAVAL-4
FE 000	N,Alpha Reac	1.4+7		EXPT	82ANTWER	156	8209	GEL	WATTECAMPS+ DIFF SIG AT 5 ANGLES.
FE 054	Evaluation	1.0-3	1.5+7	EVAL	82ANTWER	643	8209	RCN	GRUPPELAAR+ RCN-2 EVL,KEDAK-FORMAT
FE 054	Total Xsect	2.4+5	1.9+7	EXPT	82ANTWER	135	8209	GEL	CORNELIS+ GELINA, RES PARMS ANAL
FE 054	Total Xsect	5.0+5	1.0+8	THEO	82ANTWER	574	8209	BNL	PRINCE. OPT MDL FIT.
FE 054	Reson Params	1.0+3	2.0+5	EXPT	82ANTWER	127	8209	GEL	BRUSEGAN+ AREA,R-MATRIX SHAPE ANAL
FE 054	Strngth Func	5.0+5	1.0+8	THEO	82ANTWER	574	8209	BNL	PRINCE. OPT MDL FIT.
FE 054	Elastic Scat	5.0+5	1.0+8	THEO	82ANTWER	574	8209	BNL	PRINCE. OPT MDL FIT.
FE 054	Diff Elastic	5.0+5	1.0+8	THEO	82ANTWER	574	8209	BNL	PRINCE. OPT MDL FIT.
FE 054	N,Gamma	3.0+2	5.0+5	EXPT	82ANTWER	127	8209	GEL	BRUSEGAN+ LINAC,SCIN DETECTOR
FE 054	N,Gamma	1.0-3	1.5+7	EVAL	82ANTWER	643	8209	RCN	GRUPPELAAR+ RCN-2 EVL,GRPH.
FE 054	N,Proton	FISS		EXPT	82ANTWER	418	8209	KOS	DEZSO+ (CF252) REL IN115(N,N')IN115M
FE 054	N,Proton	FISS		EXPT	82ANTWER	425	8209	NBS	LAMAZE+ CF-252,U-235 N-FIELDS.
FE 054	N,Proton	1.4+7	1.5+7	EXPT	82ANTWER	406	8209	MOR	VIENNOT+ 0-120DEGR,ACTIV,REL AL XSS
FE 054	N,Proton	FAST		REVW	82ANTWER	397	8209	RCN	ZIJP+ INTEGRAL DATA CHECK,DOSIMETRY
FE 056	Total Xsect	2.4+5	1.9+7	EXPT	82ANTWER	135	8209	GEL	CORNELIS+ LINAC,TOF,400M FLIGHT PATH
FE 056	Total Xsect	5.0+5	1.0+8	THEO	82ANTWER	574	8209	BNL	PRINCE. OPT MDL FIT.
FE 056	Total Xsect	0.0+0	3.0+7	THEO	82ANTWER	552	8209	IRK	STROHMAIER+OPT,CMPD NUCL,EXCITON MDL
FE 056	Reson Params	2.4+5	8.5+5	EXPT	82ANTWER	135	8209	GEL	CORNELIS+ LINAC,TOF,400M FLIGHT PATH
FE 056	Reson Params	1.0+3	2.6+4	EXPT	82ANTWER	131	8209	GEL	CORVI+ RES-PARMS, R-MATRIX ANAL.
FE 056	Reson Params	2.8+4	1.9+5	THEO	82ANTWER	755	8209	KFK	MENGONI+ (+BOL) VALENCE MDL.
FE 056	Strngth Func	5.0+5	1.0+8	THEO	82ANTWER	574	8209	BNL	PRINCE. OPT MDL FIT.
FE 056	Elastic Scat	5.0+5	1.0+8	THEO	82ANTWER	574	8209	BNL	PRINCE. OPT MDL FIT.
FE 056	Diff Elastic	5.0+5	1.0+8	THEO	82ANTWER	574	8209	BNL	PRINCE. OPT MDL FIT.
FE 056	Diff Elastic	0.0+0	3.0+7	THEO	82ANTWER	552	8209	IRK	STROHMAIER+OPT,CMPD NUCL,EXCITON MDL
FE 056	Diff Inelast	2.0+6	4.5+6	THEO	82ANTWER	597	8209	SWR	RAMSTROEM. LVL WIDTH FLUCT CALC.
FE 056	Diff Inelast	1.5+7		THEO	82ANTWER	603	8209	AEP	ZIYANG+ EXCITON MDL CALC.
FE 056	N,Gamma	1.0+3	2.6+4	EXPT	82ANTWER	131	8209	GEL	CORVI+ RES-PARMS, R-MATRIX ANAL.
FE 056	N,Gamma	0.0+0	3.0+7	THEO	82ANTWER	552	8209	IRK	STROHMAIER+OPT,CMPD NUCL,EXCITON MDL

ELEMENT Z A	QUANTITY	ENERGY (EV) MIN	MAX	TYPE	DOCUMENTATION REF	PAGE	DATE	LAB	COMMENTS
FE 056	N,Proton	FISS		EXPT	82ANTWER	418	8209	KOS	DEZSO+ (CF252) REL IN115(N,N')IN11SM
FE 056	N,Proton	0.0+0	3.0+7	THEO	82ANTWER	552	8209	IRK	STROHMAIER+OPT,CMPD NUCL,EXCITON MDL
FE 056	N,Proton	1.4+7	1.5+7	EXPT	82ANTWER	406	8209	MOR	VIENNOT+0-120DEGR,ACTV,REL AL(NA/NP)
FE 056	N,Proton	FAST		REVW	82ANTWER	397	8209	RCN	ZIJP+ INTEGRAL DATA CHECK,DOSIMETRY
FE 056	N,Alpha Reac	0.0+0	3.0+7	THEO	82ANTWER	552	8209	IRK	STROHMAIER+OPT,CMPD NUCL,EXCITON MDL
FE 057	Total Xsect	5.0+5	1.0+8	THEO	82ANTWER	574	8209	BNL	PRINCE. OPT MDL FIT.
FE 057	Total Xsect	1.0+3	1.0+5	EXPT	82ANTWER	139	8209	GEL	ROHR+ NAI-DET,TOF,LINAC.
FE 057	Reson Params	1.0+3	2.0+5	EXPT	82ANTWER	139	8209	GEL	ROHR+ C6D6-DET,TOF,LINAC.
FE 057	Strngth Func	5.0+5	1.0+8	THEO	82ANTWER	574	8209	BNL	PRINCE. OPT MDL FIT.
FE 057	Elastic Scat	5.0+5	1.0+8	THEO	82ANTWER	574	8209	BNL	PRINCE. OPT MDL FIT.
FE 057	Diff Elastic	5.0+5	1.0+8	THEO	82ANTWER	574	8209	BNL	PRINCE. OPT MDL FIT.
FE 057	N,Gamma	1.0+3	6.0+5	EXPT	82ANTWER	139	8209	GEL	ROHR+ C6D6-DET,TOF,LINAC.
FE 057	N,Proton	1.4+7	1.5+7	EXPT	82ANTWER	406	8209	MOR	VIENNOT+0-120DEGR,ACTV,REL AL(NA/NP)
FE 058	Total Xsect	5.0+5	1.0+8	THEO	82ANTWER	574	8209	BNL	PRINCE. OPT MDL FIT.
FE 058	Reson Params	1.0+4	9.3+4	THEO	82ANTWER	755	8209	KFK	MENGONI+ (+BOL) VALENCE MDL.
FE 058	Strngth Func	5.0+5	1.0+8	THEO	82ANTWER	574	8209	BNL	PRINCE. OPT MDL FIT.
FE 058	Elastic Scat	5.0+5	1.0+8	THEO	82ANTWER	574	8209	BNL	PRINCE. OPT MDL FIT.
FE 058	Diff Elastic	5.0+5	1.0+8	THEO	82ANTWER	574	8209	BNL	PRINCE. OPT MDL FIT.
FE 058	N,Gamma	FAST		REVW	82ANTWER	397	8209	RCN	ZIJP+ INTEGRAL DATA CHECK,DOSIMETRY
CO 058	Evaluation	1.0-3	1.5+7	EVAL	82ANTWER	643	8209	RCN	GRUPPELAAR+ RCN-2 EVL,G+M STATE
CO 058	Total Xsect	1.0-3	1.5+7	EVAL	82ANTWER	643	8209	RCN	GRUPPELAAR+ RCN-2 EVL,G+M STATE
CO 058	Absorption	1.0-3	1.5+7	EVAL	82ANTWER	643	8209	RCN	GRUPPELAAR+ RCN-2 EVL,G+M STATE
CO 058	Res Int Abs	1.0-3	1.5+7	EVAL	82ANTWER	643	8209	RCN	GRUPPELAAR+ RCN-2 EVL,G+M STATE
CO 058	N,Gamma	1.0-3	1.5+7	EVAL	82ANTWER	643	8209	RCN	GRUPPELAAR+ RCN-2 EVL,G+M STATE
CO 059	Evaluation	MAXW	FAST	EVAL	82ANTWER	202	8209	KFK	WIESE+ KEDAK-4 AND ENDF-B4 CALC.
CO 059	Diff Inelast	2.0+6	4.5+6	THEO	82ANTWER	597	8209	SWR	RAMSTROEM. LVL WIDTH FLUCT CALC.
CO 059	Diff Inelast	1.5+7		THEO	82ANTWER	603	8209	AEP	ZIYANG+ EXCITON MDL CALC.
CO 059	N,2N	FISS		EXPT	82ANTWER	429	8209	PTB	MANNHART. CF252,CFD ENDFB-V
CO 059	N,2N	MAXW	FAST	EVAL	82ANTWER	202	8209	KFK	WIESE+ KEDAK-4 AND ENDF-B4 CALC.
CO 059	N,2N	FAST		REVW	82ANTWER	397	8209	RCN	ZIJP+ INTEGRAL DATA CHECK,DOSIMETRY
CO 059	N,Gamma	FAST		EXPT	82ANTWER	110	8209	CAD	DARROUZET+ ERMINE,COMP CARNAVAL-4
CO 059	N,Gamma	MAXW		EXPT	82ANTWER	681	8209	LON	JEFFERIES+ (+NPL) ACT DATA.
CO 059	N,Gamma	MAXW	FAST	EVAL	82ANTWER	202	8209	KFK	WIESE+ KEDAK-4 AND ENDF-B4 CALC.
CO 059	N,Gamma	FAST		REVW	82ANTWER	397	8209	RCN	ZIJP+ INTEGRAL DATA CHECK,DOSIMETRY
CO 059	N,Proton	FISS		EXPT	82ANTWER	429	8209	PTB	MANNHART. CF252,CFD ENDFB-V
CO 059	N,Proton	FAST		REVW	82ANTWER	397	8209	RCN	ZIJP+ INTEGRAL DATA CHECK,DOSIMETRY
CO 059	N,Alpha Reac	FISS		EXPT	82ANTWER	429	8209	PTB	MANNHART. CF252,CFD ENDFB-V
CO 059	N,Alpha Reac	FAST		REVW	82ANTWER	397	8209	RCN	ZIJP+ INTEGRAL DATA CHECK,DOSIMETRY
NI 000	Neut Emissn	1.4+7		EXPT	82ANTWER	360	8209	OSA	TAKAHASHI+ DIFF IN ANG,E' TOF
NI 000	Inelst Gamma	1.4+7		EXPT	82ANTWER	373	8209	AEP	XIAMIN+ GELI, G PROD XS,PRE-EQ MDL

ELEMENT Z A	QUANTITY	ENERGY (EV) MIN	MAX	TYPE	DOCUMENTATION REF	PAGE	DATE	LAB	COMMENTS
NI 000	N,Gamma	FAST		EXPT	82ANTWER	106	8209	BOL	AZZONI+ RB-2 REACT VALUES.
NI 000	N,Gamma	FAST		EXPT	82ANTWER	110	8209	CAD	DARROUZET+ ERMINE,COMP CARNAVAL-4
NI 000	N,Alpha Reac	1.4+7		EXPT	82ANTWER	156	8209	GEL	WATTECAMPS+ DIFF SIG AT 5 ANGLES.
NI 058	Evaluation	1.0-3	1.5+7	EVAL	82ANTWER	643	8209	RCN	GRUPPELAAR+ RCN-2 EVL,KEDAK-FORMAT
NI 058	Evaluation	MAXW	FAST	EVAL	82ANTWER	202	8209	KFK	WIESE+ KEDAK-4 AND ENDF-B4 CALC.
NI 058	Total Xsect	5.0+5	1.0+8	THEO	82ANTWER	574	8209	BNL	PRINCE. OPT MDL FIT.
NI 058	Total Xsect	0.0+0	3.0+7	THEO	82ANTWER	552	8209	IRK	STROHMAIER+OPT,CMPD NUCL,EXCITON MDL
NI 058	Reson Params	1.5+4	1.5+5	THEO	82ANTWER	755	8209	KFK	MENGONI+ (+BOL) VALENCE MDL.
NI 058	Strngth Func	5.0+5	1.0+8	THEO	82ANTWER	574	8209	BNL	PRINCE. OPT MDL FIT.
NI 058	Elastic Scat	5.0+5	1.0+8	THEO	82ANTWER	574	8209	BNL	PRINCE. OPT MDL FIT.
NI 058	Diff Elastic	5.0+5	1.0+8	THEO	82ANTWER	574	8209	BNL	PRINCE. OPT MDL FIT.
NI 058	Diff Elastic	0.0+0	3.0+7	THEO	82ANTWER	552	8209	IRK	STROHMAIER+OPT,CMPD NUCL,EXCITON MDL
NI 058	Diff Inelast	1.5+7		THEO	82ANTWER	603	8209	AEP	ZIYANG+ EXCITON MDL CALC.
NI 058	N,2N	1.4+7	1.5+7	EXPT	82ANTWER	414	8209	KOS	CSIKAI. ACT, GELI OR NATL DETECTOR
NI 058	N,2N	1.2+7	1.8+7	EXPT	82ANTWER	411	8209	AEP	HANLIN+ ACT REL AL27(NA) FE56(NP)
NI 058	N,2N	FISS		EXPT	82ANTWER	429	8209	PTB	MANNHART. CF252,CFD ENDFB-V
NI 058	N,2N	TR	2.0+7	EVAL	82ANTWER	400	8209	IRK	WINKLER+GRPH,TBL,GROUP-SIG EVL.
NI 058	N,2N	TR	2.0+7	EXPT	82ANTWER	400	8209	IRK	WINKLER+SIG ACT,NAJ WELL,REL AL27-NA
NI 058	N,2N	FAST		REVW	82ANTWER	397	8209	RCN	ZIJP+ INTEGRAL DATA CHECK,DOSIMETRY
NI 058	N,Gamma	1.0-3	1.5+7	EVAL	82ANTWER	643	8209	RCN	GRUPPELAAR+ RCN-2 EVL,KEDAK-FORMAT
NI 058	N,Gamma	0.0+0	3.0+7	THEO	82ANTWER	552	8209	IRK	STROHMAIER+OPT,CMPD NUCL,EXCITON MDL
NI 058	N,Proton	FISS		EXPT	82ANTWER	421	8209	KOS	BENABDALLAH+ (MOR) (CF-252)
NI 058	N,Proton	FISS		EXPT	82ANTWER	418	8209	KOS	DEZSO+ (CF252) REL IN115(N,N')IN115M
NI 058	N,Proton	FISS		EXPT	82ANTWER	425	8209	NBS	LAMAZE+ CF-252,U-235 N-FIELDS.
NI 058	N,Proton	0.0+0	3.0+7	THEO	82ANTWER	552	8209	IRK	STROHMAIER+OPT,CMPD NUCL,EXCITON MDL
NI 058	N,Proton	1.4+7	1.5+7	EXPT	82ANTWER	406	8209	MOR	VIENNOT+ 0-120DEGR,ACTIV,REL AL XSS
NI 058	N,Proton	MAXW	FAST	EVAL	82ANTWER	202	8209	KFK	WIESE+ KEDAK-4 AND ENDF-B4 CALC.
NI 058	N,Proton	FAST		REVW	82ANTWER	397	8209	RCN	ZIJP+ INTEGRAL DATA CHECK,DOSIMETRY
NI 058	N,Alpha Reac	5.0+6	1.0+7	EXPT	82ANTWER	356	8209	JUL	QAIM+ CYCLOTRON,DD-TARGET,ACTIV
NI 058	N,Alpha Reac	0.0+0	3.0+7	THEO	82ANTWER	552	8209	IRK	STROHMAIER+OPT,CMPD NUCL,EXCITON MDL
NI 060	Evaluation	MAXW	FAST	EVAL	82ANTWER	202	8209	KFK	WIESE+ KEDAK-4 AND ENDF-B4 CALC.
NI 060	Total Xsect	5.0+5	1.0+8	THEO	82ANTWER	574	8209	BNL	PRINCE. OPT MDL FIT.
NI 060	Total Xsect	0.0+0	3.0+7	THEO	82ANTWER	552	8209	IRK	STROHMAIER+OPT,CMPD NUCL,EXCITON MDL
NI 060	Reson Params	1.2+4	1.1+5	THEO	82ANTWER	755	8209	KFK	MENGONI+ (+BOL) VALENCE MDL.
NI 060	Strngth Func	5.0+5	1.0+8	THEO	82ANTWER	574	8209	BNL	PRINCE. OPT MDL FIT.
NI 060	Elastic Scat	5.0+5	1.0+8	THEO	82ANTWER	574	8209	BNL	PRINCE. OPT MDL FIT.
NI 060	Diff Elastic	5.0+5	1.0+8	THEO	82ANTWER	574	8209	BNL	PRINCE. OPT MDL FIT.
NI 060	Diff Elastic	0.0+0	3.0+7	THEO	82ANTWER	552	8209	IRK	STROHMAIER+OPT,CMPD NUCL,EXCITON MDL
NI 060	N,Gamma	0.0+0	3.0+7	THEO	82ANTWER	552	8209	IRK	STROHMAIER+OPT,CMPD NUCL,EXCITON MDL
NI 060	N,Proton	0.0+0	3.0+7	THEO	82ANTWER	552	8209	IRK	STROHMAIER+OPT,CMPD NUCL,EXCITON MDL

ELEMENT Z A	QUANTITY	ENERGY (EV) MIN	MAX	TYPE	DOCUMENTATION REF	PAGE	DATE	LAB	COMMENTS
NI 060	N,Proton	1.4+7	1.5+7	EXPT	82ANTWER	406	8209	MOR	VIENNOT+0-120DEGR,ACTV,REL AL(NA/NP)
NI 060	N,Proton	MAXW	FAST	EVAL	82ANTWER	202	8209	KFK	WIESE+ KEDAK-4 AND ENDF-B4 CALC.
NI 060	N,Alpha Reac	0.0+0	3.0+7	THEO	82ANTWER	552	8209	IRK	STROHMAIER+OPT,CMPD NUCL,EXCITON MDL
NI 061	Total Xsect	5.0+5	1.0+8	THEO	82ANTWER	574	8209	BNL	PRINCE. OPT MDL FIT.
NI 061	Strngth Func	5.0+5	1.0+8	THEO	82ANTWER	574	8209	BNL	PRINCE. OPT MDL FIT.
NI 061	Elastic Scat	5.0+5	1.0+8	THEO	82ANTWER	574	8209	BNL	PRINCE. OPT MDL FIT.
NI 061	Diff Elastic	5.0+5	1.0+8	THEO	82ANTWER	574	8209	BNL	PRINCE. OPT MDL FIT.
NI 061	N,Proton	5.0+6	1.0+7	EXPT	82ANTWER	356	8209	JUL	QAIM+ CYCLOTRON,DD-TARGET,ACTIV
NI 061	N,Proton	1.4+7	1.5+7	EXPT	82ANTWER	406	8209	MOR	VIENNOT+0-120DEGR,ACTV,REL AL(NA/NP)
NI 062	Evaluation	1.0-3	1.5+7	EVAL	82ANTWER	643	8209	RCN	GRUPPELAAR+ RCN-2 EVL,KEDAK-FORMAT
NI 062	Total Xsect	5.0+5	1.0+8	THEO	82ANTWER	574	8209	BNL	PRINCE. OPT MDL FIT.
NI 062	Strngth Func	5.0+5	1.0+8	THEO	82ANTWER	574	8209	BNL	PRINCE. OPT MDL FIT.
NI 062	Elastic Scat	5.0+5	1.0+8	THEO	82ANTWER	574	8209	BNL	PRINCE. OPT MDL FIT.
NI 062	Diff Elastic	5.0+5	1.0+8	THEO	82ANTWER	574	8209	BNL	PRINCE. OPT MDL FIT.
NI 062	N,Gamma	1.0-3	1.5+7	EVAL	82ANTWER	643	8209	RCN	GRUPPELAAR+ RCN-2 EVL,GRPH.
NI 062	N,Proton	4.0+6	9.0+6	EXPT	82ANTWER	356	8209	JUL	QAIM+ CYCLOTRON,DD-TARGET,ACTIV
NI 062	N,Proton	1.4+7	1.5+7	EXPT	82ANTWER	406	8209	MOR	VIENNOT+0-120DEGR,ACTV,REL AL(NA/NP)
NI 062	N,Alpha Reac	4.0+6	9.0+6	EXPT	82ANTWER	356	8209	JUL	QAIM+ CYCLOTRON,DD-TARGET,ACTIV
NI 064	Evaluation	1.0-3	1.5+7	EVAL	82ANTWER	643	8209	RCN	GRUPPELAAR+ RCN-2 EVL,KEDAK-FORMAT
NI 064	Total Xsect	5.0+5	1.0+8	THEO	82ANTWER	574	8209	BNL	PRINCE. OPT MDL FIT.
NI 064	Strngth Func	5.0+5	1.0+8	THEO	82ANTWER	574	8209	BNL	PRINCE. OPT MDL FIT.
NI 064	Elastic Scat	5.0+5	1.0+8	THEO	82ANTWER	574	8209	BNL	PRINCE. OPT MDL FIT.
NI 064	Diff Elastic	5.0+5	1.0+8	THEO	82ANTWER	574	8209	BNL	PRINCE. OPT MDL FIT.
NI 064	N,Gamma	1.0-3	1.5+7	EVAL	82ANTWER	643	8209	RCN	GRUPPELAAR+ RCN-2 EVL,GRPH.
CU 000	Total Xsect	2.0-3	5.0+0	EXPT	82ANTWER	961	8209	ORL	HARVEY+ (+NEG) POLYCR+POWDER SAMPLE
CU 000	Diff Elastic	7.8+6		EXPT	82ANTWER	783	8209	THS	BULSKI+ BE-9 N-SOURCE,FOR OPTMDL
CU 000	Polarization	7.8+6		EXPT	82ANTWER	783	8209	THS	BULSKI+ BE-9 N-SOURCE,FOR OPTMDL
CU 000	Neut Emissn	1.4+7		EXPT	82ANTWER	360	8209	OSA	TAKAHASHI+ DIFF IN ANG,E' TOF
CU 000	Inelst Gamma	1.4+7		EXPT	82ANTWER	373	8209	AEP	XIAMIN+ GELI, G PROD XS,PRE-EQ MDL
CU 000	N,Gamma	5.0+5	3.0+6	EXPT	82ANTWER	759	8209	BRC	VOIGNIER+ INTEGRATED SPECTRUM METH.
CU 063	Diff Inelast	2.0+6	4.5+6	THEO	82ANTWER	597	8209	SWR	RAMSTROEM. LVL WIDTH FLUCT CALC.
CU 063	Diff Inelast	1.5+7		THEO	82ANTWER	603	8209	AEP	ZIYANG+ EXCITON MDL CALC.
CU 063	N,2N	1.4+7		EXPT	82ANTWER	404	8209	MOR	CHIADLI+ REL AL-27(N,P)
CU 063	N,2N	FISS		EXPT	82ANTWER	429	8209	PTB	MANNHART. CF252,CFD ENDFB-V
CU 063	N,2N	FAST		REVW	82ANTWER	397	8209	RCN	ZIJP+ INTEGRAL DATA CHECK,DOSIMETRY
CU 063	N,Gamma	FISS		EXPT	82ANTWER	429	8209	PTB	MANNHART. CF252,CFD ENDFB-V
CU 063	N,Gamma	5.0+5	3.0+6	EXPT	82ANTWER	759	8209	BRC	VOIGNIER+ INTEGRATED SPECTRUM METH.
CU 063	N,Gamma	FAST		REVW	82ANTWER	397	8209	RCN	ZIJP+ INTEGRAL DATA CHECK,DOSIMETRY
CU 063	N,Alpha Reac	PILE		EXPT	82ANTWER	409	8209	GEL	LISKIEN+ REL TO AL(N,A),BE9(ND) SPEC
CU 063	N,Alpha Reac	FISS		EXPT	82ANTWER	429	8209	PTB	MANNHART. CF252,CFD ENDFB-V

ELEMENT Z	A	QUANTITY	ENERGY (EV) MIN	MAX	TYPE	DOCUMENTATION REF	PAGE	DATE	LAB	COMMENTS
CU	063	N,Alpha Reac	FAST		REVW	82ANTWER	397	8209	RCN	ZIJP+ INTEGRAL DATA CHECK,DOSIMETRY
CU	065	N,2N	1.4+7	1.5+7	EXPT	82ANTWER	414	8209	KOS	CSIKAI. ACT, GELI OR NATL DETECTOR
CU	065	N,Gamma	MAXW		EXPT	82ANTWER	681	8209	LON	JEFFERIES+ (+NPL) ACT DATA.
CU	065	N,Gamma	5.0+5	3.0+6	EXPT	82ANTWER	759	8209	BRC	VOIGNIER+ INTEGRATED SPECTRUM METH.
ZN	064	Evaluation	1.0-3	1.5+7	EVAL	82ANTWER	643	8209	RCN	GRUPPELAAR+ RCN-2 EVL,KEDAK-FORMAT
ZN	064	Diff Inelast	1.5+7		THEO	82ANTWER	603	8209	AEP	ZIYANG+ EXCITON MDL CALC.
ZN	064	N,Gamma	1.0-3	1.5+7	EVAL	82ANTWER	643	8209	RCN	GRUPPELAAR+ RCN-2 EVL,GRPH.
ZN	064	N,Proton	FISS		EXPT	82ANTWER	421	8209	KOS	BENABDALLAH+ (MOR) (CF-252)
ZN	064	N,Proton	FISS		EXPT	82ANTWER	418	8209	KOS	DEZSO+ (CF252) REL IN115(N,N')IN115M
ZN	064	N,Proton	FAST		REVW	82ANTWER	397	8209	RCN	ZIJP+ INTEGRAL DATA CHECK,DOSIMETRY
ZN	065	N,Alpha Reac	MAXW		EXPT	82ANTWER	147	8209	ILL	D'HONDT+ HIGH FLUX REAC
ZN	068	N,Gamma	FISS		EXPT	82ANTWER	421	8209	KOS	BENABDALLAH+ (MOR) (CF-252)
GA	069	Diff Inelast	1.5+7		THEO	82ANTWER	603	8209	AEP	ZIYANG+ EXCITON MDL CALC.
AS	075	N,Gamma	FAST		REVW	82ANTWER	397	8209	RCN	ZIJP+ INTEGRAL DATA CHECK,DOSIMETRY
SE	080	Diff Inelast	1.5+7		THEO	82ANTWER	603	8209	AEP	ZIYANG+ EXCITON MDL CALC.
BR	079	Tot Inelastc	1.5+7		EXPT	82ANTWER	859	8209	KIG	ANDERS+ (+HAM) ACT,THRESHOLD METH
BR	079	Diff Inelast	1.5+7		THEO	82ANTWER	603	8209	AEP	ZIYANG+ EXCITON MDL CALC.
BR	081	Total Xsect	1.3+0	5.5+0	EXPT	82ANTWER	773	8209	DUB	ALFIMENKOV+ PARITY VIOLATION EFFECTS
BR	081	Reson Params	1.3+0	5.5+0	EXPT	82ANTWER	773	8209	DUB	ALFIMENKOV+ PARITY VIOLATION EFFECTS
BR	081	Polarization	7.5-1	5.5+0	EXPT	82ANTWER	773	8209	DUB	ALFIMENKOV+ PARITY VIOLATION EFFECTS
SR	084	N,Gamma	FISS		EXPT	82ANTWER	418	8209	KOS	DEZSO+ (CF252) REL IN115(N,N')IN115M
SR	086	N,Gamma	FISS		EXPT	82ANTWER	421	8209	KOS	BENABDALLAH+ (MOR) (CF-252)+SR087DIN
SR	087	Diff Inelast	FISS		EXPT	82ANTWER	421	8209	KOS	BENABDALLAH+ (MOR) (CF-252)+SR086NG
Y	089	Diff Inelast	2.0+6	4.5+6	THEO	82ANTWER	597	8209	SWR	RAMSTROEM. LVL WIDTH FLUCT CALC.
Y	089	N,2N	TR	2.8+7	THEO	82ANTWER	547	8209	TAT	JHINGAN+ (+TRM) EQUIL STATMDI
Y	089	N,xN x>2	TR	2.8+7	THEO	82ANTWER	547	8209	TAT	JHINGAN+ (+TRM) EQUIL STATMDL
ZR	000	Total Xsect	1.0-3	3.0-1	EXPT	82ANTWER	143	8209	JAP	AIZAWA+ (MUSASHI INST) COMP ENDFB4.
ZR	000	Total Xsect	2.4+4		EXPT	82ANTWER	143	8209	JAP	AIZAWA+ (MUSASHI INST) COMP ENDFB4.
ZR	090	Diff Inelast	1.5+7		THEO	82ANTWER	603	8209	AEP	ZIYANG+ EXCITON MDL CALC.
ZR	090	N,2N	1.4+7		EXPT	82ANTWER	404	8209	MOR	CHIADLI+ REL NB-93(N,2N)
ZR	090	N,2N	1.4+7	1.5+7	EXPT	82ANTWER	414	8209	KOS	CSIKAI. ACT, GELI OR NATL DETECTOR
ZR	090	N,2N	FISS		EXPT	82ANTWER	429	8209	PTB	MANNHART. CF252,CFD ENDFB-V
ZR	090	N,2N	TR	2.0+7	EVAL	82ANTWER	400	8209	IRK	WINKLER+GRPH,TBL,GROUP-SIG EVL.
ZR	090	N,2N	TR	2.0+7	EXPT	82ANTWER	400	8209	IRK	WINKLER+SIG ACT,NAJ WELL,REL AL27-NA
ZR	092	N,Deuteron	1.5+7		EXPT	82ANTWER	356	8209	JUL	QAIM+ ACTIV, ND+NNP SIG
ZR	092	N,N Proton	1.5+7		EXPT	82ANTWER	356	8209	JUL	QAIM+ ACTIV, ND+NNP SIG
NB	093	Total Xsect	1.0-3	3.0-1	EXPT	82ANTWER	143	8209	JAP	AIZAWA+ (MUSASHI INST) COMP ENDFB4.
NB	093	Total Xsect	2.4+4		EXPT	82ANTWER	143	8209	JAP	AIZAWA+ (MUSASHI INST) COMP ENDFB4.
NB	093	Neut Emissn	1.4+7		EXPT	82ANTWER	360	8209	OSA	TAKAHASHI+ DIFF IN ANG,E' TOF
NB	093	Tot Inelastc	FAST		REVW	82ANTWER	397	8209	RCN	ZIJP+ INTEGRAL DATA CHECK,DOSIMETRY

ELEMENT Z A	QUANTITY	ENERGY (EV) MIN	MAX	TYPE	DOCUMENTATION REF	PAGE	DATE	LAB	COMMENTS
NB 093	Diff Inelast	1.5+7		THEO	82ANTWER	543	8209	TRM	GUPTA+ ANG DISTR. NEW PREEQUILIB MDL
NB 093	Diff Inelast	1.5+7		THEO	82ANTWER	603	8209	AEP	ZIYANG+ EXCITON MDL CALC.
NB 093	N,2N	1.4+7		EXPT	82ANTWER	404	8209	MOR	CHIADLI+ REL ZR-90(N,2N)
NB 093	N,2N	1.4+7	1.5+7	EXPT	82ANTWER	414	8209	KOS	CSIKAI. ACT, GELI OR NATL DETECTOR
NB 093	N,2N	1.2+7	1.8+7	EXPT	82ANTWER	411	8209	AEP	HANLIN+ ACT REL AL27(NA) FE56(NP)
NB 093	N,2N	FAST		REVW	82ANTWER	397	8209	RCN	ZIJP+ INTEGRAL DATA CHECK,DOSIMETRY
NB 093	Spect N,Gamm	4.0+5		EXPT	82ANTWER	152	8209	TIT	YAMAMURO+ NAI(TL).
NB 093	N,Alpha Reac	FISS		EXPT	82ANTWER	418	8209	KOS	DEZSO+ (CF252) REL IN115(N,N')IN115M
MO 000	Total Xsect	2.4+4		EXPT	82ANTWER	143	8209	JAP	AIZAWA+ (MUSASHI INST) COMP ENDFB4.
MO 000	Neut Emissn	1.4+7		EXPT	82ANTWER	360	8209	OSA	TAKAHASHI+ DIFF IN ANG,E' TOF
MO 000	N,Gamma	FAST		EXPT	82ANTWER	110	8209	CAD	DARROUZET+ ERMINE,COMP CARNAVAL-4
MO 000	Spect N,Gamm	4.0+5		EXPT	82ANTWER	152	8209	TIT	YAMAMURO+ NAI(TL).
MO 092	Diff Elastic	1.5+6	7.0+6	EXPT	82ANTWER	159	8209	IFU	KORZH+ TOF, CMP SPHERICAL OPT MDL.
MO 092	Diff Inelast	1.5+6	7.0+6	EXPT	82ANTWER	159	8209	IFU	KORZH+ TOF, CMP SPHERICAL OPT MDL.
MO 095	N,Gamma	FAST		EXPT	82ANTWER	175	8209	ITU	CRICCHIO+ ACTIV, RAPSODIE-REACTOR.
MO 095	N,Alpha Reac	MAXW		EXPT	82ANTWER	147	8209	ILL	D'HONDT+ HIGH FLUX REAC
MO 097	N,Gamma	FAST		EXPT	82ANTWER	175	8209	ITU	CRICCHIO+ ACTIV, RAPSODIE-REACTOR.
MO 097	N,Deuteron	1.5+7		EXPT	82ANTWER	356	8209	JUL	QAIM+ ACTIV, ND+NNP SIG
MO 097	N,Alpha Reac	MAXW		EXPT	82ANTWER	147	8209	ILL	D'HONDT+ HIGH FLUX REAC
MO 097	N,N Proton	1.5+7		EXPT	82ANTWER	356	8209	JUL	QAIM+ ACTIV, ND+NNP SIG
MO 098	N,Gamma	FAST		EXPT	82ANTWER	175	8209	ITU	CRICCHIO+ ACTIV, RAPSODIE-REACTOR.
MO 098	N,Gamma	FAST		EXPT	82ANTWER	233	8209	CAD	MARTIN DEIDIER+SPECTR PHENIX+SUDERPM
MO 098	N,Gamma	FAST		REVW	82ANTWER	397	8209	RCN	ZIJP+ INTEGRAL DATA CHECK,DOSIMETRY
MO 098	N,Deuteron	1.5+7		EXPT	82ANTWER	356	8209	JUL	QAIM+ ACTIV, ND+NNP SIG
MO 098	N,N Proton	1.5+7		EXPT	82ANTWER	356	8209	JUL	QAIM+ ACTIV, ND+NNP SIG
MO 100	N,Gamma	FAST		EXPT	82ANTWER	175	8209	ITU	CRICCHIO+ ACTIV, RAPSODIE-REACTOR.
MO 100	N,Gamma	FAST		EXPT	82ANTWER	233	8209	CAD	MARTIN DEIDIER+SPECTR PHENIX+SUDERPM
MO 100	N,Gamma	FAST		REVW	82ANTWER	397	8209	RCN	ZIJP+ INTEGRAL DATA CHECK,DOSIMETRY
RU 101	N,Alpha Reac	MAXW		EXPT	82ANTWER	147	8209	ILL	D'HONDT+ HIGH FLUX REAC
RU 102	N,Gamma	FAST		EXPT	82ANTWER	233	8209	CAD	MARTIN DEIDIER+SPECTR PHENIX+SUDERPM
RU 104	N,Gamma	FAST		EXPT	82ANTWER	233	8209	CAD	MARTIN DEIDIER+SPECTR PHENIX+SUDERPM
RH 103	Tot Inelastc	FAST		REVW	82ANTWER	397	8209	RCN	ZIJP+ INTEGRAL DATA CHECK,DOSIMETRY
PD 104	N,Gamma	3.0+3	3.0+5	EXPT	82ANTWER	222	8209	GEL	CORNELIS+ TOF,LINAC.
PD 105	N,Gamma	3.0+3	3.0+5	EXPT	82ANTWER	222	8209	GEL	CORNELIS+ TOF,LINAC.
PD 105	N,Alpha Reac	MAXW		EXPT	82ANTWER	147	8209	ILL	D'HONDT+ HIGH FLUX REAC
PD 106	N,2N	2.3+7		THEO	82ANTWER	822	8209	LBZ	ZADRO+ N-N FSI CONTRIBUTIONS.
PD 106	N,Gamma	3.0+3	3.0+5	EXPT	82ANTWER	222	8209	GEL	CORNELIS+ TOF,LINAC.
PD 106	N,Gamma	FAST		EXPT	82ANTWER	175	8209	ITU	CRICCHIO+ ACTIV, RAPSODIE-REACTOR.
PD 108	N,Gamma	3.0+3	3.0+5	EXPT	82ANTWER	222	8209	GEL	CORNELIS+ TOF,LINAC.
PD 108	N,Gamma	FAST		EXPT	82ANTWER	175	8209	ITU	CRICCHIO+ ACTIV, RAPSODIE-REACTOR.

ELEMENT Z A	QUANTITY	ENERGY (EV) MIN	MAX	TYPE	DOCUMENTATION REF	PAGE	DATE	LAB	COMMENTS
PD 108	N,Gamma	FAST		EXPT	82ANTWER	233	8209	CAD	MARTIN DEIDIER+SPECTR PHENIX+SUDERPM
PD 110	N,Gamma	3.0+3	3.0+5	EXPT	82ANTWER	222	8209	GEL	CORNELIS+ TOF,LINAC.
PD 110	N,Gamma	FAST		EXPT	82ANTWER	175	8209	ITU	CRICCHIO+ ACTIV, RAPSODIE-REACTOR.
AG 107	Total Xsect	3.2+3	7.0+5	EXPT	82ANTWER	226	8209	JAE	MIZUMOTO+ RESPARS EN,WN
AG 107	Reson Params	3.2+3	7.0+4	EXPT	82ANTWER	226	8209	JAE	MIZUMOTO+ RESPARS EN,WN
AG 107	Strngth Func	3.2+3	7.0+5	EXPT	82ANTWER	226	8209	JAE	MIZUMOTO+ RESPARS EN,WN
AG 107	N,Gamma	3.2+3	7.0+5	EXPT	82ANTWER	226	8209	JAE	MIZUMOTO+ RESPARS EN,WN
AG 107	LvL Den Law	3.2+3	7.0+4	EXPT	82ANTWER	226	8209	JAE	MIZUMOTO+ RESPARS EN,WN
AG 109	Total Xsect	3.2+3	7.0+5	EXPT	82ANTWER	226	8209	JAE	MIZUMOTO+ RESPARS EN,WN
AG 109	Reson Params	3.2+3	7.0+4	EXPT	82ANTWER	226	8209	JAE	MIZUMOTO+ RESPARS EN,WN
AG 109	Strngth Func	3.2+3	7.0+5	EXPT	82ANTWER	226	8209	JAE	MIZUMOTO+ RESPARS EN,WN
AG 109	N,Gamma	3.2+3	7.0+5	EXPT	82ANTWER	226	8209	JAE	MIZUMOTO+ RESPARS EN,WN
AG 109	LvL Den Law	3.2+3	7.0+4	EXPT	82ANTWER	226	8209	JAE	MIZUMOTO+ RESPARS EN,WN
CD 110	N,Gamma	FISS		EXPT	82ANTWER	421	8209	KOS	BENABDALLAH+ (MOR) (CF-252) +CD110NG
CD 111	Total Xsect	1.3+0	5.5+0	EXPT	82ANTWER	773	8209	DUB	ALFIMENKOV+ PARITY VIOLATION EFFECTS
CD 111	Reson Params	1.3+0	5.5+0	EXPT	82ANTWER	773	8209	DUB	ALFIMENKOV+ PARITY VIOLATION EFFECTS
CD 111	Polarization	7.5-1	5.5+0	EXPT	82ANTWER	773	8209	DUB	ALFIMENKOV+ PARITY VIOLATION EFFECTS
CD 111	Diff Inelast	FISS		EXPT	82ANTWER	421	8209	KOS	BENABDALLAH+ (MOR) (CF-252)
CD 113	N,Alpha Reac	MAXW		EXPT	82ANTWER	147	8209	ILL	D'HONDT+ HIGH FLUX REAC
CD 114	Diff Inelast	1.5+7		THEO	82ANTWER	603	8209	AEP	ZIYANG+ EXCITON MDL CALC.
IN 113	Diff Inelast	FISS		EXPT	82ANTWER	421	8209	KOS	BENABDALLAH+ (MOR) (CF-252)
IN 113	Diff Inelast	FISS		EXPT	82ANTWER	418	8209	KOS	DEZSO+ (CF252) REL IN115(N,N')IN11SM
IN 113	N,2N	1.5+7		EXPT	82ANTWER	873	8209	MOR	REGGOUG+ MET,GND, REL TO AL-27(N,A).
IN 115	Tot Inelastc	FISS		EXPT	82ANTWER	425	8209	NBS	LAMAZE+ CF-252,U-235 N-FIELDS.
IN 115	Tot Inelastc	FAST		REVW	82ANTWER	397	8209	RCN	ZIJP+ INTEGRAL DATA CHECK,DOSIMETRY
IN 115	Diff Inelast	FISS		EXPT	82ANTWER	421	8209	KOS	BENABDALLAH+ (MOR) (CF-252)
IN 115	Diff Inelast	1.5+7		THEO	82ANTWER	603	8209	AEP	ZIYANG+ EXCITON MDL CALC.
IN 115	N,2N	1.5+7		EXPT	82ANTWER	873	8209	MOR	REGGOUG+ REL TO AL-27(N,A).
IN 115	N,Gamma	2.0+6	4.5+6	EXPT	82ANTWER	866	8209	LND	ANDERSSON+ ACT,REL IN-115(N,N').
IN 115	N,Gamma	FISS		EXPT	82ANTWER	421	8209	KOS	BENABDALLAH+ (MOR) (CF-252)
IN 115	N,Gamma	MAXW		EXPT	82ANTWER	681	8209	LON	JEFFERIES+ (+NPL) ACT DATA.
SN 000	Spect N,Gamm	4.0+5		EXPT	82ANTWER	152	8209	TIT	YAMAMURO+ NAI(TL).
SN 112	Evaluation	1.0-3	1.5+7	EVAL	82ANTWER	643	8209	RCN	GRUPPELAAR+ RCN-2 EVL,KEDAK-FORMAT
SN 112	N,Gamma	1.0-3	1.5+7	EVAL	82ANTWER	643	8209	RCN	GRUPPELAAR+ RCN-2 EVL,GRPH.
SN 115	N,Alpha Reac	MAXW		EXPT	82ANTWER	147	8209	ILL	D'HONDT+ HIGH FLUX REAC
SN 116	Strngth Func	8.0+6	1.7+7	THEO	82ANTWER	802	8209	BRC	DELAROCHE+ (TNL,TUE) CPLD CHAN CALC
SN 116	Strngth Func	1.0+3	2.0+5	EXPT	82ANTWER	781	8209	DUB	NIKOLENKO+ LEGENDRE FIT.
SN 116	Diff Elastic	8.0+6	1.7+7	THEO	82ANTWER	802	8209	BRC	DELAROCHE+ (TNL,TUE) CPLD CHAN CALC
SN 116	Diff Elastic	1.0+3	2.0+5	EXPT	82ANTWER	781	8209	DUB	NIKOLENKO+ LEGENDRE FIT.
SN 116	Diff Inelast	8.0+6	1.7+7	THEO	82ANTWER	802	8209	BRC	DELAROCHE+ (TNL,TUE) CPLD CHAN CALC

ELEMENT Z A	QUANTITY	ENERGY (EV) MIN	MAX	TYPE	DOCUMENTATION REF	PAGE	DATE	LAB	COMMENTS
SN 117	Total Xsect	1.3+0	5.5+0	EXPT	82ANTWER	773	8209	DUB	ALFIMENKOV+ PARITY VIOLATION EFFECTS
SN 117	Reson Params	1.3+0	5.5+0	EXPT	82ANTWER	773	8209	DUB	ALFIMENKOV+ PARITY VIOLATION EFFECTS
SN 117	Polarization	7.5-1	5.5+0	EXPT	82ANTWER	773	8209	DUB	ALFIMENKOV+ PARITY VIOLATION EFFECTS
SN 118	Strngth Func	1.0+3	2.0+5	EXPT	82ANTWER	781	8209	DUB	NIKOLENKO+ LEGENDRE FIT.
SN 118	Diff Elastic	1.0+3	2.0+5	EXPT	82ANTWER	781	8209	DUB	NIKOLENKO+ LEGENDRE FIT.
SN 119	Strngth Func	1.0+3	2.0+5	EXPT	82ANTWER	781	8209	DUB	NIKOLENKO+ LEGENDRE FIT.
SN 119	Diff Elastic	1.0+3	2.0+5	EXPT	82ANTWER	781	8209	DUB	NIKOLENKO+ LEGENDRE FIT.
SN 120	Strngth Func	8.0+6	1.7+7	THEO	82ANTWER	802	8209	BRC	DELAROCHE+ (TNL,TUE) CPLD CHAN CALC
SN 120	Strngth Func	1.0+3	2.0+5	EXPT	82ANTWER	781	8209	DUB	NIKOLENKO+ LEGENDRE FIT.
SN 120	Diff Elastic	8.0+6	1.7+7	THEO	82ANTWER	802	8209	BRC	DELAROCHE+ (TNL,TUE) CPLD CHAN CALC
SN 120	Diff Elastic	1.0+3	2.0+5	EXPT	82ANTWER	781	8209	DUB	NIKOLENKO+ LEGENDRE FIT.
SN 120	Diff Inelast	8.0+6	1.7+7	THEO	82ANTWER	802	8209	BRC	DELAROCHE+ (TNL,TUE) CPLD CHAN CALC
SN 120	Diff Inelast	1.5+7		THEO	82ANTWER	603	8209	AEP	ZIYANG+ EXCITON MDL CALC.
SN 122	Strngth Func	1.0+3	2.0+5	EXPT	82ANTWER	781	8209	DUB	NIKOLENKO+ LEGENDRE FIT.
SN 122	Diff Elastic	1.0+3	2.0+5	EXPT	82ANTWER	781	8209	DUB	NIKOLENKO+ LEGENDRE FIT.
SN 124	Strngth Func	1.0+3	2.0+5	EXPT	82ANTWER	781	8209	DUB	NIKOLENKO+ LEGENDRE FIT.
SN 124	Diff Elastic	1.0+3	2.0+5	EXPT	82ANTWER	781	8209	DUB	NIKOLENKO+ LEGENDRE FIT.
SB 121	Diff Inelast	1.5+7		THEO	82ANTWER	603	8209	AEP	ZIYANG+ EXCITON MDL CALC.
I 127	Total Xsect	1.0+1	2.0+2	EXPT	82ANTWER	758	8209	DUB	ZO IN OK+ SHAPE ANAL,WG,WN.
I 127	Reson Params	1.0+1	2.0+2	EXPT	82ANTWER	758	8209	DUB	ZO IN OK+ SHAPE ANAL,WG,WN.
I 127	Diff Inelast	1.5+7		THEO	82ANTWER	603	8209	AEP	ZIYANG+ EXCITON MDL CALC.
I 127	N,2N	FAST		REVW	82ANTWER	397	8209	RCN	ZIJP+ INTEGRAL DATA CHECK,DOSIMETRY
CS 133	Total Xsect	1.0+1	2.0+2	EXPT	82ANTWER	758	8209	DUB	ZO IN OK+ SHAPE ANAL,WG,WN.
CS 133	Reson Params	1.0+1	2.0+2	EXPT	82ANTWER	758	8209	DUB	ZO IN OK+ SHAPE ANAL,WG,WN.
CS 133	N,Gamma	FAST		EXPT	82ANTWER	175	8209	ITU	CRICCHIO+ ACTIV, RAPSODIE-REACTOR.
BA 134	N,Gamma	FISS		EXPT	82ANTWER	421	8209	KOS	BENABDALLAH+ (MOR) (CF-252)+BA135DIN
BA 135	Diff Inelast	FISS		EXPT	82ANTWER	421	8209	KOS	BENABDALLAH+ (MOR) (CF-252) +BA134NG
BA 138	N,Gamma	FISS		EXPT	82ANTWER	421	8209	KOS	BENABDALLAH+ (MOR) (CF-252)
LA 139	Total Xsect	7.5-1	5.5+0	EXPT	82ANTWER	773	8209	DUB	ALFIMENKOV+ PARITY VIOLATION EFFECTS
LA 139	Reson Params	7.5-1	5.5+0	EXPT	82ANTWER	773	8209	DUB	ALFIMENKOV+ PARITY VIOLATION EFFECTS
LA 139	Diff Elastic	7.8+6		EXPT	82ANTWER	783	8209	THS	BULSKI+ BE-9 N-SOURCE,FOR OPTMDL
LA 139	Polarization	7.5-1	5.5+0	EXPT	82ANTWER	773	8209	DUB	ALFIMENKOV+ PARITY VIOLATION EFFECTS
LA 139	Polarization	7.8+6		EXPT	82ANTWER	783	8209	THS	BULSKI+ BE-9 N-SOURCE,FOR OPTMDL
LA 139	N,Gamma	FAST		EXPT	82ANTWER	175	8209	ITU	CRICCHIO+ ACTIV, RAPSODIE-REACTOR.
LA 139	N,Gamma	FAST		EXPT	82ANTWER	233	8209	CAD	MARTIN DEIDIER+SPECTR PHENIX+SUDERPM
LA 139	N,Gamma	5.0+5	3.0+6	EXPT	82ANTWER	759	8209	BRC	VOIGNIER+ INTEGRATED SPECTRUM METH.
CE 142	N,Gamma	FAST		EXPT	82ANTWER	233	8209	CAD	MARTIN DEIDIER+SPECTR PHENIX+SUDERPM
PR 141	N,Gamma	FAST		EXPT	82ANTWER	175	8209	ITU	CRICCHIO+ ACTIV, RAPSODIE-REACTOR.
PR 141	N,Gamma	FAST		EXPT	82ANTWER	233	8209	CAD	MARTIN DEIDIER+SPECTR PHENIX+SUDERPM
ND 143	N,Gamma	FAST		EXPT	82ANTWER	175	8209	ITU	CRICCHIO+ ACTIV, RAPSODIE-REACTOR.

ELEMENT Z A	QUANTITY	ENERGY (EV) MIN	MAX	TYPE	DOCUMENTATION REF	PAGE	DATE	LAB	COMMENTS
ND 144	N,Gamma	FAST		EXPT	82ANTWER	175	8209	ITU	CRICCHIO+ ACTIV, RAPSODIE-REACTOR.
ND 146	N,Gamma	FAST		EXPT	82ANTWER	175	8209	ITU	CRICCHIO+ ACTIV, RAPSODIE-REACTOR.
ND 146	N,Gamma	FAST		EXPT	82ANTWER	233	8209	CAD	MARTIN DEIDIER+SPECTR PHENIX+SUDERPM
ND 148	N,Gamma	FAST		EXPT	82ANTWER	175	8209	ITU	CRICCHIO+ ACTIV, RAPSODIE-REACTOR.
ND 148	N,Gamma	FAST		EXPT	82ANTWER	233	8209	CAD	MARTIN DEIDIER+SPECTR PHENIX+SUDERPM
ND 150	N,Gamma	FAST		EXPT	82ANTWER	233	8209	CAD	MARTIN DEIDIER+SPECTR PHENIX+SUDERPM
SM 149	N,Gamma	FAST		EXPT	82ANTWER	175	8209	ITU	CRICCHIO+ ACTIV, RAPSODIE-REACTOR.
SM 152	N,Gamma	FAST		EXPT	82ANTWER	233	8209	CAD	MARTIN DEIDIER+SPECTR PHENIX+SUDERPM
EU 152	Total Xsect	2.5-2		EXPT	82ANTWER	230	8209	IJI	PSHENICHNIY+ XS OF T1/2=9.3H ISOMER
EU 152	N,Gamma	2.5-2		EXPT	82ANTWER	230	8209	IJI	PSHENICHNIY+ XS OF T1/2=9.3H ISOMER
HO 165	Diff Elastic	1.0-2	2.0+7	THEO	82ANTWER	792	8209	LAS	YOUNG+ OPT MDL PARAMS (+BRC).
ER 167	Tot Inelastc	1.5+7		EXPT	82ANTWER	859	8209	KIG	ANDERS+ (+HAM) ACT,THRESHOLD METH
ER 168	N,2N	1.5+7		EXPT	82ANTWER	859	8209	KIG	ANDERS+ (+HAM) ACT,THRESHOLD METH
TM 169	Total Xsect	1.0-2	2.0+7	THEO	82ANTWER	792	8209	LAS	YOUNG+ OPT MDL PARAMS (+BRC).
TM 169	Diff Elastic	1.0-2	2.0+7	THEO	82ANTWER	792	8209	LAS	YOUNG+ OPT MDL PARAMS (+BRC).
TM 169	Diff Inelast	1.0-2	2.0+7	THEO	82ANTWER	792	8209	LAS	YOUNG+ OPT MDL PARAMS (+BRC).
TM 169	N,2N	TR	2.8+7	THEO	82ANTWER	547	8209	TAT	JHINGAN+ (+TRM) EQUIL STATMDL
TM 169	N,2N	1.0-2	2.0+7	THEO	82ANTWER	792	8209	LAS	YOUNG+ OPT MDL PARAMS (+BRC).
TM 169	N,xN x>2	TR	2.8+7	THEO	82ANTWER	547	8209	TAT	JHINGAN+ (+TRM) EQUIL STATMDL
TM 169	N,xN x>2	1.0-2	2.0+7	THEO	82ANTWER	792	8209	LAS	YOUNG+ OPT MDL PARAMS (+BRC).
TM 169	N,Gamma	1.0+5	1.5+6	EXPT	82ANTWER	462	8209	AEP	YING+ ACTIVATION,IMITATION SRCE METH
TM 169	N,Gamma	1.0-2	2.0+7	THEO	82ANTWER	792	8209	LAS	YOUNG+ OPT MDL PARAMS (+BRC).
YB 174	N,Proton	1.5+7		EXPT	82ANTWER	859	8209	KIG	ANDERS+ (+HAM) ACT,THRESHOLD METH
YB 176	Tot Inelastc	1.5+7		EXPT	82ANTWER	859	8209	KIG	ANDERS+ (+HAM) ACT,THRESHOLD METH
HF 178	Reson Params	2.6+3	2.0+6	EXPT	82ANTWER	945	8209	ORL	BEER+ (+KFK) WG, SHAPE ANALYSIS
HF 178	Strngth Func	2.6+3	2.0+6	EXPT	82ANTWER	945	8209	ORL	BEER+ (+KFK) SHAPE ANALYSIS
HF 178	N,Gamma	2.6+3	2.0+6	EXPT	82ANTWER	945	8209	ORL	BEER+ (+KFK) FOR ANAL S-,R-PROCESS
HF 179	Reson Params	2.6+3	2.0+6	EXPT	82ANTWER	945	8209	ORL	BEER+ (+KFK) WG, SHAPE ANALYSIS
HF 179	Strngth Func	2.6+3	2.0+6	EXPT	82ANTWER	945	8209	ORL	BEER+ (+KFK) SHAPE ANALYSIS
HF 179	N,Gamma	2.6+3	2.0+6	EXPT	82ANTWER	945	8209	ORL	BEER+ (+KFK) FOR ANAL S-,R-PROCESS
HF 180	Reson Params	2.6+3	2.0+6	EXPT	82ANTWER	945	8209	ORL	BEER+ (+KFK) WG, SHAPE ANALYSIS
HF 180	Strngth Func	2.6+3	2.0+6	EXPT	82ANTWER	945	8209	ORL	BEER+ (+KFK) SHAPE ANALYSIS
HF 180	N,Gamma	2.6+3	2.0+6	EXPT	82ANTWER	945	8209	ORL	BEER+ (+KFK) FOR ANAL S-,R-PROCESS
TA 101	N,Gamma	FAST		REVW	82ANTWER	397	8209	RCN	ZIJP+ INTEGRAL DATA CHECK,DOSIMETRY
TA 180	Total Xsect	4.3-1		EXPT	82ANTWER	961	8209	ORL	HARVEY+ (+NEG) TA205 SAMPLE.
TA 181	Total Xsect	2.0-3	5.0+0	EXPT	82ANTWER	961	8209	ORL	HARVEY+ (+NEG) POLYCR SAMPLE
TA 181	Diff Inelast	1.5+7		THEO	82ANTWER	603	8209	AEP	ZIYANG+ EXCITON MDL CALC.
TA 181	N,2N	1.4+7	1.5+7	EXPT	82ANTWER	414	8209	KOS	CSIKAI. ACT, GELI OR NATL DETECTOR
TA 181	N,2N	1.2+7	1.8+7	EXPT	82ANTWER	411	8209	AEP	HANLIN+ ACT REL AL27(NA) FE56(NP)
TA 181	N,Gamma	FISS		EXPT	82ANTWER	418	8209	KOS	DEZSO+ (CF252) REL IN115(N,N')IN115M

ELEMENT Z A	QUANTITY	ENERGY (EV) MIN	MAX	TYPE	DOCUMENTATION REF	PAGE	DATE	LAB	COMMENTS
W 182	Diff Elastic	2.7+6	4.0+6	THEO	82ANTWER	580	8209	BRC	DELAROCHE. B- AND G- BAND LEVELS
W 182	Diff Inelast	2.7+6	4.0+6	THEO	82ANTWER	580	8209	BRC	DELAROCHE. B- AND G- BAND LEVELS
W 183	Tot Inelastc	1.5+7		EXPT	82ANTWER	859	8209	KIG	ANDERS+ (+HAM) ACT,THRESHOLD METH
W 184	Diff Elastic	2.7+6	4.0+6	THEO	82ANTWER	580	8209	BRC	DELAROCHE. B- AND G- BAND LEVELS
W 184	Tot Inelastc	1.5+7		EXPT	82ANTWER	859	8209	KIG	ANDERS+ (+HAM) ACT,THRESHOLD METH
W 184	Diff Inelast	2.7+6	4.0+6	THEO	82ANTWER	580	8209	BRC	DELAROCHE. B- AND G- BAND LEVELS
W 184	Diff Inelast	1.5+7		THEO	82ANTWER	603	8209	AEP	ZIYANG+ EXCITON MDL CALC.
W 184	N,2N	1.5+7		EXPT	82ANTWER	859	8209	KIG	ANDERS+ (+HAM) ACT,THRESHOLD METH
W 186	Diff Elastic	2.7+6	4.0+6	THEO	82ANTWER	580	8209	BRC	DELAROCHE. B- AND G- BAND LEVELS
W 186	Diff Inelast	2.7+6	4.0+6	THEO	82ANTWER	580	8209	BRC	DELAROCHE. B- AND G- BAND LEVELS
W 186	Diff Inelast	1.8+6	2.8+6	THEO	82ANTWER	582	8209	CUB	FERNANDEZ DIAZ+ DAVYDOV-FILIPPOV MDL
W 186	N,Gamma	MAXW		EXPT	82ANTWER	681	8209	LON	JEFFERIES+ (+NPL) ACT DATA.
OS 187	Total Xsect	2.7+1	5.0+2	EXPT	82ANTWER	943	8209	ORL	WINTERS+ TOF,FOR NUCLEOSYNTHESIS.
OS 187	Reson Params	2.7+1	5.0+2	EXPT	82ANTWER	943	8209	ORL	WINTERS+ TOF,FOR NUCLEOSYNTHESIS.
OS 187	Strngth Func	2.7+1	5.0+2	EXPT	82ANTWER	943	8209	ORL	WINTERS+ TOF,FOR NUCLEOSYNTHESIS.
OS 190	Tot Inelastc	1.5+7		EXPT	82ANTWER	859	8209	KIG	ANDERS+ (+HAM) ACT,THRESHOLD METH
OS 192	Tot Inelastc	1.5+7		EXPT	82ANTWER	859	8209	KIG	ANDERS+ (+HAM) ACT,THRESHOLD METH
IR 000	Total Xsect	1.0+1	2.0+2	EXPT	82ANTWER	758	8209	DUB	ZO IN OK+ SHAPE ANAL,WG,WN.
IR 000	Reson Params	1.0+1	2.0+2	EXPT	82ANTWER	758	8209	DUB	ZO IN OK+ SHAPE ANAL,WG,WN.
IR 191	Tot Inelastc	1.5+7		EXPT	82ANTWER	859	8209	KIG	ANDERS+ (+HAM) ACT,THRESHOLD METH
AU 197	Evaluation	2.5-2		EVAL	82ANTWER	689	8209	AEP	HANRONG. NG, 98.67+-0.09 B.
AU 197	Evaluation	1.0+5	1.5+7	EVAL	82ANTWER	639	8209	KYU	UENOHARA+ NG, LEAST SQ FIT
AU 197	Diff Inelast	1.5+7		THEO	82ANTWER	543	8209	TRM	GUPTA+ ANG DISTR. NEW PREEQUILIB MDL
AU 197	Diff Inelast	1.5+7		THEO	82ANTWER	603	8209	AEP	ZIYANG+ EXCITON MDL CALC.
AU 197	N,2N	1.4+7	1.5+7	EXPT	82ANTWER	414	8209	KOS	CSIKAI. ACT, GELI OR NATL DETECTOR
AU 197	N,2N	1.2+7	1.8+7	EXPT	82ANTWER	411	8209	AEP	HANLIN+ ACT REL AL27(NA) FE56(NP)
AU 197	N,2N	TR	2.8+7	THEO	82ANTWER	547	8209	TAT	JHINGAN+ (+TRM) EQUIL STATMDL
AU 197	N,2N	FAST		REVW	82ANTWER	397	8209	RCN	ZIJP+ INTEGRAL DATA CHECK,DOSIMETRY
AU 197	N,xN x>2	TR	2.8+7	THEO	82ANTWER	547	8209	TAT	JHINGAN+ (+TRM) EQUIL STATMDL
AU 197	N,Gamma	FISS		EXPT	82ANTWER	421	8209	KOS	BENABDALLAH+ (MOR) (CF-252)
AU 197	N,Gamma	FISS		EXPT	82ANTWER	418	8209	KOS	DEZSO+ (CF252) REL IN115(N,N')IN115M
AU 197	N,Gamma	2.5-2		EVAL	82ANTWER	689	8209	AEP	HANRONG. EVL,98.67+-0.09 B.
AU 197	N,Gamma	MAXW		EXPT	82ANTWER	681	8209	LON	JEFFERIES+ (+NPL) ACT DATA.
AU 197	N,Gamma	1.0+5	1.5+7	EVAL	82ANTWER	639	8209	KYU	UENOHARA+ EVL, LEAST SQ FIT
AU 197	N,Gamma	1.0+5	1.5+6	EXPT	82ANTWER	462	8209	AEP	YING+ ACTIVATION,IMITATION SRCE METH
HG 202	Diff Inelast	1.5+7		THEO	82ANTWER	603	8209	AEP	ZIYANG+ EXCITON MDL CALC.
PB 000	Diff Elastic	4.5+7	7.5+7	EXPT	82ANTWER	800	8209	LVN	BOL+ 3-10 DEGR LAB-SYS, PLAST SCINT
PB 000	Diff Elastic	7.8+6		EXPT	82ANTWER	783	8209	THS	BULSKI+ BE-9 N-SOURCE,FOR OPTMDL
PB 000	Polarization	7.8+6		EXPT	82ANTWER	783	8209	THS	BULSKI+ BE-9 N-SOURCE,FOR OPTMDL
PB 000	Neut Emissn	1.4+7		EXPT	82ANTWER	360	8209	OSA	TAKAHASHI+ DIFF IN ANG,E' TOF

ELEMENT Z A	QUANTITY	ENERGY (EV) MIN	MAX	TYPE	DOCUMENTATION REF	PAGE	DATE	LAB	COMMENTS
PB 000	Inelst Gamma	1.4+7		EXPT	82ANTWER	373	8209	AEP	XIAMIN+ GELI, G PROD XS,PRE-EQ MDL
PB 206	Diff Inelast	2.0+6	4.5+6	THEO	82ANTWER	597	8209	SWR	RAMSTROEM. LVL WIDTH FLUCT CALC.
PB 207	Tot Inelastc	1.5+7		EXPT	82ANTWER	859	8209	KIG	ANDERS+ (+HAM) ACT,THRESHOLD METH
PB 207	Diff Inelast	2.0+6	4.5+6	THEO	82ANTWER	597	8209	SWR	RAMSTROEM. LVL WIDTH FLUCT CALC.
PB 207	N,2N	1.5+7		EXPT	82ANTWER	859	8209	KIG	ANDERS+ (+HAM) ACT,THRESHOLD METH
PB 208	Strngth Func	8.0+6	1.7+7	THEO	82ANTWER	802	8209	BRC	DELAROCHE+ (TNL,TUE) CPLD CHAN CALC
PB 208	Diff Elastic	8.0+6	1.7+7	THEO	82ANTWER	802	8209	BRC	DELAROCHE+ (TNL,TUE) CPLD CHAN CALC
PB 208	Diff Inelast	8.0+6	1.7+7	THEO	82ANTWER	802	8209	BRC	DELAROCHE+ (TNL,TUE) CPLD CHAN CALC
PB 208	Diff Inelast	1.5+7		THEO	82ANTWER	603	8209	AEP	ZIYANG+ EXCITON MDL CALC.
BI 209	Evaluation	1.0-5	2.0+7	EVAL	82ANTWER	665	8209	BRC	BERSILLON+ENDF FORM OPTMDL,TOT,NG,EL
BI 209	Total Xsect	2.4+4		EXPT	82ANTWER	143	8209	JAP	AIZAWA+ (MUSASHI INST) COMP ENDFB4.
BI 209	Total Xsect	1.0-5	2.0+7	EVAL	82ANTWER	665	8209	BRC	BERSILLON+ENDF FORM OPTMDL,EVL
BI 209	Total Xsect	2.0-3	5.0+0	EXPT	82ANTWER	961	8209	ORL	HARVEY+ (+NEG) POLYCR SAMPLE
BI 209	Elastic Scat	1.0-5	2.0+7	EVAL	82ANTWER	665	8209	BRC	BERSILLON+ENDF FORM OPTMDL,EVL
BI 209	Diff Elastic	1.0-5	2.0+7	EVAL	82ANTWER	665	8209	BRC	BERSILLON+ENDF FORM OPTMDL,EVL
BI 209	Diff Elastic	7.8+6		EXPT	82ANTWER	783	8209	THS	BULSKI+ BE-9 N-SOURCE,FOR OPTMDL
BI 209	Polarization	7.8+6		EXPT	82ANTWER	783	8209	THS	BULSKI+ BE-9 N-SOURCE,FOR OPTMDL
BI 209	Diff Inelast	2.0+6	4.5+6	THEO	82ANTWER	597	8209	SWR	RAMSTROEM. LVL WIDTH FLUCT CALC.
BI 209	Diff Inelast	1.5+7		THEO	82ANTWER	603	8209	AEP	ZIYANG+ EXCITON MDL CALC.
BI 209	Inelst Gamma	1.4+7		EXPT	82ANTWER	373	8209	AEP	XIAMIN+ GELI, G PROD XS,PRE-EQ MDL
BI 209	N,Gamma	1.0-5	2.0+7	EVAL	82ANTWER	665	8209	BRC	BERSILLON+ENDF FORM OPTMDL,EVL
BI 209	N,Gamma	5.0+5	3.0+6	EXPT	82ANTWER	759	8209	BRC	VOIGNIER+ INTEGRATED SPECTRUM METH.
TH 230	Nu,Nubar	TR		EXTH	82ANTWER	733	8209	BRC	TROCHON+ (AUA) FUNCTION OF EXC-E
TH 230	Fiss Yield	7.1+5		THEO	82ANTWER	751	8209	AUA	LANG. TRIPLE-HUMPED BARRIER.
TH 230	Fiss Yield	TR		EXTH	82ANTWER	733	8209	BRC	TROCHON+ (AUA) FUNCTION OF EXC-E
TH 230	Frag Spectra	7.1+5		THEO	82ANTWER	751	8209	AUA	LANG. TRIPLE-HUMPED BARRIER.
TH 230	Frag Spectra	TR		EXTH	82ANTWER	733	8209	BRC	TROCHON+ (AUA) FUNCTION OF EXC-E
TH 232	Evaluation	5.0+4	2.0+7	EVAL	82ANTWER	657	8209	TRM	JAIN+ RENORMALIZED TO ENDF/B-V
TH 232	Neut Emissn	1.0+6	3.5+6	EXPT	82ANTWER	39	8209	ANL	SMITH+ COMP EVAL, ANGLE-INTEGR XS
TH 232	Diff Inelast	9.0+5	1.5+6	EXPT	82ANTWER	45	8209	LTI	COUCHELL+ TOF,LVLS BETW 680-1500KEV
TH 232	Diff Inelast	8.0+5	2.5+6	THEO	82ANTWER	518	8209	LTI	SHELDON. UNIFIED S-MATRIX FORMALISM.
TH 232	Diff Inelast	1.0+6	3.5+6	EXPT	82ANTWER	39	8209	ANL	SMITH+ COMP EVAL, ANGLE-INTEGR XS
TH 232	Fission	5.0+4	2.0+7	EVAL	82ANTWER	657	8209	TRM	JAIN+ RENORMALIZED TO ENDF/B-V
TH 232	Fission	FAST		REVW	82ANTWER	397	8209	RCN	ZIJP+ INTEGRAL DATA CHECK,DOSIMETRY
TH 232	Nu,Nubar	1.0+6	1.5+7	EXPT	82ANTWER	78	8209	BRC	FREHAUT+ LIQ SCINT.
TH 232	Nu,Nubar	TR		EXTH	82ANTWER	733	8209	BRC	TROCHON+ (AUA) FUNCTION OF EXC-E
TH 232	Spect Fiss G	1.0+6	1.5+7	EXPT	82ANTWER	78	8209	BRC	FREHAUT+ LIQ SCINT.
TH 232	Fiss Yield	TR		EXTH	82ANTWER	733	8209	BRC	TROCHON+ (AUA) FUNCTION OF EXC-E
TH 232	Frag Spectra	TR		EXTH	82ANTWER	733	8209	BRC	TROCHON+ (AUA) FUNCTION OF EXC-E
TH 232	Fiss Prod Gs	PILE		EXPT	82ANTWER	863	8209	SUR	ADESANMI+ FP GAMMA SPECT,HALF LIVES.

ELEMENT Z A	QUANTITY	ENERGY (EV) MIN	MAX	TYPE	DOCUMENTATION REF	PAGE	DATE	LAB	COMMENTS
TH 232	N,Gamma	5.0+4	2.0+7	EVAL	82ANTWER	657	8209	TRM	JAIN+ RENORMALIZED TO ENDF/B-V
TH 232	N,Gamma	FAST		REVW	82ANTWER	397	8209	RCN	ZIJP+ INTEGRAL DATA CHECK,DOSIMETRY
TH 232	N,Alpha Reac	MAXW		EXPT	82ANTWER	147	8209	ILL	D'HONDT+ HIGH FLUX REAC
U 000	Total Xsect	2.0-3	2.0+0	EXPT	82ANTWER	965	8209	CAI	ADIB+ TOF + NEUTRON DIFF SPCTRMTR
U 000	Total Xsect	FAST		EXPT	82ANTWER	968	8209	TUD	MEISTER+ (+DUB) DOPPLER BROADENING
U 233	Evaluation	2.5-2		EVAL	82ANTWER	685	8209	BNL	STEHN+ SIG,NUBAR, CFD ENDFB-V
U 233	Neut Emissn	1.0+6	3.5+6	EXPT	82ANTWER	39	8209	ANL	SMITH+ COMP EVAL, ANGLE-INTEGR XS
U 233	Diff Inelast	1.0+6	3.5+6	EXPT	82ANTWER	39	8209	ANL	SMITH+ COMP EVAL, ANGLE-INTEGR XS
U 233	Fission	FAST		EXPT	82ANTWER	175	8209	ITU	CRICCHIO+ ACTIV, RAPSODIE-REACTOR.
U 233	Fission	FAST		REVW	82ANTWER	397	8209	RCN	ZIJP+ INTEGRAL DATA CHECK,DOSIMETRY
U 233	Fiss Yield	MAXW		EXPT	82ANTWER	147	8209	ILL	D'HONDT+ REAC, TER A YIELDS.
U 233	Fiss Yield	9.0+7	1.1+8	THEO	82ANTWER	719	8209	THD	THEOBALD. SCISSION POINT MDL.
U 233	Absorption	FAST		EXPT	82ANTWER	175	8209	ITU	CRICCHIO+ ACTIV, RAPSODIE-REACTOR.
U 233	N,Gamma	FAST		EXPT	82ANTWER	175	8209	ITU	CRICCHIO+ ACTIV, RAPSODIE-REACTOR.
U 233	N,Alpha Reac	MAXW		EXPT	82ANTWER	147	8209	ILL	D'HONDT+ HIGH FLUX REAC
U 235	Evaluation	2.5-2		EVAL	82ANTWER	685	8209	BNL	STEHN+ SIG,NUBAR, CFD ENDFB-V
U 235	Evaluation	1.0+5	1.5+7	EVAL	82ANTWER	639	8209	KYU	UENOHARA+ NF, LEAST SQ FIT.
U 235	Total Xsect	1.0+3	2.0+7	THEO	82ANTWER	589	8209	AEP	QINGBIAO+ OPTMDL
U 235	Diff Elastic	1.0+3	2.0+7	THEO	82ANTWER	589	8209	AEP	QINGBIAO+ OPTMDL
U 235	Neut Emissn	1.0+6	3.5+6	EXPT	82ANTWER	39	8209	ANL	SMITH+ COMP EVAL, ANGLE-INTEGR XS
U 235	Diff Inelast	1.0+6	3.5+6	EXPT	82ANTWER	39	8209	ANL	SMITH+ COMP EVAL, ANGLE-INTEGR XS
U 235	Fission	3.0+5	3.0+6	EXPT	82ANTWER	456	8209	NBS	CARLSON+ LINAC BLACK N-DETECTOR.
U 235	Fission	1.5+7		EXPT	82ANTWER	55	8209	AEP	JINGWEN+ ABSOLUTE. ASS PART METH
U 235	Fission	1.5+7		EXPT	82ANTWER	58	8209	MHG	MAHDAVI+(+NMX)+ANGDIST,TRACK-DET.
U 235	Fission	1.0+5	1.5+7	EVAL	82ANTWER	639	8209	KYU	UENOHARA+ LEAST SQ FIT.
U 235	Fission	FAST		REVW	82ANTWER	397	8209	RCN	ZIJP+ INTEGRAL DATA CHECK,DOSIMETRY
U 235	Alpha	2.0+0	3.2+1	EXPT	82ANTWER	730	8209	KUR	ADAMCHUKI+ALF AT U-235 RESONANCES.
U 235	Nu,Nubar	1.0+6	1.5+7	EXPT	82ANTWER	78	8209	BRC	FREHAUT+ LIQ SCINT.
U 235	Neut Delay	MAXW		EXPT	82ANTWER	268	8209	LON	SYNETOS. AVERAGE ENERGY.
U 235	Neut Delay	5.0+5	7.5+6	EXPT	82ANTWER	265	8209	BIR	WALKER+ SPC MEASRD, COV ERR ANAL
U 235	Neut Delay	MAXW		EXPT	82ANTWER	261	8209	BNL	YEH+ (+COR) +TRISTAN,HFR
U 235	Spect Fiss G	1.0+6	1.5+7	EXPT	82ANTWER	78	8209	BRC	FREHAUT+ LIQ SCINT.
U 235	Fiss Yield	MAXW		EXPT	82ANTWER	147	8209	ILL	D'HONDT+ REAC, TER P,T,A YIELDS.
U 235	Frag Spectra	0.0+0	2.3+7	COMP	82ANTWER	460	8209	TRM	KAPOOR+ FOR EVALUATION. ANGDIST ANIS
U 235	Frag Spectra	1.8+5	8.8+6	EXPT	82ANTWER	740	8209	ANL	MEADOWS+ ANGDIST+KE-TOT.
U 235	Photo-Fissn	1.2+7	3.0+7	EXPT	82ANTWER	748	8209	GHT	DE FRENNE+ISOT YLDS,12,15,20,30MEV
U 235	Fiss Prod Gs	PILE		EXPT	82ANTWER	863	8209	SUR	ADESANMI+ FP GAMMA SPECT,HALF LIVES.
U 235	N,Alpha Reac	MAXW		EXPT	82ANTWER	147	8209	ILL	D'HONDT+ HIGH FLUX REAC
U 238	Evaluation	1.0+5	1.5+7	EVAL	82ANTWER	639	8209	KYU	UENOHARA+ NG, LEAST SQ FIT
U 238	Total Xsect	1.0+1	2.2+4	EXPT	82ANTWER	62	8209	DUB	BAKALOV+ TRANSMISSION (+FEI)

ELEMENT Z	A	QUANTITY	ENERGY (EV) MIN	MAX	TYPE	DOCUMENTATION REF	PAGE	DATE	LAB	COMMENTS
U	238	Total Xsect	1.0+0	2.0+7	REVW	82ANTWER	9	8209	ORL	DE SAUSSURE+ RES+UNRES REG.,PROBLEMS
U	238	Total Xsect	1.0+3	2.0+7	THEO	82ANTWER	589	8209	AEP	QINGBIAO+ OPTMDL
U	238	Total Xsect	2.4+4	1.0+6	EXPT	82ANTWER	65	8209	JAE	TSUBONE+ TOF, IRON FILTERED BEAM
U	238	Reson Params	1.0+1	2.2+4	EXPT	82ANTWER	62	8209	DUB	BAKALOV+ TRANSMISSION (+FEI)
U	238	Reson Params	1.0+0	2.0+7	REVW	82ANTWER	9	8209	ORL	DE SAUSSURE+ RES+UNRES REG.,WN,WG
U	238	Reson Params	2.4+4	1.0+6	EXPT	82ANTWER	65	8209	JAE	TSUBONE+ TOF, IRON FILTERED BEAM
U	238	Strngth Func	2.4+4	1.0+6	EXPT	82ANTWER	65	8209	JAE	TSUBONE+ TOF, IRON FILTERED BEAM
U	238	Diff Elastic	7.8+6		EXPT	82ANTWER	783	8209	THS	BULSKI+ BE-9 N-SOURCE,FOR OPTMDL
U	238	Diff Elastic	1.0+0	2.0+7	REVW	82ANTWER	9	8209	ORL	DE SAUSSURE+ RES+UNRES REG.,PROBLEMS
U	238	Diff Elastic	1.0+3	2.0+7	THEO	82ANTWER	589	8209	AEP	QINGBIAO+ OPTMDL
U	238	Polarization	7.8+6		EXPT	82ANTWER	783	8209	THS	BULSKI+ BE-9 N-SOURCE,FOR OPTMDL
U	238	Neut Emissn	1.0+6	3.5+6	EXPT	82ANTWER	39	8209	ANL	SMITH+ COMP EVAL, ANGLE-INTEGR XS
U	238	Diff Inelast	9.0+5	1.5+6	EXPT	82ANTWER	45	8209	LTI	COUCHELL+ TOF,LVLS BETW 680-1500KEV
U	238	Diff Inelast	1.0+0	2.0+7	REVW	82ANTWER	9	8209	ORL	DE SAUSSURE+ RES+UNRES REG.,PROBLEMS
U	238	Diff Inelast	8.0+5	2.5+6	THEO	82ANTWER	518	8209	LTI	SHELDON. UNIFIED S-MATRIX FORMALISM.
U	238	Diff Inelast	1.0+6	3.5+6	EXPT	82ANTWER	39	8209	ANL	SMITH+ COMP EVAL, ANGLE-INTEGR XS
U	238	N,2N	1.0+0	2.0+7	REVW	82ANTWER	9	8209	ORL	DE SAUSSURE+ RES+UNRES REG.,PROBLEMS
U	238	N,2N	TR	2.8+7	THEO	82ANTWER	547	8209	TAT	JHINGAN+ (+TRM) EQUIL STATMDL
U	238	N,2N	TR	1.9+7	EVAL	82ANTWER	679	8209	CJD	KORNILOV+ EVALUATED EXCITATION FUNCT
U	238	N,xN x>2	1.0+0	2.0+7	REVW	82ANTWER	9	8209	ORL	DE SAUSSURE+ RES+UNRES REG.,PROBLEMS
U	238	N,xN x>2	TR	2.8+7	THEO	82ANTWER	547	8209	TAT	JHINGAN+ (+TRM) EQUIL STATMDL
U	238	Fission	1.0+0	2.0+7	REVW	82ANTWER	9	8209	ORL	DE SAUSSURE+ RES+UNRES REG.,PROBLEMS
U	238	Fission	1.0+5	1.5+7	EVAL	82ANTWER	639	8209	KYU	UENOHARA+ EVL,LEAST SQ FIT.
U	238	Fission	FAST		REVW	82ANTWER	397	8209	RCN	ZIJP+ INTEGRAL DATA CHECK,DOSIMETRY
U	238	Nu,Nubar	1.0+0	2.0+7	REVW	82ANTWER	9	8209	ORL	DE SAUSSURE+ RES+UNRES REG.,PROBLEMS
U	238	Neut Delay	1.0+0	2.0+7	REVW	82ANTWER	9	8209	ORL	DE SAUSSURE+ RES+UNRES REG.,PROBLEMS
U	238	Fiss Yield	1.0+0	2.0+7	REVW	82ANTWER	9	8209	ORL	DE SAUSSURE+ RES+UNRES REG.,PROBLEMS
U	238	Photo-Fissn	1.2+7	3.0+7	EXPT	82ANTWER	748	8209	GHT	DE FRENNE+ISOT YLDS,12,15,20,30MEV
U	238	N,Gamma	1.0+0	2.0+7	REVW	82ANTWER	9	8209	ORL	DE SAUSSURE+ RES+UNRES REG.,PROBLEMS
U	238	N,Gamma	1.0+5	1.5+7	EVAL	82ANTWER	639	8209	KYU	UENOHARA+ EVL, LEAST SQ FIT
U	238	N,Gamma	FAST		REVW	82ANTWER	397	8209	RCN	ZIJP+ INTEGRAL DATA CHECK,DOSIMETRY
U	238	N,Alpha Reac	MAXW		EXPT	82ANTWER	147	8209	ILL	D'HONDT+ HIGH FLUX REAC
U	OXI	Total Xsect	2.0-3	2.0+0	EXPT	82ANTWER	965	8209	CAI	ADIB+ TOF + NEUTRON DIFF SPCTRMTR
U	OXI	Total Xsect	FAST		EXPT	82ANTWER	968	8209	TUD	MEISTER+ (+DUB) UO3,VF6.DOPPLER BR.
NP	237	Evaluation	MAXW	FAST	EVAL	82ANTWER	202	8209	KFK	WIESE+ KEDAK-4 AND ENDF-B4 CALC.
NP	237	N,2N	5.0+6	1.6+7	EVAL	82ANTWER	673	8209	CAD	FORT+ COMPARED TO INTEGRAL DATA.
NP	237	N,2N	MAXW	FAST	EVAL	82ANTWER	202	8209	KFK	WIESE+ KEDAK-4 AND ENDF-B4 CALC.
NP	237	Fission	2.5+6		EXPT	82ANTWER	51	8209	BRC	CANCE+ HYBRID DET,CMP PREV.RESULTS.
NP	237	Fission	FAST		EXPT	82ANTWER	175	8209	ITU	CRICCHIO+ ACTIV, RAPSODIE-REACTOR.
NP	237	Fission	FAST		REVW	82ANTWER	397	8209	RCN	ZIJP+ INTEGRAL DATA CHECK,DOSIMETRY

ELEMENT Z A	QUANTITY	ENERGY (EV) MIN	MAX	TYPE	DOCUMENTATION REF	PAGE	DATE	LAB	COMMENTS
NP 237	Nu,Nubar	1.0+6	1.5+7	EXPT	82ANTWER	78	8209	BRC	FREHAUT+ LIQ SCINT.
NP 237	Spect Fiss G	1.0+6	1.5+7	EXPT	82ANTWER	78	8209	BRC	FREHAUT+ LIQ SCINT.
NP 237	Fiss Yield	MAXW		EXPT	82ANTWER	147	8209	ILL	D'HONDT+ REAC, TER A YIELDS.
NP 237	Absorption	FAST		EXPT	82ANTWER	175	8209	ITU	CRICCHIO+ ACTIV, RAPSODIE-REACTOR.
NP 237	N,Gamma	FAST		EXPT	82ANTWER	175	8209	ITU	CRICCHIO+ ACTIV, RAPSODIE-REACTOR.
PU 238	Evaluation	1.0-5	1.6+7	EVAL	82ANTWER	669	8209	CAD	DERRIEN. TOT,NG
PU 238	Fission	5.0+0	1.0+7	EXPT	82ANTWER	206	8209	GEL	BUDTZ-JORGENSEN+ GELINA + VDG.
PU 238	Fission	0.0+0	1.0+7	THEO	82ANTWER	744	8209	GEL	KNITTER+ 2ND CHANCE FISSION SIG.
PU 239	Evaluation	2.5-2		EVAL	82ANTWER	685	8209	BNL	STEHN+ SIG,NUBAR, CFD ENDFB-V
PU 239	Evaluation	1.0+5	1.5+7	EVAL	82ANTWER	639	8209	KYU	UENOHARA+ NF, LEAST SQ FIT
PU 239	Total Xsect	1.0+1	2.2+4	EXPT	82ANTWER	62	8209	DUB	BAKALOV+ TRANSMISSION (+FEI)
PU 239	Total Xsect	1.0+3	2.0+7	THEO	82ANTWER	589	8209	AEP	QINGBIAO+ OPTMDL
PU 239	Reson Params	1.0+1	2.2+4	EXPT	82ANTWER	62	8209	DUB	BAKALOV+ TRANSMISSION (+FEI)
PU 239	Diff Elastic	1.0+3	5.0+6	THEO	82ANTWER	556	8209	LAS	ARTHUR. CPLD CHNNL+ HAUSER-FESCHBACH
PU 239	Diff Elastic	1.0+3	2.0+7	THEO	82ANTWER	589	8209	AEP	QINGBIAO+ OPTMDL
PU 239	Neut Emissn	1.0+6	3.5+6	EXPT	82ANTWER	39	8209	ANL	SMITH+ COMP EVAL, ANGLE-INTEGR XS
PU 239	Diff Inelast	1.0+3	5.0+6	THEO	82ANTWER	556	8209	LAS	ARTHUR. CPLD CHNNL+ HAUSER-FESCHBACH
PU 239	Diff Inelast	1.0+6	3.5+6	EXPT	82ANTWER	39	8209	ANL	SMITH+ COMP EVAL, ANGLE-INTEGR XS
PU 239	Fission	1.0+3	5.0+6	THEO	82ANTWER	556	8209	LAS	ARTHUR. CPLD CHNNL+ HAUSER-FESCHBACH
PU 239	Fission	1.5+7		EXPT	82ANTWER	55	8209	AEP	JINGWEN+ ABSOLUTE. ASS PART METH
PU 239	Fission	0.0+0	1.0+7	THEO	82ANTWER	744	8209	GEL	KNITTER+ 2ND CHANCE FISSION SIG.
PU 239	Fission	1.5+7		EXPT	82ANTWER	58	8209	MHG	MAHDAVI+(+NMX)+ANGDIST,TRACK-DET.
PU 239	Fission	1.0+5	1.5+7	EVAL	82ANTWER	639	8209	KYU	UENOHARA+ EVL, LEAST SQ FIT
PU 239	Fission	1.0+6	5.6+6	EXPT	82ANTWER	36	8209	AEP	XIANJIAN+ IONIZ CH, VDG.
PU 239	Fission	FAST		REVW	82ANTWER	397	8209	RCN	ZIJP+ INTEGRAL DATA CHECK,DOSIMETRY
PU 239	Neut Delay	5.0+5	7.5+6	EXPT	82ANTWER	265	8209	BIR	WALKER+ SPC MEASRD, COV ERR ANAL
PU 239	Fiss Yield	MAXW		EXPT	82ANTWER	147	8209	ILL	D'HONDT+ REAC, TER A YIELDS.
PU 239	N,Alpha Reac	MAXW		EXPT	82ANTWER	147	8209	ILL	D'HONDT+ HIGH FLUX REAC
PU 240	Total Xsect	1.1+0		EXPT	82ANTWER	961	8209	ORL	HARVEY+ (+NEG) AL-PU SAMPLE.
PU 240	Total Xsect	1.0+3	2.0+7	THEO	82ANTWER	589	8209	AEP	QINGBIAO+ OPTMDL
PU 240	Diff Elastic	1.0+3	2.0+7	THEO	82ANTWER	589	8209	AEP	QINGBIAO+ OPTMDL
PU 240	Neut Emissn	1.0+6	3.5+6	EXPT	82ANTWER	39	8209	ANL	SMITH+ COMP EVAL, ANGLE-INTEGR XS
PU 240	Diff Inelast	8.0+5	2.5+6	THEO	82ANTWER	518	8209	LTI	SHELDON. UNIFIED S-MATRIX FORMALISM.
PU 240	Diff Inelast	1.0+6	3.5+6	EXPT	82ANTWER	39	8209	ANL	SMITH+ COMP EVAL, ANGLE-INTEGR XS
PU 240	Fission	2.5+6		EXPT	82ANTWER	51	8209	BRC	CANCE+ HYBRID DET,CMP PREV.RESULTS.
PU 240	Fission	0.0+0	1.0+7	THEO	82ANTWER	744	8209	GEL	KNITTER+ 2ND CHANCE FISSION SIG.
PU 240	N,Gamma	PILE		EXPT	82ANTWER	181	8209	CAD	DARROUZET+ MELUSINE-REACT (GRE)
PU 241	Evaluation	2.5-2		EVAL	82ANTWER	685	8209	BNL	STEHN+ SIG,NUBAR, CFD ENDFB-V
PU 241	Fission	0.0+0	1.0+7	THEO	82ANTWER	744	8209	GEL	KNITTER+ 2ND CHANCE FISSION SIG.
PU 241	Fission	1.0-2	1.0+5	EXPT	82ANTWER	69	8209	GEL	WAGEMANS+ CFD PREVIOUS RESULTS.

ELEMENT Z A	QUANTITY	ENERGY (EV) MIN	MAX	TYPE	DOCUMENTATION REF	PAGE	DATE	LAB	COMMENTS
PU 241	Fiss Yield	2.5-2		EXPT	82ANTWER	737	8209	GEL	ALLAERT+ (+MOL) FRAG MASS+ENERG DIST
PU 241	Frag Spectra	2.5-2		EXPT	82ANTWER	737	8209	GEL	ALLAERT+ (+MOL) FRAG MASS+ENERG DIST
PU 241	N,Gamma	PILE		EXPT	82ANTWER	181	8209	CAD	DARROUZET+ MELUSINE-REACT (GRE)
PU 242	Diff Inelast	8.0+5	2.5+6	THEO	82ANTWER	518	8209	LTI	SHELDON. UNIFIED S-MATRIX FORMALISM.
PU 242	Fission	2.5+6		EXPT	82ANTWER	51	8209	BRC	CANCE+ HYBRID DET,CMP PREV.RESULTS.
PU 242	Fission	0.0+0	1.0+7	THEO	82ANTWER	744	8209	GEL	KNITTER+ 2ND CHANCE FISSION SIG.
PU 242	Fiss Yield	SPON		EXPT	82ANTWER	737	8209	GEL	ALLAERT+ (+MOL) FRAG MASS+ENERG DIST
PU 242	Fiss Yield	PILE		EXPT	82ANTWER	185	8209	MOL	DE RAEDT+ (+GEL) CHAIN YLDS.
PU 242	Frag Spectra	SPON		EXPT	82ANTWER	737	8209	GEL	ALLAERT+ (+MOL) FRAG MASS+ENERG DIST
PU 242	N,Gamma	PILE		EXPT	82ANTWER	181	8209	CAD	DARROUZET+ MELUSINE-REACT (GRE)
PU 244	Fission	0.0+0	1.0+7	THEO	82ANTWER	744	8209	GEL	KNITTER+ 2ND CHANCE FISSION SIG.
PU 244	Fission	1.0+3	8.0+6	EXPT	82ANTWER	74	8209	GEL	MOORE+ REL TO U-235, GELINA,TOF
PU 244	Fiss Yield	SPON		EXPT	82ANTWER	737	8209	GEL	ALLAERT+ (+MOL) FRAG MASS+ENERG DIST
PU 244	Frag Spectra	SPON		EXPT	82ANTWER	737	8209	GEL	ALLAERT+ (+MOL) FRAG MASS+ENERG DIST
AM 241	Evaluation	1.0-3	1.5+7	EVAL	82ANTWER	211	8209	KFK	FROEHNER+ KEDAK-4.
AM 241	Evaluation	3.0-4		EVAL	82ANTWER	196	8209	KFK	GOEL+ KEDAK-4.
AM 241	Fission	FAST		EXPT	82ANTWER	175	8209	ITU	CRICCHIO+ ACTIV, RAPSODIE-REACTOR.
AM 241	Res Int Fiss	3.0-4		EVAL	82ANTWER	196	8209	KFK	GOEL+ SENSITIVITY TO CUT-OFF ENERGY.
AM 241	Absorption	FAST		EXPT	82ANTWER	175	8209	ITU	CRICCHIO+ ACTIV, RAPSODIE-REACTOR.
AM 241	Res Int Capt	3.0-4		EVAL	82ANTWER	196	8209	KFK	GOEL+ SENSITIVITY TO CUT-OFF ENERGY.
AM 241	N,Gamma	FAST		EXPT	82ANTWER	175	8209	ITU	CRICCHIO+ ACTIV, RAPSODIE-REACTOR.
AM 241	N,Gamma	PILE		EXPT	82ANTWER	181	8209	CAD	DARROUZET+ MELUSINE-REACT (GRE)
AM 241	N,Gamma	FAST		EXPT	82ANTWER	178	8209	HAR	STEVENSON+ IRRAD IN ZEBRA.
AM 242	Evaluation	1.0-3	1.5+7	EVAL	82ANTWER	211	8209	KFK	FROEHNER+ KEDAK-4.
AM 242	Evaluation	3.0-4		EVAL	82ANTWER	196	8209	KFK	GOEL+ KEDAK-4.
AM 242	Reson Params	1.0-3	2.0+1	EXPT	82ANTWER	218	8209	LRL	WHITE+ BOHR CHNNL THEORY,AM242M
AM 242	Fission	1.0-3	2.0+7	EXPT	82ANTWER	218	8209	LRL	WHITE+ REL U235, RESPAR ANAL, AM242M
AM 242	Res Int Fiss	3.0-4		EVAL	82ANTWER	196	8209	KFK	GOEL+ SENSITIVITY TO CUT-OFF ENERGY.
AM 242	Res Int Capt	3.0-4		EVAL	82ANTWER	196	8209	KFK	GOEL+ SENSITIVITY TO CUT-OFF ENERGY.
AM 242	Lvl Den Law	1.0-3	2.0+1	EXPT	82ANTWER	218	8209	LRL	WHITE+ BOHR CHNNL THEORY,AM242M
AM 243	Evaluation	1.0-3	1.5+7	EVAL	82ANTWER	211	8209	KFK	FROEHNER+ KEDAK-4.
AM 243	Evaluation	3.0-4		EVAL	82ANTWER	196	8209	KFK	GOEL+ KEDAK-4.
AM 243	Fission	FAST		EXPT	82ANTWER	175	8209	ITU	CRICCHIO+ ACTIV, RAPSODIE-REACTOR.
AM 243	Res Int Fiss	3.0-4		EVAL	82ANTWER	196	8209	KFK	GOEL+ SENSITIVITY TO CUT-OFF ENERGY.
AM 243	Absorption	FAST		EXPT	82ANTWER	175	8209	ITU	CRICCHIO+ ACTIV, RAPSODIE-REACTOR.
AM 243	Res Int Capt	3.0-4		EVAL	82ANTWER	196	8209	KFK	GOEL+ SENSITIVITY TO CUT-OFF ENERGY.
AM 243	N,Gamma	FAST		EXPT	82ANTWER	175	8209	ITU	CRICCHIO+ ACTIV, RAPSODIE-REACTOR.
AM 243	N,Gamma	PILE		EXPT	82ANTWER	181	8209	CAD	DARROUZET+ MELUSINE-REACT (GRE)
AM 243	N,Gamma	FAST		EXPT	82ANTWER	178	8209	HAR	STEVENSON+ IRRAD IN ZEBRA.
AM 243	N,Gamma	5.0+4	2.5+5	EXPT	82ANTWER	215	8209	KFK	WISSHAK+ REL TO AU-197,VDG.

ELEMENT Z A	QUANTITY	ENERGY (EV) MIN	MAX	TYPE	DOCUMENTATION REF	PAGE	DATE	LAB	COMMENTS
CM 244	Evaluation	1.0-3	1.5+7	EVAL	82ANTWER	211	8209	KFK	FROEHNER+ KEDAK-4.
CM 244	Evaluation	3.0-4		EVAL	82ANTWER	196	8209	KFK	GOEL+ KEDAK-4.
CM 244	Res Int Fiss	3.0-4		EVAL	82ANTWER	196	8209	KFK	GOEL+ SENSITIVITY TO CUT-OFF ENERGY.
CM 244	Res Int Capt	3.0-4		EVAL	82ANTWER	196	8209	KFK	GOEL+ SENSITIVITY TO CUT-OFF ENERGY.
CM 244	N,Gamma	PILE		EXPT	82ANTWER	181	8209	CAD	DARROUZET+ MELUSINE-REACT (GRE)
CM 245	Reson Params	1.0-3	3.2+1	EXPT	82ANTWER	218	8209	LRL	WHITE+ BOHR CHNNL THEORY
CM 245	Fission	1.0-3	2.0+7	EXPT	82ANTWER	218	8209	LRL	WHITE+ REL U235, RESPAR ANAL
CM 245	Fiss Yield	PILE		EXPT	82ANTWER	185	8209	MOL	DE RAEDT+ (+GEL) CHAIN YLDS.
CM 245	LvL Den Law	1.0-3	3.2+1	EXPT	82ANTWER	218	8209	LRL	WHITE+ BOHR CHNNL THEORY
CF 252	Spect Fiss N	SPON		EXPT	82ANTWER	479	8209	RI	BLINOV+ MAXWELLIAN FIT.
CF 252	Spect Fiss N	SPON		EXPT	82ANTWER	484	8209	PTB	BOETTGER+ (+IRK) E=2-14MEV ANALYSED.
CF 252	Spect Fiss N	SPON		THEO	82ANTWER	473	8209	LAS	MADLAND+ PROMPT N SPEC, MULTIPLICITY
CF 252	Spect Fiss N	SPON		EXPT	82ANTWER	488	8209	TUD	MAERTEN+ TO 28MEV,TOF+P-RECOIL.
CF 252	Spect Fiss N	SPON		EXPT	82ANTWER	465	8209	ANL	POENITZ+ BLACK N-DETECTOR.
MA 000	Evaluation	NDG		EVAL	82ANTWER	612	8209	AEP	DELIN+(RES) FOR CENDL.
MA 000	Evaluation	MAXW		EVAL	82ANTWER	681	8209	LON	JEFFERIES+ (+NPL) ACT DATA.
MA 000	Evaluation	1.0-5	2.0+7	EVAL	82ANTWER	615	8209	JAE	KIKUCHI+ JENDL-2 BENCHMARK TESTS.
MA 000	Total Xsect	0.0+0	1.0+7	THEO	82ANTWER	562	8209	IFU	FEDEROV. GENERALIZED OPTMDL.
MA 000	Total Xsect	0.0+0	3.0+7	THEO	82ANTWER	565	8209	AEP	QINGBIAO+ OPTMDL WITH SKYRME FORCES
MA 000	Reson Params	NDG		THEO	82ANTWER	809	8209	PAR	BOHIGAS+ GAUSS ORTH ENSEMBLE TESTED.
MA 000	Reson Params	NDG		THEO	82ANTWER	630	8209	MIF	BONDARENKO+ FIT OF WG. (+DUB)
MA 000	Reson Params	NDG		EVAL	82ANTWER	612	8209	AEP	DELIN+ EVL FOR CENDL.
MA 000	Reson Params	NDG		COMP	82ANTWER	632	8209	CHP	YOUXIANG+ WG. 208 ELEMENTS.
MA 000	Reson Params	NDG		THEO	82ANTWER	560	8209	AEP	ZHIXIANG. NR OF DEGR FREEDOM WN-DIS
MA 000	Strngth Func	0.0+0	1.0+7	THEO	82ANTWER	562	8209	IFU	FEDEROV. GENERALIZED OPTMDL
MA 000	Diff Elastic	0.0+0	3.0+7	THEO	82ANTWER	565	8209	AEP	QINGBIAO+ OPTMDL WITH SKYRME FORCES
MA 000	N,Gamma	2.5-3		EXPT	82ANTWER	824	8209	GHT	DE CORTE+ (+KFI) COMPILATION,70 NUCL
MA 000	LvL Den Law	0.0+0	7.0+7	THEO	82ANTWER	534	8209	KFK	ANZALDO MENESES. ANAL NUMBER THEORY.
MA 000	LvL Den Law	0.0+0	7.0+7	THEO	82ANTWER	534	8209	KFK	ANZALDO MENESES. ANAL NUMBER THEORY.